P9-DIW-490

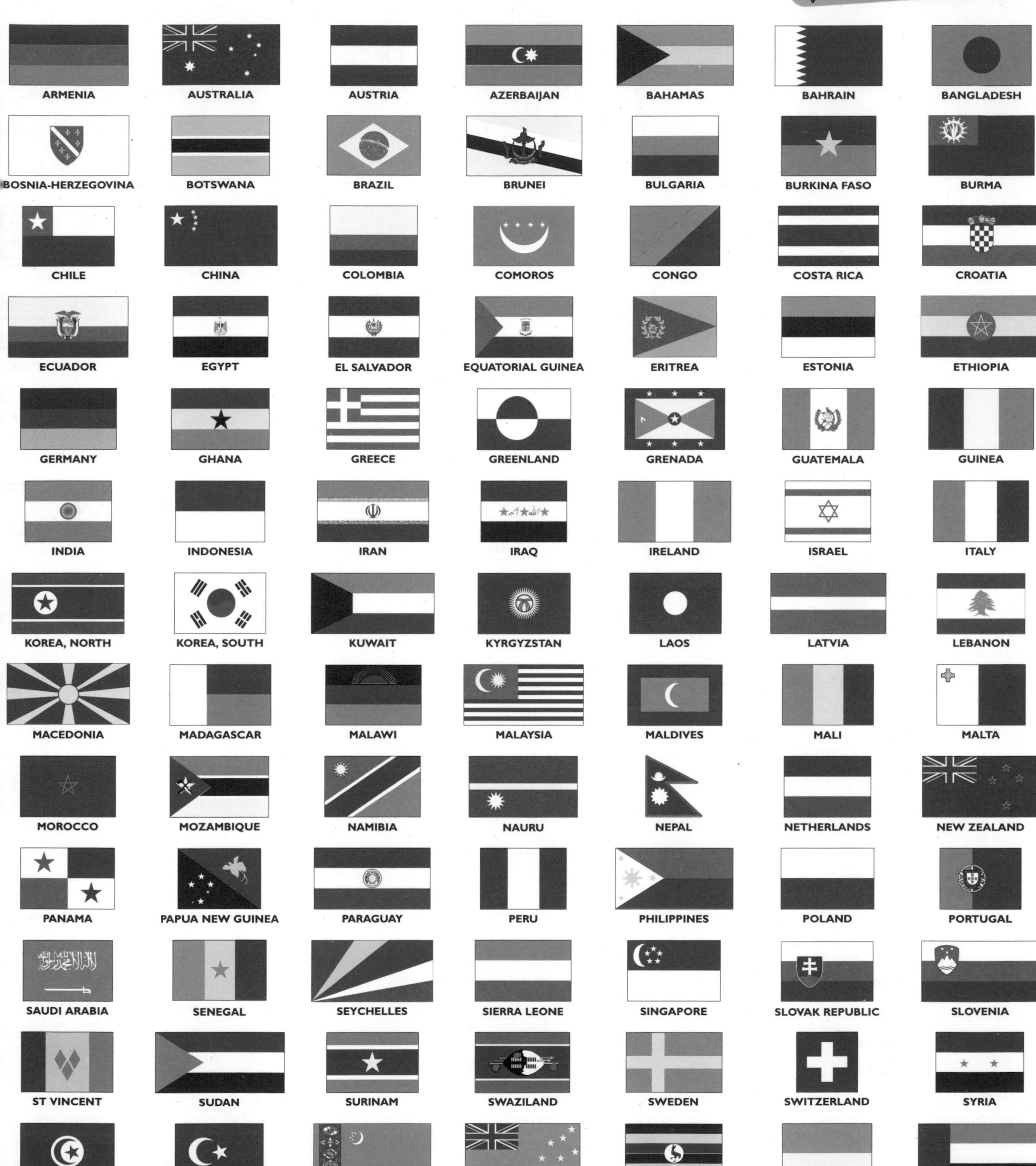
ARMENIA
AUSTRALIA
AUSTRIA
AZERBAIJAN
BAHAMAS
BAHRAIN
BANGLADESH
BOSNIA-HERZEGOVINA
BOTSWANA
BRAZIL
BRUNEI
BULGARIA
BURKINA FASO
BURMA
CHILE
CHINA
COLOMBIA
COMOROS
CONGO
COSTA RICA
CROATIA
ECUADOR
EGYPT
EL SALVADOR
EQUATORIAL GUINEA
ERITREA
ESTONIA
ETHIOPIA
GERMANY
GHANA
GREECE
GREENLAND
GRENADA
GUATEMALA
GUINEA
INDIA
INDONESIA
IRAN
IRAQ
IRELAND
ISRAEL
ITALY
KOREA, NORTH
KOREA, SOUTH
KUWAIT
KYRGYZSTAN
LAOS
LATVIA
LEBANON
MACEDONIA
MADAGASCAR
MALAWI
MALAYSIA
MALDIVES
MALI
MALTA
MOROCCO
MOZAMBIQUE
NAMIBIA
NAURU
NEPAL
NETHERLANDS
NEW ZEALAND
PANAMA
PAPUA NEW GUINEA
PARAGUAY
PERU
PHILIPPINES
POLAND
PORTUGAL
SAUDI ARABIA
SENEGAL
SEYCHELLES
SIERRA LEONE
SINGAPORE
SLOVAK REPUBLIC
SLOVENIA
ST VINCENT
SUDAN
SURINAM
SWAZILAND
SWEDEN
SWITZERLAND
SYRIA
TUNISIA
TURKEY
TURKMENISTAN
TUVALU
UGANDA
UKRAINE
UNITED ARAB EMIRATES
VIETNAM
WESTERN SAMOA
YEMEN
YUGOSLAVIA
ZAÏRE
ZAMBIA
ZIMBABWE

OXFORD

CONCISE ATLAS OF THE WORLD

FOURTH EDITION

CONTENTS

WORLD STATISTICS AND USER GUIDE

iv	**Countries**
v	**Physical Dimensions**
vi–vii	**Map Projections**
viii	**User Guide**

Picture Acknowledgements
Page 14 Science Photo Library/NOAA

Introduction to World Geography
Cartography by Philip's

Illustrations
Stefan Chabluk

CONSULTANTS

The editors are grateful to the following people who acted as specialist geography consultants on the "Introduction to World Geography" front section:

Professor D. Brunsden, Kings College, University of London, UK
Dr C. Clarke, Oxford University, UK
Dr. I. S. Evans, Durham University, UK
Professor P. Haggett, University of Bristol, UK
Professor K. McLachlan, University of London, UK
Professor M. Monmonier, Syracuse University, New York, USA
..ofessor M-L. Hsu, University of Minnesota, Minnesota, USA
..or M. J. Tooley, University of St Andrews, UK
..n, Royal Holloway, University of London, UK

..imited

..RB,

UNITED STATES MAPS

1	**Map Symbols**
2–3	**North America:**
	Physical 1:30 000 000
	Structure 1:70 000 000
	Geomorphology 1:70 000 000
4	**North America:**
	January Temperature 1:70 000 000
	July Temperature 1:70 000 000
	Rainfall – November to April 1:70 000 000
	Rainfall – May to October 1:70 000 000
5	**North America:**
	Natural Vegetation 1:32 000 000
6–7	**Continental United States:**
	Administrative 1:12 000 000
	Hawaii 1:10 000 000
8–9	**Northeastern United States** 1:2 500 000
10–11	**Eastern United States** 1:6 000 000
12–13	**Middle United States** 1:6 000 000
14–15	**Chicago and the Midwest** 1:2 500 000
16–17	**Western United States** 1:6 000 000
18–19	**Central and Southern California and Western Washington** 1:2 500 000
20–21	**New York** 1:250 000
22	**Chicago** 1:250 000
23	**Boston** 1:250 000
24	**Philadelphia** 1:250 000
25	**Washington / Baltimore** 1:250 000
26	**Detroit** 1:250 000
27	**Cleveland / Pittsburgh** 1:250 000
28	**San Francisco** 1:250 000
29	**Los Angeles** 1:250 000
30	**Alaska** 1:12 000 000 / **Hawaii** 1:5 000 000
	Puerto Rico and Virgin Islands 1:5 000 000
	Aleutian Islands 1:15 000 000
31	**United States Outlying Areas** 1:110 000 000
	Guam 1:1 000 000
	Saipan and Tinian 1:1 000 000
	..tuila 1:1 000 000
	..a 1:1 000 000
	..ia 1:20 000 000

..D GEOGRAPHY

8–9 **Oceans**
10–11 **Climate**
12–13 **Water and Vegetation**
14–15 **Environment**
16–17 **Population**
18–19 **The Human Family**
20–21 **Wealth**
22–23 **Quality of Life**
24–25 **Energy**
26–27 **Production**
28–29 **Trade**
30–31 **Travel and Tourism**
32 **Index**

WORLD MAPS

1 **Map Symbols**
2–3 **World: Political** 1:80 000 000
4 **Arctic Ocean** 1:35 000 000
5 **Antarctica** 1:35 000 000
6 **Europe: Physical** 1:20 000 000
7 **Europe: Political** 1:20 000 000
8–9 **Scandinavia** 1:5 000 000 / **Iceland** 1:5 000 000
Færoe Islands 1:5 000 000
10–11 **England and Wales** 1:2 000 000
12 **Scotland** 1:2 000 000
13 **Ireland** 1:2 000 000
14 **British Isles** 1:5 000 000
15 **Netherlands, Belgium and Luxembourg** 1:2 500 000
16–17 **Middle Europe** 1:5 000 000
18 **France** 1:5 000 000
19 **Spain and Portugal** 1:5 000 000
20–21 **Italy and the Balkan States** 1:5 000 000
22 **Balearic Islands** 1:1 000 000
Canary Islands 1:2 000 000
Madeira 1:1 000 000
23 **Malta** 1:1 000 000 / **Crete** 1:1 300 000
Corfu 1:1 000 000 / **Rhodes** 1:1 000 000
Cyprus 1:1 300 000
24–25 **Eastern Europe and Turkey** 1:10 000 000
26–27 **Russia and Central Asia** 1:20 000 000
28 **Asia: Physical** 1:50 000 000
29 **Asia: Political** 1:50 000 000
30–31 **Japan** 1:5 000 000
32–33 **China and Korea** 1:15 000 000
34–35 **Northern China and Korea** 1:6 000 000
36–37 **Indonesia and the Philippines** 1:12 500 000
38–39 **Mainland Southeast Asia** 1:6 000 000
40–41 **South Asia** 1:10 000 000
42–43 **The Indo-Gangetic Plain** 1:6 000 000
Jammu and Kashmir 1:6 000 000
44–45 **The Middle East** 1:7 000 000
46 **Arabia and the Horn of Africa** 1:15 000 000
47 **The Near East** 1:2 500 000
48 **Africa: Physical** 1:42 000 000
49 **Africa: Political** 1:42 000 000
50–51 **Northern Africa** 1:15 000 000
52–53 **Central and Southern Africa** 1:15 000 000
54–55 **East Africa** 1:8 000 000
56–57 **Southern Africa** 1:8 000 000
Madagascar 1:8 000 000
58 **Australia and Oceania:**
Physical and Political 1:50 000 000
59 **New Zealand** 1:6 000 000
Samoa Islands 1:12 000 000
Fiji and Tonga Islands 1:12 000 000
New Zealand and Southwest Pacific 1:60 000 000
60–61 **Western Australia** 1:8 000 000
62–63 **Eastern Australia** 1:8 000 000
64–65 **Pacific Ocean** 1:54 000 000
66 **North America: Physical** 1:35 000 000
67 **North America: Political** 1:35 000 000
68–69 **Canada** 1:15 000 000 / **Alaska** 1:30 000 000
70–71 **Eastern Canada** 1:7 000 000
72–73 **Western Canada** 1:7 000 000
74–75 **United States** 1:12 000 000
Hawaii 1:10 000 000
76–77 **Eastern United States** 1:6 000 000
78–79 **Northeastern United States** 1:2 500 000
80–81 **Middle United States** 1:6 000 000
82–83 **Western United States** 1:6 000 000
84–85 **Central and Southern California and Washington** 1:2 500 000
86–87 **Mexico** 1:8 000 000
88–89 **Central America and the West Indies** 1:8 000 000
90 **South America: Physical** 1:35 000 000
91 **South America: Political** 1:35 000 000
92–93 **South America – North** 1:16 000 000
94–95 **Central South America** 1:8 000 000
96 **South America – South** 1:16 000 000
97–176 **Index**
177 **Regions in the News**

World Statistics: Countries

This alphabetical list includes all the countries and territories of the world. If a territory is not completely independent, then the country it is associated with is named. The area figures give the total area of land, inland water and ice. The population figures are the latest available estimates. The annual income is the Gross National Product per capita in US dollars for 1995.

Country/Territory	Area km^2 Thousands	Area $miles^2$ Thousands	Population Thousands	Capital	Annual Income US $
Adélie Land (Fr.)	432	167	0.03	–	–
Afghanistan	652	252	19,509	Kabul	220
Albania	28.8	11.1	3,458	Tirana	340
Algeria	2,382	920	27,936	Algiers	1,650
American Samoa (US)	0.20	0.08	58	Pago Pago	2,600
Andorra	0.45	0.17	65	Andorra La Vella	14,000
Angola	1,247	481	10,844	Luanda	600
Anguilla (UK)	0.1	0.04	8	The Valley	6,800
Antigua & Barbuda	0.44	0.17	67	St John's	6,390
Argentina	2,767	1,068	34,663	Buenos Aires	7,290
Armenia	29.8	11.5	3,603	Yerevan	660
Aruba (Neths)	0.19	0.07	71	Oranjestad	17,500
Ascension Is. (UK)	0.09	0.03	1.5	Georgetown	–
Australia	7,687	2,968	18,107	Canberra	17,510
Austria	83.9	32.4	8,004	Vienna	23,120
Azerbaijan	86.6	33.4	7,559	Baku	730
Azores (Port.)	2.2	0.87	240	Ponta Delgada	4,500
Bahamas	13.9	5.4	277	Nassau	11,500
Bahrain	0.68	0.26	558	Manama	7,870
Bangladesh	144	56	118,342	Dhaka	220
Barbados	0.43	0.17	263	Bridgetown	6,240
Belarus	207.6	80.1	10,500	Minsk	2,930
Belgium	30.5	11.8	10,140	Brussels	21,210
Belize	23	8.9	216	Belmopan	2,440
Benin	113	43	5,381	Porto-Novo	420
Bermuda (UK)	0.05	0.02	64	Hamilton	27,000
Bhutan	47	18.1	1,639	Thimphu	170
Bolivia	1,099	424	7,900	La Paz/Sucre	770
Bosnia-Herzegovina	51	20	4,400	Sarajevo	2,500
Botswana	582	225	1,481	Gaborone	2,590
Brazil	8,512	3,286	161,416	Brasilia	3,020
British Indian Ocean Terr. (UK)	0.08	0.03	0	–	–
Brunei	5.8	2.2	284	Bandar Seri Begawan	9,000
Bulgaria	111	43	8,771	Sofia	1,160
Burkina Faso	274	106	10,326	Ouagadougou	300
Burma (Myanmar)	677	261	46,580	Rangoon	950
Burundi	27.8	10.7	6,412	Bujumbura	180
Cambodia	181	70	10,452	Phnom Penh	600
Cameroon	475	184	13,232	Yaoundé	770
Canada	9,976	3,852	29,972	Ottawa	20,670
Canary Is. (Spain)	7.3	2.8	1,700	Las Palmas/Santa Cruz	7,900
Cape Verde Is.	4	1.6	386	Praia	870
Cayman Is. (UK)	0.26	0.10	31	George Town	20,000
Central African Republic	623	241	3,294	Bangui	390
Chad	1,284	496	6,314	Ndjaména	200
Chatham Is. (NZ)	0.96	0.37	0.05	Waitangi	–
Chile	757	292	14,271	Santiago	3,070
China	9,597	3,705	1,226,944	Beijing	490
Christmas Is. (Aus.)	0.14	0.05	2	The Settlement	–
Cocos (Keeling) Is. (Aus.)	0.01	0.005	0.6	West Island	–
Colombia	1,139	440	34,948	Bogotá	1,400
Comoros	2.2	0.86	654	Moroni	520
Congo	342	132	2,593	Brazzaville	920
Cook Is. (NZ)	0.24	0.09	19	Avarua	900
Costa Rica	51.1	19.7	3,436	San José	2,160
Croatia	56.5	21.8	4,900	Zagreb	4,500
Cuba	111	43	11,050	Havana	1,250
Cyprus	9.3	3.6	742	Nicosia	10,380
Czech Republic	78.9	30.4	10,500	Prague	2,730
Denmark	43.1	16.6	5,229	Copenhagen	26,510
Djibouti	23.2	9	603	Djibouti	780
Dominica	0.75	0.29	89	Roseau	2,680
Dominican Republic	48.7	18.8	7,818	Santo Domingo	1,080
Ecuador	284	109	11,384	Quito	1,170
Egypt	1,001	387	64,100	Cairo	660
El Salvador	21	8.1	5,743	San Salvador	1,320
Equatorial Guinea	28.1	10.8	400	Malabo	360
Eritrea	94	36	3,850	Asmara	500
Estonia	44.7	17.3	1,531	Tallinn	3,040
Ethiopia	1,128	436	51,600	Addis Ababa	100
Falkland Is. (UK)	12.2	4.7	2	Stanley	–
Faroe Is. (Den.)	1.4	0.54	47	Tórshavn	23,660
Fiji	18.3	7.1	773	Suva	2,140
Finland	338	131	5,125	Helsinki	18,970
France	552	213	58,286	Paris	22,360
French Guiana (Fr.)	90	34.7	154	Cayenne	5,000
French Polynesia (Fr.)	4	1.5	217	Papeete	7,000
Gabon	268	103	1,316	Libreville	4,050
Gambia, The	11.3	4.4	1,144	Banjul	360
Georgia	69.7	26.9	5,448	Tbilisi	560
Germany	357	138	82,000	Berlin/Bonn	23,560
Ghana	239	92	17,462	Accra	430
Gibraltar (UK)	0.007	0.003	28	Gibraltar Town	5,000
Greece	132	51	10,510	Athens	7,390
Greenland (Den.)	2,176	840	59	Godthåb (Nuuk)	9,000
Grenada	0.34	0.13	96	St George's	2,410
Guadeloupe (Fr.)	1.7	0.66	443	Basse-Terre	9,000
Guam (US)	0.55	0.21	155	Agana	6,000
Guatemala	109	42	10,624	Guatemala City	1,110
Guinea	246	95	6,702	Conakry	510
Guinea-Bissau	36.1	13.9	1,073	Bissau	220
Guyana	215	83	832	Georgetown	350
Haiti	27.8	10.7	7,180	Port-au-Prince	800
Honduras	112	43	5,940	Tegucigalpa	580
Hong Kong (China)	1.1	0.40	6,000	–	17,860
Hungary	93	35.9	10,500	Budapest	3,330
Iceland	103	40	269	Reykjavik	23,620
India	3,288	1,269	942,989	New Delhi	290
Indonesia	1,905	735	198,644	Jakarta	730
Iran	1,648	636	68,885	Tehran	4,750
Iraq	438	169	20,184	Baghdad	2,000
Ireland	70.3	27.1	3,589	Dublin	12,580
Israel	27	10.3	5,696	Jerusalem	13,760
Italy	301	116	57,181	Rome	19,620
Ivory Coast	322	125	14,271	Yamoussoukro	630
Jamaica	11	4.2	2,700	Kingston	1,390
Jan Mayen Is. (Nor.)	0.38	0.15	0.06	–	–
Japan	378	146	125,156	Tokyo	31,450
Johnston Is. (US)	0.002	0.0009	1	–	–
Jordan	89.2	34.4	5,547	Amman	1,190
Kazakstan	2,717	1,049	17,099	Alma-Ata	1,540
Kenya	580	224	28,240	Nairobi	270
Kerguelen Is. (Fr.)	7.2	2.8	0.7	–	–
Kermadec Is. (NZ)	0.03	0.01	0.1	–	–
Kiribati	0.72	0.28	80	Tarawa	710
Korea, North	121	47	23,931	Pyŏngyang	1,100
Korea, South	99	38.2	45,088	Seoul	7,670
Kuwait	17.8	6.9	1,668	Kuwait City	23,350
Kyrgyzstan	198.5	76.6	4,738	Bishkek	830
Laos	237	91	4,906	Vientiane	290
Latvia	65	25	2,558	Riga	2,030
Lebanon	10.4	4	2,971	Beirut	1,750
Lesotho	30.4	11.7	2,064	Maseru	660
Liberia	111	43	3,092	Monrovia	800
Libya	1,760	679	5,410	Tripoli	6,500
Liechtenstein	0.16	0.06	31	Vaduz	33,510
Lithuania	65.2	25.2	3,735	Vilnius	1,310
Luxembourg	2.6	1	408	Luxembourg	35,850
Macau (Port.)	0.02	0.006	490	Macau	7,500
Macedonia	25.7	9.9	2,173	Skopje	730
Madagascar	587	227	15,206	Antananarivo	240
Madeira (Port.)	0.81	0.31	300	Funchal	4,500
Malawi	118	46	9,800	Lilongwe	220
Malaysia	330	127	20,174	Kuala Lumpur	3,160
Maldives	0.30	0.12	254	Malé	820
Mali	1,240	479	10,700	Bamako	300
Malta	0.32	0.12	370	Valletta	6,800
Marshall Is.	0.18	0.07	55	Dalap-Uliga-Darrit	1,500
Martinique (Fr.)	1.1	0.42	384	Fort-de-France	3,500
Mauritania	1,030	412	2,268	Nouakchott	510
Mauritius	2.0	0.72	1,112	Port Louis	2,980
Mayotte (Fr.)	0.37	0.14	101	Mamoundzou	1,430
Mexico	1,958	756	93,342	Mexico City	3,750
Micronesia, Fed. States of	0.70	0.27	125	Palikir	1,560
Midway Is. (US)	0.005	0.002	2	–	–
Moldova	33.7	13	4,434	Chişinău	1,180
Monaco	0.002	0.0001	32	Monaco	16,000
Mongolia	1,567	605	2,408	Ulan Bator	400
Montserrat (UK)	0.10	0.04	11	Plymouth	4,500
Morocco	447	172	26,857	Rabat	1,030
Mozambique	802	309	17,800	Maputo	80
Namibia	825	318	1,610	Windhoek	1,660
Nauru	0.02	0.008	12	Yaren District	10,000
Nepal	141	54	21,953	Katmandu	160
Netherlands	41.5	16	15,495	Amsterdam/The Hague	20,710
Neths Antilles (Neths)	0.99	0.38	202	Willemstad	9,700
New Caledonia (Fr.)	19	7.2	181	Nouméa	6,000
New Zealand	269	104	3,567	Wellington	12,900
Nicaragua	130	50	4,544	Managua	360
Niger	1,267	489	9,149	Niamey	270
Nigeria	924	357	88,515	Abuja	310
Niue (NZ)	0.26	0.10	2	Alofi	–
Norfolk Is. (Aus.)	0.03	0.01	2	Kingston	–
Northern Mariana Is. (US)	0.48	0.18	50	Saipan	11,500
Norway	324	125	4,361	Oslo	26,340
Oman	212	82	2,252	Muscat	5,600
Pakistan	796	307	143,595	Islamabad	430
Palau	0.46	0.18	18	Koror	2,260
Panama	77.1	29.8	2,629	Panama City	2,580
Papua New Guinea	463	179	4,292	Port Moresby	1,120
Paraguay	407	157	4,979	Asunción	1,500
Peru	1,285	496	23,588	Lima	1,490
Philippines	300	116	67,167	Manila	830
Pitcairn Is. (UK)	0.03	0.01	0.05	Adamstown	–
Poland	313	121	38,587	Warsaw	2,270
Portugal	92.4	35.7	10,600	Lisbon	7,890
Puerto Rico (US)	9	3.5	3,689	San Juan	7,020
Qatar	11	4.2	594	Doha	15,140
Queen Maud Land (Nor.)	2,800	1,081	0	–	–
Réunion (Fr.)	2.5	0.97	655	Saint-Denis	3,900
Romania	238	92	22,863	Bucharest	1,120
Russia	17,075	6,592	148,385	Moscow	2,350
Rwanda	26.3	10.2	7,899	Kigali	200
St Helena (UK)	0.12	0.05	6	Jamestown	–
St Kitts & Nevis	0.36	0.14	45	Basseterre	4,470
St Lucia	0.62	0.24	147	Castries	3,040
St Pierre & Miquelon (Fr.)	0.24	0.09	6	Saint Pierre	–
St Vincent & Grenadines	0.39	0.15	111	Kingstown	1,730
San Marino	0.06	0.02	26	San Marino	20,000
São Tomé & Principe	0.96	0.37	133	São Tomé	330
Saudi Arabia	2,150	830	18,395	Riyadh	8,000
Senegal	197	76	8,308	Dakar	730
Seychelles	0.46	0.18	75	Victoria	6,370
Sierra Leone	71.7	27.7	4,467	Freetown	140
Singapore	0.62	0.24	2,990	Singapore	19,310
Slovak Republic	49	18.9	5,400	Bratislava	1,900
Slovenia	20.3	7.8	2,000	Ljubljana	6,310
Solomon Is.	28.9	11.2	378	Honiara	750
Somalia	638	246	9,180	Mogadishu	500
South Africa	1,220	471	44,000	C. Town/Pretoria/Bloem.	2,900
South Georgia (UK)	3.8	1.4	0.05	–	–
Spain	505	195	39,664	Madrid	13,650
Sri Lanka	65.6	25.3	18,359	Colombo	600
Sudan	2,506	967	29,980	Khartoum	750
Surinam	163	63	421	Paramaribo	1,210
Svalbard (Nor.)	62.9	24.3	4	Longyearbyen	–
Swaziland	17.4	6.7	849	Mbabane	1,050
Sweden	450	174	8,893	Stockholm	24,830
Switzerland	41.3	15.9	7,268	Bern	36,410
Syria	185	71	14,614	Damascus	5,700
Taiwan	36	13.9	21,100	Taipei	11,000
Tajikistan	143.1	55.2	6,102	Dushanbe	470
Tanzania	945	365	29,710	Dodoma	100
Thailand	513	198	58,432	Bangkok	2,040
Togo	56.8	21.9	4,140	Lomé	330
Tokelau (NZ)	0.01	0.005	2	Nukunonu	–
Tonga	0.75	0.29	107	Nuku'alofa	1,610
Trinidad & Tobago	5.1	2	1,295	Port of Spain	3,730
Tristan da Cunha (UK)	0.11	0.04	0.33	Edinburgh	–
Tunisia	164	63	8,906	Tunis	1,780
Turkey	779	301	61,303	Ankara	2,120
Turkmenistan	488.1	188.5	4,100	Ashkhabad	1,400
Turks & Caicos Is. (UK)	0.43	0.17	15	Cockburn Town	5,000
Tuvalu	0.03	0.01	10	Fongafale	600
Uganda	236	91	21,466	Kampala	190
Ukraine	603.7	233.1	52,027	Kiev	1,910
United Arab Emirates	83.6	32.3	2,800	Abu Dhabi	22,470
United Kingdom	243.3	94	58,306	London	17,970
United States of America	9,373	3,619	263,563	Washington, DC	24,750
Uruguay	177	68	3,186	Montevideo	3,910
Uzbekistan	447.4	172.7	22,833	Tashkent	960
Vanuatu	12.2	4.7	167	Port-Vila	1,230
Vatican City	0.0004	0.0002	1	–	–
Venezuela	912	352	21,810	Caracas	2,840
Vietnam	332	127	74,580	Hanoi	170
Virgin Is. (UK)	0.15	0.06	20	Road Town	–
Virgin Is. (US)	0.34	0.13	102	Charlotte Amalie	12,000
Wake Is.	0.008	0.003	0.30	–	–
Wallis & Futuna Is. (Fr.)	0.20	0.08	13	Mata-Utu	–
Western Sahara	266	103	220	El Aaiún	300
Western Samoa	2.8	1.1	169	Apia	980
Yemen	528	204	14,609	Sana	800
Yugoslavia	102.3	39.5	10,881	Belgrade	1,000
Zaïre	2,345	905	44,504	Kinshasa	500
Zambia	753	291	9,500	Lusaka	370
Zimbabwe	391	151	11,453	Harare	540

World Statistics: Physical Dimensions

Each topic list is divided into continents and within a continent the items are listed in order of size. The order of the continents is as in the atlas. The bottom part of many of the lists is selective in order to give examples from as many different countries as possible. The figures are rounded as appropriate, and both metric and imperial measurements are given.

World, Continents, Oceans

	km^2	$miles^2$	%
The World	509,450,000	196,672,000	–
Land	149,450,000	57,688,000	29.3
Water	360,000,000	138,984,000	70.7
Asia	44,500,000	17,177,000	29.8
Africa	30,302,000	11,697,000	20.3
North America	24,241,000	9,357,000	16.2
South America	17,793,000	6,868,000	11.9
Antarctica	14,100,000	5,443,000	9.4
Europe	9,957,000	3,843,000	6.7
Australia & Oceania	8,557,000	3,303,000	5.7
Pacific Ocean	179,679,000	69,356,000	49.9
Atlantic Ocean	92,373,000	35,657,000	25.7
Indian Ocean	73,917,000	28,532,000	20.5
Arctic Ocean	14,090,000	5,439,000	3.9

Ocean Depths

Atlantic Ocean

	m	ft
Puerto Rico (Milwaukee) Deep	9,220	30,249
Cayman Trench	7,680	25,197
Gulf of Mexico	5,203	17,070
Mediterranean Sea	5,121	16,801
Black Sea	2,211	7,254
North Sea	660	2,165

Indian Ocean

	m	ft
Java Trench	7,450	24,442
Red Sea	2,635	8,454

Pacific Ocean

	m	ft
Mariana Trench	11,022	36,161
Tonga Trench	10,882	35,702
Japan Trench	10,554	34,626
Kuril Trench	10,542	34,587

Arctic Ocean

	m	ft
Molloy Deep	5,608	18,399

Mountains

Europe

		m	ft
Mont Blanc	*France/Italy*	4,807	15,771
Monte Rosa	*Italy/Switzerland*	4,634	15,203
Dom	*Switzerland*	4,545	14,911
Liskamm	*Switzerland*	4,527	14,852
Weisshorn	*Switzerland*	4,505	14,780
Taschorn	*Switzerland*	4,490	14,730
Matterhorn/Cervino	*Italy/Switzerland*	4,478	14,691
Mont Maudit	*France/Italy*	4,465	14,649
Dent Blanche	*Switzerland*	4,356	14,291
Nadelhorn	*Switzerland*	4,327	14,196
Grandes Jorasses	*France/Italy*	4,208	13,806
Jungfrau	*Switzerland*	4,158	13,642
Grossglockner	*Austria*	3,797	12,457
Mulhacén	*Spain*	3,478	11,411
Zugspitze	*Germany*	2,962	9,718
Olympus	*Greece*	2,917	9,570
Triglav	*Slovenia*	2,863	9,393
Gerlachovka	*Slovak Republic*	2,655	8,711
Galdhöpiggen	*Norway*	2,468	8,100
Kebnekaise	*Sweden*	2,117	6,946
Ben Nevis	*UK*	1,343	4,406

Asia

		m	ft
Everest	*China/Nepal*	8,848	29,029
K2 (Godwin Austen)	*China/Kashmir*	8,611	28,251
Kanchenjunga	*India/Nepal*	8,598	28,208
Lhotse	*China/Nepal*	8,516	27,939
Makalu	*China/Nepal*	8,481	27,824
Cho Oyu	*China/Nepal*	8,201	26,906
Dhaulagiri	*Nepal*	8,172	26,811
Manaslu	*Nepal*	8,156	26,758
Nanga Parbat	*Kashmir*	8,126	26,660
Annapurna	*Nepal*	8,078	26,502
Gasherbrum	*China/Kashmir*	8,068	26,469
Broad Peak	*China/Kashmir*	8,051	26,414
Xixabangma	*China*	8,012	26,286
Kangbachen	*India/Nepal*	7,902	25,925
Trivor	*Pakistan*	7,720	25,328
Pik Kommunizma	*Tajikistan*	7,495	24,590
Elbrus	*Russia*	5,642	18,510
Demavend	*Iran*	5,604	18,386
Ararat	*Turkey*	5,165	16,945
Gunong Kinabalu	*Malaysia (Borneo)*	4,101	13,455
Fuji-San	*Japan*	3,776	12,388

Africa

		m	ft
Kilimanjaro	*Tanzania*	5,895	19,340
Mt Kenya	*Kenya*	5,199	17,057
Ruwenzori (Margherita)	*Uganda/Zaïre*	5,109	16,762
Ras Dashan	*Ethiopia*	4,620	15,157
Meru	*Tanzania*	4,565	14,977
Karisimbi	*Rwanda/Zaïre*	4,507	14,787
Mt Elgon	*Kenya/Uganda*	4,321	14,176
Batu	*Ethiopia*	4,307	14,130
Toubkal	*Morocco*	4,165	13,665
Mt Cameroon	*Cameroon*	4,070	13,353

Oceania

		m	ft
Puncak Jaya	*Indonesia*	5,029	16,499
Puncak Trikora	*Indonesia*	4,750	15,584
Puncak Mandala	*Indonesia*	4,702	15,427
Mt Wilhelm	*Papua New Guinea*	4,508	14,790
Mauna Kea	*USA (Hawaii)*	4,205	13,796
Mauna Loa	*USA (Hawaii)*	4,170	13,681
Mt Cook	*New Zealand*	3,753	12,313
Mt Kosciusko	*Australia*	2,237	7,339

North America

		m	ft
Mt McKinley (Denali)	*USA (Alaska)*	6,194	20,321
Mt Logan	*Canada*	5,959	19,551
Citlaltepetl	*Mexico*	5,700	18,701
Mt St Elias	*USA/Canada*	5,489	18,008
Popocatepetl	*Mexico*	5,452	17,887
Mt Foraker	*USA (Alaska)*	5,304	17,401
Ixtaccihuatl	*Mexico*	5,286	17,342
Lucania	*Canada*	5,227	17,149
Mt Steele	*Canada*	5,073	16,644
Mt Bona	*USA (Alaska)*	5,005	16,420
Mt Whitney	*USA*	4,418	14,495
Tajumulco	*Guatemala*	4,220	13,845
Chirripó Grande	*Costa Rica*	3,837	12,589
Pico Duarte	*Dominican Rep.*	3,175	10,417

South America

		m	ft
Aconcagua	*Argentina*	6,960	22,834
Bonete	*Argentina*	6,872	22,546
Ojos del Salado	*Argentina/Chile*	6,863	22,516
Pissis	*Argentina*	6,779	22,241
Mercedario	*Argentina/Chile*	6,770	22,211
Huascaran	*Peru*	6,768	22,204
Llullaillaco	*Argentina/Chile*	6,723	22,057
Nudo de Cachi	*Argentina*	6,720	22,047
Yerupaja	*Peru*	6,632	21,758
Sajama	*Bolivia*	6,542	21,463
Chimborazo	*Ecuador*	6,267	20,561
Pico Colon	*Colombia*	5,800	19,029
Pico Bolivar	*Venezuela*	5,007	16,427

Antarctica

	m	ft
Vinson Massif	4,897	16,066
Mt Kirkpatrick	4,528	14,855

Rivers

Europe

		km	miles
Volga	*Caspian Sea*	3,700	2,300
Danube	*Black Sea*	2,850	1,770
Ural	*Caspian Sea*	2,535	1,575
Dnepr (Dnipro)	*Volga*	2,285	1,420
Kama	*Volga*	2,030	1,260
Don	*Volga*	1,990	1,240
Petchora	*Arctic Ocean*	1,790	1,110
Oka	*Volga*	1,480	920
Dnister (Dniester)	*Black Sea*	1,400	870
Vyatka	*Kama*	1,370	850
Rhine	*North Sea*	1,320	820
N. Dvina	*Arctic Ocean*	1,290	800
Elbe	*North Sea*	1,145	710

Asia

		km	miles
Yangtze	*Pacific Ocean*	6,380	3,960
Yenisey–Angara	*Arctic Ocean*	5,550	3,445
Huang He	*Pacific Ocean*	5,464	3,395
Ob–Irtysh	*Arctic Ocean*	5,410	3,360
Mekong	*Pacific Ocean*	4,500	2,795
Amur	*Pacific Ocean*	4,400	2,730
Lena	*Arctic Ocean*	4,400	2,730
Irtysh	*Ob*	4,250	2,640
Yenisey	*Arctic Ocean*	4,090	2,540
Ob	*Arctic Ocean*	3,680	2,285
Indus	*Indian Ocean*	3,100	1,925
Brahmaputra	*Indian Ocean*	2,900	1,800
Syrdarya	*Aral Sea*	2,860	1,775
Salween	*Indian Ocean*	2,800	1,740
Euphrates	*Indian Ocean*	2,700	1,675
Amudarya	*Aral Sea*	2,540	1,575

Africa

		km	miles
Nile	*Mediterranean*	6,670	4,140
Zaïre/Congo	*Atlantic Ocean*	4,670	2,900
Niger	*Atlantic Ocean*	4,180	2,595
Zambezi	*Indian Ocean*	3,540	2,200
Oubangi/Uele	*Zaïre*	2,250	1,400
Kasai	*Zaïre*	1,950	1,210
Shaballe	*Indian Ocean*	1,930	1,200
Orange	*Atlantic Ocean*	1,860	1,155
Cubango	*Okavango Swamps*	1,800	1,120
Limpopo	*Indian Ocean*	1,600	995
Senegal	*Atlantic Ocean*	1,600	995

Australia

		km	miles
Murray–Darling	*Indian Ocean*	3,750	2,330
Darling	*Murray*	3,070	1,905
Murray	*Indian Ocean*	2,575	1,600
Murrumbidgee	*Murray*	1,690	1,050

North America

		km	miles
Mississippi–Missouri	*Gulf of Mexico*	6,020	3,740
Mackenzie	*Arctic Ocean*	4,240	2,630
Mississippi	*Gulf of Mexico*	3,780	2,350
Missouri	*Mississippi*	3,780	2,350
Yukon	*Pacific Ocean*	3,185	1,980
Rio Grande	*Gulf of Mexico*	3,030	1,880
Arkansas	*Mississippi*	2,340	1,450
Colorado	*Pacific Ocean*	2,330	1,445
Red	*Mississippi*	2,040	1,270
Columbia	*Pacific Ocean*	1,950	1,210
Saskatchewan	*Lake Winnipeg*	1,940	1,205

South America

		km	miles
Amazon	*Atlantic Ocean*	6,450	4,010
Paraná–Plate	*Atlantic Ocean*	4,500	2,800
Purus	*Amazon*	3,350	2,080
Madeira	*Amazon*	3,200	1,990
São Francisco	*Atlantic Ocean*	2,900	1,800
Paraná	*Plate*	2,800	1,740
Tocantins	*Atlantic Ocean*	2,750	1,710
Paraguay	*Paraná*	2,550	1,580
Orinoco	*Atlantic Ocean*	2,500	1,550
Pilcomayo	*Paraná*	2,500	1,550
Araguaia	*Tocantins*	2,250	1,400

Lakes

Europe

		km^2	$miles^2$
Lake Ladoga	*Russia*	17,700	6,800
Lake Onega	*Russia*	9,700	3,700
Saimaa system	*Finland*	8,000	3,100
Vänern	*Sweden*	5,500	2,100

Asia

		km^2	$miles^2$
Caspian Sea	*Asia*	371,800	143,550
Aral Sea	*Kazakstan/Uzbekistan*	33,640	13,000
Lake Baykal	*Russia*	30,500	11,780
Tonlé Sap	*Cambodia*	20,000	7,700
Lake Balqash	*Kazakstan*	18,500	7,100

Africa

		km^2	$miles^2$
Lake Victoria	*East Africa*	68,000	26,000
Lake Tanganyika	*Central Africa*	33,000	13,000
Lake Malawi/Nyasa	*East Africa*	29,600	11,430
Lake Chad	*Central Africa*	25,000	9,700
Lake Turkana	*Ethiopia/Kenya*	8,500	3,300
Lake Volta	*Ghana*	8,500	3,300

Australia

		km^2	$miles^2$
Lake Eyre	*Australia*	8,900	3,400
Lake Torrens	*Australia*	5,800	2,200
Lake Gairdner	*Australia*	4,800	1,900

North America

		km^2	$miles^2$
Lake Superior	*Canada/USA*	82,350	31,800
Lake Huron	*Canada/USA*	59,600	23,010
Lake Michigan	*USA*	58,000	22,400
Great Bear Lake	*Canada*	31,800	12,280
Great Slave Lake	*Canada*	28,500	11,000
Lake Erie	*Canada/USA*	25,700	9,900
Lake Winnipeg	*Canada*	24,400	9,400
Lake Ontario	*Canada/USA*	19,500	7,500
Lake Nicaragua	*Nicaragua*	8,200	3,200

South America

		km^2	$miles^2$
Lake Titicaca	*Bolivia/Peru*	8,300	3,200
Lake Poopo	*Peru*	2,800	1,100

Islands

Europe

		km^2	$miles^2$
Great Britain	*UK*	229,880	88,700
Iceland	*Atlantic Ocean*	103,000	39,800
Ireland	*Ireland/UK*	84,400	32,600
Novaya Zemlya (N.)	*Russia*	48,200	18,600
Sicily	*Italy*	25,500	9,800
Corsica	*France*	8,700	3,400

Asia

		km^2	$miles^2$
Borneo	*Southeast Asia*	744,360	287,400
Sumatra	*Indonesia*	473,600	182,860
Honshu	*Japan*	230,500	88,980
Celebes	*Indonesia*	189,000	73,000
Java	*Indonesia*	126,700	48,900
Luzon	*Philippines*	104,700	40,400
Hokkaido	*Japan*	78,400	30,300

Africa

		km^2	$miles^2$
Madagascar	*Indian Ocean*	587,040	226,660
Socotra	*Indian Ocean*	3,600	1,400
Réunion	*Indian Ocean*	2,500	965

Oceania

		km^2	$miles^2$
New Guinea	*Indonesia/Papua NG*	821,030	317,000
New Zealand (S.)	*Pacific Ocean*	150,500	58,100
New Zealand (N.)	*Pacific Ocean*	114,700	44,300
Tasmania	*Australia*	67,800	26,200
Hawaii	*Pacific Ocean*	10,450	4,000

North America

		km^2	$miles^2$
Greenland	*Atlantic Ocean*	2,175,600	839,800
Baffin Is.	*Canada*	508,000	196,100
Victoria Is.	*Canada*	212,200	81,900
Ellesmere Is.	*Canada*	212,000	81,800
Cuba	*Caribbean Sea*	110,860	42,800
Hispaniola	*Dominican Rep./Haiti*	76,200	29,400
Jamaica	*Caribbean Sea*	11,400	4,400
Puerto Rico	*Atlantic Ocean*	8,900	3,400

South America

		km^2	$miles^2$
Tierra del Fuego	*Argentina/Chile*	47,000	18,100
Falkland Is. (E.)	*Atlantic Ocean*	6,800	2,600

Map Projections

MAP PROJECTIONS

A map projection is the systematic depiction on a plane surface of the imaginary lines of latitude or longitude from a globe of the earth. This network of lines is called the graticule and forms the framework upon which an accurate depiction of the earth is made. The map graticule, which is the basis of any map, is constructed sometimes by graphical means, but often by using mathematical formulae to give the intersections of the graticule plotted as x and y co-ordinates. The choice between projections is based upon which properties the cartographer wishes the map to possess, the map scale and also the extent of the area to be mapped. Since the globe is three dimensional, it is not possible to depict its surface on a two dimensional plane without distortion. Preservation of one of the basic properties listed below can only be secured at the expense of the others and the choice of projection is often a compromise solution.

Correct Area

In these projections the areas from the globe are to scale on the map. For example, if you look at the diagram at the top right, areas of 10° x 10° are shown from the equator to the poles. The proportion of this area at the extremities are approximately 11:1. An equal area projection will retain that proportion in its portrayal of those areas. This is particularly useful in the mapping of densities and distributions. Projections with this property are termed **Equal Area, Equivalent or Homolographic.**

Correct Distance

In these projections the scale is correct along the meridians, or in the case of the Azimuthal Equidistant scale is true along any line drawn from the centre of the projection. They are called **Equidistant**.

Correct Shape

This property can only be true within small areas as it is achieved only by having a uniform scale distortion along both x and y axes of the projection. The projections are called **Conformal** or **Orthomorphic.**

In order to minimise the distortions at the edges of some projections, central portions of them are often selected for atlas maps. Below are listed some of the major types of projection.

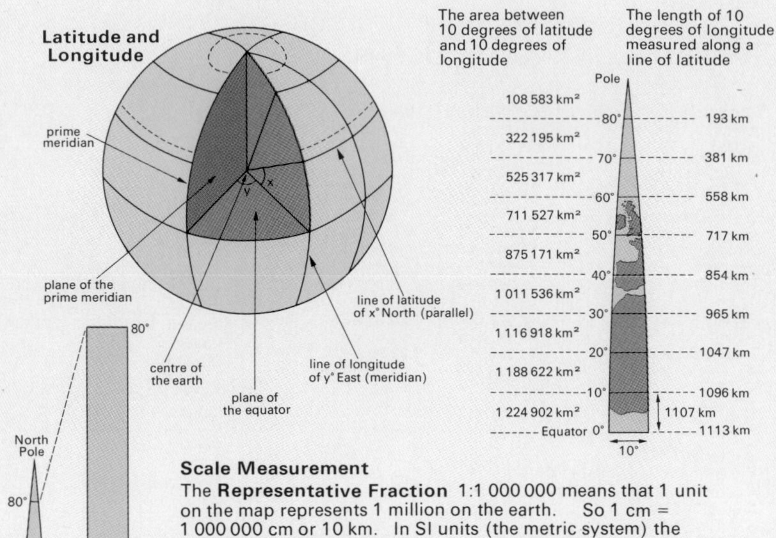

Scale Measurement

The **Representative Fraction** 1:1 000 000 means that 1 unit on the map represents 1 million on the earth. So 1 cm = 1 000 000 cm or 10 km. In SI units (the metric system) the Representative Fraction can be converted to the scale of 1 cm by moving the decimal point through five places, usually deleting the last five zeros thus :- 1: 20 000 000 – 1cm = 200 km. Scale on map projections can only be correct along specific meridians and parallels or from one or two specific points. To take an extreme example of scale distortion, the diagram on the left shows how a portion of the earth's surface, when unwrapped from around a globe (**a gore**) compares with the same area at the same equatorial scale but on Mercator's Projection. The enlargement of scale away from the equator is considerable and for this reason a variable scale for each line of latitude is given (see below).

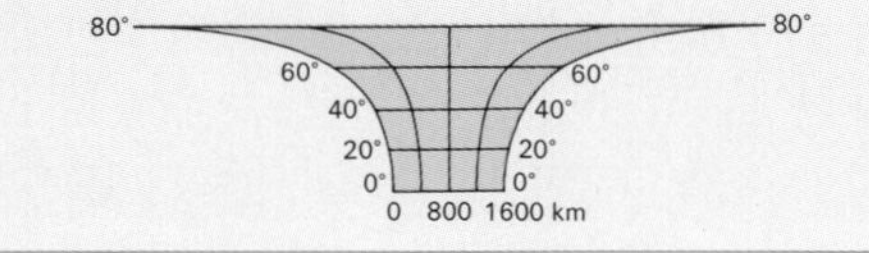

AZIMUTHAL OR ZENITHAL PROJECTIONS

These are constructed by the projection of part of the graticule from the globe onto a plane tangential to any single point on it. This plane may be tangential to the equator (**equatorial case**), the poles (**polar case**) or any other point (**oblique case**). Any straight line drawn from the point at which the plane touches the globe is the shortest distance from that point and is known as a **great circle**. In its **Gnomonic** construction *any* straight line on the map is a great circle, but there is great exaggeration towards the edges and this reduces its general uses. There are five different ways of transferring the graticule onto the plane and these are shown on the right. The central diagram below shows how the graticules vary, using the polar case as the example.

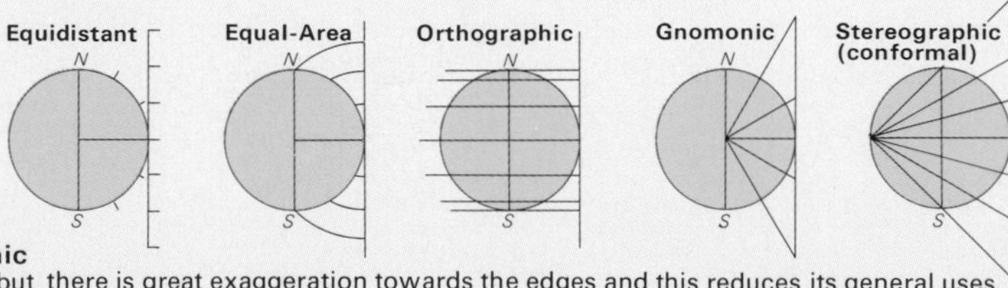

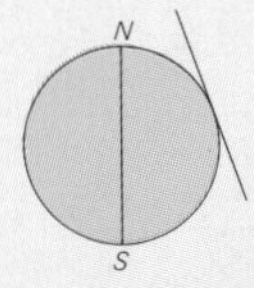

Oblique Case

The plane touches the globe at any point between the equator and poles. The oblique orthographic uses the distortion in azimuthal projections away from the centre to give a graphic depiction of the earth as seen from any desired point in space. It can also be used in both Polar and Equatorial cases. It is used not only for the earth but also for the moon and planets.

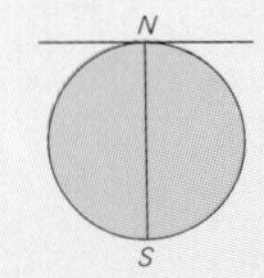

Polar Case

The polar case is the simplest to construct and the diagram below shows the differing effects of all five methods of construction comparing their coverage, distortion etc., using North America as the example.

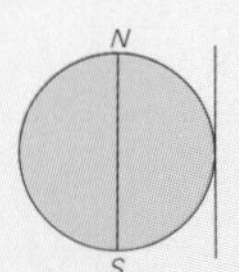

Equatorial Case

The example shown here is Lambert's Equivalent Azimuthal. It is the only projection which is both equal area and where bearing is true from the centre.

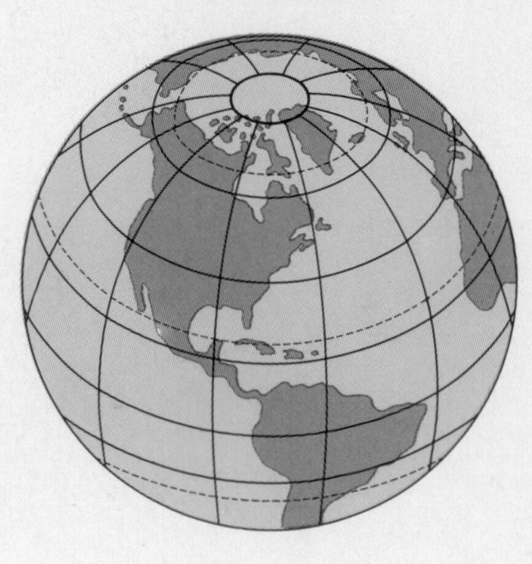

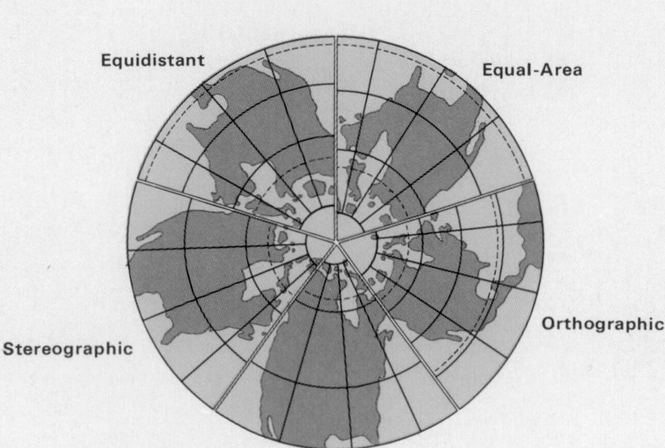

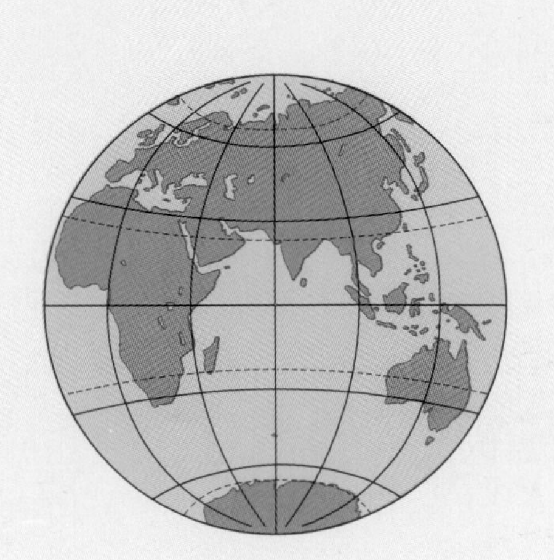

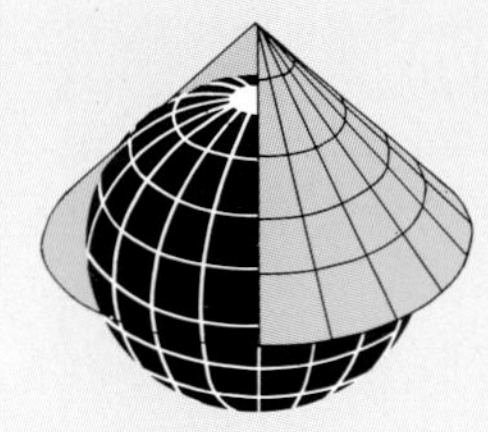

CONICAL PROJECTIONS

These use the projection of the graticule from the globe onto a cone which is tangential to a line of latitude (termed the **standard parallel**). This line is always an arc and scale is always true along it. Because of its method of construction it is used mainly for depicting the temperate latitudes around the standard parallel i.e. where there is least distortion. To reduce the distortion and include a larger range of latitudes, the projection may be constructed with the cone bisecting the surface of the globe so that there are two standard parallels each of which is true to scale. The distortion is thus spread more evenly between the two chosen parallels.

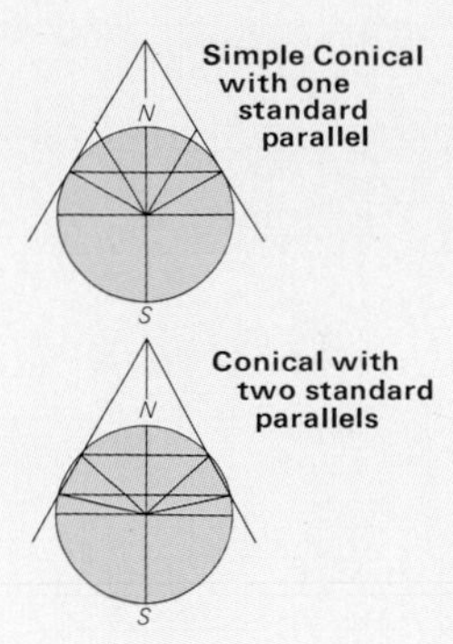

Bonne

This is a modification of the simple conic whereby the true scale along the meridians is sacrificed to enable the accurate representation of areas. However scale is true along each parallel but shapes are distorted at the edges.

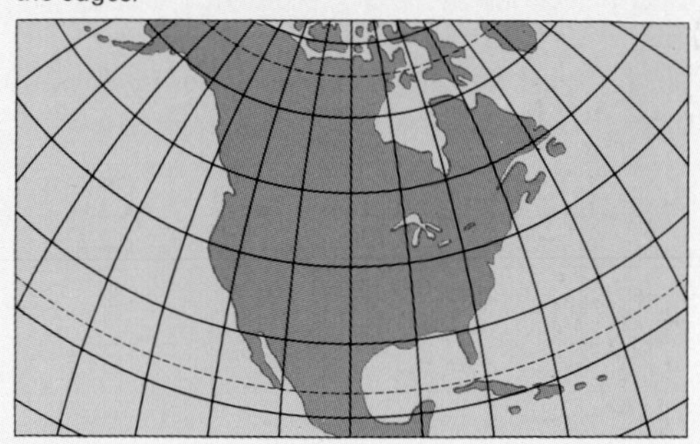

Simple Conic

Scale is correct not only along the standard parallel but also along all meridians. The selection of the standard parallel used is crucial because of the distortion away from it. The projection is usually used to portray regions or continents at small scales.

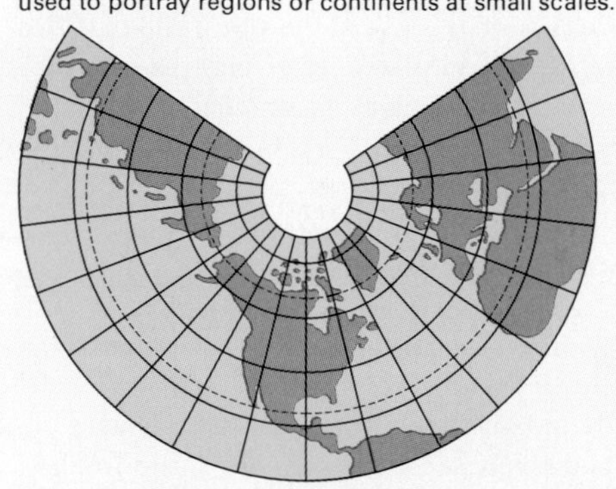

Lambert's Conformal Conic

This projection uses two standard parallels but instead of being equal area as Albers, it is Conformal. Because it has comparatively small distortion, direction and distances can be readily measured and it is therefore used for some navigational charts.

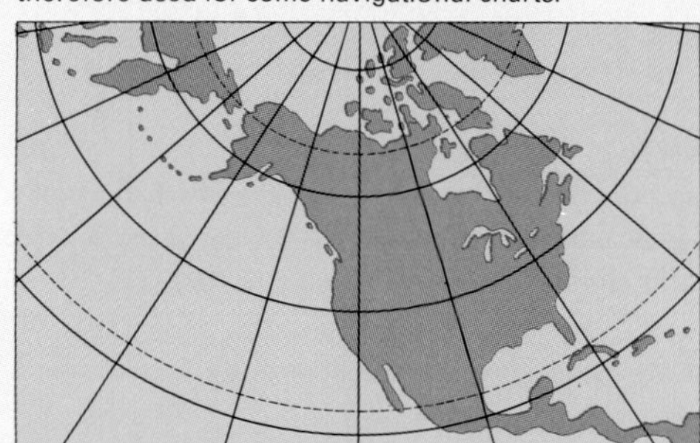

Albers Conical Equal Area

This projection uses two standard parallels and once again the selection of the two specific ones relative to the land area to be mapped is very important. It is equal area and is especially useful for large land masses oriented East-West, for example the U.S.A.

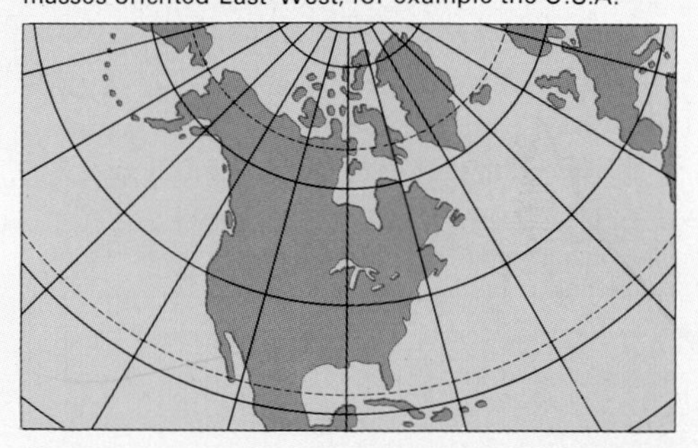

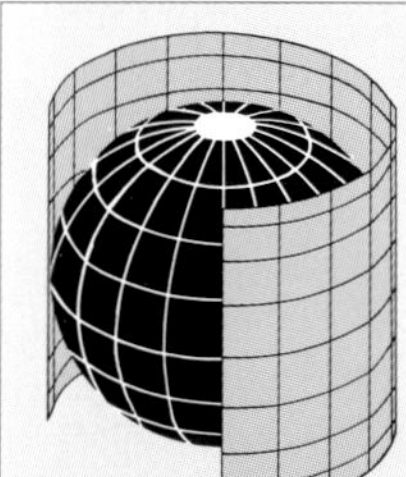

CYLINDRICAL AND OTHER WORLD PROJECTIONS

This group of projections are those which permit the whole of the Earth's surface to be depicted on one map. They are a very large group of projections and the following are only a few of them. Cylindrical projections are constructed by the projection of the graticule from the globe onto a cylinder tangential to the globe. In the examples shown here the cylinder touches the equator, but it can be moved through 90° so it touches the poles - this is called the **Transverse Aspect**. If the cylinder is twisted so that it touches anywhere between the equator and poles it is called the **Oblique Aspect.** Although cylindrical projections can depict all the main land masses, there is considerable distortion of shape and area towards the poles. One cylindrical projection, **Mercator** overcomes this shortcoming by possessing the unique navigational property that any straight drawn on it is a line of constant bearing (**loxodrome**), i.e. a straight line route on the globe crosses the parallels and meridians on the map at the same angles as on the globe. It is used for maps and charts between 15° either side of the equator. Beyond this enlargement of area is a serious drawback, although it is used for navigational charts at all latitudes.

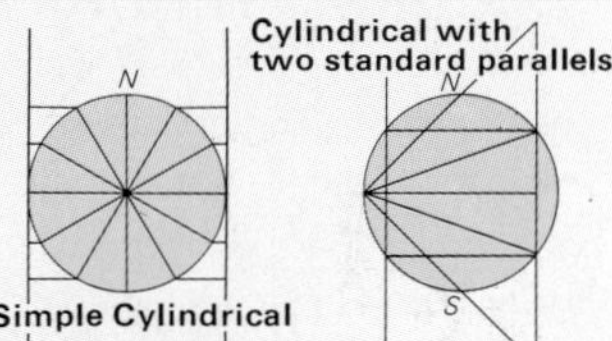

Mercator

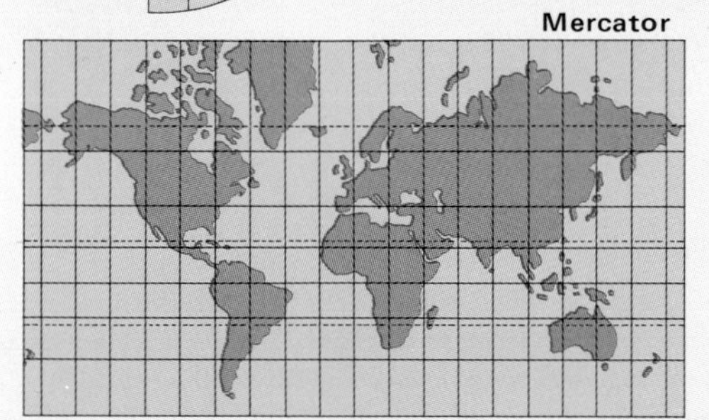

Mollweide

Sanson-Flamsteed

Mollweide and Sanson-Flamsteed

Both of these projections are termed **pseudo-cylindrical**. They are basically cylindrical projections where parallels have been progressively shortened and drawn to scale towards the poles. This allows them to overcome the gross distortions exhibited by the ordinary cylindrical projections and they are in fact Equal Area, Mollweide's giving a slightly better shape. To improve the shape of the continents still further they, like some other projections can be **Interrupted** as can be seen below, but this is at the expense of contiguous sea areas. These projections can have any central meridian and so can be 'centred' on the Atlantic, Pacific, Asia, America etc. In this form both projections are suitable for any form of mapping statistical distributions.

Hammer

This is not a cylindrical projection, but is developed from the Lambert Azimuthal Equal Area by doubling all the East-West distances along the parallels from the central meridian. Like both Sanson–Flamsteed and Mollweide it is distorted towards its edges but has curved parallels to lessen the distortion.

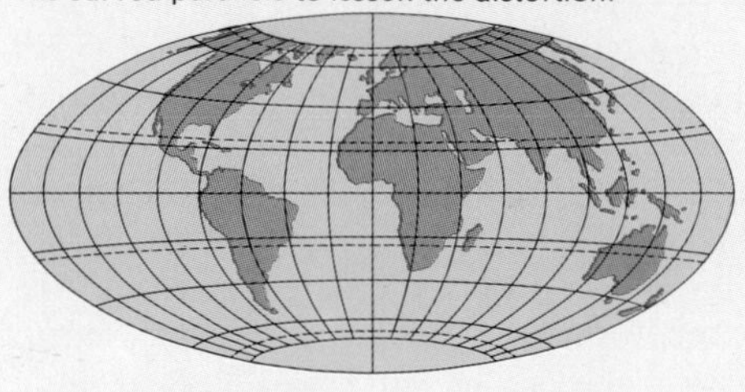

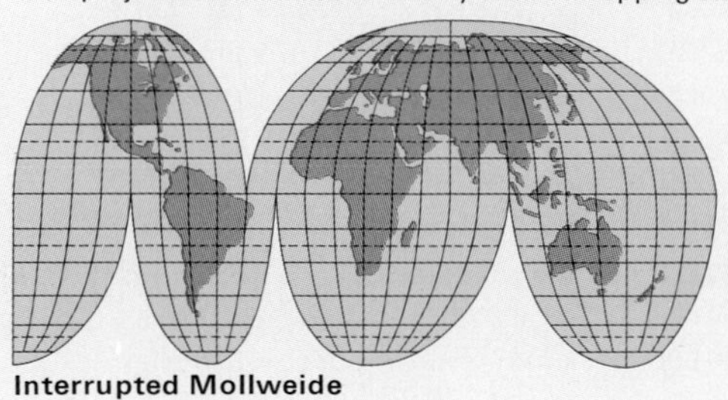

Interrupted Mollweide

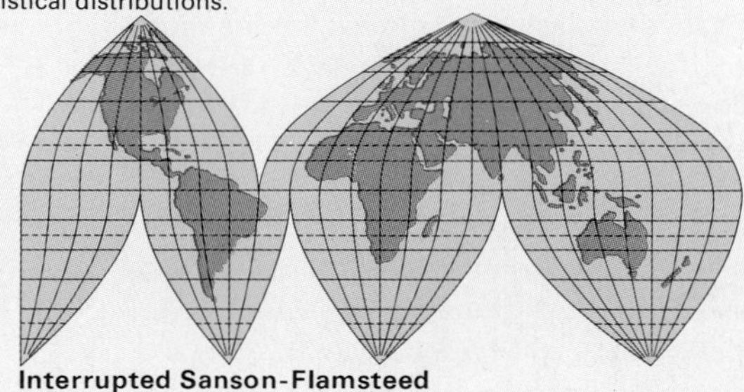

Interrupted Sanson-Flamsteed

User Guide

Organization of the atlas

Prepared in accordance with the highest standards in cartography to provide accurate and detailed representation of the earth, the atlas is made up of four separate sections and is organized with ease of use in mind.

The first section of the atlas consists of up-to-date geographical and demographical statistics for all the countries in the world, graphics on map projections intended to help the reader understand how cartographers create and use map projections, and this user guide.

The second section of the atlas, the 32-page United States Maps section, has blue page borders and offers comprehensive coverage of the United States and its outlying areas, with climate and agricultural maps, politically colored maps with some topographical detail, maps of major urban areas, and a 16-page index with latitude and longitude coordinates.

The third section of the atlas, the informative 32-page Introduction to World Geography section, consists of thematic maps, graphs, tables, and charts on a wide range of geographical and demographical topics, followed by a subject index.

The fourth and final section of the atlas, the 96-page World Maps section, has gray page borders and covers the earth continent by continent in the classic sequence adopted by cartographers since the 16th century. This section begins with Europe, then Asia, Africa, Australia and Oceania, North America, and South America. For each continent, there are maps at a variety of scales: first, physical relief maps and political maps of the whole continent, then large scale maps of the most important or densely populated areas.

The governing principle is that by turning the pages of the World Maps section, the reader moves steadily from north to south through each continent, with each map overlapping its neighbors. Immediately following the maps in the World Maps section is the comprehensive index to the maps, which contains 44,000 entries of both place names and geographical features. The index provides the latitude and longitude coordinates as well as letters and numbers, so that locating any site can be accomplished with speed and accuracy.

Map presentation

All of the maps in the atlas are drawn with north at the top (except for two maps: the map of the Arctic Ocean and the map of Antarctica). The maps in the United States Maps section and the World Maps section all contain the following information in their borders: the map title; scale; the projection used; the degrees of latitude and longitude; and on the physical relief maps, a height and depth reference panel identifying the colors used for each layer of contouring. In addition to this information, the maps in the World Maps section also contain locator diagrams which show the area covered, the page numbers for adjacent maps, and the letters and numbers used in the index for locating place names and geographical features.

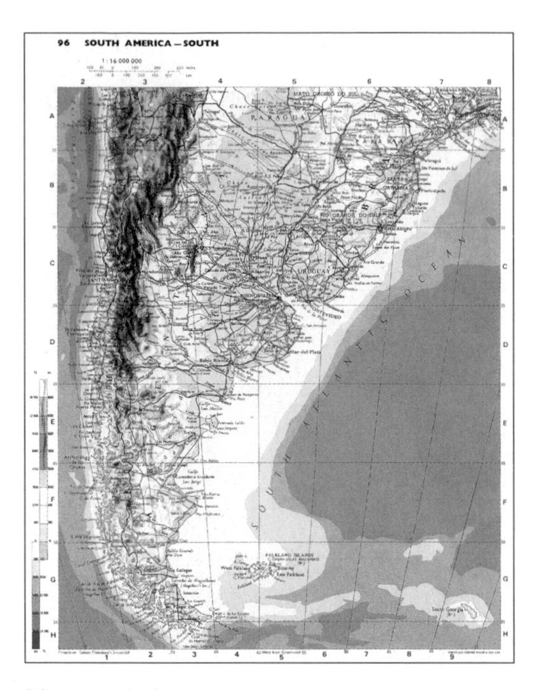

Map symbols

Each map contains a vast amount of detail which is conveyed clearly and accurately by the use of symbols. Points and circles of varying sizes locate and identify the relative importance of towns and cities; different styles of type are employed for administrative, geographical and regional place names. A variety of pictorial symbols denote landscape features such as glaciers, marshes and reefs, and man-made structures including roads, railroads, airports, canals, and dams. International borders are shown by red lines. Where neighboring countries are in dispute, the maps show the *de facto* boundary between nations, regardless of the legal or historical situation. The symbols are explained on the first page of each of the map sections.

Map scales

The scale of each map is given in the numerical form known as the representative fraction. The first figure is always one, signifying one unit of distance on the map; the second figure, usually in millions, is the number by which the map unit must be multiplied to give the equivalent distance on the earth's surface. Calculations can easily be made in centimeters and kilometers, by dividing the earth units figure by 100 000 (i.e. deleting the last five 0s).

LARGE SCALE

1:1 000 000	1 cm = 10 km	1 inch = 16 miles
1:2 500 000	1 cm = 25 km	1 inch = 39.5 miles
1:5 000 000	1 cm = 50 km	1 inch = 79 miles
1:6 000 000	1 cm = 60 km	1 inch = 95 miles
1:8 000 000	1 cm = 80 km	1 inch = 126 miles
1:10 000 000	1 cm = 100km	1 inch = 158 miles
1:15 000 000	1 cm = 150 km	1 inch = 237 miles
1:20 000 000	1 cm = 200 km	1 inch = 316 miles
1:50 000 000	1 cm = 500 km	1 inch = 790 miles

SMALL SCALE

Thus 1:1 000 000 means 1 cm = 10 km. The calculation for inches and miles is more laborious, but 1 000 000 divided by 63 360 (the number of inches in a mile) shows that 1:1 000 000 means about 1 inch = 16 miles.

Measuring distances

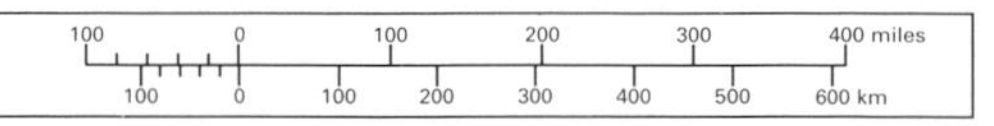

Although each map is accompanied by a scale bar, distances cannot always be measured with confidence because of the distortions involved in portraying the curved surface of the earth on a flat page. As a general rule, the larger the map scale (i.e. the lower the number of earth units in the representative fraction), the more accurate and reliable will be the distance measured. On small scale maps such as those of the world and of entire continents, measurement may only be accurate along the standard parallels, or central axes, and should not be attempted without considering the map projection.

Latitude and longitude

Accurate positioning of individual points on the earth's surface is made possible by reference to the geometrical system of latitude and longitude. Latitude parallels are drawn west–east around the earth and numbered by degrees north and south of the Equator, which is designated 0° of latitude. Longitude meridians are drawn north–south and numbered by degrees east and west of the Prime Meridian, 0° of longitude, which passes through Greenwich in England. By referring to these co-ordinates and their subdivisions of minutes (1/60th of a degree) and seconds (1/60th of a minute), any place on earth can be located to within a few hundred meters. Latitude and longitude are indicated by blue lines on the maps; they are straight or curved according to the projection employed. Reference to these lines is the easiest way of determining the relative positions of places on different maps, and for plotting compass directions.

Name forms

For ease of reference, both English and local name forms appear in the atlas. Oceans, seas and countries are shown in English throughout the atlas; country names may be abbreviated to their commonly accepted form. English conventional forms are also used for place names on the continental maps. However, local name forms are used on all large scale and regional maps, with the English form given in brackets only for important cities – the large-scale map of Russia and Central Asia thus shows Moskva (Moscow). For countries which do not use a Roman script, place names have been transcribed according to the systems adopted by the British and US Geographic Names Authorities. For China, the Pin Yin system has been used, with some more widely known forms appearing in brackets, as with Beijing (Peking). Both English and local names appear in the index.

UNITED STATES MAPS

SETTLEMENTS

WASHINGTON D.C. Tampa Fresno Waterloo *Ventura* *Barstow* *Blythe* *Hope*

Settlement symbols and type styles vary according to the scale of each map and indicate the importance of towns on the map rather than specific population figures

ADMINISTRATION

International Boundaries

Internal Boundaries

National Parks, Recreation Areas and Monuments

Country Names

CANADA

Administrative Area Names

MICHIGAN

COMMUNICATIONS

Major Highways

Other Principal Roads

Passes

Airports and Airfields

Principal Railroads

Railroads Under Construction

Other Railroads

Railroad Tunnels

Principal Canals

PHYSICAL FEATURES

Perennial Streams

Intermittent Streams

Perennial Lakes and Reservoirs

Intermittent Lakes and Salt Flats

Swamps and Marshes

Permanent Ice and Glaciers

8848 Elevations in meters

8050 Sea Depths in meters

1134 Height of Lake Surface Above Sea Level in meters

1 meter is approx. 3.3 feet

CITY MAPS

In addition to, or instead of, the symbols explained above, the following symbols are used on the city maps between pages 20-29

Urban Areas

Limited Access Roads

Woodland and Parks

Secondary Roads

Ferry Routes

Airports

Canals

Aqueducts

State Boundaries

County Boundaries

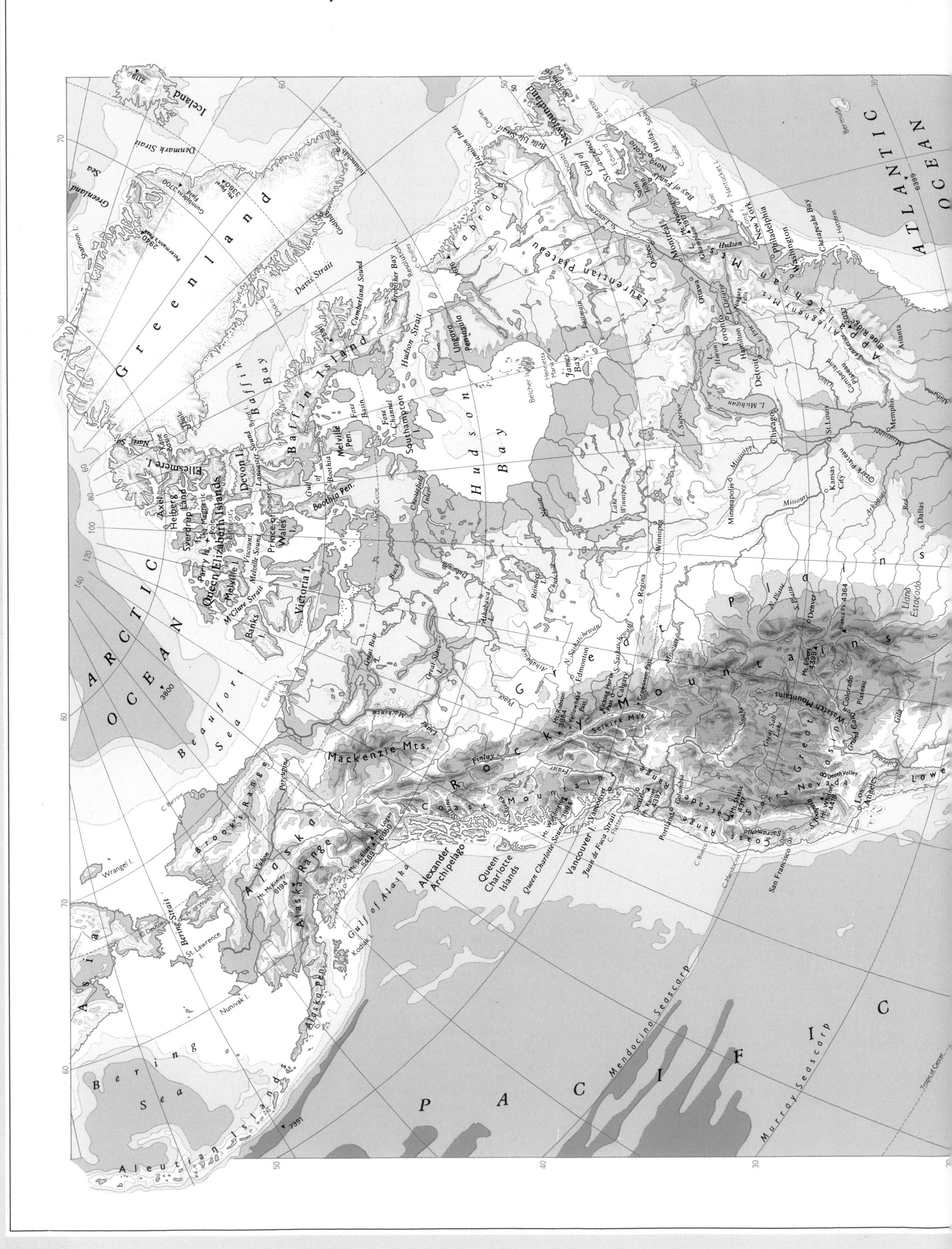
Greenland
Baffin Bay
Baffin Island
Hudson Bay
Labrador
Laurentian Plateau
ARCTIC OCEAN
Beaufort Sea
Mackenzie Mts.
Alaska Range
Brooks Range
Rocky Mountains
Great Plains
Great Basin
Coast Mountains
Cascade Range
Sierra Nevada
Appalachian Mts.
ATLANTIC OCEAN
PACIFIC
Bering Sea
Aleutian Islands
Gulf of Alaska
Mendocino Seascarp
Murray Seascarp
Victoria I.
Banks I.
Queen Elizabeth Islands
Ellesmere I.
Devon I.
Southampton I.
Iceland
Newfoundland
James Bay
Hudson Strait
Davis Strait
Alexander Archipelago
Queen Charlotte Islands
Vancouver I.
Chicago
St. Louis
New York
Philadelphia
Washington
Denver
Los Angeles
San Francisco
Winnipeg
Edmonton
Calgary
Toronto
Montreal
Quebec

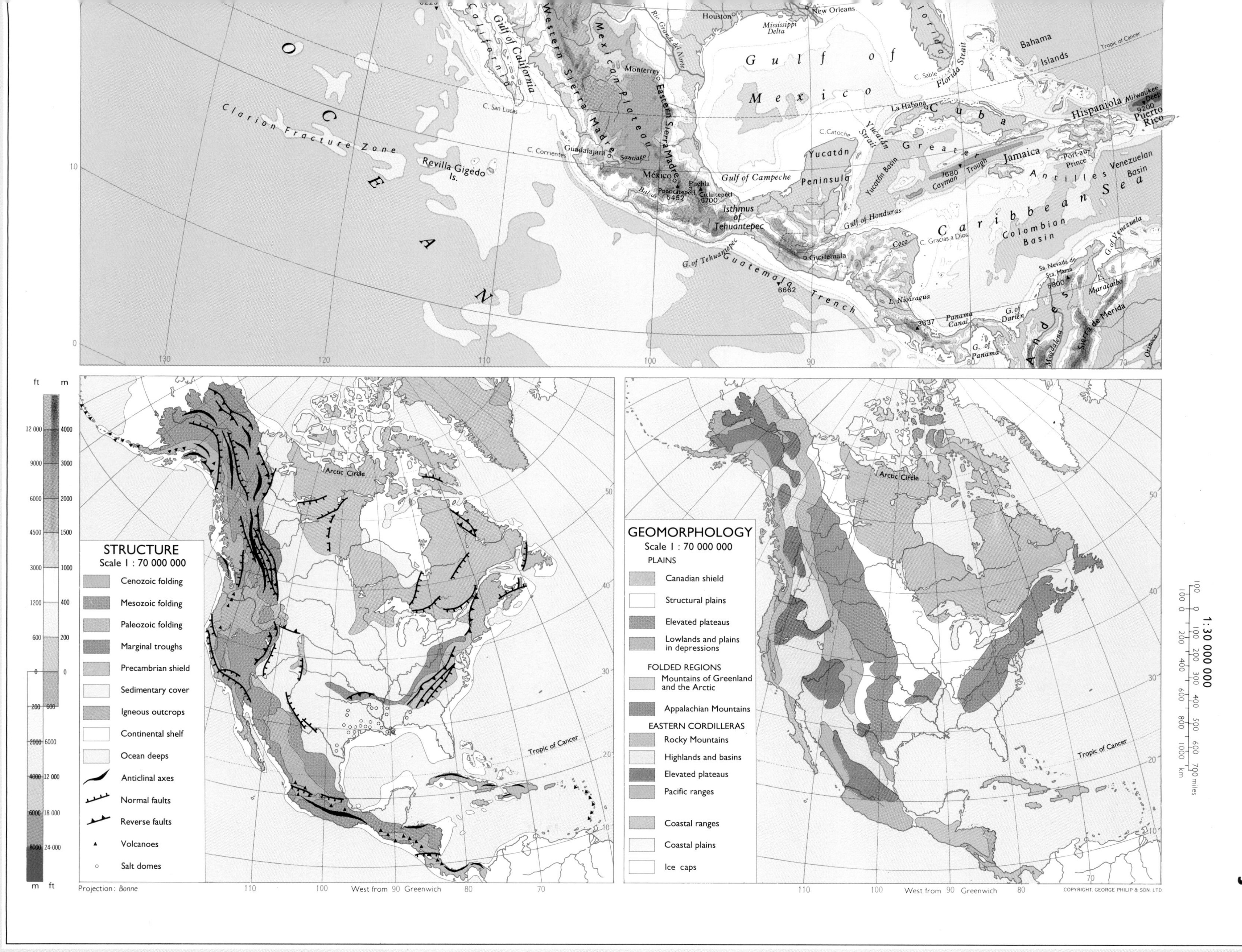

1:30 000 000
miles 100 0 100 200 300 400 500 600 700
km 100 0 200 400 600 800 1000
Gulf of Mexico
Houston
New Orleans
Mississippi Delta
Rio Grande del Norte
Florida Strait
C. Sable
Bahama Islands
Tropic of Cancer
Gulf of California
California
C. San Lucas
Western Sierra Madre
Mexican Plateau
Eastern Sierra Madre
Monterrey
Guadalajara
Santiago
México
Balsas
Puebla
Popocatepetl 5452
Citlaltepetl 5700
C. Corrientes
Revilla Gigedo Is.
Clarion Fracture Zone
OCEAN
Gulf of Campeche
Isthmus of Tehuantepec
G. of Tehuantepec
Guatemala Trench
6662
Guatemala
Yucatán Peninsula
C. Catoche
Yucatán Strait
Yucatán Basin
Gulf of Honduras
La Habana
Cuba
Greater Antilles
Cayman Trough
7680
Jamaica
Port-au-Prince
Hispaniola
Milwaukee Deep
9200
Puerto Rico
Venezuelan Basin
Caribbean Sea
Colombian Basin
C. Gracias a Dios
Coco
L. Nicaragua
3837
Panama Canal
G. of Panama
G. of Darién
Sa. Nevada de Sta. Marta
5800
G. of Venezuela
Maracaibo
Sierra de Merida
Andes
Magdalena
Orinoco
STRUCTURE
Scale 1 : 70 000 000
Cenozoic folding
Mesozoic folding
Paleozoic folding
Marginal troughs
Precambrian shield
Sedimentary cover
Igneous outcrops
Continental shelf
Ocean deeps
Anticlinal axes
Normal faults
Reverse faults
Volcanoes
Salt domes
Arctic Circle
Projection: Bonne
West from 90 Greenwich
GEOMORPHOLOGY
Scale 1 : 70 000 000
PLAINS
Canadian shield
Structural plains
Elevated plateaus
Lowlands and plains in depressions
FOLDED REGIONS
Mountains of Greenland and the Arctic
Appalachian Mountains
EASTERN CORDILLERAS
Rocky Mountains
Highlands and basins
Elevated plateaus
Pacific ranges
Coastal ranges
Coastal plains
Ice caps
ft m
12 000 4000
9000 3000
6000 2000
4500 1500
3000 1000
1200 400
600 200
0 0
200 600
2000 6000
4000 12 000
6000 18 000
8000 24 000
m ft
COPYRIGHT GEORGE PHILIP & SON LTD

NORTH AMERICA : Climate

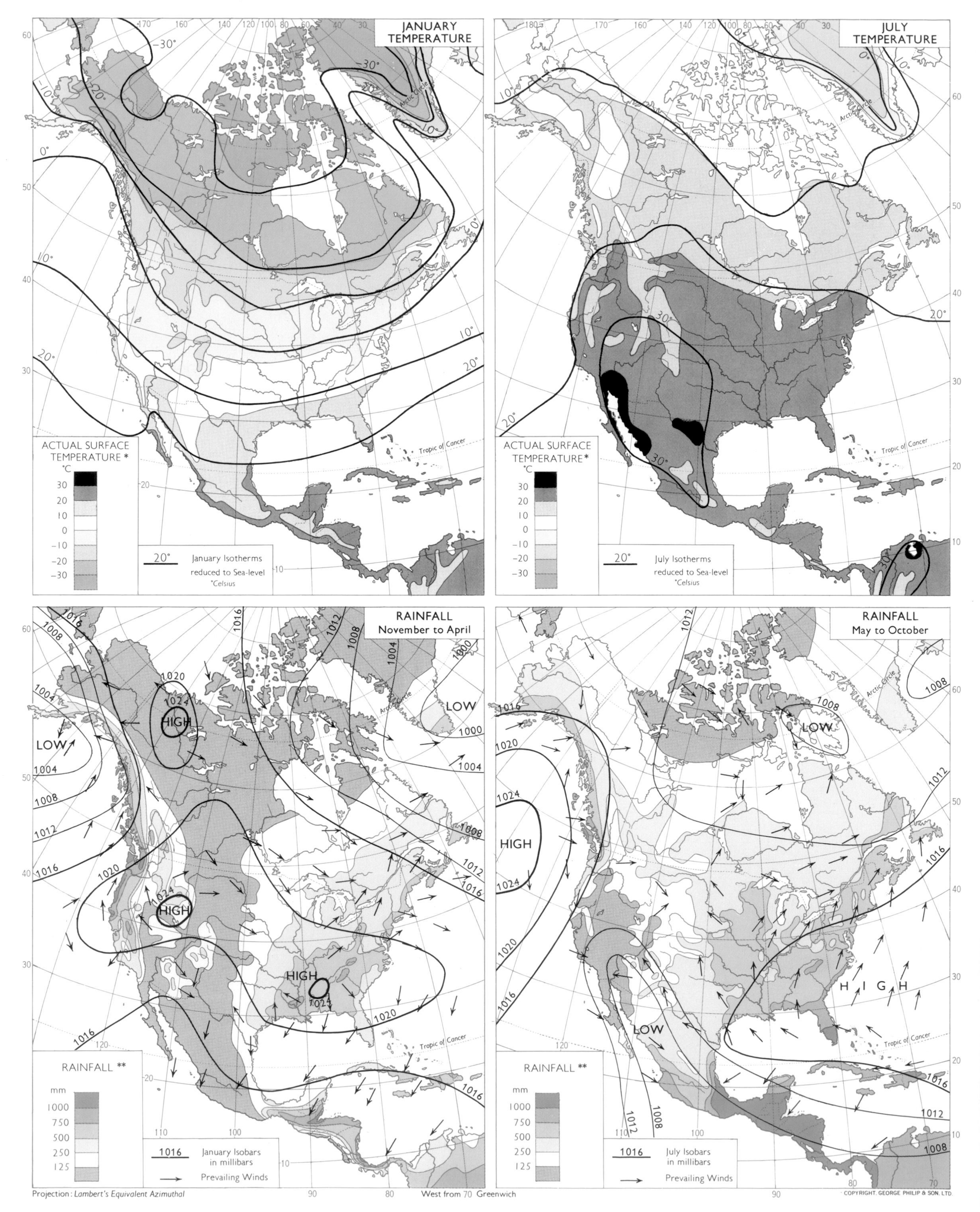

*To convert °C to °F, multiply by 1.8, then add 32 **1 in equals 25.4mm

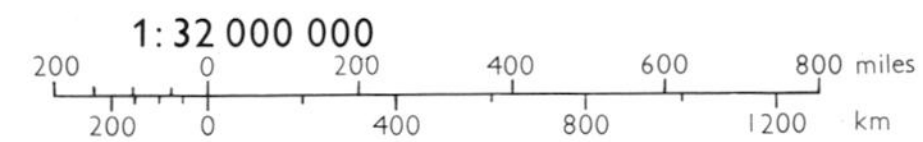
1: 32 000 000
200 0 200 400 600 800 miles
200 0 400 800 1200 km

Coast Range
Rocky Mountains
Great Plains
Prairies
Laurentian Plateau
Cascade Range
Sierra Nevada
Great Basin
Appalachian Mts
Piedmont Plateau
Ozark Plateau
Llano Estacado
Mojave Des.
Colorado Des.
Sierra Madre
ATLANTIC OCEAN
Gulf of Mexico
Tropic of Cancer
PACIFIC OCEAN
CARIBBEAN SEA
West from Greenwich
NATURAL VEGETATION
after Harschberger, Shantz, Zon, Fernow and others
FOREST VEGETATION
Northern Coniferous Forest
Sub-Arctic and Northern Forest (pine, spruce, fir, tamarack, balsam, poplar, larch; willow and birch undergrowth)
North-East Coniferous Forest (white, jack and red pines, spruce, balsam, poplar, tamarack, birch)
Central and Eastern Hardwoods
Central (oak, hickory)
Alleghanian (oak, chestnut, yellow poplar)
Piedmont (oak, pine)
North-Eastern (beech, birch, maple, hemlock)
Appalachian Mountain Forest
Broad-leaved Forest (beech, chestnut, maple, oak)
Coniferous Forest (hemlock, pine, fir, spruce)
Atlantic Pine Barrens
South-Eastern Pine Forest (longleaf and loblolly pines)
South-Eastern Swamp Forest (cypress, magnolia, white cedar)
Pacific Coniferous Forest
Northern Zone (spruce, hemlock)
Central Zone (Douglas fir, hemlock)
Southern Zone (sequoia (redwood), cypress, Douglas fir, oak)
Cordilleran and Rocky Mountain Coniferous Forest
Yellow Pine and Douglas Fir
Lodgepole, Yellow and Sugar Pine Forest
Pinon-Juniper Coniferous Woodland
Californian Chaparral (broad-leaved Woodland)
Mexican and Central American Pine and Oak Forest
Sub-tropical and Tropical Forest (palms, bamboo, tree-ferns, lianas, orchids, etc.)
Sub-tropical and Tropical Chaparral
Northern Limit of Douglas Fir
Limit of White Pine
Limit of Sugar Maple
Limit of Yucca
Northern Limit of Coastal Mangrove Swamps
GRASS VEGETATION
Temperate Grasslands
Sub-tropical and Tropical Grasslands and Savanna
Semi-desert Mesquite Grasslands
Semi-desert Mesquite Savanna
Swamp and Marsh Vegetation
STEPPE, SCRUB AND DESERT VEGETATION
Sage Brush
Creosote Shrub (yucca)
Mexican Plateau Shrub (yucca, agave, cactus)
Salt Desert Shrub (greasewood)
Ice Desert, Tundra (moss, lichen, heather bogs, dwarf willow, birch and alder, etc.). Alpine (above timber line)
Seas and Lakes frozen in Winter
Projection: Polyconic

CANADA
BRITISH COLUMBIA
ALBERTA
SASKATCHEWAN
MANITOBA
Vancouver
Vancouver I.
Calgary
Saskatoon
Regina
Lethbridge
Winnipegosis
WASHINGTON
Seattle
Tacoma
Olympia
Spokane
Mt. Rainier 4392
OREGON
Portland
Salem
Eugene
Mt. Hood 3427
IDAHO
Boise
Pocatello
Idaho Falls
Twin Falls
MONTANA
Great Falls
Helena
Butte
Billings
WYOMING
Casper
Cheyenne
Yellowstone National Park
NORTH DAKOTA
Bismarck
SOUTH DAKOTA
Pierre
NEBRASKA
KANSAS
CALIFORNIA
Sacramento
SAN FRANCISCO
Oakland
San Jose
Fresno
Bakersfield
LOS ANGELES
Long Beach
San Diego
Santa Barbara
NEVADA
Reno
Carson City
Las Vegas
Great Basin
UTAH
Salt Lake City
Ogden
Provo
Great Salt Lake
COLORADO
Denver
Colorado Springs
Pueblo
Mt. Elbert 4399
ARIZONA
Phoenix
Tucson
Grand Canyon Nat. Park
NEW MEXICO
Santa Fe
Albuquerque
El Paso
TEXAS
Amarillo
Lubbock
San Antonio
Austin
Laredo
MEXICO
SONORA
CHIHUAHUA
COAHUILA
DURANGO
LOWER CALIFORNIA
Gulf of California
Rocky Mountains
Sierra Nevada
Coast Ranges
Sierra Madre
PACIFIC OCEAN
HAWAII
1:10 000 000
Kauai
Niihau
Oahu
Honolulu
Molokai
Maui
Lanai
Hawaii
Hilo
Hawaiian Islands
PACIFIC OCEAN
Projection: Albers' Equal Area with two standard parallels
West from Greenwich
National Capital ★
State Capital

1:12 000 000

50 0 50 100 150 200 250 300 miles

50 0 50 100 150 200 250 300 350 400 450 km

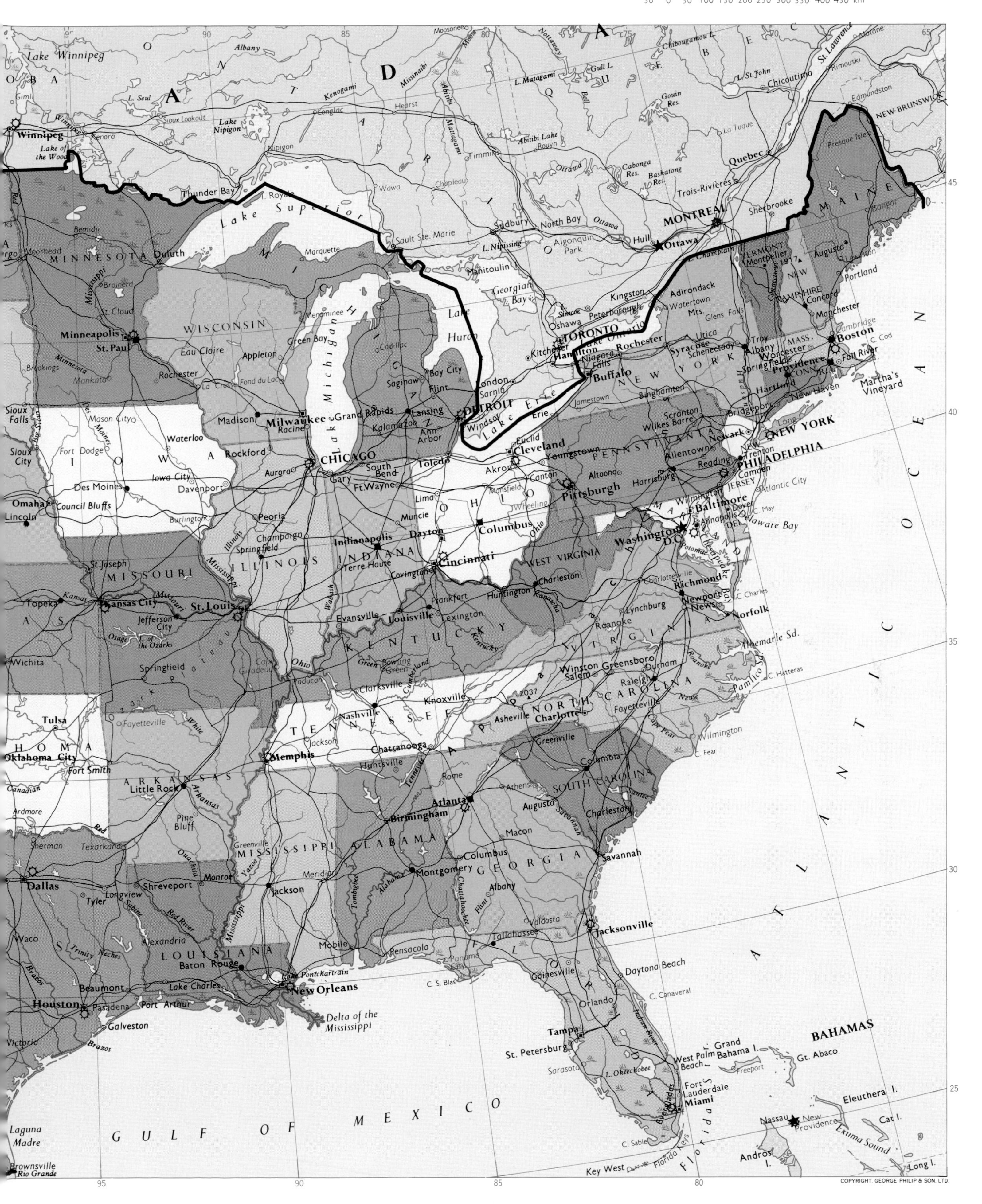

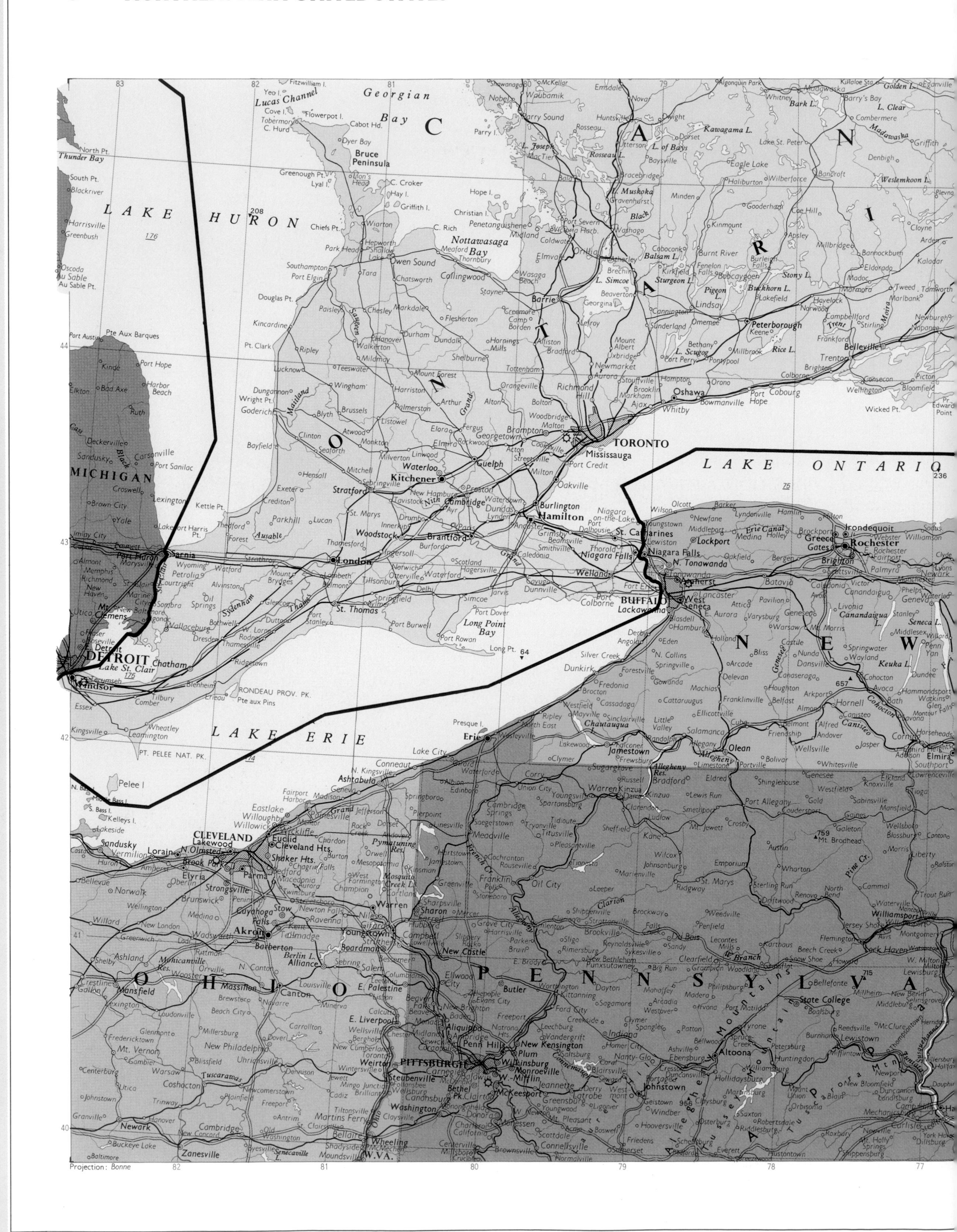
Georgian Bay
LAKE HURON
Bruce Peninsula
Nottawasaga Bay
LAKE ONTARIO
LAKE ERIE
L. Simcoe
Long Point Bay
Lake St. Clair
Chautauqua
MICHIGAN
ONTARIO
NEW YORK
OHIO
PENNSYLVANIA
W.VA.
TORONTO
Mississauga
Hamilton
Kitchener
Waterloo
Guelph
London
Brantford
Barrie
Peterborough
Oshawa
St. Catharines
Niagara Falls
Sarnia
Windsor
Chatham
St. Thomas
Owen Sound
Stratford
Belleville
DETROIT
BUFFALO
Rochester
Erie
Jamestown
Olean
Elmira
CLEVELAND
Lakewood
Euclid
Parma
Akron
Canton
Youngstown
Warren
Massillon
Mansfield
Newark
Zanesville
Wheeling
PITTSBURGH
Altoona
Johnstown
State College
Williamsport
Allegheny Mountains
Tuscarora Mtn.
Projection: Bonne

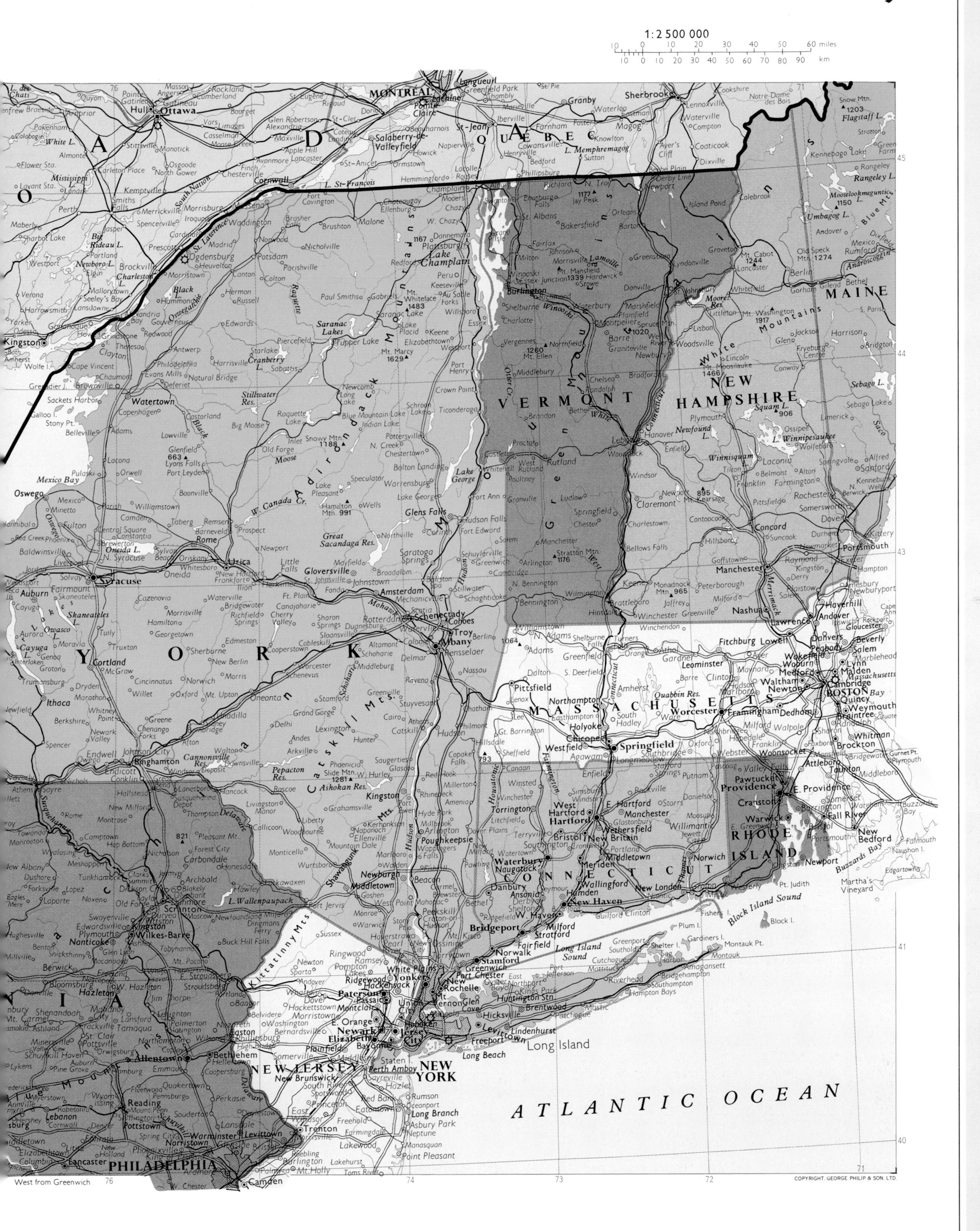
1:2 500 000
10 0 10 20 30 40 50 60 miles
10 0 10 20 30 40 50 60 70 80 90 km
MONTREAL
Ottawa
Hull
Sherbrooke
Granby
Cornwall
Kingston
Q U E B E C
O
A
D
A
St-Jean
Salaberry-de-Valleyfield
Magog
L. Memphremagog
L. St-François
Brockville
Ogdensburg
Plattsburg
Lake Champlain
Burlington
Montpelier
Barre
VERMONT
Green Mts.
St. Albans
Rutland
Brattleboro
Bennington
Mt. Mansfield 1339
Jay Peak 1177
NEW HAMPSHIRE
White Mountains
Mt. Washington 1917
Concord
Manchester
Nashua
Portsmouth
L. Winnipesaukee
MAINE
Connecticut
Lake Ontario
Oswego
Syracuse
Utica
Rome
Watertown
Mexico Bay
Adirondack Mountains
Mt. Marcy 1629
Saranac Lakes
Lake George
Glens Falls
Saratoga Springs
Great Sacandaga Res.
Amsterdam
Schenectady
Albany
Troy
Gloversville
Auburn
Cortland
Ithaca
Binghamton
Elmira
Y
O
R
K
Catskill Mts.
Kingston
Poughkeepsie
Newburgh
Middletown
Hudson
Ashokan Res.
Pepacton Res.
Cannonsville Res.
MASSACHUSETTS
Pittsfield
Springfield
Holyoke
Worcester
Framingham
Lowell
Lawrence
Haverhill
Fitchburg
Leominster
BOSTON
Cambridge
Quincy
Brockton
Salem
Lynn
Quabbin Res.
Fall River
New Bedford
Cape Ann
RHODE ISLAND
Providence
Pawtucket
Warwick
Cranston
Newport
Woonsocket
CONNECTICUT
Hartford
West Hartford
New Britain
Waterbury
Meriden
Middletown
New Haven
Bridgeport
Stamford
Norwalk
New London
Norwich
Danbury
Block Island Sound
Long Island Sound
Long Island
Montauk Pt.
Martha's Vineyard
Buzzards Bay
NEW YORK
Yonkers
White Plains
New Rochelle
Jersey City
Newark
Paterson
Elizabeth
Bayonne
Staten I.
NEW JERSEY
Perth Amboy
New Brunswick
Trenton
Camden
Asbury Park
Long Branch
Point Pleasant
Kittatinny Mts.
PHILADELPHIA
Allentown
Bethlehem
Reading
Scranton
Wilkes-Barre
Hazleton
Lancaster
Pottstown
Norristown
L. Wallenpaupack
ATLANTIC OCEAN
45
44
43
42
41
40
76
74
73
72
71
West from Greenwich
COPYRIGHT, GEORGE PHILIP & SON, LTD.

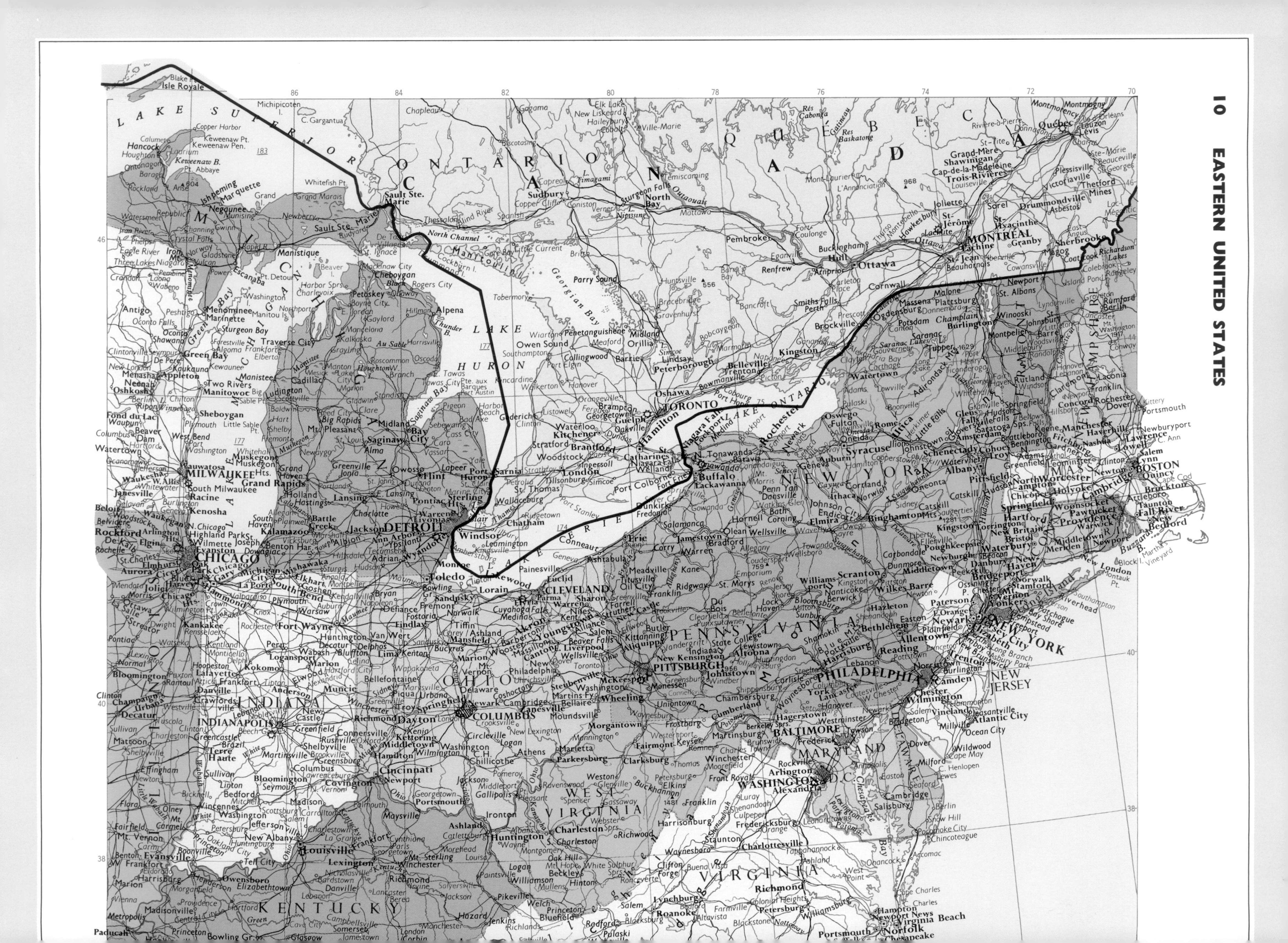
LAKE SUPERIOR
LAKE HURON
LAKE ERIE
LAKE ONTARIO
Georgian Bay
ONTARIO
QUEBEC
CANADA
MICHIGAN
NEW YORK
PENNSYLVANIA
OHIO
INDIANA
KENTUCKY
VIRGINIA
WEST VIRGINIA
MARYLAND
NEW JERSEY
DELAWARE
CHICAGO
MILWAUKEE
DETROIT
TORONTO
MONTREAL
Ottawa
BOSTON
NEW YORK
PHILADELPHIA
PITTSBURGH
CLEVELAND
BALTIMORE
WASHINGTON D.C.
INDIANAPOLIS
COLUMBUS

1:6 000 000
50 0 50 100 150 miles
50 0 50 100 150 200 km
TENNESSEE
NORTH CAROLINA
SOUTH CAROLINA
GEORGIA
ALABAMA
FLORIDA
MISSISSIPPI
MAINE
NEW HAMPSHIRE
CANADA
BAHAMAS
ATLANTIC OCEAN
GULF OF MEXICO
NASHVILLE
ATLANTA
JACKSONVILLE
Birmingham
Montgomery
Chattanooga
Knoxville
Charlotte
Raleigh
Columbia
Charleston
Savannah
Augusta
Macon
Columbus
Mobile
Pensacola
Tallahassee
Orlando
Tampa
St. Petersburg
Miami
Ft. Lauderdale
West Palm Beach
Daytona Beach
St. Augustine
Wilmington
Greensboro
Durham
Winston-Salem
Portland
Bangor
Augusta
Lewiston
Pamlico Sound
C. Hatteras
C. Fear
C. Canaveral
L. Okeechobee
Big Cypress Swamp
EVERGLADES NAT. PARK
Grand Bahama I.
Great Abaco I.
Continuation Eastwards On same scale.
Projection: Alber's Equal Area with two standard parallels
West from Greenwich
COPYRIGHT. GEORGE PHILIP & SON. LTD.

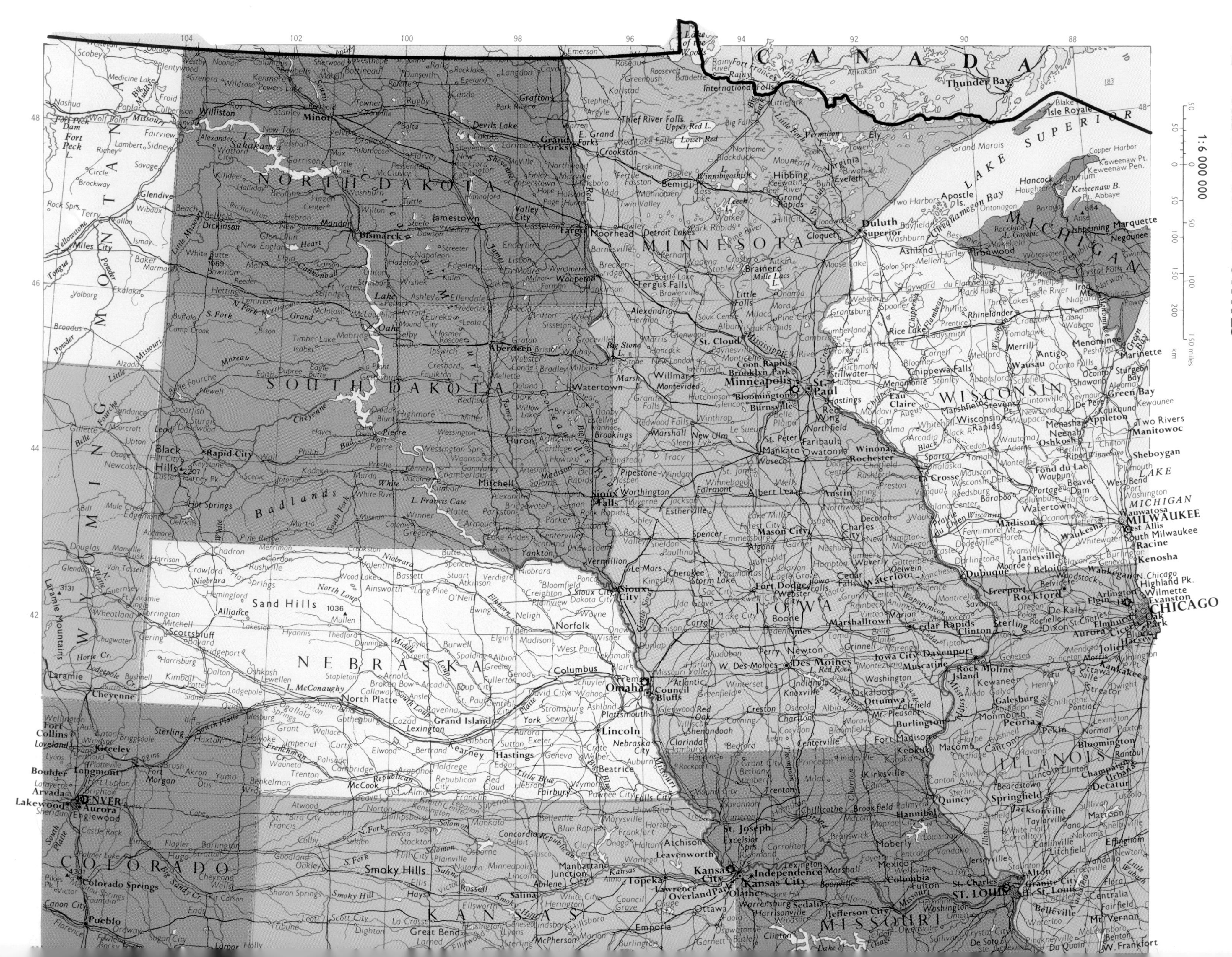
1:6 000 000
50 0 50 100 150 miles
50 0 50 100 150 200 km
CANADA
LAKE SUPERIOR
LAKE MICHIGAN
MICHIGAN
MINNESOTA
WISCONSIN
NORTH DAKOTA
SOUTH DAKOTA
IOWA
ILLINOIS
NEBRASKA
MISSOURI
KANSAS
COLORADO
WYOMING
MONTANA
CHICAGO
MILWAUKEE
ST. LOUIS
DENVER
Minneapolis
St. Paul
Des Moines
Omaha
Kansas City
Lincoln
Duluth
Thunder Bay
Bismarck
Sand Hills
Laramie Mountains
Badlands

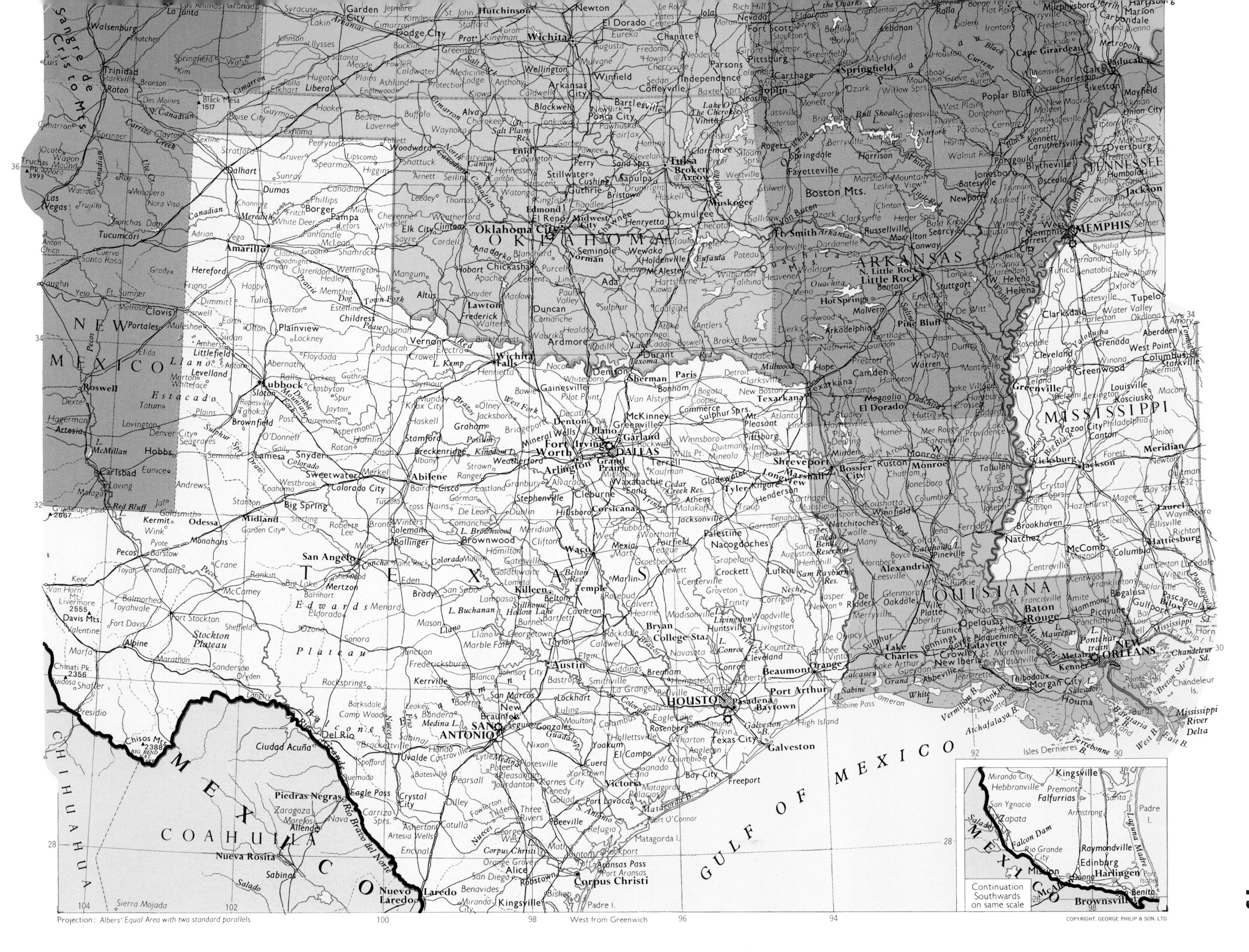
OKLAHOMA
ARKANSAS
MISSISSIPPI
LOUISIANA
TEXAS
NEW MEXICO
TENNESSEE
MEXICO
COAHUILA
CHIHUAHUA
GULF OF MEXICO
Oklahoma City
Tulsa
Wichita
DALLAS
Fort Worth
HOUSTON
SAN ANTONIO
Austin
Corpus Christi
Galveston
Beaumont
Shreveport
Little Rock
MEMPHIS
NEW ORLEANS
Baton Rouge
Jackson
Amarillo
Lubbock
Abilene
Wichita Falls
Laredo
Brownsville
Springfield
Ft. Smith
Sangre de Cristo Mts.
Edwards Plateau
Llano Estacado
Boston Mts.
Mississippi River Delta
Continuation Southwards on same scale
Projection: Albers' Equal Area with two standard parallels
West from Greenwich

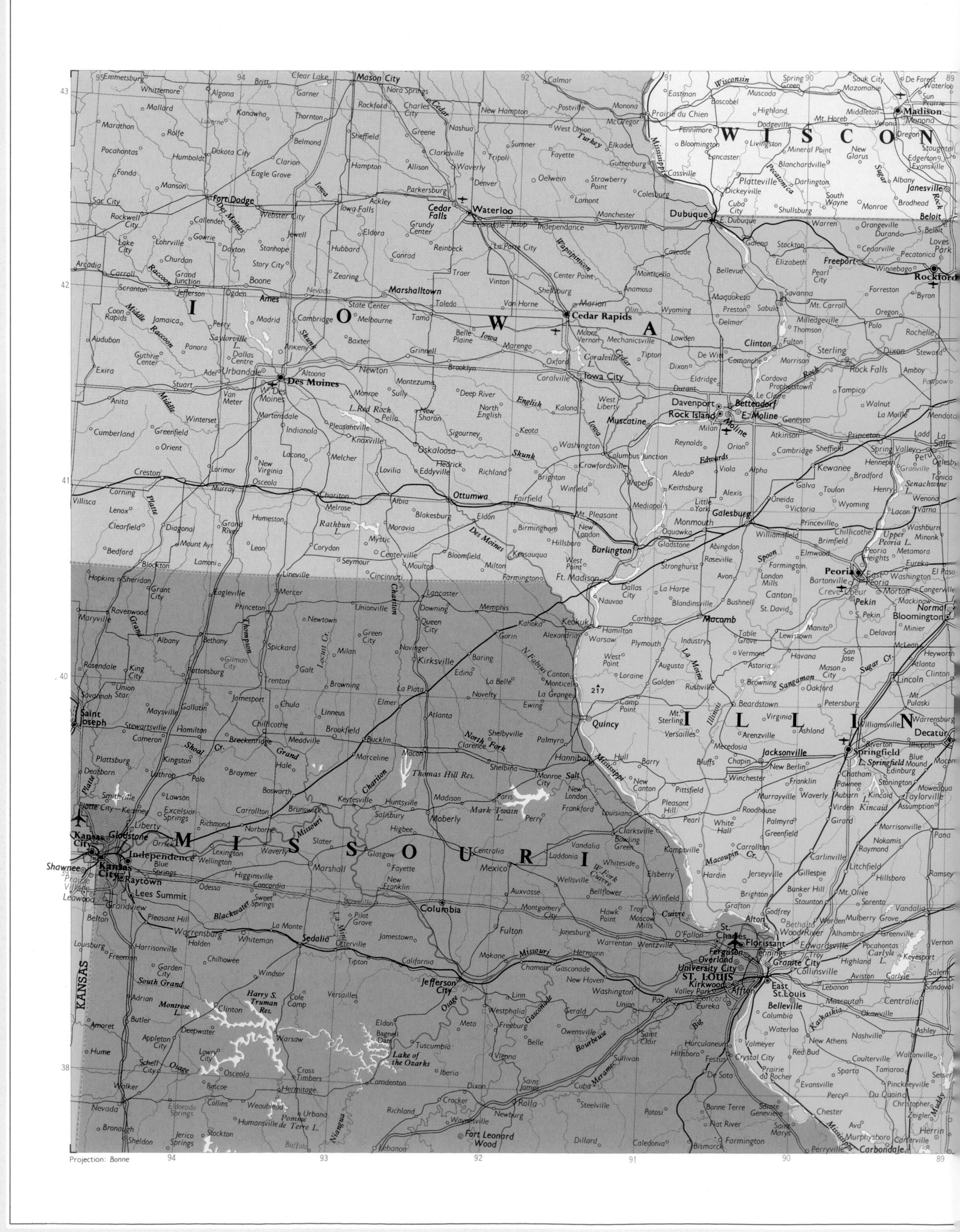
I O W A
M I S S O U R I
I L L I N
W I S C O N
KANSAS
Des Moines
Cedar Rapids
Waterloo
Cedar Falls
Dubuque
Davenport
Bettendorf
Rock Island
Moline
E. Moline
Iowa City
Marshalltown
Mason City
Fort Dodge
Ames
Ottumwa
Burlington
Muscatine
Clinton
Galesburg
Peoria
Pekin
Bloomington
Springfield
Decatur
Jacksonville
Quincy
Hannibal
Keokuk
Kirksville
Columbia
Jefferson City
Sedalia
Kansas City
Independence
Raytown
Lees Summit
Saint Joseph
St. Louis
St. Charles
Florissant
University City
Kirkwood
East St. Louis
Belleville
Granite City
Alton
Fort Leonard Wood
Rolla
Madison
Janesville
Beloit
Rockford
Freeport
Sterling
Dixon
Macomb
Carbondale
Lake of the Ozarks
Harry S. Truman Res.
Mark Twain L.
Thomas Hill Res.
Mississippi
Missouri
Des Moines
Illinois
Projection: Bonne

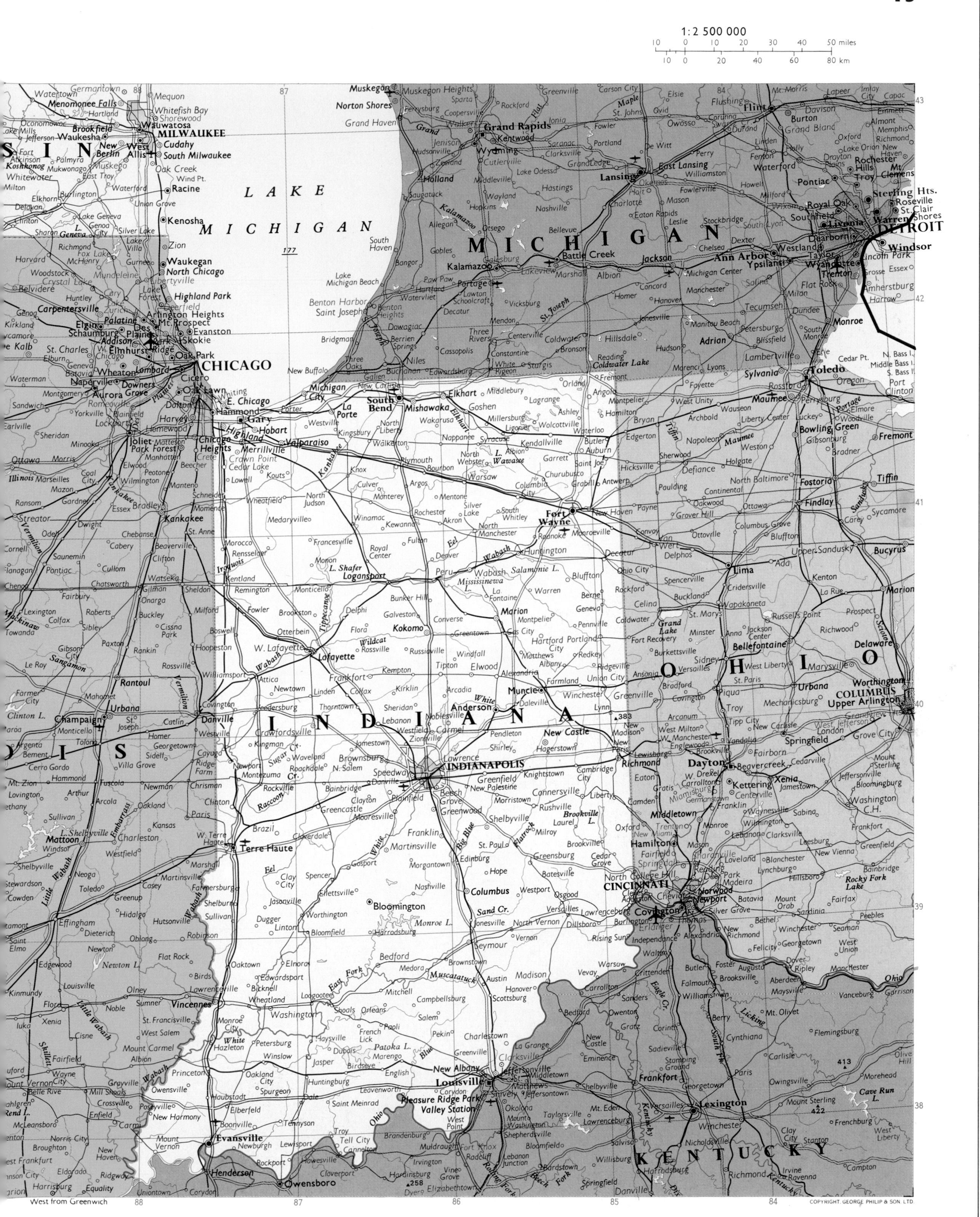

1:2 500 000
10 0 10 20 30 40 50 miles
10 0 20 40 60 80 km
LAKE MICHIGAN
MICHIGAN
INDIANA
OHIO
KENTUCKY
CHICAGO
MILWAUKEE
DETROIT
INDIANAPOLIS
COLUMBUS
CINCINNATI
Windsor
Grand Rapids
Lansing
Kalamazoo
Battle Creek
Jackson
Ann Arbor
Flint
Toledo
South Bend
Fort Wayne
Gary
Hammond
Lafayette
Kokomo
Muncie
Anderson
Terre Haute
Bloomington
Columbus
Richmond
Dayton
Springfield
Lima
Hamilton
Evansville
Louisville
Lexington
Frankfort
Owensboro
Henderson
Vincennes
Champaign
Urbana
Danville
Kankakee
Joliet
Waukegan
Racine
Kenosha
Elkhart
Michigan City
Muskegon
Holland
Monroe
Findlay
Marion
West from Greenwich
88
87
86
85
84
43
42
41
40
39
38
COPYRIGHT, GEORGE PHILIP & SON, LTD.

BRITISH COLUMBIA
ALBERTA
SASKATCHEWAN
CANADA
WASHINGTON
OREGON
IDAHO
MONTANA
WYOMING
NEVADA
UTAH
CALIFORNIA
Vancouver
Victoria
Seattle
Tacoma
Spokane
Portland
Salem
Boise
Helena
Butte
Billings
Great Falls
Missoula
Pocatello
Idaho Falls
Ogden
Salt Lake City
Provo
Reno
Sacramento
Casper
Laramie
Rock Springs
Olympic Mts.
Blue Mountains
Salmon River Mountains
Clearwater Mountains
Bitterroot Range
Lemhi Range
Lewis Range
Little Belt Mts.
Big Belt Mts.
Crazy Mts.
Bighorn Mountains
Wind River Range
Medicine Bow Mts.
Uinta Mountains
Great Salt Lake Desert
Snake River Plain
Great Sandy Desert
Harney Basin
Klamath Mts.

1:6 000 000

50 0 50 100 miles

50 0 50 100 150 km

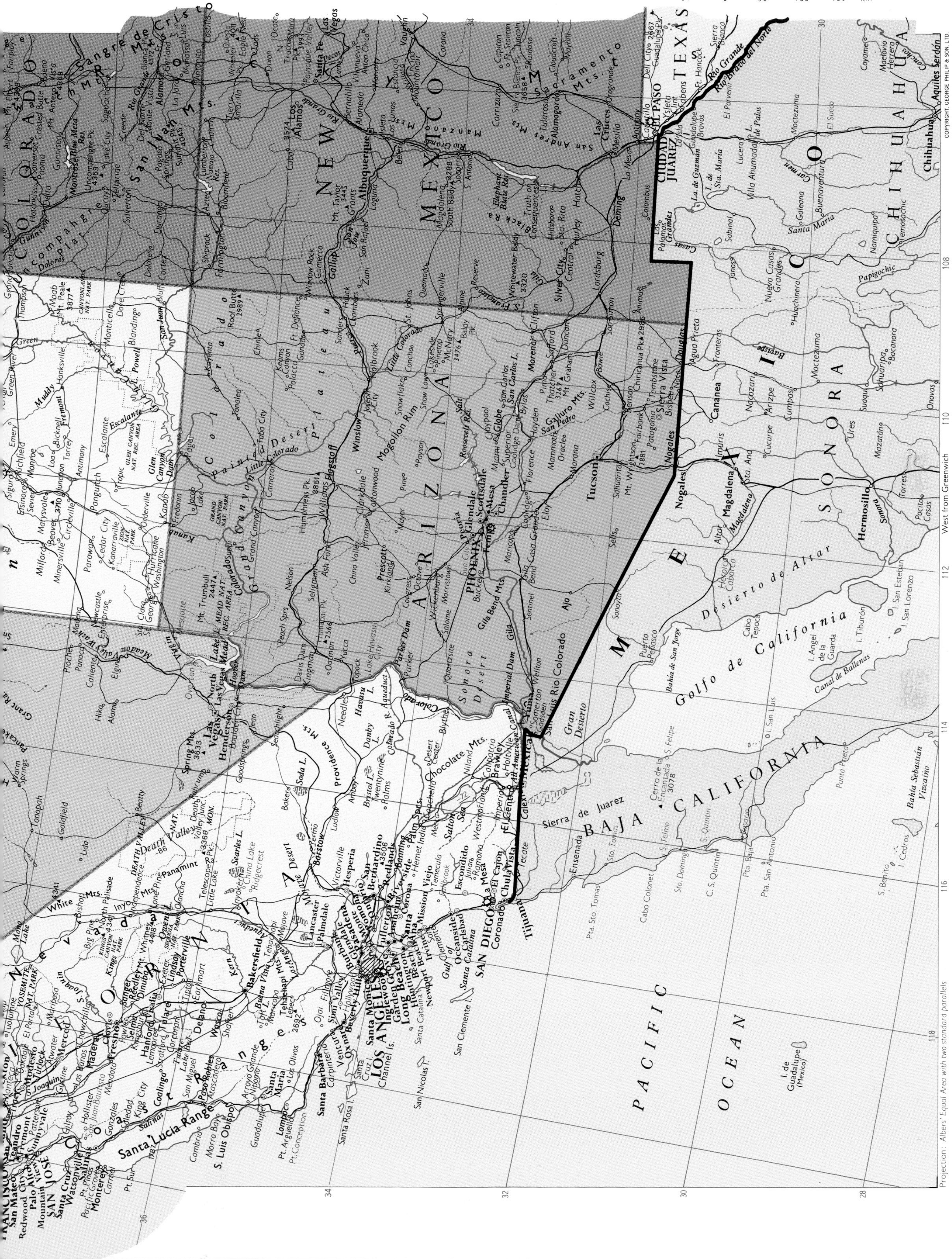

COLORADO

NEW MEXICO

ARIZONA

TEXAS

MEXICO

CHIHUAHUA

SONORA

BAJA CALIFORNIA

PACIFIC OCEAN

Golfo de California

Desierto de Altar

Projection: Albers' Equal Area with two standard parallels

West from Greenwich

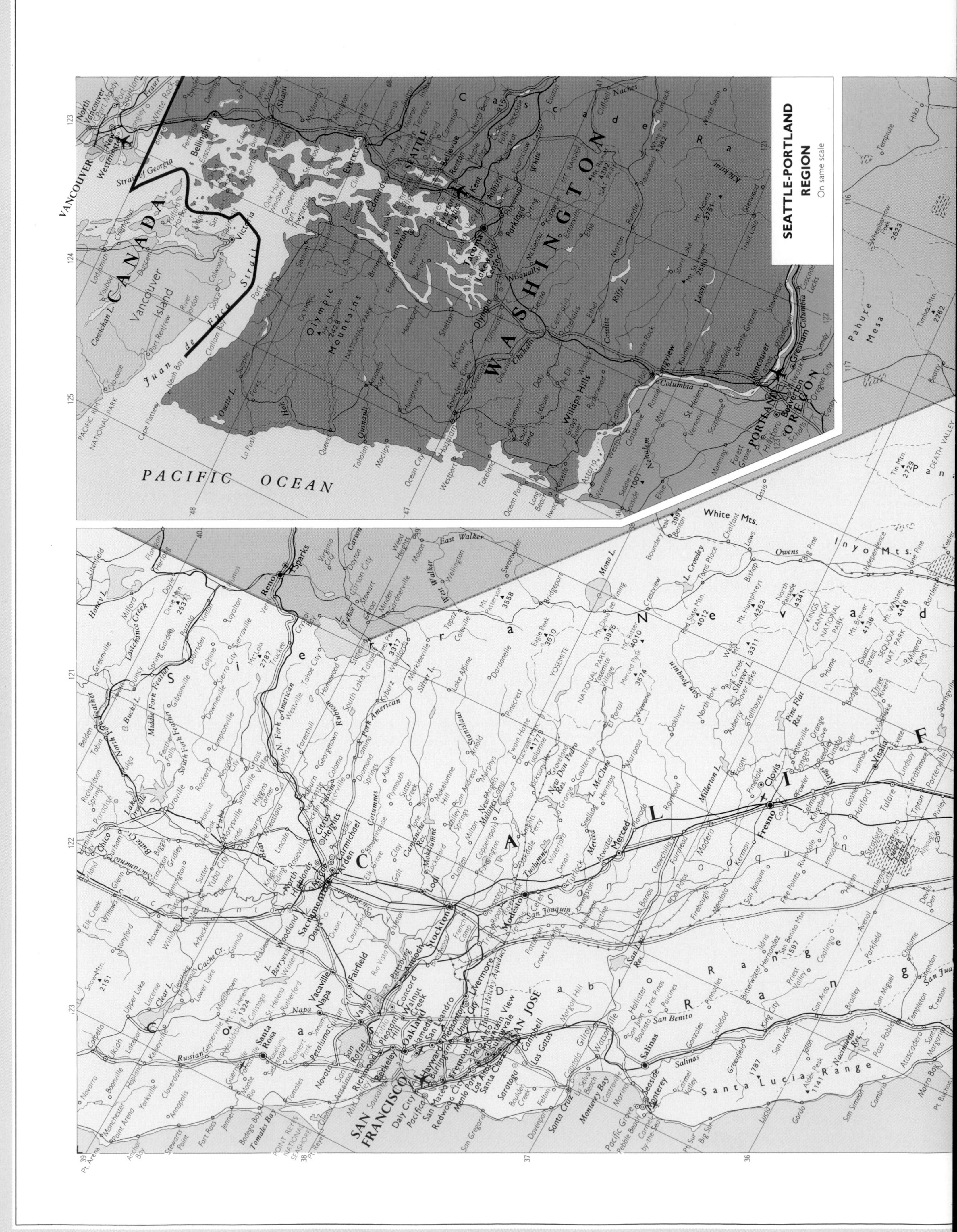
SEATTLE-PORTLAND REGION
On same scale
PACIFIC OCEAN
CANADA
WASHINGTON
OREGON
Vancouver Island
Olympic Mountains
White Mts.
SAN FRANCISCO
SAN JOSE
SEATTLE
PORTLAND

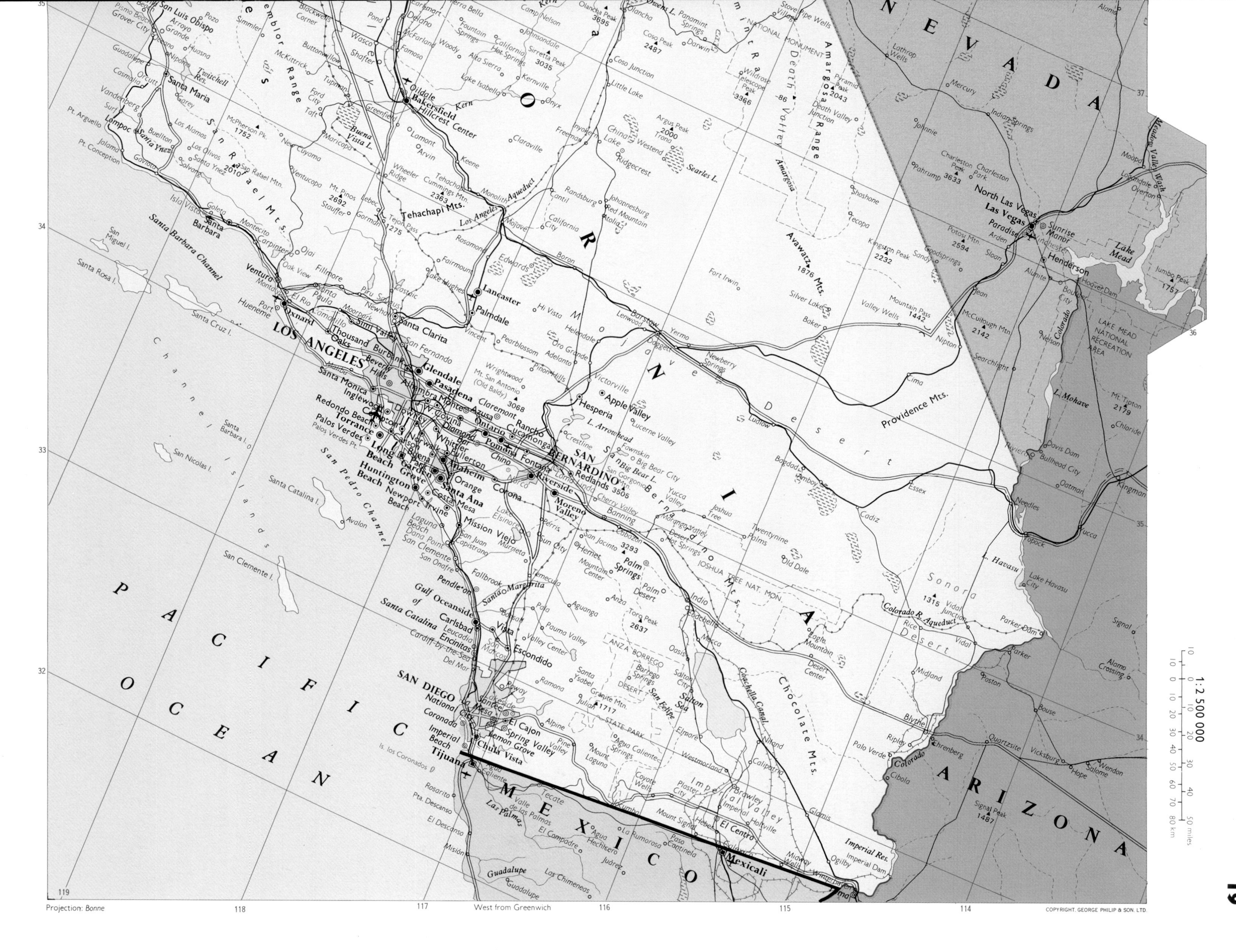
1:2 500 000
10 0 10 20 30 40 50 miles
10 0 10 20 30 40 50 60 70 80 km
COPYRIGHT GEORGE PHILIP & SON, LTD
N E V A D A
A R I Z O N A
M E X I C O
P A C I F I C
O C E A N
Lake Mead
LAKE MEAD NATIONAL RECREATION AREA
L. Mohave
L. Havasu
Las Vegas
North Las Vegas
Henderson
Boulder City
Needles
Amargosa Range
Death Valley
Avawatz Mts.
Providence Mts.
Chocolate Mts.
Coachella Canal
Colorado R. Aqueduct
Imperial Valley
Imperial Res.
Salton Sea
Mexicali
El Centro
Palm Springs
Indio
SAN BERNARDINO
Riverside
Redlands
Apple Valley
Hesperia
Victorville
Barstow
Lancaster
Palmdale
Bakersfield
Tehachapi Mts.
Santa Clarita
Glendale
Pasadena
LOS ANGELES
Long Beach
Santa Ana
Anaheim
Huntington Beach
Newport Beach
Torrance
Santa Monica
Thousand Oaks
Oxnard
Ventura
Santa Barbara
Santa Barbara Channel
San Pedro Channel
Channel Islands
Santa Catalina I.
San Clemente I.
San Nicolas I.
Santa Cruz I.
Santa Rosa I.
San Miguel I.
Gulf of Santa Catalina
Oceanside
Carlsbad
Escondido
SAN DIEGO
Chula Vista
Tijuana
Santa Maria
Lompoc
Pt. Conception
34
33
32
114
115
116
117
118
119
West from Greenwich
Projection: Bonne

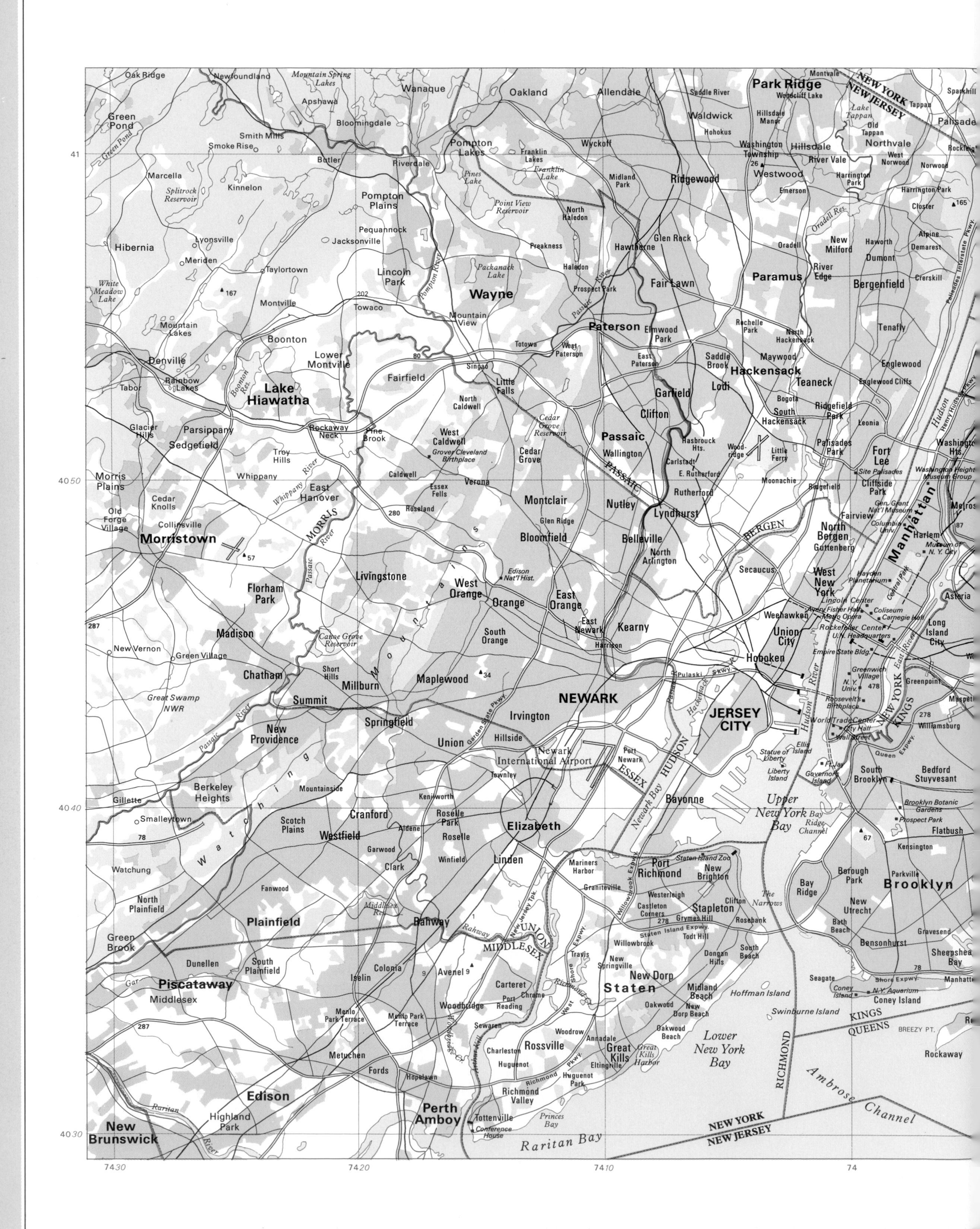

Oak Ridge
Newfoundland
Mountain Spring Lakes
Wanaque
Oakland
Allendale
Saddle River
Park Ridge
Woodcliff Lake
Montvale
NEW YORK
NEW JERSEY
Tappan
Sparkhill
Palisades
Green Pond
Apshawa
Bloomingdale
Waldwick
Hillsdale Manor
Lake Tappan
Old Tappan
Northvale
Smith Mills
Smoke Rise
Pompton Lakes
Franklin Lakes
Wyckoff
Hohokus
Washington Township
Hillsdale
River Vale
West Norwood
Norwood
Rockleigh
Butler
Riverdale
Franklin Lake
Midland Park
Ridgewood
Westwood
Harrington Park
Marcella
Kinnelon
Splitrock Reservoir
Pines Lake
Pompton Plains
Point View Reservoir
North Haledon
Emerson
Closter
Oradell Res.
Pequannock
Jacksonville
Glen Rock
Hawthorne
Oradell
New Milford
Haworth
Alpine
Demarest
Hibernia
Lyonsville
Meriden
Preakness
Dumont
Taylortown
Lincoln Park
Pompton River
Packanack Lake
Haledon
Fair Lawn
Paramus
River Edge
Bergenfield
Cresskill
White Meadow Lake
Wayne
Prospect Park
Montville
Towaco
Mountain View
Mountain Lakes
Paterson
Elmwood Park
Rochelle Park
North Hackensack
Tenafly
Boonton
Totowa
West Paterson
East Paterson
Saddle Brook
Maywood
Englewood
Denville
Lower Montville
Singac
Hackensack
Teaneck
Englewood Cliffs
Rainbow Lakes
Tabor
Boonton Res.
Lake Hiawatha
Fairfield
Little Falls
Garfield
Lodi
Bogota
Ridgefield Park
North Caldwell
Clifton
South Hackensack
Leonia
Hudson
Henry Hudson Pkwy.
Glacier Hills
Parsippany
Rockaway Neck
Pine Brook
Cedar Grove Reservoir
West Caldwell
Grover Cleveland Birthplace
Passaic
Wallington
Hasbrouck Hts.
Wood-ridge
Little Ferry
Palisades Park
Fort Lee
Washington Hts.
Sedgefield
Troy Hills
Cedar Grove
Carlstadt
E. Rutherford
Site Palisades
Washington Heights Museum Group
Whippany
Whippany River
Caldwell
Verona
PASSAIC
Moonachie
Ridgefield
Cliffside Park
Morris Plains
East Hanover
Essex Fells
Rutherford
Cedar Knolls
Montclair
Nutley
Lyndhurst
Gen. Grant Nat'l Museum
Old Forge Village
Roseland
Glen Ridge
Fairview
Columbia Univ.
Collinsville
MORRIS
BERGEN
North Bergen
Manhattan
Harlem
Morristown
Bloomfield
Belleville
Gottenberg
Museum of N. Y. City
North Arlington
Passaic River
Edison Nat'l Hist.
Secaucus
West New York
Hayden Planetarium
Livingstone
Florham Park
West Orange
Central Park
Astoria
Lincoln Center
East Orange
Orange
Avery Fisher Hall
Metro Opera
Coliseum
Carnegie Hall
Weehawken
Long Island City
East Newark
Kearny
Rockefeller Center
U.N. Headquarters
Madison
South Orange
Union City
Harrison
Empire State Bldg.
New Vernon
Green Village
Canoe Grove Reservoir
Hoboken
Greenwich Village
Greenpoint
Chatham
Short Hills
Maplewood
Pulaski Skwy.
N.Y. Univ.
NEW YORK
Maspeth
Millburn
Mountains
Hackensack River
Hudson River
East River
Great Swamp NWR
Summit
NEWARK
Roosevelt's Birthplace
KINGS
Irvington
JERSEY CITY
World Trade Center
City Hall
Wall Street
Williamsburg
Springfield
New Providence
Garden State Pkwy.
Hillside
Union
Newark International Airport
Port Newark
Statue of Liberty
Ellis Island
Queen Expwy.
HUDSON
ESSEX
Liberty Island
Ft. Jay
Governors Island
South Brooklyn
Bedford Stuyvesant
Berkeley Heights
Mountainside
Townley
Gillette
Kenilworth
Newark Bay
Bayonne
Upper New York Bay
Brooklyn Botanic Gardens
Prospect Park
Smalleytown
Scotch Plains
Cranford
Roselle Park
Elizabeth
Bay Ridge Channel
Westfield
Aldene
Roselle
Flatbush
Watchung
Garwood
Linden
Mariners Harbor
Port Richmond
Staten Island Zoo
Kensington
Clark
Winfield
New Brighton
Graniteville
Willowbrook Expwy.
Bay Ridge
Borough Park
Parkville
Brooklyn
Fanwood
Westerleigh
The Narrows
North Plainfield
Middlesex Res.
Castleton Corners
Stapleton
Clifton
New Utrecht
Plainfield
Rahway
Grymes Hill
Rosebank
Rahway River
New Jersey Tpk.
Staten Island Expwy.
Todd Hill
Bath Beach
Green Brook
UNION
MIDDLESEX
Willowbrook
Bensonhurst
Gravesend
Dunellen
South Plainfield
Dongan Hills
South Beach
Travis
Sheepshead Bay
Colonia
Avenel
New Springville
Seagate
Shore Expwy.
Manhattan
Piscataway
Iselin
Carteret
New Dorp
Staten
Midland Beach
Coney Island
N.Y. Aquarium
Middlesex
Richmond Creek
Chrome
Port Reading
Hoffman Island
Oakwood
New Dorp Beach
Swinburne Island
Menlo Park Terrace
Woodbridge
Sewaren
West Shore Expwy.
KINGS
QUEENS
BREEZY PT.
Metuchen
Woodrow
Oakwood Beach
Lower New York Bay
Charleston
Rossville
Annadale
Great Kills
Great Kills Harbor
Rockaway
Huguenot
Eltingville
RICHMOND
Fords
Hopelawn
Arthur Kill
Richmond Pkwy.
Huguenot Park
Ambrose Channel
Edison
Richmond Valley
Perth Amboy
Tottenville
Princes Bay
NEW YORK
NEW JERSEY
New Brunswick
Highland Park
Raritan River
Conference House
Raritan Bay
41
40 50
40 40
40 30
74 30
74 20
74 10
74
287
202
80
280
78
278
26
167
57
34
9
165
67

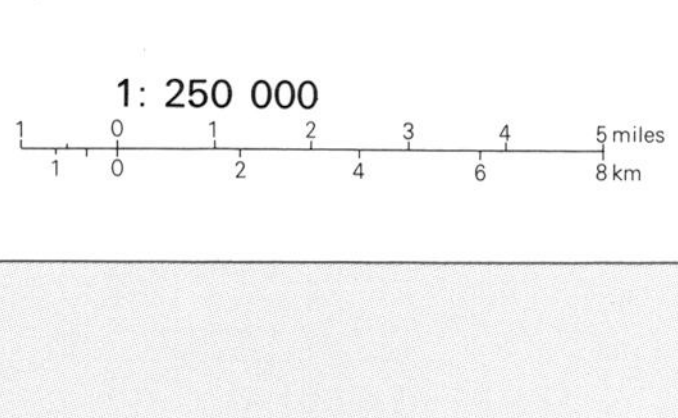
1: 250 000
1 0 1 2 3 4 5 miles
1 0 2 4 6 8 km

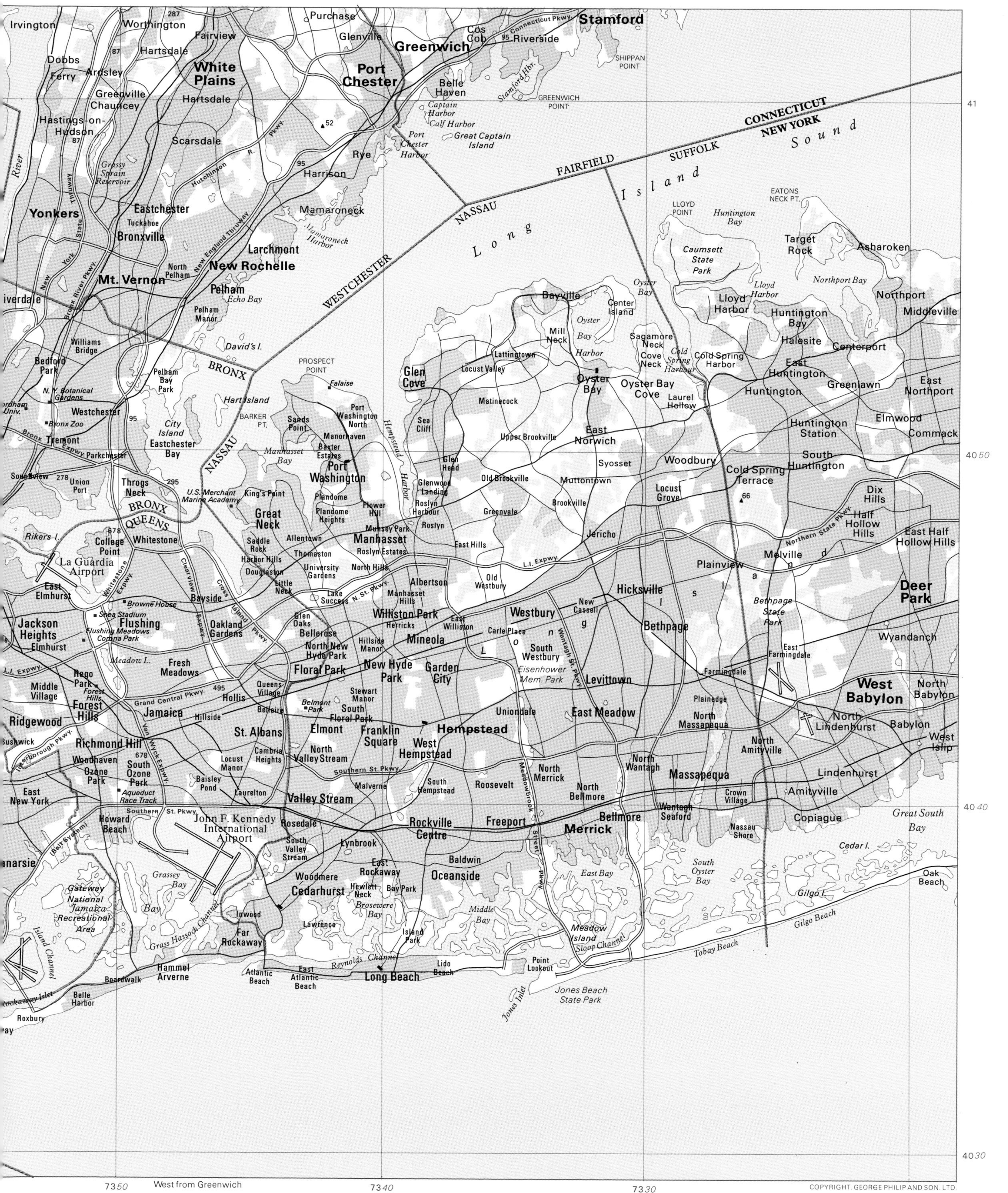
Irvington
Worthington
287
Purchase
Stamford
Connecticut Pkwy.
Cos Cob
95 Riverside
Glenville
Greenwich
Fairview
Hartsdale
87
Dobbs Ferry
Ardsley
White Plains
Port Chester
SHIPPAN POINT
Belle Haven
Greenville
Chauncey
Hartsdale
Stamford Hbr.
GREENWICH POINT
Captain Harbor
Calf Harbor
Great Captain Island
Hastings-on-Hudson
87
52
41
River
Scarsdale
Port Chester Harbor
Rye
CONNECTICUT
NEW YORK
Sound
Grassy Sprain Reservoir
Hutchinson R. Pkwy.
95
Harrison
SUFFOLK
Thruway
FAIRFIELD
Island
Yonkers
Eastchester
Tuckahoe
Mamaroneck
NASSAU
LLOYD POINT
Huntington Bay
EATONS NECK PT.
State
Bronxville
Mamaroneck Harbor
Long
Target Rock
Asharoken
New York
New England Thruway
Larchmont
Caumsett State Park
New Rochelle
North Pelham
New
Bronx River Pkwy.
Mt. Vernon
Pelham
Echo Bay
WESTCHESTER
Oyster Bay
Lloyd Harbor
Northport Bay
Riverdale
Bayville
Center Island
Lloyd Harbor
Northport
Middleville
Huntington Bay
Pelham Manor
Oyster Bay
Williams Bridge
David's I.
Mill Neck
Sagamore Neck
Halesite
Centerport
Bedford Park
PROSPECT POINT
BRONX
Lattingtown
Harbor
Cove Neck
Cold Spring Harbour
Cold Spring Harbor
East Huntington
Glen Cove
Locust Valley
Oyster Bay
Oyster Bay Cove
East Northport
N.Y. Botanical Gardens
Pelham Bay Park
Falaise
Huntington
Greenlawn
Fordham Univ.
Westchester
Hart Island
Matinecock
Laurel Hollow
Elmwood
Bronx Zoo
95
BARKER PT.
Sands Point
Port Washington North
Sea Cliff
Hempstead Harbor
Huntington Station
Commack
Tremont
City Island
Manorhaven
Upper Brookville
East Norwich
Bronx Expwy.
Eastchester Bay
Baxter Estates
South Huntington
Parkchester
Manhasset Bay
NASSAU
Glen Head
Syosset
Woodbury
40 50
Soundview
278
Union Port
Throgs Neck
295
Port Washington
Cold Spring Terrace
Glenwood Landing
Old Brookville
Muttontown
Locust Grove
Dix Hills
BRONX
U.S. Merchant Marine Academy
King's Point
Plandome
66
Roslyn Harbour
Brookville
QUEENS
Great Neck
Plandome Heights
Flower Hill
Greenvale
Half Hollow Hills
East Half Hollow Hills
678
Rikers I.
Roslyn
Jericho
Northern State Pkwy.
College Point
Whitestone
Allentown
Munsey Park
Manhasset
Saddle Rock
Thomaston
Roslyn Estates
East Hills
Melville
La Guardia Airport
Harbor Hills
L.I. Expwy.
Plainview
Douglaston
University Gardens
North Hills
Whitestone Expwy.
Clearview Expwy.
Old Westbury
Deer Park
Albertson
Hicksville
East Elmhurst
Little Neck
Manhasset Hills
Bethpage State Park
Lake Success
N. St. Pkwy.
Browne House
Bayside
New Cassel
Cross Island Pkwy.
Shea Stadium
Williston Park
Westbury
Jackson Heights
Flushing
Glen Oaks
East Williston
Herricks
Bethpage
Oakland Gardens
Wyandanch
Flushing Meadows Corona Park
Bellerose
Carle Place
Elmhurst
North New Hyde Park
Hillside Manor
Mineola
South Westbury
Wantagh St. Pkwy.
East Farmingdale
Meadow L.
L.I. Expwy.
Fresh Meadows
New Hyde Park
Garden City
Eisenhower Mem. Park
Rego Park
Floral Park
Levittown
Farmingdale
West Babylon
North Babylon
Middle Village
Forest Hills
495
Queens Village
Stewart Manor
Grand Central Pkwy.
Hollis
Belmont Park
Plainedge
Ridgewood
Jamaica
Bellaire
South Floral Park
Uniondale
East Meadow
North Lindenhurst
Babylon
Hillside
North Massapequa
Bushwick
Van Wyck Expwy.
St. Albans
Franklin Square
Hempstead
Elmont
Richmond Hill
West Hempstead
North Amityville
West Islip
Interborough Pkwy.
678
Woodhaven
South Ozone Park
Locust Manor
Cambria Heights
North Valley Stream
Ozone Park
Southern St. Pkwy.
North Wantagh
Massapequa
Lindenhurst
Meadowbrook
North Merrick
Baisley Pond
Malverne
South Hempstead
Roosevelt
North Bellmore
Crown Village
Amityville
East New York
Aqueduct Race Track
Laurelton
Valley Stream
40 40
Southern St. Pkwy.
Howard Beach
John F. Kennedy International Airport
Rosedale
Rockville Centre
Freeport
Bellmore
Wantagh
Seaford
Copiague
Great South Bay
Merrick
Street
South Valley Stream
Lynbrook
Nassau Shore
Canarsie
Belt System
Baldwin
Cedar I.
East Rockaway
Oceanside
South Oyster Bay
Woodmere
Grassey Bay
Hewlett Neck
Bay Park
East Bay
Gateway National Jamaica Recreational Area
Cedarhurst
Pkwy.
Oak Beach
Broswere Bay
Gilgo I.
Bay
Middle Bay
Inwood
Grass Hassock Channel
Lawrence
Gilgo Beach
Island Channel
Far Rockaway
Island Park
Meadow Island
Sloop Channel
Tobay Beach
Reynolds Channel
Point Lookout
Boardwalk
Hammel Arverne
Lido Beach
Atlantic Beach
East Atlantic Beach
Long Beach
Belle Harbor
Rockaway Inlet
Jones Beach State Park
Jones Inlet
Roxbury
40 30
73 50
West from Greenwich
73 40
73 30

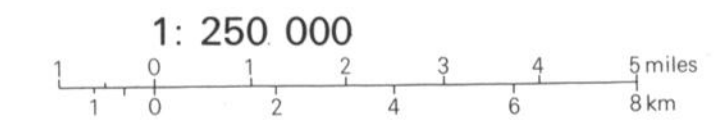
1: 250 000
5 miles
8 km

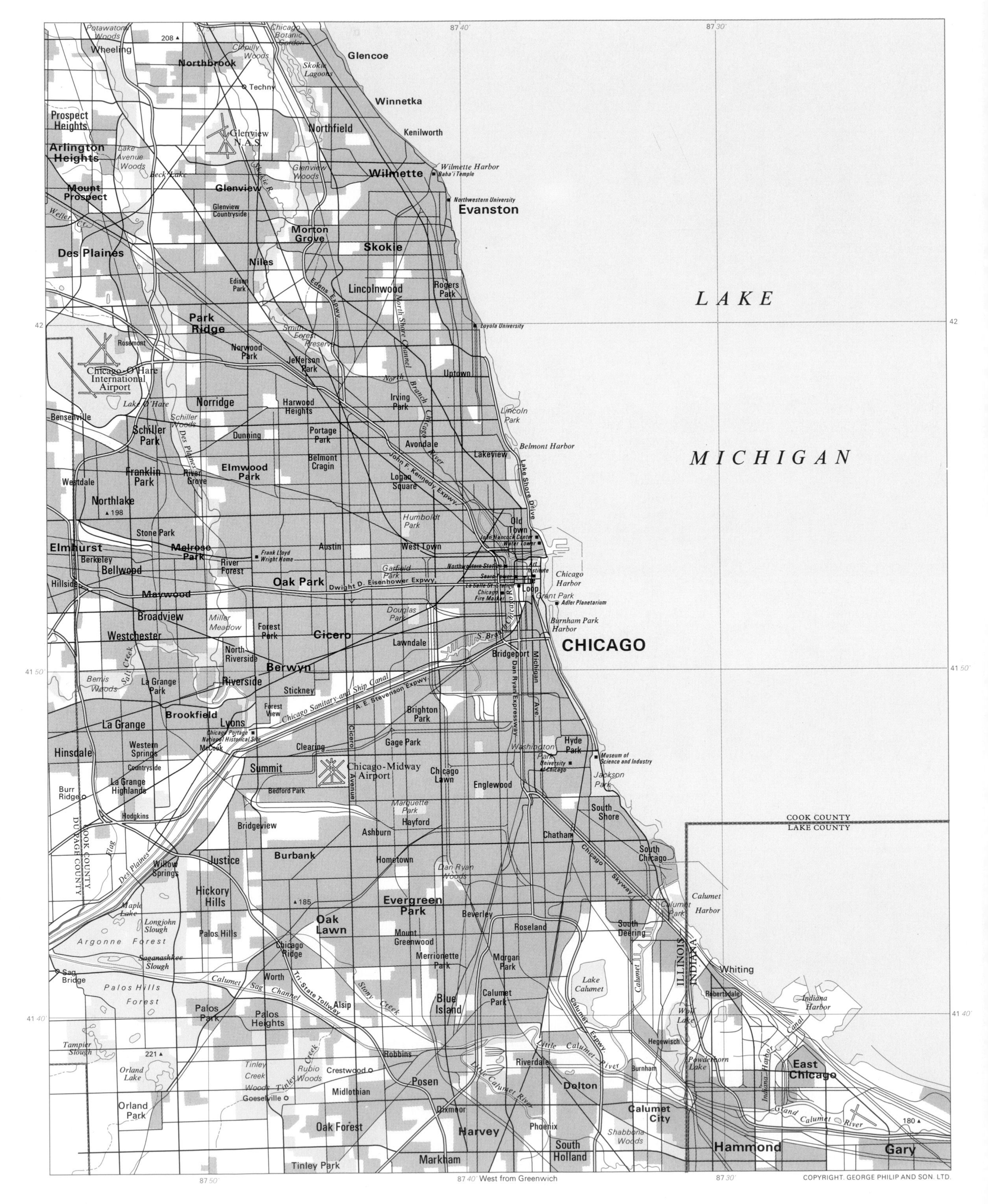
LAKE
MICHIGAN
CHICAGO
Glencoe
Winnetka
Kenilworth
Northbrook
Wheeling
Northfield
Wilmette
Wilmette Harbor
Evanston
Northwestern University
Prospect Heights
Arlington Heights
Mount Prospect
Glenview
Glenview N.A.S.
Glenview Countryside
Morton Grove
Skokie
Des Plaines
Niles
Lincolnwood
Rogers Park
Loyola University
Park Ridge
Rosemont
Chicago-O'Hare International Airport
Norwood Park
Jefferson Park
Uptown
Norridge
Harwood Heights
Irving Park
Lincoln Park
Bensenville
Schiller Park
Dunning
Portage Park
Avondale
Belmont Harbor
Lakeview
Franklin Park
River Grove
Elmwood Park
Belmont Cragin
Logan Square
Westdale
Northlake
Stone Park
Humboldt Park
Old Town
Elmhurst
Melrose Park
Austin
West Town
Berkeley
Bellwood
River Forest
Hillside
Oak Park
Chicago Harbor
Loop
Maywood
Broadview
Forest Park
Cicero
Lawndale
Burnham Park Harbor
Westchester
North Riverside
Bridgeport
Berwyn
Riverside
La Grange Park
Stickney
Brookfield
Lyons
Brighton Park
La Grange
Hinsdale
Western Springs
Clearing
Gage Park
Hyde Park
Summit
Chicago-Midway Airport
Chicago Lawn
Englewood
Burr Ridge
La Grange Highlands
Bedford Park
Hodgkins
Bridgeview
Ashburn
Hayford
Chatham
South Shore
South Chicago
Justice
Burbank
Hometown
Willow Springs
Hickory Hills
Evergreen Park
Beverley
Roseland
South Deering
Calumet Harbor
Oak Lawn
Palos Hills
Mount Greenwood
Merrionette Park
Morgan Park
Whiting
Chicago Ridge
Worth
Alsip
Blue Island
Calumet Park
Lake Calumet
Robertsdale
Indiana Harbor
Palos Park
Palos Heights
Hegewisch
Robbins
Riverdale
Burnham
East Chicago
Crestwood
Posen
Dolton
Midlothian
Dixmoor
Calumet City
Oak Forest
Harvey
Phoenix
South Holland
Hammond
Gary
Markham
Tinley Park
Orland Park
Orland Lake
ILLINOIS
INDIANA
COOK COUNTY
LAKE COUNTY
DU PAGE COUNTY
87 40' West from Greenwich
COPYRIGHT. GEORGE PHILIP AND SON. LTD.

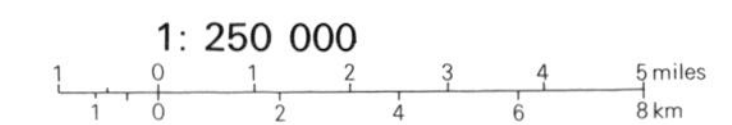

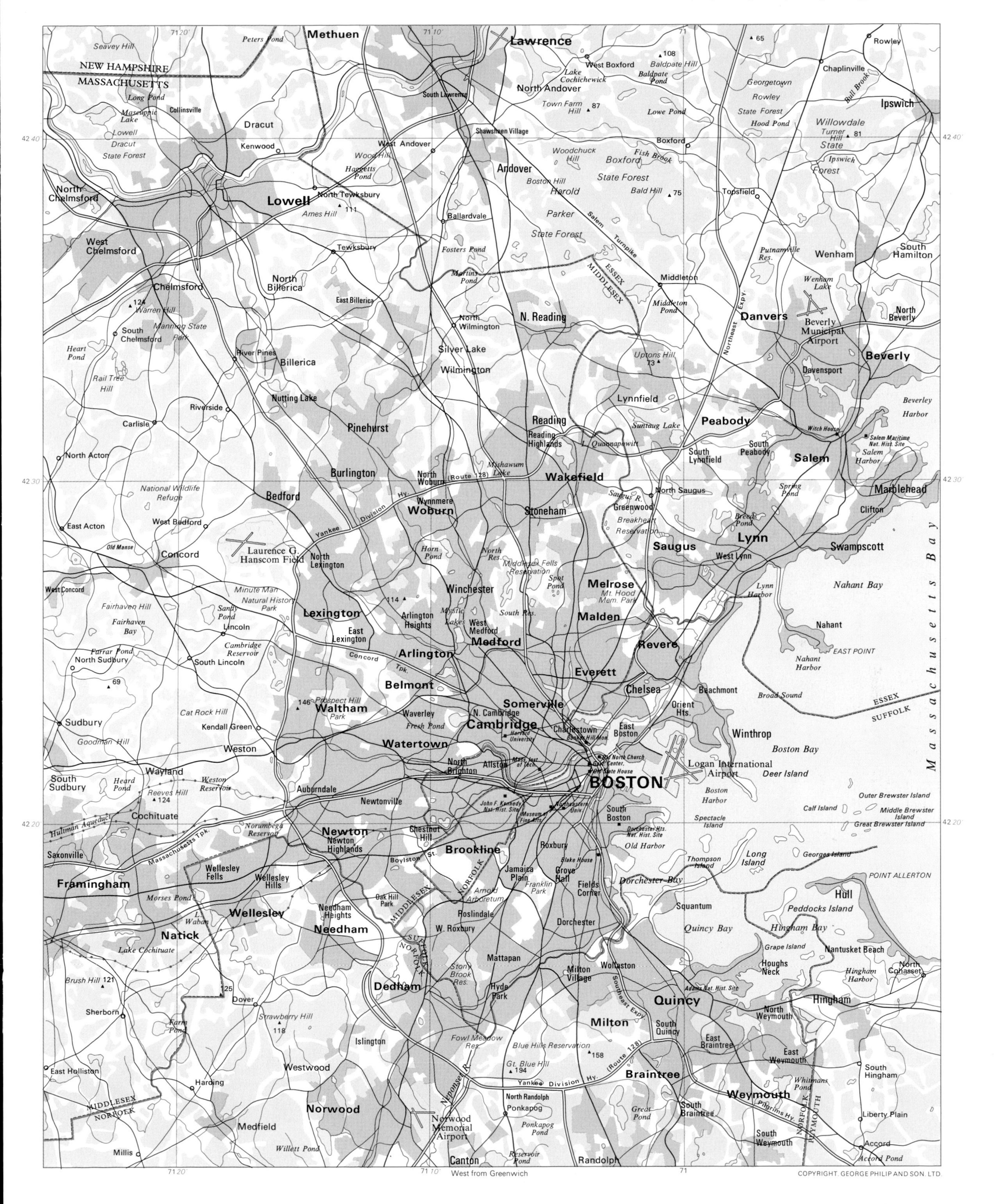

West from Greenwich

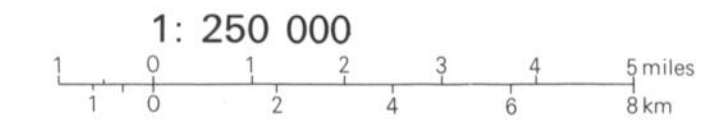
1: 250 000
1 0 1 2 3 4 5 miles
1 0 2 4 6 8 km

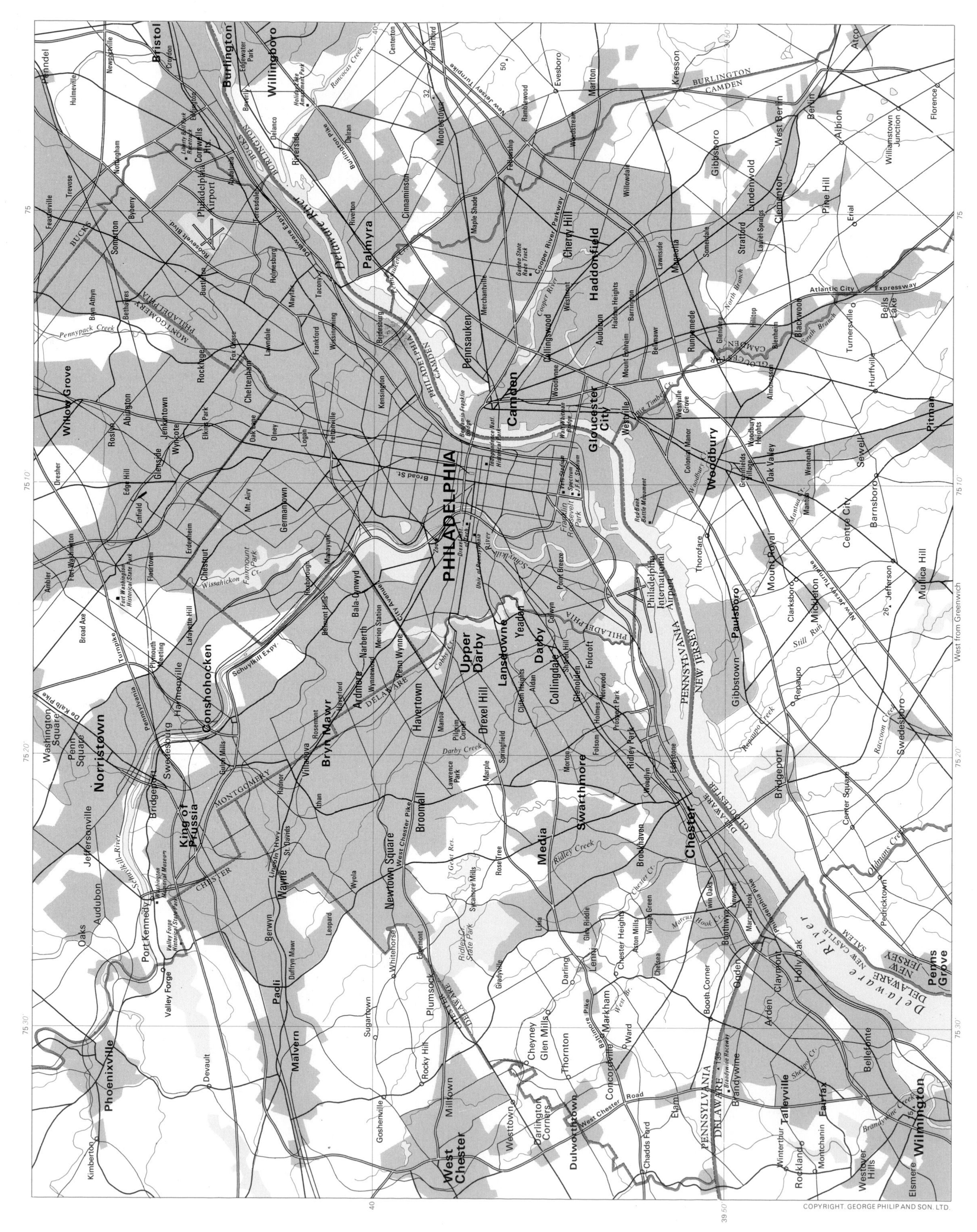
PHILADELPHIA
Camden
Chester
Wilmington
Norristown
Conshohocken
King of Prussia
Phoenixville
West Chester
Bryn Mawr
Ardmore
Upper Darby
Lansdowne
Darby
Media
Swarthmore
Haddonfield
Gloucester City
Woodbury
Burlington
Bristol
Willingboro
Palmyra
Pennsauken
Cherry Hill
Willow Grove
Jenkintown
Germantown
Chestnut Hill
Paoli
Malvern
Wayne
Havertown
Drexel Hill
Collingdale
Paulsboro
Bridgeport
Penns Grove
Delaware River
Schuylkill River
Philadelphia International Airport
Philadelphia Airport
PENNSYLVANIA
NEW JERSEY
DELAWARE
Atlantic City Expressway
New Jersey Turnpike
Schuylkill Expy.
West from Greenwich

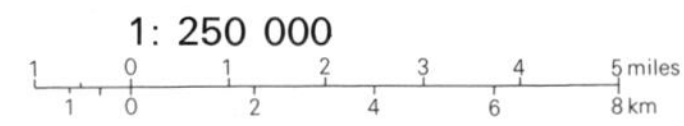
1: 250 000
0 1 2 3 4 5 miles
0 2 4 6 8 km

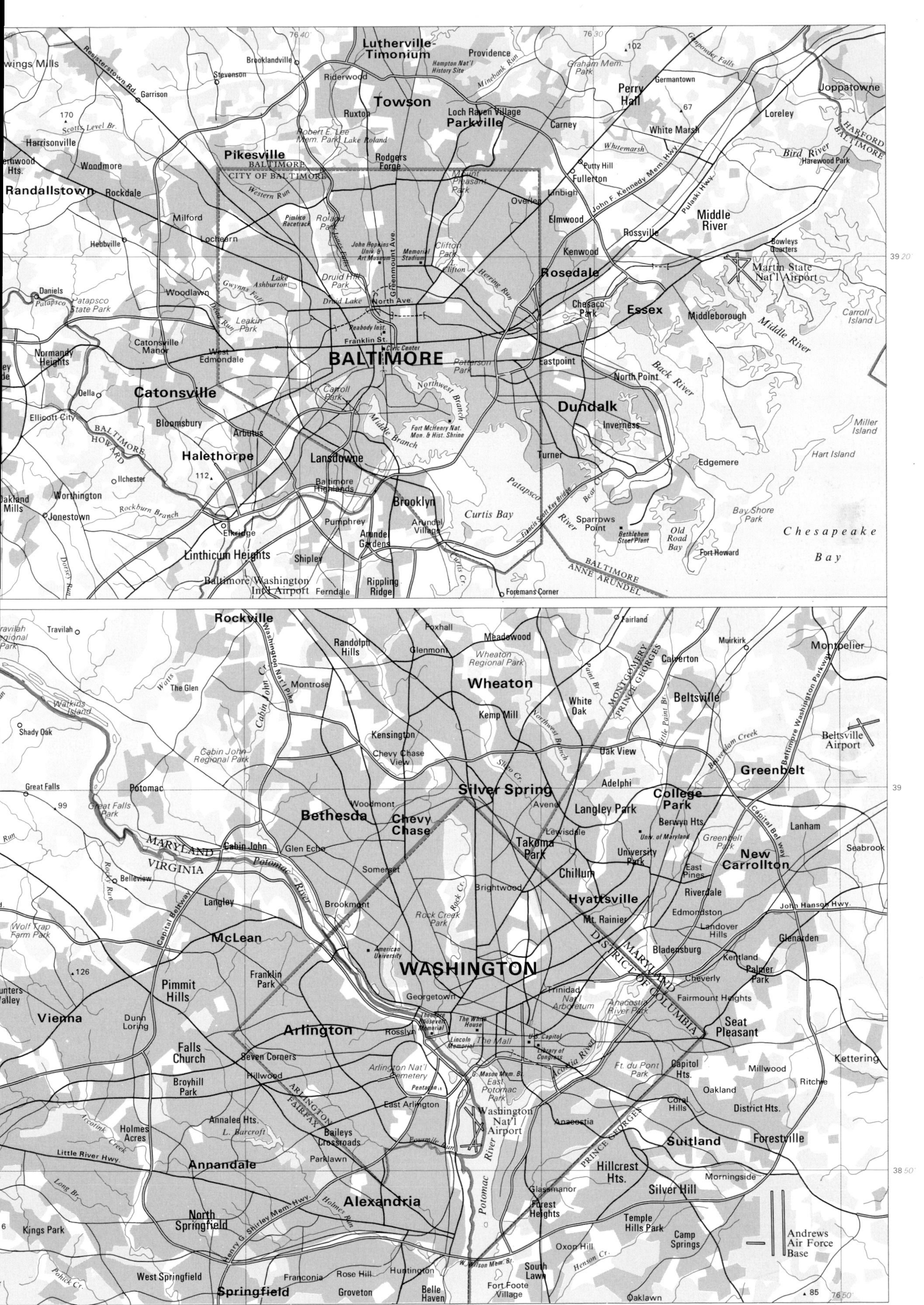
Lutherville-Timonium
Providence
Towson
Riderwood
Ruxton
Loch Raven Village
Parkville
Carney
Perry Hall
Germantown
Joppatowne
Loreley
White Marsh
Pikesville
Brooklandville
Stevenson
Garrison
Harrisonville
Woodmore
Randallstown
Rockdale
Milford
Hebbville
Lochearn
Rodgers Forge
CITY OF BALTIMORE
BALTIMORE
Pimlico Racetrack
Roland Park
Mount Pleasant Park
Overlea
Limbigh
Elmwood
Fullerton
Putty Hill
Rossville
Middle River
Kenwood
Rosedale
Bowleys Quarters
Martin State Nat'l Airport
Essex
Chesaco Park
Middleborough
Middle River
Back River
North Point
Carroll Island
Daniels
Patapsco State Park
Woodlawn
Lake Ashburton
Druid Hill Park
Druid Lake
North Ave.
Clifton Park
Herring Run
John Hopkins Univ. & Art Museum
Memorial Stadium
Greenmount Ave.
Gwynns Falls
Leakin Park
Catonsville Manor
West Edmondale
Normandy Heights
Peabody Inst.
Franklin St.
Civic Center
BALTIMORE
Patterson Park
Eastpoint
Catonsville
Oella
Ellicott City
Carroll Park
Northwest Branch
Middle Branch
Fort McHenry Nat. Mon. & Hist. Shrine
Dundalk
Inverness
Miller Island
Hart Island
Edgemere
Bloomsbury
Arbutus
Halethorpe
Lansdowne
Turner
Ilchester
Baltimore Highlands
Brooklyn
Curtis Bay
Patapsco River
Francis Scott Key Bridge
Sparrows Point
Bethlehem Steel Plant
Old Road Bay
Bay Shore Park
Chesapeake Bay
Worthington
Jonestown
Rockburn Branch
Elkridge
Pumphrey
Arundel Village
Arundel Gardens
Linthicum Heights
Shipley
Fort Howard
Baltimore/Washington Int'l Airport
Ferndale
Rippling Ridge
Foremans Corner
BALTIMORE
ANNE ARUNDEL
HOWARD
HARFORD
Rockville
Travilah
Randolph Hills
Foxhall
Glenmont
Meadowood
Wheaton Regional Park
Fairland
Muirkirk
Montpelier
Calverton
The Glen
Montrose
Wheaton
White Oak
Beltsville
Kemp Mill
MONTGOMERY
PRINCE GEORGES
Baltimore Washington Parkway
Shady Oak
Watkins Island
Kensington
Chevy Chase View
Cabin John Regional Park
Oak View
Beltsville Airport
Greenbelt
Great Falls
Potomac
Great Falls Park
Silver Spring
Adelphi
College Park
Woodmont
Bethesda
Chevy Chase
Avenel
Langley Park
Berwyn Hts.
Lanham
Lewisdale
MARYLAND
VIRGINIA
Cabin John
Glen Echo
Takoma Park
Univ. of Maryland
University Park
Greenbelt Park
New Carrollton
Seabrook
Potomac River
Belleview
Somerset
Chillum
East Pines
Brightwood
Riverdale
Langley
Brookmont
Hyattsville
John Hanson Hwy.
Wolf Trap Farm Park
Rock Creek Park
Edmondston
Mt. Rainier
Landover Hills
McLean
American University
Capital Beltway
Bladensburg
Glenarden
Kentland
Palmer Park
WASHINGTON
DISTRICT OF COLUMBIA
Franklin Park
Pimmit Hills
Georgetown
Trinidad
Nat'l Arboretum
Cheverly
Fairmount Heights
Vienna
Dunn Loring
Theodore Roosevelt Memorial
The White House
Anacostia River Park
Seat Pleasant
Arlington
Rosslyn
Lincoln Memorial
The Mall
U.S. Capitol
Library of Congress
Kettering
Falls Church
Seven Corners
Arlington Nat'l Cemetery
Ft. du Pont Park
Capitol Hts.
Millwood
Anacostia River
Hillwood
G. Mason Mem. Br.
East Potomac Park
Pentagon
Oakland
Ritchie
Broyhill Park
East Arlington
Coral Hills
District Hts.
ARLINGTON
FAIRFAX
Washington Nat'l Airport
Annalee Hts.
L. Barcroft
Baileys Crossroads
Anacostia
Holmes Acres
Accotink Creek
PRINCE GEORGES
Suitland
Forestville
Little River Hwy.
Parklawn
Four Mile Run
Annandale
Hillcrest Hts.
Morningside
Silver Hill
Glassmanor
Potomac River
Long Br.
Alexandria
Holmes Run
Forest Heights
Temple Hills Park
North Springfield
Henry G. Shirley Mem. Hwy.
Camp Springs
Andrews Air Force Base
Kings Park
Oxon Hill
Henson Cr.
Pohick Cr.
West Springfield
Franconia
Rose Hill
Huntington
W. Wilson Mem. Br.
South Lawn
Springfield
Groveton
Belle Haven
Fort Foote Village
Oaklawn
West from Greenwich
COPYRIGHT GEORGE PHILIP AND SON LTD

DETROIT

1: 250 000

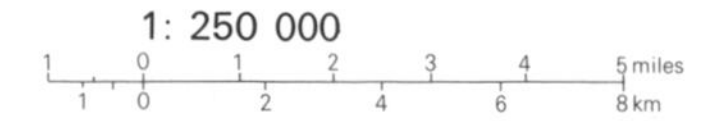

1: 250 000
miles
km

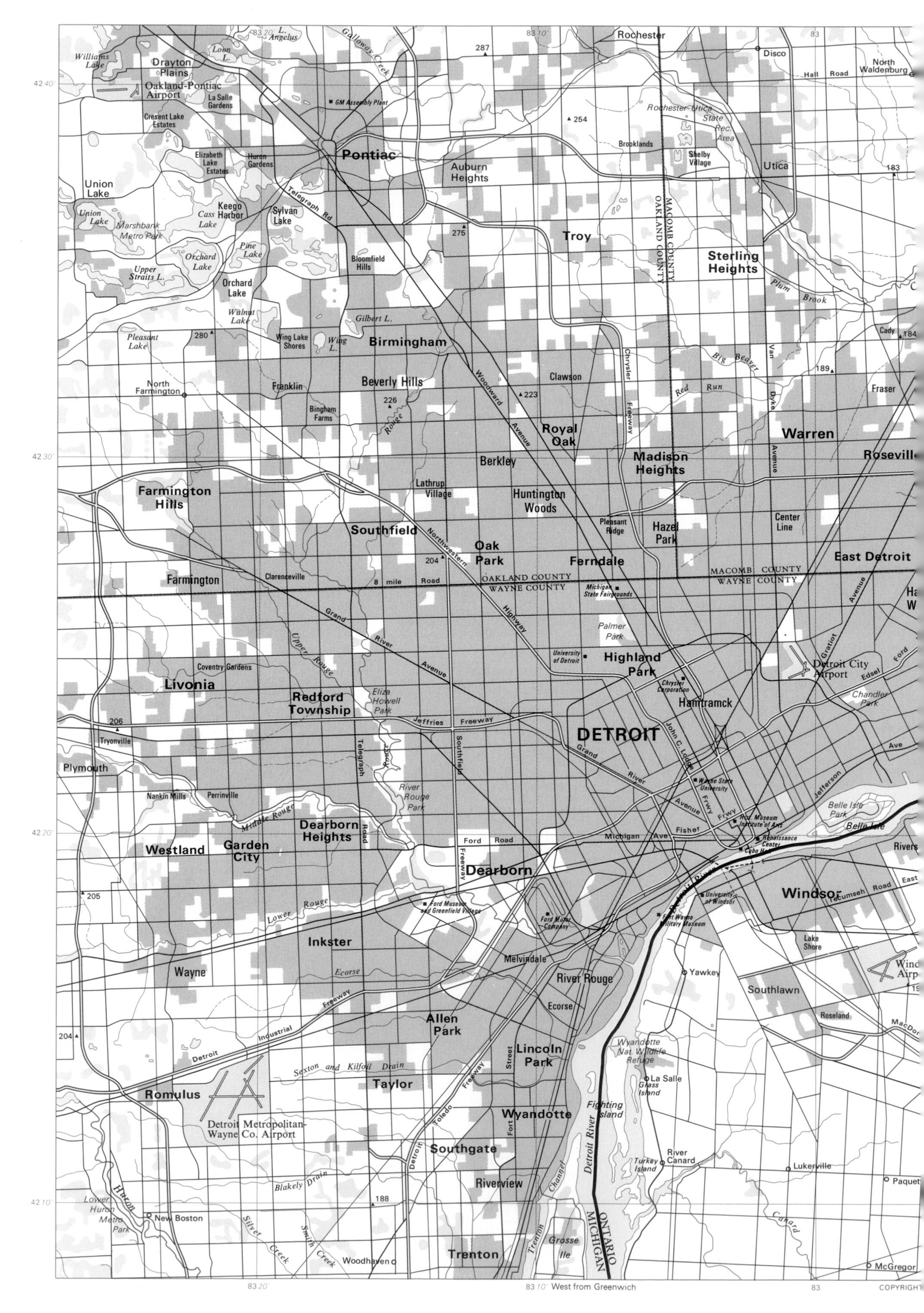

Rochester
Disco
North Waldenburg
Hall Road
Drayton Plains
Oakland-Pontiac Airport
La Salle Gardens
Crescent Lake Estates
GM Assembly Plant
Pontiac
Auburn Heights
Rochester-Utica State Rec. Area
Brooklands
Shelby Village
Utica
Union Lake
Elizabeth Lake Estates
Huron Gardens
Keego Harbor
Cass Lake
Sylvan Lake
Telegraph Rd
Marshbank Metro Park
Orchard Lake
Pine Lake
Upper Straits L.
Bloomfield Hills
Troy
OAKLAND COUNTY
MACOMB COUNTY
Sterling Heights
Plum Brook
Walnut Lake
Gilbert L.
Pleasant Lake
Wing Lake Shores
Birmingham
Clawson
Chrysler Freeway
Big Beaver
Van Dyke Avenue
Cady
Fraser
North Farmington
Franklin
Beverly Hills
Bingham Farms
Woodward Avenue
Royal Oak
Red Run
Warren
Berkley
Madison Heights
Roseville
Farmington Hills
Lathrup Village
Huntington Woods
Southfield
Pleasant Ridge
Hazel Park
Center Line
Oak Park
Ferndale
East Detroit
Northwestern Highway
Farmington
Clarenceville
8 mile Road
OAKLAND COUNTY
WAYNE COUNTY
MACOMB COUNTY
WAYNE COUNTY
Michigan State Fairgrounds
Grand River Avenue
Upper Rouge
Palmer Park
University of Detroit
Highland Park
Chrysler Corporation
Gratiot
Detroit City Airport
Coventry Gardens
Livonia
Redford Township
Eliza Howell Park
Hamtramck
Chandler Park
Jeffries Freeway
Tryonville
DETROIT
John C. Lodge Frwy
Plymouth
Telegraph
Southfield
River Rouge Park
Wayne State University
Nankin Mills
Perrinville
Middle Rouge
Dearborn Heights
Jefferson
Belle Isle Park
Belle Isle
Fisher Frwy
Michigan Ave
Detroit Inst. Museum Institute of Arts
Renaissance Center
Cobo Hall
Westland
Garden City
Ford Road
Dearborn
Windsor
Tecumseh Road
Ford Museum and Greenfield Village
University of Windsor
Lower Rouge
Ford Motor Company
Fort Wayne Military Museum
Inkster
Melvindale
Lake Shore
Wayne
Ecorse
River Rouge
Yawkey
Southlawn
Ecorse
Allen Park
Roseland
Industrial Freeway
Detroit
Wyandotte Nat. Wildlife Refuge
Lincoln Park
Sexton and Kilfoil Drain
Taylor
La Salle
Grass Island
Romulus
Fighting Island
Wyandotte
Detroit Metropolitan-Wayne Co. Airport
Toledo Freeway
Fort Street
Detroit River
Southgate
Turkey Island
River Canard
Lukerville
Paquet
Riverview
Blakely Drain
Canard
Lower Huron Metro Park
New Boston
Huron
Silver Creek
Smith Creek
Woodhaven
Trenton
Grosse Ile
ONTARIO
MICHIGAN
McGregor
West from Greenwich

1: 250 000

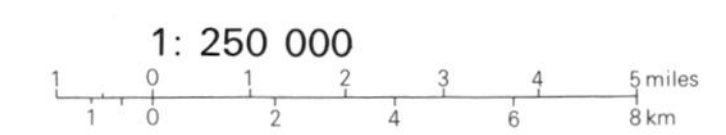
1 0 1 2 3 4 5 miles
1 0 2 4 6 8 km

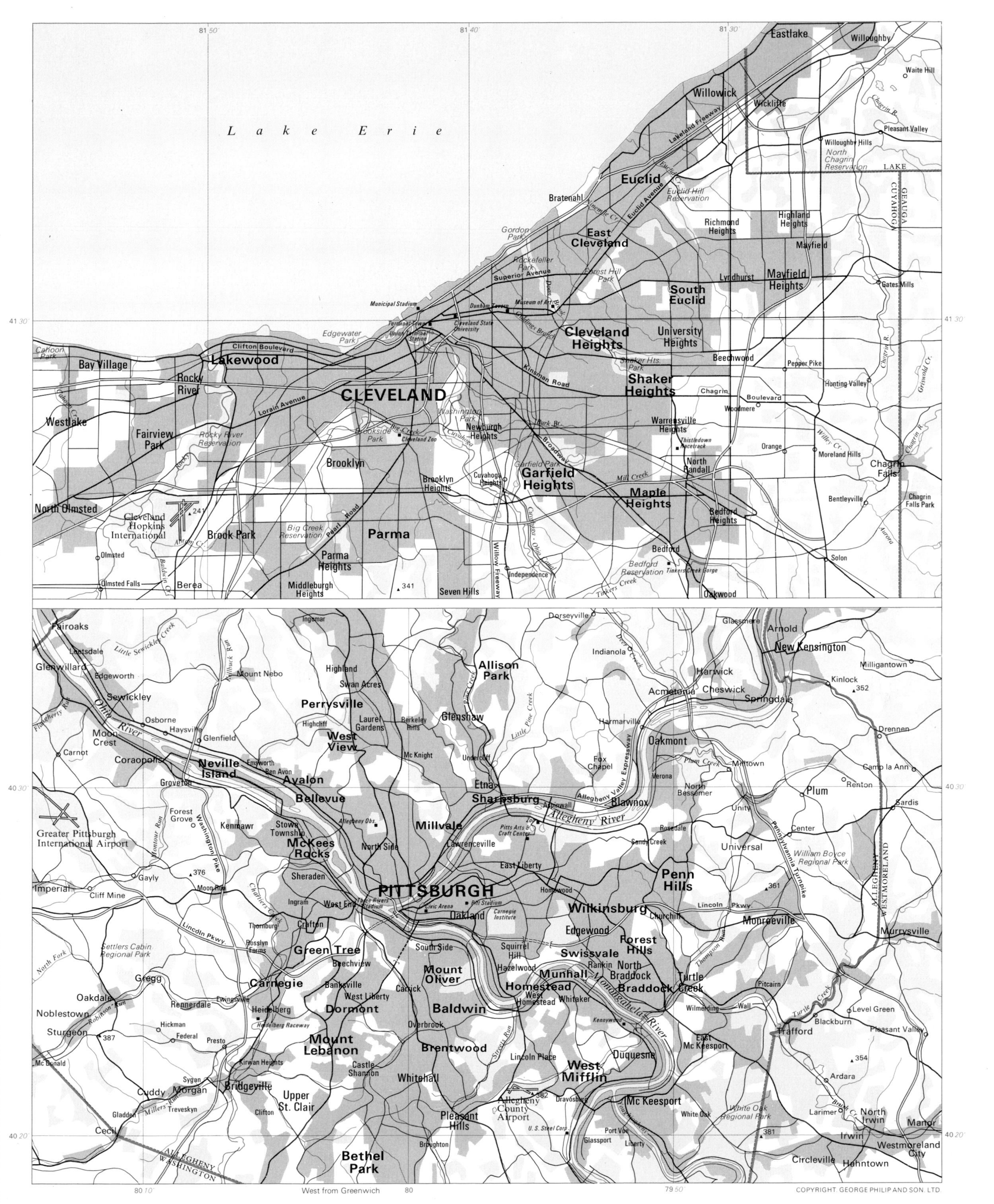
Lake Erie
Eastlake
Willoughby
Waite Hill
Willowick
Wickliffe
Lakeland Freeway
Pleasant Valley
Willoughby Hills
North Chagrin Reservation
LAKE
GEAUGA
CUYAHOGA
Euclid
Euclid Avenue
Euclid Hill Reservation
Bratenahl
Richmond Heights
Highland Heights
Mayfield
Gordon Park
East Cleveland
Rockefeller Park
Superior Avenue
Forest Hill Park
Lyndhurst
Mayfield Heights
Gates Mills
South Euclid
Municipal Stadium
Museum of Art
Cleveland State University
Edgewater Park
Terminal Tower
Union Terminal Station
Cleveland Heights
University Heights
Clifton Boulevard
Lakewood
Bay Village
Cahoon Park
Rocky River
Beechwood
Pepper Pike
Kinsman Road
Shaker Hts. Park
Shaker Heights
Hunting Valley
Chagrin
Boulevard
Woodmere
CLEVELAND
Lorain Avenue
Westlake
Fairview Park
Rocky River Reservation
Washington Park
Brookside Park
Big Creek
Cleveland Zoo
Newburgh Heights
Warrensville Heights
Thistledown Racetrack
Orange
Moreland Hills
Broadway
Chagrin Falls
Brooklyn
Brooklyn Heights
Cuyahoga Heights
Garfield Park
Garfield Heights
Mill Creek
North Randall
Maple Heights
Bentleyville
Chagrin Falls Park
North Olmsted
Cleveland Hopkins International
Brook Park
Big Creek Reservation
Pearl Road
Parma
Bedford Heights
Aurora
Olmsted
Parma Heights
Bedford
Bedford Reservation
Tinkers Creek Gorge
Solon
Independence
Olmsted Falls
Berea
Middleburgh Heights
Seven Hills
Willow Freeway
Tinkers Creek
Oakwood
81 50′
81 40′
81 30′
41 30′
Dorseyville
Glassmere
Arnold
New Kensington
Fairoaks
Ingomar
Indianola
Leetsdale
Glenwillard
Edgeworth
Mount Nebo
Highland
Swan Acres
Allison Park
Harwick
Milligantown
Kinlock
Sewickley
Perrysville
Acmetonia
Cheswick
Springdale
Glenshaw
Osborne
Haysville
Moon Crest
Glenfield
Highcliff
Laurel Gardens
Berkeley Hills
West View
Harmarville
Oakmont
Drennen
Carnot
Coraopolis
Neville Island
Emsworth
Ben Avon
Mc Knight
Fox Chapel
Undercliff
Milltown
Camp la Ann
Groveton
Avalon
Verona
Etna
Sharpsburg
North Bessemer
Renton
Plum
Bellevue
Aspinwall
Blawnox
Allegheny River
Unity
Sardis
Forest Grove
Kenmawr
Stowe Township
Allegheny Obs.
Millvale
Pitts Arts & Craft Center
Rosedale
Center
Greater Pittsburgh International Airport
McKees Rocks
North Side
Lawrenceville
Sandy Creek
Universal
William Boyce Regional Park
East Liberty
Penn Hills
Imperial
Gayly
Moon Run
Sheraden
PITTSBURGH
Pitt Stadium
Homewood
Cliff Mine
Ingram
West End
Three Rivers Stadium
Civic Arena
Oakland
Carnegie Institute
Wilkinsburg
Lincoln Pkwy
Churchill
Monroeville
Thornburg
Crafton
Edgewood
Murrysville
Settlers Cabin Regional Park
Rosslyn Farms
Green Tree
South Side
Squirrel Hill
Forest Hills
Swissvale
ALLEGHENY
WESTMORELAND
Gregg
Beechview
Hazelwood
Rankin
North Braddock
Oakdale
Carnegie
Mount Oliver
Munhall
Homestead
Braddock
Turtle Creek
Pitcairn
Renerdale
Banksville
West Liberty
Carrick
West Homestead
Whitaker
Noblestown
Dormont
Baldwin
Monongahela River
Wilmerding
Wall
Level Green
Heidelberg
Heidelberg Raceway
Overbrook
Kennywood
Blackburn
Pleasant Valley
Sturgeon
Hickman
Federal
Presto
Mount Lebanon
Brentwood
East Mc Keesport
Trafford
Mc Donald
Kirwan Heights
Castle Shannon
Lincoln Place
West Mifflin
Duquesne
Cuddy
Sygan
Morgan
Bridgeville
Upper St. Clair
Whitehall
Ardara
Allegheny County Airport
Dravosburg
Mc Keesport
Gladden
Millers Run
Treveskyn
Clifton
White Oak
White Oak Regional Park
Larimer
North Irwin
Pleasant Hills
U. S. Steel Corp.
Port Vue
Irwin
Manor
Cecil
Glassport
Liberty
Broughton
Westmoreland City
Bethel Park
Circleville
Hahntown
ALLEGHENY
WASHINGTON
Pennsylvania Turnpike
Ohio River
40 30′
40 20′
80 10′
West from Greenwich
80
79 50′

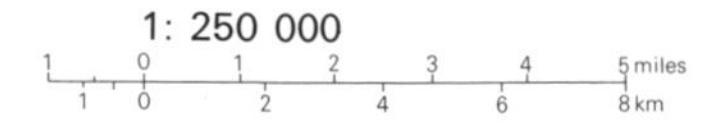
1: 250 000
miles
km

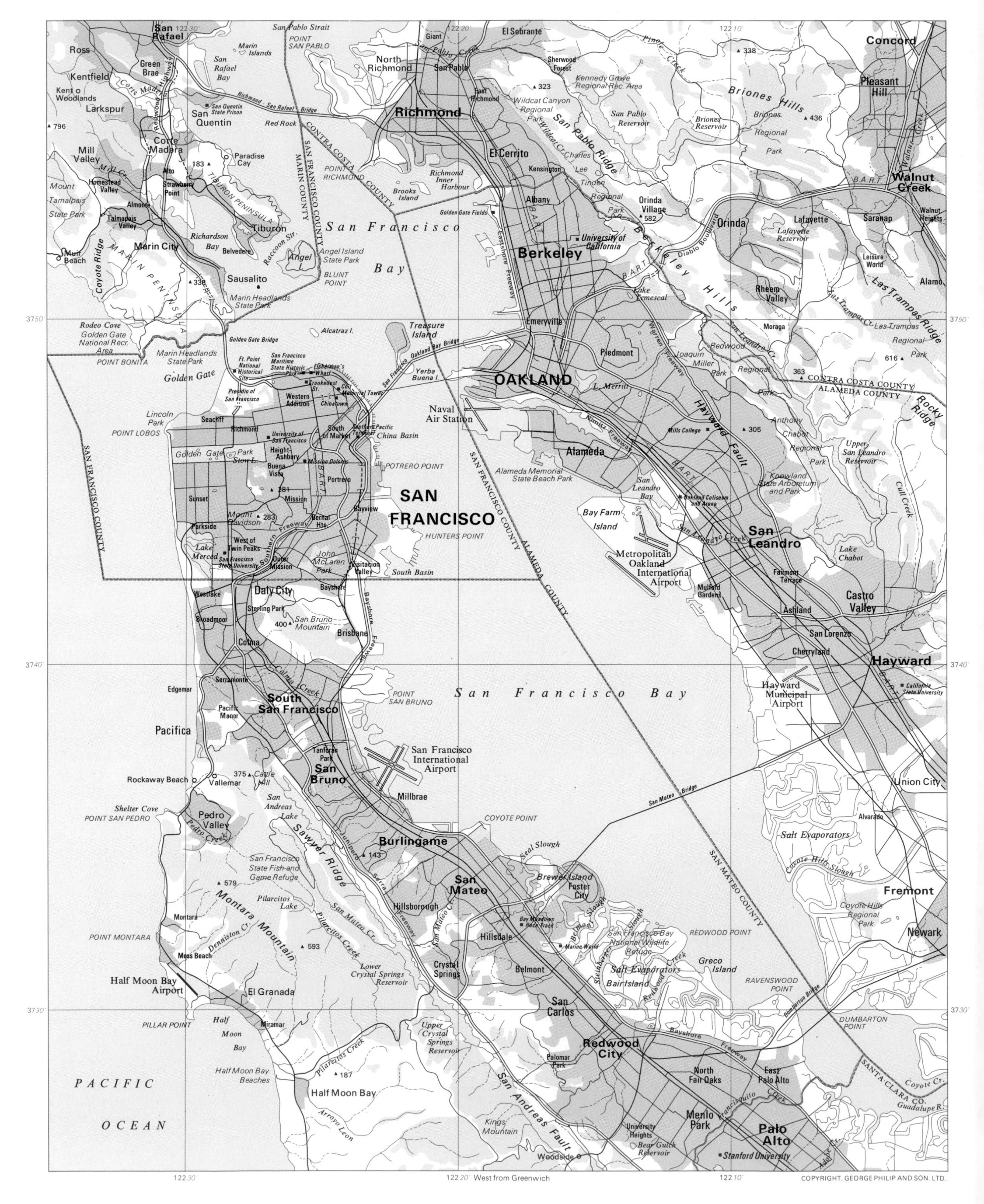
San Rafael
Ross
Kentfield
Green Brae
Larkspur
San Quentin
Mill Valley
Corte Madera
Tiburon
Marin City
Sausalito
Richmond
El Cerrito
Berkeley
Albany
Emeryville
Piedmont
OAKLAND
Alameda
SAN FRANCISCO
Daly City
South San Francisco
Pacifica
San Bruno
Millbrae
Burlingame
San Mateo
Hillsborough
Belmont
San Carlos
Redwood City
Menlo Park
Palo Alto
San Leandro
Hayward
Castro Valley
Concord
Pleasant Hill
Walnut Creek
Orinda
Lafayette
Fremont
Newark
Union City
Half Moon Bay
San Francisco Bay
PACIFIC OCEAN
San Francisco International Airport
Metropolitan Oakland International Airport
COPYRIGHT, GEORGE PHILIP AND SON, LTD.

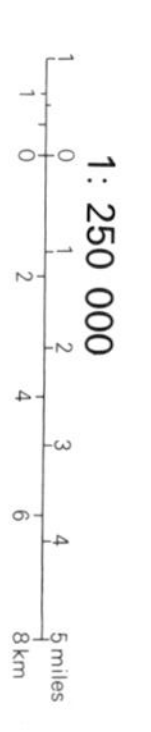
1: 250 000
1 0 1 2 3 4 5 miles
1 0 2 4 6 8 km

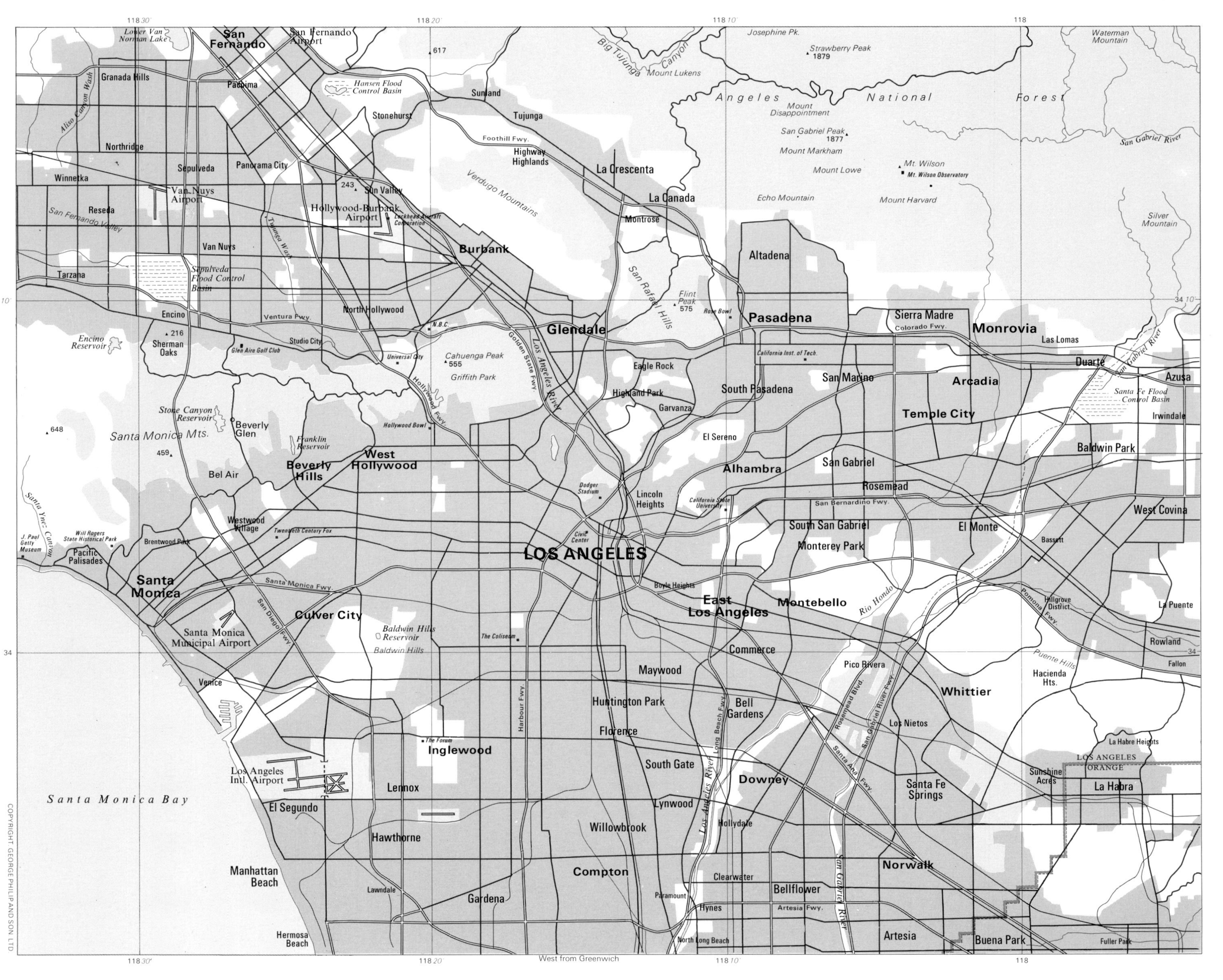
Lower Van Norman Lake
San Fernando
San Fernando Airport
617
Josephine Pk.
Strawberry Peak 1879
Waterman Mountain
Big Tujunga Canyon
Mount Lukens
Granada Hills
Pacoima
Hansen Flood Control Basin
Sunland
Angeles National Forest
Mount Disappointment
Aliso Canyon Wash
Stonehurst
Tujunga
San Gabriel Peak 1877
Mount Markham
San Gabriel River
Northridge
Foothill Fwy.
Highway Highlands
Mt. Wilson
Mt. Wilson Observatory
Winnetka
Sepulveda
Panorama City
Verdugo Mountains
La Crescenta
Mount Lowe
243
Sun Valley
Van Nuys Airport
Reseda
Hollywood-Burbank Airport
Lockheed Aircraft Corporation
La Canada
Echo Mountain
Mount Harvard
Silver Mountain
San Fernando Valley
Tujunga Wash
Montrose
Van Nuys
Burbank
Altadena
San Rafael Hills
Tarzana
Sepulveda Flood Control Basin
Flint Peak 575
Rose Bowl
Encino
Ventura Fwy.
North Hollywood
Pasadena
Sierra Madre
Colorado Fwy.
Monrovia
Las Lomas
Encino Reservoir
216
Sherman Oaks
N.B.C.
Glendale
Golden State Fwy.
Los Angeles River
California Inst. of Tech.
Duarte
Azusa
Glen Aire Golf Club
Studio City
Universal City
Cahuenga Peak 555
Griffith Park
Eagle Rock
San Marino
Arcadia
Santa Fe Flood Control Basin
Irwindale
Highland Park
South Pasadena
Stone Canyon Reservoir
Garvanza
Temple City
648
Beverly Glen
Hollywood Fwy.
Santa Monica Mts.
Hollywood Bowl
El Sereno
Franklin Reservoir
459
West Hollywood
Beverly Hills
Bel Air
Alhambra
San Gabriel
Baldwin Park
Dodger Stadium
Lincoln Heights
California State University
Rosemead
San Bernardino Fwy.
West Covina
Santa Ynez Canyon
Westwood Village
Twentieth Century Fox
South San Gabriel
El Monte
J. Paul Getty Museum
Will Rogers State Historical Park
Brentwood Park
Civic Center
Bassett
Pacific Palisades
Monterey Park
LOS ANGELES
Santa Monica
Santa Monica Fwy.
Boyle Heights
Hillgrove District
Pomona Fwy.
East Los Angeles
Montebello
Rio Hondo
La Puente
San Diego Fwy.
Culver City
Santa Monica Municipal Airport
Baldwin Hills Reservoir
Baldwin Hills
The Coliseum
Commerce
Rowland
Puente Hills
Pico Rivera
Fallon
Venice
Maywood
Hacienda Hts.
Whittier
Huntington Park
Bell Gardens
Rosemead Blvd.
Long Beach Fwy.
San Gabriel River Fwy.
Harbour Fwy.
Florence
Los Nietos
La Habre Heights
The Forum
Inglewood
LOS ANGELES
ORANGE
South Gate
Los Angeles Intl. Airport
Santa Ana Fwy.
Sunshine Acres
Downey
Santa Fe Springs
La Habra
Lennox
Santa Monica Bay
El Segundo
Lynwood
Los Angeles River
Hollydale
Willowbrook
Hawthorne
Norwalk
Manhattan Beach
Compton
Clearwater
San Gabriel River
Lawndale
Paramount
Bellflower
Gardena
Hynes
Artesia Fwy.
Artesia
Buena Park
Hermosa Beach
North Long Beach
Fuller Park
118 30'
118 20'
118 10'
118
34 10'
34
West from Greenwich

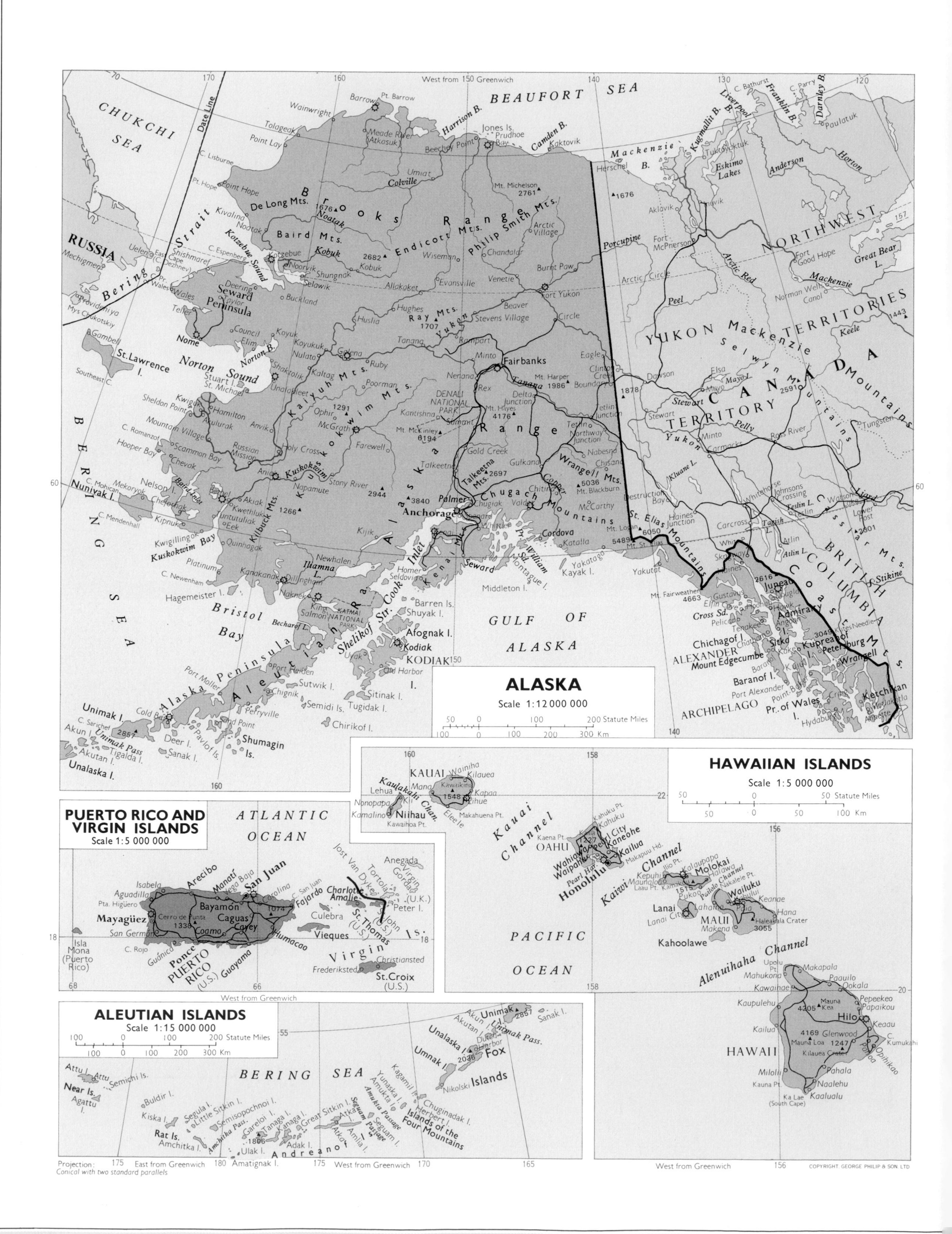
ALASKA
Scale 1:12 000 000
Statute Miles
Km
West from 150 Greenwich
BEAUFORT SEA
CHUKCHI SEA
Date Line
RUSSIA
Bering Strait
BERING SEA
Brooks Range
De Long Mts.
Baird Mts.
Endicott Mts.
Philip Smith Mts.
Seward Peninsula
Norton Sound
St. Lawrence I.
Nome
Kotzebue
Kotzebue Sound
Barrow
Pt. Barrow
Wainwright
Point Lay
Point Hope
C. Lisburne
Prudhoe Bay
Kaktovik
Colville
Umiat
Mt. Michelson 2761
Kobuk
Noatak
Arctic Village
Fort Yukon
Circle
Yukon
Ray Mts.
Kaiyuh Mts.
Kuskokwim Mts.
Kuskokwim Bay
Kuskokwim
Fairbanks
Tanana
Mt. Harper 1986
DENALI NATIONAL PARK
Mt. McKinley 6194
Alaska Range
Talkeetna Mts. 2697
Wrangell Mts.
Chugach Mountains
Mt. Blackburn
Mt. McCarthy
Anchorage
Palmer
Valdez
Cordova
Seward
Kenai Mts.
Cook Inlet
Iliamna L.
Bristol Bay
Dillingham
Naknek
KATMAI NATIONAL PARK
Alaska Peninsula
Aleutian Ra.
Shelikof Str.
Kodiak
KODIAK I.
Afognak I.
Barren Is.
Shuyak I.
Sutwik I.
Chignik
Semidi Is.
Sitinak I.
Tugidak I.
Chirikof I.
Shumagin Is.
Unimak I.
Unimak Pass
Unalaska I.
Sanak I.
GULF OF ALASKA
Middleton I.
Kayak I.
Yakutat
Mt. St. Elias
Mt. Logan 6050
St. Elias Mountains
Mt. Fairweather 4663
Juneau
Sitka
Petersburg
Wrangell
Ketchikan
Chichagof I.
ALEXANDER ARCHIPELAGO
Baranof I.
Admiralty I.
Kupreanof I.
Pr. of Wales I.
Mount Edgecumbe
Cross Sd.
Coast Mts.
Cassiar Mts.
BRITISH COLUMBIA
YUKON TERRITORY
NORTHWEST TERRITORIES
CANADA
Mackenzie Mountains
Selwyn Mountains
Mackenzie B.
Herschel
Inuvik
Aklavik
Tuktoyaktuk
Eskimo Lakes
Fort McPherson
Porcupine
Arctic Red
Peel
Dawson
Whitehorse
Carcross
Atlin
Atlin L.
Teslin L.
Watson Lake
Liard
Stikine
Kluane L.
Haines
Haines Junction
Great Bear L.
Norman Wells
Fort Good Hope
Mackenzie
Paulatuk
Franklin B.
Liverpool B.
Darnley B.
Anderson
Horton
Keele
Pelly
Ross River
Tungsten
Mayo
Elsa
Stewart
Carmacks
Minto
HAWAIIAN ISLANDS
Scale 1:5 000 000
Statute Miles
Km
KAUAI
Niihau
Lehua
Kaulakahi Chan.
Kauai Channel
OAHU
Honolulu
Pearl Har.
Kaneohe
Kailua
Kaiwi Channel
Molokai
Lanai
MAUI
Wailuku
Haleakala Crater 3055
Kahoolawe
Alenuihaha Channel
HAWAII
Mauna Kea 4205
Mauna Loa 4169
Kilauea Crater 1247
Hilo
Kailua
Ka Lae (South Cape)
PACIFIC OCEAN
PUERTO RICO AND VIRGIN ISLANDS
Scale 1:5 000 000
ATLANTIC OCEAN
San Juan
Arecibo
Mayagüez
Ponce
Bayamón
Caguas
Cerro de Punta 1338
PUERTO RICO (U.S.)
Culebra
Vieques
Charlotte Amalie
St. Thomas
St. John
Tortola
Anegada
Virgin Is.
St. Croix (U.S.)
Christiansted
Frederiksted
Isla Mona (Puerto Rico)
West from Greenwich
ALEUTIAN ISLANDS
Scale 1:15 000 000
Statute Miles
Km
BERING SEA
Attu I.
Near Is.
Agattu I.
Semichi Is.
Buldir I.
Kiska I.
Rat Is.
Amchitka I.
Amchitka Pass.
Andreanof Is.
Adak I.
Atka I.
Islands of the Four Mountains
Umnak I.
Unalaska I.
Fox Islands
Unimak I.
Unimak Pass.
Akutan I.
Akun I.
Sanak I.
Projection: Conical with two standard parallels
East from Greenwich
West from Greenwich
COPYRIGHT GEORGE PHILIP & SON LTD

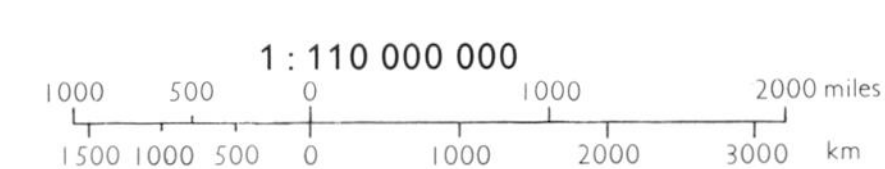
1 : 110 000 000
1000 500 0 1000 2000 miles
1500 1000 500 0 1000 2000 3000 km

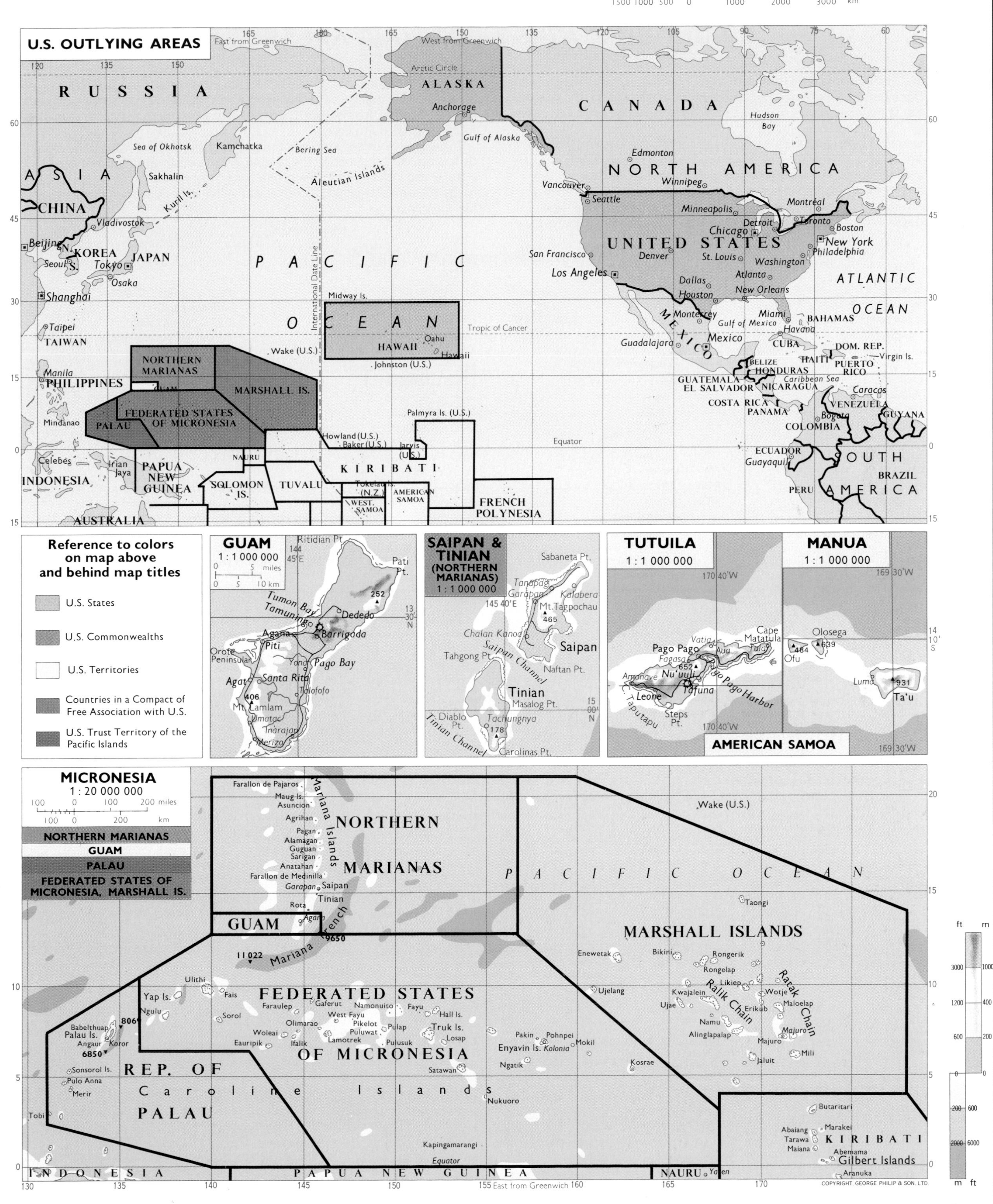
U.S. OUTLYING AREAS
East from Greenwich
West from Greenwich
RUSSIA
ASIA
CHINA
Beijing
N. KOREA
S.
Seoul
JAPAN
Tokyo
Osaka
Shanghai
Taipei
TAIWAN
Manila
PHILIPPINES
Mindanao
Celebes
INDONESIA
Irian Jaya
PAPUA NEW GUINEA
AUSTRALIA
Sea of Okhotsk
Kamchatka
Sakhalin
Vladivostok
Kuril Is.
Bering Sea
Aleutian Islands
Arctic Circle
ALASKA
Anchorage
Gulf of Alaska
CANADA
Hudson Bay
Edmonton
NORTH AMERICA
Winnipeg
Vancouver
Seattle
Montreal
Minneapolis
Detroit
Toronto
Boston
Chicago
New York
Philadelphia
UNITED STATES
Denver
St. Louis
Washington
San Francisco
Los Angeles
Atlanta
Dallas
Houston
New Orleans
ATLANTIC OCEAN
PACIFIC OCEAN
International Date Line
Midway Is.
Tropic of Cancer
HAWAII
Oahu
Hawaii
Wake (U.S.)
Johnston (U.S.)
MEXICO
Monterrey
Gulf of Mexico
Miami
BAHAMAS
Havana
Mexico
Guadalajara
CUBA
DOM. REP.
HAITI
PUERTO RICO
Virgin Is.
BELIZE
HONDURAS
Caribbean Sea
GUATEMALA
EL SALVADOR
NICARAGUA
Caracas
COSTA RICA
PANAMA
VENEZUELA
Bogotá
GUYANA
COLOMBIA
ECUADOR
Guayaquil
SOUTH AMERICA
BRAZIL
PERU
NORTHERN MARIANAS
GUAM
MARSHALL IS.
FEDERATED STATES OF MICRONESIA
PALAU
Palmyra Is. (U.S.)
Howland (U.S.)
Baker (U.S.)
Jarvis (U.S.)
Equator
NAURU
KIRIBATI
SOLOMON IS.
TUVALU
Tokelau Is. (N.Z.)
AMERICAN SAMOA
WEST. SAMOA
FRENCH POLYNESIA
Reference to colors on map above and behind map titles
U.S. States
U.S. Commonwealths
U.S. Territories
Countries in a Compact of Free Association with U.S.
U.S. Trust Territory of the Pacific Islands
GUAM
1 : 1 000 000
0 5 miles
0 5 10 km
Ritidian Pt.
144 45'E
Pati Pt.
252
13 30' N
Tumon Bay
Tamuning
Dededo
Agana
Barrigada
Piti
Orote Peninsula
Yona
Pago Bay
Agat
Santa Rita
Talofofo
406
Mt. Lamlam
Umatac
Inarajan
Merizo
SAIPAN & TINIAN (NORTHERN MARIANAS)
1 : 1 000 000
Sabaneta Pt.
Tanapag
Garapan
Kalabera
145 40'E
Mt. Tagpochau
465
Chalan Kanoa
Saipan
Tahgong Pt.
Saipan Channel
Naftan Pt.
Tinian
Masalog Pt.
15 00' N
Diablo Pt.
Tachungnya
178
Tinian Channel
Carolinas Pt.
TUTUILA
1 : 1 000 000
170 40'W
Cape Matatula
Vatia
Pago Pago
Aua
Fagasa
Tula
652
Amanave
Nu'uuli
Pago Pago Harbor
Leone
Tafuna
Taputapu
Steps Pt.
MANUA
1 : 1 000 000
169 30'W
Olosega
639
484
Ofu
Luma
931
Ta'u
14 10' S
AMERICAN SAMOA
MICRONESIA
1 : 20 000 000
100 0 100 200 miles
100 0 200 km
NORTHERN MARIANAS
GUAM
PALAU
FEDERATED STATES OF MICRONESIA, MARSHALL IS.
Farallon de Pajaros
Maug Is.
Asuncion
Agrihan
Pagan
Alamagan
Guguan
Sarigan
Anatahan
Farallon de Medinilla
Garapan
Saipan
Tinian
Rota
Agana
Mariana Islands
NORTHERN MARIANAS
GUAM
Mariana Trench
9650
11 022
Wake (U.S.)
PACIFIC OCEAN
Taongi
MARSHALL ISLANDS
Enewetak
Bikini
Rongerik
Rongelap
Ujelang
Likiep
Wotje
Kwajalein
Ujae
Ralik Chain
Ratak Chain
Erikub
Maloelap
Namu
Ailinglapalap
Majuro
Jaluit
Mili
Ulithi
Yap Is.
Fais
Ngulu
Sorol
FEDERATED STATES OF MICRONESIA
Faraulep
Gaferut
Namonuito
Fayu
Hall Is.
West Fayu
Pikelot
Olimarao
Puluwat
Pulap
Truk Is.
Losap
Woleai
Eauripik
Ifalik
Lamotrek
Pulusuk
Pakin
Pohnpei
Enyavin Is.
Kolonia
Mokil
Ngatik
Kosrae
Satawan
Babelthuap
Palau Is.
806
Angaur
Koror
6850
Sonsorol Is.
Pulo Anna
Merir
REP. OF PALAU
Caroline Islands
Nukuoro
Tobi
Kapingamarangi
Equator
Butaritari
Abaiang
Marakei
Tarawa
Maiana
KIRIBATI
Abemama
Gilbert Islands
Aranuka
INDONESIA
PAPUA NEW GUINEA
NAURU
Yaren
155 East from Greenwich 160
ft m
3000 1000
1200 400
600 200
0 0
200 600
2000 6000
m ft

INDEX

UNITED STATES & OUTLYING AREAS

This index lists all the place names which appear on the large scale maps of the United States and outlying areas (pages which precede this index). Place names for the rest of the world can be found in the World Maps Index at the end of the atlas.

The number in dark type which follows each name in the index refers to the page number on which the place or feature is located. The geographical coordinates which follow the page number give the latitude and longitude of each place. The first coordinate indicates the latitude – the distance north or south of the Equator. The second coordinate indicates the longitude – the distance east or west of the Greenwich Meridian. Both latitude and longitude are measured in degrees and minutes (there are 60 minutes in a degree). Rivers are indexed to their mouths or confluences. A solid square ■ follows the name of a country, while an open square □ signifies that the name is a state. An arrow → follows the name of a river.

The alphabetic order of names composed of two or more words is governed by the first word and then by the second. Names composed of a proper name (Alaska) and a description (Gulf of) are positioned alphabetically by the proper name. All names beginning St. are alphabetized under Saint and those beginning Mc under Mac.

Abbreviations used in the index

Ala. — Alabama
Amer. — America, American
Ariz. — Arizona
Ark. — Arkansas
B. — Bay
C. — Cape
Calif. — California
Chan. — Channel
Colo. — Colorado
Conn. — Connecticut
Cr. — Creek
D.C. — District of Columbia
Del. — Delaware
Dist. — District
E. — East, Eastern
Fla. — Florida
G. — Gulf
Ga. — Georgia
Gt. — Great
I.(s) — Island(s)
Ill. — Illinois
Ind. — Indiana
Kans. — Kansas
Ky. — Kentucky
L. — Lake
La. — Louisiana
Ld. — Land
Mass. — Massachusetts
Md. — Maryland
Mich. — Michigan
Minn. — Minnesota
Miss. — Mississippi
Mo. — Missouri
Mont. — Montana
Mt.(s) — Mountain(s)
N. — North, Northern
N.B. — New Brunswick
N.C. — North Carolina
N. Dak. — North Dakota
N.H. — New Hampshire
N.J. — New Jersey
N. Mex. — New Mexico
N.Y. — New York
Nat. Mon. — National Monument
Nat. Park — National Park
Nat. Rec. Area. — National Recreation Area
Nebr. — Nebraska
Nev. — Nevada
Okla. — Oklahoma
Oreg. — Oregon
Pa. — Pennsylvania
Pac. Oc. — Pacific Ocean
Pass. — Passage
Pen. — Peninsula
Pk. — Peak
Pt. — Point
R. — Rio, River
R.I. — Rhode Island
Ra.(s) — Range(s)
Res. — Reserve, Reservoir, Reservation
S.C. — South Carolina
S. Dak. — South Dakota
Sa. — Serra, Sierra
Sd. — Sound
St. — Saint
Ste. — Sainte
Str. — Strait
Tenn. — Tennessee
Tex. — Texas
U.S.A. — United States of America
Va. — Virginia
Vt. — Vermont
Wash. — Washington
W. — West, Western
W. Va. — West Virginia
Wis. — Wisconsin
Wyo. — Wyoming

A

Abbaye, Pt., *Mich.* **10** 46 58N 88 8W
Abbeville, *La.* **13** 29 58N 92 8W
Abbeville, *S.C.* **11** 34 11N 82 23W
Abbotsford, *Wis.* **12** 44 57N 90 19W
Aberdeen, *Ala.* **11** 33 49N 88 33W
Aberdeen, *Idaho* **16** 42 57N 112 50W
Aberdeen, *S. Dak.* **12** 45 28N 98 29W
Aberdeen, *Wash.* **18** 46 59N 123 50W
Abernathy, *Tex.* **13** 33 50N 101 51W
Abert, L., *Oreg.* **16** 42 38N 120 14W
Abilene, *Kans.* **12** 38 55N 97 13W
Abilene, *Tex.* **13** 32 28N 99 43W
Abingdon, *Ill.* **14** 40 48N 90 24W
Abingdon, *Va.* **11** 36 43N 81 59W
Absaroka Range, *Wyo.* **16** 44 45N 109 50W
Accomac, *Va.* **10** 37 43N 75 40W
Ackerman, *Miss.* **13** 33 19N 89 11W
Ada, *Minn.* **12** 47 18N 96 31W
Ada, *Okla.* **13** 34 46N 96 41W
Adams, *Mass.* **9** 42 38N 73 7W
Adams, *N.Y.* **9** 43 49N 76 1W
Adams, *Wis.* **12** 43 57N 89 49W
Adams Mt., *Wash.* **18** 46 12N 121 30W
Adel, *Ga.* **11** 31 8N 83 25W
Adelanto, *Calif.* **19** 34 35N 117 22W
Adin, *Calif.* **16** 41 12N 120 57W
Adirondack Mts., *N.Y.* **9** 44 0N 74 0W
Admiralty I., *Alaska* **30** 57 30N 134 30W
Admiralty Inlet, *Wash.* **16** 48 8N 122 58W
Adrian, *Mich.* **15** 41 54N 84 2W
Adrian, *Tex.* **13** 35 16N 102 40W
Affton, *Miss.* **14** 38 33N 90 20W
Afognak I., *Alaska* **30** 58 15N 152 30W
Afton, *N.Y.* **9** 42 14N 75 32W
Agana, *Guam* **31** 13 28N 144 45 E
Agattu I., *Alaska* **30** 52 25N 172 30 E
Agua Caliente Springs, *Calif.* **19** 32 56N 116 19W
Aguadilla, *Puerto Rico* **30** 18 26N 67 10W
Aguanga, *Calif.* **19** 33 27N 116 51W
Aiken, *S.C.* **11** 33 34N 81 43W
Ainsworth, *Nebr.* **12** 42 33N 99 52W
Aitkin, *Minn.* **12** 46 32N 93 42W
Ajo, *Ariz.* **17** 32 22N 112 52W
Akiak, *Alaska* **30** 60 55N 161 13W
Akron, *Colo.* **12** 40 10N 103 13W
Akron, *Ohio* **8** 41 5N 81 31W
Akulurak, *Alaska* **30** 62 40N 164 35W
Akun I., *Alaska* **30** 54 11N 165 32W
Akutan I., *Alaska* **30** 54 7N 165 55W
Alabama □ **11** 33 0N 87 0W
Alabama →, *Ala.* **11** 31 8N 87 57W
Alameda, *Calif.* **28** 37 46N 122 15W
Alameda, *N. Mex.* **17** 35 11N 106 37W
Alameda County, *Calif.* **28** 37 40N 122 10W
Alamo, *Nev.* **19** 36 21N 115 10W
Alamogordo, *N. Mex.* **17** 32 54N 105 57W
Alamosa, *Colo.* **17** 37 28N 105 52W
Alaska □ **30** 64 0N 154 0W
Alaska, G. of, *Pac. Oc.* **30** 58 0N 145 0W
Alaska Peninsula, *Alaska* **30** 56 0N 159 0W
Alaska Range, *Alaska* **30** 62 50N 151 0W
Alava, C., *Wash.* **16** 48 10N 124 44W
Albany, *Ga.* **11** 31 35N 84 10W
Albany, *Minn.* **12** 45 38N 94 34W
Albany, *N.Y.* **9** 42 39N 73 45W
Albany, *Oreg.* **16** 44 38N 123 6W
Albany, *Tex.* **13** 32 44N 99 18W
Albemarle, *N.C.* **11** 35 21N 80 11W
Albemarle Sd., *N.C.* **11** 36 5N 76 0W
Albert Lea, *Minn.* **12** 43 39N 93 22W
Albia, *Iowa* **14** 41 2N 92 48W
Albion, *Idaho* **16** 42 25N 113 35W
Albion, *Mich.* **15** 42 15N 84 45W
Albion, *Nebr.* **12** 41 42N 98 0W
Albuquerque, *N. Mex.* **17** 35 5N 106 39W
Alcatraz I., *Calif.* **28** 37 49N 122 25W
Alcoa, *Tenn.* **11** 35 48N 83 59W
Alcova, *Wyo.* **16** 42 34N 106 43W
Alder, *Mont.* **16** 45 19N 112 6W
Alder Pk., *Calif.* **18** 35 53N 121 22W
Aledo, *Ill.* **14** 41 12N 90 45W
Alenuihaha Channel, *Hawaii* **30** 20 30N 156 0W
Aleutian Is., *Pac. Oc.* **30** 52 0N 175 0W
Aleutian Ra., *Alaska* **30** 55 0N 155 0W
Alexander, *N. Dak.* **12** 47 51N 103 39W
Alexander Arch., *Alaska* **30** 56 0N 136 0W
Alexander City, *Ala.* **11** 32 56N 85 58W
Alexandria, *Ind.* **15** 40 16N 85 41W
Alexandria, *La.* **13** 31 18N 92 27W
Alexandria, *Minn.* **12** 45 53N 95 22W
Alexandria, *S. Dak.* **12** 43 39N 97 47W
Alexandria, *Va.* **25** 38 49N 77 6W
Alexandria Bay, *N.Y.* **9** 44 20N 75 55W
Alfred, *Maine* **9** 43 29N 70 43W
Algoma, *Wis.* **10** 44 36N 87 26W
Algona, *Iowa* **14** 43 4N 94 14W
Alhambra, *Calif.* **29** 34 5N 118 9W
Alice, *Tex.* **13** 27 45N 98 5W
Aliceville, *Ala.* **11** 33 8N 88 9W
Aliquippa, *Pa.* **8** 40 37N 80 15W
All American Canal, *Calif.* **17** 32 45N 115 15W
Allakaket, *Alaska* **30** 66 34N 152 39W
Allegan, *Mich.* **15** 42 32N 85 51W
Allegheny →, *Pa.* **8** 40 27N 80 1W
Allegheny Plateau, *Va.* **10** 38 0N 80 0W
Allen Park, *Mich.* **26** 42 14N 83 12W
Allentown, *Pa.* **9** 40 37N 75 29W
Alliance, *Nebr.* **12** 42 6N 102 52W
Alliance, *Ohio* **8** 40 55N 81 6W
Allison Park, *Pa.* **27** 40 33N 79 56W
Alma, *Ga.* **11** 31 33N 82 28W
Alma, *Kans.* **12** 39 1N 96 17W
Alma, *Mich.* **10** 43 23N 84 39W
Alma, *Nebr.* **12** 40 6N 99 22W
Alma, *Wis.* **12** 44 20N 91 55W
Almanor, L., *Calif.* **16** 40 14N 121 9W
Alpaugh, *Calif.* **18** 35 53N 119 29W
Alpena, *Mich.* **10** 45 4N 83 27W
Alpine, *Ariz.* **17** 33 51N 109 9W
Alpine, *Calif.* **19** 32 50N 116 46W
Alpine, *Tex.* **13** 30 22N 103 40W
Alta Sierra, *Calif.* **19** 35 42N 118 33W
Altadena, *Calif.* **29** 34 11N 118 8W
Altamaha →, *Ga.* **11** 31 20N 81 20W
Altamont, *N.Y.* **9** 42 43N 74 3W
Altavista, *Va.* **10** 37 6N 79 17W
Alton, *Ill.* **14** 38 53N 90 11W
Altoona, *Pa.* **8** 40 31N 78 24W
Alturas, *Calif.* **16** 41 29N 120 32W
Altus, *Okla.* **13** 34 38N 99 20W
Alva, *Okla.* **13** 36 48N 98 40W
Alvarado, *Tex.* **13** 32 24N 97 13W
Alvin, *Tex.* **13** 29 26N 95 15W
Alzada, *Mont.* **12** 45 2N 104 25W
Amargosa →, *Calif.* **19** 36 14N 116 51W
Amargosa Range, *Calif.* **19** 36 20N 116 45W
Amarillo, *Tex.* **13** 35 13N 101 50W
Amatignak I., *Alaska* **30** 51 16N 179 6W
Amboy, *Calif.* **19** 34 33N 115 45W
Amchitka I., *Alaska* **30** 51 29N 179 0 E
Amchitka Pass., *Alaska* **30** 51 30N 179 0W
American Falls, *Idaho* **16** 42 47N 112 51W
American Falls Reservoir, *Idaho* **16** 42 47N 112 52W
American Samoa ■, *Pac. Oc.* **31** 14 20S 170 40W
Americus, *Ga.* **11** 32 4N 84 14W
Ames, *Iowa* **14** 42 2N 93 37W
Amesbury, *Mass.* **9** 42 51N 70 56W
Amherst, *Mass.* **9** 42 23N 72 31W
Amherst, *Tex.* **13** 34 1N 102 25W
Amite, *La.* **13** 30 44N 90 30W
Amlia I., *Alaska* **30** 52 4N 173 30W
Amory, *Miss.* **11** 33 59N 88 29W
Amsterdam, *N.Y.* **9** 42 56N 74 11W
Amukta I., *Alaska* **30** 52 30N 171 16W
Anaconda, *Mont.* **16** 46 8N 112 57W
Anacortes, *Wash.* **18** 48 30N 122 37W
Anadarko, *Okla.* **13** 35 4N 98 15W
Anaheim, *Calif.* **19** 33 50N 117 55W
Anamoose, *N. Dak.* **12** 47 53N 100 15W
Anamosa, *Iowa* **14** 42 7N 91 17W
Anatone, *Wash.* **16** 46 8N 117 8W
Anchorage, *Alaska* **30** 61 13N 149 54W
Andalusia, *Ala.* **11** 31 18N 86 29W
Anderson, *Calif.* **16** 40 27N 122 18W
Anderson, *Ind.* **15** 40 10N 85 41W
Anderson, *Mo.* **13** 36 39N 94 27W
Anderson, *S.C.* **11** 34 31N 82 39W
Andover, *Mass.* **23** 42 39N 71 7W
Andreanof Is., *Alaska* **30** 52 0N 178 0W
Andrews, *S.C.* **11** 33 27N 79 34W
Andrews, *Tex.* **13** 32 19N 102 33W
Andrews Air Force Base, *Md.* **25** 38 48N 76 52W
Anegada I., *Virgin Is.* **30** 18 45N 64 20W
Angeles National Forest, *Calif.* **29** 34 15N 118 5W
Angels Camp, *Calif.* **18** 38 4N 120 32W
Angleton, *Tex.* **13** 29 10N 95 26W
Angola, *Ind.* **15** 41 38N 85 0W
Angoon, *Alaska* **30** 57 30N 134 35W
Aniak, *Alaska* **30** 61 35N 159 32W
Animas, *N. Mex.* **17** 31 57N 108 48W
Ann, C., *Mass.* **9** 42 38N 70 35W
Ann Arbor, *Mich.* **15** 42 17N 83 45W
Anna, *Ill.* **13** 37 28N 89 15W
Annandale, *Va.* **25** 38 50N 77 12W
Annapolis, *Md.* **10** 38 59N 76 30W
Annette, *Alaska* **30** 55 2N 131 35W
Anniston, *Ala.* **11** 33 39N 85 50W
Annville, *Pa.* **9** 40 20N 76 31W
Anoka, *Minn.* **12** 45 12N 93 23W
Ansley, *Nebr.* **12** 41 18N 99 23W
Anson, *Tex.* **13** 32 45N 99 54W
Ansonia, *Conn.* **9** 41 21N 73 5W
Antero, Mt., *Colo.* **17** 38 41N 106 15W
Anthony, *Kans.* **13** 37 9N 98 2W
Anthony, *N. Mex.* **17** 32 0N 106 36W
Antigo, *Wis.* **12** 45 9N 89 9W
Antimony, *Utah* **17** 38 7N 112 0W
Antioch, *Calif.* **18** 38 1N 121 48W

Antler, *N. Dak.* **12** 48 59N 101 17W
Antlers, *Okla.* **13** 34 14N 95 37W
Anton, *Tex.* **13** 33 49N 102 10W
Anton Chico, *N. Mex.* **17** 35 12N 105 9W
Antonito, *Colo.* **17** 37 5N 106 0W
Anvik, *Alaska* **30** 62 39N 160 13W
Anza, *Calif.* **19** 33 35N 116 39W
Apache, *Okla.* **13** 34 54N 98 22W
Apalachee B., *Fla.* **11** 30 0N 84 0W
Apalachicola, *Fla.* **11** 29 43N 84 59W
Apalachicola →, *Fla.* **11** 29 43N 84 58W
Apostle Is., *Wis.* **12** 47 0N 90 40W
Appalachian Mts., *Va.* **10** 38 0N 80 0W
Apple Valley, *Calif.* **19** 34 32N 117 14W
Appleton, *Wis.* **10** 44 16N 88 25W
Aransas Pass, *Tex.* **13** 27 55N 97 9W
Arapahoe, *Nebr.* **12** 40 18N 99 54W
Arbuckle, *Calif.* **18** 39 1N 122 3W
Arcadia, *Calif.* **29** 34 7N 118 1W
Arcadia, *Fla.* **11** 27 13N 81 52W
Arcadia, *La.* **13** 32 33N 92 55W
Arcadia, *Nebr.* **12** 41 25N 99 8W
Arcadia, *Wis.* **12** 44 15N 91 30W
Arcata, *Calif.* **16** 40 52N 124 5W
Archbald, *Pa.* **9** 41 30N 75 32W
Arco, *Idaho* **16** 43 38N 113 18W
Arctic Village, *Alaska* **30** 68 8N 145 32W
Ardmore, *Okla.* **13** 34 10N 97 8W
Ardmore, *Pa.* **24** 40 0N 75 19W
Ardmore, *S. Dak.* **12** 43 1N 103 40W
Arecibo, *Puerto Rico* **30** 18 29N 66 43W
Argonne Forest, *Ill.* **22** 41 42N 87 53W
Arguello, Pt., *Calif.* **19** 34 35N 120 39W
Argus Pk., *Calif.* **19** 35 52N 117 26W
Argyle, *Minn.* **12** 48 20N 96 49W
Arivaca, *Ariz.* **17** 31 37N 111 25W
Arizona □ **17** 34 0N 112 0W
Arkadelphia, *Ark.* **13** 34 7N 93 4W
Arkansas □ **13** 35 0N 92 30W
Arkansas →, *Ark.* **13** 33 47N 91 4W
Arkansas City, *Kans.* **13** 37 4N 97 2W
Arlee, *Mont.* **16** 47 10N 114 5W
Arlington, *Mass.* **23** 42 24N 71 9W
Arlington, *Oreg.* **16** 45 43N 120 12W
Arlington, *S. Dak.* **12** 44 22N 97 8W
Arlington, *Va.* **25** 38 53N 77 7W
Arlington, *Wash.* **18** 48 12N 122 8W
Arlington Heights, *Ill.* **22** 42 5N 87 54W
Arlington National Cemetery, *D.C.* **25** 38 52N 77 4W
Armour, *S. Dak.* **12** 43 19N 98 21W
Armstrong, *Tex.* **13** 26 56N 97 47W
Arnett, *Okla.* **13** 36 8N 99 46W
Arnold, *Calif.* **18** 38 15N 120 20W
Arnold, *Nebr.* **12** 41 26N 100 12W
Arrow Rock Res., *Idaho* **16** 43 45N 115 50W
Arrowhead, L., *Calif.* **19** 34 16N 117 10W
Arroyo Grande, *Calif.* **19** 35 7N 120 35W
Artesia, *Calif.* **29** 33 51N 118 5W
Artesia, *N. Mex.* **13** 32 51N 104 24W
Artesia Wells, *Tex.* **13** 28 17N 99 17W
Artesian, *S. Dak.* **12** 44 1N 97 55W
Arundel Gardens, *Md.* **25** 39 12N 76 37W
Arvada, *Wyo.* **16** 44 39N 106 8W
Arvin, *Calif.* **19** 35 12N 118 50W
Asbury Park, *N.J.* **9** 40 13N 74 1W
Ash Fork, *Ariz.* **17** 35 13N 112 29W
Ash Grove, *Mo.* **13** 37 19N 93 35W
Ashburn, *Ga.* **11** 31 43N 83 39W
Asheboro, *N.C.* **11** 35 43N 79 49W
Asherton, *Tex.* **13** 28 27N 99 46W
Asheville, *N.C.* **11** 35 36N 82 33W
Ashford, *Wash.* **16** 46 46N 122 2W
Ashland, *Kans.* **13** 37 11N 99 46W
Ashland, *Ky.* **10** 38 28N 82 38W
Ashland, *Maine* **11** 46 38N 68 24W
Ashland, *Mont.* **16** 45 36N 106 16W
Ashland, *Nebr.* **12** 41 3N 96 23W
Ashland, *Ohio* **8** 40 52N 82 19W
Ashland, *Oreg.* **16** 42 12N 122 43W
Ashland, *Pa.* **9** 40 45N 76 22W
Ashland, *Va.* **10** 37 46N 77 29W
Ashland, *Wis.* **12** 46 35N 90 53W
Ashley, *N. Dak.* **12** 46 2N 99 22W
Ashtabula, *Ohio* **8** 41 52N 80 47W
Ashton, *Idaho* **16** 44 4N 111 27W
Asotin, *Wash.* **16** 46 20N 117 3W
Aspen, *Colo.* **17** 39 11N 106 49W
Aspermont, *Tex.* **13** 33 8N 100 14W
Astoria, *Oreg.* **18** 46 11N 123 50W
Atascadero, *Calif.* **17** 35 32N 120 44W
Atchafalaya B., *La.* **13** 29 25N 91 25W
Atchison, *Kans.* **12** 39 34N 95 7W
Athens, *Ala.* **11** 34 48N 86 58W
Athens, *Ga.* **11** 33 57N 83 23W
Athens, *N.Y.* **9** 42 16N 73 49W
Athens, *Ohio* **10** 39 20N 82 6W
Athens, *Pa.* **9** 41 57N 76 31W
Athens, *Tenn.* **11** 35 27N 84 36W
Athens, *Tex.* **13** 32 12N 95 51W
Atka, *Alaska* **30** 52 5N 174 40W
Atkasuk, *Alaska* **30** 70 30N 157 20W
Atkinson, *Nebr.* **12** 42 32N 98 59W
Atlanta, *Ga.* **11** 33 45N 84 23W
Atlanta, *Tex.* **13** 33 7N 94 10W
Atlantic, *Iowa* **12** 41 24N 95 1W
Atlantic City, *N.J.* **10** 39 21N 74 27W
Atmore, *Ala.* **11** 31 2N 87 29W
Atoka, *Okla.* **13** 34 23N 96 8W
Atolia, *Calif.* **19** 35 19N 117 37W
Attalla, *Ala.* **11** 34 1N 86 6W
Attica, *Ind.* **15** 40 18N 87 15W
Attleboro, *Mass.* **9** 41 57N 71 17W
Attu, *Alaska* **30** 52 56N 173 15 E
Atwater, *Calif.* **18** 37 21N 120 37W
Atwood, *Kans.* **12** 39 48N 101 3W
Au Sable →, *Mich.* **10** 44 25N 83 20W
Auberry, *Calif.* **18** 37 7N 119 29W
Auburn, *Ala.* **11** 32 36N 85 29W
Auburn, *Calif.* **18** 38 54N 121 4W
Auburn, *Ind.* **15** 41 22N 85 4W
Auburn, *N.Y.* **9** 42 56N 76 34W
Auburn, *Nebr.* **12** 40 23N 95 51W
Auburndale, *Fla.* **11** 28 4N 81 48W
Audubon, *Iowa* **14** 41 43N 94 56W
Augusta, *Ark.* **13** 35 17N 91 22W
Augusta, *Ga.* **11** 33 28N 81 58W
Augusta, *Kans.* **13** 37 41N 96 59W
Augusta, *Maine* **11** 44 19N 69 47W
Augusta, *Mont.* **16** 47 30N 112 24W
Augusta, *Wis.* **12** 44 41N 91 7W
Aukum, *Calif.* **18** 38 34N 120 43W
Ault, *Colo.* **12** 40 35N 104 44W
Aurora, *Colo.* **12** 39 44N 104 52W
Aurora, *Ill.* **15** 41 45N 88 19W
Aurora, *Mo.* **13** 36 58N 93 43W
Aurora, *Nebr.* **12** 40 52N 98 0W
Austin, *Minn.* **12** 43 40N 92 58W
Austin, *Nev.* **16** 39 30N 117 4W
Austin, *Tex.* **13** 30 17N 97 45W
Avalon, *Calif.* **19** 33 21N 118 20W
Avalon, *Pa.* **27** 40 30N 80 4W
Avawatz Mts., *Calif.* **19** 35 40N 116 30W
Avenal, *Calif.* **18** 36 0N 120 8W
Avery, *Idaho* **16** 47 15N 115 49W
Avila Beach, *Calif.* **19** 35 11N 120 44W
Avon, *S. Dak.* **12** 43 0N 98 4W
Aztec, *N. Mex.* **17** 36 49N 107 59W
Azusa, *Calif.* **29** 34 7N 117 54W

B

Babb, *Mont.* **16** 48 51N 113 27W
Bad →, *S. Dak.* **12** 44 21N 100 22W
Bad Axe, *Mich.* **8** 43 48N 83 0W
Bad Lands, *S. Dak.* **12** 43 40N 102 10W
Badger, *Calif.* **18** 36 38N 119 1W
Bagdad, *Calif.* **19** 34 35N 115 53W
Baggs, *Wyo.* **16** 41 2N 107 39W
Bagley, *Minn.* **12** 47 32N 95 24W
Bainbridge, *Ga.* **11** 30 55N 84 35W
Bainbridge, *N.Y.* **9** 42 18N 75 29W
Bainville, *Mont.* **12** 48 8N 104 13W
Baird, *Tex.* **13** 32 24N 99 24W
Baird Inlet, *Alaska* **30** 60 50N 164 18W
Baird Mts., *Alaska* **30** 67 0N 160 0W
Baker, *Calif.* **19** 35 16N 116 4W
Baker, *Mont.* **12** 46 22N 104 17W
Baker, *Oreg.* **16** 44 47N 117 50W
Baker Mt., *Wash.* **16** 48 50N 121 49W
Bakersfield, *Calif.* **19** 35 23N 119 1W
Bald Knob, *Ark.* **13** 35 19N 91 34W
Baldwin, *Fla.* **11** 30 18N 81 59W
Baldwin, *Mich.* **10** 43 54N 85 51W
Baldwin, *Pa.* **27** 40 23N 79 58W
Baldwin Park, *Calif.* **29** 34 5N 117 57W
Baldwinsville, *N.Y.* **9** 43 10N 76 20W
Baldy Peak, *Ariz.* **17** 33 54N 109 34W
Ballinger, *Tex.* **13** 31 45N 99 57W
Balmorhea, *Tex.* **13** 30 59N 103 45W
Balta, *N. Dak.* **12** 48 10N 100 2W
Baltimore, *Md.* **25** 39 17N 76 37W
Baltimore Washington International Airport, *Md.* **25** 39 11N 76 41W
Bamberg, *S.C.* **11** 33 18N 81 2W
Bandera, *Tex.* **13** 29 44N 99 5W
Bangor, *Maine* **11** 44 48N 68 46W
Bangor, *Pa.* **9** 40 52N 75 13W
Banning, *Calif.* **19** 33 56N 116 53W
Bar Harbor, *Maine* **11** 44 23N 68 13W
Baraboo, *Wis.* **12** 43 28N 89 45W
Baraga, *Mich.* **12** 46 47N 88 30W
Baranof I., *Alaska* **30** 57 0N 135 0W
Barataria B., *La.* **13** 29 20N 89 55W
Barberton, *Ohio* **8** 41 0N 81 39W
Barbourville, *Ky.* **11** 36 52N 83 53W
Bardstown, *Ky.* **15** 37 49N 85 28W
Barksdale, *Tex.* **13** 29 44N 100 2W
Barnesville, *Ga.* **11** 33 3N 84 9W
Barneveld, *N.Y.* **9** 43 16N 75 14W
Barnhart, *Tex.* **13** 31 8N 101 10W
Barnsville, *Minn.* **12** 46 43N 96 28W
Barques, Pt. Aux, *Mich.* **10** 44 4N 82 58W
Barre, *Mass.* **9** 42 25N 72 6W
Barre, *Vt.* **9** 44 12N 72 30W
Barren Is., *Alaska* **30** 58 45N 152 0W
Barrington, *R.I.* **9** 41 44N 71 18W
Barrow, *Alaska* **30** 71 18N 156 47W
Barrow Pt., *Alaska* **30** 71 24N 156 29W
Barstow, *Calif.* **19** 34 54N 117 1W
Barstow, *Tex.* **13** 31 28N 103 24W
Bartlesville, *Okla.* **13** 36 45N 95 59W
Bartlett, *Calif.* **18** 36 29N 118 2W
Bartlett, *Tex.* **13** 30 48N 97 26W
Bartow, *Fla.* **11** 27 54N 81 50W
Basin, *Wyo.* **16** 44 23N 108 2W
Bassett, *Nebr.* **12** 42 35N 99 32W
Bassett, *Va.* **11** 36 46N 79 59W
Bastrop, *Tex.* **13** 30 7N 97 19W
Batavia, *N.Y.* **8** 43 0N 78 11W
Batesburg, *S.C.* **11** 33 54N 81 33W
Batesville, *Ark.* **13** 35 46N 91 39W
Batesville, *Miss.* **13** 34 19N 89 57W
Batesville, *Tex.* **13** 28 58N 99 37W
Bath, *Maine* **11** 43 55N 69 49W
Bath, *N.Y.* **8** 42 20N 77 19W
Baton Rouge, *La.* **13** 30 27N 91 11W
Battle Creek, *Mich.* **15** 42 19N 85 11W
Battle Lake, *Minn.* **12** 46 17N 95 43W
Battle Mountain, *Nev.* **16** 40 38N 116 56W
Baudette, *Minn.* **12** 48 43N 94 36W
Baxley, *Ga.* **11** 31 47N 82 21W
Baxter Springs, *Kans.* **13** 37 2N 94 44W
Bay City, *Mich.* **10** 43 36N 83 54W
Bay City, *Oreg.* **16** 45 31N 123 53W
Bay City, *Tex.* **13** 28 59N 95 58W
Bay Minette, *Ala.* **11** 30 53N 87 46W
Bay St. Louis, *Miss.* **13** 30 19N 89 20W
Bay Springs, *Miss.* **13** 31 59N 89 17W
Bay Village, *Ohio* **27** 41 29N 81 53W
Bayamón, *Puerto Rico* **30** 18 24N 66 10W
Bayard, *Nebr.* **12** 41 45N 103 20W
Bayfield, *Wis.* **12** 46 49N 90 49W
Bayonne, *N.J.* **20** 40 40N 74 6W
Baytown, *Tex.* **13** 29 43N 94 59W
Beach, *N. Dak.* **12** 46 58N 104 0W
Beacon, *N.Y.* **9** 41 30N 73 58W
Bear L., *Utah* **16** 41 59N 111 21W
Bearcreek, *Mont.* **16** 45 11N 109 6W
Beardstown, *Ill.* **14** 40 1N 90 26W
Bearpaw Mts., *Mont.* **16** 48 12N 109 30W
Beatrice, *Nebr.* **12** 40 16N 96 45W
Beatty, *Nev.* **18** 36 54N 116 46W
Beaufort, *N.C.* **11** 34 43N 76 40W
Beaufort, *S.C.* **11** 32 26N 80 40W
Beaumont, *Calif.* **19** 33 56N 116 58W
Beaumont, *Tex.* **13** 30 5N 94 6W
Beaver, *Alaska* **30** 66 22N 147 24W
Beaver, *Okla.* **13** 36 49N 100 31W
Beaver, *Utah* **17** 38 17N 112 38W
Beaver City, *Nebr.* **12** 40 8N 99 50W
Beaver Dam, *Wis.* **12** 43 28N 88 50W
Beaver Falls, *Pa.* **8** 40 46N 80 20W
Beaver I., *Mich.* **10** 45 40N 85 33W
Becharof L., *Alaska* **30** 57 56N 156 23W
Beckley, *W. Va.* **10** 37 47N 81 11W
Bedford, *Ind.* **15** 38 52N 86 29W
Bedford, *Iowa* **14** 40 40N 94 44W
Bedford, *Mass.* **23** 42 29N 71 15W
Bedford, *Ohio* **8** 41 23N 81 32W
Bedford, *Va.* **10** 37 20N 79 31W
Beech Grove, *Ind.* **15** 39 44N 86 3W
Beechey Point, *Alaska* **30** 70 27N 149 18W
Beeville, *Tex.* **13** 28 24N 97 45W
Belen, *N. Mex.* **17** 34 40N 106 46W
Belfast, *Maine* **11** 44 26N 69 1W
Belfield, *N. Dak.* **12** 46 53N 103 12W
Belfry, *Mont.* **16** 45 9N 109 1W
Belgrade, *Mont.* **16** 45 47N 111 11W
Belhaven, *N.C.* **11** 35 33N 76 37W
Bell Gardens, *Calif.* **29** 33 58N 118 9W
Bellaire, *Ohio* **8** 40 1N 80 45W
Belle Fourche, *S. Dak.* **12** 44 40N 103 51W
Belle Fourche →, *S. Dak.* **12** 44 26N 102 18W
Belle Glade, *Fla.* **11** 26 41N 80 40W
Belle Plaine, *Iowa* **14** 41 54N 92 17W
Belle Plaine, *Minn.* **12** 44 37N 93 46W
Bellefontaine, *Ohio* **15** 40 22N 83 46W
Bellefonte, *Pa.* **8** 40 55N 77 47W
Belleville, *Ill.* **14** 38 31N 89 59W
Belleville, *Kans.* **12** 39 50N 97 38W
Belleville, *N.J.* **20** 40 48N 74 9W
Belleville, *N.Y.* **9** 43 46N 76 10W
Bellevue, *Idaho* **16** 43 28N 114 16W
Bellevue, *Pa.* **27** 40 29N 80 3W
Bellflower, *Calif.* **29** 33 53N 118 8W
Bellingham, *Wash.* **18** 48 46N 122 29W
Bellmore, *N.Y.* **21** 40 39N 73 31W
Bellows Falls, *Vt.* **9** 43 8N 72 27W
Bellville, *Tex.* **13** 29 57N 96 15W
Bellwood, *Ill.* **22** 41 53N 87 53W
Belmont, *Mass.* **23** 42 24N 71 10W
Beloit, *Kans.* **12** 39 28N 98 6W
Beloit, *Wis.* **14** 42 31N 89 2W
Belton, *S.C.* **11** 34 31N 82 30W
Belton, *Tex.* **13** 31 3N 97 28W
Belton Res., *Tex.* **13** 31 8N 97 32W
Beltsville, *Md.* **25** 39 2N 76 55W
Belvidere, *Ill.* **15** 42 15N 88 50W
Belvidere, *N.J.* **9** 40 50N 75 5W
Belzoni, *Miss.* **13** 33 11N 90 29W
Bemidji, *Minn.* **12** 47 28N 94 53W
Benavides, *Tex.* **13** 27 36N 98 25W
Bend, *Oreg.* **16** 44 4N 121 19W
Benicia, *Calif.* **18** 38 3N 122 9W
Benkelman, *Nebr.* **12** 40 3N 101 32W
Bennettsville, *S.C.* **11** 34 37N 79 41W
Bennington, *N.H.* **9** 43 0N 71 55W
Benson, *Ariz.* **17** 31 58N 110 18W
Benton, *Ark.* **13** 34 34N 92 35W
Benton, *Calif.* **18** 37 48N 118 32W
Benton, *Ill.* **14** 38 0N 88 55W
Benton Harbor, *Mich.* **15** 42 6N 86 27W
Beowawe, *Nev.* **16** 40 35N 116 29W
Berea, *Ky.* **10** 37 34N 84 17W
Bergenfield, *N.J.* **20** 40 55N 73 58W
Bering Sea, *Pac. Oc.* **30** 58 0N 171 0 E
Bering Strait, *Alaska* **30** 65 30N 169 0W
Berkeley, *Calif.* **28** 37 52N 122 17W
Berkeley Springs, *W. Va.* **10** 39 38N 78 14W
Berkley, *Mich.* **26** 42 29N 83 11W
Berlin, *Md.* **10** 38 20N 75 13W
Berlin, *N.H.* **9** 44 28N 71 11W
Berlin, *Wis.* **10** 43 58N 88 57W
Bernado, *N. Mex.* **17** 34 30N 106 53W
Bernalillo, *N. Mex.* **17** 35 18N 106 33W
Berryville, *Ark.* **13** 36 22N 93 34W
Berthold, *N. Dak.* **12** 48 19N 101 44W
Berthoud, *Colo.* **12** 40 19N 105 5W
Bertrand, *Nebr.* **12** 40 32N 99 38W
Berwick, *Pa.* **9** 41 3N 76 14W
Berwyn, *Ill.* **22** 41 50N 87 47W
Bessemer, *Ala.* **11** 33 24N 86 58W
Bessemer, *Mich.* **12** 46 29N 90 3W
Bethany, *Mo.* **14** 40 16N 94 2W
Bethel, *Alaska* **30** 60 48N 161 45W
Bethel Park, *Pa.* **27** 40 19N 80 1W
Bethesda, *Md.* **25** 38 59N 77 6W
Bethlehem, *Pa.* **9** 40 37N 75 23W
Bethpage, *N.Y.* **21** 40 45N 73 29W
Beulah, *N. Dak.* **12** 47 16N 101 47W
Beverly, *Mass.* **23** 42 33N 70 52W
Beverly, *Wash.* **16** 46 50N 119 56W
Beverly Hills, *Calif.* **29** 34 5N 118 24W
Beverly Hills, *Mich.* **26** 42 31N 83 15W
Bicknell, *Ind.* **15** 38 47N 87 19W
Bicknell, *Utah* **17** 38 20N 111 33W
Biddeford, *Maine* **11** 43 30N 70 28W
Bieber, *Calif.* **16** 41 7N 121 8W
Big Bear City, *Calif.* **19** 34 16N 116 51W
Big Bear Lake, *Calif.* **19** 34 15N 116 56W
Big Belt Mts., *Mont.* **16** 46 30N 111 25W
Big Bend National Park, *Tex.* **13** 29 20N 103 5W
Big Black →, *Miss.* **13** 32 3N 91 4W
Big Blue →, *Kans.* **12** 39 35N 96 34W
Big Creek, *Calif.* **18** 37 11N 119 14W
Big Cypress Swamp, *Fla.* **11** 26 12N 81 10W
Big Falls, *Minn.* **12** 48 12N 93 48W
Big Fork →, *Minn.* **12** 48 31N 93 43W
Big Horn Mts. = Bighorn Mts., *Wyo.* **16** 44 30N 107 30W
Big Lake, *Tex.* **13** 31 12N 101 28W
Big Moose, *N.Y.* **9** 43 49N 74 58W
Big Muddy Cr. →, *Mont.* **12** 48 8N 104 36W
Big Pine, *Calif.* **18** 37 10N 118 17W
Big Piney, *Wyo.* **16** 42 32N 110 7W
Big Rapids, *Mich.* **10** 43 42N 85 29W
Big Sable Pt., *Mich.* **10** 44 3N 86 1W
Big Sandy, *Mont.* **16** 48 11N 110 7W
Big Sandy Cr. →, *Colo.* **12** 38 7N 102 29W
Big Sioux →, *S. Dak.* **12** 42 29N 96 27W
Big Spring, *Tex.* **13** 32 15N 101 28W
Big Springs, *Nebr.* **12** 41 4N 102 5W
Big Stone City, *S. Dak.* **12** 45 18N 96 28W
Big Stone Gap, *Va.* **11** 36 52N 82 47W
Big Stone L., *Minn.* **12** 45 30N 96 35W
Big Sur, *Calif.* **18** 36 15N 121 48W
Big Timber, *Mont.* **16** 45 50N 109 57W
Bigfork, *Mont.* **16** 48 4N 114 4W
Bighorn, *Mont.* **16** 46 10N 107 27W
Bighorn →, *Mont.* **16** 46 10N 107 28W
Bighorn Mts., *Wyo.* **16** 44 30N 107 30W
Bikini Atoll, *Pac. Oc.* **31** 12 0N 167 30 E
Bill, *Wyo.* **12** 43 14N 105 16W
Billerica, *Mass.* **23** 42 33N 71 15W
Billings, *Mont.* **16** 45 47N 108 30W
Biloxi, *Miss.* **13** 30 24N 88 53W
Bingham, *Maine* **11** 45 3N 69 53W
Bingham Canyon, *Utah* **16** 40 32N 112 9W
Binghamton, *N.Y.* **9** 42 6N 75 55W
Bird City, *Kans.* **12** 39 45N 101 32W
Birmingham, *Ala.* **11** 33 31N 86 48W
Birmingham, *Mich.* **26** 42 33N 83 13W
Bisbee, *Ariz.* **17** 31 27N 109 55W
Biscayne B., *Fla.* **11** 25 40N 80 12W
Bishop, *Calif.* **18** 37 22N 118 24W
Bishop, *Tex.* **13** 27 35N 97 48W
Bismarck, *N. Dak.* **12** 46 48N 100 47W
Bison, *S. Dak.* **12** 45 31N 102 28W
Bitter Creek, *Wyo.* **16** 41 33N 108 33W
Bitterroot →, *Mont.* **16** 46 52N 114 7W
Bitterroot Range, *Idaho* **16** 46 0N 114 20W
Bitterwater, *Calif.* **18** 36 23N 121 0W
Biwabik, *Minn.* **12** 47 32N 92 21W
Black →, *Ark.* **13** 35 38N 91 20W
Black →, *Wis.* **12** 43 57N 91 22W
Black Hills, *S. Dak.* **12** 44 0N 103 45W
Black L., *Mich.* **10** 45 28N 84 16W
Black Mesa, *Okla.* **13** 36 58N 102 58W
Black Range, *N. Mex.* **17** 33 15N 107 50W
Black River Falls, *Wis.* **12** 44 18N 90 51W
Black Warrior →, *Ala.* **11** 32 32N 87 51W
Blackburn, Mt., *Alaska* **30** 61 44N 143 26W
Blackduck, *Minn.* **12** 47 44N 94 33W
Blackfoot, *Idaho* **16** 43 11N 112 21W
Blackfoot →, *Mont.* **16** 46 52N 113 53W

Blackfoot River Reservoir, Idaho 16 43 0N 111 43W
Blacksburg, Va. 10 37 14N 80 25W
Blackstone, Va. 10 37 4N 78 0W
Blackwell, Okla. 13 36 48N 97 17W
Blackwells Corner, Calif. 19 35 37N 119 47W
Blackwood, N.J. 24 39 48N 75 4W
Blaine, Wash. 18 48 59N 122 45W
Blair, Nebr. 12 41 33N 96 8W
Blake Pt., Mich. 12 48 11N 88 25W
Blakely, Ga. 11 31 23N 84 56W
Blanca Peak, Colo. 17 37 35N 105 29W
Blanchard, Okla. 13 35 8N 97 39W
Blanco, Tex. 13 30 6N 98 25W
Blanco, C., Oreg. 16 42 51N 124 34W
Blanding, Utah 17 37 37N 109 29W
Block I., R.I. 9 41 11N 71 35W
Bloomer, Wis. 12 45 6N 91 29W
Bloomfield, Iowa 14 40 45N 92 25W
Bloomfield, N.J. 20 40 48N 74 12W
Bloomfield, N. Mex. 17 36 43N 107 59W
Bloomfield, Nebr. 12 42 36N 97 39W
Bloomingdale, N.J. 20 41 0N 74 15W
Bloomington, Ill. 14 40 28N 89 0W
Bloomington, Ind. 15 39 10N 86 32W
Bloomsburg, Pa. 9 41 0N 76 27W
Blossburg, Pa. 8 41 41N 77 4W
Blountstown, Fla. 11 30 27N 85 3W
Blue Island, Ill. 10 41 40N 87 40W
Blue Island, Ill. 22 41 40N 87 40W
Blue Lake, Calif. 16 40 53N 123 59W
Blue Mesa Reservoir, Colo. 17 38 28N 107 20W
Blue Mts., Oreg. 16 45 15N 119 0W
Blue Mts., Pa. 9 40 30N 76 30W
Blue Rapids, Kans. 12 39 41N 96 39W
Blue Ridge Mts., N.C. 11 36 30N 80 15W
Bluefield, Va. 10 37 15N 81 17W
Bluff, Utah 17 37 17N 109 33W
Bluffton, Ind. 15 40 44N 85 11W
Blunt, S. Dak. 12 44 31N 99 59W
Bly, Oreg. 16 42 24N 121 3W
Blythe, Calif. 19 33 37N 114 36W
Boca Raton, Fla. 11 26 21N 80 5W
Boerne, Tex. 13 29 47N 98 44W
Bogalusa, La. 13 30 47N 89 52W
Bogata, Tex. 13 33 28N 95 13W
Boise, Idaho 16 43 37N 116 13W
Boise City, Okla. 13 36 44N 102 31W
Bolivar, Mo. 13 37 37N 93 25W
Bolivar, Tenn. 13 35 12N 89 0W
Bonham, Tex. 13 33 35N 96 11W
Bonne Terre, Mo. 14 37 55N 90 33W
Bonners Ferry, Idaho 16 48 42N 116 19W
Bonsall, Calif. 19 33 16N 117 14W
Booker, Tex. 13 36 27N 100 32W
Boone, Iowa 14 42 4N 93 53W
Boone, N.C. 11 36 13N 81 41W
Booneville, Ark. 13 35 8N 93 55W
Booneville, Miss. 13 34 39N 88 34W
Boonville, Ind. 15 38 3N 87 16W
Boonville, Mo. 14 38 58N 92 44W
Boonville, N.Y. 9 43 29N 75 20W
Borah Peak, Idaho 16 44 8N 113 47W
Borger, Tex. 13 35 39N 101 24W
Boron, Calif. 19 35 0N 117 39W
Borrego Springs, Calif. 19 33 15N 116 23W
Bossier City, La. 13 32 31N 93 44W
Boston, Mass. 23 42 21N 71 3W
Boswell, Okla. 13 34 2N 95 52W
Bottineau, N. Dak. 12 48 50N 100 27W
Boulder, Colo. 12 40 1N 105 17W
Boulder, Mont. 16 46 14N 112 7W
Boulder City, Nev. 19 35 59N 114 50W
Boulder Creek, Calif. 18 37 7N 122 7W
Boulder Dam = Hoover Dam, Ariz. 19 36 1N 114 44W
Boundary, Alaska 30 64 4N 141 6W
Boundary Peak, Nev. 18 37 51N 118 21W
Bountiful, Utah 16 40 53N 111 53W
Bovill, Idaho 16 46 51N 116 24W
Bowbells, N. Dak. 12 48 48N 102 15W
Bowdle, S. Dak. 12 45 27N 99 39W
Bowie, Ariz. 17 32 19N 109 29W
Bowie, Tex. 13 33 34N 97 51W
Bowling Green, Ky. 10 36 59N 86 27W
Bowling Green, Ohio 15 41 23N 83 39W
Bowman, N. Dak. 12 46 11N 103 24W
Boyce, La. 13 31 23N 92 40W
Boyne City, Mich. 10 45 13N 85 1W
Boynton Beach, Fla. 11 26 32N 80 4W
Bozeman, Mont. 16 45 41N 111 2W
Brackettville, Tex. 13 29 19N 100 25W
Braddock, Pa. 27 40 24N 79 51W
Bradenton, Fla. 11 27 30N 82 34W
Bradford, Pa. 8 41 58N 78 38W
Bradley, Ark. 13 33 6N 93 39W
Bradley, Calif. 18 35 52N 120 48W
Bradley, S. Dak. 12 45 5N 97 39W
Brady, Tex. 13 31 9N 99 20W
Brainerd, Minn. 12 46 22N 94 12W
Braintree, Mass. 23 42 12N 71 0W
Brandywine, Del. 24 39 49N 75 32W
Branford, Conn. 9 41 17N 72 49W
Branson, Colo. 13 37 1N 103 53W
Branson, Mo. 13 36 39N 93 13W
Brasstown Bald, Ga. 11 34 53N 83 49W
Brattleboro, Vt. 9 42 51N 72 34W
Brawley, Calif. 19 32 59N 115 31W
Brazil, Ind. 15 39 32N 87 8W
Brazos →, Tex. 13 28 53N 95 23W
Breckenridge, Colo. 16 39 29N 106 3W
Breckenridge, Minn. 12 46 16N 96 35W
Breckenridge, Tex. 13 32 45N 98 54W
Bremerton, Wash. 18 47 34N 122 38W
Brenham, Tex. 13 30 10N 96 24W
Brentwood, Pa. 27 40 22N 79 59W
Breton Sd., La. 13 29 35N 89 15W
Brevard, N.C. 11 35 14N 82 44W
Brewer, Maine 11 44 48N 68 46W
Brewer, Mt., Calif. 18 36 44N 118 28W
Brewster, N.Y. 9 41 23N 73 37W
Brewster, Wash. 16 48 6N 119 47W
Brewton, Ala. 11 31 7N 87 4W
Bridgehampton, N.Y. 9 40 56N 72 19W
Bridgeport, Calif. 18 38 15N 119 14W
Bridgeport, Conn. 9 41 11N 73 12W
Bridgeport, Nebr. 12 41 40N 103 6W
Bridgeport, Pa. 24 39 48N 75 21W
Bridgeport, Tex. 13 33 13N 97 45W
Bridger, Mont. 16 45 18N 108 55W
Bridgeton, N.J. 10 39 26N 75 14W
Bridgeville, Pa. 27 21 0N 80 6W
Bridgewater, Mass. 9 41 59N 70 58W
Bridgewater, S. Dak. 12 43 33N 97 30W
Briggsdale, Colo. 12 40 38N 104 20W
Brigham City, Utah 16 41 31N 112 1W
Brighton, Colo. 12 39 59N 104 49W
Brinkley, Ark. 13 34 53N 91 12W
Bristol, Conn. 9 41 40N 72 57W
Bristol, Pa. 24 40 6N 74 53W
Bristol, R.I. 9 41 40N 71 16W
Bristol, S. Dak. 12 45 21N 97 45W
Bristol, Tenn. 11 36 36N 82 11W
Bristol B., Alaska 30 58 0N 160 0W
Bristol L., Calif. 17 34 23N 116 50W
Bristow, Okla. 13 35 50N 96 23W
Britton, S. Dak. 12 45 48N 97 45W
Broad →, S.C. 11 34 1N 81 4W
Broadus, Mont. 12 45 27N 105 25W
Broadview, Ill. 22 41 51N 87 52W
Brockport, N.Y. 8 43 13N 77 56W
Brockton, Mass. 9 42 5N 71 1W
Brockway, Mont. 12 47 18N 105 45W
Brogan, Oreg. 16 44 15N 117 31W
Broken Bow, Nebr. 12 41 24N 99 38W
Broken Bow, Okla. 13 34 2N 94 44W
Bronte, Tex. 13 31 53N 100 18W
Bronx, N.Y. 21 40 50N 73 52W
Bronxville, N.Y. 21 40 56N 73 49W
Brook Park, Ohio 27 41 24N 81 48W
Brookfield, Ill. 22 41 48N 87 50W
Brookfield, Mo. 14 39 47N 93 4W
Brookhaven, Miss. 13 31 35N 90 26W
Brookings, Oreg. 16 42 3N 124 17W
Brookings, S. Dak. 12 44 19N 96 48W
Brookline, Mass. 23 42 19N 71 7W
Brooklyn, Md. 25 39 13N 76 35W
Brooklyn, N.Y. 20 40 37N 73 57W
Brooklyn, Ohio 27 41 26N 81 44W
Brooks Ra., Alaska 30 68 40N 147 0W
Brooksville, Fla. 11 28 33N 82 23W
Brookville, Ind. 15 39 25N 85 1W
Broomall, Pa. 24 39 58N 75 22W
Brothers, Oreg. 16 43 49N 120 36W
Browerville, Minn. 12 46 5N 94 52W
Brownfield, Tex. 13 33 11N 102 17W
Browning, Mont. 16 48 34N 113 1W
Brownsville, Oreg. 16 44 24N 122 59W
Brownsville, Tenn. 13 35 36N 89 16W
Brownsville, Tex. 13 25 54N 97 30W
Brownwood, Tex. 13 31 43N 98 59W
Brownwood, L., Tex. 13 31 51N 98 35W
Brundidge, Ala. 11 31 43N 85 49W
Bruneau, Idaho 16 42 53N 115 48W
Bruneau →, Idaho 16 42 56N 115 57W
Brunswick, Ga. 11 31 10N 81 30W
Brunswick, Maine 11 43 55N 69 58W
Brunswick, Md. 10 39 19N 77 38W
Brunswick, Mo. 14 39 26N 93 8W
Brush, Colo. 12 40 15N 103 37W
Bryan, Ohio 15 41 28N 84 33W
Bryan, Tex. 13 30 40N 96 22W
Bryant, S. Dak. 12 44 35N 97 28W
Bryn Mawr, Pa. 24 40 1N 75 19W
Bryson City, N.C. 11 35 26N 83 27W
Buchanan, L., Tex. 13 30 45N 98 25W
Buchon, Pt., Calif. 18 35 15N 120 54W
Buckeye, Ariz. 17 33 22N 112 35W
Buckhannon, W. Va. 10 39 0N 80 8W
Buckland, Alaska 30 65 59N 161 8W
Buckley, Wash. 16 47 10N 122 2W
Bucklin, Kans. 13 37 33N 99 38W
Bucyrus, Ohio 15 40 48N 82 59W
Buellton, Calif. 19 34 37N 120 12W
Buena Park, Calif. 29 33 51N 118 1W
Buena Vista, Colo. 17 38 51N 106 8W
Buena Vista, Va. 10 37 44N 79 21W
Buena Vista L., Calif. 19 35 12N 119 18W
Buffalo, Mo. 14 37 39N 93 6W
Buffalo, N.Y. 8 42 53N 78 53W
Buffalo, Okla. 13 36 50N 99 38W
Buffalo, S. Dak. 12 45 35N 103 33W
Buffalo, Wyo. 16 44 21N 106 42W
Buford, Ga. 11 34 10N 84 0W
Buhl, Idaho 16 42 36N 114 46W
Buhl, Minn. 12 47 30N 92 46W
Buldir I., Alaska 30 52 21N 175 56 E
Bull Shoals L., Ark. 13 36 22N 92 35W
Bunker Hill Monument, Mass. 23 42 21N 71 3W
Bunkie, La. 13 30 57N 92 11W
Bunnell, Fla. 11 29 28N 81 16W
Buras, La. 13 29 22N 89 32W
Burbank, Calif. 29 34 11N 118 18W
Burbank, Ill. 22 41 44N 87 47W
Burkburnett, Tex. 13 34 6N 98 34W
Burke, Idaho 16 47 31N 115 49W
Burley, Idaho 16 42 32N 113 48W
Burlingame, Calif. 28 37 35N 122 22W
Burlington, Colo. 12 39 18N 102 16W
Burlington, Iowa 14 40 49N 91 14W
Burlington, Kans. 12 38 12N 95 45W
Burlington, Mass. 23 42 30N 71 13W
Burlington, N.C. 11 36 6N 79 26W
Burlington, N.J. 24 40 4N 74 54W
Burlington, Vt. 9 44 29N 73 12W
Burlington, Wash. 18 48 28N 122 20W
Burlington, Wis. 10 42 41N 88 17W
Burnet, Tex. 13 30 45N 98 14W
Burney, Calif. 16 40 53N 121 40W
Burns, Oreg. 16 43 35N 119 3W
Burns, Wyo. 12 41 12N 104 21W
Burnt Paw, Alaska 30 67 2N 142 43W
Burwell, Nebr. 12 41 47N 99 8W
Bushnell, Ill. 12 40 33N 90 31W
Bushnell, Nebr. 12 41 14N 103 54W
Butler, Mo. 14 38 16N 94 20W
Butler, Pa. 8 40 52N 79 54W
Butte, Mont. 16 46 0N 112 32W
Butte, Nebr. 12 42 58N 98 51W
Buttonwillow, Calif. 19 35 24N 119 28W
Buzzards Bay, Mass. 9 41 45N 70 37W
Byers, Colo. 12 39 43N 104 14W
Byhalia, Miss. 13 34 52N 89 41W
Bylas, Ariz. 17 33 8N 110 7W

C

Cabazon, Calif. 19 33 55N 116 47W
Cabinet Mts., Mont. 16 48 0N 115 30W
Cabool, Mo. 13 37 7N 92 6W
Caddo, Okla. 13 34 7N 96 16W
Cadillac, Mich. 10 44 15N 85 24W
Caguas, Puerto Rico 30 18 14N 66 2W
Cairo, Ga. 11 30 52N 84 13W
Cairo, Ill. 13 37 0N 89 11W
Calais, Maine 11 45 11N 67 17W
Calcasieu L., La. 13 29 55N 93 18W
Caldwell, Idaho 16 43 40N 116 41W
Caldwell, Kans. 13 37 2N 97 37W
Caldwell, Tex. 13 30 32N 96 42W
Calexico, Calif. 19 32 40N 115 30W
Calhoun, Ga. 11 34 30N 84 57W
Caliente, Nev. 17 37 37N 114 31W
California, Mo. 14 38 38N 92 34W
California □ 17 37 30N 119 30W
California, University of, Calif. 28 37 52N 122 15W
California City, Calif. 19 35 10N 117 55W
California Hot Springs, Calif. 19 35 51N 118 41W
Calipatria, Calif. 19 33 8N 115 31W
Calistoga, Calif. 18 38 35N 122 35W
Callaway, Nebr. 12 41 18N 99 56W
Calumet, Mich. 10 47 14N 88 27W
Calumet City, Ill. 22 41 37N 87 32W
Calvert, Tex. 13 30 59N 96 40W
Calwa, Calif. 18 36 42N 119 46W
Camanche Reservoir, Calif. 18 38 14N 121 1W
Camarillo, Calif. 19 34 13N 119 2W
Camas, Wash. 18 45 35N 122 24W
Camas Valley, Oreg. 16 43 2N 123 40W
Cambria, Calif. 18 35 34N 121 5W
Cambridge, Idaho 16 44 34N 116 41W
Cambridge, Mass. 23 42 22N 71 6W
Cambridge, Md. 10 38 34N 76 5W
Cambridge, Minn. 12 45 34N 93 13W
Cambridge, N.Y. 9 43 2N 73 22W
Cambridge, Nebr. 12 40 17N 100 10W
Cambridge, Ohio 8 40 2N 81 35W
Camden, Ala. 11 31 59N 87 17W
Camden, Ark. 13 33 35N 92 50W
Camden, Maine 11 44 13N 69 4W
Camden, N.J. 24 39 56N 75 7W
Camden, S.C. 11 34 16N 80 36W
Camden, B., Alaska 30 70 30N 145 0W
Camdenton, Mo. 14 38 1N 92 45W
Cameron, Ariz. 17 35 53N 111 25W
Cameron, La. 13 29 48N 93 20W
Cameron, Mo. 14 39 44N 94 14W
Cameron, Tex. 13 30 51N 96 59W
Camino, Calif. 18 38 44N 120 41W
Camp Crook, S. Dak. 12 45 33N 103 59W
Camp Nelson, Calif. 19 36 8N 118 39W
Camp Wood, Tex. 13 29 40N 100 1W
Campbell, Calif. 18 37 17N 121 57W
Campbellsville, Ky. 10 37 21N 85 20W
Canadian, Tex. 13 35 55N 100 23W
Canadian →, Okla. 13 35 28N 95 3W
Canandaigua, N.Y. 8 42 54N 77 17W
Canarsie, N.Y. 21 40 38N 73 53W
Canaveral, C., Fla. 11 28 27N 80 32W
Canby, Calif. 16 41 27N 120 52W
Canby, Minn. 12 44 43N 96 16W
Canby, Oreg. 18 45 16N 122 42W
Cando, N. Dak. 12 48 32N 99 12W
Cannon Ball →, N. Dak. 12 46 20N 100 38W
Canon City, Colo. 12 38 27N 105 14W
Cantil, Calif. 19 35 18N 117 58W
Canton, Ga. 11 34 14N 84 29W
Canton, Ill. 14 40 33N 90 2W
Canton, Miss. 13 32 37N 90 2W
Canton, Mo. 14 40 8N 91 32W
Canton, N.Y. 9 44 36N 75 10W
Canton, Ohio 8 40 48N 81 23W
Canton, Okla. 13 36 3N 98 35W
Canton, S. Dak. 12 43 18N 96 35W
Canton L., Okla. 13 36 6N 98 35W
Canutillo, Tex. 17 31 55N 106 36W
Canyon, Tex. 13 34 59N 101 55W
Canyon, Wyo. 16 44 43N 110 36W
Canyonlands National Park, Utah 17 38 15N 110 0W
Canyonville, Oreg. 16 42 56N 123 17W
Cape Charles, Va. 10 37 16N 76 1W
Cape Fear →, N.C. 11 33 53N 78 1W
Cape Girardeau, Mo. 13 37 19N 89 32W
Cape May, N.J. 10 38 56N 74 56W
Capitan, N. Mex. 17 33 35N 105 35W
Capitola, Calif. 18 36 59N 121 57W
Carbondale, Colo. 16 39 24N 107 13W
Carbondale, Ill. 14 37 44N 89 13W
Carbondale, Pa. 9 41 35N 75 30W
Cardiff-by-the-Sea, Calif. 19 33 1N 117 17W
Carey, Idaho 16 43 19N 113 57W
Carey, Ohio 15 40 57N 83 23W
Caribou, Maine 11 46 52N 68 1W
Carlin, Nev. 16 40 43N 116 7W
Carlinville, Ill. 14 39 17N 89 53W
Carlisle, Pa. 8 40 12N 77 12W
Carlsbad, Calif. 19 33 10N 117 21W
Carlsbad, N. Mex. 13 32 25N 104 14W
Carlyle, Ill. 12 38 37N 89 22W
Carmel, N.Y. 9 41 26N 73 41W
Carmel-by-the-Sea, Calif. 18 36 33N 121 55W
Carmel Valley, Calif. 18 36 29N 121 43W
Carmi, Ill. 15 38 5N 88 10W
Carmichael, Calif. 18 38 38N 121 19W
Carnegie, Pa. 27 40 24N 80 5W
Caro, Mich. 10 43 29N 83 24W
Carol City, Fla. 11 25 56N 80 16W
Caroline Is., Pac. Oc. 31 8 0N 150 0 E
Carpinteria, Calif. 19 34 24N 119 31W
Carrabelle, Fla. 11 29 51N 84 40W
Carrington, N. Dak. 12 47 27N 99 8W
Carrizo Cr. →, N. Mex. 13 36 55N 103 55W
Carrizo Springs, Tex. 13 28 31N 99 52W
Carrizozo, N. Mex. 17 33 38N 105 53W
Carroll, Iowa 14 42 4N 94 52W
Carrollton, Ga. 11 33 35N 85 5W
Carrollton, Ill. 12 39 18N 90 24W
Carrollton, Ky. 15 38 41N 85 11W
Carrollton, Mo. 14 39 22N 93 30W
Carson, N. Dak. 12 46 25N 101 34W
Carson City, Nev. 18 39 10N 119 46W
Carson Sink, Nev. 16 39 50N 118 25W
Cartersville, Ga. 11 34 10N 84 48W
Carthage, Ark. 13 34 4N 92 33W
Carthage, Ill. 14 40 25N 91 8W
Carthage, Mo. 13 37 11N 94 19W
Carthage, S. Dak. 12 44 10N 97 43W
Carthage, Tex. 13 32 9N 94 20W
Caruthersville, Mo. 13 36 11N 89 39W
Casa Grande, Ariz. 17 32 53N 111 45W
Cascade, Idaho 16 44 31N 116 2W
Cascade, Mont. 16 47 16N 111 42W
Cascade Locks, Oreg. 18 45 40N 121 54W
Cascade Ra., Wash. 18 47 0N 121 30W
Cashmere, Wash. 16 47 31N 120 28W
Casmalia, Calif. 19 34 50N 120 32W
Casper, Wyo. 16 42 51N 106 19W
Cass City, Mich. 10 43 36N 83 11W
Cass Lake, Minn. 12 47 23N 94 37W
Casselton, N. Dak. 12 46 54N 97 13W
Cassville, Mo. 13 36 41N 93 52W
Castaic, Calif. 19 34 30N 118 38W
Castle Dale, Utah 16 39 13N 111 1W
Castle Rock, Colo. 12 39 22N 104 51W
Castle Rock, Wash. 18 46 17N 122 54W
Castro Valley, Calif. 28 37 42N 122 4W
Castroville, Calif. 18 36 46N 121 45W
Castroville, Tex. 13 29 21N 98 53W
Cat I., Miss. 13 30 14N 89 6W
Catahoula L., La. 13 31 31N 92 7W
Cathlamet, Wash. 18 46 12N 123 23W
Catlettsburg, Ky. 10 38 25N 82 36W
Catonsville, Md. 25 39 16N 76 43W
Catskill, N.Y. 9 42 14N 73 52W
Catskill Mts., N.Y. 9 42 10N 74 25W
Cavalier, N. Dak. 12 48 48N 97 37W
Cave City, Ky. 10 37 8N 85 58W
Cayey, Puerto Rico 30 18 7N 66 10W
Cayuga L., N.Y. 9 42 41N 76 41W
Cedar →, Iowa 14 41 17N 91 21W
Cedar City, Utah 17 37 41N 113 4W
Cedar Creek Reservoir, Tex. 13 32 11N 96 4W

Cedar Falls, *Iowa* **14** 42 32N 92 27W
Cedar Key, *Fla.* **11** 29 8N 83 2W
Cedar Rapids, *Iowa* **14** 41 59N 91 40W
Cedarhurst, *N.Y.* **21** 40 37N 73 42W
Cedartown, *Ga.* **11** 34 1N 85 15W
Cedarville, *Calif.* **16** 41 32N 120 10W
Celina, *Ohio* **15** 40 33N 84 35W
Cement, *Okla.* **13** 34 56N 98 8W
Center, *N. Dak.* **12** 47 7N 101 18W
Center, *Tex.* **13** 31 48N 94 11W
Centerfield, *Utah* **17** 39 8N 111 49W
Centerville, *Calif.* **18** 36 44N 119 30W
Centerville, *Iowa* **14** 40 44N 92 52W
Centerville, *S. Dak.* **12** 43 7N 96 58W
Centerville, *Tenn.* **11** 35 47N 87 28W
Centerville, *Tex.* **13** 31 16N 95 59W
Central, *N. Mex.* **17** 32 47N 108 9W
Central City, *Ky.* **10** 37 18N 87 7W
Central City, *Nebr.* **12** 41 7N 98 0W
Centralia, *Ill.* **14** 38 32N 89 8W
Centralia, *Mo.* **14** 39 13N 92 8W
Centralia, *Wash.* **18** 46 43N 122 58W
Centreville, *Ala.* **11** 32 57N 87 8W
Centreville, *Miss.* **13** 31 5N 91 4W
Ceres, *Calif.* **18** 37 35N 120 57W
Cerro de Punta, Mt., *Puerto Rico* **30** 18 10N 67 0W
Chadron, *Nebr.* **12** 42 50N 103 0W
Chalfant, *Calif.* **18** 37 32N 118 21W
Challis, *Idaho* **16** 44 30N 114 14W
Chama, *N. Mex.* **17** 36 54N 106 35W
Chamberlain, *S. Dak.* **12** 43 49N 99 20W
Chambers, *Ariz.* **17** 35 11N 109 26W
Chambersburg, *Pa.* **10** 39 56N 77 40W
Champaign, *Ill.* **15** 40 7N 88 15W
Champlain, *N.Y.* **9** 44 59N 73 27W
Champlain, L., *N.Y.* **9** 44 40N 73 20W
Chandalar, *Alaska* **30** 67 30N 148 35W
Chandeleur Is., *La.* **13** 29 55N 88 57W
Chandeleur Sd., *La.* **13** 29 55N 89 0W
Chandler, *Ariz.* **17** 33 18N 111 50W
Chandler, *Okla.* **13** 35 42N 96 53W
Channel Is., *Calif.* **19** 33 40N 119 15W
Channing, *Mich.* **10** 46 9N 88 5W
Channing, *Tex.* **13** 35 41N 102 20W
Chanute, *Kans.* **13** 37 41N 95 27W
Chapel Hill, *N.C.* **11** 35 55N 79 4W
Chariton →, *Mo.* **14** 39 19N 92 58W
Charles, C., *Va.* **10** 37 7N 75 58W
Charles City, *Iowa* **14** 43 4N 92 41W
Charles Town, *W. Va.* **10** 39 17N 77 52W
Charleston, *Ill.* **10** 39 30N 88 10W
Charleston, *Miss.* **13** 34 1N 90 4W
Charleston, *Mo.* **13** 36 55N 89 21W
Charleston, *S.C.* **11** 32 46N 79 56W
Charleston, *W. Va.* **10** 38 21N 81 38W
Charlestown, *Ind.* **15** 38 27N 85 40W
Charlevoix, *Mich.* **10** 45 19N 85 16W
Charlotte, *Mich.* **15** 42 34N 84 50W
Charlotte, *N.C.* **11** 35 13N 80 51W
Charlotte Amalie, *Virgin Is.* **30** 18 21N 64 56W
Charlotte Harbor, *Fla.* **11** 26 50N 82 10W
Charlottesville, *Va.* **10** 38 2N 78 30W
Charlton, *Iowa* **12** 40 59N 93 20W
Chatfield, *Minn.* **12** 43 51N 92 11W
Chatham, *Alaska* **30** 57 30N 135 0W
Chatham, *La.* **13** 32 18N 92 27W
Chatham, *N.J.* **20** 40 44N 74 23W
Chatham, *N.Y.* **9** 42 21N 73 36W
Chattahoochee →, *Ga.* **11** 30 54N 84 57W
Chattanooga, *Tenn.* **11** 35 3N 85 19W
Cheboygan, *Mich.* **10** 45 39N 84 29W
Checotah, *Okla.* **13** 35 28N 95 31W
Chefornak, *Alaska* **30** 60 13N 164 12W
Chehalis, *Wash.* **18** 46 40N 122 58W
Chelan, *Wash.* **16** 47 51N 120 1W
Chelan, L., *Wash.* **16** 48 11N 120 30W
Chelmsford, *Mass.* **23** 42 35N 71 20W
Chelsea, *Mass.* **23** 42 23N 71 1W
Chelsea, *Okla.* **13** 36 32N 95 26W
Cheltenham, *Pa.* **24** 40 3N 75 6W
Chemult, *Oreg.* **16** 43 14N 121 47W
Chenango Forks, *N.Y.* **9** 42 15N 75 51W
Cheney, *Wash.* **16** 47 30N 117 35W
Chequamegon B., *Mich.* **12** 46 40N 90 30W
Cheraw, *S.C.* **11** 34 42N 79 53W
Cherokee, *Iowa* **12** 42 45N 95 33W
Cherokee, *Okla.* **13** 36 45N 98 21W
Cherokees, Lake O' The, *Okla.* **13** 36 28N 95 2W
Cherry Creek, *Nev.* **16** 39 54N 114 53W
Cherry Hill, *N.J.* **24** 39 54N 75 1W
Cherry Valley, *Calif.* **19** 33 59N 116 57W
Cherryvale, *Kans.* **13** 37 16N 95 33W
Chesapeake, *Va.* **10** 36 50N 76 17W
Chesapeake B., *Va.* **10** 38 0N 76 10W
Chester, *Calif.* **16** 40 19N 121 14W
Chester, *Ill.* **14** 37 55N 89 49W
Chester, *Mont.* **16** 48 31N 110 58W
Chester, *Pa.* **24** 39 51N 75 22W
Chester, *S.C.* **11** 34 43N 81 12W
Chesuncook L., *Maine* **11** 46 0N 69 21W
Chevy Chase, *Md.* **25** 38 59N 77 4W
Chewelah, *Wash.* **16** 48 17N 117 43W
Cheyenne, *Okla.* **13** 35 37N 99 40W
Cheyenne, *Wyo.* **12** 41 8N 104 49W
Cheyenne →, *S. Dak.* **12** 44 41N 101 18W
Cheyenne Wells, *Colo.* **12** 38 49N 102 21W
Cheyney, *Pa.* **24** 39 55N 75 32W
Chicago, *Ill.* **22** 41 52N 87 38W
Chicago, University of, *Ill.* **22** 41 47N 87 35W
Chicago Heights, *Ill.* **15** 41 30N 87 38W
Chicago Midway Airport, *Ill.* **22** 41 46N 87 44W
Chicago O'Hare International Airport, *Ill.* **22** 41 58N 87 54W
Chickasha, *Okla.* **13** 35 3N 97 58W
Chico, *Calif.* **18** 39 44N 121 50W
Chicopee, *Mass.* **9** 42 9N 72 37W
Chignik, *Alaska* **30** 56 18N 158 24W
Childress, *Tex.* **13** 34 25N 100 13W
Chillicothe, *Ill.* **14** 40 55N 89 29W
Chillicothe, *Mo.* **14** 39 48N 93 33W
Chillicothe, *Ohio* **10** 39 20N 82 59W
Chillum, *Md.* **25** 38 57N 76 58W
Chilton, *Wis.* **10** 44 2N 88 10W
China Lake, *Calif.* **19** 35 44N 117 37W
Chinati Peak, *Tex.* **13** 29 57N 104 29W
Chincoteague, *Va.* **10** 37 56N 75 23W
Chinle, *Ariz.* **17** 36 9N 109 33W
Chino, *Calif.* **19** 34 1N 117 41W
Chino Valley, *Ariz.* **17** 34 45N 112 27W
Chinook, *Mont.* **16** 48 35N 109 14W
Chipley, *Fla.* **11** 30 47N 85 32W
Chippewa →, *Wis.* **12** 44 25N 92 5W
Chippewa Falls, *Wis.* **12** 44 56N 91 24W
Chiricahua Peak, *Ariz.* **17** 31 51N 109 18W
Chirikof I., *Alaska* **30** 55 50N 155 40W
Chisos Mts., *Tex.* **13** 29 5N 103 15W
Chitina, *Alaska* **30** 61 31N 144 26W
Cholame, *Calif.* **18** 35 44N 120 18W
Choteau, *Mont.* **16** 47 49N 112 11W
Chowchilla, *Calif.* **18** 37 7N 120 16W
Chugach Mts., *Alaska* **30** 60 45N 147 0W
Chugiak, *Alaska* **30** 61 24N 149 29W
Chuginadak I., *Alaska* **30** 52 50N 169 45W
Chugwater, *Wyo.* **12** 41 46N 104 50W
Chula Vista, *Calif.* **19** 32 39N 117 5W
Cicero, *Ill.* **22** 41 51N 87 45W
Cimarron, *Kans.* **13** 37 48N 100 21W
Cimarron, *N. Mex.* **13** 36 31N 104 55W
Cimarron →, *Okla.* **13** 36 10N 96 17W
Cincinnati, *Ohio* **15** 39 6N 84 31W
Circle, *Alaska* **30** 65 50N 144 4W
Circle, *Mont.* **12** 47 25N 105 35W
Circleville, *Ohio* **10** 39 36N 82 57W
Circleville, *Utah* **17** 38 10N 112 16W
Cisco, *Tex.* **13** 32 23N 98 59W
Clairemont, *Tex.* **13** 33 9N 100 44W
Clanton, *Ala.* **11** 32 51N 86 38W
Claraville, *Calif.* **19** 35 24N 118 20W
Clare, *Mich.* **10** 43 49N 84 46W
Claremont, *Calif.* **19** 34 6N 117 43W
Claremont, *N.H.* **9** 43 23N 72 20W
Claremore, *Okla.* **13** 36 19N 95 36W
Clarendon, *Ark.* **13** 34 42N 91 19W
Clarendon, *Tex.* **13** 34 56N 100 53W
Clarinda, *Iowa* **12** 40 44N 95 2W
Clarion, *Iowa* **14** 42 44N 93 44W
Clark, *S. Dak.* **12** 44 53N 97 44W
Clark Fork, *Idaho* **16** 48 9N 116 11W
Clark Fork →, *Idaho* **16** 48 9N 116 15W
Clark Hill Res., *Ga.* **11** 33 45N 82 20W
Clarkdale, *Ariz.* **17** 34 46N 112 3W
Clark's Fork →, *Wyo.* **16** 45 39N 108 43W
Clarks Summit, *Pa.* **9** 41 30N 75 42W
Clarksburg, *W. Va.* **10** 39 17N 80 30W
Clarksdale, *Miss.* **13** 34 12N 90 35W
Clarkston, *Wash.* **16** 46 25N 117 3W
Clarksville, *Ark.* **13** 35 28N 93 28W
Clarksville, *Tenn.* **11** 36 32N 87 21W
Clarksville, *Tex.* **13** 33 37N 95 3W
Clatskanie, *Oreg.* **18** 46 6N 123 12W
Claude, *Tex.* **13** 35 7N 101 22W
Clay, *Calif.* **18** 38 17N 121 10W
Clay Center, *Kans.* **12** 39 23N 97 8W
Claymont, *Del.* **24** 39 48N 75 28W
Claypool, *Ariz.* **17** 33 25N 110 51W
Clayton, *Idaho* **16** 44 16N 114 24W
Clayton, *N. Mex.* **13** 36 27N 103 11W
Cle Elum, *Wash.* **16** 47 12N 120 56W
Clear L., *Calif.* **18** 39 2N 122 47W
Clear Lake, *S. Dak.* **12** 44 45N 96 41W
Clear Lake, *Wash.* **16** 48 27N 122 15W
Clear Lake Reservoir, *Calif.* **16** 41 56N 121 5W
Clearfield, *Pa.* **10** 41 2N 78 27W
Clearfield, *Utah* **16** 41 7N 112 2W
Clearmont, *Wyo.* **16** 44 38N 106 23W
Clearwater, *Fla.* **11** 27 58N 82 48W
Clearwater Mts., *Idaho* **16** 46 5N 115 20W
Cleburne, *Tex.* **13** 32 21N 97 23W
Cleveland, *Miss.* **13** 33 45N 90 43W
Cleveland, *Ohio* **27** 41 29N 81 42W
Cleveland, *Okla.* **13** 36 19N 96 28W
Cleveland, *Tenn.* **11** 35 10N 84 53W
Cleveland, *Tex.* **13** 30 21N 95 5W
Cleveland Heights, *Ohio* **27** 41 29N 81 35W
Cleveland Hopkins International Airport, *Ohio* **27** 41 24N 81 51W
Clewiston, *Fla.* **11** 26 45N 80 56W
Clifton, *Ariz.* **17** 33 3N 109 18W
Clifton, *N.J.* **20** 40 52N 74 8W
Clifton, *Tex.* **13** 31 47N 97 35W
Clifton Forge, *Va.* **10** 37 49N 79 50W
Clinch →, *Tenn.* **11** 35 53N 84 29W
Clingmans Dome, *Tenn.* **11** 35 34N 83 30W
Clint, *Tex.* **17** 31 35N 106 14W
Clinton, *Ark.* **13** 35 36N 92 28W
Clinton, *Ill.* **12** 40 9N 88 57W
Clinton, *Ind.* **15** 39 40N 87 24W
Clinton, *Iowa* **14** 41 51N 90 12W
Clinton, *Mass.* **9** 42 25N 71 41W
Clinton, *Mo.* **14** 38 22N 93 46W
Clinton, *N.C.* **11** 35 0N 78 22W
Clinton, *Okla.* **13** 35 31N 98 58W
Clinton, *S.C.* **11** 34 29N 81 53W
Clinton, *Tenn.* **11** 36 6N 84 8W
Clintonville, *Wis.* **12** 44 37N 88 46W
Cloquet, *Minn.* **12** 46 43N 92 28W
Cloud Peak, *Wyo.* **16** 44 23N 107 11W
Cloudcroft, *N. Mex.* **17** 32 58N 105 45W
Cloverdale, *Calif.* **18** 38 48N 123 1W
Clovis, *Calif.* **18** 36 49N 119 42W
Clovis, *N. Mex.* **13** 34 24N 103 12W
Coachella, *Calif.* **19** 33 41N 116 10W
Coahoma, *Tex.* **13** 32 18N 101 18W
Coalgate, *Okla.* **13** 34 32N 96 13W
Coalinga, *Calif.* **18** 36 9N 120 21W
Coalville, *Utah* **16** 40 55N 111 24W
Coamo, *Puerto Rico* **30** 18 5N 66 22W
Coast Ranges, *Calif.* **18** 39 0N 123 0W
Coatesville, *Pa.* **10** 39 59N 75 50W
Cobleskill, *N.Y.* **9** 42 41N 74 29W
Cobre, *Nev.* **16** 41 7N 114 24W
Cochise, *Ariz.* **17** 32 7N 109 55W
Cochran, *Ga.* **11** 32 23N 83 21W
Cocoa, *Fla.* **11** 28 21N 80 44W
Cody, *Wyo.* **16** 44 32N 109 3W
Cœur d'Alene, *Idaho* **16** 47 45N 116 51W
Cœur d'Alene L., *Idaho* **16** 47 32N 116 48W
Coffeyville, *Kans.* **13** 37 2N 95 37W
Cohagen, *Mont.* **16** 47 3N 106 37W
Cohoes, *N.Y.* **9** 42 46N 73 42W
Cokeville, *Wyo.* **16** 42 5N 110 57W
Colby, *Kans.* **12** 39 24N 101 3W
Coldwater, *Kans.* **13** 37 16N 99 20W
Colebrook, *N.H.* **9** 44 54N 71 30W
Coleman, *Tex.* **13** 31 50N 99 26W
Coleville, *Calif.* **18** 38 34N 119 30W
Colfax, *La.* **13** 31 31N 92 42W
Colfax, *Wash.* **16** 46 53N 117 22W
Collbran, *Colo.* **17** 39 14N 107 58W
College Park, *Ga.* **11** 33 40N 84 27W
College Park, *Md.* **25** 38 59N 76 55W
Collingdale, *Pa.* **24** 39 54N 75 16W
Colome, *S. Dak.* **12** 43 16N 99 43W
Colonial Heights, *Va.* **10** 37 15N 77 25W
Colorado □ **17** 39 30N 105 30W
Colorado →, *N. Amer.* **17** 31 45N 114 40W
Colorado →, *Tex.* **13** 28 36N 95 59W
Colorado City, *Tex.* **13** 32 24N 100 52W
Colorado Plateau, *Ariz.* **17** 37 0N 111 0W
Colorado River Aqueduct, *Calif.* **19** 34 17N 114 10W
Colorado Springs, *Colo.* **12** 38 50N 104 49W
Colton, *Calif.* **19** 34 4N 117 20W
Colton, *Wash.* **16** 46 34N 117 8W
Columbia, *La.* **13** 32 6N 92 5W
Columbia, *Md.* **25** 39 12N 76 50W
Columbia, *Miss.* **13** 31 15N 89 50W
Columbia, *Mo.* **14** 38 57N 92 20W
Columbia, *Pa.* **9** 40 2N 76 30W
Columbia, *S.C.* **11** 34 0N 81 2W
Columbia, *Tenn.* **11** 35 37N 87 2W
Columbia →, *Oreg.* **16** 46 15N 124 5W
Columbia, District of □ **10** 38 55N 77 0W
Columbia Basin, *Wash.* **16** 46 45N 119 5W
Columbia Falls, *Mont.* **16** 48 23N 114 11W
Columbia Heights, *Minn.* **12** 45 3N 93 15W
Columbus, *Ga.* **11** 32 28N 84 59W
Columbus, *Ind.* **15** 39 13N 85 55W
Columbus, *Kans.* **13** 37 10N 94 50W
Columbus, *Miss.* **11** 33 30N 88 25W
Columbus, *Mont.* **16** 45 38N 109 15W
Columbus, *N. Dak.* **12** 48 54N 102 47W
Columbus, *N. Mex.* **17** 31 50N 107 38W
Columbus, *Nebr.* **12** 41 26N 97 22W
Columbus, *Ohio* **15** 39 58N 83 0W
Columbus, *Tex.* **13** 29 42N 96 33W
Columbus, *Wis.* **12** 43 21N 89 1W
Colusa, *Calif.* **18** 39 13N 122 1W
Colville, *Wash.* **16** 48 33N 117 54W
Colville →, *Alaska* **30** 70 25N 150 30W
Comanche, *Okla.* **13** 34 22N 97 58W
Comanche, *Tex.* **13** 31 54N 98 36W
Combahee →, *S.C.* **11** 32 30N 80 31W
Commerce, *Calif.* **29** 34 0N 118 9W
Commerce, *Ga.* **11** 34 12N 83 28W
Commerce, *Tex.* **13** 33 15N 95 54W
Compton, *Calif.* **29** 33 53N 118 14W
Conception, Pt., *Calif.* **19** 34 27N 120 28W
Conchas Dam, *N. Mex.* **13** 35 22N 104 11W
Concho, *Ariz.* **17** 34 28N 109 36W
Concho →, *Tex.* **13** 31 34N 99 43W
Concord, *Calif.* **28** 37 57N 122 3W
Concord, *Mass.* **23** 42 27N 71 20W
Concord, *N.C.* **11** 35 25N 80 35W
Concord, *N.H.* **9** 43 12N 71 32W
Concordia, *Kans.* **12** 39 34N 97 40W
Concordville, *Pa.* **24** 39 53N 75 31W
Concrete, *Wash.* **16** 48 32N 121 45W
Conde, *S. Dak.* **12** 45 9N 98 6W
Condon, *Oreg.* **16** 45 14N 120 11W
Congress, *Ariz.* **17** 34 9N 112 51W
Conneaut, *Ohio* **8** 41 57N 80 34W
Connecticut □ **9** 41 30N 72 45W
Connecticut →, *Conn.* **9** 41 16N 72 20W
Connell, *Wash.* **16** 46 40N 118 52W
Connellsville, *Pa.* **8** 40 1N 79 35W
Connersville, *Ind.* **15** 39 39N 85 8W
Conrad, *Mont.* **16** 48 10N 111 57W
Conroe, *Tex.* **13** 30 19N 95 27W
Conshohocken, *Pa.* **24** 40 4N 75 18W
Contact, *Nev.* **16** 41 46N 114 45W
Contoocook, *N.H.* **9** 43 13N 71 45W
Conway, *Ark.* **13** 35 5N 92 26W
Conway, *N.H.* **9** 43 59N 71 7W
Conway, *S.C.* **11** 33 51N 79 3W
Cook, *Minn.* **12** 47 49N 92 39W
Cook Inlet, *Alaska* **30** 60 0N 152 0W
Cookeville, *Tenn.* **11** 36 10N 85 30W
Coolidge, *Ariz.* **17** 32 59N 111 31W
Coolidge Dam, *Ariz.* **17** 33 0N 110 20W
Cooper, *Tex.* **13** 33 23N 95 42W
Cooper →, *S.C.* **11** 32 50N 79 56W
Cooperstown, *N. Dak.* **12** 47 27N 98 8W
Cooperstown, *N.Y.* **9** 42 42N 74 56W
Coos Bay, *Oreg.* **16** 43 22N 124 13W
Cope, *Colo.* **12** 39 40N 102 51W
Copper →, *Alaska* **30** 60 18N 145 3W
Copper Center, *Alaska* **30** 61 58N 145 18W
Copper Harbor, *Mich.* **10** 47 28N 87 53W
Copperopolis, *Calif.* **18** 37 58N 120 38W
Coquille, *Oreg.* **16** 43 11N 124 11W
Coral Gables, *Fla.* **11** 25 45N 80 16W
Corbin, *Ky.* **10** 36 57N 84 6W
Corcoran, *Calif.* **18** 36 6N 119 33W
Cordele, *Ga.* **11** 31 58N 83 47W
Cordell, *Okla.* **13** 35 17N 98 59W
Cordova, *Ala.* **11** 33 46N 87 11W
Cordova, *Alaska* **30** 60 33N 145 45W
Corinth, *Miss.* **11** 34 56N 88 31W
Corinth, *N.Y.* **9** 43 15N 73 49W
Cornell, *Wis.* **12** 45 10N 91 9W
Corning, *Ark.* **13** 36 25N 90 35W
Corning, *Calif.* **16** 39 56N 122 11W
Corning, *Iowa* **14** 40 59N 94 44W
Corning, *N.Y.* **8** 42 9N 77 3W
Corona, *Calif.* **19** 33 53N 117 34W
Corona, *N. Mex.* **17** 34 15N 105 36W
Coronado, *Calif.* **19** 32 41N 117 11W
Corpus Christi, *Tex.* **13** 27 47N 97 24W
Corpus Christi, L., *Tex.* **13** 28 2N 97 52W
Corrigan, *Tex.* **13** 31 0N 94 52W
Corry, *Pa.* **8** 41 55N 79 39W
Corsicana, *Tex.* **13** 32 6N 96 28W
Corte Madera, *Calif.* **28** 37 55N 122 30W
Cortez, *Colo.* **17** 37 21N 108 35W
Cortland, *N.Y.* **9** 42 36N 76 11W
Corvallis, *Oreg.* **16** 44 34N 123 16W
Corydon, *Iowa* **14** 40 46N 93 19W
Coshocton, *Ohio* **8** 40 16N 81 51W
Coso Junction, *Calif.* **19** 36 3N 117 57W
Coso Pk., *Calif.* **19** 36 13N 117 44W
Costa Mesa, *Calif.* **19** 33 38N 117 55W
Costilla, *N. Mex.* **17** 36 59N 105 32W
Cosumnes →, *Calif.* **18** 38 16N 121 26W
Coteau des Prairies, *S. Dak.* **12** 45 20N 97 50W
Coteau du Missouri, *N. Dak.* **12** 47 0N 100 0W
Cottage Grove, *Oreg.* **16** 43 48N 123 3W
Cottonwood, *Ariz.* **17** 34 45N 112 1W
Cotulla, *Tex.* **13** 28 26N 99 14W
Coudersport, *Pa.* **8** 41 46N 78 1W
Coulee City, *Wash.* **16** 47 37N 119 17W
Coulterville, *Calif.* **18** 37 43N 120 12W
Council, *Alaska* **30** 64 55N 163 45W
Council, *Idaho* **16** 44 44N 116 26W
Council Bluffs, *Iowa* **12** 41 16N 95 52W
Council Grove, *Kans.* **12** 38 40N 96 29W
Courtland, *Calif.* **18** 38 20N 121 34W
Coushatta, *La.* **13** 32 1N 93 21W
Covington, *Ga.* **11** 33 36N 83 51W
Covington, *Ky.* **15** 39 5N 84 31W
Covington, *Okla.* **13** 36 18N 97 35W
Covington, *Tenn.* **13** 35 34N 89 39W
Cozad, *Nebr.* **12** 40 52N 99 59W
Craig, *Alaska* **30** 55 29N 133 9W
Craig, *Colo.* **16** 40 31N 107 33W
Crandon, *Wis.* **12** 45 34N 88 54W
Crane, *Oreg.* **16** 43 25N 118 35W
Crane, *Tex.* **13** 31 24N 102 21W
Cranford, *N.J.* **20** 40 39N 74 19W
Cranston, *R.I.* **9** 41 47N 71 26W
Crater L., *Oreg.* **16** 42 56N 122 6W
Crawford, *Nebr.* **12** 42 41N 103 25W
Crawfordsville, *Ind.* **15** 40 2N 86 54W
Crazy Mts., *Mont.* **16** 46 12N 110 20W
Creede, *Colo.* **17** 37 51N 106 56W
Creighton, *Nebr.* **12** 42 28N 97 54W
Cresbard, *S. Dak.* **12** 45 10N 98 57W
Crescent, *Okla.* **13** 35 57N 97 36W
Crescent, *Oreg.* **16** 43 28N 121 42W
Crescent City, *Calif.* **16** 41 45N 124 12W
Crested Butte, *Colo.* **17** 38 52N 106 59W
Crestline, *Calif.* **19** 34 14N 117 18W

Creston, *Calif.* **18** 35 32N 120 33W
Creston, *Iowa* **14** 41 4N 94 22W
Creston, *Wash.* **16** 47 46N 118 31W
Crestview, *Calif.* **18** 37 46N 118 58W
Crestview, *Fla.* **11** 30 46N 86 34W
Crete, *Nebr.* **12** 40 38N 96 58W
Crockett, *Tex.* **13** 31 19N 95 27W
Crooked →, *Oreg.* **16** 44 32N 121 16W
Crookston, *Minn.* **12** 47 47N 96 37W
Crookston, *Nebr.* **12** 42 56N 100 45W
Crooksville, *Ohio* **10** 39 46N 82 6W
Crosby, *Minn.* **12** 46 29N 93 58W
Crosbyton, *Tex.* **13** 33 40N 101 14W
Cross City, *Fla.* **11** 29 38N 83 7W
Cross Plains, *Tex.* **13** 32 8N 99 11W
Cross Sound, *Alaska* **30** 58 0N 135 0W
Crossett, *Ark.* **13** 33 8N 91 58W
Croton-on-Hudson, *N.Y.* **9** 41 12N 73 55W
Crow Agency, *Mont.* **16** 45 36N 107 28W
Crowell, *Tex.* **13** 33 59N 99 43W
Crowley, *La.* **13** 30 13N 92 22W
Crowley, L., *Calif.* **18** 37 35N 118 42W
Crown Point, *Ind.* **15** 41 25N 87 22W
Crows Landing, *Calif.* **18** 37 23N 121 6W
Crystal City, *Mo.* **14** 38 13N 90 23W
Crystal City, *Tex.* **13** 28 41N 99 50W
Crystal Falls, *Mich.* **10** 46 5N 88 20W
Crystal River, *Fla.* **11** 28 54N 82 35W
Crystal Springs, *Miss.* **13** 31 59N 90 21W
Cuba, *N. Mex.* **17** 36 1N 107 4W
Cudahy, *Wis.* **15** 42 58N 87 52W
Cuero, *Tex.* **13** 29 6N 97 17W
Cuervo, *N. Mex.* **13** 35 2N 104 25W
Culbertson, *Mont.* **12** 48 9N 104 31W
Culebra, Isla de, *Puerto Rico* **30** 18 19N 65 18W
Cullman, *Ala.* **11** 34 11N 86 51W
Culpeper, *Va.* **10** 38 30N 78 0W
Culver City, *Calif.* **29** 34 1N 118 23W
Cumberland, *Md.* **10** 39 39N 78 46W
Cumberland, *Wis.* **12** 45 32N 92 1W
Cumberland →, *Tenn.* **11** 36 15N 87 0W
Cumberland I., *Ga.* **11** 30 50N 81 25W
Cumberland Plateau, *Tenn.* **11** 36 0N 85 0W
Cummings Mt., *Calif.* **19** 35 2N 118 34W
Currant, *Nev.* **16** 38 51N 115 32W
Current →, *Ark.* **13** 36 15N 90 55W
Currie, *Nev.* **16** 40 16N 114 45W
Currituck Sd., *N.C.* **11** 36 20N 75 52W
Curtis, *Nebr.* **12** 40 38N 100 31W
Cushing, *Okla.* **13** 35 59N 96 46W
Custer, *S. Dak.* **12** 43 46N 103 36W
Cut Bank, *Mont.* **16** 48 38N 112 20W
Cuthbert, *Ga.* **11** 31 46N 84 48W
Cutler, *Calif.* **18** 36 31N 119 17W
Cuyahoga Falls, *Ohio* **8** 41 8N 81 29W
Cynthiana, *Ky.* **15** 38 23N 84 18W

D

Dade City, *Fla.* **11** 28 22N 82 11W
Daggett, *Calif.* **19** 34 52N 116 52W
Dahlonega, *Ga.* **11** 34 32N 83 59W
Dakota City, *Nebr.* **12** 42 25N 96 25W
Dalhart, *Tex.* **13** 36 4N 102 31W
Dallas, *Oreg.* **16** 44 55N 123 19W
Dallas, *Tex.* **13** 32 47N 96 49W
Dalton, *Ga.* **11** 34 46N 84 58W
Dalton, *Mass.* **9** 42 28N 73 11W
Dalton, *Nebr.* **12** 41 25N 102 58W
Daly City, *Calif.* **28** 37 42N 122 26W
Dana, Mt., *Calif.* **18** 37 54N 119 12W
Danbury, *Conn.* **9** 41 24N 73 28W
Danby L., *Calif.* **17** 34 13N 115 5W
Danforth, *Maine* **11** 45 40N 67 52W
Daniel, *Wyo.* **16** 42 52N 110 4W
Danielson, *Conn.* **9** 41 48N 71 53W
Dannemora, *N.Y.* **9** 44 43N 73 44W
Dansville, *N.Y.* **8** 42 34N 77 42W
Danvers, *Mass.* **23** 42 34N 70 56W
Danville, *Ill.* **15** 40 8N 87 37W
Danville, *Ky.* **15** 37 39N 84 46W
Danville, *Va.* **11** 36 36N 79 23W
Darby, *Mont.* **16** 46 1N 114 11W
Darby, *Pa.* **24** 39 55N 75 16W
Dardanelle, *Ark.* **13** 35 13N 93 9W
Dardanelle, *Calif.* **18** 38 20N 119 50W
Darlington, *S.C.* **11** 34 18N 79 52W
Darlington, *Wis.* **14** 42 41N 90 7W
Darrington, *Wash.* **16** 48 15N 121 36W
Darwin, *Calif.* **19** 36 15N 117 35W
Dauphin I., *Ala.* **11** 30 15N 88 11W
Davenport, *Calif.* **18** 37 1N 122 12W
Davenport, *Iowa* **14** 41 32N 90 35W
Davenport, *Wash.* **16** 47 39N 118 9W
David City, *Nebr.* **12** 41 15N 97 8W
Davis, *Calif.* **18** 38 33N 121 44W
Davis Dam, *Ariz.* **19** 35 11N 114 34W
Davis Mts., *Tex.* **13** 30 50N 103 55W
Dawson, *Ga.* **11** 31 46N 84 27W
Dawson, *N. Dak.* **12** 46 52N 99 45W
Dayton, *Ohio* **10** 39 45N 84 12W
Dayton, *Tenn.* **11** 35 30N 85 1W
Dayton, *Wash.* **16** 46 19N 117 59W
Daytona Beach, *Fla.* **11** 29 13N 81 1W
Dayville, *Oreg.* **16** 44 28N 119 32W
De Funiak Springs, *Fla.* **11** 30 43N 86 7W
De Kalb, *Ill.* **15** 41 56N 88 46W
De Land, *Fla.* **11** 29 2N 81 18W
De Leon, *Tex.* **13** 32 7N 98 32W
De Long Mts., *Alaska* **30** 68 30N 163 0W
De Pere, *Wis.* **10** 44 27N 88 4W
De Queen, *Ark.* **13** 34 2N 94 21W
De Quincy, *La.* **13** 30 27N 93 26W
De Ridder, *La.* **13** 30 51N 93 17W
De Smet, *S. Dak.* **12** 44 23N 97 33W
De Soto, *Mo.* **14** 38 8N 90 34W
De Tour Village, *Mich.* **10** 46 0N 83 56W
De Witt, *Ark.* **13** 34 18N 91 20W
Deadwood, *S. Dak.* **12** 44 23N 103 44W
Dearborn, *Mich.* **26** 42 19N 83 10W
Dearborn Heights, *Mich.* **26** 42 20N 83 17W
Death Valley, *Calif.* **19** 36 15N 116 50W
Death Valley Junction, *Calif.* **19** 36 20N 116 25W
Death Valley National Monument, *Calif.* **19** 36 45N 117 15W
Decatur, *Ala.* **11** 34 36N 86 59W
Decatur, *Ga.* **11** 33 47N 84 18W
Decatur, *Ill.* **14** 39 51N 88 57W
Decatur, *Ind.* **15** 40 50N 84 56W
Decatur, *Tex.* **13** 33 14N 97 35W
Decorah, *Iowa* **12** 43 18N 91 48W
Dedham, *Mass.* **23** 42 15N 71 10W
Deer I., *Alaska* **30** 54 55N 162 18W
Deer Lodge, *Mont.* **16** 46 24N 112 44W
Deer Park, *N.Y.* **21** 40 46N 73 19W
Deer Park, *Wash.* **16** 47 57N 117 28W
Deer River, *Minn.* **12** 47 20N 93 48W
Deering, *Alaska* **30** 66 4N 162 42W
Defiance, *Ohio* **15** 41 17N 84 22W
Del Mar, *Calif.* **19** 32 58N 117 16W
Del Norte, *Colo.* **17** 37 41N 106 21W
Del Rio, *Tex.* **13** 29 22N 100 54W
Delano, *Calif.* **19** 35 46N 119 15W
Delavan, *Wis.* **15** 42 38N 88 39W
Delaware, *Ohio* **15** 40 18N 83 4W
Delaware □ **10** 39 0N 75 20W
Delaware →, *Del.* **10** 39 15N 75 20W
Delhi, *N.Y.* **9** 42 17N 74 55W
Dell City, *Tex.* **17** 31 56N 105 12W
Dell Rapids, *S. Dak.* **12** 43 50N 96 43W
Delphi, *Ind.* **15** 40 36N 86 41W
Delphos, *Ohio* **15** 40 51N 84 21W
Delray Beach, *Fla.* **11** 26 28N 80 4W
Delta, *Colo.* **17** 38 44N 108 4W
Delta, *Utah* **16** 39 21N 112 35W
Deming, *N. Mex.* **17** 32 16N 107 46W
Demopolis, *Ala.* **11** 32 31N 87 50W
Denair, *Calif.* **18** 37 32N 120 48W
Denison, *Iowa* **12** 42 1N 95 21W
Denison, *Tex.* **13** 33 45N 96 33W
Denton, *Mont.* **16** 47 19N 109 57W
Denton, *Tex.* **13** 33 13N 97 8W
Denver, *Colo.* **12** 39 44N 104 59W
Denver City, *Tex.* **13** 32 58N 102 50W
Deposit, *N.Y.* **9** 42 4N 75 25W
Derby, *Conn.* **9** 41 19N 73 5W
Dernieres, Isles, *La.* **13** 29 2N 90 50W
Des Moines, *Iowa* **14** 41 35N 93 37W
Des Moines, *N. Mex.* **13** 36 46N 103 50W
Des Moines →, *Iowa* **12** 40 23N 91 25W
Des Plaines, *Ill.* **22** 42 2N 87 54W
Deschutes →, *Oreg.* **16** 45 38N 120 55W
Desert Center, *Calif.* **19** 33 43N 115 24W
Desert Hot Springs, *Calif.* **19** 33 58N 116 30W
Detour, Pt., *Mich.* **10** 45 40N 86 40W
Detroit, *Mich.* **26** 42 20N 83 2W
Detroit, *Tex.* **13** 33 40N 95 16W
Detroit City Airport, *Mich.* **26** 42 24N 83 0W
Detroit Lakes, *Minn.* **12** 46 49N 95 51W
Detroit-Wayne Airport, *Mich.* **26** 42 13N 83 20W
Devils Den, *Calif.* **18** 35 46N 119 58W
Devils Lake, *N. Dak.* **12** 48 7N 98 52W
Dexter, *Mo.* **13** 36 48N 89 57W
Dexter, *N. Mex.* **13** 33 12N 104 22W
Diablo, Mt., *Calif.* **18** 37 53N 121 56W
Diablo Range, *Calif.* **18** 37 20N 121 25W
Diamond Mts., *Nev.* **16** 39 50N 115 30W
Diamond Springs, *Calif.* **18** 38 42N 120 49W
Diamondville, *Wyo.* **16** 41 47N 110 32W
Dickinson, *N. Dak.* **12** 46 53N 102 47W
Dickson, *Tenn.* **11** 36 5N 87 23W
Dickson City, *Pa.* **9** 41 29N 75 40W
Dierks, *Ark.* **13** 34 7N 94 1W
Dighton, *Kans.* **12** 38 29N 100 28W
Dilley, *Tex.* **13** 28 40N 99 10W
Dillingham, *Alaska* **30** 59 3N 158 28W
Dillon, *Mont.* **16** 45 13N 112 38W
Dillon, *S.C.* **11** 34 25N 79 22W
Dimmitt, *Tex.* **13** 34 33N 102 19W
Dingmans Ferry, *Pa.* **9** 41 13N 74 55W
Dinosaur National Monument, *Colo.* **16** 40 30N 108 45W
Dinuba, *Calif.* **18** 36 32N 119 23W
Disappointment, C., *Wash.* **16** 46 18N 124 5W
Divide, *Mont.* **16** 45 45N 112 45W
Dixon, *Calif.* **18** 38 27N 121 49W
Dixon, *Ill.* **14** 41 50N 89 29W
Dixon, *Mont.* **16** 47 19N 114 19W
Dixon, *N. Mex.* **17** 36 12N 105 53W
Dodge Center, *Minn.* **12** 44 2N 92 52W
Dodge City, *Kans.* **13** 37 45N 100 1W
Dodgeville, *Wis.* **14** 42 58N 90 8W
Dodson, *Mont.* **16** 48 24N 108 15W
Doland, *S. Dak.* **12** 44 54N 98 6W
Dolores, *Colo.* **17** 37 28N 108 30W
Dolores →, *Utah* **17** 38 49N 109 17W
Dolton, *Ill.* **22** 41 37N 87 35W
Donaldsonville, *La.* **13** 30 6N 90 59W
Donalsonville, *Ga.* **11** 31 3N 84 53W
Doniphan, *Mo.* **13** 36 37N 90 50W
Donna, *Tex.* **13** 26 9N 98 4W
Dormont, *Pa.* **27** 40 23N 80 2W
Dorris, *Calif.* **16** 41 58N 121 55W
Dos Palos, *Calif.* **18** 36 59N 120 37W
Dothan, *Ala.* **11** 31 13N 85 24W
Douglas, *Alaska* **30** 58 17N 134 24W
Douglas, *Ariz.* **17** 31 21N 109 33W
Douglas, *Ga.* **11** 31 31N 82 51W
Douglas, *Wyo.* **12** 42 45N 105 24W
Douglasville, *Ga.* **11** 33 45N 84 45W
Dove Creek, *Colo.* **17** 37 46N 108 54W
Dover, *Del.* **10** 39 10N 75 32W
Dover, *N.H.* **9** 43 12N 70 56W
Dover, *N.J.* **9** 40 53N 74 34W
Dover, *Ohio* **8** 40 32N 81 29W
Dover-Foxcroft, *Maine* **11** 45 11N 69 13W
Dover Plains, *N.Y.* **9** 41 43N 73 35W
Dowagiac, *Mich.* **15** 41 59N 86 6W
Downey, *Calif.* **29** 33 56N 118 8W
Downey, *Idaho* **16** 42 26N 112 7W
Downieville, *Calif.* **18** 39 34N 120 50W
Doylestown, *Pa.* **9** 40 21N 75 10W
Drain, *Oreg.* **16** 43 40N 123 19W
Drake, *N. Dak.* **12** 47 55N 100 23W
Drexel Hill, *Pa.* **24** 39 56N 75 18W
Driggs, *Idaho* **16** 43 44N 111 6W
Drummond, *Mont.* **16** 46 40N 113 9W
Drumright, *Okla.* **13** 35 59N 96 36W
Dryden, *Tex.* **13** 30 3N 102 7W
Du Bois, *Pa.* **8** 41 8N 78 46W
Du Quoin, *Ill.* **14** 38 1N 89 14W
Duanesburg, *N.Y.* **9** 42 45N 74 11W
Duarte, *Calif.* **29** 34 8N 117 57W
Dublin, *Ga.* **11** 32 32N 82 54W
Dublin, *Tex.* **13** 32 5N 98 21W
Dubois, *Idaho* **16** 44 10N 112 14W
Dubuque, *Iowa* **14** 42 30N 90 41W
Duchesne, *Utah* **16** 40 10N 110 24W
Duckwall, Mt., *Calif.* **18** 37 58N 120 7W
Duluth, *Minn.* **12** 46 47N 92 6W
Dulworthtown, *Pa.* **24** 39 54N 75 33W
Dumas, *Ark.* **13** 33 53N 91 29W
Dumas, *Tex.* **13** 35 52N 101 58W
Duncan, *Ariz.* **17** 32 43N 109 6W
Duncan, *Okla.* **13** 34 30N 97 57W
Dundalk, *Md.* **25** 39 16N 76 30W
Dunedin, *Fla.* **11** 28 1N 82 47W
Dunkirk, *N.Y.* **8** 42 29N 79 20W
Dunlap, *Iowa* **12** 41 51N 95 36W
Dunmore, *Pa.* **9** 41 25N 75 38W
Dunn, *N.C.* **11** 35 19N 78 37W
Dunnellon, *Fla.* **11** 29 3N 82 28W
Dunning, *Nebr.* **12** 41 50N 100 6W
Dunseith, *N. Dak.* **12** 48 50N 100 3W
Dunsmuir, *Calif.* **16** 41 13N 122 16W
Dupree, *S. Dak.* **12** 45 4N 101 35W
Dupuyer, *Mont.* **16** 48 13N 112 30W
Duquesne, *Pa.* **27** 40 22N 79 52W
Durand, *Mich.* **15** 42 55N 83 59W
Durango, *Colo.* **17** 37 16N 107 53W
Durant, *Okla.* **13** 33 59N 96 25W
Durham, *N.C.* **11** 35 59N 78 54W
Duryea, *Pa.* **9** 41 20N 75 45W
Dutch Harbor, *Alaska* **30** 53 53N 166 32W
Dwight, *Ill.* **15** 41 5N 88 26W
Dyersburg, *Tenn.* **13** 36 3N 89 23W

E

Eads, *Colo.* **12** 38 29N 102 47W
Eagle, *Alaska* **30** 64 47N 141 12W
Eagle, *Colo.* **16** 39 39N 106 50W
Eagle Butte, *S. Dak.* **12** 45 0N 101 10W
Eagle Grove, *Iowa* **14** 42 40N 93 54W
Eagle L., *Calif.* **16** 40 39N 120 45W
Eagle L., *Maine* **11** 46 20N 69 22W
Eagle Lake, *Tex.* **13** 29 35N 96 20W
Eagle Nest, *N. Mex.* **17** 36 33N 105 16W
Eagle Pass, *Tex.* **13** 28 43N 100 30W
Eagle Pk., *Calif.* **18** 38 10N 119 25W
Eagle River, *Wis.* **12** 45 55N 89 15W
Earle, *Ark.* **13** 35 16N 90 28W
Earlimart, *Calif.* **19** 35 53N 119 16W
Earth, *Tex.* **13** 34 14N 102 24W
Easley, *S.C.* **11** 34 50N 82 36W
East B., *La.* **13** 29 0N 89 15W
East Chicago, *Ind.* **22** 41 38N 87 26W
East Cleveland, *Ohio* **27** 41 32N 81 35W
East Detroit, *Mich.* **26** 42 27N 82 58W
East Grand Forks, *Minn.* **12** 47 56N 97 1W
East Greenwich, *R.I.* **9** 41 40N 71 27W
East Hartford, *Conn.* **9** 41 46N 72 39W
East Helena, *Mont.* **16** 46 35N 111 56W
East Jordan, *Mich.* **10** 45 10N 85 7W
East Lansing, *Mich.* **15** 42 44N 84 29W
East Liverpool, *Ohio* **8** 40 37N 80 35W
East Los Angeles, *Calif.* **29** 34 1N 118 10W
East Meadow, *N.Y.* **21** 40 42N 73 31W
East Orange, *N.J.* **20** 40 46N 74 11W
East Point, *Ga.* **11** 33 41N 84 27W
East Providence, *R.I.* **9** 41 49N 71 23W
East St. Louis, *Ill.* **14** 38 37N 90 9W
East Stroudsburg, *Pa.* **9** 41 1N 75 11W
East Tawas, *Mich.* **10** 44 17N 83 29W
East Walker →, *Nev.* **18** 38 52N 119 10W
Eastchester, *N.Y.* **21** 40 57N 73 49W
Eastlake, *Ohio* **27** 41 38N 81 28W
Eastland, *Tex.* **13** 32 24N 98 49W
Eastman, *Ga.* **11** 32 12N 83 11W
Easton, *Md.* **10** 38 47N 76 5W
Easton, *Pa.* **9** 40 41N 75 13W
Easton, *Wash.* **18** 47 14N 121 11W
Eastport, *Maine* **11** 44 56N 67 0W
Eaton, *Colo.* **12** 40 32N 104 42W
Eatonton, *Ga.* **11** 33 20N 83 23W
Eatontown, *N.J.* **9** 40 19N 74 4W
Eau Claire, *Wis.* **12** 44 49N 91 30W
Eden, *N.C.* **11** 36 29N 79 53W
Eden, *Tex.* **13** 31 13N 99 51W
Eden, *Wyo.* **16** 42 3N 109 26W
Edenton, *N.C.* **11** 36 4N 76 39W
Edgar, *Nebr.* **12** 40 22N 97 58W
Edgartown, *Mass.* **9** 41 23N 70 31W
Edgefield, *S.C.* **11** 33 47N 81 56W
Edgeley, *N. Dak.* **12** 46 22N 98 43W
Edgemont, *S. Dak.* **12** 43 18N 103 50W
Edina, *Mo.* **14** 40 10N 92 11W
Edinburg, *Tex.* **13** 26 18N 98 10W
Edison, *N.J.* **20** 40 31N 74 22W
Edmeston, *N.Y.* **9** 42 42N 75 15W
Edmond, *Okla.* **13** 35 39N 97 29W
Edmonds, *Wash.* **18** 47 49N 122 23W
Edna, *Tex.* **13** 28 59N 96 39W
Edwards, *Calif.* **19** 34 55N 117 51W
Edwards Plateau, *Tex.* **13** 30 45N 101 20W
Edwardsville, *Pa.* **9** 41 15N 75 56W
Eek, *Alaska* **30** 60 14N 162 2W
Effingham, *Ill.* **15** 39 7N 88 33W
Egeland, *N. Dak.* **12** 48 38N 99 6W
Ekalaka, *Mont.* **12** 45 53N 104 33W
El Cajon, *Calif.* **19** 32 48N 116 58W
El Campo, *Tex.* **13** 29 12N 96 16W
El Centro, *Calif.* **19** 32 48N 115 34W
El Cerrito, *Calif.* **28** 37 54N 122 18W
El Dorado, *Ark.* **13** 33 12N 92 40W
El Dorado, *Kans.* **13** 37 49N 96 52W
El Granada, *Calif.* **28** 37 30N 122 28W
El Monte, *Calif.* **29** 34 3N 118 1W
El Paso, *Tex.* **17** 31 45N 106 29W
El Paso Robles, *Calif.* **18** 35 38N 120 41W
El Portal, *Calif.* **18** 37 41N 119 47W
El Reno, *Okla.* **13** 35 32N 97 57W
El Rio, *Calif.* **19** 34 14N 119 10W
El Segundo, *Calif.* **29** 33 55N 118 24W
Elba, *Ala.* **11** 31 25N 86 4W
Elbert, Mt., *Colo.* **17** 39 7N 106 27W
Elberta, *Mich.* **10** 44 37N 86 14W
Elberton, *Ga.* **11** 34 7N 82 52W
Eldon, *Mo.* **14** 38 21N 92 35W
Eldora, *Iowa* **14** 42 22N 93 5W
Eldorado, *Ill.* **15** 37 49N 88 26W
Eldorado, *Tex.* **13** 30 52N 100 36W
Eldorado Springs, *Mo.* **14** 37 52N 94 1W
Electra, *Tex.* **13** 34 2N 98 55W
Eleele, *Hawaii* **30** 21 54N 159 35W
Elephant Butte Reservoir, *N. Mex.* **17** 33 9N 107 11W
Elfin Cove, *Alaska* **30** 58 12N 136 22W
Elgin, *Ill.* **15** 42 2N 88 17W
Elgin, *N. Dak.* **12** 46 24N 101 51W
Elgin, *Nebr.* **12** 41 59N 98 5W
Elgin, *Nev.* **17** 37 21N 114 32W
Elgin, *Oreg.* **16** 45 34N 117 55W
Elgin, *Tex.* **13** 30 21N 97 22W
Elida, *N. Mex.* **13** 33 57N 103 39W
Elim, *Alaska* **30** 64 37N 162 15W
Elizabeth, *N.J.* **20** 40 39N 74 12W
Elizabeth City, *N.C.* **11** 36 18N 76 14W
Elizabethton, *Tenn.* **11** 36 21N 82 13W
Elizabethtown, *Ky.* **10** 37 42N 85 52W
Elizabethtown, *Pa.* **9** 40 9N 76 36W
Elk City, *Okla.* **13** 35 25N 99 25W
Elk Grove, *Calif.* **18** 38 25N 121 22W
Elk River, *Idaho* **16** 46 47N 116 11W
Elk River, *Minn.* **12** 45 18N 93 35W
Elkhart, *Ind.* **15** 41 41N 85 58W
Elkhart, *Kans.* **13** 37 0N 101 54W
Elkhorn →, *Nebr.* **12** 41 8N 96 19W
Elkin, *N.C.* **11** 36 15N 80 51W
Elkins, *W. Va.* **10** 38 55N 79 51W
Elko, *Nev.* **16** 40 50N 115 46W
Ellendale, *N. Dak.* **12** 46 0N 98 32W
Ellensburg, *Wash.* **16** 46 59N 120 34W
Ellenville, *N.Y.* **9** 41 43N 74 24W
Ellinwood, *Kans.* **12** 38 21N 98 35W
Ellis, *Kans.* **12** 38 56N 99 34W
Ellisville, *Miss.* **13** 31 36N 89 12W
Ellsworth, *Kans.* **12** 38 44N 98 14W
Ellwood City, *Pa.* **8** 40 52N 80 17W
Elma, *Wash.* **18** 47 0N 123 25W

Elmhurst, *Ill.* **22** 41 53N 87 55W
Elmira, *N.Y.* **8** 42 6N 76 48W
Elmont, *N.Y.* **21** 40 42N 73 42W
Elmwood Park, *Ill.* **22** 41 55N 87 48W
Eloy, *Ariz.* **17** 32 45N 111 33W
Elsinore, *Utah* **17** 38 41N 112 9W
Elwood, *Ind.* **15** 40 17N 85 50W
Elwood, *Nebr.* **12** 40 36N 99 52W
Ely, *Minn.* **12** 47 55N 91 51W
Ely, *Nev.* **16** 39 15N 114 54W
Elyria, *Ohio* **8** 41 22N 82 7W
Emery, *Utah* **17** 38 55N 111 15W
Emmetsburg, *Iowa* **14** 43 7N 94 41W
Emmett, *Idaho* **16** 43 52N 116 30W
Empire State Building, *N.Y.* **20** 40 44N 73 59W
Emporia, *Kans.* **12** 38 25N 96 11W
Emporia, *Va.* **11** 36 42N 77 32W
Emporium, *Pa.* **8** 41 31N 78 14W
Encinal, *Tex.* **13** 28 2N 99 21W
Encinitas, *Calif.* **19** 33 3N 117 17W
Encino, *N. Mex.* **17** 34 39N 105 28W
Enderlin, *N. Dak.* **12** 46 38N 97 36W
Endicott, *N.Y.* **9** 42 6N 76 4W
Endicott, *Wash.* **16** 46 56N 117 41W
Endicott Mts., *Alaska* **30** 68 0N 152 0W
England, *Ark.* **13** 34 33N 91 58W
Englewood, *Colo.* **12** 39 39N 104 59W
Englewood, *Kans.* **13** 37 2N 99 59W
Enid, *Okla.* **13** 36 24N 97 53W
Ennis, *Mont.* **16** 45 21N 111 44W
Ennis, *Tex.* **13** 32 20N 96 38W
Enterprise, *Oreg.* **16** 45 25N 117 17W
Enterprise, *Utah* **17** 37 34N 113 43W
Enumclaw, *Wash.* **18** 47 12N 121 59W
Ephraim, *Utah* **16** 39 22N 111 35W
Ephrata, *Wash.* **16** 47 19N 119 33W
Erie, *Pa.* **8** 42 8N 80 5W
Erie, L., *N. Amer.* **8** 42 15N 81 0W
Erskine, *Minn.* **12** 47 40N 96 0W
Erwin, *Tenn.* **11** 36 9N 82 25W
Escalante, *Utah* **17** 37 47N 111 36W
Escalante →, *Utah* **17** 37 24N 110 57W
Escambia →, *Fla.* **11** 30 32N 87 11W
Escanaba, *Mich.* **10** 45 45N 87 4W
Escondido, *Calif.* **19** 33 7N 117 5W
Espenberg, C., *Alaska* **30** 66 33N 163 36W
Essex, *Md.* **25** 39 18N 76 28W
Estancia, *N. Mex.* **17** 34 46N 106 4W
Estelline, *S. Dak.* **12** 44 35N 96 54W
Estelline, *Tex.* **13** 34 33N 100 26W
Estherville, *Iowa* **12** 43 24N 94 50W
Etawah →, *Ga.* **11** 34 20N 84 15W
Etowah, *Tenn.* **11** 35 20N 84 32W
Euclid, *Ohio* **27** 41 34N 81 33W
Eudora, *Ark.* **13** 33 7N 91 16W
Eufaula, *Ala.* **11** 31 54N 85 9W
Eufaula, *Okla.* **13** 35 17N 95 35W
Eufaula L., *Okla.* **13** 35 18N 95 21W
Eugene, *Oreg.* **16** 44 5N 123 4W
Eunice, *La.* **13** 30 30N 92 25W
Eunice, *N. Mex.* **13** 32 26N 103 10W
Eureka, *Calif.* **16** 40 47N 124 9W
Eureka, *Kans.* **13** 37 49N 96 17W
Eureka, *Mont.* **16** 48 53N 115 3W
Eureka, *Nev.* **16** 39 31N 115 58W
Eureka, *S. Dak.* **12** 45 46N 99 38W
Eureka, *Utah* **16** 39 58N 112 7W
Eustis, *Fla.* **11** 28 51N 81 41W
Evans, *Colo.* **12** 40 23N 104 41W
Evanston, *Ill.* **22** 42 3N 87 41W
Evanston, *Wyo.* **16** 41 16N 110 58W
Evansville, *Ind.* **15** 37 58N 87 35W
Evansville, *Wis.* **14** 42 47N 89 18W
Eveleth, *Minn.* **12** 47 28N 92 32W
Everett, *Mass.* **23** 42 24N 71 3W
Everett, *Wash.* **18** 47 59N 122 12W
Everglades, The, *Fla.* **11** 25 50N 81 0W
Everglades City, *Fla.* **11** 25 52N 81 23W
Everglades National Park, *Fla.* **11** 25 30N 81 0W
Evergreen, *Ala.* **11** 31 26N 86 57W
Evergreen Park, *Ill.* **22** 41 43N 87 42W
Everson, *Wash.* **16** 48 57N 122 22W
Evesboro, *N.J.* **24** 39 54N 74 56W
Ewing, *Nebr.* **12** 42 16N 98 21W
Excelsior Springs, *Mo.* **14** 39 20N 94 13W
Exeter, *Calif.* **18** 36 18N 119 9W
Exeter, *N.H.* **9** 42 59N 70 57W
Exeter, *Nebr.* **12** 40 39N 97 27W

F

Fabens, *Tex.* **17** 31 30N 106 10W
Fagatogo, *Amer. Samoa* **31** 14 17S 170 41W
Fair Lawn, *N.J.* **20** 40 56N 74 7W
Fair Oaks, *Calif.* **18** 38 39N 121 16W
Fairbank, *Ariz.* **17** 31 43N 110 11W
Fairbanks, *Alaska* **30** 64 51N 147 43W
Fairbury, *Nebr.* **12** 40 8N 97 11W
Fairfax, *Del.* **24** 39 47N 75 32W
Fairfax, *Okla.* **13** 36 34N 96 42W
Fairfax, *Va.* **25** 38 50N 77 19W
Fairfield, *Ala.* **11** 33 29N 86 55W
Fairfield, *Calif.* **18** 38 15N 122 3W
Fairfield, *Conn.* **9** 41 9N 73 16W
Fairfield, *Idaho* **16** 43 21N 114 44W
Fairfield, *Ill.* **15** 38 23N 88 22W
Fairfield, *Iowa* **14** 40 56N 91 57W
Fairfield, *Mont.* **16** 47 37N 111 59W
Fairfield, *Tex.* **13** 31 44N 96 10W
Fairhope, *Ala.* **11** 30 31N 87 54W
Fairmead, *Calif.* **18** 37 5N 120 10W
Fairmont, *Minn.* **12** 43 39N 94 28W
Fairmont, *W. Va.* **10** 39 29N 80 9W
Fairmount, *Calif.* **19** 34 45N 118 26W
Fairplay, *Colo.* **17** 39 15N 106 2W
Fairport, *N.Y.* **8** 43 6N 77 27W
Fairview, *Mont.* **12** 47 51N 104 3W
Fairview, *Okla.* **13** 36 16N 98 29W
Fairview, *Utah* **16** 39 50N 111 0W
Fairview Park, *Ohio* **27** 41 26N 81 52W
Fairweather, Mt., *Alaska* **30** 58 55N 137 32W
Faith, *S. Dak.* **12** 45 2N 102 2W
Fajardo, *Puerto Rico* **30** 18 20N 65 39W
Falcon Dam, *Tex.* **13** 26 50N 99 20W
Falfurrias, *Tex.* **13** 27 14N 98 9W
Fall River, *Mass.* **9** 41 43N 71 10W
Fall River Mills, *Calif.* **16** 41 3N 121 26W
Fallbrook, *Calif.* **17** 33 25N 117 12W
Fallon, *Mont.* **12** 46 50N 105 8W
Fallon, *Nev.* **16** 39 28N 118 47W
Falls Church, *Va.* **25** 38 53N 77 11W
Falls City, *Nebr.* **12** 40 3N 95 36W
Falls City, *Oreg.* **16** 44 52N 123 26W
Falmouth, *Ky.* **15** 38 41N 84 20W
Famoso, *Calif.* **19** 35 37N 119 12W
Farewell, *Alaska* **30** 62 31N 153 54W
Fargo, *N. Dak.* **12** 46 53N 96 48W
Faribault, *Minn.* **12** 44 18N 93 16W
Farmerville, *La.* **13** 32 47N 92 24W
Farmington, *Calif.* **18** 37 55N 120 59W
Farmington, *Mich.* **26** 42 26N 83 22W
Farmington, *N. Mex.* **17** 36 44N 108 12W
Farmington, *Utah* **16** 41 0N 111 12W
Farmington →, *Conn.* **9** 41 51N 72 38W
Farmington Hills, *Mich.* **26** 42 29N 83 23W
Farmville, *Va.* **10** 37 18N 78 24W
Farrell, *Pa.* **8** 41 13N 80 30W
Farwell, *Tex.* **13** 34 23N 103 2W
Faulkton, *S. Dak.* **12** 45 2N 99 8W
Fawnskin, *Calif.* **19** 34 16N 116 56W
Fayette, *Ala.* **11** 33 41N 87 50W
Fayette, *Mo.* **14** 39 9N 92 41W
Fayetteville, *Ark.* **13** 36 4N 94 10W
Fayetteville, *N.C.* **11** 35 3N 78 53W
Fayetteville, *Tenn.* **11** 35 9N 86 34W
Fear, C., *N.C.* **11** 33 50N 77 58W
Feather →, *Calif.* **16** 38 47N 121 36W
Felton, *Calif.* **18** 37 3N 122 4W
Fennimore, *Wis.* **14** 42 59N 90 39W
Fenton, *Mich.* **15** 42 48N 83 42W
Fergus Falls, *Minn.* **12** 46 17N 96 4W
Fernandina Beach, *Fla.* **11** 30 40N 81 27W
Ferndale, *Calif.* **16** 40 35N 124 16W
Ferndale, *Mich.* **26** 42 27N 83 7W
Ferndale, *Wash.* **18** 48 51N 122 36W
Fernley, *Nev.* **16** 39 36N 119 15W
Ferriday, *La.* **13** 31 38N 91 33W
Ferron, *Utah* **17** 39 5N 111 8W
Fertile, *Minn.* **12** 47 32N 96 17W
Fessenden, *N. Dak.* **12** 47 39N 99 38W
Filer, *Idaho* **16** 42 34N 114 37W
Fillmore, *Calif.* **19** 34 24N 118 55W
Fillmore, *Utah* **17** 38 58N 112 20W
Findlay, *Ohio* **15** 41 2N 83 39W
Finley, *N. Dak.* **12** 47 31N 97 50W
Firebaugh, *Calif.* **18** 36 52N 120 27W
Fitchburg, *Mass.* **9** 42 35N 71 48W
Fitzgerald, *Ga.* **11** 31 43N 83 15W
Five Points, *Calif.* **18** 36 26N 120 6W
Flagler, *Colo.* **12** 39 18N 103 4W
Flagstaff, *Ariz.* **17** 35 12N 111 39W
Flambeau →, *Wis.* **12** 45 18N 91 14W
Flaming Gorge Dam, *Utah* **16** 40 55N 109 25W
Flaming Gorge Reservoir, *Wyo.* **16** 41 10N 109 25W
Flandreau, *S. Dak.* **12** 44 3N 96 36W
Flat River, *Mo.* **13** 37 51N 90 31W
Flathead L., *Mont.* **16** 47 51N 114 8W
Flattery, C., *Wash.* **18** 48 23N 124 29W
Flaxton, *N. Dak.* **12** 48 54N 102 24W
Flint, *Mich.* **15** 43 1N 83 41W
Flint →, *Ga.* **11** 30 57N 84 34W
Floodwood, *Minn.* **12** 46 55N 92 55W
Flora, *Ill.* **10** 38 40N 88 29W
Floral Park, *N.Y.* **21** 40 43N 73 42W
Florala, *Ala.* **11** 31 0N 86 20W
Florence, *Ala.* **11** 34 48N 87 41W
Florence, *Ariz.* **17** 33 2N 111 23W
Florence, *Calif.* **29** 33 57N 118 13W
Florence, *Colo.* **12** 38 23N 105 8W
Florence, *Oreg.* **16** 43 58N 124 7W
Florence, *S.C.* **11** 34 12N 79 46W
Floresville, *Tex.* **13** 29 8N 98 10W
Florham Park, *N.J.* **20** 40 46N 74 23W
Florida □ **11** 28 0N 82 0W
Floydada, *Tex.* **13** 33 59N 101 20W
Flushing, *N.Y.* **21** 40 45N 73 49W
Folkston, *Ga.* **11** 30 50N 82 0W
Follett, *Tex.* **13** 36 26N 100 8W
Fond du Lac, *Wis.* **12** 43 47N 88 27W
Ford City, *Calif.* **19** 35 9N 119 27W
Fordyce, *Ark.* **13** 33 49N 92 25W
Forest, *Miss.* **13** 32 22N 89 29W
Forest City, *Iowa* **12** 43 16N 93 39W
Forest City, *N.C.* **11** 35 20N 81 52W
Forest Grove, *Oreg.* **18** 45 31N 123 7W
Forest Hills, *N.Y.* **21** 40 42N 73 51W
Forest Hills, *Pa.* **27** 40 25N 79 51W
Forestville, *Md.* **25** 38 50N 76 52W
Forestville, *Wis.* **10** 44 41N 87 29W
Forks, *Wash.* **18** 47 57N 124 23W
Forman, *N. Dak.* **12** 46 7N 97 38W
Forrest City, *Ark.* **13** 35 1N 90 47W
Forsyth, *Ga.* **11** 33 2N 83 56W
Forsyth, *Mont.* **16** 46 16N 106 41W
Fort Apache, *Ariz.* **17** 33 50N 110 0W
Fort Benton, *Mont.* **16** 47 49N 110 40W
Fort Bragg, *Calif.* **16** 39 26N 123 48W
Fort Bridger, *Wyo.* **16** 41 19N 110 23W
Fort Collins, *Colo.* **12** 40 35N 105 5W
Fort Davis, *Tex.* **13** 30 35N 103 54W
Fort Defiance, *Ariz.* **17** 35 45N 109 5W
Fort Dodge, *Iowa* **12** 42 30N 94 11W
Fort Garland, *Colo.* **17** 37 26N 105 26W
Fort Hancock, *Tex.* **17** 31 18N 105 51W
Fort Irwin, *Calif.* **19** 35 16N 116 34W
Fort Kent, *Maine* **11** 47 15N 68 36W
Fort Klamath, *Oreg.* **16** 42 42N 122 0W
Fort Laramie, *Wyo.* **12** 42 13N 104 31W
Fort Lauderdale, *Fla.* **11** 26 7N 80 8W
Fort Lee, *N.J.* **20** 40 50N 73 58W
Fort Lupton, *Colo.* **12** 40 5N 104 49W
Fort Madison, *Iowa* **14** 40 38N 91 27W
Fort Meade, *Fla.* **11** 27 45N 81 48W
Fort Morgan, *Colo.* **12** 40 15N 103 48W
Fort Myers, *Fla.* **11** 26 39N 81 52W
Fort Payne, *Ala.* **11** 34 26N 85 43W
Fort Peck, *Mont.* **16** 48 1N 106 27W
Fort Peck Dam, *Mont.* **16** 48 0N 106 26W
Fort Peck L., *Mont.* **16** 48 0N 106 26W
Fort Pierce, *Fla.* **11** 27 27N 80 20W
Fort Pierre, *S. Dak.* **12** 44 21N 100 22W
Fort Scott, *Kans.* **13** 37 50N 94 42W
Fort Smith, *Ark.* **13** 35 23N 94 25W
Fort Stanton, *N. Mex.* **17** 33 30N 105 31W
Fort Stockton, *Tex.* **13** 30 53N 102 53W
Fort Sumner, *N. Mex.* **13** 34 28N 104 15W
Fort Valley, *Ga.* **11** 32 33N 83 53W
Fort Walton Beach, *Fla.* **11** 30 25N 86 36W
Fort Wayne, *Ind.* **15** 41 4N 85 9W
Fort Worth, *Tex.* **13** 32 45N 97 18W
Fort Yates, *N. Dak.* **12** 46 5N 100 38W
Fort Yukon, *Alaska* **30** 66 34N 145 16W
Fortuna, *Calif.* **16** 40 36N 124 9W
Fortuna, *N. Dak.* **12** 48 55N 103 47W
Fossil, *Oreg.* **16** 45 0N 120 9W
Fosston, *Minn.* **12** 47 35N 95 45W
Fostoria, *Ohio* **15** 41 10N 83 25W
Fountain, *Colo.* **12** 38 41N 104 42W
Fountain, *Utah* **16** 39 41N 111 37W
Fountain Springs, *Calif.* **19** 35 54N 118 51W
Four Mountains, Is. of, *Alaska* **30** 53 0N 170 0W
Fowler, *Calif.* **18** 36 38N 119 41W
Fowler, *Colo.* **12** 38 8N 104 2W
Fowler, *Kans.* **13** 37 23N 100 12W
Fowlerton, *Tex.* **13** 28 28N 98 48W
Fox Is., *Alaska* **30** 52 30N 166 0W
Foxpark, *Wyo.* **16** 41 5N 106 9W
Frackville, *Pa.* **9** 40 47N 76 14W
Framingham, *Mass.* **23** 42 18N 71 23W
Frankfort, *Ind.* **15** 40 17N 86 31W
Frankfort, *Kans.* **12** 39 42N 96 25W
Frankfort, *Ky.* **15** 38 12N 84 52W
Frankfort, *Mich.* **10** 44 38N 86 14W
Franklin, *Ky.* **15** 36 43N 86 35W
Franklin, *La.* **13** 29 48N 91 30W
Franklin, *Mass.* **9** 42 5N 71 24W
Franklin, *N.H.* **9** 43 27N 71 39W
Franklin, *Nebr.* **12** 40 6N 98 57W
Franklin, *Pa.* **8** 41 24N 79 50W
Franklin, *Tenn.* **11** 35 55N 86 52W
Franklin, *Va.* **11** 36 41N 76 56W
Franklin, *W. Va.* **10** 38 39N 79 20W
Franklin D. Roosevelt L., *Wash.* **16** 48 18N 118 9W
Franklin L., *Nev.* **16** 40 25N 115 22W
Franklin Park, *Ill.* **22** 41 55N 87 52W
Franklin Square, *N.Y.* **21** 40 41N 73 40W
Franklinton, *La.* **13** 30 51N 90 9W
Franks Pk., *Wyo.* **16** 43 58N 109 18W
Frederick, *Md.* **10** 39 25N 77 25W
Frederick, *Okla.* **13** 34 23N 99 1W
Frederick, *S. Dak.* **12** 45 50N 98 31W
Fredericksburg, *Tex.* **13** 30 16N 98 52W
Fredericksburg, *Va.* **10** 38 18N 77 28W
Fredericktown, *Mo.* **13** 37 34N 90 18W
Fredonia, *Ariz.* **17** 36 57N 112 32W
Fredonia, *Kans.* **13** 37 32N 95 49W
Fredonia, *N.Y.* **8** 42 26N 79 20W
Freehold, *N.J.* **9** 40 16N 74 17W
Freel Peak, *Nev.* **18** 38 52N 119 54W
Freeland, *Pa.* **9** 41 1N 75 54W
Freeman, *Calif.* **19** 35 35N 117 53W
Freeman, *S. Dak.* **12** 43 21N 97 26W
Freeport, *Ill.* **14** 42 17N 89 36W
Freeport, *N.Y.* **21** 40 39N 73 35W
Freeport, *Tex.* **13** 28 57N 95 21W
Fremont, *Calif.* **28** 37 33N 122 2W
Fremont, *Mich.* **10** 43 28N 85 57W
Fremont, *Nebr.* **12** 41 26N 96 30W
Fremont, *Ohio* **15** 41 21N 83 7W
Fremont →, *Utah* **17** 38 24N 110 42W
Fremont L., *Wyo.* **16** 42 57N 109 48W
French Camp, *Calif.* **18** 37 53N 121 16W
French Creek →, *Pa.* **8** 41 24N 79 50W
Frenchglen, *Oreg.* **16** 42 50N 118 55W
Frenchman Cr. →, *Mont.* **16** 48 31N 107 10W
Frenchman Cr. →, *Nebr.* **12** 40 14N 100 50W
Fresno, *Calif.* **18** 36 44N 119 47W
Fresno Reservoir, *Mont.* **16** 48 36N 109 57W
Friant, *Calif.* **18** 36 59N 119 43W
Frio →, *Tex.* **13** 28 26N 98 11W
Friona, *Tex.* **13** 34 38N 102 43W
Fritch, *Tex.* **13** 35 38N 101 36W
Froid, *Mont.* **12** 48 20N 104 30W
Fromberg, *Mont.* **16** 45 24N 108 54W
Front Range, *Colo.* **16** 40 25N 105 45W
Front Royal, *Va.* **10** 38 55N 78 12W
Frostburg, *Md.* **10** 39 39N 78 56W
Fullerton, *Calif.* **19** 33 53N 117 56W
Fullerton, *Nebr.* **12** 41 22N 97 58W
Fulton, *Mo.* **14** 38 52N 91 57W
Fulton, *N.Y.* **9** 43 19N 76 25W
Fulton, *Tenn.* **11** 36 31N 88 53W

G

Gadsden, *Ala.* **11** 34 1N 86 1W
Gadsden, *Ariz.* **17** 32 33N 114 47W
Gaffney, *S.C.* **11** 35 5N 81 39W
Gail, *Tex.* **13** 32 46N 101 27W
Gainesville, *Fla.* **11** 29 40N 82 20W
Gainesville, *Ga.* **11** 34 18N 83 50W
Gainesville, *Mo.* **13** 36 36N 92 26W
Gainesville, *Tex.* **13** 33 38N 97 8W
Galax, *Va.* **11** 36 40N 80 56W
Galena, *Alaska* **30** 64 44N 156 56W
Galesburg, *Ill.* **14** 40 57N 90 22W
Galiuro Mts., *Ariz.* **17** 32 30N 110 20W
Gallatin, *Tenn.* **11** 36 24N 86 27W
Gallipolis, *Ohio* **10** 38 49N 82 12W
Gallup, *N. Mex.* **17** 35 32N 108 45W
Galt, *Calif.* **18** 38 15N 121 18W
Galva, *Ill.* **14** 41 10N 90 3W
Galveston, *Tex.* **13** 29 18N 94 48W
Galveston B., *Tex.* **13** 29 36N 94 50W
Gambell, *Alaska* **30** 63 47N 171 45W
Gamerco, *N. Mex.* **17** 35 34N 108 46W
Ganado, *Ariz.* **17** 35 43N 109 33W
Ganado, *Tex.* **13** 29 2N 96 31W
Gannett Peak, *Wyo.* **16** 43 11N 109 39W
Gannvalley, *S. Dak.* **12** 44 2N 98 59W
Garapan, *Pac. Oc.* **31** 15 12N 145 53 E
Garber, *Okla.* **13** 36 26N 97 35W
Garberville, *Calif.* **16** 40 6N 123 48W
Garden City, *Kans.* **13** 37 58N 100 53W
Garden City, *Mich.* **26** 42 20N 83 20W
Garden City, *N.Y.* **21** 40 43N 73 38W
Garden City, *Tex.* **13** 31 52N 101 29W
Garden Grove, *Calif.* **19** 33 47N 117 55W
Gardena, *Calif.* **29** 33 53N 118 17W
Gardiner, *Mont.* **16** 45 2N 110 22W
Gardiners I., *N.Y.* **9** 41 6N 72 6W
Gardner, *Mass.* **9** 42 34N 71 59W
Gardnerville, *Nev.* **18** 38 56N 119 45W
Gareloi I., *Alaska* **30** 51 48N 178 48W
Garey, *Calif.* **19** 34 53N 120 19W
Garfield, *N.J.* **20** 40 52N 74 6W
Garfield, *Wash.* **16** 47 1N 117 9W
Garfield Heights, *Ohio* **27** 41 25N 81 37W
Garland, *Utah* **16** 41 47N 112 10W
Garner, *Iowa* **14** 43 6N 93 36W
Garnett, *Kans.* **12** 38 17N 95 14W
Garrison, *Mont.* **16** 46 31N 112 49W
Garrison, *N. Dak.* **12** 47 40N 101 25W
Garrison, *Tex.* **13** 31 49N 94 30W
Garrison Res. = Sakakawea, L., *N. Dak.* **12** 47 30N 101 25W
Gary, *Ind.* **22** 41 35N 87 23W
Gassaway, *W. Va.* **10** 38 41N 80 47W
Gastonia, *N.C.* **11** 35 16N 81 11W
Gatesville, *Tex.* **13** 31 26N 97 45W
Gaviota, *Calif.* **19** 34 29N 120 13W
Gaylord, *Mich.* **10** 45 2N 84 41W
Genesee, *Idaho* **16** 46 33N 116 56W
Genesee →, *N.Y.* **8** 43 16N 77 36W
Geneseo, *Ill.* **14** 41 27N 90 9W
Geneseo, *Kans.* **12** 38 31N 98 10W
Geneva, *Ala.* **11** 31 2N 85 52W
Geneva, *N.Y.* **8** 42 52N 76 59W
Geneva, *Nebr.* **12** 40 32N 97 36W
Geneva, *Ohio* **8** 41 48N 80 57W
Geneva, L., *Wis.* **15** 42 38N 88 30W
Genoa, *N.Y.* **9** 42 40N 76 32W
Genoa, *Nebr.* **12** 41 27N 97 44W
George, L., *Fla.* **11** 29 17N 81 36W
George, L., *N.Y.* **9** 43 37N 73 33W
George West, *Tex.* **13** 28 20N 98 7W
Georgetown, *Colo.* **16** 39 42N 105 42W
Georgetown, *D.C.* **25** 38 54N 77 3W
Georgetown, *Ky.* **10** 38 13N 84 33W

Georgetown, *S.C.* **11** 33 23N 79 17W
Georgetown, *Tex.* **13** 30 38N 97 41W
Georgia □ **11** 32 50N 83 15W
Geraldine, *Mont.* **16** 47 36N 110 16W
Gering, *Nebr.* **12** 41 50N 103 40W
Gerlach, *Nev.* **16** 40 39N 119 21W
Gettysburg, *Pa.* **10** 39 50N 77 14W
Gettysburg, *S. Dak.* **12** 45 1N 99 57W
Geyser, *Mont.* **16** 47 16N 110 30W
Giant Forest, *Calif.* **18** 36 36N 118 43W
Gibbon, *Nebr.* **12** 40 45N 98 51W
Giddings, *Tex.* **13** 30 11N 96 56W
Gila →, *Ariz.* **17** 32 43N 114 33W
Gila Bend, *Ariz.* **17** 32 57N 112 43W
Gila Bend Mts., *Ariz.* ... **17** 33 10N 113 0W
Gillette, *Wyo.* **12** 44 18N 105 30W
Gilmer, *Tex.* **13** 32 44N 94 57W
Gilroy, *Calif.* **18** 37 1N 121 34W
Girard, *Kans.* **13** 37 31N 94 51W
Glacier Park, *Mont.* **16** 48 30N 113 18W
Glacier Peak, *Wash.* ... **16** 48 7N 121 7W
Gladewater, *Tex.* **13** 32 33N 94 56W
Gladstone, *Mich.* **10** 45 51N 87 1W
Gladwin, *Mich.* **10** 43 59N 84 29W
Glasco, *Kans.* **12** 39 22N 97 50W
Glasco, *N.Y.* **9** 42 3N 73 57W
Glasgow, *Ky.* **10** 37 0N 85 55W
Glasgow, *Mont.* **16** 48 12N 106 38W
Glastonbury, *Conn.* **9** 41 43N 72 37W
Glen Canyon Dam, *Ariz.* **17** 36 57N 111 29W
Glen Canyon National Recreation Area, *Utah* **17** 37 15N 111 0W
Glen Cove, *N.Y.* **21** 40 52N 73 38W
Glen Lyon, *Pa.* **9** 41 10N 76 5W
Glen Ullin, *N. Dak.* **12** 46 49N 101 50W
Glencoe, *Ill.* **22** 42 7N 87 44W
Glencoe, *Minn.* **12** 44 46N 94 9W
Glendale, *Ariz.* **17** 33 32N 112 11W
Glendale, *Calif.* **29** 34 9N 118 14W
Glendale, *Oreg.* **16** 42 44N 123 26W
Glendive, *Mont.* **12** 47 7N 104 43W
Glendo, *Wyo.* **12** 42 30N 105 2W
Glenmora, *La.* **13** 30 59N 92 35W
Glenns Ferry, *Idaho* ... **16** 42 57N 115 18W
Glenrock, *Wyo.* **16** 42 52N 105 52W
Glens Falls, *N.Y.* **9** 43 19N 73 39W
Glenshaw, *Pa.* **27** 40 32N 79 58W
Glenview, *Ill.* **22** 42 4N 87 48W
Glenville, *W. Va.* **10** 38 56N 80 50W
Glenwood, *Ark.* **13** 34 20N 93 33W
Glenwood, *Hawaii* **30** 19 29N 155 9W
Glenwood, *Iowa* **12** 41 3N 95 45W
Glenwood, *Minn.* **12** 45 39N 95 23W
Glenwood Springs, *Colo.* **16** 39 33N 107 19W
Globe, *Ariz.* **17** 33 24N 110 47W
Gloucester, *Mass.* **9** 42 37N 70 40W
Gloucester City, *N.J.* ... **24** 39 53N 75 7W
Gloversville, *N.Y.* **9** 43 3N 74 21W
Gogebic, L., *Mich.* **12** 46 30N 89 35W
Golconda, *Nev.* **16** 40 58N 117 30W
Gold Beach, *Oreg.* **16** 42 25N 124 25W
Gold Creek, *Alaska* **30** 62 46N 149 41W
Gold Hill, *Oreg.* **16** 42 26N 123 3W
Golden, *Colo.* **12** 39 42N 105 15W
Golden Gate, *Calif.* **16** 37 54N 122 30W
Golden Gate, *Calif.* **28** 37 48N 122 30W
Golden Gate Bridge, *Calif.* **28** 37 49N 122 28W
Goldendale, *Wash.* **16** 45 49N 120 50W
Goldfield, *Nev.* **17** 37 42N 117 14W
Goldsboro, *N.C.* **11** 35 23N 77 59W
Goldsmith, *Tex.* **13** 31 59N 102 37W
Goldthwaite, *Tex.* **13** 31 27N 98 34W
Goleta, *Calif.* **19** 34 27N 119 50W
Goliad, *Tex.* **13** 28 40N 97 23W
Gonzales, *Calif.* **18** 36 30N 121 26W
Gonzales, *Tex.* **13** 29 30N 97 27W
Gooding, *Idaho* **16** 42 56N 114 43W
Goodland, *Kans.* **12** 39 21N 101 43W
Goodnight, *Tex.* **13** 35 2N 101 11W
Goodsprings, *Nev.* **17** 35 50N 115 26W
Goose L., *Calif.* **16** 41 56N 120 26W
Gorda, *Calif.* **18** 35 53N 121 26W
Gordon, *Nebr.* **12** 42 48N 102 12W
Gorman, *Calif.* **19** 34 47N 118 51W
Gorman, *Tex.* **13** 32 12N 98 41W
Goshen, *Calif.* **18** 36 21N 119 25W
Goshen, *Ind.* **15** 41 35N 85 50W
Goshen, *N.Y.* **9** 41 24N 74 20W
Gothenburg, *Nebr.* **12** 40 56N 100 10W
Gowanda, *N.Y.* **8** 42 28N 78 56W
Grace, *Idaho* **16** 42 35N 111 44W
Graceville, *Minn.* **12** 45 34N 96 26W
Grady, *N. Mex.* **13** 34 49N 103 19W
Grafton, *N. Dak.* **12** 48 25N 97 25W
Graham, *N.C.* **11** 36 5N 79 25W
Graham, *Tex.* **13** 33 6N 98 35W
Graham, Mt., *Ariz.* **17** 32 42N 109 52W
Granada, *Colo.* **13** 38 4N 102 19W
Granbury, *Tex.* **13** 32 27N 97 47W
Grand →, *Mo.* **14** 39 23N 93 7W
Grand →, *S. Dak.* **12** 45 40N 100 45W
Grand Canyon, *Ariz.* ... **17** 36 3N 112 9W
Grand Canyon National Park, *Ariz.* **17** 36 15N 112 30W
Grand Coulee, *Wash.* .. **16** 47 57N 119 0W
Grand Coulee Dam, *Wash.* **16** 47 57N 118 59W
Grand Forks, *N. Dak.* ... **12** 47 55N 97 3W
Grand Haven, *Mich.* ... **15** 43 4N 86 13W
Grand I., *Mich.* **10** 46 31N 86 40W
Grand Island, *Nebr.* **12** 40 55N 98 21W
Grand Isle, *La.* **13** 29 14N 90 0W
Grand Junction, *Colo.* .. **17** 39 4N 108 33W
Grand L., *La.* **13** 29 55N 92 47W
Grand Lake, *Colo.* **16** 40 15N 105 49W
Grand Marais, *Mich.* ... **10** 46 40N 85 59W
Grand Rapids, *Mich.* ... **15** 42 58N 85 40W
Grand Rapids, *Minn.* ... **12** 47 14N 93 31W
Grand Teton, *Idaho* **16** 43 54N 111 50W
Grand Valley, *Colo.* **16** 39 27N 108 3W
Grande, Rio →, *Tex.* ... **13** 25 58N 97 9W
Grandfalls, *Tex.* **13** 31 20N 102 51W
Grandview, *Wash.* **16** 46 15N 119 54W
Granger, *Wash.* **16** 46 21N 120 11W
Granger, *Wyo.* **16** 41 35N 109 58W
Grangeville, *Idaho* **16** 45 56N 116 7W
Granite City, *Ill.* **14** 38 42N 90 9W
Granite Falls, *Minn.* ... **12** 44 49N 95 33W
Granite Mt., *Calif.* **19** 33 5N 116 28W
Granite Peak, *Mont.* ... **16** 45 10N 109 48W
Grant, *Nebr.* **12** 40 53N 101 42W
Grant, Mt., *Nev.* **16** 38 34N 118 48W
Grant City, *Mo.* **14** 40 29N 94 25W
Grant Range, *Nev.* **17** 38 30N 115 25W
Grants, *N. Mex.* **17** 35 9N 107 52W
Grants Pass, *Oreg.* **16** 42 26N 123 19W
Grantsburg, *Wis.* **12** 45 47N 92 41W
Grantsville, *Utah* **16** 40 36N 112 28W
Granville, *N. Dak.* **12** 48 16N 100 47W
Granville, *N.Y.* **10** 43 24N 73 16W
Grapeland, *Tex.* **13** 31 30N 95 29W
Grass Range, *Mont.* ... **16** 47 0N 109 0W
Grass Valley, *Calif.* **18** 39 13N 121 4W
Grass Valley, *Oreg.* **16** 45 22N 120 47W
Grayling, *Mich.* **10** 44 40N 84 43W
Grays Harbor, *Wash.* .. **16** 46 59N 124 1W
Grays L., *Idaho* **16** 43 4N 111 26W
Great Barrington, *Mass.* **9** 42 12N 73 22W
Great Basin, *Nev.* **16** 40 0N 117 0W
Great Bend, *Kans.* **12** 38 22N 98 46W
Great Bend, *Pa.* **9** 41 58N 75 45W
Great Falls, *Mont.* **16** 47 30N 111 17W
Great Kills, *N.Y.* **20** 40 32N 74 9W
Great Neck, *N.Y.* **21** 40 48N 73 44W
Great Plains, *N. Amer.* . **2** 47 0N 105 0W
Great Salt L., *Utah* **16** 41 15N 112 40W
Great Salt Lake Desert, *Utah* **16** 40 50N 113 30W
Great Salt Plains L., *Okla.* **13** 36 45N 98 8W
Great Sitkin I., *Alaska* .. **30** 52 3N 176 6W
Great Smoky Mts. Nat. Pk., *Tenn.* **11** 35 40N 83 40W
Greater Pittsburgh International Airport, *Pa.* **27** 40 29N 80 13W
Greeley, *Colo.* **12** 40 25N 104 42W
Greeley, *Nebr.* **12** 41 33N 98 32W
Green →, *Ky.* **10** 37 54N 87 30W
Green →, *Utah* **17** 38 11N 109 53W
Green B., *Wis.* **10** 45 0N 87 30W
Green Bay, *Wis.* **10** 44 31N 88 0W
Green Cove Springs, *Fla.* **11** 29 59N 81 42W
Green River, *Utah* **17** 38 59N 110 10W
Green Tree, *Pa.* **27** 40 25N 80 4W
Greenbelt, *Md.* **25** 39 0N 76 52W
Greenbush, *Minn.* **12** 48 42N 96 11W
Greencastle, *Ind.* **15** 39 38N 86 52W
Greene, *N.Y.* **9** 42 20N 75 46W
Greenfield, *Calif.* **18** 36 19N 121 15W
Greenfield, *Calif.* **19** 35 15N 119 0W
Greenfield, *Ind.* **15** 39 47N 85 46W
Greenfield, *Iowa* **14** 41 18N 94 28W
Greenfield, *Mass.* **9** 42 35N 72 36W
Greenfield, *Mo.* **13** 37 25N 93 51W
Greenport, *N.Y.* **9** 41 6N 72 22W
Greensboro, *Ga.* **11** 33 35N 83 11W
Greensboro, *N.C.* **11** 36 4N 79 48W
Greensburg, *Ind.* **15** 39 20N 85 29W
Greensburg, *Kans.* **13** 37 36N 99 18W
Greensburg, *Pa.* **8** 40 18N 79 33W
Greenville, *Ala.* **11** 31 50N 86 38W
Greenville, *Calif.* **18** 40 8N 120 57W
Greenville, *Ill.* **14** 38 53N 89 25W
Greenville, *Maine* **11** 45 28N 69 35W
Greenville, *Mich.* **15** 43 11N 85 15W
Greenville, *Miss.* **13** 33 24N 91 4W
Greenville, *N.C.* **11** 35 37N 77 23W
Greenville, *Ohio* **15** 40 6N 84 38W
Greenville, *Pa.* **8** 41 24N 80 23W
Greenville, *S.C.* **11** 34 51N 82 24W
Greenville, *Tenn.* **11** 36 13N 82 51W
Greenville, *Tex.* **13** 33 8N 96 7W
Greenwich, *Conn.* **21** 41 1N 73 38W
Greenwich, *N.Y.* **9** 43 5N 73 30W
Greenwood, *Miss.* **13** 33 31N 90 11W
Greenwood, *S.C.* **11** 34 12N 82 10W
Gregory, *S. Dak.* **12** 43 14N 99 20W
Grenada, *Miss.* **13** 33 47N 89 49W
Grenora, *N. Dak.* **12** 48 37N 103 56W
Gresham, *Oreg.* **18** 45 30N 122 26W
Greybull, *Wyo.* **16** 44 30N 108 3W
Gridley, *Calif.* **18** 39 22N 121 42W
Griffin, *Ga.* **11** 33 15N 84 16W
Grinnell, *Iowa* **14** 41 45N 92 43W
Groesbeck, *Tex.* **13** 30 48N 96 31W
Groom, *Tex.* **13** 35 12N 101 6W
Grosse Pointe, *Mich.* ... **26** 42 23N 82 54W
Groton, *Conn.* **9** 41 21N 72 5W
Groton, *S. Dak.* **12** 45 27N 98 6W
Grouse Creek, *Utah* **16** 41 42N 113 53W
Groveland, *Calif.* **18** 37 50N 120 14W
Grover City, *Calif.* **19** 35 7N 120 37W
Groveton, *N.H.* **9** 44 36N 71 31W
Groveton, *Tex.* **13** 31 4N 95 8W
Grundy Center, *Iowa* ... **14** 42 22N 92 47W
Gruver, *Tex.* **13** 36 16N 101 24W
Guadalupe, *Calif.* **19** 34 59N 120 33W
Guadalupe →, *Tex.* ... **13** 28 27N 96 47W
Guadalupe Peak, *Tex.* .. **17** 31 50N 104 52W
Guam ■, *Pac. Oc.* ..... **31** 13 27N 144 45 E
Guánica, *Puerto Rico* .. **30** 17 58N 66 55W
Guayama, *Puerto Rico* . **30** 17 59N 66 7W
Guernsey, *Wyo.* **12** 42 19N 104 45W
Gueydan, *La.* **13** 30 2N 92 31W
Guilford, *Maine* **11** 45 10N 69 23W
Gulfport, *Miss.* **13** 30 22N 89 6W
Gulkana, *Alaska* **30** 62 16N 145 23W
Gunnison, *Colo.* **17** 38 33N 106 56W
Gunnison, *Utah* **16** 39 9N 111 49W
Gunnison →, *Colo.* ... **17** 39 4N 108 35W
Guntersville, *Ala.* **11** 34 21N 86 18W
Gurdon, *Ark.* **13** 33 55N 93 9W
Gustavus, *Alaska* **30** 58 25N 136 44W
Gustine, *Calif.* **18** 37 16N 121 0W
Guthrie, *Okla.* **13** 35 53N 97 25W
Guttenberg, *Iowa* **14** 42 47N 91 6W
Guymon, *Okla.* **13** 36 41N 101 29W
Gwinn, *Mich.* **10** 46 19N 87 27W

H

Hackensack, *N.J.* **20** 40 53N 74 3W
Haddonfield, *N.J.* **24** 39 53N 75 2W
Hagemeister I., *Alaska* . **30** 58 39N 160 54W
Hagerman, *N. Mex.* **13** 33 7N 104 20W
Hagerstown, *Md.* **10** 39 39N 77 43W
Hailey, *Idaho* **16** 43 31N 114 19W
Haines, *Alaska* **30** 59 14N 135 26W
Haines, *Oreg.* **16** 44 55N 117 56W
Haines City, *Fla.* **11** 28 7N 81 38W
Halawa, *Hawaii* **30** 21 9N 156 47W
Haleakala Crater, *Hawaii* **30** 20 43N 156 16W
Halethorpe, *Md.* **25** 39 14N 76 41W
Haleyville, *Ala.* **11** 34 14N 87 37W
Half Moon B., *Calif.* **28** 37 29N 122 28W
Half Moon Bay, *Calif.* .. **28** 37 27N 122 25W
Hallettsville, *Tex.* **13** 29 27N 96 57W
Halliday, *N. Dak.* **12** 47 21N 102 20W
Hallstead, *Pa.* **9** 41 58N 75 45W
Halstad, *Minn.* **12** 47 21N 96 50W
Hamburg, *Ark.* **13** 33 14N 91 48W
Hamburg, *Iowa* **12** 40 36N 95 39W
Hamburg, *Pa.* **9** 40 33N 75 59W
Hamden, *Conn.* **9** 41 23N 72 54W
Hamilton, *Alaska* **30** 62 54N 163 53W
Hamilton, *Mo.* **12** 39 45N 93 59W
Hamilton, *Mont.* **16** 46 15N 114 10W
Hamilton, *N.Y.* **9** 42 50N 75 33W
Hamilton, *Ohio* **15** 39 24N 84 34W
Hamilton, *Tex.* **13** 31 42N 98 7W
Hamlet, *N.C.* **11** 34 53N 79 42W
Hamlin, *Tex.* **13** 32 53N 100 8W
Hammond, *Ind.* **22** 41 35N 87 29W
Hammond, *La.* **13** 30 30N 90 28W
Hammonton, *N.J.* **10** 39 39N 74 48W
Hampton, *Ark.* **13** 33 32N 92 28W
Hampton, *Iowa* **14** 42 45N 93 13W
Hampton, *N.H.* **9** 42 57N 70 50W
Hampton, *S.C.* **11** 32 52N 81 7W
Hampton, *Va.* **10** 37 2N 76 21W
Hamtramck, *Mich.* **26** 42 23N 83 4W
Hana, *Hawaii* **30** 20 45N 155 59W
Hancock, *Mich.* **12** 47 8N 88 35W
Hancock, *Minn.* **12** 45 30N 95 48W
Hancock, *N.Y.* **9** 41 57N 75 17W
Hanford, *Calif.* **18** 36 20N 119 39W
Hankinson, *N. Dak.* **12** 46 4N 96 54W
Hanksville, *Utah* **17** 38 22N 110 43W
Hannaford, *N. Dak.* **12** 47 19N 98 11W
Hannah, *N. Dak.* **12** 48 58N 98 42W
Hannibal, *Mo.* **14** 39 42N 91 22W
Hanover, *N.H.* **9** 43 42N 72 17W
Hanover, *Pa.* **10** 39 48N 76 59W
Happy, *Tex.* **13** 34 45N 101 52W
Happy Camp, *Calif.* **16** 41 48N 123 23W
Harbor Beach, *Mich.* ... **10** 43 51N 82 39W
Harbor Springs, *Mich.* . **10** 45 26N 85 0W
Hardin, *Mont.* **16** 45 44N 107 37W
Hardman, *Oreg.* **16** 45 10N 119 41W
Hardy, *Ark.* **13** 36 19N 91 29W
Harlan, *Iowa* **12** 41 39N 95 19W
Harlan, *Ky.* **11** 36 51N 83 19W
Harlem, *Mont.* **16** 48 32N 108 47W
Harlem, *N.Y.* **20** 40 48N 73 56W
Harlingen, *Tex.* **13** 26 12N 97 42W
Harlowton, *Mont.* **16** 46 26N 109 50W
Harney Basin, *Oreg.* ... **16** 43 30N 119 0W
Harney L., *Oreg.* **16** 43 14N 119 8W
Harney Peak, *S. Dak.* ... **12** 43 52N 103 32W
Harper, Mt., *Alaska* **30** 64 14N 143 51W
Harper Woods, *Mich.* .. **26** 42 26N 82 56W
Harriman, *Tenn.* **11** 35 56N 84 33W
Harrisburg, *Ill.* **15** 37 44N 88 32W
Harrisburg, *Nebr.* **12** 41 33N 103 44W
Harrisburg, *Oreg.* **16** 44 16N 123 10W
Harrisburg, *Pa.* **8** 40 16N 76 53W
Harrison, *Ark.* **13** 36 14N 93 7W
Harrison, *Idaho* **16** 47 27N 116 47W
Harrison, *Nebr.* **12** 42 41N 103 53W
Harrison Bay, *Alaska* ... **30** 70 40N 151 0W
Harrisonburg, *Va.* **10** 38 27N 78 52W
Harrisonville, *Mo.* **14** 38 39N 94 21W
Harrisville, *Mich.* **8** 44 39N 83 17W
Hart, *Mich.* **10** 43 42N 86 22W
Hartford, *Conn.* **9** 41 46N 72 41W
Hartford, *Ky.* **10** 37 27N 86 55W
Hartford, *S. Dak.* **12** 43 38N 96 57W
Hartford, *Wis.* **12** 43 19N 88 22W
Hartford City, *Ind.* **15** 40 27N 85 22W
Hartselle, *Ala.* **11** 34 27N 86 56W
Hartshorne, *Okla.* **13** 34 51N 95 34W
Hartsville, *S.C.* **11** 34 23N 80 4W
Hartwell, *Ga.* **11** 34 21N 82 56W
Harvard University, *Mass.* **23** 42 22N 71 7W
Harvey, *Ill.* **22** 41 36N 87 39W
Harvey, *N. Dak.* **12** 47 47N 99 56W
Harwood Heights, *Ill.* .. **22** 41 57N 87 47W
Haskell, *Okla.* **13** 35 50N 95 40W
Haskell, *Tex.* **13** 33 10N 99 44W
Hastings, *Mich.* **15** 42 39N 85 17W
Hastings, *Minn.* **12** 44 44N 92 51W
Hastings, *Nebr.* **12** 40 35N 98 23W
Hatch, *N. Mex.* **17** 32 40N 107 9W
Hatteras, C., *N.C.* **11** 35 14N 75 32W
Hattiesburg, *Miss.* **13** 31 20N 89 17W
Havana, *Ill.* **14** 40 18N 90 4W
Havasu, L., *Ariz.* **19** 34 18N 114 28W
Haverhill, *Mass.* **9** 42 47N 71 5W
Haverstraw, *N.Y.* **9** 41 12N 73 58W
Havertown, *Pa.* **24** 39 58N 75 18W
Havre, *Mont.* **16** 48 33N 109 41W
Haw →, *N.C.* **11** 35 36N 79 3W
Hawaii □ **30** 19 30N 156 30W
Hawaii I., *Pac. Oc.* **30** 20 0N 155 0W
Hawaiian Is., *Pac. Oc.* .. **30** 20 30N 156 0W
Hawarden, *Iowa* **12** 43 0N 96 29W
Hawkinsville, *Ga.* **11** 32 17N 83 28W
Hawley, *Minn.* **12** 46 53N 96 19W
Hawthorne, *Calif.* **29** 33 54N 118 21W
Hawthorne, *Nev.* **16** 38 32N 118 38W
Haxtun, *Colo.* **12** 40 39N 102 38W
Hay Springs, *Nebr.* **12** 42 41N 102 41W
Hayden, *Ariz.* **17** 33 0N 110 47W
Hayden, *Colo.* **16** 40 30N 107 16W
Hayes, *S. Dak.* **12** 44 23N 101 1W
Haynesville, *La.* **13** 32 58N 93 8W
Hays, *Kans.* **12** 38 53N 99 20W
Hayward, *Calif.* **28** 37 40N 122 4W
Hayward, *Wis.* **12** 46 1N 91 29W
Hazard, *Ky.* **10** 37 15N 83 12W
Hazel Park, *Mich.* **26** 42 28N 83 5W
Hazelton, *N. Dak.* **12** 46 29N 100 17W
Hazen, *N. Dak.* **12** 47 18N 101 38W
Hazen, *Nev.* **16** 39 34N 119 3W
Hazlehurst, *Ga.* **11** 31 52N 82 36W
Hazlehurst, *Miss.* **13** 31 52N 90 24W
Hazleton, *Pa.* **9** 40 57N 75 59W
Healdsburg, *Calif.* **18** 38 37N 122 52W
Healdton, *Okla.* **13** 34 14N 97 29W
Hearne, *Tex.* **13** 30 53N 96 36W
Heart →, *N. Dak.* **12** 46 46N 100 50W
Heavener, *Okla.* **13** 34 53N 94 36W
Hebbronville, *Tex.* **13** 27 18N 98 41W
Heber Springs, *Ark.* **13** 35 30N 92 2W
Hebgen L., *Mont.* **16** 44 52N 111 20W
Hebron, *N. Dak.* **12** 46 54N 102 3W
Hebron, *Nebr.* **12** 40 10N 97 35W
Hecla, *S. Dak.* **12** 45 53N 98 9W
Hedley, *Tex.* **13** 34 52N 100 39W
Helena, *Ark.* **13** 34 32N 90 36W
Helena, *Mont.* **16** 46 36N 112 2W
Helendale, *Calif.* **19** 34 44N 117 19W
Helper, *Utah* **16** 39 41N 110 51W
Hemet, *Calif.* **19** 33 45N 116 58W
Hemingford, *Nebr.* **12** 42 19N 103 4W
Hemphill, *Tex.* **13** 31 20N 93 51W
Hempstead, *N.Y.* **21** 40 42N 73 37W
Hempstead, *Tex.* **13** 30 6N 96 5W
Henderson, *Ky.* **15** 37 50N 87 35W
Henderson, *N.C.* **11** 36 20N 78 25W
Henderson, *Nev.* **19** 36 2N 114 59W
Henderson, *Tenn.* **11** 35 26N 88 38W
Henderson, *Tex.* **13** 32 9N 94 48W
Hendersonville, *N.C.* ... **11** 35 19N 82 28W
Henlopen, C., *Del.* **10** 38 48N 75 6W
Hennessey, *Okla.* **13** 36 6N 97 54W
Henrietta, *Tex.* **13** 33 49N 98 12W
Henry, *Ill.* **14** 41 7N 89 22W
Henryetta, *Okla.* **13** 35 27N 95 59W
Heppner, *Oreg.* **16** 45 21N 119 33W
Herbert I., *Alaska* **30** 52 45N 170 7W
Hereford, *Tex.* **13** 34 49N 102 24W
Herington, *Kans.* **12** 38 40N 96 57W
Herkimer, *N.Y.* **9** 43 0N 74 59W
Herman, *Minn.* **12** 45 49N 96 9W
Hermann, *Mo.* **12** 38 42N 91 27W
Hermiston, *Oreg.* **16** 45 51N 119 17W

Hernandez, *Calif.* **18** 36 24N 120 46W
Hernando, *Miss.* **13** 34 50N 90 0W
Herreid, *S. Dak.* **12** 45 50N 100 4W
Herrin, *Ill.* **14** 37 48N 89 2W
Hesperia, *Calif.* **19** 34 25N 117 18W
Hetch Hetchy Aqueduct, *Calif.* **18** 37 29N 122 19W
Hettinger, *N. Dak.* **12** 46 0N 102 42W
Hi Vista, *Calif.* **19** 34 45N 117 46W
Hialeah, *Fla.* **11** 25 50N 80 17W
Hiawatha, *Kans.* **12** 39 51N 95 32W
Hiawatha, *Utah* **16** 39 29N 111 1W
Hibbing, *Minn.* **12** 47 25N 92 56W
Hickory, *N.C.* **11** 35 44N 81 21W
Hickory Hills, *Ill.* **22** 41 43N 87 50W
Hicksville, *N.Y.* **21** 40 46N 73 30W
Higgins, *Tex.* **13** 36 7N 100 2W
High Island, *Tex.* **13** 29 34N 94 24W
High Point, *N.C.* **11** 35 57N 80 0W
High Springs, *Fla.* **11** 29 50N 82 36W
Highland Park, *Ill.* **15** 42 11N 87 48W
Highland Park, *Mich.* **26** 42 24N 83 6W
Highmore, *S. Dak.* **12** 44 31N 99 27W
Hiko, *Nev.* **18** 37 32N 115 14W
Hill City, *Idaho* **16** 43 18N 115 3W
Hill City, *Kans.* **12** 39 22N 99 51W
Hill City, *Minn.* **12** 46 59N 93 36W
Hill City, *S. Dak.* **12** 43 56N 103 35W
Hillcrest Heights, *Md.* **25** 38 50N 76 57W
Hillman, *Mich.* **10** 45 4N 83 54W
Hillsboro, *Kans.* **12** 38 21N 97 12W
Hillsboro, *N. Dak.* **12** 47 26N 97 3W
Hillsboro, *N.H.* **9** 43 7N 71 54W
Hillsboro, *N. Mex.* **17** 32 55N 107 34W
Hillsboro, *Oreg.* **18** 45 31N 122 59W
Hillsboro, *Tex.* **13** 32 1N 97 8W
Hillsdale, *Mich.* **15** 41 56N 84 38W
Hillsdale, *N.J.* **20** 41 0N 74 2W
Hillsdale, *N.Y.* **9** 42 11N 73 30W
Hilo, *Hawaii* **30** 19 44N 155 5W
Hinckley, *Utah* **16** 39 20N 112 40W
Hingham, *Mass.* **23** 42 14N 70 54W
Hingham, *Mont.* **16** 48 33N 110 25W
Hinsdale, *Ill.* **22** 41 47N 87 56W
Hinsdale, *Mont.* **16** 48 24N 107 5W
Hinton, *W. Va.* **10** 37 40N 80 54W
Hobart, *Okla.* **13** 35 1N 99 6W
Hobbs, *N. Mex.* **13** 32 42N 103 8W
Hoboken, *N.J.* **20** 40 44N 74 3W
Hogansville, *Ga.* **11** 33 10N 84 55W
Hogeland, *Mont.* **16** 48 51N 108 40W
Hohenwald, *Tenn.* **11** 35 33N 87 33W
Hoisington, *Kans.* **12** 38 31N 98 47W
Holbrook, *Ariz.* **17** 34 54N 110 10W
Holden, *Utah* **16** 39 6N 112 16W
Holdenville, *Okla.* **13** 35 5N 96 24W
Holdrege, *Nebr.* **12** 40 26N 99 23W
Holland, *Mich.* **15** 42 47N 86 7W
Hollidaysburg, *Pa.* **8** 40 26N 78 24W
Hollis, *Okla.* **13** 34 41N 99 55W
Hollister, *Calif.* **18** 36 51N 121 24W
Hollister, *Idaho* **16** 42 21N 114 35W
Holly, *Colo.* **12** 38 3N 102 7W
Holly Hill, *Fla.* **11** 29 16N 81 3W
Holly Springs, *Miss.* **13** 34 46N 89 27W
Hollywood, *Calif.* **17** 34 7N 118 25W
Hollywood, *Fla.* **11** 26 1N 80 9W
Holton, *Kans.* **12** 39 28N 95 44W
Holtville, *Calif.* **19** 32 49N 115 23W
Holy Cross, *Alaska* **30** 62 12N 159 46W
Holyoke, *Colo.* **12** 40 35N 102 18W
Holyoke, *Mass.* **9** 42 12N 72 37W
Homedale, *Idaho* **16** 43 37N 116 56W
Homer, *Alaska* **30** 59 39N 151 33W
Homer, *La.* **13** 32 48N 93 4W
Homestead, *Fla.* **11** 25 28N 80 29W
Homestead, *Oreg.* **16** 45 2N 116 51W
Homestead, *Pa.* **27** 40 24N 79 55W
Hominy, *Okla.* **13** 36 25N 96 24W
Hondo, *Tex.* **13** 29 21N 99 9W
Honey L., *Calif.* **18** 40 15N 120 19W
Honolulu, *Hawaii* **30** 21 19N 157 52W
Hood, Mt., *Oreg.* **16** 45 23N 121 42W
Hood River, *Oreg.* **16** 45 43N 121 31W
Hoodsport, *Wash.* **18** 47 24N 123 9W
Hooker, *Okla.* **13** 36 52N 101 13W
Hoonah, *Alaska* **30** 58 7N 135 27W
Hooper Bay, *Alaska* **30** 61 32N 166 6W
Hoopeston, *Ill.* **15** 40 28N 87 40W
Hoover Dam, *Ariz.* **19** 36 1N 114 44W
Hop Bottom, *Pa.* **9** 41 42N 75 46W
Hope, *Ark.* **13** 33 40N 93 36W
Hope, *N. Dak.* **12** 47 19N 97 43W
Hope, Pt., *Alaska* **30** 68 20N 166 50W
Hopkins, *Mo.* **14** 40 33N 94 49W
Hopkinsville, *Ky.* **11** 36 52N 87 29W
Hopland, *Calif.* **18** 38 58N 123 7W
Hoquiam, *Wash.* **18** 46 59N 123 53W
Horn I., *Miss.* **11** 30 14N 88 39W
Hornbeck, *La.* **13** 31 20N 93 24W
Hornbrook, *Calif.* **16** 41 55N 122 33W
Hornell, *N.Y.* **8** 42 20N 77 40W
Hornitos, *Calif.* **18** 37 30N 120 14W
Horse Creek, *Wyo.* **12** 41 57N 105 10W
Horton, *Kans.* **12** 39 40N 95 32W
Hosmer, *S. Dak.* **12** 45 34N 99 28W
Hot Creek Range, *Nev.* **16** 38 40N 116 20W
Hot Springs, *Ark.* **13** 34 31N 93 3W
Hot Springs, *S. Dak.* **12** 43 26N 103 29W
Hotchkiss, *Colo.* **17** 38 48N 107 43W
Houck, *Ariz.* **17** 35 20N 109 10W
Houghton, *Mich.* **12** 47 7N 88 34W
Houghton L., *Mich.* **10** 44 21N 84 44W
Houlton, *Maine* **11** 46 8N 67 51W
Houma, *La.* **13** 29 36N 90 43W
Houston, *Mo.* **13** 37 22N 91 58W
Houston, *Tex.* **13** 29 46N 95 22W
Howard, *Kans.* **13** 37 28N 96 16W
Howard, *S. Dak.* **12** 44 1N 97 32W
Howe, *Idaho* **16** 43 48N 113 0W
Howell, *Mich.* **15** 42 36N 83 56W
Hualapai Peak, *Ariz.* **17** 35 5N 113 54W
Huasna, *Calif.* **19** 35 6N 120 24W
Hubbard, *Tex.* **13** 31 51N 96 48W
Hudson, *Mich.* **15** 41 51N 84 21W
Hudson, *N.Y.* **9** 42 15N 73 46W
Hudson, *Wis.* **12** 44 58N 92 45W
Hudson, *Wyo.* **16** 42 54N 108 35W
Hudson →, *N.Y.* **9** 40 42N 74 2W
Hudson Falls, *N.Y.* **9** 43 18N 73 35W
Hughes, *Alaska* **30** 66 3N 154 15W
Hugo, *Colo.* **12** 39 8N 103 28W
Hugoton, *Kans.* **13** 37 11N 101 21W
Hull, *Mass.* **23** 42 18N 70 54W
Humacao, *Puerto Rico* **30** 18 9N 65 50W
Humble, *Tex.* **13** 29 59N 93 18W
Humboldt, *Iowa* **14** 42 44N 94 13W
Humboldt, *Tenn.* **13** 35 50N 88 55W
Humboldt →, *Nev.* **16** 39 59N 118 36W
Hume, *Calif.* **18** 36 48N 118 54W
Humphreys, Mt., *Calif.* **18** 37 17N 118 40W
Humphreys Peak, *Ariz.* **17** 35 21N 111 41W
Hunter, *N. Dak.* **12** 47 12N 97 13W
Hunter, *N.Y.* **9** 42 13N 74 13W
Huntingburg, *Ind.* **15** 38 18N 86 57W
Huntingdon, *Pa.* **8** 40 30N 78 1W
Huntington, *Ind.* **15** 40 53N 85 30W
Huntington, *N.Y.* **21** 40 52N 73 25W
Huntington, *Oreg.* **16** 44 21N 117 16W
Huntington, *Utah* **16** 39 20N 110 58W
Huntington, *W. Va.* **10** 38 25N 82 27W
Huntington Beach, *Calif.* **19** 33 40N 118 5W
Huntington Park, *Calif.* **29** 33 58N 118 13W
Huntington Woods, *Mich.* **26** 42 28N 83 10W
Huntsville, *Ala.* **11** 34 44N 86 35W
Huntsville, *Tex.* **13** 30 43N 95 33W
Hurley, *N. Mex.* **17** 32 42N 108 8W
Hurley, *Wis.* **12** 46 27N 90 11W
Huron, *Calif.* **18** 36 12N 120 6W
Huron, *S. Dak.* **12** 44 22N 98 13W
Huron, L., *Mich.* **8** 44 30N 82 40W
Hurricane, *Utah* **17** 37 11N 113 17W
Huslia, *Alaska* **30** 65 41N 156 24W
Hutchinson, *Kans.* **13** 38 5N 97 56W
Hutchinson, *Minn.* **12** 44 54N 94 22W
Huttig, *Ark.* **13** 33 2N 92 11W
Hyannis, *Nebr.* **12** 42 0N 101 46W
Hyattsville, *Md.* **25** 38 57N 76 58W
Hydaburg, *Alaska* **30** 55 12N 132 50W
Hyndman Peak, *Idaho* **16** 43 45N 114 8W
Hyrum, *Utah* **16** 41 38N 111 51W
Hysham, *Mont.* **16** 46 18N 107 14W

I

Ida Grove, *Iowa* **12** 42 21N 95 28W
Idabel, *Okla.* **13** 33 54N 94 50W
Idaho □ **16** 45 0N 115 0W
Idaho City, *Idaho* **16** 43 50N 115 50W
Idaho Falls, *Idaho* **16** 43 30N 112 2W
Idaho Springs, *Colo.* **16** 39 45N 105 31W
Idria, *Calif.* **18** 36 25N 120 41W
Iliamna L., *Alaska* **30** 59 30N 155 0W
Iliff, *Colo.* **12** 40 45N 103 4W
Ilio Pt., *Hawaii* **30** 21 13N 157 16W
Ilion, *N.Y.* **9** 43 1N 75 2W
Illinois □ **14** 40 15N 89 30W
Illinois →, *Ill.* **14** 38 58N 90 28W
Imbler, *Oreg.* **16** 45 28N 117 58W
Imlay, *Nev.* **16** 40 40N 118 9W
Immokalee, *Fla.* **11** 26 25N 81 25W
Imperial, *Calif.* **19** 32 51N 115 34W
Imperial, *Nebr.* **12** 40 31N 101 39W
Imperial Beach, *Calif.* **19** 32 35N 117 8W
Imperial Dam, *Ariz.* **19** 32 55N 114 25W
Independence, *Calif.* **18** 36 48N 118 12W
Independence, *Iowa* **14** 42 28N 91 54W
Independence, *Kans.* **13** 37 14N 95 42W
Independence, *Mo.* **14** 39 6N 94 25W
Independence, *Oreg.* **16** 44 51N 123 11W
Independence Mts., *Nev.* **16** 41 20N 116 0W
Indian →, *Fla.* **11** 27 59N 80 34W
Indiana, *Pa.* **8** 40 37N 79 9W
Indiana □ **15** 40 0N 86 0W
Indianapolis, *Ind.* **15** 39 46N 86 9W
Indianola, *Iowa* **14** 41 22N 93 34W
Indianola, *Miss.* **13** 33 27N 90 39W
Indio, *Calif.* **19** 33 43N 116 13W
Inglewood, *Calif.* **29** 33 57N 118 19W
Ingomar, *Mont.* **16** 46 35N 107 23W
Inkom, *Idaho* **16** 42 48N 112 15W
Inkster, *Mich.* **26** 42 17N 83 16W
Interior, *S. Dak.* **12** 43 44N 101 59W
International Falls, *Minn.* **12** 48 36N 93 25W
Inverness, *Fla.* **11** 28 50N 82 20W
Inyo Mts., *Calif.* **17** 36 40N 118 0W
Inyokern, *Calif.* **19** 35 39N 117 49W
Iola, *Kans.* **13** 37 55N 95 24W
Ione, *Calif.* **18** 38 21N 120 56W
Ione, *Wash.* **16** 48 45N 117 25W
Ionia, *Mich.* **15** 42 59N 85 4W
Iowa □ **12** 42 18N 93 30W
Iowa City, *Iowa* **14** 41 40N 91 32W
Iowa Falls, *Iowa* **14** 42 31N 93 16W
Ipswich, *Mass.* **23** 42 41N 70 50W
Ipswich, *S. Dak.* **12** 45 27N 99 2W
Iron Mountain, *Mich.* **10** 45 49N 88 4W
Iron River, *Mich.* **12** 46 6N 88 39W
Ironton, *Mo.* **13** 37 36N 90 38W
Ironton, *Ohio* **10** 38 32N 82 41W
Ironwood, *Mich.* **12** 46 27N 90 9W
Irvine, *Ky.* **15** 37 42N 83 58W
Irvington, *N.Y.* **20** 40 42N 74 13W
Isabel, *S. Dak.* **12** 45 24N 101 26W
Isabela, *Puerto Rico* **30** 18 30N 67 2W
Ishpeming, *Mich.* **10** 46 29N 87 40W
Isla Vista, *Calif.* **19** 34 25N 119 53W
Island Falls, *Maine* **11** 46 1N 68 16W
Island Pond, *Vt.* **9** 44 49N 71 53W
Isle Royale, *Mich.* **12** 48 0N 88 54W
Isleta, *N. Mex.* **17** 34 55N 106 42W
Isleton, *Calif.* **18** 38 10N 121 37W
Ismay, *Mont.* **12** 46 30N 104 48W
Istokpoga, L., *Fla.* **11** 27 23N 81 17W
Ithaca, *N.Y.* **9** 42 27N 76 30W
Ivanhoe, *Calif.* **18** 36 23N 119 13W

J

Jackman, *Maine* **11** 45 35N 70 17W
Jacksboro, *Tex.* **13** 33 14N 98 15W
Jackson, *Ala.* **11** 31 31N 87 53W
Jackson, *Calif.* **18** 38 21N 120 46W
Jackson, *Ky.* **10** 37 33N 83 23W
Jackson, *Mich.* **15** 42 15N 84 24W
Jackson, *Minn.* **12** 43 37N 95 1W
Jackson, *Miss.* **13** 32 18N 90 12W
Jackson, *Mo.* **13** 37 23N 89 40W
Jackson, *Ohio* **10** 39 3N 82 39W
Jackson, *Tenn.* **11** 35 37N 88 49W
Jackson, *Wyo.* **16** 43 29N 110 46W
Jackson Heights, *N.Y.* **21** 40 44N 73 53W
Jackson L., *Wyo.* **16** 43 52N 110 36W
Jacksonville, *Ala.* **11** 33 49N 85 46W
Jacksonville, *Calif.* **18** 37 52N 120 24W
Jacksonville, *Fla.* **11** 30 20N 81 39W
Jacksonville, *Ill.* **14** 39 44N 90 14W
Jacksonville, *N.C.* **11** 34 45N 77 26W
Jacksonville, *Oreg.* **16** 42 19N 122 57W
Jacksonville, *Tex.* **13** 31 58N 95 17W
Jacksonville Beach, *Fla.* **11** 30 17N 81 24W
Jacob Lake, *Ariz.* **17** 36 43N 112 13W
Jal, *N. Mex.* **13** 32 7N 103 12W
Jalama, *Calif.* **19** 34 29N 120 29W
Jaluit I., *Pac. Oc.* **31** 6 0N 169 30 E
Jamaica, *N.Y.* **21** 40 42N 73 48W
James →, *S. Dak.* **12** 42 52N 97 18W
Jamestown, *Ky.* **10** 36 59N 85 4W
Jamestown, *N. Dak.* **12** 46 54N 98 42W
Jamestown, *N.Y.* **8** 42 6N 79 14W
Jamestown, *Tenn.* **11** 36 26N 84 56W
Janesville, *Wis.* **14** 42 41N 89 1W
Jasper, *Ala.* **11** 33 50N 87 17W
Jasper, *Fla.* **11** 30 31N 82 57W
Jasper, *Minn.* **12** 43 51N 96 24W
Jasper, *Tex.* **13** 30 56N 94 1W
Jay, *Okla.* **13** 36 25N 94 48W
Jayton, *Tex.* **13** 33 15N 100 34W
Jean, *Nev.* **19** 35 47N 115 20W
Jeanerette, *La.* **13** 29 55N 91 40W
Jefferson, *Iowa* **14** 42 1N 94 23W
Jefferson, *Tex.* **13** 32 46N 94 21W
Jefferson, *Wis.* **15** 43 0N 88 48W
Jefferson, Mt., *Nev.* **16** 38 51N 117 0W
Jefferson, Mt., *Oreg.* **16** 44 41N 121 48W
Jefferson City, *Mo.* **14** 38 34N 92 10W
Jefferson City, *Tenn.* **11** 36 7N 83 30W
Jeffersonville, *Ind.* **15** 38 17N 85 44W
Jena, *La.* **13** 31 41N 92 8W
Jenkins, *Ky.* **10** 37 10N 82 38W
Jennings, *La.* **13** 30 13N 92 40W
Jermyn, *Pa.* **9** 41 31N 75 31W
Jerome, *Ariz.* **17** 34 45N 112 7W
Jersey City, *N.J.* **20** 40 42N 74 4W
Jersey Shore, *Pa.* **8** 41 12N 77 15W
Jerseyville, *Ill.* **14** 39 7N 90 20W
Jesup, *Ga.* **11** 31 36N 81 53W
Jetmore, *Kans.* **13** 38 4N 99 54W
Jewett, *Tex.* **13** 31 22N 96 9W
Jewett City, *Conn.* **9** 41 36N 72 0W
Johannesburg, *Calif.* **19** 35 22N 117 38W
John Day, *Oreg.* **16** 44 25N 118 57W
John Day →, *Oreg.* **16** 45 44N 120 39W
John F.Kennedy International Airport, *N.Y.* **21** 40 38N 73 46W
John H. Kerr Reservoir, *N.C.* **11** 36 36N 78 18W
Johnson, *Kans.* **13** 37 34N 101 45W
Johnson City, *N.Y.* **9** 42 7N 75 58W
Johnson City, *Tenn.* **11** 36 19N 82 21W
Johnson City, *Tex.* **13** 30 17N 98 25W
Johnsondale, *Calif.* **19** 35 58N 118 32W
Johnstown, *N.Y.* **9** 43 0N 74 22W
Johnstown, *Pa.* **8** 40 20N 78 55W
Joliet, *Ill.* **15** 41 32N 88 5W
Jolon, *Calif.* **18** 35 58N 121 9W
Jonesboro, *Ark.* **13** 35 50N 90 42W
Jonesboro, *Ill.* **13** 37 27N 89 16W
Jonesboro, *La.* **13** 32 15N 92 43W
Jonesport, *Maine* **11** 44 32N 67 37W
Joplin, *Mo.* **13** 37 6N 94 31W
Joppatowne, *Md.* **25** 39 24N 76 21W
Jordan, *Mont.* **16** 47 19N 106 55W
Jordan Valley, *Oreg.* **16** 42 59N 117 3W
Joseph, *Oreg.* **16** 45 21N 117 14W
Joseph City, *Ariz.* **17** 34 57N 110 20W
Joshua Tree, *Calif.* **19** 34 8N 116 19W
Joshua Tree National Monument, *Calif.* **19** 33 55N 116 0W
Jourdanton, *Tex.* **13** 28 55N 98 33W
Judith →, *Mont.* **16** 47 44N 109 39W
Judith, Pt., *R.I.* **9** 41 22N 71 29W
Judith Gap, *Mont.* **16** 46 41N 109 45W
Julesburg, *Colo.* **12** 40 59N 102 16W
Julian, *Calif.* **19** 33 4N 116 38W
Junction, *Tex.* **13** 30 29N 99 46W
Junction, *Utah* **17** 38 14N 112 13W
Junction City, *Kans.* **12** 39 2N 96 50W
Junction City, *Oreg.* **16** 44 13N 123 12W
Juneau, *Alaska* **30** 58 18N 134 25W
Juniata →, *Pa.* **8** 40 30N 77 40W
Juntura, *Oreg.* **16** 43 45N 118 5W
Justice, *Ill.* **22** 41 44N 87 49W

K

Ka Lae, *Hawaii* **30** 18 55N 155 41W
Kaala, *Hawaii* **30** 21 31N 158 9W
Kadoka, *S. Dak.* **12** 43 50N 101 31W
Kaena Pt., *Hawaii* **30** 21 35N 158 17W
Kagamil I., *Alaska* **30** 53 0N 169 43W
Kahoka, *Mo.* **14** 40 25N 91 44W
Kahoolawe, *Hawaii* **30** 20 33N 156 37W
Kahuku Pt., *Hawaii* **30** 21 43N 157 59W
Kahului, *Hawaii* **30** 20 54N 156 28W
Kailua Kona, *Hawaii* **30** 19 39N 155 59W
Kaiwi Channel, *Hawaii* **30** 21 15N 157 30W
Kaiyuh Mts., *Alaska* **30** 64 30N 158 0W
Kake, *Alaska* **30** 56 59N 133 57W
Kaktovik, *Alaska* **30** 70 8N 143 38W
Kalama, *Wash.* **18** 46 1N 122 51W
Kalamazoo, *Mich.* **15** 42 17N 85 35W
Kalamazoo →, *Mich.* **15** 42 40N 86 10W
Kalaupapa, *Hawaii* **30** 21 12N 156 59W
Kalispell, *Mont.* **16** 48 12N 114 19W
Kalkaska, *Mich.* **10** 44 44N 85 11W
Kamalino, *Hawaii* **30** 21 50N 160 14W
Kamiah, *Idaho* **16** 46 14N 116 2W
Kanab, *Utah* **17** 37 3N 112 32W
Kanab →, *Ariz.* **17** 36 24N 112 38W
Kanaga I., *Alaska* **30** 51 45N 177 22W
Kanakanak, *Alaska* **30** 59 0N 158 58W
Kanarraville, *Utah* **17** 37 32N 113 11W
Kanawha →, *W. Va.* **10** 38 50N 82 9W
Kane, *Pa.* **8** 41 40N 78 49W
Kaneohe, *Hawaii* **30** 21 25N 157 48W
Kankakee, *Ill.* **15** 41 7N 87 52W
Kankakee →, *Ill.* **15** 41 23N 88 15W
Kannapolis, *N.C.* **11** 35 30N 80 37W
Kansas □ **12** 38 30N 99 0W
Kansas →, *Kans.* **12** 39 7N 94 37W
Kansas City, *Kans.* **14** 39 7N 94 38W
Kansas City, *Mo.* **14** 39 6N 94 35W
Kantishna, *Alaska* **30** 63 31N 151 5W
Kapaa, *Hawaii* **30** 22 5N 159 19W
Karlstad, *Minn.* **12** 48 35N 96 31W
Karnes City, *Tex.* **13** 28 53N 97 54W
Kaskaskia →, *Ill.* **14** 37 58N 89 57W
Katalla, *Alaska* **30** 60 12N 144 31W
Katmai National Park, *Alaska* **30** 58 20N 155 0W
Kauai, *Hawaii* **30** 22 3N 159 30W
Kauai Channel, *Hawaii* **30** 21 45N 158 50W
Kaufman, *Tex.* **13** 32 35N 96 19W
Kaukauna, *Wis.* **10** 44 17N 88 17W
Kaupulehu, *Hawaii* **30** 19 43N 155 53W
Kawaihae, *Hawaii* **30** 20 3N 155 50W
Kawaihoa Pt., *Hawaii* **30** 21 47N 160 12W
Kawaikimi, *Hawaii* **30** 22 5N 159 29W
Kayak I., *Alaska* **30** 59 56N 144 23W
Kaycee, *Wyo.* **16** 43 43N 106 38W
Kayenta, *Ariz.* **17** 36 44N 110 15W
Kaysville, *Utah* **16** 41 2N 111 56W
Keaau, *Hawaii* **30** 19 37N 155 2W
Keams Canyon, *Ariz.* **17** 35 49N 110 12W
Keanae, *Hawaii* **30** 20 52N 156 9W
Kearney, *Nebr.* **12** 40 42N 99 5W
Kearny, *N.J.* **20** 40 45N 74 9W
Keeler, *Calif.* **18** 36 29N 117 52W
Keene, *Calif.* **19** 35 13N 118 33W
Keene, *N.H.* **9** 42 56N 72 17W
Keewatin, *Minn.* **12** 47 24N 93 5W

Keller, *Wash.* **16** 48 5N 118 41W
Kellogg, *Idaho* **16** 47 32N 116 7W
Kelso, *Wash.* **18** 46 9N 122 54W
Kemmerer, *Wyo.* **16** 41 48N 110 32W
Kemp, L., *Tex.* **13** 33 46N 99 9W
Kenai, *Alaska* **30** 60 33N 151 16W
Kenai Mts., *Alaska* **30** 60 0N 150 0W
Kendallville, *Ind.* **15** 41 27N 85 16W
Kendrick, *Idaho* **16** 46 37N 116 39W
Kenedy, *Tex.* **13** 28 49N 97 51W
Kenmare, *N. Dak.* **12** 48 41N 102 5W
Kennebec, *S. Dak.* **12** 43 54N 99 52W
Kennett, *Mo.* **13** 36 14N 90 3W
Kennewick, *Wash.* **16** 46 12N 119 7W
Kenosha, *Wis.* **15** 42 35N 87 49W
Kensington, *Kans.* **12** 39 46N 99 2W
Kent, *Ohio* **8** 41 9N 81 22W
Kent, *Oreg.* **16** 45 12N 120 42W
Kent, *Tex.* **13** 31 4N 104 13W
Kentfield, *Calif.* **28** 37 57N 122 33W
Kentland, *Ind.* **15** 40 46N 87 27W
Kenton, *Ohio* **15** 40 39N 83 37W
Kentucky □ **10** 37 0N 84 0W
Kentucky →, *Ky.* **15** 38 41N 85 11W
Kentucky L., *Ky.* **11** 37 1N 88 16W
Kentwood, *La.* **13** 31 0N 90 30W
Kentwood, *La.* **13** 30 56N 90 31W
Keokuk, *Iowa* **14** 40 24N 91 24W
Kepuhi, *Hawaii* **30** 21 10N 157 10W
Kerman, *Calif.* **18** 36 43N 120 4W
Kermit, *Tex.* **13** 31 52N 103 6W
Kern →, *Calif.* **19** 35 16N 119 18W
Kernville, *Calif.* **19** 35 45N 118 26W
Kerrville, *Tex.* **13** 30 3N 99 8W
Ketchikan, *Alaska* **30** 55 21N 131 39W
Ketchum, *Idaho* **16** 43 41N 114 22W
Kettle Falls, *Wash.* **16** 48 37N 118 3W
Kettleman City, *Calif.* **18** 36 1N 119 58W
Kevin, *Mont.* **16** 48 45N 111 58W
Kewanee, *Ill.* **14** 41 14N 89 56W
Kewaunee, *Wis.* **10** 44 27N 87 31W
Keweenaw B., *Mich.* **10** 47 0N 88 15W
Keweenaw Pen., *Mich.* **10** 47 30N 88 0W
Keweenaw Pt., *Mich.* **10** 47 25N 87 43W
Keyser, *W. Va.* **10** 39 26N 78 59W
Keystone, *S. Dak.* **12** 43 54N 103 25W
Kijik, *Alaska* **30** 60 20N 154 20W
Kilauea, *Hawaii* **30** 22 13N 159 25W
Kilauea Crater, *Hawaii* **30** 19 25N 155 17W
Kilbuck Mts., *Alaska* **30** 60 30N 160 0W
Kilgore, *Tex.* **13** 32 23N 94 53W
Killdeer, *N. Dak.* **12** 47 26N 102 48W
Killeen, *Tex.* **13** 31 7N 97 44W
Kim, *Colo.* **13** 37 15N 103 21W
Kimball, *Nebr.* **12** 41 14N 103 40W
Kimball, *S. Dak.* **12** 43 45N 98 57W
Kimberly, *Idaho* **16** 42 32N 114 22W
King City, *Calif.* **18** 36 13N 121 8W
King of Prussia, *Pa.* **24** 40 5N 75 22W
Kingfisher, *Okla.* **13** 35 52N 97 56W
Kingman, *Ariz.* **19** 35 12N 114 4W
Kingman, *Kans.* **13** 37 39N 98 7W
Kings →, *Calif.* **18** 36 3N 119 50W
Kings Canyon National Park, *Calif.* **18** 36 50N 118 40W
Kings Mountain, *N.C.* **11** 35 15N 81 20W
King's Peak, *Utah* **16** 40 46N 110 27W
Kingsburg, *Calif.* **18** 36 31N 119 33W
Kingsley, *Iowa* **12** 42 35N 95 58W
Kingsport, *Tenn.* **11** 36 33N 82 33W
Kingston, *N.Y.* **9** 41 56N 73 59W
Kingston, *Pa.* **9** 41 16N 75 54W
Kingston, *R.I.* **9** 41 29N 71 30W
Kingstree, *S.C.* **11** 33 40N 79 50W
Kingsville, *Tex.* **13** 27 31N 97 52W
Kinsley, *Kans.* **13** 37 55N 99 25W
Kinston, *N.C.* **11** 35 16N 77 35W
Kiowa, *Kans.* **13** 37 1N 98 29W
Kiowa, *Okla.* **13** 34 43N 95 54W
Kipnuk, *Alaska* **30** 59 56N 164 3W
Kirkland, *Ariz.* **17** 34 25N 112 43W
Kirksville, *Mo.* **14** 40 12N 92 35W
Kiska I., *Alaska* **30** 51 59N 177 30 E
Kissimmee, *Fla.* **11** 28 18N 81 24W
Kissimmee →, *Fla.* **11** 27 9N 80 52W
Kit Carson, *Colo.* **12** 38 46N 102 48W
Kittanning, *Pa.* **8** 40 49N 79 31W
Kittatinny Mts., *N.J.* **9** 41 0N 75 0W
Kittery, *Maine* **11** 43 5N 70 45W
Kivalina, *Alaska* **30** 67 44N 164 33W
Klamath →, *Calif.* **16** 41 33N 124 5W
Klamath Falls, *Oreg.* **16** 42 13N 121 46W
Klamath Mts., *Calif.* **16** 41 20N 123 0W
Klein, *Mont.* **16** 46 24N 108 33W
Klickitat, *Wash.* **16** 45 49N 121 9W
Knights Ferry, *Calif.* **18** 37 50N 120 40W
Knights Landing, *Calif.* **18** 38 48N 121 43W
Knox, *Ind.* **15** 41 18N 86 37W
Knox City, *Tex.* **13** 33 25N 99 49W
Knoxville, *Iowa* **14** 41 19N 93 6W
Knoxville, *Tenn.* **11** 35 58N 83 55W
Kobuk, *Alaska* **30** 66 55N 156 52W
Kobuk →, *Alaska* **30** 66 55N 157 0W
Kodiak, *Alaska* **30** 57 47N 152 24W
Kodiak I., *Alaska* **30** 57 30N 152 45W
Kokomo, *Ind.* **15** 40 29N 86 8W
Konawa, *Okla.* **13** 34 58N 96 45W
Kooskia, *Idaho* **16** 46 9N 115 59W
Koror, *Pac. Oc.* **31** 7 20N 134 28 E
Kosciusko, *Miss.* **13** 33 4N 89 35W
Kotzebue, *Alaska* **30** 66 53N 162 39W
Kotzebue Sound, *Alaska* **30** 66 20N 163 0W
Kountze, *Tex.* **13** 30 22N 94 19W
Koyuk, *Alaska* **30** 64 56N 161 9W
Koyukuk →, *Alaska* **30** 64 55N 157 32W
Kremmling, *Colo.* **16** 40 4N 106 24W
Kualakahi Chan, *Hawaii* **30** 22 2N 159 53W
Kuiu I., *Alaska* **30** 57 45N 134 10W
Kulm, *N. Dak.* **12** 46 18N 98 57W
Kumukahi, C., *Hawaii* **30** 19 31N 154 49W
Kupreanof I., *Alaska* **30** 56 50N 133 30W
Kuskokwim →, *Alaska* **30** 60 5N 162 25W
Kuskokwim B., *Alaska* **30** 59 45N 162 25W
Kuskokwim Mts., *Alaska* **30** 62 30N 156 0W
Kwethluk, *Alaska* **30** 60 49N 161 26W
Kwigillingok, *Alaska* **30** 59 51N 163 8W
Kwiguk, *Alaska* **30** 62 46N 164 30W
Kyburz, *Calif.* **18** 38 47N 120 18W

L

La Barge, *Wyo.* **16** 42 16N 110 12W
La Belle, *Fla.* **11** 26 46N 81 26W
La Canada, *Calif.* **29** 34 12N 118 12W
La Conner, *Wash.* **16** 48 23N 122 30W
La Crescenta, *Calif.* **29** 34 13N 118 14W
La Crosse, *Kans.* **12** 38 32N 99 18W
La Crosse, *Wis.* **12** 43 48N 91 15W
La Fayette, *Ga.* **11** 34 42N 85 17W
La Follette, *Tenn.* **11** 36 23N 84 7W
La Grande, *Oreg.* **16** 45 20N 118 5W
La Grange, *Calif.* **18** 37 42N 120 27W
La Grange, *Ga.* **11** 33 2N 85 2W
La Grange, *Ill.* **22** 41 48N 87 53W
La Grange, *Ky.* **10** 38 25N 85 23W
La Grange, *Tex.* **13** 29 54N 96 52W
La Guardia Airport, *N.Y.* **21** 40 46N 73 52W
La Habra, *Calif.* **29** 33 56N 117 57W
La Harpe, *Ill.* **14** 40 35N 90 58W
La Jara, *Colo.* **17** 37 16N 105 58W
La Junta, *Colo.* **13** 37 59N 103 33W
La Mesa, *Calif.* **19** 32 46N 117 3W
La Mesa, *N. Mex.* **17** 32 7N 106 42W
La Moure, *N. Dak.* **12** 46 21N 98 18W
La Pine, *Oreg.* **16** 43 40N 121 30W
La Plant, *S. Dak.* **12** 45 9N 100 39W
La Porte, *Ind.* **15** 41 36N 86 43W
La Push, *Wash.* **18** 47 55N 124 38W
La Salle, *Ill.* **14** 41 20N 89 6W
La Selva Beach, *Calif.* **18** 36 56N 121 51W
Laau Pt., *Hawaii* **30** 21 6N 157 19W
Lac du Flambeau, *Wis.* **12** 45 58N 89 53W
Lackawanna, *N.Y.* **8** 42 50N 78 50W
Lacona, *N.Y.* **9** 43 39N 76 10W
Laconia, *N.H.* **9** 43 32N 71 28W
Lacrosse, *Wash.* **16** 46 51N 117 58W
Ladysmith, *Wis.* **12** 45 28N 91 12W
Lafayette, *Colo.* **12** 39 58N 105 12W
Lafayette, *Ind.* **15** 40 25N 86 54W
Lafayette, *La.* **13** 30 14N 92 1W
Lafayette, *Tenn.* **11** 36 31N 86 2W
Laguna, *N. Mex.* **17** 35 2N 107 25W
Laguna Beach, *Calif.* **19** 33 33N 117 47W
Lahaina, *Hawaii* **30** 20 53N 156 41W
Lahontan Reservoir, *Nev.* **16** 39 28N 119 4W
Lake Alpine, *Calif.* **18** 38 29N 120 0W
Lake Andes, *S. Dak.* **12** 43 9N 98 32W
Lake Anse, *Mich.* **10** 46 42N 88 25W
Lake Arthur, *La.* **13** 30 5N 92 41W
Lake Charles, *La.* **13** 30 14N 93 13W
Lake City, *Colo.* **17** 38 2N 107 19W
Lake City, *Fla.* **11** 30 11N 82 38W
Lake City, *Iowa* **14** 42 16N 94 44W
Lake City, *Mich.* **10** 44 20N 85 13W
Lake City, *Minn.* **12** 44 27N 92 16W
Lake City, *S.C.* **11** 33 52N 79 45W
Lake George, *N.Y.* **9** 43 26N 73 43W
Lake Havasu City, *Ariz.* **19** 34 27N 114 22W
Lake Hiawatha, *N.J.* **20** 40 52N 74 22W
Lake Hughes, *Calif.* **19** 34 41N 118 26W
Lake Isabella, *Calif.* **19** 35 38N 118 28W
Lake Mead National Recreation Area, *Ariz.* **19** 36 15N 114 30W
Lake Mills, *Iowa* **12** 43 25N 93 32W
Lake Providence, *La.* **13** 32 48N 91 10W
Lake Village, *Ark.* **13** 33 20N 91 17W
Lake Wales, *Fla.* **11** 27 54N 81 35W
Lake Worth, *Fla.* **11** 26 37N 80 3W
Lakeland, *Fla.* **11** 28 3N 81 57W
Lakeside, *Ariz.* **17** 34 9N 109 58W
Lakeside, *Calif.* **19** 32 52N 116 55W
Lakeside, *Nebr.* **12** 42 3N 102 26W
Lakeview, *Oreg.* **16** 42 11N 120 21W
Lakewood, *Colo.* **12** 39 44N 105 5W
Lakewood, *N.J.* **9** 40 6N 74 13W
Lakewood, *Ohio* **27** 41 29N 81 49W
Lakin, *Kans.* **13** 37 57N 101 15W
Lakota, *N. Dak.* **12** 48 2N 98 21W
Lamar, *Colo.* **12** 38 5N 102 37W
Lamar, *Mo.* **13** 37 30N 94 16W
Lambert, *Mont.* **12** 47 41N 104 37W
Lame Deer, *Mont.* **16** 45 37N 106 40W
Lamesa, *Tex.* **13** 32 44N 101 58W
Lamont, *Calif.* **19** 35 15N 118 55W
Lampasas, *Tex.* **13** 31 4N 98 11W
Lamy, *N. Mex.* **17** 35 29N 105 53W
Lanai City, *Hawaii* **30** 20 50N 156 55W
Lanai I., *Hawaii* **30** 20 50N 156 55W
Lancaster, *Calif.* **19** 34 42N 118 8W
Lancaster, *Ky.* **10** 37 37N 84 35W
Lancaster, *N.H.* **9** 44 29N 71 34W
Lancaster, *Pa.* **9** 40 2N 76 19W
Lancaster, *S.C.* **11** 34 43N 80 46W
Lancaster, *Wis.* **14** 42 51N 90 43W
Lander, *Wyo.* **16** 42 50N 108 44W
Lanesboro, *Pa.* **9** 41 57N 75 34W
Lanett, *Ala.* **11** 32 52N 85 12W
Langdon, *N. Dak.* **12** 48 45N 98 22W
Langley Park, *Md.* **25** 38 59N 76 58W
Langlois, *Oreg.* **16** 42 56N 124 27W
Langtry, *Tex.* **13** 29 49N 101 34W
Lansdale, *Pa.* **9** 40 14N 75 17W
Lansdowne, *Md.* **25** 39 14N 76 39W
Lansdowne, *Pa.* **24** 39 56N 75 15W
Lansford, *Pa.* **9** 40 50N 75 53W
Lansing, *Mich.* **15** 42 44N 84 33W
Laona, *Wis.* **10** 45 34N 88 40W
Lapeer, *Mich.* **15** 43 3N 83 19W
Laporte, *Pa.* **9** 41 25N 76 30W
Laramie, *Wyo.* **12** 41 19N 105 35W
Laramie Mts., *Wyo.* **12** 42 0N 105 30W
Larchmont, *N.Y.* **21** 40 55N 73 44W
Laredo, *Tex.* **13** 27 30N 99 30W
Larimore, *N. Dak.* **12** 47 54N 97 38W
Larkspur, *Calif.* **28** 37 56N 122 32W
Larned, *Kans.* **12** 38 11N 99 6W
Las Animas, *Colo.* **12** 38 4N 103 13W
Las Cruces, *N. Mex.* **17** 32 19N 106 47W
Las Vegas, *N. Mex.* **17** 35 36N 105 13W
Las Vegas, *Nev.* **19** 36 10N 115 9W
Lassen Pk., *Wash.* **16** 40 29N 121 31W
Lathrop Wells, *Nev.* **19** 36 39N 116 24W
Laton, *Calif.* **18** 36 26N 119 41W
Laurel, *Miss.* **13** 31 41N 89 8W
Laurel, *Mont.* **16** 45 40N 108 46W
Laurens, *S.C.* **11** 34 30N 82 1W
Laurinburg, *N.C.* **11** 34 47N 79 28W
Laurium, *Mich.* **10** 47 14N 88 27W
Lava Hot Springs, *Idaho* **16** 42 37N 112 1W
Laverne, *Okla.* **13** 36 43N 99 54W
Lawrence, *Kans.* **12** 38 58N 95 14W
Lawrence, *Mass.* **23** 42 43N 71 7W
Lawrenceburg, *Ind.* **15** 39 6N 84 52W
Lawrenceburg, *Tenn.* **11** 35 14N 87 20W
Lawrenceville, *Ga.* **11** 33 57N 83 59W
Laws, *Calif.* **18** 37 24N 118 20W
Lawton, *Okla.* **13** 34 37N 98 25W
Laytonville, *Calif.* **16** 39 41N 123 29W
Le Mars, *Iowa* **12** 42 47N 96 10W
Le Roy, *Kans.* **13** 38 5N 95 38W
Le Sueur, *Minn.* **12** 44 28N 93 55W
Lead, *S. Dak.* **12** 44 21N 103 46W
Leadville, *Colo.* **17** 39 15N 106 18W
Leaf →, *Miss.* **13** 30 59N 88 44W
Leakey, *Tex.* **13** 29 44N 99 46W
Leamington, *Utah* **16** 39 32N 112 17W
Leavenworth, *Kans.* **12** 39 19N 94 55W
Leavenworth, *Wash.* **16** 47 36N 120 40W
Lebanon, *Ind.* **15** 40 3N 86 28W
Lebanon, *Kans.* **12** 39 49N 98 33W
Lebanon, *Ky.* **10** 37 34N 85 15W
Lebanon, *Mo.* **14** 37 41N 92 40W
Lebanon, *Oreg.* **16** 44 32N 122 55W
Lebanon, *Pa.* **9** 40 20N 76 26W
Lebanon, *Tenn.* **11** 36 12N 86 18W
Lebec, *Calif.* **19** 34 50N 118 52W
Lee Vining, *Calif.* **18** 37 58N 119 7W
Leech L., *Minn.* **12** 47 10N 94 24W
Leedey, *Okla.* **13** 35 52N 99 21W
Leeds, *Ala.* **11** 33 33N 86 33W
Leesburg, *Fla.* **11** 28 49N 81 53W
Leesville, *La.* **13** 31 9N 93 16W
Lefors, *Tex.* **13** 35 26N 100 48W
Lehi, *Utah* **16** 40 24N 111 51W
Lehighton, *Pa.* **9** 40 50N 75 43W
Lehua I., *Hawaii* **30** 22 1N 160 6W
Leland, *Miss.* **13** 33 24N 90 54W
Lemhi Ra., *Idaho* **16** 44 30N 113 30W
Lemmon, *S. Dak.* **12** 45 57N 102 10W
Lemon Grove, *Calif.* **19** 32 45N 117 2W
Lemoore, *Calif.* **18** 36 18N 119 46W
Lennox, *Calif.* **29** 33 56N 118 21W
Lenoir, *N.C.* **11** 35 55N 81 32W
Lenoir City, *Tenn.* **11** 35 48N 84 16W
Lenora, *Kans.* **12** 39 37N 100 0W
Lenox, *Mass.* **9** 42 22N 73 17W
Lenwood, *Calif.* **19** 34 53N 117 7W
Leola, *S. Dak.* **12** 45 43N 98 56W
Leominster, *Mass.* **9** 42 32N 71 46W
Leon, *Iowa* **14** 40 44N 93 45W
Leonardtown, *Md.* **10** 38 17N 76 38W
Leoti, *Kans.* **12** 38 29N 101 21W
Leslie, *Ark.* **13** 35 50N 92 34W
Leucadia, *Calif.* **19** 33 4N 117 18W
Levan, *Utah* **16** 39 33N 111 52W
Levelland, *Tex.* **13** 33 35N 102 23W
Levittown, *N.Y.* **21** 40 43N 73 31W
Levittown, *Pa.* **9** 40 9N 74 51W
Lewellen, *Nebr.* **12** 41 20N 102 9W
Lewes, *Del.* **10** 38 46N 75 9W
Lewis Range, *Mont.* **16** 48 5N 113 5W
Lewisburg, *Pa.* **8** 40 58N 76 54W
Lewisburg, *Tenn.* **11** 35 27N 86 48W
Lewiston, *Idaho* **16** 46 25N 117 1W
Lewiston, *Maine* **11** 44 6N 70 13W
Lewistown, *Mont.* **16** 47 4N 109 26W
Lewistown, *Pa.* **8** 40 36N 77 34W
Lexington, *Ill.* **15** 40 39N 88 47W
Lexington, *Ky.* **15** 38 3N 84 30W
Lexington, *Mass.* **23** 42 26N 71 13W
Lexington, *Miss.* **13** 33 7N 90 3W
Lexington, *Mo.* **14** 39 11N 93 52W
Lexington, *N.C.* **11** 35 49N 80 15W
Lexington, *Nebr.* **12** 40 47N 99 45W
Lexington, *Oreg.* **16** 45 27N 119 42W
Lexington, *Tenn.* **11** 35 39N 88 24W
Lexington Park, *Md.* **10** 38 16N 76 27W
Libby, *Mont.* **16** 48 23N 115 33W
Liberal, *Kans.* **13** 37 3N 100 55W
Liberal, *Mo.* **13** 37 34N 94 31W
Liberty, *Mo.* **14** 39 15N 94 25W
Liberty, *Tex.* **13** 30 3N 94 48W
Lida, *Nev.* **17** 37 28N 117 30W
Lihue, *Hawaii* **30** 21 59N 159 23W
Lima, *Mont.* **16** 44 38N 112 36W
Lima, *Ohio* **15** 40 44N 84 6W
Limon, *Colo.* **12** 39 16N 103 41W
Lincoln, *Ill.* **14** 40 9N 89 22W
Lincoln, *Kans.* **12** 39 3N 98 9W
Lincoln, *Maine* **11** 45 22N 68 30W
Lincoln, *N. Mex.* **17** 33 30N 105 23W
Lincoln, *Nebr.* **12** 40 49N 96 41W
Lincoln Park, *Mich.* **26** 42 14N 83 9W
Lincolnton, *N.C.* **11** 35 29N 81 16W
Lincolnwood, *Ill.* **22** 42 1N 87 45W
Lind, *Wash.* **16** 46 58N 118 37W
Linden, *Calif.* **18** 38 1N 121 5W
Linden, *N.J.* **20** 40 38N 74 14W
Linden, *Tex.* **13** 33 1N 94 22W
Lindsay, *Calif.* **18** 36 12N 119 5W
Lindsay, *Okla.* **13** 34 50N 97 38W
Lindsborg, *Kans.* **12** 38 35N 97 40W
Lingle, *Wyo.* **12** 42 8N 104 21W
Linthicum Heights, *Md.* **25** 39 12N 76 41W
Linton, *Ind.* **15** 39 2N 87 10W
Linton, *N. Dak.* **12** 46 16N 100 14W
Lipscomb, *Tex.* **13** 36 14N 100 16W
Lisbon, *N. Dak.* **12** 46 27N 97 41W
Lisburne, C., *Alaska* **30** 68 53N 166 13W
Litchfield, *Conn.* **9** 41 45N 73 11W
Litchfield, *Ill.* **14** 39 11N 89 39W
Litchfield, *Minn.* **12** 45 8N 94 32W
Little Belt Mts., *Mont.* **16** 46 40N 110 45W
Little Blue →, *Kans.* **13** 39 42N 96 41W
Little Colorado →, *Ariz.* **17** 36 12N 111 48W
Little Falls, *Minn.* **12** 45 59N 94 22W
Little Falls, *N.Y.* **9** 43 3N 74 51W
Little Fork →, *Minn.* **12** 48 31N 93 35W
Little Humboldt →, *Nev.* **16** 41 1N 117 43W
Little Lake, *Calif.* **19** 35 56N 117 55W
Little Missouri →, *N. Dak.* **12** 47 36N 102 25W
Little Red →, *Ark.* **13** 35 11N 91 27W
Little Rock, *Ark.* **13** 34 45N 92 17W
Little Sable Pt., *Mich.* **10** 43 38N 86 33W
Little Sioux →, *Iowa* **12** 41 48N 96 4W
Little Snake →, *Colo.* **16** 40 27N 108 26W
Little Wabash →, *Ill.* **15** 37 55N 88 5W
Littlefield, *Tex.* **13** 33 55N 102 20W
Littlefork, *Minn.* **12** 48 24N 93 34W
Littleton, *N.H.* **9** 44 18N 71 46W
Live Oak, *Fla.* **11** 30 18N 82 59W
Livermore, *Calif.* **18** 37 41N 121 47W
Livermore, Mt., *Tex.* **13** 30 38N 104 11W
Livingston, *Calif.* **18** 37 23N 120 43W
Livingston, *Mont.* **16** 45 40N 110 34W
Livingston, *N.J.* **20** 40 47N 74 18W
Livingston, *Tex.* **13** 30 43N 94 56W
Livonia, *Mich.* **26** 42 24N 83 22W
Llano, *Tex.* **13** 30 45N 98 41W
Llano →, *Tex.* **13** 30 39N 98 26W
Llano Estacado, *Tex.* **13** 33 30N 103 0W
Loa, *Utah* **17** 38 24N 111 39W
Lock Haven, *Pa.* **8** 41 8N 77 28W
Lockeford, *Calif.* **18** 38 10N 121 9W
Lockhart, *Tex.* **13** 29 53N 97 40W
Lockney, *Tex.* **13** 34 7N 101 27W
Lockport, *N.Y.* **8** 43 10N 78 42W
Lodge Grass, *Mont.* **16** 45 19N 107 22W
Lodgepole, *Nebr.* **12** 41 9N 102 38W
Lodgepole Cr. →, *Wyo.* **12** 41 20N 104 30W
Lodi, *Calif.* **18** 38 8N 121 16W
Lodi, *N.J.* **20** 40 52N 74 5W
Logan, *Kans.* **12** 39 40N 99 34W
Logan, *Ohio* **10** 39 32N 82 25W
Logan, *Utah* **16** 41 44N 111 50W
Logan, *W. Va.* **10** 37 51N 81 59W
Logan International Airport, *Mass.* **23** 42 21N 71 0W
Logansport, *Ind.* **15** 40 45N 86 22W
Logansport, *La.* **13** 31 58N 94 0W
Lolo, *Mont.* **16** 46 45N 114 5W
Loma, *Mont.* **16** 47 56N 110 30W
Loma Linda, *Calif.* **19** 34 3N 117 16W
Lometa, *Tex.* **13** 31 13N 98 24W
Lompoc, *Calif.* **19** 34 38N 120 28W
London, *Ky.* **10** 37 8N 84 5W
London, *Ohio* **15** 39 53N 83 27W
Lone Pine, *Calif.* **18** 36 36N 118 4W
Long Beach, *Calif.* **19** 33 47N 118 11W
Long Beach, *N.Y.* **21** 40 35N 73 40W
Long Beach, *Wash.* **18** 46 21N 124 3W

Long Branch, *N.J.* **9** 40 18N 74 0W
Long Creek, *Oreg.* **16** 44 43N 119 6W
Long I., *N.Y.* **9** 40 45N 73 30W
Long Island Sd., *N.Y.* . . **9** 41 10N 73 0W
Long Pine, *Nebr.* **12** 42 32N 99 42W
Longmont, *Colo.* **12** 40 10N 105 6W
Longview, *Tex.* **13** 32 30N 94 44W
Longview, *Wash.* **18** 46 8N 122 57W
Lonoke, *Ark.* **13** 34 47N 91 54W
Lookout, C., *N.C.* **11** 34 35N 76 32W
Lorain, *Ohio* **8** 41 28N 82 11W
Lordsburg, *N. Mex.* **17** 32 21N 108 43W
Los Alamos, *Calif.* **19** 34 44N 120 17W
Los Alamos, *N. Mex.* . . **17** 35 53N 106 19W
Los Altos, *Calif.* **18** 37 23N 122 7W
Los Angeles, *Calif.* **29** 34 3N 118 13W
Los Angeles Aqueduct,
Calif. **19** 35 22N 118 5W
Los Angeles
International Airport,
Calif. **29** 33 56N 118 23W
Los Banos, *Calif.* **18** 37 4N 120 51W
Los Lunas, *N. Mex.* **17** 34 48N 106 44W
Los Olivos, *Calif.* **19** 34 40N 120 7W
Loudon, *Tenn.* **11** 35 45N 84 20W
Louisa, *Ky.* **10** 38 7N 82 36W
Louisiana, *Mo.* **14** 39 27N 91 3W
Louisiana □ **13** 30 50N 92 0W
Louisville, *Ky.* **15** 38 15N 85 46W
Louisville, *Miss.* **13** 33 7N 89 3W
Loup City, *Nebr.* **12** 41 17N 98 58W
Loveland, *Colo.* **12** 40 24N 105 5W
Lovell, *Wyo.* **16** 44 50N 108 24W
Lovelock, *Nev.* **16** 40 11N 118 28W
Loving, *N. Mex.* **13** 32 17N 104 6W
Lovington, *N. Mex.* **13** 32 57N 103 21W
Lowell, *Mass.* **23** 42 38N 71 16W
Lower L., *Calif.* **16** 41 16N 120 2W
Lower Lake, *Calif.* **18** 38 55N 122 37W
Lower Red L., *Minn.* . . . **12** 47 58N 95 0W
Lowville, *N.Y.* **9** 43 47N 75 29W
Lubbock, *Tex.* **13** 33 35N 101 51W
Lucedale, *Miss.* **11** 30 56N 88 35W
Lucerne Valley, *Calif.* . . **19** 34 27N 116 57W
Ludington, *Mich.* **10** 43 57N 86 27W
Ludlow, *Calif.* **19** 34 43N 116 10W
Ludlow, *Vt.* **9** 43 24N 72 42W
Lufkin, *Tex.* **13** 31 21N 94 44W
Luling, *Tex.* **13** 29 41N 97 39W
Luma, *Amer. Samoa* . . . **31** 14 15S 169 32W
Lumberton, *Miss.* **13** 31 0N 89 27W
Lumberton, *N.C.* **11** 34 37N 79 0W
Lumberton, *N. Mex.* . . . **17** 36 56N 106 56W
Lund, *Nev.* **16** 38 52N 115 0W
Luning, *Nev.* **16** 38 30N 118 11W
Luray, *Va.* **10** 38 40N 78 28W
Lusk, *Wyo.* **12** 42 46N 104 27W
Lutherville-Timonium,
Md. **25** 39 25N 76 36W
Luverne, *Minn.* **12** 43 39N 96 13W
Lyman, *Wyo.* **16** 41 20N 110 18W
Lynchburg, *Va.* **10** 37 25N 79 9W
Lynden, *Wash.* **18** 48 57N 122 27W
Lyndhurst, *N.J.* **20** 40 49N 74 8W
Lynn, *Mass.* **23** 42 28N 70 57W
Lynwood, *Calif.* **29** 33 55N 118 12W
Lyons, *Colo.* **12** 40 14N 105 16W
Lyons, *Ga.* **11** 32 12N 82 19W
Lyons, *Ill.* **22** 41 48N 87 49W
Lyons, *Kans.* **12** 38 21N 98 12W
Lyons, *N.Y.* **8** 43 5N 77 0W
Lytle, *Tex.* **13** 29 14N 98 48W

M

Mabton, *Wash.* **16** 46 13N 120 0W
McAlester, *Okla.* **13** 34 56N 95 46W
McAllen, *Tex.* **13** 26 12N 98 14W
McCall, *Idaho* **16** 44 55N 116 6W
McCamey, *Tex.* **13** 31 8N 102 14W
McCammon, *Idaho* **16** 42 39N 112 12W
McCarthy, *Alaska* **30** 61 26N 142 56W
McCloud, *Calif.* **16** 41 15N 122 8W
McClure, L., *Calif.* **18** 37 35N 120 16W
McClusky, *N. Dak.* **12** 47 29N 100 27W
McComb, *Miss.* **13** 31 15N 90 27W
McConaughy, L., *Nebr.* . **12** 41 14N 101 40W
McCook, *Nebr.* **12** 40 12N 100 38W
McDermitt, *Nev.* **16** 41 59N 117 43W
McFarland, *Calif.* **19** 35 41N 119 14W
McGehee, *Ark.* **13** 33 38N 91 24W
McGill, *Nev.* **16** 39 23N 114 47W
McGregor, *Iowa* **14** 43 1N 91 11W
Machias, *Maine* **11** 44 43N 67 28W
McIntosh, *S. Dak.* **12** 45 55N 101 21W
Mackay, *Idaho* **16** 43 55N 113 37W
McKees Rocks, *Pa.* **27** 40 28N 80 3W
McKeesport, *Pa.* **27** 40 21N 79 51W
McKenzie, *Tenn.* **11** 36 8N 88 31W
McKenzie →, *Oreg.* . . . **16** 44 7N 123 6W
Mackinaw City, *Mich.* . . . **10** 45 47N 84 44W
McKinley, Mt., *Alaska* . . **30** 63 4N 151 0W
McKinney, *Tex.* **13** 33 12N 96 37W
McLaughlin, *S. Dak.* . . . **12** 45 49N 100 49W
McLean, *Tex.* **13** 35 14N 100 36W
McLean, *Va.* **25** 38 56N 77 10W
McLeansboro, *Ill.* **15** 38 6N 88 32W
McLoughlin, Mt., *Oreg.* . **16** 42 27N 122 19W
McMillan, L., *N. Mex.* . . . **13** 32 36N 104 21W
McMinnville, *Oreg.* **16** 45 13N 123 12W
McMinnville, *Tenn.* **11** 35 41N 85 46W
McNary, *Ariz.* **17** 34 4N 109 51W
Macomb, *Ill.* **14** 40 27N 90 40W
Macon, *Ga.* **11** 32 51N 83 38W
Macon, *Miss.* **11** 33 7N 88 34W
Macon, *Mo.* **14** 39 44N 92 28W
McPherson, *Kans.* **12** 38 22N 97 40W
McPherson Pk., *Calif.* . . **19** 34 53N 119 53W
McVille, *N. Dak.* **12** 47 46N 98 11W
Madera, *Calif.* **18** 36 57N 120 3W
Madill, *Okla.* **13** 34 6N 96 46W
Madison, *Fla.* **11** 30 28N 83 25W
Madison, *Ind.* **15** 38 44N 85 23W
Madison, *N.J.* **20** 40 45N 74 25W
Madison, *Nebr.* **12** 41 50N 97 27W
Madison, *S. Dak.* **12** 44 0N 97 7W
Madison, *Wis.* **14** 43 4N 89 24W
Madison →, *Mont.* **16** 45 56N 111 31W
Madison Heights, *Mich.* **26** 42 29N 83 6W
Madisonville, *Ky.* **10** 37 20N 87 30W
Madisonville, *Tex.* **13** 30 57N 95 55W
Madras, *Oreg.* **16** 44 38N 121 8W
Madre, Laguna, *Tex.* . . . **13** 27 0N 97 30W
Magdalena, *N. Mex.* . . . **17** 34 7N 107 15W
Magee, *Miss.* **13** 31 52N 89 44W
Magnolia, *Ark.* **13** 33 16N 93 14W
Magnolia, *Miss.* **13** 31 9N 90 28W
Mahanoy City, *Pa.* **9** 40 49N 76 9W
Mahnomen, *Minn.* **12** 47 19N 95 58W
Mahukona, *Hawaii* **30** 20 11N 155 52W
Maine □ **11** 45 20N 69 0W
Makapuu Hd., *Hawaii* . . **30** 21 19N 157 39W
Makena, *Hawaii* **30** 20 39N 156 27W
Malad City, *Idaho* **16** 42 12N 112 15W
Malaga, *N. Mex.* **13** 32 14N 104 4W
Malakoff, *Tex.* **13** 32 10N 96 1W
Malden, *Mass.* **23** 42 26N 71 3W
Malden, *Mo.* **13** 36 34N 89 57W
Malheur →, *Oreg.* **16** 44 4N 116 59W
Malheur L., *Oreg.* **16** 43 20N 118 48W
Malibu, *Calif.* **19** 34 2N 118 41W
Malone, *N.Y.* **9** 44 51N 74 18W
Malta, *Idaho* **16** 42 18N 113 22W
Malta, *Mont.* **16** 48 21N 107 52W
Malvern, *Ark.* **13** 34 22N 92 49W
Malvern, *Pa.* **24** 40 2N 75 31W
Mammoth, *Ariz.* **17** 32 43N 110 39W
Mana, *Hawaii* **30** 22 2N 159 47W
Manasquan, *N.J.* **9** 40 8N 74 3W
Manassa, *Colo.* **17** 37 11N 105 56W
Manatí, *Puerto Rico* . . . **30** 18 26N 66 29W
Mancelona, *Mich.* **10** 44 54N 85 4W
Manchester, *Conn.* **9** 41 47N 72 31W
Manchester, *Ga.* **11** 32 51N 84 37W
Manchester, *Iowa* **14** 42 29N 91 27W
Manchester, *Ky.* **10** 37 9N 83 46W
Manchester, *N.H.* **9** 42 59N 71 28W
Mandan, *N. Dak.* **12** 46 50N 100 54W
Mangum, *Okla.* **13** 34 53N 99 30W
Manhasset, *N.Y.* **21** 40 47N 73 39W
Manhattan, *Kans.* **12** 39 11N 96 35W
Manhattan, *N.Y.* **20** 40 48N 73 57W
Manhattan Beach, *Calif.* **29** 33 53N 118 25W
Manila, *Utah* **16** 40 59N 109 43W
Manistee, *Mich.* **10** 44 15N 86 19W
Manistee →, *Mich.* . . . **10** 44 15N 86 21W
Manistique, *Mich.* **10** 45 57N 86 15W
Manitou Is., *Mich.* **10** 45 8N 86 0W
Manitou Springs, *Colo.* . **12** 38 52N 104 55W
Manitowoc, *Wis.* **10** 44 5N 87 40W
Mankato, *Kans.* **12** 39 47N 98 13W
Mankato, *Minn.* **12** 44 10N 94 0W
Manning, *S.C.* **11** 33 42N 80 13W
Mannington, *W. Va.* . . . **10** 39 32N 80 21W
Mansfield, *La.* **13** 32 2N 93 43W
Mansfield, *Mass.* **9** 42 2N 71 13W
Mansfield, *Ohio* **8** 40 45N 82 31W
Mansfield, *Pa.* **8** 41 48N 77 5W
Mansfield, *Wash.* **16** 47 49N 119 38W
Manteca, *Calif.* **18** 37 48N 121 13W
Manteo, *N.C.* **11** 35 55N 75 40W
Manti, *Utah* **16** 39 16N 111 38W
Manton, *Mich.* **10** 44 25N 85 24W
Manua Is., *Amer. Samoa* **31** 14 13S 169 35W
Manville, *Wyo.* **12** 42 47N 104 37W
Many, *La.* **13** 31 34N 93 29W
Manzano Mts., *N. Mex.* **17** 34 40N 106 20W
Maple Heights, *Ohio* . . . **27** 41 25N 81 33W
Mapleton, *Oreg.* **16** 44 2N 123 52W
Maplewood, *N.J.* **20** 40 43N 74 16W
Maquoketa, *Iowa* **14** 42 4N 90 40W
Marana, *Ariz.* **17** 32 27N 111 13W
Marathon, *N.Y.* **9** 42 27N 76 2W
Marathon, *Tex.* **13** 30 12N 103 15W
Marble Falls, *Tex.* **13** 30 35N 98 16W
Marblehead, *Mass.* **23** 42 29N 70 51W
Marengo, *Iowa* **14** 41 48N 92 4W
Marfa, *Tex.* **13** 30 19N 104 1W
Mariana Trench, *Pac. Oc.* **31** 13 0N 145 0 E
Marianna, *Ark.* **13** 34 46N 90 46W
Marianna, *Fla.* **11** 30 46N 85 14W
Marias →, *Mont.* **16** 47 56N 110 30W
Maricopa, *Ariz.* **17** 33 4N 112 3W
Maricopa, *Calif.* **19** 35 4N 119 24W
Marietta, *Ga.* **11** 33 57N 84 33W
Marietta, *Ohio* **10** 39 25N 81 27W
Marin City, *Calif.* **28** 37 52N 122 30W
Marina, *Calif.* **18** 36 41N 121 48W
Marine City, *Mich.* **10** 42 43N 82 30W
Marinette, *Wis.* **10** 45 6N 87 38W
Marion, *Ala.* **11** 32 38N 87 19W
Marion, *Ill.* **14** 37 44N 88 56W
Marion, *Ind.* **15** 40 32N 85 40W
Marion, *Iowa* **14** 42 2N 91 36W
Marion, *Kans.* **12** 38 21N 97 1W
Marion, *Mich.* **10** 44 6N 85 9W
Marion, *N.C.* **11** 35 41N 82 1W
Marion, *Ohio* **15** 40 35N 83 8W
Marion, *S.C.* **11** 34 11N 79 24W
Marion, *Va.* **11** 36 50N 81 31W
Marion, L., *S.C.* **11** 33 28N 80 10W
Mariposa, *Calif.* **18** 37 29N 119 58W
Marked Tree, *Ark.* **13** 35 32N 90 25W
Markham, *Ill.* **22** 41 35N 87 41W
Markleeville, *Calif.* **18** 38 42N 119 47W
Marksville, *La.* **13** 31 8N 92 4W
Marlboro, *Mass.* **9** 42 19N 71 33W
Marlin, *Tex.* **13** 31 18N 96 54W
Marlow, *Okla.* **13** 34 39N 97 58W
Marmarth, *N. Dak.* **12** 46 18N 103 54W
Marple, *Pa.* **24** 39 56N 75 21W
Marquette, *Mich.* **10** 46 33N 87 24W
Marsh I., *La.* **13** 29 34N 91 53W
Marsh L., *Minn.* **12** 45 5N 96 0W
Marshall, *Ark.* **13** 35 55N 92 38W
Marshall, *Mich.* **15** 42 16N 84 58W
Marshall, *Minn.* **12** 44 25N 95 45W
Marshall, *Mo.* **14** 39 7N 93 12W
Marshall, *Tex.* **13** 32 33N 94 23W
Marshall Is. ■, *Pac. Oc.* **31** 9 0N 171 0 E
Marshalltown, *Iowa* . . . **14** 42 3N 92 55W
Marshfield, *Mo.* **13** 37 15N 92 54W
Marshfield, *Wis.* **12** 44 40N 90 10W
Mart, *Tex.* **13** 31 33N 96 50W
Martha's Vineyard, *Mass.* **9** 41 25N 70 38W
Martin, *S. Dak.* **12** 43 11N 101 44W
Martin, *Tenn.* **13** 36 21N 88 51W
Martin L., *Ala.* **11** 32 41N 85 55W
Martin State National
Airport, *Md.* **25** 39 19N 76 25W
Martinez, *Calif.* **18** 38 1N 122 8W
Martinsburg, *W. Va.* . . . **10** 39 27N 77 58W
Martinsville, *Ind.* **15** 39 26N 86 25W
Martinsville, *Va.* **11** 36 41N 79 52W
Maryland □ **10** 39 0N 76 30W
Marysvale, *Utah* **17** 38 27N 112 14W
Marysville, *Calif.* **18** 39 9N 121 35W
Marysville, *Kans.* **12** 39 51N 96 39W
Marysville, *Ohio* **15** 40 14N 83 22W
Maryville, *Tenn.* **11** 35 46N 83 58W
Mason, *Nev.* **18** 38 56N 119 8W
Mason, *Tex.* **13** 30 45N 99 14W
Mason City, *Iowa* **14** 43 9N 93 12W
Massachusetts □ **9** 42 30N 72 0W
Massachusetts B., *Mass.* **9** 42 20N 70 50W
Massapequa, *N.Y.* **21** 40 41N 73 28W
Massena, *N.Y.* **9** 44 56N 74 54W
Massillon, *Ohio* **8** 40 48N 81 32W
Matagorda, *Tex.* **13** 28 42N 95 58W
Matagorda B., *Tex.* **13** 28 40N 96 0W
Matagorda I., *Tex.* **13** 28 15N 96 30W
Mathis, *Tex.* **13** 28 6N 97 50W
Mattawamkeag, *Maine* . **11** 45 32N 68 21W
Mattituck, *N.Y.* **9** 40 59N 72 32W
Maui, *Hawaii* **30** 20 48N 156 20W
Maumee, *Ohio* **15** 41 34N 83 39W
Maumee →, *Ohio* **15** 41 42N 83 28W
Mauna Kea, *Hawaii* **30** 19 50N 155 28W
Mauna Loa, *Hawaii* **30** 19 30N 155 35W
Maupin, *Oreg.* **16** 45 11N 121 5W
Maurepas, L., *La.* **13** 30 15N 90 30W
Mauston, *Wis.* **12** 43 48N 90 5W
Max, *N. Dak.* **12** 47 49N 101 18W
Mayagüez, *Puerto Rico* . **30** 18 12N 67 9W
Maybell, *Colo.* **16** 40 31N 108 5W
Mayer, *Ariz.* **17** 34 24N 112 14W
Mayfield, *Ky.* **11** 36 44N 88 38W
Mayfield Heights, *Ohio* . **27** 41 31N 81 28W
Mayhill, *N. Mex.* **17** 32 53N 105 29W
Maysville, *Ky.* **15** 38 39N 83 46W
Mayville, *N. Dak.* **12** 47 30N 97 20W
Maywood, *Calif.* **29** 33 59N 118 12W
Maywood, *Ill.* **22** 41 52N 87 52W
McGrath, *Alaska* **30** 62 58N 155 40W
Mead, L., *Ariz.* **19** 36 1N 114 44W
Meade, *Kans.* **13** 37 17N 100 20W
Meade River = Atkasuk,
Alaska **30** 70 30N 157 20W
Meadow Valley
Wash →, *Nev.* **19** 36 40N 114 34W
Meadville, *Pa.* **8** 41 39N 80 9W
Meares, C., *Oreg.* **16** 45 37N 124 0W
Mecca, *Calif.* **19** 33 34N 116 5W
Mechanicsburg, *Pa.* **8** 40 13N 77 1W
Mechanicville, *N.Y.* **9** 42 54N 73 41W
Medford, *Mass.* **23** 42 25N 71 7W
Medford, *Oreg.* **16** 42 19N 122 52W
Medford, *Wis.* **12** 45 9N 90 20W
Media, *Pa.* **24** 39 55N 75 23W
Medical Lake, *Wash.* . . . **16** 47 34N 117 41W
Medicine Bow, *Wyo.* . . . **16** 41 54N 106 12W
Medicine Bow Pk., *Wyo.* **16** 41 21N 106 19W
Medicine Bow Ra., *Wyo.* **16** 41 10N 106 25W
Medicine Lake, *Mont.* . . **12** 48 30N 104 30W
Medicine Lodge, *Kans.* . **13** 37 17N 98 35W
Medina, *N. Dak.* **12** 46 54N 99 18W
Medina, *N.Y.* **8** 43 13N 78 23W
Medina, *Ohio* **8** 41 8N 81 52W
Medina →, *Tex.* **13** 29 16N 98 29W
Medina L., *Tex.* **13** 29 32N 98 56W
Meeker, *Colo.* **16** 40 2N 107 55W
Meeteetse, *Wyo.* **16** 44 9N 108 52W
Mekoryok, *Alaska* **30** 60 20N 166 20W
Melbourne, *Fla.* **11** 28 5N 80 37W
Mellen, *Wis.* **12** 46 20N 90 40W
Mellette, *S. Dak.* **12** 45 9N 98 30W
Melrose, *Mass.* **23** 42 27N 71 2W
Melrose, *N. Mex.* **13** 34 26N 103 38W
Melrose Park, *Ill.* **22** 41 53N 87 53W
Melstone, *Mont.* **16** 46 36N 107 52W
Memphis, *Tenn.* **13** 35 8N 90 3W
Memphis, *Tex.* **13** 34 44N 100 33W
Mena, *Ark.* **13** 34 35N 94 15W
Menard, *Tex.* **13** 30 55N 99 47W
Menasha, *Wis.* **10** 44 13N 88 26W
Mendenhall, C., *Alaska* . **30** 59 45N 166 10W
Mendocino, *Calif.* **16** 39 19N 123 48W
Mendocino, C., *Calif.* . . . **16** 40 26N 124 25W
Mendota, *Calif.* **18** 36 45N 120 23W
Mendota, *Ill.* **14** 41 33N 89 7W
Menlo Park, *Calif.* **28** 37 26N 122 11W
Menominee, *Mich.* **10** 45 6N 87 37W
Menominee →, *Wis.* . . **10** 45 6N 87 36W
Menomonie, *Wis.* **12** 44 53N 91 55W
Mer Rouge, *La.* **13** 32 47N 91 48W
Merced, *Calif.* **18** 37 18N 120 29W
Merced Pk., *Calif.* **18** 37 36N 119 24W
Meredith, L., *Tex.* **13** 35 43N 101 33W
Meriden, *Conn.* **9** 41 32N 72 48W
Meridian, *Idaho* **16** 43 37N 116 24W
Meridian, *Miss.* **11** 32 22N 88 42W
Meridian, *Tex.* **13** 31 56N 97 39W
Merkel, *Tex.* **13** 32 28N 100 1W
Merrick, *N.Y.* **21** 40 39N 73 32W
Merrill, *Oreg.* **16** 42 1N 121 36W
Merrill, *Wis.* **12** 45 11N 89 41W
Merriman, *Nebr.* **12** 42 55N 101 42W
Merryville, *La.* **13** 30 45N 93 33W
Mertzon, *Tex.* **13** 31 16N 100 49W
Mesa, *Ariz.* **17** 33 25N 111 50W
Meshoppen, *Pa.* **9** 41 36N 76 3W
Mesick, *Mich.* **10** 44 24N 85 43W
Mesilla, *N. Mex.* **17** 32 16N 106 48W
Mesquite, *Nev.* **17** 36 47N 114 6W
Metairie, *La.* **13** 29 58N 90 10W
Metaline Falls, *Wash.* . . **16** 48 52N 117 22W
Methuen, *Mass.* **23** 42 43N 71 12W
Metlakatla, *Alaska* **30** 55 8N 131 35W
Metropolis, *Ill.* **13** 37 9N 88 44W
Metropolitan Oakland
International Airport,
Calif. **28** 37 43N 122 13W
Mexia, *Tex.* **13** 31 41N 96 29W
Mexico, *Mo.* **14** 39 10N 91 53W
Miami, *Ariz.* **17** 33 24N 110 52W
Miami, *Fla.* **11** 25 47N 80 11W
Miami, *Tex.* **13** 35 42N 100 38W
Miami →, *Ohio* **10** 39 20N 84 40W
Miami Beach, *Fla.* **11** 25 47N 80 8W
Miamisburg, *Ohio* **15** 39 38N 84 17W
Michelson, Mt., *Alaska* . **30** 69 20N 144 20W
Michigan □ **10** 44 0N 85 0W
Michigan, L., *Mich.* **10** 44 0N 87 0W
Michigan City, *Ind.* **15** 41 43N 86 54W
Micronesia, Federated
States of ■, *Pac. Oc.* . **31** 11 0N 160 0 E
Middle Alkali L., *Calif.* . . **16** 41 27N 120 5W
Middle Loup →, *Nebr.* . **12** 41 17N 98 24W
Middle River, *Md.* **25** 39 21N 76 26W
Middleburg, *N.Y.* **9** 42 36N 74 20W
Middleport, *Ohio* **10** 39 0N 82 3W
Middlesboro, *Ky.* **11** 36 36N 83 43W
Middlesex, *N.J.* **9** 40 36N 74 30W
Middleton I., *Alaska* . . . **30** 59 26N 146 20W
Middletown, *Conn.* **9** 41 34N 72 39W
Middletown, *N.Y.* **9** 41 27N 74 25W
Middletown, *Ohio* **15** 39 31N 84 24W
Middletown, *Pa.* **9** 40 12N 76 44W
Midland, *Mich.* **10** 43 37N 84 14W
Midland, *Tex.* **13** 32 0N 102 3W
Midlothian, *Tex.* **13** 32 30N 97 0W
Midwest, *Wyo.* **16** 43 25N 106 16W
Milaca, *Minn.* **12** 45 45N 93 39W
Milan, *Mo.* **14** 40 12N 93 7W
Milan, *Tenn.* **11** 35 55N 88 46W
Milbank, *S. Dak.* **12** 45 13N 96 38W
Miles, *Tex.* **13** 31 36N 100 11W
Miles City, *Mont.* **12** 46 25N 105 51W
Milford, *Conn.* **9** 41 14N 73 3W
Milford, *Del.* **10** 38 55N 75 26W
Milford, *Mass.* **9** 42 8N 71 31W
Milford, *Pa.* **9** 41 19N 74 48W
Milford, *Utah* **17** 38 24N 113 1W
Milk →, *Mont.* **16** 48 4N 106 19W
Mill City, *Oreg.* **16** 44 45N 122 29W
Mill Valley, *Calif.* **28** 37 54N 122 32W
Millburn, *N.J.* **20** 40 43N 74 19W
Mille Lacs L., *Minn.* . . . **12** 46 15N 93 39W
Milledgeville, *Ga.* **11** 33 5N 83 14W
Millen, *Ga.* **11** 32 48N 81 57W
Miller, *S. Dak.* **12** 44 31N 98 59W
Millersburg, *Pa.* **8** 40 32N 76 58W
Millerton, *N.Y.* **9** 41 57N 73 31W

Millerton L., *Calif.* **18** 37 1N 119 41W
Millinocket, *Maine* **11** 45 39N 68 43W
Milltown, *Pa.* **24** 39 57N 75 32W
Millvale, *Pa.* **27** 40 28N 79 59W
Millville, *N.J.* **10** 39 24N 75 2W
Millwood L., *Ark.* **13** 33 42N 93 58W
Milnor, *N. Dak.* **12** 46 16N 97 27W
Milolii, *Hawaii* **30** 19 11N 155 55W
Milton, *Calif.* **18** 38 3N 120 51W
Milton, *Fla.* **11** 30 38N 87 3W
Milton, *Mass.* **23** 42 14N 71 2W
Milton, *Pa.* **8** 41 1N 76 51W
Milton-Freewater, *Oreg.* **16** 45 56N 118 23W
Milwaukee, *Wis.* **15** 43 2N 87 55W
Milwaukie, *Oreg.* **18** 45 27N 122 38W
Mina, *Nev.* **17** 38 24N 118 7W
Minden, *La.* **13** 32 37N 93 17W
Mineola, *N.Y.* **21** 40 44N 73 38W
Mineola, *Tex.* **13** 32 40N 95 29W
Mineral King, *Calif.* **18** 36 27N 118 36W
Mineral Wells, *Tex.* **13** 32 48N 98 7W
Minersville, *Pa.* **9** 40 41N 76 16W
Minersville, *Utah* **17** 38 13N 112 56W
Minetto, *N.Y.* **9** 43 24N 76 28W
Minidoka, *Idaho* **16** 42 45N 113 29W
Minneapolis, *Kans.* **12** 39 8N 97 42W
Minneapolis, *Minn.* **12** 44 59N 93 16W
Minnesota □ **12** 46 0N 94 15W
Minot, *N. Dak.* **12** 48 14N 101 18W
Minto, *Alaska* **30** 64 53N 149 11W
Minturn, *Colo.* **16** 39 35N 106 26W
Mirando City, *Tex.* **13** 27 26N 99 0W
Mishawaka, *Ind.* **15** 41 40N 86 11W
Mission, *S. Dak.* **12** 43 18N 100 39W
Mission, *Tex.* **13** 26 13N 98 20W
Mississippi □ **13** 33 0N 90 0W
Mississippi →, *La.* **13** 29 9N 89 15W
Mississippi River Delta, *La.* **13** 29 10N 89 15W
Mississippi Sd., *Miss.* **13** 30 20N 89 0W
Missoula, *Mont.* **16** 46 52N 114 1W
Missouri □ **12** 38 25N 92 30W
Missouri →, *Mo.* **12** 38 49N 90 7W
Missouri Valley, *Iowa* **12** 41 34N 95 53W
Mitchell, *Ind.* **15** 38 44N 86 28W
Mitchell, *Nebr.* **12** 41 57N 103 49W
Mitchell, *Oreg.* **16** 44 34N 120 9W
Mitchell, *S. Dak.* **12** 43 43N 98 2W
Mitchell, Mt., *N.C.* **11** 35 46N 82 16W
Moab, *Utah* **17** 38 35N 109 33W
Moberly, *Mo.* **14** 39 25N 92 26W
Mobile, *Ala.* **11** 30 41N 88 3W
Mobile B., *Ala.* **11** 30 30N 88 0W
Mobridge, *S. Dak.* **12** 45 32N 100 26W
Moclips, *Wash.* **18** 47 14N 124 13W
Modena, *Utah* **17** 37 48N 113 56W
Modesto, *Calif.* **18** 37 39N 121 0W
Mohall, *N. Dak.* **12** 48 46N 101 31W
Mohawk →, *N.Y.* **9** 42 47N 73 41W
Mohican, C., *Alaska* **30** 60 12N 167 25W
Mojave, *Calif.* **19** 35 3N 118 10W
Mojave Desert, *Calif.* **19** 35 0N 116 30W
Mokelumne →, *Calif.* **18** 38 13N 121 28W
Mokelumne Hill, *Calif.* **18** 38 18N 120 43W
Moline, *Ill.* **14** 41 30N 90 31W
Molokai, *Hawaii* **30** 21 8N 157 0W
Monahans, *Tex.* **13** 31 36N 102 54W
Mondovi, *Wis.* **12** 44 34N 91 40W
Monessen, *Pa.* **8** 40 9N 79 54W
Monett, *Mo.* **13** 36 55N 93 55W
Monmouth, *Ill.* **14** 40 55N 90 39W
Mono L., *Calif.* **18** 38 1N 119 1W
Monolith, *Calif.* **19** 35 7N 118 22W
Monroe, *Ga.* **11** 33 47N 83 43W
Monroe, *La.* **13** 32 30N 92 7W
Monroe, *Mich.* **15** 41 55N 83 24W
Monroe, *N.C.* **11** 34 59N 80 33W
Monroe, *Utah* **17** 38 38N 112 7W
Monroe, *Wis.* **14** 42 36N 89 38W
Monroe City, *Mo.* **14** 39 39N 91 44W
Monroeville, *Ala.* **11** 31 31N 87 20W
Monroeville, *Pa.* **27** 40 26N 79 46W
Monrovia, *Calif.* **29** 34 9N 118 1W
Montague, *Calif.* **16** 41 44N 122 32W
Montague I., *Alaska* **30** 60 0N 147 30W
Montalvo, *Calif.* **19** 34 15N 119 12W
Montana □ **16** 47 0N 110 0W
Montauk, *N.Y.* **9** 41 3N 71 57W
Montauk Pt., *N.Y.* **9** 41 4N 71 52W
Montclair, *N.J.* **20** 40 49N 74 12W
Monte Vista, *Colo.* **17** 37 35N 106 9W
Montebello, *Calif.* **29** 34 1N 118 8W
Montecito, *Calif.* **19** 34 26N 119 40W
Montello, *Wis.* **12** 43 48N 89 20W
Monterey, *Calif.* **18** 36 37N 121 55W
Monterey B., *Calif.* **18** 36 45N 122 0W
Monterey Park, *Calif.* **29** 34 3N 118 7W
Montesano, *Wash.* **18** 46 59N 123 36W
Montevideo, *Minn.* **12** 44 57N 95 43W
Montezuma, *Iowa* **14** 41 35N 92 32W
Montgomery, *Ala.* **11** 32 23N 86 19W
Montgomery, *W. Va.* **10** 38 11N 81 19W
Monticello, *Ark.* **13** 33 38N 91 47W
Monticello, *Fla.* **11** 30 33N 83 52W
Monticello, *Ind.* **15** 40 45N 86 46W
Monticello, *Iowa* **14** 42 15N 91 12W
Monticello, *Ky.* **11** 36 50N 84 51W
Monticello, *Minn.* **12** 45 18N 93 48W
Monticello, *Miss.* **13** 31 33N 90 7W
Monticello, *N.Y.* **9** 41 39N 74 42W
Monticello, *Utah* **17** 37 52N 109 21W
Montour Falls, *N.Y.* **8** 42 21N 76 51W
Montpelier, *Idaho* **16** 42 19N 111 18W
Montpelier, *Md.* **25** 39 3N 76 50W
Montpelier, *Ohio* **15** 41 35N 84 37W
Montpelier, *Vt.* **9** 44 16N 72 35W
Montrose, *Colo.* **17** 38 29N 107 53W
Montrose, *Pa.* **9** 41 50N 75 53W
Moorcroft, *Wyo.* **12** 44 16N 104 57W
Moorefield, *W. Va.* **10** 39 5N 78 59W
Mooresville, *N.C.* **11** 35 35N 80 48W
Moorhead, *Minn.* **12** 46 53N 96 45W
Moorpark, *Calif.* **19** 34 17N 118 53W
Moose Lake, *Minn.* **12** 46 27N 92 46W
Moosehead L., *Maine* **11** 45 38N 69 40W
Moosup, *Conn.* **9** 41 43N 71 53W
Mora, *Minn.* **12** 45 53N 93 18W
Mora, *N. Mex.* **17** 35 58N 105 20W
Moran, *Kans.* **13** 37 55N 95 10W
Moran, *Wyo.* **16** 43 53N 110 37W
Moravia, *Iowa* **14** 40 53N 92 49W
Moreau →, *S. Dak.* **12** 45 18N 100 43W
Morehead, *Ky.* **15** 38 11N 83 26W
Morehead City, *N.C.* **11** 34 43N 76 43W
Morenci, *Ariz.* **17** 33 5N 109 22W
Morgan, *Utah* **16** 41 2N 111 41W
Morgan City, *La.* **13** 29 42N 91 12W
Morgan Hill, *Calif.* **18** 37 8N 121 39W
Morganfield, *Ky.* **10** 37 41N 87 55W
Morganton, *N.C.* **11** 35 45N 81 41W
Morgantown, *W. Va.* **10** 39 38N 79 57W
Morongo Valley, *Calif.* **19** 34 3N 116 37W
Morrilton, *Ark.* **13** 35 9N 92 44W
Morris, *Ill.* **15** 41 22N 88 26W
Morris, *Minn.* **12** 45 35N 95 55W
Morrison, *Ill.* **14** 41 49N 89 58W
Morristown, *Ariz.* **17** 33 51N 112 37W
Morristown, *N.J.* **20** 40 48N 74 26W
Morristown, *S. Dak.* **12** 45 56N 101 43W
Morristown, *Tenn.* **11** 36 13N 83 18W
Morro Bay, *Calif.* **18** 35 22N 120 51W
Morton, *Tex.* **13** 33 44N 102 46W
Morton, *Wash.* **18** 46 34N 122 17W
Morton Grove, *Ill.* **22** 42 2N 87 45W
Moscow, *Idaho* **16** 46 44N 117 0W
Moses Lake, *Wash.* **16** 47 8N 119 17W
Mosquero, *N. Mex.* **13** 35 47N 103 58W
Mott, *N. Dak.* **12** 46 23N 102 20W
Moulton, *Tex.* **13** 29 35N 97 9W
Moultrie, *Ga.* **11** 31 11N 83 47W
Moultrie, L., *S.C.* **11** 33 20N 80 5W
Mound City, *Mo.* **12** 40 7N 95 14W
Mound City, *S. Dak.* **12** 45 44N 100 4W
Moundsville, *W. Va.* **8** 39 55N 80 44W
Mount Airy, *N.C.* **11** 36 31N 80 37W
Mount Angel, *Oreg.* **16** 45 4N 122 48W
Mount Carmel, *Ill.* **15** 38 25N 87 46W
Mount Clemens, *Mich.* **8** 42 35N 82 53W
Mount Clemens, *Mich.* **26** 42 35N 82 53W
Mount Desert I., *Maine* **11** 44 21N 68 20W
Mount Dora, *Fla.* **11** 28 48N 81 38W
Mount Edgecumbe, *Alaska* **30** 57 3N 135 21W
Mount Hope, *W. Va.* **10** 37 54N 81 10W
Mount Horeb, *Wis.* **14** 43 1N 89 44W
Mount Laguna, *Calif.* **19** 32 52N 116 25W
Mount Lebanon, *Pa.* **27** 40 22N 80 2W
Mount McKinley National Park, *Alaska* **30** 63 30N 150 0W
Mount Morris, *N.Y.* **8** 42 44N 77 52W
Mount Oliver, *Pa.* **27** 40 24N 79 59W
Mount Pleasant, *Iowa* **14** 40 58N 91 33W
Mount Pleasant, *Mich.* **10** 43 36N 84 46W
Mount Pleasant, *S.C.* **11** 32 47N 79 52W
Mount Pleasant, *Tenn.* **11** 35 32N 87 12W
Mount Pleasant, *Tex.* **13** 33 9N 94 58W
Mount Pleasant, *Utah* **16** 39 33N 111 27W
Mount Pocono, *Pa.* **9** 41 7N 75 22W
Mount Prospect, *Ill.* **22** 42 3N 87 55W
Mount Rainier National Park, *Wash.* **18** 46 55N 121 50W
Mount Royal, *N.J.* **24** 39 48N 75 13W
Mount Shasta, *Calif.* **16** 41 19N 122 19W
Mount Sterling, *Ill.* **14** 39 59N 90 45W
Mount Sterling, *Ky.* **15** 38 4N 83 56W
Mount Vernon, *Ind.* **15** 38 17N 88 57W
Mount Vernon, *N.Y.* **21** 40 54N 73 49W
Mount Vernon, *Ohio* **8** 40 23N 82 29W
Mount Vernon, *Wash.* **18** 48 25N 122 20W
Mount Wilson Observatory, *Calif.* **29** 34 13N 118 4W
Mountain Center, *Calif.* **19** 33 42N 116 44W
Mountain City, *Nev.* **16** 41 50N 115 58W
Mountain City, *Tenn.* **11** 36 29N 81 48W
Mountain Grove, *Mo.* **13** 37 8N 92 16W
Mountain Home, *Ark.* **13** 36 20N 92 23W
Mountain Home, *Idaho* **16** 43 8N 115 41W
Mountain Iron, *Minn.* **12** 47 32N 92 37W
Mountain View, *Ark.* **13** 35 52N 92 7W
Mountain View, *Calif.* **18** 37 23N 122 5W
Mountain Village, *Alaska* **30** 62 5N 163 43W
Mountainair, *N. Mex.* **17** 34 31N 106 15W
Muddy Cr. →, *Utah* **17** 38 24N 110 42W
Mule Creek, *Wyo.* **12** 43 19N 104 8W
Muleshoe, *Tex.* **13** 34 13N 102 43W
Mullen, *Nebr.* **12** 42 3N 101 1W
Mullens, *W. Va.* **10** 37 35N 81 23W
Mullin, *Tex.* **13** 31 33N 98 40W
Mullins, *S.C.* **11** 34 12N 79 15W
Mulvane, *Kans.* **13** 37 29N 97 15W
Muncie, *Ind.* **15** 40 12N 85 23W
Munday, *Tex.* **13** 33 27N 99 38W
Munhall, *Pa.* **27** 40 24N 79 54W
Munising, *Mich.* **10** 46 25N 86 40W
Murdo, *S. Dak.* **12** 43 53N 100 43W
Murfreesboro, *Tenn.* **11** 35 51N 86 24W
Murphy, *Idaho* **16** 43 13N 116 33W
Murphys, *Calif.* **18** 38 8N 120 28W
Murphysboro, *Ill.* **14** 37 46N 89 20W
Murray, *Ky.* **11** 36 37N 88 19W
Murray, *Utah* **16** 40 40N 111 53W
Murray, L., *S.C.* **11** 34 3N 81 13W
Murrieta, *Calif.* **19** 33 33N 117 13W
Murrysville, *Pa.* **27** 40 25N 79 41W
Muscatine, *Iowa* **14** 41 25N 91 3W
Muskegon, *Mich.* **15** 43 14N 86 16W
Muskegon →, *Mich.* **10** 43 14N 86 21W
Muskegon Heights, *Mich.* **15** 43 12N 86 16W
Muskogee, *Okla.* **13** 35 45N 95 22W
Musselshell →, *Mont.* **16** 47 21N 107 57W
Myerstown, *Pa.* **9** 40 22N 76 19W
Myrtle Beach, *S.C.* **11** 33 42N 78 53W
Myrtle Creek, *Oreg.* **16** 43 1N 123 17W
Myrtle Point, *Oreg.* **16** 43 4N 124 8W
Mystic, *Conn.* **9** 41 21N 71 58W
Myton, *Utah* **16** 40 12N 110 4W

N

Naalehu, *Hawaii* **30** 19 4N 155 35W
Nabesna, *Alaska* **30** 62 22N 143 0W
Naches, *Wash.* **16** 46 44N 120 42W
Nacimiento Reservoir, *Calif.* **18** 35 46N 120 53W
Naco, *Ariz.* **17** 31 20N 109 57W
Nacogdoches, *Tex.* **13** 31 36N 94 39W
Nakalele Pt., *Hawaii* **30** 21 2N 156 35W
Naknek, *Alaska* **30** 58 44N 157 1W
Nampa, *Idaho* **16** 43 34N 116 34W
Nanticoke, *Pa.* **9** 41 12N 76 0W
Napa, *Calif.* **18** 38 18N 122 17W
Napa →, *Calif.* **18** 38 10N 122 19W
Napamute, *Alaska* **30** 61 30N 158 45W
Napanoch, *N.Y.* **9** 41 44N 74 22W
Naples, *Fla.* **11** 26 8N 81 48W
Napoleon, *N. Dak.* **12** 46 30N 99 46W
Napoleon, *Ohio* **15** 41 23N 84 8W
Nara Visa, *N. Mex.* **13** 35 37N 103 6W
Narrows, The, *N.Y.* **20** 40 37N 74 3W
Nashua, *Iowa* **14** 42 57N 92 32W
Nashua, *Mont.* **16** 48 8N 106 22W
Nashua, *N.H.* **9** 42 45N 71 28W
Nashville, *Ark.* **13** 33 57N 93 51W
Nashville, *Ga.* **11** 31 12N 83 15W
Nashville, *Tenn.* **11** 36 10N 86 47W
Nassau, *N.Y.* **9** 42 31N 73 37W
Natchez, *Miss.* **13** 31 34N 91 24W
Natchitoches, *La.* **13** 31 46N 93 5W
Natick, *Mass.* **23** 42 16N 71 21W
National City, *Calif.* **19** 32 41N 117 6W
Natoma, *Kans.* **12** 39 11N 99 2W
Navajo Reservoir, *N. Mex.* **17** 36 48N 107 36W
Navasota, *Tex.* **13** 30 23N 96 5W
Neah Bay, *Wash.* **18** 48 22N 124 37W
Near Is., *Alaska* **30** 53 0N 172 0 E
Nebraska □ **12** 41 30N 99 30W
Nebraska City, *Nebr.* **12** 40 41N 95 52W
Necedah, *Wis.* **12** 44 2N 90 4W
Neches →, *Tex.* **13** 29 58N 93 51W
Needham, *Mass.* **23** 42 16N 71 13W
Needles, *Calif.* **19** 34 51N 114 37W
Neenah, *Wis.* **10** 44 11N 88 28W
Negaunee, *Mich.* **10** 46 30N 87 36W
Neihart, *Mont.* **16** 47 0N 110 44W
Neilton, *Wash.* **16** 47 25N 123 53W
Neligh, *Nebr.* **12** 42 8N 98 2W
Nelson, *Ariz.* **17** 35 31N 113 19W
Nelson I., *Alaska* **30** 60 40N 164 40W
Nenana, *Alaska* **30** 64 34N 149 5W
Neodesha, *Kans.* **13** 37 25N 95 41W
Neosho, *Mo.* **13** 36 52N 94 22W
Neosho →, *Okla.* **13** 36 48N 95 18W
Nephi, *Utah* **16** 39 43N 111 50W
Neptune, *N.J.* **9** 40 13N 74 2W
Neuse →, *N.C.* **11** 35 6N 76 29W
Nevada, *Mo.* **14** 37 51N 94 22W
Nevada □ **16** 39 0N 117 0W
Nevada, Sierra, *Calif.* **16** 39 0N 120 30W
Nevada City, *Calif.* **18** 39 16N 121 1W
Neville Island, *Pa.* **27** 40 30N 80 6W
New Albany, *Ind.* **15** 38 18N 85 49W
New Albany, *Miss.* **13** 34 29N 89 0W
New Albany, *Pa.* **9** 41 36N 76 27W
New Bedford, *Mass.* **9** 41 38N 70 56W
New Bern, *N.C.* **11** 35 7N 77 3W
New Boston, *Tex.* **13** 33 28N 94 25W
New Braunfels, *Tex.* **13** 29 42N 98 8W
New Britain, *Conn.* **9** 41 40N 72 47W
New Brunswick, *N.J.* **9** 40 30N 74 27W
New Carrollton, *Md.* **25** 38 58N 76 53W
New Castle, *Ind.* **15** 39 55N 85 22W
New Castle, *Pa.* **8** 41 0N 80 21W
New City, *N.Y.* **9** 41 9N 73 59W
New Cuyama, *Calif.* **19** 34 57N 119 38W
New Don Pedro Reservoir, *Calif.* **18** 37 43N 120 24W
New Dorp, *N.Y.* **20** 40 34N 74 8W
New England, *N. Dak.* **12** 46 32N 102 52W
New Hampshire □ **9** 44 0N 71 30W
New Hampton, *Iowa* **14** 43 3N 92 19W
New Haven, *Conn.* **9** 41 18N 72 55W
New Hyde Park, *N.Y.* **21** 40 43N 73 39W
New Iberia, *La.* **13** 30 1N 91 49W
New Jersey □ **9** 40 0N 74 30W
New Kensington, *Pa.* **8** 40 34N 79 46W
New Kensington, *Pa.* **27** 40 34N 79 46W
New Lexington, *Ohio* **10** 39 43N 82 13W
New London, *Conn.* **9** 41 22N 72 6W
New London, *Minn.* **12** 45 18N 94 56W
New London, *Wis.* **12** 44 23N 88 45W
New Madrid, *Mo.* **13** 36 36N 89 32W
New Meadows, *Idaho* **16** 44 58N 116 18W
New Melones L., *Calif.* **18** 37 57N 120 31W
New Mexico □ **17** 34 30N 106 0W
New Milford, *Conn.* **9** 41 35N 73 25W
New Milford, *Pa.* **9** 41 52N 75 44W
New Orleans, *La.* **13** 29 58N 90 4W
New Philadelphia, *Ohio* **8** 40 30N 81 27W
New Plymouth, *Idaho* **16** 43 58N 116 49W
New Providence, *N.J.* **20** 40 42N 74 23W
New Richmond, *Wis.* **12** 45 7N 92 32W
New Roads, *La.* **13** 30 42N 91 26W
New Rochelle, *N.Y.* **21** 40 55N 73 45W
New Rockford, *N. Dak.* **12** 47 41N 99 8W
New Salem, *N. Dak.* **12** 46 51N 101 25W
New Smyrna Beach, *Fla.* **11** 29 1N 80 56W
New Town, *N. Dak.* **12** 47 59N 102 30W
New Ulm, *Minn.* **12** 44 19N 94 28W
New York, *N.Y.* **20** 40 42N 74 0W
New York □ **9** 43 0N 75 0W
Newark, *Del.* **10** 39 41N 75 46W
Newark, *N.J.* **20** 40 43N 74 10W
Newark, *N.Y.* **8** 43 3N 77 6W
Newark, *Ohio* **8** 40 3N 82 24W
Newark International Airport, *N.J.* **20** 40 41N 74 10W
Newaygo, *Mich.* **10** 43 25N 85 48W
Newberg, *Oreg.* **16** 45 18N 122 58W
Newberry, *Mich.* **10** 46 21N 85 30W
Newberry, *S.C.* **11** 34 17N 81 37W
Newberry Springs, *Calif.* **19** 34 50N 116 41W
Newburgh, *N.Y.* **9** 41 30N 74 1W
Newburyport, *Mass.* **9** 42 49N 70 53W
Newcastle, *Wyo.* **12** 43 50N 104 11W
Newell, *S. Dak.* **12** 44 43N 103 25W
Newenham, C., *Alaska* **30** 58 39N 162 11W
Newhalen, *Alaska* **30** 59 43N 154 54W
Newhall, *Calif.* **19** 34 23N 118 32W
Newkirk, *Okla.* **13** 36 53N 97 3W
Newman, *Calif.* **18** 37 19N 121 1W
Newmarket, *N.H.* **9** 43 5N 70 56W
Newnan, *Ga.* **11** 33 23N 84 48W
Newport, *Ark.* **13** 35 37N 91 16W
Newport, *Ky.* **15** 39 5N 84 30W
Newport, *N.H.* **9** 43 22N 72 10W
Newport, *Oreg.* **16** 44 39N 124 3W
Newport, *R.I.* **9** 41 29N 71 19W
Newport, *Tenn.* **11** 35 58N 83 11W
Newport, *Vt.* **9** 44 56N 72 13W
Newport, *Wash.* **16** 48 11N 117 3W
Newport Beach, *Calif.* **19** 33 37N 117 56W
Newport News, *Va.* **10** 36 59N 76 25W
Newton, *Iowa* **14** 41 42N 93 3W
Newton, *Mass.* **23** 42 19N 71 13W
Newton, *Miss.* **13** 32 19N 89 10W
Newton, *N.C.* **11** 35 40N 81 13W
Newton, *N.J.* **9** 41 3N 74 45W
Newton, *Tex.* **13** 30 51N 93 46W
Newtown Square, *Pa.* **24** 39 59N 75 24W
Nezperce, *Idaho* **16** 46 14N 116 14W
Niagara, *Mich.* **10** 45 45N 88 0W
Niagara Falls, *N.Y.* **8** 43 5N 79 4W
Niceville, *Fla.* **11** 30 31N 86 30W
Nicholasville, *Ky.* **15** 37 53N 84 34W
Nichols, *N.Y.* **9** 42 1N 76 22W
Nicholson, *Pa.* **9** 41 37N 75 47W
Niihau, *Hawaii* **30** 21 54N 160 9W
Nikolski, *Alaska* **30** 52 56N 168 52W
Niland, *Calif.* **19** 33 14N 115 31W
Niles, *Ill.* **22** 42 1N 87 48W
Niles, *Ohio* **8** 41 11N 80 46W
Niobrara, *Nebr.* **12** 42 45N 98 2W
Niobrara →, *Nebr.* **12** 42 46N 98 3W
Nipomo, *Calif.* **19** 35 3N 120 29W
Nixon, *Tex.* **13** 29 16N 97 46W
Noatak, *Alaska* **30** 67 34N 162 58W
Noatak →, *Alaska* **30** 68 0N 161 0W
Noblesville, *Ind.* **15** 40 3N 86 1W
Nocona, *Tex.* **13** 33 47N 97 44W
Noel, *Mo.* **13** 36 33N 94 29W
Nogales, *Ariz.* **17** 31 20N 110 56W
Nome, *Alaska* **30** 64 30N 165 25W
Nonopapa, *Hawaii* **30** 21 50N 160 15W
Noonan, *N. Dak.* **12** 48 54N 103 1W
Noorvik, *Alaska* **30** 66 50N 161 3W
Norco, *Calif.* **19** 33 56N 117 33W
Norfolk, *Nebr.* **12** 42 2N 97 25W
Norfolk, *Va.* **10** 36 51N 76 17W
Norfork Res., *Ark.* **13** 36 13N 92 15W
Normal, *Ill.* **14** 40 31N 88 59W
Norman, *Okla.* **13** 35 13N 97 26W
Norridge, *Ill.* **22** 41 57N 87 49W

Norris, *Mont.* **16** 45 34N 111 41W
Norristown, *Pa.* **24** 40 7N 75 20W
North Adams, *Mass.* **9** 42 42N 73 7W
North Bend, *Oreg.* **16** 43 24N 124 14W
North Bergen, *N.J.* **20** 40 48N 74 0W
North Berwick, *Maine* **9** 43 18N 70 44W
North Billerica, *Mass.* **23** 42 35N 71 16W
North Braddock, *Pa.* **27** 40 25N 79 51W
North Canadian →, *Okla.* **13** 35 16N 95 31W
North Carolina □ **11** 35 30N 80 0W
North Chelmsford, *Mass.* **23** 42 38N 71 23W
North Chicago, *Ill.* **15** 42 19N 87 51W
North Dakota □ **12** 47 30N 100 15W
North Fork, *Calif.* **18** 37 14N 119 21W
North Las Vegas, *Nev.* **19** 36 12N 115 7W
North Loup →, *Nebr.* **12** 41 17N 98 24W
North Olmsted, *Ohio* **27** 41 24N 81 55W
North Palisade, *Calif.* **18** 37 6N 118 31W
North Platte, *Nebr.* **12** 41 8N 100 46W
North Platte →, *Nebr.* **12** 41 7N 100 42W
North Powder, *Oreg.* **16** 45 2N 117 55W
North Reading, *Mass.* **23** 42 34N 71 5W
North Richmond, *Calif.* **28** 37 57N 122 22W
North Springfield, *Va.* **25** 38 48N 77 12W
North Tonawanda, *N.Y.* **8** 43 2N 78 53W
North Truchas Pk., *N. Mex.* **17** 36 0N 105 30W
North Vernon, *Ind.* **15** 39 0N 85 38W
Northampton, *Mass.* **9** 42 19N 72 38W
Northampton, *Pa.* **9** 40 41N 75 30W
Northbridge, *Mass.* **9** 42 9N 71 39W
Northbrook, *Ill.* **22** 42 7N 87 53W
Northern Marianas ■, *Pac. Oc.* **31** 17 0N 145 0 E
Northfield, *Ill.* **22** 42 5N 87 44W
Northfield, *Minn.* **12** 44 27N 93 9W
Northlake, *Ill.* **22** 41 54N 87 53W
Northome, *Minn.* **12** 47 52N 94 17W
Northport, *Ala.* **11** 33 14N 87 35W
Northport, *Mich.* **10** 45 8N 85 37W
Northport, *Wash.* **16** 48 55N 117 48W
Northway, *Alaska* **30** 62 58N 141 56W
Northwood, *Iowa* **12** 43 27N 93 13W
Northwood, *N. Dak.* **12** 47 44N 97 34W
Norton, *Kans.* **12** 39 50N 99 53W
Norton B., *Alaska* **30** 64 45N 161 15W
Norton Sd., *Alaska* **30** 63 50N 164 0W
Norwalk, *Calif.* **29** 33 54N 118 4W
Norwalk, *Conn.* **9** 41 7N 73 22W
Norwalk, *Ohio* **8** 41 15N 82 37W
Norway, *Mich.* **10** 45 47N 87 55W
Norwich, *Conn.* **9** 41 31N 72 5W
Norwich, *N.Y.* **9** 42 32N 75 32W
Norwood, *Mass.* **23** 42 11N 71 13W
Nottoway →, *Va.* **10** 36 33N 76 55W
Novato, *Calif.* **18** 38 6N 122 35W
Noxen, *Pa.* **9** 41 25N 76 4W
Noxon, *Mont.* **16** 48 0N 115 43W
Nueces →, *Tex.* **13** 27 51N 97 30W
Nulato, *Alaska* **30** 64 43N 158 6W
Nunivak I., *Alaska* **30** 60 10N 166 30W
Nutley, *N.J.* **20** 40 49N 74 9W
Nyack, *N.Y.* **9** 41 5N 73 55W
Nyssa, *Oreg.* **16** 43 53N 117 0W

O

Oacoma, *S. Dak.* **12** 43 48N 99 24W
Oahe, L., *S. Dak.* **12** 44 27N 100 24W
Oahe Dam, *S. Dak.* **12** 44 27N 100 24W
Oahu, *Hawaii* **30** 21 28N 157 58W
Oak Creek, *Colo.* **16** 40 16N 106 57W
Oak Forest, *Ill.* **22** 41 36N 87 44W
Oak Harbor, *Wash.* **18** 48 18N 122 39W
Oak Hill, *W. Va.* **10** 37 59N 81 9W
Oak Lawn, *Ill.* **22** 41 42N 87 44W
Oak Park, *Ill.* **22** 41 52N 87 46W
Oak Park, *Mich.* **26** 42 27N 83 11W
Oak Ridge, *Tenn.* **11** 36 1N 84 16W
Oak View, *Calif.* **19** 34 24N 119 18W
Oakdale, *Calif.* **18** 37 46N 120 51W
Oakdale, *La.* **13** 30 49N 92 40W
Oakes, *N. Dak.* **12** 46 8N 98 6W
Oakesdale, *Wash.* **16** 47 8N 117 15W
Oakhurst, *Calif.* **18** 37 19N 119 40W
Oakland, *Calif.* **28** 37 48N 122 17W
Oakland, *N.J.* **20** 41 2N 74 13W
Oakland, *Oreg.* **16** 43 25N 123 18W
Oakland City, *Ind.* **15** 38 20N 87 21W
Oakland Pontiac Airport, *Mich.* **26** 42 40N 83 24W
Oakley, *Idaho* **16** 42 15N 113 53W
Oakley, *Kans.* **12** 39 8N 100 51W
Oakmont, *Pa.* **27** 40 31N 79 50W
Oakridge, *Oreg.* **16** 43 45N 122 28W
Oasis, *Calif.* **19** 33 28N 116 6W
Oasis, *Nev.* **18** 37 29N 117 55W
Oatman, *Ariz.* **19** 35 1N 114 19W
Oberlin, *Kans.* **12** 39 49N 100 32W
Oberlin, *La.* **13** 30 37N 92 46W
Ocala, *Fla.* **11** 29 11N 82 8W
Ocanomowoc, *Wis.* **12** 43 7N 88 30W
Ocate, *N. Mex.* **13** 36 11N 105 3W
Ocean City, *N.J.* **10** 39 17N 74 35W
Ocean Park, *Wash.* **18** 46 30N 124 3W
Oceano, *Calif.* **19** 35 6N 120 37W
Oceanside, *Calif.* **19** 33 12N 117 23W
Oceanside, *N.Y.* **21** 40 38N 73 37W
Ocilla, *Ga.* **11** 31 36N 83 15W
Ocmulgee →, *Ga.* **11** 31 58N 82 33W
Oconee →, *Ga.* **11** 31 58N 82 33W
Oconto, *Wis.* **10** 44 53N 87 52W
Oconto Falls, *Wis.* **10** 44 52N 88 9W
Octave, *Ariz.* **17** 34 10N 112 43W
Odessa, *Tex.* **13** 31 52N 102 23W
Odessa, *Wash.* **16** 47 20N 118 41W
O'Donnell, *Tex.* **13** 32 58N 101 50W
Oelrichs, *S. Dak.* **12** 43 11N 103 14W
Oelwein, *Iowa* **12** 42 41N 91 55W
Ofu, *Amer. Samoa* **31** 14 11 S 169 41W
Ogallala, *Nebr.* **12** 41 8N 101 43W
Ogden, *Iowa* **14** 42 2N 94 2W
Ogden, *Utah* **16** 41 13N 111 58W
Ogdensburg, *N.Y.* **9** 44 42N 75 30W
Ogeechee →, *Ga.* **11** 31 50N 81 3W
Ohio □ **10** 40 15N 82 45W
Ohio →, *Ohio* **10** 36 59N 89 8W
Oil City, *Pa.* **8** 41 26N 79 42W
Oildale, *Calif.* **19** 35 25N 119 1W
Ojai, *Calif.* **19** 34 27N 119 15W
Okanogan, *Wash.* **16** 48 22N 119 35W
Okanogan →, *Wash.* **16** 48 6N 119 44W
Okeechobee, *Fla.* **11** 27 15N 80 50W
Okeechobee, L., *Fla.* **11** 27 0N 80 50W
Okefenokee Swamp, *Ga.* **11** 30 40N 82 20W
Oklahoma □ **13** 35 20N 97 30W
Oklahoma City, *Okla.* **13** 35 30N 97 30W
Okmulgee, *Okla.* **13** 35 37N 95 58W
Okolona, *Miss.* **13** 34 0N 88 45W
Ola, *Ark.* **13** 35 2N 93 13W
Olancha, *Calif.* **19** 36 17N 118 1W
Olancha Pk., *Calif.* **19** 36 15N 118 7W
Olathe, *Kans.* **12** 38 53N 94 49W
Old Baldy Pk. = San Antonio, Mt., *Calif.* **19** 34 17N 117 38W
Old Dale, *Calif.* **19** 34 8N 115 47W
Old Forge, *N.Y.* **9** 43 43N 74 58W
Old Forge, *Pa.* **9** 41 22N 75 45W
Old Harbor, *Alaska* **30** 57 12N 153 18W
Old Town, *Maine* **11** 44 56N 68 39W
Olean, *N.Y.* **8** 42 5N 78 26W
Olema, *Calif.* **18** 38 3N 122 47W
Olney, *Ill.* **15** 38 44N 88 5W
Olney, *Tex.* **13** 33 22N 98 45W
Olosega, *Amer. Samoa* **31** 14 11 S 169 38W
Olton, *Tex.* **13** 34 11N 102 8W
Olympia, *Wash.* **18** 47 3N 122 53W
Olympic Mts., *Wash.* **18** 47 55N 123 45W
Olympic Nat. Park, *Wash.* **18** 47 48N 123 30W
Olympus, Mt., *Wash.* **18** 47 48N 123 43W
Omaha, *Nebr.* **12** 41 17N 95 58W
Omak, *Wash.* **16** 48 25N 119 31W
Onaga, *Kans.* **12** 39 29N 96 10W
Onalaska, *Wis.* **12** 43 53N 91 14W
Onamia, *Minn.* **12** 46 4N 93 40W
Onancock, *Va.* **10** 37 43N 75 45W
Onawa, *Iowa* **12** 42 2N 96 6W
Onaway, *Mich.* **10** 45 21N 84 14W
Oneida, *N.Y.* **9** 43 6N 75 39W
Oneida L., *N.Y.* **9** 43 12N 75 54W
O'Neill, *Nebr.* **12** 42 27N 98 39W
Oneonta, *Ala.* **11** 33 57N 86 28W
Oneonta, *N.Y.* **9** 42 27N 75 4W
Onida, *S. Dak.* **12** 44 42N 100 4W
Onslow B., *N.C.* **11** 34 20N 77 15W
Ontario, *Calif.* **19** 34 4N 117 39W
Ontario, *Oreg.* **16** 44 2N 116 58W
Ontario, L., *N. Amer.* **8** 43 20N 78 0W
Ontonagon, *Mich.* **12** 46 52N 89 19W
Onyx, *Calif.* **19** 35 41N 118 14W
Ookala, *Hawaii* **30** 20 1N 155 17W
Opelousas, *La.* **13** 30 32N 92 5W
Opheim, *Mont.* **16** 48 51N 106 24W
Ophir, *Alaska* **30** 63 10N 156 31W
Opp, *Ala.* **11** 31 17N 86 16W
Oracle, *Ariz.* **17** 32 37N 110 46W
Orange, *Calif.* **19** 33 47N 117 51W
Orange, *Mass.* **9** 42 35N 72 19W
Orange, *N.J.* **20** 40 46N 74 13W
Orange, *Tex.* **13** 30 6N 93 44W
Orange, *Va.* **10** 38 15N 78 7W
Orange Cove, *Calif.* **18** 36 38N 119 19W
Orange Grove, *Tex.* **13** 27 58N 97 56W
Orangeburg, *S.C.* **11** 33 30N 80 52W
Orcutt, *Calif.* **19** 34 52N 120 27W
Orderville, *Utah* **17** 37 17N 112 38W
Ordway, *Colo.* **12** 38 13N 103 46W
Oregon, *Ill.* **14** 42 1N 89 20W
Oregon □ **16** 44 0N 121 0W
Oregon City, *Oreg.* **18** 45 21N 122 36W
Orem, *Utah* **16** 40 19N 111 42W
Orinda, *Calif.* **28** 37 52N 122 10W
Orland, *Calif.* **18** 39 45N 122 12W
Orland Park, *Ill.* **22** 41 37N 87 52W
Orlando, *Fla.* **11** 28 33N 81 23W
Ormond Beach, *Fla.* **11** 29 17N 81 3W
Oro Grande, *Calif.* **19** 34 36N 117 20W
Orogrande, *N. Mex.* **17** 32 24N 106 5W
Oroville, *Calif.* **18** 39 31N 121 33W
Oroville, *Wash.* **16** 48 56N 119 26W
Osage, *Iowa* **12** 43 17N 92 49W
Osage, *Wyo.* **12** 43 59N 104 25W
Osage →, *Mo.* **14** 38 35N 91 57W
Osage City, *Kans.* **12** 38 38N 95 50W
Osawatomie, *Kans.* **12** 38 31N 94 57W
Osborne, *Kans.* **12** 39 26N 98 42W
Osceola, *Ark.* **13** 35 42N 89 58W
Osceola, *Iowa* **14** 41 2N 93 46W
Oscoda, *Mich.* **8** 44 26N 83 20W
Oshkosh, *Nebr.* **12** 41 24N 102 21W
Oshkosh, *Wis.* **12** 44 1N 88 33W
Oskaloosa, *Iowa* **14** 41 18N 92 39W
Ossabaw I., *Ga.* **11** 31 50N 81 5W
Ossining, *N.Y.* **9** 41 10N 73 55W
Oswego, *N.Y.* **9** 43 27N 76 31W
Othello, *Wash.* **16** 46 50N 119 10W
Otis, *Colo.* **12** 40 9N 102 58W
Ottawa, *Ill.* **15** 41 21N 88 51W
Ottawa, *Kans.* **12** 38 37N 95 16W
Ottumwa, *Iowa* **14** 41 1N 92 25W
Ouachita →, *La.* **13** 31 38N 91 49W
Ouachita, L., *Ark.* **13** 34 34N 93 12W
Ouachita Mts., *Ark.* **13** 34 40N 94 25W
Ouray, *Colo.* **17** 38 1N 107 40W
Outlook, *Mont.* **12** 48 53N 104 47W
Overlea, *Md.* **25** 39 21N 76 33W
Overton, *Nev.* **19** 36 33N 114 27W
Ovid, *Colo.* **12** 40 58N 102 23W
Owatonna, *Minn.* **12** 44 5N 93 14W
Owego, *N.Y.* **9** 42 6N 76 16W
Owens →, *Calif.* **18** 36 32N 117 59W
Owens L., *Calif.* **19** 36 26N 117 57W
Owensboro, *Ky.* **15** 37 46N 87 7W
Owensville, *Mo.* **14** 38 21N 91 30W
Owings Mills, *Md.* **25** 39 25N 76 48W
Owosso, *Mich.* **15** 43 0N 84 10W
Owyhee, *Nev.* **16** 41 57N 116 6W
Owyhee →, *Oreg.* **16** 43 49N 117 2W
Owyhee, L., *Oreg.* **16** 43 38N 117 14W
Oxford, *Miss.* **13** 34 22N 89 31W
Oxford, *N.C.* **11** 36 19N 78 35W
Oxford, *Ohio* **15** 39 31N 84 45W
Oxnard, *Calif.* **19** 34 12N 119 11W
Oyster Bay, *N.Y.* **21** 40 52N 73 31W
Ozark, *Ala.* **11** 31 28N 85 39W
Ozark, *Ark.* **13** 35 29N 93 50W
Ozark, *Mo.* **13** 37 1N 93 12W
Ozark Plateau, *Mo.* **13** 37 20N 91 40W
Ozarks, L. of the, *Mo.* **14** 38 12N 92 38W
Ozona, *Tex.* **13** 30 43N 101 12W

P

Paauilo, *Hawaii* **30** 20 2N 155 22W
Pacific Grove, *Calif.* **18** 36 38N 121 56W
Pacifica, *Calif.* **28** 37 38N 122 29W
Padre I., *Tex.* **13** 27 10N 97 25W
Paducah, *Ky.* **10** 37 5N 88 37W
Paducah, *Tex.* **13** 34 1N 100 18W
Page, *Ariz.* **17** 36 57N 111 27W
Page, *N. Dak.* **12** 47 10N 97 34W
Pago Pago, *Amer. Samoa* **31** 14 16 S 170 43W
Pagosa Springs, *Colo.* **17** 37 16N 107 1W
Pahala, *Hawaii* **30** 19 12N 155 29W
Pahoa, *Hawaii* **30** 19 30N 154 57W
Pahokee, *Fla.* **11** 26 50N 80 40W
Pahrump, *Nev.* **19** 36 12N 115 59W
Pahute Mesa, *Nev.* **18** 37 20N 116 45W
Paia, *Hawaii* **30** 20 54N 156 22W
Paicines, *Calif.* **18** 36 44N 121 17W
Pailolo Channel, *Hawaii* **30** 21 0N 156 40W
Painesville, *Ohio* **8** 41 43N 81 15W
Paint Rock, *Tex.* **13** 31 31N 99 55W
Painted Desert, *Ariz.* **17** 36 0N 111 0W
Paintsville, *Ky.* **10** 37 49N 82 48W
Paisley, *Oreg.* **16** 42 42N 120 32W
Pala, *Calif.* **19** 33 22N 117 5W
Palacios, *Tex.* **13** 28 42N 96 13W
Palatka, *Fla.* **11** 29 39N 81 38W
Palau ■, *Pac. Oc.* **31** 7 30N 134 30 E
Palermo, *Calif.* **16** 39 26N 121 33W
Palestine, *Tex.* **13** 31 46N 95 38W
Palisade, *Nebr.* **12** 40 21N 101 7W
Palisades, *N.Y.* **20** 41 1N 73 55W
Palm Beach, *Fla.* **11** 26 43N 80 2W
Palm Desert, *Calif.* **19** 33 43N 116 22W
Palm Springs, *Calif.* **19** 33 50N 116 33W
Palmdale, *Calif.* **19** 34 35N 118 7W
Palmer, *Alaska* **30** 61 36N 149 7W
Palmer Lake, *Colo.* **12** 39 7N 104 55W
Palmerton, *Pa.* **9** 40 48N 75 37W
Palmetto, *Fla.* **11** 27 31N 82 34W
Palmyra, *Mo.* **14** 39 48N 91 32W
Palmyra, *N.J.* **24** 40 0N 75 1W
Palo Alto, *Calif.* **28** 37 26N 122 8W
Palos Heights, *Ill.* **22** 41 40N 87 47W
Palos Hills Forest, *Ill.* **22** 41 40N 87 52W
Palos Verdes, *Calif.* **19** 33 48N 118 23W
Palos Verdes, Pt., *Calif.* **19** 33 43N 118 26W
Palouse, *Wash.* **16** 46 55N 117 4W
Pamlico →, *N.C.* **11** 35 20N 76 28W
Pamlico Sd., *N.C.* **11** 35 20N 76 0W
Pampa, *Tex.* **13** 35 32N 100 58W
Pana, *Ill.* **14** 39 23N 89 5W
Panaca, *Nev.* **17** 37 47N 114 23W
Panama City, *Fla.* **11** 30 10N 85 40W
Panamint Range, *Calif.* **19** 36 20N 117 20W
Panamint Springs, *Calif.* **19** 36 20N 117 28W
Pancake Range, *Nev.* **17** 38 30N 115 50W
Panguitch, *Utah* **17** 37 50N 112 26W
Panhandle, *Tex.* **13** 35 21N 101 23W
Paola, *Kans.* **12** 38 35N 94 53W
Paoli, *Pa.* **24** 40 2N 75 28W
Paonia, *Colo.* **17** 38 52N 107 36W
Papaikou, *Hawaii* **30** 19 47N 155 6W
Paradise, *Mont.* **16** 47 23N 114 48W
Paradise Valley, *Nev.* **16** 41 30N 117 32W
Paragould, *Ark.* **13** 36 3N 90 29W
Paramus, *N.J.* **20** 40 56N 74 2W
Paris, *Idaho* **16** 42 14N 111 24W
Paris, *Ky.* **15** 38 13N 84 15W
Paris, *Tenn.* **11** 36 18N 88 19W
Paris, *Tex.* **13** 33 40N 95 33W
Parish, *N.Y.* **9** 43 25N 76 8W
Park City, *Utah* **16** 40 39N 111 30W
Park Falls, *Wis.* **12** 45 56N 90 27W
Park Range, *Colo.* **16** 40 0N 106 30W
Park Rapids, *Minn.* **12** 46 55N 95 4W
Park Ridge, *Ill.* **22** 42 0N 87 50W
Park Ridge, *N.J.* **20** 41 2N 74 2W
Park River, *N. Dak.* **12** 48 24N 97 45W
Parker, *Ariz.* **19** 34 9N 114 17W
Parker, *S. Dak.* **12** 43 24N 97 8W
Parker Dam, *Ariz.* **19** 34 18N 114 8W
Parkersburg, *W. Va.* **10** 39 16N 81 34W
Parkfield, *Calif.* **18** 35 54N 120 26W
Parkston, *S. Dak.* **12** 43 24N 97 59W
Parkville, *Md.* **25** 39 23N 76 34W
Parma, *Idaho* **16** 43 47N 116 57W
Parma, *Ohio* **27** 41 24N 81 43W
Parma Heights, *Ohio* **27** 41 23N 81 45W
Parowan, *Utah* **17** 37 51N 112 50W
Parris I., *S.C.* **11** 32 20N 80 41W
Parshall, *N. Dak.* **12** 47 57N 102 8W
Parsons, *Kans.* **13** 37 20N 95 16W
Pasadena, *Calif.* **29** 34 9N 118 8W
Pasadena, *Tex.* **13** 29 43N 95 13W
Pascagoula, *Miss.* **13** 30 21N 88 33W
Pascagoula →, *Miss.* **13** 30 23N 88 37W
Pasco, *Wash.* **16** 46 14N 119 6W
Paso Robles, *Calif.* **17** 35 38N 120 41W
Passaic, *N.J.* **20** 40 51N 74 9W
Patagonia, *Ariz.* **17** 31 33N 110 45W
Patchogue, *N.Y.* **9** 40 46N 73 1W
Pateros, *Wash.* **16** 48 3N 119 54W
Paterson, *N.J.* **20** 40 54N 74 10W
Pathfinder Reservoir, *Wyo.* **16** 42 28N 106 51W
Patten, *Maine* **11** 46 0N 68 38W
Patterson, *Calif.* **18** 37 28N 121 8W
Patterson, *La.* **13** 29 42N 91 18W
Patterson, Mt., *Calif.* **18** 38 29N 119 20W
Paullina, *Iowa* **12** 42 59N 95 41W
Pauls Valley, *Okla.* **13** 34 44N 97 13W
Paulsboro, *N.J.* **24** 39 49N 75 14W
Pauma Valley, *Calif.* **19** 33 16N 116 58W
Pavlof Is., *Alaska* **30** 55 30N 161 30W
Pawhuska, *Okla.* **13** 36 40N 96 20W
Pawling, *N.Y.* **9** 41 34N 73 36W
Pawnee, *Okla.* **13** 36 20N 96 48W
Pawnee City, *Nebr.* **12** 40 7N 96 9W
Pawtucket, *R.I.* **9** 41 53N 71 23W
Paxton, *Ill.* **15** 40 27N 88 6W
Paxton, *Nebr.* **12** 41 7N 101 21W
Payette, *Idaho* **16** 44 5N 116 56W
Paynesville, *Minn.* **12** 45 23N 94 43W
Payson, *Ariz.* **17** 34 14N 111 20W
Payson, *Utah* **16** 40 3N 111 44W
Pe Ell, *Wash.* **18** 46 34N 123 18W
Peabody, *Mass.* **23** 42 32N 70 57W
Peach Springs, *Ariz.* **17** 35 32N 113 25W
Peale, Mt., *Utah* **17** 38 26N 109 14W
Pearblossom, *Calif.* **19** 34 30N 117 55W
Pearl →, *Miss.* **13** 30 11N 89 32W
Pearl City, *Hawaii* **30** 21 24N 157 59W
Pearl Harbor, *Hawaii* **30** 21 21N 157 57W
Pearsall, *Tex.* **13** 28 54N 99 6W
Pease →, *Tex.* **13** 34 12N 99 2W
Pebble Beach, *Calif.* **18** 36 34N 121 57W
Pecos, *Tex.* **13** 31 26N 103 30W
Pecos →, *Tex.* **13** 29 42N 101 22W
Pedro Valley, *Calif.* **28** 37 35N 122 28W
Peekskill, *N.Y.* **9** 41 17N 73 55W
Pekin, *Ill.* **14** 40 35N 89 40W
Pelham, *Ga.* **11** 31 8N 84 9W
Pelham, *N.Y.* **21** 40 54N 73 46W
Pelican, *Alaska* **30** 57 58N 136 14W
Pella, *Iowa* **14** 41 25N 92 55W
Pembina, *N. Dak.* **12** 48 58N 97 15W
Pembine, *Wis.* **10** 45 38N 87 59W
Pembroke, *Ga.* **11** 32 8N 81 37W
Pend Oreille →, *Wash.* **16** 49 4N 117 37W
Pend Oreille L., *Idaho* **16** 48 10N 116 21W
Pendleton, *Calif.* **19** 33 16N 117 23W
Pendleton, *Oreg.* **16** 45 40N 118 47W
Penn Hills, *Pa.* **27** 40 27N 79 50W
Penn Yan, *N.Y.* **8** 42 40N 77 3W
Pennsauken, *N.J.* **24** 39 57N 75 5W
Pennsylvania □ **10** 40 45N 77 30W
Pensacola, *Fla.* **11** 30 25N 87 13W
Peoria, *Ariz.* **17** 33 35N 112 14W
Peoria, *Ill.* **14** 40 42N 89 36W
Perham, *Minn.* **12** 46 36N 95 34W
Perris, *Calif.* **19** 33 47N 117 14W
Perry, *Fla.* **11** 30 7N 83 35W
Perry, *Ga.* **11** 32 28N 83 44W
Perry, *Iowa* **14** 41 51N 94 6W

Perry, *Maine* **11** 44 58N 67 5W
Perry, *Okla.* **13** 36 17N 97 14W
Perry Hall, *Md.* **25** 39 24N 76 28W
Perrysville, *Pa.* **27** 40 32N 80 1W
Perryton, *Tex.* **13** 36 24N 100 48W
Perryville, *Alaska* **30** 55 55N 159 9W
Perryville, *Mo.* **14** 37 43N 89 52W
Perth Amboy, *N.J.* **20** 40 30N 74 16W
Peru, *Ill.* **14** 41 20N 89 8W
Peru, *Ind.* **15** 40 45N 86 4W
Peshtigo, *Mich.* **10** 45 4N 87 46W
Petaluma, *Calif.* **18** 38 14N 122 39W
Peterborough, *N.H.* **9** 42 53N 71 57W
Petersburg, *Alaska* **30** 56 48N 132 58W
Petersburg, *Ind.* **15** 38 30N 87 17W
Petersburg, *Va.* **10** 37 14N 77 24W
Petersburg, *W. Va.* **10** 39 1N 79 5W
Petit Bois I., *Miss.* **11** 30 12N 88 26W
Petoskey, *Mich.* **10** 45 22N 84 57W
Phelps, *N.Y.* **8** 42 58N 77 3W
Phelps, *Wis.* **12** 46 4N 89 5W
Phenix City, *Ala.* **11** 32 28N 85 0W
Philadelphia, *Miss.* **13** 32 46N 89 7W
Philadelphia, *Pa.* **24** 39 58N 75 10W
Philadelphia Airport, *Pa.* **24** 40 4N 75 1W
Philadelphia International Airport, *Pa.* **24** 39 52N 75 14W
Philip, *S. Dak.* **12** 44 2N 101 40W
Philip Smith Mts., *Alaska* **30** 68 0N 146 0W
Philipsburg, *Mont.* **16** 46 20N 113 18W
Phillips, *Tex.* **13** 35 42N 101 22W
Phillips, *Wis.* **12** 45 42N 90 24W
Phillipsburg, *Kans.* **12** 39 45N 99 19W
Phillipsburg, *N.J.* **9** 40 42N 75 12W
Philmont, *N.Y.* **9** 42 15N 73 39W
Philomath, *Oreg.* **16** 44 32N 123 22W
Phoenix, *Ariz.* **17** 33 27N 112 4W
Phoenix, *N.Y.* **9** 43 14N 76 18W
Phoenixville, *Pa.* **24** 40 7N 75 31W
Picayune, *Miss.* **13** 30 32N 89 41W
Pico Rivera, *Calif.* **29** 33 59N 118 5W
Piedmont, *Ala.* **11** 33 55N 85 37W
Piedmont Plateau, *S.C.* **11** 34 0N 81 30W
Pierce, *Idaho* **16** 46 30N 115 48W
Pierre, *S. Dak.* **12** 44 22N 100 21W
Pigeon, *Mich.* **10** 43 50N 83 16W
Piggott, *Ark.* **13** 36 23N 90 11W
Pikes Peak, *Colo.* **12** 38 50N 105 3W
Pikesville, *Md.* **25** 39 22N 76 41W
Pikeville, *Ky.* **10** 37 29N 82 31W
Pilot Point, *Tex.* **13** 33 24N 96 58W
Pilot Rock, *Oreg.* **16** 45 29N 118 50W
Pima, *Ariz.* **17** 32 54N 109 50W
Pimmit Hills, *Va.* **25** 38 54N 77 12W
Pinckneyville, *Ill.* **14** 38 5N 89 23W
Pine, *Ariz.* **17** 34 23N 111 27W
Pine Bluff, *Ark.* **13** 34 13N 92 1W
Pine City, *Minn.* **12** 45 50N 92 59W
Pine Flat L., *Calif.* **18** 36 50N 119 20W
Pine Ridge, *S. Dak.* **12** 43 2N 102 33W
Pine River, *Minn.* **12** 46 43N 94 24W
Pine Valley, *Calif.* **19** 32 50N 116 32W
Pinecrest, *Calif.* **18** 38 12N 120 1W
Pinedale, *Calif.* **18** 36 50N 119 48W
Pinehurst, *Mass.* **23** 42 31N 71 12W
Pinetop, *Ariz.* **17** 34 8N 109 56W
Pinetree, *Wyo.* **16** 43 42N 105 52W
Pineville, *Ky.* **11** 36 46N 83 42W
Pineville, *La.* **13** 31 19N 92 26W
Pinnacles, *Calif.* **18** 36 33N 121 19W
Pinon Hills, *Calif.* **19** 34 26N 117 39W
Pinos, Mt., *Calif.* **19** 34 49N 119 8W
Pinos Pt., *Calif.* **17** 36 38N 121 57W
Pioche, *Nev.* **17** 37 56N 114 27W
Pipestone, *Minn.* **12** 44 0N 96 19W
Piqua, *Ohio* **15** 40 9N 84 15W
Piru, *Calif.* **19** 34 25N 118 48W
Piscataway, *N.J.* **20** 40 34N 74 27W
Pismo Beach, *Calif.* **19** 35 9N 120 38W
Pittsburg, *Kans.* **13** 37 25N 94 42W
Pittsburg, *Tex.* **13** 33 0N 94 59W
Pittsburgh, *Pa.* **27** 40 26N 79 59W
Pittsfield, *Ill.* **14** 39 36N 90 49W
Pittsfield, *Mass.* **9** 42 27N 73 15W
Pittsfield, *N.H.* **9** 43 18N 71 20W
Pittston, *Pa.* **9** 41 19N 75 47W
Pixley, *Calif.* **18** 35 58N 119 18W
Placerville, *Calif.* **18** 38 44N 120 48W
Plain Dealing, *La.* **13** 32 54N 93 42W
Plainfield, *N.J.* **20** 40 36N 74 24W
Plainfield, *N.J.* **9** 40 37N 74 25W
Plains, *Kans.* **13** 37 16N 100 35W
Plains, *Mont.* **16** 47 28N 114 53W
Plains, *Tex.* **13** 33 11N 102 50W
Plainview, *Nebr.* **12** 42 21N 97 47W
Plainview, *Tex.* **13** 34 11N 101 43W
Plainville, *Kans.* **12** 39 14N 99 18W
Plainwell, *Mich.* **10** 42 27N 85 38W
Planada, *Calif.* **18** 37 16N 120 19W
Plankinton, *S. Dak.* **12** 43 43N 98 29W
Plano, *Tex.* **13** 33 1N 96 42W
Plant City, *Fla.* **11** 28 1N 82 7W
Plaquemine, *La.* **13** 30 17N 91 14W
Plateau du Coteau du Missouri, *N. Dak.* **12** 47 9N 101 5W
Platinum, *Alaska* **30** 59 1N 161 49W
Platte, *S. Dak.* **12** 43 23N 98 51W
Platte →, *Mo.* **14** 39 16N 94 50W
Platteville, *Colo.* **12** 40 13N 104 49W
Plattsburgh, *N.Y.* **9** 44 42N 73 28W
Plattsmouth, *Nebr.* **12** 41 1N 95 53W
Pleasant Hill, *Calif.* **28** 37 56N 122 4W
Pleasant Hill, *Mo.* **14** 38 47N 94 16W
Pleasant Hills, *Pa.* **27** 40 20N 79 58W
Pleasanton, *Tex.* **13** 28 58N 98 29W
Pleasantville, *N.J.* **10** 39 24N 74 32W
Plentywood, *Mont.* **12** 48 47N 104 34W
Plum I., *N.Y.* **9** 41 11N 72 12W
Plummer, *Idaho* **16** 47 20N 116 53W
Plymouth, *Calif.* **18** 38 29N 120 51W
Plymouth, *Ind.* **15** 41 21N 86 19W
Plymouth, *Mass.* **9** 41 57N 70 40W
Plymouth, *N.C.* **11** 35 52N 76 43W
Plymouth, *Pa.* **9** 41 14N 75 57W
Plymouth, *Wis.* **10** 43 45N 87 59W
Plymouth Meeting, *Pa.* **24** 40 6N 75 17W
Pocahontas, *Ark.* **13** 36 16N 90 58W
Pocahontas, *Iowa* **14** 42 44N 94 40W
Pocatello, *Idaho* **16** 42 52N 112 27W
Pocomoke City, *Md.* **10** 38 5N 75 34W
Pohnpei, *Pac. Oc.* **31** 6 55N 158 10 E
Point Baker, *Alaska* **30** 56 21N 133 37W
Point Hope, *Alaska* **30** 68 21N 166 47W
Point Lay, *Alaska* **30** 69 46N 163 3W
Point Pleasant, *W. Va.* **10** 38 51N 82 8W
Pointe-à-la Hache, *La.* **13** 29 35N 89 55W
Pojoaque Valley, *N. Mex.* **17** 35 54N 106 1W
Polacca, *Ariz.* **17** 35 50N 110 23W
Pollock, *S. Dak.* **12** 45 55N 100 17W
Polo, *Ill.* **14** 41 59N 89 35W
Polson, *Mont.* **16** 47 41N 114 9W
Pomeroy, *Ohio* **10** 39 2N 82 2W
Pomeroy, *Wash.* **16** 46 28N 117 36W
Pomona, *Calif.* **19** 34 4N 117 45W
Pompano Beach, *Fla.* **11** 26 14N 80 8W
Pompeys Pillar, *Mont.* **16** 45 59N 107 57W
Pompton Plains, *N.J.* **20** 40 58N 74 18W
Ponca, *Nebr.* **12** 42 34N 96 43W
Ponca City, *Okla.* **13** 36 42N 97 5W
Ponce, *Puerto Rico* **30** 18 1N 66 37W
Ponchatoula, *La.* **13** 30 26N 90 26W
Pond, *Calif.* **19** 35 43N 119 20W
Pontchartrain L., *La.* **13** 30 5N 90 5W
Pontiac, *Ill.* **15** 40 53N 88 38W
Pontiac, *Mich.* **26** 42 38N 83 17W
Poorman, *Alaska* **30** 64 5N 155 48W
Poplar, *Mont.* **12** 48 7N 105 12W
Poplar Bluff, *Mo.* **13** 36 46N 90 24W
Poplarville, *Miss.* **13** 30 51N 89 32W
Porcupine →, *Alaska* **30** 66 34N 145 19W
Port Alexander, *Alaska* **30** 56 15N 134 38W
Port Allegany, *Pa.* **8** 41 48N 78 17W
Port Allen, *La.* **13** 30 27N 91 12W
Port Angeles, *Wash.* **18** 48 7N 123 27W
Port Aransas, *Tex.* **13** 27 50N 97 4W
Port Arthur, *Tex.* **13** 29 54N 93 56W
Port Austin, *Mich.* **8** 44 3N 83 1W
Port Chester, *N.Y.* **21** 41 0N 73 40W
Port Clinton, *Ohio* **15** 41 31N 82 56W
Port Gibson, *Miss.* **13** 31 58N 90 59W
Port Heiden, *Alaska* **30** 56 55N 158 41W
Port Henry, *N.Y.* **9** 44 3N 73 28W
Port Hueneme, *Calif.* **19** 34 7N 119 12W
Port Huron, *Mich.* **10** 42 58N 82 26W
Port Isabel, *Tex.* **13** 26 5N 97 12W
Port Jefferson, *N.Y.* **9** 40 57N 73 3W
Port Jervis, *N.Y.* **9** 41 22N 74 41W
Port Lavaca, *Tex.* **13** 28 37N 96 38W
Port O'Connor, *Tex.* **13** 28 26N 96 24W
Port Orchard, *Wash.* **18** 47 32N 122 38W
Port Orford, *Oreg.* **16** 42 45N 124 30W
Port Richmond, *N.Y.* **20** 40 38N 74 7W
Port St. Joe, *Fla.* **11** 29 49N 85 18W
Port Sanilac, *Mich.* **8** 43 26N 82 33W
Port Townsend, *Wash.* **18** 48 7N 122 45W
Port Washington, *N.Y.* **21** 40 49N 73 41W
Port Washington, *Wis.* **10** 43 23N 87 53W
Portage, *Wis.* **12** 43 33N 89 28W
Portageville, *Mo.* **13** 36 26N 89 42W
Portales, *N. Mex.* **13** 34 11N 103 20W
Porterville, *Calif.* **18** 36 4N 119 1W
Porthill, *Idaho* **16** 48 59N 116 30W
Portland, *Conn.* **9** 41 34N 72 38W
Portland, *Maine* **11** 43 39N 70 16W
Portland, *Mich.* **15** 42 52N 84 54W
Portland, *Oreg.* **18** 45 32N 122 37W
Portola, *Calif.* **18** 39 49N 120 28W
Portsmouth, *N.H.* **9** 43 5N 70 45W
Portsmouth, *Ohio* **10** 38 44N 82 57W
Portsmouth, *R.I.* **9** 41 36N 71 15W
Portsmouth, *Va.* **10** 36 50N 76 18W
Post, *Tex.* **13** 33 12N 101 23W
Post Falls, *Idaho* **16** 47 43N 116 57W
Poteau, *Okla.* **13** 35 3N 94 37W
Poteet, *Tex.* **13** 29 2N 98 35W
Potomac →, *Md.* **10** 38 0N 76 23W
Potsdam, *N.Y.* **9** 44 40N 74 59W
Potter, *Nebr.* **12** 41 13N 103 19W
Pottstown, *Pa.* **9** 40 15N 75 39W
Pottsville, *Pa.* **9** 40 41N 76 12W
Poughkeepsie, *N.Y.* **9** 41 42N 73 56W
Poulsbo, *Wash.* **18** 47 44N 122 39W
Poway, *Calif.* **19** 32 58N 117 2W
Powder →, *Mont.* **12** 46 45N 105 26W
Powder River, *Wyo.* **16** 43 2N 106 59W
Powell, *Wyo.* **16** 44 45N 108 46W
Powell L., *Utah* **17** 36 57N 111 29W
Powers, *Mich.* **10** 45 41N 87 32W
Powers, *Oreg.* **16** 42 53N 124 4W
Powers Lake, *N. Dak.* **12** 48 34N 102 39W
Pozo, *Calif.* **19** 35 20N 120 24W
Prairie →, *Tex.* **13** 34 30N 99 23W
Prairie City, *Oreg.* **16** 44 28N 118 43W
Prairie du Chien, *Wis.* **14** 43 3N 91 9W
Pratt, *Kans.* **13** 37 39N 98 44W
Prattville, *Ala.* **11** 32 28N 86 29W
Premont, *Tex.* **13** 27 22N 98 7W
Prentice, *Wis.* **12** 45 33N 90 17W
Prescott, *Ariz.* **17** 34 33N 112 28W
Prescott, *Ark.* **13** 33 48N 93 23W
Presho, *S. Dak.* **12** 43 54N 100 3W
Presidio, *Tex.* **13** 29 34N 104 22W
Presque Isle, *Maine* **11** 46 41N 68 1W
Preston, *Idaho* **16** 42 6N 111 53W
Preston, *Minn.* **12** 43 40N 92 5W
Preston, *Nev.* **16** 38 55N 115 4W
Price, *Utah* **16** 39 36N 110 49W
Prichard, *Ala.* **11** 30 44N 88 5W
Priest →, *Idaho* **16** 48 12N 116 54W
Priest L., *Idaho* **16** 48 35N 116 52W
Priest Valley, *Calif.* **18** 36 10N 120 39W
Prince of Wales, C., *Alaska* **30** 65 36N 168 5W
Prince of Wales I., *Alaska* **30** 55 47N 132 50W
Prince William Sd., *Alaska* **30** 60 40N 147 0W
Princeton, *Ill.* **14** 41 23N 89 28W
Princeton, *Ind.* **15** 38 21N 87 34W
Princeton, *Ky.* **10** 37 7N 87 53W
Princeton, *Mo.* **14** 40 24N 93 35W
Princeton, *N.J.* **9** 40 21N 74 39W
Princeton, *W. Va.* **10** 37 22N 81 6W
Prineville, *Oreg.* **16** 44 18N 120 51W
Prospect Heights, *Ill.* **22** 42 6N 87 54W
Prosser, *Wash.* **16** 46 12N 119 46W
Protection, *Kans.* **13** 37 12N 99 29W
Providence, *Ky.* **10** 37 24N 87 46W
Providence, *R.I.* **9** 41 49N 71 24W
Providence Mts., *Calif.* **17** 35 10N 115 15W
Provo, *Utah* **16** 40 14N 111 39W
Prudhoe Bay, *Alaska* **30** 70 18N 148 22W
Pryor, *Okla.* **13** 36 19N 95 19W
Pueblo, *Colo.* **12** 38 16N 104 37W
Puerco →, *N. Mex.* **17** 34 22N 107 50W
Puerto Rico ■, *W. Indies* **30** 18 15N 66 45W
Puget Sound, *Wash.* **16** 47 50N 122 30W
Pukoo, *Hawaii* **30** 21 4N 156 48W
Pulaski, *N.Y.* **9** 43 34N 76 8W
Pulaski, *Tenn.* **11** 35 12N 87 2W
Pulaski, *Va.* **10** 37 3N 80 47W
Pullman, *Wash.* **16** 46 44N 117 10W
Punta Gorda, *Fla.* **11** 26 56N 82 3W
Punxsatawney, *Pa.* **8** 40 57N 78 59W
Purcell, *Okla.* **13** 35 1N 97 22W
Putnam, *Conn.* **9** 41 55N 71 55W
Puyallup, *Wash.* **18** 47 12N 122 18W
Pyote, *Tex.* **13** 31 32N 103 8W
Pyramid L., *Nev.* **16** 40 1N 119 35W
Pyramid Pk., *Calif.* **19** 36 25N 116 37W

Q

Quakertown, *Pa.* **9** 40 26N 75 21W
Quanah, *Tex.* **13** 34 18N 99 44W
Quartzsite, *Ariz.* **19** 33 40N 114 13W
Queens, *N.Y.* **21** 40 42N 73 50W
Quemado, *N. Mex.* **17** 34 20N 108 30W
Quemado, *Tex.* **13** 28 58N 100 35W
Questa, *N. Mex.* **17** 36 42N 105 36W
Quincy, *Calif.* **18** 39 56N 120 57W
Quincy, *Fla.* **11** 30 35N 84 34W
Quincy, *Ill.* **12** 39 56N 91 23W
Quincy, *Mass.* **23** 42 14N 71 0W
Quincy, *Wash.* **16** 47 22N 119 56W
Quinhagak, *Alaska* **30** 59 45N 161 54W
Quitman, *Ga.* **11** 30 47N 83 34W
Quitman, *Miss.* **11** 32 2N 88 44W
Quitman, *Tex.* **13** 32 48N 95 27W

R

Racine, *Wis.* **15** 42 41N 87 51W
Radford, *Va.* **10** 37 8N 80 34W
Rahway, *N.J.* **20** 40 36N 74 17W
Rainier, *Wash.* **18** 46 53N 122 41W
Rainier, Mt., *Wash.* **18** 46 52N 121 46W
Raleigh, *N.C.* **11** 35 47N 78 39W
Raleigh B., *N.C.* **11** 34 50N 76 15W
Ralls, *Tex.* **13** 33 41N 101 24W
Ramona, *Calif.* **19** 33 2N 116 52W
Rampart, *Alaska* **30** 65 30N 150 10W
Ranchester, *Wyo.* **16** 44 54N 107 10W
Randallstown, *Md.* **25** 39 21N 76 46W
Randolph, *Mass.* **9** 42 10N 71 2W
Randolph, *Utah* **16** 41 40N 111 11W
Rangeley, *Maine* **9** 44 58N 70 39W
Rangely, *Colo.* **16** 40 5N 108 48W
Ranger, *Tex.* **13** 32 28N 98 41W
Rankin, *Tex.* **13** 31 13N 101 56W
Rantoul, *Ill.* **15** 40 19N 88 9W
Rapid City, *S. Dak.* **12** 44 5N 103 14W
Rapid River, *Mich.* **10** 45 55N 86 58W
Rat Islands, *Alaska* **30** 52 0N 178 0 E
Raton, *N. Mex.* **13** 36 54N 104 24W
Ravena, *N.Y.* **9** 42 28N 73 49W
Ravenna, *Nebr.* **12** 41 1N 98 55W
Ravenswood, *W. Va.* **10** 38 57N 81 46W
Rawlins, *Wyo.* **16** 41 47N 107 14W
Ray, *N. Dak.* **12** 48 21N 103 10W
Ray Mts., *Alaska* **30** 66 0N 152 0W
Raymond, *Calif.* **18** 37 13N 119 54W
Raymond, *Wash.* **18** 46 41N 123 44W
Raymondville, *Tex.* **13** 26 29N 97 47W
Rayne, *La.* **13** 30 14N 92 16W
Rayville, *La.* **13** 32 29N 91 46W
Reading, *Mass.* **23** 42 31N 71 5W
Reading, *Pa.* **9** 40 20N 75 56W
Red →, *La.* **13** 31 1N 91 45W
Red →, *N. Dak.* **12** 49 0N 97 15W
Red Bank, *N.J.* **9** 40 21N 74 5W
Red Bluff, *Calif.* **16** 40 11N 122 15W
Red Bluff L., *N. Mex.* **13** 31 54N 103 55W
Red Cloud, *Nebr.* **12** 40 5N 98 32W
Red Lake Falls, *Minn.* **12** 47 53N 96 16W
Red Lodge, *Mont.* **16** 45 11N 109 15W
Red Mountain, *Calif.* **19** 35 37N 117 38W
Red Oak, *Iowa* **12** 41 1N 95 14W
Red Rock, L., *Iowa* **14** 41 22N 92 59W
Red Slate Mt., *Calif.* **18** 37 31N 118 52W
Red Wing, *Minn.* **12** 44 34N 92 31W
Redding, *Calif.* **16** 40 35N 122 24W
Redfield, *S. Dak.* **12** 44 53N 98 31W
Redford Township, *Mich.* **26** 42 23N 83 17W
Redlands, *Calif.* **19** 34 4N 117 11W
Redmond, *Oreg.* **16** 44 17N 121 11W
Redwood City, *Calif.* **28** 37 29N 122 13W
Redwood Falls, *Minn.* **12** 44 32N 95 7W
Reed City, *Mich.* **10** 43 53N 85 31W
Reeder, *N. Dak.* **12** 46 7N 102 57W
Reedley, *Calif.* **18** 36 36N 119 27W
Reedsburg, *Wis.* **12** 43 32N 90 0W
Reedsport, *Oreg.* **16** 43 42N 124 6W
Refugio, *Tex.* **13** 28 18N 97 17W
Reidsville, *N.C.* **11** 36 21N 79 40W
Reinbeck, *Iowa* **14** 42 19N 92 36W
Reno, *Nev.* **18** 39 31N 119 48W
Renovo, *Pa.* **8** 41 20N 77 45W
Rensselaer, *Ind.* **15** 40 57N 87 9W
Rensselaer, *N.Y.* **9** 42 38N 73 45W
Renton, *Wash.* **18** 47 29N 122 12W
Republic, *Mich.* **10** 46 25N 87 59W
Republic, *Wash.* **16** 48 39N 118 44W
Republican →, *Kans.* **12** 39 4N 96 48W
Republican City, *Nebr.* **12** 40 6N 99 13W
Reserve, *N. Mex.* **17** 33 43N 108 45W
Reston, *Va.* **25** 38 57N 77 20W
Revere, *Mass.* **23** 42 25N 71 1W
Rex, *Alaska* **30** 64 10N 149 20W
Rexburg, *Idaho* **16** 43 49N 111 47W
Reyes, Pt., *Calif.* **18** 38 0N 123 0W
Rhinelander, *Wis.* **12** 45 38N 89 25W
Rhode Island □ **9** 41 40N 71 30W
Rice Lake, *Wis.* **12** 45 30N 91 44W
Rich Hill, *Mo.* **13** 38 6N 94 22W
Richardton, *N. Dak.* **12** 46 53N 102 19W
Richey, *Mont.* **12** 47 39N 105 4W
Richfield, *Idaho* **16** 43 3N 114 9W
Richfield, *Utah* **17** 38 46N 112 5W
Richland, *Ga.* **11** 32 5N 84 40W
Richland, *Oreg.* **16** 44 46N 117 10W
Richland, *Wash.* **16** 46 17N 119 18W
Richland Center, *Wis.* **12** 43 21N 90 23W
Richlands, *Va.* **10** 37 6N 81 48W
Richmond, *Calif.* **28** 37 56N 122 22W
Richmond, *Ind.* **15** 39 50N 84 53W
Richmond, *Ky.* **15** 37 45N 84 18W
Richmond, *Mo.* **12** 39 17N 93 58W
Richmond, *Tex.* **13** 29 35N 95 46W
Richmond, *Utah* **16** 41 56N 111 48W
Richmond, *Va.* **10** 37 33N 77 27W
Richmond Hill, *N.Y.* **21** 40 41N 73 50W
Richton, *Miss.* **11** 31 16N 88 56W
Richwood, *W. Va.* **10** 38 14N 80 32W
Ridgecrest, *Calif.* **19** 35 38N 117 40W
Ridgeland, *S.C.* **11** 32 29N 80 59W
Ridgewood, *N.J.* **20** 40 59N 74 6W
Ridgewood, *N.Y.* **21** 40 42N 73 52W
Ridgway, *Pa.* **8** 41 25N 78 44W
Rifle, *Colo.* **16** 39 32N 107 47W
Rigby, *Idaho* **16** 43 40N 111 55W
Riggins, *Idaho* **16** 45 25N 116 19W
Riley, *Oreg.* **16** 43 32N 119 28W
Rimrock, *Wash.* **18** 46 38N 121 10W
Ringling, *Mont.* **16** 46 16N 110 49W
Rio Grande →, *Tex.* **13** 25 57N 97 9W
Rio Grande City, *Tex.* **13** 26 23N 98 49W
Rio Vista, *Calif.* **18** 38 10N 121 42W
Ripley, *Tenn.* **13** 35 45N 89 32W
Ripon, *Calif.* **18** 37 44N 121 7W
Ripon, *Wis.* **10** 43 51N 88 50W
Rison, *Ark.* **13** 33 58N 92 11W
Ritzville, *Wash.* **16** 47 8N 118 23W
River Rouge, *Mich.* **26** 42 16N 83 8W
Riverdale, *Calif.* **18** 36 26N 119 52W
Riverdale, *N.Y.* **21** 40 54N 73 54W
Riverhead, *N.Y.* **9** 40 55N 72 40W
Riverside, *Calif.* **19** 33 59N 117 22W

Riverside, *Ill.* **22** 41 49N 87 48W
Riverside, *N.J.* **24** 40 2N 74 58W
Riverside, *Wyo.* **16** 41 13N 106 47W
Riverton, *Wyo.* **16** 43 2N 108 23W
Riverview, *Mich.* **26** 42 10N 83 11W
Roanoke, *Ala.* **11** 33 9N 85 22W
Roanoke, *Va.* **10** 37 16N 79 56W
Roanoke →, *N.C.* **11** 35 57N 76 42W
Roanoke I., *Ala.* **11** 35 55N 75 40W
Roanoke Rapids, *N.C.* **11** 36 28N 77 40W
Robert Lee, *Tex.* **13** 31 54N 100 29W
Roberts, *Idaho* **16** 43 43N 112 8W
Robstown, *Tex.* **13** 27 47N 97 40W
Rochelle, *Ill.* **14** 41 56N 89 4W
Rochester, *Ind.* **15** 41 4N 86 13W
Rochester, *Minn.* **12** 44 1N 92 28W
Rochester, *N.H.* **9** 43 18N 70 59W
Rochester, *N.Y.* **8** 43 10N 77 37W
Rock Hill, *S.C.* **11** 34 56N 81 1W
Rock Island, *Ill.* **14** 41 30N 90 34W
Rock Rapids, *Iowa* **12** 43 26N 96 10W
Rock River, *Wyo.* **16** 41 44N 105 58W
Rock Springs, *Mont.* **16** 46 49N 106 15W
Rock Springs, *Wyo.* **16** 41 35N 109 14W
Rock Valley, *Iowa* **12** 43 12N 96 18W
Rockdale, *Tex.* **13** 30 39N 97 0W
Rockford, *Ill.* **14** 42 16N 89 6W
Rocklake, *N. Dak.* **12** 48 47N 99 15W
Rockland, *Idaho* **16** 42 34N 112 53W
Rockland, *Maine* **11** 44 6N 69 7W
Rockland, *Mich.* **12** 46 44N 89 11W
Rockmart, *Ga.* **11** 34 0N 85 3W
Rockport, *Mo.* **12** 40 25N 95 31W
Rockport, *Tex.* **13** 28 2N 97 3W
Rocksprings, *Tex.* **13** 30 1N 100 13W
Rockville, *Conn.* **9** 41 52N 72 28W
Rockville, *Md.* **25** 39 4N 77 9W
Rockville Center, *N.Y.* **21** 40 39N 73 38W
Rockwall, *Tex.* **13** 32 56N 96 28W
Rockwell City, *Iowa* **14** 42 24N 94 38W
Rockwood, *Tenn.* **11** 35 52N 84 41W
Rocky Ford, *Colo.* **12** 38 3N 103 43W
Rocky Mount, *N.C.* **11** 35 57N 77 48W
Rocky Mts., *N. Amer.* **2** 39 0N 106 0W
Rocky River, *Ohio* **27** 41 28N 81 50W
Roebling, *N.J.* **9** 40 7N 74 47W
Rogers, *Ark.* **13** 36 20N 94 7W
Rogers City, *Mich.* **10** 45 25N 83 49W
Rogerson, *Idaho* **16** 42 13N 114 36W
Rogersville, *Tenn.* **11** 36 24N 83 1W
Rogue →, *Oreg.* **16** 42 26N 124 26W
Rohnert Park, *Calif.* **18** 38 16N 122 40W
Rojo, Cabo, *Puerto Rico* **30** 17 56N 67 12W
Rolette, *N. Dak.* **12** 48 40N 99 51W
Rolla, *Kans.* **13** 37 7N 101 38W
Rolla, *Mo.* **14** 37 57N 91 46W
Rolla, *N. Dak.* **12** 48 52N 99 37W
Romanzof C., *Alaska* **30** 61 49N 166 6W
Rome, *Ga.* **11** 34 15N 85 10W
Rome, *N.Y.* **9** 43 13N 75 27W
Romney, *W. Va.* **10** 39 21N 78 45W
Romulus, *Mich.* **26** 42 13N 83 23W
Ronan, *Mont.* **16** 47 32N 114 6W
Ronceverte, *W. Va.* **10** 37 45N 80 28W
Roof Butte, *Ariz.* **17** 36 28N 109 5W
Roosevelt, *Minn.* **12** 48 48N 95 6W
Roosevelt, *Utah* **16** 40 18N 109 59W
Roosevelt Res., *Ariz.* **17** 33 46N 111 0W
Ropesville, *Tex.* **13** 33 26N 102 9W
Rosalia, *Wash.* **16** 47 14N 117 22W
Rosamond, *Calif.* **19** 34 52N 118 10W
Roscoe, *S. Dak.* **12** 45 27N 99 20W
Roscommon, *Mich.* **10** 44 30N 84 35W
Roseau, *Minn.* **12** 48 51N 95 46W
Rosebud, *Tex.* **13** 31 4N 96 59W
Roseburg, *Oreg.* **16** 43 13N 123 20W
Rosedale, *Md.* **25** 39 19N 76 32W
Rosedale, *Miss.* **13** 33 51N 91 2W
Rosemead, *Calif.* **29** 34 4N 118 4W
Rosenberg, *Tex.* **13** 29 34N 95 49W
Roseville, *Calif.* **18** 38 45N 121 17W
Roseville, *Mich.* **26** 42 30N 82 57W
Ross, *Calif.* **28** 37 58N 122 33W
Ross L., *Wash.* **16** 48 44N 121 4W
Rossville, *N.Y.* **20** 40 32N 74 12W
Roswell, *N. Mex.* **13** 33 24N 104 32W
Rotan, *Tex.* **13** 32 51N 100 28W
Round Mountain, *Nev.* **16** 38 43N 117 4W
Roundup, *Mont.* **16** 46 27N 108 33W
Rouses Point, *N.Y.* **9** 44 59N 73 22W
Roxboro, *N.C.* **11** 36 24N 78 59W
Roy, *Mont.* **16** 47 20N 108 58W
Roy, *N. Mex.* **13** 35 57N 104 12W
Royal Oak, *Mich.* **26** 42 30N 83 9W
Ruby, *Alaska* **30** 64 45N 155 30W
Ruby L., *Nev.* **16** 40 10N 115 28W
Ruby Mts., *Nev.* **16** 40 30N 115 20W
Rudyard, *Mich.* **10** 46 14N 84 36W
Rugby, *N. Dak.* **12** 48 22N 100 0W
Ruidosa, *Tex.* **13** 29 59N 104 41W
Ruidoso, *N. Mex.* **17** 33 20N 105 41W
Rumford, *Maine* **9** 44 33N 70 33W
Rushford, *Minn.* **12** 43 49N 91 46W
Rushville, *Ill.* **14** 40 7N 90 34W
Rushville, *Ind.* **15** 39 37N 85 27W
Rushville, *Nebr.* **12** 42 43N 102 28W
Russell, *Kans.* **12** 38 54N 98 52W
Russellville, *Ala.* **11** 34 30N 87 44W
Russellville, *Ark.* **13** 35 17N 93 8W
Russellville, *Ky.* **11** 36 51N 86 53W
Russian Mission, *Alaska* **30** 61 47N 161 19W
Ruston, *La.* **13** 32 32N 92 38W
Ruth, *Nev.* **16** 39 17N 114 59W
Rye Patch Reservoir, *Nev.* **16** 40 28N 118 19W
Ryegate, *Mont.* **16** 46 18N 109 15W

S

Sabinal, *Tex.* **13** 29 19N 99 28W
Sabine →, *La.* **13** 29 59N 93 47W
Sabine L., *La.* **13** 29 53N 93 51W
Sabine Pass, *Tex.* **13** 29 44N 93 54W
Sac City, *Iowa* **14** 42 25N 95 0W
Saco, *Maine* **11** 43 30N 70 27W
Saco, *Mont.* **16** 48 28N 107 21W
Sacramento, *Calif.* **18** 38 35N 121 29W
Sacramento →, *Calif.* **18** 38 3N 121 56W
Sacramento Mts., *N. Mex.* **17** 32 30N 105 30W
Safford, *Ariz.* **17** 32 50N 109 43W
Sag Harbor, *N.Y.* **9** 41 0N 72 18W
Saginaw, *Mich.* **10** 43 26N 83 56W
Saginaw B., *Mich.* **10** 43 50N 83 40W
Saguache, *Colo.* **17** 38 5N 106 8W
Sahuarita, *Ariz.* **17** 31 57N 110 58W
St. Albans, *N.Y.* **21** 40 42N 73 44W
St. Albans, *Vt.* **9** 44 49N 73 5W
St. Albans, *W. Va.* **10** 38 23N 81 50W
St. Anthony, *Idaho* **16** 43 58N 111 41W
St. Augustine, *Fla.* **11** 29 54N 81 19W
St. Catherines I., *Ga.* **11** 31 40N 81 10W
St. Charles, *Ill.* **15** 41 54N 88 19W
St. Charles, *Mo.* **14** 38 47N 90 29W
St. Clair, *Pa.* **9** 40 43N 76 12W
St. Clair Shores, *Mich.* **26** 42 29N 82 54W
St. Cloud, *Fla.* **11** 28 15N 81 17W
St. Cloud, *Minn.* **12** 45 34N 94 10W
St. Croix, *Virgin Is.* **30** 17 45N 64 45W
St. Croix →, *Wis.* **12** 44 45N 92 48W
St. Croix Falls, *Wis.* **12** 45 24N 92 38W
St. Elias, Mt., *Alaska* **30** 60 18N 140 56W
St. Francis, *Kans.* **12** 39 47N 101 48W
St. Francis →, *Ark.* **13** 34 38N 90 36W
St. Francisville, *La.* **13** 30 47N 91 23W
St. George, *S.C.* **11** 33 11N 80 35W
St. George, *Utah* **17** 37 6N 113 35W
St. George, C., *Fla.* **11** 29 40N 85 5W
St. Helena, *Calif.* **16** 38 30N 122 28W
St. Helens, *Oreg.* **18** 45 52N 122 48W
St. Ignace, *Mich.* **10** 45 52N 84 44W
St. Ignatius, *Mont.* **16** 47 19N 114 6W
St. James, *Minn.* **12** 43 59N 94 38W
St. John, *Kans.* **13** 38 0N 98 46W
St. John, *N. Dak.* **12** 48 57N 99 43W
St. John →, *Maine* **11** 45 12N 66 5W
St. John I., *Virgin Is.* **30** 18 20N 64 42W
St. Johns, *Ariz.* **17** 34 30N 109 22W
St. Johns, *Mich.* **15** 43 0N 84 33W
St. Johns →, *Fla.* **11** 30 24N 81 24W
St. Johnsbury, *Vt.* **9** 44 25N 72 1W
St. Johnsville, *N.Y.* **9** 43 0N 74 43W
St. Joseph, *La.* **13** 31 55N 91 14W
St. Joseph, *Mich.* **15** 42 6N 86 29W
St. Joseph, *Mo.* **14** 39 46N 94 50W
St. Joseph →, *Mich.* **15** 42 7N 86 29W
St. Lawrence I., *Alaska* **30** 63 30N 170 30W
St. Louis, *Mich.* **10** 43 25N 84 36W
St. Louis, *Mo.* **14** 38 37N 90 12W
St. Louis →, *Minn.* **12** 47 15N 92 45W
St. Maries, *Idaho* **16** 47 19N 116 35W
St. Martinville, *La.* **13** 30 7N 91 50W
St. Marys, *Pa.* **8** 41 26N 78 34W
St. Michael, *Alaska* **30** 63 29N 162 2W
St. Paul, *Minn.* **12** 44 57N 93 6W
St. Paul, *Nebr.* **12** 41 13N 98 27W
St. Peter, *Minn.* **12** 44 20N 93 57W
St. Petersburg, *Fla.* **11** 27 46N 82 39W
St. Regis, *Mont.* **16** 47 18N 115 6W
St. Thomas I., *Virgin Is.* **30** 18 20N 64 55W
Ste. Genevieve, *Mo.* **14** 37 59N 90 2W
Saipan, *Pac. Oc.* **31** 15 12N 145 45 E
Sakakawea, L., *N. Dak.* **12** 47 30N 101 25W
Salamanca, *N.Y.* **8** 42 10N 78 43W
Salem, *Ind.* **15** 38 36N 86 6W
Salem, *Mass.* **23** 42 30N 70 55W
Salem, *Mo.* **13** 37 39N 91 32W
Salem, *N.J.* **10** 39 34N 75 28W
Salem, *Ohio* **8** 40 54N 80 52W
Salem, *Oreg.* **16** 44 56N 123 2W
Salem, *S. Dak.* **12** 43 44N 97 23W
Salem, *Va.* **10** 37 18N 80 3W
Salina, *Kans.* **12** 38 50N 97 37W
Salinas, *Calif.* **18** 36 40N 121 39W
Salinas →, *Calif.* **18** 36 45N 121 48W
Saline →, *Ark.* **13** 33 10N 92 8W
Saline →, *Kans.* **12** 38 52N 97 30W
Salisbury, *Md.* **10** 38 22N 75 36W
Salisbury, *N.C.* **11** 35 40N 80 29W
Sallisaw, *Okla.* **13** 35 28N 94 47W
Salmon, *Idaho* **16** 45 11N 113 54W
Salmon →, *Idaho* **16** 45 51N 116 47W
Salmon Falls, *Idaho* **16** 42 48N 114 59W
Salmon River Mts., *Idaho* **16** 45 0N 114 30W
Salome, *Ariz.* **19** 33 47N 113 37W
Salt →, *Ariz.* **17** 33 23N 112 19W
Salt Fork Arkansas →, *Okla.* **13** 36 36N 97 3W
Salt Lake City, *Utah* **16** 40 45N 111 53W
Salton City, *Calif.* **19** 33 29N 115 51W
Salton Sea, *Calif.* **19** 33 15N 115 45W
Saltville, *Va.* **10** 36 53N 81 46W
Saluda →, *S.C.* **11** 34 1N 81 4W
Salvador, L., *La.* **13** 29 43N 90 15W
Salyersville, *Ky.* **10** 37 45N 83 4W
Sam Rayburn Reservoir, *Tex.* **13** 31 4N 94 5W
San Andreas, *Calif.* **18** 38 12N 120 41W
San Andres Mts., *N. Mex.* **17** 33 0N 106 30W
San Angelo, *Tex.* **13** 31 28N 100 26W
San Anselmo, *Calif.* **18** 37 59N 122 34W
San Antonio, *N. Mex.* **17** 33 55N 106 52W
San Antonio, *Tex.* **13** 29 25N 98 30W
San Antonio →, *Tex.* **13** 28 30N 96 54W
San Antonio, Mt., *Calif.* **19** 34 17N 117 38W
San Ardo, *Calif.* **18** 36 1N 120 54W
San Augustine, *Tex.* **13** 31 30N 94 7W
San Benito, *Tex.* **13** 26 8N 97 38W
San Benito →, *Calif.* **18** 36 53N 121 34W
San Benito Mt., *Calif.* **18** 36 22N 120 37W
San Bernardino, *Calif.* **19** 34 7N 117 19W
San Blas, C., *Fla.* **11** 29 40N 85 21W
San Bruno, *Calif.* **28** 37 37N 122 24W
San Carlos, *Ariz.* **17** 33 21N 110 27W
San Carlos, *Calif.* **28** 37 30N 122 16W
San Carlos L., *Ariz.* **17** 33 11N 110 32W
San Clemente, *Calif.* **19** 33 26N 117 37W
San Clemente I., *Calif.* **19** 32 53N 118 29W
San Diego, *Calif.* **19** 32 43N 117 9W
San Diego, *Tex.* **13** 27 46N 98 14W
San Felipe →, *Calif.* **19** 33 12N 115 49W
San Fernando, *Calif.* **29** 34 17N 118 26W
San Francisco, *Calif.* **28** 37 46N 122 25W
San Francisco →, *Ariz.* **17** 32 59N 109 22W
San Francisco Bay, *Calif.* **28** 37 40N 122 15W
San Francisco International Airport, *Calif.* **28** 37 37N 122 22W
San Gabriel, *Calif.* **29** 34 5N 118 5W
San Germán, *Puerto Rico* **30** 18 5N 67 3W
San Gorgonio Mt., *Calif.* **19** 34 7N 116 51W
San Gregorio, *Calif.* **18** 37 20N 122 23W
San Jacinto, *Calif.* **19** 33 47N 116 57W
San Joaquin, *Calif.* **18** 36 36N 120 11W
San Joaquin →, *Calif.* **18** 38 4N 121 51W
San Joaquin Valley, *Calif.* **18** 37 20N 121 0W
San Jose, *Calif.* **18** 37 20N 121 53W
San Jose →, *N. Mex.* **17** 34 25N 106 45W
San Juan, *Dom. Rep.* **30** 18 49N 71 12W
San Juan, *Puerto Rico* **30** 18 28N 66 7W
San Juan →, *Utah* **17** 37 16N 110 26W
San Juan, C., *Puerto Rico* **30** 18 23N 65 37W
San Juan Bautista, *Calif.* **18** 36 51N 121 32W
San Juan Capistrano, *Calif.* **19** 33 30N 117 40W
San Juan Cr. →, *Calif.* **18** 35 40N 120 22W
San Juan Mts., *Colo.* **17** 37 30N 107 0W
San Leandro, *Calif.* **28** 37 44N 122 9W
San Lucas, *Calif.* **18** 36 8N 121 1W
San Luis, *Colo.* **17** 37 12N 105 25W
San Luis Obispo, *Calif.* **19** 35 17N 120 40W
San Luis Reservoir, *Calif.* **18** 37 4N 121 5W
San Marcos, *Tex.* **13** 29 53N 97 56W
San Marino, *Calif.* **29** 34 7N 118 5W
San Mateo, *Calif.* **28** 37 34N 122 20W
San Mateo Bridge, *Calif.* **28** 37 36N 122 11W
San Mateo County, *Calif.* **28** 37 30N 122 20W
San Miguel, *Calif.* **18** 35 45N 120 42W
San Miguel I., *Calif.* **19** 34 2N 120 23W
San Nicolas I., *Calif.* **19** 33 15N 119 30W
San Onofre, *Calif.* **19** 33 22N 117 34W
San Pedro →, *Ariz.* **17** 32 59N 110 47W
San Pedro Channel, *Calif.* **19** 33 30N 118 25W
San Quentin, *Calif.* **28** 37 56N 122 29W
San Rafael, *Calif.* **28** 37 58N 122 31W
San Rafael, *N. Mex.* **17** 35 7N 107 53W
San Rafael Mt., *Calif.* **19** 34 41N 119 52W
San Saba, *Tex.* **13** 31 12N 98 43W
San Simeon, *Calif.* **18** 35 39N 121 11W
San Simon, *Ariz.* **17** 32 16N 109 14W
San Ygnacio, *Tex.* **13** 27 3N 99 26W
Sanak I., *Alaska* **30** 54 25N 162 40W
Sand Point, *Alaska* **30** 55 20N 160 30W
Sand Springs, *Okla.* **13** 36 9N 96 7W
Sanders, *Ariz.* **17** 35 13N 109 20W
Sanderson, *Tex.* **13** 30 9N 102 24W
Sandpoint, *Idaho* **16** 48 17N 116 33W
Sandusky, *Mich.* **8** 43 25N 82 50W
Sandusky, *Ohio* **8** 41 27N 82 42W
Sandy Cr. →, *Wyo.* **16** 41 51N 109 47W
Sanford, *Fla.* **11** 28 48N 81 16W
Sanford, *Maine* **9** 43 27N 70 47W
Sanford, *N.C.* **11** 35 29N 79 10W
Sanger, *Calif.* **18** 36 42N 119 33W
Sangre de Cristo Mts., *N. Mex.* **13** 37 0N 105 0W
Santa Ana, *Calif.* **19** 33 46N 117 52W
Santa Barbara, *Calif.* **19** 34 25N 119 42W
Santa Barbara Channel, *Calif.* **19** 34 15N 120 0W
Santa Barbara I., *Calif.* **19** 33 29N 119 2W
Santa Catalina, Gulf of, *Calif.* **19** 33 10N 117 50W
Santa Catalina I., *Calif.* **19** 33 23N 118 25W
Santa Clara, *Calif.* **18** 37 21N 121 57W
Santa Clara, *Utah* **17** 37 8N 113 39W
Santa Cruz, *Calif.* **18** 36 58N 122 1W
Santa Cruz I., *Calif.* **19** 34 1N 119 43W
Santa Fe, *N. Mex.* **17** 35 41N 105 57W
Santa Fe Springs, *Calif.* **29** 33 56N 118 3W
Santa Lucia Range, *Calif.* **18** 36 0N 121 20W
Santa Margarita, *Calif.* **18** 35 23N 120 37W
Santa Margarita →, *Calif.* **19** 33 13N 117 23W
Santa Maria, *Calif.* **19** 34 57N 120 26W
Santa Monica, *Calif.* **29** 34 1N 118 29W
Santa Rita, *N. Mex.* **17** 32 48N 108 4W
Santa Rosa, *Calif.* **18** 38 26N 122 43W
Santa Rosa, *N. Mex.* **13** 34 57N 104 41W
Santa Rosa I., *Calif.* **19** 33 58N 120 6W
Santa Rosa I., *Fla.* **11** 30 20N 86 50W
Santa Rosa Range, *Nev.* **16** 41 45N 117 40W
Santa Ynez →, *Calif.* **19** 35 41N 120 36W
Santa Ynez Mts., *Calif.* **19** 34 30N 120 0W
Santa Ysabel, *Calif.* **19** 33 7N 116 40W
Santaquin, *Utah* **16** 39 59N 111 47W
Sapelo I., *Ga.* **11** 31 25N 81 12W
Sapulpa, *Okla.* **13** 35 59N 96 5W
Saranac Lake, *N.Y.* **9** 44 20N 74 8W
Sarasota, *Fla.* **11** 27 20N 82 32W
Saratoga, *Calif.* **18** 37 16N 122 2W
Saratoga, *Wyo.* **16** 41 27N 106 49W
Saratoga Springs, *N.Y.* **9** 43 5N 73 47W
Sargent, *Nebr.* **12** 41 39N 99 22W
Sarichef C., *Alaska* **30** 54 38N 164 59W
Sarita, *Tex.* **13** 27 13N 97 47W
Sarles, *N. Dak.* **12** 48 58N 99 0W
Satanta, *Kans.* **13** 37 26N 100 59W
Satilla →, *Ga.* **11** 30 59N 81 29W
Saugerties, *N.Y.* **9** 42 5N 73 57W
Saugus, *Mass.* **23** 42 28N 71 0W
Sauk Centre, *Minn.* **12** 45 44N 94 57W
Sauk Rapids, *Minn.* **12** 45 35N 94 10W
Sault Ste. Marie, *Mich.* **10** 46 30N 84 21W
Sausalito, *Calif.* **28** 37 51N 122 28W
Savage, *Mont.* **12** 47 27N 104 21W
Savanna, *Ill.* **14** 42 5N 90 8W
Savannah, *Ga.* **11** 32 5N 81 6W
Savannah, *Mo.* **14** 39 56N 94 50W
Savannah, *Tenn.* **11** 35 14N 88 15W
Savannah →, *Ga.* **11** 32 2N 80 53W
Sawatch Mts., *Colo.* **17** 38 30N 106 30W
Sayre, *Okla.* **13** 35 18N 99 38W
Sayre, *Pa.* **9** 41 59N 76 32W
Scammon Bay, *Alaska* **30** 61 51N 165 35W
Scenic, *S. Dak.* **12** 43 47N 102 33W
Schell Creek Ra., *Nev.* **16** 39 15N 114 30W
Schenectady, *N.Y.* **9** 42 49N 73 57W
Schiller Park, *Ill.* **22** 41 56N 87 52W
Schofield, *Wis.* **12** 44 54N 89 36W
Schurz, *Nev.* **16** 38 57N 118 49W
Schuyler, *Nebr.* **12** 41 27N 97 4W
Schuylkill Haven, *Pa.* **9** 40 37N 76 11W
Scioto →, *Ohio* **10** 38 44N 83 1W
Scobey, *Mont.* **12** 48 47N 105 25W
Scotia, *Calif.* **16** 40 29N 124 6W
Scotia, *N.Y.* **9** 42 50N 73 58W
Scotland, *S. Dak.* **12** 43 9N 97 43W
Scotland Neck, *N.C.* **11** 36 8N 77 25W
Scott City, *Kans.* **12** 38 29N 100 54W
Scottsbluff, *Nebr.* **12** 41 52N 103 40W
Scottsboro, *Ala.* **11** 34 40N 86 2W
Scottsburg, *Ind.* **15** 38 41N 85 47W
Scottsville, *Ky.* **11** 36 45N 86 11W
Scottville, *Mich.* **10** 43 58N 86 17W
Scranton, *Pa.* **9** 41 25N 75 40W
Seaford, *Del.* **10** 38 39N 75 37W
Seagraves, *Tex.* **13** 32 57N 102 34W
Sealy, *Tex.* **13** 29 47N 96 9W
Searchlight, *Nev.* **19** 35 28N 114 55W
Searcy, *Ark.* **13** 35 15N 91 44W
Searles L., *Calif.* **19** 35 44N 117 21W
Sears Tower, *Ill.* **22** 41 52N 87 37W
Seaside, *Calif.* **18** 36 37N 121 50W
Seaside, *Oreg.* **18** 46 0N 123 56W
Seat Pleasant, *Md.* **25** 38 53N 76 53W
Seattle, *Wash.* **18** 47 36N 122 20W
Sebastopol, *Calif.* **18** 38 24N 122 49W
Sebewaing, *Mich.* **10** 43 44N 83 27W
Sebring, *Fla.* **11** 27 30N 81 27W
Sedalia, *Mo.* **14** 38 42N 93 14W
Sedan, *Kans.* **13** 37 8N 96 11W
Sedro-Woolley, *Wash.* **18** 48 30N 122 14W
Seguam I., *Alaska* **30** 52 19N 172 30W
Seguam Pass, *Alaska* **30** 52 0N 172 30W
Seguin, *Tex.* **13** 29 34N 97 58W
Segula I., *Alaska* **30** 52 0N 177 50 E
Seiling, *Okla.* **13** 36 9N 98 56W
Selah, *Wash.* **16** 46 39N 120 32W
Selawik, *Alaska* **30** 66 36N 160 0W
Selby, *S. Dak.* **12** 45 31N 100 2W
Selden, *Kans.* **12** 39 33N 100 34W
Seldovia, *Alaska* **30** 59 26N 151 43W
Selfridge, *N. Dak.* **12** 46 2N 100 56W
Seligman, *Ariz.* **17** 35 20N 112 53W
Sells, *Ariz.* **17** 31 55N 111 53W
Selma, *Ala.* **11** 32 25N 87 1W
Selma, *Calif.* **18** 36 34N 119 37W
Selma, *N.C.* **11** 35 32N 78 17W

Selmer, *Tenn.* **11** 35 10N 88 36W
Seminoe Reservoir, *Wyo.* **16** 42 9N 106 55W
Seminole, *Okla.* **13** 35 14N 96 41W
Seminole, *Tex.* **13** 32 43N 102 39W
Semisopochnoi I., *Alaska* **30** 51 55N 179 36 E
Senatobia, *Miss.* **13** 34 37N 89 58W
Seneca, *Oreg.* **16** 44 8N 118 58W
Seneca, *S.C.* **11** 34 41N 82 57W
Seneca Falls, *N.Y.* **9** 42 55N 76 48W
Seneca L., *N.Y.* **8** 42 40N 76 54W
Sentinel, *Ariz.* **17** 32 52N 113 13W
Sequim, *Wash.* **18** 48 5N 123 6W
Sequoia National Park, *Calif.* **18** 36 30N 118 30W
Settlement Pt., *Bahamas* **11** 26 40N 79 0W
Sevier, *Utah* **17** 38 39N 112 11W
Sevier →, *Utah* **17** 39 4N 113 6W
Sevier L., *Utah* **16** 38 54N 113 9W
Seward, *Alaska* **30** 60 7N 149 27W
Seward, *Nebr.* **12** 40 55N 97 6W
Seward Pen., *Alaska* **30** 65 0N 164 0W
Seymour, *Conn.* **9** 41 24N 73 4W
Seymour, *Ind.* **15** 38 58N 85 53W
Seymour, *Tex.* **13** 33 35N 99 16W
Seymour, *Wis.* **10** 44 31N 88 20W
Shafter, *Calif.* **19** 35 30N 119 16W
Shafter, *Tex.* **13** 29 49N 104 18W
Shaker Heights, *Ohio* **27** 41 28N 81 33W
Shakopee, *Minn.* **12** 44 48N 93 32W
Shaktolik, *Alaska* **30** 64 30N 161 15W
Shamokin, *Pa.* **9** 40 47N 76 34W
Shamrock, *Tex.* **13** 35 13N 100 15W
Shandon, *Calif.* **18** 35 39N 120 23W
Shaniko, *Oreg.* **16** 45 0N 120 45W
Sharon, *Mass.* **9** 42 7N 71 11W
Sharon, *Pa.* **8** 41 14N 80 31W
Sharon Springs, *Kans.* **12** 38 54N 101 45W
Sharpsburg, *Pa.* **27** 40 29N 79 56W
Shasta, Mt., *Calif.* **16** 41 25N 122 12W
Shasta L., *Calif.* **16** 40 43N 122 25W
Shattuck, *Okla.* **13** 36 16N 99 53W
Shaver L., *Calif.* **18** 37 9N 119 18W
Shawano, *Wis.* **10** 44 47N 88 36W
Shawnee, *Okla.* **13** 35 20N 96 55W
Sheboygan, *Wis.* **10** 43 46N 87 45W
Sheffield, *Ala.* **11** 34 46N 87 41W
Sheffield, *Mass.* **9** 42 5N 73 21W
Sheffield, *Tex.* **13** 30 41N 101 49W
Shelburne Falls, *Mass.* **9** 42 36N 72 45W
Shelby, *Mich.* **10** 43 37N 86 22W
Shelby, *Mont.* **16** 48 30N 111 51W
Shelby, *N.C.* **11** 35 17N 81 32W
Shelbyville, *Ill.* **15** 39 24N 88 48W
Shelbyville, *Ind.* **15** 39 31N 85 47W
Shelbyville, *Tenn.* **11** 35 29N 86 28W
Sheldon, *Iowa* **12** 43 11N 95 51W
Sheldon Point, *Alaska* **30** 62 32N 164 52W
Shelikof Strait, *Alaska* **30** 57 30N 155 0W
Shelton, *Conn.* **9** 41 19N 73 5W
Shelton, *Wash.* **18** 47 13N 123 6W
Shenandoah, *Iowa* **12** 40 46N 95 22W
Shenandoah, *Pa.* **9** 40 49N 76 12W
Shenandoah, *Va.* **10** 38 29N 78 37W
Shenandoah →, *Va.* **10** 39 19N 77 44W
Sheridan, *Ark.* **13** 34 19N 92 24W
Sheridan, *Wyo.* **16** 44 48N 106 58W
Sherman, *Tex.* **13** 33 40N 96 35W
Sherwood, *N. Dak.* **12** 48 57N 101 38W
Sherwood, *Tex.* **13** 31 18N 100 45W
Sheyenne, *N. Dak.* **12** 47 50N 99 7W
Sheyenne →, *N. Dak.* **12** 47 2N 96 50W
Ship I., *Miss.* **13** 30 13N 88 55W
Shippensburg, *Pa.* **8** 40 3N 77 31W
Shiprock, *N. Mex.* **17** 36 47N 108 41W
Shishmaref, *Alaska* **30** 66 15N 166 4W
Shoshone, *Calif.* **19** 35 58N 116 16W
Shoshone, *Idaho* **16** 42 56N 114 25W
Shoshone L., *Wyo.* **16** 44 22N 110 43W
Shoshone Mts., *Nev.* **16** 39 20N 117 25W
Shoshoni, *Wyo.* **16** 43 14N 108 7W
Show Low, *Ariz.* **17** 34 15N 110 2W
Shreveport, *La.* **13** 32 31N 93 45W
Shumagin Is., *Alaska* **30** 55 7N 159 45W
Shungnak, *Alaska* **30** 66 52N 157 9W
Shuyak I., *Alaska* **30** 58 31N 152 30W
Sibley, *Iowa* **12** 43 24N 95 45W
Sibley, *La.* **13** 32 33N 93 18W
Sidney, *Mont.* **12** 47 43N 104 9W
Sidney, *N.Y.* **9** 42 19N 75 24W
Sidney, *Nebr.* **12** 41 8N 102 59W
Sidney, *Ohio* **15** 40 17N 84 9W
Sierra Blanca, *Tex.* **17** 31 11N 105 22W
Sierra Blanca Peak, *N. Mex.* **17** 33 23N 105 49W
Sierra City, *Calif.* **18** 39 34N 120 38W
Sierra Madre, *Calif.* **29** 34 9N 118 3W
Sigurd, *Utah* **17** 38 50N 111 58W
Sikeston, *Mo.* **13** 36 53N 89 35W
Siler City, *N.C.* **11** 35 44N 79 28W
Siloam Springs, *Ark.* **13** 36 11N 94 32W
Silsbee, *Tex.* **13** 30 21N 94 11W
Silver City, *N. Mex.* **17** 32 46N 108 17W
Silver City, *Nev.* **16** 39 15N 119 48W
Silver Cr. →, *Oreg.* **16** 43 16N 119 13W
Silver Creek, *N.Y.* **8** 42 33N 79 10W
Silver Hill, *Md.* **25** 38 49N 76 55W
Silver L., *Calif.* **18** 38 39N 120 6W
Silver L., *Calif.* **19** 35 21N 116 7W
Silver Lake, *Oreg.* **16** 43 8N 121 3W
Silver Spring, *Md.* **25** 39 0N 77 1W
Silverton, *Colo.* **17** 37 49N 107 40W
Silverton, *Tex.* **13** 34 28N 101 19W
Silvies →, *Oreg.* **16** 43 34N 119 2W
Simi Valley, *Calif.* **19** 34 16N 118 47W
Simmler, *Calif.* **19** 35 21N 119 59W
Sinclair, *Wyo.* **16** 41 47N 107 7W
Sinton, *Tex.* **13** 28 2N 97 31W
Sioux City, *Iowa* **12** 42 30N 96 24W
Sioux Falls, *S. Dak.* **12** 43 33N 96 44W
Sirretta Pk., *Calif.* **19** 35 56N 118 19W
Sisseton, *S. Dak.* **12** 45 40N 97 3W
Sisters, *Oreg.* **16** 44 18N 121 33W
Sitka, *Alaska* **30** 57 3N 135 20W
Skagway, *Alaska* **30** 59 28N 135 19W
Skokie, *Ill.* **22** 42 2N 87 42W
Skowhegan, *Maine* **11** 44 46N 69 43W
Skunk →, *Iowa* **14** 40 42N 91 7W
Skykomish, *Wash.* **16** 47 42N 121 22W
Slaton, *Tex.* **13** 33 26N 101 39W
Sleepy Eye, *Minn.* **12** 44 18N 94 43W
Slidell, *La.* **13** 30 17N 89 47W
Sloansville, *N.Y.* **9** 42 45N 74 22W
Sloughhouse, *Calif.* **18** 38 26N 121 12W
Smith Center, *Kans.* **12** 39 47N 98 47W
Smithfield, *N.C.* **11** 35 31N 78 21W
Smithfield, *Utah* **16** 41 50N 111 50W
Smithville, *Tex.* **13** 30 1N 97 10W
Smoky Hill →, *Kans.* **12** 39 4N 96 48W
Snake →, *Wash.* **16** 46 12N 119 2W
Snake Range, *Nev.* **16** 39 0N 114 20W
Snake River Plain, *Idaho* **16** 42 50N 114 0W
Snelling, *Calif.* **18** 37 31N 120 26W
Snohomish, *Wash.* **18** 47 55N 122 6W
Snow Hill, *Md.* **10** 38 11N 75 24W
Snowflake, *Ariz.* **17** 34 30N 110 5W
Snowshoe Pk., *Mont.* **16** 48 13N 115 41W
Snowville, *Utah* **16** 41 58N 112 43W
Snyder, *Okla.* **13** 34 40N 98 57W
Snyder, *Tex.* **13** 32 44N 100 55W
Soap Lake, *Wash.* **16** 47 23N 119 29W
Socorro, *N. Mex.* **17** 34 4N 106 54W
Soda L., *Calif.* **17** 35 10N 116 4W
Soda Springs, *Idaho* **16** 42 39N 111 36W
Sodus, *N.Y.* **8** 43 14N 77 4W
Soledad, *Calif.* **18** 36 26N 121 20W
Solomon, N. Fork →, *Kans.* **12** 39 29N 98 26W
Solomon, S. Fork →, *Kans.* **12** 39 25N 99 12W
Solon Springs, *Wis.* **12** 46 22N 91 49W
Solvang, *Calif.* **19** 34 36N 120 8W
Solvay, *N.Y.* **9** 43 3N 76 13W
Somers, *Mont.* **16** 48 5N 114 13W
Somerset, *Colo.* **17** 38 56N 107 28W
Somerset, *Ky.* **10** 37 5N 84 36W
Somerset, *Mass.* **9** 41 47N 71 8W
Somerton, *Ariz.* **17** 32 36N 114 43W
Somerville, *Mass.* **23** 42 23N 71 5W
Somerville, *N.J.* **9** 40 35N 74 38W
Sonora, *Calif.* **18** 37 59N 120 23W
Sonora, *Tex.* **13** 30 34N 100 39W
South Baldy, *N. Mex.* **17** 33 59N 107 11W
South Bend, *Ind.* **15** 41 41N 86 15W
South Bend, *Wash.* **18** 46 40N 123 48W
South Boston, *Va.* **11** 36 42N 78 54W
South C. = Ka Lae, *Hawaii* **30** 18 55N 155 41W
South Cape, *Hawaii* **30** 18 58N 155 24 E
South Carolina □ **11** 34 0N 81 0W
South Charleston, *W. Va.* **10** 38 22N 81 44W
South Dakota □ **12** 44 15N 100 0W
South Euclid, *Ohio* **27** 41 31N 81 32W
South Fork →, *Mont.* **16** 47 54N 113 15W
South Fork, American →, *Calif.* **18** 38 45N 121 5W
South Gate, *Calif.* **29** 33 56N 118 12W
South Haven, *Mich.* **15** 42 24N 86 16W
South Holland, *Ill.* **22** 41 36N 87 36W
South Loup →, *Nebr.* **12** 41 4N 98 39W
South Milwaukee, *Wis.* **15** 42 55N 87 52W
South Pasadena, *Calif.* **29** 34 7N 118 8W
South Pass, *Wyo.* **16** 42 20N 108 58W
South Pittsburg, *Tenn.* **11** 35 1N 85 42W
South Platte →, *Nebr.* **12** 41 7N 100 42W
South River, *N.J.* **9** 40 27N 74 23W
South San Francisco, *Calif.* **28** 37 39N 122 24W
South Sioux City, *Nebr.* **12** 42 28N 96 24W
Southampton, *N.Y.* **9** 40 53N 72 23W
Southbridge, *Mass.* **9** 42 5N 72 2W
Southeast C., *Alaska* **30** 62 56N 169 39W
Southern Pines, *N.C.* **11** 35 11N 79 24W
Southfield, *Mich.* **26** 42 28N 83 15W
Southgate, *Mich.* **26** 42 11N 83 12W
Southington, *Conn.* **9** 41 36N 72 53W
Southold, *N.Y.* **9** 41 4N 72 26W
Southport, *N.C.* **11** 33 55N 78 1W
Spalding, *Nebr.* **12** 41 42N 98 22W
Spanish Fork, *Utah* **16** 40 7N 111 39W
Sparks, *Nev.* **18** 39 32N 119 45W
Sparta, *Ga.* **11** 33 17N 82 58W
Sparta, *Wis.* **12** 43 56N 90 49W
Spartanburg, *S.C.* **11** 34 56N 81 57W
Spearfish, *S. Dak.* **12** 44 30N 103 52W
Spearman, *Tex.* **13** 36 12N 101 12W
Spenard, *Alaska* **30** 61 11N 149 55W
Spencer, *Idaho* **16** 44 22N 112 11W
Spencer, *Iowa* **12** 43 9N 95 9W
Spencer, *N.Y.* **9** 42 13N 76 30W
Spencer, *Nebr.* **12** 42 53N 98 42W
Spencer, *W. Va.* **10** 38 48N 81 21W
Spirit Lake, *Idaho* **16** 47 58N 116 52W
Spofford, *Tex.* **13** 29 10N 100 25W
Spokane, *Wash.* **16** 47 40N 117 24W
Spooner, *Wis.* **12** 45 50N 91 53W
Sprague, *Wash.* **16** 47 18N 117 59W
Sprague River, *Oreg.* **16** 42 27N 121 30W
Spray, *Oreg.* **16** 44 50N 119 48W
Spring City, *Utah* **16** 39 29N 111 30W
Spring Mts., *Nev.* **17** 36 0N 115 45W
Spring Valley, *Minn.* **12** 43 41N 92 23W
Springdale, *Ark.* **13** 36 11N 94 8W
Springdale, *Wash.* **16** 48 4N 117 45W
Springer, *N. Mex.* **13** 36 22N 104 36W
Springerville, *Ariz.* **17** 34 8N 109 17W
Springfield, *Colo.* **13** 37 24N 102 37W
Springfield, *Ill.* **14** 39 48N 89 39W
Springfield, *Mass.* **9** 42 6N 72 35W
Springfield, *Mo.* **13** 37 13N 93 17W
Springfield, *N.J.* **20** 40 42N 74 18W
Springfield, *Ohio* **15** 39 55N 83 49W
Springfield, *Oreg.* **16** 44 3N 123 1W
Springfield, *Tenn.* **11** 36 31N 86 53W
Springfield, *Va.* **25** 38 46N 77 10W
Springfield, *Vt.* **9** 43 18N 72 29W
Springvale, *Maine* **9** 43 28N 70 48W
Springville, *Calif.* **18** 36 8N 118 49W
Springville, *N.Y.* **8** 42 31N 78 40W
Springville, *Utah* **16** 40 10N 111 37W
Spur, *Tex.* **13** 33 28N 100 52W
Stafford, *Kans.* **13** 37 58N 98 36W
Stafford Springs, *Conn.* **9** 41 57N 72 18W
Stamford, *Conn.* **9** 41 3N 73 32W
Stamford, *Tex.* **13** 32 57N 99 48W
Stamps, *Ark.* **13** 33 22N 93 30W
Stanberry, *Mo.* **12** 40 13N 94 35W
Standish, *Mich.* **10** 43 59N 83 57W
Stanford, *Mont.* **16** 47 9N 110 13W
Stanislaus →, *Calif.* **18** 37 40N 121 14W
Stanley, *Idaho* **16** 44 13N 114 56W
Stanley, *N. Dak.* **12** 48 19N 102 23W
Stanley, *Wis.* **12** 44 58N 90 56W
Stanton, *Tex.* **13** 32 8N 101 48W
Staples, *Minn.* **12** 46 21N 94 48W
Stapleton, *N.Y.* **20** 40 36N 74 5W
Stapleton, *Nebr.* **12** 41 29N 100 31W
Starke, *Fla.* **11** 29 57N 82 7W
Starkville, *Colo.* **13** 37 8N 104 30W
Starkville, *Miss.* **11** 33 28N 88 49W
State College, *Pa.* **8** 40 48N 77 52W
Staten Island, *N.Y.* **20** 40 34N 74 9W
Statesboro, *Ga.* **11** 32 27N 81 47W
Statesville, *N.C.* **11** 35 47N 80 53W
Statue of Liberty, *N.J.* **20** 40 41N 74 2W
Stauffer, *Calif.* **19** 34 45N 119 3W
Staunton, *Ill.* **14** 39 1N 89 47W
Staunton, *Va.* **10** 38 9N 79 4W
Steamboat Springs, *Colo.* **16** 40 29N 106 50W
Steele, *N. Dak.* **12** 46 51N 99 55W
Steelton, *Pa.* **8** 40 14N 76 50W
Steelville, *Mo.* **14** 37 58N 91 22W
Stephen, *Minn.* **12** 48 27N 96 53W
Stephenville, *Tex.* **13** 32 13N 98 12W
Sterling, *Colo.* **12** 40 37N 103 13W
Sterling, *Ill.* **14** 41 48N 89 42W
Sterling, *Kans.* **12** 38 13N 98 12W
Sterling City, *Tex.* **13** 31 51N 101 0W
Sterling Heights, *Mich.* **26** 42 35N 83 3W
Steubenville, *Ohio* **8** 40 22N 80 37W
Stevens Point, *Wis.* **12** 44 31N 89 34W
Stevens Village, *Alaska* **30** 66 1N 149 6W
Stigler, *Okla.* **13** 35 15N 95 8W
Stillwater, *Minn.* **12** 45 3N 92 49W
Stillwater, *N.Y.* **9** 42 55N 73 41W
Stillwater, *Okla.* **13** 36 7N 97 4W
Stillwater Range, *Nev.* **16** 39 50N 118 5W
Stilwell, *Okla.* **13** 35 49N 94 38W
Stockett, *Mont.* **16** 47 21N 111 10W
Stockton, *Calif.* **18** 37 58N 121 17W
Stockton, *Kans.* **12** 39 26N 99 16W
Stockton, *Mo.* **14** 37 42N 93 48W
Stoneham, *Mass.* **23** 42 29N 71 5W
Stony River, *Alaska* **30** 61 47N 156 35W
Storm Lake, *Iowa* **12** 42 39N 95 13W
Stove Pipe Wells Village, *Calif.* **19** 36 35N 117 11W
Strasburg, *N. Dak.* **12** 46 8N 100 10W
Stratford, *Calif.* **18** 36 11N 119 49W
Stratford, *Conn.* **9** 41 12N 73 8W
Stratford, *Tex.* **13** 36 20N 102 4W
Strathmore, *Calif.* **18** 36 9N 119 4W
Stratton, *Colo.* **12** 39 19N 102 36W
Strawberry Reservoir, *Utah* **16** 40 8N 111 9W
Strawn, *Tex.* **13** 32 33N 98 30W
Streator, *Ill.* **15** 41 8N 88 50W
Streeter, *N. Dak.* **12** 46 39N 99 21W
Stromsburg, *Iowa* **12** 41 7N 97 36W
Stroudsburg, *Pa.* **9** 40 59N 75 12W
Struthers, *Ohio* **8** 41 4N 80 39W
Stryker, *Mont.* **16** 48 41N 114 46W
Stuart, *Fla.* **11** 27 12N 80 15W
Stuart, *Nebr.* **12** 42 36N 99 8W
Stuart I., *Alaska* **30** 63 55N 164 50W
Sturgeon Bay, *Wis.* **10** 44 50N 87 23W
Sturgis, *Mich.* **15** 41 48N 85 25W
Sturgis, *S. Dak.* **12** 44 25N 103 31W
Stuttgart, *Ark.* **13** 34 30N 91 33W
Stuyvesant, *N.Y.* **9** 42 23N 73 45W
Sudan, *Tex.* **13** 34 4N 102 32W
Suffolk, *Va.* **10** 36 44N 76 35W
Sugar City, *Colo.* **12** 38 14N 103 40W
Suitland, *Md.* **25** 38 50N 76 55W
Sullivan, *Ill.* **15** 39 36N 88 37W
Sullivan, *Ind.* **15** 39 6N 87 24W
Sullivan, *Mo.* **14** 38 13N 91 10W
Sulphur, *La.* **13** 30 14N 93 23W
Sulphur, *Okla.* **13** 34 31N 96 58W
Sulphur Springs, *Tex.* **13** 33 8N 95 36W
Sulphur Springs Draw →, *Tex.* **13** 32 12N 101 36W
Sumatra, *Mont.* **16** 46 37N 107 33W
Summer L., *Oreg.* **16** 42 50N 120 45W
Summerville, *Ga.* **11** 34 29N 85 21W
Summerville, *S.C.* **11** 33 1N 80 11W
Summit, *Alaska* **30** 63 20N 149 7W
Summit, *Ill.* **22** 41 47N 87 47W
Summit, *N.J.* **20** 40 43N 74 21W
Summit Peak, *Colo.* **17** 37 21N 106 42W
Sumner, *Iowa* **14** 42 51N 92 6W
Sumter, *S.C.* **11** 33 55N 80 21W
Sun City, *Ariz.* **17** 33 36N 112 17W
Sun City, *Calif.* **19** 33 42N 117 11W
Sunburst, *Mont.* **16** 48 53N 111 55W
Sunbury, *Pa.* **9** 40 52N 76 48W
Suncook, *N.H.* **9** 43 8N 71 27W
Sundance, *Wyo.* **12** 44 24N 104 23W
Sunnyside, *Utah* **16** 39 34N 110 23W
Sunnyside, *Wash.* **16** 46 20N 120 0W
Sunnyvale, *Calif.* **18** 37 23N 122 2W
Sunray, *Tex.* **13** 36 1N 101 49W
Sunshine Acres, *Calif.* **29** 33 56N 117 59W
Supai, *Ariz.* **17** 36 15N 112 41W
Superior, *Ariz.* **17** 33 18N 111 6W
Superior, *Mont.* **16** 47 12N 114 53W
Superior, *Nebr.* **12** 40 1N 98 4W
Superior, *Wis.* **12** 46 44N 92 6W
Superior, L., *N. Amer.* **10** 47 0N 87 0W
Sur, Pt., *Calif.* **18** 36 18N 121 54W
Surf, *Calif.* **19** 34 41N 120 36W
Susanville, *Calif.* **16** 40 25N 120 39W
Susquehanna →, *Pa.* **9** 39 33N 76 5W
Susquehanna Depot, *Pa.* **9** 41 57N 75 36W
Sussex, *N.J.* **9** 41 13N 74 37W
Sutherland, *Nebr.* **12** 41 10N 101 8W
Sutherlin, *Oreg.* **16** 43 23N 123 19W
Sutter Creek, *Calif.* **18** 38 24N 120 48W
Sutton, *Nebr.* **12** 40 36N 97 52W
Sutwik I., *Alaska* **30** 56 34N 157 12W
Suwannee →, *Fla.* **11** 29 17N 83 10W
Swainsboro, *Ga.* **11** 32 36N 82 20W
Swampscott, *Mass.* **23** 42 28N 70 53W
Swarthmore, *Pa.* **24** 39 54N 75 20W
Sweet Home, *Oreg.* **16** 44 24N 122 44W
Sweetwater, *Nev.* **18** 38 27N 119 9W
Sweetwater, *Tex.* **13** 32 28N 100 25W
Sweetwater →, *Wyo.* **16** 42 31N 107 2W
Swissvale, *Pa.* **27** 40 25N 79 52W
Sylacauga, *Ala.* **11** 33 10N 86 15W
Sylvania, *Ga.* **11** 32 45N 81 38W
Sylvester, *Ga.* **11** 31 32N 83 50W
Syracuse, *Kans.* **13** 37 59N 101 45W
Syracuse, *N.Y.* **9** 43 3N 76 9W

T

Tacoma, *Wash.* **18** 47 14N 122 26W
Taft, *Calif.* **19** 35 8N 119 28W
Taft, *Tex.* **13** 27 59N 97 24W
Tahoe, L., *Nev.* **18** 39 6N 120 2W
Tahoe City, *Calif.* **18** 39 10N 120 9W
Takoma Park, *Md.* **25** 38 58N 77 0W
Talihina, *Okla.* **13** 34 45N 95 3W
Talkeetna, *Alaska* **30** 62 20N 150 6W
Talkeetna Mts., *Alaska* **30** 62 20N 149 0W
Talladega, *Ala.* **11** 33 26N 86 6W
Tallahassee, *Fla.* **11** 30 27N 84 17W
Talleyville, *Del.* **24** 39 48N 75 32W
Tallulah, *La.* **13** 32 25N 91 11W
Tama, *Iowa* **14** 41 58N 92 35W
Tamaqua, *Pa.* **9** 40 48N 75 58W
Tampa, *Fla.* **11** 27 57N 82 27W
Tampa B., *Fla.* **11** 27 50N 82 30W
Tanana, *Alaska* **30** 65 10N 152 4W
Tanana →, *Alaska* **30** 65 10N 151 58W
Taos, *N. Mex.* **17** 36 24N 105 35W
Tappahannock, *Va.* **10** 37 56N 76 52W
Tarboro, *N.C.* **11** 35 54N 77 32W
Tarpon Springs, *Fla.* **11** 28 9N 82 45W
Tarrytown, *N.Y.* **9** 41 4N 73 52W
Tatum, *N. Mex.* **13** 33 16N 103 19W
Tau, *W. Samoa* **31** 14 15S 169 30W
Taunton, *Mass.* **9** 41 54N 71 6W
Tawas City, *Mich.* **10** 44 16N 83 31W
Taylor, *Alaska* **30** 65 40N 164 50W
Taylor, *Mich.* **26** 42 13N 83 15W
Taylor, *Nebr.* **12** 41 46N 99 23W
Taylor, *Pa.* **9** 41 23N 75 43W
Taylor, *Tex.* **13** 30 34N 97 25W
Taylor, Mt., *N. Mex.* **17** 35 14N 107 37W
Taylortown, *N.J.* **20** 40 56N 74 23W
Taylorville, *Ill.* **14** 39 33N 89 18W
Teague, *Tex.* **13** 31 38N 96 17W

Teaneck, N.J. 20 40 52N 74 1W
Tecopa, Calif. 19 35 51N 116 13W
Tecumseh, Mich. 15 42 0N 83 57W
Tehachapi, Calif. 19 35 8N 118 27W
Tehachapi Mts., Calif. 19 35 0N 118 30W
Tejon Pass, Calif. 19 34 49N 118 53W
Tekamah, Nebr. 12 41 47N 96 13W
Tekoa, Wash. 16 47 14N 117 4W
Telescope Pk., Calif. 19 36 10N 117 5W
Tell City, Ind. 15 37 57N 86 46W
Teller, Alaska 30 65 16N 166 22W
Telluride, Colo. 17 37 56N 107 49W
Temblor Range, Calif. 19 35 20N 119 50W
Temecula, Calif. 19 33 30N 117 9W
Tempe, Ariz. 17 33 25N 111 56W
Temple, Tex. 13 31 6N 97 21W
Temple City, Calif. 29 34 6N 118 2W
Templeton, Calif. 18 35 33N 120 42W
Tenaha, Tex. 13 31 57N 94 15W
Tennessee □ 11 36 0N 86 30W
Tennessee →, Tenn. 10 37 4N 88 34W
Tennille, Ga. 11 32 56N 82 48W
Terra Bella, Calif. 19 35 58N 119 3W
Terre Haute, Ind. 15 39 28N 87 25W
Terrebonne B., La. 13 29 5N 90 35W
Terrell, Tex. 13 32 44N 96 17W
Terry, Mont. 12 46 47N 105 19W
Tetlin, Alaska 30 63 8N 142 31W
Tetlin Junction, Alaska 30 63 29N 142 55W
Teton →, Mont. 16 47 56N 110 31W
Texarkana, Ark. 13 33 26N 94 2W
Texarkana, Tex. 13 33 26N 94 3W
Texas □ 13 31 40N 98 30W
Texas City, Tex. 13 29 24N 94 54W
Texhoma, Okla. 13 36 30N 101 47W
Texline, Tex. 13 36 23N 103 2W
Texoma, L., Tex. 13 33 50N 96 34W
Thames →, Conn. 9 41 18N 72 5W
Thatcher, Ariz. 17 32 51N 109 46W
Thatcher, Colo. 13 37 33N 104 7W
Thayer, Mo. 13 36 31N 91 33W
The Dalles, Oreg. 16 45 36N 121 10W
Thedford, Nebr. 12 41 59N 100 35W
Thermopolis, Wyo. 16 43 39N 108 13W
Thibodaux, La. 13 29 48N 90 49W
Thief River Falls, Minn. 12 48 7N 96 10W
Thomas, Okla. 13 35 45N 98 45W
Thomas, W. Va. 10 39 9N 79 30W
Thomaston, Ga. 11 32 53N 84 20W
Thomasville, Ala. 11 31 55N 87 44W
Thomasville, Ga. 11 30 50N 83 59W
Thomasville, N.C. 11 35 53N 80 5W
Thompson, Utah 17 38 58N 109 43W
Thompson →, Mo. 12 39 46N 93 37W
Thompson Falls, Mont. 16 47 36N 115 21W
Thompson Pk., Calif. 16 41 0N 123 0W
Thousand Oaks, Calif. 19 34 10N 118 50W
Three Forks, Mont. 16 45 54N 111 33W
Three Lakes, Wis. 12 45 48N 89 10W
Three Rivers, Calif. 18 36 26N 118 54W
Three Rivers, Tex. 13 28 28N 98 11W
Three Sisters, Oreg. 16 44 4N 121 51W
Thunder B., Mich. 8 45 0N 83 20W
Tiber Reservoir, Mont. 16 48 19N 111 6W
Tiburon, Calif. 28 37 52N 122 27W
Ticonderoga, N.Y. 9 43 51N 73 26W
Tierra Amarilla, N. Mex. 17 36 42N 106 33W
Tiffin, Ohio 15 41 7N 83 11W
Tifton, Ga. 11 31 27N 83 31W
Tigalda I., Alaska 30 54 6N 165 5W
Tilden, Nebr. 12 42 3N 97 50W
Tilden, Tex. 13 28 28N 98 33W
Tillamook, Oreg. 16 45 27N 123 51W
Tilton, N.H. 9 43 27N 71 36W
Timber Lake, S. Dak. 12 45 26N 101 5W
Timber Mt., Nev. 18 37 6N 116 28W
Tin Mt., Calif. 18 36 50N 117 10W
Tinian, Pac. Oc. 31 15 0N 145 38 E
Tioga, Pa. 8 41 55N 77 8W
Tipton, Calif. 18 36 4N 119 19W
Tipton, Ind. 15 40 17N 86 2W
Tipton, Iowa 14 41 46N 91 8W
Tiptonville, Tenn. 13 36 23N 89 29W
Tishomingo, Okla. 13 34 14N 96 41W
Titusville, Fla. 11 28 37N 80 49W
Titusville, Pa. 8 41 38N 79 41W
Tobyhanna, Pa. 9 41 11N 75 25W
Toccoa, Ga. 11 34 35N 83 19W
Tolageak, Alaska 30 70 2N 162 50W
Toledo, Ohio 15 41 39N 83 33W
Toledo, Oreg. 16 44 37N 123 56W
Toledo, Wash. 16 46 26N 122 51W
Tolleson, Ariz. 17 33 27N 112 16W
Tollhouse, Calif. 18 37 1N 119 24W
Tomah, Wis. 12 43 59N 90 30W
Tomahawk, Wis. 12 45 28N 89 44W
Tomales, Calif. 18 38 15N 122 53W
Tomales B., Calif. 18 38 15N 123 58W
Tombigbee →, Ala. 11 31 8N 87 57W
Tombstone, Ariz. 17 31 43N 110 4W
Toms Place, Calif. 18 37 34N 118 41W
Toms River, N.J. 9 39 58N 74 12W
Tonalea, Ariz. 17 36 19N 110 56W
Tonasket, Wash. 16 48 42N 119 26W
Tonawanda, N.Y. 8 43 1N 78 53W
Tongue →, Mont. 12 46 25N 105 52W
Tonkawa, Okla. 13 36 41N 97 18W
Tonopah, Nev. 17 38 4N 117 14W
Tooele, Utah 16 40 32N 112 18W
Topaz, Calif. 18 38 41N 119 30W
Topeka, Kans. 12 39 3N 95 40W
Topock, Calif. 19 34 46N 114 29W
Toppenish, Wash. 16 46 23N 120 19W
Toro Pk., Calif. 19 33 34N 116 24W
Toronto, Ohio 8 40 28N 80 36W
Torrance, Calif. 19 33 50N 118 19W
Torrey, Utah 17 38 18N 111 25W
Torrington, Conn. 9 41 48N 73 7W
Torrington, Wyo. 12 42 4N 104 11W
Tortola, Virgin Is. 30 18 19N 64 45W
Towanda, Pa. 9 41 46N 76 27W
Tower, Minn. 12 47 48N 92 17W
Towner, N. Dak. 12 48 21N 100 25W
Townsend, Mont. 16 46 19N 111 31W
Towson, Md. 25 39 24N 76 36W
Toyah, Tex. 13 31 19N 103 48W
Toyahvale, Tex. 13 30 57N 103 47W
Tracy, Calif. 18 37 44N 121 26W
Tracy, Minn. 12 44 14N 95 37W
Trapper Pk., Mont. 16 45 54N 114 18W
Traverse City, Mich. 10 44 46N 85 38W
Tremonton, Utah 16 41 43N 112 10W
Trenton, Mo. 14 40 5N 93 37W
Trenton, N.J. 9 40 14N 74 46W
Trenton, Nebr. 12 40 11N 101 1W
Trenton, Tenn. 13 35 59N 88 56W
Tres Pinos, Calif. 18 36 48N 121 19W
Tribune, Kans. 12 38 28N 101 45W
Trinidad, Colo. 13 37 10N 104 31W
Trinity, Tex. 13 30 57N 95 22W
Trinity →, Calif. 16 41 11N 123 42W
Trinity →, Tex. 13 29 45N 94 43W
Trinity Range, Nev. 16 40 15N 118 45W
Trion, Ga. 11 34 33N 85 19W
Tripp, S. Dak. 12 43 13N 97 58W
Trona, Calif. 19 35 46N 117 23W
Tropic, Utah 17 37 37N 112 5W
Troup, Tex. 13 32 9N 95 7W
Troy, Ala. 11 31 48N 85 58W
Troy, Idaho 16 46 44N 116 46W
Troy, Kans. 12 39 47N 95 5W
Troy, Mich. 26 42 35N 83 9W
Troy, Mo. 14 38 59N 90 59W
Troy, Mont. 16 48 28N 115 53W
Troy, N.Y. 9 42 44N 73 41W
Troy, Ohio 15 40 2N 84 12W
Truckee, Calif. 18 39 20N 120 11W
Trujillo, N. Mex. 13 35 32N 104 42W
Truk,
U.S. Pac. Is. Trust Terr. 31 7 25N 151 46 E
Trumann, Ark. 13 35 41N 90 31W
Trumbull, Mt., Ariz. 17 36 25N 113 8W
Truth or Consequences,
N. Mex. 17 33 8N 107 15W
Tryon, N.C. 11 35 13N 82 14W
Tuba City, Ariz. 17 36 8N 111 14W
Tucson, Ariz. 17 32 13N 110 58W
Tucumcari, N. Mex. 13 35 10N 103 44W
Tugidak I., Alaska 30 56 30N 154 40W
Tulare, Calif. 18 36 13N 119 21W
Tulare Lake Bed, Calif. 18 36 0N 119 48W
Tularosa, N. Mex. 17 33 5N 106 1W
Tulia, Tex. 13 34 32N 101 46W
Tullahoma, Tenn. 11 35 22N 86 13W
Tulsa, Okla. 13 36 10N 95 55W
Tumwater, Wash. 16 47 1N 122 54W
Tunica, Miss. 13 34 41N 90 23W
Tunkhannock, Pa. 9 41 32N 75 57W
Tuntutuliak, Alaska 30 60 22N 162 38W
Tuolumne, Calif. 18 37 58N 120 15W
Tuolumne →, Calif. 18 37 36N 121 13W
Tupelo, Miss. 11 34 16N 88 43W
Tupman, Calif. 19 35 18N 119 21W
Tupper Lake, N.Y. 9 44 14N 74 28W
Turlock, Calif. 18 37 30N 120 51W
Turner, Mont. 16 48 51N 108 24W
Turners Falls, Mass. 9 42 36N 72 33W
Turon, Kans. 13 37 48N 98 26W
Turtle Creek, Pa. 27 40 24N 79 49W
Turtle Lake, N. Dak. 12 47 31N 100 53W
Turtle Lake, Wis. 12 45 24N 92 8W
Tuscaloosa, Ala. 11 33 12N 87 34W
Tuscola, Ill. 15 39 48N 88 17W
Tuscola, Tex. 13 32 12N 99 48W
Tuscumbia, Ala. 11 34 44N 87 42W
Tuskegee, Ala. 11 32 25N 85 42W
Tuttle, N. Dak. 12 47 9N 100 0W
Tutuila, Amer. Samoa 31 14 19S 170 50W
Twain Harte, Calif. 18 38 2N 120 14W
Twentynine Palms, Calif. 19 34 8N 116 3W
Twin Bridges, Mont. 16 45 33N 112 20W
Twin Falls, Idaho 16 42 34N 114 28W
Twin Valley, Minn. 12 47 16N 96 16W
Twisp, Wash. 16 48 22N 120 7W
Two Harbors, Minn. 12 47 2N 91 40W
Two Rivers, Wis. 10 44 9N 87 34W
Tyler, Minn. 12 44 18N 96 8W
Tyler, Tex. 13 32 21N 95 18W

U

U.S.A. = United States
of America ■,
N. Amer. 6 37 0N 96 0W
Uhrichsville, Ohio 8 40 24N 81 21W
Uinta Mts., Utah 16 40 45N 110 30W
Ukiah, Calif. 18 39 9N 123 13W
Ulak I., Alaska 30 51 22N 178 57W
Ulysses, Kans. 13 37 35N 101 22W
Umatilla, Oreg. 16 45 55N 119 21W
Umiat, Alaska 30 69 22N 152 8W
Uminak I., Alaska 30 53 20N 168 20W
Umnak I., Alaska 30 53 15N 168 20W
Umpqua →, Oreg. 16 43 40N 124 12W
Unadilla, N.Y. 9 42 20N 75 19W
Unalaska, Alaska 30 53 53N 166 32W
Uncompahgre Peak,
Colo. 17 38 4N 107 28W
Unimak I., Alaska 30 54 45N 164 0W
Unimak Pass, Alaska 30 54 15N 164 30W
Union, Miss. 13 32 34N 89 7W
Union, Mo. 14 38 27N 91 0W
Union, N.J. 20 40 42N 74 15W
Union, S.C. 11 34 43N 81 37W
Union, Mt., Ariz. 17 34 34N 112 21W
Union City, Calif. 28 37 36N 122 3W
Union City, N.J. 20 40 45N 74 2W
Union City, Pa. 8 41 54N 79 51W
Union City, Tenn. 13 36 26N 89 3W
Union Gap, Wash. 16 46 33N 120 28W
Union Springs, Ala. 11 32 9N 85 43W
Uniontown, Pa. 10 39 54N 79 44W
Unionville, Mo. 14 40 29N 93 1W
United States of
America ■, N. Amer. 6 37 0N 96 0W
University Heights, Ohio 27 41 29N 81 31W
Upolu Pt., Hawaii 30 20 16N 155 52W
Upper Alkali Lake, Calif. 16 41 47N 120 8W
Upper Darby, Pa. 24 39 57N 75 16W
Upper Klamath L., Oreg. 16 42 25N 121 55W
Upper Lake, Calif. 18 39 10N 122 54W
Upper Red L., Minn. 12 48 8N 94 45W
Upper Sandusky, Ohio 15 40 50N 83 17W
Upper St. Clair, Pa. 27 40 21N 80 3W
Upton, Wyo. 12 44 6N 104 38W
Urbana, Ill. 15 40 7N 88 12W
Urbana, Ohio 15 40 7N 83 45W
Utah □ 16 39 20N 111 30W
Utah, L., Utah 16 40 10N 111 58W
Ute Creek →, N. Mex. 13 35 21N 103 50W
Utica, N.Y. 9 43 6N 75 14W
Uvalde, Tex. 13 29 13N 99 47W

V

Vacaville, Calif. 18 38 21N 121 59W
Valdez, Alaska 30 61 7N 146 16W
Valdosta, Ga. 11 30 50N 83 17W
Vale, Oreg. 16 43 59N 117 15W
Valentine, Nebr. 12 42 52N 100 33W
Valentine, Tex. 13 30 35N 104 30W
Valier, Mont. 16 48 18N 112 16W
Vallejo, Calif. 18 38 7N 122 14W
Valley Center, Calif. 19 33 13N 117 2W
Valley City, N. Dak. 12 46 55N 98 0W
Valley Falls, Oreg. 16 42 29N 120 17W
Valley Springs, Calif. 18 38 12N 120 50W
Valley Stream, N.Y. 21 40 40N 73 42W
Valparaiso, Ind. 15 41 28N 87 4W
Van Alstyne, Tex. 13 33 25N 96 35W
Van Buren, Ark. 13 35 26N 94 21W
Van Buren, Maine 11 47 10N 67 58W
Van Buren, Mo. 13 37 0N 91 1W
Van Horn, Tex. 13 31 3N 104 50W
Van Tassell, Wyo. 12 42 40N 104 5W
Van Wert, Ohio 15 40 52N 84 35W
Vancouver, Wash. 18 45 38N 122 40W
Vandalia, Ill. 14 38 58N 89 6W
Vandalia, Mo. 14 39 19N 91 29W
Vandenburg, Calif. 19 34 35N 120 33W
Vandergrift, Pa. 8 40 36N 79 34W
Variadero, N. Mex. 13 35 43N 104 17W
Vassar, Mich. 10 43 22N 83 35W
Vaughn, Mont. 16 47 33N 111 33W
Vaughn, N. Mex. 17 34 36N 105 13W
Vega, Tex. 13 35 15N 102 26W
Vega Baja, Puerto Rico 30 18 27N 66 23W
Velva, N. Dak. 12 48 4N 100 56W
Venetie, Alaska 30 67 1N 146 25W
Ventucopa, Calif. 19 34 50N 119 29W
Ventura, Calif. 19 34 17N 119 18W
Verdigre, Nebr. 12 42 36N 98 2W
Vermilion, B., La. 13 29 45N 91 55W
Vermilion L., Minn. 12 47 53N 92 26W
Vermillion, S. Dak. 12 42 47N 96 56W
Vermont □ 9 44 0N 73 0W
Vernal, Utah 16 40 27N 109 32W
Vernalis, Calif. 18 37 36N 121 17W
Vernon, Tex. 13 34 9N 99 17W
Vero Beach, Fla. 11 27 38N 80 24W
Vicksburg, Mich. 15 42 7N 85 32W
Vicksburg, Miss. 13 32 21N 90 53W
Victor, Colo. 12 38 43N 105 9W
Victoria, Kans. 12 38 52N 99 9W
Victoria, Tex. 13 28 48N 97 0W
Victorville, Calif. 19 34 32N 117 18W
Vidalia, Ga. 11 32 13N 82 25W
Vienna, Ill. 13 37 25N 88 54W
Vienna, Va. 25 38 54N 77 17W
Vieques, Isla de,
Puerto Rico 30 18 8N 65 25W
Villanueva, N. Mex. 17 35 16N 105 22W
Ville Platte, La. 13 30 41N 92 17W
Villisca, Iowa 14 40 56N 94 59W
Vincennes, Ind. 15 38 41N 87 32W
Vincent, Calif. 19 34 33N 118 11W
Vineland, N.J. 10 39 29N 75 2W
Vinita, Okla. 13 36 39N 95 9W
Vinton, Iowa 14 42 10N 92 1W
Vinton, La. 13 30 11N 93 35W
Virgin →, Nev. 17 36 28N 114 21W
Virgin Gorda, Virgin Is. 30 18 30N 64 26W
Virgin Is. (British) ■,
W. Indies 30 18 30N 64 30W
Virginia, Minn. 12 47 31N 92 32W
Virginia □ 10 37 30N 78 45W
Virginia Beach, Va. 10 36 51N 75 59W
Virginia City, Mont. 16 45 18N 111 56W
Virginia City, Nev. 18 39 19N 119 39W
Viroqua, Wis. 12 43 34N 90 53W
Visalia, Calif. 18 36 20N 119 18W
Vista, Calif. 19 33 12N 117 14W
Volborg, Mont. 12 45 51N 105 41W
Vulcan, Mich. 10 45 47N 87 53W

W

Wabash, Ind. 15 40 48N 85 49W
Wabash →, Ill. 10 37 48N 88 2W
Wabeno, Wis. 10 45 26N 88 39W
Wabuska, Nev. 16 39 9N 119 11W
Waco, Tex. 13 31 33N 97 9W
Wadena, Minn. 12 46 26N 95 8W
Wadesboro, N.C. 11 34 58N 80 5W
Wadsworth, Nev. 16 39 38N 119 17W
Wagon Mound, N. Mex. 13 36 1N 104 42W
Wagoner, Okla. 13 35 58N 95 22W
Wahiawa, Hawaii 30 21 30N 158 2W
Wahoo, Nebr. 12 41 13N 96 37W
Wahpeton, N. Dak. 12 46 16N 96 36W
Wailuku, Hawaii 30 20 53N 156 30W
Wainiha, Hawaii 30 22 9N 159 34W
Wainwright, Alaska 30 70 38N 160 2W
Waipahu, Hawaii 30 21 23N 158 1W
Waitsburg, Wash. 16 46 16N 118 9W
Wake Forest, N.C. 11 35 59N 78 30W
Wake I., Pac. Oc. 31 19 18N 166 36 E
Wakefield, Mass. 23 42 30N 71 4W
Wakefield, Mich. 12 46 29N 89 56W
Walcott, Wyo. 16 41 46N 106 51W
Walden, Colo. 16 40 44N 106 17W
Walden, N.Y. 9 41 34N 74 11W
Waldport, Oreg. 16 44 26N 124 4W
Waldron, Ark. 13 34 54N 94 5W
Wales, Alaska 30 65 37N 168 5W
Walker, Minn. 12 47 6N 94 35W
Walker L., Nev. 18 38 42N 118 43W
Wall, S. Dak. 12 44 0N 102 8W
Wall Street, N.Y. 20 40 42N 74 0W
Walla Walla, Wash. 16 46 4N 118 20W
Wallace, Idaho 16 47 28N 115 56W
Wallace, N.C. 11 34 44N 77 59W
Wallace, Nebr. 12 40 50N 101 10W
Wallowa, Oreg. 16 45 34N 117 32W
Wallowa Mts., Oreg. 16 45 20N 117 30W
Wallula, Wash. 16 46 5N 118 54W
Walnut Creek, Calif. 28 37 53N 122 3W
Walnut Ridge, Ark. 13 36 4N 90 57W
Walsenburg, Colo. 13 37 38N 104 47W
Walsh, Colo. 13 37 23N 102 17W
Walterboro, S.C. 11 32 55N 80 40W
Walters, Okla. 13 34 22N 98 19W
Waltham, Mass. 23 42 23N 71 13W
Waltman, Wyo. 16 43 4N 107 12W
Walton, N.Y. 9 42 10N 75 8W
Wamego, Kans. 12 39 12N 96 18W
Wapakoneta, Ohio 15 40 34N 84 12W
Wapato, Wash. 16 46 27N 120 25W
Wappingers Falls, N.Y. 9 41 36N 73 55W
Wapsipinicon →, Iowa 14 41 44N 90 19W
Ward Mt., Calif. 18 37 12N 118 54W
Ware, Mass. 9 42 16N 72 14W
Wareham, Mass. 9 41 46N 70 43W
Warm Springs, Nev. 17 38 10N 116 20W
Warner Mts., Calif. 16 41 40N 120 15W
Warner Robins, Ga. 11 32 37N 83 36W
Warren, Ark. 13 33 37N 92 4W
Warren, Mich. 26 42 31N 83 0W
Warren, Minn. 12 48 12N 96 46W
Warren, Ohio 8 41 14N 80 49W
Warren, Pa. 8 41 51N 79 9W
Warrensburg, Mo. 12 38 46N 93 44W
Warrenton, Oreg. 18 46 10N 123 56W
Warrington, Fla. 11 30 23N 87 17W
Warroad, Minn. 12 48 54N 95 19W
Warsaw, Ind. 15 41 14N 85 51W
Warwick, R.I. 9 41 42N 71 28W
Wasatch Ra., Utah 16 40 30N 111 15W
Wasco, Calif. 19 35 36N 119 20W
Wasco, Oreg. 16 45 36N 120 42W
Waseca, Minn. 12 44 5N 93 30W
Washburn, N. Dak. 12 47 17N 101 2W
Washburn, Wis. 12 46 40N 90 54W
Washington, D.C. 25 38 53N 77 2W
Washington, Ga. 11 33 44N 82 44W
Washington, Ind. 15 38 40N 87 10W
Washington, Iowa 14 41 18N 91 42W
Washington, Mo. 14 38 33N 91 1W
Washington, N.C. 11 35 33N 77 3W

Washington, *N.J.* **9** 40 46N 74 59W
Washington, *Pa.* **8** 40 10N 80 15W
Washington, *Utah* **17** 37 8N 113 31W
Washington □ **16** 47 30N 120 30W
Washington, Mt., *N.H.* **9** 44 16N 71 18W
Washington Heights, *N.Y.* **20** 40 50N 73 55W
Washington I., *Wis.* **10** 45 23N 86 54W
Washington National Airport, *D.C.* **25** 38 51N 77 2W
Watching Mountains, *N.J.* **20** 40 42N 74 20W
Water Valley, *Miss.* **13** 34 10N 89 38W
Waterbury, *Conn.* **9** 41 33N 73 3W
Waterford, *Calif.* **18** 37 38N 120 46W
Waterloo, *Ill.* **14** 38 20N 90 9W
Waterloo, *Iowa* **14** 42 30N 92 21W
Waterloo, *N.Y.* **8** 42 54N 76 52W
Watersmeet, *Mich.* **12** 46 16N 89 11W
Waterton-Glacier International Peace Park, *Mont.* **16** 48 45N 115 0W
Watertown, *Conn.* **9** 41 36N 73 7W
Watertown, *Mass.* **23** 42 22N 71 10W
Watertown, *N.Y.* **9** 43 59N 75 55W
Watertown, *S. Dak.* **12** 44 54N 97 7W
Watertown, *Wis.* **15** 43 12N 88 43W
Waterville, *Maine* **11** 44 33N 69 38W
Waterville, *N.Y.* **9** 42 56N 75 23W
Waterville, *Wash.* **16** 47 39N 120 4W
Watervliet, *N.Y.* **9** 42 44N 73 42W
Watford City, *N. Dak.* **12** 47 48N 103 17W
Watkins Glen, *N.Y.* **8** 42 23N 76 52W
Watonga, *Okla.* **13** 35 51N 98 25W
Watrous, *N. Mex.* **13** 35 48N 104 59W
Watseka, *Ill.* **15** 40 47N 87 44W
Watsonville, *Calif.* **18** 36 55N 121 45W
Waubay, *S. Dak.* **12** 45 20N 97 18W
Wauchula, *Fla.* **11** 27 33N 81 49W
Waukegan, *Ill.* **15** 42 22N 87 50W
Waukesha, *Wis.* **15** 43 1N 88 14W
Waukon, *Iowa* **12** 43 16N 91 29W
Wauneta, *Nebr.* **12** 40 25N 101 23W
Waupaca, *Wis.* **12** 44 21N 89 5W
Waupun, *Wis.* **12** 43 38N 88 44W
Waurika, *Okla.* **13** 34 10N 98 0W
Wausau, *Wis.* **12** 44 58N 89 38W
Wautoma, *Wis.* **12** 44 4N 89 18W
Wauwatosa, *Wis.* **15** 43 3N 88 0W
Waverly, *Iowa* **14** 42 44N 92 29W
Waverly, *N.Y.* **9** 42 1N 76 32W
Wawona, *Calif.* **18** 37 32N 119 39W
Waxahachie, *Tex.* **13** 32 24N 96 51W
Waycross, *Ga.* **11** 31 13N 82 21W
Wayland, *Mass.* **23** 42 21N 71 20W
Wayne, *Mich.* **26** 42 16N 83 22W
Wayne, *N.J.* **20** 40 55N 74 14W
Wayne, *Nebr.* **12** 42 14N 97 1W
Wayne, *Pa.* **24** 40 2N 75 24W
Wayne, *W. Va.* **10** 38 13N 82 27W
Waynesboro, *Ga.* **11** 33 6N 82 1W
Waynesboro, *Miss.* **11** 31 40N 88 39W
Waynesboro, *Pa.* **10** 39 45N 77 35W
Waynesboro, *Va.* **10** 38 4N 78 53W
Waynesburg, *Pa.* **10** 39 54N 80 11W
Waynesville, *N.C.* **11** 35 28N 82 58W
Waynoka, *Okla.* **13** 36 35N 98 53W
Weatherford, *Okla.* **13** 35 32N 98 43W
Weatherford, *Tex.* **13** 32 46N 97 48W
Weaverville, *Calif.* **16** 40 44N 122 56W
Webb City, *Mo.* **13** 37 9N 94 28W
Webster, *Mass.* **9** 42 3N 71 53W
Webster, *S. Dak.* **12** 45 20N 97 31W
Webster, *Wis.* **12** 45 53N 92 22W
Webster City, *Iowa* **14** 42 28N 93 49W
Webster Green, *Mo.* **12** 38 38N 90 20W
Webster Springs, *W. Va.* **10** 38 29N 80 25W
Weed, *Calif.* **16** 41 25N 122 23W
Weedsport, *N.Y.* **9** 43 3N 76 35W
Weiser, *Idaho* **16** 44 10N 117 0W
Welch, *W. Va.* **10** 37 26N 81 35W
Wellesley, *Mass.* **23** 42 17N 71 17W
Wellington, *Colo.* **12** 40 42N 105 0W
Wellington, *Kans.* **13** 37 16N 97 24W
Wellington, *Nev.* **18** 38 45N 119 23W
Wellington, *Tex.* **13** 34 51N 100 13W
Wells, *Maine* **9** 43 20N 70 35W
Wells, *Minn.* **12** 43 45N 93 44W
Wells, *Nev.* **16** 41 7N 114 58W
Wellsboro, *Pa.* **8** 41 45N 77 18W
Wellsville, *Mo.* **14** 39 4N 91 34W
Wellsville, *N.Y.* **8** 42 7N 77 57W
Wellsville, *Ohio* **8** 40 36N 80 39W
Wellsville, *Utah* **16** 41 38N 111 56W
Wellton, *Ariz.* **17** 32 40N 114 8W
Wenatchee, *Wash.* **16** 47 25N 120 19W
Wendell, *Idaho* **16** 42 47N 114 42W
Wendover, *Utah* **16** 40 44N 114 2W
Weott, *Calif.* **16** 40 20N 123 55W
Wesley Vale, *N. Mex.* **17** 35 3N 106 2W
Wessington, *S. Dak.* **12** 44 27N 98 42W
Wessington Springs, *S. Dak.* **12** 44 5N 98 34W
West, *Tex.* **13** 31 48N 97 6W
West B., *La.* **13** 29 3N 89 22W
West Babylon, *N.Y.* **21** 40 43N 73 21W
West Bend, *Wis.* **10** 43 25N 88 11W
West Branch, *Mich.* **10** 44 17N 84 14W
West Chelmsford, *Mass.* **23** 42 36N 71 22W
West Chester, *Pa.* **10** 39 58N 75 36W
West Chester, *Pa.* **24** 39 57N 75 35W
West Columbia, *Tex.* **13** 29 9N 95 39W
West Covina, *Calif.* **29** 34 4N 117 55W
West Des Moines, *Iowa* **14** 41 35N 93 43W
West Frankfort, *Ill.* **14** 37 54N 88 55W
West Hartford, *Conn.* **9** 41 45N 72 44W
West Haven, *Conn.* **9** 41 17N 72 57W
West Helena, *Ark.* **13** 34 33N 90 38W
West Hempstead, *N.Y.* **21** 40 41N 73 38W
West Hollywood, *Calif.* **29** 34 5N 118 21W
West Memphis, *Ark.* **13** 35 9N 90 11W
West Mifflin, *Pa.* **27** 40 21N 79 53W
West Monroe, *La.* **13** 32 31N 92 9W
West New York, *N.J.* **20** 40 46N 74 0W
West Orange, *N.J.* **20** 40 46N 74 15W
West Palm Beach, *Fla.* **11** 26 43N 80 3W
West Plains, *Mo.* **13** 36 44N 91 51W
West Point, *Ga.* **11** 32 53N 85 11W
West Point, *Miss.* **11** 33 36N 88 39W
West Point, *Nebr.* **12** 41 51N 96 43W
West Point, *Va.* **10** 37 32N 76 48W
West Rutland, *Vt.* **9** 43 38N 73 5W
West View, *Pa.* **27** 40 31N 80 2W
West Virginia □ **10** 38 45N 80 30W
West Walker →, *Nev.* **18** 38 54N 119 9W
West Yellowstone, *Mont.* **16** 44 40N 111 6W
Westbrook, *Maine* **11** 43 41N 70 22W
Westbrook, *Tex.* **13** 32 21N 101 1W
Westbury, *N.Y.* **21** 40 45N 73 35W
Westby, *Mont.* **12** 48 52N 104 3W
Westchester, *Ill.* **22** 41 51N 87 53W
Westend, *Calif.* **19** 35 42N 117 24W
Westernport, *Md.* **10** 39 29N 79 3W
Westfield, *Mass.* **9** 42 7N 72 45W
Westfield, *N.J.* **20** 40 39N 74 20W
Westhope, *N. Dak.* **12** 48 55N 101 1W
Westlake, *Ohio* **27** 41 27N 81 54W
Westland, *Mich.* **26** 42 19N 83 22W
Westminster, *Md.* **10** 39 34N 76 59W
Westmorland, *Calif.* **17** 33 2N 115 37W
Weston, *Mass.* **23** 42 22N 71 17W
Weston, *Oreg.* **16** 45 49N 118 26W
Weston, *W. Va.* **10** 39 2N 80 28W
Westport, *Wash.* **16** 46 53N 124 6W
Westville, *Ill.* **15** 40 2N 87 38W
Westville, *Okla.* **13** 35 58N 94 40W
Westwood, *Calif.* **16** 40 18N 121 0W
Westwood, *Mass.* **23** 42 12N 71 13W
Wethersfield, *Conn.* **9** 41 42N 72 40W
Wewoka, *Okla.* **13** 35 9N 96 30W
Weymouth, *Mass.* **23** 42 12N 70 57W
Wharton, *N.J.* **9** 40 54N 74 35W
Wharton, *Tex.* **13** 29 19N 96 6W
Wheatland, *Wyo.* **12** 42 3N 104 58W
Wheaton, *Md.* **25** 39 2N 77 1W
Wheaton, *Minn.* **12** 45 48N 96 30W
Wheeler, *Oreg.* **16** 45 41N 123 53W
Wheeler, *Tex.* **13** 35 27N 100 16W
Wheeler Pk., *N. Mex.* **17** 36 34N 105 25W
Wheeler Pk., *Nev.* **17** 38 57N 114 15W
Wheeler Ridge, *Calif.* **19** 35 0N 118 57W
Wheeling, *W. Va.* **8** 40 4N 80 43W
White →, *Ark.* **13** 33 57N 91 5W
White →, *Ind.* **15** 38 25N 87 45W
White →, *S. Dak.* **12** 43 42N 99 27W
White →, *Utah* **16** 40 4N 109 41W
White Bird, *Idaho* **16** 45 46N 116 18W
White Butte, *N. Dak.* **12** 46 23N 103 18W
White City, *Kans.* **12** 38 48N 96 44W
White Deer, *Tex.* **13** 35 26N 101 10W
White Hall, *Ill.* **14** 39 26N 90 24W
White Haven, *Pa.* **9** 41 4N 75 47W
White House, The, *D.C.* **25** 38 53N 77 2W
White L., *La.* **13** 29 44N 92 30W
White Mts., *Calif.* **18** 37 30N 118 15W
White Mts., *N.H.* **9** 44 15N 71 15W
White Plains, *N.Y.* **21** 41 0N 73 46W
White River, *S. Dak.* **12** 43 34N 100 45W
White Sulphur Springs, *Mont.* **16** 46 33N 110 54W
White Sulphur Springs, *W. Va.* **10** 37 48N 80 18W
Whiteface, *Tex.* **13** 33 36N 102 37W
Whitefish, *Mont.* **16** 48 25N 114 20W
Whitefish Point, *Mich.* **10** 46 45N 84 59W
Whitehall, *Mich.* **10** 43 24N 86 21W
Whitehall, *Mont.* **16** 45 52N 112 6W
Whitehall, *N.Y.* **9** 43 33N 73 24W
Whitehall, *Pa.* **27** 40 21N 80 0W
Whitehall, *Wis.* **12** 44 22N 91 19W
Whitesboro, *N.Y.* **9** 43 7N 75 18W
Whitesboro, *Tex.* **13** 33 39N 96 54W
Whitetail, *Mont.* **12** 48 54N 105 10W
Whiteville, *N.C.* **11** 34 20N 78 42W
Whitewater, *Wis.* **15** 42 50N 88 44W
Whitewater Baldy, *N. Mex.* **17** 33 20N 108 39W
Whiting, *Ind.* **22** 41 41N 87 30W
Whitman, *Mass.* **9** 42 5N 70 56W
Whitmire, *S.C.* **11** 34 30N 81 37W
Whitney, Mt., *Calif.* **18** 36 35N 118 18W
Whitney Point, *N.Y.* **9** 42 20N 75 58W
Whittier, *Alaska* **30** 60 47N 148 41W
Whittier, *Calif.* **29** 33 58N 118 2W
Whitwell, *Tenn.* **11** 35 12N 85 31W
Wibaux, *Mont.* **12** 46 59N 104 11W
Wichita, *Kans.* **13** 37 42N 97 20W
Wichita Falls, *Tex.* **13** 33 54N 98 30W
Wickenburg, *Ariz.* **17** 33 58N 112 44W
Wiggins, *Colo.* **12** 40 14N 104 4W
Wiggins, *Miss.* **13** 30 51N 89 8W
Wilber, *Nebr.* **12** 40 29N 96 58W
Wilburton, *Okla.* **13** 34 55N 95 19W
Wildrose, *Calif.* **19** 36 14N 117 11W
Wildrose, *N. Dak.* **12** 48 38N 103 11W
Wildwood, *N.J.* **10** 38 59N 74 50W
Wilkes-Barre, *Pa.* **9** 41 15N 75 53W
Wilkesboro, *N.C.* **11** 36 9N 81 10W
Wilkinsburg, *Pa.* **27** 40 26N 79 52W
Willamina, *Oreg.* **16** 45 5N 123 29W
Willapa B., *Wash.* **16** 46 40N 124 0W
Willard, *N. Mex.* **17** 34 36N 106 2W
Willard, *Utah* **16** 41 25N 112 2W
Willcox, *Ariz.* **17** 32 15N 109 50W
Williams, *Ariz.* **17** 35 15N 112 11W
Williamsburg, *Ky.* **11** 36 44N 84 10W
Williamsburg, *Va.* **10** 37 17N 76 44W
Williamson, *W. Va.* **10** 37 41N 82 17W
Williamsport, *Pa.* **8** 41 15N 77 0W
Williamston, *N.C.* **11** 35 51N 77 4W
Williamstown, *N.Y.* **9** 43 26N 75 53W
Williamsville, *Mo.* **13** 36 58N 90 33W
Willimantic, *Conn.* **9** 41 43N 72 13W
Willingboro, *N.J.* **24** 40 3N 74 54W
Williston, *Fla.* **11** 29 23N 82 27W
Williston, *N. Dak.* **12** 48 9N 103 37W
Williston Park, *N.Y.* **21** 40 45N 73 39W
Willits, *Calif.* **16** 39 25N 123 21W
Willmar, *Minn.* **12** 45 7N 95 3W
Willow Brook, *Calif.* **29** 33 55N 118 13W
Willow Grove, *Pa.* **24** 40 8N 75 7W
Willow Lake, *S. Dak.* **12** 44 38N 97 38W
Willow Springs, *Mo.* **13** 37 0N 91 58W
Willowick, *Ohio* **27** 41 36N 81 30W
Willows, *Calif.* **18** 39 31N 122 12W
Wills Point, *Tex.* **13** 32 43N 96 1W
Wilmette, *Ill.* **10** 42 5N 87 42W
Wilmette, *Ill.* **22** 42 4N 87 42W
Wilmington, *Del.* **10** 39 45N 75 33W
Wilmington, *Del.* **24** 39 44N 75 32W
Wilmington, *Ill.* **15** 41 18N 88 9W
Wilmington, *N.C.* **11** 34 14N 77 55W
Wilmington, *Ohio* **15** 39 27N 83 50W
Wilsall, *Mont.* **16** 45 59N 110 38W
Wilson, *N.C.* **11** 35 44N 77 55W
Wilton, *N. Dak.* **12** 47 10N 100 47W
Winchendon, *Mass.* **9** 42 41N 72 3W
Winchester, *Conn.* **9** 41 53N 73 9W
Winchester, *Idaho* **16** 46 14N 116 38W
Winchester, *Ind.* **15** 40 10N 84 59W
Winchester, *Ky.* **15** 38 0N 84 11W
Winchester, *Mass.* **23** 42 26N 71 8W
Winchester, *N.H.* **9** 42 46N 72 23W
Winchester, *Tenn.* **11** 35 11N 86 7W
Winchester, *Va.* **10** 39 11N 78 10W
Wind →, *Wyo.* **16** 43 12N 108 12W
Wind River Range, *Wyo.* **16** 43 0N 109 30W
Windber, *Pa.* **8** 40 14N 78 50W
Windom, *Minn.* **12** 43 52N 95 7W
Window Rock, *Ariz.* **17** 35 41N 109 3W
Windsor, *Colo.* **12** 40 29N 104 54W
Windsor, *Conn.* **9** 41 50N 72 39W
Windsor, *Mich.* **26** 42 18N 83 0W
Windsor, *Mo.* **14** 38 32N 93 31W
Windsor, *N.Y.* **9** 42 5N 75 37W
Windsor, *Vt.* **9** 43 29N 72 24W
Windsor Airport, *Mich.* **26** 42 16N 82 57W
Winfield, *Kans.* **13** 37 15N 96 59W
Winifred, *Mont.* **16** 47 34N 109 23W
Wink, *Tex.* **13** 31 45N 103 9W
Winlock, *Wash.* **18** 46 30N 122 56W
Winnebago, *Minn.* **12** 43 46N 94 10W
Winnebago, L., *Wis.* **10** 44 0N 88 26W
Winnemucca, *Nev.* **16** 40 58N 117 44W
Winnemucca L., *Nev.* **16** 40 7N 119 21W
Winner, *S. Dak.* **12** 43 22N 99 52W
Winnetka, *Ill.* **22** 42 6N 87 43W
Winnett, *Mont.* **16** 47 0N 108 21W
Winnfield, *La.* **13** 31 56N 92 38W
Winnibigoshish, L., *Minn.* **12** 47 27N 94 13W
Winnipesaukee, L., *N.H.* **9** 43 38N 71 21W
Winnsboro, *La.* **13** 32 10N 91 43W
Winnsboro, *S.C.* **11** 34 23N 81 5W
Winnsboro, *Tex.* **13** 32 58N 95 17W
Winona, *Minn.* **12** 44 3N 91 39W
Winona, *Miss.* **13** 33 29N 89 44W
Winooski, *Vt.* **9** 44 29N 73 11W
Winslow, *Ariz.* **17** 35 2N 110 42W
Winsted, *Conn.* **9** 41 55N 73 4W
Winston-Salem, *N.C.* **11** 36 6N 80 15W
Winter Garden, *Fla.* **11** 28 34N 81 35W
Winter Haven, *Fla.* **11** 28 1N 81 44W
Winter Park, *Fla.* **11** 28 36N 81 20W
Winters, *Tex.* **13** 31 58N 99 58W
Winterset, *Iowa* **14** 41 20N 94 1W
Winthrop, *Mass.* **23** 42 22N 70 58W
Winthrop, *Minn.* **12** 44 32N 94 22W
Winthrop, *Wash.* **16** 48 28N 120 10W
Winton, *N.C.* **11** 36 24N 76 56W
Wisconsin □ **12** 44 45N 89 30W
Wisconsin →, *Wis.* **12** 43 0N 91 15W
Wisconsin Dells, *Wis.* **12** 43 38N 89 46W
Wisconsin Rapids, *Wis.* **12** 44 23N 89 49W
Wisdom, *Mont.* **16** 45 37N 113 27W
Wiseman, *Alaska* **30** 67 25N 150 6W
Wishek, *N. Dak.* **12** 46 16N 99 33W
Wisner, *Nebr.* **12** 41 59N 96 55W
Woburn, *Mass.* **23** 42 29N 71 10W
Wolf Creek, *Mont.* **16** 47 0N 112 4W
Wolf Point, *Mont.* **12** 48 5N 105 39W
Wood Lake, *Nebr.* **12** 42 38N 100 14W
Woodbury, *N.J.* **24** 39 50N 75 9W
Woodbury, *N.Y.* **21** 40 49N 73 27W
Woodfords, *Calif.* **18** 38 47N 119 50W
Woodlake, *Calif.* **18** 36 25N 119 6W
Woodland, *Calif.* **18** 38 41N 121 46W
Woodlawn, *Md.* **25** 39 19N 76 44W
Woodruff, *Ariz.* **17** 34 51N 110 1W
Woodruff, *Utah* **16** 41 31N 111 10W
Woodstock, *Ill.* **15** 42 19N 88 27W
Woodstock, *Vt.* **9** 43 37N 72 31W
Woodsville, *N.H.* **9** 44 9N 72 2W
Woodville, *Tex.* **13** 30 47N 94 25W
Woodward, *Okla.* **13** 36 26N 99 24W
Woody, *Calif.* **19** 35 42N 118 50W
Woonsocket, *R.I.* **9** 42 0N 71 31W
Woonsocket, *S. Dak.* **12** 44 3N 98 17W
Wooster, *Ohio* **8** 40 48N 81 56W
Worcester, *Mass.* **9** 42 16N 71 48W
Worcester, *N.Y.* **9** 42 36N 74 45W
Worland, *Wyo.* **16** 44 1N 107 57W
Wortham, *Tex.* **13** 31 47N 96 28W
Worthington, *Minn.* **12** 43 37N 95 36W
Wrangell, *Alaska* **30** 56 28N 132 23W
Wrangell Mts., *Alaska* **30** 61 30N 142 0W
Wray, *Colo.* **12** 40 5N 102 13W
Wrens, *Ga.* **11** 33 12N 82 23W
Wrightson Mt., *Ariz.* **17** 31 42N 110 51W
Wrightwood, *Calif.* **19** 34 21N 117 38W
Wyalusing, *Pa.* **9** 41 40N 76 16W
Wyandotte, *Mich.* **26** 42 12N 83 9W
Wymore, *Nebr.* **12** 40 7N 96 40W
Wyndmere, *N. Dak.* **12** 46 16N 97 8W
Wynne, *Ark.* **13** 35 14N 90 47W
Wyoming □ **16** 43 0N 107 30W
Wytheville, *Va.* **10** 36 57N 81 5W

X

Xenia, *Ohio* **15** 39 41N 83 56W

Yadkin →, *N.C.* **11** 35 29N 80 9W
Yakataga, *Alaska* **30** 60 5N 142 32W
Yakima, *Wash.* **16** 46 36N 120 31W
Yakima →, *Wash.* **16** 47 0N 120 30W
Yakutat, *Alaska* **30** 59 33N 139 44W
Yalobusha →, *Miss.* **13** 33 33N 90 10W
Yampa →, *Colo.* **16** 40 32N 108 59W
Yankton, *S. Dak.* **12** 42 53N 97 23W
Yap Is., *U.S. Pac. Is. Trust Terr.* **31** 9 30N 138 10 E
Yates Center, *Kans.* **13** 37 53N 95 44W
Yazoo →, *Miss.* **13** 32 22N 90 54W
Yazoo City, *Miss.* **13** 32 51N 90 25W
Yellowstone →, *Mont.* **12** 47 59N 103 59W
Yellowstone L., *Wyo.* **16** 44 27N 110 22W
Yellowstone National Park, *Wyo.* **16** 44 40N 110 30W
Yellowtail Res., *Wyo.* **16** 45 6N 108 8W
Yermo, *Calif.* **19** 34 54N 116 50W
Yeso, *N. Mex.* **13** 34 26N 104 37W
Yoakum, *Tex.* **13** 29 17N 97 9W
Yonkers, *N.Y.* **21** 40 57N 73 52W
York, *Ala.* **11** 32 29N 88 18W
York, *Nebr.* **12** 40 52N 97 36W
York, *Pa.* **10** 39 58N 76 44W
Yorktown, *Tex.* **13** 28 59N 97 30W
Yorkville, *Ill.* **15** 41 38N 88 27W
Yosemite National Park, *Calif.* **18** 37 45N 119 40W
Yosemite Village, *Calif.* **18** 37 45N 119 35W
Youngstown, *N.Y.* **8** 43 15N 79 3W
Youngstown, *Ohio* **8** 41 6N 80 39W
Ypsilanti, *Mich.* **15** 42 14N 83 37W
Yreka, *Calif.* **16** 41 44N 122 38W
Ysleta, *N. Mex.* **17** 31 45N 106 24W
Yuba City, *Calif.* **18** 39 8N 121 37W
Yucca, *Ariz.* **19** 34 52N 114 9W
Yucca Valley, *Calif.* **19** 34 8N 116 27W
Yukon →, *Alaska* **30** 62 32N 163 54W
Yuma, *Ariz.* **19** 32 43N 114 37W
Yuma, *Colo.* **12** 40 8N 102 43W
Yunaska I., *Alaska* **30** 52 38N 170 40W

Z

Zanesville, *Ohio* **8** 39 56N 82 1W
Zapata, *Tex.* **13** 26 55N 99 16W
Zion National Park, *Utah* **17** 37 15N 113 5W
Zrenton, *Mich.* **26** 42 8N 83 12W
Zuni, *N. Mex.* **17** 35 4N 108 51W
Zwolle, *La.* **13** 31 38N 93 39W

INTRODUCTION TO WORLD GEOGRAPHY

Planet Earth 2
Restless Earth 4
Landforms 6
Oceans 8
Climate 10
Water & Vegetation 12
Environment 14
Population 16
The Human Family 18
Wealth 20
Quality of Life 22
Energy 24
Production 26
Trade 28
Travel & Tourism 30

PLANET EARTH

THE SOLAR SYSTEM

A minute part of one of the billions of galaxies (collections of stars) that comprises the Universe, the Solar System lies some 27,000 light-years from the center of our own galaxy, the "Milky Way." Thought to be over 4,700 million years old, it consists of a central sun with nine planets and their moons revolving around it, attracted by its gravitational pull. The planets orbit the Sun in the same direction – counterclockwise when viewed from the Northern Heavens – and almost in the same plane. Their orbital paths, however, vary enormously.

The Sun's diameter is 109 times that of Earth, and the temperature at its core – caused by continuous thermonuclear fusions of hydrogen into helium – is estimated to be 27 million degrees Fahrenheit. It is the Solar System's only source of light and heat.

PROFILE OF THE PLANETS

	Mean distance from Sun (million miles)	Mass (Earth = 1)	Period of orbit (Earth years)	Period of rotation (Earth days)	Equatorial diameter (miles)	Number of known satellites
Mercury	36.4	0.055	0.24 years	58.67	3,031	0
Venus	66.9	0.815	0.62 years	243.00	7,521	0
Earth	93.0	1.0	1.00 years	1.00	7,926	1
Mars	141.2	0.107	1.88 years	1.03	4,217	2
Jupiter	483.4	317.8	11.86 years	0.41	88,730	16
Saturn	886.8	95.2	29.46 years	0.43	74,500	20
Uranus	1,784.8	14.5	84.01 years	0.75	31,763	15
Neptune	2,797.8	17.1	164.80 years	0.80	30,775	8
Pluto	3,662.5	0.002	248.50 years	6.39	1,450	1

All planetary orbits are elliptical in form, but only Pluto and Mercury follow paths that deviate noticeably from a circular one. Near perihelion – its closest approach to the Sun – Pluto actually passes inside the orbit of Neptune, an event that last occurred in 1983. Pluto will not regain its station as outermost planet until February 1999.

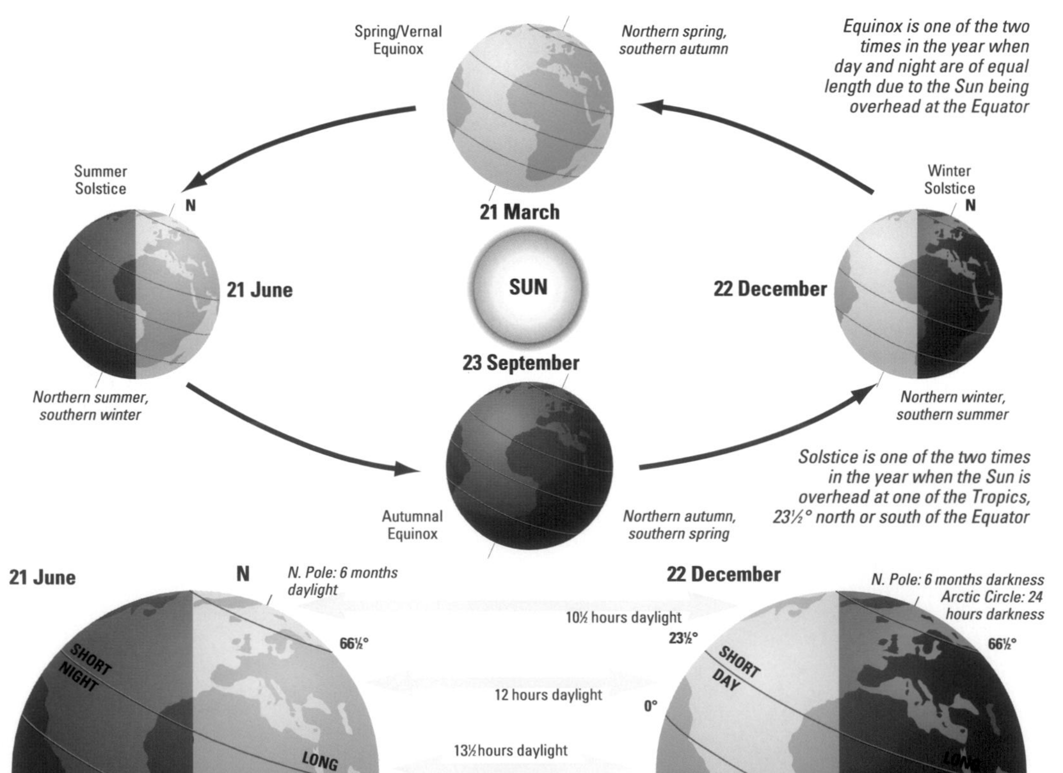

THE SEASONS

The Earth revolves around the Sun once a year in a counterclockwise direction, tilted at a constant angle of 23½°. In June, the northern hemisphere is tilted toward the Sun: as a result it receives more hours of sunshine in a day and therefore has its warmest season, summer. By December, the Earth has rotated halfway round the Sun so that the southern hemisphere is tilted toward the Sun and has its summer; the hemisphere that is tilted away from the Sun has winter. On 21 June the Sun is directly overhead at the Tropic of Cancer (23½° N), and this is midsummer in the northern hemisphere. Midsummer in the southern hemisphere occurs on 21 December, when the Sun is overhead at the Tropic of Capricorn (23½° S).

DAY AND NIGHT

The Sun appears to rise in the east, reach its highest point at noon, and then set in the west, to be followed by night. In reality it is not the Sun that is moving but the Earth revolving from west to east. Due to the tilting of the Earth the length of day and night varies from place to place and month to month, as shown on the diagram on the left.

At the summer solstice in the northern hemisphere (21 June), the Arctic has total daylight and the Antarctic total darkness. The opposite occurs at the winter solstice (21 December). At the Equator, the length of day and night are almost equal all year, at latitude 30° the length of day varies from about 14 hours to 10 hours, and at latitude 50° from about 16 hours to about 8 hours.

TIME

Year: The time taken by the Earth to revolve around the Sun, or 365.24 days.
Leap Year: A calendar year of 366 days, 29 February being the additional day. It offsets the difference between the calendar and the solar year.
Month: The approximate time taken by the Moon to revolve around the Earth. The 12 months of the year in fact vary from 28 (29 in a Leap Year) to 31 days.
Week: An artificial period of 7 days, not based on astronomical time.
Day: The time taken by the Earth to complete one rotation on its axis.
Hour: 24 hours make one day. Usually the day is divided into hours AM (ante meridiem or before noon) and PM (post meridiem or after noon), although most timetables now use the 24-hour system, from midnight to midnight.

SUNRISE

Spring Equinox — Autumnal Equinox
Hours AM: 2–9; Latitude: 60°N, 40°N, 20°N, 0°(Equator), 20°S, 40°S, 60°S
J F M A M J J A S O N D — Months of the year

SUNSET

Spring Equinox — Autumnal Equinox
Hours PM: 14–21; Latitude: 60°S, 40°S, 20°S, 0°(Equator), 20°N, 40°N, 60°N
J F M A M J J A S O N D — Months of the year

THE MOON

Distance from Earth: 221,463 mi – 252,710 mi; Mean diameter: 2,160 mi; Mass: approx. 1/81 that of Earth; Surface gravity: one-sixth of Earth's; Daily range of temperature at lunar equator: 360°F; Average orbital speed: 2,300 mph

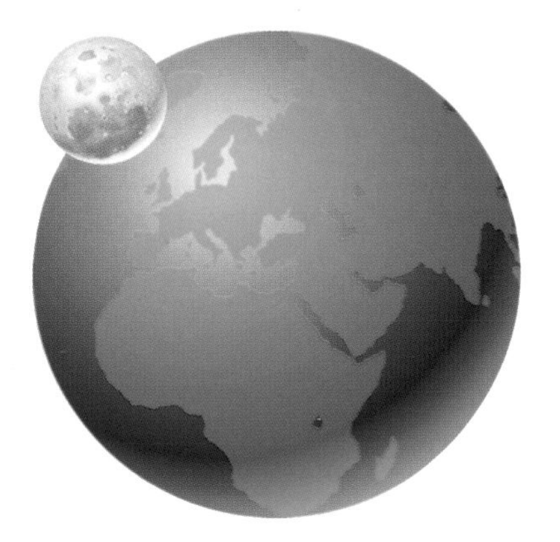

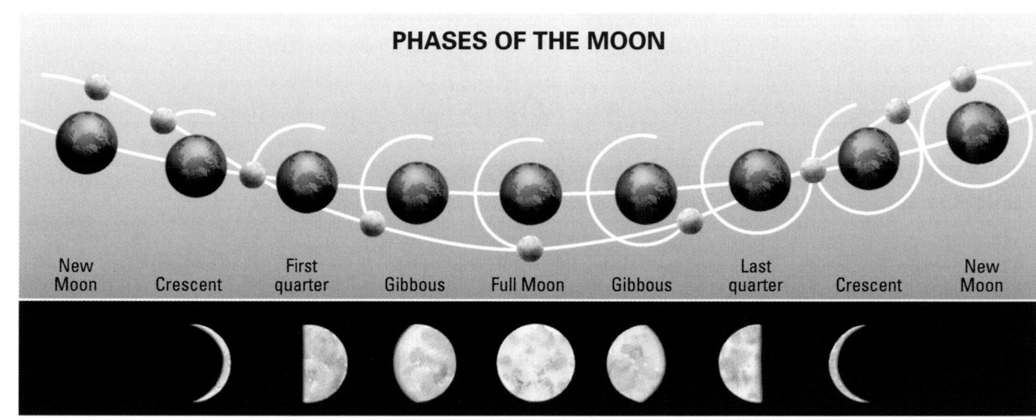

The Moon rotates more slowly than the Earth, making one complete turn on its axis in just over 27 days. Since this corresponds to its period of revolution around the Earth, the Moon always presents the same hemisphere or face to us, and we never see "the dark side." The interval between one full Moon and the next (and between new Moons) is about 29½ days – a lunar month. The apparent changes in the shape of the Moon are caused by its changing position in relation to the Earth; like the planets, it produces no light of its own and shines only by reflecting the rays of the Sun.

Partial eclipse (1)

P P P

Solar eclipse

Total eclipse (2)

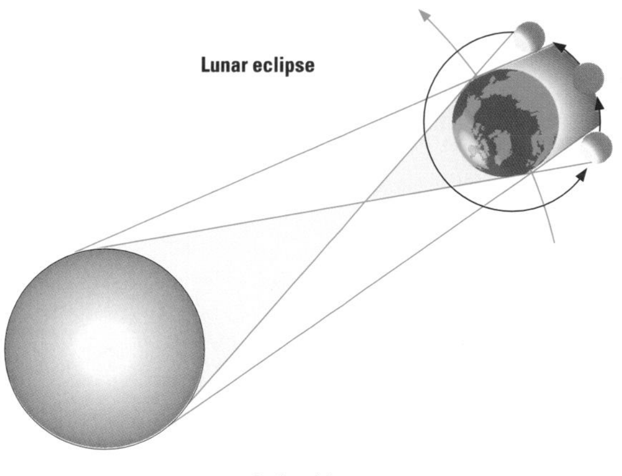

ECLIPSES

When the Moon passes between the Sun and the Earth it causes a partial eclipse of the Sun (1) if the Earth passes through the Moon's outer shadow (P), or a total eclipse (2) if the inner cone shadow crosses the Earth's surface. In a lunar eclipse, the Earth's shadow crosses the Moon and, again, provides either a partial or total eclipse. Eclipses of the Sun and the Moon do not occur every month because of the 5° difference between the plane of the Moon's orbit and the plane in which the Earth moves. In the 1990s only 14 lunar eclipses are possible, for example, seven partial and seven total; each is visible only from certain, and variable, parts of the world. The same period witnesses 13 solar eclipses – six partial (or annular) and seven total.

TIDES

The daily rise and fall of the ocean's tides are the result of the gravitational pull of the Moon and that of the Sun, though the effect of the latter is only 46.6% as strong as that of the Moon. This effect is greatest on the hemisphere facing the Moon and causes a tidal "bulge." When the Sun, Earth and Moon are in line, tide-raising forces are at a maximum and Spring tides occur: high tide reaches the highest values, and low tide falls to low levels. When lunar and solar forces are least coincidental with the Sun and Moon at an angle (near the Moon's first and third quarters), Neap tides occur, which have a small tidal range.

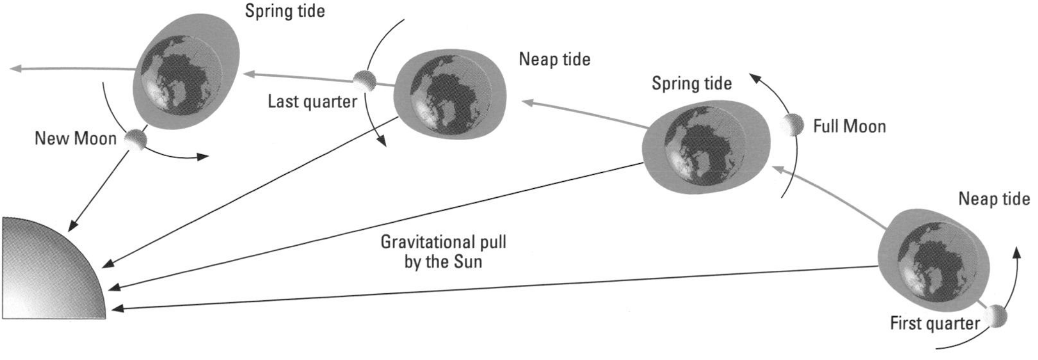

Restless Earth

THE EARTH'S STRUCTURE

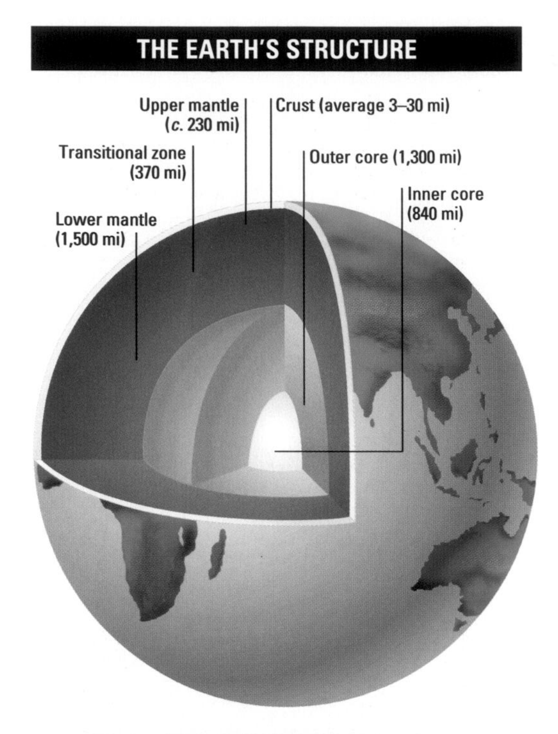

CONTINENTAL DRIFT

About 200 million years ago the original Pangaea land mass began to split into two continental groups, which further separated over time to produce the present-day configuration.

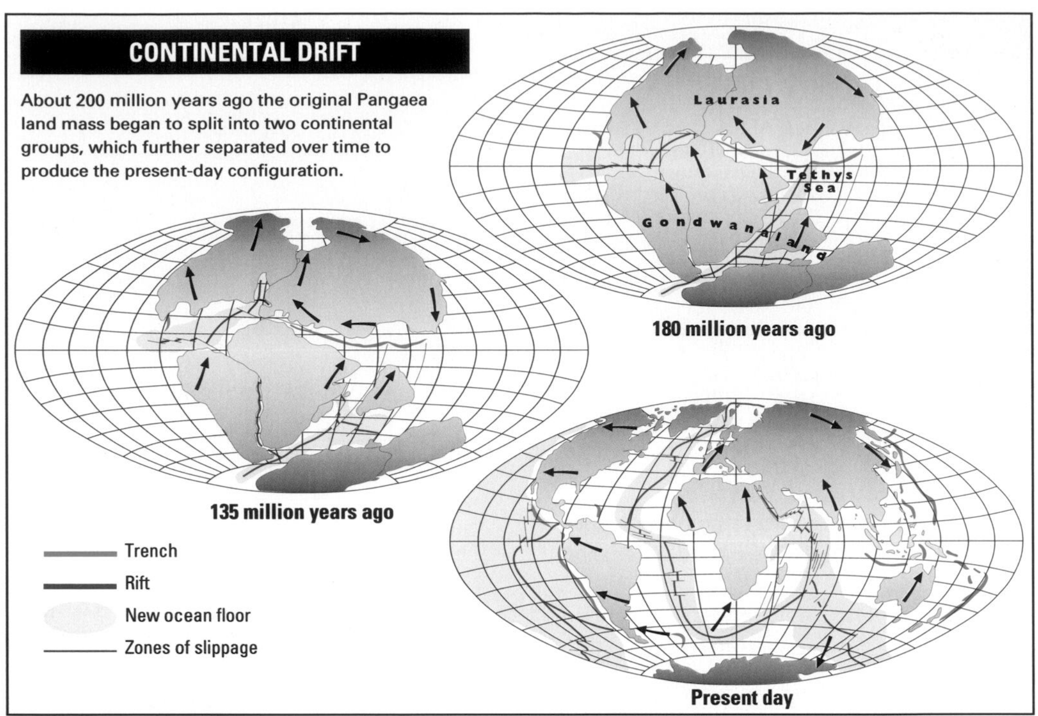

EARTHQUAKES

Earthquake magnitude is usually rated according to either the Richter or the Modified Mercalli scale, both devised by seismologists in the 1930s. The Richter scale measures absolute earthquake power with mathematical precision: each step upward represents a tenfold increase in shockwave amplitude. Theoretically, there is no upper limit, but the largest earthquakes measured have been rated at between 8.8 and 8.9. The 12-point Mercalli scale, based on observed effects, is often more meaningful, ranging from I (earthquakes noticed only by seismographs) to XII (total destruction); intermediate points include V (people awakened at night; unstable objects overturned), VII (collapse of ordinary buildings; chimneys and monuments fall) and IX (conspicuous cracks in ground; serious damage to reservoirs).

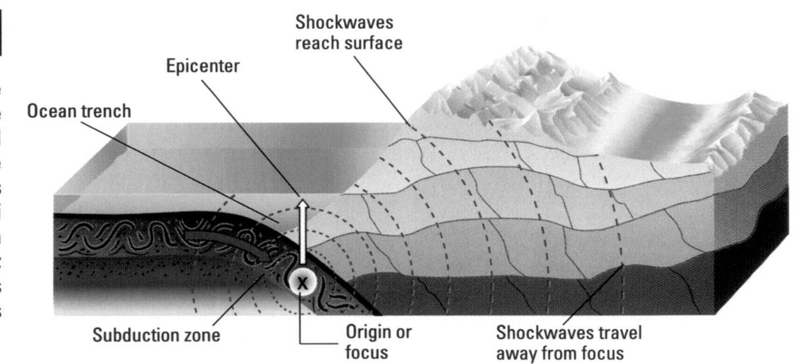

NOTABLE EARTHQUAKES SINCE 1900

Year	Location	Richter Scale	Deaths
1906	San Francisco, USA	8.3	503
1906	Valparaiso, Chile	8.6	22,000
1908	Messina, Italy	7.5	83,000
1915	Avezzano, Italy	7.5	30,000
1920	Gansu (Kansu), China	8.6	180,000
1923	Yokohama, Japan	8.3	143,000
1927	Nan Shan, China	8.3	200,000
1932	Gansu (Kansu), China	7.6	70,000
1934	Bihar, India/Nepal	8.4	10,700
1935	Quetta, India (now Pakistan)	7.5	60,000
1939	Chillan, Chile	8.3	28,000
1939	Erzincan, Turkey	7.9	30,000
1960	Agadir, Morocco	5.8	12,000
1962	Khorasan, Iran	7.1	12,230
1968	N.E. Iran	7.4	12,000
1970	N. Peru	7.7	66,794
1972	Managua, Nicaragua	6.2	5,000
1974	N. Pakistan	6.3	5,200
1976	Guatemala	7.5	22,778
1976	Tangshan, China	8.2	255,000
1978	Tabas, Iran	7.7	25,000
1980	El Asnam, Algeria	7.3	20,000
1980	S. Italy	7.2	4,800
1985	Mexico City, Mexico	8.1	4,200
1988	N.W. Armenia	6.8	55,000
1990	N. Iran	7.7	36,000
1993	Maharashtra, India	6.4	30,000
1994	Los Angeles, USA	6.6	61
1995	Kobe, Japan	7.2	5,000
1995	Sakhalin Is., Russia	7.5	2,000
1997	N.W. Iran	6.1	965

The highest magnitude recorded on the Richter scale is 8.9 in Japan on 2 March 1933 which killed 2,990 people. The most devastating earthquake ever was at Shaanxi (Shenshi) province, central China, on 3 January 1556, when an estimated 830,000 people were killed.

STRUCTURE AND EARTHQUAKES

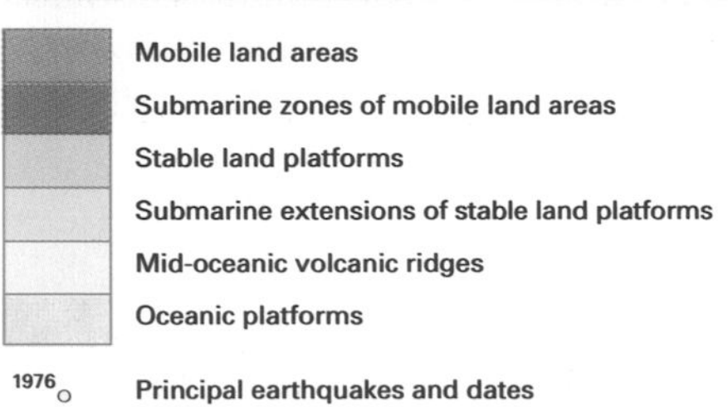

Earthquakes are a series of rapid vibrations originating from the slipping or faulting of parts of the Earth's crust when stresses within build up to breaking point. They usually happen at depths varying from 5 to 20 miles. Severe earthquakes cause extensive damage when they take place in populated areas, destroying structures and severing communications. Most initial loss of life occurs due to secondary causes such as falling masonry, fires and flooding.

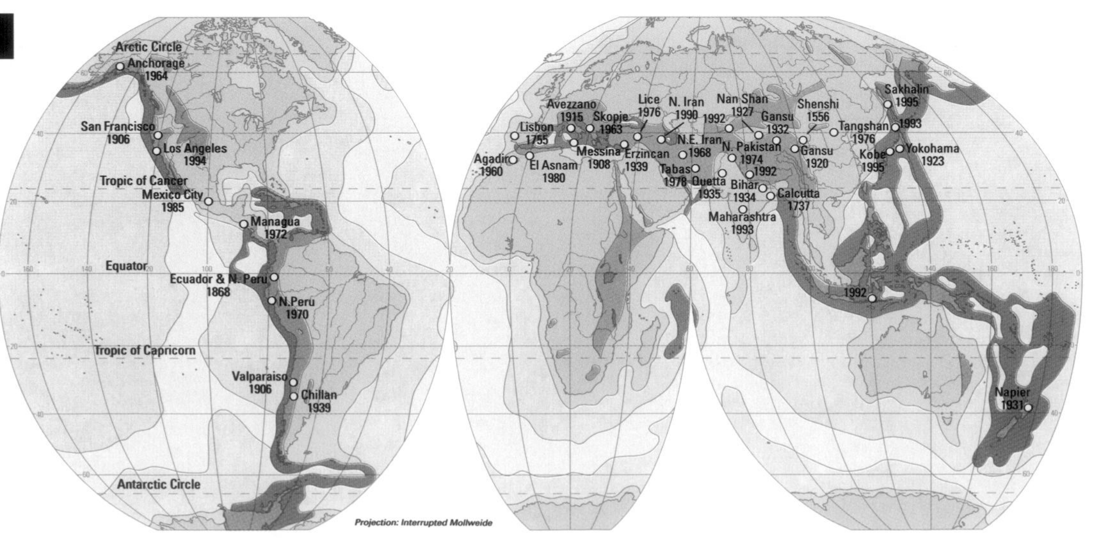

PLATE TECTONICS

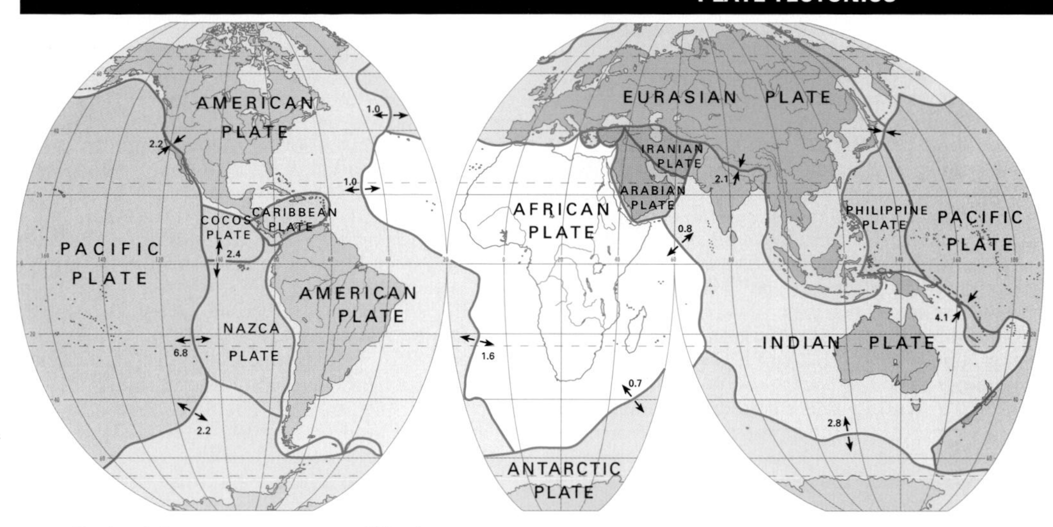

The drifting of the continents is a feature that is unique to Planet Earth. The complementary, almost jigsaw-puzzle fit of the coastlines on each side of the Atlantic Ocean inspired Alfred Wegener's theory of continental drift in 1915. The theory suggested that the ancient supercontinent, which Wegener named Pangaea, incorporated all of the Earth's land masses and gradually split up to form today's continents.

The original debate about continental drift was a prelude to a more radical idea: plate tectonics. The basic theory is that the Earth's crust is made up of a series of rigid plates which float on a soft layer of the mantle and are moved about by continental convection currents within the Earth's interior. These plates diverge and converge along margins marked by seismic activity. Plates diverge from mid-ocean ridges where molten lava pushes upward and forces the plates apart at rates of up to 1.6 in [40 mm] a year.

The three diagrams, right, give some examples of plate boundaries from around the world. Diagram a) shows sea-floor spreading at the Mid-Atlantic Ridge as the American and African plates slowly diverge. The same thing is happening in b) where sea-floor spreading at the Mid-Indian Ocean Ridge is forcing the Indian plate to collide into the Eurasian plate. In c) oceanic crust (sima) is being subducted beneath lighter continental crust (sial).

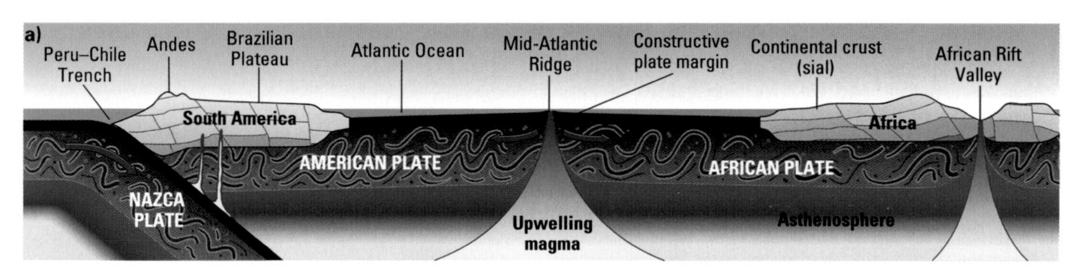

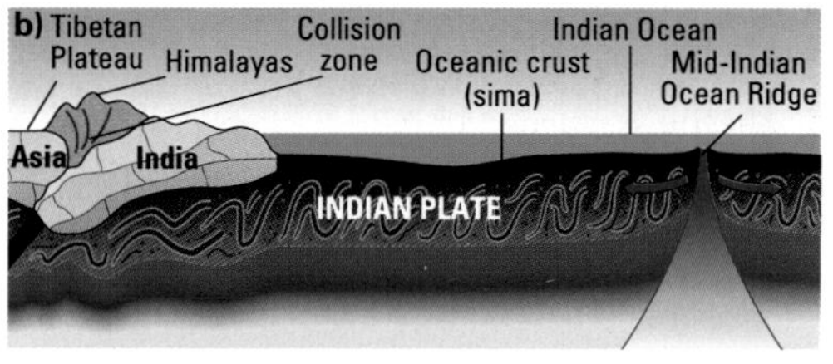

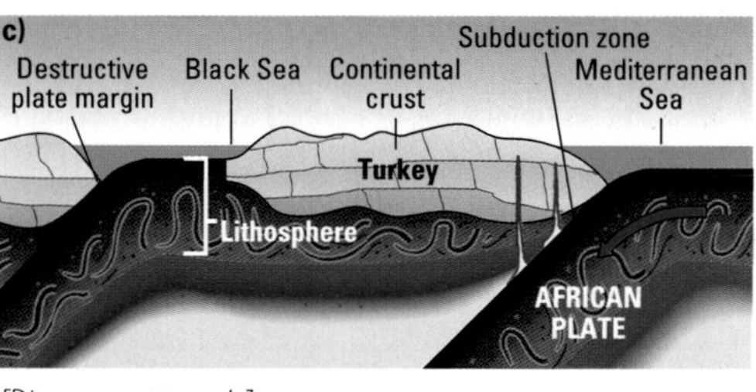

[Diagrams not to scale]

VOLCANOES

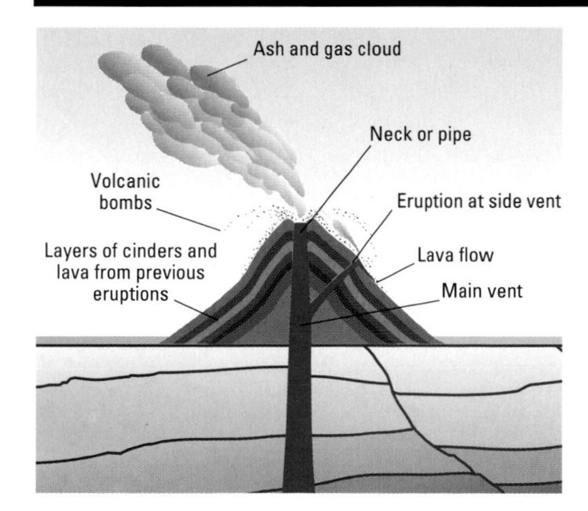

Volcanoes occur when hot liquefied rock beneath the Earth's crust is pushed up by pressure to the surface as molten lava. Some volcanoes erupt in an explosive way, throwing out rocks and ash, whilst others are effusive and lava flows out of the vent. There are volcanoes which are both, such as Mount Fuji. An accumulation of lava and cinders creates cones of variable size and shape. As a result of many eruptions over centuries Mount Etna in Sicily has a circumference of more than 75 miles [120 km].

Climatologists believe that volcanic ash, if ejected high into the atmosphere, can influence temperature and weather for several years afterward. The eruption of Mount Pinatubo in the Philippines ejected more than 20 million tons of dust and ash 20 miles [32 km] into the atmosphere and is believed to have accelerated ozone depletion over a large part of the globe.

DISTRIBUTION OF VOLCANOES

Today volcanoes may be the subject of considerable scientific study but they remain both dramatic and unpredictable, if not exactly supernatural: in 1991 Mount Pinatubo, 62 miles [100 km] north of the Philippines capital Manila, suddenly burst into life after lying dormant for more than six centuries. Most of the world's active volcanoes occur in a belt around the Pacific Ocean, on the edge of the Pacific plate, called the "ring of fire." Indonesia has the greatest concentration with 90 volcanoes, 12 of which are active. The most famous, Krakatoa, erupted in 1883 with such force that the resulting tidal wave killed 36,000 people and tremors were felt as far away as Australia.

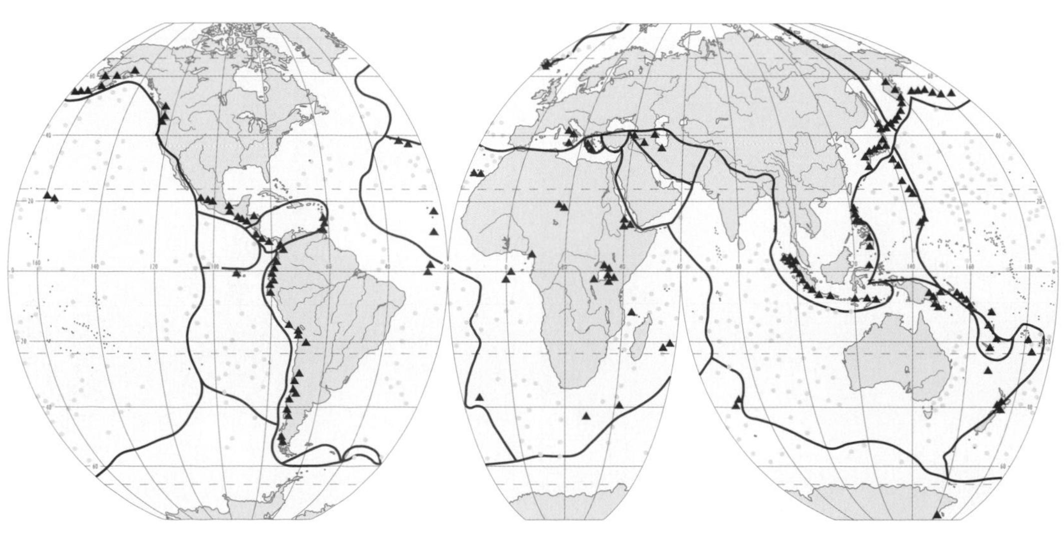

LANDFORMS

THE ROCK CYCLE

James Hutton first proposed the rock cycle in the late 1700s after he observed the slow but steady effects of erosion.

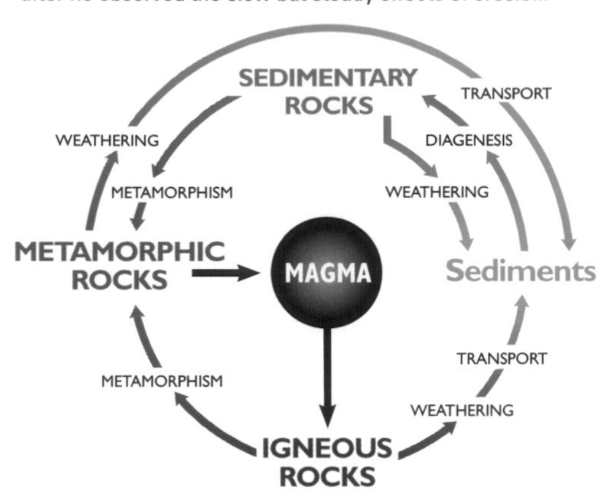

Above and below the surface of the oceans, the features of the Earth's crust are constantly changing. The phenomenal forces generated by convection currents in the molten core of our planet carry the vast segments or "plates" of the crust across the globe in an endless cycle of creation and destruction. A continent may travel little more than 1 in [25 mm] per year, yet in the vast span of geological time this process throws up giant mountain ranges and creates new land.

Destruction of the landscape, however, begins as soon as it is formed. Wind, water, ice and sea, the main agents of erosion, mount a constant assault that even the most resistant rocks cannot withstand. Mountain peaks may dwindle by as little as a fraction of an inch each year, but if they are not uplifted by further movements of the crust they will eventually be reduced to rubble and transported away. Water is the most powerful agent of erosion – it has been estimated that 100 billion tons of sediment are washed into the oceans every year. Three Asian rivers account for 20% of this total, the Huang He, in China, and the Brahmaputra and Ganges in Bangladesh.

Rivers and glaciers, like the sea itself, generate much of their effect through abrasion – pounding the land with the debris they carry with them. But as well as destroying they also create new landforms, many of them spectacular: vast deltas like those of the Mississippi and the Nile, or the deep fjords cut by glaciers in British Columbia, Norway and New Zealand.

Geologists once considered that landscapes evolved from "young," newly uplifted mountainous areas, through a "mature" hilly stage, to an "old age" stage when the land was reduced to an almost flat plain, or peneplain. This theory, called the "cycle of erosion," fell into disuse when it became evident that so many factors, including the effects of plate tectonics and climatic change, constantly interrupt the cycle, which takes no account of the highly complex interactions that shape the surface of our planet.

MOUNTAIN BUILDING

Mountains are formed when pressures on the Earth's crust caused by continental drift become so intense that the surface buckles or cracks. This happens where oceanic crust is subducted by continental crust or, more dramatically, where two tectonic plates collide: the Rockies, Andes, Alps, Urals and Himalayas resulted from such impacts. These are all known as fold mountains because they were formed by the compression of the rocks, forcing the surface to bend and fold like a crumpled rug. The Himalayas are formed from the folded former sediments of the Tethys Sea which was trapped in the collision zone between the Indian and Eurasian plates.

The other main mountain-building process occurs when the crust fractures to create faults, allowing rock to be forced upward in large blocks; or when the pressure of magma within the crust forces the surface to bulge into a dome, or erupts to form a volcano. Large mountain ranges may reveal a combination of those features; the Alps, for example, have been compressed so violently that the folds are fragmented by numerous faults and intrusions of molten igneous rock.

Over millions of years, even the greatest mountain ranges can be reduced by the agents of erosion (especially rivers) to a low rugged landscape known as a peneplain.

Types of faults: Faults occur where the crust is being stretched or compressed so violently that the rock strata break in a horizontal or vertical movement. They are classified by the direction in which the blocks of rock have moved. A normal fault results when a vertical movement causes the surface to break apart; compression causes a reverse fault. Horizontal movement causes shearing, known as a strike-slip fault. When the rock breaks in two places, the central block may be pushed up in a horst fault, or sink (creating a rift valley) in a graben fault.

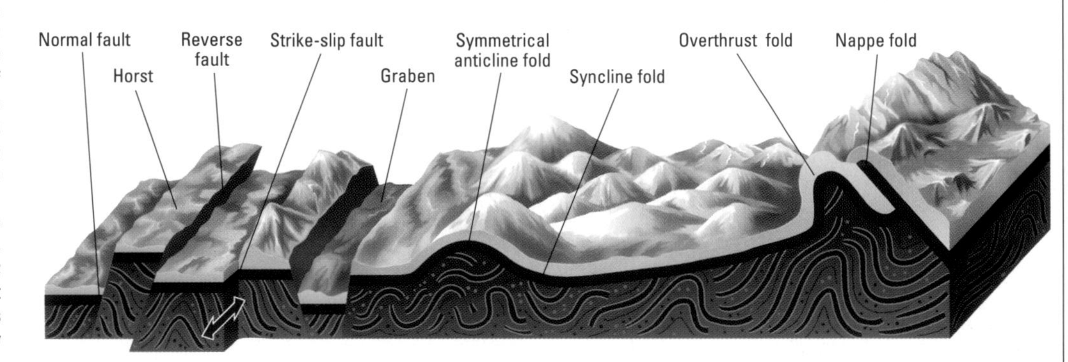

Types of fold: Folds occur when rock strata are squeezed and compressed. They are common therefore at destructive plate margins and where plates have collided, forcing the rocks to buckle into mountain ranges. Geographers give different names to the degrees of fold that result from continuing pressure on the rock. A simple fold may be symmetric, with even slopes on either side, but as the pressure builds up, one slope becomes steeper and the fold becomes asymmetric. Later, the ridge or "anticline" at the top of the fold may slide over the lower ground or "syncline" to form a recumbent fold. Eventually, the rock strata may break under the pressure to form an overthrust and finally a nappe fold.

CONTINENTAL GLACIATION

Ice sheets were at their greatest extent about 200,000 years ago. The maximum advance of the last Ice Age was about 18,000 years ago, when ice covered virtually all of Canada and reached as far south as the Bristol Channel in Britain.

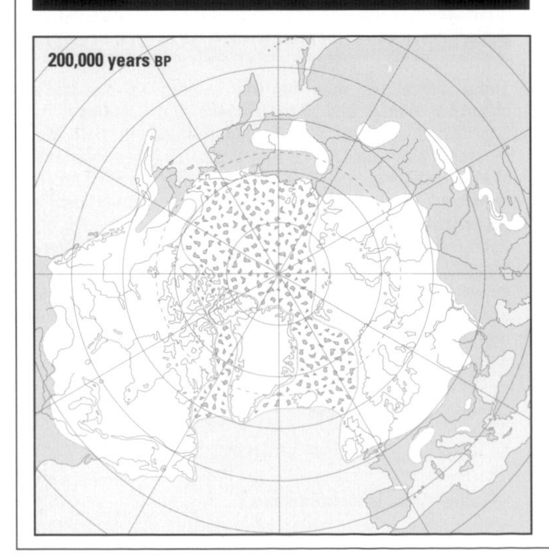

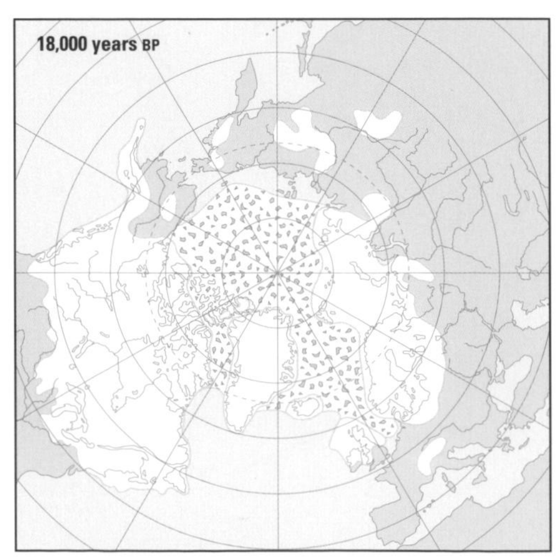

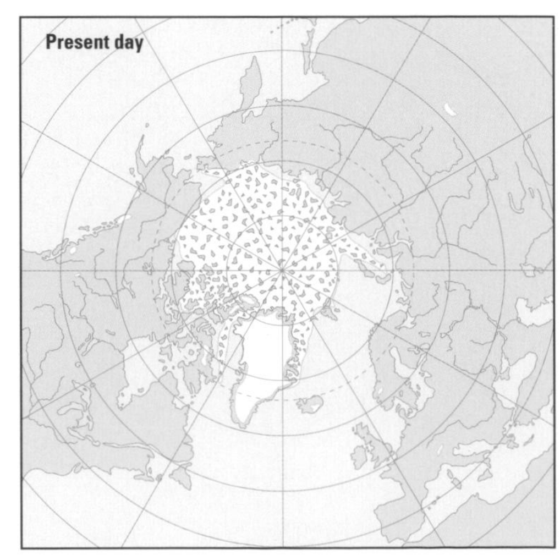

NATURAL LANDFORMS

A stylized diagram to show a selection of landforms found in the mid-latitudes.

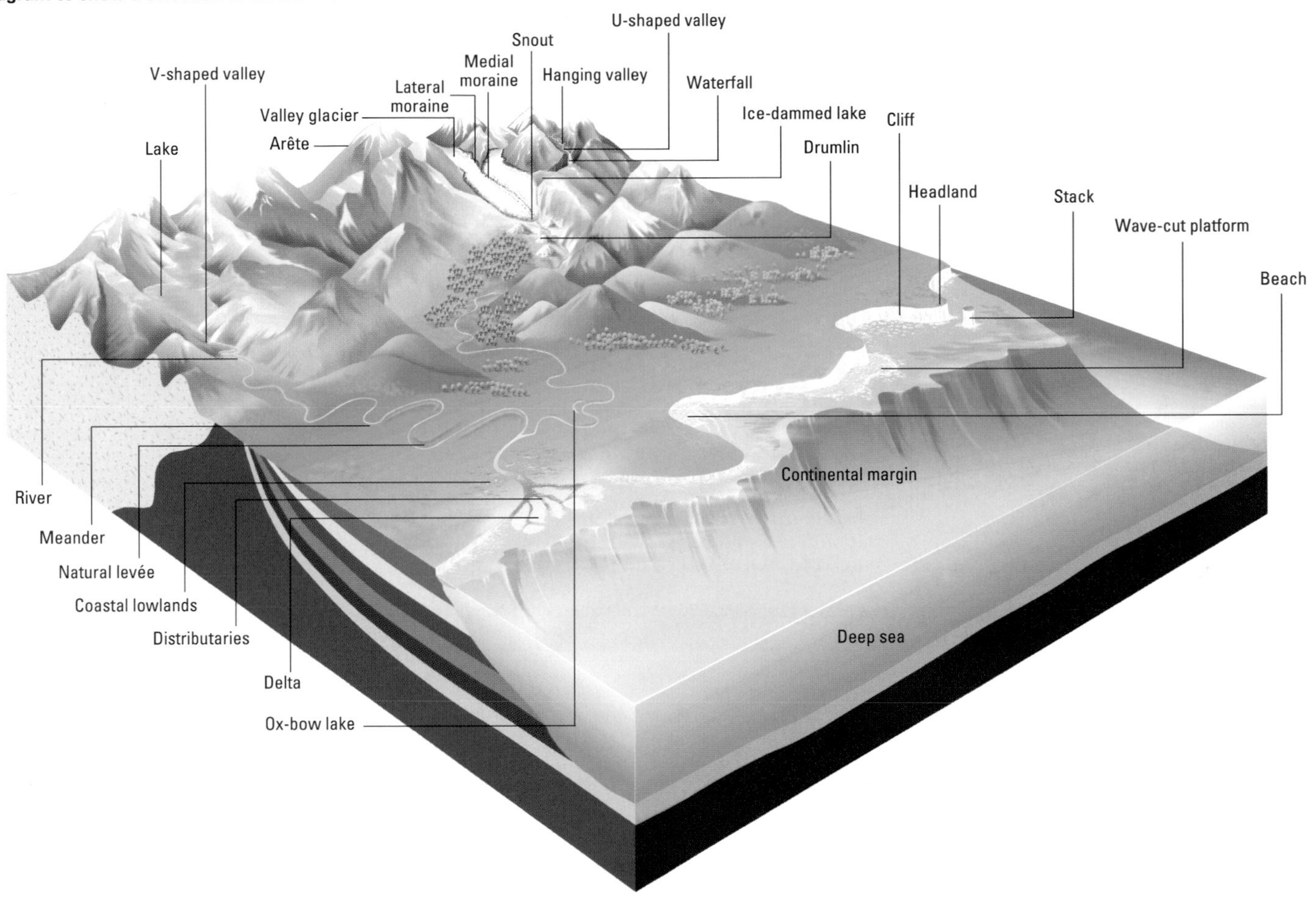

DESERT LANDSCAPES

The popular image that deserts are all huge expanses of sand is wrong. Despite harsh conditions, deserts contain some of the most varied and interesting landscapes in the world. They are also one of the most extensive environments – the hot and cold deserts together cover almost 40% of the Earth's surface.

The three types of hot desert are known by their Arabic names: sand desert, called *erg*, covers only about one-fifth of the world's desert; the rest is divided between *hammada* (areas of bare rock) and *reg* (broad plains covered by loose gravel or pebbles).

In areas of *erg*, such as the Namib Desert, the shape of the dunes reflects the character of local winds. Where winds are constant in direction, crescent-shaped *barchan* dunes form. In areas of bare rock, wind-blown sand is a major agent of erosion. The erosion is mainly confined to within six and a half feet of the surface, producing characteristic, mushroom-shaped rocks.

SURFACE PROCESSES

Catastrophic changes to natural landforms are periodically caused by such phenomena as avalanches, landslides and volcanic eruptions, but most of the processes that shape the Earth's surface operate extremely slowly in human terms. One estimate, based on a study in the United States, suggested that three feet of land was removed from the entire surface of the country, on average, every 29,500 years. However, the time scale varies from 1,300 years to 154,200 years depending on the terrain and climate.

In hot, dry climates, mechanical weathering, a result of rapid temperature changes, causes the outer layers of rock to peel away, while in cold mountainous regions, boulders are prised apart when water freezes in cracks in rocks. Chemical weathering, at its greatest in warm, humid regions, is responsible for hollowing out limestone caves and decomposing granites.

The erosion of soil and rock is greatest on sloping land and the steeper the slope, the greater the tendency for mass wasting – the movement of soil and rock downhill under the influence of gravity. The mechanisms of mass wasting (ranging from very slow to very rapid) vary with the type of material but the presence of water as a lubricant is usually an important factor.

Running water is the world's leading agent of erosion and transportation. The energy of a river depends on several factors, including its velocity and volume, and its erosive power is at its peak when it is in full flood. Sea waves also exert tremendous erosive power during storms when they hurl pebbles against the shore, undercutting cliffs and hollowing out caves.

Glacier ice forms in mountain hollows and spills out to form valley glaciers, which transport rocks shattered by frost action. As glaciers move, rocks embedded into the ice erode steep-sided, U-shaped valleys. Evidence of glaciation in mountain regions includes cirques, knife-edged ridges, or arêtes, and pyramidal peaks.

OCEANS

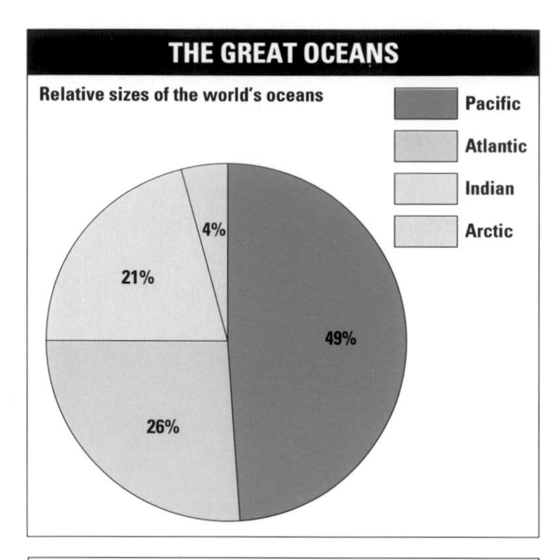

In a strict geographical sense there are only three true oceans – the Atlantic, Indian and Pacific. The legendary "Seven Seas" would require these to be divided at the Equator and the addition of the Arctic Ocean – which accounts for less than 4% of the total sea area. The International Hydrographic Bureau does not recognize the Antarctic Ocean (even less the "Southern Ocean") as a separate entity.

The Earth is a watery planet: more than 70% of its surface – over 140,000,000 square miles – is covered by the oceans and seas. The mighty Pacific alone accounts for nearly 36% of the total, and 49% of the sea area. Gravity holds in around 320 million cubic miles of water, of which over 97% is saline.

The vast underwater world starts in the shallows of the seaside and plunges to depths of more than 36,000 feet. The continental shelf, the underwater part of the land mass, drops gently to around 600 feet; here the seabed falls away suddenly at an angle of 3° to 6° – the continental slope. The third stage, called the continental rise, is more gradual with gradients varying from 1 in 100 to 1 in 700. At an average depth of 16,000 feet there begins the aptly-named abyssal plain – massive submarine depths where sunlight fails to penetrate and few creatures can survive.

From these plains rise volcanoes which, taken from base to top, rival and even surpass the biggest continental mountains in height. Mount Kea, on Hawaii, reaches a total of 33,400 feet, some 4,500 feet more than Mount Everest, though scarcely 40% is visible above sea level.

In addition there are underwater mountain chains up to 600 miles across, whose peaks sometimes appear above sea level as islands such as Iceland and Tristan da Cunha.

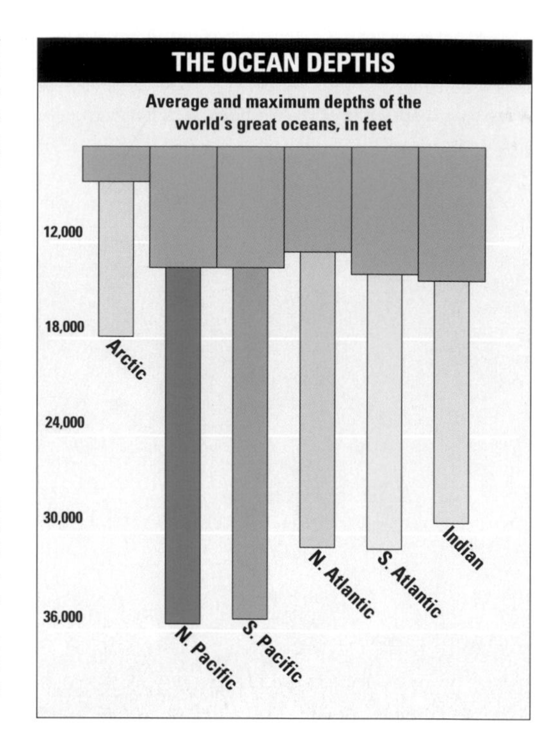

OCEAN CURRENTS

January temperatures and ocean currents

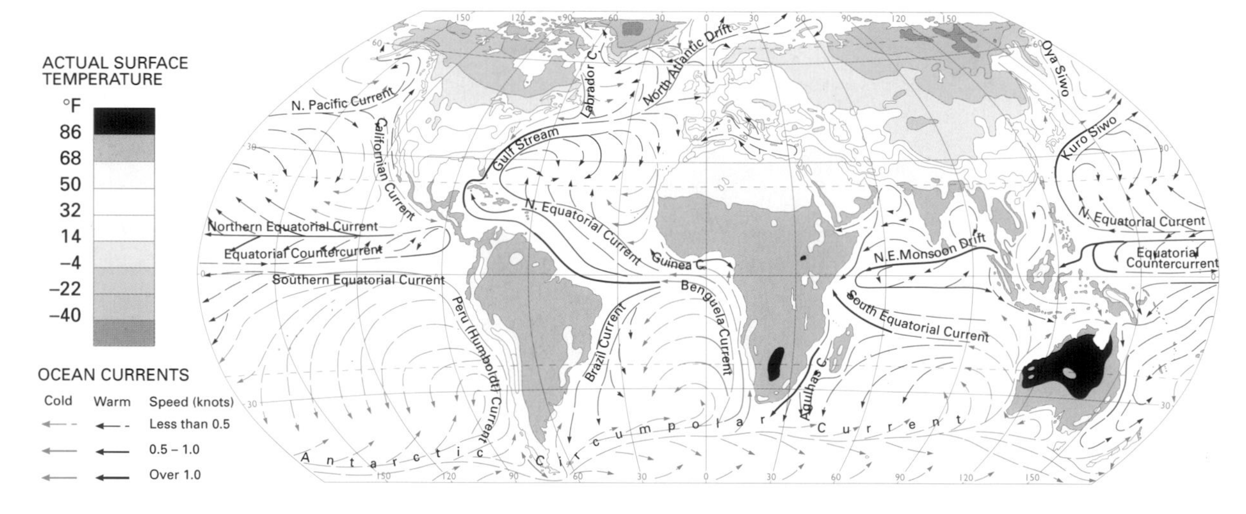

July temperatures and ocean currents

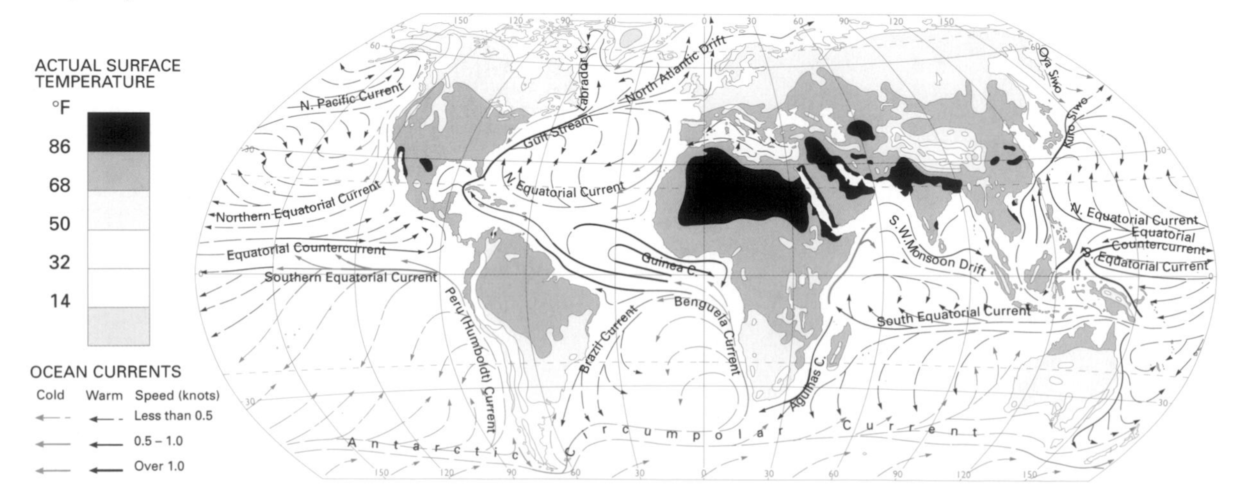

Moving immense quantities of energy as well as billions of tons of water every hour, the ocean currents are a vital part of the great heat engine that drives the Earth's climate. They themselves are produced by a twofold mechanism. At the surface, winds push huge masses of water before them; in the deep ocean, below an abrupt temperature gradient that separates the churning surface waters from the still depths, density variations cause slow vertical movements.

The pattern of circulation of the great surface currents is determined by the displacement known as the Coriolis effect. As the Earth turns beneath a moving object – whether it is a tennis ball or a vast mass of water – it appears to be deflected to one side. The deflection is most obvious near the Equator, where the Earth's surface is spinning eastward at 1,050 mph; currents moving poleward are curved clockwise in the northern hemisphere and counterclockwise in the southern.

The result is a system of spinning circles known as gyres. The Coriolis effect piles up water on the left of each gyre, creating a narrow, fast-moving stream that is matched by a slower, broader returning current on the right. North and south of the Equator, the fastest currents are located in the west and in the east respectively. In each case, warm water moves from the Equator and cold water returns to it. Cold currents often bring an upwelling of nutrients with them, supporting the world's most economically important fisheries.

Depending on the prevailing winds, some currents on or near the Equator may reverse their direction in the course of the year – a seasonal variation on which Asian monsoon rains depend, and whose occasional failure can bring disaster to millions.

WORLD FISHING AREAS

Main commercial fishing areas (numbered FAO regions)

Catch by top marine fishing areas, thousand tons (1992)

1. Pacific, NW	[61]	26,667	29.3%
2. Pacific, SE	[87]	15,317	16.8%
3. Atlantic, NE	[27]	12,202	13.4%
4. Pacific, WC	[71]	8,496	9.3%
5. Indian, W	[51]	4,129	4.5%
6. Indian, E	[57]	3,595	4.0%
7. Atlantic, EC	[34]	3,591	3.9%
8. Pacific, NE	[67]	3,470	3.8%

Principal fishing areas

Leading fishing nations

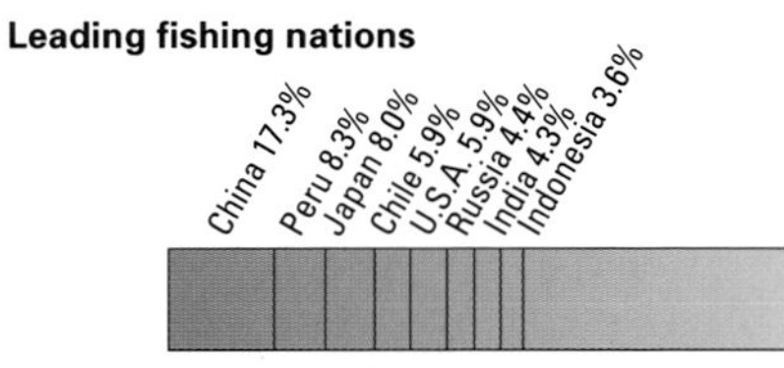

World total (1993): 111,762,080 tons
(Marine catch 83.1% Inland catch 16.9%)

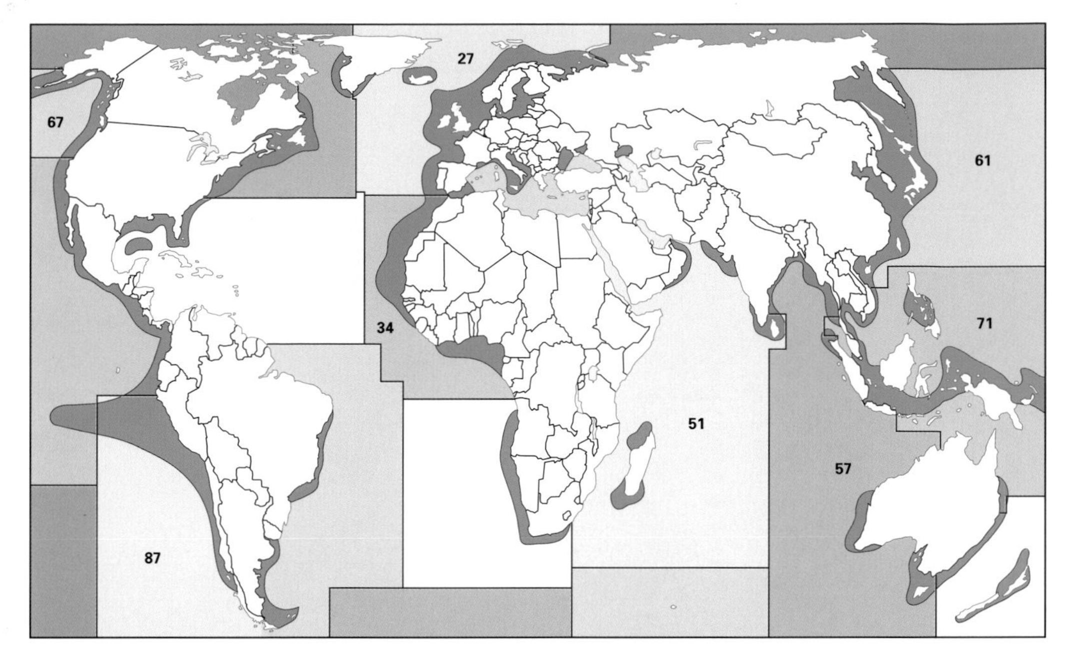

MARINE POLLUTION

Sources of marine oil pollution (latest available year)

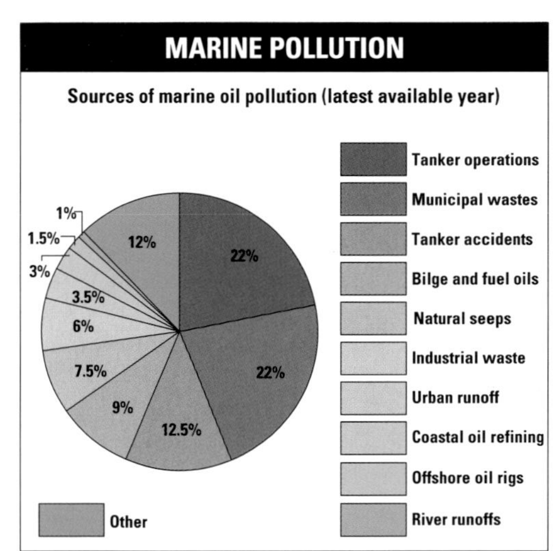

OIL SPILLS

Major oil spills from tankers and combined carriers

Year	Vessel	Location	Spill (barrels)**	Cause
1979	Atlantic Empress	West Indies	1,890,000	collision
1983	Castillo De Bellver	South Africa	1,760,000	fire
1978	Amoco Cadiz	France	1,628,000	grounding
1991	Haven	Italy	1,029,000	explosion
1988	Odyssey	Canada	1,000,000	fire
1967	Torrey Canyon	UK	909,000	grounding
1972	Sea Star	Gulf of Oman	902,250	collision
1977	Hawaiian Patriot	Hawaiian Is.	742,500	fire
1979	Independenta	Turkey	696,350	collision
1993	Braer	UK	625,000	grounding
1996	Sea Empress	UK	515,000	grounding
Other sources of major oil spills				
1983	Nowruz oilfield	The Gulf	4,250,000†	war
1979	Ixtoc 1 oilwell	Gulf of Mexico	4,200,000	blow out
1991	Kuwait	The Gulf	2,500,000†	war

** 1 barrel = 0.15 tons/159 lit./35 Imperial gal./42 US gal. † estimated

RIVER POLLUTION

Sources of river pollution, USA (latest available year)

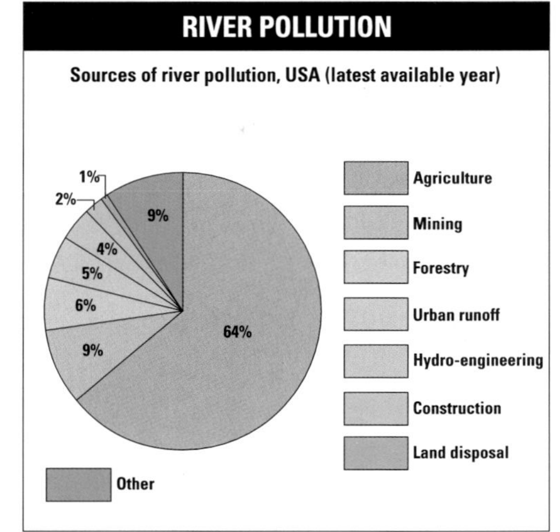

WATER POLLUTION

Severely polluted sea areas and lakes

Polluted sea areas and lakes

Areas of frequent oil pollution by shipping

Major oil tanker spills

Major oil rig blow outs

Offshore dumpsites for industrial and municipal waste

Severely polluted rivers and estuaries

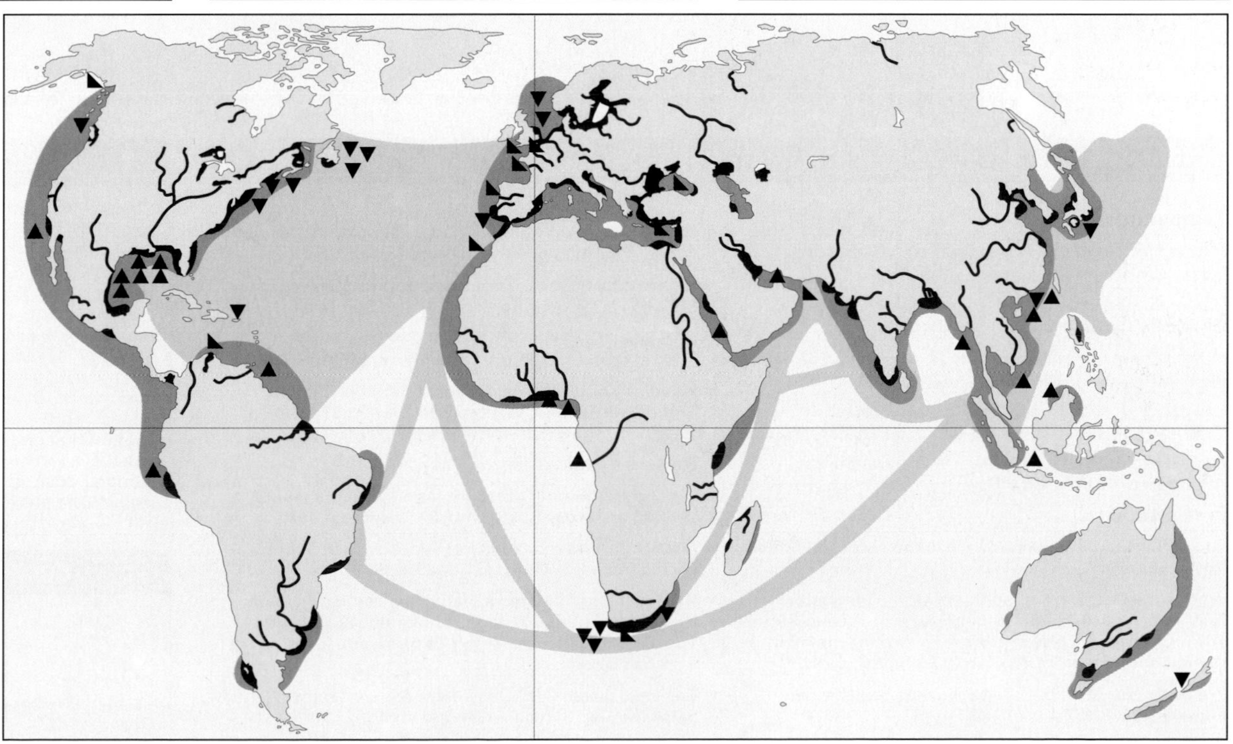

The most notorious tanker spillage of the 1980s occurred when the *Exxon Valdez* ran aground in Prince William Sound, Alaska, in 1989, spilling 267,000 barrels of crude oil close to shore in a sensitive ecological area. This rates as the world's 28th worst spill in terms of volume.

CLIMATE

CLIMATIC REGIONS

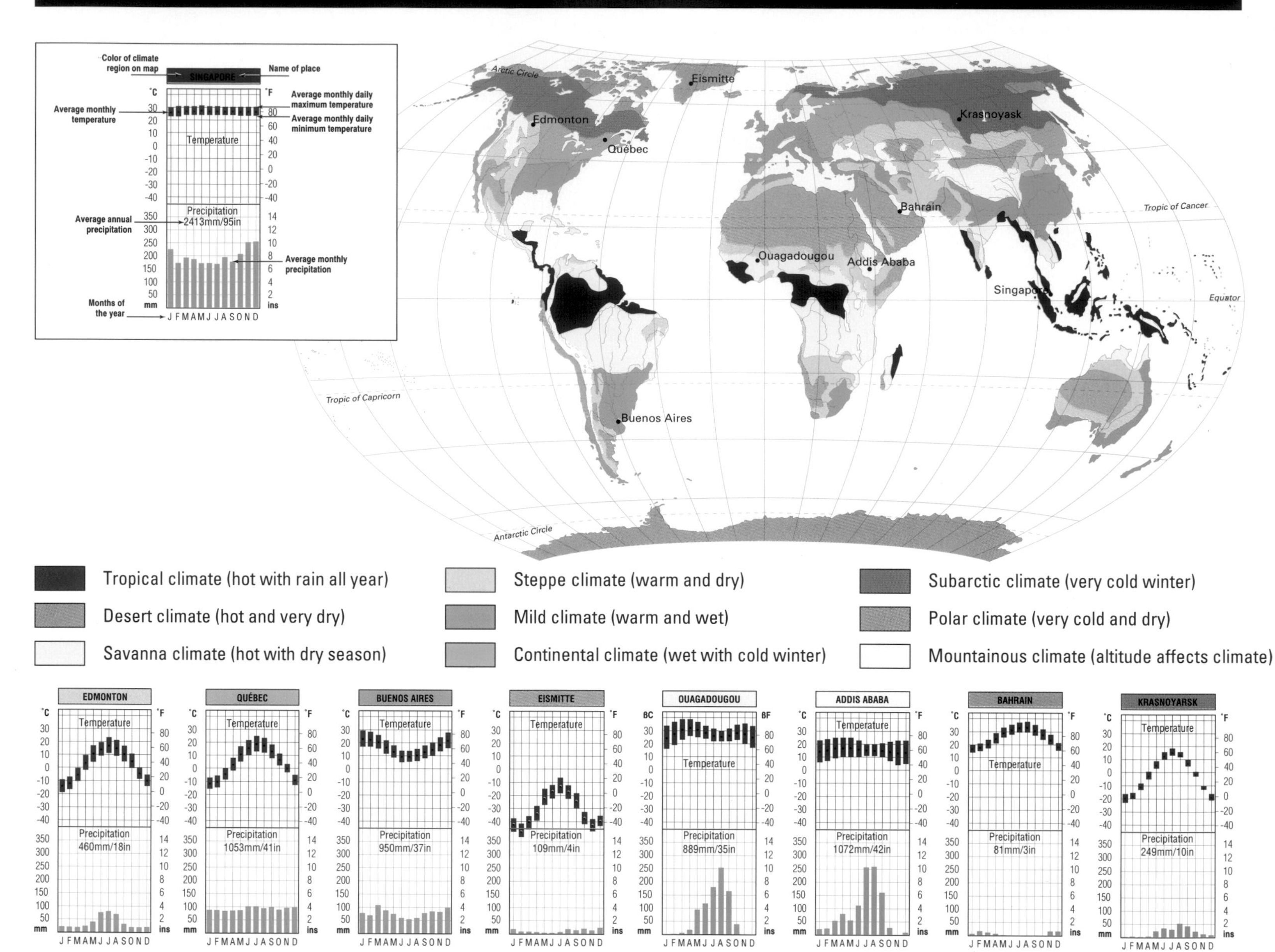

CLIMATE RECORDS

Temperature

Highest recorded shade temperature: Al Aziziyah, Libya, 136.4°F [58°C], 13 September 1922.

Highest mean annual temperature: Dallol, Ethiopia, 94°F [34.4°C], 1960–66.

Longest heatwave: Marble Bar, W. Australia, 162 days over 100°F [38°C], 23 October 1923 to 7 April 1924.

Lowest recorded temperature (outside poles): Verkhoyansk, Siberia, –90°F [–68°C], 6 February 1933.

Lowest mean annual temperature: Plateau Station, Antarctica, –72.0°F [–56.6°C]

Precipitation

Longest drought: Calama, N. Chile, no recorded rainfall in 400 years to 1971.

Wettest place (12 months): Cherrapunji, Meghalaya, N. E. India, 1,040 in [26,470 mm], August 1860 to August 1861. Cherrapunji also holds the record for the most rainfall in one month: 115.4 in [2,930 mm], July 1861.

Wettest place (average): Mawsynram, India, mean annual rainfall 467.4 in [11,873 mm].

Wettest place (24 hours): Cilaos, Réunion, Indian Ocean, 73.6 in [1,870 mm], 15–16 March 1952.

Heaviest hailstones: Gopalganj, Bangladesh, up to 2.25 lb [1.02 kg], 14 April 1986 (killed 92 people).

Heaviest snowfall (continuous): Bessans, Savoie, France, 68 in [1,730 mm] in 19 hours, 5–6 April 1969.

Heaviest snowfall (season/year): Paradise Ranger Station, Mt Rainier, Washington, USA, 1,224.5 in [31,102 mm], 19 February 1971 to 18 February 1972.

Pressure and winds

Highest barometric pressure: Agata, Siberia (at 862 ft [262 m] altitude), 1,083.8 mb, 31 December 1968.

Lowest barometric pressure: Typhoon Tip, Guam, Pacific Ocean, 870 mb, 12 October 1979.

Highest recorded wind speed: Mt Washington, New Hampshire, USA, 231 mph [371 km/h], 12 April 1934. This is three times as strong as hurricane force on the Beaufort Scale.

Windiest place: Commonwealth Bay, Antarctica, where gales frequently reach over 200 mph [320 km/h].

CLIMATE

Climate is weather in the long term: the seasonal pattern of hot and cold, wet and dry, averaged over time (usually 30 years). At the simplest level, it is caused by the uneven heating of the Earth. Surplus heat at the Equator passes toward the poles, leveling out the energy differential. Its passage is marked by a ceaseless churning of the atmosphere and the oceans, further agitated by the Earth's diurnal spin and the motion it imparts to moving air and water. The heat's means of transport – by winds and ocean currents, by the continual evaporation and recondensation of water molecules – is the weather itself. There are four basic types of climate, each of which can be further subdivided: tropical, desert (dry), temperate and polar.

COMPOSITION OF DRY AIR

Nitrogen	78.09%	Sulfur dioxide	trace
Oxygen	20.95%	Nitrogen oxide	trace
Argon	0.93%	Methane	trace
Water vapor	0.2–4.0%	Dust	trace
Carbon dioxide	0.03%	Helium	trace
Ozone	0.00006%	Neon	trace

EL NIÑO

In a normal year, southeasterly trade winds drive surface waters westward off the coast of South America, drawing cold, nutrient-rich water up from below. In an El Niño year (which occurs every 2 to 7 years), warm water from the west Pacific suppresses up-welling in the east depriving the region of nutrients. The water is warmed by as much as 12°F, disturbing the tropical atmospheric circulation. During an intense El Niño, the southeast trade winds change direction and become equatorial westerlies resulting in climatic extremes in many regions of the world, such as drought in parts of Australia and India, and heavy rainfall in southeastern USA. The UK experiences exceptionally mild and wet winters.

Normal year

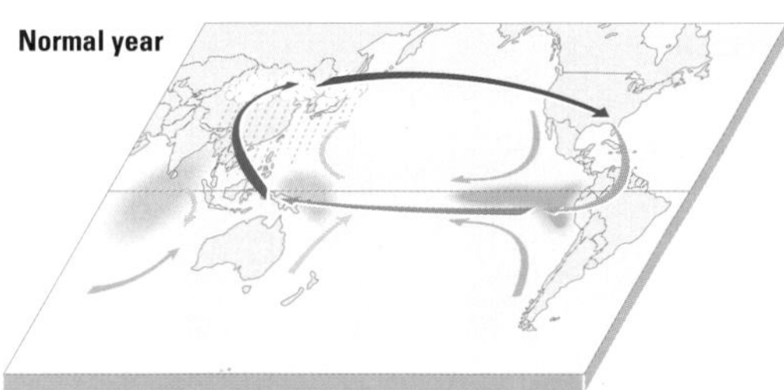

El Niño event

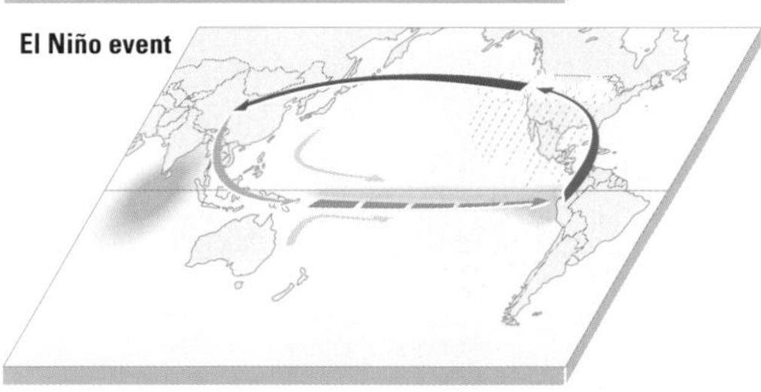

BEAUFORT WIND SCALE

Named after the 19th-century British naval officer who devised it, the Beaufort Scale assesses wind speed according to its effects. It was originally designed as an aid for sailors, but has since been adapted for use on the land.

Scale	Wind speed km/h	Wind speed mph	Effect
0	0–1	0–1	**Calm** Smoke rises vertically
1	1–5	1–3	**Light air** Wind direction shown only by smoke drift
2	6–11	4–7	**Light breeze** Wind felt on face; leaves rustle; vanes moved by wind
3	12–19	8–12	**Gentle breeze** Leaves and small twigs in constant motion; wind extends small flag
4	20–28	13–18	**Moderate** Raises dust and loose paper; small branches move
5	29–38	19–24	**Fresh** Small trees in leaf sway; wavelets on inland waters
6	39–49	25–31	**Strong** Large branches move; difficult to use umbrellas
7	50–61	32–38	**Near gale** Whole trees in motion; difficult to walk against wind
8	62–74	39–46	**Gale** Twigs break from trees; walking very difficult
9	75–88	47–54	**Strong gale** Slight structural damage
10	89–102	55–63	**Storm** Trees uprooted; serious structural damage
11	103–117	64–72	**Violent storm** Widespread damage
12	118+	73+	**Hurricane**

Conversions

°C = (°F –32) x 5/9; °F = (°C x 9/5) + 32; 0°C = 32°F

1 in = 25.4 mm; 1 mm = 0.0394 in; 100 mm = 3.94 in

TEMPERATURE

Average temperature in January

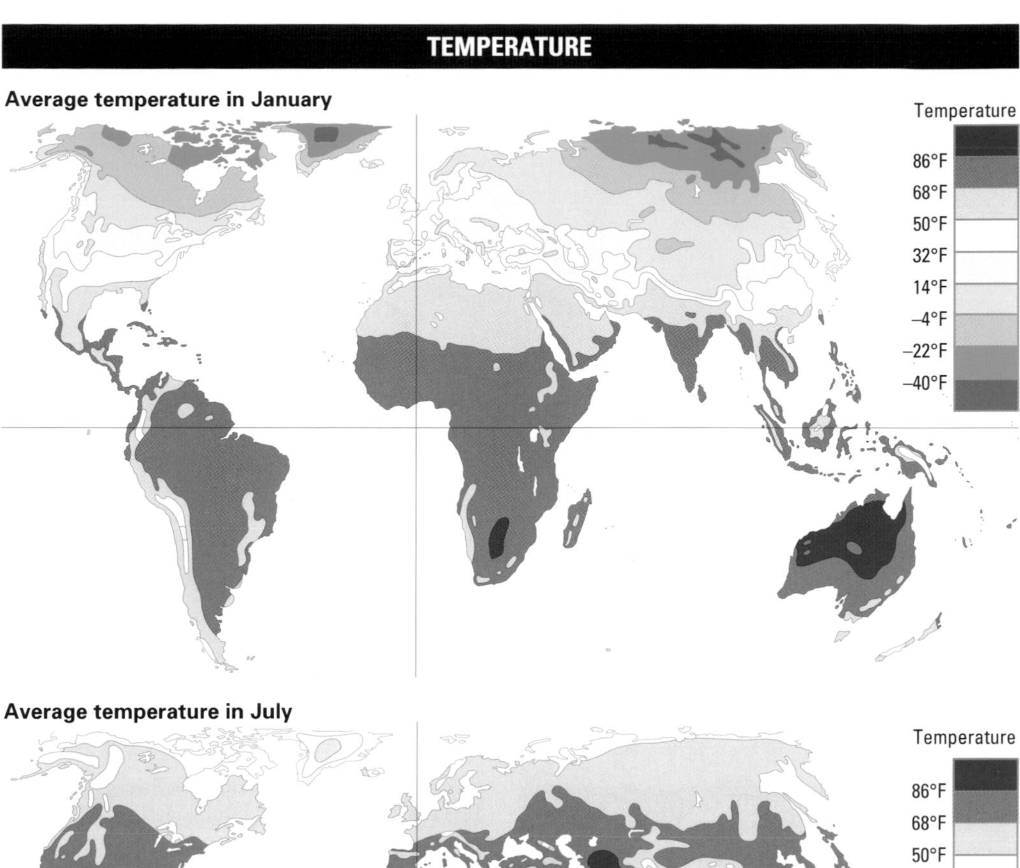

Average temperature in July

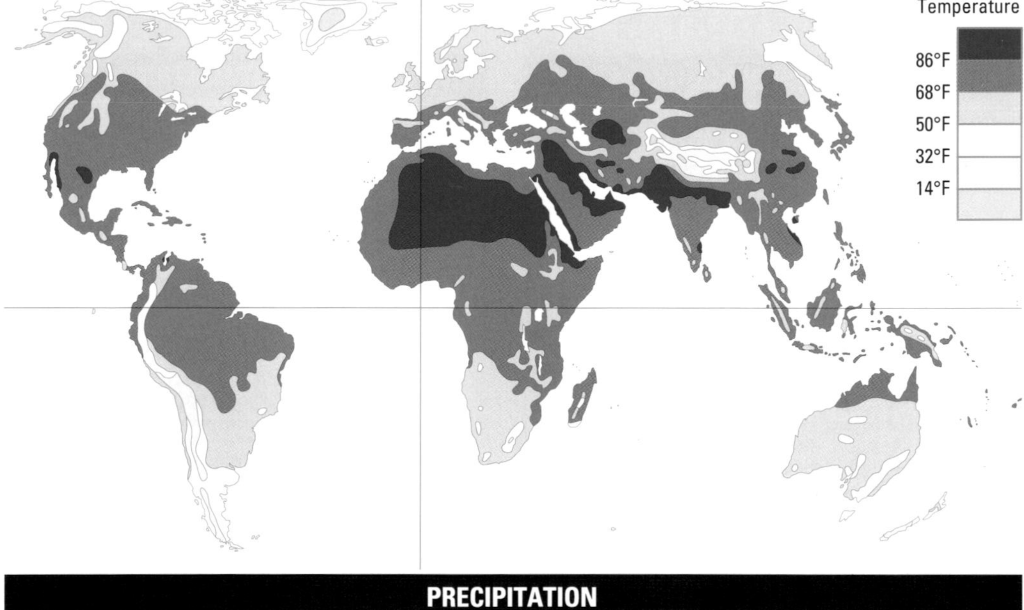

PRECIPITATION

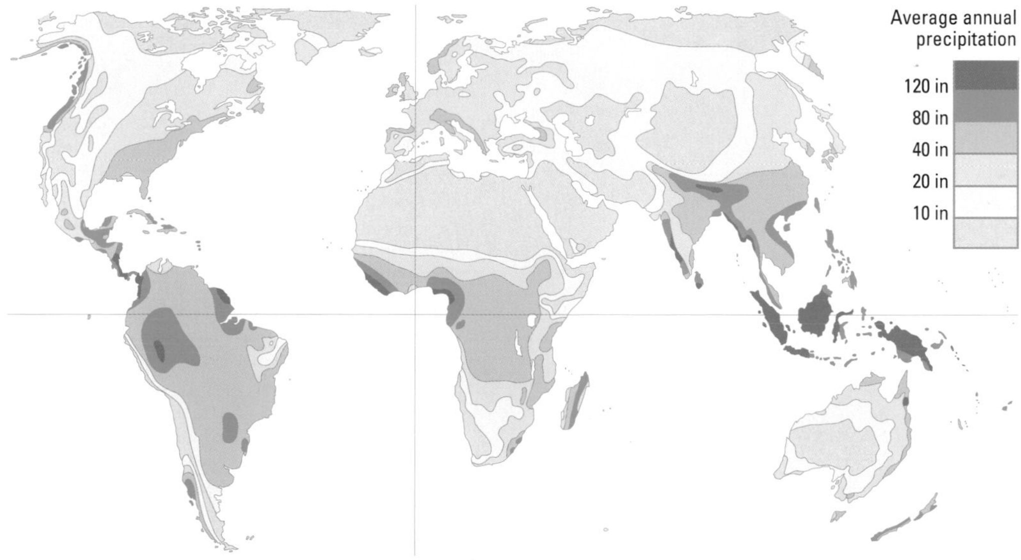

WATER AND VEGETATION

THE HYDROLOGICAL CYCLE

The world's water balance is regulated by the constant recycling of water between the oceans, atmosphere and land. The movement of water between these three reservoirs is known as the hydrological cycle. The oceans play a vital role in the hydrological cycle: 74% of the total precipitation falls over the oceans and 84% of the total evaporation comes from the oceans.

Transfer of water vapor
Evaporation from oceans
Evapotranspiration
Precipitation
Precipitation
Surface runoff
Runoff
Surface storage
Infiltration
Groundwater flow

WATER DISTRIBUTION

The distribution of planetary water, by percentage. Oceans and ice caps together account for more than 99% of the total; the breakdown of the remainder is estimated.

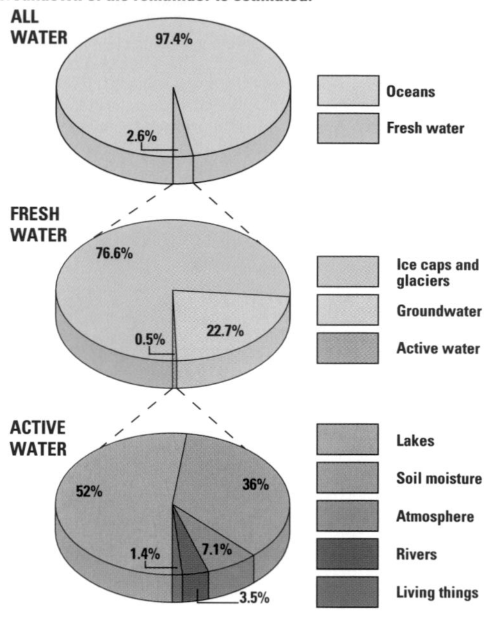

WATER USAGE

Almost all the world's water is 3,000 million years old, and all of it cycles endlessly through the hydrosphere, though at different rates. Water vapor circulates over days, even hours, deep ocean water circulates over millennia, and ice-cap water remains solid for millions of years.

Fresh water is essential to all terrestrial life. Humans cannot survive more than a few days without it, and even the hardiest desert plants and animals could not exist without some water. Agriculture requires huge quantities of fresh water: without large-scale irrigation most of the world's people would starve. In the USA, agriculture uses 43% and industry 38% of all water withdrawals.

The United States is one of the heaviest users of water in the world. According to the latest figures the average American uses 380 liters a day and the average household uses 415,000 liters a year. This is two to four times more than in Western Europe.

WATER UTILIZATION

The percentage breakdown of water usage by sector, selected countries (latest available year)

0 20 40 60 80 100

Algeria
Australia
CIS
Egypt
France
Ghana
India
Mexico
Poland
Saudi Arabia
UK
USA

0 20 40 60 80 100

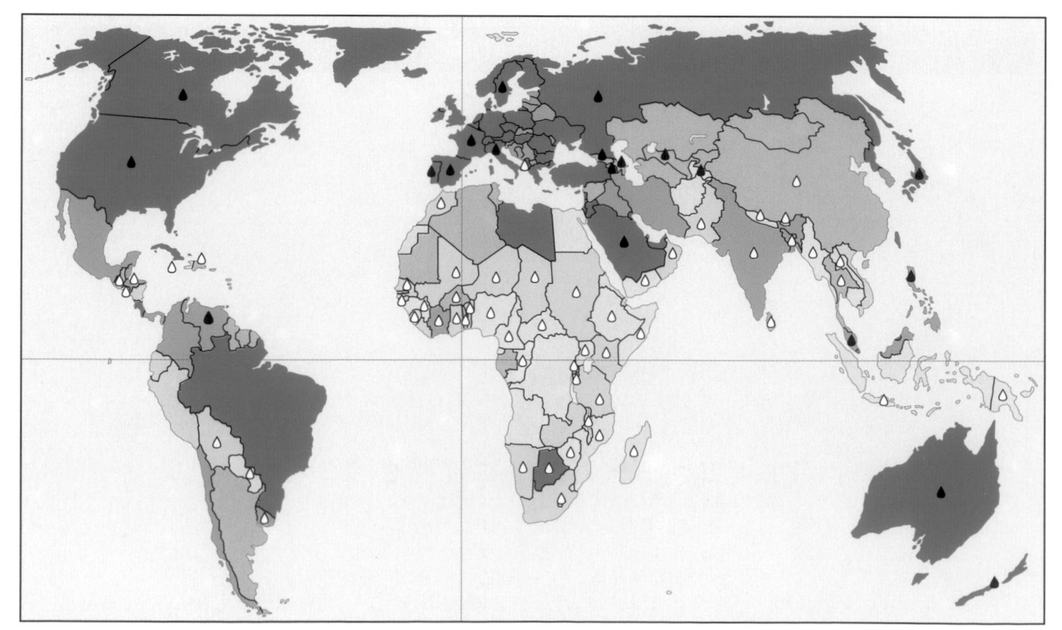

WATER SUPPLY

Percentage of total population with access to safe drinking water (latest available year)

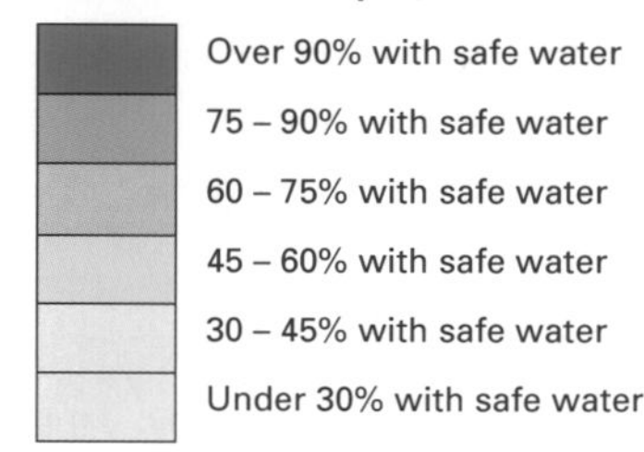

Under 80 liters per person per day domestic water consumption

Over 320 liters per person per day

Least well-provided countries

Central African Rep.	12%	Madagascar	23%
Uganda	15%	Guinea-Bissau	25%
Ethiopia	18%	Laos	28%
Mozambique	22%	Swaziland	30%
Afghanistan	23%	Tajikistan	30%

NATURAL VEGETATION

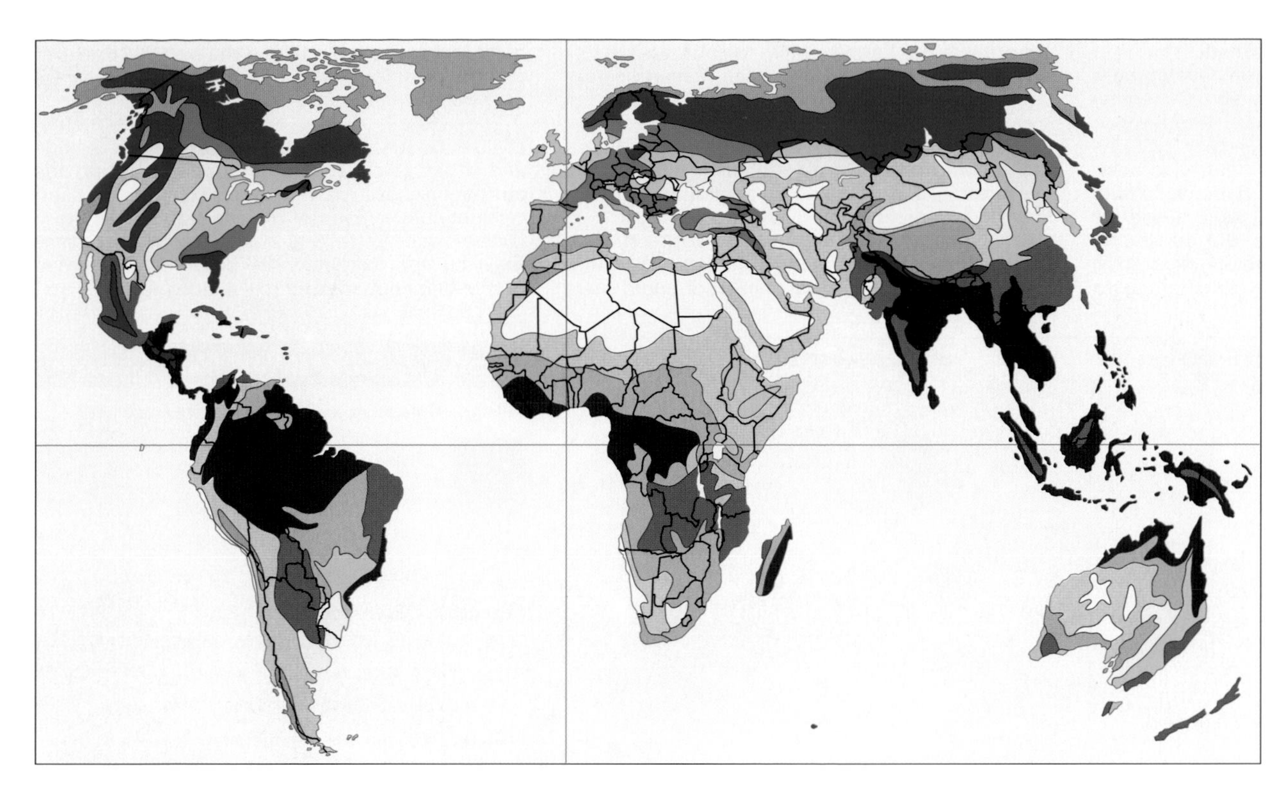

Regional variation in vegetation

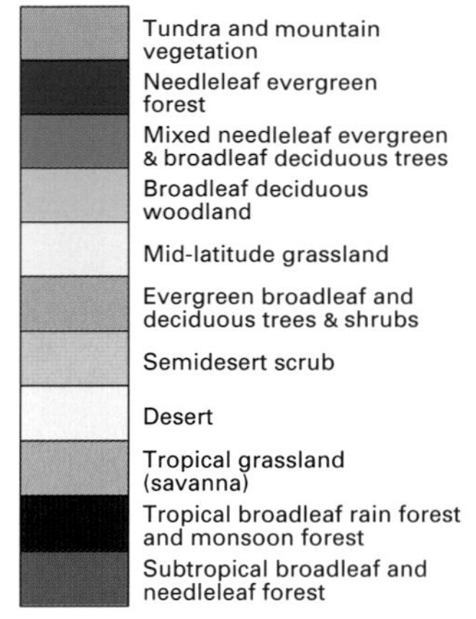

The map shows the natural "climax vegetation" of regions, as dictated by climate and topography. In most cases, however, agricultural activity has drastically altered the vegetation pattern. Western Europe, for example, lost most of its broadleaf forest many centuries ago, while irrigation has turned some natural semidesert into productive land.

LAND USE BY CONTINENT

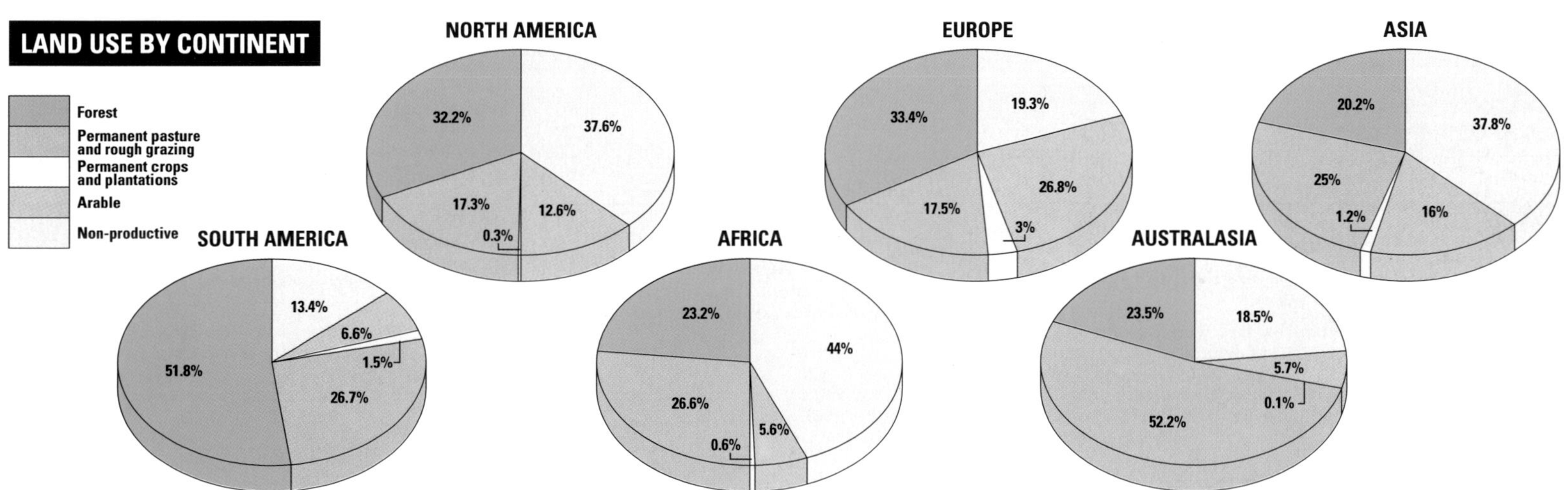

FORESTRY: PRODUCTION

	Forest & woodland (million acres)	Annual production (1993, million cubic yards) Fuelwood & charcoal	Industrial roundwood*
World	***9,854.1***	***2,453.5***	***1,999.3***
CIS	2,045.5	67.4	226.2
S. America	2,049.2	324.1	159.6
N. & C. America	1,753.9	205.0	767.4
Africa	1,691.6	645.6	77.8
Asia	1,211.3	1,133.3	363.8
Europe	388.7	66.6	356.0
Australasia	388.4	11.4	48.3

PAPER AND BOARD

Top producers (1993)**		Top exporters (1993)**	
USA	85,130	Canada	14,211
Japan	30,596	Finland	9,396
China	26,245	USA	7,875
Canada	19,348	Sweden	7,723
Germany	14,363	Germany	5,249

* roundwood is timber as it is felled
** in thousand tons

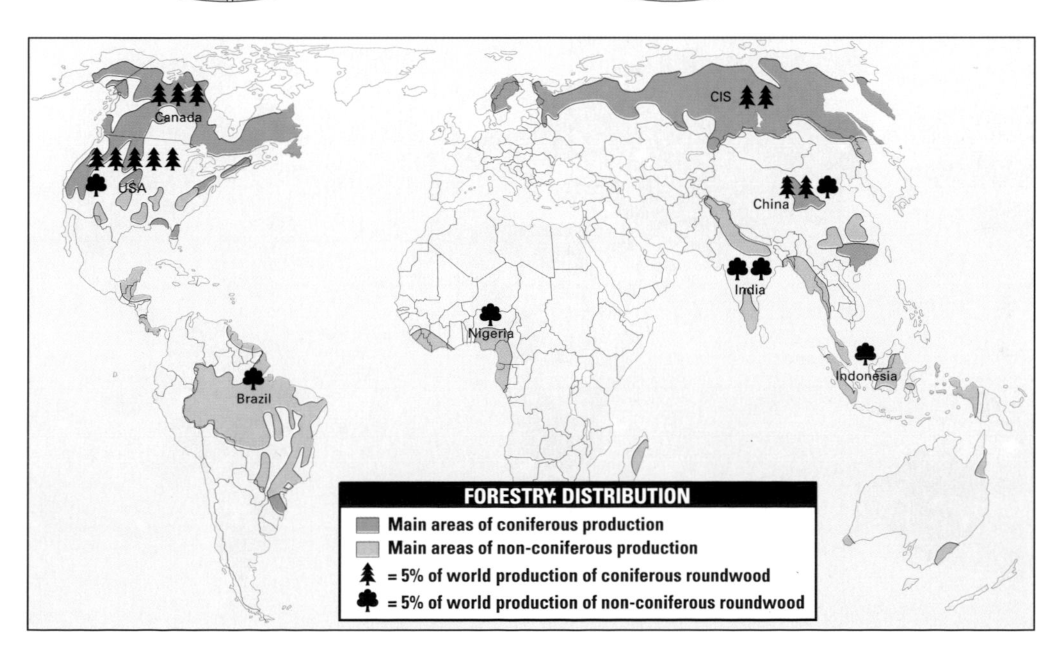

Environment

Humans have always had a dramatic effect on their environment, at least since the development of agriculture almost 10,000 years ago. Generally, the Earth has accepted human interference without obvious ill effects: the complex systems that regulate the global environment have been able to absorb substantial damage while maintaining a stable and comfortable home for the planet's trillions of lifeforms. But advancing human technology and the rapidly-expanding populations it supports are now threatening to overwhelm the Earth's ability to compensate.

Industrial wastes, acid rainfall, desertification and large-scale deforestation all combine to create environmental change at a rate far faster than the great slow cycles of planetary evolution can accommodate. As a result of overcultivation, overgrazing and overcutting of groundcover for firewood, desertification is affecting as much as 60% of the world's croplands. In addition, with fire and chainsaws, humans are destroying more forest in a day than their ancestors could have done in a century, upsetting the balance between plant and animal, carbon dioxide and oxygen, on which all life ultimately depends.

The fossil fuels that power industrial civilization have pumped enough carbon dioxide and other so-called greenhouse gases into the atmosphere to make climatic change a near-certainty. As a result of the combination of these factors, the Earth's average temperature has risen by approximately 1°F since the beginning of the 20th century, and it is still rising.

GLOBAL WARMING

Carbon dioxide emissions in tons per person per year (1992)

- Over 10 tons of CO_2
- 5 – 10 tons of CO_2
- 1 – 5 tons of CO_2
- Under 1 ton of CO_2

Changes in CO_2 emissions 1980–90

- ▲ Over 100% increase in emissions
- ▴ 50–100% increase in emissions
- ▽ Reduction in emissions
- — Coastal areas in danger of flooding from rising sea levels caused by global warming

High atmospheric concentrations of heat-absorbing gases, especially carbon dioxide, appear to be causing a steady rise in average temperatures worldwide – up to 3°F by the year 2020, according to some estimates. Global warming is likely to bring with it a rise in sea levels that may flood some of the Earth's most densely populated coastal areas.

GREENHOUSE POWER

Relative contributions to the Greenhouse Effect by the major heat-absorbing gases in the atmosphere.

The chart combines greenhouse potency and volume. Carbon dioxide has a greenhouse potential of only 1, but its concentration of 350 parts per million makes it predominate. CFC 12, with 25,000 times the absorption capacity of CO_2, is present only as 0.00044 ppm.

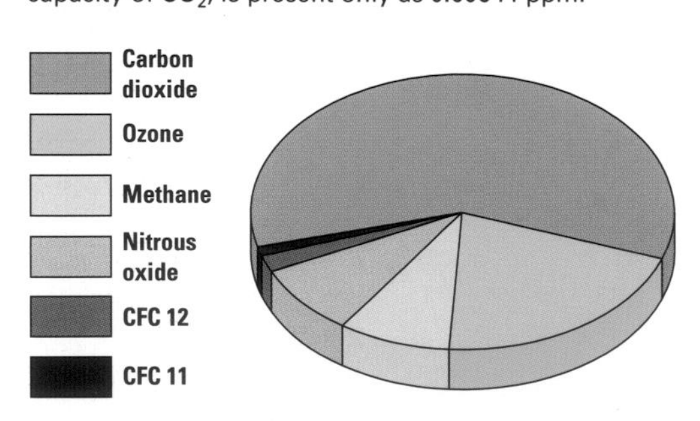

Carbon dioxide
Ozone
Methane
Nitrous oxide
CFC 12
CFC 11

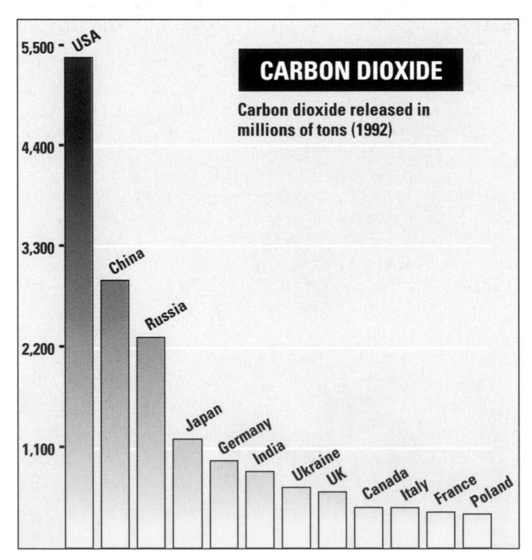

5,500
4,400
3,300
2,200
1,100
USA
China
Russia
Japan
Germany
India
Ukraine
UK
Canada
Italy
France
Poland

CARBON DIOXIDE

Carbon dioxide released in millions of tons (1992)

OZONE LAYER

The ozone "hole" over the northern hemisphere on 12 March 1995.

The colors represent Dobson Units (DU). The ozone hole is seen as the dark blue and purple patch in the center, where ozone values are around 120 DU or lower. Normal levels are around 280 DU. The ozone hole over Antarctica is much larger.

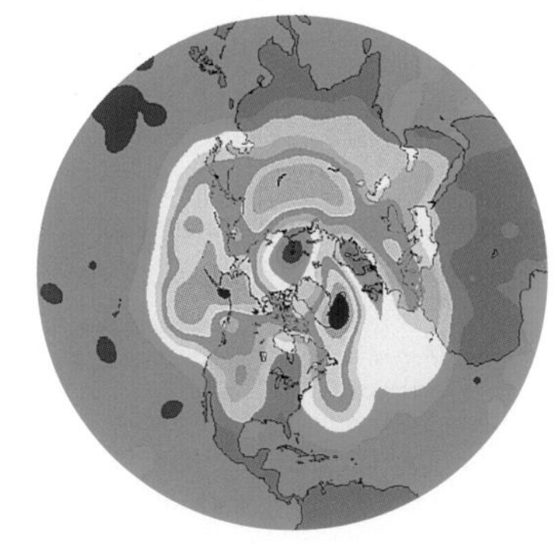

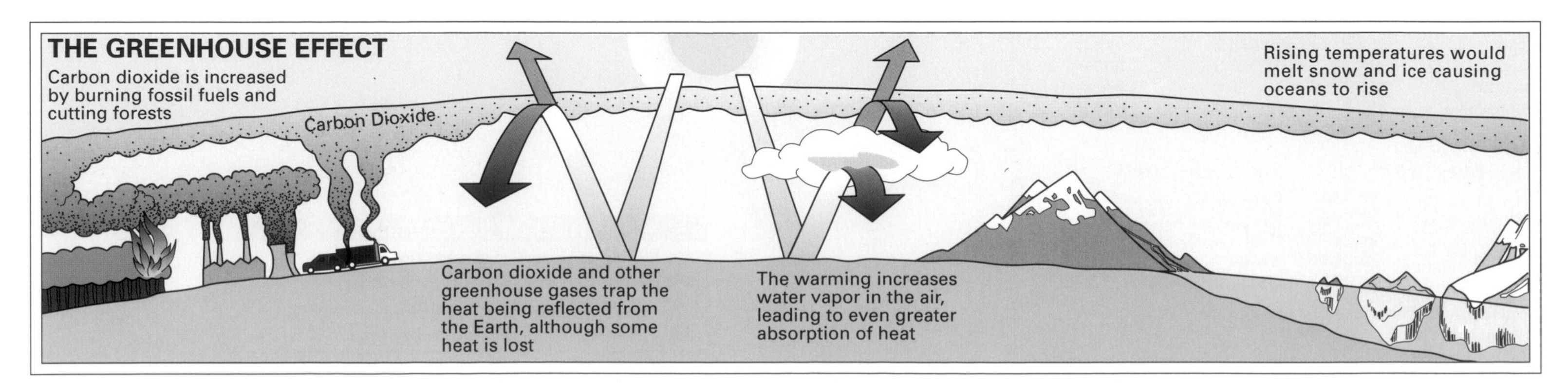

THE GREENHOUSE EFFECT
Carbon dioxide is increased by burning fossil fuels and cutting forests
Carbon Dioxide
Rising temperatures would melt snow and ice causing oceans to rise
Carbon dioxide and other greenhouse gases trap the heat being reflected from the Earth, although some heat is lost
The warming increases water vapor in the air, leading to even greater absorption of heat

DESERTIFICATION

DEFORESTATION

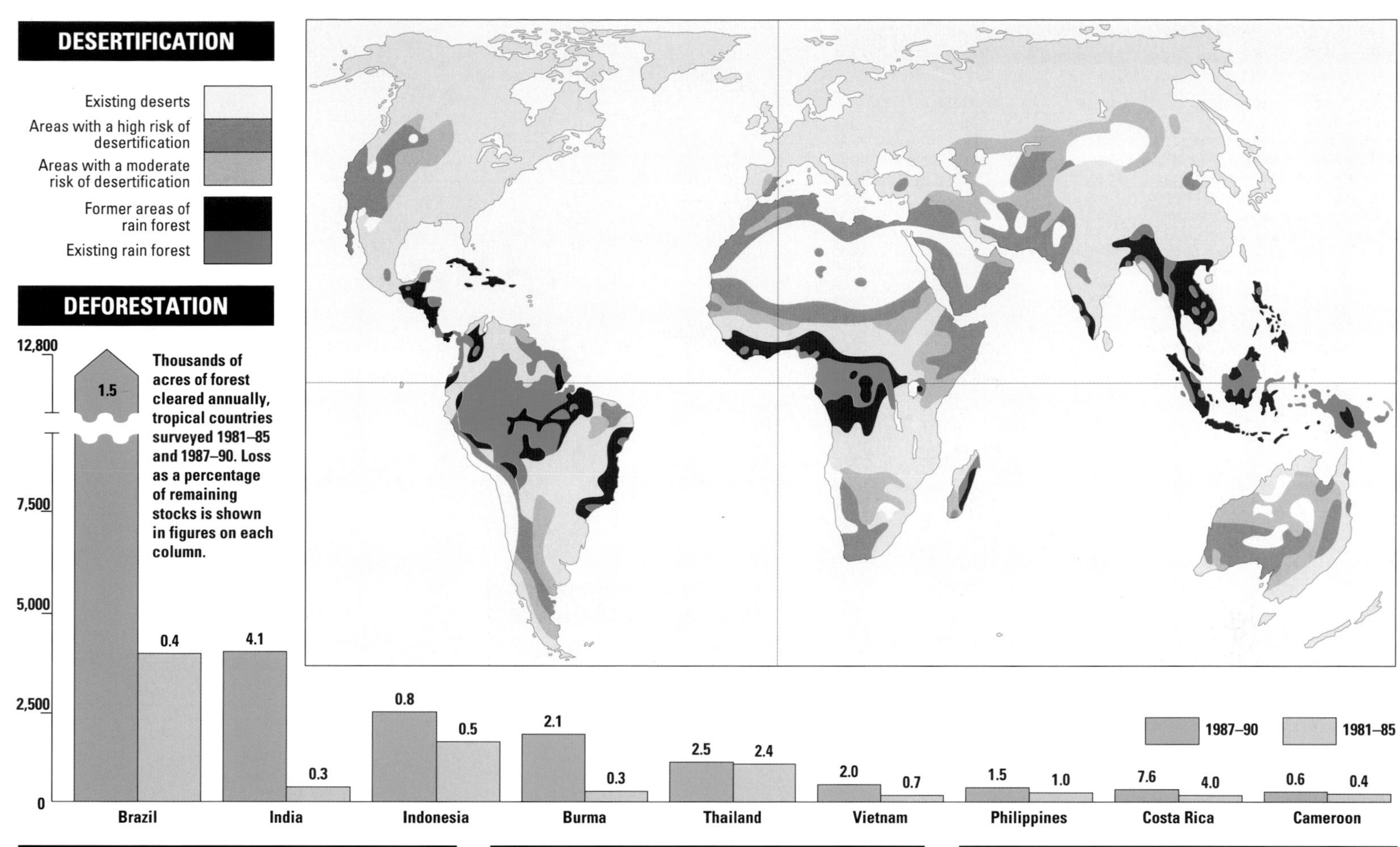

OZONE DEPLETION

The ozone layer (15–18 miles above sea level) acts as a barrier to most of the Sun's harmful ultra-violet radiation, protecting us from the ionizing radiation that can cause skin cancer and cataracts. In recent years, however, two holes in the ozone layer have been observed during winter; one over the Arctic and the other, the size of the USA, over Antarctica. By 1996, ozone had been reduced to around a half of its 1970 amount. The ozone (O_3) is broken down by chlorine released into the atmosphere as CFCs (chlorofluorocarbons) – chemicals used in refrigerators, packaging and aerosols.

DEFORESTATION

The Earth's remaining forests are under attack from three directions: expanding agriculture, logging, and growing consumption of fuelwood, often in combination. Sometimes deforestation is the direct result of government policy, as in the efforts made to resettle the urban poor in some parts of Brazil; just as often, it comes about despite state attempts at conservation. Loggers, licensed or unlicensed, blaze a trail into virgin forest, often destroying twice as many trees as they harvest. Landless farmers follow, burning away most of what remains to plant their crops, completing the destruction.

ACID RAIN

Killing trees, poisoning lakes and rivers and eating away buildings, acid rain is mostly produced by sulfur dioxide emissions from industry and volcanic eruptions. By the mid 1990s, acid rain had sterilized 4,000 or more of Sweden's lakes and left 45% of Switzerland's alpine conifers dead or dying, while the monuments of Greece were dissolving in Athens' smog. Prevailing wind patterns mean that the acids often fall many hundred miles from where the original pollutants were discharged. In parts of Europe acid deposition has slightly decreased, following reductions in emissions, but not by enough.

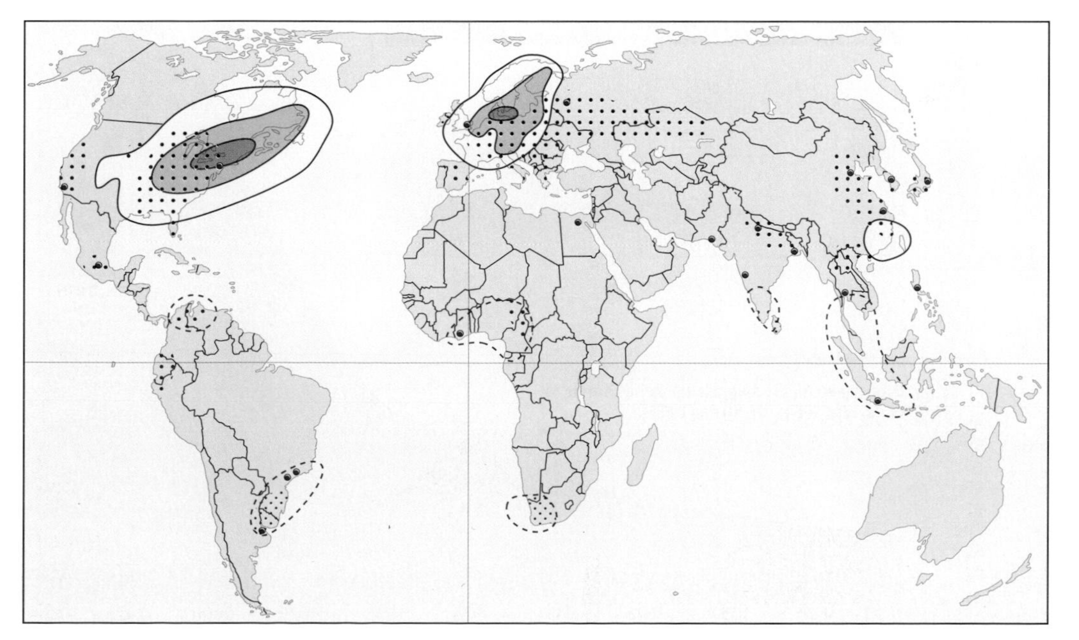

WORLD POLLUTION

Acid rain and sources of acidic emissions (latest available year)

Acid rain is caused by high levels of sulfur and nitrogen in the atmosphere. They combine with water vapor and oxygen to form acids (H_2SO_4 and HNO_3) which fall as precipitation.

Regions where sulfur and nitrogen oxides are released in high concentrations, mainly from fossil fuel combustion

Major cities with high levels of air pollution (including nitrogen and sulfur emissions)

Areas of heavy acid deposition

pH numbers indicate acidity, decreasing from a neutral 7. Normal rain, slightly acid from dissolved carbon dioxide, never exceeds a pH of 5.6.

pH less than 4.0 (most acidic)

pH 4.0 to 4.5

pH 4.5 to 5.0

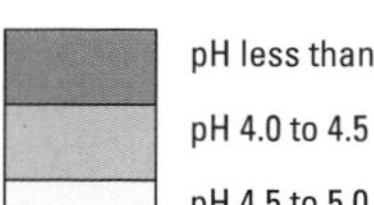

Areas where acid rain is a potential problem

POPULATION

DEMOGRAPHIC PROFILES

Developed nations such as the UK have populations evenly spread across the age groups and, usually, a growing proportion of elderly people. The great majority of the people in developing nations, however, are in the younger age groups, about to enter their most fertile years. In time, these population profiles should resemble the world profile (even Kenya has made recent progress with reducing its birth rate), but the transition will come about only after a few more generations of rapid population growth.

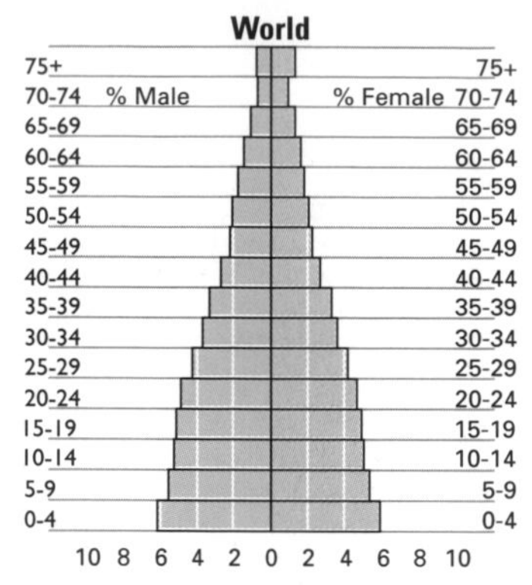

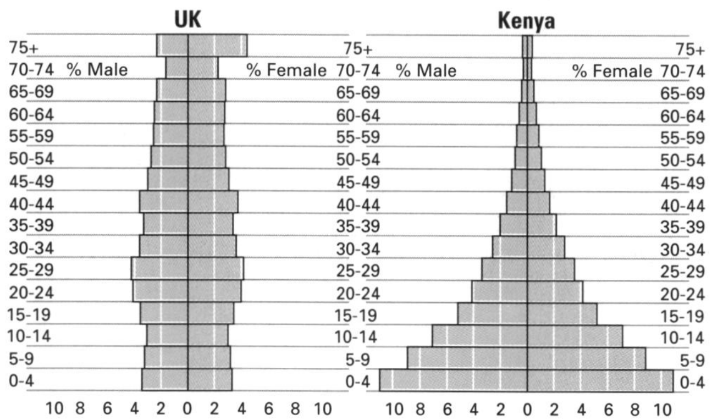

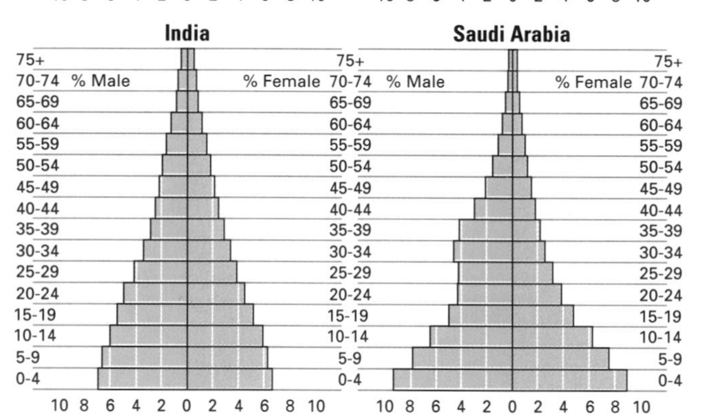

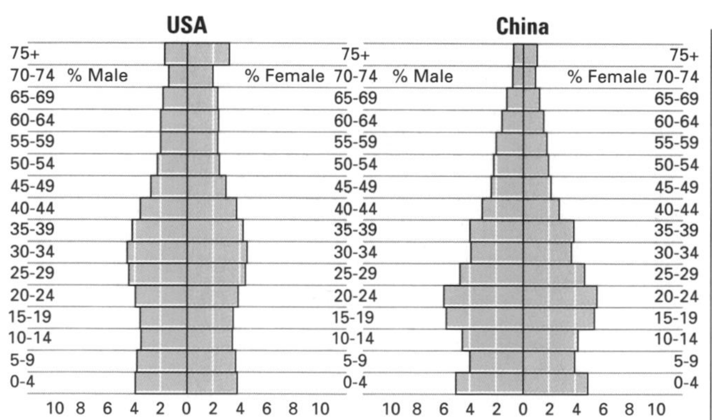

MOST POPULOUS NATIONS [in millions (1995)]

1. China	1,227	9. Bangladesh	118	17. Turkey	61
2. India	943	10. Mexico	93	18. Thailand	58
3. USA	264	11. Nigeria	89	19. UK	58
4. Indonesia	199	12. Germany	82	20. France	58
5. Brazil	161	13. Vietnam	75	21. Italy	57
6. Russia	148	14. Iran	69	22. Ukraine	52
7. Pakistan	144	15. Philippines	67	23. Ethiopia	52
8. Japan	125	16. Egypt	64	24. Burma	47

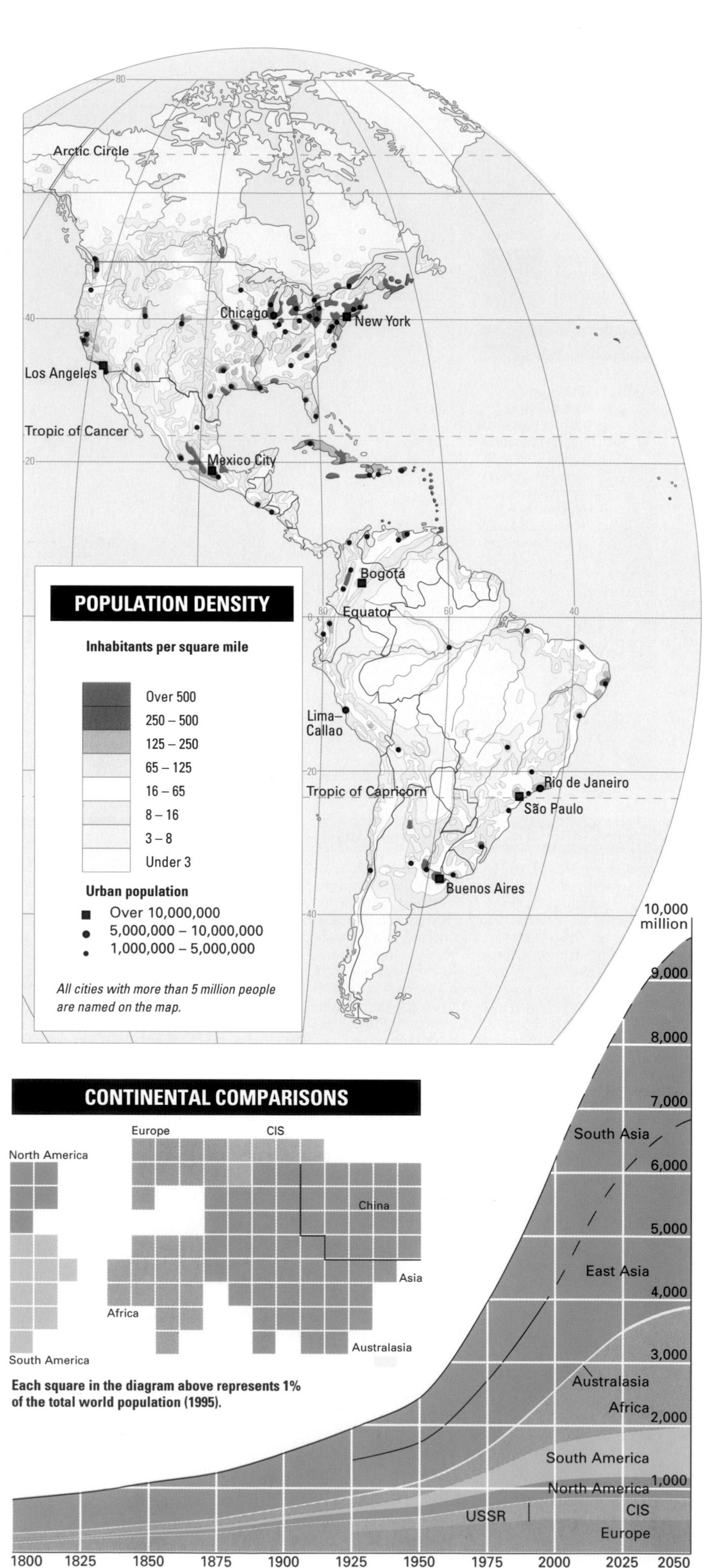

Each square in the diagram above represents 1% of the total world population (1995).

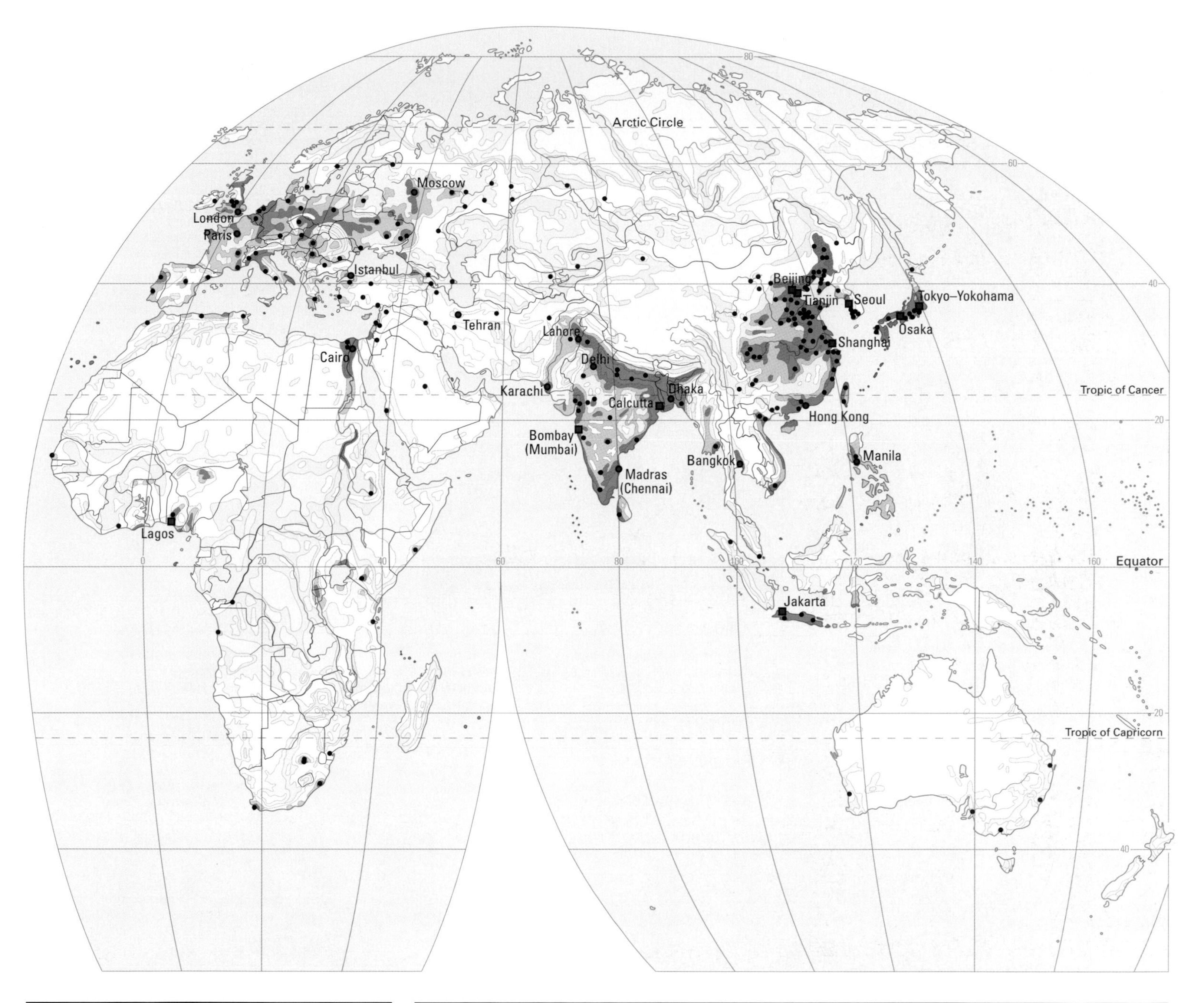

URBAN POPULATION

Percentage of total population living in towns and cities (1993)

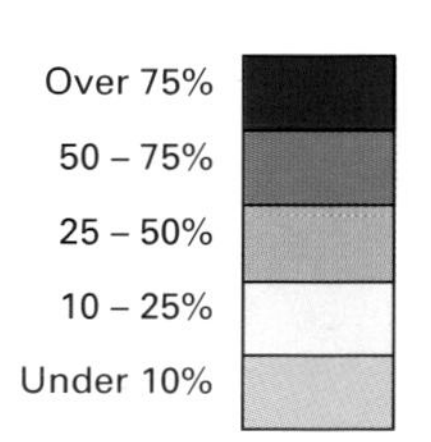

Most urbanized		**Least urbanized**	
Singapore	100%	Bhutan	6%
Belgium	97%	Rwanda	6%
Kuwait	96%	Burundi	7%
Venezuela	92%	Nepal	12%
Israel	91%	Uganda	12%

[USA 76%]

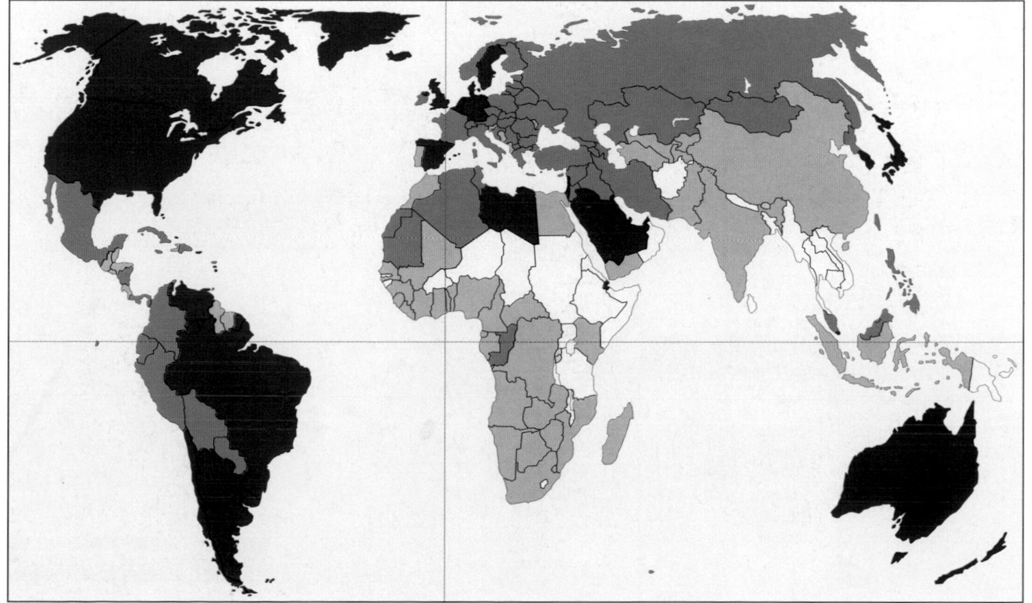

The Human Family

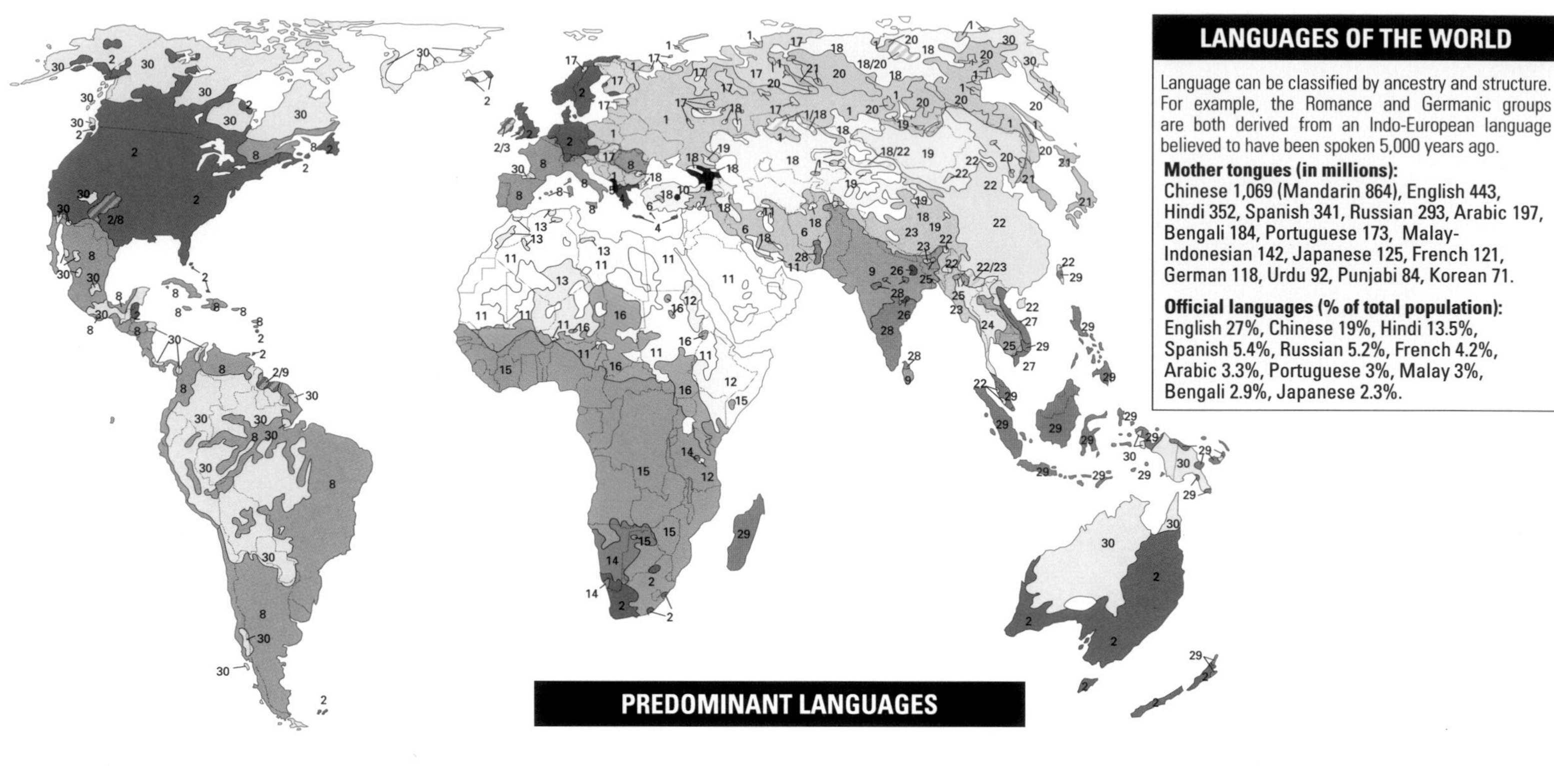

LANGUAGES OF THE WORLD

Language can be classified by ancestry and structure. For example, the Romance and Germanic groups are both derived from an Indo-European language believed to have been spoken 5,000 years ago.

Mother tongues (in millions):
Chinese 1,069 (Mandarin 864), English 443, Hindi 352, Spanish 341, Russian 293, Arabic 197, Bengali 184, Portuguese 173, Malay-Indonesian 142, Japanese 125, French 121, German 118, Urdu 92, Punjabi 84, Korean 71.

Official languages (% of total population):
English 27%, Chinese 19%, Hindi 13.5%, Spanish 5.4%, Russian 5.2%, French 4.2%, Arabic 3.3%, Portuguese 3%, Malay 3%, Bengali 2.9%, Japanese 2.3%.

INDO-EUROPEAN FAMILY

- 1 Balto-Slavic group (incl. Russian, Ukrainian)
- 2 Germanic group (incl. English, German)
- 3 Celtic group
- 4 Greek
- 5 Albanian
- 6 Iranian group
- 7 Armenian
- 8 Romance group (incl. Spanish, Portuguese, French, Italian)
- 9 Indo-Aryan group (incl. Hindi, Bengali, Urdu, Punjabi, Marathi)
- 10 CAUCASIAN FAMILY

AFRO-ASIATIC FAMILY

- 11 Semitic group (incl. Arabic)
- 12 Kushitic group
- 13 Berber group
- 14 KHOISAN FAMILY
- 15 NIGER-CONGO FAMILY
- 16 NILO-SAHARAN FAMILY
- 17 URALIC FAMILY

ALTAIC FAMILY

- 18 Turkic group
- 19 Mongolian group
- 20 Tungus-Manchu group
- 21 Japanese and Korean

SINO-TIBETAN FAMILY

- 22 Sinitic (Chinese) languages
- 23 Tibetic-Burmic languages
- 24 TAI FAMILY

AUSTRO-ASIATIC FAMILY

- 25 Mon-Khmer group
- 26 Munda group
- 27 Vietnamese
- 28 DRAVIDIAN FAMILY (incl. Telugu, Tamil)
- 29 AUSTRONESIAN FAMILY (incl. Malay-Indonesian)
- 30 OTHER LANGUAGES

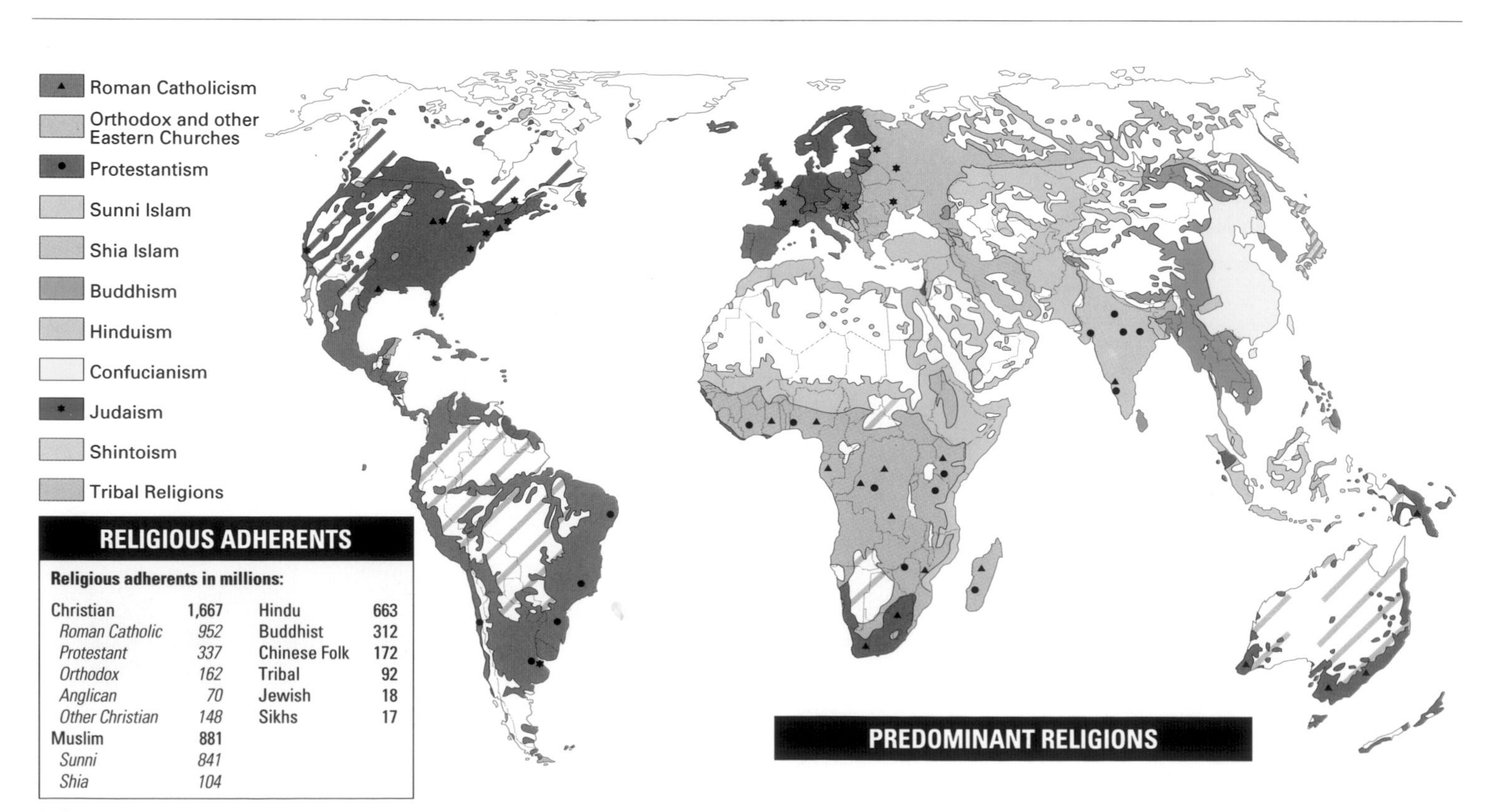

RELIGIOUS ADHERENTS

Religious adherents in millions:

Christian	1,667	Hindu	663
Roman Catholic	*952*	Buddhist	312
Protestant	*337*	Chinese Folk	172
Orthodox	*162*	Tribal	92
Anglican	*70*	Jewish	18
Other Christian	*148*	Sikhs	17
Muslim	881		
Sunni	*841*		
Shia	*104*		

UNITED NATIONS

Created in 1945 to promote peace and cooperation and based in New York, the United Nations is the world's largest international organization, with 185 members and an annual budget of US $2.6 billion (1996–97). Each member of the General Assembly has one vote, while the permanent members of the 15-nation Security Council – USA, Russia, China, UK and France – hold a veto. The Secretariat is the UN's principal administrative arm. The 54 members of the Economic and Social Council are responsible for economic, social, cultural, educational, health, and related matters. The UN has 16 specialized agencies – based in Canada, France, Switzerland and Italy, as well as the USA – which help members in fields such as education (UNESCO), agriculture (FAO), medicine (WHO) and finance (IFC). By the end of 1994, all the original 11 trust territories of The Trusteeship Council had become independent.

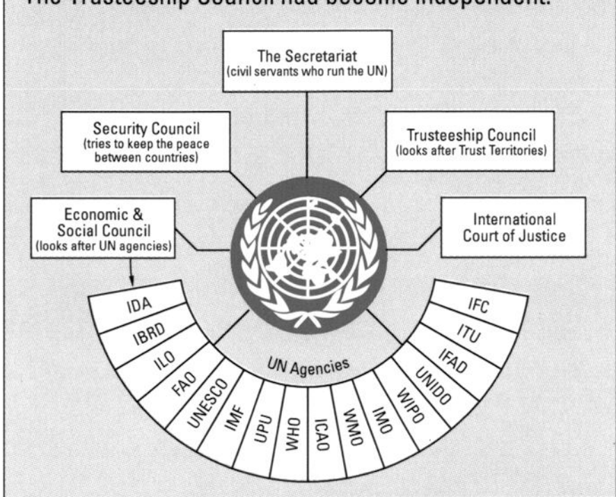

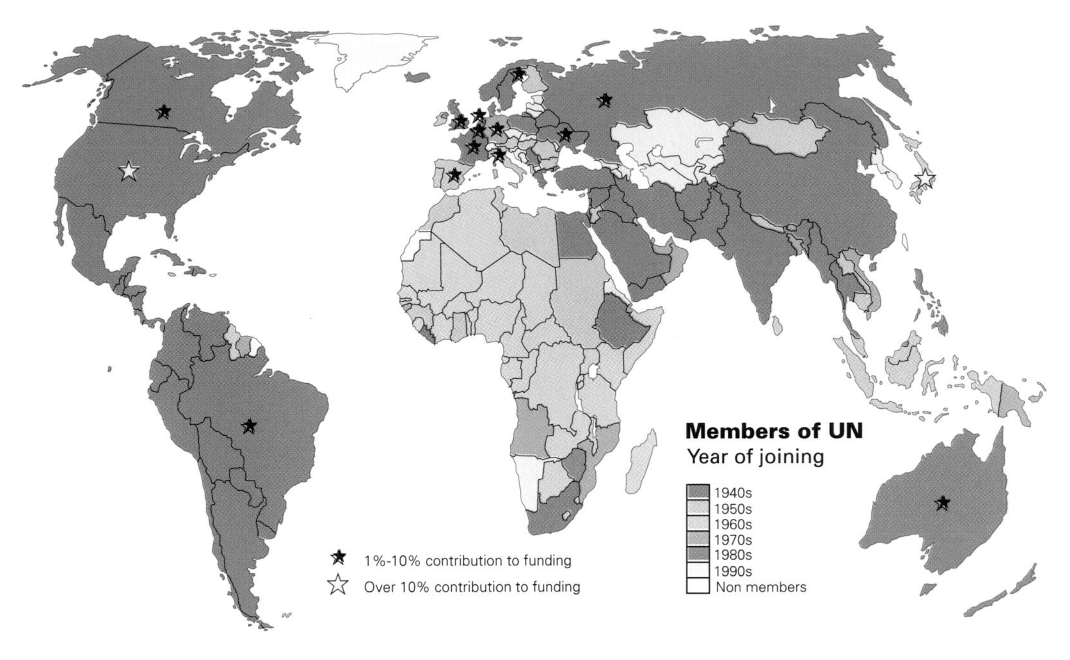

MEMBERSHIP OF THE UN In 1945 there were 51 members; by December 1994 membership had increased to 185 following the admission of Palau. There are 7 independent states which are not members of the UN – Kiribati, Nauru, Switzerland, Taiwan, Tonga, Tuvalu and the Vatican City. All the successor states of the former USSR had joined by the end of 1992. The official languages of the UN are Chinese, English, French, Russian, Spanish and Arabic.

FUNDING The UN budget for 1996–97 was US $2.6 billion. Contributions are assessed by the members' ability to pay, with the maximum 25% of the total, the minimum 0.01%. Contributions for 1996 were: USA 25.0%, Japan 15.4%, Germany 9.0%, France 6.4%, UK 5.3%, Italy 5.2%, Russia 4.5%, Canada 3.1%, Spain 2.4%, Brazil 1.6%, Netherlands 1.6%, Australia 1.5%, Sweden 1.2%, Ukraine 1.1%, Belgium 1.0%.

INTERNATIONAL ORGANIZATIONS

EU European Union (evolved from the European Community in 1993). The 15 members – Austria, Belgium, Denmark, Finland, France, Germany, Greece, Ireland, Italy, Luxembourg, Netherlands, Portugal, Spain, Sweden and the UK – aim to integrate economies, coordinate social developments and bring about political union. These members of what is now the world's biggest market share agricultural and industrial policies and tariffs on trade. The original body, the European Coal and Steel Community (ECSC), was created in 1951 following the signing of the Treaty of Paris.
EFTA European Free Trade Association (formed in 1960). Portugal left the original "Seven" in 1989 to join what was then the EC, followed by Austria, Finland and Sweden in 1995. Only 4 members remain: Norway, Iceland, Switzerland and Liechtenstein.
ACP African-Caribbean-Pacific (formed in 1963). Members have economic ties with the EU.
NATO North Atlantic Treaty Organization (formed in 1949). It continues after 1991 despite the winding up of the Warsaw Pact. There are 16 member nations.
OAS Organization of American States (formed in 1948). It aims to promote social and economic co-operation between developed countries of North America and developing nations of Latin America.
ASEAN Association of Southeast Asian Nations (formed in 1967). Vietnam joined in July 1995.
OAU Organization of African Unity (formed in 1963). Its 53 members represent over 94% of Africa's population. Arabic, French, Portuguese and English are recognized as working languages.
LAIA Latin American Integration Association (1980). Its aim is to promote freer regional trade.
OECD Organization for Economic Cooperation and Development (formed in 1961). It comprises the 29 major Western free-market economies. Poland, Hungary and South Korea joined in 1996. "G7" is its "inner group" comprising the USA, Canada, Japan, UK, Germany, Italy, and France.
COMMONWEALTH The Commonwealth of Nations evolved from the British Empire; it comprises 16 Queen's realms, 32 republics and 5 indigenous monarchies, giving a total of 53.
OPEC Organization of Petroleum Exporting Countries (formed in 1960). It controls about three-quarters of the world's oil supply. Gabon left the organization in 1996.

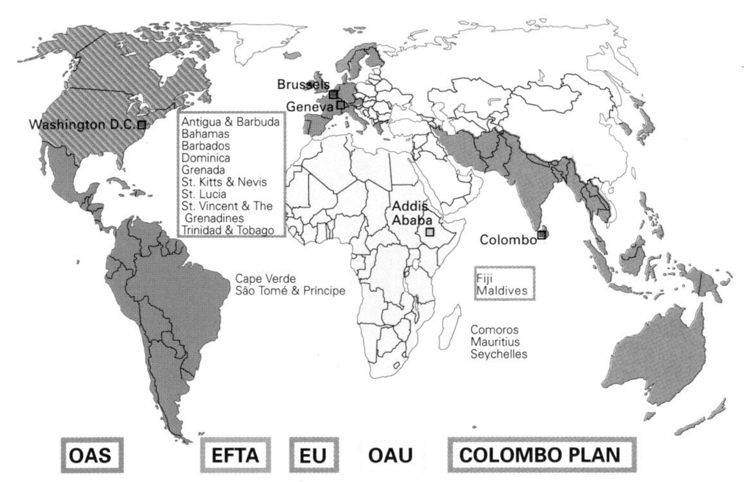

ARAB LEAGUE (formed in 1945). The League's aim is to promote economic, social, political and military cooperation. There are 21 member nations.
COLOMBO PLAN (formed in 1951). Its 26 members aim to promote economic and social development in Asia and the Pacific.

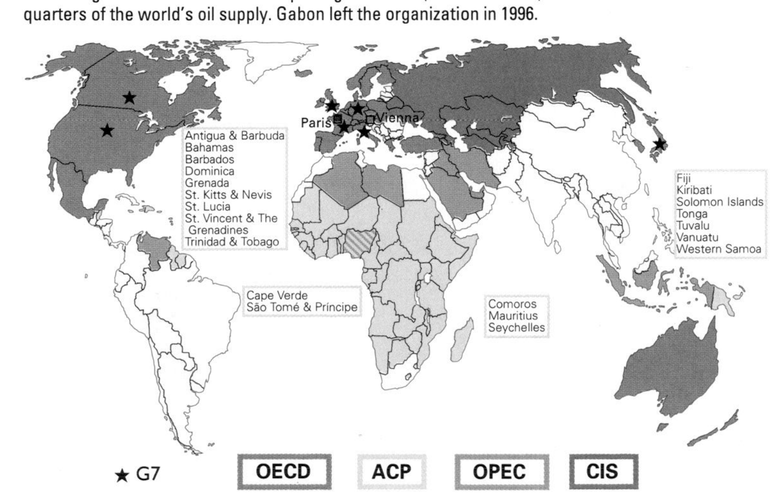

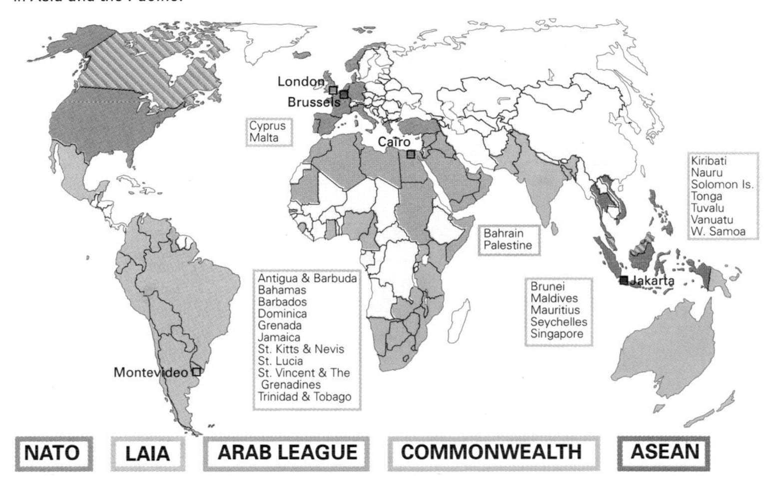

WEALTH

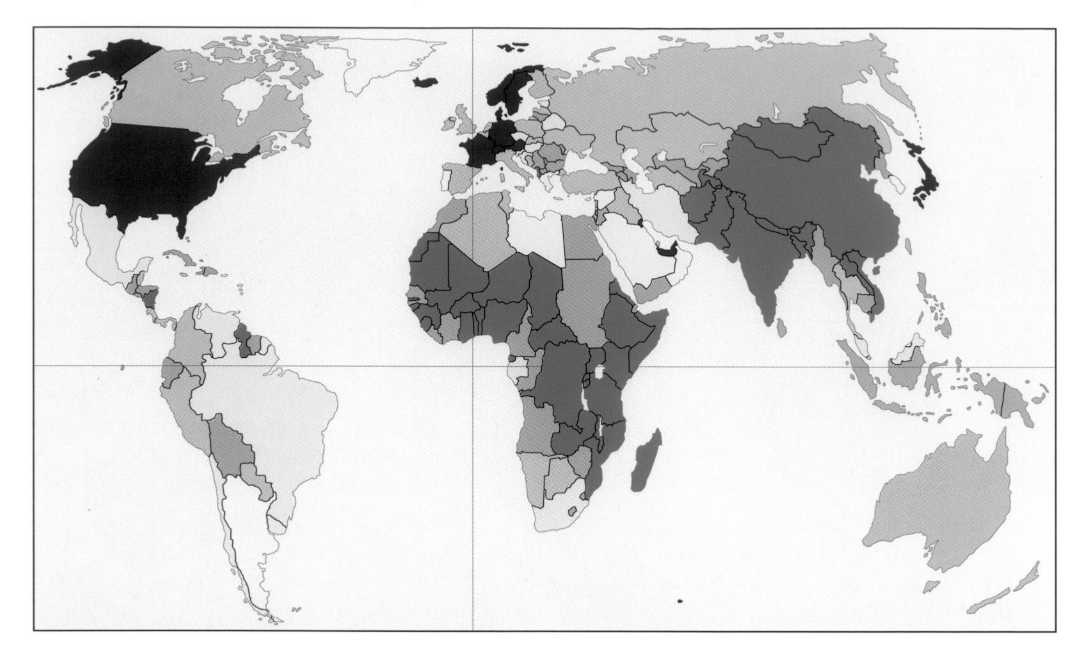

LEVELS OF INCOME

Gross National Product per capita: the value of total production divided by the population (1993)

Over 400% of world average

200 – 400% of world average

100 – 200% of world average

[World average wealth per person US $5,359]

50 – 100% of world average

25 – 50% of world average

10 – 25% of world average

Under 10% of world average

GNP per capita growth rate (%), selected countries, 1985–94

Thailand	8.2	Brazil	–0.4
Chile	6.9	Zimbabwe	–0.6
Japan	3.2	USA	–1.3
Germany	1.9	UK	–1.4
Australia	1.2	Armenia	–12.9

WEALTH CREATION

The Gross National Product (GNP) of the world's largest economies, US $ million (1994)

1.	USA	6,737,367	23.	Indonesia	167,632
2.	Japan	4,321,136	24.	Turkey	149,002
3.	Germany	2,075,452	25.	Denmark	145,384
4.	France	1,355,039	26.	Thailand	129,864
5.	Italy	1,101,258	27.	Saudi Arabia	126,597
6.	UK	1,069,457	28.	South Africa	125,225
7.	China	630,202	29.	Norway	114,328
8.	Canada	569,949	30.	Finland	95,817
9.	Brazil	536,309	31.	Poland	94,613
10.	Spain	525,334	32.	Portugal	92,124
11.	Russia	392,496	33.	Ukraine	80,921
12.	Mexico	368,679	34.	Greece	80,194
13.	South Korea	366,484	35.	Syria	80,120
14.	Netherlands	338,144	36.	Israel	78,113
15.	Australia	320,705	37.	Malaysia	68,674
16.	India	278,739	38.	Singapore	65,842
17.	Argentina	275,657	39.	Philippines	63,311
18.	Switzerland	264,974	40.	Venezuela	59,025
19.	Belgium	231,051	41.	Colombia	58,935
20.	Taiwan	228,000	42.	Pakistan	55,565
21.	Sweden	206,419	43.	Chile	50,051
22.	Austria	197,475	44.	Ireland	48,275

THE WEALTH GAP

The world's richest and poorest countries, by Gross National Product per capita in US $ (1994)

1.	Luxembourg	39,850	1.	Rwanda	80
2.	Switzerland	37,180	2.	Mozambique	80
3.	Japan	34,630	3.	Ethiopia	130
4.	Liechtenstein	33,510	4.	Tanzania	140
5.	Denmark	28,110	5.	Malawi	140
6.	Norway	26,480	6.	Sierra Leone	150
7.	USA	25,860	7.	Burundi	150
8.	Germany	25,580	8.	Chad	190
9.	Austria	24,950	9.	Vietnam	190
10.	Iceland	24,590	10.	Nepal	200
11.	Sweden	23,630	11.	Uganda	200
12.	France	23,470	12.	Haiti	220
13.	Singapore	23,360	13.	Afghanistan	220
14.	UAE	23,000	14.	Madagascar	230
15.	Belgium	22,920	15.	Bangladesh	230
16.	Netherlands	21,970	16.	Guinea-Bissau	240
17.	Hong Kong	21,650	17.	Mali	250
18.	Canada	19,570	18.	São Tomé & P.	250
19.	Italy	19,270	19.	Kenya	260
20.	Kuwait	19,040	20.	Yemen	280

GNP per capita is calculated by dividing a country's Gross National Product by its total population.

CONTINENTAL SHARES

Shares of population and of wealth (GNP) by continent

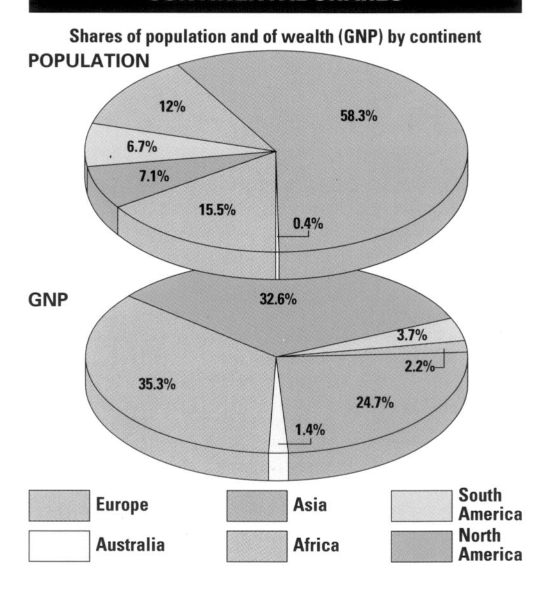

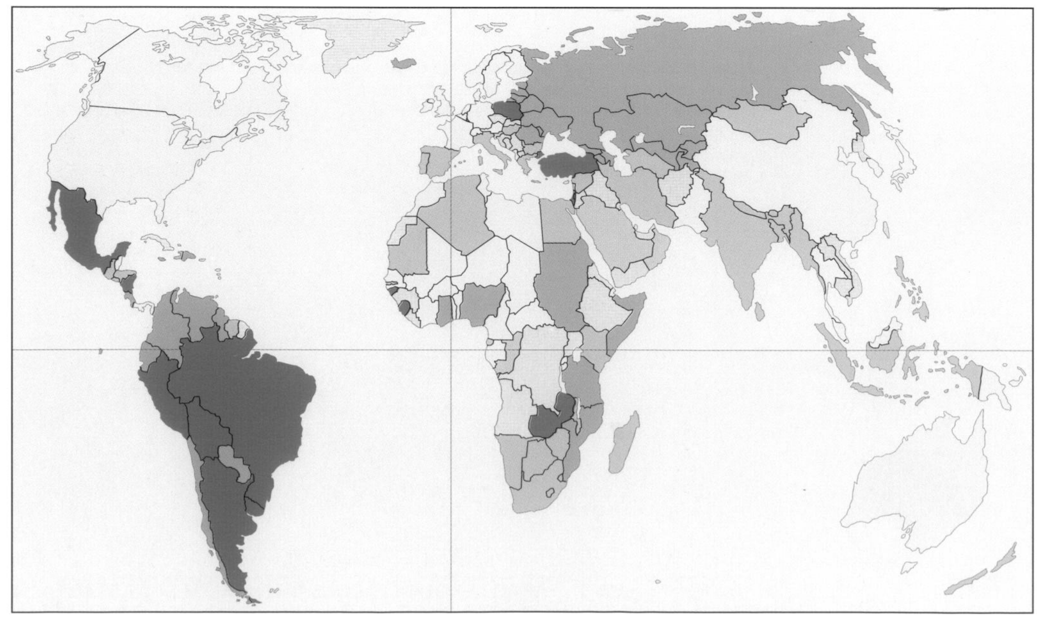

INFLATION

Average annual rate of inflation (1980–93)

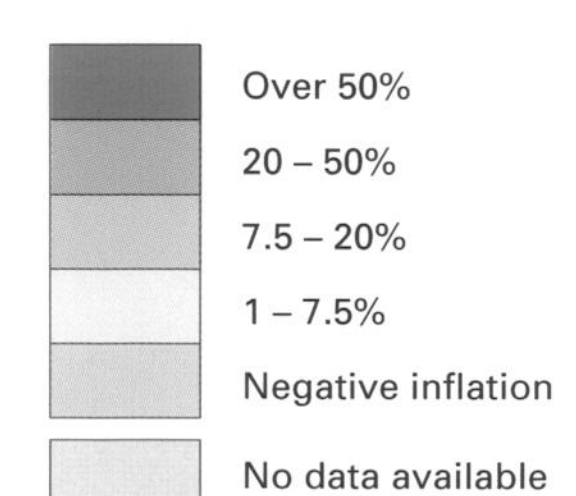

Over 50%

20 – 50%

7.5 – 20%

1 – 7.5%

Negative inflation

No data available

Highest average inflation		Lowest average inflation	
Nicaragua	665%	Brunei	–5.1%
Brazil	423%	Oman	–2.3%
Argentina	374%	Saudi Arabia	–2.1%
Peru	316%	Equatorial Guinea	–0.6%
Bolivia	187%	Congo	–0.6%
Israel	70%	Bahrain	–0.3%
Poland	69%	Libya	0.2%

INTERNATIONAL AID

Aid provided or received, divided by the total population, in US $ (1994)

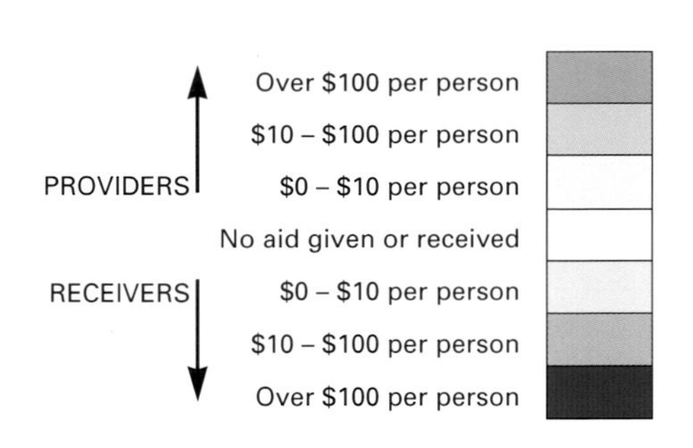

Top 5 providers per capita		Top 5 receivers per capita	
France	$279	São Tomé & P.	$378
Denmark	$260	Cape Verde	$314
Norway	$247	Djibouti	$235
Sweden	$201	Surinam	$198
Germany	$166	Mauritania	$153

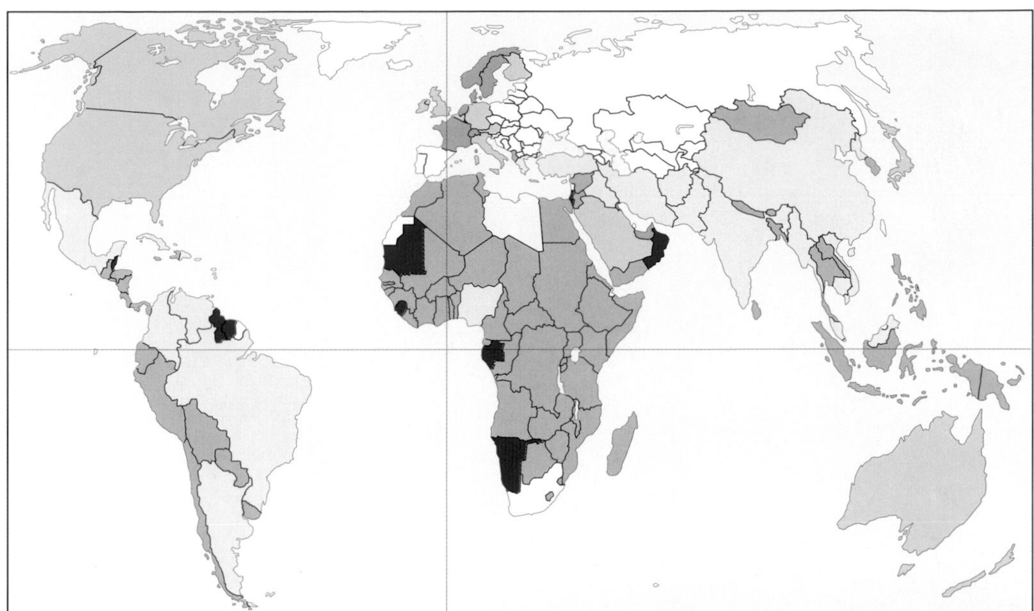

DEBT AND AID

International debtors and the aid they receive (1993)

Although aid grants make a vital contribution to many of the world's poorer countries, they are usually dwarfed by the burden of debt that the developing economies are expected to repay. In 1992, they had to pay US $160,000 million in debt service charges alone – more than two and a half times the amount of Official Development Assistance (ODA) the developing countries were receiving, and US $60,000 million more than total private flows of aid in the same year. In 1990, the debts of Mozambique, one of the world's poorest countries, were estimated to be 75 times its entire earnings from exports.

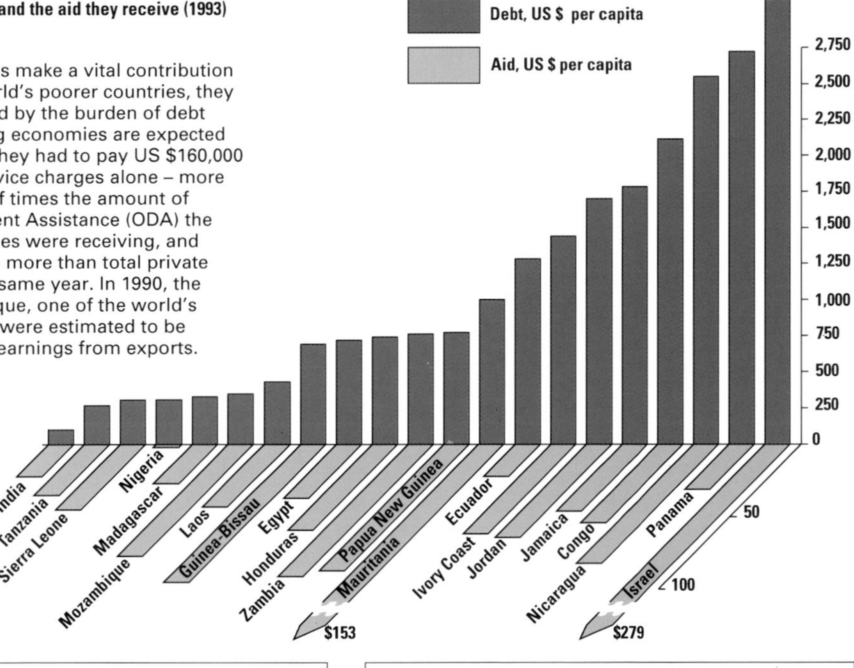

DISTRIBUTION OF SPENDING

Percentage share of household spending, selected countries

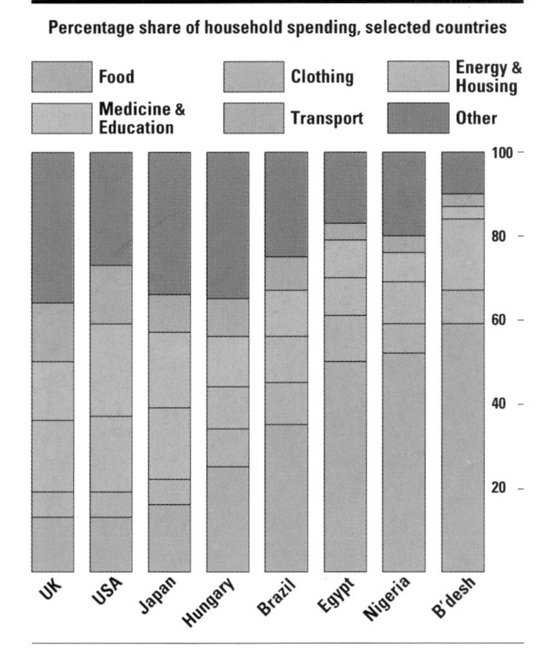

HIGH INCOME

Number of cars, televisions and telephones for each 10,000 people, selected high income countries (1993)

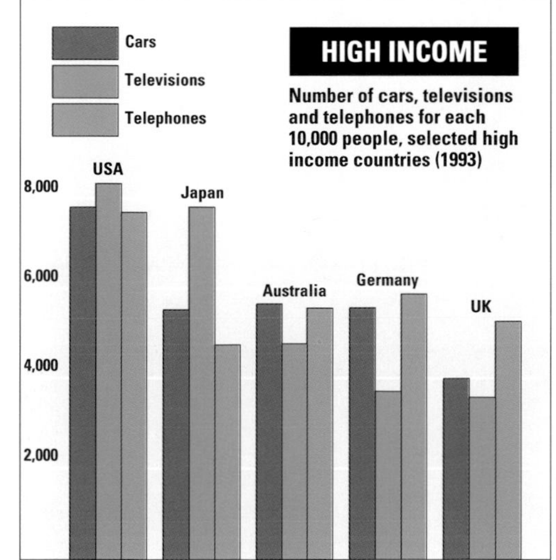

MIDDLE INCOME

Number of cars, televisions and telephones for each 10,000 people, selected middle income countries (1993)

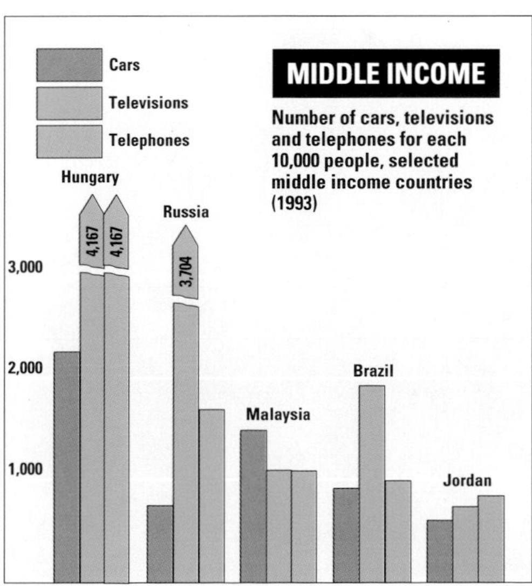

LOW INCOME

Number of cars, televisions and telephones for each 10,000 people, selected low income countries (1993)

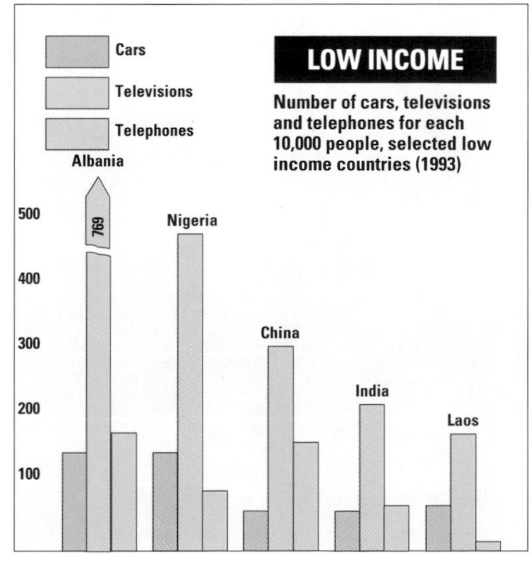

QUALITY OF LIFE

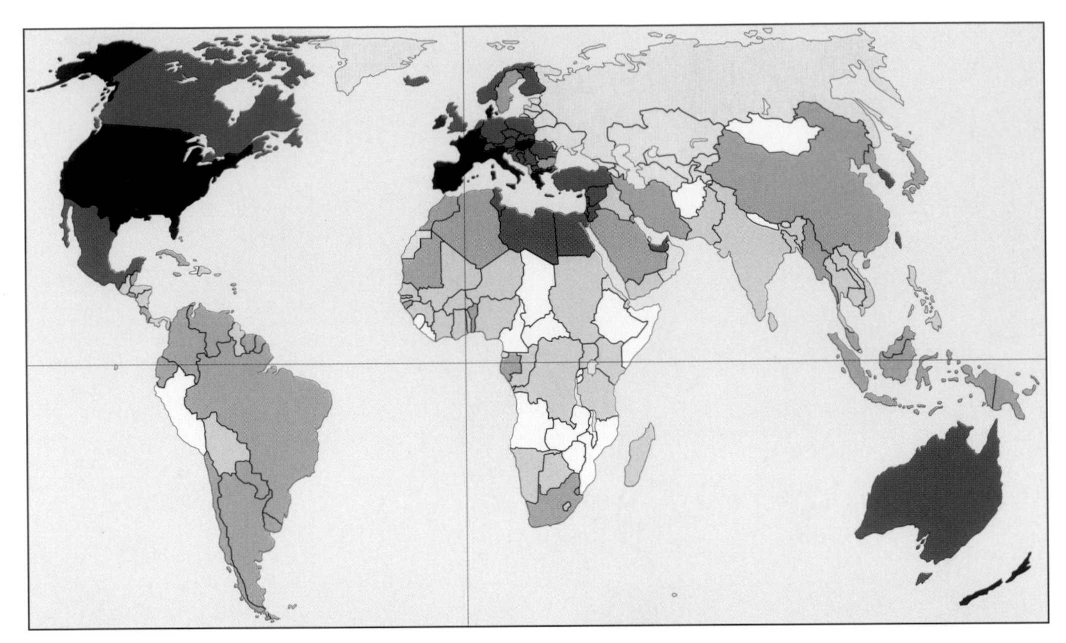

DAILY FOOD CONSUMPTION

Average daily food intake in calories per person (latest available year)

Top 5 countries		Bottom 5 countries	
Ireland	3,847 cal.	Mozambique	1,680 cal.
Greece	3,815 cal.	Liberia	1,640 cal.
Cyprus	3,779 cal.	Ethiopia	1,610 cal.
USA	3,732 cal.	Afghanistan	1,523 cal.
Spain	3,708 cal.	Somalia	1,499 cal.

HOSPITAL CAPACITY

Hospital beds available for each 1,000 people (1993)

Highest capacity		Lowest capacity	
Japan	13.6	Bangladesh	0.2
Kazakstan	13.5	Ethiopia	0.2
Ukraine	13.5	Nepal	0.3
Russia	13.5	Burkina Faso	0.4
Latvia	13.5	Afghanistan	0.5
North Korea	13.5	Pakistan	0.6
Moldova	12.8	Niger	0.6
Belarus	12.7	Mali	0.6
Finland	12.3	Indonesia	0.6
France	12.2	Guinea	0.6

[USA 4.6]

Although the ratio of people to hospital beds gives a good approximation of a country's health provision, it is not an absolute indicator. Raw numbers may mask inefficiency and other weaknesses: the high availability of beds in Kazakstan, for example, has not prevented infant mortality rates over three times as high as in the United Kingdom and the United States.

LIFE EXPECTANCY

Years of life expectancy at birth, selected countries (1990–95)

The chart shows combined data for both sexes. On average, women live longer than men worldwide, even in developing countries with high maternal mortality rates. Overall, life expectancy is steadily rising, though the difference between rich and poor nations remains dramatic.

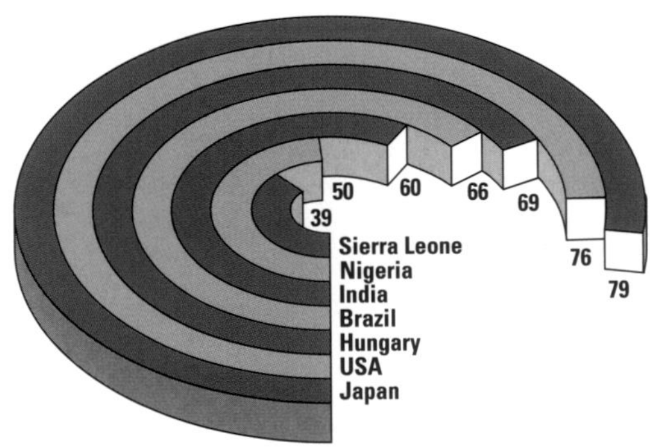

CAUSES OF DEATH

Causes of death for selected countries by % (1992–94)

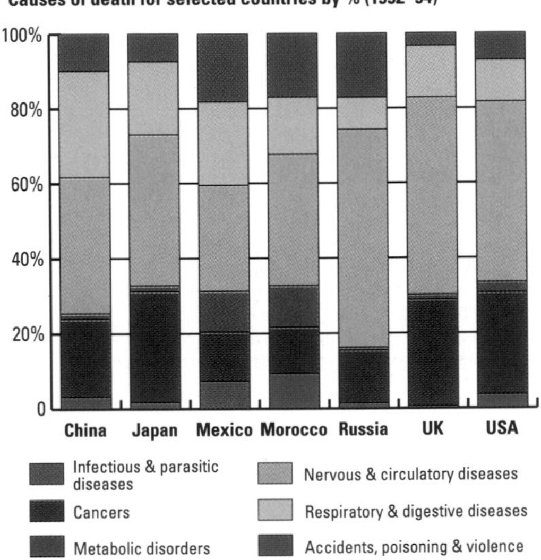

CHILD MORTALITY

Number of babies who will die under the age of one, per 1,000 births (average 1990–95)

Highest child mortality		Lowest child mortality	
Afghanistan	162	Hong Kong	6
Mali	159	Denmark	6
Sierra Leone	143	Japan	5
Guinea-Bissau	140	Iceland	5
Malawi	138	Finland	5

[USA 8 deaths]

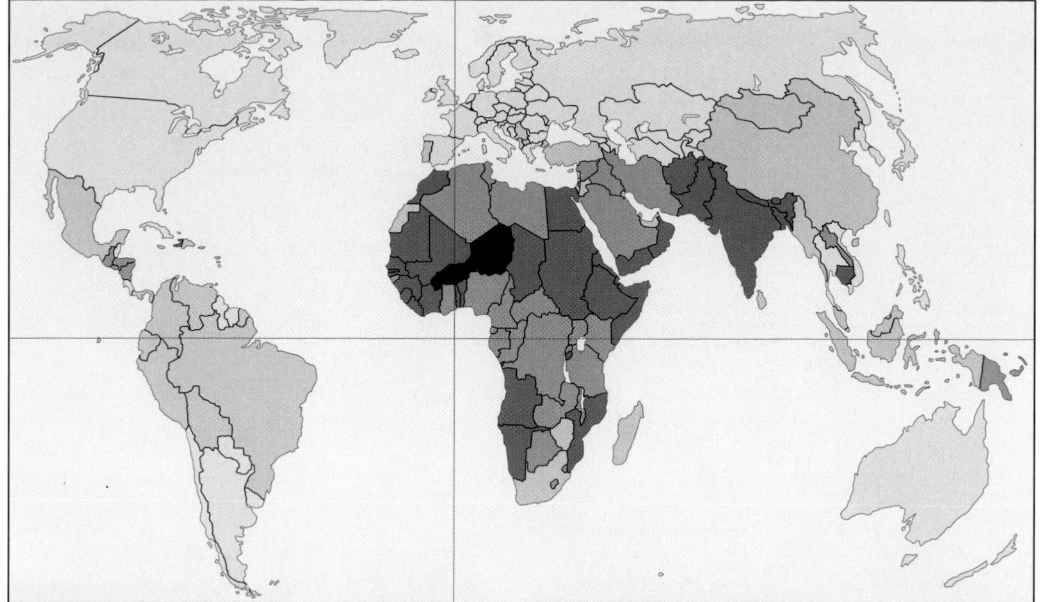

ILLITERACY

Percentage of the total population unable to read or write (latest available year)

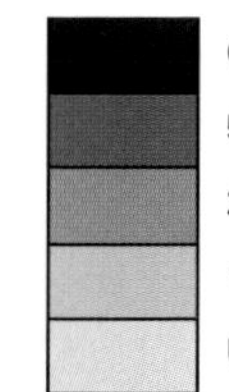

Over 75% of population illiterate

50 – 75% of population illiterate

25 – 50% of population illiterate

10 – 15% of population illiterate

Under 10% of population illiterate

Educational expenditure per person (latest available year)

Top 5 countries		Bottom 5 countries	
Sweden	$997	Chad	$2
Qatar	$989	Bangladesh	$3
Canada	$983	Ethiopia	$3
Norway	$971	Nepal	$4
Switzerland	$796	Somalia	$4

LIVING STANDARDS

At first sight, most international contrasts in living standards are swamped by differences in wealth. The rich not only have more money, they have more of everything, including years of life. Those with only a little money are obliged to spend most of it on food and clothing, the basic maintenance costs of their existence; air travel and tourism are unlikely to feature on their expenditure lists. However, poverty and wealth are both relative: slum dwellers living on social security payments in an affluent industrial country have far more resources at their disposal than an average African peasant, but feel their own poverty nonetheless. A middle-class Indian lawyer cannot command a fraction of the earnings of a counterpart living in New York, London or Rome; nevertheless, he rightly sees himself as prosperous.

The rich not only live longer, on average, than the poor, they also die from different causes. Infectious and parasitic diseases, all but eliminated in the developed world, remain a scourge in the developing nations. On the other hand, more than two-thirds of the populations of OECD nations eventually succumb to cancer or circulatory disease.

FERTILITY AND EDUCATION

Fertility rates compared with female education, selected countries (1992–95)

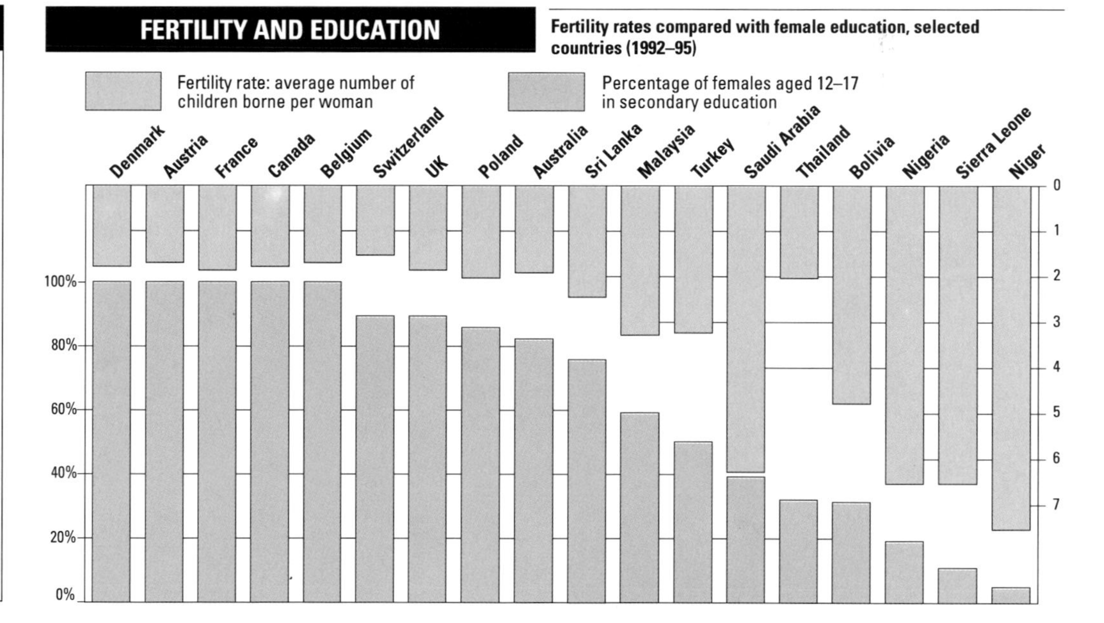

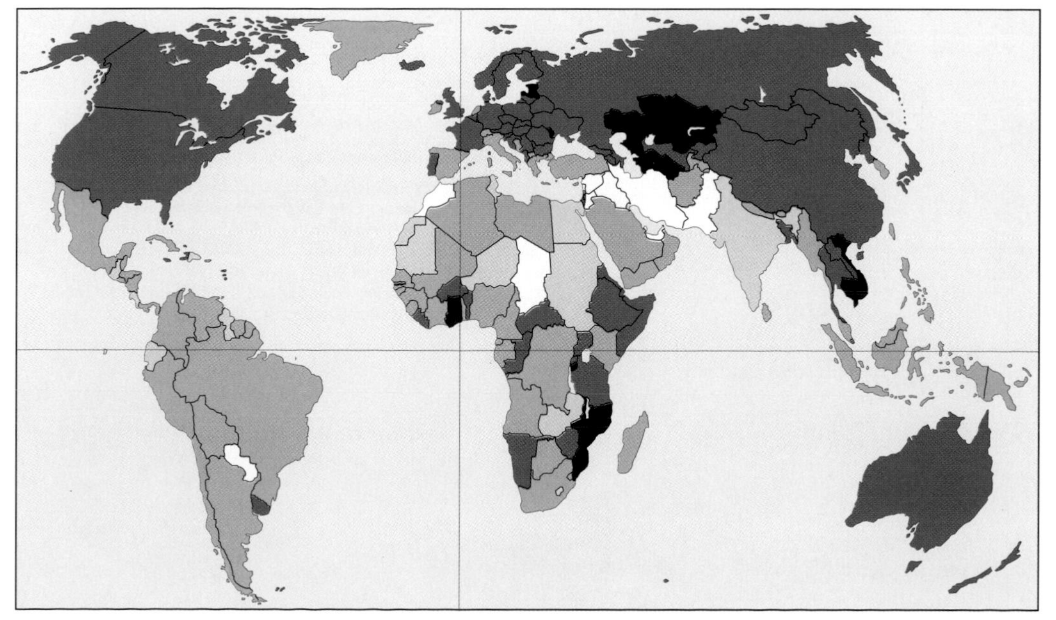

WOMEN IN THE WORK FORCE

Women in paid employment as a percentage of the total work force (latest available year)

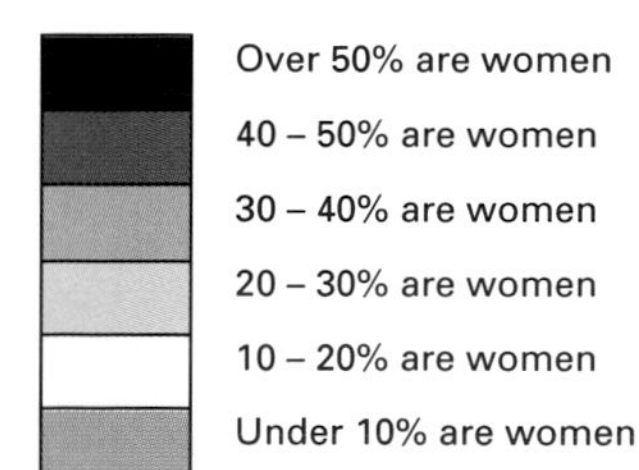

Over 50% are women

40 – 50% are women

30 – 40% are women

20 – 30% are women

10 – 20% are women

Under 10% are women

Most women in the work force		Fewest women in the work force	
Cambodia	56%	Saudi Arabia	4%
Kazakstan	54%	Oman	6%
Burundi	53%	Afghanistan	8%
Mozambique	53%	Algeria	9%
Turkmenistan	52%	Libya	9%

[USA 45]

Energy

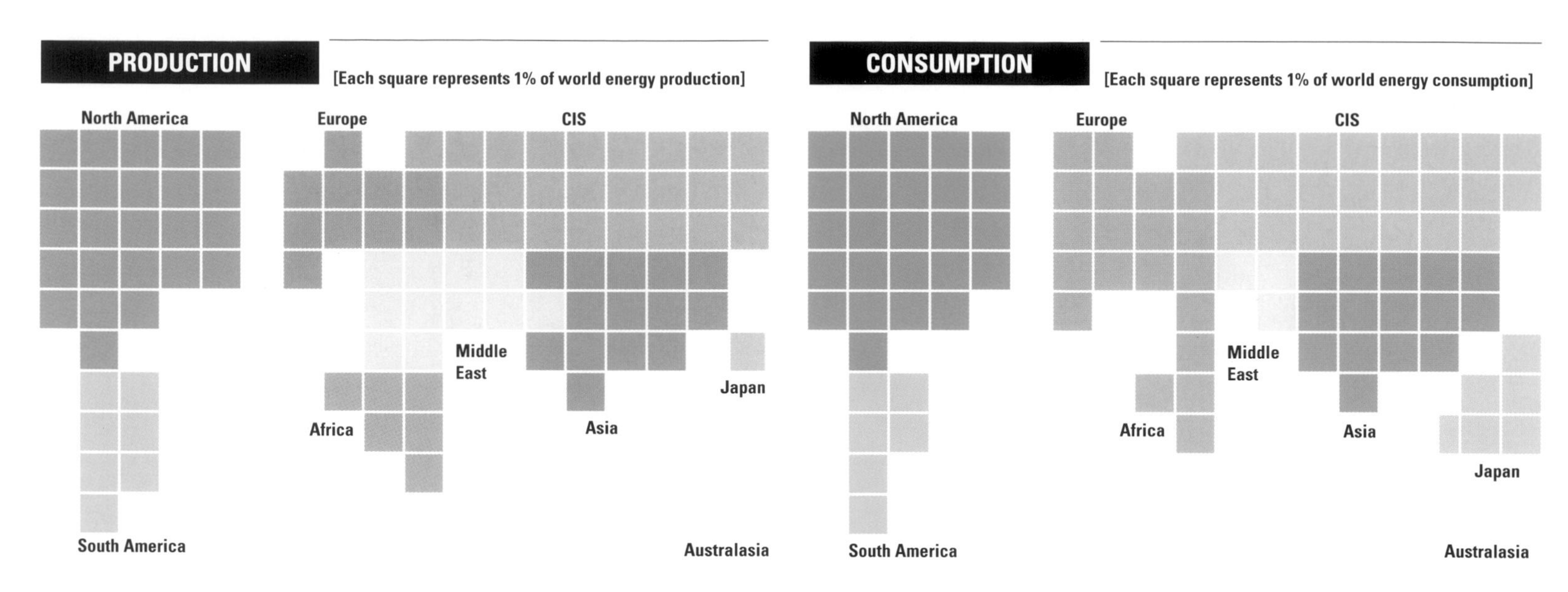

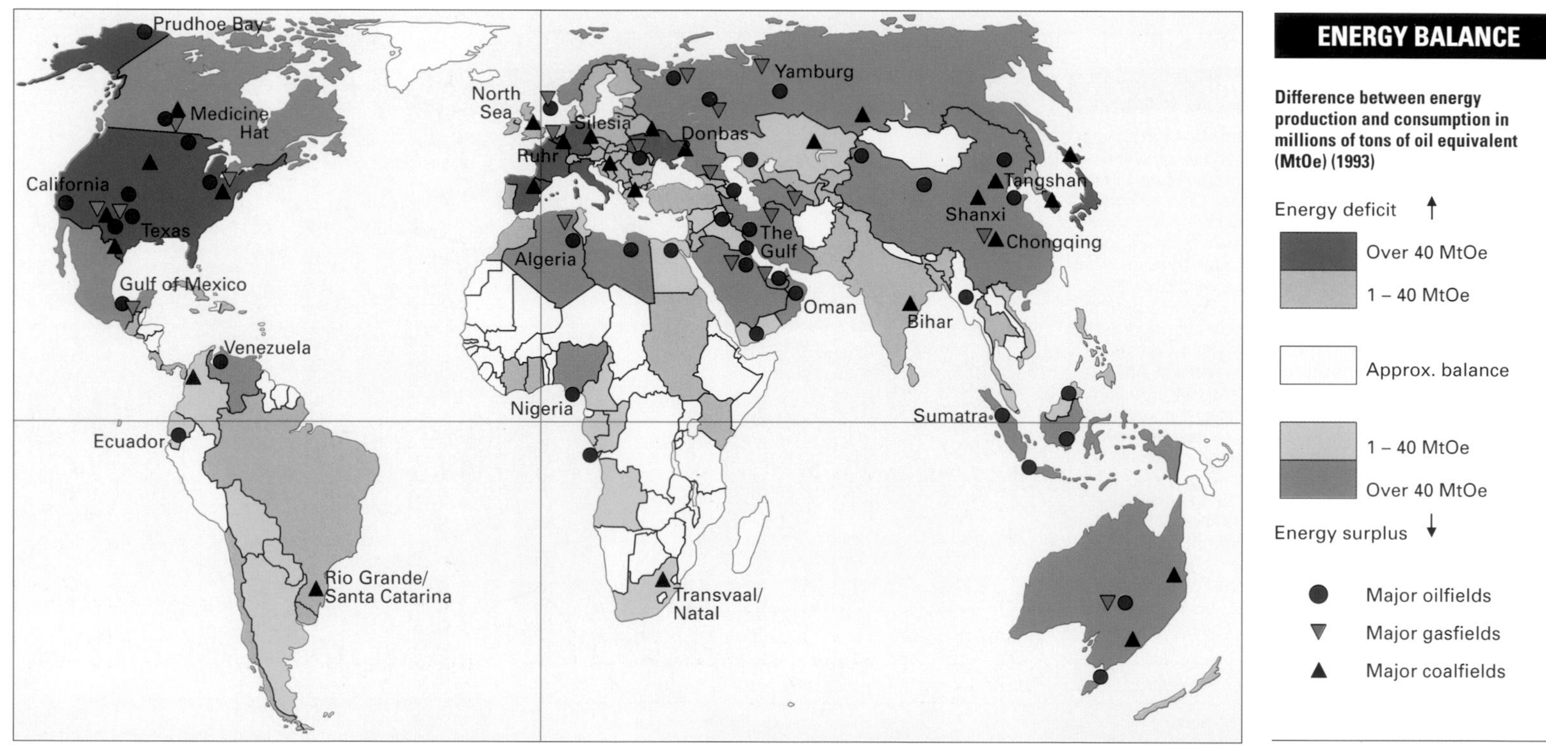

ENERGY BALANCE

Difference between energy production and consumption in millions of tons of oil equivalent (MtOe) (1993)

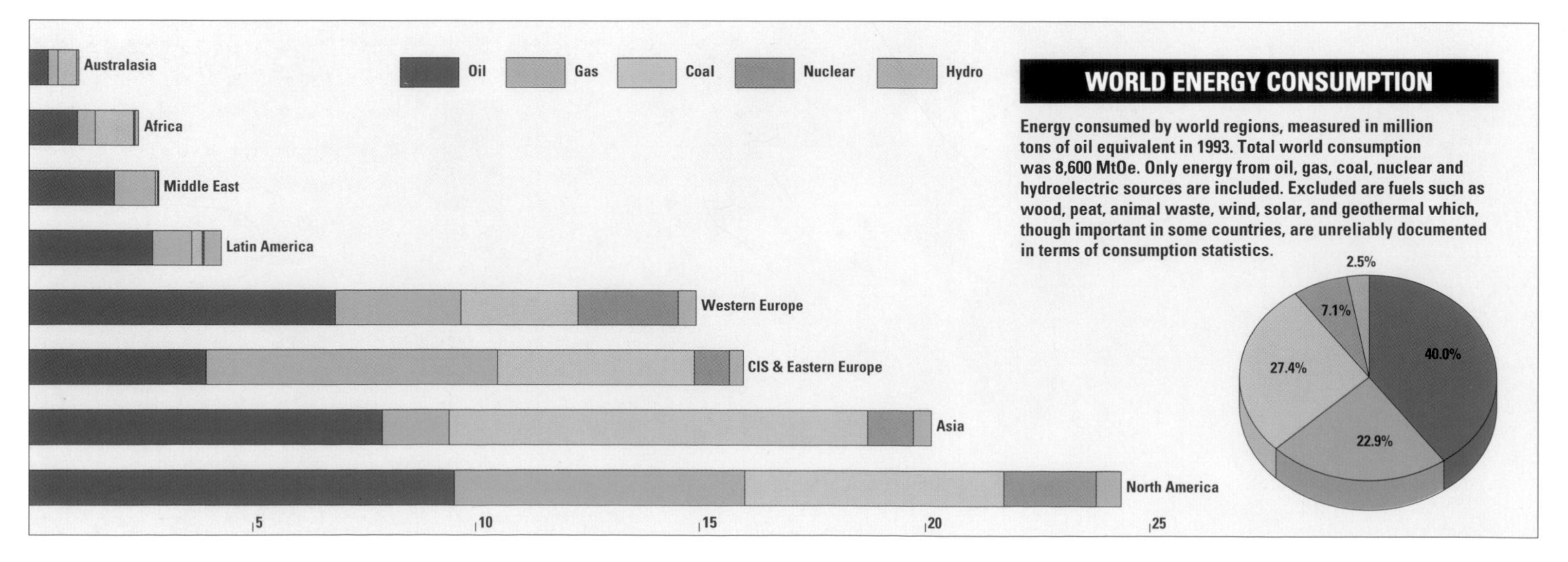

WORLD ENERGY CONSUMPTION

Energy consumed by world regions, measured in million tons of oil equivalent in 1993. Total world consumption was 8,600 MtOe. Only energy from oil, gas, coal, nuclear and hydroelectric sources are included. Excluded are fuels such as wood, peat, animal waste, wind, solar, and geothermal which, though important in some countries, are unreliably documented in terms of consumption statistics.

ENERGY

Energy is used to keep us warm or cool, fuel our industries and our transport systems, and even feed us; high-intensity agriculture, with its use of fertilizers, pesticides and machinery, is heavily energy-dependent. Although we live in a high-energy society, there are vast discrepancies between rich and poor; for example, a North American consumes 13 times as much energy as a Chinese person. But even developing nations have more power at their disposal than was imaginable a century ago.

The distribution of energy supplies, most importantly fossil fuels (coal, oil and natural gas), is very uneven. In addition, the diagrams and map opposite show that the largest producers of energy are not necessarily the largest consumers. The movement of energy supplies around the world is therefore an important component of international trade. In 1995, total world movements in oil amounted to 2,000 million tons.

As the finite reserves of fossil fuels are depleted, renewable energy sources, such as solar, hydrothermal, wind, tidal, and biomass, will become increasingly important around the world.

NUCLEAR POWER

Percentage of electricity generated by nuclear power stations, leading nations (1994)

1. Lithuania	76%	11. Spain	35%
2. France	75%	12. Taiwan	32%
3. Belgium	56%	13. Finland	30%
4. Sweden	51%	14. Germany	29%
5. Slovak Rep.	49%	15. Ukraine	29%
6. Bulgaria	46%	16. Czech Rep.	28%
7. Hungary	44%	17. Japan	27%
8. Slovenia	38%	18. UK	26%
9. Switzerland	37%	19. USA	22%
10. South Korea	36%	20. Canada	19%

Although the 1980s were a bad time for the nuclear power industry (major projects ran over budget, and fears of long-term environmental damage were heavily reinforced by the 1986 disaster at Chernobyl), the industry picked up in the early 1990s. However, whilst the number of reactors is still increasing, orders for new plants have shrunk. This is partly due to the increasingly difficult task of disposing of nuclear waste.

HYDROELECTRICITY

Percentage of electricity generated by hydroelectric power stations, leading nations (1993)

1. Paraguay	99.9%	11. Zaïre	97.3%
2. Norway	99.6%	12. Cameroon	97.1%
3. Bhutan	99.6%	13. Tajikistan	96.5%
4. Zambia	99.5%	14. Albania	96.5%
5. Ghana	99.4%	15. Sri Lanka	95.4%
6. Congo	99.3%	16. Laos	95.2%
7. Uganda	99.1%	17. Iceland	94.4%
8. Rwanda	98.3%	18. Nepal	93.5%
9. Buruni	98.3%	19. Brazil	93.3%
10. Malawi	98.0%	20. Honduras	91.9%

Countries heavily reliant on hydroelectricity are usually small and non industrial: a high proportion of hydroelectric power more often reflects a modest energy budget than vast hydroelectric resources. The USA, for instance, produces only 9% of power requirements from hydroelectricity; yet that 9% amounts to more than three times the hydro power generated by all of Africa.

FUEL EXPORTS

Fuels as a percentage of total value of exports (1990–94)

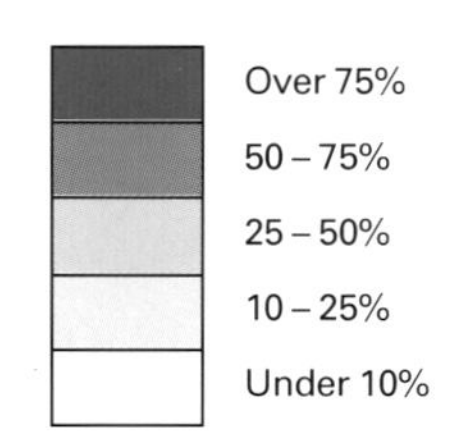

MEASUREMENTS

For historical reasons, oil is traded in "barrels." The weight and volume equivalents (shown right) are all based on average-density "Arabian light" crude oil.

The energy equivalents given for a ton of oil are also somewhat imprecise: oil and coal of different qualities will have varying energy contents, a fact usually reflected in their price on world markets.

CONVERSION RATES

1 barrel = 0.15 tons or 159 liters or 35 Imperial gallons or 42 US gallons

1 ton = 6.67 barrels or 1,075 liters or 233 Imperial gallons or 280 US gallons

1 ton oil = 1.5 tons hard coal or 3.0 tons lignite or 10,900 kWh

1 Imperial gallon = 1.201 US gallons or 4.546 liters or 277.4 cubic inches.

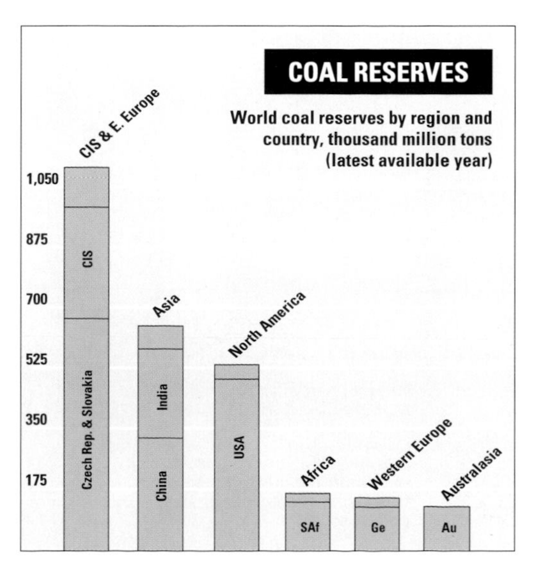

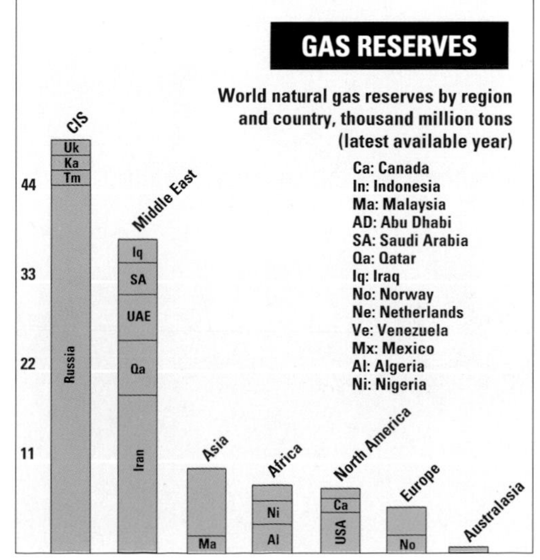

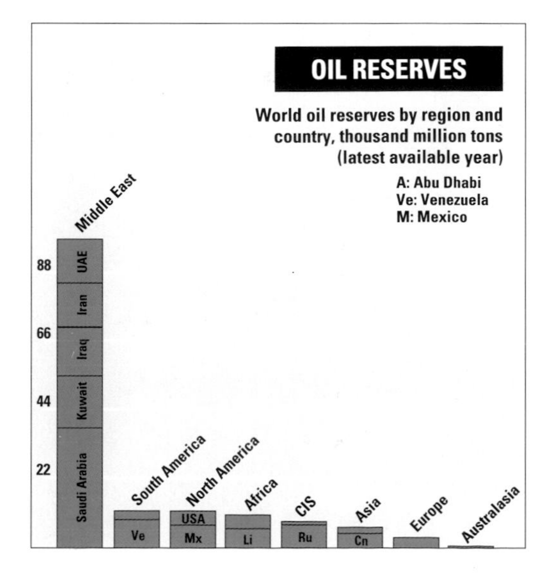

PRODUCTION

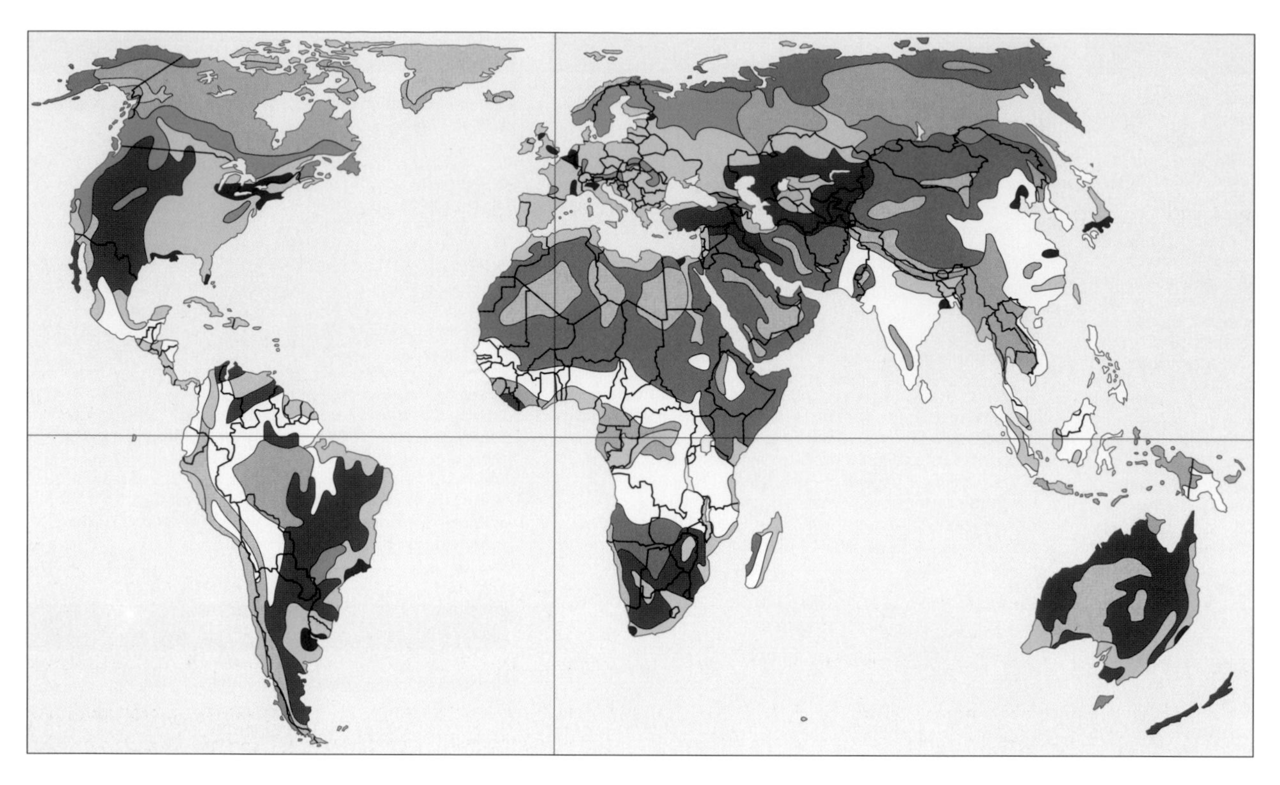

AGRICULTURE

Predominant type of farming or land use.

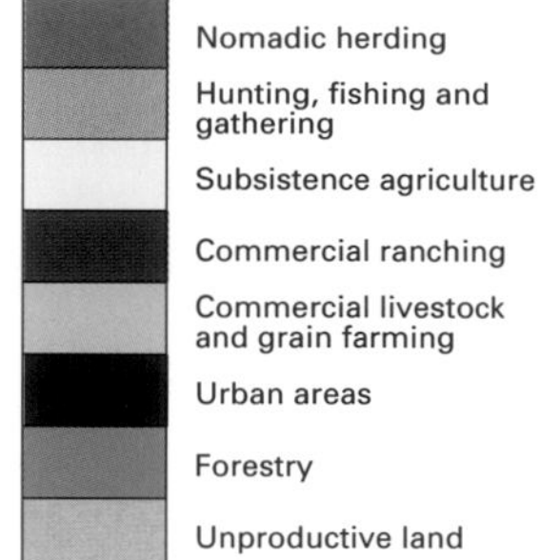

The development of agriculture transformed human existence more than any other. The whole business of farming is constantly developing: due mainly to new varieties of rice and wheat, world grain production has increased by over 70% since 1965. New machinery and modern agricultural techniques enable relatively few farmers to produce enough food for the world's 5,800 million people.

STAPLE CROPS

Wheat

China 18.6%, USA 11.6%, India 10.1%, Russia 7.5%, France 5.2%, Canada 4.9%

World total (1993): 620,902,700 tons

Rice

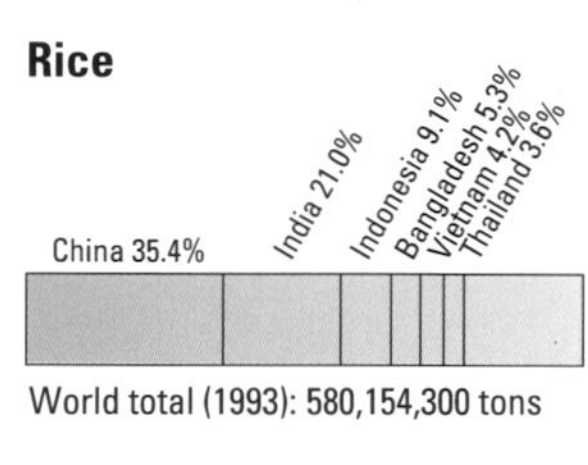

World total (1993): 580,154,300 tons

Maize

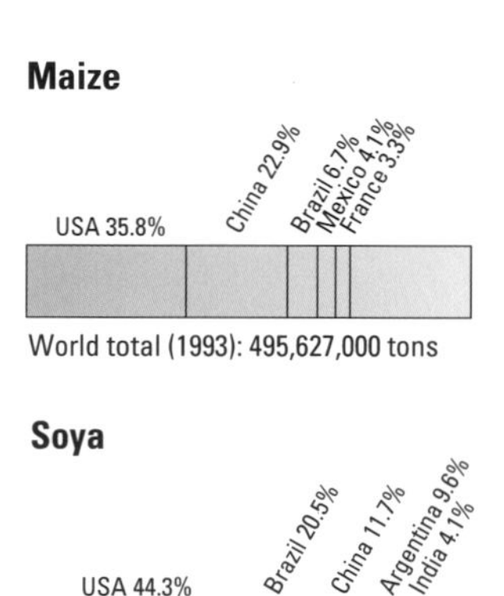

World total (1993): 495,627,000 tons

Potatoes

Russia 13.2%, Poland 12.6%, China 12.2%, Ukraine 7.3%, USA 6.6%, India 5.5%

World total (1993): 317,001,300 tons

Millet

India 37.8%, China 15.0%, Nigeria 14.4%, Niger 5.4%, Russia 4.2%

World total (1993): 29,086,200 tons

Rye

Russia 34.9%, Poland 19.0%, Germany 11.2%, Belarus 10.7%, Ukraine 4.5%

World total (1993): 28,820,000 tons

Soya

USA 44.3%, Brazil 20.5%, China 11.7%, Argentina 9.6%, India 4.1%

World total (1993): 122,112,100 tons

Cassava

Brazil 14.1%, Nigeria 13.7%, Zaïre 13.6%, Thailand 12.8%, Indonesia 10.6%, Tanzania 4.4%

World total (1993): 168,990,800 tons

SUGARS

Sugarcane

Brazil 24.2%, India 22.2%, China 6.6%, Cuba 4.2%, Mexico 4.0%, Pakistan 3.7%

World total (1993): 1,144,660,000 tons

Sugar beet

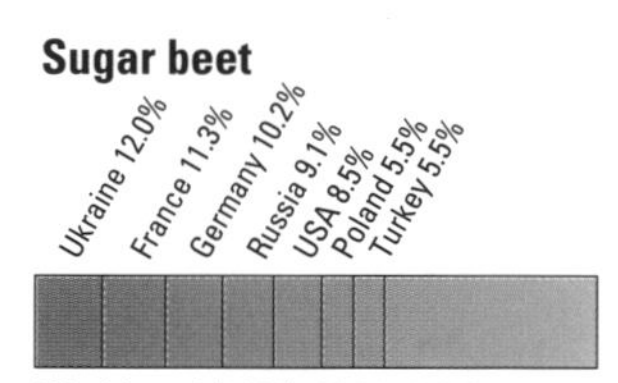

World total (1993): 309,850,200 tons

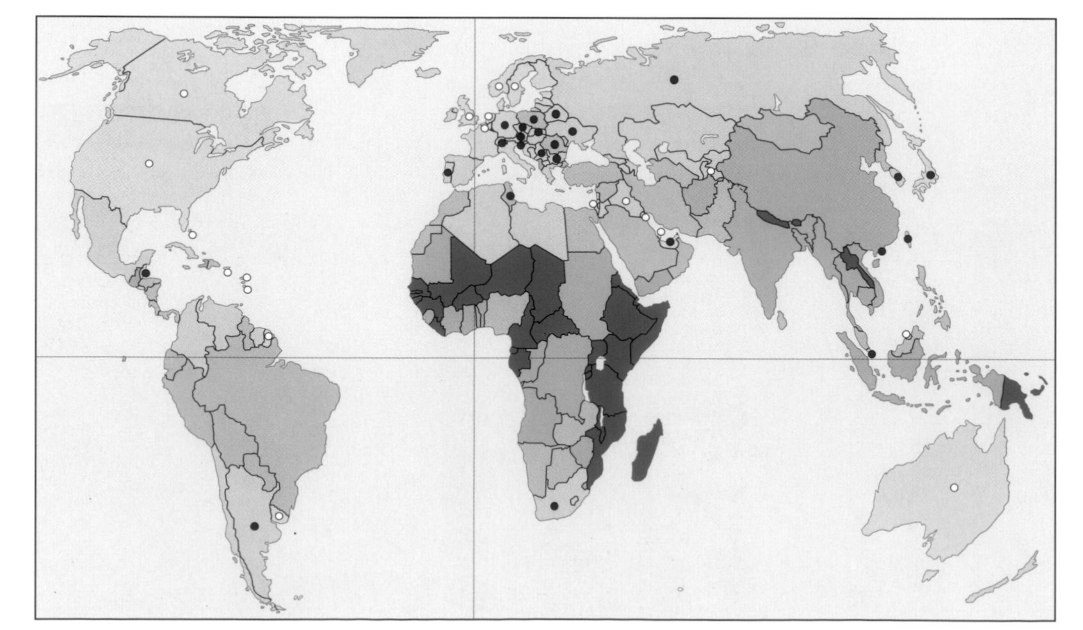

BALANCE OF EMPLOYMENT

Percentage of total work force employed in agriculture, including forestry and fishing (1990–92)

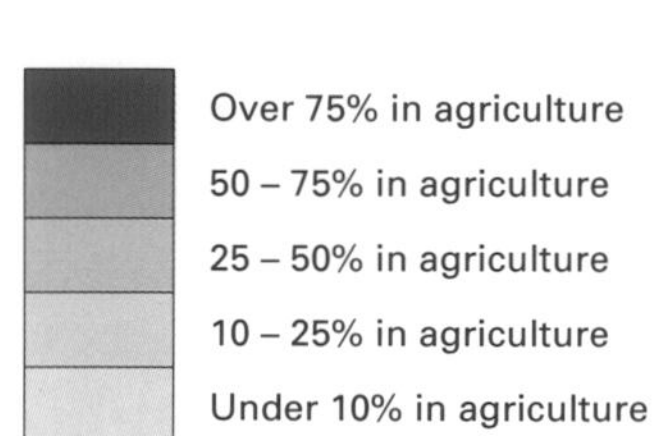

Employment in industry and services

- ● Over a third of total work force employed in manufacturing
- ○ Over two-thirds of total work force employed in service industries (work in offices, shops, tourism, transport, construction, and government)

MINERAL PRODUCTION

***Figures for aluminum are for refined metal; all other figures refer to ore production.**

Copper

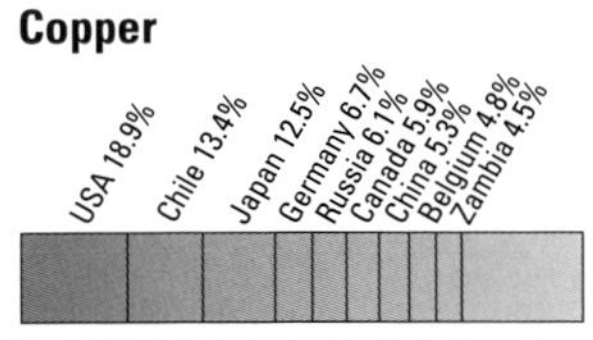

World total (1993): 10,400,000 tons*

Iron

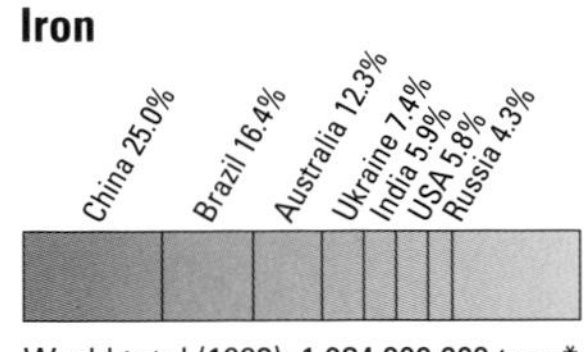

World total (1993): 1,034,000,000 tons*

Chromium

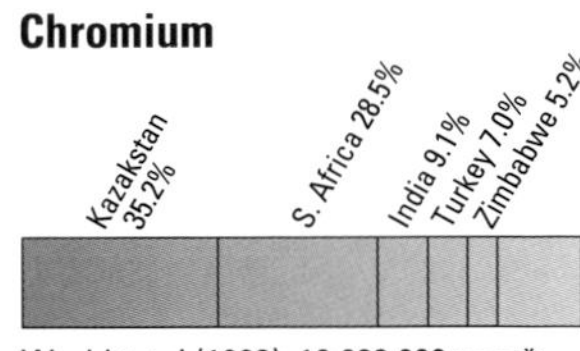

World total (1993): 10,923,000 tons*

Gold

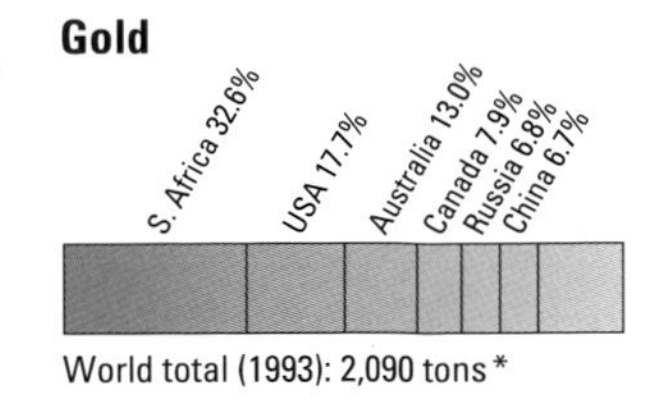

World total (1993): 2,090 tons*

Uranium

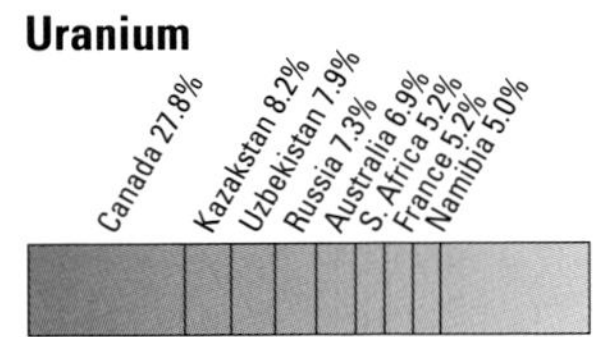

World total (1993): 36,300 tons*

Lead

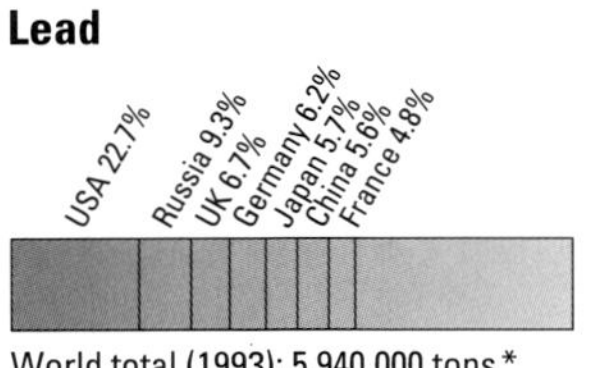

World total (1993): 5,940,000 tons*

Tin

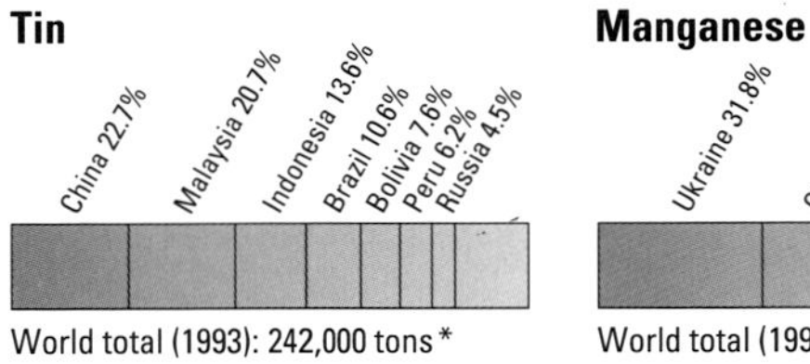

World total (1993): 242,000 tons*

Manganese

World total (1993): 24,200,000 tons*

Silver

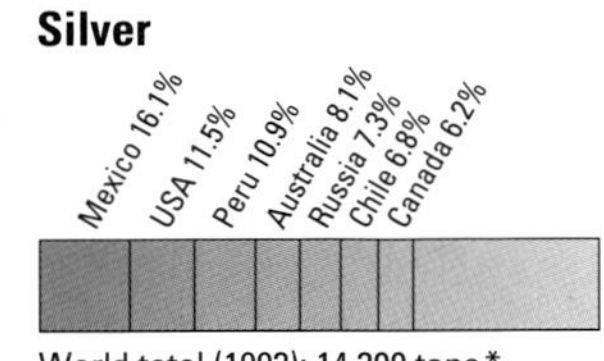

World total (1993): 14,300 tons*

Aluminum

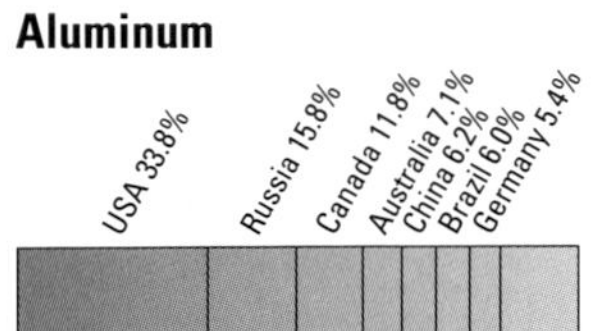

World total (1993): 21,569,900 tons*

Mercury

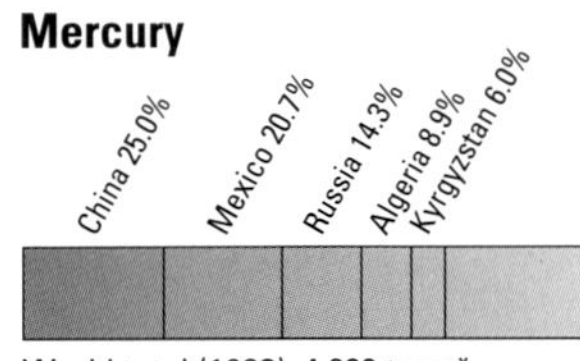

World total (1993): 4,620 tons*

Zinc

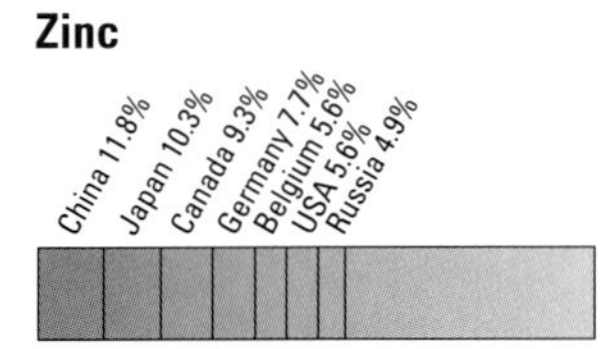

World total (1993): 7,839,700 tons*

Nickel

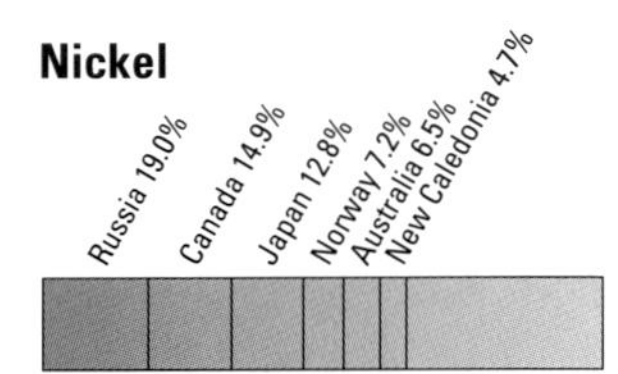

World total (1993): 869,000 tons*

Diamonds

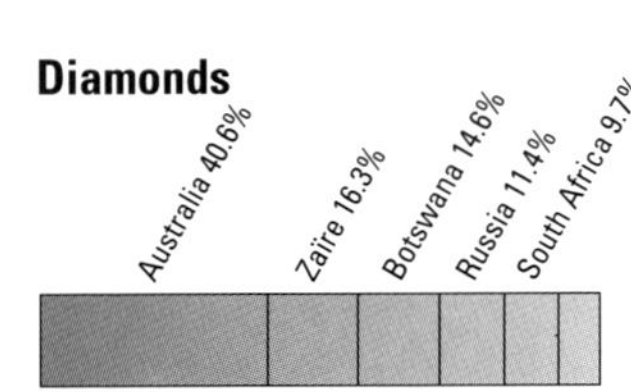

World total (1993): 100,850,000 carats

MINERAL DISTRIBUTION

The map shows the richest sources of the most important minerals. Major mineral locations are named.

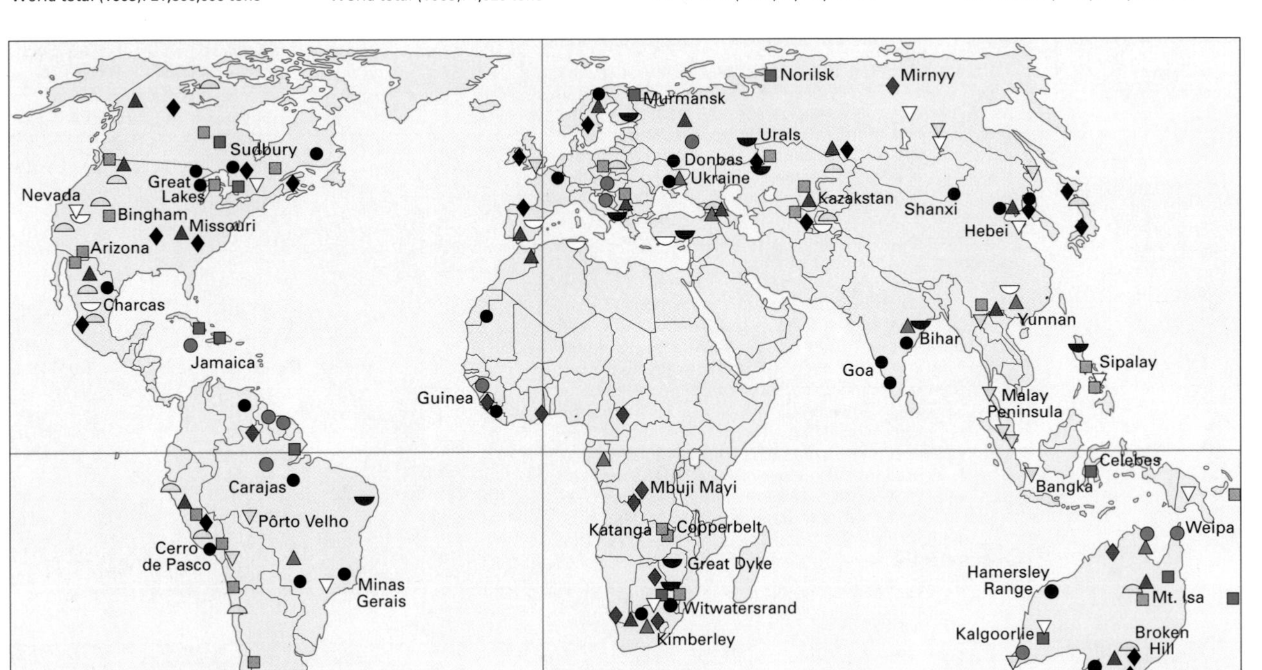

Light metals
- Bauxite

Base metals
- Copper
- Lead
- Mercury
- Tin
- Zinc

Iron and ferro-alloys
- Iron
- Chrome
- Manganese
- Nickel

Precious metals
- Gold
- Silver

Precious stones
- Diamonds

The map does not show undersea deposits, most of which are considered inaccessible.

STEEL PRODUCTION

Steel output in thousand tons (top ten countries, 1993)

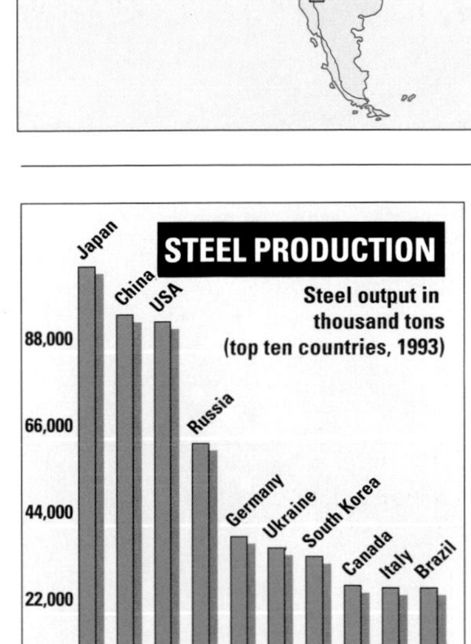

SHIPBUILDING

Merchant vessels launched, in thousand gross registered tons (top ten countries, 1993)

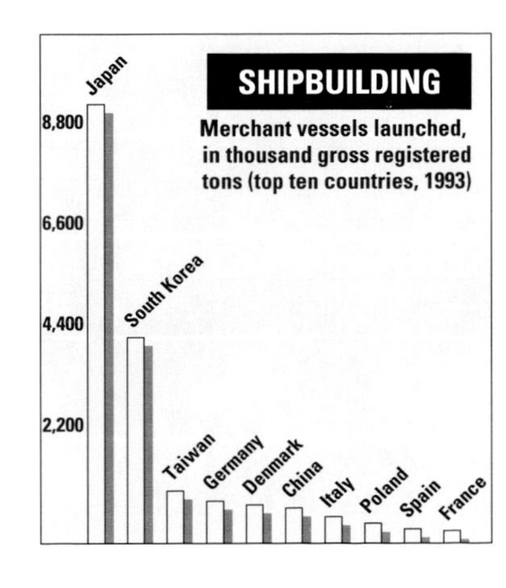

AUTOMOBILES

Production of passenger cars in thousands (top ten countries, 1993)

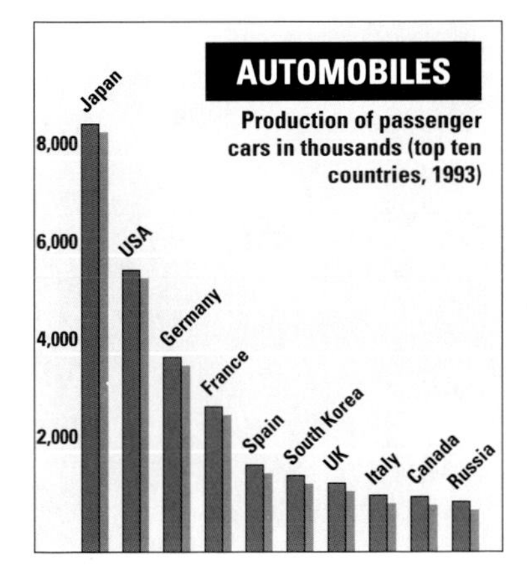

COMMERCIAL VEHICLES

Trucks, buses and coaches produced by the top ten manufacturing countries, in thousands (1993)

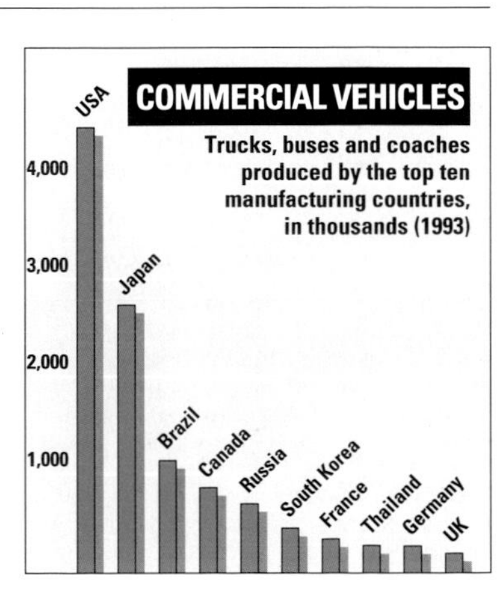

TRADE

SHARE OF WORLD TRADE

Percentage share of total world exports by value (1993)

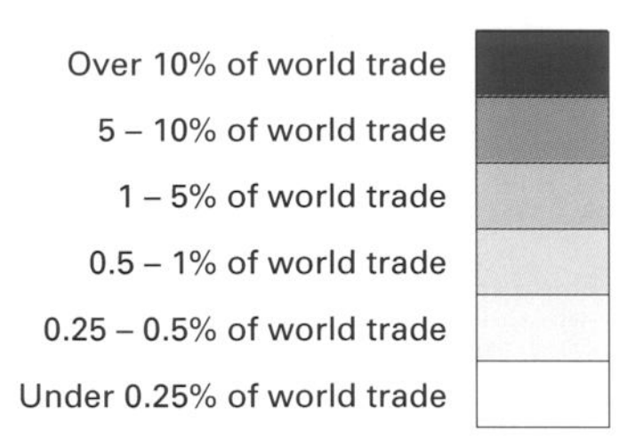

International trade is dominated by a handful of powerful maritime nations. The members of "G7," the inner circle of OECD (see page 19), and the top seven countries listed in the diagram below, account for more than half the total. The majority of nations – including all but four in Africa – contribute less than one quarter of 1% to the worldwide total of exports; the EU countries account for 40%, the Pacific Rim nations over 35%.

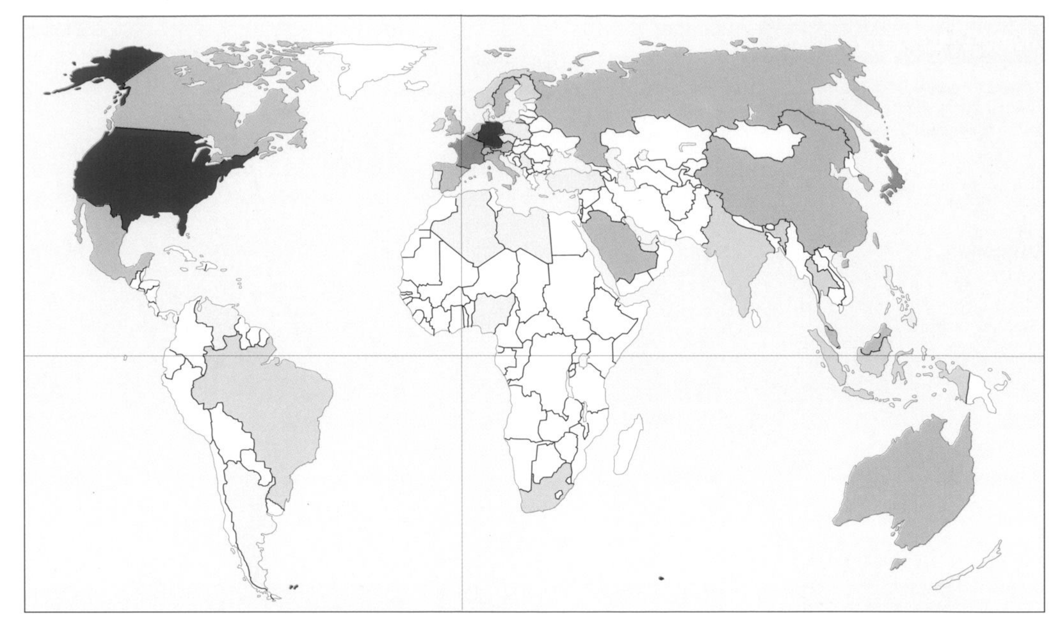

THE MAIN TRADING NATIONS

The imports and exports of the top ten trading nations as a percentage of world trade (1994). Each country's trade in manufactured goods is shown in dark blue.

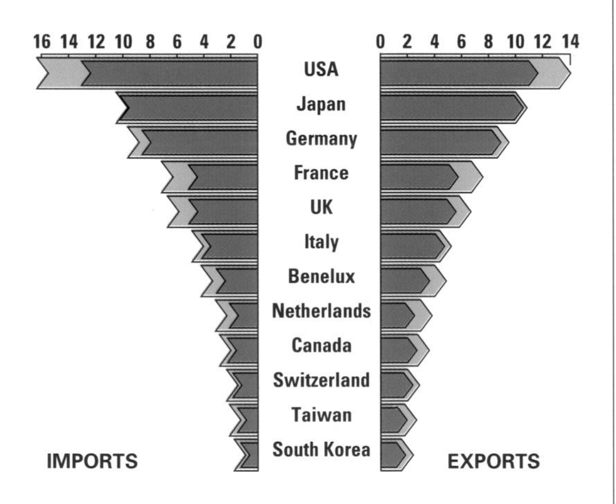

PATTERNS OF TRADE

Thriving international trade is the outward sign of a healthy world economy, the obvious indicator that some countries have goods to sell and others the means to buy them. Global exports expanded to an estimated US $3.92 trillion in 1994, an increase due partly to economic recovery in industrial nations but also to export-led growth strategies in many developing nations and lowered regional trade barriers. International trade remains dominated, however, by the rich, industrialized countries of the Organization for Economic Development: between them, OECD members account for almost 75% of world imports and exports in most years. However, continued rapid economic growth in some developing countries is altering global trade patterns. The "tiger economies" of Southeast Asia are particularly vibrant, averaging more than 8% growth between 1992 and 1994. The size of the largest trading economies means that imports and exports usually represent only a small percentage of their total wealth. In export-concious Japan, for example, trade in goods and services amounts to less than 18% of GDP. In poorer countries, trade – often in a single commodity – may amount to 50% of GDP.

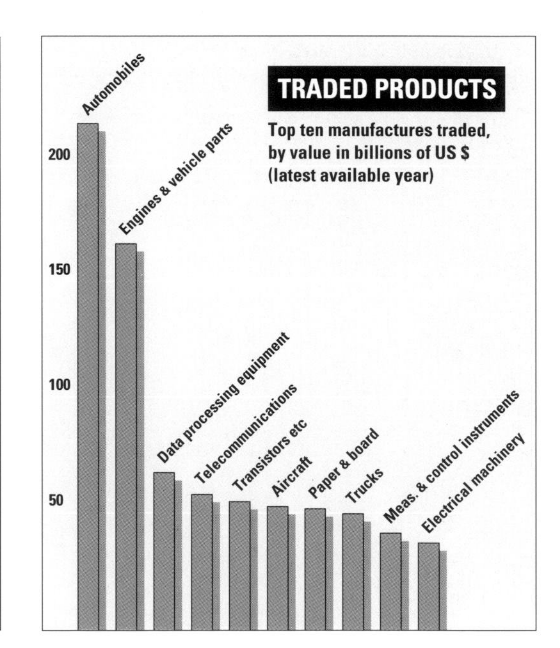

BALANCE OF TRADE

Value of exports in proportion to the value of imports (1993)

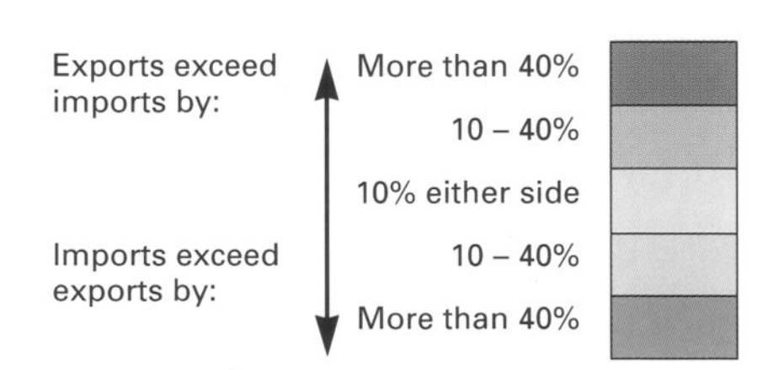

The total world trade balance should amount to zero, since exports must equal imports on a global scale. In practice, at least $100 billion in exports go unrecorded, leaving the world with an apparent deficit and many countries in a better position than public accounting reveals. However, a favorable trade balance is not necessarily a sign of prosperity: many poorer countries must maintain a high surplus in order to service debts, and do so by restricting imports below the levels needed to sustain successful economies.

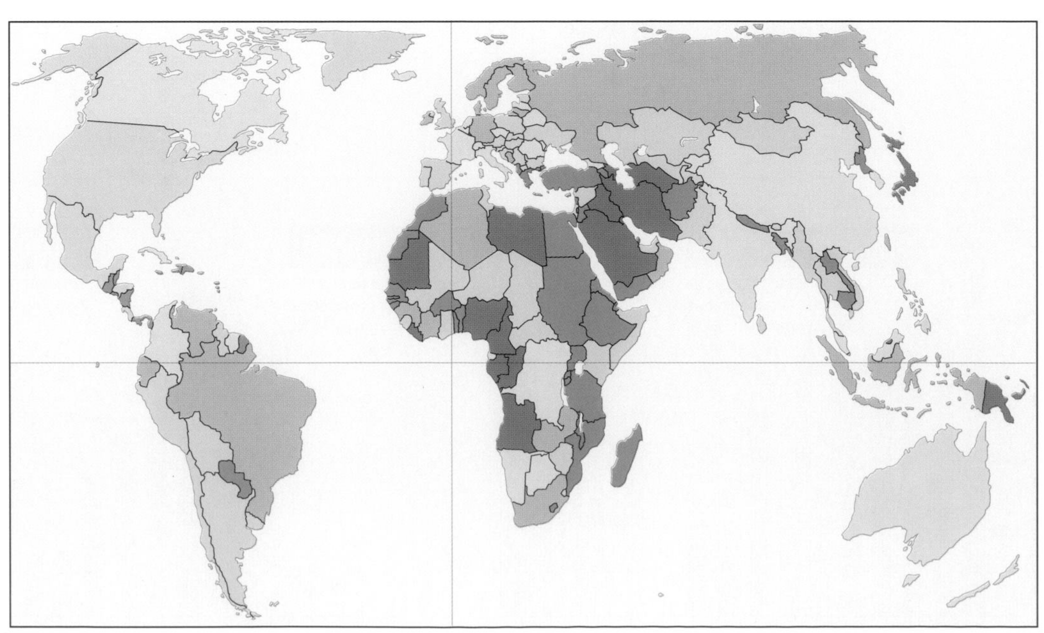

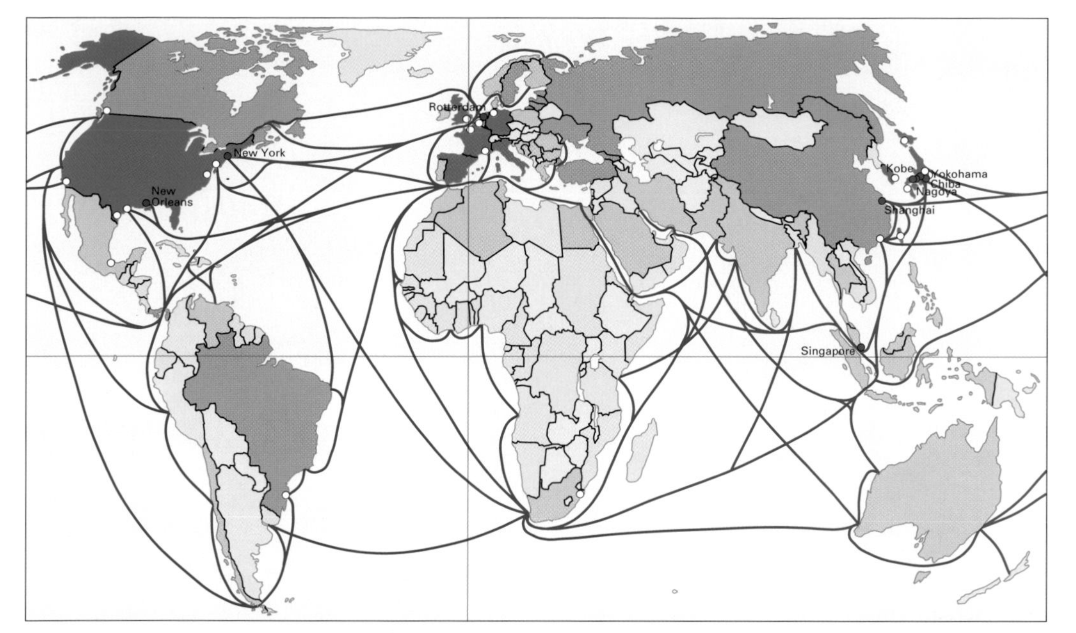

FREIGHT

Freight unloaded in millions of tons (latest available year)

CARGOES

Type of seaborne freight

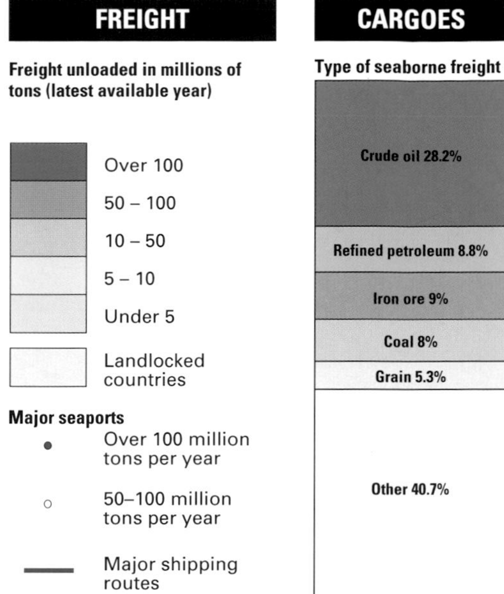

MERCHANT FLEETS

Merchant fleets in thousand gross tonnage (1994). A large number of vessels are registered in Liberia and Panama but they are not part of the national fleet.

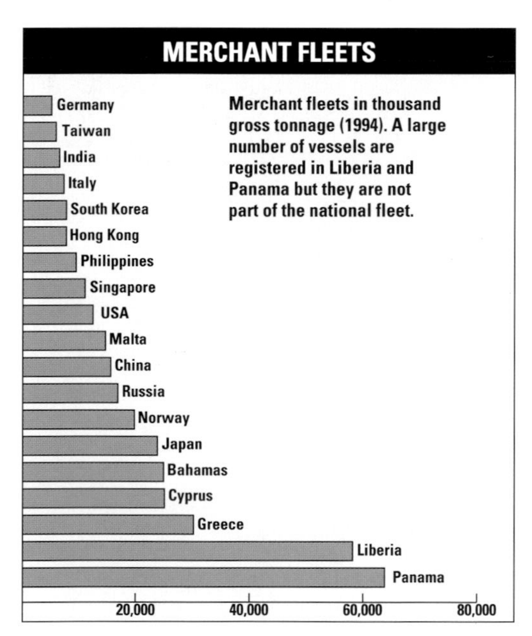

WORLD SHIPPING

World merchant fleet by type of vessel and deadweight tonnage (latest available year)

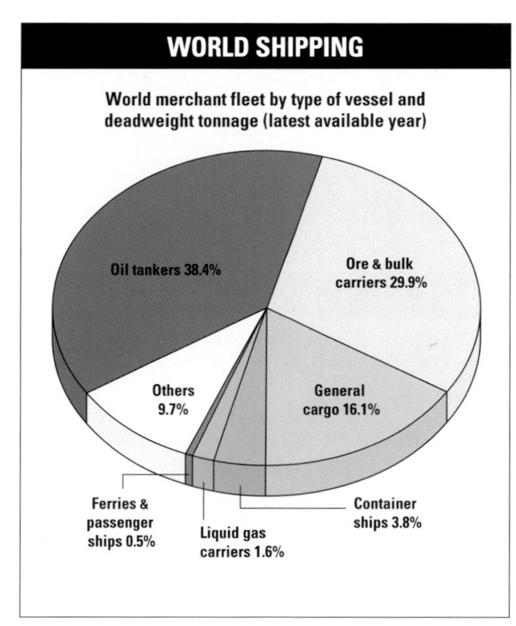

THE GREAT PORTS

The world's ten busiest ports by million tons of shipping arrivals (1992)

The busiest ports in the USA, by tonnage handled, (1993):

1. South Louisiana 193.8 million tons
2. Houston 141.5 million tons
3. New York 116.7 million tons
4. Valdez (Alaska) 85.7 million tons
5. Baton Rouge 85.1 million tons

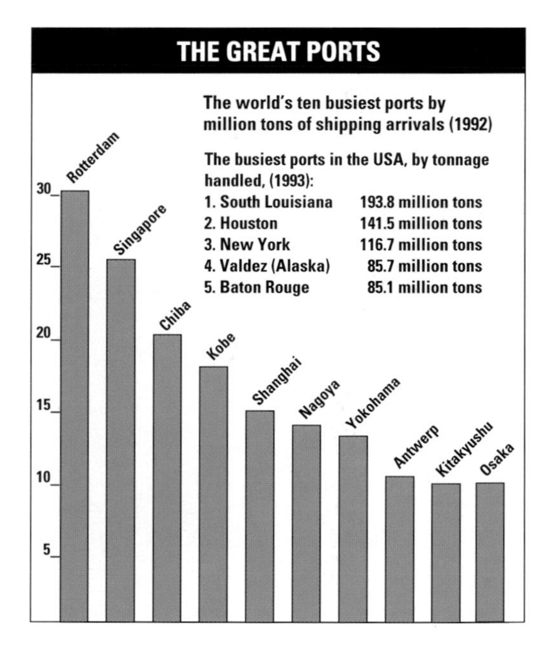

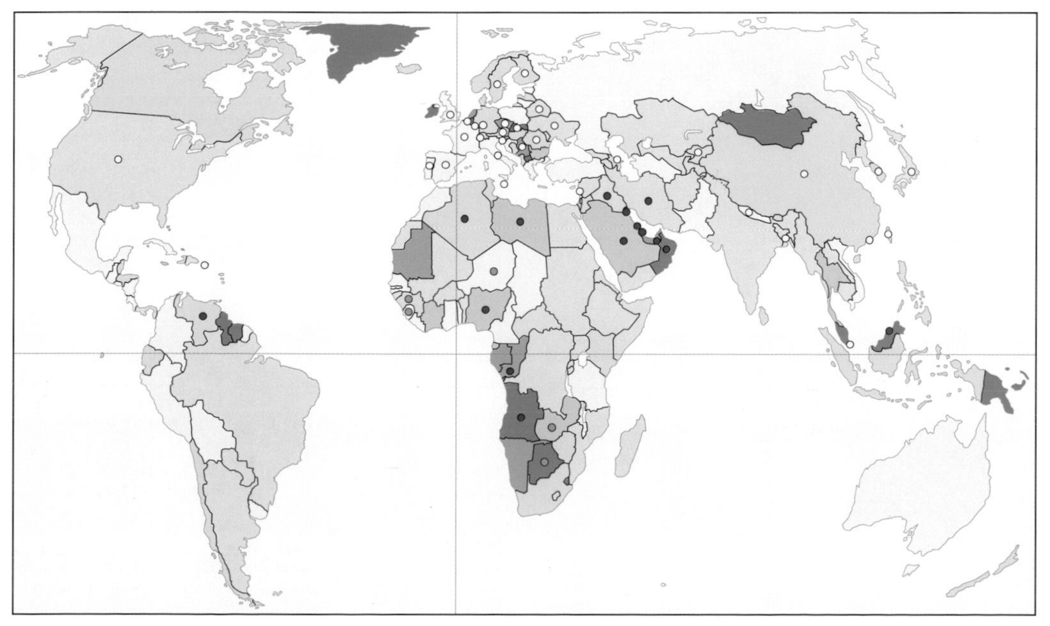

DEPENDENCE ON TRADE

Value of exports as a percentage of Gross Domestic Product (1993)

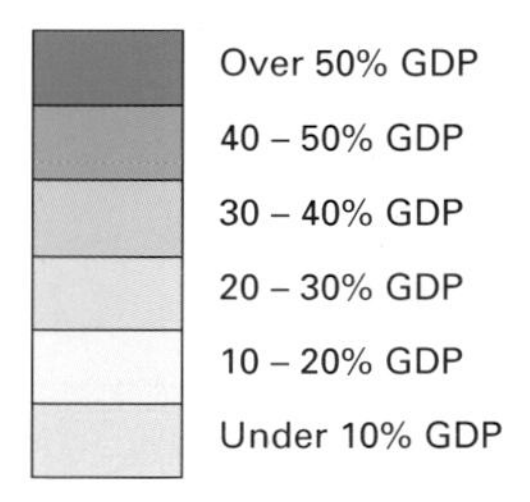

Most dependent on industrial exports (over 75% of total exports)

Most dependent on fuel exports (over 75% of total exports)

Most dependent on mineral and metal exports (over 75% of total exports)

TRAVEL AND TOURISM

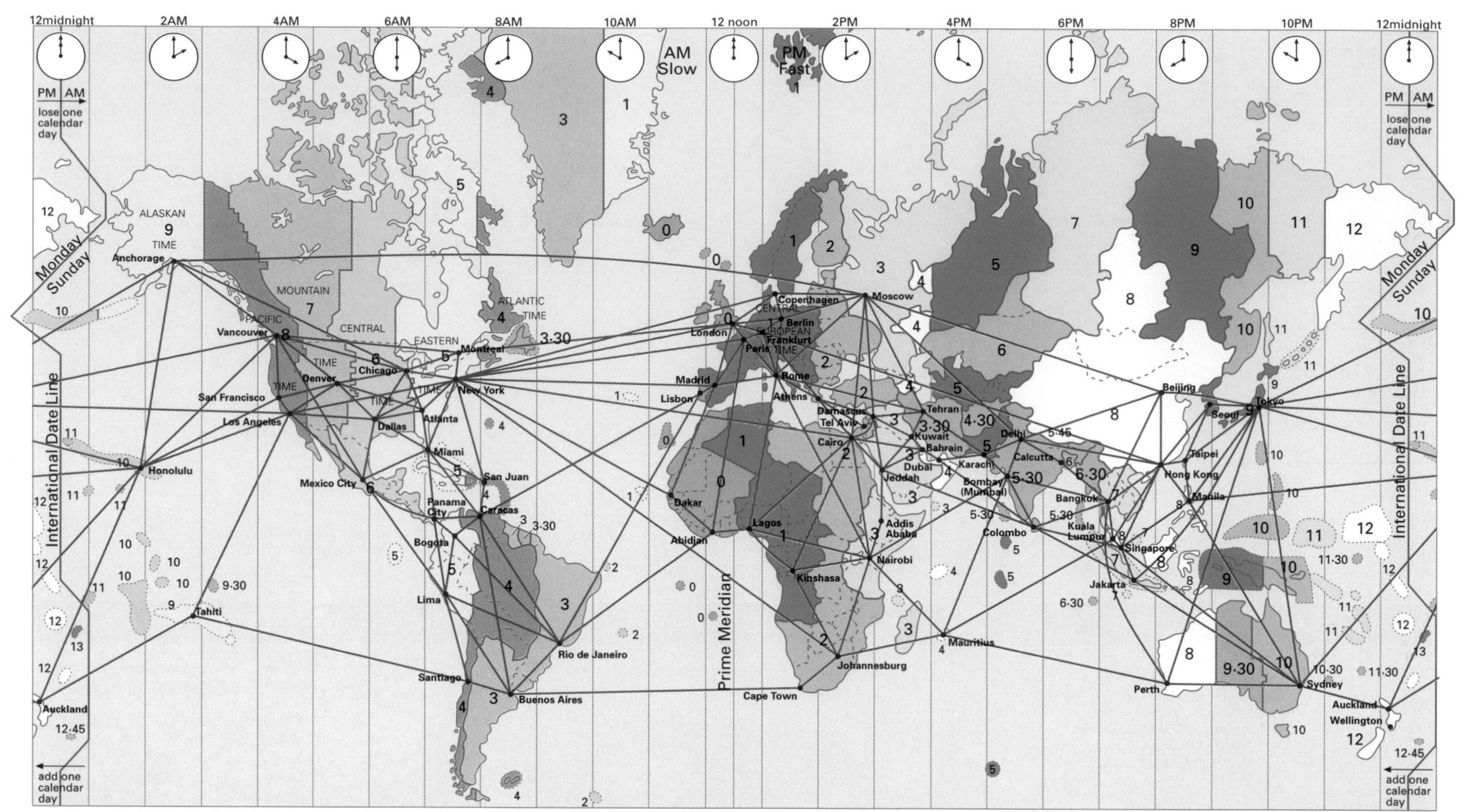

TIME ZONES

- Zones using GMT
- Zones slow of GMT
- International boundaries
- 10 — Hours slow or fast of GMT
- Zones fast of GMT
- Half-hour zones
- Time zone boundaries
- International Date Line
- Selected air routes

The world is divided into 24 time zones, each centered on meridians at 15° intervals, which is the longitudinal distance the sun travels every hour. The meridian running through Greenwich, England, passes through the middle of the first zone.

Certain time zones are affected by the incidence of "summer time" in countries where it is adopted.

Actual Solar Time, when it is noon at Greenwich, is shown along the top of the map.

RAIL AND ROAD: THE LEADING NATIONS

	Total rail network ('000 miles)	Passenger miles per head per year	Total road network ('000 miles)	Vehicle miles per head per year	Number of vehicles per mile of roads
1.	USA 148.9	Japan 1,253	USA 3,898.6	USA 7,766	Hong Kong 176
2.	Russia 54.3	Belarus 1,167	India 1,839.7	Luxembourg 4,961	Taiwan 131
3.	India 38.8	Russia 1,134	Brazil 1,133.0	Kuwait 4,503	Singapore 94
4.	China 33.5	Switzerland 1,099	Japan 702.3	France 4,435	Kuwait 87
5.	Germany 25.1	Ukraine 904	China 646.5	Sweden 4,341	Brunei 60
6.	Australia 22.2	Austria 725	Russia 549.0	Germany 4,227	Italy 57
7.	Argentina 21.2	France 628	Canada 527.5	Denmark 4,200	Israel 54
8.	France 20.2	Netherlands 617	France 504.0	Austria 4,048	Thailand 45
9.	Mexico 16.5	Latvia 570	Australia 503.2	Netherlands 3,716	Ukraine 45
10.	Poland 15.5	Denmark 549	Germany 395.1	UK 3,563	UK 42
11.	South Africa 14.7	Slovak Rep. 535	Romania 286.8	Canada 3,411	Netherlands 41
12.	Ukraine 14.0	Romania 528	Turkey 241.0	Italy 3,013	Germany 39

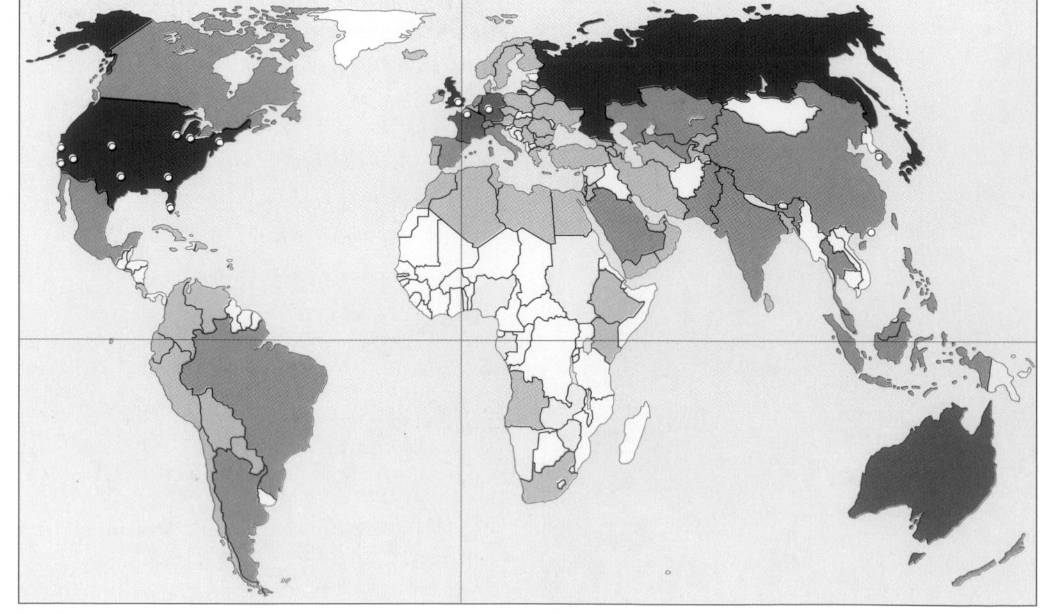

AIR TRAVEL

Passenger miles (the number of passengers – international and and domestic – multiplied by the distance flown by each passenger from the airport of origin) (1994)

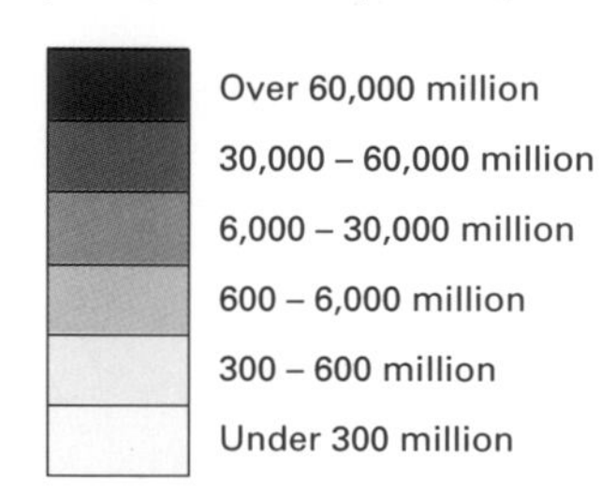

○ Major airports (handling over 25 million passengers in 1994)

World's busiest airports (total passengers)

1. Chicago (O'Hare)
2. Atlanta (Hatsfield)
3. Dallas (Dallas/Ft Worth)
4. London (Heathrow)
5. Los Angeles (Intern'l)

World's busiest airports (international passengers)

1. London (Heathrow)
2. London (Gatwick)
3. Frankfurt (International)
4. New York (Kennedy)
5. Paris (De Gaulle)

DESTINATIONS

- Cultural & historical centers
- Coastal resorts
- Ski resorts
- Centers of entertainment
- Places of pilgrimage
- Places of great natural beauty
- Popular holiday cruise routes

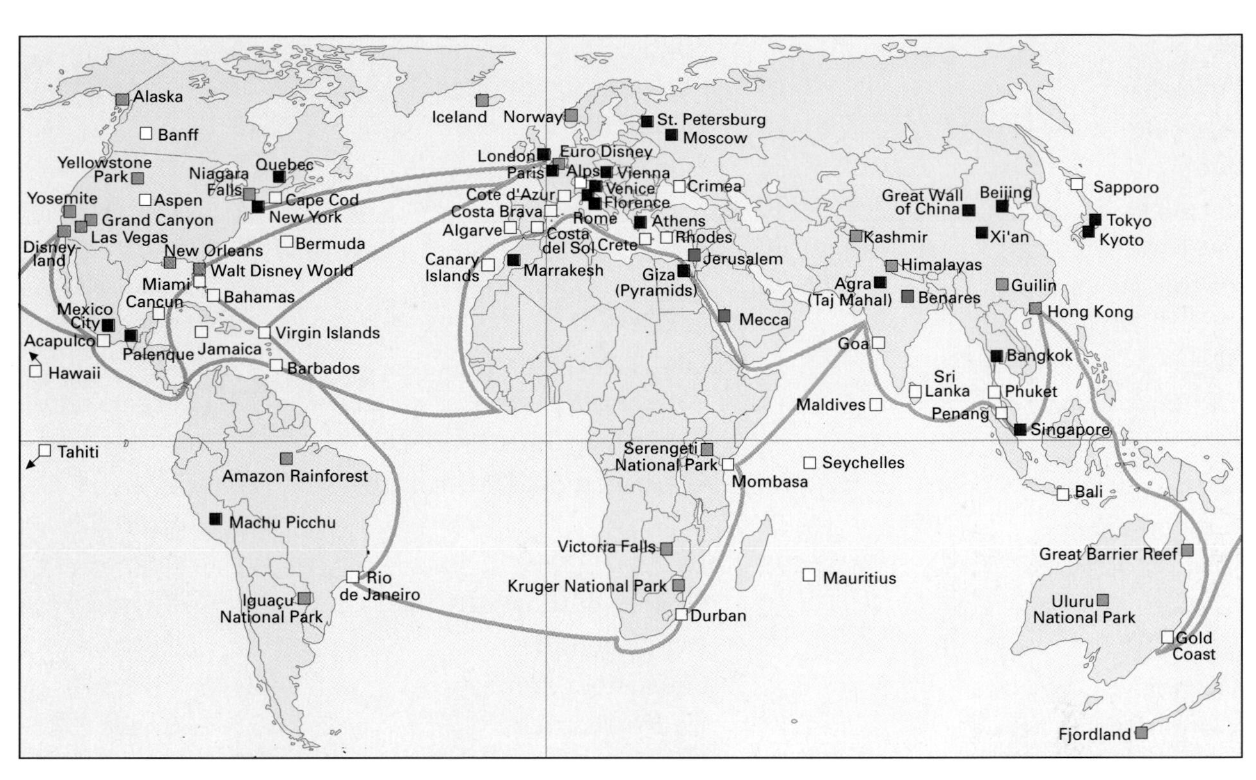

VISITORS TO THE USA

Overseas travelers to the USA, thousands (1997 projections)

1. Canada 13,900
2. Mexico 12,370
3. Japan 4,640
4. UK 3,350
5. Germany 1,990
6. France 1,030
7. Taiwan 885
8. Venezuela 860
9. South Korea 800
10. Brazil 785

In 1996, the USA earned the most from tourism, with receipts of more than US $64 billion.

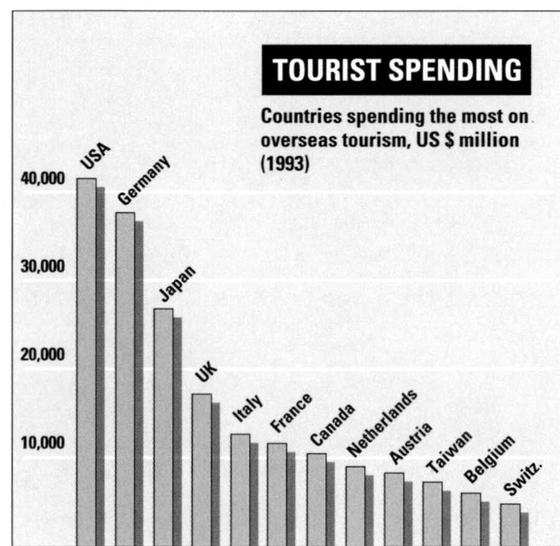

TOURIST SPENDING

Countries spending the most on overseas tourism, US $ million (1993)

IMPORTANCE OF TOURISM

	Arrivals from abroad (1995)	% of world total (1995)
1. France	60,584,000	10.68%
2. Spain	45,125,000	7.96%
3. USA	44,730,000	7.89%
4. Italy	29,184,000	5.15%
5. China	23,368,000	4.12%
6. UK	22,700,000	4.00%
7. Hungary	22,087,000	3.90%
8. Mexico	19,870,000	3.50%
9. Poland	19,225,000	3.39%
10. Austria	17,750,000	3.13%
11. Canada	16,854,000	2.97%
12. Czech Republic	16,600,000	2.93%

The latest figures reveal a 4.6% rise in the total number of people traveling abroad in 1996, to 593 million. Small economies in attractive areas are often completely dominated by tourism: in some West Indian islands, for example, tourist spending provides over 90% of total income.

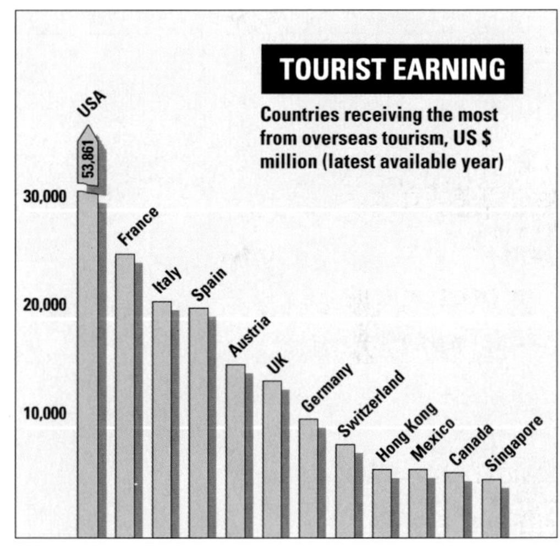

TOURIST EARNING

Countries receiving the most from overseas tourism, US $ million (latest available year)

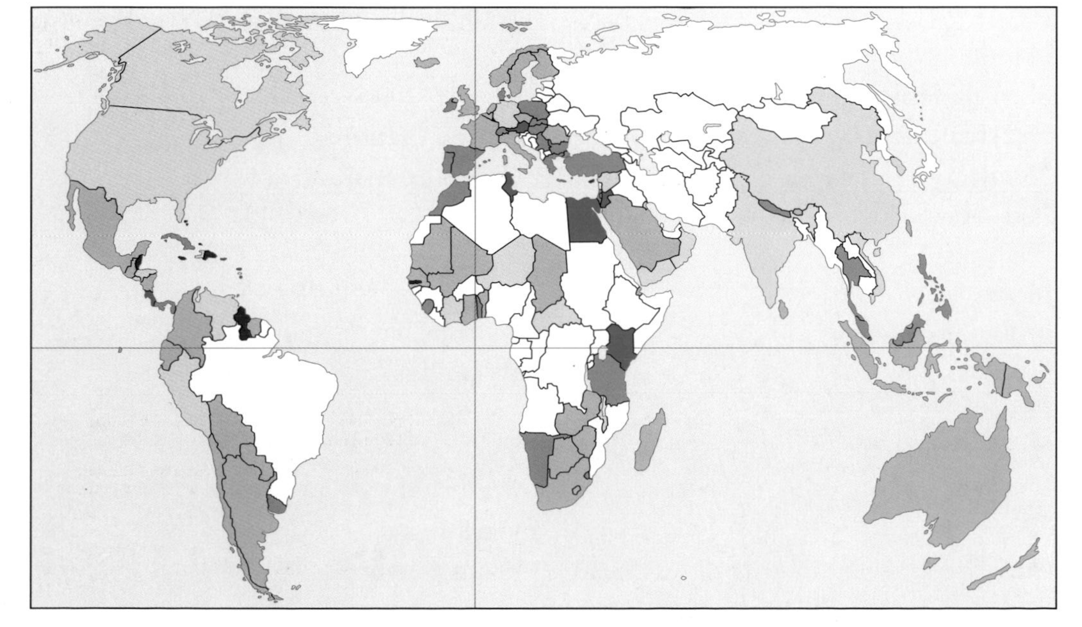

TOURISM

Tourism receipts as a percentage of Gross National Product (1994)

- Over 10% of GNP from tourism
- 5 – 10% of GNP from tourism
- 2.5 – 5% of GNP from tourism
- 1 – 2.5% of GNP from tourism
- 0.5 – 1% of GNP from tourism
- Under 0.5% of GNP from tourism

Countries spending the most on promoting tourism, millions of US $ (1996)

Australia	88
Spain	79
UK	79
France	73
Singapore	54

Fastest growing tourist destinations, % change in receipts (1994–5)

South Korea	49%
Czech Republic	27%
India	21%
Russia	19%
Philippines	18%

Introduction to World Geography: Index

A
Acid rain 15
Agriculture 26
Aid 21
Airports 30
Air travel 30
Automobiles 27–28

B
Beaufort Wind Scale 11

C
Carbon dioxide 10, 14
Child mortality 22
Climate regions 10
Coal 25
Commercial vehicles 27
Continental drift 4
Crops 26
Crust 4–5
Currents 8

D
Day and night 2
Debt 21
Deforestation 15
Desertification 15

E
Earth's structure 4
Earthquakes 4
Eclipses 3
Education 23
El Niño 11
Employment 23, 26
Energy balance 24
Epicenter 4
Equinox 2–3
Erosion 7
European Union 19
Exports 28

F
Faults 6
Female employment 23
Fertility 23
Fishing 9
Folds 6
Food consumption 22
Forestry 13
Fuel exports 25

G
Geothermal energy 25
Glaciers 7
Global warming 14
Greenhouse effect 14
Gross Domestic Product 29
Gross National Product 20

H
Holiday destinations 31
Hospital beds 22
Household spending 21
Hydroelectricity 25
Hydrological cycle 12

I
Illiteracy 23
Imports 28
Infectious disease 22
Inflation 20
International organizations 19

L
Land use 13
Language 18
Life expectancy 22

M
Mantle 4–5
Mercalli Scale 4
Merchant fleets 29
Metals 27
Mid-ocean ridge 5
Minerals 27
Moon 3
Mountain building 6

N
Nuclear power 25

O
OECD 19, 28
Oil 24–25
Ozone 14–15

P
Planets 2
Plate tectonics 5
Pollution 9
Population density 16–17
Ports 29
Precipitation 10–12

R
Railroads 30
Rain forest 13, 15
Richter Scale 4
Religion 18
Rivers 7
Roads 30

S
Sea-floor spreading 5
Seasons 2
Shipbuilding 27
Shipping 29
Solar System 2
Subduction zone 6

T
Temperature 11
Tides 3
Time zones 30
Tourism 31
Traded products 28

U
United Nations 19
Urban population 17

V
Vegetation 13
Vehicles 30
Volcanoes 5

W
Water 12
Wealth gap 20
Wind power 25

WORLD MAPS

SETTLEMENTS

PARIS Berne Livorno Brugge *Algeciras* *Fréjus* *Oberammergau* *Thira*

Settlement symbols and type styles vary according to the scale of each map and indicate the importance of towns on the map rather than specific population figures

∴ Ruins or Archæological Sites

Wells in Desert

ADMINISTRATION

International Boundaries

International Boundaries (Undefined or Disputed)

Internal Boundaries

National Parks

Country Names

NICARAGUA

Administrative Area Names

KENT

CALABRIA

International boundaries show the *de facto* situation where there are rival claims to territory

COMMUNICATIONS

Principal Roads

Other Roads

Trails and Seasonal Roads

Passes

Airfields

Principal Railroads

Railroads Under Construction

Other Railroads

Railroad Tunnels

Principal Canals

PHYSICAL FEATURES

Perennial Streams

Intermittent Streams

Perennial Lakes

Intermittent Lakes

Swamps and Marshes

Permanent Ice and Glaciers

8848 Elevations (m)

8050 Sea Depths (m)

Height of Lake Surface Above Sea Level (m)

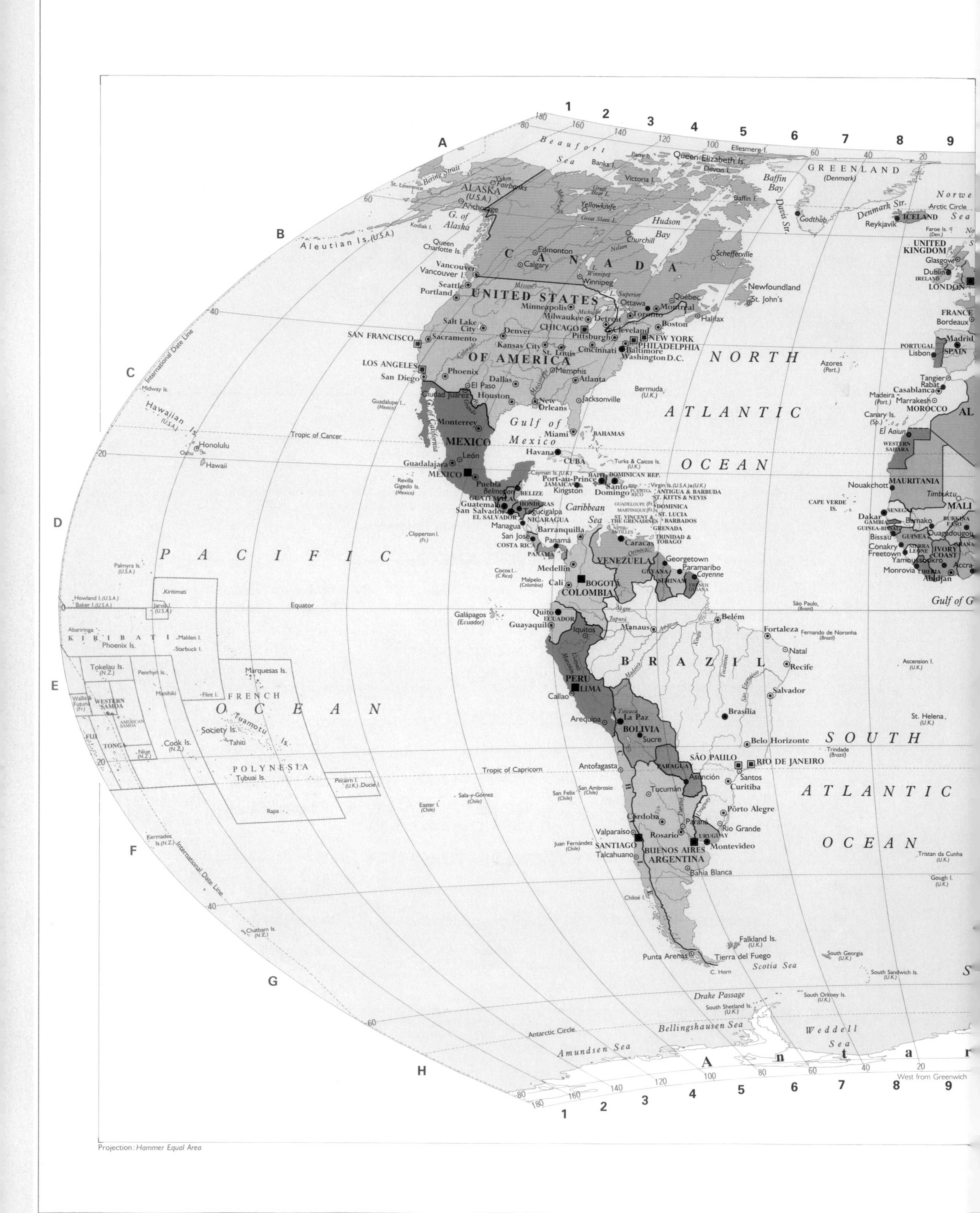
Beaufort Sea
Banks I.
Parry Is.
Queen Elizabeth Is.
Ellesmere I.
Devon I.
Victoria I.
Baffin Bay
GREENLAND (Denmark)
Norwegian Sea
Arctic Circle
Bering Strait
St. Lawrence I.
ALASKA (U.S.A.)
Yukon
Fairbanks
Anchorage
Great Bear L.
Yellowknife
Great Slave L.
Baffin I.
Davis Str.
Godthåb
Denmark Str.
ICELAND
Reykjavík
G. of Alaska
Kodiak I.
Hudson Bay
Churchill
Faroe Is. (Den.)
UNITED KINGDOM
Glasgow
Dublin
IRELAND
LONDON
Aleutian Is. (U.S.A.)
Queen Charlotte Is.
Edmonton
Calgary
CANADA
Nelson
Schefferville
Vancouver
Vancouver I.
Seattle
Portland
L. Winnipeg
Winnipeg
Missouri
UNITED STATES OF AMERICA
L. Superior
Newfoundland
St. John's
Québec
Ottawa
Montreal
Minneapolis
Milwaukee
Detroit
Toronto
Halifax
Salt Lake City
Denver
CHICAGO
Boston
SAN FRANCISCO
Sacramento
Kansas City
St. Louis
Pittsburgh
Cleveland
NEW YORK
PHILADELPHIA
Baltimore
Washington D.C.
Cincinnati
Colorado
LOS ANGELES
San Diego
Phoenix
Dallas
Memphis
Atlanta
El Paso
Houston
New Orleans
Jacksonville
Bermuda (U.K.)
Azores (Port.)
NORTH ATLANTIC OCEAN
FRANCE
Bordeaux
Madrid
SPAIN
PORTUGAL
Lisbon
Tangier
Rabat
Casablanca
Marrakesh
MOROCCO
Madeira (Port.)
Canary Is. (Sp.)
El Aaiun
WESTERN SAHARA
International Date Line
Midway Is.
Hawaiian Is. (U.S.A.)
Honolulu
Oahu
Hawaii
Guadalupe I. (Mexico)
Ciudad Juárez
G. of California
Monterrey
Gulf of Mexico
Miami
BAHAMAS
Tropic of Cancer
MEXICO
León
Guadalajara
MÉXICO
Havana
CUBA
Turks & Caicos Is. (U.K.)
Revilla Gigedo Is. (Mexico)
Cayman Is. (U.K.)
JAMAICA
Kingston
HAITI
Port-au-Prince
DOMINICAN REP.
Santo Domingo
PUERTO RICO
Virgin Is. (U.S.A.) & (U.K.)
ANTIGUA & BARBUDA
ST. KITTS & NEVIS
GUADELOUPE (Fr.)
DOMINICA
MARTINIQUE (Fr.)
ST. LUCIA
ST. VINCENT & THE GRENADINES
BARBADOS
GRENADA
TRINIDAD & TOBAGO
NETH. ANTILLES
Puebla
Belmopan
BELIZE
GUATEMALA
Guatemala
HONDURAS
Tegucigalpa
San Salvador
EL SALVADOR
NICARAGUA
Managua
Caribbean Sea
Barranquilla
San José
COSTA RICA
Panamá
PANAMA
Caracas
VENEZUELA
Georgetown
Paramaribo
Cayenne
GUYANA
SURINAM
FRENCH GUIANA
CAPE VERDE IS.
MAURITANIA
Nouakchott
Timbuktu
MALI
Dakar
GAMBIA
SENEGAL
GUINEA-BISSAU
Bissau
GUINEA
Conakry
Freetown
SIERRA LEONE
Bamako
BURKINA FASO
Ouagadougou
IVORY COAST
Yamoussoukro
Monrovia
LIBERIA
Abidjan
Accra
Gulf of Guinea
Clipperton I. (Fr.)
Medellín
Cali
BOGOTÁ
COLOMBIA
Cocos I. (C. Rica)
Malpelo (Colombia)
PACIFIC OCEAN
Palmyra Is. (U.S.A.)
Kiritimati
Howland I. (U.S.A.)
Baker I. (U.S.A.)
Jarvis I. (U.S.A.)
Equator
São Paulo (Brazil)
Galápagos (Ecuador)
Quito
ECUADOR
Guayaquil
Iquitos
Negro
Japurá
Amazon
Manaus
Belém
Fortaleza
Fernando de Noronha (Brazil)
Natal
Recife
Abariringa
KIRIBATI
Phoenix Is.
Malden I.
Starbuck I.
Tokelau Is. (N.Z.)
Penrhyn Is.
Marquesas Is.
Manihiki
Flint I.
FRENCH POLYNESIA
Tuamotu Is.
Society Is.
Tahiti
Wallis & Futuna (Fr.)
WESTERN SAMOA
AMERICAN SAMOA
FIJI
TONGA
Niue (N.Z.)
Cook Is. (N.Z.)
Tubuai Is.
Pitcairn I. (U.K.)
Ducie I.
Rapa
Ucayali
Marañón
Madeira
Xingu
Tocantins
São Francisco
BRAZIL
Ascension I. (U.K.)
PERU
LIMA
Callao
Salvador
L. Titicaca
La Paz
BOLIVIA
Sucre
Arequipa
Brasília
St. Helena (U.K.)
Belo Horizonte
SOUTH ATLANTIC OCEAN
Trindade (Brazil)
SÃO PAULO
RIO DE JANEIRO
PARAGUAY
Asunción
Santos
Curitiba
Antofagasta
Tropic of Capricorn
San Ambrosio (Chile)
San Felix (Chile)
Tucumán
Easter I. (Chile)
Sala-y-Gómez (Chile)
Pôrto Alegre
Paraná
Uruguay
Rio Grande
Córdoba
Rosario
Valparaíso
SANTIAGO
URUGUAY
Montevideo
BUENOS AIRES
ARGENTINA
Juan Fernández (Chile)
Talcahuano
Bahía Blanca
Tristan da Cunha (U.K.)
Gough I. (U.K.)
Kermadec Is. (N.Z.)
Chiloé I.
Chatham Is. (N.Z.)
Falkland Is. (U.K.)
Punta Arenas
Tierra del Fuego
C. Horn
Scotia Sea
South Georgia (U.K.)
South Sandwich Is. (U.K.)
Drake Passage
South Orkney Is. (U.K.)
South Shetland Is. (U.K.)
Bellingshausen Sea
Weddell Sea
Antarctic Circle
Amundsen Sea
Antarctica
West from Greenwich
Projection: Hammer Equal Area

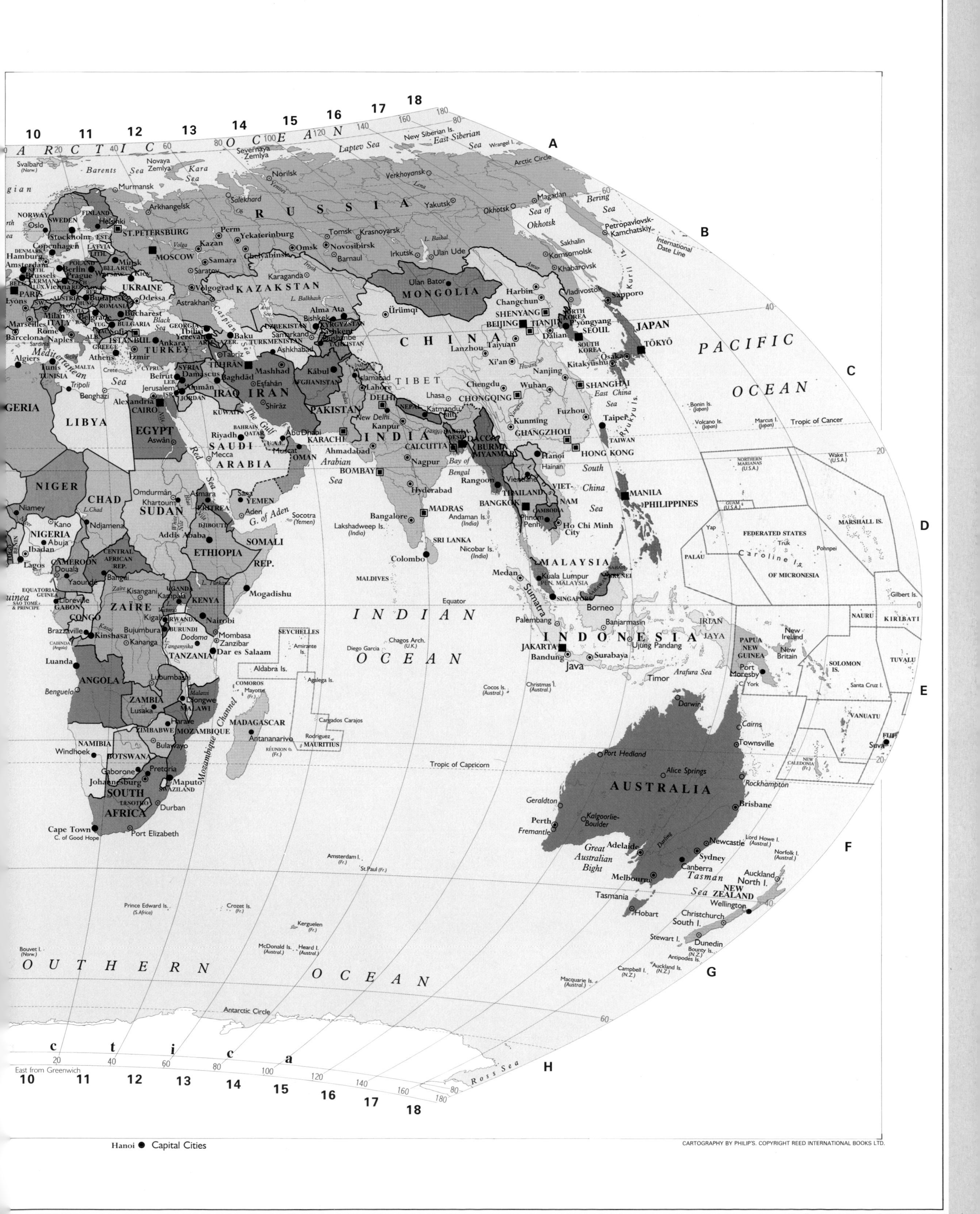

Hanoi ● Capital Cities

ARCTIC OCEAN

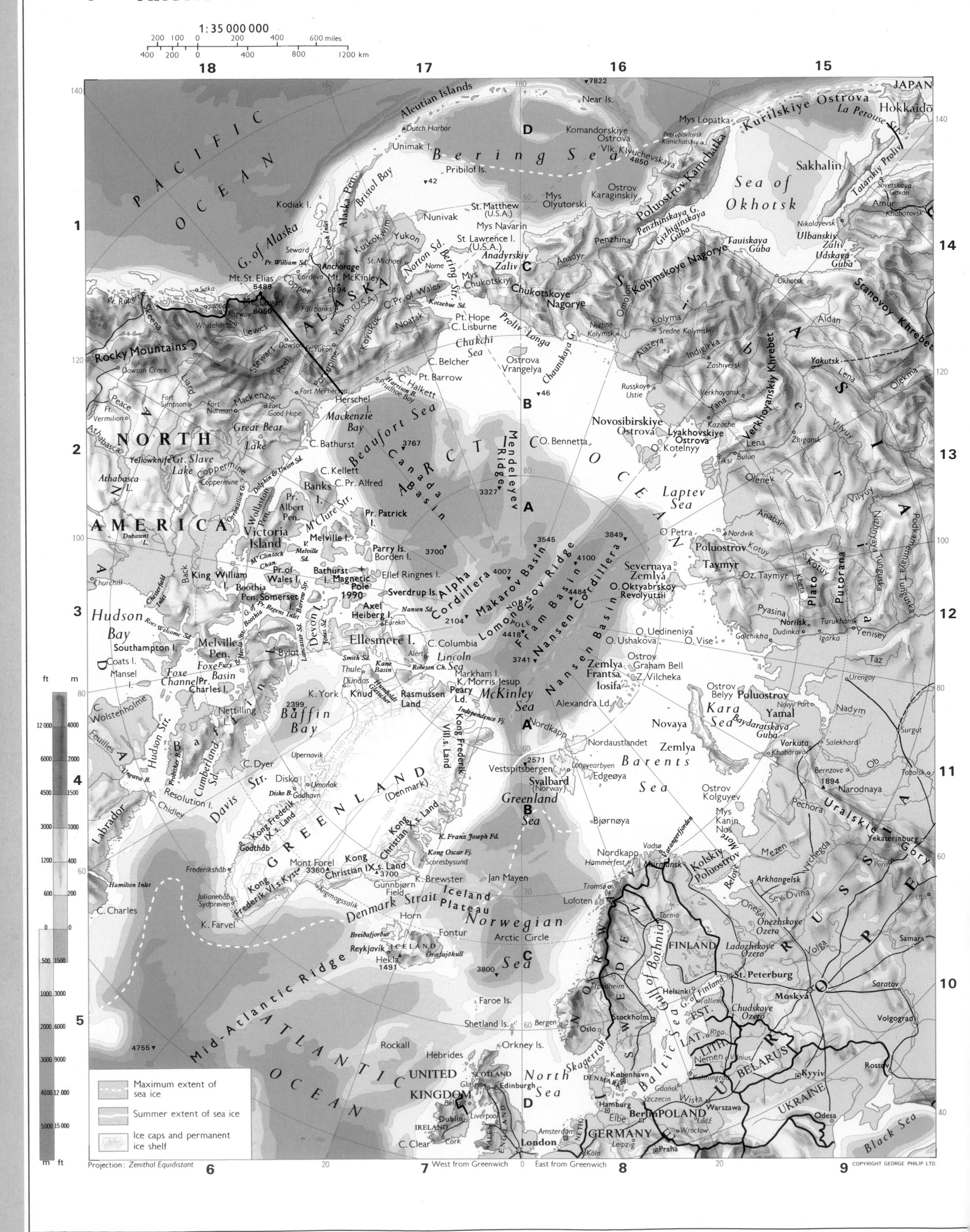

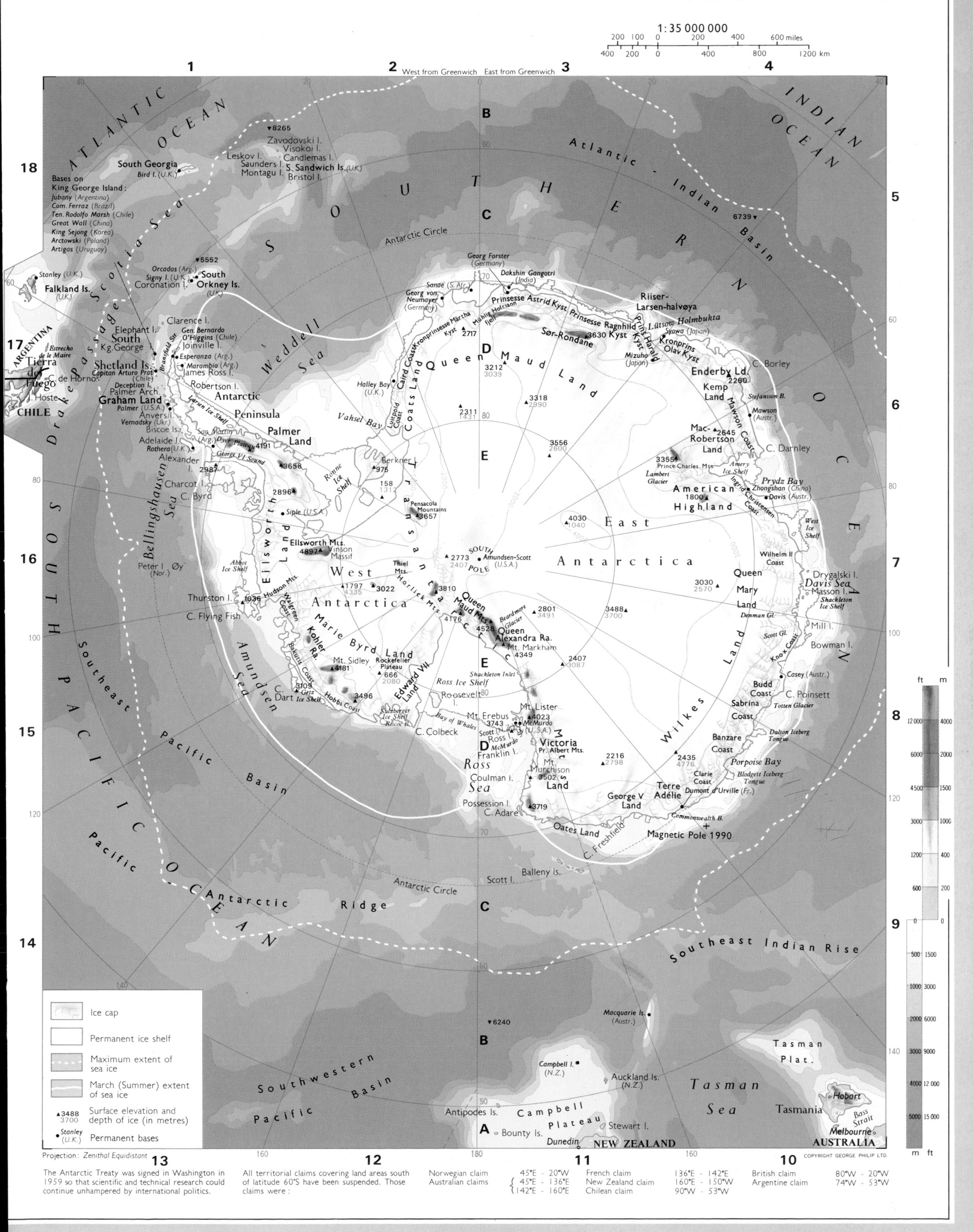

Projection: *Zenithal Equidistant*

The Antarctic Treaty was signed in Washington in 1959 so that scientific and technical research could continue unhampered by international politics.

All territorial claims covering land areas south of latitude 60°S have been suspended. Those claims were:

Norwegian claim	45°E - 20°W
Australian claims	45°E - 136°E; 142°E - 160°E
French claim	136°E - 142°E
New Zealand claim	160°E - 150°W
Chilean claim	90°W - 53°W
British claim	80°W - 20°W
Argentine claim	74°W - 53°W

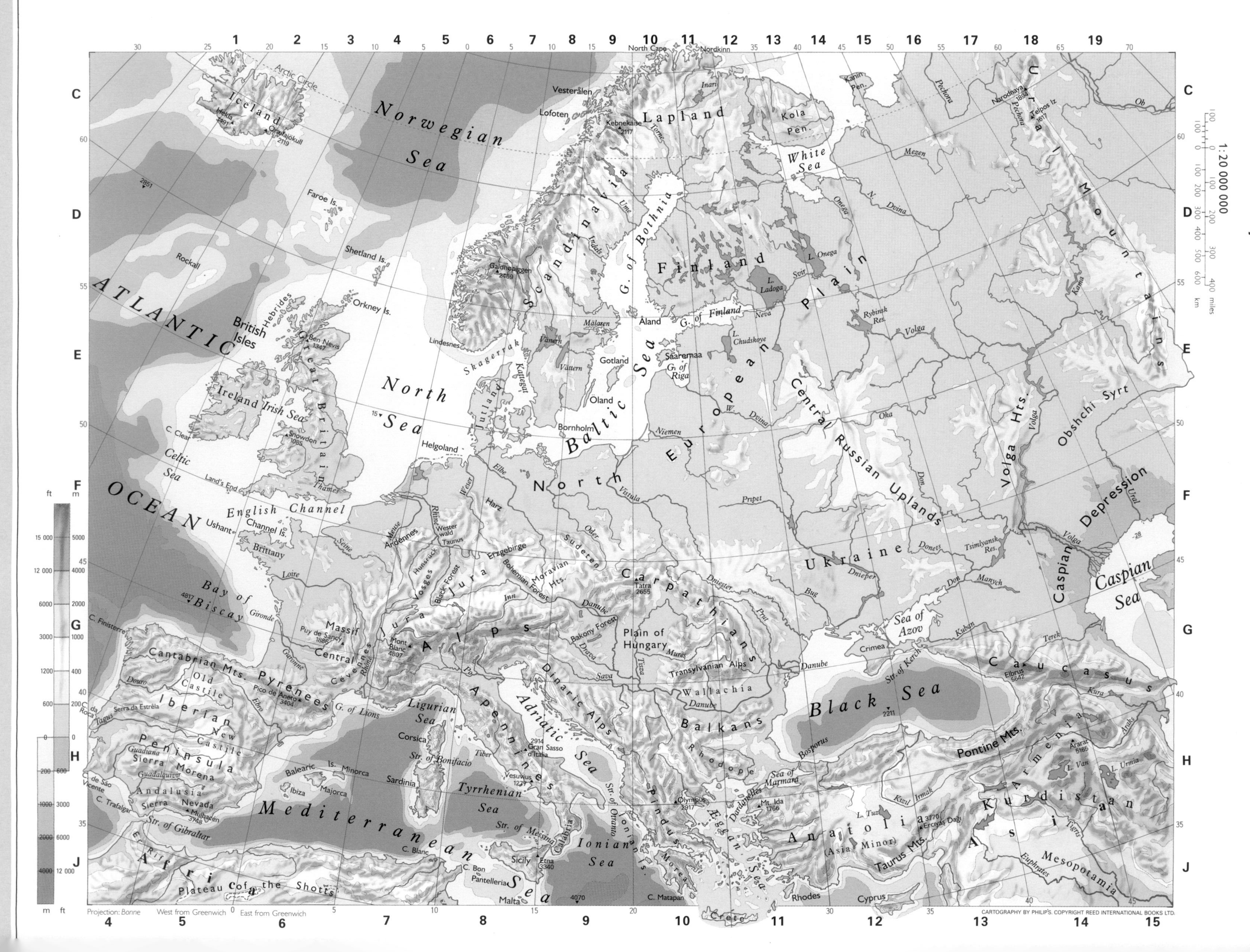
1:20 000 000
miles
km
Norwegian Sea
North Sea
ATLANTIC OCEAN
Bay of Biscay
English Channel
Celtic Sea
Irish Sea
British Isles
Great Britain
Ireland
Iceland
Faroe Is.
Shetland Is.
Orkney Is.
Hebrides
Rockall
Arctic Circle
Lofoten
Vesterålen
North Cape
Nordkinn
Lapland
Scandinavia
G. of Bothnia
Finland
G. of Finland
G. of Riga
Baltic Sea
Gotland
Öland
Bornholm
Åland
Saaremaa
Kattegat
Skagerrak
Jutland
Lindesnes
Helgoland
White Sea
Kola Pen.
Kanin Pen.
L. Ladoga
L. Onega
L. Chudskoye
Rybinsk Res.
Ural Mountains
Obshchi Syrt
Volga Hts.
Caspian Depression
Caspian Sea
North European Plain
Central Russian Uplands
Ukraine
Sea of Azov
Crimea
Str. of Kerch
Black Sea
Caucasus
Elbrus 5642
Pontine Mts.
Armenia
Kurdistan
Mesopotamia
Anatolia (Asia Minor)
Taurus Mts.
Cyprus
Rhodes
Crete
Ægean Sea
Sea of Marmara
Bosporus
Dardanelles
Balkans
Rhodope
Wallachia
Transylvanian Alps
Carpathians
Plain of Hungary
Dinaric Alps
Adriatic Sea
Apennines
Alps
Jura
Vosges
Black Forest
Bohemian Forest
Erzgebirge
Sudeten
Harz
Massif Central
Pyrenees
Iberian Peninsula
Cantabrian Mts.
Old Castile
New Castile
Sierra Morena
Sierra Nevada
Andalusia
Str. of Gibraltar
Balearic Is.
Majorca
Minorca
Ibiza
Corsica
Sardinia
Str. of Bonifacio
Ligurian Sea
Tyrrhenian Sea
Sicily
Malta
Pantelleria
Str. of Messina
Str. of Otranto
Ionian Sea
Pindus
Morea
C. Matapan
Mediterranean Sea
Africa
Plateau of the Shotts
C. Bon
C. Blanc
G. of Lions
C. Finisterre
C. Trafalgar
Ushant
Land's End
C. Clear
Channel Is.
Brittany
Volga
Don
Dnieper
Danube
Vistula
Oder
Elbe
Rhine
Rhône
Loire
Seine
Thames
Garonne
Ebro
Tagus
Douro
Po
Tiber
Tisza
Sava
Drava
Dniester
Prut
Bug
Pripet
Niemen
W. Dvina
N. Dvina
Pechora
Mezen
Onega
Kama
Ural
Terek
Kura
Euphrates
Tigris
Ob
Manych
Tsimlyansk Res.
Projection: Bonne
West from Greenwich
East from Greenwich
CARTOGRAPHY BY PHILIPS. COPYRIGHT REED INTERNATIONAL BOOKS LTD.

1 : 20 000 000

ICELAND
Reykjavik
Norwegian Sea
Faroe Is. (Den.)
Shetland Is.
Orkney Is.
Hebrides
UNITED KINGDOM
SCOTLAND
N. IRELAND
IRELAND
WALES
ENGLAND
LONDON
ATLANTIC OCEAN
North Sea
English Channel
Bay of Biscay
NORWAY
SWEDEN
FINLAND
DENMARK
ESTONIA
LATVIA
LITHUANIA
BELARUS
POLAND
GERMANY
NETHERLANDS
BELGIUM
LUX.
FRANCE
PARIS
SPAIN
PORTUGAL
ANDORRA
MONACO
SWITZERLAND
LIECH.
AUSTRIA
CZECH REP.
SLOVAK REP.
HUNGARY
SLOVENIA
CROATIA
BOSNIA-HERZ.
YUGOSLAVIA
SERBIA
MONTENEGRO
ALBANIA
MACEDONIA
GREECE
BULGARIA
ROMANIA
MOLDOVA
UKRAINE
RUSSIA
MOSCOW
ST. PETERSBURG
KAZAKSTAN
GEORGIA
ARMENIA
AZERBAIJAN
TURKEY
ISTANBUL
SYRIA
IRAQ
IRAN
CYPRUS
ITALY
SAN MARINO
MALTA
TUNISIA
ALGERIA
MOROCCO
Africa
Baltic Sea
G. of Bothnia
White Sea
Black Sea
Caspian Sea
Mediterranean Sea
Adriatic Sea
Tyrrhenian Sea
Ionian Sea
Ægean Sea
Arctic Circle
Corsica
Sardinia
Sicily
Crete
Balearic Is.
LONDON Capital Cities 9
Projection: Bonne
West from Greenwich
East from Greenwich

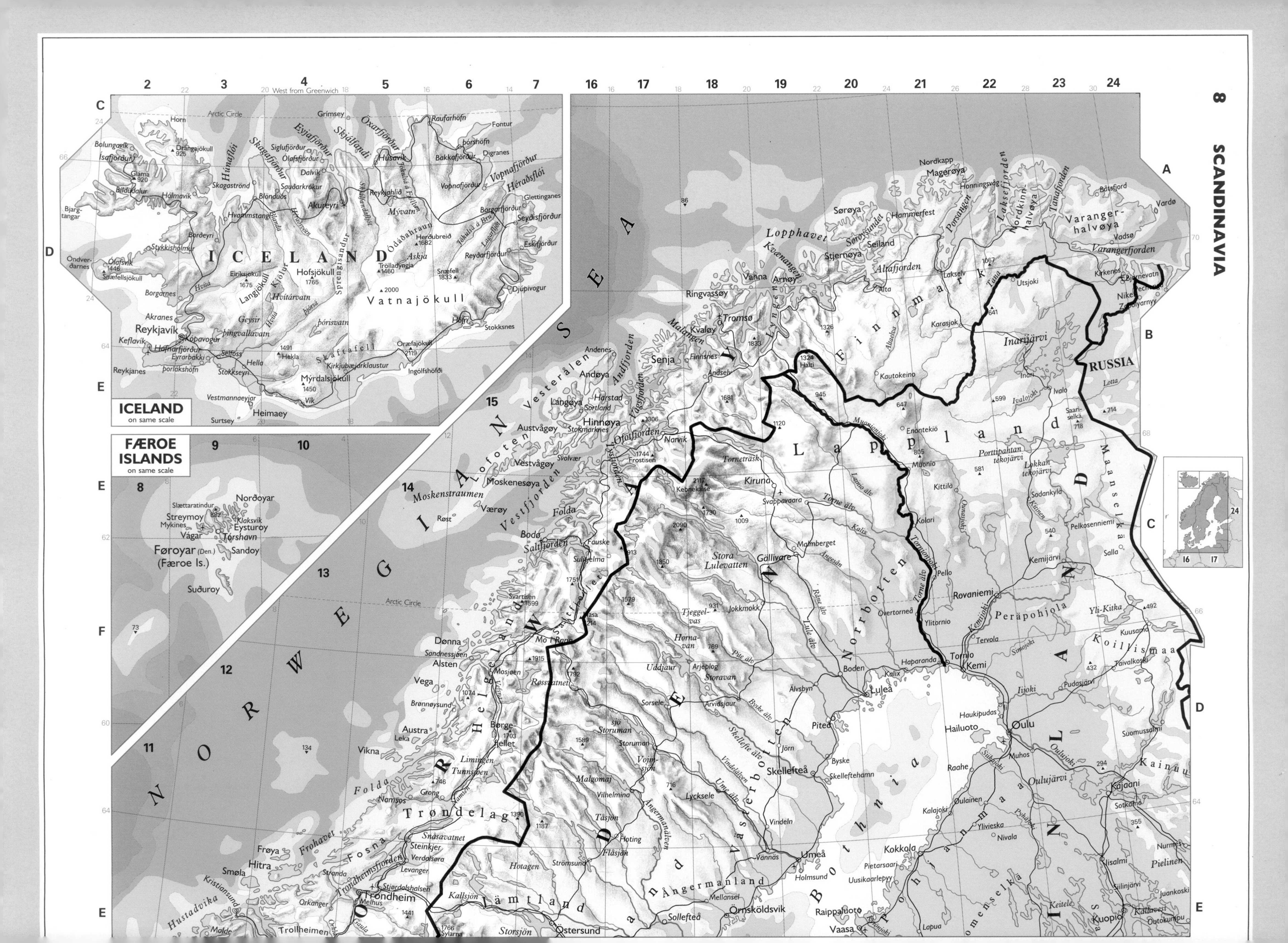

ICELAND
on same scale
FÆROE ISLANDS
on same scale
West from Greenwich
Arctic Circle
ICELAND
Vatnajökull
Hofsjökull
Langjökull
Mýrdalsjökull
Reykjavík
Akureyri
Húsavík
Heimaey
Vestmannaeyjar
Føroyar (Den.)
(Færoe Is.)
Tórshavn
Streymoy
Eysturoy
Sandoy
Suðuroy
Norðoyar
NORWEGIAN SEA
RUSSIA
Lappland
Finnmark
Nordkapp
Magerøya
Hammerfest
Varanger-halvøya
Varangerfjorden
Tromsø
Senja
Narvik
Lofoten
Vesterålen
Vestfjorden
Bodø
Kiruna
Gällivare
Norrbotten
Västerbotten
Ångermanland
Jämtland
Trøndelag
Trondheim
Östersund
Luleå
Umeå
Skellefteå
Oulu
Kemi
Tornio
Rovaniemi
Inarijärvi
Maanselkä
Peräpohjola
Koillismaa
Kainuu
Kajaani
Kuopio
Kokkola
Vaasa
Örnsköldsvik
Bothnia

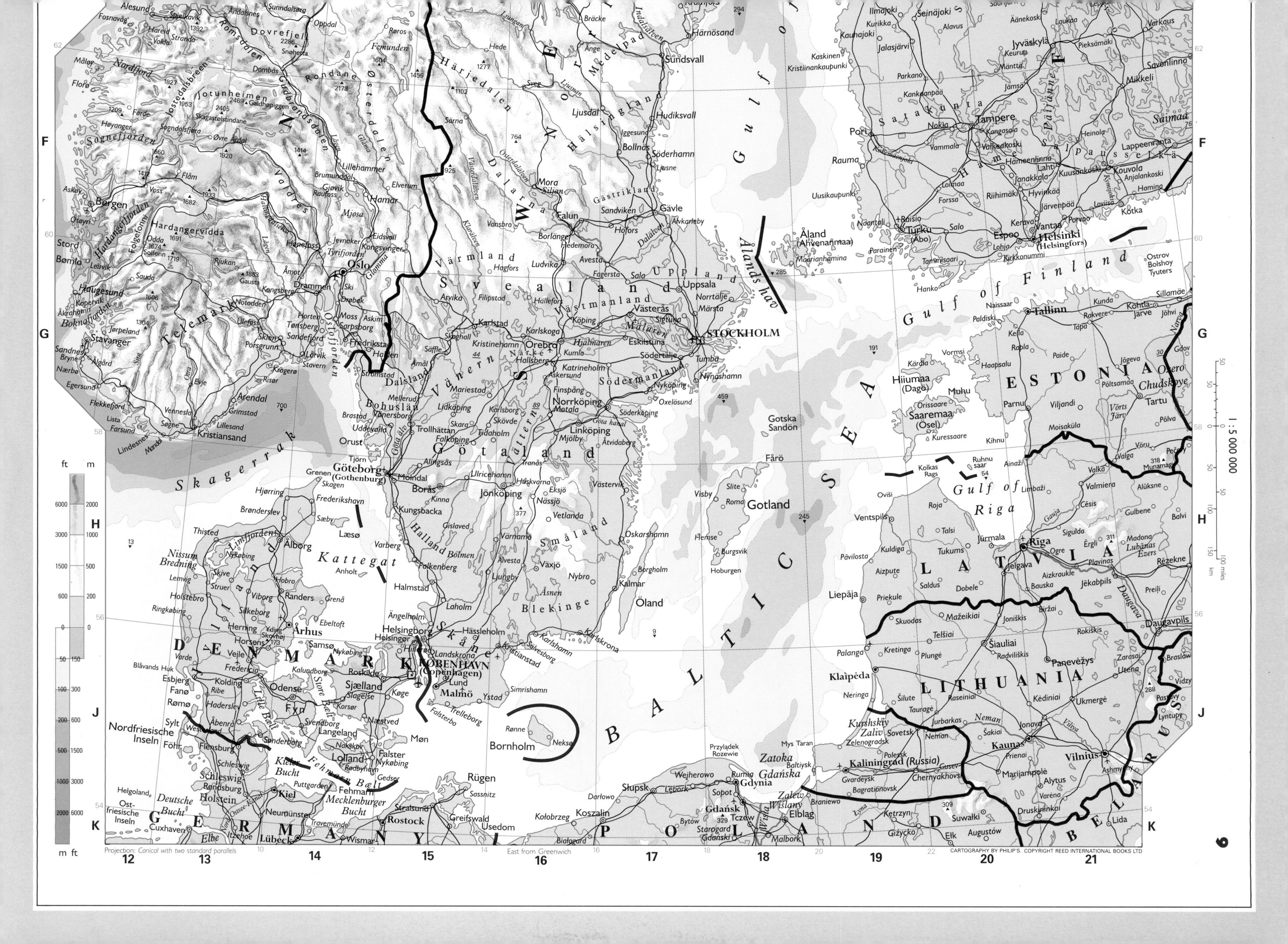

9
1 : 5 000 000
CARTOGRAPHY BY PHILIP'S. COPYRIGHT REED INTERNATIONAL BOOKS LTD
Projection: Conical with two standard parallels
East from Greenwich
BALTIC SEA
Gulf of Finland
Gulf of Riga
Ålands hav
Skagerrak
Kattegat
Bornholm
Gotland
Öland
Gotska Sandön
Fårö
Hiiumaa (Dagö)
Saaremaa (Ösel)
Åland (Ahvenanmaa)
Zatoka Gdańska
Mecklenburger Bucht
Kieler Bucht
Fehmarn Bælt
Deutsche Bucht
Nordfriesische Inseln
Ost-friesische Inseln
Lille Bælt
Store Bælt
Oslofjorden
Sognefjorden
Hardangerfjorden
Nordfjord
Svealand
Götaland
Värmland
Dalarna
Västmanland
Uppland
Södermanland
Småland
Blekinge
Skåne
Halland
Bohuslän
Dalsland
Hälsingland
Härjedalen
Telemark
Hardangervidda
Jotunheimen
Dovrefjell
Rondane
Østerdalen
Gudbrandsdalen
Valdres
Vänern
Vättern
Mälaren
Hjälmaren
Satakunta
Salpausselkä
Päijänne
Saimaa
ESTONIA
LATVIA
LITHUANIA
DENMARK
GERMANY
POLAND
BELARUS
STOCKHOLM
KØBENHAVN (Copenhagen)
Helsinki (Helsingfors)
Tallinn
Riga
Vilnius
Kaliningrad (Russia)
Oslo
Göteborg (Gothenburg)
Malmö
Uppsala
Västerås
Örebro
Norrköping
Linköping
Jönköping
Gävle
Sundsvall
Härnösand
Turku (Åbo)
Tampere
Jyväskylä
Pori
Bergen
Stavanger
Kristiansand
Arendal
Drammen
Ålborg
Århus
Odense
Esbjerg
Kiel
Lübeck
Rostock
Gdańsk
Gdynia
Klaipėda
Kaunas
Liepāja
Ventspils
Tartu
Pärnu
Daugavpils
Šiauliai
Panevėžys
Flensburg
Rügen
Usedom
Koszalin
Słupsk
Elbląg
Sjælland
Fyn
Lolland
Falster
Møn
Jylland

English Unitary Authorities
(from April 1996)
12. Hartlepool
13. Stockton-on-Tees
14. Middlesbrough
15. Redcar and Cleveland
16. Darlington
17. City of York
18. Kingston upon Hull
19. Stoke-on-Trent
20. Derby City
21. Leicester City
22. Rutland
23. Milton Keynes
24. Luton
25. Thamesdown
26. South Gloucester
27. City and County of Bristol
28. North Somerset
29. Bath and N.E. Somerset
30. Southampton
31. Portsmouth
32. Brighton and Hove
33. Bournemouth
34. Poole
Welsh Unitary Authorities
(from April 1996)
1. Neath Port Talbot
2. Rhondda Cynon Taff
3. Bridgend
4. Merthyr Tydfil
5. Caerphilly
6. Vale of Glamorgan
7. Cardiff
8. Blaenau Gwent
9. Torfaen
10. Newport
11. Monmouthshire
NORTH SEA
IRISH SEA
North Channel
SCOTLAND
Southern Uplands
NORTHUMBERLAND
DURHAM
CUMBRIA
LANCASHIRE
NORTH YORKSHIRE
EAST RIDING OF YORKSHIRE
SOUTH YORKSHIRE
W. YORKSHIRE
TYNE & WEAR
MERSEYSIDE
CHESHIRE
DERBY
NOTTS
LINCOLN
NORFOLK
STAFFORD
ISLE OF MAN
ANGLESEY
Cumbrian Mts.
N. York Moors
Edinburgh
Glasgow
Newcastle
Sunderland
Carlisle
Leeds
Bradford
Manchester
Liverpool
Sheffield
Nottingham
Derby
Lincoln
Kingston upon Hull
Blackpool
Preston
York
Middlesbrough
Douglas
Belfast
Dumfries
Stirling
Kilmarnock
Ayr
Paisley

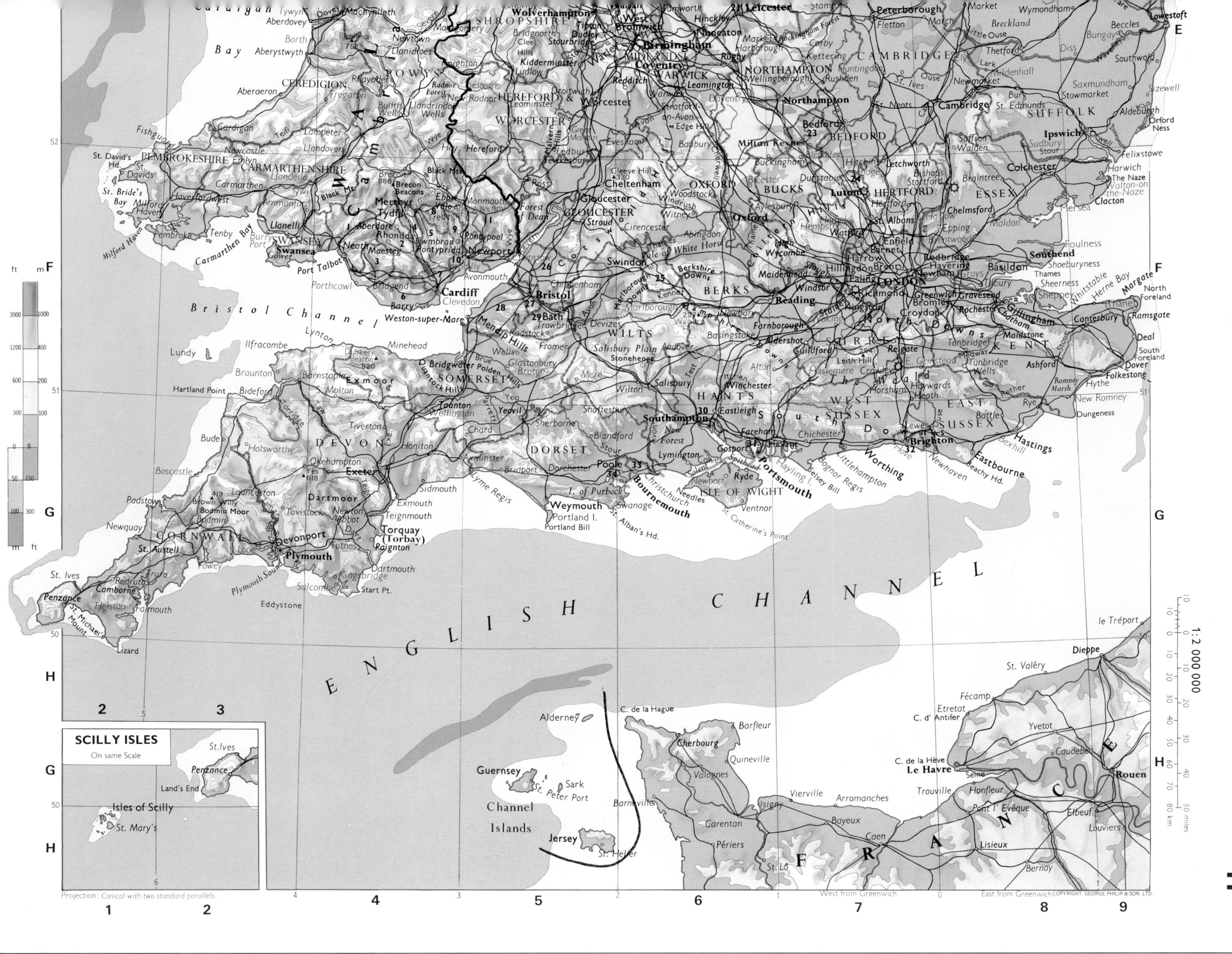

1:2 000 000
10 0 10 20 30 40 50 miles
10 0 10 20 30 40 50 60 70 80 km
ENGLISH CHANNEL
Bristol Channel
FRANCE
SCILLY ISLES
On same Scale
Isles of Scilly
St. Mary's
Land's End
Penzance
St. Ives
Channel Islands
Guernsey
Sark
St. Peter Port
Alderney
Jersey
St. Helier
C. de la Hague
Cherbourg
Barfleur
Quineville
Valognes
Carentan
Périers
St. Lô
Isigny
Vierville
Arromanches
Bayeux
Caen
Lisieux
Bernay
Honfleur
Pont l' Evêque
Trouville
Le Havre
C. de la Hève
Fécamp
Etretat
C. d' Antifer
Yvetot
Caudebec
Rouen
Elbeuf
Louviers
Dieppe
St. Valéry
le Tréport
CORNWALL
DEVON
DORSET
SOMERSET
WILTS
HANTS
BERKS
OXFORD
BUCKS
HERTFORD
ESSEX
SUFFOLK
KENT
EAST SUSSEX
WEST SUSSEX
ISLE OF WIGHT
GLOUCESTER
WORCESTER
HEREFORD
SHROPSHIRE
WARWICK
NORTHAMPTON
BEDFORD
CAMBRIDGE
CEREDIGION
CARMARTHENSHIRE
PEMBROKESHIRE
POWYS
LONDON
Plymouth
Exeter
Dartmoor
Exmoor
Torquay (Torbay)
Weymouth
Bournemouth
Southampton
Portsmouth
Brighton
Worthing
Eastbourne
Hastings
Dover
Folkestone
Canterbury
Margate
Ramsgate
Bristol
Bath
Cardiff
Swansea
Newport
Gloucester
Cheltenham
Oxford
Reading
Swindon
Birmingham
Coventry
Northampton
Cambridge
Ipswich
Colchester
Chelmsford
Southend
Luton
Bedford
Peterborough
Leicester
Lowestoft
Lundy
Eddystone
Projection: Conical with two standard parallels.
West from Greenwich
East from Greenwich
COPYRIGHT, GEORGE PHILIP & SON, LTD.

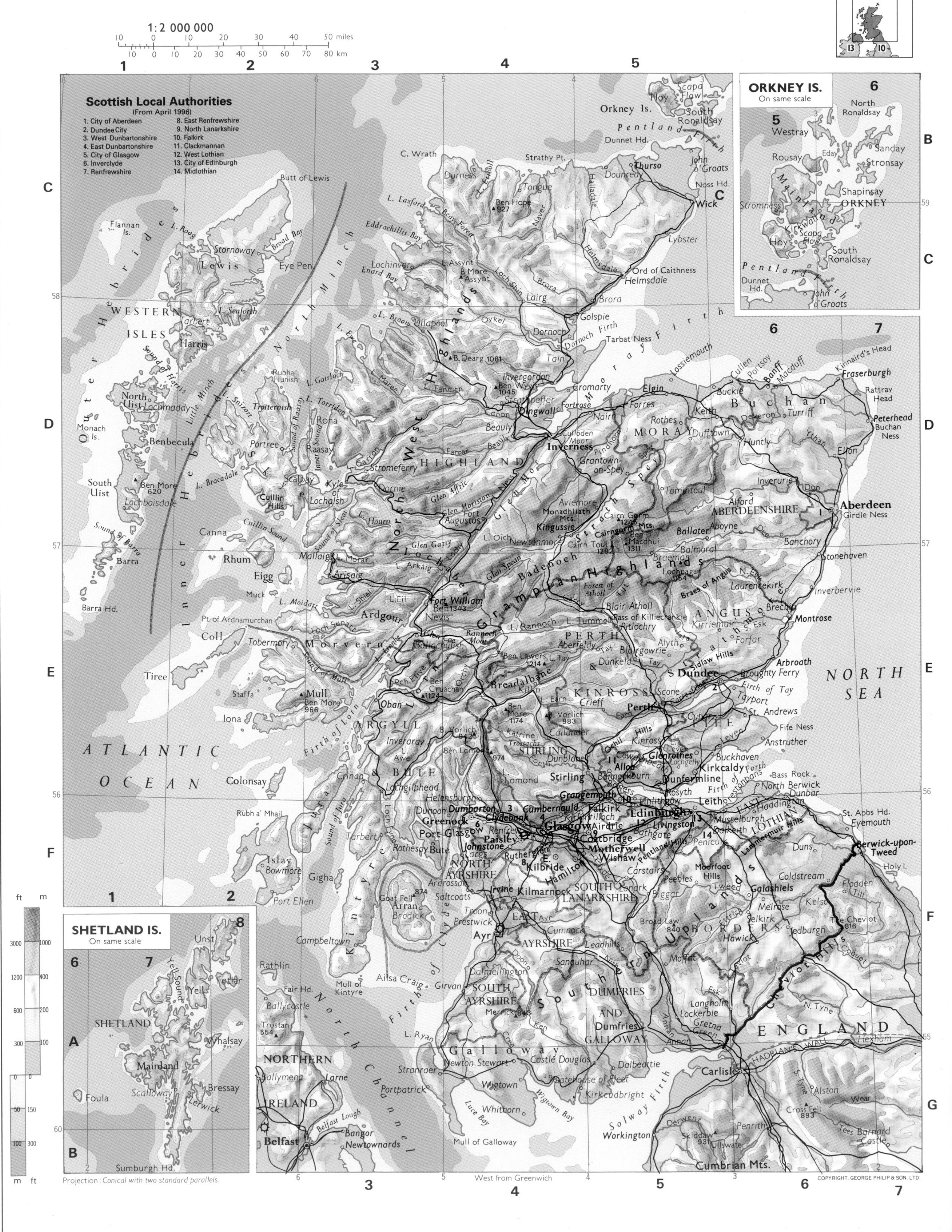
1:2 000 000
10 0 10 20 30 40 50 miles
10 0 10 20 30 40 50 60 70 80 km
Scottish Local Authorities
(From April 1996)
1. City of Aberdeen
2. Dundee City
3. West Dunbartonshire
4. East Dunbartonshire
5. City of Glasgow
6. Inverclyde
7. Renfrewshire
8. East Renfrewshire
9. North Lanarkshire
10. Falkirk
11. Clackmannan
12. West Lothian
13. City of Edinburgh
14. Midlothian
ORKNEY IS.
On same scale
SHETLAND IS.
On same scale
ATLANTIC OCEAN
NORTH SEA
WESTERN ISLES
HIGHLAND
MORAY
ABERDEENSHIRE
ANGUS
PERTH & KINROSS
FIFE
STIRLING
ARGYLL & BUTE
NORTH AYRSHIRE
EAST AYRSHIRE
SOUTH AYRSHIRE
SOUTH LANARKSHIRE
DUMFRIES AND GALLOWAY
BORDERS
EAST LOTHIAN
SHETLAND
ORKNEY
ENGLAND
NORTHERN IRELAND
Grampian Highlands
Southern Uplands
North West Highlands
Outer Hebrides
Inner Hebrides
The Minch
Little Minch
North Minch
North Channel
Pentland Firth
Moray Firth
Firth of Forth
Firth of Tay
Firth of Clyde
Firth of Lorn
Solway Firth
Aberdeen
Inverness
Dundee
Perth
Edinburgh
Glasgow
Stirling
Fort William
Oban
Ayr
Dumfries
Carlisle
Belfast
Thurso
Wick
Stornoway
Ben Nevis 1343
Cairn Gorm 1245
Ben Macdhui 1311
Cairn Toul 1292
Lochnagar 1154
Ben Lawers 1214
Ben More 1174
Ben Hope 927
The Cheviot 816
Cross Fell 893
Skiddaw 931
Cumbrian Mts.
Hadrian's Wall
Projection: Conical with two standard parallels.
West from Greenwich
COPYRIGHT. GEORGE PHILIP & SON. LTD.
ft m
3000 1000
1200 400
600 200
300 100
0 0
50 150
100 300
m ft

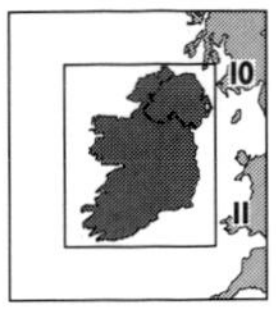

1:2 000 000

Towns underlined in Northern Ireland give their names to the Districts in which they stand

The remaining Districts are:—

1	Fermanagh	5	Castlereagh
2	Moyle	6	Ards
3	Newtownabbey	7	Down
4	North Down	8	Newry & Mourne

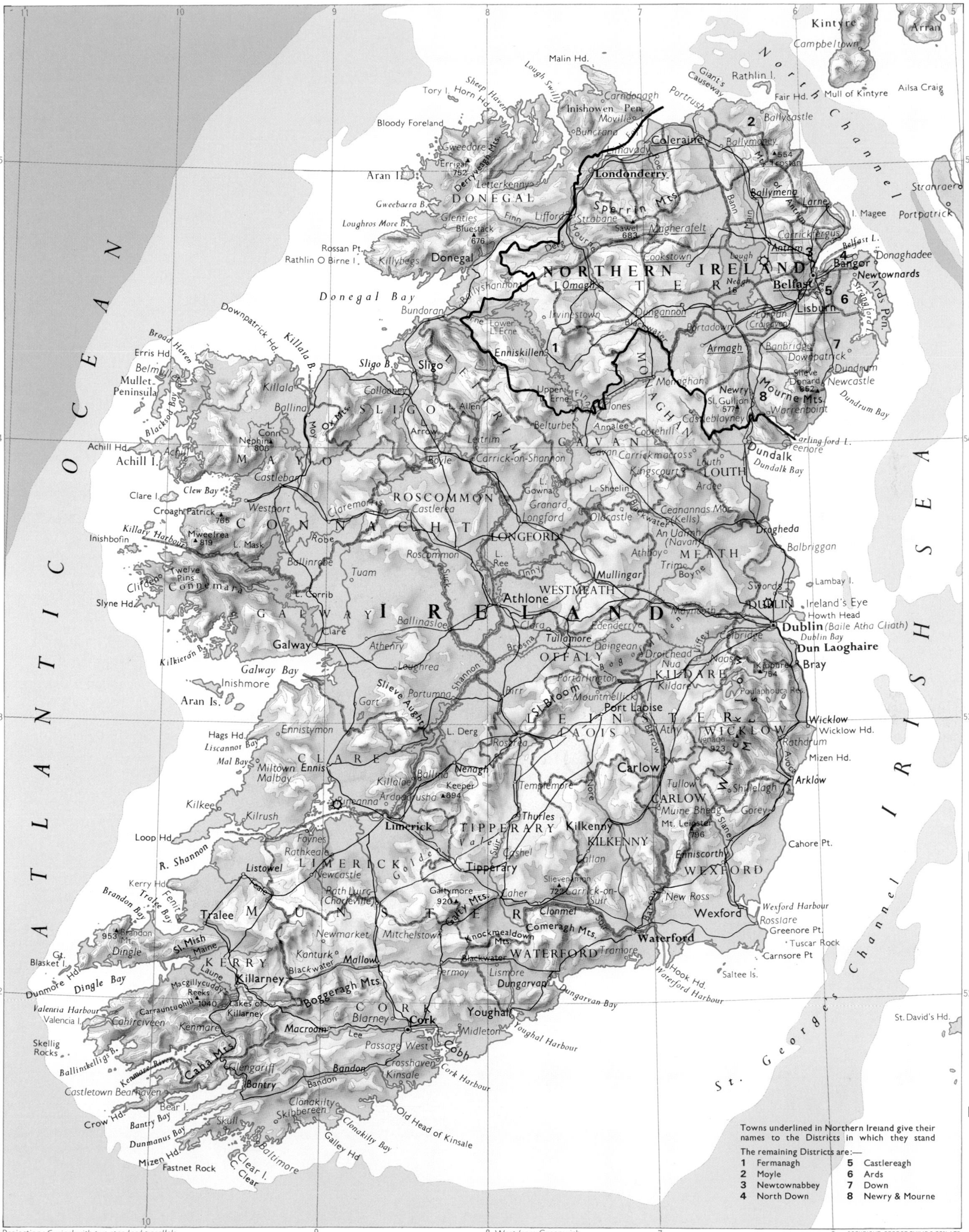

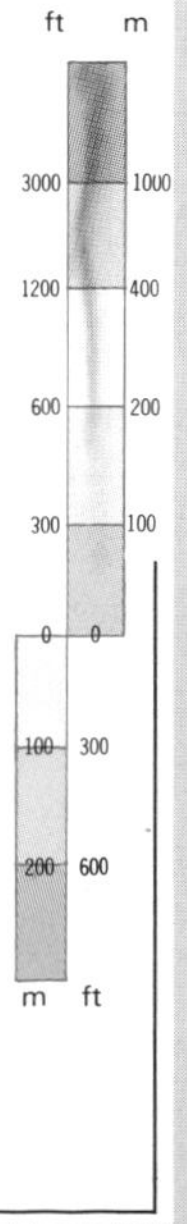

Projection: *Conical with two standard parallels.*

8 West from Greenwich

1 : 5 000 000
miles
km
Projection: Conical with two standard parallels
East from Greenwich
West from Greenwich
CARTOGRAPHY BY PHILIP'S.
COPYRIGHT REED INTERNATIONAL BOOKS LTD
ATLANTIC OCEAN
NORTH SEA
CELTIC SEA
IRISH SEA
English Channel
Bristol Channel
St. George's Channel
North Channel
UNITED KINGDOM
NORTHERN IRELAND
SCOTLAND
ENGLAND
WALES
IRELAND
NORWAY
NETHERLANDS
BELGIUM
FRANCE
Shetland Is.
Orkney Is.
Outer Hebrides
Inner Hebrides
Pentland Firth
Moray Firth
North Minch
North West Highlands
Grampian Mts.
Southern Uplands
Cheviot Hills
Pennines
Cumbrian Mts.
Cambrian Mts.
Cotswold Hills
Exmoor
Dartmoor
The Wash
Str. of Dover
Firth of Clyde
Wicklow Mts.
I. of Man
Anglesey
Cardigan Bay
Isle of Wight
Isles of Scilly
Land's End
Channel Is.
Guernsey
Jersey
Alderney
Sark
Lerwick
Kirkwall
Lewis
Harris
North Uist
South Uist
Skye
Mull
Islay
Jura
Arran
Aberdeen
Inverness
Dundee
Perth
Edinburgh
Glasgow
Stirling
Belfast
Londonderry
Dublin
Cork
Limerick
Galway
Waterford
Newcastle-upon-Tyne
Sunderland
Middlesbrough
Carlisle
Leeds
York
Bradford
Manchester
Liverpool
Sheffield
Kingston upon Hull
Nottingham
Leicester
Derby
Stoke on Trent
BIRMINGHAM
Coventry
Wolverhampton
Norwich
Cambridge
Oxford
Bristol
Cardiff
Swansea
Newport
LONDON
Southampton
Portsmouth
Brighton
Plymouth
Exeter
Bournemouth
Bergen
Stavanger
Haugesund
Texel
Den Helder
Haarlem
's-Gravenhage (Den Haag)
ROTTERDAM
Vlissingen
Oostende
Dunkerque
Calais
Boulogne-sur-Mer
Lille
Dieppe
Le Havre
Rouen
Cherbourg
Caen
Amiens

1:2 500 000
miles
km
NORTH SEA
ENGLAND
NETHERLANDS
BELGIUM
GERMANY
FRANCE
LUXEMBOURG
WADDENEILANDEN
OSTFRIESISCHE INSELN
IJsselmeer
Waddenzee
Noordoost-Polder
Flevoland-polder
FRIESLAND
DRENTHE
OVERIJSSEL
GELDERLAND
UTRECHT
ZEELAND
ANTWERPEN
LIMBURG
HAINAUT
ARDENNES
LORRAINE
SAARLAND
RHEINLAND-PFALZ
Ostfriesland
Plaine de Champagne
AMSTERDAM
ROTTERDAM
'S-GRAVENHAGE
(The Hague)
Hoek van Holland
Haarlem
Leiden
Utrecht
Delft
Dordrecht
Groningen
Leeuwarden
Den Helder
Alkmaar
Hilversum
Amersfoort
Apeldoorn
Zwolle
Deventer
Arnhem
Nijmegen
Enschede
Hengelo
Emmen
Assen
Eindhoven
Tilburg
Breda
's-Hertogenbosch
Maastricht
Vlissingen
Middelburg
Texel
Vlieland
Terschelling
Ameland
Schiermonnikoog
BRUSSEL
(Bruxelles)
Antwerpen
Gent
Brugge
Oostende
Mechelen
Leuven
Liège
Namur
Charleroi
Mons
Kortrijk
Hasselt
Luxembourg
Esch
Lille
Roubaix
Tourcoing
Calais
Dunkerque
Boulogne-sur-Mer
Abbeville
Amiens
Arras
Cambrai
Valenciennes
Douai
St-Quentin
Laon
Reims
Soissons
Compiègne
Beauvais
Épernay
Châlons-en-Champagne
Charleville-Mézières
Sedan
Verdun
Metz
Nancy
Thionville
Strasbourg
PARIS
St-Denis
Versailles
Bremerhaven
Wilhelmshaven
Oldenburg
Emden
Osnabrück
Münster
Bielefeld
Dortmund
Bochum
Essen
Duisburg
Gelsenkirchen
Oberhausen
Recklinghausen
Hagen
Wuppertal
Düsseldorf
Krefeld
Mönchengladbach
Solingen
Remscheid
Köln
Leverkusen
Bonn
Aachen
Düren
Koblenz
Trier
Wiesbaden
Mainz
Kaiserslautern
Saarbrücken
Neunkirchen
Pirmasens
Great Yarmouth
Lowestoft
Ramsgate
Dover
Rhein
Maas
Meuse
Moselle
Schelde
Marne
Aisne
Oise
Somme
Ems
Projection: Conical with two standard parallels
East from Greenwich
COPYRIGHT, GEORGE PHILIP & SON, LTD.

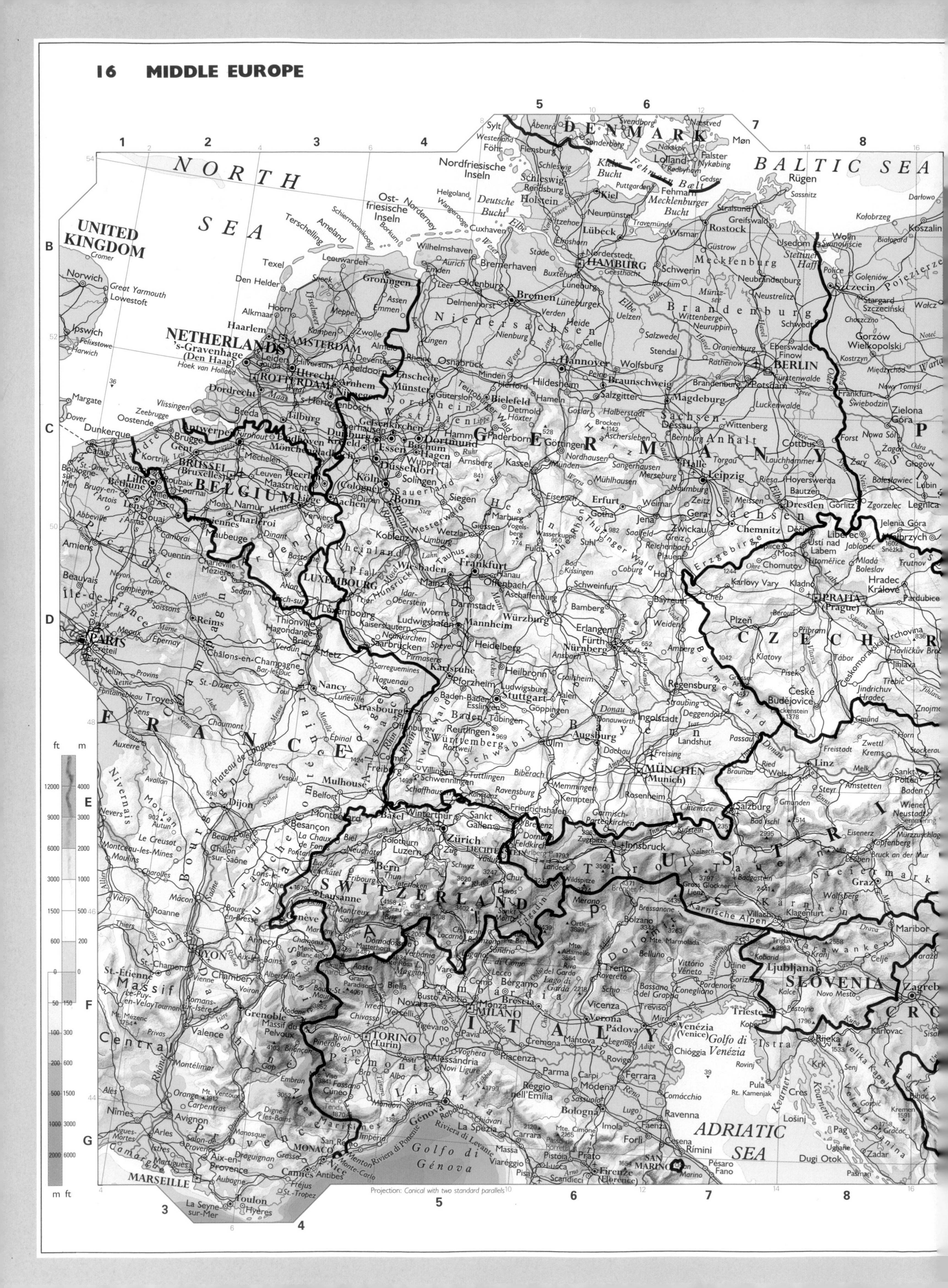
NORTH SEA
BALTIC SEA
ADRIATIC SEA
UNITED KINGDOM
NETHERLANDS
BELGIUM
LUXEMBOURG
GERMANY
DENMARK
FRANCE
SWITZERLAND
LIECHTENSTEIN
AUSTRIA
CZECH R
ITALY
SLOVENIA
MONACO
SAN MARINO
BERLIN
HAMBURG
MÜNCHEN (Munich)
PRAHA (Prague)
PARIS
AMSTERDAM
BRUSSEL (Bruxelles)
LUXEMBOURG
MILANO
TORINO (Turin)
Golfo di Génova
Golfo di Venézia
Deutsche Bucht
Kieler Bucht
Mecklenburger Bucht
Projection: Conical with two standard parallels
ft
m
12000
4000
9000
3000
6000
2000
3000
1000
1500
500
600
200
0
0
50
150
100
300
200
600
500
1500
1000
3000
2000
6000
m
ft
1
2
3
4
5
6
7
8
B
C
D
E
F
G

1:5 000 000

50 0 50 100 miles

50 0 50 100 150 km

9 10 11 12 13 14 15 16

East from Greenwich

9 10 11 12 13 14 15

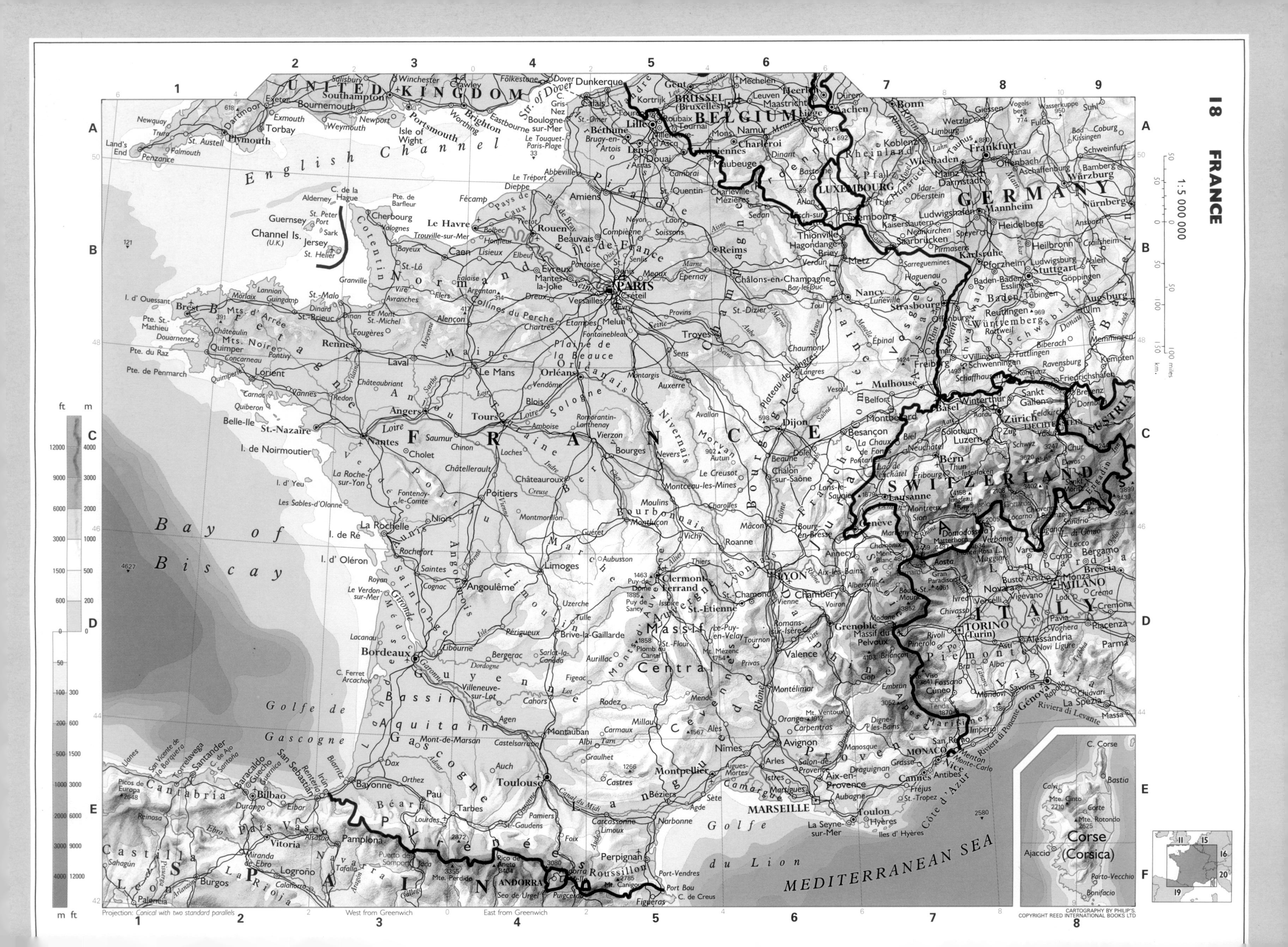
1:5 000 000
50 0 50 100 miles
50 0 50 100 150 km.
UNITED KINGDOM
BELGIUM
LUXEMBOURG
GERMANY
SWITZERLAND
AUSTRIA
ITALY
FRANCE
MONACO
ANDORRA
SPAIN
English Channel
Str. of Dover
Channel Is. (U.K.)
Bay of Biscay
Golfe de Gascogne
Golfe du Lion
MEDITERRANEAN SEA
Côte d'Azur
Corse (Corsica)
PARIS
BRUSSEL (Bruxelles)
LYON
MARSEILLE
MILANO
TORINO (Turin)
Bordeaux
Toulouse
Nantes
Strasbourg
Massif Central
Bassin Aquitain
Normandie
Bretagne
Provence
Pyrénées
Nice
Montpellier
Dijon
Grenoble
Genève
Bern
Zürich
Limoges
Orléans
Rennes
Brest
Le Havre
Rouen
Lille
Reims
Metz
Nancy
Clermont-Ferrand
St.-Étienne
Perpignan
Bilbao
San Sebastián
Pamplona
Ajaccio
Bastia
Projection: Conical with two standard parallels
West from Greenwich
East from Greenwich
CARTOGRAPHY BY PHILIP'S.
COPYRIGHT REED INTERNATIONAL BOOKS LTD

SPAIN AND PORTUGAL
19
1:5 000 000
50 0 50 100 miles
50 0 50 100 150 km
SPAIN
PORTUGAL
FRANCE
ANDORRA
MOROCCO
ALGERIA
MADRID
LISBOA
BARCELONA
ALGER
Bay of Biscay
Golfe du Lion
ATLANTIC OCEAN
MEDITERRANEAN SEA
Islas Baleares
Mallorca
Menorca
Ibiza
Formentera
Golfo de Valencia
Str. of Gibraltar
G. de Cádiz
Pyrénées
Cordillera Cantábrica
Sierra Morena
Sierra Nevada
Sierra de Gredos
Sierra de Guadarrama
Castilla y León
La Mancha
Andalucía
Zaragoza
Sevilla
Málaga
Granada
Córdoba
Bilbao
Valladolid
Valencia
Murcia
Alicante
Porto
Coimbra
Tanger
Oran
Toulouse
Montpellier
Projection: Conical with two standard parallels
West from Greenwich
East from Greenwich
CARTOGRAPHY BY PHILIP'S.
COPYRIGHT REED INTERNATIONAL BOOKS LTD
ft m
6000 2000
3000 1000
1500 500
600 200
0 0
50
100 300
200 600
500 1500
1000 3000
2000 6000
3000 9000
4000 12000
m ft

SWITZERLAND
AUSTRIA
SLOVENIA
CROATIA
FRANCE
ITALIA
ALGERIA
TUNISIA
MALTA
SAN MARINO
VATICAN CITY
LIGURIAN SEA
TYRRHENIAN SEA
ADRIATIC SEA
MEDITER
Golfo di Génova
Golfo di Venézia
Golfo di Táranto
Golfe de Tunis
Golfe de Hammamet
Corse
Sardegna
Sicilia
Ísole Eólie
Ísole Égadi
Ísole Ponziane
Ísole Pelagie (Italy)
Pantelleria (Italy)
Ustica (Italy)
Is. de la Galite (Tunisia)
Elba
Capraia
Pianosa
Montecristo
Giglio
Stromboli
Lipari
Vulcano
Salina
Lampedusa
Lampione
I. Linosa
Gozo
Valletta
Rabat
LYON
Grenoble
Valence
Avignon
MARSEILLE
Toulon
Cannes
Nice
MONACO
Aix-en-Provence
TORINO (Turin)
MILANO
Génova
Bérgamo
Bréscia
Verona
Vicenza
Pádova
Venézia (Venice)
Trieste
Trento
Bolzano
Udine
Parma
Mòdena
Bologna
Ferrara
Ravenna
Rímini
Firenze (Florence)
Pisa
Livorno
La Spezia
Siena
Perúgia
Ancona
Pescara
L'Àquila
ROMA
Latina
NÁPOLI
Salerno
Foggia
Bari
Táranto
Potenza
Cosenza
Catanzaro
Crotone
Règgio di Calábria
Messina
Palermo
Catánia
Siracusa
Agrigento
Marsala
Cágliari
Sassari
Nuoro
Oristano
Ajaccio
Bastia
Ljubljana
Zagreb
Rijeka
Karlovac
Maribor
Klagenfurt
Graz
Villach
Split
Zadar
Šibenik
Banja Luka
Tunis
Bizerte
Sousse
Kairouan
Constantine
Annaba
Skikda
Tébessa
Projection: Conical with two standard parallels

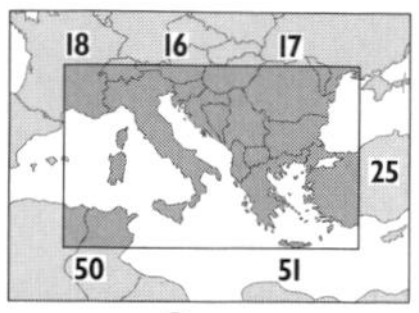
18
16
17
25
50
51

1:5 000 000

50 0 50 100 miles

50 0 50 100 150 km

HUNGARY
ROMANIA
UKRAINE
YUGOSLAVIA
SERBIA
MONTENEGRO
BOSNIA-
ZEGOVINA
ALBANIA
MACEDONIA
BULGARIA
GREECE
TURKEY
BLACK SEA
AEGEAN SEA
IONIAN SEA
RANEAN SEA
Str. of Otranto
Marmara Denizi (Sea of Marmara)
Karadeniz Boğazı (Bosporus)
Çanakkale Boğazı (Dardanelles)
Transilvania
Carpații Meridionali
Vojvodina
Kosovo
Stara Planina
Rhodopi Planina
Makedhonia
Thrace
Walachia
Dobruja
Pindhos
Pelopónnisos
Kikládhes
Dhodhekánisos
Kríti
Mysia
Lydia
Caria
BUCUREȘTI (Bucharest)
BEOGRAD (Belgrade)
SOFIYA
ATHÍNAI (Athens)
İSTANBUL
Skopje
Tirana
Sarajevo
Podgorica
Thessaloníki
İZMİR (Smyrna)
Bursa
Constanța
Varna
Burgas
Plovdiv
Craiova
Timișoara
Novi Sad
Niš
Pátrai
Lárisa
Iráklion
Ródhos (Rhodes)
Kérkira (Corfu)
Lésvos
Khíos
Sámos
Límnos
Thásos
Évvoia
Kíthira
Kárpathos
Zákinthos
Kefallinía
Thermaïkós Kólpos
Saronikós Kólpos
Korinthiakós Kólpos
Kiparissiakós Kólpos
Argolikós Kólpos
Lakonikós Kólpos
Messiniakós Kólpos
Saros Körfezi
Kerme Körfezi
Edremit Körfezi
Dunărea (Danube)
East from Greenwich

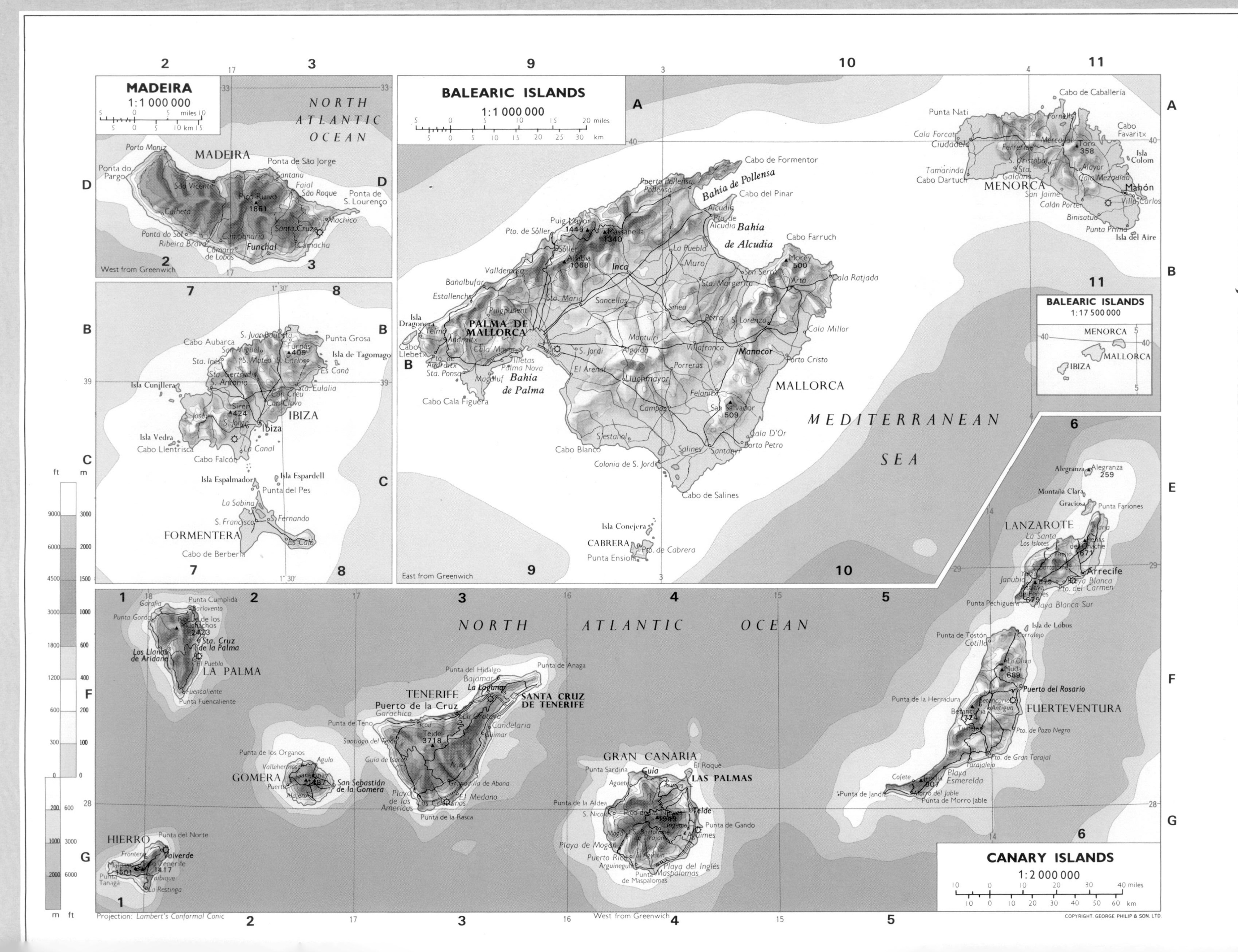
MADEIRA
1:1 000 000
NORTH ATLANTIC OCEAN
Porto Moniz
Ponta do Pargo
São Vicente
Pico Ruivo
1861
Ponta de São Jorge
Santana
Faial
São Roque
Ponta de S. Lourenço
Machico
Calheta
Ponta do Sol
Ribeira Brava
Câmara de Lobos
Funchal
Santa Cruz
Camacha
West from Greenwich
BALEARIC ISLANDS
1:1 000 000
Cabo de Formentor
Bahía de Pollensa
Cabo del Pinar
Puerto Pollensa
Pollensa
Alcudia
Bahía de Alcudia
Cabo Farruch
Puig Mayor
1445
Massanella
1340
Pto. de Sóller
Sóller
Alfabia
1068
Inca
Valldemosa
Bañalbufar
Estallenchs
Isla Dragonera
PALMA DE MALLORCA
Bahía de Palma
Andraitx
Sta. Ponsa
Cabo Cala Figuera
S. Jordi
El Arenal
Lluchmayor
Manacor
Porto Cristo
Cala Millor
Cala Ratjada
Artá
Morey
500
MALLORCA
San Salvador
509
Felanitx
Campos
Santañy
Cala D'Or
Porto Petro
Colonia de S. Jordi
Cabo de Salines
Cabo Blanco
Isla Conejera
CABRERA
Pto. de Cabrera
Punta Ensiola
MEDITERRANEAN SEA
East from Greenwich
Cabo de Caballería
Punta Nati
Cala Forcat
Ciudadela
Cabo Dartuch
MENORCA
Toro
358
Alayor
Cabo Favaritx
Isla Colom
Mahón
Villa Carlos
Punta Prima
Isla del Aire
BALEARIC ISLANDS
1:17 500 000
MENORCA
MALLORCA
IBIZA
Cabo Aubarca
Punta Grosa
Isla de Tagomago
S. Juan Bautista
Furnás
409
Sta. Eulalia
Isla Cunillera
IBIZA
Ibiza
Isla Vedra
Cabo Llentrisca
Cabo Falcón
Isla Espalmador
Isla Espardell
Punta del Pes
La Sabina
FORMENTERA
Cabo de Berbería
NORTH ATLANTIC OCEAN
LA PALMA
Punta Cumplida
Roque de los Muchachos
2423
Sta. Cruz de la Palma
Los Llanos de Aridane
Punta Fuencaliente
HIERRO
Punta del Norte
Valverde
1501
TENERIFE
Puerto de la Cruz
La Laguna
SANTA CRUZ DE TENERIFE
Punta de Anaga
Punta de Teno
Teide
3718
Candelaria
El Medano
Punta de la Rasca
GOMERA
San Sebastián de la Gomera
Punta de los Organos
GRAN CANARIA
LAS PALMAS
Guía
Telde
Punta de Gando
Playa del Inglés
Punta Maspalomas
LANZAROTE
Alegranza
259
Montaña Clara
Graciosa
Punta Fariones
Arrecife
Playa Blanca Sur
Isla de Lobos
FUERTEVENTURA
Puerto del Rosario
Corralejo
Punta de la Herradura
Pto. de Pozo Negro
Pto. de Gran Tarajal
Playa Esmerelda
Punta de Morro Jable
Punta de Jandía
CANARY ISLANDS
1:2 000 000
West from Greenwich
Projection: Lambert's Conformal Conic
COPYRIGHT. GEORGE PHILIP & SON. LTD.

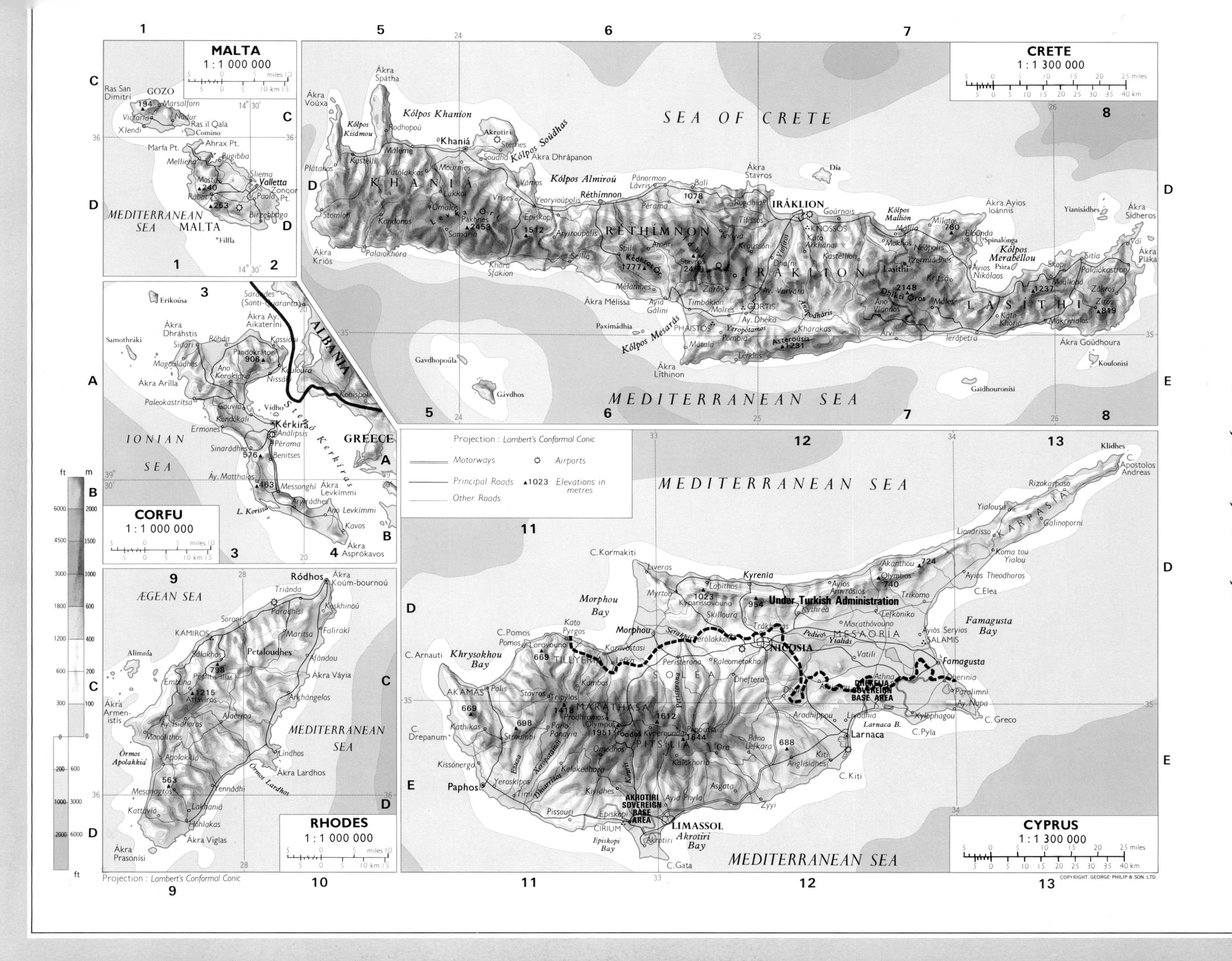

MALTA
1 : 1 000 000
GOZO
MEDITERRANEAN SEA
Valletta
CRETE
1 : 1 300 000
SEA OF CRETE
MEDITERRANEAN SEA
KHANIÁ
RÉTHÍMNON
IRÁKLION
LASÍTHI
Khaniá
Réthimnon
Iráklion
CORFU
1 : 1 000 000
IONIAN SEA
ALBANIA
GREECE
Kérkira
RHODES
1 : 1 000 000
ÆGEAN SEA
MEDITERRANEAN SEA
Ródhos
CYPRUS
1 : 1 300 000
MEDITERRANEAN SEA
Under Turkish Administration
NICOSIA
Kyrenia
Famagusta
Larnaca
LIMASSOL
Paphos
Morphou
AKROTIRI SOVEREIGN BASE AREA
DHEKELIA SOVEREIGN BASE AREA
Projection : Lambert's Conformal Conic
Motorways
Principal Roads
Other Roads
Airports
Elevations in metres
COPYRIGHT. GEORGE PHILIP & SON, LTD.

P-ov. Rybachiy
Kolguyev
M. Kanin Nos
P-ov. Kanin
Cheshskaya Guba
Ostrov Morzhovets
Pechorskaya Guba
Malozemelskaya Tundra
Bolshezemelskaya Tundra
Arctic Circle
Timanskiy Kryazh
Kolskiy Poluostrov
Kandalakshskiy Zaliv
Beloye More (White Sea)
Dvinskaya Guba
Onezhskaya Guba
Lappland
SWEDEN
NORWAY
FINLAND
KARELIA
Gulf of Bothnia
G. of Finland
Gulf of Riga
Hiiumaa (Dagö)
Saaremaa (Ösel)
Åland (Ahvenanmaa)
ESTONIA
LATVIA
LITHUANIA
BELARUS
POLAND
RUSSIA
Severnyye Uvaly
Valdayskaya Vozvyshennost
UDMURTIA
MARI EL
CHUVASHIA
TATARSTAN
MORDVINIA
BASHKORTOSTAN
Murmansk
Severomorsk
Kola
Monchegorsk
Oz. Imandra
Apatity
Kirovsk
Kandalaksha
Arkhangelsk
Severodvinsk (Molotovsk)
Onega
Mezen
Kamenka
Leshukonskoye
Naryan-Mar
Ust Tsilma
Pechora
Ukhta
Sosnogorsk
Vorkuta
Inta
Syktyvkar
Kotlas
Petrozavodsk
Ladozhskoye Ozero
Onezhskoye Ozero
Oz. Beloye
Vyborg
Kronshtadt
ST. PETERBURG (Leningrad)
Kolpino
Novgorod
Oz. Ilmen
Pskov
Tartu
Narva
Tallinn
Helsinki
Stockholm
Turku
Tampere
Pori
Oulu
Luleå
Kemi
Tornio
Riga
Daugavpils
Liepaja
Klaipeda
Kaliningrad (Russia)
Kaunas
Vilnius
Minsk
Barysaw
Vitsyebsk
Mahilyow
Homyel
Brest
Pinsk
Hrodna
WARSZAWA (Warsaw)
Bialystok
Lublin
Smolensk
Bryansk
Orel
Kursk
Voronezh
Lipetsk
Tambov
Tula
Kaluga
Novomoskovsk
Ryazan
Kolomna
MOSKVA (Moscow)
Tver
Rybinsk
Yaroslavl
Kostroma
Ivanovo
Vladimir
Vologda
Cherepovets
Rybinskoye Vdkhr.
Nizhniy Novgorod (Gorkiy)
Dzerzhinsk
Cheboksary
Yoshkar Ola
Kazan
Naberezhnyye Chelny
Nizhnekamsk
Izhevsk (Ustinov)
Votkinsk
Kirov (Vyatka)
Glazov
Perm
Solikamsk
Berezniki
Nizhniy Tagil
Yekaterinburg
Pervouralsk
Zlatoust
Miass
Ufa
Sterlitamak
Salavat
Magnitogorsk
Beloretsk
Orenburg
Orsk
Novotroitsk
Samara (Kuybyshev)
Tolyatti
Novokuybyshevsk
Syzran
Simbirsk
Saransk
Penza
Saratov
Engels
Balakovo
Arzamas
Murom
Kovrov
Serpukhov
Podolsk
Zelenograd
Sergiyev Posad
Kimry
Vyshniy Volochek
Rzhev
Velikiye Luki
Polatsk
Orsha
Kamskoye Vdkhr.
Gorkovskoye Vdkhr.
Kuybyshevskoye Vdkhr.
Votkinskoye Vdkhr.

1:10 000 000
100 50 0 50 100 150 200 miles
100 0 100 200 300 km
UKRAINE
ROMANIA
MOLDOVA
BULGARIA
TURKEY
GEORGIA
ARMENIA
AZERBAIJAN
KAZAKSTAN
TURKMENISTAN
KALMYKIA
ABKHAZIA
DAGESTAN
SYRIA
IRAQ
IRAN
LEBANON
CYPRUS
BLACK SEA
CASPIAN SEA
MEDITERRANEAN SEA
Levant
Sea of Azov
Caspian Depression
Kirgiziya Steppe
Caucasus Mountains
Anadolu Dağlari
Kuzey Anadolu Dağlari
Toros Dağlari
Kurdistan
Alborz
Bādiyat ash Sham
Gara-bogazköl Aylagy
KYYIV (Kiev)
KHARKIV (Kharkov)
Volgograd (Stalingrad)
Astrakhan
Rostov
Donetsk
Dnipropetrovsk
Zaporizhzhya
Odesa
Chişinău (Kishinev)
BUCUREŞTI (Bucharest)
Simferopol
Sevastopol
Krasnodar
Stavropol
Groznyy
Makhachkala
Tbilisi
Yerevan
BAKI (Baku)
Istanbul
Ankara
Izmir (Smyrna)
Adana
Halab
Dimashq (Damascus)
Bayrūt (Beirut)
Baghdād
Al Mawşil
Tabrīz
TEHRĀN
Elbrus 5642
Kazbek 5047
Ararat 5165
Damāvand 5604
1. Crimea (Ukr.)
2. Adygea (Russ.)
3. Karachey-Cherkessia (Russ.)
4. Kabardino-Balkaria (Russ.)
5. North Ossetia (Russ.)
6. Ingushetia (Russ.)
7. Chechenia (Russ.)
8. Naxçivan (Azer.)
ft m
12 000 4000
6000 2000
3000 1000
1200 400
600 200
0 0
200–600
1000–3000
2000–6000
4000–12 000
m ft
Projection: Conical with two standard parallels
Division between Greeks and Turks in Cyprus; Turks to the North.
East from Greenwich
COPYRIGHT, GEORGE PHILIP & SON, LTD.

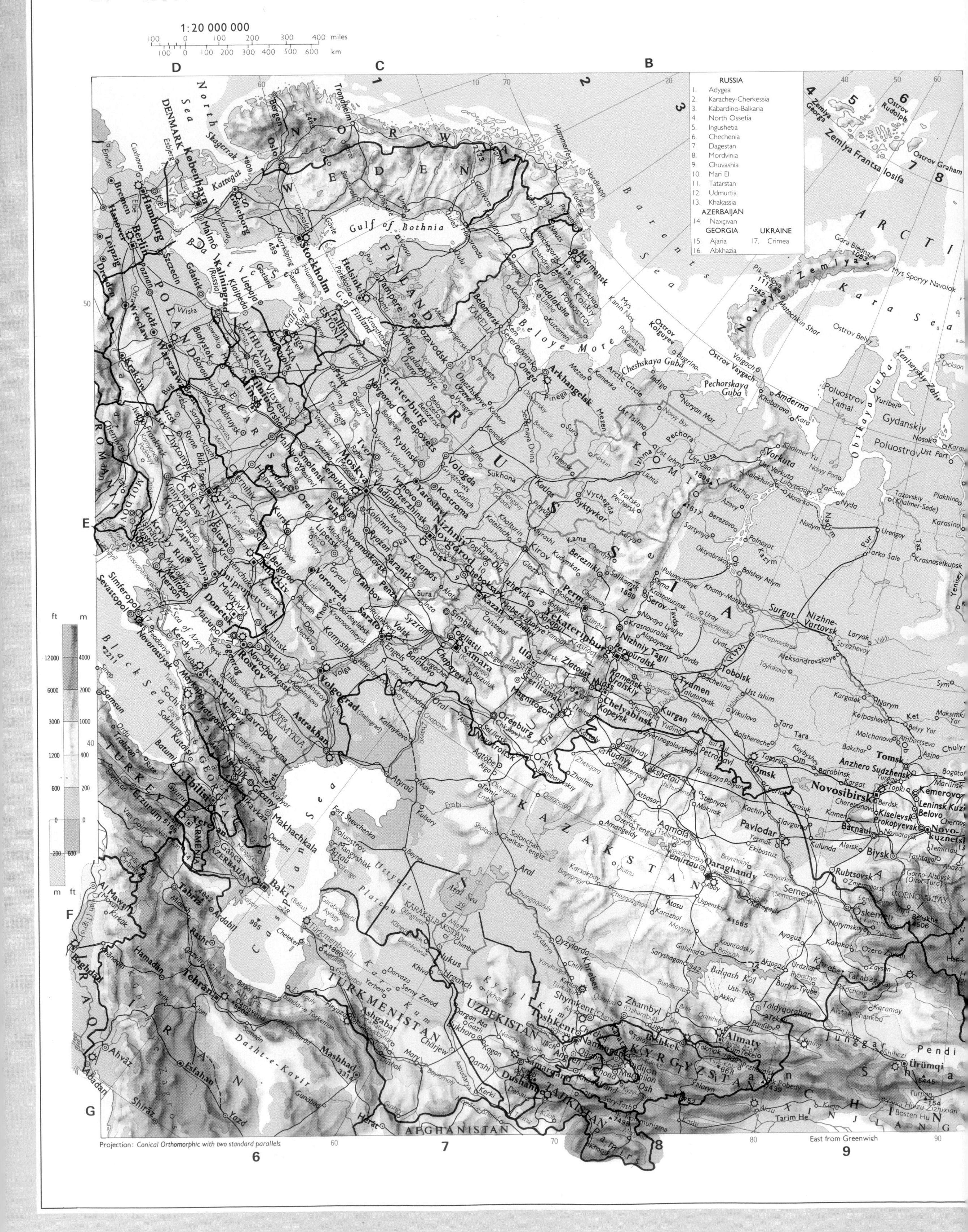
1:20 000 000
miles
km
RUSSIA
1. Adygea
2. Karachey-Cherkessia
3. Kabardino-Balkaria
4. North Ossetia
5. Ingushetia
6. Chechenia
7. Dagestan
8. Mordvinia
9. Chuvashia
10. Mari El
11. Tatarstan
12. Udmurtia
13. Khakassia
AZERBAIJAN
14. Naxçivan
GEORGIA
15. Ajaria
16. Abkhazia
UKRAINE
17. Crimea
Zemlya Frantsa Iosifa
Barents Sea
Kara Sea
Gulf of Bothnia
North Sea
Black Sea
Caspian Sea
Aral Sea
Beloye More
NORWAY
SWEDEN
FINLAND
ESTONIA
LATVIA
LITHUANIA
POLAND
BELARUS
UKRAINE
MOLDOVA
ROMANIA
TURKEY
GEORGIA
ARMENIA
AZERBAIJAN
IRAN
IRAQ
TURKMENISTAN
UZBEKISTAN
KAZAKSTAN
KYRGYZSTAN
TAJIKISTAN
AFGHANISTAN
RUSSIA
Moskva
St. Peterburg
Stockholm
Helsinki
Oslo
København
Warszawa
Kyiv
Minsk
Murmansk
Arkhangelsk
Novosibirsk
Omsk
Yekaterinburg
Volgograd
Astrakhan
Baku
Tbilisi
Tehran
Ashgabat
Toshkent
Almaty
Bishkek
Dushanbe
Qaraghandy
Projection: Conical Orthomorphic with two standard parallels
East from Greenwich

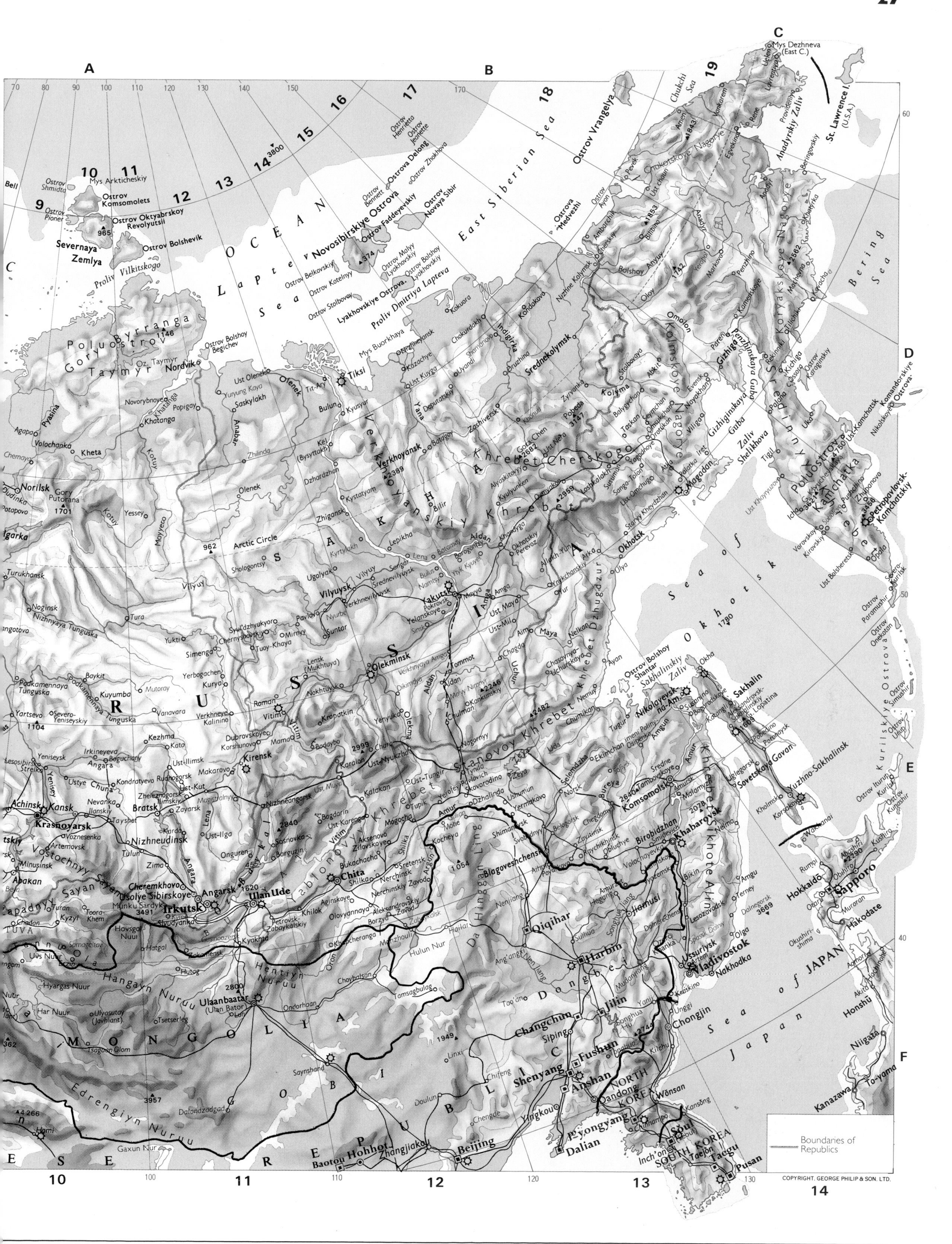
A
B
C
D
E
F
9
10
11
12
13
14
15
16
17
18
19
OCEAN
Laptev Sea
East Siberian Sea
Chukchi Sea
Bering Sea
Sea of Okhotsk
Sea of JAPAN
Japan
Severnaya Zemlya
Ostrov Komsomolets
Ostrov Oktyabrskoy Revolyutsii
Ostrov Bolshevik
Proliv Vilkitskogo
Poluostrov Taymyr
Gory Byrranga
Novosibirskiye Ostrova
Ostrov Kotelnyy
Ostrov Faddeyevskiy
Ostrov Novaya Sibir
Lyakhovskiye Ostrova
Proliv Dmitriya Lapteva
Ostrov Vrangelya
Ostrova Medvezhi
Mys Dezhneva (East C.)
St. Lawrence I. (U.S.A.)
Anadyrskiy Zaliv
Chukotskoye Nagorye
Koryakskoye Nagorye
Kolymskoye Nagorye
Khrebet Cherskogo
Verkhoyanskiy Khrebet
Poluostrov Kamchatka
Petropavlovsk-Kamchatskiy
Komandorskiye Ostrova
Kurilskiye Ostrova
Sakhalin
Zaliv Shelikhova
Penzhinskaya Guba
Gizhiginskaya Guba
Magadan
Okhotsk
Yakutsk
Verkhoyansk
Tiksi
Norilsk
Vilyuysk
Olekminsk
Arctic Circle
Stanovoy Khrebet
Khrebet Dzhugdzhur
Sikhote Alin
Khabarovsk
Komsomolsk
Nikolayevsk-na-Amure
Vladivostok
Nakhodka
Sovetskaya Gavan
Yuzhno-Sakhalinsk
Hokkaido
Sapporo
Hakodate
Honshu
Niigata
Kanazawa
Irkutsk
Angarsk
Ulan Ude
Chita
Bratsk
Kirensk
Vitim
Krasnoyarsk
Achinsk
Kansk
Nizhneudinsk
Vostochnyy Sayan
Zapadnyy Sayan
TUVA
Amur
Lena
Blagoveshchensk
Qiqihar
Harbin
Jiamusi
Jilin
Changchun
Shenyang
Fushun
Anshan
Dalian
Beijing
Baotou
Hohhot
Zhangjiakou
Ulaanbaatar (Ulan Bator)
MONGOLIA
Hangayn Nuruu
Hentiyn Nuruu
Edrengiyn Nuruu
Da Hinggan Ling
Dongbei
NORTH KOREA
SOUTH KOREA
P'yongyang
Soul
Taegu
Pusan
Chongjin
RUSSIA
SAKHA
REPUBLIC
GOBI
Boundaries of Republics
70
80
90
100
110
120
130
140
150
170
40
50
60
10
11
12
13
14
COPYRIGHT, GEORGE PHILIP & SON, LTD.

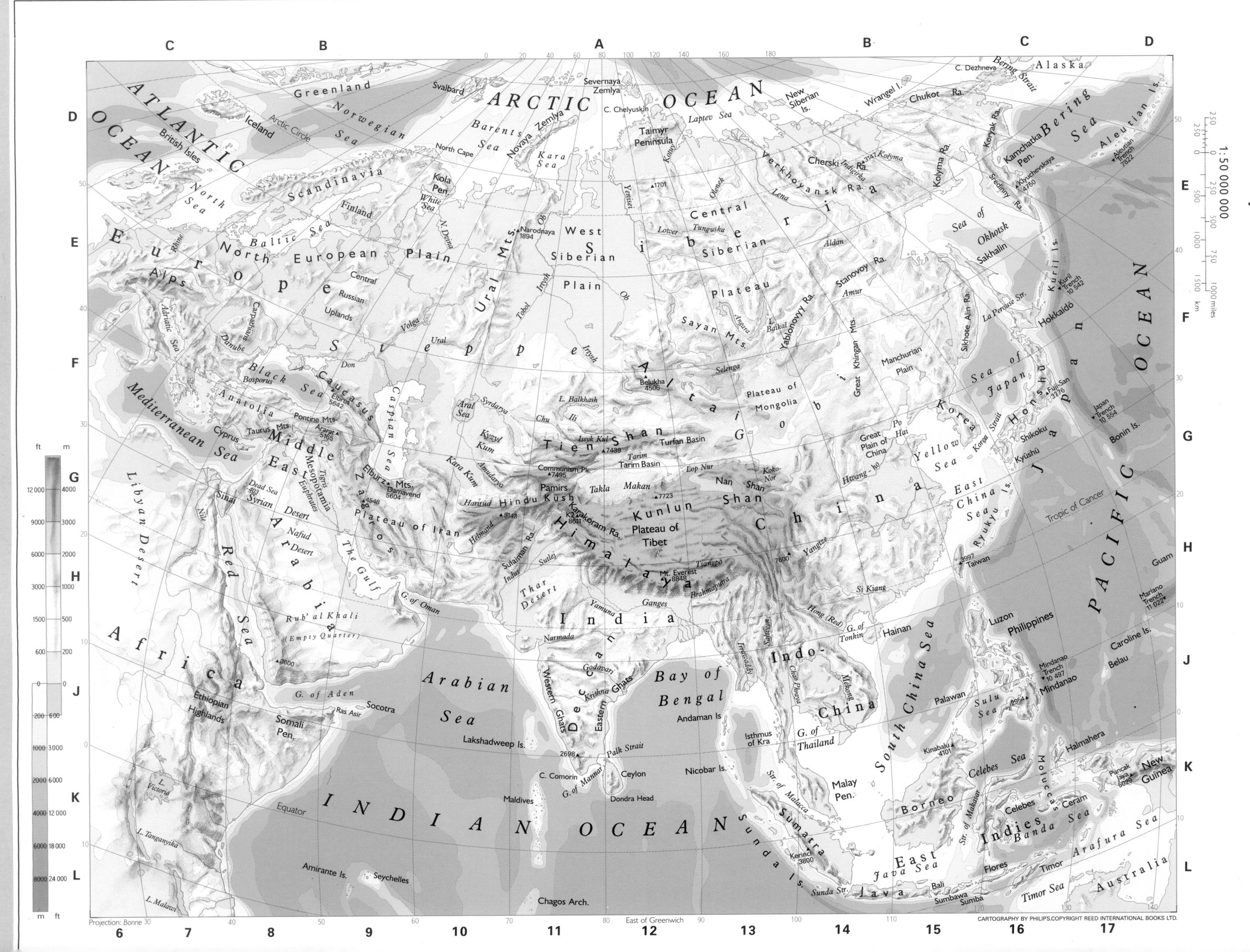
1:50 000 000
250 0 250 500 750 1000 miles
250 0 500 1000 1500 km
ARCTIC OCEAN
PACIFIC OCEAN
INDIAN OCEAN
ATLANTIC OCEAN
Bering Sea
Sea of Okhotsk
Sea of Japan
Yellow Sea
East China Sea
South China Sea
Bay of Bengal
Arabian Sea
Caspian Sea
Black Sea
Mediterranean Sea
Red Sea
Barents Sea
Kara Sea
Laptev Sea
Norwegian Sea
North Sea
Baltic Sea
Adriatic Sea
Sulu Sea
Celebes Sea
Banda Sea
Arafura Sea
Java Sea
Timor Sea
Molucca Sea
The Gulf
G. of Oman
G. of Aden
G. of Thailand
G. of Tonkin
G. of Mannar
Str. of Malacca
Str. of Makasar
Palk Strait
Bering Strait
Korea Strait
La Perouse Str.
Sunda Str.
Siberia
West Siberian Plain
Central Siberian Plateau
Ural Mts.
Narodnaya 1894
Verkhoyansk Ra.
Cherski Ra.
Kolyma Ra.
Chukot Ra.
Koryak Ra.
Sredinny Ra.
Klyuchevskaya 4750
Stanovoy Ra.
Yablonovyy Ra.
Sayan Mts.
Altai
Belukha 4506
Great Khingan Mts.
Sikhote Alin Ra.
Plateau of Mongolia
Gobi
Manchurian Plain
Great Plain of China
China
Tien Shan
Kunlun Shan
Nan Shan
Plateau of Tibet
Himalaya
Mt. Everest 8848
Karakoram Ra.
K2 8611
Hindu Kush
Pamirs
Communism Pk. 7495
Takla Makan
Tarim Basin
Turfan Basin
Lop Nur
Koko Nor
Kara Kum
Kyzyl Kum
Aral Sea
L. Balkhash
Issyk Kul
Steppe
Elburz Mts.
Damavend 5604
Plateau of Iran
Zagros
Caucasus
Elbrus 5642
Ararat 5165
Pontine Mts.
Taurus Mts.
Anatolia
Bosporus
Middle East
Mesopotamia
Dead Sea 403
Syrian Desert
Nafud Desert
Arabia
Rub' al Khali (Empty Quarter)
India
Thar Desert
Sulaiman Ra.
Western Ghats
Eastern Ghats
Deccan
Ceylon
Dondra Head
Indo-China
Malay Pen.
Isthmus of Kra
Borneo
Kinabalu 4101
Sumatra
Kerinci 3800
Java
Bali
Sumbawa
Sumba
Flores
Timor
Celebes
Ceram
Halmahera
East Indies
Sunda Is.
Philippines
Luzon
Mindanao
Palawan
Mindanao Trench 10 497
Taiwan
Hainan
Korea
Japan
Honshu
Fuji-San 3776
Hokkaido
Kyushu
Shikoku
Sakhalin
Kamchatka Pen.
Kuril Is.
Kuril Trench 10 542
Japan Trench 10 554
Aleutian Is.
Aleutian Trench 7822
Ryukyu Is.
Bonin Is.
Guam
Mariano Trench 11 022
Caroline Is.
Belau
New Guinea
Puncak Jaya 5029
Australia
Alaska
Europe
Africa
North European Plain
Central Russian Uplands
Scandinavia
Finland
Kola Pen.
White Sea
North Cape
Greenland
Iceland
British Isles
Alps
Carpathians
Cyprus
Sinai
Nile
Libyan Desert
Ethiopian Highlands
Somali Pen.
Ras Asir
L. Victoria
L. Tanganyika
L. Malawi
Maldives
Lakshadweep Is.
Chagos Arch.
Seychelles
Amirante Is.
Socotra
Andaman Is.
Nicobar Is.
C. Comorin
Novaya Zemlya
Severnaya Zemlya
New Siberian Is.
Wrangel I.
Taimyr Peninsula
C. Chelyuskin
C. Dezhneva
Svalbard
Ob
Yenisei
Lena
Irtysh
Tobol
Amur
Indus
Ganges
Yamuna
Narmada
Godavari
Krishna
Brahmaputra
Tsangpo
Yangtze
Hwang-ho
Si Kiang
Hong (Red)
Mekong
Chao Phraya
Salween
Irrawaddy
Tigris
Euphrates
Volga
Ural
Don
Danube
Rhine
Arctic Circle
Tropic of Cancer
Equator
East of Greenwich
Projection: Bonne
CARTOGRAPHY BY PHILIPS.COPYRIGHT REED INTERNATIONAL BOOKS LTD.
ft m
12 000 4000
9000 3000
6000 2000
3000 1000
1500 500
600 200
0 0
200 600
1000 3000
2000 6000
4000 12 000
6000 18 000
8000 24 000
m ft

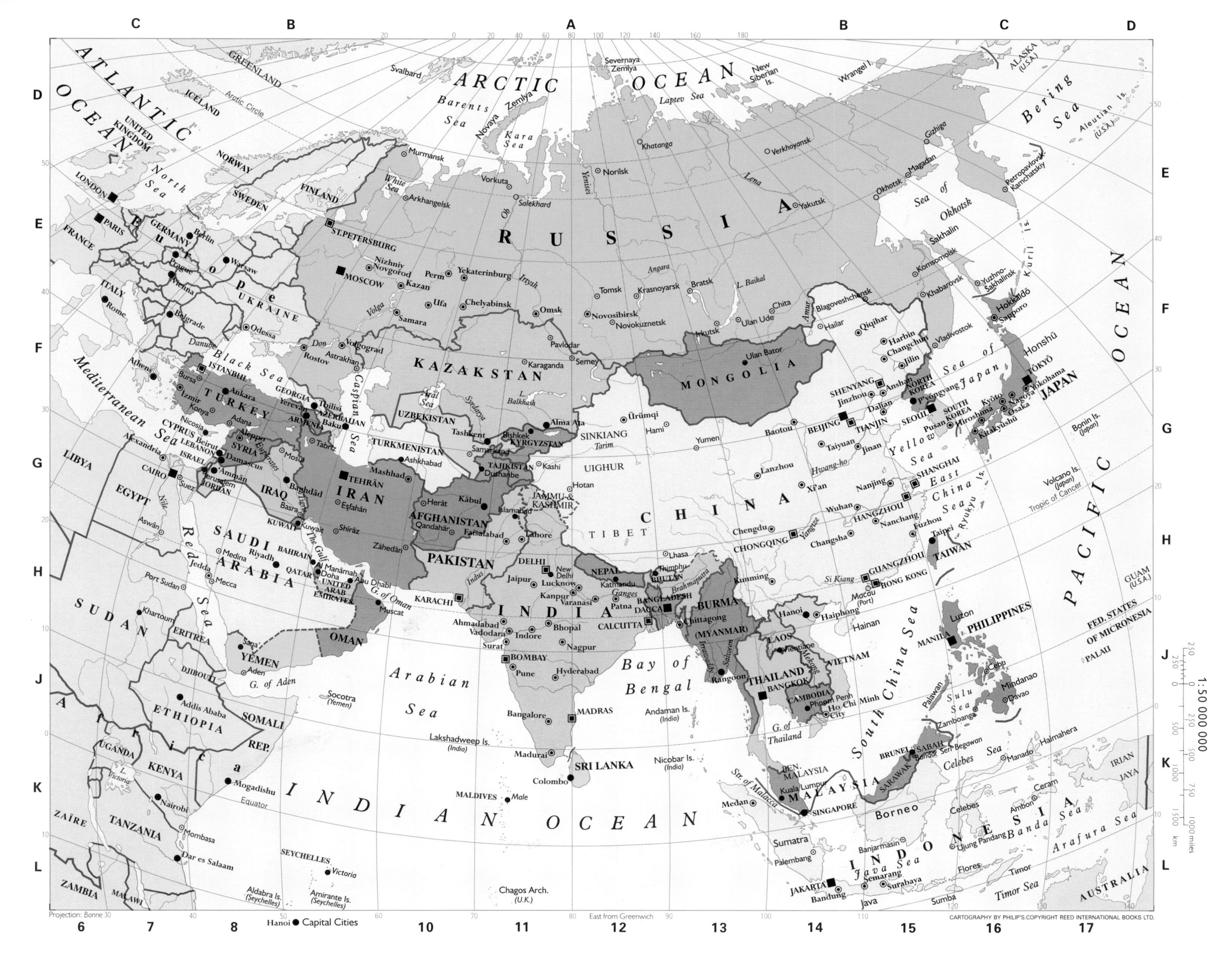
1:50 000 000
250 0 250 500 750 1000 miles
250 0 500 1000 1500 km
Projection: Bonne 30
Hanoi ● Capital Cities
East from Greenwich
CARTOGRAPHY BY PHILIP'S.COPYRIGHT REED INTERNATIONAL BOOKS LTD.
ARCTIC OCEAN
PACIFIC OCEAN
INDIAN OCEAN
ATLANTIC OCEAN
RUSSIA
CHINA
MONGOLIA
KAZAKSTAN
INDIA
PAKISTAN
AFGHANISTAN
IRAN
IRAQ
SAUDI ARABIA
TURKEY
JAPAN
INDONESIA
PHILIPPINES
THAILAND
BURMA (MYANMAR)
Bay of Bengal
Arabian Sea
South China Sea
Caspian Sea
Black Sea
Mediterranean Sea
Red Sea

SEA OF OKHOTSK
SEA OF JAPAN
RUSSIA
CHINA
NORTH KOREA
HOKKAIDO
TŌHOKU
CHŪBU
Sakhalin
La Pérouse Strait
(Sōya-Kaikyō)
Sōya-Misaki
Wakkanai
Rebun-Tō
Rishiri-Tō
Teshio
Esashi
Ōmu
Mombetsu
Abashiri-Wan
Shiretoko-Misaki
Nemuro-Kaikyō
Ostrov Kunashir
Nemuro
Kushiro
Akkeshi
Kitami
Asahigawa
Rumoi
Ishikari-Wan
(Otaru-Wan)
Otaru
SAPPORO
Iwamizawa
Obihiro
Tomakomai
Hiroo
Erimo-Misaki
Muroran
Uchiura-Wan
Okushiri-Tō
Hakodate
Tsugaru-Kaikyō
Mutsu-Wan
Aomori
Hachinohe
Hirosaki
Noshiro
Oga-Hantō
Akita
Morioka
Miyako
Kamaishi
Kesennuma
Sakata
Tsuruoka
Ishinomaki
Shiogama
Sendai
Sendai-Wan
Yamagata
Sado
Niigata
Fukushima
Svetlaya
Amgu
Terney
Plastun
Dalnegorsk
Kavalerovo
Olga
Bikin
Dalnerechensk
Lesozavodsk
Ozero Khanka
Spassk-Dalniy
Pogranichnyy
Ussuriysk
Artem
Vladivostok
Nakhodka
Preobrazheniye
Zaliv Petra Velikogo
Jiamusi
Shuangyashan
Najin
Chongjin

1:5 000 000

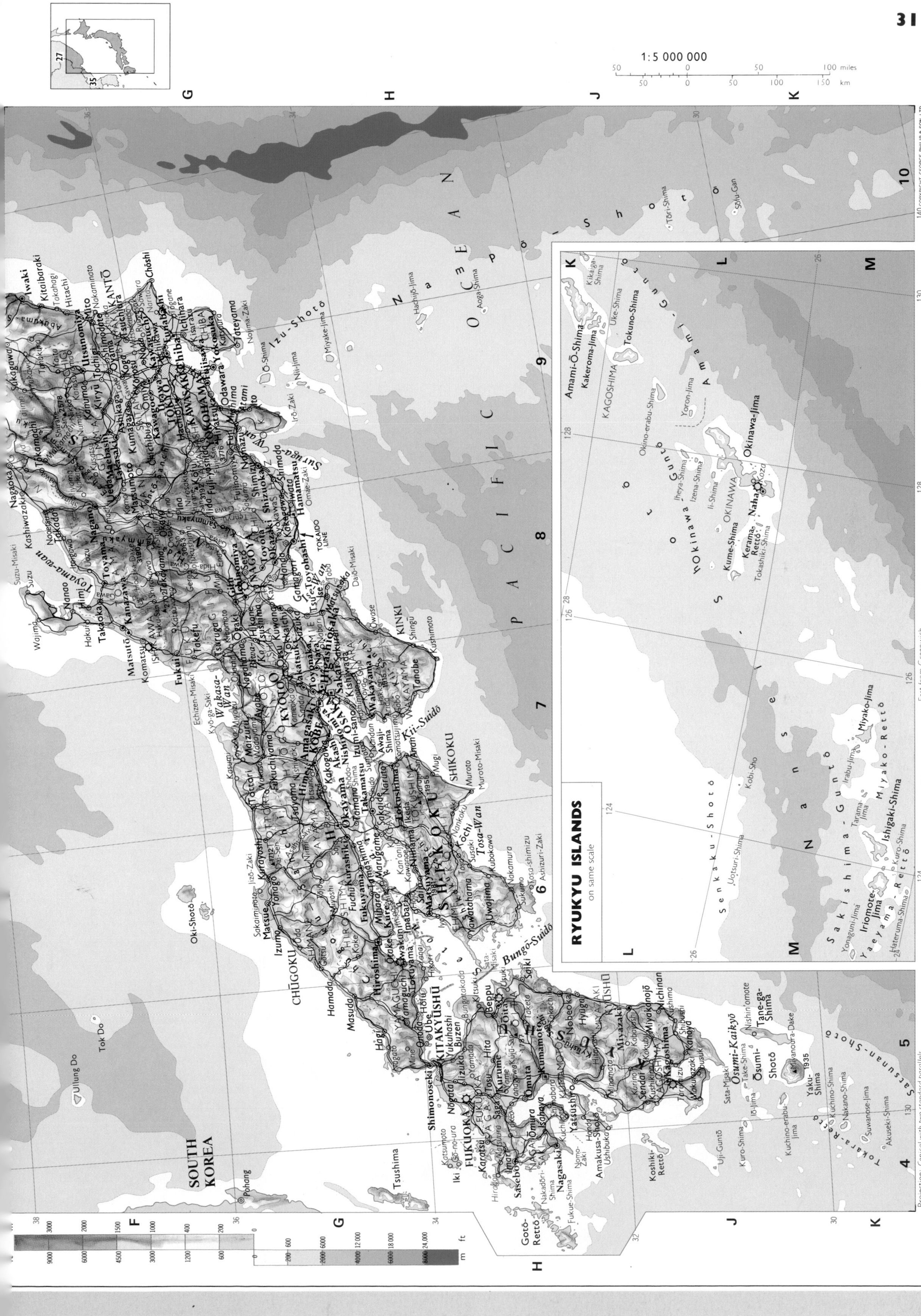

50 0 50 100 miles

50 0 50 100 150 km

m ft

9000 3000
6000 2000
4500 1500
3000 1000
1200 400
600 200
0 0
200 600
2000 6000
4000 12 000
6000 18 000
8000 24 000

SOUTH KOREA

Ullung Do

Tok Do

Oki-Shotō

Tsushima

CHŪGOKU

SHIKOKU

KINKI

KANTŌ

KYŪSHŪ

PACIFIC OCEAN

Nansei-shotō

Izu-Shotō

Nampō-shotō

Bungo-Suidō

Kii-Suidō

Tosa-Wan

Suruga-Wan

Ise-Wan

Wakasa-Wan

Toyama-wan

Ōsumi-Kaikyō

Ōsumi-Shotō

Tokara-Rettō

Satsunan-Shotō

TOKYO

YOKOHAMA

KAWASAKI

NAGOYA

KYOTO

OSAKA

KOBE

KITAKYŪSHŪ

FUKUOKA

Hiroshima

Okayama

Kagoshima

Nagasaki

Kumamoto

Matsuyama

TOKAIDO LINE

RYUKYU ISLANDS

on same scale

Amami-Ō-Shima

Kakeroma-Jima

Tokuno-Shima

Okinawa-Jima

Naha

Kume-Shima

Kerama-Rettō

Okinawa-Guntō

Amami-Guntō

Senkaku-Shotō

Sakishima-Guntō

Miyako-Jima

Miyako-Rettō

Ishigaki-Shima

Iriomote-Jima

Yaeyama-Rettō

Projection: *Conical with two standard parallels*

East from Greenwich

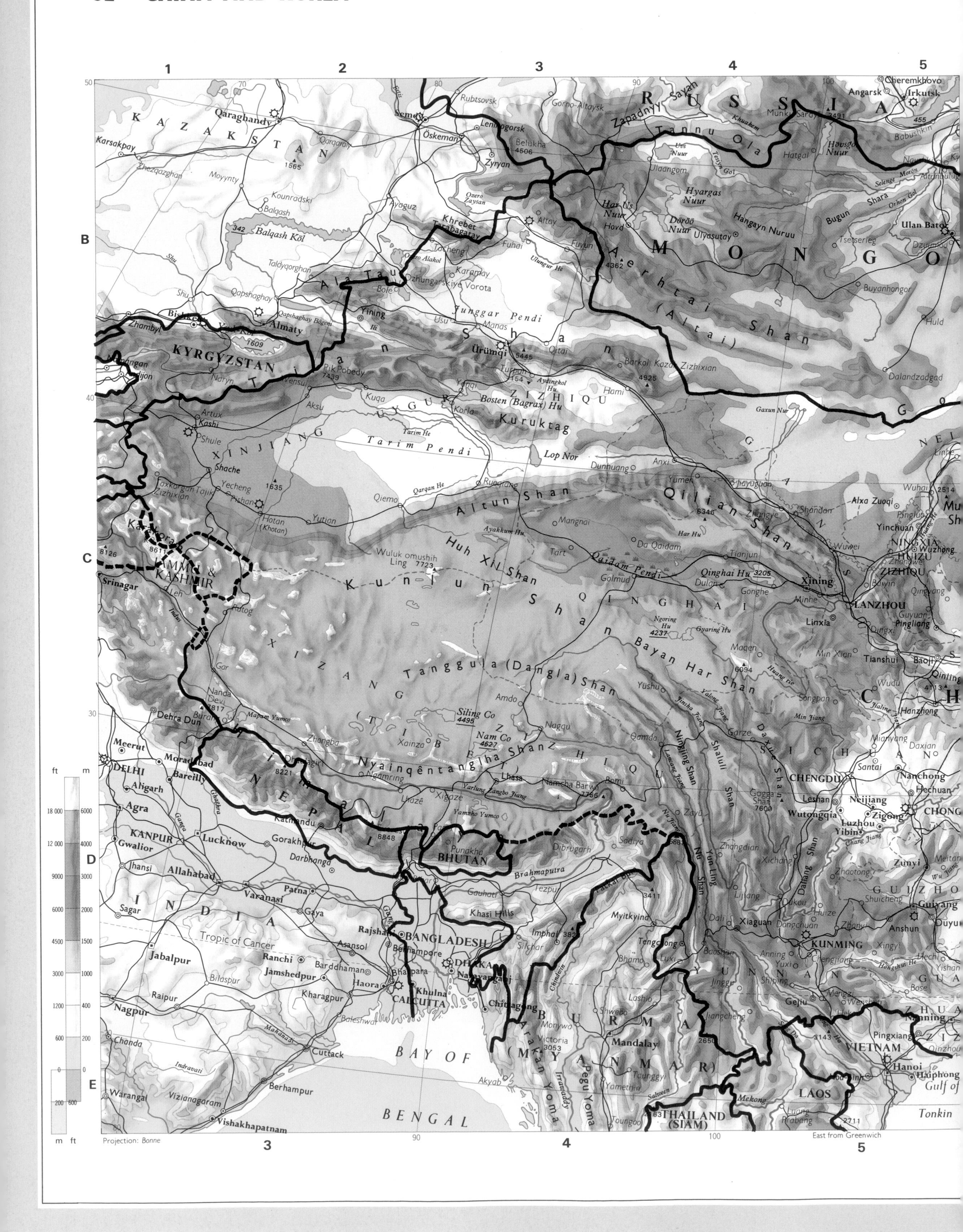

1
2
3
4
5
B
C
D
E
KAZAKSTAN
Qaraghandy
Karsakpay
Zhezqazghan
Moyynty
Kounradski
Balqash
Balqash Köl
Taldyqorghan
Qapshaghay
Shu
Bishkek
Zhambyl
Almaty
Qapshaghay Bögeni
KYRGYZSTAN
Andijon
Naryn
Semey
Öskeman
Rubtsovsk
Leninogorsk
Zyryan
Belukha
4506
Ayaguz
Ozero Zaysan
Khrebet Tarabagatay
Tacheng
Ozero Alakol
Ala Tau
Bole
Karamay
Dzhungarskiye Vorota
Yining
Junggar Pendi
Usu
Manas
Ürümqi
5445
Turpan
Ayding Hu
Fuhai
Altay
Ulungur He
Fuyun
Gorno-Altaysk
Tannu Ola
Zapadnyy Sayan
RUSSIA
Uvs Nuur
Ulaangom
Hyargas Nuur
Har Us Nuur
Hovd
Döröö Nuur
Uliastay
Hangayn Nuruu
MONGOLIA
Bugun
Ulan Bator
Tsetserleg
Dzuunmod
Bayanhongor
Huld
Dalandzadgad
Aerhtai Shan (Altai)
Angarsk
Irkutsk
Ulan-Ude
Hövsgöl Nuur
Hatgal
Selenge Mörön
Orhon Gol
Tian Shan
Pik Pobedy
7439
Aksu
Kuqa
Yanqi
Korla
Bosten (Bagrax) Hu
Kuruktag
Qitai
Barkol Kazak Zizhixian
Hami
4925
XINJIANG UYGUR ZIZHIQU
Artux
Kashi
Shule
Shache
Yecheng
Taxkorgan Tajik Zizhixian
Pishan
Hotan (Khotan)
Yutian
Qiemo
Qarqan He
Tarim He
Tarim Pendi
Ruoqiang
Lop Nor
Dunhuang
Anxi
Yumen
Jiayuguan
Gaxun Nur
Altun Shan
Qilian Shan
Mangnai
6346
Ayakkum Hu
Zhangye
Shandan
Wuwei
Wuhai
Alxa Zuoqi
Yinchuan
NINGXIA HUIZU ZIZHIQU
Wuzhong
Karakoram
8611
8126
JAMMU & KASHMIR
Srinagar
Leh
Indus
Rutog
Wuluk omushih Ling
7723
Kunlun Shan
Huh Xil Shan
Tart
Da Qaidam
Qaidam Pendi
Golmud
Tianjun
Qinghai Hu
3205
Dulan
Xining
Gonghe
Minhe
LANZHOU
Linxia
Guyuan
Pingliang
QINGHAI
Ngoring Hu
4237
Gyaring Hu
Maqen
6094
Bayan Har Shan
Huang He
Min Xian
Tianshui
Baoji
Qinling
XIZANG ZIZHIQU
Tanggula (Dangla) Shan
(TIBET)
Gar
Nanda Devi
7817
Burang
Mapam Yumco
Dehra Dun
Amdo
Siling Co
4495
Nam Co
4627
Nagqu
Zhongba
Xainza
Nyainqêntanglha Shan
Qamdo
Yushu
Jinsha Jiang
Yalong Jiang
Ningjing Shan
Lancang Jiang
Shaluli Shan
Garze
Daxue Shan
Songpan
Min Jiang
Hanzhong
Mianyang
SICHUAN
CHENGDU
Santai
Nanchong
Daxian
Leshan
Neijiang
Zigong
Gongga Shan
7600
Wutongqiao
Luzhou
Yibin
Chang Jiang
Meerut
DELHI
Moradabad
Bareilly
Aligarh
Agra
KANPUR
Lucknow
Gwalior
Jhansi
Allahabad
Varanasi
Patna
Gaya
Sagar
INDIA
Tropic of Cancer
Jabalpur
Bilaspur
Raipur
Nagpur
Chanda
Indravati
Mahanadi
Cuttack
Warangal
Vizianagaram
Berhampur
Vishakhapatnam
Ghaghra
Ganga
Gorakhpur
Darbhanga
NEPAL
Himalaya
Dhaulagiri
8221
Kathmandu
8848
Ngamring
Lhazê
Xigazê
Yamzho Yumco
Lhasa
Yarlung Zangbo Jiang
Namcha Barwa
7756
Bomi
BHUTAN
Punakha
Brahmaputra
Dibrugarh
Sadiya
Tezpur
Gauhati
Khasi Hills
Patkai Bum
3411
Imphal
Silchar
Rajshahi
BANGLADESH
Berhampore
DHAKA
Narayanganj
Chittagong
Asansol
Ranchi
Barddhaman
Jamshedpur
Haora
Bhatpara
Khulna
CALCUTTA
Kharagpur
Baleshwar
BAY OF BENGAL
Akyab
Myitkyina
Chindwin
BURMA (MYANMAR)
Monywa
Victoria
3053
Mandalay
Arakan Yoma
Irrawaddy
Pegu Yoma
Shwebo
Bhamo
Lashio
Taunggyi
Yamethin
Toungoo
Salween
Tengchong
Luxi
Nu Jiang
Nu Shan
Yun Ling
Zayü
5881
Zhongdian
Lijiang
Dali
Xiaguan
Baoshan
YUNNAN
KUNMING
Anning
Yuxi
Dongchuan
Huize
Chengjiang
Shiping
Gejiu
Jinggu
Jiangcheng
2650
Daliang Shan
Xichang
Zhaotong
Shuicheng
GUIZHOU
Guiyang
Anshun
Duyun
Zunyi
Xingyi
Hongshui He
Hechi
Yishan
Bose
Mengzi
Wenshan
Nanning
Pingxiang
Qinzhou
3143
VIETNAM
Hanoi
Haiphong
Gulf of Tonkin
LAOS
Luang Prabang
Mekong
2711
THAILAND (SIAM)
2563
CHONGQING
Projection: Bonne
East from Greenwich

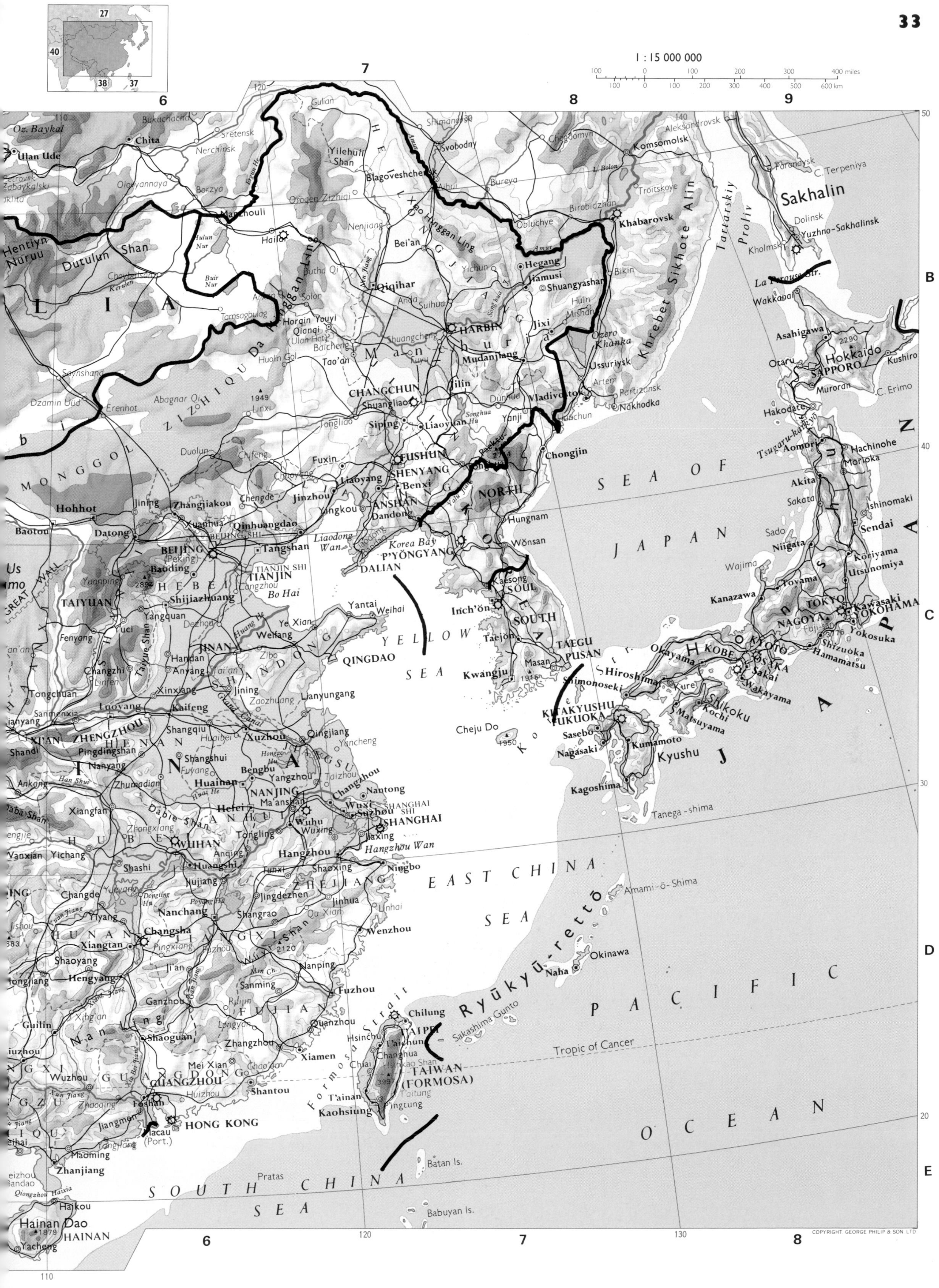

1 : 15 000 000

100 0 100 200 300 400 miles
100 0 100 200 300 400 500 600 km

Oz. Baykal
Ulan Ude
Chita
Bukachacha
Sretensk
Nerchinsk
Olovyannaya
Borzya
Manzhouli
Hailar
Hulun Nur
Buir Nur
Hentiyn Nuruu
Dutulun Shan
Choybalsan
Kerulen
Tamsagbulag
Saynshand
Dzamïn Üüd
Erenhot
Abagnar Qi
Linxi
Duolun
Chifeng
Gulian
Shimanovsk
Svobodny
Yilehuli Shan
Blagoveshchensk
Aihui
Bureya
Nenjiang
Bei'an
Qiqihar
Hegang
Jiamusi
Shuangyashan
Jixi
HARBIN
Mudanjiang
Jilin
CHANGCHUN
Siping
Liaoyuan
FUSHUN
SHENYANG
Benxi
Anshan
Dandong
Yingkou
Jinzhou
Fuxin
Liaoyang
Komsomolsk
Aleksandrovsk
Khabarovsk
Birobidzhan
Obluchye
Bikin
Ussuriysk
Vladivostok
Artem
Partizansk
Nakhodka
Sikhote Alin
Tartarskiy Proliv
Sakhalin
C. Terpeniya
Dolinsk
Yuzhno-Sakhalinsk
Kholmsk
La Perouse Str.
Wakkanai
Asahigawa
Hokkaido
SAPPORO
Otaru
Kushiro
Muroran
C. Erimo
Hakodate
Aomori
Hachinohe
Morioka
Akita
Sakata
Ishinomaki
Sendai
Niigata
Sado
Koriyama
Utsunomiya
Wajima
Toyama
Kanazawa
TOKYO
Kawasaki
YOKOHAMA
NAGOYA
Yokosuka
Shizuoka
Hamamatsu
KYOTO
KOBE
OSAKA
Sakai
Wakayama
Okayama
Hiroshima
Kure
Shikoku
Kochi
Matsuyama
Shimonoseki
KITAKYUSHU
FUKUOKA
Sasebo
Nagasaki
Kumamoto
Kyushu
Kagoshima
Tanega-shima
SEA OF JAPAN
JAPAN
Chongjin
NORTH KOREA
Hungnam
Wŏnsan
P'YŎNGYANG
Korea Bay
Kaesong
SOUL
Inch'ŏn
SOUTH KOREA
Taejŏn
TAEGU
PUSAN
Kwangju
Masan
Cheju Do
Korea Str.
Hohhot
Baotou
Datong
Jining
Zhangjiakou
Chengde
Qinhuangdao
BEIJING
Tangshan
TIANJIN
Baoding
Shijiazhuang
Cangzhou
Bo Hai
Liaodong Wan
DALIAN
TAIYUAN
Yangquan
Dezhou
Fenyang
JINAN
Zibo
Weifang
Yantai
Weihai
QINGDAO
YELLOW SEA
Handan
Anyang
Changzhi
Xinxiang
Jining
Lianyungang
Zaozhuang
Kaifeng
Luoyang
ZHENGZHOU
Pingdingshan
Nanyang
Xuzhou
Qingjiang
Yancheng
Bengbu
Huainan
Yangzhou
Shangqiu
Shangshui
Fuyang
NANJING
Ma'anshan
Hefei
Changzhou
Nantong
Wuxi
Suzhou
SHANGHAI
Wuhu
Tongling
Anqing
Jiaxing
Hangzhou
Hangzhou Wan
Shaoxing
Ningbo
ZHEJIANG
Jinhua
Jingdezhen
Shangrao
Wenzhou
Nanchang
Jiujiang
Huangshi
WUHAN
Shashi
Yichang
Xiangfan
Changde
Yiyang
Changsha
Xiangtan
Hengyang
Shaoyang
Ji'an
Ganzhou
Nanping
Fuzhou
Sanming
Quanzhou
Zhangzhou
Xiamen
Longyan
Shaoguan
Mei Xian
Shantou
GUANGZHOU
Huizhou
Foshan
HONG KONG
Macau (Port.)
Jiangmen
Maoming
Zhanjiang
Wuzhou
Guilin
Haikou
Hainan Dao
HAINAN
Yacheng
EAST CHINA SEA
Amami-ō-Shima
Ryūkyū-rettō
Okinawa
Naha
Sakashima Gunto
Formosa Strait
Chilung
TAIPEI
Hsinchu
T'aichung
Changhua
Chiai
T'ainan
Kaohsiung
P'ingtung
T'aitung
TAIWAN (FORMOSA)
Tropic of Cancer
Batan Is.
Babuyan Is.
Pratas
SOUTH CHINA SEA
PACIFIC OCEAN

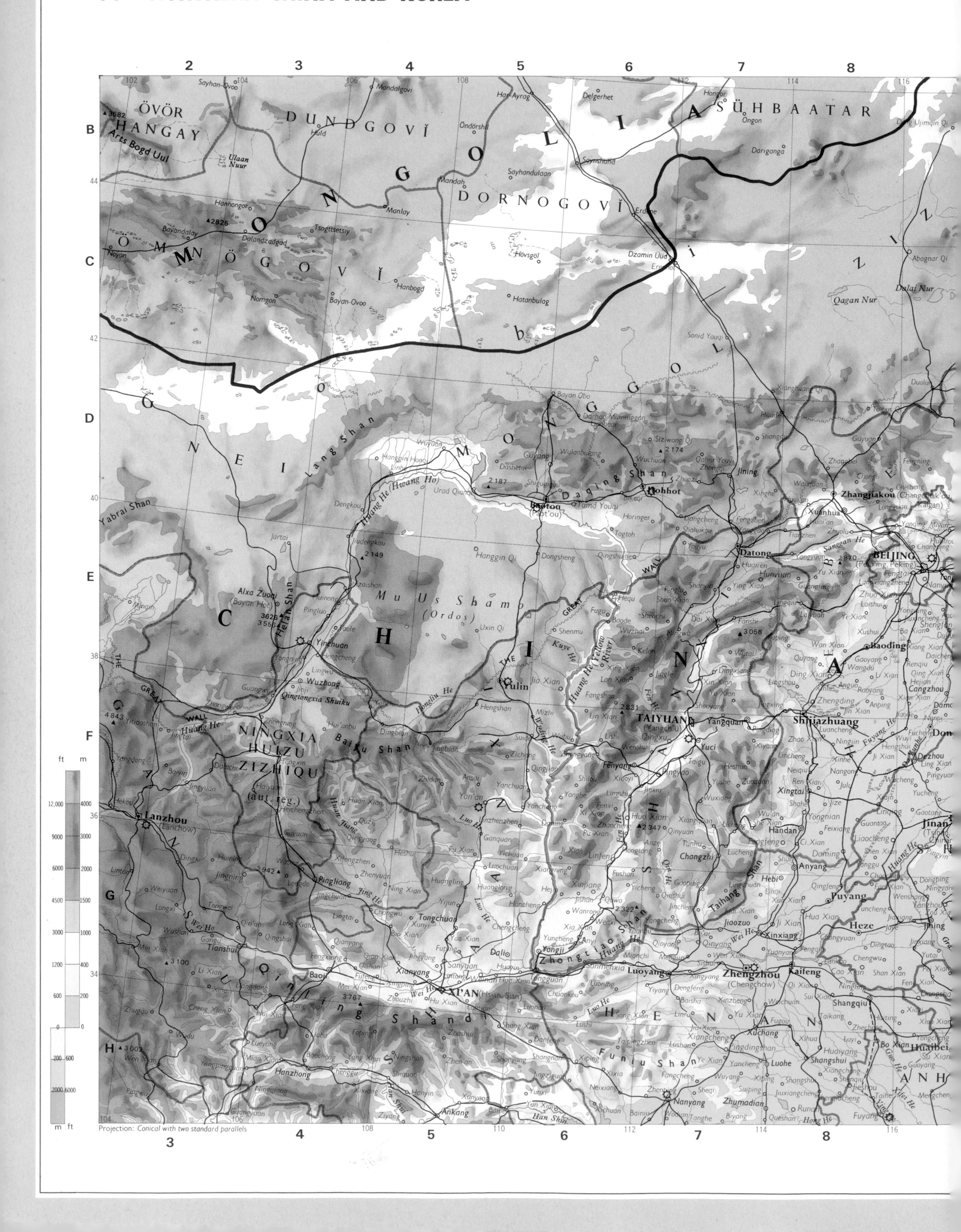
MONGOLIA
ÖVÖR HANGAY
DUNDGOVĬ
SÜHBAATAR
DORNOGOVĬ
ÖMNÖGOVĬ
NEI MONGOL
NINGXIA HUIZU ZIZHIQU (aut. reg.)
CHINA
SHANXI
HEBEI
HENAN
BEIJING
TAIYUAN
Hohhot
Baotou
Datong
Zhangjiakou
Shijiazhuang
Baoding
Yinchuan
Lanzhou
Xi'an
Zhengzhou
Luoyang
Kaifeng
Yulin
Yan'an
Mu Us Shamo (Ordos)
Huang He (Hwang Ho)
Qinling Shandi
Funiu Shan
Daqing Shan
Lang Shan
Helan Shan
GREAT WALL
Projection: Conical with two standard parallels
ft
m
12,000
9000
6000
4500
3000
1200
600
0
200
600
2000
6000
4000
3000
2000
1500
1000
400
200
0
m ft

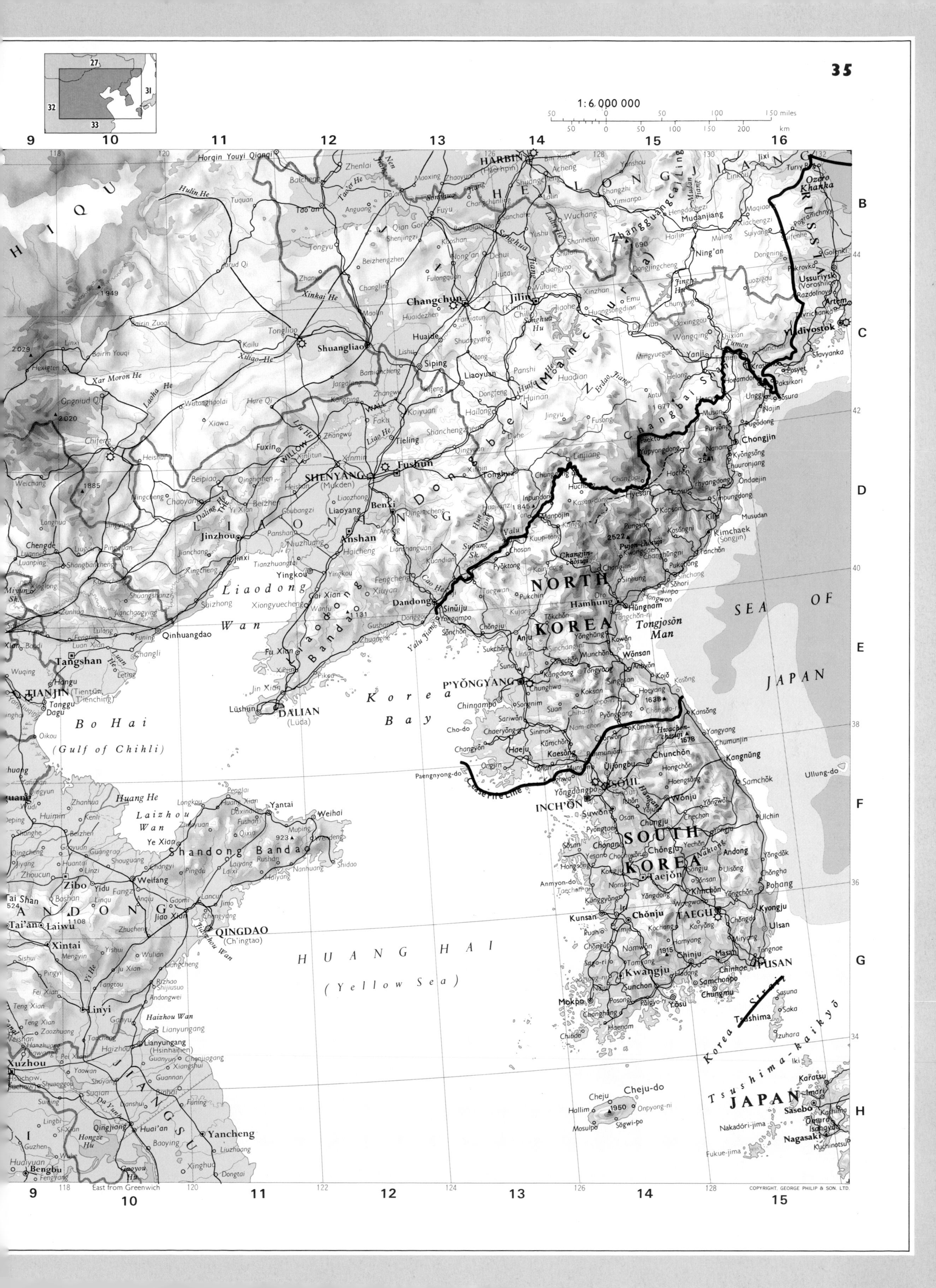
1:6 000 000
50 0 50 100 150 miles
50 0 50 100 150 200 km
HEILONGJIANG
JILIN
LIAONING
NORTH KOREA
SOUTH KOREA
RUSSIA
JAPAN
SHANDONG
JIANGSU
HARBIN (Haerhpin)
Changchun
Jilin
Shuangliao
Siping
Liaoyuan
Mudanjiang
Vladivostok
Ussuriysk
SHENYANG (Mukden)
Fushun
Benxi
Anshan
Liaoyang
Fuxin
Jinzhou
Yingkou
Dandong
Qinhuangdao
Tangshan
TIANJIN (Tientsin T'ienching)
DALIAN (Lüda)
Lüshun
P'YONGYANG
Chinnampo
Hamhung
Hungnam
Wonsan
Chongjin
Kimchaek (Songjin)
Sinuiju
Kaesong
SEOUL
INCH'ON
Suwon
Ch'unch'on
Kangnung
Taejon
TAEGU
PUSAN
Kwangju
Chonju
Kunsan
Mokpo
Ulsan
Pohang
Yantai
Weihai
Weifang
Zibo
QINGDAO (Ch'ingtao)
Lianyungang (Hsinhaihsien)
Xuzhou
Linyi
Yancheng
Huai'an
Bengbu
Cheju-do
Cheju
Tsushima
Sasebo
Nagasaki
Liaodong Wan
Bo Hai (Gulf of Chihli)
Korea Bay
HUANG HAI (Yellow Sea)
Laizhou Wan
Shandong Bandao
Liaodong Bandao
Tongjoson Man
SEA OF JAPAN
Korea Strait
Tsushima-kaikyo
Cease Fire Line
Changbai Shan
Yalu Jiang
Ozero Khanka
Paengnyong-do
Ullung-do
East from Greenwich
COPYRIGHT. GEORGE PHILIP & SON, LTD.

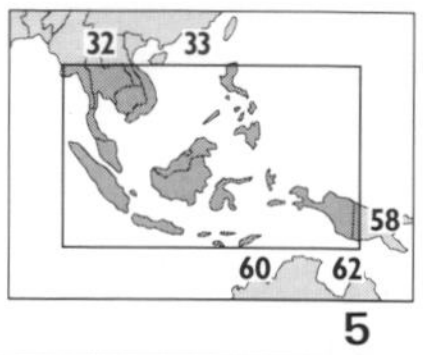

INDONESIA AND THE PHILIPPINES

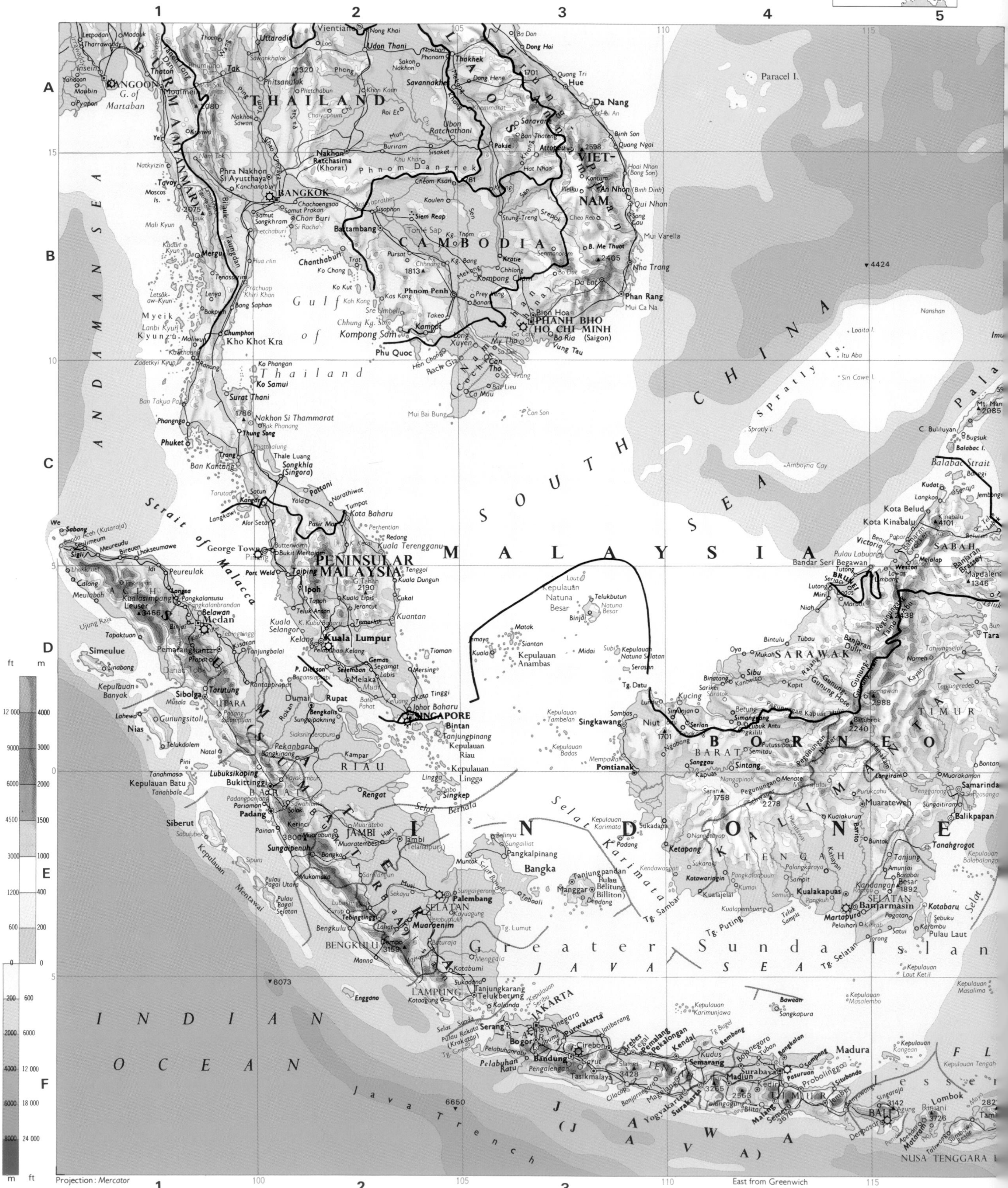

Projection: Mercator

East from Greenwich

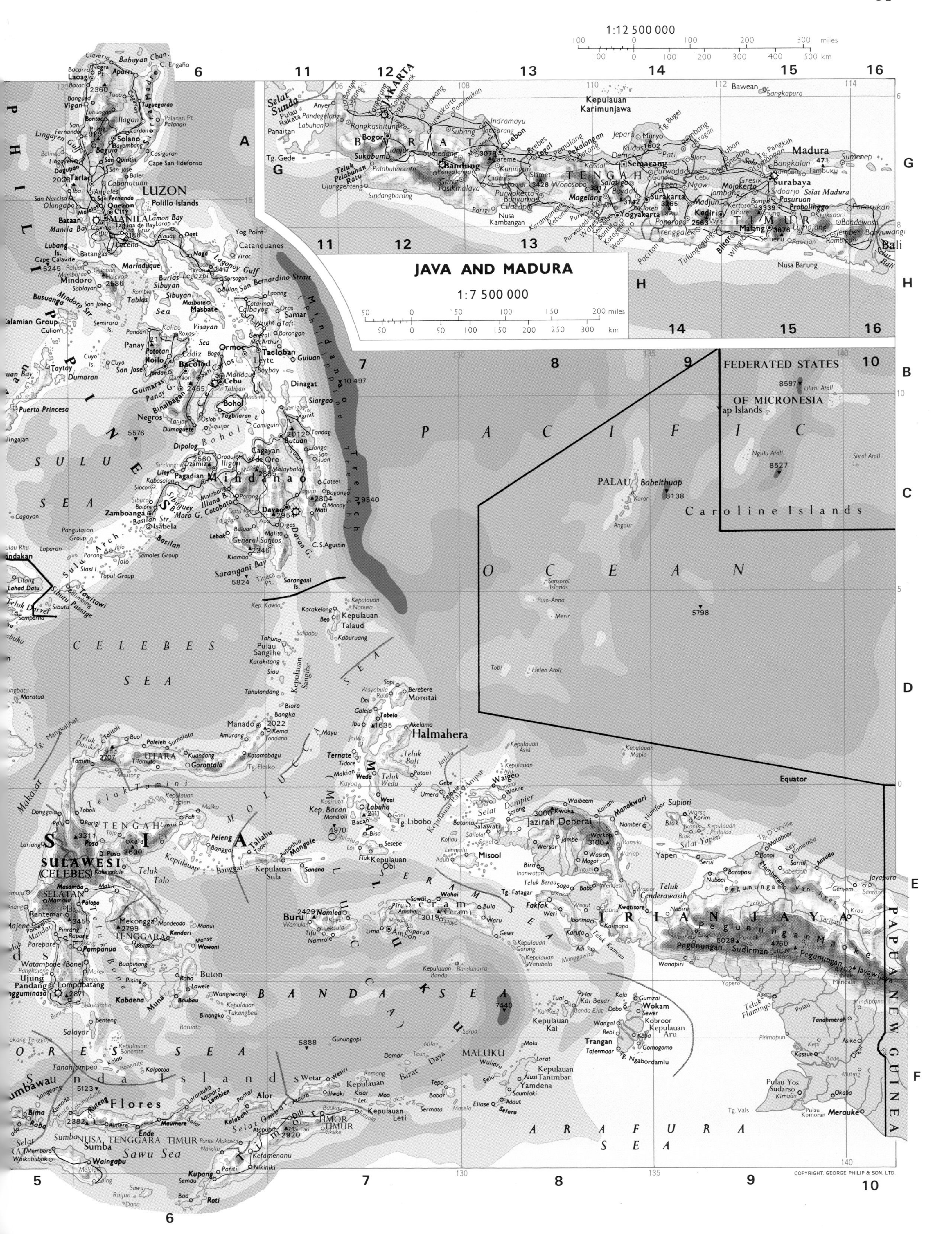
1:12 500 000
JAVA AND MADURA
1:7 500 000
PACIFIC OCEAN
CELEBES SEA
SULU SEA
BANDA SEA
ARAFURA SEA
MOLUCCA SEA
FEDERATED STATES OF MICRONESIA
Caroline Islands
PALAU
LUZON
Mindanao
Halmahera
SULAWESI (CELEBES)
IRIAN JAYA
PAPUA NEW GUINEA
MALUKU
Flores
Sumba
Sawu Sea
Equator
Mindanao Trench
COPYRIGHT, GEORGE PHILIP & SON, LTD.

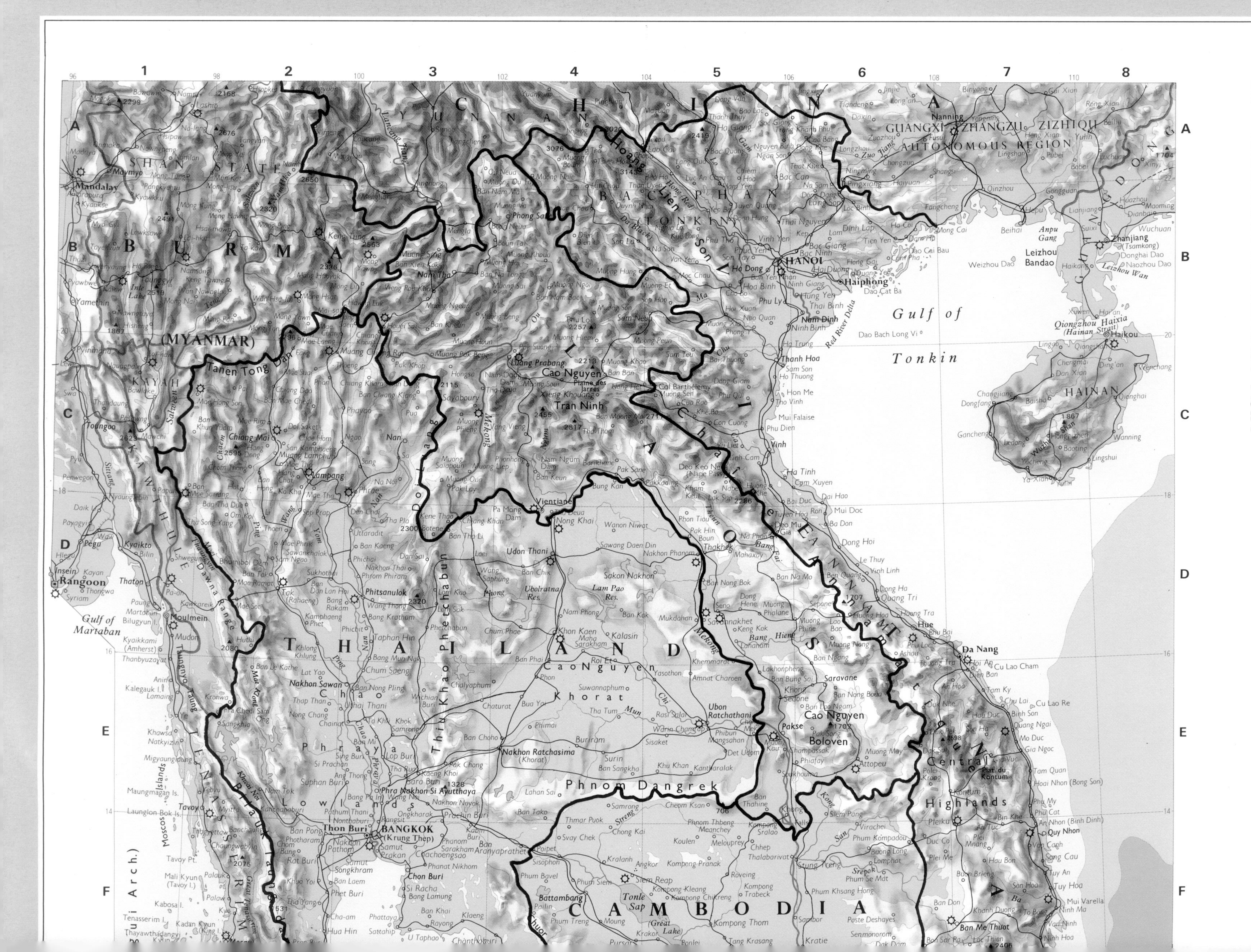

BURMA
(MYANMAR)
THAILAND
CAMBODIA
Gulf of Tonkin
HAINAN
HANOI
Haiphong
Da Nang
Vientiane
BANGKOK (Krung Thep)
Rangoon
Mandalay
Gulf of Martaban
Mekong
Khorat
Phnom Dangrek
Dawna Range
Tanen Tong Dan
Central Highlands
GUANGXI ZHANGZU ZIZHIQU AUTONOMOUS REGION

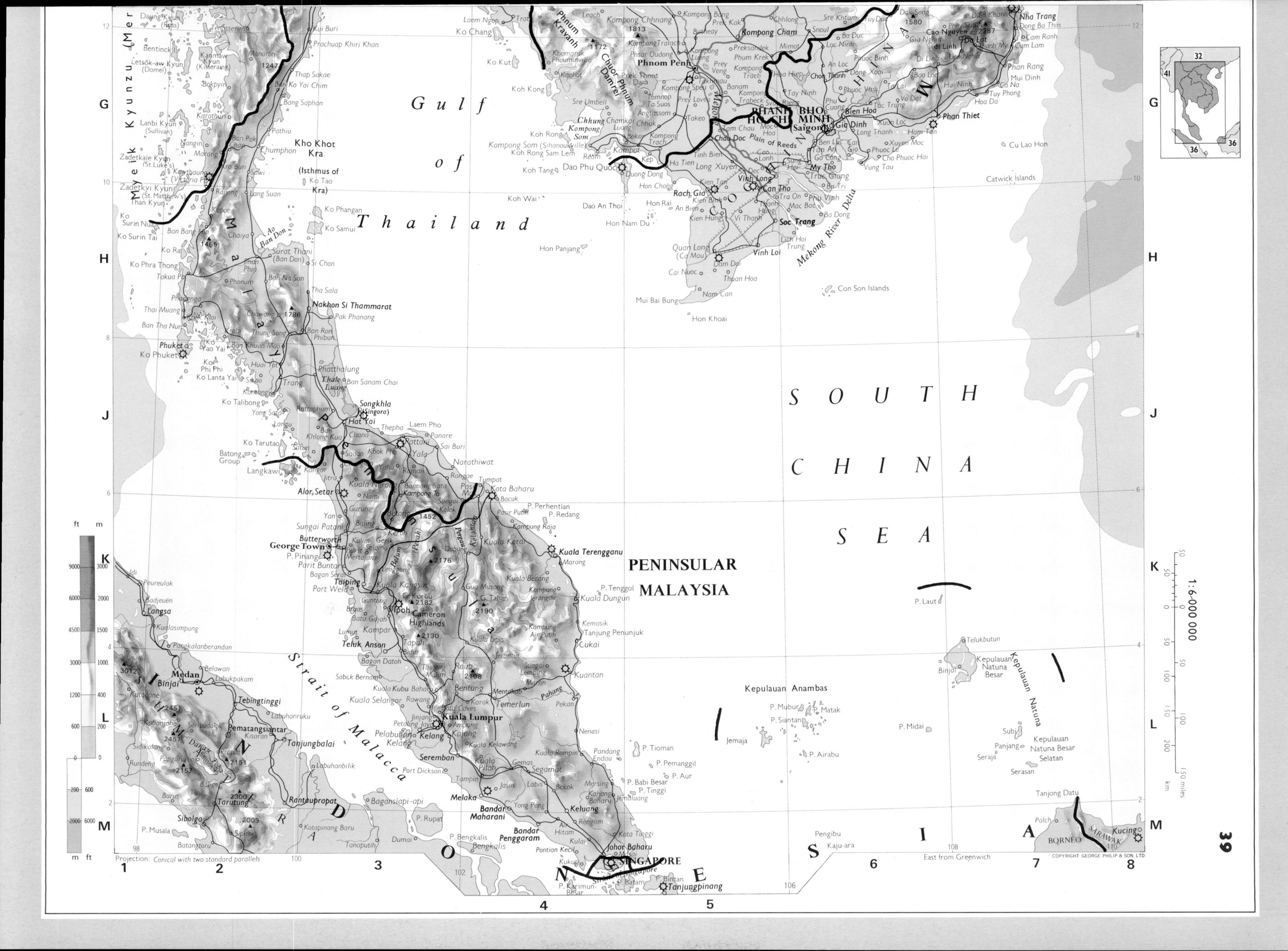
1:6 000 000
50 0 50 100 150 miles
50 0 50 100 150 200 km
Gulf of Thailand
SOUTH CHINA SEA
PENINSULAR MALAYSIA
Strait of Malacca
Mekong River Delta
Plain of Reeds
Isthmus of Kra
Phnom Penh
Kompong Cham
Kompong Chhnang
Phanh Bho Ho Chi Minh (Saigon)
Bien Hoa
Gia Dinh
My Tho
Can Tho
Vinh Long
Rach Gia
Soc Trang
Vinh Loi
Dao Phu Quoc
Con Son Islands
Nha Trang
Da Lat
Phan Thiet
Phan Rang
Cu Lao Hon
Catwick Islands
Kho Khot Kra
Ko Phangan
Ko Samui
Surat Thani
Nakhon Si Thammarat
Phuket
Ko Phuket
Songkhla
Hat Yai
Pattani
Narathiwat
Langkawi
Alor Setar
Kota Baharu
George Town
Butterworth
Taiping
Ipoh
Cameron Highlands
Kuala Terengganu
Kuantan
Teluk Anson
Kuala Lumpur
Kelang
Seremban
Melaka
Keluang
Johor Baharu
SINGAPORE
Tanjungpinang
Medan
Binjai
Tebingtinggi
Pematangsiantar
Tanjungbalai
Rantauprapat
Sibolga
Tarutung
Kepulauan Anambas
Kepulauan Natuna
Natuna Besar
Kepulauan Natuna Besar Selatan
P. Laut
P. Tioman
Tanjong Datu
Kucing
SARAWAK
BORNEO
Projection: Conical with two standard parallels
East from Greenwich
COPYRIGHT GEORGE PHILIP & SON LTD.

ARABIAN SEA
KARACHI
Hyderabad
Quetta
Kabul
Peshawar
Rawalpindi
Islamabad
Srinagar
Lahore
Amritsar
Faisalabad
Multan
Sukkur
DELHI
Agra
Jaipur
Ahmadabad
BOMBAY
Pune (Poona)
Hyderabad
Nagpur
Bhopal
Indore
Madras
Bangalore
Mysore
Colombo
SRI LANKA (CEYLON)
Kandy
Galle
Trincomalee
Jaffna
Madurai
Tiruchchirappalli
Coimbatore
Trivandrum
Mangalore
Calicut (Kozhikode)
Cochin
Pondicherry
Cape Comorin
Palk Bay
Gulf of Mannar
Gulf of Kachchh
Gulf of Khambhat
Rann of Kachchh
Mouths of the Indus
Toba Kakar
Central Makran Range
Makran Coast Range
Siahan Range
Kirthar Range
Karakoram
Thar or Great Indian Desert
RAJASTHAN
MAHARASHTRA
MADHYA
ANDHRA PRADESH
KARNATAKA
TAMIL NADU
KERALA
HARYANA
HIMACHAL PRADESH
PUNJAB
SIND
JAMMU AND KASHMIR
NORTH WEST FRONTIER
AFGHANISTAN
IRAN
TURKMEN.
HELMAND
QANDAHAR
GHAZNI
ORUZGAN
GHOWR
HERAT
FARAH
NIMRUZ
BAMIAN
Continuation Southwards on same scale
Projection: Conical with two standard parallels
Tropic of Cancer
ft m
18 000 6000
12 000 4000
9000 3000
6000 2000
4500 1500
3000 1000
1200 400
600 200
0 0
200 600
m ft

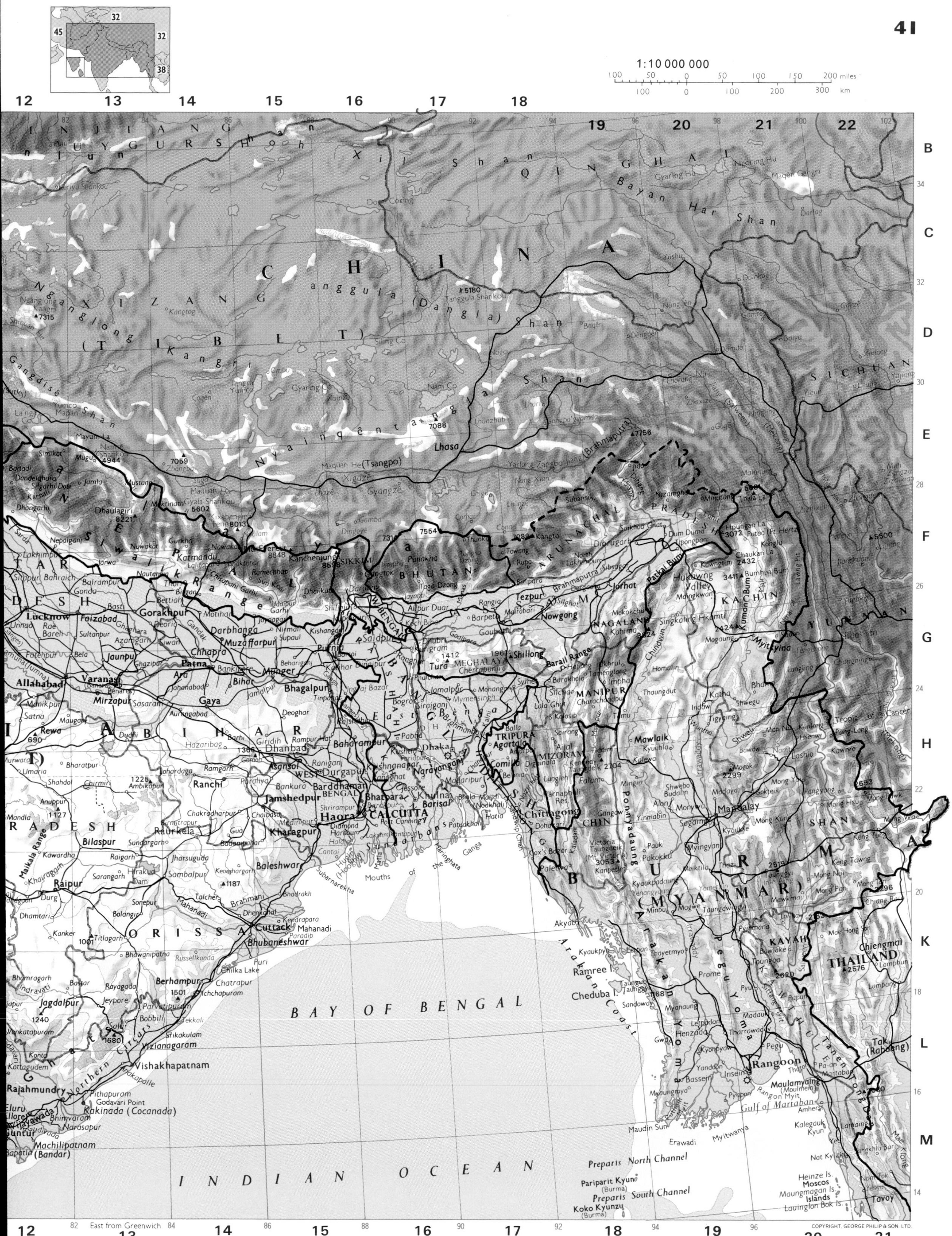

1:10 000 000

100 50 0 50 100 150 200 miles

100 0 100 200 300 km

12 13 14 15 16 17 18 19 20 21 22

B C D E F G H K L M

CHINA
XIZANG (TIBET)
QINGHAI
SICHUAN
YUNNAN
Xinjiang Uygur
Kunlun Shan
Hoh Xil Shan
Bayan Har Shan
Tanggula Shan (Dangla Shan)
Nganglong Kangri
Gangdisê Shan
Nyainqêntanglha Shan
Lhasa
Yarlung Zangbo Jiang (Brahmaputra)
Maquan He (Tsangpo)
Xigazê
Gyangzê
Nam Co
Siling Co
Gyaring Co
Dorsoidong Co
Qamdo
Nu Jiang (Salween)
Lancang Jiang (Mekong)
Jinsha Jiang
Yushu
Gyaring Hu
Ngoring Hu
Maqen Gangri
Tanggula Shankou
NEPAL
Katmandu
Everest 8848
Kanchenjunga 8598
Dhaulagiri 8221
Siwalik Range
SIKKIM
Gangtok
BHUTAN
Punakha
ARUNACHAL PRADESH
ASSAM
NAGALAND
MANIPUR
MIZORAM
TRIPURA
MEGHALAYA
Shillong
Cherrapunji
Tezpur
Jorhat
Dibrugarh
Nowgong
Imphal
Kohima
Agartala
BANGLADESH
Dhaka
Chittagong
Khulna
Barisal
Comilla
Mymensingh
Rajshahi
Cox's Bazar
WEST BENGAL
CALCUTTA
Haora
Durgapur
Asansol
Barddhaman
Kharagpur
BIHAR
Patna
Gaya
Bhagalpur
Muzaffarpur
Darbhanga
Ranchi
Jamshedpur
Dhanbad
UTTAR PRADESH
Lucknow
Allahabad
Varanasi (Benares)
Gorakhpur
Faizabad
Mirzapur
Jaunpur
MADHYA PRADESH
Rewa
Raipur
Bilaspur
ORISSA
Cuttack
Bhubaneshwar
Puri
Sambalpur
Rourkela
Baleshwar
Berhampur
Chilka Lake
Mahanadi
ANDHRA PRADESH
Vishakhapatnam
Vizianagaram
Rajahmundry
Kakinada (Cocanada)
Machilipatnam (Bandar)
Guntur
Vijayawada
Eluru
Northern Circars
Jagdalpur
BAY OF BENGAL
Mouths of the Ganga
Sundarbans
Gulf of Martaban
INDIAN OCEAN
BURMA (MYANMAR)
KACHIN
Myitkyina
Putao
CHIN
SHAN
KAYAH
Mandalay
Rangoon
Pegu
Bassein
Moulmein
Akyab
Sittwe
Ramree I.
Cheduba I.
Arakan Coast
Arakan Yoma
Pegu Yoma
Chindwin
Irrawaddy
Sagaing
Magwe
Prome
Henzada
Tavoy
Preparis North Channel
Preparis South Channel
Coco Is.
Heinze Is.
Moscos Islands
THAILAND
Chiengmai
Tak (Raheng)
Tropic of Cancer

12 82 East from Greenwich 84 13 14 86 15 88 16 90 17 92 18 94 19 96 20 21

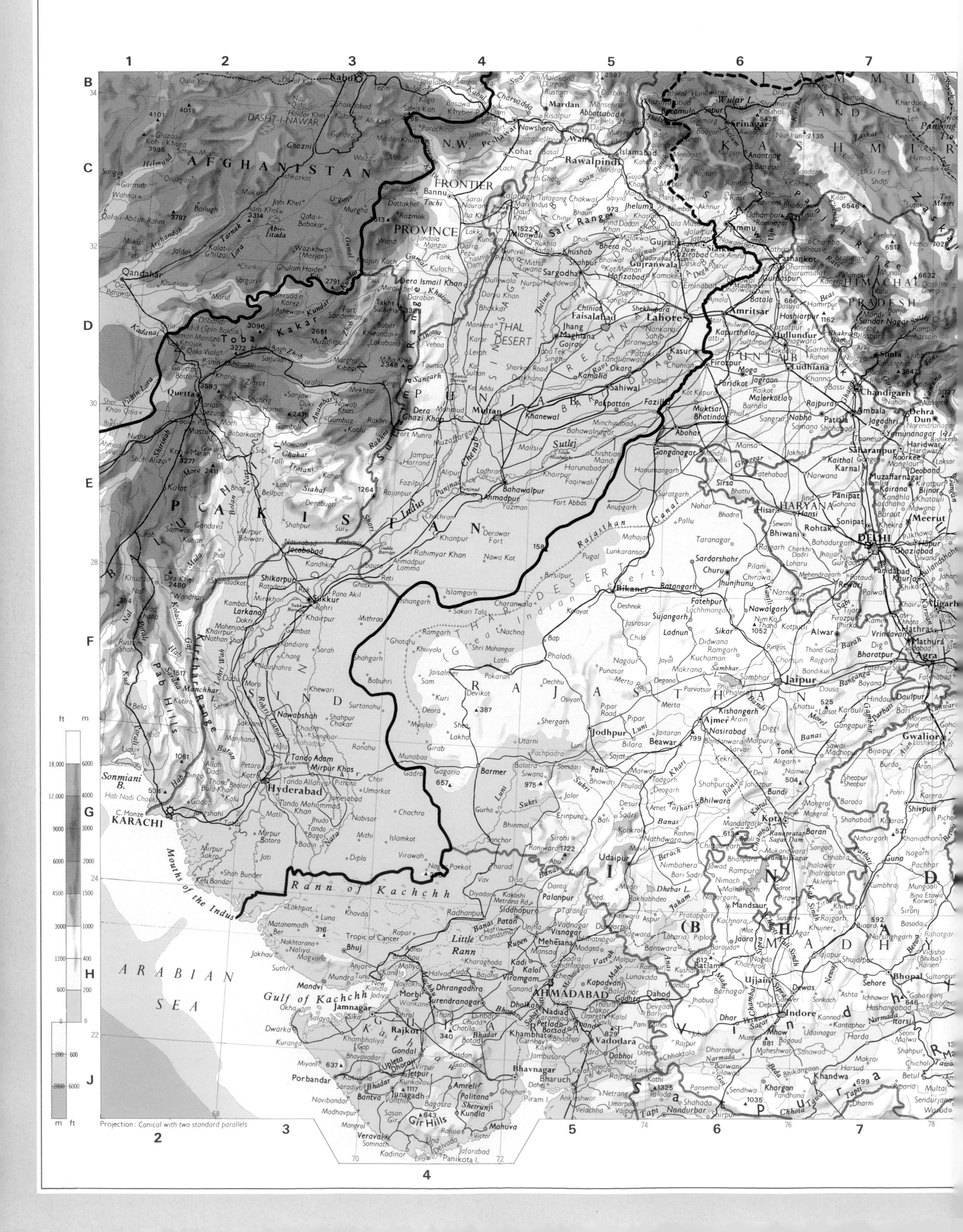

AFGHANISTAN
PAKISTAN
BALUCHISTAN
N.W. FRONTIER PROVINCE
PUNJAB
SIND
JAMMU AND KASHMIR
HIMACHAL PRADESH
HARYANA
RAJASTHAN
MADHYA PRADESH
GUJARAT
DELHI
THAL DESERT
THAR DESERT (Great Indian Desert)
Kabul
Qandahar
Ghazni
Quetta
Peshawar
Rawalpindi
Islamabad
Srinagar
Lahore
Amritsar
Multan
Faisalabad
Hyderabad
KARACHI
Jaipur
Jodhpur
Bikaner
Ajmer
Agra
AHMADABAD
Vadodara
Rajkot
Indore
Bhopal
Ujjain
Gwalior
Simla
Chandigarh
Ludhiana
Jullundur
Meerut
Bahawalpur
Sukkur
Jacobabad
Rann of Kachchh
Little Rann
Gulf of Kachchh
ARABIAN SEA
Mouths of the Indus
Tropic of Cancer
Salt Range
Sulaiman Range
Kirthar Range
Pab Hills
Gir Hills
Indus
Chenab
Jhelum
Ravi
Sutlej
Beas
Narmada
Tapi
Chambal
Luni
Banas
Projection: Conical with two standard parallels
B C D E F G H J
1 2 3 4 5 6 7
ft m
18,000 6000
12,000 4000
9000 3000
6000 2000
4500 1500
3000 1000
1200 400
600 200
0 0
200 600
2000 6000
m ft

JAMMU AND KASHMIR
On same scale as Main Map
1:6 000 000
miles
km
East from Greenwich
CHINA
XIZANG
(TIBET)
NEPAL
SIKKIM
BHUTAN
BANGLADESH
ASSAM
UTTAR PRADESH
BIHAR
MADHYA PRADESH
WEST BENGAL
N.W. FRONTIER PROVINCE
PUNJAB
HIMACHAL PRADESH
Karakoram Range
Ladakh Range
Zaskar Mountains
Deosai Mountains
Kunlun Shan
Gangdisê Shan
Nganglong Kangri
Siwalik Range
Mahabharat Lekh
Kaimur Hills
Maikala Range
Rajmahal Hills
Garo Hills
Mouths of the Ganga
The Sandheads
Sundarbans
Srinagar
Jammu
Leh
Rawalpindi
Islamabad
Gilgit
Lucknow
Kanpur
Allahabad
Varanasi
Patna
Gorakhpur
Calcutta
Dhaka
Kathmandu
Jabalpur
Bilaspur
Jamshedpur
Haora
Bhagalpur
Mt. Everest
Kanchenjunga
Annapurna
Dhaulagiri
Nanda Devi
Nanga Parbat
COPYRIGHT, GEORGE PHILIP & SON, LTD.

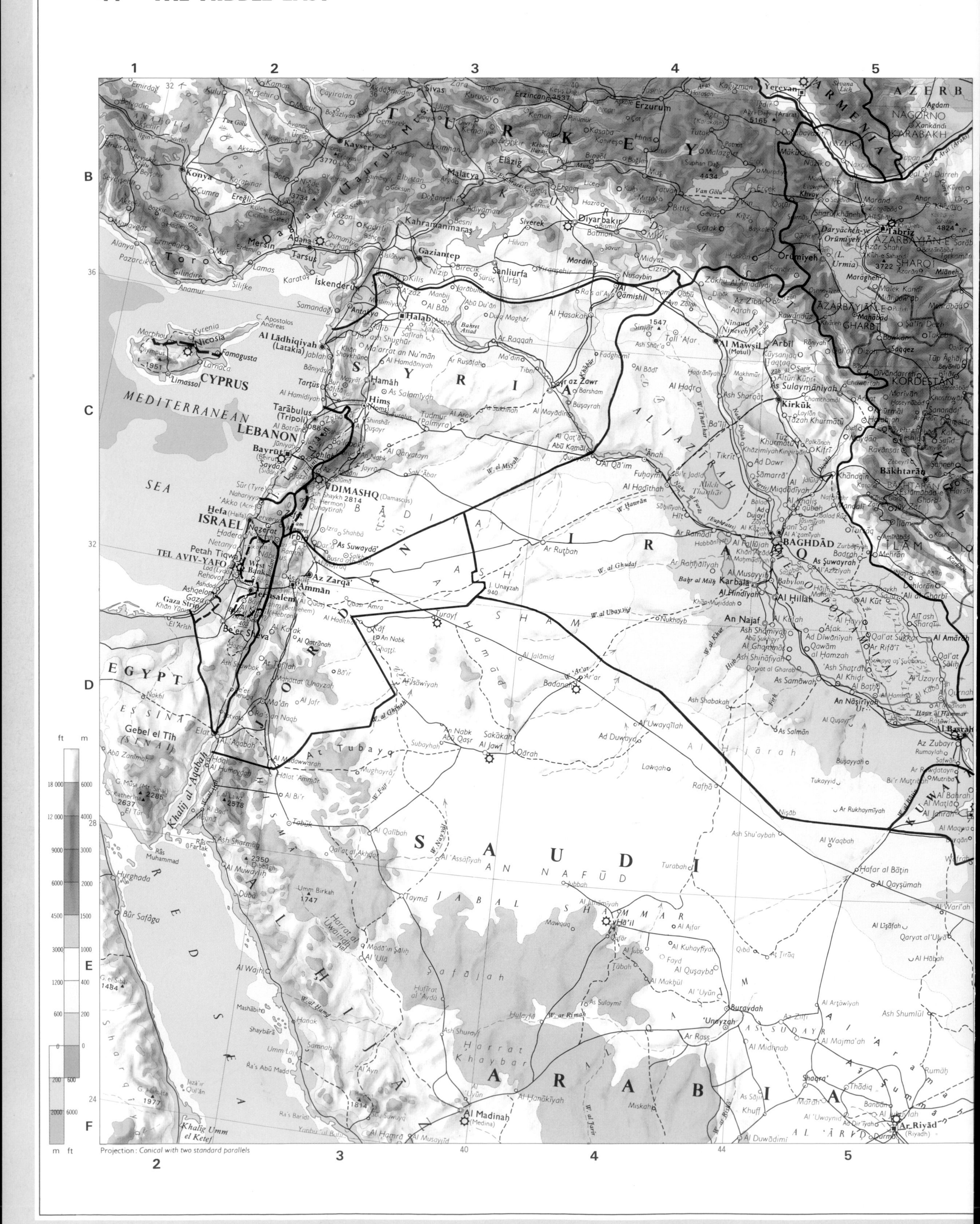

TURKEY
SYRIA
IRAQ
SAUDI ARABIA
JORDAN
LEBANON
ISRAEL
CYPRUS
EGYPT
ARMENIA
KUWAIT
MEDITERRANEAN SEA
RED SEA
AZARBAYJAN-E SHARQI
AZARBAYJAN-E GHARBI
KORDESTAN
ILAM
NAGORNO KARABAKH
Konya
Kayseri
Sivas
Erzurum
Malatya
Elazığ
Diyarbakır
Gaziantep
Sanliurfa (Urfa)
Adana
Mersin
Tarsus
İskenderun
Antakya
Mardin
Nusaybin
Yerevan
Tabriz
Orumiyeh
Nicosia
Famagusta
Limassol
Larnaca
Kyrenia
Halab (Aleppo)
Al Lādhiqiyah (Latakia)
Hamāh
Hims (Homs)
Tarābulus (Tripoli)
Bayrūt (Beirut)
Saydā (Sidon)
DIMASHQ (Damascus)
Jabal ash Shaykh (Mt. Hermon) 2814
Hefa (Haifa)
TEL AVIV-YAFO
Jerusalem
Gaza Strip
West Bank
Be'er Sheva
'Ammān
Az Zarqā'
Elat
Al 'Aqabah
Ma'ān
Tadmur (Palmyra)
Dayr az Zawr
Ar Raqqah
Al Ḥasakah
Al Qāmishlī
Al Mawṣil (Mosul)
Arbīl
Kirkūk
As Sulaymānīyah
BAGHDĀD
Karbalā'
An Najaf
Al Ḥillah
Al Kūt
Al 'Amārah
An Nāṣirīyah
Al Baṣrah
Ar Ruṭbah
Tabūk
Ḥā'il
Al Madīnah (Medina)
Buraydah
'Unayzah
Ar Riyāḍ (Riyadh)
Ḥafar al Bāṭin
Al Jawf
Sakākah
'Ar'ar
Turayf
AN NAFŪD
Gebel el Tih (SINAI)
Khalīj al 'Aqabah
Khalig Umm el Kebef
Hurghada
Bûr Safâga
Projection: Conical with two standard parallels

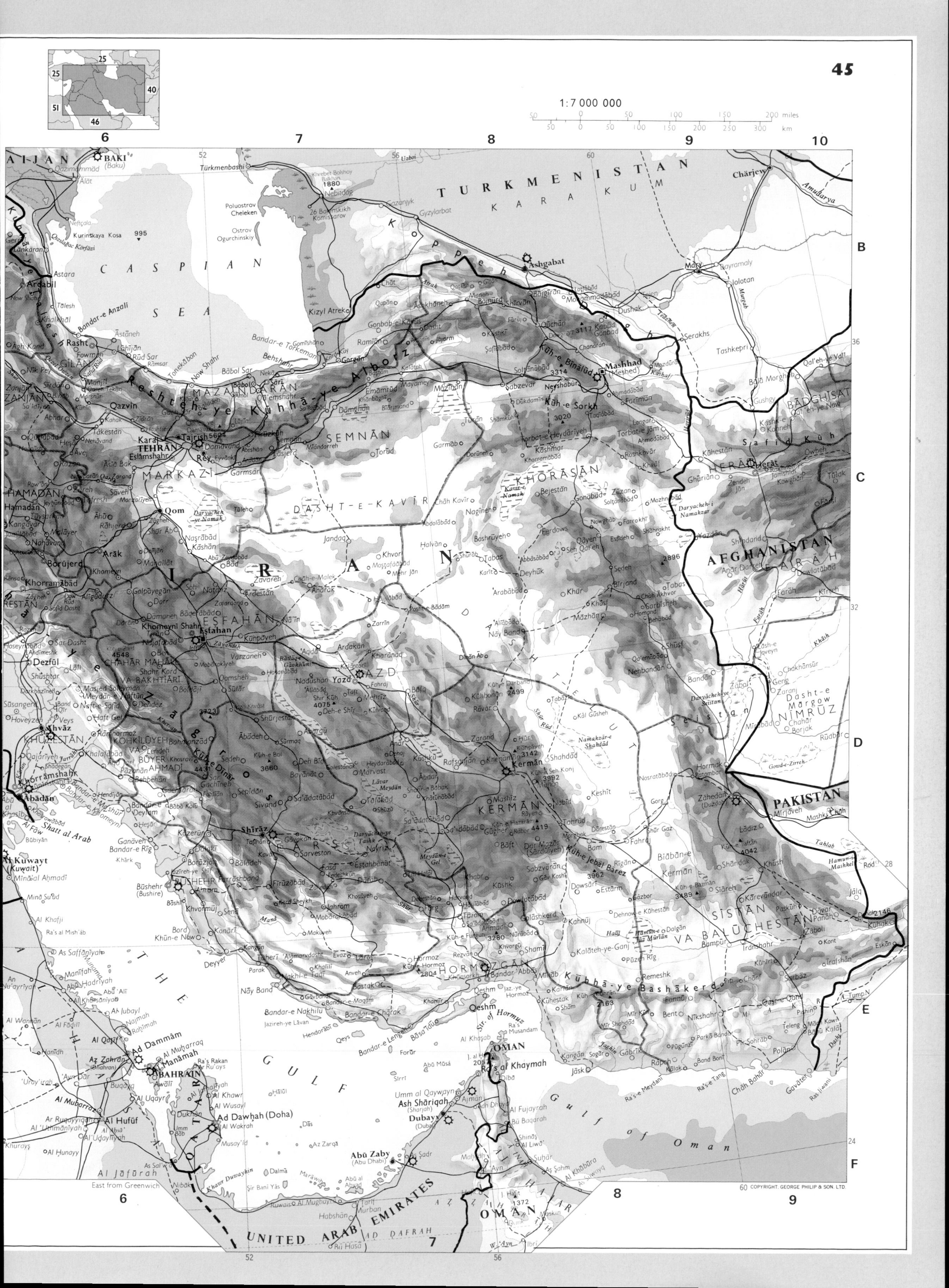
1:7 000 000
miles
km
TURKMENISTAN
KARA KUM
CASPIAN SEA
Türkmenbashi
Ashgabat
Mary
Chärjew
Amudarya
Kopet Dag
BAKI (Baku)
Astara
Ardabil
Rasht
GĪLĀN
ZANJĀN
Qazvin
MAZANDARAN
Reshteh-ye Kūhhā-ye Alborz
TEHRĀN
Karaj
Tajrīsh 5604
Damāvand
SEMNĀN
Semnān
Gorgān
Mashhad (Meshed)
Kūh-e Bīnālūd 3314
Neyshābūr
Sabzevār
Torbat-e Heydārīyeh
KHORĀSĀN
DASHT-E-KAVĪR
Qom
MARKAZĪ
Arāk
HAMADĀN
Hamadān
Borūjerd
Khorramābād
Kāshān
IRAN
ESFAHĀN
Eşfahān
Yazd
YAZD
Tabas
Bīrjand
DASHT-E LŪT
CHAHĀR MAHĀL VA BAKHTĪARĪ
Dezfūl
Ahvāz
KHŪZESTĀN
KOHKĪLŪYEH VA BŪYER AHMADĪ
Kūh-e Dīnār 4431
Zāgros
Khorramshahr
Ābādān
Shatt al Arab
Shīrāz
FĀRS
Kermān
KERMĀN
Kūh-e Hazārān 4419
Kūh-e Jebāl Bārez
Bam
Zāhedān
SĪSTĀN VA BALŪCHESTĀN
Kūh-e Taftān 4042
HORMOZGĀN
Bandar-e 'Abbās
Kūhhā-ye Bashākerd
Chāh Bahār
Jāsk
BŪSHEHR
Būshehr (Bushire)
Al Kuwayt (Kuwait)
Ad Dammām
BAHRAIN
Al Manāmah
QATAR
Ad Dawḩah (Doha)
Al Hufūf
THE GULF
Str. of Hormuz
OMAN
Ra's al Khaymah
Ash Shāriqah (Sharjah)
Dubayy (Dubai)
Abū Zaby (Abu Dhabi)
UNITED ARAB EMIRATES
Gulf of Oman
AFGHANISTAN
Herāt
HERĀT
Safīd Kūh
FARĀH
NĪMRŪZ
Dasht-e Mārgow
PAKISTAN
East from Greenwich
COPYRIGHT, GEORGE PHILIP & SON, LTD.

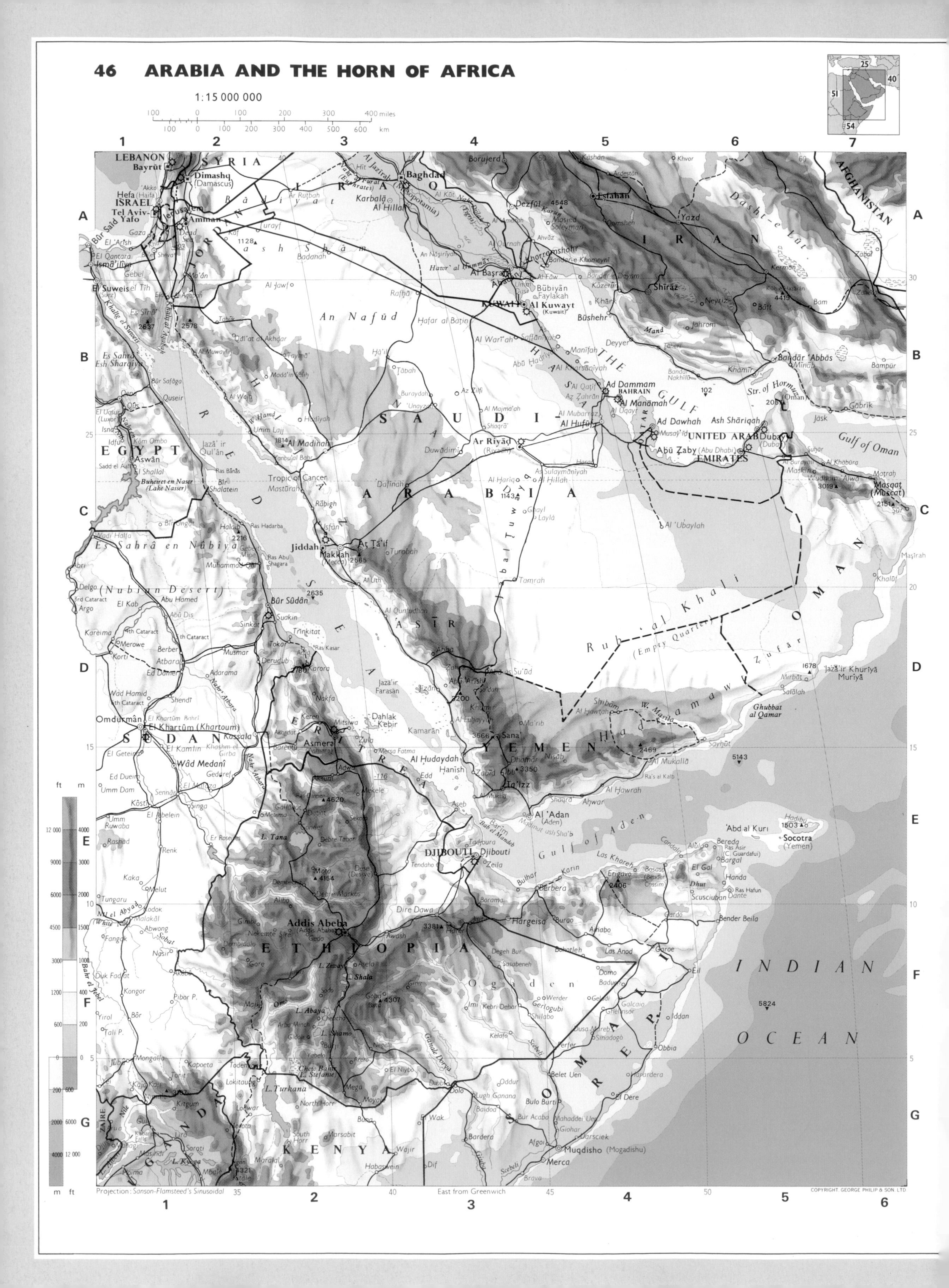
46 ARABIA AND THE HORN OF AFRICA
1:15 000 000
miles
km
LEBANON
Bayrūt
SYRIA
Dimashq
(Damascus)
ISRAEL
Tel Aviv-Yafo
Amman
JORDAN
Baghdad
IRAQ
Karbalā
Al Hillah
Al Başrah
KUWAIT
Al Kuwayt
(Kuwait)
Eşfahān
Yazd
IRAN
Dasht-e Lūt
Kermān
Shīrāz
Bandar 'Abbās
AFGHANISTAN
Dezfūl
Khorramshahr
Būshehr
An Nafūd
Ḥā'il
SAUDI ARABIA
Ar Riyāḍ
(Riyadh)
Al Madīnah
Makkah
(Mecca)
Jiddah
Aṭ Ṭā'if
Ad Dammam
BAHRAIN
Al Manāmah
QATAR
Ad Dawhah
UNITED ARAB EMIRATES
Abū Ẓaby (Abu Dhabi)
Dubayy
Ash Shāriqah
Str. of Hormuz
Gulf of Oman
Masqat
(Muscat)
OMAN
THE GULF
Rub' al Khali
(Empty Quarter)
Tropic of Cancer
EGYPT
El Suweis
Aswân
Buheiret en Naser
(Lake Nasser)
Es Sahrâ en Nûbîya
(Nubian Desert)
RED SEA
Bûr Sûdân
SUDAN
Omdurmân
El Khartûm (Khartoum)
Kassala
Wâd Medanî
Asmera
ERITREA
Dahlak Kebir
YEMEN
Sana'
Ta'izz
Al Ḥudaydah
Al Mukallā
Al 'Adan
(Aden)
Hadramawt
Ghubbat al Qamar
Gulf of Aden
DJIBOUTI
Djibouti
Socotra
(Yemen)
'Abd al Kuri
ETHIOPIA
Addis Abeba
L. Tana
Dire Dawa
Hargeisa
Berbera
Ogaden
SOMALIA
Muqdisho (Mogadishu)
L. Turkana
KENYA
UGANDA
ZAIRE
INDIAN OCEAN
Projection: Sanson-Flamsteed's Sinusoidal
East from Greenwich
COPYRIGHT GEORGE PHILIP & SON LTD

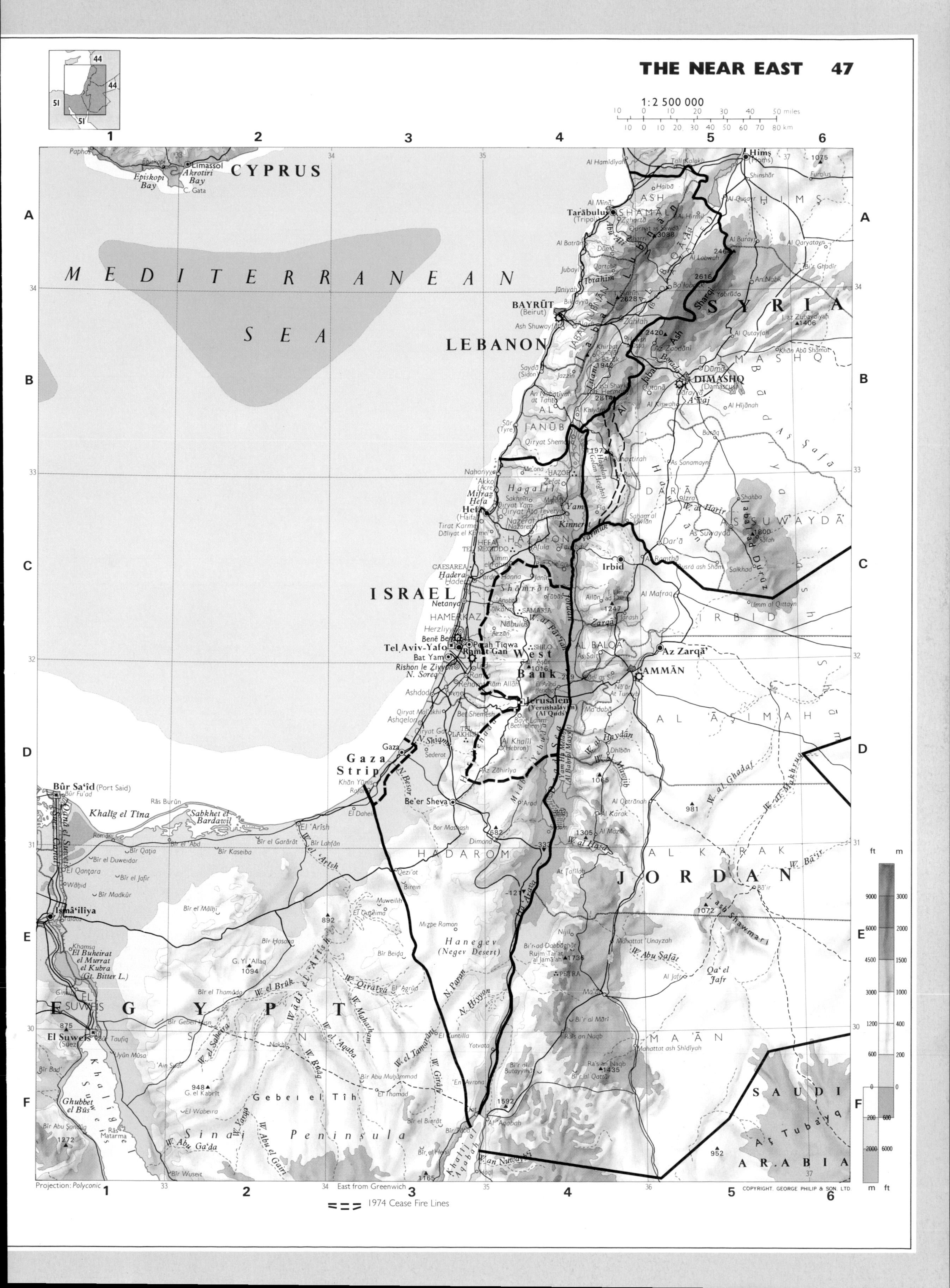
1:2 500 000
miles
km
MEDITERRANEAN
SEA
CYPRUS
LEBANON
SYRIA
ISRAEL
West Bank
Gaza Strip
JORDAN
EGYPT
SAUDI ARABIA
Sinai Peninsula
Hanegev (Negev Desert)
BAYRŪT (Beirut)
DIMASHQ (Damascus)
Jerusalem (Yerushalayim) (Al Quds)
Tel Aviv-Yafo
AMMĀN
Hefa (Haifa)
Tarābulus (Tripoli)
Ḥimṣ (Homs)
Irbid
Az Zarqā'
Gaza
Be'er Sheva
Bûr Sa'îd (Port Said)
Ismâ'ilîya
El Suweis (Suez)
Limassol
Projection: Polyconic
East from Greenwich
1974 Cease Fire Lines
ft
m

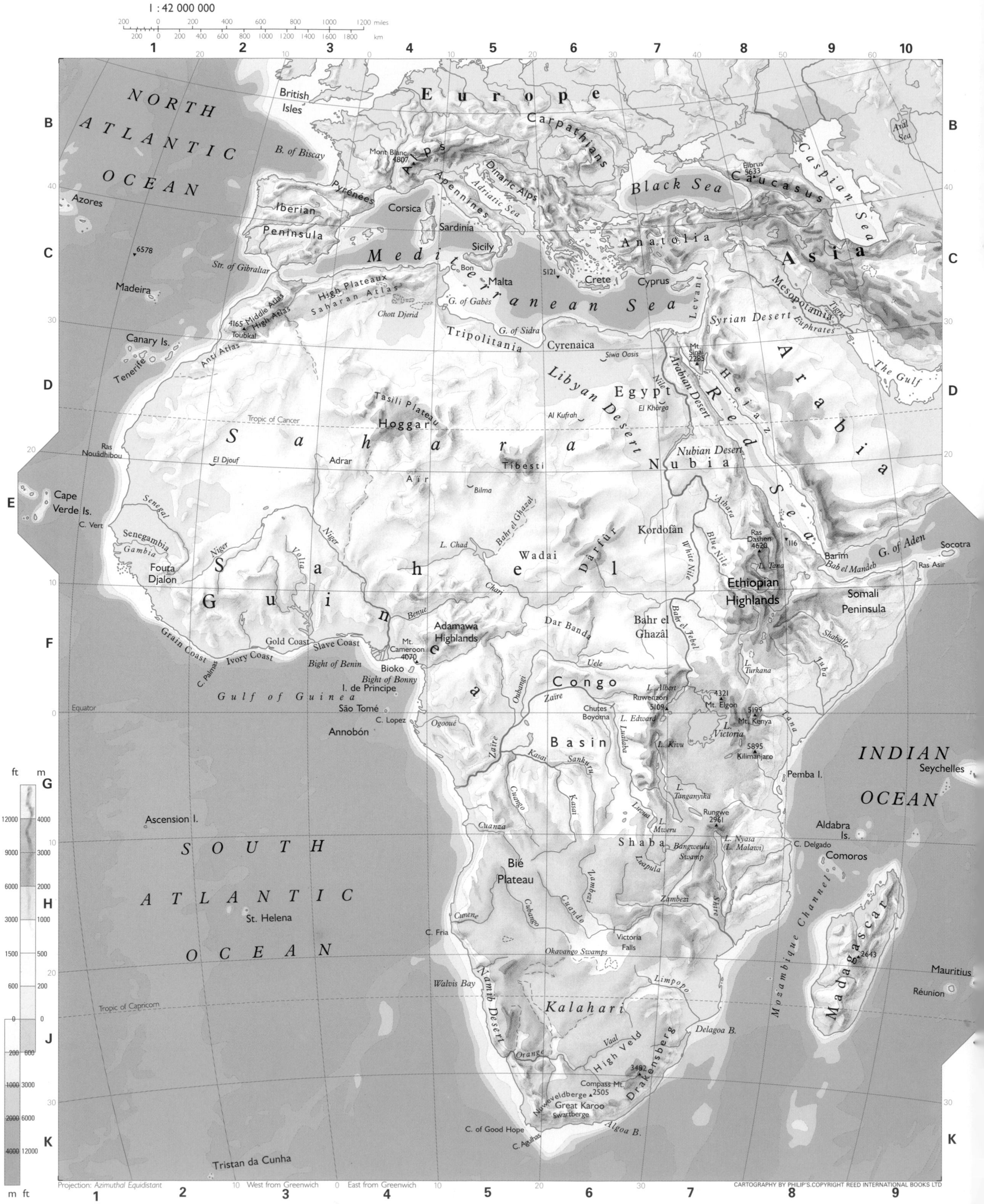
1 : 42 000 000
miles
km
Projection: Azimuthal Equidistant
West from Greenwich
East from Greenwich
CARTOGRAPHY BY PHILIP'S.COPYRIGHT REED INTERNATIONAL BOOKS LTD
NORTH ATLANTIC OCEAN
SOUTH ATLANTIC OCEAN
INDIAN OCEAN
Europe
Asia
Arabia
British Isles
B. of Biscay
Mont Blanc 4807
Alps
Carpathians
Dinaric Alps
Apennines
Adriatic Sea
Pyrénées
Iberian Peninsula
Corsica
Sardinia
Sicily
Black Sea
Elbrus 5633
Caucasus
Caspian Sea
Aral Sea
Anatolia
Azores
6578
Str. of Gibraltar
Mediterranean Sea
C. Bon
Malta
5121
Crete
Cyprus
Levant
Mesopotamia
Tigris
Euphrates
Syrian Desert
The Gulf
Madeira
High Plateaux
Saharan Atlas
Middle Atlas
High Atlas
4165 Toubkal
Anti Atlas
Chott Djerid
G. of Gabès
G. of Sidra
Tripolitania
Cyrenaica
Siwa Oasis
Mt. Sinai 2285
Canary Is.
Tenerife
Libyan Desert
Egypt
Nile
Arabian Desert
Hejaz
Red Sea
El Khârga
Al Kufrah
Tasili Plateau
Hoggar
Tropic of Cancer
Sahara
Ras Nouâdhibou
El Djouf
Adrar
Tibesti
Nubian Desert
Nubia
Aïr
Bilma
Cape Verde Is.
C. Vert
Senegal
Senegambia
Gambia
Niger
Volta
Bahr el Ghazal
L. Chad
Wadai
Dârfûr
Kordofân
White Nile
Blue Nile
Atbara
Ras Dashen 4620
116
L. Tana
Barim
Bab el Mandeb
G. of Aden
Socotra
Ras Asir
Fouta Djalon
Sahel
Guinea
Chari
Ethiopian Highlands
Somali Peninsula
Benue
Adamawa Highlands
Dar Banda
Bahr el Ghazâl
Bahr el Jebel
Shaballe
Juba
Grain Coast
Ivory Coast
Gold Coast
Slave Coast
Mt. Cameroon 4070
Bight of Benin
C. Palmas
Bioko
Bight of Bonny
I. de Principe
Gulf of Guinea
São Tomé
Equator
C. Lopez
Annobón
Uele
Congo
Oubangi
Zaïre
L. Albert
Ruwenzori 5109
4321 Mt. Elgon
L. Turkana
5199 Mt. Kenya
Tana
Chutes Boyoma
L. Edward
Lualaba
L. Victoria
Ogooué
Basin
L. Kivu
5895 Kilimanjaro
Kasai
Sankuru
Pemba I.
Seychelles
Cuango
L. Tanganyika
Lirewa
Rungwe 2961
Ascension I.
L. Mweru
L. Nyasa (L. Malawi)
Aldabra Is.
Cuanza
Shaba
Bangweulu Swamp
Luapula
C. Delgado
Comoros
Bié Plateau
Zambezi
Cuando
Shire
Mozambique Channel
Madagascar
2643
Cubango
Cunene
St. Helena
C. Fria
Okavango Swamps
Victoria Falls
Mauritius
Réunion
Walvis Bay
Namib Desert
Limpopo
Kalahari
Tropic of Capricorn
Delagoa B.
Vaal
High Veld
Orange
Drakensberg
3482
Compass Mt. 2505
Nuweveldberge
Great Karoo
Swartberge
C. of Good Hope
C. Agulhas
Algoa B.
Tristan da Cunha
ft m
12000 4000
9000 3000
6000 2000
3000 1000
1500 500
600 200
0 0
200 600
1000 3000
2000 6000
4000 12000
m ft

1 : 42 000 000

NORTH ATLANTIC OCEAN
SOUTH ATLANTIC OCEAN
INDIAN OCEAN
Mediterranean Sea
Black Sea
Red Sea
Caspian Sea
Gulf of Guinea
Bight of Benin
Mozambique Channel
G. of Aden
The Gulf
Sahara
Tropic of Cancer
Equator
Tropic of Capricorn
MOROCCO
ALGERIA
TUNISIA
LIBYA
EGYPT
WESTERN SAHARA
MAURITANIA
MALI
NIGER
CHAD
SUDAN
ERITREA
DJIBOUTI
ETHIOPIA
SOMALI REP.
SENEGAL
GAMBIA
GUINEA-BISSAU
GUINEA
SIERRA LEONE
LIBERIA
IVORY COAST
BURKINA FASO
GHANA
TOGO
BENIN
NIGERIA
CAMEROON
CENTRAL AFRICAN REP.
EQUATORIAL GUINEA
SÃO TOMÉ & PRINCIPE
GABON
CONGO
ZAÏRE
UGANDA
KENYA
RWANDA
BURUNDI
TANZANIA
ANGOLA
ZAMBIA
MALAWI
MOZAMBIQUE
ZIMBABWE
NAMIBIA
BOTSWANA
SOUTH AFRICA
LESOTHO
SWAZ.
MADAGASCAR
SAUDI ARABIA
YEMEN
IRAN
TURKEY
SYRIA
ISRAEL
JORDAN
CYPRUS
KUWAIT
SPAIN
PORTUGAL
FRANCE
ITALY
GREECE
RUSSIA
UKRAINE
KAZAKSTAN
Cairo
Capital Cities
Projection: Azimuthal Equidistant
West from Greenwich
East from Greenwich

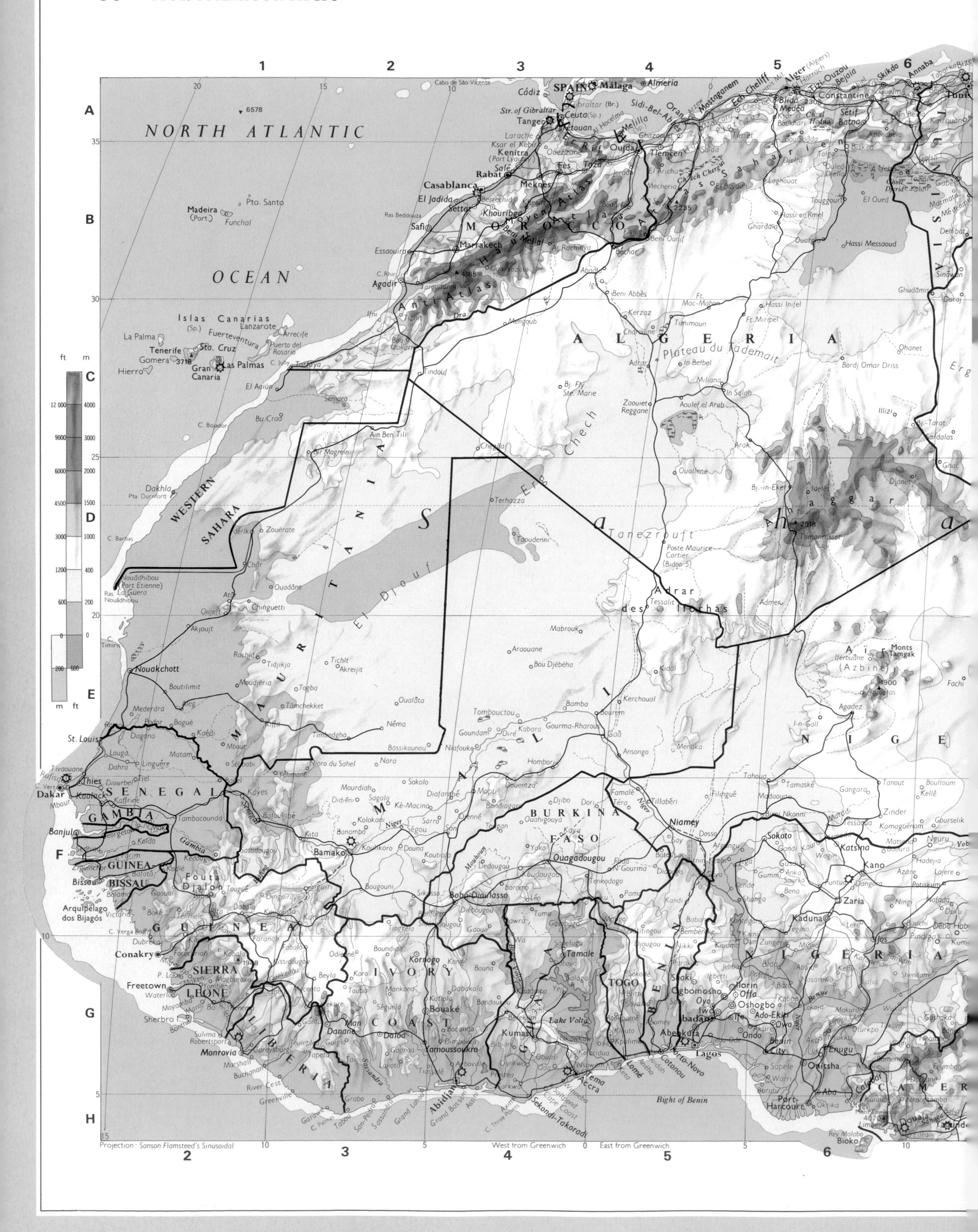

NORTH ATLANTIC
OCEAN
Madeira (Port.)
Funchal
Pto. Santo
Islas Canarias (Sp.)
La Palma
Tenerife
Gomera
Hierro
Gran Canaria
Las Palmas
Sta. Cruz
Lanzarote
Fuerteventura
Arrecife
SPAIN
Málaga
Almería
Cádiz
Str. of Gibraltar
Tanger
Ceuta (Sp.)
Tetouan
Melilla
Oran
Alger (Algiers)
Constantine
Annaba
Tunis
M O R O C C O
Rabat
Casablanca
El Jadida
Safi
Essaouira
Agadir
Marrakech
Fès
Meknès
Oujda
Kenitra (Port Lyautey)
Salé
Anti Atlas
A L G E R I A
Plateau du Tademait
Hassi Messaoud
Ghardaïa
In Salah
Adrar
Tindouf
Béchar
Ahaggar
Tamanrasset
Djanet
Illizi
Tanezrouft
Adrar des Iforhas
WESTERN SAHARA
El Aaiún
Dakhla
Nouâdhibou (Port Etienne)
M A U R I T A N I A
Nouakchott
Atar
Zouérate
El Djouf
Chech
S a h a r a
M A L I
Tombouctou
Gao
Bamako
Mopti
Ségou
Kidal
Taoudenni
N I G E R
Niamey
Agadez
Aïr (Azbine)
Monts Tamgak
S E N E G A L
Dakar
St. Louis
Thiès
Kaolack
GAMBIA
Banjul
GUINEA-BISSAU
Bissau
Arquipélago dos Bijagós
G U I N E A
Conakry
Fouta Djalon
SIERRA LEONE
Freetown
L I B E R I A
Monrovia
IVORY COAST
Abidjan
Yamoussoukro
Bouaké
G H A N A
Accra
Kumasi
Tamale
Lake Volta
Sekondi-Takoradi
TOGO
Lomé
BENIN
Porto-Novo
Cotonou
BURKINA FASO
Ouagadougou
Bobo-Dioulasso
N I G E R I A
Lagos
Ibadan
Kano
Kaduna
Zaria
Sokoto
Port Harcourt
Enugu
Bight of Benin
Bioko
Douala
C A M E R
Projection: Sanson Flamsteed's Sinusoidal
West from Greenwich
East from Greenwich

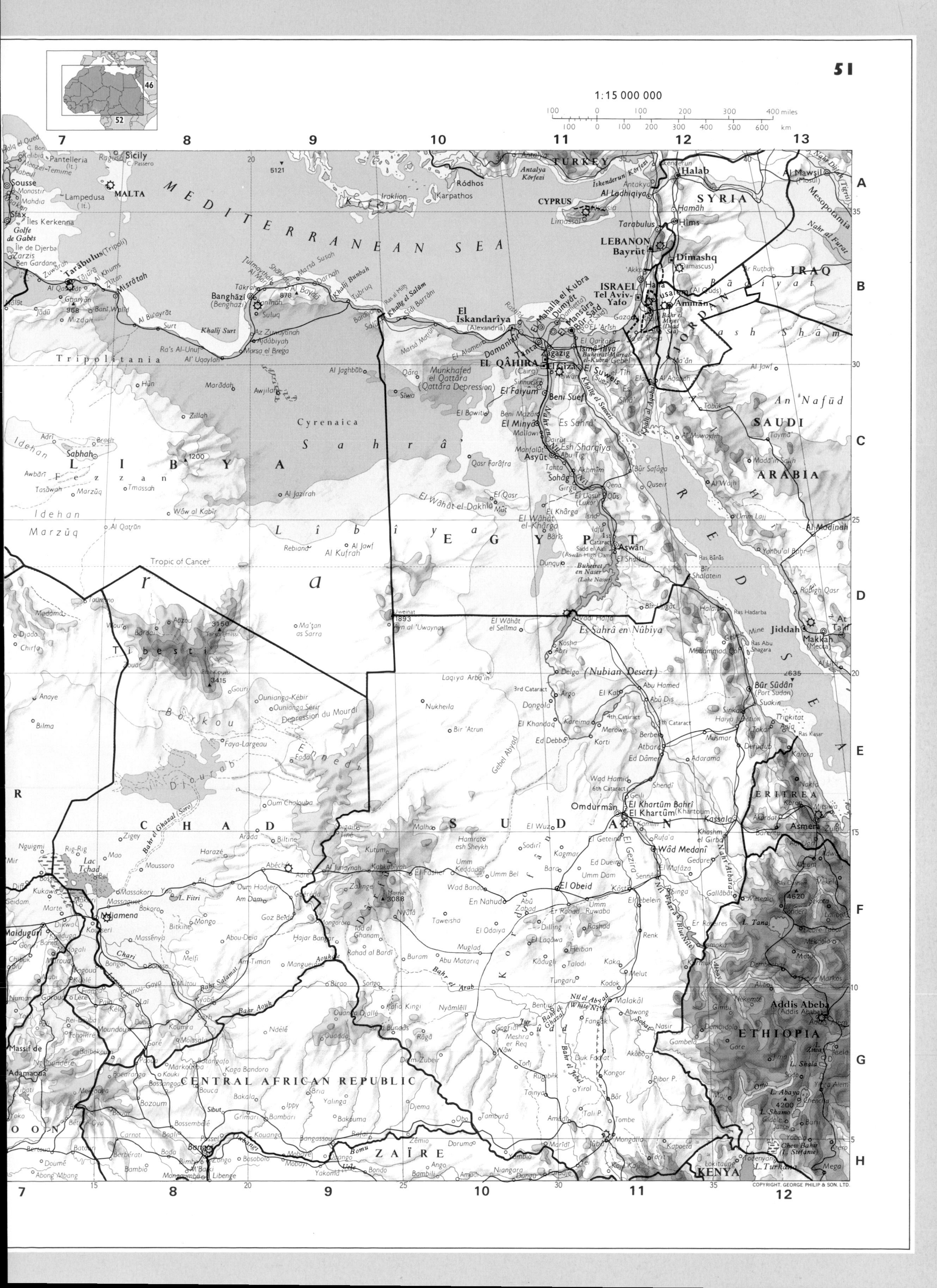
1:15 000 000
MEDITERRANEAN SEA
LIBYA
EGYPT
SUDAN
CHAD
CENTRAL AFRICAN REPUBLIC
ZAÏRE
KENYA
ETHIOPIA
ERITREA
SAUDI ARABIA
SYRIA
LEBANON
ISRAEL
JORDAN
IRAQ
TURKEY
CYPRUS
MALTA
RED SEA
El Qâhira
(Cairo)
El Iskandarîya
(Alexandria)
El Khartûm
(Khartoum)
Addis Abeba
(Addis Ababa)
Tropic of Cancer
COPYRIGHT, GEORGE PHILIP & SON, LTD.

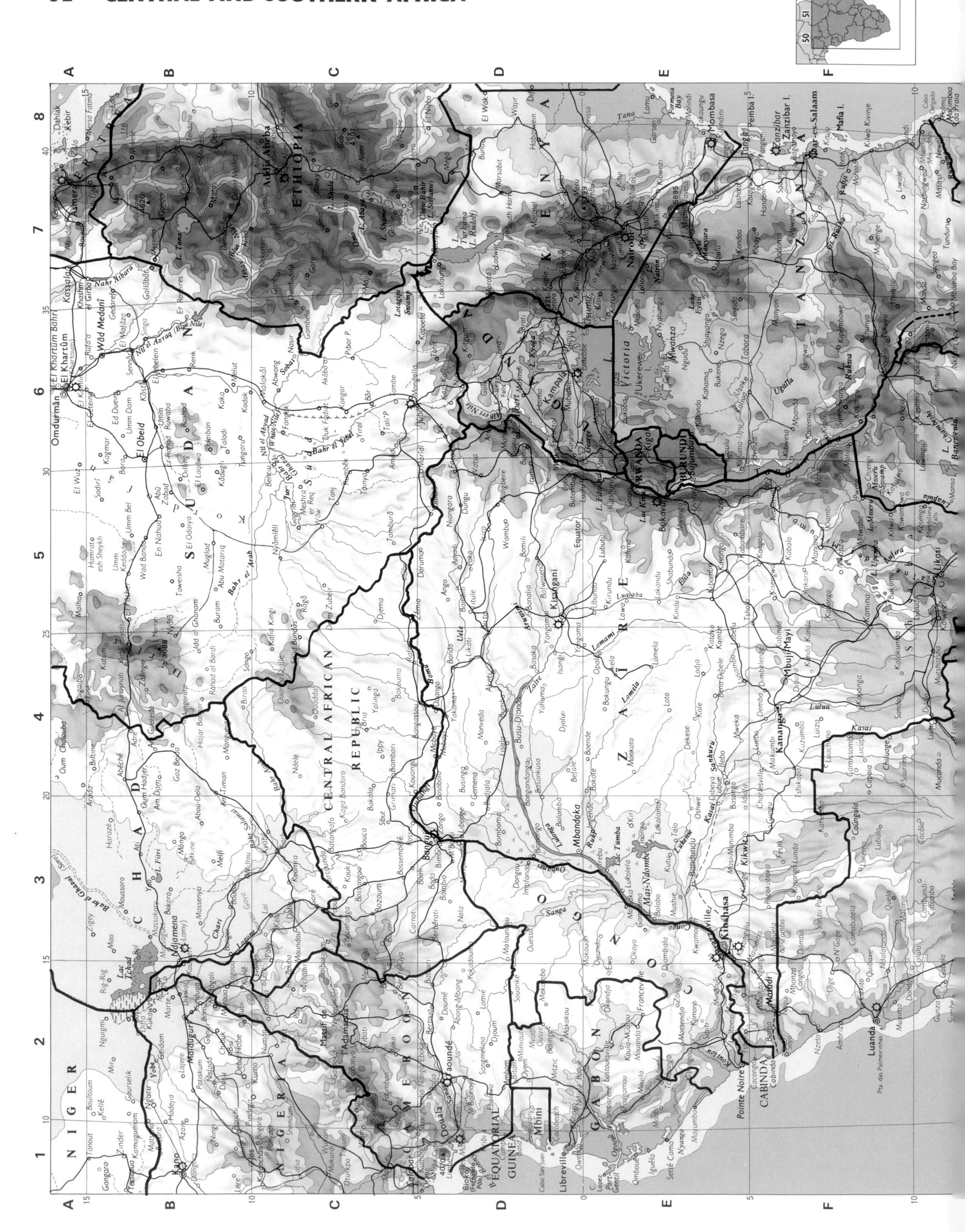
ETHIOPIA
CENTRAL AFRICAN REPUBLIC
RWANDA
BURUNDI
CABINDA
EQUATORIAL GUINEA
Addis Ababa
Nairobi
Mombasa
Kampala
Kinshasa
Kisangani
Bangui
Libreville
Luanda
Yaoundé
El Khartûm
Omdurmân
Victoria
Zanzibar
Dar-es-Salaam

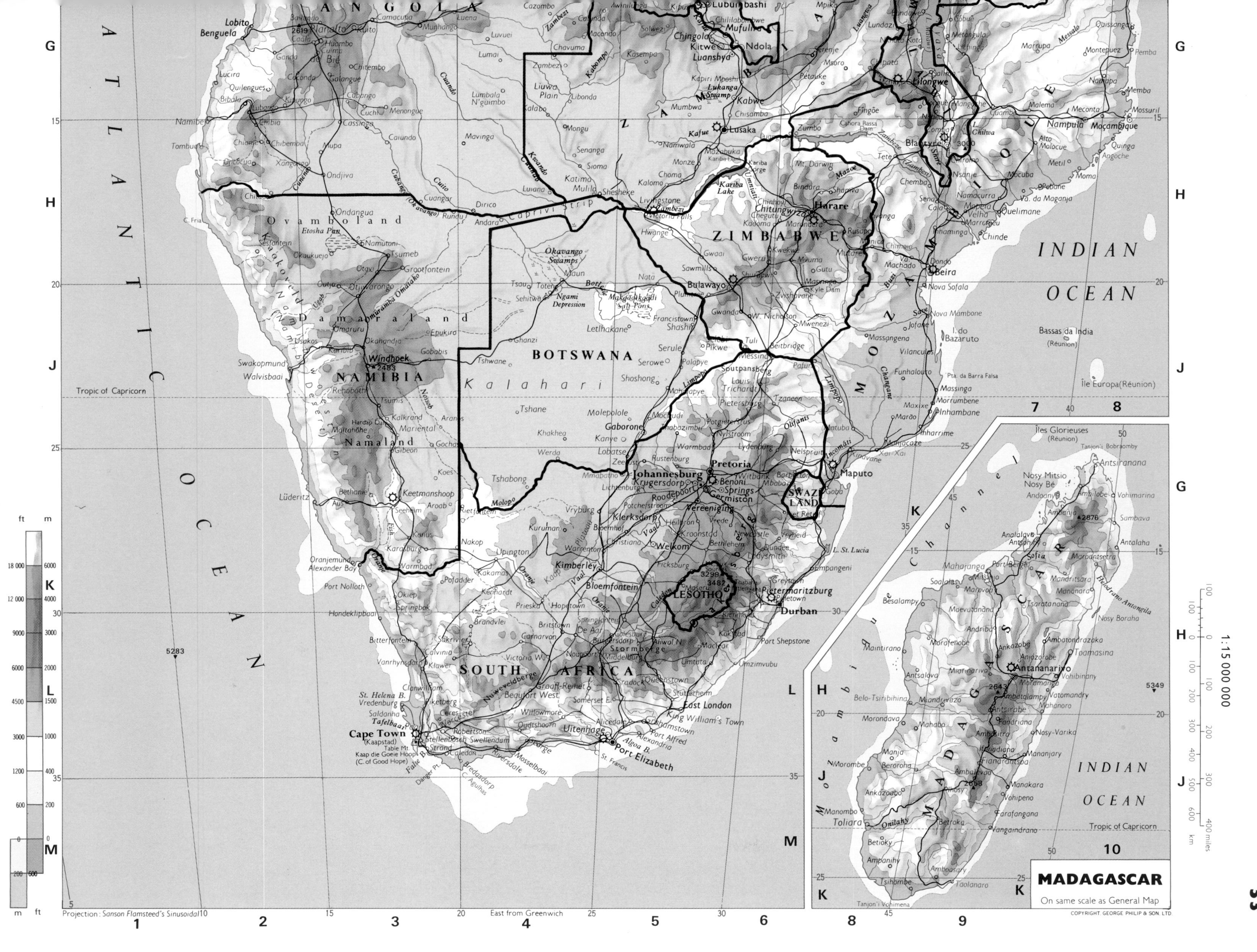
ATLANTIC OCEAN
INDIAN OCEAN
ANGOLA
ZAMBIA
ZIMBABWE
MOZAMBIQUE
BOTSWANA
NAMIBIA
SOUTH AFRICA
LESOTHO
SWAZILAND
MADAGASCAR
Kalahari
Ovamboland
Damaraland
Namaland
Namib Desert
Caprivi Strip
Okavango Swamps
Makgadikgadi Salt Pans
Ngami Depression
Etosha Pan
Mozambique Channel
Lobito
Benguela
Namibe
Lusaka
Harare
Bulawayo
Beira
Maputo
Windhoek
Gaborone
Pretoria
Johannesburg
Bloemfontein
Kimberley
Pietermaritzburg
Durban
East London
Port Elizabeth
Cape Town
(Kaapstad)
Antananarivo
Toliara
Tropic of Capricorn
1:15 000 000
Projection: Sanson Flamsteed's Sinusoidal
East from Greenwich
MADAGASCAR
On same scale as General Map
COPYRIGHT GEORGE PHILIP & SON LTD

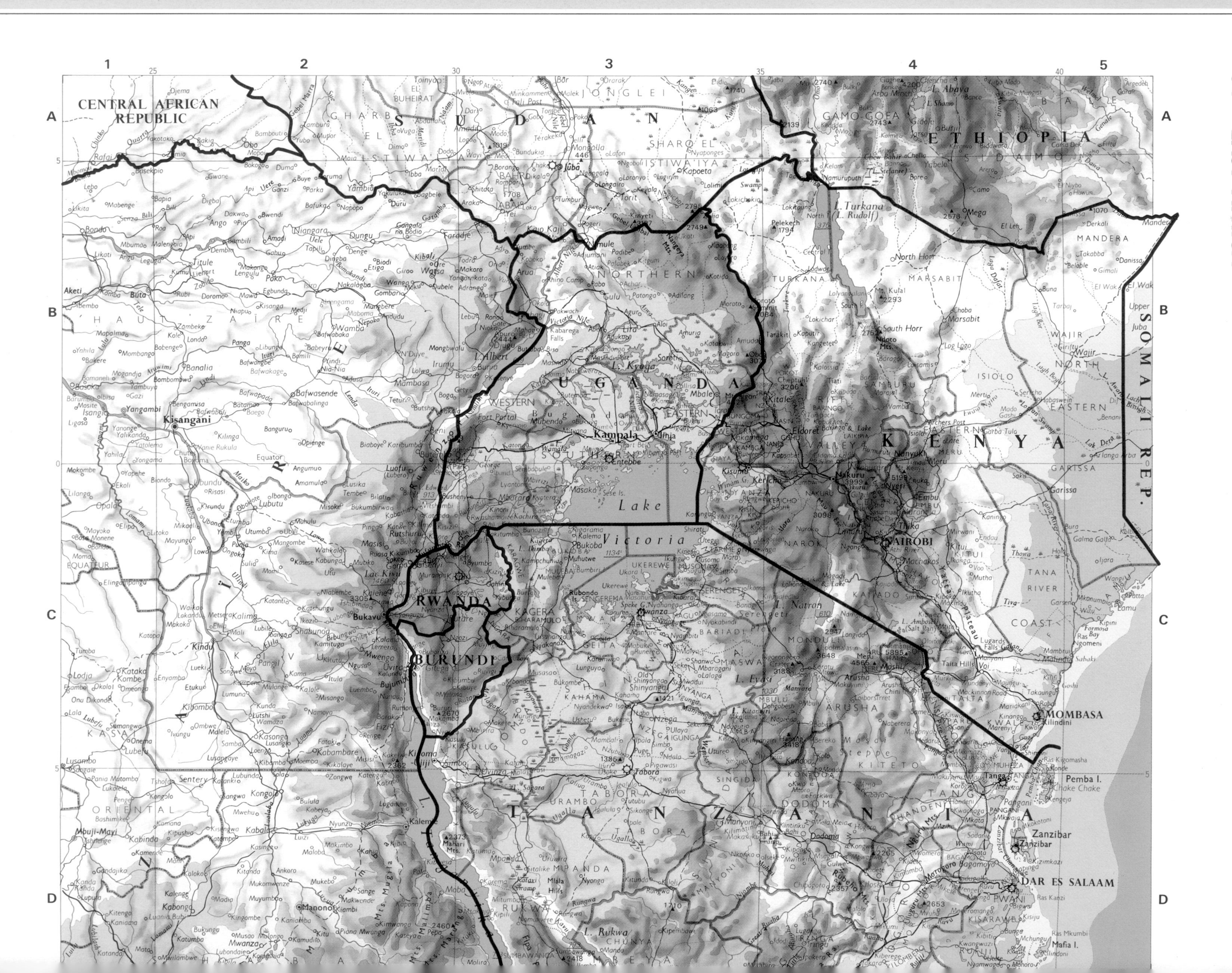
CENTRAL AFRICAN REPUBLIC
SUDAN
ETHIOPIA
SOMALI REP.
UGANDA
KENYA
RWANDA
BURUNDI
TANZANIA
Lake Victoria
L. Turkana (L. Rudolf)
L. Albert
L. Kyoga
L. Natron
L. Eyasi
L. Rukwa
Kisangani
Juba
Kampala
Entebbe
Jinja
NAIROBI
MOMBASA
DAR ES SALAAM
Zanzibar
Pemba I.
Mafia I.
Tabora
Mwanza
Bukavu
Tanga
Kisumu
Nakuru
Eldoret
Kitale
Arusha
Moshi
Dodoma
A
B
C
D
1
2
3
4
5

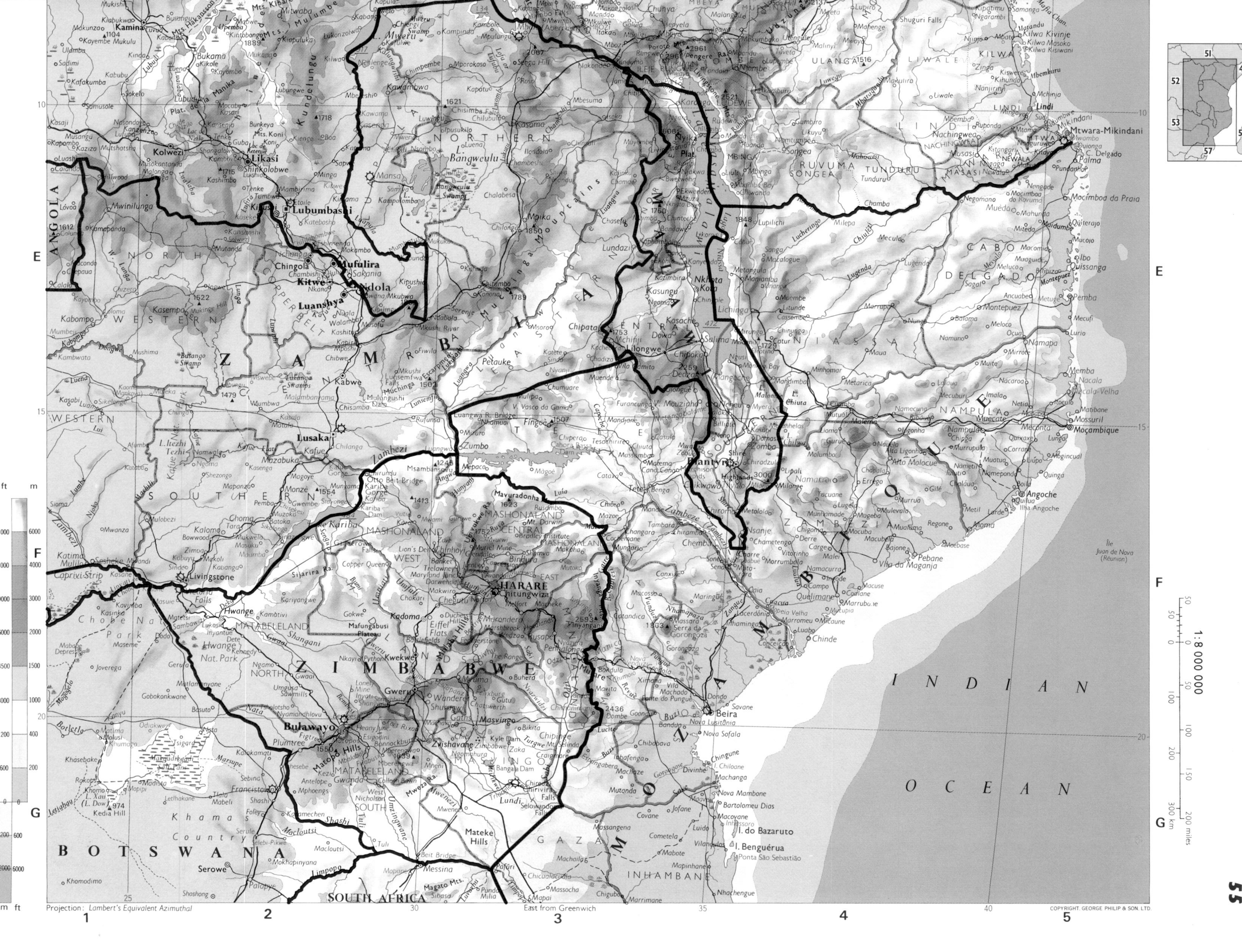

1:8 000 000
INDIAN OCEAN
ZAMBIA
ZIMBABWE
MALAWI
MOZAMBIQUE
BOTSWANA
SOUTH AFRICA
ANGOLA
HARARE
Lusaka
Bulawayo
Lilongwe
Blantyre
Beira
Moçambique
Mtwara-Mikindani
Lindi
Kitwe
Ndola
Lubumbashi
Livingstone
Zambeze
Limpopo
East from Greenwich
Projection: Lambert's Equivalent Azimuthal

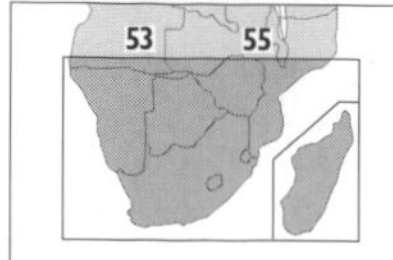
53
55

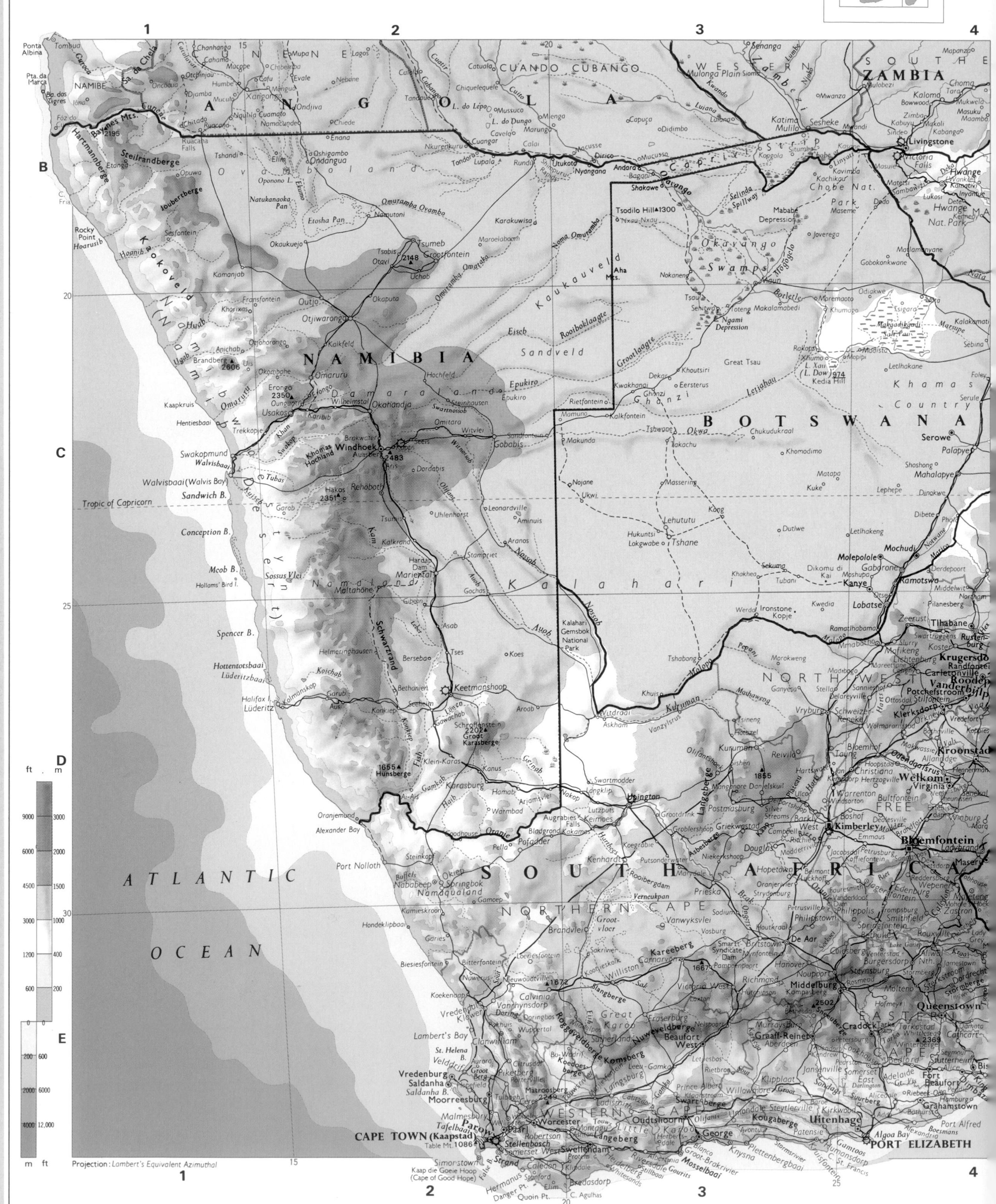
ANGOLA
NAMIBIA
BOTSWANA
SOUTH AFRICA
ZAMBIA
ATLANTIC OCEAN
CUANDO CUBANGO
NAMIBE
NORTH WEST
NORTHERN CAPE
WESTERN CAPE
EASTERN CAPE
FREE
Kalahari
Sandveld
Kaokoveld
Damaraland
Namaland
Ovamboland
Caprivi Strip
Okavango Swamps
Great Karoo
Little Karoo
Khamas Country
Chobe Nat. Park
Hwange Nat. Park
Kalahari Gemsbok National Park
Makgadikgadi Salt Pan
Etosha Pan
Ngami Depression
Tropic of Capricorn
Windhoek
Walvisbaai (Walvis Bay)
Swakopmund
Lüderitz
Keetmanshoop
Rehoboth
Okahandja
Otjiwarongo
Tsumeb
Grootfontein
Outjo
Omaruru
Gobabis
Mariental
Oranjemund
Alexander Bay
Port Nolloth
Springbok
Upington
Kimberley
Bloemfontein
Mafikeng
Gaborone
Lobatse
Kanye
Molepolole
Mochudi
Serowe
Palapye
Mahalapye
Francistown
Livingstone
Victoria Falls
Hwange
Maun
Ghanzi
Tshabong
Vryburg
Welkom
Klerksdorp
Potchefstroom
De Aar
Beaufort West
Graaff-Reinet
Cradock
Queenstown
Middelburg
Uitenhage
PORT ELIZABETH
Grahamstown
Port Alfred
George
Oudtshoorn
Mosselbaai
Knysna
Worcester
Paarl
Stellenbosch
CAPE TOWN (Kaapstad)
Table Mt. 1086
Saldanha
Vredenburg
Moorreesburg
Malmesbury
Kaap die Goeie Hoop (Cape of Good Hope)
C. Agulhas
Bredasdorp
Swellendam
Hermanus
Simonstown
Calvinia
Clanwilliam
Brandberg 2606
Erongo 2350
2483
2148
Hakos 2351
Groot Karasberge 2202
Hunsberge 1655
Schwarzrand
Langeberg
Roggeveldberge
Nuweveldberge
Kamsberg
Sneeuberge
Kareeberg
2502
2369
1672
1667
Tsodilo Hill 1300
Aha Mts.
Baynes Mts. 2195
Steilrandberge
Joubertberge
Hartmannberge
Orange
Molopo
Nossob
Auob
Fish
Cunene
Okavango
Zambezi
Kuruman
Vaal
Sandwich B.
Conception B.
Hottentotsbaai
Spencer B.
Lambert's Bay
St. Helena B.
Saldanha B.
Algoa Bay
Plettenbergbaai
Projection: Lambert's Equivalent Azimuthal
ft m
9000 3000
6000 2000
4500 1500
3000 1000
1200 400
600 200
0 0
200 600
2000 6000
4000 12,000
m ft
1 2 3 4
B C D E

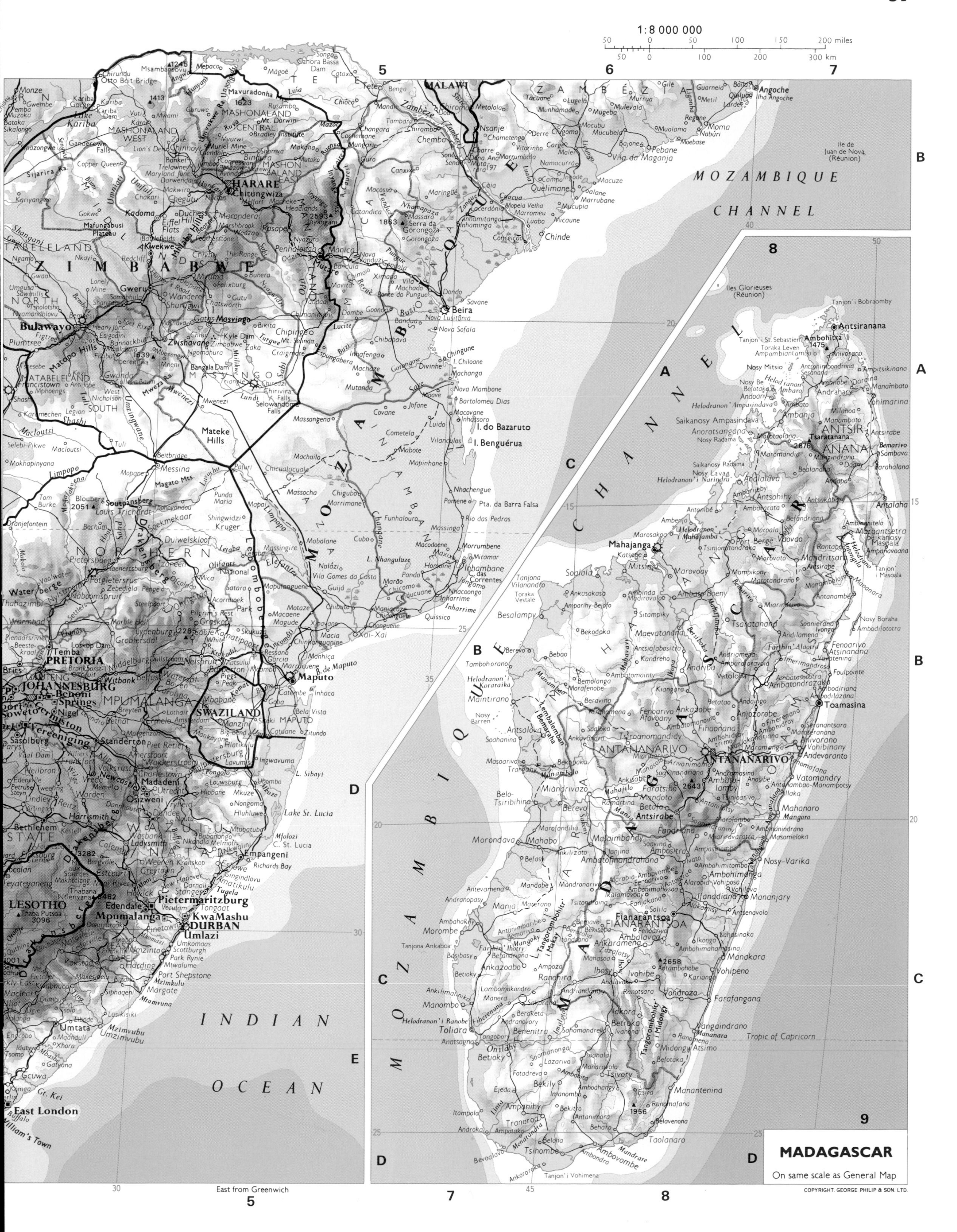
1:8 000 000
50 0 50 100 150 200 miles
50 0 100 200 300 km
MADAGASCAR
On same scale as General Map
MOZAMBIQUE CHANNEL
INDIAN OCEAN
ZIMBABWE
MOZAMBIQUE
MALAWI
SWAZILAND
LESOTHO
ZAMBÉZIA
TETE
MASHONALAND WEST
MASHONALAND CENTRAL
MASVINGO
NORTHERN
MPUMALANGA
KWAZULU NATAL
HARARE
Chitungwiza
Bulawayo
Gweru
Masvingo
Mutare
Beira
Maputo
PRETORIA
JOHANNESBURG
Soweto
Pietermaritzburg
DURBAN
Umlazi
East London
Quelimane
Chinde
Inhambane
Xai-Xai
Antsiranana
Mahajanga
Toamasina
ANTANANARIVO
Antsirabe
Fianarantsoa
Toliara
Morondava
Taolanaro
Tropic of Capricorn
Iles Glorieuses (Réunion)
Ile de Juan de Nova (Réunion)
I. do Bazaruto
I. Benguérua
Lake St. Lucia
East from Greenwich
COPYRIGHT. GEORGE PHILIP & SON. LTD.

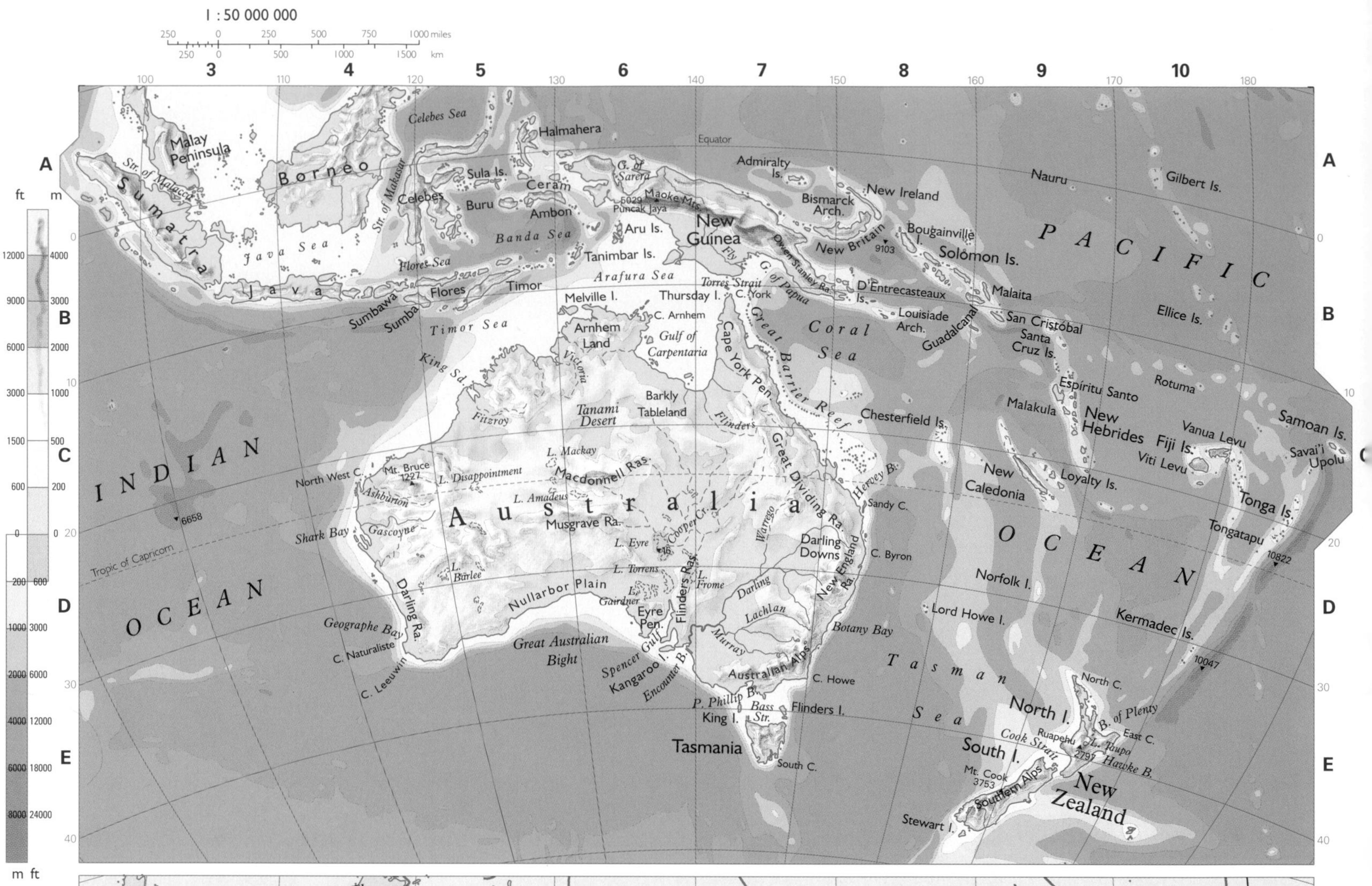
1 : 50 000 000
miles
km
INDIAN OCEAN
PACIFIC OCEAN
Malay Peninsula
Sumatra
Str. of Malacca
Borneo
Celebes Sea
Celebes
Str. of Makasar
Java Sea
Java
Halmahera
Sula Is.
Buru
Ceram
Ambon
Banda Sea
Flores Sea
Sumbawa
Sumba
Flores
Timor
Timor Sea
Aru Is.
Tanimbar Is.
Arafura Sea
Maoke Mts.
Puncak Jaya
New Guinea
Fly
Torres Strait
Thursday I.
C. York
G. of Papua
Owen Stanley Ra.
Admiralty Is.
Bismarck Arch.
New Ireland
New Britain
Bougainville
Solomon Is.
D'Entrecasteaux Is.
Louisiade Arch.
Guadalcanal
Malaita
San Cristóbal
Santa Cruz Is.
Nauru
Gilbert Is.
Ellice Is.
Melville I.
C. Arnhem
Arnhem Land
Gulf of Carpentaria
Victoria
King Sd.
Fitzroy
Tanami Desert
Barkly Tableland
Cape York Pen.
Great Barrier Reef
Coral Sea
Flinders
Mt. Bruce 1227
North West C.
Ashburton
L. Disappointment
L. Mackay
Macdonnell Ras.
L. Amadeus
Musgrave Ra.
Australia
Gascoyne
Shark Bay
L. Barlee
L. Eyre
L. Torrens
L. Gairdner
L. Frome
Cooper Cr.
Warrego
Great Dividing Ra.
Darling Downs
Darling
Lachlan
Murray
New England Ra.
Hervey B.
Sandy C.
C. Byron
Botany Bay
Nullarbor Plain
Eyre Pen.
Flinders Ras.
Spencer Gulf
Kangaroo I.
Encounter B.
Great Australian Bight
Darling Ra.
Geographe Bay
C. Naturaliste
C. Leeuwin
Australian Alps
C. Howe
P. Phillip B.
Bass Str.
King I.
Flinders I.
Tasmania
South C.
Chesterfield Is.
Espiritu Santo
Malakula
New Hebrides
New Caledonia
Loyalty Is.
Rotuma
Fiji Is.
Vanua Levu
Viti Levu
Samoan Is.
Savai'i
Upolu
Tonga Is.
Tongatapu
Norfolk I.
Lord Howe I.
Kermadec Is.
Tasman Sea
North I.
North C.
B. of Plenty
East C.
L. Taupo
Ruapehu
Hawke B.
Cook Strait
South I.
Mt. Cook 3753
Southern Alps
New Zealand
Stewart I.
Equator
Tropic of Capricorn
ft m
m ft

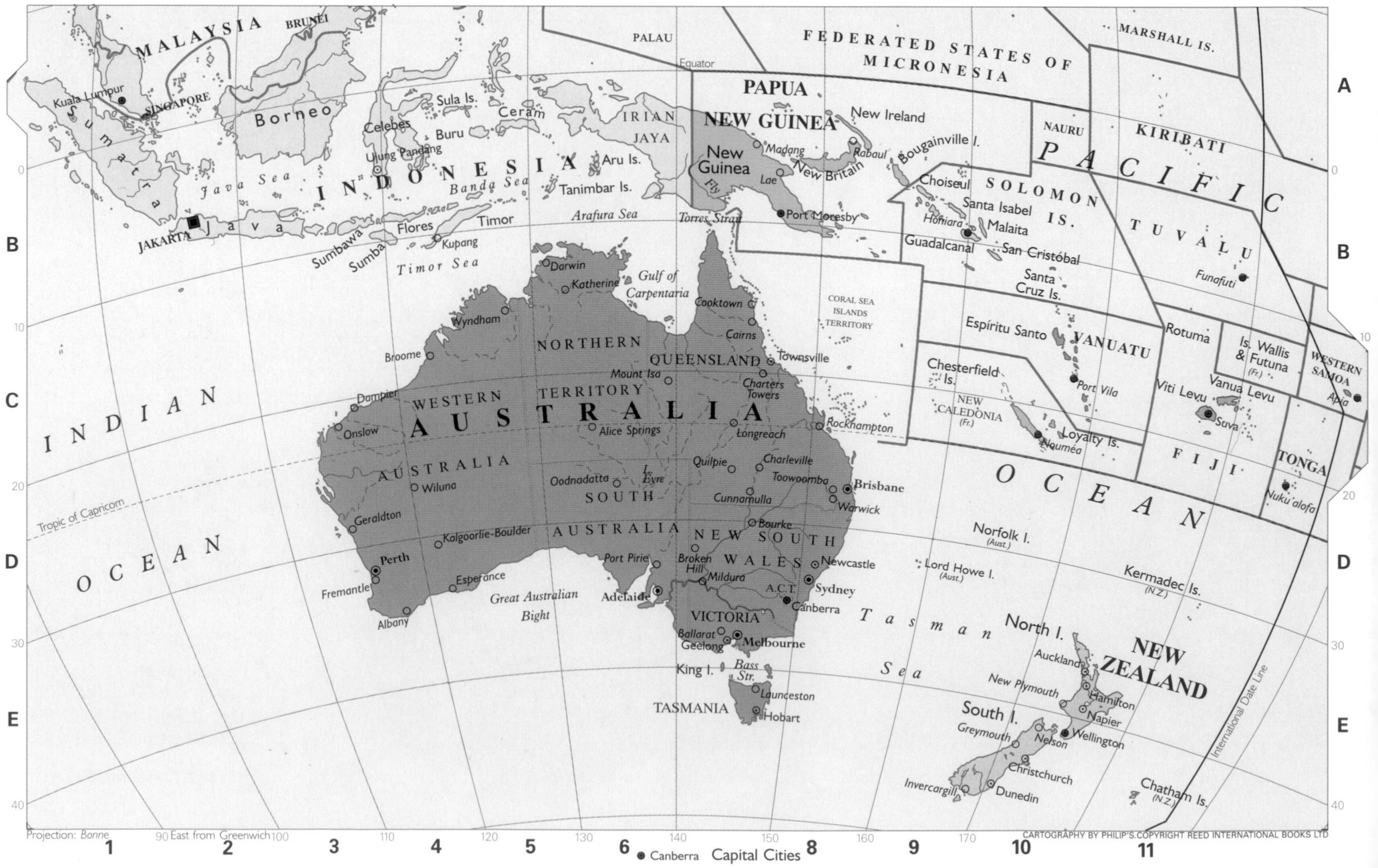
MALAYSIA
BRUNEI
Kuala Lumpur
SINGAPORE
Borneo
Sumatra
JAKARTA
Java
Java Sea
Celebes
Ujung Pandang
Sula Is.
Buru
Ceram
IRIAN JAYA
INDONESIA
Banda Sea
Aru Is.
Tanimbar Is.
Sumbawa
Sumba
Flores
Kupang
Timor
Timor Sea
Arafura Sea
PALAU
FEDERATED STATES OF MICRONESIA
MARSHALL IS.
PAPUA NEW GUINEA
New Guinea
Madang
Lae
Fly
Torres Strait
Port Moresby
New Ireland
Rabaul
New Britain
Bougainville I.
Choiseul
SOLOMON IS.
Santa Isabel
Honiara
Malaita
Guadalcanal
San Cristóbal
Santa Cruz Is.
NAURU
KIRIBATI
TUVALU
Funafuti
PACIFIC OCEAN
Darwin
Katherine
Gulf of Carpentaria
Wyndham
Cooktown
Cairns
Townsville
CORAL SEA ISLANDS TERRITORY
NORTHERN TERRITORY
QUEENSLAND
Mount Isa
Charters Towers
Broome
Dampier
Onslow
WESTERN AUSTRALIA
AUSTRALIA
Alice Springs
Longreach
Rockhampton
Wiluna
Oodnadatta
L. Eyre
Quilpie
Charleville
Toowoomba
Brisbane
Warwick
Cunnamulla
SOUTH AUSTRALIA
Geraldton
Kalgoorlie-Boulder
Bourke
NEW SOUTH WALES
Perth
Fremantle
Esperance
Albany
Port Pirie
Broken Hill
Mildura
Newcastle
Sydney
A.C.T.
Canberra
Adelaide
Great Australian Bight
VICTORIA
Ballarat
Geelong
Melbourne
King I.
Bass Str.
Launceston
TASMANIA
Hobart
Espiritu Santo
VANUATU
Port Vila
Chesterfield Is.
NEW CALEDONIA (Fr.)
Loyalty Is.
Nouméa
Rotuma
Is. Wallis & Futuna (Fr.)
WESTERN SAMOA
Apia
Viti Levu
Vanua Levu
Suva
FIJI
TONGA
Nuku'alofa
Norfolk I. (Aust.)
Lord Howe I. (Aust.)
Kermadec Is. (N.Z.)
Tasman Sea
North I.
Auckland
NEW ZEALAND
New Plymouth
Hamilton
Napier
Wellington
South I.
Greymouth
Nelson
Christchurch
Invercargill
Dunedin
Chatham Is. (N.Z.)
International Date Line
INDIAN OCEAN
Tropic of Capricorn
Equator
Projection: Bonne
East from Greenwich
Canberra Capital Cities

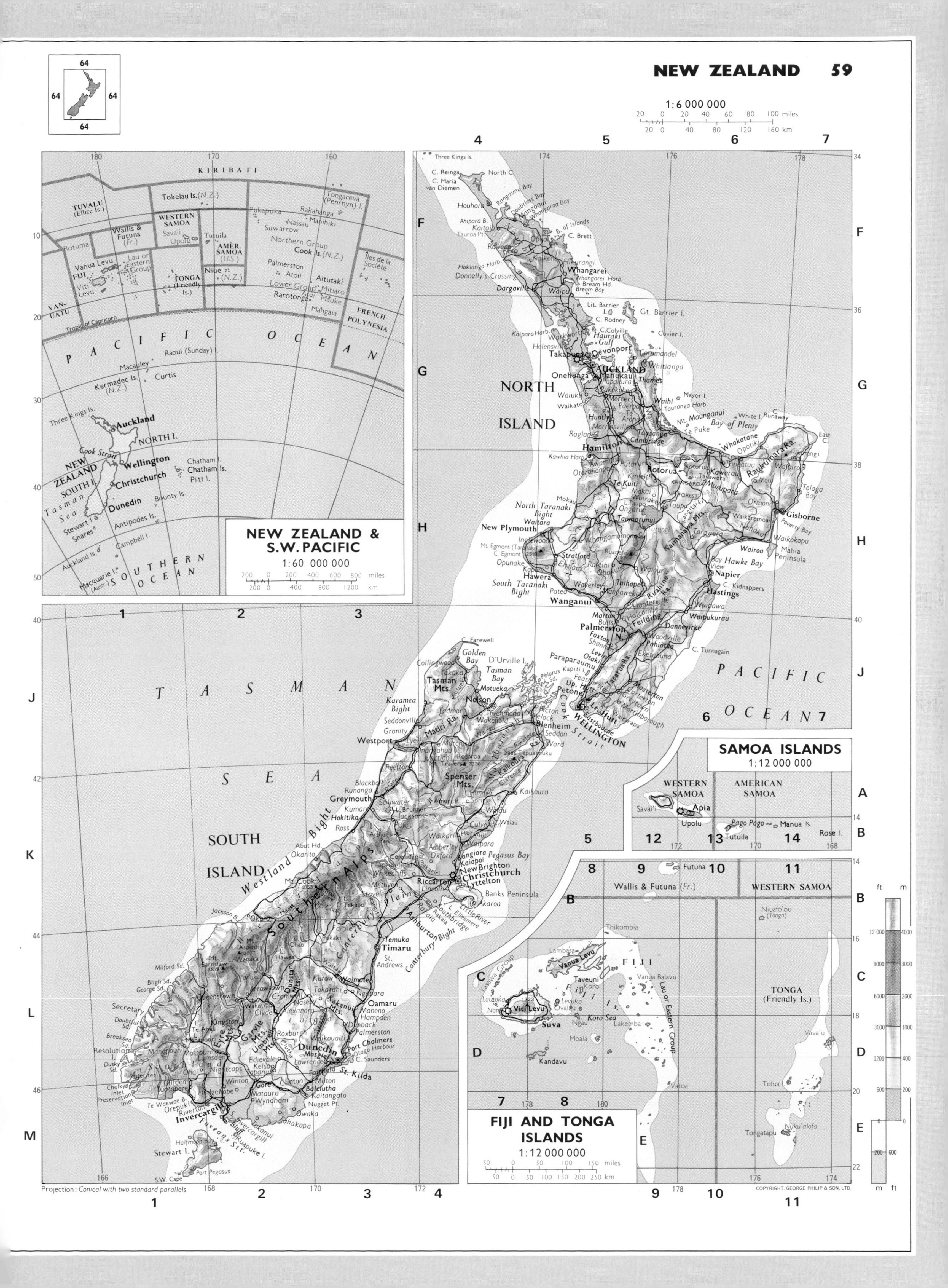
1:6 000 000
NEW ZEALAND & S.W. PACIFIC
1:60 000 000
SAMOA ISLANDS
1:12 000 000
FIJI AND TONGA ISLANDS
1:12 000 000
NORTH ISLAND
SOUTH ISLAND
TASMAN SEA
PACIFIC OCEAN
Auckland
Wellington
Christchurch
Dunedin
Hamilton
Napier
Hastings
Gisborne
New Plymouth
Wanganui
Palmerston N.
Nelson
Blenheim
Westport
Greymouth
Timaru
Oamaru
Invercargill
Stewart I.
Cook Strait
Southern Alps
WESTERN SAMOA
AMERICAN SAMOA
Apia
Pago Pago
FIJI
Suva
Viti Levu
Vanua Levu
TONGA (Friendly Is.)
Wallis & Futuna (Fr.)
Projection: Conical with two standard parallels
COPYRIGHT, GEORGE PHILIP & SON, LTD.

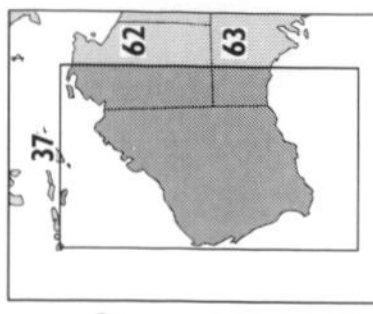
62
63
37

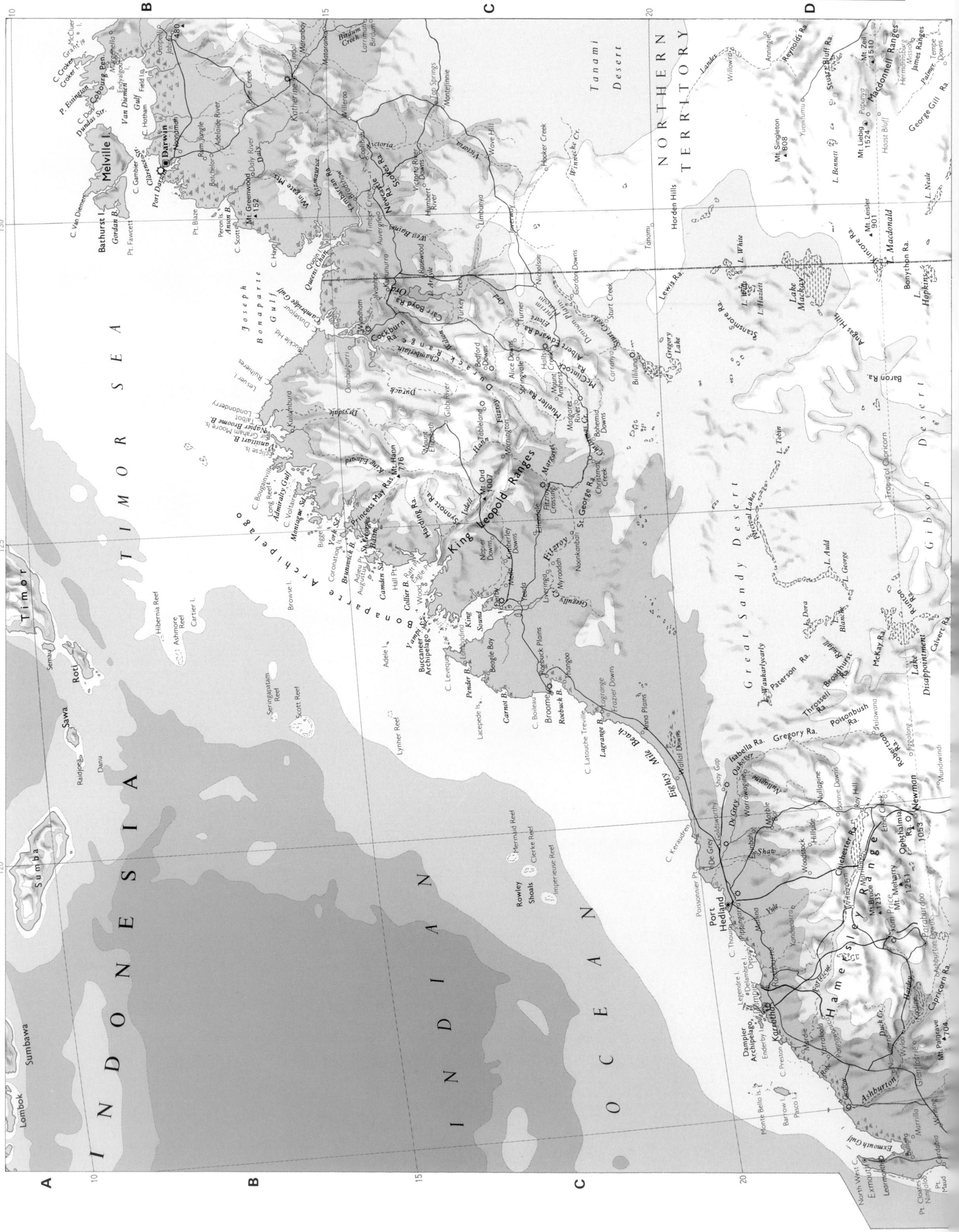
INDONESIA
TIMOR SEA
INDIAN OCEAN
NORTHERN TERRITORY
Tanami Desert
Great Sandy Desert
Gibson Desert
Joseph Bonaparte Gulf
Bonaparte Archipelago
King Leopold Ranges
Hamersley Range
Eighty Mile Beach
Dampier Archipelago
Exmouth Gulf
Macdonnell Ranges
Lake Mackay
Lake Disappointment
Melville I.
Bathurst I.
Darwin
Katherine
Wyndham
Kununurra
Derby
Broome
Port Hedland
Karratha
Newman
Timor
Roti
Sawa
Sumba
Sumbawa
Lombok

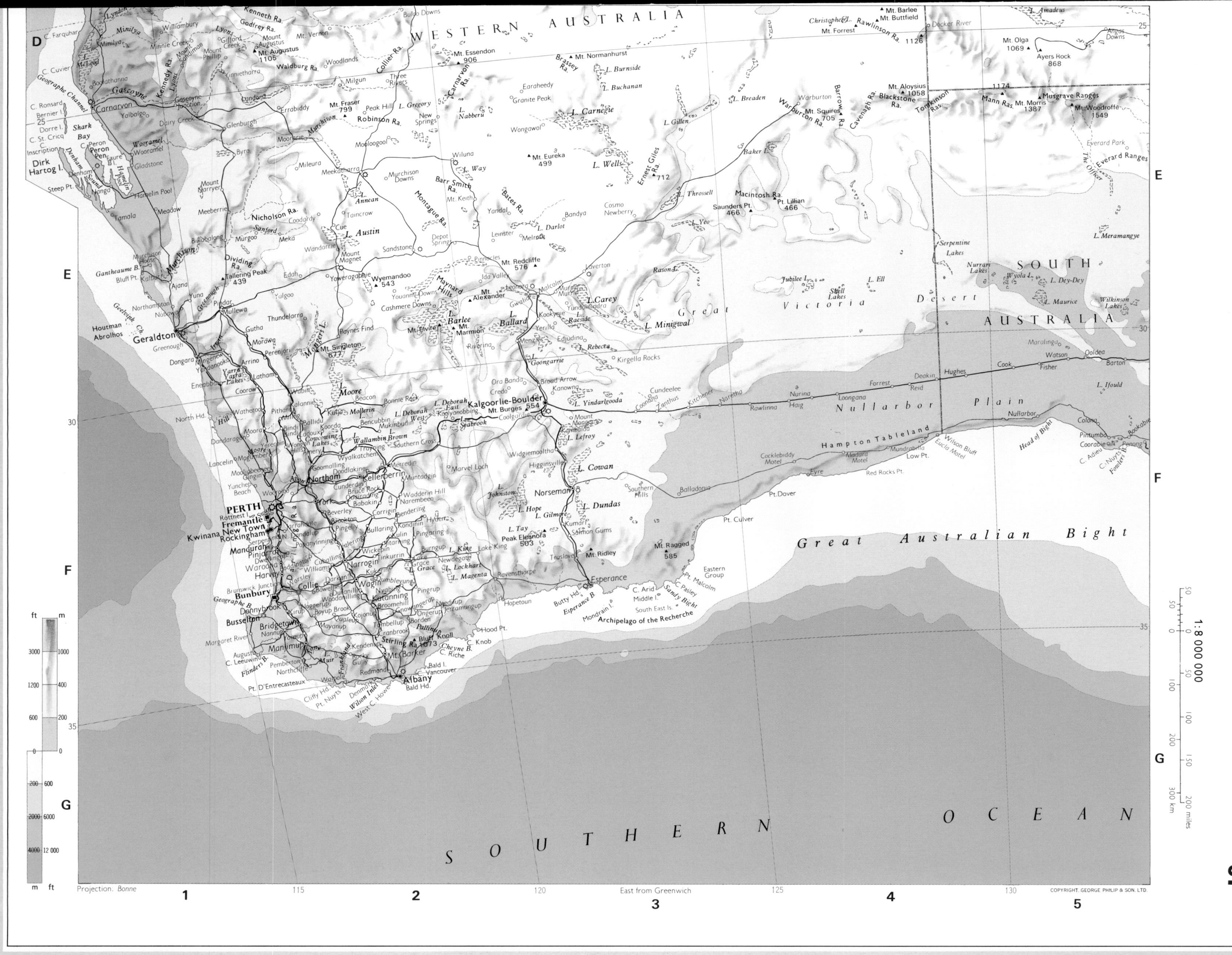
WESTERN AUSTRALIA
SOUTH AUSTRALIA
Great Victoria Desert
Nullarbor Plain
Hampton Tableland
Great Australian Bight
SOUTHERN OCEAN
PERTH
Fremantle
Kwinana New Town
Rockingham
Mandurah
Geraldton
Carnarvon
Bunbury
Busselton
Albany
Esperance
Kalgoorlie-Boulder
Norseman
Northam
Narrogin
Katanning
Collie
Archipelago of the Recherche
Shark Bay
Dirk Hartog I.
Houtman Abrolhos
L. Carnegie
L. Wells
L. Austin
L. Moore
L. Barlee
L. Ballard
L. Carey
L. Cowan
L. Dundas
L. Minigwal
Mt. Olga 1069
Ayers Rock 868
Musgrave Ranges
Mt. Woodroffe 1549
Everard Ranges
Warburton Ra.
Rawlinson Ra.
Robinson Ra.
Stirling Ra.
Head of Bight
Wilson Bluff
Eucla Motel
Forrest
Cook
Ooldea
Hughes
Deakin
Rawlinna
Zanthus
Kennedy Ra.
Collier Ra.
Ernest Giles Ra.
Macintosh Ra.
D
E
F
G
1
2
3
4
5
25
30
35
115
120
125
130
East from Greenwich
Projection. Bonne
1:8 000 000
50 0 50 100 150 200 miles
50 0 100 200 300 km
ft m
3000 1000
1200 400
600 200
0 0
200 600
2000 6000
4000 12 000
m ft
COPYRIGHT, GEORGE PHILIP & SON, LTD.

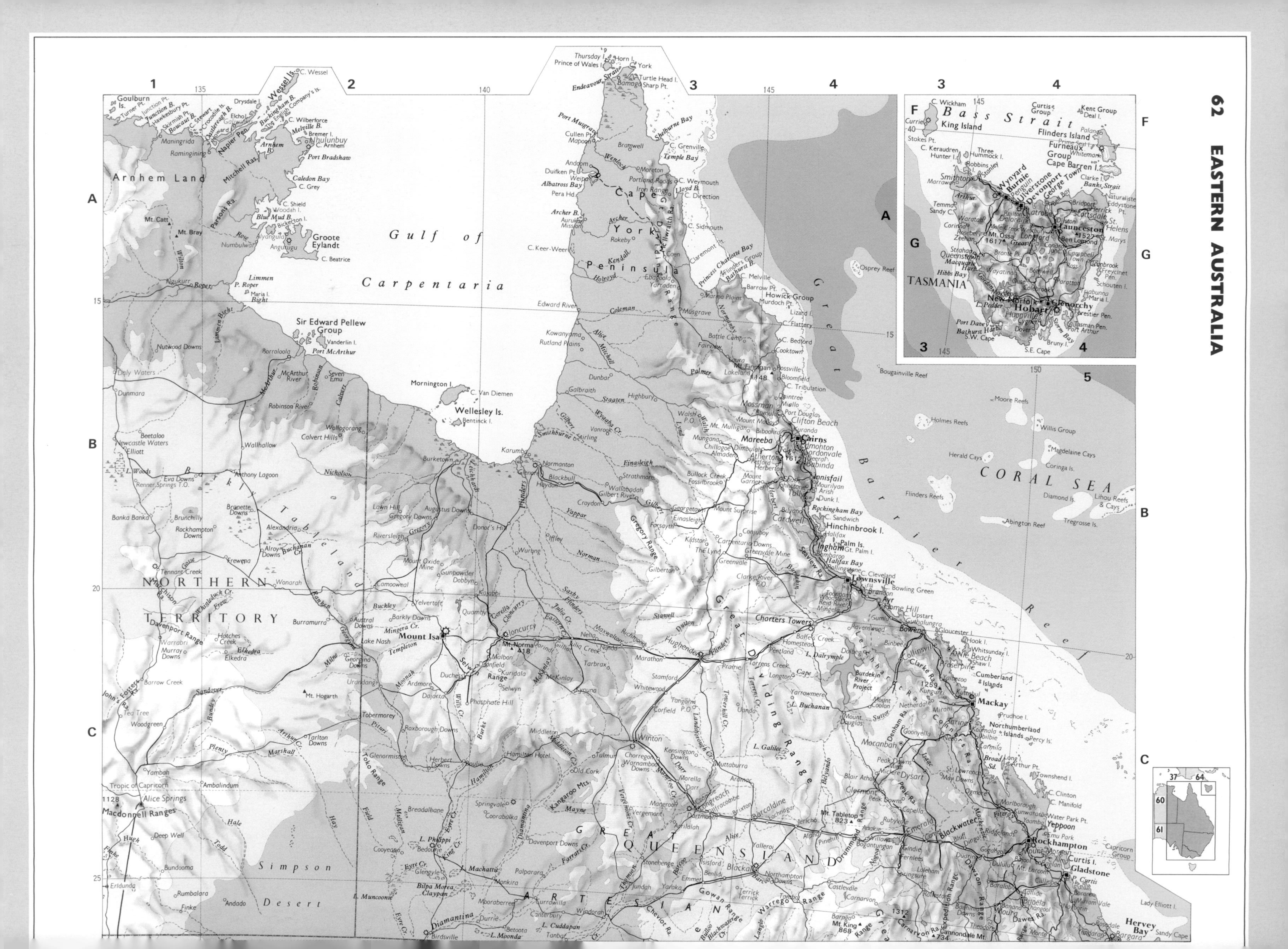

Gulf of Carpentaria
Great Barrier Reef
CORAL SEA
Bass Strait
TASMANIA
NORTHERN TERRITORY
GREAT ARTESIAN BASIN
QUEENSLAND
Great Dividing Range
Cape York Peninsula
Arnhem Land
Simpson Desert
Barkly Tableland
Groote Eylandt
Wellesley Is.
Mornington I.
Sir Edward Pellew Group
Cairns
Townsville
Mackay
Rockhampton
Gladstone
Mount Isa
Cloncurry
Charters Towers
Longreach
Winton
Alice Springs
Hobart
Launceston
King Island
Flinders Island
Hinchinbrook I.
Thursday I.
Macdonnell Ranges
Gregory Range

1:8 000 000
SOUTH AUSTRALIA
NEW SOUTH WALES
TASMAN SEA
Bass Strait
SYDNEY
CANBERRA
MELBOURNE
ADELAIDE
BRISBANE
Broken Hill
Port Augusta
Port Pirie
Whyalla
Port Lincoln
Kangaroo I.
Spencer Gulf
Lake Eyre (North)
Lake Torrens
Lake Frome
Lake Gairdner
Newcastle
Wollongong
Bendigo
Ballarat
Geelong
Mildura
Wagga Wagga
Albury
Dubbo
Tamworth
Bourke
Toowoomba
Coffs Harbour
Mount Gambier
King Island
Flinders Island
Furneaux Group
Cape Barren I.
Darling
Stuart Range
Projection: Bonne
East from Greenwich
COPYRIGHT, GEORGE PHILIP & SON, LTD.

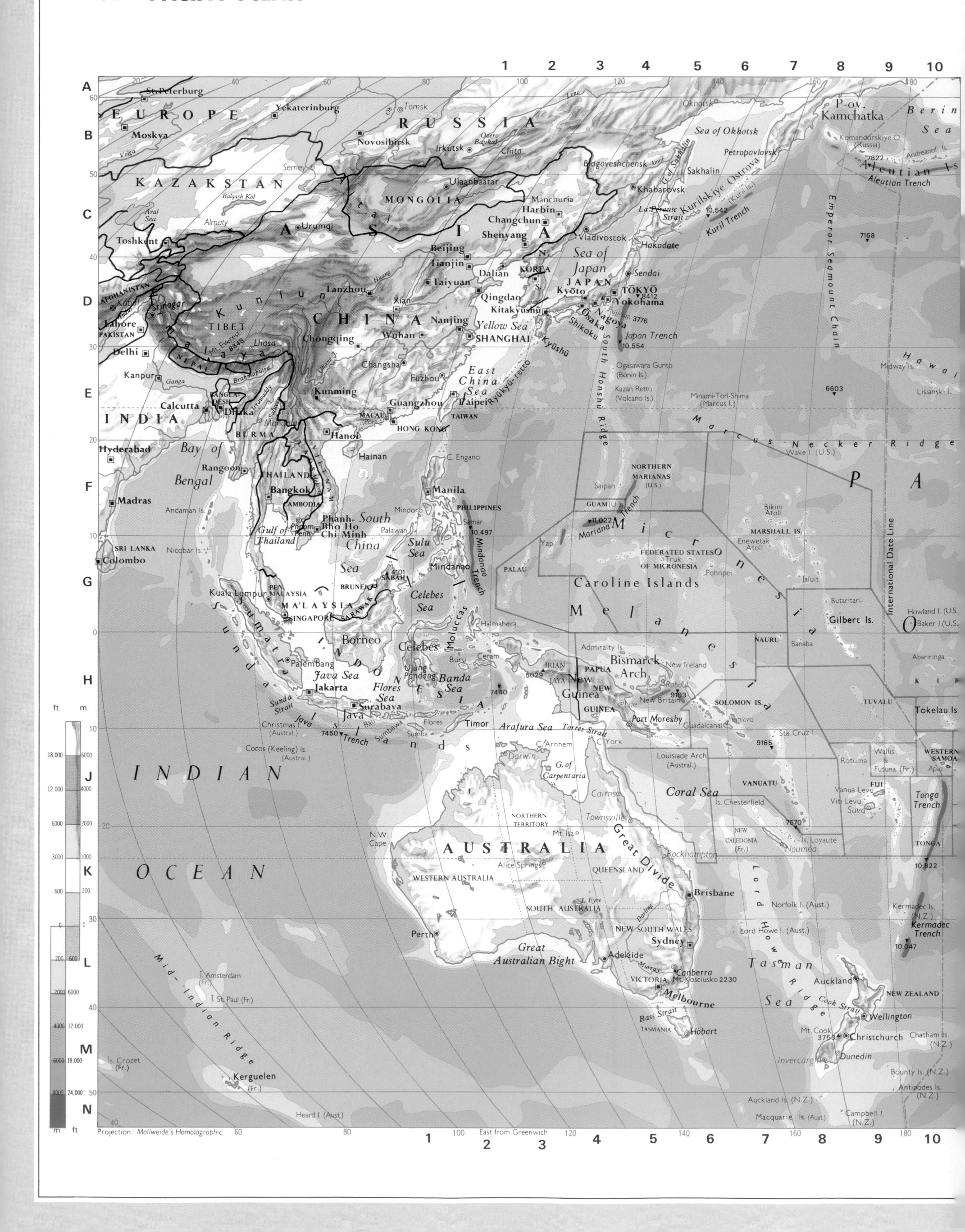
EUROPE
RUSSIA
St.Peterburg
Moskva
Yekaterinburg
Tomsk
Novosibirsk
Irkutsk
Ozero Baykal
Chita
Okhotsk
Sea of Okhotsk
P-ov. Kamchatka
Bering Sea
Komandorskiye O. (Russia)
Petropavlovsk
Aleutian Is.
Aleutian Trench
Andreanof Is.
KAZAKSTAN
Balqash Köl
Aral Sea
Almaty
Semey
Toshkent
MONGOLIA
Ulaanbaatar
Blagoveshchensk
Khabarovsk
Sakhalin
G. of Sakhalin
Kuril'skiye Ostrova
Kuril Trench
La Pérouse Strait
Manchuria
Harbin
Changchun
Shenyang
ASIA
Urumqi
Vladivostok
Hakodate
Beijing
Tianjin
Dalian
KOREA
Sea of Japan
JAPAN
Sendai
TŌKYŌ
Yokohama
Nagoya
Ōsaka
Kyōto
Shikoku
Kyūshū
Kitakyūshū
Qingdao
Taiyuan
Lanzhou
Xi'an
Kunlun
CHINA
TIBET
Nanjing
Yellow Sea
SHANGHAI
Wuhan
Chongqing
Changsha
Fuzhou
East China Sea
Taipei
TAIWAN
Ryūkyū-retto
Japan Trench
10,554
3776
8412
South Honshū Ridge
Ogasawara Gunto (Bonin Is.)
Kazan Retto (Volcano Is.)
Minami-Tori-Shima (Marcus I.)
Emperor Seamount Chain
7168
6603
Midway Is.
Hawaiian
Lisianski I.
Marcus Necker Ridge
Wake I. (U.S.)
AFGHANISTAN
Kabul
Srinagar
Lahore
PAKISTAN
Delhi
Himalaya
Mt. Everest 8848
Lhasa
NEPAL
Kanpur
Ganga
Brahmaputra
BANGLADESH
Dhaka
Calcutta
INDIA
Irrawaddy
Mandalay
BURMA
Kunming
Guangzhou
MACAU (Port.)
HONG KONG
Hanoi
Hainan
Hyderabad
Bay of Bengal
Rangoon
THAILAND
Bangkok
CAMBODIA
VIETNAM
LAOS
Phnom Penh
Phanh-Bho Ho Chi Minh
Gulf of Thailand
Madras
Andaman Is.
Nicobar Is.
SRI LANKA
Colombo
C. Engano
Manila
PHILIPPINES
Mindoro
Palawan
Samar
Mindanao
Mindanao Trench
10,497
South China Sea
Sulu Sea
Celebes Sea
BRUNEI
SABAH
SARAWAK
4101
PEN. MALAYSIA
MALAYSIA
Kuala Lumpur
SINGAPORE
Sumatra
Borneo
Celebes
Moluccas
Halmahera
INDONESIA
Palembang
Java Sea
Jakarta
Sunda Strait
Java
Surabaya
Flores Sea
Banda Sea
Ujung Pandang
Buru
Ceram
Bali
Sumbawa
Flores
Sumba
Timor
Java Trench
7450
Sunda Islands
Christmas I. (Austral.)
Cocos (Keeling) Is. (Austral.)
7440
NORTHERN MARIANAS (U.S.)
Saipan
GUAM (U.S.)
Mariana Trench
11,022
Micronesia
Yap
PALAU
FEDERATED STATES OF MICRONESIA
Truk
Pohnpei
Caroline Islands
Bikini Atoll
Enewetak Atoll
MARSHALL IS.
Jaluit
Butaritari
Gilbert Is.
Howland I. (U.S.)
Baker I. (U.S.)
International Date Line
PACIFIC OCEAN
Melanesia
NAURU
Banaba
Abariringa
KIRIBATI
Admiralty Is.
Bismarck Arch.
New Ireland
IRIAN JAYA
5029
PAPUA NEW GUINEA
New Guinea
Rabaul
New Britain
9103
Lae
SOLOMON IS.
Honiara
Guadalcanal
Port Moresby
Torres Strait
Arafura Sea
C. Arnhem
Darwin
C. York
G. of Carpentaria
Sta. Cruz I.
9165
Louisiade Arch. (Austral.)
TUVALU
Tokelau Is.
Rotuma
Wallis & Futuna (Fr.)
WESTERN SAMOA
Apia
VANUATU
FIJI
Vanua Levu
Viti Levu
Suva
Coral Sea
Cairns
Is. Chesterfield
Townsville
7570
NEW CALEDONIA (Fr.)
Nouméa
Is. Loyauté
Tonga Trench
TONGA
10,822
NORTHERN TERRITORY
Mt. Isa
Great Divide
Rockhampton
N.W. Cape
AUSTRALIA
Alice Springs
WESTERN AUSTRALIA
QUEENSLAND
Brisbane
L. Eyre
SOUTH AUSTRALIA
Darling
NEW SOUTH WALES
Sydney
Perth
Great Australian Bight
Adelaide
Murray
Canberra
Mt. Kosciusko 2230
VICTORIA
Melbourne
Bass Strait
TASMANIA
Hobart
Lord Howe Ridge
Norfolk I. (Aust.)
Lord Howe I. (Aust.)
Tasman Sea
Kermadec Is. (N.Z.)
Kermadec Trench
10,047
Auckland
NEW ZEALAND
Cook Strait
Wellington
Chatham Is. (N.Z.)
Mt. Cook 3753
Christchurch
Invercargill
Dunedin
Bounty Is. (N.Z.)
Antipodes Is. (N.Z.)
Auckland Is. (N.Z.)
Macquarie Is. (Aust.)
Campbell I. (N.Z.)
INDIAN OCEAN
Mid-Indian Ridge
I. Amsterdam (Fr.)
I. St. Paul (Fr.)
Is. Crozet (Fr.)
Kerguelen (Fr.)
Heard I. (Aust.)
Projection: Mollweide's Homolographic
East from Greenwich
ft
m
18,000
12 000
6000
3000
600
0
200
2000
4000
6000
8000
6000
4000
2000
1000
200
600
12 000
18,000
24,000

1:54 000 000

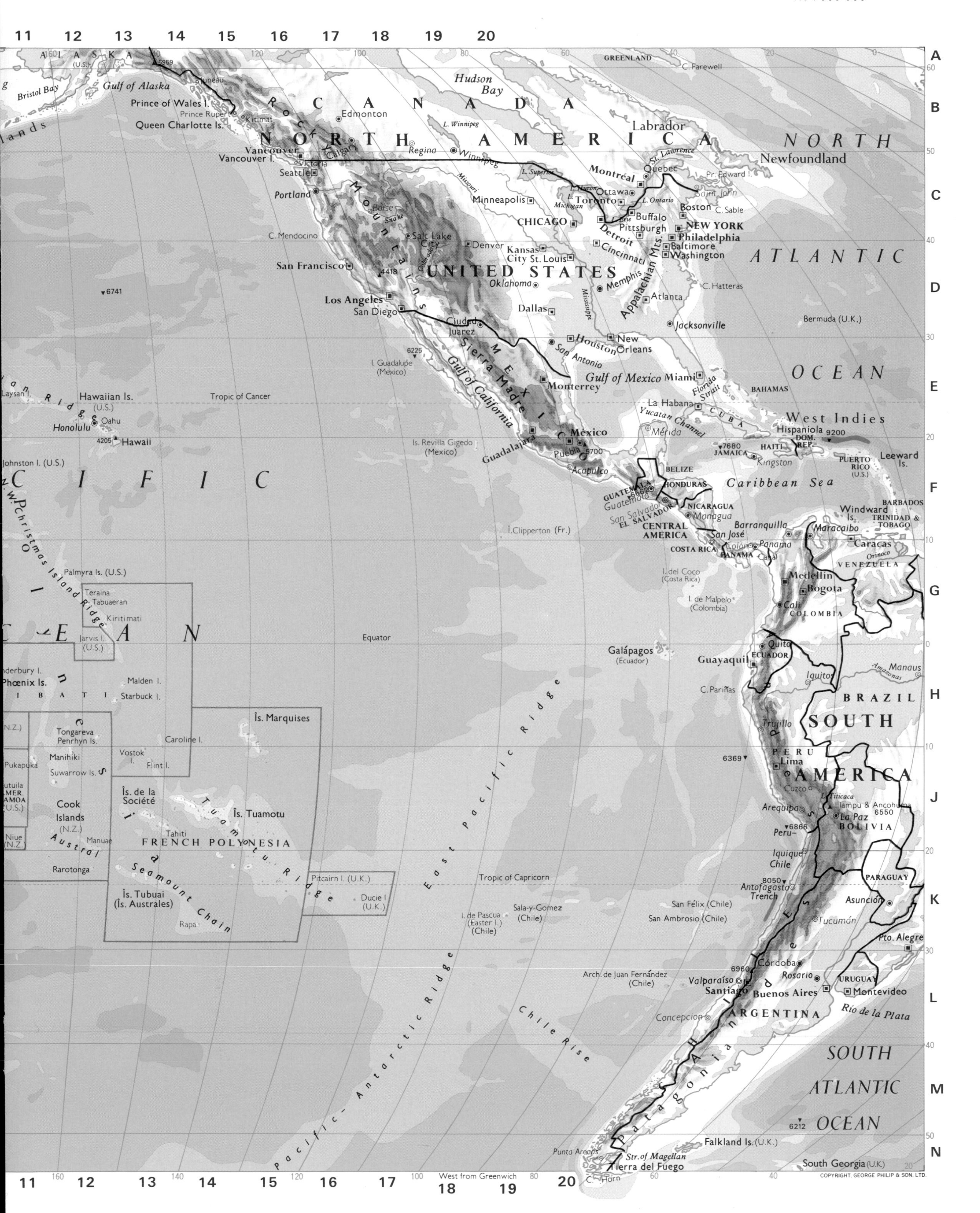

Hudson Bay
C A N A D A
N O R T H A M E R I C A
UNITED STATES
Rocky Mountains
Gulf of Alaska
Prince of Wales I.
Queen Charlotte Is.
Vancouver
Vancouver I.
Seattle
Portland
Edmonton
Calgary
Regina
Winnipeg
Minneapolis
CHICAGO
Detroit
Toronto
Ottawa
Montréal
Quebec
Boston
NEW YORK
Philadelphia
Baltimore
Washington
Buffalo
Pittsburgh
Cincinnati
Kansas City
St. Louis
Denver
Salt Lake City
San Francisco
Los Angeles
San Diego
Oklahoma
Dallas
Memphis
Atlanta
Appalachian Mts.
Houston
San Antonio
New Orleans
Jacksonville
Miami
Labrador
Newfoundland
GREENLAND
C. Farewell
C. Mendocino
C. Hatteras
ATLANTIC
OCEAN
NORTH
Bermuda (U.K.)
Gulf of Mexico
Gulf of California
Sierra Madre
M E X I C O
Monterrey
Ciudad Juarez
Guadalajara
México
Puebla
Acapulco
La Habana
CUBA
BAHAMAS
West Indies
Hispaniola
JAMAICA
Kingston
HAITI
DOM. REP.
PUERTO RICO (U.S.)
Leeward Is.
Windward Is.
BARBADOS
TRINIDAD & TOBAGO
Caribbean Sea
BELIZE
GUATEMALA
Guatemala
HONDURAS
EL SALVADOR
San Salvador
NICARAGUA
Managua
CENTRAL AMERICA
COSTA RICA
San José
PANAMA
Panama
Barranquilla
Maracaibo
Caracas
VENEZUELA
Medellín
Bogota
Cali
COLOMBIA
Quito
ECUADOR
Guayaquil
Galápagos (Ecuador)
Iquitos
Manaus
BRAZIL
SOUTH AMERICA
PERU
Trujillo
Lima
Cuzco
Arequipa
La Paz
BOLIVIA
Iquique
Antofagasta
PARAGUAY
Asunción
Tucumán
Córdoba
Rosario
Valparaíso
Santiago
Buenos Aires
URUGUAY
Montevideo
Río de la Plata
Pto. Alegre
ARGENTINA
Concepción
Patagonia
Andes
SOUTH ATLANTIC OCEAN
Falkland Is. (U.K.)
South Georgia (U.K.)
Punta Arenas
Str. of Magellan
Tierra del Fuego
C. Horn
Hawaiian Is. (U.S.)
Oahu
Honolulu
Hawaii
Johnston I. (U.S.)
Tropic of Cancer
Equator
Tropic of Capricorn
P A C I F I C
O C E A N
Palmyra Is. (U.S.)
Christmas Island Ridge
Kiritimati
Jarvis I. (U.S.)
Phoenix Is.
Malden I.
Starbuck I.
Cook Islands (N.Z.)
Rarotonga
Îs. Marquises
Îs. Tuamotu
Îs. de la Société
Tahiti
FRENCH POLYNESIA
Îs. Tubuai (Îs. Australes)
Tuamotu Ridge
Austral Seamount Chain
Pitcairn I. (U.K.)
Ducie I (U.K.)
East Pacific Ridge
Pacific-Antarctic Ridge
Chile Rise
I. de Pascua (Easter I.) (Chile)
Sala-y-Gomez (Chile)
San Félix (Chile)
San Ambrosio (Chile)
Arch. de Juan Fernández (Chile)
I. Clipperton (Fr.)
I. del Coco (Costa Rica)
I. de Malpelo (Colombia)
Is. Revilla Gigedo (Mexico)
I. Guadalupe (Mexico)
West from Greenwich
COPYRIGHT, GEORGE PHILIP & SON, LTD.

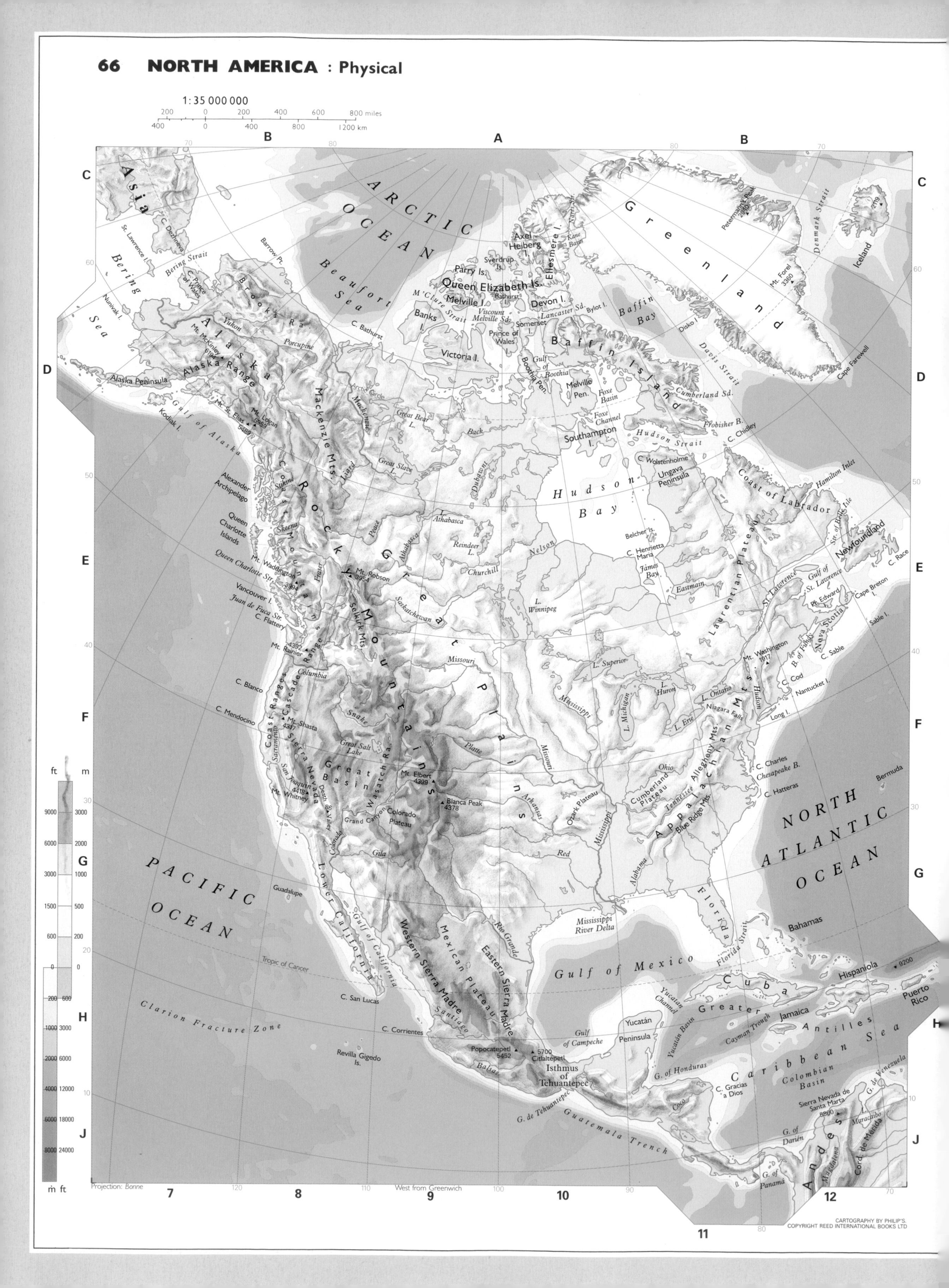
66 NORTH AMERICA : Physical
1:35 000 000
200 0 200 400 600 800 miles
400 0 400 800 1200 km
ARCTIC OCEAN
Beaufort Sea
Bering Sea
Bering Strait
Asia
C. Dezhneva
St. Lawrence I.
Nunivak I.
C. Prince of Wales
Barrow Pt.
Brooks Ra.
Yukon
Porcupine
Alaska
Alaska Range
Mt. McKinley 6194
Alaska Peninsula
Kodiak I.
Gulf of Alaska
Mt. St. Elias 5489
Mt. Logan 5959
Mackenzie Mts.
Mackenzie
Arctic Circle
Great Bear L.
Great Slave L.
Liard
Back
C. Bathurst
Banks I.
M'Clure Strait
Viscount Melville Sd.
Melville I.
Parry Is.
Queen Elizabeth Is.
Sverdrup Is.
Axel Heiberg I.
Ellesmere I.
Kane Basin
Bathurst I.
Devon I.
Lancaster Sd.
Bylot I.
Somerset I.
Prince of Wales I.
Victoria I.
Gulf of Boothia
Boothia Pen.
Melville Pen.
Foxe Basin
Foxe Channel
Southampton I.
Baffin Island
Baffin Bay
Cumberland Sd.
Frobisher B.
Hudson Strait
C. Chidley
C. Wolstenholme
Ungava Peninsula
Hudson Bay
Belcher Is.
C. Henrietta Maria
James Bay
Eastmain
Greenland
Petermann Peak 2940
Mt. Forel 3360
Disko I.
Davis Strait
Denmark Strait
Iceland
2119
C. Farewell
Coast of Labrador
Hamilton Inlet
Str. of Belle Isle
Newfoundland
C. Race
Gulf of St. Lawrence
St. Lawrence
Pr. Edward I.
Cape Breton I.
Nova Scotia
Sable I.
B. of Fundy
C. Sable
Laurentian Plateau
Alexander Archipelago
Stikine
Coast Mountains
Skeena
Queen Charlotte Islands
Queen Charlotte Str.
Mt. Waddington 3994
Vancouver I.
Juan de Fuca Str.
C. Flattery
Fraser
Rocky Mountains
Mt. Robson 3954
Selkirk Mts.
Peace
Athabasca
L. Athabasca
Dubawnt
Reindeer L.
Churchill
Nelson
Saskatchewan
L. Winnipeg
Great Plains
Missouri
Mt. Rainier 4392
Cascade Range
Columbia
C. Blanco
C. Mendocino
Coast Ranges
Mt. Shasta 4317
Sacramento
Snake
Great Salt Lake
Great Basin
Wasatch Ra.
Sierra Nevada
San Joaquin
Mt. Whitney 4418
Death Valley -86
Mt. Elbert 4399
Blanca Peak 4378
Platte
Colorado Plateau
Grand Canyon
Colorado
Gila
Arkansas
Red
Ozark Plateau
Mississippi
L. Superior
L. Michigan
L. Huron
L. Ontario
L. Erie
Niagara Falls
Mt. Washington 1917
Hudson
C. Cod
Nantucket I.
Long I.
Adirondack Mts.
Allegheny Mts.
Appalachian Mts.
Blue Ridge Mts.
Ohio
Cumberland Plateau
Tennessee
Alabama
C. Charles
Chesapeake B.
C. Hatteras
Bermuda
NORTH ATLANTIC OCEAN
Florida
Florida Strait
Bahamas
Mississippi River Delta
Gulf of Mexico
Rio Grande
PACIFIC OCEAN
Guadalupe
Lower California
Gulf of California
Tropic of Cancer
C. San Lucas
Western Sierra Madre
Mexican Plateau
Eastern Sierra Madre
Santiago
C. Corrientes
Revilla Gigedo Is.
Clarion Fracture Zone
Balsas
Popocatepetl 5452
Citlaltepetl 5700
Isthmus of Tehuantepec
G. de Tehuantepec
Guatemala Trench
Gulf of Campeche
Yucatán Peninsula
Yucatan Channel
Yucatan Basin
Cuba
Greater Antilles
Cayman Trough
Jamaica
Hispaniola
9200
Puerto Rico
Caribbean Sea
Colombian Basin
G. of Honduras
C. Gracias a Dios
Coco
G. of Darién
G. of Panamá
Sierra Nevada de Santa Marta 5800
G. de Venezuela
Maracaibo
Cord. de Mérida
Magdalena
Andes
ft m
9000 3000
6000 2000
3000 1000
1500 500
600 200
0 0
200 600
1000 3000
2000 6000
4000 12000
6000 18000
8000 24000
m ft
Projection: Bonne
West from Greenwich
CARTOGRAPHY BY PHILIP'S.
COPYRIGHT REED INTERNATIONAL BOOKS LTD

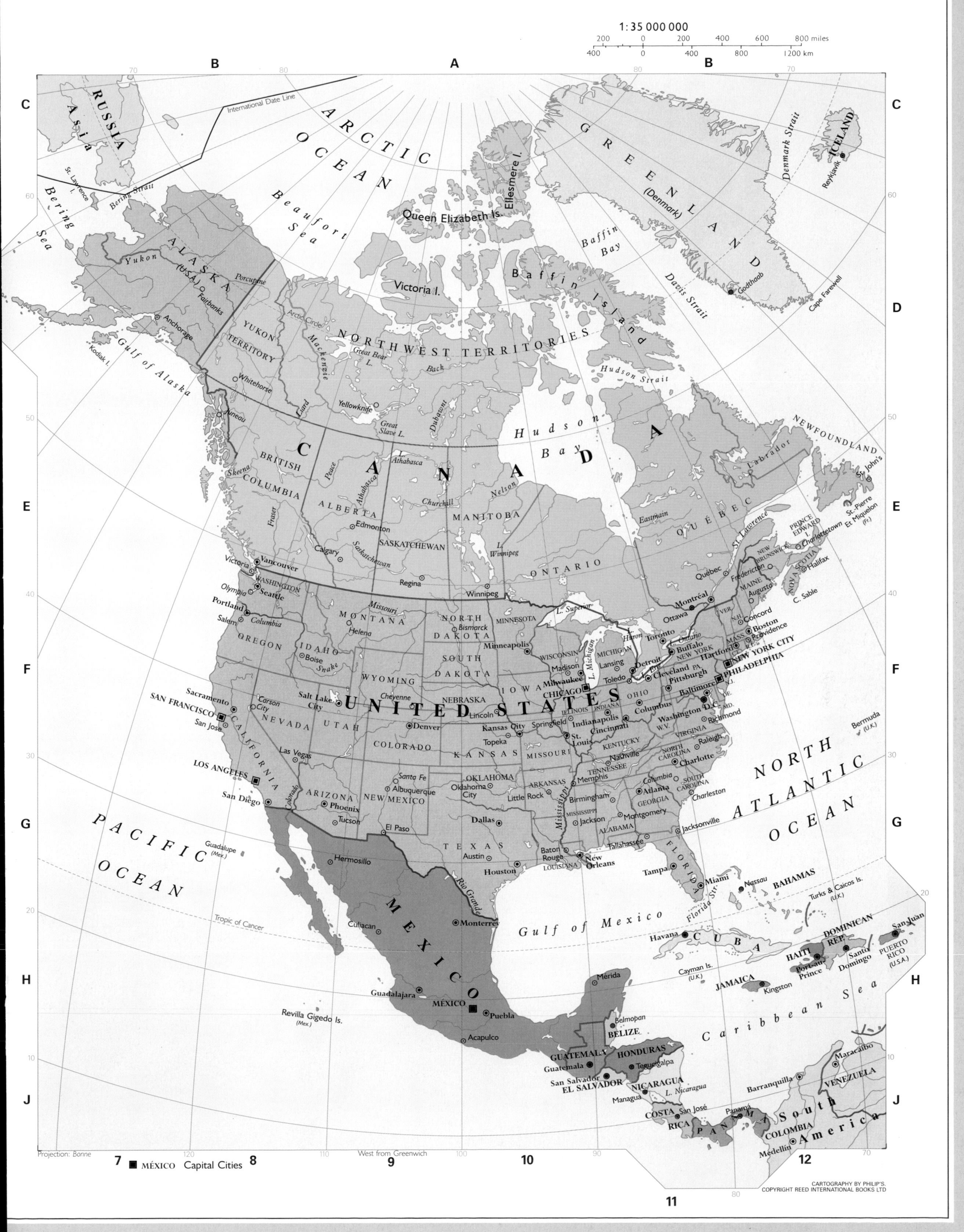

1:35 000 000
ARCTIC OCEAN
GREENLAND (Denmark)
ICELAND
RUSSIA
Asia
Bering Sea
Beaufort Sea
Queen Elizabeth Is.
Ellesmere I.
Baffin Bay
Victoria I.
Baffin Island
ALASKA (U.S.A.)
Fairbanks
Anchorage
Gulf of Alaska
YUKON TERRITORY
NORTHWEST TERRITORIES
Hudson Strait
Davis Strait
Godthaab
Cape Farewell
Denmark Strait
Reykjavik
Whitehorse
Juneau
Yellowknife
Great Slave L.
Great Bear L.
Hudson Bay
CANADA
BRITISH COLUMBIA
ALBERTA
SASKATCHEWAN
MANITOBA
ONTARIO
QUÉBEC
NEWFOUNDLAND
Labrador
St John's
Edmonton
Calgary
Regina
Winnipeg
Vancouver
Victoria
Québec
Montréal
Ottawa
Toronto
Halifax
Fredericton
Charlottetown
PRINCE EDWARD I.
NEW BRUNSWICK
NOVA SCOTIA
St. Pierre Et Miquelon (Fr.)
UNITED STATES
WASHINGTON
OREGON
IDAHO
MONTANA
WYOMING
NORTH DAKOTA
SOUTH DAKOTA
NEBRASKA
KANSAS
COLORADO
UTAH
NEVADA
CALIFORNIA
ARIZONA
NEW MEXICO
TEXAS
OKLAHOMA
MINNESOTA
IOWA
MISSOURI
ARKANSAS
WISCONSIN
MICHIGAN
ILLINOIS
INDIANA
OHIO
KENTUCKY
TENNESSEE
MISSISSIPPI
ALABAMA
GEORGIA
FLORIDA
LOUISIANA
NEW YORK
NEW YORK CITY
PHILADELPHIA
CHICAGO
LOS ANGELES
SAN FRANCISCO
Seattle
Portland
Salem
Olympia
Boise
Helena
Bismarck
Minneapolis
Salt Lake City
Cheyenne
Denver
Lincoln
Kansas City
Topeka
Sacramento
San Jose
Carson City
Las Vegas
San Diego
Phoenix
Tucson
Santa Fe
Albuquerque
El Paso
Oklahoma City
Dallas
Austin
Houston
Little Rock
Memphis
Nashville
St. Louis
Springfield
Indianapolis
Cincinnati
Columbus
Cleveland
Detroit
Lansing
Milwaukee
Madison
Toledo
Pittsburgh
Buffalo
Hartford
Boston
Providence
Concord
Augusta
Baltimore
Washington D.C.
Richmond
Raleigh
Charlotte
Columbia
Atlanta
Birmingham
Montgomery
Jackson
Baton Rouge
New Orleans
Tallahassee
Jacksonville
Tampa
Miami
Charleston
NORTH ATLANTIC OCEAN
PACIFIC OCEAN
Bermuda (U.K.)
Guadalupe (Mex.)
Gulf of Mexico
Tropic of Cancer
MEXICO
MÉXICO
Hermosillo
Culiacan
Monterrey
Guadalajara
Puebla
Acapulco
Mérida
Revilla Gigedo Is. (Mex.)
BAHAMAS
Nassau
CUBA
Havana
Turks & Caicos Is. (U.K.)
Cayman Is. (U.K.)
JAMAICA
Kingston
HAITI
Port-au-Prince
DOMINICAN REP.
Santo Domingo
PUERTO RICO (U.S.A.)
San Juan
Caribbean Sea
BELIZE
Belmopan
GUATEMALA
Guatemala
HONDURAS
Tegucigalpa
EL SALVADOR
San Salvador
NICARAGUA
Managua
L. Nicaragua
COSTA RICA
San José
PANAMA
Panamá
COLOMBIA
Barranquilla
Medellín
VENEZUELA
Maracaibo
South America
Projection: Bonne
West from Greenwich
■ MÉXICO Capital Cities

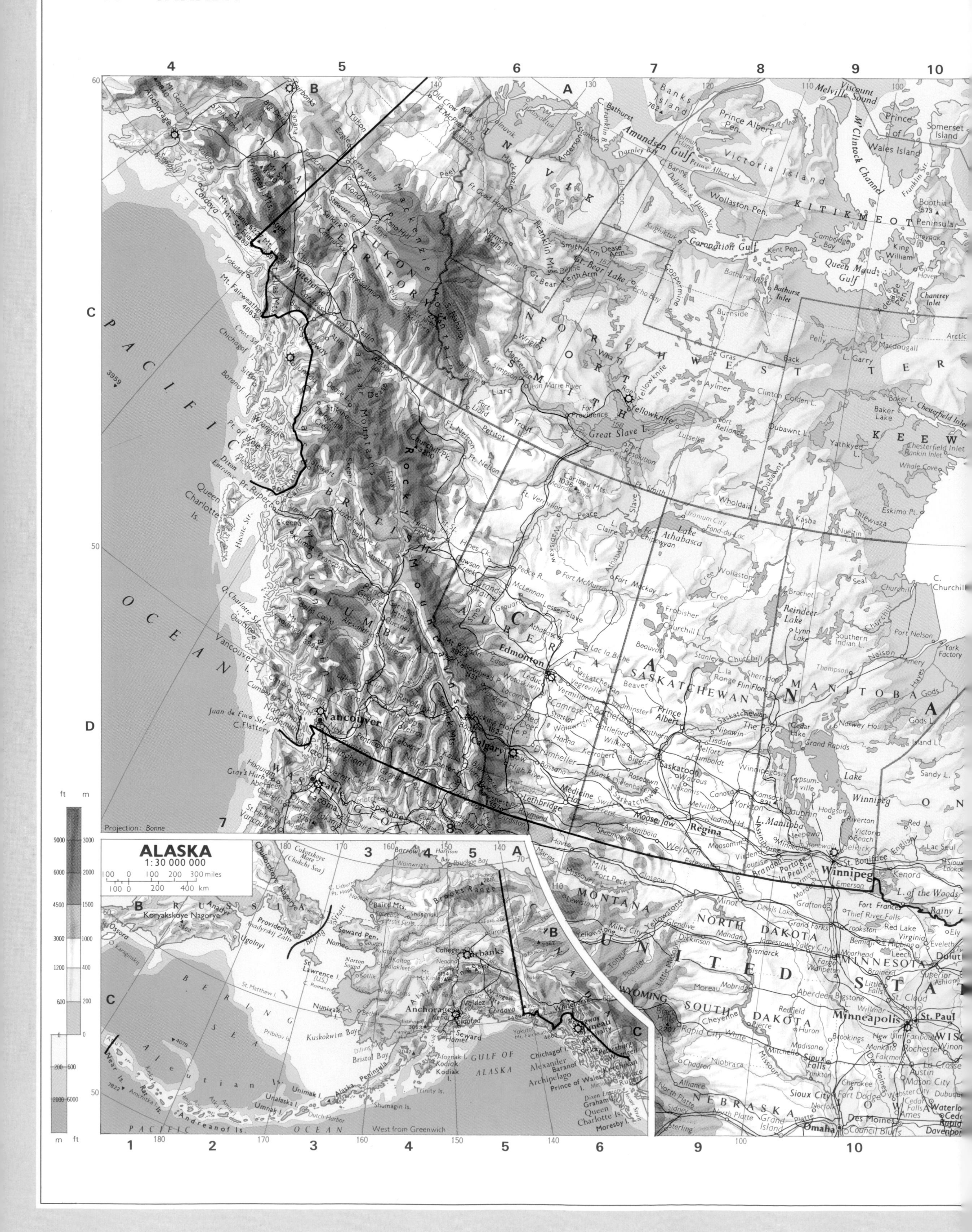
PACIFIC OCEAN
YUKON TERRITORY
BRITISH COLUMBIA
ALBERTA
SASKATCHEWAN
MANITOBA
NORTHWEST TERRITORIES
Great Slave L.
Lake Athabasca
Vancouver
Edmonton
Calgary
Regina
Winnipeg
Yellowknife
Saskatoon
Seattle
Victoria Island
Amundsen Gulf
Coronation Gulf
Queen Maud Gulf
Banks Island
Rocky Mountains
Mackenzie Mountains
MONTANA
NORTH DAKOTA
SOUTH DAKOTA
MINNESOTA
NEBRASKA
WYOMING
Minneapolis
St. Paul
Omaha
Projection: Bonne
ALASKA
1:30 000 000
Anchorage
Fairbanks
Juneau
Nome
Bering Sea
Gulf of Alaska
Aleutian Is.
Brooks Range
West from Greenwich

1:15 000 000
miles
km
GREENLAND
Baffin Bay
Davis Strait
Hudson Strait
Hudson Bay
Ungava Bay
Ungava Peninsula
Coast of Labrador
Newfoundland
Gulf of St. Lawrence
ATLANTIC OCEAN
James Bay
Foxe Basin
Cumberland Sd.
Frobisher Bay
QUEBEC
NEW BRUNSWICK
NOVA SCOTIA
PR. EDWARD I.
MONTREAL
Québec
Ottawa
TORONTO
Halifax
St. John's
Boston
NEW YORK
CHICAGO
DETROIT
Cleveland
Buffalo
West from Greenwich
COPYRIGHT, GEORGE PHILIP & SON, LTD.

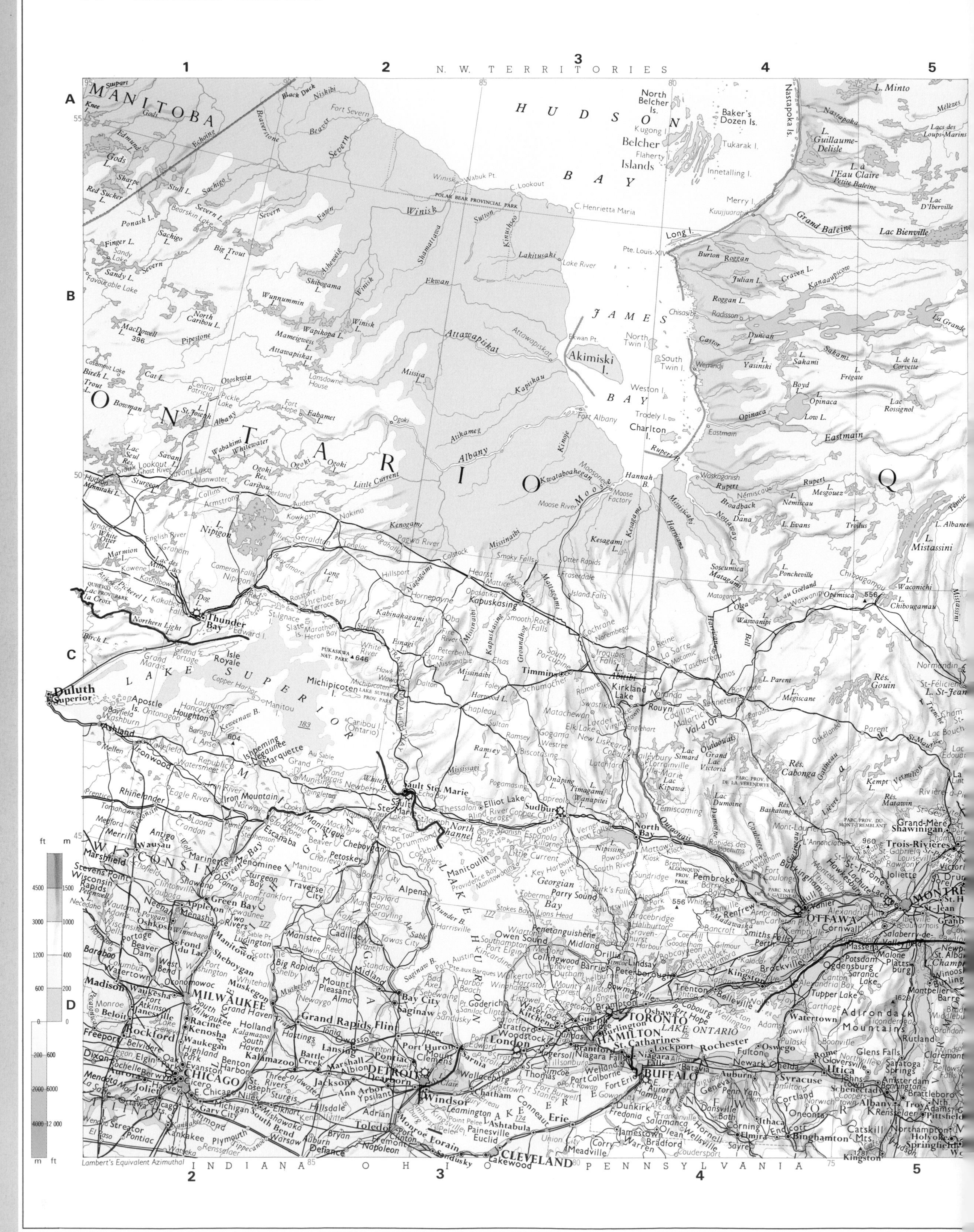

HUDSON BAY
JAMES BAY
ONTARIO
MANITOBA
LAKE SUPERIOR
LAKE HURON
LAKE ONTARIO
LAKE ERIE
Thunder Bay
Duluth
Sault Ste. Marie
Sudbury
North Bay
Timmins
Kirkland Lake
OTTAWA
MONTREAL
TORONTO
HAMILTON
BUFFALO
DETROIT
Windsor
London
CLEVELAND
CHICAGO
MILWAUKEE
Toledo
Rochester
Syracuse
Trois-Rivières
Akimiski I.
Belcher Islands
Lake Nipigon
Lake Nipissing
Georgian Bay
Manitoulin I.
Isle Royale
Adirondack Mountains
N. W. TERRITORIES
INDIANA
OHIO
PENNSYLVANIA
Lambert's Equivalent Azimuthal

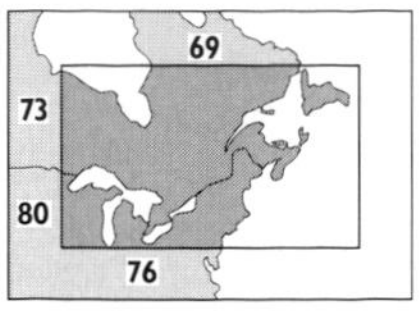

69
73
80
76

1:7 000 000

50 0 50 100 150 200 miles
50 0 50 100 150 200 250 300 km

6 7 8 9

A B C D

NEWFOUNDLAND
COAST OF LABRADOR
QUEBEC
NEW BRUNSWICK
NOVA SCOTIA
MAINE
PRINCE EDWARD ISLAND
GULF OF ST. LAWRENCE
ATLANTIC OCEAN
Cabot Strait
Str. of Belle Isle
Bay of Fundy
Î. d'Anticosti
NEWFOUNDLAND
Avalon Peninsula
Cape Breton Island
Sable I. (Nova Scotia)
SAINT-PIERRE ET MIQUELON (Fr.)
Îs. de la Madeleine (Quebec)
Mts. Chic-Chocs
Pén. de Gaspé
Mealy Mts.
Long Range Mts.
Smallwood Res.
Rés. Manicouagan
Kaniapiskau Lake
Happy Valley-Goose Bay
Labrador City
Sept-Îles
Corner Brook
Gander
St. John's
Sydney
Glace Bay
Halifax
Dartmouth
Saint John
Fredericton
Moncton
Charlottetown
Truro
New Glasgow
Yarmouth
Campbellton
Dalhousie
Bathurst
Chatham
Edmundston
Rimouski
Rivière-du-Loup
Québec
Lévis
Sherbrooke
Thetford Mines
Chicoutimi
Arvida
Jonquière
Alma
Bangor
Augusta
Portland
Lewiston
Auburn
Brunswick
Bath
Rockland
Belfast
Biddeford
Saco
Sanford
Rochester
Dover
Portsmouth
Concord
Manchester
Nashua
Lowell
Lawrence
Haverhill
Lynn
Gloucester
Boston
Brockton

West from Greenwich

70 65 60 55 50

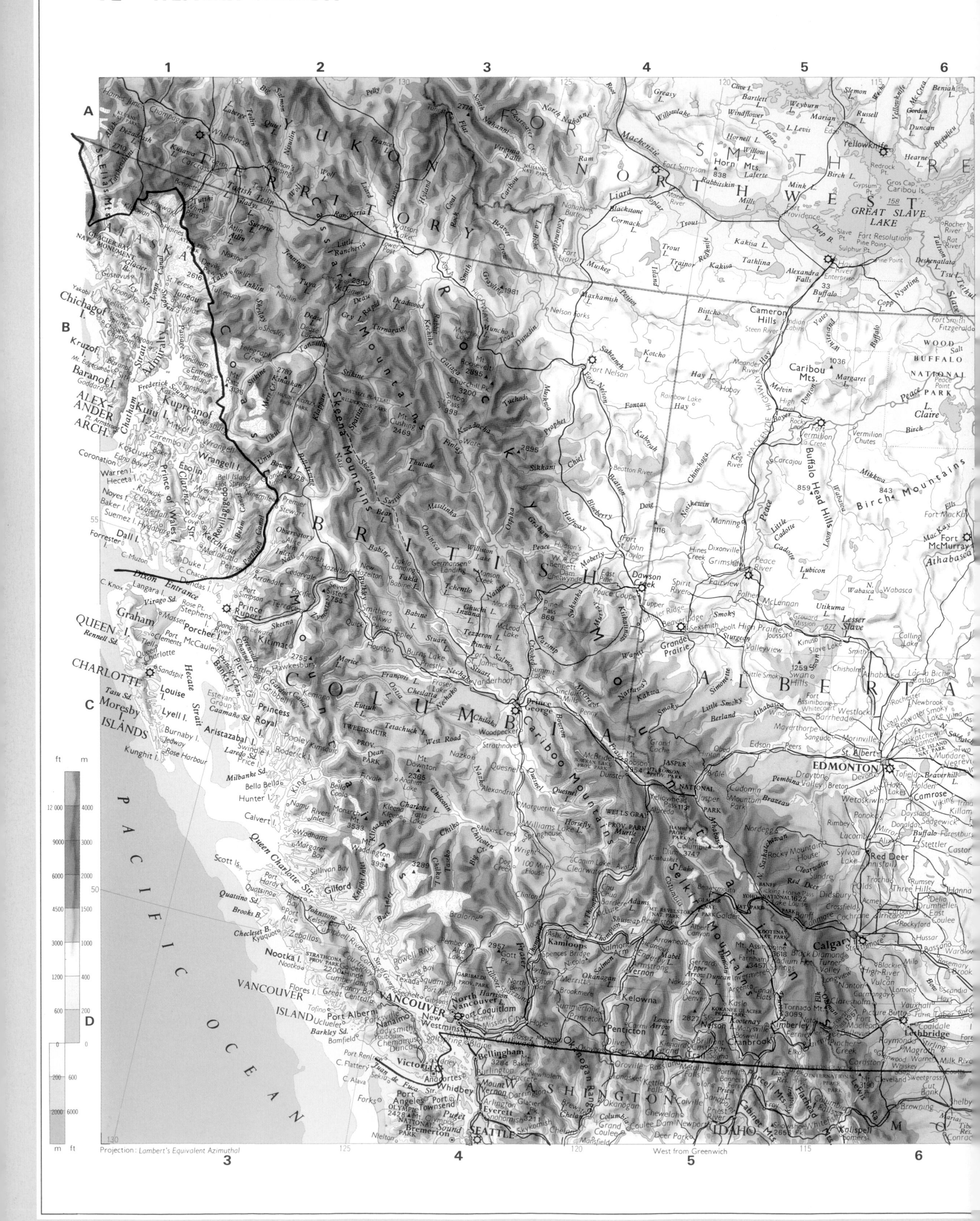
YUKON TERRITORY
NORTHWEST TERRITORIES
BRITISH COLUMBIA
ALBERTA
WASHINGTON
ROCKY MOUNTAINS
COAST MOUNTAINS
PACIFIC OCEAN
GREAT SLAVE LAKE
QUEEN CHARLOTTE ISLANDS
VANCOUVER ISLAND
ALEXANDER ARCH.
WOOD BUFFALO NATIONAL PARK
Birch Mountains
Buffalo Head Hills
Caribou Mts.
Cameron Hills
Skeena Mountains
Cariboo Mountains
Selkirk Mountains
Yellowknife
Fort Simpson
Fort Nelson
Fort St. John
Dawson Creek
Grande Prairie
Peace River
Prince Rupert
Prince George
Kamloops
Kelowna
Penticton
Vernon
VANCOUVER
New Westminster
Victoria
Nanaimo
Bellingham
SEATTLE
EDMONTON
Red Deer
Calgary
Lethbridge
Camrose
Whitehorse
Juneau
Chichagof I.
Baranof I.
Kruzof I.
Kupreanof I.
Graham I.
Moresby I.
Nootka I.
Dixon Entrance
Hecate Strait
Queen Charlotte Str.
Mt. Robson 3954
JASPER NATIONAL PARK
Ft. McMurray
Athabasca
Lesser Slave L.
Williston Lake
TWEEDSMUIR PROV. PARK
Projection: Lambert's Equivalent Azimuthal
West from Greenwich

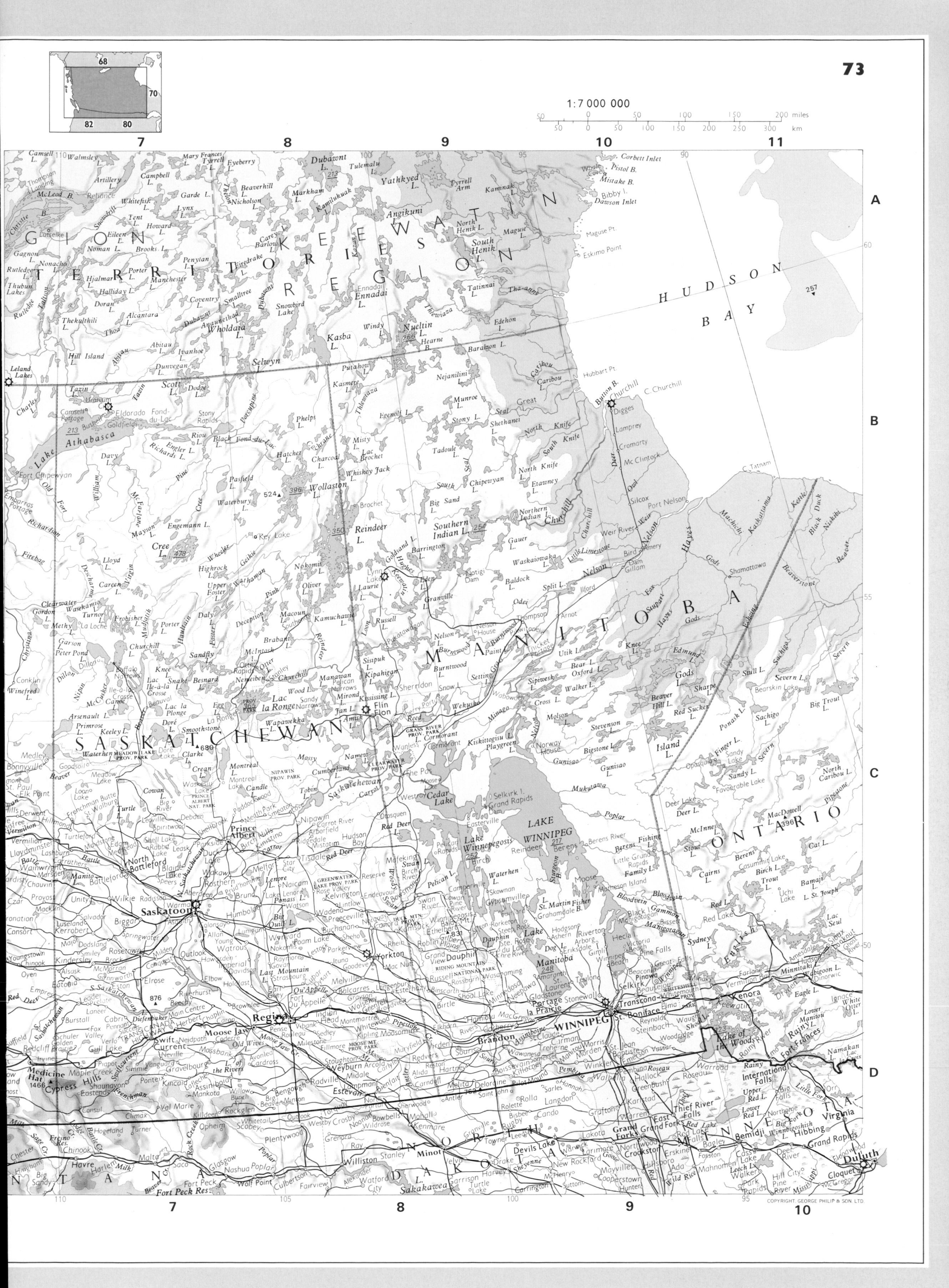

1:7 000 000
miles
km
KEEWATIN REGION
TERRITORIES
HUDSON BAY
MANITOBA
SASKATCHEWAN
ONTARIO
NORTH DAKOTA
MINNESOTA
MONTANA
Lake Athabasca
Wollaston L.
Reindeer L.
Southern Indian L.
Nueltin L.
Dubawnt L.
LAKE WINNIPEG
Lake Winnipegosis
Lake Manitoba
Cedar Lake
Lake of the Woods
Churchill
Flin Flon
Saskatoon
Prince Albert
North Battleford
Regina
Moose Jaw
Swift Current
Medicine Hat
Yorkton
Brandon
WINNIPEG
St. Boniface
Transcona
Portage la Prairie
Kenora
Fort Frances
International Falls
Duluth
Grand Forks
Minot
Devils Lake
Williston
Cypress Hills
COPYRIGHT, GEORGE PHILIP & SON, LTD.

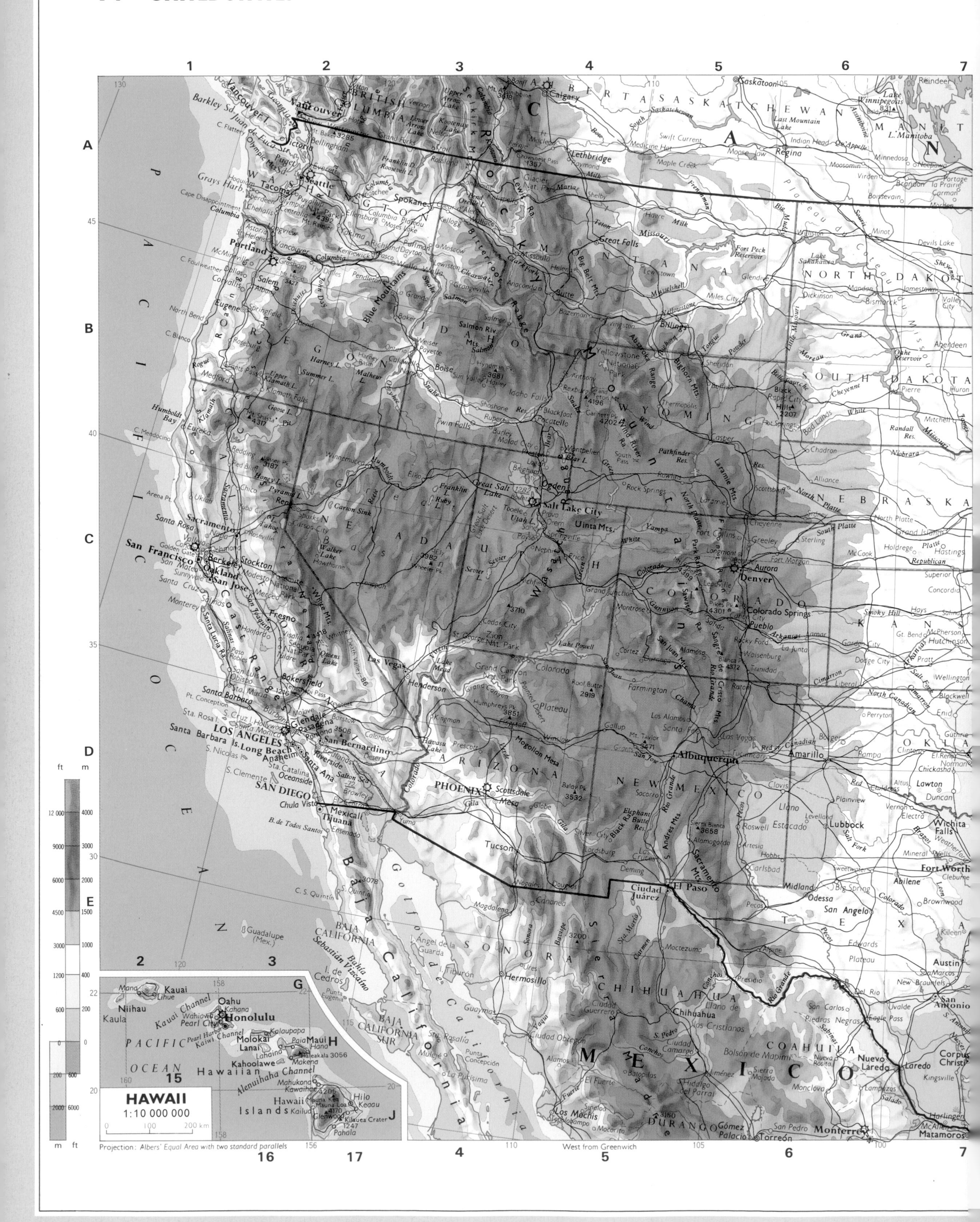

1 2 3 4 5 6 7
A B C D E
CANADA
BRITISH COLUMBIA
ALBERTA
SASKATCHEWAN
MANITOBA
PACIFIC OCEAN
WASHINGTON
OREGON
CALIFORNIA
NEVADA
IDAHO
MONTANA
WYOMING
UTAH
COLORADO
ARIZONA
NEW MEXICO
NORTH DAKOTA
SOUTH DAKOTA
NEBRASKA
KANSAS
OKLAHOMA
TEXAS
MEXICO
SONORA
CHIHUAHUA
COAHUILA
DURANGO
BAJA CALIFORNIA
BAJA CALIFORNIA SUR
Golfo de California
Vancouver
Victoria
Seattle
Tacoma
Spokane
Portland
Salem
Eugene
Boise
Great Falls
Billings
Calgary
Lethbridge
Regina
Saskatoon
Sacramento
San Francisco
Oakland
Berkeley
San Jose
Stockton
Fresno
Bakersfield
LOS ANGELES
Long Beach
Glendale
Pasadena
San Bernardino
Riverside
Anaheim
Santa Ana
Oceanside
SAN DIEGO
Chula Vista
Tijuana
Mexicali
Las Vegas
Henderson
Reno
Salt Lake City
Ogden
Provo
PHOENIX
Scottsdale
Mesa
Tucson
Denver
Aurora
Colorado Springs
Pueblo
Albuquerque
El Paso
Ciudad Juárez
Amarillo
Lubbock
Midland
Odessa
San Angelo
Abilene
Wichita Falls
Fort Worth
Austin
San Antonio
Laredo
Nuevo Laredo
Corpus Christi
Chihuahua
Hermosillo
Monterrey
Torreón
Los Mochis
Rocky Mts.
Blue Mountains
Coast Ranges
Sierra Nevada
Wasatch Ra.
Uinta Mts.
Bighorn Mts.
Great Salt Lake
Yellowstone National Park
Grand Canyon
Plateau du Coteau du Missouri
Mt. Rainier
Mt. Whitney
HAWAII
1:10 000 000
0 100 200 km
Kauai
Niihau
Oahu
Honolulu
Molokai
Lanai
Maui
Kahoolawe
Hawaii
Kauai Channel
Kaiwi Channel
Alenuihaha Channel
Hawaiian Islands
PACIFIC OCEAN
Hilo
Kailua
Mauna Kea 4205
Mauna Loa 4170
Haleakala 3056
ft m
12 000 4000
9000 3000
6000 2000
4500 1500
3000 1000
1200 400
600 200
0 0
200 600
2000 6000
m ft
Projection: Albers' Equal Area with two standard parallels
West from Greenwich

68 69 86 87

1:12 000 000

0 50 100 150 200 250 300 miles

0 50 100 150 200 250 300 350 400 450 km

8 9 10 11 12 13

A B C D E F

C A N A D A

Lake Winnipeg

Lake Superior

Lake Michigan

Lake Huron

Lake Erie

Lake Ontario

MINNESOTA

WISCONSIN

MICHIGAN

IOWA

ILLINOIS

INDIANA

OHIO

MISSOURI

KENTUCKY

TENNESSEE

WEST VIRGINIA

VIRGINIA

NORTH CAROLINA

SOUTH CAROLINA

GEORGIA

ALABAMA

MISSISSIPPI

LOUISIANA

ARKANSAS

FLORIDA

PENNSYLVANIA

NEW YORK

MAINE

NEW JERSEY

MONTREAL

TORONTO

CHICAGO

DETROIT

NEW YORK

PHILADELPHIA

HOUSTON

DALLAS

BAHAMAS

A T L A N T I C   O C E A N

G U L F   O F   M E X I C O

95 90 85 80

8 9 10 11 12

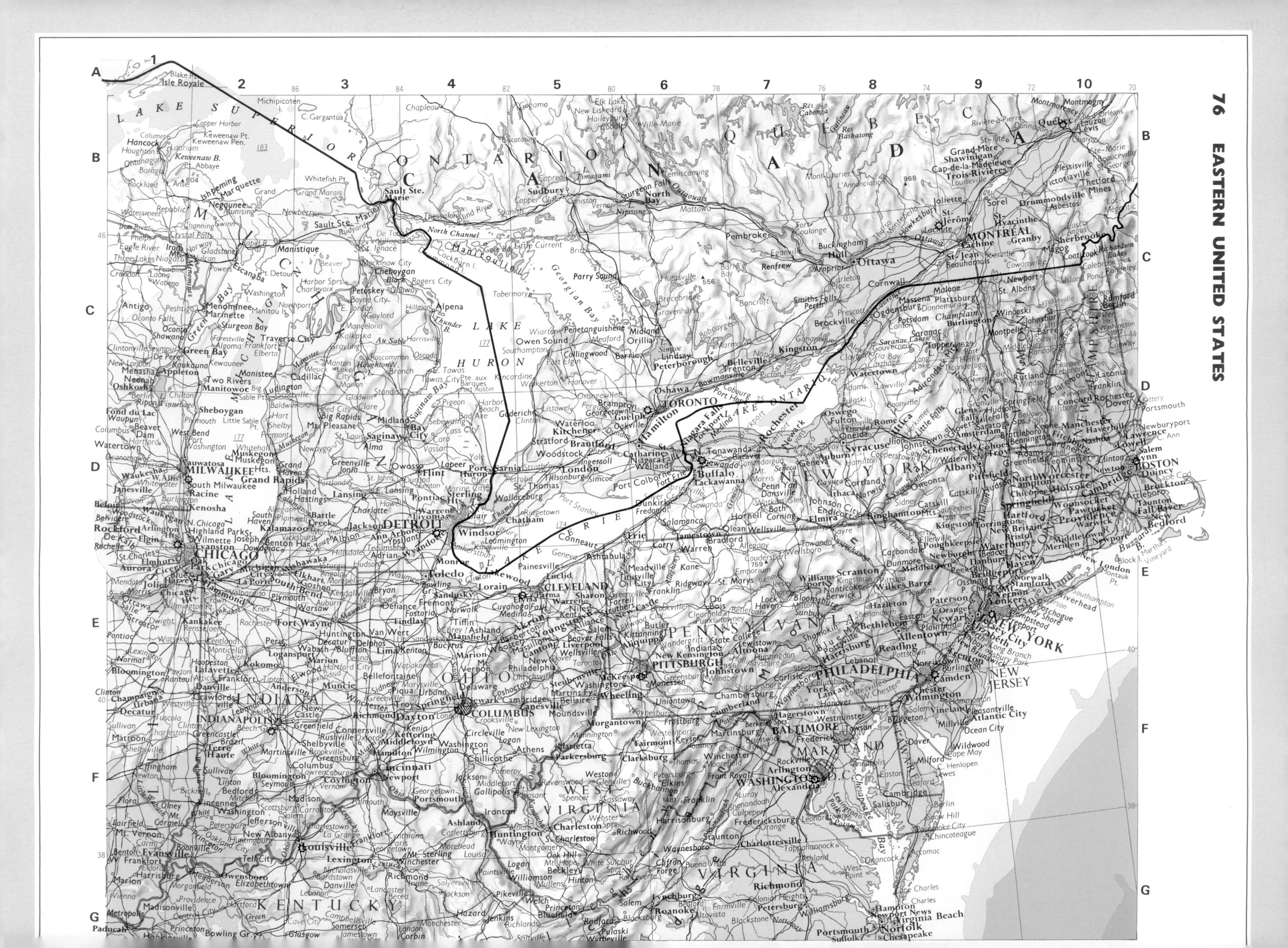
LAKE SUPERIOR
LAKE HURON
LAKE ERIE
LAKE ONTARIO
Georgian Bay
ONTARIO
QUÉBEC
CANADA
MICHIGAN
NEW YORK
PENNSYLVANIA
OHIO
INDIANA
KENTUCKY
VIRGINIA
WEST VIRGINIA
MARYLAND
NEW JERSEY
CHICAGO
DETROIT
CLEVELAND
TORONTO
MONTREAL
Ottawa
BOSTON
NEW YORK
PHILADELPHIA
BALTIMORE
WASHINGTON D.C.
PITTSBURGH
COLUMBUS
INDIANAPOLIS
Cincinnati
Louisville
MILWAUKEE
Buffalo
Rochester
Syracuse
Albany
Québec
Richmond
Norfolk
Toledo
Dayton
Lansing
Grand Rapids
Sault Ste. Marie
Chesapeake

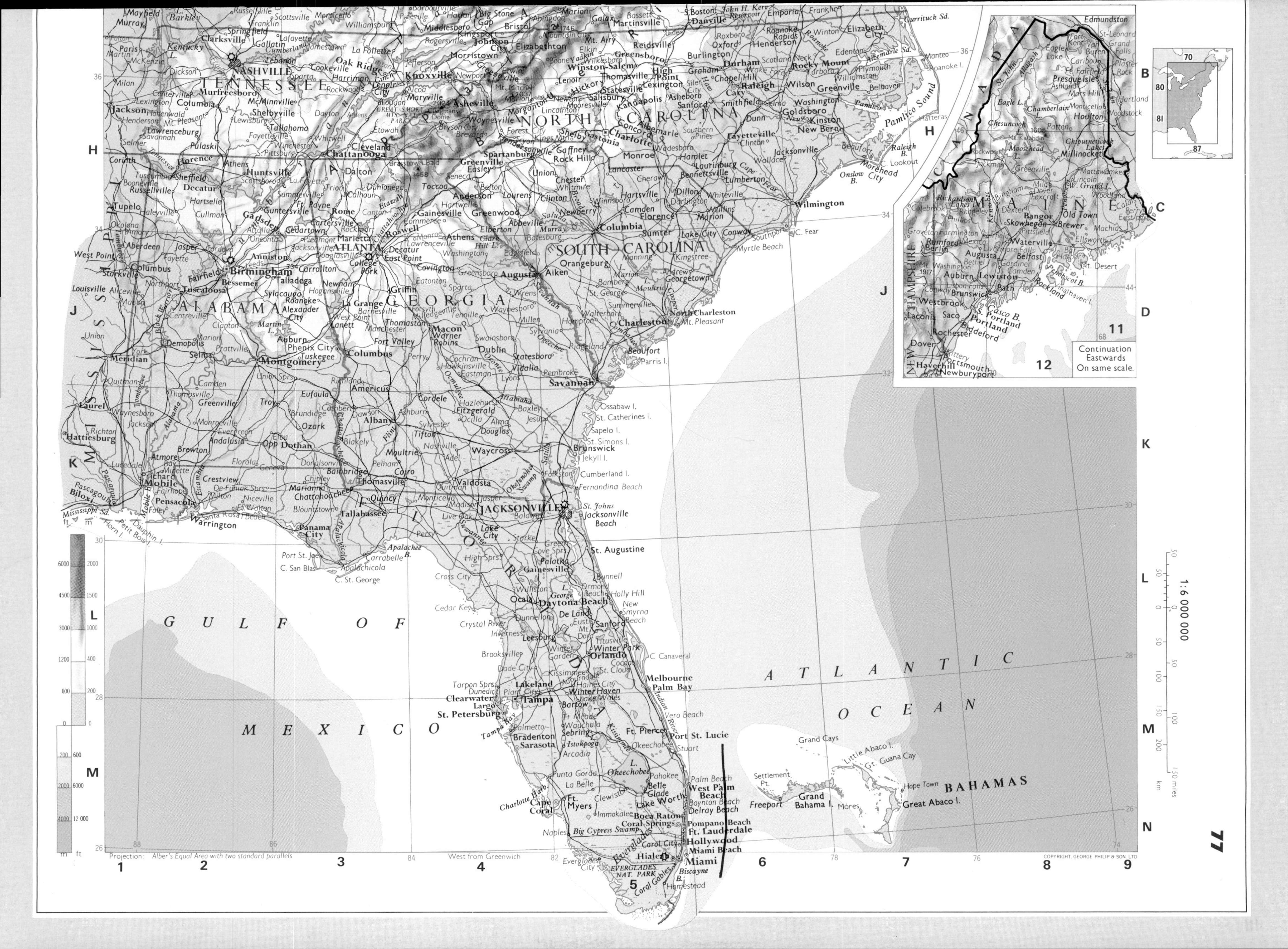
1:6 000 000
TENNESSEE
NORTH CAROLINA
SOUTH CAROLINA
GEORGIA
ALABAMA
FLORIDA
MAINE
NEW HAMPSHIRE
CANADA
BAHAMAS
ATLANTIC OCEAN
GULF OF MEXICO
Nashville
Knoxville
Chattanooga
Asheville
Charlotte
Raleigh
Durham
Greensboro
Winston-Salem
Wilmington
Columbia
Charleston
Savannah
Augusta
Atlanta
Macon
Columbus
Albany
Birmingham
Montgomery
Mobile
Pensacola
Tallahassee
Jacksonville
Orlando
Tampa
St. Petersburg
Miami
Ft. Lauderdale
West Palm Beach
Daytona Beach
Gainesville
Portland
Bangor
Augusta
Continuation Eastwards On same scale.
Projection: Alber's Equal Area with two standard parallels
West from Greenwich
COPYRIGHT GEORGE PHILIP & SON LTD

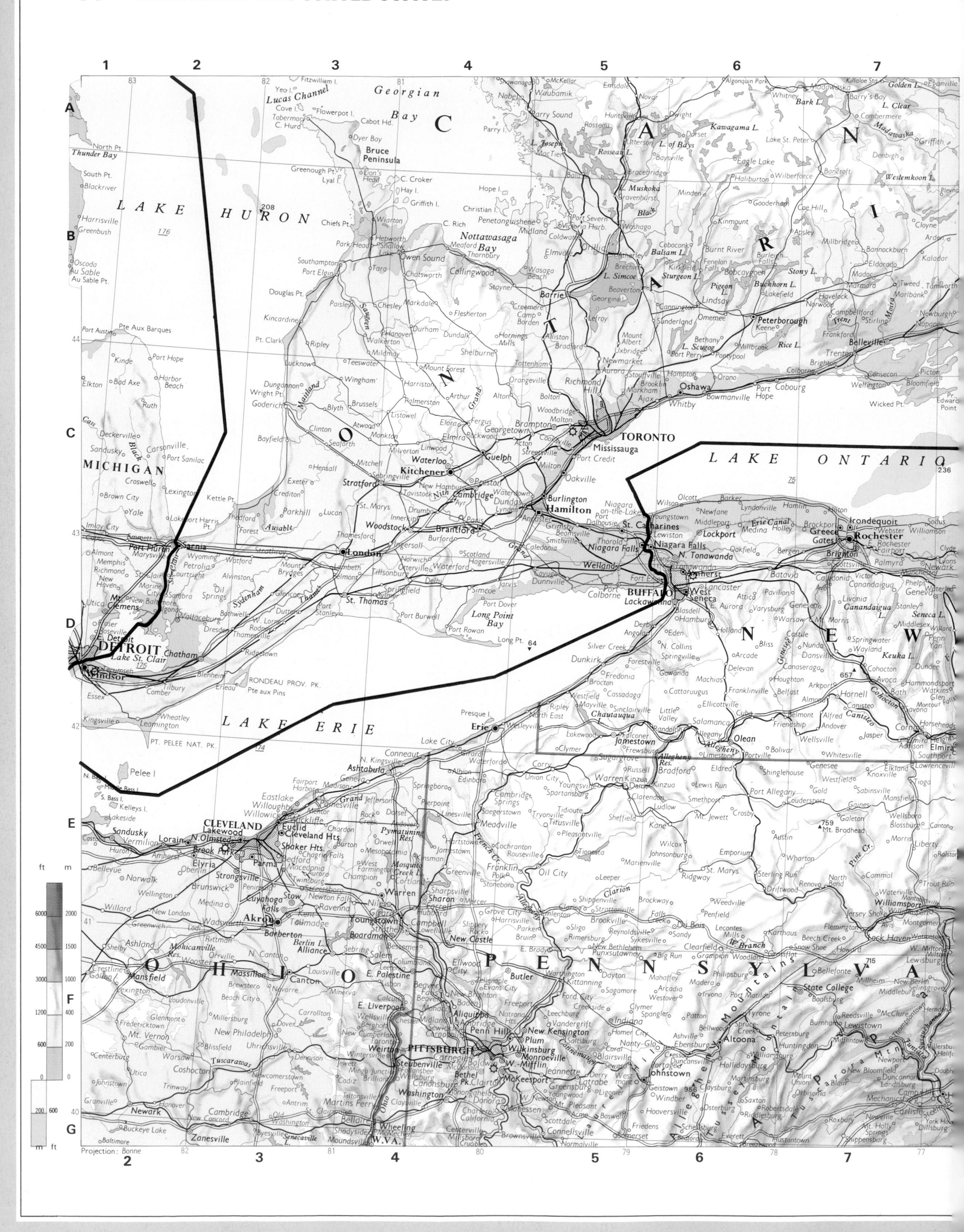

LAKE HURON
Georgian Bay
Bruce Peninsula
Nottawasaga Bay
LAKE ONTARIO
LAKE ERIE
Lake St. Clair
Long Point Bay
L. Simcoe
CANADA
ONTARIO
MICHIGAN
NEW
OHIO
PENNSYLVANIA
TORONTO
Mississauga
Hamilton
Burlington
Kitchener
Waterloo
Guelph
Cambridge
Brantford
London
St. Thomas
Sarnia
Port Huron
Windsor
DETROIT
Chatham
St. Catharines
Niagara Falls
Welland
BUFFALO
Lackawanna
West Seneca
Amherst
N. Tonawanda
Lockport
Rochester
Irondequoit
Greece
Brighton
Oshawa
Peterborough
Belleville
Barrie
Owen Sound
Erie
Jamestown
Olean
Bradford
CLEVELAND
Lakewood
Euclid
Cleveland Hts.
Shaker Hts.
Parma
Lorain
Elyria
Akron
Barberton
Canton
Massillon
Youngstown
Warren
New Castle
PITTSBURGH
Penn Hills
McKeesport
Wilkinsburg
Johnstown
Altoona
State College
Williamsport
Wheeling
Steubenville
Weirton
Mansfield
Allegheny Mountains
Tuscarora Mtn
Projection: Bonne

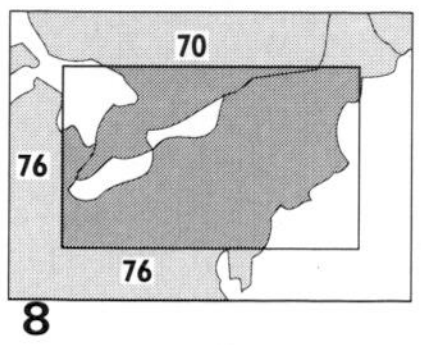

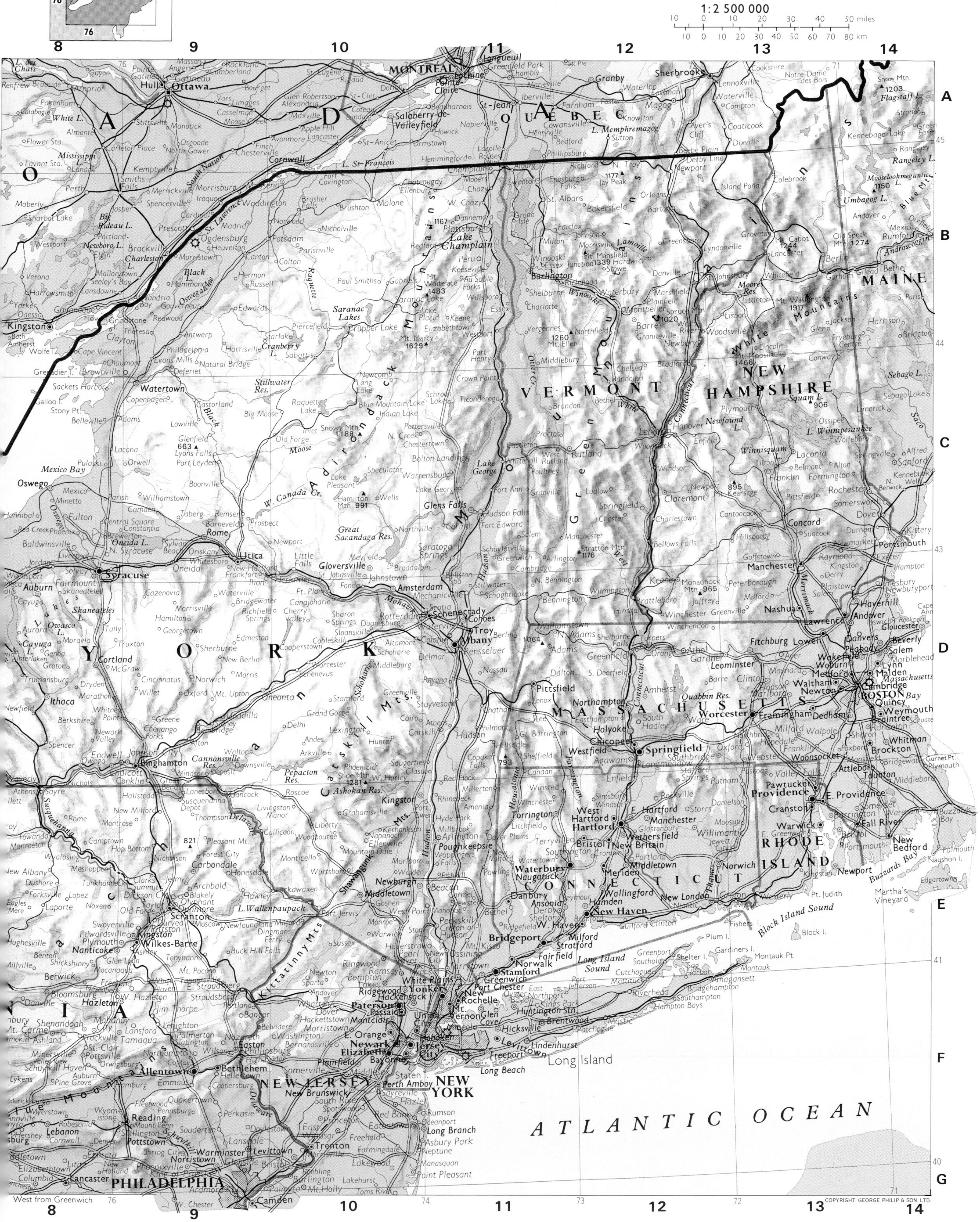

1:2 500 000

10 0 10 20 30 40 50 miles

10 0 10 20 30 40 50 60 70 80 km

8 9 10 11 12 13 14

A B C D E F G

CANADA

ONTARIO

QUEBEC

NEW YORK

PENNSYLVANIA

VERMONT

NEW HAMPSHIRE

MAINE

MASSACHUSETTS

CONNECTICUT

RHODE ISLAND

NEW JERSEY

ATLANTIC OCEAN

MONTREAL

Ottawa

Hull

Sherbrooke

Kingston

Syracuse

Utica

Albany

Schenectady

Troy

Burlington

Springfield

Hartford

New Haven

Bridgeport

Stamford

Providence

BOSTON

Worcester

Manchester

Concord

Portsmouth

Allentown

Scranton

Wilkes-Barre

Binghamton

Trenton

Newark

Paterson

Jersey City

NEW YORK

PHILADELPHIA

Camden

Long Island

Lake Champlain

L. Memphremagog

L. Winnipesaukee

Adirondack Mountains

Green Mountains

White Mountains

Catskill Mts.

Long Island Sound

Block Island Sound

Buzzards Bay

Massachusetts Bay

Mt. Marcy 1629

Mt. Washington 1917

Mt. Mansfield 1339

Slide Mtn. 1281

West from Greenwich

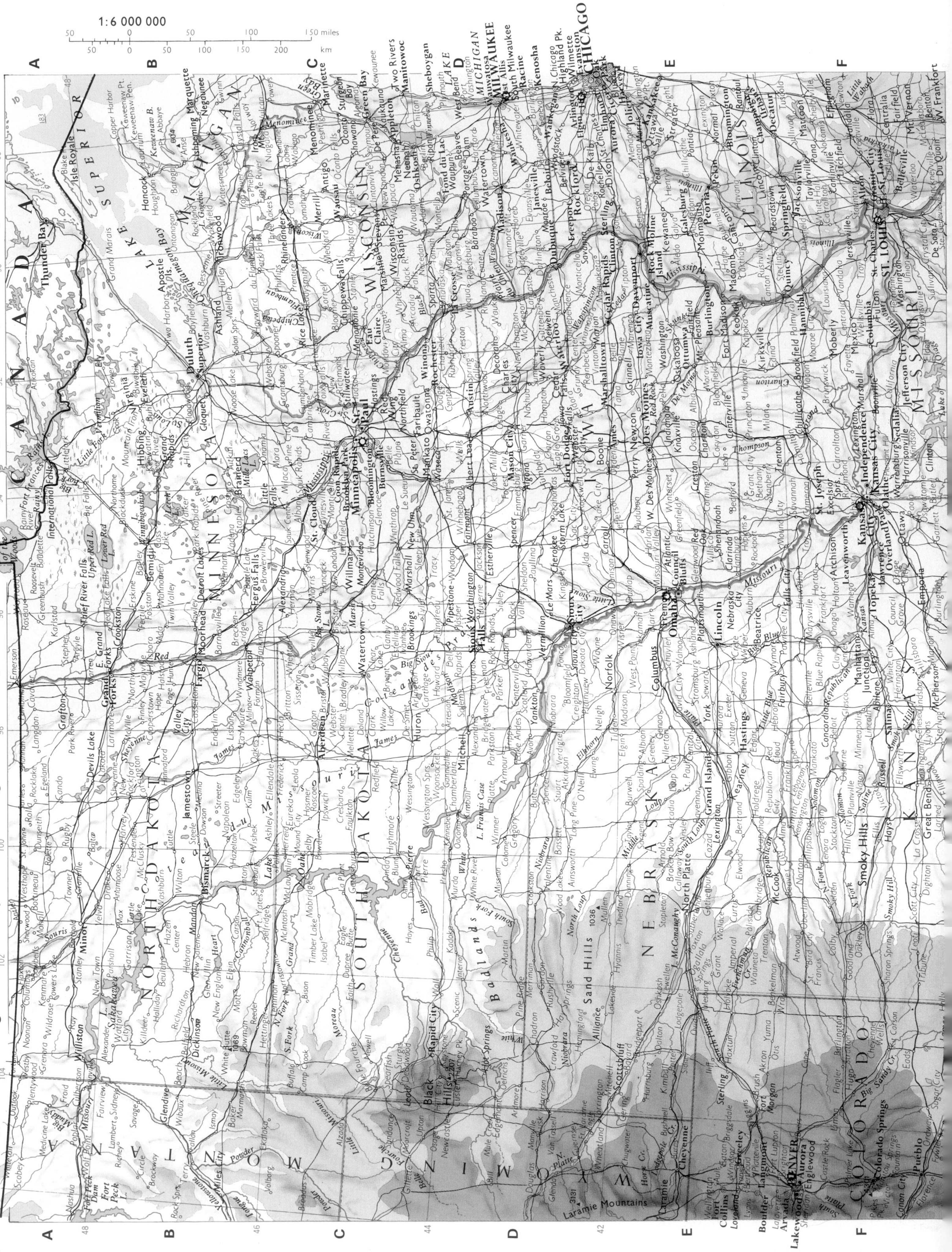
1:6 000 000
50 0 50 100 150 miles
50 0 50 100 150 200 km
CANADA
LAKE SUPERIOR
LAKE MICHIGAN
MICHIGAN
MINNESOTA
WISCONSIN
NORTH DAKOTA
SOUTH DAKOTA
IOWA
ILLINOIS
NEBRASKA
MISSOURI
KANSAS
COLORADO
WYOMING
MONTANA
Sand Hills
Badlands
Black Hills
Laramie Mountains
CHICAGO
MILWAUKEE
Minneapolis
St. Paul
Des Moines
Omaha
Kansas City
ST. LOUIS
DENVER
Duluth
Bismarck
Pierre
Lincoln
Topeka
Cheyenne
Madison
Springfield

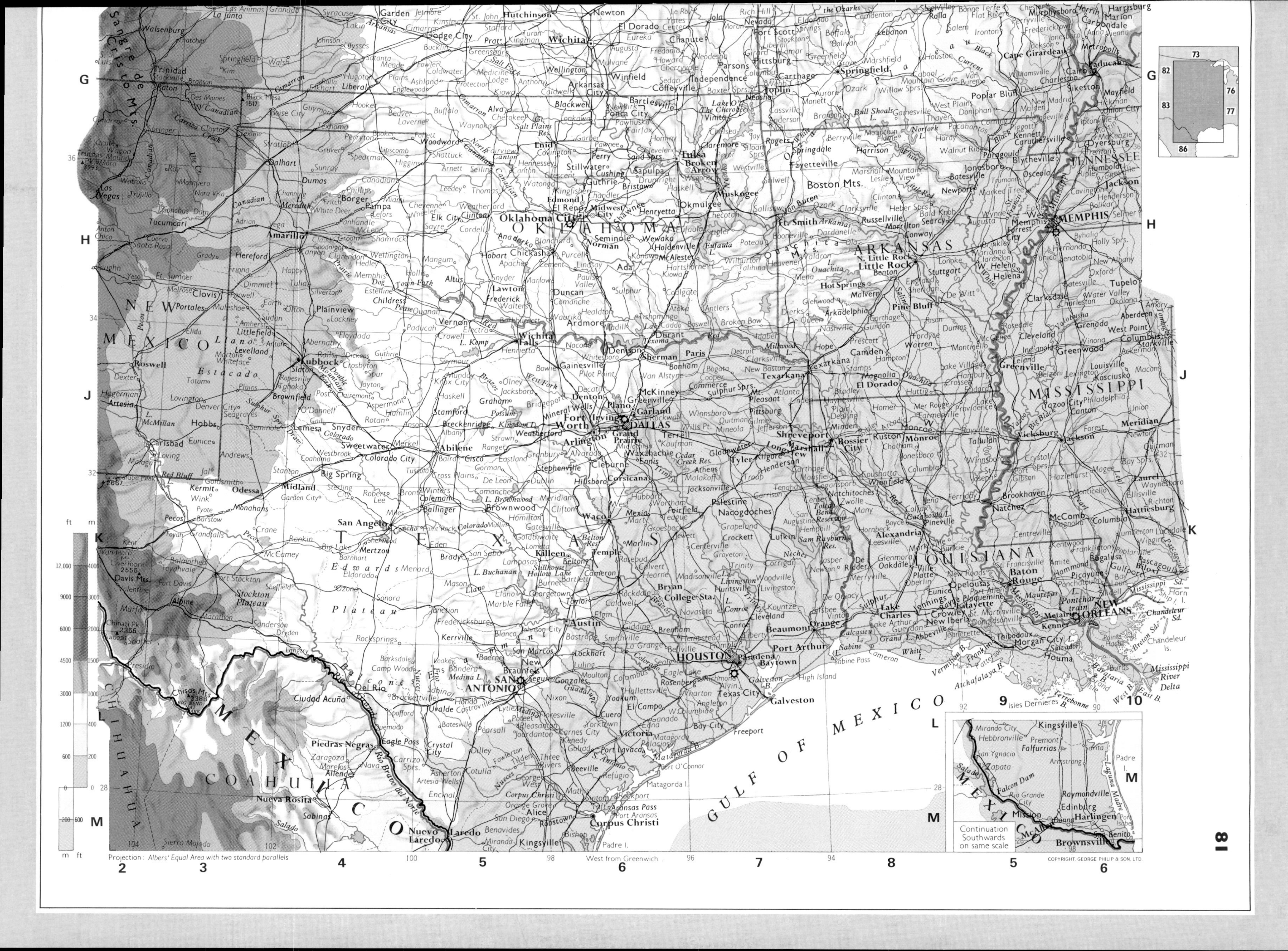
NEW MEXICO
OKLAHOMA
TEXAS
ARKANSAS
LOUISIANA
MISSISSIPPI
TENNESSEE
MEXICO
COAHUILA
CHIHUAHUA
GULF OF MEXICO
Sangre de Cristo Mts.
Llano Estacado
Edwards Plateau
Boston Mts.
Stockton Plateau
Oklahoma City
Tulsa
Wichita
Amarillo
Lubbock
Dallas
Fort Worth
Houston
San Antonio
Austin
Corpus Christi
El Paso
Shreveport
Little Rock
Memphis
Jackson
Baton Rouge
New Orleans
Galveston
Beaumont
Port Arthur
Waco
Abilene
San Angelo
Midland
Odessa
Laredo
Nuevo Laredo
Brownsville
Continuation Southwards on same scale
Projection: Albers' Equal Area with two standard parallels
West from Greenwich
COPYRIGHT. GEORGE PHILIP & SON. LTD

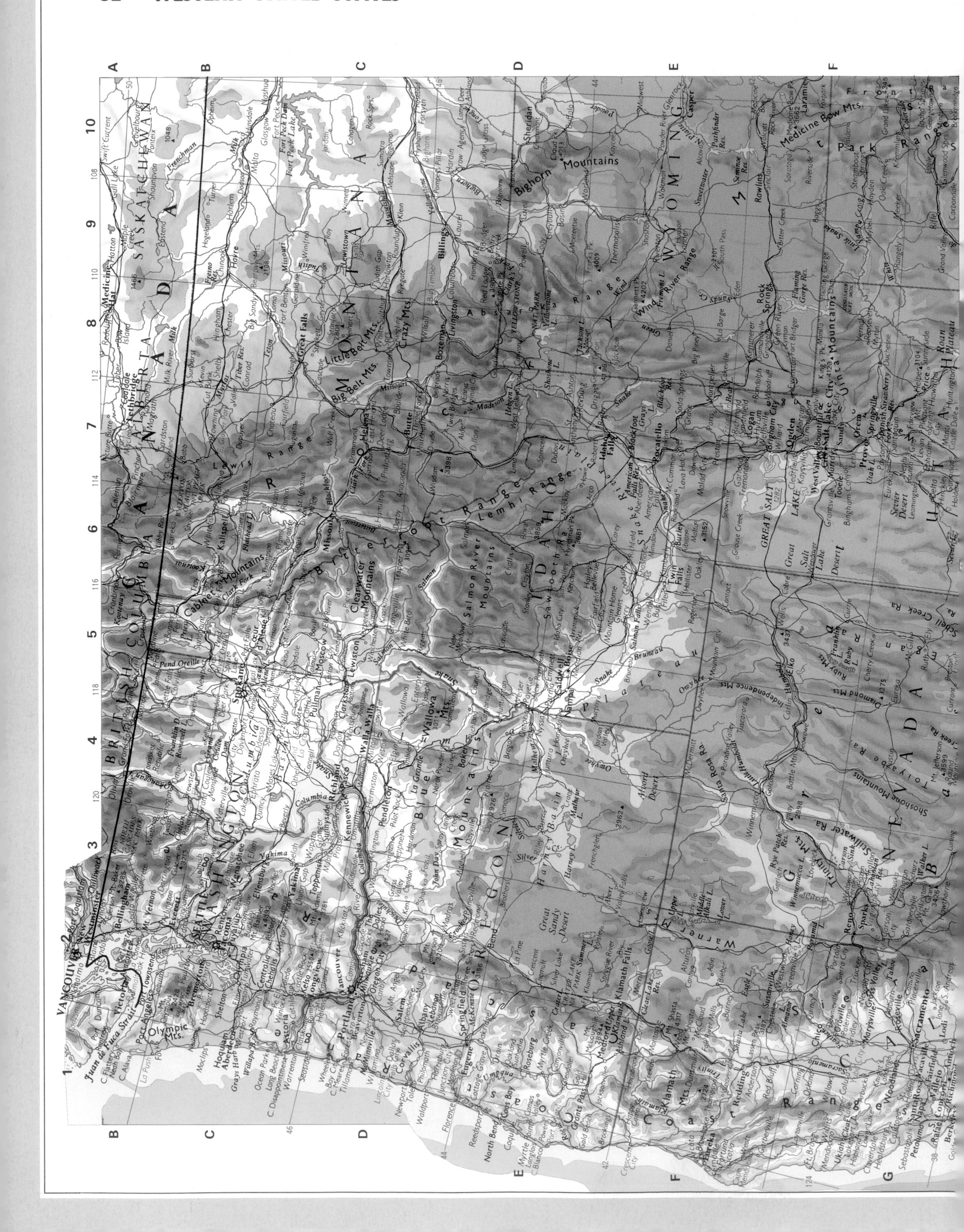
BRITISH COLUMBIA
ALBERTA
SASKATCHEWAN
CANADA
WASHINGTON
OREGON
IDAHO
MONTANA
WYOMING
NEVADA
UTAH
CALIFORNIA
Olympic Mts.
Cascade Range
Blue Mountains
Salmon River Mountains
Clearwater Mountains
Bitterroot Range
Lemhi Range
Sawtooth Ra.
Lewis Range
Little Belt Mts.
Big Belt Mts.
Crazy Mts.
Bighorn Mountains
Wind River Range
Absaroka Range
Medicine Bow Mts.
Uinta Mountains
Wasatch Range
Snake River Plain
Great Basin
Great Salt Lake
Great Salt Lake Desert
Harney Basin
Great Sandy Desert
Coast Ranges
Klamath Mts.
Sierra Nevada
Shoshone Mountains
Toiyabe Ra.
Ruby Mts.
Independence Mts.
Santa Rosa Ra.
Warner Mts.
Seattle
Tacoma
Spokane
Portland
Salem
Eugene
Boise
Pocatello
Idaho Falls
Helena
Butte
Great Falls
Billings
Missoula
Casper
Ogden
Salt Lake City
Provo
Reno
Sacramento
Vancouver
Victoria
Juan de Fuca Strait

1:6 000 000
miles
km
PACIFIC OCEAN
Golfo de California
Desierto de Altar
BAJA CALIFORNIA
SONORA
CHIHUAHUA
ARIZONA
NEW MEXICO
TEXAS
COLORADO
MEXICO
Sangre de Cristo Mts.
San Juan Mts.
Sacramento Mts.
Sonora Desert
Mojave Desert
Death Valley
Gran Desierto
Sierra de Juarez
Santa Lucia Range
Colorado Plateau
Grand Canyon
Mogollon Rim
Painted Desert
SAN FRANCISCO
Oakland
SAN JOSE
Stockton
Fresno
Bakersfield
Santa Barbara
LOS ANGELES
SAN DIEGO
Tijuana
Ensenada
Mexicali
Yuma
PHOENIX
Tucson
Nogales
Hermosillo
Flagstaff
Albuquerque
Santa Fe
Las Cruces
EL PASO
CIUDAD JUAREZ
Chihuahua
Las Vegas
Gallup
Prescott
Santa Catalina
San Clemente I.
I. de Guadalupe (Mexico)
I. Tiburón
I. Angel de la Guarda
West from Greenwich
Projection: Albers' Equal Area with two standard parallels
COPYRIGHT GEORGE PHILIP & SON, LTD.

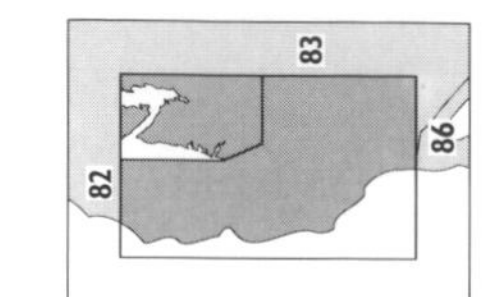

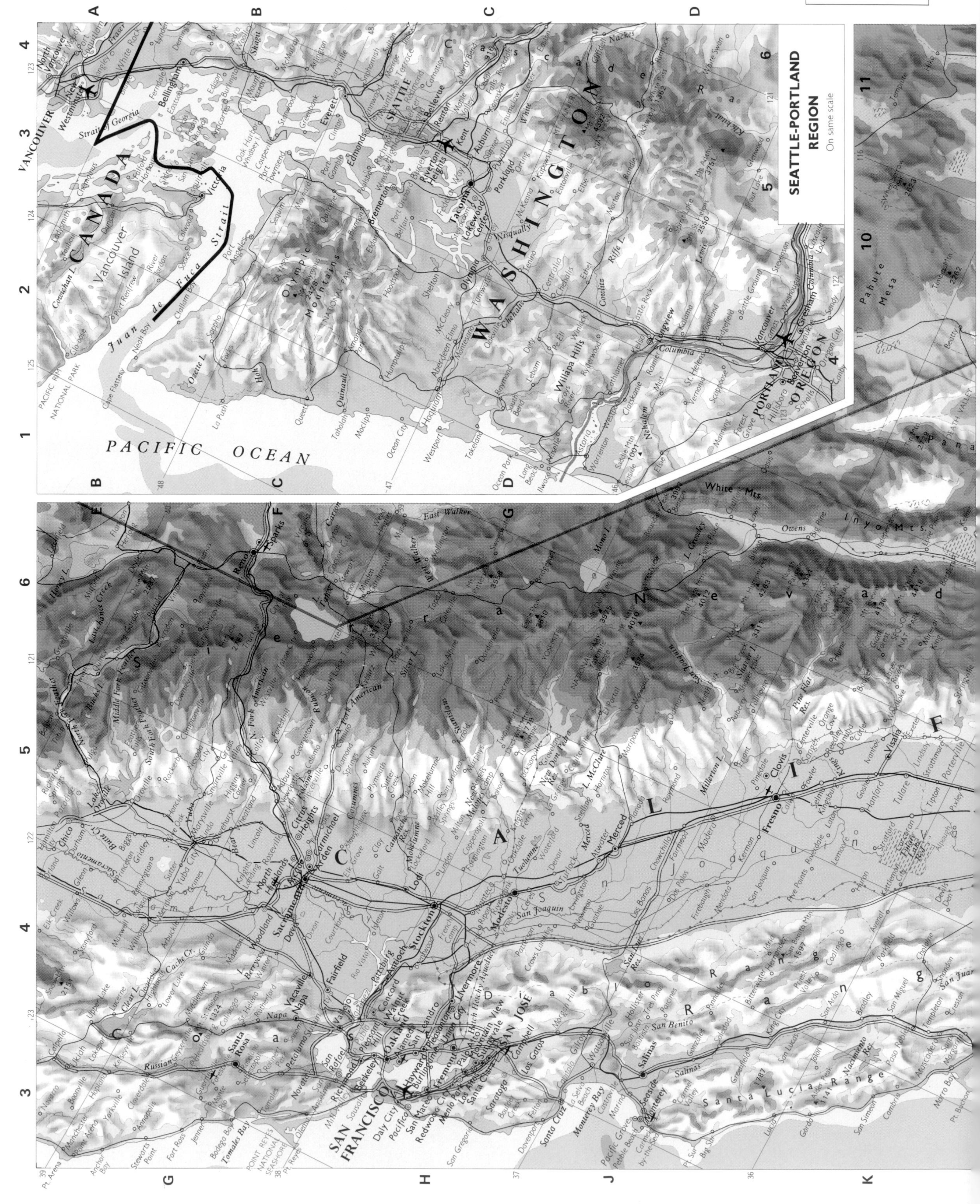

SEATTLE-PORTLAND REGION
On same scale
PACIFIC OCEAN
CANADA
Vancouver Island
WASHINGTON
OREGON
SEATTLE
PORTLAND
Tacoma
Olympia
Vancouver
Victoria
White Mts.
Inyo Mts.
Pahute Mesa
SAN FRANCISCO
SAN JOSE
Sacramento
Stockton
Fresno
Santa Lucia Range

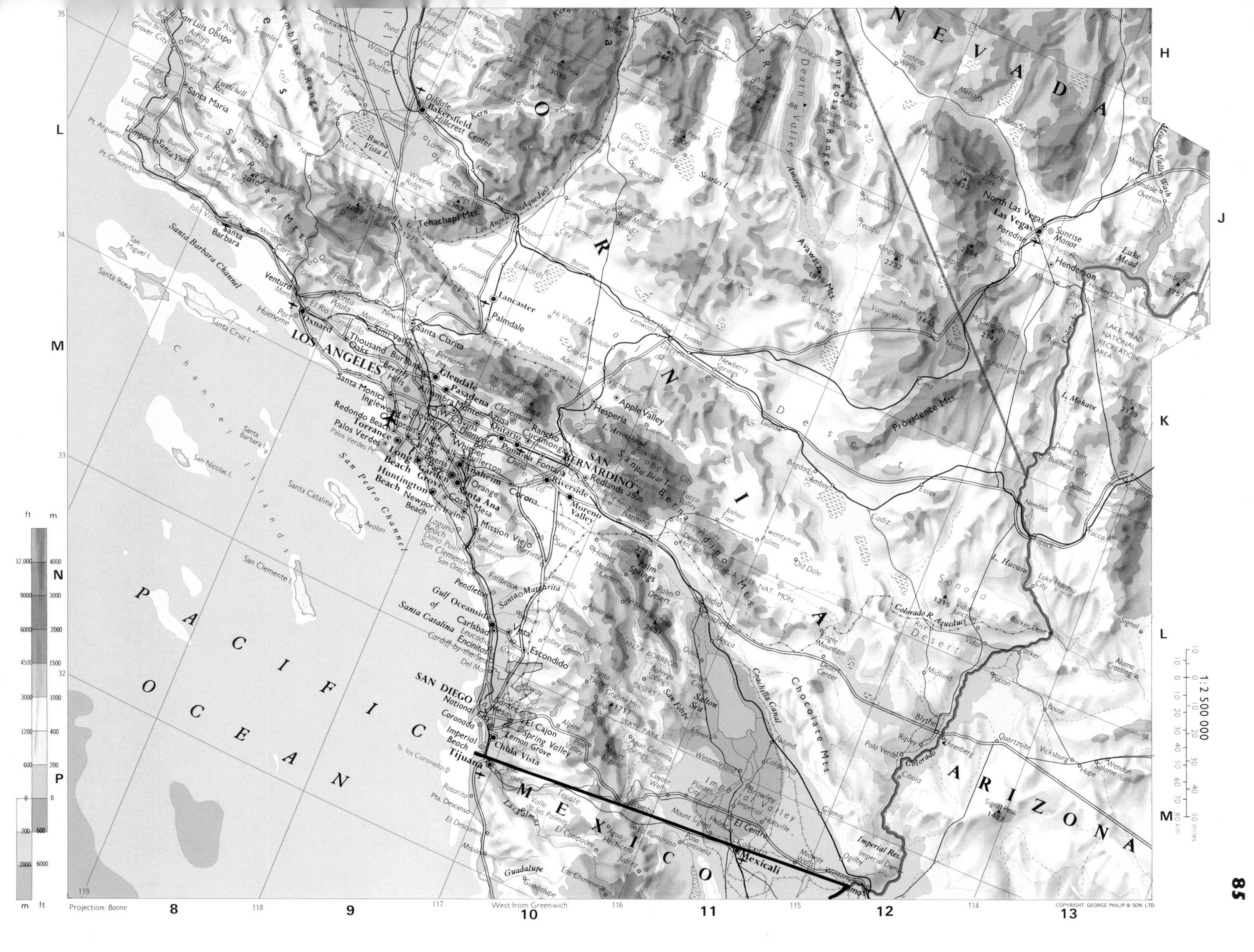
1:2 500 000
NEVADA
ARIZONA
CALIFORNIA
MEXICO
PACIFIC OCEAN
LOS ANGELES
SAN DIEGO
SAN BERNARDINO
Las Vegas
North Las Vegas
Henderson
Lake Mead
LAKE MEAD NATIONAL RECREATION AREA
Death Valley
Amargosa Range
Mojave Desert
Sonora Desert
Chocolate Mts.
Providence Mts.
Salton Sea
Coachella Canal
Colorado R. Aqueduct
Imperial Valley
Mexicali
Tijuana
Santa Barbara Channel
San Pedro Channel
Channel Islands
Gulf of Santa Catalina
Santa Catalina I.
San Clemente I.
Bakersfield
Lancaster
Palmdale
Riverside
Palm Springs
Escondido
Oceanside
Chula Vista
Ventura
Oxnard
Santa Barbara
Santa Maria
San Luis Obispo
Temblor Range
Tehachapi Mts.
West from Greenwich
Projection: Bonne
COPYRIGHT GEORGE PHILIP & SON LTD.

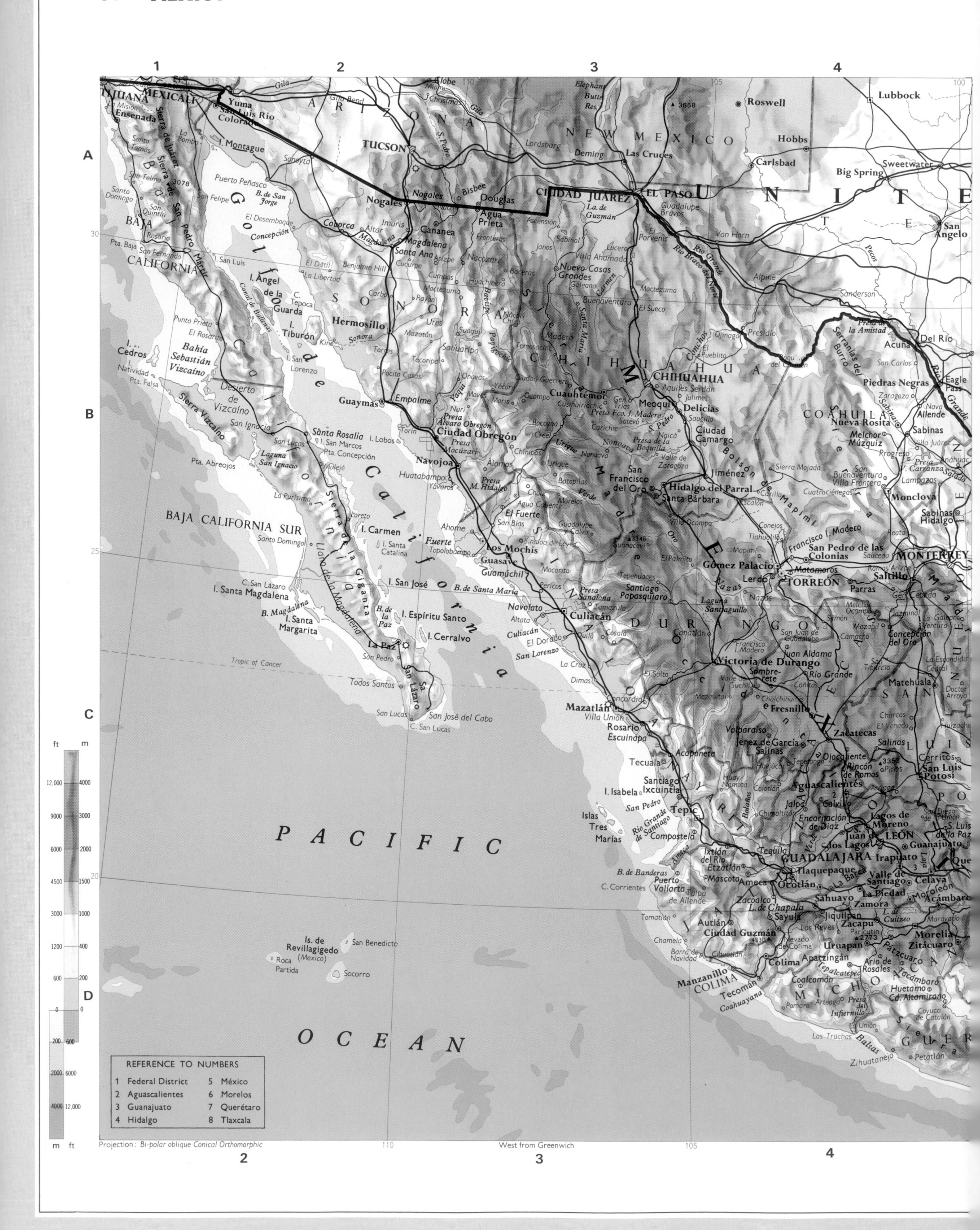

1
2
3
4
A
B
C
D
TIJUANA
MEXICALI
Ensenada
Yuma
San Luis Rio Colorado
I. Montague
Puerto Peñasco
B. de San Jorge
San Felipe
BAJA CALIFORNIA
Golfo de California
Sierra de Juárez
Sierra de San Pedro Mártir
Pta. Baja
San Fernando
I. San Luis
I. Ángel de la Guarda
Canal de Ballenas
Punta Prieta
El Rosarito
Bahía Sebastián Vizcaíno
I. Cedros
Natividad
Pta. Falsa
Desierto de Vizcaíno
Sierra Vizcaíno
San Ignacio
Laguna San Ignacio
Pta. Abreojos
Santa Rosalía
I. San Marcos
Pta. Concepción
Mulejé
La Purísima
Loreto
I. Carmen
I. Santa Catalina
BAJA CALIFORNIA SUR
Santo Domingo
Sierra de la Giganta
Llano de la Magdalena
C. San Lázaro
I. Santa Magdalena
B. Magdalena
I. Santa Margarita
I. San José
B. de la Paz
I. Espíritu Santo
I. Cerralvo
La Paz
San Pedro
Tropic of Cancer
Todos Santos
San Lázaro
San Lucas
San José del Cabo
C. San Lucas
ARIZONA
TUCSON
Gila
Gila Bend
Globe
Christmas
Sonoyta
Nogales
Bisbee
Douglas
Agua Prieta
Caborca
Altar
Imuris
Cananea
Magdalena
Santa Ana
Cucurpe
Arizpe
Nacozari
El Desemboque
Concepción
El Dátil
La Libertad
Benjamin Hill
SONORA
C. Tepoca
I. Tiburón
Kino
Hermosillo
Ures
Sonora
Mazatán
Torres
Sahuaripa
Tecoripa
Pocito Casas
I. San Lorenzo
Guaymas
Empalme
Yaqui
Nuri
Presa Alvaro Obregón
Ciudad Obregón
Presa Mocúzari
I. Lobos
Navojoa
Huatabampo
Yávaros
Presa M. Hidalgo
Álamos
El Fuerte
San Blas
Ahome
Fuerte
Topolobampo
Los Mochis
Guasave
Guamúchil
B. de Santa María
Mocorito
Navolato
Culiacán
Altata
El Dorado
San Lorenzo
La Cruz
Dimas
Concordia
Mazatlán
Villa Unión
Rosario
Escuinapa
Tecuala
Acaponeta
Santiago Ixcuintla
I. Isabela
San Pedro
Islas Tres Marías
Río Grande de Santiago
Tepic
Compostela
NAYARIT
B. de Banderas
C. Corrientes
Puerto Vallarta
Talpa de Allende
Tomatlán
Chamela
Barra de Navidad
Manzanillo
COLIMA
Tecomán
Coahuayana
Colima
Ixtlán del Río
Etzatlán
Mascota
Ameca
Tequila
GUADALAJARA
Tlaquepaque
Ocotlán
L. de Chapala
Zacoalco
Sayula
Autlán
Ciudad Guzmán
Nevado de Colima
Sahuayo
Jiquilpan
Zamora
La Piedad
Zacapu
Uruapan
Apatzingán
Coalcomán
Ario de Rosales
Pátzcuaro
Morelia
Zitácuaro
Huetamo
Cd. Altamirano
Infiernillo
Las Truchas
Balsas
Zihuatanejo
Petatlán
MICHOACAN
Acámbaro
Celaya
Irapuato
Valle de Santiago
Guanajuato
León
San Juan de los Lagos
Lagos de Moreno
Encarnación de Díaz
Aguascalientes
Calvillo
Jalpa
Rincón de Romos
Ojocaliente
Zacatecas
Fresnillo
Jerez de García Salinas
Valparaíso
Sombrerete
Victoria de Durango
Juan Aldama
Río Grande
DURANGO
Santiago Papasquiaro
Tepehuanes
Laguna Santiaguillo
Lerdo
Gómez Palacio
TORREÓN
San Pedro de las Colonias
Francisco I. Madero
Parras
Saltillo
MONTERREY
Ramos Arizpe
Concepción del Oro
Matehuala
Cerritos
San Luis Potosí
Salinas
Charcas
SAN LUIS POTOSÍ
Hidalgo del Parral
Santa Bárbara
Jiménez
San Francisco del Oro
Bolsón de Mapimí
CHIHUAHUA
Cuauhtémoc
Delicias
Meoqui
Ciudad Camargo
Ciudad Guerrero
Ojinaga
Presidio
CIUDAD JUÁREZ
EL PASO
Nuevo Casas Grandes
Villa Ahumada
El Sueco
Sierra Madre Occidental
Sierra Madre Oriental
NEW MEXICO
Roswell
Lordsburg
Deming
Las Cruces
Carlsbad
Hobbs
Lubbock
Sweetwater
Big Spring
San Angelo
Van Horn
Alpine
Sanderson
Pecos
Rio Grande
UNITED STATES
Del Rio
Eagle Pass
Presa de la Amistad
Acuña
Piedras Negras
COAHUILA
Allende
Nueva Rosita
Sabinas
Melchor Múzquiz
Monclova
Sabinas Hidalgo
Cuatrociénegas
Sierra Mojada
PACIFIC OCEAN
Is. de Revillagigedo (Mexico)
San Benedicto
Roca Partida
Socorro
REFERENCE TO NUMBERS
1 Federal District
2 Aguascalientes
3 Guanajuato
4 Hidalgo
5 México
6 Morelos
7 Querétaro
8 Tlaxcala
ft
m
Projection: Bi-polar oblique Conical Orthomorphic
West from Greenwich
110
105

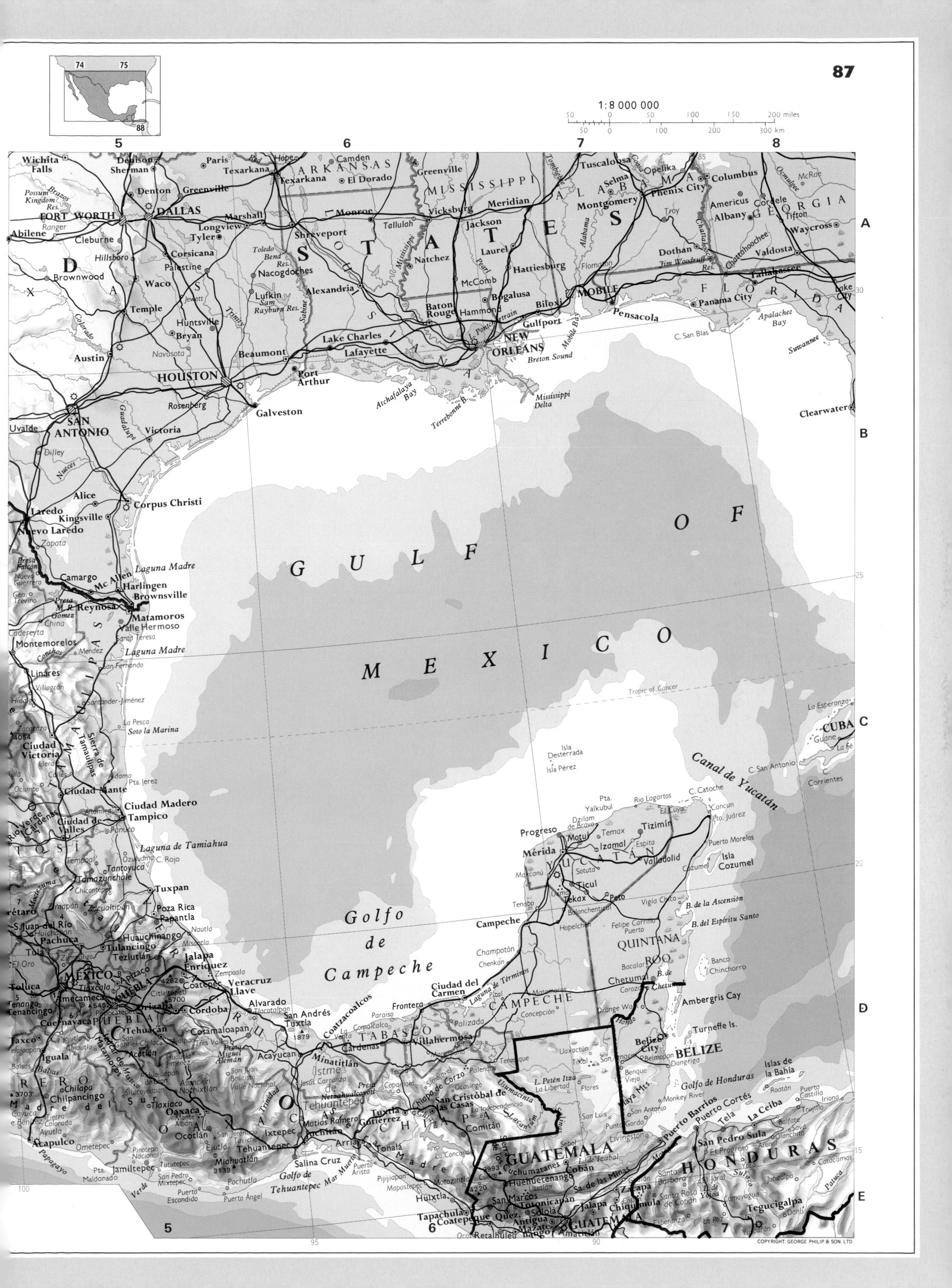

1:8 000 000
50 0 50 100 150 200 miles
50 0 100 200 300 km
74
75
88
5
6
7
8
A
B
C
D
E
GULF OF MEXICO
Golfo de Campeche
Canal de Yucatán
Tropic of Cancer
UNITED STATES
TEXAS
ARKANSAS
MISSISSIPPI
ALABAMA
GEORGIA
FLORIDA
LOUISIANA
CUBA
BELIZE
GUATEMALA
HONDURAS
YUCATÁN
QUINTANA ROO
CAMPECHE
TABASCO
CHIAPAS
OAXACA
VERACRUZ
PUEBLA
TAMAULIPAS
Wichita Falls
Denison
Sherman
Paris
Texarkana
Hope
Camden
El Dorado
Greenville
Tuscaloosa
Opelika
Columbus
Phenix City
Selma
Montgomery
Meridian
Vicksburg
Jackson
Monroe
Tallulah
Shreveport
Marshall
Longview
Tyler
Denton
DALLAS
FORT WORTH
Abilene
Cleburne
Hillsboro
Corsicana
Palestine
Nacogdoches
Brownwood
Waco
Temple
Lufkin
Sam Rayburn Res.
Toledo Bend Res.
Alexandria
Natchez
Laurel
Hattiesburg
McComb
Bogalusa
Baton Rouge
Hammond
Biloxi
Gulfport
MOBILE
Pensacola
Dothan
Americus
Albany
Cordele
Tifton
Waycross
Valdosta
Tallahassee
Panama City
Apalachee Bay
C. San Blas
Lake City
Huntsville
Bryan
Austin
Navasota
Beaumont
Lake Charles
Lafayette
NEW ORLEANS
Breton Sound
Mississippi Delta
Atchafalaya Bay
Terrebonne B.
HOUSTON
Port Arthur
Galveston
Rosenberg
SAN ANTONIO
Uvalde
Victoria
Dilley
Alice
Corpus Christi
Kingsville
Laredo
Nuevo Laredo
Zapata
Laguna Madre
Camargo
Mc Allen
Harlingen
Brownsville
Reynosa
Matamoros
Valle Hermoso
Montemorelos
Linares
San Fernando
Soto la Marina
La Pesca
Sierra de Tamaulipas
Ciudad Victoria
Ciudad Mante
Ciudad Madero
Tampico
Ciudad de Valles
Laguna de Tamiahua
C. Rojo
Tuxpan
Poza Rica
Papantla
Nautla
Huauchinango
Tulancingo
Pachuca
Teziutlán
Jalapa Enríquez
Veracruz Llave
Coatepec
Córdoba
Orizaba
MÉXICO
Toluca
Cuernavaca
PUEBLA
Tehuacán
Alvarado
Cosamaloapan
San Andrés Tuxtla
Coatzacoalcos
Minatitlán
Acayucan
Cárdenas
Villahermosa
Frontera
Ciudad del Carmen
Laguna de Términos
Champotón
Campeche
Hopelchén
Progreso
Mérida
Motul
Izamal
Tizimín
Valladolid
Ticul
Tekax
Peto
Felipe Carrillo Puerto
Chetumal
Bacalar
B. de la Ascensión
B. del Espíritu Santo
Isla Cozumel
Cancún
Pto. Juárez
Puerto Morelos
C. Catoche
Río Lagartos
Isla Desterrada
Isla Pérez
Banco Chinchorro
Ambergris Cay
Turneffe Is.
Belize City
Belmopan
Dangriga
Golfo de Honduras
Islas de la Bahía
Roatán
Puerto Castilla
Trujillo
La Ceiba
Tela
Puerto Cortés
Puerto Barrios
San Pedro Sula
Tegucigalpa
L. Petén Itzá
Flores
Cobán
Huehuetenango
Zacapa
Jalapa
Chiquimula
GUATEMALA
San Marcos
Tapachula
Quezaltenango
Retalhuleu
Antigua
Tuxtla Gutiérrez
San Cristóbal de las Casas
Comitán
Tonalá
Arriaga
Juchitán
Salina Cruz
Golfo de Tehuantepec
Istmo de Tehuantepec
Tehuantepec
Ixtepec
Oaxaca
Miahuatlán
Ocotlán
Pochutla
Puerto Ángel
Puerto Escondido
Jamiltepec
Pinotepa Nacional
Acapulco
Chilpancingo
Chilapa
Iguala
Taxco
Madre del Sur
Sierra Madre
Pta. Maldonado
Cuba
Lo Esperanza
Guane
C. San Antonio
C. Corrientes
Clearwater
Suwannee
100
95
90
85
30
25
20
15
COPYRIGHT, GEORGE PHILIP & SON, LTD

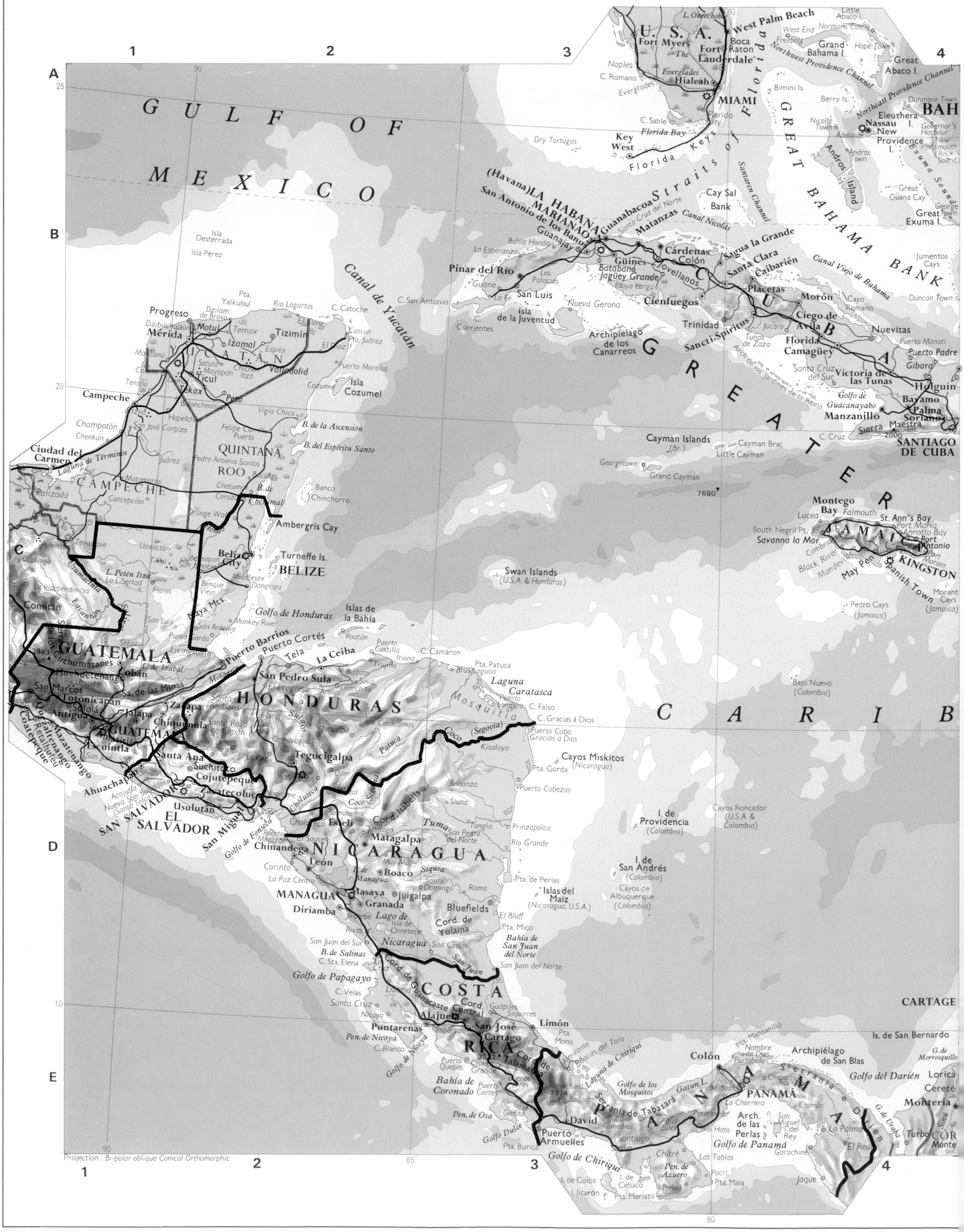

1
2
3
4
A
B
C
D
E
90
85
80
25
20
10
GULF OF MEXICO
U.S.A.
L. Okeechobee
West Palm Beach
Boca Raton
Fort Myers
Fort Lauderdale
Naples
C. Romano
Everglades
Hialeah
MIAMI
C. Sable
Florida Bay
Key West
Florida Keys
Dry Tortugas
Straits of Florida
Little Abaco I.
West End
Grand Bahama I.
Northwest Providence Channel
Great Abaco I.
Northeast Providence Channel
Dunmore Town
BAH
Bimini Is.
Berry Is.
Eleuthera I.
Nicolls Town
Nassau
New Providence I.
Governor's Harbour
Adelaide
Andros Town
Andros Island
New Portsmouth (Rock Sound)
Exuma Sound
GREAT BAHAMA BANK
Great Guana Cay
George Town
Great Exuma I.
Cay Sal Bank
Santaren Channel
Canal Nicolás
Jumentos Cays
Canal Viejo de Bahama
Duncan Town
(Havana) LA HABANA
MARIANAO
San Antonio de los Baños
Guanabacoa
Santa Cruz del Norte
Matanzas
Guanajay
Bahía Honda
La Esperanza
Pinar del Río
Los Palacios
San Luis
Guane
La Fé
Güines
Batabanó
Jagüey Grande
Cárdenas
Colón
Jovellanos
Sagua la Grande
Santa Clara
Caibarién
Playa Larga
Cienfuegos
Placetas
Morón
Cayo Romano
Nueva Gerona
Isla de la Juventud
C. San Antonio
C. Corrientes
Archipiélago de los Canarreos
Trinidad
Sancti-Spíritus
Tunas de Zaza
Júcaro
Ciego de Avila
Florida
Camagüey
Nuevitas
Puerto Manatí
Puerto Padre
Gibara
Santa Cruz del Sur
Victoria de las Tunas
Holguín
Arch. de los Jardines de la Reina
Golfo de Guacanayabo
Bayamo
Palma Soriano
Manzanillo
Sierra Maestra
2000
C. Cruz
SANTIAGO DE CUBA
CUBA
GREATER
Cayman Islands (Br.)
Cayman Brac
Little Cayman
Georgetown
Grand Cayman
7680
Montego Bay
Falmouth
Lucea
St. Ann's Bay
Port Maria
Annotto Bay
Port Antonio
South Negril Pt.
Savanna la Mar
Black River
Mandeville
May Pen
Spanish Town
KINGSTON
JAMAICA
Morant Cays (Jamaica)
Pedro Cays (Jamaica)
Swan Islands (U.S.A. & Honduras)
Bajo Nuevo (Colombia)
CARIB
Isla Desterrada
Isla Pérez
Canal de Yucatán
Pta. Yalkubul
Río Lagartos
C. Catoche
Progreso
Dzilam de Bravo
Dzibilchaltún
Motul
Temax
Tizimín
Mérida
YUCATÁN
Izamal
Cancun
Pto. Juárez
El Díaz
Sotuta
Mayapán
Chichén Itzá
Valladolid
Puerto Morelos
Ticul
Isla Cozumel
Cozumel
Campeche
Tenabo
Tekax
Peto
Bolonchenticul
Hopelchén
Vigía Chico
Champotón
San José Carpizo
Felipe Carrillo Puerto
B. de la Ascensión
Chenkán
B. del Espíritu Santo
Ciudad del Carmen
Laguna de Términos
QUINTANA ROO
Juárez
Pedro Antonio Santos
Bacalar
Matamoros
Chetumal
B. de Chetumal
Banco Chinchorro
CAMPECHE
Palizada
Concepción
Corozal
Orange Walk
Ambergris Cay
Tenosique
Palenque
Uaxactún
Belize City
Turneffe Is.
BELIZE
Cosimaga
Tikal
L. Petén Itzá
La Libertad
Flores
Belmopan
San Ignacio
Benque Viejo
Middlesex
Dangriga
Maya Mts.
Usumacinta
La Independencia
Comitán
Lacantún
San Luis
Punta Gorda
San Antonio
Monkey River
Golfo de Honduras
Islas de la Bahía
Roatán
Livingston
Puerto Barrios
Puerto Cortés
Tela
La Ceiba
Puerto Castilla
Trujillo
Iriona
C. Camarón
Pta. Patuca
Brus Laguna
Laguna Caratasca
Puerto Lempira
C. Falso
C. Gracias á Dios
Puerto Cabo Gracias á Dios
Mosquitia
Kisalaya
Cayos Miskitos (Nicaragua)
Pta. Gorda
Puerto Cabezas
GUATEMALA
3993
Cuchumatanes
Huehuetenango
Cobán
L. de Izabal
San Marcos
Totonicapán
Sa. de las Minas
Motagua
Sololá
Antigua
Jalapa
Zacapa
Chiquimula
GUATEMALA
Quezaltenango
Mazatenango
Retalhuleu
Coatepeque
Escuintla
San Pedro Sula
El Progreso
HONDURAS
Santa Rosa de Copán
Yoro
Sulaco
Juticalpa
Catacamas
Comayagua
La Paz
Tegucigalpa
Yuscarán
Patuca
Coco (Segovia)
Santa Ana
Suchitoto
Cojutepeque
Ahuachapán
Acajutla
Nueva San Salvador (Santa Tecla)
SAN SALVADOR
Zacatecoluca
Usulután
EL SALVADOR
San Miguel
Golfo de Fonseca
Choluteca
Somoto
Estelí
Chalteca
El Sauce
Puerto Morazán
Chinandega
Jinotega
Matagalpa
NICARAGUA
León
Corinto
La Paz Centro
Boaco
Siquia
Muy Muy
Tuma
Cord. Isabella
Bonanza
Siuna
Tungla
San Pedro del Norte
Prinzapolca
Río Grande
Pta. de Perlas
Islas del Maiz (Nicaragua, U.S.A.)
MANAGUA
Masaya
Granada
Juigalpa
Santo Domingo
Rama
Diriamba
Bluefields
Lago de Nicaragua
Isla de Ometepe
Rivas
Cord. de Yolaina
El Bluff
Pta. Mico
San Juan del Sur
B. de Salinas
C. Sta. Elena
San Carlos
San Juan
Bahía de San Juan del Norte
San Juan del Norte
Golfo de Papagayo
C. Velas
Liberia
Santa Cruz
Nicoya
Cord. de Guanacaste
Cord. Central
COSTA RICA
Guápiles
Siquirres
Limón
Alajuela
San José
Cartago
Puntarenas
Esparza
Pen. de Nicoya
C. Blanco
Golfo de Nicoya
Puerto Quepos
Chirripó Grande
3820
Talamanca
Bahía de Coronado
Puerto Cortés
Golfito
Pen. de Osa
Golfo Dulce
Pta. Burica
Puerto Armuelles
Pta. Mona
Almirante
Bocas del Toro
Laguna de Chiriquí
3374
Volcán Barú
Boquete
David
Remedios
Golfo de los Mosquitos
Serranía de Tabasará
Santiago
Golfo de Chiriquí
I. de Coiba
I. Jicarón
I. de Cebaco
Pta. Mariato
Chitré
Pen. de Azuero
Tonosí
Las Tablas
Pocrí
Pta. Mala
Pta. Manzanillo
Nombre de Dios
Portobelo
Colón
Gatún L.
Balboa
La Chorrera
Penonomé
Antón
Río Hato
Arch. de las Perlas
San Miguel
I. del Rey
Golfo de Panamá
Garachiné
Jaque
PANAMÁ
PANAMA
Chepo
Chimán
Serranía del Darién
La Palma
El Real
Archipiélago de San Blas
Golfo del Darién
G. de Urabá
I. de Providencia (Colombia)
Cayos Roncador (U.S.A. & Colombia)
I. de San Andrés (Colombia)
Cayos de Albuquerque (Colombia)
CARTAGE
Is. de San Bernardo
G. de Morrosquillo
Lorica
Cereté
Montería
Sinú
Turbo
COR
Monte

Projection: Bi-polar oblique Conical Orthomorphic

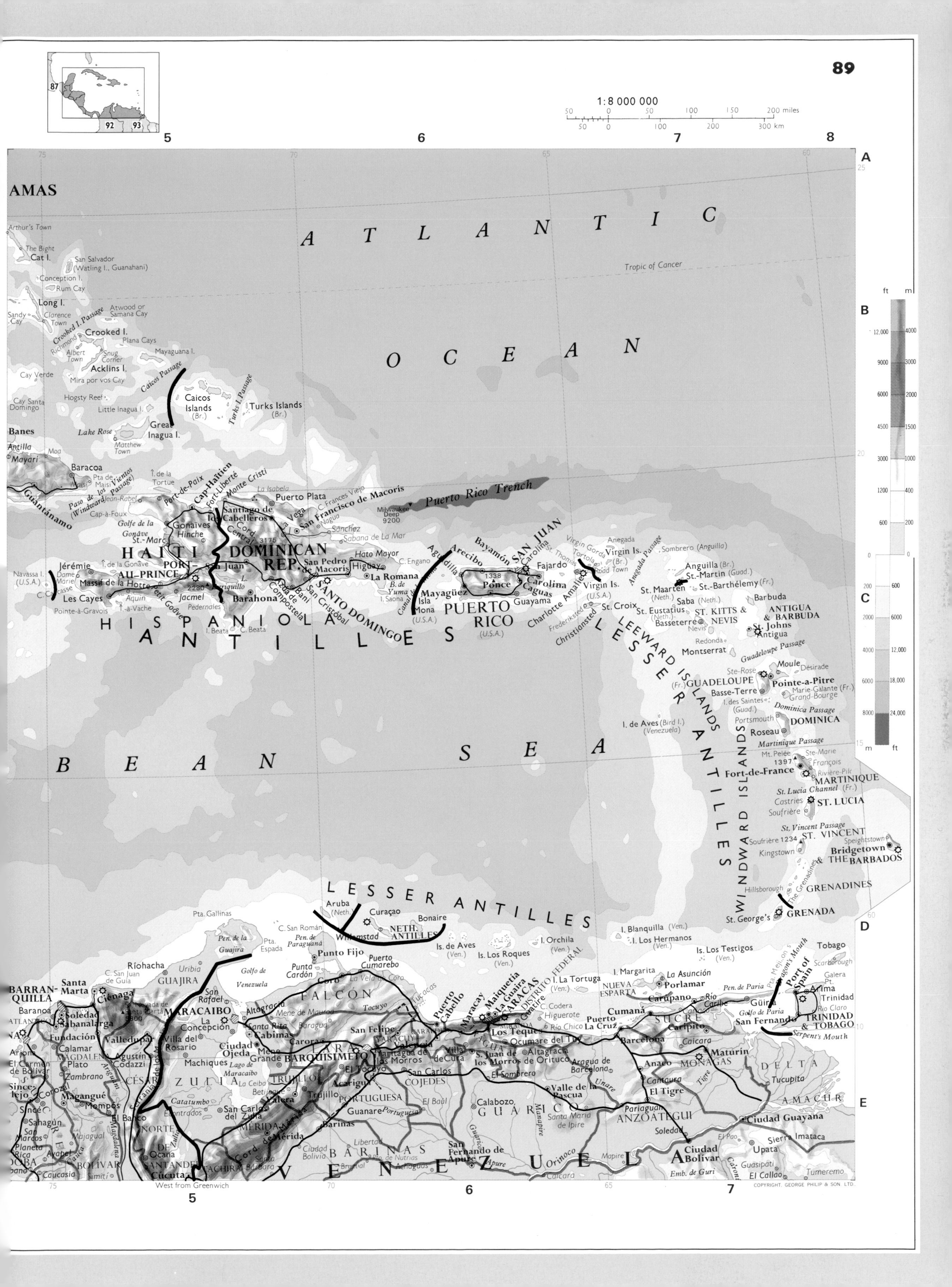
1:8 000 000
ATLANTIC OCEAN
CARIBBEAN SEA
Tropic of Cancer
Puerto Rico Trench
Milwaukee Deep 9200
HAITI
DOMINICAN REP.
HISPANIOLA
ANTILLES
PUERTO RICO (U.S.A.)
SAN JUAN
PORT-AU-PRINCE
SANTO DOMINGO
Cap-Haïtien
Santiago de los Caballeros
Puerto Plata
San Francisco de Macoris
La Romana
Mayagüez
Ponce
Arecibo
Caicos Islands (Br.)
Turks Islands (Br.)
Great Inagua I.
Acklins I.
Crooked I.
Long I.
Cat I.
Baracoa
Guantánamo
Virgin Is.
Anguilla (Br.)
St.-Martin (Guad.)
Barbuda
ANTIGUA & BARBUDA
St. Johns
ST. KITTS & NEVIS
Basseterre
Montserrat
GUADELOUPE
Pointe-a-Pitre
Basse-Terre
DOMINICA
Roseau
MARTINIQUE
Fort-de-France
ST. LUCIA
Castries
ST. VINCENT
Kingstown
Bridgetown
BARBADOS
GRENADINES
GRENADA
St. George's
LEEWARD ISLANDS
WINDWARD ISLANDS
LESSER ANTILLES
I. de Aves (Bird I.) (Venezuela)
Aruba (Neth.)
Curaçao
Bonaire
Willemstad
NETH. ANTILLES
Tobago
TRINIDAD & TOBAGO
Port of Spain
San Fernando
Serpent's Mouth
Dragon's Mouth
I. Margarita
Porlamar
Cumaná
Barcelona
Maturín
CARACAS
Valencia
Maracay
BARQUISIMETO
MARACAIBO
Golfo de Venezuela
Lago de Maracaibo
Punto Fijo
Coro
FALCÓN
ZULIA
Mérida
BARINAS
Barinas
GUÁRICO
ANZOÁTEGUI
MONAGAS
DELTA AMACURO
Ciudad Guayana
Ciudad Bolívar
VENEZUELA
Orinoco
BARRANQUILLA
Santa Marta
Riohacha
GUAJIRA
Cúcuta
Valledupar
West from Greenwich
COPYRIGHT, GEORGE PHILIP & SON, LTD.

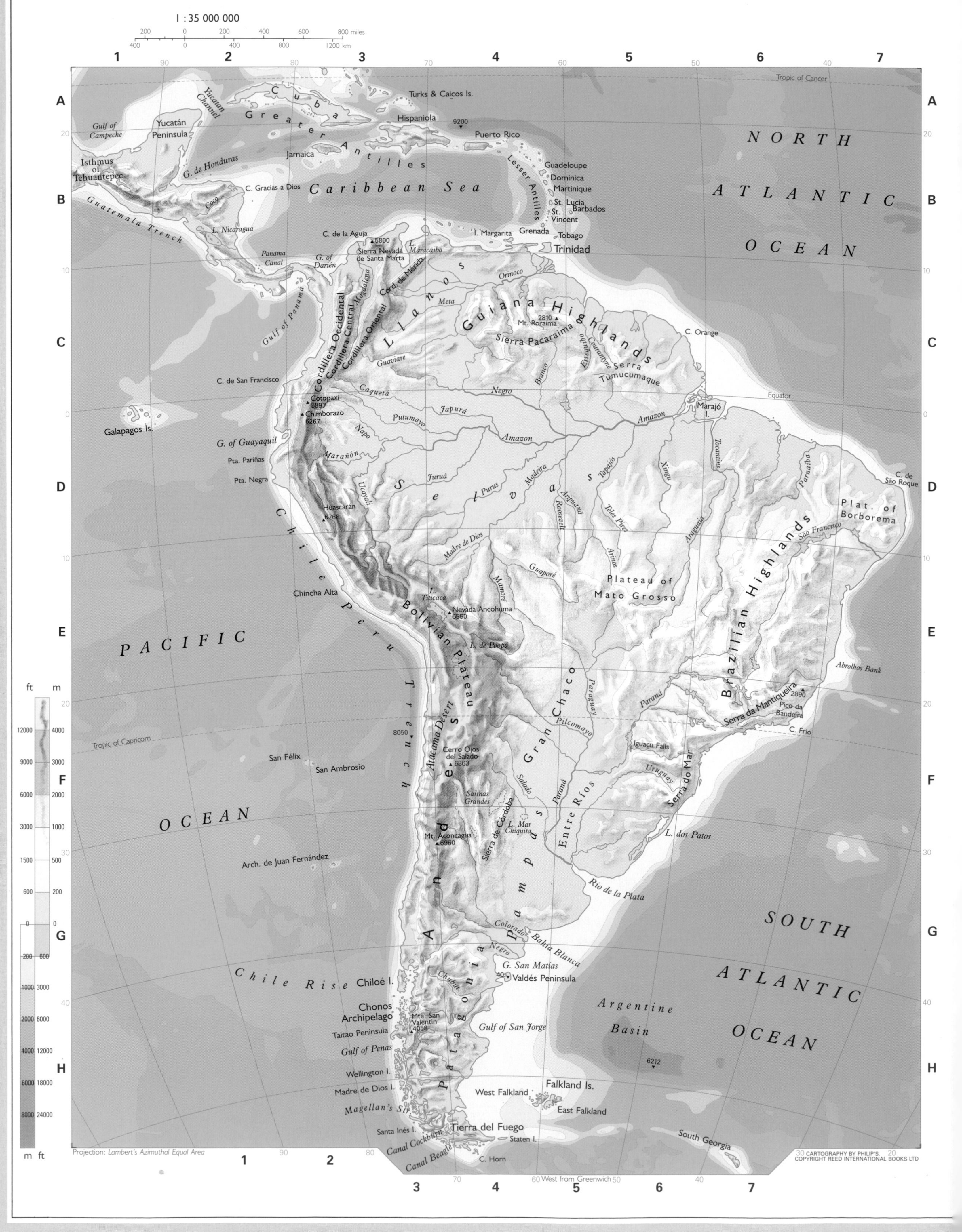
1 : 35 000 000
Tropic of Cancer
NORTH ATLANTIC OCEAN
Caribbean Sea
Greater Antilles
Lesser Antilles
Cuba
Turks & Caicos Is.
Hispaniola
Puerto Rico
Jamaica
Gulf of Campeche
Yucatán Peninsula
Yucatan Channel
Isthmus of Tehuantepec
G. de Honduras
C. Gracias a Dios
Guatemala Trench
L. Nicaragua
Panama Canal
G. of Darién
Gulf of Panama
Guadeloupe
Dominica
Martinique
St. Lucia
Barbados
St. Vincent
Grenada
Tobago
Trinidad
I. Margarita
C. de la Aguja
Sierra Nevada de Santa Marta
L. Maracaibo
Cordillera Occidental
Cordillera Central
Cordillera Oriental
Cord. de Merida
Magdalena
Llanos
Orinoco
Meta
Guiana Highlands
Mt. Roraima
Sierra Pacaraima
Serra Tumucumaque
C. Orange
Guaviare
C. de San Francisco
Caquetá
Negro
Branco
Essequibo
Courantyne
Equator
Marajó I.
Amazon
Japurá
Putumayo
Cotopaxi 5897
Chimborazo 6267
Galapagos Is.
G. of Guayaquil
Pta. Pariñas
Pta. Negra
Napo
Marañón
Selvas
Juruá
Purus
Madeira
Tapajós
Xingu
Tocantins
Parnaíba
C. de São Roque
Plat. of Borborema
Ucayali
Huascarán 6768
Aripuanã
Roosevelt
Teles Pires
Araguaia
São Francisco
Madre de Dios
Guaporé
Arinos
Mamoré
Plateau of Mato Grosso
Brazilian Highlands
Chincha Alta
Peru Chile Trench
L. Titicaca
Nevada Ancohuma 6550
Bolivian Plateau
L. de Poopó
PACIFIC OCEAN
Abrolhos Bank
Gran Chaco
Paraguay
Paraná
Serra da Mantiqueira
2890
Pico da Bandeira
C. Frio
Tropic of Capricorn
8050
Pilcomayo
Atacama Desert
Cerro Ojos del Salado 6863
Iguaçu Falls
San Félix
San Ambrosio
Uruguay
Serra do Mar
Salado
Entre Rios
Salinas Grandes
Sierra de Córdoba
L. Mar Chiquita
L. dos Patos
Mt. Aconcagua 6960
Andes
Arch. de Juan Fernández
Pampas
Río de la Plata
SOUTH ATLANTIC OCEAN
Colorado
Bahía Blanca
Negro
G. San Matías
Valdés Peninsula
Chile Rise
Chiloé I.
Chubut
Patagonia
Chonos Archipelago
Mte. San Valentin 4058
Taitao Peninsula
Gulf of San Jorge
Argentine Basin
Gulf of Penas
6212
Wellington I.
Madre de Dios I.
West Falkland
Falkland Is.
East Falkland
Magellan's Str.
Santa Inés I.
Tierra del Fuego
Canal Cockburn
Staten I.
Canal Beagle
C. Horn
South Georgia
West from Greenwich
ft m
Projection: Lambert's Azimuthal Equal Area

1 : 35 000 000
NORTH ATLANTIC OCEAN
Caribbean Sea
PACIFIC OCEAN
SOUTH ATLANTIC OCEAN
MEXICO
BELIZE
GUATEMALA
HONDURAS
NICARAGUA
EL SALVADOR
COSTA RICA
PANAMA
CUBA
HAITI
DOMINICAN REP.
JAMAICA
BAHAMAS
VENEZUELA
COLOMBIA
GUYANA
SURINAM
FRENCH GUIANA
ECUADOR
PERU
BOLIVIA
BRAZIL
PARAGUAY
CHILE
ARGENTINA
URUGUAY
LIMA
Capital Cities
Projection: Lambert's Azimuthal Equal Area
CARTOGRAPHY BY PHILIP'S. COPYRIGHT REED INTERNATIONAL BOOKS LTD

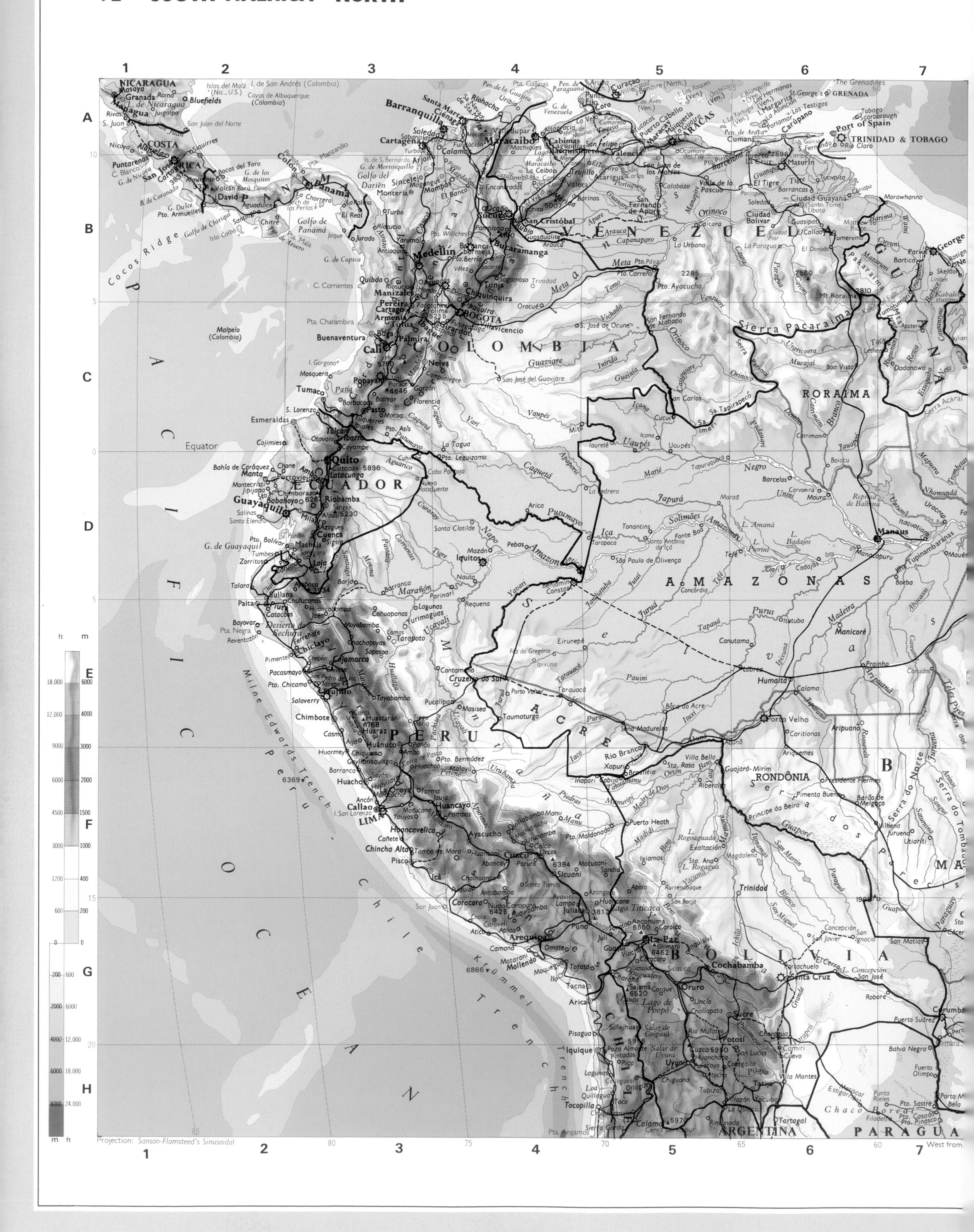

NICARAGUA
COSTA RICA
PANAMA
COLOMBIA
VENEZUELA
ECUADOR
PERU
BOLIVIA
CHILE
ARGENTINA
PARAGUAY
GUYANA
TRINIDAD & TOBAGO
GRENADA
AMAZONAS
RORAIMA
RONDÔNIA
ACRE
PACIFIC OCEAN
Equator
Cocos Ridge
Milne Edwards Trench
Peru Chile Trench
Krümmel Trench
Golfo de Panamá
Golfo de Guayaquil
Golfo del Darién
Lago de Maracaibo
Lago Titicaca
Lago de Poopó
Salar de Uyuni
Sierra Pacaraima
Managua
Bluefields
San José
Panamá
Colón
Barranquilla
Cartagena
Santa Marta
Maracaibo
Caracas
Valencia
Barquisimeto
Cumaná
Maturín
Ciudad Bolívar
Ciudad Guayana
Port of Spain
Georgetown
Medellín
Bucaramanga
Cúcuta
Bogotá
Cali
Manizales
Pereira
Ibagué
Neiva
Popayán
Pasto
Villavicencio
Buenaventura
Quito
Guayaquil
Cuenca
Riobamba
Manta
Esmeraldas
Loja
Iquitos
Piura
Chiclayo
Trujillo
Cajamarca
Chimbote
Callao
LIMA
Huancayo
Ayacucho
Cuzco
Arequipa
Juliaca
Puno
Tacna
Arica
Iquique
Antofagasta
La Paz
Cochabamba
Oruro
Sucre
Potosí
Santa Cruz
Trinidad
Tarija
Manaus
Porto Velho
Rio Branco
Boa Vista
Humaitá
Manicoré
Cruzeiro do Sul
Orinoco
Amazonas
Solimões
Negro
Japurá
Putumayo
Caquetá
Napo
Marañón
Ucayali
Juruá
Purus
Madeira
Guaporé
Mamoré
Beni
Meta
Guaviare
Vaupés
Branco
Magdalena
Cauca
Chimborazo 6267
Cotopaxi 5896
Huascarán 6768
Illimani 6462
Mt. Roraima 2810
ft
m
18,000
12,000
9000
6000
4500
3000
1200
600
0
6000
4000
3000
2000
1500
1000
400
200
200
2000
4000
6000
8000
600
6000
12,000
18,000
24,000
Projection: Sanson-Flamsteed's Sinusoidal
West from

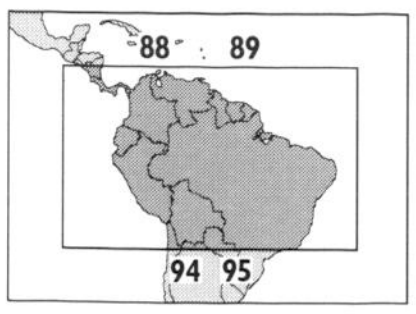

88 89
94 95

1:16 000 000

100 50 0 100 200 300 miles

100 0 100 200 300 400 km

8 9 10 11 12 13

A B C D E F G H

ATLANTIC OCEAN

Equator

SURINAM
FR. GUIANA
Paramaribo
Cayenne
C. Orange
Oiapoque
AMAPÁ
Macapá
Ilha de Maracá
C. do Norte
Estuario do Rio Amazonas
Ilha Caviana
Ilha Mexiana
C. Maguarinho
Ilha de Marajó
Belém
Bragança
Amazonas (Amazon)
Santarém
Óbidos
PARÁ
Altamira
Tucuruí
Marabá
São Luís
Fortaleza
Rocas
Fernando de Noronha (Braz.)
MARANHÃO
Imperatriz
Teresina
CEARÁ
Natal
C. de São Roque
RIO GRANDE DO NORTE
PARAÍBA
João Pessoa
Campina Grande
Olinda
RECIFE
PERNAMBUCO
Maceió
ALAGOAS
SERGIPE
Aracaju
PIAUÍ
TOCANTINS
BRAZIL
BAHIA
Feira de Santana
Salvador
Vitória da Conquista
Ilhéus
Represa de Sobradinho
6059
MATO GROSSO
Planalto do Mato Grosso
DIST. FED.
Brasília
Goiânia
GOIÁS
MATO GROSSO DO SUL
Campo Grande
MINAS GERAIS
Belo Horizonte
Montes Claros
Uberlândia
Uberaba
Vitória
ESPÍRITO SANTO
Juiz de Fora
Campos
SÃO PAULO
Campinas
Petrópolis
Niterói
RIO DE JANEIRO
Cabo Frio
Trindade (Braz.)

Greenwich 55 50 45 40 35 30

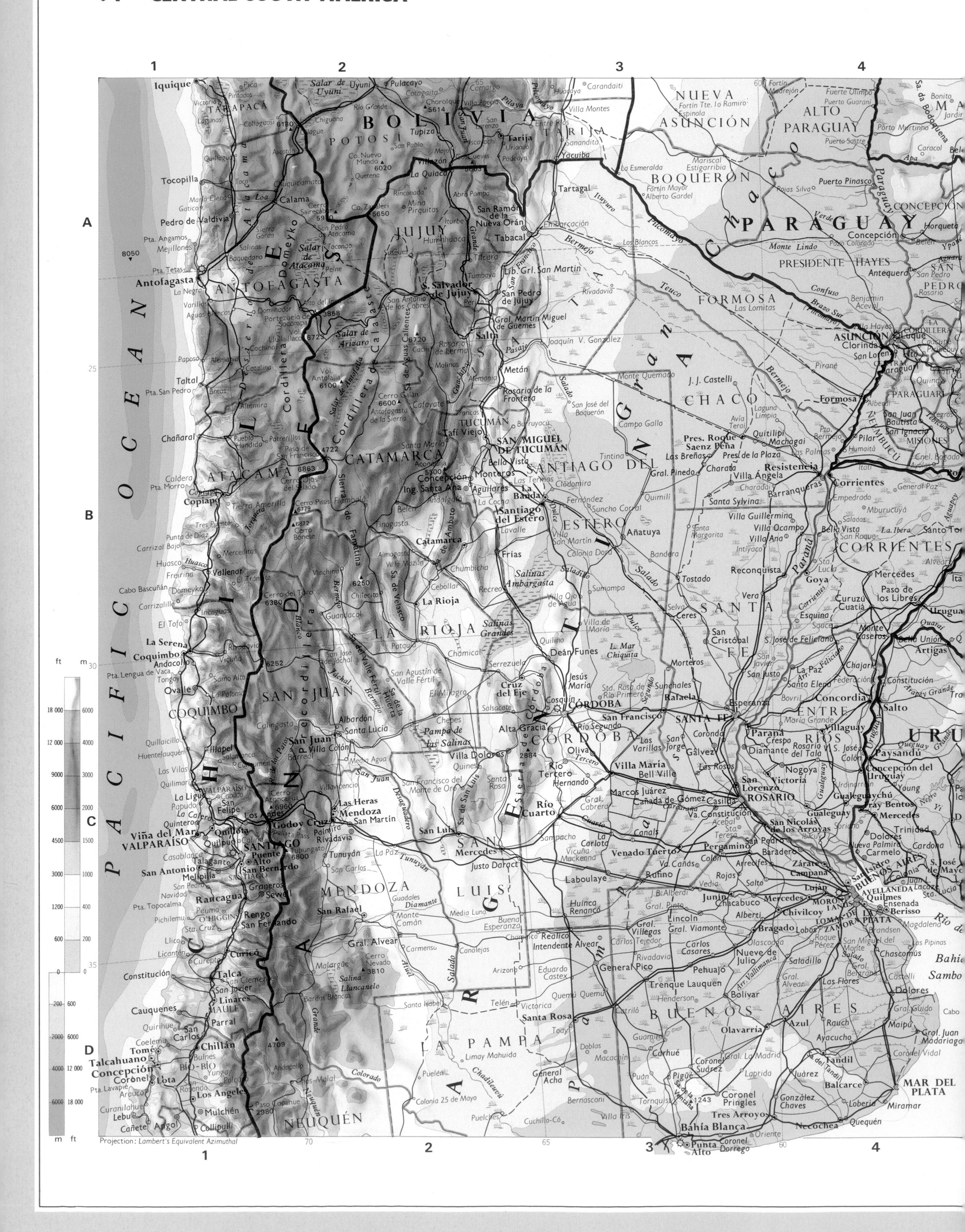
1
2
3
4
A
B
C
D
BOLIVIA
PARAGUAY
PACIFIC OCEAN
CHILE
ANDES
NUEVA ASUNCIÓN
ALTO PARAGUAY
BOQUERÓN
PRESIDENTE HAYES
FORMOSA
CHACO
SANTIAGO DEL ESTERO
SANTA FE
CORRIENTES
ENTRE RÍOS
BUENOS AIRES
LA PAMPA
MENDOZA
SAN LUIS
SAN JUAN
LA RIOJA
CATAMARCA
TUCUMÁN
JUJUY
SALTA
CÓRDOBA
NEUQUÉN
TARAPACÁ
ANTOFAGASTA
ATACAMA
COQUIMBO
POTOSÍ
TARIJA
Iquique
Tocopilla
Calama
Pedro de Valdivia
Antofagasta
Taltal
Chañaral
Copiapó
Vallenar
La Serena
Coquimbo
Ovalle
Viña del Mar
VALPARAÍSO
SANTIAGO
San Antonio
Rancagua
San Fernando
Curicó
Talca
Linares
Cauquenes
Chillán
Tomé
Talcahuano
Concepción
Coronel
Lota
Los Angeles
Mulchén
Lebu
Uyuni
Tupiza
Villazón
Tarija
Yacuiba
La Quiaca
Tartagal
San Ramón de la Nueva Orán
Tabacal
S. Salvador de Jujuy
Salta
Metán
SAN MIGUEL DE TUCUMÁN
Catamarca
La Rioja
Santiago del Estero
La Banda
Añatuya
Deán Funes
CÓRDOBA
Río Cuarto
Villa María
Bell Ville
Rosario
SANTA FE
Paraná
Rafaela
Reconquista
Resistencia
Corrientes
Formosa
ASUNCIÓN
Concepción
Puerto Pinasco
Goya
Mercedes
Concordia
Paysandú
Salto
Gualeguaychú
San Nicolás de los Arroyos
Pergamino
Venado Tuerto
Junín
Chivilcoy
Mercedes
Luján
BUENOS AIRES
LA PLATA
Quilmes
Avellaneda
Bragado
Lincoln
General Pico
Santa Rosa
Trenque Lauquen
Olavarría
Azul
Tandil
Balcarce
MAR DEL PLATA
Necochea
Tres Arroyos
Bahía Blanca
Punta Alto
Coronel Pringles
San Luis
Mercedes
Mendoza
Godoy Cruz
San Martín
San Juan
San Rafael
Gral. Alvear
Cerro Aconcagua 6960
Salar de Uyuni
Salar de Atacama
Salinas Grandes
Pilcomayo
Bermejo
Paraná
Uruguay
Río de la Plata
Colorado
Salado
Dulce
Projection: Lambert's Equivalent Azimuthal
ft
m
18 000
12 000
9000
6000
4500
3000
1200
600
6000
4000
3000
2000
1500
1000
400
200
0
25
30
35
70
65
60

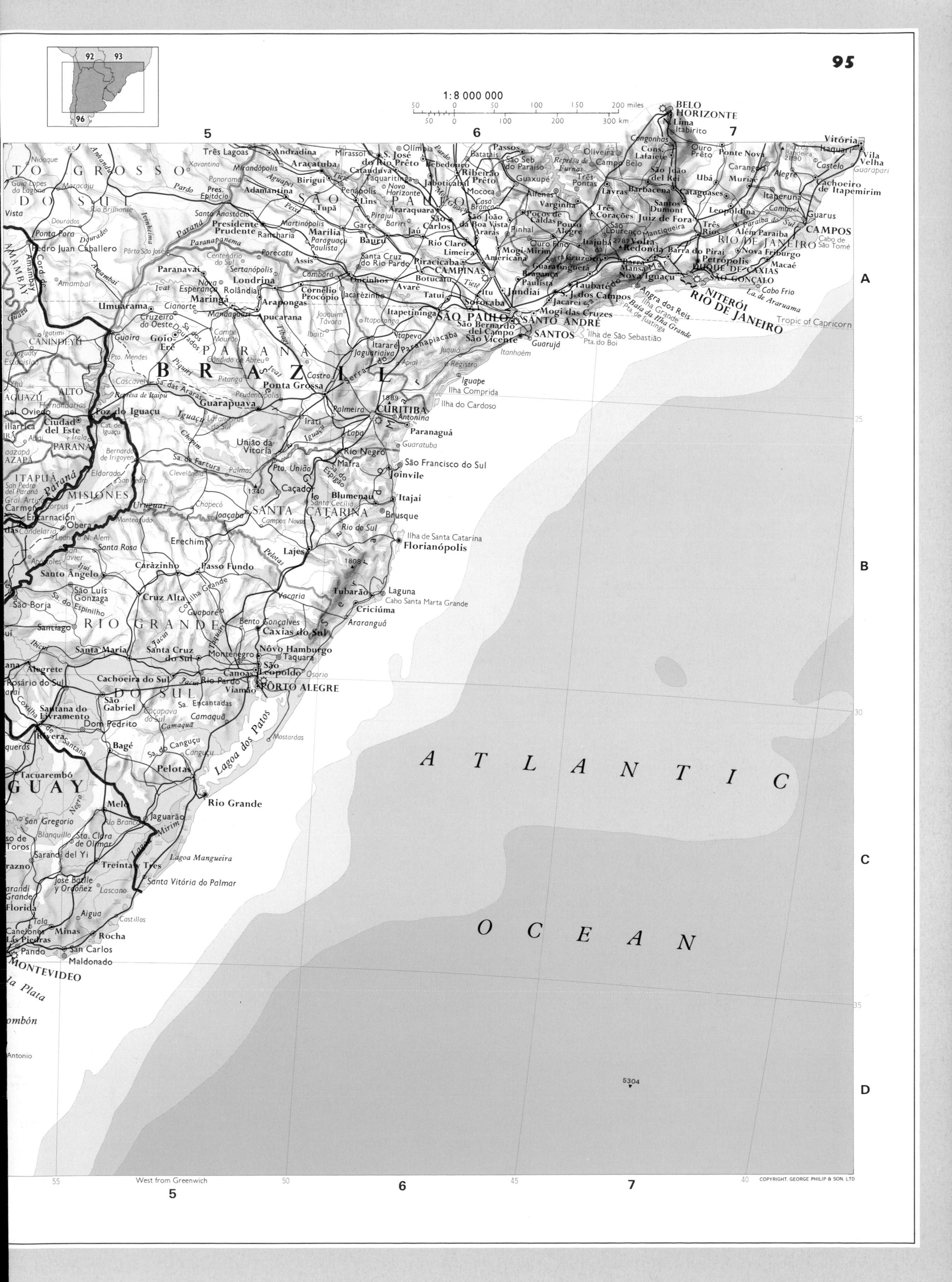
95
92
93
96
1:8 000 000
50 0 50 100 150 200 miles
50 0 100 200 300 km
5
6
7
A
B
C
D
BELO HORIZONTE
Lima
Itabirito
Congonhas
Cons. Lafaiete
Ouro Prêto
Ponte Nova
Vitória
Itaquari
Vila Velha
Guarapari
Castelo
Cachoeiro de Itapemirim
Alegre
Carangola
Muriaé
Ubá
São João del Rei
Barbacena
Cataguases
Santos Dumont
Juiz de Fora
Leopoldina
Itaperuna
Guarus
CAMPOS
Cabo de São Tomé
Três Rios
Além Paraíba
RIO DE JANEIRO
Nova Friburgo
Petrópolis
Macaé
DUQUE DE CAXIAS
SÃO GONÇALO
Cabo Frio
NITERÓI
RIO DE JANEIRO
Tropic of Capricorn
Angra dos Reis
Ilha Grande
Baía da Ilha Grande
Volta Redonda
Barra Mansa
Barra do Piraí
Nova Iguaçu
Lavras
Três Corações
Varginha
Alfenas
Oliveira
Campo Belo
Três Pontas
Poços de Caldas
Pouso Alegre
Itajubá
Guaratinguetá
Cruzeiro
Taubaté
S. J. dos Campos
Jacareí
Mogi das Cruzes
SÃO PAULO
SANTO ANDRÉ
São Bernardo do Campo
São Vicente
SANTOS
Guarujá
Ilha de São Sebastião
Pta. do Boi
Itanhaém
CAMPINAS
Jundiaí
Bragança Paulista
Mogi-Mirim
Americana
Limeira
Rio Claro
Piracicaba
Araras
São Carlos
Araraquara
Ribeirão Prêto
Batatais
Passos
Franca
São João da Boa Vista
Mococa
Guaxupé
Jaboticabal
Bebedouro
Olímpia
S. José do Rio Prêto
Mirassol
Catanduva
Taquaritinga
Novo Horizonte
Lins
Penápolis
Birigui
Araçatuba
Andradina
Três Lagoas
Mirandópolis
Adamantina
Tupã
Marília
Garça
Bauru
Jaú
Botucatu
Tietê
Itu
Sorocaba
Tatuí
Avaré
Itapetininga
Ourinhos
Santa Cruz do Rio Pardo
Assis
Presidente Prudente
Pres. Epitácio
Rancharia
Martinópolis
Paraguaçu Paulista
Cornélio Procópio
Jacarèzinho
Londrina
Arapongas
Apucarana
Maringá
Rolândia
Paranavaí
Umuarama
Cianorte
Cruzeiro do Oeste
Goio-Erê
Guaíra
Itapeva
Itararé
Jaguariaíva
Castro
Ponta Grossa
Guarapuava
Foz do Iguaçu
Ciudad del Este
CURITIBA
Antonina
Paranaguá
Guaratuba
Iguape
Ilha Comprida
Ilha do Cardoso
Rio Negro
Mafra
São Francisco do Sul
Joinvile
Itajaí
Blumenau
Brusque
Ilha de Santa Catarina
Florianópolis
Lajes
Tubarão
Laguna
Cabo Santa Marta Grande
Criciúma
Araranguá
Caxias do Sul
Bento Gonçalves
Nôvo Hamburgo
Taquara
São Leopoldo
Canoas
PÔRTO ALEGRE
Viamão
Osório
Lagoa dos Patos
Mostardas
Pelotas
Rio Grande
Jaguarão
Lagoa Mirim
Lagoa Mangueira
Santa Vitória do Palmar
Bagé
Santa Maria
Santa Cruz do Sul
Cachoeira do Sul
São Gabriel
Santana do Livramento
Rivera
Alegrete
Uruguaiana
Santiago
São Borja
São Luís Gonzaga
Santo Ângelo
Cruz Alta
Passo Fundo
Erechim
Carazinho
Vacaria
Joaçaba
Caçador
União da Vitória
Irati
Palmeira
Lapa
Chapecó
Posadas
Encarnación
Oberá
Tacuarembó
Melo
Treinta y Tres
Minas
Rocha
San Carlos
Maldonado
Pando
MONTEVIDEO
Las Piedras
Canelones
Florida
MATO GROSSO DO SUL
SÃO PAULO
PARANÁ
SANTA CATARINA
RIO GRANDE DO SUL
BRAZIL
MISIONES
ITAPUÁ
ALTO PARANÁ
AMAMBAY
CANINDEYÚ
URUGUAY
Pedro Juan Caballero
Ponta Porã
Dourados
Maracaju
ATLANTIC
OCEAN
5304
1808
1889
2787
1340
Rio de la Plata
West from Greenwich
55
50
45
40
25
30
35
COPYRIGHT. GEORGE PHILIP & SON. LTD

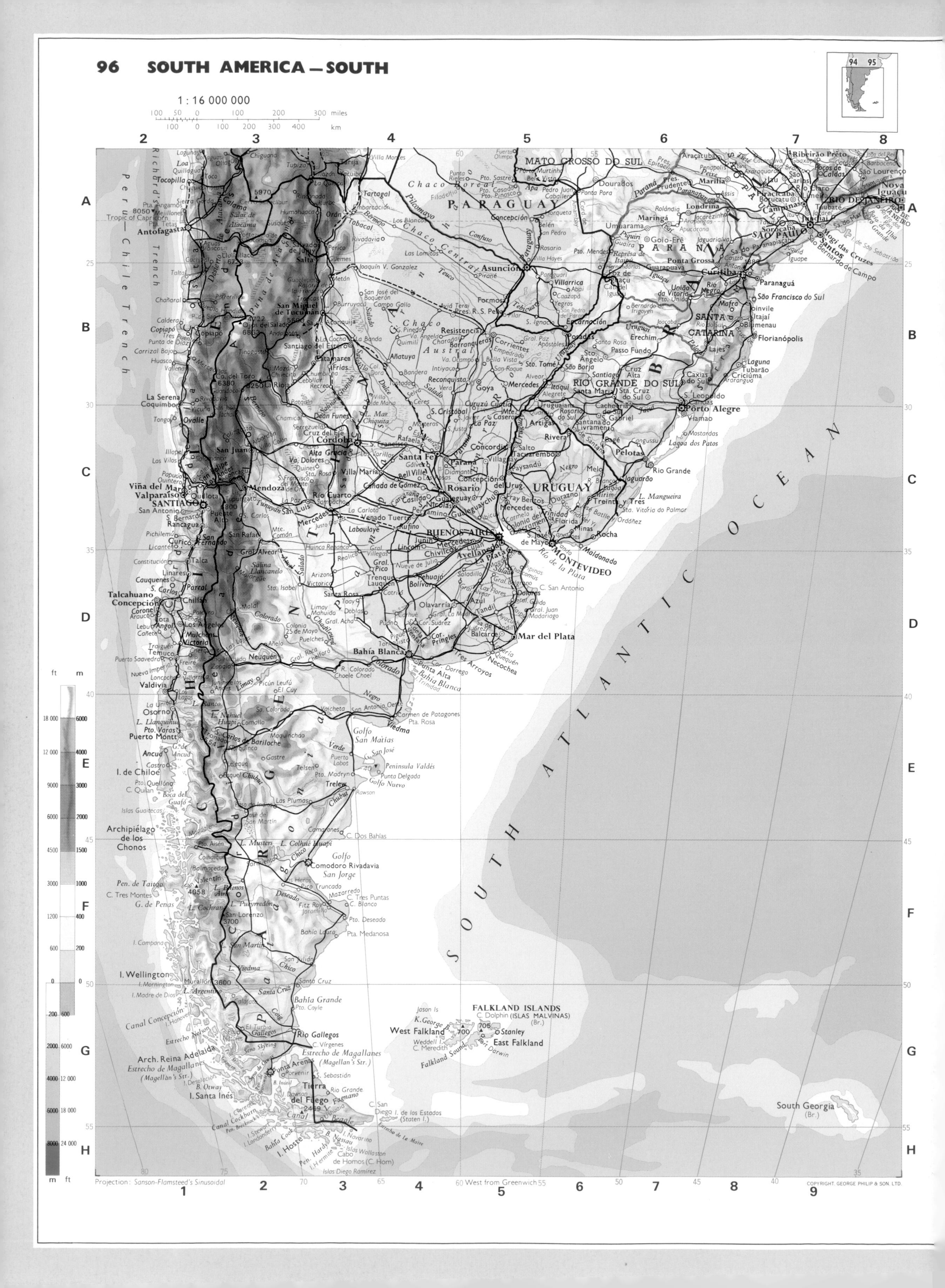
1 : 16 000 000
miles
km
PARAGUAY
URUGUAY
MATO GROSSO DO SUL
PARANÁ
SANTA CATARINA
RIO GRANDE DO SUL
SÃO PAULO
RIO DE JANEIRO
Asunción
BUENOS AIRES
MONTEVIDEO
SANTIAGO
Valparaíso
Viña del Mar
Mendoza
Córdoba
Rosario
Santa Fe
Paraná
Mar del Plata
Bahía Blanca
Neuquén
Comodoro Rivadavia
Rio Gallegos
Punta Arenas
Tierra del Fuego
Antofagasta
Tocopilla
La Serena
Coquimbo
Talcahuano
Concepción
Temuco
Valdivia
Osorno
Puerto Montt
I. de Chiloé
Archipiélago de los Chonos
I. Wellington
Arch. Reina Adelaida
I. Santa Inés
Estrecho de Magallanes
(Magellan's Str.)
Curitiba
Porto Alegre
Florianópolis
Paranaguá
Santos
Campinas
Pelotas
Rio Grande
Golfo San Matías
Península Valdés
Golfo San Jorge
Golfo Nuevo
FALKLAND ISLANDS
(ISLAS MALVINAS)
(Br.)
West Falkland
East Falkland
Stanley
South Georgia
(Br.)
SOUTH ATLANTIC OCEAN
Peru—Chile Trench
Richards Trench
Tropic of Capricorn
Río de la Plata
Cabo de Hornos (C. Horn)
Islas Diego Ramírez
I. de los Estados
(Staten I.)
Chaco Boreal
Chaco Central
Chaco Austral
ft
m
Projection: Sanson-Flamsteed's Sinusoidal
West from Greenwich
COPYRIGHT. GEORGE PHILIP & SON. LTD.

INDEX

The index contains the names of all the principal places and features shown on the World Maps. Each name is followed by an additional entry in italics giving the country or region within which it is located. The alphabetical order of names composed of two or more words is governed primarily by the first word and then by the second. This is an example of the rule:

Mīr Kūh, *Iran*	**45**	**E8**	26 22 N	58 55 E
Mīr Shahdād, *Iran*	**45**	**E8**	26 15 N	58 29 E
Miraj, *India*	**40**	**L9**	16 50 N	74 45 E
Miram Shah, *Pakistan*	**42**	**C4**	33 0 N	70 2 E
Miramar, *Mozam.*	**57**	**C6**	23 50 S	35 35 E

Physical features composed of a proper name (Erie) and a description (Lake) are positioned alphabetically by the proper name. The description is positioned after the proper name and is usually abbreviated:

Erie, L., *N. Amer.*	**78**	**D3**	42 15 N	81 0 W

Where a description forms part of a settlement or administrative name, however, it is always written in full and put in its true alphabetic position:

Mount Morris, *U.S.A.*	**78**	**D7**	42 44 N	77 52 W

Names beginning with M' and Mc are indexed as if they were spelled Mac. Names beginning St. are alphabetized under Saint, but Sankt, Sint, Sant', Santa and San are all spelled in full and are alphabetized accordingly. If the same place name occurs two or more times in the index and all are in the same country, each is followed by the name of the administrative subdivision in which it is located. The names are placed in the alphabetical order of the subdivisions. For example:

Jackson, *Ky., U.S.A.*	**76**	**G4**	37 33 N	83 23 W
Jackson, *Mich., U.S.A.*	**76**	**D3**	42 15 N	84 24 W
Jackson, *Minn., U.S.A.*	**80**	**D7**	43 37 N	95 1 W

The number in bold type which follows each name in the index refers to the number of the map page where that feature or place will be found. This is usually the largest scale at which the place or feature appears.

The letter and figure which are in bold type immediately after the page number give the grid square on the map page, within which the feature is situated. The letter represents the latitude and the figure the longitude.

In some cases the feature itself may fall within the specified square, while the name is outside. This is usually the case only with features which are larger than a grid square.

For a more precise location the geographical coordinates which follow the letter/figure references give the latitude and the longitude of each place. The first set of figures represent the latitude which is the distance north or south of the Equator measured as an angle at the centre of the Earth. The Equator is latitude 0°, the North Pole is 90°N, and the South Pole 90°S.

The second set of figures represent the longitude, which is the distance East or West of the prime meridian, which runs through Greenwich, England. Longitude is also measured as an angle at the centre of the earth and is given East or West of the prime meridian, from 0° to 180° in either direction.

The unit of measurement for latitude and longitude is the degree, which is subdivided into 60 minutes. Each index entry states the position of a place in degrees and minutes,
a space being left between the degrees and the minutes.

The latitude is followed by N(orth) or S(outh) and the longitude by E(ast) or W(est).

Rivers are indexed to their mouths or confluences, and carry the symbol → after their names. A solid square ■ follows the name of a country while, an open square □ refers to a first order administrative area.

Abbreviations used in the index

A.C.T. — Australian Capital Territory
Afghan. — Afghanistan
Ala. — Alabama
Alta. — Alberta
Amer. — America(n)
Arch. — Archipelago
Ariz. — Arizona
Ark. — Arkansas
Atl. Oc. — Atlantic Ocean
B. — Baie, Bahía, Bay, Bucht, Bugt
B.C. — British Columbia
Bangla. — Bangladesh
Barr. — Barrage
Bos. & H. — Bosnia and Herzegovina
C. — Cabo, Cap, Cape, Coast
C.A.R. — Central African Republic
C. Prov. — Cape Province
Calif. — California
Cent. — Central
Chan. — Channel
Colo. — Colorado
Conn. — Connecticut
Cord. — Cordillera
Cr. — Creek
Czech. — Czech Republic
D.C. — District of Columbia
Del. — Delaware
Dep. — Dependency
Des. — Desert
Dist. — District
Dj. — Djebel
Domin. — Dominica
Dom. Rep. — Dominican Republic
E. — East
El Salv. — El Salvador
Eq. Guin. — Equatorial Guinea
Fla. — Florida
Falk. Is. — Falkland Is.
G. — Golfe, Golfo, Gulf, Guba, Gebel
Ga. — Georgia
Gt. — Great, Greater
Guinea-Biss. — Guinea-Bissau
H.K. — Hong Kong
H.P. — Himachal Pradesh
Hants. — Hampshire
Harb. — Harbor, Harbour
Hd. — Head
Hts. — Heights
I.(s). — Île, Ilha, Insel, Isla, Island, Isle
Ill. — Illinois
Ind. — Indiana
Ind. Oc. — Indian Ocean
Ivory C. — Ivory Coast
J. — Jabal, Jebel, Jazira
Junc. — Junction
K. — Kap, Kapp
Kans. — Kansas
Kep. — Kepulauan
Ky. — Kentucky
L. — Lac, Lacul, Lago, Lagoa, Lake, Limni, Loch, Lough
La. — Louisiana
Liech. — Liechtenstein
Lux. — Luxembourg
Mad. P. — Madhya Pradesh
Madag. — Madagascar
Man. — Manitoba
Mass. — Massachusetts
Md. — Maryland
Me. — Maine
Medit. S. — Mediterranean Sea
Mich. — Michigan
Minn. — Minnesota
Miss. — Mississippi
Mo. — Missouri
Mont. — Montana
Mozam. — Mozambique
Mt.(e). — Mont, Monte, Monti, Montaña, Mountain
N. — Nord, Norte, North, Northern, Nouveau
N.B. — New Brunswick
N.C. — North Carolina
N. Cal. — New Caledonia
N. Dak. — North Dakota
N.H. — New Hampshire
N.I. — North Island
N.J. — New Jersey
N. Mex. — New Mexico
N.S. — Nova Scotia
N.S.W. — New South Wales
N.W.T. — North West Territory
N.Y. — New York
N.Z. — New Zealand
Nebr. — Nebraska
Neths. — Netherlands
Nev. — Nevada
Nfld. — Newfoundland
Nic. — Nicaragua
O. — Oued, Ouadi
Occ. — Occidentale
Okla. — Oklahoma
Ont. — Ontario
Or. — Orientale
Oreg. — Oregon
Os. — Ostrov
Oz. — Ozero
P. — Pass, Passo, Pasul, Pulau
P.E.I. — Prince Edward Island
Pa. — Pennsylvania
Pac. Oc. — Pacific Ocean
Papua N.G. — Papua New Guinea
Pass. — Passage
Pen. — Peninsula, Péninsule
Phil. — Philippines
Pk. — Park, Peak
Plat. — Plateau
P-ov. — Poluostrov
Prov. — Province, Provincial
Pt. — Point
Pta. — Ponta, Punta
Pte. — Pointe
Qué. — Québec
Queens. — Queensland
R. — Rio, River
R.I. — Rhode Island
Ra.(s). — Range(s)
Raj. — Rajasthan
Reg. — Region
Rep. — Republic
Res. — Reserve, Reservoir
S. — San, South, Sea
Si. Arabia — Saudi Arabia
S.C. — South Carolina
S. Dak. — South Dakota
S.I. — South Island
S. Leone — Sierra Leone
Sa. — Serra, Sierra
Sask. — Saskatchewan
Scot. — Scotland
Sd. — Sound
Sev. — Severnaya
Sib. — Siberia
Sprs. — Springs
St. — Saint, Sankt, Sint
Sta. — Santa, Station
Ste. — Sainte
Sto. — Santo
Str. — Strait, Stretto
Switz. — Switzerland
Tas. — Tasmania
Tenn. — Tennessee
Tex. — Texas
Tg. — Tanjung
Trin. & Tob. — Trinidad & Tobago
U.A.E. — United Arab Emirates
U.K. — United Kingdom
U.S.A. — United States of America
Ut. P. — Uttar Pradesh
Va. — Virginia
Vdkhr. — Vodokhranilishche
Vf. — Vîrful
Vic. — Victoria
Vol. — Volcano
Vt. — Vermont
W. — Wadi, West
W. Va. — West Virginia
Wash. — Washington
Wis. — Wisconsin
Wlkp. — Wielkopolski
Wyo. — Wyoming
Yorks. — Yorkshire

A

A Coruña = La Coruña, *Spain* **19 A1** 43 20N 8 25W
Aachen, *Germany* **16 C4** 50 45N 6 6 E
Aalborg = Ålborg, *Denmark* **9 H13** 57 2N 9 54 E
Aalen, *Germany* **16 D6** 48 51N 10 6 E
Aalsmeer, *Neths.* **15 B4** 52 17N 4 43 E
Aalst, *Belgium* **15 D4** 50 56N 4 2 E
Aalten, *Neths.* **15 C6** 51 56N 6 35 E
Äänekoski, *Finland* **9 E21** 62 36N 25 44 E
Aarau, *Switz.* **16 E5** 47 23N 8 4 E
Aare →, *Switz.* **16 E5** 47 33N 8 14 E
Aarhus = Århus, *Denmark* **9 H14** 56 8N 10 11 E
Aarschot, *Belgium* **15 D4** 50 59N 4 49 E
Aba, *Nigeria* **50 G6** 5 10N 7 19 E
Aba, *Zaïre* **54 B3** 3 58N 30 17 E
Ābādān, *Iran* **45 D6** 30 22N 48 20 E
Ābādeh, *Iran* **45 D7** 31 8N 52 40 E
Abadla, *Algeria* **50 B4** 31 2N 2 45W
Abaetetuba, *Brazil* **93 D9** 1 40S 48 50W
Abagnar Qi, *China* **34 C9** 43 52N 116 2 E
Abai, *Paraguay* **95 B4** 25 58S 55 54W
Abakan, *Russia* **27 D10** 53 40N 91 10 E
Abancay, *Peru* **92 F4** 13 35S 72 55W
Abariringa, *Kiribati* **64 H10** 2 50S 171 40W
Abarqū, *Iran* **45 D7** 31 10N 53 20 E
Abashiri, *Japan* **30 B12** 44 0N 144 15 E
Abashiri-Wan, *Japan* **30 B12** 44 0N 144 30 E
Abay, *Kazakstan* **26 E8** 49 38N 72 53 E
Abaya, L., *Ethiopia* **51 G12** 6 30N 37 50 E
Abaza, *Russia* **26 D10** 52 39N 90 6 E
'Abbāsābād, *Iran* **45 C8** 33 34N 58 23 E
Abbay = Nîl el Azraq →, *Sudan* **51 E11** 15 38N 32 31 E
Abbaye, Pt., *U.S.A.* **76 B1** 46 58N 88 8W
Abbeville, *France* **18 A4** 50 6N 1 49 E
Abbeville, *La., U.S.A.* **81 K8** 29 58N 92 8W
Abbeville, *S.C., U.S.A.* **77 H4** 34 11N 82 23W
Abbieglassie, *Australia* **63 D4** 27 15S 147 28 E
Abbot Ice Shelf, *Antarctica* **5 D16** 73 0S 92 0W
Abbotsford, *Canada* **72 D4** 49 5N 122 20W
Abbotsford, *U.S.A.* **80 C9** 44 57N 90 19W
Abbottabad, *Pakistan* **42 B5** 34 10N 73 15 E
Abd al Kūrī, *Ind. Oc.* **46 E5** 12 5N 52 20 E
Ābdar, *Iran* **45 D7** 30 16N 55 19 E
'Abdolābād, *Iran* **45 C8** 34 12N 56 30 E
Abéché, *Chad* **51 F9** 13 50N 20 35 E
Åbenrå, *Denmark* **9 J13** 55 3N 9 25 E
Abeokuta, *Nigeria* **50 G5** 7 3N 3 19 E
Aber, *Uganda* **54 B3** 2 12N 32 25 E
Aberaeron, *U.K.* **11 E3** 52 15N 4 15W
Aberayron = Aberaeron, *U.K.* **11 E3** 52 15N 4 15W
Aberconwy & Colwyn □, *U.K.* **10 D4** 53 10N 3 44W
Abercorn = Mbala, *Zambia* **55 D3** 8 46S 31 24 E
Abercorn, *Australia* **63 D5** 25 12S 151 5 E
Aberdare, *U.K.* **11 F4** 51 43N 3 27W
Aberdare Ra., *Kenya* **54 C4** 0 15S 36 50 E
Aberdeen, *Australia* **63 E5** 32 9S 150 56 E
Aberdeen, *Canada* **73 C7** 52 20N 106 8W
Aberdeen, *S. Africa* **56 E3** 32 28S 24 2 E
Aberdeen, *U.K.* **12 D6** 57 9N 2 5W
Aberdeen, *Ala., U.S.A.* **77 J1** 33 49N 88 33W
Aberdeen, *Idaho, U.S.A.* **82 E7** 42 57N 112 50W
Aberdeen, *S. Dak., U.S.A.* **80 C5** 45 28N 98 29W
Aberdeen, *Wash., U.S.A.* **84 D3** 46 59N 123 50W
Aberdeenshire □, *U.K.* **12 D6** 57 17N 2 36W
Aberdovey = Aberdyfi, *U.K.* **11 E3** 52 33N 4 3W
Aberdyfi, *U.K.* **11 E3** 52 33N 4 3W
Aberfeldy, *U.K.* **12 E5** 56 37N 3 51W
Abergavenny, *U.K.* **11 F4** 51 49N 3 1W
Abernathy, *U.S.A.* **81 J4** 33 50N 101 51W
Abert, L., *U.S.A.* **82 E3** 42 38N 120 14W
Aberystwyth, *U.K.* **11 E3** 52 25N 4 5W
Abhar, *Iran* **45 B6** 36 9N 49 13 E
Abhayapuri, *India* **43 F14** 26 24N 90 38 E
Abidjan, *Ivory C.* **50 G4** 5 26N 3 58W
Abilene, *Kans., U.S.A.* **80 F6** 38 55N 97 13W
Abilene, *Tex., U.S.A.* **81 J5** 32 28N 99 43W
Abingdon, *U.K.* **11 F6** 51 40N 1 17W
Abingdon, *Ill., U.S.A.* **80 E9** 40 48N 90 24W
Abingdon, *Va., U.S.A.* **77 G5** 36 43N 81 59W
Abington Reef, *Australia* **62 B4** 18 0S 149 35 E
Abitau →, *Canada* **73 B7** 59 53N 109 3W
Abitau L., *Canada* **73 A7** 60 27N 107 15W
Abitibi L., *Canada* **70 C4** 48 40N 79 40W
Abkhaz Republic □ = Abkhazia □, *Georgia* **25 F7** 43 12N 41 5 E
Abkhazia □, *Georgia* **25 F7** 43 12N 41 5 E
Abkit, *Russia* **27 C16** 64 10N 157 10 E
Abminga, *Australia* **63 D1** 26 8S 134 51 E
Åbo = Turku, *Finland* **9 F20** 60 30N 22 19 E
Abohar, *India* **42 D6** 30 10N 74 10 E
Abomey, *Benin* **50 G5** 7 10N 2 5 E
Abong-Mbang, *Cameroon* **52 D2** 4 0N 13 8 E
Abou-Deïa, *Chad* **51 F8** 11 20N 19 20 E
Aboyne, *U.K.* **12 D6** 57 4N 2 47W
Abra Pampa, *Argentina* **94 A2** 22 43S 65 42W
Abreojos, Pta., *Mexico* **86 B2** 26 50N 113 40W
Abri, *Sudan* **51 D11** 20 50N 30 27 E
Abrolhos, Banka, *Brazil* **93 G11** 18 0S 38 0W
Abrud, *Romania* **17 E12** 46 19N 23 5 E
Absaroka Range, *U.S.A.* **82 D9** 44 45N 109 50W
Abū al Khaṣīb, *Iraq* **45 D6** 30 25N 48 0 E
Abū 'Alī, *Si. Arabia* **45 E6** 27 20N 49 27 E
Abū 'Alī →, *Lebanon* **47 A4** 34 25N 35 50 E
Abū 'Arīsh, *Si. Arabia* **46 D3** 16 53N 42 48 E
Abu Dhabi = Abū Ẓāby, *U.A.E.* **45 E7** 24 28N 54 22 E
Abû Dis, *Sudan* **51 E11** 19 12N 33 38 E
Abū Du'ān, *Syria* **44 B3** 36 25N 38 15 E
Abu el Gairi, W. →, *Egypt* **47 F2** 29 35N 33 30 E
Abu Ga'da, W. →, *Egypt* **47 F1** 29 15N 32 53 E
Abū Ḥadrīyah, *Si. Arabia* **45 E6** 27 20N 48 58 E
Abu Hamed, *Sudan* **51 E11** 19 32N 33 13 E
Abū Kamāl, *Syria* **44 C4** 34 30N 41 0 E
Abū Madd, Ra's, *Si. Arabia* **44 E3** 24 50N 37 7 E
Abu Matariq, *Sudan* **51 F10** 10 59N 26 9 E
Abu Şafāt, W. →, *Jordan* **47 E5** 30 24N 36 7 E
Abū Şukhayr, *Iraq* **44 D5** 31 54N 44 30 E
Abu Tig, *Egypt* **51 C11** 27 4N 31 15 E
Abū Zabad, *Sudan* **51 F10** 12 25N 29 10 E
Abū Ẓāby, *U.A.E.* **45 E7** 24 28N 54 22 E
Abū Zeydābād, *Iran* **45 C6** 33 54N 51 45 E
Abuja, *Nigeria* **50 G6** 9 16N 7 2 E
Abukuma-Gawa →, *Japan* **30 E10** 38 6N 140 52 E
Abukuma-Sammyaku, *Japan* **30 F10** 37 30N 140 45 E
Abunã, *Brazil* **92 E5** 9 40S 65 20W
Abunã →, *Brazil* **92 E5** 9 41S 65 20W
Aburo, *Zaïre* **54 B3** 2 4N 30 53 E
Abut Hd., *N.Z.* **59 K3** 43 7S 170 15 E
Abwong, *Sudan* **51 G11** 9 2N 32 14 E
Acajutla, *El Salv.* **88 D2** 13 36N 89 50W
Acámbaro, *Mexico* **86 C4** 20 0N 100 40W
Acaponeta, *Mexico* **86 C3** 22 30N 105 20W
Acapulco, *Mexico* **87 D5** 16 51N 99 56W
Acarigua, *Venezuela* **92 B5** 9 33N 69 12W
Acatlán, *Mexico* **87 D5** 18 10N 98 3W
Acayucan, *Mexico* **87 D6** 17 59N 94 58W
Accomac, *U.S.A.* **76 G8** 37 43N 75 40W
Accra, *Ghana* **50 G4** 5 35N 0 6W
Accrington, *U.K.* **10 D5** 53 45N 2 22W
Acebal, *Argentina* **94 C3** 33 20S 60 50W
Aceh □, *Indonesia* **36 D1** 4 15N 97 30 E
Achalpur, *India* **40 J10** 21 22N 77 32 E
Acheng, *China* **35 B14** 45 30N 126 58 E
Acher, *India* **42 H5** 23 10N 72 32 E
Achill, *Ireland* **13 C2** 53 56N 9 55W
Achill Hd., *Ireland* **13 C1** 53 58N 10 15W
Achill I., *Ireland* **13 C1** 53 58N 10 1W
Achill Sd., *Ireland* **13 C2** 53 54N 9 56W
Achinsk, *Russia* **27 D10** 56 20N 90 20 E
Acireale, *Italy* **20 F6** 37 37N 15 10 E
Ackerman, *U.S.A.* **81 J10** 33 19N 89 11W
Acklins I., *Bahamas* **89 B5** 22 30N 74 0W
Acme, *Canada* **72 C6** 51 33N 113 30W
Aconcagua, Cerro, *Argentina* **94 C2** 32 39S 70 0W
Aconquija, Mt., *Argentina* **94 B2** 27 0S 66 0W
Açores, Is. dos = Azores, *Atl. Oc.* **48 C1** 38 44N 29 0W
Acraman, L., *Australia* **63 E2** 32 2S 135 23 E
Acre = 'Akko, *Israel* **47 C4** 32 55N 35 4 E
Acre □, *Brazil* **92 E4** 9 1S 71 0W
Acre →, *Brazil* **92 E5** 8 45S 67 22W
Acton, *Canada* **78 C4** 43 38N 80 3W
Ad Dammām, *Si. Arabia* **45 E6** 26 20N 50 5 E
Ad Dawhah, *Qatar* **45 E6** 25 15N 51 35 E
Ad Dawr, *Iraq* **44 C4** 34 27N 43 47 E
Ad Dir'iyah, *Si. Arabia* **44 E5** 24 44N 46 35 E
Ad Dīwānīyah, *Iraq* **44 D5** 32 0N 45 0 E
Ad Dujayl, *Iraq* **44 C5** 33 51N 44 14 E
Ad Durūz, J., *Jordan* **47 C5** 32 35N 36 40 E
Ada, *Minn., U.S.A.* **80 B6** 47 18N 96 31W
Ada, *Okla., U.S.A.* **81 H6** 34 46N 96 41W
Adaja →, *Spain* **19 B3** 41 32N 4 52W
Adamaoua, Massif de l', *Cameroon* **51 G7** 7 20N 12 20 E
Adamawa Highlands = Adamaoua, Massif de l', *Cameroon* **51 G7** 7 20N 12 20 E
Adamello, Mte., *Italy* **20 A4** 46 9N 10 30 E
Adaminaby, *Australia* **63 F4** 36 0S 148 45 E
Adams, *Mass., U.S.A.* **79 D11** 42 38N 73 7W
Adams, *N.Y., U.S.A.* **79 C8** 43 49N 76 1W
Adams, *Wis., U.S.A.* **80 D10** 43 57N 89 49W
Adam's Bridge, *Sri Lanka* **40 Q11** 9 15N 79 40 E
Adams L., *Canada* **72 C5** 51 10N 119 40W
Adams Mt., *U.S.A.* **84 D5** 46 12N 121 30W
Adam's Peak, *Sri Lanka* **40 R12** 6 48N 80 30 E
Adana, *Turkey* **25 G6** 37 0N 35 16 E
Adapazarı, *Turkey* **25 F5** 40 48N 30 25 E
Adarama, *Sudan* **51 E11** 17 10N 34 52 E
Adare, C., *Antarctica* **5 D11** 71 0S 171 0 E
Adaut, *Indonesia* **37 F8** 8 8S 131 7 E
Adavale, *Australia* **63 D3** 25 52S 144 32 E
Adda →, *Italy* **20 B3** 45 8N 9 53 E
Addis Ababa = Addis Abeba, *Ethiopia* **51 G12** 9 2N 38 42 E
Addis Abeba, *Ethiopia* **51 G12** 9 2N 38 42 E
Addis Alem, *Ethiopia* **51 G12** 9 0N 38 17 E
Addison, *U.S.A.* **78 D7** 42 1N 77 14W
Addo, *S. Africa* **56 E4** 33 32S 25 45 E
Ādeh, *Iran* **44 B5** 37 42N 45 11 E
Adel, *U.S.A.* **77 K4** 31 8N 83 25W
Adelaide, *Australia* **63 E2** 34 52S 138 30 E
Adelaide, *Bahamas* **88 A4** 25 4N 77 31W
Adelaide, *S. Africa* **56 E4** 32 42S 26 20 E
Adelaide I., *Antarctica* **5 C17** 67 15S 68 30W
Adelaide Pen., *Canada* **68 B10** 68 15N 97 30W
Adelaide River, *Australia* **60 B5** 13 15S 131 7 E
Adelanto, *U.S.A.* **85 L9** 34 35N 117 22W
Adele I., *Australia* **60 C3** 15 32S 123 9 E
Adélie, Terre, *Antarctica* **5 C10** 68 0S 140 0 E
Adélie Land = Adélie, Terre, *Antarctica* **5 C10** 68 0S 140 0 E
Aden = Al 'Adan, *Yemen* **46 E4** 12 45N 45 0 E
Aden, G. of, *Asia* **46 E4** 12 30N 47 30 E
Adendorp, *S. Africa* **56 E3** 32 15S 24 30 E
Adh Dhayd, *U.A.E.* **45 E7** 25 17N 55 53 E
Adhoi, *India* **42 H4** 23 26N 70 32 E
Adi, *Indonesia* **37 E8** 4 15S 133 30 E
Adi Ugri, *Eritrea* **51 F12** 14 58N 38 48 E
Adieu, C., *Australia* **61 F5** 32 0S 132 10 E
Adieu Pt., *Australia* **60 C3** 15 14S 124 35 E
Adige →, *Italy* **20 B5** 45 9N 12 20 E
Adilabad, *India* **40 K11** 19 33N 78 20 E
Adin, *U.S.A.* **82 F3** 41 12N 120 57W
Adin Khel, *Afghan.* **40 C6** 32 45N 68 5 E
Adirondack Mts., *U.S.A.* **79 C10** 44 0N 74 0W
Adjumani, *Uganda* **54 B3** 3 20N 31 50 E
Adlavik Is., *Canada* **71 B8** 55 2N 57 45W
Admer, *Algeria* **50 D6** 20 21N 5 27 E
Admiralty G., *Australia* **60 B4** 14 20S 125 55 E
Admiralty I., *U.S.A.* **68 C6** 57 30N 134 30W
Admiralty Inlet, *U.S.A.* **82 C2** 48 8N 122 58W
Admiralty Is., *Papua N. G.* **64 H6** 2 0S 147 0 E
Ado-Ekiti, *Nigeria* **50 G6** 7 38N 5 12 E
Adonara, *Indonesia* **37 F6** 8 15S 123 5 E
Adoni, *India* **40 M10** 15 33N 77 18 E
Adour →, *France* **18 E3** 43 32N 1 32W
Adra, *India* **43 H12** 23 30N 86 42 E
Adra, *Spain* **19 D4** 36 43N 3 3W
Adrano, *Italy* **20 F6** 37 40N 14 50 E
Adrar, *Algeria* **50 C4** 27 51N 0 11W
Adré, *Chad* **51 F9** 13 40N 22 20 E
Adrī, *Libya* **51 C7** 27 32N 13 2 E
Adrian, *Mich., U.S.A.* **76 E3** 41 54N 84 2W
Adrian, *Tex., U.S.A.* **81 H3** 35 16N 102 40W
Adriatic Sea, *Medit. S.* **20 C6** 43 0N 16 0 E
Adua, *Indonesia* **37 E7** 1 45S 129 50 E
Adwa, *Ethiopia* **51 F12** 14 15N 38 52 E
Adzhar Republic □ = Ajaria □, *Georgia* **25 F7** 41 30N 42 0 E
Ægean Sea, *Medit. S.* **21 E11** 38 30N 25 0 E
Aerhtai Shan, *Mongolia* **32 B4** 46 40N 92 45 E
'Afak, *Iraq* **44 C5** 32 4N 45 15 E
Afándou, *Greece* **23 C10** 36 18N 28 12 E
Afghanistan ■, *Asia* **40 C4** 33 0N 65 0 E
Afgoi, *Somali Rep.* **46 G3** 2 7N 44 59 E
Afognak I., *U.S.A.* **68 C4** 58 15N 152 30W
'Afrīn, *Syria* **44 B3** 36 32N 36 50 E
Afton, *U.S.A.* **79 D9** 42 14N 75 32W
Afuá, *Brazil* **93 D8** 0 15S 50 20W
Afula, *Israel* **47 C4** 32 37N 35 17 E
Afyonkarahisar, *Turkey* **25 G5** 38 45N 30 33 E
Agadès = Agadez, *Niger* **50 E6** 16 58N 7 59 E
Agadez, *Niger* **50 E6** 16 58N 7 59 E
Agadir, *Morocco* **50 B3** 30 28N 9 55W
Agaete, *Canary Is.* **22 F4** 28 6N 15 43W
Agapa, *Russia* **27 B9** 71 27N 89 15 E
Agar, *India* **42 H7** 23 40N 76 2 E
Agartala, *India* **41 H17** 23 50N 91 23 E
Agassiz, *Canada* **72 D4** 49 14N 121 46W
Agats, *Indonesia* **37 F9** 5 33S 138 0 E
Agboville, *Ivory C.* **50 G4** 5 55N 4 15W
Agde, *France* **18 E5** 43 19N 3 28 E
Agen, *France* **18 D4** 44 12N 0 38 E
Āgh Kand, *Iran* **45 B6** 37 15N 48 4 E
Aginskoye, *Russia* **27 D12** 51 6N 114 32 E
Agra, *India* **42 F7** 27 17N 77 58 E
Agri →, *Italy* **20 D7** 40 13N 16 44 E
Ağrı Dağı, *Turkey* **25 G7** 39 50N 44 15 E
Ağrı Karakose, *Turkey* **25 G7** 39 44N 43 3 E
Agrigento, *Italy* **20 F5** 37 19N 13 34 E
Agrinion, *Greece* **21 E9** 38 37N 21 27 E
Agua Caliente, *Baja Calif., Mexico* **85 N10** 32 29N 116 59W
Agua Caliente, *Sinaloa, Mexico* **86 B3** 26 30N 108 20W
Agua Caliente Springs, *U.S.A.* **85 N10** 32 56N 116 19W
Água Clara, *Brazil* **93 H8** 20 25S 52 45W
Agua Hechicero, *Mexico* **85 N10** 32 26N 116 14W
Agua Prieta, *Mexico* **86 A3** 31 20N 109 32W
Aguadas, *Colombia* **92 B3** 5 40N 75 38W
Aguadilla, *Puerto Rico* **89 C6** 18 26N 67 10W
Aguadulce, *Panama* **88 E3** 8 15N 80 32W
Aguanga, *U.S.A.* **85 M10** 33 27N 116 51W
Aguanish, *Canada* **71 B7** 50 14N 62 2W
Aguanus →, *Canada* **71 B7** 50 13N 62 5W
Aguapey →, *Argentina* **94 B4** 29 7S 56 36W
Aguaray Guazú →, *Paraguay* **94 A4** 24 47S 57 19W
Aguarico →, *Ecuador* **92 D3** 0 59S 75 11W
Aguas Blancas, *Chile* **94 A2** 24 15S 69 55W
Aguas Calientes, Sierra de, *Argentina* **94 B2** 25 26S 66 40W
Aguascalientes, *Mexico* **86 C4** 21 53N 102 12W
Aguascalientes □, *Mexico* **86 C4** 22 0N 102 20W
Aguilares, *Argentina* **94 B2** 27 26S 65 35W
Aguilas, *Spain* **19 D5** 37 23N 1 35W
Agüimes, *Canary Is.* **22 G4** 27 58N 15 27W
Aguja, C. de la, *Colombia* **90 B3** 11 18N 74 12W
Agulhas, C., *S. Africa* **56 E3** 34 52S 20 0 E
Agulo, *Canary Is.* **22 F2** 28 11N 17 12W
Agung, *Indonesia* **36 F5** 8 20S 115 28 E
Agur, *Uganda* **54 B3** 2 28N 32 55 E
Agusan →, *Phil.* **37 C7** 9 0N 125 30 E
Aha Mts., *Botswana* **56 B3** 19 45S 21 0 E
Ahaggar, *Algeria* **50 D6** 23 0N 6 30 E
Ahar, *Iran* **44 B5** 38 35N 47 0 E
Ahipara B., *N.Z.* **59 F4** 35 5S 173 5 E
Ahiri, *India* **40 K12** 19 30N 80 0 E
Ahmad Wal, *Pakistan* **42 E1** 29 18N 65 58 E
Ahmadabad, *India* **42 H5** 23 0N 72 40 E
Aḥmadābād, *Khorāsān, Iran* **45 C9** 35 3N 60 50 E
Aḥmadābād, *Khorāsān, Iran* **45 C8** 35 49N 59 42 E
Aḥmadī, *Iran* **45 E8** 27 56N 56 42 E
Ahmadnagar, *India* **40 K9** 19 7N 74 46 E
Ahmadpur, *Pakistan* **42 E4** 29 12N 71 10 E
Ahmedabad = Ahmadabad, *India* **42 H5** 23 0N 72 40 E
Ahmednagar = Ahmadnagar, *India* **40 K9** 19 7N 74 46 E
Ahome, *Mexico* **86 B3** 25 55N 109 11W
Ahram, *Iran* **45 D6** 28 52N 51 16 E
Ahrax Pt., *Malta* **23 D1** 35 59N 14 22 E
Āhū, *Iran* **45 C6** 34 33N 50 2 E
Ahuachapán, *El Salv.* **88 D2** 13 54N 89 52W
Ahvāz, *Iran* **45 D6** 31 20N 48 40 E
Ahvenanmaa = Åland, *Finland* **9 F19** 60 15N 20 0 E
Aḥwar, *Yemen* **46 E4** 13 30N 46 40 E
Aichi □, *Japan* **31 G8** 35 0N 137 15 E
Aigua, *Uruguay* **95 C5** 34 13S 54 46W
Aigues-Mortes, *France* **18 E6** 43 35N 4 12 E
Aihui, *China* **33 A7** 50 10N 127 30 E
Aija, *Peru* **92 E3** 9 50S 77 45W
Aikawa, *Japan* **30 E9** 38 2N 138 15 E
Aiken, *U.S.A.* **77 J5** 33 34N 81 43W
Aillik, *Canada* **71 A8** 55 11N 59 18W
Ailsa Craig, *U.K.* **12 F3** 55 15N 5 6W
'Ailūn, *Jordan* **47 C4** 32 18N 35 47 E
Aim, *Russia* **27 D14** 59 0N 133 55 E
Aimere, *Indonesia* **37 F6** 8 45S 121 3 E
Aimogasta, *Argentina* **94 B2** 28 33S 66 50W
Aimorés, *Brazil* **93 G10** 19 30S 41 4W
Aïn Beïda, *Algeria* **50 A6** 35 50N 7 29 E
Aïn Ben Tili, *Mauritania* **50 C3** 25 59N 9 27W
Aïn-Sefra, *Algeria* **50 B4** 32 47N 0 37W
'Aïn Sudr, *Egypt* **47 F2** 29 50N 33 6 E
Ainabo, *Somali Rep.* **46 F4** 9 0N 46 25 E
Ainaži, *Latvia* **9 H21** 57 50N 24 24 E
Ainsworth, *U.S.A.* **80 D5** 42 33N 99 52W
Aïr, *Niger* **50 E6** 18 30N 8 0 E
Air Hitam, *Malaysia* **39 M4** 1 55N 103 11 E
Airdrie, *U.K.* **12 F5** 55 52N 3 57W
Aire →, *U.K.* **10 D7** 53 43N 0 55W
Aire, I. del, *Spain* **22 B11** 39 48N 4 16 E
Airlie Beach, *Australia* **62 C4** 20 16S 148 43 E
Aisne →, *France* **18 B5** 49 26N 2 50 E
Aitkin, *U.S.A.* **80 B8** 46 32N 93 42W
Aiud, *Romania* **17 E12** 46 19N 23 44 E
Aix-en-Provence, *France* **18 E6** 43 32N 5 27 E
Aix-la-Chapelle = Aachen, *Germany* **16 C4** 50 45N 6 6 E
Aix-les-Bains, *France* **18 D6** 45 41N 5 53 E
Aiyansh, *Canada* **72 B3** 55 17N 129 2W
Aiyion, *Greece* **21 E10** 38 15N 22 5 E
Aizawl, *India* **41 H18** 23 40N 92 44 E
Aizkraukle, *Latvia* **9 H21** 56 36N 25 11 E
Aizpute, *Latvia* **9 H19** 56 43N 21 40 E
Aizuwakamatsu, *Japan* **30 F9** 37 30N 139 56 E
Ajaccio, *France* **18 F8** 41 55N 8 40 E
Ajalpan, *Mexico* **87 D5** 18 22N 97 15W
Ajanta Ra., *India* **40 J9** 20 28N 75 50 E
Ajari Rep. = Ajaria □, *Georgia* **25 F7** 41 30N 42 0 E
Ajaria □, *Georgia* **25 F7** 41 30N 42 0 E
Ajax, *Canada* **78 C5** 43 50N 79 1W
Ajdābiyah, *Libya* **51 B9** 30 54N 20 4 E
Ajka, *Hungary* **17 E9** 47 4N 17 31 E
'Ajmān, *U.A.E.* **45 E7** 25 25N 55 30 E
Ajmer, *India* **42 F6** 26 28N 74 37 E
Ajo, *U.S.A.* **83 K7** 32 22N 112 52W
Ajo, C. de, *Spain* **19 A4** 43 31N 3 35W
Akabira, *Japan* **30 C11** 43 33N 142 5 E
Akamas □, *Cyprus* **23 D11** 35 3N 32 18 E
Akanthou, *Cyprus* **23 D12** 35 22N 33 45 E
Akaroa, *N.Z.* **59 K4** 43 49S 172 59 E
Akashi, *Japan* **31 G7** 34 45N 134 58 E
Akelamo, *Indonesia* **37 D7** 1 35N 129 40 E
Aketi, *Zaïre* **52 D4** 2 38N 23 47 E
Akharnaí, *Greece* **21 E10** 38 5N 23 44 E
Akhelóös →, *Greece* **21 E9** 38 19N 21 7 E
Akhisar, *Turkey* **21 E12** 38 56N 27 48 E
Akhmîm, *Egypt* **51 C11** 26 31N 31 47 E
Akhnur, *India* **43 C6** 32 52N 74 45 E
Aki, *Japan* **31 H6** 33 30N 133 54 E
Akimiski I., *Canada* **70 B3** 52 50N 81 30W
Akita, *Japan* **30 E10** 39 45N 140 7 E
Akita □, *Japan* **30 E10** 39 40N 140 30 E
Akjoujt, *Mauritania* **50 E2** 19 45N 14 15W
Akkeshi, *Japan* **30 C12** 43 2N 144 51 E
'Akko, *Israel* **47 C4** 32 55N 35 4 E
Akkol, *Kazakstan* **26 E8** 45 0N 75 39 E
Aklavik, *Canada* **68 B6** 68 12N 135 0W
Akmolinsk = Aqmola, *Kazakstan* **26 D8** 51 10N 71 30 E
Akō, *Japan* **31 G7** 34 45N 134 24 E
Akobo →, *Ethiopia* **51 G11** 7 48N 33 3 E
Akola, *India* **40 J10** 20 42N 77 2 E
Akordat, *Eritrea* **51 E12** 15 30N 37 40 E
Akpatok I., *Canada* **69 B13** 60 25N 68 8W
Åkrahamn, *Norway* **9 G11** 59 15N 5 10 E
Akranes, *Iceland* **8 D2** 64 19N 22 5W
Akreïjit, *Mauritania* **50 E3** 18 19N 9 11W
Akron, *Colo., U.S.A.* **80 E3** 40 10N 103 13W
Akron, *Ohio, U.S.A.* **78 E3** 41 5N 81 31W
Akrotiri, *Cyprus* **23 E11** 34 36N 32 57 E
Akrotiri Bay, *Cyprus* **23 E12** 34 35N 33 10 E
Aksai Chin, *India* **43 B8** 35 15N 79 55 E
Aksarka, *Russia* **26 C7** 66 31N 67 50 E
Aksay, *Kazakstan* **24 D9** 51 11N 53 0 E
Aksenovo Zilovskoye, *Russia* **27 D12** 53 20N 117 40 E
Aksu, *China* **32 B3** 41 5N 80 10 E
Aksum, *Ethiopia* **51 F12** 14 5N 38 40 E
Aktogay, *Kazakstan* **26 E8** 46 57N 79 40 E
Aktsyabrski, *Belarus* **17 B15** 52 38N 28 53 E
Aktyubinsk = Aqtöbe, *Kazakstan* **25 D10** 50 17N 57 10 E
Aku, *Nigeria* **50 G6** 6 40N 7 18 E
Akure, *Nigeria* **50 G6** 7 15N 5 5 E
Akureyri, *Iceland* **8 D4** 65 40N 18 6W
Akuseki-Shima, *Japan* **31 K4** 29 27N 129 37 E
Akyab = Sittwe, *Burma* **41 J18** 20 18N 92 45 E
Al 'Adan, *Yemen* **46 E4** 12 45N 45 0 E
Al Aḩsā, *Si. Arabia* **45 E6** 25 50N 49 0 E
Al Ajfar, *Si. Arabia* **44 E4** 27 26N 43 0 E
Al Amādīyah, *Iraq* **44 B4** 37 5N 43 30 E
Al Amārah, *Iraq* **44 D5** 31 55N 47 15 E
Al 'Aqabah, *Jordan* **47 F4** 29 31N 35 0 E
Al Arak, *Syria* **44 C3** 34 38N 38 35 E
Al 'Aramah, *Si. Arabia* **44 E5** 25 30N 46 0 E
Al Arṭāwīyah, *Si. Arabia* **44 E5** 26 31N 45 20 E
Al 'Āṣimah □, *Jordan* **47 D5** 31 40N 36 30 E
Al' Assāfīyah, *Si. Arabia* **44 D3** 28 17N 38 59 E
Al 'Ayn, *Oman* **45 E7** 24 15N 55 45 E
Al 'Ayn, *Si. Arabia* **44 E3** 25 4N 38 6 E
Al A'ẓamīyah, *Iraq* **44 C5** 33 22N 44 22 E
Al 'Azīzīyah, *Iraq* **44 C5** 32 54N 45 4 E
Al Bāb, *Syria* **44 B3** 36 23N 37 29 E
Al Bad', *Si. Arabia* **44 D2** 28 28N 35 1 E
Al Bādī, *Iraq* **44 C4** 35 56N 41 32 E
Al Baḩrah, *Kuwait* **44 D5** 29 40N 47 52 E
Al Balqā' □, *Jordan* **47 C4** 32 5N 35 45 E
Al Bārūk, J., *Lebanon* **47 B4** 33 39N 35 40 E
Al Başrah, *Iraq* **44 D5** 30 30N 47 50 E
Al Baṭḥā, *Iraq* **44 D5** 31 6N 45 53 E
Al Batrūn, *Lebanon* **47 A4** 34 15N 35 40 E
Al Bayḍā, *Libya* **51 B9** 32 50N 21 44 E
Al Biqā □, *Lebanon* **47 A5** 34 10N 36 10 E
Al Bi'r, *Si. Arabia* **44 D3** 28 51N 36 16 E
Al Bu'ayrāt al Ḩasūn, *Libya* **51 B8** 31 24N 15 44 E
Al Burayj, *Syria* **47 A5** 34 15N 36 46 E
Al Fallūjah, *Iraq* **44 C4** 33 20N 43 55 E
Al Fāw, *Iraq* **45 D6** 30 0N 48 30 E
Al Fujayrah, *U.A.E.* **45 E8** 25 7N 56 18 E
Al Ghadaf, W. →, *Jordan* **47 D5** 31 26N 36 43 E
Al Ghammās, *Iraq* **44 D5** 31 45N 44 37 E

Al Hābah, Si. Arabia 44 E5 27 10N 47 0 E
Al Ḥadīthah, Iraq 44 C4 34 0N 41 13 E
Al Ḥadīthah, Si. Arabia 44 D3 31 28N 37 8 E
Al Ḥājānah, Syria 47 B5 33 20N 36 33 E
Al Ḥāmad, Si. Arabia 44 D3 31 30N 39 30 E
Al Hamdānīyah, Syria 44 C3 35 25N 36 50 E
Al Ḥamīdīyah, Syria 47 A4 34 42N 35 57 E
Al Ḥammār, Iraq 44 D5 30 57N 46 51 E
Al Ḥarīr, W. →, Syria 47 C4 32 44N 35 59 E
Al Ḥasā, W. →, Jordan 47 D4 31 4N 35 29 E
Al Ḥasakah, Syria 44 B4 36 35N 40 45 E
Al Ḥawrah, Yemen 46 E4 13 50N 47 35 E
Al Ḥaydān, W. →, Jordan 47 D4 31 29N 35 34 E
Al Ḥayy, Iraq 44 C5 32 5N 46 5 E
Al Ḥijāz, Si. Arabia 46 B2 26 0N 37 30 E
Al Ḥillah, Iraq 44 C5 32 30N 44 25 E
Al Ḥillah, Si. Arabia 46 C4 23 35N 46 50 E
Al Hirmil, Lebanon 47 A5 34 26N 36 24 E
Al Hoceïma, Morocco 50 A4 35 8N 3 58W
Al Ḥudaydah, Yemen 46 E3 14 50N 43 0 E
Al Hufūf, Si. Arabia 45 E6 25 25N 49 45 E
Al Ḥumayḍah, Si. Arabia 44 D2 29 14N 34 56 E
Al Ḥunayy, Si. Arabia 45 E6 25 58N 48 45 E
Al Īsāwīyah, Si. Arabia 44 D3 30 43N 37 59 E
Al Ittihad = Madīnat ash Sha'b, Yemen 46 E3 12 50N 45 0 E
Al Jafr, Jordan 47 E5 30 18N 36 14 E
Al Jaghbūb, Libya 51 C9 29 42N 24 38 E
Al Jahrah, Kuwait 44 D5 29 25N 47 40 E
Al Jalāmīd, Si. Arabia 44 D3 31 20N 39 45 E
Al Jamalīyah, Qatar 45 E6 25 37N 51 5 E
Al Janūb □, Lebanon 47 B4 33 20N 35 20 E
Al Jawf, Libya 51 D9 24 10N 23 24 E
Al Jawf, Si. Arabia 44 D3 29 55N 39 40 E
Al Jazirah, Iraq 44 C5 33 30N 44 0 E
Al Jazirah, Libya 51 C9 26 10N 21 20 E
Al Jithāmīyah, Si. Arabia 44 E4 27 41N 41 43 E
Al Jubayl, Si. Arabia 45 E6 27 0N 49 50 E
Al Jubaylah, Si. Arabia 44 E5 24 55N 46 25 E
Al Jubb, Si. Arabia 44 E4 27 11N 42 17 E
Al Junaynah, Sudan 51 F9 13 27N 22 45 E
Al Kabā'ish, Iraq 44 D5 30 58N 47 0 E
Al Karak, Jordan 47 D4 31 11N 35 42 E
Al Karak □, Jordan 47 E5 31 0N 36 0 E
Al Kāẓim Tyah, Iraq 44 C5 33 22N 44 12 E
Al Khalīl, West Bank 47 D4 31 32N 35 6 E
Al Khawr, Qatar 45 E6 25 41N 51 30 E
Al Khiḍr, Iraq 44 D5 31 12N 45 33 E
Al Khiyām, Lebanon 47 B4 33 20N 35 36 E
Al Kiswah, Syria 47 B5 33 23N 36 14 E
Al Kufrah, Libya 51 D9 24 17N 23 15 E
Al Kuhayfīyah, Si. Arabia 44 E4 27 12N 43 3 E
Al Kūt, Iraq 44 C5 32 30N 46 0 E
Al Kuwayt, Kuwait 44 D5 29 30N 48 0 E
Al Labwah, Lebanon 47 A5 34 11N 36 20 E
Al Lādhiqīyah, Syria 44 C2 35 30N 35 45 E
Al Liwā', Oman 45 E8 24 31N 56 36 E
Al Luḥayyah, Yemen 46 D3 15 45N 42 40 E
Al Madīnah, Iraq 44 D5 30 57N 47 16 E
Al Madīnah, Si. Arabia 46 C2 24 35N 39 52 E
Al-Mafraq, Jordan 47 C5 32 17N 36 14 E
Al Maḥmūdīyah, Iraq 44 C5 33 3N 44 21 E
Al Majma'ah, Si. Arabia 44 E5 25 57N 45 22 E
Al Makhruq, W. →, Jordan 47 D6 31 28N 37 0 E
Al Makḥūl, Si. Arabia 44 E4 26 37N 42 39 E
Al Manāmah, Bahrain 45 E6 26 10N 50 30 E
Al Maqwa', Kuwait 44 D5 29 10N 47 59 E
Al Marj, Libya 51 B9 32 25N 20 30 E
Al Maṭlā, Kuwait 44 D5 29 24N 47 40 E
Al Mawjib, W. →, Jordan 47 D4 31 28N 35 36 E
Al Mawṣil, Iraq 44 B4 36 15N 43 5 E
Al Mayādin, Syria 44 C4 35 1N 40 27 E
Al Mazār, Jordan 47 D4 31 4N 35 41 E
Al Midhnab, Si. Arabia 44 E5 25 50N 44 18 E
Al Minā', Lebanon 47 A4 34 24N 35 49 E
Al Miqdādīyah, Iraq 44 C5 34 0N 45 0 E
Al Mubarraz, Si. Arabia 45 E6 25 30N 49 40 E
Al Mughayrā', U.A.E. 45 E7 24 5N 53 32 E
Al Muḥarraq, Bahrain 45 E6 26 15N 50 40 E
Al Mukallā, Yemen 46 E4 14 33N 49 2 E
Al Mukhā, Yemen 46 E3 13 18N 43 15 E
Al Musayjīd, Si. Arabia 44 E3 24 5N 39 5 E
Al Musayyib, Iraq 44 C5 32 49N 44 20 E
Al Muwayliḥ, Si. Arabia 44 E2 27 40N 35 30 E
Al Qā'im, Iraq 44 C4 34 21N 41 7 E
Al Qalībah, Si. Arabia 44 D3 28 24N 37 42 E
Al Qaryatayn, Syria 47 A6 34 12N 37 13 E
Al Qaṣabát, Libya 51 B7 32 39N 14 1 E
Al Qaṭ'ā, Syria 44 C4 34 40N 40 48 E
Al Qaṭīf, Si. Arabia 45 E6 26 35N 50 0 E
Al Qatrānah, Jordan 47 D5 31 12N 36 6 E
Al Qaṭrūn, Libya 51 D8 24 56N 15 3 E
Al Qayṣūmah, Si. Arabia 44 D5 28 20N 46 7 E
Al Quds = Jerusalem, Israel 47 D4 31 47N 35 10 E
Al Qunaytirah, Syria 47 C4 32 55N 35 45 E
Al Qurnah, Iraq 44 D5 31 1N 47 25 E
Al Quṣayr, Iraq 44 D5 30 39N 45 50 E
Al Quṣayr, Syria 47 A5 34 31N 36 34 E
Al Quṭayfah, Syria 47 B5 33 44N 36 36 E
Al' Uḍaylīyah, Si. Arabia 45 E6 25 8N 49 18 E
Al 'Ulā, Si. Arabia 44 E3 26 35N 38 0 E
Al Uqaylah ash Sharqīgah, Libya 51 B8 30 12N 19 10 E
Al Uqayr, Si. Arabia 45 E6 25 40N 50 15 E
Al 'Uwaynid, Si. Arabia 44 E5 24 50N 46 0 E
Al 'Uwayqīlah, Si. Arabia 44 D4 30 30N 42 10 E
Al 'Uyūn, Si. Arabia 44 E4 26 30N 43 50 E
Al 'Uyūn, Si. Arabia 44 E3 24 33N 39 35 E
Al Wajh, Si. Arabia 44 E3 26 10N 36 30 E
Al Wakrah, Qatar 45 E6 25 10N 51 40 E
Al Wannān, Si. Arabia 45 E6 26 55N 48 24 E
Al Waqbah, Si. Arabia 44 D5 28 48N 45 33 E
Al Wari'āh, Si. Arabia 44 E5 27 51N 47 25 E
Al Wusayl, Qatar 45 E6 25 29N 51 29 E
Ala Tau Shankou = Dzhungarskiye Vorota, Kazakstan 32 B3 45 0N 82 0 E
Alabama □, U.S.A. 77 J2 33 0N 87 0W
Alabama →, U.S.A. 77 K2 31 8N 87 57W
Alaçam Dağlan, Turkey 21 E13 39 18N 28 49 E
Alaérma, Greece 23 C9 36 9N 27 57 E
Alagoa Grande, Brazil 93 E11 7 3S 35 35W
Alagoas □, Brazil 93 E11 9 0S 36 0W
Alagoinhas, Brazil 93 F11 12 7S 38 20W
Alajero, Canary Is. 22 F2 28 3N 17 13W
Alajuela, Costa Rica 88 D3 10 2N 84 8W
Alakamisy, Madag. 57 C8 21 19S 47 14 E
Alakurtti, Russia 24 A5 67 0N 30 30 E
Alameda, Calif., U.S.A. 84 H4 37 46N 122 15W
Alameda, N. Mex., U.S.A. 83 J10 35 11N 106 37W
Alamo, U.S.A. 85 J11 36 21N 115 10W
Alamo Crossing, U.S.A. 85 L13 34 16N 113 33W
Alamogordo, U.S.A. 83 K11 32 54N 105 57W
Alamos, Mexico 86 B3 27 0N 109 0W
Alamosa, U.S.A. 83 H11 37 28N 105 52W
Åland, Finland 9 F19 60 15N 20 0 E
Ålands hav, Sweden 9 F18 60 0N 19 30 E
Alandur, India 40 N12 13 0N 80 15 E
Alania = North Ossetia □, Russia 25 F7 43 30N 44 30 E
Alanya, Turkey 25 G5 36 38N 32 0 E
Alaotra, Farihin', Madag. 57 B8 17 30S 48 30 E
Alapayevsk, Russia 26 D7 57 52N 61 42 E
Alaşehir, Turkey 21 E13 38 23N 28 30 E
Alaska □, U.S.A. 68 B5 64 0N 154 0W
Alaska, G. of, Pac. Oc. 68 C5 58 0N 145 0W
Alaska Highway, Canada 72 B3 60 0N 130 0W
Alaska Peninsula, U.S.A. 68 C4 56 0N 159 0W
Alaska Range, U.S.A. 68 B4 62 50N 151 0W
Älät, Azerbaijan 25 G8 39 58N 49 25 E
Alatyr, Russia 24 D8 54 55N 46 35 E
Alausi, Ecuador 92 D3 2 0S 78 50W
Alava, C., U.S.A. 82 B1 48 10N 124 44W
Alavus, Finland 9 E20 62 35N 23 36 E
Alawoona, Australia 63 E3 34 45S 140 30 E
'Alayh, Lebanon 47 B4 33 46N 35 33 E
Alayor, Spain 22 B11 39 57N 4 8 E
Alba, Italy 20 B3 44 42N 8 2 E
Alba-Iulia, Romania 17 E12 46 8N 23 39 E
Albacete, Spain 19 C5 39 0N 1 50W
Albacutya, L., Australia 63 F3 35 45S 141 58 E
Albania ■, Europe 21 D9 41 0N 20 0 E
Albany, Australia 61 G2 35 1S 117 58 E
Albany, Ga., U.S.A. 77 K3 31 35N 84 10W
Albany, Minn., U.S.A. 80 C7 45 38N 94 34W
Albany, N.Y., U.S.A. 79 D11 42 39N 73 45W
Albany, Oreg., U.S.A. 82 D2 44 38N 123 6W
Albany, Tex., U.S.A. 81 J5 32 44N 99 18W
Albany →, Canada 70 B3 52 17N 81 31W
Albardón, Argentina 94 C2 31 20S 68 30W
Albatross B., Australia 62 A3 12 45S 141 30 E
Albemarle, U.S.A. 77 H5 35 21N 80 11W
Albemarle Sd., U.S.A. 77 H7 36 5N 76 0W
Alberche →, Spain 19 C3 39 58N 4 46W
Alberdi, Paraguay 94 B4 26 14S 58 20W
Albert, L., Australia 63 F2 35 30S 139 10 E
Albert Canyon, Canada 72 C5 51 8N 117 41W
Albert Edward Ra., Australia 60 C4 18 17S 127 57 E
Albert L., Africa 54 B3 1 30N 31 0 E
Albert Lea, U.S.A. 80 D8 43 39N 93 22W
Albert Nile →, Uganda 54 B3 3 36N 32 2 E
Albert Town, Bahamas 89 B5 22 37N 74 33W
Alberta □, Canada 72 C6 54 40N 115 0W
Alberti, Argentina 94 D3 35 1S 60 16W
Albertinia, S. Africa 56 E3 34 11S 21 34 E
Alberton, Canada 71 C7 46 50N 64 0W
Albertville = Kalemie, Zaïre 54 D2 5 55S 29 9 E
Albertville, France 18 D7 45 40N 6 22 E
Albi, France 18 E5 43 56N 2 9 E
Albia, U.S.A. 80 E8 41 2N 92 48W
Albina, Surinam 93 B8 5 37N 54 15W
Albina, Ponta, Angola 56 B1 15 52S 11 44 E
Albion, Idaho, U.S.A. 82 E7 42 25N 113 35W
Albion, Mich., U.S.A. 76 D3 42 15N 84 45W
Albion, Nebr., U.S.A. 80 E5 41 42N 98 0W
Albion, Pa., U.S.A. 78 E4 41 53N 80 22W
Alborán, Medit. S. 19 E4 35 57N 3 0W
Ålborg, Denmark 9 H13 57 2N 9 54 E
Alborz, Reshteh-ye Kūhhā-ye, Iran 45 C7 36 0N 52 0 E
Albreda, Canada 72 C5 52 35N 119 10W
Albuquerque, U.S.A. 83 J10 35 5N 106 39W
Albuquerque, Cayos de, Caribbean 88 D3 12 10N 81 50W
Alburg, U.S.A. 79 B11 44 59N 73 18W
Albury, Australia 63 F4 36 3S 146 56 E
Alcalá de Henares, Spain 19 B4 40 28N 3 22W
Alcalá la Real, Spain 19 D4 37 27N 3 57W
Alcamo, Italy 20 F5 37 59N 12 55 E
Alcaníz, Spain 19 B5 41 2N 0 8W
Alcântara, Brazil 93 D10 2 20S 44 30W
Alcántara, Embalse de, Spain 19 C2 39 44N 6 50W
Alcantara L., Canada 73 A7 60 57N 108 9W
Alcantarilla, Spain 19 D5 37 59N 1 12W
Alcaraz, Sierra de, Spain 19 C4 38 40N 2 20W
Alcaudete, Spain 19 D3 37 35N 4 5W
Alcázar de San Juan, Spain 19 C4 39 24N 3 12W
Alchevsk, Ukraine 25 E6 48 30N 38 45 E
Alcira, Spain 19 C5 39 9N 0 30W
Alcoa, U.S.A. 77 H4 35 48N 83 59W
Alcova, U.S.A. 82 E10 42 34N 106 43W
Alcoy, Spain 19 C5 38 43N 0 30W
Alcudia, Spain 22 B10 39 51N 3 7 E
Alcudia, B. de, Spain 22 B10 39 47N 3 15 E
Aldabra Is., Seychelles 49 G8 9 22S 46 28 E
Aldama, Mexico 87 C5 23 0N 98 4W
Aldan, Russia 27 D13 58 40N 125 30 E
Aldan →, Russia 27 C13 63 28N 129 35 E
Aldea, Pta. de la, Canary Is. 22 G4 28 0N 15 50W
Aldeburgh, U.K. 11 E9 52 10N 1 37 E
Alder, U.S.A. 82 D7 45 19N 112 6W
Alder Pk., U.S.A. 84 K5 35 53N 121 22W
Alderney, U.K. 11 H5 49 42N 2 11W
Aldershot, U.K. 11 F7 51 15N 0 44W
Aledo, U.S.A. 80 E9 41 12N 90 45W
Aleg, Mauritania 50 E2 17 3N 13 55W
Alegranza, Canary Is. 22 E6 29 23N 13 32W
Alegranza, I., Canary Is. 22 E6 29 23N 13 32W
Alegre, Brazil 95 A7 20 50S 41 30W
Alegrete, Brazil 95 B4 29 40S 56 0W
Aleisk, Russia 26 D9 52 40N 83 0 E
Aleksandriya = Oleksandriya, Ukraine 17 C14 50 37N 26 19 E
Aleksandrovsk-Sakhalinskiy, Russia 27 D15 50 50N 142 20 E
Aleksandrovskiy Zavod, Russia 27 D12 50 40N 117 50 E
Aleksandrovskoye, Russia 26 C8 60 35N 77 50 E
Além Paraíba, Brazil 95 A7 21 52S 42 41W
Alemania, Argentina 94 B2 25 40S 65 30W
Alemania, Chile 94 B2 25 10S 69 55W
Alençon, France 18 B4 48 27N 0 4 E
Alenuihaha Channel, U.S.A. 74 H17 20 30N 156 0W
Aleppo = Ḥalab, Syria 44 B3 36 10N 37 15 E
Alert Bay, Canada 72 C3 50 30N 126 55W
Alès, France 18 D6 44 9N 4 5 E
Alessándria, Italy 20 B3 44 54N 8 37 E
Ålesund, Norway 9 E12 62 28N 6 12 E
Aleutian Is., Pac. Oc. 68 C2 52 0N 175 0W
Aleutian Trench, Pac. Oc. 64 B10 48 0N 180 0 E
Alexander, U.S.A. 80 B3 47 51N 103 39W
Alexander, Mt., Australia 61 E3 28 58S 120 16 E
Alexander Arch., U.S.A. 72 B2 56 0N 136 0W
Alexander Bay, S. Africa 56 D2 28 40S 16 30 E
Alexander City, U.S.A. 77 J3 32 56N 85 58W
Alexander I., Antarctica 5 C17 69 0S 70 0W
Alexandra, Australia 63 F4 37 8S 145 40 E
Alexandra, N.Z. 59 L2 45 14S 169 25 E
Alexandra Falls, Canada 72 A5 60 29N 116 18W
Alexandria = El Iskandarîya, Egypt 51 B10 31 13N 29 58 E
Alexandria, Australia 62 B2 19 5S 136 40 E
Alexandria, B.C., Canada 72 C4 52 35N 122 27W
Alexandria, Ont., Canada 70 C5 45 19N 74 38W
Alexandria, Romania 17 G13 43 57N 25 24 E
Alexandria, S. Africa 56 E4 33 38S 26 28 E
Alexandria, Ind., U.S.A. 76 E3 40 16N 85 41W
Alexandria, La., U.S.A. 81 K8 31 18N 92 27W
Alexandria, Minn., U.S.A. 80 C7 45 53N 95 22W
Alexandria, S. Dak., U.S.A. 80 D6 43 39N 97 47W
Alexandria, Va., U.S.A. 76 F7 38 48N 77 3W
Alexandria Bay, U.S.A. 79 B9 44 20N 75 55W
Alexandrina, L., Australia 63 F2 35 25S 139 10 E
Alexandroúpolis, Greece 21 D11 40 50N 25 54 E
Alexis →, Canada 71 B8 52 33N 56 8W
Alexis Creek, Canada 72 C4 52 10N 123 20W
Alfabia, Spain 22 B9 39 44N 2 44 E
Alfenas, Brazil 95 A6 21 20S 46 10W
Alford, U.K. 12 D6 57 14N 2 41W
Alfred, Maine, U.S.A. 79 C14 43 29N 70 43W
Alfred, N.Y., U.S.A. 78 D7 42 16N 77 48W
Alfreton, U.K. 10 D6 53 6N 1 24W
Alga, Kazakstan 25 E10 49 53N 57 20 E
Algaida, Spain 22 B9 39 33N 2 53 E
Algård, Norway 9 G11 58 46N 5 53 E
Algarve, Portugal 19 D1 36 58N 8 20W
Algeciras, Spain 19 D3 36 9N 5 28W
Algemesí, Spain 19 C5 39 11N 0 27W
Alger, Algeria 50 A5 36 42N 3 8 E
Algeria ■, Africa 50 C5 28 30N 2 0 E
Alghero, Italy 20 D3 40 33N 8 19 E
Algiers = Alger, Algeria 50 A5 36 42N 3 8 E
Algoa B., S. Africa 56 E4 33 50S 25 45 E
Algoma, U.S.A. 76 C2 44 36N 87 26W
Algona, U.S.A. 80 D7 43 4N 94 14W
Algonac, U.S.A. 78 D2 42 37N 82 32W
Alhambra, U.S.A. 74 D3 34 8N 118 6W
Alhucemas = Al Hoceïma, Morocco 50 A4 35 8N 3 58W
'Alī al Gharbī, Iraq 44 C5 32 30N 46 45 E
Alī ash Sharqī, Iraq 44 C5 32 7N 46 44 E
'Alī Khēl, Afghan. 42 C3 33 57N 69 43 E
Alī Shāh, Iran 44 B5 38 9N 45 50 E
'Alīābād, Khorāsān, Iran 45 C8 32 30N 57 30 E
'Alīābād, Kordestān, Iran 44 C5 35 4N 46 58 E
'Alīābād, Yazd, Iran 45 D7 31 41N 53 49 E
Aliağa, Turkey 21 E12 38 47N 26 59 E
Aliákmon →, Greece 21 D10 40 30N 22 36 E
Alibo, Ethiopia 51 G12 9 52N 37 5 E
Alicante, Spain 19 C5 38 23N 0 30W
Alice, S. Africa 56 E4 32 48S 26 55 E
Alice, U.S.A. 81 M5 27 45N 98 5W
Alice →, Queens., Australia 62 C3 24 2S 144 50 E
Alice →, Queens., Australia 62 B3 15 35S 142 20 E
Alice Arm, Canada 72 B3 55 29N 129 31W
Alice Downs, Australia 60 C4 17 45S 127 56 E
Alice Springs, Australia 62 C1 23 40S 133 50 E
Alicedale, S. Africa 56 E4 33 15S 26 4 E
Aliceville, U.S.A. 77 J1 33 8N 88 9W
Alick Cr. →, Australia 62 C3 20 55S 142 20 E
Alida, Canada 73 D8 49 25N 101 55W
Aligarh, Raj., India 42 G7 25 55N 76 15 E
Aligarh, Ut. P., India 42 F8 27 55N 78 10 E
Alīgūdarz, Iran 45 C6 33 25N 49 45 E
Alimnía, Greece 23 C9 36 16N 27 43 E
Alingsås, Sweden 9 H15 57 56N 12 31 E
Alipur, Pakistan 42 E4 29 25N 70 55 E
Alipur Duar, India 41 F16 26 30N 89 35 E
Aliquippa, U.S.A. 78 F4 40 37N 80 15W
Alitus = Alytus, Lithuania 9 J21 54 24N 24 3 E
Aliwal North, S. Africa 56 E4 30 45S 26 45 E
Alix, Canada 72 C6 52 24N 113 11W
Aljustrel, Portugal 19 D1 37 55N 8 10W
Alkmaar, Neths. 15 B4 52 37N 4 45 E
All American Canal, U.S.A. 83 K6 32 45N 115 15W
Allah Dad, Pakistan 42 G2 25 38N 67 34 E
Allahabad, India 43 G9 25 25N 81 58 E
Allakh-Yun, Russia 27 C14 60 50N 137 5 E
Allan, Canada 73 C7 51 53N 106 4W
Allanmyo, Burma 41 K19 19 30N 95 17 E
Allanridge, S. Africa 56 D4 27 45S 26 40 E
Allanwater, Canada 70 B1 50 14N 90 10W
Allegan, U.S.A. 76 D3 42 32N 85 51W
Allegany, U.S.A. 78 D6 42 6N 78 30W
Allegheny →, U.S.A. 78 F5 40 27N 80 1W
Allegheny Mts., U.S.A. 66 F11 38 15N 80 10W
Allegheny Plateau, U.S.A. 76 G6 38 0N 80 0W
Allegheny Reservoir, U.S.A. 78 E6 41 50N 79 0W
Allen, Bog of, Ireland 13 C4 53 15N 7 0W
Allen, L., Ireland 13 B3 54 8N 8 4W
Allende, Mexico 86 B4 28 20N 100 50W
Allentown, U.S.A. 79 F9 40 37N 75 29W
Alleppey, India 40 Q10 9 30N 76 28 E
Aller →, Germany 16 B5 52 56N 9 12 E
Alliance, Nebr., U.S.A. 80 D3 42 6N 102 52W
Alliance, Ohio, U.S.A. 78 F3 40 55N 81 6W
Allier →, France 18 C5 46 57N 3 4 E
Alliston, Canada 70 D4 44 9N 79 52W
Alloa, U.K. 12 E5 56 7N 3 47W
Allora, Australia 63 D5 28 2S 152 0 E
Alluitsup Paa = Sydprøven, Greenland 4 C5 60 30N 45 35W
Alma, Canada 71 C5 48 35N 71 40W
Alma, Ga., U.S.A. 77 K4 31 33N 82 28W
Alma, Kans., U.S.A. 80 F6 39 1N 96 17W
Alma, Mich., U.S.A. 76 D3 43 23N 84 39W
Alma, Nebr., U.S.A. 80 E5 40 6N 99 22W
Alma, Wis., U.S.A. 80 C9 44 20N 91 55W
Alma Ata = Almaty, Kazakstan 26 E8 43 15N 76 57 E
Almada, Portugal 19 C1 38 40N 9 9W
Almaden, Australia 62 B3 17 22S 144 40 E
Almadén, Spain 19 C3 38 49N 4 52W
Almanor, L., U.S.A. 82 F3 40 14N 121 9W
Almansa, Spain 19 C5 38 51N 1 5W
Almanzor, Pico del Moro, Spain 19 B3 40 15N 5 18W
Almanzora →, Spain 19 D5 37 14N 1 46W
Almaty, Kazakstan 26 E8 43 15N 76 57 E
Almazán, Spain 19 B4 41 30N 2 30W
Almeirim, Brazil 93 D8 1 30S 52 34W
Almelo, Neths. 15 B6 52 22N 6 42 E
Almendralejo, Spain 19 C2 38 41N 6 26W
Almería, Spain 19 D4 36 52N 2 27W
Almirante, Panama 88 E3 9 10N 82 30W
Almiroú, Kólpos, Greece 23 D6 35 23N 24 20 E
Almont, U.S.A. 78 D1 42 55N 83 3W
Almonte, Canada 79 A8 45 14N 76 12W
Almora, India 43 E8 29 38N 79 40 E
Alnwick, U.K. 10 B6 55 24N 1 42W
Aloi, Uganda 54 B3 2 16N 33 10 E
Alon, Burma 41 H19 22 12N 95 5 E
Alor, Indonesia 37 F6 8 15S 124 30 E
Alor Setar, Malaysia 39 J3 6 7N 100 22 E
Aloysius, Mt., Australia 61 E4 26 0S 128 38 E
Alpaugh, U.S.A. 84 K7 35 53N 119 29W
Alpena, U.S.A. 76 C4 45 4N 83 27W
Alpha, Australia 62 C4 23 39S 146 37 E
Alpine, Ariz., U.S.A. 83 K9 33 51N 109 9W
Alpine, Calif., U.S.A. 85 N10 32 50N 116 46W
Alpine, Tex., U.S.A. 81 K3 30 22N 103 40W
Alps, Europe 16 E5 46 30N 9 30 E
Alroy Downs, Australia 62 B2 19 20S 136 5 E
Alsace, France 18 B7 48 15N 7 25 E
Alsask, Canada 73 C7 51 21N 109 59W
Alsásua, Spain 19 A4 42 54N 2 10W
Alsten, Norway 8 D15 65 58N 12 40 E
Alta, Norway 8 B20 69 57N 23 10 E
Alta Gracia, Argentina 94 C3 31 40S 64 30W
Alta Lake, Canada 72 C4 50 10N 123 0W
Alta Sierra, U.S.A. 85 K8 35 42N 118 33W
Altaelva →, Norway 8 B20 69 54N 23 17 E
Altafjorden, Norway 8 A20 70 5N 23 5 E
Altagracia, Venezuela 92 A4 10 45N 71 30W
Altai = Aerhtai Shan, Mongolia 32 B4 46 40N 92 45 E
Altamaha →, U.S.A. 77 K5 31 20N 81 20W
Altamira, Brazil 93 D8 3 12S 52 10W
Altamira, Chile 94 B2 25 47S 69 51W
Altamira, Mexico 87 C5 22 24N 97 55W
Altamont, U.S.A. 79 D10 42 43N 74 3W
Altamura, Italy 20 D7 40 49N 16 33 E
Altanbulag, Mongolia 32 A5 50 16N 106 30 E
Altar, Mexico 86 A2 30 40N 111 50W
Altata, Mexico 86 C3 24 30N 108 0W
Altavista, U.S.A. 76 G6 37 6N 79 17W
Altay, China 32 B3 47 48N 88 10 E
Altea, Spain 19 C5 38 38N 0 2W
Alto Araguaia, Brazil 93 G8 17 15S 53 20W
Alto Cuchumatanes = Cuchumatanes, Sierra de los, Guatemala 88 C1 15 35N 91 25W
Alto del Inca, Chile 94 A2 24 10S 68 10W
Alto Ligonha, Mozam. 55 F4 15 30S 38 11 E
Alto Molocue, Mozam. 55 F4 15 50S 37 35 E
Alto Paraguay □, Paraguay 94 A4 21 0S 58 30W
Alto Paraná □, Paraguay 95 B5 25 30S 54 50W
Alton, Canada 78 C4 43 54N 80 5W
Alton, U.S.A. 80 F9 38 53N 90 11W
Alton Downs, Australia 63 D2 26 7S 138 57 E
Altoona, U.S.A. 78 F6 40 31N 78 24W
Altūn Kūprī, Iraq 44 C5 35 45N 44 9 E
Altun Shan, China 32 C3 38 30N 88 0 E
Alturas, U.S.A. 82 F3 41 29N 120 32W
Altus, U.S.A. 81 H5 34 38N 99 20W
Alūksne, Latvia 9 H22 57 24N 27 3 E
Alùla, Somali Rep. 46 E5 11 50N 50 45 E
Alunite, U.S.A. 85 K12 35 59N 114 55W
Alusi, Indonesia 37 F8 7 35S 131 40 E
Al'Uzayr, Iraq 44 D5 31 19N 47 25 E
Alva, U.S.A. 81 G5 36 48N 98 40W
Alvarado, Mexico 87 D5 18 40N 95 50W
Alvarado, U.S.A. 81 J6 32 24N 97 13W
Alvaro Obregón, Presa, Mexico 86 B3 27 55N 109 52W
Alvear, Argentina 94 B4 29 5S 56 30W
Alvesta, Sweden 9 H16 56 54N 14 35 E
Alvie, Australia 63 F3 38 14S 143 30 E
Alvin, U.S.A. 81 L7 29 26N 95 15W
Alvinston, Canada 78 D3 42 49N 81 52W
Älvkarleby, Sweden 9 F17 60 34N 17 26 E
Älvsbyn, Sweden 8 D19 65 40N 21 0 E
Alwar, India 42 F7 27 38N 76 34 E
Alxa Zuoqi, China 34 E3 38 50N 105 40 E
Alyaskitovyy, Russia 27 C15 64 45N 141 30 E
Alyata = Älät, Azerbaijan 25 G8 39 58N 49 25 E
Alyth, U.K. 12 E5 56 38N 3 13W
Alytus, Lithuania 9 J21 54 24N 24 3 E
Alzada, U.S.A. 80 C2 45 2N 104 25W
Am Dam, Chad 51 F9 12 40N 20 35 E
Am-Timan, Chad 51 F9 11 0N 20 10 E
Amadeus, L., Australia 61 D5 24 54S 131 0 E
Amâdi, Sudan 51 G11 5 29N 30 25 E
Amadi, Zaïre 54 B2 3 40N 26 40 E

Amadjuak, Canada 69 B12 64 0N 72 39W
Amadjuak L., Canada ... 69 B12 65 0N 71 8W
Amagasaki, Japan ...... 31 G7 34 42N 135 20 E
Amakusa-Shotō, Japan .. 31 H5 32 15N 130 10 E
Åmål, Sweden .... 9 G15 59 3N 12 42 E
Amaliás, Greece .... 21 F9 37 47N 21 22 E
Amalner, India .... 40 J9 21 5N 75 5 E
Amambaí, Brazil 95 A4 23 5S 55 13W
Amambaí →, Brazil 95 A5 23 22S 53 56W
Amambay □, Paraguay .. 95 A4 23 0S 56 0W
Amambay, Cordillera de,
S. Amer. 95 A4 23 0S 55 45W
Amami-Guntō, Japan ... 31 L4 27 16N 129 21 E
Amami-Ō-Shima, Japan . 31 L4 28 0N 129 0 E
Amanda Park, U.S.A. 84 C3 47 28N 123 55W
Amangeldy, Kazakstan .. 26 D7 50 10N 65 10 E
Amapá, Brazil 93 C8 2 5N 50 50W
Amapá □, Brazil 93 C8 1 40N 52 0W
Amarante, Brazil 93 E10 6 14S 42 50W
Amaranth, Canada 73 C9 50 36N 98 43W
Amargosa, Brazil 93 F11 13 2S 39 36W
Amargosa →, U.S.A. ... 85 J10 36 14N 116 51W
Amargosa Range, U.S.A. 85 J10 36 20N 116 45W
Amári, Greece .... 23 D6 35 13N 24 40 E
Amarillo, U.S.A. 81 H4 35 13N 101 50W
Amaro, Mte., Italy ...... 20 C6 42 5N 14 5 E
Amarpur, India .... 43 G12 25 5N 87 0 E
Amatikulu, S. Africa .... 57 D5 29 3S 31 33 E
Amatitlán, Guatemala ... 88 D1 14 29N 90 38W
Amazon = Amazonas →,
S. Amer. 93 C9 0 5S 50 0W
Amazonas □, Brazil 92 E6 5 0S 65 0W
Amazonas →, S. Amer. 93 C9 0 5S 50 0W
Ambahakily, Madag. .... 57 C7 21 36S 43 41 E
Ambala, India .... 42 D7 30 23N 76 56 E
Ambalavao, Madag. .... 57 C8 21 50S 46 56 E
Ambalindum, Australia .. 62 C2 23 23S 135 0 E
Ambam, Cameroon ..... 52 D2 2 20N 11 15 E
Ambanja, Madag. ...... 57 A8 13 40S 48 27 E
Ambarchik, Russia ...... 27 C17 69 40N 162 20 E
Ambarijeby, Madag. .... 57 A8 14 56S 47 41 E
Ambaro, Helodranon',
Madag. .... 57 A8 13 23S 48 38 E
Ambartsevo, Russia .... 26 D9 57 30N 83 52 E
Ambato, Ecuador 92 D3 1 5S 78 42W
Ambato, Sierra de,
Argentina 94 B2 28 25S 66 10W
Ambato Boeny, Madag. . 57 B8 16 28S 46 43 E
Ambatofinandrahana,
Madag. .... 57 C8 20 33S 46 48 E
Ambatolampy, Madag. .. 57 B8 19 20S 47 35 E
Ambatondrazaka, Madag. 57 B8 17 55S 48 28 E
Ambatosoratra, Madag. . 57 B8 17 37S 48 31 E
Ambenja, Madag. ...... 57 B8 15 17S 46 58 E
Amberg, Germany ...... 16 D6 49 26N 11 52 E
Ambergris Cay, Belize ... 87 D7 18 0N 88 0W
Amberley, N.Z. .... 59 K4 43 9S 172 44 E
Ambikapur, India .... 43 H10 23 15N 83 15 E
Ambilobé, Madag. ...... 57 A8 13 10S 49 3 E
Ambinanindrano, Madag. 57 C8 20 5S 48 23 E
Ambleside, U.K. 10 C5 54 26N 2 58W
Ambo, Peru 92 F3 10 5S 76 10W
Ambodifototra, Madag. .. 57 B8 16 59S 49 52 E
Ambodilazana, Madag. .. 57 B8 18 6S 49 10 E
Ambohimahasoa, Madag. 57 C8 21 7S 47 13 E
Ambohimanga, Madag. . 57 C8 20 52S 47 36 E
Ambohitra, Madag. ..... 57 A8 12 30S 49 10 E
Amboise, France .... 18 C4 47 24N 1 2 E
Ambon, Indonesia ...... 37 E7 3 35S 128 20 E
Amboseli L., Kenya ..... 54 C4 2 40S 37 10 E
Ambositra, Madag. ..... 57 C8 20 31S 47 25 E
Ambovombé, Madag. .... 57 D8 25 11S 46 5 E
Amboy, U.S.A. 85 L11 34 33N 115 45W
Amboyna Cay,
S. China Sea .... 36 C4 7 50N 112 50 E
Ambridge, U.S.A. 78 F4 40 36N 80 14W
Ambriz, Angola .... 52 F2 7 48S 13 8 E
Amby, Australia .... 63 D4 26 30S 148 11 E
Amchitka I., U.S.A. ..... 68 C1 51 32N 179 0 E
Amderma, Russia ...... 26 C7 69 45N 61 30 E
Ameca, Mexico 86 C4 20 30N 104 0W
Ameca →, Mexico 86 C3 20 40N 105 15W
Amecameca, Mexico 87 D5 19 7N 98 46W
Ameland, Neths. .... 15 A5 53 27N 5 45 E
Amen, Russia .... 27 C18 68 45N 180 0 E
American Falls, U.S.A. .. 82 E7 42 47N 112 51W
American Falls Reservoir,
U.S.A. 82 E7 42 47N 112 52W
American Highland,
Antarctica .... 5 D6 73 0S 75 0 E
American Samoa ■,
Pac. Oc. 59 B13 14 20S 170 40W
Americana, Brazil 95 A6 22 45S 47 20W
Americus, U.S.A. 77 J3 32 4N 84 14W
Amersfoort, Neths. ..... 15 B5 52 9N 5 23 E
Amersfoort, S. Africa ... 57 D4 26 59S 29 53 E
Amery, Australia .... 61 F2 31 9S 117 5 E
Amery, Canada 73 B10 56 34N 94 3W
Amery Ice Shelf,
Antarctica .... 5 C6 69 30S 72 0 E
Ames, U.S.A. 80 E8 42 2N 93 37W
Amesbury, U.S.A. 79 D14 42 51N 70 56W
Amga, Russia .... 27 C14 60 50N 132 0 E
Amga →, Russia ...... 27 C14 62 38N 134 32 E
Amgu, Russia .... 27 E14 45 45N 137 15 E
Amgun →, Russia ..... 27 D14 52 56N 139 38 E
Amherst, Burma .... 41 L20 16 2N 97 20 E
Amherst, Canada 71 C7 45 48N 64 8W
Amherst, Mass., U.S.A. .. 79 D12 42 23N 72 31W
Amherst, N.Y., U.S.A. ... 78 D6 42 59N 78 48W
Amherst, Ohio, U.S.A. ... 78 E2 41 24N 82 14W
Amherst, Tex., U.S.A. ... 81 H3 34 1N 102 25W
Amherst I., Canada 79 B8 44 8N 76 43W
Amherstburg, Canada ... 70 D3 42 6N 83 6W
Amiata, Mte., Italy ..... 20 C4 42 53N 11 37 E
Amiens, France .... 18 B5 49 54N 2 16 E
Amīrābād, Iran .... 44 C5 33 20N 46 16 E
Amirante Is., Seychelles . 28 K9 6 0S 53 0 E
Amisk L., Canada 73 C8 54 35N 102 15W
Amistad, Presa de la,
Mexico 86 B4 29 24N 101 0W
Amite, U.S.A. 81 K9 30 44N 90 30W
Amlwch, U.K. 10 D3 53 24N 4 20W
'Ammān, Jordan .... 47 D4 31 57N 35 52 E
Ammanford, U.K. 11 F3 51 48N 3 59W
Ammassalik =
Angmagssalik,
Greenland 4 C6 65 40N 37 20W
Amnat Charoen, Thailand 38 E5 15 51N 104 38 E
Āmol, Iran .... 45 B7 36 23N 52 20 E
Amorgós, Greece .... 21 F11 36 50N 25 57 E
Amory, U.S.A. 77 J1 33 59N 88 29W
Amos, Canada 70 C4 48 35N 78 5W
Åmot, Norway .... 9 G13 59 57N 9 54 E
Amoy = Xiamen, China . 33 D6 24 25N 118 4 E
Ampang, Malaysia ...... 39 L3 3 8N 101 45 E
Ampanihy, Madag. ..... 57 C7 24 40S 44 45 E
Ampasindava,
Helodranon', Madag. .. 57 A8 13 40S 48 15 E
Ampasindava, Saikanosy,
Madag. .... 57 A8 13 42S 47 55 E
Ampenan, Indonesia .... 36 F5 8 35S 116 13 E
Amper →, Germany ... 16 D6 48 29N 11 35 E
Ampotaka, Madag. ..... 57 D7 25 3S 44 41 E
Ampoza, Madag. .... 57 C7 22 20S 44 44 E
Amqui, Canada 71 C6 48 28N 67 27W
Amravati, India .... 40 J10 20 55N 77 45 E
Amreli, India .... 42 J4 21 35N 71 17 E
Amritsar, India .... 42 D6 31 35N 74 57 E
Amroha, India .... 43 E8 28 53N 78 30 E
Amsterdam, Neths. ..... 15 B4 52 23N 4 54 E
Amsterdam, U.S.A. 79 D10 42 56N 74 11W
Amsterdam, I., Ind. Oc. .. 3 F13 38 30S 77 30 E
Amstetten, Austria ..... 16 D8 48 7N 14 51 E
Amudarya →, Uzbekistan 26 E6 43 58N 59 34 E
Amundsen Gulf, Canada . 68 A7 71 0N 124 0W
Amundsen Sea, Antarctica 5 D15 72 0S 115 0W
Amuntai, Indonesia ..... 36 E5 2 28S 115 25 E
Amur →, Russia ...... 27 D15 52 56N 141 10 E
Amurang, Indonesia .... 37 D6 1 5N 124 40 E
Amuri Pass, N.Z. .... 59 K4 42 31S 172 11 E
Amursk, Russia .... 27 D14 50 14N 136 54 E
Amurzet, Russia .... 27 E14 47 50N 131 5 E
Amyderya =
Amudarya →,
Uzbekistan .... 26 E6 43 58N 59 34 E
An Bien, Vietnam ...... 39 H5 9 45N 105 0 E
An Hoa, Vietnam ...... 38 E7 15 40N 108 5 E
An Khe, Vietnam ...... 38 F7 13 57N 108 39 E
An Nabatīyah at Tahta,
Lebanon .... 47 B4 33 23N 35 27 E
An Nabk, Si. Arabia .... 44 D3 31 20N 37 20 E
An Nabk, Syria .... 47 A5 34 2N 36 44 E
An Nabk Abū Qaşr,
Si. Arabia .... 44 D3 30 21N 38 34 E
An Nafūd, Si. Arabia ... 44 D4 28 15N 41 0 E
An Najaf, Iraq .... 44 C5 32 3N 44 15 E
An Nāşirīyah, Iraq ..... 44 D5 31 0N 46 15 E
An Nhon, Vietnam ..... 38 F7 13 55N 109 7 E
An Nu'ayrīyah, Si. Arabia 45 E6 27 30N 48 30 E
An Nuwayb'ī, W. →,
Si. Arabia .... 47 F3 29 18N 34 57 E
An Thoi, Dao, Vietnam .. 39 H4 9 58N 104 0 E
An Uaimh, Ireland 13 C5 53 39N 6 41W
Anabar →, Russia .... 27 B12 73 8N 113 36 E
'Anabtā, West Bank .... 47 C4 32 19N 35 7 E
Anaconda, U.S.A. 82 C7 46 8N 112 57W
Anacortes, U.S.A. 84 B4 48 30N 122 37W
Anadarko, U.S.A. 81 H5 35 4N 98 15W
Anadolu, Turkey .... 25 G5 39 0N 30 0 E
Anadyr, Russia .... 27 C18 64 35N 177 20 E
Anadyr →, Russia ..... 27 C18 64 55N 176 5 E
Anadyrskiy Zaliv, Russia . 27 C19 64 0N 180 0 E
Anaga, Pta. de, Canary Is. 22 F3 28 34N 16 9W
'Ānah, Iraq .... 44 C4 34 25N 42 0 E
Anaheim, U.S.A. 85 M9 33 50N 117 55W
Anahim Lake, Canada ... 72 C3 52 28N 125 18W
Anáhuac, Mexico 86 B4 27 14N 100 9W
Anakapalle, India ...... 41 L13 17 42N 83 6 E
Anakie, Australia ...... 62 C4 23 32S 147 45 E
Analalava, Madag. ..... 57 A8 14 35S 48 0 E
Análipsis, Greece ...... 23 A3 39 36N 19 55 E
Anambar →, Pakistan .. 42 D3 30 15N 68 50 E
Anambas, Kepulauan,
Indonesia .... 36 D3 3 20N 106 30 E
Anambas Is. = Anambas,
Kepulauan, Indonesia . 36 D3 3 20N 106 30 E
Anamoose, U.S.A. 80 B4 47 53N 100 15W
Anamosa, U.S.A. 80 D9 42 7N 91 17W
Anamur, Turkey .... 25 G5 36 8N 32 58 E
Anan, Japan .... 31 H7 33 54N 134 40 E
Anand, India .... 42 H5 22 32N 72 59 E
Anantnag, India .... 43 C6 33 45N 75 10 E
Ananyiv, Ukraine ...... 17 E15 47 44N 29 58 E
Anapodháris →, Greece 23 E7 34 59N 25 20 E
Anápolis, Brazil 93 G9 16 15S 48 50W
Anār, Iran .... 45 D7 30 55N 55 13 E
Anārak, Iran .... 45 C7 33 25N 53 40 E
Anatolia = Anadolu,
Turkey .... 25 G5 39 0N 30 0 E
Anatone, U.S.A. 82 C5 46 8N 117 8W
Anatsogno, Madag. ..... 57 C7 23 33S 43 46 E
Añatuya, Argentina 94 B3 28 20S 62 50W
Anaunethad L., Canada .. 73 A8 60 55N 104 25W
Anaye, Niger .... 51 E7 19 15N 12 50 E
Anbyŏn, N. Korea ...... 35 E14 39 1N 127 35 E
Anchor Bay, U.S.A. 84 G3 38 48N 123 34W
Anchorage, U.S.A. 68 B5 61 13N 149 54W
Anci, China .... 34 E9 39 20N 116 40 E
Ancohuma, Nevada,
Bolivia 92 G5 16 0S 68 50W
Ancón, Peru 92 F3 11 50S 77 10W
Ancona, Italy .... 20 C5 43 38N 13 30 E
Ancud, Chile 96 E2 42 0S 73 50W
Ancud, G. de, Chile 96 E2 42 0S 73 0W
Anda, China .... 33 B7 46 24N 125 19 E
Andacollo, Argentina ... 94 D1 37 10S 70 42W
Andacollo, Chile 94 C1 30 5S 71 10W
Andado, Australia ..... 62 D2 25 25S 135 15 E
Andalgalá, Argentina ... 94 B2 27 40S 66 30W
Åndalsnes, Norway .... 9 E12 62 35N 7 43 E
Andalucía □, Spain 19 D3 37 35N 5 0W
Andalusia, U.S.A. 77 K2 31 18N 86 29W
Andalusia □ =
Andalucía □, Spain ... 19 D3 37 35N 5 0W
Andaman Is., Ind. Oc. .. 28 H13 12 30N 92 30 E
Andaman Sea, Ind. Oc. .. 36 B1 13 0N 96 0 E
Andara, Namibia ...... 56 B3 18 2S 21 9 E
Andenes, Norway ...... 8 B17 69 19N 16 18 E
Andenne, Belgium ...... 15 D5 50 28N 5 5 E
Anderson, Calif., U.S.A. . 82 F2 40 27N 122 18W
Anderson, Ind., U.S.A. ... 76 E3 40 10N 85 41W
Anderson, Mo., U.S.A. .. 81 G7 36 39N 94 27W
Anderson, S.C., U.S.A. .. 77 H4 34 31N 82 39W
Anderson →, Canada .. 68 B7 69 42N 129 0W
Andes = Andes, Cord. de
los, S. Amer. 92 F4 20 0S 68 0W
Andes, S. Amer. 92 F3 10 0S 75 53W
Andes, Cord. de los,
S. Amer. 92 F4 20 0S 68 0W
Andfjorden, Norway .... 8 B17 69 10N 16 20 E
Andhra Pradesh □, India 40 L11 18 0N 79 0 E
Andijon, Uzbekistan .... 26 E8 41 10N 72 15 E
Andikíthira, Greece .... 21 G10 35 52N 23 15 E
Andīmeshk, Iran .... 45 C6 32 27N 48 21 E
Andizhan = Andijon,
Uzbekistan .... 26 E8 41 10N 72 15 E
Andoany, Madag. ...... 57 A8 13 25S 48 16 E
Andong, S. Korea ...... 35 F15 36 40N 128 43 E
Andongwei, China ..... 35 G10 35 6N 119 20 E
Andorra ■, Europe ..... 19 A6 42 30N 1 30 E
Andorra La Vella, Andorra 19 A6 42 31N 1 32 E
Andover, U.K. 11 F6 51 12N 1 29W
Andover, Mass., U.S.A. . 79 D13 42 40N 71 8W
Andover, N.Y., U.S.A. ... 78 D7 42 10N 77 48W
Andover, Ohio, U.S.A. ... 78 E4 41 36N 80 34W
Andøya, Norway .... 8 B16 69 10N 15 50 E
Andrahary, Mt., Madag. . 57 A8 13 37S 49 17 E
Andraitx, Spain .... 22 B9 39 39N 2 25 E
Andramasina, Madag. ... 57 B8 19 11S 47 35 E
Andranopasy, Madag. ... 57 C7 21 17S 43 44 E
Andreanof Is., U.S.A. ... 68 C2 52 0N 178 0W
Andrewilla, Australia ... 63 D2 26 31S 139 17 E
Andrews, S.C., U.S.A. ... 77 J6 33 27N 79 34W
Andrews, Tex., U.S.A. ... 81 J3 32 19N 102 33W
Ándria, Italy .... 20 D7 41 13N 16 17 E
Andriba, Madag. .... 57 B8 17 30S 46 58 E
Androka, Madag. .... 57 C7 24 58S 44 2 E
Andropov = Rybinsk,
Russia .... 24 C6 58 5N 38 50 E
Ándros, Greece .... 21 F11 37 50N 24 57 E
Andros I., Bahamas 88 B4 24 30N 78 0W
Andros Town, Bahamas .. 88 B4 24 43N 77 47W
Andselv, Norway .... 8 B18 69 4N 18 34 E
Andújar, Spain 19 C3 38 3N 4 5W
Andulo, Angola .... 52 G3 11 25S 16 45 E
Anegada I., Virgin Is. 89 C7 18 45N 64 20W
Anegada Passage,
W. Indies 89 C7 18 15N 63 45W
Anécho, Togo .... 50 G5 6 12N 1 34 E
Aneto, Pico de, Spain ... 19 A6 42 37N 0 40 E
Ang Thong, Thailand .... 38 E3 14 35N 100 31 E
Angamos, Punta, Chile .. 94 A1 23 1S 70 32W
Angara →, Russia ..... 27 D10 58 5N 94 20 E
Angarsk, Russia .... 27 D11 52 30N 104 0 E
Angas Downs, Australia . 61 E5 25 2S 132 14 E
Angas Hills, Australia ... 60 D4 23 0S 127 50 E
Angaston, Australia .... 63 E2 34 30S 139 8 E
Ånge, Sweden .... 9 E16 62 31N 15 35 E
Ángel de la Guarda, I.,
Mexico 86 B2 29 30N 113 30W
Angeles, Phil. .... 37 A6 15 9N 120 33 E
Ängelholm, Sweden .... 9 H15 56 15N 12 58 E
Angellala, Australia .... 63 D4 26 24S 146 54 E
Angels Camp, U.S.A. 83 G3 38 4N 120 32W
Ångermanälven →,
Sweden .... 8 E17 62 40N 18 0 E
Ångermanland, Sweden . 8 E18 63 36N 17 45 E
Angers, Canada 79 A9 45 31N 75 29W
Angers, France 18 C3 47 30N 0 35W
Ängesån →, Sweden ... 8 C20 66 16N 22 47 E
Angikuni L., Canada 73 A9 62 0N 100 0W
Angkor, Cambodia ..... 38 F4 13 22N 103 50 E
Anglesey, U.K. 10 D3 53 17N 4 20W
Anglesey □, U.K. 10 D3 53 16N 4 18W
Angleton, U.S.A. 81 L7 29 10N 95 26W
Anglisidhes, Cyprus .... 23 E12 34 51N 33 27 E
Angmagssalik, Greenland 4 C6 65 40N 37 20W
Ango, Zaïre .... 54 B2 4 10N 26 5 E
Angoche, Mozam. ...... 55 F4 16 8S 39 55 E
Angoche, I., Mozam. .... 55 F4 16 20S 39 50 E
Angol, Chile 94 D1 37 56S 72 45W
Angola, Ind., U.S.A. 76 E3 41 38N 85 0W
Angola, N.Y., U.S.A. 78 D5 42 38N 79 2W
Angola ■, Africa .... 53 G3 12 0S 18 0 E
Angoon, U.S.A. 72 B2 57 30N 134 35W
Angoulême, France .... 18 D4 45 39N 0 10 E
Angoumois, France .... 18 D3 45 50N 0 25 E
Angra dos Reis, Brazil ... 95 A7 23 0S 44 10W
Angren, Uzbekistan .... 26 E8 41 1N 70 12 E
Angtassom, Cambodia .. 39 G5 11 1N 104 41 E
Angu, Zaïre .... 54 B1 3 25N 24 28 E
Anguang, China .... 35 B12 45 15N 123 45 E
Anguilla ■, W. Indies ... 89 C7 18 14N 63 5W
Anguo, China .... 34 E8 38 28N 115 15 E
Angurugu, Australia .... 62 A2 14 0S 136 25 E
Angus □, U.K. 12 E6 56 46N 2 56W
Angus, Braes of, U.K. ... 12 E5 56 51N 3 10W
Anhanduí →, Brazil 95 A5 21 46S 52 9W
Anholt, Denmark ...... 9 H14 56 42N 11 33 E
Anhui □, China .... 33 C6 32 0N 117 0 E
Anhwei □ = Anhui □,
China .... 33 C6 32 0N 117 0 E
Anichab, Namibia ..... 56 C1 21 0S 14 46 E
Animas, U.S.A. 83 L9 31 57N 108 48W
Anivorano, Madag. ..... 57 B8 18 44S 48 58 E
Anjalankoski, Finland ... 9 F22 60 45N 26 51 E
Anjar, India .... 42 H4 23 6N 70 10 E
Anjidiv I., India .... 40 M9 14 40N 74 10 E
Anjou, France 18 C3 47 20N 0 15W
Anjozorobe, Madag. .... 57 B8 18 22S 47 52 E
Anju, N. Korea .... 35 E13 39 36N 125 40 E
Anka, Nigeria .... 50 F6 12 13N 5 58 E
Ankaboa, Tanjona, Madag. 57 C7 21 58S 43 20 E
Ankang, China .... 34 H5 32 40N 109 1 E
Ankara, Turkey .... 25 G5 39 57N 32 54 E
Ankaramena, Madag. ... 57 C8 21 57S 46 39 E
Ankazoabo, Madag. .... 57 C7 22 18S 44 31 E
Ankazobe, Madag. ..... 57 B8 18 20S 47 10 E
Ankisabe, Madag. ..... 57 B8 19 17S 46 29 E
Ankoro, Zaïre .... 54 D2 6 45S 26 55 E
Anmyŏn-do, S. Korea ... 35 F14 36 25N 126 25 E
Ann, C., U.S.A. 79 D14 42 38N 70 35W
Ann Arbor, U.S.A. 76 D4 42 17N 83 45W
Anna, U.S.A. 81 G10 37 28N 89 15W
Anna Plains, Australia ... 60 C3 19 17S 121 37 E
Annaba, Algeria .... 50 A6 36 50N 7 46 E
Annalee →, Ireland 13 B4 54 2N 7 24W
Annam = Trung-Phan,
Vietnam .... 38 E7 16 0N 108 0 E
Annamitique, Chaîne, Asia 38 D6 17 0N 106 0 E
Annan, U.K. 12 G5 54 59N 3 16W
Annan →, U.K. 12 G5 54 58N 3 16W
Annapolis, U.S.A. 76 F7 38 59N 76 30W
Annapolis Royal, Canada 71 D6 44 44N 65 32W
Annapurna, Nepal ..... 43 E10 28 34N 83 50 E
Annean, L., Australia ... 61 E2 26 54S 118 14 E
Annecy, France .... 18 D7 45 55N 6 8 E
Anning, China .... 32 D5 24 55N 102 26 E
Anningie, Australia .... 60 D5 21 50S 133 7 E
Anniston, U.S.A. 77 J3 33 39N 85 50W
Annobón, Atl. Oc. ...... 49 G4 1 25S 5 36 E
Annotto Bay, Jamaica ... 88 C4 18 17N 76 45W
Annuello, Australia .... 63 E3 34 53S 142 55 E
Annville, U.S.A. 79 F8 40 20N 76 31W
Áno Viánnos, Greece ... 23 D7 35 2N 25 21 E
Anoka, U.S.A. 75 A8 45 12N 93 23W
Anorotsangana, Madag. . 57 A8 13 56S 47 55 E
Anóyia, Greece .... 23 D6 35 16N 24 52 E
Anping, Hebei, China ... 34 E8 38 15N 115 30 E
Anping, Liaoning, China . 35 D12 41 5N 123 30 E
Anqing, China .... 33 C6 30 30N 117 3 E
Anqiu, China .... 35 F10 36 25N 119 10 E
Ansai, China .... 34 F5 36 50N 109 20 E
Ansbach, Germany ..... 16 D6 49 28N 10 34 E
Anshan, China .... 35 D12 41 5N 122 58 E
Anshun, China .... 32 D5 26 18N 105 57 E
Ansirabe, Madag. ...... 57 B8 19 55S 47 2 E
Ansley, U.S.A. 80 E5 41 18N 99 23W
Anson, U.S.A. 81 J5 32 45N 99 54W
Anson B., Australia .... 60 B5 13 20S 130 6 E
Ansongo, Mali .... 50 E5 15 25N 0 35 E
Ansonia, U.S.A. 79 E11 41 21N 73 5W
Anstruther, U.K. 12 E6 56 14N 2 41W
Ansudu, Indonesia ..... 37 E9 2 11S 139 22 E
Antabamba, Peru 92 F4 14 40S 73 0W
Antakya, Turkey .... 25 G6 36 14N 36 10 E
Antalaha, Madag. ...... 57 A9 14 57S 50 20 E
Antalya, Turkey .... 25 G5 36 52N 30 45 E
Antalya Körfezi, Turkey .. 25 G5 36 15N 31 30 E
Antananarivo, Madag. ... 57 B8 18 55S 47 31 E
Antananarivo □, Madag. . 57 B8 19 0S 47 0 E
Antanimbaribe, Madag. . 57 C7 21 30S 44 48 E
Antarctic Pen., Antarctica 5 C18 67 0S 60 0W
Antarctica .... 5 E3 90 0S 0 0 E
Antelope, Zimbabwe .... 55 G2 21 2S 28 31 E
Antequera, Paraguay 94 A4 24 8S 57 7W
Antequera, Spain 19 D3 37 5N 4 33W
Antero, Mt., U.S.A. 83 G10 38 41N 106 15W
Anthony, Kans., U.S.A. .. 81 G5 37 9N 98 2W
Anthony, N. Mex., U.S.A. 83 K10 32 0N 106 36W
Anthony Lagoon, Australia 62 B2 18 0S 135 30 E
Anti Atlas, Morocco 50 C3 30 0N 8 30W
Anti-Lebanon = Ash
Sharqi, Al Jabal,
Lebanon .... 47 B5 33 40N 36 10 E
Antibes, France .... 18 E7 43 34N 7 6 E
Anticosti, I. d', Canada .. 71 C7 49 30N 63 0W
Antigo, U.S.A. 80 C10 45 9N 89 9W
Antigonish, Canada 71 C7 45 38N 61 58W
Antigua, Canary Is. 22 F5 28 24N 14 1W
Antigua, Guatemala 88 D1 14 34N 90 41W
Antigua, W. Indies 89 C7 17 0N 61 50W
Antigua & Barbuda ■,
W. Indies 89 C7 17 20N 61 48W
Antilla, Cuba 88 B4 20 40N 75 50W
Antimony, U.S.A. 83 G8 38 7N 112 0W
Antioch, U.S.A. 84 G5 38 1N 121 48W
Antioquia, Colombia ... 92 B3 6 40N 75 55W
Antipodes Is., Pac. Oc. .. 64 M9 49 45S 178 40 E
Antler, U.S.A. 80 A4 48 59N 101 17W
Antler →, Canada 73 D8 49 8N 101 0W
Antlers, U.S.A. 81 H7 34 14N 95 37W
Antofagasta, Chile 94 A1 23 50S 70 30W
Antofagasta □, Chile ... 94 A2 24 0S 69 0W
Antofagasta de la Sierra,
Argentina 94 B2 26 5S 67 20W
Antofalla, Argentina ... 94 B2 25 30S 68 5W
Antofalla, Salar de,
Argentina 94 B2 25 40S 67 45W
Anton, U.S.A. 81 J3 33 49N 102 10W
Anton Chico, U.S.A. ... 83 J11 35 12N 105 9W
Antongila, Helodrano,
Madag. .... 57 B8 15 30S 49 50 E
Antonibé, Madag. ..... 57 B8 15 7S 47 24 E
Antonibé, Presq'île d',
Madag. .... 57 A8 14 55S 47 20 E
Antonina, Brazil 95 B6 25 26S 48 42W
Antonito, U.S.A. 83 H10 37 5N 106 0W
Antrim, U.K. 13 B5 54 43N 6 14W
Antrim □, U.K. 13 B5 54 56N 6 25W
Antrim, Mts. of, U.K. ... 13 B5 55 3N 6 14W
Antrim Plateau, Australia 60 C4 18 8S 128 20 E
Antsalova, Madag. ..... 57 B7 18 40S 44 37 E
Antsiranana, Madag. ... 57 A8 12 25S 49 20 E
Antsohihy, Madag. ..... 57 A8 14 50S 47 59 E
Antsohimbondrona
Seranana, Madag. .... 57 A8 13 7S 48 48 E
Antu, China .... 35 C15 42 30N 128 20 E
Antwerp = Antwerpen,
Belgium .... 15 C4 51 13N 4 25 E
Antwerp, U.S.A. 79 B9 44 12N 75 37W
Antwerpen, Belgium ... 15 C4 51 13N 4 25 E
Antwerpen □, Belgium .. 15 C4 51 15N 4 40 E
Anupgarh, India ...... 42 E5 29 10N 73 10 E
Anuradhapura, Sri Lanka 40 Q12 8 22N 80 28 E
Anveh, Iran .... 45 E7 27 23N 54 11 E
Anvers = Antwerpen,
Belgium .... 15 C4 51 13N 4 25 E
Anvers I., Antarctica 5 C17 64 30S 63 40W
Anxi, China .... 32 B4 40 30N 95 43 E
Anxious B., Australia ... 63 E1 33 24S 134 45 E
Anyang, China .... 34 F8 36 5N 114 21 E
Anyi, China .... 34 G6 35 2N 111 2 E
Anza, U.S.A. 85 M10 33 35N 116 39W
Anze, China .... 34 F7 36 10N 112 12 E

Anzhero-Sudzhensk, *Russia* **26 D9** 56 10N 86 0 E
Ánzio, *Italy* **20 D5** 41 27N 12 37 E
Aoga-Shima, *Japan* **31 H9** 32 28N 139 46 E
Aomori, *Japan* **30 D10** 40 45N 140 45 E
Aomori □, *Japan* **30 D10** 40 45N 140 40 E
Aonla, *India* **43 E8** 28 16N 79 11 E
Aosta, *Italy* **20 B2** 45 45N 7 20 E
Aouderas, *Niger* **50 E6** 17 45N 8 20 E
Aoulef el Arab, *Algeria* **50 C5** 26 55N 1 2 E
Apa →, *S. Amer.* **94 A4** 22 6S 58 2W
Apache, *U.S.A.* **81 H5** 34 54N 98 22W
Apalachee B., *U.S.A.* **77 L3** 30 0N 84 0W
Apalachicola, *U.S.A.* **77 L3** 29 43N 84 59W
Apalachicola →, *U.S.A.* **77 L3** 29 43N 84 58W
Apaporis →, *Colombia* **92 D5** 1 23S 69 25W
Aparri, *Phil.* **37 A6** 18 22N 121 38 E
Apatity, *Russia* **24 A5** 67 34N 33 22 E
Apatzingán, *Mexico* **86 D4** 19 0N 102 20W
Apeldoorn, *Neths.* **15 B5** 52 13N 5 57 E
Apennines = Appennini, *Italy* **20 B4** 44 0N 10 0 E
Apia, *W. Samoa* **59 A13** 13 50S 171 50W
Apiacás, Serra dos, *Brazil* **92 E7** 9 50S 57 0W
Apizaco, *Mexico* **87 D5** 19 26N 98 9W
Aplao, *Peru* **92 G4** 16 0S 72 40W
Apo, Mt., *Phil.* **37 C7** 6 53N 125 14 E
Apolakkiá, *Greece* **23 C9** 36 5N 27 48 E
Apolakkiá, Órmos, *Greece* **23 C9** 36 5N 27 45 E
Apollonia = Marsá Susah, *Libya* **51 B9** 32 52N 21 59 E
Apolo, *Bolivia* **92 F5** 14 30S 68 30W
Apostle Is., *U.S.A.* **80 B9** 47 0N 90 40W
Apóstoles, *Argentina* **95 B4** 28 0S 56 0W
Apostolos Andreas, C., *Cyprus* **23 D13** 35 42N 34 35 E
Apoteri, *Guyana* **92 C7** 4 2N 58 32W
Appalachian Mts., *U.S.A.* **76 G6** 38 0N 80 0W
Appennini, *Italy* **20 B4** 44 0N 10 0 E
Apple Hill, *Canada* **79 A10** 45 13N 74 46W
Apple Valley, *U.S.A.* **85 L9** 34 32N 117 14W
Appleby-in-Westmorland, *U.K.* **10 C5** 54 35N 2 29W
Appleton, *U.S.A.* **76 C1** 44 16N 88 25W
Approuague, *Fr. Guiana* **93 C8** 4 20N 52 0W
Aprília, *Italy* **20 D5** 41 36N 12 39 E
Apucarana, *Brazil* **95 A5** 23 55S 51 33W
Apure →, *Venezuela* **92 B5** 7 37N 66 25W
Apurimac →, *Peru* **92 F4** 12 17S 73 56W
Aqabah = Al 'Aqabah, *Jordan* **47 F4** 29 31N 35 0 E
'Aqabah, Khalīj al, *Red Sea* **44 D2** 28 15N 33 20 E
'Aqdā, *Iran* **45 C7** 32 26N 53 37 E
Aqīq, *Sudan* **51 E12** 18 14N 38 12 E
Aqmola, *Kazakstan* **26 D8** 51 10N 71 30 E
Aqrah, *Iraq* **44 B4** 36 46N 43 45 E
Aqtöbe, *Kazakstan* **25 D10** 50 17N 57 10 E
Aquidauana, *Brazil* **93 H7** 20 30S 55 50W
Aquiles Serdán, *Mexico* **86 B3** 28 37N 105 54W
Aquin, *Haiti* **89 C5** 18 16N 73 24W
Aquitain, Bassin, *France* **18 D3** 44 0N 0 30W
Ar Rachidiya, *Morocco* **50 B4** 31 58N 4 20W
Ar Rafid, *Syria* **47 C4** 32 57N 35 52 E
Ar Raḥḥālīyah, *Iraq* **44 C4** 32 44N 43 23 E
Ar Ramādī, *Iraq* **44 C4** 33 25N 43 20 E
Ar Ramthā, *Jordan* **47 C5** 32 34N 36 0 E
Ar Raqqah, *Syria* **44 C3** 35 59N 39 8 E
Ar Rass, *Si. Arabia* **44 E4** 25 50N 43 40 E
Ar Rifā'ī, *Iraq* **44 D5** 31 50N 46 10 E
Ar Riyāḍ, *Si. Arabia* **46 C4** 24 41N 46 42 E
Ar Ru'ays, *Qatar* **45 E6** 26 8N 51 12 E
Ar Rukhaymīyah, *Iraq* **44 D5** 29 22N 45 38 E
Ar Ruqayyidah, *Si. Arabia* **45 E6** 25 21N 49 34 E
Ar Ruṣāfah, *Syria* **44 C3** 35 45N 38 49 E
Ar Ruṭbah, *Iraq* **44 C4** 33 0N 40 15 E
Ara, *India* **43 G11** 25 35N 84 32 E
'Arab, Bahr el →, *Sudan* **51 G10** 9 0N 29 30 E
'Arabābād, *Iran* **45 C8** 33 2N 57 41 E
Arabia, *Asia* **46 C4** 25 0N 45 0 E
Arabian Desert = Es Sahrâ' Esh Sharqîya, *Egypt* **51 C11** 27 30N 32 30 E
Arabian Gulf = Gulf, The, *Asia* **45 E6** 27 0N 50 0 E
Arabian Sea, *Ind. Oc.* **29 H10** 16 0N 65 0 E
Aracaju, *Brazil* **93 F11** 10 55S 37 4W
Aracataca, *Colombia* **92 A4** 10 38N 74 9W
Aracati, *Brazil* **93 D11** 4 30S 37 44W
Araçatuba, *Brazil* **95 A5** 21 10S 50 30W
Aracena, *Spain* **19 D2** 37 53N 6 38W
Araçuaí, *Brazil* **93 G10** 16 52S 42 4W
'Arad, *Israel* **47 D4** 31 15N 35 12 E
Arad, *Romania* **17 E11** 46 10N 21 20 E
Arada, *Chad* **51 F9** 15 0N 20 20 E
Aradhippou, *Cyprus* **23 E12** 34 57N 33 36 E
Arafura Sea, *E. Indies* **37 F8** 9 0S 135 0 E
Aragón □, *Spain* **19 B5** 41 25N 0 40W
Aragón →, *Spain* **19 A5** 42 13N 1 44W
Araguacema, *Brazil* **93 E9** 8 50S 49 20W
Araguaia →, *Brazil* **93 E9** 5 21S 48 41W
Araguari, *Brazil* **93 G9** 18 38S 48 11W
Araguari →, *Brazil* **93 C9** 1 15N 49 55W
Arak, *Algeria* **50 C5** 25 20N 3 45 E
Arāk, *Iran* **45 C6** 34 0N 49 40 E
Arakan Coast, *Burma* **41 K19** 19 0N 94 0 E
Arakan Yoma, *Burma* **41 K19** 20 0N 94 40 E
Araks = Aras, Rūd-e →, *Azerbaijan* **44 B5** 40 5N 48 29 E
Aral, *Kazakstan* **26 E7** 46 41N 61 45 E
Aral Sea, *Asia* **26 E7** 44 30N 60 0 E
Aral Tengizi = Aral Sea, *Asia* **26 E7** 44 30N 60 0 E
Aralsk = Aral, *Kazakstan* **26 E7** 46 41N 61 45 E
Aralskoye More = Aral Sea, *Asia* **26 E7** 44 30N 60 0 E
Aramac, *Australia* **62 C4** 22 58S 145 14 E
Arambag, *India* **43 H12** 22 53N 87 48 E
Aran I., *Ireland* **13 B3** 55 0N 8 30W
Aran Is., *Ireland* **13 C2** 53 6N 9 38W
Aranda de Duero, *Spain* **19 B4** 41 39N 3 42W
Arandān, *Iran* **44 C5** 35 23N 46 55 E
Aranjuez, *Spain* **19 B4** 40 1N 3 40W
Aranos, *Namibia* **56 C2** 24 9S 19 7 E
Aransas Pass, *U.S.A.* **81 M6** 27 55N 97 9W
Araouane, *Mali* **50 E4** 18 55N 3 30W
Arapahoe, *U.S.A.* **80 E5** 40 18N 99 54W
Arapey Grande →, *Uruguay* **94 C4** 30 55S 57 49W
Arapiraca, *Brazil* **93 E11** 9 45S 36 39W
Arapongas, *Brazil* **95 A5** 23 29S 51 28W
Ar'ar, *Si. Arabia* **44 D4** 30 59N 41 2 E
Araranguá, *Brazil* **95 B6** 29 0S 49 30W
Araraquara, *Brazil* **93 H9** 21 50S 48 0W
Ararás, Serra das, *Brazil* **95 B5** 25 0S 53 10W
Ararat, *Australia* **63 F3** 37 16S 143 0 E
Ararat, Mt. = Ağrı Dağı, *Turkey* **25 G7** 39 50N 44 15 E
Araria, *India* **43 F12** 26 9N 87 33 E
Araripe, Chapada do, *Brazil* **93 E11** 7 20S 40 0W
Araruama, L. de, *Brazil* **95 A7** 22 53S 42 12W
Aras, Rūd-e →, *Azerbaijan* **44 B5** 40 5N 48 29 E
Arauca, *Colombia* **92 B4** 7 0N 70 40W
Arauca →, *Venezuela* **92 B5** 7 24N 66 35W
Arauco, *Chile* **94 D1** 37 16S 73 25W
Arauco □, *Chile* **94 D1** 37 40S 73 25W
Araxá, *Brazil* **93 G9** 19 35S 46 55W
Araya, Pen. de, *Venezuela* **92 A6** 10 40N 64 0W
Arbat, *Iraq* **44 C5** 35 25N 45 35 E
Arbatax, *Italy* **20 E3** 39 56N 9 42 E
Arbīl, *Iraq* **44 B5** 36 15N 44 5 E
Arborfield, *Canada* **73 C8** 53 6N 103 39W
Arborg, *Canada* **73 C9** 50 54N 97 13W
Arbroath, *U.K.* **12 E6** 56 34N 2 35W
Arbuckle, *U.S.A.* **84 F4** 39 1N 122 3W
Arcachon, *France* **18 D3** 44 40N 1 10W
Arcade, *U.S.A.* **78 D6** 42 32N 78 25W
Arcadia, *Fla., U.S.A.* **77 M5** 27 13N 81 52W
Arcadia, *La., U.S.A.* **81 J8** 32 33N 92 55W
Arcadia, *Nebr., U.S.A.* **80 E5** 41 25N 99 8W
Arcadia, *Pa., U.S.A.* **78 F6** 40 47N 78 51W
Arcadia, *Wis., U.S.A.* **80 C9** 44 15N 91 30W
Arcata, *U.S.A.* **82 F1** 40 52N 124 5W
Archangel = Arkhangelsk, *Russia* **24 B7** 64 38N 40 36 E
Archbald, *U.S.A.* **79 E9** 41 30N 75 32W
Archer →, *Australia* **62 A3** 13 28S 141 41 E
Archer B., *Australia* **62 A3** 13 20S 141 30 E
Archers Post, *Kenya* **54 B4** 0 35N 37 35 E
Arcila = Asilah, *Morocco* **50 A3** 35 29N 6 0W
Arckaringa, *Australia* **63 D1** 27 56S 134 45 E
Arckaringa Cr. →, *Australia* **63 D2** 28 10S 135 22 E
Arco, *U.S.A.* **82 E7** 43 38N 113 18W
Arcola, *Canada* **73 D8** 49 40N 102 30W
Arcos de la Frontera, *Spain* **19 D3** 36 45N 5 49W
Arcot, *India* **40 N11** 12 53N 79 20 E
Arcoverde, *Brazil* **93 E11** 8 25S 37 4W
Arctic Bay, *Canada* **69 A11** 73 1N 85 7W
Arctic Ocean, *Arctic* **4 B18** 78 0N 160 0W
Arctic Red River, *Canada* **68 B6** 67 15N 134 0W
Arda →, *Bulgaria* **21 D12** 41 40N 26 29 E
Ardabīl, *Iran* **45 B6** 38 15N 48 18 E
Ardakān = Sepīdān, *Iran* **45 D7** 30 20N 52 5 E
Ardee, *Ireland* **13 C5** 53 52N 6 33W
Arden, *Canada* **78 B8** 44 43N 76 56W
Arden, *Calif., U.S.A.* **84 G5** 38 36N 121 33W
Arden, *Nev., U.S.A.* **85 J11** 36 1N 115 14W
Ardenne, *Belgium* **15 E5** 49 50N 5 5 E
Ardennes = Ardenne, *Belgium* **15 E5** 49 50N 5 5 E
Ardestān, *Iran* **45 C7** 33 20N 52 25 E
Ardgour, *U.K.* **12 E3** 56 45N 5 25W
Ardlethan, *Australia* **63 E4** 34 22S 146 53 E
Ardmore, *Australia* **62 C2** 21 39S 139 11 E
Ardmore, *Okla., U.S.A.* **81 H6** 34 10N 97 8W
Ardmore, *Pa., U.S.A.* **79 G9** 39 58N 75 18W
Ardmore, *S. Dak., U.S.A.* **80 D3** 43 1N 103 40W
Ardnacrusha, *Ireland* **13 D3** 52 43N 8 38W
Ardnamurchan, Pt. of, *U.K.* **12 E2** 56 43N 6 14W
Ardrossan, *Australia* **63 E2** 34 26S 137 53 E
Ardrossan, *U.K.* **12 F4** 55 39N 4 49W
Ards □, *U.K.* **13 B6** 54 35N 5 30W
Ards Pen., *U.K.* **13 B6** 54 33N 5 34W
Arecibo, *Puerto Rico* **89 C6** 18 29N 66 43W
Areia Branca, *Brazil* **93 D11** 5 0S 37 0W
Arena, Pt., *U.S.A.* **84 G3** 38 57N 123 44W
Arendal, *Norway* **9 G13** 58 28N 8 46 E
Arequipa, *Peru* **92 G4** 16 20S 71 30W
Arero, *Ethiopia* **51 H12** 4 41N 38 50 E
Arévalo, *Spain* **19 B3** 41 3N 4 43W
Arezzo, *Italy* **20 C4** 43 25N 11 53 E
Argamakmur, *Indonesia* **36 E2** 3 35S 102 0 E
Arganda, *Spain* **19 B4** 40 19N 3 26W
Argentan, *France* **18 B3** 48 45N 0 1W
Argentário, Mte., *Italy* **20 C4** 42 24N 11 9 E
Argentia, *Canada* **71 C9** 47 18N 53 58W
Argentina ■, *S. Amer.* **96 D3** 35 0S 66 0W
Argentina Is., *Antarctica* **5 C17** 66 0S 64 0W
Argentino, L., *Argentina* **96 G2** 50 10S 73 0W
Argeş →, *Romania* **17 F14** 44 11N 26 25 E
Arghandab →, *Afghan.* **42 D1** 31 30N 64 15 E
Argo, *Sudan* **51 E11** 19 28N 30 30 E
Argolikós Kólpos, *Greece* **21 F10** 37 20N 22 52 E
Árgos, *Greece* **21 F10** 37 40N 22 43 E
Argostólion, *Greece* **21 E9** 38 12N 20 33 E
Arguello, Pt., *U.S.A.* **85 L6** 34 35N 120 39W
Arguineguín, *Canary Is.* **22 G4** 27 46N 15 41W
Argun →, *Russia* **27 D13** 53 20N 121 28 E
Argungu, *Nigeria* **50 F5** 12 40N 4 31 E
Argus Pk., *U.S.A.* **85 K9** 35 52N 117 26W
Argyle, *U.S.A.* **80 A6** 48 20N 96 49W
Argyle, L., *Australia* **60 C4** 16 20S 128 40 E
Argyll & Bute □, *U.K.* **12 E3** 56 13N 5 28W
Århus, *Denmark* **9 H14** 56 8N 10 11 E
Ariadnoye, *Russia* **30 B7** 45 8N 134 25 E
Ariamsvlei, *Namibia* **56 D2** 28 9S 19 51 E
Arica, *Chile* **92 G4** 18 32S 70 20W
Arica, *Colombia* **92 D4** 2 0S 71 50W
Arico, *Canary Is.* **22 F3** 28 9N 16 29W
Arid, C., *Australia* **61 F3** 34 1S 123 10 E
Arida, *Japan* **31 G7** 34 5N 135 8 E
Arīḥā, *Syria* **44 C3** 35 49N 36 35 E
Arilla, Ákra, *Greece* **23 A3** 39 43N 19 39 E
Arima, *Trin. & Tob.* **89 D7** 10 38N 61 17W
Arinos →, *Brazil* **92 F7** 10 25S 58 20W
Ario de Rosales, *Mexico* **86 D4** 19 12N 102 0W
Aripuanã, *Brazil* **92 E6** 9 25S 60 30W
Aripuanã →, *Brazil* **92 E6** 5 7S 60 25W
Ariquemes, *Brazil* **92 E6** 9 55S 63 6W
Arisaig, *U.K.* **12 E3** 56 55N 5 51W
Aristazabal I., *Canada* **72 C3** 52 40N 129 10W
Arivaca, *U.S.A.* **83 L8** 31 37N 111 25W
Arivonimamo, *Madag.* **57 B8** 19 1S 47 11 E
Arizaro, Salar de, *Argentina* **94 A2** 24 40S 67 50W
Arizona, *Argentina* **94 D2** 35 45S 65 25W
Arizona □, *U.S.A.* **83 J8** 34 0N 112 0W
Arizpe, *Mexico* **86 A2** 30 20N 110 11W
Arjeplog, *Sweden* **8 D18** 66 3N 18 2 E
Arjona, *Colombia* **92 A3** 10 14N 75 22W
Arjuno, *Indonesia* **37 G15** 7 49S 112 34 E
Arka, *Russia* **27 C15** 60 15N 142 0 E
Arkadelphia, *U.S.A.* **81 H8** 34 7N 93 4W
Arkaig, L., *U.K.* **12 E3** 56 59N 5 10W
Arkalyk = Arqalyk, *Kazakstan* **26 D7** 50 13N 66 50 E
Arkansas □, *U.S.A.* **81 H8** 35 0N 92 30W
Arkansas →, *U.S.A.* **81 J9** 33 47N 91 4W
Arkansas City, *U.S.A.* **81 G6** 37 4N 97 2W
Arkhángelos, *Greece* **23 C10** 36 13N 28 7 E
Arkhangelsk, *Russia* **24 B7** 64 38N 40 36 E
Arklow, *Ireland* **13 D5** 52 48N 6 10W
Arkticheskiy, Mys, *Russia* **27 A10** 81 10N 95 0 E
Arlanzón →, *Spain* **19 A3** 42 3N 4 17W
Arlberg P., *Austria* **16 E6** 47 9N 10 12 E
Arlee, *U.S.A.* **82 C6** 47 10N 114 5W
Arles, *France* **18 E6** 43 41N 4 40 E
Arlington, *S. Africa* **57 D4** 28 1S 27 53 E
Arlington, *Oreg., U.S.A.* **82 D3** 45 43N 120 12W
Arlington, *S. Dak., U.S.A.* **80 C6** 44 22N 97 8W
Arlington, *Va., U.S.A.* **76 F7** 38 53N 77 7W
Arlington, *Wash., U.S.A.* **84 B4** 48 12N 122 8W
Arlington Heights, *U.S.A.* **76 D2** 42 5N 87 59W
Arlon, *Belgium* **15 E5** 49 42N 5 49 E
Armagh, *U.K.* **13 B5** 54 21N 6 39W
Armagh □, *U.K.* **13 B5** 54 18N 6 37W
Armavir, *Russia* **25 E7** 45 2N 41 7 E
Armenia, *Colombia* **92 C3** 4 35N 75 45W
Armenia ■, *Asia* **25 F7** 40 20N 45 0 E
Armenistís, Ákra, *Greece* **23 C9** 36 8N 27 42 E
Armidale, *Australia* **63 E5** 30 30S 151 40 E
Armour, *U.S.A.* **80 D5** 43 19N 98 21W
Armstrong, *B.C., Canada* **72 C5** 50 25N 119 10W
Armstrong, *Ont., Canada* **70 B2** 50 18N 89 4W
Armstrong, *U.S.A.* **81 M6** 26 56N 97 47W
Armstrong →, *Australia* **60 C5** 16 35S 131 40 E
Arnarfjörður, *Iceland* **8 D2** 65 48N 23 40W
Arnaud →, *Canada* **69 B12** 60 0N 70 0W
Arnauti, C., *Cyprus* **23 D11** 35 6N 32 17 E
Arnett, *U.S.A.* **81 G5** 36 8N 99 46W
Arnhem, *Neths.* **15 C5** 51 58N 5 55 E
Arnhem, C., *Australia* **62 A2** 12 20S 137 30 E
Arnhem B., *Australia* **62 A2** 12 20S 136 10 E
Arnhem Land, *Australia* **62 A1** 13 10S 134 30 E
Arno →, *Italy* **20 C4** 43 41N 10 17 E
Arno Bay, *Australia* **63 E2** 33 54S 136 34 E
Arnold, *Calif., U.S.A.* **84 G6** 38 15N 120 20W
Arnold, *Nebr., U.S.A.* **80 E4** 41 26N 100 12W
Arnot, *Canada* **73 B9** 55 56N 96 41W
Arnøy, *Norway* **8 A19** 70 9N 20 40 E
Arnprior, *Canada* **70 C4** 45 26N 76 21W
Arnsberg, *Germany* **16 C5** 51 24N 8 5 E
Aroab, *Namibia* **56 D2** 26 41S 19 39 E
Arqalyk, *Kazakstan* **26 D7** 50 13N 66 50 E
Arrabury, *Australia* **63 D3** 26 45S 141 0 E
Arrah = Ara, *India* **43 G11** 25 35N 84 32 E
Arran, *U.K.* **12 F3** 55 34N 5 12W
Arrandale, *Canada* **72 C3** 54 57N 130 0W
Arras, *France* **18 A5** 50 17N 2 46 E
Arrecife, *Canary Is.* **22 F6** 28 57N 13 37W
Arrecifes, *Argentina* **94 C3** 34 6S 60 9W
Arrée, Mts. d', *France* **18 B2** 48 26N 3 55W
Arriaga, *Chiapas, Mexico* **87 D6** 16 15N 93 52W
Arriaga, *San Luis Potosí, Mexico* **86 C4** 21 55N 101 23W
Arrilalah P.O., *Australia* **62 C3** 23 43S 143 54 E
Arrino, *Australia* **61 E2** 29 30S 115 40 E
Arrow, L., *Ireland* **13 B3** 54 3N 8 19W
Arrow Rock Res., *U.S.A.* **82 E6** 43 45N 115 50W
Arrowhead, *Canada* **72 C5** 50 40N 117 55W
Arrowhead, L., *U.S.A.* **85 L9** 34 16N 117 10W
Arrowtown, *N.Z.* **59 L2** 44 57S 168 50 E
Arroyo Grande, *U.S.A.* **85 K6** 35 7N 120 35W
Ars, *Iran* **44 B5** 37 9N 47 46 E
Arsenault L., *Canada* **73 B7** 55 6N 108 32W
Arsenev, *Russia* **30 B6** 44 10N 133 15 E
Árta, *Greece* **21 E9** 39 8N 21 2 E
Artá, *Spain* **22 B10** 39 41N 3 21 E
Arteaga, *Mexico* **86 D4** 18 50N 102 20W
Artem, *Russia* **30 C6** 43 22N 132 13 E
Artemovsk, *Russia* **27 D10** 54 45N 93 35 E
Artesia = Mosomane, *Botswana* **56 C4** 24 2S 26 19 E
Artesia, *U.S.A.* **81 J2** 32 51N 104 24W
Artesia Wells, *U.S.A.* **81 L5** 28 17N 99 17W
Artesian, *U.S.A.* **80 C6** 44 1N 97 55W
Arthur →, *Australia* **62 G3** 41 2S 144 40 E
Arthur Cr. →, *Australia* **62 C2** 22 30S 136 25 E
Arthur Pt., *Australia* **62 C5** 22 7S 150 3 E
Arthur's Pass, *N.Z.* **59 K3** 42 54S 171 35 E
Arthur's Town, *Bahamas* **89 B4** 24 38N 75 42W
Artigas, *Uruguay* **94 C4** 30 20S 56 30W
Artillery L., *Canada* **73 A7** 63 9N 107 52W
Artois, *France* **18 A5** 50 20N 2 30 E
Artsyz, *Ukraine* **17 E15** 46 4N 29 26 E
Artvin, *Turkey* **25 F7** 41 14N 41 44 E
Aru, Kepulauan, *Indonesia* **37 F8** 6 0S 134 30 E
Aru Is. = Aru, Kepulauan, *Indonesia* **37 F8** 6 0S 134 30 E
Aru Meru □, *Tanzania* **54 C4** 3 20S 36 50 E
Arua, *Uganda* **54 B3** 3 1N 30 58 E
Aruanã, *Brazil* **93 F8** 14 54S 51 10W
Aruba ■, *W. Indies* **89 D6** 12 30N 70 0W
Arucas, *Canary Is.* **22 F4** 28 7N 15 32W
Arumpo, *Australia* **63 E3** 33 48S 142 55 E
Arun →, *Nepal* **43 F12** 26 55N 87 10 E
Arunachal Pradesh □, *India* **41 E19** 28 0N 95 0 E
Arusha, *Tanzania* **54 C4** 3 20S 36 40 E
Arusha □, *Tanzania* **54 C4** 4 0S 36 30 E
Arusha Chini, *Tanzania* **54 C4** 3 32S 37 20 E
Aruwimi →, *Zaïre* **54 B1** 1 13N 23 36 E
Arvada, *U.S.A.* **82 D10** 44 39N 106 8W
Árvi, *Greece* **23 E7** 34 59N 25 28 E
Arvida, *Canada* **71 C5** 48 25N 71 14W
Arvidsjaur, *Sweden* **8 D18** 65 35N 19 10 E
Arvika, *Sweden* **9 G15** 59 40N 12 36 E
Arvin, *U.S.A.* **85 K8** 35 12N 118 50W
Arxan, *China* **33 B6** 47 11N 119 57 E
Aryirádhes, *Greece* **23 B3** 39 27N 19 58 E
Aryiroúpolis, *Greece* **23 D6** 35 17N 24 20 E
Arys, *Kazakstan* **26 E7** 42 26N 68 48 E
Arzamas, *Russia* **24 C7** 55 27N 43 55 E
Arzew, *Algeria* **50 A4** 35 50N 0 23W
Aş Şadr, *U.A.E.* **45 E7** 24 40N 54 41 E
Aş Şafā, *Syria* **47 B6** 33 10N 37 0 E
'As Saffānīyah, *Si. Arabia* **45 D6** 28 5N 48 50 E
As Safīrah, *Syria* **44 B3** 36 5N 37 21 E
Aş Şahm, *Oman* **45 E8** 24 10N 56 53 E
As Sājir, *Si. Arabia* **44 E5** 25 11N 44 36 E
As Salamīyah, *Syria* **44 C3** 35 1N 37 2 E
As Salṭ, *Jordan* **47 C4** 32 2N 35 43 E
As Sal'w'a, *Qatar* **45 E6** 24 23N 50 50 E
As Samāwah, *Iraq* **44 D5** 31 15N 45 15 E
As Sanamayn, *Syria* **47 B5** 33 3N 36 10 E
As Sukhnah, *Syria* **44 C3** 34 52N 38 52 E
As Sulaymānīyah, *Iraq* **44 C5** 35 35N 45 29 E
As Sulaymī, *Si. Arabia* **44 E4** 26 17N 41 21 E
As Summān, *Si. Arabia* **44 E5** 25 0N 47 0 E
As Suwaydā', *Syria* **47 C5** 32 40N 36 30 E
As Suwaydā' □, *Syria* **47 C5** 32 45N 36 45 E
As Şuwayrah, *Iraq* **44 C5** 32 55N 45 0 E
Asab, *Namibia* **56 D2** 25 30S 18 0 E
Asahi-Gawa →, *Japan* **31 G6** 34 36N 133 58 E
Asahigawa, *Japan* **30 C11** 43 46N 142 22 E
Asansol, *India* **43 H12** 23 40N 87 1 E
Asbesberge, *S. Africa* **56 D3** 29 0S 23 0 E
Asbestos, *Canada* **71 C5** 45 47N 71 58W
Asbury Park, *U.S.A.* **79 F10** 40 13N 74 1W
Ascensión, *Mexico* **86 A3** 31 6N 107 59W
Ascensión, B. de la, *Mexico* **87 D7** 19 50N 87 20W
Ascension I., *Atl. Oc.* **49 G2** 8 0S 14 15W
Aschaffenburg, *Germany* **16 D5** 49 58N 9 6 E
Aschersleben, *Germany* **16 C6** 51 45N 11 29 E
Áscoli Piceno, *Italy* **20 C5** 42 51N 13 34 E
Ascope, *Peru* **92 E3** 7 46S 79 8W
Ascotán, *Chile* **94 A2** 21 45S 68 17W
Aseb, *Eritrea* **46 E3** 13 0N 42 40 E
Asela, *Ethiopia* **51 G12** 8 0N 39 0 E
Asenovgrad, *Bulgaria* **21 C11** 42 1N 24 51 E
Asgata, *Cyprus* **23 E12** 34 46N 33 15 E
Ash Fork, *U.S.A.* **83 J7** 35 13N 112 29W
Ash Grove, *U.S.A.* **81 G8** 37 19N 93 35W
Ash Shām, Bādiyat, *Asia* **28 F7** 32 0N 40 0 E
Ash Shamāl □, *Lebanon* **47 A5** 34 25N 36 0 E
Ash Shāmīyah, *Iraq* **44 D5** 31 55N 44 35 E
Ash Shāriqah, *U.A.E.* **45 E7** 25 23N 55 26 E
Ash Sharmah, *Si. Arabia* **44 D2** 28 1N 35 16 E
Ash Sharqāṭ, *Iraq* **44 C4** 35 27N 43 16 E
Ash Sharqi, Al Jabal, *Lebanon* **47 B5** 33 40N 36 10 E
Ash Shaṭrah, *Iraq* **44 D5** 31 30N 46 10 E
Ash Shawbak, *Jordan* **44 D2** 30 32N 35 34 E
Ash Shawmari, J., *Jordan* **47 E5** 30 35N 36 35 E
Ash Shaykh, J., *Lebanon* **47 B4** 33 25N 35 50 E
Ash Shināfiyah, *Iraq* **44 D5** 31 35N 44 39 E
Ash Shu'aybah, *Si. Arabia* **44 E5** 27 53N 44 43 E
Ash Shumlūl, *Si. Arabia* **44 E5** 26 31N 47 20 E
Ash Shūr'a, *Iraq* **44 C4** 35 58N 43 13 E
Ash Shurayf, *Si. Arabia* **44 E3** 25 43N 39 14 E
Ash Shuwayfāt, *Lebanon* **47 B4** 33 45N 35 30 E
Asha, *Russia* **24 D10** 55 0N 57 16 E
Ashau, *Vietnam* **38 D6** 16 6N 107 22 E
Ashburn, *U.S.A.* **77 K4** 31 43N 83 39W
Ashburton, *N.Z.* **59 K3** 43 53S 171 48 E
Ashburton →, *Australia* **60 D1** 21 40S 114 56 E
Ashburton Downs, *Australia* **60 D2** 23 25S 117 4 E
Ashby de la Zouch, *U.K.* **10 E6** 52 46N 1 29W
Ashcroft, *Canada* **72 C4** 50 40N 121 20W
Ashdod, *Israel* **47 D3** 31 49N 34 35 E
Asheboro, *U.S.A.* **77 H6** 35 43N 79 49W
Asherton, *U.S.A.* **81 L5** 28 27N 99 46W
Asheville, *U.S.A.* **77 H4** 35 36N 82 33W
Asheweig →, *Canada* **70 B2** 54 17N 87 12W
Ashford, *Australia* **63 D5** 29 15S 151 3 E
Ashford, *U.K.* **11 F8** 51 8N 0 53 E
Ashford, *U.S.A.* **82 C2** 46 46N 122 2W
Ashgabat, *Turkmenistan* **26 F6** 38 0N 57 50 E
Ashibetsu, *Japan* **30 C11** 43 31N 142 11 E
Ashikaga, *Japan* **31 F9** 36 28N 139 29 E
Ashizuri-Zaki, *Japan* **31 H6** 32 44N 133 0 E
Ashkarkot, *Afghan.* **42 C2** 33 3N 67 58 E
Ashkhabad = Ashgabat, *Turkmenistan* **26 F6** 38 0N 57 50 E
Ashland, *Kans., U.S.A.* **81 G5** 37 11N 99 46W
Ashland, *Ky., U.S.A.* **76 F4** 38 28N 82 38W
Ashland, *Maine, U.S.A.* **71 C6** 46 38N 68 24W
Ashland, *Mont., U.S.A.* **82 D10** 45 36N 106 16W
Ashland, *Nebr., U.S.A.* **80 E6** 41 3N 96 23W
Ashland, *Ohio, U.S.A.* **78 F2** 40 52N 82 19W
Ashland, *Oreg., U.S.A.* **82 E2** 42 12N 122 43W
Ashland, *Pa., U.S.A.* **79 F8** 40 45N 76 22W
Ashland, *Va., U.S.A.* **76 G7** 37 46N 77 29W
Ashland, *Wis., U.S.A.* **80 B9** 46 35N 90 53W
Ashley, *N. Dak., U.S.A.* **80 B5** 46 2N 99 22W
Ashley, *Pa., U.S.A.* **79 E9** 41 12N 75 55W
Ashmont, *Canada* **72 C6** 54 7N 111 35W
Ashmore Reef, *Australia* **60 B3** 12 14S 123 5 E
Ashmyany, *Belarus* **9 J21** 54 26N 25 52 E
Ashqelon, *Israel* **47 D3** 31 42N 34 35 E
Ashtabula, *U.S.A.* **78 E4** 41 52N 80 47W
Ashton, *S. Africa* **56 E3** 33 50S 20 5 E
Ashton, *U.S.A.* **82 D8** 44 4N 111 27W
Ashton under Lyne, *U.K.* **10 D5** 53 29N 2 6W
Ashuanipi, L., *Canada* **71 B6** 52 45N 66 15W
Asia, Kepulauan, *Indonesia* **37 D8** 1 0N 131 13 E
Āsīā Bak, *Iran* **45 C6** 35 19N 50 30 E
Asifabad, *India* **40 K11** 19 20N 79 24 E
Asike, *Indonesia* **37 F10** 6 39S 140 24 E

Asilah, *Morocco* **50 A3** 35 29N 6 0W
Asinara, *Italy* **20 D3** 41 4N 8 16 E
Asinara, G. dell', *Italy* **20 D3** 41 0N 8 30 E
Asino, *Russia* **26 D9** 57 0N 86 0 E
Asipovichy, *Belarus* **17 B15** 53 19N 28 33 E
'Asīr □, *Si. Arabia* **46 D3** 18 40N 42 30 E
Asir, Ras, *Somali Rep.* **46 E5** 11 55N 51 10 E
Askersund, *Sweden* **9 G16** 58 53N 14 55 E
Askham, *S. Africa* **56 D3** 26 59S 20 47 E
Askim, *Norway* **9 G14** 59 35N 11 10 E
Askja, *Iceland* **8 D5** 65 3N 16 48W
Askøy, *Norway* **9 F11** 60 29N 5 10 E
Asmara = Asmera, *Eritrea* **51 E12** 15 19N 38 55 E
Asmera, *Eritrea* **51 E12** 15 19N 38 55 E
Åsnen, *Sweden* **9 H16** 56 37N 14 45 E
Asotin, *U.S.A.* **82 C5** 46 20N 117 3W
Aspen, *U.S.A.* **83 G10** 39 11N 106 49W
Aspermont, *U.S.A.* **81 J4** 33 8N 100 14W
Aspiring, Mt., *N.Z.* **59 L2** 44 23S 168 46 E
Asprókavos, Ákra, *Greece* **23 B4** 39 21N 20 6 E
Aspur, *India* **42 H6** 23 58N 74 7 E
Asquith, *Canada* **73 C7** 52 8N 107 13W
Assad, Bahret, *Syria* **44 C3** 36 0N 38 15 E
Assam □, *India* **41 F18** 26 0N 93 0 E
Asse, *Belgium* **15 D4** 50 24N 4 10 E
Assen, *Neths.* **15 B6** 53 0N 6 35 E
Assini, *Ivory C.* **50 G4** 5 9N 3 17W
Assiniboia, *Canada* **73 D7** 49 40N 105 59W
Assiniboine →, *Canada* **73 D9** 49 53N 97 8W
Assis, *Brazil* **95 A5** 22 40S 50 20W
Assisi, *Italy* **20 C5** 43 4N 12 37 E
Assynt, L., *U.K.* **12 C3** 58 10N 5 3W
Astara, *Azerbaijan* **25 G8** 38 30N 48 50 E
Asteroúsia, *Greece* **23 E7** 34 59N 25 3 E
Asti, *Italy* **20 B3** 44 54N 8 12 E
Astipálaia, *Greece* **21 F12** 36 32N 26 22 E
Astorga, *Spain* **19 A2** 42 29N 6 8W
Astoria, *U.S.A.* **84 D3** 46 11N 123 50W
Astrakhan, *Russia* **25 E8** 46 25N 48 5 E
Astrakhan-Bazàr, *Azerbaijan* **25 G8** 39 14N 48 30 E
Asturias □, *Spain* **19 A3** 43 15N 6 0W
Asunción, *Paraguay* **94 B4** 25 10S 57 30W
Asunción Nochixtlán, *Mexico* **87 D5** 17 28N 97 14W
Aswa →, *Uganda* **54 B3** 3 43N 31 55 E
Aswân, *Egypt* **51 D11** 24 4N 32 57 E
Aswân High Dam = Sadd el Aali, *Egypt* **51 D11** 23 54N 32 54 E
Asyût, *Egypt* **51 C11** 27 11N 31 4 E
At Ţafilah, *Jordan* **47 E4** 30 45N 35 30 E
At Tā'if, *Si. Arabia* **46 C3** 21 5N 40 27 E
Aţ Ţirāq, *Si. Arabia* **44 E5** 27 19N 44 33 E
Atacama □, *Chile* **94 B2** 27 30S 70 0W
Atacama, Desierto de, *Chile* **94 A2** 24 0S 69 20W
Atacama, Salar de, *Chile* **94 A2** 23 30S 68 20W
Atakpamé, *Togo* **50 G5** 7 31N 1 13 E
Atalaya, *Peru* **92 F4** 10 45S 73 50W
Atalaya de Femes, *Canary Is.* **22 F6** 28 56N 13 47W
Atami, *Japan* **31 G9** 35 5N 139 4 E
Atapupu, *Indonesia* **37 F6** 9 0S 124 51 E
Atâr, *Mauritania* **50 D2** 20 30N 13 5W
Atascadero, *U.S.A.* **83 J3** 35 29N 120 40W
Atasu, *Kazakstan* **26 E8** 48 30N 71 0 E
Atauro, *Indonesia* **37 F7** 8 10S 125 30 E
Atbara, *Sudan* **51 E11** 17 42N 33 59 E
'Atbara →, *Sudan* **51 E11** 17 40N 33 56 E
Atbasar, *Kazakstan* **26 D7** 51 48N 68 20 E
Atchafalaya B., *U.S.A.* **81 L9** 29 25N 91 25W
Atchison, *U.S.A.* **80 F7** 39 34N 95 7W
Ath, *Belgium* **15 D3** 50 38N 3 47 E
Athabasca, *Canada* **72 C6** 54 45N 113 20W
Athabasca →, *Canada* **73 B6** 58 40N 110 50W
Athabasca, L., *Canada* **73 B7** 59 15N 109 15W
Athboy, *Ireland* **13 C5** 53 37N 6 56W
Athenry, *Ireland* **13 C3** 53 18N 8 44W
Athens = Athínai, *Greece* **21 F10** 37 58N 23 46 E
Athens, *Ala., U.S.A.* **77 H2** 34 48N 86 58W
Athens, *Ga., U.S.A.* **77 J4** 33 57N 83 23W
Athens, *N.Y., U.S.A.* **79 D11** 42 16N 73 49W
Athens, *Ohio, U.S.A.* **76 F4** 39 20N 82 6W
Athens, *Pa., U.S.A.* **79 E8** 41 57N 76 31W
Athens, *Tenn., U.S.A.* **77 H3** 35 27N 84 36W
Athens, *Tex., U.S.A.* **81 J7** 32 12N 95 51W
Atherley, *Canada* **78 B5** 44 37N 79 20W
Atherton, *Australia* **62 B4** 17 17S 145 30 E
Athienou, *Cyprus* **23 D12** 35 3N 33 32 E
Athínai, *Greece* **21 F10** 37 58N 23 46 E
Athlone, *Ireland* **13 C4** 53 25N 7 56W
Athna, *Cyprus* **23 D12** 35 3N 33 47 E
Atholl, Forest of, *U.K.* **12 E5** 56 51N 3 50W
Atholville, *Canada* **71 C6** 47 59N 66 43W
Áthos, *Greece* **21 D11** 40 9N 24 22 E
Athy, *Ireland* **13 D5** 53 0N 7 0W
Ati, *Chad* **51 F8** 13 13N 18 20 E
Atiak, *Uganda* **54 B3** 3 12N 32 2 E
Atico, *Peru* **92 G4** 16 14S 73 40W
Atikokan, *Canada* **70 C1** 48 45N 91 37W
Atikonak L., *Canada* **71 B7** 52 40N 64 32W
Atka, *Russia* **27 C16** 60 50N 151 48 E
Atkinson, *U.S.A.* **80 D5** 42 32N 98 59W
Atlanta, *Ga., U.S.A.* **77 J3** 33 45N 84 23W
Atlanta, *Tex., U.S.A.* **81 J7** 33 7N 94 10W
Atlantic, *U.S.A.* **80 E7** 41 24N 95 1W
Atlantic City, *U.S.A.* **76 F8** 39 21N 74 27W
Atlantic Ocean **2 E9** 0 0 20 0W
Atlas Mts. = Haut Atlas, *Morocco* **50 B3** 32 30N 5 0W
Atlin, *Canada* **72 B2** 59 31N 133 41W
Atlin, L., *Canada* **72 B2** 59 26N 133 45W
Atmore, *U.S.A.* **77 K2** 31 2N 87 29W
Atoka, *U.S.A.* **81 H6** 34 23N 96 8W
Atolia, *U.S.A.* **85 K9** 35 19N 117 37W
Atoyac →, *Mexico* **87 D5** 16 30N 97 31W
Atrak = Atrek →, *Turkmenistan* **45 B8** 37 35N 53 58 E
Atrauli, *India* **42 E8** 28 2N 78 20 E
Atrek →, *Turkmenistan* **45 B8** 37 35N 53 58 E
Atsuta, *Japan* **30 C10** 43 24N 141 26 E
Attalla, *U.S.A.* **77 H2** 34 1N 86 6W
Attáviros, *Greece* **23 C9** 36 12N 27 50 E
Attawapiskat, *Canada* **70 B3** 52 56N 82 24W
Attawapiskat →, *Canada* **70 B3** 52 57N 82 18W
Attawapiskat, L., *Canada* **70 B2** 52 18N 87 54W
Attica, *U.S.A.* **76 E2** 40 18N 87 15W
Attikamagen L., *Canada* **71 A6** 55 0N 66 30W
Attleboro, *U.S.A.* **79 E13** 41 57N 71 17W
Attock, *Pakistan* **42 C5** 33 52N 72 20 E
Attopeu, *Laos* **38 E6** 14 48N 106 50 E
Attur, *India* **40 P11** 11 35N 78 30 E
Atuel →, *Argentina* **94 D2** 36 17S 66 50W
Åtvidaberg, *Sweden* **9 G17** 58 12N 16 0 E
Atwater, *U.S.A.* **83 H3** 37 21N 120 37W
Atwood, *Canada* **78 C3** 43 40N 81 1W
Atwood, *U.S.A.* **80 F4** 39 48N 101 3W
Atyraū, *Kazakstan* **25 E9** 47 5N 52 0 E
Au Sable →, *U.S.A.* **76 C4** 44 25N 83 20W
Au Sable Pt., *U.S.A.* **70 C2** 46 40N 86 10W
Aubagne, *France* **18 E6** 43 17N 5 37 E
Aubarca, C., *Spain* **22 B7** 39 4N 1 22 E
Aube →, *France* **18 B5** 48 34N 3 43 E
Auberry, *U.S.A.* **84 H7** 37 7N 119 29W
Auburn, *Ala., U.S.A.* **77 J3** 32 36N 85 29W
Auburn, *Calif., U.S.A.* **84 G5** 38 54N 121 4W
Auburn, *Ind., U.S.A.* **76 E3** 41 22N 85 4W
Auburn, *N.Y., U.S.A.* **79 D8** 42 56N 76 34W
Auburn, *Nebr., U.S.A.* **80 E7** 40 23N 95 51W
Auburn, *Wash., U.S.A.* **84 C4** 47 18N 122 14W
Auburn Ra., *Australia* **63 D5** 25 15S 150 30 E
Auburndale, *U.S.A.* **77 L5** 28 4N 81 48W
Aubusson, *France* **18 D5** 45 57N 2 11 E
Auch, *France* **18 E4** 43 39N 0 36 E
Auckland, *N.Z.* **59 G5** 36 52S 174 46 E
Auckland Is., *Pac. Oc.* **64 N8** 50 40S 166 5 E
Aude →, *France* **18 E5** 43 13N 3 14 E
Auden, *Canada* **70 B2** 50 14N 87 53W
Audubon, *U.S.A.* **80 E7** 41 43N 94 56W
Augathella, *Australia* **63 D4** 25 48S 146 35 E
Augrabies Falls, *S. Africa* **56 D3** 28 35S 20 20 E
Augsburg, *Germany* **16 D6** 48 25N 10 52 E
Augusta, *Italy* **20 F6** 37 13N 15 13 E
Augusta, *Ark., U.S.A.* **81 H9** 35 17N 91 22W
Augusta, *Ga., U.S.A.* **77 J5** 33 28N 81 58W
Augusta, *Kans., U.S.A.* **81 G6** 37 41N 96 59W
Augusta, *Maine, U.S.A.* **71 D6** 44 19N 69 47W
Augusta, *Mont., U.S.A.* **82 C7** 47 30N 112 24W
Augusta, *Wis., U.S.A.* **80 C9** 44 41N 91 7W
Augustów, *Poland* **17 B12** 53 51N 23 0 E
Augustus, Mt., *Australia* **60 D2** 24 20S 116 50 E
Augustus Downs, *Australia* **62 B2** 18 35S 139 55 E
Augustus I., *Australia* **60 C3** 15 20S 124 30 E
Aukum, *U.S.A.* **84 G6** 38 34N 120 43W
Auld, L., *Australia* **60 D3** 22 25S 123 50 E
Ault, *U.S.A.* **80 E2** 40 35N 104 44W
Aunis, *France* **18 C3** 46 5N 0 50W
Auponhia, *Indonesia* **37 E7** 1 58S 125 27 E
Aur, P., *Malaysia* **39 L5** 2 35N 104 10 E
Auraiya, *India* **43 F8** 26 28N 79 33 E
Aurangabad, *Bihar, India* **43 G11** 24 45N 84 18 E
Aurangabad, *Maharashtra, India* **40 K9** 19 50N 75 23 E
Aurich, *Germany* **16 B4** 53 28N 7 28 E
Aurillac, *France* **18 D5** 44 55N 2 26 E
Aurora, *Canada* **78 C5** 44 0N 79 28W
Aurora, *S. Africa* **56 E2** 32 40S 18 29 E
Aurora, *Colo., U.S.A.* **80 F2** 39 44N 104 52W
Aurora, *Ill., U.S.A.* **76 E1** 41 45N 88 19W
Aurora, *Mo., U.S.A.* **81 G8** 36 58N 93 43W
Aurora, *Nebr., U.S.A.* **80 E6** 40 52N 98 0W
Aurora, *Ohio, U.S.A.* **78 E3** 41 21N 81 20W
Aurukun Mission, *Australia* **62 A3** 13 20S 141 45 E
Aus, *Namibia* **56 D2** 26 35S 16 12 E
Auschwitz = Oświęcim, *Poland* **17 C10** 50 2N 19 11 E
Austin, *Minn., U.S.A.* **80 D8** 43 40N 92 58W
Austin, *Nev., U.S.A.* **82 G5** 39 30N 117 4W
Austin, *Pa., U.S.A.* **78 E6** 41 38N 78 6W
Austin, *Tex., U.S.A.* **81 K6** 30 17N 97 45W
Austin, L., *Australia* **61 E2** 27 40S 118 0 E
Austra, *Norway* **8 D14** 65 8N 11 55 E
Austral Downs, *Australia* **62 C2** 20 30S 137 45 E
Austral Is. = Tubuai Is., *Pac. Oc.* **65 K12** 25 0S 150 0W
Austral Seamount Chain, *Pac. Oc.* **65 K13** 24 0S 150 0W
Australia ■, *Oceania* **64 K5** 23 0S 135 0 E
Australian Alps, *Australia* **63 F4** 36 30S 148 30 E
Australian Capital Territory □, *Australia* **63 F4** 35 30S 149 0 E
Austria ■, *Europe* **16 E8** 47 0N 14 0 E
Austvågøy, *Norway* **8 B16** 68 20N 14 40 E
Autlán, *Mexico* **86 D4** 19 40N 104 30W
Autun, *France* **18 C6** 46 58N 4 17 E
Auvergne, *Australia* **60 C5** 15 39S 130 1 E
Auvergne, *France* **18 D5** 45 20N 3 15 E
Auvergne, Mts. d', *France* **18 D5** 45 20N 2 55 E
Auxerre, *France* **18 C5** 47 48N 3 32 E
Avallon, *France* **18 C5** 47 30N 3 53 E
Avalon, *U.S.A.* **85 M8** 33 21N 118 20W
Avalon Pen., *Canada* **71 C9** 47 30N 53 20W
Avaré, *Brazil* **95 A6** 23 4S 48 58W
Avawatz Mts., *U.S.A.* **85 K10** 35 40N 116 30W
Aveiro, *Brazil* **93 D7** 3 10S 55 5W
Aveiro, *Portugal* **19 B1** 40 37N 8 38W
Āvej, *Iran* **45 C6** 35 40N 49 15 E
Avellaneda, *Argentina* **94 C4** 34 50S 58 10W
Avellino, *Italy* **20 D6** 40 54N 14 47 E
Avenal, *U.S.A.* **84 K6** 36 0N 120 8W
Aversa, *Italy* **20 D6** 40 58N 14 12 E
Avery, *U.S.A.* **82 C6** 47 15N 115 49W
Aves, I. de, *W. Indies* **89 C7** 15 45N 63 55W
Aves, Is. de, *Venezuela* **89 D6** 12 0N 67 30W
Avesta, *Sweden* **9 F17** 60 9N 16 10 E
Avezzano, *Italy* **20 C5** 42 2N 13 25 E
Aviá Terai, *Argentina* **94 B3** 26 45S 60 50W
Avignon, *France* **18 E6** 43 57N 4 50 E
Ávila, *Spain* **19 B3** 40 39N 4 43W
Avila Beach, *U.S.A.* **85 K6** 35 11N 120 44W
Avilés, *Spain* **19 A3** 43 35N 5 57W
Avoca, *Ireland* **13 D5** 52 51N 6 13W
Avoca, *U.S.A.* **78 D7** 42 25N 77 25W
Avoca →, *Australia* **63 F3** 35 40S 143 43 E
Avola, *Canada* **72 C5** 51 45N 119 19W
Avola, *Italy* **20 F6** 36 56N 15 7 E
Avon, *N.Y., U.S.A.* **78 D7** 42 55N 77 45W
Avon, *S. Dak., U.S.A.* **80 D5** 43 0N 98 4W
Avon →, *Australia* **61 F2** 31 40S 116 7 E
Avon →, *Bristol, U.K.* **11 F5** 51 29N 2 41W
Avon →, *Dorset, U.K.* **11 G6** 50 44N 1 46W
Avon →, *Warks., U.K.* **11 F5** 52 10N 1 51W
Avondale, *Zimbabwe* **55 F3** 17 43S 30 58 E
Avonlea, *Canada* **73 D7** 50 0N 105 0W
Avonmore, *Canada* **79 A10** 45 10N 74 58W
Avonmouth, *U.K.* **11 F5** 51 30N 2 42W
Avranches, *France* **18 B3** 48 40N 1 20W
A'waj →, *Syria* **47 B5** 33 23N 36 20 E
Awaji-Shima, *Japan* **31 G7** 34 30N 134 50 E
'Awālī, *Bahrain* **45 E6** 26 0N 50 30 E
Awantipur, *India* **43 C6** 33 55N 75 3 E
Awash, *Ethiopia* **46 F3** 9 1N 40 10 E
Awatere →, *N.Z.* **59 J5** 41 37S 174 10 E
Awbārī, *Libya* **51 C7** 26 46N 12 57 E
Awe, L., *U.K.* **12 E3** 56 17N 5 16W
Awjilah, *Libya* **51 C9** 29 8N 21 7 E
Axel Heiberg I., *Canada* **4 B3** 80 0N 90 0W
Axim, *Ghana* **50 H4** 4 51N 2 15W
Axiós →, *Greece* **21 D10** 40 57N 22 35 E
Axminster, *U.K.* **11 G4** 50 46N 3 0W
Ayabaca, *Peru* **92 D3** 4 40S 79 53W
Ayabe, *Japan* **31 G7** 35 20N 135 20 E
Ayacucho, *Argentina* **94 D4** 37 5S 58 20W
Ayacucho, *Peru* **92 F4** 13 0S 74 0W
Ayaguz, *Kazakstan* **26 E9** 48 10N 80 10 E
Ayamonte, *Spain* **19 D2** 37 12N 7 24W
Ayan, *Russia* **27 D14** 56 30N 138 16 E
Ayaviri, *Peru* **92 F4** 14 50S 70 35W
Aydın, *Turkey* **21 F12** 37 51N 27 51 E
Ayer's Cliff, *Canada* **79 A12** 45 10N 72 3W
Ayers Rock, *Australia* **61 E5** 25 23S 131 5 E
Ayia Aikateríni, Ákra, *Greece* **23 A3** 39 50N 19 50 E
Ayia Dhéka, *Greece* **23 D6** 35 3N 24 58 E
Ayia Gálini, *Greece* **23 D6** 35 6N 24 41 E
Ayia Napa, *Cyprus* **23 E13** 34 59N 34 0 E
Ayia Phyla, *Cyprus* **23 E12** 34 43N 33 1 E
Ayia Varvára, *Greece* **23 D7** 35 8N 25 1 E
Áyios Amvrósios, *Cyprus* **23 D12** 35 20N 33 35 E
Áyios Evstrátios, *Greece* **21 E11** 39 34N 24 58 E
Áyios Ioánnis, Ákra, *Greece* **23 D7** 35 20N 25 40 E
Áyios Isidhoros, *Greece* **23 C9** 36 9N 27 51 E
Áyios Matthaios, *Greece* **23 B3** 39 30N 19 47 E
Áyios Nikólaos, *Greece* **23 D7** 35 11N 25 41 E
Áyios Seryios, *Cyprus* **23 D12** 35 12N 33 53 E
Ayios Theodhoros, *Cyprus* **23 D13** 35 22N 34 1 E
Aykino, *Russia* **24 B8** 62 15N 49 56 E
Aylesbury, *U.K.* **11 F7** 51 49N 0 49W
Aylmer, *Canada* **78 D4** 42 46N 80 59W
Aylmer, L., *Canada* **68 B8** 64 0N 110 8W
Ayolas, *Paraguay* **94 B4** 27 10S 56 59W
Ayon, Ostrov, *Russia* **27 C17** 69 50N 169 0 E
Ayr, *Australia* **62 B4** 19 35S 147 25 E
Ayr, *U.K.* **12 F4** 55 28N 4 38W
Ayr →, *U.K.* **12 F4** 55 28N 4 38W
Ayre, Pt. of, *U.K.* **10 C3** 54 25N 4 21W
Aytos, *Bulgaria* **21 C12** 42 42N 27 16 E
Ayu, Kepulauan, *Indonesia* **37 D8** 0 35N 131 5 E
Ayutla, *Guatemala* **88 D1** 14 40N 92 10W
Ayutla, *Mexico* **87 D5** 16 58N 99 17W
Ayvacık, *Turkey* **21 E12** 39 36N 26 24 E
Ayvalık, *Turkey* **21 E12** 39 20N 26 46 E
Az Zabdānī, *Syria* **47 B5** 33 43N 36 5 E
Az Ẓāhirīyah, *West Bank* **47 D3** 31 25N 34 58 E
Az Ẓahrān, *Si. Arabia* **45 E6** 26 10N 50 7 E
Az Zarqā, *Jordan* **47 C5** 32 5N 36 4 E
Az Zībār, *Iraq* **44 B5** 36 52N 44 4 E
Az-Zilfī, *Si. Arabia* **44 E5** 26 12N 44 52 E
Az Zubayr, *Iraq* **44 D5** 30 26N 47 40 E
Az Zuwaytīnah, *Libya* **51 B9** 30 58N 20 7 E
Azamgarh, *India* **43 F10** 26 5N 83 13 E
Āzār Shahr, *Iran* **44 B5** 37 45N 45 59 E
Azärbayjan = Azerbaijan ■, *Asia* **25 F8** 40 20N 48 0 E
Āzarbāyjān-e Gharbī □, *Iran* **44 B5** 37 0N 44 30 E
Āzarbāyjān-e Sharqī □, *Iran* **44 B5** 37 20N 47 0 E
Azare, *Nigeria* **50 F7** 11 55N 10 10 E
A'zāz, *Syria* **44 B3** 36 36N 37 4 E
Azbine = Aïr, *Niger* **50 E6** 18 30N 8 0 E
Azerbaijan ■, *Asia* **25 F8** 40 20N 48 0 E
Azerbaijchan = Azerbaijan ■, *Asia* **25 F8** 40 20N 48 0 E
Azimganj, *India* **43 G13** 24 14N 88 16 E
Azogues, *Ecuador* **92 D3** 2 35S 78 0W
Azores, *Atl. Oc.* **48 C1** 38 44N 29 0W
Azov, *Russia* **25 E6** 47 3N 39 25 E
Azov, Sea of, *Europe* **25 E6** 46 0N 36 30 E
Azovskoye More = Azov, Sea of, *Europe* **25 E6** 46 0N 36 30 E
Azovy, *Russia* **26 C7** 64 55N 65 1 E
Aztec, *U.S.A.* **83 H10** 36 49N 107 59W
Azúa, *Dom. Rep.* **89 C5** 18 25N 70 44W
Azuaga, *Spain* **19 C3** 38 16N 5 39W
Azuero, Pen. de, *Panama* **88 E3** 7 30N 80 30W
Azul, *Argentina* **94 D4** 36 42S 59 43W
Azusa, *U.S.A.* **85 L9** 34 8N 117 52W
'Azzūn, *West Bank* **47 C4** 32 10N 35 2 E

B

Ba Don, *Vietnam* **38 D6** 17 45N 106 26 E
Ba Dong, *Vietnam* **39 H6** 9 40N 106 33 E
Ba Ngoi = Cam Lam, *Vietnam* **39 G7** 11 54N 109 10 E
Ba Ria, *Vietnam* **39 G6** 10 30N 107 10 E
Ba Tri, *Vietnam* **39 G6** 10 2N 106 36 E
Ba Xian, *China* **34 E9** 39 8N 116 22 E
Baa, *Indonesia* **37 F6** 10 50S 123 0 E
Baarle Nassau, *Belgium* **15 C4** 51 27N 4 56 E
Baarn, *Neths.* **15 B5** 52 12N 5 17 E
Bab el Mandeb, *Red Sea* **46 E3** 12 35N 43 25 E
Baba Burnu, *Turkey* **21 E12** 39 29N 26 2 E
Bābā Kalū, *Iran* **45 D6** 30 7N 50 49 E
Babadag, *Romania* **17 F15** 44 53N 28 44 E
Babadayhan, *Turkmenistan* **26 F7** 37 42N 60 23 E
Babaeski, *Turkey* **21 D12** 41 26N 27 6 E
Babahoyo, *Ecuador* **92 D3** 1 40S 79 30W
Babakin, *Australia* **61 F2** 32 7S 118 1 E
Babana, *Nigeria* **50 F5** 10 31N 3 46 E
Babar, *Indonesia* **37 F7** 8 0S 129 30 E
Babar, *Pakistan* **42 D3** 31 7N 69 32 E
Babarkach, *Pakistan* **42 E2** 29 45N 68 0 E
Babb, *U.S.A.* **82 B7** 48 51N 113 27W
Babi Besar, P., *Malaysia* **39 L4** 2 25N 103 59 E
Babinda, *Australia* **62 B4** 17 20S 145 56 E
Babine, *Canada* **72 B3** 55 22N 126 37W
Babine →, *Canada* **72 B3** 55 45N 127 44W
Babine L., *Canada* **72 C3** 54 48N 126 0W
Babo, *Indonesia* **37 E8** 2 30S 133 30 E
Bābol, *Iran* **45 B7** 36 40N 52 50 E
Bābol Sar, *Iran* **45 B7** 36 45N 52 45 E
Baboua, *C.A.R.* **52 C2** 5 49N 14 58 E
Babruysk, *Belarus* **17 B15** 53 10N 29 15 E
Babura, *Nigeria* **50 F6** 12 51N 8 59 E
Babusar Pass, *Pakistan* **43 B5** 35 12N 73 59 E
Babuyan Chan., *Phil.* **37 A6** 18 40N 121 30 E
Babylon, *Iraq* **44 C5** 32 34N 44 22 E
Bac Can, *Vietnam* **38 A5** 22 8N 105 49 E
Bac Giang, *Vietnam* **38 B6** 21 16N 106 11 E
Bac Ninh, *Vietnam* **38 B6** 21 13N 106 4 E
Bac Phan, *Vietnam* **38 B5** 22 0N 105 0 E
Bac Quang, *Vietnam* **38 A5** 22 30N 104 48 E
Bacabal, *Brazil* **93 D10** 4 15S 44 45W
Bacalar, *Mexico* **87 D7** 18 50N 87 27W
Bacan, Kepulauan, *Indonesia* **37 E7** 0 35S 127 30 E
Bacan, Pulau, *Indonesia* **37 E7** 0 50S 127 30 E
Bacarra, *Phil.* **37 A6** 18 15N 120 37 E
Bacău, *Romania* **17 E14** 46 35N 26 55 E
Bacerac, *Mexico* **86 A3** 30 18N 108 50W
Bach Long Vi, Dao, *Vietnam* **38 B6** 20 10N 107 40 E
Bachelina, *Russia* **26 D7** 57 45N 67 20 E
Back →, *Canada* **68 B9** 65 10N 104 0W
Backstairs Passage, *Australia* **63 F2** 35 40S 138 5 E
Bacolod, *Phil.* **37 B6** 10 40N 122 57 E
Bacuk, *Malaysia* **39 J4** 6 4N 102 25 E
Bād, *Iran* **45 C7** 33 41N 52 1 E
Bad →, *U.S.A.* **80 C4** 44 21N 100 22W
Bad Axe, *U.S.A.* **78 C2** 43 48N 83 0W
Bad Ischl, *Austria* **16 E7** 47 44N 13 38 E
Bad Kissingen, *Germany* **16 C6** 50 11N 10 4 E
Bad Lands, *U.S.A.* **80 D3** 43 40N 102 10W
Badagara, *India* **40 P9** 11 35N 75 40 E
Badajoz, *Spain* **19 C2** 38 50N 6 59W
Badalona, *Spain* **19 B7** 41 26N 2 15 E
Badalzai, *Afghan.* **42 E1** 29 50N 65 35 E
Badampahar, *India* **41 H15** 22 10N 86 10 E
Badanah, *Si. Arabia* **44 D4** 30 58N 41 30 E
Badarinath, *India* **43 D8** 30 45N 79 30 E
Badas, *Brunei* **36 D4** 4 33N 114 25 E
Badas, Kepulauan, *Indonesia* **36 D3** 0 45N 107 5 E
Baddo →, *Pakistan* **40 F4** 28 0N 64 20 E
Bade, *Indonesia* **37 F9** 7 10S 139 35 E
Baden, *Austria* **16 D9** 48 1N 16 13 E
Baden-Baden, *Germany* **16 D5** 48 44N 8 13 E
Baden-Württemberg □, *Germany* **16 D5** 48 20N 8 40 E
Badenoch, *U.K.* **12 E4** 56 59N 4 15W
Badgastein, *Austria* **16 E7** 47 7N 13 9 E
Badger, *Canada* **71 C8** 49 0N 56 4W
Badger, *U.S.A.* **84 J7** 36 38N 119 1W
Bādghīsāt □, *Afghan.* **40 B3** 35 0N 63 0 E
Badgom, *India* **43 B6** 34 1N 74 45 E
Badin, *Pakistan* **42 G3** 24 38N 68 54 E
Baduen, *Somali Rep.* **46 F4** 7 15N 47 40 E
Badulla, *Sri Lanka* **40 R12** 7 1N 81 7 E
Baena, *Spain* **19 D3** 37 37N 4 20W
Baeza, *Spain* **19 D4** 37 57N 3 25W
Bafatá, *Guinea-Biss.* **50 F2** 12 8N 14 40W
Baffin B., *Canada* **4 B4** 72 0N 64 0W
Baffin I., *Canada* **69 B12** 68 0N 75 0W
Bafia, *Cameroon* **50 H7** 4 40N 11 10 E
Bafing →, *Mali* **50 F2** 13 49N 10 50W
Bafliyūn, *Syria* **44 B3** 36 37N 36 59 E
Bafoulabé, *Mali* **50 F2** 13 50N 10 55W
Bāfq, *Iran* **45 D7** 31 40N 55 25 E
Bāft, *Iran* **45 D8** 29 15N 56 38 E
Bafwasende, *Zaïre* **54 B2** 1 3N 27 5 E
Bagamoyo, *Tanzania* **54 D4** 6 28S 38 55 E
Bagamoyo □, *Tanzania* **54 D4** 6 20S 38 30 E
Bagan Datoh, *Malaysia* **39 L3** 3 59N 100 47 E
Bagan Serai, *Malaysia* **39 K3** 5 1N 100 32 E
Baganga, *Phil.* **37 C7** 7 34N 126 33 E
Bagani, *Namibia* **56 B3** 18 7S 21 41 E
Bagansiapiapi, *Indonesia* **36 D2** 2 12N 100 50 E
Bagasra, *India* **42 J4** 21 30N 71 0 E
Bagdad, *U.S.A.* **85 L11** 34 35N 115 53W
Bagdarin, *Russia* **27 D12** 54 26N 113 36 E
Bagé, *Brazil* **95 C5** 31 20S 54 15W
Bagenalstown = Muine Bheag, *Ireland* **13 D5** 52 42N 6 58W
Baggs, *U.S.A.* **82 F10** 41 2N 107 39W
Bagh, *Pakistan* **43 C5** 33 59N 73 45 E
Baghdād, *Iraq* **44 C5** 33 20N 44 30 E
Bagheria, *Italy* **20 E5** 38 5N 13 30 E
Baghlān, *Afghan.* **40 A6** 36 12N 69 0 E
Bagley, *U.S.A.* **80 B7** 47 32N 95 24W
Bagotville, *Canada* **71 C5** 48 22N 70 54W
Bagrationovsk, *Russia* **9 J19** 54 23N 20 39 E
Baguio, *Phil.* **37 A6** 16 26N 120 34 E
Bahadurgarh, *India* **42 E7** 28 40N 76 57 E
Bahama, Canal Viejo de, *W. Indies* **88 B4** 22 10N 77 30W
Bahamas ■, *N. Amer.* **89 B5** 24 0N 75 0W
Baharampur, *India* **43 G13** 24 2N 88 27 E
Bahau, *Malaysia* **39 L4** 2 48N 102 26 E
Bahawalnagar, *Pakistan* **42 D5** 30 0N 73 15 E
Bahawalpur, *Pakistan* **42 E4** 29 24N 71 40 E
Baheri, *India* **43 E8** 28 45N 79 34 E
Bahi, *Tanzania* **54 D4** 5 58S 35 21 E
Bahi Swamp, *Tanzania* **54 D4** 6 10S 35 0 E
Bahía = Salvador, *Brazil* **93 F11** 13 0S 38 30W
Bahía □, *Brazil* **93 F10** 12 0S 42 0W
Bahía, Is. de la, *Honduras* **88 C2** 16 45N 86 15W

Bahía Blanca, *Argentina* **94 D3** 38 35S 62 13W
Bahía de Caráquez, *Ecuador* **92 D2** 0 40S 80 27W
Bahía Honda, *Cuba* **88 B3** 22 54N 83 10W
Bahía Laura, *Argentina* **96 F3** 48 10S 66 30W
Bahía Negra, *Paraguay* **92 H7** 20 5S 58 5W
Bahmanzād, *Iran* **45 D6** 31 15N 51 47 E
Bahr Aouk →, *C.A.R.* **52 C3** 8 40N 19 0 E
Bahr el Ghazâl □, *Sudan* **48 F6** 7 0N 28 0 E
Bahr Salamat →, *Chad* **51 G8** 9 20N 18 0 E
Bahraich, *India* **43 F9** 27 38N 81 37 E
Bahrain ■, *Asia* **45 E6** 26 0N 50 35 E
Bahror, *India* **42 F7** 27 51N 76 20 E
Bāhū Kalāt, *Iran* **45 E9** 25 43N 61 25 E
Bai Bung, Mui, *Vietnam* **39 H5** 8 38N 104 44 E
Bai Duc, *Vietnam* **38 C5** 18 3N 105 49 E
Bai Thuong, *Vietnam* **38 C5** 19 54N 105 23 E
Baia Mare, *Romania* **17 E12** 47 40N 23 35 E
Baïbokoum, *Chad* **51 G8** 7 46N 15 43 E
Baicheng, *China* **35 B12** 45 38N 122 42 E
Baidoa, *Somali Rep.* **46 G3** 3 8N 43 30 E
Baie Comeau, *Canada* **71 C6** 49 12N 68 10W
Baie-St-Paul, *Canada* **71 C5** 47 28N 70 32W
Baie Trinité, *Canada* **71 C6** 49 25N 67 20W
Baie Verte, *Canada* **71 C8** 49 55N 56 12W
Baihe, *China* **34 H6** 32 50N 110 5 E
Ba'ījī, *Iraq* **44 C4** 35 0N 43 30 E
Baikal, L. = Baykal, Oz., *Russia* **27 D11** 53 0N 108 0 E
Baile Atha Cliath = Dublin, *Ireland* **13 C5** 53 21N 6 15W
Băileşti, *Romania* **17 F12** 44 1N 23 20 E
Bailundo, *Angola* **53 G3** 12 10S 15 50 E
Bainbridge, *Ga., U.S.A.* **77 K3** 30 55N 84 35W
Bainbridge, *N.Y., U.S.A.* **79 D9** 42 18N 75 29W
Baing, *Indonesia* **37 F6** 10 14S 120 34 E
Bainiu, *China* **34 H7** 32 50N 112 15 E
Bainville, *U.S.A.* **80 A2** 48 8N 104 13W
Bā'ir, *Jordan* **47 E5** 30 45N 36 55 E
Baird, *U.S.A.* **81 J5** 32 24N 99 24W
Baird Mts., *U.S.A.* **68 B3** 67 0N 160 0W
Bairin Youqi, *China* **35 C10** 43 30N 118 35 E
Bairin Zuoqi, *China* **35 C10** 43 58N 119 15 E
Bairnsdale, *Australia* **63 F4** 37 48S 147 36 E
Baisha, *China* **34 G7** 34 20N 112 32 E
Baitadi, *Nepal* **43 E9** 29 35N 80 25 E
Baiyin, *China* **34 F3** 36 45N 104 14 E
Baiyu Shan, *China* **34 F4** 37 15N 107 30 E
Baj Baj, *India* **43 H13** 22 30N 88 5 E
Baja, *Hungary* **17 E10** 46 12N 18 59 E
Baja, Pta., *Mexico* **86 B1** 29 50N 116 0W
Baja California, *Mexico* **86 A1** 31 10N 115 12W
Baja California □, *Mexico* **86 B2** 30 0N 115 0W
Baja California Sur □, *Mexico* **86 B2** 25 50N 111 50W
Bajamar, *Canary Is.* **22 F3** 28 33N 16 20W
Bajana, *India* **42 H4** 23 7N 71 49 E
Bājgīrān, *Iran* **45 B8** 37 36N 58 24 E
Bajimba, Mt., *Australia* **63 D5** 29 17S 152 6 E
Bajo Nuevo, *Caribbean* **88 C4** 15 40N 78 50W
Bajool, *Australia* **62 C5** 23 40S 150 35 E
Bakala, *C.A.R.* **51 G9** 6 15N 20 20 E
Bakchar, *Russia* **26 D9** 57 1N 82 5 E
Bakel, *Senegal* **50 F2** 14 56N 12 20W
Baker, *Calif., U.S.A.* **85 K10** 35 16N 116 4W
Baker, *Mont., U.S.A.* **80 B2** 46 22N 104 17W
Baker, *Oreg., U.S.A.* **82 D5** 44 47N 117 50W
Baker, L., *Canada* **68 B10** 64 0N 96 0W
Baker I., *Pac. Oc.* **64 G10** 0 10N 176 35W
Baker L., *Australia* **61 E4** 26 54S 126 5 E
Baker Lake, *Canada* **68 B10** 64 20N 96 3W
Baker Mt., *U.S.A.* **82 B3** 48 50N 121 49W
Bakers Creek, *Australia* **62 C4** 21 13S 149 7 E
Baker's Dozen Is., *Canada* **70 A4** 56 45N 78 45W
Bakersfield, *Calif., U.S.A.* **85 K7** 35 23N 119 1W
Bakersfield, *Vt., U.S.A.* **79 B12** 44 45N 72 48W
Bākhtarān, *Iran* **44 C5** 34 23N 47 0 E
Bākhtarān □, *Iran* **44 C5** 34 0N 46 30 E
Bakı, *Azerbaijan* **25 F8** 40 29N 49 56 E
Bakkafjörður, *Iceland* **8 C6** 66 2N 14 48W
Bakony Forest = Bakony Hegyseg, *Hungary* **17 E9** 47 10N 17 30 E
Bakony Hegyseg, *Hungary* **17 E9** 47 10N 17 30 E
Bakouma, *C.A.R.* **51 G9** 5 40N 22 56 E
Baku = Bakı, *Azerbaijan* **25 F8** 40 29N 49 56 E
Bakutis Coast, *Antarctica* **5 D15** 74 0S 120 0W
Baky = Bakı, *Azerbaijan* **25 F8** 40 29N 49 56 E
Bala, *Canada* **78 A5** 45 1N 79 37W
Bala, L., *U.K.* **10 E4** 52 53N 3 37W
Balabac I., *Phil.* **36 C5** 8 0N 117 0 E
Balabac Str., *E. Indies* **36 C5** 7 53N 117 5 E
Balabagh, *Afghan.* **42 B4** 34 25N 70 12 E
Ba'labakk, *Lebanon* **47 A5** 34 0N 36 10 E
Balabalangan, Kepulauan, *Indonesia* **36 E5** 2 20S 117 30 E
Balad, *Iraq* **44 C5** 34 1N 44 9 E
Balad Rūz, *Iraq* **44 C5** 33 42N 45 5 E
Bālādeh, *Fārs, Iran* **45 D6** 29 17N 51 56 E
Bālādeh, *Māzandaran, Iran* **45 B6** 36 12N 51 48 E
Balaghat, *India* **40 J12** 21 49N 80 12 E
Balaghat Ra., *India* **40 K10** 18 50N 76 30 E
Balaguer, *Spain* **19 B6** 41 50N 0 50 E
Balaklava, *Australia* **63 E2** 34 7S 138 22 E
Balaklava, *Ukraine* **25 F5** 44 30N 33 30 E
Balakovo, *Russia* **24 D8** 52 4N 47 55 E
Balancán, *Mexico* **87 D6** 17 48N 91 32W
Balashov, *Russia* **24 D7** 51 30N 43 10 E
Balasinor, *India* **42 H5** 22 57N 73 23 E
Balasore = Baleshwar, *India* **41 J15** 21 35N 87 3 E
Balaton, *Hungary* **17 E9** 46 50N 17 40 E
Balbina, Represa de, *Brazil* **92 D7** 2 0S 59 30W
Balboa, *Panama* **88 E4** 8 57N 79 34W
Balbriggan, *Ireland* **13 C5** 53 37N 6 11W
Balcarce, *Argentina* **94 D4** 38 0S 58 10W
Balcarres, *Canada* **73 C8** 50 50N 103 35W
Balchik, *Bulgaria* **21 C13** 43 28N 28 11 E
Balclutha, *N.Z.* **59 M2** 46 15S 169 45 E
Bald Hd., *Australia* **61 G2** 35 6S 118 1 E
Bald I., *Australia* **61 F2** 34 57S 118 27 E
Bald Knob, *U.S.A.* **81 H9** 35 19N 91 34W
Baldock L., *Canada* **73 B9** 56 33N 97 57W
Baldwin, *Fla., U.S.A.* **77 K4** 30 18N 81 59W
Baldwin, *Mich., U.S.A.* **76 D3** 43 54N 85 51W
Baldwinsville, *U.S.A.* **79 C8** 43 10N 76 20W
Baldy Peak, *U.S.A.* **83 K9** 33 54N 109 34W
Baleares, Is., *Spain* **22 B10** 39 30N 3 0 E
Balearic Is. = Baleares, Is., *Spain* **22 B10** 39 30N 3 0 E
Baler, *Phil.* **37 A6** 15 46N 121 34 E
Baleshwar, *India* **41 J15** 21 35N 87 3 E
Balfate, *Honduras* **88 C2** 15 48N 86 25W
Balfe's Creek, *Australia* **62 C4** 20 12S 145 55 E
Bali, *Cameroon* **50 G7** 5 54N 10 0 E
Balí, *Greece* **23 D6** 35 25N 24 47 E
Bali, *Indonesia* **36 F5** 8 20S 115 0 E
Bali □, *Indonesia* **36 F5** 8 20S 115 0 E
Bali, Selat, *Indonesia* **37 H16** 8 18S 114 25 E
Balikeşir, *Turkey* **21 E12** 39 35N 27 58 E
Balikpapan, *Indonesia* **36 E5** 1 10S 116 55 E
Balimbing, *Phil.* **37 C5** 5 5N 119 58 E
Baling, *Malaysia* **39 K3** 5 41N 100 55 E
Balipara, *India* **41 F18** 26 50N 92 45 E
Baliza, *Brazil* **93 G8** 16 0S 52 20W
Balkan Mts. = Stara Planina, *Bulgaria* **21 C10** 43 15N 23 0 E
Balkhash = Balqash, *Kazakstan* **26 E8** 46 50N 74 50 E
Balkhash, Ozero = Balqash Köl, *Kazakstan* **26 E8** 46 0N 74 50 E
Balla, *Bangla.* **41 G17** 24 10N 91 35 E
Ballachulish, *U.K.* **12 E3** 56 41N 5 8W
Balladonia, *Australia* **61 F3** 32 27S 123 51 E
Ballarat, *Australia* **63 F3** 37 33S 143 50 E
Ballard, L., *Australia* **61 E3** 29 20S 120 40 E
Ballater, *U.K.* **12 D5** 57 3N 3 3W
Ballenas, Canal de, *Mexico* **86 B2** 29 10N 113 45W
Balleny Is., *Antarctica* **5 C11** 66 30S 163 0 E
Ballia, *India* **43 G11** 25 46N 84 12 E
Ballidu, *Australia* **61 F2** 30 35S 116 45 E
Ballina, *Australia* **63 D5** 28 50S 153 31 E
Ballina, *Mayo, Ireland* **13 B2** 54 7N 9 9W
Ballina, *Tipp., Ireland* **13 D3** 52 49N 8 26W
Ballinasloe, *Ireland* **13 C3** 53 20N 8 13W
Ballinger, *U.S.A.* **81 K5** 31 45N 99 57W
Ballinrobe, *Ireland* **13 C2** 53 38N 9 13W
Ballinskelligs B., *Ireland* **13 E1** 51 48N 10 13W
Ballycastle, *U.K.* **13 A5** 55 12N 6 15W
Ballymena, *U.K.* **13 B5** 54 52N 6 17W
Ballymena □, *U.K.* **13 B5** 54 53N 6 18W
Ballymoney, *U.K.* **13 A5** 55 5N 6 31W
Ballymoney □, *U.K.* **13 A5** 55 5N 6 23W
Ballyshannon, *Ireland* **13 B3** 54 30N 8 11W
Balmaceda, *Chile* **96 F2** 46 0S 71 50W
Balmoral, *Australia* **63 F3** 37 15S 141 48 E
Balmoral, *U.K.* **12 D5** 57 3N 3 13W
Balmorhea, *U.S.A.* **81 K3** 30 59N 103 45W
Balonne →, *Australia* **63 D4** 28 47S 147 56 E
Balqash, *Kazakstan* **26 E8** 46 50N 74 50 E
Balqash Köl, *Kazakstan* **26 E8** 46 0N 74 50 E
Balrampur, *India* **43 F10** 27 30N 82 20 E
Balranald, *Australia* **63 E3** 34 38S 143 33 E
Balsas, *Mexico* **87 D5** 18 0N 99 40W
Balsas →, *Mexico* **86 D4** 17 55N 102 10W
Balston Spa, *U.S.A.* **79 D11** 43 0N 73 52W
Balta, *Ukraine* **17 D15** 48 2N 29 45 E
Balta, *U.S.A.* **80 A4** 48 10N 100 2W
Bălţi, *Moldova* **17 E14** 47 48N 28 0 E
Baltic Sea, *Europe* **9 H18** 57 0N 19 0 E
Baltimore, *Ireland* **13 E2** 51 29N 9 22W
Baltimore, *U.S.A.* **76 F7** 39 17N 76 37W
Baltit, *Pakistan* **43 A6** 36 15N 74 40 E
Baltiysk, *Russia* **9 J18** 54 41N 19 58 E
Baluchistan □, *Pakistan* **40 F4** 27 30N 65 0 E
Balurghat, *India* **43 G13** 25 15N 88 44 E
Balvi, *Latvia* **9 H22** 57 8N 27 15 E
Balya, *Turkey* **21 E12** 39 44N 27 35 E
Balygychan, *Russia* **27 C16** 63 56N 154 12 E
Bam, *Iran* **45 D8** 29 7N 58 14 E
Bama, *Nigeria* **51 F7** 11 33N 13 41 E
Bamako, *Mali* **50 F3** 12 34N 7 55W
Bamba, *Mali* **50 E4** 17 5N 1 24W
Bambari, *C.A.R.* **51 G9** 5 40N 20 35 E
Bambaroo, *Australia* **62 B4** 18 50S 146 10 E
Bamberg, *Germany* **16 D6** 49 54N 10 54 E
Bamberg, *U.S.A.* **77 J5** 33 18N 81 2W
Bambili, *Zaïre* **54 B2** 3 40N 26 0 E
Bamenda, *Cameroon* **50 G7** 5 57N 10 11 E
Bamfield, *Canada* **72 D3** 48 45N 125 10W
Bāmīān □, *Afghan.* **40 B5** 35 0N 67 0 E
Bamiancheng, *China* **35 C13** 43 15N 124 2 E
Bampūr, *Iran* **45 E9** 27 15N 60 21 E
Ban Aranyaprathet, *Thailand* **38 F4** 13 41N 102 30 E
Ban Ban, *Laos* **38 C4** 19 31N 103 30 E
Ban Bang Hin, *Thailand* **39 H2** 9 32N 98 35 E
Ban Chiang Klang, *Thailand* **38 C3** 19 25N 100 55 E
Ban Chik, *Laos* **38 D4** 17 15N 102 22 E
Ban Choho, *Thailand* **38 E4** 15 2N 102 9 E
Ban Dan Lan Hoi, *Thailand* **38 D2** 17 0N 99 35 E
Ban Don = Surat Thani, *Thailand* **39 H2** 9 6N 99 20 E
Ban Don, *Vietnam* **38 F6** 12 53N 107 48 E
Ban Don, Ao, *Thailand* **39 H2** 9 20N 99 25 E
Ban Dong, *Thailand* **38 C3** 19 30N 100 59 E
Ban Hong, *Thailand* **38 C2** 18 18N 98 50 E
Ban Kaeng, *Thailand* **38 D3** 17 29N 100 7 E
Ban Keun, *Laos* **38 C4** 18 22N 102 35 E
Ban Khai, *Thailand* **38 F3** 12 46N 101 18 E
Ban Kheun, *Laos* **38 B3** 20 13N 101 7 E
Ban Khlong Kua, *Thailand* **39 J3** 6 57N 100 8 E
Ban Khuan Mao, *Thailand* **39 J2** 7 50N 99 37 E
Ban Khun Yuam, *Thailand* **38 C1** 18 49N 97 57 E
Ban Ko Yai Chim, *Thailand* **39 G2** 11 17N 99 26 E
Ban Kok, *Thailand* **38 D4** 16 40N 103 40 E
Ban Laem, *Thailand* **38 F2** 13 13N 99 59 E
Ban Lao Ngam, *Laos* **38 E6** 15 28N 106 10 E
Ban Le Kathe, *Thailand* **38 E2** 15 49N 98 53 E
Ban Mae Chedi, *Thailand* **38 C2** 19 11N 99 31 E
Ban Mae Laeng, *Thailand* **38 B2** 20 1N 99 17 E
Ban Mae Sariang, *Thailand* **38 C1** 18 10N 97 56 E
Ban Mê Thuôt = Buon Me Thuot, *Vietnam* **38 F7** 12 40N 108 3 E
Ban Mi, *Thailand* **38 E3** 15 3N 100 32 E
Ban Muong Mo, *Laos* **38 C4** 19 4N 103 58 E
Ban Na Mo, *Laos* **38 D5** 17 7N 105 40 E
Ban Na San, *Thailand* **39 H2** 8 53N 99 52 E
Ban Na Tong, *Laos* **38 B3** 20 56N 101 47 E
Ban Nam Bac, *Laos* **38 B4** 20 38N 102 20 E
Ban Nam Ma, *Laos* **38 A3** 22 2N 101 37 E
Ban Ngang, *Laos* **38 E6** 15 59N 106 11 E
Ban Nong Bok, *Laos* **38 D5** 17 5N 104 48 E
Ban Nong Boua, *Laos* **38 E6** 15 40N 106 33 E
Ban Nong Pling, *Thailand* **38 E3** 15 40N 100 10 E
Ban Pak Chan, *Thailand* **39 G2** 10 32N 98 51 E
Ban Phai, *Thailand* **38 D4** 16 4N 102 44 E
Ban Pong, *Thailand* **38 F2** 13 50N 99 55 E
Ban Ron Phibun, *Thailand* **39 H2** 8 9N 99 51 E
Ban Sanam Chai, *Thailand* **39 J3** 7 33N 100 25 E
Ban Sangkha, *Thailand* **38 E4** 14 37N 103 52 E
Ban Tak, *Thailand* **38 D2** 17 2N 99 4 E
Ban Tako, *Thailand* **38 E4** 14 5N 102 40 E
Ban Tha Dua, *Thailand* **38 D2** 17 59N 98 39 E
Ban Tha Li, *Thailand* **38 D3** 17 37N 101 25 E
Ban Tha Nun, *Thailand* **39 H2** 8 12N 98 18 E
Ban Thahine, *Laos* **38 E5** 14 12N 105 33 E
Ban Xien Kok, *Laos* **38 B3** 20 54N 100 39 E
Ban Yen Nhan, *Vietnam* **38 B6** 20 57N 106 2 E
Banaba, *Kiribati* **64 H8** 0 45S 169 50 E
Bañalbufar, *Spain* **22 B9** 39 42N 2 31 E
Banalia, *Zaïre* **54 B2** 1 32N 25 5 E
Banam, *Cambodia* **39 G5** 11 20N 105 17 E
Banamba, *Mali* **50 F3** 13 29N 7 22W
Banana, *Australia* **62 C5** 24 28S 150 8 E
Bananal, I. do, *Brazil* **93 F8** 11 30S 50 30W
Banaras = Varanasi, *India* **43 G10** 25 22N 83 0 E
Banas →, *Gujarat, India* **42 H4** 23 45N 71 25 E
Banas →, *Mad. P., India* **43 G9** 24 15N 81 30 E
Banbān, *Si. Arabia* **44 E5** 25 1N 46 35 E
Banbridge, *U.K.* **13 B5** 54 22N 6 16W
Banbridge □, *U.K.* **13 B5** 54 21N 6 16W
Banbury, *U.K.* **11 E6** 52 4N 1 20W
Banchory, *U.K.* **12 D6** 57 3N 2 29W
Bancroft, *Canada* **70 C4** 45 3N 77 51W
Band Bonī, *Iran* **45 E8** 25 30N 59 33 E
Band Qīr, *Iran* **45 D6** 31 39N 48 53 E
Banda, *India* **43 G9** 25 30N 80 26 E
Banda, Kepulauan, *Indonesia* **37 E7** 4 37S 129 50 E
Banda Aceh, *Indonesia* **36 C1** 5 35N 95 20 E
Banda Banda, Mt., *Australia* **63 E5** 31 10S 152 28 E
Banda Elat, *Indonesia* **37 F8** 5 40S 133 5 E
Banda Is. = Banda, Kepulauan, *Indonesia* **37 E7** 4 37S 129 50 E
Banda Sea, *Indonesia* **37 F7** 6 0S 130 0 E
Bandai-San, *Japan* **30 F10** 37 36N 140 4 E
Bandān, *Iran* **45 D9** 31 23N 60 44 E
Bandanaira, *Indonesia* **37 E7** 4 32S 129 54 E
Bandanwara, *India* **42 F6** 26 9N 74 38 E
Bandar = Machilipatnam, *India* **41 L12** 16 12N 81 8 E
Bandār 'Abbās, *Iran* **45 E8** 27 15N 56 15 E
Bandar-e Anzalī, *Iran* **45 B6** 37 30N 49 30 E
Bandar-e Bushehr = Būshehr, *Iran* **45 D6** 28 55N 50 55 E
Bandar-e Chārak, *Iran* **45 E7** 26 45N 54 20 E
Bandar-e Deylam, *Iran* **45 D6** 30 5N 50 10 E
Bandar-e Khomeyni, *Iran* **45 D6** 30 30N 49 5 E
Bandar-e Lengeh, *Iran* **45 E7** 26 35N 54 58 E
Bandar-e Maqām, *Iran* **45 E7** 26 56N 53 29 E
Bandar-e Ma'shur, *Iran* **45 D6** 30 35N 49 10 E
Bandar-e Nakhīlū, *Iran* **45 E7** 26 58N 53 30 E
Bandar-e Rīg, *Iran* **45 D6** 29 29N 50 38 E
Bandar-e Torkeman, *Iran* **45 B7** 37 0N 54 10 E
Bandar Maharani = Muar, *Malaysia* **39 L4** 2 3N 102 34 E
Bandar Penggaram = Batu Pahat, *Malaysia* **39 M4** 1 50N 102 56 E
Bandar Seri Begawan, *Brunei* **36 C4** 4 52N 115 0 E
Bandawe, *Malawi* **55 E3** 11 58S 34 5 E
Bandeira, Pico da, *Brazil* **95 A7** 20 26S 41 47W
Bandera, *Argentina* **94 B3** 28 55S 62 20W
Bandera, *U.S.A.* **81 L5** 29 44N 99 5W
Banderas, B. de, *Mexico* **86 C3** 20 40N 105 30W
Bandiagara, *Mali* **50 F4** 14 12N 3 29W
Bandırma, *Turkey* **21 D13** 40 20N 28 0 E
Bandon, *Ireland* **13 E3** 51 44N 8 44W
Bandon →, *Ireland* **13 E3** 51 43N 8 37W
Bandula, *Mozam.* **55 F3** 19 0S 33 7 E
Bandundu, *Zaïre* **52 E3** 3 15S 17 22 E
Bandung, *Indonesia* **37 G12** 6 54S 107 36 E
Bandya, *Australia* **61 E3** 27 40S 122 5 E
Bāneh, *Iran* **44 C5** 35 59N 45 53 E
Banes, *Cuba* **89 B4** 21 0N 75 42W
Banff, *Canada* **72 C5** 51 10N 115 34W
Banff, *U.K.* **12 D6** 57 40N 2 33W
Banff Nat. Park, *Canada* **72 C5** 51 30N 116 15W
Banfora, *Burkina Faso* **50 F4** 10 40N 4 40W
Bang Fai →, *Laos* **38 D5** 16 57N 104 45 E
Bang Hieng →, *Laos* **38 D5** 16 10N 105 10 E
Bang Krathum, *Thailand* **38 D3** 16 34N 100 18 E
Bang Lamung, *Thailand* **38 F3** 13 3N 100 56 E
Bang Mun Nak, *Thailand* **38 D3** 16 2N 100 23 E
Bang Pa In, *Thailand* **38 E3** 14 14N 100 35 E
Bang Rakam, *Thailand* **38 D3** 16 45N 100 7 E
Bang Saphan, *Thailand* **39 G2** 11 14N 99 28 E
Bangala Dam, *Zimbabwe* **55 G3** 21 7S 31 25 E
Bangalore, *India* **40 N10** 12 59N 77 40 E
Bangaon, *India* **43 H13** 23 0N 88 47 E
Bangassou, *C.A.R.* **52 D4** 4 55N 23 7 E
Banggai, Kepulauan, *Indonesia* **37 E6** 1 40S 123 30 E
Banggi, P., *Malaysia* **36 C5** 7 17N 117 12 E
Banghāzī, *Libya* **51 B9** 32 11N 20 3 E
Bangil, *Indonesia* **37 G15** 7 36S 112 50 E
Bangka, P., *Sulawesi, Indonesia* **37 D7** 1 50N 125 5 E
Bangka, P., *Sumatera, Indonesia* **36 E3** 2 0S 105 50 E
Bangka, Selat, *Indonesia* **36 E3** 2 30S 105 30 E
Bangkalan, *Indonesia* **37 G15** 7 2S 112 46 E
Bangkinang, *Indonesia* **36 D2** 0 18N 101 5 E
Bangko, *Indonesia* **36 E2** 2 5S 102 9 E
Bangkok, *Thailand* **38 F3** 13 45N 100 35 E
Bangladesh ■, *Asia* **41 H17** 24 0N 90 0 E
Bangong Co, *India* **43 B8** 35 50N 79 20 E
Bangor, *Down, U.K.* **13 B6** 54 40N 5 40W
Bangor, *Gwynedd, U.K.* **10 D3** 53 14N 4 8W
Bangor, *Maine, U.S.A.* **71 D6** 44 48N 68 46W
Bangor, *Pa., U.S.A.* **79 F9** 40 52N 75 13W
Bangued, *Phil.* **37 A6** 17 40N 120 37 E
Bangui, *C.A.R.* **52 D3** 4 23N 18 35 E
Banguru, *Zaïre* **54 B2** 0 30N 27 10 E
Bangweulu, L., *Zambia* **55 E3** 11 0S 30 0 E
Bangweulu Swamp, *Zambia* **55 E3** 11 20S 30 15 E
Bani, *Dom. Rep.* **89 C5** 18 16N 70 22W
Banī Sa'd, *Iraq* **44 C5** 33 34N 44 32 E
Banī Walīd, *Libya* **51 B7** 31 36N 13 53 E
Banihal Pass, *India* **43 C6** 33 30N 75 12 E
Banīnah, *Libya* **51 B9** 32 0N 20 12 E
Bāniyās, *Syria* **44 C3** 35 10N 36 0 E
Banja Luka, *Bos.-H.* **20 B7** 44 49N 17 11 E
Banjar, *Indonesia* **37 G13** 7 24S 108 30 E
Banjarmasin, *Indonesia* **36 E4** 3 20S 114 35 E
Banjarnegara, *Indonesia* **37 G13** 7 24S 109 42 E
Banjul, *Gambia* **50 F1** 13 28N 16 40W
Banka Banka, *Australia* **62 B1** 18 50S 134 0 E
Banket, *Zimbabwe* **55 F3** 17 27S 30 19 E
Bankipore, *India* **43 G11** 25 35N 85 10 E
Banks I., *B.C., Canada* **72 C3** 53 20N 130 0W
Banks I., *N.W.T., Canada* **68 A7** 73 15N 121 30W
Banks Pen., *N.Z.* **59 K4** 43 45S 173 15 E
Banks Str., *Australia* **62 G4** 40 40S 148 10 E
Bankura, *India* **43 H12** 23 11N 87 18 E
Bann →, *Arm., U.K.* **13 B5** 54 30N 6 31W
Bann →, *L'derry., U.K.* **13 A5** 55 8N 6 41W
Bannang Sata, *Thailand* **39 J3** 6 16N 101 16 E
Banning, *U.S.A.* **85 M10** 33 56N 116 53W
Banningville = Bandundu, *Zaïre* **52 E3** 3 15S 17 22 E
Bannockburn, *Canada* **78 B7** 44 39N 77 33W
Bannockburn, *U.K.* **12 E5** 56 5N 3 55W
Bannockburn, *Zimbabwe* **55 G2** 20 17S 29 48 E
Bannu, *Pakistan* **40 C7** 33 0N 70 18 E
Banská Bystrica, *Slovak Rep.* **17 D10** 48 46N 19 14 E
Banswara, *India* **42 H6** 23 32N 74 24 E
Banten, *Indonesia* **37 G12** 6 5S 106 8 E
Bantry, *Ireland* **13 E2** 51 41N 9 27W
Bantry B., *Ireland* **13 E2** 51 38N 9 44W
Bantul, *Indonesia* **37 G14** 7 55S 110 19 E
Bantva, *India* **42 J4** 21 29N 70 12 E
Banu, *Afghan.* **40 B6** 35 35N 69 5 E
Banyak, Kepulauan, *Indonesia* **36 D1** 2 10N 97 10 E
Banyo, *Cameroon* **50 G7** 6 52N 11 45 E
Banyumas, *Indonesia* **37 G13** 7 32S 109 18 E
Banyuwangi, *Indonesia* **37 H16** 8 13S 114 21 E
Banzare Coast, *Antarctica* **5 C9** 68 0S 125 0 E
Banzyville = Mobayi, *Zaïre* **52 D4** 4 15N 21 8 E
Bao Ha, *Vietnam* **38 A5** 22 11N 104 21 E
Bao Lac, *Vietnam* **38 A5** 22 57N 105 40 E
Bao Loc, *Vietnam* **39 G6** 11 32N 107 48 E
Baocheng, *China* **34 H4** 33 12N 106 56 E
Baode, *China* **34 E6** 39 1N 111 5 E
Baodi, *China* **35 E9** 39 38N 117 20 E
Baoding, *China* **34 E8** 38 50N 115 28 E
Baoji, *China* **34 G4** 34 20N 107 5 E
Baoshan, *China* **32 D4** 25 10N 99 5 E
Baotou, *China* **34 D6** 40 32N 110 2 E
Baoying, *China* **35 H10** 33 17N 119 20 E
Bap, *India* **42 F5** 27 23N 72 18 E
Bapatla, *India* **41 M12** 15 55N 80 30 E
Bāqerābād, *Iran* **45 C6** 33 2N 51 58 E
Ba'qūbah, *Iraq* **44 C5** 33 45N 44 50 E
Baquedano, *Chile* **94 A2** 23 20S 69 52W
Bar, *Montenegro, Yug.* **21 C8** 42 8N 19 8 E
Bar, *Ukraine* **17 D14** 49 4N 27 40 E
Bar Bigha, *India* **43 G11** 25 21N 85 47 E
Bar Harbor, *U.S.A.* **71 D6** 44 23N 68 13W
Bar-le-Duc, *France* **18 B6** 48 47N 5 10 E
Barabai, *Indonesia* **36 E5** 2 32S 115 34 E
Barabinsk, *Russia* **26 D8** 55 20N 78 20 E
Baraboo, *U.S.A.* **80 D10** 43 28N 89 45W
Baracaldo, *Spain* **19 A4** 43 18N 2 59W
Baracoa, *Cuba* **89 B5** 20 20N 74 30W
Baradero, *Argentina* **94 C4** 33 52S 59 29W
Baraga, *U.S.A.* **80 B10** 46 47N 88 30W
Barahona, *Dom. Rep.* **89 C5** 18 13N 71 7W
Barail Range, *India* **41 G18** 25 15N 93 20 E
Barakhola, *India* **41 G18** 25 0N 92 45 E
Barakot, *India* **43 J11** 21 33N 84 59 E
Barakpur, *India* **43 H13** 22 44N 88 30 E
Barakula, *Australia* **63 D5** 26 30S 150 33 E
Baralaba, *Australia* **62 C4** 24 13S 149 50 E
Baralzon L., *Canada* **73 B9** 60 0N 98 3W
Baramula, *India* **43 B6** 34 15N 74 20 E
Baran, *India* **42 G7** 25 9N 76 40 E
Baranavichy, *Belarus* **17 B14** 53 10N 26 0 E
Baranof I., *U.S.A.* **72 B1** 57 0N 135 0W
Barão de Melgaço, *Brazil* **92 F6** 11 50S 60 45W
Barapasi, *Indonesia* **37 E9** 2 15S 137 5 E
Barasat, *India* **43 H13** 22 46N 88 31 E
Barat Daya, Kepulauan, *Indonesia* **37 F7** 7 30S 128 0 E
Barataria B., *U.S.A.* **81 L10** 29 20N 89 55W
Baraut, *India* **42 E7** 29 13N 77 7 E
Barbacena, *Brazil* **95 A7** 21 15S 43 56W
Barbacoas, *Colombia* **92 C3** 1 45N 78 0W
Barbados ■, *W. Indies* **89 D8** 13 10N 59 30W
Barbastro, *Spain* **19 A6** 42 2N 0 5 E
Barberton, *S. Africa* **57 D5** 25 42S 31 2 E
Barberton, *U.S.A.* **78 E3** 41 0N 81 39W
Barbourville, *U.S.A.* **77 G4** 36 52N 83 53W
Barbuda, *W. Indies* **89 C7** 17 30N 61 40W
Barcaldine, *Australia* **62 C4** 23 43S 145 6 E
Barcellona Pozzo di Gotto, *Italy* **20 E6** 38 9N 15 13 E
Barcelona, *Spain* **19 B7** 41 21N 2 10 E
Barcelona, *Venezuela* **92 A6** 10 10N 64 40W
Barcelos, *Brazil* **92 D6** 1 0S 63 0W
Barcoo →, *Australia* **62 D3** 25 30S 142 50 E
Bardai, *Chad* **51 D8** 21 25N 17 0 E
Bardas Blancas, *Argentina* **94 D2** 35 49S 69 45W
Barddhaman, *India* **43 H12** 23 14N 87 39 E
Bardejov, *Slovak Rep.* **17 D11** 49 18N 21 15 E
Bardera, *Somali Rep.* **46 G3** 2 20N 42 27 E
Bardīyah, *Libya* **51 B9** 31 45N 25 5 E
Bardsey I., *U.K.* **10 E3** 52 45N 4 47W
Bardstown, *U.S.A.* **76 G3** 37 49N 85 28W
Bareilly, *India* **43 E8** 28 22N 79 27 E

Barents Sea, *Arctic* **4 B9** 73 0N 39 0 E
Barentu, *Eritrea* **51 E12** 15 2N 37 35 E
Barfleur, Pte. de, *France* **18 B3** 49 42N 1 16W
Bargal, *Somali Rep.* **46 E5** 11 25N 51 0 E
Bargara, *Australia* **62 C5** 24 50S 152 25 E
Barguzin, *Russia* **27 D11** 53 37N 109 37 E
Barh, *India* **43 G11** 25 29N 85 46 E
Barhaj, *India* **43 F10** 26 18N 83 44 E
Barhi, *India* **43 G11** 24 15N 85 25 E
Bari, *India* **42 F7** 26 39N 77 39 E
Bari, *Italy* **20 D7** 41 8N 16 51 E
Bari Doab, *Pakistan* **42 D5** 30 20N 73 0 E
Bariadi □, *Tanzania* **54 C3** 2 45S 34 40 E
Barīm, *Yemen* **46 E3** 12 39N 43 25 E
Barinas, *Venezuela* **92 B4** 8 36N 70 15W
Baring, C., *Canada* **68 B8** 70 0N 117 30W
Baringo, *Kenya* **54 B4** 0 47N 36 16 E
Baringo □, *Kenya* **54 B4** 0 55N 36 0 E
Baringo, L., *Kenya* **54 B4** 0 47N 36 16 E
Bârîs, *Egypt* **51 D11** 24 42N 30 31 E
Barisal, *Bangla.* **41 H17** 22 45N 90 20 E
Barisan, Bukit, *Indonesia* **36 E2** 3 30S 102 15 E
Barito →, *Indonesia* **36 E4** 4 0S 114 50 E
Bark L., *Canada* **78 A7** 45 27N 77 51W
Barker, *U.S.A.* **78 C6** 43 20N 78 33W
Barkley Sound, *Canada* **72 D3** 48 50N 125 10W
Barkly Downs, *Australia* **62 C2** 20 30S 138 30 E
Barkly East, *S. Africa* **56 E4** 30 58S 27 33 E
Barkly Tableland, *Australia* **62 B2** 17 50S 136 40 E
Barkly West, *S. Africa* **56 D3** 28 5S 24 31 E
Barkol, *China* **32 B4** 43 37N 93 2 E
Barksdale, *U.S.A.* **81 L4** 29 44N 100 2W
Barlee, L., *Australia* **61 E2** 29 15S 119 30 E
Barlee, Mt., *Australia* **61 D4** 24 38S 128 13 E
Barletta, *Italy* **20 D7** 41 19N 16 17 E
Barlovento, *Canary Is.* **22 F2** 28 48N 17 48W
Barlow L., *Canada* **73 A8** 62 0N 103 0W
Barmedman, *Australia* **63 E4** 34 9S 147 21 E
Barmer, *India* **42 G4** 25 45N 71 20 E
Barmera, *Australia* **63 E3** 34 15S 140 28 E
Barmouth, *U.K.* **10 E3** 52 44N 4 4W
Barnagar, *India* **42 H6** 23 7N 75 19 E
Barnard Castle, *U.K.* **10 C6** 54 33N 1 55W
Barnato, *Australia* **63 E3** 31 38S 145 0 E
Barnaul, *Russia* **26 D9** 53 20N 83 40 E
Barnesville, *U.S.A.* **77 J3** 33 3N 84 9W
Barnet, *U.K.* **11 F7** 51 38N 0 9W
Barneveld, *Neths.* **15 B5** 52 7N 5 36 E
Barneveld, *U.S.A.* **79 C9** 43 16N 75 14W
Barngo, *Australia* **62 D4** 25 3S 147 20 E
Barnhart, *U.S.A.* **81 K4** 31 8N 101 10W
Barnsley, *U.K.* **10 D6** 53 34N 1 27W
Barnstaple, *U.K.* **11 F3** 51 5N 4 4W
Barnsville, *U.S.A.* **80 B6** 46 43N 96 28W
Baro, *Nigeria* **50 G6** 8 35N 6 18 E
Baroda = Vadodara, *India* **42 H5** 22 20N 73 10 E
Baroda, *India* **42 G7** 25 29N 76 35 E
Baroe, *S. Africa* **56 E3** 33 13S 24 33 E
Baron Ra., *Australia* **60 D4** 23 30S 127 45 E
Barpeta, *India* **41 F17** 26 20N 91 10 E
Barques, Pt. Aux, *U.S.A.* **76 C4** 44 4N 82 58W
Barquísimeto, *Venezuela* **92 A5** 10 4N 69 19W
Barra, *Brazil* **93 F10** 11 5S 43 10W
Barra, *U.K.* **12 E1** 57 0N 7 29W
Barra, Sd. of, *U.K.* **12 D1** 57 4N 7 25W
Barra de Navidad, *Mexico* **86 D4** 19 12N 104 41W
Barra do Corda, *Brazil* **93 E9** 5 30S 45 10W
Barra do Piraí, *Brazil* **95 A7** 22 30S 43 50W
Barra Falsa, Pta. da, *Mozam.* **57 C6** 22 58S 35 37 E
Barra Hd., *U.K.* **12 E1** 56 47N 7 40W
Barra Mansa, *Brazil* **95 A7** 22 35S 44 12W
Barraba, *Australia* **63 E5** 30 21S 150 35 E
Barrackpur = Barakpur, *India* **43 H13** 22 44N 88 30 E
Barraigh = Barra, *U.K.* **12 E1** 57 0N 7 29W
Barranca, *Lima, Peru* **92 F3** 10 45S 77 50W
Barranca, *Loreto, Peru* **92 D3** 4 50S 76 50W
Barrancabermeja, *Colombia* **92 B4** 7 0N 73 50W
Barrancas, *Venezuela* **92 B6** 8 55N 62 5W
Barrancos, *Portugal* **19 C2** 38 10N 6 58W
Barranqueras, *Argentina* **94 B4** 27 30S 59 0W
Barranquilla, *Colombia* **92 A4** 11 0N 74 50W
Barras, *Brazil* **93 D10** 4 15S 42 18W
Barraute, *Canada* **70 C4** 48 26N 77 38W
Barre, *Mass., U.S.A.* **79 D12** 42 25N 72 6W
Barre, *Vt., U.S.A.* **79 B12** 44 12N 72 30W
Barreal, *Argentina* **94 C2** 31 33S 69 28W
Barreiras, *Brazil* **93 F10** 12 8S 45 0W
Barreirinhas, *Brazil* **93 D10** 2 30S 42 50W
Barreiro, *Portugal* **19 C1** 38 40N 9 6W
Barreiros, *Brazil* **93 E11** 8 49S 35 12W
Barren, Nosy, *Madag.* **57 B7** 18 25S 43 40 E
Barretos, *Brazil* **93 H9** 20 30S 48 35W
Barrhead, *Canada* **72 C6** 54 10N 114 24W
Barrie, *Canada* **70 D4** 44 24N 79 40W
Barrier Ra., *Australia* **63 E3** 31 0S 141 30 E
Barrière, *Canada* **72 C4** 51 12N 120 7W
Barrington, *U.S.A.* **79 E13** 41 44N 71 18W
Barrington L., *Canada* **73 B8** 56 55N 100 15W
Barrington Tops, *Australia* **63 E5** 32 6S 151 28 E
Barringun, *Australia* **63 D4** 29 1S 145 41 E
Barrow, *U.S.A.* **68 A4** 71 18N 156 47W
Barrow →, *Ireland* **13 D4** 52 25N 6 58W
Barrow Creek, *Australia* **62 C1** 21 30S 133 55 E
Barrow I., *Australia* **60 D2** 20 45S 115 20 E
Barrow-in-Furness, *U.K.* **10 C4** 54 7N 3 14W
Barrow Pt., *Australia* **62 A3** 14 20S 144 40 E
Barrow Pt., *U.S.A.* **66 B4** 71 24N 156 29W
Barrow Ra., *Australia* **61 E4** 26 0S 127 40 E
Barrow Str., *Canada* **4 B3** 74 20N 95 0W
Barry, *U.K.* **11 F4** 51 24N 3 16W
Barry's Bay, *Canada* **70 C4** 45 29N 77 41W
Barsat, *Pakistan* **43 A5** 36 10N 72 45 E
Barsham, *Syria* **44 C4** 35 21N 40 33 E
Barsi, *India* **40 K9** 18 10N 75 50 E
Barsoi, *India* **41 G15** 25 48N 87 57 E
Barstow, *Calif., U.S.A.* **85 L9** 34 54N 117 1W
Barstow, *Tex., U.S.A.* **81 K3** 31 28N 103 24W
Barthélemy, Col, *Vietnam* **38 C5** 19 26N 104 6 E
Bartica, *Guyana* **92 B7** 6 25N 58 40W
Bartlesville, *U.S.A.* **81 G7** 36 45N 95 59W
Bartlett, *Calif., U.S.A.* **84 J8** 36 29N 118 2W
Bartlett, *Tex., U.S.A.* **81 K6** 30 48N 97 26W
Bartlett, L., *Canada* **72 A5** 63 5N 118 20W
Bartolomeu Dias, *Mozam.* **55 G4** 21 10S 35 8 E
Barton, *Australia* **61 F5** 30 31S 132 39 E
Barton upon Humber, *U.K.* **10 D7** 53 41N 0 25W
Bartow, *U.S.A.* **77 M5** 27 54N 81 50W
Barú, Volcan, *Panama* **88 E3** 8 55N 82 35W
Barumba, *Zaïre* **54 B1** 1 3N 23 37 E
Barwani, *India* **42 H6** 22 2N 74 57 E
Barysaw, *Belarus* **17 A15** 54 17N 28 28 E
Barzān, *Iraq* **44 B5** 36 55N 44 3 E
Bāsa'idū, *Iran* **45 E7** 26 35N 55 20 E
Basal, *Pakistan* **42 C5** 33 33N 72 13 E
Basankusa, *Zaïre* **52 D3** 1 5N 19 50 E
Basarabeasca, *Moldova* **17 E15** 46 21N 28 58 E
Basawa, *Afghan.* **42 B4** 34 15N 70 50 E
Bascuñán, C., *Chile* **94 B1** 28 52S 71 35W
Basel, *Switz.* **16 E4** 47 35N 7 35 E
Bāshī, *Iran* **45 D6** 28 41N 51 4 E
Bashkir Republic = Bashkortostan □, *Russia* **24 D10** 54 0N 57 0 E
Bashkortostan □, *Russia* **24 D10** 54 0N 57 0 E
Basilan, *Phil.* **37 C6** 6 35N 122 0 E
Basilan Str., *Phil.* **37 C6** 6 50N 122 0 E
Basildon, *U.K.* **11 F8** 51 34N 0 28 E
Basim = Washim, *India* **40 J10** 20 3N 77 0 E
Basin, *U.S.A.* **82 D9** 44 23N 108 2W
Basingstoke, *U.K.* **11 F6** 51 15N 1 5W
Baskatong, Rés., *Canada* **70 C4** 46 46N 75 50W
Basle = Basel, *Switz.* **16 E4** 47 35N 7 35 E
Basoda, *India* **42 H7** 23 52N 77 54 E
Basoka, *Zaïre* **54 B1** 1 16N 23 40 E
Basongo, *Zaïre* **52 E4** 4 15S 20 20 E
Basque Provinces = Pais Vasco □, *Spain* **19 A4** 42 50N 2 45W
Basra = Al Başrah, *Iraq* **44 D5** 30 30N 47 50 E
Bass Rock, *U.K.* **12 E6** 56 5N 2 38W
Bass Str., *Australia* **62 F4** 39 15S 146 30 E
Bassano, *Canada* **72 C6** 50 48N 112 20W
Bassano del Grappa, *Italy* **20 B4** 45 46N 11 44 E
Bassas da India, *Ind. Oc.* **53 J7** 22 0S 39 0 E
Basse-Terre, *Guadeloupe* **89 C7** 16 0N 61 44W
Bassein, *Burma* **41 L19** 16 45N 94 30 E
Basseterre, *St. Kitts & Nevis* **89 C7** 17 17N 62 43W
Bassett, *Nebr., U.S.A.* **80 D5** 42 35N 99 32W
Bassett, *Va., U.S.A.* **77 G6** 36 46N 79 59W
Bassi, *India* **42 D7** 30 44N 76 21 E
Bassikounou, *Mauritania* **50 E3** 15 55N 6 1W
Bastak, *Iran* **45 E7** 27 15N 54 25 E
Baştām, *Iran* **45 B7** 36 29N 55 4 E
Bastar, *India* **41 K12** 19 15N 81 40 E
Basti, *India* **43 F10** 26 52N 82 55 E
Bastia, *France* **18 E8** 42 40N 9 30 E
Bastogne, *Belgium* **15 D5** 50 1N 5 43 E
Bastrop, *U.S.A.* **81 K6** 30 7N 97 19W
Bat Yam, *Israel* **47 C3** 32 2N 34 44 E
Bata, *Eq. Guin.* **52 D1** 1 57N 9 50 E
Bataan, *Phil.* **37 B6** 14 40N 120 25 E
Batabanó, *Cuba* **88 B3** 22 40N 82 20W
Batabanó, G. de, *Cuba* **88 B3** 22 30N 82 30W
Batac, *Phil.* **37 A6** 18 3N 120 34 E
Batagoy, *Russia* **27 C14** 67 38N 134 38 E
Batama, *Zaïre* **54 B2** 0 58N 26 33 E
Batamay, *Russia* **27 C13** 63 30N 129 15 E
Batang, *Indonesia* **37 G13** 6 55S 109 45 E
Batangafo, *C.A.R.* **51 G8** 7 25N 18 20 E
Batangas, *Phil.* **37 B6** 13 35N 121 10 E
Batanta, *Indonesia* **37 E8** 0 55S 130 40 E
Batatais, *Brazil* **95 A6** 20 54S 47 37W
Batavia, *U.S.A.* **78 D6** 43 0N 78 11W
Batchelor, *Australia* **60 B5** 13 4S 131 1 E
Bateman's B., *Australia* **63 F5** 35 40S 150 12 E
Batemans Bay, *Australia* **63 F5** 35 44S 150 11 E
Bates Ra., *Australia* **61 E3** 27 27S 121 5 E
Batesburg, *U.S.A.* **77 J5** 33 54N 81 33W
Batesville, *Ark., U.S.A.* **81 H9** 35 46N 91 39W
Batesville, *Miss., U.S.A.* **81 H10** 34 19N 89 57W
Batesville, *Tex., U.S.A.* **81 L5** 28 58N 99 37W
Bath, *U.K.* **11 F5** 51 23N 2 22W
Bath, *Maine, U.S.A.* **71 D6** 43 55N 69 49W
Bath, *N.Y., U.S.A.* **78 D7** 42 20N 77 19W
Bath & North East Somerset □, *U.K.* **11 F5** 51 21N 2 27W
Batheay, *Cambodia* **39 G5** 11 59N 104 57 E
Bathgate, *U.K.* **12 F5** 55 54N 3 39W
Bathurst = Banjul, *Gambia* **50 F1** 13 28N 16 40W
Bathurst, *Australia* **63 E4** 33 25S 149 31 E
Bathurst, *Canada* **71 C6** 47 37N 65 43W
Bathurst, *S. Africa* **56 E4** 33 30S 26 50 E
Bathurst, C., *Canada* **68 A7** 70 34N 128 0W
Bathurst B., *Australia* **62 A3** 14 16S 144 25 E
Bathurst Harb., *Australia* **62 G4** 43 15S 146 10 E
Bathurst I., *Australia* **60 B5** 11 30S 130 10 E
Bathurst I., *Canada* **4 B2** 76 0N 100 30W
Bathurst Inlet, *Canada* **68 B9** 66 50N 108 1W
Batlow, *Australia* **63 F4** 35 31S 148 9 E
Batna, *Algeria* **50 A6** 35 34N 6 15 E
Batoka, *Zambia* **55 F2** 16 45S 27 15 E
Baton Rouge, *U.S.A.* **81 K9** 30 27N 91 11W
Batong, Ko, *Thailand* **39 J2** 6 32N 99 12 E
Batopilas, *Mexico* **86 B3** 27 0N 107 45W
Batouri, *Cameroon* **52 D2** 4 30N 14 25 E
Båtsfjord, *Norway* **8 A23** 70 38N 29 39 E
Battambang, *Cambodia* **38 F4** 13 7N 103 12 E
Batticaloa, *Sri Lanka* **40 R12** 7 43N 81 45 E
Battipáglia, *Italy* **20 D6** 40 37N 14 58 E
Battle, *U.K.* **11 G8** 50 55N 0 30 E
Battle →, *Canada* **73 C7** 52 43N 108 15W
Battle Camp, *Australia* **62 B3** 15 20S 144 40 E
Battle Creek, *U.S.A.* **76 D3** 42 19N 85 11W
Battle Ground, *U.S.A.* **84 E4** 45 47N 122 32W
Battle Harbour, *Canada* **71 B8** 52 16N 55 35W
Battle Lake, *U.S.A.* **80 B7** 46 17N 95 43W
Battle Mountain, *U.S.A.* **82 F5** 40 38N 116 56W
Battlefields, *Zimbabwe* **55 F2** 18 37S 29 47 E
Battleford, *Canada* **73 C7** 52 45N 108 15W
Batu, *Ethiopia* **46 F2** 6 55N 39 45 E
Batu, Kepulauan, *Indonesia* **36 E1** 0 30S 98 25 E
Batu Caves, *Malaysia* **39 L3** 3 15N 101 40 E
Batu Gajah, *Malaysia* **39 K3** 4 28N 101 3 E
Batu Is. = Batu, Kepulauan, *Indonesia* **36 E1** 0 30S 98 25 E
Batu Pahat, *Malaysia* **39 M4** 1 50N 102 56 E
Batuata, *Indonesia* **37 F6** 6 12S 122 42 E
Batumi, *Georgia* **25 F7** 41 39N 41 44 E
Baturaja, *Indonesia* **36 E2** 4 11S 104 15 E
Baturité, *Brazil* **93 D11** 4 28S 38 45W
Bau, *Malaysia* **36 D4** 1 25N 110 9 E
Baubau, *Indonesia* **37 F6** 5 25S 122 38 E
Bauchi, *Nigeria* **50 F6** 10 22N 9 48 E
Baudette, *U.S.A.* **80 A7** 48 43N 94 36W
Bauer, C., *Australia* **63 E1** 32 44S 134 4 E
Bauhinia Downs, *Australia* **62 C4** 24 35S 149 18 E
Baukau, *Indonesia* **37 F7** 8 27S 126 27 E
Bauru, *Brazil* **95 A6** 22 10S 49 0W
Baús, *Brazil* **93 G8** 18 22S 52 47W
Bauska, *Latvia* **9 H21** 56 24N 24 15 E
Bautzen, *Germany* **16 C8** 51 10N 14 26 E
Bavānāt, *Iran* **45 D7** 30 28N 53 27 E
Bavaria = Bayern □, *Germany* **16 D6** 48 50N 12 0 E
Bavi Sadri, *India* **42 G6** 24 28N 74 30 E
Bavispe →, *Mexico* **86 B3** 29 30N 109 11W
Bawdwin, *Burma* **41 H20** 23 5N 97 20 E
Bawean, *Indonesia* **36 F4** 5 46S 112 35 E
Bawku, *Ghana* **50 F4** 11 3N 0 19W
Bawlake, *Burma* **41 K20** 19 11N 97 21 E
Baxley, *U.S.A.* **77 K4** 31 47N 82 21W
Baxter Springs, *U.S.A.* **81 G7** 37 2N 94 44W
Bay, L. de, *Phil.* **37 B6** 14 20N 121 11 E
Bay Bulls, *Canada* **71 C9** 47 19N 52 50W
Bay City, *Mich., U.S.A.* **76 D4** 43 36N 83 54W
Bay City, *Oreg., U.S.A.* **82 D2** 45 31N 123 53W
Bay City, *Tex., U.S.A.* **81 L7** 28 59N 95 58W
Bay de Verde, *Canada* **71 C9** 48 5N 52 54W
Bay Minette, *U.S.A.* **77 K2** 30 53N 87 46W
Bay St. Louis, *U.S.A.* **81 K10** 30 19N 89 20W
Bay Springs, *U.S.A.* **81 K10** 31 59N 89 17W
Bay View, *N.Z.* **59 H6** 39 25S 176 50 E
Baya, *Zaïre* **55 E2** 11 53S 27 25 E
Bayamo, *Cuba* **88 B4** 20 20N 76 40W
Bayamón, *Puerto Rico* **89 C6** 18 24N 66 10W
Bayan Har Shan, *China* **32 C4** 34 0N 98 0 E
Bayan Hot = Alxa Zuoqi, *China* **34 E3** 38 50N 105 40 E
Bayan Obo, *China* **34 D5** 41 52N 109 59 E
Bayan-Ovoo, *Mongolia* **34 C4** 42 55N 106 5 E
Bayana, *India* **42 F7** 26 55N 77 18 E
Bayanaūyl, *Kazakstan* **26 D8** 50 45N 75 45 E
Bayandalay, *Mongolia* **34 C2** 43 30N 103 29 E
Bayanhongor, *Mongolia* **32 B5** 46 8N 102 43 E
Bayard, *U.S.A.* **80 E3** 41 45N 103 20W
Baybay, *Phil.* **37 B6** 10 40N 124 55 E
Bayern □, *Germany* **16 D6** 48 50N 12 0 E
Bayeux, *France* **18 B3** 49 17N 0 42W
Bayfield, *Canada* **78 C3** 43 34N 81 42W
Bayfield, *U.S.A.* **80 B9** 46 49N 90 49W
Bayındır, *Turkey* **21 E12** 38 13N 27 39 E
Baykal, Oz., *Russia* **27 D11** 53 0N 108 0 E
Baykit, *Russia* **27 C10** 61 50N 95 50 E
Baykonur = Bayqongyr, *Kazakstan* **26 E7** 47 48N 65 50 E
Baymak, *Russia* **24 D10** 52 36N 58 19 E
Baynes Mts., *Namibia* **56 B1** 17 15S 13 0 E
Bayombong, *Phil.* **37 A6** 16 30N 121 10 E
Bayonne, *France* **18 E3** 43 30N 1 28W
Bayonne, *U.S.A.* **79 F10** 40 40N 74 7W
Bayovar, *Peru* **92 E2** 5 50S 81 0W
Bayqongyr, *Kazakstan* **26 E7** 47 48N 65 50 E
Bayram-Ali = Bayramaly, *Turkmenistan* **26 F7** 37 37N 62 10 E
Bayramaly, *Turkmenistan* **26 F7** 37 37N 62 10 E
Bayramiç, *Turkey* **21 E12** 39 48N 26 36 E
Bayreuth, *Germany* **16 D6** 49 56N 11 35 E
Bayrūt, *Lebanon* **47 B4** 33 53N 35 31 E
Bayt Lahm, *West Bank* **47 D4** 31 43N 35 12 E
Baytown, *U.S.A.* **81 L7** 29 43N 94 59W
Baza, *Spain* **19 D4** 37 30N 2 47W
Bazaruto, I. do, *Mozam.* **57 C6** 21 40S 35 28 E
Bazmān, Kūh-e, *Iran* **45 D9** 28 4N 60 1 E
Beach, *U.S.A.* **80 B3** 46 58N 104 0W
Beach City, *U.S.A.* **78 F3** 40 39N 81 35W
Beachport, *Australia* **63 F2** 37 29S 140 0 E
Beachy Hd., *U.K.* **11 G8** 50 44N 0 15 E
Beacon, *Australia* **61 F2** 30 26S 117 52 E
Beacon, *U.S.A.* **79 E11** 41 30N 73 58W
Beaconia, *Canada* **73 C9** 50 25N 96 31W
Beagle, Canal, *S. Amer.* **96 G3** 55 0S 68 30W
Beagle Bay, *Australia* **60 C3** 16 58S 122 40 E
Bealanana, *Madag.* **57 A8** 14 33S 48 44 E
Beamsville, *Canada* **78 C5** 43 12N 79 28W
Bear →, *U.S.A.* **84 G5** 38 56N 121 36W
Bear I., *Ireland* **13 E2** 51 38N 9 50W
Bear L., *B.C., Canada* **72 B3** 56 10N 126 52W
Bear L., *Man., Canada* **73 B9** 55 8N 96 0W
Bear L., *U.S.A.* **82 E8** 41 59N 111 21W
Bearcreek, *U.S.A.* **82 D9** 45 11N 109 6W
Beardmore, *Canada* **70 C2** 49 36N 87 57W
Beardmore Glacier, *Antarctica* **5 E11** 84 30S 170 0 E
Beardstown, *U.S.A.* **80 F9** 40 1N 90 26W
Béarn, *France* **18 E3** 43 20N 0 30W
Bearpaw Mts., *U.S.A.* **82 B9** 48 12N 109 30W
Bearskin Lake, *Canada* **70 B1** 53 58N 91 2W
Beata, C., *Dom. Rep.* **89 C5** 17 40N 71 30W
Beata, I., *Dom. Rep.* **89 C5** 17 34N 71 31W
Beatrice, *U.S.A.* **80 E6** 40 16N 96 45W
Beatrice, *Zimbabwe* **55 F3** 18 15S 30 55 E
Beatrice, C., *Australia* **62 A2** 14 20S 136 55 E
Beatton →, *Canada* **72 B4** 56 15N 120 45W
Beatton River, *Canada* **72 B4** 57 26N 121 20W
Beatty, *U.S.A.* **83 H5** 36 54N 116 46W
Beauce, Plaine de la, *France* **18 B4** 48 10N 1 45 E
Beauceville, *Canada* **71 C5** 46 13N 70 46W
Beaudesert, *Australia* **63 D5** 27 59S 153 0 E
Beaufort, *Malaysia* **36 C5** 5 30N 115 40 E
Beaufort, *N.C., U.S.A.* **77 H7** 34 43N 76 40W
Beaufort, *S.C., U.S.A.* **77 J5** 32 26N 80 40W
Beaufort Sea, *Arctic* **4 B1** 72 0N 140 0W
Beaufort West, *S. Africa* **56 E3** 32 18S 22 36 E
Beauharnois, *Canada* **70 C5** 45 20N 73 52W
Beaulieu →, *Canada* **72 A6** 62 3N 113 11W
Beauly, *U.K.* **12 D4** 57 30N 4 28W
Beauly →, *U.K.* **12 D4** 57 29N 4 27W
Beaumaris, *U.K.* **10 D3** 53 16N 4 6W
Beaumont, *U.S.A.* **81 K7** 30 5N 94 6W
Beaune, *France* **18 C6** 47 2N 4 50 E
Beauséjour, *Canada* **73 C9** 50 5N 96 35W
Beauvais, *France* **18 B5** 49 25N 2 8 E
Beauval, *Canada* **73 B7** 55 9N 107 37W
Beaver, *Alaska, U.S.A.* **68 B5** 66 22N 147 24W
Beaver, *Okla., U.S.A.* **81 G4** 36 49N 100 31W
Beaver, *Pa., U.S.A.* **78 F4** 40 42N 80 19W
Beaver, *Utah, U.S.A.* **83 G7** 38 17N 112 38W
Beaver →, *B.C., Canada* **72 B4** 59 52N 124 20W
Beaver →, *Ont., Canada* **70 A2** 55 55N 87 48W
Beaver →, *Sask., Canada* **73 B7** 55 26N 107 45W
Beaver City, *U.S.A.* **80 E5** 40 8N 99 50W
Beaver Dam, *U.S.A.* **80 D10** 43 28N 88 50W
Beaver Falls, *U.S.A.* **78 F4** 40 46N 80 20W
Beaver Hill L., *Canada* **73 C10** 54 5N 94 50W
Beaver I., *U.S.A.* **76 C3** 45 40N 85 33W
Beaverhill L., *Alta., Canada* **72 C6** 53 27N 112 32W
Beaverhill L., *N.W.T., Canada* **73 A8** 63 2N 104 22W
Beaverlodge, *Canada* **72 B5** 55 11N 119 29W
Beavermouth, *Canada* **72 C5** 51 32N 117 23W
Beaverstone →, *Canada* **70 B2** 54 59N 89 25W
Beaverton, *Canada* **78 B5** 44 26N 79 9W
Beaverton, *U.S.A.* **84 E4** 45 29N 122 48W
Beawar, *India* **42 F6** 26 3N 74 18 E
Bebedouro, *Brazil* **95 A6** 21 0S 48 25W
Beboa, *Madag.* **57 B7** 17 22S 44 33 E
Beccles, *U.K.* **11 E9** 52 27N 1 35 E
Bečej, *Serbia, Yug.* **21 B9** 45 36N 20 3 E
Béchar, *Algeria* **50 B4** 31 38N 2 18W
Beckley, *U.S.A.* **76 G5** 37 47N 81 11W
Bedford, *Canada* **70 C5** 45 7N 72 59W
Bedford, *S. Africa* **56 E4** 32 40S 26 10 E
Bedford, *U.K.* **11 E7** 52 8N 0 28W
Bedford, *Ind., U.S.A.* **76 F2** 38 52N 86 29W
Bedford, *Iowa, U.S.A.* **80 E7** 40 40N 94 44W
Bedford, *Ohio, U.S.A.* **78 E3** 41 23N 81 32W
Bedford, *Pa., U.S.A.* **78 F6** 40 1N 78 30W
Bedford, *Va., U.S.A.* **76 G6** 37 20N 79 31W
Bedford, C., *Australia* **62 B4** 15 14S 145 21 E
Bedford Downs, *Australia* **60 C4** 17 19S 127 20 E
Bedfordshire □, *U.K.* **11 E7** 52 4N 0 28W
Bedourie, *Australia* **62 C2** 24 30S 139 30 E
Beech Grove, *U.S.A.* **76 F2** 39 44N 86 3W
Beechy, *Canada* **73 C7** 50 53N 107 24W
Beenleigh, *Australia* **63 D5** 27 43S 153 10 E
Be'er Menuha, *Israel* **44 D2** 30 19N 35 8 E
Be'er Sheva, *Israel* **47 D3** 31 15N 34 48 E
Beersheba = Be'er Sheva, *Israel* **47 D3** 31 15N 34 48 E
Beeston, *U.K.* **10 E6** 52 56N 1 14W
Beetaloo, *Australia* **62 B1** 17 15S 133 50 E
Beeville, *U.S.A.* **81 L6** 28 24N 97 45W
Befale, *Zaïre* **52 D4** 0 25N 20 45 E
Befandriana, *Madag.* **57 C7** 21 55S 44 0 E
Befotaka, *Madag.* **57 C8** 23 49S 47 0 E
Bega, *Australia* **63 F4** 36 41S 149 51 E
Begusarai, *India* **43 G12** 25 24N 86 9 E
Behābād, *Iran* **45 C8** 32 24N 59 47 E
Behara, *Madag.* **57 C8** 24 55S 46 20 E
Behbehān, *Iran* **45 D6** 30 30N 50 15 E
Behshahr, *Iran* **45 B7** 36 45N 53 35 E
Bei Jiang →, *China* **33 D6** 23 2N 112 58 E
Bei'an, *China* **33 B7** 48 10N 126 20 E
Beihai, *China* **33 D5** 21 28N 109 6 E
Beijing, *China* **34 E9** 39 55N 116 20 E
Beijing □, *China* **34 E9** 39 55N 116 20 E
Beilen, *Neths.* **15 B6** 52 52N 6 27 E
Beilpajah, *Australia* **63 E3** 32 54S 143 52 E
Beinn na Faoghla = Benbecula, *U.K.* **12 D1** 57 26N 7 21W
Beipiao, *China* **35 D11** 41 52N 120 32 E
Beira, *Mozam.* **55 F3** 19 50S 34 52 E
Beirut = Bayrūt, *Lebanon* **47 B4** 33 53N 35 31 E
Beitaolaizhao, *China* **35 B13** 44 58N 125 58 E
Beitbridge, *Zimbabwe* **55 G3** 22 12S 30 0 E
Beizhen, *Liaoning, China* **35 D11** 41 38N 121 54 E
Beizhen, *Shandong, China* **35 F10** 37 20N 118 2 E
Beizhengzhen, *China* **35 B12** 44 31N 123 30 E
Beja, *Portugal* **19 C2** 38 2N 7 53W
Béja, *Tunisia* **50 A6** 36 43N 9 12 E
Bejaia, *Algeria* **50 A6** 36 42N 5 2 E
Béjar, *Spain* **19 B3** 40 23N 5 46W
Bejestān, *Iran* **45 C8** 34 30N 58 5 E
Bekaa Valley = Al Biqā □, *Lebanon* **47 A5** 34 10N 36 10 E
Bekasi, *Indonesia* **37 G12** 6 14S 106 59 E
Békéscsaba, *Hungary* **17 E11** 46 40N 21 5 E
Bekily, *Madag.* **57 C8** 24 13S 45 19 E
Bekok, *Malaysia* **39 L4** 2 20N 103 7 E
Bela, *India* **43 G9** 25 50N 82 0 E
Bela, *Pakistan* **42 F2** 26 12N 66 20 E
Bela Crkva, *Serbia, Yug.* **21 B9** 44 55N 21 27 E
Bela Vista, *Brazil* **94 A4** 22 12S 56 20W
Bela Vista, *Mozam.* **57 D5** 26 10S 32 44 E
Belarus ■, *Europe* **17 B14** 53 30N 27 0 E
Belau = Palau ■, *Pac. Oc.* **28 J17** 7 30N 134 30 E
Belavenona, *Madag.* **57 C8** 24 50S 47 4 E
Belawan, *Indonesia* **36 D1** 3 33N 98 32 E
Belaya →, *Russia* **24 C9** 54 40N 56 0 E
Belaya Tserkov = Bila Tserkva, *Ukraine* **17 D16** 49 45N 30 10 E
Belcher Is., *Canada* **69 C12** 56 15N 78 45W
Belden, *U.S.A.* **84 E5** 40 2N 121 17W
Belebey, *Russia* **24 D9** 54 7N 54 7 E
Belém, *Brazil* **93 D9** 1 20S 48 30W
Belén, *Argentina* **94 B2** 27 40S 67 5W
Belén, *Paraguay* **94 A4** 23 30S 57 6W
Belen, *U.S.A.* **83 J10** 34 40N 106 46W
Belet Uen, *Somali Rep.* **46 G4** 4 30N 45 5 E
Belev, *Russia* **24 D6** 53 50N 36 5 E
Belfair, *U.S.A.* **84 C4** 47 27N 122 50W
Belfast, *S. Africa* **57 D5** 25 42S 30 2 E
Belfast, *U.K.* **13 B6** 54 37N 5 56W
Belfast, *Maine, U.S.A.* **71 D6** 44 26N 69 1W
Belfast, *N.Y., U.S.A.* **78 D6** 42 21N 78 7W
Belfast □, *U.K.* **13 B6** 54 35N 5 56W
Belfast L., *U.K.* **13 B6** 54 40N 5 50W
Belfield, *U.S.A.* **80 B3** 46 53N 103 12W
Belfort, *France* **18 C7** 47 38N 6 50 E
Belfry, *U.S.A.* **82 D9** 45 9N 109 1W
Belgaum, *India* **40 M9** 15 55N 74 35 E
Belgium ■, *Europe* **15 D5** 50 30N 5 0 E

Belgorod, Russia 25 D6 50 35N 36 35 E
Belgorod-Dnestrovskiy = Bilhorod-Dnistrovskyy, Ukraine 25 E5 46 11N 30 23 E
Belgrade = Beograd, Serbia, Yug. 21 B9 44 50N 20 37 E
Belgrade, U.S.A. 82 D8 45 47N 111 11W
Belhaven, U.S.A. 77 H7 35 33N 76 37W
Beli Drim →, Europe 21 C9 42 6N 20 25 E
Belinga, Gabon 52 D2 1 10N 13 2 E
Belinyu, Indonesia 36 E3 1 35S 105 50 E
Beliton Is. = Belitung, Indonesia 36 E3 3 10S 107 50 E
Belitung, Indonesia 36 E3 3 10S 107 50 E
Belize ■, Cent. Amer. 87 D7 17 0N 88 30W
Belize City, Belize 87 D7 17 25N 88 0W
Belkovskiy, Ostrov, Russia 27 B14 75 32N 135 44 E
Bell →, Canada 70 C4 49 48N 77 38W
Bell Bay, Australia 62 G4 41 6S 146 53 E
Bell I., Canada 71 B8 50 46N 55 35W
Bell-Irving →, Canada 72 B3 56 12N 129 5W
Bell Peninsula, Canada 69 B11 63 50N 82 0W
Bell Ville, Argentina 94 C3 32 40S 62 40W
Bella Bella, Canada 72 C3 52 10N 128 10W
Bella Coola, Canada 72 C3 52 25N 126 40W
Bella Unión, Uruguay 94 C4 30 15S 57 40W
Bella Vista, Corrientes, Argentina 94 B4 28 33S 59 0W
Bella Vista, Tucuman, Argentina 94 B2 27 10S 65 25W
Bellaire, U.S.A. 78 F4 40 1N 80 45W
Bellary, India 40 M10 15 10N 76 56 E
Bellata, Australia 63 D4 29 53S 149 46 E
Belle Fourche, U.S.A. 80 C3 44 40N 103 51W
Belle Fourche →, U.S.A. 80 C3 44 26N 102 18W
Belle Glade, U.S.A. 77 M5 26 41N 80 40W
Belle-Ile, France 18 C2 47 20N 3 10W
Belle Isle, Canada 71 B8 51 57N 55 25W
Belle Isle, Str. of, Canada 71 B8 51 30N 56 30W
Belle Plaine, Iowa, U.S.A. 80 E8 41 54N 92 17W
Belle Plaine, Minn., U.S.A. 80 C8 44 37N 93 46W
Belledune, Canada 71 C6 47 55N 65 50W
Bellefontaine, U.S.A. 76 E4 40 22N 83 46W
Bellefonte, U.S.A. 78 F7 40 55N 77 47W
Belleoram, Canada 71 C8 47 31N 55 25W
Belleville, Canada 70 D4 44 10N 77 23W
Belleville, Ill., U.S.A. 80 F10 38 31N 89 59W
Belleville, Kans., U.S.A. 80 F6 39 50N 97 38W
Belleville, N.Y., U.S.A. 79 C8 43 46N 76 10W
Bellevue, Canada 72 D6 49 35N 114 22W
Bellevue, Idaho, U.S.A. 82 E6 43 28N 114 16W
Bellevue, Ohio, U.S.A. 78 E2 41 17N 82 51W
Bellevue, Wash., U.S.A. 84 C4 47 37N 122 12W
Bellin = Kangirsuk, Canada 69 B13 60 0N 70 0W
Bellingen, Australia 63 E5 30 25S 152 50 E
Bellingham, U.S.A. 84 B4 48 46N 122 29W
Bellingshausen Sea, Antarctica 5 C17 66 0S 80 0W
Bellinzona, Switz. 16 E5 46 11N 9 1 E
Bellows Falls, U.S.A. 79 C12 43 8N 72 27W
Bellpat, Pakistan 42 E3 29 0N 68 5 E
Belluno, Italy 20 A5 46 9N 12 13 E
Bellville, U.S.A. 81 L6 29 57N 96 15W
Bellwood, U.S.A. 78 F6 40 36N 78 20W
Belmont, Australia 63 E5 33 4S 151 42 E
Belmont, Canada 78 D3 42 53N 81 5W
Belmont, S. Africa 56 D3 29 28S 24 22 E
Belmont, U.S.A. 78 D6 42 14N 78 2W
Belmonte, Brazil 93 G11 16 0S 39 0W
Belmopan, Belize 87 D7 17 18N 88 30W
Belmullet, Ireland 13 B2 54 14N 9 58W
Belo Horizonte, Brazil 93 G10 19 55S 43 56W
Belo-sur-Mer, Madag. 57 C7 20 42S 44 0 E
Belo-Tsiribihina, Madag. 57 B7 19 40S 44 30 E
Belogorsk, Russia 27 D13 51 0N 128 20 E
Beloha, Madag. 57 D8 25 10S 45 3 E
Beloit, Kans., U.S.A. 80 F5 39 28N 98 6W
Beloit, Wis., U.S.A. 80 D10 42 31N 89 2W
Belokorovichi, Ukraine 17 C15 51 7N 28 2 E
Belomorsk, Russia 24 B5 64 35N 34 54 E
Belonia, India 41 H17 23 15N 91 30 E
Beloretsk, Russia 24 D10 53 58N 58 24 E
Belorussia ■ = Belarus ■, Europe 17 B14 53 30N 27 0 E
Belovo, Russia 26 D9 54 30N 86 0 E
Beloye, Ozero, Russia 24 B6 60 10N 37 35 E
Beloye More, Russia 24 A6 66 30N 38 0 E
Belozersk, Russia 24 B6 60 1N 37 45 E
Beltana, Australia 63 E2 30 48S 138 25 E
Belterra, Brazil 93 D8 2 45S 55 0W
Belton, S.C., U.S.A. 77 H4 34 31N 82 30W
Belton, Tex., U.S.A. 81 K6 31 3N 97 28W
Belton Res., U.S.A. 81 K6 31 8N 97 32W
Beltsy = Bălţi, Moldova 17 E14 47 48N 28 0 E
Belturbet, Ireland 13 B4 54 6N 7 26W
Belukha, Russia 26 E9 49 50N 86 50 E
Beluran, Malaysia 36 C5 5 48N 117 35 E
Belvidere, Ill., U.S.A. 80 D10 42 15N 88 50W
Belvidere, N.J., U.S.A. 79 F9 40 50N 75 5W
Belyando →, Australia 62 C4 21 38S 146 50 E
Belyy, Ostrov, Russia 26 B8 73 30N 71 0 E
Belyy Yar, Russia 26 D9 58 26N 84 39 E
Belzoni, U.S.A. 81 J9 33 11N 90 29W
Bemaraha, Lembalemban' i, Madag. 57 B7 18 40S 44 45 E
Bemarivo, Madag. 57 C7 21 45S 44 45 E
Bemarivo →, Madag. 57 B8 15 27S 47 40 E
Bemavo, Madag. 57 C8 21 33S 45 25 E
Bembéréke, Benin 50 F5 10 11N 2 43 E
Bembesi, Zimbabwe 55 F2 20 0S 28 58 E
Bembesi →, Zimbabwe 55 F2 18 57S 27 47 E
Bemidji, U.S.A. 80 B7 47 28N 94 53W
Ben, Iran 45 C6 32 32N 50 45 E
Ben Cruachan, U.K. 12 E3 56 26N 5 8W
Ben Dearg, U.K. 12 D4 57 47N 4 56W
Ben Gardane, Tunisia 51 B7 33 11N 11 11 E
Ben Hope, U.K. 12 C4 58 25N 4 36W
Ben Lawers, U.K. 12 E4 56 32N 4 14W
Ben Lomond, N.S.W., Australia 63 E5 30 1S 151 43 E
Ben Lomond, Tas., Australia 62 G4 41 38S 147 42 E
Ben Lomond, U.K. 12 E4 56 11N 4 38W
Ben Luc, Vietnam 39 G6 10 39N 106 29 E
Ben Macdhui, U.K. 12 D5 57 4N 3 40W
Ben Mhor, U.K. 12 D1 57 15N 7 18W
Ben More, Arg. & Bute, U.K. 12 E2 56 26N 6 1W
Ben More, Stirl., U.K. 12 E4 56 23N 4 32W
Ben More Assynt, U.K. 12 C4 58 8N 4 52W
Ben Nevis, U.K. 12 E4 56 48N 5 1W
Ben Quang, Vietnam 38 D6 17 3N 106 55 E
Ben Tre, Vietnam 39 G6 10 3N 106 36 E
Ben Vorlich, U.K. 12 E4 56 21N 4 14W
Ben Wyvis, U.K. 12 D4 57 40N 4 35W
Bena, Nigeria 50 F6 11 20N 5 50 E
Bena Dibele, Zaïre 52 E4 4 4S 22 50 E
Benagerie, Australia 63 E3 31 25S 140 22 E
Benalla, Australia 63 F4 36 30S 146 0 E
Benambra, Mt., Australia 63 F4 36 31S 147 34 E
Benares = Varanasi, India 43 G10 25 22N 83 0 E
Benavente, Spain 19 A3 42 2N 5 43W
Benavides, U.S.A. 81 M5 27 36N 98 25W
Benbecula, U.K. 12 D1 57 26N 7 21W
Benbonyathe, Australia 63 E2 30 25S 139 11 E
Bencubbin, Australia 61 F2 30 48S 117 52 E
Bend, U.S.A. 82 D3 44 4N 121 19W
Bender Beila, Somali Rep. 46 F5 9 30N 50 48 E
Bendering, Australia 61 F2 32 23S 118 18 E
Bendery = Tighina, Moldova 17 E15 46 50N 29 30 E
Bendigo, Australia 63 F3 36 40S 144 15 E
Benē Beraq, Israel 47 C3 32 6N 34 51 E
Benenitra, Madag. 57 C8 23 27S 45 5 E
Benevento, Italy 20 D6 41 8N 14 45 E
Benga, Mozam. 55 F3 16 11S 33 40 E
Bengal, Bay of, Ind. Oc. 41 K16 15 0N 90 0 E
Bengbu, China 35 H9 32 58N 117 20 E
Benghazi = Banghāzī, Libya 51 B9 32 11N 20 3 E
Bengkalis, Indonesia 36 D2 1 30N 102 10 E
Bengkulu, Indonesia 36 E2 3 50S 102 12 E
Bengkulu □, Indonesia 36 E2 3 48S 102 16 E
Bengough, Canada 73 D7 49 25N 105 10W
Benguela, Angola 53 G2 12 37S 13 25 E
Benguérua, I., Mozam. 57 C6 21 58S 35 28 E
Beni, Zaïre 54 B2 0 30N 29 27 E
Beni →, Bolivia 92 F5 10 23S 65 24W
Beni Abbès, Algeria 50 B4 30 5N 2 5W
Beni Mazâr, Egypt 51 C11 28 32N 30 44 E
Beni Mellal, Morocco 50 B3 32 21N 6 21W
Beni Ounif, Algeria 50 B4 32 0N 1 10W
Beni Suef, Egypt 51 C11 29 5N 31 6 E
Beniah L., Canada 72 A6 63 23N 112 17W
Benicia, U.S.A. 84 G4 38 3N 122 9W
Benidorm, Spain 19 C5 38 33N 0 9W
Benin ■, Africa 50 G5 10 0N 2 0 E
Benin, Bight of, W. Afr. 50 H5 5 0N 3 0 E
Benin City, Nigeria 50 G6 6 20N 5 31 E
Benitses, Greece 23 A3 39 32N 19 55 E
Benjamin Aceval, Paraguay 94 A4 24 58S 57 34W
Benjamin Constant, Brazil 92 D4 4 40S 70 15W
Benjamin Hill, Mexico 86 A2 30 10N 111 10W
Benkelman, U.S.A. 80 E4 40 3N 101 32W
Benlidi, Australia 62 C3 24 35S 144 50 E
Bennett, Canada 72 B2 59 56N 134 53W
Bennett, L., Australia 60 D5 22 50S 131 2 E
Bennett, Ostrov, Russia 27 B15 76 21N 148 56 E
Bennettsville, U.S.A. 77 H6 34 37N 79 41W
Bennington, U.S.A. 79 D11 43 0N 71 55W
Benoni, S. Africa 57 D4 26 11S 28 18 E
Benque Viejo, Belize 87 D7 17 5N 89 8W
Benson, U.S.A. 83 L8 31 58N 110 18W
Bent, Iran 45 E8 26 20N 59 31 E
Benteng, Indonesia 37 F6 6 10S 120 30 E
Bentinck I., Australia 62 B2 17 3S 139 35 E
Bento Gonçalves, Brazil 95 B5 29 10S 51 31W
Benton, Ark., U.S.A. 81 H8 34 34N 92 35W
Benton, Calif., U.S.A. 84 H8 37 48N 118 32W
Benton, Ill., U.S.A. 80 F10 38 0N 88 55W
Benton Harbor, U.S.A. 76 D2 42 6N 86 27W
Bentung, Malaysia 39 L3 3 31N 101 55 E
Benue →, Nigeria 50 G6 7 48N 6 46 E
Benxi, China 35 D12 41 20N 123 48 E
Beo, Indonesia 37 D7 4 25N 126 50 E
Beograd, Serbia, Yug. 21 B9 44 50N 20 37 E
Beowawe, U.S.A. 82 F5 40 35N 116 29W
Beppu, Japan 31 H5 33 15N 131 30 E
Beqaa Valley = Al Biqā □, Lebanon 47 A5 34 10N 36 10 E
Berati, Albania 21 D8 40 43N 19 59 E
Berau, Teluk, Indonesia 37 E8 2 30S 132 30 E
Berber, Sudan 51 E11 18 0N 34 0 E
Berbera, Somali Rep. 46 E4 10 30N 45 2 E
Berbérati, C.A.R. 52 D3 4 15N 15 40 E
Berberia, C. del, Spain 22 C7 38 39N 1 24 E
Berbice →, Guyana 92 B7 6 20N 57 32W
Berdichev = Berdychiv, Ukraine 17 D15 49 57N 28 30 E
Berdsk, Russia 26 D9 54 47N 83 2 E
Berdyansk, Ukraine 25 E6 46 45N 36 50 E
Berdychiv, Ukraine 17 D15 49 57N 28 30 E
Berea, U.S.A. 76 G3 37 34N 84 17W
Berebere, Indonesia 37 D7 2 25N 128 45 E
Bereda, Somali Rep. 46 E5 11 45N 51 0 E
Berehove, Ukraine 17 D12 48 15N 22 35 E
Berekum, Ghana 50 G4 7 29N 2 34W
Berens →, Canada 73 C9 52 25N 97 2W
Berens I., Canada 73 C9 52 18N 97 18W
Berens River, Canada 73 C9 52 25N 97 0W
Berestechko, Ukraine 17 C13 50 22N 25 5 E
Berevo, Mahajanga, Madag. 57 B7 17 14S 44 17 E
Berevo, Toliara, Madag. 57 B7 19 44S 44 58 E
Bereza, Belarus 17 B13 52 31N 24 51 E
Berezhany, Ukraine 17 D13 49 26N 24 58 E
Berezina = Byarezina →, Belarus 17 B16 52 33N 30 14 E
Berezniki, Russia 26 D6 59 24N 56 46 E
Berezovo, Russia 24 B11 64 0N 65 0 E
Berga, Spain 19 A6 42 6N 1 48 E
Bergama, Turkey 21 E12 39 8N 27 15 E
Bérgamo, Italy 20 B3 45 41N 9 43 E
Bergen, Neths. 15 B4 52 40N 4 43 E
Bergen, Norway 9 F11 60 20N 5 20 E
Bergen, U.S.A. 78 C7 43 5N 77 57W
Bergen-op-Zoom, Neths. 15 C4 51 28N 4 18 E
Bergerac, France 18 D4 44 51N 0 30 E
Bergum, Neths. 15 A5 53 13N 5 59 E
Bergville, S. Africa 57 D4 28 52S 29 18 E
Berhala, Selat, Indonesia 36 E2 1 0S 104 15 E
Berhampore = Baharampur, India 43 G13 24 2N 88 27 E
Berhampur, India 41 K14 19 15N 84 54 E
Bering Sea, Pac. Oc. 68 C1 58 0N 171 0 E
Bering Strait, U.S.A. 68 B3 65 30N 169 0W
Beringen, Belgium 15 C5 51 3N 5 14 E
Beringovskiy, Russia 27 C18 63 3N 179 19 E
Berisso, Argentina 94 C4 34 56S 57 50W
Berja, Spain 19 D4 36 50N 2 56W
Berkeley, U.K. 11 F5 51 41N 2 27W
Berkeley, U.S.A. 84 H4 37 52N 122 16W
Berkeley Springs, U.S.A. 76 F6 39 38N 78 14W
Berkner I., Antarctica 5 D18 79 30S 50 0W
Berkshire □, U.K. 11 F6 51 25N 1 17W
Berland →, Canada 72 C5 54 0N 116 50W
Berlin, Germany 16 B7 52 30N 13 25 E
Berlin, Md., U.S.A. 76 F8 38 20N 75 13W
Berlin, N.H., U.S.A. 79 B13 44 28N 71 11W
Berlin, Wis., U.S.A. 76 D1 43 58N 88 57W
Bermejo →, Formosa, Argentina 94 B4 26 51S 58 23W
Bermejo →, San Juan, Argentina 94 C2 32 30S 67 30W
Bermuda ■, Atl. Oc. 66 F13 32 45N 65 0W
Bern, Switz. 16 E4 46 57N 7 28 E
Bernado, U.S.A. 83 J10 34 30N 106 53W
Bernalillo, U.S.A. 83 J10 35 18N 106 33W
Bernardo de Irigoyen, Argentina 95 B5 26 15S 53 40W
Bernardo O'Higgins □, Chile 94 C1 34 15S 70 45W
Bernasconi, Argentina 94 D3 37 55S 63 44W
Bernburg, Germany 16 C6 51 47N 11 44 E
Berne = Bern, Switz. 16 E4 46 57N 7 28 E
Bernier I., Australia 61 D1 24 50S 113 12 E
Bernina, Piz, Switz. 16 E5 46 20N 9 54 E
Beroroha, Madag. 57 C8 21 40S 45 10 E
Beroun, Czech. 16 D8 49 57N 14 5 E
Berrechid, Morocco 50 B3 33 18N 7 36W
Berri, Australia 63 E3 34 14S 140 35 E
Berry, Australia 63 E5 34 46S 150 43 E
Berry, France 18 C5 46 50N 2 0 E
Berry Is., Bahamas 88 A4 25 40N 77 50W
Berryessa L., U.S.A. 84 G4 38 31N 122 6W
Berryville, U.S.A. 81 G8 36 22N 93 34W
Bershad, Ukraine 17 D15 48 22N 29 31 E
Berthold, U.S.A. 80 A4 48 19N 101 44W
Berthoud, U.S.A. 80 E2 40 19N 105 5W
Bertoua, Cameroon 52 D2 4 30N 13 45 E
Bertrand, U.S.A. 80 E5 40 32N 99 38W
Berwick, U.S.A. 79 E8 41 3N 76 14W
Berwick-upon-Tweed, U.K. 10 B5 55 46N 2 0W
Berwyn Mts., U.K. 10 E4 52 54N 3 26W
Besal, Pakistan 43 B5 35 4N 73 56 E
Besalampy, Madag. 57 B7 16 43S 44 29 E
Besançon, France 18 C7 47 15N 6 2 E
Besar, Indonesia 36 E5 2 40S 116 0 E
Besnard L., Canada 73 B7 55 25N 106 0W
Besor, N. →, Egypt 47 D3 31 28N 34 22 E
Bessarabiya, Moldova 17 E15 47 0N 28 10 E
Bessarabka = Basarabeasca, Moldova 17 E15 46 21N 28 58 E
Bessemer, Ala., U.S.A. 77 J2 33 24N 86 58W
Bessemer, Mich., U.S.A. 80 B9 46 29N 90 3W
Bet She'an, Israel 47 C4 32 30N 35 30 E
Bet Shemesh, Israel 47 D3 31 44N 35 0 E
Betafo, Madag. 57 B8 19 50S 46 51 E
Betancuria, Canary Is. 22 F5 28 25N 14 3W
Betanzos, Spain 19 A1 43 15N 8 12W
Bétaré Oya, Cameroon 52 C2 5 40N 14 5 E
Bethal, S. Africa 57 D4 26 27S 29 28 E
Bethanien, Namibia 56 D2 26 31S 17 8 E
Bethany, U.S.A. 80 E7 40 16N 94 2W
Bethel, Alaska, U.S.A. 68 B3 60 48N 161 45W
Bethel, Vt., U.S.A. 79 C12 43 50N 72 38W
Bethel Park, U.S.A. 78 F4 40 20N 80 1W
Bethlehem = Bayt Lahm, West Bank 47 D4 31 43N 35 12 E
Bethlehem, S. Africa 57 D4 28 14S 28 18 E
Bethlehem, U.S.A. 79 F9 40 37N 75 23W
Bethulie, S. Africa 56 E4 30 30S 25 59 E
Béthune, France 18 A5 50 30N 2 38 E
Bethungra, Australia 63 E4 34 45S 147 51 E
Betioky, Madag. 57 C7 23 48S 44 20 E
Betong, Thailand 39 K3 5 45N 101 5 E
Betoota, Australia 62 D3 25 45S 140 42 E
Betroka, Madag. 57 C8 23 16S 46 0 E
Betsiamites, Canada 71 C6 48 56N 68 40W
Betsiamites →, Canada 71 C6 48 56N 68 38W
Betsiboka →, Madag. 57 B8 16 3S 46 36 E
Bettiah, India 43 F11 26 48N 84 33 E
Betul, India 40 J10 21 58N 77 59 E
Betung, Malaysia 36 D4 1 24N 111 31 E
Beulah, U.S.A. 80 B4 47 16N 101 47W
Beverley, Australia 61 F2 32 9S 116 56 E
Beverley, U.K. 10 D7 53 51N 0 26W
Beverly, Mass., U.S.A. 79 D14 42 33N 70 53W
Beverly, Wash., U.S.A. 82 C4 46 50N 119 56W
Beverly Hills, U.S.A. 85 L8 34 4N 118 25W
Beverwijk, Neths. 15 B4 52 28N 4 38 E
Beya, Russia 27 D10 52 40N 92 30 E
Beyānlū, Iran 44 C5 36 0N 47 51 E
Beyla, Guinea 50 G3 8 30N 8 38W
Beyneu, Kazakstan 25 E10 45 18N 55 9 E
Beypazarı, Turkey 25 F5 40 10N 31 56 E
Beyşehir Gölü, Turkey 25 G5 37 41N 31 33 E
Bezhitsa, Russia 24 D5 53 19N 34 17 E
Béziers, France 18 E5 43 20N 3 12 E
Bezwada = Vijayawada, India 41 L12 16 31N 80 39 E
Bhachau, India 40 H7 23 20N 70 16 E
Bhadarwah, India 43 C6 32 58N 75 46 E
Bhadrakh, India 41 J15 21 10N 86 30 E
Bhadravati, India 40 N9 13 49N 75 40 E
Bhagalpur, India 43 G12 25 10N 87 0 E
Bhakkar, Pakistan 42 D4 31 40N 71 5 E
Bhakra Dam, India 42 D7 31 30N 76 45 E
Bhamo, Burma 41 G20 24 15N 97 15 E
Bhandara, India 40 J11 21 5N 79 42 E
Bhanrer Ra., India 42 H8 23 40N 79 45 E
Bharat = India ■, Asia 40 K11 20 0N 78 0 E
Bharatpur, India 42 F7 27 15N 77 30 E
Bhatinda, India 42 D6 30 15N 74 57 E
Bhatpara, India 43 H13 22 50N 88 25 E
Bhaun, Pakistan 42 C5 32 55N 72 40 E
Bhaunagar = Bhavnagar, India 42 J5 21 45N 72 10 E
Bhavnagar, India 42 J5 21 45N 72 10 E
Bhawanipatna, India 41 K12 19 55N 80 10 E
Bhera, Pakistan 42 C5 32 29N 72 57 E
Bhilsa = Vidisha, India 42 H7 23 28N 77 53 E
Bhilwara, India 42 G6 25 25N 74 38 E
Bhima →, India 40 L10 16 25N 77 17 E
Bhimavaram, India 41 L12 16 30N 81 30 E
Bhimbar, Pakistan 43 C6 32 59N 74 3 E
Bhind, India 43 F8 26 30N 78 46 E
Bhiwandi, India 40 K8 19 20N 73 0 E
Bhiwani, India 42 E7 28 50N 76 9 E
Bhola, Bangla. 41 H17 22 45N 90 35 E
Bhopal, India 42 H7 23 20N 77 30 E
Bhubaneshwar, India 41 J14 20 15N 85 50 E
Bhuj, India 42 H3 23 15N 69 49 E
Bhumiphol Dam = Phumiphon, Khuan, Thailand 38 D2 17 15N 98 58 E
Bhusaval, India 40 J9 21 3N 75 46 E
Bhutan ■, Asia 41 F17 27 25N 90 30 E
Biafra, B. of = Bonny, Bight of, Africa 52 D1 3 30N 9 20 E
Biak, Indonesia 37 E9 1 10S 136 6 E
Biała Podlaska, Poland 17 B12 52 4N 23 6 E
Białogard, Poland 16 A8 54 2N 15 58 E
Białystok, Poland 17 B12 53 10N 23 10 E
Bīārjmand, Iran 45 B7 36 6N 55 53 E
Biaro, Indonesia 37 D7 2 5N 125 26 E
Biarritz, France 18 E3 43 29N 1 33W
Bibai, Japan 30 C10 43 19N 141 52 E
Bibala, Angola 53 G2 14 44S 13 24 E
Bibby I., Canada 73 A10 61 55N 93 0W
Biberach, Germany 16 D5 48 5N 9 47 E
Bibiani, Ghana 50 G4 6 30N 2 8W
Biboohra, Australia 62 B4 16 56S 145 25 E
Bibungwa, Zaïre 54 C2 2 40S 28 15 E
Bic, Canada 71 C6 48 20N 68 41W
Bickerton I., Australia 62 A2 13 45S 136 10 E
Bicknell, Ind., U.S.A. 76 F2 38 47N 87 19W
Bicknell, Utah, U.S.A. 83 G8 38 20N 111 33W
Bida, Nigeria 50 G6 9 3N 5 58 E
Bidar, India 40 L10 17 55N 77 35 E
Biddeford, U.S.A. 71 D5 43 30N 70 28W
Bideford, U.K. 11 F3 51 1N 4 13W
Bidon 5 = Poste Maurice Cortier, Algeria 50 D5 22 14N 1 2 E
Bidor, Malaysia 39 K3 4 6N 101 15 E
Bié, Planalto de, Angola 53 G3 12 0S 16 0 E
Bieber, U.S.A. 82 F3 41 7N 121 8W
Biel, Switz. 16 E4 47 8N 7 14 E
Bielé Karpaty, Europe 17 D9 49 5N 18 0 E
Bielefeld, Germany 16 B5 52 1N 8 33 E
Biella, Italy 20 B3 45 34N 8 3 E
Bielsk Podlaski, Poland 17 B12 52 47N 23 12 E
Bielsko-Biała, Poland 17 D10 49 50N 19 2 E
Bien Hoa, Vietnam 39 G6 10 57N 106 49 E
Bienfait, Canada 73 D8 49 10N 102 50W
Bienne = Biel, Switz. 16 E4 47 8N 7 14 E
Bienville, L., Canada 70 A5 55 5N 72 40W
Biesiesfontein, S. Africa 56 E2 30 57S 17 58 E
Big →, Canada 71 B8 54 50N 58 55W
Big B., Canada 71 A7 55 43N 60 35W
Big Bear City, U.S.A. 85 L10 34 16N 116 51W
Big Bear Lake, U.S.A. 85 L10 34 15N 116 56W
Big Beaver, Canada 73 D7 49 10N 105 10W
Big Belt Mts., U.S.A. 82 C8 46 30N 111 25W
Big Bend, Swaziland 57 D5 26 50S 31 58 E
Big Bend National Park, U.S.A. 81 L3 29 20N 103 5W
Big Black →, U.S.A. 81 J9 32 3N 91 4W
Big Blue →, U.S.A. 80 F6 39 35N 96 34W
Big Cr. →, Canada 72 C4 51 42N 122 41W
Big Creek, U.S.A. 84 H7 37 11N 119 14W
Big Cypress Swamp, U.S.A. 77 M5 26 12N 81 10W
Big Falls, U.S.A. 80 A8 48 12N 93 48W
Big Fork →, U.S.A. 80 A8 48 31N 93 43W
Big Horn Mts. = Bighorn Mts., U.S.A. 82 D10 44 30N 107 30W
Big Lake, U.S.A. 81 K4 31 12N 101 28W
Big Moose, U.S.A. 79 C10 43 49N 74 58W
Big Muddy Cr. →, U.S.A. 80 A2 48 8N 104 36W
Big Pine, U.S.A. 83 H4 37 10N 118 17W
Big Piney, U.S.A. 82 E8 42 32N 110 7W
Big Quill L., Canada 73 C8 51 55N 104 50W
Big Rapids, U.S.A. 76 D3 43 42N 85 29W
Big River, Canada 73 C7 53 50N 107 0W
Big Run, U.S.A. 78 F6 40 57N 78 55W
Big Sable Pt., U.S.A. 76 C2 44 3N 86 1W
Big Sand L., Canada 73 B9 57 45N 99 45W
Big Sandy, U.S.A. 82 B8 48 11N 110 7W
Big Sandy Cr. →, U.S.A. 80 F3 38 7N 102 29W
Big Sioux →, U.S.A. 80 D6 42 29N 96 27W
Big Spring, U.S.A. 81 J4 32 15N 101 28W
Big Springs, U.S.A. 80 E3 41 4N 102 5W
Big Stone City, U.S.A. 80 C6 45 18N 96 28W
Big Stone Gap, U.S.A. 77 G4 36 52N 82 47W
Big Stone L., U.S.A. 80 C6 45 30N 96 35W
Big Sur, U.S.A. 84 J5 36 15N 121 48W
Big Timber, U.S.A. 82 D9 45 50N 109 57W
Big Trout L., Canada 70 B1 53 40N 90 0W
Biğa, Turkey 21 D12 40 13N 27 14 E
Bigadiç, Turkey 21 E13 39 22N 28 7 E
Bigfork, U.S.A. 82 B6 48 4N 114 4W
Biggar, Canada 73 C7 52 4N 108 0W
Biggar, U.K. 12 F5 55 38N 3 32W
Bigge I., Australia 60 B4 14 35S 125 10 E
Biggenden, Australia 63 D5 25 31S 152 4 E
Biggs, U.S.A. 84 F5 39 25N 121 43W
Bighorn, U.S.A. 82 C10 46 10N 107 27W
Bighorn →, U.S.A. 82 C10 46 10N 107 28W
Bighorn Mts., U.S.A. 82 D10 44 30N 107 30W
Bigstone L., Canada 73 C9 53 42N 95 44W
Bigwa, Tanzania 54 D4 7 10S 39 10 E
Bihać, Bos.-H. 16 F8 44 49N 15 57 E
Bihar, India 43 G11 25 5N 85 40 E

Bihar □, India 43 G11 25 0N 86 0 E
Biharamulo, Tanzania 54 C3 2 25S 31 25 E
Biharamulo □, Tanzania 54 C3 2 30S 31 20 E
Bihor, Munţii, Romania 17 E12 46 29N 22 47 E
Bijagós, Arquipélago dos, Guinea-Biss. 50 F1 11 15N 16 10W
Bijaipur, India 42 F7 26 2N 77 20 E
Bijapur, Karnataka, India 40 L9 16 50N 75 55 E
Bijapur, Mad. P., India 41 K12 18 50N 80 50 E
Bījār, Iran 44 C5 35 52N 47 35 E
Bijeljina, Bos.-H. 21 B8 44 46N 19 17 E
Bijnor, India 42 E8 29 27N 78 11 E
Bikaner, India 42 E5 28 2N 73 18 E
Bikapur, India 43 F10 26 30N 82 7 E
Bikeqi, China 34 D6 40 43N 111 20 E
Bikfayyā, Lebanon 47 B4 33 55N 35 41 E
Bikin, Russia 27 E14 46 50N 134 20 E
Bikin →, Russia 30 A7 46 51N 134 2 E
Bikini Atoll, Pac. Oc. 64 F8 12 0N 167 30 E
Bila Tserkva, Ukraine 17 D16 49 45N 30 10 E
Bilara, India 42 F5 26 14N 73 53 E
Bilaspur, Mad. P., India 43 H10 22 2N 82 15 E
Bilaspur, Punjab, India 42 D7 31 19N 76 50 E
Bilauk Taungdan, Thailand 38 F2 13 0N 99 0 E
Bilbao, Spain 19 A4 43 16N 2 56W
Bilbo = Bilbao, Spain 19 A4 43 16N 2 56W
Bildudalur, Iceland 8 D2 65 41N 23 36W
Bilecik, Turkey 25 F5 40 5N 30 5 E
Bilhorod-Dnistrovskyy, Ukraine 25 E5 46 11N 30 23 E
Bilibino, Russia 27 C17 68 3N 166 20 E
Bilibiza, Mozam. 55 E5 12 30S 40 20 E
Bilir, Russia 27 C14 65 40N 131 20 E
Bill, U.S.A. 80 D2 43 14N 105 16W
Billabalong, Australia 61 E2 27 25S 115 49 E
Billiluna, Australia 60 C4 19 37S 127 41 E
Billingham, U.K. 10 C6 54 36N 1 17W
Billings, U.S.A. 82 D9 45 47N 108 30W
Billiton Is. = Belitung, Indonesia 36 E3 3 10S 107 50 E
Bilma, Niger 51 E7 18 50N 13 30 E
Biloela, Australia 62 C5 24 24S 150 31 E
Biloxi, U.S.A. 81 K10 30 24N 88 53W
Bilpa Morea Claypan, Australia 62 D2 25 0S 140 0 E
Biltine, Chad 51 F9 14 40N 20 50 E
Bilyana, Australia 62 B4 18 5S 145 50 E
Bima, Indonesia 37 F5 8 22S 118 49 E
Bimbo, C.A.R. 52 D3 4 15N 18 33 E
Bimini Is., Bahamas 88 A4 25 42N 79 25W
Bin Xian, Heilongjiang, China 35 B14 45 42N 127 32 E
Bin Xian, Shaanxi, China 34 G5 35 2N 108 4 E
Bina-Etawah, India 42 G8 24 13N 78 14 E
Bināb, Iran 45 B6 36 35N 48 41 E
Binalbagan, Phil. 37 B6 10 12N 122 50 E
Binalong, Australia 63 E4 34 40S 148 39 E
Bīnālūd, Kūh-e, Iran 45 B8 36 30N 58 30 E
Binatang, Malaysia 36 D4 2 10N 111 40 E
Binbee, Australia 62 C4 20 19S 147 56 E
Binche, Belgium 15 D4 50 26N 4 10 E
Binda, Australia 63 D4 27 52S 147 21 E
Bindle, Australia 63 D4 27 40S 148 45 E
Bindura, Zimbabwe 55 F3 17 18S 31 18 E
Bingara, N.S.W., Australia 63 D5 29 52S 150 36 E
Bingara, Queens., Australia 63 D3 28 10S 144 37 E
Bingham, U.S.A. 71 C6 45 3N 69 53W
Bingham Canyon, U.S.A. 82 F7 40 32N 112 9W
Binghamton, U.S.A. 79 D9 42 6N 75 55W
Binh Dinh = An Nhon, Vietnam 38 F7 13 55N 109 7 E
Binh Khe, Vietnam 38 F7 13 57N 108 51 E
Binh Son, Vietnam 38 E7 15 20N 108 40 E
Binhai, China 35 G10 34 2N 119 49 E
Binisatua, Spain 22 B11 39 50N 4 11 E
Binjai, Indonesia 36 D1 3 20N 98 30 E
Binnaway, Australia 63 E4 31 28S 149 24 E
Binongko, Indonesia 37 F6 5 55S 123 55 E
Binscarth, Canada 73 C8 50 37N 101 17W
Bintan, Indonesia 36 D2 1 0N 104 0 E
Bintulu, Malaysia 36 D4 3 10N 113 0 E
Bintuni, Indonesia 37 E8 2 7S 133 32 E
Binzert = Bizerte, Tunisia 50 A6 37 15N 9 50 E
Bío Bío □, Chile 94 D1 37 35S 72 0W
Bioko, Eq. Guin. 50 H6 3 30N 8 40 E
Bir, India 40 K9 19 0N 75 54 E
Bîr Abu Muḥammad, Egypt 47 F3 29 44N 34 14 E
Bi'r ad Dabbāghāt, Jordan 47 E4 30 26N 35 32 E
Bi'r al Butayyihāt, Jordan 47 F4 29 47N 35 20 E
Bi'r al Mārī, Jordan 47 E4 30 4N 35 33 E
Bi'r al Qattār, Jordan 47 F4 29 47N 35 32 E
Bir Autrun, Sudan 51 E10 18 15N 26 40 E
Bîr Beiḍa, Egypt 47 E3 30 25N 34 29 E
Bîr el 'Abd, Egypt 47 D2 31 2N 33 0 E
Bîr el Biarât, Egypt 47 F3 29 30N 34 43 E
Bîr el Duweidar, Egypt 47 E1 30 56N 32 32 E
Bîr el Garârât, Egypt 47 D2 31 3N 33 34 E
Bîr el Heisi, Egypt 47 F3 29 22N 34 36 E
Bîr el Jafir, Egypt 47 E1 30 50N 32 41 E
Bîr el Mâlḥi, Egypt 47 E2 30 38N 33 19 E
Bîr el Thamâda, Egypt 47 E2 30 12N 33 27 E
Bîr Gebeil Ḥiṣn, Egypt 47 E2 30 2N 33 18 E
Bi'r Ghadīr, Syria 47 A6 34 6N 37 3 E
Bîr Ḥasana, Egypt 47 E2 30 29N 33 46 E
Bi'r Jadīd, Iraq 44 C4 34 1N 42 54 E
Bîr Kaseiba, Egypt 47 E2 31 0N 33 17 E
Bîr Lahfân, Egypt 47 D2 31 0N 33 51 E
Bîr Madkûr, Egypt 47 E1 30 44N 32 33 E
Bir Mogrein, Mauritania 50 C2 25 10N 11 25W
Bi'r Muṭribah, Kuwait 44 D5 29 54N 47 17 E
Bîr Qaṭia, Egypt 47 E1 30 58N 32 45 E
Bîr Ungât, Egypt 51 D11 22 8N 33 48 E
Bira, Indonesia 37 E8 2 3S 132 2 E
Birao, C.A.R. 51 F9 10 20N 22 47 E
Birawa, Zaïre 54 C2 2 20S 28 48 E
Birch Hills, Canada 73 C7 52 59N 105 25W
Birch I., Canada 73 C9 52 26N 99 54W
Birch L., N.W.T., Canada 72 A5 62 4N 116 33W
Birch L., Ont., Canada 70 B1 51 23N 92 18W
Birch L., U.S.A. 70 C1 47 45N 91 51W
Birch Mts., Canada 72 B6 57 30N 113 10W
Birch River, Canada 73 C8 52 24N 101 6W
Birchip, Australia 63 F3 35 56S 142 55 E
Bird, Canada 73 B10 56 30N 94 13W
Bird City, U.S.A. 80 F4 39 45N 101 32W
Bird I. = Aves, I. de, W. Indies 89 C7 15 45N 63 55W
Birdlip, U.K. 11 F5 51 50N 2 5W
Birdsville, Australia 62 D2 25 51S 139 20 E
Birdum, Australia 60 C5 15 39S 133 13 E
Birein, Israel 47 E3 30 50N 34 28 E
Bireuen, Indonesia 36 C1 5 14N 96 39 E
Birigui, Brazil 95 A5 21 18S 50 16W
Birkenhead, U.K. 10 D4 53 23N 3 2W
Bîrlad, Romania 17 E14 46 15N 27 38 E
Bîrlad →, Romania 17 F14 45 38N 27 32 E
Birmingham, U.K. 11 E6 52 29N 1 52W
Birmingham, U.S.A. 77 J2 33 31N 86 48W
Birmitrapur, India 41 H14 22 24N 84 46 E
Birni Nkonni, Niger 50 F6 13 55N 5 15 E
Birnin Kebbi, Nigeria 50 F5 12 32N 4 12 E
Birobidzhan, Russia 27 E14 48 50N 132 50 E
Birr, Ireland 13 C4 53 6N 7 54W
Birrie →, Australia 63 D4 29 43S 146 37 E
Birsilpur, India 42 E5 28 11N 72 15 E
Birsk, Russia 24 C10 55 25N 55 30 E
Birtle, Canada 73 C8 50 30N 101 5W
Birur, India 40 N9 13 30N 75 55 E
Biržai, Lithuania 9 H21 56 11N 24 45 E
Birzebugga, Malta 23 D2 35 49N 14 32 E
Bisa, Indonesia 37 E7 1 15S 127 28 E
Bisalpur, India 43 E8 28 14N 79 48 E
Bisbee, U.S.A. 83 L9 31 27N 109 55W
Biscay, B. of, Atl. Oc. 18 D1 45 0N 2 0W
Biscayne B., U.S.A. 77 N5 25 40N 80 12W
Biscoe Bay, Antarctica 5 D13 77 0S 152 0W
Biscoe Is., Antarctica 5 C17 66 0S 67 0W
Biscostasing, Canada 70 C3 47 18N 82 9W
Bishkek, Kyrgyzstan 26 E8 42 54N 74 46 E
Bishnupur, India 43 H12 23 8N 87 20 E
Bisho, S. Africa 57 E4 32 50S 27 23 E
Bishop, Calif., U.S.A. 83 H4 37 22N 118 24W
Bishop, Tex., U.S.A. 81 M6 27 35N 97 48W
Bishop Auckland, U.K. 10 C6 54 39N 1 40W
Bishop's Falls, Canada 71 C8 49 2N 55 30W
Bishop's Stortford, U.K. 11 F8 51 52N 0 10 E
Bisina, L., Uganda 54 B3 1 38N 33 56 E
Biskra, Algeria 50 B6 34 50N 5 44 E
Bislig, Phil. 37 C7 8 15N 126 27 E
Bismarck, U.S.A. 80 B4 46 48N 100 47W
Bismarck Arch., Papua N. G. 64 H6 2 30S 150 0 E
Biso, Uganda 54 B3 1 44N 31 26 E
Bison, U.S.A. 80 C3 45 31N 102 28W
Bīsotūn, Iran 44 C5 34 23N 47 26 E
Bissagos = Bijagós, Arquipélago dos, Guinea-Biss. 50 F1 11 15N 16 10W
Bissau, Guinea-Biss. 50 F1 11 45N 15 45W
Bissett, Canada 73 C9 51 2N 95 41W
Bistcho L., Canada 72 B5 59 45N 118 50W
Bistriţa, Romania 17 E13 47 9N 24 35 E
Bistriţa →, Romania 17 E14 46 30N 26 57 E
Biswan, India 43 F9 27 29N 81 2 E
Bitam, Gabon 52 D2 2 5N 11 25 E
Bitkine, Chad 51 F8 11 59N 18 13 E
Bitlis, Turkey 25 G7 38 20N 42 3 E
Bitola, Macedonia 21 D9 41 5N 21 10 E
Bitolj = Bitola, Macedonia 21 D9 41 5N 21 10 E
Bitter Creek, U.S.A. 82 F9 41 33N 108 33W
Bitter L. = Buheirat-Murrat-el-Kubra, Egypt 51 B11 30 18N 32 26 E
Bitterfontein, S. Africa 56 E2 31 1S 18 32 E
Bitterroot →, U.S.A. 82 C6 46 52N 114 7W
Bitterroot Range, U.S.A. 82 D6 46 0N 114 20W
Bitterwater, U.S.A. 84 J6 36 23N 121 0W
Biu, Nigeria 51 F7 10 40N 12 3 E
Biwa-Ko, Japan 31 G8 35 15N 136 10 E
Biwabik, U.S.A. 80 B8 47 32N 92 21W
Biyang, China 34 H7 32 38N 113 21 E
Biysk, Russia 26 D9 52 40N 85 0 E
Bizana, S. Africa 57 E4 30 50S 29 52 E
Bizen, Japan 31 G7 34 43N 134 8 E
Bizerte, Tunisia 50 A6 37 15N 9 50 E
Bjargtangar, Iceland 8 D1 65 30N 24 30W
Bjelovar, Croatia 20 B7 45 56N 16 49 E
Bjørnevatn, Norway 8 B23 69 40N 30 0 E
Bjørnøya, Arctic 4 B8 74 30N 19 0 E
Black = Da →, Vietnam 38 B5 21 15N 105 20 E
Black →, Canada 78 B5 44 42N 79 19W
Black →, Ark., U.S.A. 81 H9 35 38N 91 20W
Black →, N.Y., U.S.A. 79 C8 43 59N 76 4W
Black →, Wis., U.S.A. 80 D9 43 57N 91 22W
Black Diamond, Canada 72 C6 50 45N 114 14W
Black Forest = Schwarzwald, Germany 16 D5 48 30N 8 20 E
Black Hills, U.S.A. 80 C3 44 0N 103 45W
Black I., Canada 73 C9 51 12N 96 30W
Black L., Canada 73 B7 59 12N 105 15W
Black L., U.S.A. 76 C3 45 28N 84 16W
Black Mesa, U.S.A. 81 G3 36 58N 102 58W
Black Mt. = Mynydd Du, U.K. 11 F4 51 52N 3 50W
Black Mts., U.K. 11 F4 51 55N 3 7W
Black Range, U.S.A. 83 K10 33 15N 107 50W
Black River, Jamaica 88 C4 18 0N 77 50W
Black River Falls, U.S.A. 80 C9 44 18N 90 51W
Black Sea, Eurasia 25 F6 43 30N 35 0 E
Black Volta →, Africa 50 G4 8 41N 1 33W
Black Warrior →, U.S.A. 77 J2 32 32N 87 51W
Blackall, Australia 62 C4 24 25S 145 45 E
Blackball, N.Z. 59 K3 42 22S 171 26 E
Blackbull, Australia 62 B3 17 55S 141 45 E
Blackburn, U.K. 10 D5 53 45N 2 29W
Blackduck, U.S.A. 80 B7 47 44N 94 33W
Blackfoot, U.S.A. 82 E7 43 11N 112 21W
Blackfoot →, U.S.A. 82 C7 46 52N 113 53W
Blackfoot River Reservoir, U.S.A. 82 E8 43 0N 111 43W
Blackie, Canada 72 C6 50 36N 113 37W
Blackpool, U.K. 10 D4 53 49N 3 3W
Blackriver, U.S.A. 78 B1 44 46N 83 17W
Blacks Harbour, Canada 71 C6 45 3N 66 49W
Blacksburg, U.S.A. 76 G5 37 14N 80 25W
Blacksod B., Ireland 13 B2 54 6N 10 0W
Blackstone, U.S.A. 76 G6 37 4N 78 0W
Blackstone →, Canada 72 A4 61 5N 122 55W
Blackstone Ra., Australia 61 E4 26 0S 128 30 E
Blackville, Canada 71 C6 46 44N 65 50W
Blackwater, Australia 62 C4 23 35S 148 53 E
Blackwater →, Ireland 13 E4 52 4N 7 52W
Blackwater →, U.K. 13 B5 54 31N 6 35W
Blackwater Cr. →, Australia 63 D3 25 56S 144 30 E
Blackwell, U.S.A. 81 G6 36 48N 97 17W
Blackwells Corner, U.S.A. 85 K7 35 37N 119 47W
Blaenau Ffestiniog, U.K. 10 E4 53 0N 3 56W
Blaenau Gwent □, U.K. 11 F4 51 48N 3 12W
Blagodarnoye = Blagodarnyy, Russia 25 E7 45 7N 43 37 E
Blagodarnyy, Russia 25 E7 45 7N 43 37 E
Blagoevgrad, Bulgaria 21 C10 42 2N 23 5 E
Blagoveshchensk, Russia 27 D13 50 20N 127 30 E
Blaine, U.S.A. 84 B4 48 59N 122 45W
Blaine Lake, Canada 73 C7 52 51N 106 52W
Blair, U.S.A. 80 E6 41 33N 96 8W
Blair Athol, Australia 62 C4 22 42S 147 31 E
Blair Atholl, U.K. 12 E5 56 46N 3 50W
Blairgowrie, U.K. 12 E5 56 35N 3 21W
Blairmore, Canada 72 D6 49 40N 114 25W
Blairsden, U.S.A. 84 F6 39 47N 120 37W
Blairsville, U.S.A. 78 F5 40 26N 79 16W
Blake Pt., U.S.A. 80 A10 48 11N 88 25W
Blakely, U.S.A. 77 K3 31 23N 84 56W
Blanc, Mont, Alps 18 D7 45 48N 6 50 E
Blanca, B., Argentina 96 D4 39 10S 61 30W
Blanca Peak, U.S.A. 83 H11 37 35N 105 29W
Blanchard, U.S.A. 81 H6 35 8N 97 39W
Blanche, C., Australia 63 E1 33 1S 134 9 E
Blanche, L., S. Austral., Australia 63 D2 29 15S 139 40 E
Blanche, L., W. Austral., Australia 60 D3 22 25S 123 17 E
Blanco, S. Africa 56 E3 33 55S 22 23 E
Blanco, U.S.A. 81 K5 30 6N 98 25W
Blanco →, Argentina 94 C2 30 20S 68 42W
Blanco, C., Costa Rica 88 E2 9 34N 85 8W
Blanco, C., Spain 22 B9 39 21N 2 51 E
Blanco, C., U.S.A. 82 E1 42 51N 124 34W
Blanda →, Iceland 8 D3 65 37N 20 9W
Blandford Forum, U.K. 11 G5 50 51N 2 9W
Blanding, U.S.A. 83 H9 37 37N 109 29W
Blanes, Spain 19 B7 41 40N 2 48 E
Blankenberge, Belgium 15 C3 51 20N 3 9 E
Blanquillo, Uruguay 95 C4 32 53S 55 37W
Blantyre, Malawi 55 F4 15 45S 35 0 E
Blarney, Ireland 13 E3 51 56N 8 33W
Blåvands Huk, Denmark 9 J13 55 33N 8 4 E
Blaydon, U.K. 10 C6 54 58N 1 42W
Blayney, Australia 63 E4 33 32S 149 14 E
Blaze, Pt., Australia 60 B5 12 56S 130 11 E
Blednaya, Gora, Russia 26 B7 76 20N 65 0 E
Blekinge, Sweden 9 H16 56 25N 15 20 E
Blenheim, Canada 78 D2 42 20N 82 0W
Blenheim, N.Z. 59 J4 41 38S 173 57 E
Bletchley, U.K. 11 F7 51 59N 0 44W
Blida, Algeria 50 A5 36 30N 2 49 E
Bligh Sound, N.Z. 59 L1 44 47S 167 32 E
Blind River, Canada 70 C3 46 10N 82 58W
Blitar, Indonesia 37 H15 8 5S 112 11 E
Blitta, Togo 50 G5 8 23N 1 6 E
Block I., U.S.A. 79 E13 41 11N 71 35W
Block Island Sd., U.S.A. 79 E13 41 15N 71 40W
Blodgett Iceberg Tongue, Antarctica 5 C9 66 8S 130 35 E
Bloemfontein, S. Africa 56 D4 29 6S 26 7 E
Bloemhof, S. Africa 56 D4 27 38S 25 32 E
Blois, France 18 C4 47 35N 1 20 E
Blönduós, Iceland 8 D3 65 40N 20 12W
Bloodvein →, Canada 73 C9 51 47N 96 43W
Bloody Foreland, Ireland 13 A3 55 10N 8 17W
Bloomer, U.S.A. 80 C9 45 6N 91 29W
Bloomfield, Australia 62 B4 15 56S 145 22 E
Bloomfield, Canada 78 C7 43 59N 77 14W
Bloomfield, Iowa, U.S.A. 80 E8 40 45N 92 25W
Bloomfield, N. Mex., U.S.A. 83 H10 36 43N 107 59W
Bloomfield, Nebr., U.S.A. 80 D6 42 36N 97 39W
Bloomington, Ill., U.S.A. 80 E10 40 28N 89 0W
Bloomington, Ind., U.S.A. 76 F2 39 10N 86 32W
Bloomington, Minn., U.S.A. 80 C8 44 50N 93 17W
Bloomsburg, U.S.A. 79 F8 41 0N 76 27W
Blora, Indonesia 37 G14 6 57S 111 25 E
Blossburg, U.S.A. 78 E7 41 41N 77 4W
Blouberg, S. Africa 57 C4 23 8S 28 59 E
Blountstown, U.S.A. 77 K3 30 27N 85 3W
Blue Island, U.S.A. 76 E2 41 40N 87 40W
Blue Lake, U.S.A. 82 F2 40 53N 123 59W
Blue Mesa Reservoir, U.S.A. 83 G10 38 28N 107 20W
Blue Mts., Oreg., U.S.A. 82 D4 45 15N 119 0W
Blue Mts., Pa., U.S.A. 79 F8 40 30N 76 30W
Blue Mud B., Australia 62 A2 13 30S 136 0 E
Blue Nile = Nîl el Azraq →, Sudan 51 E11 15 38N 32 31 E
Blue Rapids, U.S.A. 80 F6 39 41N 96 39W
Blue Ridge Mts., U.S.A. 77 G5 36 30N 80 15W
Blue Stack Mts., Ireland 13 B3 54 42N 8 12W
Blueberry →, Canada 72 B4 56 45N 120 49W
Bluefield, U.S.A. 76 G5 37 15N 81 17W
Bluefields, Nic. 88 D3 12 20N 83 50W
Bluff, Australia 62 C4 23 35S 149 4 E
Bluff, N.Z. 59 M2 46 37S 168 20 E
Bluff, U.S.A. 83 H9 37 17N 109 33W
Bluff Knoll, Australia 61 F2 34 24S 118 15 E
Bluff Pt., Australia 61 E1 27 50S 114 5 E
Bluffton, U.S.A. 76 E3 40 44N 85 11W
Blumenau, Brazil 95 B6 27 0S 49 0W
Blunt, U.S.A. 80 C4 44 31N 99 59W
Bly, U.S.A. 82 E3 42 24N 121 3W
Blyth, Canada 78 C3 43 44N 81 26W
Blyth, U.K. 10 B6 55 8N 1 31W
Blythe, U.S.A. 85 M12 33 37N 114 36W
Blythe Bridge, U.K. 10 E5 52 59N 2 4W
Bo, S. Leone 50 G2 7 55N 11 50W
Bo Duc, Vietnam 39 G6 11 58N 106 50 E
Bo Hai, China 35 E10 39 0N 119 0 E
Bo Xian, China 34 H8 33 55N 115 41 E
Boa Vista, Brazil 92 C6 2 48N 60 30W
Boaco, Nic. 88 D2 12 29N 85 35W
Bo'ai, China 34 G7 35 10N 113 3 E
Boardman, U.S.A. 78 E4 41 2N 80 40W
Boatman, Australia 63 D4 27 16S 146 55 E
Bobadah, Australia 63 E4 32 19S 146 41 E
Bobbili, India 41 K13 18 35N 83 30 E
Bobcaygeon, Canada 70 D4 44 33N 78 33W
Bobo-Dioulasso, Burkina Faso 50 F4 11 8N 4 13W
Bóbr →, Poland 16 B8 52 4N 15 4 E
Bobraomby, Tanjon' i, Madag. 57 A8 12 40S 49 10 E
Bobruysk = Babruysk, Belarus 17 B15 53 10N 29 15 E
Bôca do Acre, Brazil 92 E5 8 50S 67 27W
Boca Raton, U.S.A. 77 M5 26 21N 80 5W
Bocaiúva, Brazil 93 G10 17 7S 43 49W
Bocanda, Ivory C. 50 G4 7 5N 4 31W
Bocaranga, C.A.R. 51 G8 7 0N 15 35 E
Bocas del Toro, Panama 88 E3 9 15N 82 20W
Bochnia, Poland 17 D11 49 58N 20 27 E
Bochum, Germany 16 C4 51 28N 7 13 E
Bocoyna, Mexico 86 B3 27 52N 107 35W
Boda, C.A.R. 52 D3 4 19N 17 26 E
Bodaybo, Russia 27 D12 57 50N 114 0 E
Boddington, Australia 61 F2 32 50S 116 30 E
Bodega Bay, U.S.A. 84 G3 38 20N 123 3W
Boden, Sweden 8 D19 65 50N 21 42 E
Bodensee, Europe 16 E5 47 35N 9 25 E
Bodhan, India 40 K10 18 40N 77 44 E
Bodmin, U.K. 11 G3 50 28N 4 43W
Bodmin Moor, U.K. 11 G3 50 33N 4 36W
Bodø, Norway 8 C16 67 17N 14 24 E
Bodrog →, Hungary 17 D11 48 11N 21 22 E
Bodrum, Turkey 21 F12 37 5N 27 30 E
Boende, Zaïre 52 E4 0 24S 21 12 E
Boerne, U.S.A. 81 L5 29 47N 98 44W
Boffa, Guinea 50 F2 10 16N 14 3W
Bogalusa, U.S.A. 81 K10 30 47N 89 52W
Bogan Gate, Australia 63 E4 33 7S 147 49 E
Bogantungan, Australia 62 C4 23 41S 147 17 E
Bogata, U.S.A. 81 J7 33 28N 95 13W
Boggabilla, Australia 63 D5 28 36S 150 24 E
Boggabri, Australia 63 E5 30 45S 150 5 E
Boggeragh Mts., Ireland 13 D3 52 2N 8 55W
Bognor Regis, U.K. 11 G7 50 47N 0 40W
Bogo, Phil. 37 B6 11 3N 124 0 E
Bogong, Mt., Australia 63 F4 36 47S 147 17 E
Bogor, Indonesia 37 G12 6 36S 106 48 E
Bogorodskoye, Russia 27 D15 52 22N 140 30 E
Bogotá, Colombia 92 C4 4 34N 74 0W
Bogotol, Russia 26 D9 56 15N 89 50 E
Bogra, Bangla. 41 G16 24 51N 89 22 E
Boguchany, Russia 27 D10 58 40N 97 30 E
Bogué, Mauritania 50 E2 16 45N 14 10W
Bohemia Downs, Australia 60 C4 18 53S 126 14 E
Bohemian Forest = Böhmerwald, Germany 16 D7 49 8N 13 14 E
Bohena Cr. →, Australia 63 E4 30 17S 149 42 E
Böhmerwald, Germany 16 D7 49 8N 13 14 E
Bohol, Phil. 37 C6 9 50N 124 10 E
Bohol Sea, Phil. 37 C6 9 0N 124 0 E
Bohotleh, Somali Rep. 46 F4 8 20N 46 25 E
Bohuslän, Sweden 9 G14 58 25N 12 0 E
Boi, Pta. de, Brazil 95 A6 23 55S 45 15W
Boileau, C., Australia 60 C3 17 40S 122 7 E
Boise, U.S.A. 82 E5 43 37N 116 13W
Boise City, U.S.A. 81 G3 36 44N 102 31W
Boissevain, Canada 73 D8 49 15N 100 5W
Bojador C., W. Sahara 50 C2 26 0N 14 30W
Bojana →, Albania 21 D8 41 52N 19 22 E
Bojnūrd, Iran 45 B8 37 30N 57 20 E
Bojonegoro, Indonesia 37 G14 7 11S 111 54 E
Boké, Guinea 50 F2 10 56N 14 17W
Bokhara →, Australia 63 D4 29 55S 146 42 E
Boknafjorden, Norway 9 G11 59 14N 5 40 E
Bokoro, Chad 51 F8 12 25N 17 14 E
Bokote, Zaïre 52 E4 0 12S 21 8 E
Bokungu, Zaïre 52 E4 0 35S 22 50 E
Bol, Chad 51 F7 13 30N 14 40 E
Bolama, Guinea-Biss. 50 F1 11 30N 15 30W
Bolan Pass, Pakistan 40 E5 29 50N 67 20 E
Bolaños →, Mexico 86 C4 21 14N 104 8W
Bolbec, France 18 B4 49 30N 0 30 E
Boldājī, Iran 45 D6 31 56N 51 3 E
Bole, China 32 B3 45 11N 81 37 E
Bolekhiv, Ukraine 17 D12 49 0N 23 57 E
Bolekhov = Bolekhiv, Ukraine 17 D12 49 0N 23 57 E
Bolesławiec, Poland 16 C8 51 17N 15 37 E
Bolgrad = Bolhrad, Ukraine 17 F15 45 40N 28 32 E
Bolhrad, Ukraine 17 F15 45 40N 28 32 E
Bolinao C., Phil. 37 A5 16 23N 119 55 E
Bolívar, Argentina 94 D3 36 15S 60 53W
Bolívar, Colombia 92 C3 2 0N 77 0W
Bolivar, Mo., U.S.A. 81 G8 37 37N 93 25W
Bolivar, Tenn., U.S.A. 81 H10 35 12N 89 0W
Bolivia ■, S. Amer. 92 G6 17 6S 64 0W
Bolivian Plateau, S. Amer. 90 E4 20 0S 67 30W
Bollnäs, Sweden 9 F17 61 21N 16 24 E
Bollon, Australia 63 D4 28 2S 147 29 E
Bolmen, Sweden 9 H15 56 55N 13 40 E
Bolobo, Zaïre 52 E3 2 6S 16 20 E
Bologna, Italy 20 B4 44 29N 11 20 E
Bologoye, Russia 24 C5 57 55N 34 5 E
Bolomba, Zaïre 52 D3 0 35N 19 0 E
Bolonchenticul, Mexico 87 D7 20 0N 89 49W
Bolong, Phil. 37 C6 7 6N 122 14 E
Boloven, Cao Nguyen, Laos 38 E6 15 10N 106 30 E
Bolpur, India 43 H12 23 40N 87 45 E
Bolsena, L. di, Italy 20 C4 42 36N 11 56 E
Bolshereche, Russia 26 D8 56 4N 74 45 E
Bolshevik, Ostrov, Russia 27 B11 78 30N 102 0 E
Bolshezemelskaya Tundra, Russia 24 A10 67 0N 56 0 E
Bolshoi Kavkas = Caucasus Mountains, Eurasia 25 F7 42 50N 44 0 E
Bolshoy Anyuy →, Russia 27 C17 68 30N 160 49 E
Bolshoy Atlym, Russia 26 C7 62 25N 66 50 E

Bolshoy Begichev, Ostrov, *Russia* **27 B12** 74 20N 112 30 E
Bolshoy Lyakhovskiy, Ostrov, *Russia* **27 B15** 73 35N 142 0 E
Bolshoy Tyuters, Ostrov, *Russia* **9 G22** 59 51N 27 13 E
Bolsward, *Neths.* **15 A5** 53 3N 5 32 E
Bolton, *Canada* **78 C5** 43 54N 79 45W
Bolton, *U.K.* **10 D5** 53 35N 2 26W
Bolu, *Turkey* **25 F5** 40 45N 31 35 E
Bolungavík, *Iceland* **8 C2** 66 9N 23 15W
Bolvadin, *Turkey* **25 G5** 38 45N 31 4 E
Bolzano, *Italy* **20 A4** 46 31N 11 22 E
Bom Despacho, *Brazil* **93 G9** 19 43S 45 15W
Bom Jesus da Lapa, *Brazil* **93 F10** 13 15S 43 25W
Boma, *Zaïre* **52 F2** 5 50S 13 4 E
Bomaderry, *Australia* **63 E5** 34 52S 150 37 E
Bombala, *Australia* **63 F4** 36 56S 149 15 E
Bombay, *India* **40 K8** 18 55N 72 50 E
Bomboma, *Zaïre* **52 D3** 2 25N 18 55 E
Bombombwa, *Zaïre* **54 B2** 1 40N 25 40 E
Bomili, *Zaïre* **54 B2** 1 45N 27 5 E
Bømlo, *Norway* **9 G11** 59 37N 5 13 E
Bomokandi →, *Zaïre* **54 B2** 3 39N 26 8 E
Bomongo, *Zaïre* **52 D3** 1 27N 18 21 E
Bomu →, *C.A.R.* **52 D4** 4 40N 22 30 E
Bon, C., *Tunisia* **51 A7** 37 1N 11 2 E
Bon Sar Pa, *Vietnam* **38 F6** 12 24N 107 35 E
Bonaire, *Neth. Ant.* **89 D6** 12 10N 68 15W
Bonang, *Australia* **63 F4** 37 11S 148 41 E
Bonanza, *Nic.* **88 D3** 13 54N 84 35W
Bonaparte Arch., *Australia* **60 B3** 14 0S 124 30 E
Bonaventure, *Canada* **71 C6** 48 5N 65 32W
Bonavista, *Canada* **71 C9** 48 40N 53 5W
Bonavista, C., *Canada* **71 C9** 48 42N 53 5W
Bondo, *Zaïre* **54 B1** 3 55N 23 53 E
Bondoukou, *Ivory C.* **50 G4** 8 2N 2 47W
Bondowoso, *Indonesia* **37 G15** 7 55S 113 49 E
Bone, Teluk, *Indonesia* **37 E6** 4 10S 120 50 E
Bonerate, *Indonesia* **37 F6** 7 25S 121 5 E
Bonerate, Kepulauan, *Indonesia* **37 F6** 6 30S 121 10 E
Bo'ness, *U.K.* **12 E5** 56 1N 3 37W
Bonete, Cerro, *Argentina* **94 B2** 27 55S 68 40W
Bong Son = Hoai Nhon, *Vietnam* **38 E7** 14 28N 109 1 E
Bongandanga, *Zaïre* **52 D4** 1 24N 21 3 E
Bongor, *Chad* **51 F8** 10 35N 15 20 E
Bonham, *U.S.A.* **81 J6** 33 35N 96 11W
Bonifacio, *France* **18 F8** 41 24N 9 10 E
Bonin Is. = Ogasawara Gunto, *Pac. Oc.* **28 G18** 27 0N 142 0 E
Bonn, *Germany* **16 C4** 50 46N 7 6 E
Bonne Terre, *U.S.A.* **81 G9** 37 55N 90 33W
Bonners Ferry, *U.S.A.* **82 B5** 48 42N 116 19W
Bonney, L., *Australia* **63 F3** 37 50S 140 20 E
Bonnie Downs, *Australia* **62 C3** 22 7S 143 50 E
Bonnie Rock, *Australia* **61 F2** 30 29S 118 22 E
Bonny, Bight of, *Africa* **52 D1** 3 30N 9 20 E
Bonnyville, *Canada* **73 C6** 54 20N 110 45W
Bonoi, *Indonesia* **37 E9** 1 45S 137 41 E
Bonsall, *U.S.A.* **85 M9** 33 16N 117 14W
Bontang, *Indonesia* **36 D5** 0 10N 117 30 E
Bonthain, *Indonesia* **37 F5** 5 34S 119 56 E
Bonthe, *S. Leone* **50 G2** 7 30N 12 33W
Bontoc, *Phil.* **37 A6** 17 7N 120 58 E
Bonython Ra., *Australia* **60 D4** 23 40S 128 45 E
Bookabie, *Australia* **61 F5** 31 50S 132 41 E
Booker, *U.S.A.* **81 G4** 36 27N 100 32W
Boolaboolka L., *Australia* **63 E3** 32 38S 143 10 E
Booligal, *Australia* **63 E3** 33 58S 144 53 E
Boom, *Belgium* **15 C4** 51 6N 4 20 E
Boonah, *Australia* **63 D5** 27 58S 152 41 E
Boone, *Iowa, U.S.A.* **80 D8** 42 4N 93 53W
Boone, *N.C., U.S.A.* **77 G5** 36 13N 81 41W
Booneville, *Ark., U.S.A.* **81 H8** 35 8N 93 55W
Booneville, *Miss., U.S.A.* **77 H1** 34 39N 88 34W
Boonville, *Calif., U.S.A.* **84 F3** 39 1N 123 22W
Boonville, *Ind., U.S.A.* **76 F2** 38 3N 87 16W
Boonville, *Mo., U.S.A.* **80 F8** 38 58N 92 44W
Boonville, *N.Y., U.S.A.* **79 C9** 43 29N 75 20W
Boorindal, *Australia* **63 E4** 30 22S 146 11 E
Boorowa, *Australia* **63 E4** 34 28S 148 44 E
Boothia, Gulf of, *Canada* **69 A11** 71 0N 90 0W
Boothia Pen., *Canada* **68 A10** 71 0N 94 0W
Bootle, *U.K.* **10 D4** 53 28N 3 1W
Booué, *Gabon* **52 E2** 0 5S 11 55 E
Boquete, *Panama* **88 E3** 8 46N 82 27W
Boquilla, Presa de la, *Mexico* **86 B3** 27 40N 105 30W
Boquillas del Carmen, *Mexico* **86 B4** 29 17N 102 53W
Bor, *Serbia, Yug.* **21 B10** 44 5N 22 7 E
Bôr, *Sudan* **51 G11** 6 10N 31 40 E
Bor Mashash, *Israel* **47 D3** 31 7N 34 50 E
Boradā →, *Syria* **47 B5** 33 33N 36 34 E
Borah Peak, *U.S.A.* **82 D7** 44 8N 113 47W
Borama, *Somali Rep.* **46 F3** 9 55N 43 7 E
Borås, *Sweden* **9 H15** 57 43N 12 56 E
Borāzjān, *Iran* **45 D6** 29 22N 51 10 E
Borba, *Brazil* **92 D7** 4 12S 59 34W
Borborema, Planalto da, *Brazil* **90 D7** 7 0S 37 0W
Bord Khūn-e Now, *Iran* **45 D6** 28 3N 51 28 E
Borda, C., *Australia* **63 F2** 35 45S 136 34 E
Bordeaux, *France* **18 D3** 44 50N 0 36W
Borden, *Australia* **61 F2** 34 3S 118 12 E
Borden, *Canada* **71 C7** 46 18N 63 47W
Borden I., *Canada* **4 B2** 78 30N 111 30W
Borders □, *U.K.* **12 F6** 55 35N 2 50W
Bordertown, *Australia* **63 F3** 36 19S 140 45 E
Borðeyri, *Iceland* **8 D3** 65 12N 21 6W
Bordj Fly Ste. Marie, *Algeria* **50 C4** 27 19N 2 32W
Bordj-in-Eker, *Algeria* **50 D6** 24 9N 5 3 E
Bordj Omar Driss, *Algeria* **50 C6** 28 10N 6 40 E
Bordj-Tarat, *Algeria* **50 C6** 25 55N 9 3 E
Borgå = Porvoo, *Finland* **9 F21** 60 24N 25 40 E
Borgarfjörður, *Iceland* **8 D7** 65 31N 13 49W
Borgarnes, *Iceland* **8 D3** 64 32N 21 55W
Børgefjellet, *Norway* **8 D15** 65 20N 13 45 E
Borger, *Neths.* **15 B6** 52 54N 6 44 E
Borger, *U.S.A.* **81 H4** 35 39N 101 24W
Borgholm, *Sweden* **9 H17** 56 52N 16 39 E
Borikhane, *Laos* **38 C4** 18 33N 103 43 E
Borisoglebsk, *Russia* **25 D7** 51 27N 42 5 E
Borisov = Barysaw, *Belarus* **17 A15** 54 17N 28 28 E
Borja, *Peru* **92 D3** 4 20S 77 40W
Borkou, *Chad* **51 E8** 18 15N 18 50 E
Borkum, *Germany* **16 B4** 53 34N 6 40 E
Borlänge, *Sweden* **9 F16** 60 29N 15 26 E
Borley, C., *Antarctica* **5 C5** 66 15S 52 30 E
Borneo, *E. Indies* **36 D5** 1 0N 115 0 E
Bornholm, *Denmark* **9 J16** 55 10N 15 0 E
Borobudur, *Indonesia* **37 G14** 7 36S 110 13 E
Borogontsy, *Russia* **27 C14** 62 42N 131 8 E
Boromo, *Burkina Faso* **50 F4** 11 45N 2 58W
Boron, *U.S.A.* **85 L9** 35 0N 117 39W
Borongan, *Phil.* **37 B7** 11 37N 125 26 E
Bororen, *Australia* **62 C5** 24 13S 151 33 E
Borovichi, *Russia* **24 C5** 58 25N 33 55 E
Borrego Springs, *U.S.A.* **85 M10** 33 15N 116 23W
Borroloola, *Australia* **62 B2** 16 4S 136 17 E
Borşa, *Romania* **17 E13** 47 41N 24 50 E
Borth, *U.K.* **11 E3** 52 29N 4 2W
Borūjerd, *Iran* **45 C6** 33 55N 48 50 E
Boryslav, *Ukraine* **17 D12** 49 18N 23 28 E
Borzya, *Russia* **27 D12** 50 24N 116 31 E
Bosa, *Italy* **20 D3** 40 18N 8 30 E
Bosanska Gradiška, *Bos.-H.* **20 B7** 45 10N 17 15 E
Bosaso, *Somali Rep.* **46 E4** 11 12N 49 18 E
Boscastle, *U.K.* **11 G3** 50 41N 4 42W
Boshan, *China* **35 F9** 36 28N 117 49 E
Boshof, *S. Africa* **56 D4** 28 31S 25 13 E
Boshrūyeh, *Iran* **45 C8** 33 50N 57 30 E
Bosna →, *Bos.-H.* **21 B8** 45 4N 18 29 E
Bosna i Hercegovina = Bosnia-Herzegovina ■, *Europe* **20 B7** 44 0N 17 0 E
Bosnia-Herzegovina ■, *Europe* **20 B7** 44 0N 17 0 E
Bosnik, *Indonesia* **37 E9** 1 5S 136 10 E
Bosobolo, *Zaïre* **52 D3** 4 15N 19 50 E
Bosporus = Karadeniz Boğazı, *Turkey* **21 D13** 41 10N 29 10 E
Bossangoa, *C.A.R.* **51 G8** 6 35N 17 30 E
Bossembélé, *C.A.R.* **51 G8** 5 25N 17 40 E
Bossier City, *U.S.A.* **81 J8** 32 31N 93 44W
Bosso, *Niger* **51 F7** 13 43N 13 19 E
Bostānābād, *Iran* **44 B5** 37 50N 46 50 E
Bosten Hu, *China* **32 B3** 41 55N 87 40 E
Boston, *U.K.* **10 E7** 52 59N 0 2W
Boston, *U.S.A.* **79 D13** 42 22N 71 4W
Boston Bar, *Canada* **72 D4** 49 52N 121 30W
Boswell, *Canada* **72 D5** 49 28N 116 45W
Boswell, *Okla., U.S.A.* **81 H7** 34 2N 95 52W
Boswell, *Pa., U.S.A.* **78 F5** 40 10N 79 2W
Botad, *India* **42 H4** 22 15N 71 40 E
Botany B., *Australia* **63 E5** 34 0S 151 14 E
Botene, *Laos* **38 D3** 17 35N 101 12 E
Bothaville, *S. Africa* **56 D4** 27 23S 26 34 E
Bothnia, G. of, *Europe* **8 E19** 63 0N 20 15 E
Bothwell, *Australia* **62 G4** 42 20S 147 1 E
Bothwell, *Canada* **78 D3** 42 38N 81 52W
Botletle →, *Botswana* **56 C3** 20 10S 23 15 E
Botoşani, *Romania* **17 E14** 47 42N 26 41 E
Botswana ■, *Africa* **56 C3** 22 0S 24 0 E
Bottineau, *U.S.A.* **80 A4** 48 50N 100 27W
Bottrop, *Germany* **15 C6** 51 31N 6 58 E
Botucatu, *Brazil* **95 A6** 22 55S 48 30W
Botwood, *Canada* **71 C8** 49 6N 55 23W
Bou Djébéha, *Mali* **50 E4** 18 25N 2 45W
Bou Izakarn, *Morocco* **50 C3** 29 12N 9 46W
Bouaké, *Ivory C.* **50 G3** 7 40N 5 2W
Bouar, *C.A.R.* **52 C3** 6 0N 15 40 E
Bouârfa, *Morocco* **50 B4** 32 32N 1 58W
Bouca, *C.A.R.* **51 G8** 6 45N 18 25 E
Boucaut B., *Australia* **62 A1** 12 0S 134 25 E
Bougainville C., *Australia* **60 B4** 13 57S 126 4 E
Bougainville Reef, *Australia* **62 B4** 15 30S 147 5 E
Bougie = Bejaia, *Algeria* **50 A6** 36 42N 5 2 E
Bougouni, *Mali* **50 F3** 11 30N 7 20W
Bouillon, *Belgium* **15 E5** 49 44N 5 3 E
Boulder, *Colo., U.S.A.* **80 E2** 40 1N 105 17W
Boulder, *Mont., U.S.A.* **82 C7** 46 14N 112 7W
Boulder City, *U.S.A.* **85 K12** 35 59N 114 50W
Boulder Creek, *U.S.A.* **84 H4** 37 7N 122 7W
Boulder Dam = Hoover Dam, *U.S.A.* **85 K12** 36 1N 114 44W
Boulia, *Australia* **62 C2** 22 52S 139 51 E
Boulogne-sur-Mer, *France* **18 A4** 50 42N 1 36 E
Boultoum, *Niger* **50 F7** 14 45N 10 25 E
Boun Neua, *Laos* **38 B3** 21 38N 101 54 E
Boun Tai, *Laos* **38 B3** 21 23N 101 58 E
Bouna, *Ivory C.* **50 G4** 9 10N 3 0W
Boundary Peak, *U.S.A.* **84 H8** 37 51N 118 21W
Boundiali, *Ivory C.* **50 G3** 9 30N 6 20W
Bountiful, *U.S.A.* **82 F8** 40 53N 111 53W
Bounty Is., *Pac. Oc.* **64 M9** 48 0S 178 30 E
Bourbonnais, *France* **18 C5** 46 28N 3 0 E
Bourem, *Mali* **50 E4** 17 0N 0 24W
Bourg-en-Bresse, *France* **18 C6** 46 13N 5 12 E
Bourg-St.-Maurice, *France* **18 D7** 45 35N 6 46 E
Bourges, *France* **18 C5** 47 9N 2 25 E
Bourget, *Canada* **79 A9** 45 26N 75 9W
Bourgogne, *France* **18 C6** 47 0N 4 50 E
Bourke, *Australia* **63 E4** 30 8S 145 55 E
Bournemouth, *U.K.* **11 G6** 50 43N 1 52W
Bouse, *U.S.A.* **85 M13** 33 56N 114 0W
Bousso, *Chad* **51 F8** 10 34N 16 52 E
Boutilimit, *Mauritania* **50 E2** 17 45N 14 40W
Bouvet I. = Bouvetøya, *Antarctica* **3 G10** 54 26S 3 24 E
Bouvetøya, *Antarctica* **3 G10** 54 26S 3 24 E
Bovigny, *Belgium* **15 D5** 50 12N 5 55 E
Bovill, *U.S.A.* **82 C5** 46 51N 116 24W
Bow Island, *Canada* **72 D6** 49 50N 111 23W
Bowbells, *U.S.A.* **80 A3** 48 48N 102 15W
Bowdle, *U.S.A.* **80 C5** 45 27N 99 39W
Bowelling, *Australia* **61 F2** 33 25S 116 30 E
Bowen, *Australia* **62 C4** 20 0S 148 16 E
Bowen Mts., *Australia* **63 F4** 37 0S 147 50 E
Bowie, *Ariz., U.S.A.* **83 K9** 32 19N 109 29W
Bowie, *Tex., U.S.A.* **81 J6** 33 34N 97 51W
Bowkān, *Iran* **44 B5** 36 31N 46 12 E
Bowland, Forest of, *U.K.* **10 D5** 54 0N 2 30W
Bowling Green, *Ky., U.S.A.* **76 G2** 36 59N 86 27W
Bowling Green, *Ohio, U.S.A.* **76 E4** 41 23N 83 39W
Bowling Green, C., *Australia* **62 B4** 19 19S 147 25 E
Bowman, *U.S.A.* **80 B3** 46 11N 103 24W
Bowman I., *Antarctica* **5 C8** 65 0S 104 0 E
Bowmans, *Australia* **63 E2** 34 10S 138 17 E
Bowmanville, *Canada* **70 D4** 43 55N 78 41W
Bowmore, *U.K.* **12 F2** 55 45N 6 17W
Bowral, *Australia* **63 E5** 34 26S 150 27 E
Bowraville, *Australia* **63 E5** 30 37S 152 52 E
Bowron →, *Canada* **72 C4** 54 3N 121 50W
Bowser L., *Canada* **72 B3** 56 30N 129 30W
Bowsman, *Canada* **73 C8** 52 14N 101 12W
Bowwood, *Zambia* **55 F2** 17 5S 26 20 E
Boxtel, *Neths.* **15 C5** 51 36N 5 20 E
Boyce, *U.S.A.* **81 K8** 31 23N 92 40W
Boyer →, *Canada* **72 B5** 58 27N 115 57W
Boyle, *Ireland* **13 C3** 53 59N 8 18W
Boyne →, *Ireland* **13 C5** 53 43N 6 15W
Boyne City, *U.S.A.* **76 C3** 45 13N 85 1W
Boynton Beach, *U.S.A.* **77 M5** 26 32N 80 4W
Boyoma, Chutes, *Zaïre* **54 B2** 0 35N 25 23 E
Boyup Brook, *Australia* **61 F2** 33 50S 116 23 E
Boz Dağları, *Turkey* **21 E13** 38 20N 28 0 E
Bozburun, *Turkey* **21 F13** 36 43N 28 8 E
Bozcaada, *Turkey* **21 E12** 39 49N 26 3 E
Bozdoğan, *Turkey* **21 F13** 37 40N 28 17 E
Bozeman, *U.S.A.* **82 D8** 45 41N 111 2W
Bozen = Bolzano, *Italy* **20 A4** 46 31N 11 22 E
Bozoum, *C.A.R.* **51 G8** 6 25N 16 35 E
Bra, *Italy* **20 B2** 44 42N 7 51 E
Brabant □, *Belgium* **15 D4** 50 46N 4 30 E
Brabant L., *Canada* **73 B8** 55 58N 103 43W
Brač, *Croatia* **20 C7** 43 20N 16 40 E
Bracadale, L., *U.K.* **12 D2** 57 20N 6 30W
Bracciano, L. di, *Italy* **20 C5** 42 7N 12 14 E
Bracebridge, *Canada* **70 C4** 45 2N 79 19W
Brach, *Libya* **51 C7** 27 31N 14 20 E
Bräcke, *Sweden* **9 E16** 62 45N 15 26 E
Brackettville, *U.S.A.* **81 L4** 29 19N 100 25W
Brad, *Romania* **17 E12** 46 10N 22 50 E
Bradenton, *U.S.A.* **77 M4** 27 30N 82 34W
Bradford, *Canada* **78 B5** 44 7N 79 34W
Bradford, *U.K.* **10 D6** 53 47N 1 45W
Bradford, *Pa., U.S.A.* **78 E6** 41 58N 78 38W
Bradford, *Vt., U.S.A.* **79 C12** 43 59N 72 9W
Bradley, *Ark., U.S.A.* **81 J8** 33 6N 93 39W
Bradley, *Calif., U.S.A.* **84 K6** 35 52N 120 48W
Bradley, *S. Dak., U.S.A.* **80 C6** 45 5N 97 39W
Bradley Institute, *Zimbabwe* **55 F3** 17 7S 31 25 E
Bradore Bay, *Canada* **71 B8** 51 27N 57 18W
Bradshaw, *Australia* **60 C5** 15 21S 130 16 E
Brady, *U.S.A.* **81 K5** 31 9N 99 20W
Braemar, *Australia* **63 E2** 33 12S 139 35 E
Braeside, *Canada* **79 A8** 45 28N 76 24W
Braga, *Portugal* **19 B1** 41 35N 8 25W
Bragado, *Argentina* **94 D3** 35 2S 60 27W
Bragança, *Brazil* **93 D9** 1 0S 47 2W
Bragança, *Portugal* **19 B2** 41 48N 6 50W
Bragança Paulista, *Brazil* **95 A6** 22 55S 46 32W
Brahmanbaria, *Bangla.* **41 H17** 23 58N 91 15 E
Brahmani →, *India* **41 J15** 20 39N 86 46 E
Brahmaputra →, *India* **43 G13** 23 58N 89 50 E
Braich-y-pwll, *U.K.* **10 E3** 52 47N 4 46W
Braidwood, *Australia* **63 F4** 35 27S 149 49 E
Brăila, *Romania* **17 F14** 45 19N 27 59 E
Brainerd, *U.S.A.* **80 B7** 46 22N 94 12W
Braintree, *U.K.* **11 F8** 51 53N 0 34 E
Braintree, *U.S.A.* **79 D14** 42 13N 71 0W
Brak →, *S. Africa* **56 D3** 29 35S 22 55 E
Brakwater, *Namibia* **56 C2** 22 28S 17 3 E
Bralorne, *Canada* **72 C4** 50 50N 122 50W
Brampton, *Canada* **70 D4** 43 45N 79 45W
Bramwell, *Australia* **62 A3** 12 8S 142 37 E
Branco →, *Brazil* **92 D6** 1 20S 61 50W
Brandenburg = Neubrandenburg, *Germany* **16 B7** 53 33N 13 15 E
Brandenburg, *Germany* **16 B7** 52 25N 12 33 E
Brandenburg □, *Germany* **16 B6** 52 50N 13 0 E
Brandfort, *S. Africa* **56 D4** 28 40S 26 30 E
Brandon, *Canada* **73 D9** 49 50N 99 57W
Brandon, *U.S.A.* **79 C11** 43 48N 73 4W
Brandon B., *Ireland* **13 D1** 52 17N 10 8W
Brandon Mt., *Ireland* **13 D1** 52 15N 10 15W
Brandsen, *Argentina* **94 D4** 35 10S 58 15W
Brandvlei, *S. Africa* **56 E3** 30 25S 20 30 E
Branford, *U.S.A.* **79 E12** 41 17N 72 49W
Braniewo, *Poland* **17 A10** 54 25N 19 50 E
Bransfield Str., *Antarctica* **5 C18** 63 0S 59 0W
Branson, *Colo., U.S.A.* **81 G3** 37 1N 103 53W
Branson, *Mo., U.S.A.* **81 G8** 36 39N 93 13W
Brantford, *Canada* **70 D3** 43 10N 80 15W
Branxholme, *Australia* **63 F3** 37 52S 141 49 E
Bras d'Or, L., *Canada* **71 C7** 45 50N 60 50W
Brasil, Planalto, *Brazil* **90 E6** 18 0S 46 30W
Brasiléia, *Brazil* **92 F5** 11 0S 68 45W
Brasília, *Brazil* **93 G9** 15 47S 47 55W
Braslaw, *Belarus* **9 J22** 55 38N 27 0 E
Braşov, *Romania* **17 F13** 45 38N 25 35 E
Brasschaat, *Belgium* **15 C4** 51 19N 4 27 E
Brassey, Banjaran, *Malaysia* **36 D5** 5 0N 117 15 E
Brassey Ra., *Australia* **61 E3** 25 8S 122 15 E
Brasstown Bald, *U.S.A.* **77 H4** 34 53N 83 49W
Brastad, *Sweden* **9 G14** 58 23N 11 30 E
Bratislava, *Slovak Rep.* **17 D9** 48 10N 17 7 E
Bratsk, *Russia* **27 D11** 56 10N 101 30 E
Brattleboro, *U.S.A.* **79 D12** 42 51N 72 34W
Braunau, *Austria* **16 D7** 48 15N 13 3 E
Braunschweig, *Germany* **16 B6** 52 15N 10 31 E
Braunton, *U.K.* **11 F3** 51 7N 4 10W
Brava, *Somali Rep.* **46 G3** 1 20N 44 8 E
Bravo del Norte →, *Mexico* **86 B5** 25 57N 97 9W
Bravo del Norte, R. → = Grande, Rio →, *U.S.A.* **81 N6** 25 58N 97 9W
Brawley, *U.S.A.* **85 N11** 32 59N 115 31W
Bray, *Ireland* **13 C5** 53 13N 6 7W
Bray, Mt., *Australia* **62 A1** 14 0S 134 30 E
Bray, Pays de, *France* **18 B4** 49 46N 1 26 E
Brazeau →, *Canada* **72 C5** 52 55N 115 14W
Brazil, *U.S.A.* **76 F2** 39 32N 87 8W
Brazil ■, *S. Amer.* **93 F9** 12 0S 50 0W
Brazilian Highlands = Brasil, Planalto, *Brazil* **90 E6** 18 0S 46 30W
Brazo Sur →, *S. Amer.* **94 B4** 25 21S 57 42W
Brazos →, *U.S.A.* **81 L7** 28 53N 95 23W
Brazzaville, *Congo* **52 E3** 4 9S 15 12 E
Brčko, *Bos.-H.* **21 B8** 44 54N 18 46 E
Breadalbane, *Australia* **62 C2** 23 50S 139 35 E
Breadalbane, *U.K.* **12 E4** 56 30N 4 15W
Breaden, L., *Australia* **61 E4** 25 51S 125 28 E
Breaksea Sd., *N.Z.* **59 L1** 45 35S 166 35 E
Bream B., *N.Z.* **59 F5** 35 56S 174 28 E
Bream Hd., *N.Z.* **59 F5** 35 51S 174 36 E
Breas, *Chile* **94 B1** 25 29S 70 24W
Brebes, *Indonesia* **37 G13** 6 52S 109 3 E
Brechin, *Canada* **78 B5** 44 32N 79 10W
Brechin, *U.K.* **12 E6** 56 44N 2 39W
Breckenridge, *Colo., U.S.A.* **82 G10** 39 29N 106 3W
Breckenridge, *Minn., U.S.A.* **80 B6** 46 16N 96 35W
Breckenridge, *Tex., U.S.A.* **81 J5** 32 45N 98 54W
Breckland, *U.K.* **11 E8** 52 30N 0 40 E
Brecon, *U.K.* **11 F4** 51 57N 3 23W
Brecon Beacons, *U.K.* **11 F4** 51 53N 3 26W
Breda, *Neths.* **15 C4** 51 35N 4 45 E
Bredasdorp, *S. Africa* **56 E3** 34 33S 20 2 E
Bredbo, *Australia* **63 F4** 35 58S 149 10 E
Bregenz, *Austria* **16 E5** 47 30N 9 45 E
Breiðafjörður, *Iceland* **8 D2** 65 15N 23 15W
Brejo, *Brazil* **93 D10** 3 41S 42 47W
Bremen, *Germany* **16 B5** 53 4N 8 47 E
Bremer I., *Australia* **62 A2** 12 5S 136 45 E
Bremerhaven, *Germany* **16 B5** 53 33N 8 36 E
Bremerton, *U.S.A.* **84 C4** 47 34N 122 38W
Brenham, *U.S.A.* **81 K6** 30 10N 96 24W
Brenner P., *Austria* **16 E6** 47 2N 11 30 E
Brent, *Canada* **70 C4** 46 2N 78 29W
Brent, *U.K.* **11 F7** 51 33N 0 16W
Brentwood, *U.K.* **11 F8** 51 37N 0 19 E
Brentwood, *U.S.A.* **79 F11** 40 47N 73 15W
Bréscia, *Italy* **20 B4** 45 33N 10 15 E
Breskens, *Neths.* **15 C3** 51 23N 3 33 E
Breslau = Wrocław, *Poland* **17 C9** 51 5N 17 5 E
Bressanone, *Italy* **20 A4** 46 43N 11 39 E
Bressay, *U.K.* **12 A7** 60 9N 1 6W
Brest, *Belarus* **17 B12** 52 10N 23 40 E
Brest, *France* **18 B1** 48 24N 4 31W
Brest-Litovsk = Brest, *Belarus* **17 B12** 52 10N 23 40 E
Bretagne, *France* **18 B2** 48 10N 3 0W
Breton, *Canada* **72 C6** 53 7N 114 28W
Breton Sd., *U.S.A.* **81 L10** 29 35N 89 15W
Brett, C., *N.Z.* **59 F5** 35 10S 174 20 E
Brevard, *U.S.A.* **77 H4** 35 14N 82 44W
Brewarrina, *Australia* **63 D4** 30 0S 146 51 E
Brewer, *U.S.A.* **71 D6** 44 48N 68 46W
Brewer, Mt., *U.S.A.* **84 J8** 36 44N 118 28W
Brewster, *N.Y., U.S.A.* **79 E11** 41 23N 73 37W
Brewster, *Wash., U.S.A.* **82 B4** 48 6N 119 47W
Brewster, Kap, *Greenland* **4 B6** 70 7N 22 0W
Brewton, *U.S.A.* **77 K2** 31 7N 87 4W
Breyten, *S. Africa* **57 D4** 26 16S 30 0 E
Brezhnev = Naberezhnyye Chelny, *Russia* **24 C9** 55 42N 52 19 E
Bria, *C.A.R.* **51 G9** 6 30N 21 58 E
Briançon, *France* **18 D7** 44 54N 6 39 E
Bribie I., *Australia* **63 D5** 27 0S 153 10 E
Bridgehampton, *U.S.A.* **79 F12** 40 56N 72 19W
Bridgend, *U.K.* **11 F4** 51 30N 3 34W
Bridgend □, *U.K.* **11 F4** 51 36N 3 36W
Bridgeport, *Calif., U.S.A.* **83 G4** 38 15N 119 14W
Bridgeport, *Conn., U.S.A.* **79 E11** 41 11N 73 12W
Bridgeport, *Nebr., U.S.A.* **80 E3** 41 40N 103 6W
Bridgeport, *Tex., U.S.A.* **81 J6** 33 13N 97 45W
Bridger, *U.S.A.* **82 D9** 45 18N 108 55W
Bridgeton, *U.S.A.* **76 F8** 39 26N 75 14W
Bridgetown, *Australia* **61 F2** 33 58S 116 7 E
Bridgetown, *Barbados* **89 D8** 13 5N 59 30W
Bridgetown, *Canada* **71 D6** 44 55N 65 18W
Bridgewater, *Canada* **71 D7** 44 25N 64 31W
Bridgewater, *Mass., U.S.A.* **79 E14** 41 59N 70 58W
Bridgewater, *S. Dak., U.S.A.* **80 D6** 43 33N 97 30W
Bridgewater, C., *Australia* **63 F3** 38 23S 141 23 E
Bridgnorth, *U.K.* **11 E5** 52 32N 2 25W
Bridgton, *U.S.A.* **79 B14** 44 3N 70 42W
Bridgwater, *U.K.* **11 F4** 51 8N 2 59W
Bridlington, *U.K.* **10 C7** 54 5N 0 12W
Bridport, *Australia* **62 G4** 40 59S 147 23 E
Bridport, *U.K.* **11 G5** 50 44N 2 45W
Brig, *Switz.* **16 E4** 46 18N 7 59 E
Brigg, *U.K.* **10 D7** 53 34N 0 28W
Briggsdale, *U.S.A.* **80 E2** 40 38N 104 20W
Brigham City, *U.S.A.* **82 F7** 41 31N 112 1W
Bright, *Australia* **63 F4** 36 42S 146 56 E
Brighton, *Australia* **63 F2** 35 5S 138 30 E
Brighton, *Canada* **70 D4** 44 2N 77 44W
Brighton, *U.K.* **11 G7** 50 49N 0 7W
Brighton, *U.S.A.* **80 F2** 39 59N 104 49W
Brilliant, *Canada* **72 D5** 49 19N 117 38W
Brilliant, *U.S.A.* **78 F4** 40 15N 80 39W
Bríndisi, *Italy* **21 D7** 40 39N 17 55 E
Brinkley, *U.S.A.* **81 H9** 34 53N 91 12W
Brinkworth, *Australia* **63 E2** 33 42S 138 26 E
Brinnon, *U.S.A.* **84 C4** 47 41N 122 54W
Brion, I., *Canada* **71 C7** 47 46N 61 26W
Brisbane, *Australia* **63 D5** 27 25S 153 2 E
Brisbane →, *Australia* **63 D5** 27 24S 153 9 E
Bristol, *U.K.* **11 F5** 51 26N 2 35W
Bristol, *Conn., U.S.A.* **79 E12** 41 40N 72 57W
Bristol, *Pa., U.S.A.* **79 F10** 40 6N 74 51W
Bristol, *R.I., U.S.A.* **79 E13** 41 40N 71 16W
Bristol, *S. Dak., U.S.A.* **80 C6** 45 21N 97 45W
Bristol, *Tenn., U.S.A.* **77 G4** 36 36N 82 11W
Bristol □, *U.K.* **11 F5** 51 27N 2 36W
Bristol B., *U.S.A.* **68 C4** 58 0N 160 0W
Bristol Channel, *U.K.* **11 F3** 51 18N 4 30W
Bristol I., *Antarctica* **5 B1** 58 45S 28 0W
Bristol L., *U.S.A.* **83 J5** 34 23N 116 50W

Bristow, U.S.A. 81 H6 35 50N 96 23W
British Columbia □, Canada 72 C3 55 0N 125 15W
British Isles, Europe 6 E5 54 0N 4 0W
Brits, S. Africa 57 D4 25 37S 27 48 E
Britstown, S. Africa 56 E3 30 37S 23 30 E
Britt, Canada 70 C3 45 46N 80 34W
Brittany = Bretagne, France 18 B2 48 10N 3 0W
Britton, U.S.A. 80 C6 45 48N 97 45W
Brive-la-Gaillarde, France 18 D4 45 10N 1 32 E
Brixen = Bressanone, Italy 20 A4 46 43N 11 39 E
Brixton, Australia 62 C3 23 32S 144 57 E
Brlik, Kazakstan 26 E8 43 40N 73 49 E
Brno, Czech. 17 D9 49 10N 16 35 E
Broad —►, U.S.A. 77 J5 34 1N 81 4W
Broad Arrow, Australia 61 F3 30 23S 121 15 E
Broad B., U.K. 12 C2 58 14N 6 18W
Broad Haven, Ireland 13 B2 54 20N 9 55W
Broad Law, U.K. 12 F5 55 30N 3 21W
Broad Sd., Australia 62 C4 22 0S 149 45 E
Broadhurst Ra., Australia 60 D3 22 30S 122 30 E
Broads, The, U.K. 10 E9 52 45N 1 30 E
Broadus, U.S.A. 80 C2 45 27N 105 25W
Broadview, Canada 73 C8 50 22N 102 35W
Brochet, Canada 73 B8 57 53N 101 40W
Brochet, L., Canada 73 B8 58 36N 101 35W
Brock, Canada 73 C7 51 26N 108 43W
Brocken, Germany 16 C6 51 47N 10 37 E
Brockport, U.S.A. 78 C7 43 13N 77 56W
Brockton, U.S.A. 79 D13 42 5N 71 1W
Brockville, Canada 70 D4 44 35N 75 41W
Brockway, Mont., U.S.A. 80 B2 47 18N 105 45W
Brockway, Pa., U.S.A. 78 E6 41 15N 78 47W
Brocton, U.S.A. 78 D5 42 23N 79 26W
Brodeur Pen., Canada 69 A11 72 30N 88 10W
Brodick, U.K. 12 F3 55 35N 5 9W
Brodnica, Poland 17 B10 53 15N 19 25 E
Brody, Ukraine 17 C13 50 5N 25 10 E
Brogan, U.S.A. 82 D5 44 15N 117 31W
Broken Arrow, U.S.A. 81 G7 36 3N 95 48W
Broken Bow, Nebr., U.S.A. 80 E5 41 24N 99 38W
Broken Bow, Okla., U.S.A. 81 H7 34 2N 94 44W
Broken Hill = Kabwe, Zambia 55 E2 14 30S 28 29 E
Broken Hill, Australia 63 E3 31 58S 141 29 E
Bromfield, U.K. 11 E5 52 24N 2 45W
Bromley, U.K. 11 F8 51 24N 0 2 E
Brønderslev, Denmark 9 H13 57 16N 9 57 E
Bronkhorstspruit, S. Africa 57 D4 25 46S 28 45 E
Brønnøysund, Norway 8 D15 65 28N 12 14 E
Bronte, U.S.A. 81 K4 31 53N 100 18W
Bronte Park, Australia 62 G4 42 8S 146 30 E
Brook Park, U.S.A. 78 E4 41 24N 80 51W
Brookfield, U.S.A. 80 F8 39 47N 93 4W
Brookhaven, U.S.A. 81 K9 31 35N 90 26W
Brookings, Oreg., U.S.A. 82 E1 42 3N 124 17W
Brookings, S. Dak., U.S.A. 80 C6 44 19N 96 48W
Brooklin, Canada 78 C6 43 55N 78 55W
Brooklyn Park, U.S.A. 80 C8 45 6N 93 23W
Brookmere, Canada 72 D4 49 52N 120 53W
Brooks, Canada 72 C6 50 35N 111 55W
Brooks B., Canada 72 C3 50 15N 127 55W
Brooks L., Canada 73 A7 61 55N 106 35W
Brooks Ra., U.S.A. 68 B5 68 40N 147 0W
Brooksville, U.S.A. 77 L4 28 33N 82 23W
Brookville, U.S.A. 76 F3 39 25N 85 1W
Brooloo, Australia 63 D5 26 30S 152 43 E
Broom, L., U.K. 12 D3 57 55N 5 15W
Broome, Australia 60 C3 18 0S 122 15 E
Broomehill, Australia 61 F2 33 51S 117 39 E
Brora, U.K. 12 C5 58 0N 3 52W
Brora —►, U.K. 12 C5 58 0N 3 51W
Brosna —►, Ireland 13 C4 53 14N 7 58W
Brothers, U.S.A. 82 E3 43 49N 120 36W
Brough, U.K. 10 C5 54 32N 2 18W
Broughton Island, Canada 69 B13 67 33N 63 0W
Broughty Ferry, U.K. 12 E6 56 29N 2 51W
Brouwershaven, Neths. 15 C3 51 45N 3 55 E
Browerville, U.S.A. 80 B7 46 5N 94 52W
Brown, Pt., Australia 63 E1 32 32S 133 50 E
Brown Willy, U.K. 11 G3 50 35N 4 37W
Brownfield, U.S.A. 81 J3 33 11N 102 17W
Browning, U.S.A. 82 B7 48 34N 113 1W
Brownlee, Canada 73 C7 50 43N 106 1W
Brownsville, Oreg., U.S.A. 82 D2 44 24N 122 59W
Brownsville, Tenn., U.S.A. 81 H10 35 36N 89 16W
Brownsville, Tex., U.S.A. 81 N6 25 54N 97 30W
Brownwood, U.S.A. 81 K5 31 43N 98 59W
Brownwood, L., U.S.A. 81 K5 31 51N 98 35W
Browse I., Australia 60 B3 14 7S 123 33 E
Bruas, Malaysia 39 K3 4 30N 100 47 E
Bruay-en-Artois, France 18 A5 50 29N 2 33 E
Bruce, Mt., Australia 60 D2 22 37S 118 8 E
Bruce Pen., Canada 78 A3 45 0N 81 30W
Bruce Rock, Australia 61 F2 31 52S 118 8 E
Bruck an der Leitha, Austria 17 D9 48 1N 16 47 E
Bruck an der Mur, Austria 16 E8 47 24N 15 16 E
Brue —►, U.K. 11 F5 51 13N 2 59W
Bruges = Brugge, Belgium 15 C3 51 13N 3 13 E
Brugge, Belgium 15 C3 51 13N 3 13 E
Brûlé, Canada 72 C5 53 15N 117 58W
Brumado, Brazil 93 F10 14 14S 41 40W
Brumunddal, Norway 9 F14 60 53N 10 56 E
Brunchilly, Australia 62 B1 18 50S 134 30 E
Brundidge, U.S.A. 77 K3 31 43N 85 49W
Bruneau, U.S.A. 82 E6 42 53N 115 48W
Bruneau —►, U.S.A. 82 E6 42 56N 115 57W
Brunei = Bandar Seri Begawan, Brunei 36 C4 4 52N 115 0 E
Brunei ■, Asia 36 D4 4 50N 115 0 E
Brunette Downs, Australia 62 B2 18 40S 135 55 E
Brunner, L., N.Z. 59 K3 42 37S 171 27 E
Bruno, Canada 73 C7 52 20N 105 30W
Brunswick = Braunschweig, Germany 16 B6 52 15N 10 31 E
Brunswick, Ga., U.S.A. 77 K5 31 10N 81 30W
Brunswick, Maine, U.S.A. 71 D6 43 55N 69 58W
Brunswick, Md., U.S.A. 76 F7 39 19N 77 38W
Brunswick, Mo., U.S.A. 80 F8 39 26N 93 8W
Brunswick, Ohio, U.S.A. 78 E3 41 14N 81 51W
Brunswick, Pen. de, Chile 96 G2 53 30S 71 30W
Brunswick B., Australia 60 C3 15 15S 124 50 E
Brunswick Junction, Australia 61 F2 33 15S 115 50 E
Bruny I., Australia 62 G4 43 20S 147 15 E
Brus Laguna, Honduras 88 C3 15 47N 84 35W
Brush, U.S.A. 80 E3 40 15N 103 37W
Brushton, U.S.A. 79 B10 44 50N 74 31W
Brusque, Brazil 95 B6 27 5S 49 0W
Brussel, Belgium 15 D4 50 51N 4 21 E
Brussels = Brussel, Belgium 15 D4 50 51N 4 21 E
Brussels, Canada 78 C3 43 44N 81 15W
Bruthen, Australia 63 F4 37 42S 147 50 E
Bruxelles = Brussel, Belgium 15 D4 50 51N 4 21 E
Bryan, Ohio, U.S.A. 76 E3 41 28N 84 33W
Bryan, Tex., U.S.A. 81 K6 30 40N 96 22W
Bryan, Mt., Australia 63 E2 33 30S 139 0 E
Bryansk, Russia 24 D5 53 13N 34 25 E
Bryant, U.S.A. 80 C6 44 35N 97 28W
Bryne, Norway 9 G11 58 44N 5 38 E
Bryson City, U.S.A. 77 H4 35 26N 83 27W
Bsharri, Lebanon 47 A5 34 15N 36 0 E
Bū Baqarah, U.A.E. 45 E8 25 35N 56 25 E
Bu Craa, W. Sahara 50 C2 26 45N 12 50W
Bū Ḩasā, U.A.E. 45 F7 23 30N 53 20 E
Bua Yai, Thailand 38 E4 15 33N 102 26 E
Buapinang, Indonesia 37 E6 4 40S 121 30 E
Buayan, Phil. 37 C7 6 3N 125 6 E
Bubanza, Burundi 54 C2 3 6S 29 23 E
Būbiyān, Kuwait 45 D6 29 45N 48 15 E
Bucaramanga, Colombia 92 B4 7 0N 73 0W
Buccaneer Arch., Australia 60 C3 16 7S 123 20 E
Buchach, Ukraine 17 D13 49 5N 25 25 E
Buchan, U.K. 12 D6 57 32N 2 21W
Buchan Ness, U.K. 12 D7 57 29N 1 46W
Buchanan, Canada 73 C8 51 40N 102 45W
Buchanan, Liberia 50 G2 5 57N 10 2W
Buchanan, L., Queens., Australia 62 C4 21 35S 145 52 E
Buchanan, L., W. Austral., Australia 61 E3 25 33S 123 2 E
Buchanan, L., U.S.A. 81 K5 30 45N 98 25W
Buchanan Cr. —►, Australia 62 B2 19 13S 136 33 E
Buchans, Canada 71 C8 48 50N 56 52W
Bucharest = Bucureşti, Romania 17 F14 44 27N 26 10 E
Buchon, Pt., U.S.A. 84 K6 35 15N 120 54W
Buckeye, U.S.A. 83 K7 33 22N 112 35W
Buckhannon, U.S.A. 76 F5 39 0N 80 8W
Buckhaven, U.K. 12 E5 56 11N 3 3W
Buckie, U.K. 12 D6 57 41N 2 58W
Buckingham, Canada 70 C4 45 37N 75 24W
Buckingham, U.K. 11 F7 51 59N 0 57W
Buckingham B., Australia 62 A2 12 10S 135 40 E
Buckinghamshire □, U.K. 11 F7 51 53N 0 55W
Buckle Hd., Australia 60 B4 14 26S 127 52 E
Buckleboo, Australia 63 E2 32 54S 136 12 E
Buckley, U.S.A. 82 C2 47 10N 122 2W
Buckley —►, Australia 62 C2 20 10S 138 49 E
Bucklin, U.S.A. 81 G5 37 33N 99 38W
Bucks L., U.S.A. 84 F5 39 54N 121 12W
Buctouche, Canada 71 C7 46 30N 64 45W
Bucureşti, Romania 17 F14 44 27N 26 10 E
Bucyrus, U.S.A. 76 E4 40 48N 82 59W
Budalin, Burma 41 H19 22 20N 95 10 E
Budapest, Hungary 17 E10 47 29N 19 5 E
Budaun, India 43 E8 28 5N 79 10 E
Budd Coast, Antarctica 5 C8 68 0S 112 0 E
Bude, U.K. 11 G3 50 49N 4 34W
Budennovsk, Russia 25 F7 44 50N 44 10 E
Budge Budge = Baj Baj, India 43 H13 22 30N 88 5 E
Budgewoi, Australia 63 E5 33 13S 151 34 E
Budjala, Zaïre 52 D3 2 50N 19 40 E
Buellton, U.S.A. 85 L6 34 37N 120 12W
Buena Park, U.S.A. 85 M9 33 52N 117 59W
Buena Vista, Colo., U.S.A. 83 G10 38 51N 106 8W
Buena Vista, Va., U.S.A. 76 G6 37 44N 79 21W
Buena Vista L., U.S.A. 85 K7 35 12N 119 18W
Buenaventura, Colombia 92 C3 3 53N 77 4W
Buenaventura, Mexico 86 B3 29 50N 107 30W
Buenos Aires, Argentina 94 C4 34 30S 58 20W
Buenos Aires, Costa Rica 88 E3 9 10N 83 20W
Buenos Aires □, Argentina 94 D4 36 30S 60 0W
Buenos Aires, L., Chile 96 F2 46 35S 72 30W
Buffalo, Mo., U.S.A. 81 G8 37 39N 93 6W
Buffalo, N.Y., U.S.A. 78 D6 42 53N 78 53W
Buffalo, Okla., U.S.A. 81 G5 36 50N 99 38W
Buffalo, S. Dak., U.S.A. 80 C3 45 35N 103 33W
Buffalo, Wyo., U.S.A. 82 D10 44 21N 106 42W
Buffalo —►, Canada 72 A5 60 5N 115 5W
Buffalo Head Hills, Canada 72 B5 57 25N 115 55W
Buffalo L., Canada 72 C6 52 27N 112 54W
Buffalo Narrows, Canada 73 B7 55 51N 108 29W
Buffels —►, S. Africa 56 D2 29 36S 17 3 E
Buford, U.S.A. 77 H4 34 10N 84 0W
Bug —► = Buh —►, Ukraine 25 E5 46 59N 31 58 E
Bug —►, Poland 17 B11 52 31N 21 5 E
Buga, Colombia 92 C3 4 0N 76 15W
Buganda, Uganda 54 C3 0 0 31 30 E
Buganga, Uganda 54 C3 0 3S 32 0 E
Bugel, Tanjung, Indonesia 36 F4 6 26S 111 3 E
Bugibba, Malta 23 D1 35 57N 14 25 E
Bugsuk, Phil. 36 C5 8 15N 117 15 E
Bugulma, Russia 24 D9 54 33N 52 48 E
Bugun Shara, Mongolia 32 B5 49 0N 104 0 E
Buguruslan, Russia 24 D9 53 39N 52 26 E
Buh —►, Ukraine 25 E5 46 59N 31 58 E
Buheirat-Murrat-el-Kubra, Egypt 51 B11 30 18N 32 26 E
Buhl, Idaho, U.S.A. 82 E6 42 36N 114 46W
Buhl, Minn., U.S.A. 80 B8 47 30N 92 46W
Buick, U.S.A. 81 G9 37 38N 91 2W
Builth Wells, U.K. 11 E4 52 9N 3 25W
Buir Nur, Mongolia 33 B6 47 50N 117 42 E
Bujumbura, Burundi 54 C2 3 16S 29 18 E
Bukachacha, Russia 27 D12 52 55N 116 50 E
Bukama, Zaïre 55 D2 9 10S 25 50 E
Bukavu, Zaïre 54 C2 2 20S 28 52 E
Bukene, Tanzania 54 C3 4 15S 32 48 E
Bukhara = Bukhoro, Uzbekistan 26 F7 39 48N 64 25 E
Bukhoro, Uzbekistan 26 F7 39 48N 64 25 E
Bukima, Tanzania 54 C3 1 50S 33 25 E
Bukit Mertajam, Malaysia 39 K3 5 22N 100 28 E
Bukittinggi, Indonesia 36 E2 0 20S 100 20 E
Bukoba, Tanzania 54 C3 1 20S 31 49 E
Bukoba □, Tanzania 54 C3 1 30S 32 0 E
Bukuya, Uganda 54 B3 0 40N 31 52 E
Bula, Indonesia 37 E8 3 6S 130 30 E
Bulahdelah, Australia 63 E5 32 23S 152 13 E
Bulan, Phil. 37 B6 12 40N 123 52 E
Bulandshahr, India 42 E7 28 28N 77 51 E
Bulawayo, Zimbabwe 55 G2 20 7S 28 32 E
Buldan, Turkey 21 E13 38 2N 28 50 E
Bulgaria ■, Europe 21 C11 42 35N 25 30 E
Bulgroo, Australia 63 D3 25 47S 143 58 E
Bulgunnia, Australia 63 E1 30 10S 134 53 E
Bulhar, Somali Rep. 46 E3 10 25N 44 30 E
Buli, Teluk, Indonesia 37 D7 1 5N 128 25 E
Buliluyan, C., Phil. 36 C5 8 20N 117 15 E
Bulkley —►, Canada 72 B3 55 15N 127 40W
Bull Shoals L., U.S.A. 81 G8 36 22N 92 35W
Bullara, Australia 60 D1 22 40S 114 3 E
Bullaring, Australia 61 F2 32 30S 117 45 E
Bulli, Australia 63 E5 34 15S 150 57 E
Bullock Creek, Australia 62 B3 17 43S 144 31 E
Bulloo —►, Australia 63 D3 28 43S 142 30 E
Bulloo Downs, Queens., Australia 63 D3 28 31S 142 57 E
Bulloo Downs, W. Austral., Australia 60 D2 24 0S 119 32 E
Bulloo L., Australia 63 D3 28 43S 142 25 E
Bulls, N.Z. 59 J5 40 10S 175 24 E
Bulnes, Chile 94 D1 36 42S 72 19W
Bulo Burti, Somali Rep. 46 G4 3 50N 45 33 E
Bulsar = Valsad, India 40 J8 20 40N 72 58 E
Bultfontein, S. Africa 56 D4 28 18S 26 10 E
Bulukumba, Indonesia 37 F6 5 33S 120 11 E
Bulun, Russia 27 B13 70 37N 127 30 E
Bulus, Russia 27 C13 63 10N 129 10 E
Bumba, Zaïre 52 D4 2 13N 22 30 E
Bumbiri I., Tanzania 54 C3 1 40S 31 55 E
Bumhpa Bum, Burma 41 F20 26 51N 97 14 E
Bumi —►, Zimbabwe 55 F2 17 0S 28 20 E
Buna, Kenya 54 B4 2 58N 39 30 E
Bunazi, Tanzania 54 C3 1 3S 31 23 E
Bunbah, Khalīj, Libya 51 B9 32 20N 23 15 E
Bunbury, Australia 61 F2 33 20S 115 35 E
Buncrana, Ireland 13 A4 55 8N 7 27W
Bundaberg, Australia 63 C5 24 54S 152 22 E
Bundey —►, Australia 62 C2 21 46S 135 37 E
Bundi, India 42 G6 25 30N 75 35 E
Bundooma, Australia 62 C1 24 54S 134 16 E
Bundoran, Ireland 13 B3 54 28N 8 16W
Bung Kan, Thailand 38 C4 18 23N 103 37 E
Bungatakada, Japan 31 H5 33 35N 131 25 E
Bungil Cr. —►, Australia 62 D4 27 5S 149 5 E
Bungo-Suidō, Japan 31 H6 33 0N 132 15 E
Bungoma, Kenya 54 B3 0 34N 34 34 E
Bungu, Tanzania 54 D4 7 35S 39 0 E
Bunia, Zaïre 54 B3 1 35N 30 20 E
Bunji, Pakistan 43 B6 35 45N 74 40 E
Bunkie, U.S.A. 81 K8 30 57N 92 11W
Bunnell, U.S.A. 77 L5 29 28N 81 16W
Buntok, Indonesia 36 E4 1 40S 114 58 E
Bunyu, Indonesia 36 D5 3 35N 117 50 E
Buol, Indonesia 37 D6 1 15N 121 32 E
Buon Brieng, Vietnam 38 F7 13 9N 108 12 E
Buon Me Thuot, Vietnam 38 F7 12 40N 108 3 E
Buong Long, Cambodia 38 F6 13 44N 106 59 E
Buorkhaya, Mys, Russia 27 B14 71 50N 132 40 E
Buqayq, Si. Arabia 45 E6 26 0N 49 45 E
Bur Acaba, Somali Rep. 46 G3 3 12N 44 20 E
Bûr Safâga, Egypt 51 C11 26 43N 33 57 E
Bûr Sa'îd, Egypt 51 B11 31 16N 32 18 E
Bûr Sûdân, Sudan 51 E12 19 32N 37 9 E
Bura, Kenya 54 C4 1 4S 39 58 E
Burao, Somali Rep. 46 F4 9 32N 45 32 E
Burāq, Syria 47 B5 33 11N 36 29 E
Buras, U.S.A. 81 L10 29 22N 89 32W
Buraydah, Si. Arabia 44 E5 26 20N 44 8 E
Burbank, U.S.A. 85 L8 34 11N 118 19W
Burcher, Australia 63 E4 33 30S 147 16 E
Burdekin —►, Australia 62 B4 19 38S 147 25 E
Burdett, Canada 72 D6 49 50N 111 32W
Burdur, Turkey 25 G5 37 45N 30 17 E
Burdwan = Barddhaman, India 43 H12 23 14N 87 39 E
Bure —►, U.K. 10 E9 52 38N 1 43 E
Bureya —►, Russia 27 E13 49 27N 129 30 E
Burford, Canada 78 C4 43 7N 80 27W
Burgas, Bulgaria 21 C12 42 33N 27 29 E
Burgeo, Canada 71 C8 47 37N 57 38W
Burgersdorp, S. Africa 56 E4 31 0S 26 20 E
Burges, Mt., Australia 61 F3 30 50S 121 5 E
Burgos, Spain 19 A4 42 21N 3 41W
Burgsvik, Sweden 9 H18 57 3N 18 19 E
Burgundy = Bourgogne, France 18 C6 47 0N 4 50 E
Burhaniye, Turkey 21 E12 39 30N 26 58 E
Burhanpur, India 40 J10 21 18N 76 14 E
Burias, Phil. 37 B6 12 55N 123 5 E
Burica, Pta., Costa Rica 88 E3 8 3N 82 51W
Burigi, L., Tanzania 54 C3 2 2S 31 22 E
Burin, Canada 71 C8 47 1N 55 14W
Buriram, Thailand 38 E4 15 0N 103 0 E
Burj Sāfitā, Syria 44 C3 34 48N 36 7 E
Burji, Ethiopia 51 G12 5 29N 37 51 E
Burkburnett, U.S.A. 81 H5 34 6N 98 34W
Burke, U.S.A. 82 C6 47 31N 115 49W
Burke —►, Australia 62 C2 23 12S 139 33 E
Burketown, Australia 62 B2 17 45S 139 33 E
Burkina Faso ■, Africa 50 F4 12 0N 1 0W
Burk's Falls, Canada 70 C4 45 37N 79 24W
Burley, U.S.A. 82 E7 42 32N 113 48W
Burlingame, U.S.A. 84 H4 37 35N 122 21W
Burlington, Canada 78 C5 43 18N 79 45W
Burlington, Colo., U.S.A. 80 F3 39 18N 102 16W
Burlington, Iowa, U.S.A. 80 E9 40 49N 91 14W
Burlington, Kans., U.S.A. 80 F7 38 12N 95 45W
Burlington, N.C., U.S.A. 77 G6 36 6N 79 26W
Burlington, N.J., U.S.A. 79 F10 40 4N 74 51W
Burlington, Vt., U.S.A. 79 B11 44 29N 73 12W
Burlington, Wash., U.S.A. 84 B4 48 28N 122 20W
Burlington, Wis., U.S.A. 76 D1 42 41N 88 17W
Burlyu-Tyube, Kazakstan 26 E8 46 30N 79 10 E
Burma ■, Asia 41 J20 21 0N 96 30 E
Burnaby I., Canada 72 C2 52 25N 131 19W
Burnet, U.S.A. 81 K5 30 45N 98 14W
Burney, U.S.A. 82 F3 40 53N 121 40W
Burngup, Australia 61 F2 33 2S 118 42 E
Burnham, U.S.A. 78 F7 40 38N 77 34W
Burnie, Australia 62 G4 41 4S 145 56 E
Burnley, U.K. 10 D5 53 47N 2 14W
Burns, Oreg., U.S.A. 82 E4 43 35N 119 3W
Burns, Wyo., U.S.A. 80 E2 41 12N 104 21W
Burns Lake, Canada 72 C3 54 20N 125 45W
Burnside —►, Canada 68 B9 66 51N 108 4W
Burnside, L., Australia 61 E3 25 22S 123 0 E
Burnsville, U.S.A. 80 C8 44 47N 93 17W
Burnt River, Canada 78 B6 44 41N 78 42W
Burntwood —►, Canada 73 B9 56 8N 96 34W
Burntwood L., Canada 73 B8 55 22N 100 26W
Burqān, Kuwait 44 D5 29 0N 47 57 E
Burra, Australia 63 E2 33 40S 138 55 E
Burramurra, Australia 62 C2 20 25S 137 15 E
Burren Junction, Australia 63 E4 30 7S 148 59 E
Burrendong Dam, Australia 63 E4 32 39S 149 6 E
Burrinjuck Res., Australia 63 F4 35 0S 148 36 E
Burro, Serranías del, Mexico 86 B4 29 0N 102 0W
Burruyacú, Argentina 94 B3 26 30S 64 40W
Burry Port, U.K. 11 F3 51 41N 4 15W
Bursa, Turkey 21 D13 40 15N 29 5 E
Burstall, Canada 73 C7 50 39N 109 54W
Burton L., Canada 70 B4 54 45N 78 20W
Burton upon Trent, U.K. 10 E6 52 48N 1 38W
Burtundy, Australia 63 E3 33 45S 142 15 E
Buru, Indonesia 37 E7 3 30S 126 30 E
Burûn, Râs, Egypt 47 D2 31 14N 33 7 E
Burundi ■, Africa 54 C3 3 15S 30 0 E
Bururi, Burundi 54 C2 3 57S 29 37 E
Burutu, Nigeria 50 G6 5 20N 5 29 E
Burwell, U.S.A. 80 E5 41 47N 99 8W
Bury, U.K. 10 D5 53 35N 2 17W
Bury St. Edmunds, U.K. 11 E8 52 15N 0 43 E
Buryatia □, Russia 27 D11 53 0N 110 0 E
Busango Swamp, Zambia 55 E2 14 15S 25 45 E
Buşayrah, Syria 44 C4 35 9N 40 26 E
Buşayyah, Iraq 44 D5 30 0N 46 10 E
Būshehr, Iran 45 D6 28 55N 50 55 E
Būshehr □, Iran 45 D6 28 20N 51 45 E
Bushell, Canada 73 B7 59 31N 108 45W
Bushenyi, Uganda 54 C3 0 35S 30 10 E
Bushire = Būshehr, Iran 45 D6 28 55N 50 55 E
Bushnell, Ill., U.S.A. 80 E9 40 33N 90 31W
Bushnell, Nebr., U.S.A. 80 E3 41 14N 103 54W
Busia □, Kenya 54 B3 0 25N 34 6 E
Businga, Zaïre 52 D4 3 16N 20 59 E
Busra ash Shām, Syria 47 C5 32 30N 36 25 E
Busselton, Australia 61 F2 33 42S 115 15 E
Bussum, Neths. 15 B5 52 16N 5 10 E
Busto Arsizio, Italy 20 B3 45 37N 8 51 E
Busu-Djanoa, Zaïre 52 D4 1 43N 21 23 E
Busuanga, Phil. 37 B5 12 10N 120 0 E
Buta, Zaïre 54 B1 2 50N 24 53 E
Butare, Rwanda 54 C2 2 31S 29 52 E
Butaritari, Kiribati 64 G9 3 30N 174 0 E
Bute, U.K. 12 F3 55 48N 5 2W
Bute Inlet, Canada 72 C4 50 40N 124 53W
Butemba, Uganda 54 B3 1 9N 31 37 E
Butembo, Zaïre 54 B2 0 9N 29 18 E
Butha Qi, China 33 B7 48 0N 122 32 E
Butiaba, Uganda 54 B3 1 50N 31 20 E
Butler, Mo., U.S.A. 80 F7 38 16N 94 20W
Butler, Pa., U.S.A. 78 F5 40 52N 79 54W
Buton, Indonesia 37 E6 5 0S 122 45 E
Butte, Mont., U.S.A. 82 C7 46 0N 112 32W
Butte, Nebr., U.S.A. 80 D5 42 58N 98 51W
Butte Creek —►, U.S.A. 84 F5 39 12N 121 56W
Butterworth = Gcuwa, S. Africa 57 E4 32 20S 28 11 E
Butterworth, Malaysia 39 K3 5 24N 100 23 E
Buttfield, Mt., Australia 61 D4 24 45S 128 9 E
Button B., Canada 73 B10 58 45N 94 23W
Buttonwillow, U.S.A. 85 K7 35 24N 119 28W
Butty Hd., Australia 61 F3 33 54S 121 39 E
Butuan, Phil. 37 C7 8 57N 125 33 E
Butung = Buton, Indonesia 37 E6 5 0S 122 45 E
Buturlinovka, Russia 25 D7 50 50N 40 35 E
Buxar, India 43 G10 25 34N 83 58 E
Buxtehude, Germany 16 B5 53 28N 9 39 E
Buxton, U.K. 10 D6 53 16N 1 54W
Buy, Russia 24 C7 58 28N 41 28 E
Büyük Menderes —►, Turkey 21 F12 37 28N 27 11 E
Büyükçekmece, Turkey 21 D13 41 2N 28 35 E
Buzău, Romania 17 F14 45 10N 26 50 E
Buzău —►, Romania 17 F14 45 26N 27 44 E
Buzen, Japan 31 H5 33 35N 131 5 E
Buzi —►, Mozam. 55 F3 19 50S 34 43 E
Buzuluk, Russia 24 D9 52 48N 52 12 E
Buzzards Bay, U.S.A. 79 E14 41 45N 70 37W
Bwana Mkubwe, Zaïre 55 E2 13 8S 28 38 E
Byarezina —►, Belarus 17 B16 52 33N 30 14 E
Bydgoszcz, Poland 17 B9 53 10N 18 0 E
Byelarus = Belarus ■, Europe 17 B14 53 30N 27 0 E
Byelorussia = Belarus ■, Europe 17 B14 53 30N 27 0 E
Byers, U.S.A. 80 F2 39 43N 104 14W
Byesville, U.S.A. 78 G3 39 58N 81 32W
Byhalia, U.S.A. 81 H10 34 52N 89 41W
Bykhaw, Belarus 17 B16 53 31N 30 14 E
Bykhov = Bykhaw, Belarus 17 B16 53 31N 30 14 E
Bylas, U.S.A. 83 K8 33 8N 110 7W
Bylot I., Canada 69 A12 73 13N 78 34W
Byrd, C., Antarctica 5 C17 69 38S 76 7W
Byro, Australia 61 E2 26 5S 116 11 E
Byrock, Australia 63 E4 30 40S 146 27 E
Byron Bay, Australia 63 D5 28 43S 153 37 E
Byrranga, Gory, Russia 27 B11 75 0N 100 0 E

Byrranga Mts. = Byrranga, Gory, Russia **27 B11** 75 0N 100 0 E
Byske, Sweden **8 D19** 64 57N 21 11 E
Byske älv →, Sweden **8 D19** 64 57N 21 13 E
Bytom, Poland **17 C10** 50 25N 18 54 E
Bytów, Poland **17 A9** 54 10N 17 30 E
Byumba, Rwanda **54 C3** 1 35S 30 4 E

C

Ca →, Vietnam **38 C5** 18 45N 105 45 E
Ca Mau = Quan Long, Vietnam **39 H5** 9 7N 105 8 E
Ca Mau, Mui = Bai Bung, Mui, Vietnam **39 H5** 8 38N 104 44 E
Ca Na, Vietnam **39 G7** 11 20N 108 54 E
Caacupé, Paraguay **94 B4** 25 23S 57 5W
Caála, Angola **53 G3** 12 46S 15 30 E
Caamano Sd., Canada **72 C3** 52 55N 129 25W
Caazapá, Paraguay **94 B4** 26 8S 56 19W
Caazapá □, Paraguay **95 B4** 26 10S 56 0W
Caballeria, C. de, Spain **22 A11** 40 5N 4 5 E
Cabanatuan, Phil. **37 A6** 15 30N 120 58 E
Cabano, Canada **71 C6** 47 40N 68 56W
Cabazon, U.S.A. **85 M10** 33 55N 116 47W
Cabedelo, Brazil **93 E12** 7 0S 34 50W
Cabildo, Chile **94 C1** 32 30S 71 5W
Cabimas, Venezuela **92 A4** 10 23N 71 25W
Cabinda, Angola **52 F2** 5 33S 12 11 E
Cabinda □, Angola **52 F2** 5 0S 12 30 E
Cabinet Mts., U.S.A. **82 C6** 48 0N 115 30W
Cabo Blanco, Argentina **96 F3** 47 15S 65 47W
Cabo Frio, Brazil **95 A7** 22 51S 42 3W
Cabo Pantoja, Peru **92 D3** 1 0S 75 10W
Cabonga, Réservoir, Canada **70 C4** 47 20N 76 40W
Cabool, U.S.A. **81 G8** 37 7N 92 6W
Caboolture, Australia **63 D5** 27 5S 152 58 E
Cabora Bassa Dam = Cahora Bassa Dam, Mozam. **55 F3** 15 20S 32 50 E
Caborca, Mexico **86 A2** 30 40N 112 10W
Cabot, Mt., U.S.A. **79 B13** 44 30N 71 25W
Cabot Str., Canada **71 C8** 47 15N 59 40W
Cabra, Spain **19 D3** 37 30N 4 28W
Cabrera, Spain **22 B9** 39 8N 2 57 E
Cabri, Canada **73 C7** 50 35N 108 25W
Cabriel →, Spain **19 C5** 39 14N 1 3W
Čačak, Serbia, Yug. **21 C9** 43 54N 20 20 E
Cáceres, Brazil **92 G7** 16 5S 57 40W
Cáceres, Spain **19 C2** 39 26N 6 23W
Cache Bay, Canada **70 C4** 46 22N 80 0W
Cache Cr. →, U.S.A. **84 G5** 38 42N 121 42W
Cachi, Argentina **94 B2** 25 5S 66 10W
Cachimbo, Serra do, Brazil **93 E7** 9 30S 55 30W
Cachoeira, Brazil **93 F11** 12 30S 39 0W
Cachoeira de Itapemirim, Brazil **95 A7** 20 51S 41 7W
Cachoeira do Sul, Brazil **95 C5** 30 3S 52 53W
Cacólo, Angola **52 G3** 10 9S 19 21 E
Caconda, Angola **53 G3** 13 48S 15 8 E
Cacongo, Angola **52 F2** 5 11S 12 5 E
Caddo, U.S.A. **81 H6** 34 7N 96 16W
Cadell Cr. →, Australia **62 C3** 22 35S 141 51 E
Cader Idris, U.K. **10 E4** 52 42N 3 53W
Cadibarrawirracanna, L., Australia **63 D2** 28 52S 135 27 E
Cadillac, Canada **70 C4** 48 14N 78 23W
Cadillac, U.S.A. **76 C3** 44 15N 85 24W
Cadiz, Phil. **37 B6** 10 57N 123 15 E
Cádiz, Spain **19 D2** 36 30N 6 20W
Cadiz, U.S.A. **78 F4** 40 22N 81 0W
Cádiz, G. de, Spain **19 D2** 36 40N 7 0W
Cadney Park, Australia **63 D1** 27 55S 134 3 E
Cadomin, Canada **72 C5** 53 2N 117 20W
Cadotte →, Canada **72 B5** 56 43N 117 10W
Cadoux, Australia **61 F2** 30 46S 117 7 E
Caen, France **18 B3** 49 10N 0 22W
Caernarfon, U.K. **10 D3** 53 8N 4 16W
Caernarfon B., U.K. **10 D3** 53 4N 4 40W
Caernarvon = Caernarfon, U.K. **10 D3** 53 8N 4 16W
Caerphilly, U.K. **11 F4** 51 35N 3 13W
Caerphilly □, U.K. **11 F4** 51 37N 3 12W
Caesarea, Israel **47 C3** 32 30N 34 53 E
Caeté, Brazil **93 G10** 19 55S 43 40W
Caetité, Brazil **93 F10** 13 50S 42 32W
Cafayate, Argentina **94 B2** 26 2S 66 0W
Cafu, Angola **56 B2** 16 30S 15 8 E
Cagayan →, Phil. **37 A6** 18 25N 121 42 E
Cagayan de Oro, Phil. **37 C6** 8 30N 124 40 E
Cágliari, Italy **20 E3** 39 13N 9 7 E
Cágliari, G. di, Italy **20 E3** 39 8N 9 11 E
Caguas, Puerto Rico **89 C6** 18 14N 66 2W
Caha Mts., Ireland **13 E2** 51 45N 9 40W
Cahama, Angola **56 B1** 16 17S 14 19 E
Caher, Ireland **13 D4** 52 22N 7 56W
Caherciveen, Ireland **13 E1** 51 56N 10 14W
Cahora Bassa Dam, Mozam. **55 F3** 15 20S 32 50 E
Cahore Pt., Ireland **13 D5** 52 33N 6 12W
Cahors, France **18 D4** 44 27N 1 27 E
Cahuapanas, Peru **92 E3** 5 15S 77 0W
Cahul, Moldova **17 F15** 45 50N 28 15 E
Cai Bau, Dao, Vietnam **38 B6** 21 10N 107 27 E
Cai Nuoc, Vietnam **39 H5** 8 56N 105 1 E
Caia, Mozam. **55 F4** 17 51S 35 24 E
Caianda, Angola **55 E1** 11 2S 23 31 E
Caibarién, Cuba **88 B4** 22 30N 79 30W
Caicara, Venezuela **92 B5** 7 38N 66 10W
Caicó, Brazil **93 E11** 6 20S 37 0W
Caicos Is., W. Indies **89 B5** 21 40N 71 40W
Caicos Passage, W. Indies **89 B5** 22 45N 72 45W
Caird Coast, Antarctica **5 D1** 75 0S 25 0W
Cairn Gorm, U.K. **12 D5** 57 7N 3 39W
Cairn Toul, U.K. **12 D5** 57 3N 3 44W
Cairngorm Mts., U.K. **12 D5** 57 6N 3 42W
Cairns, Australia **62 B4** 16 57S 145 45 E
Cairo = El Qâhira, Egypt **51 B11** 30 1N 31 14 E
Cairo, Ga., U.S.A. **77 K3** 30 52N 84 13W
Cairo, Ill., U.S.A. **81 G10** 37 0N 89 11W
Caithness, Ord of, U.K. **12 C5** 58 8N 3 36W
Caiundo, Angola **53 H3** 15 50S 17 28 E
Caiza, Bolivia **92 H5** 20 2S 65 40W
Cajamarca, Peru **92 E3** 7 5S 78 28W
Cajàzeiras, Brazil **93 E11** 6 52S 38 30W
Cala d'Or, Spain **22 B10** 39 23N 3 14 E
Cala Figuera, C., Spain **22 B9** 39 27N 2 31 E
Cala Forcat, Spain **22 A10** 40 0N 3 47 E
Cala Mayor, Spain **22 B9** 39 33N 2 37 E
Cala Mezquida, Spain **22 B11** 39 55N 4 16 E
Cala Millor, Spain **22 B10** 39 35N 3 22 E
Cala Ratjada, Spain **22 B10** 39 43N 3 27 E
Calabar, Nigeria **50 H6** 4 57N 8 20 E
Calábria □, Italy **20 E7** 39 0N 16 30 E
Calafate, Argentina **96 G2** 50 19S 72 15W
Calahorra, Spain **19 A5** 42 18N 1 59W
Calais, France **18 A4** 50 57N 1 56 E
Calais, U.S.A. **71 C6** 45 11N 67 17W
Calalaste, Cord. de, Argentina **94 B2** 25 0S 67 0W
Calama, Brazil **92 E6** 8 0S 62 50W
Calama, Chile **94 A2** 22 30S 68 55W
Calamar, Bolívar, Colombia **92 A4** 10 15N 74 55W
Calamar, Vaupés, Colombia **92 C4** 1 58N 72 32W
Calamian Group, Phil. **37 B5** 11 50N 119 55 E
Calamocha, Spain **19 B5** 40 50N 1 17W
Calán Porter, Spain **22 B11** 39 52N 4 8 E
Calang, Indonesia **36 D1** 4 37N 95 37 E
Calapan, Phil. **37 B6** 13 25N 121 7 E
Călăraşi, Romania **17 F14** 44 12N 27 20 E
Calatayud, Spain **19 B5** 41 20N 1 40W
Calauag, Phil. **37 B6** 13 55N 122 15 E
Calavite, C., Phil. **37 B6** 13 26N 120 20 E
Calbayog, Phil. **37 B6** 12 4N 124 38 E
Calca, Peru **92 F4** 13 22S 72 0W
Calcasieu L., U.S.A. **81 L8** 29 55N 93 18W
Calcutta, India **43 H13** 22 36N 88 24 E
Caldas da Rainha, Portugal **19 C1** 39 24N 9 8W
Calder →, U.K. **10 D6** 53 44N 1 22W
Caldera, Chile **94 B1** 27 5S 70 55W
Caldwell, Idaho, U.S.A. **82 E5** 43 40N 116 41W
Caldwell, Kans., U.S.A. **81 G6** 37 2N 97 37W
Caldwell, Tex., U.S.A. **81 K6** 30 32N 96 42W
Caledon, S. Africa **56 E2** 34 14S 19 26 E
Caledon →, S. Africa **56 E4** 30 31S 26 5 E
Caledon B., Australia **62 A2** 12 45S 137 0 E
Caledonia, Canada **78 C5** 43 7N 79 58W
Caledonia, U.S.A. **78 D7** 42 58N 77 51W
Calemba, Angola **56 B2** 16 0S 15 44 E
Calexico, U.S.A. **85 N11** 32 40N 115 30W
Calf of Man, U.K. **10 C3** 54 3N 4 48W
Calgary, Canada **72 C6** 51 0N 114 10W
Calheta, Madeira **22 D2** 32 44N 17 11W
Calhoun, U.S.A. **77 H3** 34 30N 84 57W
Cali, Colombia **92 C3** 3 25N 76 35W
Calicut, India **40 P9** 11 15N 75 43 E
Caliente, U.S.A. **83 H6** 37 37N 114 31W
California, Mo., U.S.A. **80 F8** 38 38N 92 34W
California, Pa., U.S.A. **78 F5** 40 4N 79 54W
California □, U.S.A. **83 H4** 37 30N 119 30W
California, Baja, Mexico **86 A1** 32 10N 115 12W
California, Baja, T.N. □ = Baja California □, Mexico **86 B2** 30 0N 115 0W
California, Baja, T.S. □ = Baja California Sur □, Mexico **86 B2** 25 50N 111 50W
California, G. de, Mexico **86 B2** 27 0N 111 0W
California City, U.S.A. **85 K9** 35 10N 117 55W
California Hot Springs, U.S.A. **85 K8** 35 51N 118 41W
Calingasta, Argentina **94 C2** 31 15S 69 30W
Calipatria, U.S.A. **85 M11** 33 8N 115 31W
Calistoga, U.S.A. **84 G4** 38 35N 122 35W
Calitzdorp, S. Africa **56 E3** 33 33S 21 42 E
Callabonna, L., Australia **63 D3** 29 40S 140 5 E
Callan, Ireland **13 D4** 52 32N 7 24W
Callander, U.K. **12 E4** 56 15N 4 13W
Callao, Peru **92 F3** 12 0S 77 0W
Callaway, U.S.A. **80 E5** 41 18N 99 56W
Calles, Mexico **87 C5** 23 2N 98 42W
Callide, Australia **62 C5** 24 18S 150 28 E
Calling Lake, Canada **72 B6** 55 15N 113 12W
Calliope, Australia **62 C5** 24 0S 151 16 E
Calola, Angola **56 B2** 16 25S 17 48 E
Caloundra, Australia **63 D5** 26 45S 153 10 E
Calpella, U.S.A. **84 F3** 39 14N 123 12W
Calpine, U.S.A. **84 F6** 39 40N 120 27W
Calstock, Canada **70 C3** 49 47N 84 9W
Caltagirone, Italy **20 F6** 37 14N 14 31 E
Caltanissetta, Italy **20 F6** 37 29N 14 4 E
Calulo, Angola **52 G2** 10 1S 14 56 E
Calumet, U.S.A. **76 B1** 47 14N 88 27W
Calunda, Angola **53 G4** 12 7S 23 36 E
Calvert, U.S.A. **81 K6** 30 59N 96 40W
Calvert →, Australia **62 B2** 16 17S 137 44 E
Calvert Hills, Australia **62 B2** 17 15S 137 20 E
Calvert I., Canada **72 C3** 51 30N 128 0W
Calvert Ra., Australia **60 D3** 24 0S 122 30 E
Calvi, France **18 E8** 42 34N 8 45 E
Calviá, Spain **19 C7** 39 34N 2 31 E
Calvillo, Mexico **86 C4** 21 51N 102 43W
Calvinia, S. Africa **56 E2** 31 28S 19 45 E
Calwa, U.S.A. **84 J7** 36 42N 119 46W
Cam →, U.K. **11 E8** 52 21N 0 16 E
Cam Lam, Vietnam **39 G7** 11 54N 109 10 E
Cam Pha, Vietnam **38 B6** 21 7N 107 18 E
Cam Ranh, Vietnam **39 G7** 11 54N 109 12 E
Cam Xuyen, Vietnam **38 C6** 18 15N 106 0 E
Camabatela, Angola **52 F3** 8 20S 15 26 E
Camacha, Madeira **22 D3** 32 41N 16 49W
Camacho, Mexico **86 C4** 24 25N 102 18W
Camacupa, Angola **53 G3** 11 58S 17 22 E
Camagüey, Cuba **88 B4** 21 20N 78 0W
Camaná, Peru **92 G4** 16 30S 72 50W
Camanche Reservoir, U.S.A. **84 G6** 38 14N 121 1W
Camaquã →, Brazil **95 C5** 31 17S 51 47W
Câmara de Lobos, Madeira **22 D3** 32 39N 16 59W
Camargo, Bolivia **92 H5** 20 38S 65 15W
Camargue, France **18 E6** 43 34N 4 34 E
Camarillo, U.S.A. **85 L7** 34 13N 119 2W
Camarón, C., Honduras **88 C2** 16 0N 85 5W
Camarones, Argentina **96 E3** 44 50S 65 40W
Camas, U.S.A. **84 E4** 45 35N 122 24W
Camas Valley, U.S.A. **82 E2** 43 2N 123 40W
Cambará, Brazil **95 A5** 23 2S 50 5W
Cambay = Khambhat, India **42 H5** 22 23N 72 33 E
Cambay, G. of = Khambat, G. of, India **42 J5** 20 45N 72 30 E
Cambodia ■, Asia **38 F5** 12 15N 105 0 E
Camborne, U.K. **11 G2** 50 12N 5 19W
Cambrai, France **18 A5** 50 11N 3 14 E
Cambria, U.S.A. **83 J3** 35 34N 121 5W
Cambrian Mts., U.K. **11 E4** 52 3N 3 57W
Cambridge, Canada **70 D3** 43 23N 80 15W
Cambridge, Jamaica **88 C4** 18 18N 77 54W
Cambridge, N.Z. **59 G5** 37 54S 175 29 E
Cambridge, U.K. **11 E8** 52 12N 0 8 E
Cambridge, Idaho, U.S.A. **82 D5** 44 34N 116 41W
Cambridge, Mass., U.S.A. **79 D13** 42 22N 71 6W
Cambridge, Md., U.S.A. **76 F7** 38 34N 76 5W
Cambridge, Minn., U.S.A. **80 C8** 45 34N 93 13W
Cambridge, N.Y., U.S.A. **79 C11** 43 2N 73 22W
Cambridge, Nebr., U.S.A. **80 E4** 40 17N 100 10W
Cambridge, Ohio, U.S.A. **78 F3** 40 2N 81 35W
Cambridge Bay, Canada **68 B9** 69 10N 105 0W
Cambridge G., Australia **60 B4** 14 55S 128 15 E
Cambridge Springs, U.S.A. **78 E4** 41 48N 80 4W
Cambridgeshire □, U.K. **11 E8** 52 25N 0 7W
Cambuci, Brazil **95 A7** 21 35S 41 55W
Cambundi-Catembo, Angola **52 G3** 10 10S 17 35 E
Camden, Ala., U.S.A. **77 K2** 31 59N 87 17W
Camden, Ark., U.S.A. **81 J8** 33 35N 92 50W
Camden, Maine, U.S.A. **71 D6** 44 13N 69 4W
Camden, N.J., U.S.A. **79 G9** 39 56N 75 7W
Camden, S.C., U.S.A. **77 H5** 34 16N 80 36W
Camden Sd., Australia **60 C3** 15 27S 124 25 E
Camdenton, U.S.A. **81 F8** 38 1N 92 45W
Cameron, Ariz., U.S.A. **83 J8** 35 53N 111 25W
Cameron, La., U.S.A. **81 L8** 29 48N 93 20W
Cameron, Mo., U.S.A. **80 F7** 39 44N 94 14W
Cameron, Tex., U.S.A. **81 K6** 30 51N 96 59W
Cameron Falls, Canada **70 C2** 49 8N 88 19W
Cameron Highlands, Malaysia **39 K3** 4 27N 101 22 E
Cameron Hills, Canada **72 B5** 59 48N 118 0W
Cameroon ■, Africa **51 G7** 6 0N 12 30 E
Cameroun, Mt., Cameroon **50 H6** 4 13N 9 10 E
Cametá, Brazil **93 D9** 2 12S 49 30W
Caminha, Portugal **19 B1** 41 50N 8 50W
Camino, U.S.A. **84 G6** 38 44N 120 41W
Camira Creek, Australia **63 D5** 29 15S 152 58 E
Camissombo, Angola **52 F4** 8 7S 20 38 E
Cammal, U.S.A. **78 E7** 41 24N 77 28W
Camocim, Brazil **93 D10** 2 55S 40 50W
Camooweal, Australia **62 B2** 19 56S 138 7 E
Camopi →, Fr. Guiana **93 C8** 3 10N 52 20W
Camp Crook, U.S.A. **80 C3** 45 33N 103 59W
Camp Nelson, U.S.A. **85 J8** 36 8N 118 39W
Camp Wood, U.S.A. **81 L4** 29 40N 100 1W
Campana, Argentina **94 C4** 34 10S 58 55W
Campana, I., Chile **96 F1** 48 20S 75 20W
Campanário, Madeira **22 D2** 32 39N 17 2W
Campánia □, Italy **20 D6** 41 0N 14 30 E
Campbell, S. Africa **56 D3** 28 48S 23 44 E
Campbell, Calif., U.S.A. **84 H5** 37 17N 121 57W
Campbell, Ohio, U.S.A. **78 E4** 41 5N 80 37W
Campbell I., Pac. Oc. **64 N8** 52 30S 169 0 E
Campbell L., Canada **73 A7** 63 14N 106 55W
Campbell River, Canada **72 C3** 50 5N 125 20W
Campbell Town, Australia **62 G4** 41 52S 147 30 E
Campbellford, Canada **78 B7** 44 18N 77 48W
Campbellpur, Pakistan **42 C5** 33 46N 72 26 E
Campbellsville, U.S.A. **76 G3** 37 21N 85 20W
Campbellton, Canada **71 C6** 47 57N 66 43W
Campbelltown, Australia **63 E5** 34 4S 150 49 E
Campbeltown, U.K. **12 F3** 55 26N 5 36W
Campeche, Mexico **87 D6** 19 50N 90 32W
Campeche □, Mexico **87 D6** 19 50N 90 32W
Campeche, B. de, Mexico **87 D6** 19 30N 93 0W
Camperdown, Australia **63 F3** 38 14S 143 9 E
Camperville, Canada **73 C8** 51 59N 100 9W
Campina Grande, Brazil **93 E11** 7 20S 35 47W
Campinas, Brazil **95 A6** 22 50S 47 0W
Campo, Cameroon **52 D1** 2 22N 9 50 E
Campo Belo, Brazil **93 H9** 20 52S 45 16W
Campo Formoso, Brazil **93 F10** 10 30S 40 20W
Campo Grande, Brazil **93 H8** 20 25S 54 40W
Campo Maíor, Brazil **93 D10** 4 50S 42 12W
Campo Mourão, Brazil **95 A5** 24 3S 52 22W
Campoalegre, Colombia **92 C3** 2 41N 75 20W
Campobasso, Italy **20 D6** 41 34N 14 39 E
Campos, Brazil **95 A7** 21 50S 41 20W
Campos Belos, Brazil **93 F9** 13 10S 47 3W
Campos del Puerto, Spain **22 B10** 39 26N 3 1 E
Campos Novos, Brazil **95 B5** 27 21S 51 50W
Camptonville, U.S.A. **84 F5** 39 27N 121 3W
Campuya →, Peru **92 D4** 1 40S 73 30W
Camrose, Canada **72 C6** 53 0N 112 50W
Camsell Portage, Canada **73 B7** 59 37N 109 15W
Çan, Turkey **21 D12** 40 2N 27 3 E
Can Clavo, Spain **22 C7** 38 57N 1 27 E
Can Creu, Spain **22 C7** 38 58N 1 28 E
Can Gio, Vietnam **39 G6** 10 25N 106 58 E
Can Tho, Vietnam **39 G5** 10 2N 105 46 E
Canaan, U.S.A. **79 D11** 42 2N 73 20W
Canada ■, N. Amer. **68 C10** 60 0N 100 0W
Cañada de Gómez, Argentina **94 C3** 32 40S 61 30W
Canadian, U.S.A. **81 H4** 35 55N 100 23W
Canadian →, U.S.A. **81 H7** 35 28N 95 3W
Canadian Shield, Canada **69 C10** 53 0N 75 0W
Çanakkale, Turkey **21 D12** 40 8N 26 24 E
Çanakkale Boğazı, Turkey **21 D12** 40 17N 26 32 E
Canal Flats, Canada **72 C5** 50 10N 115 48W
Canalejas, Argentina **94 D2** 35 15S 66 34W
Canals, Argentina **94 C3** 33 35S 62 53W
Canandaigua, U.S.A. **78 D7** 42 54N 77 17W
Cananea, Mexico **86 A2** 31 0N 110 20W
Canarias, Is., Atl. Oc. **22 F4** 28 30N 16 0W
Canarreos, Arch. de los, Cuba **88 B3** 21 35N 81 40W
Canary Is. = Canarias, Is., Atl. Oc. **22 F4** 28 30N 16 0W
Canatlán, Mexico **86 C4** 24 31N 104 47W
Canaveral, C., U.S.A. **77 L5** 28 27N 80 32W
Canavieiras, Brazil **93 G11** 15 39S 39 0W
Canbelego, Australia **63 E4** 31 32S 146 18 E
Canberra, Australia **63 F4** 35 15S 149 8 E
Canby, Calif., U.S.A. **82 F3** 41 27N 120 52W
Canby, Minn., U.S.A. **80 C6** 44 43N 96 16W
Canby, Oreg., U.S.A. **84 E4** 45 16N 122 42W
Cancún, Mexico **87 C7** 21 8N 86 44W
Candala, Somali Rep. **46 E4** 11 30N 49 58 E
Candelaria, Argentina **95 B4** 27 29S 55 44W
Candelaria, Canary Is. **22 F3** 28 22N 16 22W
Candelo, Australia **63 F4** 36 47S 149 43 E
Candia = Iráklion, Greece **23 D7** 35 20N 25 12 E
Candle L., Canada **73 C7** 53 50N 105 18W
Candlemas I., Antarctica **5 B1** 57 3S 26 40W
Cando, U.S.A. **80 A5** 48 32N 99 12W
Canea = Khaniá, Greece **23 D6** 35 30N 24 4 E
Canelones, Uruguay **95 C4** 34 32S 56 17W
Cañete, Chile **94 D1** 37 50S 73 30W
Cañete, Peru **92 F3** 13 8S 76 30W
Cangas de Narcea, Spain **19 A2** 43 10N 6 32W
Canguaretama, Brazil **93 E11** 6 20S 35 5W
Canguçu, Brazil **95 C5** 31 22S 52 43W
Cangzhou, China **34 E9** 38 19N 116 52 E
Canicatti, Italy **20 F5** 37 21N 13 51 E
Canigou, Mt., France **18 E5** 42 31N 2 27 E
Canim Lake, Canada **72 C4** 51 47N 120 54W
Canindeyu □, Paraguay **95 A4** 24 10S 55 0W
Canipaan, Phil. **36 C5** 8 33N 117 15 E
Canisteo, U.S.A. **78 D7** 42 16N 77 36W
Canisteo →, U.S.A. **78 D7** 42 7N 77 8W
Cañitas, Mexico **86 C4** 23 36N 102 43W
Çankırı, Turkey **25 F5** 40 40N 33 37 E
Cankuzo, Burundi **54 C3** 3 10S 30 31 E
Canmore, Canada **72 C5** 51 7N 115 18W
Cann River, Australia **63 F4** 37 35S 149 7 E
Canna, U.K. **12 D2** 57 3N 6 33W
Cannanore, India **40 P9** 11 53N 75 27 E
Cannes, France **18 E7** 43 32N 7 1 E
Canning Town = Port Canning, India **43 H13** 22 23N 88 40 E
Cannington, Canada **78 B5** 44 20N 79 2W
Cannock, U.K. **10 E5** 52 41N 2 1W
Cannon Ball →, U.S.A. **80 B4** 46 20N 100 38W
Cannondale Mt., Australia **62 D4** 25 13S 148 57 E
Canoas, Brazil **95 B5** 29 56S 51 11W
Canoe L., Canada **73 B7** 55 10N 108 15W
Canon City, U.S.A. **80 F2** 38 27N 105 14W
Canora, Canada **73 C8** 51 40N 102 30W
Canowindra, Australia **63 E4** 33 35S 148 38 E
Canso, Canada **71 C7** 45 20N 61 0W
Cantabria □, Spain **19 A4** 43 10N 4 0W
Cantabrian Mts. = Cantábrica, Cordillera, Spain **19 A3** 43 0N 5 10W
Cantábrica, Cordillera, Spain **19 A3** 43 0N 5 10W
Cantal, Plomb du, France **18 D5** 45 3N 2 45 E
Canterbury, Australia **62 D3** 25 23S 141 53 E
Canterbury, U.K. **11 F9** 51 16N 1 6 E
Canterbury □, N.Z. **59 K3** 43 45S 171 19 E
Canterbury Bight, N.Z. **59 L3** 44 16S 171 55 E
Canterbury Plains, N.Z. **59 K3** 43 55S 171 22 E
Cantil, U.S.A. **85 K9** 35 18N 117 58W
Canton = Guangzhou, China **33 D6** 23 5N 113 10 E
Canton, Ga., U.S.A. **77 H3** 34 14N 84 29W
Canton, Ill., U.S.A. **80 E9** 40 33N 90 2W
Canton, Miss., U.S.A. **81 J9** 32 37N 90 2W
Canton, Mo., U.S.A. **80 E9** 40 8N 91 32W
Canton, N.Y., U.S.A. **79 B9** 44 36N 75 10W
Canton, Ohio, U.S.A. **78 F3** 40 48N 81 23W
Canton, Okla., U.S.A. **81 G5** 36 3N 98 35W
Canton, S. Dak., U.S.A. **80 D6** 43 18N 96 35W
Canton L., U.S.A. **81 G5** 36 6N 98 35W
Canudos, Brazil **92 E7** 7 13S 58 5W
Canutama, Brazil **92 E6** 6 30S 64 20W
Canutillo, U.S.A. **83 L10** 31 55N 106 36W
Canyon, Tex., U.S.A. **81 H4** 34 59N 101 55W
Canyon, Wyo., U.S.A. **82 D8** 44 43N 110 36W
Canyonlands National Park, U.S.A. **83 G9** 38 15N 110 0W
Canyonville, U.S.A. **82 E2** 42 56N 123 17W
Cao Bang, Vietnam **38 A6** 22 40N 106 15 E
Cao He →, China **35 D13** 40 10N 124 32 E
Cao Lanh, Vietnam **39 G5** 10 27N 105 38 E
Cao Xian, China **34 G8** 34 50N 115 35 E
Cap-aux-Meules, Canada **71 C7** 47 23N 61 52W
Cap-Chat, Canada **71 C6** 49 6N 66 40W
Cap-de-la-Madeleine, Canada **70 C5** 46 22N 72 31W
Cap-Haïtien, Haiti **89 C5** 19 40N 72 20W
Cap St.-Jacques = Vung Tau, Vietnam **39 G6** 10 21N 107 4 E
Capa, Vietnam **38 A4** 22 21N 103 50 E
Capaia, Angola **52 F4** 8 27S 20 13 E
Capanaparo →, Venezuela **92 B5** 7 1N 67 7W
Cape →, Australia **62 C4** 20 59S 146 51 E
Cape Barren I., Australia **62 G4** 40 25S 148 15 E
Cape Breton Highlands Nat. Park, Canada **71 C7** 46 50N 60 40W
Cape Breton I., Canada **71 C7** 46 0N 60 30W
Cape Charles, U.S.A. **76 G8** 37 16N 76 1W
Cape Coast, Ghana **50 G4** 5 5N 1 15W
Cape Coral, U.S.A. **77 M5** 26 33N 81 57W
Cape Dorset, Canada **69 B12** 64 14N 76 32W
Cape Dyer, Canada **69 B13** 66 30N 61 22W
Cape Fear →, U.S.A. **77 H6** 33 53N 78 1W
Cape Girardeau, U.S.A. **81 G10** 37 19N 89 32W
Cape Jervis, Australia **63 F2** 35 40S 138 5 E
Cape May, U.S.A. **76 F8** 38 56N 74 56W
Cape May Point, U.S.A. **75 C12** 38 56N 74 58W
Cape Tormentine, Canada **71 C7** 46 8N 63 47W
Cape Town, S. Africa **56 E2** 33 55S 18 22 E
Cape Verde Is. ■, Atl. Oc. **49 E1** 17 10N 25 20W
Cape Vincent, U.S.A. **79 B8** 44 8N 76 20W
Cape York Peninsula, Australia **62 A3** 12 0S 142 30 E

Capela, Brazil **93 F11** 10 30S 37 0W
Capella, Australia **62 C4** 23 2S 148 1 E
Capim →, Brazil **93 D9** 1 40S 47 47W
Capitan, U.S.A. **83 K11** 33 35N 105 35W
Capitola, U.S.A. **84 J5** 36 59N 121 57W
Capoche →, Mozam. **55 F3** 15 35S 33 0 E
Capraia, Italy **20 C3** 43 2N 9 50 E
Capreol, Canada **70 C3** 46 43N 80 56W
Capri, Italy **20 D6** 40 33N 14 14 E
Capricorn Group, Australia **62 C5** 23 30S 151 55 E
Capricorn Ra., Australia **60 D2** 23 20S 116 50 E
Caprivi Strip, Namibia **56 B3** 18 0S 23 0 E
Captainganj, India **43 F10** 26 55N 83 45 E
Captain's Flat, Australia **63 F4** 35 35S 149 27 E
Caquetá →, Colombia **92 D5** 1 15S 69 15W
Caracal, Romania **17 F13** 44 8N 24 22 E
Caracas, Venezuela **92 A5** 10 30N 66 55W
Caracol, Brazil **93 E10** 9 15S 43 22W
Caradoc, Australia **63 E3** 30 35S 143 5 E
Carajás, Serra dos, Brazil **93 E8** 6 0S 51 30W
Carangola, Brazil **95 A7** 20 44S 42 5W
Carani, Australia **61 F2** 30 57S 116 28 E
Caransebeş, Romania **17 F12** 45 28N 22 18 E
Caratasca, L., Honduras **88 C3** 15 20N 83 40W
Caratinga, Brazil **93 G10** 19 50S 42 10W
Caraúbas, Brazil **93 E11** 5 43S 37 33W
Caravaca, Spain **19 C5** 38 8N 1 52W
Caravelas, Brazil **93 G11** 17 45S 39 15W
Caraveli, Peru **92 G4** 15 45S 73 25W
Caràzinho, Brazil **95 B5** 28 16S 52 46W
Carballo, Spain **19 A1** 43 13N 8 41W
Carberry, Canada **73 D9** 49 50N 99 25W
Carbó, Mexico **86 B2** 29 42N 110 58W
Carbon, Canada **72 C6** 51 30N 113 9W
Carbonara, C., Italy **20 E3** 39 6N 9 31 E
Carbondale, Colo., U.S.A. **82 G10** 39 24N 107 13W
Carbondale, Ill., U.S.A. **81 G10** 37 44N 89 13W
Carbondale, Pa., U.S.A. **79 E9** 41 35N 75 30W
Carbonear, Canada **71 C9** 47 42N 53 13W
Carbonia, Italy **20 E3** 39 10N 8 30 E
Carcajou, Canada **72 B5** 57 47N 117 6W
Carcasse, C., Haiti **89 C5** 18 30N 74 28W
Carcassonne, France **18 E5** 43 13N 2 20 E
Carcross, Canada **68 B6** 60 13N 134 45W
Cardabia, Australia **60 D1** 23 2S 113 48 E
Cardamon Hills, India **40 Q10** 9 30N 77 15 E
Cárdenas, Cuba **88 B3** 23 0N 81 30W
Cárdenas, San Luis Potosí, Mexico **87 C5** 22 0N 99 41W
Cárdenas, Tabasco, Mexico **87 D6** 17 59N 93 21W
Cardiff, U.K. **11 F4** 51 29N 3 10W
Cardiff □, U.K. **11 F4** 51 31N 3 12W
Cardiff-by-the-Sea, U.S.A. **85 M9** 33 1N 117 17W
Cardigan, U.K. **11 E3** 52 5N 4 40W
Cardigan B., U.K. **11 E3** 52 30N 4 30W
Cardinal, Canada **79 B9** 44 47N 75 23W
Cardona, Uruguay **94 C4** 33 53S 57 18W
Cardross, Canada **73 D7** 49 50N 105 40W
Cardston, Canada **72 D6** 49 15N 113 20W
Cardwell, Australia **62 B4** 18 14S 146 2 E
Careen L., Canada **73 B7** 57 0N 108 11W
Carei, Romania **17 E12** 47 40N 22 29 E
Careme, Indonesia **37 G13** 6 55S 108 27 E
Carey, Idaho, U.S.A. **82 E7** 43 19N 113 57W
Carey, Ohio, U.S.A. **76 E4** 40 57N 83 23W
Carey, L., Australia **61 E3** 29 0S 122 15 E
Carey L., Canada **73 A8** 62 12N 102 55W
Careysburg, Liberia **50 G2** 6 34N 10 30W
Carhué, Argentina **94 D3** 37 10S 62 50W
Caria, Turkey **21 F13** 37 20N 28 10 E
Cariacica, Brazil **93 H10** 20 16S 40 25W
Caribbean Sea, W. Indies **89 C5** 15 0N 75 0W
Cariboo Mts., Canada **72 C4** 53 0N 121 0W
Caribou, U.S.A. **71 C6** 46 52N 68 1W
Caribou →, Man., Canada **73 B10** 59 20N 94 44W
Caribou →, N.W.T., Canada **72 A3** 61 27N 125 45W
Caribou I., Canada **70 C2** 47 22N 85 49W
Caribou Is., Canada **72 A6** 61 55N 113 15W
Caribou L., Man., Canada **73 B9** 59 21N 96 10W
Caribou L., Ont., Canada **70 B2** 50 25N 89 5W
Caribou Mts., Canada **72 B5** 59 12N 115 40W
Carichic, Mexico **86 B3** 27 56N 107 3W
Carillo, Mexico **86 B4** 26 50N 103 55W
Carinda, Australia **63 E4** 30 28S 147 41 E
Carinhanha, Brazil **93 F10** 14 15S 44 46W
Carinthia □ = Kärnten □, Austria **16 E8** 46 52N 13 30 E
Caripito, Venezuela **92 A6** 10 8N 63 6W
Caritianas, Brazil **92 E6** 9 20S 63 6W
Carleton Place, Canada **70 C4** 45 8N 76 9W
Carletonville, S. Africa **56 D4** 26 23S 27 22 E
Carlin, U.S.A. **82 F5** 40 43N 116 7W
Carlingford L., U.K. **13 B5** 54 3N 6 9W
Carlinville, U.S.A. **80 F10** 39 17N 89 53W
Carlisle, U.K. **10 C5** 54 54N 2 56W
Carlisle, U.S.A. **78 F7** 40 12N 77 12W
Carlos Casares, Argentina **94 D3** 35 32S 61 20W
Carlos Tejedor, Argentina **94 D3** 35 25S 62 25W
Carlow, Ireland **13 D5** 52 50N 6 56W
Carlow □, Ireland **13 D5** 52 43N 6 50W
Carlsbad, Calif., U.S.A. **85 M9** 33 10N 117 21W
Carlsbad, N. Mex., U.S.A. **81 J2** 32 25N 104 14W
Carlyle, Canada **73 D8** 49 40N 102 20W
Carlyle, U.S.A. **80 F10** 38 37N 89 22W
Carmacks, Canada **68 B6** 62 5N 136 16W
Carman, Canada **73 D9** 49 30N 98 0W
Carmangay, Canada **72 C6** 50 10N 113 10W
Carmanville, Canada **71 C9** 49 23N 54 19W
Carmarthen, U.K. **11 F3** 51 52N 4 19W
Carmarthen B., U.K. **11 F3** 51 40N 4 30W
Carmarthenshire □, U.K. **11 F3** 51 55N 4 13W
Carmaux, France **18 D5** 44 3N 2 10 E
Carmel, U.S.A. **79 E11** 41 26N 73 41W
Carmel-by-the-Sea, U.S.A. **83 H3** 36 33N 121 55W
Carmel Valley, U.S.A. **84 J5** 36 29N 121 43W
Carmelo, Uruguay **94 C4** 34 0S 58 20W
Carmen, Colombia **92 B3** 9 43N 75 8W
Carmen, Paraguay **95 B4** 27 13S 56 12W
Carmen →, Mexico **86 A3** 30 42N 106 29W
Carmen, I., Mexico **86 B2** 26 0N 111 20W
Carmen de Patagones, Argentina **96 E4** 40 50S 63 0W
Carmensa, Argentina **94 D2** 35 15S 67 40W
Carmi, U.S.A. **76 F1** 38 5N 88 10W
Carmichael, U.S.A. **84 G5** 38 38N 121 19W
Carmila, Australia **62 C4** 21 55S 149 24 E
Carmona, Spain **19 D3** 37 28N 5 42W
Carnac, France **18 C2** 47 45N 3 6W
Carnarvon, Queens., Australia **62 C4** 24 48S 147 45 E
Carnarvon, W. Austral., Australia **61 D1** 24 51S 113 42 E
Carnarvon, S. Africa **56 E3** 30 56S 22 8 E
Carnarvon Ra., Queens., Australia **62 D4** 25 15S 148 30 E
Carnarvon Ra., W. Austral., Australia **61 E3** 25 20S 120 45 E
Carnation, U.S.A. **84 C5** 47 39N 121 55W
Carndonagh, Ireland **13 A4** 55 16N 7 15W
Carnduff, Canada **73 D8** 49 10N 101 50W
Carnegie, U.S.A. **78 F4** 40 24N 80 5W
Carnegie, L., Australia **61 E3** 26 5S 122 30 E
Carnic Alps = Karnische Alpen, Europe **16 E7** 46 36N 13 0 E
Carniche Alpi = Karnische Alpen, Europe **16 E7** 46 36N 13 0 E
Carnot, C.A.R. **52 D3** 4 59N 15 56 E
Carnot, C., Australia **63 E2** 34 57S 135 38 E
Carnot B., Australia **60 C3** 17 20S 122 15 E
Carnsore Pt., Ireland **13 D5** 52 10N 6 22W
Caro, U.S.A. **76 D4** 43 29N 83 24W
Carol City, U.S.A. **77 N5** 25 56N 80 16W
Carolina, Brazil **93 E9** 7 10S 47 30W
Carolina, Puerto Rico **89 C6** 18 23N 65 58W
Carolina, S. Africa **57 D5** 26 5S 30 6 E
Caroline I., Kiribati **65 H12** 9 15S 150 3W
Caroline Is., Pac. Oc. **28 J17** 8 0N 150 0 E
Caron, Canada **73 C7** 50 30N 105 50W
Caroni →, Venezuela **92 B6** 8 21N 62 43W
Caronie = Nébrodi, Monti, Italy **20 F6** 37 54N 14 35 E
Caroona, Australia **63 E5** 31 24S 150 26 E
Carpathians, Europe **17 D11** 49 30N 21 0 E
Carpaţii Meridionali, Romania **17 F13** 45 30N 25 0 E
Carpentaria, G. of, Australia **62 A2** 14 0S 139 0 E
Carpentaria Downs, Australia **62 B3** 18 44S 144 20 E
Carpentras, France **18 D6** 44 3N 5 2 E
Carpi, Italy **20 B4** 44 47N 10 53 E
Carpinteria, U.S.A. **85 L7** 34 24N 119 31W
Carpolac = Morea, Australia **63 F3** 36 45S 141 18 E
Carr Boyd Ra., Australia **60 C4** 16 15S 128 35 E
Carrabelle, U.S.A. **77 L3** 29 51N 84 40W
Carranya, Australia **60 C4** 19 14S 127 46 E
Carrara, Italy **20 B4** 44 5N 10 6 E
Carrauntoohill, Ireland **13 E2** 52 0N 9 45W
Carrick-on-Shannon, Ireland **13 C3** 53 57N 8 5W
Carrick-on-Suir, Ireland **13 D4** 52 21N 7 24W
Carrickfergus, U.K. **13 B6** 54 43N 5 49W
Carrickfergus □, U.K. **13 B6** 54 43N 5 49W
Carrickmacross, Ireland **13 C5** 53 59N 6 43W
Carrieton, Australia **63 E2** 32 25S 138 31 E
Carrington, U.S.A. **80 B5** 47 27N 99 8W
Carrizal Bajo, Chile **94 B1** 28 5S 71 20W
Carrizalillo, Chile **94 B1** 29 5S 71 30W
Carrizo Cr. →, U.S.A. **81 G3** 36 55N 103 55W
Carrizo Springs, U.S.A. **81 L5** 28 31N 99 52W
Carrizozo, U.S.A. **83 K11** 33 38N 105 53W
Carroll, U.S.A. **80 D7** 42 4N 94 52W
Carrollton, Ga., U.S.A. **77 J3** 33 35N 85 5W
Carrollton, Ill., U.S.A. **80 F9** 39 18N 90 24W
Carrollton, Ky., U.S.A. **76 F3** 38 41N 85 11W
Carrollton, Mo., U.S.A. **80 F8** 39 22N 93 30W
Carrollton, Ohio, U.S.A. **78 F3** 40 34N 81 5W
Carron →, U.K. **12 D3** 57 19N 5 26W
Carron, L., U.K. **12 D3** 57 22N 5 35W
Carrot →, Canada **73 C8** 53 50N 101 17W
Carrot River, Canada **73 C8** 53 17N 103 35W
Carruthers, Canada **73 C7** 52 52N 109 16W
Carse of Gowrie, U.K. **12 E5** 56 30N 3 10W
Carson, Calif., U.S.A. **85 M8** 33 48N 118 17W
Carson, N. Dak., U.S.A. **80 B4** 46 25N 101 34W
Carson →, U.S.A. **84 F8** 39 45N 118 40W
Carson City, U.S.A. **84 F7** 39 10N 119 46W
Carson Sink, U.S.A. **82 G4** 39 50N 118 25W
Carstairs, U.K. **12 F5** 55 42N 3 41W
Cartagena, Colombia **92 A3** 10 25N 75 33W
Cartagena, Spain **19 D5** 37 38N 0 59W
Cartago, Colombia **92 C3** 4 45N 75 55W
Cartago, Costa Rica **88 E3** 9 50N 83 55W
Cartersville, U.S.A. **77 H3** 34 10N 84 48W
Carterton, N.Z. **59 J5** 41 2S 175 31 E
Carthage, Ark., U.S.A. **81 H8** 34 4N 92 33W
Carthage, Ill., U.S.A. **80 E9** 40 25N 91 8W
Carthage, Mo., U.S.A. **81 G7** 37 11N 94 19W
Carthage, S. Dak., U.S.A. **80 C6** 44 10N 97 43W
Carthage, Tex., U.S.A. **81 J7** 32 9N 94 20W
Cartier I., Australia **60 B3** 12 31S 123 29 E
Cartwright, Canada **71 B8** 53 41N 56 58W
Caruaru, Brazil **93 E11** 8 15S 35 55W
Carúpano, Venezuela **92 A6** 10 39N 63 15W
Caruthersville, U.S.A. **81 G10** 36 11N 89 39W
Carvoeiro, Brazil **92 D6** 1 30S 61 59W
Carvoeiro, C., Portugal **19 C1** 39 21N 9 24W
Cary, U.S.A. **77 H6** 35 47N 78 46W
Casa Grande, U.S.A. **83 K8** 32 53N 111 45W
Casablanca, Chile **94 C1** 33 20S 71 25W
Casablanca, Morocco **50 B3** 33 36N 7 36W
Casas Grandes, Mexico **86 A3** 30 22N 108 0W
Cascade, Idaho, U.S.A. **82 D5** 44 31N 116 2W
Cascade, Mont., U.S.A. **82 C8** 47 16N 111 42W
Cascade Locks, U.S.A. **84 E5** 45 40N 121 54W
Cascade Ra., U.S.A. **84 D5** 47 0N 121 30W
Cascade Range, U.S.A. **66 E7** 45 0N 121 45W
Cascais, Portugal **19 C1** 38 41N 9 25W
Cascavel, Brazil **95 A5** 24 57S 53 28W
Cáscina, Italy **20 C4** 43 41N 10 33 E
Caserta, Italy **20 D6** 41 4N 14 20 E
Cashel, Ireland **13 D4** 52 30N 7 53W
Cashmere, U.S.A. **82 C3** 47 31N 120 28W
Cashmere Downs, Australia **61 E2** 28 57S 119 35 E
Casiguran, Phil. **37 A6** 16 22N 122 7 E
Casilda, Argentina **94 C3** 33 10S 61 10W
Casino, Australia **63 D5** 28 52S 153 3 E
Casiquiare →, Venezuela **92 C5** 2 1N 67 7W
Caslan, Canada **72 C6** 54 38N 112 31W
Casma, Peru **92 E3** 9 30S 78 20W
Casmalia, U.S.A. **85 L6** 34 50N 120 32W
Caspe, Spain **19 B5** 41 14N 0 1W
Casper, U.S.A. **82 E10** 42 51N 106 19W
Caspian Depression, Eurasia **25 E8** 47 0N 48 0 E
Caspian Sea, Eurasia **25 F9** 43 0N 50 0 E
Cass City, U.S.A. **76 D4** 43 36N 83 11W
Cass Lake, U.S.A. **80 B7** 47 23N 94 37W
Casselman, Canada **79 A9** 45 19N 75 5W
Casselton, U.S.A. **80 B6** 46 54N 97 13W
Cassiar, Canada **72 B3** 59 16N 129 40W
Cassiar Mts., Canada **72 B2** 59 30N 130 30W
Cassinga, Angola **53 H3** 15 5S 16 4 E
Cassino, Italy **20 D5** 41 30N 13 49 E
Cassville, U.S.A. **81 G8** 36 41N 93 52W
Castaic, U.S.A. **85 L8** 34 30N 118 38W
Castellammare di Stábia, Italy **20 D6** 40 42N 14 29 E
Castelli, Argentina **94 D4** 36 7S 57 47W
Castellón de la Plana, Spain **19 C5** 39 58N 0 3W
Castelo, Brazil **95 A7** 20 33S 41 14W
Castelo Branco, Portugal **19 C2** 39 50N 7 31W
Castelsarrasin, France **18 E4** 44 2N 1 7 E
Castelvetrano, Italy **20 F5** 37 41N 12 47 E
Casterton, Australia **63 F3** 37 30S 141 30 E
Castilla La Mancha □, Spain **19 C4** 39 30N 3 30W
Castilla y Leon □, Spain **19 B3** 42 0N 5 0W
Castillos, Uruguay **95 C5** 34 12S 53 52W
Castle Dale, U.S.A. **82 G8** 39 13N 111 1W
Castle Douglas, U.K. **12 G5** 54 56N 3 56W
Castle Rock, Colo., U.S.A. **80 F2** 39 22N 104 51W
Castle Rock, Wash., U.S.A. **84 D4** 46 17N 122 54W
Castlebar, Ireland **13 C2** 53 52N 9 18W
Castleblaney, Ireland **13 B5** 54 7N 6 44W
Castlegar, Canada **72 D5** 49 20N 117 40W
Castlemaine, Australia **63 F3** 37 2S 144 12 E
Castlerea, Ireland **13 C3** 53 46N 8 29W
Castlereagh □, U.K. **13 B6** 54 33N 5 53W
Castlereagh →, Australia **63 E4** 30 12S 147 32 E
Castlereagh B., Australia **62 A2** 12 10S 135 10 E
Castletown, U.K. **10 C3** 54 5N 4 38W
Castletown Bearhaven, Ireland **13 E2** 51 39N 9 55W
Castlevale, Australia **62 C4** 24 30S 146 48 E
Castor, Canada **72 C6** 52 15N 111 50W
Castres, France **18 E5** 43 37N 2 13 E
Castries, St. Lucia **89 D7** 14 2N 60 58W
Castro, Brazil **95 A5** 24 45S 50 0W
Castro, Chile **96 E2** 42 30S 73 50W
Castro Alves, Brazil **93 F11** 12 46S 39 33W
Castroville, Calif., U.S.A. **84 J5** 36 46N 121 45W
Castroville, Tex., U.S.A. **81 L5** 29 21N 98 53W
Castuera, Spain **19 C3** 38 43N 5 37W
Casummit Lake, Canada **70 B1** 51 29N 92 22W
Cat Ba, Dao, Vietnam **38 B6** 20 50N 107 0 E
Cat I., Bahamas **89 B4** 24 30N 75 30W
Cat I., U.S.A. **81 K10** 30 14N 89 6W
Cat L., Canada **70 B1** 51 40N 91 50W
Catacamas, Honduras **88 D2** 14 54N 85 56W
Catacáos, Peru **92 E2** 5 20S 80 45W
Cataguases, Brazil **95 A7** 21 23S 42 39W
Catahoula L., U.S.A. **81 K8** 31 31N 92 7W
Catalão, Brazil **93 G9** 18 10S 47 57W
Çatalca, Turkey **21 D13** 41 8N 28 27 E
Catalina, Canada **71 C9** 48 31N 53 4W
Catalonia = Cataluña □, Spain **19 B6** 41 40N 1 15 E
Cataluña □, Spain **19 B6** 41 40N 1 15 E
Catamarca, Argentina **94 B2** 28 30S 65 50W
Catamarca □, Argentina **94 B2** 27 0S 65 50W
Catanduanes, Phil. **37 B6** 13 50N 124 20 E
Catanduva, Brazil **95 A6** 21 5S 48 58W
Catánia, Italy **20 F6** 37 30N 15 6 E
Catanzaro, Italy **20 E7** 38 54N 16 35 E
Catarman, Phil. **37 B6** 12 28N 124 35 E
Cateel, Phil. **37 C7** 7 47N 126 24 E
Cathcart, S. Africa **56 E4** 32 18S 27 10 E
Cathlamet, U.S.A. **84 D3** 46 12N 123 23W
Catlettsburg, U.S.A. **76 F4** 38 25N 82 36W
Catoche, C., Mexico **87 C7** 21 40N 87 8W
Catrimani, Brazil **92 C6** 0 27N 61 41W
Catskill, U.S.A. **79 D11** 42 14N 73 52W
Catskill Mts., U.S.A. **79 D10** 42 10N 74 25W
Catt, Mt., Australia **62 A1** 13 49S 134 23 E
Cattaraugus, U.S.A. **78 D6** 42 22N 78 52W
Catuala, Angola **56 B2** 16 25S 19 2 E
Catur, Mozam. **55 E4** 13 45S 35 30 E
Catwick Is., Vietnam **39 G7** 10 0N 109 0 E
Cauca →, Colombia **92 B4** 8 54N 74 28W
Caucaia, Brazil **93 D11** 3 40S 38 35W
Caucasus Mountains, Eurasia **25 F7** 42 50N 44 0 E
Caúngula, Angola **52 F3** 8 26S 18 38 E
Cauquenes, Chile **94 D1** 36 0S 72 22W
Caura →, Venezuela **92 B6** 7 38N 64 53W
Cauresi →, Mozam. **55 F3** 17 8S 33 0 E
Causapscal, Canada **71 C6** 48 19N 67 12W
Cauvery →, India **40 P11** 11 9N 78 52 E
Caux, Pays de, France **18 B4** 49 38N 0 35 E
Cavalier, U.S.A. **80 A6** 48 48N 97 37W
Cavan, Ireland **13 C4** 54 0N 7 22W
Cavan □, Ireland **13 C4** 54 1N 7 16W
Cave City, U.S.A. **76 G3** 37 8N 85 58W
Cavenagh Ra., Australia **61 E4** 26 12S 127 55 E
Cavendish, Australia **63 F3** 37 31S 142 2 E
Caviana, I., Brazil **93 C8** 0 10N 50 10W
Cavite, Phil. **37 B6** 14 29N 120 55 E
Cawndilla L., Australia **63 E3** 32 30S 142 15 E
Cawnpore = Kanpur, India **43 F9** 26 28N 80 20 E
Caxias, Brazil **93 D10** 4 55S 43 20W
Caxias do Sul, Brazil **95 B5** 29 10S 51 10W
Caxito, Angola **52 F2** 8 30S 13 30 E
Cay Sal Bank, Bahamas **88 B3** 23 45N 80 0W
Cayambe, Ecuador **92 C3** 0 3N 78 8W
Cayenne, Fr. Guiana **93 B8** 5 5N 52 18W
Cayman Brac, Cayman Is. **88 C4** 19 43N 79 49W
Cayman Is. ■, W. Indies **88 C3** 19 40N 80 30W
Cayo Romano, Cuba **89 B4** 22 0N 78 0W
Cayuga, Canada **78 D5** 42 59N 79 50W
Cayuga, U.S.A. **79 D8** 42 54N 76 44W
Cayuga L., U.S.A. **79 D8** 42 41N 76 41W
Cazombo, Angola **53 G4** 11 54S 22 56 E
Ceadâr-Lunga, Moldova **17 E15** 46 3N 28 51 E
Ceanannus Mor, Ireland **13 C5** 53 44N 6 53W
Ceará = Fortaleza, Brazil **93 D11** 3 45S 38 35W
Ceará □, Brazil **93 E11** 5 0S 40 0W
Ceará Mirim, Brazil **93 E11** 5 38S 35 25W
Cebaco, I. de, Panama **88 E3** 7 33N 81 9W
Cebollar, Argentina **94 B2** 29 10S 66 35W
Cebu, Phil. **37 B6** 10 18N 123 54 E
Cecil Plains, Australia **63 D5** 27 30S 151 11 E
Cedar →, U.S.A. **80 E9** 41 17N 91 21W
Cedar City, U.S.A. **83 H7** 37 41N 113 4W
Cedar Creek Reservoir, U.S.A. **81 J6** 32 11N 96 4W
Cedar Falls, Iowa, U.S.A. **80 D8** 42 32N 92 27W
Cedar Falls, Wash., U.S.A. **84 C5** 47 25N 121 45W
Cedar Key, U.S.A. **77 L4** 29 8N 83 2W
Cedar L., Canada **73 C8** 53 10N 100 0W
Cedar Rapids, U.S.A. **80 E9** 41 59N 91 40W
Cedartown, U.S.A. **77 H3** 34 1N 85 15W
Cedarvale, Canada **72 B3** 55 1N 128 22W
Cedarville, S. Africa **57 E4** 30 23S 29 3 E
Cedarville, U.S.A. **82 F3** 41 32N 120 10W
Cedral, Mexico **86 C4** 23 50N 100 42W
Cedro, Brazil **93 E11** 6 34S 39 3W
Cedros, I. de, Mexico **86 B1** 28 10N 115 20W
Ceduna, Australia **63 E1** 32 7S 133 46 E
Cefalù, Italy **20 E6** 38 2N 14 1 E
Cegléd, Hungary **17 E10** 47 11N 19 47 E
Celaya, Mexico **86 C4** 20 31N 100 37W
Celbridge, Ireland **13 C5** 53 20N 6 32W
Celebes = Sulawesi □, Indonesia **37 E6** 2 0S 120 0 E
Celebes Sea, Indonesia **37 D6** 3 0N 123 0 E
Celina, U.S.A. **76 E3** 40 33N 84 35W
Celje, Slovenia **16 E8** 46 16N 15 18 E
Celle, Germany **16 B6** 52 37N 10 4 E
Cement, U.S.A. **81 H5** 34 56N 98 8W
Center, N. Dak., U.S.A. **80 B4** 47 7N 101 18W
Center, Tex., U.S.A. **81 K7** 31 48N 94 11W
Centerfield, U.S.A. **83 G8** 39 8N 111 49W
Centerville, Calif., U.S.A. **84 J7** 36 44N 119 30W
Centerville, Iowa, U.S.A. **80 E8** 40 44N 92 52W
Centerville, Pa., U.S.A. **78 F5** 40 3N 79 59W
Centerville, S. Dak., U.S.A. **80 D6** 43 7N 96 58W
Centerville, Tenn., U.S.A. **77 H2** 35 47N 87 28W
Centerville, Tex., U.S.A. **81 K7** 31 16N 95 59W
Central, U.S.A. **83 K9** 32 47N 108 9W
Central □, Kenya **54 C4** 0 30S 37 30 E
Central □, Malawi **55 E3** 13 30S 33 30 E
Central □, Zambia **55 E2** 14 25S 28 50 E
Central, Cordillera, Colombia **92 C4** 5 0N 75 0W
Central, Cordillera, Costa Rica **88 D3** 10 10N 84 5W
Central, Cordillera, Dom. Rep. **89 C5** 19 15N 71 0W
Central African Rep. ■, Africa **51 G9** 7 0N 20 0 E
Central City, Ky., U.S.A. **76 G2** 37 18N 87 7W
Central City, Nebr., U.S.A. **80 E5** 41 7N 98 0W
Central I., Kenya **54 B4** 3 30N 36 0 E
Central Makran Range, Pakistan **40 F4** 26 30N 64 15 E
Central Patricia, Canada **70 B1** 51 30N 90 9W
Central Russian Uplands, Europe **6 E13** 54 0N 36 0 E
Central Siberian Plateau, Russia **28 C14** 65 0N 105 0 E
Centralia, Ill., U.S.A. **80 F10** 38 32N 89 8W
Centralia, Mo., U.S.A. **80 F8** 39 13N 92 8W
Centralia, Wash., U.S.A. **84 D4** 46 43N 122 58W
Centreville, Ala., U.S.A. **77 J2** 32 57N 87 8W
Centreville, Miss., U.S.A. **81 K9** 31 5N 91 4W
Cephalonia = Kefallinía, Greece **21 E9** 38 20N 20 30 E
Cepu, Indonesia **37 G14** 7 9S 111 35 E
Ceram = Seram, Indonesia **37 E7** 3 10S 129 0 E
Ceram Sea = Seram Sea, Indonesia **37 E7** 2 30S 128 30 E
Ceres, Argentina **94 B3** 29 55S 61 55W
Ceres, S. Africa **56 E2** 33 21S 19 18 E
Ceres, U.S.A. **84 H6** 37 35N 120 57W
Ceridigion □, U.K. **11 E3** 52 16N 4 15W
Cerignola, Italy **20 D6** 41 17N 15 53 E
Cerigo = Kíthira, Greece **21 F10** 36 8N 23 0 E
Çerkeşköy, Turkey **21 D12** 41 17N 27 59 E
Cerralvo, I., Mexico **86 C3** 24 20N 109 45W
Cerritos, Mexico **86 C4** 22 20N 100 20W
Cervera, Spain **19 B6** 41 40N 1 16 E
Cesena, Italy **20 B5** 44 8N 12 15 E
Cēsis, Latvia **9 H21** 57 18N 25 15 E
Česka Republika = Czech Rep. ■, Europe **16 D8** 50 0N 15 0 E
České Budějovice, Czech. **16 D8** 48 55N 14 25 E
Ceskomoravská Vrchovina, Czech. **16 D8** 49 30N 15 40 E
Çeşme, Turkey **21 E12** 38 20N 26 23 E
Cessnock, Australia **63 E5** 32 50S 151 21 E
Cetinje, Montenegro, Yug. **21 C8** 42 23N 18 59 E
Cetraro, Italy **20 E6** 39 31N 15 55 E
Ceuta, N. Afr. **19 E3** 35 52N 5 18W
Cévennes, France **18 D5** 44 10N 3 50 E
Ceyhan →, Turkey **25 G6** 36 38N 35 40 E
Ceylon = Sri Lanka ■, Asia **40 R12** 7 30N 80 50 E
Cha-am, Thailand **38 F2** 12 48N 99 58 E
Chacabuco, Argentina **94 C3** 34 40S 60 27W
Chachapoyas, Peru **92 E3** 6 15S 77 50W
Chachoengsao, Thailand **38 F3** 13 42N 101 5 E
Chachran, Pakistan **40 E7** 28 55N 70 30 E
Chachro, Pakistan **42 G4** 25 5N 70 15 E
Chaco □, Argentina **94 B3** 26 30S 61 0W
Chaco □, Paraguay **94 B3** 26 0S 60 0W
Chad ■, Africa **51 E8** 15 0N 17 15 E
Chad, L. = Tchad, L., Chad **51 F7** 13 30N 14 30 E

Chadan, *Russia* **27 D10** 51 17N 91 35 E
Chadileuvú →, *Argentina* **94 D2** 37 46S 66 0W
Chadiza, *Zambia* **55 E3** 14 45S 32 27 E
Chadron, *U.S.A.* **80 D3** 42 50N 103 0W
Chadyr-Lunga = Ceadâr-Lunga, *Moldova* **17 E15** 46 3N 28 51 E
Chae Hom, *Thailand* **38 C2** 18 43N 99 35 E
Chaem →, *Thailand* **38 C2** 18 11N 98 38 E
Chaeryŏng, *N. Korea* **35 E13** 38 24N 125 36 E
Chagai Hills, *Afghan.* **40 E3** 29 30N 63 0 E
Chagda, *Russia* **27 D14** 58 45N 130 38 E
Chagos Arch., *Ind. Oc.* **29 K11** 6 0S 72 0 E
Chāh Ākhvor, *Iran* **45 C8** 32 41N 59 40 E
Chāh Bahār, *Iran* **45 E9** 25 20N 60 40 E
Chāh-e-Malek, *Iran* **45 D8** 28 35N 59 7 E
Chāh Kavīr, *Iran* **45 D7** 31 45N 54 52 E
Chahar Burjak, *Afghan.* **40 D3** 30 15N 62 0 E
Chaibasa, *India* **41 H14** 22 42N 85 49 E
Chainat, *Thailand* **38 E3** 15 11N 100 8 E
Chaiya, *Thailand* **39 H2** 9 23N 99 14 E
Chaj Doab, *Pakistan* **42 C5** 32 15N 73 0 E
Chajari, *Argentina* **94 C4** 30 42S 58 0W
Chake Chake, *Tanzania* **54 D4** 5 15S 39 45 E
Chakhānsūr, *Afghan.* **40 D3** 31 10N 62 0 E
Chakonipau, L., *Canada* **71 A6** 56 18N 68 30W
Chakradharpur, *India* **43 H11** 22 45N 85 40 E
Chakwal, *Pakistan* **42 C5** 32 56N 72 53 E
Chala, *Peru* **92 G4** 15 48S 74 20W
Chalchihuites, *Mexico* **86 C4** 23 29N 103 53W
Chalcis = Khalkis, *Greece* **21 E10** 38 27N 23 42 E
Chaleur B., *Canada* **71 C6** 47 55N 65 30W
Chalfant, *U.S.A.* **84 H8** 37 32N 118 21W
Chalhuanca, *Peru* **92 F4** 14 15S 73 15W
Chalisgaon, *India* **40 J9** 20 30N 75 10 E
Chalky Inlet, *N.Z.* **59 M1** 46 3S 166 31 E
Challapata, *Bolivia* **92 G5** 18 53S 66 50W
Challis, *U.S.A.* **82 D6** 44 30N 114 14W
Chalna, *India* **43 H13** 22 36N 89 35 E
Chalon-sur-Saône, *France* **18 C6** 46 48N 4 50 E
Châlons-en-Champagne, *France* **18 B6** 48 58N 4 20 E
Chalyaphum, *Thailand* **38 E4** 15 48N 102 2 E
Cham, Cu Lao, *Vietnam* **38 E7** 15 57N 108 30 E
Chama, *U.S.A.* **83 H10** 36 54N 106 35W
Chaman, *Pakistan* **40 D5** 30 58N 66 25 E
Chamba, *India* **42 C7** 32 35N 76 10 E
Chamba, *Tanzania* **55 E4** 11 37S 37 0 E
Chambal →, *India* **43 F8** 26 29N 79 15 E
Chamberlain, *U.S.A.* **80 D5** 43 49N 99 20W
Chamberlain →, *Australia* **60 C4** 15 30S 127 54 E
Chambers, *U.S.A.* **83 J9** 35 11N 109 26W
Chambersburg, *U.S.A.* **76 F7** 39 56N 77 40W
Chambéry, *France* **18 D6** 45 34N 5 55 E
Chambly, *Canada* **79 A11** 45 27N 73 17W
Chambord, *Canada* **71 C5** 48 25N 72 6W
Chamchamal, *Iraq* **44 C5** 35 32N 44 50 E
Chamela, *Mexico* **86 D3** 19 32N 105 5W
Chamical, *Argentina* **94 C2** 30 22S 66 27W
Chamkar Luong, *Cambodia* **39 G4** 11 0N 103 45 E
Chamonix-Mont Blanc, *France* **18 D7** 45 55N 6 51 E
Champa, *India* **43 H10** 22 2N 82 43 E
Champagne, *Canada* **72 A1** 60 49N 136 30W
Champagne, *France* **18 B6** 48 40N 4 20 E
Champaign, *U.S.A.* **76 E1** 40 7N 88 15W
Champassak, *Laos* **38 E5** 14 53N 105 52 E
Champlain, *Canada* **76 B9** 46 27N 72 24W
Champlain, *U.S.A.* **79 B11** 44 59N 73 27W
Champlain, L., *U.S.A.* **79 B11** 44 40N 73 20W
Champotón, *Mexico* **87 D6** 19 20N 90 50W
Chana, *Thailand* **39 J3** 6 55N 100 44 E
Chañaral, *Chile* **94 B1** 26 23S 70 40W
Chanārān, *Iran* **45 B8** 36 39N 59 6 E
Chanasma, *India* **42 H5** 23 44N 72 5 E
Chandannagar, *India* **43 H13** 22 52N 88 24 E
Chandausi, *India* **43 E8** 28 27N 78 49 E
Chandeleur Is., *U.S.A.* **81 L10** 29 55N 88 57W
Chandeleur Sd., *U.S.A.* **81 L10** 29 55N 89 0W
Chandigarh, *India* **42 D7** 30 43N 76 47 E
Chandler, *Australia* **63 D1** 27 0S 133 19 E
Chandler, *Canada* **71 C7** 48 18N 64 46W
Chandler, *Ariz., U.S.A.* **83 K8** 33 18N 111 50W
Chandler, *Okla., U.S.A.* **81 H6** 35 42N 96 53W
Chandpur, *Bangla.* **41 H17** 23 8N 90 45 E
Chandpur, *India* **42 E8** 29 8N 78 19 E
Chandrapur, *India* **40 K11** 19 57N 79 25 E
Chānf, *Iran* **45 E9** 26 38N 60 29 E
Chang, *Pakistan* **42 F3** 26 59N 68 30 E
Chang, Ko, *Thailand* **39 F4** 12 0N 102 23 E
Ch'ang Chiang = Chang Jiang →, *China* **33 C7** 31 48N 121 10 E
Chang Jiang →, *China* **33 C7** 31 48N 121 10 E
Changa, *India* **43 C7** 33 53N 77 35 E
Changanacheri, *India* **40 Q10** 9 25N 76 31 E
Changane →, *Mozam.* **57 C5** 24 30S 33 30 E
Changbai, *China* **35 D15** 41 25N 128 5 E
Changbai Shan, *China* **35 C15** 42 20N 129 0 E
Changchiak'ou = Zhangjiakou, *China* **34 D8** 40 48N 114 55 E
Ch'angchou = Changzhou, *China* **33 C6** 31 47N 119 58 E
Changchun, *China* **35 C13** 43 57N 125 17 E
Changchunling, *China* **35 B13** 45 18N 125 27 E
Changde, *China* **33 D6** 29 4N 111 35 E
Changdo-ri, *N. Korea* **35 E14** 38 30N 127 40 E
Changhai = Shanghai, *China* **33 C7** 31 15N 121 26 E
Changhua, *Taiwan* **33 D7** 24 2N 120 30 E
Changhŭng, *S. Korea* **35 G14** 34 41N 126 52 E
Changhŭngni, *N. Korea* **35 D15** 40 24N 128 19 E
Changjiang, *China* **38 C7** 19 20N 108 55 E
Changjin, *N. Korea* **35 D14** 40 23N 127 15 E
Changjin-chŏsuji, *N. Korea* **35 D14** 40 30N 127 15 E
Changli, *China* **35 E10** 39 40N 119 13 E
Changling, *China* **35 B12** 44 20N 123 58 E
Changlun, *Malaysia* **39 J3** 6 25N 100 26 E
Changping, *China* **34 D9** 40 14N 116 12 E
Changsha, *China* **33 D6** 28 12N 113 0 E
Changwu, *China* **34 G4** 35 10N 107 45 E
Changyi, *China* **35 F10** 36 40N 119 30 E
Changyŏn, *N. Korea* **35 E13** 38 15N 125 6 E
Changyuan, *China* **34 G8** 35 15N 114 42 E
Changzhi, *China* **34 F7** 36 10N 113 6 E
Changzhou, *China* **33 C6** 31 47N 119 58 E
Chanhanga, *Angola* **56 B1** 16 0S 14 8 E
Channapatna, *India* **40 N10** 12 40N 77 15 E
Channel Is., *U.K.* **11 H5** 49 19N 2 24W
Channel Is., *U.S.A.* **85 M7** 33 40N 119 15W
Channel-Port aux Basques, *Canada* **71 C8** 47 30N 59 9W
Channing, *Mich., U.S.A.* **76 B1** 46 9N 88 5W
Channing, *Tex., U.S.A.* **81 H3** 35 41N 102 20W
Chantada, *Spain* **19 A2** 42 36N 7 46W
Chanthaburi, *Thailand* **38 F4** 12 38N 102 12 E
Chantrey Inlet, *Canada* **68 B10** 67 48N 96 20W
Chanute, *U.S.A.* **81 G7** 37 41N 95 27W
Chao Phraya →, *Thailand* **38 F3** 13 32N 100 36 E
Chao Phraya Lowlands, *Thailand* **38 E3** 15 30N 100 0 E
Chao'an, *China* **33 D6** 23 42N 116 32 E
Chaocheng, *China* **34 F8** 36 4N 115 37 E
Chaoyang, *China* **35 D11** 41 35N 120 22 E
Chapala, *Mozam.* **55 F4** 15 50S 37 35 E
Chapala, L. de, *Mexico* **86 C4** 20 10N 103 20W
Chapayev, *Kazakstan* **25 D9** 50 25N 51 10 E
Chapayevsk, *Russia* **24 D8** 53 0N 49 40 E
Chapecó, *Brazil* **95 B5** 27 14S 52 41W
Chapel Hill, *U.S.A.* **77 H6** 35 55N 79 4W
Chapleau, *Canada* **70 C3** 47 50N 83 24W
Chaplin, *Canada* **73 C7** 50 28N 106 40W
Chapra = Chhapra, *India* **43 G11** 25 48N 84 44 E
Char, *Mauritania* **50 D2** 21 32N 12 45W
Chara, *Russia* **27 D12** 56 54N 118 20 E
Charadai, *Argentina* **94 B4** 27 35S 59 55W
Charagua, *Bolivia* **92 G6** 19 45S 63 10W
Charaña, *Bolivia* **92 G5** 17 30S 69 25W
Charata, *Argentina* **94 B3** 27 13S 61 14W
Charcas, *Mexico* **86 C4** 23 10N 101 20W
Charcoal L., *Canada* **73 B8** 58 49N 102 22W
Chard, *U.K.* **11 G5** 50 52N 2 58W
Chardara, *Kazakstan* **26 E7** 41 16N 67 59 E
Chardon, *U.S.A.* **78 E3** 41 35N 81 12W
Chardzhou = Chärjew, *Turkmenistan* **26 F7** 39 6N 63 34 E
Charente →, *France* **18 D3** 45 57N 1 5W
Chari →, *Chad* **51 F7** 12 58N 14 31 E
Chārīkār, *Afghan.* **40 B6** 35 0N 69 10 E
Chariton →, *U.S.A.* **80 F8** 39 19N 92 58W
Chärjew, *Turkmenistan* **26 F7** 39 6N 63 34 E
Charkhari, *India* **43 G8** 25 24N 79 45 E
Charkhi Dadri, *India* **42 E7** 28 37N 76 17 E
Charleroi, *Belgium* **15 D4** 50 24N 4 27 E
Charleroi, *U.S.A.* **78 F5** 40 9N 79 57W
Charles, C., *U.S.A.* **76 G8** 37 7N 75 58W
Charles City, *U.S.A.* **80 D8** 43 4N 92 41W
Charles L., *Canada* **73 B6** 59 50N 110 33W
Charles Town, *U.S.A.* **76 F7** 39 17N 77 52W
Charleston, *Ill., U.S.A.* **76 F1** 39 30N 88 10W
Charleston, *Miss., U.S.A.* **81 H9** 34 1N 90 4W
Charleston, *Mo., U.S.A.* **81 G10** 36 55N 89 21W
Charleston, *S.C., U.S.A.* **77 J6** 32 46N 79 56W
Charleston, *W. Va., U.S.A.* **76 F5** 38 21N 81 38W
Charleston Peak, *U.S.A.* **85 J11** 36 16N 115 42W
Charlestown, *S. Africa* **57 D4** 27 26S 29 53 E
Charlestown, *U.S.A.* **76 F3** 38 27N 85 40W
Charlesville, *Zaïre* **52 F4** 5 27S 20 59 E
Charleville = Rath Luirc, *Ireland* **13 D3** 52 21N 8 40W
Charleville, *Australia* **63 D4** 26 24S 146 15 E
Charleville-Mézières, *France* **18 B6** 49 44N 4 40 E
Charlevoix, *U.S.A.* **76 C3** 45 19N 85 16W
Charlotte, *Mich., U.S.A.* **76 D3** 42 34N 84 50W
Charlotte, *N.C., U.S.A.* **77 H5** 35 13N 80 51W
Charlotte Amalie, *Virgin Is.* **89 C7** 18 21N 64 56W
Charlotte Harbor, *U.S.A.* **77 M4** 26 50N 82 10W
Charlottesville, *U.S.A.* **76 F6** 38 2N 78 30W
Charlottetown, *Canada* **71 C7** 46 14N 63 8W
Charlton, *Australia* **63 F3** 36 16S 143 24 E
Charlton, *U.S.A.* **80 E8** 40 59N 93 20W
Charlton I., *Canada* **70 B4** 52 0N 79 20W
Charny, *Canada* **71 C5** 46 43N 71 15W
Charolles, *France* **18 C6** 46 27N 4 16 E
Charouine, *Algeria* **50 C4** 29 0N 0 15W
Charre, *Mozam.* **55 F4** 17 13S 35 10 E
Charsadda, *Pakistan* **42 B4** 34 7N 71 45 E
Charters Towers, *Australia* **62 C4** 20 5S 146 13 E
Chartres, *France* **18 B4** 48 29N 1 30 E
Chascomús, *Argentina* **94 D4** 35 30S 58 0W
Chasefu, *Zambia* **55 E3** 11 55S 33 8 E
Chasovnya-Uchurskaya, *Russia* **27 D14** 57 15N 132 50 E
Chāt, *Iran* **45 B7** 37 59N 55 16 E
Châteaubriant, *France* **18 C3** 47 43N 1 23W
Châteaulin, *France* **18 B1** 48 11N 4 8W
Châteauroux, *France* **18 C4** 46 50N 1 40 E
Châtellerault, *France* **18 C4** 46 50N 0 30 E
Chatfield, *U.S.A.* **80 D9** 43 51N 92 11W
Chatham, *N.B., Canada* **71 C6** 47 2N 65 28W
Chatham, *Ont., Canada* **70 D3** 42 24N 82 11W
Chatham, *U.K.* **11 F8** 51 22N 0 32 E
Chatham, *La., U.S.A.* **81 J8** 32 18N 92 27W
Chatham, *N.Y., U.S.A.* **79 D11** 42 21N 73 36W
Chatham Is., *Pac. Oc.* **64 M10** 44 0S 176 40W
Chatham Str., *U.S.A.* **72 B2** 57 0N 134 40W
Chatmohar, *Bangla.* **43 G13** 24 15N 89 15 E
Chatra, *India* **43 G11** 24 12N 84 56 E
Chatrapur, *India* **41 K14** 19 22N 85 2 E
Chats, L. des, *Canada* **79 A8** 45 30N 76 20W
Chatsworth, *Canada* **78 B4** 44 27N 80 54W
Chatsworth, *Zimbabwe* **55 F3** 19 38S 31 13 E
Chattahoochee →, *U.S.A.* **77 K3** 30 54N 84 57W
Chattanooga, *U.S.A.* **77 H3** 35 3N 85 19W
Chaturat, *Thailand* **38 E3** 15 40N 101 51 E
Chau Doc, *Vietnam* **39 G5** 10 42N 105 7 E
Chauk, *Burma* **41 J19** 20 53N 94 49 E
Chaukan La, *Burma* **41 F20** 27 0N 97 15 E
Chaumont, *France* **18 B6** 48 7N 5 8 E
Chaumont, *U.S.A.* **79 B8** 44 4N 76 8W
Chautauqua L., *U.S.A.* **78 D5** 42 10N 79 24W
Chauvin, *Canada* **73 C6** 52 45N 110 10W
Chaves, *Brazil* **93 D9** 0 15S 49 55W
Chaves, *Portugal* **19 B2** 41 45N 7 32W
Chavuma, *Zambia* **53 G4** 13 4S 22 40 E
Chawang, *Thailand* **39 H2** 8 25N 99 30 E
Chaykovskiy, *Russia* **24 C9** 56 47N 54 9 E
Chazy, *U.S.A.* **79 B11** 44 53N 73 26W
Cheb, *Czech.* **16 C7** 50 9N 12 28 E
Cheboksary, *Russia* **24 C8** 56 8N 47 12 E
Cheboygan, *U.S.A.* **76 C3** 45 39N 84 29W
Chech, Erg, *Africa* **50 D4** 25 0N 2 15W
Chechenia □, *Russia* **25 F8** 43 30N 45 29 E
Checheno-Ingush Republic = Chechenia □, *Russia* **25 F8** 43 30N 45 29 E
Chechnya = Chechenia □, *Russia* **25 F8** 43 30N 45 29 E
Chechon, *S. Korea* **35 F15** 37 8N 128 12 E
Checleset B., *Canada* **72 C3** 50 5N 127 35W
Checotah, *U.S.A.* **81 H7** 35 28N 95 31W
Chedabucto B., *Canada* **71 C7** 45 25N 61 8W
Cheduba I., *Burma* **41 K18** 18 45N 93 40 E
Cheepie, *Australia* **63 D4** 26 33S 145 1 E
Chegdomyn, *Russia* **27 D14** 51 7N 133 1 E
Chegga, *Mauritania* **50 C3** 25 27N 5 40W
Chegutu, *Zimbabwe* **55 F3** 18 10S 30 14 E
Chehalis, *U.S.A.* **84 D4** 46 40N 122 58W
Cheju Do, *S. Korea* **35 H14** 33 29N 126 34 E
Chekiang = Zhejiang □, *China* **33 D7** 29 0N 120 0 E
Chela, Sa. da, *Angola* **56 B1** 16 20S 13 20 E
Chelan, *U.S.A.* **82 C4** 47 51N 120 1W
Chelan, L., *U.S.A.* **82 C3** 48 11N 120 30W
Cheleken, *Turkmenistan* **25 G9** 39 34N 53 16 E
Chelforó, *Argentina* **96 D3** 39 0S 66 33W
Chelkar = Shalqar, *Kazakstan* **26 E6** 47 48N 59 39 E
Chelkar Tengiz, Solonchak, *Kazakstan* **26 E7** 48 5N 63 7 E
Chełm, *Poland* **17 C12** 51 8N 23 30 E
Chełmno, *Poland* **17 B10** 53 20N 18 30 E
Chelmsford, *U.K.* **11 F8** 51 44N 0 29 E
Chelsea, *Okla., U.S.A.* **81 G7** 36 32N 95 26W
Chelsea, *Vt., U.S.A.* **79 C12** 43 59N 72 27W
Cheltenham, *U.K.* **11 F5** 51 54N 2 4W
Chelyabinsk, *Russia* **26 D7** 55 10N 61 24 E
Chelyuskin, C., *Russia* **28 B14** 77 30N 103 0 E
Chemainus, *Canada* **72 D4** 48 55N 123 42W
Chemnitz, *Germany* **16 C7** 50 51N 12 54 E
Chemult, *U.S.A.* **82 E3** 43 14N 121 47W
Chen, Gora, *Russia* **27 C15** 65 16N 141 50 E
Chenab →, *Pakistan* **42 D4** 30 23N 71 2 E
Chenango Forks, *U.S.A.* **79 D9** 42 15N 75 51W
Chencha, *Ethiopia* **51 G12** 6 15N 37 32 E
Cheney, *U.S.A.* **82 C5** 47 30N 117 35W
Cheng Xian, *China* **34 H3** 33 43N 105 42 E
Chengcheng, *China* **34 G5** 35 8N 109 56 E
Chengchou = Zhengzhou, *China* **34 G7** 34 45N 113 34 E
Chengde, *China* **35 D9** 40 59N 117 58 E
Chengdu, *China* **32 C5** 30 38N 104 2 E
Chenggu, *China* **34 H4** 33 10N 107 21 E
Chengjiang, *China* **32 D5** 24 39N 103 0 E
Ch'engtu = Chengdu, *China* **32 C5** 30 38N 104 2 E
Chengwu, *China* **34 G8** 34 58N 115 50 E
Chengyang, *China* **35 F11** 36 18N 120 21 E
Chenjiagang, *China* **35 G10** 34 23N 119 47 E
Chenkán, *Mexico* **87 D6** 19 8N 90 58W
Chennai = Madras, *India* **40 N12** 13 8N 80 19 E
Cheo Reo, *Vietnam* **38 F7** 13 25N 108 28 E
Cheom Ksan, *Cambodia* **38 E5** 14 13N 104 56 E
Chepén, *Peru* **92 E3** 7 15S 79 23W
Chepes, *Argentina* **94 C2** 31 20S 66 35W
Chepo, *Panama* **88 E4** 9 10N 79 6W
Cheptulil, Mt., *Kenya* **54 B4** 1 25N 35 35 E
Chequamegon B., *U.S.A.* **80 B9** 46 40N 90 30W
Cher →, *France* **18 C4** 47 21N 0 29 E
Cheraw, *U.S.A.* **77 H6** 34 42N 79 53W
Cherbourg, *France* **18 B3** 49 39N 1 40W
Cherchell, *Algeria* **50 A5** 36 35N 2 12 E
Cherdyn, *Russia* **24 B10** 60 24N 56 29 E
Cheremkhovo, *Russia* **27 D11** 53 8N 103 1 E
Cherepanovo, *Russia* **26 D9** 54 15N 83 30 E
Cherepovets, *Russia* **24 C6** 59 5N 37 55 E
Chergui, Chott ech, *Algeria* **50 B5** 34 21N 0 25 E
Cherikov = Cherykaw, *Belarus* **17 B16** 53 32N 31 20 E
Cherkasy, *Ukraine* **25 E5** 49 27N 32 4 E
Cherlak, *Russia* **26 D8** 54 15N 74 55 E
Chernaya, *Russia* **27 B9** 70 30N 89 10 E
Chernigov = Chernihiv, *Ukraine* **24 D5** 51 28N 31 20 E
Chernihiv, *Ukraine* **24 D5** 51 28N 31 20 E
Chernikovsk, *Russia* **24 D10** 54 48N 56 8 E
Chernivtsi, *Ukraine* **17 D13** 48 15N 25 52 E
Chernobyl = Chornobyl, *Ukraine* **17 C16** 51 20N 30 15 E
Chernogorsk, *Russia* **27 D10** 53 49N 91 18 E
Chernovtsy = Chernivtsi, *Ukraine* **17 D13** 48 15N 25 52 E
Chernyakhovsk, *Russia* **9 J19** 54 36N 21 48 E
Chernyshovskiy, *Russia* **27 C12** 63 0N 112 30 E
Cherokee, *Iowa, U.S.A.* **80 D7** 42 45N 95 33W
Cherokee, *Okla., U.S.A.* **81 G5** 36 45N 98 21W
Cherokees, Lake O' The, *U.S.A.* **81 G7** 36 28N 95 2W
Cherquenco, *Chile* **96 D2** 38 35S 72 0W
Cherrapunji, *India* **41 G17** 25 17N 91 47 E
Cherry Creek, *U.S.A.* **82 G6** 39 54N 114 53W
Cherry Valley, *U.S.A.* **85 M10** 33 59N 116 57W
Cherryvale, *U.S.A.* **81 G7** 37 16N 95 33W
Cherskiy, *Russia* **27 C17** 68 45N 161 18 E
Cherskogo Khrebet, *Russia* **27 C15** 65 0N 143 0 E
Cherven, *Belarus* **17 B15** 53 45N 28 28 E
Chervonohrad, *Ukraine* **17 C13** 50 25N 24 10 E
Cherwell →, *U.K.* **11 F6** 51 44N 1 14W
Cherykaw, *Belarus* **17 B16** 53 32N 31 20 E
Chesapeake, *U.S.A.* **76 G7** 36 50N 76 17W
Chesapeake B., *U.S.A.* **76 F7** 38 0N 76 10W
Cheshire □, *U.K.* **10 D5** 53 14N 2 30W
Cheshskaya Guba, *Russia* **24 A8** 67 20N 47 0 E
Cheslatta L., *Canada* **72 C3** 53 49N 125 20W
Chesley, *Canada* **78 B3** 44 17N 81 5W
Chester, *U.K.* **10 D5** 53 12N 2 53W
Chester, *Calif., U.S.A.* **82 F3** 40 19N 121 14W
Chester, *Ill., U.S.A.* **81 G10** 37 55N 89 49W
Chester, *Mont., U.S.A.* **82 B8** 48 31N 110 58W
Chester, *Pa., U.S.A.* **76 F8** 39 51N 75 22W
Chester, *S.C., U.S.A.* **77 H5** 34 43N 81 12W
Chesterfield, *U.K.* **10 D6** 53 15N 1 25W
Chesterfield, Is., *N. Cal.* **64 J7** 19 52S 158 15 E
Chesterfield Inlet, *Canada* **68 B10** 63 30N 90 45W
Chesterton Ra., *Australia* **63 D4** 25 30S 147 27 E
Chesterville, *Canada* **79 A9** 45 6N 75 14W
Chesuncook L., *U.S.A.* **71 C6** 46 0N 69 21W
Chéticamp, *Canada* **71 C7** 46 37N 60 59W
Chetumal, B. de, *Mexico* **87 D7** 18 40N 88 10W
Chetwynd, *Canada* **72 B4** 55 45N 121 36W
Cheviot, The, *U.K.* **10 B5** 55 29N 2 9W
Cheviot Hills, *U.K.* **10 B5** 55 20N 2 30W
Cheviot Ra., *Australia* **62 D3** 25 20S 143 45 E
Chew Bahir, *Ethiopia* **51 H12** 4 40N 36 50 E
Chewelah, *U.S.A.* **82 B5** 48 17N 117 43W
Cheyenne, *Okla., U.S.A.* **81 H5** 35 37N 99 40W
Cheyenne, *Wyo., U.S.A.* **80 E2** 41 8N 104 49W
Cheyenne →, *U.S.A.* **80 C4** 44 41N 101 18W
Cheyenne Wells, *U.S.A.* **80 F3** 38 49N 102 21W
Cheyne B., *Australia* **61 F2** 34 35S 118 50 E
Chhabra, *India* **42 G7** 24 40N 76 54 E
Chhapra, *India* **43 G11** 25 48N 84 44 E
Chhata, *India* **42 F7** 27 42N 77 30 E
Chhatarpur, *India* **43 G8** 24 55N 79 35 E
Chhep, *Cambodia* **38 F5** 13 45N 105 24 E
Chhindwara, *India* **43 H8** 22 2N 78 59 E
Chhlong, *Cambodia* **39 F5** 12 15N 105 58 E
Chhuk, *Cambodia* **39 G5** 10 46N 104 28 E
Chi →, *Thailand* **38 E5** 15 11N 104 43 E
Chiai, *Taiwan* **33 D7** 23 29N 120 25 E
Chiamis, *Indonesia* **37 G13** 7 20S 108 21 E
Chiamussu = Jiamusi, *China* **33 B8** 46 40N 130 26 E
Chiang Dao, *Thailand* **38 C2** 19 22N 98 58 E
Chiang Kham, *Thailand* **38 C3** 19 32N 100 18 E
Chiang Khan, *Thailand* **38 D3** 17 52N 101 36 E
Chiang Khong, *Thailand* **38 B3** 20 17N 100 24 E
Chiang Mai, *Thailand* **38 C2** 18 47N 98 59 E
Chiang Saen, *Thailand* **38 B3** 20 16N 100 5 E
Chiange, *Angola* **53 H2** 15 35S 13 40 E
Chiapa →, *Mexico* **87 D6** 16 42N 93 0W
Chiapa de Corzo, *Mexico* **87 D6** 16 42N 93 0W
Chiapas □, *Mexico* **87 D6** 17 0N 92 45W
Chiautla, *Mexico* **87 D5** 18 18N 98 34W
Chiávari, *Italy* **20 B3** 44 19N 9 19 E
Chiavenna, *Italy* **20 A3** 46 19N 9 24 E
Chiba, *Japan* **31 G10** 35 30N 140 7 E
Chibabava, *Mozam.* **57 C5** 20 17S 33 35 E
Chibatu, *Indonesia* **37 G12** 7 6S 107 59 E
Chibemba, *Cunene, Angola* **53 H2** 15 48S 14 8 E
Chibemba, *Huila, Angola* **56 B2** 16 20S 15 20 E
Chibia, *Angola* **53 H2** 15 10S 13 42 E
Chibougamau, *Canada* **70 C5** 49 56N 74 24W
Chibougamau L., *Canada* **70 C5** 49 50N 74 20W
Chibuk, *Nigeria* **51 F7** 10 52N 12 50 E
Chic-Chocs, Mts., *Canada* **71 C6** 48 55N 66 0W
Chicacole = Srikakulam, *India* **41 K13** 18 14N 83 58 E
Chicago, *U.S.A.* **76 E2** 41 53N 87 38W
Chicago Heights, *U.S.A.* **76 E2** 41 30N 87 38W
Chichagof I., *U.S.A.* **72 B1** 57 30N 135 30W
Chicheng, *China* **34 D8** 40 55N 115 55 E
Chichester, *U.K.* **11 G7** 50 50N 0 47W
Chichibu, *Japan* **31 F9** 36 5N 139 10 E
Ch'ich'ihaerh = Qiqihar, *China* **27 E13** 47 26N 124 0 E
Chickasha, *U.S.A.* **81 H5** 35 3N 97 58W
Chiclana de la Frontera, *Spain* **19 D2** 36 26N 6 9W
Chiclayo, *Peru* **92 E3** 6 42S 79 50W
Chico, *U.S.A.* **84 F5** 39 44N 121 50W
Chico →, *Chubut, Argentina* **96 E3** 44 0S 67 0W
Chico →, *Santa Cruz, Argentina* **96 G3** 50 0S 68 30W
Chicomo, *Mozam.* **57 C5** 24 31S 34 6 E
Chicontepec, *Mexico* **87 C5** 20 58N 98 10W
Chicopee, *U.S.A.* **79 D12** 42 9N 72 37W
Chicoutimi, *Canada* **71 C5** 48 28N 71 5W
Chicualacuala, *Mozam.* **57 C5** 22 6S 31 42 E
Chidambaram, *India* **40 P11** 11 20N 79 45 E
Chidenguele, *Mozam.* **57 C5** 24 55S 34 11 E
Chidley, C., *Canada* **69 B13** 60 23N 64 26W
Chiede, *Angola* **56 B2** 17 15S 16 22 E
Chiefs Pt., *Canada* **78 B3** 44 41N 81 18W
Chiem Hoa, *Vietnam* **38 A5** 22 12N 105 17 E
Chiemsee, *Germany* **16 E7** 47 53N 12 28 E
Chiengi, *Zambia* **55 D2** 8 45S 29 10 E
Chiengmai = Chiang Mai, *Thailand* **38 C2** 18 47N 98 59 E
Chiese →, *Italy* **20 B4** 45 8N 10 25 E
Chieti, *Italy* **20 C6** 42 21N 14 10 E
Chifeng, *China* **35 C10** 42 18N 118 58 E
Chignecto B., *Canada* **71 C7** 45 30N 64 40W
Chiguana, *Bolivia* **94 A2** 21 0S 67 58W
Chiha-ri, *N. Korea* **35 E14** 38 40N 126 30 E
Chihli, G. of = Bo Hai, *China* **35 E10** 39 0N 119 0 E
Chihli, G. of = Po Hai, *China* **28 F15** 38 30N 119 0 E
Chihuahua, *Mexico* **86 B3** 28 40N 106 3W
Chihuahua □, *Mexico* **86 B3** 28 40N 106 3W
Chiili, *Kazakstan* **26 E7** 44 20N 66 15 E
Chik Bollapur, *India* **40 N10** 13 25N 77 45 E
Chikmagalur, *India* **40 N9** 13 15N 75 45 E
Chikwawa, *Malawi* **55 F3** 16 2S 34 50 E
Chilac, *Mexico* **87 D5** 18 20N 97 24W
Chilako →, *Canada* **72 C4** 53 53N 122 57W
Chilam Chavki, *Pakistan* **43 B6** 35 5N 75 5 E
Chilanga, *Zambia* **55 F2** 15 33S 28 16 E
Chilapa, *Mexico* **87 D5** 17 40N 99 11W
Chilas, *Pakistan* **43 B6** 35 25N 74 5 E
Chilaw, *Sri Lanka* **40 R11** 7 30N 79 50 E
Chilcotin →, *Canada* **72 C4** 51 44N 122 23W
Childers, *Australia* **63 D5** 25 15S 152 17 E
Childress, *U.S.A.* **81 H4** 34 25N 100 13W
Chile ■, *S. Amer.* **96 D2** 35 0S 72 0W
Chile Rise, *Pac. Oc.* **65 L18** 38 0S 92 0W
Chilecito, *Argentina* **94 B2** 29 10S 67 30W
Chilete, *Peru* **92 E3** 7 10S 78 50W
Chililabombwe, *Zambia* **55 E2** 12 18S 27 43 E
Chilin = Jilin, *China* **35 C14** 43 44N 126 30 E
Chilka L., *India* **41 K14** 19 40N 85 25 E
Chilko →, *Canada* **72 C4** 52 0N 123 40W
Chilko, L., *Canada* **72 C4** 51 20N 124 10W

Chillagoe, *Australia* **62 B3** 17 7S 144 33 E
Chillán, *Chile* **94 D1** 36 40S 72 10W
Chillicothe, *Ill., U.S.A.* **80 E10** 40 55N 89 29W
Chillicothe, *Mo., U.S.A.* **80 F8** 39 48N 93 33W
Chillicothe, *Ohio, U.S.A.* **76 F4** 39 20N 82 59W
Chilliwack, *Canada* **72 D4** 49 10N 121 54W
Chilo, *India* **42 F5** 27 25N 73 32 E
Chiloane, I., *Mozam.* **57 C5** 20 40S 34 55 E
Chiloé, I. de, *Chile* **96 E2** 42 30S 73 50W
Chilpancingo, *Mexico* **87 D5** 17 30N 99 30W
Chiltern Hills, *U.K.* **11 F7** 51 40N 0 53W
Chilton, *U.S.A.* **76 C1** 44 2N 88 10W
Chiluage, *Angola* **52 F4** 9 30S 21 50 E
Chilubi, *Zambia* **55 E2** 11 5S 29 58 E
Chilubula, *Zambia* **55 E3** 10 14S 30 51 E
Chilumba, *Malawi* **55 E3** 10 28S 34 12 E
Chilung, *Taiwan* **33 D7** 25 3N 121 45 E
Chilwa, L., *Malawi* **55 F4** 15 15S 35 40 E
Chimaltitán, *Mexico* **86 C4** 21 46N 103 50W
Chimán, *Panama* **88 E4** 8 45N 78 40W
Chimay, *Belgium* **15 D4** 50 3N 4 20 E
Chimbay, *Uzbekistan* **26 E6** 42 57N 59 47 E
Chimborazo, *Ecuador* **92 D3** 1 29S 78 55W
Chimbote, *Peru* **92 E3** 9 0S 78 35W
Chimkent = Shymkent, *Kazakstan* **26 E7** 42 18N 69 36 E
Chimoio, *Mozam.* **55 F3** 19 4S 33 30 E
Chimpembe, *Zambia* **55 D2** 9 31S 29 33 E
Chin □, *Burma* **41 J18** 22 0N 93 0 E
Chin Ling Shan = Qinling Shandi, *China* **34 H5** 33 50N 108 10 E
China, *Mexico* **87 B5** 25 40N 99 20W
China ■, *Asia* **34 E3** 30 0N 110 0 E
China Lake, *U.S.A.* **85 K9** 35 44N 117 37W
Chinan = Jinan, *China* **34 F9** 36 38N 117 1 E
Chinandega, *Nic.* **88 D2** 12 35N 87 12W
Chinati Peak, *U.S.A.* **81 K2** 29 57N 104 29W
Chincha Alta, *Peru* **92 F3** 13 25S 76 7W
Chinchilla, *Australia* **63 D5** 26 45S 150 38 E
Chinchorro, Banco, *Mexico* **87 D7** 18 35N 87 20W
Chinchou = Jinzhou, *China* **35 D11** 41 5N 121 3 E
Chincoteague, *U.S.A.* **76 G8** 37 56N 75 23W
Chinde, *Mozam.* **55 F4** 18 35S 36 30 E
Chindo, *S. Korea* **35 G14** 34 28N 126 15 E
Chindwin →, *Burma* **41 J19** 21 26N 95 15 E
Chineni, *India* **43 C6** 33 2N 75 15 E
Chinga, *Mozam.* **55 F4** 15 13S 38 35 E
Chingola, *Zambia* **55 E2** 12 31S 27 53 E
Chingole, *Malawi* **55 E3** 13 4S 34 17 E
Ch'ingtao = Qingdao, *China* **35 F11** 36 5N 120 20 E
Chinguetti, *Mauritania* **50 D2** 20 25N 12 24W
Chingune, *Mozam.* **57 C5** 20 33S 34 58 E
Chinhae, *S. Korea* **35 G15** 35 9N 128 47 E
Chinhanguanine, *Mozam.* **57 D5** 25 21S 32 30 E
Chinhoyi, *Zimbabwe* **55 F3** 17 20S 30 8 E
Chiniot, *Pakistan* **42 D5** 31 45N 73 0 E
Chinipas, *Mexico* **86 B3** 27 22N 108 32W
Chinju, *S. Korea* **35 G15** 35 12N 128 2 E
Chinle, *U.S.A.* **83 H9** 36 9N 109 33W
Chinnampo, *N. Korea* **35 E13** 38 52N 125 10 E
Chino, *Japan* **31 G9** 35 59N 138 9 E
Chino, *U.S.A.* **85 L9** 34 1N 117 41W
Chino Valley, *U.S.A.* **83 J7** 34 45N 112 27W
Chinon, *France* **18 C4** 47 10N 0 15 E
Chinook, *Canada* **73 C6** 51 28N 110 59W
Chinook, *U.S.A.* **82 B9** 48 35N 109 14W
Chinsali, *Zambia* **55 E3** 10 30S 32 2 E
Chióggia, *Italy* **20 B5** 45 13N 12 17 E
Chios = Khíos, *Greece* **21 E12** 38 27N 26 9 E
Chipata, *Zambia* **55 E3** 13 38S 32 28 E
Chipewyan L., *Canada* **73 B9** 58 0N 98 27W
Chipinge, *Zimbabwe* **55 G3** 20 13S 32 28 E
Chipley, *U.S.A.* **77 K3** 30 47N 85 32W
Chipman, *Canada* **71 C6** 46 6N 65 53W
Chipoka, *Malawi* **55 E3** 13 57S 34 28 E
Chippenham, *U.K.* **11 F5** 51 27N 2 6W
Chippewa →, *U.S.A.* **80 C8** 44 25N 92 5W
Chippewa Falls, *U.S.A.* **80 C9** 44 56N 91 24W
Chiquián, *Peru* **92 F3** 10 10S 77 0W
Chiquimula, *Guatemala* **88 D2** 14 51N 89 37W
Chiquinquira, *Colombia* **92 B4** 5 37N 73 50W
Chirala, *India* **40 M12** 15 50N 80 26 E
Chiramba, *Mozam.* **55 F3** 16 55S 34 39 E
Chirawa, *India* **42 E6** 28 14N 75 42 E
Chirchiq, *Uzbekistan* **26 E7** 41 29N 69 35 E
Chiricahua Peak, *U.S.A.* **83 L9** 31 51N 109 18W
Chiriquí, G. de, *Panama* **88 E3** 8 0N 82 10W
Chiriquí, L. de, *Panama* **88 E3** 9 10N 82 0W
Chirivira Falls, *Zimbabwe* **55 G3** 21 10S 32 12 E
Chirmiri, *India* **41 H13** 23 15N 82 20 E
Chiromo, *Malawi* **53 H7** 16 30S 35 7 E
Chirripó Grande, Cerro, *Costa Rica* **88 E3** 9 29N 83 29W
Chisamba, *Zambia* **55 E2** 14 55S 28 20 E
Chisapani Garhi, *Nepal* **41 F14** 27 30N 84 2 E
Chisasibi, *Canada* **70 B4** 53 50N 79 0W
Chisholm, *Canada* **72 C6** 54 55N 114 10W
Chishtian Mandi, *Pakistan* **42 E5** 29 50N 72 55 E
Chisimaio, *Somali Rep.* **49 G8** 0 22S 42 32 E
Chisimba Falls, *Zambia* **55 E3** 10 12S 30 56 E
Chişinău, *Moldova* **17 E15** 47 0N 28 50 E
Chisos Mts., *U.S.A.* **81 L3** 29 5N 103 15W
Chistopol, *Russia* **24 C9** 55 25N 50 38 E
Chita, *Russia* **27 D12** 52 0N 113 35 E
Chitado, *Angola* **53 H2** 17 10S 14 8 E
Chitembo, *Angola* **53 G3** 13 30S 16 50 E
Chitipa, *Malawi* **55 D3** 9 41S 33 19 E
Chitose, *Japan* **30 C10** 42 49N 141 39 E
Chitral, *Pakistan* **40 B7** 35 50N 71 56 E
Chitré, *Panama* **88 E3** 7 59N 80 27W
Chittagong, *Bangla.* **41 H17** 22 19N 91 48 E
Chittagong □, *Bangla.* **41 G17** 24 5N 91 0 E
Chittaurgarh, *India* **42 G6** 24 52N 74 38 E
Chittoor, *India* **40 N11** 13 15N 79 5 E
Chitungwiza, *Zimbabwe* **55 F3** 18 0S 31 6 E
Chiusi, *Italy* **20 C4** 43 1N 11 57 E
Chivasso, *Italy* **20 B2** 45 11N 7 53 E
Chivhu, *Zimbabwe* **55 F3** 19 2S 30 52 E
Chivilcoy, *Argentina* **94 C4** 34 55S 60 0W
Chiwanda, *Tanzania* **55 E3** 11 23S 34 55 E
Chizera, *Zambia* **55 E1** 13 10S 25 0 E
Chkalov = Orenburg, *Russia* **24 D10** 51 45N 55 6 E
Chloride, *U.S.A.* **85 K12** 35 25N 114 12W
Cho Bo, *Vietnam* **38 B5** 20 46N 105 10 E
Cho-do, *N. Korea* **35 E13** 38 30N 124 40 E
Cho Phuoc Hai, *Vietnam* **39 G6** 10 26N 107 18 E
Choba, *Kenya* **54 B4** 2 30N 38 5 E
Chobe National Park, *Botswana* **56 B3** 18 0S 25 0 E
Chochiwŏn, *S. Korea* **35 F14** 36 37N 127 18 E
Choctawhatchee B., *U.S.A.* **75 D9** 30 20N 86 20W
Choele Choel, *Argentina* **96 D3** 39 11S 65 40W
Choix, *Mexico* **86 B3** 26 40N 108 23W
Chojnice, *Poland* **17 B9** 53 42N 17 32 E
Chōkai-San, *Japan* **30 E10** 39 6N 140 3 E
Chokurdakh, *Russia* **27 B15** 70 38N 147 55 E
Cholame, *U.S.A.* **84 K6** 35 44N 120 18W
Cholet, *France* **18 C3** 47 4N 0 52W
Choluteca, *Honduras* **88 D2** 13 20N 87 14W
Choluteca →, *Honduras* **88 D2** 13 0N 87 20W
Chom Bung, *Thailand* **38 F2** 13 37N 99 36 E
Chom Thong, *Thailand* **38 C2** 18 25N 98 41 E
Choma, *Zambia* **55 F2** 16 48S 26 59 E
Chomun, *India* **42 F6** 27 15N 75 40 E
Chomutov, *Czech.* **16 C7** 50 28N 13 23 E
Chon Buri, *Thailand* **38 F3** 13 21N 101 1 E
Chon Thanh, *Vietnam* **39 G6** 11 24N 106 36 E
Chonan, *S. Korea* **35 F14** 36 48N 127 9 E
Chone, *Ecuador* **92 D2** 0 40S 80 0W
Chong Kai, *Cambodia* **38 F4** 13 57N 103 35 E
Chong Mek, *Thailand* **38 E5** 15 10N 105 27 E
Chŏngdo, *S. Korea* **35 G15** 35 38N 128 42 E
Chŏngha, *S. Korea* **35 F15** 36 12N 129 21 E
Chŏngjin, *N. Korea* **35 D15** 41 47N 129 50 E
Chŏngju, *N. Korea* **35 E13** 39 40N 125 5 E
Chŏngju, *S. Korea* **35 F14** 36 39N 127 27 E
Chongli, *China* **34 D8** 40 58N 115 15 E
Chongqing, *China* **32 D5** 29 35N 106 25 E
Chŏngŭp, *S. Korea* **35 G14** 35 35N 126 50 E
Chŏnju, *S. Korea* **35 G14** 35 50N 127 4 E
Chonos, Arch. de los, *Chile* **96 F2** 45 0S 75 0W
Chop, *Ukraine* **17 D12** 48 26N 22 12 E
Chopim →, *Brazil* **95 B5** 25 35S 53 5W
Chorbat La, *India* **43 B7** 34 42N 76 37 E
Chorley, *U.K.* **10 D5** 53 39N 2 38W
Chornobyl, *Ukraine* **17 C16** 51 20N 30 15 E
Chorolque, Cerro, *Bolivia* **94 A2** 20 59S 66 5W
Chorregon, *Australia* **62 C3** 22 40S 143 32 E
Chortkiv, *Ukraine* **17 D13** 49 2N 25 46 E
Chŏrwŏn, *S. Korea* **35 E14** 38 15N 127 10 E
Chorzów, *Poland* **17 C10** 50 18N 18 57 E
Chos-Malal, *Argentina* **94 D1** 37 20S 70 15W
Chosan, *N. Korea* **35 D13** 40 50N 125 47 E
Choszczno, *Poland* **16 B8** 53 7N 15 25 E
Choteau, *U.S.A.* **82 C7** 47 49N 112 11W
Chotila, *India* **42 H4** 22 23N 71 15 E
Chowchilla, *U.S.A.* **83 H3** 37 7N 120 16W
Choybalsan, *Mongolia* **33 B6** 48 4N 114 30 E
Christchurch, *N.Z.* **59 K4** 43 33S 172 47 E
Christchurch, *U.K.* **11 G6** 50 44N 1 47W
Christian I., *Canada* **78 B4** 44 50N 80 12W
Christiana, *S. Africa* **56 D4** 27 52S 25 8 E
Christiansted, *Virgin Is.* **89 C7** 17 45N 64 42W
Christie B., *Canada* **73 A6** 62 32N 111 10W
Christina →, *Canada* **73 B6** 56 40N 111 3W
Christmas Cr. →, *Australia* **60 C4** 18 29S 125 23 E
Christmas Creek, *Australia* **60 C4** 18 29S 125 23 E
Christmas I. = Kiritimati, *Kiribati* **65 G12** 1 58N 157 27W
Christmas I., *Ind. Oc.* **64 J2** 10 30S 105 40 E
Christopher L., *Australia* **61 D4** 24 49S 127 42 E
Chtimba, *Malawi* **55 E3** 10 35S 34 13 E
Chu = Shu, *Kazakstan* **26 E8** 43 36N 73 42 E
Chu → = Shu →, *Kazakstan* **28 E10** 45 0N 67 44 E
Chu →, *Vietnam* **38 C5** 19 53N 105 45 E
Chu Chua, *Canada* **72 C4** 51 22N 120 10W
Chu Lai, *Vietnam* **38 E7** 15 28N 108 45 E
Ch'uanchou = Quanzhou, *China* **33 D6** 24 55N 118 34 E
Chuankou, *China* **34 G6** 34 20N 110 59 E
Chūbu □, *Japan* **31 F8** 36 45N 137 30 E
Chubut →, *Argentina* **96 E3** 43 20S 65 5W
Chuchi L., *Canada* **72 B4** 55 12N 124 30W
Chudskoye, Oz., *Russia* **9 G22** 58 13N 27 30 E
Chūgoku □, *Japan* **31 G6** 35 0N 133 0 E
Chūgoku-Sanchi, *Japan* **31 G6** 35 0N 133 0 E
Chugwater, *U.S.A.* **80 E2** 41 46N 104 50W
Chukchi Sea, *Russia* **27 C19** 68 0N 175 0W
Chukotskoye Nagorye, *Russia* **27 C18** 68 0N 175 0 E
Chula Vista, *U.S.A.* **85 N9** 32 39N 117 5W
Chulman, *Russia* **27 D13** 56 52N 124 52 E
Chulucanas, *Peru* **92 E2** 5 8S 80 10W
Chulym →, *Russia* **26 D9** 57 43N 83 51 E
Chum Phae, *Thailand* **38 D4** 16 40N 102 6 E
Chum Saeng, *Thailand* **38 E3** 15 55N 100 15 E
Chumar, *India* **43 C8** 32 40N 78 35 E
Chumbicha, *Argentina* **94 B2** 29 0S 66 10W
Chumikan, *Russia* **27 D14** 54 40N 135 10 E
Chumphon, *Thailand* **39 G2** 10 35N 99 14 E
Chumuare, *Mozam.* **55 E3** 14 31S 31 50 E
Chumunjin, *S. Korea* **35 F15** 37 55N 128 54 E
Chuna →, *Russia* **27 D10** 57 47N 94 37 E
Chunchŏn, *S. Korea* **35 F14** 37 58N 127 44 E
Chunchura, *India* **43 H13** 22 53N 88 27 E
Chunga, *Zambia* **55 F2** 15 0S 26 2 E
Chunggang-ŭp, *N. Korea* **35 D14** 41 48N 126 48 E
Chunghwa, *N. Korea* **35 E13** 38 52N 125 47 E
Chungju, *S. Korea* **35 F14** 36 58N 127 58 E
Chungking = Chongqing, *China* **32 D5** 29 35N 106 25 E
Chungmu, *S. Korea* **35 G15** 34 50N 128 20 E
Chungt'iaoshan = Zhongtiao Shan, *China* **34 G6** 35 0N 111 10 E
Chunian, *Pakistan* **42 D6** 30 57N 74 0 E
Chunya, *Tanzania* **55 D3** 8 30S 33 27 E
Chunya □, *Tanzania* **54 D3** 7 48S 33 0 E
Chunyang, *China* **35 C15** 43 38N 129 23 E
Chuquibamba, *Peru* **92 G4** 15 47S 72 44W
Chuquicamata, *Chile* **94 A2** 22 15S 69 0W
Chur, *Switz.* **16 E5** 46 52N 9 32 E
Churachandpur, *India* **41 G18** 24 20N 93 40 E
Churchill, *Canada* **73 B10** 58 47N 94 11W
Churchill →, *Man., Canada* **73 B10** 58 47N 94 12W
Churchill →, *Nfld., Canada* **71 B7** 53 19N 60 10W
Churchill, C., *Canada* **73 B10** 58 46N 93 12W
Churchill Falls, *Canada* **71 B7** 53 36N 64 19W
Churchill L., *Canada* **73 B7** 55 55N 108 20W
Churchill Pk., *Canada* **72 B3** 58 10N 125 10W
Churu, *India* **42 E6** 28 20N 74 50 E
Chushal, *India* **43 C8** 33 40N 78 40 E
Chusovoy, *Russia* **24 C10** 58 22N 57 50 E
Chuuronjang, *N. Korea* **35 D15** 41 35N 129 40 E
Chuvash Republic □ = Chuvashia □, *Russia* **24 C8** 55 30N 47 0 E
Chuvashia □, *Russia* **24 C8** 55 30N 47 0 E
Chuwārtah, *Iraq* **44 C5** 35 43N 45 34 E
Ci Xian, *China* **34 F8** 36 20N 114 25 E
Cianjur, *Indonesia* **37 G12** 6 49S 107 8 E
Cibadok, *Indonesia* **37 G12** 6 53S 106 47 E
Cibatu, *Indonesia* **37 G12** 7 8S 107 59 E
Cibola, *U.S.A.* **85 M12** 33 17N 114 42W
Cicero, *U.S.A.* **76 E2** 41 48N 87 48W
Ciechanów, *Poland* **17 B11** 52 52N 20 38 E
Ciego de Avila, *Cuba* **88 B4** 21 50N 78 50W
Ciénaga, *Colombia* **92 A4** 11 1N 74 15W
Cienfuegos, *Cuba* **88 B3** 22 10N 80 30W
Cieszyn, *Poland* **17 D10** 49 45N 18 35 E
Cieza, *Spain* **19 C5** 38 17N 1 23W
Cihuatlán, *Mexico* **86 D4** 19 14N 104 35W
Cijara, Pantano de, *Spain* **19 C3** 39 18N 4 52W
Cijulang, *Indonesia* **37 G13** 7 42S 108 27 E
Cikajang, *Indonesia* **37 G12** 7 25S 107 48 E
Cikampek, *Indonesia* **37 G12** 6 23S 107 28 E
Cilacap, *Indonesia* **37 G13** 7 43S 109 0 E
Cill Chainnigh = Kilkenny, *Ireland* **13 D4** 52 39N 7 15W
Cima, *U.S.A.* **85 K11** 35 14N 115 30W
Cimahi, *Indonesia* **37 G12** 6 53S 107 33 E
Cimarron, *Kans., U.S.A.* **81 G4** 37 48N 100 21W
Cimarron, *N. Mex., U.S.A.* **81 G2** 36 31N 104 55W
Cimarron →, *U.S.A.* **81 G6** 36 10N 96 17W
Cimişlia, *Moldova* **17 E15** 46 34N 28 44 E
Cimone, Mte., *Italy* **20 B4** 44 12N 10 42 E
Cîmpina, *Romania* **17 F13** 45 10N 25 45 E
Cîmpulung, *Romania* **17 F13** 45 17N 25 3 E
Cinca →, *Spain* **19 B6** 41 26N 0 21 E
Cincar, *Bos.-H.* **20 C7** 43 55N 17 5 E
Cincinnati, *U.S.A.* **76 F3** 39 6N 84 31W
Çine, *Turkey* **21 F13** 37 37N 28 2 E
Ciney, *Belgium* **15 D5** 50 18N 5 5 E
Cinto, Mte., *France* **18 E8** 42 24N 8 54 E
Circle, *Alaska, U.S.A.* **68 B5** 65 50N 144 4W
Circle, *Mont., U.S.A.* **80 B2** 47 25N 105 35W
Circleville, *Ohio, U.S.A.* **76 F4** 39 36N 82 57W
Circleville, *Utah, U.S.A.* **83 G7** 38 10N 112 16W
Cirebon, *Indonesia* **37 G13** 6 45S 108 32 E
Cirencester, *U.K.* **11 F6** 51 43N 1 57W
Cirium, *Cyprus* **23 E11** 34 40N 32 53 E
Cisco, *U.S.A.* **81 J5** 32 23N 98 59W
Citlaltépetl, *Mexico* **87 D5** 19 0N 97 20W
Citrus Heights, *U.S.A.* **84 G5** 38 42N 121 17W
Citrusdal, *S. Africa* **56 E2** 32 35S 19 0 E
Città di Castello, *Italy* **20 C5** 43 27N 12 14 E
City of Aberdeen □, *U.K.* **12 D6** 57 10N 2 10W
City of Edinburgh □, *U.K.* **12 F5** 55 57N 3 17W
City of Glasgow □, *U.K.* **12 F4** 55 51N 4 12W
Ciudad Altamirano, *Mexico* **86 D4** 18 20N 100 40W
Ciudad Bolívar, *Venezuela* **92 B6** 8 5N 63 36W
Ciudad Camargo, *Mexico* **86 B3** 27 41N 105 10W
Ciudad Chetumal, *Mexico* **87 D7** 18 30N 88 20W
Ciudad de Valles, *Mexico* **87 C5** 22 0N 99 0W
Ciudad del Carmen, *Mexico* **87 D6** 18 38N 91 50W
Ciudad del Este, *Paraguay* **95 B5** 25 30S 54 50W
Ciudad Delicias = Delicias, *Mexico* **86 B3** 28 10N 105 30W
Ciudad Guayana, *Venezuela* **92 B6** 8 0N 62 30W
Ciudad Guerrero, *Mexico* **86 B3** 28 33N 107 28W
Ciudad Guzmán, *Mexico* **86 D4** 19 40N 103 30W
Ciudad Juárez, *Mexico* **86 A3** 31 40N 106 28W
Ciudad Madero, *Mexico* **87 C5** 22 19N 97 50W
Ciudad Mante, *Mexico* **87 C5** 22 50N 99 0W
Ciudad Obregón, *Mexico* **86 B3** 27 28N 109 59W
Ciudad Real, *Spain* **19 C4** 38 59N 3 55W
Ciudad Rodrigo, *Spain* **19 B2** 40 35N 6 32W
Ciudad Trujillo = Santo Domingo, *Dom. Rep.* **89 C6** 18 30N 69 59W
Ciudad Victoria, *Mexico* **87 C5** 23 41N 99 9W
Ciudadela, *Spain* **22 B10** 40 0N 3 50 E
Civitanova Marche, *Italy* **20 C5** 43 18N 13 44 E
Civitavécchia, *Italy* **20 C4** 42 6N 11 48 E
Cizre, *Turkey* **25 G7** 37 19N 42 10 E
Clackmannan □, *U.K.* **12 E5** 56 10N 3 43W
Clacton-on-Sea, *U.K.* **11 F9** 51 47N 1 11 E
Claire, L., *Canada* **72 B6** 58 35N 112 5W
Clairemont, *U.S.A.* **81 J4** 33 9N 100 44W
Clairton, *U.S.A.* **78 F5** 40 18N 79 53W
Clallam Bay, *U.S.A.* **84 B2** 48 15N 124 16W
Clanton, *U.S.A.* **77 J2** 32 51N 86 38W
Clanwilliam, *S. Africa* **56 E2** 32 11S 18 52 E
Clara, *Ireland* **13 C4** 53 21N 7 37W
Clara →, *Australia* **62 B3** 19 8S 142 30 E
Claraville, *U.S.A.* **85 K8** 35 24N 118 20W
Clare, *Australia* **63 E2** 33 50S 138 37 E
Clare, *U.S.A.* **76 D3** 43 49N 84 46W
Clare □, *Ireland* **13 D3** 52 45N 9 0W
Clare →, *Ireland* **13 C2** 53 20N 9 2W
Clare I., *Ireland* **13 C2** 53 49N 10 0W
Claremont, *Calif., U.S.A.* **85 L9** 34 6N 117 43W
Claremont, *N.H., U.S.A.* **79 C12** 43 23N 72 20W
Claremont Pt., *Australia* **62 A3** 14 1S 143 41 E
Claremore, *U.S.A.* **81 G7** 36 19N 95 36W
Claremorris, *Ireland* **13 C3** 53 45N 9 0W
Clarence →, *Australia* **63 D5** 29 25S 153 22 E
Clarence →, *N.Z.* **59 K4** 42 10S 173 56 E
Clarence, I., *Chile* **96 G2** 54 0S 72 0W
Clarence I., *Antarctica* **5 C18** 61 10S 54 0W
Clarence Str., *Australia* **60 B5** 12 0S 131 0 E
Clarence Str., *U.S.A.* **72 B2** 55 40N 132 10W
Clarence Town, *Bahamas* **89 B5** 23 6N 74 59W
Clarendon, *Ark., U.S.A.* **81 H9** 34 42N 91 19W
Clarendon, *Tex., U.S.A.* **81 H4** 34 56N 100 53W
Clarenville, *Canada* **71 C9** 48 10N 54 1W
Claresholm, *Canada* **72 C6** 50 0N 113 33W
Clarie Coast, *Antarctica* **5 C9** 68 0S 135 0 E
Clarinda, *U.S.A.* **80 E7** 40 44N 95 2W
Clarion, *Iowa, U.S.A.* **80 D8** 42 44N 93 44W
Clarion, *Pa., U.S.A.* **78 E5** 41 13N 79 23W
Clarion →, *U.S.A.* **78 E5** 41 7N 79 41W
Clark, *U.S.A.* **80 C6** 44 53N 97 44W
Clark, Pt., *Canada* **78 B3** 44 4N 81 45W
Clark Fork, *U.S.A.* **82 B5** 48 9N 116 11W
Clark Fork →, *U.S.A.* **82 B5** 48 9N 116 15W
Clark Hill Res., *U.S.A.* **77 J4** 33 45N 82 20W
Clarkdale, *U.S.A.* **83 J7** 34 46N 112 3W
Clarke City, *Canada* **71 B6** 50 12N 66 38W
Clarke I., *Australia* **62 G4** 40 32S 148 10 E
Clarke L., *Canada* **73 C7** 54 24N 106 54W
Clarke Ra., *Australia* **62 C4** 20 40S 148 30 E
Clark's Fork →, *U.S.A.* **82 D9** 45 39N 108 43W
Clark's Harbour, *Canada* **71 D6** 43 25N 65 38W
Clarks Summit, *U.S.A.* **79 E9** 41 30N 75 42W
Clarksburg, *U.S.A.* **76 F5** 39 17N 80 30W
Clarksdale, *U.S.A.* **81 H9** 34 12N 90 35W
Clarkston, *U.S.A.* **82 C5** 46 25N 117 3W
Clarksville, *Ark., U.S.A.* **81 H8** 35 28N 93 28W
Clarksville, *Tenn., U.S.A.* **77 G2** 36 32N 87 21W
Clarksville, *Tex., U.S.A.* **81 J7** 33 37N 95 3W
Clatskanie, *U.S.A.* **84 D3** 46 6N 123 12W
Claude, *U.S.A.* **81 H4** 35 7N 101 22W
Claveria, *Phil.* **37 A6** 18 37N 121 4 E
Clay, *U.S.A.* **84 G5** 38 17N 121 10W
Clay Center, *U.S.A.* **80 F6** 39 23N 97 8W
Claypool, *U.S.A.* **83 K8** 33 25N 110 51W
Claysville, *U.S.A.* **78 F4** 40 7N 80 25W
Clayton, *Idaho, U.S.A.* **82 D6** 44 16N 114 24W
Clayton, *N. Mex., U.S.A.* **81 G3** 36 27N 103 11W
Cle Elum, *U.S.A.* **82 C3** 47 12N 120 56W
Clear, C., *Ireland* **13 E2** 51 25N 9 32W
Clear I., *Ireland* **13 E2** 51 26N 9 30W
Clear L., *U.S.A.* **84 F4** 39 2N 122 47W
Clear Lake, *S. Dak., U.S.A.* **80 C6** 44 45N 96 41W
Clear Lake, *Wash., U.S.A.* **82 B2** 48 27N 122 15W
Clear Lake Reservoir, *U.S.A.* **82 F3** 41 56N 121 5W
Clearfield, *Pa., U.S.A.* **76 E6** 41 2N 78 27W
Clearfield, *Utah, U.S.A.* **82 F7** 41 7N 112 2W
Clearlake Highlands, *U.S.A.* **84 G4** 38 57N 122 38W
Clearmont, *U.S.A.* **82 D10** 44 38N 106 23W
Clearwater, *Canada* **72 C4** 51 38N 120 2W
Clearwater, *U.S.A.* **77 M4** 27 58N 82 48W
Clearwater →, *Alta., Canada* **72 C6** 52 22N 114 57W
Clearwater →, *Alta., Canada* **73 B6** 56 44N 111 23W
Clearwater Cr. →, *Canada* **72 A3** 61 36N 125 30W
Clearwater Mts., *U.S.A.* **82 C6** 46 5N 115 20W
Clearwater Prov. Park, *Canada* **73 C8** 54 0N 101 0W
Cleburne, *U.S.A.* **81 J6** 32 21N 97 23W
Cleethorpes, *U.K.* **10 D7** 53 33N 0 3W
Cleeve Hill, *U.K.* **11 F6** 51 54N 2 0W
Clerke Reef, *Australia* **60 C2** 17 22S 119 20 E
Clermont, *Australia* **62 C4** 22 49S 147 39 E
Clermont-Ferrand, *France* **18 D5** 45 46N 3 4 E
Clervaux, *Lux.* **15 D6** 50 4N 6 2 E
Cleveland, *Australia* **63 D5** 27 30S 153 15 E
Cleveland, *Miss., U.S.A.* **81 J9** 33 45N 90 43W
Cleveland, *Ohio, U.S.A.* **78 E3** 41 30N 81 42W
Cleveland, *Okla., U.S.A.* **81 G6** 36 19N 96 28W
Cleveland, *Tenn., U.S.A.* **77 H3** 35 10N 84 53W
Cleveland, *Tex., U.S.A.* **81 K7** 30 21N 95 5W
Cleveland, C., *Australia* **62 B4** 19 11S 147 1 E
Cleveland Heights, *U.S.A.* **78 E3** 41 30N 81 34W
Clevelândia, *Brazil* **95 B5** 26 24S 52 23W
Clew B., *Ireland* **13 C2** 53 50N 9 49W
Clewiston, *U.S.A.* **77 M5** 26 45N 80 56W
Clifden, *Ireland* **13 C1** 53 29N 10 1W
Clifden, *N.Z.* **59 M1** 46 1S 167 42 E
Cliffdell, *U.S.A.* **84 D5** 46 56N 121 5W
Clifton, *Australia* **63 D5** 27 59S 151 53 E
Clifton, *Ariz., U.S.A.* **83 K9** 33 3N 109 18W
Clifton, *Tex., U.S.A.* **81 K6** 31 47N 97 35W
Clifton Beach, *Australia* **62 B4** 16 46S 145 39 E
Clifton Forge, *U.S.A.* **76 G6** 37 49N 79 50W
Clifton Hills, *Australia* **63 D2** 27 1S 138 54 E
Climax, *Canada* **73 D7** 49 10N 108 20W
Clinch →, *U.S.A.* **77 H3** 35 53N 84 29W
Clingmans Dome, *U.S.A.* **77 H4** 35 34N 83 30W
Clint, *U.S.A.* **83 L10** 31 35N 106 14W
Clinton, *B.C., Canada* **72 C4** 51 6N 121 35W
Clinton, *Ont., Canada* **70 D3** 43 37N 81 32W
Clinton, *N.Z.* **59 M2** 46 12S 169 23 E
Clinton, *Ark., U.S.A.* **81 H8** 35 36N 92 28W
Clinton, *Ill., U.S.A.* **80 E10** 40 9N 88 57W
Clinton, *Ind., U.S.A.* **76 F2** 39 40N 87 24W
Clinton, *Iowa, U.S.A.* **80 E9** 41 51N 90 12W
Clinton, *Mass., U.S.A.* **79 D13** 42 25N 71 41W
Clinton, *Mo., U.S.A.* **80 F8** 38 22N 93 46W
Clinton, *N.C., U.S.A.* **77 H6** 35 0N 78 22W
Clinton, *Okla., U.S.A.* **81 H5** 35 31N 98 58W
Clinton, *S.C., U.S.A.* **77 H5** 34 29N 81 53W
Clinton, *Tenn., U.S.A.* **77 G3** 36 6N 84 8W
Clinton, *Wash., U.S.A.* **84 C4** 47 59N 122 21W
Clinton C., *Australia* **62 C5** 22 30S 150 45 E
Clinton Colden L., *Canada* **68 B9** 63 58N 107 27W
Clintonville, *U.S.A.* **80 C10** 44 37N 88 46W
Clipperton, I., *Pac. Oc.* **65 F17** 10 18N 109 13W
Clive L., *Canada* **72 A5** 63 13N 118 54W
Cloates, Pt., *Australia* **60 D1** 22 43S 113 40 E
Clocolan, *S. Africa* **57 D4** 28 55S 27 34 E
Clodomira, *Argentina* **94 B3** 27 35S 64 14W
Clonakilty, *Ireland* **13 E3** 51 37N 8 53W
Clonakilty B., *Ireland* **13 E3** 51 35N 8 51W
Cloncurry, *Australia* **62 C3** 20 40S 140 28 E
Cloncurry →, *Australia* **62 B3** 18 37S 140 40 E
Clones, *Ireland* **13 B4** 54 11N 7 15W
Clonmel, *Ireland* **13 D4** 52 21N 7 42W
Cloquet, *U.S.A.* **80 B8** 46 43N 92 28W
Clorinda, *Argentina* **94 B4** 25 16S 57 45W
Cloud Peak, *U.S.A.* **82 D10** 44 23N 107 11W
Cloudcroft, *U.S.A.* **83 K11** 32 58N 105 45W

Cloverdale, U.S.A. 84 G4 38 48N 123 1W
Clovis, Calif., U.S.A. 83 H4 36 49N 119 42W
Clovis, N. Mex., U.S.A. 81 H3 34 24N 103 12W
Cluj-Napoca, Romania 17 E12 46 47N 23 38 E
Clunes, Australia 63 F3 37 20S 143 45 E
Clutha →, N.Z. 59 M2 46 20S 169 49 E
Clwyd →, U.K. 10 D4 53 19N 3 31W
Clyde, N.Z. 59 L2 45 12S 169 20 E
Clyde, U.S.A. 78 C8 43 5N 76 52W
Clyde →, U.K. 12 F4 55 55N 4 30W
Clyde, Firth of, U.K. 12 F4 55 22N 5 1W
Clyde River, Canada 69 A13 70 30N 68 30W
Clydebank, U.K. 12 F4 55 54N 4 23W
Clymer, U.S.A. 78 D5 40 40N 79 1W
Coachella, U.S.A. 85 M10 33 41N 116 10W
Coachella Canal, U.S.A. 85 N12 32 43N 114 57W
Coahoma, U.S.A. 81 J4 32 18N 101 18W
Coahuayana →, Mexico 86 D4 18 41N 103 45W
Coahuayutla, Mexico 86 D4 18 19N 101 42W
Coahuila □, Mexico 86 B4 27 0N 103 0W
Coal →, Canada 72 B3 59 39N 126 57W
Coalane, Mozam. 55 F4 17 48S 37 2 E
Coalcomán, Mexico 86 D4 18 40N 103 10W
Coaldale, Canada 72 D6 49 45N 112 35W
Coalgate, U.S.A. 81 H6 34 32N 96 13W
Coalinga, U.S.A. 83 H3 36 9N 120 21W
Coalville, U.K. 10 E6 52 44N 1 23W
Coalville, U.S.A. 82 F8 40 55N 111 24W
Coari, Brazil 92 D6 4 8S 63 7W
Coast □, Kenya 54 C4 2 40S 39 45 E
Coast Mts., Canada 72 C3 55 0N 129 20W
Coast Ranges, U.S.A. 84 G4 39 0N 123 0W
Coatbridge, U.K. 12 F4 55 52N 4 6W
Coatepec, Mexico 87 D5 19 27N 96 58W
Coatepeque, Guatemala 88 D1 14 46N 91 55W
Coatesville, U.S.A. 76 F8 39 59N 75 50W
Coaticook, Canada 71 C5 45 10N 71 46W
Coats I., Canada 69 B11 62 30N 83 0W
Coats Land, Antarctica 5 D1 77 0S 25 0W
Coatzacoalcos, Mexico 87 D6 18 7N 94 25W
Cobalt, Canada 70 C4 47 25N 79 42W
Cobán, Guatemala 88 C1 15 30N 90 21W
Cobar, Australia 63 E4 31 27S 145 48 E
Cóbh, Ireland 13 E3 51 51N 8 17W
Cobham, Australia 63 E3 30 18S 142 7 E
Cobija, Bolivia 92 F5 11 0S 68 50W
Cobleskill, U.S.A. 79 D10 42 41N 74 29W
Coboconk, Canada 78 B6 44 39N 78 48W
Cobourg, Canada 70 D4 43 58N 78 10W
Cobourg Pen., Australia 60 B5 11 20S 132 15 E
Cobram, Australia 63 F4 35 54S 145 40 E
Cobre, U.S.A. 82 F6 41 7N 114 24W
Cóbué, Mozam. 55 E3 12 0S 34 58 E
Coburg, Germany 16 C6 50 15N 10 58 E
Cocanada = Kakinada, India 41 L13 16 57N 82 11 E
Cochabamba, Bolivia 92 G5 17 26S 66 10W
Cochemane, Mozam. 55 F3 17 0S 32 54 E
Cochin, India 40 Q10 9 59N 76 22 E
Cochin China = Nam-Phan, Vietnam 39 G6 10 30N 106 0 E
Cochise, U.S.A. 83 K9 32 7N 109 55W
Cochran, U.S.A. 77 J4 32 23N 83 21W
Cochrane, Alta., Canada 72 C6 51 11N 114 30W
Cochrane, Ont., Canada 70 C3 49 0N 81 0W
Cochrane →, Canada 73 B8 59 0N 103 40W
Cochrane, L., Chile 96 F2 47 10S 72 0W
Cockburn, Australia 63 E3 32 5S 141 0 E
Cockburn, Canal, Chile 96 G2 54 30S 72 0W
Cockburn I., Canada 70 C3 45 55N 83 22W
Cockburn Ra., Australia 60 C4 15 46S 128 0 E
Cocklebiddy Motel, Australia 61 F4 32 0S 126 3 E
Coco →, Cent. Amer. 88 D3 15 0N 83 8W
Cocoa, U.S.A. 77 L5 28 21N 80 44W
Cocobeach, Gabon 52 D1 0 59N 9 34 E
Cocos, I. del, Pac. Oc. 65 G19 5 25N 87 55W
Cocos Is., Ind. Oc. 64 J1 12 10S 96 55 E
Cod, C., U.S.A. 75 B13 42 5N 70 10W
Codajás, Brazil 92 D6 3 55S 62 0W
Coderre, Canada 73 C7 50 11N 106 31W
Codó, Brazil 93 D10 4 30S 43 55W
Cody, U.S.A. 82 D9 44 32N 109 3W
Coe Hill, Canada 70 D4 44 52N 77 50W
Coelemu, Chile 94 D1 36 30S 72 48W
Coen, Australia 62 A3 13 52S 143 12 E
Cœur d'Alene, U.S.A. 82 C5 47 45N 116 51W
Cœur d'Alene L., U.S.A. 82 C5 47 32N 116 48W
Coevorden, Neths. 15 B6 52 40N 6 44 E
Cofete, Canary Is. 22 F5 28 6N 14 23W
Coffeyville, U.S.A. 81 G7 37 2N 95 37W
Coffin B., Australia 63 E2 34 38S 135 28 E
Coffin Bay Peninsula, Australia 63 E2 34 32S 135 15 E
Coffs Harbour, Australia 63 E5 30 16S 153 5 E
Cognac, France 18 D3 45 41N 0 20W
Cohagen, U.S.A. 82 C10 47 3N 106 37W
Cohoes, U.S.A. 79 D11 42 46N 73 42W
Cohuna, Australia 63 F3 35 45S 144 15 E
Coiba, I., Panama 88 E3 7 30N 81 40W
Coig →, Argentina 96 G3 51 0S 69 10W
Coihaique, Chile 96 F2 45 30S 71 45W
Coimbatore, India 40 P10 11 2N 76 59 E
Coimbra, Brazil 92 G7 19 55S 57 48W
Coimbra, Portugal 19 B1 40 15N 8 27W
Coín, Spain 19 D3 36 40N 4 48W
Cojimies, Ecuador 92 C2 0 20N 80 0W
Cojutepequé, El Salv. 88 D2 13 41N 88 54W
Cokeville, U.S.A. 82 E8 42 5N 110 57W
Colac, Australia 63 F3 38 21S 143 35 E
Colatina, Brazil 93 G10 19 32S 40 37W
Colbeck, C., Antarctica 5 D13 77 6S 157 48W
Colbinabbin, Australia 63 F3 36 38S 144 48 E
Colborne, Canada 78 B7 44 0N 77 53W
Colby, U.S.A. 80 F4 39 24N 101 3W
Colchagua □, Chile 94 C1 34 30S 71 0W
Colchester, U.K. 11 F8 51 54N 0 55 E
Coldstream, U.K. 12 F6 55 39N 2 15W
Coldwater, Canada 78 B5 44 42N 79 40W
Coldwater, U.S.A. 81 G5 37 16N 99 20W
Colebrook, Australia 62 G4 42 31S 147 21 E
Colebrook, U.S.A. 79 B13 44 54N 71 30W
Coleman, Canada 72 D6 49 40N 114 30W
Coleman, U.S.A. 81 K5 31 50N 99 26W
Coleman →, Australia 62 B3 15 6S 141 38 E
Colenso, S. Africa 57 D4 28 44S 29 50 E
Coleraine, Australia 63 F3 37 36S 141 40 E
Coleraine, U.K. 13 A5 55 8N 6 41W
Coleraine □, U.K. 13 A5 55 8N 6 40W
Coleridge, L., N.Z. 59 K3 43 17S 171 30 E
Colesberg, S. Africa 56 E4 30 45S 25 5 E
Coleville, U.S.A. 84 G7 38 34N 119 30W
Colfax, Calif., U.S.A. 84 F6 39 6N 120 57W
Colfax, La., U.S.A. 81 K8 31 31N 92 42W
Colfax, Wash., U.S.A. 82 C5 46 53N 117 22W
Colhué Huapi, L., Argentina 96 F3 45 30S 69 0W
Coligny, S. Africa 57 D4 26 17S 26 15 E
Colima, Mexico 86 D4 19 14N 103 43W
Colima □, Mexico 86 D4 19 10N 103 40W
Colima, Nevado de, Mexico 86 D4 19 35N 103 45W
Colina, Chile 94 C1 33 13S 70 45W
Colinas, Brazil 93 E10 6 0S 44 10W
Coll, U.K. 12 E2 56 39N 6 34W
Collaguasi, Chile 94 A2 21 5S 68 45W
Collarenebri, Australia 63 D4 29 33S 148 34 E
Collbran, U.S.A. 83 G10 39 14N 107 58W
Colleen Bawn, Zimbabwe 55 G2 21 0S 29 12 E
College Park, U.S.A. 77 J3 33 40N 84 27W
College Station, U.S.A. 81 K6 30 37N 96 21W
Collette, Canada 71 C6 46 40N 65 30W
Collie, Australia 61 F2 33 22S 116 8 E
Collier B., Australia 60 C3 16 10S 124 15 E
Collier Ra., Australia 60 D2 24 45S 119 10 E
Collingwood, Canada 78 B4 44 29N 80 13W
Collingwood, N.Z. 59 J4 40 41S 172 40 E
Collins, Canada 70 B2 50 17N 89 27W
Collinsville, Australia 62 C4 20 30S 147 56 E
Collipulli, Chile 94 D1 37 55S 72 30W
Collooney, Ireland 13 B3 54 11N 8 29W
Colmar, France 18 B7 48 5N 7 20 E
Colne, U.K. 10 D5 53 51N 2 9W
Colo →, Australia 63 E5 33 25S 150 52 E
Cologne = Köln, Germany 16 C4 50 56N 6 57 E
Colom, I., Spain 22 B11 39 58N 4 16 E
Coloma, U.S.A. 84 G6 38 48N 120 53W
Colomb-Béchar = Béchar, Algeria 50 B4 31 38N 2 18W
Colômbia, Brazil 93 H9 20 10S 48 40W
Colombia ■, S. Amer. 92 C4 3 45N 73 0W
Colombian Basin, S. Amer. 66 H12 14 0N 76 0W
Colombo, Sri Lanka 40 R11 6 56N 79 58 E
Colome, U.S.A. 80 D5 43 16N 99 43W
Colón, Argentina 94 C4 32 12S 58 10W
Colón, Cuba 88 B3 22 42N 80 54W
Colón, Panama 88 E4 9 20N 79 54W
Colona, Australia 61 F5 31 38S 132 4 E
Colonia, Uruguay 94 C4 34 25S 57 50W
Colonia de San Jordi, Spain 22 B9 39 19N 2 59 E
Colonia Dora, Argentina 94 B3 28 34S 62 59W
Colonial Heights, U.S.A. 76 G7 37 15N 77 25W
Colonsay, Canada 73 C7 51 59N 105 52W
Colonsay, U.K. 12 E2 56 5N 6 12W
Colorado □, U.S.A. 83 G10 39 30N 105 30W
Colorado →, Argentina 96 D4 39 50S 62 8W
Colorado →, N. Amer. 83 L6 31 45N 114 40W
Colorado →, U.S.A. 81 L7 28 36N 95 59W
Colorado City, U.S.A. 81 J4 32 24N 100 52W
Colorado Desert, U.S.A. 74 D3 34 20N 116 0W
Colorado Plateau, U.S.A. 83 H8 37 0N 111 0W
Colorado River Aqueduct, U.S.A. 85 L12 34 17N 114 10W
Colorado Springs, U.S.A. 80 F2 38 50N 104 49W
Colotlán, Mexico 86 C4 22 6N 103 16W
Colton, N.Y., U.S.A. 79 B10 44 33N 74 56W
Colton, Wash., U.S.A. 82 C5 46 34N 117 8W
Columbia, La., U.S.A. 81 J8 32 6N 92 5W
Columbia, Miss., U.S.A. 81 K10 31 15N 89 50W
Columbia, Mo., U.S.A. 80 F8 38 57N 92 20W
Columbia, Pa., U.S.A. 79 F8 40 2N 76 30W
Columbia, S.C., U.S.A. 77 H5 34 0N 81 2W
Columbia, Tenn., U.S.A. 77 H2 35 37N 87 2W
Columbia →, U.S.A. 82 C1 46 15N 124 5W
Columbia, C., Canada 4 A4 83 0N 70 0W
Columbia, District of □, U.S.A. 76 F7 38 55N 77 0W
Columbia, Mt., Canada 72 C5 52 8N 117 20W
Columbia Basin, U.S.A. 82 C4 46 45N 119 5W
Columbia Falls, U.S.A. 82 B6 48 23N 114 11W
Columbia Heights, U.S.A. 80 C8 45 3N 93 15W
Columbiana, U.S.A. 78 F4 40 53N 80 42W
Columbretes, Is., Spain 19 C6 39 50N 0 50 E
Columbus, Ga., U.S.A. 77 J3 32 28N 84 59W
Columbus, Ind., U.S.A. 76 F3 39 13N 85 55W
Columbus, Kans., U.S.A. 81 G7 37 10N 94 50W
Columbus, Miss., U.S.A. 77 J1 33 30N 88 25W
Columbus, Mont., U.S.A. 82 D9 45 38N 109 15W
Columbus, N. Dak., U.S.A. 80 A3 48 54N 102 47W
Columbus, N. Mex., U.S.A. 83 L10 31 50N 107 38W
Columbus, Nebr., U.S.A. 80 E6 41 26N 97 22W
Columbus, Ohio, U.S.A. 76 F4 39 58N 83 0W
Columbus, Tex., U.S.A. 81 L6 29 42N 96 33W
Columbus, Wis., U.S.A. 80 D10 43 21N 89 1W
Colusa, U.S.A. 84 F4 39 13N 122 1W
Colville, U.S.A. 82 B5 48 33N 117 54W
Colville →, U.S.A. 68 A4 70 25N 150 30W
Colville, C., N.Z. 59 G5 36 29S 175 21 E
Colwyn Bay, U.K. 10 D4 53 18N 3 44W
Comácchio, Italy 20 B5 44 42N 12 11 E
Comalcalco, Mexico 87 D6 18 16N 93 13W
Comallo, Argentina 96 E2 41 0S 70 5W
Comanche, Okla., U.S.A. 81 H6 34 22N 97 58W
Comanche, Tex., U.S.A. 81 K5 31 54N 98 36W
Comayagua, Honduras 88 D2 14 25N 87 37W
Combahee →, U.S.A. 77 J5 32 30N 80 31W
Comber, Canada 78 D2 42 14N 82 33W
Comblain-au-Pont, Belgium 15 D5 50 29N 5 35 E
Comeragh Mts., Ireland 13 D4 52 18N 7 34W
Comet, Australia 62 C4 23 36S 148 38 E
Comilla, Bangla. 41 H17 23 28N 91 10 E
Comino, Malta 23 C1 36 2N 14 20 E
Comino, C., Italy 20 D3 40 32N 9 49 E
Comitán, Mexico 87 D6 16 18N 92 9W
Commerce, Ga., U.S.A. 77 H4 34 12N 83 28W
Commerce, Tex., U.S.A. 81 J7 33 15N 95 54W
Committee B., Canada 69 B11 68 30N 86 30W
Commonwealth B., Antarctica 5 C10 67 0S 144 0 E
Commoron Cr. →, Australia 63 D5 28 22S 150 8 E
Communism Pk. = Kommunizma, Pik, Tajikistan 26 F8 39 0N 72 2 E
Como, Italy 20 B3 45 47N 9 5 E
Como, L. di, Italy 20 B3 46 0N 9 11 E
Comodoro Rivadavia, Argentina 96 F3 45 50S 67 40W
Comorin, C., India 40 Q10 8 3N 77 40 E
Comoro Is. = Comoros ■, Ind. Oc. 49 H8 12 10S 44 15 E
Comoros ■, Ind. Oc. 49 H8 12 10S 44 15 E
Comox, Canada 72 D4 49 42N 124 55W
Compiègne, France 18 B5 49 24N 2 50 E
Compostela, Mexico 86 C4 21 15N 104 53W
Comprida, I., Brazil 95 A6 24 50S 47 42W
Compton, U.S.A. 85 M8 33 54N 118 13W
Compton Downs, Australia 63 E4 30 28S 146 30 E
Comrat, Moldova 17 E15 46 18N 28 40 E
Con Cuong, Vietnam 38 C5 19 2N 104 54 E
Con Son, Vietnam 39 H6 8 41N 106 37 E
Conakry, Guinea 50 G2 9 29N 13 49W
Conara Junction, Australia 62 G4 41 50S 147 26 E
Concarneau, France 18 C2 47 52N 3 56W
Conceição, Mozam. 55 F4 18 47S 36 7 E
Conceição da Barra, Brazil 93 G11 18 35S 39 45W
Conceição do Araguaia, Brazil 93 E9 8 0S 49 2W
Concepción, Argentina 94 B2 27 20S 65 35W
Concepción, Bolivia 92 G6 16 15S 62 8W
Concepción, Chile 94 D1 36 50S 73 0W
Concepción, Mexico 87 D6 18 15N 90 5W
Concepción, Paraguay 94 A4 23 22S 57 26W
Concepción □, Chile 94 D1 37 0S 72 30W
Concepción →, Mexico 86 A2 30 32N 113 2W
Concepción, L., Bolivia 92 G6 17 20S 61 20W
Concepción, Punta, Mexico 86 B2 26 55N 111 59W
Concepción del Oro, Mexico 86 C4 24 40N 101 30W
Concepción del Uruguay, Argentina 94 C4 32 35S 58 20W
Conception, Pt., U.S.A. 85 L6 34 27N 120 28W
Conception B., Namibia 56 C1 23 55S 14 22 E
Conception I., Bahamas 89 B4 23 52N 75 9W
Concession, Zimbabwe 55 F3 17 27S 30 56 E
Conchas Dam, U.S.A. 81 H2 35 22N 104 11W
Conche, Canada 71 B8 50 55N 55 58W
Concho, U.S.A. 83 J9 34 28N 109 36W
Concho →, U.S.A. 81 K5 31 34N 99 43W
Conchos →, Chihuahua, Mexico 86 B4 29 32N 105 0W
Conchos →, Tamaulipas, Mexico 87 B5 25 9N 98 35W
Concord, Calif., U.S.A. 84 H4 37 59N 122 2W
Concord, N.C., U.S.A. 77 H5 35 25N 80 35W
Concord, N.H., U.S.A. 79 C13 43 12N 71 32W
Concordia, Argentina 94 C4 31 20S 58 2W
Concórdia, Brazil 92 D5 4 36S 66 36W
Concordia, Mexico 86 C3 23 18N 106 2W
Concordia, U.S.A. 80 F6 39 34N 97 40W
Concrete, U.S.A. 82 B3 48 32N 121 45W
Condamine, Australia 63 D5 26 56S 150 9 E
Conde, U.S.A. 80 C5 45 9N 98 6W
Condeúba, Brazil 93 F10 14 52S 42 0W
Condobolin, Australia 63 E4 33 4S 147 6 E
Condon, U.S.A. 82 D3 45 14N 120 11W
Conegliano, Italy 20 B5 45 53N 12 18 E
Conejera, I., Spain 22 B9 39 11N 2 58 E
Conejos, Mexico 86 B4 26 14N 103 53W
Confuso →, Paraguay 94 B4 25 9S 57 34W
Congleton, U.K. 10 D5 53 10N 2 13W
Congo = Zaïre →, Africa 52 F2 6 4S 12 24 E
Congo (Kinshasa) = Zaïre ■, Africa 52 E4 3 0S 23 0 E
Congo ■, Africa 52 E3 1 0S 16 0 E
Congo Basin, Africa 48 G6 0 10S 24 30 E
Congonhas, Brazil 95 A7 20 30S 43 52W
Congress, U.S.A. 83 J7 34 9N 112 51W
Coniston, Canada 70 C3 46 29N 80 51W
Conjeeveram = Kanchipuram, India 40 N11 12 52N 79 45 E
Conjuboy, Australia 62 B3 18 35S 144 35 E
Conklin, Canada 73 B6 55 38N 111 5W
Conlea, Australia 63 E3 30 7S 144 35 E
Conn, L., Ireland 13 B2 54 3N 9 15W
Connacht □, Ireland 13 C3 53 43N 9 12W
Conneaut, U.S.A. 78 E4 41 57N 80 34W
Connecticut □, U.S.A. 79 E12 41 30N 72 45W
Connecticut →, U.S.A. 79 E12 41 16N 72 20W
Connell, U.S.A. 82 C4 46 40N 118 52W
Connellsville, U.S.A. 78 F5 40 1N 79 35W
Connemara, Ireland 13 C2 53 29N 9 45W
Connemaugh →, U.S.A. 78 F5 40 28N 79 19W
Connersville, U.S.A. 76 F3 39 39N 85 8W
Connors Ra., Australia 62 C4 21 40S 149 10 E
Conoble, Australia 63 E3 32 55S 144 33 E
Cononaco →, Ecuador 92 D3 1 32S 75 35W
Cononbridge, U.K. 12 D4 57 34N 4 27W
Conquest, Canada 73 C7 51 32N 107 14W
Conrad, U.S.A. 82 B8 48 10N 111 57W
Conran, C., Australia 63 F4 37 49S 148 44 E
Conroe, U.S.A. 81 K7 30 19N 95 27W
Conselheiro Lafaiete, Brazil 95 A7 20 40S 43 48W
Consort, Canada 73 C6 52 1N 110 46W
Constance = Konstanz, Germany 16 E5 47 40N 9 10 E
Constance, L. = Bodensee, Europe 16 E5 47 35N 9 25 E
Constanța, Romania 17 F15 44 14N 28 38 E
Constantine, Algeria 50 A6 36 25N 6 42 E
Constitución, Chile 94 D1 35 20S 72 30W
Constitución, Uruguay 94 C4 31 0S 57 50W
Consul, Canada 73 D7 49 20N 109 30W
Contact, U.S.A. 82 F6 41 46N 114 45W
Contai, India 43 J12 21 54N 87 46 E
Contamana, Peru 92 E4 7 19S 74 55W
Contas →, Brazil 93 F11 14 17S 39 1W
Contoocook, U.S.A. 79 C13 43 13N 71 45W
Contra Costa, Mozam. 57 D5 25 9S 33 30 E
Conway = Conwy, U.K. 10 D4 53 17N 3 50W
Conway = Conwy →, U.K. 10 D4 53 17N 3 50W
Conway, Ark., U.S.A. 81 H8 35 5N 92 26W
Conway, N.H., U.S.A. 79 C13 43 59N 71 7W
Conway, S.C., U.S.A. 77 J6 33 51N 79 3W
Conway, L., Australia 63 D2 28 17S 135 35 E
Conwy, U.K. 10 D4 53 17N 3 50W
Conwy →, U.K. 10 D4 53 17N 3 50W
Coober Pedy, Australia 63 D1 29 1S 134 43 E
Cooch Behar = Koch Bihar, India 41 F16 26 22N 89 29 E
Coodardy, Australia 61 E2 27 15S 117 39 E
Cook, Australia 61 F5 30 37S 130 25 E
Cook, U.S.A. 80 B8 47 49N 92 39W
Cook, B., Chile 96 H3 55 10S 70 0W
Cook, Mt., N.Z. 59 K3 43 36S 170 9 E
Cook Inlet, U.S.A. 68 C4 60 0N 152 0W
Cook Is., Pac. Oc. 65 J11 17 0S 160 0W
Cook Strait, N.Z. 59 J5 41 15S 174 29 E
Cookeville, U.S.A. 77 G3 36 10N 85 30W
Cookhouse, S. Africa 56 E4 32 44S 25 47 E
Cookshire, Canada 79 A13 45 25N 71 38W
Cookstown, U.K. 13 B5 54 39N 6 45W
Cookstown □, U.K. 13 B5 54 40N 6 43W
Cooksville, Canada 78 C5 43 36N 79 35W
Cooktown, Australia 62 B4 15 30S 145 16 E
Coolabah, Australia 63 E4 31 1S 146 43 E
Cooladdi, Australia 63 D4 26 37S 145 23 E
Coolah, Australia 63 E4 31 48S 149 41 E
Coolamon, Australia 63 E4 34 46S 147 8 E
Coolangatta, Australia 63 D5 28 11S 153 29 E
Coolgardie, Australia 61 F3 30 55S 121 8 E
Coolibah, Australia 60 C5 15 33S 130 56 E
Coolidge, U.S.A. 83 K8 32 59N 111 31W
Coolidge Dam, U.S.A. 83 K8 33 0N 110 20W
Cooma, Australia 63 F4 36 12S 149 8 E
Coon Rapids, U.S.A. 80 C8 45 9N 93 19W
Coonabarabran, Australia 63 E4 31 14S 149 18 E
Coonamble, Australia 63 E4 30 56S 148 27 E
Coonana, Australia 61 F3 31 0S 123 0 E
Coondapoor, India 40 N9 13 42N 74 40 E
Coongie, Australia 63 D3 27 9S 140 8 E
Coongoola, Australia 63 D4 27 43S 145 51 E
Cooninie, L., Australia 63 D2 26 4S 139 59 E
Cooper, U.S.A. 81 J7 33 23N 95 42W
Cooper →, U.S.A. 77 J6 32 50N 79 56W
Cooper Cr. →, Australia 63 D2 28 29S 137 46 E
Cooperstown, N. Dak., U.S.A. 80 B5 47 27N 98 8W
Cooperstown, N.Y., U.S.A. 79 D10 42 42N 74 56W
Coorabie, Australia 61 F5 31 54S 132 18 E
Coorabulka, Australia 62 C3 23 41S 140 20 E
Coorow, Australia 61 E2 29 53S 116 2 E
Cooroy, Australia 63 D5 26 22S 152 54 E
Coos Bay, U.S.A. 82 E1 43 22N 124 13W
Cootamundra, Australia 63 E4 34 36S 148 1 E
Cootehill, Ireland 13 B4 54 4N 7 5W
Cooyar, Australia 63 D5 26 59S 151 51 E
Cooyeana, Australia 62 C2 24 29S 138 45 E
Copahue Paso, Argentina 94 D1 37 49S 71 8W
Copainalá, Mexico 87 D6 17 8N 93 11W
Copán, Honduras 88 D2 14 50N 89 9W
Cope, U.S.A. 80 F3 39 40N 102 51W
Copenhagen = København, Denmark 9 J15 55 41N 12 34 E
Copiapó, Chile 94 B1 27 30S 70 20W
Copiapó →, Chile 94 B1 27 19S 70 56W
Copley, Australia 63 E2 30 36S 138 26 E
Copp L., Canada 72 A6 60 14N 114 40W
Copper Center, U.S.A. 68 B5 61 58N 145 18W
Copper Cliff, Canada 70 C3 46 28N 81 4W
Copper Harbor, U.S.A. 76 B2 47 28N 87 53W
Copper Queen, Zimbabwe 55 F2 17 29S 29 18 E
Copperbelt □, Zambia 55 E2 13 15S 27 30 E
Coppermine = Kuglugtuk, Canada 68 B8 67 50N 115 5W
Coppermine →, Canada 68 B8 67 49N 116 4W
Copperopolis, U.S.A. 84 H6 37 58N 120 38W
Coquet →, U.K. 10 B6 55 20N 1 32W
Coquilhatville = Mbandaka, Zaïre 52 D3 0 1N 18 18 E
Coquille, U.S.A. 82 E1 43 11N 124 11W
Coquimbo, Chile 94 B1 30 0S 71 20W
Coquimbo □, Chile 94 C1 31 0S 71 0W
Corabia, Romania 17 G13 43 48N 24 30 E
Coracora, Peru 92 G4 15 5S 73 45W
Coral Gables, U.S.A. 77 N5 25 45N 80 16W
Coral Harbour, Canada 69 B11 64 8N 83 10W
Coral Sea, Pac. Oc. 64 J7 15 0S 150 0 E
Coral Springs, U.S.A. 77 M5 26 16N 80 13W
Coraopolis, U.S.A. 78 F4 40 31N 80 10W
Corato, Italy 20 D7 41 9N 16 25 E
Corbin, U.S.A. 76 G3 36 57N 84 6W
Corby, U.K. 11 E7 52 30N 0 41W
Corby Glen, U.K. 11 E7 52 49N 0 30W
Corcaigh = Cork, Ireland 13 E3 51 54N 8 29W
Corcoran, U.S.A. 83 H4 36 6N 119 33W
Corcubión, Spain 19 A1 42 56N 9 12W
Cordele, U.S.A. 77 K4 31 58N 83 47W
Cordell, U.S.A. 81 H5 35 17N 98 59W
Córdoba, Argentina 94 C3 31 20S 64 10W
Córdoba, Mexico 87 D5 18 50N 97 0W
Córdoba, Spain 19 D3 37 50N 4 50W
Córdoba □, Argentina 94 C3 31 22S 64 15W
Córdoba, Sierra de, Argentina 94 C3 31 10S 64 25W
Cordon, Phil. 37 A6 16 42N 121 32 E
Cordova, Ala., U.S.A. 77 J2 33 46N 87 11W
Cordova, Alaska, U.S.A. 68 B5 60 33N 145 45W
Corella →, Australia 62 B3 19 34S 140 47 E
Corfield, Australia 62 C3 21 40S 143 21 E
Corfu = Kérkira, Greece 23 A3 39 38N 19 50 E
Corfu, Str of, Greece 23 A4 39 34N 20 0 E
Coria, Spain 19 C2 39 58N 6 33W
Corigliano Cálabro, Italy 20 E7 39 36N 16 31 E
Coringa Is., Australia 62 B4 16 58S 149 58 E
Corinna, Australia 62 G4 41 35S 145 10 E
Corinth = Kórinthos, Greece 21 F10 37 56N 22 55 E
Corinth, Miss., U.S.A. 77 H1 34 56N 88 31W
Corinth, N.Y., U.S.A. 79 C11 43 15N 73 49W

Corinth, G. of = Korinthiakós Kólpos, Greece 21 E10 38 16N 22 30 E
Corinto, Brazil 93 G10 18 20S 44 30W
Corinto, Nic. 88 D2 12 30N 87 10W
Cork, Ireland 13 E3 51 54N 8 29W
Cork □, Ireland 13 E3 51 57N 8 40W
Cork Harbour, Ireland 13 E3 51 47N 8 16W
Çorlu, Turkey 21 D12 41 11N 27 49 E
Cormack L., Canada 72 A4 60 56N 121 37W
Cormorant, Canada 73 C8 54 14N 100 35W
Cormorant L., Canada 73 C8 54 15N 100 50W
Corn Is. = Maíz, Is. del, Nic. 88 D3 12 15N 83 4W
Cornélio Procópio, Brazil 95 A5 23 7S 50 40W
Cornell, U.S.A. 80 C9 45 10N 91 9W
Corner Brook, Canada 71 C8 48 57N 57 58W
Corneşti, Moldova 17 E15 47 21N 28 1 E
Corning, Ark., U.S.A. 81 G9 36 25N 90 35W
Corning, Calif., U.S.A. 82 G2 39 56N 122 11W
Corning, Iowa, U.S.A. 80 E7 40 59N 94 44W
Corning, N.Y., U.S.A. 78 D7 42 9N 77 3W
Cornwall, Canada 70 C5 45 2N 74 44W
Cornwall □, U.K. 11 G3 50 26N 4 40W
Corny Pt., Australia 63 E2 34 55S 137 0 E
Coro, Venezuela 92 A5 11 25N 69 41W
Coroatá, Brazil 93 D10 4 8S 44 0W
Corocoro, Bolivia 92 G5 17 15S 68 28W
Coroico, Bolivia 92 G5 16 0S 67 50W
Coromandel, N.Z. 59 G5 36 45S 175 31 E
Coromandel Coast, India 40 N12 12 30N 81 0 E
Corona, Australia 63 E3 31 16S 141 24 E
Corona, Calif., U.S.A. 85 M9 33 53N 117 34W
Corona, N. Mex., U.S.A. 83 J11 34 15N 105 36W
Coronado, U.S.A. 85 N9 32 41N 117 11W
Coronado, B. de, Costa Rica 88 E3 9 0N 83 40W
Coronados, Is. los, U.S.A. 85 N9 32 25N 117 15W
Coronation, Canada 72 C6 52 5N 111 27W
Coronation Gulf, Canada 68 B8 68 25N 110 0W
Coronation I., Antarctica 5 C18 60 45S 46 0W
Coronation I., U.S.A. 72 B2 55 52N 134 20W
Coronation Is., Australia 60 B3 14 57S 124 55 E
Coronda, Argentina 94 C3 31 58S 60 56W
Coronel, Chile 94 D1 37 0S 73 10W
Coronel Bogado, Paraguay 94 B4 27 11S 56 18W
Coronel Dorrego, Argentina 94 D3 38 40S 61 10W
Coronel Oviedo, Paraguay 94 B4 25 24S 56 30W
Coronel Pringles, Argentina 94 D3 38 0S 61 30W
Coronel Suárez, Argentina 94 D3 37 30S 61 52W
Coronel Vidal, Argentina 94 D4 37 28S 57 45W
Corowa, Australia 63 F4 35 58S 146 21 E
Corozal, Belize 87 D7 18 23N 88 23W
Corpus, Argentina 95 B4 27 10S 55 30W
Corpus Christi, U.S.A. 81 M6 27 47N 97 24W
Corpus Christi, L., U.S.A. 81 L6 28 2N 97 52W
Corque, Bolivia 92 G5 18 20S 67 41W
Corralejo, Canary Is. 22 F6 28 43N 13 53W
Correntes, C. das, Mozam. 57 C6 24 6S 35 34 E
Corrib, L., Ireland 13 C2 53 27N 9 16W
Corrientes, Argentina 94 B4 27 30S 58 45W
Corrientes □, Argentina 94 B4 28 0S 57 0W
Corrientes →, Argentina 94 C4 30 42S 59 38W
Corrientes →, Peru 92 D4 3 43S 74 35W
Corrientes, C., Colombia 92 B3 5 30N 77 34W
Corrientes, C., Cuba 88 B3 21 43N 84 30W
Corrientes, C., Mexico 86 C3 20 25N 105 42W
Corrigan, U.S.A. 81 K7 31 0N 94 52W
Corrigin, Australia 61 F2 32 20S 117 53 E
Corry, U.S.A. 78 E5 41 55N 79 39W
Corse, France 18 F8 42 0N 9 0 E
Corse, C., France 18 E8 43 1N 9 25 E
Corsica = Corse, France 18 F8 42 0N 9 0 E
Corsicana, U.S.A. 81 J6 32 6N 96 28W
Corte, France 18 E8 42 19N 9 11 E
Cortez, U.S.A. 83 H9 37 21N 108 35W
Cortland, U.S.A. 79 D8 42 36N 76 11W
Çorum, Turkey 25 F5 40 30N 34 57 E
Corumbá, Brazil 92 G7 19 0S 57 30W
Corumbá de Goiás, Brazil 93 G9 16 0S 48 50W
Corunna = La Coruña, Spain 19 A1 43 20N 8 25W
Corvallis, U.S.A. 82 D2 44 34N 123 16W
Corvette, L. de la, Canada 70 B5 53 25N 74 3W
Corydon, U.S.A. 80 E8 40 46N 93 19W
Cosalá, Mexico 86 C3 24 28N 106 40W
Cosamaloapan, Mexico 87 D5 18 23N 95 50W
Cosenza, Italy 20 E7 39 18N 16 15 E
Coshocton, U.S.A. 78 F3 40 16N 81 51W
Cosmo Newberry, Australia 61 E3 28 0S 122 54 E
Coso Junction, U.S.A. 85 J9 36 3N 117 57W
Coso Pk., U.S.A. 85 J9 36 13N 117 44W
Cosquín, Argentina 94 C3 31 15S 64 30W
Costa Blanca, Spain 19 C5 38 25N 0 10W
Costa Brava, Spain 19 B7 41 30N 3 0 E
Costa del Sol, Spain 19 D3 36 30N 4 30W
Costa Dorada, Spain 19 B6 41 12N 1 15 E
Costa Mesa, U.S.A. 85 M9 33 38N 117 55W
Costa Rica ■, Cent. Amer. 88 D3 10 0N 84 0W
Costilla, U.S.A. 83 H11 36 59N 105 32W
Cosumnes →, U.S.A. 84 G5 38 16N 121 26W
Cotabato, Phil. 37 C6 7 14N 124 15 E
Cotagaita, Bolivia 94 A2 20 45S 65 40W
Côte d'Azur, France 18 E7 43 25N 7 10 E
Côte-d'Ivoire ■ = Ivory Coast ■, Africa 50 G3 7 30N 5 0W
Coteau des Prairies, U.S.A. 80 C6 45 20N 97 50W
Coteau du Missouri, U.S.A. 80 B4 47 0N 100 0W
Coteau Landing, Canada 79 A10 45 15N 74 13W
Cotentin, France 18 B3 49 15N 1 30W
Cotillo, Canary Is. 22 F5 28 41N 14 1W
Cotonou, Benin 50 G5 6 20N 2 25 E
Cotopaxi, Ecuador 92 D3 0 40S 78 30W
Cotswold Hills, U.K. 11 F5 51 42N 2 10W
Cottage Grove, U.S.A. 82 E2 43 48N 123 3W
Cottbus, Germany 16 C8 51 45N 14 20 E
Cottingham, U.K. 10 C5 53 47N 0 23W
Cottonwood, U.S.A. 83 J7 34 45N 112 1W
Cotulla, U.S.A. 81 L5 28 26N 99 14W
Coudersport, U.S.A. 78 E6 41 46N 78 1W
Couedic, C. du, Australia 63 F2 36 5S 136 40 E
Coulee City, U.S.A. 82 C4 47 37N 119 17W
Coulman I., Antarctica 5 D11 73 35S 170 0 E
Coulonge →, Canada 70 C4 45 52N 76 46W
Coulterville, U.S.A. 84 H6 37 43N 120 12W
Council, Alaska, U.S.A. 68 B3 64 55N 163 45W
Council, Idaho, U.S.A. 82 D5 44 44N 116 26W
Council Bluffs, U.S.A. 80 E7 41 16N 95 52W
Council Grove, U.S.A. 80 F6 38 40N 96 29W
Coupeville, U.S.A. 84 B4 48 13N 122 41W
Courantyne →, S. Amer. 92 B7 5 55N 57 5W
Courtenay, Canada 72 D3 49 45N 125 0W
Courtland, U.S.A. 84 G5 38 20N 121 34W
Courtrai = Kortrijk, Belgium 15 D3 50 50N 3 17 E
Courtright, Canada 78 D2 42 49N 82 28W
Coushatta, U.S.A. 81 J8 32 1N 93 21W
Coutts, Canada 72 D6 49 0N 111 57W
Coventry, U.K. 11 E6 52 25N 1 28W
Coventry L., Canada 73 A7 61 15N 106 15W
Covilhã, Portugal 19 B2 40 17N 7 31W
Covington, Ga., U.S.A. 77 J4 33 36N 83 51W
Covington, Ky., U.S.A. 76 F3 39 5N 84 31W
Covington, Okla., U.S.A. 81 G6 36 18N 97 35W
Covington, Tenn., U.S.A. 81 H10 35 34N 89 39W
Cowal, L., Australia 63 E4 33 40S 147 25 E
Cowan, Canada 73 C8 52 5N 100 45W
Cowan, L., Australia 61 F3 31 45S 121 45 E
Cowan L., Canada 73 C7 54 0N 107 15W
Cowangie, Australia 63 F3 35 12S 141 26 E
Cowansville, Canada 79 A12 45 14N 72 46W
Cowarie, Australia 63 D2 27 45S 138 15 E
Cowcowing Lakes, Australia 61 F2 30 55S 117 20 E
Cowdenbeath, U.K. 12 E5 56 7N 3 21W
Cowell, Australia 63 E2 33 39S 136 56 E
Cowes, U.K. 11 G6 50 45N 1 18W
Cowlitz →, U.S.A. 84 D4 46 6N 122 55W
Cowra, Australia 63 E4 33 49S 148 42 E
Coxilha Grande, Brazil 95 B5 28 18S 51 30W
Coxim, Brazil 93 G8 18 30S 54 55W
Cox's Bazar, Bangla. 41 J17 21 26N 91 59 E
Cox's Cove, Canada 71 C8 49 7N 58 5W
Coyame, Mexico 86 B3 29 28N 105 6W
Coyote Wells, U.S.A. 85 N11 32 44N 115 58W
Coyuca de Benítez, Mexico 87 D4 17 1N 100 8W
Coyuca de Catalan, Mexico 86 D4 18 18N 100 41W
Cozad, U.S.A. 80 E5 40 52N 99 59W
Cozumel, Mexico 87 C7 20 31N 86 55W
Cozumel, I. de, Mexico 87 C7 20 30N 86 40W
Craboon, Australia 63 E4 32 3S 149 30 E
Cracow = Kraków, Poland 17 C10 50 4N 19 57 E
Cracow, Australia 63 D5 25 17S 150 17 E
Cradock, S. Africa 56 E4 32 8S 25 36 E
Craig, Alaska, U.S.A. 72 B2 55 29N 133 9W
Craig, Colo., U.S.A. 82 F10 40 31N 107 33W
Craigavon, U.K. 13 B5 54 27N 6 23W
Craigmore, Zimbabwe 55 G3 20 28S 32 50 E
Crailsheim, Germany 16 D6 49 8N 10 5 E
Craiova, Romania 17 F12 44 21N 23 48 E
Cramsie, Australia 62 C3 23 20S 144 15 E
Cranberry Portage, Canada 73 C8 54 35N 101 23W
Cranbrook, Tas., Australia 62 G4 42 0S 148 5 E
Cranbrook, W. Austral., Australia 61 F2 34 18S 117 33 E
Cranbrook, Canada 72 D5 49 30N 115 46W
Crandon, U.S.A. 80 C10 45 34N 88 54W
Crane, Oreg., U.S.A. 82 E4 43 25N 118 35W
Crane, Tex., U.S.A. 81 K3 31 24N 102 21W
Cranston, U.S.A. 79 E13 41 47N 71 26W
Crater L., U.S.A. 82 E2 42 56N 122 6W
Crateús, Brazil 93 E10 5 10S 40 39W
Crato, Brazil 93 E11 7 10S 39 25W
Crawford, U.S.A. 80 D3 42 41N 103 25W
Crawfordsville, U.S.A. 76 E2 40 2N 86 54W
Crawley, U.K. 11 F7 51 7N 0 11W
Crazy Mts., U.S.A. 82 C8 46 12N 110 20W
Crean L., Canada 73 C7 54 5N 106 9W
Crediton, Canada 78 C3 43 17N 81 33W
Credo, Australia 61 F3 30 28S 120 45 E
Cree →, Canada 73 B7 58 57N 105 47W
Cree →, U.K. 12 G4 54 55N 4 25W
Cree L., Canada 73 B7 57 30N 106 30W
Creede, U.S.A. 83 H10 37 51N 106 56W
Creel, Mexico 86 B3 27 45N 107 38W
Creighton, U.S.A. 80 D6 42 28N 97 54W
Crema, Italy 20 B3 45 22N 9 41 E
Cremona, Italy 20 B4 45 7N 10 2 E
Cres, Croatia 16 F8 44 58N 14 25 E
Cresbard, U.S.A. 80 C5 45 10N 98 57W
Crescent, Okla., U.S.A. 81 H6 35 57N 97 36W
Crescent, Oreg., U.S.A. 82 E3 43 28N 121 42W
Crescent City, U.S.A. 82 F1 41 45N 124 12W
Crespo, Argentina 94 C3 32 2S 60 19W
Cressy, Australia 63 F3 38 2S 143 40 E
Crested Butte, U.S.A. 83 G10 38 52N 106 59W
Crestline, Calif., U.S.A. 85 L9 34 14N 117 18W
Crestline, Ohio, U.S.A. 78 F2 40 47N 82 44W
Creston, Canada 72 D5 49 10N 116 31W
Creston, Calif., U.S.A. 84 K6 35 32N 120 33W
Creston, Iowa, U.S.A. 80 E7 41 4N 94 22W
Creston, Wash., U.S.A. 82 C4 47 46N 118 31W
Crestview, Calif., U.S.A. 84 H8 37 46N 118 58W
Crestview, Fla., U.S.A. 77 K2 30 46N 86 34W
Crete = Kríti, Greece 23 D7 35 15N 25 0 E
Crete, U.S.A. 80 E6 40 38N 96 58W
Créteil, France 18 B5 48 47N 2 28 E
Creus, C. de, Spain 19 A7 42 20N 3 19 E
Creuse →, France 18 C4 47 0N 0 34 E
Crewe, U.K. 10 D5 53 6N 2 26W
Criciúma, Brazil 95 B6 28 40S 49 23W
Crieff, U.K. 12 E5 56 22N 3 50W
Crimean Pen. = Krymskyy Pivostriv, Ukraine 25 E5 45 0N 34 0 E
Crişul Alb →, Romania 17 E11 46 42N 21 17 E
Crişul Negru →, Romania 17 E11 46 42N 21 16 E
Crna Gora = Montenegro □, Yugoslavia 21 C8 42 40N 19 20 E
Crna Gora, Serbia, Yug. 21 C9 42 10N 21 30 E
Crna Reka →, Macedonia 21 D9 41 33N 21 59 E
Croagh Patrick, Ireland 13 C2 53 46N 9 40W
Croatia ■, Europe 16 F9 45 20N 16 0 E
Crocker, Banjaran, Malaysia 36 C5 5 40N 116 30 E
Crockett, U.S.A. 81 K7 31 19N 95 27W
Crocodile = Krokodil →, Mozam. 57 D5 25 14S 32 18 E
Crocodile Is., Australia 62 A1 12 3S 134 58 E
Croix, L. La, Canada 70 C1 48 20N 92 15W
Croker, C., Australia 60 B5 10 58S 132 35 E
Croker I., Australia 60 B5 11 12S 132 32 E
Cromarty, Canada 73 B10 58 3N 94 9W
Cromarty, U.K. 12 D4 57 40N 4 2W
Cromer, U.K. 10 E9 52 56N 1 17 E
Cromwell, N.Z. 59 L2 45 3S 169 14 E
Cronulla, Australia 63 E5 34 3S 151 8 E
Crooked →, Canada 72 C4 54 50N 122 54W
Crooked →, U.S.A. 82 D3 44 32N 121 16W
Crooked I., Bahamas 89 B5 22 50N 74 10W
Crooked Island Passage, Bahamas 89 B5 23 0N 74 30W
Crookston, Minn., U.S.A. 80 B6 47 47N 96 37W
Crookston, Nebr., U.S.A. 80 D4 42 56N 100 45W
Crooksville, U.S.A. 76 F4 39 46N 82 6W
Crookwell, Australia 63 E4 34 28S 149 24 E
Crosby, Minn., U.S.A. 80 B8 46 29N 93 58W
Crosby, N. Dak., U.S.A. 73 D8 48 55N 103 18W
Crosby, Pa., U.S.A. 78 E6 41 45N 78 23W
Crosbyton, U.S.A. 81 J4 33 40N 101 14W
Cross City, U.S.A. 77 L4 29 38N 83 7W
Cross Fell, U.K. 10 C5 54 43N 2 28W
Cross L., Canada 73 C9 54 45N 97 30W
Cross Plains, U.S.A. 81 J5 32 8N 99 11W
Cross Sound, U.S.A. 68 C6 58 0N 135 0W
Crossett, U.S.A. 81 J9 33 8N 91 58W
Crossfield, Canada 72 C6 51 25N 114 0W
Crosshaven, Ireland 13 E3 51 47N 8 17W
Croton-on-Hudson, U.S.A. 79 E11 41 12N 73 55W
Crotone, Italy 20 E7 39 5N 17 8 E
Crow →, Canada 72 B4 59 41N 124 20W
Crow Agency, U.S.A. 82 D10 45 36N 107 28W
Crow Hd., Ireland 13 E1 51 35N 10 9W
Crowell, U.S.A. 81 J5 33 59N 99 43W
Crowley, U.S.A. 81 K8 30 13N 92 22W
Crowley, L., U.S.A. 84 H8 37 35N 118 42W
Crown Point, U.S.A. 76 E2 41 25N 87 22W
Crows Landing, U.S.A. 84 H5 37 23N 121 6W
Crows Nest, Australia 63 D5 27 16S 152 4 E
Crowsnest Pass, Canada 72 D6 49 40N 114 40W
Croydon, Australia 62 B3 18 13S 142 14 E
Croydon, U.K. 11 F7 51 22N 0 5W
Crozet Is., Ind. Oc. 3 G12 46 27S 52 0 E
Cruz, C., Cuba 88 C4 19 50N 77 50W
Cruz Alta, Brazil 95 B5 28 45S 53 40W
Cruz del Eje, Argentina 94 C3 30 45S 64 50W
Cruzeiro, Brazil 95 A7 22 33S 45 0W
Cruzeiro do Oeste, Brazil 95 A5 23 46S 53 4W
Cruzeiro do Sul, Brazil 92 E4 7 35S 72 35W
Cry L., Canada 72 B3 58 45N 129 0W
Crystal Bay, U.S.A. 84 F7 39 15N 120 0W
Crystal Brook, Australia 63 E2 33 21S 138 12 E
Crystal City, Mo., U.S.A. 80 F9 38 13N 90 23W
Crystal City, Tex., U.S.A. 81 L5 28 41N 99 50W
Crystal Falls, U.S.A. 76 B1 46 5N 88 20W
Crystal River, U.S.A. 77 L4 28 54N 82 35W
Crystal Springs, U.S.A. 81 K9 31 59N 90 21W
Csongrád, Hungary 17 E11 46 43N 20 12 E
Cu Lao Hon, Vietnam 39 G7 10 54N 108 18 E
Cua Rao, Vietnam 38 C5 19 16N 104 27 E
Cuácua →, Mozam. 55 F4 17 54S 37 0 E
Cuamato, Angola 56 B2 17 2S 15 7 E
Cuamba, Mozam. 55 E4 14 45S 36 22 E
Cuando →, Angola 53 H4 17 30S 23 15 E
Cuando Cubango □, Angola 56 B3 16 25S 20 0 E
Cuangar, Angola 56 B2 17 36S 18 39 E
Cuanza →, Angola 48 G5 9 2S 13 30 E
Cuarto →, Argentina 94 C3 33 25S 63 2W
Cuatrociénegas, Mexico 86 B4 26 59N 102 5W
Cuauhtémoc, Mexico 86 B3 28 25N 106 52W
Cuba, N. Mex., U.S.A. 83 J10 36 1N 107 4W
Cuba, N.Y., U.S.A. 78 D6 42 13N 78 17W
Cuba ■, W. Indies 88 B4 22 0N 79 0W
Cuballing, Australia 61 F2 32 50S 117 10 E
Cubango →, Africa 56 B3 18 50S 22 25 E
Cuchi, Angola 53 G3 14 37S 16 58 E
Cuchumatanes, Sierra de los, Guatemala 88 C1 15 35N 91 25W
Cucurpe, Mexico 86 A2 30 20N 110 43W
Cúcuta, Colombia 92 B4 7 54N 72 31W
Cuddalore, India 40 P11 11 46N 79 45 E
Cuddapah, India 40 M11 14 30N 78 47 E
Cuddapan, L., Australia 62 D3 25 45S 141 26 E
Cudgewa, Australia 63 F4 36 10S 147 42 E
Cue, Australia 61 E2 27 25S 117 54 E
Cuenca, Ecuador 92 D3 2 50S 79 9W
Cuenca, Spain 19 B4 40 5N 2 10W
Cuenca, Serranía de, Spain 19 C5 39 55N 1 50W
Cuernavaca, Mexico 87 D5 18 55N 99 15W
Cuero, U.S.A. 81 L6 29 6N 97 17W
Cuervo, U.S.A. 81 H2 35 2N 104 25W
Cuevas del Almanzora, Spain 19 D5 37 18N 1 58W
Cuevo, Bolivia 92 H6 20 15S 63 30W
Cuiabá, Brazil 93 G7 15 30S 56 0W
Cuiabá →, Brazil 93 G7 17 5S 56 36W
Cuilco, Guatemala 88 C1 15 24N 91 58W
Cuillin Hills, U.K. 12 D2 57 13N 6 15W
Cuillin Sd., U.K. 12 D2 57 4N 6 20W
Cuima, Angola 53 G3 13 25S 15 45 E
Cuito →, Angola 56 B3 18 1S 20 48 E
Cuitzeo, L. de, Mexico 86 D4 19 55N 101 5W
Cukai, Malaysia 39 K4 4 13N 103 25 E
Culbertson, U.S.A. 80 A2 48 9N 104 31W
Culcairn, Australia 63 F4 35 41S 147 3 E
Culgoa →, Australia 63 D4 29 56S 146 20 E
Culiacán, Mexico 86 C3 24 50N 107 23W
Culiacán →, Mexico 86 C3 24 30N 107 42W
Culion, Phil. 37 B6 11 54N 120 1 E
Cullarin Ra., Australia 63 E4 34 30S 149 30 E
Cullen, U.K. 12 D6 57 42N 2 49W
Cullen Pt., Australia 62 A3 11 57S 141 54 E
Cullera, Spain 19 C5 39 9N 0 17W
Cullman, U.S.A. 77 H2 34 11N 86 51W
Culloden, U.K. 12 D4 57 30N 4 9W
Culpeper, U.S.A. 76 F7 38 30N 78 0W
Culuene →, Brazil 93 F8 12 56S 52 51W
Culver, Pt., Australia 61 F3 32 54S 124 43 E
Culverden, N.Z. 59 K4 42 47S 172 49 E
Cumaná, Venezuela 92 A6 10 30N 64 5W
Cumberland, Canada 72 D3 49 40N 125 0W
Cumberland, Md., U.S.A. 76 F6 39 39N 78 46W
Cumberland, Wis., U.S.A. 80 C8 45 32N 92 1W
Cumberland →, U.S.A. 77 G2 36 15N 87 0W
Cumberland I., U.S.A. 77 K5 30 50N 81 25W
Cumberland Is., Australia 62 C4 20 35S 149 10 E
Cumberland L., Canada 73 C8 54 3N 102 18W
Cumberland Pen., Canada 69 B13 67 0N 64 0W
Cumberland Plateau, U.S.A. 77 H3 36 0N 85 0W
Cumberland Sd., Canada 69 B13 65 30N 66 0W
Cumborah, Australia 63 D4 29 40S 147 45 E
Cumbria □, U.K. 10 C5 54 42N 2 52W
Cumbrian Mts., U.K. 10 C5 54 30N 3 0W
Cumbum, India 40 M11 15 40N 79 10 E
Cummings Mt., U.S.A. 85 K8 35 2N 118 34W
Cummins, Australia 63 E2 34 16S 135 43 E
Cumnock, Australia 63 E4 32 59S 148 46 E
Cumnock, U.K. 12 F4 55 28N 4 17W
Cumpas, Mexico 86 A3 30 0N 109 48W
Cumplida, Pta., Canary Is. 22 F2 28 50N 17 48W
Cuncumén, Chile 94 C1 31 53S 70 38W
Cundeelee, Australia 61 F3 30 43S 123 26 E
Cunderdin, Australia 61 F2 31 37S 117 12 E
Cunene →, Angola 56 B1 17 20S 11 50 E
Cúneo, Italy 20 B2 44 23N 7 32 E
Cunillera, I., Spain 22 C7 38 59N 1 13 E
Cunnamulla, Australia 63 D4 28 2S 145 38 E
Cupar, Canada 73 C8 50 57N 104 10W
Cupar, U.K. 12 E5 56 19N 3 1W
Cupica, G. de, Colombia 92 B3 6 25N 77 30W
Curaçao, Neth. Ant. 89 D6 12 10N 69 0W
Curanilahue, Chile 94 D1 37 29S 73 28W
Curaray →, Peru 92 D4 2 20S 74 5W
Curepto, Chile 94 D1 35 8S 72 1W
Curiapo, Venezuela 92 B6 8 33N 61 5W
Curicó, Chile 94 C1 34 55S 71 20W
Curicó □, Chile 94 C1 34 50S 71 15W
Curitiba, Brazil 95 B6 25 20S 49 10W
Currabubula, Australia 63 E5 31 16S 150 44 E
Currais Novos, Brazil 93 E11 6 13S 36 30W
Curralinho, Brazil 93 D9 1 45S 49 46W
Currant, U.S.A. 82 G6 38 51N 115 32W
Curraweena, Australia 63 E4 30 47S 145 54 E
Currawilla, Australia 62 D3 25 10S 141 20 E
Current →, U.S.A. 81 G9 36 15N 90 55W
Currie, Australia 62 F3 39 56S 143 53 E
Currie, U.S.A. 82 F6 40 16N 114 45W
Currituck Sd., U.S.A. 77 G8 36 20N 75 52W
Curtea de Argeş, Romania 17 F13 45 12N 24 42 E
Curtis, U.S.A. 80 E4 40 38N 100 31W
Curtis Group, Australia 62 F4 39 30S 146 37 E
Curtis I., Australia 62 C5 23 35S 151 10 E
Curuápanema →, Brazil 93 D7 2 25S 55 2W
Curuçá, Brazil 93 D9 0 43S 47 50W
Curuguaty, Paraguay 95 A4 24 31S 55 42W
Çürüksu Çayi →, Turkey 25 G4 37 27N 27 11 E
Curup, Indonesia 36 E2 4 26S 102 13 E
Cururupu, Brazil 93 D10 1 50S 44 50W
Curuzú Cuatiá, Argentina 94 B4 29 50S 58 5W
Cushing, U.S.A. 81 H6 35 59N 96 46W
Cushing, Mt., Canada 72 B3 57 35N 126 57W
Cusihuiriáchic, Mexico 86 B3 28 10N 106 50W
Custer, U.S.A. 80 D3 43 46N 103 36W
Cut Bank, U.S.A. 82 B7 48 38N 112 20W
Cuthbert, U.S.A. 77 K3 31 46N 84 48W
Cutler, U.S.A. 84 J7 36 31N 119 17W
Cuttaburra →, Australia 63 D3 29 43S 144 22 E
Cuttack, India 41 J14 20 25N 85 57 E
Cuvier, C., Australia 61 D1 23 14S 113 22 E
Cuvier I., N.Z. 59 G5 36 27S 175 50 E
Cuxhaven, Germany 16 B5 53 51N 8 41 E
Cuyahoga Falls, U.S.A. 78 E3 41 8N 81 29W
Cuyo, Phil. 37 B6 10 50N 121 5 E
Cuzco, Bolivia 92 H5 20 0S 66 50W
Cuzco, Peru 92 F4 13 32S 72 0W
Cwmbran, U.K. 11 F4 51 39N 3 2W
Cyangugu, Rwanda 54 C2 2 29S 28 54 E
Cyclades = Kikládhes, Greece 21 F11 37 20N 24 30 E
Cygnet, Australia 62 G4 43 8S 147 1 E
Cynthiana, U.S.A. 76 F3 38 23N 84 18W
Cypress Hills, Canada 73 D7 49 40N 109 30W
Cyprus ■, Asia 23 E12 35 0N 33 0 E
Cyrenaica, Libya 51 C9 27 0N 23 0 E
Cyrene = Shaḥḥāt, Libya 51 B9 32 48N 21 54 E
Czar, Canada 73 C6 52 27N 110 50W
Czech Rep. ■, Europe 16 D8 50 0N 15 0 E
Częstochowa, Poland 17 C10 50 49N 19 7 E

D

Da →, Vietnam 38 B5 21 15N 105 20 E
Da Hinggan Ling, China 33 B7 48 0N 121 0 E
Da Lat, Vietnam 39 G7 11 56N 108 25 E
Da Nang, Vietnam 38 D7 16 4N 108 13 E
Da Qaidam, China 32 C4 37 50N 95 15 E
Da Yunhe →, China 35 G11 34 25N 120 5 E
Da'an, China 35 B13 45 30N 124 7 E
Daba Shan, China 33 C5 32 0N 109 0 E
Dabakala, Ivory C. 50 G4 8 15N 4 20W
Dabhoi, India 42 H5 22 10N 73 20 E
Dabo, Indonesia 36 E2 0 30S 104 33 E
Dabola, Guinea 50 F2 10 50N 11 5W
Daboya, Ghana 50 G4 9 30N 1 20W
Dabung, Malaysia 39 K4 5 23N 102 1 E
Dacca = Dhaka, Bangla. 43 H14 23 43N 90 26 E
Dacca = Dhaka □, Bangla. 43 G14 24 25N 90 25 E
Dachau, Germany 16 D6 48 15N 11 26 E
Dadanawa, Guyana 92 C7 2 50N 59 30W
Dade City, U.S.A. 77 L4 28 22N 82 11W
Dadra and Nagar Haveli □, India 40 J8 20 5N 73 0 E

Dadri = Charkhi Dadri, India **42 E7** 28 37N 76 17 E
Dadu, Pakistan **42 F2** 26 45N 67 45 E
Daet, Phil. **37 B6** 14 2N 122 55 E
Dagana, Senegal **50 E1** 16 30N 15 35W
Dagestan □, Russia **25 F8** 42 30N 47 0 E
Daggett, U.S.A. **85 L10** 34 52N 116 52W
Daghestan Republic = Dagestan □, Russia **25 F8** 42 30N 47 0 E
Dagö = Hiiumaa, Estonia **9 G20** 58 50N 22 45 E
Dagu, China **35 E9** 38 59N 117 40 E
Dagupan, Phil. **37 A6** 16 3N 120 20 E
Dahlak Kebir, Eritrea **46 D3** 15 50N 40 10 E
Dahlonega, U.S.A. **77 H4** 34 32N 83 59W
Dahod, India **42 H6** 22 50N 74 15 E
Dahomey = Benin ■, Africa **50 G5** 10 0N 2 0 E
Dahra, Senegal **50 E1** 15 22N 15 30W
Dai Hao, Vietnam **38 C6** 18 1N 106 25 E
Dai-Sen, Japan **31 G6** 35 22N 133 32 E
Dai Xian, China **34 E7** 39 4N 112 58 E
Daicheng, China **34 E9** 38 42N 116 38 E
Daingean, Ireland **13 C4** 53 18N 7 17W
Daintree, Australia **62 B4** 16 20S 145 20 E
Daiō-Misaki, Japan **31 G8** 34 15N 136 45 E
Dairût, Egypt **51 C11** 27 34N 30 43 E
Daisetsu-Zan, Japan **30 C11** 43 30N 142 57 E
Dajarra, Australia **62 C2** 21 42S 139 30 E
Dak Dam, Cambodia **38 F6** 12 20N 107 21 E
Dak Nhe, Vietnam **38 E6** 15 28N 107 48 E
Dak Pek, Vietnam **38 E6** 15 4N 107 44 E
Dak Song, Vietnam **39 F6** 12 19N 107 35 E
Dak Sui, Vietnam **38 E6** 14 55N 107 43 E
Dakar, Senegal **50 F1** 14 34N 17 29W
Dakhla, W. Sahara **50 D1** 23 50N 15 53W
Dakhla, El Wâhât el-, Egypt **51 C10** 25 30N 28 50 E
Dakhovskaya, Russia **25 F7** 44 13N 40 13 E
Dakor, India **42 H5** 22 45N 73 11 E
Dakota City, U.S.A. **80 D6** 42 25N 96 25W
Đakovica, Serbia, Yug. **21 C9** 42 22N 20 26 E
Dalachi, China **34 F3** 36 48N 105 0 E
Dalai Nur, China **34 C9** 43 20N 116 45 E
Dālakī, Iran **45 D6** 29 26N 51 17 E
Dalälven, Sweden **9 F17** 60 12N 16 43 E
Dalaman →, Turkey **21 F13** 36 41N 28 43 E
Dalandzadgad, Mongolia **34 C3** 43 27N 104 30 E
Dalarna, Sweden **9 F16** 61 0N 14 0 E
Dālbandīn, Pakistan **40 E4** 29 0N 64 23 E
Dalbeattie, U.K. **12 G5** 54 56N 3 50W
Dalby, Australia **63 D5** 27 10S 151 17 E
Dalgān, Iran **45 E8** 27 31N 59 19 E
Dalhart, U.S.A. **81 G3** 36 4N 102 31W
Dalhousie, Canada **71 C6** 48 5N 66 26W
Dalhousie, India **42 C6** 32 38N 75 58 E
Dali, Shaanxi, China **34 G5** 34 48N 109 58 E
Dali, Yunnan, China **32 D5** 25 40N 100 10 E
Dalian, China **35 E11** 38 50N 121 40 E
Daliang Shan, China **32 D5** 28 0N 102 45 E
Daling He →, China **35 D11** 40 55N 121 40 E
Dāliyat el Karmel, Israel **47 C4** 32 43N 35 2 E
Dalkeith, U.K. **12 F5** 55 54N 3 4W
Dall I., U.S.A. **72 C2** 54 59N 133 25W
Dallarnil, Australia **63 D5** 25 19S 152 2 E
Dallas, Oreg., U.S.A. **82 D2** 44 55N 123 19W
Dallas, Tex., U.S.A. **81 J6** 32 47N 96 49W
Dalmacija, Croatia **20 C7** 43 20N 17 0 E
Dalmatia = Dalmacija, Croatia **20 C7** 43 20N 17 0 E
Dalmellington, U.K. **12 F4** 55 19N 4 23W
Dalnegorsk, Russia **27 E14** 44 32N 135 33 E
Dalnerechensk, Russia **27 E14** 45 50N 133 40 E
Daloa, Ivory C. **50 G3** 7 0N 6 30W
Dalsland, Sweden **9 G14** 58 50N 12 15 E
Daltenganj, India **43 G11** 24 0N 84 4 E
Dalton, Canada **70 C3** 48 11N 84 1W
Dalton, Ga., U.S.A. **77 H3** 34 46N 84 58W
Dalton, Mass., U.S.A. **79 D11** 42 28N 73 11W
Dalton, Nebr., U.S.A. **80 E3** 41 25N 102 58W
Dalton Iceberg Tongue, Antarctica **5 C9** 66 15S 121 30 E
Dalvik, Iceland **8 D4** 65 58N 18 32W
Daly →, Australia **60 B5** 13 35S 130 19 E
Daly City, U.S.A. **84 H4** 37 42N 122 28W
Daly L., Canada **73 B7** 56 32N 105 39W
Daly Waters, Australia **62 B1** 16 15S 133 24 E
Dam Doi, Vietnam **39 H5** 8 50N 105 12 E
Dam Ha, Vietnam **38 B6** 21 21N 107 36 E
Daman, India **40 J8** 20 25N 72 57 E
Dāmaneh, Iran **45 C6** 33 1N 50 29 E
Damanhûr, Egypt **51 B11** 31 0N 30 30 E
Damanzhuang, China **34 E9** 38 5N 116 35 E
Damar, Indonesia **37 F7** 7 7S 128 40 E
Damaraland, Namibia **56 C2** 21 0S 17 0 E
Damascus = Dimashq, Syria **47 B5** 33 30N 36 18 E
Damāvand, Iran **45 C7** 35 47N 52 0 E
Damāvand, Qolleh-ye, Iran **45 C7** 35 56N 52 10 E
Damba, Angola **52 F3** 6 44S 15 20 E
Dame Marie, Haiti **89 C5** 18 36N 74 26W
Dāmghān, Iran **45 B7** 36 10N 54 17 E
Damiel, Spain **19 C4** 39 4N 3 37W
Damietta = Dumyât, Egypt **51 B11** 31 24N 31 48 E
Daming, China **34 F8** 36 15N 115 6 E
Damīr Qābū, Syria **44 B4** 36 58N 41 51 E
Dammam = Ad Dammām, Si. Arabia **45 E6** 26 20N 50 5 E
Damodar →, India **43 H12** 23 17N 87 35 E
Damoh, India **43 H8** 23 50N 79 28 E
Dampier, Australia **60 D2** 20 41S 116 42 E
Dampier, Selat, Indonesia **37 E8** 0 40S 131 0 E
Dampier Arch., Australia **60 D2** 20 38S 116 32 E
Damrei, Chuor Phnum, Cambodia **39 G4** 11 30N 103 0 E
Dana, Indonesia **37 F6** 11 0S 122 52 E
Dana, L., Canada **70 B4** 50 53N 77 20W
Dana, Mt., U.S.A. **84 H7** 37 54N 119 12W
Danbury, U.S.A. **79 E11** 41 24N 73 28W
Danby L., U.S.A. **83 J6** 34 13N 115 5W
Dand, Afghan. **42 D1** 31 28N 65 32 E
Dandaragan, Australia **61 F2** 30 40S 115 40 E
Dandeldhura, Nepal **43 E9** 29 20N 80 35 E
Dandeli, India **40 M9** 15 5N 74 30 E
Dandenong, Australia **63 F4** 38 0S 145 15 E
Dandong, China **35 D13** 40 10N 124 20 E
Danfeng, China **34 H6** 33 45N 110 25 E
Danforth, U.S.A. **71 C6** 45 40N 67 52W
Danger Is. = Pukapuka, Cook Is. **65 J11** 10 53S 165 49W
Danger Pt., S. Africa **56 E2** 34 40S 19 17 E
Dangora, Nigeria **50 F6** 11 30N 8 7 E
Dangrek, Phnom, Thailand **38 E5** 14 15N 105 0 E
Dangriga, Belize **87 D7** 17 0N 88 13W
Dangshan, China **34 G9** 34 27N 116 22 E
Daniel, U.S.A. **82 E8** 42 52N 110 4W
Daniel's Harbour, Canada **71 B8** 50 13N 57 35W
Danielskuil, S. Africa **56 D3** 28 11S 23 33 E
Danielson, U.S.A. **79 E13** 41 48N 71 53W
Danilov, Russia **24 C7** 58 16N 40 13 E
Daning, China **34 F6** 36 28N 110 45 E
Danissa, Kenya **54 B5** 3 15N 40 58 E
Dankhar Gompa, India **42 C11** 32 10N 78 10 E
Danlí, Honduras **88 D2** 14 4N 86 35W
Dannemora, U.S.A. **79 B11** 44 43N 73 44W
Dannevirke, N.Z. **59 J6** 40 12S 176 8 E
Dannhauser, S. Africa **57 D5** 28 0S 30 3 E
Dansville, U.S.A. **78 D7** 42 34N 77 42W
Dantan, India **43 J12** 21 57N 87 20 E
Dante, Somali Rep. **46 E5** 10 25N 51 16 E
Danube = Dunărea →, Europe **17 F15** 45 20N 29 40 E
Danube →, Europe **6 F11** 45 20N 29 40 E
Danvers, U.S.A. **79 D14** 42 34N 70 56W
Danville, Ill., U.S.A. **76 E2** 40 8N 87 37W
Danville, Ky., U.S.A. **76 G3** 37 39N 84 46W
Danville, Va., U.S.A. **77 G6** 36 36N 79 23W
Danzig = Gdańsk, Poland **17 A10** 54 22N 18 40 E
Dao, Phil. **37 B6** 10 30N 121 57 E
Daoud = Aïn Beïda, Algeria **50 A6** 35 50N 7 29 E
Daqing Shan, China **34 D6** 40 40N 111 0 E
Dar Banda, Africa **48 F6** 8 0N 23 0 E
Dar el Beida = Casablanca, Morocco **50 B3** 33 36N 7 36W
Dar es Salaam, Tanzania **54 D4** 6 50S 39 12 E
Dar Mazār, Iran **45 D8** 29 14N 57 20 E
Dar'ā, Syria **47 C5** 32 36N 36 7 E
Dar'ā □, Syria **47 C5** 32 55N 36 10 E
Dārāb, Iran **45 D7** 28 50N 54 30 E
Daraj, Libya **50 B7** 30 10N 10 28 E
Dārān, Iran **45 C6** 32 59N 50 24 E
Dārayyā, Syria **47 B5** 33 28N 36 15 E
Darband, Pakistan **42 B5** 34 20N 72 50 E
Darband, Kūh-e, Iran **45 D8** 31 34N 57 8 E
Darbhanga, India **43 F11** 26 15N 85 55 E
Darby, U.S.A. **82 C6** 46 1N 114 11W
Dardanelle, Ark., U.S.A. **81 H8** 35 13N 93 9W
Dardanelle, Calif., U.S.A. **84 G7** 38 20N 119 50W
Dardanelles = Çanakkale Boğazı, Turkey **21 D12** 40 17N 26 32 E
Dārestān, Iran **45 D8** 29 9N 58 42 E
Dârfûr, Sudan **51 F9** 13 40N 24 0 E
Dargai, Pakistan **42 B4** 34 25N 71 55 E
Dargan Ata, Uzbekistan **26 E7** 40 29N 62 10 E
Dargaville, N.Z. **59 F4** 35 57S 173 52 E
Darhan Muminggan Lianheqi, China **34 D6** 41 40N 110 28 E
Darica, Turkey **21 D13** 40 45N 29 23 E
Darién, G. del, Colombia **92 B3** 9 0N 77 0W
Dariganga, Mongolia **34 B7** 45 21N 113 45 E
Darjeeling = Darjiling, India **43 F13** 27 3N 88 18 E
Darjiling, India **43 F13** 27 3N 88 18 E
Dark Cove, Canada **71 C9** 48 47N 54 13W
Darkan, Australia **61 F2** 33 20S 116 43 E
Darkhazineh, Iran **45 D6** 31 54N 48 39 E
Darkot Pass, Pakistan **43 A5** 36 45N 73 26 E
Darling →, Australia **63 E3** 34 4S 141 54 E
Darling Downs, Australia **63 D5** 27 30S 150 30 E
Darling Ra., Australia **61 F2** 32 30S 116 0 E
Darlington, U.K. **10 C6** 54 32N 1 33W
Darlington, S.C., U.S.A. **77 H6** 34 18N 79 52W
Darlington, Wis., U.S.A. **80 D9** 42 41N 90 7W
Darlington, L., S. Africa **56 E4** 33 10S 25 9 E
Darlot, L., Australia **61 E3** 27 48S 121 35 E
Darłowo, Poland **16 A9** 54 25N 16 25 E
Darmstadt, Germany **16 D5** 49 51N 8 39 E
Darnah, Libya **51 B9** 32 45N 22 45 E
Darnall, S. Africa **57 D5** 29 23S 31 18 E
Darnley, C., Antarctica **5 C6** 68 0S 69 0 E
Darnley B., Canada **68 B7** 69 30N 123 30W
Darr, Australia **62 C3** 23 13S 144 7 E
Darr →, Australia **62 C3** 23 39S 143 50 E
Darrington, U.S.A. **82 B3** 48 15N 121 36W
Dart →, U.K. **11 G4** 50 24N 3 39W
Dart, C., Antarctica **5 D14** 73 6S 126 20W
Dartmoor, U.K. **11 G4** 50 38N 3 57W
Dartmouth, Australia **62 C3** 23 31S 144 44 E
Dartmouth, Canada **71 D7** 44 40N 63 30W
Dartmouth, U.K. **11 G4** 50 21N 3 36W
Dartmouth, L., Australia **63 D4** 26 4S 145 18 E
Dartuch, C., Spain **22 B10** 39 55N 3 49 E
Darvaza, Turkmenistan **26 E6** 40 11N 58 24 E
Darvel, Teluk, Malaysia **37 D5** 4 50N 118 20 E
Darwha, India **40 J10** 20 15N 77 45 E
Darwin, Australia **60 B5** 12 25S 130 51 E
Darwin, U.S.A. **85 J9** 36 15N 117 35W
Darwin River, Australia **60 B5** 12 50S 130 58 E
Daryoi Amu = Amudarya →, Uzbekistan **26 E6** 43 58N 59 34 E
Dās, U.A.E. **45 E7** 25 20N 53 30 E
Dashetai, China **34 D5** 41 0N 109 5 E
Dashhowuz, Turkmenistan **26 E6** 41 49N 59 58 E
Dasht, Iran **45 B8** 37 17N 56 7 E
Dasht →, Pakistan **40 G2** 25 10N 61 40 E
Dasht-e Mārgow, Afghan. **40 D3** 30 40N 62 30 E
Dasht-i-Nawar, Afghan. **42 C3** 33 52N 68 0 E
Daska, Pakistan **42 C6** 32 20N 74 20 E
Datça, Turkey **21 F12** 36 46N 27 40 E
Datia, India **43 G8** 25 39N 78 27 E
Datong, China **34 D7** 40 6N 113 18 E
Datu, Tanjung, Indonesia **36 D3** 2 5N 109 39 E
Datu Piang, Phil. **37 C6** 7 2N 124 30 E
Daugava →, Latvia **9 H21** 57 4N 24 3 E
Daugavpils, Latvia **9 J22** 55 53N 26 32 E
Daulpur, India **42 F7** 26 45N 77 59 E
Dauphin, Canada **73 C8** 51 9N 100 5W
Dauphin I., U.S.A. **77 K1** 30 15N 88 11W
Dauphin L., Canada **73 C9** 51 20N 99 45W
Dauphiné, France **18 D6** 45 15N 5 25 E
Dausa, India **42 F7** 26 52N 76 20 E
Davangere, India **40 M9** 14 25N 75 55 E
Davao, Phil. **37 C7** 7 0N 125 40 E
Davao, G. of, Phil. **37 C7** 6 30N 125 48 E
Dāvar Panāh, Iran **45 E9** 27 25N 62 15 E
Davenport, Calif., U.S.A. **84 H4** 37 1N 122 12W
Davenport, Iowa, U.S.A. **80 E9** 41 32N 90 35W
Davenport, Wash., U.S.A. **82 C4** 47 39N 118 9W
Davenport Downs, Australia **62 C3** 24 8S 141 7 E
Davenport Ra., Australia **62 C1** 20 28S 134 0 E
David, Panama **88 E3** 8 30N 82 30W
David City, U.S.A. **80 E6** 41 15N 97 8W
David Gorodok = Davyd Haradok, Belarus **17 B14** 52 4N 27 8 E
Davidson, Canada **73 C7** 51 16N 105 59W
Davis, U.S.A. **84 G5** 38 33N 121 44W
Davis Dam, U.S.A. **85 K12** 35 11N 114 34W
Davis Inlet, Canada **71 A7** 55 50N 60 59W
Davis Mts., U.S.A. **81 K2** 30 50N 103 55W
Davis Sea, Antarctica **5 C7** 66 0S 92 0 E
Davis Str., N. Amer. **69 B14** 65 0N 58 0W
Davos, Switz. **16 E5** 46 48N 9 49 E
Davy L., Canada **73 B7** 58 53N 108 18W
Davyd Haradok, Belarus **17 B14** 52 4N 27 8 E
Dawes Ra., Australia **62 C5** 24 40S 150 40 E
Dawson, Canada **68 B6** 64 10N 139 30W
Dawson, Ga., U.S.A. **77 K3** 31 46N 84 27W
Dawson, N. Dak., U.S.A. **80 B5** 46 52N 99 45W
Dawson, I., Chile **96 G2** 53 50S 70 50W
Dawson Creek, Canada **72 B4** 55 45N 120 15W
Dawson Inlet, Canada **73 A10** 61 50N 93 25W
Dawson Ra., Australia **62 C4** 24 30S 149 48 E
Dax, France **18 E3** 43 44N 1 3W
Daxian, China **32 C5** 31 15N 107 23 E
Daxindian, China **35 F11** 37 30N 120 50 E
Daxinggou, China **35 C15** 43 25N 129 40 E
Daxue Shan, China **32 C5** 30 30N 101 30 E
Daylesford, Australia **63 F3** 37 21S 144 9 E
Dayr az Zawr, Syria **44 C4** 35 20N 40 5 E
Daysland, Canada **72 C6** 52 50N 112 20W
Dayton, Nev., U.S.A. **84 F7** 39 14N 119 36W
Dayton, Ohio, U.S.A. **76 F3** 39 45N 84 12W
Dayton, Pa., U.S.A. **78 F5** 40 53N 79 15W
Dayton, Tenn., U.S.A. **77 H3** 35 30N 85 1W
Dayton, Wash., U.S.A. **82 C4** 46 19N 117 59W
Daytona Beach, U.S.A. **77 L5** 29 13N 81 1W
Dayville, U.S.A. **82 D4** 44 28N 119 32W
De Aar, S. Africa **56 E3** 30 39S 24 0 E
De Funiak Springs, U.S.A. **77 K2** 30 43N 86 7W
De Grey, Australia **60 D2** 20 12S 119 12 E
De Grey →, Australia **60 D2** 20 12S 119 13 E
De Kalb, U.S.A. **80 E10** 41 56N 88 46W
De Land, U.S.A. **77 L5** 29 2N 81 18W
De Leon, U.S.A. **81 J5** 32 7N 98 32W
De Pere, U.S.A. **76 C1** 44 27N 88 4W
De Queen, U.S.A. **81 H7** 34 2N 94 21W
De Quincy, U.S.A. **81 K8** 30 27N 93 26W
De Ridder, U.S.A. **81 K8** 30 51N 93 17W
De Smet, U.S.A. **80 C6** 44 23N 97 33W
De Soto, U.S.A. **80 F9** 38 8N 90 34W
De Tour Village, U.S.A. **76 C4** 46 0N 83 56W
De Witt, U.S.A. **81 H9** 34 18N 91 20W
Dead Sea, Asia **47 D4** 31 30N 35 30 E
Deadwood, U.S.A. **80 C3** 44 23N 103 44W
Deadwood L., Canada **72 B3** 59 10N 128 30W
Deakin, Australia **61 F4** 30 46S 128 58 E
Deal, U.K. **11 F9** 51 13N 1 25 E
Deal I., Australia **62 F4** 39 30S 147 20 E
Dealesville, S. Africa **56 D4** 28 41S 25 44 E
Dean, Forest of, U.K. **11 F5** 51 45N 2 33W
Deán Funes, Argentina **94 C3** 30 20S 64 20W
Dearborn, U.S.A. **70 D3** 42 19N 83 11W
Dease →, Canada **72 B3** 59 56N 128 32W
Dease L., Canada **72 B2** 58 40N 130 5W
Dease Lake, Canada **72 B2** 58 25N 130 6W
Death Valley, U.S.A. **85 J10** 36 15N 116 50W
Death Valley Junction, U.S.A. **85 J10** 36 20N 116 25W
Death Valley National Monument, U.S.A. **85 J10** 36 45N 117 15W
Deba Habe, Nigeria **50 F7** 10 14N 11 20 E
Debar, Macedonia **21 D9** 41 31N 20 30 E
Debden, Canada **73 C7** 53 30N 106 50W
Dębica, Poland **17 C11** 50 2N 21 25 E
Debolt, Canada **72 B5** 55 12N 118 1W
Deborah East, L., Australia **61 F2** 30 45S 119 0 E
Deborah West, L., Australia **61 F2** 30 45S 118 50 E
Debre Markos, Ethiopia **51 F12** 10 20N 37 40 E
Debre Tabor, Ethiopia **51 F12** 11 50N 38 26 E
Debrecen, Hungary **17 E11** 47 33N 21 42 E
Decatur, Ala., U.S.A. **77 H2** 34 36N 86 59W
Decatur, Ga., U.S.A. **77 J3** 33 47N 84 18W
Decatur, Ill., U.S.A. **80 F10** 39 51N 88 57W
Decatur, Ind., U.S.A. **76 E3** 40 50N 84 56W
Decatur, Tex., U.S.A. **81 J6** 33 14N 97 35W
Deccan, India **40 M10** 18 0N 79 0 E
Deception L., Canada **73 B8** 56 33N 104 13W
Děčín, Czech. **16 C8** 50 47N 14 12 E
Deckerville, U.S.A. **78 C2** 43 32N 82 44W
Decorah, U.S.A. **80 D9** 43 18N 91 48W
Dedéagach = Alexandroúpolis, Greece **21 D11** 40 50N 25 54 E
Dedham, U.S.A. **79 D13** 42 15N 71 10W
Dédougou, Burkina Faso **50 F4** 12 30N 3 25W
Dedza, Malawi **55 E3** 14 20S 34 20 E
Dee →, C. of Aberd., U.K. **12 D6** 57 9N 2 5W
Dee →, Wales, U.K. **10 D4** 53 22N 3 17W
Deep B., Canada **72 A5** 61 15N 116 35W
Deep Well, Australia **62 C1** 24 20S 134 0 E
Deepwater, Australia **63 D5** 29 25S 151 51 E
Deer →, Canada **73 B10** 58 23N 94 13W
Deer Lake, Nfld., Canada **71 C8** 49 11N 57 27W
Deer Lake, Ont., Canada **73 C10** 52 36N 94 20W
Deer Lodge, U.S.A. **82 C7** 46 24N 112 44W
Deer Park, U.S.A. **82 C5** 47 57N 117 28W
Deer River, U.S.A. **80 B8** 47 20N 93 48W
Deeral, Australia **62 B4** 17 14S 145 55 E
Deerdepoort, S. Africa **56 C4** 24 37S 26 27 E
Deferiet, U.S.A. **79 B9** 44 2N 75 41W
Defiance, U.S.A. **76 E3** 41 17N 84 22W
Degeh Bur, Ethiopia **46 F3** 8 11N 43 31 E
Deggendorf, Germany **16 D7** 48 50N 12 57 E
Deh Bīd, Iran **45 D7** 30 39N 53 11 E
Deh-e Shīr, Iran **45 D7** 31 29N 53 45 E
Dehaj, Iran **45 D7** 30 42N 54 53 E
Dehdez, Iran **45 D6** 31 43N 50 17 E
Dehestān, Iran **45 D7** 28 30N 55 35 E
Dehgolān, Iran **44 C5** 35 17N 47 25 E
Dehi Titan, Afghan. **40 C3** 33 45N 63 50 E
Dehibat, Tunisia **50 B7** 32 0N 10 47 E
Dehlorān, Iran **44 C5** 32 41N 47 16 E
Dehnow-e Kūhestān, Iran **45 E8** 27 58N 58 32 E
Dehra Dun, India **42 D8** 30 20N 78 4 E
Dehri, India **43 G11** 24 50N 84 15 E
Dehui, China **35 B13** 44 30N 125 40 E
Deinze, Belgium **15 D3** 50 59N 3 32 E
Dej, Romania **17 E12** 47 10N 23 52 E
Dekese, Zaïre **52 E4** 3 24S 21 24 E
Del Mar, U.S.A. **85 N9** 32 58N 117 16W
Del Norte, U.S.A. **83 H10** 37 41N 106 21W
Del Rio, U.S.A. **81 L4** 29 22N 100 54W
Delano, U.S.A. **85 K7** 35 46N 119 15W
Delareyville, S. Africa **56 D4** 26 41S 25 26 E
Delavan, U.S.A. **80 D10** 42 38N 88 39W
Delaware, U.S.A. **76 E4** 40 18N 83 4W
Delaware □, U.S.A. **76 F8** 39 0N 75 20W
Delaware →, U.S.A. **76 F8** 39 15N 75 20W
Delaware B., U.S.A. **75 C12** 39 0N 75 10W
Delegate, Australia **63 F4** 37 4S 148 56 E
Delft, Neths. **15 B4** 52 1N 4 22 E
Delfzijl, Neths. **15 A6** 53 20N 6 55 E
Delgado, C., Mozam. **55 E5** 10 45S 40 40 E
Delgerhet, Mongolia **34 B6** 45 50N 110 30 E
Delgo, Sudan **51 D11** 20 6N 30 40 E
Delhi, Canada **78 D4** 42 51N 80 30W
Delhi, India **42 E7** 28 38N 77 17 E
Delhi, U.S.A. **79 D10** 42 17N 74 55W
Delia, Canada **72 C6** 51 38N 112 23W
Delice →, Turkey **25 G5** 39 45N 34 15 E
Delicias, Mexico **86 B3** 28 10N 105 30W
Delījān, Iran **45 C6** 33 59N 50 40 E
Déline, Canada **68 B7** 65 10N 123 30W
Dell City, U.S.A. **83 L11** 31 56N 105 12W
Dell Rapids, U.S.A. **80 D6** 43 50N 96 43W
Delmar, U.S.A. **79 D11** 42 37N 73 47W
Delmenhorst, Germany **16 B5** 53 3N 8 37 E
Delmiro Gouveia, Brazil **93 E11** 9 24S 38 6W
Delong, Ostrova, Russia **27 B15** 76 40N 149 20 E
Deloraine, Australia **62 G4** 41 30S 146 40 E
Deloraine, Canada **73 D8** 49 15N 100 29W
Delphi, U.S.A. **76 E2** 40 36N 86 41W
Delphos, U.S.A. **76 E3** 40 51N 84 21W
Delportshoop, S. Africa **56 D3** 28 22S 24 20 E
Delray Beach, U.S.A. **77 M5** 26 28N 80 4W
Delta, Colo., U.S.A. **83 G9** 38 44N 108 4W
Delta, Utah, U.S.A. **82 G7** 39 21N 112 35W
Delungra, Australia **63 D5** 29 39S 150 51 E
Delvinë, Albania **21 E9** 39 59N 20 4 E
Demanda, Sierra de la, Spain **19 A4** 42 15N 3 0W
Demavand = Damāvand, Iran **45 C7** 35 47N 52 0 E
Demba, Zaïre **52 F4** 5 28S 22 15 E
Dembecha, Ethiopia **51 F12** 10 32N 37 30 E
Dembia, Zaïre **54 B2** 3 33N 25 48 E
Dembidolo, Ethiopia **51 G11** 8 34N 34 50 E
Demer →, Belgium **15 D4** 50 57N 4 42 E
Deming, N. Mex., U.S.A. **83 K10** 32 16N 107 46W
Deming, Wash., U.S.A. **84 B4** 48 50N 122 13W
Demini →, Brazil **92 D6** 0 46S 62 56W
Demirci, Turkey **21 E13** 39 2N 28 38 E
Demirköy, Turkey **21 D12** 41 49N 27 45 E
Demopolis, U.S.A. **77 J2** 32 31N 87 50W
Dempo, Indonesia **36 E2** 4 2S 103 15 E
Den Burg, Neths. **15 A4** 53 3N 4 47 E
Den Chai, Thailand **38 D3** 17 59N 100 4 E
Den Haag = 's-Gravenhage, Neths. **15 B4** 52 7N 4 17 E
Den Helder, Neths. **15 B4** 52 57N 4 45 E
Den Oever, Neths. **15 B5** 52 56N 5 2 E
Denain, France **15 D3** 50 20N 3 22 E
Denair, U.S.A. **84 H6** 37 32N 120 48W
Denau, Uzbekistan **26 F7** 38 16N 67 54 E
Denbigh, U.K. **10 D4** 53 12N 3 25W
Denbighshire □, U.K. **10 D4** 53 8N 3 22W
Dendang, Indonesia **36 E3** 3 7S 107 56 E
Dendermonde, Belgium **15 C4** 51 2N 4 5 E
Dengfeng, China **34 G7** 34 25N 113 2 E
Dengkou, China **34 D4** 40 18N 106 55 E
Denham, Australia **61 E1** 25 56S 113 31 E
Denham Ra., Australia **62 C4** 21 55S 147 46 E
Denham Sd., Australia **61 E1** 25 45S 113 15 E
Denia, Spain **19 C6** 38 49N 0 8 E
Denial B., Australia **63 E1** 32 14S 133 32 E
Deniliquin, Australia **63 F3** 35 30S 144 58 E
Denison, Iowa, U.S.A. **80 D7** 42 1N 95 21W
Denison, Tex., U.S.A. **81 J6** 33 45N 96 33W
Denison Plains, Australia **60 C4** 18 35S 128 0 E
Denizli, Turkey **25 G4** 37 42N 29 2 E
Denman Glacier, Antarctica **5 C7** 66 45S 99 25 E
Denmark, Australia **61 F2** 34 59S 117 25 E
Denmark ■, Europe **9 J13** 55 30N 9 0 E
Denmark Str., Atl. Oc. **4 C6** 66 0N 30 0W
Dennison, U.S.A. **78 F3** 40 24N 81 19W
Denpasar, Indonesia **36 F5** 8 45S 115 14 E
Denton, Mont., U.S.A. **82 C9** 47 19N 109 57W
Denton, Tex., U.S.A. **81 J6** 33 13N 97 8W
D'Entrecasteaux, Pt., Australia **61 F2** 34 50S 115 57 E
Denver, U.S.A. **80 F2** 39 44N 104 59W
Denver City, U.S.A. **81 J3** 32 58N 102 50W
Deoband, India **42 E7** 29 42N 77 43 E
Deoghar, India **43 G12** 24 30N 86 42 E
Deolali, India **40 K8** 19 58N 73 50 E
Deoli = Devli, India **42 G6** 25 50N 75 20 E
Deoria, India **43 F10** 26 31N 83 48 E
Deosai Mts., Pakistan **43 B6** 35 40N 75 0 E
Deping, China **35 F9** 37 25N 116 58 E
Deposit, U.S.A. **79 D9** 42 4N 75 25W
Depot Springs, Australia **61 E3** 27 55S 120 3 E
Deputatskiy, Russia **27 C14** 69 18N 139 54 E
Dera Ghazi Khan, Pakistan **42 D4** 30 5N 70 43 E

Dera Ismail Khan, Pakistan 42 D4 31 50N 70 50 E
Derbent, Russia 25 F8 42 5N 48 15 E
Derby, Australia 60 C3 17 18S 123 38 E
Derby, U.K. 10 E6 52 56N 1 28W
Derby, Conn., U.S.A. 79 E11 41 19N 73 5W
Derby, N.Y., U.S.A. 78 D6 42 41N 78 58W
Derbyshire □, U.K. 10 E6 53 11N 1 38W
Derg →, U.K. 13 B4 54 44N 7 26W
Derg, L., Ireland 13 D3 53 0N 8 20W
Dergaon, India 41 F19 26 45N 94 0 E
Dernieres, Isles, U.S.A. 81 L9 29 2N 90 50W
Derry = Londonderry, U.K. 13 B4 55 0N 7 20W
Derryveagh Mts., Ireland 13 B3 54 56N 8 11W
Derudub, Sudan 51 E12 17 31N 36 7 E
Derwent, Canada 73 C6 53 41N 110 58W
Derwent →, Derby, U.K. 10 E6 52 57N 1 28W
Derwent →, N. Yorks., U.K. 10 D7 53 45N 0 58W
Derwent Water, U.K. 10 C4 54 35N 3 9W
Des Moines, Iowa, U.S.A. 80 E8 41 35N 93 37W
Des Moines, N. Mex., U.S.A. 81 G3 36 46N 103 50W
Des Moines →, U.S.A. 80 E9 40 23N 91 25W
Desaguadero →, Argentina 94 C2 34 30S 66 46W
Desaguadero →, Bolivia 92 G5 16 35S 69 5W
Descanso, Pta., Mexico 85 N9 32 21N 117 3W
Deschaillons, Canada 71 C5 46 32N 72 7W
Descharme →, Canada 73 B7 56 51N 109 13W
Deschutes →, U.S.A. 82 D3 45 38N 120 55W
Dese, Ethiopia 46 E2 11 5N 39 40 E
Desert Center, U.S.A. 85 M11 33 43N 115 24W
Desert Hot Springs, U.S.A. 85 M10 33 58N 116 30W
Désirade, I., Guadeloupe 89 C7 16 18N 61 3W
Deskenatlata L., Canada 72 A6 60 55N 112 3W
Desna →, Ukraine 17 C16 50 33N 30 32 E
Desolación, I., Chile 96 G2 53 0S 74 0W
Despeñaperros, Paso, Spain 19 C4 38 24N 3 30W
Dessau, Germany 16 C7 51 51N 12 14 E
Dessye = Dese, Ethiopia 46 E2 11 5N 39 40 E
D'Estrees B., Australia 63 F2 35 55S 137 45 E
Desuri, India 42 G5 25 18N 73 35 E
Det Udom, Thailand 38 E5 14 54N 105 5 E
Dete, Zimbabwe 55 F2 18 38S 26 50 E
Detmold, Germany 16 C5 51 56N 8 52 E
Detour, Pt., U.S.A. 76 C2 45 40N 86 40W
Detroit, Mich., U.S.A. 75 B10 42 20N 83 3W
Detroit, Tex., U.S.A. 81 J7 33 40N 95 16W
Detroit Lakes, U.S.A. 80 B7 46 49N 95 51W
Deurne, Belgium 15 C4 51 12N 4 24 E
Deurne, Neths. 15 C5 51 27N 5 49 E
Deutsche Bucht, Germany 16 A5 54 15N 8 0 E
Deva, Romania 17 F12 45 53N 22 55 E
Devakottai, India 40 Q11 9 55N 78 45 E
Devaprayag, India 43 D8 30 13N 78 35 E
Deventer, Neths. 15 B6 52 15N 6 10 E
Deveron →, U.K. 12 D6 57 41N 2 32W
Devgadh Bariya, India 42 H5 22 40N 73 55 E
Devils Den, U.S.A. 84 K7 35 46N 119 58W
Devils Lake, U.S.A. 80 A5 48 7N 98 52W
Devils Paw, Canada 72 B2 58 47N 134 0W
Devizes, U.K. 11 F6 51 22N 1 58W
Devli, India 42 G6 25 50N 75 20 E
Devon, Canada 72 C6 53 24N 113 44W
Devon □, U.K. 11 G4 50 50N 3 40W
Devon I., Canada 4 B3 75 10N 85 0W
Devonport, Australia 62 G4 41 10S 146 22 E
Devonport, N.Z. 59 G5 36 49S 174 49 E
Devonport, U.K. 11 G3 50 22N 4 11W
Dewas, India 42 H7 22 59N 76 3 E
Dewetsdorp, S. Africa 56 D4 29 33S 26 39 E
Dewsbury, U.K. 10 D6 53 42N 1 37W
Dexter, Mo., U.S.A. 81 G9 36 48N 89 57W
Dexter, N. Mex., U.S.A. 81 J2 33 12N 104 22W
Dey-Dey, L., Australia 61 E5 29 12S 131 4 E
Deyhūk, Iran 45 C8 33 15N 57 30 E
Deyyer, Iran 45 E6 27 55N 51 55 E
Dezadeash L., Canada 72 A1 60 28N 136 58W
Dezfūl, Iran 45 C6 32 20N 48 30 E
Dezhneva, Mys, Russia 27 C19 66 5N 169 40W
Dezhou, China 34 F9 37 26N 116 18 E
Dháfni, Greece 23 D7 35 13N 25 3 E
Dhahiriya = Az Zāhirīyah, West Bank 47 D3 31 25N 34 58 E
Dhahran = Az Zahrān, Si. Arabia 45 E6 26 10N 50 7 E
Dhaka, Bangla. 43 H14 23 43N 90 26 E
Dhaka □, Bangla. 43 G14 24 25N 90 25 E
Dhali, Cyprus 23 D12 35 1N 33 25 E
Dhamar, Yemen 46 E3 14 30N 44 20 E
Dhampur, India 43 E8 29 19N 78 33 E
Dhamtari, India 41 J12 20 42N 81 35 E
Dhanbad, India 43 H12 23 50N 86 30 E
Dhangarhi, Nepal 41 E12 28 55N 80 40 E
Dhankuta, Nepal 43 F12 26 55N 87 40 E
Dhar, India 42 H6 22 35N 75 26 E
Dharampur, India 42 H6 22 13N 75 18 E
Dharamsala = Dharmsala, India 42 C7 32 16N 76 23 E
Dharmapuri, India 40 N11 12 10N 78 10 E
Dharmsala, India 42 C7 32 16N 76 23 E
Dharwad, India 40 M9 15 22N 75 15 E
Dhaulagiri, Nepal 43 E10 28 39N 83 28 E
Dhebar, L., India 42 G6 24 10N 74 0 E
Dheftera, Cyprus 23 D12 35 5N 33 16 E
Dhenkanal, India 41 J14 20 45N 85 35 E
Dherinia, Cyprus 23 D12 35 3N 33 57 E
Dhiarrizos →, Cyprus 23 E11 34 41N 32 34 E
Dhībān, Jordan 47 D4 31 30N 35 46 E
Dhikti Óros, Greece 23 D7 35 8N 25 22 E
Dhírfis, Greece 21 E10 38 40N 23 54 E
Dhodhekánisos, Greece 21 F12 36 35N 27 0 E
Dholka, India 42 H5 22 44N 72 29 E
Dhoraji, India 42 J4 21 45N 70 37 E
Dhráhstis, Ákra, Greece 23 A3 39 48N 19 40 E
Dhrangadhra, India 42 H4 22 59N 71 31 E
Dhrápanon, Ákra, Greece 23 D6 35 28N 24 14 E
Dhrol, India 42 H4 22 33N 70 25 E
Dhuburi, India 41 F16 26 2N 89 59 E
Dhule, India 40 J9 20 58N 74 50 E
Dhut →, Somali Rep. 46 E5 10 30N 50 0 E
Di Linh, Vietnam 39 G7 11 35N 108 4 E
Di Linh, Cao Nguyen, Vietnam 39 G7 11 30N 108 0 E
Día, Greece 23 D7 35 28N 25 14 E
Diablo, Mt., U.S.A. 84 H5 37 53N 121 56W
Diablo Range, U.S.A. 84 J5 37 20N 121 25W
Diafarabé, Mali 50 F4 14 9N 4 57W
Diamante, Argentina 94 C3 32 5S 60 40W
Diamante →, Argentina 94 C2 34 30S 66 46W
Diamantina, Brazil 93 G10 18 17S 43 40W
Diamantina →, Australia 63 D2 26 45S 139 10 E
Diamantino, Brazil 93 F7 14 30S 56 30W
Diamond Bar, U.S.A. 85 L9 34 1N 117 48W
Diamond Harbour, India 43 H13 22 11N 88 14 E
Diamond Is., Australia 62 B5 17 25S 151 5 E
Diamond Mts., U.S.A. 82 G6 39 50N 115 30W
Diamond Springs, U.S.A. 84 G6 38 42N 120 49W
Diamondville, U.S.A. 82 F8 41 47N 110 32W
Diapaga, Burkina Faso 50 F5 12 5N 1 46 E
Dibā, Oman 45 E8 25 45N 56 16 E
Dibaya, Zaïre 52 F4 6 30S 22 57 E
Dibaya-Lubue, Zaïre 52 E3 4 12S 19 54 E
Dibbi, Ethiopia 46 G3 4 10N 41 52 E
Dibete, Botswana 56 C4 23 45S 26 32 E
Dibrugarh, India 41 F19 27 29N 94 55 E
Dickinson, U.S.A. 80 B3 46 53N 102 47W
Dickson, Russia 26 B9 73 40N 80 5 E
Dickson, U.S.A. 77 G2 36 5N 87 23W
Dickson City, U.S.A. 79 E9 41 29N 75 40W
Didiéni, Mali 50 F3 13 53N 8 6W
Didsbury, Canada 72 C6 51 35N 114 10W
Didwana, India 42 F6 27 23N 74 36 E
Diébougou, Burkina Faso 50 F4 11 0N 3 15W
Diefenbaker L., Canada 73 C7 51 0N 106 55W
Diego Garcia, Ind. Oc. 3 E13 7 50S 72 50 E
Diekirch, Lux. 15 E6 49 52N 6 10 E
Dien Ban, Vietnam 38 E7 15 53N 108 16 E
Dien Bien, Vietnam 38 B4 21 20N 103 0 E
Dien Khanh, Vietnam 39 F7 12 15N 109 6 E
Dieppe, France 18 B4 49 54N 1 4 E
Dieren, Neths. 15 B6 52 3N 6 6 E
Dierks, U.S.A. 81 H7 34 7N 94 1W
Diest, Belgium 15 D5 50 58N 5 4 E
Differdange, Lux. 15 E5 49 31N 5 54 E
Dig, India 42 F7 27 28N 77 20 E
Digba, Zaïre 54 B2 4 25N 25 48 E
Digby, Canada 71 D6 44 38N 65 50W
Digges, Canada 73 B10 58 40N 94 0W
Digges Is., Canada 69 B12 62 40N 77 50W
Dighinala, Bangla. 41 H18 23 15N 92 5 E
Dighton, U.S.A. 80 F4 38 29N 100 28W
Digne-les-Bains, France 18 D7 44 5N 6 12 E
Digos, Phil. 37 C7 6 45N 125 20 E
Digranes, Iceland 8 C6 66 4N 14 44W
Digul →, Indonesia 37 F9 7 7S 138 42 E
Dihang →, India 41 F19 27 48N 95 30 E
Dīhōk, Iraq 44 B3 36 50N 43 1 E
Dijlah, Nahr →, Asia 44 D5 31 0N 47 25 E
Dijon, France 18 C6 47 20N 5 3 E
Dikimdya, Russia 27 D13 59 1N 121 47 E
Dikomu di Kai, Botswana 56 C3 24 58S 24 36 E
Diksmuide, Belgium 15 C2 51 2N 2 52 E
Dikson = Dickson, Russia 26 B9 73 40N 80 5 E
Dikwa, Nigeria 51 F7 12 4N 13 30 E
Dili, Indonesia 37 F7 8 39S 125 34 E
Dilley, U.S.A. 81 L5 28 40N 99 10W
Dilling, Sudan 51 F10 12 3N 29 35 E
Dillingham, U.S.A. 68 C4 59 3N 158 28W
Dillon, Canada 73 B7 55 56N 108 35W
Dillon, Mont., U.S.A. 82 D7 45 13N 112 38W
Dillon, S.C., U.S.A. 77 H6 34 25N 79 22W
Dillon →, Canada 73 B7 55 56N 108 56W
Dilolo, Zaïre 52 G4 10 28S 22 18 E
Dilston, Australia 62 G4 41 22S 147 10 E
Dimas, Mexico 86 C3 23 43N 106 47W
Dimashq, Syria 47 B5 33 30N 36 18 E
Dimashq □, Syria 47 B5 33 30N 36 30 E
Dimbaza, S. Africa 57 E4 32 50S 27 14 E
Dimbokro, Ivory C. 50 G4 6 45N 4 46W
Dimboola, Australia 63 F3 36 28S 142 7 E
Dîmboviţa →, Romania 17 F14 44 5N 26 35 E
Dimbulah, Australia 62 B4 17 8S 145 4 E
Dimitrovgrad, Bulgaria 21 C11 42 5N 25 35 E
Dimitrovgrad, Russia 24 D8 54 14N 49 39 E
Dimitrovo = Pernik, Bulgaria 21 C10 42 35N 23 2 E
Dimmitt, U.S.A. 81 H3 34 33N 102 19W
Dimona, Israel 47 D4 31 2N 35 1 E
Dinagat, Phil. 37 B7 10 10N 125 40 E
Dinajpur, Bangla. 41 G16 25 33N 88 43 E
Dinan, France 18 B2 48 28N 2 2W
Dīnān Āb, Iran 45 C8 32 4N 56 49 E
Dinant, Belgium 15 D4 50 16N 4 55 E
Dinapur, India 43 G11 25 38N 85 5 E
Dīnār, Kūh-e, Iran 45 D6 30 42N 51 46 E
Dinara Planina, Croatia 20 C7 44 0N 16 30 E
Dinard, France 18 B2 48 38N 2 6W
Dinaric Alps = Dinara Planina, Croatia 20 C7 44 0N 16 30 E
Dindigul, India 40 P11 10 25N 78 0 E
Ding Xian, China 34 E8 38 30N 114 59 E
Dingbian, China 34 F4 37 35N 107 32 E
Dingle, Ireland 13 D1 52 9N 10 17W
Dingle B., Ireland 13 D1 52 3N 10 20W
Dingmans Ferry, U.S.A. 79 E10 41 13N 74 55W
Dingo, Australia 62 C4 23 38S 149 19 E
Dingtao, China 34 G8 35 5N 115 35 E
Dinguiraye, Guinea 50 F2 11 18N 10 49W
Dingwall, U.K. 12 D4 57 36N 4 26W
Dingxi, China 34 G3 35 30N 104 33 E
Dingxiang, China 34 E7 38 30N 112 58 E
Dinh, Mui, Vietnam 39 G7 11 22N 109 1 E
Dinh Lap, Vietnam 38 B6 21 33N 107 6 E
Dinokwe, Botswana 56 C4 23 29S 26 37 E
Dinosaur National Monument, U.S.A. 82 F9 40 30N 108 45W
Dinuba, U.S.A. 83 H4 36 32N 119 23W
Diourbel, Senegal 50 F1 14 39N 16 12W
Diplo, Pakistan 42 G3 24 35N 69 35 E
Dipolog, Phil. 37 C6 8 36N 123 20 E
Dir, Pakistan 40 B7 35 8N 71 59 E
Diré, Mali 50 E4 16 20N 3 25W
Dire Dawa, Ethiopia 46 F3 9 35N 41 45 E
Diriamba, Nic. 88 D2 11 51N 86 19W
Dirico, Angola 53 H4 17 50S 20 42 E
Dirk Hartog I., Australia 61 E1 25 50S 113 5 E
Dirranbandi, Australia 63 D4 28 33S 148 17 E
Disa, India 42 G5 24 18N 72 10 E
Disappointment, C., U.S.A. 82 C1 46 18N 124 5W
Disappointment, L., Australia 60 D3 23 20S 122 40 E
Disaster B., Australia 63 F4 37 15S 149 58 E
Discovery B., Australia 63 F3 38 10S 140 40 E
Disko, Greenland 4 C5 69 45N 53 30W
Disko Bugt, Greenland 4 C5 69 10N 52 0W
Disteghil Sar, Pakistan 43 A6 36 20N 75 12 E
Distrito Federal □, Brazil 93 G9 15 45S 47 45W
Diu, India 42 J4 20 45N 70 58 E
Dīvāndarreh, Iran 44 C5 35 55N 47 2 E
Divide, U.S.A. 82 D7 45 45N 112 45W
Dividing Ra., Australia 61 E2 27 45S 116 0 E
Divinópolis, Brazil 93 H10 20 10S 44 54W
Divnoye, Russia 25 E7 45 55N 43 21 E
Diwāl Kol, Afghan. 42 B2 34 23N 67 52 E
Dixie Mt., U.S.A. 84 F6 39 55N 120 16W
Dixon, Calif., U.S.A. 84 G5 38 27N 121 49W
Dixon, Ill., U.S.A. 80 E10 41 50N 89 29W
Dixon, Mont., U.S.A. 82 C6 47 19N 114 19W
Dixon, N. Mex., U.S.A. 83 H11 36 12N 105 53W
Dixon Entrance, U.S.A. 72 C2 54 30N 132 0W
Dixonville, Canada 72 B5 56 32N 117 40W
Diyarbakır, Turkey 25 G7 37 55N 40 18 E
Djado, Niger 51 D7 21 4N 12 14 E
Djakarta = Jakarta, Indonesia 37 G12 6 9S 106 49 E
Djamba, Angola 56 B1 16 45S 13 58 E
Djambala, Congo 52 E2 2 32S 14 30 E
Djanet, Algeria 50 D6 24 35N 9 32 E
Djawa = Jawa, Indonesia 37 G14 7 0S 110 0 E
Djelfa, Algeria 50 B5 34 40N 3 15 E
Djema, C.A.R. 54 A2 6 3N 25 15 E
Djenné, Mali 50 F4 14 0N 4 30W
Djerba, I. de, Tunisia 51 B7 33 50N 10 48 E
Djerid, Chott, Tunisia 50 B6 33 42N 8 30 E
Djibo, Burkina Faso 50 F4 14 9N 1 35W
Djibouti, Djibouti 46 E3 11 30N 43 5 E
Djibouti ■, Africa 46 E3 12 0N 43 0 E
Djolu, Zaïre 52 D4 0 35N 22 5 E
Djougou, Benin 50 G5 9 40N 1 45 E
Djoum, Cameroon 52 D2 2 41N 12 35 E
Djourab, Chad 51 E8 16 40N 18 50 E
Djugu, Zaïre 54 B3 1 55N 30 35 E
Djúpivogur, Iceland 8 D6 64 39N 14 17W
Dmitriya Lapteva, Proliv, Russia 27 B15 73 0N 140 0 E
Dnepr → = Dnipro →, Ukraine 25 E5 46 30N 32 18 E
Dneprodzerzhinsk = Dniprodzerzhynsk, Ukraine 25 E5 48 32N 34 37 E
Dnepropetrovsk = Dnipropetrovsk, Ukraine 25 E5 48 30N 35 0 E
Dnestr → = Dnister →, Europe 17 E16 46 18N 30 17 E
Dnestrovski = Belgorod, Russia 25 D6 50 35N 36 35 E
Dnieper = Dnipro →, Ukraine 25 E5 46 30N 32 18 E
Dniester = Dnister →, Europe 17 E16 46 18N 30 17 E
Dnipro →, Ukraine 25 E5 46 30N 32 18 E
Dniprodzerzhynsk, Ukraine 25 E5 48 32N 34 37 E
Dnipropetrovsk, Ukraine 25 E5 48 30N 35 0 E
Dnister →, Europe 17 E16 46 18N 30 17 E
Dnistrovskyy Lyman, Ukraine 17 E16 46 15N 30 17 E
Dnyapro = Dnipro →, Ukraine 25 E5 46 30N 32 18 E
Doan Hung, Vietnam 38 B5 21 30N 105 10 E
Doba, Chad 51 G8 8 40N 16 50 E
Dobbyn, Australia 62 B3 19 44S 140 2 E
Dobele, Latvia 9 H20 56 37N 23 16 E
Doberai, Jazirah, Indonesia 37 E8 1 25S 133 0 E
Doblas, Argentina 94 D3 37 5S 64 0W
Dobo, Indonesia 37 F8 5 45S 134 15 E
Doboj, Bos.-H. 21 B8 44 46N 18 6 E
Dobreta-Turnu-Severin, Romania 17 F12 44 39N 22 41 E
Dobrich, Bulgaria 21 C12 43 37N 27 49 E
Dobruja, Romania 17 F15 44 30N 28 15 E
Dobrush, Belarus 17 B16 52 25N 31 22 E
Doc, Mui, Vietnam 38 D6 17 58N 106 30 E
Doda, India 43 C6 33 10N 75 34 E
Dodecanese = Dhodhekánisos, Greece 21 F12 36 35N 27 0 E
Dodge Center, U.S.A. 80 C8 44 2N 92 52W
Dodge City, U.S.A. 81 G5 37 45N 100 1W
Dodge L., Canada 73 B7 59 50N 105 36W
Dodgeville, U.S.A. 80 D9 42 58N 90 8W
Dodoma, Tanzania 54 D4 6 8S 35 45 E
Dodoma □, Tanzania 54 D4 6 0S 36 0 E
Dodsland, Canada 73 C7 51 50N 108 45W
Dodson, U.S.A. 82 B9 48 24N 108 15W
Doetinchem, Neths. 15 C6 51 59N 6 18 E
Dog Creek, Canada 72 C4 51 35N 122 14W
Dog L., Man., Canada 73 C9 51 2N 98 31W
Dog L., Ont., Canada 70 C2 48 48N 89 30W
Dogi, Afghan. 40 C3 32 20N 62 50 E
Dogran, Pakistan 42 D5 31 48N 73 35 E
Doha = Ad Dawhah, Qatar 45 E6 25 15N 51 35 E
Dohazari, Bangla. 41 H18 22 10N 92 5 E
Doi, Indonesia 37 D7 2 14N 127 49 E
Doi Luang, Thailand 38 C3 18 30N 101 0 E
Doi Saket, Thailand 38 C2 18 52N 99 9 E
Doig →, Canada 72 B4 56 25N 120 40W
Dois Irmãos, Sa., Brazil 93 E10 9 0S 42 30W
Dokkum, Neths. 15 A5 53 20N 5 59 E
Dokri, Pakistan 42 F3 27 25N 68 7 E
Doland, U.S.A. 80 C5 44 54N 98 6W
Dolbeau, Canada 71 C5 48 53N 72 18W
Dole, France 18 C6 47 7N 5 31 E
Dolgellau, U.K. 10 E4 52 45N 3 53W
Dolgelley = Dolgellau, U.K. 10 E4 52 45N 3 53W
Dollart, Neths. 15 A7 53 20N 7 10 E
Dolomites = Dolomiti, Italy 20 A4 46 23N 11 51 E
Dolomiti, Italy 20 A4 46 23N 11 51 E
Dolores, Argentina 94 D4 36 20S 57 40W
Dolores, Uruguay 94 C4 33 34S 58 15W
Dolores, U.S.A. 83 H9 37 28N 108 30W
Dolores →, U.S.A. 83 G9 38 49N 109 17W
Dolphin, C., Falk. Is. 96 G5 51 10S 59 0W
Dolphin and Union Str., Canada 68 B8 69 5N 114 45W
Dom Pedrito, Brazil 95 C5 31 0S 54 40W
Domasi, Malawi 55 F4 15 15S 35 22 E
Dombarovskiy, Russia 26 D6 50 46N 59 32 E
Dombås, Norway 9 E13 62 4N 9 8 E
Domburg, Neths. 15 C3 51 34N 3 30 E
Domeyko, Chile 94 B1 29 0S 71 0W
Domeyko, Cordillera, Chile 94 A2 24 30S 69 0W
Dominador, Chile 94 A2 24 21S 69 20W
Dominica ■, W. Indies 89 C7 15 20N 61 20W
Dominica Passage, W. Indies 89 C7 15 10N 61 20W
Dominican Rep. ■, W. Indies 89 C5 19 0N 70 30W
Domo, Ethiopia 46 F4 7 50N 47 10 E
Domodóssola, Italy 20 A3 46 7N 8 17 E
Domville, Mt., Australia 63 D5 28 1S 151 15 E
Don →, Russia 25 E6 47 4N 39 18 E
Don →, C. of Aberd., U.K. 12 D6 57 11N 2 5W
Don →, S. Yorks., U.K. 10 D7 53 41N 0 52W
Don, C., Australia 60 B5 11 18S 131 46 E
Don Benito, Spain 19 C3 38 53N 5 51W
Don Duong, Vietnam 39 G7 11 51N 108 35 E
Don Martín, Presa de, Mexico 86 B4 27 30N 100 50W
Dona Ana = Nhamaabué, Mozam. 55 F4 17 25S 35 5 E
Donaghadee, U.K. 13 B6 54 39N 5 33W
Donald, Australia 63 F3 36 23S 143 0 E
Donalda, Canada 72 C6 52 35N 112 34W
Donaldsonville, U.S.A. 81 K9 30 6N 90 59W
Donalsonville, U.S.A. 77 K3 31 3N 84 53W
Donau = Dunărea →, Europe 17 F15 45 20N 29 40 E
Donauwörth, Germany 16 D6 48 43N 10 47 E
Doncaster, U.K. 10 D6 53 32N 1 6W
Dondo, Angola 52 F2 9 45S 14 25 E
Dondo, Mozam. 55 F3 19 33S 34 46 E
Dondo, Teluk, Indonesia 37 D6 0 29N 120 30 E
Dondra Head, Sri Lanka 40 S12 5 55N 80 40 E
Donegal, Ireland 13 B3 54 39N 8 5W
Donegal □, Ireland 13 B4 54 53N 8 0W
Donegal B., Ireland 13 B3 54 31N 8 49W
Donets →, Russia 25 E7 47 33N 40 55 E
Donetsk, Ukraine 25 E6 48 0N 37 45 E
Dong Ba Thin, Vietnam 39 F7 12 8N 109 13 E
Dong Dang, Vietnam 38 B6 21 54N 106 42 E
Dong Giam, Vietnam 38 C5 19 25N 105 31 E
Dong Ha, Vietnam 38 D6 16 55N 107 8 E
Dong Hene, Laos 38 D5 16 40N 105 18 E
Dong Hoi, Vietnam 38 D6 17 29N 106 36 E
Dong Khe, Vietnam 38 A6 22 26N 106 27 E
Dong Ujimqin Qi, China 34 B9 45 32N 116 55 E
Dong Van, Vietnam 38 A5 23 16N 105 22 E
Dong Xoai, Vietnam 39 G6 11 32N 106 55 E
Dongara, Australia 61 E1 29 14S 114 57 E
Dongbei, China 35 D13 42 0N 125 0 E
Dongchuan, China 32 D5 26 8N 103 1 E
Dongfang, China 38 C7 18 50N 108 33 E
Dongfeng, China 35 C13 42 40N 125 34 E
Donggala, Indonesia 37 E5 0 30S 119 40 E
Donggou, China 35 E13 39 52N 124 10 E
Dongguang, China 34 F9 37 50N 116 30 E
Dongjingcheng, China 35 B15 44 5N 129 10 E
Dongning, China 35 B16 44 2N 131 5 E
Dongola, Sudan 51 E11 19 9N 30 22 E
Dongou, Congo 52 D3 2 0N 18 5 E
Dongping, China 34 G9 35 55N 116 20 E
Dongsheng, China 34 E6 39 50N 110 0 E
Dongtai, China 35 H11 32 51N 120 21 E
Dongting Hu, China 33 D6 29 18N 112 45 E
Donington, C., Australia 63 E2 34 45S 136 0 E
Doniphan, U.S.A. 81 G9 36 37N 90 50W
Dønna, Norway 8 C15 66 6N 12 30 E
Donna, U.S.A. 81 M5 26 9N 98 4W
Donnaconna, Canada 71 C5 46 41N 71 41W
Donnelly's Crossing, N.Z. 59 F4 35 42S 173 38 E
Donnybrook, Australia 61 F2 33 34S 115 48 E
Donnybrook, S. Africa 57 D4 29 59S 29 48 E
Donora, U.S.A. 78 F5 40 11N 79 52W
Donor's Hill, Australia 62 B3 18 42S 140 33 E
Donostia = San Sebastián, Spain 19 A5 43 17N 1 58W
Doon →, U.K. 12 F4 55 27N 4 39W
Dora, L., Australia 60 D3 22 0S 123 0 E
Dora Báltea →, Italy 20 B3 45 11N 8 3 E
Doran L., Canada 73 A7 61 13N 108 6W
Dorchester, U.K. 11 G5 50 42N 2 27W
Dorchester, C., Canada 69 B12 65 27N 77 27W
Dordogne →, France 18 D3 45 2N 0 36W
Dordrecht, Neths. 15 C4 51 48N 4 39 E
Dordrecht, S. Africa 56 E4 31 20S 27 3 E
Doré L., Canada 73 C7 54 46N 107 17W
Doré Lake, Canada 73 C7 54 38N 107 36W
Dori, Burkina Faso 50 F4 14 3N 0 2W
Doring →, S. Africa 56 E2 31 54S 18 39 E
Doringbos, S. Africa 56 E2 31 59S 19 16 E
Dorion, Canada 70 C5 45 23N 74 3W
Dornbirn, Austria 16 E5 47 25N 9 45 E
Dornoch, U.K. 12 D4 57 53N 4 2W
Dornoch Firth, U.K. 12 D4 57 51N 4 4W
Dornogovĭ □, Mongolia 34 B6 44 0N 110 0 E
Dorohoi, Romania 17 E14 47 56N 26 30 E
Döröö Nuur, Mongolia 32 B4 48 0N 93 0 E
Dorr, Iran 45 C6 33 17N 50 38 E
Dorre I., Australia 61 E1 25 13S 113 12 E
Dorrigo, Australia 63 E5 30 20S 152 44 E
Dorris, U.S.A. 82 F3 41 58N 121 55W
Dorset, Canada 78 A6 45 14N 78 54W
Dorset, U.S.A. 78 E4 41 4N 80 40W
Dorset □, U.K. 11 G5 50 45N 2 26W
Dortmund, Germany 16 C4 51 30N 7 28 E
Doruma, Zaïre 54 B2 4 42N 27 33 E
Dorūneh, Iran 45 C8 35 10N 57 18 E
Dos Bahías, C., Argentina 96 E3 44 58S 65 32W
Dos Hermanas, Spain 19 D3 37 16N 5 55W
Dos Palos, U.S.A. 84 J6 36 59N 120 37W

Dosso, Niger 50 F5 13 0N 3 13 E
Dothan, U.S.A. 77 K3 31 13N 85 24W
Doty, U.S.A. 84 D3 46 38N 123 17W
Douai, France 18 A5 50 21N 3 4 E
Douala, Cameroon 50 H6 4 0N 9 45 E
Douarnenez, France 18 B1 48 6N 4 21W
Double Island Pt., Australia 63 D5 25 56S 153 11 E
Doubs →, France 18 C6 46 53N 5 1 E
Doubtful Sd., N.Z. 59 L1 45 20S 166 49 E
Doubtless B., N.Z. 59 F4 34 55S 173 26 E
Douentza, Mali 50 F4 14 58N 2 48W
Douglas, S. Africa 56 D3 29 4S 23 46 E
Douglas, U.K. 10 C3 54 10N 4 28W
Douglas, Alaska, U.S.A. 72 B2 58 17N 134 24W
Douglas, Ariz., U.S.A. 83 L9 31 21N 109 33W
Douglas, Ga., U.S.A. 77 K4 31 31N 82 51W
Douglas, Wyo., U.S.A. 80 D2 42 45N 105 24W
Douglastown, Canada 71 C7 48 46N 64 24W
Douglasville, U.S.A. 77 J3 33 45N 84 45W
Doumé, Cameroon 52 D2 4 15N 13 25 E
Dounreay, U.K. 12 C5 58 35N 3 44W
Dourados, Brazil 95 A5 22 9S 54 50W
Dourados →, Brazil 95 A5 21 58S 54 18W
Douro →, Europe 19 B1 41 8N 8 40W
Dove →, U.K. 10 E6 52 51N 1 36W
Dove Creek, U.S.A. 83 H9 37 46N 108 54W
Dover, Australia 62 G4 43 18S 147 2 E
Dover, U.K. 11 F9 51 7N 1 19 E
Dover, Del., U.S.A. 76 F8 39 10N 75 32W
Dover, N.H., U.S.A. 79 C14 43 12N 70 56W
Dover, N.J., U.S.A. 79 F10 40 53N 74 34W
Dover, Ohio, U.S.A. 78 F3 40 32N 81 29W
Dover, Pt., Australia 61 F4 32 32S 125 32 E
Dover, Str. of, Europe 18 A4 51 0N 1 30 E
Dover-Foxcroft, U.S.A. 71 C6 45 11N 69 13W
Dover Plains, U.S.A. 79 E11 41 43N 73 35W
Dovey = Dyfi →, U.K. 11 E4 52 32N 4 3W
Dovrefjell, Norway 9 E13 62 15N 9 33 E
Dow Rūd, Iran 45 C6 33 28N 49 4 E
Dowa, Malawi 55 E3 13 38S 33 58 E
Dowagiac, U.S.A. 76 E2 41 59N 86 6W
Dowgha'i, Iran 45 B8 36 54N 58 32 E
Dowlatābād, Iran 45 D8 28 20N 56 40 E
Down □, U.K. 13 B6 54 23N 6 2W
Downey, Calif., U.S.A. 85 M8 33 56N 118 7W
Downey, Idaho, U.S.A. 82 E7 42 26N 112 7W
Downham Market, U.K. 11 E8 52 37N 0 23 E
Downieville, U.S.A. 84 F6 39 34N 120 50W
Downpatrick, U.K. 13 B6 54 20N 5 43W
Downpatrick Hd., Ireland 13 B2 54 20N 9 21W
Dowsārī, Iran 45 D8 28 25N 57 59 E
Doyle, U.S.A. 84 E6 40 2N 120 6W
Doylestown, U.S.A. 79 F9 40 21N 75 10W
Draa, Oued →, Morocco 50 C2 28 40N 11 10W
Drachten, Neths. 15 A6 53 7N 6 5 E
Drăgăşani, Romania 17 F13 44 39N 24 17 E
Dragichyn, Belarus 17 B13 52 15N 25 8 E
Dragoman, Prokhod, Bulgaria 21 C10 42 58N 22 53 E
Dragonera, I., Spain 22 B9 39 35N 2 19 E
Draguignan, France 18 E7 43 32N 6 27 E
Drain, U.S.A. 82 E2 43 40N 123 19W
Drake, Australia 63 D5 28 55S 152 25 E
Drake, U.S.A. 80 B4 47 55N 100 23W
Drake Passage, S. Ocean 5 B17 58 0S 68 0W
Drakensberg, S. Africa 57 E4 31 0S 28 0 E
Dráma, Greece 21 D11 41 9N 24 10 E
Drammen, Norway 9 G14 59 42N 10 12 E
Drangajökull, Iceland 8 C2 66 9N 22 15W
Dras, India 43 B6 34 25N 75 48 E
Drau = Drava →, Croatia 21 B8 45 33N 18 55 E
Drava →, Croatia 21 B8 45 33N 18 55 E
Drayton Valley, Canada 72 C6 53 12N 114 58W
Drenthe □, Neths. 15 B6 52 52N 6 40 E
Drepanum, C., Cyprus 23 E11 34 54N 32 19 E
Dresden, Canada 78 D2 42 35N 82 11W
Dresden, Germany 16 C7 51 3N 13 44 E
Dreux, France 18 B4 48 44N 1 23 E
Driffield, U.K. 10 C7 54 0N 0 26W
Driftwood, U.S.A. 78 E6 41 20N 78 8W
Driggs, U.S.A. 82 E8 43 44N 111 6W
Drina →, Bos.-H. 21 B8 44 53N 19 21 E
Drini →, Albania 21 C8 42 1N 19 38 E
Drøbak, Norway 9 G14 59 39N 10 39 E
Drochia, Moldova 17 D14 48 2N 27 48 E
Drogheda, Ireland 13 C5 53 43N 6 22W
Drogichin = Dragichyn, Belarus 17 B13 52 15N 25 8 E
Drogobych = Drohobych, Ukraine 17 D12 49 20N 23 30 E
Drohobych, Ukraine 17 D12 49 20N 23 30 E
Droichead Atha = Drogheda, Ireland 13 C5 53 43N 6 22W
Droichead Nua, Ireland 13 C5 53 11N 6 48W
Droitwich, U.K. 11 E5 52 16N 2 8W
Dromedary, C., Australia 63 F5 36 17S 150 10 E
Dronfield, Australia 62 C3 21 12S 140 3 E
Drumbo, Canada 78 C4 43 16N 80 35W
Drumheller, Canada 72 C6 51 25N 112 40W
Drummond, U.S.A. 82 C7 46 40N 113 9W
Drummond I., U.S.A. 70 C3 46 1N 83 39W
Drummond Pt., Australia 63 E2 34 9S 135 16 E
Drummond Ra., Australia 62 C4 23 45S 147 10 E
Drummondville, Canada 70 C5 45 55N 72 25W
Drumright, U.S.A. 81 H6 35 59N 96 36W
Druskininkai, Lithuania 9 J20 54 3N 23 58 E
Drut →, Belarus 17 B16 53 8N 30 5 E
Druzhina, Russia 27 C15 68 14N 145 18 E
Dry Tortugas, U.S.A. 88 B3 24 38N 82 55W
Dryden, Canada 73 D10 49 47N 92 50W
Dryden, U.S.A. 81 K3 30 3N 102 7W
Drygalski I., Antarctica 5 C7 66 0S 92 0 E
Drysdale →, Australia 60 B4 13 59S 126 51 E
Drysdale I., Australia 62 A2 11 41S 136 0 E
Dschang, Cameroon 50 G7 5 32N 10 3 E
Du Bois, U.S.A. 78 E6 41 8N 78 46W
Du Quoin, U.S.A. 80 G10 38 1N 89 14W
Duanesburg, U.S.A. 79 D10 42 45N 74 11W
Duaringa, Australia 62 C4 23 42S 149 42 E
Dubā, Si. Arabia 44 E2 27 10N 35 40 E
Dubai = Dubayy, U.A.E. 45 E7 25 18N 55 20 E
Dubăsari, Moldova 17 E15 47 15N 29 10 E
Dubăsari Vdkhr., Moldova 17 E15 47 30N 29 0 E
Dubawnt →, Canada 73 A8 64 33N 100 6W
Dubawnt, L., Canada 73 A8 63 4N 101 42W
Dubayy, U.A.E. 45 E7 25 18N 55 20 E
Dubbo, Australia 63 E4 32 11S 148 35 E
Dubele, Zaïre 54 B2 2 56N 29 35 E
Dublin, Ireland 13 C5 53 21N 6 15W
Dublin, Ga., U.S.A. 77 J4 32 32N 82 54W
Dublin, Tex., U.S.A. 81 J5 32 5N 98 21W
Dublin □, Ireland 13 C5 53 24N 6 20W
Dublin B., Ireland 13 C5 53 18N 6 5W
Dubno, Ukraine 17 C13 50 25N 25 45 E
Dubois, U.S.A. 82 D7 44 10N 112 14W
Dubossary = Dubăsari, Moldova 17 E15 47 15N 29 10 E
Dubossary Vdkhr. = Dubăsari Vdkhr., Moldova 17 E15 47 30N 29 0 E
Dubovka, Russia 25 E7 49 5N 44 50 E
Dubrajpur, India 43 H12 23 48N 87 25 E
Dubréka, Guinea 50 G2 9 46N 13 31W
Dubrovitsa = Dubrovytsya, Ukraine 17 C14 51 31N 26 35 E
Dubrovnik, Croatia 21 C8 42 39N 18 6 E
Dubrovskoye, Russia 27 D12 58 55N 111 10 E
Dubrovytsya, Ukraine 17 C14 51 31N 26 35 E
Dubuque, U.S.A. 80 D9 42 30N 90 41W
Duchesne, U.S.A. 82 F8 40 10N 110 24W
Duchess, Australia 62 C2 21 20S 139 50 E
Ducie I., Pac. Oc. 65 K15 24 40S 124 48W
Duck Cr. →, Australia 60 D2 22 37S 116 53 E
Duck Lake, Canada 73 C7 52 50N 106 16W
Duck Mountain Prov. Park, Canada 73 C8 51 45N 101 0W
Duckwall, Mt., U.S.A. 84 H6 37 58N 120 7W
Dudhi, India 41 G13 24 15N 83 10 E
Dudinka, Russia 27 C9 69 30N 86 13 E
Dudley, U.K. 11 E5 52 31N 2 5W
Duero = Douro →, Europe 19 B1 41 8N 8 40W
Dufftown, U.K. 12 D5 57 27N 3 8W
Dugi Otok, Croatia 16 G8 44 0N 15 3 E
Duifken Pt., Australia 62 A3 12 33S 141 38 E
Duisburg, Germany 16 C4 51 26N 6 45 E
Duiwelskloof, S. Africa 57 C5 23 42S 30 10 E
Dūkdamīn, Iran 45 C8 35 59N 57 43 E
Duke I., U.S.A. 72 C2 54 50N 131 20W
Dukelský Průsmyk, Slovak Rep. 17 D11 49 25N 21 42 E
Dukhān, Qatar 45 E6 25 25N 50 50 E
Duki, Pakistan 40 D6 30 14N 68 25 E
Duku, Nigeria 50 F7 10 43N 10 43 E
Dulce →, Argentina 94 C3 30 32S 62 33W
Dulce, G., Costa Rica 88 E3 8 40N 83 20W
Dulf, Iraq 44 C5 35 7N 45 51 E
Dulit, Banjaran, Malaysia 36 D4 3 15N 114 30 E
Duliu, China 34 E9 39 2N 116 55 E
Dullewala, Pakistan 42 D4 31 50N 71 25 E
Dulq Maghār, Syria 44 B3 36 22N 38 39 E
Dululu, Australia 62 C5 23 48S 150 15 E
Duluth, U.S.A. 80 B8 46 47N 92 6W
Dum Dum, India 43 H13 22 39N 88 33 E
Dum Duma, India 41 F19 27 40N 95 40 E
Dum Hadjer, Chad 51 F8 13 18N 19 41 E
Dūmā, Lebanon 47 A4 34 12N 35 50 E
Dūmā, Syria 47 B5 33 34N 36 24 E
Dumaguete, Phil. 37 C6 9 17N 123 15 E
Dumai, Indonesia 36 D2 1 35N 101 28 E
Dumaran, Phil. 37 B5 10 33N 119 50 E
Dumas, Ark., U.S.A. 81 J9 33 53N 91 29W
Dumas, Tex., U.S.A. 81 H4 35 52N 101 58W
Dumbarton, U.K. 12 F4 55 57N 4 33W
Dumbleyung, Australia 61 F2 33 17S 117 42 E
Dumfries, U.K. 12 F5 55 4N 3 37W
Dumfries & Galloway □, U.K. 12 F5 55 9N 3 58W
Dumka, India 43 G12 24 12N 87 15 E
Dumoine →, Canada 70 C4 46 13N 77 51W
Dumoine L., Canada 70 C4 46 55N 77 55W
Dumraon, India 43 G11 25 33N 84 8 E
Dumyât, Egypt 51 B11 31 24N 31 48 E
Dún Dealgan = Dundalk, Ireland 13 B5 54 1N 6 24W
Dun Laoghaire, Ireland 13 C5 53 17N 6 8W
Duna = Dunărea →, Europe 17 F15 45 20N 29 40 E
Dunaj = Dunărea →, Europe 17 F15 45 20N 29 40 E
Dunakeszi, Hungary 17 E10 47 37N 19 8 E
Dunărea →, Europe 17 F15 45 20N 29 40 E
Dunaújváros, Hungary 17 E10 47 0N 18 57 E
Dunav = Dunărea →, Europe 17 F15 45 20N 29 40 E
Dunay, Russia 30 C6 42 52N 132 22 E
Dunback, N.Z. 59 L3 45 23S 170 36 E
Dunbar, Australia 62 B3 16 0S 142 22 E
Dunbar, U.K. 12 E6 56 0N 2 31W
Dunblane, U.K. 12 E5 56 11N 3 58W
Duncan, Canada 72 D4 48 45N 123 40W
Duncan, Ariz., U.S.A. 83 K9 32 43N 109 6W
Duncan, Okla., U.S.A. 81 H6 34 30N 97 57W
Duncan, L., Canada 70 B4 53 29N 77 58W
Duncan L., Canada 72 A6 62 51N 113 58W
Duncan Town, Bahamas 88 B4 22 15N 75 45W
Duncannon, U.S.A. 78 F7 40 23N 77 2W
Dundalk, Canada 78 B4 44 10N 80 24W
Dundalk, Ireland 13 B5 54 1N 6 24W
Dundalk Bay, Ireland 13 C5 53 55N 6 15W
Dundas, Canada 70 D4 43 17N 79 59W
Dundas, L., Australia 61 F3 32 35S 121 50 E
Dundas I., Canada 72 C2 54 30N 130 50W
Dundas Str., Australia 60 B5 11 15S 131 35 E
Dundee, S. Africa 57 D5 28 11S 30 15 E
Dundee, U.K. 12 E6 56 28N 2 59W
Dundee City □, U.K. 12 E6 56 30N 2 58W
Dundgovĭ □, Mongolia 34 B4 45 10N 106 0 E
Dundoo, Australia 63 D3 27 40S 144 37 E
Dundrum, U.K. 13 B6 54 16N 5 52W
Dundrum B., U.K. 13 B6 54 13N 5 47W
Dundwara, India 43 F8 27 48N 79 9 E
Dunedin, N.Z. 59 L3 45 50S 170 33 E
Dunedin, U.S.A. 77 L4 28 1N 82 47W
Dunedin →, Canada 72 B4 59 30N 124 5W
Dunfermline, U.K. 12 E5 56 5N 3 27W
Dungannon, Canada 78 C3 43 51N 81 36W
Dungannon, U.K. 13 B5 54 31N 6 46W
Dungannon □, U.K. 13 B5 54 30N 6 55W
Dungarpur, India 42 H5 23 52N 73 45 E
Dungarvan, Ireland 13 D4 52 5N 7 37W
Dungarvan Harbour, Ireland 13 D4 52 4N 7 35W
Dungeness, U.K. 11 G8 50 54N 0 59 E
Dungo, L. do, Angola 56 B2 17 15S 19 0 E
Dungog, Australia 63 E5 32 22S 151 46 E
Dungu, Zaïre 54 B2 3 40N 28 32 E
Dunhua, China 35 C15 43 20N 128 14 E
Dunhuang, China 32 B4 40 8N 94 36 E
Dunk I., Australia 62 B4 17 59S 146 29 E
Dunkeld, U.K. 12 E5 56 34N 3 35W
Dunkerque, France 18 A5 51 2N 2 20 E
Dunkery Beacon, U.K. 11 F4 51 9N 3 36W
Dunkirk = Dunkerque, France 18 A5 51 2N 2 20 E
Dunkirk, U.S.A. 78 D5 42 29N 79 20W
Dunkwa, Ghana 50 G4 6 0N 1 47W
Dunlap, U.S.A. 80 E7 41 51N 95 36W
Dúnleary = Dun Laoghaire, Ireland 13 C5 53 17N 6 8W
Dunmanus B., Ireland 13 E2 51 31N 9 50W
Dunmara, Australia 62 B1 16 42S 133 25 E
Dunmore, U.S.A. 79 E9 41 25N 75 38W
Dunmore Hd., Ireland 13 D1 52 10N 10 35W
Dunmore Town, Bahamas 88 A4 25 30N 76 39W
Dunn, U.S.A. 77 H6 35 19N 78 37W
Dunnellon, U.S.A. 77 L4 29 3N 82 28W
Dunnet Hd., U.K. 12 C5 58 40N 3 21W
Dunning, U.S.A. 80 E4 41 50N 100 6W
Dunnville, Canada 78 D5 42 54N 79 36W
Dunolly, Australia 63 F3 36 51S 143 44 E
Dunoon, U.K. 12 F4 55 57N 4 56W
Dunqul, Egypt 51 D11 23 26N 31 37 E
Duns, U.K. 12 F6 55 47N 2 20W
Dunseith, U.S.A. 80 A4 48 50N 100 3W
Dunsmuir, U.S.A. 82 F2 41 13N 122 16W
Dunstable, U.K. 11 F7 51 53N 0 32W
Dunstan Mts., N.Z. 59 L2 44 53S 169 35 E
Dunster, Canada 72 C5 53 8N 119 50W
Dunvegan L., Canada 73 A7 60 8N 107 10W
Duolun, China 34 C9 42 12N 116 28 E
Duong Dong, Vietnam 39 G4 10 13N 103 58 E
Dupree, U.S.A. 80 C4 45 4N 101 35W
Dupuyer, U.S.A. 82 B7 48 13N 112 30W
Duque de Caxias, Brazil 95 A7 22 45S 43 19W
Durack →, Australia 60 C4 15 33S 127 52 E
Durack Ra., Australia 60 C4 16 50S 127 40 E
Durance →, France 18 E6 43 55N 4 45 E
Durand, U.S.A. 76 D4 42 55N 83 59W
Durango = Victoria de Durango, Mexico 86 C4 24 3N 104 39W
Durango, Spain 19 A4 43 13N 2 40W
Durango, U.S.A. 83 H10 37 16N 107 53W
Durango □, Mexico 86 C4 25 0N 105 0W
Duranillin, Australia 61 F2 33 30S 116 45 E
Durant, U.S.A. 81 J6 33 59N 96 25W
Durazno, Uruguay 94 C4 33 25S 56 31W
Durazzo = Durrësi, Albania 21 D8 41 19N 19 28 E
Durban, S. Africa 57 D5 29 49S 31 1 E
Düren, Germany 16 C4 50 48N 6 29 E
Durg, India 41 J12 21 15N 81 22 E
Durgapur, India 43 H12 23 30N 87 20 E
Durham, Canada 78 B4 44 10N 80 49W
Durham, U.K. 10 C6 54 47N 1 34W
Durham, Calif., U.S.A. 84 F5 39 39N 121 48W
Durham, N.C., U.S.A. 77 G6 35 59N 78 54W
Durham □, U.K. 10 C6 54 42N 1 45W
Durham Downs, Australia 63 D4 26 6S 149 5 E
Durmitor, Montenegro, Yug. 21 C8 43 10N 19 0 E
Durness, U.K. 12 C4 58 34N 4 45W
Durrësi, Albania 21 D8 41 19N 19 28 E
Durrie, Australia 62 D3 25 40S 140 15 E
Dursunbey, Turkey 21 E13 39 35N 28 37 E
Duru, Zaïre 54 B2 4 14N 28 50 E
D'Urville, Tanjung, Indonesia 37 E9 1 28S 137 54 E
D'Urville I., N.Z. 59 J4 40 50S 173 55 E
Duryea, U.S.A. 79 E9 41 20N 75 45W
Dusa Mareb, Somali Rep. 46 F4 5 30N 46 15 E
Dushak, Turkmenistan 26 F7 37 13N 60 1 E
Dushanbe, Tajikistan 26 F7 38 33N 68 48 E
Dusky Sd., N.Z. 59 L1 45 47S 166 30 E
Dussejour, C., Australia 60 B4 14 45S 128 13 E
Düsseldorf, Germany 16 C4 51 14N 6 47 E
Dutch Harbor, U.S.A. 68 C3 53 53N 166 32W
Dutlwe, Botswana 56 C3 23 58S 23 46 E
Dutton, Canada 78 D3 42 39N 81 30W
Dutton →, Australia 62 C3 20 44S 143 10 E
Duyun, China 32 D5 26 18N 107 29 E
Duzdab = Zāhedān, Iran 45 D9 29 30N 60 50 E
Dvina, Severnaya →, Russia 24 B7 64 32N 40 30 E
Dvinsk = Daugavpils, Latvia 9 J22 55 53N 26 32 E
Dvinskaya Guba, Russia 24 B6 65 0N 39 0 E
Dwarka, India 42 H3 22 18N 69 8 E
Dwellingup, Australia 61 F2 32 43S 116 4 E
Dwight, Canada 78 A5 45 20N 79 1W
Dwight, U.S.A. 76 E1 41 5N 88 26W
Dyatlovo = Dzyatlava, Belarus 17 B13 53 28N 25 28 E
Dyer, C., Canada 69 B13 66 40N 61 0W
Dyer Plateau, Antarctica 5 D17 70 45S 65 30W
Dyersburg, U.S.A. 81 G10 36 3N 89 23W
Dyfi →, U.K. 11 E4 52 32N 4 3W
Dymer, Ukraine 17 C16 50 47N 30 18 E
Dynevor Downs, Australia 63 D3 28 10S 144 20 E
Dysart, Canada 73 C8 50 57N 104 2W
Dzamin Üüd, Mongolia 34 C6 43 50N 111 58 E
Dzerzhinsk, Russia 24 C7 56 14N 43 30 E
Dzhalinda, Russia 27 D13 53 26N 124 0 E
Dzhambul = Zhambyl, Kazakstan 26 E8 42 54N 71 22 E
Dzhankoy, Ukraine 25 E5 45 40N 34 20 E
Dzhardzhan, Russia 27 C13 68 10N 124 10 E
Dzhetygara = Zhetiqara, Kazakstan 26 D7 52 11N 61 12 E
Dzhezkazgan = Zhezqazghan, Kazakstan 26 E7 47 44N 67 40 E
Dzhizak = Jizzakh, Uzbekistan 26 E7 40 6N 67 50 E
Dzhugdzur, Khrebet, Russia 27 D14 57 30N 138 0 E
Dzhungarskiye Vorota, Kazakstan 32 B3 45 0N 82 0 E
Działdowa, Poland 17 B11 53 15N 20 15 E
Dzierzoniów, Poland 17 C9 50 45N 16 39 E
Dzilam de Bravo, Mexico 87 C7 21 24N 88 53W
Dzungaria = Junggar Pendi, China 32 B3 44 30N 86 0 E
Dzungarian Gates = Dzhungarskiye Vorota, Kazakstan 32 B3 45 0N 82 0 E
Dzuumod, Mongolia 32 B5 47 45N 106 58 E
Dzyarzhynsk, Belarus 17 B14 53 40N 27 1 E
Dzyatlava, Belarus 17 B13 53 28N 25 28 E

E

Eabamet, L., Canada 70 B2 51 30N 87 46W
Eads, U.S.A. 80 F3 38 29N 102 47W
Eagle, U.S.A. 82 G10 39 39N 106 50W
Eagle →, Canada 71 B8 53 36N 57 26W
Eagle Butte, U.S.A. 80 C4 45 0N 101 10W
Eagle Grove, U.S.A. 80 D8 42 40N 93 54W
Eagle L., Calif., U.S.A. 82 F3 40 39N 120 45W
Eagle L., Maine, U.S.A. 71 C6 46 20N 69 22W
Eagle Lake, U.S.A. 81 L6 29 35N 96 20W
Eagle Mountain, U.S.A. 85 M11 33 49N 115 27W
Eagle Nest, U.S.A. 83 H11 36 33N 105 16W
Eagle Pass, U.S.A. 81 L4 28 43N 100 30W
Eagle Pk., U.S.A. 84 G7 38 10N 119 25W
Eagle Pt., Australia 60 C3 16 11S 124 23 E
Eagle River, U.S.A. 80 C10 45 55N 89 15W
Ealing, U.K. 11 F7 51 31N 0 20W
Earaheedy, Australia 61 E3 25 34S 121 29 E
Earl Grey, Canada 73 C8 50 57N 104 43W
Earle, U.S.A. 81 H9 35 16N 90 28W
Earlimart, U.S.A. 85 K7 35 53N 119 16W
Earn →, U.K. 12 E5 56 21N 3 18W
Earn, L., U.K. 12 E4 56 23N 4 13W
Earnslaw, Mt., N.Z. 59 L2 44 32S 168 27 E
Earth, U.S.A. 81 H3 34 14N 102 24W
Easley, U.S.A. 77 H4 34 50N 82 36W
East Angus, Canada 71 C5 45 30N 71 40W
East Aurora, U.S.A. 78 D6 42 46N 78 37W
East Ayrshire □, U.K. 12 F4 55 26N 4 11W
East B., U.S.A. 81 L10 29 0N 89 15W
East Bengal, Bangla. 41 G17 24 0N 90 0 E
East Beskids = Vychodné Beskydy, Europe 17 D11 49 20N 22 0 E
East Brady, U.S.A. 78 F5 40 59N 79 36W
East C., N.Z. 59 G7 37 42S 178 35 E
East Chicago, U.S.A. 76 E2 41 38N 87 27W
East China Sea, Asia 33 C7 30 5N 126 0 E
East Coulee, Canada 72 C6 51 23N 112 27W
East Dunbartonshire □, U.K. 12 F4 55 57N 4 13W
East Falkland, Falk. Is. 96 G5 51 30S 58 30W
East Grand Forks, U.S.A. 80 B6 47 56N 97 1W
East Greenwich, U.S.A. 79 E13 41 40N 71 27W
East Hartford, U.S.A. 79 E12 41 46N 72 39W
East Helena, U.S.A. 82 C8 46 35N 111 56W
East Indies, Asia 37 E6 0 0 120 0 E
East Jordan, U.S.A. 76 C3 45 10N 85 7W
East Lansing, U.S.A. 76 D3 42 44N 84 29W
East Liverpool, U.S.A. 78 F4 40 37N 80 35W
East London, S. Africa 57 E4 33 0S 27 55 E
East Lothian □, U.K. 12 F6 55 58N 2 44W
East Main = Eastmain, Canada 70 B4 52 10N 78 30W
East Orange, U.S.A. 79 F10 40 46N 74 13W
East Pacific Ridge, Pac. Oc. 65 J17 15 0S 110 0W
East Palestine, U.S.A. 78 F4 40 50N 80 33W
East Pine, Canada 72 B4 55 48N 120 12W
East Point, U.S.A. 77 J3 33 41N 84 27W
East Providence, U.S.A. 79 E13 41 49N 71 23W
East Pt., Canada 71 C7 46 27N 61 58W
East Renfrewshire □, U.K. 12 F4 55 46N 4 21W
East Retford = Retford, U.K. 10 D7 53 19N 0 56W
East Riding □, U.K. 10 D7 53 55N 0 30W
East St. Louis, U.S.A. 80 F9 38 37N 90 9W
East Schelde → = Oosterschelde, Neths. 15 C4 51 33N 4 0 E
East Siberian Sea, Russia 27 B17 73 0N 160 0 E
East Stroudsburg, U.S.A. 79 E9 41 1N 75 11W
East Sussex □, U.K. 11 G8 50 56N 0 19 E
East Tawas, U.S.A. 76 C4 44 17N 83 29W
East Toorale, Australia 63 E4 30 27S 145 28 E
East Walker →, U.S.A. 84 G7 38 52N 119 10W
Eastbourne, N.Z. 59 J5 41 19S 174 55 E
Eastbourne, U.K. 11 G8 50 46N 0 18 E
Eastend, Canada 73 D7 49 32N 108 50W
Easter Islands = Pascua, I. de, Pac. Oc. 65 K17 27 0S 109 0W
Eastern □, Kenya 54 B4 0 0 38 30 E
Eastern □, Uganda 54 B3 1 50N 33 45 E
Eastern Cape □, S. Africa 56 E4 32 0S 26 0 E
Eastern Cr. →, Australia 62 C3 20 40S 141 35 E
Eastern Ghats, India 40 N11 14 0N 78 50 E
Eastern Group = Lau Group, Fiji 59 C9 17 0S 178 30W
Eastern Group, Australia 61 F3 33 30S 124 30 E
Eastern Transvaal = Mpumalanga □, S. Africa 57 B5 26 0S 30 0 E
Easterville, Canada 73 C9 53 8N 99 49W
Easthampton, U.S.A. 79 D12 42 16N 72 40W
Eastland, U.S.A. 81 J5 32 24N 98 49W
Eastleigh, U.K. 11 G6 50 58N 1 21W
Eastmain, Canada 70 B4 52 10N 78 30W
Eastmain →, Canada 70 B4 52 27N 78 26W
Eastman, Canada 79 A12 45 18N 72 19W
Eastman, U.S.A. 77 J4 32 12N 83 11W
Easton, Md., U.S.A. 76 F7 38 47N 76 5W
Easton, Pa., U.S.A. 79 F9 40 41N 75 13W
Easton, Wash., U.S.A. 84 C5 47 14N 121 11W
Eastport, U.S.A. 71 D6 44 56N 67 0W

Eastsound, *U.S.A.* **84 B4** 48 42N 122 55W
Eaton, *U.S.A.* **80 E2** 40 32N 104 42W
Eatonia, *Canada* **73 C7** 51 13N 109 25W
Eatonton, *U.S.A.* **77 J4** 33 20N 83 23W
Eatontown, *U.S.A.* **79 F10** 40 19N 74 4W
Eatonville, *U.S.A.* **84 D4** 46 52N 122 16W
Eau Claire, *U.S.A.* **80 C9** 44 49N 91 30W
Ebagoola, *Australia* ..... **62 A3** 14 15S 143 12 E
Ebbw Vale, *U.K.* **11 F4** 51 46N 3 12W
Ebeltoft, *Denmark* ...... **9 H14** 56 12N 10 41 E
Ebensburg, *U.S.A.* **78 F6** 40 29N 78 44W
Eberswalde-Finow, *Germany* .... **16 B7** 52 50N 13 49 E
Ebetsu, *Japan* .... **30 C10** 43 7N 141 34 E
Ebolowa, *Cameroon* .... **52 D2** 2 55N 11 10 E
Ebro —►, *Spain* .... **19 B6** 40 43N 0 54 E
Eceabat, *Turkey* .... **21 D12** 40 11N 26 21 E
Ech Cheliff, *Algeria* ..... **50 A5** 36 10N 1 20 E
Echigo-Sammyaku, *Japan* **31 F9** 36 50N 139 50 E
Echizen-Misaki, *Japan* ... **31 G7** 35 59N 135 57 E
Echo Bay, *N.W.T., Canada* **68 B8** 66 5N 117 55W
Echo Bay, *Ont., Canada* . **70 C3** 46 29N 84 4W
Echoing —►, *Canada* **73 B10** 55 51N 92 5W
Echternach, *Lux.* .... **15 E6** 49 49N 6 25 E
Echuca, *Australia* .... **63 F3** 36 10S 144 20 E
Ecija, *Spain* **19 D3** 37 30N 5 10W
Eclipse Is., *Australia* .... **60 B4** 13 54S 126 19 E
Ecuador ■, *S. Amer.* **92 D3** 2 0S 78 0W
Ed Dâmer, *Sudan* .... **51 E11** 17 27N 34 0 E
Ed Debba, *Sudan* .... **51 E11** 18 0N 30 51 E
Ed Dueim, *Sudan* .... **51 F11** 14 0N 32 10 E
Edah, *Australia* .... **61 E2** 28 16S 117 10 E
Edam, *Canada* **73 C7** 53 11N 108 46W
Edam, *Neths.* .... **15 B5** 52 31N 5 3 E
Eday, *U.K.* **12 B6** 59 11N 2 47W
Edd, *Eritrea* .... **46 E3** 14 0N 41 38 E
Eddrachillis B., *U.K.* **12 C3** 58 17N 5 14W
Eddystone, *U.K.* **11 G3** 50 11N 4 16W
Eddystone Pt., *Australia* . **62 G4** 40 59S 148 20 E
Ede, *Neths.* .... **15 B5** 52 4N 5 40 E
Édea, *Cameroon* .... **50 H7** 3 51N 10 9 E
Edehon L., *Canada* **73 A9** 60 25N 97 15W
Eden, *Australia* .... **63 F4** 37 3S 149 55 E
Eden, *N.C., U.S.A.* **77 G6** 36 29N 79 53W
Eden, *N.Y., U.S.A.* **78 D6** 42 39N 78 55W
Eden, *Tex., U.S.A.* **81 K5** 31 13N 99 51W
Eden, *Wyo., U.S.A.* **82 E9** 42 3N 109 26W
Eden —►, *U.K.* **10 C4** 54 57N 3 1W
Eden L., *Canada* **73 B8** 56 38N 100 15W
Edenburg, *S. Africa* ..... **56 D4** 29 43S 25 58 E
Edendale, *S. Africa* ..... **57 D5** 29 39S 30 18 E
Edenderry, *Ireland* **13 C4** 53 21N 7 4W
Edenton, *U.S.A.* **77 G7** 36 4N 76 39W
Edenville, *S. Africa* ..... **57 D4** 27 37S 27 34 E
Eder —►, *Germany* ...... **16 C5** 51 12N 9 28 E
Edgar, *U.S.A.* **80 E5** 40 22N 97 58W
Edgartown, *U.S.A.* **79 E14** 41 23N 70 31W
Edge Hill, *U.K.* **11 E6** 52 8N 1 26W
Edgefield, *U.S.A.* **77 J5** 33 47N 81 56W
Edgeley, *U.S.A.* **80 B5** 46 22N 98 43W
Edgemont, *U.S.A.* **80 D3** 43 18N 103 50W
Edgeøya, *Svalbard* ...... **4 B9** 77 45N 22 30 E
Édhessa, *Greece* .... **21 D10** 40 48N 22 5 E
Edievale, *N.Z.* .... **59 L2** 45 49S 169 22 E
Edina, *U.S.A.* **80 E8** 40 10N 92 11W
Edinburg, *U.S.A.* **81 M5** 26 18N 98 10W
Edinburgh, *U.K.* **12 F5** 55 57N 3 13W
Edinița, *Moldova* .... **17 D14** 48 9N 27 18 E
Edirne, *Turkey* .... **21 D12** 41 40N 26 34 E
Edison, *U.S.A.* **84 B4** 48 33N 122 27W
Edithburgh, *Australia* ... **63 F2** 35 5S 137 43 E
Edjudina, *Australia* ..... **61 E3** 29 48S 122 23 E
Edmeston, *U.S.A.* **79 D9** 42 42N 75 15W
Edmond, *U.S.A.* **81 H6** 35 39N 97 29W
Edmonds, *U.S.A.* **84 C4** 47 49N 122 23W
Edmonton, *Australia* .... **62 B4** 17 2S 145 46 E
Edmonton, *Canada* **72 C6** 53 30N 113 30W
Edmund L., *Canada* **73 C10** 54 45N 93 17W
Edmundston, *Canada* ... **71 C6** 47 23N 68 20W
Edna, *U.S.A.* **81 L6** 28 59N 96 39W
Edna Bay, *U.S.A.* **72 B2** 55 55N 133 40W
Edremit, *Turkey* .... **21 E12** 39 34N 27 0 E
Edremit Körfezi, *Turkey* .. **21 E12** 39 30N 26 45 E
Edson, *Canada* **72 C5** 53 35N 116 28W
Eduardo Castex, *Argentina* **94 D3** 35 50S 64 18W
Edward —►, *Australia* ... **63 F3** 35 5S 143 30 E
Edward, L., *Africa* ...... **54 C2** 0 25S 29 40 E
Edward I., *Canada* **70 C2** 48 22N 88 37W
Edward River, *Australia* . **62 A3** 14 59S 141 26 E
Edward VII Land, *Antarctica* **5 E13** 80 0S 150 0W
Edwards, *U.S.A.* **85 L9** 34 55N 117 51W
Edwards Plateau, *U.S.A.* . **81 K4** 30 45N 101 20W
Edwardsville, *U.S.A.* **79 E9** 41 15N 75 56W
Edzo, *Canada* **72 A5** 62 49N 116 4W
Eekloo, *Belgium* .... **15 C3** 51 11N 3 33 E
Effingham, *U.S.A.* **76 F1** 39 7N 88 33W
Égadi, Ísole, *Italy* ...... **20 F5** 37 55N 12 16 E
Eganville, *Canada* **70 C4** 45 32N 77 5W
Egeland, *U.S.A.* **80 A5** 48 38N 99 6W
Egenolf L., *Canada* **73 B9** 59 3N 100 0W
Eger = Cheb, *Czech.* .... **16 C7** 50 9N 12 28 E
Eger, *Hungary* .... **17 E11** 47 53N 20 27 E
Egersund, *Norway* ...... **9 G12** 58 26N 6 1 E
Egg L., *Canada* **73 B7** 55 5N 105 30W
Eginbah, *Australia* ...... **60 D2** 20 53S 119 47 E
Egmont, C., *N.Z.* .... **59 H4** 39 16S 173 45 E
Egmont, Mt., *N.Z.* ...... **59 H5** 39 17S 174 5 E
Eğridir, *Turkey* .... **25 G5** 37 52N 30 51 E
Eğridir Gölü, *Turkey* .... **25 G5** 37 53N 30 50 E
Egvekinot, *Russia* **27 C19** 66 19N 179 50W
Egypt ■, *Africa* .... **51 C11** 28 0N 31 0 E
Ehime □, *Japan* .... **31 H6** 33 30N 132 40 E
Ehrenberg, *U.S.A.* **85 M12** 33 36N 114 31W
Eibar, *Spain* **19 A4** 43 11N 2 28W
Eidsvold, *Australia* ..... **63 D5** 25 25S 151 12 E
Eidsvoll, *Norway* .... **9 F14** 60 19N 11 14 E
Eifel, *Germany* .... **16 C4** 50 15N 6 50 E
Eiffel Flats, *Zimbabwe* ... **55 F3** 18 20S 30 0 E
Eigg, *U.K.* **12 E2** 56 54N 6 10W
Eighty Mile Beach, *Australia* .... **60 C3** 19 30S 120 40 E
Eil, *Somali Rep.* .... **46 F4** 8 0N 49 50 E
Eil, L., *U.K.* **12 E3** 56 51N 5 16W
Eildon, L., *Australia* .... **63 F4** 37 10S 146 0 E
Eileen L., *Canada* **73 A7** 62 16N 107 37W
Einasleigh, *Australia* .... **62 B3** 18 32S 144 5 E
Einasleigh —►, *Australia* **62 B3** 17 30S 142 17 E
Eindhoven, *Neths.* ...... **15 C5** 51 26N 5 28 E
Eire = Ireland ■, *Europe* **13 D4** 53 50N 7 52W
Eiriksjökull, *Iceland* **8 D3** 64 46N 20 24W
Eirunepé, *Brazil* **92 E5** 6 35S 69 53W
Eisenach, *Germany* ..... **16 C6** 50 58N 10 19 E
Eisenerz, *Austria* ...... **16 E8** 47 32N 14 54 E
Eivissa = Ibiza, *Spain* ... **22 C7** 38 54N 1 26 E
Ejutla, *Mexico* **87 D5** 16 34N 96 44W
Ekalaka, *U.S.A.* **80 C2** 45 53N 104 33W
Eketahuna, *N.Z.* .... **59 J5** 40 38S 175 43 E
Ekibastuz, *Kazakstan* .... **26 D8** 51 50N 75 10 E
Ekimchan, *Russia* ...... **27 D14** 53 0N 133 0 E
Ekoli, *Zaïre* .... **54 C1** 0 23S 24 13 E
Eksjö, *Sweden* .... **9 H16** 57 40N 14 58 E
Ekwan —►, *Canada* **70 B3** 53 12N 82 15W
Ekwan Pt., *Canada* **70 B3** 53 16N 82 7W
El Aaiún, *W. Sahara* **50 C2** 27 9N 13 12W
El 'Agrûd, *Egypt* .... **47 E3** 30 14N 34 24 E
El Alamein, *Egypt* ...... **51 B10** 30 48N 28 58 E
El 'Aqaba, W. —►, *Egypt* **47 E2** 30 7N 33 54 E
El Arenal, *Spain* .... **22 B9** 39 30N 2 45 E
El Aricha, *Algeria* **50 B4** 34 13N 1 10W
El Arīhā, *West Bank* .... **47 D4** 31 52N 35 27 E
El Arish, *Australia* ..... **62 B4** 17 35S 146 1 E
El 'Arîsh, *Egypt* .... **47 D2** 31 8N 33 50 E
El 'Arîsh, W. —►, *Egypt* . **47 D2** 31 8N 33 47 E
El Asnam = Ech Cheliff, *Algeria* .... **50 A5** 36 10N 1 20 E
El Bawiti, *Egypt* .... **51 C10** 28 25N 28 45 E
El Bayadh, *Algeria* ..... **50 B5** 33 40N 1 1 E
El Bluff, *Nic.* **88 D3** 11 59N 83 40W
El Brûk, W. —►, *Egypt* . **47 E2** 30 15N 33 50 E
El Cajon, *U.S.A.* **85 N10** 32 48N 116 58W
El Callao, *Venezuela* **92 B6** 7 18N 61 50W
El Campo, *U.S.A.* **81 L6** 29 12N 96 16W
El Centro, *U.S.A.* **85 N11** 32 48N 115 34W
El Cerro, *Bolivia* **92 G6** 17 30S 61 40W
El Compadre, *Mexico* ... **85 N10** 32 20N 116 14W
El Cuy, *Argentina* **96 D3** 39 55S 68 25W
El Cuyo, *Mexico* **87 C7** 21 30N 87 40W
El Daheir, *Egypt* .... **47 D3** 31 13N 34 10 E
El Dere, *Somali Rep.* ... **46 G4** 3 50N 47 8 E
El Descanso, *Mexico* ... **85 N10** 32 12N 116 58W
El Desemboque, *Mexico* . **86 A2** 30 30N 112 57W
El Diviso, *Colombia* **92 C3** 1 22N 78 14W
El Djouf, *Mauritania* **50 E3** 20 0N 9 0W
El Dorado, *Ark., U.S.A.* .. **81 J8** 33 12N 92 40W
El Dorado, *Kans., U.S.A.* . **81 G6** 37 49N 96 52W
El Dorado, *Venezuela* ... **92 B6** 6 55N 61 37W
El Escorial, *Spain* **19 B3** 40 35N 4 7W
El Faiyûm, *Egypt* ...... **51 C11** 29 19N 30 50 E
El Fâsher, *Sudan* ...... **51 F10** 13 33N 25 26 E
El Ferrol, *Spain* **19 A1** 43 29N 8 15W
El Fuerte, *Mexico* **86 B3** 26 30N 108 40W
El Gal, *Somali Rep.* .... **46 E5** 10 58N 50 20 E
El Geneina = Al Junaynah, *Sudan* ..... **51 F9** 13 27N 22 45 E
El Geteina, *Sudan* ..... **51 F11** 14 50N 32 27 E
El Gîza, *Egypt* .... **51 C11** 30 0N 31 10 E
El Iskandarîya, *Egypt* ... **51 B10** 31 13N 29 58 E
El Jadida, *Morocco* **50 B3** 33 11N 8 17W
El Jebelein, *Sudan* ..... **51 F11** 12 40N 32 55 E
El Kab, *Sudan* .... **51 E11** 19 27N 32 46 E
El Kabrît, G., *Egypt* .... **47 F2** 29 42N 33 16 E
El Kala, *Algeria* .... **50 A6** 36 50N 8 30 E
El Kamlin, *Sudan* ...... **51 E11** 15 3N 33 11 E
El Kef, *Tunisia* .... **50 A6** 36 12N 8 47 E
El Khandaq, *Sudan* ..... **51 E11** 18 30N 30 30 E
El Khârga, *Egypt* ...... **51 C11** 25 30N 30 33 E
El Khartûm, *Sudan* ..... **51 E11** 15 31N 32 35 E
El Khartûm Bahrî, *Sudan* **51 E11** 15 40N 32 31 E
El Kuntilla, *Egypt* ...... **47 E3** 30 1N 34 45 E
El Laqâwa, *Sudan* ..... **51 F10** 11 25N 29 1 E
El Mafâza, *Sudan* ...... **51 F11** 13 38N 34 30 E
El Mahalla el Kubra, *Egypt* **51 B11** 31 0N 31 0 E
El Mansûra, *Egypt* ..... **51 B11** 31 0N 31 19 E
El Medano, *Canary Is.* .. **22 F3** 28 3N 16 32W
El Milagro, *Argentina* ... **94 C2** 30 59S 65 59W
El Minyâ, *Egypt* .... **51 C11** 28 7N 30 33 E
El Monte, *U.S.A.* **85 L8** 34 4N 118 1W
El Obeid, *Sudan* .... **51 F11** 13 8N 30 10 E
El Odaiya, *Sudan* ...... **51 F10** 12 8N 28 12 E
El Oro, *Mexico* **87 D4** 19 48N 100 8W
El Oued, *Algeria* ...... **50 B6** 33 20N 6 58 E
El Palmito, Presa, *Mexico* **86 B3** 25 40N 105 30W
El Paso, *U.S.A.* **83 L10** 31 45N 106 29W
El Paso Robles, *U.S.A.* .. **84 K6** 35 38N 120 41W
El Portal, *U.S.A.* **83 H4** 37 41N 119 47W
El Porvenir, *Mexico* **86 A3** 31 15N 105 51W
El Prat de Llobregat, *Spain* **19 B7** 41 18N 2 3 E
El Progreso, *Honduras* .. **88 C2** 15 26N 87 51W
El Pueblito, *Mexico* **86 B3** 29 3N 105 4W
El Pueblo, *Canary Is.* ... **22 F2** 28 36N 17 47W
El Puerto de Santa María, *Spain* **19 D2** 36 36N 6 13W
El Qâhira, *Egypt* ...... **51 B11** 30 1N 31 14 E
El Qantara, *Egypt* ..... **47 E1** 30 51N 32 20 E
El Qasr, *Egypt* .... **51 C10** 25 44N 28 42 E
El Quseima, *Egypt* ..... **47 E3** 30 40N 34 15 E
El Reno, *U.S.A.* **81 H6** 35 32N 97 57W
El Rio, *U.S.A.* **85 L7** 34 14N 119 10W
El Roque, Pta., *Canary Is.* **22 F4** 28 10N 15 25W
El Rosarito, *Mexico* **86 B2** 28 38N 114 4W
El Saheira, W. —►, *Egypt* **47 E2** 30 5N 33 25 E
El Salto, *Mexico* **86 C3** 23 47N 105 22W
El Salvador ■, *Cent. Amer.* **88 D2** 13 50N 89 0W
El Sauce, *Nic.* **88 D2** 13 0N 86 40W
El Shallal, *Egypt* ...... **51 D11** 24 0N 32 53 E
El Suweis, *Egypt* ...... **51 C11** 29 58N 32 31 E
El Tamarâni, W. —►, *Egypt* .... **47 E3** 30 7N 34 43 E
El Thamad, *Egypt* ...... **47 F3** 29 40N 34 28 E
El Tigre, *Venezuela* **92 B6** 8 44N 64 15W
El Tîh, G., *Egypt* ...... **47 F2** 29 40N 33 50 E
El Tîna, Khalîg, *Egypt* .. **47 D1** 31 10N 32 40 E
El Tocuyo, *Venezuela* ... **92 B5** 9 47N 69 48W
El Tofo, *Chile* **94 B1** 29 22S 71 18W
El Tránsito, *Chile* **94 B1** 28 52S 70 17W
El Turbio, *Argentina* **96 G2** 51 45S 72 5W
El Uqsur, *Egypt* .... **51 C11** 25 41N 32 38 E
El Venado, *Mexico* **86 C4** 22 56N 101 10W
El Vigía, *Venezuela* **92 B4** 8 38N 71 39W
El Wabeira, *Egypt* ..... **47 F2** 29 34N 33 6 E
El Wak, *Kenya* .... **54 B5** 2 49N 40 56 E
El Wuz, *Sudan* .... **51 E11** 15 5N 30 7 E
Elat, *Israel* .... **47 F3** 29 30N 34 56 E
Elâzığ, *Turkey* .... **25 G6** 38 37N 39 14 E
Elba, *Italy* .... **20 C4** 42 46N 10 17 E
Elba, *U.S.A.* **77 K2** 31 25N 86 4W
Elbasani, *Albania* ..... **21 D9** 41 9N 20 9 E
Elbe, *U.S.A.* **84 D4** 46 45N 122 10W
Elbe —►, *Europe* ...... **16 B5** 53 50N 9 0 E
Elbert, Mt., *U.S.A.* **83 G10** 39 7N 106 27W
Elberta, *U.S.A.* **76 C2** 44 37N 86 14W
Elberton, *U.S.A.* **77 H4** 34 7N 82 52W
Elbeuf, *France* .... **18 B4** 49 17N 1 2 E
Elbing = Elbląg, *Poland* . **17 A10** 54 10N 19 25 E
Elbląg, *Poland* .... **17 A10** 54 10N 19 25 E
Elbow, *Canada* **73 C7** 51 7N 106 35W
Elbrus, *Asia* .... **25 F7** 43 21N 42 30 E
Elburg, *Neths.* .... **15 B5** 52 26N 5 50 E
Elburz Mts. = Alborz, Reshteh-ye Kūhhā-ye, *Iran* .... **45 C7** 36 0N 52 0 E
Elche, *Spain* **19 C5** 38 15N 0 42W
Elcho I., *Australia* ..... **62 A2** 11 55S 135 45 E
Elda, *Spain* **19 C5** 38 29N 0 47W
Elde —►, *Germany* .... **16 B6** 53 7N 11 15 E
Eldon, *Mo., U.S.A.* **80 F8** 38 21N 92 35W
Eldon, *Wash., U.S.A.* ... **84 C3** 47 33N 123 3W
Eldora, *U.S.A.* **80 D8** 42 22N 93 5W
Eldorado, *Argentina* ... **95 B5** 26 28S 54 43W
Eldorado, *Canada* **73 B7** 59 35N 108 30W
Eldorado, *Mexico* **86 C3** 24 20N 107 22W
Eldorado, *Ill., U.S.A.* ... **76 G1** 37 49N 88 26W
Eldorado, *Tex., U.S.A.* .. **81 K4** 30 52N 100 36W
Eldorado Springs, *U.S.A.* **81 G8** 37 52N 94 1W
Eldoret, *Kenya* .... **54 B4** 0 30N 35 17 E
Eldred, *U.S.A.* **78 E6** 41 58N 78 23W
Elea, C., *Cyprus* ...... **23 D13** 35 19N 34 4 E
Electra, *U.S.A.* **81 H5** 34 2N 98 55W
Elefantes —►, *Mozam.* . **57 C5** 24 10S 32 40 E
Elektrostal, *Russia* .... **24 C6** 55 41N 38 32 E
Elephant Butte Reservoir, *U.S.A.* **83 K10** 33 9N 107 11W
Elephant I., *Antarctica* .. **5 C18** 61 0S 55 0W
Eleuthera, *Bahamas* **88 A4** 25 0N 76 20W
Elgeyo-Marakwet □, *Kenya* .... **54 B4** 0 45N 35 30 E
Elgin, *N.B., Canada* **71 C6** 45 48N 65 10W
Elgin, *Ont., Canada* **79 B8** 44 36N 76 13W
Elgin, *U.K.* **12 D5** 57 39N 3 19W
Elgin, *Ill., U.S.A.* **76 D1** 42 2N 88 17W
Elgin, *N. Dak., U.S.A.* .. **80 B4** 46 24N 101 51W
Elgin, *Nebr., U.S.A.* ... **80 E5** 41 59N 98 5W
Elgin, *Nev., U.S.A.* **83 H6** 37 21N 114 32W
Elgin, *Oreg., U.S.A.* ... **82 D5** 45 34N 117 55W
Elgin, *Tex., U.S.A.* **81 K6** 30 21N 97 22W
Elgon, Mt., *Africa* ..... **54 B3** 1 10N 34 30 E
Eliase, *Indonesia* ..... **37 F8** 8 21S 130 48 E
Elida, *U.S.A.* **81 J3** 33 57N 103 39W
Elim, *S. Africa* .... **56 E2** 34 35S 19 45 E
Elisabethville = Lubumbashi, *Zaïre* .... **55 E2** 11 40S 27 28 E
Elista, *Russia* .... **25 E7** 46 16N 44 14 E
Elizabeth, *Australia* ... **63 E2** 34 42S 138 41 E
Elizabeth, *U.S.A.* **79 F10** 40 40N 74 13W
Elizabeth City, *U.S.A.* .. **77 G7** 36 18N 76 14W
Elizabethton, *U.S.A.* ... **77 G4** 36 21N 82 13W
Elizabethtown, *Ky., U.S.A.* **76 G3** 37 42N 85 52W
Elizabethtown, *N.Y., U.S.A.* **79 B11** 44 13N 73 36W
Elizabethtown, *Pa., U.S.A.* **79 F8** 40 9N 76 36W
Ełk, *Poland* .... **17 B12** 53 50N 22 21 E
Elk City, *U.S.A.* **81 H5** 35 25N 99 25W
Elk Creek, *U.S.A.* **84 F4** 39 36N 122 32W
Elk Grove, *U.S.A.* **84 G5** 38 25N 121 22W
Elk Island Nat. Park, *Canada* **72 C6** 53 35N 112 59W
Elk Lake, *Canada* **70 C3** 47 40N 80 25W
Elk Point, *Canada* **73 C6** 53 54N 110 55W
Elk River, *Idaho, U.S.A.* . **82 C5** 46 47N 116 11W
Elk River, *Minn., U.S.A.* . **80 C8** 45 18N 93 35W
Elkedra, *Australia* .... **62 C2** 21 9S 135 33 E
Elkedra —►, *Australia* .. **62 C2** 21 8S 136 22 E
Elkhart, *Ind., U.S.A.* ... **76 E3** 41 41N 85 58W
Elkhart, *Kans., U.S.A.* .. **81 G4** 37 0N 101 54W
Elkhorn, *Canada* **73 D8** 49 59N 101 14W
Elkhorn —►, *U.S.A.* **80 E6** 41 8N 96 19W
Elkhovo, *Bulgaria* .... **21 C12** 42 10N 26 40 E
Elkin, *U.S.A.* **77 G5** 36 15N 80 51W
Elkins, *U.S.A.* **76 F6** 38 55N 79 51W
Elko, *Canada* **72 D5** 49 20N 115 10W
Elko, *U.S.A.* **82 F6** 40 50N 115 46W
Ell, L., *Australia* ..... **61 E4** 29 13S 127 46 E
Ellef Ringnes I., *Canada* . **4 B2** 78 30N 102 2W
Ellendale, *Australia* ... **60 C3** 17 56S 124 48 E
Ellendale, *U.S.A.* **80 B5** 46 0N 98 32W
Ellensburg, *U.S.A.* ... **82 C3** 46 59N 120 34W
Ellenville, *U.S.A.* **79 E10** 41 43N 74 24W
Ellery, Mt., *Australia* .. **63 F4** 37 28S 148 47 E
Ellesmere, L., *N.Z.* .... **59 M4** 47 47S 172 28 E
Ellesmere I., *Canada* .. **4 B4** 79 30N 80 0W
Ellesmere Port, *U.K.* .. **10 D5** 53 17N 2 54W
Ellice Is. = Tuvalu ■, *Pac. Oc.* .... **64 H9** 8 0S 178 0 E
Ellinwood, *U.S.A.* **80 F5** 38 21N 98 35W
Elliot, *Australia* ...... **62 B1** 17 33S 133 32 E
Elliot, *S. Africa* ...... **57 E4** 31 22S 27 48 E
Elliot Lake, *Canada* ... **70 C3** 46 25N 82 35W
Elliotdale = Xhora, *S. Africa* .... **57 E4** 31 55S 28 38 E
Ellis, *U.S.A.* **80 F5** 38 56N 99 34W
Elliston, *Australia* .... **63 E1** 33 39S 134 53 E
Ellisville, *U.S.A.* **81 K10** 31 36N 89 12W
Ellon, *U.K.* **12 D6** 57 22N 2 4W
Ellore = Eluru, *India* ... **41 L12** 16 48N 81 8 E
Ells —►, *Canada* **72 B6** 57 18N 111 40W
Ellsworth, *U.S.A.* **80 F5** 38 44N 98 14W
Ellsworth Land, *Antarctica* **5 D16** 76 0S 89 0W
Ellsworth Mts., *Antarctica* **5 D16** 78 30S 85 0W
Ellwood City, *U.S.A.* .. **78 F4** 40 52N 80 17W
Elma, *Canada* **73 D9** 49 52N 95 55W
Elma, *U.S.A.* **84 D3** 47 0N 123 25W
Elmalı, *Turkey* .... **25 G4** 36 44N 29 56 E
Elmhurst, *U.S.A.* **76 E2** 41 53N 87 56W
Elmira, *Canada* **78 C4** 43 36N 80 33W
Elmira, *U.S.A.* **78 D8** 42 6N 76 48W
Elmore, *Australia* .... **63 F3** 36 30S 144 37 E
Elmore, *U.S.A.* **85 M11** 33 7N 115 49W
Elmshorn, *Germany* ... **16 B5** 53 43N 9 40 E
Elmvale, *Canada* **78 B5** 44 35N 79 52W
Elora, *Canada* **78 C4** 43 41N 80 26W
Eloúnda, *Greece* ..... **23 D7** 35 16N 25 42 E
Eloy, *U.S.A.* **83 K8** 32 45N 111 33W
Elrose, *Canada* **73 C7** 51 12N 108 0W
Elsas, *Canada* **70 C3** 48 32N 82 55W
Elsie, *U.S.A.* **84 E3** 45 52N 123 36W
Elsinore = Helsingør, *Denmark* .... **9 H15** 56 2N 12 35 E
Elsinore, *U.S.A.* **83 G7** 38 41N 112 9W
Eltham, *N.Z.* .... **59 H5** 39 26S 174 19 E
Eluru, *India* .... **41 L12** 16 48N 81 8 E
Elvas, *Portugal* **19 C2** 38 50N 7 10W
Elverum, *Norway* ..... **9 F14** 60 53N 11 34 E
Elvire —►, *Australia* ... **60 C4** 17 51S 128 11 E
Elwood, *Ind., U.S.A.* .. **76 E3** 40 17N 85 50W
Elwood, *Nebr., U.S.A.* .. **80 E5** 40 36N 99 52W
Elx = Elche, *Spain* ... **19 C5** 38 15N 0 42W
Ely, *U.K.* .... **11 E8** 52 24N 0 16 E
Ely, *Minn., U.S.A.* **80 B9** 47 55N 91 51W
Ely, *Nev., U.S.A.* **82 G6** 39 15N 114 54W
Elyria, *U.S.A.* **78 E2** 41 22N 82 7W
Emāmrūd, *Iran* ...... **45 B7** 36 30N 55 0 E
Emba = Embi, *Kazakstan* **26 E6** 48 50N 58 8 E
Emba —► = Embi —►, *Kazakstan* .... **25 E9** 46 55N 53 28 E
Embarcación, *Argentina* . **94 A3** 23 10S 64 0W
Embarras Portage, *Canada* **73 B6** 58 27N 111 28W
Embetsu, *Japan* ...... **30 B10** 44 44N 141 47 E
Embi, *Kazakstan* ..... **26 E6** 48 50N 58 8 E
Embi —►, *Kazakstan* ... **25 E9** 46 55N 53 28 E
Embóna, *Greece* ..... **23 C9** 36 13N 27 51 E
Embrun, *France* ...... **18 D7** 44 34N 6 30 E
Embu, *Kenya* .... **54 C4** 0 32S 37 38 E
Embu □, *Kenya* ..... **54 C4** 0 30S 37 35 E
Emden, *Germany* .... **16 B4** 53 21N 7 12 E
Emerald, *Australia* ... **62 C4** 23 32S 148 10 E
Emerson, *Canada* **73 D9** 49 0N 97 10W
Emery, *U.S.A.* **83 G8** 38 55N 111 15W
Emet, *Turkey* .... **21 E13** 39 20N 29 15 E
Emi Koussi, *Chad* .... **51 E8** 19 45N 18 55 E
Eminabad, *Pakistan* ... **42 C6** 32 2N 74 8 E
Emine, Nos, *Bulgaria* .. **21 C12** 42 40N 27 56 E
Emlenton, *U.S.A.* **78 E5** 41 11N 79 43W
Emmeloord, *Neths.* ... **15 B5** 52 44N 5 46 E
Emmen, *Neths.* ...... **15 B6** 52 48N 6 57 E
Emmet, *Australia* .... **62 C3** 24 45S 144 30 E
Emmetsburg, *U.S.A.* .. **80 D7** 43 7N 94 41W
Emmett, *U.S.A.* **82 E5** 43 52N 116 30W
Empalme, *Mexico* **86 B2** 28 1N 110 49W
Empangeni, *S. Africa* .. **57 D5** 28 50S 31 52 E
Empedrado, *Argentina* . **94 B4** 28 0S 58 46W
Emperor Seamount Chain, *Pac. Oc.* .... **64 D9** 40 0N 170 0 E
Emporia, *Kans., U.S.A.* . **80 F6** 38 25N 96 11W
Emporia, *Va., U.S.A.* .. **77 G7** 36 42N 77 32W
Emporium, *U.S.A.* ... **78 E6** 41 31N 78 14W
Empress, *Canada* **73 C6** 50 57N 110 0W
Empty Quarter = Rub' al Khali, *Si. Arabia* ..... **46 D4** 18 0N 48 0 E
Ems —►, *Germany* ... **16 B4** 53 20N 7 12 E
Emsdale, *Canada* **78 A5** 45 32N 79 19W
Emu, *China* .... **35 C15** 43 40N 128 6 E
Emu Park, *Australia* .. **62 C5** 23 13S 150 50 E
'En 'Avrona, *Israel* ... **47 F3** 29 43N 35 0 E
En Nahud, *Sudan* .... **51 F10** 12 45N 28 25 E
Ena, *Japan* .... **31 G8** 35 25N 137 25 E
Enana, *Namibia* ..... **56 B2** 17 30S 16 23 E
Enaratoli, *Indonesia* .. **37 E9** 3 55S 136 21 E
Enard B., *U.K.* **12 C3** 58 5N 5 20W
Enare = Inarijärvi, *Finland* **8 B22** 69 0N 28 0 E
Encantadas, Serra, *Brazil* **95 C5** 30 40S 53 0W
Encanto, C., *Phil.* ... **37 A6** 15 45N 121 38 E
Encarnación, *Paraguay* . **95 B4** 27 15S 55 50W
Encarnación de Diaz, *Mexico* **86 C4** 21 30N 102 13W
Encinal, *U.S.A.* **81 L5** 28 2N 99 21W
Encinitas, *U.S.A.* ... **85 M9** 33 3N 117 17W
Encino, *U.S.A.* **83 J11** 34 39N 105 28W
Encounter B., *Australia* . **63 F2** 35 45S 138 45 E
Ende, *Indonesia* .... **37 F6** 8 45S 121 40 E
Endeavour, *Canada* .. **73 C8** 52 10N 102 39W
Endeavour Str., *Australia* **62 A3** 10 45S 142 0 E
Enderbury I., *Kiribati* .. **64 H10** 3 8S 171 5W
Enderby, *Canada* **72 C5** 50 35N 119 10W
Enderby I., *Australia* . **60 D2** 20 35S 116 30 E
Enderby Land, *Antarctica* **5 C5** 66 0S 53 0 E
Enderlin, *U.S.A.* **80 B6** 46 38N 97 36W
Endicott, *N.Y., U.S.A.* . **79 D8** 42 6N 76 4W
Endicott, *Wash., U.S.A.* . **82 C5** 46 56N 117 41W
Endyalgout I., *Australia* . **60 B5** 11 40S 132 35 E
Enewetak Atoll, *Pac. Oc.* . **64 F8** 11 30N 162 15 E
Enez, *Turkey* .... **21 D12** 40 45N 26 5 E
Enfield, *U.K.* **11 F7** 51 38N 0 5W
Engadin, *Switz.* .... **16 E6** 46 45N 10 10 E
Engaño, C., *Dom. Rep.* . **89 C6** 18 30N 68 20W
Engaño, C., *Phil.* .... **37 A6** 18 35N 122 23 E
Engcobo, *S. Africa* ... **57 E4** 31 37S 28 0 E
Engels, *Russia* ...... **24 D8** 51 28N 46 6 E
Engemann L., *Canada* . **73 B7** 58 0N 106 55W
Enggano, *Indonesia* .. **36 F2** 5 20S 102 40 E
Enghien, *Belgium* .... **15 D4** 50 37N 4 2 E
Engkilili, *Malaysia* ... **36 D4** 1 3N 111 42 E
England, *U.S.A.* **81 H9** 34 33N 91 58W
England □, *U.K.* **7 E5** 53 0N 2 0W
Englee, *Canada* **71 B8** 50 45N 56 5W
Englehart, *Canada* ... **70 C4** 47 49N 79 52W
Engler L., *Canada* ... **73 B7** 59 8N 106 52W
Englewood, *Colo., U.S.A.* **80 F2** 39 39N 104 59W
Englewood, *Kans., U.S.A.* **81 G5** 37 2N 99 59W
English —►, *Canada* ... **73 C10** 50 35N 93 30W
English Bazar = Ingraj Bazar, *India* .... **43 G13** 24 58N 88 10 E
English Channel, *Europe* . **18 A3** 50 0N 2 0W
English River, *Canada* . **70 C1** 49 14N 91 0W
Enid, *U.S.A.* **81 G6** 36 24N 97 53W

Enkhuizen, *Neths.* ...... **15 B5** 52 42N 5 17 E
Enna, *Italy* .... **20 F6** 37 34N 14 16 E
Ennadai, *Canada* **73 A8** 61 8N 100 53W
Ennadai L., *Canada* **73 A8** 61 0N 101 0W
Ennedi, *Chad* .... **51 E9** 17 15N 22 0 E
Enngonia, *Australia* ..... **63 D4** 29 21S 145 50 E
Ennis, *Ireland* **13 D3** 52 51N 8 59W
Ennis, *Mont., U.S.A.* **82 D8** 45 21N 111 44W
Ennis, *Tex., U.S.A.* **81 J6** 32 20N 96 38W
Enniscorthy, *Ireland* **13 D5** 52 30N 6 34W
Enniskillen, *U.K.* **13 B4** 54 21N 7 39W
Ennistimon, *Ireland* **13 D2** 52 57N 9 17W
Enns →, *Austria* .... **16 D8** 48 14N 14 32 E
Enontekiö, *Finland* ...... **8 B20** 68 23N 23 37 E
Enriquillo, L., *Dom. Rep.* . **89 C5** 18 20N 72 5W
Enschede, *Neths.* .... **15 B6** 52 13N 6 53 E
Ensenada, *Argentina* **94 C4** 34 55S 57 55W
Ensenada, *Mexico* **86 A1** 31 50N 116 50W
Ensiola, Pta., *Spain* ..... **22 B9** 39 7N 2 55 E
Entebbe, *Uganda* .... **54 B3** 0 4N 32 28 E
Enterprise, *Canada* **72 A5** 60 47N 115 45W
Enterprise, *Oreg., U.S.A.* . **82 D5** 45 25N 117 17W
Enterprise, *Utah, U.S.A.* . **83 H7** 37 34N 113 43W
Entre Ríos, *Bolivia* **94 A3** 21 30S 64 25W
Entre Ríos □, *Argentina* . **94 C4** 30 30S 58 30W
Entroncamento, *Portugal* **19 C1** 39 28N 8 28W
Enugu, *Nigeria* .... **50 G6** 6 20N 7 30 E
Enugu Ezike, *Nigeria* .... **50 G6** 7 0N 7 29 E
Enumclaw, *U.S.A.* **84 C5** 47 12N 121 59W
Eólie, Ís., *Italy* .... **20 E6** 38 30N 14 57 E
Epe, *Neths.* .... **15 B5** 52 21N 5 59 E
Épernay, *France* .... **18 B5** 49 3N 3 56 E
Ephesus, *Turkey* .... **21 F12** 37 55N 27 22 E
Ephraim, *U.S.A.* **82 G8** 39 22N 111 35W
Ephrata, *U.S.A.* **82 C4** 47 19N 119 33W
Épinal, *France* .... **18 B7** 48 10N 6 27 E
Episkopi, *Cyprus* .... **23 E11** 34 40N 32 54 E
Episkopí, *Greece* .... **23 D6** 35 20N 24 20 E
Episkopi Bay, *Cyprus* .... **23 E11** 34 35N 32 50 E
Epping, *U.K.* .... **11 F8** 51 41N 0 7 E
Epukiro, *Namibia* .... **56 C2** 21 40S 19 9 E
Equatorial Guinea ■,
Africa .... **52 D1** 2 0N 8 0 E
Er Rahad, *Sudan* .... **51 F11** 12 45N 30 32 E
Er Rif, *Morocco* **50 A4** 35 1N 4 1W
Er Roseires, *Sudan* ..... **51 F11** 11 55N 34 30 E
Erâwadī Myit =
Irrawaddy →, *Burma* . **41 M19** 15 50N 95 6 E
Erbil = Arbīl, *Iraq* .... **44 B5** 36 15N 44 5 E
Erciyaş Dağı, *Turkey* .... **25 G6** 38 30N 35 30 E
Érd, *Hungary* .... **17 E10** 47 22N 18 56 E
Erdao Jiang →, *China* .. **35 C14** 43 0N 127 0 E
Erdek, *Turkey* .... **21 D12** 40 23N 27 47 E
Erdene, *Mongolia* .... **34 B6** 44 13N 111 10 E
Erebus, Mt., *Antarctica* ... **5 D11** 77 35S 167 0 E
Erechim, *Brazil* **95 B5** 27 35S 52 15W
Ereğli, *Konya, Turkey* .... **25 G5** 37 31N 34 4 E
Ereğli, *Zonguldak, Turkey* **25 F5** 41 15N 31 24 E
Erenhot, *China* .... **34 C7** 43 48N 112 2 E
Eresma →, *Spain* **19 B3** 41 26N 4 45W
Erewadi Myitwanya,
Burma .... **41 M19** 15 30N 95 0 E
Erfenisdam, *S. Africa* .... **56 D4** 28 30S 26 50 E
Erfurt, *Germany* .... **16 C6** 50 58N 11 2 E
Ergeni Vozvyshennost,
Russia .... **25 E7** 47 0N 44 0 E
Ērgli, *Latvia* .... **9 H21** 56 54N 25 38 E
Eriboll, L., *U.K.* **12 C4** 58 30N 4 42W
Érice, *Italy* .... **20 E5** 38 2N 12 35 E
Erie, *U.S.A.* **78 D4** 42 8N 80 5W
Erie, L., *N. Amer.* **78 D3** 42 15N 81 0W
Erie Canal, *U.S.A.* **78 C6** 43 5N 78 43W
Erieau, *Canada* **78 D3** 42 16N 81 57W
Erigavo, *Somali Rep.* .... **46 E4** 10 35N 47 20 E
Erikoúsa, *Greece* .... **23 A3** 39 53N 19 34 E
Eriksdale, *Canada* **73 C9** 50 52N 98 7W
Erímanthos, *Greece* ..... **21 F9** 37 57N 21 50 E
Erimo-misaki, *Japan* .... **30 D11** 41 50N 143 15 E
Eritrea ■, *Africa* .... **51 F12** 14 0N 38 30 E
Erlangen, *Germany* ..... **16 D6** 49 36N 11 0 E
Erldunda, *Australia* ..... **62 D1** 25 14S 133 12 E
Ermelo, *Neths.* .... **15 B5** 52 18N 5 35 E
Ermelo, *S. Africa* .... **57 D4** 26 31S 29 59 E
Ermones, *Greece* .... **23 A3** 39 37N 19 46 E
Ermoúpolis = Síros,
Greece .... **21 F11** 37 28N 24 57 E
Ernakulam = Cochin, *India* **40 Q10** 9 59N 76 22 E
Erne →, *Ireland* **13 B3** 54 30N 8 16W
Erne, Lower L., *U.K.* **13 B4** 54 28N 7 47W
Erne, Upper L., *U.K.* **13 B4** 54 14N 7 32W
Ernest Giles Ra., *Australia* **61 E3** 27 0S 123 45 E
Erode, *India* .... **40 P10** 11 24N 77 45 E
Eromanga, *Australia* .... **63 D3** 26 40S 143 11 E
Erongo, *Namibia* .... **56 C2** 21 39S 15 58 E
Errabiddy, *Australia* .... **61 E2** 25 25S 117 5 E
Erramala Hills, *India* .... **40 M11** 15 30N 78 15 E
Errigal, *Ireland* **13 A3** 55 2N 8 6W
Erris Hd., *Ireland* **13 B1** 54 19N 10 0W
Erskine, *U.S.A.* **80 B7** 47 40N 96 0W
Ertis → = Irtysh →,
Russia .... **26 C7** 61 4N 68 52 E
Erwin, *U.S.A.* **77 G4** 36 9N 82 25W
Erzgebirge, *Germany* .... **16 C7** 50 27N 12 55 E
Erzin, *Russia* .... **27 D10** 50 15N 95 10 E
Erzincan, *Turkey* .... **25 G6** 39 46N 39 30 E
Erzurum, *Turkey* .... **25 G7** 39 57N 41 15 E
Es Caló, *Spain* .... **22 C8** 38 40N 1 30 E
Es Caná, *Spain* .... **22 B8** 39 2N 1 36 E
Es Sahrâ' Esh Sharqîya,
Egypt .... **51 C11** 27 30N 32 30 E
Es Sînâ', *Egypt* .... **51 C11** 29 0N 34 0 E
Esambo, *Zaïre* .... **54 C1** 3 48S 23 30 E
Esan-Misaki, *Japan* ..... **30 D10** 41 40N 141 10 E
Esashi, *Hokkaidō, Japan* . **30 B11** 44 56N 142 35 E
Esashi, *Hokkaidō, Japan* . **30 D10** 41 52N 140 7 E
Esbjerg, *Denmark* ...... **9 J13** 55 29N 8 29 E
Escalante, *U.S.A.* **83 H8** 37 47N 111 36W
Escalante →, *U.S.A.* ... **83 H8** 37 24N 110 57W
Escalón, *Mexico* **86 B4** 26 46N 104 20W
Escambia →, *U.S.A.* ... **77 K2** 30 32N 87 11W
Escanaba, *U.S.A.* **76 C2** 45 45N 87 4W
Esch-sur-Alzette, *Lux.* ... **18 B6** 49 32N 6 0 E
Escondido, *U.S.A.* **85 M9** 33 7N 117 5W
Escuinapa, *Mexico* **86 C3** 22 50N 105 50W
Escuintla, *Guatemala* **88 D1** 14 20N 90 48W
Esenguly, *Turkmenistan* . **26 F6** 37 37N 53 59 E
Eşfahān, *Iran* .... **45 C6** 32 39N 51 43 E
Esfideh, *Iran* .... **45 C8** 33 39N 59 46 E
Esh Sham = Dimashq,
Syria .... **47 B5** 33 30N 36 18 E
Eshowe, *S. Africa* .... **57 D5** 28 50S 31 30 E
Esil → = Ishim →,
Russia .... **26 D8** 57 45N 71 10 E
Esk →, *Cumb., U.K.* **12 G5** 54 58N 3 2W
Esk →, *N. Yorks., U.K.* . **10 C7** 54 30N 0 37W
Eskifjörður, *Iceland* **8 D7** 65 3N 13 55W
Eskilstuna, *Sweden* ..... **9 G17** 59 22N 16 32 E
Eskimo Pt., *Canada* **73 A10** 61 10N 94 15W
Eskişehir, *Turkey* .... **25 G5** 39 50N 30 35 E
Esla →, *Spain* **19 B2** 41 29N 6 3W
Eslāmābād-e Gharb, *Iran* **44 C5** 34 10N 46 30 E
Eşme, *Turkey* .... **21 E13** 38 23N 28 58 E
Esmeraldas, *Ecuador* **92 C3** 1 0N 79 40W
Espalmador, I., *Spain* ... **22 C7** 38 47N 1 26 E
Espanola, *Canada* **70 C3** 46 15N 81 46W
Espardell, I. del, *Spain* .. **22 C7** 38 48N 1 29 E
Esparta, *Costa Rica* **88 E3** 9 59N 84 40W
Esperance, *Australia* .... **61 F3** 33 45S 121 55 E
Esperance B., *Australia* . **61 F3** 33 48S 121 55 E
Esperanza, *Argentina* ... **94 C3** 31 29S 61 3W
Espichel, C., *Portugal* ... **19 C1** 38 22N 9 16W
Espigão, Serra do, *Brazil* . **95 B5** 26 35S 50 30W
Espinal, *Colombia* **92 C4** 4 9N 74 53W
Espinazo, Sierra del =
Espinhaço, Serra do,
Brazil **93 G10** 17 30S 43 30W
Espinhaço, Serra do, *Brazil* **93 G10** 17 30S 43 30W
Espinilho, Serra do, *Brazil* **95 B5** 28 30S 55 0W
Espírito Santo □, *Brazil* .. **93 G10** 20 0S 40 45W
Espiritu Santo, B. del,
Mexico **87 D7** 19 15N 87 0W
Espiritu Santo, I., *Mexico* **86 C2** 24 30N 110 23W
Espita, *Mexico* **87 C7** 21 1N 88 19W
Espoo, *Finland* .... **9 F21** 60 12N 24 40 E
Espungabera, *Mozam.* ... **57 C5** 20 29S 32 45 E
Esquel, *Argentina* **96 E2** 42 55S 71 20W
Esquina, *Argentina* **94 B4** 30 0S 59 30W
Essaouira, *Morocco* **50 B3** 31 32N 9 42W
Essebie, *Zaïre* .... **54 B3** 2 58N 30 40 E
Essen, *Belgium* .... **15 C4** 51 28N 4 28 E
Essen, *Germany* .... **16 C4** 51 28N 7 0 E
Essendon, Mt., *Australia* . **61 E3** 25 0S 120 29 E
Essequibo →, *Guyana* .. **92 B7** 6 50N 58 30W
Essex, *Canada* **78 D2** 42 10N 82 49W
Essex, *Calif., U.S.A.* **85 L11** 34 44N 115 15W
Essex, *N.Y., U.S.A.* **79 B11** 44 19N 73 21W
Essex □, *U.K.* .... **11 F8** 51 54N 0 27 E
Esslingen, *Germany* ..... **16 D5** 48 44N 9 18 E
Estados, I. de Los,
Argentina **96 G4** 54 40S 64 30W
Eşţahbānāt, *Iran* .... **45 D7** 29 8N 54 4 E
Estallenchs, *Spain* ...... **22 B9** 39 39N 2 29 E
Estância, *Brazil* **93 F11** 11 16S 37 26W
Estancia, *U.S.A.* **83 J10** 34 46N 106 4W
Estārm, *Iran* .... **45 D8** 28 21N 58 21 E
Estcourt, *S. Africa* ...... **57 D4** 29 0S 29 53 E
Estelí, *Nic.* **88 D2** 13 9N 86 22W
Estelline, *S. Dak., U.S.A.* . **80 C6** 44 35N 96 54W
Estelline, *Tex., U.S.A.* ... **81 H4** 34 33N 100 26W
Esterhazy, *Canada* **73 C8** 50 37N 102 5W
Estevan, *Canada* **73 D8** 49 10N 102 59W
Estevan Group, *Canada* . **72 C3** 53 3N 129 38W
Estherville, *U.S.A.* **80 D7** 43 24N 94 50W
Eston, *Canada* **73 C7** 51 8N 108 40W
Estonia ■, *Europe* ..... **9 G21** 58 30N 25 30 E
Estrêla, Serra da, *Portugal* **19 B2** 40 10N 7 45W
Estremoz, *Portugal* **19 C2** 38 51N 7 39W
Estrondo, Serra do, *Brazil* **93 E9** 7 20S 48 0W
Esztergom, *Hungary* .... **17 E10** 47 47N 18 44 E
Etadunna, *Australia* ..... **63 D2** 28 43S 138 38 E
Etah, *India* .... **43 F8** 27 35N 78 40 E
Etamamu, *Canada* **71 B8** 50 18N 59 59W
Étampes, *France* .... **18 B5** 48 26N 2 10 E
Etanga, *Namibia* .... **56 B1** 17 55S 13 0 E
Etawah, *India* .... **43 F8** 26 48N 79 6 E
Etawah →, *U.S.A.* **77 H3** 34 20N 84 15W
Etawney L., *Canada* **73 B9** 57 50N 96 50W
Ethel, *U.S.A.* **84 D4** 46 32N 122 46W
Ethel Creek, *Australia* ... **60 D3** 22 55S 120 11 E
Ethelbert, *Canada* **73 C8** 51 32N 100 25W
Ethiopia ■, *Africa* ...... **46 F3** 8 0N 40 0 E
Ethiopian Highlands,
Ethiopia .... **28 J7** 10 0N 37 0 E
Etive, L., *U.K.* **12 E3** 56 29N 5 10W
Etna, *Italy* .... **20 F6** 37 50N 14 55 E
Etoile, *Zaïre* .... **55 E2** 11 33S 27 30 E
Etolin I., *U.S.A.* **72 B2** 56 5N 132 20W
Etosha Pan, *Namibia* .... **56 B2** 18 40S 16 30 E
Etowah, *U.S.A.* **77 H3** 35 20N 84 32W
Ettrick Water →, *U.K.* .. **12 F6** 55 31N 2 55W
Etuku, *Zaïre* .... **54 C2** 3 42S 25 45 E
Etzatlán, *Mexico* **86 C4** 20 48N 104 5W
Euboea = Évvoia, *Greece* **21 E11** 38 30N 24 0 E
Eucla Motel, *Australia* ... **61 F4** 31 41S 128 52 E
Euclid, *U.S.A.* **78 E3** 41 34N 81 32W
Eucumbene, L., *Australia* **63 F4** 36 2S 148 40 E
Eudora, *U.S.A.* **81 J9** 33 7N 91 16W
Eufaula, *Ala., U.S.A.* **77 K3** 31 54N 85 9W
Eufaula, *Okla., U.S.A.* **81 H7** 35 17N 95 35W
Eufaula L., *U.S.A.* **81 H7** 35 18N 95 21W
Eugene, *U.S.A.* **82 E2** 44 5N 123 4W
Eugowra, *Australia* ..... **63 E4** 33 22S 148 24 E
Eulo, *Australia* .... **63 D4** 28 10S 145 3 E
Eunice, *La., U.S.A.* **81 K8** 30 30N 92 25W
Eunice, *N. Mex., U.S.A.* .. **81 J3** 32 26N 103 10W
Eupen, *Belgium* .... **15 D6** 50 37N 6 3 E
Euphrates = Furāt, Nahr
al →, *Asia* .... **44 D5** 31 0N 47 25 E
Eureka, *Canada* **4 B3** 80 0N 85 56W
Eureka, *Calif., U.S.A.* **82 F1** 40 47N 124 9W
Eureka, *Kans., U.S.A.* **81 G6** 37 49N 96 17W
Eureka, *Mont., U.S.A.* **82 B6** 48 53N 115 3W
Eureka, *Nev., U.S.A.* **82 G5** 39 31N 115 58W
Eureka, *S. Dak., U.S.A.* .. **80 C5** 45 46N 99 38W
Eureka, *Utah, U.S.A.* **82 G7** 39 58N 112 7W
Eureka, Mt., *Australia* ... **61 E3** 26 35S 121 35 E
Euroa, *Australia* .... **63 F4** 36 44S 145 35 E
Europa, I., *Ind. Oc.* ...... **53 J8** 22 20S 40 22 E
Europa, Picos de, *Spain* . **19 A3** 43 10N 4 49W
Europa, Pta. de, *Gib.* **19 D3** 36 3N 5 21W
Europa Pt. = Europa, Pta.
de, *Gib.* **19 D3** 36 3N 5 21W
Europoort, *Neths.* ...... **15 C4** 51 57N 4 10 E
Eustis, *U.S.A.* **77 L5** 28 51N 81 41W
Eutsuk L., *Canada* **72 C3** 53 20N 126 45W
Eva Downs, *Australia* ... **62 B1** 18 1S 134 52 E
Evale, *Angola* .... **56 B2** 16 33S 15 44 E
Evans, *U.S.A.* **80 E2** 40 23N 104 41W
Evans Head, *Australia* ... **63 D5** 29 7S 153 27 E
Evans L., *Canada* **70 B4** 50 50N 77 0W
Evans Mills, *U.S.A.* **79 B9** 44 6N 75 48W
Evanston, *Ill., U.S.A.* **76 D2** 42 3N 87 41W
Evanston, *Wyo., U.S.A.* .. **82 F8** 41 16N 110 58W
Evansville, *Ind., U.S.A.* .. **76 F2** 37 58N 87 35W
Evansville, *Wis., U.S.A.* .. **80 D10** 42 47N 89 18W
Evaz, *Iran* .... **45 E7** 27 46N 53 59 E
Eveleth, *U.S.A.* **80 B8** 47 28N 92 32W
Evensk, *Russia* .... **27 C16** 62 12N 159 30 E
Everard, L., *Australia* .... **63 E1** 31 30S 135 0 E
Everard Park, *Australia* .. **61 E5** 27 1S 132 43 E
Everard Ras., *Australia* .. **61 E5** 27 5S 132 28 E
Everest, Mt., *Nepal* ..... **43 E12** 28 5N 86 58 E
Everett, *Pa., U.S.A.* **78 F6** 40 1N 78 23W
Everett, *Wash., U.S.A.* ... **84 C4** 47 59N 122 12W
Everglades, The, *U.S.A.* . **77 N5** 25 50N 81 0W
Everglades City, *U.S.A.* .. **77 N5** 25 52N 81 23W
Everglades National Park,
U.S.A. **77 N5** 25 30N 81 0W
Evergreen, *U.S.A.* **77 K2** 31 26N 86 57W
Everson, *U.S.A.* **82 B2** 48 57N 122 22W
Evesham, *U.K.* **11 E6** 52 6N 1 56W
Evinayong, *Eq. Guin.* .... **52 D2** 1 26N 10 35 E
Evje, *Norway* .... **9 G12** 58 36N 7 51 E
Évora, *Portugal* **19 C2** 38 33N 7 57W
Evowghlī, *Iran* .... **44 B5** 38 43N 45 13 E
Évreux, *France* .... **18 B4** 49 3N 1 8 E
Évros →, *Bulgaria* ..... **21 D12** 41 40N 26 34 E
Evry, *France* .... **18 B5** 48 38N 2 27 E
Évvoia, *Greece* .... **21 E11** 38 30N 24 0 E
Ewe, L., *U.K.* **12 D3** 57 49N 5 38W
Ewing, *U.S.A.* **80 D5** 42 16N 98 21W
Ewo, *Congo* .... **52 E2** 0 48S 14 45 E
Exaltación, *Bolivia* **92 F5** 13 10S 65 20W
Excelsior Springs, *U.S.A.* **80 F7** 39 20N 94 13W
Exe →, *U.K.* **11 G4** 50 41N 3 29W
Exeter, *Canada* **78 C3** 43 21N 81 29W
Exeter, *U.K.* **11 G4** 50 43N 3 31W
Exeter, *Calif., U.S.A.* **83 H4** 36 18N 119 9W
Exeter, *N.H., U.S.A.* **79 D14** 42 59N 70 57W
Exeter, *Nebr., U.S.A.* **80 E6** 40 39N 97 27W
Exmoor, *U.K.* **11 F4** 51 12N 3 45W
Exmouth, *Australia* ..... **60 D1** 21 54S 114 10 E
Exmouth, *U.K.* **11 G4** 50 37N 3 25W
Exmouth G., *Australia* ... **60 D1** 22 15S 114 15 E
Expedition Ra., *Australia* . **62 C4** 24 30S 149 12 E
Extremadura □, *Spain* ... **19 C2** 39 30N 6 5W
Exuma Sound, *Bahamas* . **88 B4** 24 30N 76 20W
Eyasi, L., *Tanzania* ..... **54 C4** 3 30S 35 0 E
Eyeberry L., *Canada* **73 A8** 63 8N 104 43W
Eyemouth, *U.K.* **12 F6** 55 52N 2 5W
Eyjafjörður, *Iceland* **8 C4** 66 15N 18 30W
Eyre, *Australia* .... **61 F4** 32 15S 126 18 E
Eyre (North), L., *Australia* **63 D2** 28 30S 137 20 E
Eyre (South), L., *Australia* **63 D2** 29 18S 137 25 E
Eyre Cr. →, *Australia* .. **63 D2** 26 40S 139 0 E
Eyre Mts., *N.Z.* .... **59 L2** 45 25S 168 25 E
Eyre Pen., *Australia* ..... **63 E2** 33 30S 136 17 E
Eysturoy, *Færoe Is.* **8 E9** 62 13N 6 54W
Eyvānkī, *Iran* .... **45 C6** 35 24N 51 56 E
Ezine, *Turkey* .... **21 E12** 39 48N 26 20 E
Ezouza →, *Cyprus* ..... **23 E11** 34 44N 32 27 E

F

F.Y.R.O.M. =
Macedonia ■, *Europe* . **21 D9** 41 53N 21 40 E
Fabens, *U.S.A.* **83 L10** 31 30N 106 10W
Fabriano, *Italy* .... **20 C5** 43 20N 12 54 E
Facatativá, *Colombia* ... **92 C4** 4 49N 74 22W
Fachi, *Niger* .... **50 E7** 18 6N 11 34 E
Fada, *Chad* .... **51 E9** 17 13N 21 34 E
Fada-n-Gourma,
Burkina Faso .... **50 F5** 12 10N 0 30 E
Faddeyevskiy, Ostrov,
Russia .... **27 B15** 76 0N 144 0 E
Fadghāmī, *Syria* .... **44 C4** 35 53N 40 52 E
Faenza, *Italy* .... **20 B4** 44 17N 11 53 E
Færoe Is. = Føroyar,
Atl. Oc. **8 F9** 62 0N 7 0W
Făgăras, *Romania* ...... **17 F13** 45 48N 24 58 E
Fagersta, *Sweden* ...... **9 F16** 60 1N 15 46 E
Fagnano, L., *Argentina* .. **96 G3** 54 30S 68 0W
Fahlīān, *Iran* .... **45 D6** 30 11N 51 28 E
Fahraj, *Kermān, Iran* .... **45 D8** 29 0N 59 0 E
Fahraj, *Yazd, Iran* ...... **45 D7** 31 46N 54 36 E
Faial, *Madeira* **22 D3** 32 47N 16 53W
Fair Hd., *U.K.* **13 A5** 55 14N 6 9W
Fair Oaks, *U.S.A.* **84 G5** 38 39N 121 16W
Fairbank, *U.S.A.* **83 L8** 31 43N 110 11W
Fairbanks, *U.S.A.* **68 B5** 64 51N 147 43W
Fairbury, *U.S.A.* **80 E6** 40 8N 97 11W
Fairfax, *U.S.A.* **81 G6** 36 34N 96 42W
Fairfield, *Ala., U.S.A.* **77 J2** 33 29N 86 55W
Fairfield, *Calif., U.S.A.* ... **84 G4** 38 15N 122 3W
Fairfield, *Conn., U.S.A.* .. **79 E11** 41 9N 73 16W
Fairfield, *Idaho, U.S.A.* .. **82 E6** 43 21N 114 44W
Fairfield, *Ill., U.S.A.* **76 F1** 38 23N 88 22W
Fairfield, *Iowa, U.S.A.* ... **80 E9** 40 56N 91 57W
Fairfield, *Mont., U.S.A.* .. **82 C8** 47 37N 111 59W
Fairfield, *Tex., U.S.A.* **81 K7** 31 44N 96 10W
Fairford, *Canada* **73 C9** 51 37N 98 38W
Fairhope, *U.S.A.* **77 K2** 30 31N 87 54W
Fairlie, *N.Z.* .... **59 L3** 44 5S 170 49 E
Fairmead, *U.S.A.* **84 H6** 37 5N 120 10W
Fairmont, *Minn., U.S.A.* .. **80 D7** 43 39N 94 28W
Fairmont, *W. Va., U.S.A.* . **76 F5** 39 29N 80 9W
Fairplay, *U.S.A.* **83 G11** 39 15N 106 2W
Fairport, *U.S.A.* **78 C7** 43 6N 77 27W
Fairport Harbor, *U.S.A.* .. **78 E3** 41 45N 81 17W
Fairview, *Australia* ..... **62 B3** 15 31S 144 17 E
Fairview, *Canada* **72 B5** 56 5N 118 25W
Fairview, *Mont., U.S.A.* .. **80 B2** 47 51N 104 3W
Fairview, *Okla., U.S.A.* ... **81 G5** 36 16N 98 29W
Fairview, *Utah, U.S.A.* ... **82 G8** 39 50N 111 0W
Fairweather, Mt., *U.S.A.* . **68 C6** 58 55N 137 32W
Faisalabad, *Pakistan* .... **42 D5** 31 30N 73 5 E
Faith, *U.S.A.* **80 C3** 45 2N 102 2W
Faizabad, *India* .... **43 F10** 26 45N 82 10 E
Fajardo, *Puerto Rico* **89 C6** 18 20N 65 39W
Fakfak, *Indonesia* ...... **37 E8** 3 0S 132 15 E
Faku, *China* .... **35 C12** 42 32N 123 21 E
Falaise, *France* **18 B3** 48 54N 0 12W
Falaise, Mui, *Vietnam* ... **38 C5** 19 6N 105 45 E
Falam, *Burma* .... **41 H18** 23 0N 93 45 E
Falcón, C., *Spain* ...... **22 C7** 38 50N 1 23 E
Falcon Dam, *U.S.A.* **81 M5** 26 50N 99 20W
Falconara Marittima, *Italy* **20 C5** 43 37N 13 24 E
Falcone, C., *Italy* .... **20 D3** 40 58N 8 12 E
Falconer, *U.S.A.* **78 D5** 42 7N 79 13W
Faleshty = Fălești,
Moldova .... **17 E14** 47 32N 27 44 E
Fălești, *Moldova* .... **17 E14** 47 32N 27 44 E
Falfurrias, *U.S.A.* **81 M5** 27 14N 98 9W
Falher, *Canada* **72 B5** 55 44N 117 15W
Faliraki, *Greece* .... **23 C10** 36 22N 28 12 E
Falkenberg, *Sweden* .... **9 H15** 56 54N 12 30 E
Falkirk, *U.K.* **12 F5** 56 0N 3 47W
Falkirk □, *U.K.* **12 F5** 55 58N 3 49W
Falkland Is. □, *Atl. Oc.* ... **96 G5** 51 30S 59 0W
Falkland Sd., *Falk. Is.* **96 G5** 52 0S 60 0W
Falköping, *Sweden* ..... **9 G15** 58 12N 13 33 E
Fall River, *U.S.A.* **79 E13** 41 43N 71 10W
Fall River Mills, *U.S.A.* .. **82 F3** 41 3N 121 26W
Fallbrook, *U.S.A.* **83 K5** 33 25N 117 12W
Fallbrook, *Calif., U.S.A.* .. **85 M9** 33 23N 117 15W
Fallon, *Mont., U.S.A.* **80 B2** 46 50N 105 8W
Fallon, *Nev., U.S.A.* **82 G4** 39 28N 118 47W
Falls City, *Nebr., U.S.A.* .. **80 E7** 40 3N 95 36W
Falls City, *Oreg., U.S.A.* .. **82 D2** 44 52N 123 26W
Falls Creek, *U.S.A.* **78 E6** 41 9N 78 48W
Falmouth, *Jamaica* **88 C4** 18 30N 77 40W
Falmouth, *U.K.* **11 G2** 50 9N 5 5W
Falmouth, *U.S.A.* **76 F3** 38 41N 84 20W
False B., *S. Africa* ...... **56 E2** 34 15S 18 40 E
Falso, C., *Honduras* **88 C3** 15 12N 83 21W
Falster, *Denmark* ...... **9 J14** 54 45N 11 55 E
Falsterbo, *Sweden* ..... **9 J15** 55 23N 12 50 E
Fălticeni, *Romania* ..... **17 E14** 47 21N 26 20 E
Falun, *Sweden* .... **9 F16** 60 37N 15 37 E
Famagusta, *Cyprus* .... **23 D12** 35 8N 33 55 E
Famagusta Bay, *Cyprus* . **23 D13** 35 15N 34 0 E
Famatina, Sierra de,
Argentina **94 B2** 27 30S 68 0W
Family L., *Canada* **73 C9** 51 54N 95 27W
Famoso, *U.S.A.* **85 K7** 35 37N 119 12W
Fan Xian, *China* .... **34 G8** 35 55N 115 38 E
Fandriana, *Madag.* ..... **57 C8** 20 14S 47 21 E
Fang, *Thailand* .... **38 C2** 19 55N 99 13 E
Fangcheng, *China* ...... **34 H7** 33 18N 112 59 E
Fangshan, *China* .... **34 E6** 38 3N 111 25 E
Fangzi, *China* .... **35 F10** 36 33N 119 10 E
Fanjiatun, *China* .... **35 C13** 43 40N 125 15 E
Fannich, L., *U.K.* **12 D4** 57 38N 4 59W
Fannūj, *Iran* .... **45 E8** 26 35N 59 38 E
Fanny Bay, *Canada* **72 D4** 49 37N 124 48W
Fanø, *Denmark* .... **9 J13** 55 25N 8 25 E
Fano, *Italy* .... **20 C5** 43 50N 13 1 E
Fanshaw, *U.S.A.* **72 B2** 57 11N 133 30W
Fanshi, *China* .... **34 E7** 39 12N 113 20 E
Fao = Al Fāw, *Iraq* ..... **45 D6** 30 0N 48 30 E
Faqirwali, *Pakistan* ..... **42 E5** 29 27N 73 0 E
Faradje, *Zaïre* .... **54 B2** 3 50N 29 45 E
Farafangana, *Madag.* .... **57 C8** 22 49S 47 50 E
Farāh, *Afghan.* .... **40 C3** 32 20N 62 7 E
Farāh □, *Afghan.* .... **40 C3** 32 25N 62 10 E
Farahalana, *Madag.* ..... **57 A9** 14 26S 50 10 E
Faranah, *Guinea* **50 F2** 10 3N 10 45W
Farasān, Jazā'ir, *Si. Arabia* **46 D3** 16 45N 41 55 E
Farasan Is. = Farasān,
Jazā'ir, *Si. Arabia* .... **46 D3** 16 45N 41 55 E
Faratsiho, *Madag.* ...... **57 B8** 19 24S 46 57 E
Fareham, *U.K.* **11 G6** 50 51N 1 11W
Farewell, C., *N.Z.* .... **59 J4** 40 29S 172 43 E
Farewell C. = Farvel, Kap,
Greenland **4 D5** 59 48N 43 55W
Farghona, *Uzbekistan* ... **26 E8** 40 23N 71 19 E
Fargo, *U.S.A.* **80 B6** 46 53N 96 48W
Fār'iah, W. al →,
West Bank .... **47 C4** 32 12N 35 27 E
Faribault, *U.S.A.* **80 C8** 44 18N 93 16W
Faridkot, *India* .... **42 D6** 30 44N 74 45 E
Faridpur, *Bangla.* ...... **43 H13** 23 15N 89 55 E
Farim, *Guinea-Biss.* **50 F1** 12 27N 15 9W
Farīmān, *Iran* .... **45 C8** 35 40N 59 49 E
Farina, *Australia* .... **63 E2** 30 3S 138 15 E
Fariones, Pta., *Canary Is.* **22 E6** 29 13N 13 28W
Farmerville, *U.S.A.* **81 J8** 32 47N 92 24W
Farmington, *Calif., U.S.A.* **84 H6** 37 55N 120 59W
Farmington, *N.H., U.S.A.* **79 C13** 43 24N 71 4W
Farmington, *N. Mex.,
U.S.A.* **83 H9** 36 44N 108 12W
Farmington, *Utah, U.S.A.* **82 F8** 41 0N 111 12W
Farmington →, *U.S.A.* .. **79 E12** 41 51N 72 38W
Farmville, *U.S.A.* **76 G6** 37 18N 78 24W
Farnborough, *U.K.* **11 F7** 51 16N 0 45W
Farne Is., *U.K.* **10 B6** 55 38N 1 37W
Farnham, *Canada* **79 A12** 45 17N 72 59W
Faro, *Brazil* **93 D7** 2 10S 56 39W
Faro, *Portugal* **19 D2** 37 2N 7 55W
Fårö, *Sweden* .... **9 H18** 57 55N 19 5 E
Farquhar, C., *Australia* .. **61 D1** 23 50S 113 36 E
Farrars Cr. →, *Australia* **62 D3** 25 35S 140 43 E
Farrāshband, *Iran* ...... **45 D7** 28 57N 52 5 E
Farrell, *U.S.A.* **78 E4** 41 13N 80 30W
Farrell Flat, *Australia* ... **63 E2** 33 48S 138 48 E
Farrokhī, *Iran* .... **45 C8** 33 50N 59 31 E
Farruch, C., *Spain* ...... **22 B10** 39 47N 3 21 E
Farrukhabad-cum-
Fatehgarh, *India* ..... **43 F8** 27 30N 79 32 E
Fārs □, *Iran* .... **45 D7** 29 30N 55 0 E
Fársala, *Greece* .... **21 E10** 39 17N 22 23 E
Farsund, *Norway* .... **9 G12** 58 5N 6 55 E

Fartak, Râs, *Si. Arabia* ... **44 D2** 28 5N 34 34 E
Fartura, Serra da, *Brazil* . **95 B5** 26 21S 52 52W
Fârûj, *Iran* .... **45 B8** 37 14N 58 14 E
Farvel, Kap, *Greenland* .. **4 D5** 59 48N 43 55W
Farwell, *U.S.A.* **81 H3** 34 23N 103 2W
Fasā, *Iran* .... **45 D7** 29 0N 53 39 E
Fasano, *Italy* .... **20 D7** 40 50N 17 22 E
Fastiv, *Ukraine* .... **17 C15** 50 7N 29 57 E
Fastnet Rock, *Ireland* ... **13 E2** 51 22N 9 37W
Fastov = Fastiv, *Ukraine* . **17 C15** 50 7N 29 57 E
Fatagar, Tanjung, *Indonesia* .... **37 E8** 2 46S 131 57 E
Fatehgarh, *India* .... **43 F8** 27 25N 79 35 E
Fatehpur, *Raj., India* .... **42 F6** 28 0N 74 40 E
Fatehpur, *Ut. P., India* ... **43 G9** 25 56N 81 13 E
Fatima, *Canada* **71 C7** 47 24N 61 53W
Faulkton, *U.S.A.* **80 C5** 45 2N 99 8W
Faure I., *Australia* .... **61 E1** 25 52S 113 50 E
Fauresmith, *S. Africa* .... **56 D4** 29 44S 25 17 E
Fauske, *Norway* .... **8 C16** 67 17N 15 25 E
Favara, *Italy* .... **20 F5** 37 19N 13 39 E
Favaritx, C., *Spain* ...... **22 A11** 40 0N 4 15 E
Favignana, *Italy* .... **20 F5** 37 56N 12 20 E
Favourable Lake, *Canada* **70 B1** 52 50N 93 39W
Fawn →, *Canada* **70 A2** 55 20N 87 35W
Fawnskin, *U.S.A.* **85 L10** 34 16N 116 56W
Faxaflói, *Iceland* **8 D2** 64 29N 23 0W
Faya-Largeau, *Chad* ..... **51 E8** 17 58N 19 6 E
Fayd, *Si. Arabia* .... **44 E4** 27 1N 42 52 E
Fayette, *Ala., U.S.A.* **77 J2** 33 41N 87 50W
Fayette, *Mo., U.S.A.* **80 F8** 39 9N 92 41W
Fayetteville, *Ark., U.S.A.* . **81 G7** 36 4N 94 10W
Fayetteville, *N.C., U.S.A.* . **77 H6** 35 3N 78 53W
Fayetteville, *Tenn., U.S.A.* **77 H2** 35 9N 86 34W
Fazilka, *India* .... **42 D6** 30 27N 74 2 E
Fazilpur, *Pakistan* .... **42 E4** 29 18N 70 29 E
Fdérik, *Mauritania* **50 D2** 22 40N 12 45W
Feale →, *Ireland* **13 D2** 52 27N 9 37W
Fear, C., *U.S.A.* **77 J7** 33 50N 77 58W
Feather →, *U.S.A.* **82 G3** 38 47N 121 36W
Feather Falls, *U.S.A.* **84 F5** 39 36N 121 16W
Featherston, *N.Z.* .... **59 J5** 41 6S 175 20 E
Featherstone, *Zimbabwe* **55 F3** 18 42S 30 55 E
Fécamp, *France* .... **18 B4** 49 45N 0 22 E
Federación, *Argentina* ... **94 C4** 31 0S 57 55W
Fedeshkūh, *Iran* .... **45 D7** 28 49N 53 50 E
Fehmarn, *Germany* ...... **16 A6** 54 27N 11 7 E
Fehmarn Bælt, *Europe* ... **9 J14** 54 35N 11 20 E
Fei Xian, *China* .... **35 G9** 35 18N 117 59 E
Feilding, *N.Z.* .... **59 J5** 40 13S 175 35 E
Feira de Santana, *Brazil* . **93 F11** 12 15S 38 57W
Feixiang, *China* .... **34 F8** 36 30N 114 45 E
Felanitx, *Spain* .... **22 B10** 39 28N 3 9 E
Feldkirch, *Austria* .... **16 E5** 47 15N 9 37 E
Felipe Carrillo Puerto, *Mexico* **87 D7** 19 38N 88 3W
Felixstowe, *U.K.* .... **11 F9** 51 58N 1 23 E
Felton, *U.K.* **10 B6** 55 18N 1 42W
Felton, *U.S.A.* **84 H4** 37 3N 122 4W
Femunden, *Norway* ...... **9 E14** 62 10N 11 53 E
Fen He →, *China* ...... **34 G6** 35 36N 110 42 E
Fenelon Falls, *Canada* **78 B6** 44 32N 78 45W
Feng Xian, *Jiangsu, China* **34 G9** 34 43N 116 35 E
Feng Xian, *Shaanxi, China* **34 H4** 33 54N 106 40 E
Fengcheng, *China* ...... **35 D13** 40 28N 124 5 E
Fengfeng, *China* .... **34 F8** 36 28N 114 8 E
Fengjie, *China* .... **33 C5** 31 5N 109 36 E
Fengning, *China* .... **34 D9** 41 10N 116 33 E
Fengqiu, *China* .... **34 G8** 35 2N 114 25 E
Fengrun, *China* .... **35 E10** 39 48N 118 8 E
Fengtai, *China* .... **34 E9** 39 50N 116 18 E
Fengxiang, *China* .... **34 G4** 34 29N 107 25 E
Fengyang, *China* .... **35 H9** 32 51N 117 29 E
Fengzhen, *China* .... **34 D7** 40 25N 113 2 E
Fenit, *Ireland* **13 D2** 52 17N 9 51W
Fennimore, *U.S.A.* **80 D9** 42 59N 90 39W
Fenoarivo Afovoany, *Madag.* .... **57 B8** 18 26S 46 34 E
Fenoarivo Atsinanana, *Madag.* .... **57 B8** 17 22S 49 25 E
Fens, The, *U.K.* **10 E8** 52 38N 0 2W
Fenton, *U.S.A.* **76 D4** 42 48N 83 42W
Fenxi, *China* .... **34 F6** 36 40N 111 31 E
Fenyang, *China* .... **34 F6** 37 18N 111 48 E
Feodosiya, *Ukraine* ...... **25 E6** 45 2N 35 16 E
Ferdows, *Iran* .... **45 C8** 33 58N 58 2 E
Ferfer, *Somali Rep.* ...... **46 F4** 5 4N 45 9 E
Fergana = Farghona, *Uzbekistan* .... **26 E8** 40 23N 71 19 E
Fergus, *Canada* **70 D3** 43 43N 80 24W
Fergus Falls, *U.S.A.* **80 B6** 46 17N 96 4W
Ferland, *Canada* **70 B2** 50 19N 88 27W
Fermanagh □, *U.K.* **13 B4** 54 21N 7 40W
Fermo, *Italy* .... **20 C5** 43 9N 13 43 E
Fermoy, *Ireland* **13 D3** 52 9N 8 16W
Fernández, *Argentina* **94 B3** 27 55S 63 50W
Fernandina Beach, *U.S.A.* **77 K5** 30 40N 81 27W
Fernando de Noronha, *Brazil* **93 D12** 4 0S 33 10W
Fernando Póo = Bioko, *Eq. Guin.* .... **50 H6** 3 30N 8 40 E
Ferndale, *Calif., U.S.A.* ... **82 F1** 40 35N 124 16W
Ferndale, *Wash., U.S.A.* ... **84 B4** 48 51N 122 36W
Fernie, *Canada* **72 D5** 49 30N 115 5W
Fernlees, *Australia* ...... **62 C4** 23 51S 148 7 E
Fernley, *U.S.A.* **82 G4** 39 36N 119 15W
Ferozepore = Firozpur, *India* .... **42 D6** 30 55N 74 40 E
Ferrara, *Italy* .... **20 B4** 44 50N 11 35 E
Ferreñafe, *Peru* **92 E3** 6 42S 79 50W
Ferrerías, *Spain* .... **22 B11** 39 59N 4 1 E
Ferret, C., *France* **18 D3** 44 38N 1 15W
Ferriday, *U.S.A.* **81 K9** 31 38N 91 33W
Ferrol = El Ferrol, *Spain* . **19 A1** 43 29N 8 15W
Ferron, *U.S.A.* **83 G8** 39 5N 111 8W
Ferryland, *Canada* **71 C9** 47 2N 52 53W
Fertile, *U.S.A.* **80 B6** 47 32N 96 17W
Fès, *Morocco* **50 B4** 34 0N 5 0W
Feshi, *Zaïre* .... **52 F3** 6 8S 18 10 E
Fessenden, *U.S.A.* **80 B5** 47 39N 99 38W
Feteşti, *Romania* .... **17 F14** 44 22N 27 51 E
Fetlar, *U.K.* **12 A8** 60 36N 0 52W
Feuilles →, *Canada* **69 C12** 58 47N 70 4W
Fezzan, *Libya* .... **51 C8** 27 0N 15 0 E
Ffestiniog, *U.K.* **10 E4** 52 57N 3 55W
Fiambalá, *Argentina* **94 B2** 27 45S 67 37W
Fianarantsoa, *Madag.* ... **57 C8** 21 26S 47 5 E
Fianarantsoa □, *Madag.* . **57 B8** 19 30S 47 0 E
Fianga, *Cameroon* ...... **51 G8** 9 55N 15 9 E
Ficksburg, *S. Africa* ..... **57 D4** 28 51S 27 53 E
Field, *Canada* **70 C3** 46 31N 80 1W
Field →, *Australia* ..... **62 C2** 23 48S 138 0 E
Field I., *Australia* .... **60 B5** 12 5S 132 23 E
Fieri, *Albania* .... **21 D8** 40 43N 19 33 E
Fife □, *U.K.* **12 E5** 56 16N 3 1W
Fife Ness, *U.K.* **12 E6** 56 17N 2 35W
Figeac, *France* .... **18 D5** 44 37N 2 2 E
Figtree, *Zimbabwe* ...... **55 G2** 20 22S 28 20 E
Figueira da Foz, *Portugal* **19 B1** 40 7N 8 54W
Figueras, *Spain* .... **19 A7** 42 18N 2 58 E
Figuig, *Morocco* **50 B4** 32 5N 1 11W
Fihaonana, *Madag.* ...... **57 B8** 18 36S 47 12 E
Fiherenana, *Madag.* ..... **57 B8** 18 29S 48 24 E
Fiherenana →, *Madag.* . **57 C7** 23 19S 43 37 E
Fiji ■, *Pac. Oc.* .... **59 C8** 17 20S 179 0 E
Filer, *U.S.A.* **82 E6** 42 34N 114 37W
Filey, *U.K.* **10 C7** 54 12N 0 18W
Filfla, *Malta* .... **23 D1** 35 47N 14 24 E
Filiatrá, *Greece* .... **21 F9** 37 9N 21 35 E
Filipstad, *Sweden* .... **9 G16** 59 43N 14 9 E
Fillmore, *Canada* **73 D8** 49 50N 103 25W
Fillmore, *Calif., U.S.A.* ... **85 L8** 34 24N 118 55W
Fillmore, *Utah, U.S.A.* ... **83 G7** 38 58N 112 20W
Finch, *Canada* **79 A9** 45 11N 75 7W
Findhorn →, *U.K.* **12 D5** 57 38N 3 38W
Findlay, *U.S.A.* **76 E4** 41 2N 83 39W
Finger L., *Canada* **73 C10** 53 33N 93 30W
Fingõe, *Mozam.* .... **55 E3** 14 55S 31 50 E
Finisterre, C., *Spain* **19 A1** 42 50N 9 19W
Finke, *Australia* .... **62 D1** 25 34S 134 35 E
Finke →, *Australia* ..... **63 D2** 27 0S 136 10 E
Finland ■, *Europe* ...... **8 E22** 63 0N 27 0 E
Finland, G. of, *Europe* ... **9 G21** 60 0N 26 0 E
Finlay →, *Canada* **72 B3** 57 0N 125 10W
Finley, *Australia* .... **63 F4** 35 38S 145 35 E
Finley, *U.S.A.* **80 B6** 47 31N 97 50W
Finn →, *Ireland* **13 B4** 54 51N 7 28W
Finnigan, Mt., *Australia* .. **62 B4** 15 49S 145 17 E
Finniss, C., *Australia* .... **63 E1** 33 8S 134 51 E
Finnmark, *Norway* ...... **8 B20** 69 37N 23 57 E
Finnsnes, *Norway* ...... **8 B18** 69 14N 18 0 E
Finspång, *Sweden* ...... **9 G16** 58 43N 15 47 E
Fiora →, *Italy* .... **20 C4** 42 20N 11 34 E
Fiq, *Syria* .... **47 C4** 32 46N 35 41 E
Firat = Furāt, Nahr al →, *Asia* .... **44 D5** 31 0N 47 25 E
Fire River, *Canada* **70 C3** 48 47N 83 21W
Firebag →, *Canada* **73 B6** 57 45N 111 21W
Firebaugh, *U.S.A.* **84 J6** 36 52N 120 27W
Firedrake L., *Canada* **73 A8** 61 25N 104 30W
Firenze, *Italy* .... **20 C4** 43 46N 11 15 E
Firk →, *Iraq* .... **44 D5** 30 59N 44 34 E
Firozabad, *India* .... **43 F8** 27 10N 78 25 E
Firozpur, *India* .... **42 D6** 30 55N 74 40 E
Fīrūzābād, *Iran* .... **45 D7** 28 52N 52 35 E
Fīrūzkūh, *Iran* .... **45 C7** 35 50N 52 50 E
Firvale, *Canada* **72 C3** 52 27N 126 13W
Fish →, *Namibia* ...... **56 D2** 28 7S 17 10 E
Fish →, *S. Africa* ..... **56 E3** 31 30S 20 16 E
Fisher, *Australia* .... **61 F5** 30 30S 131 0 E
Fisher B., *Canada* **73 C9** 51 35N 97 13W
Fishguard, *U.K.* **11 F3** 52 0N 5 0W
Fishing L., *Canada* **73 C9** 52 10N 95 24W
Fitchburg, *U.S.A.* **79 D13** 42 35N 71 48W
Fitri, L., *Chad* .... **51 F8** 12 50N 17 28 E
Fitz Roy, *Argentina* **96 F3** 47 0S 67 0W
Fitzgerald, *Canada* **72 B6** 59 51N 111 36W
Fitzgerald, *U.S.A.* **77 K4** 31 43N 83 15W
Fitzmaurice →, *Australia* **60 B5** 14 45S 130 5 E
Fitzroy →, *Queens., Australia* .... **62 C5** 23 32S 150 52 E
Fitzroy →, *W. Austral., Australia* .... **60 C3** 17 31S 123 35 E
Fitzroy Crossing, *Australia* **60 C4** 18 9S 125 38 E
Fitzwilliam I., *Canada* **78 A3** 45 30N 81 45W
Fiume = Rijeka, *Croatia* . **16 F8** 45 20N 14 21 E
Five Points, *U.S.A.* **84 J6** 36 26N 120 6W
Fizi, *Zaïre* .... **54 C2** 4 17S 28 55 E
Flagler, *U.S.A.* **80 F3** 39 18N 103 4W
Flagstaff, *U.S.A.* **83 J8** 35 12N 111 39W
Flaherty I., *Canada* **70 A4** 56 15N 79 15W
Flåm, *Norway* .... **9 F12** 60 50N 7 7 E
Flambeau →, *U.S.A.* ... **80 C9** 45 18N 91 14W
Flamborough Hd., *U.K.* .. **10 C7** 54 7N 0 5W
Flaming Gorge Dam, *U.S.A.* **82 F9** 40 55N 109 25W
Flaming Gorge Reservoir, *U.S.A.* **82 F9** 41 10N 109 25W
Flamingo, Teluk, *Indonesia* **37 F9** 5 30S 138 0 E
Flanders = West-Vlaanderen □, *Belgium* **15 D3** 51 0N 3 0 E
Flandre, *Europe* .... **16 C2** 51 0N 3 0 E
Flandre Occidentale = West-Vlaanderen □, *Belgium* .... **15 D3** 51 0N 3 0 E
Flandre Orientale = Oost-Vlaanderen □, *Belgium* **15 C3** 51 5N 3 50 E
Flandreau, *U.S.A.* **80 C6** 44 3N 96 36W
Flanigan, *U.S.A.* **84 E7** 40 10N 119 53W
Flannan Is., *U.K.* **12 C1** 58 9N 7 52W
Flåsjön, *Sweden* .... **8 D16** 64 5N 15 40 E
Flat →, *Canada* **72 A3** 61 33N 125 18W
Flat River, *U.S.A.* **81 G9** 37 51N 90 31W
Flathead L., *U.S.A.* **82 C6** 47 51N 114 8W
Flattery, C., *Australia* ... **62 A4** 14 58S 145 21 E
Flattery, C., *U.S.A.* **84 B2** 48 23N 124 29W
Flaxton, *U.S.A.* **80 A3** 48 54N 102 24W
Fleetwood, *U.K.* **10 D4** 53 55N 3 1W
Flekkefjord, *Norway* ..... **9 G12** 58 18N 6 39 E
Flemington, *U.S.A.* **78 E7** 41 7N 77 28W
Flensburg, *Germany* ..... **16 A5** 54 47N 9 27 E
Flers, *France* **18 B3** 48 47N 0 33W
Flesherton, *Canada* **78 B4** 44 16N 80 33W
Flesko, Tanjung, *Indonesia* **37 D6** 0 29N 124 30 E
Flevoland □, *Neths.* ..... **15 B5** 52 30N 5 30 E
Flin Flon, *Canada* **73 C8** 54 46N 101 53W
Flinders →, *Australia* .. **62 B3** 17 36S 140 36 E
Flinders B., *Australia* .... **61 F2** 34 19S 115 19 E
Flinders Group, *Australia* **62 A3** 14 11S 144 15 E
Flinders I., *Australia* ..... **62 F4** 40 0S 148 0 E
Flinders Ranges, *Australia* **63 E2** 31 30S 138 30 E
Flinders Reefs, *Australia* . **62 B4** 17 37S 148 31 E
Flint, *U.K.* **10 D4** 53 15N 3 8W
Flint, *U.S.A.* **76 D4** 43 1N 83 41W
Flint →, *U.S.A.* **77 K3** 30 57N 84 34W
Flint I., *Kiribati* **65 J12** 11 26S 151 48W
Flinton, *Australia* .... **63 D4** 27 55S 149 32 E
Flintshire □, *U.K.* **10 D4** 53 17N 3 17W
Flodden, *U.K.* **10 B5** 55 37N 2 8W
Floodwood, *U.S.A.* **80 B8** 46 55N 92 55W
Flora, *U.S.A.* **76 F1** 38 40N 88 29W
Floral a, *U.S.A.* **77 K2** 31 0N 86 20W
Florence = Firenze, *Italy* . **20 C4** 43 46N 11 15 E
Florence, *Ala., U.S.A.* ... **77 H2** 34 48N 87 41W
Florence, *Ariz., U.S.A.* ... **83 K8** 33 2N 111 23W
Florence, *Colo., U.S.A.* ... **80 F2** 38 23N 105 8W
Florence, *Oreg., U.S.A.* .. **82 E1** 43 58N 124 7W
Florence, *S.C., U.S.A.* ... **77 H6** 34 12N 79 46W
Florence, L., *Australia* ... **63 D2** 28 53S 138 9 E
Florennes, *Belgium* ..... **15 D4** 50 15N 4 35 E
Florenville, *Belgium* ..... **15 E5** 49 40N 5 19 E
Flores, *Guatemala* **88 C2** 16 59N 89 50W
Flores, *Indonesia* .... **37 F6** 8 35S 121 0 E
Flores I., *Canada* **72 D3** 49 20N 126 10W
Flores Sea, *Indonesia* ... **37 F6** 6 30S 120 0 E
Floreşti, *Moldova* .... **17 E15** 47 53N 28 17 E
Floresville, *U.S.A.* **81 L5** 29 8N 98 10W
Floriano, *Brazil* **93 E10** 6 50S 43 0W
Florianópolis, *Brazil* **95 B6** 27 30S 48 30W
Florida, *Cuba* **88 B4** 21 32N 78 14W
Florida, *Uruguay* **95 C4** 34 7S 56 10W
Florida □, *U.S.A.* **77 L5** 28 0N 82 0W
Florida, Straits of, *U.S.A.* **88 B3** 25 0N 80 0W
Florida B., *U.S.A.* **88 A3** 25 0N 80 45W
Florida Keys, *U.S.A.* **75 F10** 24 40N 81 0W
Flórina, *Greece* .... **21 D9** 40 48N 21 26 E
Florø, *Norway* .... **9 F11** 61 35N 5 1 E
Flower Station, *Canada* .. **79 A8** 45 10N 76 41W
Flower's Cove, *Canada* .. **71 B8** 51 14N 56 46W
Floydada, *U.S.A.* **81 J4** 33 59N 101 20W
Fluk, *Indonesia* .... **37 E7** 1 42S 127 44 E
Flushing = Vlissingen, *Neths.* .... **15 C3** 51 26N 3 34 E
Flying Fish, C., *Antarctica* **5 D15** 72 6S 102 29W
Foam Lake, *Canada* **73 C8** 51 40N 103 32W
Foça, *Turkey* .... **21 E12** 38 39N 26 46 E
Focşani, *Romania* ...... **17 F14** 45 41N 27 15 E
Fóggia, *Italy* .... **20 D6** 41 27N 15 34 E
Fogo, *Canada* **71 C9** 49 43N 54 17W
Fogo I., *Canada* **71 C9** 49 40N 54 5W
Föhr, *Germany* .... **16 A5** 54 43N 8 30 E
Foix, *France* .... **18 E4** 42 58N 1 38 E
Folda, *Nord-Trøndelag, Norway* .... **8 D14** 64 32N 10 30 E
Folda, *Nordland, Norway* . **8 C16** 67 38N 14 50 E
Foleyet, *Canada* **70 C3** 48 15N 82 25W
Folgefonni, *Norway* ..... **9 F12** 60 3N 6 23 E
Foligno, *Italy* .... **20 C5** 42 57N 12 42 E
Folkestone, *U.K.* .... **11 F9** 51 5N 1 12 E
Folkston, *U.S.A.* **77 K5** 30 50N 82 0W
Follett, *U.S.A.* **81 G4** 36 26N 100 8W
Folsom Res., *U.S.A.* **84 G5** 38 42N 121 9W
Fond-du-Lac, *Canada* ... **73 B7** 59 19N 107 12W
Fond du Lac, *U.S.A.* **80 D10** 43 47N 88 27W
Fond-du-Lac →, *Canada* **73 B7** 59 17N 106 0W
Fonda, *U.S.A.* **79 D10** 42 57N 74 22W
Fondi, *Italy* .... **20 D5** 41 21N 13 25 E
Fonsagrada, *Spain* **19 A2** 43 8N 7 4W
Fonseca, G. de, *Cent. Amer.* **88 D2** 13 10N 87 40W
Fontainebleau, *France* ... **18 B5** 48 24N 2 40 E
Fontana, *U.S.A.* **85 L9** 34 6N 117 26W
Fontas →, *Canada* **72 B4** 58 14N 121 48W
Fonte Boa, *Brazil* **92 D5** 2 33S 66 0W
Fontenay-le-Comte, *France* **18 C3** 46 28N 0 48W
Fontur, *Iceland* **8 C6** 66 23N 14 32W
Foochow = Fuzhou, *China* **33 D6** 26 5N 119 16 E
Foping, *China* .... **34 H4** 33 41N 108 0 E
Forbes, *Australia* .... **63 E4** 33 22S 148 0 E
Forbesganj, *India* .... **43 F12** 26 17N 87 18 E
Ford City, *Calif., U.S.A.* .. **85 K7** 35 9N 119 27W
Ford City, *Pa., U.S.A.* **78 F5** 40 46N 79 32W
Førde, *Norway* .... **9 F11** 61 27N 5 53 E
Ford's Bridge, *Australia* .. **63 D4** 29 41S 145 29 E
Fordyce, *U.S.A.* **81 J8** 33 49N 92 25W
Forécariah, *Guinea* **50 G2** 9 28N 13 10W
Forel, Mt., *Greenland* **4 C6** 66 52N 36 55W
Foremost, *Canada* **72 D6** 49 26N 111 34W
Forest, *Canada* **78 C3** 43 6N 82 0W
Forest, *U.S.A.* **81 J10** 32 22N 89 29W
Forest City, *Iowa, U.S.A.* . **80 D8** 43 16N 93 39W
Forest City, *N.C., U.S.A.* . **77 H5** 35 20N 81 52W
Forest City, *Pa., U.S.A.* ... **79 E9** 41 39N 75 28W
Forest Grove, *U.S.A.* **84 E3** 45 31N 123 7W
Forestburg, *Canada* **72 C6** 52 35N 112 1W
Foresthill, *U.S.A.* **84 F6** 39 1N 120 49W
Forestier Pen., *Australia* . **62 G4** 43 0S 148 0 E
Forestville, *Canada* **71 C6** 48 48N 69 2W
Forestville, *Calif., U.S.A.* . **84 G4** 38 28N 122 54W
Forestville, *Wis., U.S.A.* .. **76 C2** 44 41N 87 29W
Forfar, *U.K.* **12 E6** 56 39N 2 53W
Forks, *U.S.A.* **84 C2** 47 57N 124 23W
Forlì, *Italy* .... **20 B5** 44 13N 12 3 E
Forman, *U.S.A.* **80 B6** 46 7N 97 38W
Formby Pt., *U.K.* **10 D4** 53 33N 3 6W
Formentera, *Spain* ...... **22 C7** 38 43N 1 27 E
Formentor, C. de, *Spain* . **22 B10** 39 58N 3 13 E
Former Yugoslav Republic of Macedonia = Macedonia ■, *Europe* . **21 D9** 41 53N 21 40 E
Fórmia, *Italy* .... **20 D5** 41 15N 13 37 E
Formosa = Taiwan ■, *Asia* .... **33 D7** 23 30N 121 0 E
Formosa, *Argentina* **94 B4** 26 15S 58 10W
Formosa □, *Argentina* ... **94 B3** 25 0S 60 0W
Formosa, Serra, *Brazil* ... **93 F8** 12 0S 55 0W
Formosa Bay, *Kenya* ..... **54 C5** 2 40S 40 20 E
Fornells, *Spain* .... **22 A11** 40 3N 4 7 E
Føroyar, *Atl. Oc.* **8 F9** 62 0N 7 0W
Forres, *U.K.* **12 D5** 57 37N 3 37W
Forrest, *Vic., Australia* ... **63 F3** 38 33S 143 47 E
Forrest, *W. Austral., Australia* .... **61 F4** 30 51S 128 6 E
Forrest, Mt., *Australia* ... **61 D4** 24 48S 127 45 E
Forrest City, *U.S.A.* **81 H9** 35 1N 90 47W
Forsayth, *Australia* ...... **62 B3** 18 33S 143 34 E
Forssa, *Finland* .... **9 F20** 60 49N 23 38 E
Forst, *Germany* .... **16 C8** 51 45N 14 37 E
Forster, *Australia* .... **63 E5** 32 12S 152 31 E
Forsyth, *Ga., U.S.A.* **77 J4** 33 2N 83 56W
Forsyth, *Mont., U.S.A.* ... **82 C10** 46 16N 106 41W
Fort Albany, *Canada* **70 B3** 52 15N 81 35W
Fort Apache, *U.S.A.* **83 K9** 33 50N 110 0W
Fort Assiniboine, *Canada* **72 C6** 54 20N 114 45W
Fort Augustus, *U.K.* **12 D4** 57 9N 4 42W
Fort Beaufort, *S. Africa* .. **56 E4** 32 46S 26 40 E
Fort Benton, *U.S.A.* **82 C8** 47 49N 110 40W
Fort Bragg, *U.S.A.* **82 G2** 39 26N 123 48W
Fort Bridger, *U.S.A.* **82 F8** 41 19N 110 23W
Fort Chipewyan, *Canada* . **73 B6** 58 42N 111 8W
Fort Collins, *U.S.A.* **80 E2** 40 35N 105 5W
Fort-Coulonge, *Canada* .. **70 C4** 45 50N 76 45W
Fort Davis, *U.S.A.* **81 K3** 30 35N 103 54W
Fort-de-France, *Martinique* **89 D7** 14 36N 61 2W
Fort de Possel = Possel, *C.A.R.* .... **52 C3** 5 5N 19 10 E
Fort Defiance, *U.S.A.* ... **83 J9** 35 45N 109 5W
Fort Dodge, *U.S.A.* **80 D7** 42 30N 94 11W
Fort Edward, *U.S.A.* **79 C11** 43 16N 73 35W
Fort Frances, *Canada* **73 D10** 48 36N 93 24W
Fort Garland, *U.S.A.* **83 H11** 37 26N 105 26W
Fort George = Chisasibi, *Canada* **70 B4** 53 50N 79 0W
Fort Good-Hope, *Canada* **68 B7** 66 14N 128 40W
Fort Hancock, *U.S.A.* **83 L11** 31 18N 105 51W
Fort Hertz = Putao, *Burma* **41 F20** 27 28N 97 30 E
Fort Hope, *Canada* **70 B2** 51 30N 88 0W
Fort Irwin, *U.S.A.* **85 K10** 35 16N 116 34W
Fort Jameson = Chipata, *Zambia* .... **55 E3** 13 38S 32 28 E
Fort Kent, *U.S.A.* **71 C6** 47 15N 68 36W
Fort Klamath, *U.S.A.* **82 E3** 42 42N 122 0W
Fort-Lamy = Ndjamena, *Chad* .... **51 F7** 12 10N 14 59 E
Fort Laramie, *U.S.A.* **80 D2** 42 13N 104 31W
Fort Lauderdale, *U.S.A.* .. **77 M5** 26 7N 80 8W
Fort Liard, *Canada* **72 A4** 60 14N 123 30W
Fort Liberté, *Haiti* **89 C5** 19 42N 71 51W
Fort Lupton, *U.S.A.* **80 E2** 40 5N 104 49W
Fort Mackay, *Canada* **72 B6** 57 12N 111 41W
Fort McKenzie, *Canada* .. **71 A6** 57 20N 69 0W
Fort Macleod, *Canada* ... **72 D6** 49 45N 113 30W
Fort MacMahon, *Algeria* . **50 C5** 29 43N 1 45 E
Fort McMurray, *Canada* .. **72 B6** 56 44N 111 7W
Fort McPherson, *Canada* . **68 B6** 67 30N 134 55W
Fort Madison, *U.S.A.* **80 E9** 40 38N 91 27W
Fort Meade, *U.S.A.* **77 M5** 27 45N 81 48W
Fort Miribel, *Algeria* ..... **50 C5** 29 25N 2 55 E
Fort Morgan, *U.S.A.* **80 E3** 40 15N 103 48W
Fort Myers, *U.S.A.* **77 M5** 26 39N 81 52W
Fort Nelson, *Canada* **72 B4** 58 50N 122 44W
Fort Nelson →, *Canada* **72 B4** 59 32N 124 0W
Fort Norman = Tulita, *Canada* **68 B7** 64 57N 125 30W
Fort Payne, *U.S.A.* **77 H3** 34 26N 85 43W
Fort Peck, *U.S.A.* **82 B10** 48 1N 106 27W
Fort Peck Dam, *U.S.A.* ... **82 C10** 48 0N 106 26W
Fort Peck L., *U.S.A.* **82 C10** 48 0N 106 26W
Fort Pierce, *U.S.A.* **77 M5** 27 27N 80 20W
Fort Pierre, *U.S.A.* **80 C4** 44 21N 100 22W
Fort Plain, *U.S.A.* **79 D10** 42 56N 74 37W
Fort Portal, *Uganda* ..... **54 B3** 0 40N 30 20 E
Fort Providence, *Canada* . **72 A5** 61 3N 117 40W
Fort Qu'Appelle, *Canada* . **73 C8** 50 45N 103 50W
Fort Resolution, *Canada* . **72 A6** 61 10N 113 40W
Fort Rixon, *Zimbabwe* ... **55 G2** 20 2S 29 17 E
Fort Rosebery = Mansa, *Zambia* .... **55 E2** 11 13S 28 55 E
Fort Ross, *U.S.A.* **84 G3** 38 32N 123 13W
Fort Rupert = Waskaganish, *Canada* .. **70 B4** 51 30N 78 40W
Fort St. James, *Canada* .. **72 C4** 54 30N 124 10W
Fort St. John, *Canada* .. **72 B4** 56 15N 120 50W
Fort Sandeman, *Pakistan* **42 D3** 31 20N 69 31 E
Fort Saskatchewan, *Canada* **72 C6** 53 40N 113 15W
Fort Scott, *U.S.A.* **81 G7** 37 50N 94 42W
Fort Severn, *Canada* **70 A2** 56 0N 87 40W
Fort Shevchenko, *Kazakstan* .... **25 F9** 44 35N 50 23 E
Fort Simpson, *Canada* ... **72 A4** 61 45N 121 15W
Fort Smith, *Canada* **72 B6** 60 0N 111 51W
Fort Smith, *U.S.A.* **81 H7** 35 23N 94 25W
Fort Stanton, *U.S.A.* **83 K11** 33 30N 105 31W
Fort Stockton, *U.S.A.* ... **81 K3** 30 53N 102 53W
Fort Sumner, *U.S.A.* **81 H2** 34 28N 104 15W
Fort Trinquet = Bir Mogrein, *Mauritania* .. **50 C2** 25 10N 11 25W
Fort Valley, *U.S.A.* **77 J4** 32 33N 83 53W
Fort Vermilion, *Canada* .. **72 B5** 58 24N 116 0W
Fort Walton Beach, *U.S.A.* **77 K2** 30 25N 86 36W
Fort Wayne, *U.S.A.* **76 E3** 41 4N 85 9W
Fort William, *U.K.* **12 E3** 56 49N 5 7W
Fort Worth, *U.S.A.* **81 J6** 32 45N 97 18W
Fort Yates, *U.S.A.* **80 B4** 46 5N 100 38W
Fort Yukon, *U.S.A.* **68 B5** 66 34N 145 16W
Fortaleza, *Brazil* **93 D11** 3 45S 38 35W
Forteau, *Canada* **71 B8** 51 28N 56 58W
Forth →, *U.K.* **12 E5** 56 9N 3 50W
Forth, Firth of, *U.K.* **12 E6** 56 5N 2 55W
Fortrose, *U.K.* **12 D4** 57 35N 4 9W
Fortuna, *Calif., U.S.A.* ... **82 F1** 40 36N 124 9W
Fortuna, *N. Dak., U.S.A.* . **80 A3** 48 55N 103 47W
Fortune B., *Canada* **71 C8** 47 30N 55 22W
Foshan, *China* .... **33 D6** 23 4N 113 5 E
Fosna, *Norway* .... **8 E14** 63 50N 10 20 E
Fosnavåg, *Norway* ...... **9 E11** 62 22N 5 38 E
Fossano, *Italy* .... **20 B2** 44 33N 7 43 E
Fossil, *U.S.A.* **82 D3** 45 0N 120 9W
Fossilbrook, *Australia* ... **62 B3** 17 47S 144 29 E
Fosston, *U.S.A.* **80 B7** 47 35N 95 45W
Foster, *Canada* **79 A12** 45 17N 72 30W
Foster →, *Canada* **73 B7** 55 47N 105 49W
Fosters Ra., *Australia* ... **62 C1** 21 35S 133 48 E
Fostoria, *U.S.A.* **76 E4** 41 10N 83 25W

Fougamou, *Gabon* **52 E2** 1 16S 10 30 E
Fougères, *France* **18 B3** 48 21N 1 14W
Foul Pt., *Sri Lanka* **40 Q12** 8 35N 81 18 E
Foula, *U.K.* **12 A6** 60 10N 2 5W
Foulness I., *U.K.* **11 F8** 51 36N 0 55 E
Foulpointe, *Madag.* **57 B8** 17 41S 49 31 E
Foumban, *Cameroon* **50 G7** 5 45N 10 50 E
Fountain, *Colo., U.S.A.* **80 F2** 38 41N 104 42W
Fountain, *Utah, U.S.A.* **82 G8** 39 41N 111 37W
Fountain Springs, *U.S.A.* **85 K8** 35 54N 118 51W
Fourchu, *Canada* **71 C7** 45 43N 60 17W
Fouriesburg, *S. Africa* **56 D4** 28 38S 28 14 E
Foúrnoi, *Greece* **21 F12** 37 36N 26 32 E
Fouta Djalon, *Guinea* **50 F2** 11 20N 12 10W
Foux, Cap-à-, *Haiti* **89 C5** 19 43N 73 27W
Foveaux Str., *N.Z.* **59 M2** 46 42S 168 10 E
Fowey, *U.K.* **11 G3** 50 20N 4 39W
Fowler, *Calif., U.S.A.* **83 H4** 36 38N 119 41W
Fowler, *Colo., U.S.A.* **80 F2** 38 8N 104 2W
Fowler, *Kans., U.S.A.* **81 G4** 37 23N 100 12W
Fowlers B., *Australia* **61 F5** 31 59S 132 34 E
Fowlerton, *U.S.A.* **81 L5** 28 28N 98 48W
Fox →, *Canada* **73 B10** 56 3N 93 18W
Fox Valley, *Canada* **73 C7** 50 30N 109 25W
Foxe Basin, *Canada* **69 B12** 66 0N 77 0W
Foxe Chan., *Canada* **69 B11** 65 0N 80 0W
Foxe Pen., *Canada* **69 B12** 65 0N 76 0W
Foxpark, *U.S.A.* **82 F10** 41 5N 106 9W
Foxton, *N.Z.* **59 J5** 40 29S 175 18 E
Foyle, Lough, *U.K.* **13 A4** 55 7N 7 4W
Foynes, *Ireland* **13 D2** 52 37N 9 7W
Fóz do Cunene, *Angola* **56 B1** 17 15S 11 48 E
Foz do Gregório, *Brazil* **92 E4** 6 47S 70 44W
Foz do Iguaçu, *Brazil* **95 B5** 25 30S 54 30W
Frackville, *U.S.A.* **79 F8** 40 47N 76 14W
Framingham, *U.S.A.* **79 D13** 42 17N 71 25W
Franca, *Brazil* **93 H9** 20 33S 47 30W
Francavilla Fontana, *Italy* **21 D7** 40 32N 17 35 E
France ■, *Europe* **18 C5** 47 0N 3 0 E
Frances, *Australia* **63 F3** 36 41S 140 55 E
Frances →, *Canada* **72 A3** 60 16N 129 10W
Frances L., *Canada* **72 A3** 61 23N 129 30W
Francés Viejo, C., *Dom. Rep.* **89 C6** 19 40N 69 55W
Franceville, *Gabon* **52 E2** 1 40S 13 32 E
Franche-Comté, *France* **18 C6** 46 50N 5 55 E
Francisco I. Madero, *Coahuila, Mexico* **86 B4** 25 48N 103 18W
Francisco I. Madero, *Durango, Mexico* **86 C4** 24 32N 104 22W
Francistown, *Botswana* **57 C4** 21 7S 27 33 E
François, *Canada* **71 C8** 47 35N 56 45W
François L., *Canada* **72 C3** 54 0N 125 30W
Franeker, *Neths.* **15 A5** 53 12N 5 33 E
Frankfort, *S. Africa* **57 D4** 27 17S 28 30 E
Frankfort, *Ind., U.S.A.* **76 E2** 40 17N 86 31W
Frankfort, *Kans., U.S.A.* **80 F6** 39 42N 96 25W
Frankfort, *Ky., U.S.A.* **76 F3** 38 12N 84 52W
Frankfort, *Mich., U.S.A.* **76 C2** 44 38N 86 14W
Frankfurt, Brandenburg, *Germany* **16 B8** 52 20N 14 32 E
Frankfurt, Hessen, *Germany* **16 C5** 50 7N 8 41 E
Fränkische Alb, *Germany* **16 D6** 49 10N 11 23 E
Frankland →, *Australia* **61 G2** 35 0S 116 48 E
Franklin, *Ky., U.S.A.* **77 G2** 36 43N 86 35W
Franklin, *La., U.S.A.* **81 L9** 29 48N 91 30W
Franklin, *Mass., U.S.A.* **79 D13** 42 5N 71 24W
Franklin, *N.H., U.S.A.* **79 C13** 43 27N 71 39W
Franklin, *Nebr., U.S.A.* **80 E5** 40 6N 98 57W
Franklin, *Pa., U.S.A.* **78 E5** 41 24N 79 50W
Franklin, *Tenn., U.S.A.* **77 H2** 35 55N 86 52W
Franklin, *Va., U.S.A.* **77 G7** 36 41N 76 56W
Franklin, *W. Va., U.S.A.* **76 F6** 38 39N 79 20W
Franklin B., *Canada* **68 B7** 69 45N 126 0W
Franklin D. Roosevelt L., *U.S.A.* **82 B4** 48 18N 118 9W
Franklin I., *Antarctica* **5 D11** 76 10S 168 30 E
Franklin L., *U.S.A.* **82 F6** 40 25N 115 22W
Franklin Mts., *Canada* **68 B7** 65 0N 125 0W
Franklin Str., *Canada* **68 A10** 72 0N 96 0W
Franklinton, *U.S.A.* **81 K9** 30 51N 90 9W
Franklinville, *U.S.A.* **78 D6** 42 20N 78 27W
Franks Pk., *U.S.A.* **82 E9** 43 58N 109 18W
Frankston, *Australia* **63 F4** 38 8S 145 8 E
Frantsa Iosifa, Zemlya, *Russia* **26 A6** 82 0N 55 0 E
Franz, *Canada* **70 C3** 48 25N 84 30W
Franz Josef Land = Frantsa Iosifa, Zemlya, *Russia* **26 A6** 82 0N 55 0 E
Fraser →, *B.C., Canada* **72 D4** 49 7N 123 11W
Fraser →, *Nfld., Canada* **71 A7** 56 39N 62 10W
Fraser, Mt., *Australia* **61 E2** 25 35S 118 20 E
Fraser I., *Australia* **63 D5** 25 15S 153 10 E
Fraser Lake, *Canada* **72 C4** 54 0N 124 50W
Fraserburg, *S. Africa* **56 E3** 31 55S 21 30 E
Fraserburgh, *U.K.* **12 D6** 57 42N 2 1W
Fraserdale, *Canada* **70 C3** 49 55N 81 37W
Fray Bentos, *Uruguay* **94 C4** 33 10S 58 15W
Frazier Downs, *Australia* **60 C3** 18 48S 121 42 E
Fredericia, *Denmark* **9 J13** 55 34N 9 45 E
Frederick, *Md., U.S.A.* **76 F7** 39 25N 77 25W
Frederick, *Okla., U.S.A.* **81 H5** 34 23N 99 1W
Frederick, *S. Dak., U.S.A.* **80 C5** 45 50N 98 31W
Frederick Sd., *U.S.A.* **72 B2** 57 10N 134 0W
Fredericksburg, *Tex., U.S.A.* **81 K5** 30 16N 98 52W
Fredericksburg, *Va., U.S.A.* **76 F7** 38 18N 77 28W
Fredericktown, *U.S.A.* **81 G9** 37 34N 90 18W
Frederico I. Madero, Presa, *Mexico* **86 B3** 28 7N 105 40W
Fredericton, *Canada* **71 C6** 45 57N 66 40W
Fredericton Junc., *Canada* **71 C6** 45 41N 66 40W
Frederikshåb, *Greenland* **4 C5** 62 0N 49 43W
Frederikshavn, *Denmark* **9 H14** 57 28N 10 31 E
Frederiksted, *Virgin Is.* **89 C7** 17 43N 64 53W
Fredonia, *Ariz., U.S.A.* **83 H7** 36 57N 112 32W
Fredonia, *Kans., U.S.A.* **81 G7** 37 32N 95 49W
Fredonia, *N.Y., U.S.A.* **78 D5** 42 26N 79 20W
Fredrikstad, *Norway* **9 G14** 59 13N 10 57 E
Free State □, *S. Africa* **56 D4** 28 30S 27 0 E
Freehold, *U.S.A.* **79 F10** 40 16N 74 17W
Freel Peak, *U.S.A.* **84 G7** 38 52N 119 54W
Freeland, *U.S.A.* **79 E9** 41 1N 75 54W
Freels, C., *Canada* **71 C9** 49 15N 53 30W
Freeman, *Calif., U.S.A.* **85 K9** 35 35N 117 53W
Freeman, *S. Dak., U.S.A.* **80 D6** 43 21N 97 26W
Freeport, *Bahamas* **88 A4** 26 30N 78 47W
Freeport, *Canada* **71 D6** 44 15N 66 20W
Freeport, *Ill., U.S.A.* **80 D10** 42 17N 89 36W
Freeport, *N.Y., U.S.A.* **79 F11** 40 39N 73 35W
Freeport, *Tex., U.S.A.* **81 L7** 28 57N 95 21W
Freetown, *S. Leone* **50 G2** 8 30N 13 17W
Frégate, L., *Canada* **70 B5** 53 15N 74 45W
Fregenal de la Sierra, *Spain* **19 C2** 38 10N 6 39W
Freibourg = Fribourg, *Switz.* **16 E4** 46 49N 7 9 E
Freiburg, *Germany* **16 E4** 47 59N 7 51 E
Freire, *Chile* **96 D2** 38 54S 72 38W
Freirina, *Chile* **94 B1** 28 30S 71 10W
Freising, *Germany* **16 D6** 48 24N 11 45 E
Freistadt, *Austria* **16 D8** 48 30N 14 30 E
Fréjus, *France* **18 E7** 43 25N 6 44 E
Fremantle, *Australia* **61 F2** 32 7S 115 47 E
Fremont, *Calif., U.S.A.* **83 H2** 37 32N 121 57W
Fremont, *Mich., U.S.A.* **76 D3** 43 28N 85 57W
Fremont, *Nebr., U.S.A.* **80 E6** 41 26N 96 30W
Fremont, *Ohio, U.S.A.* **76 E4** 41 21N 83 7W
Fremont →, *U.S.A.* **83 G8** 38 24N 110 42W
Fremont L., *U.S.A.* **82 E9** 42 57N 109 48W
French Camp, *U.S.A.* **84 H5** 37 53N 121 16W
French Creek →, *U.S.A.* **78 E5** 41 24N 79 50W
French Guiana ■, *S. Amer.* **93 C8** 4 0N 53 0W
French Pass, *N.Z.* **59 J4** 40 55S 173 55 E
French Polynesia ■, *Pac. Oc.* **65 J13** 20 0S 145 0W
Frenchglen, *U.S.A.* **82 E4** 42 50N 118 55W
Frenchman Butte, *Canada* **73 C7** 53 35N 109 38W
Frenchman Cr. →, *Mont., U.S.A.* **82 B10** 48 31N 107 10W
Frenchman Cr. →, *Nebr., U.S.A.* **80 E4** 40 14N 100 50W
Fresco →, *Brazil* **93 E8** 7 15S 51 30W
Freshfield, C., *Antarctica* **5 C10** 68 25S 151 10 E
Fresnillo, *Mexico* **86 C4** 23 10N 103 0W
Fresno, *U.S.A.* **83 H4** 36 44N 119 47W
Fresno Reservoir, *U.S.A.* **82 B9** 48 36N 109 57W
Frew →, *Australia* **62 C2** 20 0S 135 38 E
Frewena, *Australia* **62 B2** 19 25S 135 25 E
Freycinet Pen., *Australia* **62 G4** 42 10S 148 25 E
Fria, C., *Namibia* **56 B1** 18 0S 12 0 E
Friant, *U.S.A.* **84 J7** 36 59N 119 43W
Frías, *Argentina* **94 B2** 28 40S 65 5W
Fribourg, *Switz.* **16 E4** 46 49N 7 9 E
Friday Harbor, *U.S.A.* **84 B3** 48 32N 123 1W
Friedrichshafen, *Germany* **16 E5** 47 39N 9 30 E
Friendly Is. = Tonga ■, *Pac. Oc.* **59 D11** 19 50S 174 30W
Friesland □, *Neths.* **15 A5** 53 5N 5 50 E
Frio →, *U.S.A.* **81 L5** 28 26N 98 11W
Frio, C., *Brazil* **90 F6** 22 50S 41 50W
Friona, *U.S.A.* **81 H3** 34 38N 102 43W
Fritch, *U.S.A.* **81 H4** 35 38N 101 36W
Frobisher B., *Canada* **69 B13** 62 30N 66 0W
Frobisher Bay = Iqaluit, *Canada* **69 B13** 63 44N 68 31W
Frobisher L., *Canada* **73 B7** 56 20N 108 15W
Frohavet, *Norway* **8 E13** 64 0N 9 30 E
Froid, *U.S.A.* **80 A2** 48 20N 104 30W
Fromberg, *U.S.A.* **82 D9** 45 24N 108 54W
Frome, *U.K.* **11 F5** 51 14N 2 19W
Frome, L., *Australia* **63 E2** 30 45S 139 45 E
Frome Downs, *Australia* **63 E2** 31 13S 139 45 E
Front Range, *U.S.A.* **82 G11** 40 25N 105 45W
Front Royal, *U.S.A.* **76 F6** 38 55N 78 12W
Frontera, *Canary Is.* **22 G2** 27 47N 17 59W
Frontera, *Mexico* **87 D6** 18 30N 92 40W
Frosinone, *Italy* **20 D5** 41 38N 13 19 E
Frostburg, *U.S.A.* **76 F6** 39 39N 78 56W
Frostisen, *Norway* **8 B17** 68 14N 17 10 E
Frøya, *Norway* **8 E13** 63 43N 8 40 E
Frunze = Bishkek, *Kyrgyzstan* **26 E8** 42 54N 74 46 E
Frutal, *Brazil* **93 G9** 20 0S 49 0W
Frýdek-Místek, *Czech.* **17 D10** 49 40N 18 20 E
Fu Xian, *Liaoning, China* **35 E11** 39 38N 121 58 E
Fu Xian, *Shaanxi, China* **34 F5** 36 0N 109 20 E
Fucheng, *China* **34 F9** 37 50N 116 10 E
Fuchou = Fuzhou, *China* **33 D6** 26 5N 119 16 E
Fuchū, *Japan* **31 G6** 34 34N 133 14 E
Fuencaliente, *Canary Is.* **22 F2** 28 28N 17 50W
Fuencaliente, Pta., *Canary Is.* **22 F2** 28 27N 17 51W
Fuengirola, *Spain* **19 D3** 36 32N 4 41W
Fuentes de Oñoro, *Spain* **19 B2** 40 33N 6 52W
Fuerte →, *Mexico* **86 B3** 25 50N 109 25W
Fuerte Olimpo, *Paraguay* **94 A4** 21 0S 57 51W
Fuerteventura, *Canary Is.* **22 F6** 28 30N 14 0W
Fufeng, *China* **34 G4** 34 22N 108 0 E
Fugou, *China* **34 G8** 34 3N 114 25 E
Fugu, *China* **34 E6** 39 2N 111 3 E
Fuhai, *China* **32 B3** 47 2N 87 25 E
Fuḩaymī, *Iraq* **44 C4** 34 16N 42 10 E
Fuji, *Japan* **31 G9** 35 9N 138 39 E
Fuji-San, *Japan* **31 G9** 35 22N 138 44 E
Fuji-yoshida, *Japan* **31 G9** 35 30N 138 46 E
Fujian □, *China* **33 D6** 26 0N 118 0 E
Fujinomiya, *Japan* **31 G9** 35 10N 138 40 E
Fujisawa, *Japan* **31 G9** 35 22N 139 29 E
Fukien = Fujian □, *China* **33 D6** 26 0N 118 0 E
Fukuchiyama, *Japan* **31 G7** 35 19N 135 9 E
Fukue-Shima, *Japan* **31 H4** 32 40N 128 45 E
Fukui, *Japan* **31 F8** 36 5N 136 10 E
Fukui □, *Japan* **31 G8** 36 0N 136 12 E
Fukuoka, *Japan* **31 H5** 33 39N 130 21 E
Fukuoka □, *Japan* **31 H5** 33 30N 131 0 E
Fukushima, *Japan* **30 F10** 37 44N 140 28 E
Fukushima □, *Japan* **30 F10** 37 30N 140 15 E
Fukuyama, *Japan* **31 G6** 34 35N 133 20 E
Fulda, *Germany* **16 C5** 50 32N 9 40 E
Fulda →, *Germany* **16 C5** 51 25N 9 39 E
Fullerton, *Calif., U.S.A.* **85 M9** 33 53N 117 56W
Fullerton, *Nebr., U.S.A.* **80 E6** 41 22N 97 58W
Fulongquan, *China* **35 B13** 44 20N 124 42 E
Fulton, *Mo., U.S.A.* **80 F9** 38 52N 91 57W
Fulton, *N.Y., U.S.A.* **79 C8** 43 19N 76 25W
Fulton, *Tenn., U.S.A.* **77 G1** 36 31N 88 53W
Funabashi, *Japan* **31 G10** 35 45N 140 0 E
Funchal, *Madeira* **22 D3** 32 38N 16 54W
Fundación, *Colombia* **92 A4** 10 31N 74 11W
Fundão, *Portugal* **19 B2** 40 8N 7 30W
Fundy, B. of, *Canada* **71 D6** 45 0N 66 0W
Funing, *Hebei, China* **35 E10** 39 53N 119 12 E
Funing, *Jiangsu, China* **35 H10** 33 45N 119 50 E
Funiu Shan, *China* **34 H7** 33 30N 112 20 E
Funtua, *Nigeria* **50 F6** 11 30N 7 18 E
Fuping, *Hebei, China* **34 E8** 38 48N 114 12 E
Fuping, *Shaanxi, China* **34 G5** 34 42N 109 10 E
Furano, *Japan* **30 C11** 43 21N 142 23 E
Furāt, Nahr al →, *Asia* **44 D5** 31 0N 47 25 E
Fūrg, *Iran* **45 D7** 28 18N 55 13 E
Furnás, *Spain* **22 B8** 39 3N 1 32 E
Furnas, Reprêsa de, *Brazil* **95 A6** 20 50S 45 30W
Furneaux Group, *Australia* **62 G4** 40 10S 147 50 E
Furness, *U.K.* **10 C4** 54 14N 3 8W
Furqlus, *Syria* **47 A6** 34 36N 37 8 E
Fürstenwalde, *Germany* **16 B8** 52 22N 14 3 E
Fürth, *Germany* **16 D6** 49 28N 10 59 E
Furukawa, *Japan* **30 E10** 38 34N 140 58 E
Fury and Hecla Str., *Canada* **69 B11** 69 56N 84 0W
Fusagasuga, *Colombia* **92 C4** 4 21N 74 22W
Fushan, *Shandong, China* **35 F11** 37 30N 121 15 E
Fushan, *Shanxi, China* **34 G6** 35 58N 111 51 E
Fushun, *China* **35 D12** 41 50N 123 56 E
Fusong, *China* **35 C14** 42 20N 127 15 E
Futuna, *Wall. & F. Is.* **59 B8** 14 25S 178 20 E
Fuxin, *China* **35 C11** 42 5N 121 48 E
Fuyang, *China* **34 H8** 33 0N 115 48 E
Fuyang He →, *China* **34 E9** 38 12N 117 0 E
Fuyu, *China* **35 B13** 45 12N 124 43 E
Fuzhou, *China* **33 D6** 26 5N 119 16 E
Fylde, *U.K.* **10 D5** 53 50N 2 58W
Fyn, *Denmark* **9 J14** 55 20N 10 30 E
Fyne, L., *U.K.* **12 F3** 55 59N 5 23W

G

Gabela, *Angola* **52 G2** 11 0S 14 24 E
Gabès, *Tunisia* **50 B7** 33 53N 10 2 E
Gabès, G. de, *Tunisia* **51 B7** 34 0N 10 30 E
Gabon ■, *Africa* **52 E2** 0 10S 10 0 E
Gaborone, *Botswana* **56 C4** 24 45S 25 57 E
Gabriels, *U.S.A.* **79 B10** 44 26N 74 12W
Gābrīk, *Iran* **45 E8** 25 44N 58 28 E
Gabrovo, *Bulgaria* **21 C11** 42 52N 25 19 E
Gāch Sār, *Iran* **45 B6** 36 7N 51 19 E
Gachsārān, *Iran* **45 D6** 30 15N 50 45 E
Gadag, *India* **40 M9** 15 30N 75 45 E
Gadap, *Pakistan* **42 G2** 25 5N 67 28 E
Gadarwara, *India* **43 H8** 22 50N 78 50 E
Gadhada, *India* **42 J4** 22 0N 71 35 E
Gadsden, *Ala., U.S.A.* **77 H2** 34 1N 86 1W
Gadsden, *Ariz., U.S.A.* **83 K6** 32 33N 114 47W
Gadwal, *India* **40 L10** 16 10N 77 50 E
Gaffney, *U.S.A.* **77 H5** 35 5N 81 39W
Gafsa, *Tunisia* **50 B6** 34 24N 8 43 E
Gagetown, *Canada* **71 C6** 45 46N 66 10W
Gagnoa, *Ivory C.* **50 G3** 6 56N 5 16W
Gagnon, *Canada* **71 B6** 51 50N 68 5W
Gagnon, L., *Canada* **73 A6** 62 3N 110 27W
Gahini, *Rwanda* **54 C3** 1 50S 30 30 E
Gahmar, *India* **43 G10** 25 27N 83 49 E
Gai Xian, *China* **35 D12** 40 22N 122 20 E
Gaïdhouronísi, *Greece* **23 E7** 34 53N 25 41 E
Gail, *U.S.A.* **81 J4** 32 46N 101 27W
Gaillimh = Galway, *Ireland* **13 C2** 53 17N 9 3W
Gaines, *U.S.A.* **78 E7** 41 46N 77 35W
Gainesville, *Fla., U.S.A.* **77 L4** 29 40N 82 20W
Gainesville, *Ga., U.S.A.* **77 H4** 34 18N 83 50W
Gainesville, *Mo., U.S.A.* **81 G8** 36 36N 92 26W
Gainesville, *Tex., U.S.A.* **81 J6** 33 38N 97 8W
Gainsborough, *U.K.* **10 D7** 53 24N 0 46W
Gairdner, L., *Australia* **63 E2** 31 30S 136 0 E
Gairloch, L., *U.K.* **12 D3** 57 43N 5 45W
Gakuch, *Pakistan* **43 A5** 36 7N 73 45 E
Galán, Cerro, *Argentina* **94 B2** 25 55S 66 52W
Galana →, *Kenya* **54 C5** 3 9S 40 8 E
Galangue, *Angola* **53 G3** 13 42S 16 9 E
Galápagos, *Pac. Oc.* **90 D1** 0 0 91 0W
Galashiels, *U.K.* **12 F6** 55 37N 2 49W
Galaţi, *Romania* **17 F15** 45 27N 28 2 E
Galatina, *Italy* **21 D8** 40 10N 18 10 E
Galax, *U.S.A.* **77 G5** 36 40N 80 56W
Galbraith, *Australia* **62 B3** 16 25S 141 30 E
Galcaio, *Somali Rep.* **46 F4** 6 30N 47 30 E
Galdhøpiggen, *Norway* **9 F12** 61 38N 8 18 E
Galeana, *Mexico* **86 C4** 24 50N 100 4W
Galela, *Indonesia* **37 D7** 1 50N 127 49 E
Galera Point, *Trin. & Tob.* **89 D7** 10 8N 61 0W
Galesburg, *U.S.A.* **80 E9** 40 57N 90 22W
Galeton, *U.S.A.* **78 E7** 41 44N 77 39W
Galich, *Russia* **24 C7** 58 22N 42 24 E
Galicia □, *Spain* **19 A2** 42 43N 7 45W
Galilee = Hagalil, *Israel* **47 C4** 32 53N 35 18 E
Galilee, L., *Australia* **62 C4** 22 20S 145 50 E
Galilee, Sea of = Yam Kinneret, *Israel* **47 C4** 32 45N 35 35 E
Galinoporni, *Cyprus* **23 D13** 35 31N 34 18 E
Galion, *U.S.A.* **78 F2** 40 44N 82 47W
Galiuro Mts., *U.S.A.* **83 K8** 32 30N 110 20W
Gallabat, *Sudan* **51 F12** 12 58N 36 11 E
Gallatin, *U.S.A.* **77 G2** 36 24N 86 27W
Galle, *Sri Lanka* **40 R12** 6 5N 80 10 E
Gállego →, *Spain* **19 B5** 41 39N 0 51W
Gallegos →, *Argentina* **96 G3** 51 35S 69 0W
Galley Hd., *Ireland* **13 E3** 51 32N 8 55W
Gallinas, Pta., *Colombia* **92 A4** 12 28N 71 40W
Gallipoli = Gelibolu, *Turkey* **21 D12** 40 28N 26 43 E
Gallipoli, *Italy* **21 D8** 40 3N 17 58 E
Gallipolis, *U.S.A.* **76 F4** 38 49N 82 12W
Gällivare, *Sweden* **8 C19** 67 9N 20 40 E
Galloway, *U.K.* **12 G4** 55 1N 4 29W
Galloway, Mull of, *U.K.* **12 G4** 54 39N 4 52W
Gallup, *U.S.A.* **83 J9** 35 32N 108 45W
Galong, *Australia* **63 E4** 34 37S 148 34 E
Galoya, *Sri Lanka* **40 Q12** 8 10N 80 55 E
Galt, *U.S.A.* **84 G5** 38 15N 121 18W
Galty Mts., *Ireland* **13 D3** 52 22N 8 10W
Galtymore, *Ireland* **13 D3** 52 21N 8 11W
Galva, *U.S.A.* **80 E9** 41 10N 90 3W
Galveston, *U.S.A.* **81 L7** 29 18N 94 48W
Galveston B., *U.S.A.* **81 L7** 29 36N 94 50W
Gálvez, *Argentina* **94 C3** 32 0S 61 14W
Galway, *Ireland* **13 C2** 53 17N 9 3W
Galway □, *Ireland* **13 C2** 53 22N 9 1W
Galway B., *Ireland* **13 C2** 53 13N 9 10W
Gam →, *Vietnam* **38 B5** 21 55N 105 12 E
Gamagori, *Japan* **31 G8** 34 50N 137 14 E
Gambaga, *Ghana* **50 F4** 10 30N 0 28W
Gambat, *Pakistan* **42 F3** 27 17N 68 26 E
Gambela, *Ethiopia* **51 G11** 8 14N 34 38 E
Gambia ■, *W. Afr.* **50 F1** 13 25N 16 0W
Gambia →, *W. Afr.* **50 F1** 13 28N 16 34W
Gambier, C., *Australia* **60 B5** 11 56S 130 57 E
Gambier Is., *Australia* **63 F2** 35 3S 136 30 E
Gamboli, *Pakistan* **42 E3** 29 53N 68 24 E
Gamboma, *Congo* **52 E3** 1 55S 15 52 E
Gamerco, *U.S.A.* **83 J9** 35 34N 108 46W
Gamlakarleby = Kokkola, *Finland* **8 E20** 63 50N 23 8 E
Gammon →, *Canada* **73 C9** 51 24N 95 44W
Gan Jiang →, *China* **33 D6** 29 15N 116 0 E
Ganado, *Ariz., U.S.A.* **83 J9** 35 43N 109 33W
Ganado, *Tex., U.S.A.* **81 L6** 29 2N 96 31W
Gananoque, *Canada* **70 D4** 44 20N 76 10W
Ganāveh, *Iran* **45 D6** 29 35N 50 35 E
Gäncä, *Azerbaijan* **25 F8** 40 45N 46 20 E
Gand = Gent, *Belgium* **15 C3** 51 2N 3 42 E
Ganda, *Angola* **53 G2** 13 3S 14 35 E
Gandak →, *India* **43 G11** 25 39N 85 13 E
Gandava, *Pakistan* **42 E2** 28 32N 67 32 E
Gander, *Canada* **71 C9** 48 58N 54 35W
Gander L., *Canada* **71 C9** 48 58N 54 35W
Ganderowe Falls, *Zimbabwe* **55 F2** 17 20S 29 10 E
Gandhi Sagar, *India* **42 G6** 24 40N 75 40 E
Gandi, *Nigeria* **50 F6** 12 55N 5 49 E
Gandía, *Spain* **19 C5** 38 58N 0 9W
Gando, Pta., *Canary Is.* **22 G4** 27 55N 15 22W
Ganedidalem = Gani, *Indonesia* **37 E7** 0 48S 128 14 E
Ganga →, *India* **43 H14** 23 20N 90 30 E
Ganga, Mouths of the, *India* **43 J13** 21 30N 90 0 E
Ganganagar, *India* **42 E5** 29 56N 73 56 E
Gangapur, *India* **42 F7** 26 32N 76 49 E
Gangara, *Niger* **50 F6** 14 35N 8 29 E
Gangaw, *Burma* **41 H19** 22 5N 94 5 E
Gangdisê Shan, *China* **41 D12** 31 20N 81 0 E
Ganges = Ganga →, *India* **43 H14** 23 20N 90 30 E
Gangoh, *India* **42 E7** 29 46N 77 18 E
Gangtok, *India* **41 F16** 27 20N 88 37 E
Gangu, *China* **34 G3** 34 40N 105 15 E
Gangyao, *China* **35 B14** 44 12N 126 37 E
Gani, *Indonesia* **37 E7** 0 48S 128 14 E
Ganj, *India* **43 F8** 27 45N 78 57 E
Gannett Peak, *U.S.A.* **82 E9** 43 11N 109 39W
Gannvalley, *U.S.A.* **80 C5** 44 2N 98 59W
Ganquan, *China* **34 F5** 36 20N 109 20 E
Gansu □, *China* **34 G3** 36 0N 104 0 E
Ganta, *Liberia* **50 G3** 7 15N 8 59W
Gantheaume, C., *Australia* **63 F2** 36 4S 137 32 E
Gantheaume B., *Australia* **61 E1** 27 40S 114 10 E
Gantsevichi = Hantsavichy, *Belarus* **17 B14** 52 49N 26 30 E
Ganyem, *Indonesia* **37 E10** 2 46S 140 12 E
Ganyu, *China* **35 G10** 34 50N 119 8 E
Ganzhou, *China* **33 D6** 25 51N 114 56 E
Gaomi, *China* **35 F10** 36 20N 119 42 E
Gaoping, *China* **34 G7** 35 45N 112 55 E
Gaotang, *China* **34 F9** 36 50N 116 15 E
Gaoua, *Burkina Faso* **50 F4** 10 20N 3 8W
Gaoual, *Guinea* **50 F2** 11 45N 13 25W
Gaoxiong = Kaohsiung, *Taiwan* **33 D7** 22 35N 120 16 E
Gaoyang, *China* **34 E8** 38 40N 115 45 E
Gaoyou Hu, *China* **35 H10** 32 45N 119 20 E
Gaoyuan, *China* **35 F9** 37 8N 117 58 E
Gap, *France* **18 D7** 44 33N 6 5 E
Gar, *China* **32 C2** 32 10N 79 58 E
Garabogazköl Aylagy, *Turkmenistan* **25 F9** 41 0N 53 30 E
Garachico, *Canary Is.* **22 F3** 28 22N 16 46W
Garachiné, *Panama* **88 E4** 8 0N 78 12W
Garafia, *Canary Is.* **22 F2** 28 48N 17 57W
Garajonay, *Canary Is.* **22 F2** 28 7N 17 14W
Garanhuns, *Brazil* **93 E11** 8 50S 36 30W
Garawe, *Liberia* **50 H3** 4 35N 8 0W
Garba Tula, *Kenya* **54 B4** 0 30N 38 32 E
Garber, *U.S.A.* **81 G6** 36 26N 97 35W
Garberville, *U.S.A.* **82 F2** 40 6N 123 48W
Gard, *Somali Rep.* **46 F4** 9 30N 49 6 E
Garda, L. di, *Italy* **20 B4** 45 40N 10 41 E
Garde L., *Canada* **73 A7** 62 50N 106 13W
Garden City, *Kans., U.S.A.* **81 G4** 37 58N 100 53W
Garden City, *Tex., U.S.A.* **81 K4** 31 52N 101 29W
Garden Grove, *U.S.A.* **85 M9** 33 47N 117 55W
Gardēz, *Afghan.* **42 C3** 33 37N 69 9 E
Gardiner, *U.S.A.* **82 D8** 45 2N 110 22W
Gardiners I., *U.S.A.* **79 E12** 41 6N 72 6W
Gardner, *U.S.A.* **79 D13** 42 34N 71 59W
Gardner Canal, *Canada* **72 C3** 53 27N 128 8W
Gardnerville, *U.S.A.* **84 G7** 38 56N 119 45W
Garey, *U.S.A.* **85 L6** 34 53N 120 19W
Garfield, *U.S.A.* **82 C5** 47 1N 117 9W
Gargano, Mte., *Italy* **20 D6** 41 43N 15 43 E
Garhshankar, *India* **42 D7** 31 13N 76 11 E
Garibaldi Prov. Park, *Canada* **72 D4** 49 50N 122 40W
Garies, *S. Africa* **56 E2** 30 32S 17 59 E
Garigliano →, *Italy* **20 D5** 41 13N 13 45 E
Garissa, *Kenya* **54 C4** 0 25S 39 40 E
Garissa □, *Kenya* **54 C5** 0 20S 40 0 E
Garland, *Tex., U.S.A.* **81 J6** 32 55N 96 38W
Garland, *Utah, U.S.A.* **82 F7** 41 47N 112 10W
Garm, *Tajikistan* **26 F8** 39 0N 70 20 E

Garmāb, *Iran* 45 C8 35 25N 56 45 E
Garmisch-Partenkirchen, *Germany* 16 E6 47 30N 11 6 E
Garmsār, *Iran* 45 C7 35 20N 52 25 E
Garner, *U.S.A.* 80 D8 43 6N 93 36W
Garnett, *U.S.A.* 80 F7 38 17N 95 14W
Garo Hills, *India* 43 G14 25 30N 90 30 E
Garoe, *Somali Rep.* 46 F4 8 25N 48 33 E
Garonne →, *France* 18 D3 45 2N 0 36W
Garoua, *Cameroon* 51 G7 9 19N 13 21 E
Garrison, *Mont., U.S.A.* 82 C7 46 31N 112 49W
Garrison, *N. Dak., U.S.A.* 80 B4 47 40N 101 25W
Garrison, *Tex., U.S.A.* 81 K7 31 49N 94 30W
Garrison Res. = Sakakawea, L., *U.S.A.* 80 B3 47 30N 101 25W
Garry →, *U.K.* 12 E5 56 44N 3 47W
Garry, L., *Canada* 68 B9 65 58N 100 18W
Garsen, *Kenya* 54 C5 2 20S 40 5 E
Garson L., *Canada* 73 B6 56 19N 110 2W
Garub, *Namibia* 56 D2 26 37S 16 0 E
Garut, *Indonesia* 37 G12 7 14S 107 53 E
Garvie Mts., *N.Z.* 59 L2 45 30S 168 50 E
Garwa = Garoua, *Cameroon* 51 G7 9 19N 13 21 E
Garwa, *India* 43 G10 24 11N 83 47 E
Gary, *U.S.A.* 76 E2 41 36N 87 20W
Garzê, *China* 32 C5 31 38N 100 1 E
Garzón, *Colombia* 92 C3 2 10N 75 40W
Gas-San, *Japan* 30 E10 38 32N 140 1 E
Gasan Kuli = Esenguly, *Turkmenistan* 26 F6 37 37N 53 59 E
Gascogne, *France* 18 E4 43 45N 0 20 E
Gascogne, G. de, *Europe* 18 D2 44 0N 2 0W
Gascony = Gascogne, *France* 18 E4 43 45N 0 20 E
Gascoyne →, *Australia* 61 D1 24 52S 113 37 E
Gascoyne Junc. T.O., *Australia* 61 E2 25 2S 115 17 E
Gashaka, *Nigeria* 50 G7 7 20N 11 29 E
Gasherbrum, *Pakistan* 43 B7 35 40N 76 40 E
Gaspé, *Canada* 71 C7 48 52N 64 30W
Gaspé, C. de, *Canada* 71 C7 48 48N 64 7W
Gaspé, Pén. de, *Canada* 71 C6 48 45N 65 40W
Gaspésie, Parc Prov. de la, *Canada* 71 C6 48 55N 65 50W
Gassaway, *U.S.A.* 76 F5 38 41N 80 47W
Gasteiz = Vitoria, *Spain* 19 A4 42 50N 2 41W
Gastonia, *U.S.A.* 77 H5 35 16N 81 11W
Gastre, *Argentina* 96 E3 42 20S 69 15W
Gata, C., *Cyprus* 23 E12 34 34N 33 2 E
Gata, C. de, *Spain* 19 D4 36 41N 2 13W
Gata, Sierra de, *Spain* 19 B2 40 20N 6 45W
Gataga →, *Canada* 72 B3 58 35N 126 59W
Gates, *U.S.A.* 78 C7 43 9N 77 42W
Gateshead, *U.K.* 10 C6 54 57N 1 35W
Gatesville, *U.S.A.* 81 K6 31 26N 97 45W
Gaths, *Zimbabwe* 55 G3 20 2S 30 32 E
Gatico, *Chile* 94 A1 22 29S 70 20W
Gatineau →, *Canada* 70 C4 45 27N 75 42W
Gatineau, Parc de la, *Canada* 70 C4 45 40N 76 0W
Gatun, L., *Panama* 88 E4 9 7N 79 56W
Gatyana, *S. Africa* 57 E4 32 16S 28 31 E
Gau, *Fiji* 59 D8 18 2S 179 18 E
Gauer L., *Canada* 73 B9 57 0N 97 50W
Gauhati, *India* 43 F14 26 10N 91 45 E
Gauja →, *Latvia* 9 H21 57 10N 24 16 E
Gaula →, *Norway* 8 E14 63 21N 10 14 E
Gausta, *Norway* 9 G13 59 48N 8 40 E
Gauteng □, *S. Africa* 57 D4 26 0S 28 0 E
Gāv Koshī, *Iran* 45 D8 28 38N 57 12 E
Gāvakān, *Iran* 45 D7 29 37N 53 10 E
Gavāter, *Iran* 45 E9 25 10N 61 31 E
Gāvbandī, *Iran* 45 E7 27 12N 53 4 E
Gavdhopoúla, *Greece* 23 E6 34 56N 24 0 E
Gávdhos, *Greece* 23 E6 34 50N 24 5 E
Gaviota, *U.S.A.* 85 L6 34 29N 120 13W
Gävle, *Sweden* 9 F17 60 40N 17 9 E
Gawachab, *Namibia* 56 D2 27 4S 17 55 E
Gawilgarh Hills, *India* 40 J10 21 15N 76 45 E
Gawler, *Australia* 63 E2 34 30S 138 42 E
Gaxun Nur, *China* 32 B5 42 22N 100 30 E
Gay, *Russia* 24 D10 51 27N 58 27 E
Gaya, *India* 43 G11 24 47N 85 4 E
Gaya, *Niger* 50 F5 11 52N 3 28 E
Gaylord, *U.S.A.* 76 C3 45 2N 84 41W
Gayndah, *Australia* 63 D5 25 35S 151 32 E
Gaysin = Haysyn, *Ukraine* 17 D15 48 57N 29 25 E
Gayvoron = Hayvoron, *Ukraine* 17 D15 48 22N 29 52 E
Gaza, *Gaza Strip* 47 D3 31 30N 34 28 E
Gaza □, *Mozam.* 57 C5 23 10S 32 45 E
Gaza Strip □, *Asia* 47 D3 31 29N 34 25 E
Gāzbor, *Iran* 45 D8 28 5N 58 51 E
Gazi, *Zaïre* 54 B1 1 3N 24 30 E
Gaziantep, *Turkey* 25 G6 37 6N 37 23 E
Gazli, *Uzbekistan* 26 E7 40 14N 63 24 E
Gcuwa, *S. Africa* 57 E4 32 20S 28 11 E
Gdańsk, *Poland* 17 A10 54 22N 18 40 E
Gdańska, Zatoka, *Poland* 17 A10 54 30N 19 20 E
Gdov, *Russia* 9 G22 58 48N 27 55 E
Gdynia, *Poland* 17 A10 54 35N 18 33 E
Gebe, *Indonesia* 37 D7 0 5N 129 25 E
Gebeit Mine, *Sudan* 51 D12 21 3N 36 29 E
Gebze, *Turkey* 21 D13 40 47N 29 25 E
Gedaref, *Sudan* 51 F12 14 2N 35 28 E
Gede, Tanjung, *Indonesia* 36 F3 6 46S 105 12 E
Gediz →, *Turkey* 21 E12 38 35N 26 48 E
Gedser, *Denmark* 9 J14 54 35N 11 55 E
Geegully Cr. →, *Australia* 60 C3 18 32S 123 41 E
Geelong, *Australia* 63 F3 38 10S 144 22 E
Geelvink Chan., *Australia* 61 E1 28 30S 114 0 E
Geesthacht, *Germany* 16 B6 53 26N 10 22 E
Geidam, *Nigeria* 51 F7 12 57N 11 57 E
Geikie →, *Canada* 73 B8 57 45N 103 52W
Geili, *Sudan* 51 E11 16 1N 32 37 E
Geita, *Tanzania* 54 C3 2 48S 32 12 E
Geita □, *Tanzania* 54 C3 2 50S 32 10 E
Gejiu, *China* 32 D5 23 20N 103 10 E
Gela, *Italy* 20 F6 37 4N 14 15 E
Geladi, *Ethiopia* 46 F4 6 59N 46 30 E
Gelderland □, *Neths.* 15 B6 52 5N 6 10 E
Geldermalsen, *Neths.* 15 C5 51 53N 5 17 E
Geldrop, *Neths.* 15 C5 51 25N 5 32 E
Geleen, *Neths.* 15 D5 50 57N 5 49 E
Gelehun, *S. Leone* 50 G2 8 20N 11 40W
Gelibolu, *Turkey* 21 D12 40 28N 26 43 E
Gelsenkirchen, *Germany* 16 C4 51 32N 7 1 E
Gemas, *Malaysia* 39 L4 2 37N 102 36 E
Gembloux, *Belgium* 15 D4 50 34N 4 43 E
Gemena, *Zaïre* 52 D3 3 13N 19 48 E
Gemlik, *Turkey* 21 D13 40 26N 29 9 E
Gendringen, *Neths.* 15 C6 51 52N 6 21 E
General Acha, *Argentina* 94 D3 37 20S 64 38W
General Alvear, *Buenos Aires, Argentina* 94 D3 36 0S 60 0W
General Alvear, *Mendoza, Argentina* 94 D2 35 0S 67 40W
General Artigas, *Paraguay* 94 B4 26 52S 56 16W
General Belgrano, *Argentina* 94 D4 36 35S 58 47W
General Cabrera, *Argentina* 94 C3 32 53S 63 52W
General Cepeda, *Mexico* 86 B4 25 23N 101 27W
General Guido, *Argentina* 94 D4 36 40S 57 50W
General Juan Madariaga, *Argentina* 94 D4 37 0S 57 0W
General La Madrid, *Argentina* 94 D3 37 17S 61 20W
General MacArthur, *Phil.* 37 B7 11 18N 125 28 E
General Martin Miguel de Güemes, *Argentina* 94 A3 24 50S 65 0W
General Paz, *Argentina* 94 B4 27 45S 57 36W
General Pico, *Argentina* 94 D3 35 45S 63 50W
General Pinedo, *Argentina* 94 B3 27 15S 61 20W
General Pinto, *Argentina* 94 C3 34 45S 61 50W
General Santos, *Phil.* 37 C7 6 5N 125 14 E
General Trevino, *Mexico* 87 B5 26 14N 99 29W
General Trías, *Mexico* 86 B3 28 21N 106 22W
General Viamonte, *Argentina* 94 D3 35 1S 61 3W
General Villegas, *Argentina* 94 D3 35 5S 63 0W
Genesee, *Idaho, U.S.A.* 82 C5 46 33N 116 56W
Genesee, *Pa., U.S.A.* 78 E7 41 59N 77 54W
Genesee →, *U.S.A.* 78 C7 43 16N 77 36W
Geneseo, *Ill., U.S.A.* 80 E9 41 27N 90 9W
Geneseo, *Kans., U.S.A.* 80 F5 38 31N 98 10W
Geneseo, *N.Y., U.S.A.* 78 D7 42 48N 77 49W
Geneva = Genève, *Switz.* 16 E4 46 12N 6 9 E
Geneva, *Ala., U.S.A.* 77 K3 31 2N 85 52W
Geneva, *N.Y., U.S.A.* 78 D7 42 52N 76 59W
Geneva, *Nebr., U.S.A.* 80 E6 40 32N 97 36W
Geneva, *Ohio, U.S.A.* 78 E4 41 48N 80 57W
Geneva, L. = Léman, L., *Europe* 16 E4 46 26N 6 30 E
Geneva, L., *U.S.A.* 76 D1 42 38N 88 30W
Genève, *Switz.* 16 E4 46 12N 6 9 E
Genil →, *Spain* 19 D3 37 42N 5 19W
Genk, *Belgium* 15 D5 50 58N 5 32 E
Gennargentu, Mti. del, *Italy* 20 D3 40 1N 9 19 E
Gennep, *Neths.* 15 C5 51 41N 5 59 E
Genoa = Génova, *Italy* 20 B3 44 25N 8 57 E
Genoa, *Australia* 63 F4 37 29S 149 35 E
Genoa, *N.Y., U.S.A.* 79 D8 42 40N 76 32W
Genoa, *Nebr., U.S.A.* 80 E6 41 27N 97 44W
Genoa, *Nev., U.S.A.* 84 F7 39 2N 119 50W
Génova, *Italy* 20 B3 44 25N 8 57 E
Génova, G. di, *Italy* 20 C3 44 0N 9 0 E
Gent, *Belgium* 15 C3 51 2N 3 42 E
Geographe B., *Australia* 61 F2 33 30S 115 15 E
Geographe Chan., *Australia* 61 D1 24 30S 113 0 E
Georga, Zemlya, *Russia* 26 A5 80 30N 49 0 E
George, *S. Africa* 56 E3 33 58S 22 29 E
George →, *Canada* 71 A6 58 49N 66 10W
George, L., *N.S.W., Australia* 63 F4 35 10S 149 25 E
George, L., *S. Austral., Australia* 63 F3 37 25S 140 0 E
George, L., *W. Austral., Australia* 60 D3 22 45S 123 40 E
George, L., *Uganda* 54 B3 0 5N 30 10 E
George, L., *Fla., U.S.A.* 77 L5 29 17N 81 36W
George, L., *N.Y., U.S.A.* 79 C11 43 37N 73 33W
George Gill Ra., *Australia* 60 D5 24 22S 131 45 E
George River = Kangiqsualujjuaq, *Canada* 69 C13 58 30N 65 59W
George Sound, *N.Z.* 59 L1 44 52S 167 25 E
George Town, *Bahamas* 88 B4 23 33N 75 47W
George Town, *Malaysia* 39 K3 5 25N 100 15 E
George V Land, *Antarctica* 5 C10 69 0S 148 0 E
George VI Sound, *Antarctica* 5 D17 71 0S 68 0W
George West, *U.S.A.* 81 L5 28 20N 98 7W
Georgetown, *Australia* 62 B3 18 17S 143 33 E
Georgetown, *Ont., Canada* 70 D4 43 40N 79 56W
Georgetown, *P.E.I., Canada* 71 C7 46 13N 62 24W
Georgetown, *Cayman Is.* 88 C3 19 20N 81 24W
Georgetown, *Gambia* 50 F2 13 30N 14 47W
Georgetown, *Guyana* 92 B7 6 50N 58 12W
Georgetown, *Calif., U.S.A.* 84 G6 38 54N 120 50W
Georgetown, *Colo., U.S.A.* 82 G11 39 42N 105 42W
Georgetown, *Ky., U.S.A.* 76 F3 38 13N 84 33W
Georgetown, *S.C., U.S.A.* 77 J6 33 23N 79 17W
Georgetown, *Tex., U.S.A.* 81 K6 30 38N 97 41W
Georgia □, *U.S.A.* 77 J4 32 50N 83 15W
Georgia ■, *Asia* 25 F7 42 0N 43 0 E
Georgia, Str. of, *Canada* 72 D4 49 25N 124 0W
Georgian B., *Canada* 70 C3 45 15N 81 0W
Georgina →, *Australia* 62 C2 23 30S 139 47 E
Georgina Downs, *Australia* 62 C2 21 10S 137 40 E
Georgiu-Dezh = Liski, *Russia* 25 D6 51 3N 39 30 E
Georgiyevsk, *Russia* 25 F7 44 12N 43 28 E
Gera, *Germany* 16 C7 50 53N 12 4 E
Geraardsbergen, *Belgium* 15 D3 50 45N 3 53 E
Geral, Serra, *Brazil* 95 B6 26 25S 50 0W
Geral de Goiás, Serra, *Brazil* 93 F9 12 0S 46 0W
Geraldine, *U.S.A.* 82 C8 47 36N 110 16W
Geraldton, *Australia* 61 E1 28 48S 114 32 E
Geraldton, *Canada* 70 C2 49 44N 86 59W
Gereshk, *Afghan.* 40 D4 31 47N 64 35 E
Gerik, *Malaysia* 39 K3 5 50N 101 15 E
Gering, *U.S.A.* 80 E3 41 50N 103 40W
Gerlach, *U.S.A.* 82 F4 40 39N 119 21W
Gerlogubi, *Ethiopia* 46 F4 6 53N 45 3 E
Germansen Landing, *Canada* 72 B4 55 43N 124 40W
Germany ■, *Europe* 16 C6 51 0N 10 0 E
Germiston, *S. Africa* 57 D4 26 15S 28 10 E
Gero, *Japan* 31 G8 35 48N 137 14 E
Gerona, *Spain* 19 B7 41 58N 2 46 E
Gerrard, *Canada* 72 C5 50 30N 117 17W
Geser, *Indonesia* 37 E8 3 50S 130 54 E
Getafe, *Spain* 19 B4 40 18N 3 44W
Gethsémani, *Canada* 71 B7 50 13N 60 40W
Gettysburg, *Pa., U.S.A.* 76 F7 39 50N 77 14W
Gettysburg, *S. Dak., U.S.A.* 80 C5 45 1N 99 57W
Getz Ice Shelf, *Antarctica* 5 D14 75 0S 130 0W
Geyser, *U.S.A.* 82 C8 47 16N 110 30W
Geyserville, *U.S.A.* 84 G4 38 42N 122 54W
Ghaghara →, *India* 43 G11 25 45N 84 40 E
Ghana ■, *W. Afr.* 50 G4 8 0N 1 0W
Ghansor, *India* 43 H9 22 39N 80 1 E
Ghanzi, *Botswana* 56 C3 21 50S 21 34 E
Ghanzi □, *Botswana* 56 C3 21 50S 21 45 E
Ghardaïa, *Algeria* 50 B5 32 20N 3 37 E
Gharyān, *Libya* 51 B7 32 10N 13 0 E
Ghat, *Libya* 50 D7 24 59N 10 11 E
Ghatal, *India* 43 H12 22 40N 87 46 E
Ghatampur, *India* 43 F9 26 8N 80 13 E
Ghaṭṭī, *Si. Arabia* 44 D3 31 16N 37 31 E
Ghawdex = Gozo, *Malta* 23 C1 36 3N 14 13 E
Ghazal, Bahr el →, *Chad* 51 F8 13 0N 15 47 E
Ghazâl, Bahr el →, *Sudan* 51 G11 9 31N 30 25 E
Ghazaouet, *Algeria* 50 A4 35 8N 1 50W
Ghaziabad, *India* 42 E7 28 42N 77 26 E
Ghazipur, *India* 43 G10 25 38N 83 35 E
Ghaznī, *Afghan.* 42 C3 33 30N 68 28 E
Ghaznī □, *Afghan.* 40 C6 32 10N 68 20 E
Ghèlinsor, *Somali Rep.* 46 F4 6 28N 46 39 E
Ghent = Gent, *Belgium* 15 C3 51 2N 3 42 E
Ghizao, *Afghan.* 42 C1 33 20N 65 44 E
Ghizar →, *Pakistan* 43 A5 36 15N 73 43 E
Ghogha, *India* 42 J5 21 40N 72 20 E
Ghotaru, *India* 42 F4 27 20N 70 1 E
Ghotki, *Pakistan* 42 E3 28 5N 69 21 E
Ghowr □, *Afghan.* 40 C4 34 0N 64 20 E
Ghudaf, W. al →, *Iraq* 44 C4 32 56N 43 30 E
Ghudāmis, *Libya* 49 C4 30 11N 9 29 E
Ghughri, *India* 43 H9 22 39N 80 41 E
Ghugus, *India* 40 K11 19 58N 79 12 E
Ghulam Mohammad Barrage, *Pakistan* 42 G3 25 30N 68 20 E
Ghūriān, *Afghan.* 40 B2 34 17N 61 25 E
Gia Dinh, *Vietnam* 39 G6 10 49N 106 42 E
Gia Lai = Pleiku, *Vietnam* 38 F7 13 57N 108 0 E
Gia Nghia, *Vietnam* 39 G6 11 58N 107 42 E
Gia Ngoc, *Vietnam* 38 E7 14 50N 108 58 E
Gia Vuc, *Vietnam* 38 E7 14 42N 108 34 E
Gian, *Phil.* 37 C7 5 45N 125 20 E
Giant Forest, *U.S.A.* 84 J8 36 36N 118 43W
Giants Causeway, *U.K.* 13 A5 55 16N 6 29W
Giarabub = Al Jaghbūb, *Libya* 51 C9 29 42N 24 38 E
Giarre, *Italy* 20 F6 37 43N 15 11 E
Gibara, *Cuba* 88 B4 21 9N 76 11W
Gibb River, *Australia* 60 C4 16 26S 126 26 E
Gibbon, *U.S.A.* 80 E5 40 45N 98 51W
Gibraltar ■, *Europe* 19 D3 36 7N 5 22W
Gibraltar, Str. of, *Medit. S.* 19 E3 35 55N 5 40W
Gibson Desert, *Australia* 60 D4 24 0S 126 0 E
Gibsons, *Canada* 72 D4 49 24N 123 32W
Gibsonville, *U.S.A.* 84 F6 39 46N 120 54W
Giddings, *U.S.A.* 81 K6 30 11N 96 56W
Gidole, *Ethiopia* 51 G12 5 40N 37 25 E
Giessen, *Germany* 16 C5 50 34N 8 41 E
Gīfān, *Iran* 45 B8 37 54N 57 28 E
Gifford Creek, *Australia* 60 D2 24 3S 116 16 E
Gifu, *Japan* 31 G8 35 30N 136 45 E
Gifu □, *Japan* 31 G8 35 40N 137 0 E
Giganta, Sa. de la, *Mexico* 86 B2 25 30N 111 30W
Gigha, *U.K.* 12 F3 55 42N 5 44W
Giglio, *Italy* 20 C4 42 20N 10 52 E
Gijón, *Spain* 19 A3 43 32N 5 42W
Gil I., *Canada* 72 C3 53 12N 129 15W
Gila →, *U.S.A.* 83 K6 32 43N 114 33W
Gila Bend, *U.S.A.* 83 K7 32 57N 112 43W
Gila Bend Mts., *U.S.A.* 83 K7 33 10N 113 0W
Gīlān □, *Iran* 45 B6 37 0N 50 0 E
Gilbert →, *Australia* 62 B3 16 35S 141 15 E
Gilbert Is., *Kiribati* 64 G9 1 0N 172 0 E
Gilbert Plains, *Canada* 73 C8 51 9N 100 28W
Gilbert River, *Australia* 62 B3 18 9S 142 52 E
Gilberton, *Australia* 62 B3 19 16S 143 35 E
Gilford I., *Canada* 72 C3 50 40N 126 30W
Gilgandra, *Australia* 63 E4 31 43S 148 39 E
Gilgil, *Kenya* 54 C4 0 30S 36 20 E
Gilgit, *India* 43 B6 35 50N 74 15 E
Gilgit →, *Pakistan* 43 B6 35 44N 74 37 E
Gillam, *Canada* 73 B10 56 20N 94 40W
Gillen, L., *Australia* 61 E3 26 11S 124 38 E
Gilles, L., *Australia* 63 E2 32 50S 136 45 E
Gillette, *U.S.A.* 80 C2 44 18N 105 30W
Gilliat, *Australia* 62 C3 20 40S 141 28 E
Gillingham, *U.K.* 11 F8 51 23N 0 33 E
Gilmer, *U.S.A.* 81 J7 32 44N 94 57W
Gilmore, *Australia* 63 F4 35 20S 148 12 E
Gilmore, L., *Australia* 61 F3 32 29S 121 37 E
Gilmour, *Canada* 70 D4 44 48N 77 37W
Gilroy, *U.S.A.* 83 H3 37 1N 121 34W
Gimbi, *Ethiopia* 51 G12 9 3N 35 42 E
Gimli, *Canada* 73 C9 50 40N 97 0W
Gin Gin, *Australia* 63 D5 25 0S 151 58 E
Gindie, *Australia* 62 C4 23 44S 148 8 E
Gingin, *Australia* 61 F2 31 22S 115 54 E
Ginir, *Ethiopia* 46 F3 7 6N 40 40 E
Giohar, *Somali Rep.* 46 G4 2 48N 45 30 E
Gióna, Óros, *Greece* 21 E10 38 38N 22 14 E
Gir Hills, *India* 42 J4 21 0N 71 0 E
Girab, *India* 42 F4 26 2N 70 38 E
Girâfi, W. →, *Egypt* 47 F3 29 58N 34 39 E
Girard, *Kans., U.S.A.* 81 G7 37 31N 94 51W
Girard, *Ohio, U.S.A.* 78 E4 41 9N 80 42W
Girard, *Pa., U.S.A.* 78 D4 42 0N 80 19W
Girardot, *Colombia* 92 C4 4 18N 74 48W
Girdle Ness, *U.K.* 12 D6 57 9N 2 3W
Giresun, *Turkey* 25 F6 40 55N 38 30 E
Girga, *Egypt* 51 C11 26 17N 31 55 E
Giridih, *India* 43 G12 24 10N 86 21 E
Girilambone, *Australia* 63 E4 31 16S 146 57 E
Girne = Kyrenia, *Cyprus* 23 D12 35 20N 33 20 E
Girona = Gerona, *Spain* 19 B7 41 58N 2 46 E
Gironde →, *France* 18 D3 45 32N 1 7W
Giru, *Australia* 62 B4 19 30S 147 5 E
Girvan, *U.K.* 12 F4 55 14N 4 51W
Gisborne, *N.Z.* 59 H7 38 39S 178 5 E
Gisenyi, *Rwanda* 54 C2 1 41S 29 15 E
Gislaved, *Sweden* 9 H15 57 19N 13 32 E
Gitega, *Burundi* 54 C2 3 26S 29 56 E
Giuba →, *Somali Rep.* 46 G3 1 30N 42 35 E
Giurgiu, *Romania* 17 G13 43 52N 25 57 E
Giza = El Gîza, *Egypt* 51 C11 30 0N 31 10 E
Gizhiga, *Russia* 27 C17 62 3N 160 30 E
Gizhiginskaya Guba, *Russia* 27 C16 61 0N 158 0 E
Giżycko, *Poland* 17 A11 54 2N 21 48 E
Gjirokastra, *Albania* 21 D9 40 7N 20 10 E
Gjoa Haven, *Canada* 68 B10 68 20N 96 8W
Gjøvik, *Norway* 9 F14 60 47N 10 43 E
Glace Bay, *Canada* 71 C8 46 11N 59 58W
Glacier Bay, *U.S.A.* 72 B1 58 40N 136 0W
Glacier Nat. Park, *Canada* 72 C5 51 15N 117 30W
Glacier Park, *U.S.A.* 82 B7 48 30N 113 18W
Glacier Peak, *U.S.A.* 82 B3 48 7N 121 7W
Gladewater, *U.S.A.* 81 J7 32 33N 94 56W
Gladstone, *Queens., Australia* 62 C5 23 52S 151 16 E
Gladstone, *S. Austral., Australia* 63 E2 33 15S 138 22 E
Gladstone, *W. Austral., Australia* 61 E1 25 57S 114 17 E
Gladstone, *Canada* 73 C9 50 13N 98 57W
Gladstone, *U.S.A.* 76 C2 45 51N 87 1W
Gladwin, *U.S.A.* 76 D3 43 59N 84 29W
Gladys L., *Canada* 72 B2 59 50N 133 0W
Glåma = Glomma →, *Norway* 9 G14 59 12N 10 57 E
Gláma, *Iceland* 8 D2 65 48N 23 0W
Glamis, *U.S.A.* 85 N11 32 55N 115 5W
Glasco, *Kans., U.S.A.* 80 F6 39 22N 97 50W
Glasco, *N.Y., U.S.A.* 79 D11 42 3N 73 57W
Glasgow, *U.K.* 12 F4 55 51N 4 15W
Glasgow, *Ky., U.S.A.* 76 G3 37 0N 85 55W
Glasgow, *Mont., U.S.A.* 82 B10 48 12N 106 38W
Glastonbury, *U.K.* 11 F5 51 9N 2 43W
Glastonbury, *U.S.A.* 79 E12 41 43N 72 37W
Glazov, *Russia* 24 C9 58 9N 52 40 E
Gleiwitz = Gliwice, *Poland* 17 C10 50 22N 18 41 E
Glen, *U.S.A.* 79 B13 44 7N 71 11W
Glen Affric, *U.K.* 12 D4 57 17N 5 1W
Glen Canyon Dam, *U.S.A.* 83 H8 36 57N 111 29W
Glen Canyon National Recreation Area, *U.S.A.* 83 H8 37 15N 111 0W
Glen Coe, *U.K.* 12 E4 56 40N 5 0W
Glen Cove, *U.S.A.* 79 F11 40 52N 73 38W
Glen Garry, *U.K.* 12 D3 57 3N 5 7W
Glen Innes, *Australia* 63 D5 29 44S 151 44 E
Glen Lyon, *U.S.A.* 79 E8 41 10N 76 5W
Glen Mor, *U.K.* 12 D4 57 9N 4 37W
Glen Moriston, *U.K.* 12 D4 57 11N 4 52W
Glen Orchy, *U.K.* 12 E4 56 27N 4 52W
Glen Spean, *U.K.* 12 E4 56 53N 4 40W
Glen Ullin, *U.S.A.* 80 B4 46 49N 101 50W
Glenburgh, *Australia* 61 E2 25 26S 116 6 E
Glencoe, *Canada* 78 D3 42 45N 81 43W
Glencoe, *S. Africa* 57 D5 28 11S 30 11 E
Glencoe, *U.S.A.* 80 C7 44 46N 94 9W
Glendale, *Ariz., U.S.A.* 83 K7 33 32N 112 11W
Glendale, *Calif., U.S.A.* 85 L8 34 9N 118 15W
Glendale, *Oreg., U.S.A.* 82 E2 42 44N 123 26W
Glendale, *Zimbabwe* 55 F3 17 22S 31 5 E
Glendive, *U.S.A.* 80 B2 47 7N 104 43W
Glendo, *U.S.A.* 80 D2 42 30N 105 2W
Glenelg, *Australia* 63 E2 34 58S 138 31 E
Glenelg →, *Australia* 63 F3 38 4S 140 59 E
Glenflorrie, *Australia* 60 D2 22 55S 115 59 E
Glengarriff, *Ireland* 13 E2 51 45N 9 34W
Glengyle, *Australia* 62 C2 24 48S 139 37 E
Glenmora, *U.S.A.* 81 K8 30 59N 92 35W
Glenmorgan, *Australia* 63 D4 27 14S 149 42 E
Glenn, *U.S.A.* 84 F4 39 31N 122 1W
Glenns Ferry, *U.S.A.* 82 E6 42 57N 115 18W
Glenorchy, *Australia* 62 G4 42 49S 147 18 E
Glenore, *Australia* 62 B3 17 50S 141 12 E
Glenormiston, *Australia* 62 C2 22 55S 138 50 E
Glenreagh, *Australia* 63 E5 30 2S 153 1 E
Glenrock, *U.S.A.* 82 E11 42 52N 105 52W
Glenrothes, *U.K.* 12 E5 56 12N 3 10W
Glens Falls, *U.S.A.* 79 C11 43 19N 73 39W
Glenties, *Ireland* 13 B3 54 49N 8 16W
Glenville, *U.S.A.* 76 F5 38 56N 80 50W
Glenwood, *Alta., Canada* 72 D6 49 21N 113 31W
Glenwood, *Nfld., Canada* 71 C9 49 0N 54 58W
Glenwood, *Ark., U.S.A.* 81 H8 34 20N 93 33W
Glenwood, *Hawaii, U.S.A.* 74 J17 19 29N 155 9W
Glenwood, *Iowa, U.S.A.* 80 E7 41 3N 95 45W
Glenwood, *Minn., U.S.A.* 80 C7 45 39N 95 23W
Glenwood, *Wash., U.S.A.* 84 D5 46 1N 121 17W
Glenwood Springs, *U.S.A.* 82 G10 39 33N 107 19W
Glettinganes, *Iceland* 8 D7 65 30N 13 37W
Gliwice, *Poland* 17 C10 50 22N 18 41 E
Globe, *U.S.A.* 83 K8 33 24N 110 47W
Głogów, *Poland* 16 C9 51 37N 16 5 E
Glomma →, *Norway* 9 G14 59 12N 10 57 E
Glorieuses, Is., *Ind. Oc.* 57 A8 11 30S 47 20 E
Glossop, *U.K.* 10 D6 53 27N 1 56W
Gloucester, *Australia* 63 E5 32 0S 151 59 E
Gloucester, *U.K.* 11 F5 51 53N 2 15W
Gloucester, *U.S.A.* 79 D14 42 37N 70 40W
Gloucester I., *Australia* 62 B4 20 0S 148 30 E
Gloucestershire □, *U.K.* 11 F5 51 46N 2 15W
Gloversville, *U.S.A.* 79 C10 43 3N 74 21W
Glovertown, *Canada* 71 C9 48 40N 54 3W
Glusk, *Belarus* 17 B15 52 53N 28 41 E
Gmünd, *Austria* 16 D8 48 45N 15 0 E
Gmunden, *Austria* 16 E7 47 55N 13 48 E
Gniezno, *Poland* 17 B9 52 30N 17 35 E
Gnowangerup, *Australia* 61 F2 33 58S 117 59 E
Go Cong, *Vietnam* 39 G6 10 22N 106 40 E

Gô-no-ura, Japan **31 H4** 33 44N 129 40 E
Go Quao, Vietnam **39 H5** 9 43N 105 17 E
Goa, India **40 M8** 15 33N 73 59 E
Goa □, India **40 M8** 15 33N 73 59 E
Goalen Hd., Australia **63 F5** 36 33S 150 4 E
Goalpara, India **41 F17** 26 10N 90 40 E
Goalundo Ghat, Bangla. **43 H13** 23 50N 89 47 E
Goat Fell, U.K. **12 F3** 55 38N 5 11W
Goba, Ethiopia **46 F2** 7 1N 39 59 E
Goba, Mozam. **57 D5** 26 15S 32 13 E
Gobabis, Namibia **56 C2** 22 30S 19 0 E
Gobi, Asia **34 C5** 44 0N 111 0 E
Gobō, Japan **31 H7** 33 53N 135 10 E
Gochas, Namibia **56 C2** 24 59S 18 55 E
Godavari →, India **41 L13** 16 25N 82 18 E
Godavari Point, India **41 L13** 17 0N 82 20 E
Godbout, Canada **71 C6** 49 20N 67 38W
Godda, India **43 G12** 24 50N 87 13 E
Goderich, Canada **70 D3** 43 45N 81 41W
Godhavn, Greenland **4 C5** 69 15N 53 38W
Godhra, India **42 H5** 22 49N 73 40 E
Godoy Cruz, Argentina **94 C2** 32 56S 68 52W
Gods →, Canada **73 B10** 56 22N 92 51W
Gods L., Canada **73 C10** 54 40N 94 15W
Godthåb, Greenland **69 B14** 64 10N 51 35W
Godwin Austen = K2, Pakistan **43 B7** 35 58N 76 32 E
Goeie Hoop, Kaap die = Good Hope, C. of, S. Africa **56 E2** 34 24S 18 30 E
Goéland, L. au, Canada **70 C4** 49 50N 76 48W
Goeree, Neths. **15 C3** 51 50N 4 0 E
Goes, Neths. **15 C3** 51 30N 3 55 E
Gogama, Canada **70 C3** 47 35N 81 43W
Gogango, Australia **62 C5** 23 40S 150 2 E
Gogebic, L., U.S.A. **80 B10** 46 30N 89 35W
Gogra = Ghaghara →, India **43 G11** 25 45N 84 40 E
Goiânia, Brazil **93 G9** 16 43S 49 20W
Goiás, Brazil **93 G8** 15 55S 50 10W
Goiás □, Brazil **93 F9** 12 10S 48 0W
Goio-Ere, Brazil **95 A5** 24 12S 53 1W
Gojō, Japan **31 G7** 34 21N 135 42 E
Gojra, Pakistan **42 D5** 31 10N 72 40 E
Gokarannath, India **43 F9** 27 57N 80 39 E
Gökçeada, Turkey **21 D11** 40 10N 25 50 E
Gokteik, Burma **41 H20** 22 26N 97 0 E
Gokurt, Pakistan **42 E2** 29 40N 67 26 E
Gola, India **43 E9** 28 3N 80 32 E
Golakganj, India **43 F13** 26 8N 89 52 E
Golan Heights = Hagolan, Syria **47 B4** 33 0N 35 45 E
Golāshkerd, Iran **45 E8** 27 59N 57 16 E
Golchikha, Russia **4 B12** 71 45N 83 30 E
Golconda, U.S.A. **82 F5** 40 58N 117 30W
Gold Beach, U.S.A. **82 E1** 42 25N 124 25W
Gold Coast, Australia **63 D5** 28 0S 153 25 E
Gold Coast, W. Afr. **48 F3** 4 0N 1 40W
Gold Hill, U.S.A. **82 E2** 42 26N 123 3W
Golden, Canada **72 C5** 51 20N 116 59W
Golden, U.S.A. **80 F2** 39 42N 105 15W
Golden B., N.Z. **59 J4** 40 40S 172 50 E
Golden Gate, U.S.A. **82 H2** 37 54N 122 30W
Golden Hinde, Canada **72 D3** 49 40N 125 44W
Golden Lake, Canada **78 A7** 45 34N 77 21W
Golden Prairie, Canada **73 C7** 50 13N 109 37W
Golden Vale, Ireland **13 D3** 52 33N 8 17W
Goldendale, U.S.A. **82 D3** 45 49N 120 50W
Goldfield, U.S.A. **83 H5** 37 42N 117 14W
Goldfields, Canada **73 B7** 59 28N 108 29W
Goldsand L., Canada **73 B8** 57 2N 101 8W
Goldsboro, U.S.A. **77 H7** 35 23N 77 59W
Goldsmith, U.S.A. **81 K3** 31 59N 102 37W
Goldsworthy, Australia **60 D2** 20 21S 119 30 E
Goldthwaite, U.S.A. **81 K5** 31 27N 98 34W
Goleniów, Poland **16 B8** 53 35N 14 50 E
Golestānak, Iran **45 D7** 30 36N 54 14 E
Goleta, U.S.A. **85 L7** 34 27N 119 50W
Golfito, Costa Rica **88 E3** 8 41N 83 5W
Golfo Aranci, Italy **20 D3** 40 59N 9 38 E
Goliad, U.S.A. **81 L6** 28 40N 97 23W
Golpāyegān, Iran **45 C6** 33 27N 50 18 E
Golra, Pakistan **42 C5** 33 37N 72 56 E
Golspie, U.K. **12 D5** 57 58N 3 59W
Goma, Rwanda **54 C2** 2 11S 29 18 E
Goma, Zaïre **54 C2** 1 37S 29 10 E
Gomati →, India **43 G10** 25 32N 83 11 E
Gombari, Zaïre **54 B2** 2 45N 29 3 E
Gombe →, Tanzania **54 C3** 4 38S 31 40 E
Gomel = Homyel, Belarus **17 B16** 52 28N 31 0 E
Gomera, Canary Is. **22 F2** 28 7N 17 14W
Gómez Palacio, Mexico **86 B4** 25 40N 104 0W
Gomīshān, Iran **45 B7** 37 4N 54 6 E
Gomogomo, Indonesia **37 F8** 6 39S 134 43 E
Gomoh, India **41 H15** 23 52N 86 10 E
Gompa = Ganta, Liberia **50 G3** 7 15N 8 59W
Gonābād, Iran **45 C8** 34 15N 58 45 E
Gonaïves, Haiti **89 C5** 19 20N 72 42W
Gonâve, G. de la, Haiti **89 C5** 19 29N 72 42W
Gonâve, I. de la, Haiti **89 C5** 18 45N 73 0W
Gonbab-e Kāvūs, Iran **45 B7** 37 20N 55 25 E
Gonda, India **43 F9** 27 9N 81 58 E
Gondal, India **42 J4** 21 58N 70 52 E
Gonder, Ethiopia **51 F12** 12 39N 37 30 E
Gondia, India **40 J12** 21 23N 80 10 E
Gondola, Mozam. **55 F3** 19 10S 33 37 E
Gönen, Turkey **21 D12** 40 6N 27 39 E
Gonghe, China **32 C5** 36 18N 100 32 E
Gongolgon, Australia **63 E4** 30 21S 146 54 E
Goniri, Nigeria **51 F7** 11 30N 12 15 E
Gonzales, Calif., U.S.A. **83 H3** 36 30N 121 26W
Gonzales, Tex., U.S.A. **81 L6** 29 30N 97 27W
González Chaves, Argentina **94 D3** 38 2S 60 5W
Good Hope, C. of, S. Africa **56 E2** 34 24S 18 30 E
Gooderham, Canada **70 D4** 44 54N 78 21W
Goodeve, Canada **73 C8** 51 4N 103 10W
Gooding, U.S.A. **82 E6** 42 56N 114 43W
Goodland, U.S.A. **80 F4** 39 21N 101 43W
Goodnight, U.S.A. **81 H4** 35 2N 101 11W
Goodooga, Australia **63 D4** 29 3S 147 28 E
Goodsoil, Canada **73 C7** 54 24N 109 13W
Goodsprings, U.S.A. **83 J6** 35 50N 115 26W
Goole, U.K. **10 D7** 53 42N 0 53W
Goolgowi, Australia **63 E4** 33 58S 145 41 E
Goomalling, Australia **61 F2** 31 15S 116 49 E
Goombalie, Australia **63 D4** 29 59S 145 26 E
Goonda, Mozam. **55 F3** 19 48S 33 57 E
Goondiwindi, Australia **63 D5** 28 30S 150 21 E
Goongarrie, L., Australia **61 F3** 30 3S 121 9 E
Goonyella, Australia **62 C4** 21 47S 147 58 E
Goor, Neths. **15 B6** 52 13N 6 33 E
Gooray, Australia **63 D5** 28 25S 150 2 E
Goose →, Canada **71 B7** 53 20N 60 35W
Goose L., U.S.A. **82 F3** 41 56N 120 26W
Gop, India **40 H6** 22 5N 69 50 E
Gopalganj, India **43 F11** 26 28N 84 30 E
Göppingen, Germany **16 D5** 48 42N 9 39 E
Gorakhpur, India **43 F10** 26 47N 83 23 E
Goražde, Bos.-H. **21 C8** 43 38N 18 58 E
Gorda, U.S.A. **84 K5** 35 53N 121 26W
Gorda, Pta., Canary Is. **22 F2** 28 45N 18 0W
Gorda, Pta., Nic. **88 D3** 14 20N 83 10W
Gordan B., Australia **60 B5** 11 35S 130 10 E
Gordon, U.S.A. **80 D3** 42 48N 102 12W
Gordon →, Australia **62 G4** 42 27S 145 30 E
Gordon Downs, Australia **60 C4** 18 48S 128 33 E
Gordon L., Alta., Canada **73 B6** 56 30N 110 25W
Gordon L., N.W.T., Canada **72 A6** 63 5N 113 11W
Gordonvale, Australia **62 B4** 17 5S 145 50 E
Gore, Australia **63 D5** 28 17S 151 30 E
Goré, Chad **51 G8** 7 59N 16 31 E
Gore, Ethiopia **51 G12** 8 12N 35 32 E
Gore, N.Z. **59 M2** 46 5S 168 58 E
Gore Bay, Canada **70 C3** 45 57N 82 28W
Gorey, Ireland **13 D5** 52 41N 6 18W
Gorg, Iran **45 D8** 29 29N 59 43 E
Gorgān, Iran **45 B7** 36 50N 54 29 E
Gorgona, I., Colombia **92 C3** 3 0N 78 10W
Gorham, U.S.A. **79 B13** 44 23N 71 10W
Gorinchem, Neths. **15 C4** 51 50N 4 59 E
Gorízia, Italy **20 B5** 45 56N 13 37 E
Gorki = Nizhniy Novgorod, Russia **24 C7** 56 20N 44 0 E
Gorkiy = Nizhniy Novgorod, Russia **24 C7** 56 20N 44 0 E
Gorkovskoye Vdkhr., Russia **24 C7** 57 2N 43 4 E
Görlitz, Germany **16 C8** 51 9N 14 58 E
Gorlovka = Horlivka, Ukraine **25 E6** 48 19N 38 5 E
Gorman, Calif., U.S.A. **85 L8** 34 47N 118 51W
Gorman, Tex., U.S.A. **81 J5** 32 12N 98 41W
Gorna Dzhumayo = Blagoevgrad, Bulgaria **21 C10** 42 2N 23 5 E
Gorna Oryakhovitsa, Bulgaria **21 C11** 43 7N 25 40 E
Gorno-Altay □, Russia **26 D9** 51 0N 86 0 E
Gorno-Altaysk, Russia **26 D9** 51 50N 86 5 E
Gorno Slinkino = Gornopravdinsk, Russia **26 C8** 60 5N 70 0 E
Gornopravdinsk, Russia **26 C8** 60 5N 70 0 E
Gornyatski, Russia **24 A11** 67 32N 64 3 E
Gornyi, Russia **30 B6** 44 57N 133 59 E
Gorodenka = Horodenka, Ukraine **17 D13** 48 41N 25 29 E
Gorodok = Horodok, Ukraine **17 D12** 49 46N 23 32 E
Gorokhov = Horokhiv, Ukraine **17 C13** 50 30N 24 45 E
Goromonzi, Zimbabwe **55 F3** 17 52S 31 22 E
Gorongose →, Mozam. **57 C5** 20 30S 34 40 E
Gorongoza, Mozam. **55 F3** 18 44S 34 2 E
Gorongoza, Sa. da, Mozam. **55 F3** 18 27S 34 2 E
Gorontalo, Indonesia **37 D6** 0 35N 123 5 E
Gort, Ireland **13 C3** 53 3N 8 49W
Gortis, Greece **23 D6** 35 4N 24 58 E
Gorzów Wielkopolski, Poland **16 B8** 52 43N 15 15 E
Gosford, Australia **63 E5** 33 23S 151 18 E
Goshen, Calif., U.S.A. **84 J7** 36 21N 119 25W
Goshen, Ind., U.S.A. **76 E3** 41 35N 85 50W
Goshen, N.Y., U.S.A. **79 E10** 41 24N 74 20W
Goshogawara, Japan **30 D10** 40 48N 140 27 E
Goslar, Germany **16 C6** 51 54N 10 25 E
Gospič, Croatia **16 F8** 44 35N 15 23 E
Gosport, U.K. **11 G6** 50 48N 1 9W
Gosse →, Australia **62 B1** 19 32S 134 37 E
Göta älv →, Sweden **9 H14** 57 42N 11 54 E
Göta kanal, Sweden **9 G16** 58 30N 15 58 E
Götaland, Sweden **9 G15** 57 30N 14 30 E
Göteborg, Sweden **9 H14** 57 43N 11 59 E
Gotha, Germany **16 C6** 50 56N 10 42 E
Gothenburg = Göteborg, Sweden **9 H14** 57 43N 11 59 E
Gothenburg, U.S.A. **80 E4** 40 56N 100 10W
Gotland, Sweden **9 H18** 57 30N 18 33 E
Gotska Sandön, Sweden **9 G18** 58 24N 19 15 E
Gōtsu, Japan **31 G6** 35 0N 132 14 E
Göttingen, Germany **16 C5** 51 31N 9 55 E
Gottwaldov = Zlin, Czech. **17 D9** 49 14N 17 40 E
Goubangzi, China **35 D11** 41 20N 121 52 E
Gouda, Neths. **15 B4** 52 1N 4 42 E
Goúdhoura, Ákra, Greece **23 E8** 34 59N 26 6 E
Gough I., Atl. Oc. **2 G9** 40 10S 9 45W
Gouin, Rés., Canada **70 C5** 48 35N 74 40W
Goulburn, Australia **63 E4** 34 44S 149 44 E
Goulburn Is., Australia **62 A1** 11 40S 133 20 E
Gounou-Gaya, Chad **51 G8** 9 38N 15 31 E
Gouri, Chad **51 E8** 19 36N 19 36 E
Gourits →, S. Africa **56 E3** 34 21S 21 52 E
Gourma Rharous, Mali **50 E4** 16 55N 1 50W
Goúrnais, Greece **23 D7** 35 19N 25 16 E
Gourock Ra., Australia **63 F4** 36 0S 149 25 E
Gouverneur, U.S.A. **79 B9** 44 20N 75 28W
Gouviá, Greece **23 A3** 39 39N 19 50 E
Govan, Canada **73 C8** 51 20N 105 0W
Governador Valadares, Brazil **93 G10** 18 15S 41 57W
Governor's Harbour, Bahamas **88 A4** 25 10N 76 14W
Gowan Ra., Australia **62 C4** 25 0S 145 0 E
Gowanda, U.S.A. **78 D6** 42 28N 78 56W
Gowd-e Zirreh, Afghan. **40 E3** 29 45N 62 0 E
Gower, U.K. **11 F3** 51 35N 4 10W
Gowna, L., Ireland **13 C4** 53 51N 7 34W
Goya, Argentina **94 B4** 29 10S 59 10W
Goyder Lagoon, Australia **63 D2** 27 3S 138 58 E
Goyllarisquisga, Peru **92 F3** 10 31S 76 24W
Goz Beïda, Chad **51 F9** 12 10N 21 20 E
Gozo, Malta **23 C1** 36 3N 14 13 E
Graaff-Reinet, S. Africa **56 E3** 32 13S 24 32 E
Gračac, Croatia **16 F8** 44 18N 15 57 E
Grace, U.S.A. **82 E8** 42 35N 111 44W
Graceville, U.S.A. **80 C6** 45 34N 96 26W
Gracias a Dios, C., Honduras **88 C3** 15 0N 83 10W
Graciosa, I., Canary Is. **22 E6** 29 15N 13 32W
Grado, Spain **19 A2** 43 23N 6 4W
Gradule, Australia **63 D4** 28 32S 149 15 E
Grady, U.S.A. **81 H3** 34 49N 103 19W
Grafton, Australia **63 D5** 29 38S 152 58 E
Grafton, U.S.A. **80 A6** 48 25N 97 25W
Graham, Canada **70 C1** 49 20N 90 30W
Graham, N.C., U.S.A. **77 G6** 36 5N 79 25W
Graham, Tex., U.S.A. **81 J5** 33 6N 98 35W
Graham →, Canada **72 B4** 56 31N 122 17W
Graham, Mt., U.S.A. **83 K9** 32 42N 109 52W
Graham Bell, Os., Russia **26 A7** 81 0N 62 0 E
Graham I., Canada **72 C2** 53 40N 132 30W
Graham Land, Antarctica **5 C17** 65 0S 64 0W
Grahamdale, Canada **73 C9** 51 23N 98 30W
Grahamstown, S. Africa **56 E4** 33 19S 26 31 E
Grain Coast, W. Afr. **48 F2** 4 20N 10 0W
Grajaú, Brazil **93 E9** 5 50S 46 4W
Grajaú →, Brazil **93 D10** 3 41S 44 48W
Grampian Highlands = Grampian Mts., U.K. **12 E5** 56 50N 4 0W
Grampian Mts., U.K. **12 E5** 56 50N 4 0W
Gran Canaria, Canary Is. **22 F4** 27 55N 15 35W
Gran Chaco, S. Amer. **94 B3** 25 0S 61 0W
Gran Paradiso, Italy **20 B2** 45 33N 7 17 E
Gran Sasso d'Italia, Italy **20 C5** 42 27N 13 42 E
Granada, Nic. **88 D2** 11 58N 86 0W
Granada, Spain **19 D4** 37 10N 3 35W
Granada, U.S.A. **81 F3** 38 4N 102 19W
Granadilla de Abona, Canary Is. **22 F3** 28 7N 16 33W
Granard, Ireland **13 C4** 53 47N 7 30W
Granbury, U.S.A. **81 J6** 32 27N 97 47W
Granby, Canada **70 C5** 45 25N 72 45W
Grand →, Mo., U.S.A. **80 F8** 39 23N 93 7W
Grand →, S. Dak., U.S.A. **80 C4** 45 40N 100 45W
Grand Bahama, Bahamas **88 A4** 26 40N 78 30W
Grand Bank, Canada **71 C8** 47 6N 55 48W
Grand Bassam, Ivory C. **50 G4** 5 10N 3 49W
Grand-Bourg, Guadeloupe **89 C7** 15 53N 61 19W
Grand Canal = Yun Ho →, China **35 E9** 39 10N 117 10 E
Grand Canyon, U.S.A. **83 H7** 36 3N 112 9W
Grand Canyon National Park, U.S.A. **83 H7** 36 15N 112 30W
Grand Cayman, Cayman Is. **88 C3** 19 20N 81 20W
Grand Coulee, U.S.A. **82 C4** 47 57N 119 0W
Grand Coulee Dam, U.S.A. **82 C4** 47 57N 118 59W
Grand Falls, Canada **71 C8** 48 56N 55 40W
Grand Forks, Canada **72 D5** 49 0N 118 30W
Grand Forks, U.S.A. **80 B6** 47 55N 97 3W
Grand Haven, U.S.A. **76 D2** 43 4N 86 13W
Grand I., U.S.A. **76 B2** 46 31N 86 40W
Grand Island, U.S.A. **80 E5** 40 55N 98 21W
Grand Isle, U.S.A. **81 L10** 29 14N 90 0W
Grand Junction, U.S.A. **83 G9** 39 4N 108 33W
Grand L., N.B., Canada **71 C6** 45 57N 66 7W
Grand L., Nfld., Canada **71 C8** 49 0N 57 30W
Grand L., Nfld., Canada **71 B7** 53 40N 60 30W
Grand L., U.S.A. **81 L8** 29 55N 92 47W
Grand Lac Victoria, Canada **70 C4** 47 35N 77 35W
Grand Lahou, Ivory C. **50 G3** 5 10N 5 5W
Grand Lake, U.S.A. **82 F11** 40 15N 105 49W
Grand Manan I., Canada **71 D6** 44 45N 66 52W
Grand Marais, Canada **80 B9** 47 45N 90 25W
Grand Marais, U.S.A. **76 B3** 46 40N 85 59W
Grand-Mère, Canada **70 C5** 46 36N 72 40W
Grand Portage, U.S.A. **70 C2** 47 58N 89 41W
Grand Prairie, U.S.A. **81 J6** 32 47N 97 0W
Grand Rapids, Canada **73 C9** 53 12N 99 19W
Grand Rapids, Mich., U.S.A. **76 D2** 42 58N 85 40W
Grand Rapids, Minn., U.S.A. **80 B8** 47 14N 93 31W
Grand St.-Bernard, Col du, Europe **16 F4** 45 50N 7 10 E
Grand Teton, U.S.A. **82 E8** 43 54N 111 50W
Grand Valley, U.S.A. **82 G9** 39 27N 108 3W
Grand View, Canada **73 C8** 51 10N 100 42W
Grande →, Jujuy, Argentina **94 A2** 24 20S 65 2W
Grande →, Mendoza, Argentina **94 D2** 36 52S 69 45W
Grande →, Bolivia **92 G6** 15 51S 64 39W
Grande →, Bahia, Brazil **93 F10** 11 30S 44 30W
Grande →, Minas Gerais, Brazil **93 H8** 20 6S 51 4W
Grande, B., Argentina **96 G3** 50 30S 68 20W
Grande, Rio →, U.S.A. **81 N6** 25 58N 97 9W
Grande Baie, Canada **71 C5** 48 19N 70 52W
Grande Baleine, R. de la →, Canada **70 A4** 55 16N 77 47W
Grande Cache, Canada **72 C5** 53 53N 119 8W
Grande de Santiago →, Mexico **86 C3** 21 36N 105 26W
Grande-Entrée, Canada **71 C7** 47 30N 61 40W
Grande Prairie, Canada **72 B5** 55 10N 118 50W
Grande-Rivière, Canada **71 C7** 48 26N 64 30W
Grande-Vallée, Canada **71 C6** 49 14N 65 8W
Grandes-Bergeronnes, Canada **71 C6** 48 16N 69 35W
Grandfalls, U.S.A. **81 K3** 31 20N 102 51W
Grandoe Mines, Canada **72 B3** 56 29N 129 54W
Grandview, U.S.A. **82 C4** 46 15N 119 54W
Graneros, Chile **94 C1** 34 5S 70 45W
Grangemouth, U.K. **12 E5** 56 1N 3 42W
Granger, Wash., U.S.A. **82 C3** 46 21N 120 11W
Granger, Wyo., U.S.A. **82 F9** 41 35N 109 58W
Grangeville, U.S.A. **82 D5** 45 56N 116 7W
Granite City, U.S.A. **80 F9** 38 42N 90 9W
Granite Falls, U.S.A. **80 C7** 44 49N 95 33W
Granite Mt., U.S.A. **85 M10** 33 5N 116 28W
Granite Peak, Australia **61 E3** 25 40S 121 20 E
Granite Peak, U.S.A. **82 D9** 45 10N 109 48W
Granity, N.Z. **59 J3** 41 39S 171 51 E
Granja, Brazil **93 D10** 3 7S 40 50W
Granollers, Spain **19 B7** 41 39N 2 18 E
Grant, U.S.A. **80 E4** 40 53N 101 42W
Grant, Mt., U.S.A. **82 G4** 38 34N 118 48W
Grant City, U.S.A. **80 E7** 40 29N 94 25W
Grant I., Australia **60 B5** 11 10S 132 52 E
Grant Range, U.S.A. **83 G6** 38 30N 115 25W
Grantham, U.K. **10 E7** 52 55N 0 38W
Grantown-on-Spey, U.K. **12 D5** 57 20N 3 36W
Grants, U.S.A. **83 J10** 35 9N 107 52W
Grants Pass, U.S.A. **82 E2** 42 26N 123 19W
Grantsburg, U.S.A. **80 C8** 45 47N 92 41W
Grantsville, U.S.A. **82 F7** 40 36N 112 28W
Granville, France **18 B3** 48 50N 1 35W
Granville, N. Dak., U.S.A. **80 A4** 48 16N 100 47W
Granville, N.Y., U.S.A. **76 D9** 43 24N 73 16W
Granville L., Canada **73 B8** 56 18N 100 30W
Grapeland, U.S.A. **81 K7** 31 30N 95 29W
Gras, L. de, Canada **68 B8** 64 30N 110 30W
Graskop, S. Africa **57 C5** 24 56S 30 49 E
Grass →, Canada **73 B9** 56 3N 96 33W
Grass Range, U.S.A. **82 C9** 47 0N 109 0W
Grass River Prov. Park, Canada **73 C8** 54 40N 100 50W
Grass Valley, Calif., U.S.A. **84 F6** 39 13N 121 4W
Grass Valley, Oreg., U.S.A. **82 D3** 45 22N 120 47W
Grasse, France **18 E7** 43 38N 6 56 E
Grassmere, Australia **63 E3** 31 24S 142 38 E
Graulhet, France **18 E4** 43 45N 1 59 E
Gravelbourg, Canada **73 D7** 49 50N 106 35W
's-Gravenhage, Neths. **15 B4** 52 7N 4 17 E
Gravenhurst, Canada **78 B5** 44 52N 79 20W
Gravesend, Australia **63 D5** 29 35S 150 20 E
Gravesend, U.K. **11 F8** 51 26N 0 22 E
Gravois, Pointe-à-, Haiti **89 C5** 18 15N 73 56W
Grayling, U.S.A. **76 C3** 44 40N 84 43W
Grayling →, Canada **72 B3** 59 21N 125 0W
Grays Harbor, U.S.A. **82 C1** 46 59N 124 1W
Grays L., U.S.A. **82 E8** 43 4N 111 26W
Grays River, U.S.A. **84 D3** 46 21N 123 37W
Grayson, Canada **73 C8** 50 45N 102 40W
Graz, Austria **16 E8** 47 4N 15 27 E
Greasy L., Canada **72 A4** 62 55N 122 12W
Great Abaco I., Bahamas **88 A4** 26 25N 77 10W
Great Artesian Basin, Australia **62 C3** 23 0S 144 0 E
Great Australian Bight, Australia **61 F5** 33 30S 130 0 E
Great Bahama Bank, Bahamas **88 B4** 23 15N 78 0W
Great Barrier I., N.Z. **59 G5** 36 11S 175 25 E
Great Barrier Reef, Australia **62 B4** 18 0S 146 50 E
Great Barrington, U.S.A. **79 D11** 42 12N 73 22W
Great Basin, U.S.A. **82 G5** 40 0N 117 0W
Great Bear →, Canada **68 B7** 65 0N 124 0W
Great Bear L., Canada **68 B7** 65 30N 120 0W
Great Belt = Store Bælt, Denmark **9 J14** 55 20N 11 0 E
Great Bend, Kans., U.S.A. **80 F5** 38 22N 98 46W
Great Bend, Pa., U.S.A. **79 E9** 41 58N 75 45W
Great Blasket I., Ireland **13 D1** 52 6N 10 32W
Great Britain, Europe **6 E5** 54 0N 2 15W
Great Central, Canada **72 D3** 49 20N 125 10W
Great Dividing Ra., Australia **62 C4** 23 0S 146 0 E
Great Driffield = Driffield, U.K. **10 C7** 54 0N 0 26W
Great Exuma I., Bahamas **88 B4** 23 30N 75 50W
Great Falls, Canada **73 C9** 50 27N 96 1W
Great Falls, U.S.A. **82 C8** 47 30N 111 17W
Great Fish = Groot Vis →, S. Africa **56 E4** 33 28S 27 5 E
Great Guana Cay, Bahamas **88 B4** 24 0N 76 20W
Great Harbour Deep, Canada **71 B8** 50 25N 56 32W
Great I., Canada **73 B9** 58 53N 96 35W
Great Inagua I., Bahamas **89 B5** 21 0N 73 20W
Great Indian Desert = Thar Desert, India **42 F4** 28 0N 72 0 E
Great Karoo, S. Africa **56 E3** 31 55S 21 0 E
Great Lake, Australia **62 G4** 41 50S 146 40 E
Great Malvern, U.K. **11 E5** 52 7N 2 18W
Great Ormes Head, U.K. **10 D4** 53 20N 3 52W
Great Ouse →, U.K. **10 E8** 52 48N 0 21 E
Great Palm I., Australia **62 B4** 18 45S 146 40 E
Great Plains, N. Amer. **74 A6** 47 0N 105 0W
Great Ruaha →, Tanzania **54 D4** 7 56S 37 52 E
Great Saint Bernard P. = Grand St.-Bernard, Col du, Europe **16 F4** 45 50N 7 10 E
Great Salt L., U.S.A. **82 F7** 41 15N 112 40W
Great Salt Lake Desert, U.S.A. **82 F7** 40 50N 113 30W
Great Salt Plains L., U.S.A. **81 G5** 36 45N 98 8W
Great Sandy Desert, Australia **60 D3** 21 0S 124 0 E
Great Sangi = Sangihe, P., Indonesia **37 D7** 3 45N 125 30 E
Great Slave L., Canada **72 A5** 61 23N 115 38W
Great Smoky Mts. Nat. Pk., U.S.A. **77 H4** 35 40N 83 40W
Great Stour = Stour →, U.K. **11 F9** 51 18N 1 22 E
Great Victoria Desert, Australia **61 E4** 29 30S 126 30 E
Great Wall, China **34 E5** 38 30N 109 30 E
Great Whernside, U.K. **10 C6** 54 10N 1 58W
Great Yarmouth, U.K. **10 E9** 52 37N 1 44 E
Greater Antilles, W. Indies **89 C5** 17 40N 74 0W
Greater London □, U.K. **11 F7** 51 30N 0 6W
Greater Manchester □, U.K. **10 D5** 53 30N 2 15W
Greater Sunda Is., Indonesia **36 F4** 7 0S 112 0 E
Greco, C., Cyprus **23 E13** 34 57N 34 5 E
Gredos, Sierra de, Spain **19 B3** 40 20N 5 0W
Greece, U.S.A. **78 C7** 43 13N 77 41W
Greece ■, Europe **21 E9** 40 0N 23 0 E

Greeley, *Colo., U.S.A.* **80 E2** 40 25N 104 42W
Greeley, *Nebr., U.S.A.* **80 E5** 41 33N 98 32W
Green —►, *Ky., U.S.A.* **76 G2** 37 54N 87 30W
Green —►, *Utah, U.S.A.* **83 G9** 38 11N 109 53W
Green B., *U.S.A.* **76 C2** 45 0N 87 30W
Green Bay, *U.S.A.* **76 C2** 44 31N 88 0W
Green C., *Australia* **63 F5** 37 13S 150 1 E
Green Cove Springs, *U.S.A.* **77 L5** 29 59N 81 42W
Green River, *U.S.A.* **83 G8** 38 59N 110 10W
Greenbank, *U.S.A.* **84 B4** 48 6N 122 34W
Greenbush, *Mich., U.S.A.* **78 B1** 44 35N 83 19W
Greenbush, *Minn., U.S.A.* **80 A6** 48 42N 96 11W
Greencastle, *U.S.A.* **76 F2** 39 38N 86 52W
Greene, *U.S.A.* **79 D9** 42 20N 75 46W
Greenfield, *Calif., U.S.A.* **84 J5** 36 19N 121 15W
Greenfield, *Calif., U.S.A.* **85 K8** 35 15N 119 0W
Greenfield, *Ind., U.S.A.* **76 F3** 39 47N 85 46W
Greenfield, *Iowa, U.S.A.* **80 E7** 41 18N 94 28W
Greenfield, *Mass., U.S.A.* **79 D12** 42 35N 72 36W
Greenfield, *Mo., U.S.A.* **81 G8** 37 25N 93 51W
Greenfield Park, *Canada* **79 A11** 45 29N 73 29W
Greenland ■, *N. Amer.* **4 C5** 66 0N 45 0W
Greenland Sea, *Arctic* **4 B7** 73 0N 10 0W
Greenock, *U.K.* **12 F4** 55 57N 4 46W
Greenore, *Ireland* **13 B5** 54 2N 6 8W
Greenore Pt., *Ireland* **13 D5** 52 14N 6 19W
Greenough —►, *Australia* **61 E1** 28 51S 114 38 E
Greenport, *U.S.A.* **79 E12** 41 6N 72 22W
Greensboro, *Ga., U.S.A.* **77 J4** 33 35N 83 11W
Greensboro, *N.C., U.S.A.* **77 G6** 36 4N 79 48W
Greensburg, *Ind., U.S.A.* **76 F3** 39 20N 85 29W
Greensburg, *Kans., U.S.A.* **81 G5** 37 36N 99 18W
Greensburg, *Pa., U.S.A.* **78 F5** 40 18N 79 33W
Greenville, *Liberia* **50 G3** 5 1N 9 6W
Greenville, *Ala., U.S.A.* **77 K2** 31 50N 86 38W
Greenville, *Calif., U.S.A.* **84 E6** 40 8N 120 57W
Greenville, *Ill., U.S.A.* **80 F10** 38 53N 89 25W
Greenville, *Maine, U.S.A.* **71 C6** 45 28N 69 35W
Greenville, *Mich., U.S.A.* **76 D3** 43 11N 85 15W
Greenville, *Miss., U.S.A.* **81 J9** 33 24N 91 4W
Greenville, *N.C., U.S.A.* **77 H7** 35 37N 77 23W
Greenville, *Ohio, U.S.A.* **76 E3** 40 6N 84 38W
Greenville, *Pa., U.S.A.* **78 E4** 41 24N 80 23W
Greenville, *S.C., U.S.A.* **77 H4** 34 51N 82 24W
Greenville, *Tenn., U.S.A.* **77 G4** 36 13N 82 51W
Greenville, *Tex., U.S.A.* **81 J6** 33 8N 96 7W
Greenwater Lake Prov. Park, *Canada* **73 C8** 52 32N 103 30W
Greenwich, *U.K.* **11 F8** 51 29N 0 1 E
Greenwich, *Conn., U.S.A.* **79 E11** 41 2N 73 38W
Greenwich, *N.Y., U.S.A.* **79 C11** 43 5N 73 30W
Greenwich, *Ohio, U.S.A.* **78 E2** 41 2N 82 31W
Greenwood, *Canada* **72 D5** 49 10N 118 40W
Greenwood, *Miss., U.S.A.* **81 J9** 33 31N 90 11W
Greenwood, *S.C., U.S.A.* **77 H4** 34 12N 82 10W
Greenwood, Mt., *Australia* **60 B5** 13 48S 130 4 E
Gregory, *U.S.A.* **80 D5** 43 14N 99 20W
Gregory —►, *Australia* **62 B2** 17 53S 139 17 E
Gregory, L., *S. Austral., Australia* **63 D2** 28 55S 139 0 E
Gregory, L., *W. Austral., Australia* **61 E2** 25 38S 119 58 E
Gregory Downs, *Australia* **62 B2** 18 35S 138 45 E
Gregory L., *Australia* **60 D4** 20 0S 127 40 E
Gregory Ra., *Queens., Australia* **62 B3** 19 30S 143 40 E
Gregory Ra., *W. Austral., Australia* **60 D3** 21 20S 121 12 E
Greifswald, *Germany* **16 A7** 54 5N 13 23 E
Greiz, *Germany* **16 C7** 50 39N 12 10 E
Gremikha, *Russia* **24 A6** 67 59N 39 47 E
Grenå, *Denmark* **9 H14** 56 25N 10 53 E
Grenada, *U.S.A.* **81 J10** 33 47N 89 49W
Grenada ■, *W. Indies* **89 D7** 12 10N 61 40W
Grenadines, *W. Indies* **89 D7** 12 40N 61 20W
Grenen, *Denmark* **9 H14** 57 44N 10 40 E
Grenfell, *Australia* **63 E4** 33 52S 148 8 E
Grenfell, *Canada* **73 C8** 50 30N 102 56W
Grenoble, *France* **18 D6** 45 12N 5 42 E
Grenora, *U.S.A.* **80 A3** 48 37N 103 56W
Grenville, C., *Australia* **62 A3** 12 0S 143 13 E
Grenville Chan., *Canada* **72 C3** 53 40N 129 46W
Gresham, *U.S.A.* **84 E4** 45 30N 122 26W
Gresik, *Indonesia* **37 G15** 7 13S 112 38 E
Gretna Green, *U.K.* **12 F5** 55 1N 3 3W
Grevenmacher, *Lux.* **15 E6** 49 41N 6 26 E
Grey —►, *N.Z.* **59 K3** 42 27S 171 12 E
Grey, C., *Australia* **62 A2** 13 0S 136 35 E
Grey Ra., *Australia* **63 D3** 27 0S 143 30 E
Grey Res., *Canada* **71 C8** 48 20N 56 30W
Greybull, *U.S.A.* **82 D9** 44 30N 108 3W
Greymouth, *N.Z.* **59 K3** 42 29S 171 13 E
Greytown, *N.Z.* **59 J5** 41 5S 175 29 E
Greytown, *S. Africa* **57 D5** 29 1S 30 36 E
Gribbell I., *Canada* **72 C3** 53 23N 129 0W
Gridley, *U.S.A.* **84 F5** 39 22N 121 42W
Griekwastad, *S. Africa* **56 D3** 28 49S 23 15 E
Griffin, *U.S.A.* **77 J3** 33 15N 84 16W
Griffith, *Australia* **63 E4** 34 18S 146 2 E
Grimari, *C.A.R.* **51 G9** 5 43N 20 6 E
Grimaylov = Hrymayliv, *Ukraine* **17 D14** 49 20N 26 5 E
Grimes, *U.S.A.* **84 F5** 39 4N 121 54W
Grimsby, *Canada* **78 C5** 43 12N 79 34W
Grimsby, *U.K.* **10 D7** 53 34N 0 5W
Grímsey, *Iceland* **8 C5** 66 33N 17 58W
Grimshaw, *Canada* **72 B5** 56 10N 117 40W
Grimstad, *Norway* **9 G13** 58 20N 8 35 E
Grinnell, *U.S.A.* **80 E8** 41 45N 92 43W
Gris-Nez, C., *France* **18 A4** 50 52N 1 35 E
Groais I., *Canada* **71 B8** 50 55N 55 35W
Groblersdal, *S. Africa* **57 D4** 25 15S 29 25 E
Grodno = Hrodna, *Belarus* **17 B12** 53 42N 23 52 E
Grodzyanka = Hrodzyanka, *Belarus* **17 B15** 53 31N 28 42 E
Groesbeck, *U.S.A.* **81 K6** 30 48N 96 31W
Grójec, *Poland* **17 C11** 51 50N 20 58 E
Grong, *Norway* **8 D15** 64 25N 12 8 E
Groningen, *Neths.* **15 A6** 53 15N 6 35 E
Groningen □, *Neths.* **15 A6** 53 16N 6 40 E
Groom, *U.S.A.* **81 H4** 35 12N 101 6W
Groot —►, *S. Africa* **56 E3** 33 45S 24 36 E
Groot Berg —►, *S. Africa* **56 E2** 32 47S 18 8 E
Groot-Brakrivier, *S. Africa* **56 E3** 34 2S 22 18 E
Groot-Kei —►, *S. Africa* **57 E4** 32 41S 28 22 E
Groot Vis —►, *S. Africa* **56 E4** 33 28S 27 5 E
Groote Eylandt, *Australia* **62 A2** 14 0S 136 40 E
Grootfontein, *Namibia* **56 B2** 19 31S 18 6 E
Grootlaagte —►, *Africa* **56 C3** 20 55S 21 27 E
Grootvloer, *S. Africa* **56 E3** 30 0S 20 40 E
Gros C., *Canada* **72 A6** 61 59N 113 32W
Grosa, Pta., *Spain* **22 B8** 39 6N 1 36 E
Gross Glockner, *Austria* **16 E7** 47 5N 12 40 E
Grosser Arber, *Germany* **16 D7** 49 3N 13 8 E
Grosseto, *Italy* **20 C4** 42 46N 11 8 E
Groswater B., *Canada* **71 B8** 54 20N 57 40W
Groton, *Conn., U.S.A.* **79 E12** 41 21N 72 5W
Groton, *S. Dak., U.S.A.* **80 C5** 45 27N 98 6W
Grouard Mission, *Canada* **72 B5** 55 33N 116 9W
Groundhog —►, *Canada* **70 C3** 48 45N 82 58W
Grouse Creek, *U.S.A.* **82 F7** 41 42N 113 53W
Grove City, *U.S.A.* **78 E4** 41 10N 80 5W
Groveland, *U.S.A.* **84 H6** 37 50N 120 14W
Grover City, *U.S.A.* **85 K6** 35 7N 120 37W
Groveton, *N.H., U.S.A.* **79 B13** 44 36N 71 31W
Groveton, *Tex., U.S.A.* **81 K7** 31 4N 95 8W
Groznyy, *Russia* **25 F8** 43 20N 45 45 E
Grudziądz, *Poland* **17 B10** 53 30N 18 47 E
Grundy Center, *U.S.A.* **80 D8** 42 22N 92 47W
Gruver, *U.S.A.* **81 G4** 36 16N 101 24W
Gryazi, *Russia* **24 D6** 52 30N 39 58 E
Gryazovets, *Russia* **26 D5** 58 50N 40 10 E
Gua, *India* **41 H14** 22 18N 85 20 E
Gua Musang, *Malaysia* **39 K3** 4 53N 101 58 E
Guacanayabo, G. de, *Cuba* **88 B4** 20 40N 77 20W
Guachípas —►, *Argentina* **94 B2** 25 40S 65 30W
Guadalajara, *Mexico* **86 C4** 20 40N 103 20W
Guadalajara, *Spain* **19 B4** 40 37N 3 12W
Guadalcanal, *Solomon Is.* **64 H8** 9 32S 160 12 E
Guadales, *Argentina* **94 C2** 34 30S 67 55W
Guadalete —►, *Spain* **19 D2** 36 35N 6 13W
Guadalquivir —►, *Spain* **19 D2** 36 47N 6 22W
Guadalupe = Guadeloupe ■, *W. Indies* **89 C7** 16 20N 61 40W
Guadalupe, *Mexico* **85 N10** 32 4N 116 32W
Guadalupe, *U.S.A.* **85 L6** 34 59N 120 33W
Guadalupe —►, *Mexico* **85 N10** 32 6N 116 51W
Guadalupe —►, *U.S.A.* **81 L6** 28 27N 96 47W
Guadalupe, Sierra de, *Spain* **19 C3** 39 28N 5 30W
Guadalupe Bravos, *Mexico* **86 A3** 31 20N 106 10W
Guadalupe I., *Pac. Oc.* **66 G8** 29 0N 118 50W
Guadalupe Peak, *U.S.A.* **83 L11** 31 50N 104 52W
Guadalupe y Calvo, *Mexico* **86 B3** 26 6N 106 58W
Guadarrama, Sierra de, *Spain* **19 B4** 41 0N 4 0W
Guadeloupe ■, *W. Indies* **89 C7** 16 20N 61 40W
Guadeloupe Passage, *W. Indies* **89 C7** 16 50N 62 15W
Guadiana —►, *Portugal* **19 D2** 37 14N 7 22W
Guadix, *Spain* **19 D4** 37 18N 3 11W
Guafo, Boca del, *Chile* **96 E2** 43 35S 74 0W
Guaíra, *Brazil* **95 A5** 24 5S 54 10W
Guaitecas, Is., *Chile* **96 E2** 44 0S 74 30W
Guajará-Mirim, *Brazil* **92 F5** 10 50S 65 20W
Guajira, Pen. de la, *Colombia* **92 A4** 12 0N 72 0W
Gualán, *Guatemala* **88 C2** 15 8N 89 22W
Gualeguay, *Argentina* **94 C4** 33 10S 59 14W
Gualeguaychú, *Argentina* **94 C4** 33 3S 59 31W
Guam ■, *Pac. Oc.* **64 F6** 13 27N 144 45 E
Guaminí, *Argentina* **94 D3** 37 1S 62 28W
Guamúchil, *Mexico* **86 B3** 25 25N 108 3W
Guanabacoa, *Cuba* **88 B3** 23 8N 82 18W
Guanacaste, Cordillera del, *Costa Rica* **88 D2** 10 40N 85 4W
Guanacevi, *Mexico* **86 B3** 25 40N 106 0W
Guanahani = San Salvador, *Bahamas* **89 B5** 24 0N 74 40W
Guanajay, *Cuba* **88 B3** 22 56N 82 42W
Guanajuato, *Mexico* **86 C4** 21 0N 101 20W
Guanajuato □, *Mexico* **86 C4** 20 40N 101 20W
Guandacol, *Argentina* **94 B2** 29 30S 68 40W
Guane, *Cuba* **88 B3** 22 10N 84 7W
Guangdong □, *China* **33 D6** 23 0N 113 0 E
Guangling, *China* **34 E8** 39 47N 114 22 E
Guangrao, *China* **35 F10** 37 5N 118 25 E
Guangwu, *China* **34 F3** 37 48N 105 57 E
Guangxi Zhuangzu Zizhiqu □, *China* **33 D5** 24 0N 109 0 E
Guangzhou, *China* **33 D6** 23 5N 113 10 E
Guanipa —►, *Venezuela* **92 B6** 9 56N 62 26W
Guannan, *China* **35 G10** 34 8N 119 21 E
Guantánamo, *Cuba* **89 B4** 20 10N 75 14W
Guantao, *China* **34 F8** 36 42N 115 25 E
Guanyun, *China* **35 G10** 34 20N 119 18 E
Guápiles, *Costa Rica* **88 D3** 10 10N 83 46W
Guaporé —►, *Brazil* **92 F5** 11 55S 65 4W
Guaqui, *Bolivia* **92 G5** 16 41S 68 54W
Guarapari, *Brazil* **95 A7** 20 40S 40 30W
Guarapuava, *Brazil* **95 B5** 25 20S 51 30W
Guaratinguetá, *Brazil* **95 A6** 22 49S 45 9W
Guaratuba, *Brazil* **95 B6** 25 53S 48 38W
Guarda, *Portugal* **19 B2** 40 32N 7 20W
Guardafui, C. = Asir, Ras, *Somali Rep.* **46 E5** 11 55N 51 10 E
Guaria □, *Paraguay* **94 B4** 25 45S 56 30W
Guarujá, *Brazil* **95 A6** 24 2S 46 25W
Guarus, *Brazil* **95 A7** 21 44S 41 20W
Guasave, *Mexico* **86 B3** 25 34N 108 27W
Guasdualito, *Venezuela* **92 B4** 7 15N 70 44W
Guasipati, *Venezuela* **92 B6** 7 28N 61 54W
Guatemala, *Guatemala* **88 D1** 14 40N 90 22W
Guatemala ■, *Cent. Amer.* **88 C1** 15 40N 90 30W
Guatire, *Venezuela* **92 A5** 10 28N 66 32W
Guaviare □, *Colombia* **92 C4** 2 0N 72 30W
Guaviare —►, *Colombia* **92 C5** 4 3N 67 44W
Guaxupé, *Brazil* **95 A6** 21 10S 47 5W
Guayama, *Puerto Rico* **89 C6** 17 59N 66 7W
Guayaquil, *Ecuador* **92 D3** 2 15S 79 52W
Guayaquil, G. de, *Ecuador* **92 D2** 3 10S 81 0W
Guaymas, *Mexico* **86 B2** 27 59N 110 54W
Guba, *Zaïre* **55 E2** 10 38S 26 27 E
Gudbrandsdalen, *Norway* **9 F14** 61 33N 10 10 E
Guddu Barrage, *Pakistan* **40 E6** 28 30N 69 50 E
Gudivada, *India* **41 L12** 16 30N 81 3 E
Gudur, *India* **40 M11** 14 12N 79 55 E
Guecho, *Spain* **19 A4** 43 21N 2 59W
Guékédou, *Guinea* **50 G2** 8 40N 10 5W
Guelma, *Algeria* **50 A6** 36 25N 7 29 E
Guelph, *Canada* **70 D3** 43 35N 80 20W
Guéréda, *Chad* **51 F9** 14 31N 22 5 E
Guéret, *France* **18 C4** 46 11N 1 51 E
Guerneville, *U.S.A.* **84 G4** 38 30N 123 0W
Guernica, *Spain* **19 A4** 43 19N 2 40W
Guernsey, *U.K.* **11 H5** 49 26N 2 35W
Guernsey, *U.S.A.* **80 D2** 42 19N 104 45W
Guerrero □, *Mexico* **87 D5** 17 30N 100 0W
Gueydan, *U.S.A.* **81 K8** 30 2N 92 31W
Gügher, *Iran* **45 D8** 29 28N 56 27 E
Guia, *Canary Is.* **22 F4** 28 8N 15 38W
Guia de Isora, *Canary Is.* **22 F3** 28 12N 16 46W
Guia Lopes da Laguna, *Brazil* **95 A4** 21 26S 56 7W
Guiana, *S. Amer.* **90 C4** 5 10N 60 40W
Guidónia-Montecélio, *Italy* **20 C5** 42 1N 12 45 E
Guiglo, *Ivory C.* **50 G3** 6 45N 7 30W
Guijá, *Mozam.* **57 C5** 24 27S 33 0 E
Guildford, *U.K.* **11 F7** 51 14N 0 34W
Guilford, *U.S.A.* **71 C6** 45 10N 69 23W
Guilin, *China* **33 D6** 25 18N 110 15 E
Güímar, *Canary Is.* **22 F3** 28 18N 16 24W
Guimarães, *Brazil* **93 D10** 2 9S 44 42W
Guimarães, *Portugal* **19 B1** 41 28N 8 24W
Guimaras, *Phil.* **37 B6** 10 35N 122 37 E
Guinda, *U.S.A.* **84 G4** 38 50N 122 12W
Guinea, *Africa* **48 F4** 8 0N 8 0 E
Guinea ■, *W. Afr.* **50 F2** 10 20N 11 30W
Guinea, Gulf of, *Atl. Oc.* **48 F4** 3 0N 2 30 E
Guinea-Bissau ■, *Africa* **50 F2** 12 0N 15 0W
Güines, *Cuba* **88 B3** 22 50N 82 0W
Guingamp, *France* **18 B2** 48 34N 3 10W
Güiria, *Venezuela* **92 A6** 10 32N 62 18W
Guiuan, *Phil.* **37 B7** 11 5N 125 55 E
Guiyang, *China* **32 D5** 26 32N 106 40 E
Guizhou □, *China* **32 D5** 27 0N 107 0 E
Gujarat □, *India* **42 H4** 23 20N 71 0 E
Gujranwala, *Pakistan* **42 C6** 32 10N 74 12 E
Gujrat, *Pakistan* **42 C6** 32 40N 74 2 E
Gulbarga, *India* **40 L10** 17 20N 76 50 E
Gulbene, *Latvia* **9 H22** 57 8N 26 52 E
Gulf, The, *Asia* **45 E6** 27 0N 50 0 E
Gulfport, *U.S.A.* **81 K10** 30 22N 89 6W
Gulgong, *Australia* **63 E4** 32 20S 149 49 E
Gulistan, *Pakistan* **42 D2** 30 30N 66 35 E
Gull Lake, *Canada* **73 C7** 50 10N 108 29W
Güllük, *Turkey* **21 F12** 37 14N 27 35 E
Gulmarg, *India* **43 B6** 34 3N 74 25 E
Gulshad, *Kazakstan* **26 E8** 46 45N 74 25 E
Gulu, *Uganda* **54 B3** 2 48N 32 17 E
Gulwe, *Tanzania* **54 D4** 6 30S 36 25 E
Gum Lake, *Australia* **63 E3** 32 42S 143 9 E
Gumal —►, *Pakistan* **42 D4** 31 40N 71 50 E
Gumbaz, *Pakistan* **42 D3** 30 2N 69 0 E
Gumlu, *Australia* **62 B4** 19 53S 147 41 E
Gumma □, *Japan* **31 F9** 36 30N 138 20 E
Gummi, *Nigeria* **50 F6** 12 4N 5 9 E
Gumzai, *Indonesia* **37 F8** 5 28S 134 42 E
Guna, *India* **42 G7** 24 40N 77 19 E
Gundagai, *Australia* **63 F4** 35 3S 148 6 E
Gundih, *Indonesia* **37 G14** 7 10S 110 56 E
Gungu, *Zaïre* **52 F3** 5 43S 19 20 E
Gunisao —►, *Canada* **73 C9** 53 56N 97 53W
Gunisao L., *Canada* **73 C9** 53 33N 96 15W
Gunnbjørn Fjeld, *Greenland* **4 C6** 68 55N 29 47W
Gunnedah, *Australia* **63 E5** 30 59S 150 15 E
Gunningbar Cr. —►, *Australia* **63 E4** 31 14S 147 6 E
Gunnison, *Colo., U.S.A.* **83 G10** 38 33N 106 56W
Gunnison, *Utah, U.S.A.* **82 G8** 39 9N 111 49W
Gunnison —►, *U.S.A.* **83 G9** 39 4N 108 35W
Gunpowder, *Australia* **62 B2** 19 42S 139 22 E
Guntakal, *India* **40 M10** 15 11N 77 27 E
Guntersville, *U.S.A.* **77 H2** 34 21N 86 18W
Guntong, *Malaysia* **39 K3** 4 36N 101 3 E
Guntur, *India* **41 L12** 16 23N 80 30 E
Gunungapi, *Indonesia* **37 F7** 6 45S 126 30 E
Gunungsitoli, *Indonesia* **36 D1** 1 15N 97 30 E
Gunza, *Angola* **52 G2** 10 50S 13 50 E
Guo He —►, *China* **35 H9** 32 59N 117 10 E
Guoyang, *China* **34 H9** 33 32N 116 12 E
Gupis, *Pakistan* **43 A5** 36 15N 73 20 E
Gurdaspur, *India* **42 C6** 32 5N 75 31 E
Gurdon, *U.S.A.* **81 J8** 33 55N 93 9W
Gurgaon, *India* **42 E7** 28 27N 77 1 E
Gurha, *India* **42 G4** 25 12N 71 39 E
Gurkha, *Nepal* **43 E11** 28 5N 84 40 E
Gurley, *Australia* **63 D4** 29 45S 149 48 E
Gurué, *Mozam.* **55 F4** 15 25S 36 58 E
Gurun, *Malaysia* **39 K3** 5 49N 100 27 E
Gurupá, *Brazil* **93 D8** 1 25S 51 35W
Gurupá, I. Grande de, *Brazil* **93 D8** 1 25S 51 45W
Gurupi —►, *Brazil* **93 D9** 1 13S 46 6W
Guryev = Atyraū, *Kazakstan* **25 E9** 47 5N 52 0 E
Gusau, *Nigeria* **50 F6** 12 12N 6 40 E
Gusev, *Russia* **9 J20** 54 35N 22 10 E
Gushan, *China* **35 E12** 39 50N 123 35 E
Gushgy, *Turkmenistan* **26 F7** 35 20N 62 18 E
Gusinoozersk, *Russia* **27 D11** 51 16N 106 27 E
Gustine, *U.S.A.* **83 H3** 37 16N 121 0W
Güstrow, *Germany* **16 B7** 53 47N 12 10 E
Gütersloh, *Germany* **16 C5** 51 54N 8 24 E
Gutha, *Australia* **61 E2** 28 58S 115 55 E
Guthalongra, *Australia* **62 B4** 19 52S 147 50 E
Guthrie, *U.S.A.* **81 H6** 35 53N 97 25W
Guttenberg, *U.S.A.* **80 D9** 42 47N 91 6W
Guyana ■, *S. Amer.* **92 C7** 5 0N 59 0W
Guyane française ■ = French Guiana ■, *S. Amer.* **93 C8** 4 0N 53 0W
Guyang, *China* **34 D6** 41 0N 110 5 E
Guyenne, *France* **18 D4** 44 30N 0 40 E
Guymon, *U.S.A.* **81 G4** 36 41N 101 29W
Guyra, *Australia* **63 E5** 30 15S 151 40 E
Guyuan, *Hebei, China* **34 D8** 41 37N 115 40 E
Guyuan, *Ningxia Huizu, China* **34 F4** 36 0N 106 20 E
Guzhen, *China* **35 H9** 33 22N 117 18 E
Guzmán, L. de, *Mexico* **86 A3** 31 25N 107 25W
Gvardeysk, *Russia* **9 J19** 54 39N 21 5 E
Gwa, *Burma* **41 L19** 17 36N 94 34 E
Gwaai, *Zimbabwe* **55 F2** 19 15S 27 45 E
Gwabegar, *Australia* **63 E4** 30 31S 149 0 E
Gwādar, *Pakistan* **40 G3** 25 10N 62 18 E
Gwalia, *Australia* **61 E3** 28 54S 121 20 E
Gwalior, *India* **42 F8** 26 12N 78 10 E
Gwanda, *Zimbabwe* **55 G2** 20 55S 29 0 E
Gwane, *Zaïre* **54 B2** 4 45N 25 48 E
Gweebarra B., *Ireland* **13 B3** 54 51N 8 23W
Gweedore, *Ireland* **13 A3** 55 3N 8 13W
Gweru, *Zimbabwe* **55 F2** 19 28S 29 45 E
Gwinn, *U.S.A.* **76 B2** 46 19N 87 27W
Gwydir —►, *Australia* **63 D4** 29 27S 149 48 E
Gwynedd □, *U.K.* **10 E3** 52 52N 4 10W
Gyandzha = Gäncä, *Azerbaijan* **25 F8** 40 45N 46 20 E
Gyaring Hu, *China* **32 C4** 34 50N 97 40 E
Gydanskiy P-ov., *Russia* **26 C8** 70 0N 78 0 E
Gympie, *Australia* **63 D5** 26 11S 152 38 E
Gyöngyös, *Hungary* **17 E10** 47 48N 19 56 E
Győr, *Hungary* **17 E9** 47 41N 17 40 E
Gypsum Pt., *Canada* **72 A6** 61 53N 114 35W
Gypsumville, *Canada* **73 C9** 51 45N 98 40W
Gyula, *Hungary* **17 E11** 46 38N 21 17 E
Gyumri, *Armenia* **25 F7** 40 47N 43 50 E
Gyzylarbat, *Turkmenistan* **26 F6** 39 4N 56 23 E

H

Ha 'Arava —►, *Israel* **47 E4** 30 50N 35 20 E
Ha Coi, *Vietnam* **38 B6** 21 26N 107 46 E
Ha Dong, *Vietnam* **38 B5** 20 58N 105 46 E
Ha Giang, *Vietnam* **38 A5** 22 50N 104 59 E
Ha Tien, *Vietnam* **39 G5** 10 23N 104 29 E
Ha Tinh, *Vietnam* **38 C5** 18 20N 105 54 E
Ha Trung, *Vietnam* **38 C5** 19 58N 105 50 E
Haapsalu, *Estonia* **9 G20** 58 56N 23 30 E
Haarlem, *Neths.* **15 B4** 52 23N 4 39 E
Haast —►, *N.Z.* **59 K2** 43 50S 169 2 E
Haast Bluff, *Australia* **60 D5** 23 22S 132 0 E
Hab Nadi Chauki, *Pakistan* **42 G2** 25 0N 66 50 E
Habaswein, *Kenya* **54 B4** 1 2N 39 30 E
Habay, *Canada* **72 B5** 58 50N 118 44W
Ḥabbānīyah, *Iraq* **44 C4** 33 17N 43 29 E
Haboro, *Japan* **30 B10** 44 22N 141 42 E
Hachijō-Jima, *Japan* **31 H9** 33 5N 139 45 E
Hachinohe, *Japan* **30 D10** 40 30N 141 29 E
Hachiōji, *Japan* **31 G9** 35 40N 139 20 E
Hachŏn, *N. Korea* **35 D15** 41 29N 129 2 E
Hackensack, *U.S.A.* **79 F10** 40 53N 74 3W
Hadali, *Pakistan* **42 C5** 32 16N 72 11 E
Hadarba, Ras, *Sudan* **51 D12** 22 4N 36 51 E
Hadarom □, *Israel* **47 E3** 31 0N 35 0 E
Haddington, *U.K.* **12 F6** 55 57N 2 47W
Hadejia, *Nigeria* **50 F7** 12 30N 10 5 E
Haden, *Australia* **63 D5** 27 13S 151 54 E
Hadera, *Israel* **47 C3** 32 27N 34 55 E
Hadera, N. —►, *Israel* **47 C3** 32 28N 34 52 E
Haderslev, *Denmark* **9 J13** 55 15N 9 30 E
Hadhramaut = Hadramawt, *Yemen* **46 D4** 15 30N 49 30 E
Hadong, *S. Korea* **35 G14** 35 5N 127 44 E
Hadramawt, *Yemen* **46 D4** 15 30N 49 30 E
Ḥadrānīyah, *Iraq* **44 C4** 35 38N 43 14 E
Hadrian's Wall, *U.K.* **10 C5** 55 0N 2 30W
Haeju, *N. Korea* **35 E13** 38 3N 125 45 E
Haenam, *S. Korea* **35 G14** 34 34N 126 35 E
Haerhpin = Harbin, *China* **35 B14** 45 48N 126 40 E
Hafar al Bāṭin, *Si. Arabia* **44 D5** 28 32N 45 52 E
Ḥafirat al 'Aydā, *Si. Arabia* **44 E3** 26 26N 39 12 E
Hafizabad, *Pakistan* **42 C5** 32 5N 73 40 E
Haflong, *India* **41 G18** 25 10N 93 5 E
Hafnarfjörður, *Iceland* **8 D3** 64 4N 21 57W
Hafun, Ras, *Somali Rep.* **46 E6** 10 29N 51 30 E
Hagalil, *Israel* **47 C4** 32 53N 35 18 E
Hagen, *Germany* **16 C4** 51 21N 7 27 E
Hagerman, *U.S.A.* **81 J2** 33 7N 104 20W
Hagerstown, *U.S.A.* **76 F7** 39 39N 77 43W
Hagfors, *Sweden* **9 F15** 60 3N 13 45 E
Hagi, *Japan* **31 G5** 34 30N 131 22 E
Hagolan, *Syria* **47 B4** 33 0N 35 45 E
Hagondange-Briey, *France* **18 B7** 49 16N 6 11 E
Hags Hd., *Ireland* **13 D2** 52 57N 9 28W
Hague, C. de la, *France* **18 B3** 49 44N 1 56W
Hague, The = 's-Gravenhage, *Neths.* **15 B4** 52 7N 4 17 E
Haguenau, *France* **18 B7** 48 49N 7 47 E
Hai □, *Tanzania* **54 C4** 3 10S 37 10 E
Hai Duong, *Vietnam* **38 B6** 20 56N 106 19 E
Haicheng, *China* **35 D12** 40 50N 122 45 E
Haidar Khel, *Afghan.* **42 C3** 33 58N 68 38 E
Haifa = Hefa, *Israel* **47 C3** 32 46N 35 0 E
Haig, *Australia* **61 F4** 30 55S 126 10 E
Haikou, *China* **33 D6** 20 1N 110 16 E
Ḥā'il, *Si. Arabia* **44 E4** 27 28N 41 45 E
Hailar, *China* **33 B6** 49 10N 119 38 E
Hailey, *U.S.A.* **82 E6** 43 31N 114 19W
Haileybury, *Canada* **70 C4** 47 30N 79 38W
Hailin, *China* **35 B15** 44 37N 129 30 E
Hailong, *China* **35 C13** 42 32N 125 40 E
Hailuoto, *Finland* **8 D21** 65 3N 24 45 E
Hainan □, *China* **33 E5** 19 0N 109 30 E
Hainaut □, *Belgium* **15 D4** 50 30N 4 0 E
Haines, *U.S.A.* **82 D5** 44 55N 117 56W
Haines City, *U.S.A.* **77 L5** 28 7N 81 38W
Haines Junction, *Canada* **72 A1** 60 45N 137 30W
Haiphong, *Vietnam* **32 D5** 20 47N 106 41 E
Haiti ■, *W. Indies* **89 C5** 19 0N 72 30W
Haiya Junction, *Sudan* **51 E12** 18 20N 36 21 E
Haiyang, *China* **35 F11** 36 47N 121 9 E
Haiyuan, *China* **34 F3** 36 35N 105 52 E
Haizhou, *China* **35 G10** 34 37N 119 7 E
Haizhou Wan, *China* **35 G10** 34 50N 119 20 E
Hajar Bangar, *Sudan* **51 F9** 10 40N 22 45 E
Hajdúböszörmény, *Hungary* **17 E11** 47 40N 21 30 E

Hajipur, India 43 G11 25 45N 85 13 E
Hājjī Muḩsin, Iraq 44 C5 32 35N 45 29 E
Ḩājjīābād, Eşfahan, Iran 45 C7 33 41N 54 50 E
Ḩājjīābād, Hormozgān, Iran 45 D7 28 19N 55 55 E
Hajnówka, Poland 17 B12 52 47N 23 35 E
Hakansson, Mts., Zaïre 55 D2 8 40S 25 45 E
Hakken-Zan, Japan 31 G7 34 10N 135 54 E
Hakodate, Japan 30 D10 41 45N 140 44 E
Haku-San, Japan 31 F8 36 9N 136 46 E
Hakui, Japan 31 F8 36 53N 136 47 E
Hala, Pakistan 40 G6 25 43N 68 20 E
Ḩalab, Syria 44 B3 36 10N 37 15 E
Ḩalabjah, Iraq 44 C5 35 10N 45 58 E
Halaib, Sudan 51 D12 22 12N 36 30 E
Hālat 'Ammār, Si. Arabia 44 D3 29 10N 36 4 E
Halbā, Lebanon 47 A5 34 34N 36 6 E
Halberstadt, Germany 16 C6 51 54N 11 3 E
Halcombe, N.Z. 59 J5 40 8S 175 30 E
Halcon, Mt., Phil. 37 B6 13 0N 121 30 E
Halden, Norway 9 G14 59 9N 11 23 E
Haldia, India 41 H16 22 5N 88 3 E
Haldwani, India 43 E8 29 31N 79 30 E
Hale →, Australia 62 C2 24 56S 135 53 E
Haleakala Crater, U.S.A. 74 H16 20 43N 156 16W
Haleyville, U.S.A. 77 H2 34 14N 87 37W
Halfway →, Canada 72 B4 56 12N 121 32W
Haliburton, Canada 70 C4 45 3N 78 30W
Halifax, Australia 62 B4 18 32S 146 22 E
Halifax, Canada 71 D7 44 38N 63 35W
Halifax, U.K. 10 D6 53 43N 1 52W
Halifax B., Australia 62 B4 18 50S 147 0 E
Halifax I., Namibia 56 D2 26 38S 15 4 E
Halīl →, Iran 45 E8 27 40N 58 30 E
Hall Beach, Canada 69 B11 68 46N 81 12W
Hall Pt., Australia 60 C3 15 40S 124 23 E
Halland, Sweden 9 H15 57 8N 12 47 E
Halle, Belgium 15 D4 50 44N 4 13 E
Halle, Germany 16 C6 51 30N 11 56 E
Hällefors, Sweden 9 G16 59 47N 14 31 E
Hallett, Australia 63 E2 33 25S 138 55 E
Hallettsville, U.S.A. 81 L6 29 27N 96 57W
Halliday, U.S.A. 80 B3 47 21N 102 20W
Halliday L., Canada 73 A7 61 21N 108 56W
Hallim, S. Korea 35 H14 33 24N 126 15 E
Hallingdalselva →, Norway 9 F13 60 40N 8 50 E
Hallock, U.S.A. 73 D9 48 47N 96 57W
Halls Creek, Australia 60 C4 18 16S 127 38 E
Hallsberg, Sweden 9 G16 59 5N 15 7 E
Hallstead, U.S.A. 79 E9 41 58N 75 45W
Halmahera, Indonesia 37 D7 0 40N 128 0 E
Halmstad, Sweden 9 H15 56 41N 12 52 E
Halq el Oued, Tunisia 51 A7 36 53N 10 18 E
Hälsingborg = Helsingborg, Sweden 9 H15 56 3N 12 42 E
Hälsingland, Sweden 9 F16 61 40N 16 5 E
Halstad, U.S.A. 80 B6 47 21N 96 50W
Halti, Finland 8 B19 69 17N 21 18 E
Halul, Qatar 45 E7 25 40N 52 40 E
Ḩalvān, Iran 45 C8 33 57N 56 15 E
Ham Tan, Vietnam 39 G6 10 40N 107 45 E
Ham Yen, Vietnam 38 A5 22 4N 105 3 E
Hamab, Namibia 56 D2 28 7S 19 16 E
Hamada, Japan 31 G6 34 56N 132 4 E
Hamadān, Iran 45 C6 34 52N 48 32 E
Hamadān □, Iran 45 C6 35 0N 49 0 E
Ḩamāh, Syria 44 C3 35 5N 36 40 E
Hamamatsu, Japan 31 G8 34 45N 137 45 E
Hamar, Norway 9 F14 60 48N 11 7 E
Hambantota, Sri Lanka 40 R12 6 10N 81 10 E
Hamber Prov. Park, Canada 72 C5 52 20N 118 0W
Hamburg, Germany 16 B5 53 33N 9 59 E
Hamburg, Ark., U.S.A. 81 J9 33 14N 91 48W
Hamburg, Iowa, U.S.A. 80 E7 40 36N 95 39W
Hamburg, N.Y., U.S.A. 78 D6 42 43N 78 50W
Hamburg, Pa., U.S.A. 79 F9 40 33N 75 59W
Ḩamḑ, W. al →, Si. Arabia 44 E3 24 55N 36 20 E
Hamden, U.S.A. 79 E12 41 23N 72 54W
Häme, Finland 9 F20 61 38N 25 10 E
Hämeenlinna, Finland 9 F21 61 0N 24 28 E
Hamelin Pool, Australia 61 E1 26 22S 114 20 E
Hameln, Germany 16 B5 52 6N 9 21 E
Hamerkaz □, Israel 47 C3 32 15N 34 55 E
Hamersley Ra., Australia 60 D2 22 0S 117 45 E
Hamhung, N. Korea 35 E14 39 54N 127 30 E
Hami, China 32 B4 42 55N 93 25 E
Hamilton, Australia 63 F3 37 45S 142 2 E
Hamilton, Canada 70 D4 43 15N 79 50W
Hamilton, N.Z. 59 G5 37 47S 175 19 E
Hamilton, U.K. 12 F4 55 46N 4 2W
Hamilton, Mo., U.S.A. 80 F8 39 45N 93 59W
Hamilton, Mont., U.S.A. 82 C6 46 15N 114 10W
Hamilton, N.Y., U.S.A. 79 D9 42 50N 75 33W
Hamilton, Ohio, U.S.A. 76 F3 39 24N 84 34W
Hamilton, Tex., U.S.A. 81 K5 31 42N 98 7W
Hamilton →, Australia 62 C2 23 30S 139 47 E
Hamilton City, U.S.A. 84 F4 39 45N 122 1W
Hamilton Hotel, Australia 62 C3 22 45S 140 40 E
Hamilton Inlet, Canada 71 B8 54 0N 57 30W
Hamina, Finland 9 F22 60 34N 27 12 E
Hamiota, Canada 73 C8 50 11N 100 38W
Hamlet, U.S.A. 77 H6 34 53N 79 42W
Hamley Bridge, Australia 63 E2 34 17S 138 35 E
Hamlin = Hameln, Germany 16 B5 52 6N 9 21 E
Hamlin, N.Y., U.S.A. 78 C7 43 17N 77 55W
Hamlin, Tex., U.S.A. 81 J4 32 53N 100 8W
Hamm, Germany 16 C4 51 40N 7 50 E
Hammerfest, Norway 8 A20 70 39N 23 41 E
Hammond, Ind., U.S.A. 76 E2 41 38N 87 30W
Hammond, La., U.S.A. 81 K9 30 30N 90 28W
Hammonton, U.S.A. 76 F8 39 39N 74 48W
Hampden, N.Z. 59 L3 45 18S 170 50 E
Hampshire □, U.K. 11 F6 51 7N 1 23W
Hampshire Downs, U.K. 11 F6 51 15N 1 10W
Hampton, Ark., U.S.A. 81 J8 33 32N 92 28W
Hampton, Iowa, U.S.A. 80 D8 42 45N 93 13W
Hampton, N.H., U.S.A. 79 D14 42 57N 70 50W
Hampton, S.C., U.S.A. 77 J5 32 52N 81 7W
Hampton, Va., U.S.A. 76 G7 37 2N 76 21W
Hampton Tableland, Australia 61 F4 32 0S 127 0 E
Hamrat esh Sheykh, Sudan 51 F10 14 38N 27 55 E
Hamyang, S. Korea 35 G14 35 32N 127 42 E
Han Pijesak, Bos.-H. 21 B8 44 5N 18 57 E
Hana, U.S.A. 74 H17 20 45N 155 59W
Hanak, Si. Arabia 44 E3 25 32N 37 0 E
Hanamaki, Japan 30 E10 39 23N 141 7 E
Hanang, Tanzania 54 C4 4 30S 35 25 E
Hanau, Germany 16 C5 50 7N 8 56 E
Hanbogd, Mongolia 34 C4 43 11N 107 10 E
Hancheng, China 34 G6 35 31N 110 25 E
Hancock, Mich., U.S.A. 80 B10 47 8N 88 35W
Hancock, Minn., U.S.A. 80 C7 45 30N 95 48W
Hancock, N.Y., U.S.A. 79 E9 41 57N 75 17W
Handa, Japan 31 G8 34 53N 136 55 E
Handa, Somali Rep. 46 E5 10 37N 51 2 E
Handan, China 34 F8 36 35N 114 28 E
Handeni, Tanzania 54 D4 5 25S 38 2 E
Handeni □, Tanzania 54 D4 5 30S 38 0 E
Handwara, India 43 B6 34 21N 74 20 E
Hanegev, Israel 47 E3 30 50N 35 0 E
Haney, Canada 72 D4 49 12N 122 40W
Hanford, U.S.A. 83 H4 36 20N 119 39W
Hang Chat, Thailand 38 C2 18 20N 99 21 E
Hang Dong, Thailand 38 C2 18 41N 98 55 E
Hangang →, S. Korea 35 F14 37 50N 126 30 E
Hangayn Nuruu, Mongolia 32 B4 47 30N 99 0 E
Hangchou = Hangzhou, China 33 C7 30 18N 120 11 E
Hanggin Houqi, China 34 D4 40 58N 107 4 E
Hanggin Qi, China 34 E5 39 52N 108 50 E
Hangu, China 35 E9 39 18N 117 53 E
Hangzhou, China 33 C7 30 18N 120 11 E
Hangzhou Wan, China 33 C7 30 15N 120 45 E
Hanhongor, Mongolia 34 C3 43 55N 104 28 E
Ḩanīdh, Si. Arabia 45 E6 26 35N 48 38 E
Ḩanīsh, Yemen 46 E3 13 45N 42 46 E
Hankinson, U.S.A. 80 B6 46 4N 96 54W
Hanko, Finland 9 G20 59 50N 22 57 E
Hanksville, U.S.A. 83 G8 38 22N 110 43W
Hanle, India 43 C8 32 42N 79 4 E
Hanmer Springs, N.Z. 59 K4 42 32S 172 50 E
Hann →, Australia 60 C4 17 26S 126 17 E
Hann, Mt., Australia 60 C4 15 45S 126 0 E
Hanna, Canada 72 C6 51 40N 111 54W
Hannaford, U.S.A. 80 B5 47 19N 98 11W
Hannah, U.S.A. 80 A5 48 58N 98 42W
Hannah B., Canada 70 B4 51 40N 80 0W
Hannibal, U.S.A. 80 F9 39 42N 91 22W
Hannover, Germany 16 B5 52 22N 9 46 E
Hanoi, Vietnam 32 D5 21 5N 105 55 E
Hanover = Hannover, Germany 16 B5 52 22N 9 46 E
Hanover, Canada 78 B3 44 9N 81 2W
Hanover, S. Africa 56 E3 31 4S 24 29 E
Hanover, N.H., U.S.A. 79 C12 43 42N 72 17W
Hanover, Ohio, U.S.A. 78 F2 40 4N 82 16W
Hanover, Pa., U.S.A. 76 F7 39 48N 76 59W
Hanover, I., Chile 96 G2 51 0S 74 50W
Hansi, India 42 E6 29 10N 75 57 E
Hanson, L., Australia 63 E2 31 0S 136 15 E
Hantsavichy, Belarus 17 B14 52 49N 26 30 E
Hanzhong, China 34 H4 33 10N 107 1 E
Hanzhuang, China 35 G9 34 33N 117 23 E
Haora, India 43 H13 22 37N 88 20 E
Haparanda, Sweden 8 D21 65 52N 24 8 E
Happy, U.S.A. 81 H4 34 45N 101 52W
Happy Camp, U.S.A. 82 F2 41 48N 123 23W
Happy Valley-Goose Bay, Canada 71 B7 53 15N 60 20W
Hapsu, N. Korea 35 D15 41 13N 128 51 E
Hapur, India 42 E7 28 45N 77 45 E
Ḩaql, Si. Arabia 47 F3 29 10N 34 58 E
Har, Indonesia 37 F8 5 16S 133 14 E
Har-Ayrag, Mongolia 34 B5 45 47N 109 16 E
Har Hu, China 32 C4 38 20N 97 38 E
Har Us Nuur, Mongolia 32 B4 48 0N 92 0 E
Har Yehuda, Israel 47 D3 31 35N 34 57 E
Ḩaraḑ, Si. Arabia 46 C4 24 22N 49 0 E
Haranomachi, Japan 30 F10 37 38N 140 58 E
Harardera, Somali Rep. 46 G4 4 33N 47 38 E
Harare, Zimbabwe 55 F3 17 43S 31 2 E
Harazé, Chad 51 F8 14 20N 19 12 E
Harbin, China 35 B14 45 48N 126 40 E
Harbor Beach, U.S.A. 76 D4 43 51N 82 39W
Harbor Springs, U.S.A. 76 C3 45 26N 85 0W
Harbour Breton, Canada 71 C8 47 29N 55 50W
Harbour Grace, Canada 71 C9 47 40N 53 22W
Harda, India 42 H7 22 27N 77 5 E
Hardangerfjorden, Norway 9 F12 60 5N 6 0 E
Hardangervidda, Norway 9 F12 60 7N 7 20 E
Hardap Dam, Namibia 56 C2 24 32S 17 50 E
Hardenberg, Neths. 15 B6 52 34N 6 37 E
Harderwijk, Neths. 15 B5 52 21N 5 38 E
Hardey →, Australia 60 D2 22 45S 116 8 E
Hardin, U.S.A. 82 D10 45 44N 107 37W
Harding, S. Africa 57 E4 30 35S 29 55 E
Harding Ra., Australia 60 C3 16 17S 124 55 E
Hardisty, Canada 72 C6 52 40N 111 18W
Hardman, U.S.A. 82 D4 45 10N 119 41W
Hardoi, India 43 F9 27 26N 80 6 E
Hardwar = Haridwar, India 42 E8 29 58N 78 9 E
Hardwick, U.S.A. 79 B12 44 30N 72 22W
Hardy, U.S.A. 81 G9 36 19N 91 29W
Hardy, Pen., Chile 96 H3 55 30S 68 20W
Hare B., Canada 71 B8 51 15N 55 45W
Hareid, Norway 9 E12 62 22N 6 1 E
Harer, Ethiopia 46 F3 9 20N 42 8 E
Hargeisa, Somali Rep. 46 F3 9 30N 44 2 E
Hari →, Indonesia 36 E2 1 16S 104 5 E
Haria, Canary Is. 22 E6 29 8N 13 32W
Haridwar, India 42 E8 29 58N 78 9 E
Haringhata →, Bangla. 41 J16 22 0N 89 58 E
Harīrūd →, Asia 40 A2 37 24N 60 38 E
Härjedalen, Sweden 9 E15 62 22N 13 5 E
Harlan, Iowa, U.S.A. 80 E7 41 39N 95 19W
Harlan, Ky., U.S.A. 77 G4 36 51N 83 19W
Harlech, U.K. 10 E3 52 52N 4 6W
Harlem, U.S.A. 82 B9 48 32N 108 47W
Harlingen, Neths. 15 A5 53 11N 5 25 E
Harlingen, U.S.A. 81 M6 26 12N 97 42W
Harlowton, U.S.A. 82 C9 46 26N 109 50W
Harney Basin, U.S.A. 82 E4 43 30N 119 0W
Harney L., U.S.A. 82 E4 43 14N 119 8W
Harney Peak, U.S.A. 80 D3 43 52N 103 32W
Härnösand, Sweden 9 E17 62 38N 17 55 E
Harp L., Canada 71 A7 55 5N 61 50W
Harrand, Pakistan 42 E4 29 28N 70 3 E
Harriman, U.S.A. 77 H3 35 56N 84 33W
Harrington Harbour, Canada 71 B8 50 31N 59 30W
Harris, U.K. 12 D2 57 50N 6 55W
Harris, Sd. of, U.K. 12 D1 57 44N 7 6W
Harris L., Australia 63 E2 31 10S 135 10 E
Harrisburg, Ill., U.S.A. 81 G10 37 44N 88 32W
Harrisburg, Nebr., U.S.A. 80 E3 41 33N 103 44W
Harrisburg, Oreg., U.S.A. 82 D2 44 16N 123 10W
Harrisburg, Pa., U.S.A. 78 F8 40 16N 76 53W
Harrismith, S. Africa 57 D4 28 15S 29 8 E
Harrison, Ark., U.S.A. 81 G8 36 14N 93 7W
Harrison, Idaho, U.S.A. 82 C5 47 27N 116 47W
Harrison, Nebr., U.S.A. 80 D3 42 41N 103 53W
Harrison, C., Canada 71 B8 54 55N 57 55W
Harrison Bay, U.S.A. 68 A4 70 40N 151 0W
Harrison L., Canada 72 D4 49 33N 121 50W
Harrisonburg, U.S.A. 76 F6 38 27N 78 52W
Harrisonville, U.S.A. 80 F7 38 39N 94 21W
Harriston, Canada 70 D3 43 57N 80 53W
Harrisville, U.S.A. 78 B1 44 39N 83 17W
Harrogate, U.K. 10 D6 54 0N 1 33W
Harrow, U.K. 11 F7 51 35N 0 21W
Harsin, Iran 44 C5 34 18N 47 33 E
Harstad, Norway 8 B17 68 48N 16 30 E
Hart, U.S.A. 76 D2 43 42N 86 22W
Hart, L., Australia 63 E2 31 10S 136 25 E
Hartbees →, S. Africa 56 D3 28 45S 20 32 E
Hartford, Conn., U.S.A. 79 E12 41 46N 72 41W
Hartford, Ky., U.S.A. 76 G2 37 27N 86 55W
Hartford, S. Dak., U.S.A. 80 D6 43 38N 96 57W
Hartford, Wis., U.S.A. 80 D10 43 19N 88 22W
Hartford City, U.S.A. 76 E3 40 27N 85 22W
Hartland, Canada 71 C6 46 20N 67 32W
Hartland Pt., U.K. 11 F3 51 1N 4 32W
Hartlepool, U.K. 10 C6 54 42N 1 13W
Hartlepool □, U.K. 10 C6 54 42N 1 17W
Hartley Bay, Canada 72 C3 53 25N 129 15W
Hartmannberge, Namibia 56 B1 17 0S 13 0 E
Hartney, Canada 73 D8 49 30N 100 35W
Harts →, S. Africa 56 D3 28 24S 24 17 E
Hartselle, U.S.A. 77 H2 34 27N 86 56W
Hartshorne, U.S.A. 81 H7 34 51N 95 34W
Hartsville, U.S.A. 77 H5 34 23N 80 4W
Hartwell, U.S.A. 77 H4 34 21N 82 56W
Harunabad, Pakistan 42 E5 29 35N 73 8 E
Harvand, Iran 45 D7 28 25N 55 43 E
Harvey, Australia 61 F2 33 5S 115 54 E
Harvey, Ill., U.S.A. 76 E2 41 36N 87 50W
Harvey, N. Dak., U.S.A. 80 B5 47 47N 99 56W
Harwich, U.K. 11 F9 51 56N 1 17 E
Haryana □, India 42 E7 29 0N 76 10 E
Haryn →, Belarus 17 B14 52 7N 27 17 E
Harz, Germany 16 C6 51 38N 10 44 E
Hasan Kīādeh, Iran 45 B6 37 24N 49 58 E
Ḩasanābād, Iran 45 C7 32 8N 52 44 E
Hasanpur, India 42 E8 28 43N 78 17 E
Hashimoto, Japan 31 G7 34 19N 135 37 E
Hashtjerd, Iran 45 C6 35 52N 50 40 E
Haskell, Okla., U.S.A. 81 H7 35 50N 95 40W
Haskell, Tex., U.S.A. 81 J5 33 10N 99 44W
Hasselt, Belgium 15 D5 50 56N 5 21 E
Hassi Inifel, Algeria 50 C5 29 50N 3 41 E
Hassi Messaoud, Algeria 50 B6 31 51N 6 1 E
Hässleholm, Sweden 9 H15 56 10N 13 46 E
Hastings, N.Z. 59 H6 39 39S 176 52 E
Hastings, U.K. 11 G8 50 51N 0 35 E
Hastings, Mich., U.S.A. 76 D3 42 39N 85 17W
Hastings, Minn., U.S.A. 80 C8 44 44N 92 51W
Hastings, Nebr., U.S.A. 80 E5 40 35N 98 23W
Hastings Ra., Australia 63 E5 31 15S 152 14 E
Hat Yai, Thailand 39 J3 7 1N 100 27 E
Hatanbulag, Mongolia 34 C5 43 8N 109 5 E
Hatay = Antalya, Turkey 25 G5 36 52N 30 45 E
Hatch, U.S.A. 83 K10 32 40N 107 9W
Hatches Creek, Australia 62 C2 20 56S 135 12 E
Hatchet L., Canada 73 B8 58 36N 103 40W
Hateruma-Shima, Japan 31 M1 24 3N 123 47 E
Hatfield P.O., Australia 63 E3 33 54S 143 49 E
Hatgal, Mongolia 32 A5 50 26N 100 9 E
Hathras, India 42 F8 27 36N 78 6 E
Hatia, Bangla. 41 H17 22 30N 91 5 E
Hato Mayor, Dom. Rep. 89 C6 18 46N 69 15W
Hattah, Australia 63 E3 34 48S 142 17 E
Hatteras, C., U.S.A. 77 H8 35 14N 75 32W
Hattiesburg, U.S.A. 81 K10 31 20N 89 17W
Hatvan, Hungary 17 E10 47 40N 19 45 E
Hau Bon = Cheo Reo, Vietnam 38 F7 13 25N 108 28 E
Hau Duc, Vietnam 38 E7 15 20N 108 13 E
Haugesund, Norway 9 G11 59 23N 5 13 E
Haukipudas, Finland 8 D21 65 12N 25 20 E
Haultain →, Canada 73 B7 55 51N 106 46W
Hauraki G., N.Z. 59 G5 36 35S 175 5 E
Haut Atlas, Morocco 50 B3 32 30N 5 0W
Haut Zaïre □, Zaïre 54 B2 2 20N 26 0 E
Hauterive, Canada 71 C6 49 10N 68 16W
Hautes Fagnes = Hohe Venn, Belgium 15 D6 50 30N 6 5 E
Hauts Plateaux, Algeria 50 A5 35 0N 1 0 E
Havana = La Habana, Cuba 88 B3 23 8N 82 22W
Havana, U.S.A. 80 E9 40 18N 90 4W
Havant, U.K. 11 G7 50 51N 0 58W
Havasu, L., U.S.A. 85 L12 34 18N 114 28W
Havel →, Germany 16 B7 52 50N 12 3 E
Havelange, Belgium 15 D5 50 23N 5 15 E
Havelian, Pakistan 42 B5 34 2N 73 10 E
Havelock, N.B., Canada 71 C6 46 2N 65 24W
Havelock, Ont., Canada 70 D4 44 26N 77 53W
Havelock, N.Z. 59 J4 41 17S 173 48 E
Haverfordwest, U.K. 11 F3 51 48N 4 58W
Haverhill, U.S.A. 79 D13 42 47N 71 5W
Havering, U.K. 11 F8 51 34N 0 13 E
Haverstraw, U.S.A. 79 E11 41 12N 73 58W
Havířov, Czech Rep. 17 D10 49 46N 18 20 E
Havlíčkův Brod, Czech. 16 D8 49 36N 15 33 E
Havre, U.S.A. 82 B9 48 33N 109 41W
Havre-Aubert, Canada 71 C7 47 12N 61 56W
Havre-St.-Pierre, Canada 71 B7 50 18N 63 33W
Haw →, U.S.A. 77 H6 35 36N 79 3W
Hawaii □, U.S.A. 74 H16 19 30N 156 30W
Hawaii I., Pac. Oc. 74 J17 20 0N 155 0W
Hawaiian Is., Pac. Oc. 74 H17 20 30N 156 0W
Hawaiian Ridge, Pac. Oc. 65 E11 24 0N 165 0W
Hawarden, Canada 73 C7 51 25N 106 36W
Hawarden, U.S.A. 80 D6 43 0N 96 29W
Hawea, L., N.Z. 59 L2 44 28S 169 19 E
Hawera, N.Z. 59 H5 39 35S 174 19 E
Hawick, U.K. 12 F6 55 26N 2 47W
Hawk Junction, Canada 70 C3 48 5N 84 38W
Hawke B., N.Z. 59 H6 39 25S 177 20 E
Hawker, Australia 63 E2 31 59S 138 22 E
Hawkesbury, Canada 70 C5 45 37N 74 37W
Hawkesbury I., Canada 72 C3 53 37N 129 3W
Hawkesbury Pt., Australia 62 A1 11 55S 134 5 E
Hawkinsville, U.S.A. 77 J4 32 17N 83 28W
Hawkwood, Australia 63 D5 25 45S 150 50 E
Hawley, U.S.A. 80 B6 46 53N 96 19W
Ḩawrān, Syria 47 C5 32 45N 36 15 E
Hawsh Müssá, Lebanon 47 B4 33 45N 35 55 E
Hawthorne, U.S.A. 82 G4 38 32N 118 38W
Haxtun, U.S.A. 80 E3 40 39N 102 38W
Hay, Australia 63 E3 34 30S 144 51 E
Hay →, Australia 62 C2 24 50S 138 0 E
Hay →, Canada 72 A5 60 50N 116 26W
Hay, C., Australia 60 B4 14 5S 129 29 E
Hay L., Canada 72 B5 58 50N 118 50W
Hay Lakes, Canada 72 C6 53 12N 113 2W
Hay-on-Wye, U.K. 11 E4 52 5N 3 8W
Hay River, Canada 72 A5 60 51N 115 44W
Hay Springs, U.S.A. 80 D3 42 41N 102 41W
Haya, Indonesia 37 E7 3 19S 129 37 E
Hayachine-San, Japan 30 E10 39 34N 141 29 E
Hayden, Ariz., U.S.A. 83 K8 33 0N 110 47W
Hayden, Colo., U.S.A. 82 F10 40 30N 107 16W
Haydon, Australia 62 B3 18 0S 141 30 E
Hayes, U.S.A. 80 C4 44 23N 101 1W
Hayes →, Canada 73 B10 57 3N 92 12W
Haynesville, U.S.A. 81 J8 32 58N 93 8W
Hayrabolu, Turkey 21 D12 41 12N 27 5 E
Hays, Canada 72 C6 50 6N 111 48W
Hays, U.S.A. 80 F5 38 53N 99 20W
Haysyn, Ukraine 17 D15 48 57N 29 25 E
Hayvoron, Ukraine 17 D15 48 22N 29 52 E
Hayward, Calif., U.S.A. 84 H4 37 40N 122 5W
Hayward, Wis., U.S.A. 80 B9 46 1N 91 29W
Haywards Heath, U.K. 11 F7 51 0N 0 5W
Hazafon □, Israel 47 C4 32 40N 35 20 E
Hazarām, Kūh-e, Iran 45 D8 29 30N 57 18 E
Hazard, U.S.A. 76 G4 37 15N 83 12W
Hazaribag, India 43 H11 23 58N 85 26 E
Hazaribag Road, India 43 G11 24 12N 85 57 E
Hazelton, Canada 72 B3 55 20N 127 42W
Hazelton, U.S.A. 80 B4 46 29N 100 17W
Hazen, N. Dak., U.S.A. 80 B4 47 18N 101 38W
Hazen, Nev., U.S.A. 82 G4 39 34N 119 3W
Hazlehurst, Ga., U.S.A. 77 K4 31 52N 82 36W
Hazlehurst, Miss., U.S.A. 81 K9 31 52N 90 24W
Hazleton, U.S.A. 79 F9 40 57N 75 59W
Hazlett, L., Australia 60 D4 21 30S 128 48 E
Hazor, Israel 47 B4 33 2N 35 32 E
Head of Bight, Australia 61 F5 31 30S 131 25 E
Headlands, Zimbabwe 55 F3 18 15S 32 2 E
Healdsburg, U.S.A. 84 G4 38 37N 122 52W
Healdton, U.S.A. 81 H6 34 14N 97 29W
Healesville, Australia 63 F4 37 35S 145 30 E
Heanor, U.K. 10 D6 53 1N 1 21W
Heard I., Ind. Oc. 3 G13 53 0S 74 0 E
Hearne, U.S.A. 81 K6 30 53N 96 36W
Hearne B., Canada 73 A9 60 10N 99 10W
Hearne L., Canada 72 A6 62 20N 113 10W
Hearst, Canada 70 C3 49 40N 83 41W
Heart →, U.S.A. 80 B4 46 46N 100 50W
Heart's Content, Canada 71 C9 47 54N 53 27W
Heath Pt., Canada 71 C7 49 8N 61 40W
Heath Steele, Canada 71 C6 47 17N 66 5W
Heavener, U.S.A. 81 H7 34 53N 94 36W
Hebbronville, U.S.A. 81 M5 27 18N 98 41W
Hebei □, China 34 E9 39 0N 116 0 E
Hebel, Australia 63 D4 28 58S 147 47 E
Heber, U.S.A. 85 N11 32 44N 115 32W
Heber Springs, U.S.A. 81 H9 35 30N 92 2W
Hebert, Canada 73 C7 50 30N 107 10W
Hebgen L., U.S.A. 82 D8 44 52N 111 20W
Hebi, China 34 G8 35 57N 114 7 E
Hebrides, U.K. 12 D1 57 30N 7 0W
Hebron = Al Khalīl, West Bank 47 D4 31 32N 35 6 E
Hebron, Canada 69 C13 58 5N 62 30W
Hebron, N. Dak., U.S.A. 80 B3 46 54N 102 3W
Hebron, Nebr., U.S.A. 80 E6 40 10N 97 35W
Hecate Str., Canada 72 C2 53 10N 130 30W
Hechi, China 32 D5 24 40N 108 2 E
Hechuan, China 32 C5 30 2N 106 12 E
Hecla, U.S.A. 80 C5 45 53N 98 9W
Hecla I., Canada 73 C9 51 10N 96 43W
Hede, Sweden 9 E15 62 23N 13 30 E
Hedemora, Sweden 9 F16 60 18N 15 58 E
Hedley, U.S.A. 81 H4 34 52N 100 39W
Heemstede, Neths. 15 B4 52 22N 4 37 E
Heerde, Neths. 15 B6 52 24N 6 2 E
Heerenveen, Neths. 15 B5 52 57N 5 55 E
Heerlen, Neths. 18 A6 50 55N 5 58 E
Hefa, Israel 47 C3 32 46N 35 0 E
Hefa □, Israel 47 C4 32 40N 35 0 E
Hefei, China 33 C6 31 52N 117 18 E
Hegang, China 33 B8 47 20N 130 19 E
Heichengzhen, China 34 F4 36 24N 106 3 E
Heidelberg, Germany 16 D5 49 24N 8 42 E
Heidelberg, S. Africa 56 E3 34 6S 20 59 E
Heilbron, S. Africa 57 D4 27 16S 27 59 E
Heilbronn, Germany 16 D5 49 9N 9 13 E
Heilongjiang □, China 35 B14 48 0N 126 0 E
Heilunkiang = Heilongjiang □, China 35 B14 48 0N 126 0 E
Heimaey, Iceland 8 E3 63 26N 20 17W
Heinola, Finland 9 F22 61 13N 26 2 E
Heinze Is., Burma 41 M20 14 25N 97 45 E

Heishan, China 35 D12 41 40N 122 5 E
Heishui, China 35 C10 42 8N 119 30 E
Hejaz = Al Ḥijāz, Si. Arabia 46 B2 26 0N 37 30 E
Hejian, China 34 E9 38 25N 116 5 E
Hejin, China 34 G6 35 35N 110 42 E
Hekla, Iceland 8 E4 63 56N 19 35W
Hekou, Gansu, China 34 F2 36 10N 103 28 E
Hekou, Yunnan, China 32 D5 22 30N 103 59 E
Helan Shan, China 34 E3 38 30N 105 55 E
Helena, Ark., U.S.A. 81 H9 34 32N 90 36W
Helena, Mont., U.S.A. 82 C7 46 36N 112 2W
Helendale, U.S.A. 85 L9 34 44N 117 19W
Helensburgh, U.K. 12 E4 56 1N 4 43W
Helensville, N.Z. 59 G5 36 41S 174 29 E
Helgeland, Norway 8 C15 66 7N 13 29 E
Helgoland, Germany 16 A4 54 10N 7 53 E
Heligoland = Helgoland, Germany 16 A4 54 10N 7 53 E
Heligoland B. = Deutsche Bucht, Germany 16 A5 54 15N 8 0 E
Hella, Iceland 8 E3 63 50N 20 24W
Hellendoorn, Neths. 15 B6 52 24N 6 27 E
Hellevoetsluis, Neths. 15 C4 51 50N 4 8 E
Hellín, Spain 19 C5 38 31N 1 40W
Helmand □, Afghan. 40 D4 31 20N 64 0 E
Helmand →, Afghan. 40 D2 31 12N 61 34 E
Helmond, Neths. 15 C5 51 29N 5 41 E
Helmsdale, U.K. 12 C5 58 7N 3 39W
Helong, China 35 C15 42 40N 129 0 E
Helper, U.S.A. 82 G8 39 41N 110 51W
Helsingborg, Sweden 9 H15 56 3N 12 42 E
Helsingfors = Helsinki, Finland 9 F21 60 15N 25 3 E
Helsingør, Denmark 9 H15 56 2N 12 35 E
Helsinki, Finland 9 F21 60 15N 25 3 E
Helston, U.K. 11 G2 50 6N 5 17W
Helvellyn, U.K. 10 C4 54 32N 3 1W
Helwân, Egypt 51 C11 29 50N 31 20 E
Hemet, U.S.A. 85 M10 33 45N 116 58W
Hemingford, U.S.A. 80 D3 42 19N 103 4W
Hemphill, U.S.A. 81 K8 31 20N 93 51W
Hempstead, U.S.A. 81 K6 30 6N 96 5W
Hemse, Sweden 9 H18 57 15N 18 22 E
Henan □, China 34 G8 34 0N 114 0 E
Henares →, Spain 19 B4 40 24N 3 30W
Henashi-Misaki, Japan 30 D9 40 37N 139 51 E
Henderson, Argentina 94 D3 36 18S 61 43W
Henderson, Ky., U.S.A. 76 G2 37 50N 87 35W
Henderson, N.C., U.S.A. 77 G6 36 20N 78 25W
Henderson, Nev., U.S.A. 85 J12 36 2N 114 59W
Henderson, Tenn., U.S.A. 77 H1 35 26N 88 38W
Henderson, Tex., U.S.A. 81 J7 32 9N 94 48W
Hendersonville, U.S.A. 77 H4 35 19N 82 28W
Hendījān, Iran 45 D6 30 14N 49 43 E
Hendon, Australia 63 D5 28 5S 151 50 E
Hengcheng, China 34 E4 38 18N 106 28 E
Hengdaohezi, China 35 B15 44 52N 129 0 E
Hengelo, Neths. 15 B6 52 3N 6 19 E
Hengshan, China 34 F5 37 58N 109 5 E
Hengshui, China 34 F8 37 41N 115 40 E
Hengyang, China 33 D6 26 52N 112 33 E
Henlopen, C., U.S.A. 76 F8 38 48N 75 6W
Hennenman, S. Africa 56 D4 27 59S 27 1 E
Hennessey, U.S.A. 81 G6 36 6N 97 54W
Henrietta, U.S.A. 81 J5 33 49N 98 12W
Henrietta, Ostrov, Russia 27 B16 77 6N 156 30 E
Henrietta Maria, C., Canada 70 A3 55 9N 82 20W
Henry, U.S.A. 80 E10 41 7N 89 22W
Henryetta, U.S.A. 81 H6 35 27N 95 59W
Hensall, Canada 78 C3 43 26N 81 30W
Hentiyn Nuruu, Mongolia 33 B5 48 30N 108 30 E
Henty, Australia 63 F4 35 30S 147 0 E
Henzada, Burma 41 L19 17 38N 95 26 E
Heppner, U.S.A. 82 D4 45 21N 119 33W
Hepworth, Canada 78 B3 44 37N 81 9W
Hequ, China 34 E6 39 20N 111 15 E
Héraðsflói, Iceland 8 D6 65 42N 14 12W
Héraðsvötn →, Iceland 8 D4 65 45N 19 25W
Herald Cays, Australia 62 B4 16 58S 149 9 E
Herāt, Afghan. 40 B3 34 20N 62 7 E
Herāt □, Afghan. 40 B3 35 0N 62 0 E
Herbert →, Australia 62 B4 18 31S 146 17 E
Herbert Downs, Australia 62 C2 23 7S 139 9 E
Herberton, Australia 62 B4 17 20S 145 25 E
Hercegnovi, Montenegro, Yug. 21 C8 42 30N 18 33 E
Herðubreið, Iceland 8 D5 65 11N 16 21W
Hereford, U.K. 11 E5 52 4N 2 43W
Hereford, U.S.A. 81 H3 34 49N 102 24W
Hereford and Worcester □, U.K. 11 E5 52 10N 2 30W
Herentals, Belgium 15 C4 51 12N 4 51 E
Herford, Germany 16 B5 52 7N 8 39 E
Herington, U.S.A. 80 F6 38 40N 96 57W
Herkimer, U.S.A. 79 D10 43 0N 74 59W
Herlong, U.S.A. 84 E6 40 8N 120 8W
Herman, U.S.A. 80 C6 45 49N 96 9W
Hermann, U.S.A. 80 F9 38 42N 91 27W
Hermannsburg Mission, Australia 60 D5 23 57S 132 45 E
Hermanus, S. Africa 56 E2 34 27S 19 12 E
Hermidale, Australia 63 E4 31 30S 146 42 E
Hermiston, U.S.A. 82 D4 45 51N 119 17W
Hermitage, N.Z. 59 K3 43 44S 170 5 E
Hermite, I., Chile 96 H3 55 50S 68 0W
Hermon, Mt. = Ash Shaykh, J., Lebanon 47 B4 33 25N 35 50 E
Hermosillo, Mexico 86 B2 29 10N 111 0W
Hernád →, Hungary 17 D11 47 56N 21 8 E
Hernandarias, Paraguay 95 B5 25 20S 54 40W
Hernandez, U.S.A. 84 J6 36 24N 120 46W
Hernando, Argentina 94 C3 32 28S 63 40W
Hernando, U.S.A. 81 H10 34 50N 90 0W
Herne, Germany 15 C7 51 32N 7 14 E
Herne Bay, U.K. 11 F9 51 21N 1 8 E
Herning, Denmark 9 H13 56 8N 8 58 E
Heroica = Caborca, Mexico 86 A2 30 40N 112 10W
Heroica Nogales = Nogales, Mexico 86 A2 31 20N 110 56W
Heron Bay, Canada 70 C2 48 40N 86 25W
Herradura, Pta. de la, Canary Is. 22 F5 28 26N 14 8W
Herreid, U.S.A. 80 C4 45 50N 100 4W
Herrick, Australia 62 G4 41 5S 147 55 E
Herrin, U.S.A. 81 G10 37 48N 89 2W
Hersonissos, Greece 23 D7 35 18N 25 22 E
Herstal, Belgium 15 D5 50 40N 5 38 E
Hertford, U.K. 11 F7 51 48N 0 4W
Hertfordshire □, U.K. 11 F7 51 51N 0 5W
's-Hertogenbosch, Neths. 15 C5 51 42N 5 17 E
Hertzogville, S. Africa 56 D4 28 9S 25 30 E
Herzliyya, Israel 47 C3 32 10N 34 50 E
Ḥeşār, Fārs, Iran 45 D6 29 52N 50 16 E
Ḥeşār, Markazī, Iran 45 C6 35 50N 49 12 E
Heshui, China 34 G5 36 0N 108 0 E
Heshun, China 34 F7 37 22N 113 32 E
Hesperia, U.S.A. 85 L9 34 25N 117 18W
Hesse = Hessen □, Germany 16 C5 50 30N 9 0 E
Hessen □, Germany 16 C5 50 30N 9 0 E
Hetch Hetchy Aqueduct, U.S.A. 84 H5 37 29N 122 19W
Hettinger, U.S.A. 80 C3 46 0N 102 42W
Hewett, C., Canada 69 A13 70 16N 67 45W
Hexham, U.K. 10 C5 54 58N 2 4W
Hexigten Qi, China 35 C9 43 18N 117 30 E
Ḥeydarābād, Iran 45 D7 30 33N 55 38 E
Heyfield, Australia 63 F4 37 59S 146 47 E
Heysham, U.K. 10 C5 54 3N 2 53W
Heywood, Australia 63 F3 38 8S 141 37 E
Heze, China 34 G8 35 14N 115 20 E
Hi Vista, U.S.A. 85 L9 34 45N 117 46W
Hialeah, U.S.A. 77 N5 25 50N 80 17W
Hiawatha, Kans., U.S.A. 80 F7 39 51N 95 32W
Hiawatha, Utah, U.S.A. 82 G8 39 29N 111 1W
Hibbing, U.S.A. 80 B8 47 25N 92 56W
Hibbs B., Australia 62 G4 42 35S 145 15 E
Hibernia Reef, Australia 60 B3 12 0S 123 23 E
Hickory, U.S.A. 77 H5 35 44N 81 21W
Hicks, Pt., Australia 63 F4 37 49S 149 17 E
Hicksville, U.S.A. 79 F11 40 46N 73 32W
Hida-Gawa →, Japan 31 G8 35 26N 137 3 E
Hida-Sammyaku, Japan 31 F8 36 30N 137 40 E
Hidaka-Sammyaku, Japan 30 C11 42 35N 142 45 E
Hidalgo, Mexico 87 C5 24 15N 99 26W
Hidalgo □, Mexico 87 C5 20 30N 99 10W
Hidalgo, Presa M., Mexico 86 B3 26 30N 108 35W
Hidalgo, Pta. del, Canary Is. 22 F3 28 33N 16 19W
Hidalgo del Parral, Mexico 86 B3 26 58N 105 40W
Hierro, Canary Is. 22 G1 27 44N 18 0W
Higashiajima-San, Japan 30 F10 37 40N 140 10 E
Higashiōsaka, Japan 31 G7 34 40N 135 37 E
Higgins, U.S.A. 81 G4 36 7N 100 2W
Higgins Corner, U.S.A. 84 F5 39 2N 121 5W
Higginsville, Australia 61 F3 31 42S 121 38 E
High Atlas = Haut Atlas, Morocco 50 B3 32 30N 5 0W
High I., Canada 71 A7 56 40N 61 10W
High Island, U.S.A. 81 L7 29 34N 94 24W
High Level, Canada 72 B5 58 31N 117 8W
High Point, U.S.A. 77 H6 35 57N 80 0W
High Prairie, Canada 72 B5 55 30N 116 30W
High River, Canada 72 C6 50 30N 113 50W
High Springs, U.S.A. 77 L4 29 50N 82 36W
High Tatra = Tatry, Slovak Rep. 17 D11 49 20N 20 0 E
High Veld, Africa 48 J6 27 0S 27 0 E
High Wycombe, U.K. 11 F7 51 37N 0 45W
Highbury, Australia 62 B3 16 25S 143 9 E
Highland □, U.K. 12 D4 57 17N 4 21W
Highland Park, U.S.A. 76 D2 42 11N 87 48W
Highmore, U.S.A. 80 C5 44 31N 99 27W
Highrock L., Canada 73 B7 57 5N 105 32W
Higüay, Dom. Rep. 89 C6 18 37N 68 42W
Hiiumaa, Estonia 9 G20 58 50N 22 45 E
Ḥijāz □, Si. Arabia 46 C2 24 0N 40 0 E
Hijo = Tagum, Phil. 37 C7 7 33N 125 53 E
Hikari, Japan 31 H5 33 58N 131 58 E
Hiko, U.S.A. 83 H6 37 32N 115 14W
Hikone, Japan 31 G8 35 15N 136 10 E
Hikurangi, N.Z. 59 F5 35 36S 174 17 E
Hikurangi, Mt., N.Z. 59 H6 38 21S 176 52 E
Hildesheim, Germany 16 B5 52 9N 9 56 E
Hill →, Australia 61 F2 30 23S 115 3 E
Hill City, Idaho, U.S.A. 82 E6 43 18N 115 3W
Hill City, Kans., U.S.A. 80 F5 39 22N 99 51W
Hill City, Minn., U.S.A. 80 B8 46 59N 93 36W
Hill City, S. Dak., U.S.A. 80 D3 43 56N 103 35W
Hill Island L., Canada 73 A7 60 30N 109 50W
Hillcrest Center, U.S.A. 85 K8 35 23N 118 57W
Hillegom, Neths. 15 B4 52 18N 4 35 E
Hillerød, Denmark 9 J15 55 56N 12 19 E
Hillingdon, U.K. 11 F7 51 32N 0 27W
Hillman, U.S.A. 76 C4 45 4N 83 54W
Hillmond, Canada 73 C7 53 26N 109 41W
Hillsboro, Kans., U.S.A. 80 F6 38 21N 97 12W
Hillsboro, N. Dak., U.S.A. 80 B6 47 26N 97 3W
Hillsboro, N.H., U.S.A. 79 C13 43 7N 71 54W
Hillsboro, N. Mex., U.S.A. 83 K10 32 55N 107 34W
Hillsboro, Oreg., U.S.A. 84 E4 45 31N 122 59W
Hillsboro, Tex., U.S.A. 81 J6 32 1N 97 8W
Hillsborough, Grenada 89 D7 12 28N 61 28W
Hillsdale, Mich., U.S.A. 76 E3 41 56N 84 38W
Hillsdale, N.Y., U.S.A. 79 D11 42 11N 73 30W
Hillside, Australia 60 D2 21 45S 119 23 E
Hillsport, Canada 70 C2 49 27N 85 34W
Hillston, Australia 63 E4 33 30S 145 31 E
Hilo, U.S.A. 74 J17 19 44N 155 5W
Hilton, U.S.A. 78 C7 43 17N 77 48W
Hilversum, Neths. 15 B5 52 14N 5 10 E
Himachal Pradesh □, India 42 D7 31 30N 77 0 E
Himalaya, Asia 43 E11 29 0N 84 0 E
Himatnagar, India 40 H8 23 37N 72 57 E
Himeji, Japan 31 G7 34 50N 134 40 E
Himi, Japan 31 F8 36 50N 136 55 E
Ḥimş, Syria 47 A5 34 40N 36 45 E
Ḥims □, Syria 47 A5 34 30N 37 0 E
Hinche, Haiti 89 C5 19 9N 72 1W
Hinchinbrook I., Australia 62 B4 18 20S 146 15 E
Hinckley, U.K. 11 E6 52 33N 1 22W
Hinckley, U.S.A. 82 G7 39 20N 112 40W
Hindaun, India 42 F7 26 44N 77 5 E
Hindmarsh, L., Australia 63 F3 36 5S 141 55 E
Hindu Bagh, Pakistan 42 D2 30 56N 67 50 E
Hindu Kush, Asia 40 B7 36 0N 71 0 E
Hindubagh, Pakistan 40 D5 30 56N 67 57 E
Hindupur, India 40 N10 13 49N 77 32 E
Hines Creek, Canada 72 B5 56 20N 118 40W
Hinganghat, India 40 J11 20 30N 78 52 E
Hingham, U.S.A. 82 B8 48 33N 110 25W
Hingoli, India 40 K10 19 41N 77 15 E
Hinna = Imi, Ethiopia 46 F3 6 28N 42 10 E
Hinnøya, Norway 8 B16 68 35N 15 50 E
Hinojosa del Duque, Spain 19 C3 38 30N 5 9W
Hinsdale, U.S.A. 82 B10 48 24N 107 5W
Hinton, Canada 72 C5 53 26N 117 34W
Hinton, U.S.A. 76 G5 37 40N 80 54W
Hippolytushoef, Neths. 15 B4 52 54N 4 58 E
Hirado, Japan 31 H4 33 22N 129 33 E
Hirakud Dam, India 41 J13 21 32N 83 45 E
Hiratsuka, Japan 31 G9 35 19N 139 21 E
Hiroo, Japan 30 C11 42 17N 143 19 E
Hirosaki, Japan 30 D10 40 34N 140 28 E
Hiroshima, Japan 31 G6 34 24N 132 30 E
Hiroshima □, Japan 31 G6 34 50N 133 0 E
Hisar, India 42 E6 29 12N 75 45 E
Hisb →, Iraq 44 D5 31 45N 44 17 E
Ḥismá, Si. Arabia 44 D3 28 30N 36 0 E
Hispaniola, W. Indies 89 C5 19 0N 71 0W
Hīt, Iraq 44 C4 33 38N 42 49 E
Hita, Japan 31 H5 33 20N 130 58 E
Hitachi, Japan 31 F10 36 36N 140 39 E
Hitchin, U.K. 11 F7 51 58N 0 16W
Hitoyoshi, Japan 31 H5 32 13N 130 45 E
Hitra, Norway 8 E13 63 30N 8 45 E
Ḥiyyon, N. →, Israel 47 E4 30 25N 35 10 E
Hjalmar L., Canada 73 A7 61 33N 109 25W
Hjälmaren, Sweden 9 G16 59 18N 15 40 E
Hjørring, Denmark 9 H13 57 29N 9 59 E
Hluhluwe, S. Africa 57 D5 28 1S 32 15 E
Hlyboka, Ukraine 17 D13 48 5N 25 56 E
Ho, Ghana 50 G5 6 37N 0 27 E
Ho Chi Minh City = Phanh Bho Ho Chi Minh, Vietnam 39 G6 10 58N 106 40 E
Ho Thuong, Vietnam 38 C5 19 32N 105 48 E
Hoa Binh, Vietnam 38 B5 20 50N 105 20 E
Hoa Da, Vietnam 39 G7 11 16N 108 40 E
Hoa Hiep, Vietnam 39 G5 11 34N 105 51 E
Hoai Nhon, Vietnam 38 E7 14 28N 109 1 E
Hoang Lien Son, Vietnam 38 A4 22 0N 104 0 E
Hoare B., Canada 69 B13 65 17N 62 30W
Hobart, Australia 62 G4 42 50S 147 21 E
Hobart, U.S.A. 81 H5 35 1N 99 6W
Hobbs, U.S.A. 81 J3 32 42N 103 8W
Hobbs Coast, Antarctica 5 D14 74 50S 131 0W
Hoboken, Belgium 15 C4 51 11N 4 21 E
Hoboken, U.S.A. 79 F10 40 45N 74 4W
Hobro, Denmark 9 H13 56 39N 9 46 E
Hoburgen, Sweden 9 H18 56 55N 18 7 E
Hodaka-Dake, Japan 31 F8 36 17N 137 39 E
Hodgson, Canada 73 C9 51 13N 97 36W
Hódmezövásárhely, Hungary 17 E11 46 28N 20 22 E
Hodna, Chott el, Algeria 50 A5 35 26N 4 43 E
Hodonín, Czech. 17 D9 48 50N 17 10 E
Hoeamdong, N. Korea 35 C16 42 30N 130 16 E
Hoek van Holland, Neths. 15 C4 52 0N 4 7 E
Hoengsŏng, S. Korea 35 F14 37 29N 127 59 E
Hoeryong, N. Korea 35 C15 42 30N 129 45 E
Hoeyang, N. Korea 35 E14 38 43N 127 36 E
Hof, Germany 16 C6 50 19N 11 55 E
Hofmeyr, S. Africa 56 E4 31 39S 25 50 E
Höfn, Iceland 8 D6 64 15N 15 13W
Hofors, Sweden 9 F17 60 31N 16 15 E
Hofsjökull, Iceland 8 D4 64 49N 18 48W
Hōfu, Japan 31 G5 34 3N 131 34 E
Hogan Group, Australia 62 F4 39 13S 147 1 E
Hogansville, U.S.A. 77 J3 33 10N 84 55W
Hogeland, U.S.A. 82 B9 48 51N 108 40W
Hoggar = Ahaggar, Algeria 50 D6 23 0N 6 30 E
Hogsty Reef, Bahamas 89 B5 21 41N 73 48W
Hoh →, U.S.A. 84 C2 47 45N 124 29W
Hohe Rhön, Germany 16 C5 50 24N 9 58 E
Hohe Venn, Belgium 15 D6 50 30N 6 5 E
Hohenwald, U.S.A. 77 H2 35 33N 87 33W
Hohhot, China 34 D6 40 52N 111 40 E
Hóhlakas, Greece 23 D9 35 57N 27 53 E
Hoi An, Vietnam 38 E7 15 30N 108 19 E
Hoi Xuan, Vietnam 38 B5 20 25N 105 9 E
Hoisington, U.S.A. 80 F5 38 31N 98 47W
Höjö, Japan 31 H6 33 58N 132 46 E
Hokianga Harbour, N.Z. 59 F4 35 31S 173 22 E
Hokitika, N.Z. 59 K3 42 42S 171 0 E
Hokkaidō □, Japan 30 C11 43 30N 143 0 E
Holbrook, Australia 63 F4 35 42S 147 18 E
Holbrook, U.S.A. 83 J8 34 54N 110 10W
Holden, Canada 72 C6 53 13N 112 11W
Holden, U.S.A. 82 G7 39 6N 112 16W
Holdenville, U.S.A. 81 H6 35 5N 96 24W
Holderness, U.K. 10 D7 53 45N 0 5W
Holdfast, Canada 73 C7 50 58N 105 25W
Holdrege, U.S.A. 80 E5 40 26N 99 23W
Holguín, Cuba 88 B4 20 50N 76 20W
Hollams Bird I., Namibia 56 C1 24 40S 14 30 E
Holland, U.S.A. 76 D2 42 47N 86 7W
Hollandia = Jayapura, Indonesia 37 E10 2 28S 140 38 E
Hollidaysburg, U.S.A. 78 F6 40 26N 78 24W
Hollis, U.S.A. 81 H5 34 41N 99 55W
Hollister, Calif., U.S.A. 83 H3 36 51N 121 24W
Hollister, Idaho, U.S.A. 82 E6 42 21N 114 35W
Holly, U.S.A. 80 F3 38 3N 102 7W
Holly Hill, U.S.A. 77 L5 29 16N 81 3W
Holly Springs, U.S.A. 81 H10 34 46N 89 27W
Hollywood, Calif., U.S.A. 83 J4 34 7N 118 25W
Hollywood, Fla., U.S.A. 77 N5 26 1N 80 9W
Holman Island, Canada 68 A8 70 42N 117 41W
Hólmavík, Iceland 8 D3 65 42N 21 40W
Holmes Reefs, Australia 62 B4 16 27S 148 0 E
Holmsund, Sweden 8 E19 63 41N 20 20 E
Holroyd →, Australia 62 A3 14 10S 141 36 E
Holstebro, Denmark 9 H13 56 22N 8 37 E
Holsworthy, U.K. 11 G3 50 48N 4 22W
Holton, Canada 71 B8 54 31N 57 12W
Holton, U.S.A. 80 F7 39 28N 95 44W
Holtville, U.S.A. 85 N11 32 49N 115 23W
Holwerd, Neths. 15 A5 53 22N 5 54 E
Holy Cross, U.S.A. 68 B4 62 12N 159 46W
Holy I., Angl., U.K. 10 D3 53 17N 4 37W
Holy I., Northumb., U.K. 10 B6 55 40N 1 47W
Holyhead, U.K. 10 D3 53 18N 4 38W
Holyoke, Colo., U.S.A. 80 E3 40 35N 102 18W
Holyoke, Mass., U.S.A. 79 D12 42 12N 72 37W
Holyrood, Canada 71 C9 47 27N 53 8W
Homa Bay, Kenya 54 C3 0 36S 34 30 E
Homa Bay □, Kenya 54 C3 0 50S 34 30 E
Homalin, Burma 41 G19 24 55N 95 0 E
Homand, Iran 45 C8 32 28N 59 37 E
Hombori, Mali 50 E4 15 20N 1 38W
Home B., Canada 69 B13 68 40N 67 10W
Home Hill, Australia 62 B4 19 43S 147 25 E
Homedale, U.S.A. 82 E5 43 37N 116 56W
Homer, Alaska, U.S.A. 68 C4 59 39N 151 33W
Homer, La., U.S.A. 81 J8 32 48N 93 4W
Homestead, Australia 62 C4 20 20S 145 40 E
Homestead, Fla., U.S.A. 77 N5 25 28N 80 29W
Homestead, Oreg., U.S.A. 82 D5 45 2N 116 51W
Homewood, U.S.A. 84 F6 39 4N 120 8W
Hominy, U.S.A. 81 G6 36 25N 96 24W
Homoine, Mozam. 57 C6 23 55S 35 8 E
Homs = Ḥimş, Syria 47 A5 34 40N 36 45 E
Homyel, Belarus 17 B16 52 28N 31 0 E
Hon Chong, Vietnam 39 G5 10 25N 104 30 E
Hon Me, Vietnam 38 C5 19 23N 105 56 E
Hon Quan, Vietnam 39 G6 11 40N 106 50 E
Honan = Henan □, China 34 G8 34 0N 114 0 E
Honbetsu, Japan 30 C11 43 7N 143 37 E
Honcut, U.S.A. 84 F5 39 20N 121 32W
Honda, Colombia 92 B4 5 12N 74 45W
Hondeklipbaai, S. Africa 56 E2 30 19S 17 17 E
Hondo, Japan 31 H5 32 27N 130 12 E
Hondo, U.S.A. 81 L5 29 21N 99 9W
Hondo →, Belize 87 D7 18 25N 88 21W
Honduras ■, Cent. Amer. 88 D2 14 40N 86 30W
Honduras, G. de, Caribbean 88 C2 16 50N 87 0W
Hønefoss, Norway 9 F14 60 10N 10 18 E
Honesdale, U.S.A. 79 E9 41 34N 75 16W
Honey L., U.S.A. 84 E6 40 15N 120 19W
Honfleur, France 18 B4 49 25N 0 13 E
Hong Gai, Vietnam 38 B6 20 57N 107 5 E
Hong He →, China 34 H8 32 25N 115 35 E
Hong Kong, Asia 33 D6 22 11N 114 14 E
Hongchŏn, S. Korea 35 F14 37 44N 127 53 E
Hongha →, Vietnam 32 D5 22 0N 104 0 E
Hongjiang, China 33 D5 27 7N 109 59 E
Hongliu He →, China 34 F5 38 0N 109 50 E
Hongor, Mongolia 34 B7 45 45N 112 50 E
Hongsa, Laos 38 C3 19 43N 101 20 E
Hongshui He →, China 33 D5 23 48N 109 30 E
Hongsŏng, S. Korea 35 F14 36 37N 126 38 E
Hongtong, China 34 F6 36 16N 111 40 E
Honguedo, Détroit d', Canada 71 C7 49 15N 64 0W
Hongwon, N. Korea 35 E14 40 0N 127 56 E
Hongze Hu, China 35 H10 33 15N 118 35 E
Honiara, Solomon Is. 64 H7 9 27S 159 57 E
Honiton, U.K. 11 G4 50 47N 3 11W
Honjō, Japan 30 E10 39 23N 140 3 E
Honningsvåg, Norway 8 A21 70 59N 25 59 E
Honolulu, U.S.A. 74 H16 21 19N 157 52W
Honshū, Japan 31 G9 36 0N 138 0 E
Hood, Mt., U.S.A. 82 D3 45 23N 121 42W
Hood, Pt., Australia 61 F2 34 23S 119 34 E
Hood River, U.S.A. 82 D3 45 43N 121 31W
Hoodsport, U.S.A. 84 C3 47 24N 123 9W
Hoogeveen, Neths. 15 B6 52 44N 6 28 E
Hoogezand, Neths. 15 A6 53 11N 6 45 E
Hooghly → = Hugli →, India 43 J13 21 56N 88 4 E
Hooghly-Chinsura = Chunchura, India 43 H13 22 53N 88 27 E
Hook Hd., Ireland 13 D5 52 7N 6 56W
Hook I., Australia 62 C4 20 4S 149 0 E
Hook of Holland = Hoek van Holland, Neths. 15 C4 52 0N 4 7 E
Hooker, U.S.A. 81 G4 36 52N 101 13W
Hooker Creek, Australia 60 C5 18 23S 130 38 E
Hoopeston, U.S.A. 76 E2 40 28N 87 40W
Hoopstad, S. Africa 56 D4 27 50S 25 55 E
Hoorn, Neths. 15 B5 52 38N 5 4 E
Hoover Dam, U.S.A. 85 K12 36 1N 114 44W
Hooversville, U.S.A. 78 F6 40 9N 78 55W
Hop Bottom, U.S.A. 79 E9 41 42N 75 46W
Hope, Canada 72 D4 49 25N 121 25W
Hope, Ariz., U.S.A. 85 M13 33 43N 113 42W
Hope, Ark., U.S.A. 81 J8 33 40N 93 36W
Hope, N. Dak., U.S.A. 80 B6 47 19N 97 43W
Hope, L., Australia 63 D2 28 24S 139 18 E
Hope, Pt., U.S.A. 68 B3 68 20N 166 50W
Hope Town, Bahamas 88 A4 26 35N 76 57W
Hopedale, Canada 71 A7 55 28N 60 13W
Hopefield, S. Africa 56 E2 33 3S 18 22 E
Hopei = Hebei □, China 34 E9 39 0N 116 0 E
Hopelchén, Mexico 87 D7 19 46N 89 50W
Hopetoun, Vic., Australia 63 F3 35 42S 142 22 E
Hopetoun, W. Austral., Australia 61 F3 33 57S 120 7 E
Hopetown, S. Africa 56 D3 29 34S 24 3 E
Hopkins, U.S.A. 80 E7 40 33N 94 49W
Hopkins, L., Australia 60 D4 24 15S 128 35 E
Hopkinsville, U.S.A. 77 G2 36 52N 87 29W
Hopland, U.S.A. 84 G3 38 58N 123 7W
Hoquiam, U.S.A. 84 D3 46 59N 123 53W
Horden Hills, Australia 60 D5 20 15S 130 0 E
Horinger, China 34 D6 40 28N 111 48 E
Horlick Mts., Antarctica 5 E15 84 0S 102 0W
Horlivka, Ukraine 25 E6 48 19N 38 5 E
Hormoz, Iran 45 E7 27 35N 55 0 E
Hormoz, Jaz. ye, Iran 45 E8 27 8N 56 28 E
Hormuz, Str. of, The Gulf 45 E8 26 30N 56 30 E
Horn, Austria 16 D8 48 39N 15 40 E
Horn, Iceland 8 C2 66 28N 22 28W
Horn →, Canada 72 A5 61 30N 118 1W
Horn, Cape = Hornos, C. de, Chile 96 H3 55 50S 67 30W

Horn Head, *Ireland* **13 A3** 55 14N 8 0W
Horn I., *Australia* **62 A3** 10 37S 142 17 E
Horn I., *U.S.A.* **77 K1** 30 14N 88 39W
Horn Mts., *Canada* **72 A5** 62 15N 119 15W
Hornavan, *Sweden* **8 C17** 66 15N 17 30 E
Hornbeck, *U.S.A.* **81 K8** 31 20N 93 24W
Hornbrook, *U.S.A.* **82 F2** 41 55N 122 33W
Horncastle, *U.K.* **10 D7** 53 13N 0 7W
Hornell, *U.S.A.* **78 D7** 42 20N 77 40W
Hornell L., *Canada* **72 A5** 62 20N 119 25W
Hornepayne, *Canada* **70 C3** 49 14N 84 48W
Hornitos, *U.S.A.* **84 H6** 37 30N 120 14W
Hornos, C. de, *Chile* **96 H3** 55 50S 67 30W
Hornsby, *Australia* **63 E5** 33 42S 151 2 E
Hornsea, *U.K.* **10 D7** 53 55N 0 11W
Horobetsu, *Japan* **30 C10** 42 24N 141 6 E
Horodenka, *Ukraine* **17 D13** 48 41N 25 29 E
Horodok, *Khmelnytskyy, Ukraine* **17 D14** 49 10N 26 34 E
Horodok, *Lviv, Ukraine* **17 D12** 49 46N 23 32 E
Horokhiv, *Ukraine* **17 C13** 50 30N 24 45 E
Horqin Youyi Qianqi, *China* **35 A12** 46 5N 122 3 E
Horqueta, *Paraguay* **94 A4** 23 15S 56 55W
Horse Creek, *U.S.A.* **80 E3** 41 57N 105 10W
Horse Is., *Canada* **71 B8** 50 15N 55 50W
Horsefly L., *Canada* **72 C4** 52 25N 121 0W
Horsens, *Denmark* **9 J13** 55 52N 9 51 E
Horsham, *Australia* **63 F3** 36 44S 142 13 E
Horsham, *U.K.* **11 F7** 51 4N 0 20W
Horten, *Norway* **9 G14** 59 25N 10 32 E
Horton, *U.S.A.* **80 F7** 39 40N 95 32W
Horton →, *Canada* **68 B7** 69 56N 126 52W
Horwood, L., *Canada* **70 C3** 48 5N 82 20W
Hose, Gunung-Gunung, *Malaysia* **36 D4** 2 5N 114 6 E
Ḥoseynābād, *Khuzestān, Iran* **45 C6** 32 45N 48 20 E
Ḥoseynābād, *Kordestān, Iran* **44 C5** 35 33N 47 8 E
Hoshangabad, *India* **42 H7** 22 45N 77 45 E
Hoshiarpur, *India* **42 D6** 31 30N 75 58 E
Hosmer, *U.S.A.* **80 C5** 45 34N 99 28W
Hospet, *India* **40 M10** 15 15N 76 20 E
Hospitalet de Llobregat, *Spain* **19 B7** 41 21N 2 6 E
Hoste, I., *Chile* **96 H3** 55 0S 69 0W
Hot, *Thailand* **38 C2** 18 8N 98 29 E
Hot Creek Range, *U.S.A.* **82 G5** 38 40N 116 20W
Hot Springs, *Ark., U.S.A.* **81 H8** 34 31N 93 3W
Hot Springs, *S. Dak., U.S.A.* **80 D3** 43 26N 103 29W
Hotagen, *Sweden* **8 E16** 63 50N 14 30 E
Hotan, *China* **32 C2** 37 25N 79 55 E
Hotazel, *S. Africa* **56 D3** 27 17S 22 58 E
Hotchkiss, *U.S.A.* **83 G10** 38 48N 107 43W
Hotham, C., *Australia* **60 B5** 12 2S 131 18 E
Hoting, *Sweden* **8 D17** 64 8N 16 15 E
Hotte, Massif de la, *Haiti* **89 C5** 18 30N 73 45W
Hottentotsbaai, *Namibia* **56 D1** 26 8S 14 59 E
Houck, *U.S.A.* **83 J9** 35 20N 109 10W
Houei Sai, *Laos* **38 B3** 20 18N 100 26 E
Houffalize, *Belgium* **15 D5** 50 8N 5 48 E
Houghton, *U.S.A.* **80 B10** 47 7N 88 34W
Houghton L., *U.S.A.* **76 C3** 44 21N 84 44W
Houghton-le-Spring, *U.K.* **10 C6** 54 51N 1 28W
Houhora Heads, *N.Z.* **59 F4** 34 49S 173 9 E
Houlton, *U.S.A.* **71 C6** 46 8N 67 51W
Houma, *U.S.A.* **81 L9** 29 36N 90 43W
Houston, *Canada* **72 C3** 54 25N 126 39W
Houston, *Mo., U.S.A.* **81 G9** 37 22N 91 58W
Houston, *Tex., U.S.A.* **81 L7** 29 46N 95 22W
Houtman Abrolhos, *Australia* **61 E1** 28 43S 113 48 E
Hovd, *Mongolia* **32 B4** 48 2N 91 37 E
Hove, *U.K.* **11 G7** 50 50N 0 10W
Hoveyzeh, *Iran* **45 D6** 31 27N 48 4 E
Hövsgöl, *Mongolia* **34 C5** 43 37N 109 39 E
Hövsgöl Nuur, *Mongolia* **32 A5** 51 0N 100 30 E
Howard, *Australia* **63 D5** 25 16S 152 32 E
Howard, *Kans., U.S.A.* **81 G6** 37 28N 96 16W
Howard, *Pa., U.S.A.* **78 E7** 41 1N 77 40W
Howard, *S. Dak., U.S.A.* **80 C6** 44 1N 97 32W
Howard I., *Australia* **62 A2** 12 10S 135 24 E
Howard L., *Canada* **73 A7** 62 15N 105 57W
Howe, *U.S.A.* **82 E7** 43 48N 113 0W
Howe, C., *Australia* **63 F5** 37 30S 150 0 E
Howell, *U.S.A.* **76 D4** 42 36N 83 56W
Howick, *Canada* **79 A11** 45 11N 73 51W
Howick, *S. Africa* **57 D5** 29 28S 30 14 E
Howick Group, *Australia* **62 A4** 14 20S 145 30 E
Howitt, L., *Australia* **63 D2** 27 40S 138 40 E
Howland I., *Pac. Oc.* **64 G10** 0 48N 176 38W
Howley, *Canada* **71 C8** 49 12N 57 2W
Howrah = Haora, *India* **43 H13** 22 37N 88 20 E
Howth Hd., *Ireland* **13 C5** 53 22N 6 3W
Höxter, *Germany* **16 C5** 51 46N 9 22 E
Hoy, *U.K.* **12 C5** 58 50N 3 15W
Høyanger, *Norway* **9 F12** 61 13N 6 4 E
Hoyerswerda, *Germany* **16 C8** 51 26N 14 14 E
Hpungan Pass, *Burma* **41 F20** 27 30N 96 55 E
Hradec Králové, *Czech.* **16 C8** 50 15N 15 50 E
Hrodna, *Belarus* **17 B12** 53 42N 23 52 E
Hrodzyanka, *Belarus* **17 B15** 53 31N 28 42 E
Hron →, *Slovak Rep.* **17 E10** 47 49N 18 45 E
Hrvatska = Croatia ■, *Europe* **16 F9** 45 20N 16 0 E
Hrymayliv, *Ukraine* **17 D14** 49 20N 26 5 E
Hsenwi, *Burma* **41 H20** 23 22N 97 55 E
Hsiamen = Xiamen, *China* **33 D6** 24 25N 118 4 E
Hsian = Xi'an, *China* **34 G5** 34 15N 109 0 E
Hsinchu, *Taiwan* **33 D7** 24 48N 120 58 E
Hsinhailien = Lianyungang, *China* **35 G10** 34 40N 119 11 E
Hsüchou = Xuzhou, *China* **35 G9** 34 18N 117 10 E
Hu Xian, *China* **34 G5** 34 8N 108 42 E
Hua Hin, *Thailand* **38 F2** 12 34N 99 58 E
Hua Xian, *Henan, China* **34 G8** 35 30N 114 30 E
Hua Xian, *Shaanxi, China* **34 G5** 34 30N 109 48 E
Huachinera, *Mexico* **86 A3** 30 9N 108 55W
Huacho, *Peru* **92 F3** 11 10S 77 35W
Huachón, *Peru* **92 F3** 10 35S 76 0W
Huade, *China* **34 D7** 41 55N 113 59 E
Huadian, *China* **35 C14** 43 0N 126 40 E
Huai He →, *China* **33 C6** 33 0N 118 30 E
Huai Yot, *Thailand* **39 J2** 7 45N 99 37 E
Huai'an, *Hebei, China* **34 D8** 40 30N 114 20 E
Huai'an, *Jiangsu, China* **35 H10** 33 30N 119 10 E
Huaibei, *China* **34 G9** 34 0N 116 48 E
Huaide, *China* **35 C13** 43 30N 124 40 E
Huaidezhen, *China* **35 C13** 43 48N 124 50 E
Huainan, *China* **33 C6** 32 38N 116 58 E
Huairen, *China* **34 E7** 39 48N 113 20 E
Huairou, *China* **34 D9** 40 20N 116 35 E
Huaiyang, *China* **34 H8** 33 40N 114 52 E
Huaiyuan, *China* **35 H9** 32 55N 117 10 E
Huajianzi, *China* **35 D13** 41 23N 125 20 E
Huajuapan de Leon, *Mexico* **87 D5** 17 50N 97 48W
Hualapai Peak, *U.S.A.* **83 J7** 35 5N 113 54W
Huallaga →, *Peru* **92 E3** 5 15S 75 30W
Huambo, *Angola* **53 G3** 12 42S 15 54 E
Huan Jiang →, *China* **34 G5** 34 28N 109 0 E
Huan Xian, *China* **34 F4** 36 33N 107 7 E
Huancabamba, *Peru* **92 E3** 5 10S 79 15W
Huancane, *Peru* **92 G5** 15 10S 69 44W
Huancapi, *Peru* **92 F4** 13 40S 74 0W
Huancavelica, *Peru* **92 F3** 12 50S 75 5W
Huancayo, *Peru* **92 F3** 12 5S 75 12W
Huang Hai = Yellow Sea, *China* **35 G12** 35 0N 123 0 E
Huang He →, *China* **35 F10** 37 55N 118 50 E
Huang Xian, *China* **35 F11** 37 38N 120 30 E
Huangling, *China* **34 G5** 35 34N 109 15 E
Huanglong, *China* **34 G5** 35 30N 109 59 E
Huangshi, *China* **33 C6** 30 10N 115 3 E
Huangsongdian, *China* **35 C14** 43 45N 127 25 E
Huantai, *China* **35 F9** 36 58N 117 56 E
Huánuco, *Peru* **92 E3** 9 55S 76 15W
Huaraz, *Peru* **92 E3** 9 30S 77 32W
Huarmey, *Peru* **92 F3** 10 5S 78 5W
Huascarán, *Peru* **92 E3** 9 8S 77 36W
Huasco, *Chile* **94 B1** 28 30S 71 15W
Huasco →, *Chile* **94 B1** 28 27S 71 13W
Huasna, *U.S.A.* **85 K6** 35 6N 120 24W
Huatabampo, *Mexico* **86 B3** 26 50N 109 50W
Huauchinango, *Mexico* **87 C5** 20 11N 98 3W
Huautla de Jiménez, *Mexico* **87 D5** 18 8N 96 51W
Huay Namota, *Mexico* **86 C4** 21 56N 104 30W
Huayin, *China* **34 G6** 34 35N 110 5 E
Huayllay, *Peru* **92 F3** 11 3S 76 21W
Hubbard, *U.S.A.* **81 K6** 31 51N 96 48W
Hubbart Pt., *Canada* **73 B10** 59 21N 94 41W
Hubei □, *China* **33 C6** 31 0N 112 0 E
Hubli-Dharwad = Dharwad, *India* **40 M9** 15 22N 75 15 E
Huchang, *N. Korea* **35 D14** 41 25N 127 2 E
Huddersfield, *U.K.* **10 D6** 53 39N 1 47W
Hudiksvall, *Sweden* **9 F17** 61 43N 17 10 E
Hudson, *Canada* **73 C10** 50 6N 92 9W
Hudson, *Mass., U.S.A.* **79 D13** 42 23N 71 34W
Hudson, *Mich., U.S.A.* **76 E3** 41 51N 84 21W
Hudson, *N.Y., U.S.A.* **79 D11** 42 15N 73 46W
Hudson, *Wis., U.S.A.* **80 C8** 44 58N 92 45W
Hudson, *Wyo., U.S.A.* **82 E9** 42 54N 108 35W
Hudson →, *U.S.A.* **79 F10** 40 42N 74 2W
Hudson Bay, *N.W.T., Canada* **69 C11** 60 0N 86 0W
Hudson Bay, *Sask., Canada* **73 C8** 52 51N 102 23W
Hudson Falls, *U.S.A.* **79 C11** 43 18N 73 35W
Hudson Mts., *Antarctica* **5 D16** 74 32S 99 20W
Hudson Str., *Canada* **69 B13** 62 0N 70 0W
Hudson's Hope, *Canada* **72 B4** 56 0N 121 54W
Hue, *Vietnam* **38 D6** 16 30N 107 35 E
Huehuetenango, *Guatemala* **88 C1** 15 20N 91 28W
Huejúcar, *Mexico* **86 C4** 22 21N 103 13W
Huelva, *Spain* **19 D2** 37 18N 6 57W
Huentelauquén, *Chile* **94 C1** 31 38S 71 33W
Huerta, Sa. de la, *Argentina* **94 C2** 31 10S 67 30W
Huesca, *Spain* **19 A5** 42 8N 0 25W
Huetamo, *Mexico* **86 D4** 18 36N 100 54W
Hugh →, *Australia* **62 D1** 25 1S 134 1 E
Hughenden, *Australia* **62 C3** 20 52S 144 10 E
Hughes, *Australia* **61 F4** 30 42S 129 31 E
Hugli →, *India* **43 J13** 21 56N 88 4 E
Hugo, *U.S.A.* **80 F3** 39 8N 103 28W
Hugoton, *U.S.A.* **81 G4** 37 11N 101 21W
Hui Xian, *Gansu, China* **34 H4** 33 50N 106 4 E
Hui Xian, *Henan, China* **34 G7** 35 27N 113 12 E
Hui'anbu, *China* **34 F4** 37 28N 106 38 E
Huichapán, *Mexico* **87 C5** 20 24N 99 40W
Huifa He →, *China* **35 C14** 43 0N 127 50 E
Huila, Nevado del, *Colombia* **92 C3** 3 0N 76 0W
Huimin, *China* **35 F9** 37 27N 117 28 E
Huinan, *China* **35 C14** 42 40N 126 2 E
Huinca Renancó, *Argentina* **94 C3** 34 51S 64 22W
Huining, *China* **34 G3** 35 38N 105 0 E
Huinong, *China* **34 E4** 39 5N 106 35 E
Huiting, *China* **34 G9** 34 5N 116 5 E
Huixtla, *Mexico* **87 D6** 15 9N 92 28W
Huize, *China* **32 D5** 26 24N 103 15 E
Hukawng Valley, *Burma* **41 F20** 26 30N 96 30 E
Hukuntsi, *Botswana* **56 C3** 23 58S 21 45 E
Ḥulayfā', *Si. Arabia* **44 E4** 25 58N 40 45 E
Huld, *Mongolia* **34 B3** 45 5N 105 30 E
Hulin He →, *China* **35 B12** 45 0N 122 10 E
Hull = Kingston upon Hull, *U.K.* **10 D7** 53 45N 0 21W
Hull, *Canada* **70 C4** 45 25N 75 44W
Hull →, *U.K.* **10 D7** 53 44N 0 20W
Hulst, *Neths.* **15 C4** 51 17N 4 2 E
Hulun Nur, *China* **33 B6** 49 0N 117 30 E
Humahuaca, *Argentina* **94 A2** 23 10S 65 25W
Humaitá, *Brazil* **92 E6** 7 35S 63 1W
Humaitá, *Paraguay* **94 B4** 27 2S 58 31W
Humansdorp, *S. Africa* **56 E3** 34 2S 24 46 E
Humbe, *Angola* **56 B1** 16 40S 14 55 E
Humber →, *U.K.* **10 D7** 53 42N 0 27W
Humbert River, *Australia* **60 C5** 16 30S 130 45 E
Humble, *U.S.A.* **81 L8** 29 59N 93 18W
Humboldt, *Canada* **73 C7** 52 15N 105 9W
Humboldt, *Iowa, U.S.A.* **80 D7** 42 44N 94 13W
Humboldt, *Tenn., U.S.A.* **81 H10** 35 50N 88 55W
Humboldt →, *U.S.A.* **82 F4** 39 59N 118 36W
Humboldt Gletscher, *Greenland* **4 B4** 79 30N 62 0W
Hume, *U.S.A.* **84 J8** 36 48N 118 54W
Hume, L., *Australia* **63 F4** 36 0S 147 5 E
Humenné, *Slovak Rep.* **17 D11** 48 55N 21 50 E
Humphreys, Mt., *U.S.A.* **84 H8** 37 17N 118 40W
Humphreys Peak, *U.S.A.* **83 J8** 35 21N 111 41W
Humptulips, *U.S.A.* **84 C3** 47 14N 123 57W
Hūn, *Libya* **51 C8** 29 2N 16 0 E
Hun Jiang →, *China* **35 D13** 40 50N 125 38 E
Húnaflói, *Iceland* **8 D3** 65 50N 20 50W
Hunan □, *China* **33 D6** 27 30N 112 0 E
Hunchun, *China* **35 C16** 42 52N 130 28 E
Hundred Mile House, *Canada* **72 C4** 51 38N 121 18W
Hunedoara, *Romania* **17 F12** 45 40N 22 50 E
Hung Yen, *Vietnam* **38 B6** 20 39N 106 4 E
Hungary ■, *Europe* **17 E10** 47 20N 19 20 E
Hungary, Plain of, *Europe* **6 F10** 47 0N 20 0 E
Hungerford, *Australia* **63 D3** 28 58S 144 24 E
Hŭngnam, *N. Korea* **35 E14** 39 49N 127 45 E
Hunsberge, *Namibia* **56 D2** 27 45S 17 12 E
Hunsrück, *Germany* **16 D4** 49 56N 7 27 E
Hunstanton, *U.K.* **10 E8** 52 56N 0 29 E
Hunter, *N. Dak., U.S.A.* **80 B6** 47 12N 97 13W
Hunter, *N.Y., U.S.A.* **79 D10** 42 13N 74 13W
Hunter I., *Australia* **62 G3** 40 30S 144 45 E
Hunter I., *Canada* **72 C3** 51 55N 128 0W
Hunter Ra., *Australia* **63 E5** 32 45S 150 15 E
Hunters Road, *Zimbabwe* **55 F2** 19 9S 29 49 E
Hunterville, *N.Z.* **59 H5** 39 56S 175 35 E
Huntingburg, *U.S.A.* **76 F2** 38 18N 86 57W
Huntingdon, *Canada* **70 C5** 45 6N 74 10W
Huntingdon, *U.K.* **11 E7** 52 20N 0 11W
Huntingdon, *U.S.A.* **78 F6** 40 30N 78 1W
Huntington, *Ind., U.S.A.* **76 E3** 40 53N 85 30W
Huntington, *N.Y., U.S.A.* **79 F11** 40 52N 73 26W
Huntington, *Oreg., U.S.A.* **82 D5** 44 21N 117 16W
Huntington, *Utah, U.S.A.* **82 G8** 39 20N 110 58W
Huntington, *W. Va., U.S.A.* **76 F4** 38 25N 82 27W
Huntington Beach, *U.S.A.* **85 M8** 33 40N 118 5W
Huntington Park, *U.S.A.* **83 K4** 33 58N 118 15W
Huntly, *N.Z.* **59 G5** 37 34S 175 11 E
Huntly, *U.K.* **12 D6** 57 27N 2 47W
Huntsville, *Canada* **70 C4** 45 20N 79 14W
Huntsville, *Ala., U.S.A.* **77 H2** 34 44N 86 35W
Huntsville, *Tex., U.S.A.* **81 K7** 30 43N 95 33W
Hunyani →, *Zimbabwe* **55 F3** 15 57S 30 39 E
Hunyuan, *China* **34 E7** 39 42N 113 42 E
Hunza →, *India* **43 B6** 35 54N 74 20 E
Huo Xian, *China* **34 F6** 36 36N 111 42 E
Huong Hoa, *Vietnam* **38 D6** 16 37N 106 45 E
Huong Khe, *Vietnam* **38 C5** 18 13N 105 41 E
Huonville, *Australia* **62 G4** 43 0S 147 5 E
Hupeh = Hubei □, *China* **33 C6** 31 0N 112 0 E
Ḥūr, *Iran* **45 D8** 30 50N 57 7 E
Hure Qi, *China* **35 C11** 42 45N 121 45 E
Hurley, *N. Mex., U.S.A.* **83 K9** 32 42N 108 8W
Hurley, *Wis., U.S.A.* **80 B9** 46 27N 90 11W
Huron, *Calif., U.S.A.* **84 J6** 36 12N 120 6W
Huron, *Ohio, U.S.A.* **78 E2** 41 24N 82 33W
Huron, *S. Dak., U.S.A.* **80 C5** 44 22N 98 13W
Huron, L., *U.S.A.* **78 B2** 44 30N 82 40W
Hurricane, *U.S.A.* **83 H7** 37 11N 113 17W
Hurunui →, *N.Z.* **59 K4** 42 54S 173 18 E
Húsavík, *Iceland* **8 C5** 66 3N 17 21W
Huşi, *Romania* **17 E15** 46 41N 28 7 E
Huskvarna, *Sweden* **9 H16** 57 47N 14 15 E
Hussar, *Canada* **72 C6** 51 3N 112 41W
Hustadvika, *Norway* **8 E12** 63 0N 7 0 E
Hutchinson, *Kans., U.S.A.* **81 F6** 38 5N 97 56W
Hutchinson, *Minn., U.S.A.* **80 C7** 44 54N 94 22W
Huttig, *U.S.A.* **81 J8** 33 2N 92 11W
Hutton, Mt., *Australia* **63 D4** 25 51S 148 20 E
Huy, *Belgium* **15 D5** 50 31N 5 15 E
Hvammstangi, *Iceland* **8 D3** 65 24N 20 57W
Hvar, *Croatia* **20 C7** 43 11N 16 28 E
Hvítá →, *Iceland* **8 D3** 64 30N 21 58W
Hwachon-chosuji, *S. Korea* **35 E14** 38 5N 127 50 E
Hwang Ho = Huang He →, *China* **35 F10** 37 55N 118 50 E
Hwange, *Zimbabwe* **55 F2** 18 18S 26 30 E
Hwange Nat. Park, *Zimbabwe* **56 B4** 19 0S 26 30 E
Hyannis, *U.S.A.* **80 E4** 42 0N 101 46W
Hyargas Nuur, *Mongolia* **32 B4** 49 0N 93 0 E
Hyden, *Australia* **61 F2** 32 24S 118 53 E
Hyderabad, *India* **40 L11** 17 22N 78 29 E
Hyderabad, *Pakistan* **42 G3** 25 23N 68 24 E
Hyères, *France* **18 E7** 43 8N 6 9 E
Hyères, Is. d', *France* **18 E7** 43 0N 6 20 E
Hyesan, *N. Korea* **35 D15** 41 20N 128 10 E
Hyland →, *Canada* **72 B3** 59 52N 128 12W
Hymia, *India* **43 C8** 33 40N 78 2 E
Hyndman Peak, *U.S.A.* **82 E6** 43 45N 114 8W
Hyōgo □, *Japan* **31 G7** 35 15N 134 50 E
Hyrum, *U.S.A.* **82 F8** 41 38N 111 51W
Hysham, *U.S.A.* **82 C10** 46 18N 107 14W
Hythe, *U.K.* **11 F9** 51 4N 1 5 E
Hyūga, *Japan* **31 H5** 32 25N 131 35 E
Hyvinge = Hyvinkää, *Finland* **9 F21** 60 38N 24 50 E
Hyvinkää, *Finland* **9 F21** 60 38N 24 50 E

I

I-n-Gall, *Niger* **50 E6** 16 51N 7 1 E
Iaco →, *Brazil* **92 E5** 9 3S 68 34W
Iakora, *Madag.* **57 C8** 23 6S 46 40 E
Ialomiţa →, *Romania* **17 F14** 44 42N 27 51 E
Iaşi, *Romania* **17 E14** 47 10N 27 40 E
Iba, *Phil.* **37 A6** 15 22N 120 0 E
Ibadan, *Nigeria* **50 G5** 7 22N 3 58 E
Ibagué, *Colombia* **92 C3** 4 20N 75 20W
Ibar →, *Serbia, Yug.* **21 C9** 43 43N 20 45 E
Ibaraki □, *Japan* **31 F10** 36 10N 140 10 E
Ibarra, *Ecuador* **92 C3** 0 21N 78 7W
Ibembo, *Zaïre* **54 B1** 2 35N 23 35 E
Ibera, L., *Argentina* **94 B4** 28 30S 57 9W
Iberian Peninsula, *Europe* **6 H5** 40 0N 5 0W
Iberville, *Canada* **70 C5** 45 19N 73 17W
Iberville, Lac d', *Canada* **70 A5** 55 55N 73 15W
Ibi, *Nigeria* **50 G6** 8 15N 9 44 E
Ibiá, *Brazil* **93 G9** 19 30S 46 30W
Ibicuy, *Argentina* **94 C4** 33 55S 59 10W
Ibioapaba, Sa. da, *Brazil* **93 D10** 4 0S 41 30W
Ibiza, *Spain* **22 C7** 38 54N 1 26 E
Ibo, *Mozam.* **55 E5** 12 22S 40 40 E
Ibonma, *Indonesia* **37 E8** 3 29S 133 31 E
Ibotirama, *Brazil* **93 F10** 12 13S 43 12W
Ibrāhīm →, *Lebanon* **47 A4** 34 4N 35 38 E
Ibu, *Indonesia* **37 D7** 1 35N 127 33 E
Ibusuki, *Japan* **31 J5** 31 12N 130 40 E
Icá, *Peru* **92 F3** 14 0S 75 48W
Iça →, *Brazil* **92 D5** 2 55S 67 58W
Içana, *Brazil* **92 C5** 0 21N 67 19W
Içel = Mersin, *Turkey* **25 G5** 36 51N 34 36 E
Iceland ■, *Europe* **8 D4** 64 45N 19 0W
Icha, *Russia* **27 D16** 55 30N 156 0 E
Ich'ang = Yichang, *China* **33 C6** 30 40N 111 20 E
Ichchapuram, *India* **41 K14** 19 10N 84 40 E
Ichihara, *Japan* **31 G10** 35 28N 140 5 E
Ichikawa, *Japan* **31 G9** 35 44N 139 55 E
Ichilo →, *Bolivia* **92 G6** 15 57S 64 50W
Ichinohe, *Japan* **30 D10** 40 13N 141 17 E
Ichinomiya, *Japan* **31 G8** 35 18N 136 48 E
Ichinoseki, *Japan* **30 E10** 38 55N 141 8 E
Ichŏn, *S. Korea* **35 F14** 37 17N 127 27 E
Icod, *Canary Is.* **22 F3** 28 22N 16 43W
Icy Str., *U.S.A.* **72 B1** 58 20N 135 30W
Ida Grove, *U.S.A.* **80 D7** 42 21N 95 28W
Ida Valley, *Australia* **61 E3** 28 42S 120 29 E
Idabel, *U.S.A.* **81 J7** 33 54N 94 50W
Idaho □, *U.S.A.* **82 D6** 45 0N 115 0W
Idaho City, *U.S.A.* **82 E6** 43 50N 115 50W
Idaho Falls, *U.S.A.* **82 E7** 43 30N 112 2W
Idaho Springs, *U.S.A.* **82 G11** 39 45N 105 31W
Idar-Oberstein, *Germany* **16 D4** 49 43N 7 16 E
Idd el Ghanam, *Sudan* **51 F9** 11 30N 24 19 E
Iddan, *Somali Rep.* **46 F4** 6 10N 48 55 E
Idehan, *Libya* **51 C7** 27 10N 11 30 E
Idehan Marzūq, *Libya* **51 D7** 24 50N 13 51 E
Idelès, *Algeria* **50 D6** 23 50N 5 53 E
Idfû, *Egypt* **51 D11** 24 55N 32 49 E
Ídhi Óros, *Greece* **23 D6** 35 15N 24 45 E
Ídhra, *Greece* **21 F10** 37 20N 23 28 E
Idi, *Indonesia* **36 C1** 5 2N 97 37 E
Idiofa, *Zaïre* **52 E3** 4 55S 19 42 E
Idlib, *Syria* **44 C3** 35 55N 36 36 E
Idria, *U.S.A.* **84 J6** 36 25N 120 41W
Idutywa, *S. Africa* **57 E4** 32 8S 28 18 E
Ieper, *Belgium* **15 D2** 50 51N 2 53 E
Ierápetra, *Greece* **23 E7** 35 0N 25 44 E
Iesi, *Italy* **20 C5** 43 31N 13 14 E
'Ifāl, W. al →, *Si. Arabia* **44 D2** 28 7N 35 3 E
Ifanadiana, *Madag.* **57 C8** 21 19S 47 39 E
Ife, *Nigeria* **50 G5** 7 30N 4 31 E
Iffley, *Australia* **62 B3** 18 53S 141 12 E
Ifni, *Morocco* **50 C2** 29 29N 10 12W
Iforas, Adrar des, *Mali* **50 E5** 19 40N 1 40 E
Ifould, L., *Australia* **61 F5** 30 52S 132 6 E
Iganga, *Uganda* **54 B3** 0 37N 33 28 E
Igarapava, *Brazil* **93 H9** 20 3S 47 47W
Igarapé Açu, *Brazil* **93 D9** 1 4S 47 33W
Igarka, *Russia* **26 C9** 67 30N 86 33 E
Igatimi, *Paraguay* **95 A4** 24 5S 55 40W
Igbetti, *Nigeria* **50 G5** 8 44N 4 8 E
Iggesund, *Sweden* **9 F17** 61 39N 17 10 E
Iglésias, *Italy* **20 E3** 39 19N 8 32 E
Igli, *Algeria* **50 B4** 30 25N 2 19W
Igloolik, *Canada* **69 B11** 69 20N 81 49W
Ignace, *Canada* **70 C1** 49 30N 91 40W
İğneada Burnu, *Turkey* **21 D13** 41 53N 28 2 E
Igoumenítsa, *Greece* **21 E9** 39 32N 20 18 E
Iguaçu →, *Brazil* **95 B5** 25 36S 54 36W
Iguaçu, Cat. del, *Brazil* **95 B5** 25 41S 54 26W
Iguaçu Falls = Iguaçu, Cat. del, *Brazil* **95 B5** 25 41S 54 26W
Iguala, *Mexico* **87 D5** 18 20N 99 40W
Igualada, *Spain* **19 B6** 41 37N 1 37 E
Iguassu = Iguaçu →, *Brazil* **95 B5** 25 36S 54 36W
Iguatu, *Brazil* **93 E11** 6 20S 39 18W
Iguéla, *Gabon* **52 E1** 2 0S 9 16 E
Igunga □, *Tanzania* **54 C3** 4 20S 33 45 E
Iheya-Shima, *Japan* **31 L3** 27 4N 127 58 E
Ihosy, *Madag.* **57 C8** 22 24S 46 8 E
Ihotry, L., *Madag.* **57 C7** 21 56S 43 41 E
Ii, *Finland* **8 D21** 65 19N 25 22 E
Ii-Shima, *Japan* **31 L3** 26 43N 127 47 E
Iida, *Japan* **31 G8** 35 35N 137 50 E
Iijoki →, *Finland* **8 D21** 65 20N 25 20 E
Iisalmi, *Finland* **8 E22** 63 32N 27 10 E
Iiyama, *Japan* **31 F9** 36 51N 138 22 E
Iizuka, *Japan* **31 H5** 33 38N 130 42 E
Ijebu-Ode, *Nigeria* **50 G5** 6 47N 3 58 E
IJmuiden, *Neths.* **15 B4** 52 28N 4 35 E
IJssel →, *Neths.* **15 B5** 52 35N 5 50 E
IJsselmeer, *Neths.* **15 B5** 52 45N 5 20 E
Ijuí →, *Brazil* **95 B4** 27 58S 55 20W
Ikaría, *Greece* **21 F12** 37 35N 26 10 E
Ikeda, *Japan* **31 G6** 34 1N 133 48 E
Ikela, *Zaïre* **52 E4** 1 6S 23 6 E
Iki, *Japan* **31 H4** 33 45N 129 42 E
Ikimba L., *Tanzania* **54 C3** 1 30S 31 20 E
Ikopa →, *Madag.* **57 B8** 16 45S 46 40 E
Ikungu, *Tanzania* **54 C3** 1 33S 33 42 E
Ilagan, *Phil.* **37 A6** 17 7N 121 53 E
Īlām, *Iran* **44 C5** 33 36N 46 36 E
Ilam, *Nepal* **43 F12** 26 58N 87 58 E
Ilanskiy, *Russia* **27 D10** 56 14N 96 3 E
Iława, *Poland* **17 B10** 53 36N 19 34 E
Ilbilbie, *Australia* **62 C4** 21 45S 149 20 E
Île-à-la-Crosse, *Canada* **73 B7** 55 27N 107 53W
Île-à-la-Crosse, Lac, *Canada* **73 B7** 55 40N 107 45W
Île-de-France, *France* **18 B5** 49 0N 2 20 E
Ilebo, *Zaïre* **52 E4** 4 17S 20 55 E
Ileje □, *Tanzania* **55 D3** 9 30S 33 25 E

Ilek, *Russia* **26 D6** 51 32N 53 21 E
Ilek →, *Russia* **24 D9** 51 30N 53 22 E
Ilford, *Canada* **73 B9** 56 4N 95 35W
Ilfracombe, *Australia* **62 C3** 23 30S 144 30 E
Ilfracombe, *U.K.* **11 F3** 51 12N 4 8W
Ilhéus, *Brazil* **93 F11** 14 49S 39 2W
Ili →, *Kazakstan* **26 E8** 45 53N 77 10 E
Ilich, *Kazakstan* **26 E7** 40 50N 68 27 E
Iliff, *U.S.A.* **80 E3** 40 45N 103 4W
Iligan, *Phil.* **37 C6** 8 12N 124 13 E
Ilion, *U.S.A.* **79 D9** 43 1N 75 2W
Ilkeston, *U.K.* **10 E6** 52 58N 1 19W
Illampu = Ancohuma, Nevada, *Bolivia* **92 G5** 16 0S 68 50W
Illana B., *Phil.* **37 C6** 7 35N 123 45 E
Illapel, *Chile* **94 C1** 32 0S 71 10W
Iller →, *Germany* **16 D6** 48 23N 9 58 E
Illetas, *Spain* **22 B9** 39 32N 2 35 E
Illimani, *Bolivia* **92 G5** 16 30S 67 50W
Illinois □, *U.S.A.* **75 C9** 40 15N 89 30W
Illinois →, *U.S.A.* **75 C8** 38 58N 90 28W
Illium = Troy, *Turkey* **21 E12** 39 57N 26 12 E
Ilmajoki, *Finland* **9 E20** 62 44N 22 34 E
Ilmen, Ozero, *Russia* **24 C5** 58 15N 31 10 E
Ilo, *Peru* **92 G4** 17 40S 71 20W
Iloilo, *Phil.* **37 B6** 10 45N 122 33 E
Ilorin, *Nigeria* **50 G5** 8 30N 4 35 E
Ilwaco, *U.S.A.* **84 D2** 46 19N 124 3W
Ilwaki, *Indonesia* **37 F7** 7 55S 126 30 E
Imabari, *Japan* **31 G6** 34 4N 133 0 E
Imaloto →, *Madag.* **57 C8** 23 27S 45 13 E
Imandra, Ozero, *Russia* **24 A5** 67 30N 33 0 E
Imari, *Japan* **31 H4** 33 15N 129 52 E
Imbler, *U.S.A.* **82 D5** 45 28N 117 58W
imeni 26 Bakinskikh Komissarov = Neftçala, *Azerbaijan* **25 G8** 39 19N 49 12 E
imeni 26 Bakinskikh Komissarov, *Turkmenistan* **25 G9** 39 22N 54 10 E
Imeni Poliny Osipenko, *Russia* **27 D14** 52 30N 136 29 E
Imeri, Serra, *Brazil* **92 C5** 0 50N 65 25W
Imerimandroso, *Madag.* **57 B8** 17 26S 48 35 E
Imi, *Ethiopia* **46 F3** 6 28N 42 10 E
Imlay, *U.S.A.* **82 F4** 40 40N 118 9W
Imlay City, *U.S.A.* **78 C1** 43 2N 83 5W
Immingham, *U.K.* **10 D7** 53 37N 0 13W
Immokalee, *U.S.A.* **77 M5** 26 25N 81 25W
Imola, *Italy* **20 B4** 44 20N 11 42 E
Imperatriz, *Brazil* **93 E9** 5 30S 47 29W
Impéria, *Italy* **20 C3** 43 53N 8 3 E
Imperial, *Canada* **73 C7** 51 21N 105 28W
Imperial, *Calif., U.S.A.* **85 N11** 32 51N 115 34W
Imperial, *Nebr., U.S.A.* **80 E4** 40 31N 101 39W
Imperial Beach, *U.S.A.* **85 N9** 32 35N 117 8W
Imperial Dam, *U.S.A.* **85 N12** 32 55N 114 25W
Imperial Reservoir, *U.S.A.* **85 N12** 32 53N 114 28W
Imperial Valley, *U.S.A.* **85 N11** 33 0N 115 30W
Imperieuse Reef, *Australia* **60 C2** 17 36S 118 50 E
Impfondo, *Congo* **52 D3** 1 40N 18 0 E
Imphal, *India* **41 G18** 24 48N 93 56 E
İmroz = Gökçeada, *Turkey* **21 D11** 40 10N 25 50 E
Imuruan B., *Phil.* **37 B5** 10 40N 119 10 E
In Belbel, *Algeria* **50 C5** 27 55N 1 12 E
In Salah, *Algeria* **50 C5** 27 10N 2 32 E
Ina, *Japan* **31 G8** 35 50N 137 55 E
Inangahua Junction, *N.Z.* **59 J3** 41 52S 171 59 E
Inanwatan, *Indonesia* **37 E8** 2 10S 132 14 E
Iñapari, *Peru* **92 F5** 11 0S 69 40W
Inari, *Finland* **8 B22** 68 54N 27 5 E
Inarijärvi, *Finland* **8 B22** 69 0N 28 0 E
Inawashiro-Ko, *Japan* **30 F10** 37 29N 140 6 E
Inca, *Spain* **22 B9** 39 43N 2 54 E
Incaguasi, *Chile* **94 B1** 29 12S 71 5W
İnce Burun, *Turkey* **25 F5** 42 7N 34 56 E
Inchon, *S. Korea* **35 F14** 37 27N 126 40 E
İncirliova, *Turkey* **21 F12** 37 50N 27 41 E
Incomáti →, *Mozam.* **57 D5** 25 46S 32 43 E
Indalsälven →, *Sweden* **9 E17** 62 36N 17 30 E
Indaw, *Burma* **41 G20** 24 15N 96 5 E
Independence, *Calif., U.S.A.* **83 H4** 36 48N 118 12W
Independence, *Iowa, U.S.A.* **80 D9** 42 28N 91 54W
Independence, *Kans., U.S.A.* **81 G7** 37 14N 95 42W
Independence, *Mo., U.S.A.* **80 F7** 39 6N 94 25W
Independence, *Oreg., U.S.A.* **82 D2** 44 51N 123 11W
Independence Fjord, *Greenland* **4 A6** 82 10N 29 0W
Independence Mts., *U.S.A.* **82 F5** 41 20N 116 0W
Index, *U.S.A.* **84 C5** 47 50N 121 33W
India ■, *Asia* **40 K11** 20 0N 78 0 E
Indian →, *U.S.A.* **77 M5** 27 59N 80 34W
Indian Cabins, *Canada* **72 B5** 59 52N 117 40W
Indian Harbour, *Canada* **71 B8** 54 27N 57 13W
Indian Head, *Canada* **73 C8** 50 30N 103 41W
Indian Ocean **28 K11** 5 0S 75 0 E
Indian Springs, *U.S.A.* **85 J11** 36 35N 115 40W
Indiana, *U.S.A.* **78 F5** 40 37N 79 9W
Indiana □, *U.S.A.* **76 E3** 40 0N 86 0W
Indianapolis, *U.S.A.* **76 F2** 39 46N 86 9W
Indianola, *Iowa, U.S.A.* **80 E8** 41 22N 93 34W
Indianola, *Miss., U.S.A.* **81 J9** 33 27N 90 39W
Indiga, *Russia* **24 A8** 67 38N 49 9 E
Indigirka →, *Russia* **27 B15** 70 48N 148 54 E
Indio, *U.S.A.* **85 M10** 33 43N 116 13W
Indonesia ■, *Asia* **36 F5** 5 0S 115 0 E
Indore, *India* **42 H6** 22 42N 75 53 E
Indramayu, *Indonesia* **37 G13** 6 20S 108 19 E
Indravati →, *India* **41 K12** 19 20N 80 20 E
Indre →, *France* **18 C4** 47 16N 0 11 E
Indus →, *Pakistan* **42 G2** 24 20N 67 47 E
Indus, Mouth of the, *Pakistan* **42 H2** 24 0N 68 0 E
İnebolu, *Turkey* **25 F5** 41 55N 33 40 E
Infiernillo, Presa del, *Mexico* **86 D4** 18 9N 102 0W
Ingende, *Zaïre* **52 E3** 0 12S 18 57 E
Ingenio, *Canary Is.* **22 G4** 27 55N 15 26W
Ingenio Santa Ana, *Argentina* **94 B2** 27 25S 65 40W
Ingersoll, *Canada* **78 C4** 43 4N 80 55W
Ingham, *Australia* **62 B4** 18 43S 146 10 E
Ingleborough, *U.K.* **10 C5** 54 10N 2 22W
Inglewood, *Queens., Australia* **63 D5** 28 25S 151 2 E
Inglewood, *Vic., Australia* **63 F3** 36 29S 143 53 E
Inglewood, *N.Z.* **59 H5** 39 9S 174 14 E
Inglewood, *U.S.A.* **85 M8** 33 58N 118 21W
Ingólfshöfði, *Iceland* **8 E5** 63 48N 16 39W
Ingolstadt, *Germany* **16 D6** 48 46N 11 26 E
Ingomar, *U.S.A.* **82 C10** 46 35N 107 23W
Ingonish, *Canada* **71 C7** 46 42N 60 18W
Ingraj Bazar, *India* **43 G13** 24 58N 88 10 E
Ingrid Christensen Coast, *Antarctica* **5 C6** 69 30S 76 0 E
Ingulec = Inhulec, *Ukraine* **25 E5** 47 42N 33 14 E
Ingwavuma, *S. Africa* **57 D5** 27 9S 31 59 E
Inhaca, I., *Mozam.* **57 D5** 26 1S 32 57 E
Inhafenga, *Mozam.* **57 C5** 20 36S 33 53 E
Inhambane, *Mozam.* **57 C6** 23 54S 35 30 E
Inhambane □, *Mozam.* **57 C5** 22 30S 34 20 E
Inhaminga, *Mozam.* **55 F4** 18 26S 35 0 E
Inharrime, *Mozam.* **57 C6** 24 30S 35 0 E
Inharrime →, *Mozam.* **57 C6** 24 30S 35 0 E
Inhulec, *Ukraine* **25 E5** 47 42N 33 14 E
Ining = Yining, *China* **26 E9** 43 58N 81 10 E
Inírida →, *Colombia* **92 C5** 3 55N 67 52W
Inishbofin, *Ireland* **13 C1** 53 37N 10 13W
Inishmore, *Ireland* **13 C2** 53 8N 9 45W
Inishowen Pen., *Ireland* **13 A4** 55 14N 7 15W
Injune, *Australia* **63 D4** 25 53S 148 32 E
Inklin, *Canada* **72 B2** 58 56N 133 5W
Inklin →, *Canada* **72 B2** 58 50N 133 10W
Inkom, *U.S.A.* **82 E7** 42 48N 112 15W
Inle L., *Burma* **41 J20** 20 30N 96 58 E
Inn →, *Austria* **16 D7** 48 35N 13 28 E
Innamincka, *Australia* **63 D3** 27 44S 140 46 E
Inner Hebrides, *U.K.* **12 D2** 57 0N 6 30W
Inner Mongolia = Nei Monggol Zizhiqu □, *China* **34 C6** 42 0N 112 0 E
Inner Sound, *U.K.* **12 D3** 57 30N 5 55W
Innerkip, *Canada* **78 C4** 43 13N 80 42W
Innetalling I., *Canada* **70 A4** 56 0N 79 0W
Innisfail, *Australia* **62 B4** 17 33S 146 5 E
Innisfail, *Canada* **72 C6** 52 0N 113 57W
In'no-shima, *Japan* **31 G6** 34 19N 133 10 E
Innsbruck, *Austria* **16 E6** 47 16N 11 23 E
Inny →, *Ireland* **13 C4** 53 30N 7 50W
Inongo, *Zaïre* **52 E3** 1 55S 18 30 E
Inoucdjouac = Inukjuak, *Canada* **69 C12** 58 25N 78 15W
Inowrocław, *Poland* **17 B10** 52 50N 18 12 E
Inpundong, *N. Korea* **35 D14** 41 25N 126 34 E
Inquisivi, *Bolivia* **92 G5** 16 50S 67 10W
Inscription, C., *Australia* **61 E1** 25 29S 112 59 E
Insein, *Burma* **41 L20** 16 50N 96 5 E
Inta, *Russia* **24 A11** 66 5N 60 8 E
Intendente Alvear, *Argentina* **94 D3** 35 12S 63 32W
Interior, *U.S.A.* **80 D4** 43 44N 101 59W
Interlaken, *Switz.* **16 E4** 46 41N 7 50 E
International Falls, *U.S.A.* **80 A8** 48 36N 93 25W
Intiyaco, *Argentina* **94 B3** 28 43S 60 5W
Inukjuak, *Canada* **69 C12** 58 25N 78 15W
Inútil, B., *Chile* **96 G2** 53 30S 70 15W
Inuvik, *Canada* **68 B6** 68 16N 133 40W
Inveraray, *U.K.* **12 E3** 56 14N 5 5W
Inverbervie, *U.K.* **12 E6** 56 51N 2 17W
Invercargill, *N.Z.* **59 M2** 46 24S 168 24 E
Inverclyde □, *U.K.* **12 F4** 55 55N 4 49W
Inverell, *Australia* **63 D5** 29 45S 151 8 E
Invergordon, *U.K.* **12 D4** 57 41N 4 10W
Invermere, *Canada* **72 C5** 50 30N 116 2W
Inverness, *Canada* **71 C7** 46 15N 61 19W
Inverness, *U.K.* **12 D4** 57 29N 4 13W
Inverness, *U.S.A.* **77 L4** 28 50N 82 20W
Inverurie, *U.K.* **12 D6** 57 17N 2 23W
Inverway, *Australia* **60 C4** 17 50S 129 38 E
Investigator Group, *Australia* **63 E1** 34 45S 134 20 E
Investigator Str., *Australia* **63 F2** 35 30S 137 0 E
Inya, *Russia* **26 D9** 50 28N 86 37 E
Inyanga, *Zimbabwe* **55 F3** 18 12S 32 40 E
Inyangani, *Zimbabwe* **55 F3** 18 5S 32 50 E
Inyantue, *Zimbabwe* **55 F2** 18 30S 26 40 E
Inyo Mts., *U.S.A.* **83 H5** 36 40N 118 0W
Inyokern, *U.S.A.* **85 K9** 35 39N 117 49W
Inza, *Russia* **24 D8** 53 55N 46 25 E
Iō-Jima, *Japan* **31 J5** 30 48N 130 18 E
Ioánnina, *Greece* **21 E9** 39 42N 20 47 E
Iola, *U.S.A.* **81 G7** 37 55N 95 24W
Iona, *U.K.* **12 E2** 56 20N 6 25W
Ione, *Calif., U.S.A.* **84 G6** 38 21N 120 56W
Ione, *Wash., U.S.A.* **82 B5** 48 45N 117 25W
Ionia, *U.S.A.* **76 D3** 42 59N 85 4W
Ionian Is. = Iónioi Nísoi, *Greece* **21 E9** 38 40N 20 0 E
Ionian Sea, *Medit. S.* **21 E7** 37 30N 17 30 E
Iónioi Nísoi, *Greece* **21 E9** 38 40N 20 0 E
Íos, *Greece* **21 F11** 36 41N 25 20 E
Iowa □, *U.S.A.* **80 D8** 42 18N 93 30W
Iowa City, *U.S.A.* **80 E9** 41 40N 91 32W
Iowa Falls, *U.S.A.* **80 D8** 42 31N 93 16W
Ipala, *Tanzania* **54 C3** 4 30S 32 52 E
Ipameri, *Brazil* **93 G9** 17 44S 48 9W
Ipatinga, *Brazil* **93 G10** 19 32S 42 30W
Ipiales, *Colombia* **92 C3** 0 50N 77 37W
Ipin = Yibin, *China* **32 D5** 28 45N 104 32 E
Ipixuna, *Brazil* **92 E4** 7 0S 71 40W
Ipoh, *Malaysia* **39 K3** 4 35N 101 5 E
Ippy, *C.A.R.* **51 G9** 6 5N 21 7 E
İpsala, *Turkey* **21 D12** 40 55N 26 23 E
Ipswich, *Australia* **63 D5** 27 35S 152 40 E
Ipswich, *U.K.* **11 E9** 52 4N 1 10 E
Ipswich, *Mass., U.S.A.* **79 D14** 42 41N 70 50W
Ipswich, *S. Dak., U.S.A.* **80 C5** 45 27N 99 2W
Ipu, *Brazil* **93 D10** 4 23S 40 44W
Iqaluit, *Canada* **69 B13** 63 44N 68 31W
Iquique, *Chile* **92 H4** 20 19S 70 5W
Iquitos, *Peru* **92 D4** 3 45S 73 10W
Irabu-Jima, *Japan* **31 M2** 24 50N 125 10 E
Iracoubo, *Fr. Guiana* **93 B8** 5 30N 53 10W
Īrafshān, *Iran* **45 E9** 26 42N 61 56 E
Iráklion, *Greece* **23 D7** 35 20N 25 12 E
Iráklion □, *Greece* **23 D7** 35 10N 25 10 E
Irala, *Paraguay* **95 B5** 25 55S 54 35W
Iramba □, *Tanzania* **54 C3** 4 30S 34 30 E
Iran ■, *Asia* **45 C7** 33 0N 53 0 E
Iran, Gunung-Gunung, *Malaysia* **36 D4** 2 20N 114 50 E
Iran, Plateau of, *Asia* **28 F9** 32 0N 55 0 E
Iran Ra. = Iran, Gunung-Gunung, *Malaysia* **36 D4** 2 20N 114 50 E
Īrānshahr, *Iran* **45 E9** 27 15N 60 40 E
Irapuato, *Mexico* **86 C4** 20 40N 101 30W
Iraq ■, *Asia* **44 C5** 33 0N 44 0 E
Irati, *Brazil* **95 B5** 25 25S 50 38W
Irbid, *Jordan* **47 C4** 32 35N 35 48 E
Irbid □, *Jordan* **47 C5** 32 15N 36 35 E
Irebu, *Zaïre* **52 E3** 0 40S 17 46 E
Ireland ■, *Europe* **13 D4** 53 50N 7 52W
Ireland's Eye, *Ireland* **13 C5** 53 24N 6 4W
Iret, *Russia* **27 C16** 60 3N 154 20 E
Irhyangdong, *N. Korea* **35 D15** 41 15N 129 30 E
Iri, *S. Korea* **35 G14** 35 59N 127 0 E
Irian Jaya □, *Indonesia* **37 E9** 4 0S 137 0 E
Iringa, *Tanzania* **54 D4** 7 48S 35 43 E
Iringa □, *Tanzania* **54 D4** 7 48S 35 43 E
Iriomote-Jima, *Japan* **31 M1** 24 19N 123 48 E
Iriona, *Honduras* **88 C2** 15 57N 85 11W
Iriri →, *Brazil* **93 D8** 3 52S 52 37W
Irish Republic ■, *Europe* **13 D4** 53 0N 8 0W
Irish Sea, *U.K.* **10 D3** 53 38N 4 48W
Irkineyeva, *Russia* **27 D10** 58 30N 96 49 E
Irkutsk, *Russia* **27 D11** 52 18N 104 20 E
Irma, *Canada* **73 C6** 52 55N 111 14W
Irō-Zaki, *Japan* **31 G9** 34 36N 138 51 E
Iron Baron, *Australia* **63 E2** 32 58S 137 11 E
Iron Gate = Portile de Fier, *Europe* **17 F12** 44 42N 22 30 E
Iron Knob, *Australia* **63 E2** 32 46S 137 8 E
Iron Mountain, *U.S.A.* **76 C1** 45 49N 88 4W
Iron Ra., *Australia* **62 A3** 12 46S 143 16 E
Iron River, *U.S.A.* **80 B10** 46 6N 88 39W
Ironbridge, *U.K.* **11 E5** 52 38N 2 30W
Irondequoit, *U.S.A.* **78 C7** 43 13N 77 35W
Ironstone Kopje, *Botswana* **56 D3** 25 17S 24 5 E
Ironton, *Mo., U.S.A.* **81 G9** 37 36N 90 38W
Ironton, *Ohio, U.S.A.* **76 F4** 38 32N 82 41W
Ironwood, *U.S.A.* **80 B9** 46 27N 90 9W
Iroquois Falls, *Canada* **70 C3** 48 46N 80 41W
Irpin, *Ukraine* **17 C16** 50 30N 30 15 E
Irrara Cr. →, *Australia* **63 D4** 29 35S 145 31 E
Irrawaddy □, *Burma* **41 L19** 17 0N 95 0 E
Irrawaddy →, *Burma* **41 M19** 15 50N 95 6 E
Irtysh →, *Russia* **26 C7** 61 4N 68 52 E
Irumu, *Zaïre* **54 B2** 1 32N 29 53 E
Irún, *Spain* **19 A5** 43 20N 1 52W
Irunea = Pamplona, *Spain* **19 A5** 42 48N 1 38W
Irvine, *Canada* **73 D6** 49 57N 110 16W
Irvine, *U.K.* **12 F4** 55 37N 4 41W
Irvine, *Calif., U.S.A.* **85 M9** 33 41N 117 46W
Irvine, *Ky., U.S.A.* **76 G4** 37 42N 83 58W
Irvinestown, *U.K.* **13 B4** 54 28N 7 39W
Irving, *U.S.A.* **81 J6** 32 49N 96 56W
Irvona, *U.S.A.* **78 F6** 40 46N 78 33W
Irwin →, *Australia* **61 E1** 29 15S 114 54 E
Irymple, *Australia* **63 E3** 34 14S 142 8 E
Isaac →, *Australia* **62 C4** 22 55S 149 20 E
Isabel, *U.S.A.* **80 C4** 45 24N 101 26W
Isabela, I., *Mexico* **86 C3** 21 51N 105 55W
Isabella, *Phil.* **37 C6** 6 40N 122 10 E
Isabella, Cord., *Nic.* **88 D2** 13 30N 85 25W
Isabella Ra., *Australia* **60 D3** 21 0S 121 4 E
Ísafjarðardjúp, *Iceland* **8 C2** 66 10N 23 0W
Ísafjörður, *Iceland* **8 C2** 66 5N 23 9W
Isagarh, *India* **42 G7** 24 48N 77 51 E
Isahaya, *Japan* **31 H5** 32 52N 130 2 E
Isaka, *Tanzania* **54 C3** 3 56S 32 59 E
Isangi, *Zaïre* **52 D4** 0 52N 24 10 E
Isar →, *Germany* **16 D7** 48 48N 12 57 E
Íschia, *Italy* **20 D5** 40 44N 13 57 E
Isdell →, *Australia* **60 C3** 16 27S 124 51 E
Ise, *Japan* **31 G8** 34 25N 136 45 E
Ise-Wan, *Japan* **31 G8** 34 43N 136 43 E
Iseramagazi, *Tanzania* **54 C3** 4 37S 32 10 E
Isère →, *France* **18 D6** 44 59N 4 51 E
Isérnia, *Italy* **20 D6** 41 36N 14 14 E
Ishigaki-Shima, *Japan* **31 M2** 24 20N 124 10 E
Ishikari-Gawa →, *Japan* **30 C10** 43 15N 141 23 E
Ishikari-Sammyaku, *Japan* **30 C11** 43 30N 143 0 E
Ishikari-Wan, *Japan* **30 C10** 43 25N 141 1 E
Ishikawa □, *Japan* **31 F8** 36 30N 136 30 E
Ishim, *Russia* **26 D7** 56 10N 69 30 E
Ishim →, *Russia* **26 D8** 57 45N 71 10 E
Ishinomaki, *Japan* **30 E10** 38 32N 141 20 E
Ishioka, *Japan* **31 F10** 36 11N 140 16 E
Ishkuman, *Pakistan* **43 A5** 36 30N 73 50 E
Ishpeming, *U.S.A.* **76 B2** 46 29N 87 40W
Isil Kul, *Russia* **26 D8** 54 55N 71 16 E
Isiolo, *Kenya* **54 B4** 0 24N 37 33 E
Isiolo □, *Kenya* **54 B4** 2 30N 37 30 E
Isiro, *Zaïre* **54 B2** 2 53N 27 40 E
Isisford, *Australia* **62 C3** 24 15S 144 21 E
İskenderun, *Turkey* **25 G6** 36 32N 36 10 E
İskenderun Körfezi, *Turkey* **25 G6** 36 40N 35 50 E
Iskŭr →, *Bulgaria* **21 C11** 43 45N 24 25 E
Iskut →, *Canada* **72 B2** 56 45N 131 49W
Isla →, *U.K.* **12 E5** 56 32N 3 20W
Isla Vista, *U.S.A.* **85 L7** 34 25N 119 53W
Islamabad, *Pakistan* **42 C5** 33 40N 73 10 E
Islamkot, *Pakistan* **42 G4** 24 42N 70 13 E
Island →, *Canada* **72 A4** 60 25N 121 12W
Island Falls, *Canada* **70 C3** 49 35N 81 20W
Island Falls, *U.S.A.* **71 C6** 46 1N 68 16W
Island L., *Canada* **73 C10** 53 47N 94 25W
Island Lagoon, *Australia* **63 E2** 31 30S 136 40 E
Island Pond, *U.S.A.* **79 B13** 44 49N 71 53W
Islands, B. of, *Canada* **71 C8** 49 11N 58 15W
Islay, *U.K.* **12 F2** 55 46N 6 10W
Isle →, *France* **18 D3** 44 55N 0 15W
Isle aux Morts, *Canada* **71 C8** 47 35N 59 0W
Isle of Wight □, *U.K.* **11 G6** 50 41N 1 17W
Isle Royale, *U.S.A.* **80 A10** 48 0N 88 54W
Isleta, *U.S.A.* **83 J10** 34 55N 106 42W
Isleton, *U.S.A.* **84 G5** 38 10N 121 37W
Ismail = Izmayil, *Ukraine* **17 F15** 45 22N 28 46 E
Ismâ'ilîya, *Egypt* **51 B11** 30 37N 32 18 E
Ismay, *U.S.A.* **80 B2** 46 30N 104 48W
Isna, *Egypt* **51 C11** 25 17N 32 30 E
Isogstalo, *India* **43 B8** 34 15N 78 46 E
İsparta, *Turkey* **25 G5** 37 47N 30 30 E
Íspica, *Italy* **20 F6** 36 47N 14 55 E
Israel ■, *Asia* **47 D3** 32 0N 34 50 E
Issoire, *France* **18 D5** 45 32N 3 15 E
Issyk-Kul = Ysyk-Köl, *Kyrgyzstan* **28 E11** 42 26N 76 12 E
Issyk-Kul, Ozero = Ysyk-Köl, Ozero, *Kyrgyzstan* **26 E8** 42 25N 77 15 E
Istaihah, *U.A.E.* **45 F7** 23 19N 54 4 E
İstanbul, *Turkey* **21 D13** 41 0N 29 0 E
Istiaía, *Greece* **21 E10** 38 57N 23 9 E
Istokpoga, L., *U.S.A.* **77 M5** 27 23N 81 17W
Istra, *Croatia* **16 F7** 45 10N 14 0 E
İstranca Dağlan, *Turkey* **21 D12** 41 48N 27 36 E
Istres, *France* **18 E6** 43 31N 4 59 E
Istria = Istra, *Croatia* **16 F7** 45 10N 14 0 E
Itá, *Paraguay* **94 B4** 25 29S 57 21W
Itabaiana, *Brazil* **93 E11** 7 18S 35 19W
Itaberaba, *Brazil* **93 F10** 12 32S 40 18W
Itabira, *Brazil* **93 G10** 19 37S 43 13W
Itabirito, *Brazil* **95 A7** 20 15S 43 48W
Itabuna, *Brazil* **93 F11** 14 48S 39 16W
Itaipú, Represa de, *Brazil* **95 B5** 25 30S 54 30W
Itaituba, *Brazil* **93 D7** 4 10S 55 50W
Itajaí, *Brazil* **95 B6** 27 50S 48 39W
Itajubá, *Brazil* **95 A6** 22 24S 45 30W
Itaka, *Tanzania* **55 D3** 8 50S 32 49 E
Italy ■, *Europe* **20 C5** 42 0N 13 0 E
Itampolo, *Madag.* **57 C7** 24 41S 43 57 E
Itapecuru-Mirim, *Brazil* **93 D10** 3 24S 44 20W
Itaperuna, *Brazil* **95 A7** 21 10S 41 54W
Itapetininga, *Brazil* **95 A6** 23 36S 48 7W
Itapeva, *Brazil* **95 A6** 23 59S 48 59W
Itapicuru →, *Bahia, Brazil* **93 F11** 11 47S 37 32W
Itapicuru →, *Maranhão, Brazil* **93 D10** 2 52S 44 12W
Itapipoca, *Brazil* **93 D11** 3 30S 39 35W
Itapuá □, *Paraguay* **95 B4** 26 40S 55 40W
Itaquari, *Brazil* **95 A7** 20 20S 40 25W
Itaquatiara, *Brazil* **92 D7** 2 58S 58 30W
Itaquí, *Brazil* **94 B4** 29 8S 56 30W
Itararé, *Brazil* **95 A6** 24 6S 49 23W
Itarsi, *India* **42 H7** 22 36N 77 51 E
Itati, *Argentina* **94 B4** 27 16S 58 15W
Itatuba, *Brazil* **92 E6** 5 46S 63 20W
Itchen →, *U.K.* **11 G6** 50 55N 1 22W
Itezhi Tezhi, L., *Zambia* **55 F2** 15 30S 25 30 E
Ithaca = Itháki, *Greece* **21 E9** 38 25N 20 40 E
Ithaca, *U.S.A.* **79 D8** 42 27N 76 30W
Itháki, *Greece* **21 E9** 38 25N 20 40 E
Ito, *Japan* **31 G9** 34 58N 139 5 E
Itoigawa, *Japan* **31 F8** 37 2N 137 51 E
Itonamas →, *Bolivia* **92 F6** 12 28S 64 24W
Ittoqqortoormiit = Scoresbysund, *Greenland* **4 B6** 70 20N 23 0W
Itu, *Brazil* **95 A6** 23 17S 47 15W
Ituaçu, *Brazil* **93 F10** 13 50S 41 18W
Ituiutaba, *Brazil* **93 G9** 19 0S 49 25W
Itumbiara, *Brazil* **93 G9** 18 20S 49 10W
Ituna, *Canada* **73 C8** 51 10N 103 24W
Itunge Port, *Tanzania* **55 D3** 9 40S 33 55 E
Iturbe, *Argentina* **94 A2** 23 0S 65 25W
Ituri →, *Zaïre* **54 B2** 1 40N 27 1 E
Iturup, Ostrov, *Russia* **27 E15** 45 0N 148 0 E
Ituyuro →, *Argentina* **94 A3** 22 40S 63 50W
Itzehoe, *Germany* **16 B5** 53 55N 9 31 E
Ivaí →, *Brazil* **95 A5** 23 18S 53 42W
Ivalo, *Finland* **8 B22** 68 38N 27 35 E
Ivalojoki →, *Finland* **8 B22** 68 40N 27 40 E
Ivanava, *Belarus* **17 B13** 52 7N 25 29 E
Ivanhoe, *N.S.W., Australia* **63 E3** 32 56S 144 20 E
Ivanhoe, *W. Austral., Australia* **60 C4** 15 41S 128 41 E
Ivanhoe, *U.S.A.* **84 J7** 36 23N 119 13W
Ivanhoe L., *Canada* **73 A7** 60 25N 106 30W
Ivano-Frankivsk, *Ukraine* **17 D13** 48 40N 24 40 E
Ivano-Frankovsk = Ivano-Frankivsk, *Ukraine* **17 D13** 48 40N 24 40 E
Ivanovo = Ivanava, *Belarus* **17 B13** 52 7N 25 29 E
Ivanovo, *Russia* **24 C7** 57 5N 41 0 E
Ivato, *Madag.* **57 C8** 20 37S 47 10 E
Ivatsevichy, *Belarus* **17 B13** 52 43N 25 21 E
Ivdel, *Russia* **24 B11** 60 42N 60 24 E
Ivinheima →, *Brazil* **95 A5** 23 14S 53 42W
Ivohibe, *Madag.* **57 C8** 22 31S 46 57 E
Ivory Coast ■, *Africa* **50 G3** 7 30N 5 0W
Ivrea, *Italy* **20 B2** 45 28N 7 52 E
Ivujivik, *Canada* **69 B12** 62 24N 77 55W
Iwahig, *Phil.* **36 C5** 8 36N 117 32 E
Iwaizumi, *Japan* **30 E10** 39 50N 141 45 E
Iwaki, *Japan* **31 F10** 37 3N 140 55 E
Iwakuni, *Japan* **31 G6** 34 15N 132 8 E
Iwamizawa, *Japan* **30 C10** 43 12N 141 46 E
Iwanai, *Japan* **30 C10** 42 58N 140 30 E
Iwata, *Japan* **31 G8** 34 42N 137 51 E
Iwate □, *Japan* **30 E10** 39 30N 141 30 E
Iwate-San, *Japan* **30 E10** 39 51N 141 0 E
Iwo, *Nigeria* **50 G5** 7 39N 4 9 E
Ixiamas, *Bolivia* **92 F5** 13 50S 68 5W
Ixopo, *S. Africa* **57 E5** 30 11S 30 5 E
Ixtepec, *Mexico* **87 D5** 16 32N 95 10W
Ixtlán del Río, *Mexico* **86 C4** 21 5N 104 21W
Iyo, *Japan* **31 H6** 33 45N 132 45 E
Izabal, L. de, *Guatemala* **88 C2** 15 30N 89 10W
Izamal, *Mexico* **87 C7** 20 56N 89 1W
Izegem, *Belgium* **15 D3** 50 55N 3 12 E
Izena-Shima, *Japan* **31 L3** 26 56N 127 56 E
Izhevsk, *Russia* **24 C9** 56 51N 53 14 E
Izmayil, *Ukraine* **17 F15** 45 22N 28 46 E
İzmir, *Turkey* **21 E12** 38 25N 27 8 E
İzmit, *Turkey* **25 F4** 40 45N 29 50 E
İznik Gölü, *Turkey* **21 D13** 40 27N 29 30 E
Izra, *Syria* **47 C5** 32 51N 36 15 E
Izu-Shotō, *Japan* **31 G10** 34 30N 140 0 E
Izumi-sano, *Japan* **31 G7** 34 23N 135 18 E
Izumo, *Japan* **31 G6** 35 20N 132 46 E
Izyaslav, *Ukraine* **17 C14** 50 5N 26 50 E

J

Jabal Lubnān, Lebanon 47 B4 33 45N 35 40 E
Jabalpur, India 43 H8 23 9N 79 58 E
Jabbūl, Syria 44 B3 36 4N 37 30 E
Jablah, Syria 44 C3 35 20N 36 0 E
Jablanica, Macedonia 21 D9 41 15N 20 30 E
Jablonec, Czech. 16 C8 50 43N 15 10 E
Jaboatão, Brazil 93 E11 8 7S 35 1W
Jaboticabal, Brazil 95 A6 21 15S 48 17W
Jaburu, Brazil 92 E6 5 30S 64 0W
Jaca, Spain 19 A5 42 35N 0 33W
Jacareí, Brazil 95 A6 23 20S 46 0W
Jacarèzinho, Brazil 95 A6 23 5S 49 58W
Jackman, U.S.A. 71 C5 45 35N 70 17W
Jacksboro, U.S.A. 81 J5 33 14N 98 15W
Jackson, Australia 63 D4 26 39S 149 39 E
Jackson, Ala., U.S.A. 77 K2 31 31N 87 53W
Jackson, Calif., U.S.A. 84 G6 38 21N 120 46W
Jackson, Ky., U.S.A. 76 G4 37 33N 83 23W
Jackson, Mich., U.S.A. 76 D3 42 15N 84 24W
Jackson, Minn., U.S.A. 80 D7 43 37N 95 1W
Jackson, Miss., U.S.A. 81 J9 32 18N 90 12W
Jackson, Mo., U.S.A. 81 G10 37 23N 89 40W
Jackson, Ohio, U.S.A. 76 F4 39 3N 82 39W
Jackson, Tenn., U.S.A. 77 H1 35 37N 88 49W
Jackson, Wyo., U.S.A. 82 E8 43 29N 110 46W
Jackson B., N.Z. 59 K2 43 58S 168 42 E
Jackson L., U.S.A. 82 E8 43 52N 110 36W
Jacksons, N.Z. 59 K3 42 46S 171 32 E
Jacksonville, Ala., U.S.A. 77 J3 33 49N 85 46W
Jacksonville, Calif., U.S.A. 84 H6 37 52N 120 24W
Jacksonville, Fla., U.S.A. 77 K5 30 20N 81 39W
Jacksonville, Ill., U.S.A. 80 F9 39 44N 90 14W
Jacksonville, N.C., U.S.A. 77 H7 34 45N 77 26W
Jacksonville, Oreg., U.S.A. 82 E2 42 19N 122 57W
Jacksonville, Tex., U.S.A. 81 K7 31 58N 95 17W
Jacksonville Beach, U.S.A. 77 K5 30 17N 81 24W
Jacmel, Haiti 89 C5 18 14N 72 32W
Jacob Lake, U.S.A. 83 H7 36 43N 112 13W
Jacobabad, Pakistan 42 E3 28 20N 68 29 E
Jacobina, Brazil 93 F10 11 11S 40 30W
Jacques-Cartier, Mt., Canada 71 C6 48 57N 66 0W
Jacuí →, Brazil 95 C5 30 2S 51 15W
Jacumba, U.S.A. 85 N10 32 37N 116 11W
Jacundá →, Brazil 93 D8 1 57S 50 26W
Jadotville = Likasi, Zaïre 55 E2 10 55S 26 48 E
Jādū, Libya 51 B7 32 0N 12 0 E
Jaén, Peru 92 E3 5 25S 78 40W
Jaén, Spain 19 D4 37 44N 3 43W
Jaffa = Tel Aviv-Yafo, Israel 47 C3 32 4N 34 48 E
Jaffa, C., Australia 63 F2 36 58S 139 40 E
Jaffna, Sri Lanka 40 Q12 9 45N 80 2 E
Jagadhri, India 42 D7 30 10N 77 20 E
Jagadishpur, India 43 G11 25 30N 84 21 E
Jagdalpur, India 41 K12 19 3N 82 0 E
Jagersfontein, S. Africa 56 D4 29 44S 25 27 E
Jagraon, India 40 D9 30 50N 75 25 E
Jagtial, India 40 K11 18 50N 79 0 E
Jaguariaíva, Brazil 95 A6 24 10S 49 50W
Jaguaribe →, Brazil 93 D11 4 25S 37 45W
Jagüey Grande, Cuba 88 B3 22 35N 81 7W
Jahangirabad, India 42 E8 28 19N 78 4 E
Jahrom, Iran 45 D7 28 30N 53 31 E
Jailolo, Indonesia 37 D7 1 5N 127 30 E
Jailolo, Selat, Indonesia 37 D7 0 5N 129 5 E
Jaipur, India 42 F6 27 0N 75 50 E
Jājarm, Iran 45 B8 36 58N 56 27 E
Jakarta, Indonesia 37 G12 6 9S 106 49 E
Jakobstad = Pietarsaari, Finland 8 E20 63 40N 22 43 E
Jal, U.S.A. 81 J3 32 7N 103 12W
Jalalabad, Afghan. 42 B4 34 30N 70 29 E
Jalalabad, India 43 F8 27 41N 79 42 E
Jalalpur Jattan, Pakistan 42 C6 32 38N 74 11 E
Jalama, U.S.A. 85 L6 34 29N 120 29W
Jalapa, Guatemala 88 D2 14 39N 89 59W
Jalapa Enríquez, Mexico 87 D5 19 32N 96 55W
Jalasjärvi, Finland 9 E20 62 29N 22 47 E
Jalaun, India 43 F8 26 8N 79 25 E
Jaleswar, Nepal 43 F11 26 38N 85 48 E
Jalgaon, Maharashtra, India 40 J10 21 2N 76 31 E
Jalgaon, Maharashtra, India 40 J9 21 0N 75 42 E
Jalībah, Iraq 44 D5 30 35N 46 32 E
Jalisco □, Mexico 86 C4 20 0N 104 0W
Jalkot, Pakistan 43 B5 35 14N 73 24 E
Jalna, India 40 K9 19 48N 75 38 E
Jalón →, Spain 19 B5 41 47N 1 4W
Jalpa, Mexico 86 C4 21 38N 102 58W
Jalpaiguri, India 41 F16 26 32N 88 46 E
Jaluit I., Pac. Oc. 64 G8 6 0N 169 30 E
Jalūlā, Iraq 44 C5 34 16N 45 10 E
Jamaica ■, W. Indies 88 C4 18 10N 77 30W
Jamalpur, Bangla. 41 G16 24 52N 89 56 E
Jamalpur, India 43 G12 25 18N 86 28 E
Jamalpurganj, India 43 H13 23 2N 88 1 E
Jamanxim →, Brazil 93 D7 4 43S 56 18W
Jambe, Indonesia 37 E8 1 15S 132 10 E
Jambi, Indonesia 36 E2 1 38S 103 30 E
Jambi □, Indonesia 36 E2 1 30S 102 30 E
Jambusar, India 42 H5 22 3N 72 51 E
James →, U.S.A. 80 D6 42 52N 97 18W
James B., Canada 69 C11 51 30N 80 0W
James Ras., Australia 60 D5 24 10S 132 30 E
James Ross I., Antarctica 5 C18 63 58S 57 50W
Jamestown, Australia 63 E2 33 10S 138 32 E
Jamestown, S. Africa 56 E4 31 6S 26 45 E
Jamestown, Ky., U.S.A. 76 G3 36 59N 85 4W
Jamestown, N. Dak., U.S.A. 80 B5 46 54N 98 42W
Jamestown, N.Y., U.S.A. 78 D5 42 6N 79 14W
Jamestown, Pa., U.S.A. 78 E4 41 29N 80 27W
Jamestown, Tenn., U.S.A. 77 G3 36 26N 84 56W
Jamīlābād, Iran 45 C6 34 24N 48 28 E
Jamiltepec, Mexico 87 D5 16 17N 97 49W
Jamkhandi, India 40 L9 16 30N 75 15 E
Jammu, India 42 C6 32 43N 74 54 E
Jammu & Kashmir □, India 43 B7 34 25N 77 0 E
Jamnagar, India 42 H4 22 30N 70 6 E
Jampur, Pakistan 42 E4 29 39N 70 40 E
Jamrud, Pakistan 42 C4 33 59N 71 24 E
Jämsä, Finland 9 F21 61 53N 25 10 E
Jamshedpur, India 43 H12 22 44N 86 12 E
Jamtara, India 43 H12 23 59N 86 49 E
Jämtland, Sweden 8 E15 63 31N 14 0 E
Jan L., Canada 73 C8 54 56N 102 55W
Jan Mayen, Arctic 4 B7 71 0N 9 0W
Janakkala, Finland 9 F21 60 54N 24 36 E
Jand, Pakistan 42 C5 33 30N 72 6 E
Jandaq, Iran 45 C7 34 3N 54 22 E
Jandia, Canary Is. 22 F5 28 6N 14 21W
Jandia, Pta. de, Canary Is. 22 F5 28 3N 14 31W
Jandola, Pakistan 42 C4 32 20N 70 9 E
Jandowae, Australia 63 D5 26 45S 151 7 E
Janesville, U.S.A. 80 D10 42 41N 89 1W
Janīn, West Bank 47 C4 32 28N 35 18 E
Janos, Mexico 86 A3 30 45N 108 10W
Januária, Brazil 93 G10 15 25S 44 25W
Janubio, Canary Is. 22 F6 28 56N 13 50W
Jaora, India 42 H6 23 40N 75 10 E
Japan ■, Asia 31 G8 36 0N 136 0 E
Japan, Sea of, Asia 30 E7 40 0N 135 0 E
Japan Trench, Pac. Oc. 28 F18 32 0N 142 0 E
Japen = Yapen, Indonesia 37 E9 1 50S 136 0 E
Japurá →, Brazil 92 D5 3 8S 65 46W
Jaque, Panama 92 B3 7 27N 78 8W
Jarābulus, Syria 44 B3 36 49N 38 1 E
Jarama →, Spain 19 B4 40 24N 3 32W
Jaranwala, Pakistan 42 D5 31 15N 73 26 E
Jarash, Jordan 47 C4 32 17N 35 54 E
Jardim, Brazil 94 A4 21 28S 56 2W
Jardines de la Reina, Is., Cuba 88 B4 20 50N 78 50W
Jargalang, China 35 C12 43 5N 122 55 E
Jargalant = Hovd, Mongolia 32 B4 48 2N 91 37 E
Jarīr, W. al →, Si. Arabia 44 E4 25 38N 42 30 E
Jarosław, Poland 17 C12 50 2N 22 42 E
Jarrahdale, Australia 61 F2 32 24S 116 5 E
Jarres, Plaine des, Laos 38 C4 19 27N 103 10 E
Jarso, Ethiopia 51 G12 5 15N 37 30 E
Jartai, China 34 E3 39 45N 105 48 E
Jarud Qi, China 35 B11 44 28N 120 50 E
Järvenpää, Finland 9 F21 60 29N 25 5 E
Jarvis, Canada 78 D4 42 53N 80 6W
Jarvis I., Pac. Oc. 65 H12 0 15S 159 55W
Jarwa, India 43 F10 27 38N 82 30 E
Jāsimīyah, Iraq 44 C5 33 45N 44 41 E
Jasin, Malaysia 39 L4 2 20N 102 26 E
Jāsk, Iran 45 E8 25 38N 57 45 E
Jasło, Poland 17 D11 49 45N 21 30 E
Jasper, Alta., Canada 72 C5 52 55N 118 5W
Jasper, Ont., Canada 79 B9 44 52N 75 57W
Jasper, Ala., U.S.A. 77 J2 33 50N 87 17W
Jasper, Fla., U.S.A. 77 K4 30 31N 82 57W
Jasper, Minn., U.S.A. 80 D6 43 51N 96 24W
Jasper, Tex., U.S.A. 81 K8 30 56N 94 1W
Jasper Nat. Park, Canada 72 C5 52 50N 118 8W
Jászberény, Hungary 17 E10 47 30N 19 55 E
Jataí, Brazil 93 G8 17 58S 51 48W
Jati, Pakistan 42 G3 24 20N 68 19 E
Jatibarang, Indonesia 37 G13 6 28S 108 18 E
Jatinegara, Indonesia 37 G12 6 13S 106 52 E
Játiva, Spain 19 C5 39 0N 0 32W
Jaú, Brazil 95 A6 22 10S 48 30W
Jauja, Peru 92 F3 11 45S 75 15W
Jaunpur, India 43 G10 25 46N 82 44 E
Java = Jawa, Indonesia 37 G14 7 0S 110 0 E
Java Sea, Indonesia 36 E3 4 35S 107 15 E
Java Trench, Ind. Oc. 64 H2 9 0S 105 0 E
Javhlant = Ulyasutay, Mongolia 32 B4 47 56N 97 28 E
Jawa, Indonesia 37 G14 7 0S 110 0 E
Jay, U.S.A. 81 G7 36 25N 94 48W
Jaya, Puncak, Indonesia 37 E9 3 57S 137 17 E
Jayanti, India 41 F16 26 45N 89 40 E
Jayapura, Indonesia 37 E10 2 28S 140 38 E
Jayawijaya, Pegunungan, Indonesia 37 E9 5 0S 139 0 E
Jaynagar, India 41 F15 26 43N 86 9 E
Jayrūd, Syria 44 C3 33 49N 36 44 E
Jayton, U.S.A. 81 J4 33 15N 100 34W
Jazīreh-ye Shīf, Iran 45 D6 29 4N 50 54 E
Jazminal, Mexico 86 C4 24 56N 101 25W
Jazzīn, Lebanon 47 B4 33 31N 35 35 E
Jean, U.S.A. 85 K11 35 47N 115 20W
Jean Marie River, Canada 72 A4 61 32N 120 38W
Jean Rabel, Haiti 89 C5 19 50N 73 5W
Jeanerette, U.S.A. 81 L9 29 55N 91 40W
Jeanette, Ostrov, Russia 27 B16 76 43N 158 0 E
Jeannette, U.S.A. 78 F5 40 20N 79 36W
Jebba, Nigeria 50 G5 9 9N 4 48 E
Jebel, Bahr el →, Sudan 51 G11 9 30N 30 25 E
Jedburgh, U.K. 12 F6 55 29N 2 33W
Jedda = Jiddah, Si. Arabia 46 C2 21 29N 39 10 E
Jędrzejów, Poland 17 C11 50 35N 20 15 E
Jedway, Canada 72 C2 52 17N 131 14W
Jefferson, Iowa, U.S.A. 80 D7 42 1N 94 23W
Jefferson, Ohio, U.S.A. 78 E4 41 44N 80 46W
Jefferson, Tex., U.S.A. 81 J7 32 46N 94 21W
Jefferson, Wis., U.S.A. 80 D10 43 0N 88 48W
Jefferson, Mt., Nev., U.S.A. 82 G5 38 51N 117 0W
Jefferson, Mt., Oreg., U.S.A. 82 D3 44 41N 121 48W
Jefferson City, Mo., U.S.A. 80 F8 38 34N 92 10W
Jefferson City, Tenn., U.S.A. 77 G4 36 7N 83 30W
Jeffersonville, U.S.A. 76 F3 38 17N 85 44W
Jega, Nigeria 50 F5 12 15N 4 23 E
Jēkabpils, Latvia 9 H21 56 29N 25 57 E
Jelenia Góra, Poland 16 C8 50 50N 15 45 E
Jelgava, Latvia 9 H20 56 41N 23 49 E
Jellicoe, Canada 70 C2 49 40N 87 30W
Jemaja, Indonesia 36 D3 3 5N 105 45 E
Jemaluang, Malaysia 39 L4 2 16N 103 52 E
Jember, Indonesia 37 H15 8 11S 113 41 E
Jembongan, Malaysia 36 C5 6 45N 117 20 E
Jemeppe, Belgium 15 D5 50 37N 5 30 E
Jena, Germany 16 C6 50 54N 11 35 E
Jena, U.S.A. 81 K8 31 41N 92 8W
Jenkins, U.S.A. 76 G4 37 10N 82 38W
Jenner, U.S.A. 84 G3 38 27N 123 7W
Jennings, U.S.A. 81 K8 30 13N 92 40W
Jennings →, Canada 72 B2 59 38N 132 5W
Jeparit, Australia 63 F3 36 8S 142 1 E
Jequié, Brazil 93 F10 13 51S 40 5W
Jequitinhonha, Brazil 93 G10 16 30S 41 0W
Jequitinhonha →, Brazil 93 G11 15 51S 38 53W
Jerada, Morocco 50 B4 34 17N 2 10W
Jerantut, Malaysia 39 L4 3 56N 102 22 E
Jérémie, Haiti 89 C5 18 40N 74 10W
Jerez, Punta, Mexico 87 C5 22 58N 97 40W
Jerez de García Salinas, Mexico 86 C4 22 39N 103 0W
Jerez de la Frontera, Spain 19 D2 36 41N 6 7W
Jerez de los Caballeros, Spain 19 C2 38 20N 6 45W
Jericho = Arīḥā, Syria 44 C3 35 49N 36 35 E
Jericho = El Arīḥā, West Bank 47 D4 31 52N 35 27 E
Jericho, Australia 62 C4 23 38S 146 6 E
Jerilderie, Australia 63 F4 35 20S 145 41 E
Jermyn, U.S.A. 79 E9 41 31N 75 31W
Jerome, U.S.A. 83 J8 34 45N 112 7W
Jersey, U.K. 11 H5 49 11N 2 7W
Jersey City, U.S.A. 79 F10 40 44N 74 4W
Jersey Shore, U.S.A. 78 E7 41 12N 77 15W
Jerseyville, U.S.A. 80 F9 39 7N 90 20W
Jerusalem, Israel 47 D4 31 47N 35 10 E
Jervis B., Australia 63 F5 35 8S 150 46 E
Jesselton = Kota Kinabalu, Malaysia 36 C5 6 0N 116 4 E
Jessore, Bangla. 41 H16 23 10N 89 10 E
Jesup, U.S.A. 77 K5 31 36N 81 53W
Jesús Carranza, Mexico 87 D5 17 28N 95 1W
Jesús María, Argentina 94 C3 30 59S 64 5W
Jetmore, U.S.A. 81 F5 38 4N 99 54W
Jetpur, India 42 J4 21 45N 70 10 E
Jevnaker, Norway 9 F14 60 15N 10 26 E
Jewett, Ohio, U.S.A. 78 F3 40 22N 81 2W
Jewett, Tex., U.S.A. 81 K6 31 22N 96 9W
Jewett City, U.S.A. 79 E13 41 36N 72 0W
Jeyḥūnābād, Iran 45 C6 34 58N 48 59 E
Jeypore, India 41 K13 18 50N 82 38 E
Jhajjar, India 42 E7 28 37N 76 42 E
Jhal Jhao, Pakistan 40 F4 26 20N 65 35 E
Jhalawar, India 42 G7 24 40N 76 10 E
Jhang Maghiana, Pakistan 42 D5 31 15N 72 22 E
Jhansi, India 43 G8 25 30N 78 36 E
Jharia, India 43 H12 23 45N 86 26 E
Jharsuguda, India 41 J14 21 56N 84 5 E
Jhelum, Pakistan 42 C5 33 0N 73 45 E
Jhelum →, Pakistan 42 D5 31 20N 72 10 E
Jhunjhunu, India 42 E6 28 10N 75 30 E
Ji Xian, Hebei, China 34 F8 37 35N 115 30 E
Ji Xian, Henan, China 34 G8 35 22N 114 5 E
Ji Xian, Shanxi, China 34 F6 36 7N 110 40 E
Jia Xian, Henan, China 34 H7 33 59N 113 12 E
Jia Xian, Shaanxi, China 34 E6 38 12N 110 28 E
Jiamusi, China 33 B8 46 40N 130 26 E
Ji'an, Jiangxi, China 33 D6 27 6N 114 59 E
Ji'an, Jilin, China 35 D14 41 5N 126 10 E
Jianchang, China 35 D11 40 55N 120 35 E
Jianchangying, China 35 D10 40 10N 118 50 E
Jiangcheng, China 32 D5 22 36N 101 52 E
Jiangmen, China 33 D6 22 32N 113 0 E
Jiangsu □, China 35 H10 33 0N 120 0 E
Jiangxi □, China 33 D6 27 30N 116 0 E
Jiao Xian, China 35 F11 36 18N 120 1 E
Jiaohe, Hebei, China 34 E9 38 2N 116 20 E
Jiaohe, Jilin, China 35 C14 43 40N 127 22 E
Jiaozhou Wan, China 35 F11 36 5N 120 10 E
Jiaozuo, China 34 G7 35 16N 113 12 E
Jiawang, China 35 G9 34 28N 117 26 E
Jiaxiang, China 34 G9 35 25N 116 20 E
Jiaxing, China 33 C7 30 49N 120 45 E
Jiayi = Chiai, Taiwan 33 D7 23 29N 120 25 E
Jibuti = Djibouti ■, Africa 46 E3 12 0N 43 0 E
Jicarón, I., Panama 88 E3 7 10N 81 50W
Jiddah, Si. Arabia 46 C2 21 29N 39 10 E
Jido, India 41 E19 29 2N 94 58 E
Jieshou, China 34 H8 33 18N 115 22 E
Jiexiu, China 34 F6 37 2N 111 55 E
Jiggalong, Australia 60 D3 23 21S 120 47 E
Jihlava, Czech. 16 D8 49 28N 15 35 E
Jihlava →, Czech. 17 D9 48 55N 16 36 E
Jijel, Algeria 50 A6 36 52N 5 50 E
Jijiga, Ethiopia 46 F3 9 20N 42 50 E
Jilin, China 35 C14 43 44N 126 30 E
Jilin □, China 35 C13 44 0N 127 0 E
Jilong = Chilung, Taiwan 33 D7 25 3N 121 45 E
Jima, Ethiopia 51 G12 7 40N 36 47 E
Jiménez, Mexico 86 B4 27 10N 104 54W
Jimo, China 35 F11 36 23N 120 30 E
Jin Xian, Hebei, China 34 E8 38 2N 115 2 E
Jin Xian, Liaoning, China 35 E11 38 55N 121 42 E
Jinan, China 34 F9 36 38N 117 1 E
Jincheng, China 34 G7 35 29N 112 50 E
Jind, India 42 E7 29 19N 76 22 E
Jindabyne, Australia 63 F4 36 25S 148 35 E
Jindrichuv Hradeç, Czech. 16 D8 49 10N 15 2 E
Jing He →, China 34 G5 34 27N 109 4 E
Jingbian, China 34 F5 37 20N 108 30 E
Jingchuan, China 34 G4 35 20N 107 20 E
Jingdezhen, China 33 D6 29 20N 117 11 E
Jinggu, China 32 D5 23 35N 100 41 E
Jinghai, China 34 E9 38 55N 116 55 E
Jingle, China 34 E6 38 20N 111 55 E
Jingning, China 34 G3 35 30N 105 43 E
Jingpo Hu, China 35 C15 43 55N 128 55 E
Jingtai, China 34 F3 37 10N 104 6 E
Jingxing, China 34 E8 38 2N 114 8 E
Jingyang, China 34 G5 34 30N 108 50 E
Jingyu, China 35 C14 42 25N 126 45 E
Jingyuan, China 34 F3 36 30N 104 40 E
Jingziguan, China 34 H6 33 15N 111 0 E
Jinhua, China 33 D6 29 8N 119 38 E
Jining, Nei Mongol Zizhiqu, China 34 D7 41 5N 113 0 E
Jining, Shandong, China 34 G9 35 22N 116 34 E
Jinja, Uganda 54 B3 0 25N 33 12 E
Jinjang, Malaysia 39 L3 3 13N 101 39 E
Jinji, China 34 F4 37 58N 106 8 E
Jinnah Barrage, Pakistan 40 C7 32 58N 71 33 E
Jinotega, Nic. 88 D2 13 6N 85 59W
Jinotepe, Nic. 88 D2 11 50N 86 10W
Jinsha Jiang →, China 32 D5 28 50N 104 36 E
Jinxi, China 35 D11 40 52N 120 50 E
Jinxiang, China 34 G9 35 5N 116 22 E
Jinzhou, China 35 D11 41 5N 121 3 E
Jiparaná →, Brazil 92 E6 8 3S 62 52W
Jipijapa, Ecuador 92 D2 1 0S 80 40W
Jiquilpan, Mexico 86 D4 19 57N 102 42W
Jishan, China 34 G6 35 34N 110 58 E
Jisr ash Shughūr, Syria 44 C3 35 49N 36 18 E
Jitarning, Australia 61 F2 32 48S 117 57 E
Jitra, Malaysia 39 J3 6 16N 100 25 E
Jiu →, Romania 17 F12 43 47N 23 48 E
Jiudengkou, China 34 E4 39 56N 106 40 E
Jiujiang, China 33 D6 29 42N 115 58 E
Jiutai, China 35 B13 44 10N 125 50 E
Jiuxiangcheng, China 34 H8 33 12N 114 50 E
Jiuxincheng, China 34 E8 39 17N 115 59 E
Jixi, China 35 B16 45 20N 130 50 E
Jiyang, China 35 F9 37 0N 117 12 E
Jīzān, Si. Arabia 46 D3 17 0N 42 20 E
Jize, China 34 F8 36 54N 114 56 E
Jizō-Zaki, Japan 31 G6 35 34N 133 20 E
Jizzakh, Uzbekistan 26 E7 40 6N 67 50 E
Joaçaba, Brazil 95 B5 27 5S 51 31W
João Pessoa, Brazil 93 E12 7 10S 34 52W
Joaquín V. González, Argentina 94 B3 25 10S 64 0W
Jodhpur, India 42 F5 26 23N 73 8 E
Joensuu, Finland 24 B4 62 37N 29 49 E
Jofane, Mozam. 57 C5 21 15S 34 18 E
Jõgeva, Estonia 9 G22 58 45N 26 24 E
Joggins, Canada 71 C7 45 42N 64 27W
Jogjakarta = Yogyakarta, Indonesia 37 G14 7 49S 110 22 E
Johannesburg, S. Africa 57 D4 26 10S 28 2 E
Johannesburg, U.S.A. 85 K9 35 22N 117 38W
John Day, U.S.A. 82 D4 44 25N 118 57W
John Day →, U.S.A. 82 D3 45 44N 120 39W
John H. Kerr Reservoir, U.S.A. 77 G6 36 36N 78 18W
John o' Groats, U.K. 12 C5 58 38N 3 4W
Johnnie, U.S.A. 85 J10 36 25N 116 5W
John's Ra., Australia 62 C1 21 55S 133 23 E
Johnson, U.S.A. 81 G4 37 34N 101 45W
Johnson City, N.Y., U.S.A. 79 D9 42 7N 75 58W
Johnson City, Tenn., U.S.A. 77 G4 36 19N 82 21W
Johnson City, Tex., U.S.A. 81 K5 30 17N 98 25W
Johnsonburg, U.S.A. 78 E6 41 29N 78 41W
Johnsondale, U.S.A. 85 K8 35 58N 118 32W
Johnson's Crossing, Canada 72 A2 60 29N 133 18W
Johnston, L., Australia 61 F3 32 25S 120 30 E
Johnston Falls = Mambilima Falls, Zambia 55 E2 10 31S 28 45 E
Johnston I., Pac. Oc. 65 F11 17 10N 169 8W
Johnstone Str., Canada 72 C3 50 28N 126 0W
Johnstown, N.Y., U.S.A. 79 C10 43 0N 74 22W
Johnstown, Pa., U.S.A. 78 F6 40 20N 78 55W
Johor Baharu, Malaysia 39 M4 1 28N 103 46 E
Jõhvi, Estonia 9 G22 59 22N 27 27 E
Joinvile, Brazil 95 B6 26 15S 48 55W
Joinville I., Antarctica 5 C18 65 0S 55 30W
Jojutla, Mexico 87 D5 18 37N 99 11W
Jokkmokk, Sweden 8 C18 66 35N 19 50 E
Jökulsá á Bru →, Iceland 8 D6 65 40N 14 16W
Jökulsá á Fjöllum →, Iceland 8 C5 66 10N 16 30W
Jolfā, Āzarbājān-e Sharqī, Iran 44 B5 38 57N 45 38 E
Jolfā, Eşfahan, Iran 45 C6 32 58N 51 37 E
Joliet, U.S.A. 76 E1 41 32N 88 5W
Joliette, Canada 70 C5 46 3N 73 24W
Jolo, Phil. 37 C6 6 0N 121 0 E
Jolon, U.S.A. 84 K5 35 58N 121 9W
Jombang, Indonesia 37 G15 7 33S 112 14 E
Jome, Indonesia 37 E7 1 16S 127 30 E
Jonava, Lithuania 9 J21 55 8N 24 12 E
Jones Sound, Canada 4 B3 76 0N 85 0W
Jonesboro, Ark., U.S.A. 81 H9 35 50N 90 42W
Jonesboro, Ill., U.S.A. 81 G10 37 27N 89 16W
Jonesboro, La., U.S.A. 81 J8 32 15N 92 43W
Jonesport, U.S.A. 71 D6 44 32N 67 37W
Joniškis, Lithuania 9 H20 56 13N 23 35 E
Jönköping, Sweden 9 H16 57 45N 14 10 E
Jonquière, Canada 71 C5 48 27N 71 14W
Joplin, U.S.A. 81 G7 37 6N 94 31W
Jordan, U.S.A. 82 C10 47 19N 106 55W
Jordan ■, Asia 47 E5 31 0N 36 0 E
Jordan →, Asia 47 D4 31 48N 35 32 E
Jordan Valley, U.S.A. 82 E5 42 59N 117 3W
Jorhat, India 41 F19 26 45N 94 12 E
Jörn, Sweden 8 D19 65 4N 20 1 E
Jorong, Indonesia 36 E4 3 58S 114 56 E
Jørpeland, Norway 9 G11 59 3N 6 1 E
Jorquera →, Chile 94 B2 28 3S 69 58W
Jos, Nigeria 50 G6 9 53N 8 51 E
José Batlle y Ordóñez, Uruguay 95 C4 33 20S 55 10W
Joseph, U.S.A. 82 D5 45 21N 117 14W
Joseph, L., Nfld., Canada 71 B6 52 45N 65 18W
Joseph, L., Ont., Canada 78 A5 45 10N 79 44W
Joseph Bonaparte G., Australia 60 B4 14 35S 128 50 E
Joseph City, U.S.A. 83 J8 34 57N 110 20W
Joshua Tree, U.S.A. 85 L10 34 8N 116 19W
Joshua Tree National Monument, U.S.A. 85 M10 33 55N 116 0W
Jostedalsbreen, Norway 9 F12 61 40N 6 59 E
Jotunheimen, Norway 9 F13 61 35N 8 25 E
Jourdanton, U.S.A. 81 L5 28 55N 98 33W
Joussard, Canada 72 B5 55 22N 115 50W
Jovellanos, Cuba 88 B3 22 40N 81 10W
Ju Xian, China 35 F10 36 35N 118 20 E
Juan Aldama, Mexico 86 C4 24 20N 103 23W
Juan Bautista Alberdi, Argentina 94 C3 34 26S 61 48W
Juan de Fuca Str., Canada 84 B2 48 15N 124 0W

Juan de Nova, Ind. Oc. .. 57 B7 17 3S 43 45 E
Juan Fernández, Arch. de, Pac. Oc. 90 G2 33 50S 80 0W
Juan José Castelli, Argentina 94 B3 25 27S 60 57W
Juan L. Lacaze, Uruguay . 94 C4 34 26S 57 25W
Juankoski, Finland ...... 8 E23 63 3N 28 19 E
Juárez, Argentina 94 D4 37 40S 59 43W
Juárez, Mexico 85 N11 32 20N 115 57W
Juárez, Sierra de, Mexico 86 A1 32 0N 116 0W
Juàzeiro, Brazil 93 E10 9 30S 40 30W
Juàzeiro do Norte, Brazil 93 E11 7 10S 39 18W
Jubayl, Lebanon .... 47 A4 34 5N 35 39 E
Jubbah, Si. Arabia ...... 44 D4 28 2N 40 56 E
Jubbulpore = Jabalpur, India .... 43 H8 23 9N 79 58 E
Jubilee L., Australia ..... 61 E4 29 0S 126 50 E
Juby, C., Morocco 50 C2 28 0N 12 59W
Júcar —▸, Spain 19 C5 39 5N 0 10W
Júcaro, Cuba 88 B4 21 37N 78 51W
Juchitán, Mexico 87 D5 16 27N 95 5W
Judaea = Har Yehuda, Israel .... 47 D3 31 35N 34 57 E
Judith —▸, U.S.A. 82 C9 47 44N 109 39W
Judith, Pt., U.S.A. 79 E13 41 22N 71 29W
Judith Gap, U.S.A. 82 C9 46 41N 109 45W
Jugoslavia = Yugoslavia ■, Europe . 21 B9 44 0N 20 0 E
Juigalpa, Nic. 88 D2 12 6N 85 26W
Juiz de Fora, Brazil 95 A7 21 43S 43 19W
Jujuy □, Argentina 94 A2 23 20S 65 40W
Julesburg, U.S.A. 80 E3 40 59N 102 16W
Juli, Peru 92 G5 16 10S 69 25W
Julia Cr. —▸, Australia .. 62 C3 20 0S 141 11 E
Julia Creek, Australia ... 62 C3 20 39S 141 44 E
Juliaca, Peru 92 G4 15 25S 70 10W
Julian, U.S.A. 85 M10 33 4N 116 38W
Julianehåb, Greenland ... 4 C5 60 43N 46 0W
Julimes, Mexico 86 B3 28 25N 105 27W
Jullundur, India .... 42 D6 31 20N 75 40 E
Julu, China .... 34 F8 37 15N 115 2 E
Jumbo, Zimbabwe .... 55 F3 17 30S 30 58 E
Jumbo Pk., U.S.A. 85 J12 36 12N 114 11W
Jumentos Cays, Bahamas 89 B4 23 0N 75 40W
Jumet, Belgium .... 15 D4 50 27N 4 25 E
Jumilla, Spain 19 C5 38 28N 1 19W
Jumla, Nepal .... 43 E10 29 15N 82 13 E
Jumna = Yamuna —▸, India .... 43 G9 25 30N 81 53 E
Junagadh, India .... 42 J4 21 30N 70 30 E
Junction, Tex., U.S.A. ... 81 K5 30 29N 99 46W
Junction, Utah, U.S.A. ... 83 G7 38 14N 112 13W
Junction B., Australia ... 62 A1 11 52S 133 55 E
Junction City, Kans., U.S.A. 80 F6 39 2N 96 50W
Junction City, Oreg., U.S.A. 82 D2 44 13N 123 12W
Junction Pt., Australia ... 62 A1 11 45S 133 50 E
Jundah, Australia .... 62 C3 24 46S 143 2 E
Jundiaí, Brazil 95 A6 24 30S 47 0W
Juneau, U.S.A. 68 C6 58 18N 134 25W
Junee, Australia .... 63 E4 34 53S 147 35 E
Jungfrau, Switz. .... 16 E4 46 32N 7 58 E
Junggar Pendi, China ... 32 B3 44 30N 86 0 E
Jungshahi, Pakistan ..... 42 G2 24 52N 67 44 E
Juniata —▸, U.S.A. 78 F7 40 30N 77 40W
Junín, Argentina 94 C3 34 33S 60 57W
Junín de los Andes, Argentina 96 D2 39 45S 71 0W
Jūniyah, Lebanon ...... 47 B4 33 59N 35 38 E
Juntura, U.S.A. 82 E4 43 45N 118 5W
Jupiter —▸, Canada 71 C7 49 29N 63 37W
Jur, Nahr el —▸, Sudan . 51 G10 8 45N 29 15 E
Jura = Jura, Mts. du, Europe .... 18 C7 46 40N 6 5 E
Jura = Schwäbische Alb, Germany .... 16 D5 48 20N 9 30 E
Jura, U.K. 12 F3 56 0N 5 50W
Jura, Mts. du, Europe ... 18 C7 46 40N 6 5 E
Jura, Sd. of, U.K. 12 F3 55 57N 5 45W
Jurado, Colombia 92 B3 7 7N 77 46W
Jurbarkas, Lithuania .... 9 J20 55 6N 22 47 E
Jūrmala, Latvia .... 9 H20 56 58N 23 34 E
Juruá —▸, Brazil 92 D5 2 37S 65 44W
Juruena —▸, Brazil 92 E7 7 20S 58 3W
Juruti, Brazil 93 D7 2 9S 56 4W
Justo Daract, Argentina . 94 C2 33 52S 65 12W
Juticalpa, Honduras 88 D2 14 40N 86 12W
Jutland = Jylland, Denmark .... 9 H13 56 25N 9 30 E
Juventud, I. de la, Cuba . 88 B3 21 40N 82 40W
Juwain, Afghan. .... 40 D2 31 45N 61 30 E
Jūy Zar, Iran .... 44 C5 33 50N 46 18 E
Juye, China .... 34 G9 35 22N 116 5 E
Jylland, Denmark .... 9 H13 56 25N 9 30 E
Jyväskylä, Finland ...... 9 E21 62 14N 25 50 E

K

K2, Pakistan .... 43 B7 35 58N 76 32 E
Kaap Plateau, S. Africa .. 56 D3 28 30S 24 0 E
Kaapkruis, Namibia ..... 56 C1 21 55S 13 57 E
Kaapstad = Cape Town, S. Africa .... 56 E2 33 55S 18 22 E
Kabaena, Indonesia ..... 37 F6 5 15S 122 0 E
Kabala, S. Leone 50 G2 9 38N 11 37W
Kabale, Uganda .... 54 C3 1 15S 30 0 E
Kabalo, Zaïre .... 54 D2 6 0S 27 0 E
Kabambare, Zaïre ...... 54 C2 4 41S 27 39 E
Kabango, Zaïre .... 55 D2 8 35S 28 30 E
Kabanjahe, Indonesia ... 36 D1 3 6N 98 30 E
Kabara, Mali 50 E4 16 40N 2 50W
Kabardino-Balkar Republic = Kabardino Balkaria □, Russia .... 25 F7 43 30N 43 30 E
Kabardino Balkaria □, Russia .... 25 F7 43 30N 43 30 E
Kabare, Indonesia ...... 37 E8 0 4S 130 58 E
Kabarega Falls, Uganda . 54 B3 2 15N 31 30 E
Kabasalan, Phil. .... 37 C6 7 47N 122 44 E
Kabba, Nigeria .... 50 G6 7 50N 6 3 E
Kabin Buri, Thailand .... 38 F3 13 57N 101 43 E
Kabinakagami L., Canada 70 C3 48 54N 84 25W
Kabīr, Zab al —▸, Iraq ... 44 C4 36 1N 43 24 E
Kabkabīyah, Sudan ..... 51 F9 13 50N 24 0 E
Kabompo, Zambia ...... 55 E1 13 36S 24 14 E
Kabompo —▸, Zambia .. 53 G4 14 10S 23 11 E
Kabondo, Zaïre .... 55 D2 8 58S 25 40 E
Kabongo, Zaïre .... 54 D2 7 22S 25 33 E
Kabra, Australia .... 62 C5 23 25S 150 25 E
Kabūd Gonbad, Iran .... 45 B8 37 5N 59 45 E
Kābul, Afghan. .... 42 B3 34 28N 69 11 E
Kābul □, Afghan. .... 40 B6 34 30N 69 0 E
Kabul —▸, Pakistan .... 42 C5 33 55N 72 14 E
Kabunga, Zaïre .... 54 C2 1 38S 28 3 E
Kaburuang, Indonesia ... 37 D7 3 50N 126 30 E
Kabwe, Zambia .... 55 E2 14 30S 28 29 E
Kachchh, Gulf of, India .. 42 H3 22 50N 69 15 E
Kachchh, Rann of, India . 42 G4 24 0N 70 0 E
Kachebera, Zambia ..... 55 E3 13 50S 32 50 E
Kachin □, Burma .... 41 F20 26 0N 97 30 E
Kachira, L., Uganda ..... 54 C3 0 40S 31 7 E
Kachiry, Kazakstan ...... 26 D8 53 10N 75 50 E
Kachot, Cambodia ...... 39 G4 11 30N 103 3 E
Kaçkar, Turkey .... 25 F7 40 45N 41 10 E
Kadan Kyun, Burma ..... 36 B1 12 30N 98 20 E
Kadanai —▸, Afghan. ... 42 D1 31 22N 65 45 E
Kadi, India .... 42 H5 23 18N 72 23 E
Kadina, Australia .... 63 E2 33 55S 137 43 E
Kadiyevka = Stakhanov, Ukraine .... 25 E6 48 35N 38 40 E
Kadoka, U.S.A. 80 D4 43 50N 101 31W
Kadoma, Zimbabwe .... 55 F2 18 20S 29 52 E
Kâdugli, Sudan .... 51 F10 11 0N 29 45 E
Kaduna, Nigeria .... 50 F6 10 30N 7 21 E
Kaédi, Mauritania 50 E2 16 9N 13 28W
Kaélé, Cameroon .... 51 F7 10 7N 14 27 E
Kaeng Khoï, Thailand ... 38 E3 14 35N 101 0 E
Kaesŏng, N. Korea ...... 35 F14 37 58N 126 35 E
Kāf, Si. Arabia .... 44 D3 31 25N 37 29 E
Kafakumba, Zaïre ...... 52 F4 9 38S 23 46 E
Kafan = Kapan, Armenia 25 G8 39 18N 46 27 E
Kafanchan, Nigeria ..... 50 G6 9 40N 8 20 E
Kaffrine, Senegal 50 F1 14 8N 15 36W
Kafia Kingi, Sudan ..... 51 G9 9 20N 24 25 E
Kafinda, Zambia .... 55 E3 12 32S 30 20 E
Kafirévs, Ákra, Greece ... 21 E11 38 9N 24 38 E
Kafue, Zambia .... 55 F2 15 46S 28 9 E
Kafue —▸, Zambia ..... 53 H5 15 30S 29 0 E
Kafue Flats, Zambia .... 55 F2 15 40S 27 25 E
Kafue Nat. Park, Zambia . 55 F2 15 0S 25 30 E
Kafulwe, Zambia .... 55 D2 9 0S 29 1 E
Kaga, Afghan. .... 42 B4 34 14N 70 10 E
Kaga Bandoro, C.A.R. ... 51 G8 7 0N 19 10 E
Kagan, Uzbekistan ...... 26 F7 39 43N 64 33 E
Kagawa □, Japan ...... 31 G6 34 15N 134 0 E
Kagera □, Tanzania ..... 54 C3 2 0S 31 30 E
Kagera —▸, Uganda ..... 54 C3 0 57S 31 47 E
Kagoshima, Japan ...... 31 J5 31 35N 130 33 E
Kagoshima □, Japan .... 31 J5 31 30N 130 30 E
Kagul = Cahul, Moldova . 17 F15 45 50N 28 15 E
Kahak, Iran .... 45 B6 36 6N 49 46 E
Kahama, Tanzania ..... 54 C3 4 8S 32 30 E
Kahama □, Tanzania .... 54 C3 3 50S 32 0 E
Kahang, Malaysia ...... 39 L4 2 12N 103 32 E
Kahayan —▸, Indonesia . 36 E4 3 40S 114 0 E
Kahe, Tanzania .... 54 C4 3 30S 37 25 E
Kahemba, Zaïre .... 52 F3 7 18S 18 55 E
Kahniah —▸, Canada ... 72 B4 58 15N 120 55W
Kahnūj, Iran .... 45 E8 27 55N 57 40 E
Kahoka, U.S.A. 80 E9 40 25N 91 44W
Kahoolawe, U.S.A. 74 H16 20 33N 156 37W
Kahramanmaraş, Turkey . 25 G6 37 37N 36 53 E
Kahuta, Pakistan .... 42 C5 33 35N 73 24 E
Kai, Kepulauan, Indonesia 37 F8 5 55S 132 45 E
Kai Besar, Indonesia .... 37 F8 5 35S 133 0 E
Kai Is. = Kai, Kepulauan, Indonesia .... 37 F8 5 55S 132 45 E
Kai Kecil, Indonesia ..... 37 F8 5 45S 132 40 E
Kaiama, Nigeria .... 50 G5 9 36N 4 1 E
Kaiapoi, N.Z. .... 59 K4 43 24S 172 40 E
Kaieteur Falls, Guyana .. 92 B7 5 1N 59 10W
Kaifeng, China .... 34 G8 34 48N 114 21 E
Kaikohe, N.Z. .... 59 F4 35 25S 173 49 E
Kaikoura, N.Z. .... 59 K4 42 25S 173 43 E
Kaikoura Ra., N.Z. ...... 59 J4 41 59S 173 41 E
Kailu, China .... 35 C11 43 38N 121 18 E
Kailua Kona, U.S.A. 74 J17 19 39N 155 59W
Kaimana, Indonesia .... 37 E8 3 39S 133 45 E
Kaimanawa Mts., N.Z. ... 59 H5 39 15S 175 56 E
Kaimganj, India .... 43 F8 27 33N 79 24 E
Kaimur Hills, India ...... 43 G9 24 30N 82 0 E
Kaingaroa Forest, N.Z. .. 59 H6 38 24S 176 30 E
Kainji Res., Nigeria ..... 50 F5 10 1N 4 40 E
Kainuu, Finland .... 8 D23 64 30N 29 7 E
Kaipara Harbour, N.Z. ... 59 G5 36 25S 174 14 E
Kaipokok B., Canada 71 B8 54 54N 59 47W
Kairana, India .... 42 E7 29 24N 77 15 E
Kaironi, Indonesia ...... 37 E8 0 47S 133 40 E
Kairouan, Tunisia ...... 50 A7 35 45N 10 5 E
Kaiserslautern, Germany 16 D4 49 26N 7 45 E
Kaitaia, N.Z. .... 59 F4 35 8S 173 17 E
Kaitangata, N.Z. .... 59 M2 46 17S 169 51 E
Kaithal, India .... 42 E7 29 48N 76 26 E
Kaitu —▸, Pakistan ..... 42 C4 33 10N 70 30 E
Kaiwi Channel, U.S.A. ... 74 H16 21 15N 157 30W
Kaiyuan, China .... 35 C13 42 28N 124 1 E
Kajaani, Finland .... 8 D22 64 17N 27 46 E
Kajabbi, Australia ...... 62 B3 20 0S 140 1 E
Kajana = Kajaani, Finland 8 D22 64 17N 27 46 E
Kajang, Malaysia .... 39 L3 2 59N 101 48 E
Kajiado, Kenya .... 54 C4 1 53S 36 48 E
Kajiado □, Kenya ...... 54 C4 2 0S 36 30 E
Kajo Kaji, Sudan .... 51 H11 3 58N 31 40 E
Kaka, Sudan .... 51 F11 10 38N 32 10 E
Kakabeka Falls, Canada .. 70 C2 48 24N 89 37W
Kakamas, S. Africa ..... 56 D3 28 45S 20 33 E
Kakamega, Kenya ...... 54 B3 0 20N 34 46 E
Kakamega □, Kenya .... 54 B3 0 20N 34 46 E
Kakanui Mts., N.Z. ..... 59 L3 45 10S 170 30 E
Kake, Japan .... 31 G6 34 36N 132 19 E
Kakegawa, Japan ...... 31 G9 34 45N 138 1 E
Kakeroma-Jima, Japan .. 31 K4 28 8N 129 14 E
Kakhovka, Ukraine ..... 25 E5 46 45N 33 30 E
Kakhovske Vdskh., Ukraine 25 E5 47 5N 34 0 E
Kakinada, India .... 41 L13 16 57N 82 11 E
Kakisa —▸, Canada 72 A5 61 3N 118 10W
Kakisa L., Canada 72 A5 60 56N 117 43W
Kakogawa, Japan ...... 31 G7 34 46N 134 51 E
Kakwa —▸, Canada 72 C5 54 37N 118 28W
Kāl Gūsheh, Iran .... 45 D8 30 59N 58 12 E
Kal Safīd, Iran .... 44 C5 34 52N 47 23 E
Kalabagh, Pakistan ..... 42 C4 33 0N 71 28 E
Kalabahi, Indonesia .... 37 F6 8 13S 124 31 E
Kalabo, Zambia .... 53 G4 14 58S 22 40 E
Kalach, Russia .... 25 D7 50 22N 41 0 E
Kaladan —▸, Burma ..... 41 J18 20 20N 93 5 E
Kaladar, Canada 78 B7 44 37N 77 5W
Kalahari, Africa .... 56 C3 24 0S 21 30 E
Kalahari Gemsbok Nat. Park, S. Africa .... 56 D3 25 30S 20 30 E
Kalajoki, Finland .... 8 D20 64 12N 24 10 E
Kālak, Iran .... 45 E8 25 29N 59 22 E
Kalakamati, Botswana ... 57 C4 20 40S 27 25 E
Kalakan, Russia .... 27 D12 55 15N 116 45 E
Kalakh, Syria .... 44 C3 34 55N 36 10 E
K'alak'unlun Shank'ou, Pakistan .... 43 B7 35 33N 77 46 E
Kalam, Pakistan .... 43 B5 35 34N 72 30 E
Kalama, U.S.A. 84 E4 46 1N 122 51W
Kalama, Zaïre .... 54 C2 2 52S 28 35 E
Kalámai, Greece .... 21 F10 37 3N 22 10 E
Kalamata = Kalámai, Greece .... 21 F10 37 3N 22 10 E
Kalamazoo, U.S.A. 76 D3 42 17N 85 35W
Kalamazoo —▸, U.S.A. ... 76 D2 42 40N 86 10W
Kalambo Falls, Tanzania . 55 D3 8 37S 31 35 E
Kalannie, Australia ..... 61 F2 30 22S 117 5 E
Kalāntarī, Iran .... 45 C7 32 10N 54 8 E
Kalao, Indonesia ...... 37 F6 7 21S 121 0 E
Kalaotoa, Indonesia .... 37 F6 7 20S 121 50 E
Kalasin, Thailand ...... 38 D4 16 26N 103 30 E
Kalat, Pakistan .... 40 E5 29 8N 66 31 E
Kalāteh, Iran .... 45 B7 36 33N 55 41 E
Kalāteh-ye-Ganj, Iran .... 45 E8 27 31N 57 55 E
Kalbarri, Australia ...... 61 E1 27 40S 114 10 E
Kalce, Slovenia .... 16 F8 45 54N 14 13 E
Kale, Turkey .... 21 F13 37 27N 28 49 E
Kalegauk Kyun, Burma .. 41 M20 15 33N 97 35 E
Kalehe, Zaïre .... 54 C2 2 6S 28 50 E
Kalema, Tanzania ...... 54 C3 1 12S 31 55 E
Kalemie, Zaïre .... 54 D2 5 55S 29 9 E
Kalewa, Burma .... 41 H19 23 10N 94 15 E
Kalgan = Zhangjiakou, China .... 34 D8 40 48N 114 55 E
Kalgoorlie-Boulder, Australia .... 61 F3 30 40S 121 22 E
Kaliakra, Nos, Bulgaria .. 21 C13 43 21N 28 30 E
Kalianda, Indonesia .... 36 F3 5 50S 105 45 E
Kalibo, Phil. .... 37 B6 11 43N 122 22 E
Kaliganj, Bangla. .... 43 H13 22 25N 89 8 E
Kalima, Zaïre .... 54 C2 2 33S 26 32 E
Kalimantan, Indonesia ... 36 E4 0 0 114 0 E
Kalimantan Barat □, Indonesia .... 36 E4 0 0 110 30 E
Kalimantan Selatan □, Indonesia .... 36 E5 2 30S 115 30 E
Kalimantan Tengah □, Indonesia .... 36 E4 2 0S 113 30 E
Kalimantan Timur □, Indonesia .... 36 D5 1 30N 116 30 E
Kálimnos, Greece .... 21 F12 37 0N 27 0 E
Kalimpong, India .... 43 F13 27 4N 88 35 E
Kalinin = Tver, Russia ... 24 C6 56 55N 35 55 E
Kaliningrad, Kaliningd., Russia .... 9 J19 54 42N 20 32 E
Kaliningrad, Moskva, Russia .... 24 C6 55 58N 37 54 E
Kalinkavichy, Belarus ... 17 B15 52 12N 29 20 E
Kalinkovichi = Kalinkavichy, Belarus .. 17 B15 52 12N 29 20 E
Kaliro, Uganda .... 54 B3 0 56N 33 30 E
Kalispell, U.S.A. 82 B6 48 12N 114 19W
Kalisz, Poland .... 17 C10 51 45N 18 8 E
Kaliua, Tanzania .... 54 D3 5 5S 31 48 E
Kalix, Sweden .... 8 D20 65 53N 23 12 E
Kalix —▸, Sweden ...... 8 D20 65 50N 23 11 E
Kalka, India .... 42 D7 30 46N 76 57 E
Kalkaska, U.S.A. 76 C3 44 44N 85 11W
Kalkfeld, Namibia ...... 56 C2 20 57S 16 14 E
Kalkfontein, Botswana .. 56 C3 22 4S 20 57 E
Kalkrand, Namibia ...... 56 C2 24 1S 17 35 E
Kallavesi, Finland ...... 8 E22 62 58N 27 30 E
Kallsjön, Sweden .... 8 E15 63 38N 13 0 E
Kalmar, Sweden .... 9 H17 56 40N 16 20 E
Kalmyk Republic = Kalmykia □, Russia .... 25 E8 46 5N 46 1 E
Kalmykia □, Russia ..... 25 E8 46 5N 46 1 E
Kalmykovo, Kazakstan ... 25 E9 49 0N 51 47 E
Kalna, India .... 43 H13 23 13N 88 25 E
Kalocsa, Hungary ...... 17 E10 46 32N 19 0 E
Kalokhorio, Cyprus ..... 23 E12 34 51N 33 2 E
Kaloko, Zaïre .... 54 D2 6 47S 25 48 E
Kalol, Gujarat, India .... 42 H5 22 37N 73 31 E
Kalol, Gujarat, India .... 42 H5 23 15N 72 33 E
Kalomo, Zambia .... 55 F2 17 0S 26 30 E
Kalpi, India .... 43 F8 26 8N 79 47 E
Kalu, Pakistan .... 42 G2 25 5N 67 39 E
Kaluga, Russia .... 24 D6 54 35N 36 10 E
Kalulushi, Zambia ...... 55 E2 12 50S 28 3 E
Kalundborg, Denmark ... 9 J14 55 41N 11 5 E
Kalush, Ukraine .... 17 D13 49 3N 24 23 E
Kalutara, Sri Lanka ..... 40 R11 6 35N 80 0 E
Kalya, Russia .... 24 B10 60 15N 59 59 E
Kama, Zaïre .... 54 C2 3 30S 27 5 E
Kama —▸, Russia .... 24 C9 55 45N 52 0 E
Kamachumu, Tanzania .. 54 C3 1 37S 31 37 E
Kamaishi, Japan .... 30 E10 39 16N 141 53 E
Kamalia, Pakistan ...... 42 D5 30 44N 72 42 E
Kamapanda, Zambia .... 55 E1 12 5S 24 0 E
Kamaran, Yemen .... 46 D3 15 21N 42 35 E
Kamativi, Zimbabwe .... 55 F2 18 15S 27 27 E
Kambalda, Australia .... 61 F3 31 10S 121 37 E
Kambar, Pakistan .... 42 F3 27 37N 68 1 E
Kambarka, Russia ...... 24 C9 56 15N 54 11 E
Kambolé, Zambia ...... 55 D3 8 47S 30 48 E
Kambos, Cyprus .... 23 D11 35 2N 32 44 E
Kambove, Zaïre .... 55 E2 10 51S 26 33 E
Kamchatka, P-ov., Russia 27 D16 57 0N 160 0 E
Kamchatka Pen. = Kamchatka, P-ov., Russia .... 27 D16 57 0N 160 0 E
Kamchiya —▸, Bulgaria . 21 C12 43 4N 27 44 E
Kamen, Russia .... 26 D9 53 50N 81 30 E
Kamen-Rybolov, Russia . 30 B6 44 46N 132 2 E
Kamenjak, Rt., Croatia ... 16 F7 44 47N 13 55 E
Kamenka, Russia ...... 24 A7 65 58N 44 0 E
Kamenka Bugskaya = Kamyanka-Buzka, Ukraine .... 17 C13 50 8N 24 16 E
Kamensk Uralskiy, Russia 26 D7 56 25N 62 2 E
Kamenskoye, Russia .... 27 C17 62 45N 165 30 E
Kameoka, Japan .... 31 G7 35 0N 135 35 E
Kamiah, U.S.A. 82 C5 46 14N 116 2W
Kamieskroon, S. Africa .. 56 E2 30 9S 17 56 E
Kamilukuak, L., Canada .. 73 A8 62 22N 101 40W
Kamin-Kashyrskyy, Ukraine .... 17 C13 51 39N 24 56 E
Kamina, Zaïre .... 55 D1 8 45S 25 0 E
Kaminak L., Canada 73 A9 62 10N 95 0W
Kaminoyama, Japan .... 30 E10 38 9N 140 17 E
Kamiros, Greece .... 23 C9 36 20N 27 56 E
Kamituga, Zaïre .... 54 C2 3 2S 28 10 E
Kamloops, Canada 72 C4 50 40N 120 20W
Kamo, Japan .... 30 F9 37 39N 139 3 E
Kamoke, Pakistan ...... 42 C6 32 4N 74 4 E
Kampala, Uganda ...... 54 B3 0 20N 32 30 E
Kampar, Malaysia ...... 39 K3 4 18N 101 9 E
Kampar —▸, Indonesia .. 36 D2 0 30N 103 8 E
Kampen, Neths. .... 15 B5 52 33N 5 53 E
Kamphaeng Phet, Thailand .... 38 D2 16 28N 99 30 E
Kampolombo, L., Zambia 55 E2 11 37S 29 42 E
Kampong To, Thailand ... 39 J3 6 3N 101 13 E
Kampot, Cambodia ..... 39 G5 10 36N 104 10 E
Kampuchea = Cambodia ■, Asia .... 38 F5 12 15N 105 0 E
Kampung —▸, Indonesia 37 F9 5 44S 138 24 E
Kampung Air Putih, Malaysia .... 39 K4 4 15N 103 10 E
Kampung Jerangau, Malaysia .... 39 K4 4 50N 103 10 E
Kampung Raja, Malaysia 39 K4 5 45N 102 35 E
Kampungbaru = Tolitoli, Indonesia .... 37 D6 1 5N 120 50 E
Kamrau, Teluk, Indonesia 37 E8 3 30S 133 36 E
Kamsack, Canada 73 C8 51 34N 101 54W
Kamskoye Vdkhr., Russia 24 C10 58 41N 56 7 E
Kamuchawie L., Canada . 73 B8 56 18N 101 59W
Kamui-Misaki, Japan .... 30 C10 43 20N 140 21 E
Kamyanets-Podilskyy, Ukraine .... 17 D14 48 45N 26 40 E
Kamyanka-Buzka, Ukraine 17 C13 50 8N 24 16 E
Kāmyārān, Iran .... 44 C5 34 47N 46 56 E
Kamyshin, Russia ...... 25 D8 50 10N 45 24 E
Kanaaupscow, Canada .. 70 B4 54 2N 76 30W
Kanab, U.S.A. 83 H7 37 3N 112 32W
Kanab —▸, U.S.A. 83 H7 36 24N 112 38W
Kanagi, Japan .... 30 D10 40 54N 140 27 E
Kanairiktok —▸, Canada . 71 A7 55 2N 60 18W
Kananga, Zaïre .... 52 F4 5 55S 22 18 E
Kanarraville, U.S.A. 83 H7 37 32N 113 11W
Kanash, Russia .... 24 C8 55 30N 47 32 E
Kanaskat, U.S.A. 84 C5 47 19N 121 54W
Kanastraíon, Ákra = Palioúrion, Ákra, Greece 21 E10 39 57N 23 45 E
Kanawha —▸, U.S.A. 76 F4 38 50N 82 9W
Kanazawa, Japan ...... 31 F8 36 30N 136 38 E
Kanchanaburi, Thailand . 38 E2 14 2N 99 31 E
Kanchenjunga, Nepal ... 43 F13 27 50N 88 10 E
Kanchipuram, India .... 40 N11 12 52N 79 45 E
Kanda Kanda, Zaïre .... 52 F4 6 52S 23 48 E
Kandahar = Qandahār, Afghan. .... 40 D4 31 32N 65 30 E
Kandalaksha, Russia .... 24 A5 67 9N 32 30 E
Kandalakshkiy Zaliv, Russia .... 24 A5 66 0N 35 0 E
Kandalu, Afghan. .... 40 E3 29 55N 63 20 E
Kandangan, Indonesia ... 36 E5 2 50S 115 20 E
Kandanos, Greece ...... 23 D5 35 19N 23 44 E
Kandhkot, Pakistan ..... 42 E3 28 16N 69 8 E
Kandhla, India .... 42 E7 29 18N 77 19 E
Kandi, Benin .... 50 F5 11 7N 2 55 E
Kandi, India .... 43 H13 23 58N 88 5 E
Kandla, India .... 42 H4 23 0N 70 10 E
Kandos, Australia ...... 63 E4 32 45S 149 58 E
Kandy, Sri Lanka .... 40 R12 7 18N 80 43 E
Kane, U.S.A. 78 E6 41 40N 78 49W
Kane Basin, Greenland .. 4 B4 79 1N 70 0W
Kangān, Fārs, Iran ..... 45 E7 27 50N 52 3 E
Kangān, Hormozgān, Iran 45 E8 25 48N 57 28 E
Kangar, Malaysia ...... 39 J3 6 27N 100 12 E
Kangaroo I., Australia ... 63 F2 35 45S 137 0 E
Kangasala, Finland ..... 9 F21 61 28N 24 4 E
Kangāvar, Iran .... 45 C6 34 40N 48 0 E
Kangdong, N. Korea .... 35 E14 39 9N 126 5 E
Kangean, Kepulauan, Indonesia .... 36 F5 6 55S 115 23 E
Kangean Is. = Kangean, Kepulauan, Indonesia . 36 F5 6 55S 115 23 E
Kanggye, N. Korea ...... 35 D14 41 0N 126 35 E
Kanggyŏng, S. Korea .... 35 F14 36 10N 127 0 E
Kanghwa, S. Korea ..... 35 F14 37 45N 126 30 E
Kangiqsualujjuaq, Canada 69 C13 58 30N 65 59W
Kangiqsujuaq, Canada .. 69 B12 61 30N 72 0W
Kangirsuk, Canada 69 B13 60 0N 70 0W
Kangnŭng, S. Korea .... 35 F15 37 45N 128 54 E
Kango, Gabon .... 52 D2 0 11N 10 5 E
Kangping, China .... 35 C12 42 43N 123 18 E
Kangto, India .... 41 F18 27 50N 92 35 E
Kaniama, Zaïre .... 54 D1 7 30S 24 12 E
Kaniapiskau —▸, Canada 71 A6 56 40N 69 30W
Kaniapiskau L., Canada .. 71 B6 54 10N 69 55W
Kanin, Poluostrov, Russia 24 A8 68 0N 45 0 E
Kanin Nos, Mys, Russia . 24 A7 68 39N 43 32 E
Kanin Pen. = Kanin, Poluostrov, Russia .... 24 A8 68 0N 45 0 E
Kaniva, Australia ...... 63 F3 36 22S 141 18 E
Kanjut Sar, Pakistan .... 43 A6 36 7N 75 25 E
Kankaanpää, Finland .... 9 F20 61 44N 22 50 E
Kankakee, U.S.A. 76 E2 41 7N 87 52W
Kankakee —▸, U.S.A. 76 E1 41 23N 88 15W
Kankan, Guinea 50 F3 10 23N 9 15W

Kankendy = Xankändi, *Azerbaijan* 25 G8 39 52N 46 49 E
Kanker, *India* 41 J12 20 10N 81 40 E
Kankunskiy, *Russia* 27 D13 57 37N 126 8 E
Kannapolis, *U.S.A.* 77 H5 35 30N 80 37W
Kannauj, *India* 43 F8 27 3N 79 56 E
Kannod, *India* 40 H10 22 45N 76 40 E
Kano, *Nigeria* 50 F6 12 2N 8 30 E
Kan'onji, *Japan* 31 G6 34 7N 133 39 E
Kanowit, *Malaysia* 36 D4 2 14N 112 20 E
Kanowna, *Australia* 61 F3 30 32S 121 31 E
Kanoya, *Japan* 31 J5 31 25N 130 50 E
Kanpetlet, *Burma* 41 J18 21 10N 93 59 E
Kanpur, *India* 43 F9 26 28N 80 20 E
Kansas □, *U.S.A.* 80 F6 38 30N 99 0W
Kansas →, *U.S.A.* 80 F7 39 7N 94 37W
Kansas City, *Kans., U.S.A.* 80 F7 39 7N 94 38W
Kansas City, *Mo., U.S.A.* 80 F7 39 6N 94 35W
Kansenia, *Zaïre* 55 E2 10 20S 26 0 E
Kansk, *Russia* 27 D10 56 20N 95 37 E
Kansŏng, *S. Korea* 35 E15 38 24N 128 30 E
Kansu = Gansu □, *China* 34 G3 36 0N 104 0 E
Kantang, *Thailand* 39 J2 7 25N 99 31 E
Kantharalak, *Thailand* 38 E5 14 39N 104 39 E
Kantō □, *Japan* 31 F9 36 15N 139 30 E
Kantō-Sanchi, *Japan* 31 G9 35 59N 138 50 E
Kanturk, *Ireland* 13 D3 52 11N 8 54W
Kanuma, *Japan* 31 F9 36 34N 139 42 E
Kanus, *Namibia* 56 D2 27 50S 18 39 E
Kanye, *Botswana* 56 C4 24 55S 25 28 E
Kanzenze, *Zaïre* 55 E2 10 30S 25 12 E
Kanzi, Ras, *Tanzania* 54 D4 7 1S 39 33 E
Kaohsiung, *Taiwan* 33 D7 22 35N 120 16 E
Kaokoveld, *Namibia* 56 B1 19 15S 14 30 E
Kaolack, *Senegal* 50 F1 14 5N 16 8W
Kaoshan, *China* 35 B13 44 38N 124 50 E
Kapadvanj, *India* 42 H5 23 5N 73 0 E
Kapan, *Armenia* 25 G8 39 18N 46 27 E
Kapanga, *Zaïre* 52 F4 8 30S 22 40 E
Kapchagai = Qapshaghay, *Kazakstan* 26 E8 43 51N 77 14 E
Kapema, *Zaïre* 55 E2 10 45S 28 22 E
Kapfenberg, *Austria* 16 E8 47 26N 15 18 E
Kapiri Mposhi, *Zambia* 55 E2 13 59S 28 43 E
Kapiskau →, *Canada* 70 B3 52 47N 81 55W
Kapit, *Malaysia* 36 D4 2 0N 112 55 E
Kapiti I., *N.Z.* 59 J5 40 50S 174 56 E
Kapoe, *Thailand* 39 H2 9 34N 98 32 E
Kapoeta, *Sudan* 51 H11 4 50N 33 35 E
Kaposvár, *Hungary* 17 E9 46 25N 17 47 E
Kapowsin, *U.S.A.* 84 D4 46 59N 122 13W
Kapps, *Namibia* 56 C2 22 32S 17 18 E
Kapsan, *N. Korea* 35 D15 41 4N 128 19 E
Kapsukas = Marijampolė, *Lithuania* 9 J20 54 33N 23 19 E
Kapuas →, *Indonesia* 36 E3 0 25S 109 20 E
Kapuas Hulu, Pegunungan, *Malaysia* 36 D4 1 30N 113 30 E
Kapuas Hulu Ra. = Kapuas Hulu, Pegunungan, *Malaysia* 36 D4 1 30N 113 30 E
Kapulo, *Zaïre* 55 D2 8 18S 29 15 E
Kapunda, *Australia* 63 E2 34 20S 138 56 E
Kapuni, *N.Z.* 59 H5 39 29S 174 8 E
Kapurthala, *India* 42 D6 31 23N 75 25 E
Kapuskasing, *Canada* 70 C3 49 25N 82 30W
Kapuskasing →, *Canada* 70 C3 49 49N 82 0W
Kaputar, *Australia* 63 E5 30 15S 150 10 E
Kaputir, *Kenya* 54 B4 2 5N 35 28 E
Kara, *Russia* 26 C7 69 10N 65 0 E
Kara Bogaz Gol, Zaliv = Garabogazköl Aylagy, *Turkmenistan* 25 F9 41 0N 53 30 E
Kara Kalpak Republic □ = Karakalpakstan □, *Uzbekistan* 26 E6 43 0N 58 0 E
Kara Kum, *Turkmenistan* 26 F6 39 30N 60 0 E
Kara Sea, *Russia* 26 B7 75 0N 70 0 E
Karabiğa, *Turkey* 21 D12 40 24N 27 18 E
Karaburun, *Turkey* 21 E12 38 41N 26 28 E
Karabutak = Qarabutaq, *Kazakstan* 26 E7 49 59N 60 14 E
Karacabey, *Turkey* 21 D13 40 12N 28 21 E
Karacasu, *Turkey* 21 F13 37 43N 28 35 E
Karachi, *Pakistan* 42 G2 24 53N 67 0 E
Karad, *India* 40 L9 17 15N 74 10 E
Karadeniz Boğazı, *Turkey* 21 D13 41 10N 29 10 E
Karaganda = Qaraghandy, *Kazakstan* 26 E8 49 50N 73 10 E
Karagayly, *Kazakstan* 26 E8 49 26N 76 0 E
Karaginskiy, Ostrov, *Russia* 27 D17 58 45N 164 0 E
Karagiye, Vpadina, *Kazakstan* 25 F9 43 27N 51 45 E
Karagiye Depression = Karagiye, Vpadina, *Kazakstan* 25 F9 43 27N 51 45 E
Karagwe □, *Tanzania* 54 C3 2 0S 31 0 E
Karaikal, *India* 40 P11 10 59N 79 50 E
Karaikkudi, *India* 40 P11 10 5N 78 45 E
Karaj, *Iran* 45 C6 35 48N 51 0 E
Karak, *Malaysia* 39 L4 3 25N 102 2 E
Karakalpakstan □, *Uzbekistan* 26 E6 43 0N 58 0 E
Karakas, *Kazakstan* 26 E9 48 20N 83 30 E
Karakelong, *Indonesia* 37 D7 4 35N 126 50 E
Karakitang, *Indonesia* 37 D7 3 14N 125 28 E
Karaklis = Vanadzor, *Armenia* 25 F7 40 48N 44 30 E
Karakoram Pass, *Pakistan* 43 B7 35 33N 77 50 E
Karakoram Ra., *Pakistan* 43 B7 35 30N 77 0 E
Karalon, *Russia* 27 D12 57 5N 115 50 E
Karaman, *Turkey* 25 G5 37 14N 33 13 E
Karamay, *China* 32 B3 45 30N 84 58 E
Karambu, *Indonesia* 36 E5 3 53S 116 6 E
Karamea Bight, *N.Z.* 59 J3 41 22S 171 40 E
Karamsad, *India* 42 H5 22 35N 72 50 E
Karand, *Iran* 44 C5 34 16N 46 15 E
Karanganyar, *Indonesia* 37 G13 7 38S 109 37 E
Karasburg, *Namibia* 56 D2 28 0S 18 44 E
Karasino, *Russia* 26 C9 66 50N 86 50 E
Karasjok, *Norway* 8 B21 69 27N 25 30 E
Karasuk, *Russia* 26 D8 53 44N 78 2 E
Karasuyama, *Japan* 31 F10 36 39N 140 9 E
Karatau = Qarataū, *Kazakstan* 26 E8 43 10N 70 28 E
Karatau, Khrebet, *Kazakstan* 26 E7 43 30N 69 30 E
Karauli, *India* 42 F7 26 30N 77 4 E
Karavostasi, *Cyprus* 23 D11 35 8N 32 50 E
Karawang, *Indonesia* 37 G12 6 30S 107 15 E
Karawanken, *Europe* 16 E8 46 30N 14 40 E
Karazhal, *Kazakstan* 26 E8 48 2N 70 49 E
Karbalā, *Iraq* 44 C5 32 36N 44 3 E
Karcag, *Hungary* 17 E11 47 19N 20 57 E
Karcha →, *Pakistan* 43 B7 34 45N 76 10 E
Karda, *Russia* 27 D11 55 0N 103 16 E
Kardhítsa, *Greece* 21 E9 39 23N 21 54 E
Kärdla, *Estonia* 9 G20 58 50N 22 40 E
Kareeberge, *S. Africa* 56 E3 30 59S 21 50 E
Karelia □, *Russia* 24 A5 65 30N 32 30 E
Karelian Republic □ = Karelia □, *Russia* 24 A5 65 30N 32 30 E
Kārevāndar, *Iran* 45 E9 27 53N 60 44 E
Kargasok, *Russia* 26 D9 59 3N 80 53 E
Kargat, *Russia* 26 D9 55 10N 80 15 E
Kargil, *India* 43 B7 34 32N 76 12 E
Kargopol, *Russia* 24 B6 61 30N 38 58 E
Karīān, *Iran* 45 E8 26 57N 57 14 E
Kariba, *Zimbabwe* 55 F2 16 28S 28 50 E
Kariba, L., *Zimbabwe* 55 F2 16 40S 28 25 E
Kariba Dam, *Zimbabwe* 55 F2 16 30S 28 35 E
Kariba Gorge, *Zambia* 55 F2 16 30S 28 50 E
Karibib, *Namibia* 56 C2 22 0S 15 56 E
Karimata, Kepulauan, *Indonesia* 36 E3 1 25S 109 0 E
Karimata, Selat, *Indonesia* 36 E3 2 0S 108 40 E
Karimata Is. = Karimata, Kepulauan, *Indonesia* 36 E3 1 25S 109 0 E
Karimnagar, *India* 40 K11 18 26N 79 10 E
Karimunjawa, Kepulauan, *Indonesia* 36 F4 5 50S 110 30 E
Karin, *Somali Rep.* 46 E4 10 50N 45 52 E
Karīt, *Iran* 45 C8 33 29N 56 55 E
Kariya, *Japan* 31 G8 34 58N 137 1 E
Karkaralinsk = Qarqaraly, *Kazakstan* 26 E8 49 26N 75 30 E
Karkinitska Zatoka, *Ukraine* 25 E5 45 56N 33 0 E
Karkinitskiy Zaliv = Karkinitska Zatoka, *Ukraine* 25 E5 45 56N 33 0 E
Karl-Marx-Stadt = Chemnitz, *Germany* 16 C7 50 51N 12 54 E
Karlovac, *Croatia* 16 F8 45 31N 15 36 E
Karlovo, *Bulgaria* 21 C11 42 38N 24 47 E
Karlovy Vary, *Czech.* 16 C7 50 13N 12 51 E
Karlsbad = Karlovy Vary, *Czech.* 16 C7 50 13N 12 51 E
Karlsborg, *Sweden* 9 G16 58 33N 14 33 E
Karlshamn, *Sweden* 9 H16 56 10N 14 51 E
Karlskoga, *Sweden* 9 G16 59 22N 14 33 E
Karlskrona, *Sweden* 9 H16 56 10N 15 35 E
Karlsruhe, *Germany* 16 D5 49 0N 8 23 E
Karlstad, *Sweden* 9 G15 59 23N 13 30 E
Karlstad, *U.S.A.* 80 A6 48 35N 96 31W
Karnal, *India* 42 E7 29 42N 77 2 E
Karnali →, *Nepal* 43 E9 28 45N 81 16 E
Karnaphuli Res., *Bangla.* 41 H18 22 40N 92 20 E
Karnataka □, *India* 40 N10 13 15N 77 0 E
Karnes City, *U.S.A.* 81 L6 28 53N 97 54W
Karnische Alpen, *Europe* 16 E7 46 36N 13 0 E
Kärnten □, *Austria* 16 E8 46 52N 13 30 E
Karoi, *Zimbabwe* 55 F2 16 48S 29 45 E
Karonga, *Malawi* 55 D3 9 57S 33 55 E
Karoonda, *Australia* 63 F2 35 1S 139 59 E
Karora, *Sudan* 51 E12 17 44N 38 15 E
Karpasia □, *Cyprus* 23 D13 35 32N 34 15 E
Kárpathos, *Greece* 21 G12 35 37N 27 10 E
Karpinsk, *Russia* 24 C11 59 45N 60 1 E
Karpogory, *Russia* 24 B7 64 0N 44 27 E
Karpuz Burnu = Apostolos Andreas, C., *Cyprus* 23 D13 35 42N 34 35 E
Kars, *Turkey* 25 F7 40 40N 43 5 E
Karsakpay, *Kazakstan* 26 E7 47 55N 66 40 E
Karshi = Qarshi, *Uzbekistan* 26 F7 38 53N 65 48 E
Karsiyang, *India* 43 F13 26 56N 88 18 E
Karsun, *Russia* 24 D8 54 14N 46 57 E
Kartaly, *Russia* 26 D7 53 3N 60 40 E
Kartapur, *India* 42 D6 31 27N 75 32 E
Karthaus, *U.S.A.* 78 E6 41 8N 78 9W
Karufa, *Indonesia* 37 E8 3 50S 133 20 E
Karumba, *Australia* 62 B3 17 31S 140 50 E
Karumo, *Tanzania* 54 C3 2 25S 32 50 E
Karumwa, *Tanzania* 54 C3 3 12S 32 38 E
Karungu, *Kenya* 54 C3 0 50S 34 10 E
Karviná, *Czech.* 17 D10 49 53N 18 25 E
Karwar, *India* 40 M9 14 55N 74 13 E
Karwi, *India* 43 G9 25 12N 80 57 E
Kasache, *Malawi* 55 E3 13 25S 34 20 E
Kasai →, *Zaïre* 52 E3 3 30S 16 10 E
Kasai Oriental □, *Zaïre* 54 C1 5 0S 24 30 E
Kasaji, *Zaïre* 55 E1 10 25S 23 27 E
Kasama, *Zambia* 55 E3 10 16S 31 9 E
Kasan-dong, *N. Korea* 35 D14 41 18N 126 55 E
Kasane, *Namibia* 56 B3 17 34S 24 50 E
Kasanga, *Tanzania* 55 D3 8 30S 31 10 E
Kasangulu, *Zaïre* 52 E3 4 33S 15 15 E
Kasaragod, *India* 40 N9 12 30N 74 58 E
Kasba L., *Canada* 73 A8 60 20N 102 10W
Kāseh Garān, *Iran* 44 C5 34 5N 46 2 E
Kasempa, *Zambia* 55 E2 13 30S 25 44 E
Kasenga, *Zaïre* 55 E2 10 20S 28 45 E
Kasese, *Uganda* 54 B3 0 13N 30 3 E
Kasewa, *Zambia* 55 E2 14 28S 28 53 E
Kasganj, *India* 43 F8 27 48N 78 42 E
Kashabowie, *Canada* 70 C1 48 40N 90 26W
Kāshān, *Iran* 45 C6 34 5N 51 30 E
Kashi, *China* 32 C2 39 30N 76 2 E
Kashimbo, *Zaïre* 55 E2 11 12S 26 19 E
Kashipur, *India* 43 E8 29 15N 79 0 E
Kashiwazaki, *Japan* 31 F9 37 22N 138 33 E
Kashk-e Kohneh, *Afghan.* 40 B3 34 55N 62 30 E
Kāshmar, *Iran* 45 C8 35 16N 58 26 E
Kashmir, *Asia* 43 C7 34 0N 76 0 E
Kashmor, *Pakistan* 42 E3 28 28N 69 32 E
Kashun Noerh = Gaxun Nur, *China* 32 B5 42 22N 100 30 E
Kasimov, *Russia* 24 D7 54 55N 41 20 E
Kasinge, *Zaïre* 54 D2 6 15S 26 58 E
Kasiruta, *Indonesia* 37 E7 0 25S 127 12 E
Kaskaskia →, *U.S.A.* 80 G10 37 58N 89 57W
Kaskattama →, *Canada* 73 B10 57 3N 90 4W
Kaskinen, *Finland* 9 E19 62 22N 21 15 E
Kaslo, *Canada* 72 D5 49 55N 116 55W
Kasmere L., *Canada* 73 B8 59 34N 101 10W
Kasongo, *Zaïre* 54 C2 4 30S 26 33 E
Kasongo Lunda, *Zaïre* 52 F3 6 35S 16 49 E
Kásos, *Greece* 21 G12 35 20N 26 55 E
Kassalâ, *Sudan* 51 E12 15 30N 36 0 E
Kassel, *Germany* 16 C5 51 18N 9 26 E
Kassiópi, *Greece* 23 A3 39 48N 19 53 E
Kassue, *Indonesia* 37 F9 6 58S 139 21 E
Kastamonu, *Turkey* 25 F5 41 25N 33 43 E
Kastélli, *Greece* 23 D5 35 29N 23 38 E
Kastéllion, *Greece* 23 D7 35 12N 25 20 E
Kastoría, *Greece* 21 D9 40 30N 21 19 E
Kasulu, *Tanzania* 54 C3 4 37S 30 5 E
Kasulu □, *Tanzania* 54 C3 4 37S 30 5 E
Kasumi, *Japan* 31 G7 35 38N 134 38 E
Kasungu, *Malawi* 55 E3 13 0S 33 29 E
Kasur, *Pakistan* 42 D6 31 5N 74 25 E
Kata, *Russia* 27 D11 58 46N 102 40 E
Kataba, *Zambia* 55 F2 16 5S 25 10 E
Katako Kombe, *Zaïre* 54 C1 3 25S 24 20 E
Katale, *Tanzania* 54 C3 4 52S 31 7 E
Katamatite, *Australia* 63 F4 36 6S 145 41 E
Katanda, *Kivu, Zaïre* 54 C2 0 55S 29 21 E
Katanda, *Shaba, Zaïre* 54 D1 7 52S 24 13 E
Katanga = Shaba □, *Zaïre* 54 D2 8 0S 25 0 E
Katangi, *India* 40 J11 21 56N 79 50 E
Katangli, *Russia* 27 D15 51 42N 143 14 E
Katavi Swamp, *Tanzania* 54 D3 6 50S 31 10 E
Kateríni, *Greece* 21 D10 40 18N 22 37 E
Katha, *Burma* 41 G20 24 10N 96 30 E
Katherine, *Australia* 60 B5 14 27S 132 20 E
Kathiawar, *India* 42 H4 22 20N 71 0 E
Kathikas, *Cyprus* 23 E11 34 55N 32 25 E
Katihar, *India* 43 G12 25 34N 87 36 E
Katima Mulilo, *Zambia* 56 B3 17 28S 24 13 E
Katimbira, *Malawi* 55 E3 12 40S 34 0 E
Katingan = Mendawai →, *Indonesia* 36 E4 3 30S 113 0 E
Katiola, *Ivory C.* 50 G3 8 10N 5 10W
Katmandu, *Nepal* 43 F11 27 45N 85 20 E
Káto Arkhánai, *Greece* 23 D7 35 15N 25 10 E
Káto Khorió, *Greece* 23 D7 35 3N 25 47 E
Kato Pyrgos, *Cyprus* 23 D11 35 11N 32 41 E
Katompe, *Zaïre* 54 D2 6 2S 26 23 E
Katonga →, *Uganda* 54 B3 0 34N 31 50 E
Katoomba, *Australia* 63 E5 33 41S 150 19 E
Katowice, *Poland* 17 C10 50 17N 19 5 E
Katrine, L., *U.K.* 12 E4 56 15N 4 30W
Katrineholm, *Sweden* 9 G17 59 9N 16 12 E
Katsepe, *Madag.* 57 B8 15 45S 46 15 E
Katsina, *Nigeria* 50 F6 13 0N 7 32 E
Katsumoto, *Japan* 31 H4 33 51N 129 42 E
Katsuura, *Japan* 31 G10 35 10N 140 20 E
Katsuyama, *Japan* 31 F8 36 3N 136 30 E
Kattaviá, *Greece* 23 D9 35 57N 27 46 E
Kattegat, *Denmark* 9 H14 57 0N 11 20 E
Katumba, *Zaïre* 54 D2 7 40S 25 17 E
Katungu, *Kenya* 54 C5 2 55S 40 3 E
Katwa, *India* 43 H13 23 30N 88 5 E
Katwijk-aan-Zee, *Neths.* 15 B4 52 12N 4 24 E
Kauai, *U.S.A.* 74 H15 22 3N 159 30W
Kauai Channel, *U.S.A.* 74 H15 21 45N 158 50W
Kaufman, *U.S.A.* 81 J6 32 35N 96 19W
Kauhajoki, *Finland* 9 E20 62 25N 22 10 E
Kaukauna, *U.S.A.* 76 C1 44 17N 88 17W
Kaukauveld, *Namibia* 56 C3 20 0S 20 15 E
Kaunas, *Lithuania* 9 J20 54 54N 23 54 E
Kaura Namoda, *Nigeria* 50 F6 12 37N 6 33 E
Kautokeino, *Norway* 8 B20 69 0N 23 4 E
Kavacha, *Russia* 27 C17 60 16N 169 51 E
Kavalerovo, *Russia* 30 B7 44 15N 135 4 E
Kavali, *India* 40 M12 14 55N 80 1 E
Kavála, *Greece* 21 D11 40 57N 24 28 E
Kavār, *Iran* 45 D7 29 11N 52 44 E
Kavos, *Greece* 23 B4 39 23N 20 3 E
Kaw, *Fr. Guiana* 93 C8 4 30N 52 15W
Kawagama L., *Canada* 78 A6 45 18N 78 45W
Kawagoe, *Japan* 31 G9 35 55N 139 29 E
Kawaguchi, *Japan* 31 G9 35 52N 139 45 E
Kawaihae, *U.S.A.* 74 H17 20 3N 155 50W
Kawambwa, *Zambia* 55 D2 9 48S 29 3 E
Kawanoe, *Japan* 31 G6 34 1N 133 34 E
Kawardha, *India* 43 J9 22 0N 81 17 E
Kawasaki, *Japan* 31 G9 35 35N 139 42 E
Kawene, *Canada* 70 C1 48 45N 91 15W
Kawerau, *N.Z.* 59 H6 38 7S 176 42 E
Kawhia Harbour, *N.Z.* 59 H5 38 5S 174 51 E
Kawio, Kepulauan, *Indonesia* 37 D7 4 30N 125 30 E
Kawnro, *Burma* 41 H21 22 48N 99 8 E
Kawthoolei = Kawthule □, *Burma* 41 L20 18 0N 97 30 E
Kawthule □, *Burma* 41 L20 18 0N 97 30 E
Kaya, *Burkina Faso* 50 F4 13 4N 1 10W
Kayah □, *Burma* 41 K20 19 15N 97 15 E
Kayan →, *Indonesia* 36 D5 2 55N 117 35 E
Kaycee, *U.S.A.* 82 E10 43 43N 106 38W
Kayeli, *Indonesia* 37 E7 3 20S 127 10 E
Kayenta, *U.S.A.* 83 H8 36 44N 110 15W
Kayes, *Mali* 50 F2 14 25N 11 30W
Kayoa, *Indonesia* 37 D7 0 1N 127 28 E
Kayomba, *Zambia* 55 E1 13 11S 24 2 E
Kayrunnera, *Australia* 63 E3 30 40S 142 30 E
Kayseri, *Turkey* 25 G6 38 45N 35 30 E
Kaysville, *U.S.A.* 82 F8 41 2N 111 56W
Kayuagung, *Indonesia* 36 E2 3 24S 104 50 E
Kazachye, *Russia* 27 B14 70 52N 135 58 E
Kazakstan ■, *Asia* 26 E7 50 0N 70 0 E
Kazan, *Russia* 24 C8 55 50N 49 10 E
Kazan-Rettō, *Pac. Oc.* 64 E6 25 0N 141 0 E
Kazanlŭk, *Bulgaria* 21 C11 42 38N 25 20 E
Kazatin = Kozyatyn, *Ukraine* 17 D15 49 45N 28 50 E
Kāzerūn, *Iran* 45 D6 29 38N 51 40 E
Kazumba, *Zaïre* 52 F4 6 25S 22 5 E
Kazuno, *Japan* 30 D10 40 10N 140 45 E
Kazym →, *Russia* 26 C7 63 54N 65 50 E
Kê-Macina, *Mali* 50 F3 13 58N 5 22W
Kéa, *Greece* 21 F11 37 35N 24 22 E
Keams Canyon, *U.S.A.* 83 J8 35 49N 110 12W
Kearney, *U.S.A.* 80 E5 40 42N 99 5W
Keban, *Turkey* 25 G6 38 50N 38 50 E
Kebnekaise, *Sweden* 8 C18 67 53N 18 33 E
Kebri Dehar, *Ethiopia* 46 F3 6 45N 44 17 E
Kebumen, *Indonesia* 37 G13 7 42S 109 40 E
Kechika →, *Canada* 72 B3 59 41N 127 12W
Kecskemét, *Hungary* 17 E10 46 57N 19 42 E
Kedgwick, *Canada* 71 C6 47 40N 67 20W
Kédhros Óros, *Greece* 23 D6 35 11N 24 37 E
Kedia Hill, *Botswana* 56 C3 21 28S 24 37 E
Kediniai, *Lithuania* 9 J21 55 15N 24 2 E
Kediri, *Indonesia* 37 G15 7 51S 112 1 E
Kédougou, *Senegal* 50 F2 12 35N 12 10W
Keeler, *U.S.A.* 84 J9 36 29N 117 52W
Keeley L., *Canada* 73 C7 54 54N 108 8W
Keeling Is. = Cocos Is., *Ind. Oc.* 64 J1 12 10S 96 55 E
Keene, *Calif., U.S.A.* 85 K8 35 13N 118 33W
Keene, *N.H., U.S.A.* 79 D12 42 56N 72 17W
Keeper Hill, *Ireland* 13 D3 52 45N 8 16W
Keer-Weer, C., *Australia* 62 A3 14 0S 141 32 E
Keeseville, *U.S.A.* 79 B11 44 29N 73 30W
Keetmanshoop, *Namibia* 56 D2 26 35S 18 8 E
Keewatin, *U.S.A.* 80 B8 47 24N 93 5W
Keewatin □, *Canada* 73 A9 63 20N 95 0W
Keewatin →, *Canada* 73 B8 56 29N 100 46W
Kefallinía, *Greece* 21 E9 38 20N 20 30 E
Kefamenanu, *Indonesia* 37 F6 9 28S 124 29 E
Keffi, *Nigeria* 50 G6 8 55N 7 43 E
Keflavík, *Iceland* 8 D2 64 2N 22 35W
Keg River, *Canada* 72 B5 57 54N 117 55W
Kegaska, *Canada* 71 B7 50 9N 61 18W
Keighley, *U.K.* 10 D6 53 52N 1 54W
Keila, *Estonia* 9 G21 59 18N 24 25 E
Keimoes, *S. Africa* 56 D3 28 41S 20 59 E
Keitele, *Finland* 8 E22 63 10N 26 20 E
Keith, *Australia* 63 F3 36 6S 140 20 E
Keith, *U.K.* 12 D6 57 32N 2 57W
Keith Arm, *Canada* 68 B7 64 20N 122 15W
Kejser Franz Joseph Fjord = Kong Franz Joseph Fd., *Greenland* 4 B6 73 30N 24 30W
Kekri, *India* 42 G6 26 0N 75 10 E
Kël, *Russia* 27 C13 69 30N 124 10 E
Kelan, *China* 34 E6 38 43N 111 31 E
Kelang, *Malaysia* 39 L3 3 2N 101 26 E
Kelantan →, *Malaysia* 39 J4 6 13N 102 14 E
Kelibia, *Tunisia* 51 A7 36 50N 11 3 E
Kellé, *Congo* 52 E2 0 8S 14 38 E
Keller, *U.S.A.* 82 B4 48 5N 118 41W
Kellerberrin, *Australia* 61 F2 31 36S 117 38 E
Kellett, C., *Canada* 4 B1 72 0N 126 0W
Kelleys I., *U.S.A.* 78 E2 41 36N 82 42W
Kellogg, *U.S.A.* 82 C5 47 32N 116 7W
Kells = Ceanannus Mor, *Ireland* 13 C5 53 44N 6 53W
Kélo, *Chad* 51 G8 9 10N 15 45 E
Kelokedhara, *Cyprus* 23 E11 34 48N 32 39 E
Kelowna, *Canada* 72 D5 49 50N 119 25W
Kelsey Bay, *Canada* 72 C3 50 25N 126 0W
Kelseyville, *U.S.A.* 84 G4 38 59N 122 50W
Kelso, *N.Z.* 59 L2 45 54S 169 15 E
Kelso, *U.K.* 12 F6 55 36N 2 26W
Kelso, *U.S.A.* 84 D4 46 9N 122 54W
Keluang, *Malaysia* 39 L4 2 3N 103 18 E
Kelvington, *Canada* 73 C8 52 10N 103 30W
Kem, *Russia* 24 B5 65 0N 34 38 E
Kem →, *Russia* 24 B5 64 57N 34 41 E
Kema, *Indonesia* 37 D7 1 22N 125 8 E
Kemano, *Canada* 72 C3 53 35N 128 0W
Kemasik, *Malaysia* 39 K4 4 25N 103 27 E
Kemerovo, *Russia* 26 D9 55 20N 86 5 E
Kemi, *Finland* 8 D21 65 44N 24 34 E
Kemi älv = Kemijoki →, *Finland* 8 D21 65 47N 24 32 E
Kemijärvi, *Finland* 8 C22 66 43N 27 22 E
Kemijoki →, *Finland* 8 D21 65 47N 24 32 E
Kemmerer, *U.S.A.* 82 F8 41 48N 110 32W
Kemmuna = Comino, *Malta* 23 C1 36 2N 14 20 E
Kemp, L., *U.S.A.* 81 J5 33 46N 99 9W
Kemp Land, *Antarctica* 5 C5 69 0S 55 0 E
Kempsey, *Australia* 63 E5 31 1S 152 50 E
Kempt, L., *Canada* 70 C5 47 25N 74 22W
Kempten, *Germany* 16 E6 47 45N 10 17 E
Kemptville, *Canada* 70 C4 45 0N 75 38W
Kendal, *Indonesia* 36 F4 6 56S 110 14 E
Kendal, *U.K.* 10 C5 54 20N 2 44W
Kendall, *Australia* 63 E5 31 35S 152 44 E
Kendall →, *Australia* 62 A3 14 4S 141 35 E
Kendallville, *U.S.A.* 76 E3 41 27N 85 16W
Kendari, *Indonesia* 37 E6 3 50S 122 30 E
Kendawangan, *Indonesia* 36 E4 2 32S 110 17 E
Kende, *Nigeria* 50 F5 11 30N 4 12 E
Kendenup, *Australia* 61 F2 34 30S 117 38 E
Kendrapara, *India* 41 J15 20 35N 86 30 E
Kendrew, *S. Africa* 56 E3 32 32S 24 30 E
Kendrick, *U.S.A.* 82 C5 46 37N 116 39W
Kene Thao, *Laos* 38 D3 17 44N 101 10 E
Kenedy, *U.S.A.* 81 L6 28 49N 97 51W
Kenema, *S. Leone* 50 G2 7 50N 11 14W
Keng Kok, *Laos* 38 D5 16 26N 105 12 E
Keng Tawng, *Burma* 41 J21 20 45N 98 18 E
Keng Tung, *Burma* 41 J21 21 0N 99 30 E
Kenge, *Zaïre* 52 E3 4 50S 17 4 E
Kengeja, *Tanzania* 54 D4 5 26S 39 45 E
Kenhardt, *S. Africa* 56 D3 29 19S 21 12 E
Kenitra, *Morocco* 50 B3 34 15N 6 40W
Kenli, *China* 35 F10 37 30N 118 20 E
Kenmare, *Ireland* 13 E2 51 53N 9 36W
Kenmare, *U.S.A.* 80 A3 48 41N 102 5W
Kenmare →, *Ireland* 13 E2 51 48N 9 51W
Kennebec, *U.S.A.* 80 D5 43 54N 99 52W
Kennedy, *Zimbabwe* 55 F2 18 52S 27 10 E
Kennedy Ra., *Australia* 61 D2 24 45S 115 10 E
Kennedy Taungdeik, *Burma* 41 H18 23 15N 93 45 E
Kenner, *U.S.A.* 81 L9 29 59N 90 15W

Kennet —►, U.K. 11 F7 51 27N 0 57W
Kenneth Ra., Australia 60 D2 23 50S 117 8 E
Kennett, U.S.A. 81 G9 36 14N 90 3W
Kennewick, U.S.A. 82 C4 46 12N 119 7W
Kénogami, Canada 71 C5 48 25N 71 15W
Kenogami —►, Canada 70 B3 51 6N 84 28W
Kenora, Canada 73 D10 49 47N 94 29W
Kenosha, U.S.A. 76 D2 42 35N 87 49W
Kensington, Canada 71 C7 46 28N 63 34W
Kensington, U.S.A. 80 F5 39 46N 99 2W
Kensington Downs, Australia 62 C3 22 31S 144 19 E
Kent, Ohio, U.S.A. 78 E3 41 9N 81 22W
Kent, Oreg., U.S.A. 82 D3 45 12N 120 42W
Kent, Tex., U.S.A. 81 K2 31 4N 104 13W
Kent, Wash., U.S.A. 84 C4 47 23N 122 14W
Kent □, U.K. 11 F8 51 12N 0 40 E
Kent Group, Australia 62 F4 39 30S 147 20 E
Kent Pen., Canada 68 B9 68 30N 107 0W
Kentau, Kazakstan 26 E7 43 32N 68 36 E
Kentland, U.S.A. 76 E2 40 46N 87 27W
Kenton, U.S.A. 76 E4 40 39N 83 37W
Kentucky □, U.S.A. 76 G3 37 0N 84 0W
Kentucky —►, U.S.A. 76 F3 38 41N 85 11W
Kentucky L., U.S.A. 77 G2 37 1N 88 16W
Kentville, Canada 71 C7 45 6N 64 29W
Kentwood, U.S.A. 81 K9 30 56N 90 31W
Kenya ■, Africa 54 B4 1 0N 38 0 E
Kenya, Mt., Kenya 54 C4 0 10S 37 18 E
Keo Neua, Deo, Vietnam 38 C5 18 23N 105 10 E
Keokuk, U.S.A. 80 E9 40 24N 91 24W
Kep, Cambodia 39 G5 10 29N 104 19 E
Kep, Vietnam 38 B6 21 24N 106 16 E
Kepi, Indonesia 37 F9 6 32S 139 19 E
Kerala □, India 40 P10 11 0N 76 15 E
Kerama-Rettō, Japan 31 L3 26 5N 127 15 E
Keran, Pakistan 43 B5 34 35N 73 59 E
Kerang, Australia 63 F3 35 40S 143 55 E
Keraudren, C., Australia 60 C2 19 58S 119 45 E
Kerava, Finland 9 F21 60 25N 25 5 E
Kerch, Ukraine 25 E6 45 20N 36 20 E
Kerchoual, Mali 50 E5 17 12N 0 20 E
Keren, Eritrea 51 E12 15 45N 38 28 E
Kerguelen, Ind. Oc. 3 G13 49 15S 69 10 E
Kericho, Kenya 54 C4 0 22S 35 15 E
Kericho □, Kenya 54 C4 0 30S 35 15 E
Kerinci, Indonesia 36 E2 1 40S 101 15 E
Kerki, Turkmenistan 26 F7 37 50N 65 12 E
Kérkira, Greece 23 A3 39 38N 19 50 E
Kerkrade, Neths. 15 D6 50 53N 6 4 E
Kermadec Is., Pac. Oc. 64 K10 30 0S 178 15W
Kermadec Trench, Pac. Oc. 64 L10 30 30S 176 0W
Kermān, Iran 45 D8 30 15N 57 1 E
Kerman, U.S.A. 84 J6 36 43N 120 4W
Kermān □, Iran 45 D8 30 0N 57 0 E
Kermānshāh = Bākhtarān, Iran 44 C5 34 23N 47 0 E
Kerme Körfezi, Turkey 21 F12 36 55N 27 50 E
Kermit, U.S.A. 81 K3 31 52N 103 6W
Kern —►, U.S.A. 85 K7 35 16N 119 18W
Kernville, U.S.A. 85 K8 35 45N 118 26W
Keroh, Malaysia 39 K3 5 43N 101 1 E
Kerrobert, Canada 73 C7 51 56N 109 8W
Kerrville, U.S.A. 81 K5 30 3N 99 8W
Kerry □, Ireland 13 D2 52 7N 9 35W
Kerry Hd., Ireland 13 D2 52 25N 9 56W
Kertosono, Indonesia 37 G15 7 38S 112 9 E
Kerulen —►, Asia 33 B6 48 48N 117 0 E
Kerzaz, Algeria 50 C4 29 29N 1 37W
Kesagami —►, Canada 70 B4 51 40N 79 45W
Kesagami L., Canada 70 B3 50 23N 80 15W
Keşan, Turkey 21 D12 40 49N 26 38 E
Kesennuma, Japan 30 E10 38 54N 141 35 E
Keshit, Iran 45 D8 29 43N 58 17 E
Kestell, S. Africa 57 D4 28 17S 28 42 E
Kestenga, Russia 24 A5 65 50N 31 45 E
Keswick, U.K. 10 C4 54 36N 3 8W
Ket —►, Russia 26 D9 58 55N 81 32 E
Keta, Ghana 50 G5 5 49N 1 0 E
Ketapang, Indonesia 36 E4 1 55S 110 0 E
Ketchikan, U.S.A. 68 C6 55 21N 131 39W
Ketchum, U.S.A. 82 E6 43 41N 114 22W
Keti Bandar, Pakistan 42 G2 24 8N 67 27 E
Ketri, India 42 E6 28 1N 75 50 E
Kętrzyn, Poland 17 A11 54 7N 21 22 E
Kettering, U.K. 11 E7 52 24N 0 43W
Kettering, U.S.A. 76 F3 39 41N 84 10W
Kettle —►, Canada 73 B11 56 40N 89 34W
Kettle Falls, U.S.A. 82 B4 48 37N 118 3W
Kettleman City, U.S.A. 84 J7 36 1N 119 58W
Keuruu, Finland 9 E21 62 16N 24 41 E
Kevin, U.S.A. 82 B8 48 45N 111 58W
Kewanee, U.S.A. 80 E10 41 14N 89 56W
Kewaunee, U.S.A. 76 C2 44 27N 87 31W
Keweenaw B., U.S.A. 76 B1 47 0N 88 15W
Keweenaw Pen., U.S.A. 76 B2 47 30N 88 0W
Keweenaw Pt., U.S.A. 76 B2 47 25N 87 43W
Key Harbour, Canada 70 C3 45 50N 80 45W
Key West, U.S.A. 75 F10 24 33N 81 48W
Keyser, U.S.A. 76 F6 39 26N 78 59W
Keystone, U.S.A. 80 D3 43 54N 103 25W
Kezhma, Russia 27 D11 58 59N 101 9 E
Khabarovo, Russia 26 C7 69 30N 60 30 E
Khabarovsk, Russia 27 E14 48 30N 135 5 E
Khabr, Iran 45 D8 28 51N 56 22 E
Khābūr —►, Syria 44 C4 35 17N 40 35 E
Khachrod, India 42 H6 23 25N 75 20 E
Khadro, Pakistan 42 F3 26 11N 68 50 E
Khadzhilyangar, India 43 B8 35 45N 79 20 E
Khagaria, India 43 G12 25 30N 86 32 E
Khaipur, Bahawalpur, Pakistan 42 E5 29 34N 72 17 E
Khaipur, Hyderabad, Pakistan 42 F3 27 32N 68 49 E
Khair, India 42 F7 27 57N 77 46 E
Khairabad, India 43 F9 27 33N 80 47 E
Khairagarh, India 43 J9 21 27N 81 2 E
Khairpur, Pakistan 40 F6 27 32N 68 49 E
Khakassia □, Russia 26 D9 53 0N 90 0 E
Khakhea, Botswana 56 C3 24 48S 23 22 E
Khalafābād, Iran 45 D6 30 54N 49 24 E
Khalilabad, India 43 F10 26 48N 83 5 E
Khalīlī, Iran 45 E7 27 38N 53 17 E
Khalkhāl, Iran 45 B6 37 37N 48 32 E
Khalkís, Greece 21 E10 38 27N 23 42 E
Khalmer-Sede = Tazovskiy, Russia 26 C8 67 30N 78 44 E
Khalmer Yu, Russia 24 A12 67 58N 65 1 E
Khalturin, Russia 24 C8 58 40N 48 50 E
Khalūf, Oman 46 C6 20 30N 58 13 E
Kham Keut, Laos 38 C5 18 15N 104 43 E
Khamas Country, Botswana 56 C4 21 45S 26 30 E
Khambat, G. of, India 42 J5 20 45N 72 30 E
Khambhaliya, India 42 H3 22 14N 69 41 E
Khambhat, India 42 H5 22 23N 72 33 E
Khamīr, Iran 45 E7 26 57N 55 36 E
Khamir, Yemen 46 D3 16 2N 44 0 E
Khamsa, Egypt 47 E1 30 27N 32 23 E
Khān Abū Shāmat, Syria 47 B5 33 39N 36 53 E
Khān Azād, Iraq 44 C5 33 7N 44 22 E
Khān Mujiddah, Iraq 44 C4 32 21N 43 48 E
Khān Shaykhūn, Syria 44 C3 35 26N 36 38 E
Khān Yūnis, Gaza Strip 47 D3 31 21N 34 18 E
Khānaqīn, Iraq 44 C5 34 23N 45 25 E
Khānbāghī, Iran 45 B7 36 10N 55 25 E
Khandwa, India 40 J10 21 49N 76 22 E
Khandyga, Russia 27 C14 62 42N 135 35 E
Khāneh, Iran 44 B5 36 41N 45 8 E
Khanewal, Pakistan 42 D4 30 20N 71 55 E
Khanh Duong, Vietnam 38 F7 12 44N 108 44 E
Khaniá, Greece 23 D6 35 30N 24 4 E
Khaniá □, Greece 23 D6 35 30N 24 0 E
Khanion, Kólpos, Greece 23 D5 35 33N 23 55 E
Khanka, Ozero, Asia 27 E14 45 0N 132 24 E
Khankendy = Xankändi, Azerbaijan 25 G8 39 52N 46 49 E
Khanna, India 42 D7 30 42N 76 16 E
Khanpur, Pakistan 42 E4 28 42N 70 35 E
Khanty-Mansiysk, Russia 26 C7 61 0N 69 0 E
Khapalu, Pakistan 43 B7 35 10N 76 20 E
Khapcheranga, Russia 27 E12 49 42N 112 24 E
Kharagpur, India 43 H12 22 20N 87 25 E
Khárakas, Greece 23 D7 35 1N 25 7 E
Kharan Kalat, Pakistan 40 E4 28 34N 65 21 E
Kharānaq, Iran 45 C7 32 20N 54 45 E
Kharda, India 40 K9 18 40N 75 34 E
Khardung La, India 43 B7 34 20N 77 43 E
Khârga, El Wâhât el, Egypt 51 C11 25 10N 30 35 E
Khargon, India 40 J9 21 45N 75 40 E
Khārk, Jazireh, Iran 45 D6 29 15N 50 28 E
Kharkiv, Ukraine 25 E6 49 58N 36 20 E
Kharkov = Kharkiv, Ukraine 25 E6 49 58N 36 20 E
Kharovsk, Russia 24 C7 59 56N 40 13 E
Kharta, Turkey 21 D13 40 55N 29 7 E
Khartoum = El Khartûm, Sudan 51 E11 15 31N 32 35 E
Khasan, Russia 30 C5 42 25N 130 40 E
Khāsh, Iran 40 E2 28 15N 61 15 E
Khashm el Girba, Sudan 51 F12 14 59N 35 58 E
Khaskovo, Bulgaria 21 D11 41 56N 25 30 E
Khatanga, Russia 27 B11 72 0N 102 20 E
Khatanga —►, Russia 27 B11 72 55N 106 0 E
Khatauli, India 42 E7 29 17N 77 43 E
Khātūnābād, Iran 45 C6 35 30N 51 40 E
Khatyrka, Russia 27 C18 62 3N 175 15 E
Khaybar, Harrat, Si. Arabia 44 E4 25 45N 40 0 E
Khāzimiyah, Iraq 44 C4 34 46N 43 37 E
Khe Bo, Vietnam 38 C5 19 8N 104 41 E
Khe Long, Vietnam 38 B5 21 29N 104 46 E
Khed Brahma, India 40 G8 24 7N 73 5 E
Khekra, India 42 E7 28 52N 77 20 E
Khemarak Phouminville, Cambodia 39 G4 11 37N 102 59 E
Khemmarat, Thailand 38 D5 16 10N 105 15 E
Khenāmān, Iran 45 D8 30 27N 56 29 E
Khenchela, Algeria 50 A6 35 28N 7 11 E
Khenifra, Morocco 50 B3 32 58N 5 46W
Kherson, Ukraine 25 E5 46 35N 32 35 E
Khersónisos Akrotíri, Greece 23 D6 35 30N 24 10 E
Kheta —►, Russia 27 B11 71 54N 102 6 E
Khilok, Russia 27 D12 51 30N 110 45 E
Khíos, Greece 21 E12 38 27N 26 9 E
Khirbat Qanāfār, Lebanon 47 B4 33 39N 35 43 E
Khiuma = Hiiumaa, Estonia 9 G20 58 50N 22 45 E
Khiva, Uzbekistan 26 E7 41 30N 60 18 E
Khīyāv, Iran 44 B5 38 30N 47 45 E
Khlong Khlung, Thailand 38 D2 16 12N 99 43 E
Khmelnik, Ukraine 17 D14 49 33N 27 58 E
Khmelnitskiy = Khmelnytskyy, Ukraine 17 D14 49 23N 27 0 E
Khmelnytskyy, Ukraine 17 D14 49 23N 27 0 E
Khmer Rep. = Cambodia ■, Asia 38 F5 12 15N 105 0 E
Khoai, Hon, Vietnam 39 H5 8 26N 104 50 E
Khodoriv, Ukraine 17 D13 49 24N 24 19 E
Khodzent = Khudzhand, Tajikistan 26 E7 40 17N 69 37 E
Khojak P., Afghan. 40 D5 30 55N 66 30 E
Khok Kloi, Thailand 39 H2 8 17N 98 19 E
Khok Pho, Thailand 39 J3 6 43N 101 6 E
Kholm, Russia 24 C5 57 10N 31 15 E
Kholmsk, Russia 27 E15 47 40N 142 5 E
Khomas Hochland, Namibia 56 C2 22 40S 16 0 E
Khomeyn, Iran 45 C6 33 40N 50 7 E
Khon Kaen, Thailand 38 D4 16 30N 102 47 E
Khong, Laos 38 E5 14 7N 105 51 E
Khong Sedone, Laos 38 E5 15 34N 105 49 E
Khonuu, Russia 27 C15 66 30N 143 12 E
Khoper —►, Russia 26 E5 49 30N 42 20 E
Khóra Sfakion, Greece 23 D6 35 15N 24 9 E
Khorāsān □, Iran 45 C8 34 0N 58 0 E
Khorat = Nakhon Ratchasima, Thailand 38 E4 14 59N 102 12 E
Khorat, Cao Nguyen, Thailand 38 E4 15 30N 102 50 E
Khorixas, Namibia 56 C1 20 16S 14 59 E
Khorramābād, Khorāsān, Iran 45 C8 35 6N 57 57 E
Khorramābād, Lorestān, Iran 45 C6 33 30N 48 25 E
Khorrāmshahr, Iran 45 D6 30 29N 48 15 E
Khorugh, Tajikistan 26 F8 37 30N 71 36 E
Khosravī, Iran 45 D6 30 48N 51 28 E
Khosrowābād, Khuzestān, Iran 45 D6 30 10N 48 25 E
Khosrowābād, Kordestān, Iran 44 C5 35 31N 47 38 E
Khosūyeh, Iran 45 D7 28 32N 54 26 E
Khotyn, Ukraine 17 D14 48 31N 26 27 E
Khouribga, Morocco 50 B3 32 58N 6 57W
Khowai, Bangla. 41 G17 24 5N 91 40 E
Khoyniki, Belarus 17 C15 51 54N 29 55 E
Khrysokhou B., Cyprus 23 D11 35 6N 32 25 E
Khu Khan, Thailand 38 E5 14 42N 104 12 E
Khudzhand, Tajikistan 26 E7 40 17N 69 37 E
Khuff, Si. Arabia 44 E5 24 55N 44 53 E
Khūgīānī, Afghan. 42 D1 31 28N 65 14 E
Khulna, Bangla. 41 H16 22 45N 89 34 E
Khulna □, Bangla. 41 H16 22 25N 89 35 E
Khumago, Botswana 56 C3 20 26S 24 32 E
Khūnsorkh, Iran 45 E8 27 9N 56 7 E
Khūr, Iran 45 C8 32 55N 58 18 E
Khurai, India 42 G8 24 3N 78 23 E
Khurayş, Si. Arabia 45 E6 25 6N 48 2 E
Khūrīyā Mūrīyā, Jazā 'ir, Oman 46 D6 17 30N 55 58 E
Khurja, India 42 E7 28 15N 77 58 E
Khūsf, Iran 45 C8 32 46N 58 53 E
Khush, Afghan. 40 C3 32 55N 62 10 E
Khushab, Pakistan 42 C5 32 20N 72 20 E
Khust, Ukraine 17 D12 48 10N 23 18 E
Khuzdar, Pakistan 42 F2 27 52N 66 30 E
Khūzestān □, Iran 45 D6 31 0N 49 0 E
Khvājeh, Iran 44 B5 38 9N 46 35 E
Khvānsār, Iran 45 D7 29 56N 54 8 E
Khvor, Iran 45 C7 33 45N 55 0 E
Khvorgū, Iran 45 E8 27 34N 56 27 E
Khvormūj, Iran 45 D6 28 40N 51 30 E
Khvoy, Iran 44 B5 38 35N 45 0 E
Khyber Pass, Afghan. 42 B4 34 10N 71 8 E
Kiabukwa, Zaïre 55 D1 8 40S 24 48 E
Kiama, Australia 63 E5 34 40S 150 50 E
Kiamba, Phil. 37 C6 6 2N 124 46 E
Kiambi, Zaïre 54 D2 7 15S 28 0 E
Kiambu, Kenya 54 C4 1 8S 36 50 E
Kiangsi = Jiangxi □, China 33 D6 27 30N 116 0 E
Kiangsu = Jiangsu □, China 35 H10 33 0N 120 0 E
Kibanga Port, Uganda 54 B3 0 10N 32 58 E
Kibangou, Congo 52 E2 3 26S 12 22 E
Kibara, Tanzania 54 C3 2 8S 33 30 E
Kibare, Mts., Zaïre 54 D2 8 25S 27 10 E
Kibombo, Zaïre 54 C2 3 57S 25 53 E
Kibondo, Tanzania 54 C3 3 35S 30 45 E
Kibondo □, Tanzania 54 C3 4 0S 30 55 E
Kibumbu, Burundi 54 C2 3 32S 29 45 E
Kibungu, Rwanda 54 C3 2 10S 30 32 E
Kibuye, Burundi 54 C2 3 39S 29 59 E
Kibuye, Rwanda 54 C2 2 3S 29 21 E
Kibwesa, Tanzania 54 D2 6 30S 29 58 E
Kibwezi, Kenya 54 C4 2 27S 37 57 E
Kichiga, Russia 27 D17 59 50N 163 5 E
Kicking Horse Pass, Canada 72 C5 51 28N 116 16W
Kidal, Mali 50 E5 18 26N 1 22 E
Kidderminster, U.K. 11 E5 52 24N 2 15W
Kidete, Tanzania 54 D4 6 25S 37 17 E
Kidnappers, C., N.Z. 59 H6 39 38S 177 5 E
Kidston, Australia 62 B3 18 52S 144 8 E
Kidugallo, Tanzania 54 D4 6 49S 38 15 E
Kiel, Germany 16 A6 54 19N 10 8 E
Kiel Canal = Nord-Ostsee-Kanal —►, Germany 16 A5 54 12N 9 32 E
Kielce, Poland 17 C11 50 52N 20 42 E
Kieler Bucht, Germany 16 A6 54 35N 10 25 E
Kien Binh, Vietnam 39 H5 9 55N 105 19 E
Kien Tan, Vietnam 39 G5 10 7N 105 17 E
Kienge, Zaïre 55 E2 10 30S 27 30 E
Kiev = Kyyiv, Ukraine 17 C16 50 30N 30 28 E
Kiffa, Mauritania 50 E2 16 37N 11 24W
Kifrī, Iraq 44 C5 34 45N 45 0 E
Kigali, Rwanda 54 C3 1 59S 30 4 E
Kigarama, Tanzania 54 C3 1 1S 31 50 E
Kigoma □, Tanzania 54 D2 5 0S 30 0 E
Kigoma-Ujiji, Tanzania 54 C2 4 55S 29 36 E
Kigomasha, Ras, Tanzania 54 C4 4 58S 38 58 E
Kihee, Australia 63 D3 27 23S 142 37 E
Kihnu, Estonia 9 G21 58 9N 24 1 E
Kii-Sanchi, Japan 31 G7 34 20N 136 0 E
Kii-Suidō, Japan 31 H7 33 40N 134 45 E
Kikaiga-Shima, Japan 31 K4 28 19N 129 59 E
Kikinda, Serbia, Yug. 21 B9 45 50N 20 30 E
Kikládhes, Greece 21 F11 37 20N 24 30 E
Kikwit, Zaïre 52 E3 5 0S 18 45 E
Kilauea Crater, U.S.A. 74 J17 19 25N 155 17W
Kilcoy, Australia 63 D5 26 59S 152 30 E
Kildare, Ireland 13 C5 53 9N 6 55W
Kildare □, Ireland 13 C5 53 10N 6 50W
Kilgore, U.S.A. 81 J7 32 23N 94 53W
Kilifi, Kenya 54 C4 3 40S 39 48 E
Kilifi □, Kenya 54 C4 3 30S 39 40 E
Kilimanjaro, Tanzania 54 C4 3 7S 37 20 E
Kilimanjaro □, Tanzania 54 C4 4 0S 38 0 E
Kilindini, Kenya 54 C4 4 4S 39 40 E
Kiliya, Ukraine 17 F15 45 28N 29 16 E
Kilju, N. Korea 35 D15 40 57N 129 25 E
Kilkee, Ireland 13 D2 52 41N 9 39W
Kilkenny, Ireland 13 D4 52 39N 7 15W
Kilkenny □, Ireland 13 D4 52 35N 7 15W
Kilkieran B., Ireland 13 C2 53 20N 9 41W
Kilkís, Greece 21 D10 40 58N 22 57 E
Killala, Ireland 13 B2 54 13N 9 12W
Killala B., Ireland 13 B2 54 16N 9 8W
Killaloe, Ireland 13 D3 52 48N 8 28W
Killaloe Sta., Canada 78 A7 45 33N 77 25W
Killam, Canada 72 C6 52 47N 111 51W
Killarney, Australia 63 D5 28 20S 152 18 E
Killarney, Canada 70 C3 45 55N 81 30W
Killarney, Ireland 13 D2 52 4N 9 30W
Killarney, Lakes of, Ireland 13 E2 52 0N 9 30W
Killary Harbour, Ireland 13 C2 53 38N 9 52W
Killdeer, Canada 73 D7 49 6N 106 22W
Killdeer, U.S.A. 80 B3 47 26N 102 48W
Killeen, U.S.A. 81 K6 31 7N 97 44W
Killiecrankie, Pass of, U.K. 12 E5 56 44N 3 46W
Killin, U.K. 12 E4 56 28N 4 19W
Killíni, Greece 21 F10 37 54N 22 25 E
Killybegs, Ireland 13 B3 54 38N 8 26W
Kilmarnock, U.K. 12 F4 55 37N 4 29W
Kilmore, Australia 63 F3 37 25S 144 53 E
Kilondo, Tanzania 55 D3 9 45S 34 20 E
Kilosa, Tanzania 54 D4 6 48S 37 0 E
Kilosa □, Tanzania 54 D4 6 48S 37 0 E
Kilrush, Ireland 13 D2 52 38N 9 29W
Kilwa □, Tanzania 55 D4 9 0S 39 0 E
Kilwa Kisiwani, Tanzania 55 D4 8 58S 39 32 E
Kilwa Kivinje, Tanzania 55 D4 8 45S 39 25 E
Kilwa Masoko, Tanzania 55 D4 8 55S 39 30 E
Kim, U.S.A. 81 G3 37 15N 103 21W
Kimaam, Indonesia 37 F9 7 58S 138 53 E
Kimamba, Tanzania 54 D4 6 45S 37 10 E
Kimba, Australia 63 E2 33 8S 136 23 E
Kimball, Nebr., U.S.A. 80 E3 41 14N 103 40W
Kimball, S. Dak., U.S.A. 80 D5 43 45N 98 57W
Kimberley, Canada 72 D5 49 40N 115 59W
Kimberley, S. Africa 56 D3 28 43S 24 46 E
Kimberley Downs, Australia 60 C3 17 24S 124 22 E
Kimberly, U.S.A. 82 E6 42 32N 114 22W
Kimchaek, N. Korea 35 D15 40 40N 129 10 E
Kimchŏn, S. Korea 35 F15 36 11N 128 4 E
Kimje, S. Korea 35 G14 35 48N 126 45 E
Kimmirut, Canada 69 B13 62 50N 69 50W
Kimry, Russia 24 C6 56 55N 37 15 E
Kimsquit, Canada 72 C3 52 45N 126 57W
Kinabalu, Gunong, Malaysia 36 C5 6 3N 116 14 E
Kinaskan L., Canada 72 B2 57 38N 130 8W
Kinbasket L., Canada 72 C5 52 0N 118 10W
Kincaid, Canada 73 D7 49 40N 107 0W
Kincardine, Canada 70 D3 44 10N 81 40W
Kinda, Zaïre 55 D2 9 18S 25 4 E
Kinder Scout, U.K. 10 D6 53 24N 1 52W
Kindersley, Canada 73 C7 51 30N 109 10W
Kindia, Guinea 50 G2 10 0N 12 52W
Kindu, Zaïre 54 C2 2 55S 25 50 E
Kineshma, Russia 24 C7 57 30N 42 5 E
Kinesi, Tanzania 54 C3 1 25S 33 50 E
King, L., Australia 61 F2 33 10S 119 35 E
King, Mt., Australia 62 D4 25 10S 147 30 E
King City, U.S.A. 83 H3 36 13N 121 8W
King Cr. —►, Australia 62 C2 24 35S 139 30 E
King Edward —►, Australia 60 B4 14 14S 126 35 E
King Frederick VI Land = Kong Frederik VI.s Kyst, Greenland 4 C5 63 0N 43 0W
King George B., Falk. Is. 96 G4 51 30S 60 30W
King George I., Antarctica 5 C18 60 0S 60 0W
King George Is., Canada 69 C11 57 20N 80 30W
King I. = Kadan Kyun, Burma 36 B1 12 30N 98 20 E
King I., Australia 62 F3 39 50S 144 0 E
King I., Canada 72 C3 52 10N 127 40W
King Leopold Ras., Australia 60 C4 17 30S 125 45 E
King Sd., Australia 60 C3 16 50S 123 20 E
King William I., Canada 68 B10 69 10N 97 25W
King William's Town, S. Africa 56 E4 32 51S 27 22 E
Kingaroy, Australia 63 D5 26 32S 151 51 E
Kingfisher, U.S.A. 81 H6 35 52N 97 56W
Kingirbān, Iraq 44 C5 34 40N 44 54 E
Kingisepp = Kuressaare, Estonia 9 G20 58 15N 22 30 E
Kingman, Ariz., U.S.A. 85 K12 35 12N 114 4W
Kingman, Kans., U.S.A. 81 G5 37 39N 98 7W
Kingoonya, Australia 63 E2 30 55S 135 19 E
Kings —►, U.S.A. 83 H4 36 3N 119 50W
Kings Canyon National Park, U.S.A. 83 H4 36 50N 118 40W
King's Lynn, U.K. 10 E8 52 45N 0 24 E
Kings Mountain, U.S.A. 77 H5 35 15N 81 20W
King's Peak, U.S.A. 82 F8 40 46N 110 27W
Kingsbridge, U.K. 11 G4 50 17N 3 47W
Kingsburg, U.S.A. 83 H4 36 31N 119 33W
Kingscote, Australia 63 F2 35 40S 137 38 E
Kingscourt, Ireland 13 C5 53 55N 6 48W
Kingsley, U.S.A. 80 D7 42 35N 95 58W
Kingsport, U.S.A. 77 G4 36 33N 82 33W
Kingston, Canada 70 D4 44 14N 76 30W
Kingston, Jamaica 88 C4 18 0N 76 50W
Kingston, N.Z. 59 L2 45 20S 168 43 E
Kingston, N.Y., U.S.A. 79 E10 41 56N 73 59W
Kingston, Pa., U.S.A. 79 E9 41 16N 75 54W
Kingston, R.I., U.S.A. 79 E13 41 29N 71 30W
Kingston Pk., U.S.A. 85 K11 35 45N 115 54W
Kingston South East, Australia 63 F2 36 51S 139 55 E
Kingston upon Hull, U.K. 10 D7 53 45N 0 21W
Kingston upon Hull □, U.K. 10 D7 53 45N 0 21W
Kingston-upon-Thames, U.K. 11 F7 51 24N 0 17W
Kingstown, St. Vincent 89 D7 13 10N 61 10W
Kingstree, U.S.A. 77 J6 33 40N 79 50W
Kingsville, Canada 70 D3 42 2N 82 45W
Kingsville, U.S.A. 81 M6 27 31N 97 52W
Kingussie, U.K. 12 D4 57 6N 4 2W
Kınık, Turkey 21 E12 39 5N 27 23 E
Kinistino, Canada 73 C7 52 57N 105 2W
Kinkala, Congo 52 E2 4 18S 14 49 E
Kinki □, Japan 31 H8 33 45N 136 0 E
Kinleith, N.Z. 59 H5 38 20S 175 56 E
Kinmount, Canada 78 B6 44 48N 78 45W
Kinna, Sweden 9 H15 57 32N 12 42 E
Kinnaird, Canada 72 D5 49 17N 117 39W
Kinnairds Hd., U.K. 12 D7 57 43N 2 1W
Kinnarodden, Norway 6 A11 71 8N 27 40 E
Kino, Mexico 86 B2 28 45N 111 59W
Kinoje —►, Canada 70 B3 52 8N 81 25W
Kinomoto, Japan 31 G8 35 30N 136 13 E
Kinoni, Uganda 54 C3 0 41S 30 28 E
Kinross, U.K. 12 E5 56 13N 3 25W
Kinsale, Ireland 13 E3 51 42N 8 31W
Kinsale, Old Hd. of, Ireland 13 E3 51 37N 8 33W
Kinsha = Chang Jiang —►, China 33 C7 31 48N 121 10 E
Kinshasa, Zaïre 52 E3 4 20S 15 15 E
Kinsley, U.S.A. 81 G5 37 55N 99 25W

Kinston, U.S.A. 77 H7 35 16N 77 35W
Kintampo, Ghana 50 G4 8 5N 1 41W
Kintap, Indonesia 36 E5 3 51S 115 13 E
Kintore Ra., Australia 60 D4 23 15S 128 47 E
Kintyre, U.K. 12 F3 55 30N 5 35W
Kintyre, Mull of, U.K. 12 F3 55 17N 5 47W
Kinushseo →, Canada 70 A3 55 15N 83 45W
Kinuso, Canada 72 B5 55 20N 115 25W
Kinyangiri, Tanzania 54 C3 4 25S 34 37 E
Kinzua, U.S.A. 78 E6 41 52N 78 58W
Kinzua Dam, U.S.A. 78 E5 41 53N 79 0W
Kiosk, Canada 70 C4 46 6N 78 53W
Kiowa, Kans., U.S.A. 81 G5 37 1N 98 29W
Kiowa, Okla., U.S.A. 81 H7 34 43N 95 54W
Kipahigan L., Canada 73 B8 55 20N 101 55W
Kipanga, Tanzania 54 D4 6 15S 35 20 E
Kiparissia, Greece 21 F9 37 15N 21 40 E
Kiparissiakós Kólpos, Greece 21 F9 37 25N 21 25 E
Kipembawe, Tanzania 54 D3 7 38S 33 27 E
Kipengere Ra., Tanzania 55 D3 9 12S 34 15 E
Kipili, Tanzania 54 D3 7 28S 30 32 E
Kipini, Kenya 54 C5 2 30S 40 32 E
Kipling, Canada 73 C8 50 6N 102 38W
Kippure, Ireland 13 C5 53 11N 6 21W
Kipushi, Zaïre 55 E2 11 48S 27 12 E
Kiratpur, India 42 E8 29 32N 78 12 E
Kirensk, Russia 27 D11 57 50N 107 55 E
Kirgella Rocks, Australia 61 F3 30 5S 122 50 E
Kirghizia ■ = Kyrgyzstan ■, Asia 26 E8 42 0N 75 0 E
Kirghizstan = Kyrgyzstan ■, Asia 26 E8 42 0N 75 0 E
Kirgiziya Steppe, Eurasia 25 D10 50 0N 55 0 E
Kiri, Zaïre 52 E3 1 29S 19 0 E
Kiribati ■, Pac. Oc. 64 H10 5 0S 180 0 E
Kırıkkale, Turkey 25 G5 39 51N 33 32 E
Kirillov, Russia 24 C6 59 49N 38 24 E
Kirin = Jilin, China 35 C14 43 44N 126 30 E
Kirin = Jilin □, China 35 C13 44 0N 127 0 E
Kiritimati, Kiribati 65 G12 1 58N 157 27W
Kirkcaldy, U.K. 12 E5 56 7N 3 9W
Kirkcudbright, U.K. 12 G4 54 50N 4 2W
Kirkee, India 40 K8 18 34N 73 56 E
Kirkenes, Norway 8 B23 69 40N 30 5 E
Kirkintilloch, U.K. 12 F4 55 56N 4 8W
Kirkjubæjarklaustur, Iceland 8 E4 63 47N 18 4W
Kirkkonummi, Finland 9 F21 60 8N 24 26 E
Kirkland, U.S.A. 83 J7 34 25N 112 43W
Kirkland Lake, Canada 70 C3 48 9N 80 2W
Kırklareli, Turkey 21 D12 41 44N 27 15 E
Kirksville, U.S.A. 80 E8 40 12N 92 35W
Kirkūk, Iraq 44 C5 35 30N 44 21 E
Kirkwall, U.K. 12 C6 58 59N 2 58W
Kirkwood, S. Africa 56 E4 33 22S 25 15 E
Kirov, Russia 24 C8 58 35N 49 40 E
Kirovabad = Gäncä, Azerbaijan 25 F8 40 45N 46 20 E
Kirovakan = Vanadzor, Armenia 25 F7 40 48N 44 30 E
Kirovograd = Kirovohrad, Ukraine 25 E5 48 35N 32 20 E
Kirovohrad, Ukraine 25 E5 48 35N 32 20 E
Kirovsk = Babadayhan, Turkmenistan 26 F7 37 42N 60 23 E
Kirovsk, Russia 24 A5 67 32N 33 41 E
Kirovskiy, Kamchatka, Russia 27 D16 54 27N 155 42 E
Kirovskiy, Primorsk, Russia 30 B6 45 7N 133 30 E
Kirriemuir, U.K. 12 E6 56 41N 3 1W
Kirsanov, Russia 24 D7 52 35N 42 40 E
Kırşehir, Turkey 25 G5 39 14N 34 5 E
Kirthar Range, Pakistan 42 F2 27 0N 67 0 E
Kiruna, Sweden 8 C19 67 52N 20 15 E
Kirundu, Zaïre 54 C2 0 50S 25 35 E
Kirup, Australia 61 F2 33 40S 115 50 E
Kiryū, Japan 31 F9 36 24N 139 20 E
Kisaga, Tanzania 54 C3 4 30S 34 23 E
Kisalaya, Nic. 88 D3 14 40N 84 3W
Kisámou, Kólpos, Greece 23 D5 35 30N 23 38 E
Kisanga, Zaïre 54 B2 2 30N 26 35 E
Kisangani, Zaïre 54 B2 0 35N 25 15 E
Kisar, Indonesia 37 F7 8 5S 127 10 E
Kisaran, Indonesia 36 D1 3 0N 99 37 E
Kisarawe, Tanzania 54 D4 6 53S 39 0 E
Kisarawe □, Tanzania 54 D4 7 3S 39 0 E
Kisarazu, Japan 31 G9 35 23N 139 55 E
Kiselevsk, Russia 26 D9 54 5N 86 39 E
Kishanganga →, Pakistan 43 B5 34 18N 73 28 E
Kishanganj, India 43 F13 26 3N 88 14 E
Kishangarh, India 42 F4 27 50N 70 30 E
Kishinev = Chişinău, Moldova 17 E15 47 0N 28 50 E
Kishiwada, Japan 31 G7 34 28N 135 22 E
Kishtwar, India 43 C6 33 20N 75 48 E
Kisii, Kenya 54 C3 0 40S 34 45 E
Kisii □, Kenya 54 C3 0 40S 34 45 E
Kisiju, Tanzania 54 D4 7 23S 39 19 E
Kisizi, Uganda 54 C2 1 0S 29 58 E
Kiska I., U.S.A. 68 C1 51 59N 177 30 E
Kiskatinaw →, Canada 72 B4 56 8N 120 10W
Kiskittogisu L., Canada 73 C9 54 13N 98 20W
Kiskörös, Hungary 17 E10 46 37N 19 20 E
Kiskunfélegyháza, Hungary 17 E10 46 42N 19 53 E
Kiskunhalas, Hungary 17 E10 46 28N 19 37 E
Kislovodsk, Russia 25 F7 43 50N 42 45 E
Kismayu = Chisimaio, Somali Rep. 49 G8 0 22S 42 32 E
Kiso-Gawa →, Japan 31 G8 35 20N 136 45 E
Kiso-Sammyaku, Japan 31 G8 35 45N 137 45 E
Kisofukushima, Japan 31 G8 35 52N 137 43 E
Kisoro, Uganda 54 C2 1 17S 29 48 E
Kissidougou, Guinea 50 G2 9 5N 10 5W
Kissimmee, U.S.A. 77 L5 28 18N 81 24W
Kissimmee →, U.S.A. 77 M5 27 9N 80 52W
Kississing L., Canada 73 B8 55 10N 101 20W
Kissónerga, Cyprus 23 E11 34 49N 32 24 E
Kisumu, Kenya 54 C3 0 3S 34 45 E
Kiswani, Tanzania 54 C4 4 5S 37 57 E
Kiswere, Tanzania 55 D4 9 27S 39 30 E
Kit Carson, U.S.A. 80 F3 38 46N 102 48W
Kita, Mali 50 F3 13 5N 9 25W
Kitab, Uzbekistan 26 F7 39 7N 66 52 E
Kitaibaraki, Japan 31 F10 36 50N 140 45 E
Kitakami, Japan 30 E10 39 20N 141 10 E
Kitakami-Gawa →, Japan 30 E10 38 25N 141 19 E
Kitakami-Sammyaku, Japan 30 E10 39 30N 141 30 E
Kitakata, Japan 30 F9 37 39N 139 52 E
Kitakyūshū, Japan 31 H5 33 50N 130 50 E
Kitale, Kenya 54 B4 1 0N 35 0 E
Kitami, Japan 30 C11 43 48N 143 54 E
Kitami-Sammyaku, Japan 30 B11 44 22N 142 43 E
Kitangiri, L., Tanzania 54 C3 4 5S 34 20 E
Kitaya, Tanzania 55 E5 10 38S 40 8 E
Kitchener, Australia 61 F3 30 55S 124 8 E
Kitchener, Canada 70 D3 43 27N 80 29W
Kitega = Gitega, Burundi 54 C2 3 26S 29 56 E
Kitengo, Zaïre 54 D1 7 26S 24 8 E
Kiteto □, Tanzania 54 C4 5 0S 37 0 E
Kitgum, Uganda 54 B3 3 17N 32 52 E
Kíthira, Greece 21 F10 36 8N 23 0 E
Kíthnos, Greece 21 F11 37 26N 24 27 E
Kiti, Cyprus 23 E12 34 50N 33 34 E
Kiti, C., Cyprus 23 E12 34 48N 33 36 E
Kitikmeot □, Canada 68 A9 70 0N 110 0W
Kitimat, Canada 72 C3 54 3N 128 38W
Kitinen →, Finland 8 C22 67 14N 27 27 E
Kitsuki, Japan 31 H5 33 25N 131 37 E
Kittakittaooloo, L., Australia 63 D2 28 3S 138 14 E
Kittanning, U.S.A. 78 F5 40 49N 79 31W
Kittatinny Mts., U.S.A. 79 F10 41 0N 75 0W
Kittery, U.S.A. 77 D10 43 5N 70 45W
Kittilä, Finland 8 C21 67 40N 24 51 E
Kitui, Kenya 54 C4 1 17S 38 0 E
Kitui □, Kenya 54 C4 1 30S 38 25 E
Kitwe, Zambia 55 E2 12 54S 28 13 E
Kivarli, India 42 G5 24 33N 72 46 E
Kivertsi, Ukraine 17 C13 50 50N 25 28 E
Kividhes, Cyprus 23 E11 34 46N 32 51 E
Kivu □, Zaïre 54 C2 3 10S 27 0 E
Kivu, L., Zaïre 54 C2 1 48S 29 0 E
Kiyev = Kyyiv, Ukraine 17 C16 50 30N 30 28 E
Kiyevskoye Vdkhr. = Kyyivske Vdskh., Ukraine 17 C16 51 0N 30 25 E
Kizel, Russia 24 C10 59 3N 57 40 E
Kiziguru, Rwanda 54 C3 1 46S 30 23 E
Kızıl Irmak →, Turkey 25 F6 41 44N 35 58 E
Kizil Jilga, India 43 B8 35 26N 78 50 E
Kizimkazi, Tanzania 54 D4 6 28S 39 30 E
Kizlyar, Russia 25 F8 43 51N 46 40 E
Kizyl-Arvat = Gyzylarbat, Turkmenistan 26 F6 39 4N 56 23 E
Kjölur, Iceland 8 D4 64 50N 19 25W
Kladno, Czech. 16 C8 50 10N 14 7 E
Klaeng, Thailand 38 F3 12 47N 101 39 E
Klagenfurt, Austria 16 E8 46 38N 14 20 E
Klaipėda, Lithuania 9 J19 55 43N 21 10 E
Klaksvík, Færoe Is. 8 E9 62 14N 6 35W
Klamath →, U.S.A. 82 F1 41 33N 124 5W
Klamath Falls, U.S.A. 82 E3 42 13N 121 46W
Klamath Mts., U.S.A. 82 F2 41 20N 123 0W
Klappan →, Canada 72 B3 58 0N 129 43W
Klarälven →, Sweden 9 G15 59 23N 13 32 E
Klaten, Indonesia 37 G14 7 43S 110 36 E
Klatovy, Czech. 16 D7 49 23N 13 18 E
Klawer, S. Africa 56 E2 31 44S 18 36 E
Klawock, U.S.A. 72 B2 55 33N 133 6W
Kleena Kleene, Canada 72 C4 52 0N 124 59W
Klein, U.S.A. 82 C9 46 24N 108 33W
Klein-Karas, Namibia 56 D2 27 33S 18 7 E
Klerksdorp, S. Africa 56 D4 26 53S 26 38 E
Kletsk = Klyetsk, Belarus 17 B14 53 5N 26 45 E
Kletskiy, Russia 26 E5 49 16N 43 11 E
Klickitat, U.S.A. 82 D3 45 49N 121 9W
Klickitat →, U.S.A. 84 E5 45 42N 121 17W
Klidhes, Cyprus 23 D13 35 42N 34 36 E
Klin, Russia 26 D4 56 20N 36 48 E
Klinaklini →, Canada 72 C3 51 21N 125 40W
Klipdale, S. Africa 56 E2 34 19S 19 57 E
Klipplaat, S. Africa 56 E3 33 1S 24 22 E
Kłodzko, Poland 17 C9 50 28N 16 38 E
Klondike, Canada 68 B6 64 0N 139 26W
Klouto, Togo 50 G5 6 57N 0 44 E
Kluane L., Canada 68 B6 61 15N 138 40W
Kluczbork, Poland 17 C10 50 58N 18 12 E
Klyetsk, Belarus 17 B14 53 5N 26 45 E
Klyuchevskaya, Gora, Russia 27 D17 55 50N 160 30 E
Knaresborough, U.K. 10 C6 54 1N 1 28W
Knee L., Man., Canada 73 B10 55 3N 94 45W
Knee L., Sask., Canada 73 B7 55 51N 107 0W
Knight Inlet, Canada 72 C3 50 45N 125 40W
Knighton, U.K. 11 E4 52 21N 3 3W
Knights Ferry, U.S.A. 84 H6 37 50N 120 40W
Knights Landing, U.S.A. 84 G5 38 48N 121 43W
Knob, C., Australia 61 F2 34 32S 119 16 E
Knockmealdown Mts., Ireland 13 D4 52 14N 7 56W
Knokke, Belgium 15 C3 51 20N 3 17 E
Knossós, Greece 23 D7 35 16N 25 10 E
Knox, U.S.A. 76 E2 41 18N 86 37W
Knox, C., Canada 72 C2 54 11N 133 5W
Knox City, U.S.A. 81 J5 33 25N 99 49W
Knox Coast, Antarctica 5 C8 66 30S 108 0 E
Knoxville, Iowa, U.S.A. 80 E8 41 19N 93 6W
Knoxville, Tenn., U.S.A. 77 H4 35 58N 83 55W
Knysna, S. Africa 56 E3 34 2S 23 2 E
Ko Kha, Thailand 38 C2 18 11N 99 24 E
Ko Tao, Thailand 39 G2 10 6N 99 48 E
Koartac = Quaqtaq, Canada 69 B13 60 55N 69 40W
Koba, Aru, Indonesia 37 F8 6 37S 134 37 E
Koba, Bangka, Indonesia 36 E3 2 26S 106 14 E
Kobarid, Slovenia 16 E7 46 15N 13 30 E
Kobayashi, Japan 31 J5 31 56N 130 59 E
Kobdo = Hovd, Mongolia 32 B4 48 2N 91 37 E
Kōbe, Japan 31 G7 34 45N 135 10 E
København, Denmark 9 J15 55 41N 12 34 E
Kōbi-Sho, Japan 31 M1 25 56N 123 41 E
Koblenz, Germany 16 C4 50 21N 7 36 E
Kobroor, Kepulauan, Indonesia 37 F8 6 10S 134 30 E
Kobryn, Belarus 17 B13 52 15N 24 22 E
Kocaeli = İzmit, Turkey 25 F4 40 45N 29 50 E
Kočani, Macedonia 21 D10 41 55N 22 25 E
Koch Bihar, India 41 F16 26 22N 89 29 E
Kochang, S. Korea 35 G14 35 41N 127 55 E
Kochas, India 43 G10 25 15N 83 56 E
Kocheya, Russia 27 D13 52 32N 120 42 E
Kōchi, Japan 31 H6 33 30N 133 35 E
Kōchi □, Japan 31 H6 33 40N 133 30 E
Kochiu = Gejiu, China 32 D5 23 20N 103 10 E
Kodiak, U.S.A. 68 C4 57 47N 152 24W
Kodiak I., U.S.A. 68 C4 57 30N 152 45W
Kodinar, India 42 J4 20 46N 70 46 E
Koes, Namibia 56 D2 26 0S 19 15 E
Koffiefontein, S. Africa 56 D4 29 30S 25 0 E
Kofiau, Indonesia 37 E7 1 11S 129 50 E
Koforidua, Ghana 50 G4 6 3N 0 17W
Kōfu, Japan 31 G9 35 40N 138 30 E
Koga, Japan 31 F9 36 11N 139 43 E
Kogaluk →, Canada 71 A7 56 12N 61 44W
Kogan, Australia 63 D5 27 2S 150 40 E
Køge, Denmark 9 J15 55 27N 12 11 E
Koh-i-Bābā, Afghan. 40 B5 34 30N 67 0 E
Koh-i-Khurd, Afghan. 42 C1 33 30N 65 59 E
Kohat, Pakistan 42 C4 33 40N 71 29 E
Kohima, India 41 G19 25 35N 94 10 E
Kohkīlūyeh va Būyer Aḥmadi □, Iran 45 D6 31 30N 50 30 E
Kohler Ra., Antarctica 5 D15 77 0S 110 0W
Kohtla-Järve, Estonia 9 G22 59 20N 27 20 E
Koillismaa, Finland 8 D23 65 44N 28 36 E
Koin-dong, N. Korea 35 D14 40 28N 126 18 E
Kojŏ, N. Korea 35 E14 38 58N 127 58 E
Kojonup, Australia 61 F2 33 48S 117 10 E
Kojūr, Iran 45 B6 36 23N 51 43 E
Kokand = Qŭqon, Uzbekistan 26 E8 40 30N 70 57 E
Kokanee Glacier Prov. Park, Canada 72 D5 49 47N 117 10W
Kokas, Indonesia 37 E8 2 42S 132 26 E
Kokchetav = Kökshetaū, Kazakstan 26 D7 53 20N 69 25 E
Kokemäenjoki →, Finland 9 F19 61 32N 21 44 E
Kokkola, Finland 8 E20 63 50N 23 8 E
Koko Kyunzu, Burma 41 M18 14 10N 93 25 E
Kokomo, U.S.A. 76 E2 40 29N 86 8W
Kokonau, Indonesia 37 E9 4 43S 136 26 E
Koksan, N. Korea 35 E14 38 46N 126 40 E
Kökshetaū, Kazakstan 26 D7 53 20N 69 25 E
Koksoak →, Canada 69 C13 58 30N 68 10W
Kokstad, S. Africa 57 E4 30 32S 29 29 E
Kokubu, Japan 31 J5 31 44N 130 46 E
Kokuora, Russia 27 B15 71 35N 144 50 E
Kola, Indonesia 37 F8 5 35S 134 30 E
Kola, Russia 24 A5 68 45N 33 8 E
Kola Pen. = Kolskiy Poluostrov, Russia 24 A6 67 30N 38 0 E
Kolahoi, India 43 B6 34 12N 75 22 E
Kolaka, Indonesia 37 E6 4 3S 121 46 E
Kolar, India 40 N11 13 12N 78 15 E
Kolar Gold Fields, India 40 N11 12 58N 78 16 E
Kolari, Finland 8 C20 67 20N 23 48 E
Kolayat, India 40 F8 27 50N 72 50 E
Kolchugino = Leninsk-Kuznetskiy, Russia 26 D9 54 44N 86 10 E
Kolda, Senegal 50 F2 12 55N 14 57W
Kolding, Denmark 9 J13 55 30N 9 29 E
Kole, Zaïre 52 E4 3 16S 22 42 E
Kolepom = Yos Sudarso, Pulau, Indonesia 37 F9 8 0S 138 30 E
Kolguyev, Ostrov, Russia 24 A8 69 20N 48 30 E
Kolhapur, India 40 L9 16 43N 74 15 E
Kolín, Czech. 16 C8 50 2N 15 9 E
Kolkas Rags, Latvia 9 H20 57 46N 22 37 E
Kolmanskop, Namibia 56 D2 26 45S 15 14 E
Köln, Germany 16 C4 50 56N 6 57 E
Koło, Poland 17 B10 52 14N 18 40 E
Kołobrzeg, Poland 16 A8 54 10N 15 35 E
Kolokani, Mali 50 F3 13 35N 7 45W
Kolomna, Russia 24 C6 55 8N 38 45 E
Kolomyya, Ukraine 17 D13 48 31N 25 2 E
Kolonodale, Indonesia 37 E6 2 3S 121 25 E
Kolosib, India 41 G18 24 15N 92 45 E
Kolpashevo, Russia 26 D9 58 20N 83 5 E
Kolpino, Russia 24 C5 59 44N 30 39 E
Kolskiy Poluostrov, Russia 24 A6 67 30N 38 0 E
Kolskiy Zaliv, Russia 24 A5 69 23N 34 0 E
Kolwezi, Zaïre 55 E2 10 40S 25 25 E
Kolyma →, Russia 27 C17 69 30N 161 0 E
Kolymskoye Nagorye, Russia 27 C16 63 0N 157 0 E
Komandorskiye Is. = Komandorskiye Ostrova, Russia 27 D17 55 0N 167 0 E
Komandorskiye Ostrova, Russia 27 D17 55 0N 167 0 E
Komárno, Slovak Rep. 17 E10 47 49N 18 5 E
Komatipoort, S. Africa 57 D5 25 25S 31 55 E
Komatou Yialou, Cyprus 23 D13 35 25N 34 8 E
Komatsu, Japan 31 F8 36 25N 136 30 E
Komatsujima, Japan 31 H7 34 0N 134 35 E
Komi □, Russia 24 B10 64 0N 55 0 E
Kommunarsk = Alchevsk, Ukraine 25 E6 48 30N 38 45 E
Kommunizma, Pik, Tajikistan 26 F8 39 0N 72 2 E
Komodo, Indonesia 37 F5 8 37S 119 20 E
Komono, Congo 52 E2 3 10S 13 20 E
Komoran, Pulau, Indonesia 37 F9 8 18S 138 45 E
Komoro, Japan 31 F9 36 19N 138 26 E
Komotini, Greece 21 D11 41 9N 25 26 E
Kompasberg, S. Africa 56 E3 31 45S 24 32 E
Kompong Bang, Cambodia 39 F5 12 24N 104 40 E
Kompong Cham, Cambodia 39 F5 12 0N 105 30 E
Kompong Chhnang, Cambodia 39 F5 12 20N 104 35 E
Kompong Chikreng, Cambodia 38 F5 13 5N 104 18 E
Kompong Kleang, Cambodia 38 F5 13 6N 104 8 E
Kompong Luong, Cambodia 39 G5 11 49N 104 48 E
Kompong Pranak, Cambodia 38 F5 13 35N 104 55 E
Kompong Som, Cambodia 39 G4 10 38N 103 30 E
Kompong Som, Chhung, Cambodia 39 G4 10 50N 103 32 E
Kompong Speu, Cambodia 39 G5 11 26N 104 32 E
Kompong Sralao, Cambodia 38 E5 14 5N 105 46 E
Kompong Thom, Cambodia 38 F5 12 35N 104 51 E
Kompong Trabeck, Cambodia 38 F5 13 6N 105 14 E
Kompong Trabeck, Cambodia 39 G5 11 9N 105 28 E
Kompong Trach, Cambodia 39 G5 11 25N 105 48 E
Kompong Tralach, Cambodia 39 G5 11 54N 104 47 E
Komrat = Comrat, Moldova 17 E15 46 18N 28 40 E
Komsberg, S. Africa 56 E3 32 40S 20 45 E
Komsomolets, Ostrov, Russia 27 A10 80 30N 95 0 E
Komsomolsk, Russia 27 D14 50 30N 137 0 E
Konarhá □, Afghan. 40 B7 35 30N 71 3 E
Konārī, Iran 45 D6 28 13N 51 36 E
Konawa, U.S.A. 81 H6 34 58N 96 45W
Konch, India 43 G8 26 0N 79 10 E
Kondakovo, Russia 27 C16 69 36N 152 0 E
Konde, Tanzania 54 C4 4 57S 39 45 E
Kondinin, Australia 61 F2 32 34S 118 8 E
Kondoa, Tanzania 54 C4 4 55S 35 50 E
Kondoa □, Tanzania 54 D4 5 0S 36 0 E
Kondókali, Greece 23 A3 39 38N 19 51 E
Kondopaga, Russia 24 B5 62 12N 34 17 E
Kondratyevo, Russia 27 D10 57 22N 98 15 E
Konduga, Nigeria 51 F7 11 35N 13 26 E
Köneürgench, Turkmenistan 26 E6 42 19N 59 10 E
Konevo, Russia 24 B6 62 8N 39 20 E
Kong, Ivory C. 50 G4 8 54N 4 36W
Kong →, Cambodia 38 F5 13 32N 105 58 E
Kong, Koh, Cambodia 39 G4 11 20N 103 0 E
Kong Christian IX.s Land, Greenland 4 C6 68 0N 36 0W
Kong Christian X.s Land, Greenland 4 B6 74 0N 29 0W
Kong Franz Joseph Fd., Greenland 4 B6 73 30N 24 30W
Kong Frederik IX.s Land, Greenland 4 C5 67 0N 52 0W
Kong Frederik VI.s Kyst, Greenland 4 C5 63 0N 43 0W
Kong Frederik VIII.s Land, Greenland 4 B6 78 30N 26 0W
Kong Oscar Fjord, Greenland 4 B6 72 20N 24 0W
Kongju, S. Korea 35 F14 36 30N 127 0 E
Konglu, Burma 41 F20 27 13N 97 57 E
Kongolo, Kasai Or., Zaïre 54 D1 5 26S 24 49 E
Kongolo, Shaba, Zaïre 54 D2 5 22S 27 0 E
Kongor, Sudan 51 G11 7 1N 31 27 E
Kongsberg, Norway 9 G13 59 39N 9 39 E
Kongsvinger, Norway 9 F15 60 12N 12 2 E
Kongwa, Tanzania 54 D4 6 11S 36 26 E
Koni, Zaïre 55 E2 10 40S 27 11 E
Koni, Mts., Zaïre 55 E2 10 36S 27 10 E
Königsberg = Kaliningrad, Russia 9 J19 54 42N 20 32 E
Konin, Poland 17 B10 52 12N 18 15 E
Konjic, Bos.-H. 21 C7 43 42N 17 58 E
Konkiep, Namibia 56 D2 26 49S 17 15 E
Konosha, Russia 24 B7 61 0N 40 5 E
Kōnosu, Japan 31 F9 36 3N 139 31 E
Konotop, Ukraine 25 D5 51 12N 33 7 E
Końskie, Poland 17 C11 51 15N 20 23 E
Konstanz, Germany 16 E5 47 40N 9 10 E
Kont, Iran 45 E9 26 55N 61 50 E
Kontagora, Nigeria 50 F6 10 23N 5 27 E
Kontum, Vietnam 38 E7 14 24N 108 0 E
Kontum, Plateau du, Vietnam 38 E7 14 30N 108 30 E
Konya, Turkey 25 G5 37 52N 32 35 E
Konza, Kenya 54 C4 1 45S 37 7 E
Kookynie, Australia 61 E3 29 17S 121 22 E
Kooline, Australia 60 D2 22 57S 116 20 E
Kooloonong, Australia 63 E3 34 48S 143 10 E
Koolyanobbing, Australia 61 F2 30 48S 119 36 E
Koondrook, Australia 63 F3 35 33S 144 8 E
Koonibba, Australia 63 E1 31 54S 133 25 E
Koorawatha, Australia 63 E4 34 2S 148 33 E
Koorda, Australia 61 F2 30 48S 117 35 E
Kooskia, U.S.A. 82 C6 46 9N 115 59W
Kootenai →, Canada 82 B5 49 15N 117 39W
Kootenay L., Canada 72 D5 49 45N 116 50W
Kootenay Nat. Park, Canada 72 C5 51 0N 116 0W
Kootjieskolk, S. Africa 56 E3 31 15S 20 21 E
Kopaonik, Serbia, Yug. 21 C9 43 10N 20 50 E
Kópavogur, Iceland 8 D3 64 6N 21 55W
Koper, Slovenia 16 F7 45 31N 13 44 E
Kopervik, Norway 9 G11 59 17N 5 17 E
Kopeysk, Russia 26 D7 55 7N 61 37 E
Kopi, Australia 63 E2 33 24S 135 40 E
Köping, Sweden 9 G17 59 31N 16 3 E
Koppies, S. Africa 57 D4 27 20S 27 30 E
Koprivnica, Croatia 20 A7 46 12N 16 45 E
Kopychyntsi, Ukraine 17 D13 49 7N 25 58 E
Korab, Macedonia 21 D9 41 44N 20 40 E
Koraküna, Greece 23 A3 39 42N 19 45 E
Korba, India 43 H10 22 20N 82 45 E
Korbu, G., Malaysia 39 K3 4 41N 101 18 E
Korça, Albania 21 D9 40 37N 20 50 E
Korce = Korça, Albania 21 D9 40 37N 20 50 E
Korčula, Croatia 20 C7 42 56N 16 57 E
Kord Kūy, Iran 45 B7 36 48N 54 7 E
Kord Sheykh, Iran 45 D7 28 31N 52 53 E
Kordestān □, Iran 44 C5 36 0N 47 0 E
Kordofân, Sudan 51 F10 13 0N 29 0 E
Korea, North ■, Asia 35 E14 40 0N 127 0 E
Korea, South ■, Asia 35 F15 36 0N 128 0 E

Korea Bay, *Korea* .... **35 E13** 39 0N 124 0 E
Korea Strait, *Asia* .... **35 G15** 34 0N 129 30 E
Korets, *Ukraine* .... **17 C14** 50 40N 27 5 E
Korhogo, *Ivory C.* **50 G3** 9 29N 5 28W
Korim, *Indonesia* .... **37 E9** 0 58S 136 10 E
Korinthiakós Kólpos,
Greece .... **21 E10** 38 16N 22 30 E
Kórinthos, *Greece* .... **21 F10** 37 56N 22 55 E
Koríssa, Límni, *Greece* .. **23 B3** 39 27N 19 53 E
Kōriyama, *Japan* .... **30 F10** 37 24N 140 23 E
Korla, *China* .... **32 B3** 41 45N 86 4 E
Kormakiti, C., *Cyprus* .... **23 D11** 35 23N 32 56 E
Korneshty = Corneşti,
Moldova .... **17 E15** 47 21N 28 1 E
Koro, *Fiji* .... **59 C8** 17 19S 179 23 E
Koro, *Ivory C.* **50 G3** 8 32N 7 30W
Koro, *Mali* **50 F4** 14 1N 2 58W
Koro Sea, *Fiji* **59 C9** 17 30S 179 45W
Korogwe, *Tanzania* ..... **54 D4** 5 5S 38 25 E
Korogwe □, *Tanzania* ... **54 D4** 5 0S 38 20 E
Koroit, *Australia* .... **63 F3** 38 18S 142 24 E
Koror, *Pac. Oc.* .... **37 C8** 7 20N 134 28 E
Körös →, *Hungary* ..... **17 E11** 46 43N 20 12 E
Korosten, *Ukraine* ...... **17 C15** 50 54N 28 36 E
Korostyshev, *Ukraine* .... **17 C15** 50 19N 29 4 E
Korraraika, Helodranon' i,
Madag. .... **57 B7** 17 45S 43 57 E
Korsakov, *Russia* .... **27 E15** 46 36N 142 42 E
Korshunovo, *Russia* ..... **27 D12** 58 37N 110 10 E
Korsør, *Denmark* .... **9 J14** 55 20N 11 9 E
Korti, *Sudan* .... **51 E11** 18 6N 31 33 E
Kortrijk, *Belgium* .... **15 D3** 50 50N 3 17 E
Korwai, *India* .... **42 G8** 24 7N 78 5 E
Koryakskoye Nagorye,
Russia .... **27 C18** 61 0N 171 0 E
Koryŏng, *S. Korea* ...... **35 G15** 35 44N 128 15 E
Kos, *Greece* .... **21 F12** 36 50N 27 15 E
Koschagyl, *Kazakstan* ... **25 E9** 46 40N 54 0 E
Kościan, *Poland* .... **17 B9** 52 5N 16 40 E
Kosciusko, *U.S.A.* **81 J10** 33 4N 89 35W
Kosciusko, Mt., *Australia* **63 F4** 36 27S 148 16 E
Kosciusko I., *U.S.A.* **72 B2** 56 0N 133 40W
Kosha, *Sudan* .... **51 D11** 20 50N 30 30 E
K'oshih = Kashi, *China* .. **32 C2** 39 30N 76 2 E
Koshiki-Rettō, *Japan* .... **31 J4** 31 45N 129 49 E
Kosi, *India* .... **42 F7** 27 48N 77 29 E
Košice, *Slovak Rep.* ..... **17 D11** 48 42N 21 15 E
Koskhinoú, *Greece* ...... **23 C10** 36 23N 28 13 E
Koslan, *Russia* .... **24 B8** 63 34N 49 14 E
Kosŏng, *N. Korea* ...... **35 E15** 38 40N 128 22 E
Kosovo □, *Serbia, Yug.* ... **21 C9** 42 30N 21 0 E
Kosovska-Mitrovica =
Titova-Mitrovica,
Serbia, Yug. .... **21 C9** 42 54N 20 52 E
Kostamuksa, *Russia* ..... **24 B5** 62 34N 32 44 E
Koster, *S. Africa* .... **56 D4** 25 52S 26 54 E
Kôstî, *Sudan* .... **51 F11** 13 8N 32 43 E
Kostopil, *Ukraine* .... **17 C14** 50 51N 26 22 E
Kostroma, *Russia* .... **24 C7** 57 50N 40 58 E
Kostrzyn, *Poland* .... **16 B8** 52 35N 14 39 E
Koszalin, *Poland* .... **16 A9** 54 11N 16 8 E
Kot Addu, *Pakistan* ..... **42 D4** 30 30N 71 0 E
Kot Moman, *Pakistan* ... **42 C5** 32 13N 73 0 E
Kota, *India* .... **42 G6** 25 14N 75 49 E
Kota Baharu, *Malaysia* .. **39 J4** 6 7N 102 14 E
Kota Belud, *Malaysia* ... **36 C5** 6 21N 116 26 E
Kota Kinabalu, *Malaysia* . **36 C5** 6 0N 116 4 E
Kota Tinggi, *Malaysia* ... **39 M4** 1 44N 103 53 E
Kotaagung, *Indonesia* ... **36 F2** 5 38S 104 29 E
Kotabaru, *Indonesia* .... **36 E5** 3 20S 116 20 E
Kotabumi, *Indonesia* .... **36 E2** 4 49S 104 54 E
Kotagede, *Indonesia* .... **37 G14** 7 54S 110 26 E
Kotamobagu, *Indonesia* .. **37 D6** 0 57N 124 31 E
Kotaneelee →, *Canada* .. **72 A4** 60 11N 123 42W
Kotawaringin, *Indonesia* . **36 E4** 2 28S 111 27 E
Kotcho L., *Canada* **72 B4** 59 7N 121 12W
Kotelnich, *Russia* ...... **24 C8** 58 22N 48 24 E
Kotelnikovo, *Russia* ..... **26 E5** 47 38N 43 8 E
Kotelnyy, Ostrov, *Russia* **27 B14** 75 10N 139 0 E
Kothi, *India* .... **43 G9** 24 45N 80 40 E
Kotiro, *Pakistan* .... **42 F2** 26 17N 67 13 E
Kotka, *Finland* .... **9 F22** 60 28N 26 58 E
Kotlas, *Russia* .... **24 B8** 61 17N 46 43 E
Kotli, *Pakistan* .... **42 C5** 33 30N 73 55 E
Kotmul, *Pakistan* .... **43 B6** 35 32N 75 10 E
Kotor, *Montenegro, Yug.* **21 C8** 42 25N 18 47 E
Kotovsk, *Ukraine* .... **17 E15** 47 45N 29 35 E
Kotputli, *India* .... **42 F7** 27 43N 76 12 E
Kotri, *Pakistan* .... **42 G3** 25 22N 68 22 E
Kottayam, *India* .... **40 Q10** 9 35N 76 33 E
Kotturu, *India* .... **40 M10** 14 45N 76 10 E
Kotuy →, *Russia* ...... **27 B11** 71 54N 102 6 E
Kotzebue, *U.S.A.* **68 B3** 66 53N 162 39W
Kouango, *C.A.R.* .... **52 C4** 5 0N 20 10 E
Koudougou, *Burkina Faso* **50 F4** 12 10N 2 20W
Koufonísi, *Greece* ...... **23 E8** 34 56N 26 8 E
Kougaberge, *S. Africa* ... **56 E3** 33 48S 23 50 E
Kouilou →, *Congo* ..... **52 E2** 4 10S 12 5 E
Kouki, *C.A.R.* .... **52 C3** 7 22N 17 3 E
Koula Moutou, *Gabon* ... **52 E2** 1 15S 12 25 E
Koulen, *Cambodia* ...... **38 F5** 13 50N 104 40 E
Koulikoro, *Mali* **50 F3** 12 40N 7 50W
Kouloúra, *Greece* .... **23 A3** 39 42N 19 54 E
Koúm-bournoú, Ákra,
Greece .... **23 C10** 36 15N 28 11 E
Koumala, *Australia* ..... **62 C4** 21 38S 149 15 E
Koumra, *Chad* .... **51 G8** 8 50N 17 35 E
Kounradskiy, *Kazakstan* . **26 E8** 46 59N 75 0 E
Kountze, *U.S.A.* **81 K7** 30 22N 94 19W
Kouris →, *Cyprus* ..... **23 E11** 34 38N 32 54 E
Kouroussa, *Guinea* **50 F3** 10 45N 9 45W
Kousseri, *Cameroon* .... **51 F7** 12 0N 14 55 E
Koutiala, *Mali* **50 F3** 12 25N 5 23W
Kouvola, *Finland* .... **9 F22** 60 52N 26 43 E
Kovdor, *Russia* .... **24 A5** 67 34N 30 24 E
Kovel, *Ukraine* .... **17 C13** 51 11N 24 38 E
Kovrov, *Russia* .... **24 C7** 56 25N 41 25 E
Kowanyama, *Australia* .. **62 B3** 15 29S 141 44 E
Kowkash, *Canada* **70 B2** 50 20N 87 12W
Kowŏn, *N. Korea* .... **35 E14** 39 26N 127 14 E
Köyceğiz, *Turkey* .... **21 F13** 36 57N 28 40 E
Koyuk, *U.S.A.* **68 B3** 64 56N 161 9W
Koyukuk →, *U.S.A.* **68 B4** 64 55N 157 32W
Koza, *Japan* .... **31 L3** 26 19N 127 46 E
Kozáni, *Greece* .... **21 D9** 40 19N 21 47 E
Kozhikode = Calicut, *India* **40 P9** 11 15N 75 43 E
Kozhva, *Russia* .... **24 A10** 65 10N 57 0 E
Kozyatyn, *Ukraine* ...... **17 D15** 49 45N 28 50 E
Kpalimé, *Togo* .... **50 G5** 6 57N 0 44 E
Kra, Isthmus of = Kra,
Kho Khot, *Thailand* ... **39 G2** 10 15N 99 30 E
Kra, Kho Khot, *Thailand* . **39 G2** 10 15N 99 30 E
Kra Buri, *Thailand* ...... **39 G2** 10 22N 98 46 E
Krabi, *Thailand* .... **39 H2** 8 4N 98 55 E
Kragan, *Indonesia* ...... **37 G14** 6 43S 111 38 E
Kragerø, *Norway* .... **9 G13** 58 52N 9 25 E
Kragujevac, *Serbia, Yug.* . **21 B9** 44 2N 20 56 E
Krajina, *Bos.-H.* .... **20 B7** 44 45N 16 35 E
Krakatau = Rakata, Pulau,
Indonesia .... **36 F3** 6 10S 105 20 E
Krakor, *Cambodia* ...... **38 F5** 12 32N 104 12 E
Kraków, *Poland* .... **17 C10** 50 4N 19 57 E
Kraksaan, *Indonesia* .... **37 G15** 7 43S 113 23 E
Kralanh, *Cambodia* ..... **38 F4** 13 35N 103 25 E
Kraljevo, *Serbia, Yug.* ... **21 C9** 43 44N 20 41 E
Kramatorsk, *Ukraine* .... **25 E6** 48 50N 37 30 E
Kramfors, *Sweden* ...... **9 E17** 62 55N 17 48 E
Kranj, *Slovenia* .... **16 E8** 46 16N 14 22 E
Krankskop, *S. Africa* .... **57 D5** 28 0S 30 47 E
Krasavino, *Russia* ...... **24 B8** 60 58N 46 29 E
Kraskino, *Russia* .... **27 E14** 42 44N 130 48 E
Kraśnik, *Poland* .... **17 C12** 50 55N 22 5 E
Krasnoarmeysk, *Russia* .. **26 D5** 51 0N 45 42 E
Krasnodar, *Russia* ...... **25 E6** 45 5N 39 0 E
Krasnokamsk, *Russia* ... **24 C10** 58 4N 55 48 E
Krasnoperekopsk, *Ukraine* **25 E5** 46 0N 33 54 E
Krasnorechenskiy, *Russia* **30 B7** 44 41N 135 14 E
Krasnoselkupsk, *Russia* .. **26 C9** 65 20N 82 10 E
Krasnoturinsk, *Russia* ... **24 C11** 59 46N 60 12 E
Krasnoufimsk, *Russia* ... **24 C10** 56 36N 57 38 E
Krasnouralsk, *Russia* ... **24 C11** 58 21N 60 3 E
Krasnovishersk, *Russia* .. **24 B10** 60 23N 57 3 E
Krasnovodsk =
Türkmenbashi,
Turkmenistan .... **25 F9** 40 5N 53 5 E
Krasnoyarsk, *Russia* .... **27 D10** 56 8N 93 0 E
Krasnyy Luch, *Ukraine* .. **25 E6** 48 13N 39 0 E
Krasnyy Yar, *Russia* .... **25 E8** 46 43N 48 23 E
Kratie, *Cambodia* .... **38 F6** 12 32N 106 10 E
Krau, *Indonesia* .... **37 E10** 3 19S 140 5 E
Kravanh, Chuor Phnum,
Cambodia .... **39 G4** 12 0N 103 32 E
Krefeld, *Germany* .... **16 C4** 51 20N 6 33 E
Kremen, *Croatia* .... **16 F8** 44 28N 15 53 E
Kremenchug =
Kremenchuk, *Ukraine* . **25 E5** 49 5N 33 25 E
Kremenchuk, *Ukraine* ... **25 E5** 49 5N 33 25 E
Kremenchuksk Vdskh.,
Ukraine .... **25 E5** 49 20N 32 30 E
Kremenets, *Ukraine* ..... **17 C13** 50 8N 25 43 E
Kremmling, *U.S.A.* **82 F10** 40 4N 106 24W
Krems, *Austria* .... **16 D8** 48 25N 15 36 E
Kretinga, *Lithuania* ..... **9 J19** 55 53N 21 15 E
Kribi, *Cameroon* .... **52 D1** 2 57N 9 56 E
Krichev = Krychaw,
Belarus .... **17 B16** 53 40N 31 41 E
Kriós, Ákra, *Greece* ..... **23 D5** 35 13N 23 34 E
Krishna →, *India* ...... **41 M12** 15 57N 80 59 E
Krishnanagar, *India* ..... **43 H13** 23 24N 88 33 E
Kristiansand, *Norway* ... **9 G13** 58 8N 8 1 E
Kristianstad, *Sweden* ... **9 H16** 56 2N 14 9 E
Kristiansund, *Norway* ... **8 E12** 63 7N 7 45 E
Kristiinankaupunki,
Finland .... **9 E19** 62 16N 21 21 E
Kristinehamn, *Sweden* .. **9 G16** 59 18N 14 13 E
Kristinestad =
Kristiinankaupunki,
Finland .... **9 E19** 62 16N 21 21 E
Kriti, *Greece* .... **23 D7** 35 15N 25 0 E
Kritsá, *Greece* .... **23 D7** 35 10N 25 41 E
Krivoy Rog = Kryvyy Rih,
Ukraine .... **25 E5** 47 51N 33 20 E
Krk, *Croatia* .... **16 F8** 45 8N 14 40 E
Krokodil →, *Mozam.* ... **57 D5** 25 14S 32 18 E
Kronprins Olav Kyst,
Antarctica .... **5 C5** 69 0S 42 0 E
Kronshtadt, *Russia* ..... **24 B4** 59 57N 29 51 E
Kroonstad, *S. Africa* .... **56 D4** 27 43S 27 19 E
Kropotkin, *Irkutsk, Russia* **27 D12** 59 0N 115 30 E
Kropotkin, *Krasnodar,*
Russia .... **25 E7** 45 28N 40 28 E
Krosno, *Poland* .... **17 D11** 49 42N 21 46 E
Krotoszyn, *Poland* ...... **17 C9** 51 42N 17 23 E
Kroussón, *Greece* ...... **23 D6** 35 13N 24 59 E
Kruger Nat. Park, *S. Africa* **57 C5** 23 30S 31 40 E
Krugersdorp, *S. Africa* ... **57 D4** 26 5S 27 46 E
Kruisfontein, *S. Africa* ... **56 E3** 33 59S 24 43 E
Krung Thep = Bangkok,
Thailand .... **38 F3** 13 45N 100 35 E
Krupki, *Belarus* .... **17 A15** 54 19N 29 8 E
Kruševac, *Serbia, Yug.* .. **21 C9** 43 35N 21 28 E
Kruzof I., *U.S.A.* **72 B1** 57 10N 135 40W
Krychaw, *Belarus* .... **17 B16** 53 40N 31 41 E
Krymskiy Poluostrov =
Krymskyy Pivostriv,
Ukraine .... **25 E5** 45 0N 34 0 E
Krymskyy Pivostriv,
Ukraine .... **25 E5** 45 0N 34 0 E
Kryvyy Rih, *Ukraine* .... **25 E5** 47 51N 33 20 E
Ksar el Boukhari, *Algeria* **50 A5** 35 51N 2 52 E
Ksar el Kebir, *Morocco* .. **50 B3** 35 0N 6 0W
Ksar es Souk = Ar
Rachidiya, *Morocco* ... **50 B4** 31 58N 4 20W
Kuala, *Indonesia* .... **36 D3** 2 55N 105 47 E
Kuala Berang, *Malaysia* . **39 K4** 5 5N 103 1 E
Kuala Dungun, *Malaysia* . **39 K4** 4 45N 103 25 E
Kuala Kangsar, *Malaysia* . **39 K3** 4 46N 100 56 E
Kuala Kelawang, *Malaysia* **39 L4** 2 56N 102 5 E
Kuala Kerai, *Malaysia* ... **39 K4** 5 30N 102 12 E
Kuala Kubu Baharu,
Malaysia .... **39 L3** 3 34N 101 39 E
Kuala Lipis, *Malaysia* ... **39 K4** 4 10N 102 3 E
Kuala Lumpur, *Malaysia* . **39 L3** 3 9N 101 41 E
Kuala Nerang, *Malaysia* . **39 J3** 6 16N 100 37 E
Kuala Pilah, *Malaysia* ... **39 L4** 2 45N 102 15 E
Kuala Rompin, *Malaysia* . **39 L4** 2 49N 103 29 E
Kuala Selangor, *Malaysia* **39 L3** 3 20N 101 15 E
Kuala Terengganu,
Malaysia .... **39 K4** 5 20N 103 8 E
Kualajelai, *Indonesia* .... **36 E4** 2 58S 110 46 E
Kualakapuas, *Indonesia* . **36 E4** 2 55S 114 20 E
Kualakurun, *Indonesia* ... **36 E4** 1 10S 113 50 E
Kualapembuang,
Indonesia .... **36 E4** 3 14S 112 38 E
Kualasimpang, *Indonesia* **36 D1** 4 17N 98 3 E
Kuancheng, *China* ...... **35 D10** 40 37N 118 30 E
Kuandang, *Indonesia* ... **37 D6** 0 56N 123 1 E
Kuandian, *China* .... **35 D13** 40 45N 124 45 E
Kuangchou = Guangzhou,
China .... **33 D6** 23 5N 113 10 E
Kuantan, *Malaysia* ...... **39 L4** 3 49N 103 20 E
Kuba = Quba, *Azerbaijan* **25 F8** 41 21N 48 32 E
Kuban →, *Russia* ...... **25 E6** 45 20N 37 30 E
Kubokawa, *Japan* .... **31 H6** 33 12N 133 8 E
Kucha Gompa, *India* .... **43 B7** 34 25N 76 56 E
Kuchaman, *India* .... **42 F6** 27 13N 74 47 E
Kuchino-eruba-Jima,
Japan .... **31 J5** 30 28N 130 12 E
Kuchino-Shima, *Japan* .. **31 K4** 29 57N 129 55 E
Kuchinotsu, *Japan* ...... **31 H5** 32 36N 130 11 E
Kucing, *Malaysia* .... **36 D4** 1 33N 110 25 E
Kud →, *Pakistan* ...... **42 F2** 26 5N 66 20 E
Kuda, *India* .... **40 H7** 23 10N 71 15 E
Kudat, *Malaysia* .... **36 C5** 6 55N 116 55 E
Kudus, *Indonesia* .... **37 G14** 6 48S 110 51 E
Kudymkar, *Russia* ...... **26 D6** 59 1N 54 39 E
Kueiyang = Guiyang,
China .... **32 D5** 26 32N 106 40 E
Kufra Oasis = Al Kufrah,
Libya .... **51 D9** 24 17N 23 15 E
Kufstein, *Austria* .... **16 E7** 47 35N 12 11 E
Kuglugtuk, *Canada* **68 B8** 67 50N 115 5W
Kugong I., *Canada* **70 A4** 56 18N 79 50W
Kūh-e-Hazārām, *Iran* .... **45 D8** 29 35N 57 20 E
Kūhak, *Iran* .... **40 F3** 27 12N 63 10 E
Kūhbonān, *Iran* .... **45 D8** 31 23N 56 19 E
Kūhestak, *Iran* .... **45 E8** 26 47N 57 2 E
Kūhīn, *Iran* .... **45 C6** 35 13N 48 25 E
Kūhīrī, *Iran* .... **45 E9** 26 55N 61 2 E
Kūhpāyeh, *Eşfahan, Iran* . **45 C7** 32 44N 52 20 E
Kūhpāyeh, *Kermān, Iran* . **45 D8** 30 35N 57 15 E
Kui Buri, *Thailand* ...... **39 F2** 12 3N 99 52 E
Kuito, *Angola* .... **53 G3** 12 22S 16 55 E
Kujang, *N. Korea* .... **35 E14** 39 57N 126 1 E
Kuji, *Japan* .... **30 D10** 40 11N 141 46 E
Kujū-San, *Japan* .... **31 H5** 33 5N 131 15 E
Kukawa, *Nigeria* .... **51 F7** 12 58N 13 27 E
Kukerin, *Australia* ...... **61 F2** 33 13S 118 0 E
Kukësi, *Albania* .... **21 C9** 42 5N 20 20 E
Kukup, *Malaysia* .... **39 M4** 1 20N 103 27 E
Kula, *Turkey* .... **21 E13** 38 32N 28 40 E
Kulai, *Malaysia* .... **39 M4** 1 44N 103 35 E
Kulal, Mt., *Kenya* ...... **54 B4** 2 42N 36 57 E
Kulasekarappattinam,
India .... **40 Q11** 8 20N 78 5 E
Kuldiga, *Latvia* .... **9 H19** 56 58N 21 59 E
Kuldja = Yining, *China* .. **26 E9** 43 58N 81 10 E
Kulgam, *India* .... **43 C6** 33 36N 75 2 E
Kulim, *Malaysia* .... **39 K3** 5 22N 100 34 E
Kulin, *Australia* .... **61 F2** 32 40S 118 2 E
Kulja, *Australia* .... **61 F2** 30 28S 117 18 E
Kulm, *U.S.A.* **80 B5** 46 18N 98 57W
Kŭlob, *Tajikistan* .... **26 F7** 37 55N 69 50 E
Kulsary, *Kazakstan* ..... **25 E9** 46 59N 54 1 E
Kulti, *India* .... **43 H12** 23 43N 86 50 E
Kulumbura, *Australia* ... **60 B4** 13 55S 126 35 E
Kulunda, *Russia* .... **26 D8** 52 35N 78 57 E
Kulungar, *Afghan.* ...... **42 C3** 34 0N 69 2 E
Kūlvand, *Iran* .... **45 D7** 31 21N 54 35 E
Kulwin, *Australia* .... **63 F3** 35 0S 142 42 E
Kulyab = Kŭlob, *Tajikistan* **26 F7** 37 55N 69 50 E
Kum Tekei, *Kazakstan* ... **26 E8** 43 10N 79 30 E
Kuma →, *Russia* .... **25 F8** 44 55N 47 0 E
Kumagaya, *Japan* ...... **31 F9** 36 9N 139 22 E
Kumai, *Indonesia* .... **36 E4** 2 44S 111 43 E
Kumamba, Kepulauan,
Indonesia .... **37 E9** 1 36S 138 45 E
Kumamoto, *Japan* ...... **31 H5** 32 45N 130 45 E
Kumamoto □, *Japan* .... **31 H5** 32 55N 130 55 E
Kumanovo, *Macedonia* .. **21 C9** 42 9N 21 42 E
Kumara, *N.Z.* .... **59 K3** 42 37S 171 12 E
Kumarl, *Australia* ...... **61 F3** 32 47S 121 33 E
Kumasi, *Ghana* **50 G4** 6 41N 1 38W
Kumayri = Gyumri,
Armenia .... **25 F7** 40 47N 43 50 E
Kumba, *Cameroon* ..... **50 H6** 4 36N 9 24 E
Kumbakonam, *India* .... **40 P11** 10 58N 79 25 E
Kumbarilla, *Australia* ... **63 D5** 27 15S 150 55 E
Kŭmchŏn, *N. Korea* .... **35 E14** 38 10N 126 29 E
Kumdok, *India* .... **43 C8** 33 32N 78 10 E
Kume-Shima, *Japan* .... **31 L3** 26 20N 126 47 E
Kumertau, *Russia* ...... **24 D10** 52 45N 55 57 E
Kŭmhwa, *S. Korea* ..... **35 E14** 38 17N 127 28 E
Kumi, *Uganda* .... **54 B3** 1 30N 33 58 E
Kumla, *Sweden* .... **9 G16** 59 8N 15 10 E
Kumo, *Nigeria* .... **50 F7** 10 1N 11 12 E
Kumon Bum, *Burma* .... **41 F20** 26 30N 97 15 E
Kunama, *Australia* ..... **63 F4** 35 35S 148 4 E
Kunashir, Ostrov, *Russia* **27 E15** 44 0N 146 0 E
Kunda, *Estonia* .... **9 G22** 59 30N 26 34 E
Kundla, *India* .... **42 J4** 21 21N 71 25 E
Kungala, *Australia* ..... **63 D5** 29 58S 153 7 E
Kunghit I., *Canada* **72 C2** 52 6N 131 3W
Kungrad = Qŭnghirot,
Uzbekistan .... **26 E6** 43 6N 58 54 E
Kungsbacka, *Sweden* ... **9 H15** 57 30N 12 5 E
Kungur, *Russia* .... **24 C10** 57 25N 56 57 E
Kungurri, *Australia* .... **62 C4** 21 3S 148 46 E
Kunhar →, *Pakistan* ... **43 B5** 34 20N 73 30 E
Kuningan, *Indonesia* ... **37 G13** 6 59S 108 29 E
Kunlong, *Burma* .... **41 H21** 23 20N 98 50 E
Kunlun Shan, *Asia* ..... **32 C3** 36 0N 86 30 E
Kunming, *China* .... **32 D5** 25 1N 102 41 E
Kunsan, *S. Korea* ...... **35 G14** 35 59N 126 45 E
Kununurra, *Australia* ... **60 C4** 15 40S 128 50 E
Kunwarara, *Australia* ... **62 C5** 22 55S 150 9 E
Kunya-Urgench =
Köneürgench,
Turkmenistan .... **26 E6** 42 19N 59 10 E
Kuopio, *Finland* .... **8 E22** 62 53N 27 35 E
Kupa →, *Croatia* ...... **16 F9** 45 28N 16 24 E
Kupang, *Indonesia* ..... **37 F6** 10 19S 123 39 E
Kupyansk, *Ukraine* ..... **26 E4** 49 52N 37 35 E
Kuqa, *China* .... **32 B3** 41 35N 82 30 E
Kür →, *Azerbaijan* ..... **25 G8** 39 29N 49 15 E
Kura = Kür →,
Azerbaijan .... **25 G8** 39 29N 49 15 E
Kuranda, *Australia* ..... **62 B4** 16 48S 145 35 E
Kurashiki, *Japan* .... **31 G6** 34 40N 133 50 E
Kurayoshi, *Japan* ...... **31 G6** 35 26N 133 50 E
Kŭrdzhali, *Bulgaria* .... **21 D11** 41 38N 25 21 E
Kure, *Japan* .... **31 G6** 34 14N 132 32 E
Kuressaare, *Estonia* .... **9 G20** 58 15N 22 30 E
Kurgaldzhinskiy, *Kazakstan* **26 D8** 50 35N 70 20 E
Kurgan, *Russia* .... **26 D7** 55 26N 65 18 E
Kuria Maria Is. = Khūrīyā
Mūrīyā, Jazā 'ir, *Oman* **46 D6** 17 30N 55 58 E
Kuridala, *Australia* ..... **62 C3** 21 16S 140 29 E
Kurigram, *Bangla.* ...... **41 G16** 25 49N 89 39 E
Kurikka, *Finland* .... **9 E20** 62 36N 22 24 E
Kuril Is. = Kurilskiye
Ostrova, *Russia* ...... **27 E15** 45 0N 150 0 E
Kuril Trench, *Pac. Oc.* ... **28 E19** 44 0N 153 0 E
Kurilsk, *Russia* .... **27 E15** 45 14N 147 53 E
Kurilskiye Ostrova, *Russia* **27 E15** 45 0N 150 0 E
Kurino, *Japan* .... **31 J5** 31 57N 130 43 E
Kurmuk, *Sudan* .... **51 F11** 10 33N 34 21 E
Kurnool, *India* .... **40 M10** 15 45N 78 0 E
Kuro-Shima, *Kagoshima,*
Japan .... **31 J4** 30 50N 129 57 E
Kuro-Shima, *Okinawa,*
Japan .... **31 M2** 24 14N 124 1 E
Kurow, *N.Z.* .... **59 L3** 44 44S 170 29 E
Kurrajong, *Australia* .... **63 E5** 33 33S 150 42 E
Kurram →, *Pakistan* ... **42 C4** 32 36N 71 20 E
Kurri Kurri, *Australia* ... **63 E5** 32 50S 151 28 E
Kurshskiy Zaliv, *Russia* .. **9 J19** 55 9N 21 6 E
Kursk, *Russia* .... **24 D6** 51 42N 36 11 E
Kuruktag, *China* .... **32 B3** 41 0N 89 0 E
Kuruman, *S. Africa* ..... **56 D3** 27 28S 23 28 E
Kuruman →, *S. Africa* .. **56 D3** 26 56S 20 39 E
Kurume, *Japan* .... **31 H5** 33 15N 130 30 E
Kurunegala, *Sri Lanka* ... **40 R12** 7 30N 80 23 E
Kurya, *Russia* .... **27 C11** 61 15N 108 10 E
Kus Gölü, *Turkey* .... **21 D12** 40 10N 27 55 E
Kuşadası, *Turkey* .... **21 F12** 37 52N 27 15 E
Kusatsu, *Japan* .... **31 F9** 36 37N 138 36 E
Kusawa L., *Canada* **72 A1** 60 20N 136 13W
Kushikino, *Japan* .... **31 J5** 31 44N 130 16 E
Kushima, *Japan* .... **31 J5** 31 29N 131 14 E
Kushimoto, *Japan* ...... **31 H7** 33 28N 135 47 E
Kushiro, *Japan* .... **30 C12** 43 0N 144 25 E
Kushiro →, *Japan* ..... **30 C12** 42 59N 144 23 E
Kūshk, *Iran* .... **45 D8** 28 46N 56 51 E
Kushka = Gushgy,
Turkmenistan .... **26 F7** 35 20N 62 18 E
Kūshki, *Īlām, Iran* .... **44 C5** 33 31N 47 13 E
Kūshki, *Khorāsān, Iran* .. **45 B8** 37 2N 57 26 E
Kūshkū, *Iran* .... **45 E7** 27 19N 53 28 E
Kushol, *India* .... **43 C7** 33 40N 76 36 E
Kushtia, *Bangla.* .... **41 H16** 23 55N 89 5 E
Kushva, *Russia* .... **24 C10** 58 18N 59 45 E
Kuskokwim →, *U.S.A.* .. **68 B3** 60 5N 162 25W
Kuskokwim B., *U.S.A.* ... **68 C3** 59 45N 162 25W
Kussharo-Ko, *Japan* .... **30 C12** 43 38N 144 21 E
Kustanay = Qostanay,
Kazakstan .... **26 D7** 53 10N 63 35 E
Kut, Ko, *Thailand* .... **39 G4** 11 40N 102 35 E
Kütahya, *Turkey* .... **25 G5** 39 30N 30 2 E
Kutaisi, *Georgia* .... **25 F7** 42 19N 42 40 E
Kutaraja = Banda Aceh,
Indonesia .... **36 C1** 5 35N 95 20 E
Kutch, Gulf of = Kachchh,
Gulf of, *India* .... **42 H3** 22 50N 69 15 E
Kutch, Rann of =
Kachchh, Rann of, *India* **42 G4** 24 0N 70 0 E
Kutiyana, *India* .... **42 J4** 21 36N 70 2 E
Kutno, *Poland* .... **17 B10** 52 15N 19 23 E
Kuttabul, *Australia* ..... **62 C4** 21 5S 148 48 E
Kutu, *Zaïre* .... **52 E3** 2 40S 18 11 E
Kutum, *Sudan* .... **51 F9** 14 10N 24 40 E
Kuujjuaq, *Canada* **69 C13** 58 6N 68 15W
Kuŭp-tong, *N. Korea* ... **35 D14** 40 45N 126 1 E
Kuusamo, *Finland* ...... **8 D23** 65 57N 29 8 E
Kuusankoski, *Finland* ... **9 F22** 60 55N 26 38 E
Kuwait = Al Kuwayt,
Kuwait .... **44 D5** 29 30N 48 0 E
Kuwait ■, *Asia* .... **44 D5** 29 30N 47 30 E
Kuwana, *Japan* .... **31 G8** 35 5N 136 43 E
Kuybyshev = Samara,
Russia .... **24 D9** 53 8N 50 6 E
Kuybyshev, *Russia* ..... **26 D8** 55 27N 78 19 E
Kuybyshevskoye Vdkhr.,
Russia .... **24 C8** 55 2N 49 30 E
Kuye He →, *China* ..... **34 E6** 38 23N 110 46 E
Kūyeh, *Iran* .... **44 B5** 38 45N 47 57 E
Kūysanjaq, *Iraq* .... **44 B5** 36 5N 44 38 E
Kuyto, Ozero, *Russia* ... **24 B5** 65 6N 31 20 E
Kuyumba, *Russia* .... **27 C10** 60 58N 96 59 E
Kuzey Anadolu Dağlan,
Turkey .... **25 F6** 41 30N 35 0 E
Kuznetsk, *Russia* .... **24 D8** 53 12N 46 40 E
Kuzomen, *Russia* .... **24 A6** 66 22N 36 50 E
Kvænangen, *Norway* .... **8 A19** 70 5N 21 15 E
Kvaløy, *Norway* .... **8 B18** 69 40N 18 30 E
Kvarner, *Croatia* .... **16 F8** 44 50N 14 10 E
Kvarnerič, *Croatia* ..... **16 F8** 44 43N 14 37 E
Kwabhaca, *S. Africa* .... **57 E4** 30 51S 29 0 E
Kwadacha →, *Canada* .. **72 B3** 57 28N 125 38W
Kwakhanai, *Botswana* ... **56 C3** 21 39S 21 16 E
Kwakoegron, *Surinam* ... **93 B7** 5 12N 55 25W
Kwale, *Kenya* .... **54 C4** 4 15S 39 31 E
Kwale □, *Kenya* .... **54 C4** 4 15S 39 10 E
KwaMashu, *S. Africa* ... **57 D5** 29 45S 30 58 E
Kwamouth, *Zaïre* .... **52 E3** 3 9S 16 12 E
Kwando →, *Africa* ..... **56 B3** 18 27S 23 32 E
Kwangdaeri, *N. Korea* ... **35 D14** 40 31N 127 32 E
Kwangju, *S. Korea* ..... **35 G14** 35 9N 126 54 E
Kwango →, *Zaïre* ...... **49 G5** 3 14S 17 22 E
Kwangsi-Chuang =
Guangxi Zhuangzu
Zizhiqu □, *China* ..... **33 D5** 24 0N 109 0 E
Kwangtung =
Guangdong □, *China* .. **33 D6** 23 0N 113 0 E

Kwataboahegan —►, *Canada* **70 B3** 51 9N 80 50W
Kwatisore, *Indonesia* **37 E8** 3 18S 134 50 E
KwaZulu Natal □, *S. Africa* **57 D5** 29 0S 30 0 E
Kweichow = Guizhou □, *China* **32 D5** 27 0N 107 0 E
Kwekwe, *Zimbabwe* **55 F2** 18 58S 29 48 E
Kwidzyn, *Poland* **17 B10** 53 44N 18 55 E
Kwimba □, *Tanzania* **54 C3** 3 0S 33 0 E
Kwinana New Town, *Australia* **61 F2** 32 15S 115 47 E
Kwoka, *Indonesia* **37 E8** 0 31S 132 27 E
Kyabé, *Chad* **51 G8** 9 30N 19 0 E
Kyabra Cr. —►, *Australia* **63 D3** 25 36S 142 55 E
Kyabram, *Australia* **63 F4** 36 19S 145 4 E
Kyaikto, *Burma* **38 D1** 17 20N 97 3 E
Kyakhta, *Russia* **27 D11** 50 30N 106 25 E
Kyancutta, *Australia* **63 E2** 33 8S 135 33 E
Kyangin, *Burma* **41 K19** 18 20N 95 20 E
Kyaukpadaung, *Burma* **41 J19** 20 52N 95 8 E
Kyaukpyu, *Burma* **41 K18** 19 28N 93 30 E
Kyaukse, *Burma* **41 J20** 21 36N 96 10 E
Kyburz, *U.S.A.* **84 G6** 38 47N 120 18W
Kyenjojo, *Uganda* **54 B3** 0 40N 30 37 E
Kyle Dam, *Zimbabwe* **55 G3** 20 15S 31 0 E
Kyle of Lochalsh, *U.K.* **12 D3** 57 17N 5 44W
Kymijoki —►, *Finland* **9 F22** 60 30N 26 55 E
Kyneton, *Australia* **63 F3** 37 10S 144 29 E
Kynuna, *Australia* **62 C3** 21 37S 141 55 E
Kyō-ga-Saki, *Japan* **31 G7** 35 45N 135 15 E
Kyoga, L., *Uganda* **54 B3** 1 35N 33 0 E
Kyogle, *Australia* **63 D5** 28 40S 153 0 E
Kyongju, *S. Korea* **35 G15** 35 51N 129 14 E
Kyongpyaw, *Burma* **41 L19** 17 12N 95 10 E
Kyŏngsŏng, *N. Korea* **35 D15** 41 35N 129 36 E
Kyōto, *Japan* **31 G7** 35 0N 135 45 E
Kyōto □, *Japan* **31 G7** 35 15N 135 45 E
Kyparissovouno, *Cyprus* **23 D12** 35 19N 33 10 E
Kyperounda, *Cyprus* **23 E11** 34 56N 32 58 E
Kyren, *Russia* **27 D11** 51 45N 101 45 E
Kyrenia, *Cyprus* **23 D12** 35 20N 33 20 E
Kyrgyzstan ■, *Asia* **26 E8** 42 0N 75 0 E
Kyrönjoki —►, *Finland* **8 E19** 63 14N 21 45 E
Kyrtylakh, *Russia* **27 C13** 65 30N 123 40 E
Kystatyam, *Russia* **27 C13** 67 20N 123 10 E
Kythréa, *Cyprus* **23 D12** 35 15N 33 29 E
Kyulyunken, *Russia* **27 C14** 64 10N 137 5 E
Kyunhla, *Burma* **41 H19** 23 25N 95 15 E
Kyuquot, *Canada* **72 C3** 50 3N 127 25W
Kyūshū, *Japan* **31 H5** 33 0N 131 0 E
Kyūshū □, *Japan* **31 H5** 33 0N 131 0 E
Kyūshū-Sanchi, *Japan* **31 H5** 32 35N 131 17 E
Kyustendil, *Bulgaria* **21 C10** 42 16N 22 41 E
Kyusyur, *Russia* **27 B13** 70 19N 127 30 E
Kywong, *Australia* **63 E4** 34 58S 146 44 E
Kyyiv, *Ukraine* **17 C16** 50 30N 30 28 E
Kyyivske Vdskh., *Ukraine* **17 C16** 51 0N 30 25 E
Kyzyl, *Russia* **27 D10** 51 50N 94 30 E
Kyzyl Kum, *Uzbekistan* **26 E7** 42 30N 65 0 E
Kyzyl-Kyya, *Kyrgyzstan* **26 E8** 40 16N 72 8 E
Kzyl-Orda = Qyzylorda, *Kazakstan* **26 E7** 44 48N 65 28 E

L

La Albufera, *Spain* **19 C5** 39 20N 0 27W
La Alcarria, *Spain* **19 B4** 40 31N 2 45W
La Asunción, *Venezuela* **92 A6** 11 2N 63 53W
La Banda, *Argentina* **94 B3** 27 45S 64 10W
La Barca, *Mexico* **86 C4** 20 20N 102 40W
La Barge, *U.S.A.* **82 E8** 42 16N 110 12W
La Belle, *U.S.A.* **77 M5** 26 46N 81 26W
La Biche —►, *Canada* **72 B4** 59 57N 123 50W
La Bomba, *Mexico* **86 A1** 31 53N 115 2W
La Calera, *Chile* **94 C1** 32 50S 71 10W
La Canal, *Spain* **22 C7** 38 51N 1 23 E
La Carlota, *Argentina* **94 C3** 33 30S 63 20W
La Ceiba, *Honduras* **88 C2** 15 40N 86 50W
La Chaux de Fonds, *Switz.* **16 E4** 47 7N 6 50 E
La Cocha, *Argentina* **94 B2** 27 50S 65 40W
La Concordia, *Mexico* **87 D6** 16 8N 92 38W
La Conner, *U.S.A.* **82 B2** 48 23N 122 30W
La Coruña, *Spain* **19 A1** 43 20N 8 25W
La Crete, *Canada* **72 B5** 58 11N 116 24W
La Crosse, *Kans., U.S.A.* **80 F5** 38 32N 99 18W
La Crosse, *Wis., U.S.A.* **80 D9** 43 48N 91 15W
La Cruz, *Costa Rica* **88 D2** 11 4N 85 39W
La Cruz, *Mexico* **86 C3** 23 55N 106 54W
La Dorada, *Colombia* **92 B4** 5 30N 74 40W
La Escondida, *Mexico* **86 C5** 24 6N 99 55W
La Esmeralda, *Paraguay* **94 A3** 22 16S 62 33W
La Esperanza, *Cuba* **88 B3** 22 46N 83 44W
La Esperanza, *Honduras* **88 D2** 14 15N 88 10W
La Estrada, *Spain* **19 A1** 42 43N 8 27W
La Fayette, *U.S.A.* **77 H3** 34 42N 85 17W
La Fé, *Cuba* **88 B3** 22 2N 84 15W
La Follette, *U.S.A.* **77 G3** 36 23N 84 7W
La Grande, *U.S.A.* **82 D4** 45 20N 118 5W
La Grange, *Calif., U.S.A.* **84 H6** 37 42N 120 27W
La Grange, *Ga., U.S.A.* **77 J3** 33 2N 85 2W
La Grange, *Ky., U.S.A.* **76 F3** 38 25N 85 23W
La Grange, *Tex., U.S.A.* **81 L6** 29 54N 96 52W
La Guaira, *Venezuela* **92 A5** 10 36N 66 56W
La Güera, *Mauritania* **50 D1** 20 51N 17 0W
La Habana, *Cuba* **88 B3** 23 8N 82 22W
La Harpe, *U.S.A.* **80 E9** 40 35N 90 58W
La Independencia, *Mexico* **87 D6** 16 31N 91 47W
La Isabela, *Dom. Rep.* **89 C5** 19 58N 71 2W
La Jara, *U.S.A.* **83 H11** 37 16N 105 58W
La Junta, *U.S.A.* **81 F3** 37 59N 103 33W
La Laguna, *Canary Is.* **22 F3** 28 28N 16 18W
La Libertad, *Guatemala* **88 C1** 16 47N 90 7W
La Libertad, *Mexico* **86 B2** 29 55N 112 41W
La Ligua, *Chile* **94 C1** 32 30S 71 16W
La Línea de la Concepción, *Spain* **19 D3** 36 15N 5 23W
La Loche, *Canada* **73 B7** 56 29N 109 26W
La Louvière, *Belgium* **15 D4** 50 27N 4 10 E
La Malbaie, *Canada* **71 C5** 47 40N 70 10W
La Mancha, *Spain* **19 C4** 39 10N 2 54W
La Mesa, *Calif., U.S.A.* **85 N9** 32 46N 117 3W
La Mesa, *N. Mex., U.S.A.* **83 K10** 32 7N 106 42W
La Misión, *Mexico* **86 A1** 32 5N 116 50W
La Moure, *U.S.A.* **80 B5** 46 21N 98 18W
La Negra, *Chile* **94 A1** 23 46S 70 18W
La Oliva, *Canary Is.* **22 F6** 28 36N 13 57W
La Orotava, *Canary Is.* **22 F3** 28 22N 16 31W
La Palma, *Canary Is.* **22 F2** 28 40N 17 50W
La Palma, *Panama* **88 E4** 8 15N 78 0W
La Palma del Condado, *Spain* **19 D2** 37 21N 6 38W
La Paloma, *Chile* **94 C1** 30 35S 71 0W
La Pampa □, *Argentina* **94 D2** 36 50S 66 0W
La Paragua, *Venezuela* **92 B6** 6 50N 63 20W
La Paz, *Entre Ríos, Argentina* **94 C4** 30 50S 59 45W
La Paz, *San Luis, Argentina* **94 C2** 33 30S 67 20W
La Paz, *Bolivia* **92 G5** 16 20S 68 10W
La Paz, *Honduras* **88 D2** 14 20N 87 47W
La Paz, *Mexico* **86 C2** 24 10N 110 20W
La Paz Centro, *Nic.* **88 D2** 12 20N 86 41W
La Pedrera, *Colombia* **92 D5** 1 18S 69 43W
La Perouse Str., *Asia* **30 B11** 45 40N 142 0 E
La Pesca, *Mexico* **87 C5** 23 46N 97 47W
La Piedad, *Mexico* **86 C4** 20 20N 102 1W
La Pine, *U.S.A.* **82 E3** 43 40N 121 30W
La Plant, *U.S.A.* **80 C4** 45 9N 100 39W
La Plata, *Argentina* **94 D4** 35 0S 57 55W
La Porte, *U.S.A.* **76 E2** 41 36N 86 43W
La Purísima, *Mexico* **86 B2** 26 10N 112 4W
La Push, *U.S.A.* **84 C2** 47 55N 124 38W
La Quiaca, *Argentina* **94 A2** 22 5S 65 35W
La Reine, *Canada* **70 C4** 48 50N 79 30W
La Restinga, *Canary Is.* **22 G2** 27 38N 17 59W
La Rioja, *Argentina* **94 B2** 29 20S 67 0W
La Rioja □, *Argentina* **94 B2** 29 30S 67 0W
La Rioja □, *Spain* **19 A4** 42 20N 2 20W
La Robla, *Spain* **19 A3** 42 50N 5 41W
La Roche-sur-Yon, *France* **18 C3** 46 40N 1 25W
La Rochelle, *France* **18 C3** 46 10N 1 9W
La Roda, *Spain* **19 C4** 39 13N 2 15W
La Romana, *Dom. Rep.* **89 C6** 18 27N 68 57W
La Ronge, *Canada* **73 B7** 55 5N 105 20W
La Rumorosa, *Mexico* **85 N10** 32 33N 116 4W
La Sabina, *Spain* **22 C7** 38 44N 1 25 E
La Salle, *U.S.A.* **80 E10** 41 20N 89 6W
La Santa, *Canary Is.* **22 E6** 29 5N 13 40W
La Sarre, *Canada* **70 C4** 48 45N 79 15W
La Scie, *Canada* **71 C8** 49 57N 55 36W
La Selva Beach, *U.S.A.* **84 J5** 36 56N 121 51W
La Serena, *Chile* **94 B1** 29 55S 71 10W
La Seyne-sur-Mer, *France* **18 E6** 43 7N 5 52 E
La Spézia, *Italy* **20 B3** 44 7N 9 50 E
La Tortuga, *Venezuela* **89 D6** 11 0N 65 22W
La Tuque, *Canada* **70 C5** 47 30N 72 50W
La Unión, *Chile* **96 E2** 40 10S 73 0W
La Unión, *El Salv.* **88 D2** 13 20N 87 50W
La Union, *Mexico* **86 D4** 17 58N 101 49W
La Urbana, *Venezuela* **92 B5** 7 8N 66 56W
La Vega, *Dom. Rep.* **89 C5** 19 20N 70 30W
La Venta, *Mexico* **87 D6** 18 8N 94 3W
La Ventura, *Mexico* **86 C4** 24 38N 100 54W
Labe = Elbe —►, *Europe* **16 B5** 53 50N 9 0 E
Labé, *Guinea* **50 F2** 11 24N 12 16W
Laberge, L., *Canada* **72 A1** 61 11N 135 12W
Labis, *Malaysia* **39 L4** 2 22N 103 2 E
Laboulaye, *Argentina* **94 C3** 34 10S 63 30W
Labrador, Coast of □, *Canada* **71 B7** 53 20N 61 0W
Labrador City, *Canada* **71 B6** 52 57N 66 55W
Lábrea, *Brazil* **92 E6** 7 15S 64 51W
Labuan, Pulau, *Malaysia* **36 C5** 5 21N 115 13 E
Labuha, *Indonesia* **37 E7** 0 30S 127 30 E
Labuhan, *Indonesia* **37 G11** 6 22S 105 50 E
Labuhanbajo, *Indonesia* **37 F6** 8 28S 120 1 E
Labuk, Telok, *Malaysia* **36 C5** 6 10N 117 50 E
Labyrinth, L., *Australia* **63 E2** 30 40S 135 11 E
Labytnangi, *Russia* **24 A12** 66 39N 66 21 E
Lac Allard, *Canada* **71 B7** 50 33N 63 24W
Lac Bouchette, *Canada* **71 C5** 48 16N 72 11W
Lac du Flambeau, *U.S.A.* **80 B10** 45 58N 89 53W
Lac Édouard, *Canada* **70 C5** 47 40N 72 16W
Lac La Biche, *Canada* **72 C6** 54 45N 111 58W
Lac la Martre = Wha Ti, *Canada* **68 B8** 63 8N 117 16W
Lac-Mégantic, *Canada* **71 C5** 45 35N 70 53W
Lac Seul, Res., *Canada* **70 B1** 50 25N 92 30W
Lac Thien, *Vietnam* **38 F7** 12 25N 108 11 E
Lacanau, *France* **18 D3** 44 58N 1 5W
Lacantúm —►, *Mexico* **87 D6** 16 36N 90 40W
Laccadive Is. = Lakshadweep Is., *Ind. Oc.* **28 H11** 10 0N 72 30 E
Lacepede B., *Australia* **63 F2** 36 40S 139 40 E
Lacepede Is., *Australia* **60 C3** 16 55S 122 0 E
Lacerdónia, *Mozam.* **55 F4** 18 3S 35 35 E
Lacey, *U.S.A.* **84 C4** 47 7N 122 49W
Lachhmangarh, *India* **42 F6** 27 50N 75 4 E
Lachi, *Pakistan* **42 C4** 33 25N 71 20 E
Lachine, *Canada* **70 C5** 45 30N 73 40W
Lachlan —►, *Australia* **63 E3** 34 22S 143 55 E
Lachute, *Canada* **70 C5** 45 39N 74 21W
Lackawanna, *U.S.A.* **78 D6** 42 50N 78 50W
Lacolle, *Canada* **79 A11** 45 5N 73 22W
Lacombe, *Canada* **72 C6** 52 30N 113 44W
Lacona, *U.S.A.* **79 C8** 43 39N 76 10W
Laconia, *U.S.A.* **79 C13** 43 32N 71 28W
Lacrosse, *U.S.A.* **82 C5** 46 51N 117 58W
Ladakh Ra., *India* **43 B8** 34 0N 78 0 E
Ladismith, *S. Africa* **56 E3** 33 28S 21 15 E
Lādīz, *Iran* **45 D9** 28 55N 61 15 E
Ladnun, *India* **42 F6** 27 38N 74 25 E
Ladoga, L. = Ladozhskoye Ozero, *Russia* **24 B5** 61 15N 30 30 E
Ladozhskoye Ozero, *Russia* **24 B5** 61 15N 30 30 E
Lady Grey, *S. Africa* **56 E4** 30 43S 27 13 E
Ladybrand, *S. Africa* **56 D4** 29 9S 27 29 E
Ladysmith, *Canada* **72 D4** 49 0N 123 49W
Ladysmith, *S. Africa* **57 D4** 28 32S 29 46 E
Ladysmith, *U.S.A.* **80 C9** 45 28N 91 12W
Lae, *Papua N. G.* **64 H6** 6 40S 147 2 E
Laem Ngop, *Thailand* **39 F4** 12 10N 102 26 E
Laem Pho, *Thailand* **39 J3** 6 55N 101 19 E
Læsø, *Denmark* **9 H14** 57 15N 10 53 E
Lafayette, *Colo., U.S.A.* **80 F2** 39 58N 105 12W
Lafayette, *Ind., U.S.A.* **76 E2** 40 25N 86 54W
Lafayette, *La., U.S.A.* **81 K9** 30 14N 92 1W
Lafayette, *Tenn., U.S.A.* **77 G3** 36 31N 86 2W
Laferte —►, *Canada* **72 A5** 61 53N 117 44W
Lafia, *Nigeria* **50 G6** 8 30N 8 34 E
Lafleche, *Canada* **73 D7** 49 45N 106 40W
Lagan —►, *U.K.* **13 B6** 54 36N 5 55W
Lagarfljót —►, *Iceland* **8 D6** 65 40N 14 18W
Lågen —►, *Oppland, Norway* **9 F14** 61 8N 10 25 E
Lågen —►, *Vestfold, Norway* **9 G14** 59 3N 10 3 E
Laghouat, *Algeria* **50 B5** 33 50N 2 59 E
Lagonoy Gulf, *Phil.* **37 B6** 13 50N 123 50 E
Lagos, *Nigeria* **50 G5** 6 25N 3 27 E
Lagos, *Portugal* **19 D1** 37 5N 8 41W
Lagos de Moreno, *Mexico* **86 C4** 21 21N 101 55W
Lagrange, *Australia* **60 C3** 18 45S 121 43 E
Lagrange B., *Australia* **60 C3** 18 38S 121 42 E
Laguna, *Brazil* **95 B6** 28 30S 48 50W
Laguna, *U.S.A.* **83 J10** 35 2N 107 25W
Laguna Beach, *U.S.A.* **85 M9** 33 33N 117 47W
Laguna Limpia, *Argentina* **94 B4** 26 32S 59 45W
Laguna Madre, *U.S.A.* **87 B5** 27 0N 97 20W
Lagunas, *Chile* **94 A2** 21 0S 69 45W
Lagunas, *Peru* **92 E3** 5 10S 75 35W
Lahad Datu, *Malaysia* **37 C5** 5 0N 118 20 E
Lahan Sai, *Thailand* **38 E4** 14 25N 102 52 E
Lahanam, *Laos* **38 D5** 16 16N 105 16 E
Laharpur, *India* **43 F9** 27 43N 80 56 E
Lahat, *Indonesia* **36 E2** 3 45S 103 30 E
Lahewa, *Indonesia* **36 D1** 1 22N 97 12 E
Lāhījān, *Iran* **45 B6** 37 10N 50 6 E
Lahn —►, *Germany* **16 C4** 50 19N 7 37 E
Laholm, *Sweden* **9 H15** 56 30N 13 2 E
Lahontan Reservoir, *U.S.A.* **82 G4** 39 28N 119 4W
Lahore, *Pakistan* **42 D6** 31 32N 74 22 E
Lahti, *Finland* **9 F21** 60 58N 25 40 E
Lahtis = Lahti, *Finland* **9 F21** 60 58N 25 40 E
Laï, *Chad* **51 G8** 9 25N 16 18 E
Lai Chau, *Vietnam* **38 A4** 22 5N 103 3 E
Laidley, *Australia* **63 D5** 27 39S 152 20 E
Laikipia □, *Kenya* **54 B4** 0 30N 36 30 E
Laingsburg, *S. Africa* **56 E3** 33 9S 20 52 E
Lainio älv —►, *Sweden* **8 C20** 67 35N 22 40 E
Lairg, *U.K.* **12 C4** 58 2N 4 24W
Laishui, *China* **34 E8** 39 23N 115 45 E
Laiwu, *China* **35 F9** 36 15N 117 40 E
Laixi, *China* **35 F11** 36 50N 120 31 E
Laiyang, *China* **35 F11** 36 59N 120 45 E
Laiyuan, *China* **34 E8** 39 20N 114 40 E
Laizhou Wan, *China* **35 F10** 37 30N 119 30 E
Laja —►, *Mexico* **86 C4** 20 55N 100 46W
Lajere, *Nigeria* **50 F7** 12 10N 11 25 E
Lajes, *Brazil* **95 B5** 27 48S 50 20W
Lak Sao, *Laos* **38 C5** 18 11N 104 59 E
Lakaband, *Pakistan* **42 D3** 31 2N 69 15 E
Lake Alpine, *U.S.A.* **84 G7** 38 29N 120 0W
Lake Andes, *U.S.A.* **80 D5** 43 9N 98 32W
Lake Anse, *U.S.A.* **76 B1** 46 42N 88 25W
Lake Arthur, *U.S.A.* **81 K8** 30 5N 92 41W
Lake Cargelligo, *Australia* **63 E4** 33 15S 146 22 E
Lake Charles, *U.S.A.* **81 K8** 30 14N 93 13W
Lake City, *Colo., U.S.A.* **83 G10** 38 2N 107 19W
Lake City, *Fla., U.S.A.* **77 K4** 30 11N 82 38W
Lake City, *Iowa, U.S.A.* **80 D7** 42 16N 94 44W
Lake City, *Mich., U.S.A.* **76 C3** 44 20N 85 13W
Lake City, *Minn., U.S.A.* **80 C8** 44 27N 92 16W
Lake City, *Pa., U.S.A.* **78 D4** 42 1N 80 21W
Lake City, *S.C., U.S.A.* **77 J6** 33 52N 79 45W
Lake George, *U.S.A.* **79 C11** 43 26N 73 43W
Lake Grace, *Australia* **61 F2** 33 7S 118 28 E
Lake Harbour = Kimmirut, *Canada* **69 B13** 62 50N 69 50W
Lake Havasu City, *U.S.A.* **85 L12** 34 27N 114 22W
Lake Hughes, *U.S.A.* **85 L8** 34 41N 118 26W
Lake Isabella, *U.S.A.* **85 K8** 35 38N 118 28W
Lake King, *Australia* **61 F2** 33 5S 119 45 E
Lake Lenore, *Canada* **73 C8** 52 24N 104 59W
Lake Louise, *Canada* **72 C5** 51 30N 116 10W
Lake Mead National Recreation Area, *U.S.A.* **85 K12** 36 15N 114 30W
Lake Mills, *U.S.A.* **80 D8** 43 25N 93 32W
Lake Nash, *Australia* **62 C2** 20 57S 138 0 E
Lake Providence, *U.S.A.* **81 J9** 32 48N 91 10W
Lake River, *Canada* **70 B3** 54 30N 82 31W
Lake Superior Prov. Park, *Canada* **70 C3** 47 45N 84 45W
Lake Village, *U.S.A.* **81 J9** 33 20N 91 17W
Lake Wales, *U.S.A.* **77 M5** 27 54N 81 35W
Lake Worth, *U.S.A.* **77 M5** 26 37N 80 3W
Lakefield, *Canada* **70 D4** 44 25N 78 16W
Lakeland, *Australia* **62 B3** 15 49S 144 57 E
Lakeland, *U.S.A.* **77 L5** 28 3N 81 57W
Lakemba, *Fiji* **59 D9** 18 13S 178 47W
Lakeport, *U.S.A.* **84 F4** 39 3N 122 55W
Lakes Entrance, *Australia* **63 F4** 37 50S 148 0 E
Lakeside, *Ariz., U.S.A.* **83 J9** 34 9N 109 58W
Lakeside, *Calif., U.S.A.* **85 N10** 32 52N 116 55W
Lakeside, *Nebr., U.S.A.* **80 D3** 42 3N 102 26W
Lakeview, *U.S.A.* **82 E3** 42 11N 120 21W
Lakewood, *Colo., U.S.A.* **80 F2** 39 44N 105 5W
Lakewood, *N.J., U.S.A.* **79 F10** 40 6N 74 13W
Lakewood, *Ohio, U.S.A.* **78 E3** 41 29N 81 48W
Lakewood Center, *U.S.A.* **84 C4** 47 11N 122 32W
Lakhaniá, *Greece* **23 D9** 35 58N 27 54 E
Lakhonpheng, *Laos* **38 E5** 15 54N 105 34 E
Lakhpat, *India* **42 H3** 23 48N 68 47 E
Lakin, *U.S.A.* **81 G4** 37 57N 101 15W
Lakitusaki —►, *Canada* **70 B3** 54 21N 82 25W
Lákkoi, *Greece* **23 D5** 35 24N 23 57 E
Lakonikós Kólpos, *Greece* **21 F10** 36 40N 22 40 E
Lakor, *Indonesia* **37 F7** 8 15S 128 17 E
Lakota, *Ivory C.* **50 G3** 5 50N 5 30W
Lakota, *U.S.A.* **80 A5** 48 2N 98 21W
Laksefjorden, *Norway* **8 A22** 70 45N 26 50 E
Lakselv, *Norway* **8 A21** 70 2N 25 0 E
Lakshadweep Is., *Ind. Oc.* **28 H11** 10 0N 72 30 E
Lakshmikantapur, *India* **43 H13** 22 5N 88 20 E
Lala Ghat, *India* **41 G18** 24 30N 92 40 E
Lala Musa, *Pakistan* **42 C5** 32 40N 73 57 E
Lalago, *Tanzania* **54 C3** 3 28S 33 58 E
Lalapanzi, *Zimbabwe* **55 F3** 19 20S 30 15 E
Lalganj, *India* **43 G11** 25 52N 85 13 E
Lalibela, *Ethiopia* **51 F12** 12 2N 39 2 E
Lalin, *China* **35 B14** 45 12N 127 0 E
Lalín, *Spain* **19 A1** 42 40N 8 5W
Lalin He —►, *China* **35 B13** 45 32N 125 40 E
Lalitapur = Patan, *Nepal* **41 F14** 27 40N 85 20 E
Lalitpur, *India* **43 G8** 24 42N 78 28 E
Lam, *Vietnam* **38 B6** 21 21N 106 31 E
Lam Pao Res., *Thailand* **38 D4** 16 50N 103 15 E
Lamaing, *Burma* **41 M20** 15 25N 97 53 E
Lamar, *Colo., U.S.A.* **80 F3** 38 5N 102 37W
Lamar, *Mo., U.S.A.* **81 G7** 37 30N 94 16W
Lamas, *Peru* **92 E3** 6 28S 76 31W
Lambaréné, *Gabon* **52 E2** 0 41S 10 12 E
Lambasa, *Fiji* **59 C8** 16 30S 179 10 E
Lambay I., *Ireland* **13 C5** 53 29N 6 1W
Lambert, *U.S.A.* **80 B2** 47 41N 104 37W
Lambert Glacier, *Antarctica* **5 D6** 71 0S 70 0 E
Lamberts Bay, *S. Africa* **56 E2** 32 5S 18 17 E
Lame, *Nigeria* **50 F6** 10 30N 9 20 E
Lame Deer, *U.S.A.* **82 D10** 45 37N 106 40W
Lamego, *Portugal* **19 B2** 41 5N 7 52W
Lamèque, *Canada* **71 C7** 47 45N 64 38W
Lameroo, *Australia* **63 F3** 35 19S 140 33 E
Lamesa, *U.S.A.* **81 J4** 32 44N 101 58W
Lamía, *Greece* **21 E10** 38 55N 22 26 E
Lammermuir Hills, *U.K.* **12 F6** 55 50N 2 40W
Lamon Bay, *Phil.* **37 B6** 14 30N 122 20 E
Lamont, *Canada* **72 C6** 53 46N 112 50W
Lamont, *U.S.A.* **85 K8** 35 15N 118 55W
Lampa, *Peru* **92 G4** 15 22S 70 22W
Lampang, *Thailand* **38 C2** 18 16N 99 32 E
Lampasas, *U.S.A.* **81 K5** 31 4N 98 11W
Lampazos de Naranjo, *Mexico* **86 B4** 27 2N 100 32W
Lampedusa, *Medit. S.* **20 G5** 35 36N 12 40 E
Lampeter, *U.K.* **11 E3** 52 7N 4 4W
Lampione, *Medit. S.* **20 G5** 35 33N 12 20 E
Lampman, *Canada* **73 D8** 49 25N 102 50W
Lamprey, *Canada* **73 B10** 58 33N 94 8W
Lampung □, *Indonesia* **36 F2** 5 30S 104 30 E
Lamu, *Kenya* **54 C5** 2 16S 40 55 E
Lamu □, *Kenya* **54 C5** 2 0S 40 45 E
Lamy, *U.S.A.* **83 J11** 35 29N 105 53W
Lan Xian, *China* **34 E6** 38 15N 111 35 E
Lanai I., *U.S.A.* **74 H16** 20 50N 156 55W
Lanak La, *India* **43 B8** 34 27N 79 32 E
Lanak'o Shank'ou = Lanak La, *India* **43 B8** 34 27N 79 32 E
Lanao, L., *Phil.* **37 C6** 7 52N 124 15 E
Lanark, *Canada* **79 A8** 45 1N 76 22W
Lanark, *U.K.* **12 F5** 55 40N 3 47W
Lancang Jiang —►, *China* **32 D5** 21 40N 101 10 E
Lancashire □, *U.K.* **10 D5** 53 50N 2 48W
Lancaster, *Canada* **79 A10** 45 10N 74 30W
Lancaster, *U.K.* **10 C5** 54 3N 2 48W
Lancaster, *Calif., U.S.A.* **85 L8** 34 42N 118 8W
Lancaster, *Ky., U.S.A.* **76 G3** 37 37N 84 35W
Lancaster, *N.H., U.S.A.* **79 B13** 44 29N 71 34W
Lancaster, *N.Y., U.S.A.* **78 D6** 42 54N 78 40W
Lancaster, *Pa., U.S.A.* **79 F8** 40 2N 76 19W
Lancaster, *S.C., U.S.A.* **77 H5** 34 43N 80 46W
Lancaster, *Wis., U.S.A.* **80 D9** 42 51N 90 43W
Lancaster Sd., *Canada* **69 A11** 74 13N 84 0W
Lancer, *Canada* **73 C7** 50 48N 108 53W
Lanchow = Lanzhou, *China* **34 F2** 36 1N 103 52 E
Lanciano, *Italy* **20 C6** 42 14N 14 23 E
Lancun, *China* **35 F11** 36 25N 120 10 E
Landeck, *Austria* **16 E6** 47 9N 10 34 E
Landen, *Belgium* **15 D5** 50 45N 5 3 E
Lander, *U.S.A.* **82 E9** 42 50N 108 44W
Lander —►, *Australia* **60 D5** 22 0S 132 0 E
Landes, *France* **18 D3** 44 0N 1 0W
Landi Kotal, *Pakistan* **42 B4** 34 7N 71 6 E
Landor, *Australia* **61 E2** 25 10S 116 54 E
Land's End, *U.K.* **11 G2** 50 4N 5 44W
Landsborough Cr. —►, *Australia* **62 C3** 22 28S 144 35 E
Landshut, *Germany* **16 D7** 48 34N 12 8 E
Landskrona, *Sweden* **9 J15** 55 53N 12 50 E
Lanesboro, *U.S.A.* **79 E9** 41 57N 75 34W
Lanett, *U.S.A.* **77 J3** 32 52N 85 12W
Lang Bay, *Canada* **72 D4** 49 45N 124 21W
Lang Qua, *Vietnam* **38 A5** 22 16N 104 27 E
Lang Shan, *China* **34 D4** 41 0N 106 30 E
Lang Son, *Vietnam* **38 B6** 21 52N 106 42 E
Lang Suan, *Thailand* **39 H2** 9 57N 99 4 E
La'nga Co, *China* **41 D12** 30 45N 81 15 E
Langar, *Iran* **45 C9** 35 23N 60 25 E
Langara I., *Canada* **72 C2** 54 14N 133 1W
Langdon, *U.S.A.* **80 A5** 48 45N 98 22W
Langeberg, *S. Africa* **56 E3** 33 55S 21 0 E
Langeberge, *S. Africa* **56 D3** 28 15S 22 33 E
Langeland, *Denmark* **9 J14** 54 56N 10 48 E
Langenburg, *Canada* **73 C8** 50 51N 101 43W
Langholm, *U.K.* **12 F6** 55 9N 3 0W
Langjökull, *Iceland* **8 D3** 64 39N 20 12W
Langkawi, P., *Malaysia* **39 J2** 6 25N 99 45 E
Langklip, *S. Africa* **56 D3** 28 12S 20 20 E
Langkon, *Malaysia* **36 C5** 6 30N 116 40 E
Langlade, *St- P. & M.* **71 C8** 46 50N 56 20W
Langlois, *U.S.A.* **82 E1** 42 56N 124 27W
Langøya, *Norway* **8 B16** 68 45N 14 50 E
Langres, *France* **18 C6** 47 52N 5 20 E
Langres, Plateau de, *France* **18 C6** 47 45N 5 3 E
Langsa, *Indonesia* **36 D1** 4 30N 97 57 E
Langtry, *U.S.A.* **81 L4** 29 49N 101 34W
Langu, *Thailand* **39 J2** 6 53N 99 47 E
Languedoc, *France* **18 E5** 43 58N 3 55 E
Langxiangzhen, *China* **34 E9** 39 43N 116 8 E
Lanigan, *Canada* **73 C7** 51 51N 105 2W
Lankao, *China* **34 G8** 34 48N 114 50 E
Länkäran, *Azerbaijan* **25 G8** 38 48N 48 52 E
Lannion, *France* **18 B2** 48 46N 3 29W
L'Annonciation, *Canada* **70 C5** 46 25N 74 55W
Lansdale, *U.S.A.* **79 F9** 40 14N 75 17W
Lansdowne, *Australia* **63 E5** 31 48S 152 30 E
Lansdowne, *Canada* **79 B8** 44 24N 76 1W

Lansdowne House, Canada 70 B2 52 14N 87 53W
L'Anse, U.S.A. 70 C2 46 45N 88 27W
L'Anse au Loup, Canada 71 B8 51 32N 56 50W
Lansford, U.S.A. 79 F9 40 50N 75 53W
Lansing, U.S.A. 76 D3 42 44N 84 33W
Lanta Yai, Ko, Thailand 39 J2 7 35N 99 3 E
Lantian, China 34 G5 34 11N 109 20 E
Lanus, Argentina 94 C4 34 44S 58 27W
Lanusei, Italy 20 E3 39 52N 9 34 E
Lanzarote, Canary Is. 22 E6 29 0N 13 40W
Lanzhou, China 34 F2 36 1N 103 52 E
Lao Bao, Laos 38 D6 16 35N 106 30 E
Lao Cai, Vietnam 38 A4 22 30N 103 57 E
Laoag, Phil. 37 A6 18 7N 120 34 E
Laoang, Phil. 37 B7 12 32N 125 8 E
Laoha He →, China 35 C11 43 25N 120 35 E
Laois □, Ireland 13 D4 52 57N 7 36W
Laon, France 18 B5 49 33N 3 35 E
Laona, U.S.A. 76 C1 45 34N 88 40W
Laos ■, Asia 38 D5 17 45N 105 0 E
Lapa, Brazil 95 B6 25 46S 49 44W
Laparan, Phil. 37 C6 6 0N 120 0 E
Lapeer, U.S.A. 76 D4 43 3N 83 19W
Lapithos, Cyprus 23 D12 35 21N 33 11 E
Lapland = Lappland, Europe 8 B21 68 7N 24 0 E
Laporte, U.S.A. 79 E8 41 25N 76 30W
Lappeenranta, Finland 9 F23 61 3N 28 12 E
Lappland, Europe 8 B21 68 7N 24 0 E
Laprida, Argentina 94 D3 37 34S 60 45W
Lapseki, Turkey 21 D12 40 20N 26 41 E
Laptev Sea, Russia 27 B13 76 0N 125 0 E
Lapua, Finland 8 E20 62 58N 23 0 E
L'Áquila, Italy 20 C5 42 22N 13 22 E
Lār, Āzarbājān-e Sharqī, Iran 44 B5 38 30N 47 52 E
Lār, Fārs, Iran 45 E7 27 40N 54 14 E
Larache, Morocco 50 A3 35 10N 6 5W
Laramie, U.S.A. 80 E2 41 19N 105 35W
Laramie Mts., U.S.A. 80 E2 42 0N 105 30W
Laranjeiras do Sul, Brazil 95 B5 25 23S 52 23W
Larantuka, Indonesia 37 F6 8 21S 122 55 E
Larap, Phil. 37 B6 14 18N 122 39 E
Larat, Indonesia 37 F8 7 0S 132 0 E
Larde, Mozam. 55 F4 16 28S 39 43 E
Larder Lake, Canada 70 C4 48 5N 79 40W
Lardhos, Ákra, Greece 23 C10 36 4N 28 10 E
Lardhos, Órmos, Greece 23 C10 36 4N 28 2 E
Laredo, U.S.A. 81 M5 27 30N 99 30W
Laredo Sd., Canada 72 C3 52 30N 128 53W
Largo, U.S.A. 77 M4 27 55N 82 47W
Largs, U.K. 12 F4 55 47N 4 52W
Lariang, Indonesia 37 E5 1 26S 119 17 E
Larimore, U.S.A. 80 B6 47 54N 97 38W
Lārīn, Iran 45 C7 35 55N 52 19 E
Lárisa, Greece 21 E10 39 36N 22 27 E
Larkana, Pakistan 42 F3 27 32N 68 18 E
Larnaca, Cyprus 23 E12 34 55N 33 38 E
Larnaca Bay, Cyprus 23 E12 34 53N 33 45 E
Larne, U.K. 13 B6 54 51N 5 51W
Larned, U.S.A. 80 F5 38 11N 99 6W
Larrimah, Australia 60 C5 15 35S 133 12 E
Larsen Ice Shelf, Antarctica 5 C17 67 0S 62 0W
Larvik, Norway 9 G14 59 4N 10 0 E
Laryak, Russia 26 C8 61 15N 80 0 E
Las Animas, U.S.A. 80 F3 38 4N 103 13W
Las Anod, Somali Rep. 46 F4 8 26N 47 19 E
Las Brenãs, Argentina 94 B3 27 5S 61 7W
Las Chimeneas, Mexico 85 N10 32 8N 116 5W
Las Cruces, U.S.A. 83 K10 32 19N 106 47W
Las Flores, Argentina 94 D4 36 10S 59 7W
Las Heras, Argentina 94 C2 32 51S 68 49W
Las Khoreh, Somali Rep. 46 E4 11 10N 48 20 E
Las Lajas, Argentina 96 D2 38 30S 70 25W
Las Lomitas, Argentina 94 A3 24 43S 60 35W
Las Palmas, Argentina 94 B4 27 8S 58 45W
Las Palmas, Canary Is. 22 F4 28 7N 15 26W
Las Palmas →, Mexico 85 N10 32 26N 116 54W
Las Piedras, Uruguay 95 C4 34 44S 56 14W
Las Pipinas, Argentina 94 D4 35 30S 57 19W
Las Plumas, Argentina 96 E3 43 40S 67 15W
Las Rosas, Argentina 94 C3 32 30S 61 35W
Las Tablas, Panama 88 E3 7 49N 80 14W
Las Termas, Argentina 94 B3 27 29S 64 52W
Las Truchas, Mexico 86 D4 17 57N 102 13W
Las Varillas, Argentina 94 C3 31 50S 62 50W
Las Vegas, N. Mex., U.S.A. 83 J11 35 36N 105 13W
Las Vegas, Nev., U.S.A. 85 J11 36 10N 115 9W
Lascano, Uruguay 95 C5 33 35S 54 12W
Lashburn, Canada 73 C7 53 10N 109 40W
Lashio, Burma 41 H20 22 56N 97 45 E
Lashkar, India 42 F8 26 10N 78 10 E
Lasíthi, Greece 23 D7 35 11N 25 31 E
Lasíthi □, Greece 23 D7 35 5N 25 50 E
Lassen Pk., U.S.A. 82 F3 40 29N 121 31W
Last Mountain L., Canada 73 C7 51 5N 105 14W
Lastchance Cr. →, U.S.A. 84 E5 40 2N 121 15W
Lastoursville, Gabon 52 E2 0 55S 12 38 E
Lastovo, Croatia 20 C7 42 46N 16 55 E
Lat Yao, Thailand 38 E2 15 45N 99 48 E
Latacunga, Ecuador 92 D3 0 50S 78 35W
Latakia = Al Lādhiqīyah, Syria 44 C2 35 30N 35 45 E
Latchford, Canada 70 C4 47 20N 79 50W
Latham, Australia 61 E2 29 44S 116 20 E
Lathrop Wells, U.S.A. 85 J10 36 39N 116 24W
Latina, Italy 20 D5 41 28N 12 52 E
Latium = Lazio □, Italy 20 C5 42 10N 12 30 E
Laton, U.S.A. 84 J7 36 26N 119 41W
Latouche Treville, C., Australia 60 C3 18 27S 121 49 E
Latrobe, Australia 62 G4 41 14S 146 30 E
Latrobe, U.S.A. 78 F5 40 19N 79 23W
Latvia ■, Europe 9 H20 56 50N 24 0 E
Lau Group, Fiji 59 C9 17 0S 178 30W
Lauchhammer, Germany 16 C7 51 29N 13 47 E
Laukaa, Finland 9 E21 62 24N 25 56 E
Launceston, Australia 62 G4 41 24S 147 8 E
Launceston, U.K. 11 G3 50 38N 4 22W
Laune →, Ireland 13 D2 52 7N 9 47W
Laura, Australia 62 B3 15 32S 144 32 E
Laurel, Miss., U.S.A. 81 K10 31 41N 89 8W
Laurel, Mont., U.S.A. 82 D9 45 40N 108 46W
Laurencekirk, U.K. 12 E6 56 50N 2 28W
Laurens, U.S.A. 77 H4 34 30N 82 1W
Laurentian Plateau, Canada 71 B6 52 0N 70 0W
Laurentides, Parc Prov. des, Canada 71 C5 47 45N 71 15W
Lauria, Italy 20 E6 40 2N 15 50 E
Laurie L., Canada 73 B8 56 35N 101 57W
Laurinburg, U.S.A. 77 H6 34 47N 79 28W
Laurium, U.S.A. 76 B1 47 14N 88 27W
Lausanne, Switz. 16 E4 46 32N 6 38 E
Laut, Indonesia 36 D3 4 45N 108 0 E
Laut Kecil, Kepulauan, Indonesia 36 E5 4 45S 115 40 E
Lautoka, Fiji 59 C7 17 37S 177 27 E
Lauzon, Canada 71 C5 46 48N 71 10W
Lava Hot Springs, U.S.A. 82 E7 42 37N 112 1W
Laval, France 18 B3 48 4N 0 48W
Lavalle, Argentina 94 B2 28 15S 65 15W
Laverne, U.S.A. 81 G5 36 43N 99 54W
Laverton, Australia 61 E3 28 44S 122 29 E
Lavras, Brazil 95 A7 21 20S 45 0W
Lavrentiya, Russia 27 C19 65 35N 171 0W
Lávrion, Greece 21 F11 37 40N 24 4 E
Lávris, Greece 23 D6 35 25N 24 40 E
Lavumisa, Swaziland 57 D5 27 20S 31 55 E
Lawas, Malaysia 36 D5 4 55N 115 25 E
Lawele, Indonesia 37 F6 5 16S 123 3 E
Lawn Hill, Australia 62 B2 18 36S 138 33 E
Lawng Pit, Burma 41 G20 25 30N 97 25 E
Lawqah, Si. Arabia 44 D4 29 49N 42 45 E
Lawrence, N.Z. 59 L2 45 55S 169 41 E
Lawrence, Kans., U.S.A. 80 F7 38 58N 95 14W
Lawrence, Mass., U.S.A. 79 D13 42 43N 71 10W
Lawrenceburg, Ind., U.S.A. 76 F3 39 6N 84 52W
Lawrenceburg, Tenn., U.S.A. 77 H2 35 14N 87 20W
Lawrenceville, U.S.A. 77 J4 33 57N 83 59W
Laws, U.S.A. 84 H8 37 24N 118 20W
Lawton, U.S.A. 81 H5 34 37N 98 25W
Lawu, Indonesia 37 G14 7 40S 111 13 E
Laxford, L., U.K. 12 C3 58 24N 5 6W
Laylān, Iraq 44 C5 35 18N 44 31 E
Laysan I., Pac. Oc. 65 E11 25 30N 167 0W
Laytonville, U.S.A. 82 G2 39 41N 123 29W
Lazio □, Italy 20 C5 42 10N 12 30 E
Lazo, Russia 30 C6 43 25N 133 55 E
Le Creusot, France 18 C6 46 48N 4 24 E
Le François, Martinique 89 D7 14 38N 60 57W
Le Havre, France 18 B4 49 30N 0 5 E
Le Mans, France 18 C4 48 0N 0 10 E
Le Mars, U.S.A. 80 D6 42 47N 96 10W
Le Mont-St.-Michel, France 18 B3 48 40N 1 30W
Le Moule, Guadeloupe 89 C7 16 20N 61 22W
Le Puy-en-Velay, France 18 D5 45 3N 3 52 E
Le Roy, U.S.A. 81 F7 38 5N 95 38W
Le Sueur, U.S.A. 80 C8 44 28N 93 55W
Le Thuy, Vietnam 38 D6 17 14N 106 49 E
Le Touquet-Paris-Plage, France 18 A4 50 30N 1 36 E
Le Tréport, France 18 A4 50 3N 1 20 E
Le Verdon-sur-Mer, France 18 D3 45 33N 1 4W
Lea →, U.K. 11 F7 51 31N 0 1 E
Leach, Cambodia 39 F4 12 21N 103 46 E
Lead, U.S.A. 80 C3 44 21N 103 46W
Leader, Canada 73 C7 50 50N 109 30W
Leadhills, U.K. 12 F5 55 25N 3 45W
Leadville, U.S.A. 83 G10 39 15N 106 18W
Leaf →, U.S.A. 81 K10 30 59N 88 44W
Leakey, U.S.A. 81 L5 29 44N 99 46W
Leamington, Canada 70 D3 42 3N 82 36W
Leamington, U.S.A. 82 G7 39 32N 112 17W
Leamington Spa = Royal Leamington Spa, U.K. 11 E6 52 18N 1 31W
Leandro Norte Alem, Argentina 95 B4 27 34S 55 15W
Learmonth, Australia 60 D1 22 13S 114 10 E
Leask, Canada 73 C7 53 5N 106 45W
Leavenworth, Kans., U.S.A. 80 F7 39 19N 94 55W
Leavenworth, Wash., U.S.A. 82 C3 47 36N 120 40W
Lebak, Phil. 37 C6 6 32N 124 5 E
Lebam, U.S.A. 84 D3 46 34N 123 33W
Lebanon, Ind., U.S.A. 76 E2 40 3N 86 28W
Lebanon, Kans., U.S.A. 80 F5 39 49N 98 33W
Lebanon, Ky., U.S.A. 76 G3 37 34N 85 15W
Lebanon, Mo., U.S.A. 81 G8 37 41N 92 40W
Lebanon, Oreg., U.S.A. 82 D2 44 32N 122 55W
Lebanon, Pa., U.S.A. 79 F8 40 20N 76 26W
Lebanon, Tenn., U.S.A. 77 G2 36 12N 86 18W
Lebanon ■, Asia 47 B4 34 0N 36 0 E
Lebec, U.S.A. 85 L8 34 50N 118 52W
Lebomboberge, S. Africa 57 C5 24 30S 32 0 E
Lębork, Poland 17 A9 54 33N 17 46 E
Lebrija, Spain 19 D2 36 53N 6 5W
Lebu, Chile 94 D1 37 40S 73 47W
Lecce, Italy 21 D8 40 23N 18 11 E
Lecco, Italy 20 B3 45 51N 9 23 E
Lech →, Germany 16 D6 48 43N 10 56 E
Łęczyca, Poland 17 B10 52 5N 19 15 E
Ledbury, U.K. 11 E5 52 2N 2 25W
Ledong, China 38 C7 18 41N 109 5 E
Leduc, Canada 72 C6 53 15N 113 30W
Lee, U.S.A. 79 D11 42 19N 73 15W
Lee →, Ireland 13 E3 51 53N 8 56W
Lee Vining, U.S.A. 84 H7 37 58N 119 7W
Leech L., U.S.A. 80 B7 47 10N 94 24W
Leedey, U.S.A. 81 H5 35 52N 99 21W
Leeds, U.K. 10 D6 53 48N 1 33W
Leeds, U.S.A. 77 J2 33 33N 86 33W
Leek, U.K. 10 D5 53 7N 2 1W
Leer, Germany 16 B4 53 13N 7 26 E
Leesburg, U.S.A. 77 L5 28 49N 81 53W
Leesville, U.S.A. 81 K8 31 9N 93 16W
Leeton, Australia 63 E4 34 33S 146 23 E
Leetonia, U.S.A. 78 F4 40 53N 80 45W
Leeu Gamka, S. Africa 56 E3 32 47S 21 59 E
Leeuwarden, Neths. 15 A5 53 15N 5 48 E
Leeuwin, C., Australia 61 F2 34 20S 115 9 E
Leeward Is., Atl. Oc. 89 C7 16 30N 63 30W
Lefka, Cyprus 23 D11 35 6N 32 51 E
Lefkoniko, Cyprus 23 D12 35 18N 33 44 E
Lefors, U.S.A. 81 H4 35 26N 100 48W
Lefroy, L., Australia 61 F3 31 21S 121 40 E
Legal, Canada 72 C6 53 55N 113 35W
Leganés, Spain 19 B4 40 19N 3 45W
Legazpi, Phil. 37 B6 13 10N 123 45 E
Legendre I., Australia 60 D2 20 22S 116 55 E
Leghorn = Livorno, Italy 20 C4 43 33N 10 19 E
Legionowo, Poland 17 B11 52 25N 20 50 E
Legnago, Italy 20 B4 45 11N 11 18 E
Legnica, Poland 16 C9 51 12N 16 10 E
Legume, Australia 63 D5 28 20S 152 19 E
Leh, India 43 B7 34 9N 77 35 E
Lehi, U.S.A. 82 F8 40 24N 111 51W
Lehighton, U.S.A. 79 F9 40 50N 75 43W
Lehututu, Botswana 56 C3 23 54S 21 55 E
Leiah, Pakistan 42 D4 30 58N 70 58 E
Leicester, U.K. 11 E6 52 38N 1 8W
Leicestershire □, U.K. 11 E6 52 41N 1 17W
Leichhardt →, Australia 62 B2 17 35S 139 48 E
Leichhardt Ra., Australia 62 C4 20 46S 147 40 E
Leiden, Neths. 15 B4 52 9N 4 30 E
Leie →, Belgium 15 C3 51 2N 3 45 E
Leine →, Germany 16 B5 52 43N 9 36 E
Leinster, Australia 61 E3 27 51S 120 36 E
Leinster □, Ireland 13 C4 53 3N 7 8W
Leinster, Mt., Ireland 13 D5 52 37N 6 46W
Leipzig, Germany 16 C7 51 18N 12 22 E
Leiria, Portugal 19 C1 39 46N 8 53W
Leirvik, Norway 9 G11 59 47N 5 28 E
Leisler, Mt., Australia 60 D4 23 23S 129 20 E
Leith, U.K. 12 F5 55 59N 3 11W
Leith Hill, U.K. 11 F7 51 11N 0 22W
Leitrim, Ireland 13 B3 54 0N 8 5W
Leitrim □, Ireland 13 B3 54 8N 8 0W
Leizhou Bandao, China 33 D6 21 0N 110 0 E
Lek →, Neths. 15 C4 51 54N 4 35 E
Leka, Norway 8 D14 65 5N 11 35 E
Leksula, Indonesia 37 E7 3 46S 126 31 E
Lékva Ori, Greece 23 D6 35 18N 24 3 E
Leland, U.S.A. 81 J9 33 24N 90 54W
Leland Lakes, Canada 73 A6 60 0N 110 59W
Leleque, Argentina 96 E2 42 28S 71 0W
Lelystad, Neths. 15 B5 52 30N 5 25 E
Léman, L., Europe 16 E4 46 26N 6 30 E
Lemera, Zaïre 54 C2 3 0S 28 55 E
Lemhi Ra., U.S.A. 82 D7 44 30N 113 30W
Lemmer, Neths. 15 B5 52 51N 5 43 E
Lemmon, U.S.A. 80 C3 45 57N 102 10W
Lemon Grove, U.S.A. 85 N9 32 45N 117 2W
Lemoore, U.S.A. 83 H4 36 18N 119 46W
Lemvig, Denmark 9 H13 56 33N 8 20 E
Lena →, Russia 27 B13 72 52N 126 40 E
Léndas, Greece 23 E6 34 56N 24 56 E
Lendeh, Iran 45 D6 30 58N 50 25 E
Lenggong, Malaysia 39 K3 5 6N 100 58 E
Lengua de Vaca, Pta., Chile 94 C1 30 14S 71 38W
Leninabad = Khudzhand, Tajikistan 26 E7 40 17N 69 37 E
Leninakan = Gyumri, Armenia 25 F7 40 47N 43 50 E
Leningrad = Sankt-Peterburg, Russia 24 C5 59 55N 30 20 E
Leninogorsk, Kazakstan 26 D9 50 20N 83 30 E
Leninsk, Russia 25 E8 48 40N 45 15 E
Leninsk-Kuznetskiy, Russia 26 D9 54 44N 86 10 E
Leninskoye, Russia 27 E14 47 56N 132 38 E
Lenkoran = Länkäran, Azerbaijan 25 G8 38 48N 48 52 E
Lenmalu, Indonesia 37 E8 1 45S 130 15 E
Lennoxville, Canada 79 A13 45 22N 71 51W
Lenoir, U.S.A. 77 H5 35 55N 81 32W
Lenoir City, U.S.A. 77 H3 35 48N 84 16W
Lenora, U.S.A. 80 F4 39 37N 100 0W
Lenore L., Canada 73 C8 52 30N 104 59W
Lenox, U.S.A. 79 D11 42 22N 73 17W
Lens, France 18 A5 50 26N 2 50 E
Lensk, Russia 27 C12 60 48N 114 55 E
Lentini, Italy 20 F6 37 17N 15 0 E
Lenwood, U.S.A. 85 L9 34 53N 117 7W
Leoben, Austria 16 E8 47 22N 15 5 E
Leodhas = Lewis, U.K. 12 C2 58 9N 6 40W
Leola, U.S.A. 80 C5 45 43N 98 56W
Leominster, U.K. 11 E5 52 14N 2 43W
Leominster, U.S.A. 79 D13 42 32N 71 46W
León, Mexico 86 C4 21 7N 101 40W
León, Nic. 88 D2 12 20N 86 51W
León, Spain 19 A3 42 38N 5 34W
Leon, U.S.A. 80 E8 40 44N 93 45W
León, Montañas de, Spain 19 A2 42 30N 6 18W
Leonardtown, U.S.A. 76 F7 38 17N 76 38W
Leongatha, Australia 63 F4 38 30S 145 58 E
Leonora, Australia 61 E3 28 49S 121 19 E
Léopold II, Lac = Mai-Ndombe, L., Zaïre 52 E3 2 0S 18 20 E
Leopoldina, Brazil 95 A7 21 28S 42 40W
Leopoldsburg, Belgium 15 C5 51 7N 5 13 E
Léopoldville = Kinshasa, Zaïre 52 E3 4 20S 15 15 E
Leoti, U.S.A. 80 F4 38 29N 101 21W
Leova, Moldova 17 E15 46 28N 28 15 E
Leoville, Canada 73 C7 53 39N 107 33W
Lépa, L. do, Angola 56 B2 17 0S 19 0 E
Lepel = Lyepyel, Belarus 24 D4 54 50N 28 40 E
Lepikha, Russia 27 C13 64 45N 125 55 E
Leppävirta, Finland 9 E22 62 29N 27 46 E
Lerdo, Mexico 86 B4 25 32N 103 32W
Léré, Chad 51 G7 9 39N 14 13 E
Leribe, Lesotho 57 D4 28 51S 28 3 E
Lérida, Spain 19 B6 41 37N 0 39 E
Lerwick, U.K. 12 A7 60 9N 1 9W
Les Cayes, Haiti 89 C5 18 15N 73 46W
Les Étroits, Canada 71 C6 47 24N 68 54W
Les Sables-d'Olonne, France 18 C3 46 30N 1 45W
Lesbos = Lésvos, Greece 21 E12 39 10N 26 20 E
Leshan, China 32 D5 29 33N 103 41 E
Leshukonskoye, Russia 24 B8 64 54N 45 46 E
Leskov I., Antarctica 5 B1 56 0S 28 0W
Leskovac, Serbia, Yug. 21 C9 43 0N 21 58 E
Leslie, U.S.A. 81 H8 35 50N 92 34W
Lesopilnoye, Russia 30 A7 46 44N 134 20 E
Lesotho ■, Africa 57 D4 29 40S 28 0 E
Lesozavodsk, Russia 27 E14 45 30N 133 29 E
Lesse →, Belgium 15 D4 50 15N 4 54 E
Lesser Antilles, W. Indies 89 C7 15 0N 61 0W
Lesser Slave L., Canada 72 B5 55 30N 115 25W
Lesser Sunda Is., Indonesia 37 F6 7 0S 120 0 E
Lessines, Belgium 15 D3 50 42N 3 50 E
Lester, U.S.A. 84 C5 47 12N 121 29W
Lestock, Canada 73 C8 51 19N 103 59W
Lesuer I., Australia 60 B4 13 50S 127 17 E
Lésvos, Greece 21 E12 39 10N 26 20 E
Leszno, Poland 17 C9 51 50N 16 30 E
Letchworth, U.K. 11 F7 51 59N 0 13W
Lethbridge, Canada 72 D6 49 45N 112 45W
Leti, Kepulauan, Indonesia 37 F7 8 10S 128 0 E
Leti Is. = Leti, Kepulauan, Indonesia 37 F7 8 10S 128 0 E
Letiahau →, Botswana 56 C3 21 16S 24 0 E
Leticia, Colombia 92 D4 4 9S 70 0W
Leting, China 35 E10 39 23N 118 55 E
Letjiesbos, S. Africa 56 E3 32 34S 22 16 E
Letlhakeng, Botswana 56 C3 24 0S 24 59 E
Letpadan, Burma 41 L19 17 45N 95 45 E
Letpan, Burma 41 K19 19 28N 94 10 E
Letterkenny, Ireland 13 B4 54 57N 7 45W
Leucadia, U.S.A. 85 M9 33 4N 117 18W
Leuser, G., Indonesia 36 D1 3 46N 97 12 E
Leuven, Belgium 15 D4 50 52N 4 42 E
Leuze, Hainaut, Belgium 15 D3 50 36N 3 37 E
Leuze, Namur, Belgium 15 D4 50 33N 4 54 E
Levádhia, Greece 21 E10 38 27N 22 54 E
Levan, U.S.A. 82 G8 39 33N 111 52W
Levanger, Norway 8 E14 63 45N 11 19 E
Levelland, U.S.A. 81 J3 33 35N 102 23W
Leven, U.K. 12 E6 56 12N 3 0W
Leven, L., U.K. 12 E5 56 12N 3 22W
Leven, Toraka, Madag. 57 A8 12 30S 47 45 E
Leveque C., Australia 60 C3 16 20S 123 0 E
Levice, Slovak Rep. 17 D10 48 13N 18 35 E
Levin, N.Z. 59 J5 40 37S 175 18 E
Lévis, Canada 71 C5 46 48N 71 9W
Levis, L., Canada 72 A5 62 37N 117 58W
Levittown, N.Y., U.S.A. 79 F11 40 44N 73 31W
Levittown, Pa., U.S.A. 79 F10 40 9N 74 51W
Levkás, Greece 21 E9 38 40N 20 43 E
Levkímmi, Greece 23 B4 39 25N 20 3 E
Levkímmi, Ákra, Greece 23 B4 39 29N 20 4 E
Levkôsia = Nicosia, Cyprus 23 D12 35 10N 33 25 E
Levskigrad = Karlovo, Bulgaria 21 C11 42 38N 24 47 E
Lewellen, U.S.A. 80 E3 41 20N 102 9W
Lewes, U.K. 11 G8 50 52N 0 1 E
Lewes, U.S.A. 76 F8 38 46N 75 9W
Lewis, U.K. 12 C2 58 9N 6 40W
Lewis →, U.S.A. 84 E4 45 51N 122 48W
Lewis, Butt of, U.K. 12 C2 58 31N 6 16W
Lewis Ra., Australia 60 D4 20 3S 128 50 E
Lewis Range, U.S.A. 82 C7 48 5N 113 5W
Lewisburg, Pa., U.S.A. 78 F8 40 58N 76 54W
Lewisburg, Tenn., U.S.A. 77 H2 35 27N 86 48W
Lewisporte, Canada 71 C8 49 15N 55 3W
Lewiston, Idaho, U.S.A. 82 C5 46 25N 117 1W
Lewiston, Maine, U.S.A. 77 C11 44 6N 70 13W
Lewistown, Mont., U.S.A. 82 C9 47 4N 109 26W
Lewistown, Pa., U.S.A. 78 F7 40 36N 77 34W
Lexington, Ill., U.S.A. 80 E10 40 39N 88 47W
Lexington, Ky., U.S.A. 76 F3 38 3N 84 30W
Lexington, Miss., U.S.A. 81 J9 33 7N 90 3W
Lexington, Mo., U.S.A. 80 F8 39 11N 93 52W
Lexington, N.C., U.S.A. 77 H5 35 49N 80 15W
Lexington, Nebr., U.S.A. 80 E5 40 47N 99 45W
Lexington, Ohio, U.S.A. 78 F2 40 41N 82 35W
Lexington, Oreg., U.S.A. 82 D4 45 27N 119 42W
Lexington, Tenn., U.S.A. 77 H1 35 39N 88 24W
Lexington Park, U.S.A. 76 F7 38 16N 76 27W
Leyte, Phil. 37 B6 11 0N 125 0 E
Lezha, Albania 21 D8 41 47N 19 42 E
Lhasa, China 32 D4 29 25N 90 58 E
Lhazê, China 32 D3 29 5N 87 38 E
Lhokkruet, Indonesia 36 D1 4 55N 95 24 E
Lhokseumawe, Indonesia 36 C1 5 10N 97 10 E
Lhuntsi Dzong, India 41 F17 27 39N 91 10 E
Li, Thailand 38 D2 17 48N 98 57 E
Li Xian, Gansu, China 34 G3 34 10N 105 5 E
Li Xian, Hebei, China 34 E8 38 30N 115 35 E
Lianga, Phil. 37 C7 8 38N 126 6 E
Liangcheng, Nei Mongol Zizhiqu, China 34 D7 40 28N 112 25 E
Liangcheng, Shandong, China 35 G10 35 32N 119 37 E
Liangdang, China 34 H4 33 56N 106 18 E
Lianshanguan, China 35 D12 40 53N 123 43 E
Lianshui, China 35 H10 33 42N 119 20 E
Lianyungang, China 35 G10 34 40N 119 11 E
Liao He →, China 35 D11 41 0N 121 50 E
Liaocheng, China 34 F8 36 28N 115 58 E
Liaodong Bandao, China 35 E12 40 0N 122 30 E
Liaodong Wan, China 35 D11 40 20N 121 10 E
Liaoning □, China 35 D12 41 40N 122 30 E
Liaoyang, China 35 D12 41 15N 122 58 E
Liaoyuan, China 35 C13 42 58N 125 2 E
Liaozhong, China 35 D12 41 23N 122 50 E
Liard →, Canada 72 A4 61 51N 121 18W
Liari, Pakistan 42 G2 25 37N 66 30 E
Libau = Liepāja, Latvia 9 H19 56 30N 21 0 E
Libby, U.S.A. 82 B6 48 23N 115 33W
Libenge, Zaïre 52 D3 3 40N 18 55 E
Liberal, Kans., U.S.A. 81 G4 37 3N 100 55W
Liberal, Mo., U.S.A. 81 G7 37 34N 94 31W
Liberec, Czech. 16 C8 50 47N 15 7 E
Liberia, Costa Rica 88 D2 10 40N 85 30W
Liberia ■, W. Afr. 50 G3 6 30N 9 30W
Liberty, Mo., U.S.A. 80 F7 39 15N 94 25W
Liberty, Tex., U.S.A. 81 K7 30 3N 94 48W
Lībīya, Sahrā', Africa 51 C9 25 0N 25 0 E
Libobo, Tanjung, Indonesia 37 E7 0 54S 128 28 E
Libode, S. Africa 57 E4 31 33S 29 2 E
Libonda, Zambia 53 G4 14 28S 23 12 E
Libourne, France 18 D3 44 55N 0 14W
Libramont, Belgium 15 E5 49 55N 5 23 E
Libreville, Gabon 52 D1 0 25N 9 26 E
Libya ■, N. Afr. 51 C8 27 0N 17 0 E

Libyan Desert = Lîbîya, Sahrâ', Africa 51 C9 25 0N 25 0 E
Licantén, Chile 94 D1 35 55S 72 0W
Licata, Italy 20 F5 37 6N 13 56 E
Licheng, China 34 F7 36 28N 113 20 E
Lichfield, U.K. 10 E6 52 41N 1 49W
Lichinga, Mozam. 55 E4 13 13S 35 11 E
Lichtenburg, S. Africa 56 D4 26 8S 26 8 E
Lida, Belarus 9 K21 53 53N 25 15 E
Lida, U.S.A. 83 H5 37 28N 117 30W
Lidköping, Sweden 9 G15 58 31N 13 14 E
Liebig, Mt., Australia 60 D5 23 18S 131 22 E
Liechtenstein ■, Europe 16 E5 47 8N 9 35 E
Liège, Belgium 15 D5 50 38N 5 35 E
Liège □, Belgium 15 D5 50 32N 5 35 E
Liegnitz = Legnica, Poland 16 C9 51 12N 16 10 E
Lienart, Zaïre 54 B2 3 3N 25 31 E
Lienyünchiangshih = Lianyungang, China 35 G10 34 40N 119 11 E
Lienz, Austria 16 E7 46 50N 12 46 E
Liepāja, Latvia 9 H19 56 30N 21 0 E
Lier, Belgium 15 C4 51 7N 4 34 E
Lièvre →, Canada 70 C4 45 31N 75 26W
Liffey →, Ireland 13 C5 53 21N 6 13W
Lifford, Ireland 13 B4 54 51N 7 29W
Lifudzin, Russia 30 B7 44 21N 134 58 E
Lightning Ridge, Australia 63 D4 29 22S 148 0 E
Liguria □, Italy 20 B3 44 30N 8 50 E
Ligurian Sea, Medit. S. 20 C3 43 20N 9 0 E
Lihou Reefs and Cays, Australia 62 B5 17 25S 151 40 E
Lihue, U.S.A. 74 H15 21 59N 159 23W
Lijiang, China 32 D5 26 55N 100 20 E
Likasi, Zaïre 55 E2 10 55S 26 48 E
Likati, Zaïre 52 D4 3 20N 24 0 E
Likoma I., Malawi 55 E3 12 3S 34 45 E
Likumburu, Tanzania 55 D4 9 43S 35 8 E
Lille, France 18 A5 50 38N 3 3 E
Lille Bælt, Denmark 9 J13 55 20N 9 45 E
Lillehammer, Norway 9 F14 61 8N 10 30 E
Lillesand, Norway 9 G13 58 15N 8 23 E
Lilleshall, U.K. 11 E5 52 44N 2 23W
Lillian Point, Mt., Australia 61 E4 27 40S 126 6 E
Lillooet →, Canada 72 D4 49 15N 121 57W
Lilongwe, Malawi 55 E3 14 0S 33 48 E
Liloy, Phil. 37 C6 8 4N 122 39 E
Lim →, Bos.-H. 21 C8 43 45N 19 15 E
Lima, Indonesia 37 E7 3 37S 128 4 E
Lima, Peru 92 F3 12 0S 77 0W
Lima, Mont., U.S.A. 82 D7 44 38N 112 36W
Lima, Ohio, U.S.A. 76 E3 40 44N 84 6W
Lima →, Portugal 19 B1 41 41N 8 50W
Limages, Canada 79 A9 45 20N 75 16W
Limassol, Cyprus 23 E12 34 42N 33 1 E
Limavady, U.K. 13 A5 55 3N 6 56W
Limavady □, U.K. 13 B5 55 0N 6 55W
Limay →, Argentina 96 D3 39 0S 68 0W
Limay Mahuida, Argentina 94 D2 37 10S 66 45W
Limbang, Brunei 36 D5 4 42N 115 6 E
Limbaži, Latvia 9 H21 57 31N 24 42 E
Limbdi, India 42 H4 22 34N 71 51 E
Limbe, Cameroon 50 H6 4 1N 9 10 E
Limbri, Australia 63 E5 31 3S 151 5 E
Limbunya, Australia 60 C4 17 14S 129 50 E
Limburg, Germany 16 C5 50 22N 8 4 E
Limburg □, Belgium 15 C5 51 2N 5 25 E
Limburg □, Neths. 15 C5 51 20N 5 55 E
Limeira, Brazil 95 A6 22 35S 47 28W
Limerick, Ireland 13 D3 52 40N 8 37W
Limerick □, Ireland 13 D3 52 30N 8 50W
Limestone, U.S.A. 78 D6 42 2N 78 38W
Limestone →, Canada 73 B10 56 31N 94 7W
Limfjorden, Denmark 9 H13 56 55N 9 0 E
Limia = Lima →, Portugal 19 B1 41 41N 8 50W
Limingen, Norway 8 D15 64 48N 13 35 E
Limmen Bight, Australia 62 A2 14 40S 135 35 E
Limmen Bight →, Australia 62 B2 15 7S 135 44 E
Límnos, Greece 21 E11 39 50N 25 5 E
Limoeiro do Norte, Brazil 93 E11 5 5S 38 0W
Limoges, France 18 D4 45 50N 1 15 E
Limón, Costa Rica 88 D3 10 0N 83 2W
Limon, U.S.A. 80 F3 39 16N 103 41W
Limousin, France 18 D4 45 30N 1 30 E
Limoux, France 18 E5 43 4N 2 12 E
Limpopo →, Africa 57 D5 25 5S 33 30 E
Limuru, Kenya 54 C4 1 2S 36 35 E
Lin Xian, China 34 F6 37 57N 110 58 E
Linares, Chile 94 D1 35 50S 71 40W
Linares, Mexico 87 C5 24 50N 99 40W
Linares, Spain 19 C4 38 10N 3 40W
Linares □, Chile 94 D1 36 0S 71 0W
Lincheng, China 34 F8 37 25N 114 30 E
Lincoln, Argentina 94 C3 34 55S 61 30W
Lincoln, N.Z. 59 K4 43 38S 172 30 E
Lincoln, U.K. 10 D7 53 14N 0 32W
Lincoln, Calif., U.S.A. 84 G5 38 54N 121 17W
Lincoln, Ill., U.S.A. 80 E10 40 9N 89 22W
Lincoln, Kans., U.S.A. 80 F5 39 3N 98 9W
Lincoln, Maine, U.S.A. 71 C6 45 22N 68 30W
Lincoln, N.H., U.S.A. 79 B13 44 3N 71 40W
Lincoln, N. Mex., U.S.A. 83 K11 33 30N 105 23W
Lincoln, Nebr., U.S.A. 80 E6 40 49N 96 41W
Lincoln Hav = Lincoln Sea, Arctic 4 A5 84 0N 55 0W
Lincoln Sea, Arctic 4 A5 84 0N 55 0W
Lincolnshire □, U.K. 10 D7 53 14N 0 32W
Lincolnshire Wolds, U.K. 10 D7 53 26N 0 13W
Lincolnton, U.S.A. 77 H5 35 29N 81 16W
Lind, U.S.A. 82 C4 46 58N 118 37W
Linda, U.S.A. 84 F5 39 8N 121 34W
Linden, Guyana 92 B7 6 0N 58 10W
Linden, Calif., U.S.A. 84 G5 38 1N 121 5W
Linden, Tex., U.S.A. 81 J7 33 1N 94 22W
Lindenhurst, U.S.A. 79 F11 40 41N 73 23W
Lindesnes, Norway 9 H12 57 58N 7 3 E
Líndhos, Greece 23 C10 36 6N 28 4 E
Lindi, Tanzania 55 D4 9 58S 39 38 E
Lindi □, Tanzania 55 D4 9 40S 38 30 E
Lindi →, Zaïre 54 B2 0 33N 25 5 E
Lindsay, Canada 70 D4 44 22N 78 43W
Lindsay, Calif., U.S.A. 83 H4 36 12N 119 5W
Lindsay, Okla., U.S.A. 81 H6 34 50N 97 38W
Lindsborg, U.S.A. 80 F6 38 35N 97 40W
Linfen, China 34 F6 36 3N 111 30 E
Ling Xian, China 34 F9 37 22N 116 30 E
Lingao, China 38 C7 19 56N 109 42 E
Lingayen, Phil. 37 A6 16 1N 120 14 E
Lingayen G., Phil. 37 A6 16 10N 120 15 E
Lingbi, China 35 H9 33 33N 117 33 E
Lingchuan, China 34 G7 35 45N 113 12 E
Lingen, Germany 16 B4 52 31N 7 19 E
Lingga, Indonesia 36 E2 0 12S 104 37 E
Lingga, Kepulauan, Indonesia 36 E2 0 10S 104 30 E
Lingga Arch. = Lingga, Kepulauan, Indonesia 36 E2 0 10S 104 30 E
Lingle, U.S.A. 80 D2 42 8N 104 21W
Lingqiu, China 34 E8 39 28N 114 22 E
Lingshi, China 34 F6 36 48N 111 48 E
Lingshou, China 34 E8 38 20N 114 20 E
Lingshui, China 38 C8 18 27N 110 0 E
Lingtai, China 34 G4 35 0N 107 40 E
Linguéré, Senegal 50 E1 15 25N 15 5W
Lingwu, China 34 E4 38 6N 106 20 E
Lingyuan, China 35 D10 41 10N 119 15 E
Linh Cam, Vietnam 38 C5 18 31N 105 31 E
Linhai, China 33 D7 28 50N 121 8 E
Linhares, Brazil 93 G10 19 25S 40 4W
Linhe, China 34 D4 40 48N 107 20 E
Linjiang, China 35 D14 41 50N 127 0 E
Linköping, Sweden 9 G16 58 28N 15 36 E
Linkou, China 35 B16 45 15N 130 18 E
Linlithgow, U.K. 12 F5 55 58N 3 37W
Linnhe, L., U.K. 12 E3 56 36N 5 25W
Linosa, I., Medit. S. 20 G5 35 51N 12 50 E
Linqi, China 34 G7 35 45N 113 52 E
Linqing, China 34 F8 36 50N 115 42 E
Linqu, China 35 F10 36 25N 118 30 E
Linru, China 34 G7 34 11N 112 52 E
Lins, Brazil 95 A6 21 40S 49 44W
Lintao, China 34 G2 35 18N 103 52 E
Lintlaw, Canada 73 C8 52 4N 103 14W
Linton, Canada 71 C5 47 15N 72 16W
Linton, Ind., U.S.A. 76 F2 39 2N 87 10W
Linton, N. Dak., U.S.A. 80 B4 46 16N 100 14W
Lintong, China 34 G5 34 20N 109 10 E
Linville, Australia 63 D5 26 50S 152 11 E
Linwood, Canada 78 C4 43 35N 80 43W
Linxi, China 35 C10 43 36N 118 2 E
Linxia, China 32 C5 35 36N 103 10 E
Linyanti →, Africa 56 B4 17 50S 25 5 E
Linyi, China 35 G10 35 5N 118 21 E
Linz, Austria 16 D8 48 18N 14 18 E
Linzhenzhen, China 34 F5 36 30N 109 59 E
Linzi, China 35 F10 36 50N 118 20 E
Lion, G. du, France 18 E6 43 10N 4 0 E
Lionárisso, Cyprus 23 D13 35 28N 34 8 E
Lions, G. of = Lion, G. du, France 18 E6 43 10N 4 0 E
Lion's Den, Zimbabwe 55 F3 17 15S 30 5 E
Lion's Head, Canada 70 D3 44 58N 81 15W
Lipa, Phil. 37 B6 13 57N 121 10 E
Lipali, Mozam. 55 F4 15 50S 35 50 E
Lípari, Italy 20 E6 38 26N 14 58 E
Lípari, Is. = Eólie, Ís., Italy 20 E6 38 30N 14 57 E
Lipcani, Moldova 17 D14 48 14N 26 48 E
Lipetsk, Russia 24 D6 52 37N 39 35 E
Lipkany = Lipcani, Moldova 17 D14 48 14N 26 48 E
Lipovcy Manzovka, Russia 30 B6 44 12N 132 26 E
Lipovets, Ukraine 17 D15 49 12N 29 1 E
Lippe →, Germany 16 C4 51 39N 6 36 E
Lipscomb, U.S.A. 81 G4 36 14N 100 16W
Liptrap C., Australia 63 F4 38 50S 145 55 E
Lira, Uganda 54 B3 2 17N 32 57 E
Liria, Spain 19 C5 39 37N 0 35W
Lisala, Zaïre 52 D4 2 12N 21 38 E
Lisboa, Portugal 19 C1 38 42N 9 10W
Lisbon = Lisboa, Portugal 19 C1 38 42N 9 10W
Lisbon, N. Dak., U.S.A. 80 B6 46 27N 97 41W
Lisbon, N.H., U.S.A. 79 B13 44 13N 71 55W
Lisbon, Ohio, U.S.A. 78 F4 40 46N 80 46W
Lisburn, U.K. 13 B5 54 31N 6 3W
Lisburne, C., U.S.A. 68 B3 68 53N 166 13W
Liscannor, B., Ireland 13 D2 52 55N 9 24W
Lishi, China 34 F6 37 31N 111 8 E
Lishu, China 35 C13 43 20N 124 18 E
Lisianski I., Pac. Oc. 64 E10 26 2N 174 0W
Lisichansk = Lysychansk, Ukraine 25 E6 48 55N 38 30 E
Lisieux, France 18 B4 49 10N 0 12 E
Liski, Russia 25 D6 51 3N 39 30 E
Lismore, Australia 63 D5 28 44S 153 21 E
Lismore, Ireland 13 D4 52 8N 7 55W
Lisse, Neths. 15 B4 52 16N 4 33 E
Lista, Norway 9 G12 58 7N 6 39 E
Lister, Mt., Antarctica 5 D11 78 0S 162 0 E
Liston, Australia 63 D5 28 39S 152 6 E
Listowel, Canada 70 D3 43 44N 80 58W
Listowel, Ireland 13 D2 52 27N 9 29W
Litang, Malaysia 37 C5 5 27N 118 31 E
Litani →, Lebanon 47 B4 33 20N 35 15 E
Litchfield, Calif., U.S.A. 84 E6 40 24N 120 23W
Litchfield, Conn., U.S.A. 79 E11 41 45N 73 11W
Litchfield, Ill., U.S.A. 80 F10 39 11N 89 39W
Litchfield, Minn., U.S.A. 80 C7 45 8N 94 32W
Lithgow, Australia 63 E5 33 25S 150 8 E
Líthinon, Ákra, Greece 23 E6 34 55N 24 44 E
Lithuania ■, Europe 9 J20 55 30N 24 0 E
Litoměřice, Czech. 16 C8 50 33N 14 10 E
Little Abaco I., Bahamas 88 A4 26 50N 77 30W
Little Barrier I., N.Z. 59 G5 36 12S 175 8 E
Little Belt Mts., U.S.A. 82 C8 46 40N 110 45W
Little Blue →, U.S.A. 80 F6 39 42N 96 41W
Little Cadotte →, Canada 72 B5 56 41N 117 6W
Little Cayman, I., Cayman Is. 88 C3 19 41N 80 3W
Little Churchill →, Canada 73 B9 57 30N 95 22W
Little Colorado →, U.S.A. 83 H8 36 12N 111 48W
Little Current, Canada 70 C3 45 55N 82 0W
Little Current →, Canada 70 B3 50 57N 84 36W
Little Falls, Minn., U.S.A. 80 C7 45 59N 94 22W
Little Falls, N.Y., U.S.A. 79 C10 43 3N 74 51W
Little Fork →, U.S.A. 80 A8 48 31N 93 35W
Little Grand Rapids, Canada 73 C9 52 0N 95 29W
Little Humboldt →, U.S.A. 82 F5 41 1N 117 43W
Little Inagua I., Bahamas 89 B5 21 40N 73 50W
Little Karoo, S. Africa 56 E3 33 45S 21 0 E
Little Lake, U.S.A. 85 K9 35 56N 117 55W
Little Laut Is. = Laut Kecil, Kepulauan, Indonesia 36 E5 4 45S 115 40 E
Little Minch, U.K. 12 D2 57 35N 6 45W
Little Missouri →, U.S.A. 80 B3 47 36N 102 25W
Little Ouse →, U.K. 11 E8 52 22N 1 12 E
Little Rann, India 42 H4 23 25N 71 25 E
Little Red →, U.S.A. 81 H9 35 11N 91 27W
Little River, N.Z. 59 K4 43 45S 172 49 E
Little Rock, U.S.A. 81 H8 34 45N 92 17W
Little Ruaha →, Tanzania 54 D4 7 57S 37 53 E
Little Sable Pt., U.S.A. 76 D2 43 38N 86 33W
Little Sioux →, U.S.A. 80 E6 41 48N 96 4W
Little Smoky →, Canada 72 C5 54 44N 117 11W
Little Snake →, U.S.A. 82 F9 40 27N 108 26W
Little Valley, U.S.A. 78 D6 42 15N 78 48W
Little Wabash →, U.S.A. 76 G1 37 55N 88 5W
Littlefield, U.S.A. 81 J3 33 55N 102 20W
Littlefork, U.S.A. 80 A8 48 24N 93 34W
Littlehampton, U.K. 11 G7 50 49N 0 32W
Littleton, U.S.A. 79 B13 44 18N 71 46W
Liu He →, China 35 D11 40 55N 121 35 E
Liuba, China 34 H4 33 38N 106 55 E
Liugou, China 35 D10 40 57N 118 15 E
Liuhe, China 35 C13 42 17N 125 43 E
Liukang Tenggaja, Indonesia 37 F5 6 45S 118 50 E
Liuli, Tanzania 55 E3 11 3S 34 38 E
Liuwa Plain, Zambia 53 G4 14 20S 22 30 E
Liuzhou, China 33 D5 24 22N 109 22 E
Liuzhuang, China 35 H11 33 12N 120 18 E
Livadhia, Cyprus 23 E12 34 57N 33 38 E
Live Oak, Calif., U.S.A. 84 F5 39 17N 121 40W
Live Oak, Fla., U.S.A. 77 K4 30 18N 82 59W
Liveras, Cyprus 23 D11 35 23N 32 57 E
Liveringa, Australia 60 C3 18 3S 124 10 E
Livermore, U.S.A. 84 H5 37 41N 121 47W
Livermore, Mt., U.S.A. 81 K2 30 38N 104 11W
Liverpool, Australia 63 E5 33 54S 150 58 E
Liverpool, Canada 71 D7 44 5N 64 41W
Liverpool, U.K. 10 D5 53 25N 3 0W
Liverpool Plains, Australia 63 E5 31 15S 150 15 E
Liverpool Ra., Australia 63 E5 31 50S 150 30 E
Livingston, Guatemala 88 C2 15 50N 88 50W
Livingston, Calif., U.S.A. 84 H6 37 23N 120 43W
Livingston, Mont., U.S.A. 82 D8 45 40N 110 34W
Livingston, Tex., U.S.A. 81 K7 30 43N 94 56W
Livingstone, Zambia 55 F2 17 46S 25 52 E
Livingstone Mts., Tanzania 55 D3 9 40S 34 20 E
Livingstonia, Malawi 55 E3 10 38S 34 5 E
Livny, Russia 24 D6 52 30N 37 30 E
Livonia, U.S.A. 76 D4 42 23N 83 23W
Livorno, Italy 20 C4 43 33N 10 19 E
Livramento, Brazil 95 C4 30 55S 55 30W
Liwale, Tanzania 55 D4 9 48S 37 58 E
Liwale □, Tanzania 55 D4 9 0S 38 0 E
Lizard I., Australia 62 A4 14 42S 145 30 E
Lizard Pt., U.K. 11 H2 49 57N 5 13W
Ljubljana, Slovenia 16 E8 46 4N 14 33 E
Ljungan →, Sweden 9 E17 62 18N 17 23 E
Ljungby, Sweden 9 H15 56 49N 13 55 E
Ljusdal, Sweden 9 F17 61 46N 16 3 E
Ljusnan →, Sweden 9 F17 61 12N 17 8 E
Ljusne, Sweden 9 F17 61 13N 17 7 E
Llancanelo, Salina, Argentina 94 D2 35 40S 69 8W
Llandeilo, U.K. 11 F3 51 53N 3 59W
Llandovery, U.K. 11 F4 51 59N 3 48W
Llandrindod Wells, U.K. 11 E4 52 14N 3 22W
Llandudno, U.K. 10 D4 53 19N 3 50W
Llanelli, U.K. 11 F3 51 41N 4 10W
Llanes, Spain 19 A3 43 25N 4 50W
Llangollen, U.K. 10 E4 52 58N 3 11W
Llanidloes, U.K. 11 E4 52 27N 3 31W
Llano, U.S.A. 81 K5 30 45N 98 41W
Llano →, U.S.A. 81 K5 30 39N 98 26W
Llano Estacado, U.S.A. 81 J3 33 30N 103 0W
Llanos, S. Amer. 92 B4 5 0N 71 35W
Llebetx, C., Spain 22 B9 39 33N 2 18 E
Lleida = Lérida, Spain 19 B6 41 37N 0 39 E
Llentrisca, C., Spain 22 C7 38 52N 1 15 E
Llera, Mexico 87 C5 23 19N 99 1W
Llico, Chile 94 C1 34 46S 72 5W
Llobregat →, Spain 19 B7 41 19N 2 9 E
Lloret de Mar, Spain 19 B7 41 41N 2 53 E
Lloyd B., Australia 62 A3 12 45S 143 27 E
Lloyd L., Canada 73 B7 57 22N 108 57W
Lloydminster, Canada 73 C6 53 17N 110 0W
Lluchmayor, Spain 22 B9 39 29N 2 53 E
Llullaillaco, Volcán, S. Amer. 94 A2 24 43S 68 30W
Lo →, Vietnam 38 B5 21 18N 105 25 E
Loa, U.S.A. 83 G8 38 24N 111 39W
Loa →, Chile 94 A1 21 26S 70 41W
Lobatse, Botswana 56 D4 25 12S 25 40 E
Lobería, Argentina 94 D4 38 10S 58 40W
Lobito, Angola 53 G2 12 18S 13 35 E
Lobos, Argentina 94 D4 35 10S 59 0W
Lobos, I., Mexico 86 B2 27 15N 110 30W
Lobos, I. de, Canary Is. 22 F6 28 45N 13 50W
Loc Binh, Vietnam 38 B6 21 46N 106 54 E
Loc Ninh, Vietnam 39 G6 11 50N 106 34 E
Locarno, Switz. 16 E5 46 10N 8 47 E
Loch Garman = Wexford, Ireland 13 D5 52 20N 6 28W
Lochaber, U.K. 12 E4 56 59N 5 1W
Lochcarron, U.K. 12 D3 57 25N 5 31W
Lochem, Neths. 15 B6 52 9N 6 26 E
Loches, France 18 C4 47 7N 1 0 E
Lochgelly, U.K. 12 E5 56 7N 3 19W
Lochgilphead, U.K. 12 E3 56 2N 5 26W
Lochinver, U.K. 12 C3 58 9N 5 14W
Lochnagar, Australia 62 C4 23 33S 145 38 E
Lochnagar, U.K. 12 E5 56 57N 3 15W
Lochy →, U.K. 12 E3 56 52N 5 3W
Lock, Australia 63 E2 33 34S 135 46 E
Lock Haven, U.S.A. 78 E7 41 8N 77 28W
Lockeford, U.S.A. 84 G5 38 10N 121 9W
Lockeport, Canada 71 D6 43 47N 65 4W
Lockerbie, U.K. 12 F5 55 7N 3 21W
Lockhart, U.S.A. 81 L6 29 53N 97 40W
Lockhart, L., Australia 61 F2 33 15S 119 3 E
Lockney, U.S.A. 81 H4 34 7N 101 27W
Lockport, U.S.A. 78 C6 43 10N 78 42W
Lod, Israel 47 D3 31 57N 34 54 E
Lodeinoye Pole, Russia 24 B5 60 44N 33 33 E
Lodge Grass, U.S.A. 82 D10 45 19N 107 22W
Lodgepole, U.S.A. 80 E3 41 9N 102 38W
Lodgepole Cr. →, U.S.A. 80 E2 41 20N 104 30W
Lodhran, Pakistan 42 E4 29 32N 71 30 E
Lodi, Italy 20 B3 45 19N 9 30 E
Lodi, U.S.A. 84 G5 38 8N 121 16W
Lodja, Zaïre 54 C1 3 30S 23 23 E
Lodwar, Kenya 54 B4 3 10N 35 40 E
Łódź, Poland 17 C10 51 45N 19 27 E
Loei, Thailand 38 D3 17 29N 101 35 E
Loengo, Zaïre 54 C2 4 48S 26 30 E
Loeriesfontein, S. Africa 56 E2 31 0S 19 26 E
Lofoten, Norway 8 B15 68 30N 14 0 E
Logan, Kans., U.S.A. 80 F5 39 40N 99 34W
Logan, Ohio, U.S.A. 76 F4 39 32N 82 25W
Logan, Utah, U.S.A. 82 F8 41 44N 111 50W
Logan, W. Va., U.S.A. 76 G5 37 51N 81 59W
Logan, Mt., Canada 68 B5 60 31N 140 22W
Logan Pass, U.S.A. 72 D6 48 41N 113 44W
Logandale, U.S.A. 85 J12 36 36N 114 29W
Logansport, Ind., U.S.A. 76 E2 40 45N 86 22W
Logansport, La., U.S.A. 81 K8 31 58N 94 0W
Logone →, Chad 51 F8 12 6N 15 2 E
Logroño, Spain 19 A4 42 28N 2 27W
Lohardaga, India 43 H11 23 27N 84 45 E
Lohja, Finland 9 F21 60 12N 24 5 E
Loi-kaw, Burma 41 K20 19 40N 97 17 E
Loimaa, Finland 9 F20 60 50N 23 5 E
Loir →, France 18 C3 47 33N 0 32W
Loire →, France 18 C2 47 16N 2 10W
Loja, Ecuador 92 D3 3 59S 79 16W
Loja, Spain 19 D3 37 10N 4 10W
Loji, Indonesia 37 E7 1 38S 127 28 E
Lokandu, Zaïre 54 C2 2 30S 25 45 E
Lokeren, Belgium 15 C3 51 6N 3 59 E
Lokichokio, Kenya 54 B3 4 19N 34 13 E
Lokitaung, Kenya 54 B4 4 12N 35 48 E
Lokkan tekojärvi, Finland 8 C22 67 55N 27 35 E
Lokoja, Nigeria 50 G6 7 47N 6 45 E
Lokolama, Zaïre 52 E3 2 35S 19 50 E
Lola, Mt., U.S.A. 84 F6 39 26N 120 22W
Loliondo, Tanzania 54 C4 2 2S 35 39 E
Lolland, Denmark 9 J14 54 45N 11 30 E
Lolo, U.S.A. 82 C6 46 45N 114 5W
Lom, Bulgaria 21 C10 43 48N 23 12 E
Lom Kao, Thailand 38 D3 16 53N 101 14 E
Lom Sak, Thailand 38 D3 16 47N 101 15 E
Loma, U.S.A. 82 C8 47 56N 110 30W
Loma Linda, U.S.A. 85 L9 34 3N 117 16W
Lomami →, Zaïre 54 B1 0 46N 24 16 E
Lomas de Zamóra, Argentina 94 C4 34 45S 58 25W
Lombadina, Australia 60 C3 16 31S 122 54 E
Lombárdia □, Italy 20 B3 45 40N 9 30 E
Lombardy = Lombárdia □, Italy 20 B3 45 40N 9 30 E
Lomblen, Indonesia 37 F6 8 30S 123 32 E
Lombok, Indonesia 36 F5 8 45S 116 30 E
Lomé, Togo 50 G5 6 9N 1 20 E
Lomela, Zaïre 52 E4 2 19S 23 15 E
Lomela →, Zaïre 52 E4 0 15S 20 40 E
Lometa, U.S.A. 81 K5 31 13N 98 24W
Lomié, Cameroon 52 D2 3 13N 13 38 E
Lomond, Canada 72 C6 50 24N 112 36W
Lomond, L., U.K. 12 E4 56 8N 4 38W
Lomphat, Cambodia 38 F6 13 30N 106 59 E
Lompobatang, Indonesia 37 F5 5 24S 119 56 E
Lompoc, U.S.A. 85 L6 34 38N 120 28W
Łomża, Poland 17 B12 53 10N 22 2 E
Loncoche, Chile 96 D2 39 20S 72 50W
Londa, India 40 M9 15 30N 74 30 E
Londiani, Kenya 54 C4 0 10S 35 33 E
London, Canada 70 D3 42 59N 81 15W
London, U.K. 11 F7 51 30N 0 3W
London, Ky., U.S.A. 76 G3 37 8N 84 5W
London, Ohio, U.S.A. 76 F4 39 53N 83 27W
London, Greater □, U.K. 11 F7 51 36N 0 5W
Londonderry, U.K. 13 B4 55 0N 7 20W
Londonderry □, U.K. 13 B4 55 0N 7 20W
Londonderry, C., Australia 60 B4 13 45S 126 55 E
Londonderry, I., Chile 96 H2 55 0S 71 0W
Londrina, Brazil 95 A5 23 18S 51 10W
Lone Pine, U.S.A. 83 H4 36 36N 118 4W
Long Beach, Calif., U.S.A. 85 M8 33 47N 118 11W
Long Beach, N.Y., U.S.A. 79 F11 40 35N 73 39W
Long Beach, Wash., U.S.A. 84 D2 46 21N 124 3W
Long Branch, U.S.A. 79 F11 40 18N 74 0W
Long Creek, U.S.A. 82 D4 44 43N 119 6W
Long Eaton, U.K. 10 E6 52 53N 1 15W
Long I., Australia 62 C4 22 8S 149 53 E
Long I., Bahamas 89 B4 23 20N 75 10W
Long I., U.S.A. 79 F11 40 45N 73 30W
Long Island Sd., U.S.A. 79 E12 41 10N 73 0W
Long L., Canada 70 C2 49 30N 86 50W
Long Lake, U.S.A. 79 C10 43 58N 74 25W
Long Pine, U.S.A. 80 D5 42 32N 99 42W
Long Point B., Canada 78 D4 42 40N 80 10W
Long Pt., Nfld., Canada 71 C8 48 47N 58 46W
Long Pt., Ont., Canada 78 D4 42 35N 80 2W
Long Range Mts., Canada 71 C8 49 30N 57 30W
Long Reef, Australia 60 B4 14 1S 125 48 E
Long Str. = Longa, Proliv, Russia 4 C16 70 0N 175 0 E
Long Thanh, Vietnam 39 G6 10 47N 106 57 E
Long Xian, China 34 G4 34 55N 106 55 E
Long Xuyen, Vietnam 39 G5 10 19N 105 28 E
Longa, Proliv, Russia 4 C16 70 0N 175 0 E
Longde, China 34 G4 35 30N 106 20 E
Longford, Australia 62 G4 41 32S 147 3 E
Longford, Ireland 13 C4 53 43N 7 49W
Longford □, Ireland 13 C4 53 42N 7 45W
Longguan, China 34 D8 40 45N 115 30 E
Longhua, China 35 D9 41 18N 117 45 E
Longido, Tanzania 54 C4 2 43S 36 42 E
Longiram, Indonesia 36 E5 0 5S 115 45 E
Longkou, China 35 F11 37 40N 120 18 E

Longlac, Canada 70 C2 49 45N 86 25W
Longmont, U.S.A. 80 E2 40 10N 105 6W
Longnawan, Indonesia 36 D4 1 51N 114 55 E
Longreach, Australia 62 C3 23 28S 144 14 E
Longton, Australia 62 C4 20 58S 145 55 E
Longtown, U.K. 11 F5 51 58N 2 59W
Longueuil, Canada 79 A11 45 32N 73 28W
Longview, Canada 72 C6 50 32N 114 10W
Longview, Tex., U.S.A. 81 J7 32 30N 94 44W
Longview, Wash., U.S.A. 84 D4 46 8N 122 57W
Longxi, China 34 G3 34 53N 104 40 E
Lonoke, U.S.A. 81 H9 34 47N 91 54W
Lons-le-Saunier, France 18 C6 46 40N 5 31 E
Lookout, C., Canada 70 A3 55 18N 83 56W
Lookout, C., U.S.A. 77 H7 34 35N 76 32W
Loolmalasin, Tanzania 54 C4 3 0S 35 53 E
Loon →, Alta., Canada 72 B5 57 8N 115 3W
Loon →, Man., Canada 73 B8 55 53N 101 59W
Loon Lake, Canada 73 C7 54 2N 109 10W
Loongana, Australia 61 F4 30 52S 127 5 E
Loop Hd., Ireland 13 D2 52 34N 9 56W
Lop Buri, Thailand 38 E3 14 48N 100 37 E
Lop Nor = Lop Nur, China 32 B4 40 20N 90 10 E
Lop Nur, China 32 B4 40 20N 90 10 E
Lopatina, G., Russia 27 D15 50 47N 143 10 E
Lopez, C., Gabon 52 E1 0 47S 8 40 E
Lopphavet, Norway 8 A19 70 27N 21 15 E
Lora →, Afghan. 40 D4 31 35N 65 50 E
Lora, Hamun-i-, Pakistan 40 E4 29 38N 64 58 E
Lora Cr. →, Australia 63 D2 28 10S 135 22 E
Lora del Río, Spain 19 D3 37 39N 5 33W
Lorain, U.S.A. 78 E2 41 28N 82 11W
Loralai, Pakistan 42 D3 30 20N 68 41 E
Lorca, Spain 19 D5 37 41N 1 42W
Lord Howe I., Pac. Oc. 64 L7 31 33S 159 6 E
Lord Howe Ridge, Pac. Oc. 64 L8 30 0S 162 30 E
Lordsburg, U.S.A. 83 K9 32 21N 108 43W
Loreto, Brazil 93 E9 7 5S 45 10W
Loreto, Mexico 86 B2 26 1N 111 21W
Lorient, France 18 C2 47 45N 3 23W
Lorn, U.K. 12 E3 56 26N 5 10W
Lorn, Firth of, U.K. 12 E3 56 20N 5 40W
Lorne, Australia 63 F3 38 33S 143 59 E
Lorovouno, Cyprus 23 D11 35 8N 32 36 E
Lorraine, France 18 B7 48 53N 6 0 E
Lorrainville, Canada 70 C4 47 21N 79 23W
Los Alamos, Calif., U.S.A. 85 L6 34 44N 120 17W
Los Alamos, N. Mex., U.S.A. 83 J10 35 53N 106 19W
Los Altos, U.S.A. 84 H4 37 23N 122 7W
Los Andes, Chile 94 C1 32 50S 70 40W
Los Angeles, Chile 94 D1 37 28S 72 23W
Los Angeles, U.S.A. 85 L8 34 4N 118 15W
Los Angeles Aqueduct, U.S.A. 85 K9 35 22N 118 5W
Los Banos, U.S.A. 83 H3 37 4N 120 51W
Los Blancos, Argentina 94 A3 23 40S 62 30W
Los Cristianos, Canary Is. 22 F3 28 3N 16 42W
Los Gatos, U.S.A. 84 H5 37 14N 121 59W
Los Hermanos, Venezuela 89 D7 11 45N 64 25W
Los Islotes, Canary Is. 22 E6 29 4N 13 44W
Los Llanos de Aridane, Canary Is. 22 F2 28 38N 17 54W
Los Lunas, U.S.A. 83 J10 34 48N 106 44W
Los Mochis, Mexico 86 B3 25 45N 108 57W
Los Olivos, U.S.A. 85 L6 34 40N 120 7W
Los Palacios, Cuba 88 B3 22 35N 83 15W
Los Reyes, Mexico 86 D4 19 34N 102 30W
Los Roques, Venezuela 92 A5 11 50N 66 45W
Los Testigos, Venezuela 92 A6 11 23N 63 6W
Los Vilos, Chile 94 C1 32 10S 71 30W
Loshkalakh, Russia 27 C15 62 45N 147 20 E
Lošinj, Croatia 16 F8 44 30N 14 30 E
Lossiemouth, U.K. 12 D5 57 42N 3 17W
Lot →, France 18 D4 44 18N 0 20 E
Lota, Chile 94 D1 37 5S 73 10W
Loṭfābād, Iran 45 B8 37 32N 59 20 E
Lothair, S. Africa 57 D5 26 22S 30 27 E
Loubomo, Congo 52 E2 4 9S 12 47 E
Loudon, U.S.A. 77 H3 35 45N 84 20W
Loudonville, U.S.A. 78 F2 40 38N 82 14W
Louga, Senegal 50 E1 15 45N 16 5W
Loughborough, U.K. 10 E6 52 47N 1 11W
Loughrea, Ireland 13 C3 53 12N 8 33W
Loughros More B., Ireland 13 B3 54 48N 8 32W
Louis Trichardt, S. Africa 57 C4 23 1S 29 43 E
Louis XIV, Pte., Canada 70 B4 54 37N 79 45W
Louisa, U.S.A. 76 F4 38 7N 82 36W
Louisbourg, Canada 71 C8 45 55N 60 0W
Louise I., Canada 72 C2 52 55N 131 50W
Louiseville, Canada 70 C5 46 20N 72 56W
Louisiade Arch., Papua N. G. 64 J7 11 10S 153 0 E
Louisiana, U.S.A. 80 F9 39 27N 91 3W
Louisiana □, U.S.A. 81 K9 30 50N 92 0W
Louisville, Ky., U.S.A. 76 F3 38 15N 85 46W
Louisville, Miss., U.S.A. 81 J10 33 7N 89 3W
Loulé, Portugal 19 D1 37 9N 8 0W
Loup City, U.S.A. 80 E5 41 17N 98 58W
Lourdes, France 18 E3 43 6N 0 3W
Lourdes-du-Blanc-Sablon, Canada 71 B8 51 24N 57 12W
Lourenço-Marques = Maputo, Mozam. 57 D5 25 58S 32 32 E
Louth, Australia 63 E4 30 30S 145 8 E
Louth, Ireland 13 C5 53 58N 6 32W
Louth, U.K. 10 D7 53 22N 0 1W
Louth □, Ireland 13 C5 53 56N 6 34W
Louvain = Leuven, Belgium 15 D4 50 52N 4 42 E
Louwsburg, S. Africa 57 D5 27 37S 31 7 E
Love, Canada 73 C8 53 29N 104 10W
Lovech, Bulgaria 21 C11 43 8N 24 42 E
Loveland, U.S.A. 80 E2 40 24N 105 5W
Lovell, U.S.A. 82 D9 44 50N 108 24W
Lovelock, U.S.A. 82 F4 40 11N 118 28W
Loviisa, Finland 9 F22 60 28N 26 12 E
Loving, U.S.A. 81 J2 32 17N 104 6W
Lovington, U.S.A. 81 J3 32 57N 103 21W
Lovisa = Loviisa, Finland 9 F22 60 28N 26 12 E
Low Pt., Australia 61 F4 32 25S 127 25 E
Low Tatra = Nizké Tatry, Slovak Rep. 17 D10 48 55N 19 30 E
Lowa, Zaïre 54 C2 1 25S 25 47 E
Lowa →, Zaïre 54 C2 1 24S 25 51 E
Lowell, U.S.A. 79 D13 42 38N 71 19W
Lower Arrow L., Canada 72 D5 49 40N 118 5W
Lower California = Baja California, Mexico 86 A1 31 10N 115 12W
Lower Hutt, N.Z. 59 J5 41 10S 174 55 E
Lower L., U.S.A. 82 F3 41 16N 120 2W
Lower Lake, U.S.A. 84 G4 38 55N 122 37W
Lower Post, Canada 72 B3 59 58N 128 30W
Lower Red L., U.S.A. 80 B7 47 58N 95 0W
Lower Saxony = Niedersachsen □, Germany 16 B5 53 8N 9 0 E
Lower Tunguska = Tunguska, Nizhnyaya →, Russia 27 C9 65 48N 88 4 E
Lowestoft, U.K. 11 E9 52 29N 1 45 E
Łowicz, Poland 17 B10 52 6N 19 55 E
Lowville, U.S.A. 79 C9 43 47N 75 29W
Loxton, Australia 63 E3 34 28S 140 31 E
Loxton, S. Africa 56 E3 31 30S 22 22 E
Loyalton, U.S.A. 84 F6 39 41N 120 14W
Loyalty Is. = Loyauté, Is., N. Cal. 64 K8 20 50S 166 30 E
Loyang = Luoyang, China 34 G7 34 40N 112 26 E
Loyauté, Is., N. Cal. 64 K8 20 50S 166 30 E
Loyev = Loyew, Belarus 17 C16 51 56N 30 46 E
Loyew, Belarus 17 C16 51 56N 30 46 E
Loyoro, Uganda 54 B3 3 22N 34 14 E
Luachimo, Angola 52 F4 7 23S 20 48 E
Luacono, Angola 52 G4 11 15S 21 37 E
Lualaba →, Zaïre 54 B5 0 26N 25 20 E
Luampa, Zambia 55 F1 15 4S 24 20 E
Luan Chau, Vietnam 38 B4 21 38N 103 24 E
Luan He →, China 35 E10 39 20N 119 5 E
Luan Xian, China 35 E10 39 40N 118 40 E
Luancheng, China 34 F8 37 53N 114 40 E
Luanda, Angola 52 F2 8 50S 13 15 E
Luang Prabang, Laos 38 C4 19 52N 102 10 E
Luang Thale, Thailand 39 J3 7 30N 100 15 E
Luangwa, Zambia 55 F3 15 35S 30 16 E
Luangwa →, Zambia 55 E3 14 25S 30 25 E
Luangwa Valley, Zambia 55 E3 13 30S 31 30 E
Luanne, China 35 D9 40 55N 117 40 E
Luanping, China 35 D9 40 53N 117 23 E
Luanshya, Zambia 55 E2 13 3S 28 28 E
Luapula □, Zambia 55 E2 11 0S 29 0 E
Luapula →, Africa 55 D2 9 26S 28 33 E
Luarca, Spain 19 A2 43 32N 6 32W
Luashi, Zaïre 55 E1 10 50S 23 36 E
Luau, Angola 52 G4 10 40S 22 10 E
Lubalo, Angola 52 F3 9 10S 19 15 E
Lubana, Ozero = Lubānas Ezers, Latvia 9 H22 56 45N 27 0 E
Lubānas Ezers, Latvia 9 H22 56 45N 27 0 E
Lubang Is., Phil. 37 B6 13 50N 120 12 E
Lubbock, U.S.A. 81 J4 33 35N 101 51W
Lübeck, Germany 16 B6 53 52N 10 40 E
Lubefu, Zaïre 54 C1 4 47S 24 27 E
Lubefu →, Zaïre 54 C1 4 10S 23 0 E
Lubero = Luofu, Zaïre 54 C2 0 10S 29 15 E
Lubicon L., Canada 72 B5 56 23N 115 56W
Lubin, Poland 16 C9 51 24N 16 11 E
Lublin, Poland 17 C12 51 12N 22 38 E
Lubnān, J., Lebanon 47 B4 33 50N 35 45 E
Lubny, Ukraine 26 D4 50 3N 32 58 E
Lubongola, Zaïre 54 C2 2 35S 27 50 E
Lubuagan, Phil. 37 A6 17 21N 121 10 E
Lubudi →, Zaïre 55 D2 9 0S 25 35 E
Lubuk Antu, Malaysia 36 D4 1 3N 111 50 E
Lubuklinggau, Indonesia 36 E2 3 15S 102 55 E
Lubuksikaping, Indonesia 36 D2 0 10N 100 15 E
Lubumbashi, Zaïre 55 E2 11 40S 27 28 E
Lubunda, Zaïre 54 D2 5 12S 26 41 E
Lubungu, Zambia 55 E2 14 35S 26 24 E
Lubutu, Zaïre 54 C2 0 45S 26 30 E
Luc An Chau, Vietnam 38 A5 22 6N 104 43 E
Lucan, Canada 78 C3 43 11N 81 24W
Lucca, Italy 20 C4 43 50N 10 29 E
Luce Bay, U.K. 12 G4 54 45N 4 48W
Lucea, Jamaica 88 C4 18 25N 78 10W
Lucedale, U.S.A. 77 K1 30 56N 88 35W
Lucena, Phil. 37 B6 13 56N 121 37 E
Lucena, Spain 19 D3 37 27N 4 31W
Lučenec, Slovak Rep. 17 D10 48 18N 19 42 E
Lucerne = Luzern, Switz. 16 E5 47 3N 8 18 E
Lucerne, U.S.A. 84 F4 39 6N 122 48W
Lucerne Valley, U.S.A. 85 L10 34 27N 116 57W
Lucero, Mexico 86 A3 30 49N 106 30W
Lucheng, China 34 F7 36 20N 113 11 E
Lucheringo →, Mozam. 55 E4 11 43S 36 17 E
Lucira, Angola 53 G2 14 0S 12 35 E
Luckenwalde, Germany 16 B7 52 5N 13 10 E
Lucknow, India 43 F9 26 50N 81 0 E
Lüda = Dalian, China 35 E11 38 50N 121 40 E
Lüderitz, Namibia 56 D2 26 41S 15 8 E
Ludewe □, Tanzania 55 D3 10 0S 34 50 E
Ludhiana, India 42 D6 30 57N 75 56 E
Ludington, U.S.A. 76 D2 43 57N 86 27W
Ludlow, U.K. 11 E5 52 22N 2 42W
Ludlow, Calif., U.S.A. 85 L10 34 43N 116 10W
Ludlow, Vt., U.S.A. 79 C12 43 24N 72 42W
Ludvika, Sweden 9 F16 60 8N 15 14 E
Ludwigsburg, Germany 16 D5 48 53N 9 11 E
Ludwigshafen, Germany 16 D5 49 29N 8 26 E
Luebo, Zaïre 52 F4 5 21S 21 23 E
Lueki, Zaïre 54 C2 3 20S 25 48 E
Luena, Zaïre 55 D2 9 28S 25 43 E
Luena, Zambia 55 E3 10 40S 30 25 E
Lüeyang, China 34 H4 33 22N 106 10 E
Lufira →, Zaïre 55 D2 9 30S 27 0 E
Lufkin, U.S.A. 81 K7 31 21N 94 44W
Lufupa, Zaïre 55 E1 10 37S 24 56 E
Luga, Russia 24 C4 58 40N 29 55 E
Lugano, Switz. 16 E5 46 0N 8 57 E
Lugansk = Luhansk, Ukraine 25 E6 48 38N 39 15 E
Lugard's Falls, Kenya 54 C4 3 6S 38 41 E
Lugela, Mozam. 55 F4 16 25S 36 43 E
Lugenda →, Mozam. 55 E4 11 25S 38 33 E
Lugh Ganana, Somali Rep. 46 G3 3 48N 42 34 E
Lugnaquilla, Ireland 13 D5 52 58N 6 28W
Lugo, Italy 20 B4 44 25N 11 54 E
Lugo, Spain 19 A2 43 2N 7 35W
Lugoj, Romania 17 F11 45 42N 21 57 E
Lugovoy, Kazakstan 26 E8 42 55N 72 43 E
Luhansk, Ukraine 25 E6 48 38N 39 15 E
Luiana, Angola 56 B3 17 25S 22 59 E
Luimneach = Limerick, Ireland 13 D3 52 40N 8 37W
Luís Correia, Brazil 93 D10 3 0S 41 35W
Luitpold Coast, Antarctica 5 D1 78 30S 32 0W
Luiza, Zaïre 52 F4 7 40S 22 30 E
Luizi, Zaïre 54 D2 6 0S 27 25 E
Luján, Argentina 94 C4 34 45S 59 5W
Lukanga Swamp, Zambia 55 E2 14 30S 27 40 E
Lukenie →, Zaïre 52 E3 3 0S 18 50 E
Lukhisaral, India 43 G12 25 11N 86 5 E
Lukolela, Equateur, Zaïre 52 E3 1 10S 17 12 E
Lukolela, Kasai Or., Zaïre 54 D1 5 23S 24 32 E
Lukosi, Zimbabwe 55 F2 18 30S 26 30 E
Łuków, Poland 17 C12 51 55N 22 23 E
Lule älv →, Sweden 8 D19 65 35N 22 10 E
Luleå, Sweden 8 D20 65 35N 22 10 E
Lüleburgaz, Turkey 21 D12 41 23N 27 22 E
Luling, U.S.A. 81 L6 29 41N 97 39W
Lulong, China 35 E10 39 53N 118 51 E
Lulonga →, Zaïre 52 D3 1 0N 18 10 E
Lulua →, Zaïre 52 E4 4 30S 20 30 E
Luluabourg = Kananga, Zaïre 52 F4 5 55S 22 18 E
Lumai, Angola 53 G4 13 13S 21 25 E
Lumajang, Indonesia 37 H15 8 8S 113 13 E
Lumbala N'guimbo, Angola 53 G4 14 18S 21 18 E
Lumberton, Miss., U.S.A. 81 K10 31 0N 89 27W
Lumberton, N.C., U.S.A. 77 H6 34 37N 79 0W
Lumberton, N. Mex., U.S.A. 83 H10 36 56N 106 56W
Lumbwa, Kenya 54 C4 0 12S 35 28 E
Lumsden, N.Z. 59 L2 45 44S 168 27 E
Lumut, Malaysia 39 K3 4 13N 100 37 E
Lumut, Tg., Indonesia 36 E3 3 50S 105 58 E
Lunavada, India 42 H5 23 8N 73 37 E
Lund, Sweden 9 J15 55 44N 13 12 E
Lund, U.S.A. 82 G6 38 52N 115 0W
Lundazi, Zambia 55 E3 12 20S 33 7 E
Lundi →, Zimbabwe 55 G3 21 43S 32 34 E
Lundu, Malaysia 36 D3 1 40N 109 50 E
Lundy, U.K. 11 F3 51 10N 4 41W
Lune →, U.K. 10 C5 54 0N 2 51W
Lüneburg, Germany 16 B6 53 15N 10 24 E
Lüneburg Heath = Lüneburger Heide, Germany 16 B6 53 10N 10 12 E
Lüneburger Heide, Germany 16 B6 53 10N 10 12 E
Lunenburg, Canada 71 D7 44 22N 64 18W
Lunéville, France 18 B7 48 36N 6 30 E
Lunga →, Zambia 55 E2 14 34S 26 25 E
Lunglei, India 41 H18 22 55N 92 45 E
Luni, India 42 F5 26 0N 73 6 E
Luni →, India 42 G4 24 41N 71 14 E
Luninets = Luninyets, Belarus 17 B14 52 15N 26 50 E
Luning, U.S.A. 82 G4 38 30N 118 11W
Luninyets, Belarus 17 B14 52 15N 26 50 E
Lunsemfwa →, Zambia 55 E3 14 54S 30 12 E
Lunsemfwa Falls, Zambia 55 E2 14 30S 29 6 E
Luo He →, China 34 G6 34 35N 110 20 E
Luochuan, China 34 G5 35 45N 109 26 E
Luofu, Zaïre 54 C2 0 10S 29 15 E
Luohe, China 34 H8 33 32N 114 2 E
Luonan, China 34 G6 34 5N 110 10 E
Luoning, China 34 G6 34 35N 111 40 E
Luoyang, China 34 G7 34 40N 112 26 E
Luozi, Zaïre 52 E2 4 54S 14 0 E
Luozigou, China 35 C16 43 42N 130 18 E
Lupanshui, China 32 D5 26 38N 104 48 E
Lupilichi, Mozam. 55 E4 11 47S 35 13 E
Luque, Paraguay 94 B4 25 19S 57 25W
Luray, U.S.A. 76 F6 38 40N 78 28W
Luremo, Angola 52 F3 8 30S 17 50 E
Lurgan, U.K. 13 B5 54 28N 6 19W
Lusaka, Zambia 55 F2 15 28S 28 16 E
Lusambo, Zaïre 54 C1 4 58S 23 28 E
Lusangaye, Zaïre 54 C2 4 54S 26 0 E
Luseland, Canada 73 C7 52 5N 109 24W
Lushan, China 34 H7 33 45N 112 55 E
Lushi, China 34 G6 34 3N 111 3 E
Lushnja, Albania 21 D8 40 55N 19 41 E
Lushoto, Tanzania 54 C4 4 47S 38 20 E
Lushoto □, Tanzania 54 C4 4 45S 38 20 E
Lüshun, China 35 E11 38 45N 121 15 E
Lusk, U.S.A. 80 D2 42 46N 104 27W
Luta = Dalian, China 35 E11 38 50N 121 40 E
Luton, U.K. 11 F7 51 53N 0 24W
Lutong, Malaysia 36 D4 4 28N 114 0 E
Lutselke, Canada 73 A6 62 24N 110 44W
Lutsk, Ukraine 17 C13 50 50N 25 15 E
Lützow Holmbukta, Antarctica 5 C4 69 10S 37 30 E
Lutzputs, S. Africa 56 D3 28 3S 20 40 E
Luverne, U.S.A. 80 D6 43 39N 96 13W
Luvua, Zaïre 55 D2 8 48S 25 17 E
Luvua →, Zaïre 54 D2 6 50S 27 30 E
Luwegu →, Tanzania 55 D4 8 31S 37 23 E
Luwuk, Indonesia 37 E6 0 56S 122 47 E
Luxembourg, Lux. 18 B7 49 37N 6 9 E
Luxembourg □, Belgium 15 E5 49 58N 5 30 E
Luxembourg ■, Europe 18 B7 49 45N 6 0 E
Luxi, China 32 D4 24 27N 98 36 E
Luxor = El Uqsur, Egypt 51 C11 25 41N 32 38 E
Luyi, China 34 H8 33 50N 115 35 E
Luza, Russia 24 B8 60 39N 47 10 E
Luzern, Switz. 16 E5 47 3N 8 18 E
Luzhou, China 32 D5 28 52N 105 20 E
Luziânia, Brazil 93 G9 16 20S 48 0W
Luzon, Phil. 37 A6 16 0N 121 0 E
Lviv, Ukraine 17 D13 49 50N 24 0 E
Lvov = Lviv, Ukraine 17 D13 49 50N 24 0 E
Lyakhavichy, Belarus 17 B14 53 2N 26 32 E
Lyakhovskiye, Ostrova, Russia 27 B15 73 40N 141 0 E
Lyallpur = Faisalabad, Pakistan 42 D5 31 30N 73 5 E
Lycksele, Sweden 8 D18 64 38N 18 40 E
Lydda = Lod, Israel 47 D3 31 57N 34 54 E
Lydenburg, S. Africa 57 D5 25 10S 30 29 E
Lydia, Turkey 21 E13 38 48N 28 19 E
Lyell, N.Z. 59 J4 41 48S 172 4 E
Lyell I., Canada 72 C2 52 40N 131 35W
Lyepyel, Belarus 24 D4 54 50N 28 40 E
Lyman, U.S.A. 82 F8 41 20N 110 18W
Lyme Regis, U.K. 11 G5 50 43N 2 57W
Lymington, U.K. 11 G6 50 45N 1 32W
Łyna →, Poland 9 J19 54 37N 21 14 E
Lynchburg, U.S.A. 76 G6 37 25N 79 9W
Lynd →, Australia 62 B3 16 28S 143 18 E
Lynd Ra., Australia 63 D4 25 30S 149 20 E
Lynden, Canada 78 C4 43 14N 80 9W
Lynden, U.S.A. 84 B4 48 57N 122 27W
Lyndhurst, Queens., Australia 62 B3 19 12S 144 20 E
Lyndhurst, S. Austral., Australia 63 E2 30 15S 138 18 E
Lyndon →, Australia 61 D1 23 29S 114 6 E
Lyndonville, N.Y., U.S.A. 78 C6 43 20N 78 23W
Lyndonville, Vt., U.S.A. 79 B12 44 31N 72 1W
Lyngen, Norway 8 B19 69 45N 20 30 E
Lynher Reef, Australia 60 C3 15 27S 121 55 E
Lynn, U.S.A. 79 D14 42 28N 70 57W
Lynn Canal, U.S.A. 72 B1 58 50N 135 15W
Lynn Lake, Canada 73 B8 56 51N 101 3W
Lynnwood, U.S.A. 84 C4 47 49N 122 19W
Lynton, U.K. 11 F4 51 13N 3 50W
Lyntupy, Belarus 9 J22 55 4N 26 23 E
Lynx L., Canada 73 A7 62 25N 106 15W
Lyon, France 18 D6 45 46N 4 50 E
Lyonnais, France 18 D6 45 45N 4 15 E
Lyons = Lyon, France 18 D6 45 46N 4 50 E
Lyons, Colo., U.S.A. 80 E2 40 14N 105 16W
Lyons, Ga., U.S.A. 77 J4 32 12N 82 19W
Lyons, Kans., U.S.A. 80 F5 38 21N 98 12W
Lyons, N.Y., U.S.A. 78 C8 43 5N 77 0W
Lys = Leie →, Belgium 15 C3 51 2N 3 45 E
Lysva, Russia 24 C10 58 7N 57 49 E
Lysychansk, Ukraine 25 E6 48 55N 38 30 E
Lytle, U.S.A. 81 L5 29 14N 98 48W
Lyttelton, N.Z. 59 K4 43 35S 172 44 E
Lytton, Canada 72 C4 50 13N 121 31W
Lyubertsy, Russia 24 C6 55 39N 37 50 E
Lyuboml, Ukraine 17 C13 51 11N 24 4 E

M

Ma →, Vietnam 38 C5 19 47N 105 56 E
Ma'adaba, Jordan 47 E4 30 43N 35 47 E
Maamba, Zambia 56 B4 17 17S 26 28 E
Ma'ān, Jordan 47 E4 30 12N 35 44 E
Ma'ān □, Jordan 47 F5 30 0N 36 0 E
Maanselkä, Finland 8 C23 63 52N 28 32 E
Ma'anshan, China 33 C6 31 44N 118 29 E
Maarianhamina, Finland 9 F18 60 5N 19 55 E
Ma'arrat an Nu'mān, Syria 44 C3 35 43N 36 43 E
Maas →, Neths. 15 C4 51 45N 4 32 E
Maaseik, Belgium 15 C5 51 6N 5 45 E
Maassluis, Neths. 15 C4 51 56N 4 16 E
Maastricht, Neths. 18 A6 50 50N 5 40 E
Maave, Mozam. 57 C5 21 4S 34 47 E
Mabel L., Canada 72 C5 50 35N 118 43W
Mabenge, Zaïre 54 B1 4 15N 24 12 E
Mablethorpe, U.K. 10 D8 53 20N 0 15 E
Maboma, Zaïre 54 B2 2 30N 28 10 E
Mabrouk, Mali 50 E4 19 29N 1 15W
Mabton, U.S.A. 82 C3 46 13N 120 0W
Mac Bac, Vietnam 39 H6 9 46N 106 7 E
Macachín, Argentina 94 D3 37 10S 63 43W
Macaé, Brazil 95 A7 22 20S 41 43W
McAlester, U.S.A. 81 H7 34 56N 95 46W
McAllen, U.S.A. 81 M5 26 12N 98 14W
Macamic, Canada 70 C4 48 45N 79 0W
Macao = Macau ■, China 33 D6 22 16N 113 35 E
Macapá, Brazil 93 C8 0 5N 51 4W
McArthur →, Australia 62 B2 15 54S 136 40 E
McArthur, Port, Australia 62 B2 16 4S 136 23 E
McArthur River, Australia 62 B2 16 27S 136 7 E
Macau, Brazil 93 E11 5 15S 36 40W
Macau ■, China 33 D6 22 16N 113 35 E
McBride, Canada 72 C4 53 20N 120 19W
McCall, U.S.A. 82 D5 44 55N 116 6W
McCamey, U.S.A. 81 K3 31 8N 102 14W
McCammon, U.S.A. 82 E7 42 39N 112 12W
McCauley I., Canada 72 C2 53 40N 130 15W
McCleary, U.S.A. 84 C3 47 3N 123 16W
Macclesfield, U.K. 10 D5 53 15N 2 8W
McClintock, Canada 73 B10 57 50N 94 10W
M'Clintock Chan., Canada 68 A9 72 0N 102 0W
McClintock Ra., Australia 60 C4 18 44S 127 38 E
McCloud, U.S.A. 82 F2 41 15N 122 8W
McCluer I., Australia 60 B5 11 5S 133 0 E
McClure, U.S.A. 78 F7 40 42N 77 19W
McClure, L., U.S.A. 84 H6 37 35N 120 16W
M'Clure Str., Canada 4 B2 75 0N 119 0W
McClusky, U.S.A. 80 B4 47 29N 100 27W
McComb, U.S.A. 81 K9 31 15N 90 27W
McConaughy, L., U.S.A. 80 E4 41 14N 101 40W
McCook, U.S.A. 80 E4 40 12N 100 38W
McCullough Mt., U.S.A. 85 K11 35 35N 115 13W
McCusker →, Canada 73 B7 55 32N 108 39W
McDame, Canada 72 B3 59 44N 128 59W
McDermitt, U.S.A. 82 F5 41 59N 117 43W
Macdonald, L., Australia 60 D4 23 30S 129 0 E
McDonald Is., Ind. Oc. 3 G13 53 0S 73 0 E
Macdonnell Ranges, Australia 60 D5 23 40S 133 0 E
McDouall Peak, Australia 63 D1 29 51S 134 55 E
Macdougall L., Canada 68 B10 66 0N 98 27W
MacDowell L., Canada 70 B1 52 15N 92 45W
Macduff, U.K. 12 D6 57 40N 2 31W
Macedonia = Makedhonia □, Greece 21 D10 40 39N 22 0 E
Macedonia ■, Europe 21 D9 41 53N 21 40 E
Maceió, Brazil 93 E11 9 40S 35 41W
Macenta, Guinea 50 G3 8 35N 9 32W
Macerata, Italy 20 C5 43 18N 13 27 E
McFarland, U.S.A. 85 K7 35 41N 119 14W
McFarlane →, Canada 73 B7 59 12N 107 58W
Macfarlane, L., Australia 63 E2 32 0S 136 40 E

McGehee, *U.S.A.* **81 J9** 33 38N 91 24W
McGill, *U.S.A.* **82 G6** 39 23N 114 47W
Macgillycuddy's Reeks, *Ireland* **13 D2** 51 58N 9 45W
MacGregor, *Canada* **73 D9** 49 57N 98 48W
McGregor, *U.S.A.* **80 D9** 43 1N 91 11W
McGregor —▸, *Canada* **72 B4** 55 10N 122 0W
McGregor Ra., *Australia* **63 D3** 27 0S 142 45 E
Mach, *Pakistan* **40 E5** 29 50N 67 20 E
Māch Kowr, *Iran* **45 E9** 25 48N 61 28 E
Machado = Jiparaná —▸, *Brazil* **92 E6** 8 3S 62 52W
Machagai, *Argentina* **94 B3** 26 56S 60 2W
Machakos, *Kenya* **54 C4** 1 30S 37 15 E
Machakos □, *Kenya* **54 C4** 1 30S 37 15 E
Machala, *Ecuador* **92 D3** 3 20S 79 57W
Machanga, *Mozam.* **57 C6** 20 59S 35 0 E
Machattie, L., *Australia* **62 C2** 24 50S 139 48 E
Machava, *Mozam.* **57 D5** 25 54S 32 28 E
Machece, *Mozam.* **55 F4** 19 15S 35 32 E
Machevna, *Russia* **27 C18** 61 20N 172 20 E
Machias, *U.S.A.* **71 D6** 44 43N 67 28W
Machichi —▸, *Canada* **73 B10** 57 3N 92 6W
Machico, *Madeira* **22 D3** 32 43N 16 44W
Machilipatnam, *India* **41 L12** 16 12N 81 8 E
Machiques, *Venezuela* **92 A4** 10 4N 72 34W
Machupicchu, *Peru* **92 F4** 13 8S 72 30W
Machynlleth, *U.K.* **11 E4** 52 35N 3 50W
McIlwraith Ra., *Australia* **62 A3** 13 50S 143 20 E
McIntosh, *U.S.A.* **80 C4** 45 55N 101 21W
McIntosh L., *Canada* **73 B8** 55 45N 105 0W
Macintosh Ra., *Australia* **61 E4** 27 39S 125 32 E
Macintyre —▸, *Australia* **63 D5** 28 37S 150 47 E
Mackay, *Australia* **62 C4** 21 8S 149 11 E
Mackay, *U.S.A.* **82 E7** 43 55N 113 37W
MacKay —▸, *Canada* **72 B6** 57 10N 111 38W
Mackay, L., *Australia* **60 D4** 22 30S 129 0 E
McKay Ra., *Australia* **60 D3** 23 0S 122 30 E
McKeesport, *U.S.A.* **78 F5** 40 21N 79 52W
McKenna, *U.S.A.* **84 D4** 46 56N 122 33W
Mackenzie, *Canada* **72 B4** 55 20N 123 5W
McKenzie, *U.S.A.* **77 G1** 36 8N 88 31W
Mackenzie —▸, *Australia* **62 C4** 23 38S 149 46 E
Mackenzie —▸, *Canada* **68 B6** 69 10N 134 20W
McKenzie —▸, *U.S.A.* **82 D2** 44 7N 123 6W
Mackenzie Bay, *Canada* **4 B1** 69 0N 137 30W
Mackenzie City = Linden, *Guyana* **92 B7** 6 0N 58 10W
Mackenzie Highway, *Canada* **72 B5** 58 0N 117 15W
Mackenzie Mts., *Canada* **68 B6** 64 0N 130 0W
Mackinaw City, *U.S.A.* **76 C3** 45 47N 84 44W
McKinlay, *Australia* **62 C3** 21 16S 141 18 E
McKinlay —▸, *Australia* **62 C3** 20 50S 141 28 E
McKinley, Mt., *U.S.A.* **68 B4** 63 4N 151 0W
McKinley Sea, *Arctic* **4 A7** 82 0N 0 0 E
McKinney, *U.S.A.* **81 J6** 33 12N 96 37W
Mackinnon Road, *Kenya* **54 C4** 3 40S 39 1 E
Macksville, *Australia* **63 E5** 30 40S 152 56 E
McLaughlin, *U.S.A.* **80 C4** 45 49N 100 49W
Maclean, *Australia* **63 D5** 29 26S 153 16 E
McLean, *U.S.A.* **81 H4** 35 14N 100 36W
McLeansboro, *U.S.A.* **80 F10** 38 6N 88 32W
Maclear, *S. Africa* **57 E4** 31 2S 28 23 E
Macleay —▸, *Australia* **63 E5** 30 56S 153 0 E
McLennan, *Canada* **72 B5** 55 42N 116 50W
MacLeod, B., *Canada* **73 A7** 62 53N 110 0W
McLeod, L., *Australia* **61 D1** 24 9S 113 47 E
MacLeod Lake, *Canada* **72 C4** 54 58N 123 0W
McLoughlin, Mt., *U.S.A.* **82 E2** 42 27N 122 19W
McLure, *Canada* **72 C4** 51 2N 120 13W
McMechen, *U.S.A.* **78 G4** 39 57N 80 44W
McMillan, L., *U.S.A.* **81 J2** 32 36N 104 21W
McMinnville, *Oreg., U.S.A.* **82 D2** 45 13N 123 12W
McMinnville, *Tenn., U.S.A.* **77 H3** 35 41N 85 46W
McMorran, *Canada* **73 C7** 51 19N 108 42W
McMurdo Sd., *Antarctica* **5 D11** 77 0S 170 0 E
McMurray = Fort McMurray, *Canada* **72 B6** 56 44N 111 7W
McMurray, *U.S.A.* **84 B4** 48 19N 122 14W
McNary, *U.S.A.* **83 J9** 34 4N 109 51W
MacNutt, *Canada* **73 C8** 51 5N 101 36W
Macodoene, *Mozam.* **57 C6** 23 32S 35 5 E
Macomb, *U.S.A.* **80 E9** 40 27N 90 40W
Mâcon, *France* **18 C6** 46 19N 4 50 E
Macon, *Ga., U.S.A.* **77 J4** 32 51N 83 38W
Macon, *Miss., U.S.A.* **77 J1** 33 7N 88 34W
Macon, *Mo., U.S.A.* **80 F8** 39 44N 92 28W
Macondo, *Angola* **53 G4** 12 37S 23 46 E
Macossa, *Mozam.* **55 F3** 17 55S 33 56 E
Macoun L., *Canada* **73 B8** 56 32N 103 40W
Macovane, *Mozam.* **57 C6** 21 30S 35 2 E
McPherson, *U.S.A.* **80 F6** 38 22N 97 40W
McPherson Pk., *U.S.A.* **85 L7** 34 53N 119 53W
McPherson Ra., *Australia* **63 D5** 28 15S 153 15 E
Macquarie Harbour, *Australia* **62 G4** 42 15S 145 23 E
Macquarie Is., *Pac. Oc.* **64 N7** 54 36S 158 55 E
MacRobertson Land, *Antarctica* **5 D6** 71 0S 64 0 E
Macroom, *Ireland* **13 E3** 51 54N 8 57W
Macroy, *Australia* **60 D2** 20 53S 118 2 E
MacTier, *Canada* **78 A5** 45 9N 79 46W
Macubela, *Mozam.* **55 F4** 16 53S 37 49 E
Macuiza, *Mozam.* **55 F3** 18 7S 34 29 E
Macuse, *Mozam.* **55 F4** 17 45S 37 10 E
Macuspana, *Mexico* **87 D6** 17 46N 92 36W
Macusse, *Angola* **56 B3** 17 48S 20 23 E
McVille, *U.S.A.* **80 B5** 47 46N 98 11W
Madadeni, *S. Africa* **57 D5** 27 43S 30 3 E
Madagali, *Nigeria* **51 F7** 10 56N 13 33 E
Madagascar ■, *Africa* **57 C8** 20 0S 47 0 E
Madā'in Sālih, *Si. Arabia* **44 E3** 26 46N 37 57 E
Madama, *Niger* **51 D7** 22 0N 13 40 E
Madame I., *Canada* **71 C7** 45 30N 60 58W
Madaoua, *Niger* **50 F6** 14 5N 6 27 E
Madaripur, *Bangla.* **41 H17** 23 19N 90 15 E
Madauk, *Burma* **41 L20** 17 56N 96 52 E
Madawaska, *Canada* **78 A7** 45 30N 78 0W
Madawaska —▸, *Canada* **78 A8** 45 27N 76 21W
Madaya, *Burma* **41 H20** 22 12N 96 10 E
Maddalena, *Italy* **20 D3** 41 16N 9 23 E
Madeira, *Atl. Oc.* **22 D3** 32 50N 17 0W
Madeira —▸, *Brazil* **92 D7** 3 22S 58 45W
Madeleine, Is. de la, *Canada* **71 C7** 47 30N 61 40W
Madera, *U.S.A.* **83 H3** 36 57N 120 3W
Madha, *India* **40 L9** 18 0N 75 30 E
Madhubani, *India* **43 F12** 26 21N 86 7 E
Madhya Pradesh □, *India* **42 J7** 22 50N 78 0 E
Madikeri, *India* **40 N9** 12 30N 75 45 E
Madill, *U.S.A.* **81 H6** 34 6N 96 46W
Madimba, *Zaïre* **52 E3** 4 58S 15 5 E
Ma'din, *Syria* **44 C3** 35 45N 39 36 E
Madīnat ash Sha'b, *Yemen* **46 E3** 12 50N 45 0 E
Madingou, *Congo* **52 E2** 4 10S 13 33 E
Madirovalo, *Madag.* **57 B8** 16 26S 46 32 E
Madison, *Calif., U.S.A.* **84 G5** 38 41N 121 59W
Madison, *Fla., U.S.A.* **77 K4** 30 28N 83 25W
Madison, *Ind., U.S.A.* **76 F3** 38 44N 85 23W
Madison, *Nebr., U.S.A.* **80 E6** 41 50N 97 27W
Madison, *Ohio, U.S.A.* **78 E3** 41 46N 81 3W
Madison, *S. Dak., U.S.A.* **80 D6** 44 0N 97 7W
Madison, *Wis., U.S.A.* **80 D10** 43 4N 89 24W
Madison —▸, *U.S.A.* **82 D8** 45 56N 111 31W
Madisonville, *Ky., U.S.A.* **76 G2** 37 20N 87 30W
Madisonville, *Tex., U.S.A.* **81 K7** 30 57N 95 55W
Madista, *Botswana* **56 C4** 21 15S 25 6 E
Madiun, *Indonesia* **37 G14** 7 38S 111 32 E
Madley, *U.K.* **11 E5** 52 2N 2 51W
Madona, *Latvia* **9 H22** 56 53N 26 5 E
Madras = Tamil Nadu □, *India* **40 P10** 11 0N 77 0 E
Madras, *India* **40 N12** 13 8N 80 19 E
Madras, *U.S.A.* **82 D3** 44 38N 121 8W
Madre, L., *Mexico* **87 B5** 25 0N 97 30W
Madre, Laguna, *U.S.A.* **81 M6** 27 0N 97 30W
Madre, Sierra, *Phil.* **37 A6** 17 0N 122 0 E
Madre de Dios —▸, *Bolivia* **92 F5** 10 59S 66 8W
Madre de Dios, I., *Chile* **96 G1** 50 20S 75 10W
Madre del Sur, Sierra, *Mexico* **87 D5** 17 30N 100 0W
Madre Occidental, Sierra, *Mexico* **86 B3** 27 0N 107 0W
Madre Oriental, Sierra, *Mexico* **86 C4** 25 0N 100 0W
Madri, *India* **42 G5** 24 16N 73 32 E
Madrid, *Spain* **19 B4** 40 25N 3 45W
Madura, Selat, *Indonesia* **37 G15** 7 30S 113 20 E
Madura Motel, *Australia* **61 F4** 31 55S 127 0 E
Madurai, *India* **40 Q11** 9 55N 78 10 E
Madurantakam, *India* **40 N11** 12 30N 79 50 E
Mae Chan, *Thailand* **38 B2** 20 9N 99 52 E
Mae Hong Son, *Thailand* **38 C2** 19 16N 98 1 E
Mae Khlong —▸, *Thailand* **38 F3** 13 24N 100 0 E
Mae Phrik, *Thailand* **38 D2** 17 27N 99 7 E
Mae Ramat, *Thailand* **38 D2** 16 58N 98 31 E
Mae Rim, *Thailand* **38 C2** 18 54N 98 57 E
Mae Sot, *Thailand* **38 D2** 16 43N 98 34 E
Mae Suai, *Thailand* **38 C2** 19 39N 99 33 E
Mae Tha, *Thailand* **38 C2** 18 28N 99 8 E
Maebashi, *Japan* **31 F9** 36 24N 139 4 E
Maesteg, *U.K.* **11 F4** 51 36N 3 40W
Maestra, Sierra, *Cuba* **88 B4** 20 15N 77 0W
Maestrazgo, Mts. del, *Spain* **19 B5** 40 30N 0 25W
Maevatanana, *Madag.* **57 B8** 16 56S 46 49 E
Mafeking = Mafikeng, *S. Africa* **56 D4** 25 50S 25 38 E
Mafeking, *Canada* **73 C8** 52 40N 101 10W
Mafeteng, *Lesotho* **56 D4** 29 51S 27 15 E
Maffra, *Australia* **63 F4** 37 53S 146 58 E
Mafia I., *Tanzania* **54 D4** 7 45S 39 50 E
Mafikeng, *S. Africa* **56 D4** 25 50S 25 38 E
Mafra, *Brazil* **95 B6** 26 10S 49 55W
Mafra, *Portugal* **19 C1** 38 55N 9 20W
Mafungabusi Plateau, *Zimbabwe* **55 F2** 18 30S 29 8 E
Magadan, *Russia* **27 D16** 59 38N 150 50 E
Magadi, *Kenya* **54 C4** 1 54S 36 19 E
Magadi, L., *Kenya* **54 C4** 1 54S 36 19 E
Magaliesburg, *S. Africa* **57 D4** 26 0S 27 32 E
Magallanes, Estrecho de, *Chile* **96 G2** 52 30S 75 0W
Magangué, *Colombia* **92 B4** 9 14N 74 45W
Magburaka, *S. Leone* **50 G2** 8 47N 12 0W
Magdalen Is. = Madeleine, Is. de la, *Canada* **71 C7** 47 30N 61 40W
Magdalena, *Argentina* **94 D4** 35 5S 57 30W
Magdalena, *Bolivia* **92 F6** 13 13S 63 57W
Magdalena, *Malaysia* **36 D5** 4 25N 117 55 E
Magdalena, *Mexico* **86 A2** 30 50N 112 0W
Magdalena, *U.S.A.* **83 J10** 34 7N 107 15W
Magdalena —▸, *Colombia* **92 A4** 11 6N 74 51W
Magdalena —▸, *Mexico* **86 A2** 30 40N 112 25W
Magdalena, B., *Mexico* **86 C2** 24 30N 112 10W
Magdalena, Llano de la, *Mexico* **86 C2** 25 0N 111 30W
Magdeburg, *Germany* **16 B6** 52 7N 11 38 E
Magdelaine Cays, *Australia* **62 B5** 16 33S 150 18 E
Magee, *U.S.A.* **81 K10** 31 52N 89 44W
Magee, I., *U.K.* **13 B6** 54 48N 5 43W
Magelang, *Indonesia* **37 G14** 7 29S 110 13 E
Magellan's Str. = Magallanes, Estrecho de, *Chile* **96 G2** 52 30S 75 0W
Magenta, L., *Australia* **61 F2** 33 30S 119 2 E
Magerøya, *Norway* **8 A21** 71 3N 25 40 E
Maggiore, L., *Italy* **20 B3** 45 57N 8 39 E
Magherafelt, *U.K.* **13 B5** 54 45N 6 37W
Magistralnyy, *Russia* **27 D11** 56 16N 107 36 E
Magnetic Pole (North) = North Magnetic Pole, *Canada* **4 B2** 77 58N 102 8W
Magnetic Pole (South) = South Magnetic Pole, *Antarctica* **5 C9** 64 8S 138 8 E
Magnitogorsk, *Russia* **24 D10** 53 27N 59 4 E
Magnolia, *Ark., U.S.A.* **81 J8** 33 16N 93 14W
Magnolia, *Miss., U.S.A.* **81 K9** 31 9N 90 28W
Magog, *Canada* **71 C5** 45 18N 72 9W
Magoro, *Uganda* **54 B3** 1 45N 34 12 E
Magosa = Famagusta, *Cyprus* **23 D12** 35 8N 33 55 E
Magouládhes, *Greece* **23 A3** 39 45N 19 42 E
Magoye, *Zambia* **55 F2** 16 1S 27 30 E
Magpie L., *Canada* **71 B7** 51 0N 64 41W
Magrath, *Canada* **72 D6** 49 25N 112 50W
Magu □, *Tanzania* **54 C3** 2 31S 33 28 E
Maguarinho, C., *Brazil* **93 D9** 0 15S 48 30W
Mağusa = Famagusta, *Cyprus* **23 D12** 35 8N 33 55 E
Maguse L., *Canada* **73 A9** 61 40N 95 10W
Maguse Pt., *Canada* **73 A10** 61 20N 93 50W
Magwe, *Burma* **41 J19** 20 10N 95 0 E
Maha Sarakham, *Thailand* **38 D4** 16 12N 103 16 E
Mahābād, *Iran* **44 B5** 36 50N 45 45 E
Mahabharat Lekh, *Nepal* **43 E9** 28 30N 82 0 E
Mahabo, *Madag.* **57 C7** 20 23S 44 40 E
Mahadeo Hills, *India* **42 H8** 22 20N 78 30 E
Mahagi, *Zaïre* **54 B3** 2 20N 31 0 E
Mahajamba —▸, *Madag.* **57 B8** 15 33S 47 8 E
Mahajamba, Helodranon' i, *Madag.* **57 B8** 15 24S 47 5 E
Mahajan, *India* **42 E5** 28 48N 73 56 E
Mahajanga, *Madag.* **57 B8** 15 40S 46 25 E
Mahajanga □, *Madag.* **57 B8** 17 0S 47 0 E
Mahajilo —▸, *Madag.* **57 B8** 19 42S 45 22 E
Mahakam —▸, *Indonesia* **36 E5** 0 35S 117 17 E
Mahalapye, *Botswana* **56 C4** 23 1S 26 51 E
Maḩallāt, *Iran* **45 C6** 33 55N 50 30 E
Māhān, *Iran* **45 D8** 30 5N 57 18 E
Mahanadi —▸, *India* **41 J15** 20 20N 86 25 E
Mahanoro, *Madag.* **57 B8** 19 54S 48 48 E
Mahanoy City, *U.S.A.* **79 F8** 40 49N 76 9W
Maharashtra □, *India* **40 J9** 20 30N 75 30 E
Mahari Mts., *Tanzania* **54 D2** 6 20S 30 0 E
Mahasham, W. —▸, *Egypt* **47 E3** 30 15N 34 10 E
Mahasolo, *Madag.* **57 B8** 19 7S 46 22 E
Mahattat ash Shīdīyah, *Jordan* **47 F4** 29 55N 35 55 E
Mahattat 'Unayzah, *Jordan* **47 E4** 30 30N 35 47 E
Mahaxay, *Laos* **38 D5** 17 22N 105 12 E
Mahbubnagar, *India* **40 L10** 16 45N 77 59 E
Maḩḑah, *Oman* **45 E7** 24 24N 55 59 E
Mahdia, *Tunisia* **51 A7** 35 28N 11 0 E
Mahe, *India* **43 C8** 33 10N 78 32 E
Mahenge, *Tanzania* **55 D4** 8 45S 36 41 E
Maheno, *N.Z.* **59 L3** 45 10S 170 50 E
Mahesana, *India* **42 H5** 23 39N 72 26 E
Mahia Pen., *N.Z.* **59 H6** 39 9S 177 55 E
Mahilyow, *Belarus* **17 B16** 53 55N 30 18 E
Mahmud Kot, *Pakistan* **42 D4** 30 16N 71 0 E
Mahnomen, *U.S.A.* **80 B7** 47 19N 95 58W
Mahoba, *India* **43 G8** 25 15N 79 55 E
Mahón, *Spain* **22 B11** 39 53N 4 16 E
Mahone Bay, *Canada* **71 D7** 44 30N 64 20W
Mai-Ndombe, L., *Zaïre* **52 E3** 2 0S 18 20 E
Mai-Sai, *Thailand* **38 B2** 20 20N 99 55 E
Maicurú —▸, *Brazil* **93 D8** 2 14S 54 17W
Maidan Khula, *Afghan.* **42 C3** 33 36N 69 50 E
Maidenhead, *U.K.* **11 F7** 51 31N 0 42W
Maidstone, *Canada* **73 C7** 53 5N 109 20W
Maidstone, *U.K.* **11 F8** 51 16N 0 32 E
Maiduguri, *Nigeria* **51 F7** 12 0N 13 20 E
Maijdi, *Bangla.* **41 H17** 22 48N 91 10 E
Maikala Ra., *India* **41 J12** 22 0N 81 0 E
Mailsi, *Pakistan* **42 E5** 29 48N 72 15 E
Main —▸, *Germany* **16 C5** 50 0N 8 18 E
Main —▸, *U.K.* **13 B5** 54 48N 6 18W
Main Centre, *Canada* **73 C7** 50 35N 107 21W
Maine, *France* **18 C3** 47 55N 0 25W
Maine □, *U.S.A.* **71 C6** 45 20N 69 0W
Maine —▸, *Ireland* **13 D2** 52 9N 9 45W
Maingkwan, *Burma* **41 F20** 26 15N 96 37 E
Mainit, L., *Phil.* **37 C7** 9 31N 125 30 E
Mainland, *Orkney, U.K.* **12 C5** 58 59N 3 8W
Mainland, *Shet., U.K.* **12 A7** 60 15N 1 22W
Mainpuri, *India* **43 F8** 27 18N 79 4 E
Maintirano, *Madag.* **57 B7** 18 3S 44 1 E
Mainz, *Germany* **16 C5** 50 1N 8 14 E
Maipú, *Argentina* **94 D4** 36 52S 57 50W
Maiquetía, *Venezuela* **92 A5** 10 36N 66 57W
Mairabari, *India* **41 F18** 26 30N 92 22 E
Maisi, *Cuba* **89 B5** 20 17N 74 9W
Maisi, Pta. de, *Cuba* **89 B5** 20 10N 74 10W
Maitland, *N.S.W., Australia* **63 E5** 32 33S 151 36 E
Maitland, *S. Austral., Australia* **63 E2** 34 23S 137 40 E
Maitland —▸, *Canada* **78 C3** 43 45N 81 43W
Maiz, Is. del, *Nic.* **88 D3** 12 15N 83 4W
Maizuru, *Japan* **31 G7** 35 25N 135 22 E
Majalengka, *Indonesia* **37 G13** 6 50S 108 13 E
Majene, *Indonesia* **37 E5** 3 38S 118 57 E
Maji, *Ethiopia* **51 G12** 6 12N 35 30 E
Major, *Canada* **73 C7** 51 52N 109 37W
Majorca = Mallorca, *Spain* **22 B10** 39 30N 3 0 E
Maka, *Senegal* **50 F2** 13 40N 14 10W
Makale, *Indonesia* **37 E5** 3 6S 119 51 E
Makamba, *Burundi* **54 C2** 4 8S 29 49 E
Makari, *Cameroon* **52 B2** 12 35N 14 28 E
Makarikari = Makgadikgadi Salt Pans, *Botswana* **56 C4** 20 40S 25 45 E
Makarovo, *Russia* **27 D11** 57 40N 107 45 E
Makasar = Ujung Pandang, *Indonesia* **37 F5** 5 10S 119 20 E
Makasar, Selat, *Indonesia* **37 E5** 1 0S 118 20 E
Makasar, Str. of = Makasar, Selat, *Indonesia* **37 E5** 1 0S 118 20 E
Makat, *Kazakstan* **25 E9** 47 39N 53 19 E
Makedhonía □, *Greece* **21 D10** 40 39N 22 0 E
Makedonija = Macedonia ■, *Europe* **21 D9** 41 53N 21 40 E
Makena, *U.S.A.* **74 H16** 20 39N 156 27W
Makeni, *S. Leone* **50 G2** 8 55N 12 5W
Makeyevka = Makiyivka, *Ukraine* **25 E6** 48 0N 38 0 E
Makgadikgadi Salt Pans, *Botswana* **56 C4** 20 40S 25 45 E
Makhachkala, *Russia* **25 F8** 43 0N 47 30 E
Makhmūr, *Iraq* **44 C4** 35 46N 43 35 E
Makian, *Indonesia* **37 D7** 0 20N 127 20 E
Makindu, *Kenya* **54 C4** 2 18S 37 50 E
Makinsk, *Kazakstan* **26 D8** 52 37N 70 26 E
Makiyivka, *Ukraine* **25 E6** 48 0N 38 0 E
Makkah, *Si. Arabia* **46 C2** 21 30N 39 54 E
Makkovik, *Canada* **71 A8** 55 10N 59 10W
Makó, *Hungary* **17 E11** 46 14N 20 33 E
Makokou, *Gabon* **52 D2** 0 40N 12 50 E
Makongo, *Zaïre* **54 B2** 3 25N 26 17 E
Makoro, *Zaïre* **54 B2** 3 10N 29 59 E
Makoua, *Congo* **52 E3** 0 5S 15 50 E
Makrai, *India* **40 H10** 22 2N 77 0 E
Makran Coast Range, *Pakistan* **40 G4** 25 40N 64 0 E
Makrana, *India* **42 F6** 27 2N 74 46 E
Makriyialos, *Greece* **23 D7** 35 2N 25 59 E
Maksimkin Yar, *Russia* **26 D9** 58 42N 86 50 E
Mākū, *Iran* **44 B5** 39 15N 44 31 E
Makumbi, *Zaïre* **52 F4** 5 50S 20 43 E
Makunda, *Botswana* **56 C3** 22 30S 20 7 E
Makurazaki, *Japan* **31 J5** 31 15N 130 20 E
Makurdi, *Nigeria* **50 G6** 7 43N 8 35 E
Makūyeh, *Iran* **45 D7** 28 7N 53 9 E
Makwassie, *S. Africa* **56 D4** 27 17S 26 0 E
Mal B., *Ireland* **13 D2** 52 50N 9 30W
Mala, Pta., *Panama* **88 E3** 7 28N 80 2W
Malabang, *Phil.* **37 C6** 7 36N 124 3 E
Malabar Coast, *India* **40 P9** 11 0N 75 0 E
Malabo = Rey Malabo, *Eq. Guin.* **50 H6** 3 45N 8 50 E
Malacca, Str. of, *Indonesia* **39 L3** 3 0N 101 0 E
Malad City, *U.S.A.* **82 E7** 42 12N 112 15W
Maladzyechna, *Belarus* **17 A14** 54 20N 26 50 E
Málaga, *Spain* **19 D3** 36 43N 4 23W
Malaga, *U.S.A.* **81 J2** 32 14N 104 4W
Malagarasi, *Tanzania* **54 D3** 5 5S 30 50 E
Malagarasi —▸, *Tanzania* **54 D2** 5 12S 29 47 E
Malaimbandy, *Madag.* **57 C8** 20 20S 45 36 E
Malakâl, *Sudan* **51 G11** 9 33N 31 40 E
Malakand, *Pakistan* **42 B4** 34 40N 71 55 E
Malakoff, *U.S.A.* **81 J7** 32 10N 96 1W
Malamyzh, *Russia* **27 E14** 49 50N 136 50 E
Malang, *Indonesia* **37 G15** 7 59S 112 45 E
Malangen, *Norway* **8 B18** 69 24N 18 37 E
Malanje, *Angola* **52 F3** 9 36S 16 17 E
Mälaren, *Sweden* **9 G17** 59 30N 17 10 E
Malargüe, *Argentina* **94 D2** 35 32S 69 30W
Malartic, *Canada* **70 C4** 48 9N 78 9W
Malaryta, *Belarus* **17 C13** 51 50N 24 3 E
Malatya, *Turkey* **25 G6** 38 25N 38 20 E
Malawi ■, *Africa* **55 E3** 11 55S 34 0 E
Malawi, L., *Africa* **55 E3** 12 30S 34 30 E
Malay Pen., *Asia* **39 J3** 7 25N 100 0 E
Malaybalay, *Phil.* **37 C7** 8 5N 125 7 E
Malāyer, *Iran* **45 C6** 34 19N 48 51 E
Malaysia ■, *Asia* **36 D4** 5 0N 110 0 E
Malazgirt, *Turkey* **25 G7** 39 10N 42 33 E
Malbon, *Australia* **62 C3** 21 5S 140 17 E
Malbooma, *Australia* **63 E1** 30 41S 134 11 E
Malbork, *Poland* **17 B10** 54 3N 19 1 E
Malcolm, *Australia* **61 E3** 28 51S 121 25 E
Malcolm, Pt., *Australia* **61 F3** 33 48S 123 45 E
Maldegem, *Belgium* **15 C3** 51 14N 3 26 E
Malden, *Mass., U.S.A.* **79 D13** 42 26N 71 4W
Malden, *Mo., U.S.A.* **81 G10** 36 34N 89 57W
Malden I., *Kiribati* **65 H12** 4 3S 155 1W
Maldives ■, *Ind. Oc.* **29 J11** 5 0N 73 0 E
Maldonado, *Uruguay* **95 C5** 34 59S 55 0W
Maldonado, Punta, *Mexico* **87 D5** 16 19N 98 35W
Malé Karpaty, *Slovak Rep.* **17 D9** 48 30N 17 20 E
Maléa, Ákra, *Greece* **21 F10** 36 28N 23 7 E
Malegaon, *India* **40 J9** 20 30N 74 38 E
Malei, *Mozam.* **55 F4** 17 12S 36 58 E
Malek Kandī, *Iran* **44 B5** 37 9N 46 6 E
Malela, *Zaïre* **54 C2** 4 22S 26 8 E
Malema, *Mozam.* **55 E4** 14 57S 37 20 E
Máleme, *Greece* **23 D5** 35 31N 23 49 E
Malerkotla, *India* **42 D6** 30 32N 75 58 E
Máles, *Greece* **23 D7** 35 6N 25 35 E
Malgomaj, *Sweden* **8 D17** 64 40N 16 30 E
Malha, *Sudan* **51 E10** 15 8N 25 10 E
Malheur —▸, *U.S.A.* **82 D5** 44 4N 116 59W
Malheur L., *U.S.A.* **82 E4** 43 20N 118 48W
Mali ■, *Africa* **50 E4** 17 0N 3 0W
Mali —▸, *Burma* **41 G20** 25 40N 97 40 E
Malibu, *U.S.A.* **85 L8** 34 2N 118 41W
Malik, *Indonesia* **37 E6** 0 39S 123 16 E
Malili, *Indonesia* **37 E6** 2 42S 121 6 E
Malimba, Mts., *Zaïre* **54 D2** 7 30S 29 30 E
Malin Hd., *Ireland* **13 A4** 55 23N 7 23W
Malindi, *Kenya* **54 C5** 3 12S 40 5 E
Malines = Mechelen, *Belgium* **15 C4** 51 2N 4 29 E
Malino, *Indonesia* **37 D6** 1 0N 121 0 E
Malinyi, *Tanzania* **55 D4** 8 56S 36 0 E
Malita, *Phil.* **37 C7** 6 19N 125 39 E
Malkara, *Turkey* **21 D12** 40 53N 26 53 E
Mallacoota, *Australia* **63 F4** 37 40S 149 40 E
Mallacoota Inlet, *Australia* **63 F4** 37 34S 149 40 E
Mallaig, *U.K.* **12 E3** 57 0N 5 50W
Mallawan, *India* **43 F9** 27 4N 80 12 E
Mallawi, *Egypt* **51 C11** 27 44N 30 44 E
Mállia, *Greece* **23 D7** 35 17N 25 27 E
Mallión, Kólpos, *Greece* **23 D7** 35 19N 25 27 E
Mallorca, *Spain* **22 B10** 39 30N 3 0 E
Mallorytown, *Canada* **79 B9** 44 29N 75 53W
Mallow, *Ireland* **13 D3** 52 8N 8 39W
Malmberget, *Sweden* **8 C19** 67 11N 20 40 E
Malmédy, *Belgium* **15 D6** 50 25N 6 2 E
Malmesbury, *S. Africa* **56 E2** 33 28S 18 41 E
Malmö, *Sweden* **9 J15** 55 36N 12 59 E
Malolos, *Phil.* **37 B6** 14 50N 120 49 E
Malombe L., *Malawi* **55 E4** 14 40S 35 15 E
Malone, *U.S.A.* **79 B10** 44 51N 74 18W
Måløy, *Norway* **9 F11** 61 57N 5 6 E
Malozemelskaya Tundra, *Russia* **24 A9** 67 0N 50 0 E
Malpaso, *Canary Is.* **22 G1** 27 43N 18 3W
Malpelo, *Colombia* **92 C2** 4 3N 81 35W
Malta, *Idaho, U.S.A.* **82 E7** 42 18N 113 22W
Malta, *Mont., U.S.A.* **82 B10** 48 21N 107 52W
Malta ■, *Europe* **23 D1** 35 50N 14 30 E
Maltahöhe, *Namibia* **56 C2** 24 55S 17 0 E
Malton, *Canada* **78 C5** 43 42N 79 38W
Malton, *U.K.* **10 C7** 54 8N 0 49W
Maluku, *Indonesia* **37 E7** 1 0S 127 0 E
Maluku □, *Indonesia* **37 E7** 3 0S 128 0 E

Maluku Sea = Molucca Sea, Indonesia 37 E6 2 0S 124 0 E
Malvan, India 40 L8 16 2N 73 30 E
Malvern, U.S.A. 81 H8 34 22N 92 49W
Malvern Hills, U.K. 11 E5 52 0N 2 19W
Malvinas, Is. = Falkland Is. □, Atl. Oc. 96 G5 51 30S 59 0W
Malya, Tanzania 54 C3 3 5S 33 38 E
Malyn, Ukraine 17 C15 50 46N 29 3 E
Malyy Lyakhovskiy, Ostrov, Russia 27 B15 74 7N 140 36 E
Malyy Nimnyr, Russia 27 D13 57 50N 125 10 E
Mama, Russia 27 D12 58 18N 112 54 E
Mamanguape, Brazil 93 E11 6 50S 35 4W
Mamasa, Indonesia 37 E5 2 55S 119 20 E
Mambasa, Zaïre 54 B2 1 22N 29 3 E
Mamberamo →, Indonesia 37 E9 2 0S 137 50 E
Mambilima Falls, Zambia 55 E2 10 31S 28 45 E
Mambirima, Zaïre 55 E2 11 25S 27 33 E
Mambo, Tanzania 54 C4 4 52S 38 22 E
Mambrui, Kenya 54 C5 3 5S 40 5 E
Mamburao, Phil. 37 B6 13 13N 120 39 E
Mameigwess L., Canada 70 B2 52 35N 87 50W
Mamfe, Cameroon 50 G6 5 50N 9 15 E
Mammoth, U.S.A. 83 K8 32 43N 110 39W
Mamoré →, Bolivia 92 F5 10 23S 65 53W
Mamou, Guinea 50 F2 10 15N 12 0W
Mamuju, Indonesia 37 E5 2 41S 118 50 E
Man, Ivory C. 50 G3 7 30N 7 40W
Man, I. of, U.K. 10 C3 54 15N 4 30W
Man Na, Burma 41 H20 23 27N 97 19 E
Mana, Fr. Guiana 93 B8 5 45N 53 55W
Manaar, G. of = Mannar, G. of, Asia 40 Q11 8 30N 79 0 E
Manacapuru, Brazil 92 D6 3 16S 60 37W
Manacor, Spain 22 B10 39 34N 3 13 E
Manado, Indonesia 37 D6 1 29N 124 51 E
Managua, Nic. 88 D2 12 6N 86 20W
Managua, L., Nic. 88 D2 12 20N 86 30W
Manakara, Madag. 57 C8 22 8S 48 1 E
Manama = Al Manāmah, Bahrain 45 E6 26 10N 50 30 E
Manambao →, Madag. 57 B7 17 35S 44 0 E
Manambato, Madag. 57 A8 13 43S 49 7 E
Manambolo →, Madag. 57 B7 19 18S 44 22 E
Manambolosy, Madag. 57 B8 16 2S 49 40 E
Mananara, Madag. 57 B8 16 10S 49 46 E
Mananara →, Madag. 57 C8 23 21S 47 42 E
Mananjary, Madag. 57 C8 21 13S 48 20 E
Manantenina, Madag. 57 C8 24 17S 47 19 E
Manaos = Manaus, Brazil 92 D7 3 0S 60 0W
Manapouri, N.Z. 59 L1 45 34S 167 39 E
Manapouri, L., N.Z. 59 L1 45 32S 167 32 E
Manas, China 32 B3 44 17N 85 56 E
Manas →, India 41 F17 26 12N 90 40 E
Manaslu, Nepal 43 E11 28 33N 84 33 E
Manasquan, U.S.A. 79 F10 40 8N 74 3W
Manassa, U.S.A. 83 H11 37 11N 105 56W
Manaung, Burma 41 K18 18 45N 93 40 E
Manaus, Brazil 92 D7 3 0S 60 0W
Manawan L., Canada 73 B8 55 24N 103 14W
Manay, Phil. 37 C7 7 17N 126 33 E
Manbij, Syria 44 B3 36 31N 37 57 E
Mancelona, U.S.A. 76 C3 44 54N 85 4W
Manchester, U.K. 10 D5 53 29N 2 12W
Manchester, Calif., U.S.A. 84 G3 38 58N 123 41W
Manchester, Conn., U.S.A. 79 E12 41 47N 72 31W
Manchester, Ga., U.S.A. 77 J3 32 51N 84 37W
Manchester, Iowa, U.S.A. 80 D9 42 29N 91 27W
Manchester, Ky., U.S.A. 76 G4 37 9N 83 46W
Manchester, N.H., U.S.A. 79 D13 42 59N 71 28W
Manchester, N.Y., U.S.A. 78 D7 42 56N 77 16W
Manchester, Vt., U.S.A. 79 C11 43 10N 73 5W
Manchester L., Canada 73 A7 61 28N 107 29W
Manchuria = Dongbei, China 35 D13 42 0N 125 0 E
Manchurian Plain, China 28 E16 47 0N 124 0 E
Mand →, Iran 45 D7 28 20N 52 30 E
Manda, Chunya, Tanzania 54 D3 6 51S 32 29 E
Manda, Ludewe, Tanzania 55 E3 10 30S 34 40 E
Mandabé, Madag. 57 C7 21 0S 44 55 E
Mandaguari, Brazil 95 A5 23 32S 51 42W
Mandah, Mongolia 34 B5 44 27N 108 2 E
Mandal, Norway 9 G12 58 2N 7 25 E
Mandalay, Burma 41 J20 22 0N 96 4 E
Mandale = Mandalay, Burma 41 J20 22 0N 96 4 E
Mandalgovi, Mongolia 34 B4 45 45N 106 10 E
Mandalī, Iraq 44 C5 33 43N 45 28 E
Mandan, U.S.A. 80 B4 46 50N 100 54W
Mandar, Teluk, Indonesia 37 E5 3 35S 119 15 E
Mandaue, Phil. 37 B6 10 20N 123 56 E
Mandera, Kenya 54 B5 3 55N 41 53 E
Mandera □, Kenya 54 B5 3 30N 41 0 E
Mandi, India 42 D7 31 39N 76 58 E
Mandimba, Mozam. 55 E4 14 20S 35 40 E
Mandioli, Indonesia 37 E7 0 40S 127 20 E
Mandla, India 43 H9 22 39N 80 30 E
Mandoto, Madag. 57 B8 19 34S 46 17 E
Mandra, Pakistan 42 C5 33 23N 73 12 E
Mandrare →, Madag. 57 D8 25 10S 46 30 E
Mandritsara, Madag. 57 B8 15 50S 48 49 E
Mandsaur, India 42 G6 24 3N 75 8 E
Mandurah, Australia 61 F2 32 36S 115 48 E
Mandvi, India 42 H3 22 51N 69 22 E
Mandya, India 40 N10 12 30N 77 0 E
Mandzai, Pakistan 42 D2 30 55N 67 6 E
Maneh, Iran 45 B8 37 39N 57 7 E
Maneroo, Australia 62 C3 23 22S 143 53 E
Maneroo Cr. →, Australia 62 C3 23 21S 143 53 E
Manfalût, Egypt 51 C11 27 20N 30 52 E
Manfred, Australia 63 E3 33 19S 143 45 E
Manfredónia, Italy 20 D6 41 38N 15 55 E
Mangalia, Romania 17 G15 43 50N 28 35 E
Mangalore, India 40 N9 12 55N 74 47 E
Mangaweka, N.Z. 59 H5 39 48S 175 47 E
Manggar, Indonesia 36 E3 2 50S 108 10 E
Manggawitu, Indonesia 37 E8 4 8S 133 32 E
Mangkalihat, Tanjung, Indonesia 37 D5 1 2N 118 59 E
Mangla Dam, Pakistan 43 C5 33 9N 73 44 E
Manglaur, India 42 E7 29 44N 77 49 E
Mangnai, China 32 C4 37 52N 91 43 E
Mango, Togo 50 F5 10 20N 0 30 E
Mangoche, Malawi 55 E4 14 25S 35 16 E
Mangoky →, Madag. 57 C7 21 29S 43 41 E
Mangole, Indonesia 37 E7 1 50S 125 55 E
Mangombe, Zaïre 54 C2 1 20S 26 48 E
Mangonui, N.Z. 59 F4 35 1S 173 32 E
Mangueigne, Chad 51 F9 10 30N 21 15 E
Mangueira, L. da, Brazil 95 C5 33 0S 52 50W
Mangum, U.S.A. 81 H5 34 53N 99 30W
Mangyshlak Poluostrov, Kazakstan 26 E6 44 30N 52 30 E
Manhattan, U.S.A. 80 F6 39 11N 96 35W
Manhiça, Mozam. 57 D5 25 23S 32 49 E
Manhuaçu, Brazil 93 H10 20 15S 42 2W
Mania →, Madag. 57 B8 19 42S 45 22 E
Manica, Mozam. 57 B5 18 58S 32 59 E
Manica e Sofala □, Mozam. 57 B5 19 10S 33 45 E
Manicaland □, Zimbabwe 55 F3 19 0S 32 30 E
Manicoré, Brazil 92 E6 5 48S 61 16W
Manicouagan →, Canada 71 C6 49 30N 68 30W
Manīfah, Si. Arabia 45 E6 27 44N 49 0 E
Manifold, Australia 62 C5 22 41S 150 40 E
Manifold, C., Australia 62 C5 22 41S 150 50 E
Manigotagan, Canada 73 C9 51 6N 96 18W
Manihiki, Cook Is. 65 J11 10 24S 161 1W
Manika, Plateau de la, Zaïre 55 E2 10 0S 25 5 E
Manila, Phil. 37 B6 14 40N 121 3 E
Manila, U.S.A. 82 F9 40 59N 109 43W
Manila B., Phil. 37 B6 14 40N 120 35 E
Manilla, Australia 63 E5 30 45S 150 43 E
Maningrida, Australia 62 A1 12 3S 134 13 E
Manipur □, India 41 G18 25 0N 94 0 E
Manipur →, Burma 41 H19 23 45N 94 20 E
Manisa, Turkey 21 E12 38 38N 27 30 E
Manistee, U.S.A. 76 C2 44 15N 86 19W
Manistee →, U.S.A. 76 C2 44 15N 86 21W
Manistique, U.S.A. 76 C2 45 57N 86 15W
Manito L., Canada 73 C7 52 43N 109 43W
Manitoba □, Canada 73 B9 55 30N 97 0W
Manitoba, L., Canada 73 C9 51 0N 98 45W
Manitou, Canada 73 D9 49 15N 98 32W
Manitou I., U.S.A. 70 C2 47 25N 87 37W
Manitou Is., U.S.A. 76 C2 45 8N 86 0W
Manitou L., Canada 71 B6 50 55N 65 17W
Manitou Springs, U.S.A. 80 F2 38 52N 104 55W
Manitoulin I., Canada 70 C3 45 40N 82 30W
Manitowaning, Canada 70 C3 45 46N 81 49W
Manitowoc, U.S.A. 76 C2 44 5N 87 40W
Manizales, Colombia 92 B3 5 5N 75 32W
Manja, Madag. 57 C7 21 26S 44 20 E
Manjacaze, Mozam. 57 C5 24 45S 34 0 E
Manjakandriana, Madag. 57 B8 18 55S 47 47 E
Manjhand, Pakistan 42 G3 25 50N 68 10 E
Manjil, Iran 45 B6 36 46N 49 30 E
Manjimup, Australia 61 F2 34 15S 116 6 E
Manjra →, India 40 K10 18 49N 77 52 E
Mankato, Kans., U.S.A. 80 F5 39 47N 98 13W
Mankato, Minn., U.S.A. 80 C8 44 10N 94 0W
Mankayane, Swaziland 57 D5 26 40S 31 4 E
Mankono, Ivory C. 50 G3 8 1N 6 10W
Mankota, Canada 73 D7 49 25N 107 5W
Manlay, Mongolia 34 B4 44 9N 107 0 E
Manly, Australia 63 E5 33 48S 151 17 E
Manmad, India 40 J9 20 18N 74 28 E
Mann Ras., Australia 61 E5 26 6S 130 5 E
Manna, Indonesia 36 E2 4 25S 102 55 E
Mannahill, Australia 63 E3 32 25S 140 0 E
Mannar, Sri Lanka 40 Q11 9 1N 79 54 E
Mannar, G. of, Asia 40 Q11 8 30N 79 0 E
Mannar I., Sri Lanka 40 Q11 9 5N 79 45 E
Mannheim, Germany 16 D5 49 29N 8 29 E
Manning, Canada 72 B5 56 53N 117 39W
Manning, Oreg., U.S.A. 84 E3 45 45N 123 13W
Manning, S.C., U.S.A. 77 J5 33 42N 80 13W
Manning Prov. Park, Canada 72 D4 49 5N 120 45W
Mannington, U.S.A. 76 F5 39 32N 80 21W
Mannum, Australia 63 E2 34 50S 139 20 E
Mano, S. Leone 50 G2 8 3N 12 2W
Manokwari, Indonesia 37 E8 0 54S 134 0 E
Manombo, Madag. 57 C7 22 57S 43 28 E
Manono, Zaïre 54 D2 7 15S 27 25 E
Manosque, France 18 E6 43 49N 5 47 E
Manouane, L., Canada 71 B5 50 45N 70 45W
Manpojin, N. Korea 35 D14 41 6N 126 24 E
Manresa, Spain 19 B6 41 48N 1 50 E
Mansa, Gujarat, India 42 H5 23 27N 72 45 E
Mansa, Punjab, India 42 E6 30 0N 75 27 E
Mansa, Zambia 55 E2 11 13S 28 55 E
Mansehra, Pakistan 42 B5 34 20N 73 15 E
Mansel I., Canada 69 B11 62 0N 80 0W
Mansfield, Australia 63 F4 37 4S 146 6 E
Mansfield, U.K. 10 D6 53 9N 1 11W
Mansfield, La., U.S.A. 81 J8 32 2N 93 43W
Mansfield, Mass., U.S.A. 79 D13 42 2N 71 13W
Mansfield, Ohio, U.S.A. 78 F2 40 45N 82 31W
Mansfield, Pa., U.S.A. 78 E7 41 48N 77 5W
Mansfield, Wash., U.S.A. 82 C4 47 49N 119 38W
Manson Creek, Canada 72 B4 55 37N 124 32W
Manta, Ecuador 92 D2 1 0S 80 40W
Mantalingajan, Mt., Phil. 36 C5 8 55N 117 45 E
Mantare, Tanzania 54 C3 2 42S 33 13 E
Manteca, U.S.A. 83 H3 37 48N 121 13W
Manteo, U.S.A. 77 H8 35 55N 75 40W
Mantes-la-Jolie, France 18 B4 48 58N 1 41 E
Manthani, India 40 K11 18 40N 79 35 E
Manti, U.S.A. 82 G8 39 16N 111 38W
Mantiqueira, Serra da, Brazil 95 A7 22 0S 44 0W
Manton, U.S.A. 76 C3 44 25N 85 24W
Mántova, Italy 20 B4 45 9N 10 48 E
Mänttä, Finland 9 E21 62 0N 24 40 E
Mantua = Mántova, Italy 20 B4 45 9N 10 48 E
Manu, Peru 92 F4 12 10S 70 51W
Manua Is., Amer. Samoa 59 B14 14 13S 169 35W
Manuae, Cook Is. 65 J12 19 30S 159 0W
Manuel Alves →, Brazil 93 F9 11 19S 48 28W
Manui, Indonesia 37 E6 3 35S 123 5 E
Manville, U.S.A. 80 D2 42 47N 104 37W
Many, U.S.A. 81 K8 31 34N 93 29W
Manyara, L., Tanzania 54 C4 3 35S 35 50 E
Manych-Gudilo, Ozero, Russia 25 E7 46 24N 42 38 E
Manyonga →, Tanzania 54 C3 4 10S 34 15 E
Manyoni, Tanzania 54 D3 5 45S 34 55 E
Manyoni □, Tanzania 54 D3 6 30S 34 30 E
Manzai, Pakistan 42 C4 32 12N 70 15 E
Manzanares, Spain 19 C4 39 2N 3 22W
Manzanillo, Cuba 88 B4 20 20N 77 31W
Manzanillo, Mexico 86 D4 19 0N 104 20W
Manzanillo, Pta., Panama 88 E4 9 30N 79 40W
Manzano Mts., U.S.A. 83 J10 34 40N 106 20W
Manzarīyeh, Iran 45 C6 34 53N 50 50 E
Manzhouli, China 33 B6 49 35N 117 25 E
Manzini, Swaziland 57 D5 26 30S 31 25 E
Mao, Chad 51 F8 14 4N 15 19 E
Maoke, Pegunungan, Indonesia 37 E9 3 40S 137 30 E
Maolin, China 35 C12 43 58N 123 30 E
Maoming, China 33 D6 21 50N 110 54 E
Maoxing, China 35 B13 45 28N 124 40 E
Mapam Yumco, China 32 C3 30 45N 81 28 E
Mapastepec, Mexico 87 D6 15 26N 92 54W
Mapia, Kepulauan, Indonesia 37 D8 0 50N 134 20 E
Mapimí, Mexico 86 B4 25 50N 103 50W
Mapimí, Bolsón de, Mexico 86 B4 27 30N 104 15W
Mapinga, Tanzania 54 D4 6 40S 39 12 E
Mapinhane, Mozam. 57 C6 22 20S 35 0 E
Maple Creek, Canada 73 D7 49 55N 109 29W
Maple Valley, U.S.A. 84 C4 47 25N 122 3W
Mapleton, U.S.A. 82 D2 44 2N 123 52W
Mapuera →, Brazil 92 D7 1 5S 57 2W
Maputo, Mozam. 57 D5 25 58S 32 32 E
Maputo, B. de, Mozam. 57 D5 25 50S 32 45 E
Maqiaohe, China 35 B16 44 40N 130 30 E
Maqnā, Si. Arabia 44 D2 28 25N 34 50 E
Maquela do Zombo, Angola 52 F3 6 0S 15 15 E
Maquinchao, Argentina 96 E3 41 15S 68 50W
Maquoketa, U.S.A. 80 D9 42 4N 90 40W
Mar, Serra do, Brazil 95 B6 25 30S 49 0W
Mar Chiquita, L., Argentina 94 C3 30 40S 62 50W
Mar del Plata, Argentina 94 D4 38 0S 57 30W
Mar Menor, Spain 19 D5 37 40N 0 45W
Mara, Tanzania 54 C3 1 30S 34 32 E
Mara □, Tanzania 54 C3 1 45S 34 20 E
Maraã, Brazil 92 D5 1 52S 65 25W
Marabá, Brazil 93 E9 5 20S 49 5W
Maracá, I. de, Brazil 93 C8 2 10N 50 30W
Maracaibo, Venezuela 92 A4 10 40N 71 37W
Maracaibo, L. de, Venezuela 92 B4 9 40N 71 30W
Maracaju, Brazil 95 A4 21 38S 55 9W
Maracay, Venezuela 92 A5 10 15N 67 28W
Marādah, Libya 51 C8 29 15N 19 15 E
Maradi, Niger 50 F6 13 29N 7 20 E
Marāgheh, Iran 44 B5 37 30N 46 12 E
Marāh, Si. Arabia 44 E5 25 0N 45 35 E
Marajó, I. de, Brazil 93 D9 1 0S 49 30W
Marākand, Iran 44 B5 38 51N 45 16 E
Maralal, Kenya 54 B4 1 0N 36 38 E
Maralinga, Australia 61 F5 30 13S 131 32 E
Marama, Australia 63 F3 35 10S 140 10 E
Marampa, S. Leone 50 G2 8 45N 12 28W
Maran, Malaysia 39 L4 3 35N 102 45 E
Marana, U.S.A. 83 K8 32 27N 111 13W
Maranboy, Australia 60 B5 14 40S 132 39 E
Marand, Iran 44 B5 38 30N 45 45 E
Marang, Malaysia 39 K4 5 12N 103 13 E
Maranguape, Brazil 93 D11 3 55S 38 50W
Maranhão = São Luís, Brazil 93 D10 2 39S 44 15W
Maranhão □, Brazil 93 E9 5 0S 46 0W
Maranoa →, Australia 63 D4 27 50S 148 37 E
Marañón →, Peru 92 D4 4 30S 73 35W
Marão, Mozam. 57 C5 24 18S 34 2 E
Maraş = Kahramanmaraş, Turkey 25 G6 37 37N 36 53 E
Marathasa □, Cyprus 23 E11 34 59N 32 51 E
Marathon, Australia 62 C3 20 51S 143 32 E
Marathon, Canada 70 C2 48 44N 86 23W
Marathon, N.Y., U.S.A. 79 D8 42 27N 76 2W
Marathon, Tex., U.S.A. 81 K3 30 12N 103 15W
Marathóvouno, Cyprus 23 D12 35 13N 33 37 E
Maratua, Indonesia 37 D5 2 10N 118 35 E
Maravatío, Mexico 86 D4 19 51N 100 25W
Marāwih, U.A.E. 45 E7 24 18N 53 18 E
Marbella, Spain 19 D3 36 30N 4 57W
Marble Bar, Australia 60 D2 21 9S 119 44 E
Marble Falls, U.S.A. 81 K5 30 35N 98 16W
Marblehead, U.S.A. 79 D14 42 30N 70 51W
Marburg, Germany 16 C5 50 47N 8 46 E
March, U.K. 11 E8 52 33N 0 5 E
Marche, France 18 C4 46 5N 1 20 E
Marche-en-Famenne, Belgium 15 D5 50 14N 5 19 E
Marchena, Spain 19 D3 37 18N 5 23W
Marcos Juárez, Argentina 94 C3 32 42S 62 5W
Marcus I. = Minami-Tori-Shima, Pac. Oc. 64 E7 24 0N 153 45 E
Marcus Necker Ridge, Pac. Oc. 64 F9 20 0N 175 0 E
Marcy, Mt., U.S.A. 79 B11 44 7N 73 56W
Mardan, Pakistan 42 B5 34 20N 72 0 E
Mardie, Australia 60 D2 21 12S 115 59 E
Mardin, Turkey 25 G7 37 20N 40 43 E
Maree, L., U.K. 12 D3 57 40N 5 26W
Mareeba, Australia 62 B4 16 59S 145 28 E
Marek = Stanke Dimitrov, Bulgaria 21 C10 42 17N 23 9 E
Marek, Indonesia 37 E6 4 41S 120 24 E
Marengo, U.S.A. 80 E8 41 48N 92 4W
Marenyi, Kenya 54 C4 4 22S 39 8 E
Marerano, Madag. 57 C7 21 23S 44 52 E
Marfa, U.S.A. 81 K2 30 19N 104 1W
Marfa Pt., Malta 23 D1 35 59N 14 19 E
Margaret →, Australia 60 C4 18 9S 125 41 E
Margaret Bay, Canada 72 C3 51 20N 127 35W
Margaret L., Canada 72 B5 58 56N 115 25W
Margaret River, Australia 60 C4 18 38S 126 52 E
Margarita, I. de, Venezuela 92 A6 11 0N 64 0W
Margaritovo, Russia 30 C7 43 25N 134 45 E
Margate, S. Africa 57 E5 30 50S 30 20 E
Margate, U.K. 11 F9 51 23N 1 23 E
Margelan = Marghilon, Uzbekistan 26 E8 40 27N 71 42 E
Marghilon, Uzbekistan 26 E8 40 27N 71 42 E
Marguerite, Canada 72 C4 52 30N 122 25W
Mari El □, Russia 24 C8 56 30N 48 0 E
Mari Republic □ = Mari El □, Russia 24 C8 56 30N 48 0 E
María Elena, Chile 94 A2 22 18S 69 40W
María Grande, Argentina 94 C4 31 45S 59 55W
Maria I., N. Terr., Australia 62 A2 14 52S 135 45 E
Maria I., Tas., Australia 62 G4 42 35S 148 0 E
Maria van Diemen, C., N.Z. 59 F4 34 29S 172 40 E
Mariakani, Kenya 54 C4 3 50S 39 27 E
Marian L., Canada 72 A5 63 0N 116 15W
Mariana Trench, Pac. Oc. 28 H18 13 0N 145 0 E
Marianao, Cuba 88 B3 23 8N 82 24W
Marianna, Ark., U.S.A. 81 H9 34 46N 90 46W
Marianna, Fla., U.S.A. 77 K3 30 46N 85 14W
Marias →, U.S.A. 82 C8 47 56N 110 30W
Mariato, Punta, Panama 88 E3 7 12N 80 52W
Ma'rib, Yemen 46 D4 15 25N 45 21 E
Maribor, Slovenia 16 E8 46 36N 15 40 E
Marico →, Africa 56 C4 23 35S 26 57 E
Maricopa, Ariz., U.S.A. 83 K7 33 4N 112 3W
Maricopa, Calif., U.S.A. 85 K7 35 4N 119 24W
Maricourt, Canada 69 C12 56 34N 70 49W
Marîdî, Sudan 51 H10 4 55N 29 25 E
Marie Byrd Land, Antarctica 5 D14 79 30S 125 0W
Marie-Galante, Guadeloupe 89 C7 15 56N 61 16W
Mariecourt = Kangiqsujuaq, Canada 69 B12 61 30N 72 0W
Marienberg, Neths. 15 B6 52 2N 6 35 E
Marienbourg, Belgium 15 D4 50 6N 4 31 E
Mariental, Namibia 56 C2 24 36S 18 0 E
Marienville, U.S.A. 78 E5 41 28N 79 8W
Mariestad, Sweden 9 G15 58 43N 13 50 E
Marietta, Ga., U.S.A. 77 J3 33 57N 84 33W
Marietta, Ohio, U.S.A. 76 F5 39 25N 81 27W
Marieville, Canada 79 A11 45 26N 73 10W
Mariinsk, Russia 26 D9 56 10N 87 20 E
Marijampole, Lithuania 9 J20 54 33N 23 19 E
Marília, Brazil 95 A5 22 13S 50 0W
Marillana, Australia 60 D2 22 37S 119 16 E
Marín, Spain 19 A1 42 23N 8 42W
Marina, U.S.A. 84 J5 36 41N 121 48W
Marina Plains, Australia 62 A3 14 37S 143 57 E
Marinduque, Phil. 37 B6 13 25N 122 0 E
Marine City, U.S.A. 76 D4 42 43N 82 30W
Marinette, U.S.A. 76 C2 45 6N 87 38W
Maringá, Brazil 95 A5 23 26S 52 2W
Marion, Ala., U.S.A. 77 J2 32 38N 87 19W
Marion, Ill., U.S.A. 81 G10 37 44N 88 56W
Marion, Ind., U.S.A. 76 E3 40 32N 85 40W
Marion, Iowa, U.S.A. 80 D9 42 2N 91 36W
Marion, Kans., U.S.A. 80 F6 38 21N 97 1W
Marion, Mich., U.S.A. 76 C3 44 6N 85 9W
Marion, N.C., U.S.A. 77 H4 35 41N 82 1W
Marion, Ohio, U.S.A. 76 E4 40 35N 83 8W
Marion, S.C., U.S.A. 77 H6 34 11N 79 24W
Marion, Va., U.S.A. 77 G5 36 50N 81 31W
Marion, L., U.S.A. 77 J5 33 28N 80 10W
Mariposa, U.S.A. 83 H4 37 29N 119 58W
Mariscal Estigarribia, Paraguay 94 A3 22 3S 60 40W
Maritime Alps = Maritimes, Alpes, Europe 16 F4 44 10N 7 10 E
Maritimes, Alpes, Europe 16 F4 44 10N 7 10 E
Maritsa = Évros →, Bulgaria 21 D12 41 40N 26 34 E
Maritsa, Greece 23 C10 36 22N 28 10 E
Mariupol, Ukraine 25 E6 47 5N 37 31 E
Marīvān, Iran 44 C5 35 30N 46 25 E
Markazī □, Iran 45 C6 35 0N 49 30 E
Markdale, Canada 78 B4 44 19N 80 39W
Marked Tree, U.S.A. 81 H9 35 32N 90 25W
Marken, Neths. 15 B5 52 26N 5 12 E
Market Drayton, U.K. 10 E5 52 54N 2 29W
Market Harborough, U.K. 11 E7 52 29N 0 55W
Markham, Canada 78 C5 43 52N 79 16W
Markham, Mt., Antarctica 5 E11 83 0S 164 0 E
Markham L., Canada 73 A8 62 30N 102 35W
Markleeville, U.S.A. 84 G7 38 42N 119 47W
Markovo, Russia 27 C17 64 40N 169 40 E
Marks, Russia 24 D8 51 45N 46 50 E
Marksville, U.S.A. 81 K8 31 8N 92 4W
Marla, Australia 63 D1 27 19S 133 33 E
Marlboro, U.S.A. 79 D13 42 19N 71 33W
Marlborough, Australia 62 C4 22 46S 149 52 E
Marlborough Downs, U.K. 11 F6 51 27N 1 53W
Marlin, U.S.A. 81 K6 31 18N 96 54W
Marlow, U.S.A. 81 H6 34 39N 97 58W
Marmagao, India 40 M8 15 25N 73 56 E
Marmara, Turkey 21 D12 40 35N 27 38 E
Marmara, Sea of = Marmara Denizi, Turkey 21 D13 40 45N 28 15 E
Marmara Denizi, Turkey 21 D13 40 45N 28 15 E
Marmaris, Turkey 21 F13 36 50N 28 14 E
Marmarth, U.S.A. 80 B3 46 18N 103 54W
Marmion, Mt., Australia 61 E2 29 16S 119 50 E
Marmion L., Canada 70 C1 48 55N 91 20W
Marmolada, Mte., Italy 20 A4 46 26N 11 51 E
Marmora, Canada 70 D4 44 28N 77 41W
Marne →, France 18 B5 48 48N 2 24 E
Maroala, Madag. 57 B8 15 23S 47 59 E
Maroantsetra, Madag. 57 B8 15 26S 49 44 E
Maromandia, Madag. 57 A8 14 13S 48 5 E
Marondera, Zimbabwe 55 F3 18 5S 31 42 E
Maroni →, Fr. Guiana 93 B8 5 30N 54 0W
Maroochydore, Australia 63 D5 26 29S 153 5 E
Maroona, Australia 63 F3 37 27S 142 54 E
Marosakoa, Madag. 57 B8 15 26S 46 38 E
Maroua, Cameroon 51 F7 10 40N 14 20 E
Marovoay, Madag. 57 B8 16 6S 46 39 E
Marquard, S. Africa 56 D4 28 40S 27 28 E
Marquesas Is. = Marquises, Is., Pac. Oc. 65 H14 9 30S 140 0W
Marquette, U.S.A. 76 B2 46 33N 87 24W
Marquises, Is., Pac. Oc. 65 H14 9 30S 140 0W

Marracuene, Mozam. 57 D5 25 45S 32 35 E
Marrakech, Morocco 50 B3 31 9N 8 0W
Marrawah, Australia 62 G3 40 55S 144 42 E
Marree, Australia 63 D2 29 39S 138 1 E
Marrilla, Australia 60 D1 22 31S 114 25 E
Marrimane, Mozam. 57 C5 22 58S 33 34 E
Marromeu, Mozam. 57 B6 18 15S 36 25 E
Marrowie Cr. →, Australia 63 E4 33 23S 145 40 E
Marrubane, Mozam. 55 F4 18 0S 37 0 E
Marrupa, Mozam. 55 E4 13 8S 37 30 E
Marsá Matrûh, Egypt 51 B10 31 19N 27 9 E
Marsá Susah, Libya 51 B9 32 52N 21 59 E
Marsabit, Kenya 54 B4 2 18N 38 0 E
Marsabit □, Kenya 54 B4 2 45N 37 45 E
Marsala, Italy 20 F5 37 48N 12 26 E
Marsalforn, Malta 23 C1 36 4N 14 15 E
Marsden, Australia 63 E4 33 47S 147 32 E
Marseille, France 18 E6 43 18N 5 23 E
Marseilles = Marseille, France 18 E6 43 18N 5 23 E
Marsh I., U.S.A. 81 L9 29 34N 91 53W
Marsh L., U.S.A. 80 C6 45 5N 96 0W
Marshall, Liberia 50 G2 6 8N 10 22W
Marshall, Ark., U.S.A. 81 H8 35 55N 92 38W
Marshall, Mich., U.S.A. 76 D3 42 16N 84 58W
Marshall, Minn., U.S.A. 80 C7 44 25N 95 45W
Marshall, Mo., U.S.A. 80 F8 39 7N 93 12W
Marshall, Tex., U.S.A. 81 J7 32 33N 94 23W
Marshall →, Australia 62 C2 22 59S 136 59 E
Marshall Is. ■, Pac. Oc. 64 G9 9 0N 171 0 E
Marshalltown, U.S.A. 80 D8 42 3N 92 55W
Marshfield, Mo., U.S.A. 81 G8 37 15N 92 54W
Marshfield, Wis., U.S.A. 80 C9 44 40N 90 10W
Marshūn, Iran 45 B6 36 19N 49 23 E
Märsta, Sweden 9 G17 59 37N 17 52 E
Mart, U.S.A. 81 K6 31 33N 96 50W
Martaban, Burma 41 L20 16 30N 97 35 E
Martaban, G. of, Burma 41 L20 16 5N 96 30 E
Martapura, Kalimantan, Indonesia 36 E4 3 22S 114 47 E
Martapura, Sumatera, Indonesia 36 E2 4 19S 104 22 E
Marte, Nigeria 51 F7 12 23N 13 46 E
Martelange, Belgium 15 E5 49 49N 5 43 E
Martha's Vineyard, U.S.A. 79 E14 41 25N 70 38W
Martigny, Switz. 16 E4 46 6N 7 3 E
Martigues, France 18 E6 43 24N 5 4 E
Martin, Slovak Rep. 17 D10 49 6N 18 48 E
Martin, S. Dak., U.S.A. 80 D4 43 11N 101 44W
Martin, Tenn., U.S.A. 81 G10 36 21N 88 51W
Martin L., U.S.A. 77 J3 32 41N 85 55W
Martina Franca, Italy 20 D7 40 42N 17 20 E
Martinborough, N.Z. 59 J5 41 14S 175 29 E
Martinez, U.S.A. 84 G4 38 1N 122 8W
Martinique ■, W. Indies 89 D7 14 40N 61 0W
Martinique Passage, W. Indies 89 C7 15 15N 61 0W
Martinópolis, Brazil 95 A5 22 11S 51 12W
Martins Ferry, U.S.A. 78 F4 40 6N 80 44W
Martinsburg, Pa., U.S.A. 78 F6 40 19N 78 20W
Martinsburg, W. Va., U.S.A. 76 F7 39 27N 77 58W
Martinsville, Ind., U.S.A. 76 F2 39 26N 86 25W
Martinsville, Va., U.S.A. 77 G6 36 41N 79 52W
Marton, N.Z. 59 J5 40 4S 175 23 E
Martos, Spain 19 D4 37 44N 3 58W
Marudi, Malaysia 36 D4 4 11N 114 19 E
Ma'ruf, Afghan. 40 D5 31 30N 67 6 E
Marugame, Japan 31 G6 34 15N 133 40 E
Marulan, Australia 63 E5 34 43S 150 3 E
Marunga, Angola 56 B3 17 28S 20 2 E
Marungu, Mts., Zaïre 54 D2 7 30S 30 0 E
Marvast, Iran 45 D7 30 30N 54 15 E
Marwar, India 42 G5 25 43N 73 45 E
Mary, Turkmenistan 26 F7 37 40N 61 50 E
Mary Frances L., Canada 73 A7 63 19N 106 13W
Mary Kathleen, Australia 62 C2 20 44S 139 48 E
Maryborough = Port Laoise, Ireland 13 C4 53 2N 7 18W
Maryborough, Queens., Australia 63 D5 25 31S 152 37 E
Maryborough, Vic., Australia 63 F3 37 0S 143 44 E
Maryfield, Canada 73 D8 49 50N 101 35W
Maryland □, U.S.A. 76 F7 39 0N 76 30W
Maryland Junction, Zimbabwe 55 F3 17 45S 30 31 E
Maryport, U.K. 10 C4 54 44N 3 28W
Mary's Harbour, Canada 71 B8 52 18N 55 51W
Marystown, Canada 71 C8 47 10N 55 10W
Marysvale, U.S.A. 83 G7 38 27N 112 14W
Marysville, Canada 72 D5 49 35N 116 0W
Marysville, Calif., U.S.A. 84 F5 39 9N 121 35W
Marysville, Kans., U.S.A. 80 F6 39 51N 96 39W
Marysville, Mich., U.S.A. 78 D2 42 54N 82 29W
Marysville, Ohio, U.S.A. 76 E4 40 14N 83 22W
Marysville, Wash., U.S.A. 84 B4 48 3N 122 11W
Maryvale, Australia 63 D5 28 4S 152 12 E
Maryville, U.S.A. 77 H4 35 46N 83 58W
Marzūq, Libya 51 C7 25 53N 13 57 E
Masahunga, Tanzania 54 C3 2 6S 33 18 E
Masai, Malaysia 39 M4 1 29N 103 55 E
Masai Steppe, Tanzania 54 C4 4 30S 36 30 E
Masaka, Uganda 54 C3 0 21S 31 45 E
Masalembo, Kepulauan, Indonesia 36 F4 5 35S 114 30 E
Masalima, Kepulauan, Indonesia 36 F5 5 4S 117 5 E
Masamba, Indonesia 37 E6 2 30S 120 15 E
Masan, S. Korea 35 G15 35 11N 128 32 E
Masasi, Tanzania 55 E4 10 45S 38 52 E
Masasi □, Tanzania 55 E4 10 45S 38 50 E
Masaya, Nic. 88 D2 12 0N 86 7W
Masbate, Phil. 37 B6 12 21N 123 36 E
Mascara, Algeria 50 A5 35 26N 0 6 E
Mascota, Mexico 86 C4 20 30N 104 50W
Masela, Indonesia 37 F7 8 9S 129 51 E
Maseru, Lesotho 56 D4 29 18S 27 30 E
Mashaba, Zimbabwe 55 G3 20 2S 30 29 E
Mashābih, Si. Arabia 44 E3 25 35N 36 30 E
Masherbrum, Pakistan 43 B7 35 38N 76 18 E
Mashhad, Iran 45 B8 36 20N 59 35 E
Mashīz, Iran 45 D8 29 56N 56 37 E
Mashkel, Hamun-i-, Pakistan 40 E3 28 30N 63 0 E
Mashki Chāh, Pakistan 40 E3 29 5N 62 30 E
Mashonaland Central □, Zimbabwe 57 B5 17 30S 31 0 E
Mashonaland East □, Zimbabwe 57 B5 18 0S 32 0 E
Mashonaland West □, Zimbabwe 57 B4 17 30S 29 30 E
Masi Manimba, Zaïre 52 E3 4 40S 17 54 E
Masindi, Uganda 54 B3 1 40N 31 43 E
Masindi Port, Uganda 54 B3 1 43N 32 2 E
Masisea, Peru 92 E4 8 35S 74 22W
Masisi, Zaïre 54 C2 1 23S 28 49 E
Masjed Soleyman, Iran 45 D6 31 55N 49 18 E
Mask, L., Ireland 13 C2 53 36N 9 22W
Masoala, Tanjon' i, Madag. 57 B9 15 59S 50 13 E
Masoarivo, Madag. 57 B7 19 3S 44 19 E
Masohi, Indonesia 37 E7 3 20S 128 55 E
Masomeloka, Madag. 57 C8 20 17S 48 37 E
Mason, Nev., U.S.A. 84 G7 38 56N 119 8W
Mason, Tex., U.S.A. 81 K5 30 45N 99 14W
Mason City, U.S.A. 80 D8 43 9N 93 12W
Maspalomas, Canary Is. 22 G4 27 46N 15 35W
Maspalomas, Pta., Canary Is. 22 G4 27 43N 15 36W
Masqat, Oman 46 C6 23 37N 58 36 E
Massa, Italy 20 B4 44 1N 10 9 E
Massachusetts □, U.S.A. 79 D12 42 30N 72 0W
Massachusetts B., U.S.A. 79 D14 42 20N 70 50W
Massaguet, Chad 51 F8 12 28N 15 26 E
Massakory, Chad 51 F8 13 0N 15 49 E
Massanella, Spain 22 B9 39 48N 2 51 E
Massangena, Mozam. 57 C5 21 34S 33 0 E
Massawa = Mitsiwa, Eritrea 51 E12 15 35N 39 25 E
Massena, U.S.A. 79 B10 44 56N 74 54W
Massénya, Chad 51 F8 11 21N 16 9 E
Masset, Canada 72 C2 54 2N 132 10W
Massif Central, France 18 D5 44 55N 3 0 E
Massillon, U.S.A. 78 F3 40 48N 81 32W
Massinga, Mozam. 57 C6 23 15S 35 22 E
Masson, Canada 79 A9 45 32N 75 25W
Masson I., Antarctica 5 C7 66 10S 93 20 E
Mastanli = Momchilgrad, Bulgaria 21 D11 41 33N 25 23 E
Masterton, N.Z. 59 J5 40 56S 175 39 E
Mastuj, Pakistan 43 A5 36 20N 72 36 E
Mastung, Pakistan 40 E5 29 50N 66 56 E
Masty, Belarus 17 B13 53 27N 24 38 E
Masuda, Japan 31 G5 34 40N 131 51 E
Masvingo, Zimbabwe 55 G3 20 8S 30 49 E
Masvingo □, Zimbabwe 55 G3 21 0S 31 30 E
Maswa □, Tanzania 54 C3 3 30S 34 0 E
Maşyāf, Syria 44 C3 35 4N 36 20 E
Matabeleland North □, Zimbabwe 55 F2 19 0S 28 0 E
Matabeleland South □, Zimbabwe 55 G2 21 0S 29 0 E
Mataboor, Indonesia 37 E9 1 41S 138 3 E
Matachewan, Canada 70 C3 47 56N 80 39W
Matadi, Zaïre 52 F2 5 52S 13 31 E
Matagalpa, Nic. 88 D2 13 0N 85 58W
Matagami, Canada 70 C4 49 45N 77 34W
Matagami, L., Canada 70 C4 49 50N 77 40W
Matagorda, U.S.A. 81 L7 28 42N 95 58W
Matagorda B., U.S.A. 81 L6 28 40N 96 0W
Matagorda I., U.S.A. 81 L6 28 15N 96 30W
Matak, P., Indonesia 39 L6 3 18N 106 16 E
Matakana, Australia 63 E4 32 59S 145 54 E
Mátala, Greece 23 E6 34 59N 24 45 E
Matam, Senegal 50 E2 15 34N 13 17W
Matamoros, Campeche, Mexico 87 D6 18 50N 90 50W
Matamoros, Coahuila, Mexico 86 B4 25 33N 103 15W
Matamoros, Puebla, Mexico 87 D5 18 2N 98 17W
Matamoros, Tamaulipas, Mexico 87 B5 25 50N 97 30W
Ma'ṭan as Sarra, Libya 51 D9 21 45N 22 0 E
Matandu →, Tanzania 55 D3 8 45S 34 19 E
Matane, Canada 71 C6 48 50N 67 33W
Matanzas, Cuba 88 B3 23 0N 81 40W
Matapan, C. = Taínaron, Ákra, Greece 21 F10 36 22N 22 27 E
Matapédia, Canada 71 C6 48 0N 66 59W
Matara, Sri Lanka 40 S12 5 58N 80 30 E
Mataram, Indonesia 36 F5 8 41S 116 10 E
Matarani, Peru 92 G4 17 0S 72 10W
Mataranka, Australia 60 B5 14 55S 133 4 E
Matarma, Râs, Egypt 47 E1 30 27N 32 44 E
Mataró, Spain 19 B7 41 32N 2 29 E
Matatiele, S. Africa 57 E4 30 20S 28 49 E
Mataura, N.Z. 59 M2 46 11S 168 51 E
Matehuala, Mexico 86 C4 23 40N 100 40W
Mateke Hills, Zimbabwe 55 G3 21 48S 31 0 E
Matera, Italy 20 D7 40 40N 16 36 E
Matetsi, Zimbabwe 55 F2 18 12S 26 0 E
Matheson Island, Canada 73 C9 51 45N 96 56W
Mathis, U.S.A. 81 L6 28 6N 97 50W
Mathura, India 42 F7 27 30N 77 40 E
Mati, Phil. 37 C7 6 55N 126 15 E
Matías Romero, Mexico 87 D5 16 53N 95 2W
Matibane, Mozam. 55 E5 14 49S 40 45 E
Matima, Botswana 56 C3 20 15S 24 26 E
Matiri Ra., N.Z. 59 J4 41 38S 172 20 E
Matlock, U.K. 10 D6 53 9N 1 33W
Matmata, Tunisia 50 B6 33 37N 9 59 E
Mato Grosso □, Brazil 93 F8 14 0S 55 0W
Mato Grosso, Planalto do, Brazil 93 G8 15 0S 55 0W
Mato Grosso, Plateau of, Brazil 90 E5 15 0S 54 0W
Mato Grosso do Sul □, Brazil 93 G8 18 0S 55 0W
Matochkin Shar, Russia 26 B6 73 10N 56 40 E
Matopo Hills, Zimbabwe 55 G2 20 36S 28 20 E
Matopos, Zimbabwe 55 G2 20 20S 28 29 E
Matosinhos, Portugal 19 B1 41 11N 8 42W
Matsue, Japan 31 G6 35 25N 133 10 E
Matsumae, Japan 30 D10 41 26N 140 7 E
Matsumoto, Japan 31 F9 36 15N 138 0 E
Matsusaka, Japan 31 G8 34 34N 136 32 E
Matsuura, Japan 31 H4 33 20N 129 49 E
Matsuyama, Japan 31 H6 33 45N 132 45 E
Mattagami →, Canada 70 B3 50 43N 81 29W
Mattancheri, India 40 Q10 9 50N 76 15 E
Mattawa, Canada 70 C4 46 20N 78 45W
Mattawamkeag, U.S.A. 71 C6 45 32N 68 21W
Matterhorn, Switz. 16 F4 45 58N 7 39 E
Matthew Town, Bahamas 89 B5 20 57N 73 40W
Matthew's Ridge, Guyana 92 B6 7 37N 60 10W
Mattice, Canada 70 C3 49 40N 83 20W
Mattituck, U.S.A. 79 F12 40 59N 72 32W
Matuba, Mozam. 57 C5 24 28S 32 49 E
Matucana, Peru 92 F3 11 55S 76 25W
Matun, Afghan. 42 C3 33 22N 69 58 E
Maturín, Venezuela 92 B6 9 45N 63 11W
Mau, India 43 G10 25 56N 83 33 E
Mau Escarpment, Kenya 54 C4 0 40S 36 0 E
Mau Ranipur, India 43 G8 25 16N 79 8 E
Maubeuge, France 18 A6 50 17N 3 57 E
Maud, Pt., Australia 60 D1 23 6S 113 45 E
Maude, Australia 63 E3 34 29S 144 18 E
Maudin Sun, Burma 41 M19 16 0N 94 30 E
Maués, Brazil 92 D7 3 20S 57 45W
Mauganj, India 41 G12 24 50N 81 55 E
Maui, U.S.A. 74 H16 20 48N 156 20W
Maulamyaing = Moulmein, Burma 41 L20 16 30N 97 40 E
Maule □, Chile 94 D1 36 5S 72 30W
Maumee, U.S.A. 76 E4 41 34N 83 39W
Maumee →, U.S.A. 76 E4 41 42N 83 28W
Maumere, Indonesia 37 F6 8 38S 122 13 E
Maun, Botswana 56 B3 20 0S 23 26 E
Mauna Kea, U.S.A. 74 J17 19 50N 155 28W
Mauna Loa, U.S.A. 74 J17 19 30N 155 35W
Maungmagan Kyunzu, Burma 41 M20 14 0N 97 48 E
Maupin, U.S.A. 82 D3 45 11N 121 5W
Maurepas, L., U.S.A. 81 K9 30 15N 90 30W
Maurice, L., Australia 61 E5 29 30S 131 0 E
Mauritania ■, Africa 50 D3 20 50N 10 0W
Mauritius ■, Ind. Oc. 49 J9 20 0S 57 0 E
Mauston, U.S.A. 80 D9 43 48N 90 5W
Mavinga, Angola 53 H4 15 50S 20 21 E
Mavli, India 42 G5 24 45N 73 55 E
Mavuradonha Mts., Zimbabwe 55 F3 16 30S 31 30 E
Mawa, Zaïre 54 B2 2 45N 26 40 E
Mawana, India 42 E7 29 6N 77 58 E
Mawand, Pakistan 42 E3 29 33N 68 38 E
Mawk Mai, Burma 41 J20 20 14N 97 37 E
Mawlaik, Burma 41 H19 23 40N 94 26 E
Mawquq, Si. Arabia 44 E4 27 25N 41 8 E
Mawson Coast, Antarctica 5 C6 68 30S 63 0 E
Max, U.S.A. 80 B4 47 49N 101 18W
Maxcanú, Mexico 87 C6 20 40N 92 0W
Maxesibeni, S. Africa 57 E4 30 49S 29 23 E
Maxhamish L., Canada 72 B4 59 50N 123 17W
Maxixe, Mozam. 57 C6 23 54S 35 17 E
Maxville, Canada 79 A10 45 17N 74 51W
Maxwell, U.S.A. 84 F4 39 17N 122 11W
Maxwelton, Australia 62 C3 20 43S 142 41 E
May Downs, Australia 62 C4 22 38S 148 55 E
May Pen, Jamaica 88 C4 17 58N 77 15W
Maya →, Russia 27 D14 60 28N 134 28 E
Maya Mts., Belize 87 D7 16 30N 89 0W
Mayaguana, Bahamas 89 B5 22 30N 72 44W
Mayagüez, Puerto Rico 89 C6 18 12N 67 9W
Mayāmey, Iran 45 B7 36 24N 55 42 E
Mayarí, Cuba 89 B4 20 40N 75 41W
Maybell, U.S.A. 82 F9 40 31N 108 5W
Maydān, Iraq 44 C5 34 55N 45 37 E
Maydena, Australia 62 G4 42 45S 146 30 E
Mayenne →, France 18 C3 47 30N 0 32W
Mayer, U.S.A. 83 J7 34 24N 112 14W
Mayerthorpe, Canada 72 C5 53 57N 115 8W
Mayfield, U.S.A. 77 G1 36 44N 88 38W
Mayhill, U.S.A. 83 K11 32 53N 105 29W
Maykop, Russia 25 F7 44 35N 40 10 E
Maymyo, Burma 38 A1 22 2N 96 28 E
Maynard, U.S.A. 84 C4 47 59N 122 55W
Maynard Hills, Australia 61 E2 28 28S 119 49 E
Mayne →, Australia 62 C3 23 40S 141 55 E
Maynooth, Ireland 13 C5 53 23N 6 34W
Mayo, Canada 68 B6 63 38N 135 57W
Mayo □, Ireland 13 C2 53 53N 9 3W
Mayo L., Canada 68 B6 63 45N 135 0W
Mayon Volcano, Phil. 37 B6 13 15N 123 41 E
Mayor I., N.Z. 59 G6 37 16S 176 17 E
Mayotte, I., Mayotte 49 H8 12 50S 45 10 E
Mayson L., Canada 73 B7 57 55N 107 10W
Maysville, U.S.A. 76 F4 38 39N 83 46W
Mayu, Indonesia 37 D7 1 30N 126 30 E
Mayville, N. Dak., U.S.A. 80 B6 47 30N 97 20W
Mayville, N.Y., U.S.A. 78 D5 42 15N 79 30W
Mayya, Russia 27 C14 61 44N 130 18 E
Mazabuka, Zambia 55 F2 15 52S 27 44 E
Mazagán = El Jadida, Morocco 50 B3 33 11N 8 17W
Mazagão, Brazil 93 D8 0 7S 51 16W
Mazán, Peru 92 D4 3 30S 73 0W
Māzandarān □, Iran 45 B7 36 30N 52 0 E
Mazapil, Mexico 86 C4 24 38N 101 34W
Mazara del Vallo, Italy 20 F5 37 39N 12 35 E
Mazarredo, Argentina 96 F3 47 10S 66 50W
Mazarrón, Spain 19 D5 37 38N 1 19W
Mazaruni →, Guyana 92 B7 6 25N 58 35W
Mazatán, Mexico 86 B2 29 0N 110 8W
Mazatenango, Guatemala 88 D1 14 35N 91 30W
Mazatlán, Mexico 86 C3 23 13N 106 25W
Mažeikiai, Lithuania 9 H20 56 20N 22 20 E
Māzhān, Iran 45 C8 32 30N 59 0 E
Mazīnān, Iran 45 B8 36 19N 56 56 E
Mazoe, Mozam. 55 F3 16 42S 33 7 E
Mazoe →, Mozam. 55 F3 16 20S 33 30 E
Mazowe, Zimbabwe 55 F3 17 28S 30 58 E
Mazurian Lakes = Mazurski, Pojezierze, Poland 17 B11 53 50N 21 0 E
Mazurski, Pojezierze, Poland 17 B11 53 50N 21 0 E
Mazyr, Belarus 17 B15 51 59N 29 15 E
Mbabane, Swaziland 57 D5 26 18S 31 6 E
Mbaïki, C.A.R. 52 D3 3 53N 18 1 E
Mbala, Zambia 55 D3 8 46S 31 24 E
Mbale, Uganda 54 B3 1 8N 34 12 E
Mbalmayo, Cameroon 52 D2 3 33N 11 33 E
Mbamba Bay, Tanzania 55 E3 11 13S 34 49 E
Mbandaka, Zaïre 52 D3 0 1N 18 18 E
Mbanza Congo, Angola 52 F2 6 18S 14 16 E
Mbanza Ngungu, Zaïre 52 F2 5 12S 14 53 E
Mbarara, Uganda 54 C3 0 35S 30 40 E
Mbashe →, S. Africa 57 E4 32 15S 28 54 E
Mbenkuru →, Tanzania 55 D4 9 25S 39 50 E
Mberengwa, Zimbabwe 55 G2 20 29S 29 57 E
Mberengwa, Mt., Zimbabwe 55 G2 20 37S 29 55 E
Mbesuma, Zambia 55 D3 10 0S 32 2 E
Mbeya, Tanzania 55 D3 8 54S 33 29 E
Mbeya □, Tanzania 54 D3 8 15S 33 30 E
Mbinga, Tanzania 55 E4 10 50S 35 0 E
Mbinga □, Tanzania 55 E3 10 50S 35 0 E
Mbini □, Eq. Guin. 52 D2 1 30N 10 0 E
Mbour, Senegal 50 F1 14 22N 16 54W
Mbout, Mauritania 50 E2 16 1N 12 38W
Mbozi □, Tanzania 55 D3 9 0S 32 50 E
Mbuji-Mayi, Zaïre 54 D1 6 9S 23 40 E
Mbulu, Tanzania 54 C4 3 45S 35 30 E
Mbulu □, Tanzania 54 C4 3 52S 35 33 E
Mburucuyá, Argentina 94 B4 28 1S 58 14W
Mchinja, Tanzania 55 D4 9 44S 39 45 E
Mchinji, Malawi 55 E3 13 47S 32 58 E
Mead, L., U.S.A. 85 J12 36 1N 114 44W
Meade, U.S.A. 81 G4 37 17N 100 20W
Meadow, Australia 61 E1 26 35S 114 40 E
Meadow Lake, Canada 73 C7 54 10N 108 26W
Meadow Lake Prov. Park, Canada 73 C7 54 27N 109 0W
Meadow Valley Wash →, U.S.A. 85 J12 36 40N 114 34W
Meadville, U.S.A. 78 E4 41 39N 80 9W
Meaford, Canada 70 D3 44 36N 80 35W
Mealy Mts., Canada 71 B8 53 10N 58 0W
Meander River, Canada 72 B5 59 2N 117 42W
Meares, C., U.S.A. 82 D2 45 37N 124 0W
Mearim →, Brazil 93 D10 3 4S 44 35W
Meath □, Ireland 13 C5 53 40N 6 57W
Meath Park, Canada 73 C7 53 27N 105 22W
Meaux, France 18 B5 48 58N 2 50 E
Mebechi-Gawa →, Japan 30 D10 40 31N 141 31 E
Mecanhelas, Mozam. 55 F4 15 12S 35 54 E
Mecca = Makkah, Si. Arabia 46 C2 21 30N 39 54 E
Mecca, U.S.A. 85 M10 33 34N 116 5W
Mechanicsburg, U.S.A. 78 F8 40 13N 77 1W
Mechanicville, U.S.A. 79 D11 42 54N 73 41W
Mechelen, Belgium 15 C4 51 2N 4 29 E
Mecheria, Algeria 50 B4 33 35N 0 18W
Mecklenburg, Germany 16 B6 53 33N 11 40 E
Mecklenburger Bucht, Germany 16 A6 54 20N 11 40 E
Meconta, Mozam. 55 E4 14 59S 39 50 E
Meda, Australia 60 C3 17 22S 123 59 E
Medan, Indonesia 36 D1 3 40N 98 38 E
Medanosa, Pta., Argentina 96 F3 48 8S 66 0W
Médéa, Algeria 50 A5 36 12N 2 50 E
Medellín, Colombia 92 B3 6 15N 75 35W
Medelpad, Sweden 9 E17 62 33N 16 30 E
Medemblik, Neths. 15 B5 52 46N 5 8 E
Mederdra, Mauritania 50 E1 17 0N 15 38W
Medford, Mass., U.S.A. 79 D13 42 25N 71 7W
Medford, Oreg., U.S.A. 82 E2 42 19N 122 52W
Medford, Wis., U.S.A. 80 C9 45 9N 90 20W
Medgidia, Romania 17 F15 44 15N 28 19 E
Media Agua, Argentina 94 C2 31 58S 68 25W
Media Luna, Argentina 94 C2 34 45S 66 44W
Mediaş, Romania 17 E13 46 9N 24 22 E
Medical Lake, U.S.A. 82 C5 47 34N 117 41W
Medicine Bow, U.S.A. 82 F10 41 54N 106 12W
Medicine Bow Pk., U.S.A. 82 F10 41 21N 106 19W
Medicine Bow Ra., U.S.A. 82 F10 41 10N 106 25W
Medicine Hat, Canada 73 D6 50 0N 110 45W
Medicine Lake, U.S.A. 80 A2 48 30N 104 30W
Medicine Lodge, U.S.A. 81 G5 37 17N 98 35W
Medina = Al Madīnah, Si. Arabia 46 C2 24 35N 39 52 E
Medina, N. Dak., U.S.A. 80 B5 46 54N 99 18W
Medina, N.Y., U.S.A. 78 C6 43 13N 78 23W
Medina, Ohio, U.S.A. 78 E3 41 8N 81 52W
Medina →, U.S.A. 81 L5 29 16N 98 29W
Medina del Campo, Spain 19 B3 41 18N 4 55W
Medina L., U.S.A. 81 L5 29 32N 98 56W
Medina-Sidonia, Spain 19 D3 36 28N 5 57W
Medinipur, India 43 H12 22 25N 87 21 E
Mediterranean Sea, Europe 6 H7 35 0N 15 0 E
Medley, Canada 73 C6 54 25N 110 16W
Médoc, France 18 D3 45 10N 0 50W
Medstead, Canada 73 C7 53 19N 108 5W
Medveditsa →, Russia 25 E7 49 35N 42 41 E
Medvezhi, Ostrava, Russia 27 B17 71 0N 161 0 E
Medvezhyegorsk, Russia 24 B5 63 0N 34 25 E
Medway →, U.K. 11 F8 51 27N 0 46 E
Meeberrie, Australia 61 E2 26 57S 115 51 E
Meekatharra, Australia 61 E2 26 32S 118 29 E
Meeker, U.S.A. 82 F10 40 2N 107 55W
Meerut, India 42 E7 29 1N 77 42 E
Meeteetse, U.S.A. 82 D9 44 9N 108 52W
Mega, Ethiopia 51 H12 3 57N 38 19 E
Mégara, Greece 21 F10 37 58N 23 22 E
Meghalaya □, India 41 G17 25 50N 91 0 E
Mégiscane, L., Canada 70 C4 48 35N 75 55W
Mehndawal, India 43 F10 26 58N 83 5 E
Mehr Jān, Iran 45 C7 33 50N 55 6 E
Mehrābād, Iran 44 B5 36 53N 47 55 E
Mehrān, Iran 44 C5 33 7N 46 10 E
Mehrīz, Iran 45 D7 31 35N 54 28 E
Mei Xian, Guangdong, China 33 D6 24 16N 116 6 E
Mei Xian, Shaanxi, China 34 G4 34 18N 107 55 E
Meiganga, Cameroon 52 C2 6 30N 14 25 E
Meiktila, Burma 41 J19 20 53N 95 54 E
Meissen, Germany 16 C7 51 9N 13 29 E
Mejillones, Chile 94 A1 23 10S 70 30W
Meka, Australia 61 E2 27 25S 116 48 E
Mékambo, Gabon 52 D2 1 2N 13 50 E
Mekdela, Ethiopia 51 F12 11 24N 39 10 E
Mekhtar, Pakistan 40 D6 30 30N 69 15 E

Meknès, *Morocco* **50 B3** 33 57N 5 33W
Mekong →, *Asia* **39 H6** 9 30N 106 15 E
Mekongga, *Indonesia* **37 E6** 3 39S 121 15 E
Mekvari = Kür →, *Azerbaijan* **25 G8** 39 29N 49 15 E
Melagiri Hills, *India* **40 N10** 12 20N 77 30 E
Melaka, *Malaysia* **39 L4** 2 15N 102 15 E
Melalap, *Malaysia* **36 C5** 5 10N 116 5 E
Mélambes, *Greece* **23 D6** 35 8N 24 40 E
Melanesia, *Pac. Oc.* **64 H7** 4 0S 155 0 E
Melbourne, *Australia* **63 F3** 37 50S 145 0 E
Melbourne, *U.S.A.* **77 L5** 28 5N 80 37W
Melchor Múzquiz, *Mexico* **86 B4** 27 50N 101 30W
Melchor Ocampo, *Mexico* **86 C4** 24 52N 101 40W
Mélèzes →, *Canada* **69 C12** 57 30N 71 0W
Melfi, *Chad* **51 F8** 11 0N 17 59 E
Melfort, *Canada* **73 C8** 52 50N 104 37W
Melfort, *Zimbabwe* **55 F3** 18 0S 31 25 E
Melhus, *Norway* **8 E14** 63 17N 10 18 E
Melilla, *N. Afr.* **19 E4** 35 21N 2 57W
Melipilla, *Chile* **94 C1** 33 42S 71 15W
Mélissa, Ákra, *Greece* **23 D6** 35 6N 24 33 E
Melita, *Canada* **73 D8** 49 15N 101 0W
Melitopol, *Ukraine* **25 E6** 46 50N 35 22 E
Melk, *Austria* **16 D8** 48 13N 15 20 E
Mellansel, *Sweden* **8 E18** 63 25N 18 17 E
Mellen, *U.S.A.* **80 B9** 46 20N 90 40W
Mellerud, *Sweden* **9 G15** 58 41N 12 28 E
Mellette, *U.S.A.* **80 C5** 45 9N 98 30W
Mellieha, *Malta* **23 D1** 35 57N 14 21 E
Melo, *Uruguay* **95 C5** 32 20S 54 10W
Melolo, *Indonesia* **37 F6** 9 53S 120 40 E
Melouprey, *Cambodia* **38 F5** 13 48N 105 16 E
Melrose, *N.S.W., Australia* **63 E4** 32 42S 146 57 E
Melrose, *W. Austral., Australia* **61 E3** 27 50S 121 15 E
Melrose, *U.K.* **12 F6** 55 36N 2 43W
Melrose, *U.S.A.* **81 H3** 34 26N 103 38W
Melstone, *U.S.A.* **82 C10** 46 36N 107 52W
Melton Mowbray, *U.K.* **10 E7** 52 47N 0 54W
Melun, *France* **18 B5** 48 32N 2 39 E
Melut, *Sudan* **51 F11** 10 30N 32 13 E
Melville, *Canada* **73 C8** 50 55N 102 50W
Melville, C., *Australia* **62 A3** 14 11S 144 30 E
Melville, L., *Canada* **71 B8** 53 30N 60 0W
Melville B., *Australia* **62 A2** 12 0S 136 45 E
Melville I., *Australia* **60 B5** 11 30S 131 0 E
Melville I., *Canada* **4 B2** 75 30N 112 0W
Melville Pen., *Canada* **69 B11** 68 0N 84 0W
Melvin →, *Canada* **72 B5** 59 11N 117 31W
Memba, *Mozam.* **55 E5** 14 11S 40 30 E
Memboro, *Indonesia* **37 F5** 9 30S 119 30 E
Memel = Klaipėda, *Lithuania* **9 J19** 55 43N 21 10 E
Memel, *S. Africa* **57 D4** 27 38S 29 36 E
Memmingen, *Germany* **16 E6** 47 58N 10 10 E
Mempawah, *Indonesia* **36 D3** 0 30N 109 5 E
Memphis, *Tenn., U.S.A.* **81 H10** 35 8N 90 3W
Memphis, *Tex., U.S.A.* **81 H4** 34 44N 100 33W
Mena, *U.S.A.* **81 H7** 34 35N 94 15W
Menai Strait, *U.K.* **10 D3** 53 11N 4 13W
Ménaka, *Mali* **50 E5** 15 59N 2 18 E
Menan = Chao Phraya →, *Thailand* **38 F3** 13 32N 100 36 E
Menarandra →, *Madag.* **57 D7** 25 17S 44 30 E
Menard, *U.S.A.* **81 K5** 30 55N 99 47W
Menasha, *U.S.A.* **76 C1** 44 13N 88 26W
Menate, *Indonesia* **36 E4** 0 12S 113 3 E
Mendawai →, *Indonesia* **36 E4** 3 30S 113 0 E
Mende, *France* **18 D5** 44 31N 3 30 E
Mendez, *Mexico* **87 B5** 25 7N 98 34W
Mendhar, *India* **43 C6** 33 35N 74 10 E
Mendip Hills, *U.K.* **11 F5** 51 17N 2 40W
Mendocino, *U.S.A.* **82 G2** 39 19N 123 48W
Mendocino, C., *U.S.A.* **82 F1** 40 26N 124 25W
Mendota, *Calif., U.S.A.* **83 H3** 36 45N 120 23W
Mendota, *Ill., U.S.A.* **80 E10** 41 33N 89 7W
Mendoza, *Argentina* **94 C2** 32 50S 68 52W
Mendoza □, *Argentina* **94 C2** 33 0S 69 0W
Mene Grande, *Venezuela* **92 B4** 9 49N 70 56W
Menemen, *Turkey* **21 E12** 38 34N 27 3 E
Menen, *Belgium* **15 D3** 50 47N 3 7 E
Menggala, *Indonesia* **36 E3** 4 30S 105 15 E
Mengjin, *China* **34 G7** 34 55N 112 45 E
Mengyin, *China* **35 G9** 35 40N 117 58 E
Mengzi, *China* **32 D5** 23 20N 103 22 E
Menihek L., *Canada* **71 B6** 54 0N 67 0W
Menin = Menen, *Belgium* **15 D3** 50 47N 3 7 E
Menindee, *Australia* **63 E3** 32 20S 142 25 E
Menindee L., *Australia* **63 E3** 32 20S 142 25 E
Meningie, *Australia* **63 F2** 35 50S 139 18 E
Menlo Park, *U.S.A.* **84 H4** 37 27N 122 12W
Menominee, *U.S.A.* **76 C2** 45 6N 87 37W
Menominee →, *U.S.A.* **76 C2** 45 6N 87 36W
Menomonie, *U.S.A.* **80 C9** 44 53N 91 55W
Menongue, *Angola* **53 G3** 14 48S 17 52 E
Menorca, *Spain* **22 B11** 40 0N 4 0 E
Mentakab, *Malaysia* **39 L4** 3 29N 102 21 E
Mentawai, Kepulauan, *Indonesia* **36 E1** 2 0S 99 0 E
Menton, *France* **18 E7** 43 50N 7 29 E
Mentor, *U.S.A.* **78 E3** 41 40N 81 21W
Menzelinsk, *Russia* **24 C9** 55 47N 53 11 E
Menzies, *Australia* **61 E3** 29 40S 121 2 E
Me'ona, *Israel* **47 B4** 33 1N 35 15 E
Meoqui, *Mexico* **86 B3** 28 17N 105 29W
Mepaco, *Mozam.* **55 F3** 15 57S 30 48 E
Meppel, *Neths.* **15 B6** 52 42N 6 12 E
Mer Rouge, *U.S.A.* **81 J9** 32 47N 91 48W
Merabéllou, Kólpos, *Greece* **23 D7** 35 10N 25 50 E
Meramangye, L., *Australia* **61 E5** 28 25S 132 13 E
Meran = Merano, *Italy* **20 A4** 46 40N 11 9 E
Merano, *Italy* **20 A4** 46 40N 11 9 E
Merauke, *Indonesia* **37 F10** 8 29S 140 24 E
Merbabu, *Indonesia* **37 G14** 7 30S 110 40 E
Merbein, *Australia* **63 E3** 34 10S 142 2 E
Merca, *Somali Rep.* **46 G3** 1 48N 44 50 E
Mercadal, *Spain* **22 B11** 39 59N 4 5 E
Merced, *U.S.A.* **83 H3** 37 18N 120 29W
Merced Pk., *U.S.A.* **84 H7** 37 36N 119 24W
Mercedes, *Buenos Aires, Argentina* **94 C4** 34 40S 59 30W
Mercedes, *Corrientes, Argentina* **94 B4** 29 10S 58 5W
Mercedes, *San Luis, Argentina* **94 C2** 33 40S 65 21W
Mercedes, *Uruguay* **94 C4** 33 12S 58 0W
Merceditas, *Chile* **94 B1** 28 20S 70 35W
Mercer, *N.Z.* **59 G5** 37 16S 175 5 E
Mercer, *U.S.A.* **78 E4** 41 14N 80 15W
Mercury, *U.S.A.* **85 J11** 36 40N 115 58W
Mercy C., *Canada* **69 B13** 65 0N 63 30W
Meredith, C., *Falk. Is.* **96 G4** 52 15S 60 40W
Meredith, L., *U.S.A.* **81 H4** 35 43N 101 33W
Merga = Nukheila, *Sudan* **51 E10** 19 1N 26 21 E
Mergui Arch. = Myeik Kyunzu, *Burma* **39 G1** 11 30N 97 30 E
Mérida, *Mexico* **87 C7** 20 58N 89 37W
Mérida, *Spain* **19 C2** 38 55N 6 25W
Mérida, *Venezuela* **92 B4** 8 24N 71 8W
Mérida, Cord. de, *Venezuela* **90 C3** 9 0N 71 0W
Meriden, *U.S.A.* **79 E12** 41 32N 72 48W
Meridian, *Calif., U.S.A.* **84 F5** 39 9N 121 55W
Meridian, *Idaho, U.S.A.* **82 E5** 43 37N 116 24W
Meridian, *Miss., U.S.A.* **77 J1** 32 22N 88 42W
Meridian, *Tex., U.S.A.* **81 K6** 31 56N 97 39W
Meriruma, *Brazil* **93 C8** 1 15N 54 50W
Merkel, *U.S.A.* **81 J4** 32 28N 100 1W
Merksem, *Belgium* **15 C4** 51 16N 4 25 E
Mermaid Reef, *Australia* **60 C2** 17 6S 119 36 E
Merowe, *Sudan* **51 E11** 18 29N 31 46 E
Merredin, *Australia* **61 F2** 31 28S 118 18 E
Merrick, *U.K.* **12 F4** 55 8N 4 28W
Merrickville, *Canada* **79 B9** 44 55N 75 50W
Merrill, *Oreg., U.S.A.* **82 E3** 42 1N 121 36W
Merrill, *Wis., U.S.A.* **80 C10** 45 11N 89 41W
Merriman, *U.S.A.* **80 D4** 42 55N 101 42W
Merritt, *Canada* **72 C4** 50 10N 120 45W
Merriwa, *Australia* **63 E5** 32 6S 150 22 E
Merriwagga, *Australia* **63 E4** 33 47S 145 43 E
Merry I., *Canada* **70 A4** 55 29N 77 31W
Merrygoen, *Australia* **63 E4** 31 51S 149 12 E
Merryville, *U.S.A.* **81 K8** 30 45N 93 33W
Mersa Fatma, *Eritrea* **46 E3** 14 57N 40 17 E
Mersch, *Lux.* **15 E6** 49 44N 6 7 E
Merseburg, *Germany* **16 C6** 51 22N 11 59 E
Mersey →, *U.K.* **10 D5** 53 25N 3 1W
Merseyside □, *U.K.* **10 D5** 53 31N 3 2W
Mersin, *Turkey* **25 G5** 36 51N 34 36 E
Mersing, *Malaysia* **39 L4** 2 25N 103 50 E
Merta, *India* **42 F6** 26 39N 74 4 E
Merthyr Tydfil, *U.K.* **11 F4** 51 45N 3 22W
Merthyr Tydfil □, *U.K.* **11 F4** 51 46N 3 21W
Mértola, *Portugal* **19 D2** 37 40N 7 40W
Mertzon, *U.S.A.* **81 K4** 31 16N 100 49W
Meru, *Kenya* **54 B4** 0 3N 37 40 E
Meru, *Tanzania* **54 C4** 3 15S 36 46 E
Meru □, *Kenya* **54 B4** 0 3N 37 46 E
Mesa, *U.S.A.* **83 K8** 33 25N 111 50W
Mesanagrós, *Greece* **23 C9** 36 1N 27 49 E
Mesaoría □, *Cyprus* **23 D12** 35 12N 33 14 E
Mesarás, Kólpos, *Greece* **23 D6** 35 6N 24 47 E
Mesgouez, L., *Canada* **70 B4** 51 20N 75 0W
Meshed = Mashhad, *Iran* **45 B8** 36 20N 59 35 E
Meshoppen, *U.S.A.* **79 E8** 41 36N 76 3W
Meshra er Req, *Sudan* **51 G10** 8 25N 29 18 E
Mesick, *U.S.A.* **76 C3** 44 24N 85 43W
Mesilinka →, *Canada* **72 B4** 56 6N 124 30W
Mesilla, *U.S.A.* **83 K10** 32 16N 106 48W
Mesolóngion, *Greece* **21 E9** 38 21N 21 28 E
Mesopotamia = Al Jazirah, *Iraq* **44 C5** 33 30N 44 0 E
Mesquite, *U.S.A.* **83 H6** 36 47N 114 6W
Mess Cr. →, *Canada* **72 B2** 57 55N 131 14W
Messalo →, *Mozam.* **55 E4** 12 25S 39 15 E
Messina, *Italy* **20 E6** 38 11N 15 34 E
Messina, *S. Africa* **57 C5** 22 20S 30 5 E
Messina, Str. di, *Italy* **20 F6** 38 15N 15 35 E
Messíni, *Greece* **21 F10** 37 4N 22 1 E
Messiniakós Kólpos, *Greece* **21 F10** 36 45N 22 5 E
Messonghi, *Greece* **23 B3** 39 29N 19 56 E
Mesta →, *Bulgaria* **21 D11** 40 54N 24 49 E
Meta →, *S. Amer.* **92 B5** 6 12N 67 28W
Metairie, *U.S.A.* **81 L9** 29 58N 90 10W
Metaline Falls, *U.S.A.* **82 B5** 48 52N 117 22W
Metán, *Argentina* **94 B3** 25 30S 65 0W
Metangula, *Mozam.* **55 E3** 12 40S 34 50 E
Metema, *Ethiopia* **51 F12** 12 56N 36 13 E
Metengobalame, *Mozam.* **55 E3** 14 49S 34 30 E
Methven, *N.Z.* **59 K3** 43 38S 171 40 E
Methy L., *Canada* **73 B7** 56 28N 109 30W
Metil, *Mozam.* **55 F4** 16 24S 39 0 E
Metlakatla, *U.S.A.* **72 B2** 55 8N 131 35W
Metropolis, *U.S.A.* **81 G10** 37 9N 88 44W
Mettur Dam, *India* **40 P10** 11 45N 77 45 E
Metz, *France* **18 B7** 49 8N 6 10 E
Meulaboh, *Indonesia* **36 D1** 4 11N 96 3 E
Meureudu, *Indonesia* **36 C1** 5 19N 96 10 E
Meuse →, *Europe* **18 A6** 50 45N 5 41 E
Mexborough, *U.K.* **10 D6** 53 30N 1 15W
Mexia, *U.S.A.* **81 K6** 31 41N 96 29W
Mexiana, I., *Brazil* **93 C9** 0 0 49 30W
Mexicali, *Mexico* **86 A1** 32 40N 115 30W
Mexican Plateau, *Mexico* **66 G9** 25 0N 104 0W
México, *Mexico* **87 D5** 19 20N 99 10W
Mexico, *Maine, U.S.A.* **79 B14** 44 34N 70 33W
Mexico, *Mo., U.S.A.* **80 F9** 39 10N 91 53W
México □, *Mexico* **86 D5** 19 20N 99 10W
Mexico ■, *Cent. Amer.* **86 C4** 25 0N 105 0W
Mexico, G. of, *Cent. Amer.* **87 C7** 25 0N 90 0W
Meymaneh, *Afghan.* **40 B4** 35 53N 64 38 E
Mezen, *Russia* **24 A7** 65 50N 44 20 E
Mezen →, *Russia* **24 A7** 65 44N 44 22 E
Mézene, *France* **18 D6** 44 54N 4 11 E
Mezőkövesd, *Hungary* **17 E11** 47 49N 20 35 E
Mezőtúr, *Hungary* **17 E11** 46 58N 20 41 E
Mezquital, *Mexico* **86 C4** 23 29N 104 23W
Mgeta, *Tanzania* **55 D4** 8 22S 36 6 E
Mhlaba Hills, *Zimbabwe* **55 F3** 18 30S 30 30 E
Mhow, *India* **42 H6** 22 33N 75 50 E
Miahuatlán, *Mexico* **87 D5** 16 21N 96 36W
Miallo, *Australia* **62 B4** 16 28S 145 22 E
Miami, *Ariz., U.S.A.* **83 K8** 33 24N 110 52W
Miami, *Fla., U.S.A.* **77 N5** 25 47N 80 11W
Miami, *Tex., U.S.A.* **81 H4** 35 42N 100 38W
Miami →, *U.S.A.* **76 F3** 39 20N 84 40W
Miami Beach, *U.S.A.* **77 N5** 25 47N 80 8W
Mian Xian, *China* **34 H4** 33 10N 106 32 E
Mianchi, *China* **34 G6** 34 48N 111 48 E
Mīāndowāb, *Iran* **44 B5** 37 0N 46 5 E
Miandrivazo, *Madag.* **57 B8** 19 31S 45 29 E
Mīāneh, *Iran* **44 B5** 37 30N 47 40 E
Mianwali, *Pakistan* **42 C4** 32 38N 71 28 E
Miarinarivo, *Madag.* **57 B8** 18 57S 46 55 E
Miass, *Russia* **24 D11** 54 59N 60 6 E
Michalovce, *Slovak Rep.* **17 D11** 48 47N 21 58 E
Michigan □, *U.S.A.* **76 C3** 44 0N 85 0W
Michigan, L., *U.S.A.* **76 C2** 44 0N 87 0W
Michigan City, *U.S.A.* **76 E2** 41 43N 86 54W
Michikamau L., *Canada* **71 B7** 54 20N 63 10W
Michipicoten, *Canada* **70 C3** 47 55N 84 55W
Michipicoten I., *Canada* **70 C2** 47 40N 85 40W
Michoacan □, *Mexico* **86 D4** 19 0N 102 0W
Michurin, *Bulgaria* **21 C12** 42 9N 27 51 E
Michurinsk, *Russia* **24 D7** 52 58N 40 27 E
Miclere, *Australia* **62 C4** 22 34S 147 32 E
Mico, Pta., *Nic.* **88 D3** 12 0N 83 30W
Micronesia, Federated States of ■, *Pac. Oc.* **64 G7** 9 0N 150 0 E
Midai, P., *Indonesia* **39 L6** 3 0N 107 47 E
Midale, *Canada* **73 D8** 49 25N 103 20W
Middelburg, *Neths.* **15 C3** 51 30N 3 36 E
Middelburg, *Eastern Cape, S. Africa* **56 E3** 31 30S 25 0 E
Middelburg, *Mpumalanga, S. Africa* **57 D4** 25 49S 29 28 E
Middelwit, *S. Africa* **56 C4** 24 51S 27 3 E
Middle Alkali L., *U.S.A.* **82 F3** 41 27N 120 5W
Middle Fork Feather →, *U.S.A.* **84 F5** 38 33N 121 30W
Middle I., *Australia* **61 F3** 34 6S 123 11 E
Middle Loup →, *U.S.A.* **80 E5** 41 17N 98 24W
Middleboro, *U.S.A.* **79 E14** 41 54N 70 55W
Middleburg, *N.Y., U.S.A.* **79 D10** 42 36N 74 20W
Middleburg, *Pa., U.S.A.* **78 F7** 40 47N 77 3W
Middlebury, *U.S.A.* **79 B11** 44 1N 73 10W
Middleport, *U.S.A.* **76 F4** 39 0N 82 3W
Middlesboro, *U.S.A.* **77 G4** 36 36N 83 43W
Middlesbrough, *U.K.* **10 C6** 54 35N 1 13W
Middlesbrough □, *U.K.* **10 C6** 54 28N 1 13W
Middlesex, *Belize* **88 C2** 17 2N 88 31W
Middlesex, *U.S.A.* **79 F10** 40 36N 74 30W
Middleton, *Australia* **62 C3** 22 22S 141 32 E
Middleton, *Canada* **71 D6** 44 57N 65 4W
Middletown, *Calif., U.S.A.* **84 G4** 38 45N 122 37W
Middletown, *Conn., U.S.A.* **79 E12** 41 34N 72 39W
Middletown, *N.Y., U.S.A.* **79 E10** 41 27N 74 25W
Middletown, *Ohio, U.S.A.* **76 F3** 39 31N 84 24W
Middletown, *Pa., U.S.A.* **79 F8** 40 12N 76 44W
Midi, Canal du →, *France* **18 E4** 43 45N 1 21 E
Midland, *Canada* **70 D4** 44 45N 79 50W
Midland, *Calif., U.S.A.* **85 M12** 33 52N 114 48W
Midland, *Mich., U.S.A.* **76 D3** 43 37N 84 14W
Midland, *Pa., U.S.A.* **78 F4** 40 39N 80 27W
Midland, *Tex., U.S.A.* **81 K3** 32 0N 102 3W
Midlands □, *Zimbabwe* **55 F2** 19 40S 29 0 E
Midleton, *Ireland* **13 E3** 51 55N 8 10W
Midlothian, *U.S.A.* **81 J6** 32 30N 97 0W
Midlothian □, *U.K.* **12 F5** 55 51N 3 5W
Midongy, Tangorombohitr' i, *Madag.* **57 C8** 23 30S 47 0 E
Midongy Atsimo, *Madag.* **57 C8** 23 35S 47 1 E
Midway Is., *Pac. Oc.* **64 E10** 28 13N 177 22W
Midway Wells, *U.S.A.* **85 N11** 32 41N 115 7W
Midwest, *U.S.A.* **75 B9** 42 0N 90 0W
Midwest, *Wyo., U.S.A.* **82 E10** 43 25N 106 16W
Midwest City, *U.S.A.* **81 H6** 35 27N 97 24W
Midžor, *Bulgaria* **21 C10** 43 24N 22 40 E
Mie □, *Japan* **31 G8** 34 30N 136 10 E
Międzychód, *Poland* **16 B8** 52 35N 15 53 E
Międzyrzec Podlaski, *Poland* **17 C12** 51 58N 22 45 E
Mielec, *Poland* **17 C11** 50 15N 21 25 E
Mienga, *Angola* **56 B2** 17 12S 19 48 E
Miercurea Ciuc, *Romania* **17 E13** 46 21N 25 48 E
Mieres, *Spain* **19 A3** 43 18N 5 48W
Mifflintown, *U.S.A.* **78 F7** 40 34N 77 24W
Mifraz Hefa, *Israel* **47 C4** 32 52N 35 0 E
Migdāl, *Israel* **47 C4** 32 51N 35 30 E
Miguel Alemán, Presa, *Mexico* **87 D5** 18 15N 96 40W
Miguel Alves, *Brazil* **93 D10** 4 11S 42 55W
Mihara, *Japan* **31 G6** 34 24N 133 5 E
Mikese, *Tanzania* **54 D4** 6 48S 37 55 E
Mikhaylovgrad, *Bulgaria* **21 C10** 43 27N 23 16 E
Mikkeli, *Finland* **9 F22** 61 43N 27 15 E
Mikkwa →, *Canada* **72 B6** 58 25N 114 46W
Míkonos, *Greece* **21 F11** 37 30N 25 25 E
Mikumi, *Tanzania* **54 D4** 7 26S 37 0 E
Mikun, *Russia* **24 B9** 62 20N 50 0 E
Milaca, *U.S.A.* **80 C8** 45 45N 93 39W
Milagro, *Ecuador* **92 D3** 2 11S 79 36W
Milan = Milano, *Italy* **20 B3** 45 28N 9 12 E
Milan, *Mo., U.S.A.* **80 E8** 40 12N 93 7W
Milan, *Tenn., U.S.A.* **77 H1** 35 55N 88 46W
Milang, *Australia* **63 E2** 32 2S 139 10 E
Milange, *Mozam.* **55 F4** 16 3S 35 45 E
Milano, *Italy* **20 B3** 45 28N 9 12 E
Milâs, *Turkey* **21 F12** 37 20N 27 50 E
Milatos, *Greece* **23 D7** 35 18N 25 34 E
Milazzo, *Italy* **20 E6** 38 13N 15 15 E
Milbank, *U.S.A.* **80 C6** 45 13N 96 38W
Milden, *Canada* **73 C7** 51 29N 107 32W
Mildmay, *Canada* **78 B3** 44 3N 81 7W
Mildura, *Australia* **63 E3** 34 13S 142 9 E
Miles, *Australia* **63 D5** 26 40S 150 9 E
Miles, *U.S.A.* **81 K4** 31 36N 100 11W
Miles City, *U.S.A.* **80 B2** 46 25N 105 51W
Milestone, *Canada* **73 D8** 49 59N 104 31W
Miletus, *Turkey* **21 F12** 37 30N 27 18 E
Mileura, *Australia* **61 E2** 26 22S 117 20 E
Milford, *Calif., U.S.A.* **84 E6** 40 10N 120 22W
Milford, *Conn., U.S.A.* **79 E11** 41 14N 73 3W
Milford, *Del., U.S.A.* **76 F8** 38 55N 75 26W
Milford, *Mass., U.S.A.* **79 D13** 42 8N 71 31W
Milford, *Pa., U.S.A.* **79 E10** 41 19N 74 48W
Milford, *Utah, U.S.A.* **83 G7** 38 24N 113 1W
Milford Haven, *U.K.* **11 F2** 51 42N 5 7W
Milford Sd., *N.Z.* **59 L1** 44 41S 167 47 E
Milgun, *Australia* **61 D2** 24 56S 118 18 E
Milh, Bahr al, *Iraq* **44 C4** 32 40N 43 35 E
Miliana, *Algeria* **50 C5** 27 20N 2 32 E
Miling, *Australia* **61 F2** 30 30S 116 17 E
Milk →, *U.S.A.* **82 B10** 48 4N 106 19W
Milk River, *Canada* **72 D6** 49 10N 112 5W
Mill City, *U.S.A.* **82 D2** 44 45N 122 29W
Mill I., *Antarctica* **5 C8** 66 0S 101 30 E
Mill Valley, *U.S.A.* **84 H4** 37 54N 122 32W
Millau, *France* **18 D5** 44 8N 3 4 E
Millbridge, *Canada* **78 B7** 44 41N 77 36W
Millbrook, *Canada* **78 B6** 44 10N 78 29W
Mille Lacs, L. des, *Canada* **70 C1** 48 45N 90 35W
Mille Lacs L., *U.S.A.* **80 B8** 46 15N 93 39W
Milledgeville, *U.S.A.* **77 J4** 33 5N 83 14W
Millen, *U.S.A.* **77 J5** 32 48N 81 57W
Miller, *U.S.A.* **80 C5** 44 31N 98 59W
Millersburg, *Ohio, U.S.A.* **78 F3** 40 33N 81 55W
Millersburg, *Pa., U.S.A.* **78 F8** 40 32N 76 58W
Millerton, *U.S.A.* **79 E11** 41 57N 73 31W
Millerton L., *U.S.A.* **84 J7** 37 1N 119 41W
Millicent, *Australia* **63 F3** 37 34S 140 21 E
Millinocket, *U.S.A.* **71 C6** 45 39N 68 43W
Millmerran, *Australia* **63 D5** 27 53S 151 16 E
Mills L., *Canada* **72 A5** 61 30N 118 20W
Millsboro, *U.S.A.* **78 G4** 40 0N 80 0W
Milltown Malbay, *Ireland* **13 D2** 52 52N 9 24W
Millville, *U.S.A.* **76 F8** 39 24N 75 2W
Millwood L., *U.S.A.* **81 J8** 33 42N 93 58W
Milne →, *Australia* **62 C2** 21 10S 137 33 E
Milne Inlet, *Canada* **69 A11** 72 30N 80 0W
Milnor, *U.S.A.* **80 B6** 46 16N 97 27W
Milo, *Canada* **72 C6** 50 34N 112 53W
Mílos, *Greece* **21 F11** 36 44N 24 25 E
Milparinka P.O., *Australia* **63 D3** 29 46S 141 57 E
Milton, *Canada* **78 C5** 43 31N 79 53W
Milton, *N.Z.* **59 M2** 46 7S 169 59 E
Milton, *U.K.* **12 D4** 57 18N 4 32W
Milton, *Calif., U.S.A.* **84 G6** 38 3N 120 51W
Milton, *Fla., U.S.A.* **77 K2** 30 38N 87 3W
Milton, *Pa., U.S.A.* **78 F8** 41 1N 76 51W
Milton-Freewater, *U.S.A.* **82 D4** 45 56N 118 23W
Milton Keynes, *U.K.* **11 E7** 52 1N 0 44W
Miltou, *Chad* **51 F8** 10 14N 17 26 E
Milverton, *Canada* **78 C4** 43 34N 80 55W
Milwaukee, *U.S.A.* **76 D2** 43 2N 87 55W
Milwaukee Deep, *Atl. Oc.* **89 C6** 19 50N 68 0W
Milwaukie, *U.S.A.* **84 E4** 45 27N 122 38W
Min Chiang →, *China* **33 D6** 26 0N 119 35 E
Min Jiang →, *China* **32 D5** 28 45N 104 40 E
Min Xian, *China* **34 G3** 34 25N 104 5 E
Mina, *U.S.A.* **83 G4** 38 24N 118 7W
Mina Pirquitas, *Argentina* **94 A2** 22 40S 66 30W
Mīnā Su'ud, *Si. Arabia* **45 D6** 28 45N 48 28 E
Mīnā'al Aḥmadī, *Kuwait* **45 D6** 29 5N 48 10 E
Mīnāb, *Iran* **45 E8** 27 10N 57 1 E
Minago →, *Canada* **73 C9** 54 33N 98 59W
Minaki, *Canada* **73 D10** 49 59N 94 40W
Minamata, *Japan* **31 H5** 32 10N 130 30 E
Minami-Tori-Shima, *Pac. Oc.* **64 E7** 24 0N 153 45 E
Minas, *Uruguay* **95 C4** 34 20S 55 10W
Minas, Sierra de las, *Guatemala* **88 C2** 15 9N 89 31W
Minas Basin, *Canada* **71 C7** 45 20N 64 12W
Minas Gerais □, *Brazil* **93 G9** 18 50S 46 0W
Minatitlán, *Mexico* **87 D6** 17 59N 94 31W
Minbu, *Burma* **41 J19** 20 10N 94 52 E
Mindanao, *Phil.* **37 C6** 8 0N 125 0 E
Mindanao Sea = Bohol Sea, *Phil.* **37 C6** 9 0N 124 0 E
Mindanao Trench, *Pac. Oc.* **37 B7** 12 0N 126 6 E
Minden, *Canada* **78 B6** 44 55N 78 43W
Minden, *Germany* **16 B5** 52 17N 8 55 E
Minden, *La., U.S.A.* **81 J8** 32 37N 93 17W
Minden, *Nev., U.S.A.* **84 G7** 38 57N 119 46W
Mindiptana, *Indonesia* **37 F10** 5 55S 140 22 E
Mindoro, *Phil.* **37 B6** 13 0N 121 0 E
Mindoro Str., *Phil.* **37 B6** 12 30N 120 30 E
Mindouli, *Congo* **52 E2** 4 12S 14 28 E
Mine, *Japan* **31 G5** 34 12N 131 7 E
Minehead, *U.K.* **11 F4** 51 12N 3 29W
Mineola, *U.S.A.* **81 J7** 32 40N 95 29W
Mineral King, *U.S.A.* **84 J8** 36 27N 118 36W
Mineral Wells, *U.S.A.* **81 J5** 32 48N 98 7W
Minersville, *Pa., U.S.A.* **79 F8** 40 41N 76 16W
Minersville, *Utah, U.S.A.* **83 G7** 38 13N 112 56W
Minerva, *U.S.A.* **78 F3** 40 44N 81 6W
Minetto, *U.S.A.* **79 C8** 43 24N 76 28W
Mingäçevir Su Anbarı, *Azerbaijan* **25 F8** 40 57N 46 50 E
Mingan, *Canada* **71 B7** 50 20N 64 0W
Mingechaurskoye Vdkhr. = Mingäçevir Su Anbarı, *Azerbaijan* **25 F8** 40 57N 46 50 E
Mingela, *Australia* **62 B4** 19 52S 146 38 E
Mingenew, *Australia* **61 E2** 29 12S 115 21 E
Mingera Cr. →, *Australia* **62 C2** 20 38S 137 45 E
Mingin, *Burma* **41 H19** 22 50N 94 30 E
Mingt'iehkaitafan = Mintaka Pass, *Pakistan* **43 A6** 37 0N 74 58 E
Mingyuegue, *China* **35 C15** 43 2N 128 50 E
Minho = Miño →, *Spain* **19 A2** 41 52N 8 40W
Minho, *Portugal* **19 B1** 41 25N 8 20W
Minidoka, *U.S.A.* **82 E7** 42 45N 113 29W
Minigwal, L., *Australia* **61 E3** 29 31S 123 14 E
Minilya, *Australia* **61 D1** 23 55S 114 0 E
Minilya →, *Australia* **61 D1** 23 45S 114 0 E
Minipi, L., *Canada* **71 B7** 52 25N 60 45W
Mink L., *Canada* **72 A5** 61 54N 117 40W
Minna, *Nigeria* **50 G6** 9 37N 6 30 E
Minneapolis, *Kans., U.S.A.* **80 F6** 39 8N 97 42W
Minneapolis, *Minn., U.S.A.* **80 C8** 44 59N 93 16W
Minnedosa, *Canada* **73 C9** 50 14N 99 50W
Minnesota □, *U.S.A.* **80 B7** 46 0N 94 15W
Minnie Creek, *Australia* **61 D2** 24 3S 115 42 E
Minnipa, *Australia* **63 E2** 32 51S 135 9 E
Minnitaki L., *Canada* **70 C1** 49 57N 92 10W
Mino, *Japan* **31 G8** 35 32N 136 55 E
Miño →, *Spain* **19 A2** 41 52N 8 40W
Minorca = Menorca, *Spain* **22 B11** 40 0N 4 0 E

Minore, Australia 63 E4 32 14S 148 27 E
Minot, U.S.A. 80 A4 48 14N 101 18W
Minqin, China 34 E2 38 38N 103 20 E
Minsk, Belarus 17 B14 53 52N 27 30 E
Mińsk Mazowiecki, Poland 17 B11 52 10N 21 33 E
Mintaka Pass, Pakistan 43 A6 37 0N 74 58 E
Minto, U.S.A. 68 B5 64 53N 149 11W
Minton, Canada 73 D8 49 10N 104 35W
Minturn, U.S.A. 82 G10 39 35N 106 26W
Minusinsk, Russia 27 D10 53 50N 91 20 E
Minutang, India 41 E20 28 15N 96 30 E
Minvoul, Gabon 52 D2 2 9N 12 8 E
Mir, Niger 51 F7 14 5N 11 59 E
Mīr Kūh, Iran 45 E8 26 22N 58 55 E
Mīr Shahdād, Iran 45 E8 26 15N 58 29 E
Mira, Italy 20 B5 45 26N 12 8 E
Mira por vos Cay, Bahamas 89 B5 22 9N 74 30W
Miraj, India 40 L9 16 50N 74 45 E
Miram Shah, Pakistan 42 C4 33 0N 70 2 E
Miramar, Argentina 94 D4 38 15S 57 50W
Miramar, Mozam. 57 C6 23 50S 35 35 E
Miramichi B., Canada 71 C7 47 15N 65 0W
Miranda, Brazil 93 H7 20 10S 56 15W
Miranda de Ebro, Spain 19 A4 42 41N 2 57W
Miranda do Douro, Portugal 19 B2 41 30N 6 16W
Mirando City, U.S.A. 81 M5 27 26N 99 0W
Mirandópolis, Brazil 95 A5 21 9S 51 6W
Mirango, Malawi 55 E3 13 32S 34 58 E
Mirani, Australia 62 C4 21 9S 148 53 E
Mirassol, Brazil 95 A6 20 46S 49 28W
Mirbāţ, Oman 46 D5 17 0N 54 45 E
Miri, Malaysia 36 D4 4 23N 113 59 E
Miriam Vale, Australia 62 C5 24 20S 151 33 E
Mirim, L., S. Amer. 95 C5 32 45S 52 50W
Mirnyy, Russia 27 C12 62 33N 113 53 E
Mirond L., Canada 73 B8 55 6N 102 47W
Mirpur, Pakistan 43 C5 33 32N 73 56 E
Mirpur Bibiwari, Pakistan 42 E2 28 33N 67 44 E
Mirpur Khas, Pakistan 42 G3 25 30N 69 0 E
Mirpur Sakro, Pakistan 42 G2 24 33N 67 41 E
Mirror, Canada 72 C6 52 30N 113 7W
Miryang, S. Korea 35 G15 35 31N 128 44 E
Mirzapur, India 43 G10 25 10N 82 34 E
Mirzapur-cum-Vindhyachal = Mirzapur, India 43 G10 25 10N 82 34 E
Misantla, Mexico 87 D5 19 56N 96 50W
Misawa, Japan 30 D10 40 41N 141 24 E
Miscou I., Canada 71 C7 47 57N 64 31W
Mish'āb, Ra'as al, Si. Arabia 45 D6 28 15N 48 43 E
Mishan, China 33 B8 45 37N 131 48 E
Mishawaka, U.S.A. 76 E2 41 40N 86 11W
Mishima, Japan 31 G9 35 10N 138 52 E
Misión, Mexico 85 N10 32 6N 116 53W
Misiones □, Argentina 95 B5 27 0S 55 0W
Misiones □, Paraguay 94 B4 27 0S 56 0W
Miskah, Si. Arabia 44 E4 24 49N 42 56 E
Miskitos, Cayos, Nic. 88 D3 14 26N 82 50W
Miskolc, Hungary 17 D11 48 7N 20 50 E
Misoke, Zaïre 54 C2 0 42S 28 2 E
Misool, Indonesia 37 E8 1 52S 130 10 E
Misrātah, Libya 51 B8 32 24N 15 3 E
Missanabie, Canada 70 C3 48 20N 84 6W
Missinaibi →, Canada 70 B3 50 43N 81 29W
Missinaibi L., Canada 70 C3 48 23N 83 40W
Mission, S. Dak., U.S.A. 80 D4 43 18N 100 39W
Mission, Tex., U.S.A. 81 M5 26 13N 98 20W
Mission City, Canada 72 D4 49 10N 122 15W
Mission Viejo, U.S.A. 85 M9 33 36N 117 40W
Missisa L., Canada 70 B2 52 20N 85 7W
Mississagi →, Canada 70 C3 46 15N 83 9W
Mississippi □, U.S.A. 81 J10 33 0N 90 0W
Mississippi →, U.S.A. 81 L10 29 9N 89 15W
Mississippi L., Canada 79 A8 45 5N 76 10W
Mississippi River Delta, U.S.A. 81 L9 29 10N 89 15W
Mississippi Sd., U.S.A. 81 K10 30 20N 89 0W
Missoula, U.S.A. 82 C6 46 52N 114 1W
Missouri □, U.S.A. 80 F8 38 25N 92 30W
Missouri →, U.S.A. 80 F9 38 49N 90 7W
Missouri Valley, U.S.A. 80 E7 41 34N 95 53W
Mist, U.S.A. 84 E3 45 59N 123 15W
Mistake B., Canada 73 A10 62 8N 93 0W
Mistassini →, Canada 71 C5 48 42N 72 20W
Mistassini L., Canada 70 B5 51 0N 73 30W
Mistastin L., Canada 71 A7 55 57N 63 20W
Mistatim, Canada 73 C8 52 52N 103 22W
Misty L., Canada 73 B8 58 53N 101 40W
Misurata = Misrātah, Libya 51 B8 32 24N 15 3 E
Mitchell, Australia 63 D4 26 29S 147 58 E
Mitchell, Canada 78 C3 43 28N 81 12W
Mitchell, Ind., U.S.A. 76 F2 38 44N 86 28W
Mitchell, Nebr., U.S.A. 80 E3 41 57N 103 49W
Mitchell, Oreg., U.S.A. 82 D3 44 34N 120 9W
Mitchell, S. Dak., U.S.A. 80 D5 43 43N 98 2W
Mitchell →, Australia 62 B3 15 12S 141 35 E
Mitchell, Mt., U.S.A. 77 H4 35 46N 82 16W
Mitchell Ras., Australia 62 A2 12 49S 135 36 E
Mitchelstown, Ireland 13 D3 52 15N 8 16W
Mitha Tiwana, Pakistan 42 C5 32 13N 72 6 E
Mitilíni, Greece 21 E12 39 6N 26 35 E
Mito, Japan 31 F10 36 20N 140 30 E
Mitrovica = Titova-Mitrovica, Serbia, Yug. 21 C9 42 54N 20 52 E
Mitsinjo, Madag. 57 B8 16 1S 45 52 E
Mitsiwa, Eritrea 51 E12 15 35N 39 25 E
Mitsukaidō, Japan 31 F9 36 1N 139 59 E
Mittagong, Australia 63 E5 34 28S 150 29 E
Mitú, Colombia 92 C4 1 8N 70 3W
Mitumba, Tanzania 54 D3 7 8S 31 2 E
Mitumba, Chaîne des, Zaïre 54 D2 7 0S 27 30 E
Mitumba Mts. = Mitumba, Chaîne des, Zaïre 54 D2 7 0S 27 30 E
Mitwaba, Zaïre 55 D2 8 2S 27 17 E
Mityana, Uganda 54 B3 0 23N 32 2 E
Mitzic, Gabon 52 D2 0 45N 11 40 E
Mixteco →, Mexico 87 D5 18 11N 98 30W
Miyagi □, Japan 30 E10 38 15N 140 45 E
Miyah, W. el →, Syria 44 C3 34 44N 39 57 E
Miyake-Jima, Japan 31 G9 34 5N 139 30 E
Miyako, Japan 30 E10 39 40N 141 59 E
Miyako-Jima, Japan 31 M2 24 45N 125 20 E
Miyako-Rettō, Japan 31 M2 24 24N 125 0 E
Miyakonojō, Japan 31 J5 31 40N 131 5 E
Miyanoura-Dake, Japan 31 J5 30 20N 130 31 E
Miyazaki, Japan 31 J5 31 56N 131 30 E
Miyazaki □, Japan 31 H5 32 30N 131 30 E
Miyazu, Japan 31 G7 35 35N 135 10 E
Miyet, Bahr el = Dead Sea, Asia 47 D4 31 30N 35 30 E
Miyoshi, Japan 31 G6 34 48N 132 51 E
Miyun, China 34 D9 40 28N 116 50 E
Miyun Shuiku, China 35 D9 40 30N 117 0 E
Mizdah, Libya 51 B7 31 30N 13 0 E
Mizen Hd., Cork, Ireland 13 E2 51 27N 9 50W
Mizen Hd., Wick., Ireland 13 D5 52 51N 6 4W
Mizhi, China 34 F6 37 47N 110 12 E
Mizoram □, India 41 H18 23 30N 92 40 E
Mizpe Ramon, Israel 47 E3 30 34N 34 49 E
Mizusawa, Japan 30 E10 39 8N 141 8 E
Mjölby, Sweden 9 G16 58 20N 15 10 E
Mjøsa, Norway 9 F14 60 40N 11 0 E
Mkata, Tanzania 54 D4 5 45S 38 20 E
Mkokotoni, Tanzania 54 D4 5 55S 39 15 E
Mkomazi, Tanzania 54 C4 4 40S 38 7 E
Mkomazi →, S. Africa 57 E5 30 12S 30 50 E
Mkulwe, Tanzania 55 D3 8 37S 32 20 E
Mkumbi, Ras, Tanzania 54 D4 7 38S 39 55 E
Mkushi, Zambia 55 E2 14 25S 29 15 E
Mkushi River, Zambia 55 E2 13 32S 29 45 E
Mkuze, S. Africa 57 D5 27 10S 32 0 E
Mladá Boleslav, Czech. 16 C8 50 27N 14 53 E
Mlala Hills, Tanzania 54 D3 6 50S 31 40 E
Mlange, Malawi 55 F4 16 2S 35 33 E
Mława, Poland 17 B11 53 9N 20 25 E
Mljet, Croatia 20 C7 42 43N 17 30 E
Mmabatho, S. Africa 56 D4 25 49S 25 30 E
Mo i Rana, Norway 8 C16 66 20N 14 7 E
Moa, Indonesia 37 F7 8 0S 128 0 E
Moab, U.S.A. 83 G9 38 35N 109 33W
Moabi, Gabon 52 E2 2 24S 10 59 E
Moala, Fiji 59 D8 18 36S 179 53 E
Moalie Park, Australia 63 D3 29 42S 143 3 E
Moba, Zaïre 54 D2 7 0S 29 48 E
Mobārakābād, Iran 45 D7 28 24N 53 20 E
Mobārakīyeh, Iran 45 C6 32 23N 51 37 E
Mobaye, C.A.R. 52 D4 4 25N 21 5 E
Mobayi, Zaïre 52 D4 4 15N 21 8 E
Moberly, U.S.A. 80 F8 39 25N 92 26W
Moberly →, Canada 72 B4 56 12N 120 55W
Mobile, U.S.A. 77 K1 30 41N 88 3W
Mobile B., U.S.A. 77 K2 30 30N 88 0W
Mobridge, U.S.A. 80 C4 45 32N 100 26W
Mobutu Sese Seko, L. = Albert L., Africa 54 B3 1 30N 31 0 E
Moc Chau, Vietnam 38 B5 20 50N 104 38 E
Moc Hoa, Vietnam 39 G5 10 46N 105 56 E
Mocabe Kasari, Zaïre 55 D2 9 58S 26 12 E
Moçambique, Mozam. 55 F5 15 3S 40 42 E
Moçâmedes = Namibe, Angola 53 H2 15 7S 12 11 E
Mochudi, Botswana 56 C4 24 27S 26 7 E
Mocimboa da Praia, Mozam. 55 E5 11 25S 40 20 E
Moclips, U.S.A. 84 C2 47 14N 124 13W
Mocoa, Colombia 92 C3 1 7N 76 35W
Mococa, Brazil 95 A6 21 28S 47 0W
Mocorito, Mexico 86 B3 25 30N 107 53W
Moctezuma, Mexico 86 B3 29 50N 109 0W
Moctezuma →, Mexico 87 C5 21 59N 98 34W
Mocuba, Mozam. 55 F4 16 54S 36 57 E
Mocúzari, Presa, Mexico 86 B3 27 10N 109 10W
Modane, France 18 D7 45 12N 6 40 E
Modasa, India 42 H5 23 30N 73 21 E
Modder →, S. Africa 56 D3 29 2S 24 37 E
Modderrivier, S. Africa 56 D3 29 2S 24 38 E
Módena, Italy 20 B4 44 40N 10 55 E
Modena, U.S.A. 83 H7 37 48N 113 56W
Modesto, U.S.A. 83 H3 37 39N 121 0W
Módica, Italy 20 F6 36 52N 14 46 E
Moe, Australia 63 F4 38 12S 146 19 E
Moebase, Mozam. 55 F4 17 3S 38 41 E
Moengo, Surinam 93 B8 5 45N 54 20W
Moffat, U.K. 12 F5 55 21N 3 27W
Moga, India 42 D6 30 48N 75 8 E
Mogadishu = Muqdisho, Somali Rep. 46 G4 2 2N 45 25 E
Mogador = Essaouira, Morocco 50 B3 31 32N 9 42W
Mogalakwena →, S. Africa 57 C4 22 38S 28 40 E
Mogami →, Japan 30 E10 38 45N 140 0 E
Mogán, Canary Is. 22 G4 27 53N 15 43W
Mogaung, Burma 41 G20 25 20N 97 0 E
Mogi das Cruzes, Brazil 95 A6 23 31S 46 11W
Mogi-Guaçu →, Brazil 95 A6 20 53S 48 10W
Mogi-Mirim, Brazil 95 A6 22 29S 47 0W
Mogilev = Mahilyow, Belarus 17 B16 53 55N 30 18 E
Mogilev-Podolskiy = Mohyliv-Podilskyy, Ukraine 17 D14 48 26N 27 48 E
Mogincual, Mozam. 55 F5 15 35S 40 25 E
Mogocha, Russia 27 D12 53 40N 119 50 E
Mogoi, Indonesia 37 E8 1 55S 133 10 E
Mogok, Burma 41 H20 23 0N 96 40 E
Mogumber, Australia 61 F2 31 2S 116 3 E
Mohács, Hungary 17 F10 45 58N 18 41 E
Mohales Hoek, Lesotho 56 E4 30 7S 27 26 E
Mohall, U.S.A. 80 A4 48 46N 101 31W
Moḩammadābād, Iran 45 B8 37 52N 59 5 E
Mohave, L., U.S.A. 85 K12 35 12N 114 34W
Mohawk →, U.S.A. 79 D11 42 47N 73 41W
Mohoro, Tanzania 54 D4 8 6S 39 8 E
Mohyliv-Podilskyy, Ukraine 17 D14 48 26N 27 48 E
Moidart, L., U.K. 12 E3 56 47N 5 52W
Moires, Greece 23 D6 35 4N 24 56 E
Moisaküla, Estonia 9 G21 58 3N 25 12 E
Moisie, Canada 71 B6 50 12N 66 1W
Moisie →, Canada 71 B6 50 14N 66 5W
Moïssala, Chad 51 G8 8 21N 17 46 E
Mojave, U.S.A. 85 K8 35 3N 118 10W
Mojave Desert, U.S.A. 85 L10 35 0N 116 30W
Mojo, Bolivia 94 A2 21 48S 65 33W
Mojokerto, Indonesia 37 G15 7 28S 112 26 E
Mokai, N.Z. 59 H5 38 32S 175 56 E
Mokambo, Zaïre 55 E2 12 25S 28 20 E
Mokameh, India 43 G11 25 24N 85 55 E
Mokelumne →, U.S.A. 84 G5 38 13N 121 28W
Mokelumne Hill, U.S.A. 84 G6 38 18N 120 43W
Mokhós, Greece 23 D7 35 16N 25 27 E
Mokhotlong, Lesotho 57 D4 29 22S 29 2 E
Mokokchung, India 41 F19 26 15N 94 30 E
Mokra Gora, Serbia, Yug. 21 C9 42 50N 20 30 E
Mol, Belgium 15 C5 51 11N 5 5 E
Molchanovo, Russia 26 D9 57 40N 83 50 E
Mold, U.K. 10 D4 53 9N 3 8W
Moldavia ■ = Moldova ■, Europe 17 E15 47 0N 28 0 E
Molde, Norway 8 E12 62 45N 7 9 E
Moldova ■, Europe 17 E15 47 0N 28 0 E
Moldoveana, Romania 17 F13 45 36N 24 45 E
Molepolole, Botswana 56 C4 24 28S 25 28 E
Molfetta, Italy 20 D7 41 12N 16 36 E
Moline, U.S.A. 80 E9 41 30N 90 31W
Molinos, Argentina 94 B2 25 28S 66 15W
Moliro, Zaïre 54 D3 8 12S 30 30 E
Mollahat, Bangla. 43 H13 22 56N 89 48 E
Mollendo, Peru 92 G4 17 0S 72 0W
Mollerin, L., Australia 61 F2 30 30S 117 35 E
Mölndal, Sweden 9 H15 57 40N 12 3 E
Molodechno = Maladzyechna, Belarus 17 A14 54 20N 26 50 E
Molokai, U.S.A. 74 H16 21 8N 157 0W
Molong, Australia 63 E4 33 5S 148 54 E
Molopo →, Africa 56 D3 27 30S 20 13 E
Molotov = Perm, Russia 24 C10 58 0N 56 10 E
Moloundou, Cameroon 52 D3 2 8N 15 15 E
Molson L., Canada 73 C9 54 22N 96 40W
Molteno, S. Africa 56 E4 31 22S 26 22 E
Molu, Indonesia 37 F8 6 45S 131 40 E
Molucca Sea, Indonesia 37 E6 2 0S 124 0 E
Moluccas = Maluku, Indonesia 37 E7 1 0S 127 0 E
Moma, Mozam. 55 F4 16 47S 39 4 E
Moma, Zaïre 54 C1 1 35S 23 52 E
Mombasa, Kenya 54 C4 4 2S 39 43 E
Mombetsu, Japan 30 B11 44 21N 143 22 E
Momchilgrad, Bulgaria 21 D11 41 33N 25 23 E
Momi, Zaïre 54 C2 1 42S 27 0 E
Mompós, Colombia 92 B4 9 14N 74 26W
Møn, Denmark 9 J15 54 57N 12 15 E
Mon →, Burma 41 J19 20 25N 94 30 E
Mona, Canal de la, W. Indies 89 C6 18 30N 67 45W
Mona, Isla, Puerto Rico 89 C6 18 5N 67 54W
Mona, Pta., Costa Rica 88 E3 9 37N 82 36W
Monach Is., U.K. 12 D1 57 32N 7 40W
Monaco ■, Europe 18 E7 43 46N 7 23 E
Monadhliath Mts., U.K. 12 D4 57 10N 4 4W
Monaghan, Ireland 13 B5 54 15N 6 57W
Monaghan □, Ireland 13 B5 54 11N 6 56W
Monahans, U.S.A. 81 K3 31 36N 102 54W
Monapo, Mozam. 55 E5 14 56S 40 19 E
Monarch Mt., Canada 72 C3 51 55N 125 57W
Monastir = Bitola, Macedonia 21 D9 41 5N 21 10 E
Monastir, Tunisia 51 A7 35 50N 10 49 E
Moncayo, Sierra del, Spain 19 B5 41 48N 1 50W
Monchegorsk, Russia 24 A5 67 54N 32 58 E
Mönchengladbach, Germany 16 C4 51 11N 6 27 E
Monchique, Portugal 19 D1 37 19N 8 38W
Monclova, Mexico 86 B4 26 50N 101 30W
Moncton, Canada 71 C7 46 7N 64 51W
Mondego →, Portugal 19 B1 40 9N 8 52W
Mondeodo, Indonesia 37 E6 3 34S 122 9 E
Mondovì, Italy 20 B2 44 23N 7 49 E
Mondovi, U.S.A. 80 C9 44 34N 91 40W
Mondrain I., Australia 61 F3 34 9S 122 14 E
Monduli □, Tanzania 54 C4 3 0S 36 0 E
Monessen, U.S.A. 78 F5 40 9N 79 54W
Monett, U.S.A. 81 G8 36 55N 93 55W
Monforte de Lemos, Spain 19 A2 42 31N 7 33W
Mong Hsu, Burma 41 J21 21 54N 98 30 E
Mong Kung, Burma 41 J20 21 35N 97 35 E
Mong Nai, Burma 41 J20 20 32N 97 46 E
Mong Pawk, Burma 41 H21 22 4N 99 16 E
Mong Ton, Burma 41 J21 20 17N 98 45 E
Mong Wa, Burma 41 J22 21 26N 100 27 E
Mong Yai, Burma 41 H21 22 21N 98 3 E
Mongalla, Sudan 51 G11 5 8N 31 42 E
Mongers, L., Australia 61 E2 29 25S 117 5 E
Monghyr = Munger, India 43 G12 25 23N 86 30 E
Mongibello = Etna, Italy 20 F6 37 50N 14 55 E
Mongo, Chad 51 F8 12 14N 18 43 E
Mongolia ■, Asia 27 E10 47 0N 103 0 E
Mongororo, Chad 51 F9 12 3N 22 26 E
Mongu, Zambia 53 H4 15 16S 23 12 E
Môngua, Angola 56 B2 16 43S 15 20 E
Monkey Bay, Malawi 55 E4 14 7S 35 1 E
Monkey River, Belize 87 D7 16 22N 88 29W
Monkira, Australia 62 C3 24 46S 140 30 E
Monkoto, Zaïre 52 E4 1 38S 20 35 E
Monmouth, U.K. 11 F5 51 48N 2 42W
Monmouth, U.S.A. 80 E9 40 55N 90 39W
Monmouthshire □, U.K. 11 F5 51 48N 2 54W
Mono L., U.S.A. 83 H4 38 1N 119 1W
Monolith, U.S.A. 85 K8 35 7N 118 22W
Monólithos, Greece 23 C9 36 7N 27 45 E
Monongahela, U.S.A. 78 F5 40 12N 79 56W
Monópoli, Italy 20 D7 40 57N 17 18 E
Monqoumba, C.A.R. 52 D3 3 33N 18 40 E
Monroe, Ga., U.S.A. 77 J4 33 47N 83 43W
Monroe, La., U.S.A. 81 J8 32 30N 92 7W
Monroe, Mich., U.S.A. 76 E4 41 55N 83 24W
Monroe, N.C., U.S.A. 77 H5 34 59N 80 33W
Monroe, N.Y., U.S.A. 79 E10 41 20N 74 11W
Monroe, Utah, U.S.A. 83 G7 38 38N 112 7W
Monroe, Wash., U.S.A. 84 C5 47 51N 121 58W
Monroe, Wis., U.S.A. 80 D10 42 36N 89 38W
Monroe City, U.S.A. 80 F9 39 39N 91 44W
Monroeville, Ala., U.S.A. 77 K2 31 31N 87 20W
Monroeville, Pa., U.S.A. 78 F5 40 26N 79 45W
Monrovia, Liberia 50 G2 6 18N 10 47W
Mons, Belgium 15 D3 50 27N 3 58 E
Monse, Indonesia 37 E6 4 0S 123 10 E
Mont-de-Marsan, France 18 E3 43 54N 0 31W
Mont-Joli, Canada 71 C6 48 37N 68 10W
Mont-Laurier, Canada 70 C4 46 35N 75 30W
Mont-St.-Michel, Le = Le Mont-St.-Michel, France 18 B3 48 40N 1 30W
Mont Tremblant Prov. Park, Canada 70 C5 46 30N 74 30W
Montagu, S. Africa 56 E3 33 45S 20 8 E
Montagu I., Antarctica 5 B1 58 25S 26 20W
Montague, Canada 71 C7 46 10N 62 39W
Montague, U.S.A. 82 F2 41 44N 122 32W
Montague, I., Mexico 86 A2 31 40N 114 56W
Montague Ra., Australia 61 E2 27 15S 119 30 E
Montague Sd., Australia 60 B4 14 28S 125 20 E
Montalbán, Spain 19 B5 40 50N 0 45W
Montalvo, U.S.A. 85 L7 34 15N 119 12W
Montaña, Peru 92 E4 6 0S 73 0W
Montana □, U.S.A. 82 C9 47 0N 110 0W
Montaña Clara, I., Canary Is. 22 E6 29 17N 13 33W
Montargis, France 18 C5 47 59N 2 43 E
Montauban, France 18 D4 44 2N 1 21 E
Montauk, U.S.A. 79 E13 41 3N 71 57W
Montauk Pt., U.S.A. 79 E13 41 4N 71 52W
Montbéliard, France 18 C7 47 31N 6 48 E
Montceau-les-Mines, France 18 C6 46 40N 4 23 E
Montclair, U.S.A. 79 F10 40 49N 74 13W
Monte Albán, Mexico 87 D5 17 2N 96 45W
Monte Alegre, Brazil 93 D8 2 0S 54 0W
Monte Azul, Brazil 93 G10 15 9S 42 53W
Monte Bello Is., Australia 60 D2 20 30S 115 45 E
Monte-Carlo, Monaco 16 G4 43 46N 7 23 E
Monte Caseros, Argentina 94 C4 30 10S 57 50W
Monte Comán, Argentina 94 C2 34 40S 67 53W
Monte Cristi, Dom. Rep. 89 C5 19 52N 71 39W
Monte Lindo →, Paraguay 94 A4 23 56S 57 12W
Monte Quemado, Argentina 94 B3 25 53S 62 41W
Monte Rio, U.S.A. 84 G4 38 28N 123 0W
Monte Santu, C. di, Italy 20 D3 40 5N 9 44 E
Monte Vista, U.S.A. 83 H10 37 35N 106 9W
Monteagudo, Argentina 95 B5 27 14S 54 8W
Montebello, Canada 70 C5 45 40N 74 55W
Montecito, U.S.A. 85 L7 34 26N 119 40W
Montecristi, Ecuador 92 D2 1 0S 80 40W
Montecristo, Italy 20 C4 42 20N 10 19 E
Montego Bay, Jamaica 88 C4 18 30N 78 0W
Montejinnie, Australia 60 C5 16 40S 131 38 E
Montélimar, France 18 D6 44 33N 4 45 E
Montello, U.S.A. 80 D10 43 48N 89 20W
Montemorelos, Mexico 87 B5 25 11N 99 42W
Montenegro, Brazil 95 B5 29 39S 51 29W
Montenegro □, Yugoslavia 21 C8 42 40N 19 20 E
Montepuez, Mozam. 55 E4 13 8S 38 59 E
Montepuez →, Mozam. 55 E5 12 32S 40 27 E
Monterey, U.S.A. 83 H3 36 37N 121 55W
Monterey B., U.S.A. 84 J5 36 45N 122 0W
Montería, Colombia 92 B3 8 46N 75 53W
Monteros, Argentina 94 B2 27 11S 65 30W
Monterrey, Mexico 86 B4 25 40N 100 30W
Montes Claros, Brazil 93 G10 16 30S 43 50W
Montesano, U.S.A. 84 D3 46 59N 123 36W
Montesilvano Marina, Italy 20 C6 42 29N 14 8 E
Montevideo, Uruguay 95 C4 34 50S 56 11W
Montevideo, U.S.A. 80 C7 44 57N 95 43W
Montezuma, U.S.A. 80 E8 41 35N 92 32W
Montgomery = Sahiwal, Pakistan 42 D5 30 45N 73 8 E
Montgomery, U.K. 11 E4 52 34N 3 8W
Montgomery, Ala., U.S.A. 77 J2 32 23N 86 19W
Montgomery, W. Va., U.S.A. 76 F5 38 11N 81 19W
Monticello, Ark., U.S.A. 81 J9 33 38N 91 47W
Monticello, Fla., U.S.A. 77 K4 30 33N 83 52W
Monticello, Ind., U.S.A. 76 E2 40 45N 86 46W
Monticello, Iowa, U.S.A. 80 D9 42 15N 91 12W
Monticello, Ky., U.S.A. 77 G3 36 50N 84 51W
Monticello, Minn., U.S.A. 80 C8 45 18N 93 48W
Monticello, Miss., U.S.A. 81 K9 31 33N 90 7W
Monticello, N.Y., U.S.A. 79 E10 41 39N 74 42W
Monticello, Utah, U.S.A. 83 H9 37 52N 109 21W
Montijo, Portugal 19 C1 38 41N 8 54W
Montilla, Spain 19 D3 37 36N 4 40W
Montluçon, France 18 C5 46 22N 2 36 E
Montmagny, Canada 71 C5 46 58N 70 34W
Montmartre, Canada 73 C8 50 14N 103 27W
Montmorency, Canada 71 C5 46 53N 71 11W
Montmorillon, France 18 C4 46 26N 0 50 E
Monto, Australia 62 C5 24 52S 151 6 E
Montoro, Spain 19 C3 38 1N 4 27W
Montour Falls, U.S.A. 78 D8 42 21N 76 51W
Montpelier, Idaho, U.S.A. 82 E8 42 19N 111 18W
Montpelier, Ohio, U.S.A. 76 E3 41 35N 84 37W
Montpelier, Vt., U.S.A. 79 B12 44 16N 72 35W
Montpellier, France 18 E5 43 37N 3 52 E
Montréal, Canada 70 C5 45 31N 73 34W
Montreal L., Canada 73 C7 54 20N 105 45W
Montreal Lake, Canada 73 C7 54 3N 105 46W
Montreux, Switz. 16 E4 46 26N 6 55 E
Montrose, U.K. 12 E6 56 44N 2 27W
Montrose, Colo., U.S.A. 83 G10 38 29N 107 53W
Montrose, Pa., U.S.A. 79 E9 41 50N 75 53W
Monts, Pte. des, Canada 71 C6 49 20N 67 12W
Montserrat ■, W. Indies 89 C7 16 40N 62 10W
Montuiri, Spain 22 B9 39 34N 2 59 E
Monveda, Zaïre 52 D4 2 52N 21 30 E
Monywa, Burma 41 H19 22 7N 95 11 E
Monza, Italy 20 B3 45 35N 9 16 E
Monze, Zambia 55 F2 16 17S 27 29 E
Monze, C., Pakistan 42 G2 24 47N 66 37 E
Monzón, Spain 19 B6 41 52N 0 10 E
Mooi River, S. Africa 57 D4 29 13S 29 50 E
Moolawatana, Australia 63 D2 29 55S 139 45 E
Mooliabeenee, Australia 61 F2 31 20S 116 2 E
Mooloogool, Australia 61 E2 26 2S 119 5 E
Moomin Cr. →, Australia 63 D4 29 44S 149 20 E
Moonah →, Australia 62 C2 22 3S 138 33 E
Moonbeam, Canada 70 C3 49 20N 82 10W
Moonda, L., Australia 63 D2 25 52S 140 25 E
Moonie, Australia 63 D5 27 46S 150 20 E
Moonie →, Australia 63 D4 29 19S 148 43 E

Moonta, *Australia* **63 E2** 34 6S 137 32 E
Moora, *Australia* **61 F2** 30 37S 115 58 E
Mooraberree, *Australia* **62 D3** 25 13S 140 54 E
Moorarie, *Australia* **61 E2** 25 56S 117 35 E
Moorcroft, *U.S.A.* **80 C2** 44 16N 104 57W
Moore →, *Australia* **61 F2** 31 22S 115 30 E
Moore, L., *Australia* **61 E2** 29 50S 117 35 E
Moore Reefs, *Australia* **62 B4** 16 0S 149 5 E
Moorefield, *U.S.A.* **76 F6** 39 5N 78 59W
Moores Res., *U.S.A.* **79 B13** 44 45N 71 50W
Mooresville, *U.S.A.* **77 H5** 35 35N 80 48W
Moorfoot Hills, *U.K.* **12 F5** 55 44N 3 8W
Moorhead, *U.S.A.* **80 B6** 46 53N 96 45W
Mooroopna, *Australia* **63 F4** 36 25S 145 22 E
Moorpark, *U.S.A.* **85 L8** 34 17N 118 53W
Moorreesburg, *S. Africa* **56 E2** 33 6S 18 38 E
Moose →, *Canada* **70 B3** 51 20N 80 25W
Moose Factory, *Canada* **70 B3** 51 16N 80 32W
Moose I., *Canada* **73 C9** 51 42N 97 10W
Moose Jaw, *Canada* **73 C7** 50 24N 105 30W
Moose Jaw →, *Canada* **73 C7** 50 34N 105 18W
Moose Lake, *Canada* **73 C8** 53 43N 100 20W
Moose Lake, *U.S.A.* **80 B8** 46 27N 92 46W
Moose Mountain Cr. →, *Canada* **73 D8** 49 13N 102 12W
Moose Mountain Prov. Park, *Canada* **73 D8** 49 48N 102 25W
Moose River, *Canada* **70 B3** 50 48N 81 17W
Moosehead L., *U.S.A.* **71 C6** 45 38N 69 40W
Moosomin, *Canada* **73 C8** 50 9N 101 40W
Moosonee, *Canada* **70 B3** 51 17N 80 39W
Moosup, *U.S.A.* **79 E13** 41 43N 71 53W
Mopeia Velha, *Mozam.* **55 F4** 17 30S 35 40 E
Mopipi, *Botswana* **56 C3** 21 6S 24 55 E
Mopoi, *C.A.R.* **54 A2** 5 6N 26 54 E
Mopti, *Mali* **50 F4** 14 30N 4 0W
Moquegua, *Peru* **92 G4** 17 15S 70 46W
Mora, *Sweden* **9 F16** 61 2N 14 38 E
Mora, *Minn., U.S.A.* **80 C8** 45 53N 93 18W
Mora, *N. Mex., U.S.A.* **83 J11** 35 58N 105 20W
Moradabad, *India* **43 E8** 28 50N 78 50 E
Morafenobe, *Madag.* **57 B7** 17 50S 44 53 E
Moramanga, *Madag.* **57 B8** 18 56S 48 12 E
Moran, *Kans., U.S.A.* **81 G7** 37 55N 95 10W
Moran, *Wyo., U.S.A.* **82 E8** 43 53N 110 37W
Moranbah, *Australia* **62 C4** 22 1S 148 6 E
Morant Cays, *Jamaica* **88 C4** 17 22N 76 0W
Morant Pt., *Jamaica* **88 C4** 17 55N 76 12W
Morar, L., *U.K.* **12 E3** 56 57N 5 40W
Moratuwa, *Sri Lanka* **40 R11** 6 45N 79 55 E
Morava →, *Serbia, Yug.* **21 B9** 44 36N 21 4 E
Morava →, *Slovak Rep.* **17 D9** 48 10N 16 59 E
Moravia, *U.S.A.* **80 E8** 40 53N 92 49W
Moravian Hts. = Ceskomoravská Vrchovina, *Czech.* **16 D8** 49 30N 15 40 E
Morawa, *Australia* **61 E2** 29 13S 116 0 E
Morawhanna, *Guyana* **92 B7** 8 30N 59 40W
Moray □, *U.K.* **12 D5** 57 31N 3 18W
Moray Firth, *U.K.* **12 D5** 57 40N 3 52W
Morbi, *India* **42 H4** 22 50N 70 42 E
Morden, *Canada* **73 D9** 49 15N 98 10W
Mordovian Republic □ = Mordvinia □, *Russia* **24 D7** 54 20N 44 30 E
Mordvinia □, *Russia* **24 D7** 54 20N 44 30 E
Morea, *Australia* **63 F3** 36 45S 141 18 E
Morea, *Greece* **6 H10** 37 45N 22 10 E
Moreau →, *U.S.A.* **80 C4** 45 18N 100 43W
Morecambe, *U.K.* **10 C5** 54 5N 2 52W
Morecambe B., *U.K.* **10 C5** 54 7N 3 0W
Moree, *Australia* **63 D4** 29 28S 149 54 E
Morehead, *U.S.A.* **76 F4** 38 11N 83 26W
Morehead City, *U.S.A.* **77 H7** 34 43N 76 43W
Morelia, *Mexico* **86 D4** 19 42N 101 7W
Morella, *Australia* **62 C3** 23 0S 143 52 E
Morella, *Spain* **19 B5** 40 35N 0 5W
Morelos, *Mexico* **86 B3** 26 42N 107 40W
Morelos □, *Mexico* **87 D5** 18 40N 99 10W
Morena, Sierra, *Spain* **19 C3** 38 20N 4 0W
Morenci, *U.S.A.* **83 K9** 33 5N 109 22W
Moreno Valley, *U.S.A.* **85 M10** 33 56N 116 58W
Moresby I., *Canada* **72 C2** 52 30N 131 40W
Moreton, *Australia* **62 A3** 12 22S 142 40 E
Moreton I., *Australia* **63 D5** 27 10S 153 25 E
Morey, *Spain* **22 B10** 39 44N 3 20 E
Morgan, *Australia* **63 E2** 34 2S 139 35 E
Morgan, *U.S.A.* **82 F8** 41 2N 111 41W
Morgan City, *U.S.A.* **81 L9** 29 42N 91 12W
Morgan Hill, *U.S.A.* **84 H5** 37 8N 121 39W
Morganfield, *U.S.A.* **76 G2** 37 41N 87 55W
Morganton, *U.S.A.* **77 H5** 35 45N 81 41W
Morgantown, *U.S.A.* **76 F6** 39 38N 79 57W
Morgenzon, *S. Africa* **57 D4** 26 45S 29 36 E
Morghak, *Iran* **45 D8** 29 7N 57 54 E
Morice L., *Canada* **72 C3** 53 50N 127 40W
Morinville, *Canada* **72 C6** 53 49N 113 41W
Morioka, *Japan* **30 E10** 39 45N 141 8 E
Moris, *Mexico* **86 B3** 28 8N 108 32W
Morlaix, *France* **18 B2** 48 36N 3 52W
Mornington, *Vic., Australia* **63 F4** 38 15S 145 5 E
Mornington, *W. Austral., Australia* **60 C4** 17 31S 126 6 E
Mornington, I., *Chile* **96 F1** 49 50S 75 30W
Mornington I., *Australia* **62 B2** 16 30S 139 30 E
Moro G., *Phil.* **37 C6** 6 30N 123 0 E
Morocco ■, *N. Afr.* **50 B3** 32 0N 5 50W
Morococha, *Peru* **92 F3** 11 40S 76 5W
Morogoro, *Tanzania* **54 D4** 6 50S 37 40 E
Morogoro □, *Tanzania* **54 D4** 8 0S 37 0 E
Moroleón, *Mexico* **86 C4** 20 8N 101 32W
Morombe, *Madag.* **57 C7** 21 45S 43 22 E
Moron, *Argentina* **94 C4** 34 39S 58 37W
Morón, *Cuba* **88 B4** 22 8N 78 39W
Morón de la Frontera, *Spain* **19 D3** 37 6N 5 28W
Morondava, *Madag.* **57 C7** 20 17S 44 17 E
Morongo Valley, *U.S.A.* **85 L10** 34 3N 116 37W
Morotai, *Indonesia* **37 D7** 2 10N 128 30 E
Moroto, *Uganda* **54 B3** 2 28N 34 42 E
Moroto Summit, *Kenya* **54 B3** 2 30N 34 43 E
Morpeth, *U.K.* **10 B6** 55 10N 1 41W
Morphou, *Cyprus* **23 D11** 35 12N 32 59 E
Morphou Bay, *Cyprus* **23 D11** 35 15N 32 50 E
Morrilton, *U.S.A.* **81 H8** 35 9N 92 44W
Morrinhos, *Brazil* **93 G9** 17 45S 49 10W
Morrinsville, *N.Z.* **59 G5** 37 40S 175 32 E
Morris, *Canada* **73 D9** 49 25N 97 22W
Morris, *Ill., U.S.A.* **76 E1** 41 22N 88 26W
Morris, *Minn., U.S.A.* **80 C7** 45 35N 95 55W
Morris, Mt., *Australia* **61 E5** 26 9S 131 4 E
Morrisburg, *Canada* **70 D4** 44 55N 75 7W
Morrison, *U.S.A.* **80 E10** 41 49N 89 58W
Morristown, *Ariz., U.S.A.* **83 K7** 33 51N 112 37W
Morristown, *N.J., U.S.A.* **79 F10** 40 48N 74 29W
Morristown, *S. Dak., U.S.A.* **80 C4** 45 56N 101 43W
Morristown, *Tenn., U.S.A.* **77 G4** 36 13N 83 18W
Morro, Pta., *Chile* **94 B1** 27 6S 71 0W
Morro Bay, *U.S.A.* **83 J3** 35 22N 120 51W
Morro del Jable, *Canary Is.* **22 F5** 28 3N 14 23W
Morro Jable, Pta. de, *Canary Is.* **22 F5** 28 2N 14 20W
Morrosquillo, G. de, *Colombia* **88 E4** 9 35N 75 40W
Morrumbene, *Mozam.* **57 C6** 23 31S 35 16 E
Morshansk, *Russia* **24 D7** 53 28N 41 50 E
Morteros, *Argentina* **94 C3** 30 50S 62 0W
Mortes, R. das →, *Brazil* **93 F8** 11 45S 50 44W
Mortlake, *Australia* **63 F3** 38 5S 142 50 E
Morton, *Tex., U.S.A.* **81 J3** 33 44N 102 46W
Morton, *Wash., U.S.A.* **84 D4** 46 34N 122 17W
Morundah, *Australia* **63 E4** 34 57S 146 19 E
Moruya, *Australia* **63 F5** 35 58S 150 3 E
Morvan, *France* **18 C6** 47 5N 4 3 E
Morven, *Australia* **63 D4** 26 22S 147 5 E
Morvern, *U.K.* **12 E3** 56 38N 5 44W
Morwell, *Australia* **63 F4** 38 10S 146 22 E
Morzhovets, Ostrov, *Russia* **24 A7** 66 44N 42 35 E
Moscos Is., *Burma* **38 E1** 14 0N 97 30 E
Moscow = Moskva, *Russia* **24 C6** 55 45N 37 35 E
Moscow, *U.S.A.* **82 C5** 46 44N 117 0W
Mosel →, *Europe* **18 A7** 50 22N 7 36 E
Moselle = Mosel →, *Europe* **18 A7** 50 22N 7 36 E
Moses Lake, *U.S.A.* **82 C4** 47 8N 119 17W
Mosgiel, *N.Z.* **59 L3** 45 53S 170 21 E
Moshi, *Tanzania* **54 C4** 3 22S 37 18 E
Moshi □, *Tanzania* **54 C4** 3 22S 37 18 E
Moshupa, *Botswana* **56 C4** 24 46S 25 29 E
Mosjøen, *Norway* **8 D15** 65 51N 13 12 E
Moskenesøya, *Norway* **8 C15** 67 58N 13 0 E
Moskenstraumen, *Norway* **8 C15** 67 47N 12 45 E
Moskva, *Russia* **24 C6** 55 45N 37 35 E
Moskva →, *Russia* **24 C6** 55 5N 38 51 E
Mosomane, *Botswana* **56 C4** 24 2S 26 19 E
Moson-magyaróvar, *Hungary* **17 E9** 47 52N 17 18 E
Mosquera, *Colombia* **92 C3** 2 35N 78 24W
Mosquero, *U.S.A.* **81 H3** 35 47N 103 58W
Mosquitia, *Honduras* **88 C3** 15 20N 84 10W
Mosquitos, G. de los, *Panama* **88 E3** 9 15N 81 10W
Moss, *Norway* **9 G14** 59 27N 10 40 E
Moss Vale, *Australia* **63 E5** 34 32S 150 25 E
Mossaka, *Congo* **52 E3** 1 15S 16 45 E
Mossbank, *Canada* **73 D7** 49 56N 105 56W
Mossburn, *N.Z.* **59 L2** 45 41S 168 15 E
Mosselbaai, *S. Africa* **56 E3** 34 11S 22 8 E
Mossendjo, *Congo* **52 E2** 2 55S 12 42 E
Mossgiel, *Australia* **63 E3** 33 15S 144 5 E
Mossman, *Australia* **62 B4** 16 21S 145 15 E
Mossoró, *Brazil* **93 E11** 5 10S 37 15W
Mossuril, *Mozam.* **55 E5** 14 58S 40 42 E
Mossy →, *Canada* **73 C8** 54 5N 102 58W
Most, *Czech.* **16 C7** 50 31N 13 38 E
Mosta, *Malta* **23 D1** 35 54N 14 24 E
Moşţafáābād, *Iran* **45 C7** 33 39N 54 53 E
Mostaganem, *Algeria* **50 A5** 35 54N 0 5 E
Mostar, *Bos.-H.* **21 C7** 43 22N 17 50 E
Mostardas, *Brazil* **95 C5** 31 2S 50 51W
Mostiska = Mostyska, *Ukraine* **17 D12** 49 48N 23 4 E
Mosty = Masty, *Belarus* **17 B13** 53 27N 24 38 E
Mostyska, *Ukraine* **17 D12** 49 48N 23 4 E
Mosul = Al Mawşil, *Iraq* **44 B4** 36 15N 43 5 E
Mosulpo, *S. Korea* **35 H14** 33 20N 126 17 E
Motagua →, *Guatemala* **88 C2** 15 44N 88 14W
Motala, *Sweden* **9 G16** 58 32N 15 1 E
Motherwell, *U.K.* **12 F5** 55 47N 3 58W
Motihari, *India* **43 F11** 26 30N 84 55 E
Motozintla de Mendoza, *Mexico* **87 D6** 15 21N 92 14W
Motril, *Spain* **19 D4** 36 31N 3 37W
Mott, *U.S.A.* **80 B3** 46 23N 102 20W
Motueka, *N.Z.* **59 J4** 41 7S 173 1 E
Motueka →, *N.Z.* **59 J4** 41 5S 173 1 E
Motul, *Mexico* **87 C7** 21 0N 89 20W
Mouanda, *Gabon* **52 E2** 1 28S 13 7 E
Mouchalagane →, *Canada* **71 B6** 50 56N 68 41W
Moúdhros, *Greece* **21 E11** 39 50N 25 18 E
Moudjeria, *Mauritania* **50 E2** 17 50N 12 28W
Mouila, *Gabon* **52 E2** 1 50S 11 0 E
Moulamein, *Australia* **63 F3** 35 3S 144 1 E
Moulianá, *Greece* **23 D7** 35 10N 25 59 E
Moulins, *France* **18 C5** 46 35N 3 19 E
Moulmein, *Burma* **41 L20** 16 30N 97 40 E
Moulton, *U.S.A.* **81 L6** 29 35N 97 9W
Moultrie, *U.S.A.* **77 K4** 31 11N 83 47W
Moultrie, L., *U.S.A.* **77 J5** 33 20N 80 5W
Mound City, *Mo., U.S.A.* **80 E7** 40 7N 95 14W
Mound City, *S. Dak., U.S.A.* **80 C4** 45 44N 100 4W
Moundou, *Chad* **51 G8** 8 40N 16 10 E
Moundsville, *U.S.A.* **78 G4** 39 55N 80 44W
Moung, *Cambodia* **38 F4** 12 46N 103 27 E
Mount Airy, *U.S.A.* **77 G5** 36 31N 80 37W
Mount Albert, *Canada* **78 B5** 44 8N 79 19W
Mount Amherst, *Australia* **60 C4** 18 24S 126 58 E
Mount Angel, *U.S.A.* **82 D2** 45 4N 122 48W
Mount Augustus, *Australia* **60 D2** 24 20S 116 56 E
Mount Barker, *S. Austral., Australia* **63 F2** 35 5S 138 52 E
Mount Barker, *W. Austral., Australia* **61 F2** 34 38S 117 40 E
Mount Carmel, *U.S.A.* **76 F2** 38 25N 87 46W
Mount Clemens, *U.S.A.* **78 D2** 42 35N 82 53W
Mount Coolon, *Australia* **62 C4** 21 25S 147 25 E
Mount Darwin, *Zimbabwe* **55 F3** 16 47S 31 38 E
Mount Desert I., *U.S.A.* **71 D6** 44 21N 68 20W
Mount Dora, *U.S.A.* **77 L5** 28 48N 81 38W
Mount Douglas, *Australia* **62 C4** 21 35S 146 50 E
Mount Eba, *Australia* **63 E2** 30 11S 135 40 E
Mount Edgecumbe, *U.S.A.* **72 B1** 57 3N 135 21W
Mount Elizabeth, *Australia* **60 C4** 16 0S 125 50 E
Mount Fletcher, *S. Africa* **57 E4** 30 40S 28 30 E
Mount Forest, *Canada* **70 D3** 43 59N 80 43W
Mount Gambier, *Australia* **63 F3** 37 50S 140 46 E
Mount Garnet, *Australia* **62 B4** 17 37S 145 6 E
Mount Hope, *N.S.W., Australia* **63 E4** 32 51S 145 51 E
Mount Hope, *S. Austral., Australia* **63 E2** 34 7S 135 23 E
Mount Hope, *U.S.A.* **76 G5** 37 54N 81 10W
Mount Horeb, *U.S.A.* **80 D10** 43 1N 89 44W
Mount Howitt, *Australia* **63 D3** 26 31S 142 16 E
Mount Isa, *Australia* **62 C2** 20 42S 139 26 E
Mount Keith, *Australia* **61 E3** 27 15S 120 30 E
Mount Laguna, *U.S.A.* **85 N10** 32 52N 116 25W
Mount Larcom, *Australia* **62 C5** 23 48S 150 59 E
Mount Lofty Ra., *Australia* **63 E2** 34 35S 139 5 E
Mount McKinley National Park, *U.S.A.* **68 B5** 63 30N 150 0W
Mount Magnet, *Australia* **61 E2** 28 2S 117 47 E
Mount Margaret, *Australia* **63 D3** 26 54S 143 21 E
Mount Maunganui, *N.Z.* **59 G6** 37 40S 176 14 E
Mount Molloy, *Australia* **62 B4** 16 42S 145 20 E
Mount Monger, *Australia* **61 F3** 31 0S 122 0 E
Mount Morgan, *Australia* **62 C5** 23 40S 150 25 E
Mount Morris, *U.S.A.* **78 D7** 42 44N 77 52W
Mount Mulligan, *Australia* **62 B3** 16 45S 144 47 E
Mount Narryer, *Australia* **61 E2** 26 30S 115 55 E
Mount Olympus = Uludağ, *Turkey* **21 D13** 40 4N 29 13 E
Mount Oxide Mine, *Australia* **62 B2** 19 30S 139 29 E
Mount Pearl, *Canada* **71 C9** 47 31N 52 47W
Mount Perry, *Australia* **63 D5** 25 13S 151 42 E
Mount Phillips, *Australia* **60 D2** 24 25S 116 15 E
Mount Pleasant, *Iowa, U.S.A.* **80 E9** 40 58N 91 33W
Mount Pleasant, *Mich., U.S.A.* **76 D3** 43 36N 84 46W
Mount Pleasant, *Pa., U.S.A.* **78 F5** 40 9N 79 33W
Mount Pleasant, *S.C., U.S.A.* **77 J6** 32 47N 79 52W
Mount Pleasant, *Tenn., U.S.A.* **77 H2** 35 32N 87 12W
Mount Pleasant, *Tex., U.S.A.* **81 J7** 33 9N 94 58W
Mount Pleasant, *Utah, U.S.A.* **82 G8** 39 33N 111 27W
Mount Pocono, *U.S.A.* **79 E9** 41 7N 75 22W
Mount Rainier National Park, *U.S.A.* **84 D5** 46 55N 121 50W
Mount Revelstoke Nat. Park, *Canada* **72 C5** 51 5N 118 30W
Mount Robson Prov. Park, *Canada* **72 C5** 53 0N 119 0W
Mount Sandiman, *Australia* **61 D2** 24 25S 115 30 E
Mount Shasta, *U.S.A.* **82 F2** 41 19N 122 19W
Mount Signal, *U.S.A.* **85 N11** 32 39N 115 37W
Mount Sterling, *Ill., U.S.A.* **80 F9** 39 59N 90 45W
Mount Sterling, *Ky., U.S.A.* **76 F4** 38 4N 83 56W
Mount Surprise, *Australia* **62 B3** 18 10S 144 17 E
Mount Union, *U.S.A.* **78 F7** 40 23N 77 53W
Mount Vernon, *Australia* **60 D2** 24 9S 118 2 E
Mount Vernon, *Ind., U.S.A.* **80 F10** 38 17N 88 57W
Mount Vernon, *N.Y., U.S.A.* **79 F11** 40 55N 73 50W
Mount Vernon, *Ohio, U.S.A.* **78 F2** 40 23N 82 29W
Mount Vernon, *Wash., U.S.A.* **84 B4** 48 25N 122 20W
Mountain Center, *U.S.A.* **85 M10** 33 42N 116 44W
Mountain City, *Nev., U.S.A.* **82 F6** 41 50N 115 58W
Mountain City, *Tenn., U.S.A.* **77 G5** 36 29N 81 48W
Mountain Grove, *U.S.A.* **81 G8** 37 8N 92 16W
Mountain Home, *Ark., U.S.A.* **81 G8** 36 20N 92 23W
Mountain Home, *Idaho, U.S.A.* **82 E6** 43 8N 115 41W
Mountain Iron, *U.S.A.* **80 B8** 47 32N 92 37W
Mountain Park, *Canada* **72 C5** 52 50N 117 15W
Mountain Pass, *U.S.A.* **85 K11** 35 29N 115 35W
Mountain View, *Ark., U.S.A.* **81 H8** 35 52N 92 7W
Mountain View, *Calif., U.S.A.* **83 H2** 37 23N 122 5W
Mountainair, *U.S.A.* **83 J10** 34 31N 106 15W
Mountmellick, *Ireland* **13 C4** 53 7N 7 20W
Moura, *Australia* **62 C4** 24 35S 149 58 E
Moura, *Brazil* **92 D6** 1 32S 61 38W
Moura, *Portugal* **19 C2** 38 7N 7 30W
Mourdi, Dépression du, *Chad* **51 E9** 18 10N 23 0 E
Mourdiah, *Mali* **50 F3** 14 35N 7 25W
Mourilyan, *Australia* **62 B4** 17 35S 146 3 E
Mourne →, *U.K.* **13 B4** 54 52N 7 26W
Mourne Mts., *U.K.* **13 B5** 54 10N 6 0W
Mournies, *Greece* **23 D6** 35 29N 24 1 E
Mouscron, *Belgium* **15 D3** 50 45N 3 12 E
Moussoro, *Chad* **51 F8** 13 41N 16 35 E
Moutohara, *N.Z.* **59 H6** 38 27S 177 32 E
Moutong, *Indonesia* **37 D6** 0 28N 121 13 E
Movas, *Mexico* **86 B3** 28 10N 109 25W
Moville, *Ireland* **13 A4** 55 11N 7 3W
Moy →, *Ireland* **13 B3** 54 8N 9 8W
Moyale, *Kenya* **46 G2** 3 30N 39 0 E
Moyamba, *S. Leone* **50 G2** 8 4N 12 30W
Moyen Atlas, *Morocco* **50 B3** 33 0N 5 0W
Moyle □, *U.K.* **13 A5** 55 10N 6 15W
Moyo, *Indonesia* **36 F5** 8 10S 117 40 E
Moyobamba, *Peru* **92 E3** 6 0S 77 0W
Moyyero →, *Russia* **27 C11** 68 44N 103 42 E
Moyynty, *Kazakstan* **26 E8** 47 10N 73 18 E
Mozambique = Moçambique, *Mozam.* **55 F5** 15 3S 40 42 E
Mozambique ■, *Africa* **55 F4** 19 0S 35 0 E
Mozambique Chan., *Africa* **57 B7** 17 30S 42 30 E
Mozdok, *Russia* **25 F7** 43 45N 44 48 E
Mozdūrān, *Iran* **45 B9** 36 9N 60 35 E
Mozhnābād, *Iran* **45 C9** 34 7N 60 6 E
Mozyr = Mazyr, *Belarus* **17 B15** 51 59N 29 15 E
Mpanda, *Tanzania* **54 D3** 6 23S 31 1 E
Mpanda □, *Tanzania* **54 D3** 6 23S 31 40 E
Mpika, *Zambia* **55 E3** 11 51S 31 25 E
Mpulungu, *Zambia* **55 D3** 8 51S 31 5 E
Mpumalanga, *S. Africa* **57 D5** 29 50S 30 33 E
Mpumalanga □, *S. Africa* **57 B5** 26 0S 30 0 E
Mpwapwa, *Tanzania* **54 D4** 6 23S 36 30 E
Mpwapwa □, *Tanzania* **54 D4** 6 30S 36 20 E
Msaken, *Tunisia* **51 A7** 35 49N 10 33 E
Msambansovu, *Zimbabwe* **55 F3** 15 50S 30 3 E
Msoro, *Zambia* **55 E3** 13 35S 31 50 E
Mstislavl = Mstsislaw, *Belarus* **17 A16** 54 0N 31 50 E
Mstsislaw, *Belarus* **17 A16** 54 0N 31 50 E
Mtama, *Tanzania* **55 E4** 10 17S 39 21 E
Mtilikwe →, *Zimbabwe* **55 G3** 21 9S 31 30 E
Mtubatuba, *S. Africa* **57 D5** 28 30S 32 8 E
Mtwara-Mikindani, *Tanzania* **55 E5** 10 20S 40 20 E
Mu Gia, Deo, *Vietnam* **38 D5** 17 40N 105 47 E
Mu Us Shamo, *China* **34 E5** 39 0N 109 0 E
Muaná, *Brazil* **93 D9** 1 25S 49 15W
Muang Chiang Rai, *Thailand* **38 C2** 19 52N 99 50 E
Muang Lamphun, *Thailand* **38 C2** 18 40N 99 2 E
Muang Pak Beng, *Laos* **38 C3** 19 54N 101 8 E
Muar, *Malaysia* **39 L4** 2 3N 102 34 E
Muarabungo, *Indonesia* **36 E2** 1 28S 102 52 E
Muaraenim, *Indonesia* **36 E2** 3 40S 103 50 E
Muarajuloi, *Indonesia* **36 E4** 0 12S 114 3 E
Muarakaman, *Indonesia* **36 E5** 0 2S 116 45 E
Muaratebo, *Indonesia* **36 E2** 1 30S 102 26 E
Muaratembesi, *Indonesia* **36 E2** 1 42S 103 8 E
Muaratewe, *Indonesia* **36 E4** 0 58S 114 52 E
Mubarakpur, *India* **43 F10** 26 6N 83 18 E
Mubarraz = Al Mubarraz, *Si. Arabia* **45 E6** 25 30N 49 40 E
Mubende, *Uganda* **54 B3** 0 33N 31 22 E
Mubi, *Nigeria* **51 F7** 10 18N 13 16 E
Mubur, P., *Indonesia* **39 L6** 3 20N 106 12 E
Muchachos, Roque de los, *Canary Is.* **22 F2** 28 44N 17 52W
Muchinga Mts., *Zambia* **55 E3** 11 30S 31 30 E
Muck, *U.K.* **12 E2** 56 50N 6 15W
Muckadilla, *Australia* **63 D4** 26 35S 148 23 E
Muconda, *Angola* **52 G4** 10 31S 21 15 E
Mucuri, *Brazil* **93 G11** 18 0S 39 36W
Mucusso, *Angola* **56 B3** 18 1S 21 25 E
Muda, *Canary Is.* **22 F6** 28 34N 13 57W
Mudan Jiang →, *China* **35 A15** 46 20N 129 30 E
Mudanjiang, *China* **35 B15** 44 38N 129 30 E
Mudanya, *Turkey* **21 D13** 40 25N 28 50 E
Muddy Cr. →, *U.S.A.* **83 H8** 38 24N 110 42W
Mudgee, *Australia* **63 E4** 32 32S 149 31 E
Mudjatik →, *Canada* **73 B7** 56 1N 107 36W
Muecate, *Mozam.* **55 E4** 14 55S 39 40 E
Muêda, *Mozam.* **55 E4** 11 36S 39 28 E
Mueller Ra., *Australia* **60 C4** 18 18S 126 46 E
Muende, *Mozam.* **55 E3** 14 28S 33 0 E
Muerto, Mar, *Mexico* **87 D6** 16 10N 94 10W
Mufindi □, *Tanzania* **55 D4** 8 30S 35 20 E
Mufulira, *Zambia* **55 E2** 12 32S 28 15 E
Mufumbiro Range, *Africa* **54 C2** 1 25S 29 30 E
Mughayrā', *Si. Arabia* **44 D3** 29 17N 37 41 E
Mugi, *Japan* **31 H7** 33 40N 134 25 E
Mugila, Mts., *Zaïre* **54 D2** 7 0S 28 50 E
Muğla, *Turkey* **21 F13** 37 15N 28 22 E
Mugu, *Nepal* **43 E10** 29 45N 82 30 E
Muhammad Qol, *Sudan* **51 D12** 20 53N 37 9 E
Muhammadabad, *India* **43 F10** 26 4N 83 25 E
Muhesi →, *Tanzania* **54 D4** 7 0S 35 20 E
Muheza □, *Tanzania* **54 C4** 5 0S 39 0 E
Mühlhausen, *Germany* **16 C6** 51 12N 10 27 E
Mühlig Hofmann fjell, *Antarctica* **5 D3** 72 30S 5 0 E
Muhos, *Finland* **8 D22** 64 47N 25 59 E
Muhu, *Estonia* **9 G20** 58 36N 23 11 E
Muhutwe, *Tanzania* **54 C3** 1 35S 31 45 E
Muikamachi, *Japan* **31 F9** 37 15N 138 50 E
Muine Bheag, *Ireland* **13 D5** 52 42N 6 58W
Muir, L., *Australia* **61 F2** 34 30S 116 40 E
Mukacheve, *Ukraine* **17 D12** 48 27N 22 45 E
Mukachevo = Mukacheve, *Ukraine* **17 D12** 48 27N 22 45 E
Mukah, *Malaysia* **36 D4** 2 55N 112 5 E
Mukdahan, *Thailand* **38 D5** 16 32N 104 43 E
Mukden = Shenyang, *China* **35 D12** 41 48N 123 27 E
Mukhtuya = Lensk, *Russia* **27 C12** 60 48N 114 55 E
Mukinbudin, *Australia* **61 F2** 30 55S 118 5 E
Mukishi, *Zaïre* **55 D1** 8 30S 24 44 E
Mukomuko, *Indonesia* **36 E2** 2 30S 101 10 E
Mukomwenze, *Zaïre* **54 D2** 6 49S 27 15 E
Muktsar, *India* **42 D6** 30 30N 74 30 E
Mukur, *Afghan.* **42 C2** 32 50N 67 42 E
Mukutawa →, *Canada* **73 C9** 53 10N 97 24W
Mukwela, *Zambia* **55 F2** 17 0S 26 40 E
Mula, *Spain* **19 C5** 38 3N 1 33W
Mulange, *Zaïre* **54 C2** 3 40S 27 10 E
Mulchén, *Chile* **94 D1** 37 45S 72 20W
Mulde →, *Germany* **16 C7** 51 53N 12 15 E
Mule Creek, *U.S.A.* **80 D2** 43 19N 104 8W
Muleba, *Tanzania* **54 C3** 1 50S 31 37 E
Muleba □, *Tanzania* **54 C3** 2 0S 31 30 E
Muleshoe, *U.S.A.* **81 H3** 34 13N 102 43W
Mulgathing, *Australia* **63 E1** 30 15S 134 8 E
Mulgrave, *Canada* **71 C7** 45 38N 61 31W
Mulhacén, *Spain* **19 D4** 37 4N 3 20W
Mulhouse, *France* **18 C7** 47 40N 7 20 E
Muling, *China* **35 B16** 44 35N 130 10 E

Mull, *U.K.* **12 E3** 56 25N 5 56W
Mullaittvu, *Sri Lanka* **40 Q12** 9 15N 80 49 E
Mullen, *U.S.A.* **80 D4** 42 3N 101 1W
Mullengudgery, *Australia* **63 E4** 31 43S 147 23 E
Mullens, *U.S.A.* **76 G5** 37 35N 81 23W
Muller, Pegunungan, *Indonesia* **36 D4** 0 30N 113 30 E
Mullet Pen., *Ireland* **13 B1** 54 13N 10 2W
Mullewa, *Australia* **61 E2** 28 29S 115 30 E
Mulligan →, *Australia* **62 C2** 25 0S 139 0 E
Mullin, *U.S.A.* **81 K5** 31 33N 98 40W
Mullingar, *Ireland* **13 C4** 53 31N 7 21W
Mullins, *U.S.A.* **77 H6** 34 12N 79 15W
Mullumbimby, *Australia* **63 D5** 28 30S 153 30 E
Mulobezi, *Zambia* **55 F2** 16 45S 25 7 E
Multan, *Pakistan* **42 D4** 30 15N 71 36 E
Mulumbe, Mts., *Zaïre* **55 D2** 8 40S 27 30 E
Mulungushi Dam, *Zambia* **55 E2** 14 48S 28 48 E
Mulvane, *U.S.A.* **81 G6** 37 29N 97 15W
Mulwala, *Australia* **63 F4** 35 59S 146 0 E
Mumbai = Bombay, *India* **40 K8** 18 55N 72 50 E
Mumbwa, *Zambia* **55 E2** 15 0S 27 0 E
Mun →, *Thailand* **38 E5** 15 19N 105 30 E
Muna, *Indonesia* **37 F6** 5 0S 122 30 E
Munamagi, *Estonia* **9 H22** 57 43N 27 4 E
München, *Germany* **16 D6** 48 8N 11 34 E
Munchen-Gladbach = Mönchengladbach, *Germany* **16 C4** 51 11N 6 27 E
Muncho Lake, *Canada* **72 B3** 59 0N 125 50W
Munchŏn, *N. Korea* **35 E14** 39 14N 127 19 E
Muncie, *U.S.A.* **76 E3** 40 12N 85 23W
Muncoonie, L., *Australia* **62 D2** 25 12S 138 40 E
Mundala, *Indonesia* **37 E10** 4 30S 141 0 E
Mundare, *Canada* **72 C6** 53 35N 112 20W
Munday, *U.S.A.* **81 J5** 33 27N 99 38W
Münden, *Germany* **16 C5** 51 25N 9 38 E
Mundiwindi, *Australia* **60 D3** 23 47S 120 9 E
Mundo Novo, *Brazil* **93 F10** 11 50S 40 29W
Mundra, *India* **42 H3** 22 54N 69 48 E
Mundrabilla, *Australia* **61 F4** 31 52S 127 51 E
Mungallala, *Australia* **63 D4** 26 28S 147 34 E
Mungallala Cr. →, *Australia* **63 D4** 28 53S 147 5 E
Mungana, *Australia* **62 B3** 17 8S 144 27 E
Mungaoli, *India* **42 G8** 24 24N 78 7 E
Mungari, *Mozam.* **55 F3** 17 12S 33 30 E
Mungbere, *Zaïre* **54 B2** 2 36N 28 28 E
Munger, *India* **43 G12** 25 23N 86 30 E
Mungindi, *Australia* **63 D4** 28 58S 149 1 E
Munhango, *Angola* **53 G3** 12 10S 18 38 E
Munich = München, *Germany* **16 D6** 48 8N 11 34 E
Munising, *U.S.A.* **76 B2** 46 25N 86 40W
Munku-Sardyk, *Russia* **27 D11** 51 45N 100 20 E
Muñoz Gamero, Pen., *Chile* **96 G2** 52 30S 73 5W
Munroe L., *Canada* **73 B9** 59 13N 98 35W
Munsan, *S. Korea* **35 F14** 37 51N 126 48 E
Münster, *Germany* **16 C4** 51 58N 7 37 E
Munster □, *Ireland* **13 D3** 52 18N 8 44W
Muntadgin, *Australia* **61 F2** 31 45S 118 33 E
Muntok, *Indonesia* **36 E3** 2 5S 105 10 E
Munyama, *Zambia* **55 F2** 16 5S 28 31 E
Muong Beng, *Laos* **38 B3** 20 23N 101 46 E
Muong Boum, *Vietnam* **38 A4** 22 24N 102 49 E
Muong Et, *Laos* **38 B5** 20 49N 104 1 E
Muong Hai, *Laos* **38 B3** 21 3N 101 49 E
Muong Hiem, *Laos* **38 B4** 20 5N 103 22 E
Muong Houn, *Laos* **38 B3** 20 8N 101 23 E
Muong Hung, *Vietnam* **38 B4** 20 56N 103 53 E
Muong Kau, *Laos* **38 E5** 15 6N 105 47 E
Muong Khao, *Laos* **38 C4** 19 38N 103 32 E
Muong Khoua, *Laos* **38 B4** 21 5N 102 31 E
Muong Liep, *Laos* **38 C3** 18 29N 101 40 E
Muong May, *Laos* **38 E6** 14 49N 106 56 E
Muong Ngeun, *Laos* **38 B3** 20 36N 101 3 E
Muong Ngoi, *Laos* **38 B4** 20 43N 102 41 E
Muong Nhie, *Vietnam* **38 A4** 22 12N 102 28 E
Muong Nong, *Laos* **38 D6** 16 22N 106 30 E
Muong Ou Tay, *Laos* **38 A3** 22 7N 101 48 E
Muong Oua, *Laos* **38 C3** 18 18N 101 20 E
Muong Peun, *Laos* **38 B4** 20 13N 103 52 E
Muong Phalane, *Laos* **38 D5** 16 39N 105 34 E
Muong Phieng, *Laos* **38 C3** 19 6N 101 32 E
Muong Phine, *Laos* **38 D6** 16 32N 106 2 E
Muong Sai, *Laos* **38 B3** 20 42N 101 59 E
Muong Saiapoun, *Laos* **38 C3** 18 24N 101 31 E
Muong Sen, *Vietnam* **38 C5** 19 24N 104 8 E
Muong Sing, *Laos* **38 B3** 21 11N 101 9 E
Muong Son, *Laos* **38 B4** 20 27N 103 19 E
Muong Soui, *Laos* **38 C4** 19 33N 102 52 E
Muong Va, *Laos* **38 B4** 21 53N 102 19 E
Muong Xia, *Vietnam* **38 B5** 20 19N 104 50 E
Muonio, *Finland* **8 C20** 67 57N 23 40 E
Muonionjoki →, *Finland* **8 C20** 67 11N 23 34 E
Mupa, *Angola* **53 H3** 16 5S 15 50 E
Muping, *China* **35 F11** 37 22N 121 36 E
Muqdisho, *Somali Rep.* **46 G4** 2 2N 45 25 E
Mur →, *Austria* **17 E9** 46 18N 16 52 E
Murakami, *Japan* **30 E9** 38 14N 139 29 E
Murallón, Cuerro, *Chile* **96 F2** 49 48S 73 30W
Muranda, *Rwanda* **54 C2** 1 52S 29 20 E
Murang'a, *Kenya* **54 C4** 0 45S 37 9 E
Murashi, *Russia* **24 C8** 59 30N 49 0 E
Muratlı, *Turkey* **21 D12** 41 10N 27 29 E
Murayama, *Japan* **30 E10** 38 30N 140 25 E
Murban, *U.A.E.* **45 F7** 23 50N 53 45 E
Murchison →, *Australia* **61 E1** 27 45S 114 0 E
Murchison, Mt., *Antarctica* **5 D11** 73 0S 168 0 E
Murchison Falls = Kabarega Falls, *Uganda* **54 B3** 2 15N 31 30 E
Murchison House, *Australia* **61 E1** 27 39S 114 14 E
Murchison Ra., *Australia* **62 C1** 20 0S 134 10 E
Murchison Rapids, *Malawi* **55 F3** 15 55S 34 35 E
Murcia, *Spain* **19 D5** 38 5N 1 10W
Murcia □, *Spain* **19 D5** 37 50N 1 30W
Murdo, *U.S.A.* **80 D4** 43 53N 100 43W
Murdoch Pt., *Australia* **62 A3** 14 37S 144 55 E
Mureş →, *Romania* **17 E11** 46 15N 20 13 E
Mureşul = Mureş →, *Romania* **17 E11** 46 15N 20 13 E
Murfreesboro, *U.S.A.* **77 H2** 35 51N 86 24W
Murgab = Murghob, *Tajikistan* **26 F8** 38 10N 74 2 E
Murghob, *Tajikistan* **26 F8** 38 10N 74 2 E
Murgon, *Australia* **63 D5** 26 15S 151 54 E
Murgoo, *Australia* **61 E2** 27 24S 116 28 E
Muria, *Indonesia* **37 G14** 6 36S 110 53 E
Muriaé, *Brazil* **95 A7** 21 8S 42 23W
Muriel Mine, *Zimbabwe* **55 F3** 17 14S 30 40 E
Müritz-see, *Germany* **16 B7** 53 25N 12 42 E
Murka, *Kenya* **54 C4** 3 27S 38 0 E
Murmansk, *Russia* **24 A5** 68 57N 33 10 E
Muro, *Spain* **22 B10** 39 44N 3 3 E
Murom, *Russia* **24 C7** 55 35N 42 3 E
Muroran, *Japan* **30 C10** 42 25N 141 0 E
Muroto, *Japan* **31 H7** 33 18N 134 9 E
Muroto-Misaki, *Japan* **31 H7** 33 15N 134 10 E
Murphy, *U.S.A.* **82 E5** 43 13N 116 33W
Murphys, *U.S.A.* **84 G6** 38 8N 120 28W
Murphysboro, *U.S.A.* **81 G10** 37 46N 89 20W
Murray, *Ky., U.S.A.* **77 G1** 36 37N 88 19W
Murray, *Utah, U.S.A.* **82 F8** 40 40N 111 53W
Murray →, *Australia* **63 F2** 35 20S 139 22 E
Murray →, *Canada* **72 B4** 56 11N 120 45W
Murray, L., *U.S.A.* **77 H5** 34 3N 81 13W
Murray Bridge, *Australia* **63 F2** 35 6S 139 14 E
Murray Downs, *Australia* **62 C1** 21 4S 134 40 E
Murray Harbour, *Canada* **71 C7** 46 0N 62 28W
Murraysburg, *S. Africa* **56 E3** 31 58S 23 47 E
Murree, *Pakistan* **42 C5** 33 56N 73 28 E
Murrieta, *U.S.A.* **85 M9** 33 33N 117 13W
Murrin Murrin, *Australia* **61 E3** 28 58S 121 33 E
Murrumbidgee →, *Australia* **63 E3** 34 43S 143 12 E
Murrumburrah, *Australia* **63 E4** 34 32S 148 22 E
Murrurundi, *Australia* **63 E5** 31 42S 150 51 E
Murshidabad, *India* **43 G13** 24 11N 88 19 E
Murtle L., *Canada* **72 C5** 52 8N 119 38W
Murtoa, *Australia* **63 F3** 36 35S 142 28 E
Murungu, *Tanzania* **54 C3** 4 12S 31 10 E
Murwara, *India* **43 H9** 23 46N 80 28 E
Murwillumbah, *Australia* **63 D5** 28 18S 153 27 E
Mürzzuschlag, *Austria* **16 E8** 47 36N 15 41 E
Muş, *Turkey* **25 G7** 38 45N 41 30 E
Mûsa, G., *Egypt* **51 C11** 28 33N 33 59 E
Musa Khel, *Pakistan* **42 D3** 30 59N 69 52 E
Mūsá Qal'eh, *Afghan.* **40 C4** 32 20N 64 50 E
Musaffargarh, *Pakistan* **40 D7** 30 10N 71 10 E
Musala, *Bulgaria* **21 C10** 42 13N 23 37 E
Musala, *Indonesia* **36 D1** 1 41N 98 28 E
Musan, *N. Korea* **35 C15** 42 12N 129 12 E
Musangu, *Zaïre* **55 E1** 10 28S 23 55 E
Musasa, *Tanzania* **54 C3** 3 25S 31 30 E
Musay'īd, *Qatar* **45 E6** 25 0N 51 33 E
Muscat = Masqat, *Oman* **46 C6** 23 37N 58 36 E
Muscat & Oman = Oman ■, *Asia* **46 C6** 23 0N 58 0 E
Muscatine, *U.S.A.* **80 E9** 41 25N 91 3W
Musgrave, *Australia* **62 A3** 14 47S 143 30 E
Musgrave Ras., *Australia* **61 E5** 26 0S 132 0 E
Mushie, *Zaïre* **52 E3** 2 56S 16 55 E
Musi →, *Indonesia* **36 E2** 2 20S 104 56 E
Muskeg →, *Canada* **72 A4** 60 20N 123 20W
Muskegon, *U.S.A.* **76 D2** 43 14N 86 16W
Muskegon →, *U.S.A.* **76 D2** 43 14N 86 21W
Muskegon Heights, *U.S.A.* **76 D2** 43 12N 86 16W
Muskogee, *U.S.A.* **81 H7** 35 45N 95 22W
Muskwa →, *Canada* **72 B4** 58 47N 122 48W
Muslimiyah, *Syria* **44 B3** 36 19N 37 12 E
Musmar, *Sudan* **51 E12** 18 13N 35 40 E
Musofu, *Zambia* **55 E2** 13 30S 29 0 E
Musoma, *Tanzania* **54 C3** 1 30S 33 48 E
Musoma □, *Tanzania* **54 C3** 1 50S 34 30 E
Musquaro, L., *Canada* **71 B7** 50 38N 61 5W
Musquodoboit Harbour, *Canada* **71 D7** 44 50N 63 9W
Musselburgh, *U.K.* **12 F5** 55 57N 3 2W
Musselshell →, *U.S.A.* **82 C10** 47 21N 107 57W
Mussoorie, *India* **42 D8** 30 27N 78 6 E
Mussuco, *Angola* **56 B2** 17 2S 19 3 E
Mustafakemalpaşa, *Turkey* **21 D13** 40 2N 28 24 E
Mustang, *Nepal* **43 E10** 29 10N 83 55 E
Musters, L., *Argentina* **96 F3** 45 20S 69 25W
Musudan, *N. Korea* **35 D15** 40 50N 129 43 E
Muswellbrook, *Australia* **63 E5** 32 16S 150 56 E
Mût, *Egypt* **51 C10** 25 28N 28 58 E
Mutanda, *Mozam.* **57 C5** 21 0S 33 34 E
Mutanda, *Zambia* **55 E2** 12 24S 26 13 E
Mutare, *Zimbabwe* **55 F3** 18 58S 32 38 E
Muting, *Indonesia* **37 F10** 7 23S 140 20 E
Mutoray, *Russia* **27 C11** 60 56N 101 0 E
Mutshatsha, *Zaïre* **55 E1** 10 35S 24 20 E
Mutsu, *Japan* **30 D10** 41 5N 140 55 E
Mutsu-Wan, *Japan* **30 D10** 41 5N 140 55 E
Muttaburra, *Australia* **62 C3** 22 38S 144 29 E
Mutuáli, *Mozam.* **55 E4** 14 55S 37 0 E
Muweilih, *Egypt* **47 E3** 30 42N 34 19 E
Muxima, *Angola* **52 F2** 9 33S 13 58 E
Muy Muy, *Nic.* **88 D2** 12 39N 85 36W
Muyinga, *Burundi* **54 C3** 3 14S 30 33 E
Muynak, *Uzbekistan* **26 E6** 43 44N 59 10 E
Muzaffarabad, *Pakistan* **43 B5** 34 25N 73 30 E
Muzaffargarh, *Pakistan* **42 D4** 30 5N 71 14 E
Muzaffarnagar, *India* **42 E7** 29 26N 77 40 E
Muzaffarpur, *India* **43 F11** 26 7N 85 23 E
Muzhi, *Russia* **26 C7** 65 25N 64 40 E
Muzon, C., *U.S.A.* **72 C2** 54 40N 132 42W
Mvuma, *Zimbabwe* **55 F3** 19 16S 30 30 E
Mvurwi, *Zimbabwe* **55 F3** 17 0S 30 57 E
Mwadui, *Tanzania* **54 C3** 3 26S 33 32 E
Mwambo, *Tanzania* **55 E5** 10 30S 40 22 E
Mwandi, *Zambia* **55 F1** 17 30S 24 51 E
Mwanza, *Tanzania* **54 C3** 2 30S 32 58 E
Mwanza, *Zaïre* **54 D2** 7 55S 26 43 E
Mwanza, *Zambia* **55 F1** 16 58S 24 28 E
Mwanza □, *Tanzania* **54 C3** 2 0S 33 0 E
Mwaya, *Tanzania* **55 D3** 9 32S 33 55 E
Mweelrea, *Ireland* **13 C2** 53 39N 9 49W
Mweka, *Zaïre* **52 E4** 4 50S 21 34 E
Mwenezi, *Zimbabwe* **55 G3** 21 15S 30 48 E
Mwenezi →, *Mozam.* **55 G3** 22 40S 31 50 E
Mwenga, *Zaïre* **54 C2** 3 1S 28 28 E
Mweru, L., *Zambia* **55 D2** 9 0S 28 40 E
Mweza Range, *Zimbabwe* **55 G3** 21 0S 30 0 E
Mwilambwe, *Zaïre* **54 D5** 8 7S 25 5 E
Mwimbi, *Tanzania* **55 D3** 8 38S 31 39 E
Mwinilunga, *Zambia* **55 E1** 11 43S 24 25 E
My Tho, *Vietnam* **39 G6** 10 29N 106 23 E
Myajlar, *India* **42 F4** 26 15N 70 20 E
Myanaung, *Burma* **41 K19** 18 18N 95 22 E
Myanmar = Burma ■, *Asia* **41 J20** 21 0N 96 30 E
Myaungmya, *Burma* **41 L19** 16 30N 94 40 E
Mycenæ, *Greece* **21 F10** 37 39N 22 52 E
Myeik Kyunzu, *Burma* **39 G1** 11 30N 97 30 E
Myerstown, *U.S.A.* **79 F8** 40 22N 76 19W
Myingyan, *Burma* **41 J19** 21 30N 95 20 E
Myitkyina, *Burma* **41 G20** 25 24N 97 26 E
Mykines, *Færoe Is.* **8 E9** 62 7N 7 35W
Mykolayiv, *Ukraine* **25 E5** 46 58N 32 0 E
Mymensingh, *Bangla.* **41 G17** 24 45N 90 24 E
Mynydd Du, *U.K.* **11 F4** 51 52N 3 50W
Mýrdalsjökull, *Iceland* **8 E4** 63 40N 19 6W
Myroodah, *Australia* **60 C3** 18 7S 124 16 E
Myrtle Beach, *U.S.A.* **77 J6** 33 42N 78 53W
Myrtle Creek, *U.S.A.* **82 E2** 43 1N 123 17W
Myrtle Point, *U.S.A.* **82 E1** 43 4N 124 8W
Myrtou, *Cyprus* **23 D12** 35 18N 33 4 E
Mysia, *Turkey* **21 E12** 39 50N 27 0 E
Mysore = Karnataka □, *India* **40 N10** 13 15N 77 0 E
Mysore, *India* **40 N10** 12 17N 76 41 E
Mystic, *U.S.A.* **79 E13** 41 21N 71 58W
Myszków, *Poland* **17 C10** 50 45N 19 22 E
Mytishchi, *Russia* **24 C6** 55 50N 37 50 E
Myton, *U.S.A.* **82 F8** 40 12N 110 4W
Mývatn, *Iceland* **8 D5** 65 36N 17 0W
Mzimba, *Malawi* **55 E3** 11 55S 33 39 E
Mzimkulu →, *S. Africa* **57 E5** 30 44S 30 28 E
Mzimvubu →, *S. Africa* **57 E4** 31 38S 29 33 E
Mzuzu, *Malawi* **55 E3** 11 30S 33 55 E

N

Na Hearadh = Harris, *U.K.* **12 D2** 57 50N 6 55W
Na Noi, *Thailand* **38 C3** 18 19N 100 43 E
Na Phao, *Laos* **38 D5** 17 35N 105 44 E
Na Sam, *Vietnam* **38 A6** 22 3N 106 37 E
Na San, *Vietnam* **38 B5** 21 12N 104 2 E
Naab →, *Germany* **16 D6** 49 1N 12 2 E
Naantali, *Finland* **9 F19** 60 29N 22 2 E
Naas, *Ireland* **13 C5** 53 12N 6 40W
Nababiep, *S. Africa* **56 D2** 29 36S 17 46 E
Nabadwip = Navadwip, *India* **43 H13** 23 34N 88 20 E
Nabari, *Japan* **31 G8** 34 37N 136 5 E
Nabawa, *Australia* **61 E1** 28 30S 114 48 E
Nabberu, L., *Australia* **61 E3** 25 50S 120 30 E
Naberezhnyye Chelny, *Russia* **24 C9** 55 42N 52 19 E
Nabeul, *Tunisia* **51 A7** 36 30N 10 44 E
Nabha, *India* **42 D7** 30 26N 76 14 E
Nabīd, *Iran* **45 D8** 29 40N 57 38 E
Nabire, *Indonesia* **37 E9** 3 15S 135 26 E
Nabisar, *Pakistan* **42 G3** 25 8N 69 40 E
Nabisipi →, *Canada* **71 B7** 50 14N 62 13W
Nabiswera, *Uganda* **54 B3** 1 27N 32 15 E
Nablus = Nābulus, *West Bank* **47 C4** 32 14N 35 15 E
Naboomspruit, *S. Africa* **57 C4** 24 32S 28 40 E
Nābulus, *West Bank* **47 C4** 32 14N 35 15 E
Nacala, *Mozam.* **55 E5** 14 31S 40 34 E
Nacala-Velha, *Mozam.* **55 E5** 14 32S 40 34 E
Nacaome, *Honduras* **88 D2** 13 31N 87 30W
Nacaroa, *Mozam.* **55 E4** 14 22S 39 56 E
Naches, *U.S.A.* **82 C3** 46 44N 120 42W
Naches →, *U.S.A.* **84 D6** 46 38N 120 31W
Nachingwea, *Tanzania* **55 E4** 10 23S 38 49 E
Nachingwea □, *Tanzania* **55 E4** 10 30S 38 30 E
Nachna, *India* **42 F4** 27 34N 71 41 E
Nacimiento Reservoir, *U.S.A.* **84 K6** 35 46N 120 53W
Nackara, *Australia* **63 E2** 32 48S 139 12 E
Naco, *Mexico* **86 A3** 31 20N 109 56W
Naco, *U.S.A.* **83 L9** 31 20N 109 57W
Nacogdoches, *U.S.A.* **81 K7** 31 36N 94 39W
Nácori Chico, *Mexico* **86 B3** 29 39N 109 1W
Nacozari, *Mexico* **86 A3** 30 24N 109 39W
Nadiad, *India* **42 H5** 22 41N 72 56 E
Nadur, *Malta* **23 C1** 36 2N 14 17 E
Nadūshan, *Iran* **45 C7** 32 2N 53 35 E
Nadvirna, *Ukraine* **17 D13** 48 37N 24 30 E
Nadvoitsy, *Russia* **24 B5** 63 52N 34 14 E
Nadvornaya = Nadvirna, *Ukraine* **17 D13** 48 37N 24 30 E
Nadym, *Russia* **26 C8** 65 35N 72 42 E
Nadym →, *Russia* **26 C8** 66 12N 72 0 E
Nærbø, *Norway* **9 G11** 58 40N 5 39 E
Næstved, *Denmark* **9 J14** 55 13N 11 44 E
Nafada, *Nigeria* **50 F7** 11 8N 11 20 E
Naftshahr, *Iran* **44 C5** 34 0N 45 30 E
Nafud Desert = An Nafūd, *Si. Arabia* **44 D4** 28 15N 41 0 E
Naga, *Phil.* **37 B6** 13 38N 123 15 E
Nagagami →, *Canada* **70 C3** 49 40N 84 40W
Nagahama, *Japan* **31 G8** 35 23N 136 16 E
Nagai, *Japan* **30 E10** 38 6N 140 2 E
Nagaland □, *India* **41 F19** 26 0N 94 30 E
Nagano, *Japan* **31 F9** 36 40N 138 10 E
Nagano □, *Japan* **31 F9** 36 15N 138 0 E
Nagaoka, *Japan* **31 F9** 37 27N 138 51 E
Nagappattinam, *India* **40 P11** 10 46N 79 51 E
Nagar Parkar, *Pakistan* **42 G4** 24 28N 70 46 E
Nagasaki, *Japan* **31 H4** 32 47N 129 50 E
Nagasaki □, *Japan* **31 H4** 32 50N 129 40 E
Nagato, *Japan* **31 G5** 34 19N 131 5 E
Nagaur, *India* **42 F5** 27 15N 73 45 E
Nagercoil, *India* **40 Q10** 8 12N 77 26 E
Nagina, *India* **43 E8** 29 30N 78 30 E
Nagineh, *Iran* **45 C8** 34 20N 57 15 E
Nagir, *Pakistan* **43 A6** 36 12N 74 42 E
Nagoorin, *Australia* **62 C5** 24 17S 151 15 E
Nagornyy, *Russia* **27 D13** 55 58N 124 57 E
Nagoya, *Japan* **31 G8** 35 10N 136 50 E
Nagpur, *India* **40 J11** 21 8N 79 10 E
Nagua, *Dom. Rep.* **89 C6** 19 23N 69 50W
Nagykanizsa, *Hungary* **17 E9** 46 28N 17 0 E
Nagykörös, *Hungary* **17 E10** 47 5N 19 48 E
Naha, *Japan* **31 L3** 26 13N 127 42 E
Nahanni Butte, *Canada* **72 A4** 61 2N 123 31W
Nahanni Nat. Park, *Canada* **72 A3** 61 15N 125 0W
Nahariyya, *Israel* **44 C2** 33 1N 35 5 E
Nahāvand, *Iran* **45 C6** 34 10N 48 22 E
Nahlin, *Canada* **72 B2** 58 55N 131 38W
Naicá, *Mexico* **86 B3** 27 53N 105 31W
Naicam, *Canada* **73 C8** 52 30N 104 30W
Nā'ifah, *Si. Arabia* **46 D5** 19 59N 50 46 E
Nain, *Canada* **71 A7** 56 34N 61 40W
Nā'īn, *Iran* **45 C7** 32 54N 53 0 E
Naini Tal, *India* **43 E8** 29 30N 79 30 E
Nainpur, *India* **40 H12** 22 30N 80 10 E
Naira, *Indonesia* **37 E7** 4 28S 130 0 E
Nairn, *U.K.* **12 D5** 57 35N 3 53W
Nairobi, *Kenya* **54 C4** 1 17S 36 48 E
Naissaar, *Estonia* **9 G21** 59 34N 24 29 E
Naivasha, *Kenya* **54 C4** 0 40S 36 30 E
Naivasha, L., *Kenya* **54 C4** 0 48S 36 20 E
Najafābād, *Iran* **45 C6** 32 40N 51 15 E
Najibabad, *India* **42 E8** 29 40N 78 20 E
Najin, *N. Korea* **35 C16** 42 12N 130 15 E
Najmah, *Si. Arabia* **45 E6** 26 42N 50 6 E
Naju, *S. Korea* **35 G14** 35 3N 126 43 E
Nakadōri-Shima, *Japan* **31 H4** 32 57N 129 4 E
Nakalagba, *Zaïre* **54 B2** 2 50N 27 58 E
Nakaminato, *Japan* **31 F10** 36 21N 140 36 E
Nakamura, *Japan* **31 H6** 32 59N 132 56 E
Nakano, *Japan* **31 F9** 36 45N 138 22 E
Nakano-Shima, *Japan* **31 K4** 29 51N 129 52 E
Nakashibetsu, *Japan* **30 C12** 43 33N 144 59 E
Nakfa, *Eritrea* **51 E12** 16 40N 38 32 E
Nakhichevan = Naxçıvan, *Azerbaijan* **25 G8** 39 12N 45 15 E
Nakhichevan Republic □ = Naxçıvan □, *Azerbaijan* **25 G8** 39 25N 45 26 E
Nakhl, *Egypt* **47 F2** 29 55N 33 43 E
Nakhl-e Taqī, *Iran* **45 E7** 27 28N 52 36 E
Nakhodka, *Russia* **27 E14** 42 53N 132 54 E
Nakhon Nayok, *Thailand* **38 E3** 14 12N 101 13 E
Nakhon Pathom, *Thailand* **38 F3** 13 49N 100 3 E
Nakhon Phanom, *Thailand* **38 D5** 17 23N 104 43 E
Nakhon Ratchasima, *Thailand* **38 E4** 14 59N 102 12 E
Nakhon Sawan, *Thailand* **38 E3** 15 35N 100 10 E
Nakhon Si Thammarat, *Thailand* **39 H3** 8 29N 100 0 E
Nakhon Thai, *Thailand* **38 D3** 17 5N 100 44 E
Nakina, *B.C., Canada* **72 B2** 59 12N 132 52W
Nakina, *Ont., Canada* **70 B2** 50 10N 86 40W
Nakodar, *India* **42 D6** 31 8N 75 31 E
Nakskov, *Denmark* **9 J14** 54 50N 11 8 E
Naktong →, *S. Korea* **35 G15** 35 7N 128 57 E
Nakuru, *Kenya* **54 C4** 0 15S 36 4 E
Nakuru □, *Kenya* **54 C4** 0 15S 35 5 E
Nakuru, L., *Kenya* **54 C4** 0 23S 36 5 E
Nakusp, *Canada* **72 C5** 50 20N 117 45W
Nal →, *Pakistan* **42 G1** 25 20N 65 30 E
Nalchik, *Russia* **25 F7** 43 30N 43 33 E
Nalgonda, *India* **40 L11** 17 6N 79 15 E
Nalhati, *India* **43 G12** 24 17N 87 52 E
Nallamalai Hills, *India* **40 M11** 15 30N 78 50 E
Nālūt, *Libya* **51 B7** 31 54N 11 0 E
Nam Can, *Vietnam* **39 H5** 8 46N 104 59 E
Nam Co, *China* **32 C4** 30 30N 90 45 E
Nam Dinh, *Vietnam* **38 B6** 20 25N 106 5 E
Nam Du, Hon, *Vietnam* **39 H5** 9 41N 104 21 E
Nam Ngum Dam, *Laos* **38 C4** 18 35N 102 34 E
Nam-Phan, *Vietnam* **39 G6** 10 30N 106 0 E
Nam Phong, *Thailand* **38 D4** 16 42N 102 52 E
Nam Tha, *Laos* **38 B3** 20 58N 101 30 E
Nam Tok, *Thailand* **38 E2** 14 21N 99 4 E
Namacunde, *Angola* **56 B2** 17 18S 15 50 E
Namacurra, *Mozam.* **57 B6** 17 30S 36 50 E
Namak, Daryācheh-ye, *Iran* **45 C7** 34 30N 52 0 E
Namak, Kavir-e, *Iran* **45 C8** 34 30N 57 30 E
Namaland, *Namibia* **56 C2** 24 30S 17 0 E
Namangan, *Uzbekistan* **26 E8** 41 0N 71 40 E
Namapa, *Mozam.* **55 E4** 13 43S 39 50 E
Namaqualand, *S. Africa* **56 D2** 30 0S 17 25 E
Namasagali, *Uganda* **54 B3** 1 2N 33 0 E
Namber, *Indonesia* **37 E8** 1 2S 134 49 E
Nambour, *Australia* **63 D5** 26 32S 152 58 E
Nambucca Heads, *Australia* **63 E5** 30 37S 153 0 E
Namcha Barwa, *China* **32 D4** 29 40N 95 10 E
Namche Bazar, *Nepal* **43 F12** 27 51N 86 47 E
Namchonjŏm, *N. Korea* **35 E14** 38 15N 126 26 E
Namecunda, *Mozam.* **55 E4** 14 54S 37 37 E
Nameh, *Indonesia* **36 D5** 2 34N 116 21 E
Nameponda, *Mozam.* **55 F4** 15 50S 39 50 E
Nametil, *Mozam.* **55 F4** 15 40S 39 21 E
Namew L., *Canada* **73 C8** 54 14N 101 56W
Namib Desert = Namibwoestyn, *Namibia* **56 C2** 22 30S 15 0 E
Namibe, *Angola* **53 H2** 15 7S 12 11 E
Namibe □, *Angola* **56 B1** 16 35S 12 30 E
Namibia ■, *Africa* **56 C2** 22 0S 18 9 E
Namibwoestyn, *Namibia* **56 C2** 22 30S 15 0 E
Namlea, *Indonesia* **37 E7** 3 18S 127 5 E
Namoi →, *Australia* **63 E4** 30 12S 149 30 E
Nampa, *U.S.A.* **82 E5** 43 34N 116 34W
Nampō-Shotō, *Japan* **31 J10** 32 0N 140 0 E
Nampula, *Mozam.* **55 F4** 15 6S 39 15 E
Namrole, *Indonesia* **37 E7** 3 46S 126 46 E
Namse Shankou, *China* **41 E13** 30 0N 82 25 E
Namsen →, *Norway* **8 D14** 64 28N 11 37 E
Namsos, *Norway* **8 D14** 64 29N 11 30 E
Namtsy, *Russia* **27 C13** 62 43N 129 37 E
Namtu, *Burma* **41 H20** 23 5N 97 28 E
Namtumbo, *Tanzania* **55 E4** 10 30S 36 4 E
Namu, *Canada* **72 C3** 51 52N 127 50W
Namur, *Belgium* **15 D4** 50 27N 4 52 E
Namur □, *Belgium* **15 D4** 50 17N 5 0 E
Namutoni, *Namibia* **56 B2** 18 49S 16 55 E
Namwala, *Zambia* **55 F2** 15 44S 26 30 E

Namwŏn, *S. Korea* ..... **35 G14** 35 23N 127 23 E
Nan, *Thailand* .... **38 C3** 18 48N 100 46 E
Nan →, *Thailand* .... **38 E3** 15 42N 100 9 E
Nanaimo, *Canada* **72 D4** 49 10N 124 0W
Nanam, *N. Korea* .... **35 D15** 41 44N 129 40 E
Nanango, *Australia* ..... **63 D5** 26 40S 152 0 E
Nanao, *Japan* .... **31 F8** 37 0N 137 0 E
Nanchang, *China* .... **33 D6** 28 42N 115 55 E
Nanching = Nanjing, *China* .... **33 C6** 32 2N 118 47 E
Nanchong, *China* .... **32 C5** 30 43N 106 2 E
Nancy, *France* .... **18 B7** 48 42N 6 12 E
Nanda Devi, *India* ...... **43 D8** 30 23N 79 59 E
Nandan, *Japan* .... **31 G7** 34 10N 134 42 E
Nanded, *India* .... **40 K10** 19 10N 77 20 E
Nandewar Ra., *Australia* . **63 E5** 30 15S 150 35 E
Nandi, *Fiji* .... **59 C7** 17 42S 177 20 E
Nandi □, *Kenya* .... **54 B4** 0 15N 35 0 E
Nandurbar, *India* .... **40 J9** 21 20N 74 15 E
Nandyal, *India* .... **40 M11** 15 30N 78 30 E
Nanga, *Australia* .... **61 E1** 26 7S 113 45 E
Nanga-Eboko, *Cameroon* **52 D2** 4 41N 12 22 E
Nanga Parbat, *Pakistan* .. **43 B6** 35 10N 74 35 E
Nangade, *Mozam.* ...... **55 E4** 11 5S 39 36 E
Nangapinoh, *Indonesia* .. **36 E4** 0 20S 111 44 E
Nangarhár □, *Afghan.* .... **40 B7** 34 20N 70 0 E
Nangatayap, *Indonesia* .. **36 E4** 1 32S 110 34 E
Nangeya Mts., *Uganda* .. **54 B3** 3 30N 33 30 E
Nangong, *China* .... **34 F8** 37 23N 115 22 E
Nanhuang, *China* .... **35 F11** 36 58N 121 48 E
Nanjeko, *Zambia* .... **55 F1** 15 31S 23 30 E
Nanjing, *China* .... **33 C6** 32 2N 118 47 E
Nanjirinji, *Tanzania* ..... **55 D4** 9 41S 39 5 E
Nankana Sahib, *Pakistan* **42 D5** 31 27N 73 38 E
Nanking = Nanjing, *China* **33 C6** 32 2N 118 47 E
Nankoku, *Japan* .... **31 H6** 33 39N 133 44 E
Nanning, *China* .... **32 D5** 22 48N 108 20 E
Nannup, *Australia* ...... **61 F2** 33 59S 115 48 E
Nanpara, *India* .... **43 F9** 27 52N 81 33 E
Nanpi, *China* .... **34 E9** 38 2N 116 45 E
Nanping, *China* .... **33 D6** 26 38N 118 10 E
Nanripe, *Mozam.* .... **55 E4** 13 52S 38 52 E
Nansei-Shotō = Ryūkyū-rettō, *Japan* .... **31 M2** 26 0N 126 0 E
Nansen Sd., *Canada* **4 A3** 81 0N 91 0W
Nansio, *Tanzania* .... **54 C3** 2 3S 33 4 E
Nantes, *France* **18 C3** 47 12N 1 33W
Nanticoke, *U.S.A.* **79 E8** 41 12N 76 0W
Nanton, *Canada* **72 C6** 50 21N 113 46W
Nantong, *China* .... **33 C7** 32 1N 120 52 E
Nantucket I., *U.S.A.* **66 E12** 41 16N 70 5W
Nanuque, *Brazil* **93 G10** 17 50S 40 21W
Nanusa, Kepulauan, *Indonesia* .... **37 D7** 4 45N 127 1 E
Nanutarra, *Australia* .... **60 D2** 22 32S 115 30 E
Nanyang, *China* .... **34 H7** 33 11N 112 30 E
Nanyuan, *China* .... **34 E9** 39 44N 116 22 E
Nanyuki, *Kenya* .... **54 B4** 0 2N 37 4 E
Nao, C. de la, *Spain* ..... **19 C6** 38 44N 0 14 E
Naococane L., *Canada* ... **71 B5** 52 50N 70 45W
Naoetsu, *Japan* .... **31 F9** 37 12N 138 10 E
Napa, *U.S.A.* **84 G4** 38 18N 122 17W
Napa →, *U.S.A.* **84 G4** 38 10N 122 19W
Napanee, *Canada* **70 D4** 44 15N 77 0W
Napanoch, *U.S.A.* **79 E10** 41 44N 74 22W
Nape, *Laos* .... **38 C5** 18 18N 105 6 E
Nape Pass = Keo Neua, Deo, *Vietnam* .... **38 C5** 18 23N 105 10 E
Napier, *N.Z.* .... **59 H6** 39 30S 176 56 E
Napier Broome B., *Australia* .... **60 B4** 14 2S 126 37 E
Napier Downs, *Australia* . **60 C3** 17 11S 124 36 E
Napier Pen., *Australia* ... **62 A2** 12 4S 135 43 E
Naples = Nápoli, *Italy* ... **20 D6** 40 50N 14 15 E
Naples, *U.S.A.* **77 M5** 26 8N 81 48W
Napo →, *Peru* **92 D4** 3 20S 72 40W
Napoleon, *N. Dak., U.S.A.* **80 B5** 46 30N 99 46W
Napoleon, *Ohio, U.S.A.* .. **76 E3** 41 23N 84 8W
Nápoli, *Italy* .... **20 D6** 40 50N 14 15 E
Napopo, *Zaïre* .... **54 B2** 4 15N 28 0 E
Nappa Merrie, *Australia* . **63 D3** 27 36S 141 7 E
Naqqāsh, *Iran* .... **45 C6** 35 40N 49 6 E
Nara, *Japan* .... **31 G7** 34 40N 135 49 E
Nara, *Mali* **50 E3** 15 10N 7 20W
Nara □, *Japan* .... **31 G8** 34 30N 136 0 E
Nara Canal, *Pakistan* .... **42 G3** 24 30N 69 20 E
Nara Visa, *U.S.A.* **81 H3** 35 37N 103 6W
Naracoorte, *Australia* ... **63 F3** 36 58S 140 45 E
Naradhan, *Australia* .... **63 E4** 33 34S 146 17 E
Narasapur, *India* .... **41 L12** 16 26N 81 40 E
Narathiwat, *Thailand* .... **39 J3** 6 30N 101 48 E
Narayanganj, *Bangla.* ... **41 H17** 23 40N 90 33 E
Narayanpet, *India* ...... **40 L10** 16 45N 77 30 E
Narbonne, *France* ...... **18 E5** 43 11N 3 0 E
Nardīn, *Iran* .... **45 B7** 37 3N 55 59 E
Nardò, *Italy* .... **21 D8** 40 11N 18 2 E
Narembeen, *Australia* ... **61 F2** 32 7S 118 24 E
Nares Str., *Arctic* **66 A13** 80 0N 70 0W
Naretha, *Australia* ...... **61 F3** 31 0S 124 45 E
Narew →, *Poland* ...... **17 B11** 52 26N 20 41 E
Nari →, *Pakistan* ...... **42 E2** 28 0N 67 40 E
Narin, *Afghan.* .... **40 A6** 36 5N 69 0 E
Narindra, Helodranon' i, *Madag.* .... **57 A8** 14 55S 47 30 E
Narita, *Japan* .... **31 G10** 35 47N 140 19 E
Narmada →, *India* ..... **42 J5** 21 38N 72 36 E
Narmland, *Sweden* ..... **9 F15** 60 0N 13 30 E
Narnaul, *India* .... **42 E7** 28 5N 76 11 E
Narodnaya, *Russia* ..... **24 A10** 65 5N 59 58 E
Narok, *Kenya* .... **54 C4** 1 55S 35 52 E
Narok □, *Kenya* .... **54 C4** 1 20S 36 30 E
Narooma, *Australia* ..... **63 F5** 36 14S 150 4 E
Narowal, *Pakistan* ...... **42 C6** 32 6N 74 52 E
Narrabri, *Australia* ..... **63 E4** 30 19S 149 46 E
Narran →, *Australia* ... **63 D4** 28 37S 148 12 E
Narrandera, *Australia* ... **63 E4** 34 42S 146 31 E
Narraway →, *Canada* .. **72 B5** 55 44N 119 55W
Narrogin, *Australia* ..... **61 F2** 32 58S 117 14 E
Narromine, *Australia* ... **63 E4** 32 12S 148 12 E
Narsimhapur, *India* ..... **43 H8** 22 54N 79 14 E
Naruto, *Japan* .... **31 G7** 34 11N 134 37 E
Narva, *Estonia* .... **24 C4** 59 23N 28 12 E
Narva →, *Russia* ...... **9 G22** 59 27N 28 2 E
Narvik, *Norway* .... **8 B17** 68 28N 17 26 E
Narwana, *India* .... **42 E7** 29 39N 76 6 E
Naryan-Mar, *Russia* ..... **24 A9** 67 42N 53 12 E
Naryilco, *Australia* ...... **63 D3** 28 37S 141 53 E
Narym, *Russia* .... **26 D9** 59 0N 81 30 E
Narymskoye, *Kazakstan* . **26 E9** 49 10N 84 15 E
Naryn, *Kyrgyzstan* ...... **26 E8** 41 26N 75 58 E
Nasa, *Norway* .... **8 C16** 66 29N 15 23 E
Nasarawa, *Nigeria* ...... **50 G6** 8 32N 7 41 E
Naseby, *N.Z.* .... **59 L3** 45 1S 170 10 E
Naselle, *U.S.A.* **84 D3** 46 22N 123 49W
Naser, Buheirat en, *Egypt* **51 D11** 23 0N 32 30 E
Nashua, *Iowa, U.S.A.* ... **80 D8** 42 57N 92 32W
Nashua, *Mont., U.S.A.* ... **82 B10** 48 8N 106 22W
Nashua, *N.H., U.S.A.* ... **79 D13** 42 45N 71 28W
Nashville, *Ark., U.S.A.* ... **81 J8** 33 57N 93 51W
Nashville, *Ga., U.S.A.* ... **77 K4** 31 12N 83 15W
Nashville, *Tenn., U.S.A.* . **77 G2** 36 10N 86 47W
Nasik, *India* .... **40 K8** 19 58N 73 50 E
Nasirabad, *India* .... **42 F6** 26 15N 74 45 E
Naskaupi →, *Canada* ... **71 B7** 53 47N 60 51W
Naṣriān-e Pā'īn, *Iran* .... **44 C5** 32 52N 46 52 E
Nass →, *Canada* **72 B3** 55 0N 129 40W
Nassau, *Bahamas* **88 A4** 25 5N 77 20W
Nassau, *U.S.A.* **79 D11** 42 31N 73 37W
Nassau, B., *Chile* **96 H3** 55 20S 68 0W
Nasser, L. = Naser, Buheirat en, *Egypt* .... **51 D11** 23 0N 32 30 E
Nässjö, *Sweden* .... **9 H16** 57 39N 14 42 E
Nat Kyizin, *Burma* ...... **41 M20** 14 57N 97 59 E
Nata, *Botswana* .... **56 C4** 20 12S 26 12 E
Natagaima, *Colombia* ... **92 C3** 3 37N 75 6W
Natal, *Brazil* **93 E11** 5 47S 35 13W
Natal, *Canada* **72 D6** 49 43N 114 51W
Natal, *Indonesia* .... **36 D1** 0 35N 99 7 E
Naṭanz, *Iran* .... **45 C6** 33 30N 51 55 E
Natashquan, *Canada* **71 B7** 50 14N 61 46W
Natashquan →, *Canada* **71 B7** 50 7N 61 50W
Natchez, *U.S.A.* **81 K9** 31 34N 91 24W
Natchitoches, *U.S.A.* **81 K8** 31 46N 93 5W
Nathalia, *Australia* ...... **63 F4** 36 1S 145 13 E
Nathdwara, *India* .... **42 G5** 24 55N 73 50 E
Nati, Pta., *Spain* .... **22 A10** 40 3N 3 50 E
Natimuk, *Australia* ...... **63 F3** 36 42S 142 0 E
Nation →, *Canada* **72 B4** 55 30N 123 32W
National City, *U.S.A.* **85 N9** 32 41N 117 6W
Natitingou, *Benin* .... **50 F5** 10 20N 1 26 E
Natividad, I., *Mexico* **86 B1** 27 50N 115 10W
Natoma, *U.S.A.* **80 F5** 39 11N 99 2W
Natron, L., *Tanzania* .... **54 C4** 2 20S 36 0 E
Natrona Heights, *U.S.A.* . **78 F5** 40 37N 79 44W
Natuna Besar, Kepulauan, *Indonesia* .... **39 L7** 4 0N 108 15 E
Natuna Is. = Natuna Besar, Kepulauan, *Indonesia* .... **39 L7** 4 0N 108 15 E
Natuna Selatan, Kepulauan, *Indonesia* . **39 L7** 2 45N 109 0 E
Natural Bridge, *U.S.A.* ... **79 B9** 44 5N 75 30W
Naturaliste, C., *Australia* . **62 G4** 40 50S 148 15 E
Nau Qala, *Afghan.* ...... **42 B3** 34 5N 68 5 E
Naubinway, *U.S.A.* **70 C2** 46 6N 85 27W
Naugatuck, *U.S.A.* **79 E11** 41 30N 73 3W
Naumburg, *Germany* .... **16 C6** 51 9N 11 47 E
Nā'ūr at Tunayb, *Jordan* . **47 D4** 31 48N 35 57 E
Nauru ■, *Pac. Oc.* ...... **64 H8** 1 0S 166 0 E
Naushahra = Nowshera, *Pakistan* .... **40 B8** 34 0N 72 0 E
Nauta, *Peru* **92 D4** 4 31S 73 35W
Nautanwa, *India* .... **41 F13** 27 20N 83 25 E
Nautla, *Mexico* **87 C5** 20 20N 96 50W
Nava, *Mexico* **86 B4** 28 25N 100 46W
Navadwip, *India* .... **43 H13** 23 34N 88 20 E
Navahrudak, *Belarus* .... **17 B13** 53 40N 25 50 E
Navajo Reservoir, *U.S.A.* **83 H10** 36 48N 107 36W
Navalmoral de la Mata, *Spain* **19 C3** 39 52N 5 33W
Navan = An Uaimh, *Ireland* **13 C5** 53 39N 6 41W
Navarino, I., *Chile* **96 H3** 55 0S 67 40W
Navarra □, *Spain* **19 A5** 42 40N 1 40W
Navarre, *U.S.A.* **78 F3** 40 43N 81 31W
Navarro →, *U.S.A.* **84 F3** 39 11N 123 45W
Navasota, *U.S.A.* **81 K6** 30 23N 96 5W
Navassa, *W. Indies* **89 C4** 18 30N 75 0W
Naver →, *U.K.* **12 C4** 58 32N 4 14W
Navidad, *Chile* **94 C1** 33 57S 71 50W
Năvodari, *Romania* ...... **17 F15** 44 19N 28 36 E
Navoi = Nawoiy, *Uzbekistan* .... **26 E7** 40 9N 65 22 E
Navojoa, *Mexico* **86 B3** 27 0N 109 30W
Navolato, *Mexico* **86 C3** 24 47N 107 42W
Návpaktos, *Greece* ..... **21 E9** 38 23N 21 50 E
Návplion, *Greece* .... **21 F10** 37 33N 22 50 E
Navsari, *India* .... **40 J8** 20 57N 72 59 E
Nawa Kot, *Pakistan* ..... **42 E4** 28 21N 71 24 E
Nawabganj, *Ut. P., India* . **43 F9** 26 56N 81 14 E
Nawabganj, *Ut. P., India* . **43 E8** 28 32N 79 40 E
Nawabshah, *Pakistan* ... **42 F3** 26 15N 68 25 E
Nawada, *India* .... **43 G11** 24 50N 85 33 E
Nawakot, *Nepal* .... **43 F11** 27 55N 85 10 E
Nawalgarh, *India* .... **42 F6** 27 50N 75 15 E
Nawanshahr, *India* ..... **43 C6** 32 33N 74 48 E
Nawoiy, *Uzbekistan* ..... **26 E7** 40 9N 65 22 E
Naxçıvan, *Azerbaijan* ... **25 G8** 39 12N 45 15 E
Naxçıvan □, *Azerbaijan* .. **25 G8** 39 25N 45 26 E
Náxos, *Greece* .... **21 F11** 37 8N 25 25 E
Nāy Band, *Iran* .... **45 E7** 27 20N 52 40 E
Nayakhan, *Russia* ...... **27 C16** 61 56N 159 0 E
Nayarit □, *Mexico* **86 C4** 22 0N 105 0W
Nayoro, *Japan* .... **30 B11** 44 21N 142 28 E
Nayyāl, W. →, *Si. Arabia* **44 D3** 28 35N 39 4 E
Nazareth = Nazerat, *Israel* **47 C4** 32 42N 35 17 E
Nazas, *Mexico* **86 B4** 25 10N 104 6W
Nazas →, *Mexico* **86 B4** 25 35N 103 25W
Naze, The, *U.K.* .... **11 F9** 51 53N 1 18 E
Nazerat, *Israel* .... **47 C4** 32 42N 35 17 E
Nāzik, *Iran* .... **44 B5** 39 1N 45 4 E
Nazilli, *Turkey* .... **21 F13** 37 55N 28 15 E
Nazir Hat, *Bangla.* ..... **41 H17** 22 35N 91 49 E
Nazko, *Canada* **72 C4** 53 1N 123 37W
Nazko →, *Canada* **72 C4** 53 7N 123 34W
Nchanga, *Zambia* .... **55 E2** 12 30S 27 49 E
Ncheu, *Malawi* .... **55 E3** 14 50S 34 47 E
Ndala, *Tanzania* .... **54 C3** 4 45S 33 15 E
Ndalatando, *Angola* ..... **52 F2** 9 12S 14 48 E
Ndareda, *Tanzania* ...... **54 C4** 4 12S 35 30 E
Ndélé, *C.A.R.* .... **51 G9** 8 25N 20 36 E
Ndendé, *Gabon* .... **52 E2** 2 22S 11 23 E
Ndjamena, *Chad* .... **51 F7** 12 10N 14 59 E
Ndjolé, *Gabon* .... **52 E2** 0 10S 10 45 E
Ndola, *Zambia* .... **55 E2** 13 0S 28 34 E
Ndoto Mts., *Kenya* ...... **54 B4** 2 0N 37 0 E
Nduguti, *Tanzania* ...... **54 C3** 4 18S 34 41 E
Neagh, Lough, *U.K.* **13 B5** 54 37N 6 25W
Neah Bay, *U.S.A.* **84 B2** 48 22N 124 37W
Neale, L., *Australia* ..... **60 D5** 24 15S 130 0 E
Neápolis, *Greece* .... **23 D7** 35 15N 25 37 E
Near Is., *U.S.A.* .... **68 C1** 53 0N 172 0 E
Neath, *U.K.* **11 F4** 51 39N 3 48W
Neath Port Talbot □, *U.K.* **11 F4** 51 42N 3 45W
Nebine Cr. →, *Australia* **63 D4** 29 27S 146 56 E
Nebitdag, *Turkmenistan* . **25 G9** 39 30N 54 22 E
Nebraska □, *U.S.A.* **80 E5** 41 30N 99 30W
Nebraska City, *U.S.A.* ... **80 E7** 40 41N 95 52W
Nébrodi, Monti, *Italy* .... **20 F6** 37 54N 14 35 E
Necedah, *U.S.A.* **80 C9** 44 2N 90 4W
Nechako →, *Canada* ... **72 C4** 53 30N 122 44W
Neches →, *U.S.A.* **81 L8** 29 58N 93 51W
Neckar →, *Germany* .... **16 D5** 49 27N 8 29 E
Necochea, *Argentina* **94 D4** 38 30S 58 50W
Needles, *U.S.A.* **85 L12** 34 51N 114 37W
Needles, The, *U.K.* **11 G6** 50 39N 1 35W
Ñeembucú □, *Paraguay* . **94 B4** 27 0S 58 0W
Neemuch = Nimach, *India* **42 G6** 24 30N 74 56 E
Neenah, *U.S.A.* **76 C1** 44 11N 88 28W
Neepawa, *Canada* **73 C9** 50 15N 99 30W
Nefta, *Tunisia* .... **50 B6** 33 53N 7 50 E
Neftçala, *Azerbaijan* .... **25 G8** 39 19N 49 12 E
Neftyannyye Kamni, *Azerbaijan* .... **25 F9** 40 20N 50 55 E
Negapatam = Nagappattinam, *India* . **40 P11** 10 46N 79 51 E
Negaunee, *U.S.A.* **76 B2** 46 30N 87 36W
Negele, *Ethiopia* .... **46 F2** 5 20N 39 36 E
Negev Desert = Hanegev, *Israel* .... **47 E3** 30 50N 35 0 E
Negombo, *Sri Lanka* .... **40 R11** 7 12N 79 50 E
Negotin, *Serbia, Yug.* ... **21 B10** 44 16N 22 37 E
Negra, Pta., *Peru* **90 D2** 6 6S 81 10W
Negra Pt., *Phil.* .... **37 A6** 18 40N 120 50 E
Negrais, C. = Maudin Sun, *Burma* .... **41 M19** 16 0N 94 30 E
Negro →, *Argentina* ... **96 E4** 41 2S 62 47W
Negro →, *Brazil* **92 D6** 3 0S 60 0W
Negro →, *Uruguay* **95 C4** 33 24S 58 22W
Negros, *Phil.* .... **37 C6** 9 30N 122 40 E
Nehalem →, *U.S.A.* ... **84 E3** 45 40N 123 56W
Nehāvand, *Iran* .... **45 C6** 35 56N 49 31 E
Nehbandān, *Iran* .... **45 D9** 31 35N 60 5 E
Nei Monggol Zizhiqu □, *China* .... **34 C6** 42 0N 112 0 E
Neidpath, *Canada* **73 C7** 50 12N 107 20W
Neihart, *U.S.A.* **82 C8** 47 0N 110 44W
Neijiang, *China* .... **32 D5** 29 35N 104 55 E
Neilton, *U.S.A.* **82 C2** 47 25N 123 53W
Neiqiu, *China* .... **34 F8** 37 15N 114 30 E
Neiva, *Colombia* **92 C3** 2 56N 75 18W
Neixiang, *China* .... **34 H6** 33 10N 111 52 E
Nejanilini L., *Canada* **73 B9** 59 33N 97 48W
Nekā, *Iran* .... **45 B7** 36 39N 53 19 E
Nekemte, *Ethiopia* ...... **51 G12** 9 4N 36 30 E
Neksø, *Denmark* .... **9 J16** 55 4N 15 8 E
Nelia, *Australia* .... **62 C3** 20 39S 142 12 E
Neligh, *U.S.A.* **80 D5** 42 8N 98 2W
Nelkan, *Russia* .... **27 D14** 57 40N 136 4 E
Nellore, *India* .... **40 M11** 14 27N 79 59 E
Nelma, *Russia* .... **27 E14** 47 39N 139 0 E
Nelson, *Canada* **72 D5** 49 30N 117 20W
Nelson, *N.Z.* .... **59 J4** 41 18S 173 16 E
Nelson, *U.K.* **10 D5** 53 50N 2 13W
Nelson, *U.S.A.* **83 J7** 35 31N 113 19W
Nelson →, *Canada* **73 C9** 54 33N 98 2W
Nelson, C., *Australia* .... **63 F3** 38 26S 141 32 E
Nelson, Estrecho, *Chile* . **96 G2** 51 30S 75 0W
Nelson Forks, *Canada* ... **72 B4** 59 30N 124 0W
Nelson House, *Canada* .. **73 B9** 55 47N 98 51W
Nelson L., *Canada* **73 B8** 55 48N 100 7W
Nelspoort, *S. Africa* ..... **56 E3** 32 7S 23 0 E
Nelspruit, *S. Africa* ..... **57 D5** 25 29S 30 59 E
Néma, *Mauritania* **50 E3** 16 40N 7 15W
Neman, *Russia* .... **9 J20** 55 2N 22 2 E
Neman →, *Lithuania* ... **9 J19** 55 25N 21 10 E
Nemeiben L., *Canada* ... **73 B7** 55 20N 105 20W
Nemunas = Neman →, *Lithuania* .... **9 J19** 55 25N 21 10 E
Nemuro, *Japan* .... **30 C12** 43 20N 145 35 E
Nemuro-Kaikyō, *Japan* .. **30 C12** 43 30N 145 30 E
Nemuy, *Russia* .... **27 D14** 55 40N 136 9 E
Nen Jiang →, *China* ... **35 B13** 45 28N 124 30 E
Nenagh, *Ireland* **13 D3** 52 52N 8 11W
Nenana, *U.S.A.* **68 B5** 64 34N 149 5W
Nenasi, *Malaysia* .... **39 L4** 3 9N 103 23 E
Nene →, *U.K.* .... **10 E8** 52 49N 0 11 E
Nenjiang, *China* .... **33 B7** 49 10N 125 10 E
Neno, *Malawi* .... **55 F3** 15 25S 34 40 E
Neodesha, *U.S.A.* **81 G7** 37 25N 95 41W
Neosho, *U.S.A.* **81 G7** 36 52N 94 22W
Neosho →, *U.S.A.* **81 H7** 36 48N 95 18W
Nepal ■, *Asia* .... **43 F11** 28 0N 84 30 E
Nepalganj, *Nepal* .... **43 E9** 28 5N 81 40 E
Nephi, *U.S.A.* **82 G8** 39 43N 111 50W
Nephin, *Ireland* **13 B2** 54 1N 9 22W
Neptune, *U.S.A.* **79 F10** 40 13N 74 2W
Nerchinsk, *Russia* ...... **27 D12** 52 0N 116 39 E
Nerchinskiy Zavod, *Russia* **27 D12** 51 20N 119 40 E
Néret L., *Canada* **71 B5** 54 45N 70 44W
Neretva →, *Croatia* .... **21 C7** 43 1N 17 27 E
Neringa, *Lithuania* ...... **9 J19** 55 30N 21 5 E
Ness, L., *U.K.* **12 D4** 57 15N 4 32W
Nesterov, *Ukraine* ...... **17 C12** 50 4N 23 58 E
Nesvizh = Nyasvizh, *Belarus* .... **17 B14** 53 14N 26 38 E
Netanya, *Israel* .... **47 C3** 32 20N 34 51 E
Nète →, *Belgium* ...... **15 C4** 51 7N 4 14 E
Netherdale, *Australia* ... **62 C4** 21 10S 148 33 E
Netherlands ■, *Europe* .. **15 C5** 52 0N 5 30 E
Netherlands Antilles ■, *W. Indies* **92 A5** 12 15N 69 0W
Nettilling L., *Canada* **69 B12** 66 30N 71 0W
Netzahualcoyotl, Presa, *Mexico* **87 D6** 17 10N 93 30W
Neubrandenburg, *Germany* .... **16 B7** 53 33N 13 15 E
Neuchâtel, *Switz.* .... **16 E4** 47 0N 6 55 E
Neuchâtel, Lac de, *Switz.* **16 E4** 46 53N 6 50 E
Neufchâteau, *Belgium* ... **15 E5** 49 50N 5 25 E
Neumünster, *Germany* .. **16 A5** 54 4N 9 58 E
Neunkirchen, *Germany* .. **16 D4** 49 20N 7 9 E
Neuquén, *Argentina* **96 D3** 38 55S 68 0W
Neuquén □, *Argentina* .. **94 D2** 38 0S 69 50W
Neuruppin, *Germany* .... **16 B7** 52 55N 12 48 E
Neuse →, *U.S.A.* **77 H7** 35 6N 76 29W
Neusiedler See, *Austria* . **17 E9** 47 50N 16 47 E
Neuss, *Germany* .... **15 C6** 51 11N 6 42 E
Neustrelitz, *Germany* .... **16 B7** 53 21N 13 4 E
Neva →, *Russia* ...... **24 C5** 59 50N 30 30 E
Nevada, *U.S.A.* **81 G7** 37 51N 94 22W
Nevada □, *U.S.A.* **82 G5** 39 0N 117 0W
Nevada, Sierra, *Spain* ... **19 D4** 37 3N 3 15W
Nevada, Sierra, *U.S.A.* .. **82 G3** 39 0N 120 30W
Nevada City, *U.S.A.* **84 F6** 39 16N 121 1W
Nevado, Cerro, *Argentina* **94 D2** 35 30S 68 32W
Nevanka, *Russia* .... **27 D10** 56 31N 98 55 E
Nevers, *France* .... **18 C5** 47 0N 3 9 E
Nevertire, *Australia* ..... **63 E4** 31 50S 147 44 E
Neville, *Canada* **73 D7** 49 58N 107 39W
Nevinnomyssk, *Russia* .. **25 F7** 44 40N 42 0 E
Nevis, *W. Indies* **89 C7** 17 0N 62 30W
Nevyansk, *Russia* ...... **24 C11** 57 30N 60 13 E
New Albany, *Ind., U.S.A.* **76 F3** 38 18N 85 49W
New Albany, *Miss., U.S.A.* **81 H10** 34 29N 89 0W
New Albany, *Pa., U.S.A.* . **79 E8** 41 36N 76 27W
New Amsterdam, *Guyana* **92 B7** 6 15N 57 36W
New Angledool, *Australia* **63 D4** 29 5S 147 55 E
New Bedford, *U.S.A.* **79 E14** 41 38N 70 56W
New Bern, *U.S.A.* **77 H7** 35 7N 77 3W
New Bethlehem, *U.S.A.* . **78 E5** 41 0N 79 20W
New Bloomfield, *U.S.A.* .. **78 F7** 40 25N 77 11W
New Boston, *U.S.A.* **81 J7** 33 28N 94 25W
New Braunfels, *U.S.A.* ... **81 L5** 29 42N 98 8W
New Brighton, *N.Z.* ..... **59 K4** 43 29S 172 43 E
New Brighton, *U.S.A.* ... **78 F4** 40 42N 80 19W
New Britain, *Papua N. G.* **64 H7** 5 50S 150 20 E
New Britain, *U.S.A.* **79 E12** 41 40N 72 47W
New Brunswick, *U.S.A.* .. **79 F10** 40 30N 74 27W
New Brunswick □, *Canada* **71 C6** 46 50N 66 30W
New Caledonia ■, *Pac. Oc.* **64 K8** 21 0S 165 0 E
New Castle, *Ind., U.S.A.* . **76 F3** 39 55N 85 22W
New Castle, *Pa., U.S.A.* . **78 E4** 41 0N 80 21W
New City, *U.S.A.* **79 E11** 41 9N 73 59W
New Cumberland, *U.S.A.* . **78 F4** 40 30N 80 36W
New Cuyama, *U.S.A.* **85 L7** 34 57N 119 38W
New Delhi, *India* .... **42 E7** 28 37N 77 13 E
New Denver, *Canada* **72 D5** 50 0N 117 25W
New Don Pedro Reservoir, *U.S.A.* **84 H6** 37 43N 120 24W
New England, *U.S.A.* **80 B3** 46 32N 102 52W
New England Ra., *Australia* .... **63 E5** 30 20S 151 45 E
New Forest, *U.K.* **11 G6** 50 53N 1 34W
New Glasgow, *Canada* .. **71 C7** 45 35N 62 36W
New Guinea, *Oceania* ... **28 K17** 4 0S 136 0 E
New Hamburg, *Canada* .. **78 C4** 43 23N 80 42W
New Hampshire □, *U.S.A.* **79 C13** 44 0N 71 30W
New Hampton, *U.S.A.* ... **80 D8** 43 3N 92 19W
New Hanover, *S. Africa* .. **57 D5** 29 22S 30 31 E
New Haven, *Conn., U.S.A.* **79 E12** 41 18N 72 55W
New Haven, *Mich., U.S.A.* **78 D2** 42 44N 82 48W
New Hazelton, *Canada* .. **72 B3** 55 20N 127 30W
New Hebrides = Vanuatu ■, *Pac. Oc.* ... **64 J8** 15 0S 168 0 E
New Iberia, *U.S.A.* **81 K9** 30 1N 91 49W
New Ireland, *Papua N. G.* **64 H7** 3 20S 151 50 E
New Jersey □, *U.S.A.* ... **79 F10** 40 0N 74 30W
New Kensington, *U.S.A.* . **78 F5** 40 34N 79 46W
New Lexington, *U.S.A.* .. **76 F4** 39 43N 82 13W
New Liskeard, *Canada* ... **70 C4** 47 31N 79 41W
New London, *Conn., U.S.A.* **79 E12** 41 22N 72 6W
New London, *Minn., U.S.A.* **80 C7** 45 18N 94 56W
New London, *Ohio, U.S.A.* **78 E2** 41 5N 82 24W
New London, *Wis., U.S.A.* **80 C10** 44 23N 88 45W
New Madrid, *U.S.A.* **81 G10** 36 36N 89 32W
New Meadows, *U.S.A.* ... **82 D5** 44 58N 116 18W
New Melones L., *U.S.A.* .. **84 H6** 37 57N 120 31W
New Mexico □, *U.S.A.* .. **83 J10** 34 30N 106 0W
New Milford, *Conn., U.S.A.* **79 E11** 41 35N 73 25W
New Milford, *Pa., U.S.A.* . **79 E9** 41 52N 75 44W
New Norcia, *Australia* ... **61 F2** 30 57S 116 13 E
New Norfolk, *Australia* .. **62 G4** 42 46S 147 2 E
New Orleans, *U.S.A.* **81 K9** 29 58N 90 4W
New Philadelphia, *U.S.A.* **78 F3** 40 30N 81 27W
New Plymouth, *N.Z.* ..... **59 H5** 39 4S 174 5 E
New Plymouth, *U.S.A.* ... **82 E5** 43 58N 116 49W
New Providence, *Bahamas* **88 A4** 25 25N 78 35W
New Radnor, *U.K.* **11 E4** 52 15N 3 9W
New Richmond, *U.S.A.* .. **80 C8** 45 7N 92 32W
New Roads, *U.S.A.* **81 K9** 30 42N 91 26W
New Rochelle, *U.S.A.* ... **79 F11** 40 55N 73 47W
New Rockford, *U.S.A.* ... **80 B5** 47 41N 99 8W
New Ross, *Ireland* **13 D5** 52 23N 6 57W
New Salem, *U.S.A.* **80 B4** 46 51N 101 25W
New Scone, *U.K.* **12 E5** 56 25N 3 24W
New Siberian Is. = Novaya Sibir, Ostrov, *Russia* .... **27 B16** 75 10N 150 0 E
New Siberian Is. = Novosibirskiye Ostrova, *Russia* .... **27 B15** 75 0N 142 0 E
New Smyrna Beach, *U.S.A.* **77 L5** 29 1N 80 56W
New South Wales □, *Australia* .... **63 E4** 33 0S 146 0 E
New Springs, *Australia* .. **61 E3** 25 49S 120 1 E
New Town, *U.S.A.* **80 A3** 47 59N 102 30W
New Ulm, *U.S.A.* **80 C7** 44 19N 94 28W

New Waterford, Canada 71 C7 46 13N 60 4W
New Westminster, Canada 72 D4 49 13N 122 55W
New York □, U.S.A. 79 D9 43 0N 75 0W
New York City, U.S.A. 79 F11 40 45N 74 0W
New Zealand ■, Oceania 59 J5 40 0S 176 0 E
Newala, Tanzania 55 E4 10 58S 39 18 E
Newala □, Tanzania 55 E4 10 46S 39 20 E
Newark, Del., U.S.A. 76 F8 39 41N 75 46W
Newark, N.J., U.S.A. 79 F10 40 44N 74 10W
Newark, N.Y., U.S.A. 78 C7 43 3N 77 6W
Newark, Ohio, U.S.A. 78 F2 40 3N 82 24W
Newark-on-Trent, U.K. 10 D7 53 5N 0 48W
Newaygo, U.S.A. 76 D3 43 25N 85 48W
Newberg, U.S.A. 82 D2 45 18N 122 58W
Newberry, Mich., U.S.A. 76 B3 46 21N 85 30W
Newberry, S.C., U.S.A. 77 H5 34 17N 81 37W
Newberry Springs, U.S.A. 85 L10 34 50N 116 41W
Newbridge = Droichead Nua, Ireland 13 C5 53 11N 6 48W
Newbrook, Canada 72 C6 54 24N 112 57W
Newburgh, U.S.A. 79 E10 41 30N 74 1W
Newbury, U.K. 11 F6 51 24N 1 20W
Newbury, U.S.A. 79 B12 43 19N 72 3W
Newburyport, U.S.A. 79 D14 42 49N 70 53W
Newcastle, Australia 63 E5 33 0S 151 46 E
Newcastle, Canada 71 C6 47 1N 65 38W
Newcastle, S. Africa 57 D4 27 45S 29 58 E
Newcastle, U.K. 13 B6 54 13N 5 54W
Newcastle, Calif., U.S.A. 84 G5 38 53N 121 8W
Newcastle, Wyo., U.S.A. 80 D2 43 50N 104 11W
Newcastle Emlyn, U.K. 11 E3 52 2N 4 28W
Newcastle Ra., Australia 60 C5 15 45S 130 15 E
Newcastle-under-Lyme, U.K. 10 D5 53 1N 2 14W
Newcastle-upon-Tyne, U.K. 10 C6 54 58N 1 36W
Newcastle Waters, Australia 62 B1 17 30S 133 28 E
Newdegate, Australia 61 F2 33 6S 119 0 E
Newell, U.S.A. 80 C3 44 43N 103 25W
Newfoundland □, Canada 71 B8 53 0N 58 0W
Newfoundland I., N. Amer. 66 E14 49 0N 55 0W
Newhalem, U.S.A. 72 D4 48 40N 121 15W
Newhall, U.S.A. 85 L8 34 23N 118 32W
Newham, U.K. 11 F8 51 31N 0 3 E
Newhaven, U.K. 11 G8 50 47N 0 3 E
Newkirk, U.S.A. 81 G6 36 53N 97 3W
Newman, Australia 60 D2 23 18S 119 45 E
Newman, U.S.A. 84 H5 37 19N 121 1W
Newmarket, Canada 78 B5 44 3N 79 28W
Newmarket, Ireland 13 D3 52 13N 9 0W
Newmarket, U.K. 11 E8 52 15N 0 25 E
Newmarket, U.S.A. 79 C14 43 5N 70 56W
Newnan, U.S.A. 77 J3 33 23N 84 48W
Newport, I. of W., U.K. 11 G6 50 42N 1 17W
Newport, Newp., U.K. 11 F5 51 35N 3 0W
Newport, Ark., U.S.A. 81 H9 35 37N 91 16W
Newport, Ky., U.S.A. 76 F3 39 5N 84 30W
Newport, N.H., U.S.A. 79 C12 43 22N 72 10W
Newport, Oreg., U.S.A. 82 D1 44 39N 124 3W
Newport, Pa., U.S.A. 78 F7 40 29N 77 8W
Newport, R.I., U.S.A. 79 E13 41 29N 71 19W
Newport, Tenn., U.S.A. 77 H4 35 58N 83 11W
Newport, Vt., U.S.A. 79 B12 44 56N 72 13W
Newport, Wash., U.S.A. 82 B5 48 11N 117 3W
Newport □, U.K. 11 F4 51 33N 3 1W
Newport Beach, U.S.A. 85 M9 33 37N 117 56W
Newport News, U.S.A. 76 G7 36 59N 76 25W
Newquay, U.K. 11 G2 50 25N 5 6W
Newry, U.K. 13 B5 54 11N 6 21W
Newry & Mourne □, U.K. 13 B5 54 10N 6 15W
Newton, Iowa, U.S.A. 80 E8 41 42N 93 3W
Newton, Mass., U.S.A. 79 D13 42 21N 71 12W
Newton, Miss., U.S.A. 81 J10 32 19N 89 10W
Newton, N.C., U.S.A. 77 H5 35 40N 81 13W
Newton, N.J., U.S.A. 79 E10 41 3N 74 45W
Newton, Tex., U.S.A. 81 K8 30 51N 93 46W
Newton Abbot, U.K. 11 G4 50 32N 3 37W
Newton Boyd, Australia 63 D5 29 45S 152 16 E
Newton Stewart, U.K. 12 G4 54 57N 4 30W
Newtonmore, U.K. 12 D4 57 4N 4 8W
Newtown, U.K. 11 E4 52 31N 3 19W
Newtownabbey □, U.K. 13 B6 54 45N 6 0W
Newtownards, U.K. 13 B6 54 36N 5 42W
Newville, U.S.A. 78 F7 40 10N 77 24W
Neya, Russia 24 C7 58 21N 43 49 E
Neyrīz, Iran 45 D7 29 15N 54 19 E
Neyshābūr, Iran 45 B8 36 10N 58 50 E
Nezhin = Nizhyn, Ukraine 25 D5 51 5N 31 55 E
Nezperce, U.S.A. 82 C5 46 14N 116 14W
Ngabang, Indonesia 36 D3 0 23N 109 55 E
Ngabordamlu, Tanjung, Indonesia 37 F8 6 56S 134 11 E
Ngami Depression, Botswana 56 C3 20 30S 22 46 E
Ngamo, Zimbabwe 55 F2 19 3S 27 32 E
Nganglong Kangri, China 41 C12 33 0N 81 0 E
Nganjuk, Indonesia 37 G14 7 32S 111 55 E
Ngao, Thailand 38 C2 18 46N 99 59 E
Ngaoundéré, Cameroon 52 C2 7 15N 13 35 E
Ngapara, N.Z. 59 L3 44 57S 170 46 E
Ngara, Tanzania 54 C3 2 29S 30 40 E
Ngara □, Tanzania 54 C3 2 29S 30 40 E
Ngawi, Indonesia 37 G14 7 24S 111 26 E
Nghia Lo, Vietnam 38 B5 21 33N 104 28 E
Ngoma, Malawi 55 E3 13 8S 33 45 E
Ngomahura, Zimbabwe 55 G3 20 26S 30 43 E
Ngomba, Tanzania 55 D3 8 20S 32 53 E
Ngoring Hu, China 32 C4 34 55N 97 5 E
Ngorongoro, Tanzania 54 C4 3 11S 35 32 E
Ngozi, Burundi 54 C2 2 54S 29 50 E
Ngudu, Tanzania 54 C3 2 58S 33 25 E
Nguigmi, Niger 51 F7 14 20N 13 20 E
Ngukurr, Australia 62 A1 14 44S 134 44 E
Ngunga, Tanzania 54 C3 3 37S 33 37 E
Nguru, Nigeria 50 F7 12 56N 10 29 E
Nguru Mts., Tanzania 54 D4 6 0S 37 30 E
Nguyen Binh, Vietnam 38 A5 22 39N 105 56 E
Nha Trang, Vietnam 39 F7 12 16N 109 10 E
Nhacoongo, Mozam. 57 C6 24 18S 35 14 E
Nhamaabué, Mozam. 55 F4 17 25S 35 5 E
Nhangutazi, L., Mozam. 57 C5 24 0S 34 30 E
Nhill, Australia 63 F3 36 18S 141 40 E
Nho Quan, Vietnam 38 B5 20 18N 105 45 E
Nhulunbuy, Australia 62 A2 12 10S 137 20 E
Nia-nia, Zaïre 54 B2 1 30N 27 40 E
Niafounké, Mali 50 E4 16 0N 4 5W
Niagara, U.S.A. 76 C1 45 45N 88 0W
Niagara Falls, Canada 70 D4 43 7N 79 5W
Niagara Falls, U.S.A. 78 C6 43 5N 79 4W
Niagara-on-the-Lake, Canada 78 C5 43 15N 79 4W
Niah, Malaysia 36 D4 3 58N 113 46 E
Niamey, Niger 50 F5 13 27N 2 6 E
Niangara, Zaïre 54 B2 3 42N 27 50 E
Nias, Indonesia 36 D1 1 0N 97 30 E
Niassa □, Mozam. 55 E4 13 30S 36 0 E
Nicaragua ■, Cent. Amer. 88 D2 11 40N 85 30W
Nicaragua, L. de, Nic. 88 D2 12 0N 85 30W
Nicastro, Italy 20 E7 38 59N 16 19 E
Nice, France 18 E7 43 42N 7 14 E
Niceville, U.S.A. 77 K2 30 31N 86 30W
Nichinan, Japan 31 J5 31 38N 131 23 E
Nicholás, Canal, W. Indies 88 B3 23 30N 80 5W
Nicholasville, U.S.A. 76 G3 37 53N 84 34W
Nichols, U.S.A. 79 D8 42 1N 76 22W
Nicholson, Australia 60 C4 18 2S 128 54 E
Nicholson, U.S.A. 79 E9 41 37N 75 47W
Nicholson →, Australia 62 B2 17 31S 139 36 E
Nicholson Ra., Australia 61 E2 27 15S 116 45 E
Nicobar Is., Ind. Oc. 28 J13 9 0N 93 0 E
Nicola, Canada 72 C4 50 12N 120 40W
Nicolet, Canada 70 C5 46 17N 72 35W
Nicolls Town, Bahamas 88 A4 25 8N 78 0W
Nicosia, Cyprus 23 D12 35 10N 33 25 E
Nicoya, Costa Rica 88 D2 10 9N 85 27W
Nicoya, G. de, Costa Rica 88 E3 10 0N 85 0W
Nicoya, Pen. de, Costa Rica 88 E2 9 45N 85 40W
Nidd →, U.K. 10 C6 53 59N 1 23W
Niedersachsen □, Germany 16 B5 53 8N 9 0 E
Niekerkshoop, S. Africa 56 D3 29 19S 22 51 E
Niemba, Zaïre 54 D2 5 58S 28 24 E
Niemen = Neman →, Lithuania 9 J19 55 25N 21 10 E
Nienburg, Germany 16 B5 52 39N 9 13 E
Nieu Bethesda, S. Africa 56 E3 31 51S 24 34 E
Nieuw Amsterdam, Surinam 93 B7 5 53N 55 5W
Nieuw Nickerie, Surinam 93 B7 6 0N 56 59W
Nieuwoudtville, S. Africa 56 E2 31 23S 19 7 E
Nieuwpoort, Belgium 15 C2 51 8N 2 45 E
Nieves, Pico de las, Canary Is. 22 G4 27 57N 15 35W
Niğde, Turkey 25 G5 37 58N 34 40 E
Nigel, S. Africa 57 D4 26 27S 28 25 E
Niger ■, W. Afr. 50 E6 17 30N 10 0 E
Niger →, W. Afr. 50 G6 5 33N 6 33 E
Nigeria ■, W. Afr. 50 G6 8 30N 8 0 E
Nightcaps, N.Z. 59 L2 45 57S 168 2 E
Nihtaur, India 43 E8 29 20N 78 23 E
Nii-Jima, Japan 31 G9 34 20N 139 15 E
Niigata, Japan 30 F9 37 58N 139 0 E
Niigata □, Japan 31 F9 37 15N 138 45 E
Niihama, Japan 31 H6 33 55N 133 16 E
Niihau, U.S.A. 74 H14 21 54N 160 9W
Niimi, Japan 31 G6 34 59N 133 28 E
Niitsu, Japan 30 F9 37 48N 139 7 E
Nijil, Jordan 47 E4 30 32N 35 33 E
Nijkerk, Neths. 15 B5 52 13N 5 30 E
Nijmegen, Neths. 15 C5 51 50N 5 52 E
Nijverdal, Neths. 15 B6 52 22N 6 28 E
Nīk Pey, Iran 45 B6 36 50N 48 10 E
Nikiniki, Indonesia 37 F6 9 49S 124 30 E
Nikki, Benin 50 G5 9 58N 3 12 E
Nikkō, Japan 31 F9 36 45N 139 35 E
Nikolayev = Mykolayiv, Ukraine 25 E5 46 58N 32 0 E
Nikolayevsk, Russia 25 D8 50 0N 45 35 E
Nikolayevsk-na-Amur, Russia 27 D15 53 8N 140 44 E
Nikolskoye, Russia 27 D17 55 12N 166 0 E
Nikopol, Ukraine 25 E5 47 35N 34 25 E
Nikshahr, Iran 45 E9 26 15N 60 10 E
Nikšić, Montenegro, Yug. 21 C8 42 50N 18 57 E
Nîl, Nahr en →, Africa 51 B11 30 10N 31 6 E
Nîl el Abyad →, Sudan 51 E11 15 38N 32 31 E
Nîl el Azraq →, Sudan 51 E11 15 38N 32 31 E
Niland, U.S.A. 85 M11 33 14N 115 31W
Nile = Nîl, Nahr en →, Africa 51 B11 30 10N 31 6 E
Niles, U.S.A. 78 E4 41 11N 80 46W
Nimach, India 42 G6 24 30N 74 56 E
Nimbahera, India 42 G6 24 37N 74 45 E
Nîmes, France 18 E6 43 50N 4 23 E
Nimfaíon, Ákra = Pínnes, Ákra, Greece 21 D11 40 5N 24 20 E
Nimmitabel, Australia 63 F4 36 29S 149 15 E
Nimule, Sudan 52 D6 3 32N 32 3 E
Ninawá, Iraq 44 B4 36 25N 43 10 E
Nindigully, Australia 63 D4 28 21S 148 50 E
Ninemile, U.S.A. 72 B2 56 0N 130 7W
Nineveh = Ninawá, Iraq 44 B4 36 25N 43 10 E
Ning Xian, China 34 G4 35 30N 107 58 E
Ningaloo, Australia 60 D1 22 41S 113 41 E
Ning'an, China 35 B15 44 22N 129 20 E
Ningbo, China 33 D7 29 51N 121 28 E
Ningcheng, China 35 D10 41 32N 119 53 E
Ningjin, China 34 F8 37 35N 114 57 E
Ningjing Shan, China 32 C4 30 0N 98 20 E
Ningling, China 34 G8 34 25N 115 22 E
Ningpo = Ningbo, China 33 D7 29 51N 121 28 E
Ningqiang, China 34 H4 32 47N 106 15 E
Ningshan, China 34 H5 33 21N 108 21 E
Ningsia Hui A.R. = Ningxia Huizu Zizhiqu □, China 34 E3 38 0N 106 0 E
Ningwu, China 34 E7 39 0N 112 18 E
Ningxia Huizu Zizhiqu □, China 34 E3 38 0N 106 0 E
Ningyang, China 34 G9 35 47N 116 45 E
Ninh Binh, Vietnam 38 B5 20 15N 105 55 E
Ninh Giang, Vietnam 38 B6 20 44N 106 24 E
Ninh Hoa, Vietnam 38 F7 12 30N 109 7 E
Ninh Ma, Vietnam 38 F7 12 48N 109 21 E
Ninove, Belgium 15 D4 50 51N 4 2 E
Nioaque, Brazil 95 A4 21 5S 55 50W
Niobrara, U.S.A. 80 D6 42 45N 98 2W
Niobrara →, U.S.A. 80 D6 42 46N 98 3W
Nioro du Sahel, Mali 50 E3 15 15N 9 30W
Niort, France 18 C3 46 19N 0 29W
Nipawin, Canada 73 C8 53 20N 104 0W
Nipawin Prov. Park, Canada 73 C8 54 0N 104 37W
Nipigon, Canada 70 C2 49 0N 88 17W
Nipigon, L., Canada 70 C2 49 50N 88 30W
Nipin →, Canada 73 B7 55 46N 108 35W
Nipishish L., Canada 71 B7 54 12N 60 45W
Nipissing L., Canada 70 C4 46 20N 80 0W
Nipomo, U.S.A. 85 K6 35 3N 120 29W
Nipton, U.S.A. 85 K11 35 28N 115 16W
Niquelândia, Brazil 93 F9 14 33S 48 23W
Nīr, Iran 44 B5 38 2N 47 59 E
Nirasaki, Japan 31 G9 35 42N 138 27 E
Nirmal, India 40 K11 19 3N 78 20 E
Nirmali, India 43 F12 26 20N 86 35 E
Niš, Serbia, Yug. 21 C9 43 19N 21 58 E
Nişāb, Si. Arabia 44 D5 29 11N 44 43 E
Nişāb, Yemen 46 E4 14 25N 46 29 E
Nishinomiya, Japan 31 G7 34 45N 135 20 E
Nishin'omote, Japan 31 J5 30 43N 130 59 E
Nishiwaki, Japan 31 G7 34 59N 134 58 E
Niskibi →, Canada 70 A2 56 29N 88 9W
Nisqually →, U.S.A. 84 C4 47 6N 122 42W
Nissáki, Greece 23 A3 39 43N 19 52 E
Nissum Bredning, Denmark 9 H13 56 40N 8 20 E
Nistru = Dnister →, Europe 17 E16 46 18N 30 17 E
Nisutlin →, Canada 72 A2 60 14N 132 34W
Nitchequon, Canada 71 B5 53 10N 70 58W
Niterói, Brazil 95 A7 22 52S 43 0W
Nith →, U.K. 12 F5 55 14N 3 33W
Nitra, Slovak Rep. 17 D10 48 19N 18 4 E
Nitra →, Slovak Rep. 17 E10 47 46N 18 10 E
Niuafo'ou, Tonga 59 B11 15 30S 175 58W
Niue, Cook Is. 65 J11 19 2S 169 54W
Niut, Indonesia 36 D4 0 55N 110 6 E
Niuzhuang, China 35 D12 40 58N 122 28 E
Nivala, Finland 8 E21 63 56N 24 57 E
Nivelles, Belgium 15 D4 50 35N 4 20 E
Nivernais, France 18 C5 47 15N 3 30 E
Nixon, U.S.A. 81 L6 29 16N 97 46W
Nizamabad, India 40 K11 18 45N 78 7 E
Nizamghat, India 41 E19 28 20N 95 45 E
Nizhne Kolymsk, Russia 27 C17 68 34N 160 55 E
Nizhne-Vartovsk, Russia 26 C8 60 56N 76 38 E
Nizhneangarsk, Russia 27 D11 55 47N 109 30 E
Nizhnekamsk, Russia 24 C9 55 38N 51 49 E
Nizhneudinsk, Russia 27 D10 54 54N 99 3 E
Nizhneyansk, Russia 27 B14 71 26N 136 4 E
Nizhniy Novgorod, Russia 24 C7 56 20N 44 0 E
Nizhniy Tagil, Russia 24 C10 57 55N 59 57 E
Nizhyn, Ukraine 25 D5 51 5N 31 55 E
Nízké Tatry, Slovak Rep. 17 D10 48 55N 19 30 E
Njakwa, Malawi 55 E3 11 1S 33 56 E
Njanji, Zambia 55 E3 14 25S 31 46 E
Njinjo, Tanzania 55 D4 8 48S 38 54 E
Njombe, Tanzania 55 D3 9 20S 34 50 E
Njombe □, Tanzania 55 D3 9 20S 34 49 E
Njombe →, Tanzania 54 D4 6 56S 35 6 E
Nkambe, Cameroon 50 G7 6 35N 10 40 E
Nkana, Zambia 55 E2 12 50S 28 8 E
Nkawkaw, Ghana 50 G4 6 36N 0 49W
Nkayi, Zimbabwe 55 F2 19 41S 29 20 E
Nkhata Bay, Malawi 52 G6 11 33S 34 16 E
Nkhota Kota, Malawi 55 E3 12 56S 34 15 E
Nkongsamba, Cameroon 50 H6 4 55N 9 55 E
Nkurenkuru, Namibia 56 B2 17 42S 18 32 E
Nmai →, Burma 41 G20 25 30N 97 25 E
Noakhali = Maijdi, Bangla. 41 H17 22 48N 91 10 E
Noatak, U.S.A. 68 B3 67 34N 162 58W
Nobel, Canada 78 A4 45 25N 80 6W
Nobeoka, Japan 31 H5 32 36N 131 41 E
Noblesville, U.S.A. 76 E3 40 3N 86 1W
Nocera Inferiore, Italy 20 D6 40 44N 14 38 E
Nockatunga, Australia 63 D3 27 42S 142 42 E
Nocona, U.S.A. 81 J6 33 47N 97 44W
Noda, Japan 31 G9 35 56N 139 52 E
Noel, U.S.A. 81 G7 36 33N 94 29W
Nogales, Mexico 86 A2 31 20N 110 56W
Nogales, U.S.A. 83 L8 31 20N 110 56W
Nōgata, Japan 31 H5 33 48N 130 44 E
Noggerup, Australia 61 F2 33 32S 116 5 E
Noginsk, Russia 27 C10 64 30N 90 50 E
Nogoa →, Australia 62 C4 23 40S 147 55 E
Nogoyá, Argentina 94 C4 32 24S 59 48W
Nohar, India 42 E6 29 11N 74 49 E
Noire, Mts., France 18 B2 48 7N 3 28W
Noirmoutier, I. de, France 18 C2 46 58N 2 10W
Nojane, Botswana 56 C3 23 15S 20 14 E
Nojima-Zaki, Japan 31 G9 34 54N 139 53 E
Nok Kundi, Pakistan 40 E3 28 50N 62 45 E
Nokaneng, Botswana 56 B3 19 40S 22 17 E
Nokhtuysk, Russia 27 C12 60 0N 117 45 E
Nokia, Finland 9 F20 61 30N 23 30 E
Nokomis, Canada 73 C8 51 35N 105 0W
Nokomis L., Canada 73 B8 57 0N 103 0W
Nola, C.A.R. 52 D3 3 35N 16 4 E
Noma Omuramba →, Namibia 56 B3 18 52S 20 53 E
Noman L., Canada 73 A7 62 15N 108 55W
Nombre de Dios, Panama 88 E4 9 34N 79 28W
Nome, U.S.A. 68 B3 64 30N 165 25W
Nomo-Zaki, Japan 31 H4 32 35N 129 44 E
Nonacho L., Canada 73 A7 61 42N 109 40W
Nonda, Australia 62 C3 20 40S 142 28 E
Nong Chang, Thailand 38 E2 15 23N 99 51 E
Nong Het, Laos 38 C4 19 29N 103 59 E
Nong Khai, Thailand 38 D4 17 50N 102 46 E
Nong'an, China 35 B13 44 25N 125 5 E
Nongoma, S. Africa 57 D5 27 58S 31 35 E
Nonoava, Mexico 86 B3 27 28N 106 44W
Nonthaburi, Thailand 38 F3 13 51N 100 34 E
Noonamah, Australia 60 B5 12 40S 131 4 E
Noonan, U.S.A. 80 A3 48 54N 103 1W
Noondoo, Australia 63 D4 28 35S 148 30 E
Noonkanbah, Australia 60 C3 18 30S 124 50 E
Noord Brabant □, Neths. 15 C5 51 40N 5 0 E
Noord Holland □, Neths. 15 B4 52 30N 4 45 E
Noordbeveland, Neths. 15 C3 51 35N 3 50 E
Noordoostpolder, Neths. 15 B5 52 45N 5 45 E
Noordwijk aan Zee, Neths. 15 B4 52 14N 4 26 E
Nootka, Canada 72 D3 49 38N 126 38W
Nootka I., Canada 72 D3 49 32N 126 42W
Nóqui, Angola 52 F2 5 55S 13 30 E
Noranda, Canada 70 C4 48 20N 79 0W
Norco, U.S.A. 85 M9 33 56N 117 33W
Nord-Ostsee-Kanal →, Germany 16 A5 54 12N 9 32 E
Nordaustlandet, Svalbard 4 B9 79 14N 23 0 E
Nordegg, Canada 72 C5 52 29N 116 5W
Norderney, Germany 16 B4 53 42N 7 9 E
Norderstedt, Germany 16 B5 53 42N 9 58 E
Nordfjord, Norway 9 F11 61 55N 5 30 E
Nordfriesische Inseln, Germany 16 A5 54 40N 8 20 E
Nordhausen, Germany 16 C6 51 30N 10 47 E
Norðoyar, Færoe Is. 8 E9 62 17N 6 35W
Nordkapp, Norway 8 A21 71 10N 25 50 E
Nordkapp, Svalbard 4 A9 80 31N 20 0 E
Nordkinn = Kinnarodden, Norway 6 A11 71 8N 27 40 E
Nordkinn-halvøya, Norway 8 A22 70 55N 27 40 E
Nordrhein-Westfalen □, Germany 16 C4 51 45N 7 30 E
Nordvik, Russia 27 B12 74 2N 111 32 E
Norembega, Canada 70 C3 48 59N 80 43W
Norfolk, Nebr., U.S.A. 80 D6 42 2N 97 25W
Norfolk, Va., U.S.A. 76 G7 36 51N 76 17W
Norfolk □, U.K. 10 E9 52 39N 0 54 E
Norfolk Broads, U.K. 10 E9 52 30N 1 15 E
Norfolk I., Pac. Oc. 64 K8 28 58S 168 3 E
Norfork Res., U.S.A. 81 G8 36 13N 92 15W
Norilsk, Russia 27 C9 69 20N 88 6 E
Norley, Australia 63 D3 27 45S 143 48 E
Norma, Mt., Australia 62 C3 20 55S 140 42 E
Normal, U.S.A. 80 E10 40 31N 88 59W
Norman, U.S.A. 81 H6 35 13N 97 26W
Norman →, Australia 62 B3 19 18S 141 51 E
Norman Wells, Canada 68 B7 65 17N 126 51W
Normanby →, Australia 62 A3 14 23S 144 10 E
Normandie, France 18 B4 48 45N 0 10 E
Normandin, Canada 70 C5 48 49N 72 31W
Normandy = Normandie, France 18 B4 48 45N 0 10 E
Normanhurst, Mt., Australia 61 E3 25 4S 122 30 E
Normanton, Australia 62 B3 17 40S 141 10 E
Norquay, Canada 73 C8 51 53N 102 5W
Norquinco, Argentina 96 E2 41 51S 70 55W
Norrbotten □, Sweden 8 C19 66 30N 22 30 E
Norris, U.S.A. 82 D8 45 34N 111 41W
Norristown, U.S.A. 79 F9 40 7N 75 21W
Norrköping, Sweden 9 G17 58 37N 16 11 E
Norrland, Sweden 9 E16 62 15N 15 45 E
Norrtälje, Sweden 9 G18 59 46N 18 42 E
Norseman, Australia 61 F3 32 8S 121 43 E
Norsk, Russia 27 D14 52 30N 130 5 E
Norte, Pta. del, Canary Is. 22 G2 27 51N 17 57W
North Adams, U.S.A. 79 D11 42 42N 73 7W
North Ayrshire □, U.K. 12 F4 55 45N 4 44W
North Battleford, Canada 73 C7 52 50N 108 17W
North Bay, Canada 70 C4 46 20N 79 30W
North Belcher Is., Canada 70 A4 56 50N 79 50W
North Bend, Canada 72 D4 49 50N 121 27W
North Bend, Oreg., U.S.A. 82 E1 43 24N 124 14W
North Bend, Pa., U.S.A. 78 E7 41 20N 77 42W
North Bend, Wash., U.S.A. 84 C5 47 30N 121 47W
North Berwick, U.K. 12 E6 56 4N 2 42W
North Berwick, U.S.A. 79 C14 43 18N 70 44W
North C., Canada 71 C7 47 2N 60 20W
North C., N.Z. 59 F4 34 23S 173 4 E
North Canadian →, U.S.A. 81 H7 35 16N 95 31W
North Cape = Nordkapp, Norway 8 A21 71 10N 25 50 E
North Cape = Nordkapp, Svalbard 4 A9 80 31N 20 0 E
North Caribou L., Canada 70 B1 52 50N 90 40W
North Carolina □, U.S.A. 77 H5 35 30N 80 0W
North Channel, Canada 70 C3 46 0N 83 0W
North Channel, U.K. 12 G3 55 13N 5 52W
North Charleston, U.S.A. 77 J6 32 53N 79 58W
North Chicago, U.S.A. 76 D2 42 19N 87 51W
North Dakota □, U.S.A. 80 B5 47 30N 100 15W
North Dandalup, Australia 61 F2 32 30S 115 57 E
North Down □, U.K. 13 B6 54 40N 5 45W
North Downs, U.K. 11 F8 51 19N 0 21 E
North East, U.S.A. 78 D5 42 13N 79 50W
North East Frontier Agency = Arunachal Pradesh □, India 41 E19 28 0N 95 0 E
North East Lincolnshire □, U.K. 10 D7 53 34N 0 2W
North East Providence Chan., W. Indies 88 A4 26 0N 76 0W
North Eastern □, Kenya 54 B5 1 30N 40 0 E
North Esk →, U.K. 12 E6 56 46N 2 24W
North European Plain, Europe 6 E10 55 0N 25 0 E
North Foreland, U.K. 11 F9 51 22N 1 28 E
North Fork, U.S.A. 84 H7 37 14N 119 21W
North Fork American →, U.S.A. 84 G5 38 57N 120 59W
North Fork Feather →, U.S.A. 84 F5 38 33N 121 30W
North Frisian Is. = Nordfriesische Inseln, Germany 16 A5 54 40N 8 20 E
North Henik L., Canada 73 A9 61 45N 97 40W
North Highlands, U.S.A. 84 G5 38 40N 121 23W
North Horr, Kenya 54 B4 3 20N 37 8 E
North I., Kenya 54 B4 4 5N 36 5 E
North I., N.Z. 59 H5 38 0S 175 0 E
North Kingsville, U.S.A. 78 E4 41 54N 80 42W
North Knife →, Canada 73 B10 58 53N 94 45W
North Koel →, India 43 G10 24 45N 83 50 E
North Korea ■, Asia 35 E14 40 0N 127 0 E
North Lakhimpur, India 41 F19 27 14N 94 7 E
North Lanarkshire □, U.K. 12 F5 55 52N 3 56W

North Las Vegas, *U.S.A.* . **85 J11** 36 12N 115 7W
North Lincolnshire □, *U.K.* **10 D7** 53 36N 0 30W
North Little Rock, *U.S.A.* . **81 H8** 34 45N 92 16W
North Loup →, *U.S.A.* . . **80 E5** 41 17N 98 24W
North Magnetic Pole, *Canada* **4 B2** 77 58N 102 8W
North Minch, *U.K.* **12 C3** 58 5N 5 55W
North Nahanni →, *Canada* **72 A4** 62 15N 123 20W
North Olmsted, *U.S.A.* . . **78 E3** 41 25N 81 56W
North Ossetia □, *Russia* . **25 F7** 43 30N 44 30 E
North Pagai, I. = Pagai Utara, *Indonesia* . . . . . . **36 E2** 2 35S 100 0 E
North Palisade, *U.S.A.* . . . **83 H4** 37 6N 118 31W
North Platte, *U.S.A.* **80 E4** 41 8N 100 46W
North Platte →, *U.S.A.* . **80 E4** 41 7N 100 42W
North Pole, *Arctic* . . . . . . **4 A** 90 0N 0 0 E
North Portal, *Canada* **73 D8** 49 0N 102 33W
North Powder, *U.S.A.* **82 D5** 45 2N 117 55W
North Pt., *Canada* **71 C7** 47 5N 64 0W
North Rhine Westphalia □ = Nordrhein-Westfalen □, *Germany* **16 C4** 51 45N 7 30 E
North Ronaldsay, *U.K.* . . . **12 B6** 59 22N 2 26W
North Saskatchewan →, *Canada* **73 C7** 53 15N 105 5W
North Sea, *Europe* . . . . . . **6 D6** 56 0N 4 0 E
North Somerset □, *U.K.* . **11 F5** 51 24N 2 45W
North Sporades = Voríai Sporádhes, *Greece* . . . . **21 E10** 39 15N 23 30 E
North Sydney, *Canada* . . **71 C7** 46 12N 60 15W
North Taranaki Bight, *N.Z.* **59 H5** 38 50S 174 15 E
North Thompson →, *Canada* **72 C4** 50 40N 120 20W
North Tonawanda, *U.S.A.* **78 C6** 43 2N 78 53W
North Troy, *U.S.A.* **79 B12** 45 0N 72 24W
North Truchas Pk., *U.S.A.* **83 J11** 36 0N 105 30W
North Twin I., *Canada* . . . **70 B3** 53 20N 80 0W
North Tyne →, *U.K.* **10 C5** 55 0N 2 8W
North Uist, *U.K.* **12 D1** 57 40N 7 15W
North Vancouver, *Canada* **72 D4** 49 25N 123 3W
North Vernon, *U.S.A.* **76 F3** 39 0N 85 38W
North Wabasca L., *Canada* **72 B6** 56 0N 113 55W
North Walsham, *U.K.* . . . **10 E9** 52 50N 1 22 E
North-West □, *S. Africa* . **56 D4** 27 0S 25 0 E
North West C., *Australia* . **60 D1** 21 45S 114 9 E
North West Christmas I. Ridge, *Pac. Oc.* **65 G11** 6 30N 165 0W
North West Frontier □, *Pakistan* . . . . **42 C4** 34 0N 72 0 E
North West Highlands, *U.K.* **12 D3** 57 33N 4 58W
North West Providence Channel, *W. Indies* . . . **88 A4** 26 0N 78 0W
North West River, *Canada* **71 B7** 53 30N 60 10W
North West Territories □, *Canada* **68 B9** 67 0N 110 0W
North Western □, *Zambia* **55 E2** 13 30S 25 30 E
North York Moors, *U.K.* . **10 C7** 54 23N 0 53W
North Yorkshire □, *U.K.* . **10 C6** 54 15N 1 25W
Northallerton, *U.K.* **10 C6** 54 20N 1 26W
Northam, *S. Africa* . . . . . . **56 C4** 24 56S 27 18 E
Northampton, *Australia* . **61 E1** 28 27S 114 33 E
Northampton, *U.K.* **11 E7** 52 15N 0 53W
Northampton, *Mass., U.S.A.* **79 D12** 42 19N 72 38W
Northampton, *Pa., U.S.A.* **79 F9** 40 41N 75 30W
Northampton Downs, *Australia* . . . . **62 C4** 24 35S 145 48 E
Northamptonshire □, *U.K.* **11 E7** 52 16N 0 55W
Northbridge, *U.S.A.* **79 D13** 42 9N 71 39W
Northcliffe, *Australia* . . . . **61 F2** 34 39S 116 7 E
Northern □, *Malawi* . . . . . **55 E3** 11 0S 34 0 E
Northern □, *Uganda* . . . . **54 B3** 3 5N 32 30 E
Northern □, *Zambia* . . . . **55 E3** 10 30S 31 0 E
Northern Cape □, *S. Africa* **56 D3** 30 0S 20 0 E
Northern Circars, *India* . . **41 L13** 17 30N 82 30 E
Northern Indian L., *Canada* **73 B9** 57 20N 97 20W
Northern Ireland □, *U.K.* . **13 B5** 54 45N 7 0W
Northern Light, L., *Canada* **70 C1** 48 15N 90 39W
Northern Marianas ■, *Pac. Oc.* . . . . **64 F6** 17 0N 145 0 E
Northern Territory □, *Australia* . . . . **60 D5** 20 0S 133 0 E
Northern Transvaal □, *S. Africa* . . . . **57 C4** 24 0S 29 0 E
Northfield, *U.S.A.* **80 C8** 44 27N 93 9W
Northland □, *N.Z.* . . . . **59 F4** 35 30S 173 30 E
Northome, *U.S.A.* **80 B7** 47 52N 94 17W
Northport, *Ala., U.S.A.* . . **77 J2** 33 14N 87 35W
Northport, *Mich., U.S.A.* . **76 C3** 45 8N 85 37W
Northport, *Wash., U.S.A.* . **82 B5** 48 55N 117 48W
Northumberland □, *U.K.* . **10 B5** 55 12N 2 0W
Northumberland, C., *Australia* . . . . **63 F3** 38 5S 140 40 E
Northumberland Is., *Australia* . . . . **62 C4** 21 30S 149 50 E
Northumberland Str., *Canada* **71 C7** 46 20N 64 0W
Northwich, *U.K.* **10 D5** 53 15N 2 31W
Northwood, *Iowa, U.S.A.* . **80 D8** 43 27N 93 13W
Northwood, *N. Dak., U.S.A.* **80 B6** 47 44N 97 34W
Norton, *U.S.A.* **80 F5** 39 50N 99 53W
Norton, *Zimbabwe* . . . . . . **55 F3** 17 52S 30 40 E
Norton Sd., *U.S.A.* **68 B3** 63 50N 164 0W
Norwalk, *Calif., U.S.A.* . . . **85 M8** 33 54N 118 5W
Norwalk, *Conn., U.S.A.* . . **79 E11** 41 7N 73 22W
Norwalk, *Ohio, U.S.A.* . . . **78 E2** 41 15N 82 37W
Norway, *U.S.A.* **76 C2** 45 47N 87 55W
Norway ■, *Europe* . . . . . . **8 E14** 63 0N 11 0 E
Norway House, *Canada* . **73 C9** 53 59N 97 50W
Norwegian Sea, *Atl. Oc.* . **4 C8** 66 0N 1 0 E
Norwich, *Canada* **78 D4** 42 59N 80 36W
Norwich, *U.K.* . . . . **10 E9** 52 38N 1 18 E
Norwich, *Conn., U.S.A.* . . **79 E12** 41 31N 72 5W
Norwich, *N.Y., U.S.A.* . . . **79 D9** 42 32N 75 32W
Norwood, *Canada* **78 B7** 44 23N 77 59W
Noshiro, *Japan* . . . . **30 D10** 40 12N 140 0 E
Nosok, *Russia* . . . . **26 B9** 70 10N 82 20 E
Noss Hd., *U.K.* **12 C5** 58 28N 3 3W
Nossob →, *S. Africa* . . . **56 D3** 26 55S 20 45 E
Nosy Bé, *Madag.* . . . . **53 G9** 13 25S 48 15 E
Nosy Boraha, *Madag.* . . . **57 B8** 16 50S 49 55 E
Nosy Mitsio, *Madag.* . . . . **53 G9** 12 54S 48 36 E
Nosy Varika, *Madag.* . . . . **57 C8** 20 35S 48 32 E
Noteć →, *Poland* . . . . . . **16 B8** 52 44N 15 26 E
Notigi Dam, *Canada* **73 B9** 56 40N 99 10W
Notikewin →, *Canada* . . . **72 B5** 57 2N 117 38W
Notodden, *Norway* . . . . . **9 G13** 59 35N 9 17 E
Notre-Dame, *Canada* **71 C7** 46 18N 64 46W
Notre Dame B., *Canada* . . **71 C8** 49 45N 55 30W
Notre Dame de Koartac = Quaqtaq, *Canada* **69 B13** 60 55N 69 40W
Notre Dame d'Ivugivic = Ivujivik, *Canada* **69 B12** 62 24N 77 55W
Nottaway →, *Canada* . . . **70 B4** 51 22N 78 55W
Nottingham, *U.K.* **10 E6** 52 58N 1 10W
Nottinghamshire □, *U.K.* **10 D7** 53 10N 1 3W
Nottoway →, *U.S.A.* . . . **76 G7** 36 33N 76 55W
Notwane →, *Botswana* . **56 C4** 23 35S 26 58 E
Nouâdhibou, *Mauritania* . **50 D1** 20 54N 17 0W
Nouâdhibou, Ras, *Mauritania* **50 D1** 20 50N 17 0W
Nouakchott, *Mauritania* . . **50 E1** 18 9N 15 58W
Nouméa, *N. Cal.* . . . . **64 K8** 22 17S 166 30 E
Noupoort, *S. Africa* . . . . . **56 E3** 31 10S 24 57 E
Nouveau Comptoir = Wemindji, *Canada* **70 B4** 53 0N 78 49W
Nouvelle-Calédonie = New Caledonia ■, *Pac. Oc.* . . . . **64 K8** 21 0S 165 0 E
Nova Casa Nova, *Brazil* . . **93 E10** 9 25S 41 5W
Nova Cruz, *Brazil* **93 E11** 6 28S 35 25W
Nova Esperança, *Brazil* . . **95 A5** 23 8S 52 24W
Nova Friburgo, *Brazil* . . . **95 A7** 22 16S 42 30W
Nova Gaia = Cambundi-Catembo, *Angola* . . . . . **52 G3** 10 10S 17 35 E
Nova Iguaçu, *Brazil* **95 A7** 22 45S 43 28W
Nova Iorque, *Brazil* **93 E10** 7 0S 44 5W
Nova Lima, *Brazil* **95 A7** 19 59S 43 51W
Nova Lisboa = Huambo, *Angola* . . . . **53 G3** 12 42S 15 54 E
Nova Lusitânia, *Mozam.* . **55 F3** 19 50S 34 34 E
Nova Mambone, *Mozam.* . **57 C6** 21 0S 35 3 E
Nova Scotia □, *Canada* . . **71 C7** 45 10N 63 0W
Nova Sofala, *Mozam.* . . . **57 C5** 20 7S 34 42 E
Nova Venécia, *Brazil* **93 G10** 18 45S 40 24W
Nova Zagora, *Bulgaria* . . **21 C11** 42 32N 25 59 E
Novara, *Italy* . . . . **20 B3** 45 28N 8 38 E
Novato, *U.S.A.* **84 G4** 38 6N 122 35W
Novaya Ladoga, *Russia* . . **24 B5** 60 7N 32 16 E
Novaya Lyalya, *Russia* . . **26 D7** 59 4N 60 45 E
Novaya Sibir, Ostrov, *Russia* . . . . **27 B16** 75 10N 150 0 E
Novaya Zemlya, *Russia* . . **26 B6** 75 0N 56 0 E
Nové Zámky, *Slovak Rep.* **17 D10** 48 2N 18 8 E
Novgorod, *Russia* . . . . . . **24 C5** 58 30N 31 25 E
Novgorod-Severskiy = Novhorod-Siverskyy, *Ukraine* . . . . **24 D5** 52 2N 33 10 E
Novhorod-Siverskyy, *Ukraine* . . . . **24 D5** 52 2N 33 10 E
Novi Ligure, *Italy* . . . . **20 B3** 44 46N 8 47 E
Novi Pazar, *Serbia, Yug.* . . **21 C9** 43 12N 20 28 E
Novi Sad, *Serbia, Yug.* . . **21 B8** 45 18N 19 52 E
Nôvo Hamburgo, *Brazil* . . **95 B5** 29 37S 51 7W
Novo Mesto, *Slovenia* . . . **16 F8** 45 48N 15 10 E
Novo Remanso, *Brazil* . . . **93 E10** 9 41S 42 4W
Novoataysk, *Russia* . . . . **26 D9** 53 30N 84 0 E
Novocherkassk, *Russia* . . . **25 E7** 47 27N 40 15 E
Novogrudok = Navahrudak, *Belarus* . . **17 B13** 53 40N 25 50 E
Novohrad-Volynskyy, *Ukraine* . . . . **17 C14** 50 34N 27 35 E
Novokachalinsk, *Russia* . . **30 B6** 45 5N 132 0 E
Novokazalinsk = Zhangaqazaly, *Kazakstan* . . . . **26 E7** 45 48N 62 6 E
Novokuybyshevsk, *Russia* **24 D8** 53 7N 49 58 E
Novokuznetsk, *Russia* . . . **26 D9** 53 45N 87 10 E
Novomoskovsk, *Russia* . . **24 D6** 54 5N 38 15 E
Novorossiysk, *Russia* . . . **25 F6** 44 43N 37 46 E
Novorybnoye, *Russia* . . . . **27 B11** 72 50N 105 50 E
Novoselytsya, *Ukraine* . . **17 D14** 48 14N 26 15 E
Novoshakhtinsk, *Russia* . . **25 E6** 47 46N 39 58 E
Novosibirsk, *Russia* . . . . . **26 D9** 55 0N 83 5 E
Novosibirskiye Ostrova, *Russia* . . . . **27 B15** 75 0N 142 0 E
Novotroitsk, *Russia* . . . . . **26 D6** 51 10N 58 15 E
Novouzensk, *Russia* . . . . . **25 D8** 50 32N 48 17 E
Novovolynsk, *Ukraine* . . . **17 C13** 50 45N 24 4 E
Novska, *Croatia* . . . . **20 B7** 45 19N 17 0 E
Novyy Port, *Russia* . . . . . **26 C8** 67 40N 72 30 E
Now Shahr, *Iran* . . . . **45 B6** 36 40N 51 30 E
Nowa Sól, *Poland* . . . . . . **16 C8** 51 48N 15 44 E
Nowbarān, *Iran* . . . . **45 C6** 35 8N 49 42 E
Nowghāb, *Iran* . . . . **45 C8** 33 53N 59 4 E
Nowgong, *India* . . . . **41 F18** 26 20N 92 50 E
Nowra, *Australia* . . . . **63 E5** 34 53S 150 35 E
Nowshera, *Pakistan* . . . . . **40 B8** 34 0N 72 0 E
Nowy Sącz, *Poland* . . . . . **17 D11** 49 40N 20 41 E
Nowy Targ, *Poland* . . . . . **17 D11** 49 29N 20 2 E
Nowy Tomyśl, *Poland* . . . **16 B9** 52 19N 16 10 E
Noxen, *U.S.A.* **79 E8** 41 25N 76 4W
Noxon, *U.S.A.* **82 C6** 48 0N 115 43W
Noyes I., *U.S.A.* **72 B2** 55 30N 133 40W
Noyon, *France* . . . . **18 B5** 49 34N 2 59 E
Noyon, *Mongolia* . . . . **34 C2** 43 2N 102 4 E
Nsanje, *Malawi* . . . . **55 F4** 16 55S 35 12 E
Nsawam, *Ghana* **50 G4** 5 50N 0 24W
Nsomba, *Zambia* . . . . **55 E2** 10 45S 29 51 E
Nsukka, *Nigeria* . . . . **50 G6** 6 51N 7 29 E
Nu Jiang →, *China* . . . . **32 D4** 29 58N 97 25 E
Nu Shan, *China* . . . . **32 D4** 26 0N 99 20 E
Nubia, *Africa* . . . . **48 D7** 21 0N 32 0 E
Nubian Desert = Nûbîya, Es Sahrâ En, *Sudan* . . . **51 D11** 21 30N 33 30 E
Nûbîya, Es Sahrâ En, *Sudan* . . . . **51 D11** 21 30N 33 30 E
Ñuble □, *Chile* **94 D1** 37 0S 72 0W
Nuboai, *Indonesia* . . . . . . **37 E9** 2 10S 136 30 E
Nubra →, *India* . . . . **43 B7** 34 35N 77 35 E
Nueces →, *U.S.A.* **81 M6** 27 51N 97 30W
Nueltin L., *Canada* **73 A9** 60 30N 99 30W
Nueva Asunción □, *Paraguay* **94 A3** 21 0S 61 0W
Nueva Gerona, *Cuba* **88 B3** 21 53N 82 49W
Nueva Imperial, *Chile* . . . **96 D2** 38 45S 72 58W
Nueva Palmira, *Uruguay* . **94 C4** 33 52S 58 20W
Nueva Rosita, *Mexico* . . . **86 B4** 28 0N 101 11W
Nueva San Salvador, *El Salv.* **88 D2** 13 40N 89 18W
Nuéve de Julio, *Argentina* **94 D3** 35 30S 61 0W
Nuevitas, *Cuba* **88 B4** 21 30N 77 20W
Nuevo, G., *Argentina* **96 E4** 43 0S 64 30W
Nuevo Guerrero, *Mexico* . **87 B5** 26 34N 99 15W
Nuevo Laredo, *Mexico* . . . **87 B5** 27 30N 99 30W
Nuevo León □, *Mexico* . . . **86 C4** 25 0N 100 0W
Nugget Pt., *N.Z.* . . . . **59 M2** 46 27S 169 50 E
Nuhaka, *N.Z.* . . . . **59 H6** 39 3S 177 45 E
Nukey Bluff, *Australia* . . . **63 E2** 32 26S 135 29 E
Nukheila, *Sudan* . . . . **51 E10** 19 1N 26 21 E
Nuku'alofa, *Tonga* **59 E11** 21 10S 174 0W
Nukus, *Uzbekistan* . . . . . **26 E6** 42 27N 59 41 E
Nulato, *U.S.A.* **68 B4** 64 43N 158 6W
Nullagine →, *Australia* . . **60 D3** 21 20S 120 20 E
Nullarbor, *Australia* . . . . . **61 F5** 31 28S 130 55 E
Nullarbor Plain, *Australia* **61 F4** 31 10S 129 0 E
Numalla, L., *Australia* . . . **63 D3** 28 43S 144 20 E
Numan, *Nigeria* . . . . **51 G7** 9 29N 12 3 E
Numata, *Japan* . . . . **31 F9** 36 45N 139 4 E
Numazu, *Japan* . . . . **31 G9** 35 7N 138 51 E
Numbulwar, *Australia* . . . **62 A2** 14 15S 135 45 E
Numfoor, *Indonesia* . . . . . **37 E8** 1 0S 134 50 E
Numurkah, *Australia* . . . . **63 F4** 36 5S 145 26 E
Nunaksaluk I., *Canada* . . . **71 A7** 55 49N 60 20W
Nuneaton, *U.K.* **11 E6** 52 32N 1 27W
Nungo, *Mozam.* . . . . **55 E4** 13 23S 37 43 E
Nungwe, *Tanzania* . . . . . . **54 C3** 2 48S 32 2 E
Nunivak I., *U.S.A.* **68 B3** 60 10N 166 30W
Nunkun, *India* . . . . **43 C7** 33 57N 76 2 E
Nunspeet, *Neths.* . . . . **15 B5** 52 21N 5 45 E
Nùoro, *Italy* . . . . **20 D3** 40 20N 9 20 E
Nūrābād, *Iran* . . . . **45 E8** 27 47N 57 12 E
Nuremberg = Nürnberg, *Germany* . . . . **16 D6** 49 27N 11 3 E
Nuri, *Mexico* **86 B3** 28 2N 109 22W
Nurina, *Australia* . . . . **61 F4** 30 56S 126 33 E
Nuriootpa, *Australia* . . . . **63 E2** 34 27S 139 0 E
Nurmes, *Finland* . . . . **8 E23** 63 33N 29 10 E
Nürnberg, *Germany* . . . . **16 D6** 49 27N 11 3 E
Nurran, L. = Terewah, L., *Australia* . . . . **63 D4** 29 52S 147 35 E
Nurrari Lakes, *Australia* . **61 E5** 29 1S 130 5 E
Nusa Barung, *Indonesia* . **37 H15** 8 10S 113 30 E
Nusa Kambangan, *Indonesia* . . . . **37 G13** 7 40S 108 10 E
Nusa Tenggara Barat □, *Indonesia* . . . . **36 F5** 8 50S 117 30 E
Nusa Tenggara Timur □, *Indonesia* . . . . **37 F6** 9 30S 122 0 E
Nusaybin, *Turkey* . . . . **25 G7** 37 3N 41 10 E
Nushki, *Pakistan* . . . . **42 E2** 29 35N 66 0 E
Nutak, *Canada* **69 C13** 57 28N 61 59W
Nutwood Downs, *Australia* **62 B1** 15 49S 134 10 E
Nuuk = Godthåb, *Greenland* **69 B14** 64 10N 51 35W
Nuwakot, *Nepal* . . . . **43 E10** 28 10N 83 55 E
Nuweveldberge, *S. Africa* **56 E3** 32 10S 21 45 E
Nuyts, C., *Australia* . . . . . **61 F5** 32 2S 132 21 E
Nuyts Arch., *Australia* . . . **63 E1** 32 35S 133 20 E
Nxau-Nxau, *Botswana* . . . **56 B3** 18 57S 21 4 E
Nyack, *U.S.A.* **79 E11** 41 5N 73 55W
Nyah West, *Australia* . . . . **63 F3** 35 16S 143 21 E
Nyahanga, *Tanzania* . . . . . **54 C3** 2 20S 33 37 E
Nyahua, *Tanzania* . . . . **54 D3** 5 25S 33 23 E
Nyahururu, *Kenya* . . . . . . **54 B4** 0 2N 36 27 E
Nyainqentanglha Shan, *China* . . . . **32 D3** 30 0N 90 0 E
Nyakanazi, *Tanzania* . . . . **54 C3** 3 2S 31 10 E
Nyâlâ, *Sudan* . . . . **51 F9** 12 2N 24 58 E
Nyamandhlovu, *Zimbabwe* . . . . **55 F2** 19 55S 28 16 E
Nyambiti, *Tanzania* . . . . . **54 C3** 2 48S 33 27 E
Nyamwaga, *Tanzania* . . . . **54 C3** 1 27S 34 33 E
Nyandekwa, *Tanzania* . . . **54 C3** 3 57S 32 32 E
Nyandoma, *Russia* . . . . . . **24 B7** 61 40N 40 12 E
Nyangana, *Namibia* . . . . . **56 B3** 18 0S 20 40 E
Nyanguge, *Tanzania* . . . . **54 C3** 2 30S 33 12 E
Nyanza, *Burundi* . . . . **54 C2** 4 21S 29 36 E
Nyanza, *Rwanda* . . . . **54 C2** 2 20S 29 42 E
Nyanza □, *Kenya* . . . . **54 C3** 0 10S 34 15 E
Nyarling →, *Canada* **72 A6** 60 41N 113 23W
Nyasa, L. = Malawi, L., *Africa* . . . . **55 E3** 12 30S 34 30 E
Nyasvizh, *Belarus* . . . . . . **17 B14** 53 14N 26 38 E
Nyazepetrovsk, *Russia* . . **24 C10** 56 3N 59 36 E
Nyazura, *Zimbabwe* . . . . **55 F3** 18 40S 32 16 E
Nyazwidzi →, *Zimbabwe* **55 F3** 20 0S 31 17 E
Nybro, *Sweden* . . . . **9 H16** 56 44N 15 55 E
Nyda, *Russia* . . . . **26 C8** 66 40N 72 58 E
Nyeri, *Kenya* . . . . **54 C4** 0 23S 36 56 E
Nyíregyháza, *Hungary* . . . **17 E11** 47 58N 21 47 E
Nykøbing, *Storstrøm, Denmark* . . . . **9 J14** 54 56N 11 52 E
Nykøbing, *Vestsjælland, Denmark* . . . . **9 J14** 55 55N 11 40 E
Nykøbing, *Viborg, Denmark* . . . . **9 H13** 56 48N 8 51 E
Nyköping, *Sweden* . . . . . . **9 G17** 58 45N 17 0 E
Nylstroom, *S. Africa* . . . . **57 C4** 24 42S 28 22 E
Nymagee, *Australia* . . . . . **63 E4** 32 7S 146 20 E
Nynäshamn, *Sweden* . . . . **9 G17** 58 54N 17 57 E
Nyngan, *Australia* . . . . . . **63 E4** 31 30S 147 8 E
Nyoman = Neman →, *Lithuania* . . . . **9 J19** 55 25N 21 10 E
Nysa, *Poland* . . . . **17 C9** 50 30N 17 22 E
Nysa →, *Europe* . . . . **16 B8** 52 4N 14 46 E
Nyssa, *U.S.A.* **82 E5** 43 53N 117 0W
Nyunzu, *Zaïre* . . . . **54 D2** 5 57S 27 58 E
Nyurbe, *Russia* . . . . **27 C12** 63 17N 118 28 E
Nzega, *Tanzania* . . . . **54 C3** 4 10S 33 12 E
Nzega □, *Tanzania* . . . . . . **54 C3** 4 10S 33 10 E
N'Zérékoré, *Guinea* **50 G3** 7 49N 8 48W
Nzeto, *Angola* . . . . **52 F2** 7 10S 12 52 E
Nzilo, Chutes de, *Zaïre* . . **55 E2** 10 18S 25 27 E
Nzubuka, *Tanzania* . . . . . . **54 C3** 4 45S 32 50 E

O

Ō-Shima, *Nagasaki, Japan* **31 G4** 34 29N 129 33 E
Ō-Shima, *Shizuoka, Japan* **31 G9** 34 44N 139 24 E
Oacoma, *U.S.A.* **80 D5** 43 48N 99 24W
Oahe, L., *U.S.A.* **80 C4** 44 27N 100 24W
Oahe Dam, *U.S.A.* **80 C4** 44 27N 100 24W
Oahu, *U.S.A.* **74 H16** 21 28N 157 58W
Oak Creek, *U.S.A.* **82 F10** 40 16N 106 57W
Oak Harbor, *U.S.A.* **84 B4** 48 18N 122 39W
Oak Hill, *U.S.A.* **76 G5** 37 59N 81 9W
Oak Park, *U.S.A.* **76 E2** 41 53N 87 47W
Oak Ridge, *U.S.A.* **77 G3** 36 1N 84 16W
Oak View, *U.S.A.* **85 L7** 34 24N 119 18W
Oakan-Dake, *Japan* . . . . . **30 C12** 43 27N 144 10 E
Oakbank, *Australia* . . . . . **63 E3** 33 4S 140 33 E
Oakdale, *Calif., U.S.A.* . . . **83 H3** 37 46N 120 51W
Oakdale, *La., U.S.A.* **81 K8** 30 49N 92 40W
Oakengates, *U.K.* **10 E5** 52 41N 2 26W
Oakes, *U.S.A.* **80 B5** 46 8N 98 6W
Oakesdale, *U.S.A.* **82 C5** 47 8N 117 15W
Oakey, *Australia* . . . . **63 D5** 27 25S 151 43 E
Oakham, *U.K.* **10 E7** 52 40N 0 43W
Oakhurst, *U.S.A.* **84 H7** 37 19N 119 40W
Oakland, *Calif., U.S.A.* . . . **83 H2** 37 49N 122 16W
Oakland, *Oreg., U.S.A.* . . . **82 E2** 43 25N 123 18W
Oakland City, *U.S.A.* **76 F2** 38 20N 87 21W
Oakley, *Idaho, U.S.A.* **82 E7** 42 15N 113 53W
Oakley, *Kans., U.S.A.* **80 F4** 39 8N 100 51W
Oakover →, *Australia* . . . **60 D3** 21 0S 120 40 E
Oakridge, *U.S.A.* **82 E2** 43 45N 122 28W
Oakville, *U.S.A.* **84 D3** 46 51N 123 14W
Oamaru, *N.Z.* . . . . **59 L3** 45 5S 170 59 E
Oasis, *Calif., U.S.A.* **85 M10** 33 28N 116 6W
Oasis, *Nev., U.S.A.* **84 H9** 37 29N 117 55W
Oates Land, *Antarctica* . . **5 C11** 69 0S 160 0 E
Oatman, *U.S.A.* **85 K12** 35 1N 114 19W
Oaxaca, *Mexico* **87 D5** 17 2N 96 40W
Oaxaca □, *Mexico* **87 D5** 17 0N 97 0W
Ob →, *Russia* . . . . **26 C7** 66 45N 69 30 E
Oba, *Canada* **70 C3** 49 4N 84 7W
Obama, *Japan* . . . . **31 G7** 35 30N 135 45 E
Oban, *U.K.* **12 E3** 56 25N 5 29W
Obbia, *Somali Rep.* . . . . . **46 F4** 5 25N 48 30 E
Obed, *Canada* **72 C5** 53 30N 117 10W
Obera, *Argentina* **95 B4** 27 21S 55 2W
Oberhausen, *Germany* . . . **16 C4** 51 28N 6 51 E
Oberlin, *Kans., U.S.A.* . . . **80 F4** 39 49N 100 32W
Oberlin, *La., U.S.A.* **81 K8** 30 37N 92 46W
Oberlin, *Ohio, U.S.A.* **78 E2** 41 18N 82 13W
Oberon, *Australia* . . . . **63 E4** 33 45S 149 52 E
Obi, Kepulauan, *Indonesia* **37 E7** 1 23S 127 45 E
Obi Is. = Obi, Kepulauan, *Indonesia* . . . . **37 E7** 1 23S 127 45 E
Óbidos, *Brazil* **93 D7** 1 50S 55 30W
Obihiro, *Japan* . . . . **30 C11** 42 56N 143 12 E
Obilatu, *Indonesia* . . . . . . **37 E7** 1 25S 127 20 E
Obluchye, *Russia* . . . . **27 E14** 49 1N 131 4 E
Obo, *C.A.R.* . . . . **54 A2** 5 20N 26 32 E
Oboa, Mt., *Uganda* . . . . . **54 B3** 1 45N 34 45 E
Oboyan, *Russia* . . . . **26 D4** 51 15N 36 21 E
Obozerskaya = Obozerskiy, *Russia* . . . . **26 C5** 63 34N 40 21 E
Obozerskiy, *Russia* . . . . . **26 C5** 63 34N 40 21 E
Observatory Inlet, *Canada* **72 B3** 55 10N 129 54W
Obshchi Syrt, *Russia* . . . . **6 E16** 52 0N 53 0 E
Obskaya Guba, *Russia* . . **26 C8** 69 0N 73 0 E
Obuasi, *Ghana* **50 G4** 6 17N 1 40W
Ocala, *U.S.A.* **77 L4** 29 11N 82 8W
Ocampo, *Mexico* **86 B3** 28 9N 108 24W
Ocaña, *Spain* **19 C4** 39 55N 3 30W
Ocanomowoc, *U.S.A.* . . . **80 D10** 43 7N 88 30W
Ocate, *U.S.A.* **81 G2** 36 11N 105 3W
Occidental, Cordillera, *Colombia* **92 C3** 5 0N 76 0W
Ocean City, *N.J., U.S.A.* . **76 F8** 39 17N 74 35W
Ocean City, *Wash., U.S.A.* **84 C2** 47 4N 124 10W
Ocean I. = Banaba, *Kiribati* . . . . **64 H8** 0 45S 169 50 E
Ocean Park, *U.S.A.* **84 D2** 46 30N 124 3W
Oceano, *U.S.A.* **85 K6** 35 6N 120 37W
Oceanport, *U.S.A.* **79 F10** 40 19N 74 3W
Oceanside, *U.S.A.* **85 M9** 33 12N 117 23W
Ochil Hills, *U.K.* **12 E5** 56 14N 3 40W
Ochre River, *Canada* **73 C9** 51 4N 99 47W
Ocilla, *U.S.A.* **77 K4** 31 36N 83 15W
Ocmulgee →, *U.S.A.* **77 K4** 31 58N 82 33W
Ocnița, *Moldova* . . . . **17 D14** 48 25N 27 30 E
Oconee →, *U.S.A.* **77 K4** 31 58N 82 33W
Oconto, *U.S.A.* **76 C2** 44 53N 87 52W
Oconto Falls, *U.S.A.* **76 C1** 44 52N 88 9W
Ocosingo, *Mexico* **87 D6** 17 10N 92 15W
Ocotal, *Nic.* **88 D2** 13 41N 86 31W
Ocotlán, *Mexico* **86 C4** 20 21N 102 42W
Octave, *U.S.A.* **83 J7** 34 10N 112 43W
Ocumare del Tuy, *Venezuela* **92 A5** 10 7N 66 46W
Ōda, *Japan* . . . . **31 G6** 35 11N 132 30 E
Ódáðahraun, *Iceland* **8 D5** 65 5N 17 0W
Odate, *Japan* . . . . **30 D10** 40 16N 140 34 E
Odawara, *Japan* . . . . **31 G9** 35 20N 139 6 E
Odda, *Norway* . . . . **9 F12** 60 3N 6 35 E
Oddur, *Somali Rep.* . . . . . **46 G3** 4 11N 43 52 E
Odei →, *Canada* **73 B9** 56 6N 96 54W
Ödemiş, *Turkey* . . . . **21 E13** 38 15N 28 0 E
Odendaalsrus, *S. Africa* . . **56 D4** 27 48S 26 45 E
Odense, *Denmark* . . . . . . **9 J14** 55 22N 10 23 E
Oder →, *Germany* . . . . . . **16 B8** 53 33N 14 38 E
Odesa, *Ukraine* . . . . **25 E5** 46 30N 30 45 E
Odessa = Odesa, *Ukraine* **25 E5** 46 30N 30 45 E
Odessa, *Canada* **79 B8** 44 17N 76 43W
Odessa, *Tex., U.S.A.* **81 K3** 31 52N 102 23W
Odessa, *Wash., U.S.A.* . . . **82 C4** 47 20N 118 41W
Odiakwe, *Botswana* . . . . . **56 C4** 20 12S 25 17 E
Odienné, *Ivory C.* **50 G3** 9 30N 7 34W
Odintsovo, *Russia* . . . . . . **24 C6** 55 39N 37 15 E
O'Donnell, *U.S.A.* **81 J4** 32 58N 101 50W
Odorheiu Secuiesc, *Romania* . . . . **17 E13** 46 21N 25 21 E
Odra = Oder →, *Germany* . . . . **16 B8** 53 33N 14 38 E

Odzi, Zimbabwe 57 B5 19 0S 32 20 E
Oeiras, Brazil 93 E10 7 0S 42 8W
Oelrichs, U.S.A. 80 D3 43 11N 103 14W
Oelwein, U.S.A. 80 D9 42 41N 91 55W
Oenpelli, Australia 60 B5 12 20S 133 4 E
Ofanto →, Italy 20 D7 41 22N 16 13 E
Offa, Nigeria 50 G5 8 13N 4 42 E
Offaly □, Ireland 13 C4 53 15N 7 30W
Offenbach, Germany 16 C5 50 6N 8 44 E
Offenburg, Germany 16 D4 48 28N 7 56 E
Ofotfjorden, Norway 8 B17 68 27N 17 0 E
Ōfunato, Japan 30 E10 39 4N 141 43 E
Oga, Japan 30 E9 39 55N 139 50 E
Oga-Hantō, Japan 30 E9 39 58N 139 47 E
Ogahalla, Canada 70 B2 50 6N 85 51W
Ōgaki, Japan 31 G8 35 21N 136 37 E
Ogallala, U.S.A. 80 E4 41 8N 101 43W
Ogasawara Gunto, Pac. Oc. 28 G18 27 0N 142 0 E
Ogbomosho, Nigeria 50 G5 8 1N 4 11 E
Ogden, Iowa, U.S.A. 80 D8 42 2N 94 2W
Ogden, Utah, U.S.A. 82 F7 41 13N 111 58W
Ogdensburg, U.S.A. 79 B9 44 42N 75 30W
Ogeechee →, U.S.A. 77 K5 31 50N 81 3W
Ogilby, U.S.A. 85 N12 32 49N 114 50W
Oglio →, Italy 20 B4 45 2N 10 39 E
Ogmore, Australia 62 C4 22 37S 149 35 E
Ogoki →, Canada 70 B2 51 38N 85 57W
Ogoki L., Canada 70 B2 50 50N 87 10W
Ogoki Res., Canada 70 B2 50 45N 88 15W
Ogooué →, Gabon 52 E1 1 0S 9 0 E
Ogowe = Ogooué →, Gabon 52 E1 1 0S 9 0 E
Ogre, Latvia 9 H21 56 49N 24 36 E
Ohai, N.Z. 59 L2 45 55S 168 0 E
Ohakune, N.Z. 59 H5 39 24S 175 24 E
Ohanet, Algeria 50 C6 28 44N 8 46 E
Ohata, Japan 30 D10 41 24N 141 10 E
Ohau, L., N.Z. 59 L2 44 15S 169 53 E
Ohey, Belgium 15 D5 50 26N 5 8 E
Ohio □, U.S.A. 76 E3 40 15N 82 45W
Ohio →, U.S.A. 76 G1 36 59N 89 8W
Ohre →, Czech. 16 C8 50 30N 14 10 E
Ohrid, Macedonia 21 D9 41 8N 20 52 E
Ohridsko Jezero, Macedonia 21 D9 41 8N 20 52 E
Ohrigstad, S. Africa 57 C5 24 39S 30 36 E
Oikou, China 35 E9 38 35N 117 42 E
Oil City, U.S.A. 78 E5 41 26N 79 42W
Oildale, U.S.A. 85 K7 35 25N 119 1W
Oise →, France 18 B5 49 0N 2 4 E
Ōita, Japan 31 H5 33 14N 131 36 E
Ōita □, Japan 31 H5 33 15N 131 30 E
Oiticica, Brazil 93 E10 5 3S 41 5W
Ojai, U.S.A. 85 L7 34 27N 119 15W
Ojinaga, Mexico 86 B4 29 34N 104 25W
Ojiya, Japan 31 F9 37 18N 138 48 E
Ojos del Salado, Cerro, Argentina 94 B2 27 0S 68 40W
Oka →, Russia 26 D5 56 20N 43 59 E
Okaba, Indonesia 37 F9 8 6S 139 42 E
Okahandja, Namibia 56 C2 22 0S 16 59 E
Okahukura, N.Z. 59 H5 38 48S 175 14 E
Okanagan L., Canada 72 C5 50 0N 119 30W
Okandja, Gabon 52 E2 0 35S 13 45 E
Okanogan, U.S.A. 82 B4 48 22N 119 35W
Okanogan →, U.S.A. 82 B4 48 6N 119 44W
Okaputa, Namibia 56 C2 20 5S 17 0 E
Okara, Pakistan 42 D5 30 50N 73 31 E
Okarito, N.Z. 59 K3 43 15S 170 9 E
Okaukuejo, Namibia 56 B2 19 10S 16 0 E
Okavango Swamps, Botswana 56 B3 18 45S 22 45 E
Okaya, Japan 31 F9 36 5N 138 10 E
Okayama, Japan 31 G6 34 40N 133 54 E
Okayama □, Japan 31 G6 35 0N 133 50 E
Okazaki, Japan 31 G8 34 57N 137 10 E
Okeechobee, U.S.A. 77 M5 27 15N 80 50W
Okeechobee, L., U.S.A. 77 M5 27 0N 80 50W
Okefenokee Swamp, U.S.A. 77 K4 30 40N 82 20W
Okehampton, U.K. 11 G3 50 44N 4 0W
Okha, Russia 27 D15 53 40N 143 0 E
Okhotsk, Russia 27 D15 59 20N 143 10 E
Okhotsk, Sea of, Asia 27 D15 55 0N 145 0 E
Okhotskiy Perevoz, Russia 27 C14 61 52N 135 35 E
Oki-Shotō, Japan 31 F6 36 5N 133 15 E
Okiep, S. Africa 56 D2 29 39S 17 53 E
Okinawa □, Japan 31 L3 26 40N 128 0 E
Okinawa-Guntō, Japan 31 L3 26 40N 128 0 E
Okinawa-Jima, Japan 31 L4 26 32N 128 0 E
Okino-erabu-Shima, Japan 31 L4 27 21N 128 33 E
Oklahoma □, U.S.A. 81 H6 35 20N 97 30W
Oklahoma City, U.S.A. 81 H6 35 30N 97 30W
Okmulgee, U.S.A. 81 H7 35 37N 95 58W
Oknitsa = Ocniţa, Moldova 17 D14 48 25N 27 30 E
Okolo, Uganda 54 B3 2 37N 31 8 E
Okolona, U.S.A. 81 H10 34 0N 88 45W
Okrika, Nigeria 50 H6 4 40N 7 10 E
Oksovskiy, Russia 24 B6 62 33N 39 57 E
Oktabrsk = Oktyabrsk, Kazakstan 25 E10 49 28N 57 25 E
Oktyabrsk, Kazakstan 25 E10 49 28N 57 25 E
Oktyabrskiy = Aktsyabrski, Belarus 17 B15 52 38N 28 53 E
Oktyabrskiy, Russia 24 D9 54 28N 53 28 E
Oktyabrskoy Revolyutsii, Os., Russia 27 B10 79 30N 97 0 E
Oktyabrskoye, Russia 26 C7 62 28N 66 3 E
Okuru, N.Z. 59 K2 43 55S 168 55 E
Okushiri-Tō, Japan 30 C9 42 15N 139 30 E
Okwa →, Botswana 56 C3 22 30S 23 0 E
Ola, U.S.A. 81 H8 35 2N 93 13W
Ólafsfjörður, Iceland 8 C4 66 4N 18 39W
Ólafsvík, Iceland 8 D2 64 53N 23 43W
Olancha, U.S.A. 85 J8 36 17N 118 1W
Olancha Pk., U.S.A. 85 J8 36 15N 118 7W
Olanchito, Honduras 88 C2 15 30N 86 30W
Öland, Sweden 9 H17 56 45N 16 38 E
Olary, Australia 63 E3 32 18S 140 19 E
Olascoaga, Argentina 94 D3 35 15S 60 39W
Olathe, U.S.A. 80 F7 38 53N 94 49W
Olavarría, Argentina 94 D3 36 55S 60 20W
Oława, Poland 17 C9 50 57N 17 20 E
Ólbia, Italy 20 D3 40 55N 9 31 E
Old Bahama Chan. = Bahama, Canal Viejo de, W. Indies 88 B4 22 10N 77 30W
Old Baldy Pk. = San Antonio, Mt., U.S.A. 85 L9 34 17N 117 38W
Old Cork, Australia 62 C3 22 57S 141 52 E
Old Crow, Canada 68 B6 67 30N 139 55W
Old Dale, U.S.A. 85 L11 34 8N 115 47W
Old Fletton, U.K. 11 E7 52 33N 0 14W
Old Forge, N.Y., U.S.A. 79 C10 43 43N 74 58W
Old Forge, Pa., U.S.A. 79 E9 41 22N 75 45W
Old Fort →, Canada 73 B6 58 36N 110 24W
Old Shinyanga, Tanzania 54 C3 3 33S 33 27 E
Old Speck Mt., U.S.A. 79 B14 44 34N 70 57W
Old Town, U.S.A. 71 D6 44 56N 68 39W
Old Wives L., Canada 73 C7 50 5N 106 0W
Oldbury, U.K. 11 F5 51 38N 2 33W
Oldcastle, Ireland 13 C4 53 46N 7 10W
Oldeani, Tanzania 54 C4 3 22S 35 35 E
Oldenburg, Germany 16 B5 53 9N 8 13 E
Oldenzaal, Neths. 15 B6 52 19N 6 53 E
Oldham, U.K. 10 D5 53 33N 2 7W
Oldman →, Canada 72 D6 49 57N 111 42W
Olds, Canada 72 C6 51 50N 114 10W
Olean, U.S.A. 78 D6 42 5N 78 26W
Olekma →, Russia 27 C13 60 22N 120 42 E
Olekminsk, Russia 27 C13 60 25N 120 30 E
Oleksandriya, Ukraine 17 C14 50 37N 26 19 E
Olema, U.S.A. 84 G4 38 3N 122 47W
Olenegorsk, Russia 24 A5 68 9N 33 18 E
Olenek, Russia 27 C12 68 28N 112 18 E
Olenek →, Russia 27 B13 73 0N 120 10 E
Oléron, I. d', France 18 D3 45 55N 1 15W
Oleśnica, Poland 17 C9 51 13N 17 22 E
Olevsk, Ukraine 17 C14 51 12N 27 39 E
Olga, Russia 27 E14 43 50N 135 14 E
Olga, L., Canada 70 C4 49 47N 77 15W
Olga, Mt., Australia 61 E5 25 20S 130 50 E
Olhão, Portugal 19 D2 37 3N 7 48W
Olifants →, Africa 57 C5 23 57S 31 58 E
Olifantshoek, S. Africa 56 D3 27 57S 22 42 E
Ólimbos, Óros, Greece 21 D10 40 6N 22 23 E
Olímpia, Brazil 95 A6 20 44S 48 54W
Olinda, Brazil 93 E12 8 1S 34 51W
Oliva, Argentina 94 C3 32 0S 63 38W
Olivehurst, U.S.A. 84 F5 39 6N 121 34W
Oliveira, Brazil 93 H10 20 39S 44 50W
Olivenza, Spain 19 C2 38 41N 7 9W
Oliver, Canada 72 D5 49 13N 119 37W
Oliver L., Canada 73 B8 56 56N 103 22W
Ollagüe, Chile 94 A2 21 15S 68 10W
Olney, Ill., U.S.A. 76 F1 38 44N 88 5W
Olney, Tex., U.S.A. 81 J5 33 22N 98 45W
Olomane →, Canada 71 B7 50 14N 60 37W
Olomouc, Czech. 17 D9 49 38N 17 12 E
Olonets, Russia 24 B5 61 0N 32 54 E
Olongapo, Phil. 37 B6 14 50N 120 18 E
Olot, Spain 19 A7 42 11N 2 30 E
Olovyannaya, Russia 27 D12 50 58N 115 35 E
Oloy →, Russia 27 C16 66 29N 159 29 E
Olsztyn, Poland 17 B11 53 48N 20 29 E
Olt →, Romania 17 G13 43 43N 24 51 E
Olteniţa, Romania 17 F14 44 7N 26 42 E
Olton, U.S.A. 81 H3 34 11N 102 8W
Olymbos, Cyprus 23 D12 35 21N 33 45 E
Olympia, Greece 21 F9 37 39N 21 39 E
Olympia, U.S.A. 84 D4 47 3N 122 53W
Olympic Mts., U.S.A. 84 C3 47 55N 123 45W
Olympic Nat. Park, U.S.A. 84 C3 47 48N 123 30W
Olympus, Cyprus 23 E11 34 56N 32 52 E
Olympus, Mt. = Ólimbos, Óros, Greece 21 D10 40 6N 22 23 E
Olympus, Mt., U.S.A. 84 C3 47 48N 123 43W
Olyphant, U.S.A. 79 E9 41 27N 75 36W
Om →, Russia 26 D8 54 59N 73 22 E
Om Koi, Thailand 38 D2 17 48N 98 22 E
Ōma, Japan 30 D10 41 45N 141 5 E
Ōmachi, Japan 31 F8 36 30N 137 50 E
Ōmae-Zaki, Japan 31 G9 34 36N 138 14 E
Ōmagari, Japan 30 E10 39 27N 140 29 E
Omagh, U.K. 13 B4 54 36N 7 19W
Omagh □, U.K. 13 B4 54 35N 7 15W
Omaha, U.S.A. 80 E7 41 17N 95 58W
Omak, U.S.A. 82 B4 48 25N 119 31W
Omalos, Greece 23 D5 35 19N 23 55 E
Oman ■, Asia 46 C6 23 0N 58 0 E
Oman, G. of, Asia 45 E8 24 30N 58 30 E
Omaruru, Namibia 56 C2 21 26S 16 0 E
Omaruru →, Namibia 56 C1 22 7S 14 15 E
Omate, Peru 92 G4 16 45S 71 0W
Ombai, Selat, Indonesia 37 F6 8 30S 124 50 E
Omboué, Gabon 52 E1 1 35S 9 15 E
Ombrone →, Italy 20 C4 42 42N 11 5 E
Omdurmân, Sudan 51 E11 15 40N 32 28 E
Omeonga, Zaïre 54 C1 3 40S 24 22 E
Ometepe, I. de, Nic. 88 D2 11 32N 85 35W
Ometepec, Mexico 87 D5 16 39N 98 23W
Ominato, Japan 30 D10 41 17N 141 10 E
Omineca →, Canada 72 B4 56 3N 124 16W
Omitara, Namibia 56 C2 22 16S 18 2 E
Ōmiya, Japan 31 G9 35 54N 139 38 E
Ommen, Neths. 15 B6 52 31N 6 26 E
Ömnögovĭ □, Mongolia 34 C3 43 15N 104 0 E
Omo →, Ethiopia 51 G12 6 25N 36 10 E
Omodhos, Cyprus 23 E11 34 51N 32 48 E
Omolon →, Russia 27 C16 68 42N 158 36 E
Omono-Gawa →, Japan 30 E10 39 46N 140 3 E
Omsk, Russia 26 D8 55 0N 73 12 E
Omsukchan, Russia 27 C16 62 32N 155 48 E
Ōmu, Japan 30 B11 44 34N 142 58 E
Omul, Vf., Romania 17 F13 45 27N 25 29 E
Ōmura, Japan 31 H4 32 56N 129 57 E
Omuramba Omatako →, Namibia 53 H4 17 45S 20 25 E
Ōmuta, Japan 31 H5 33 5N 130 26 E
Onaga, U.S.A. 80 F6 39 29N 96 10W
Onalaska, U.S.A. 80 D9 43 53N 91 14W
Onamia, U.S.A. 80 B8 46 4N 93 40W
Onancock, U.S.A. 76 G8 37 43N 75 45W
Onang, Indonesia 37 E5 3 2S 118 49 E
Onaping L., Canada 70 C3 47 3N 81 30W
Onavas, Mexico 86 B3 28 28N 109 30W
Onawa, U.S.A. 80 D6 42 2N 96 6W
Onaway, U.S.A. 76 C3 45 21N 84 14W
Oncócua, Angola 56 B1 16 30S 13 25 E
Onda, Spain 19 C5 39 55N 0 17W
Ondaejin, N. Korea 35 D15 41 34N 129 40 E
Ondangua, Namibia 56 B2 17 57S 16 4 E
Ondjiva, Angola 56 B2 16 48S 15 50 E
Ondo, Nigeria 50 G5 7 4N 4 47 E
Öndörshil, Mongolia 34 B5 45 13N 108 5 E
Öndverðarnes, Iceland 8 D1 64 52N 24 0W
Onega, Russia 24 B6 64 0N 38 10 E
Onega →, Russia 24 B6 63 58N 38 2 E
Onega, G. of = Onezhskaya Guba, Russia 24 B6 64 24N 36 38 E
Onega, L. = Onezhskoye Ozero, Russia 24 B6 61 44N 35 22 E
Onehunga, N.Z. 59 G5 36 55S 174 48 E
Oneida, U.S.A. 79 C9 43 6N 75 39W
Oneida L., U.S.A. 79 C9 43 12N 75 54W
O'Neill, U.S.A. 80 D5 42 27N 98 39W
Onekotan, Ostrov, Russia 27 E16 49 25N 154 45 E
Onema, Zaïre 54 C1 4 35S 24 30 E
Oneonta, Ala., U.S.A. 77 J2 33 57N 86 28W
Oneonta, N.Y., U.S.A. 79 D9 42 27N 75 4W
Oneşti, Romania 17 E14 46 15N 26 45 E
Onezhskaya Guba, Russia 24 B6 64 24N 36 38 E
Onezhskoye Ozero, Russia 24 B6 61 44N 35 22 E
Ongarue, N.Z. 59 H5 38 42S 175 19 E
Ongerup, Australia 61 F2 33 58S 118 28 E
Ongjin, N. Korea 35 F13 37 56N 125 21 E
Ongkharak, Thailand 38 E3 14 8N 101 1 E
Ongniud Qi, China 35 C10 43 0N 118 38 E
Ongoka, Zaïre 54 C2 1 20S 26 0 E
Ongole, India 40 M12 15 33N 80 2 E
Ongon, Mongolia 34 B7 45 41N 113 5 E
Onguren, Russia 27 D11 53 38N 107 36 E
Onida, U.S.A. 80 C4 44 42N 100 4W
Onilahy →, Madag. 57 C7 23 34S 43 45 E
Onitsha, Nigeria 50 G6 6 6N 6 42 E
Onoda, Japan 31 G5 34 2N 131 25 E
Onpyŏng-ni, S. Korea 35 H14 33 25N 126 55 E
Onslow, Australia 60 D2 21 40S 115 12 E
Onslow B., U.S.A. 77 H7 34 20N 77 15W
Onstwedde, Neths. 15 A7 53 2N 7 4 E
Ontake-San, Japan 31 G8 35 53N 137 29 E
Ontario, Calif., U.S.A. 85 L9 34 4N 117 39W
Ontario, Oreg., U.S.A. 82 D5 44 2N 116 58W
Ontario □, Canada 70 B2 48 0N 83 0W
Ontario, L., U.S.A. 70 D4 43 20N 78 0W
Ontonagon, U.S.A. 80 B10 46 52N 89 19W
Onyx, U.S.A. 85 K8 35 41N 118 14W
Oodnadatta, Australia 63 D2 27 33S 135 30 E
Ooldea, Australia 61 F5 30 27S 131 50 E
Oombulgurri, Australia 60 C4 15 15S 127 45 E
Oona River, Canada 72 C2 53 57N 130 16W
Oorindi, Australia 62 C3 20 40S 141 1 E
Oost-Vlaanderen □, Belgium 15 C3 51 5N 3 50 E
Oostende, Belgium 15 C2 51 15N 2 54 E
Oosterhout, Neths. 15 C4 51 39N 4 47 E
Oosterschelde, Neths. 15 C4 51 33N 4 0 E
Ootacamund, India 40 P10 11 30N 76 44 E
Ootsa L., Canada 72 C3 53 50N 126 2W
Opala, Russia 27 D16 51 58N 156 30 E
Opala, Zaïre 54 C1 0 40S 24 20 E
Opanake, Sri Lanka 40 R12 6 35N 80 40 E
Opasatika, Canada 70 C3 49 30N 82 50W
Opasquia, Canada 73 C10 53 16N 93 34W
Opava, Czech. 17 D9 49 57N 17 58 E
Opelousas, U.S.A. 81 K8 30 32N 92 5W
Opémisca, L., Canada 70 C5 49 56N 74 52W
Opheim, U.S.A. 82 B10 48 51N 106 24W
Ophthalmia Ra., Australia 60 D2 23 15S 119 30 E
Opinaca →, Canada 70 B4 52 15N 78 2W
Opinaca L., Canada 70 B4 52 39N 76 20W
Opiskotish, L., Canada 71 B6 53 10N 67 50W
Opole, Poland 17 C9 50 42N 17 58 E
Oporto = Porto, Portugal 19 B1 41 8N 8 40W
Opotiki, N.Z. 59 H6 38 1S 177 19 E
Opp, U.S.A. 77 K2 31 17N 86 16W
Oppdal, Norway 9 E13 62 35N 9 41 E
Opua, N.Z. 59 F5 35 19S 174 9 E
Opunake, N.Z. 59 H4 39 26S 173 52 E
Ora, Cyprus 23 E12 34 51N 33 12 E
Ora Banda, Australia 61 F3 30 20S 121 0 E
Oracle, U.S.A. 83 K8 32 37N 110 46W
Oradea, Romania 17 E11 47 2N 21 58 E
Öræfajökull, Iceland 8 D5 64 2N 16 39W
Orai, India 43 G8 25 58N 79 30 E
Oral = Zhayyq →, Kazakstan 25 E9 47 0N 51 48 E
Oral, Kazakstan 24 D9 51 20N 51 20 E
Oran, Algeria 50 A4 35 45N 0 39W
Oran, Argentina 94 A3 23 10S 64 20W
Orange = Oranje →, S. Africa 56 D2 28 41S 16 28 E
Orange, Australia 63 E4 33 15S 149 7 E
Orange, France 18 D6 44 8N 4 47 E
Orange, Calif., U.S.A. 85 M9 33 47N 117 51W
Orange, Mass., U.S.A. 79 D12 42 35N 72 19W
Orange, Tex., U.S.A. 81 K8 30 6N 93 44W
Orange, Va., U.S.A. 76 F6 38 15N 78 7W
Orange, C., Brazil 93 C8 4 20N 51 30W
Orange Cove, U.S.A. 84 J7 36 38N 119 19W
Orange Free State □ = Free State □, S. Africa 56 D4 28 30S 27 0 E
Orange Grove, U.S.A. 81 M6 27 58N 97 56W
Orange Walk, Belize 87 D7 18 6N 88 33W
Orangeburg, U.S.A. 77 J5 33 30N 80 52W
Orangeville, Canada 70 D3 43 55N 80 5W
Oranienburg, Germany 16 B7 52 45N 13 14 E
Oranje →, S. Africa 56 D2 28 41S 16 28 E
Oranje Vrystaat □ = Free State □, S. Africa 56 D4 28 30S 27 0 E
Oranjemund, Namibia 56 D2 28 38S 16 29 E
Oranjerivier, S. Africa 56 D3 29 40S 24 12 E
Oras, Phil. 37 B7 12 9N 125 28 E
Oraşul Stalin = Braşov, Romania 17 F13 45 38N 25 35 E
Orbetello, Italy 20 C4 42 27N 11 13 E
Orbost, Australia 63 F4 37 40S 148 29 E
Orchila, I., Venezuela 92 A5 11 48N 66 10W
Orcutt, U.S.A. 85 L6 34 52N 120 27W
Ord →, Australia 60 C4 15 33S 128 15 E
Ord, Mt., Australia 60 C4 17 20S 125 34 E
Orderville, U.S.A. 83 H7 37 17N 112 38W
Ordos = Mu Us Shamo, China 34 E5 39 0N 109 0 E
Ordway, U.S.A. 80 F3 38 13N 103 46W
Ordzhonikidze = Vladikavkaz, Russia 25 F7 43 0N 44 35 E
Ore, Zaïre 54 B2 3 17N 29 30 E
Ore Mts. = Erzgebirge, Germany 16 C7 50 27N 12 55 E
Örebro, Sweden 9 G16 59 20N 15 18 E
Oregon, U.S.A. 80 D10 42 1N 89 20W
Oregon □, U.S.A. 82 E3 44 0N 121 0W
Oregon City, U.S.A. 84 E4 45 21N 122 36W
Orekhovo-Zuyevo, Russia 24 C6 55 50N 38 55 E
Orel, Russia 24 D6 52 57N 36 3 E
Orem, U.S.A. 82 F8 40 19N 111 42W
Ören, Turkey 21 F12 37 3N 27 57 E
Orenburg, Russia 24 D10 51 45N 55 6 E
Orense, Spain 19 A2 42 19N 7 55W
Orepuki, N.Z. 59 M1 46 19S 167 46 E
Orestiás, Greece 21 D12 41 30N 26 33 E
Orford Ness, U.K. 11 E9 52 5N 1 35 E
Organos, Pta. de los, Canary Is. 22 F2 28 12N 17 17W
Orgaz, Spain 19 C4 39 39N 3 53W
Orgeyev = Orhei, Moldova 17 E15 47 24N 28 50 E
Orhaneli, Turkey 21 E13 39 54N 28 59 E
Orhangazi, Turkey 21 D13 40 29N 29 18 E
Orhei, Moldova 17 E15 47 24N 28 50 E
Orhon Gol →, Mongolia 32 A5 50 21N 106 0 E
Orient, Australia 63 D3 28 7S 142 50 E
Oriental, Cordillera, Colombia 92 B4 6 0N 73 0W
Oriente, Argentina 94 D3 38 44S 60 37W
Orihuela, Spain 19 C5 38 7N 0 55W
Orinoco →, Venezuela 92 B6 9 15N 61 30W
Orissa □, India 41 K14 20 0N 84 0 E
Orissaare, Estonia 9 G20 58 34N 23 5 E
Oristano, Italy 20 E3 39 54N 8 36 E
Oristano, G. di, Italy 20 E3 39 50N 8 29 E
Orizaba, Mexico 87 D5 18 51N 97 6W
Orkanger, Norway 8 E13 63 18N 9 52 E
Orkla →, Norway 8 E13 63 18N 9 51 E
Orkney, S. Africa 56 D4 26 58S 26 40 E
Orkney □, U.K. 12 C6 59 2N 3 13W
Orkney Is., U.K. 12 C6 59 0N 3 0W
Orland, U.S.A. 84 F4 39 45N 122 12W
Orlando, U.S.A. 77 L5 28 33N 81 23W
Orléanais, France 18 C5 48 0N 2 0 E
Orléans, France 18 C4 47 54N 1 52 E
Orleans, U.S.A. 79 B12 44 49N 72 12W
Orléans, I. d', Canada 71 C5 46 54N 70 58W
Ormara, Pakistan 40 G4 25 16N 64 33 E
Ormoc, Phil. 37 B6 11 0N 124 37 E
Ormond, N.Z. 59 H6 38 33S 177 56 E
Ormond Beach, U.S.A. 77 L5 29 17N 81 3W
Ormstown, Canada 79 A11 45 8N 74 0W
Örnsköldsvik, Sweden 8 E18 63 17N 18 40 E
Oro, N. Korea 35 D14 40 1N 127 27 E
Oro →, Mexico 86 B3 25 35N 105 2W
Oro Grande, U.S.A. 85 L9 34 36N 117 20W
Orocué, Colombia 92 C4 4 48N 71 20W
Orogrande, U.S.A. 83 K10 32 24N 106 5W
Orol Dengizi = Aral Sea, Asia 26 E7 44 30N 60 0 E
Oromocto, Canada 71 C6 45 54N 66 29W
Orono, Canada 78 C6 43 59N 78 37W
Oroqen Zizhiqi, China 33 A7 50 34N 123 43 E
Oroquieta, Phil. 37 C6 8 32N 123 44 E
Orós, Brazil 93 E11 6 15S 38 55W
Orosháza, Hungary 17 E11 46 32N 20 42 E
Orotukan, Russia 27 C16 62 16N 151 42 E
Oroville, Calif., U.S.A. 84 F5 39 31N 121 33W
Oroville, Wash., U.S.A. 82 B4 48 56N 119 26W
Oroville, L., U.S.A. 84 F5 39 33N 121 29W
Orroroo, Australia 63 E2 32 43S 138 38 E
Orrville, U.S.A. 78 F3 40 50N 81 46W
Orsha, Belarus 24 D5 54 30N 30 25 E
Orsk, Russia 24 D10 51 12N 58 34 E
Orşova, Romania 17 F12 44 41N 22 25 E
Ortaca, Turkey 21 F13 36 49N 28 45 E
Ortegal, C., Spain 19 A2 43 43N 7 52W
Orthez, France 18 E3 43 29N 0 48W
Ortigueira, Spain 19 A2 43 40N 7 50W
Orting, U.S.A. 84 C4 47 6N 122 12W
Ortles, Italy 20 A4 46 31N 10 33 E
Ortón →, Bolivia 92 F5 10 50S 67 0W
Orūmīyeh, Iran 44 B5 37 40N 45 0 E
Orūmīyeh, Daryācheh-ye, Iran 44 B5 37 50N 45 30 E
Oruro, Bolivia 92 G5 18 0S 67 9W
Orust, Sweden 9 G14 58 10N 11 40 E
Oruzgān □, Afghan. 40 C5 33 30N 66 0 E
Orvieto, Italy 20 C5 42 43N 12 7 E
Orwell, U.S.A. 78 E4 41 32N 80 52W
Orwell →, U.K. 11 F9 51 59N 1 18 E
Oryakhovo, Bulgaria 21 C10 43 40N 23 57 E
Osa, Russia 24 C10 57 17N 55 26 E
Osa, Pen. de, Costa Rica 88 E3 8 0N 84 0W
Osage, Iowa, U.S.A. 80 D8 43 17N 92 49W
Osage, Wyo., U.S.A. 80 D2 43 59N 104 25W
Osage →, U.S.A. 80 F9 38 35N 91 57W
Osage City, U.S.A. 80 F7 38 38N 95 50W
Ōsaka, Japan 31 G7 34 40N 135 30 E
Osan, S. Korea 35 F14 37 11N 127 4 E
Osawatomie, U.S.A. 80 F7 38 31N 94 57W
Osborne, U.S.A. 80 F5 39 26N 98 42W
Osceola, Ark., U.S.A. 81 H10 35 42N 89 58W
Osceola, Iowa, U.S.A. 80 E8 41 2N 93 46W
Oscoda, U.S.A. 78 B1 44 26N 83 20W
Ösel = Saaremaa, Estonia 9 G20 58 30N 22 30 E
Osh, Kyrgyzstan 26 E8 40 37N 72 49 E
Oshawa, Canada 70 D4 43 50N 78 50W
Oshkosh, Nebr., U.S.A. 80 E3 41 24N 102 21W
Oshkosh, Wis., U.S.A. 80 C10 44 1N 88 33W
Oshmyany = Ashmyany, Belarus 9 J21 54 26N 25 52 E

Oshnovīyeh, *Iran* **44 B5** 37 2N 45 6 E
Oshogbo, *Nigeria* **50 G5** 7 48N 4 37 E
Oshtorīnān, *Iran* **45 C6** 34 1N 48 38 E
Oshwe, *Zaïre* **52 E3** 3 25S 19 28 E
Osijek, *Croatia* **21 B8** 45 34N 18 41 E
Osipenko = Berdyansk, *Ukraine* **25 E6** 46 45N 36 50 E
Osipovichi = Asipovichy, *Belarus* **17 B15** 53 19N 28 33 E
Osizweni, *S. Africa* **57 D5** 27 49S 30 7 E
Oskaloosa, *U.S.A.* **80 E8** 41 18N 92 39W
Oskarshamn, *Sweden* **9 H17** 57 15N 16 27 E
Oskélanéo, *Canada* **70 C4** 48 5N 75 15W
Öskemen, *Kazakstan* **26 E9** 50 0N 82 36 E
Oslo, *Norway* **9 G14** 59 55N 10 45 E
Oslob, *Phil.* **37 C6** 9 31N 123 26 E
Oslofjorden, *Norway* **9 G14** 59 20N 10 35 E
Osmanabad, *India* **40 K10** 18 5N 76 10 E
Osmaniye, *Turkey* **25 G6** 37 5N 36 10 E
Osnabrück, *Germany* **16 B5** 52 17N 8 3 E
Osorio, *Brazil* **95 B5** 29 53S 50 17W
Osorno, *Chile* **96 E2** 40 25S 73 0W
Osoyoos, *Canada* **72 D5** 49 0N 119 30W
Osøyri, *Norway* **9 F11** 60 9N 5 30 E
Ospika →, *Canada* **72 B4** 56 20N 124 0W
Osprey Reef, *Australia* **62 A4** 13 52S 146 36 E
Oss, *Neths.* **15 C5** 51 46N 5 32 E
Ossa, Mt., *Australia* **62 G4** 41 52S 146 3 E
Óssa, Oros, *Greece* **21 E10** 39 47N 22 42 E
Ossabaw I., *U.S.A.* **77 K5** 31 50N 81 5W
Ossining, *U.S.A.* **79 E11** 41 10N 73 55W
Ossipee, *U.S.A.* **79 C13** 43 41N 71 7W
Ossokmanuan L., *Canada* **71 B7** 53 25N 65 0W
Ossora, *Russia* **27 D17** 59 20N 163 13 E
Ostend = Oostende, *Belgium* **15 C2** 51 15N 2 54 E
Oster, *Ukraine* **17 C16** 50 57N 30 53 E
Österdalälven, *Sweden* **9 F16** 61 30N 13 45 E
Østerdalen, *Norway* **9 F14** 61 40N 10 50 E
Östersund, *Sweden* **8 E16** 63 10N 14 38 E
Ostfriesische Inseln, *Germany* **16 B4** 53 42N 7 0 E
Ostrava, *Czech.* **17 D10** 49 51N 18 18 E
Ostróda, *Poland* **17 B10** 53 42N 19 58 E
Ostroh, *Ukraine* **17 C14** 50 20N 26 30 E
Ostrołęka, *Poland* **17 B11** 53 4N 21 32 E
Ostrów Mazowiecka, *Poland* **17 B11** 52 50N 21 51 E
Ostrów Wielkopolski, *Poland* **17 C9** 51 36N 17 44 E
Ostrowiec-Świętokrzyski, *Poland* **17 C11** 50 55N 21 22 E
Ostuni, *Italy* **21 D7** 40 44N 17 35 E
Ōsumi-Kaikyō, *Japan* **31 J5** 30 55N 131 0 E
Ōsumi-Shotō, *Japan* **31 J5** 30 30N 130 0 E
Osuna, *Spain* **19 D3** 37 14N 5 8W
Oswego, *U.S.A.* **79 C8** 43 27N 76 31W
Oswestry, *U.K.* **10 E4** 52 52N 3 3W
Oświęcim, *Poland* **17 C10** 50 2N 19 11 E
Otago □, *N.Z.* **59 L2** 45 15S 170 0 E
Otago Harbour, *N.Z.* **59 L3** 45 47S 170 42 E
Ōtake, *Japan* **31 G6** 34 12N 132 13 E
Otaki, *N.Z.* **59 J5** 40 45S 175 10 E
Otaru, *Japan* **30 C10** 43 10N 141 0 E
Otaru-Wan = Ishikari-Wan, *Japan* **30 C10** 43 25N 141 1 E
Otavalo, *Ecuador* **92 C3** 0 13N 78 20W
Otavi, *Namibia* **56 B2** 19 40S 17 24 E
Otchinjau, *Angola* **56 B1** 16 30S 13 56 E
Othello, *U.S.A.* **82 C4** 46 50N 119 10W
Otira Gorge, *N.Z.* **59 K3** 42 53S 171 33 E
Otis, *U.S.A.* **80 E3** 40 9N 102 58W
Otjiwarongo, *Namibia* **56 C2** 20 30S 16 33 E
Otoineppu, *Japan* **30 B11** 44 44N 142 16 E
Otorohanga, *N.Z.* **59 H5** 38 12S 175 14 E
Otoskwin →, *Canada* **70 B2** 52 13N 88 6W
Otosquen, *Canada* **73 C8** 53 17N 102 1W
Otra →, *Norway* **9 G13** 58 9N 8 1 E
Otranto, *Italy* **21 D8** 40 9N 18 28 E
Otranto, C. d', *Italy* **21 D8** 40 7N 18 30 E
Otranto, Str. of, *Italy* **21 D8** 40 15N 18 40 E
Otse, *S. Africa* **56 D4** 25 2S 25 45 E
Ōtsu, *Japan* **31 G7** 35 0N 135 50 E
Ōtsuki, *Japan* **31 G9** 35 36N 138 57 E
Ottawa = Outaouais →, *Canada* **70 C5** 45 27N 74 8W
Ottawa, *Canada* **70 C4** 45 27N 75 42W
Ottawa, *Ill., U.S.A.* **80 E10** 41 21N 88 51W
Ottawa, *Kans., U.S.A.* **80 F7** 38 37N 95 16W
Ottawa Is., *Canada* **69 C11** 59 35N 80 10W
Otter L., *Canada* **73 B8** 55 35N 104 39W
Otter Rapids, *Ont., Canada* **70 B3** 50 11N 81 39W
Otter Rapids, *Sask., Canada* **73 B8** 55 38N 104 44W
Otterville, *Canada* **78 D4** 42 55N 80 36W
Otto Beit Bridge, *Zimbabwe* **55 F2** 15 59S 28 56 E
Ottosdal, *S. Africa* **56 D4** 26 46S 25 59 E
Ottumwa, *U.S.A.* **80 E8** 41 1N 92 25W
Oturkpo, *Nigeria* **50 G6** 7 16N 8 8 E
Otway, B., *Chile* **96 G2** 53 30S 74 0W
Otway, C., *Australia* **63 F3** 38 52S 143 30 E
Otwock, *Poland* **17 B11** 52 5N 21 20 E
Ou →, *Laos* **38 B4** 20 4N 102 13 E
Ou Neua, *Laos* **38 A3** 22 18N 101 48 E
Ou-Sammyaku, *Japan* **30 E10** 39 20N 140 35 E
Ouachita →, *U.S.A.* **81 K9** 31 38N 91 49W
Ouachita, L., *U.S.A.* **81 H8** 34 34N 93 12W
Ouachita Mts., *U.S.A.* **81 H7** 34 40N 94 25W
Ouadâne, *Mauritania* **50 D2** 20 50N 11 40W
Ouadda, *C.A.R.* **51 G9** 8 15N 22 20 E
Ouagadougou, *Burkina Faso* **50 F4** 12 25N 1 30W
Ouahran = Oran, *Algeria* **50 A4** 35 45N 0 39W
Ouallene, *Algeria* **50 D5** 24 41N 1 11 E
Ouanda Djallé, *C.A.R.* **51 G9** 8 55N 22 53 E
Ouango, *C.A.R.* **52 D4** 4 19N 22 30 E
Ouargla, *Algeria* **50 B6** 31 59N 5 16 E
Ouarzazate, *Morocco* **50 B3** 30 55N 6 50W
Oubangi →, *Zaïre* **52 E3** 0 30S 17 50 E
Ouddorp, *Neths.* **15 C3** 51 50N 3 57 E
Oude Rijn →, *Neths.* **15 B4** 52 12N 4 24 E
Oudenaarde, *Belgium* **15 D3** 50 50N 3 37 E
Oudtshoorn, *S. Africa* **56 E3** 33 35S 22 14 E
Ouessant, I. d', *France* **18 B1** 48 28N 5 6W
Ouesso, *Congo* **52 D3** 1 37N 16 5 E
Ouest, Pte., *Canada* **71 C7** 49 52N 64 40W
Ouezzane, *Morocco* **50 B3** 34 51N 5 35W
Ouidah, *Benin* **50 G5** 6 25N 2 0 E
Oujda, *Morocco* **50 B4** 34 41N 1 55W
Oujeft, *Mauritania* **50 D2** 20 2N 13 0W
Oulainen, *Finland* **8 D21** 64 17N 24 47 E
Ouled Djellal, *Algeria* **50 B6** 34 28N 5 2 E
Oulu, *Finland* **8 D21** 65 1N 25 29 E
Oulujärvi, *Finland* **8 D22** 64 25N 27 15 E
Oulujoki →, *Finland* **8 D21** 65 1N 25 30 E
Oum Chalouba, *Chad* **51 E9** 15 48N 20 46 E
Ounasjoki →, *Finland* **8 C21** 66 31N 25 40 E
Ounguati, *Namibia* **56 C2** 22 0S 15 46 E
Ounianga-Kébir, *Chad* **51 E9** 19 4N 20 29 E
Ounianga Sérir, *Chad* **51 E9** 18 54N 20 51 E
Our →, *Lux.* **15 E6** 49 55N 6 5 E
Ouray, *U.S.A.* **83 G10** 38 1N 107 40W
Ourense = Orense, *Spain* **19 A2** 42 19N 7 55W
Ouricuri, *Brazil* **93 E10** 7 53S 40 5W
Ourinhos, *Brazil* **95 A6** 23 0S 49 54W
Ouro Fino, *Brazil* **95 A6** 22 16S 46 25W
Ouro Prêto, *Brazil* **95 A7** 20 20S 43 30W
Ourthe →, *Belgium* **15 D5** 50 29N 5 35 E
Ouse, *Australia* **62 G4** 42 38S 146 42 E
Ouse →, *E. Susx., U.K.* **11 G8** 50 47N 0 4 E
Ouse →, *N. Yorks., U.K.* **10 C8** 53 44N 0 55W
Outaouais →, *Canada* **70 C5** 45 27N 74 8W
Outardes →, *Canada* **71 C6** 49 24N 69 30W
Outer Hebrides, *U.K.* **12 D1** 57 30N 7 40W
Outer I., *Canada* **71 B8** 51 10N 58 35W
Outjo, *Namibia* **56 C2** 20 5S 16 7 E
Outlook, *Canada* **73 C7** 51 30N 107 0W
Outlook, *U.S.A.* **80 A2** 48 53N 104 47W
Outokumpu, *Finland* **8 E23** 62 43N 29 1 E
Ouyen, *Australia* **63 F3** 35 1S 142 22 E
Ovalau, *Fiji* **59 C8** 17 40S 178 48 E
Ovalle, *Chile* **94 C1** 30 33S 71 18W
Ovamboland, *Namibia* **56 B2** 18 30S 16 0 E
Overflakkee, *Neths.* **15 C4** 51 44N 4 10 E
Overijssel □, *Neths.* **15 B6** 52 25N 6 35 E
Overland Park, *U.S.A.* **80 F7** 38 55N 94 50W
Overpelt, *Belgium* **15 C5** 51 12N 5 20 E
Overton, *U.S.A.* **85 J12** 36 33N 114 27W
Övertorneå, *Sweden* **8 C20** 66 23N 23 38 E
Ovid, *U.S.A.* **80 E3** 40 58N 102 23W
Oviedo, *Spain* **19 A3** 43 25N 5 50W
Oviši, *Latvia* **9 H19** 57 33N 21 44 E
Övör Hangay □, *Mongolia* **34 B2** 45 0N 102 30 E
Øvre Årdal, *Norway* **9 F12** 61 19N 7 48 E
Ovruch, *Ukraine* **17 C15** 51 25N 28 45 E
Owaka, *N.Z.* **59 M2** 46 27S 169 40 E
Owambo = Ovamboland, *Namibia* **56 B2** 18 30S 16 0 E
Owase, *Japan* **31 G8** 34 7N 136 12 E
Owatonna, *U.S.A.* **80 C8** 44 5N 93 14W
Owbeh, *Afghan.* **40 B3** 34 28N 63 10 E
Owego, *U.S.A.* **79 D8** 42 6N 76 16W
Owen Falls Dam, *Uganda* **54 B3** 0 30N 33 5 E
Owen Sound, *Canada* **70 D3** 44 35N 80 55W
Owendo, *Gabon* **52 D1** 0 17N 9 30 E
Owens →, *U.S.A.* **84 J9** 36 32N 117 59W
Owens L., *U.S.A.* **85 J9** 36 26N 117 57W
Owensboro, *U.S.A.* **76 G2** 37 46N 87 7W
Owensville, *U.S.A.* **80 F9** 38 21N 91 30W
Owl →, *Canada* **73 B10** 57 51N 92 44W
Owo, *Nigeria* **50 G6** 7 10N 5 39 E
Owosso, *U.S.A.* **76 D3** 43 0N 84 10W
Owyhee, *U.S.A.* **82 F5** 41 57N 116 6W
Owyhee →, *U.S.A.* **82 E5** 43 49N 117 2W
Owyhee, L., *U.S.A.* **82 E5** 43 38N 117 14W
Öxarfjörður, *Iceland* **8 C5** 66 15N 16 45W
Oxelösund, *Sweden* **9 G17** 58 43N 17 15 E
Oxford, *N.Z.* **59 K4** 43 18S 172 11 E
Oxford, *U.K.* **11 F6** 51 46N 1 15W
Oxford, *Miss., U.S.A.* **81 H10** 34 22N 89 31W
Oxford, *N.C., U.S.A.* **77 G6** 36 19N 78 35W
Oxford, *Ohio, U.S.A.* **76 F3** 39 31N 84 45W
Oxford L., *Canada* **73 C9** 54 51N 95 37W
Oxfordshire □, *U.K.* **11 F6** 51 48N 1 16W
Oxley, *Australia* **63 E3** 34 11S 144 6 E
Oxnard, *U.S.A.* **85 L7** 34 12N 119 11W
Oxus = Amudarya →, *Uzbekistan* **26 E6** 43 58N 59 34 E
Oya, *Malaysia* **36 D4** 2 55N 111 55 E
Oyama, *Japan* **31 F9** 36 18N 139 48 E
Oyem, *Gabon* **52 D2** 1 34N 11 31 E
Oyen, *Canada* **73 C6** 51 22N 110 28W
Oykel →, *U.K.* **12 D4** 57 56N 4 26W
Oymyakon, *Russia* **27 C15** 63 25N 142 44 E
Oyo, *Nigeria* **50 G5** 7 46N 3 56 E
Oyster Bay, *U.S.A.* **79 F11** 40 52N 73 32W
Ōyūbari, *Japan* **30 C11** 43 1N 142 5 E
Ozamiz, *Phil.* **37 C6** 8 15N 123 50 E
Ozark, *Ala., U.S.A.* **77 K3** 31 28N 85 39W
Ozark, *Ark., U.S.A.* **81 H8** 35 29N 93 50W
Ozark, *Mo., U.S.A.* **81 G8** 37 1N 93 12W
Ozark Plateau, *U.S.A.* **81 G9** 37 20N 91 40W
Ozarks, L. of the, *U.S.A.* **80 F8** 38 12N 92 38W
Ózd, *Hungary* **17 D11** 48 14N 20 15 E
Ozette L., *U.S.A.* **84 B2** 48 6N 124 38W
Ozona, *U.S.A.* **81 K4** 30 43N 101 12W
Ozuluama, *Mexico* **87 C5** 21 40N 97 50W

P

Pa-an, *Burma* **41 L20** 16 51N 97 40 E
Pa Mong Dam, *Thailand* **38 D4** 18 0N 102 22 E
Paamiut = Frederikshåb, *Greenland* **4 C5** 62 0N 49 43W
Paarl, *S. Africa* **56 E2** 33 45S 18 56 E
Paauilo, *U.S.A.* **74 H17** 20 2N 155 22W
Pab Hills, *Pakistan* **42 F2** 26 30N 66 45 E
Pabianice, *Poland* **17 C10** 51 40N 19 20 E
Pabna, *Bangla.* **41 G16** 24 1N 89 18 E
Pabo, *Uganda* **54 B3** 3 1N 32 10 E
Pacaja →, *Brazil* **93 D8** 1 56S 50 50W
Pacaraima, Sierra, *Venezuela* **92 C6** 4 0N 62 30W
Pacasmayo, *Peru* **92 E3** 7 20S 79 35W
Pachhar, *India* **42 G7** 24 40N 77 42 E
Pachpadra, *India* **40 G8** 25 58N 72 10 E
Pachuca, *Mexico* **87 C5** 20 10N 98 40W
Pacific, *Canada* **72 C3** 54 48N 128 28W
Pacific-Antarctic Ridge, *Pac. Oc.* **65 M16** 43 0S 115 0W
Pacific Grove, *U.S.A.* **83 H3** 36 38N 121 56W
Pacific Ocean, *Pac. Oc.* **65 G14** 10 0N 140 0W
Pacifica, *U.S.A.* **84 H4** 37 36N 122 30W
Pacitan, *Indonesia* **37 H14** 8 12S 111 7 E
Packwood, *U.S.A.* **84 D5** 46 36N 121 40W
Padaido, Kepulauan, *Indonesia* **37 E9** 1 5S 138 0 E
Padang, *Indonesia* **36 E2** 1 0S 100 20 E
Padangpanjang, *Indonesia* **36 E2** 0 40S 100 20 E
Padangsidempuan, *Indonesia* **36 D1** 1 30N 99 15 E
Paddockwood, *Canada* **73 C7** 53 30N 105 30W
Paderborn, *Germany* **16 C5** 51 42N 8 45 E
Padloping Island, *Canada* **69 B13** 67 0N 62 50W
Pádova, *Italy* **20 B4** 45 25N 11 53 E
Padra, *India* **42 H5** 22 15N 73 7 E
Padrauna, *India* **43 F10** 26 54N 83 59 E
Padre I., *U.S.A.* **81 M6** 27 10N 97 25W
Padstow, *U.K.* **11 G3** 50 33N 4 58W
Padua = Pádova, *Italy* **20 B4** 45 25N 11 53 E
Paducah, *Ky., U.S.A.* **76 G1** 37 5N 88 37W
Paducah, *Tex., U.S.A.* **81 H4** 34 1N 100 18W
Paengnyong-do, *S. Korea* **35 F13** 37 57N 124 40 E
Paeroa, *N.Z.* **59 G5** 37 23S 175 41 E
Pafúri, *Mozam.* **57 C5** 22 28S 31 17 E
Pag, *Croatia* **16 F8** 44 25N 15 3 E
Pagadian, *Phil.* **37 C6** 7 55N 123 30 E
Pagai Selatan, P., *Indonesia* **36 E2** 3 0S 100 15 E
Pagai Utara, *Indonesia* **36 E2** 2 35S 100 0 E
Pagalu = Annobón, *Atl. Oc.* **49 G4** 1 25S 5 36 E
Pagastikós Kólpos, *Greece* **21 E10** 39 15N 23 0 E
Pagatan, *Indonesia* **36 E5** 3 33S 115 59 E
Page, *Ariz., U.S.A.* **83 H8** 36 57N 111 27W
Page, *N. Dak., U.S.A.* **80 B6** 47 10N 97 34W
Pago Pago, *Amer. Samoa* **59 B13** 14 16S 170 43W
Pagosa Springs, *U.S.A.* **83 H10** 37 16N 107 1W
Pagwa River, *Canada* **70 B2** 50 2N 85 14W
Pahala, *U.S.A.* **74 J17** 19 12N 155 29W
Pahang →, *Malaysia* **39 L4** 3 30N 103 9 E
Pahiatua, *N.Z.* **59 J5** 40 27S 175 50 E
Pahokee, *U.S.A.* **77 M5** 26 50N 80 40W
Pahrump, *U.S.A.* **85 J11** 36 12N 115 59W
Pahute Mesa, *U.S.A.* **84 H10** 37 20N 116 45W
Pai, *Thailand* **38 C2** 19 19N 98 27 E
Paia, *U.S.A.* **74 H16** 20 54N 156 22W
Paicines, *U.S.A.* **84 J5** 36 44N 121 17W
Paide, *Estonia* **9 G21** 58 57N 25 31 E
Paignton, *U.K.* **11 G4** 50 26N 3 35W
Päijänne, *Finland* **9 F21** 61 30N 25 30 E
Painan, *Indonesia* **36 E2** 1 21S 100 34 E
Painesville, *U.S.A.* **78 E3** 41 43N 81 15W
Paint Hills = Wemindji, *Canada* **70 B4** 53 0N 78 49W
Paint L., *Canada* **73 B9** 55 28N 97 57W
Paint Rock, *U.S.A.* **81 K5** 31 31N 99 55W
Painted Desert, *U.S.A.* **83 J8** 36 0N 111 0W
Paintsville, *U.S.A.* **76 G4** 37 49N 82 48W
País Vasco □, *Spain* **19 A4** 42 50N 2 45W
Paisley, *Canada* **78 B3** 44 18N 81 16W
Paisley, *U.K.* **12 F4** 55 50N 4 25W
Paisley, *U.S.A.* **82 E3** 42 42N 120 32W
Paita, *Peru* **92 E2** 5 11S 81 9W
Pajares, Puerto de, *Spain* **19 A3** 42 58N 5 46W
Pak Lay, *Laos* **38 C3** 18 15N 101 27 E
Pak Phanang, *Thailand* **39 H3** 8 21N 100 12 E
Pak Sane, *Laos* **38 C4** 18 22N 103 39 E
Pak Song, *Laos* **38 E6** 15 11N 106 14 E
Pak Suong, *Laos* **38 C4** 19 58N 102 15 E
Pakaraima Mts., *Guyana* **92 B6** 6 0N 60 0W
Pákhnes, *Greece* **23 D6** 35 16N 24 4 E
Pakistan ■, *Asia* **42 E3** 30 0N 70 0 E
Pakkading, *Laos* **38 C4** 18 19N 103 59 E
Pakokku, *Burma* **41 J19** 21 20N 95 0 E
Pakpattan, *Pakistan* **42 D5** 30 25N 73 27 E
Pakse, *Laos* **38 E5** 15 5N 105 52 E
Paktīā □, *Afghan.* **40 C6** 33 0N 69 15 E
Pakwach, *Uganda* **54 B3** 2 28N 31 27 E
Pala, *Chad* **51 G8** 9 25N 15 5 E
Pala, *U.S.A.* **85 M9** 33 22N 117 5W
Pala, *Zaïre* **54 D2** 6 45S 29 30 E
Palabek, *Uganda* **54 B3** 3 22N 32 33 E
Palacios, *U.S.A.* **81 L6** 28 42N 96 13W
Palagruža, *Croatia* **20 C7** 42 24N 16 15 E
Palaiókastron, *Greece* **23 D8** 35 12N 26 15 E
Palaiokhóra, *Greece* **23 D5** 35 16N 23 39 E
Palam, *India* **40 K10** 19 0N 77 0 E
Palampur, *India* **42 C7** 32 10N 76 30 E
Palana, *Australia* **62 F4** 39 45S 147 55 E
Palana, *Russia* **27 D16** 59 10N 159 59 E
Palanan, *Phil.* **37 A6** 17 8N 122 29 E
Palanan Pt., *Phil.* **37 A6** 17 17N 122 30 E
Palandri, *Pakistan* **43 C5** 33 42N 73 40 E
Palanga, *Lithuania* **9 J19** 55 58N 21 3 E
Palangkaraya, *Indonesia* **36 E4** 2 16S 113 56 E
Palani Hills, *India* **40 P10** 10 14N 77 33 E
Palanpur, *India* **42 G5** 24 10N 72 25 E
Palanro, *Indonesia* **37 E5** 3 21S 119 23 E
Palapye, *Botswana* **56 C4** 22 30S 27 7 E
Palas, *Pakistan* **43 B5** 35 4N 73 14 E
Palatka, *Russia* **27 C16** 60 6N 150 54 E
Palatka, *U.S.A.* **77 L5** 29 39N 81 38W
Palau ■, *Pac. Oc.* **28 J17** 7 30N 134 30 E
Palawan, *Phil.* **36 C5** 9 30N 118 30 E
Palayankottai, *India* **40 Q10** 8 45N 77 45 E
Paldiski, *Estonia* **9 G21** 59 23N 24 9 E
Paleleh, *Indonesia* **37 D6** 1 10N 121 50 E
Palembang, *Indonesia* **36 E2** 3 0S 104 50 E
Palencia, *Spain* **19 A3** 42 1N 4 34W
Paleokastritsa, *Greece* **23 A3** 39 40N 19 41 E
Paleometokho, *Cyprus* **23 D12** 35 7N 33 11 E
Palermo, *Italy* **20 E5** 38 7N 13 22 E
Palermo, *U.S.A.* **82 G3** 39 26N 121 33W
Palestine, *Asia* **47 D4** 32 0N 35 0 E
Palestine, *U.S.A.* **81 K7** 31 46N 95 38W
Paletwa, *Burma* **41 J18** 21 10N 92 50 E
Palghat, *India* **40 P10** 10 46N 76 42 E
Palgrave, Mt., *Australia* **60 D2** 23 22S 115 58 E
Pali, *India* **42 G5** 25 50N 73 20 E
Palioúrion, Ákra, *Greece* **21 E10** 39 57N 23 45 E
Palisade, *U.S.A.* **80 E4** 40 21N 101 7W
Palitana, *India* **42 J4** 21 32N 71 49 E
Palizada, *Mexico* **87 D6** 18 18N 92 8W
Palk Bay, *Asia* **40 Q11** 9 30N 79 15 E
Palk Strait, *Asia* **40 Q11** 10 0N 79 45 E
Palkānah, *Iraq* **44 C5** 35 49N 44 26 E
Palla Road = Dinokwe, *Botswana* **56 C4** 23 29S 26 37 E
Pallanza = Verbánia, *Italy* **20 B3** 45 56N 8 33 E
Pallisa, *Uganda* **54 B3** 1 12N 33 43 E
Pallu, *India* **42 E6** 28 59N 74 14 E
Palm Bay, *U.S.A.* **77 L5** 28 2N 80 35W
Palm Beach, *U.S.A.* **77 M6** 26 43N 80 2W
Palm Desert, *U.S.A.* **85 M10** 33 43N 116 22W
Palm Is., *Australia* **62 B4** 18 40S 146 35 E
Palm Springs, *U.S.A.* **85 M10** 33 50N 116 33W
Palma, *Mozam.* **55 E5** 10 46S 40 29 E
Palma →, *Brazil* **93 F9** 12 33S 47 52W
Palma, B. de, *Spain* **22 B9** 39 30N 2 39 E
Palma de Mallorca, *Spain* **22 B9** 39 35N 2 39 E
Palma Soriano, *Cuba* **88 B4** 20 15N 76 0W
Palmares, *Brazil* **93 E11** 8 41S 35 28W
Palmas, *Brazil* **95 B5** 26 29S 52 0W
Palmas, C., *Liberia* **50 H3** 4 27N 7 46W
Pálmas, G. di, *Italy* **20 E3** 39 0N 8 30 E
Palmdale, *U.S.A.* **85 L8** 34 35N 118 7W
Palmeira dos Índios, *Brazil* **93 E11** 9 25S 36 37W
Palmeirinhas, Pta. das, *Angola* **52 F2** 9 2S 12 57 E
Palmer, *U.S.A.* **68 B5** 61 36N 149 7W
Palmer →, *Australia* **62 B3** 16 0S 142 26 E
Palmer Arch., *Antarctica* **5 C17** 64 15S 65 0W
Palmer Lake, *U.S.A.* **80 F2** 39 7N 104 55W
Palmer Land, *Antarctica* **5 D18** 73 0S 63 0W
Palmerston, *Canada* **78 C4** 43 50N 80 51W
Palmerston, *N.Z.* **59 L3** 45 29S 170 43 E
Palmerston North, *N.Z.* **59 J5** 40 21S 175 39 E
Palmerton, *U.S.A.* **79 F9** 40 48N 75 37W
Palmetto, *U.S.A.* **77 M4** 27 31N 82 34W
Palmi, *Italy* **20 E6** 38 21N 15 51 E
Palmira, *Argentina* **94 C2** 32 59S 68 34W
Palmira, *Colombia* **92 C3** 3 32N 76 16W
Palmyra = Tudmur, *Syria* **44 C3** 34 36N 38 15 E
Palmyra, *Mo., U.S.A.* **80 F9** 39 48N 91 32W
Palmyra, *N.Y., U.S.A.* **78 C7** 43 5N 77 18W
Palmyra Is., *Pac. Oc.* **65 G11** 5 52N 162 5W
Palo Alto, *U.S.A.* **83 H2** 37 27N 122 10W
Palo Verde, *U.S.A.* **85 M12** 33 26N 114 44W
Palopo, *Indonesia* **37 E6** 3 0S 120 16 E
Palos, C. de, *Spain* **19 D5** 37 38N 0 40W
Palos Verdes, *U.S.A.* **85 M8** 33 48N 118 23W
Palos Verdes, Pt., *U.S.A.* **85 M8** 33 43N 118 26W
Palouse, *U.S.A.* **82 C5** 46 55N 117 4W
Palparara, *Australia* **62 C3** 24 47S 141 28 E
Palu, *Indonesia* **37 E5** 1 0S 119 52 E
Palu, *Turkey* **25 G7** 38 45N 40 0 E
Paluan, *Phil.* **37 B6** 13 26N 120 29 E
Palwal, *India* **42 E7** 28 8N 77 19 E
Pama, *Burkina Faso* **50 F5** 11 19N 0 44 E
Pamanukan, *Indonesia* **37 G12** 6 16S 107 49 E
Pamekasan, *Indonesia* **37 G15** 7 10S 113 28 E
Pamiers, *France* **18 E4** 43 7N 1 39 E
Pamirs, *Tajikistan* **26 F8** 37 40N 73 0 E
Pamlico →, *U.S.A.* **77 H7** 35 20N 76 28W
Pamlico Sd., *U.S.A.* **77 H8** 35 20N 76 0W
Pampa, *U.S.A.* **81 H4** 35 32N 100 58W
Pampa de las Salinas, *Argentina* **94 C2** 32 1S 66 58W
Pampanua, *Indonesia* **37 E6** 4 16S 120 8 E
Pampas, *Argentina* **94 D3** 35 0S 63 0W
Pampas, *Peru* **92 F4** 12 20S 74 50W
Pamplona, *Colombia* **92 B4** 7 23N 72 39W
Pamplona, *Spain* **19 A5** 42 48N 1 38W
Pampoenpoort, *S. Africa* **56 E3** 31 3S 22 40 E
Pana, *U.S.A.* **80 F10** 39 23N 89 5W
Panaca, *U.S.A.* **83 H6** 37 47N 114 23W
Panaitan, *Indonesia* **37 G11** 6 36S 105 12 E
Panaji, *India* **40 M8** 15 25N 73 50 E
Panamá, *Panama* **88 E4** 9 0N 79 25W
Panama ■, *Cent. Amer.* **88 E4** 8 48N 79 55W
Panamá, G. de, *Panama* **88 E4** 8 4N 79 20W
Panama Canal, *Panama* **88 E4** 9 10N 79 37W
Panama City, *U.S.A.* **77 K3** 30 10N 85 40W
Panamint Range, *U.S.A.* **85 J9** 36 20N 117 20W
Panamint Springs, *U.S.A.* **85 J9** 36 20N 117 28W
Panão, *Peru* **92 E3** 9 55S 75 55W
Panare, *Thailand* **39 J3** 6 51N 101 30 E
Panarukan, *Indonesia* **37 G15** 7 42S 113 56 E
Panay, *Phil.* **37 B6** 11 10N 122 30 E
Panay, G., *Phil.* **37 B6** 11 0N 122 30 E
Pancake Range, *U.S.A.* **83 G6** 38 30N 115 50W
Pančevo, *Serbia, Yug.* **21 B9** 44 52N 20 41 E
Pandan, *Phil.* **37 B6** 11 45N 122 10 E
Pandegelang, *Indonesia* **37 G12** 6 25S 106 5 E
Pandharpur, *India* **40 L9** 17 41N 75 20 E
Pando, *Uruguay* **95 C4** 34 44S 56 0W
Pando, L. = Hope, L., *Australia* **63 D2** 28 24S 139 18 E
Pandokrátor, *Greece* **23 A3** 39 45N 19 50 E
Pandora, *Costa Rica* **88 E3** 9 43N 83 3W
Panevėžys, *Lithuania* **9 J21** 55 42N 24 25 E
Panfilov, *Kazakstan* **26 E8** 44 10N 80 0 E
Pang-Long, *Burma* **41 H21** 23 11N 98 45 E
Pang-Yang, *Burma* **41 H21** 22 7N 98 48 E
Panga, *Zaïre* **54 B2** 1 52N 26 18 E
Pangalanes, Canal des, *Madag.* **57 C8** 22 48S 47 50 E
Pangani, *Tanzania* **54 D4** 5 25S 38 58 E
Pangani □, *Tanzania* **54 D4** 5 25S 39 0 E
Pangani →, *Tanzania* **54 D4** 5 26S 38 58 E
Pangfou = Bengbu, *China* **35 H9** 32 58N 117 20 E
Pangil, *Zaïre* **54 C2** 3 10S 26 35 E
Pangkah, Tanjung, *Indonesia* **37 G15** 6 51S 112 33 E
Pangkajene, *Indonesia* **37 E5** 4 46S 119 34 E
Pangkalanbrandan, *Indonesia* **36 D1** 4 1N 98 20 E
Pangkalanbuun, *Indonesia* **36 E4** 2 41S 111 37 E

Pangkalansusu, *Indonesia* **36 D1** 4 2N 98 13 E
Pangkalpinang, *Indonesia* **36 E3** 2 0S 106 0 E
Pangkoh, *Indonesia* **36 E4** 3 5S 114 8 E
Pangnirtung, *Canada* **69 B13** 66 8N 65 54W
Pangrango, *Indonesia* **37 G12** 6 46S 107 1 E
Panguitch, *U.S.A.* **83 H7** 37 50N 112 26W
Pangutaran Group, *Phil.* **37 C6** 6 18N 120 34 E
Panhandle, *U.S.A.* **81 H4** 35 21N 101 23W
Pani Mines, *India* **42 H5** 22 29N 73 50 E
Pania-Mutombo, *Zaïre* **54 D1** 5 11S 23 51 E
Panipat, *India* **42 E7** 29 25N 77 2 E
Panjal Range, *India* **42 C7** 32 30N 76 50 E
Panjgur, *Pakistan* **40 F4** 27 0N 64 5 E
Panjim = Panaji, *India* **40 M8** 15 25N 73 50 E
Panjinad Barrage, *Pakistan* **40 E7** 29 22N 71 15 E
Panjwai, *Afghan.* **42 D1** 31 26N 65 27 E
Panmunjŏm, *N. Korea* **35 F14** 37 59N 126 38 E
Panna, *India* **43 G9** 24 40N 80 15 E
Panna Hills, *India* **43 G9** 24 40N 81 15 E
Pano Lefkara, *Cyprus* **23 E12** 34 53N 33 20 E
Pano Panayia, *Cyprus* **23 E11** 34 55N 32 38 E
Panorama, *Brazil* **95 A5** 21 21S 51 51W
Pánormon, *Greece* **23 D6** 35 25N 24 41 E
Panshan, *China* **35 D12** 41 3N 122 2 E
Panshi, *China* **35 C14** 42 58N 126 5 E
Pantar, *Indonesia* **37 F6** 8 28S 124 10 E
Pante Macassar, *Indonesia* **37 F6** 9 30S 123 58 E
Pantelleria, *Italy* **20 F4** 36 50N 11 57 E
Pánuco, *Mexico* **87 C5** 22 0N 98 15W
Panyam, *Nigeria* **50 G6** 9 27N 9 8 E
Paola, *Malta* **23 D2** 35 52N 14 30 E
Paola, *U.S.A.* **80 F7** 38 35N 94 53W
Paonia, *U.S.A.* **83 G10** 38 52N 107 36W
Paoting = Baoding, *China* **34 E8** 38 50N 115 28 E
Paot'ou = Baotou, *China* **34 D6** 40 32N 110 2 E
Paoua, *C.A.R.* **51 G8** 7 9N 16 20 E
Pápa, *Hungary* **17 E9** 47 22N 17 30 E
Papagayo →, *Mexico* **87 D5** 16 36N 99 43W
Papagayo, G. de, *Costa Rica* **88 D2** 10 30N 85 50W
Papakura, *N.Z.* **59 G5** 37 4S 174 59 E
Papantla, *Mexico* **87 C5** 20 30N 97 30W
Papar, *Malaysia* **36 C5** 5 45N 116 0 E
Paphos, *Cyprus* **23 E11** 34 46N 32 25 E
Papien Chiang = Da →, *Vietnam* **38 B5** 21 15N 105 20 E
Papigochic →, *Mexico* **86 B3** 29 9N 109 40W
Paposo, *Chile* **94 B1** 25 0S 70 30W
Papoutsa, *Cyprus* **23 E12** 34 54N 33 4 E
Papua New Guinea ■, *Oceania* **64 H6** 8 0S 145 0 E
Papudo, *Chile* **94 C1** 32 29S 71 27W
Papun, *Burma* **41 K20** 18 2N 97 30 E
Papunya, *Australia* **60 D5** 23 15S 131 54 E
Pará = Belém, *Brazil* **93 D9** 1 20S 48 30W
Pará □, *Brazil* **93 D8** 3 20S 52 0W
Paraburdoo, *Australia* **60 D2** 23 14S 117 32 E
Paracatu, *Brazil* **93 G9** 17 10S 46 50W
Paracel Is., *S. China Sea* **36 A4** 15 50N 112 0 E
Parachilna, *Australia* **63 E2** 31 10S 138 21 E
Parachinar, *Pakistan* **42 C4** 33 55N 70 5 E
Paradhísi, *Greece* **23 C10** 36 18N 28 7 E
Paradip, *India* **41 J15** 20 15N 86 35 E
Paradise, *Calif., U.S.A.* **84 F5** 39 46N 121 37W
Paradise, *Mont., U.S.A.* **82 C6** 47 23N 114 48W
Paradise, *Nev., U.S.A.* **85 J11** 36 9N 115 10W
Paradise →, *Canada* **71 B8** 53 27N 57 19W
Paradise Valley, *U.S.A.* **82 F5** 41 30N 117 32W
Parado, *Indonesia* **37 F5** 8 42S 118 30 E
Paragould, *U.S.A.* **81 G9** 36 3N 90 29W
Paragua →, *Venezuela* **92 B6** 6 55N 62 55W
Paraguaçu →, *Brazil* **93 F11** 12 45S 38 54W
Paraguaçu Paulista, *Brazil* **95 A5** 22 22S 50 35W
Paraguaná, Pen. de, *Venezuela* **92 A4** 12 0N 70 0W
Paraguarí, *Paraguay* **94 B4** 25 36S 57 0W
Paraguarí □, *Paraguay* **94 B4** 26 0S 57 10W
Paraguay ■, *S. Amer.* **94 A4** 23 0S 57 0W
Paraguay →, *Paraguay* **94 B4** 27 18S 58 38W
Paraíba = João Pessoa, *Brazil* **93 E12** 7 10S 34 52W
Paraíba □, *Brazil* **93 E11** 7 0S 36 0W
Paraíba do Sul →, *Brazil* **95 A7** 21 37S 41 3W
Parainen, *Finland* **9 F20** 60 18N 22 18 E
Paraiso, *Mexico* **87 D6** 18 24N 93 14W
Parak, *Iran* **45 E7** 27 38N 52 25 E
Parakou, *Benin* **50 G5** 9 25N 2 40 E
Paralimni, *Cyprus* **23 D12** 35 2N 33 58 E
Paramaribo, *Surinam* **93 B7** 5 50N 55 10W
Paramushir, Ostrov, *Russia* **27 D16** 50 24N 156 0 E
Paran →, *Israel* **47 E4** 30 20N 35 10 E
Paraná, *Argentina* **94 C3** 31 45S 60 30W
Paranã, *Brazil* **93 F9** 12 30S 47 48W
Paraná □, *Brazil* **95 A5** 24 30S 51 0W
Paraná →, *Argentina* **94 C4** 33 43S 59 15W
Paranaguá, *Brazil* **95 B6** 25 30S 48 30W
Paranaíba →, *Brazil* **93 H8** 20 6S 51 4W
Paranapanema →, *Brazil* **95 A5** 22 40S 53 9W
Paranapiacaba, Serra do, *Brazil* **95 A6** 24 31S 48 35W
Paranavaí, *Brazil* **95 A5** 23 4S 52 56W
Parang, *Jolo, Phil.* **37 C6** 5 55N 120 54 E
Parang, *Mindanao, Phil.* **37 C6** 7 23N 124 16 E
Paratinga, *Brazil* **93 F10** 12 40S 43 10W
Paratoo, *Australia* **63 E2** 32 42S 139 20 E
Parattah, *Australia* **62 G4** 42 22S 147 23 E
Parbati →, *India* **42 G7** 25 50N 76 30 E
Parbhani, *India* **40 K10** 19 8N 76 52 E
Parchim, *Germany* **16 B6** 53 26N 11 52 E
Pardes Hanna, *Israel* **47 C3** 32 28N 34 57 E
Pardo →, *Bahia, Brazil* **93 G11** 15 40S 39 0W
Pardo →, *Mato Grosso, Brazil* **95 A5** 21 46S 52 9W
Pardo →, *São Paulo, Brazil* **93 H9** 20 10S 48 38W
Pardubice, *Czech.* **16 C8** 50 3N 15 45 E
Pare, *Indonesia* **37 G15** 7 43S 112 12 E
Pare □, *Tanzania* **54 C4** 4 10S 38 0 E
Pare Mts., *Tanzania* **54 C4** 4 0S 37 45 E
Parecis, Serra dos, *Brazil* **92 F7** 13 0S 60 0W
Pareh, *Iran* **44 B5** 38 52N 45 42 E
Paren, *Russia* **27 C17** 62 30N 163 15 E
Parent, *Canada* **70 C5** 47 55N 74 35W
Parent, L., *Canada* **70 C4** 48 31N 77 1W
Parepare, *Indonesia* **37 E5** 4 0S 119 40 E
Párga, *Greece* **21 E9** 39 15N 20 29 E
Pargo, Pta. do, *Madeira* **22 D2** 32 49N 17 17W
Parguba, *Russia* **24 B5** 62 20N 34 27 E
Pariaguán, *Venezuela* **92 B6** 8 51N 64 34W
Pariaman, *Indonesia* **36 E2** 0 47S 100 11 E
Paricutín, Cerro, *Mexico* **86 D4** 19 28N 102 15W
Parigi, *Java, Indonesia* **37 G13** 7 42S 108 29 E
Parigi, *Sulawesi, Indonesia* **37 E6** 0 50S 120 5 E
Parika, *Guyana* **92 B7** 6 50N 58 20W
Parima, Serra, *Brazil* **92 C6** 2 30N 64 0W
Parinari, *Peru* **92 D4** 4 35S 74 25W
Pariñas, Pta., *S. Amer.* **90 D2** 4 30S 82 0W
Parîngul Mare, *Romania* **17 F12** 45 20N 23 37 E
Parintins, *Brazil* **93 D7** 2 40S 56 50W
Pariparit Kyun, *Burma* **41 M18** 14 55N 93 45 E
Paris, *Canada* **70 D3** 43 12N 80 25W
Paris, *France* **18 B5** 48 50N 2 20 E
Paris, *Idaho, U.S.A.* **82 E8** 42 14N 111 24W
Paris, *Ky., U.S.A.* **76 F3** 38 13N 84 15W
Paris, *Tenn., U.S.A.* **77 G1** 36 18N 88 19W
Paris, *Tex., U.S.A.* **81 J7** 33 40N 95 33W
Parish, *U.S.A.* **79 C8** 43 25N 76 8W
Pariti, *Indonesia* **37 F6** 10 15S 123 45 E
Park, *U.S.A.* **84 B4** 48 45N 122 18W
Park City, *U.S.A.* **82 F8** 40 39N 111 30W
Park Falls, *U.S.A.* **80 C9** 45 56N 90 27W
Park Range, *U.S.A.* **82 G10** 40 0N 106 30W
Park Rapids, *U.S.A.* **80 B7** 46 55N 95 4W
Park River, *U.S.A.* **80 A6** 48 24N 97 45W
Park Rynie, *S. Africa* **57 E5** 30 25S 30 45 E
Parkā Bandar, *Iran* **45 E8** 25 55N 59 35 E
Parkano, *Finland* **9 E20** 62 1N 23 0 E
Parker, *Ariz., U.S.A.* **85 L12** 34 9N 114 17W
Parker, *S. Dak., U.S.A.* **80 D6** 43 24N 97 8W
Parker Dam, *U.S.A.* **85 L12** 34 18N 114 8W
Parkersburg, *U.S.A.* **76 F5** 39 16N 81 34W
Parkerview, *Canada* **73 C8** 51 21N 103 18W
Parkes, *Australia* **63 E4** 33 9S 148 11 E
Parkfield, *U.S.A.* **84 K6** 35 54N 120 26W
Parkland, *U.S.A.* **84 C4** 47 9N 122 26W
Parkside, *Canada* **73 C7** 53 10N 106 33W
Parkston, *U.S.A.* **80 D5** 43 24N 97 59W
Parksville, *Canada* **72 D4** 49 20N 124 21W
Parla, *Spain* **19 B4** 40 14N 3 46W
Parma, *Italy* **20 B4** 44 48N 10 20 E
Parma, *Idaho, U.S.A.* **82 E5** 43 47N 116 57W
Parma, *Ohio, U.S.A.* **78 E3** 41 23N 81 43W
Parnaguá, *Brazil* **93 F10** 10 10S 44 38W
Parnaíba, *Piauí, Brazil* **93 D10** 2 54S 41 47W
Parnaíba, *São Paulo, Brazil* **93 G8** 19 34S 51 14W
Parnaíba →, *Brazil* **93 D10** 3 0S 41 50W
Parnassós, *Greece* **21 E10** 38 35N 22 30 E
Parnu, *Estonia* **9 G21** 58 28N 24 33 E
Paroo →, *Australia* **63 E3** 31 28S 143 32 E
Páros, *Greece* **21 F11** 37 5N 25 12 E
Parowan, *U.S.A.* **83 H7** 37 51N 112 50W
Parral, *Chile* **94 D1** 36 10S 71 52W
Parramatta, *Australia* **63 E5** 33 48S 151 1 E
Parras, *Mexico* **86 B4** 25 30N 102 20W
Parrett →, *U.K.* **11 F5** 51 12N 3 1W
Parris I., *U.S.A.* **77 J5** 32 20N 80 41W
Parrsboro, *Canada* **71 C7** 45 30N 64 25W
Parry Is., *Canada* **4 B2** 77 0N 110 0W
Parry Sound, *Canada* **70 C3** 45 20N 80 0W
Parshall, *U.S.A.* **80 B3** 47 57N 102 8W
Parsnip →, *Canada* **72 B4** 55 10N 123 2W
Parsons, *U.S.A.* **81 G7** 37 20N 95 16W
Parsons Ra., *Australia* **62 A2** 13 30S 135 15 E
Partinico, *Italy* **20 E5** 38 3N 13 7 E
Paru →, *Brazil* **93 D8** 1 33S 52 38W
Paruro, *Peru* **92 F4** 13 45S 71 50W
Parvān □, *Afghan.* **40 B6** 35 0N 69 0 E
Parvatipuram, *India* **41 K13** 18 50N 83 25 E
Parys, *S. Africa* **56 D4** 26 52S 27 29 E
Pasadena, *Calif., U.S.A.* **85 L8** 34 9N 118 9W
Pasadena, *Tex., U.S.A.* **81 L7** 29 43N 95 13W
Pasaje, *Ecuador* **92 D3** 3 23S 79 50W
Pasaje →, *Argentina* **94 B3** 25 39S 63 56W
Pascagoula, *U.S.A.* **81 K10** 30 21N 88 33W
Pascagoula →, *U.S.A.* **81 K10** 30 23N 88 37W
Paşcani, *Romania* **17 E14** 47 14N 26 45 E
Pasco, *U.S.A.* **82 C4** 46 14N 119 6W
Pasco, Cerro de, *Peru* **92 F3** 10 45S 76 10W
Pascua, I. de, *Pac. Oc.* **65 K17** 27 0S 109 0W
Pasfield L., *Canada* **73 B7** 58 24N 105 20W
Pashiwari, *Pakistan* **43 B6** 34 40N 75 10 E
Pashmakli = Smolyan, *Bulgaria* **21 D11** 41 36N 24 38 E
Pasirian, *Indonesia* **37 H15** 8 13S 113 8 E
Paskūh, *Iran* **45 E9** 27 34N 61 39 E
Pasley, C., *Australia* **61 F3** 33 52S 123 35 E
Pašman, *Croatia* **16 G8** 43 58N 15 20 E
Pasni, *Pakistan* **40 G3** 25 15N 63 27 E
Paso Cantinela, *Mexico* **85 N11** 32 33N 115 47W
Paso de Indios, *Argentina* **96 E3** 43 55S 69 0W
Paso de los Libres, *Argentina* **94 B4** 29 44S 57 10W
Paso de los Toros, *Uruguay* **94 C4** 32 45S 56 30W
Paso Robles, *U.S.A.* **83 J3** 35 38N 120 41W
Paspébiac, *Canada* **71 C6** 48 3N 65 17W
Pasrur, *Pakistan* **42 C6** 32 16N 74 43 E
Passage West, *Ireland* **13 E3** 51 52N 8 21W
Passaic, *U.S.A.* **79 F10** 40 51N 74 7W
Passau, *Germany* **16 D7** 48 34N 13 28 E
Passero, C., *Italy* **20 F6** 36 41N 15 10 E
Passo Fundo, *Brazil* **95 B5** 28 10S 52 20W
Passos, *Brazil* **93 H9** 20 45S 46 37W
Pastavy, *Belarus* **9 J22** 55 4N 26 50 E
Pastaza →, *Peru* **92 D3** 4 50S 76 52W
Pasto, *Colombia* **92 C3** 1 13N 77 17W
Pasuruan, *Indonesia* **37 G15** 7 40S 112 44 E
Patagonia, *Argentina* **96 F3** 45 0S 69 0W
Patagonia, *U.S.A.* **83 L8** 31 33N 110 45W
Patambar, *Iran* **45 D9** 29 45N 60 17 E
Patan, *India* **40 H8** 23 54N 72 14 E
Patan, *Maharashtra, India* **42 H5** 23 54N 72 14 E
Patan, *Nepal* **41 F14** 27 40N 85 20 E
Patani, *Indonesia* **37 D7** 0 20N 128 50 E
Pataudi, *India* **42 E7** 28 18N 76 48 E
Patchewollock, *Australia* **63 F3** 35 22S 142 12 E
Patchogue, *U.S.A.* **79 F11** 40 46N 73 1W
Patea, *N.Z.* **59 H5** 39 45S 174 30 E
Pategi, *Nigeria* **50 G6** 8 50N 5 45 E
Patensie, *S. Africa* **56 E3** 33 46S 24 49 E
Paternò, *Italy* **20 F6** 37 34N 14 54 E
Pateros, *U.S.A.* **82 B4** 48 3N 119 54W
Paterson, *U.S.A.* **79 F10** 40 55N 74 11W
Paterson Ra., *Australia* **60 D3** 21 45S 122 10 E
Pathankot, *India* **42 C6** 32 18N 75 45 E
Pathfinder Reservoir, *U.S.A.* **82 E10** 42 28N 106 51W
Pathiu, *Thailand* **39 G2** 10 42N 99 19 E
Pathum Thani, *Thailand* **38 E3** 14 1N 100 32 E
Pati, *Indonesia* **37 G14** 6 45S 111 1 E
Patiala, *India* **42 D7** 30 23N 76 26 E
Patkai Bum, *India* **41 F19** 27 0N 95 30 E
Pátmos, *Greece* **21 F12** 37 21N 26 36 E
Patna, *India* **43 G11** 25 35N 85 12 E
Patonga, *Uganda* **54 B3** 2 45N 33 15 E
Patos, L. dos, *Brazil* **95 C5** 31 20S 51 0W
Patos de Minas, *Brazil* **93 G9** 18 35S 46 32W
Patquía, *Argentina* **94 C2** 30 2S 66 55W
Pátrai, *Greece* **21 E9** 38 14N 21 47 E
Pátraikós Kólpos, *Greece* **21 E9** 38 17N 21 30 E
Patras = Pátrai, *Greece* **21 E9** 38 14N 21 47 E
Patrocínio, *Brazil* **93 G9** 18 57S 47 0W
Patta, *Kenya* **54 C5** 2 10S 41 0 E
Pattani, *Thailand* **39 J3** 6 48N 101 15 E
Patten, *U.S.A.* **71 C6** 46 0N 68 38W
Patterson, *Calif., U.S.A.* **83 H3** 37 28N 121 8W
Patterson, *La., U.S.A.* **81 L9** 29 42N 91 18W
Patterson, Mt., *U.S.A.* **84 G7** 38 29N 119 20W
Patti, *India* **42 D6** 31 17N 74 54 E
Pattoki, *Pakistan* **42 D5** 31 5N 73 52 E
Patton, *U.S.A.* **78 F6** 40 38N 78 39W
Patuakhali, *Bangla.* **41 H17** 22 20N 90 25 E
Patuca →, *Honduras* **88 C3** 15 50N 84 18W
Patuca, Punta, *Honduras* **88 C3** 15 49N 84 14W
Pátzcuaro, *Mexico* **86 D4** 19 30N 101 40W
Pau, *France* **18 E3** 43 19N 0 25W
Pauini →, *Brazil* **92 D6** 1 42S 62 50W
Pauk, *Burma* **41 J19** 21 27N 94 30 E
Paul I., *Canada* **71 A7** 56 30N 61 20W
Paulis = Isiro, *Zaïre* **54 B2** 2 53N 27 40 E
Paulistana, *Brazil* **93 E10** 8 9S 41 9W
Paullina, *U.S.A.* **80 D7** 42 59N 95 41W
Paulo Afonso, *Brazil* **93 E11** 9 21S 38 15W
Paulpietersburg, *S. Africa* **57 D5** 27 23S 30 50 E
Pauls Valley, *U.S.A.* **81 H6** 34 44N 97 13W
Pauma Valley, *U.S.A.* **85 M10** 33 16N 116 58W
Pāveh, *Iran* **44 C5** 35 3N 46 22 E
Pavia, *Italy* **20 B3** 45 7N 9 8 E
Pāvilosta, *Latvia* **9 H19** 56 53N 21 14 E
Pavlodar, *Kazakstan* **26 D8** 52 33N 77 0 E
Pavlograd = Pavlohrad, *Ukraine* **25 E6** 48 30N 35 52 E
Pavlohrad, *Ukraine* **25 E6** 48 30N 35 52 E
Pavlovo, *Oka, Russia* **24 C7** 55 58N 43 5 E
Pavlovo, *Sakha, Russia* **27 C12** 63 5N 115 25 E
Pavlovsk, *Russia* **25 D7** 50 26N 40 5 E
Pawhuska, *U.S.A.* **81 G6** 36 40N 96 20W
Pawling, *U.S.A.* **79 E11** 41 34N 73 36W
Pawnee, *U.S.A.* **81 G6** 36 20N 96 48W
Pawnee City, *U.S.A.* **80 E6** 40 7N 96 9W
Pawtucket, *U.S.A.* **79 E13** 41 53N 71 23W
Paximádhia, *Greece* **23 D6** 35 0N 24 35 E
Paxoí, *Greece* **21 E9** 39 14N 20 12 E
Paxton, *Ill., U.S.A.* **76 E1** 40 27N 88 6W
Paxton, *Nebr., U.S.A.* **80 E4** 41 7N 101 21W
Payakumbuh, *Indonesia* **36 E2** 0 20S 100 35 E
Payette, *U.S.A.* **82 D5** 44 5N 116 56W
Payne Bay = Kangirsuk, *Canada* **69 B13** 60 0N 70 0W
Paynes Find, *Australia* **61 E2** 29 15S 117 42 E
Paynesville, *U.S.A.* **80 C7** 45 23N 94 43W
Paysandú, *Uruguay* **94 C4** 32 19S 58 8W
Payson, *Ariz., U.S.A.* **83 J8** 34 14N 111 20W
Payson, *Utah, U.S.A.* **82 F8** 40 3N 111 44W
Paz →, *Guatemala* **88 D1** 13 44N 90 10W
Paz, B. la, *Mexico* **86 C2** 24 15N 110 25W
Pāzanān, *Iran* **45 D6** 30 35N 49 59 E
Pazardzhik, *Bulgaria* **21 C11** 42 12N 24 20 E
Pe Ell, *U.S.A.* **84 D3** 46 34N 123 18W
Peabody, *U.S.A.* **79 D14** 42 31N 70 56W
Peace →, *Canada* **72 B6** 59 0N 111 25W
Peace Point, *Canada* **72 B6** 59 7N 112 27W
Peace River, *Canada* **72 B5** 56 15N 117 18W
Peach Springs, *U.S.A.* **83 J7** 35 32N 113 25W
Peak, The = Kinder Scout, *U.K.* **10 D6** 53 24N 1 52W
Peak Downs, *Australia* **62 C4** 22 55S 148 5 E
Peak Downs Mine, *Australia* **62 C4** 22 17S 148 11 E
Peak Hill, *N.S.W., Australia* **63 E4** 32 47S 148 11 E
Peak Hill, *W. Austral., Australia* **61 E2** 25 35S 118 43 E
Peak Ra., *Australia* **62 C4** 22 50S 148 20 E
Peake, *Australia* **63 F2** 35 25S 139 55 E
Peake Cr. →, *Australia* **63 D2** 28 2S 136 7 E
Peale, Mt., *U.S.A.* **83 G9** 38 26N 109 14W
Pearblossom, *U.S.A.* **85 L9** 34 30N 117 55W
Pearl →, *U.S.A.* **81 K10** 30 11N 89 32W
Pearl City, *U.S.A.* **74 H16** 21 24N 157 59W
Pearsall, *U.S.A.* **81 L5** 28 54N 99 6W
Pearse I., *Canada* **72 C2** 54 52N 130 14W
Peary Land, *Greenland* **4 A6** 82 40N 33 0W
Pease →, *U.S.A.* **81 H5** 34 12N 99 2W
Pebane, *Mozam.* **55 F4** 17 10S 38 8 E
Pebas, *Peru* **92 D4** 3 10S 71 46W
Pebble Beach, *U.S.A.* **84 J5** 36 34N 121 57W
Peć, *Serbia, Yug.* **21 C9** 42 40N 20 17 E
Pechenga, *Russia* **24 A5** 69 29N 31 4 E
Pechenizhyn, *Ukraine* **17 D13** 48 30N 24 48 E
Pechiguera, Pta., *Canary Is.* **22 F6** 28 51N 13 53W
Pechora →, *Russia* **24 A9** 68 13N 54 15 E
Pechorskaya Guba, *Russia* **24 A9** 68 40N 54 0 E
Pečory, *Russia* **9 H22** 57 48N 27 40 E
Pecos, *U.S.A.* **81 K3** 31 26N 103 30W
Pecos →, *U.S.A.* **81 L3** 29 42N 101 22W
Pécs, *Hungary* **17 E10** 46 5N 18 15 E
Pedder, L., *Australia* **62 G4** 42 55S 146 10 E
Peddie, *S. Africa* **57 E4** 33 14S 27 7 E
Pédernales, *Dom. Rep.* **89 C5** 18 2N 71 44W
Pedieos →, *Cyprus* **23 D12** 35 10N 33 54 E
Pedirka, *Australia* **63 D2** 26 40S 135 14 E
Pedra Azul, *Brazil* **93 G10** 16 2S 41 17W
Pedreiras, *Brazil* **93 D10** 4 32S 44 40W
Pedro Afonso, *Brazil* **93 E9** 9 0S 48 10W
Pedro Cays, *Jamaica* **88 C4** 17 5N 77 48W
Pedro de Valdivia, *Chile* **94 A2** 22 55S 69 38W
Pedro Juan Caballero, *Paraguay* **95 A4** 22 30S 55 40W
Peebinga, *Australia* **63 E3** 34 52S 140 57 E
Peebles, *U.K.* **12 F5** 55 40N 3 11W
Peekskill, *U.S.A.* **79 E11** 41 17N 73 55W
Peel, *U.K.* **10 C3** 54 13N 4 40W
Peel →, *Australia* **63 E5** 30 50S 150 29 E
Peel →, *Canada* **68 B6** 67 0N 135 0W
Peera Peera Poolanna L., *Australia* **63 D2** 26 30S 138 0 E
Peers, *Canada* **72 C5** 53 40N 116 0W
Pegasus Bay, *N.Z.* **59 K4** 43 20S 173 10 E
Pegu, *Burma* **41 L20** 17 20N 96 29 E
Pegu Yoma, *Burma* **41 K19** 19 0N 96 0 E
Pehuajó, *Argentina* **94 D3** 35 45S 62 0W
Pei Xian, *China* **34 G9** 34 44N 116 55 E
Peine, *Chile* **94 A2** 23 45S 68 8W
Peine, *Germany* **16 B6** 52 19N 10 14 E
Peip'ing = Beijing, *China* **34 E9** 39 55N 116 20 E
Peipus, L. = Chudskoye, Oz., *Russia* **9 G22** 58 13N 27 30 E
Peixe, *Brazil* **93 F9** 12 0S 48 40W
Pekalongan, *Indonesia* **37 G13** 6 53S 109 40 E
Pekan, *Malaysia* **39 L4** 3 30N 103 25 E
Pekanbaru, *Indonesia* **36 D2** 0 30N 101 15 E
Pekin, *U.S.A.* **80 E10** 40 35N 89 40W
Peking = Beijing, *China* **34 E9** 39 55N 116 20 E
Pelabuhan Kelang, *Malaysia* **39 L3** 3 0N 101 23 E
Pelabuhan Ratu, Teluk, *Indonesia* **37 G12** 7 5S 106 30 E
Pelabuhanratu, *Indonesia* **37 G12** 7 0S 106 32 E
Pelagie, Is., *Italy* **20 G5** 35 39N 12 33 E
Pelaihari, *Indonesia* **36 E4** 3 55S 114 45 E
Peleaga, Vf., *Romania* **17 F12** 45 22N 22 55 E
Pelée, Mt., *Martinique* **89 D7** 14 48N 61 10W
Pelee, Pt., *Canada* **70 D3** 41 54N 82 31W
Pelee I., *Canada* **70 D3** 41 47N 82 40W
Pelekech, *Kenya* **54 B4** 3 52N 35 8 E
Peleng, *Indonesia* **37 E6** 1 20S 123 30 E
Pelham, *U.S.A.* **77 K3** 31 8N 84 9W
Pelican L., *Canada* **73 C8** 52 28N 100 20W
Pelican Narrows, *Canada* **73 B8** 55 10N 102 56W
Pelican Rapids, *Canada* **73 C8** 52 45N 100 42W
Pelješac, *Croatia* **20 C7** 42 55N 17 25 E
Pelkosenniemi, *Finland* **8 C22** 67 6N 27 28 E
Pella, *S. Africa* **56 D2** 29 1S 19 6 E
Pella, *U.S.A.* **80 E8** 41 25N 92 55W
Pello, *Finland* **8 C21** 66 47N 23 59 E
Pelly →, *Canada* **68 B6** 62 47N 137 19W
Pelly Bay, *Canada* **69 B11** 68 38N 89 50W
Pelly L., *Canada* **68 B9** 66 0N 102 0W
Peloponnese = Pelopónnisos □, *Greece* **21 F10** 37 10N 22 0 E
Pelopónnisos □, *Greece* **21 F10** 37 10N 22 0 E
Peloro, C., *Italy* **20 E6** 38 16N 15 39 E
Pelorus Sd., *N.Z.* **59 J4** 40 59S 173 59 E
Pelotas, *Brazil* **95 C5** 31 42S 52 23W
Pelvoux, Massif du, *France* **18 D7** 44 52N 6 20 E
Pemalang, *Indonesia* **37 G13** 6 53S 109 23 E
Pematangsiantar, *Indonesia* **36 D1** 2 57N 99 5 E
Pemba, *Mozam.* **55 E5** 12 58S 40 30 E
Pemba, *Zambia* **55 F2** 16 30S 27 28 E
Pemba Channel, *Tanzania* **54 D4** 5 0S 39 37 E
Pemba I., *Tanzania* **54 D4** 5 0S 39 45 E
Pemberton, *Australia* **61 F2** 34 30S 116 0 E
Pemberton, *Canada* **72 C4** 50 25N 122 50W
Pembina, *U.S.A.* **80 A6** 48 58N 97 15W
Pembina →, *U.S.A.* **73 D9** 48 58N 97 14W
Pembine, *U.S.A.* **76 C2** 45 38N 87 59W
Pembroke, *Canada* **70 C4** 45 50N 77 7W
Pembroke, *U.K.* **11 F3** 51 41N 4 55W
Pembroke, *U.S.A.* **77 J5** 32 8N 81 37W
Pembrokeshire □, *U.K.* **11 F3** 51 52N 4 56W
Pen-y-Ghent, *U.K.* **10 C5** 54 10N 2 14W
Penang = Pinang, *Malaysia* **39 K3** 5 25N 100 15 E
Penápolis, *Brazil* **95 A6** 21 30S 50 0W
Peñarroya-Pueblonuevo, *Spain* **19 C3** 38 19N 5 16W
Peñas, C. de, *Spain* **19 A3** 43 42N 5 52W
Penas, G. de, *Chile* **96 F2** 47 0S 75 0W
Peñas del Chache, *Canary Is.* **22 E6** 29 6N 13 33W
Pench'i = Benxi, *China* **35 D12** 41 20N 123 48 E
Pend Oreille →, *U.S.A.* **82 B5** 49 4N 117 37W
Pend Oreille L., *U.S.A.* **82 C5** 48 10N 116 21W
Pendembu, *S. Leone* **50 G2** 9 7N 12 14W
Pender B., *Australia* **60 C3** 16 45S 122 42 E
Pendleton, *Calif., U.S.A.* **85 M9** 33 16N 117 23W
Pendleton, *Oreg., U.S.A.* **82 D4** 45 40N 118 47W
Penedo, *Brazil* **93 F11** 10 15S 36 36W
Penetanguishene, *Canada* **70 D4** 44 50N 79 55W
Pengalengan, *Indonesia* **37 G12** 7 9S 107 30 E
Penge, *Kasai Or., Zaïre* **54 D1** 5 30S 24 33 E
Penge, *Kivu, Zaïre* **54 C2** 4 27S 28 25 E
Penglai, *China* **35 F11** 37 48N 120 42 E
Penguin, *Australia* **62 G4** 41 8S 146 6 E
Penhalonga, *Zimbabwe* **55 F3** 18 52S 32 40 E
Peniche, *Portugal* **19 C1** 39 19N 9 22W
Penicuik, *U.K.* **12 F5** 55 50N 3 13W
Penida, *Indonesia* **36 F5** 8 45S 115 30 E
Peninsular Malaysia □, *Malaysia* **39 L4** 4 0N 102 0 E
Penmarch, Pte. de, *France* **18 C1** 47 48N 4 22W
Penn Hills, *U.S.A.* **78 F5** 40 28N 79 52W
Penn Yan, *U.S.A.* **78 D7** 42 40N 77 3W
Pennant, *Canada* **73 C7** 50 32N 108 14W
Penner →, *India* **40 M12** 14 35N 80 10 E
Pennines, *U.K.* **10 C5** 54 45N 2 27W
Pennington, *U.S.A.* **84 F5** 39 15N 121 47W
Pennsylvania □, *U.S.A.* **76 E6** 40 45N 77 30W
Penny, *Canada* **72 C4** 53 51N 121 20W

Penola, Australia 63 F3 37 25S 140 48 E
Penong, Australia 61 F5 31 56S 133 1 E
Penonomé, Panama 88 E3 8 31N 80 21W
Penrith, Australia 63 E5 33 43S 150 38 E
Penrith, U.K. 10 C5 54 40N 2 45W
Pensacola, U.S.A. 77 K2 30 25N 87 13W
Pensacola Mts., Antarctica 5 E1 84 0S 40 0W
Pense, Canada 73 C8 50 25N 104 59W
Penshurst, Australia 63 F3 37 49S 142 20 E
Penticton, Canada 72 D5 49 30N 119 38W
Pentland, Australia 62 C4 20 32S 145 25 E
Pentland Firth, U.K. 12 C5 58 43N 3 10W
Pentland Hills, U.K. 12 F5 55 48N 3 25W
Penylan L., Canada 73 A7 61 50N 106 20W
Penza, Russia 24 D8 53 15N 45 5 E
Penzance, U.K. 11 G2 50 7N 5 33W
Penzhino, Russia 27 C17 63 30N 167 55 E
Penzhinskaya Guba, Russia 27 C17 61 30N 163 0 E
Peoria, Ariz., U.S.A. 83 K7 33 35N 112 14W
Peoria, Ill., U.S.A. 80 E10 40 42N 89 36W
Pera Hd., Australia 62 A3 12 55S 141 37 E
Perabumulih, Indonesia 36 E2 3 27S 104 15 E
Pérama, Kérkira, Greece 23 A3 39 34N 19 54 E
Pérama, Kríti, Greece 23 D6 35 20N 24 40 E
Peräpohjola, Finland 8 C22 66 16N 26 10 E
Percé, Canada 71 C7 48 31N 64 13W
Perche, Collines du, France 18 B4 48 30N 0 40 E
Percival Lakes, Australia 60 D4 21 25S 125 0 E
Percy Is., Australia 62 C5 21 39S 150 16 E
Perdido, Mte., Spain 19 A6 42 40N 0 5 E
Perdu, Mt. = Perdido, Mte., Spain 19 A6 42 40N 0 5 E
Pereira, Colombia 92 C3 4 49N 75 43W
Perekerten, Australia 63 E3 34 55S 143 40 E
Perenjori, Australia 61 E2 29 26S 116 16 E
Pereyaslav-Khmelnytskyy, Ukraine 25 D5 50 3N 31 28 E
Pérez, I., Mexico 87 C7 22 24N 89 42W
Pergamino, Argentina 94 C3 33 52S 60 30W
Pergau →, Malaysia 39 K3 5 23N 102 2 E
Perham, U.S.A. 80 B7 46 36N 95 34W
Perhentian, Kepulauan, Malaysia 39 K4 5 54N 102 42 E
Péribonca →, Canada 71 C5 48 45N 72 5W
Péribonca, L., Canada 71 B5 50 1N 71 10W
Perico, Argentina 94 A2 24 20S 65 5W
Pericos, Mexico 86 B3 25 3N 107 42W
Périgueux, France 18 D4 45 10N 0 42 E
Perijá, Sierra de, Colombia 92 B4 9 30N 73 3W
Peristerona →, Cyprus 23 D12 35 8N 33 5 E
Perlas, Arch. de las, Panama 88 E4 8 41N 79 7W
Perlas, Punta de, Nic. 88 D3 12 30N 83 30W
Perm, Russia 24 C10 58 0N 56 10 E
Pernambuco = Recife, Brazil 93 E12 8 0S 35 0W
Pernambuco □, Brazil 93 E11 8 0S 37 0W
Pernatty Lagoon, Australia 63 E2 31 30S 137 12 E
Pernik, Bulgaria 21 C10 42 35N 23 2 E
Peron, C., Australia 61 E1 25 30S 113 30 E
Peron Is., Australia 60 B5 13 9S 130 4 E
Peron Pen., Australia 61 E1 26 0S 113 10 E
Perow, Canada 72 C3 54 35N 126 10W
Perpendicular Pt., Australia 63 E5 31 37S 152 52 E
Perpignan, France 18 E5 42 42N 2 53 E
Perris, U.S.A. 85 M9 33 47N 117 14W
Perry, Fla., U.S.A. 77 K4 30 7N 83 35W
Perry, Ga., U.S.A. 77 J4 32 28N 83 44W
Perry, Iowa, U.S.A. 80 E7 41 51N 94 6W
Perry, Maine, U.S.A. 77 C12 44 58N 67 5W
Perry, Okla., U.S.A. 81 G6 36 17N 97 14W
Perryton, U.S.A. 81 G4 36 24N 100 48W
Perryville, U.S.A. 81 G10 37 43N 89 52W
Pershotravensk, Ukraine 17 C14 50 13N 27 40 E
Persia = Iran ■, Asia 45 C7 33 0N 53 0 E
Persian Gulf = Gulf, The, Asia 45 E6 27 0N 50 0 E
Perth, Australia 61 F2 31 57S 115 52 E
Perth, Canada 70 D4 44 55N 76 15W
Perth, U.K. 12 E5 56 24N 3 26W
Perth & Kinross □, U.K. 12 E5 56 45N 3 55W
Perth Amboy, U.S.A. 79 F10 40 31N 74 16W
Peru, Ill., U.S.A. 80 E10 41 20N 89 8W
Peru, Ind., U.S.A. 76 E2 40 45N 86 4W
Peru ■, S. Amer. 92 E3 4 0S 75 0W
Peru-Chile Trench, Pac. Oc. 65 K20 20 0S 72 0W
Perúgia, Italy 20 C5 43 7N 12 23 E
Pervomaysk, Ukraine 25 E5 48 10N 30 46 E
Pervouralsk, Russia 24 C10 56 59N 59 59 E
Pes, Pta. del, Spain 22 C7 38 46N 1 26 E
Pésaro, Italy 20 C5 43 54N 12 55 E
Pescara, Italy 20 C6 42 28N 14 13 E
Peshawar, Pakistan 42 B4 34 2N 71 37 E
Peshkopi, Albania 21 D9 41 41N 20 25 E
Peshtigo, U.S.A. 76 C2 45 4N 87 46W
Pesqueira, Brazil 93 E11 8 20S 36 42W
Petah Tiqwa, Israel 47 C3 32 6N 34 53 E
Petaling Jaya, Malaysia 39 L3 3 4N 101 42 E
Petaloudhes, Greece 23 C10 36 18N 28 5 E
Petaluma, U.S.A. 84 G4 38 14N 122 39W
Petange, Lux. 15 E5 49 33N 5 55 E
Petatlán, Mexico 86 D4 17 31N 101 16W
Petauke, Zambia 55 E3 14 14S 31 20 E
Petawawa, Canada 70 C4 45 54N 77 17W
Petén Itzá, L., Guatemala 88 C2 16 58N 89 50W
Peter I.s Øy, Antarctica 5 C16 69 0S 91 0W
Peter Pond L., Canada 73 B7 55 55N 108 44W
Peterbell, Canada 70 C3 48 36N 83 21W
Peterborough, Australia 63 E2 32 58S 138 51 E
Peterborough, Canada 69 D12 44 20N 78 20W
Peterborough, U.K. 11 E7 52 35N 0 15W
Peterborough, U.S.A. 79 D13 42 53N 71 57W
Peterhead, U.K. 12 D7 57 31N 1 48W
Petermann Bjerg, Greenland 66 B17 73 7N 28 25W
Petersburg, Alaska, U.S.A. 72 B2 56 48N 132 58W
Petersburg, Ind., U.S.A. 76 F2 38 30N 87 17W
Petersburg, Va., U.S.A. 76 G7 37 14N 77 24W
Petersburg, W. Va., U.S.A. 76 F6 39 1N 79 5W
Petford, Australia 62 B3 17 20S 144 58 E
Petit Bois I., U.S.A. 77 K1 30 12N 88 26W
Petit-Cap, Canada 71 C7 49 3N 64 30W
Petit Goâve, Haiti 89 C5 18 27N 72 51W
Petit Lac Manicouagan, Canada 71 B6 51 25N 67 40W
Petitcodiac, Canada 71 C6 45 57N 65 11W
Petite Baleine →, Canada 70 A4 56 0N 76 45W
Petite Saguenay, Canada 71 C5 48 15N 70 4W
Petitsikapau, L., Canada 71 B6 54 37N 66 25W
Petlad, India 42 H5 22 30N 72 45 E
Peto, Mexico 87 C7 20 10N 88 53W
Petone, N.Z. 59 J5 41 13S 174 53 E
Petoskey, U.S.A. 76 C3 45 22N 84 57W
Petra, Jordan 47 E4 30 20N 35 22 E
Petra, Spain 22 B10 39 37N 3 6 E
Petra, Ostrova, Russia 4 B13 76 15N 118 30 E
Petra Velikogo, Zaliv, Russia 30 C5 42 40N 132 0 E
Petrich, Bulgaria 21 D10 41 24N 23 13 E
Petrikov = Pyetrikaw, Belarus 17 B15 52 11N 28 29 E
Petrograd = Sankt-Peterburg, Russia 24 C5 59 55N 30 20 E
Petrolândia, Brazil 93 E11 9 5S 38 20W
Petrolia, Canada 70 D3 42 54N 82 9W
Petrolina, Brazil 93 E10 9 24S 40 30W
Petropavl, Kazakstan 26 D7 54 53N 69 13 E
Petropavlovsk = Petropavl, Kazakstan 26 D7 54 53N 69 13 E
Petropavlovsk-Kamchatskiy, Russia 27 D16 53 3N 158 43 E
Petrópolis, Brazil 95 A7 22 33S 43 9W
Petroşani, Romania 17 F12 45 28N 23 20 E
Petrovaradin, Serbia, Yug. 21 B8 45 16N 19 55 E
Petrovsk, Russia 24 D8 52 22N 45 19 E
Petrovsk-Zabaykalskiy, Russia 27 D11 51 20N 108 55 E
Petrozavodsk, Russia 24 B5 61 41N 34 20 E
Petrus Steyn, S. Africa 57 D4 27 38S 28 8 E
Petrusburg, S. Africa 56 D4 29 4S 25 26 E
Peumo, Chile 94 C1 34 21S 71 12W
Peureulak, Indonesia 36 D1 4 48N 97 45 E
Pevek, Russia 27 C18 69 41N 171 19 E
Pforzheim, Germany 16 D5 48 52N 8 41 E
Phagwara, India 40 D9 31 10N 75 40 E
Phaistós, Greece 23 D6 35 2N 24 50 E
Phala, Botswana 56 C4 23 45S 26 50 E
Phalera = Phulera, India 42 F6 26 52N 75 16 E
Phalodi, India 42 F5 27 12N 72 24 E
Phan, Thailand 38 C2 19 28N 99 43 E
Phan Rang, Vietnam 39 G7 11 34N 109 0 E
Phan Ri = Hoa Da, Vietnam 39 G7 11 16N 108 40 E
Phan Thiet, Vietnam 39 G7 11 1N 108 9 E
Phanat Nikhom, Thailand 38 F3 13 27N 101 11 E
Phangan, Ko, Thailand 39 H3 9 45N 100 0 E
Phangnga, Thailand 39 H2 8 28N 98 30 E
Phanh Bho Ho Chi Minh, Vietnam 39 G6 10 58N 106 40 E
Phanom Sarakham, Thailand 38 F3 13 45N 101 21 E
Pharenda, India 43 F10 27 5N 83 17 E
Phatthalung, Thailand 39 J3 7 39N 100 6 E
Phayao, Thailand 38 C2 19 11N 99 55 E
Phelps, N.Y., U.S.A. 78 D7 42 58N 77 3W
Phelps, Wis., U.S.A. 80 B10 46 4N 89 5W
Phelps L., Canada 73 B8 59 15N 103 15W
Phenix City, U.S.A. 77 J3 32 28N 85 0W
Phet Buri, Thailand 38 F2 13 1N 99 55 E
Phetchabun, Thailand 38 D3 16 25N 101 8 E
Phetchabun, Thiu Khao, Thailand 38 E3 16 0N 101 20 E
Phetchaburi = Phet Buri, Thailand 38 F2 13 1N 99 55 E
Phi Phi, Ko, Thailand 39 J2 7 45N 98 46 E
Phiafay, Laos 38 E6 14 48N 106 0 E
Phibun Mangsahan, Thailand 38 E5 15 14N 105 14 E
Phichai, Thailand 38 D3 17 22N 100 10 E
Phichit, Thailand 38 D3 16 26N 100 22 E
Philadelphia, Miss., U.S.A. 81 J10 32 46N 89 7W
Philadelphia, N.Y., U.S.A. 79 B9 44 9N 75 43W
Philadelphia, Pa., U.S.A. 79 F9 39 57N 75 10W
Philip, U.S.A. 80 C4 44 2N 101 40W
Philippeville, Belgium 15 D4 50 12N 4 33 E
Philippi L., Australia 62 C2 24 20S 138 55 E
Philippines ■, Asia 37 B6 12 0N 123 0 E
Philippolis, S. Africa 56 E4 30 15S 25 16 E
Philippopolis = Plovdiv, Bulgaria 21 C11 42 8N 24 44 E
Philipsburg, Mont., U.S.A. 82 C7 46 20N 113 18W
Philipsburg, Pa., U.S.A. 78 F6 40 54N 78 13W
Philipstown = Daingean, Ireland 13 C4 53 18N 7 17W
Philipstown, S. Africa 56 E3 30 28S 24 30 E
Phillip I., Australia 63 F4 38 30S 145 12 E
Phillips, Tex., U.S.A. 81 H4 35 42N 101 22W
Phillips, Wis., U.S.A. 80 C9 45 42N 90 24W
Phillipsburg, Kans., U.S.A. 80 F5 39 45N 99 19W
Phillipsburg, N.J., U.S.A. 79 F9 40 42N 75 12W
Phillott, Australia 63 D4 27 53S 145 50 E
Philmont, U.S.A. 79 D11 42 15N 73 39W
Philomath, U.S.A. 82 D2 44 32N 123 22W
Phimai, Thailand 38 E4 15 13N 102 30 E
Phitsanulok, Thailand 38 D3 16 50N 100 12 E
Phnom Dangrek, Thailand 38 E5 14 20N 104 0 E
Phnom Penh, Cambodia 39 G5 11 33N 104 55 E
Phoenix, Ariz., U.S.A. 83 K7 33 27N 112 4W
Phoenix, N.Y., U.S.A. 79 C8 43 14N 76 18W
Phoenix Is., Kiribati 64 H10 3 30S 172 0W
Phoenixville, U.S.A. 79 F9 40 8N 75 31W
Phon, Thailand 38 E4 15 49N 102 36 E
Phon Tiou, Laos 38 D5 17 53N 104 37 E
Phong →, Thailand 38 D4 16 23N 102 56 E
Phong Saly, Laos 38 B4 21 42N 102 9 E
Phong Tho, Vietnam 38 A4 22 32N 103 21 E
Phonhong, Laos 38 C4 18 30N 102 25 E
Phonum, Thailand 39 H2 8 49N 98 48 E
Phosphate Hill, Australia 62 C2 21 53S 139 58 E
Photharam, Thailand 38 F2 13 41N 99 51 E
Phra Chedi Sam Ong, Thailand 38 E2 15 16N 98 23 E
Phra Nakhon Si Ayutthaya, Thailand 38 E3 14 25N 100 30 E
Phra Thong, Ko, Thailand 39 H2 9 5N 98 17 E
Phrae, Thailand 38 C3 18 7N 100 9 E
Phrom Phiram, Thailand 38 D3 17 2N 100 12 E
Phu Dien, Vietnam 38 C5 18 58N 105 31 E
Phu Loi, Laos 38 B4 20 14N 103 14 E
Phu Ly, Vietnam 38 B5 20 35N 105 50 E
Phu Tho, Vietnam 38 B5 21 24N 105 13 E
Phuc Yen, Vietnam 38 B5 21 16N 105 45 E
Phuket, Thailand 39 J2 7 52N 98 22 E
Phuket, Ko, Thailand 39 J2 8 0N 98 22 E
Phulera, India 42 F6 26 52N 75 16 E
Phumiphon, Khuan, Thailand 38 D2 17 15N 98 58 E
Phun Phin, Thailand 39 H2 9 7N 99 12 E
Piacenza, Italy 20 B3 45 1N 9 40 E
Pialba, Australia 63 D5 25 20S 152 45 E
Pian Cr. →, Australia 63 E4 30 2S 148 12 E
Pianosa, Italy 20 C4 42 35N 10 5 E
Piapot, Canada 73 D7 49 59N 109 8W
Piatra Neamţ, Romania 17 E14 46 56N 26 21 E
Piauí □, Brazil 93 E10 7 0S 43 0W
Piave →, Italy 20 B5 45 32N 12 44 E
Pibor Post, Sudan 51 G11 6 47N 33 3 E
Pica, Chile 92 H5 20 35S 69 25W
Picardie, France 18 B5 49 50N 3 0 E
Picardy = Picardie, France 18 B5 49 50N 3 0 E
Picayune, U.S.A. 81 K10 30 32N 89 41W
Pichilemu, Chile 94 C1 34 22S 72 0W
Pickerel L., Canada 70 C1 48 40N 91 25W
Pickle Lake, Canada 70 B1 51 30N 90 12W
Pico Truncado, Argentina 96 F3 46 40S 68 0W
Picton, Australia 63 E5 34 12S 150 34 E
Picton, Canada 70 D4 44 1N 77 9W
Picton, N.Z. 59 J5 41 18S 174 3 E
Pictou, Canada 71 C7 45 41N 62 42W
Picture Butte, Canada 72 D6 49 55N 112 45W
Picún Leufú, Argentina 96 D3 39 30S 69 5W
Pidurutalagala, Sri Lanka 40 R12 7 10N 80 50 E
Piedmont = Piemonte □, Italy 20 B2 45 0N 8 0 E
Piedmont, U.S.A. 77 J3 33 55N 85 37W
Piedmont Plateau, U.S.A. 77 J5 34 0N 81 30W
Piedras, R. de las →, Peru 92 F5 12 30S 69 15W
Piedras Negras, Mexico 86 B4 28 42N 100 31W
Pieksämäki, Finland 9 E22 62 18N 27 10 E
Piemonte □, Italy 20 B2 45 0N 8 0 E
Pierce, U.S.A. 82 C6 46 30N 115 48W
Piercefield, U.S.A. 79 B10 44 13N 74 35W
Pierre, U.S.A. 80 C4 44 22N 100 21W
Piet Retief, S. Africa 57 D5 27 1S 30 50 E
Pietarsaari, Finland 8 E20 63 40N 22 43 E
Pietermaritzburg, S. Africa 57 D5 29 35S 30 25 E
Pietersburg, S. Africa 57 C4 23 54S 29 25 E
Pietrosul, Romania 17 E13 47 12N 25 8 E
Pietrosul, Romania 17 E13 47 35N 24 43 E
Pigeon, U.S.A. 76 D4 43 50N 83 16W
Piggott, U.S.A. 81 G9 36 23N 90 11W
Pigüe, Argentina 94 D3 37 36S 62 25W
Pihani, India 43 F9 27 36N 80 15 E
Pihlajavesi, Finland 9 F23 61 45N 28 45 E
Pikes Peak, U.S.A. 80 F2 38 50N 105 3W
Piketberg, S. Africa 56 E2 32 55S 18 40 E
Pikeville, U.S.A. 76 G4 37 29N 82 31W
Pikou, China 35 E12 39 18N 122 22 E
Pikwitonei, Canada 73 B9 55 35N 97 9W
Piła, Poland 17 B9 53 10N 16 48 E
Pilani, India 42 E6 28 22N 75 33 E
Pilar, Brazil 93 E11 9 36S 35 56W
Pilar, Paraguay 94 B4 26 50S 58 20W
Pilas Group, Phil. 37 C6 6 45N 121 35 E
Pilcomayo →, Paraguay 94 B4 25 21S 57 42W
Pilibhit, India 43 E8 28 40N 79 50 E
Pilica →, Poland 17 C11 51 52N 21 17 E
Pilkhawa, India 42 E7 28 43N 77 42 E
Pílos, Greece 21 F9 36 55N 21 42 E
Pilot Mound, Canada 73 D9 49 15N 98 54W
Pilot Point, U.S.A. 81 J6 33 24N 96 58W
Pilot Rock, U.S.A. 82 D4 45 29N 118 50W
Pilsen = Plzeň, Czech. 16 D7 49 45N 13 22 E
Pima, U.S.A. 83 K9 32 54N 109 50W
Pimba, Australia 63 E2 31 18S 136 46 E
Pimenta Bueno, Brazil 92 F6 11 35S 61 10W
Pimentel, Peru 92 E3 6 45S 79 55W
Pinang, Malaysia 39 K3 5 25N 100 15 E
Pinar, C. del, Spain 22 B10 39 53N 3 12 E
Pinar del Río, Cuba 88 B3 22 26N 83 40W
Pınarhisar, Turkey 21 D12 41 37N 27 30 E
Pincher Creek, Canada 72 D6 49 30N 113 57W
Pinchi L., Canada 72 C4 54 38N 124 30W
Pinckneyville, U.S.A. 80 F10 38 5N 89 23W
Pińczów, Poland 17 C11 50 32N 20 32 E
Pind Dadan Khan, Pakistan 42 C5 32 36N 73 7 E
Pindar, Australia 61 E2 28 30S 115 47 E
Pindi Gheb, Pakistan 42 C5 33 14N 72 21 E
Pindiga, Nigeria 50 G7 9 58N 10 53 E
Pindos Óros, Greece 21 E9 40 0N 21 0 E
Pindus Mts. = Pindos Óros, Greece 21 E9 40 0N 21 0 E
Pine, U.S.A. 83 J8 34 23N 111 27W
Pine →, Canada 73 B7 58 50N 105 38W
Pine, C., Canada 71 C9 46 37N 53 32W
Pine Bluff, U.S.A. 81 H8 34 13N 92 1W
Pine City, U.S.A. 80 C8 45 50N 92 59W
Pine Falls, Canada 73 C9 50 34N 96 11W
Pine Flat L., U.S.A. 84 J7 36 50N 119 20W
Pine Pass, Canada 72 B4 55 25N 122 42W
Pine Point, Canada 72 A6 60 50N 114 28W
Pine Ridge, U.S.A. 80 D3 43 2N 102 33W
Pine River, Canada 73 C8 51 45N 100 30W
Pine River, U.S.A. 80 B7 46 43N 94 24W
Pine Valley, U.S.A. 85 N10 32 50N 116 32W
Pinecrest, U.S.A. 84 G6 38 12N 120 1W
Pinedale, U.S.A. 84 J7 36 50N 119 48W
Pinega →, Russia 24 B8 64 30N 44 19 E
Pinehill, Australia 62 C4 23 38S 146 57 E
Pinerolo, Italy 20 B2 44 53N 7 21 E
Pinetop, U.S.A. 83 J9 34 8N 109 56W
Pinetown, S. Africa 57 D5 29 48S 30 54 E
Pinetree, U.S.A. 82 E11 43 42N 105 52W
Pineville, Ky., U.S.A. 77 G4 36 46N 83 42W
Pineville, La., U.S.A. 81 K8 31 19N 92 26W
Ping →, Thailand 38 E3 15 42N 100 9 E
Pingaring, Australia 61 F2 32 40S 118 32 E
Pingding, China 34 F7 37 47N 113 38 E
Pingdingshan, China 34 H7 33 43N 113 27 E
Pingdong, Taiwan 33 D7 22 39N 120 30 E
Pingdu, China 35 F10 36 42N 119 59 E
Pingelly, Australia 61 F2 32 32S 117 5 E
Pingliang, China 34 G4 35 35N 106 31 E
Pinglu, China 34 E7 39 31N 112 30 E
Pingluo, China 34 E4 38 52N 106 30 E
Pingquan, China 35 D10 41 1N 118 37 E
Pingrup, Australia 61 F2 33 32S 118 29 E
P'ingtung, Taiwan 33 D7 22 38N 120 30 E
Pingwu, China 34 H3 32 25N 104 30 E
Pingxiang, China 32 D5 22 6N 106 46 E
Pingyao, China 34 F7 37 12N 112 10 E
Pingyi, China 35 G9 35 30N 117 35 E
Pingyin, China 34 F9 36 20N 116 25 E
Pingyuan, China 34 F9 37 10N 116 22 E
Pinhal, Brazil 95 A6 22 10S 46 46W
Pinhel, Portugal 19 B2 40 50N 7 1W
Pini, Indonesia 36 D1 0 10N 98 40 E
Piniós →, Greece 21 E10 39 55N 22 41 E
Pinjarra, Australia 61 F2 32 37S 115 52 E
Pink →, Canada 73 B8 56 50N 103 50W
Pinnacles, Australia 61 E3 28 12S 120 26 E
Pinnacles, U.S.A. 84 J5 36 33N 121 19W
Pinnaroo, Australia 63 F3 35 17S 140 53 E
Pinnes, Ákra, Greece 21 D11 40 5N 24 20 E
Pinon Hills, U.S.A. 85 L9 34 26N 117 39W
Pinos, Mexico 86 C4 22 20N 101 40W
Pinos, Mt., U.S.A. 85 L7 34 49N 119 8W
Pinos Pt., U.S.A. 83 H3 36 38N 121 57W
Pinotepa Nacional, Mexico 87 D5 16 19N 98 3W
Pinrang, Indonesia 37 E5 3 46S 119 41 E
Pinsk, Belarus 17 B14 52 10N 26 1 E
Pintados, Chile 92 H5 20 35S 69 40W
Pintumba, Australia 61 F5 31 30S 132 12 E
Pinyug, Russia 24 B8 60 5N 48 0 E
Pioche, U.S.A. 83 H6 37 56N 114 27W
Piombino, Italy 20 C4 42 55N 10 32 E
Pioner, Os., Russia 27 B10 79 50N 92 0 E
Piorini, L., Brazil 92 D6 3 15S 62 35W
Piotrków Trybunalski, Poland 17 C10 51 23N 19 43 E
Pīp, Iran 45 E9 26 45N 60 10 E
Pipar, India 42 F5 26 25N 73 31 E
Piparia, India 42 H8 22 45N 78 23 E
Pipestone, U.S.A. 80 D6 44 0N 96 19W
Pipestone →, Canada 70 B2 52 53N 89 23W
Pipestone Cr. →, Canada 73 D8 49 38N 100 15W
Pipmuacan, Rés., Canada 71 C5 49 45N 70 30W
Pippingarra, Australia 60 D2 20 27S 118 42 E
Piqua, U.S.A. 76 E3 40 9N 84 15W
Piquiri →, Brazil 95 A5 24 3S 54 14W
Pīr Sohrāb, Iran 45 E9 25 44N 60 54 E
Piracicaba, Brazil 95 A6 22 45S 47 40W
Piracuruca, Brazil 93 D10 3 50S 41 50W
Piræus = Piraiévs, Greece 21 F10 37 57N 23 42 E
Piraiévs, Greece 21 F10 37 57N 23 42 E
Pirajuí, Brazil 95 A6 21 59S 49 29W
Pirané, Argentina 94 B4 25 42S 59 6W
Pirapora, Brazil 93 G10 17 20S 44 56W
Pírgos, Greece 21 F9 37 40N 21 27 E
Piribebuy, Paraguay 94 B4 25 26S 57 2W
Pirin Planina, Bulgaria 21 D10 41 40N 23 30 E
Piripiri, Brazil 93 D10 4 15S 41 46W
Pirmasens, Germany 16 D4 49 12N 7 36 E
Pirot, Serbia, Yug. 21 C10 43 9N 22 39 E
Piru, Indonesia 37 E7 3 4S 128 12 E
Piru, U.S.A. 85 L8 34 25N 118 48W
Pisa, Italy 20 C4 43 43N 10 23 E
Pisagua, Chile 92 G4 19 40S 70 15W
Pisciotta, Italy 20 D6 40 6N 15 14 E
Pisco, Peru 92 F3 13 50S 76 12W
Písek, Czech. 16 D8 49 19N 14 10 E
Pishan, China 32 C2 37 30N 78 33 E
Pishin Lora →, Pakistan 42 E1 29 9N 64 5 E
Pising, Indonesia 37 F6 5 8S 121 53 E
Pismo Beach, U.S.A. 85 K6 35 9N 120 38W
Pissis, Cerro, Argentina 94 B2 27 45S 68 48W
Pissouri, Cyprus 23 E11 34 40N 32 42 E
Pistóia, Italy 20 C4 43 55N 10 54 E
Pistol B., Canada 73 A10 62 25N 92 37W
Pisuerga →, Spain 19 B3 41 33N 4 52W
Pitarpunga, L., Australia 63 E3 34 24S 143 30 E
Pitcairn I., Pac. Oc. 65 K14 25 5S 130 5W
Pite älv →, Sweden 8 D19 65 20N 21 25 E
Piteå, Sweden 8 D19 65 20N 21 25 E
Piteşti, Romania 17 F13 44 52N 24 54 E
Pithapuram, India 41 L13 17 10N 82 15 E
Pithara, Australia 61 F2 30 20S 116 35 E
Pitlochry, U.K. 12 E5 56 42N 3 44W
Pitsilia □, Cyprus 23 E12 34 55N 33 0 E
Pitt I., Canada 72 C3 53 30N 129 50W
Pittsburg, Kans., U.S.A. 81 G7 37 25N 94 42W
Pittsburg, Tex., U.S.A. 81 J7 33 0N 94 59W
Pittsburgh, U.S.A. 78 F5 40 26N 80 1W
Pittsfield, Ill., U.S.A. 80 F9 39 36N 90 49W
Pittsfield, Mass., U.S.A. 79 D11 42 27N 73 15W
Pittsfield, N.H., U.S.A. 79 C13 43 18N 71 20W
Pittston, U.S.A. 79 E9 41 19N 75 47W
Pittsworth, Australia 63 D5 27 41S 151 37 E
Pituri →, Australia 62 C2 22 35S 138 30 E
Piura, Peru 92 E2 5 15S 80 38W
Pixley, U.S.A. 84 K7 35 58N 119 18W
Placentia, Canada 71 C9 47 20N 54 0W
Placentia B., Canada 71 C9 47 0N 54 40W
Placerville, U.S.A. 84 G6 38 44N 120 48W
Placetas, Cuba 88 B4 22 15N 79 44W
Plain Dealing, U.S.A. 81 J8 32 54N 93 42W
Plainfield, U.S.A. 79 F10 40 37N 74 25W
Plains, Kans., U.S.A. 81 G4 37 16N 100 35W
Plains, Mont., U.S.A. 82 C6 47 28N 114 53W
Plains, Tex., U.S.A. 81 J3 33 11N 102 50W
Plainview, Nebr., U.S.A. 80 D6 42 21N 97 47W
Plainview, Tex., U.S.A. 81 H4 34 11N 101 43W
Plainville, U.S.A. 80 F5 39 14N 99 18W
Plainwell, U.S.A. 76 D3 42 27N 85 38W
Pláka, Ákra, Greece 23 D8 35 11N 26 19 E
Plakhino, Russia 26 C9 67 45N 86 5 E
Plana Cays, Bahamas 89 B5 22 38N 73 30W

Planada, U.S.A. 84 H6 37 16N 120 19W
Plankinton, U.S.A. 80 D5 43 43N 98 29W
Plano, U.S.A. 81 J6 33 1N 96 42W
Plant City, U.S.A. 77 L4 28 1N 82 7W
Plaquemine, U.S.A. 81 K9 30 17N 91 14W
Plasencia, Spain 19 B2 40 3N 6 8W
Plaster City, U.S.A. 85 N11 32 47N 115 51W
Plaster Rock, Canada 71 C6 46 53N 67 22W
Plastun, Russia 30 B8 44 45N 136 19 E
Plata, Río de la, S. Amer. 94 C4 34 45S 57 30W
Platani →, Italy 20 F5 37 23N 13 16 E
Plátanos, Greece 23 D5 35 28N 23 33 E
Plateau du Coteau du Missouri, U.S.A. 80 B4 47 9N 101 5W
Plato, Colombia 92 B4 9 47N 74 47W
Platte, U.S.A. 80 D5 43 23N 98 51W
Platte →, Mo., U.S.A. 80 F7 39 16N 94 50W
Platte →, Nebr., U.S.A. 66 E10 41 4N 95 53W
Platteville, U.S.A. 80 E2 40 13N 104 49W
Plattsburgh, U.S.A. 79 B11 44 42N 73 28W
Plattsmouth, U.S.A. 80 E7 41 1N 95 53W
Plauen, Germany 16 C7 50 30N 12 8 E
Plavinas, Latvia 9 H21 56 35N 25 46 E
Playa Blanca, Canary Is. 22 F6 28 55N 13 37W
Playa Blanca Sur, Canary Is. 22 F6 28 51N 13 50W
Playa de las Americas, Canary Is. 22 F3 28 5N 16 43W
Playa de Mogán, Canary Is. 22 G4 27 48N 15 47W
Playa del Inglés, Canary Is. 22 G4 27 45N 15 33W
Playa Esmerelda, Canary Is. 22 F5 28 8N 14 16W
Playgreen L., Canada 73 C9 54 0N 98 15W
Pleasant Bay, Canada 71 C7 46 51N 60 48W
Pleasant Hill, Calif., U.S.A. 84 H4 37 57N 122 4W
Pleasant Hill, Mo., U.S.A. 80 F7 38 47N 94 16W
Pleasanton, U.S.A. 81 L5 28 58N 98 29W
Pleasantville, U.S.A. 76 F8 39 24N 74 32W
Pleiku, Vietnam 38 F7 13 57N 108 0 E
Plenty →, Australia 62 C2 23 25S 136 31 E
Plenty, B. of, N.Z. 59 G6 37 45S 177 0 E
Plentywood, U.S.A. 80 A2 48 47N 104 34W
Plesetsk, Russia 24 B7 62 43N 40 20 E
Plessisville, Canada 71 C5 46 14N 71 47W
Pletipi L., Canada 71 B5 51 44N 70 6W
Pleven, Bulgaria 21 C11 43 26N 24 37 E
Plevlja, Montenegro, Yug. 21 C8 43 21N 19 21 E
Płock, Poland 17 B10 52 32N 19 40 E
Plöckenstein, Germany 16 D7 48 46N 13 51 E
Ploieşti, Romania 17 F14 44 57N 26 5 E
Plonge, Lac la, Canada 73 B7 55 8N 107 20W
Plovdiv, Bulgaria 21 C11 42 8N 24 44 E
Plum, U.S.A. 78 F5 40 29N 79 47W
Plum I., U.S.A. 79 E12 41 11N 72 12W
Plumas, U.S.A. 84 F7 39 45N 119 4W
Plummer, U.S.A. 82 C5 47 20N 116 53W
Plumtree, Zimbabwe 55 G2 20 27S 27 55 E
Plunge, Lithuania 9 J19 55 53N 21 59 E
Plymouth, U.K. 11 G3 50 22N 4 10W
Plymouth, Calif., U.S.A. 84 G6 38 29N 120 51W
Plymouth, Ind., U.S.A. 76 E2 41 21N 86 19W
Plymouth, Mass., U.S.A. 79 E14 41 57N 70 40W
Plymouth, N.C., U.S.A. 77 H7 35 52N 76 43W
Plymouth, N.H., U.S.A. 79 C13 43 46N 71 41W
Plymouth, Pa., U.S.A. 79 E9 41 14N 75 57W
Plymouth, Wis., U.S.A. 76 D2 43 45N 87 59W
Plynlimon = Pumlumon Fawr, U.K. 11 E4 52 28N 3 46W
Plzeň, Czech. 16 D7 49 45N 13 22 E
Po →, Italy 20 B5 44 57N 12 4 E
Po Hai = Bo Hai, China 35 E10 39 0N 119 0 E
Po Hai, China 28 F15 38 30N 119 0 E
Pobeda, Russia 27 C15 65 12N 146 12 E
Pobedino, Russia 27 E15 49 51N 142 49 E
Pobedy Pik, Kyrgyzstan 26 E8 40 45N 79 58 E
Pocahontas, Ark., U.S.A. 81 G9 36 16N 90 58W
Pocahontas, Iowa, U.S.A. 80 D7 42 44N 94 40W
Pocatello, U.S.A. 82 E7 42 52N 112 27W
Pochutla, Mexico 87 D5 15 50N 96 31W
Pocito Casas, Mexico 86 B2 28 32N 111 6W
Pocomoke City, U.S.A. 76 F8 38 5N 75 34W
Poços de Caldas, Brazil 95 A6 21 50S 46 33W
Podgorica, Montenegro, Yug. 21 C8 42 30N 19 19 E
Podilska Vysochyna, Ukraine 17 D14 49 0N 28 0 E
Podkamennaya Tunguska →, Russia 27 C10 61 50N 90 13 E
Podolsk, Russia 24 C6 55 25N 37 30 E
Podor, Senegal 50 E1 16 40N 15 2W
Podporozhy, Russia 24 B5 60 55N 34 2 E
Pofadder, S. Africa 56 D2 29 10S 19 22 E
Pogamasing, Canada 70 C3 46 55N 81 50W
Pogranitšnyi, Russia 30 B5 44 25N 131 24 E
Poh, Indonesia 37 E6 0 46S 122 51 E
Pohang, S. Korea 35 F15 36 1N 129 23 E
Pohjanmaa, Finland 8 E20 62 58N 22 50 E
Pohnpei, Pac. Oc. 64 G7 6 55N 158 10 E
Poinsett, C., Antarctica 5 C8 65 42S 113 18 E
Point Edward, Canada 70 D3 43 0N 82 30W
Point Pedro, Sri Lanka 40 Q12 9 50N 80 15 E
Point Pleasant, N.J., U.S.A. 79 F10 40 5N 74 4W
Point Pleasant, W. Va., U.S.A. 76 F4 38 51N 82 8W
Pointe-à-la Hache, U.S.A. 81 L10 29 35N 89 55W
Pointe-à-Pitre, Guadeloupe 89 C7 16 10N 61 30W
Pointe Noire, Congo 52 E2 4 48S 11 53 E
Poisonbush Ra., Australia 60 D3 22 30S 121 30 E
Poitiers, France 18 C4 46 35N 0 20 E
Poitou, France 18 C3 46 40N 0 10W
Pojoaque Valley, U.S.A. 83 J11 35 54N 106 1W
Pokaran, India 40 F7 27 0N 71 50 E
Pokataroo, Australia 63 D4 29 30S 148 36 E
Poko, Zaïre 54 B2 3 7N 26 52 E
Pokrovsk = Engels, Russia 24 D8 51 28N 46 6 E
Pokrovsk, Russia 27 C13 61 29N 129 0 E
Pola = Pula, Croatia 16 F7 44 54N 13 57 E
Polacca, U.S.A. 83 J8 35 50N 110 23W
Polan, Iran 45 E9 25 30N 61 10 E
Poland ■, Europe 17 C10 52 0N 20 0 E
Polatsk, Belarus 24 C4 55 30N 28 50 E
Polcura, Chile 94 D1 37 17S 71 43W
Polden Hills, U.K. 11 F5 51 7N 2 50W
Polessk, Russia 9 J19 54 50N 21 8 E
Polesye = Pripet Marshes, Europe 17 B15 52 10N 28 10 E
Polevskoy, Russia 24 C11 56 26N 60 11 E
Pŏlgyo-ri, S. Korea 35 G14 34 51N 127 21 E
Poli, Cameroon 52 C2 8 34N 13 15 E
Police, Poland 16 B8 53 33N 14 33 E
Polillo Is., Phil. 37 B6 14 56N 122 0 E
Polis, Cyprus 23 D11 35 2N 32 26 E
Políyiros, Greece 21 D10 40 23N 23 25 E
Polk, U.S.A. 78 E5 41 22N 79 56W
Pollachi, India 40 P10 10 35N 77 0 E
Pollensa, Spain 22 B10 39 54N 3 1 E
Pollensa, B. de, Spain 22 B10 39 53N 3 8 E
Pollock, U.S.A. 80 C4 45 55N 100 17W
Polnovat, Russia 26 C7 63 50N 65 54 E
Polo, U.S.A. 80 E10 41 59N 89 35W
Polonne, Ukraine 17 C14 50 6N 27 30 E
Polonnoye = Polonne, Ukraine 17 C14 50 6N 27 30 E
Polson, U.S.A. 82 C6 47 41N 114 9W
Poltava, Ukraine 25 E5 49 35N 34 35 E
Põltsamaa, Estonia 9 G21 58 41N 25 58 E
Polunochnoye, Russia 24 B11 60 52N 60 25 E
Põlva, Estonia 9 G22 58 3N 27 3 E
Polyarny, Russia 24 A5 69 8N 33 20 E
Polynesia, Pac. Oc. 65 H11 10 0S 162 0W
Polynésie française = French Polynesia ■, Pac. Oc. 65 J13 20 0S 145 0W
Pomaro, Mexico 86 D4 18 20N 103 18W
Pombal, Brazil 93 E11 6 45S 37 50W
Pombal, Portugal 19 C1 39 55N 8 40W
Pómbia, Greece 23 D6 35 0N 24 51 E
Pomeroy, Ohio, U.S.A. 76 F4 39 2N 82 2W
Pomeroy, Wash., U.S.A. 82 C5 46 28N 117 36W
Pomézia, Italy 20 D5 41 40N 12 30 E
Pomona, U.S.A. 85 L9 34 4N 117 45W
Pomorski, Pojezierze, Poland 17 B9 53 40N 16 37 E
Pomos, Cyprus 23 D11 35 9N 32 33 E
Pomos, C., Cyprus 23 D11 35 10N 32 33 E
Pompano Beach, U.S.A. 77 M5 26 14N 80 8W
Pompeys Pillar, U.S.A. 82 D10 45 59N 107 57W
Ponape = Pohnpei, Pac. Oc. 64 G7 6 55N 158 10 E
Ponask, L., Canada 70 B1 54 0N 92 41W
Ponass L., Canada 73 C8 52 16N 103 58W
Ponca, U.S.A. 80 D6 42 34N 96 43W
Ponca City, U.S.A. 81 G6 36 42N 97 5W
Ponce, Puerto Rico 89 C6 18 1N 66 37W
Ponchatoula, U.S.A. 81 K9 30 26N 90 26W
Poncheville, L., Canada 70 B4 50 10N 76 55W
Pond, U.S.A. 85 K7 35 43N 119 20W
Pond Inlet, Canada 69 A12 72 40N 77 0W
Pondicherry, India 40 P11 11 59N 79 50 E
Ponds, I. of, Canada 71 B8 53 27N 55 52W
Ponferrada, Spain 19 A2 42 32N 6 35W
Ponnani, India 40 P9 10 45N 75 59 E
Ponnyadaung, Burma 41 J19 22 0N 94 10 E
Ponoka, Canada 72 C6 52 42N 113 40W
Ponorogo, Indonesia 37 G14 7 52S 111 27 E
Ponoy, Russia 24 A7 67 0N 41 13 E
Ponoy →, Russia 24 A7 66 59N 41 17 E
Ponta do Sol, Madeira 22 D2 32 42N 17 7W
Ponta Grossa, Brazil 95 B5 25 7S 50 10W
Ponta Pora, Brazil 95 A4 22 20S 55 35W
Pontarlier, France 18 C7 46 54N 6 20 E
Pontchartrain L., U.S.A. 81 K9 30 5N 90 5W
Ponte do Pungué, Mozam. 55 F3 19 30S 34 33 E
Ponte Nova, Brazil 95 A7 20 25S 42 54W
Pontefract, U.K. 10 D6 53 42N 1 18W
Ponteix, Canada 73 D7 49 46N 107 29W
Pontevedra, Spain 19 A1 42 26N 8 40W
Pontiac, Ill., U.S.A. 80 E10 40 53N 88 38W
Pontiac, Mich., U.S.A. 76 D4 42 38N 83 18W
Pontian Kecil, Malaysia 39 M4 1 29N 103 23 E
Pontianak, Indonesia 36 E3 0 3S 109 15 E
Pontine Is. = Ponziane, Ísole, Italy 20 D5 40 55N 12 57 E
Pontine Mts. = Kuzey Anadolu Dağlan, Turkey 25 F6 41 30N 35 0 E
Pontivy, France 18 B2 48 5N 2 58W
Pontoise, France 18 B5 49 3N 2 5 E
Ponton →, Canada 72 B5 58 27N 116 11W
Pontypool, Canada 78 B6 44 6N 78 38W
Pontypool, U.K. 11 F4 51 42N 3 2W
Pontypridd, U.K. 11 F4 51 36N 3 20W
Ponziane, Ísole, Italy 20 D5 40 55N 12 57 E
Poochera, Australia 63 E1 32 43S 134 51 E
Poole, U.K. 11 G6 50 43N 1 59W
Pooley I., Canada 72 C3 52 45N 128 15W
Poona = Pune, India 40 K8 18 29N 73 57 E
Pooncarie, Australia 63 E3 33 22S 142 31 E
Poopelloe L., Australia 63 E3 31 40S 144 0 E
Poopó, L. de, Bolivia 92 G5 18 30S 67 35W
Popanyinning, Australia 61 F2 32 40S 117 2 E
Popayán, Colombia 92 C3 2 27N 76 36W
Poperinge, Belgium 15 D2 50 51N 2 42 E
Popigay, Russia 27 B12 72 1N 110 39 E
Popilta, L., Australia 63 E3 33 10S 141 42 E
Popio L., Australia 63 E3 33 10S 141 52 E
Poplar, U.S.A. 80 A2 48 7N 105 12W
Poplar →, Man., Canada 73 C9 53 0N 97 19W
Poplar →, N.W.T., Canada 72 A4 61 22N 121 52W
Poplar Bluff, U.S.A. 81 G9 36 46N 90 24W
Poplarville, U.S.A. 81 K10 30 51N 89 32W
Popocatépetl, Volcán, Mexico 87 D5 19 2N 98 38W
Popokabaka, Zaïre 52 F3 5 41S 16 40 E
Poprád, Slovak Rep. 17 D11 49 3N 20 18 E
Porali →, Pakistan 42 G2 25 35N 66 26 E
Porbandar, India 42 J3 21 44N 69 43 E
Porcher I., Canada 72 C2 53 50N 130 30W
Porcupine →, Canada 73 B8 59 11N 104 46W
Porcupine →, U.S.A. 68 B5 66 34N 145 19W
Pordenone, Italy 20 B5 45 57N 12 39 E
Pori, Finland 9 F19 61 29N 21 48 E
Porlamar, Venezuela 92 A6 10 57N 63 51W
Poronaysk, Russia 27 E15 49 13N 143 0 E
Poroshiri-Dake, Japan 30 C11 42 41N 142 52 E
Poroto Mts., Tanzania 55 D3 9 0S 33 30 E
Porpoise B., Antarctica 5 C9 66 0S 127 0 E
Porreras, Spain 22 B10 39 31N 3 2 E
Porretta, Passo di, Italy 20 B4 44 2N 10 56 E
Porsangen, Norway 8 A21 70 40N 25 40 E
Porsgrunn, Norway 9 G13 59 10N 9 40 E
Port Adelaide, Australia 63 E2 34 46S 138 30 E
Port Alberni, Canada 72 D4 49 14N 124 50W
Port Alfred, Canada 71 C5 48 18N 70 53W
Port Alfred, S. Africa 56 E4 33 36S 26 55 E
Port Alice, Canada 72 C3 50 20N 127 25W
Port Allegany, U.S.A. 78 E6 41 48N 78 17W
Port Allen, U.S.A. 81 K9 30 27N 91 12W
Port Alma, Australia 62 C5 23 38S 150 53 E
Port Angeles, U.S.A. 84 B3 48 7N 123 27W
Port Antonio, Jamaica 88 C4 18 10N 76 30W
Port Aransas, U.S.A. 81 M6 27 50N 97 4W
Port Arthur = Lüshun, China 35 E11 38 45N 121 15 E
Port Arthur, Australia 62 G4 43 7S 147 50 E
Port Arthur, U.S.A. 81 L8 29 54N 93 56W
Port au Port B., Canada 71 C8 48 40N 58 50W
Port-au-Prince, Haiti 89 C5 18 40N 72 20W
Port Augusta, Australia 63 E2 32 30S 137 50 E
Port Augusta West, Australia 63 E2 32 29S 137 29 E
Port Austin, U.S.A. 78 B2 44 3N 83 1W
Port Bell, Uganda 54 B3 0 18N 32 35 E
Port Bergé Vaovao, Madag. 57 B8 15 33S 47 40 E
Port Blandford, Canada 71 C9 48 20N 54 10W
Port Bou, Spain 19 A7 42 25N 3 9 E
Port Bradshaw, Australia 62 A2 12 30S 137 20 E
Port Broughton, Australia 63 E2 33 37S 137 56 E
Port Burwell, Canada 70 D3 42 40N 80 48W
Port Canning, India 43 H13 22 23N 88 40 E
Port-Cartier, Canada 71 B6 50 2N 66 50W
Port Chalmers, N.Z. 59 L3 45 49S 170 30 E
Port Chester, U.S.A. 79 F11 41 0N 73 40W
Port Clements, Canada 72 C2 53 40N 132 10W
Port Clinton, U.S.A. 76 E4 41 31N 82 56W
Port Colborne, Canada 70 D4 42 50N 79 10W
Port Coquitlam, Canada 72 D4 49 15N 122 45W
Port Credit, Canada 78 C5 43 33N 79 35W
Port Curtis, Australia 62 C5 23 57S 151 20 E
Port Dalhousie, Canada 78 C5 43 13N 79 16W
Port Darwin, Australia 60 B5 12 24S 130 45 E
Port Darwin, Falk. Is. 96 G5 51 50S 59 0W
Port Davey, Australia 62 G4 43 16S 145 55 E
Port-de-Paix, Haiti 89 C5 19 50N 72 50W
Port Dickson, Malaysia 39 L3 2 30N 101 49 E
Port Douglas, Australia 62 B4 16 30S 145 30 E
Port Dover, Canada 78 D4 42 47N 80 12W
Port Edward, Canada 72 C2 54 12N 130 10W
Port Elgin, Canada 70 D3 44 25N 81 25W
Port Elizabeth, S. Africa 56 E4 33 58S 25 40 E
Port Ellen, U.K. 12 F2 55 38N 6 11W
Port Erin, U.K. 10 C3 54 5N 4 45W
Port Essington, Australia 60 B5 11 15S 132 10 E
Port Etienne = Nouâdhibou, Mauritania 50 D1 20 54N 17 0W
Port Fairy, Australia 63 F3 38 22S 142 12 E
Port Gamble, U.S.A. 84 C4 47 51N 122 35W
Port-Gentil, Gabon 52 E1 0 40S 8 50 E
Port Gibson, U.S.A. 81 K9 31 58N 90 59W
Port Glasgow, U.K. 12 F4 55 56N 4 41W
Port Harcourt, Nigeria 50 H6 4 40N 7 10 E
Port Hardy, Canada 72 C3 50 41N 127 30W
Port Harrison = Inukjuak, Canada 69 C12 58 25N 78 15W
Port Hawkesbury, Canada 71 C7 45 36N 61 22W
Port Hedland, Australia 60 D2 20 25S 118 35 E
Port Henry, U.S.A. 79 B11 44 3N 73 28W
Port Hood, Canada 71 C7 46 0N 61 32W
Port Hope, Canada 70 D4 43 56N 78 20W
Port Hueneme, U.S.A. 85 L7 34 7N 119 12W
Port Huron, U.S.A. 76 D4 42 58N 82 26W
Port Isabel, U.S.A. 81 M6 26 5N 97 12W
Port Jefferson, U.S.A. 79 F11 40 57N 73 3W
Port Jervis, U.S.A. 79 E10 41 22N 74 41W
Port Kelang = Pelabuhan Kelang, Malaysia 39 L3 3 0N 101 23 E
Port Kembla, Australia 63 E5 34 52S 150 49 E
Port Kenny, Australia 63 E1 33 10S 134 41 E
Port Lairge = Waterford, Ireland 13 D4 52 15N 7 8W
Port Laoise, Ireland 13 C4 53 2N 7 18W
Port Lavaca, U.S.A. 81 L6 28 37N 96 38W
Port Lincoln, Australia 63 E2 34 42S 135 52 E
Port Loko, S. Leone 50 G2 8 48N 12 46W
Port Lyautey = Kenitra, Morocco 50 B3 34 15N 6 40W
Port MacDonnell, Australia 63 F3 38 5S 140 48 E
Port Macquarie, Australia 63 E5 31 25S 152 25 E
Port Maria, Jamaica 88 C4 18 25N 76 55W
Port Mellon, Canada 72 D4 49 32N 123 31W
Port-Menier, Canada 71 C7 49 51N 64 15W
Port Morant, Jamaica 88 C4 17 54N 76 19W
Port Moresby, Papua N. G. 64 H6 9 24S 147 8 E
Port Mouton, Canada 71 D7 43 58N 64 50W
Port Musgrave, Australia 62 A3 11 55S 141 50 E
Port Nelson, Canada 73 B10 57 3N 92 36W
Port Nolloth, S. Africa 56 D2 29 17S 16 52 E
Port Nouveau-Québec = Kangiqsualujjuaq, Canada 69 C13 58 30N 65 59W
Port O'Connor, U.S.A. 81 L6 28 26N 96 24W
Port of Spain, Trin. & Tob. 89 D7 10 40N 61 31W
Port Orchard, U.S.A. 84 C4 47 32N 122 38W
Port Orford, U.S.A. 82 E1 42 45N 124 30W
Port Pegasus, N.Z. 59 M1 47 12S 167 41 E
Port Perry, Canada 70 D4 44 6N 78 56W
Port Phillip B., Australia 63 F3 38 10S 144 50 E
Port Pirie, Australia 63 E2 33 10S 138 1 E
Port Radium = Echo Bay, Canada 68 B8 66 5N 117 55W
Port Renfrew, Canada 72 D4 48 30N 124 20W
Port Roper, Australia 62 A2 14 45S 135 25 E
Port Rowan, Canada 78 D4 42 40N 80 30W
Port Safaga = Bûr Safâga, Egypt 51 C11 26 43N 33 57 E
Port Said = Bûr Sa'îd, Egypt 51 B11 31 16N 32 18 E
Port St. Joe, U.S.A. 77 L3 29 49N 85 18W
Port St. Johns, S. Africa 57 E4 31 38S 29 33 E
Port Sanilac, U.S.A. 78 C2 43 26N 82 33W
Port Saunders, Canada 71 B8 50 40N 57 18W
Port Severn, Canada 78 B5 44 48N 79 43W
Port Shepstone, S. Africa 57 E5 30 44S 30 28 E
Port Simpson, Canada 72 C2 54 30N 130 20W
Port Stanley = Stanley, Falk. Is. 96 G5 51 40S 59 51W
Port Stanley, Canada 70 D3 42 40N 81 10W
Port Sudan = Bûr Sûdân, Sudan 51 E12 19 32N 37 9 E
Port Talbot, U.K. 11 F4 51 35N 3 47W
Port Townsend, U.S.A. 84 B4 48 7N 122 45W
Port-Vendres, France 18 E5 42 32N 3 8 E
Port Vladimir, Russia 24 A5 69 25N 33 6 E
Port Wakefield, Australia 63 E2 34 12S 138 10 E
Port Washington, U.S.A. 76 D2 43 23N 87 53W
Port Weld, Malaysia 39 K3 4 50N 100 38 E
Porta Orientalis, Romania 17 F12 45 6N 22 18 E
Portachuelo, Bolivia 92 G6 17 10S 63 20W
Portadown, U.K. 13 B5 54 25N 6 27W
Portage, U.S.A. 80 D10 43 33N 89 28W
Portage La Prairie, Canada 73 D9 49 58N 98 18W
Portageville, U.S.A. 81 G10 36 26N 89 42W
Portalegre, Portugal 19 C2 39 19N 7 25W
Portales, U.S.A. 81 H3 34 11N 103 20W
Portarlington, Ireland 13 C4 53 9N 7 14W
Porter L., N.W.T., Canada 73 A7 61 41N 108 5W
Porter L., Sask., Canada 73 B7 56 20N 107 20W
Porterville, S. Africa 56 E2 33 0S 19 0 E
Porterville, U.S.A. 83 H4 36 4N 119 1W
Porthcawl, U.K. 11 F4 51 29N 3 42W
Porthill, U.S.A. 82 B5 48 59N 116 30W
Portile de Fier, Europe 17 F12 44 42N 22 30 E
Portimão, Portugal 19 D1 37 8N 8 32W
Portland, N.S.W., Australia 63 E4 33 20S 150 0 E
Portland, Vic., Australia 63 F3 38 20S 141 35 E
Portland, Canada 79 B8 44 42N 76 12W
Portland, Conn., U.S.A. 79 E12 41 34N 72 38W
Portland, Maine, U.S.A. 71 D5 43 39N 70 16W
Portland, Mich., U.S.A. 76 D3 42 52N 84 54W
Portland, Oreg., U.S.A. 84 E4 45 32N 122 37W
Portland, I. of, U.K. 11 G5 50 33N 2 26W
Portland B., Australia 63 F3 38 15S 141 45 E
Portland Bill, U.K. 11 G5 50 31N 2 28W
Portland Prom., Canada 69 C12 58 40N 78 33W
Portlands Roads, Australia 62 A3 12 36S 143 25 E
Portneuf, Canada 71 C5 46 43N 71 55W
Porto, Portugal 19 B1 41 8N 8 40W
Pôrto Alegre, Brazil 95 C5 30 5S 51 10W
Porto Amboim = Gunza, Angola 52 G2 10 50S 13 50 E
Porto Cristo, Spain 22 B10 39 33N 3 20 E
Pôrto de Móz, Brazil 93 D8 1 41S 52 13W
Porto Empédocle, Italy 20 F5 37 17N 13 32 E
Pôrto Esperança, Brazil 92 G7 19 37S 57 29W
Pôrto Franco, Brazil 93 E9 6 20S 47 24W
Porto Mendes, Brazil 95 A5 24 30S 54 15W
Porto Moniz, Madeira 22 D2 32 52N 17 11W
Pôrto Murtinho, Brazil 92 H7 21 45S 57 55W
Pôrto Nacional, Brazil 93 F9 10 40S 48 30W
Porto Novo, Benin 50 G5 6 23N 2 42 E
Porto Petro, Spain 22 B10 39 22N 3 13 E
Porto Santo, Madeira 50 B1 33 45N 16 25W
Pôrto São José, Brazil 95 A5 22 43S 53 10W
Pôrto Seguro, Brazil 93 G11 16 26S 39 5W
Porto Tórres, Italy 20 D3 40 50N 8 24 E
Pôrto União, Brazil 95 B5 26 10S 51 10W
Pôrto Válter, Brazil 92 E4 8 15S 72 40W
Porto-Vecchio, France 18 F8 41 35N 9 16 E
Pôrto Velho, Brazil 92 E6 8 46S 63 54W
Portobelo, Panama 88 E4 9 35N 79 42W
Portoferráio, Italy 20 C4 42 48N 10 20 E
Portola, U.S.A. 84 F6 39 49N 120 28W
Portoscuso, Italy 20 E3 39 12N 8 24 E
Portoviejo, Ecuador 92 D2 1 7S 80 28W
Portpatrick, U.K. 12 G3 54 51N 5 7W
Portree, U.K. 12 D2 57 25N 6 12W
Portrush, U.K. 13 A5 55 12N 6 40W
Portsmouth, Domin. 89 C7 15 34N 61 27W
Portsmouth, U.K. 11 G6 50 48N 1 6W
Portsmouth, N.H., U.S.A. 79 C14 43 5N 70 45W
Portsmouth, Ohio, U.S.A. 76 F4 38 44N 82 57W
Portsmouth, R.I., U.S.A. 79 E13 41 36N 71 15W
Portsmouth, Va., U.S.A. 76 G7 36 50N 76 18W
Portsoy, U.K. 12 D6 57 41N 2 41W
Porttipahtan tekojärvi, Finland 8 B22 68 5N 26 40 E
Portugal ■, Europe 19 C1 40 0N 8 0W
Portumna, Ireland 13 C3 53 6N 8 14W
Portville, U.S.A. 78 D6 42 3N 78 20W
Porvenir, Chile 96 G2 53 10S 70 16W
Porvoo, Finland 9 F21 60 24N 25 40 E
Posadas, Argentina 95 B4 27 30S 55 50W
Poshan = Boshan, China 35 F9 36 28N 117 49 E
Posht-e-Badam, Iran 45 C7 33 2N 55 23 E
Poso, Indonesia 37 E6 1 20S 120 55 E
Posong, S. Korea 35 G14 34 46N 127 5 E
Posse, Brazil 93 F9 14 4S 46 18W
Possel, C.A.R. 52 C3 5 5N 19 10 E
Possession I., Antarctica 5 D11 72 4S 172 0 E
Post, U.S.A. 81 J4 33 12N 101 23W
Post Falls, U.S.A. 82 C5 47 43N 116 57W
Postavy = Pastavy, Belarus 9 J22 55 4N 26 50 E
Poste Maurice Cortier, Algeria 50 D5 22 14N 1 2 E
Postmasburg, S. Africa 56 D3 28 18S 23 5 E
Postojna, Slovenia 16 F8 45 46N 14 12 E
Poston, U.S.A. 85 M12 34 0N 114 24W
Potchefstroom, S. Africa 56 D4 26 41S 27 7 E
Poteau, U.S.A. 81 H7 35 3N 94 37W
Poteet, U.S.A. 81 L5 29 2N 98 35W
Potenza, Italy 20 D6 40 38N 15 48 E
Poteriteri, L., N.Z. 59 M1 46 5S 167 10 E
Potgietersrus, S. Africa 57 C4 24 10S 28 55 E
Poti, Georgia 25 F7 42 10N 41 38 E
Potiskum, Nigeria 50 F7 11 39N 11 2 E
Potomac →, U.S.A. 76 F7 38 0N 76 23W
Potosí, Bolivia 92 G5 19 38S 65 50W
Potosi Mt., U.S.A. 85 K11 35 57N 115 29W
Pototan, Phil. 37 B6 10 54N 122 38 E

Potrerillos, *Chile* **94 B2** 26 30S 69 30W
Potsdam, *Germany* **16 B7** 52 25N 13 4 E
Potsdam, *U.S.A.* **79 B10** 44 40N 74 59W
Potter, *U.S.A.* **80 E3** 41 13N 103 19W
Pottstown, *U.S.A.* **79 F9** 40 15N 75 39W
Pottsville, *U.S.A.* **79 F8** 40 41N 76 12W
Pottuvil, *Sri Lanka* **40 R12** 6 55N 81 50 E
Pouce Coupé, *Canada* **72 B4** 55 40N 120 10W
Poughkeepsie, *U.S.A.* **79 E11** 41 42N 73 56W
Poulaphouca Res., *Ireland* **13 C5** 53 8N 6 30W
Poulsbo, *U.S.A.* **84 C4** 47 44N 122 39W
Pouso Alegre, *Mato Grosso, Brazil* **93 F7** 11 46S 57 16W
Pouso Alegre, *Minas Gerais, Brazil* **95 A6** 22 14S 45 57W
Považská Bystrica, *Slovak Rep.* **17 D10** 49 8N 18 27 E
Povenets, *Russia* **24 B5** 62 50N 34 50 E
Poverty B., *N.Z.* **59 H7** 38 43S 178 2 E
Póvoa de Varzim, *Portugal* **19 B1** 41 25N 8 46W
Powassan, *Canada* **70 C4** 46 5N 79 25W
Poway, *U.S.A.* **85 N9** 32 58N 117 2W
Powder —►, *U.S.A.* **80 B2** 46 45N 105 26W
Powder River, *U.S.A.* **82 E10** 43 2N 106 59W
Powell, *U.S.A.* **82 D9** 44 45N 108 46W
Powell L., *U.S.A.* **83 H8** 36 57N 111 29W
Powell River, *Canada* **72 D4** 49 50N 124 35W
Powers, *Mich., U.S.A.* **76 C2** 45 41N 87 32W
Powers, *Oreg., U.S.A.* **82 E1** 42 53N 124 4W
Powers Lake, *U.S.A.* **80 A3** 48 34N 102 39W
Powys □, *U.K.* **11 E4** 52 20N 3 20W
Poyang Hu, *China* **33 D6** 29 5N 116 20 E
Poyarkovo, *Russia* **27 E13** 49 36N 128 41 E
Poza Rica, *Mexico* **87 C5** 20 33N 97 27W
Požarevac, *Serbia, Yug.* **21 B9** 44 35N 21 18 E
Poznań, *Poland* **17 B9** 52 25N 16 55 E
Pozo, *U.S.A.* **85 K6** 35 20N 120 24W
Pozo Almonte, *Chile* **92 H5** 20 10S 69 50W
Pozo Colorado, *Paraguay* **94 A4** 23 30S 58 45W
Pozo del Dátil, *Mexico* **86 B2** 30 0N 112 15W
Pozoblanco, *Spain* **19 C3** 38 23N 4 51W
Pozzuoli, *Italy* **20 D6** 40 49N 14 7 E
Prachin Buri, *Thailand* **38 E3** 14 0N 101 25 E
Prachuap Khiri Khan, *Thailand* **39 G2** 11 49N 99 48 E
Prado, *Brazil* **93 G11** 17 20S 39 13W
Prague = Praha, *Czech.* **16 C8** 50 5N 14 22 E
Praha, *Czech.* **16 C8** 50 5N 14 22 E
Praia, *C. Verde Is.* **49 E1** 17 0N 25 0W
Prainha, *Amazonas, Brazil* **92 E6** 7 10S 60 30W
Prainha, *Pará, Brazil* **93 D8** 1 45S 53 30W
Prairie, *Australia* **62 C3** 20 50S 144 35 E
Prairie —►, *U.S.A.* **81 H5** 34 30N 99 23W
Prairie City, *U.S.A.* **82 D4** 44 28N 118 43W
Prairie du Chien, *U.S.A.* **80 D9** 43 3N 91 9W
Prairies, *Canada* **68 C9** 52 0N 108 0W
Pran Buri, *Thailand* **38 F2** 12 23N 99 55 E
Prapat, *Indonesia* **36 D1** 2 41N 98 58 E
Prasonísi, Ákra, *Greece* **23 D9** 35 42N 27 46 E
Prata, *Brazil* **93 G9** 19 25S 48 54W
Pratapgarh, *India* **42 G6** 24 2N 74 40 E
Prato, *Italy* **20 C4** 43 53N 11 6 E
Pratt, *U.S.A.* **81 G5** 37 39N 98 44W
Prattville, *U.S.A.* **77 J2** 32 28N 86 29W
Pravia, *Spain* **19 A2** 43 30N 6 12W
Praya, *Indonesia* **36 F5** 8 39S 116 17 E
Precordillera, *Argentina* **94 C2** 30 0S 69 1W
Preeceville, *Canada* **73 C8** 51 57N 102 40W
Preiļi, *Latvia* **9 H22** 56 18N 26 43 E
Prelate, *Canada* **73 C7** 50 51N 109 24W
Premier, *Canada* **72 B3** 56 4N 129 56W
Premont, *U.S.A.* **81 M5** 27 22N 98 7W
Prentice, *U.S.A.* **80 C9** 45 33N 90 17W
Preobrazheniye, *Russia* **30 C6** 42 54N 133 54 E
Preparis North Channel, *Ind. Oc.* **41 M18** 15 12N 93 40 E
Preparis South Channel, *Ind. Oc.* **41 M18** 14 36N 93 40 E
Přerov, *Czech.* **17 D9** 49 28N 17 27 E
Prescott, *Canada* **70 D4** 44 45N 75 30W
Prescott, *Ariz., U.S.A.* **83 J7** 34 33N 112 28W
Prescott, *Ark., U.S.A.* **81 J8** 33 48N 93 23W
Preservation Inlet, *N.Z.* **59 M1** 46 8S 166 35 E
Presho, *U.S.A.* **80 D4** 43 54N 100 3W
Presidencia de la Plaza, *Argentina* **94 B4** 27 0S 59 50W
Presidencia Roque Saenz Peña, *Argentina* **94 B3** 26 45S 60 30W
Presidente Epitácio, *Brazil* **93 H8** 21 56S 52 6W
Presidente Hayes □, *Paraguay* **94 A4** 24 0S 59 0W
Presidente Hermes, *Brazil* **92 F6** 11 17S 61 55W
Presidente Prudente, *Brazil* **95 A5** 22 5S 51 25W
Presidio, *Mexico* **86 B4** 29 29N 104 23W
Presidio, *U.S.A.* **81 L2** 29 34N 104 22W
Prešov, *Slovak Rep.* **17 D11** 49 0N 21 15 E
Prespa, L. = Prespansko Jezero, *Macedonia* **21 D9** 40 55N 21 0 E
Prespansko Jezero, *Macedonia* **21 D9** 40 55N 21 0 E
Presque Isle, *U.S.A.* **71 C6** 46 41N 68 1W
Prestbury, *U.K.* **11 F5** 51 54N 2 2W
Presteigne, *U.K.* **11 E5** 52 17N 3 0W
Preston, *Canada* **78 C4** 43 23N 80 21W
Preston, *U.K.* **10 D5** 53 46N 2 42W
Preston, *Idaho, U.S.A.* **82 E8** 42 6N 111 53W
Preston, *Minn., U.S.A.* **80 D8** 43 40N 92 5W
Preston, *Nev., U.S.A.* **82 G6** 38 55N 115 4W
Preston, C., *Australia* **60 D2** 20 51S 116 12 E
Prestonpans, *U.K.* **12 F6** 55 58N 2 58W
Prestwick, *U.K.* **12 F4** 55 29N 4 37W
Pretoria, *S. Africa* **57 D4** 25 44S 28 12 E
Préveza, *Greece* **21 E9** 38 57N 20 47 E
Pribilof Is., *Bering S.* **4 D17** 56 0N 170 0W
Příbram, *Czech.* **16 D8** 49 41N 14 2 E
Price, *U.S.A.* **82 G8** 39 36N 110 49W
Price I., *Canada* **72 C3** 52 23N 128 41W
Prichard, *U.S.A.* **77 K1** 30 44N 88 5W
Priekule, *Latvia* **9 H19** 56 26N 21 35 E
Prienai, *Lithuania* **9 J20** 54 40N 23 55 E
Prieska, *S. Africa* **56 D3** 29 40S 22 42 E
Priest —►, *U.S.A.* **82 B5** 48 12N 116 54W
Priest L., *U.S.A.* **82 B5** 48 35N 116 52W
Priest Valley, *U.S.A.* **84 J6** 36 10N 120 39W
Priestly, *Canada* **72 C3** 54 8N 125 20W
Prievidza, *Slovak Rep.* **17 D10** 48 46N 18 36 E
Prikaspiyskaya Nizmennost = Caspian Depression, *Eurasia* **25 E8** 47 0N 48 0 E
Prilep, *Macedonia* **21 D9** 41 21N 21 37 E
Priluki = Pryluky, *Ukraine* **25 D5** 50 30N 32 24 E
Prime Seal I., *Australia* **62 G4** 40 3S 147 43 E
Primrose L., *Canada* **73 C7** 54 55N 109 45W
Prince Albert, *Canada* **73 C7** 53 15N 105 50W
Prince Albert, *S. Africa* **56 E3** 33 12S 22 2 E
Prince Albert Mts., *Antarctica* **5 D11** 76 0S 161 30 E
Prince Albert Nat. Park, *Canada* **73 C7** 54 0N 106 25W
Prince Albert Pen., *Canada* **68 A8** 72 30N 116 0W
Prince Albert Sd., *Canada* **68 A8** 70 25N 115 0W
Prince Alfred, C., *Canada* **4 B1** 74 20N 124 40W
Prince Charles I., *Canada* **69 B12** 67 47N 76 12W
Prince Charles Mts., *Antarctica* **5 D6** 72 0S 67 0 E
Prince Edward I. □, *Canada* **71 C7** 46 20N 63 20W
Prince Edward Is., *Ind. Oc.* **3 G11** 46 35S 38 0 E
Prince George, *Canada* **72 C4** 53 55N 122 50W
Prince of Wales, C., *U.S.A.* **66 C3** 65 36N 168 5W
Prince of Wales I., *Australia* **62 A3** 10 40S 142 10 E
Prince of Wales I., *Canada* **68 A10** 73 0N 99 0W
Prince of Wales I., *U.S.A.* **72 B2** 55 47N 132 50W
Prince Patrick I., *Canada* **4 B2** 77 0N 120 0W
Prince Regent Inlet, *Canada* **4 B3** 73 0N 90 0W
Prince Rupert, *Canada* **72 C2** 54 20N 130 20W
Princess Charlotte B., *Australia* **62 A3** 14 25S 144 0 E
Princess May Ras., *Australia* **60 C4** 15 30S 125 30 E
Princess Royal I., *Canada* **72 C3** 53 0N 128 40W
Princeton, *Canada* **72 D4** 49 27N 120 30W
Princeton, *Calif., U.S.A.* **84 F4** 39 24N 122 1W
Princeton, *Ill., U.S.A.* **80 E10** 41 23N 89 28W
Princeton, *Ind., U.S.A.* **76 F2** 38 21N 87 34W
Princeton, *Ky., U.S.A.* **76 G2** 37 7N 87 53W
Princeton, *Mo., U.S.A.* **80 E8** 40 24N 93 35W
Princeton, *N.J., U.S.A.* **79 F10** 40 21N 74 39W
Princeton, *W. Va., U.S.A.* **76 G5** 37 22N 81 6W
Principe, I. de, *Atl. Oc.* **48 F4** 1 37N 7 27 E
Principe Chan., *Canada* **72 C2** 53 28N 130 0W
Principe da Beira, *Brazil* **92 F6** 12 20S 64 30W
Prineville, *U.S.A.* **82 D3** 44 18N 120 51W
Prins Harald Kyst, *Antarctica* **5 D4** 70 0S 35 1 E
Prinsesse Astrid Kyst, *Antarctica* **5 D3** 70 45S 12 30 E
Prinsesse Ragnhild Kyst, *Antarctica* **5 D4** 70 15S 27 30 E
Prinzapolca, *Nic.* **88 D3** 13 20N 83 35W
Priozersk, *Russia* **24 B5** 61 2N 30 7 E
Pripet —► = Prypyat —►, *Europe* **17 C16** 51 20N 30 15 E
Pripet Marshes, *Europe* **17 B15** 52 10N 28 10 E
Pripyat Marshes = Pripet Marshes, *Europe* **17 B15** 52 10N 28 10 E
Pripyats = Prypyat —►, *Europe* **17 C16** 51 20N 30 15 E
Priština, *Serbia, Yug.* **21 C9** 42 40N 21 13 E
Privas, *France* **18 D6** 44 45N 4 37 E
Privolzhskaya Vozvyshennost, *Russia* **25 D8** 51 0N 46 0 E
Prizren, *Serbia, Yug.* **21 C9** 42 13N 20 45 E
Probolinggo, *Indonesia* **37 G15** 7 46S 113 13 E
Proddatur, *India* **40 M11** 14 45N 78 30 E
Prodhromos, *Cyprus* **23 E11** 34 57N 32 50 E
Profítis Ilías, *Greece* **23 C9** 36 17N 27 56 E
Progreso, *Mexico* **87 C7** 21 20N 89 40W
Prokopyevsk, *Russia* **26 D9** 54 0N 86 45 E
Prokuplje, *Serbia, Yug.* **21 C9** 43 16N 21 36 E
Prome = Pyè, *Burma* **41 K19** 18 49N 95 13 E
Prophet —►, *Canada* **72 B4** 58 48N 122 40W
Propriá, *Brazil* **93 F11** 10 13S 36 51W
Proserpine, *Australia* **62 C4** 20 21S 148 36 E
Prosser, *U.S.A.* **82 C4** 46 12N 119 46W
Prostějov, *Czech.* **17 D9** 49 30N 17 9 E
Proston, *Australia* **63 D5** 26 8S 151 32 E
Protection, *U.S.A.* **81 G5** 37 12N 99 29W
Provence, *France* **18 E6** 43 40N 5 46 E
Providence, *Ky., U.S.A.* **76 G2** 37 24N 87 46W
Providence, *R.I., U.S.A.* **79 E13** 41 49N 71 24W
Providence Bay, *Canada* **70 C3** 45 41N 82 15W
Providence Mts., *U.S.A.* **83 J6** 35 10N 115 15W
Providencia, I. de, *Colombia* **88 D3** 13 25N 81 26W
Provideniya, *Russia* **27 C19** 64 23N 173 18W
Provins, *France* **18 B5** 48 33N 3 15 E
Provo, *U.S.A.* **82 F8** 40 14N 111 39W
Provost, *Canada* **73 C6** 52 25N 110 20W
Prozna —►, *Poland* **17 B9** 52 6N 17 44 E
Prud'homme, *Canada* **73 C7** 52 20N 105 54W
Pruszków, *Poland* **17 B11** 52 9N 20 49 E
Prut —►, *Romania* **17 F15** 45 28N 28 10 E
Pruzhany, *Belarus* **17 B13** 52 33N 24 28 E
Prydz B., *Antarctica* **5 C6** 69 0S 74 0 E
Pryluky, *Ukraine* **25 D5** 50 30N 32 24 E
Pryor, *U.S.A.* **81 G7** 36 19N 95 19W
Prypyat —►, *Europe* **17 C16** 51 20N 30 15 E
Przemyśl, *Poland* **17 D12** 49 50N 22 45 E
Przhevalsk, *Kyrgyzstan* **26 E8** 42 30N 78 20 E
Psará, *Greece* **21 E11** 38 37N 25 38 E
Psíra, *Greece* **23 D7** 35 12N 25 52 E
Pskov, *Russia* **24 C4** 57 50N 28 25 E
Pskovskoye, Ozero, *Russia* **9 H22** 58 0N 27 58 E
Ptich = Ptsich —►, *Belarus* **17 B15** 52 9N 28 52 E
Ptolemaís, *Greece* **21 D9** 40 30N 21 43 E
Ptsich —►, *Belarus* **17 B15** 52 9N 28 52 E
Pu Xian, *China* **34 F6** 36 24N 111 6 E
Pua, *Thailand* **38 C3** 19 11N 100 55 E
Puán, *Argentina* **94 D3** 37 30S 62 45W
Puan, *S. Korea* **35 G14** 35 44N 126 44 E
Pucallpa, *Peru* **92 E4** 8 25S 74 30W
Pudasjärvi, *Finland* **8 D22** 65 23N 26 53 E
Pudozh, *Russia* **24 B6** 61 48N 36 32 E
Pudukkottai, *India* **40 P11** 10 28N 78 47 E
Puebla, *Mexico* **87 D5** 19 3N 98 12W
Puebla □, *Mexico* **87 D5** 18 30N 98 0W
Pueblo, *U.S.A.* **80 F2** 38 16N 104 37W
Pueblo Hundido, *Chile* **94 B1** 26 20S 70 5W
Puelches, *Argentina* **94 D2** 38 5S 65 51W
Puelén, *Argentina* **94 D2** 37 32S 67 38W
Puente Alto, *Chile* **94 C1** 33 32S 70 35W
Puente-Genil, *Spain* **19 D3** 37 22N 4 47W
Puerco —►, *U.S.A.* **83 J10** 34 22N 107 50W
Puerto, *Canary Is.* **22 F2** 28 5N 17 20W
Puerto Aisén, *Chile* **96 F2** 45 27S 73 0W
Puerto Ángel, *Mexico* **87 D5** 15 40N 96 29W
Puerto Arista, *Mexico* **87 D6** 15 56N 93 48W
Puerto Armuelles, *Panama* **88 E3** 8 20N 82 51W
Puerto Ayacucho, *Venezuela* **92 B5** 5 40N 67 35W
Puerto Barrios, *Guatemala* **88 C2** 15 40N 88 32W
Puerto Bermejo, *Argentina* **94 B4** 26 55S 58 34W
Puerto Bermúdez, *Peru* **92 F4** 10 20S 74 58W
Puerto Bolívar, *Ecuador* **92 D3** 3 19S 79 55W
Puerto Cabello, *Venezuela* **92 A5** 10 28N 68 1W
Puerto Cabezas, *Nic.* **88 D3** 14 0N 83 30W
Puerto Cabo Gracias á Dios, *Nic.* **88 D3** 15 0N 83 10W
Puerto Carreño, *Colombia* **92 B5** 6 12N 67 22W
Puerto Castilla, *Honduras* **88 C2** 16 0N 86 0W
Puerto Chicama, *Peru* **92 E3** 7 45S 79 20W
Puerto Coig, *Argentina* **96 G3** 50 54S 69 15W
Puerto Cortes, *Costa Rica* **88 E3** 8 55N 84 0W
Puerto Cortés, *Honduras* **88 C2** 15 51N 88 0W
Puerto Cumarebo, *Venezuela* **92 A5** 11 29N 69 30W
Puerto de Alcudia, *Spain* **22 B10** 39 50N 3 7 E
Puerto de Andraitx, *Spain* **22 B9** 39 32N 2 23 E
Puerto de Cabrera, *Spain* **22 B9** 39 8N 2 56 E
Puerto de Gran Tarajal, *Canary Is.* **22 F5** 28 13N 14 1W
Puerto de la Cruz, *Canary Is.* **22 F3** 28 24N 16 32W
Puerto de Pozo Negro, *Canary Is.* **22 F6** 28 19N 13 55W
Puerto de Sóller, *Spain* **22 B9** 39 48N 2 42 E
Puerto del Carmen, *Canary Is.* **22 F6** 28 55N 13 38W
Puerto del Rosario, *Canary Is.* **22 F6** 28 30N 13 52W
Puerto Deseado, *Argentina* **96 F3** 47 55S 66 0W
Puerto Heath, *Bolivia* **92 F5** 12 34S 68 39W
Puerto Juárez, *Mexico* **87 C7** 21 11N 86 49W
Puerto La Cruz, *Venezuela* **92 A6** 10 13N 64 38W
Puerto Leguízamo, *Colombia* **92 D4** 0 12S 74 46W
Puerto Lobos, *Argentina* **96 E3** 42 0S 65 3W
Puerto Madryn, *Argentina* **96 E3** 42 48S 65 4W
Puerto Maldonado, *Peru* **92 F5** 12 30S 69 10W
Puerto Manotí, *Cuba* **88 B4** 21 22N 76 50W
Puerto Montt, *Chile* **96 E2** 41 28S 73 0W
Puerto Morelos, *Mexico* **87 C7** 20 49N 86 52W
Puerto Natales, *Chile* **96 G2** 51 45S 72 15W
Puerto Padre, *Cuba* **88 B4** 21 13N 76 35W
Puerto Páez, *Venezuela* **92 B5** 6 13N 67 28W
Puerto Peñasco, *Mexico* **86 A2** 31 20N 113 33W
Puerto Pinasco, *Paraguay* **94 A4** 22 36S 57 50W
Puerto Pirámides, *Argentina* **96 E4** 42 35S 64 20W
Puerto Plata, *Dom. Rep.* **89 C5** 19 48N 70 45W
Puerto Pollensa, *Spain* **22 B10** 39 54N 3 4 E
Puerto Princesa, *Phil.* **37 C5** 9 46N 118 45 E
Puerto Quellón, *Chile* **96 E2** 43 7S 73 37W
Puerto Quepos, *Costa Rica* **88 E3** 9 29N 84 6W
Puerto Rico, *Canary Is.* **22 G4** 27 47N 15 42W
Puerto Rico ■, *W. Indies* **89 C6** 18 15N 66 45W
Puerto Rico Trench, *Atl. Oc.* **89 C6** 19 50N 66 0W
Puerto Sastre, *Paraguay* **94 A4** 22 2S 57 55W
Puerto Suárez, *Bolivia* **92 G7** 18 58S 57 52W
Puerto Vallarta, *Mexico* **86 C3** 20 36N 105 15W
Puerto Wilches, *Colombia* **92 B4** 7 21N 73 54W
Puertollano, *Spain* **19 C3** 38 43N 4 7W
Pueyrredón, L., *Argentina* **96 F2** 47 20S 72 0W
Pugachev, *Russia* **24 D8** 52 0N 48 49 E
Puge, *Tanzania* **54 C3** 4 45S 33 11 E
Puget Sound, *U.S.A.* **82 C2** 47 50N 122 30W
Pugŏdong, *N. Korea* **35 C16** 42 5N 130 0 E
Pugu, *Tanzania* **54 D4** 6 55S 39 4 E
Pūgūnzī, *Iran* **45 E8** 25 49N 59 10 E
Puig Mayor, *Spain* **22 B9** 39 48N 2 47 E
Puigcerdá, *Spain* **19 A6** 42 24N 1 50 E
Puigpuñent, *Spain* **22 B9** 39 38N 2 32 E
Pujon-chosuji, *N. Korea* **35 D14** 40 35N 127 35 E
Pukaki L., *N.Z.* **59 L3** 44 4S 170 1 E
Pukapuka, *Cook Is.* **65 J11** 10 53S 165 49W
Pukatawagan, *Canada* **73 B8** 55 45N 101 20W
Pukchin, *N. Korea* **35 D13** 40 12N 125 45 E
Pukchŏng, *N. Korea* **35 D15** 40 14N 128 10 E
Pukekohe, *N.Z.* **59 G5** 37 12S 174 55 E
Pula, *Croatia* **16 F7** 44 54N 13 57 E
Pulaski, *N.Y., U.S.A.* **79 C8** 43 34N 76 8W
Pulaski, *Tenn., U.S.A.* **77 H2** 35 12N 87 2W
Pulaski, *Va., U.S.A.* **76 G5** 37 3N 80 47W
Puławy, *Poland* **17 C11** 51 23N 21 59 E
Pulga, *U.S.A.* **84 F5** 39 48N 121 29W
Pulicat, L., *India* **40 N12** 13 40N 80 15 E
Pullman, *U.S.A.* **82 C5** 46 44N 117 10W
Pulog, *Phil.* **37 A6** 16 40N 120 50 E
Pułtusk, *Poland* **17 B11** 52 43N 21 6 E
Pumlumon Fawr, *U.K.* **11 E4** 52 28N 3 46W
Puna, *Bolivia* **92 G5** 19 45S 65 28W
Puná, I., *Ecuador* **92 D2** 2 55S 80 5W
Punakha, *Bhutan* **41 F16** 27 42N 89 52 E
Punasar, *India* **42 F5** 27 6N 73 6 E
Punata, *Bolivia* **92 G5** 17 32S 65 50W
Punch, *India* **43 C6** 33 48N 74 4 E
Pune, *India* **40 K8** 18 29N 73 57 E
Pungsan, *N. Korea* **35 D15** 40 50N 128 9 E
Pungue, Ponte de, *Mozam.* **55 F3** 19 0S 34 0 E
Punjab □, *India* **42 D6** 31 0N 76 0 E
Punjab □, *Pakistan* **42 E6** 32 0N 74 30 E
Puno, *Peru* **92 G4** 15 55S 70 3W
Punta Alta, *Argentina* **96 D4** 38 53S 62 4W
Punta Arenas, *Chile* **96 G2** 53 10S 71 0W
Punta de Díaz, *Chile* **94 B1** 28 0S 70 45W
Punta Gorda, *Belize* **87 D7** 16 10N 88 45W
Punta Gorda, *U.S.A.* **77 M5** 26 56N 82 3W
Punta Prieta, *Mexico* **86 B2** 28 58N 114 17W
Punta Prima, *Spain* **22 B11** 39 48N 4 16 E
Puntabie, *Australia* **63 E1** 32 12S 134 13 E
Puntarenas, *Costa Rica* **88 E3** 10 0N 84 50W
Punto Fijo, *Venezuela* **92 A4** 11 50N 70 13W
Punxsatawney, *U.S.A.* **78 F5** 40 57N 78 59W
Puquio, *Peru* **92 F4** 14 45S 74 10W
Pur —►, *Russia* **26 C8** 67 31N 77 55 E
Purace, Vol., *Colombia* **92 C3** 2 21N 76 23W
Puralia = Puruliya, *India* **43 H12** 23 17N 86 24 E
Purbeck, Isle of, *U.K.* **11 G5** 50 39N 1 59W
Purcell, *U.S.A.* **81 H6** 35 1N 97 22W
Puri, *India* **41 K14** 19 50N 85 58 E
Purmerend, *Neths.* **15 B4** 52 32N 4 58 E
Purnia, *India* **43 G12** 25 45N 87 31 E
Purukcahu, *Indonesia* **36 E4** 0 35S 114 35 E
Puruliya, *India* **43 H12** 23 17N 86 24 E
Purus —►, *Brazil* **92 D6** 3 42S 61 28W
Purwakarta, *Indonesia* **37 G12** 6 35S 107 29 E
Purwodadi, *Jawa, Indonesia* **37 G14** 7 7S 110 55 E
Purwodadi, *Jawa, Indonesia* **37 G13** 7 51S 110 0 E
Purwokerto, *Indonesia* **37 G13** 7 25S 109 14 E
Purworejo, *Indonesia* **37 G14** 7 43S 110 2 E
Puryŏng, *N. Korea* **35 C15** 42 5N 129 43 E
Pusan, *S. Korea* **35 G15** 35 5N 129 0 E
Pushchino, *Russia* **27 D16** 54 10N 158 0 E
Pushkino, *Russia* **25 D8** 51 16N 47 0 E
Putahow L., *Canada* **73 B8** 59 54N 100 40W
Putao, *Burma* **41 F20** 27 28N 97 30 E
Putaruru, *N.Z.* **59 H5** 38 2S 175 50 E
Puthein Myit —►, *Burma* **41 M19** 15 56N 94 18 E
Putignano, *Italy* **20 D7** 40 51N 17 7 E
Puting, Tanjung, *Indonesia* **36 E4** 3 31S 111 46 E
Putnam, *U.S.A.* **79 E13** 41 55N 71 55W
Putorana, Gory, *Russia* **27 C10** 69 0N 95 0 E
Puttalam, *Sri Lanka* **40 Q11** 8 1N 79 55 E
Putten, *Neths.* **15 B5** 52 16N 5 36 E
Puttgarden, *Germany* **16 A6** 54 30N 11 10 E
Putumayo —►, *S. Amer.* **92 D5** 3 7S 67 58W
Putussibau, *Indonesia* **36 D4** 0 50N 112 56 E
Puy-de-Dôme, *France* **18 D5** 45 46N 2 57 E
Puyallup, *U.S.A.* **84 C4** 47 12N 122 18W
Puyang, *China* **34 G8** 35 40N 115 1 E
Pūzeh Rīg, *Iran* **45 E8** 27 20N 58 40 E
Pwani □, *Tanzania* **54 D4** 7 0S 39 0 E
Pweto, *Zaïre* **55 D2** 8 25S 28 51 E
Pwllheli, *U.K.* **10 E3** 52 53N 4 25W
Pya-ozero, *Russia* **24 A5** 66 5N 30 58 E
Pyapon, *Burma* **41 L19** 16 20N 95 40 E
Pyasina —►, *Russia* **27 B9** 73 30N 87 0 E
Pyatigorsk, *Russia* **25 F7** 44 2N 43 6 E
Pyè, *Burma* **41 K19** 18 49N 95 13 E
Pyetrikaw, *Belarus* **17 B15** 52 11N 28 29 E
Pyhäjoki, *Finland* **8 D21** 64 28N 24 14 E
Pyinmana, *Burma* **41 K20** 19 45N 96 12 E
Pyla, C., *Cyprus* **23 E12** 34 56N 33 51 E
Pyŏktong, *N. Korea* **35 D13** 40 50N 125 50 E
Pyŏnggang, *N. Korea* **35 E14** 38 24N 127 17 E
Pyŏngtaek, *S. Korea* **35 F14** 37 1N 127 4 E
P'yŏngyang, *N. Korea* **35 E13** 39 0N 125 30 E
Pyote, *U.S.A.* **81 K3** 31 32N 103 8W
Pyramid L., *U.S.A.* **82 G4** 40 1N 119 35W
Pyramid Pk., *U.S.A.* **85 J10** 36 25N 116 37W
Pyrénées, *Europe* **19 A6** 42 45N 0 18 E
Pyu, *Burma* **41 K20** 18 30N 96 28 E

Q

Qaanaaq = Thule, *Greenland* **4 B4** 77 40N 69 0W
Qachasnek, *S. Africa* **57 E4** 30 6S 28 42 E
Qāḑib, *Yemen* **46 E5** 12 37N 53 57 E
Qa'el Jafr, *Jordan* **47 E5** 30 20N 36 25 E
Qā'emābād, *Iran* **45 D9** 31 44N 60 2 E
Qā'emshahr, *Iran* **45 B7** 36 30N 52 53 E
Qagan Nur, *China* **34 C8** 43 30N 114 55 E
Qahar Youyi Zhongqi, *China* **34 D7** 41 12N 112 40 E
Qahremānshahr = Bākhtarān, *Iran* **44 C5** 34 23N 47 0 E
Qaidam Pendi, *China* **32 C4** 37 0N 95 0 E
Qajarīyeh, *Iran* **45 D6** 31 1N 48 22 E
Qala, Ras il, *Malta* **23 C1** 36 1N 14 20 E
Qala-i-Jadid, *Afghan.* **42 D2** 31 1N 66 25 E
Qala Yangi, *Afghan.* **42 B2** 34 20N 66 30 E
Qal'at al Akhḑar, *Si. Arabia* **44 E3** 28 0N 37 10 E
Qal'at Sukkar, *Iraq* **44 D5** 31 51N 46 5 E
Qal'eh Darreh, *Iran* **44 B5** 38 47N 47 2 E
Qal'eh Shaharak, *Afghan.* **40 B4** 34 10N 64 20 E
Qamar, Ghubbat al, *Yemen* **46 D5** 16 20N 52 30 E
Qamdo, *China* **32 C4** 31 15N 97 6 E
Qamruddin Karez, *Pakistan* **42 D3** 31 45N 68 20 E
Qandahār, *Afghan.* **40 D4** 31 32N 65 30 E
Qandahār □, *Afghan.* **40 D4** 31 0N 65 0 E
Qapān, *Iran* **45 B7** 37 40N 55 47 E
Qapshaghay, *Kazakstan* **26 E8** 43 51N 77 14 E
Qaqortoq = Julianehåb, *Greenland* **4 C5** 60 43N 46 0W
Qâra, *Egypt* **51 C10** 29 38N 26 30 E
Qara Qash —►, *India* **43 B8** 35 0N 78 30 E
Qarabutaq, *Kazakstan* **26 E7** 49 59N 60 14 E
Qaraghandy, *Kazakstan* **26 E8** 49 50N 73 10 E
Qārah, *Si. Arabia* **44 D4** 29 55N 40 3 E
Qarataū, *Kazakstan* **26 E8** 43 10N 70 28 E
Qareh —►, *Iran* **44 B5** 39 25N 47 22 E
Qareh Tekān, *Iran* **45 B6** 36 38N 49 29 E
Qarqan He —►, *China* **32 C3** 39 30N 88 30 E
Qarqaraly, *Kazakstan* **26 E8** 49 26N 75 30 E
Qarshi, *Uzbekistan* **26 F7** 38 53N 65 48 E
Qartabā, *Lebanon* **47 A4** 34 4N 35 50 E
Qaryat al Gharab, *Iraq* **44 D5** 31 27N 44 48 E
Qaryat al 'Ulyā, *Si. Arabia* **44 E5** 27 33N 47 42 E

Qasr 'Amra, Jordan 44 D3 31 48N 36 35 E
Qaşr-e Qand, Iran 45 E9 26 15N 60 45 E
Qasr Farâfra, Egypt 51 C10 27 0N 28 1 E
Qatanā, Syria 47 B5 33 26N 36 4 E
Qatar ■, Asia 45 E6 25 30N 51 15 E
Qatlīsh, Iran 45 B8 37 50N 57 19 E
Qattâra, Munkhafed el, Egypt 51 C10 29 30N 27 30 E
Qattâra Depression = Qattâra, Munkhafed el, Egypt 51 C10 29 30N 27 30 E
Qawām al Ḩamzah, Iraq 44 D5 31 43N 44 58 E
Qāyen, Iran 45 C8 33 40N 59 10 E
Qazaqstan = Kazakstan ■, Asia 26 E7 50 0N 70 0 E
Qazvin, Iran 45 B6 36 15N 50 0 E
Qena, Egypt 51 C11 26 10N 32 43 E
Qeqertarsuaq = Disko, Greenland 4 C5 69 45N 53 30W
Qeqertarsuaq = Godhavn, Greenland 4 C5 69 15N 53 38W
Qeshlāq, Iran 44 C5 34 55N 46 28 E
Qeshm, Iran 45 E8 26 55N 56 10 E
Qezi'ot, Israel 47 E3 30 52N 34 26 E
Qi Xian, China 34 G8 34 40N 114 48 E
Qian Gorlos, China 35 B13 45 5N 124 42 E
Qian Xian, China 34 G5 34 31N 108 15 E
Qianyang, China 34 G4 34 40N 107 8 E
Qibā', Si. Arabia 44 E5 27 24N 44 20 E
Qila Safed, Pakistan 40 E2 29 0N 61 30 E
Qila Saifullāh, Pakistan 42 D3 30 45N 68 17 E
Qilian Shan, China 32 C4 38 30N 96 0 E
Qin He →, China 34 G7 35 1N 113 22 E
Qin Ling = Qinling Shandi, China 34 H5 33 50N 108 10 E
Qin'an, China 34 G3 34 48N 105 40 E
Qing Xian, China 34 E9 38 35N 116 45 E
Qingcheng, China 35 F9 37 15N 117 40 E
Qingdao, China 35 F11 36 5N 120 20 E
Qingfeng, China 34 G8 35 52N 115 8 E
Qinghai □, China 32 C4 36 0N 98 0 E
Qinghai Hu, China 32 C5 36 40N 100 10 E
Qinghecheng, China 35 D13 41 15N 124 30 E
Qinghemen, China 35 D11 41 48N 121 25 E
Qingjian, China 34 F6 37 8N 110 8 E
Qingjiang, China 35 H10 33 30N 119 2 E
Qingshui, China 34 G4 34 48N 106 8 E
Qingshuihe, China 34 E6 39 55N 111 35 E
Qingtongxia Shuiku, China 34 F3 37 50N 105 58 E
Qingxu, China 34 F7 37 34N 112 22 E
Qingyang, China 34 F4 36 2N 107 55 E
Qingyuan, China 35 C13 42 10N 124 55 E
Qingyun, China 35 F9 37 45N 117 20 E
Qinhuangdao, China 35 E10 39 56N 119 30 E
Qinling Shandi, China 34 H5 33 50N 108 10 E
Qinshui, China 34 G7 35 40N 112 8 E
Qinyang, China 34 G7 35 7N 112 57 E
Qinyuan, China 34 F7 36 29N 112 20 E
Qinzhou, China 32 D5 21 58N 108 38 E
Qionghai, China 38 C8 19 15N 110 26 E
Qiongshan, China 38 C8 19 51N 110 26 E
Qiongzhou Haixia, China 38 B8 20 10N 110 15 E
Qiqihar, China 27 E13 47 26N 124 0 E
Qiraîya, W. →, Egypt 47 E3 30 27N 34 0 E
Qiryat Ata, Israel 47 C4 32 47N 35 6 E
Qiryat Gat, Israel 47 D3 31 32N 34 46 E
Qiryat Mal'akhi, Israel 47 D3 31 44N 34 44 E
Qiryat Shemona, Israel 47 B4 33 13N 35 35 E
Qiryat Yam, Israel 47 C4 32 51N 35 4 E
Qishan, China 34 G4 34 25N 107 38 E
Qitai, China 32 B3 44 2N 89 35 E
Qixia, China 35 F11 37 17N 120 52 E
Qojūr, Iran 44 B5 36 12N 47 55 E
Qom, Iran 45 C6 34 40N 51 0 E
Qomsheh, Iran 45 D6 32 0N 51 55 E
Qostanay, Kazakstan 26 D7 53 10N 63 35 E
Qu Xian, China 33 D6 28 57N 118 54 E
Quairading, Australia 61 F2 32 0S 117 21 E
Quakertown, U.S.A. 79 F9 40 26N 75 21W
Qualeup, Australia 61 F2 33 48S 116 48 E
Quambatook, Australia 63 F3 35 49S 143 34 E
Quambone, Australia 63 E4 30 57S 147 53 E
Quamby, Australia 62 C3 20 22S 140 17 E
Quan Long, Vietnam 39 H5 9 7N 105 8 E
Quanah, U.S.A. 81 H5 34 18N 99 44W
Quandialla, Australia 63 E4 34 1S 147 47 E
Quang Ngai, Vietnam 38 E7 15 13N 108 58 E
Quang Yen, Vietnam 38 B6 20 56N 106 52 E
Quantock Hills, U.K. 11 F4 51 8N 3 10W
Quanzhou, China 33 D6 24 55N 118 34 E
Quaqtaq, Canada 69 B13 60 55N 69 40W
Quaraí, Brazil 94 C4 30 15S 56 20W
Quartu Sant'Elena, Italy 20 E3 39 15N 9 10 E
Quartzsite, U.S.A. 85 M12 33 40N 114 13W
Quatsino, Canada 72 C3 50 30N 127 40W
Quatsino Sd., Canada 72 C3 50 25N 127 58W
Quba, Azerbaijan 25 F8 41 21N 48 32 E
Qūchān, Iran 45 B8 37 10N 58 27 E
Queanbeyan, Australia 63 F4 35 17S 149 14 E
Québec, Canada 71 C5 46 52N 71 13W
Québec □, Canada 71 B6 48 0N 74 0W
Queen Alexandra Ra., Antarctica 5 E11 85 0S 170 0 E
Queen Charlotte, Canada 72 C2 53 15N 132 2W
Queen Charlotte Is., Canada 72 C2 53 20N 132 10W
Queen Charlotte Str., Canada 72 C3 51 0N 128 0W
Queen Elizabeth Is., Canada 66 B10 76 0N 95 0W
Queen Elizabeth Nat. Park, Uganda 54 C3 0 0 30 0 E
Queen Mary Land, Antarctica 5 D7 70 0S 95 0 E
Queen Maud G., Canada 68 B9 68 15N 102 30W
Queen Maud Land, Antarctica 5 D3 72 30S 12 0 E
Queen Maud Mts., Antarctica 5 E13 86 0S 160 0W
Queens Chan., Australia 60 C4 15 0S 129 30 E
Queenscliff, Australia 63 F3 38 16S 144 39 E
Queensland □, Australia 62 C3 22 0S 142 0 E
Queenstown, Australia 62 G4 42 4S 145 35 E
Queenstown, N.Z. 59 L2 45 1S 168 40 E
Queenstown, S. Africa 56 E4 31 52S 26 52 E
Queets, U.S.A. 84 C2 47 32N 124 20W
Queguay Grande →, Uruguay 94 C4 32 9S 58 9W
Queimadas, Brazil 93 F11 11 0S 39 38W
Quela, Angola 52 F3 9 10S 16 56 E
Quelimane, Mozam. 55 F4 17 53S 36 58 E
Quelpart = Cheju Do, S. Korea 35 H14 33 29N 126 34 E
Quemado, N. Mex., U.S.A. 83 J9 34 20N 108 30W
Quemado, Tex., U.S.A. 81 L4 28 58N 100 35W
Quemú-Quemú, Argentina 94 D3 36 3S 63 36W
Quequén, Argentina 94 D4 38 30S 58 30W
Querétaro, Mexico 86 C4 20 36N 100 23W
Querétaro □, Mexico 86 C5 20 30N 100 0W
Queshan, China 34 H8 32 55N 114 2 E
Quesnel, Canada 72 C4 53 0N 122 30W
Quesnel →, Canada 72 C4 52 58N 122 29W
Quesnel L., Canada 72 C4 52 30N 121 20W
Questa, U.S.A. 83 H11 36 42N 105 36W
Quetico Prov. Park, Canada 70 C1 48 30N 91 45W
Quetta, Pakistan 42 D2 30 15N 66 55 E
Quezaltenango, Guatemala 88 D1 14 50N 91 30W
Quezon City, Phil. 37 B6 14 38N 121 0 E
Qufār, Si. Arabia 44 E4 27 26N 41 37 E
Qui Nhon, Vietnam 38 F7 13 40N 109 13 E
Quibaxe, Angola 52 F2 8 24S 14 27 E
Quibdo, Colombia 92 B3 5 42N 76 40W
Quiberon, France 18 C2 47 29N 3 9W
Quick, Canada 72 C3 54 36N 126 54W
Quiet L., Canada 72 A2 61 5N 133 5W
Quiindy, Paraguay 94 B4 25 58S 57 14W
Quila, Mexico 86 C3 24 23N 107 13W
Quilán, C., Chile 96 E2 43 15S 74 30W
Quilcene, U.S.A. 84 C4 47 49N 122 53W
Quilengues, Angola 53 G2 14 12S 14 12 E
Quilimarí, Chile 94 C1 32 5S 71 30W
Quilino, Argentina 94 C3 30 14S 64 29W
Quillabamba, Peru 92 F4 12 50S 72 50W
Quillagua, Chile 94 A2 21 40S 69 40W
Quillaicillo, Chile 94 C1 31 17S 71 40W
Quillota, Chile 94 C1 32 54S 71 16W
Quilmes, Argentina 94 C4 34 43S 58 15W
Quilon, India 40 Q10 8 50N 76 38 E
Quilpie, Australia 63 D3 26 35S 144 11 E
Quilpué, Chile 94 C1 33 5S 71 33W
Quilua, Mozam. 55 F4 16 17S 39 54 E
Quimilí, Argentina 94 B3 27 40S 62 30W
Quimper, France 18 B1 48 0N 4 9W
Quimperlé, France 18 C2 47 53N 3 33W
Quinault →, U.S.A. 84 C2 47 21N 124 18W
Quincy, Calif., U.S.A. 84 F6 39 56N 120 57W
Quincy, Fla., U.S.A. 77 K3 30 35N 84 34W
Quincy, Ill., U.S.A. 80 F9 39 56N 91 23W
Quincy, Mass., U.S.A. 79 D14 42 15N 71 0W
Quincy, Wash., U.S.A. 82 C4 47 22N 119 56W
Quines, Argentina 94 C2 32 13S 65 48W
Quinga, Mozam. 55 F5 15 49S 40 15 E
Quintana Roo □, Mexico 87 D7 19 0N 88 0W
Quintanar de la Orden, Spain 19 C4 39 36N 3 5W
Quintero, Chile 94 C1 32 45S 71 30W
Quinyambie, Australia 63 E3 30 15S 141 0 E
Quipungo, Angola 53 G2 14 37S 14 40 E
Quirihue, Chile 94 D1 36 15S 72 35W
Quirindi, Australia 63 E5 31 28S 150 40 E
Quissanga, Mozam. 55 E5 12 24S 40 28 E
Quitilipi, Argentina 94 B3 26 50S 60 13W
Quitman, Ga., U.S.A. 77 K4 30 47N 83 34W
Quitman, Miss., U.S.A. 77 J1 32 2N 88 44W
Quitman, Tex., U.S.A. 81 J7 32 48N 95 27W
Quito, Ecuador 92 D3 0 15S 78 35W
Quixadá, Brazil 93 D11 4 55S 39 0W
Quixaxe, Mozam. 55 F5 15 17S 40 4 E
Qumbu, S. Africa 57 E4 31 10S 28 48 E
Quneitra, Syria 47 B4 33 7N 35 48 E
Qŭnghirot, Uzbekistan 26 E6 43 6N 58 54 E
Quoin I., Australia 60 B4 14 54S 129 32 E
Quoin Pt., S. Africa 56 E2 34 46S 19 37 E
Quondong, Australia 63 E3 33 6S 140 18 E
Quorn, Australia 63 E2 32 25S 138 5 E
Qŭqon, Uzbekistan 26 E8 40 30N 70 57 E
Qurnat as Sawdā', Lebanon 47 A5 34 18N 36 6 E
Qûs, Egypt 51 C11 25 55N 32 50 E
Qusaybah, Iraq 44 C4 34 24N 40 59 E
Quseir, Egypt 51 C11 26 7N 34 16 E
Qūshchī, Iran 44 B5 37 59N 45 3 E
Quthing, Lesotho 57 E4 30 25S 27 36 E
Qūţīābād, Iran 45 C6 35 47N 48 30 E
Quwo, China 34 G6 35 38N 111 25 E
Quyang, China 34 E8 38 35N 114 40 E
Quynh Nhai, Vietnam 38 B4 21 49N 103 33 E
Quzi, China 34 F4 36 20N 107 20 E
Qyzylorda, Kazakstan 26 E7 44 48N 65 28 E

R

Ra, Ko, Thailand 39 H2 9 13N 98 16 E
Raahe, Finland 8 D21 64 40N 24 28 E
Raasay, U.K. 12 D2 57 25N 6 4W
Raasay, Sd. of, U.K. 12 D2 57 30N 6 8W
Raba, Indonesia 37 F5 8 36S 118 55 E
Rába →, Hungary 17 E9 47 38N 17 38 E
Rabai, Kenya 54 C4 3 50S 39 31 E
Rabat, Malta 23 D1 35 53N 14 25 E
Rabat, Morocco 50 B3 34 2N 6 48W
Rabaul, Papua N. G. 64 H7 4 24S 152 18 E
Rabbit →, Canada 72 B3 59 41N 127 12W
Rabbit Lake, Canada 73 C7 53 8N 107 46W
Rabbitskin →, Canada 72 A4 61 47N 120 42W
Rābor, Iran 45 D8 29 17N 56 55 E
Race, C., Canada 71 C9 46 40N 53 5W
Rach Gia, Vietnam 39 G5 10 5N 105 5 E
Racibórz, Poland 17 C10 50 7N 18 18 E
Racine, U.S.A. 76 D2 42 41N 87 51W
Rackerby, U.S.A. 84 F5 39 26N 121 22W
Radama, Nosy, Madag. 57 A8 14 0S 47 47 E
Radama, Saikanosy, Madag. 57 A8 14 16S 47 53 E
Rădăuţi, Romania 17 E13 47 50N 25 59 E
Radekhiv, Ukraine 17 C13 50 25N 24 32 E
Radekhov = Radekhiv, Ukraine 17 C13 50 25N 24 32 E
Radford, U.S.A. 76 G5 37 8N 80 34W
Radhanpur, India 42 H4 23 50N 71 38 E
Radisson, Canada 73 C7 52 30N 107 20W
Radium Hot Springs, Canada 72 C5 50 35N 116 2W
Radnor Forest, U.K. 11 E4 52 17N 3 10W
Radom, Poland 17 C11 51 23N 21 12 E
Radomsko, Poland 17 C10 51 5N 19 28 E
Radomyshl, Ukraine 17 C15 50 30N 29 12 E
Radstock, U.K. 11 F5 51 17N 2 26W
Radstock, C., Australia 63 E1 33 12S 134 20 E
Radviliškis, Lithuania 9 J20 55 49N 23 33 E
Radville, Canada 73 D8 49 30N 104 15W
Rae, Canada 72 A5 62 50N 116 3W
Rae Bareli, India 43 F9 26 18N 81 20 E
Rae Isthmus, Canada 69 B11 66 40N 87 30W
Raeren, Belgium 15 D6 50 41N 6 7 E
Raeside, L., Australia 61 E3 29 20S 122 0 E
Raetihi, N.Z. 59 H5 39 25S 175 17 E
Rafaela, Argentina 94 C3 31 10S 61 30W
Rafah, Gaza Strip 47 D3 31 18N 34 14 E
Rafai, C.A.R. 54 B1 4 59N 23 58 E
Rafḩā, Si. Arabia 44 D4 29 35N 43 35 E
Rafsanjān, Iran 45 D8 30 30N 56 5 E
Raft Pt., Australia 60 C3 16 4S 124 26 E
Ragachow, Belarus 17 B16 53 8N 30 5 E
Ragama, Sri Lanka 40 R11 7 0N 79 50 E
Ragged, Mt., Australia 61 F3 33 27S 123 25 E
Raglan, Australia 62 C5 23 42S 150 49 E
Raglan, N.Z. 59 G5 37 55S 174 55 E
Ragusa, Italy 20 F6 36 55N 14 44 E
Raha, Indonesia 37 E6 4 55S 123 0 E
Rahad al Bardī, Sudan 51 F9 11 20N 23 40 E
Rahaeng = Tak, Thailand 38 D2 16 52N 99 8 E
Raḩīmah, Si. Arabia 45 E6 26 42N 50 4 E
Rahimyar Khan, Pakistan 42 E4 28 30N 70 25 E
Rāhjerd, Iran 45 C6 34 22N 50 22 E
Raichur, India 40 L10 16 10N 77 20 E
Raiganj, India 43 G13 25 37N 88 10 E
Raigarh, India 41 J13 21 56N 83 25 E
Raijua, Indonesia 37 F6 10 37S 121 36 E
Railton, Australia 62 G4 41 25S 146 28 E
Rainbow Lake, Canada 72 B5 58 30N 119 23W
Rainier, U.S.A. 84 D4 46 53N 122 41W
Rainier, Mt., U.S.A. 84 D5 46 52N 121 46W
Rainy L., Canada 73 D10 48 42N 93 10W
Rainy River, Canada 73 D10 48 43N 94 29W
Raippaluoto, Finland 8 E19 63 13N 21 14 E
Raipur, India 41 J12 21 17N 81 45 E
Raisio, Finland 9 F20 60 28N 22 11 E
Raj Nandgaon, India 41 J12 21 5N 81 5 E
Raja, Ujung, Indonesia 36 D1 3 40N 96 25 E
Raja Ampat, Kepulauan, Indonesia 37 E7 0 30S 130 0 E
Rajahmundry, India 41 L12 17 1N 81 48 E
Rajang →, Malaysia 36 D4 2 30N 112 0 E
Rajapalaiyam, India 40 Q10 9 25N 77 35 E
Rajasthan □, India 42 F5 26 45N 73 30 E
Rajasthan Canal, India 42 E5 28 0N 72 0 E
Rajauri, India 43 C6 33 25N 74 21 E
Rajgarh, Mad. P., India 42 G7 24 2N 76 45 E
Rajgarh, Raj., India 42 E6 28 40N 75 25 E
Rajkot, India 42 H4 22 15N 70 56 E
Rajmahal Hills, India 43 G12 24 30N 87 30 E
Rajpipla, India 40 J8 21 50N 73 30 E
Rajpura, India 42 D7 30 25N 76 32 E
Rajshahi, Bangla. 41 G16 24 22N 88 39 E
Rajshahi □, Bangla. 43 G13 25 0N 89 0 E
Rakaia, N.Z. 59 K4 43 45S 172 1 E
Rakaia →, N.Z. 59 K4 43 36S 172 15 E
Rakan, Ra's, Qatar 45 E6 26 10N 51 20 E
Rakaposhi, Pakistan 43 A6 36 10N 74 25 E
Rakata, Pulau, Indonesia 36 F3 6 10S 105 20 E
Rakhiv, Ukraine 17 D13 48 3N 24 12 E
Rakhni, Pakistan 42 D3 30 4N 69 56 E
Rakitnoye, Russia 30 B7 45 36N 134 17 E
Rakops, Botswana 56 C3 21 1S 24 28 E
Rakvere, Estonia 9 G22 59 20N 26 25 E
Raleigh, U.S.A. 77 H6 35 47N 78 39W
Raleigh B., U.S.A. 77 H7 34 50N 76 15W
Ralls, U.S.A. 81 J4 33 41N 101 24W
Ram →, Canada 72 A4 62 1N 123 41W
Rām Allāh, West Bank 47 D4 31 55N 35 10 E
Ram Hd., Australia 63 F4 37 47S 149 30 E
Rama, Nic. 88 D3 12 9N 84 15W
Raman, Thailand 39 J3 6 29N 101 18 E
Ramanathapuram, India 40 Q11 9 25N 78 55 E
Ramanetaka, B. de, Madag. 57 A8 14 13S 47 52 E
Ramat Gan, Israel 47 C3 32 4N 34 48 E
Ramatlhabama, S. Africa 56 D4 25 37S 25 33 E
Ramban, India 43 C6 33 14N 75 12 E
Rambipuji, Indonesia 37 H15 8 12S 113 37 E
Ramea, Canada 71 C8 47 31N 57 23W
Ramechhap, Nepal 43 F12 27 25N 86 10 E
Ramelau, Indonesia 37 F7 8 55S 126 22 E
Ramgarh, Bihar, India 43 H11 23 40N 85 35 E
Ramgarh, Raj., India 42 F6 27 16N 75 14 E
Ramgarh, Raj., India 42 F4 27 30N 70 36 E
Rāmhormoz, Iran 45 D6 31 15N 49 35 E
Ramīān, Iran 45 B7 37 3N 55 16 E
Ramingining, Australia 62 A2 12 19S 135 3 E
Ramla, Israel 47 D3 31 55N 34 52 E
Ramnad = Ramanathapuram, India 40 Q11 9 25N 78 55 E
Ramnagar, India 43 C6 32 47N 75 18 E
Ramona, U.S.A. 85 M10 33 2N 116 52W
Ramore, Canada 70 C3 48 30N 80 25W
Ramotswa, Botswana 56 C4 24 50S 25 52 E
Rampur, H.P., India 42 D7 31 26N 77 43 E
Rampur, Mad. P., India 42 H5 23 25N 73 53 E
Rampur, Ut. P., India 43 E8 28 50N 79 5 E
Rampur Hat, India 43 G12 24 10N 87 50 E
Rampura, India 42 G6 24 30N 75 27 E
Ramree I. = Ramree Kyun, Burma 41 K18 19 0N 94 0 E
Ramree Kyun, Burma 41 K18 19 0N 94 0 E
Rāmsar, Iran 45 B6 36 53N 50 41 E
Ramsey, Canada 70 C3 47 25N 82 20W
Ramsey, U.K. 10 C3 54 20N 4 22W
Ramsgate, U.K. 11 F9 51 20N 1 25 E
Ramtek, India 40 J11 21 20N 79 15 E
Ranaghat, India 43 H13 23 15N 88 35 E
Ranahu, Pakistan 42 G3 25 55N 69 45 E
Ranau, Malaysia 36 C5 6 2N 116 40 E
Rancagua, Chile 94 C1 34 10S 70 50W
Rancheria →, Canada 72 A3 60 13N 129 7W
Ranchester, U.S.A. 82 D10 44 54N 107 10W
Ranchi, India 43 H11 23 19N 85 27 E
Rancho Cucamonga, U.S.A. 85 L9 34 10N 117 30W
Randers, Denmark 9 H14 56 29N 10 1 E
Randfontein, S. Africa 57 D4 26 8S 27 45 E
Randle, U.S.A. 84 D5 46 32N 121 57W
Randolph, Mass., U.S.A. 79 D13 42 10N 71 2W
Randolph, N.Y., U.S.A. 78 D6 42 10N 78 59W
Randolph, Utah, U.S.A. 82 F8 41 40N 111 11W
Randolph, Vt., U.S.A. 79 C12 43 55N 72 40W
Råne älv →, Sweden 8 D20 65 50N 22 20 E
Rangae, Thailand 39 J3 6 19N 101 44 E
Rangaunu B., N.Z. 59 F4 34 51S 173 15 E
Rangeley, U.S.A. 79 B14 44 58N 70 39W
Rangely, U.S.A. 82 F9 40 5N 108 48W
Ranger, U.S.A. 81 J5 32 28N 98 41W
Rangia, India 41 F17 26 28N 91 38 E
Rangiora, N.Z. 59 K4 43 19S 172 36 E
Rangitaiki →, N.Z. 59 G6 37 54S 176 49 E
Rangitata →, N.Z. 59 K3 43 45S 171 15 E
Rangkasbitung, Indonesia 37 G12 6 21S 106 15 E
Rangon →, Burma 41 L20 16 28N 96 40 E
Rangoon, Burma 41 L20 16 45N 96 20 E
Rangpur, Bangla. 41 G16 25 42N 89 22 E
Rangsit, Thailand 38 F3 13 59N 100 37 E
Ranibennur, India 40 M9 14 35N 75 30 E
Raniganj, India 43 H12 23 40N 87 5 E
Raniwara, India 40 G8 24 50N 72 10 E
Rāniyah, Iraq 44 B5 36 15N 44 53 E
Ranken →, Australia 62 C2 20 31S 137 36 E
Rankin, U.S.A. 81 K4 31 13N 101 56W
Rankin Inlet, Canada 68 B10 62 30N 93 0W
Rankins Springs, Australia 63 E4 33 49S 146 14 E
Rannoch, L., U.K. 12 E4 56 41N 4 20W
Rannoch Moor, U.K. 12 E4 56 38N 4 48W
Ranobe, Helodranon' i, Madag. 57 C7 23 3S 43 33 E
Ranohira, Madag. 57 C8 22 29S 45 24 E
Ranomafana, Toamasina, Madag. 57 B8 18 57S 48 50 E
Ranomafana, Toliara, Madag. 57 C8 24 34S 47 0 E
Ranong, Thailand 39 H2 9 56N 98 40 E
Rānsa, Iran 45 C6 33 39N 48 18 E
Ransiki, Indonesia 37 E8 1 30S 134 10 E
Rantau, Indonesia 36 E5 2 56S 115 9 E
Rantauprapat, Indonesia 36 D1 2 15N 99 50 E
Rantemario, Indonesia 37 E5 3 15S 119 57 E
Rantoul, U.S.A. 76 E1 40 19N 88 9W
Raoyang, China 34 E8 38 15N 115 45 E
Rapa, Pac. Oc. 65 K13 27 35S 144 20W
Rapallo, Italy 20 B3 44 21N 9 14 E
Rāpch, Iran 45 E8 25 40N 59 15 E
Rapid →, Canada 72 B3 59 15N 129 5W
Rapid City, U.S.A. 80 D3 44 5N 103 14W
Rapid River, U.S.A. 76 C2 45 55N 86 58W
Rapides des Joachims, Canada 70 C4 46 13N 77 43W
Rapla, Estonia 9 G21 59 1N 24 52 E
Rarotonga, Cook Is. 65 K12 21 30S 160 0W
Ra's al 'Ayn, Syria 44 B4 36 45N 40 12 E
Ra's al Khaymah, U.A.E. 45 E8 25 50N 55 59 E
Ra's al-Unuf, Libya 51 B8 30 46N 18 11 E
Ra's an Naqb, Jordan 47 F4 30 0N 35 29 E
Ras Bânâs, Egypt 51 D12 23 57N 35 59 E
Ras Dashen, Ethiopia 51 F12 13 8N 38 26 E
Râs Timirist, Mauritania 50 E1 19 21N 16 30W
Rasa, Punta, Argentina 96 E4 40 50S 62 15W
Rasca, Pta. de la, Canary Is. 22 G3 27 59N 16 41W
Raseiniai, Lithuania 9 J20 55 25N 23 5 E
Rashad, Sudan 51 F11 11 55N 31 0 E
Rashîd, Egypt 51 B11 31 21N 30 22 E
Rasht, Iran 45 B6 37 20N 49 40 E
Rasi Salai, Thailand 38 E5 15 20N 104 9 E
Rason L., Australia 61 E3 28 45S 124 25 E
Rasra, India 43 G10 25 50N 83 50 E
Rat Buri, Thailand 38 F2 13 30N 99 54 E
Rat Islands, U.S.A. 68 C1 52 0N 178 0 E
Rat River, Canada 72 A6 61 7N 112 36W
Ratangarh, India 42 E6 28 5N 74 35 E
Raţāwī, Iraq 44 D5 30 38N 47 13 E
Rath, India 43 G8 25 36N 79 37 E
Rath Luirc, Ireland 13 D3 52 21N 8 40W
Rathdrum, Ireland 13 D5 52 56N 6 14W
Rathenow, Germany 16 B7 52 37N 12 19 E
Rathkeale, Ireland 13 D3 52 32N 8 56W
Rathlin I., U.K. 13 A5 55 18N 6 14W
Rathlin O'Birne I., Ireland 13 B3 54 40N 8 49W
Ratibor = Racibórz, Poland 17 C10 50 7N 18 18 E
Ratlam, India 42 H6 23 20N 75 0 E
Ratnagiri, India 40 L8 16 57N 73 18 E
Raton, U.S.A. 81 G2 36 54N 104 24W
Rattaphum, Thailand 39 J3 7 8N 100 16 E
Rattray Hd., U.K. 12 D7 57 38N 1 50W
Ratz, Mt., Canada 72 B2 57 23N 132 12W
Raub, Malaysia 39 L3 3 47N 101 52 E
Rauch, Argentina 94 D4 36 45S 59 5W
Raufarhöfn, Iceland 8 C6 66 27N 15 57W
Raufoss, Norway 9 F14 60 44N 10 37 E
Raukumara Ra., N.Z. 59 H6 38 5S 177 55 E
Rauma, Finland 9 F19 61 10N 21 30 E
Raurkela, India 43 H11 22 14N 84 50 E
Rausu-Dake, Japan 30 B12 44 4N 145 7 E
Rava-Ruska, Ukraine 17 C12 50 15N 23 42 E
Rava Russkaya = Rava-Ruska, Ukraine 17 C12 50 15N 23 42 E
Ravānsar, Iran 44 C5 34 43N 46 40 E
Rāvar, Iran 45 D8 31 20N 56 51 E
Ravena, U.S.A. 79 D11 42 28N 73 49W

Ravenna, Italy 20 B5 44 25N 12 12 E
Ravenna, Nebr., U.S.A. 80 E5 41 1N 98 55W
Ravenna, Ohio, U.S.A. 78 E3 41 9N 81 15W
Ravensburg, Germany 16 E5 47 46N 9 36 E
Ravenshoe, Australia 62 B4 17 37S 145 29 E
Ravensthorpe, Australia 61 F3 33 35S 120 2 E
Ravenswood, Australia 62 C4 20 6S 146 54 E
Ravenswood, U.S.A. 76 F5 38 57N 81 46W
Ravi →, Pakistan 42 D4 30 35N 71 49 E
Rawalpindi, Pakistan 42 C5 33 38N 73 8 E
Rawāndūz, Iraq 44 B5 36 40N 44 30 E
Rawang, Malaysia 39 L3 3 20N 101 35 E
Rawdon, Canada 70 C5 46 3N 73 40W
Rawene, N.Z. 59 F4 35 25S 173 32 E
Rawlinna, Australia 61 F4 30 58S 125 28 E
Rawlins, U.S.A. 82 F10 41 47N 107 14W
Rawlinson Ra., Australia 61 D4 24 40S 128 30 E
Rawson, Argentina 96 E3 43 15S 65 5W
Ray, U.S.A. 80 A3 48 21N 103 10W
Ray, C., Canada 71 C8 47 33N 59 15W
Rayadurg, India 40 M10 14 40N 76 50 E
Rayagada, India 41 K13 19 15N 83 20 E
Raychikhinsk, Russia 27 E13 49 46N 129 25 E
Rāyen, Iran 45 D8 29 34N 57 26 E
Raymond, Canada 72 D6 49 30N 112 35W
Raymond, Calif., U.S.A. 84 H7 37 13N 119 54W
Raymond, Wash., U.S.A. 84 D3 46 41N 123 44W
Raymondville, U.S.A. 81 M6 26 29N 97 47W
Raymore, Canada 73 C8 51 25N 104 31W
Rayne, U.S.A. 81 K8 30 14N 92 16W
Rayón, Mexico 86 B2 29 43N 110 35W
Rayong, Thailand 38 F3 12 40N 101 20 E
Rayville, U.S.A. 81 J9 32 29N 91 46W
Raz, Pte. du, France 18 C1 48 2N 4 47W
Razan, Iran 45 C6 35 23N 49 2 E
Razdel'naya = Rozdilna, Ukraine 17 E16 46 50N 30 2 E
Razdolnoye, Russia 30 C5 43 30N 131 52 E
Razeh, Iran 45 C6 32 47N 48 9 E
Razelm, Lacul, Romania 17 F15 44 50N 29 0 E
Razgrad, Bulgaria 21 C12 43 33N 26 34 E
Razmak, Pakistan 42 C3 32 45N 69 50 E
Ré, I. de, France 18 C3 46 12N 1 30W
Reading, U.K. 11 F7 51 27N 0 58W
Reading, U.S.A. 79 F9 40 20N 75 56W
Realicó, Argentina 94 D3 35 0S 64 15W
Reata, Mexico 86 B4 26 8N 101 5W
Rebecca, L., Australia 61 F3 30 0S 122 15 E
Rebi, Indonesia 37 F8 6 23S 134 7 E
Rebiana, Libya 51 D9 24 12N 22 10 E
Rebun-Tō, Japan 30 B10 45 23N 141 2 E
Recherche, Arch. of the, Australia 61 F3 34 15S 122 50 E
Rechytsa, Belarus 17 B16 52 21N 30 24 E
Recife, Brazil 93 E12 8 0S 35 0W
Recklinghausen, Germany 15 C7 51 37N 7 12 E
Reconquista, Argentina 94 B4 29 10S 59 45W
Recreo, Argentina 94 B2 29 25S 65 10W
Red →, La., U.S.A. 81 K9 31 1N 91 45W
Red →, N. Dak., U.S.A. 80 A6 49 0N 97 15W
Red Bank, U.S.A. 79 F10 40 21N 74 5W
Red Bay, Canada 71 B8 51 44N 56 25W
Red Bluff, U.S.A. 82 F2 40 11N 122 15W
Red Bluff L., U.S.A. 81 K3 31 54N 103 55W
Red Cliffs, Australia 63 E3 34 19S 142 11 E
Red Cloud, U.S.A. 80 E5 40 5N 98 32W
Red Deer, Canada 72 C6 52 20N 113 50W
Red Deer →, Alta., Canada 73 C6 50 58N 110 0W
Red Deer →, Man., Canada 73 C8 52 53N 101 1W
Red Deer L., Canada 73 C8 52 55N 101 20W
Red Indian L., Canada 71 C8 48 35N 57 0W
Red Lake, Canada 73 C10 51 3N 93 49W
Red Lake Falls, U.S.A. 80 B6 47 53N 96 16W
Red Lodge, U.S.A. 82 D9 45 11N 109 15W
Red Mountain, U.S.A. 85 K9 35 37N 117 38W
Red Oak, U.S.A. 80 E7 41 1N 95 14W
Red Rock, Canada 70 C2 48 55N 88 15W
Red Rock, L., U.S.A. 80 E8 41 22N 92 59W
Red Rocks Pt., Australia 61 F4 32 13S 127 32 E
Red Sea, Asia 46 C2 25 0N 36 0 E
Red Slate Mt., U.S.A. 84 H8 37 31N 118 52W
Red Sucker L., Canada 73 C10 54 9N 93 40W
Red Tower Pass = Turnu Roşu, P., Romania 17 F13 45 33N 24 17 E
Red Wing, U.S.A. 80 C8 44 34N 92 31W
Redbridge, U.K. 11 F8 51 35N 0 7 E
Redcar, U.K. 10 C6 54 37N 1 4W
Redcar & Cleveland □, U.K. 10 C6 54 29N 1 0W
Redcliff, Canada 73 C6 50 10N 110 50W
Redcliffe, Australia 63 D5 27 12S 153 0 E
Redcliffe, Mt., Australia 61 E3 28 30S 121 30 E
Reddersburg, S. Africa 56 D4 29 41S 26 10 E
Redding, U.S.A. 82 F2 40 35N 122 24W
Redditch, U.K. 11 E6 52 18N 1 55W
Redfield, U.S.A. 80 C5 44 53N 98 31W
Redknife →, Canada 72 A5 61 14N 119 22W
Redlands, U.S.A. 85 M9 34 4N 117 11W
Redmond, Australia 61 F2 34 55S 117 40 E
Redmond, Oreg., U.S.A. 82 D3 44 17N 121 11W
Redmond, Wash., U.S.A. 84 C4 47 41N 122 7W
Redon, France 18 C2 47 40N 2 6W
Redonda, Antigua 89 C7 16 58N 62 19W
Redondela, Spain 19 A1 42 15N 8 38W
Redondo Beach, U.S.A. 85 M8 33 50N 118 23W
Redrock Pt., Canada 72 A5 62 11N 115 2W
Redruth, U.K. 11 G2 50 14N 5 14W
Redvers, Canada 73 D8 49 35N 101 40W
Redwater, Canada 72 C6 53 55N 113 6W
Redwood, U.S.A. 79 B9 44 18N 75 48W
Redwood City, U.S.A. 83 H2 37 30N 122 15W
Redwood Falls, U.S.A. 80 C7 44 32N 95 7W
Ree, L., Ireland 13 C4 53 35N 8 0W
Reed, L., Canada 73 C8 54 38N 100 30W
Reed City, U.S.A. 76 D3 43 53N 85 31W
Reeder, U.S.A. 80 B3 46 7N 102 57W
Reedley, U.S.A. 83 H4 36 36N 119 27W
Reedsburg, U.S.A. 80 D9 43 32N 90 0W
Reedsport, U.S.A. 82 E1 43 42N 124 6W
Reefton, N.Z. 59 K3 42 6S 171 51 E
Refugio, U.S.A. 81 L6 28 18N 97 17W
Regensburg, Germany 16 D7 49 1N 12 6 E
Réggio di Calábria, Italy 20 E6 38 6N 15 39 E
Réggio nell'Emília, Italy 20 B4 44 43N 10 36 E
Reghin, Romania 17 E13 46 46N 24 42 E
Regina, Canada 73 C8 50 27N 104 35W
Registro, Brazil 95 A6 24 29S 47 49W
Rehar →, India 43 H10 23 55N 82 40 E
Rehoboth, Namibia 56 C2 23 15S 17 4 E
Rehovot, Israel 47 D3 31 54N 34 48 E
Rei-Bouba, Cameroon 51 G7 8 40N 14 15 E
Reichenbach, Germany 16 C7 50 37N 12 17 E
Reid, Australia 61 F4 30 49S 128 26 E
Reid River, Australia 62 B4 19 40S 146 48 E
Reidsville, U.S.A. 77 G6 36 21N 79 40W
Reigate, U.K. 11 F7 51 14N 0 12W
Reims, France 18 B6 49 15N 4 1 E
Reina Adelaida, Arch., Chile 96 G2 52 20S 74 0W
Reinbeck, U.S.A. 80 D8 42 19N 92 36W
Reindeer →, Canada 73 B8 55 36N 103 11W
Reindeer I., Canada 73 C9 52 30N 98 0W
Reindeer L., Canada 73 B8 57 15N 102 15W
Reinga, C., N.Z. 59 F4 34 25S 172 43 E
Reinosa, Spain 19 A3 43 2N 4 15W
Reitz, S. Africa 57 D4 27 48S 28 29 E
Reivilo, S. Africa 56 D3 27 36S 24 8 E
Rekinniki, Russia 27 C17 60 51N 163 40 E
Reliance, Canada 73 A7 63 0N 109 20W
Remarkable, Mt., Australia 63 E2 32 48S 138 10 E
Rembang, Indonesia 37 G14 6 42S 111 21 E
Remedios, Panama 88 E3 8 15N 81 50W
Remeshk, Iran 45 E8 26 55N 58 50 E
Remich, Lux. 15 E6 49 32N 6 22 E
Ren Xian, China 34 F8 37 8N 114 40 E
Rendsburg, Germany 16 A5 54 17N 9 39 E
Rene, Russia 27 C19 66 2N 179 25W
Renfrew, Canada 70 C4 45 30N 76 40W
Renfrew, U.K. 12 F4 55 52N 4 24W
Renfrewshire □, U.K. 12 F4 55 49N 4 38W
Rengat, Indonesia 36 E2 0 30S 102 45 E
Rengo, Chile 94 C1 34 24S 70 50W
Reni, Ukraine 17 F15 45 28N 28 15 E
Renk, Sudan 51 F11 11 50N 32 50 E
Renkum, Neths. 15 C5 51 58N 5 43 E
Renmark, Australia 63 E3 34 11S 140 43 E
Rennell Sd., Canada 72 C2 53 23N 132 35W
Renner Springs T.O., Australia 62 B1 18 20S 133 47 E
Rennes, France 18 B3 48 7N 1 41W
Reno, U.S.A. 84 F7 39 31N 119 48W
Reno →, Italy 20 B5 44 38N 12 16 E
Renovo, U.S.A. 78 E7 41 20N 77 45W
Renqiu, China 34 E9 38 43N 116 5 E
Rensselaer, Ind., U.S.A. 76 E2 40 57N 87 9W
Rensselaer, N.Y., U.S.A. 79 D11 42 38N 73 45W
Rentería, Spain 19 A5 43 19N 1 54W
Renton, U.S.A. 84 C4 47 29N 122 12W
Reotipur, India 43 G10 25 33N 83 45 E
Republic, Mich., U.S.A. 76 B2 46 25N 87 59W
Republic, Wash., U.S.A. 82 B4 48 39N 118 44W
Republican →, U.S.A. 80 F6 39 4N 96 48W
Republican City, U.S.A. 80 E5 40 6N 99 13W
Repulse Bay, Canada 69 B11 66 30N 86 30W
Requena, Peru 92 E4 5 5S 73 52W
Requena, Spain 19 C5 39 30N 1 4W
Resadiye = Datça, Turkey 21 F12 36 46N 27 40 E
Reserve, Canada 73 C8 52 28N 102 39W
Reserve, U.S.A. 83 K9 33 43N 108 45W
Resht = Rasht, Iran 45 B6 37 20N 49 40 E
Resistencia, Argentina 94 B4 27 30S 59 0W
Reşiţa, Romania 17 F11 45 18N 21 53 E
Resolution I., Canada 69 B13 61 30N 65 0W
Resolution I., N.Z. 59 L1 45 40S 166 40 E
Ressano Garcia, Mozam. 57 D5 25 25S 32 0 E
Reston, Canada 73 D8 49 33N 101 6W
Retalhuleu, Guatemala 88 D1 14 33N 91 46W
Retenue, L. de, Zaïre 55 E2 11 0S 27 0 E
Retford, U.K. 10 D7 53 19N 0 56W
Réthímnon, Greece 23 D6 35 18N 24 30 E
Réthímnon □, Greece 23 D6 35 23N 24 28 E
Réunion ■, Ind. Oc. 49 J9 21 0S 56 0 E
Reus, Spain 19 B6 41 10N 1 5 E
Reutlingen, Germany 16 D5 48 29N 9 12 E
Reval = Tallinn, Estonia 9 G21 59 22N 24 48 E
Revda, Russia 24 C10 56 48N 59 57 E
Revelganj, India 43 G11 25 50N 84 40 E
Revelstoke, Canada 72 C5 51 0N 118 10W
Revilla Gigedo, Is., Pac. Oc. 66 H8 18 40N 112 0W
Revillagigedo I., U.S.A. 72 B2 55 50N 131 20W
Revuè →, Mozam. 55 F3 19 50S 34 0 E
Rewa, India 43 G9 24 33N 81 25 E
Rewari, India 42 E7 28 15N 76 40 E
Rexburg, U.S.A. 82 E8 43 49N 111 47W
Rey, Iran 45 C6 35 35N 51 25 E
Rey Malabo, Eq. Guin. 50 H6 3 45N 8 50 E
Reyðarfjörður, Iceland 8 D6 65 2N 14 13W
Reyes, Pt., U.S.A. 84 H3 38 0N 123 0W
Reykjahlið, Iceland 8 D5 65 40N 16 55W
Reykjanes, Iceland 8 E2 63 48N 22 40W
Reykjavík, Iceland 8 D3 64 10N 21 57W
Reynolds, Canada 73 D9 49 40N 95 55W
Reynolds Ra., Australia 60 D5 22 30S 133 0 E
Reynoldsville, U.S.A. 78 E6 41 5N 78 58W
Reynosa, Mexico 87 B5 26 5N 98 18W
Rēzekne, Latvia 9 H22 56 30N 27 17 E
Rezvān, Iran 45 E8 27 34N 56 6 E
Rhayader, U.K. 11 E4 52 18N 3 29W
Rheden, Neths. 15 B6 52 0N 6 3 E
Rhein, Canada 73 C8 51 25N 102 15W
Rhein →, Europe 15 C6 51 52N 6 2 E
Rhein-Main-Donau-Kanal, Germany 16 D6 49 15N 11 15 E
Rheine, Germany 16 B4 52 17N 7 26 E
Rheinland-Pfalz □, Germany 16 C4 50 0N 7 0 E
Rhin = Rhein →, Europe 15 C6 51 52N 6 2 E
Rhine = Rhein →, Europe 15 C6 51 52N 6 2 E
Rhineland-Palatinate □ = Rheinland-Pfalz □, Germany 16 C4 50 0N 7 0 E
Rhinelander, U.S.A. 80 C10 45 38N 89 25W
Rhino Camp, Uganda 54 B3 3 0N 31 22 E
Rhode Island □, U.S.A. 79 E13 41 40N 71 30W
Rhodes = Ródhos, Greece 23 C10 36 15N 28 10 E
Rhodesia = Zimbabwe ■, Africa 55 F2 19 0S 30 0 E
Rhodope Mts. = Rhodopi Planina, Bulgaria 21 D11 41 40N 24 20 E
Rhodopi Planina, Bulgaria 21 D11 41 40N 24 20 E
Rhön = Hohe Rhön, Germany 16 C5 50 24N 9 58 E
Rhondda, U.K. 11 F4 51 39N 3 31W
Rhondda Cynon Taff □, U.K. 11 F4 51 42N 3 27W
Rhône →, France 18 E6 43 28N 4 42 E
Rhum, U.K. 12 E2 57 0N 6 20W
Rhyl, U.K. 10 D4 53 20N 3 29W
Rhymney, U.K. 11 F4 51 46N 3 17W
Riachão, Brazil 93 E9 7 20S 46 37W
Riasi, India 43 C6 33 10N 74 50 E
Riau □, Indonesia 36 D2 0 0 102 35 E
Riau, Kepulauan, Indonesia 36 D2 0 30N 104 20 E
Riau Arch. = Riau, Kepulauan, Indonesia 36 D2 0 30N 104 20 E
Ribadeo, Spain 19 A2 43 35N 7 5W
Ribble →, U.K. 10 C5 53 52N 2 25W
Ribe, Denmark 9 J13 55 19N 8 44 E
Ribeira Brava, Madeira 22 D2 32 41N 17 4W
Ribeirão Prêto, Brazil 95 A6 21 10S 47 50W
Riberalta, Bolivia 92 F5 11 0S 66 0W
Rîbniţa, Moldova 17 E15 47 45N 29 0 E
Riccarton, N.Z. 59 K4 43 32S 172 37 E
Rice, U.S.A. 85 L12 34 5N 114 51W
Rice L., Canada 78 B6 44 12N 78 10W
Rice Lake, U.S.A. 80 C9 45 30N 91 44W
Rich Hill, U.S.A. 81 F7 38 6N 94 22W
Richards Bay, S. Africa 57 D5 28 48S 32 6 E
Richards L., Canada 73 B7 59 10N 107 10W
Richardson →, Canada 73 B6 58 25N 111 14W
Richardson Springs, U.S.A. 84 F5 39 51N 121 46W
Richardton, U.S.A. 80 B3 46 53N 102 19W
Riche, C., Australia 61 F2 34 36S 118 47 E
Richey, U.S.A. 80 B2 47 39N 105 4W
Richfield, Idaho, U.S.A. 82 E6 43 3N 114 9W
Richfield, Utah, U.S.A. 83 G8 38 46N 112 5W
Richford, U.S.A. 79 B12 45 0N 72 40W
Richibucto, Canada 71 C7 46 42N 64 54W
Richland, Ga., U.S.A. 77 J3 32 5N 84 40W
Richland, Oreg., U.S.A. 82 D5 44 46N 117 10W
Richland, Wash., U.S.A. 82 C4 46 17N 119 18W
Richland Center, U.S.A. 80 D9 43 21N 90 23W
Richlands, U.S.A. 76 G5 37 6N 81 48W
Richmond, N.S.W., Australia 63 E5 33 35S 150 42 E
Richmond, Queens., Australia 62 C3 20 43S 143 8 E
Richmond, N.Z. 59 J4 41 20S 173 12 E
Richmond, U.K. 10 C6 54 25N 1 43W
Richmond, Calif., U.S.A. 84 H4 37 56N 122 21W
Richmond, Ind., U.S.A. 76 F3 39 50N 84 53W
Richmond, Ky., U.S.A. 76 G3 37 45N 84 18W
Richmond, Mich., U.S.A. 78 D2 42 49N 82 45W
Richmond, Mo., U.S.A. 80 F8 39 17N 93 58W
Richmond, Tex., U.S.A. 81 L7 29 35N 95 46W
Richmond, Utah, U.S.A. 82 F8 41 56N 111 48W
Richmond, Va., U.S.A. 76 G7 37 33N 77 27W
Richmond Ra., Australia 63 D5 29 0S 152 45 E
Richmond-upon-Thames, U.K. 11 F7 51 27N 0 17W
Richton, U.S.A. 77 K1 31 16N 88 56W
Richwood, U.S.A. 76 F5 38 14N 80 32W
Ridder = Leninogorsk, Kazakstan 26 D9 50 20N 83 30 E
Ridgecrest, U.S.A. 85 K9 35 38N 117 40W
Ridgedale, Canada 73 C8 53 0N 104 10W
Ridgefield, U.S.A. 84 E4 45 49N 122 45W
Ridgeland, U.S.A. 77 J5 32 29N 80 59W
Ridgelands, Australia 62 C5 23 16S 150 17 E
Ridgetown, Canada 70 D3 42 26N 81 52W
Ridgewood, U.S.A. 79 F10 40 59N 74 7W
Ridgway, U.S.A. 78 E6 41 25N 78 44W
Riding Mountain Nat. Park, Canada 73 C8 50 50N 100 0W
Ridley, Mt., Australia 61 F3 33 12S 122 7 E
Ried, Austria 16 D7 48 14N 13 30 E
Riesa, Germany 16 C7 51 17N 13 17 E
Riet →, S. Africa 56 D3 29 0S 23 54 E
Rieti, Italy 20 C5 42 24N 12 51 E
Riffe L., U.S.A. 84 D4 46 32N 122 26W
Rifle, U.S.A. 82 G10 39 32N 107 47W
Rift Valley □, Kenya 54 B4 0 20N 36 0 E
Rig Rig, Chad 51 F7 14 13N 14 25 E
Rīga, Latvia 9 H21 56 53N 24 8 E
Riga, G. of, Latvia 9 H20 57 40N 23 45 E
Rīgān, Iran 45 D8 28 37N 58 58 E
Rigas Jūras Licis = Riga, G. of, Latvia 9 H20 57 40N 23 45 E
Rigaud, Canada 79 A10 45 29N 74 18W
Rigby, U.S.A. 82 E8 43 40N 111 55W
Rigestān □, Afghan. 40 D4 30 15N 65 0 E
Riggins, U.S.A. 82 D5 45 25N 116 19W
Rigolet, Canada 71 B8 54 10N 58 23W
Riihimäki, Finland 9 F21 60 45N 24 48 E
Riiser-Larsen-halvøya, Antarctica 5 C4 68 0S 35 0 E
Rijeka, Croatia 16 F8 45 20N 14 21 E
Rijn →, Neths. 15 B4 52 12N 4 21 E
Rijssen, Neths. 15 B6 52 19N 6 31 E
Rijswijk, Neths. 15 B4 52 4N 4 22 E
Rikuzentakada, Japan 30 E10 39 0N 141 40 E
Riley, U.S.A. 82 E4 43 32N 119 28W
Rimah, Wadi ar →, Si. Arabia 44 E4 26 5N 41 30 E
Rimbey, Canada 72 C6 52 35N 114 15W
Rímini, Italy 20 B5 44 3N 12 33 E
Rîmnicu Sărat, Romania 17 F14 45 26N 27 3 E
Rîmnicu Vîlcea, Romania 17 F13 45 9N 24 21 E
Rimouski, Canada 71 C6 48 27N 68 30W
Rimrock, U.S.A. 84 D5 46 38N 121 10W
Rinca, Indonesia 37 F5 8 45S 119 35 E
Rincón de Romos, Mexico 86 C4 22 14N 102 18W
Rinconada, Argentina 94 A2 22 26S 66 10W
Ringkøbing, Denmark 9 H13 56 5N 8 15 E
Ringling, U.S.A. 82 C8 46 16N 110 49W
Ringvassøy, Norway 8 B18 69 56N 19 15 E
Rinjani, Indonesia 36 F5 8 24S 116 28 E
Rio Branco, Brazil 92 E5 9 58S 67 49W
Río Branco, Uruguay 95 C5 32 40S 53 40W
Rio Brilhante, Brazil 95 A5 21 48S 54 33W
Rio Claro, Brazil 95 A6 22 19S 47 35W
Rio Claro, Trin. & Tob. 89 D7 10 20N 61 25W
Río Colorado, Argentina 96 D4 39 0S 64 0W
Río Cuarto, Argentina 94 C3 33 10S 64 25W
Rio das Pedras, Mozam. 57 C6 23 8S 35 28 E
Rio de Janeiro, Brazil 95 A7 23 0S 43 12W
Rio de Janeiro □, Brazil 95 A7 22 50S 43 0W
Rio do Sul, Brazil 95 B6 27 13S 49 37W
Río Gallegos, Argentina 96 G3 51 35S 69 15W
Río Grande, Argentina 96 G3 53 50S 67 45W
Rio Grande, Brazil 95 C5 32 0S 52 20W
Río Grande, Mexico 86 C4 23 50N 103 2W
Río Grande, Nic. 88 D3 12 54N 83 33W
Rio Grande →, U.S.A. 81 N6 25 57N 97 9W
Rio Grande City, U.S.A. 81 M5 26 23N 98 49W
Río Grande del Norte →, N. Amer. 75 E7 26 0N 97 0W
Rio Grande do Norte □, Brazil 93 E11 5 40S 36 0W
Rio Grande do Sul □, Brazil 95 C5 30 0S 53 0W
Río Hato, Panama 88 E3 8 22N 80 10W
Río Lagartos, Mexico 87 C7 21 36N 88 10W
Rio Largo, Brazil 93 E11 9 28S 35 50W
Río Mulatos, Bolivia 92 G5 19 40S 66 50W
Río Muni = Mbini □, Eq. Guin. 52 D2 1 30N 10 0 E
Rio Negro, Brazil 95 B6 26 0S 49 55W
Rio Pardo, Brazil 95 C5 30 0S 52 30W
Río Segundo, Argentina 94 C3 31 40S 63 59W
Río Tercero, Argentina 94 C3 32 15S 64 8W
Rio Verde, Brazil 93 G8 17 50S 51 0W
Río Verde, Mexico 87 C5 21 56N 99 59W
Rio Vista, U.S.A. 84 G5 38 10N 121 42W
Riobamba, Ecuador 92 D3 1 50S 78 45W
Ríohacha, Colombia 92 A4 11 33N 72 55W
Riosucio, Caldas, Colombia 92 B3 5 30N 75 40W
Riosucio, Choco, Colombia 92 B3 7 27N 77 7W
Riou L., Canada 73 B7 59 7N 106 25W
Ripley, Canada 78 B3 44 4N 81 35W
Ripley, Calif., U.S.A. 85 M12 33 32N 114 39W
Ripley, N.Y., U.S.A. 78 D5 42 16N 79 43W
Ripley, Tenn., U.S.A. 81 H10 35 45N 89 32W
Ripon, U.K. 10 C6 54 9N 1 31W
Ripon, Calif., U.S.A. 84 H5 37 44N 121 7W
Ripon, Wis., U.S.A. 76 D1 43 51N 88 50W
Risalpur, Pakistan 42 B4 34 3N 71 59 E
Rishā', W. ar →, Si. Arabia 44 E5 25 33N 44 5 E
Rishiri-Tō, Japan 30 B10 45 11N 141 15 E
Rishon le Ẕiyyon, Israel 47 D3 31 58N 34 48 E
Rison, U.S.A. 81 J8 33 58N 92 11W
Risør, Norway 9 G13 58 43N 9 13 E
Rittman, U.S.A. 78 F3 40 58N 81 47W
Ritzville, U.S.A. 82 C4 47 8N 118 23W
Riva del Garda, Italy 20 B4 45 53N 10 50 E
Rivadavia, Buenos Aires, Argentina 94 D3 35 29S 62 59W
Rivadavia, Mendoza, Argentina 94 C2 33 13S 68 30W
Rivadavia, Salta, Argentina 94 A3 24 5S 62 54W
Rivadavia, Chile 94 B1 29 57S 70 35W
Rivas, Nic. 88 D2 11 30N 85 50W
Rivera, Uruguay 95 C4 31 0S 55 50W
Riverdale, U.S.A. 84 J7 36 26N 119 52W
Riverhead, U.S.A. 79 F12 40 55N 72 40W
Riverhurst, Canada 73 C7 50 55N 106 50W
Riverina, Australia 61 E3 29 45S 120 40 E
Rivers, Canada 73 C8 50 2N 100 14W
Rivers, L. of the, Canada 73 D7 49 49N 105 44W
Rivers Inlet, Canada 72 C3 51 42N 127 15W
Riversdale, S. Africa 56 E3 34 7S 21 15 E
Riverside, Calif., U.S.A. 85 M9 33 59N 117 22W
Riverside, Wyo., U.S.A. 82 F10 41 13N 106 47W
Riversleigh, Australia 62 B2 19 5S 138 40 E
Riverton, Australia 63 E2 34 10S 138 46 E
Riverton, Canada 73 C9 51 1N 97 0W
Riverton, N.Z. 59 M1 46 21S 168 0 E
Riverton, U.S.A. 82 E9 43 2N 108 23W
Riverton Heights, U.S.A. 84 C4 47 28N 122 17W
Riviera di Levante, Italy 20 B3 44 15N 9 30 E
Riviera di Ponente, Italy 20 B3 44 10N 8 20 E
Rivière-à-Pierre, Canada 71 C5 46 59N 72 11W
Rivière-au-Renard, Canada 71 C7 48 59N 64 23W
Rivière-du-Loup, Canada 71 C6 47 50N 69 30W
Rivière-Pentecôte, Canada 71 C6 49 57N 67 1W
Rivière-Pilote, Martinique 89 D7 14 26N 60 53W
Rivne, Ukraine 17 C14 50 40N 26 10 E
Rívoli, Italy 20 B2 45 3N 7 31 E
Rivoli B., Australia 63 F3 37 32S 140 3 E
Riyadh = Ar Riyāḍ, Si. Arabia 46 C4 24 41N 46 42 E
Rize, Turkey 25 F7 41 0N 40 30 E
Rizhao, China 35 G10 35 25N 119 30 E
Rizokarpaso, Cyprus 23 D13 35 36N 34 23 E
Rizzuto, C., Italy 20 E7 38 53N 17 5 E
Rjukan, Norway 9 G13 59 54N 8 33 E
Road Town, Virgin Is. 89 C7 18 27N 64 37W
Roag, L., U.K. 12 C2 58 12N 6 51W
Roanne, France 18 C6 46 3N 4 4 E
Roanoke, Ala., U.S.A. 77 J3 33 9N 85 22W
Roanoke, Va., U.S.A. 76 G6 37 16N 79 56W
Roanoke →, U.S.A. 77 H7 35 57N 76 42W
Roanoke I., U.S.A. 77 H8 35 55N 75 40W
Roanoke Rapids, U.S.A. 77 G7 36 28N 77 40W
Roatán, Honduras 88 C2 16 18N 86 35W
Robbins I., Australia 62 G4 40 42S 145 0 E
Robe →, Australia 60 D2 21 42S 116 15 E
Robe →, Ireland 13 C2 53 38N 9 10W
Robert Lee, U.S.A. 81 K4 31 54N 100 29W
Roberts, U.S.A. 82 E7 43 43N 112 8W
Robertsganj, India 43 G10 24 44N 83 4 E
Robertson, S. Africa 56 E2 33 46S 19 50 E
Robertson I., Antarctica 5 C18 65 15S 59 30W

Robertson Ra., *Australia* . **60 D3** 23 15S 121 0 E
Robertsport, *Liberia* **50 G2** 6 45N 11 26W
Robertstown, *Australia* . . . **63 E2** 33 58S 139 5 E
Roberval, *Canada* **71 C5** 48 32N 72 15W
Robeson Chan., *Greenland* **4 A4** 82 0N 61 30W
Robinson ➝, *Australia* . . **62 B2** 16 3S 137 16 E
Robinson Ra., *Australia* . . **61 E2** 25 40S 119 0 E
Robinson River, *Australia* **62 B2** 16 45S 136 58 E
Robinvale, *Australia* . . . . **63 E3** 34 40S 142 45 E
Roblin, *Canada* **73 C8** 51 14N 101 21W
Roboré, *Bolivia* **92 G7** 18 10S 59 45W
Robson, Mt., *Canada* **72 C5** 53 10N 119 10W
Robstown, *U.S.A.* **81 M6** 27 47N 97 40W
Roca, C. da, *Portugal* **19 C1** 38 40N 9 31W
Roca Partida, I., *Mexico* . . **86 D2** 19 1N 112 2W
Rocas, I., *Brazil* **93 D12** 4 0S 34 1W
Rocha, *Uruguay* **95 C5** 34 30S 54 25W
Rochdale, *U.K.* **10 D5** 53 38N 2 9W
Rochefort, *Belgium* . . . . . . **15 D5** 50 9N 5 12 E
Rochefort, *France* **18 D3** 45 56N 0 57W
Rochelle, *U.S.A.* **80 E10** 41 56N 89 4W
Rocher River, *Canada* **72 A6** 61 23N 112 44W
Rochester, *Canada* **72 C6** 54 22N 113 27W
Rochester, *U.K.* . . . . **11 F8** 51 23N 0 31 E
Rochester, *Ind., U.S.A.* . . . **76 E2** 41 4N 86 13W
Rochester, *Minn., U.S.A.* . . **80 C8** 44 1N 92 28W
Rochester, *N.H., U.S.A.* . . **79 C14** 43 18N 70 59W
Rochester, *N.Y., U.S.A.* . . **78 C7** 43 10N 77 37W
Rock ➝, *Canada* **72 A3** 60 7N 127 7W
Rock Hill, *U.S.A.* **77 H5** 34 56N 81 1W
Rock Island, *U.S.A.* **80 E9** 41 30N 90 34W
Rock Rapids, *U.S.A.* **80 D6** 43 26N 96 10W
Rock River, *U.S.A.* **82 F11** 41 44N 105 58W
Rock Sound, *Bahamas* . . **88 B4** 24 54N 76 12W
Rock Springs, *Mont.,*
U.S.A. **82 C10** 46 49N 106 15W
Rock Springs, *Wyo.,*
U.S.A. **82 F9** 41 35N 109 14W
Rock Valley, *U.S.A.* **80 D6** 43 12N 96 18W
Rockall, *Atl. Oc.* **6 D3** 57 37N 13 42W
Rockdale, *Tex., U.S.A.* **81 K6** 30 39N 97 0W
Rockdale, *Wash., U.S.A.* . . **84 C5** 47 22N 121 28W
Rockefeller Plateau,
Antarctica **5 E14** 80 0S 140 0W
Rockford, *U.S.A.* **80 D10** 42 16N 89 6W
Rockglen, *Canada* **73 D7** 49 11N 105 57W
Rockhampton, *Australia* . **62 C5** 23 22S 150 32 E
Rockhampton Downs,
Australia . . . . **62 B2** 18 57S 135 10 E
Rockingham, *Australia* . . **61 F2** 32 15S 115 38 E
Rockingham B., *Australia* **62 B4** 18 5S 146 10 E
Rockingham Forest, *U.K.* **11 E7** 52 29N 0 42W
Rocklake, *U.S.A.* **80 A5** 48 47N 99 15W
Rockland, *Canada* **79 A9** 45 33N 75 17W
Rockland, *Idaho, U.S.A.* . . **82 E7** 42 34N 112 53W
Rockland, *Maine, U.S.A.* . . **71 D6** 44 6N 69 7W
Rockland, *Mich., U.S.A.* . . **80 B10** 46 44N 89 11W
Rocklin, *U.S.A.* **84 G5** 38 48N 121 14W
Rockmart, *U.S.A.* **77 H3** 34 0N 85 3W
Rockport, *Mo., U.S.A.* **80 E7** 40 25N 95 31W
Rockport, *Tex., U.S.A.* **81 L6** 28 2N 97 3W
Rocksprings, *U.S.A.* **81 K4** 30 1N 100 13W
Rockville, *Conn., U.S.A.* . . **79 E12** 41 52N 72 28W
Rockville, *Md., U.S.A.* . . . **76 F7** 39 5N 77 9W
Rockwall, *U.S.A.* **81 J6** 32 56N 96 28W
Rockwell City, *U.S.A.* **80 D7** 42 24N 94 38W
Rockwood, *U.S.A.* **77 H3** 35 52N 84 41W
Rocky Ford, *U.S.A.* **80 F3** 38 3N 103 43W
Rocky Gully, *Australia* . . . **61 F2** 34 30S 116 57 E
Rocky Lane, *Canada* **72 B5** 58 31N 116 22W
Rocky Mount, *U.S.A.* **77 H7** 35 57N 77 48W
Rocky Mountain House,
Canada **72 C6** 52 22N 114 55W
Rocky Mts., *N. Amer.* . . . **72 C4** 55 0N 121 0W
Rockyford, *Canada* **72 C6** 51 14N 113 10W
Rod, *Pakistan* . . . . **40 E3** 28 10N 63 5 E
Rødbyhavn, *Denmark* . . . . **9 J14** 54 39N 11 22 E
Roddickton, *Canada* **71 B8** 50 51N 56 8W
Roderick I., *Canada* **72 C3** 52 38N 128 22W
Rodez, *France* . . . . **18 D5** 44 21N 2 33 E
Rodhopoú, *Greece* . . . . . . **23 D5** 35 34N 23 45 E
Ródhos, *Greece* . . . . **23 C10** 36 15N 28 10 E
Rodney, *Canada* **78 D3** 42 34N 81 41W
Rodney, C., *N.Z.* . . . . **59 G5** 36 17S 174 50 E
Rodriguez, *Ind. Oc.* . . . . . **3 E13** 19 45S 63 20 E
Roe ➝, *U.K.* **13 A5** 55 6N 6 59W
Roebling, *U.S.A.* **79 F10** 40 7N 74 47W
Roebourne, *Australia* . . . . **60 D2** 20 44S 117 9 E
Roebuck B., *Australia* . . . . **60 C3** 18 5S 122 20 E
Roebuck Plains, *Australia* **60 C3** 17 56S 122 28 E
Roermond, *Neths.* . . . . **15 C5** 51 12N 6 0 E
Roes Welcome Sd.,
Canada **69 B11** 65 0N 87 0W
Roeselare, *Belgium* . . . . . **15 D3** 50 57N 3 7 E
Rogachev = Ragachow,
Belarus . . . . **17 B16** 53 8N 30 5 E
Rogagua, L., *Bolivia* **92 F5** 13 43S 66 50W
Rogatyn, *Ukraine* . . . . **17 D13** 49 24N 24 36 E
Rogdhia, *Greece* . . . . **23 D7** 35 22N 25 1 E
Rogers, *U.S.A.* **81 G7** 36 20N 94 7W
Rogers City, *U.S.A.* **76 C4** 45 25N 83 49W
Rogerson, *U.S.A.* **82 E6** 42 13N 114 36W
Rogersville, *U.S.A.* **77 G4** 36 24N 83 1W
Roggan River, *Canada* . . . **70 B4** 54 25N 79 32W
Roggeveldberge, *S. Africa* **56 E3** 32 10S 20 10 E
Rogoaguado, L., *Bolivia* . **92 F5** 13 0S 65 30W
Rogue ➝, *U.S.A.* **82 E1** 42 26N 124 26W
Róhda, *Greece* . . . . **23 A3** 39 48N 19 46 E
Rohnert Park, *U.S.A.* **84 G4** 38 16N 122 40W
Rohri, *Pakistan* . . . . **42 F3** 27 45N 68 51 E
Rohri Canal, *Pakistan* . . . **42 F3** 26 15N 68 27 E
Rohtak, *India* . . . . **42 E7** 28 55N 76 43 E
Roi Et, *Thailand* . . . . **38 D4** 16 4N 103 40 E
Roja, *Latvia* . . . . **9 H20** 57 29N 22 43 E
Rojas, *Argentina* **94 C3** 34 10S 60 45W
Rojo, C., *Mexico* **87 C5** 21 33N 97 20W
Rokan ➝, *Indonesia* . . . . **36 D2** 2 0N 100 50 E
Rokeby, *Australia* . . . . **62 A3** 13 39S 142 40 E
Rokiškis, *Lithuania* . . . . . . **9 J21** 55 55N 25 35 E
Rolândia, *Brazil* **95 A5** 23 18S 51 23W
Rolette, *U.S.A.* **80 A5** 48 40N 99 51W
Rolla, *Kans., U.S.A.* **81 G4** 37 7N 101 38W
Rolla, *Mo., U.S.A.* **81 G9** 37 57N 91 46W
Rolla, *N. Dak., U.S.A.* **80 A5** 48 52N 99 37W
Rolleston, *Australia* . . . . . **62 C4** 24 28S 148 35 E
Rollingstone, *Australia* . . **62 B4** 19 2S 146 24 E
Roma, *Australia* . . . . **63 D4** 26 32S 148 49 E
Roma, *Italy* . . . . **20 D5** 41 54N 12 29 E
Roma, *Sweden* . . . . **9 H18** 57 32N 18 26 E
Roman, *Romania* . . . . **17 E14** 46 57N 26 55 E
Roman, *Russia* . . . . **27 C12** 60 4N 112 14 E
Romang, *Indonesia* . . . . . . **37 F7** 7 30S 127 20 E
Români, *Egypt* . . . . **47 E1** 30 59N 32 38 E
Romania ■, *Europe* . . . . . **17 F12** 46 0N 25 0 E
Romano, Cayo, *Cuba* **88 B4** 22 0N 77 30W
Romanovka =
Basarabeasca, *Moldova* **17 E15** 46 21N 28 58 E
Romans-sur-Isère, *France* **18 D6** 45 3N 5 3 E
Romblon, *Phil.* . . . . **37 B6** 12 33N 122 17 E
Rombo □, *Tanzania* . . . . . **54 C4** 3 10S 37 30 E
Rome = Roma, *Italy* . . . . . **20 D5** 41 54N 12 29 E
Rome, *Ga., U.S.A.* **77 H3** 34 15N 85 10W
Rome, *N.Y., U.S.A.* **79 C9** 43 13N 75 27W
Romney, *U.S.A.* **76 F6** 39 21N 78 45W
Romney Marsh, *U.K.* . . . . **11 F8** 51 2N 0 54 E
Rømø, *Denmark* . . . . **9 J13** 55 10N 8 30 E
Romorantin-Lanthenay,
France . . . . **18 C4** 47 21N 1 45 E
Romsdalen, *Norway* . . . . **9 E12** 62 25N 7 52 E
Ron, *Vietnam* . . . . **38 D6** 17 53N 106 27 E
Rona, *U.K.* **12 D3** 57 34N 5 59W
Ronan, *U.S.A.* **82 C6** 47 32N 114 6W
Roncador, Cayos,
Caribbean **88 D3** 13 32N 80 4W
Roncador, Serra do, *Brazil* **93 F8** 12 30S 52 30W
Ronceverte, *U.S.A.* **76 G5** 37 45N 80 28W
Ronda, *Spain* **19 D3** 36 46N 5 12W
Rondane, *Norway* . . . . **9 F13** 61 57N 9 50 E
Rondônia □, *Brazil* **92 F6** 11 0S 63 0W
Rondonópolis, *Brazil* **93 G8** 16 28S 54 38W
Ronge, L. la, *Canada* **73 B7** 55 6N 105 17W
Rønne, *Denmark* . . . . **9 J16** 55 6N 14 43 E
Ronne Ice Shelf,
Antarctica **5 D18** 78 0S 60 0W
Ronsard, C., *Australia* . . . **61 D1** 24 46S 113 10 E
Ronse, *Belgium* . . . . **15 D3** 50 45N 3 35 E
Roodepoort, *S. Africa* . . . **57 D4** 26 11S 27 54 E
Roof Butte, *U.S.A.* **83 H9** 36 28N 109 5W
Roorkee, *India* . . . . **42 E7** 29 52N 77 59 E
Roosendaal, *Neths.* . . . . . **15 C4** 51 32N 4 29 E
Roosevelt, *Minn., U.S.A.* . **80 A7** 48 48N 95 6W
Roosevelt, *Utah, U.S.A.* . . **82 F8** 40 18N 109 59W
Roosevelt ➝, *Brazil* **92 E6** 7 35S 60 20W
Roosevelt, Mt., *Canada* . . **72 B3** 58 26N 125 20W
Roosevelt I., *Antarctica* . . **5 D12** 79 30S 162 0W
Roosevelt Res., *U.S.A.* . . . **83 K8** 33 46N 111 0W
Roper ➝, *Australia* . . . . **62 A2** 14 43S 135 27 E
Ropesville, *U.S.A.* **81 J3** 33 26N 102 9W
Roque Pérez, *Argentina* . . **94 D4** 35 25S 59 24W
Roquetas de Mar, *Spain* . . **19 D4** 36 46N 2 36W
Roraima □, *Brazil* **92 C6** 2 0N 61 30W
Roraima, Mt., *Venezuela* . **92 B6** 5 10N 60 40W
Rorketon, *Canada* **73 C9** 51 24N 99 35W
Røros, *Norway* . . . . **9 E14** 62 35N 11 23 E
Rosa, *Zambia* . . . . **55 D3** 9 33S 31 15 E
Rosa, Monte, *Europe* . . . . **16 F4** 45 57N 7 53 E
Rosalia, *U.S.A.* **82 C5** 47 14N 117 22W
Rosamond, *U.S.A.* **85 L8** 34 52N 118 10W
Rosario, *Argentina* **94 C3** 33 0S 60 40W
Rosário, *Brazil* **93 D10** 3 0S 44 15W
Rosario, *Baja Calif.,*
Mexico **86 A1** 30 0N 115 50W
Rosario, *Sinaloa, Mexico* **86 C3** 23 0N 105 52W
Rosario, *Paraguay* **94 A4** 24 30S 57 35W
Rosario de la Frontera,
Argentina **94 B3** 25 50S 65 0W
Rosario de Lerma,
Argentina **94 A2** 24 59S 65 35W
Rosario del Tala,
Argentina **94 C4** 32 20S 59 10W
Rosário do Sul, *Brazil* . . . **95 C5** 30 15S 54 55W
Rosarito, *Mexico* **85 N9** 32 18N 117 4W
Rosas, G. de, *Spain* . . . . . **19 A7** 42 10N 3 15 E
Roscoe, *U.S.A.* **80 C5** 45 27N 99 20W
Roscommon, *Ireland* **13 C3** 53 38N 8 11W
Roscommon, *U.S.A.* **76 C3** 44 30N 84 35W
Roscommon □, *Ireland* . . **13 C3** 53 49N 8 23W
Roscrea, *Ireland* **13 D4** 52 57N 7 49W
Rose ➝, *Australia* . . . . . **62 A2** 14 16S 135 45 E
Rose Blanche, *Canada* . . . **71 C8** 47 38N 58 45W
Rose Harbour, *Canada* . . . **72 C2** 52 15N 131 10W
Rose Pt., *Canada* **72 C2** 54 11N 131 39W
Rose Valley, *Canada* **73 C8** 52 19N 103 49W
Roseau, *Domin.* **89 C7** 15 20N 61 24W
Roseau, *U.S.A.* **80 A7** 48 51N 95 46W
Rosebery, *Australia* . . . . . **62 G4** 41 46S 145 33 E
Rosebud, *U.S.A.* **81 K6** 31 4N 96 59W
Roseburg, *U.S.A.* **82 E2** 43 13N 123 20W
Rosedale, *Australia* . . . . . **62 C5** 24 38S 151 53 E
Rosedale, *U.S.A.* **81 J9** 33 51N 91 2W
Roseland, *U.S.A.* **84 G4** 38 25N 122 43W
Rosemary, *Canada* **72 C6** 50 46N 112 5W
Rosenberg, *U.S.A.* **81 L7** 29 34N 95 49W
Rosenheim, *Germany* . . . **16 E7** 47 51N 12 7 E
Rosetown, *Canada* **73 C7** 51 35N 107 59W
Rosetta = Rashîd, *Egypt* . **51 B11** 31 21N 30 22 E
Roseville, *U.S.A.* **84 G5** 38 45N 121 17W
Rosewood, *N. Terr.,*
Australia . . . . **60 C4** 16 28S 128 58 E
Rosewood, *Queens.,*
Australia . . . . **63 D5** 27 38S 152 36 E
Roshkhvār, *Iran* . . . . **45 C8** 34 58N 59 37 E
Rosignano Marittimo, *Italy* **20 C4** 43 24N 10 28 E
Rosignol, *Guyana* **92 B7** 6 15N 57 30W
Roşiori-de-Vede, *Romania* **17 F13** 44 7N 24 59 E
Roskilde, *Denmark* . . . . . . **9 J15** 55 38N 12 3 E
Roslavl, *Russia* . . . . **24 D5** 53 57N 32 55 E
Roslyn, *Australia* . . . . **63 E4** 34 29S 149 37 E
Rosmead, *S. Africa* . . . . . . **56 E4** 31 29S 25 8 E
Ross, *Australia* . . . . **62 G4** 42 2S 147 30 E
Ross, *N.Z.* . . . . **59 K3** 42 53S 170 49 E
Ross I., *Antarctica* . . . . . . **5 D11** 77 30S 168 0 E
Ross Ice Shelf, *Antarctica* **5 E12** 80 0S 180 0 E
Ross L., *U.S.A.* **82 B3** 48 44N 121 4W
Ross-on-Wye, *U.K.* **11 F5** 51 54N 2 34W
Ross Sea, *Antarctica* . . . . **5 D11** 74 0S 178 0 E
Rossan Pt., *Ireland* **13 B3** 54 42N 8 47W
Rossano Cálabro, *Italy* . . **20 E7** 39 36N 16 39 E
Rossburn, *Canada* **73 C8** 50 40N 100 49W
Rosseau, *Canada* **78 A5** 45 16N 79 39W
Rossignol, L., *Canada* **70 B5** 52 43N 73 40W
Rossignol Res., *Canada* . . **71 D6** 44 12N 65 10W
Rossland, *Canada* **72 D5** 49 6N 117 50W
Rosslare, *Ireland* **13 D5** 52 17N 6 24W
Rosso, *Mauritania* **50 E1** 16 40N 15 45W
Rossosh, *Russia* . . . . **25 D6** 50 15N 39 28 E
Rossport, *Canada* **70 C2** 48 50N 87 30W
Røssvatnet, *Norway* . . . . . **8 D16** 65 45N 14 5 E
Rossville, *Australia* . . . . . **62 B4** 15 48S 145 15 E
Røst, *Norway* . . . . **8 C15** 67 32N 12 0 E
Rosthern, *Canada* **73 C7** 52 40N 106 20W
Rostock, *Germany* . . . . . . **16 A7** 54 5N 12 8 E
Rostov, *Don, Russia* . . . . **25 E6** 47 15N 39 45 E
Rostov, *Yarosl., Russia* . . **24 C6** 57 14N 39 25 E
Roswell, *Ga., U.S.A.* **77 H3** 34 2N 84 22W
Roswell, *N. Mex., U.S.A.* . . **81 J2** 33 24N 104 32W
Rosyth, *U.K.* **12 E5** 56 2N 3 25W
Rotan, *U.S.A.* **81 J4** 32 51N 100 28W
Rother ➝, *U.K.* . . . . **11 G8** 50 59N 0 45 E
Rotherham, *U.K.* **10 D6** 53 26N 1 20W
Rothes, *U.K.* **12 D5** 57 32N 3 13W
Rothesay, *Canada* **71 C6** 45 23N 66 0W
Rothesay, *U.K.* **12 F3** 55 50N 5 3W
Roti, *Indonesia* . . . . **37 F6** 10 50S 123 0 E
Roto, *Australia* . . . . **63 E4** 33 0S 145 30 E
Rotondo Mte., *France* . . . **18 E8** 42 14N 9 8 E
Rotoroa, L., *N.Z.* . . . . **59 J4** 41 55S 172 39 E
Rotorua, *N.Z.* . . . . **59 H6** 38 9S 176 16 E
Rotorua, L., *N.Z.* . . . . **59 H6** 38 5S 176 18 E
Rotterdam, *Neths.* . . . . . . **15 C4** 51 55N 4 30 E
Rottnest I., *Australia* . . . . **61 F2** 32 0S 115 27 E
Rottumeroog, *Neths.* . . . . **15 A6** 53 33N 6 34 E
Rottweil, *Germany* . . . . . . **16 D5** 48 9N 8 37 E
Rotuma, *Fiji* . . . . **64 J9** 12 25S 177 5 E
Roubaix, *France* . . . . **18 A5** 50 40N 3 10 E
Rouen, *France* . . . . **18 B4** 49 27N 1 4 E
Rouleau, *Canada* **73 C8** 50 10N 104 56W
Round Mountain, *U.S.A.* . . **82 G5** 38 43N 117 4W
Round Mt., *Australia* . . . . **63 E5** 30 26S 152 16 E
Roundup, *U.S.A.* **82 C9** 46 27N 108 33W
Rousay, *U.K.* **12 B5** 59 10N 3 2W
Rouses Point, *U.S.A.* **79 B11** 44 59N 73 22W
Roussillon, *France* . . . . . . **18 E5** 42 30N 2 35 E
Rouxville, *S. Africa* . . . . . **56 E4** 30 25S 26 50 E
Rouyn, *Canada* **70 C4** 48 20N 79 0W
Rovaniemi, *Finland* . . . . . **8 C21** 66 29N 25 41 E
Rovereto, *Italy* . . . . **20 B4** 45 53N 11 3 E
Rovigo, *Italy* . . . . **20 B4** 45 4N 11 47 E
Rovinj, *Croatia* . . . . **16 F7** 45 5N 13 40 E
Rovno = Rivne, *Ukraine* . . **17 C14** 50 40N 26 10 E
Rovuma ➝, *Tanzania* . . . **55 E5** 10 29S 40 28 E
Row'ān, *Iran* . . . . **45 C6** 35 8N 48 51 E
Rowena, *Australia* . . . . . . **63 D4** 29 48S 148 55 E
Rowley Shoals, *Australia* **60 C2** 17 30S 119 0 E
Roxas, *Phil.* . . . . **37 B6** 11 36N 122 49 E
Roxboro, *U.S.A.* **77 G6** 36 24N 78 59W
Roxborough Downs,
Australia . . . . **62 C2** 22 30S 138 45 E
Roxburgh, *N.Z.* . . . . **59 L2** 45 33S 169 19 E
Roy, *Mont., U.S.A.* **82 C9** 47 20N 108 58W
Roy, *N. Mex., U.S.A.* **81 H2** 35 57N 104 12W
Roy Hill, *Australia* . . . . . . **60 D2** 22 37S 119 58 E
Royal Leamington Spa,
U.K. **11 E6** 52 18N 1 31W
Royal Tunbridge Wells,
U.K. . . . . **11 F8** 51 7N 0 16 E
Royan, *France* **18 D3** 45 37N 1 2W
Rozdilna, *Ukraine* . . . . **17 E16** 46 50N 30 2 E
Rozhyshche, *Ukraine* . . . . **17 C13** 50 54N 25 15 E
Rtishchevo, *Russia* . . . . . . **24 C7** 52 18N 43 46 E
Ruacaná, *Angola* . . . . **56 B1** 17 20S 14 12 E
Ruahine Ra., *N.Z.* . . . . **59 H6** 39 55S 176 2 E
Ruapehu, *N.Z.* . . . . **59 H5** 39 17S 175 35 E
Ruapuke I., *N.Z.* . . . . **59 M2** 46 46S 168 31 E
Ruâq, W. ➝, *Egypt* . . . . **47 F2** 30 0N 33 49 E
Rub' al Khali, *Si. Arabia* . . **46 D4** 18 0N 48 0 E
Rubeho Mts., *Tanzania* . . **54 D4** 6 50S 36 25 E
Rubh a' Mhail, *U.K.* **12 F2** 55 56N 6 8W
Rubha Hunish, *U.K.* **12 D2** 57 42N 6 20W
Rubha Robhanais =
Lewis, Butt of, *U.K.* . . . **12 C2** 58 31N 6 16W
Rubicon ➝, *U.S.A.* **84 G5** 38 53N 121 4W
Rubio, *Venezuela* **92 B4** 7 43N 72 22W
Rubtsovsk, *Russia* . . . . **26 D9** 51 30N 81 10 E
Ruby L., *U.S.A.* **82 F6** 40 10N 115 28W
Ruby Mts., *U.S.A.* **82 F6** 40 30N 115 20W
Rūd Sar, *Iran* . . . . **45 B6** 37 8N 50 18 E
Rudall, *Australia* . . . . **63 E2** 33 43S 136 17 E
Rudall ➝, *Australia* . . . . . **60 D3** 22 34S 122 13 E
Rudewa, *Tanzania* . . . . . . **55 E3** 10 7S 34 40 E
Rudnichnyy, *Russia* . . . . . **24 C9** 59 38N 52 26 E
Rudnogorsk, *Russia* . . . . . **27 D11** 57 15N 103 42 E
Rudnyy, *Kazakstan* . . . . . . **26 D7** 52 57N 63 7 E
Rudolf, Ostrov, *Russia* . . . **26 A6** 81 45N 58 30 E
Rudyard, *U.S.A.* **76 B3** 46 14N 84 36W
Rufa'a, *Sudan* . . . . **51 F11** 14 44N 33 22 E
Rufiji □, *Tanzania* . . . . . . **54 D4** 8 0S 38 30 E
Rufiji ➝, *Tanzania* . . . . . **54 D4** 7 50S 39 15 E
Rufino, *Argentina* **94 C3** 34 20S 62 50W
Rufisque, *Senegal* **50 F1** 14 40N 17 15W
Rufunsa, *Zambia* . . . . **55 F2** 15 4S 29 34 E
Rugby, *U.K.* **11 E6** 52 23N 1 16W
Rugby, *U.S.A.* **80 A5** 48 22N 100 0W
Rügen, *Germany* . . . . **16 A7** 54 22N 13 24 E
Ruhengeri, *Rwanda* . . . . . **54 C2** 1 30S 29 36 E
Ruhnu saar, *Estonia* . . . . **9 H20** 57 48N 23 15 E
Ruhr ➝, *Germany* . . . . . . **16 C4** 51 27N 6 43 E
Ruhuhu ➝, *Tanzania* . . . **55 E3** 10 31S 34 34 E
Ruidosa, *U.S.A.* **81 L2** 29 59N 104 41W
Ruidoso, *U.S.A.* **83 K11** 33 20N 105 41W
Ruivo, Pico, *Madeira* **22 D3** 32 45N 16 56W
Rujm Tal'at al Jamā'ah,
Jordan . . . . **47 E4** 30 24N 35 30 E
Ruk, *Pakistan* . . . . **42 F3** 27 50N 68 42 E
Rukwa □, *Tanzania* . . . . . **54 D3** 7 0S 31 30 E
Rukwa L., *Tanzania* . . . . . **54 D3** 8 0S 32 20 E
Rulhieres, C., *Australia* . . **60 B4** 13 56S 127 22 E
Rum = Rhum, *U.K.* **12 E2** 57 0N 6 20W
Rum Cay, *Bahamas* **89 B5** 23 40N 74 58W
Rum Jungle, *Australia* . . . **60 B5** 13 0S 130 59 E
Rumāḩ, *Si. Arabia* . . . . . . **44 E5** 25 29N 47 10 E
Rumania = Romania ■,
Europe . . . . **17 F12** 46 0N 25 0 E
Rumaylah, *Iraq* . . . . **44 D5** 30 47N 47 37 E
Rumbalara, *Australia* . . . . **62 D1** 25 20S 134 29 E
Rumbêk, *Sudan* . . . . **51 G10** 6 54N 29 37 E
Rumford, *U.S.A.* **79 B14** 44 33N 70 33W
Rumia, *Poland* . . . . **17 A10** 54 37N 18 25 E
Rumoi, *Japan* . . . . **30 C10** 43 56N 141 39 E
Rumonge, *Burundi* . . . . . . **54 C2** 3 59S 29 26 E
Rumsey, *Canada* **72 C6** 51 51N 112 48W
Rumula, *Australia* . . . . **62 B4** 16 35S 145 20 E
Rumuruti, *Kenya* . . . . **54 B4** 0 17N 36 32 E
Runan, *China* . . . . **34 H8** 33 0N 114 30 E
Runanga, *N.Z.* . . . . **59 K3** 42 25S 171 15 E
Runaway, C., *N.Z.* . . . . **59 G6** 37 32S 177 59 E
Runcorn, *U.K.* **10 D5** 53 21N 2 44W
Rungwa, *Tanzania* . . . . . . **54 D3** 6 55S 33 32 E
Rungwa ➝, *Tanzania* . . . **54 D3** 7 36S 31 50 E
Rungwe, *Tanzania* . . . . . . **55 D3** 9 11S 33 32 E
Rungwe □, *Tanzania* . . . . **55 D3** 9 25S 33 32 E
Runton Ra., *Australia* . . . . **60 D3** 23 31S 123 6 E
Ruoqiang, *China* . . . . **32 C3** 38 55N 88 10 E
Rupa, *India* . . . . **41 F18** 27 15N 92 21 E
Rupar, *India* . . . . **42 D7** 31 2N 76 38 E
Rupat, *Indonesia* . . . . **36 D2** 1 45N 101 40 E
Rupert ➝, *Canada* **70 B4** 51 29N 78 45W
Rupert House =
Waskaganish, *Canada* . **70 B4** 51 30N 78 40W
Rurrenabaque, *Bolivia* . . . **92 F5** 14 30S 67 32W
Rusambo, *Zimbabwe* . . . . **55 F3** 16 30S 32 4 E
Rusape, *Zimbabwe* . . . . . . **55 F3** 18 35S 32 8 E
Ruschuk = Ruse, *Bulgaria* **21 C12** 43 48N 25 59 E
Ruse, *Bulgaria* . . . . **21 C12** 43 48N 25 59 E
Rushan, *China* . . . . **35 F11** 36 56N 121 30 E
Rushden, *U.K.* **11 E7** 52 18N 0 35W
Rushford, *U.S.A.* **80 D9** 43 49N 91 46W
Rushville, *Ill., U.S.A.* **80 E9** 40 7N 90 34W
Rushville, *Ind., U.S.A.* **76 F3** 39 37N 85 27W
Rushville, *Nebr., U.S.A.* . . **80 D3** 42 43N 102 28W
Rushworth, *Australia* . . . . **63 F4** 36 32S 145 1 E
Russas, *Brazil* **93 D11** 4 55S 37 50W
Russell, *Canada* **73 C8** 50 50N 101 20W
Russell, *U.S.A.* **80 F5** 38 54N 98 52W
Russell L., *Man., Canada* . **73 B8** 56 15N 101 30W
Russell L., *N.W.T., Canada* **72 A5** 63 5N 115 44W
Russellkonda, *India* . . . . . **41 K14** 19 57N 84 42 E
Russellville, *Ala., U.S.A.* . . **77 H2** 34 30N 87 44W
Russellville, *Ark., U.S.A.* . . **81 H8** 35 17N 93 8W
Russellville, *Ky., U.S.A.* . . **77 G2** 36 51N 86 53W
Russia ■, *Eurasia* . . . . . . **27 C11** 62 0N 105 0 E
Russian ➝, *U.S.A.* **84 G3** 38 27N 123 8W
Russkaya Polyana,
Kazakstan . . . . **26 D8** 53 47N 73 53 E
Russkoye Ustie, *Russia* . . **4 B15** 71 0N 149 0 E
Rustam, *Pakistan* . . . . **42 B5** 34 25N 72 13 E
Rustam Shahr, *Pakistan* . **42 F2** 26 58N 66 6 E
Rustavi, *Georgia* . . . . **25 F8** 41 30N 45 0 E
Rustenburg, *S. Africa* . . . . **56 D4** 25 41S 27 14 E
Ruston, *U.S.A.* **81 J8** 32 32N 92 38W
Rutana, *Burundi* . . . . **54 C2** 3 55S 30 0 E
Ruteng, *Indonesia* . . . . . . **37 F6** 8 35S 120 30 E
Ruth, *Mich., U.S.A.* **78 C2** 43 42N 82 45W
Ruth, *Nev., U.S.A.* **82 G6** 39 17N 114 59W
Rutherford, *U.S.A.* **84 G4** 38 26N 122 24W
Rutherglen, *U.K.* **12 F4** 55 49N 4 13W
Rutland Plains, *Australia* . **62 B3** 15 38S 141 43 E
Rutland □, *U.K.* **10 E7** 52 38N 0 40W
Rutledge ➝, *Canada* **73 A6** 61 4N 112 0W
Rutledge L., *Canada* **73 A6** 61 33N 110 47W
Rutshuru, *Zaïre* . . . . **54 C2** 1 13S 29 25 E
Ruurlo, *Neths.* . . . . **15 B6** 52 5N 6 24 E
Ruvu, *Tanzania* . . . . **54 D4** 6 49S 38 43 E
Ruvu ➝, *Tanzania* . . . . . . **54 D4** 6 23S 38 52 E
Ruvuma □, *Tanzania* . . . . **55 E4** 10 20S 36 0 E
Ruwais, *U.A.E.* . . . . **45 E7** 24 5N 52 50 E
Ruwenzori, *Africa* . . . . **54 B2** 0 30N 29 55 E
Ruyigi, *Burundi* . . . . **54 C3** 3 29S 30 15 E
Ružomberok, *Slovak Rep.* **17 D10** 49 3N 19 17 E
Rwanda ■, *Africa* . . . . **54 C3** 2 0S 30 0 E
Ryan, L., *U.K.* **12 G3** 55 0N 5 2W
Ryazan, *Russia* . . . . **24 D6** 54 40N 39 40 E
Ryazhsk, *Russia* . . . . **24 D7** 53 45N 40 3 E
Rybache = Rybachye,
Kazakstan . . . . **26 E9** 46 40N 81 20 E
Rybachiy Poluostrov,
Russia . . . . **24 A5** 69 43N 32 0 E
Rybachye = Ysyk-Köl,
Kyrgyzstan . . . . **28 E11** 42 26N 76 12 E
Rybachye, *Kazakstan* . . . . **26 E9** 46 40N 81 20 E
Rybinsk, *Russia* . . . . **24 C6** 58 5N 38 50 E
Rybinskoye Vdkhr., *Russia* **24 C6** 58 30N 38 25 E
Rybnitsa = Rîbniţa,
Moldova . . . . **17 E15** 47 45N 29 0 E
Ryde, *U.K.* **11 G6** 50 43N 1 9W
Ryderwood, *U.S.A.* **84 D3** 46 23N 123 3W
Rye, *U.K.* . . . . **11 G8** 50 57N 0 45 E
Rye ➝, *U.K.* **10 C7** 54 11N 0 44W
Rye Patch Reservoir,
U.S.A. **82 F4** 40 28N 118 19W
Ryegate, *U.S.A.* **82 C9** 46 18N 109 15W
Rylstone, *Australia* . . . . . . **63 E4** 32 46S 149 58 E
Ryōthu, *Japan* . . . . **30 E9** 38 5N 138 26 E
Rypin, *Poland* . . . . **17 B10** 53 3N 19 25 E
Ryūgasaki, *Japan* . . . . **31 G10** 35 54N 140 11 E
Ryūkyū Is. = Ryūkyū-rettō,
Japan . . . . **31 M2** 26 0N 126 0 E
Ryūkyū-rettō, *Japan* . . . . **31 M2** 26 0N 126 0 E
Rzeszów, *Poland* . . . . **17 C11** 50 5N 21 58 E
Rzhev, *Russia* . . . . **24 C5** 56 20N 34 20 E

S

Sa, *Thailand* . . . . **38 C3** 18 34N 100 45 E
Sa Dec, *Vietnam* . . . . **39 G5** 10 20N 105 46 E
Sa'ādatābād, *Fārs, Iran* . . **45 D7** 30 10N 53 5 E
Sa'ādatābād, *Kermān, Iran* **45 D7** 28 3N 55 53 E
Saale ➝, *Germany* . . . . . **16 C6** 51 56N 11 54 E
Saalfeld, *Germany* . . . . . . **16 C6** 50 38N 11 21 E
Saar ➝, *Europe* . . . . **16 D4** 49 41N 6 32 E

Saarbrücken, *Germany* **16 D4** 49 14N 6 59 E
Saaremaa, *Estonia* **9 G20** 58 30N 22 30 E
Saarijärvi, *Finland* **9 E21** 62 43N 25 16 E
Saariselkä, *Finland* **8 B23** 68 16N 28 15 E
Saarland □, *Germany* **15 E7** 49 20N 7 0 E
Sab 'Âbar, *Syria* **44 C3** 33 46N 37 41 E
Saba, *W. Indies* **89 C7** 17 42N 63 26W
Šabac, *Serbia, Yug.* **21 B8** 44 48N 19 42 E
Sabadell, *Spain* **19 B7** 41 28N 2 7 E
Sabah □, *Malaysia* **36 C5** 6 0N 117 0 E
Sabak Bernam, *Malaysia* **39 L3** 3 46N 100 58 E
Sábana de la Mar, *Dom. Rep.* **89 C6** 19 7N 69 24W
Sábanalarga, *Colombia* **92 A4** 10 38N 74 55W
Sabang, *Indonesia* **36 C1** 5 50N 95 15 E
Sabará, *Brazil* **93 G10** 19 55S 43 46W
Sabattis, *U.S.A.* **79 B10** 44 6N 74 40W
Saberania, *Indonesia* **37 E9** 2 5S 138 18 E
Sabhah, *Libya* **51 C7** 27 9N 14 29 E
Sabie, *S. Africa* **57 D5** 25 10S 30 48 E
Sabinal, *Mexico* **86 A3** 30 58N 107 25W
Sabinal, *U.S.A.* **81 L5** 29 19N 99 28W
Sabinas, *Mexico* **86 B4** 27 50N 101 10W
Sabinas →, *Mexico* **86 B4** 27 37N 100 42W
Sabinas Hidalgo, *Mexico* **86 B4** 26 33N 100 10W
Sabine →, *U.S.A.* **81 L8** 29 59N 93 47W
Sabine L., *U.S.A.* **81 L8** 29 53N 93 51W
Sabine Pass, *U.S.A.* **81 L8** 29 44N 93 54W
Sabkhet el Bardawîl, *Egypt* **47 D2** 31 10N 33 15 E
Sablayan, *Phil.* **37 B6** 12 50N 120 50 E
Sable, C., *Canada* **71 D6** 43 29N 65 38W
Sable, C., *U.S.A.* **75 E10** 25 9N 81 8W
Sable I., *Canada* **71 D8** 44 0N 60 0W
Sabrina Coast, *Antarctica* **5 C9** 68 0S 120 0 E
Sabulubek, *Indonesia* **36 E1** 1 36S 98 40 E
Sabzevār, *Iran* **45 B8** 36 15N 57 40 E
Sabzvārān, *Iran* **45 D8** 28 45N 57 50 E
Sac City, *U.S.A.* **80 D7** 42 25N 95 0W
Săcele, *Romania* **17 F13** 45 37N 25 41 E
Sachigo →, *Canada* **70 A2** 55 6N 88 58W
Sachigo, L., *Canada* **70 B1** 53 50N 92 12W
Sachsen □, *Germany* **16 C7** 50 55N 13 10 E
Sachsen-Anhalt □, *Germany* **16 C7** 52 0N 12 0 E
Sackets Harbor, *U.S.A.* **79 C8** 43 57N 76 7W
Saco, *Maine, U.S.A.* **77 D10** 43 30N 70 27W
Saco, *Mont., U.S.A.* **82 B10** 48 28N 107 21W
Sacramento, *U.S.A.* **84 G5** 38 35N 121 29W
Sacramento →, *U.S.A.* **84 G5** 38 3N 121 56W
Sacramento Mts., *U.S.A.* **83 K11** 32 30N 105 30W
Sacramento Valley, *U.S.A.* **84 G5** 39 30N 122 0W
Sadani, *Tanzania* **54 D4** 5 58S 38 35 E
Sadao, *Thailand* **39 J3** 6 38N 100 26 E
Sadd el Aali, *Egypt* **51 D11** 23 54N 32 54 E
Saddle Mt., *U.S.A.* **84 E3** 45 58N 123 41W
Sadimi, *Zaïre* **55 D1** 9 25S 23 32 E
Sado, *Japan* **30 E9** 38 0N 138 25 E
Sadon, *Burma* **41 G20** 25 28N 97 55 E
Sæby, *Denmark* **9 H14** 57 21N 10 30 E
Saegerstown, *U.S.A.* **78 E4** 41 43N 80 9W
Şafājah, *Si. Arabia* **44 E3** 26 25N 39 0 E
Säffle, *Sweden* **9 G15** 59 8N 12 55 E
Safford, *U.S.A.* **83 K9** 32 50N 109 43W
Saffron Walden, *U.K.* **11 E8** 52 1N 0 16 E
Safi, *Morocco* **50 B3** 32 18N 9 20W
Şafīābād, *Iran* **45 B8** 36 45N 57 58 E
Safīd Dasht, *Iran* **45 C6** 33 27N 48 11 E
Safīd Kūh, *Afghan.* **40 B3** 34 45N 63 0 E
Safwān, *Iraq* **44 D5** 30 7N 47 43 E
Sag Harbor, *U.S.A.* **79 F12** 41 0N 72 18W
Saga, *Indonesia* **37 E8** 2 40S 132 55 E
Saga, *Japan* **31 H5** 33 15N 130 16 E
Saga □, *Japan* **31 H5** 33 15N 130 20 E
Sagae, *Japan* **30 E10** 38 22N 140 17 E
Sagala, *Mali* **50 F3** 14 9N 6 38W
Sagar, *India* **40 M9** 14 14N 75 6 E
Sagara, L., *Tanzania* **54 D3** 5 20S 31 0 E
Saginaw, *U.S.A.* **76 D4** 43 26N 83 56W
Saginaw B., *U.S.A.* **76 D4** 43 50N 83 40W
Şağīr, Zāb aş →, *Iraq* **44 C4** 35 17N 43 29 E
Saglouc = Salluit, *Canada* **69 B12** 62 14N 75 38W
Sagŏ-ri, *S. Korea* **35 G14** 35 25N 126 49 E
Sagua la Grande, *Cuba* **88 B3** 22 50N 80 10W
Saguache, *U.S.A.* **83 G10** 38 5N 106 8W
Saguenay →, *Canada* **71 C5** 48 22N 71 0W
Sagunto, *Spain* **19 C5** 39 42N 0 18W
Sahagún, *Spain* **19 A3** 42 18N 5 2W
Saḩam al Jawlān, *Syria* **47 C4** 32 45N 35 55 E
Sahand, Kūh-e, *Iran* **44 B5** 37 44N 46 27 E
Sahara, *Africa* **50 D5** 23 0N 5 0 E
Saharan Atlas = Saharien, Atlas, *Algeria* **50 B5** 33 30N 1 0 E
Saharanpur, *India* **42 E7** 29 58N 77 33 E
Saharien, Atlas, *Algeria* **50 B5** 33 30N 1 0 E
Sahasinaka, *Madag.* **57 C8** 21 49S 47 49 E
Sahaswan, *India* **43 E8** 28 5N 78 45 E
Sahibganj, *India* **43 G12** 25 12N 87 40 E
Sāḩilīyah, *Iraq* **44 C4** 33 43N 42 42 E
Sahiwal, *Pakistan* **42 D5** 30 45N 73 8 E
Şaḩneh, *Iran* **44 C5** 34 29N 47 41 E
Sahtaneh →, *Canada* **72 B4** 59 2N 122 28W
Sahuaripa, *Mexico* **86 B3** 29 0N 109 13W
Sahuarita, *U.S.A.* **83 L8** 31 57N 110 58W
Sahuayo, *Mexico* **86 C4** 20 4N 102 43W
Sai Buri, *Thailand* **39 J3** 6 43N 101 45 E
Sa'id Bundas, *Sudan* **51 G9** 8 24N 24 48 E
Saïda, *Algeria* **50 B5** 34 50N 0 11 E
Sa'īdābād, *Kermān, Iran* **45 D7** 29 30N 55 45 E
Sa'īdābād, *Semnān, Iran* **45 B7** 36 8N 54 11 E
Sa'īdīyeh, *Iran* **45 B6** 36 20N 48 55 E
Saidpur, *Bangla.* **41 G16** 25 48N 89 0 E
Saidu, *Pakistan* **43 B5** 34 43N 72 24 E
Saigon = Phanh Bho Ho Chi Minh, *Vietnam* **39 G6** 10 58N 106 40 E
Saijō, *Japan* **31 H6** 33 55N 133 11 E
Saikhoa Ghat, *India* **41 F19** 27 50N 95 40 E
Saiki, *Japan* **31 H5** 32 58N 131 51 E
Sailolof, *Indonesia* **37 E8** 1 7S 130 46 E
Saimaa, *Finland* **9 F23** 61 15N 28 15 E
Şa'īn Dezh, *Iran* **44 B5** 36 40N 46 25 E
St. Abb's Head, *U.K.* **12 F6** 55 55N 2 8W
St. Alban's, *Canada* **71 C8** 47 51N 55 50W
St. Albans, *U.K.* **11 F7** 51 45N 0 19W
St. Albans, *Vt., U.S.A.* **79 B11** 44 49N 73 5W
St. Albans, *W. Va., U.S.A.* **76 F5** 38 23N 81 50W
St. Alban's Head, *U.K.* **11 G5** 50 34N 2 4W
St. Albert, *Canada* **72 C6** 53 37N 113 32W
St. Andrew's, *Canada* **71 C8** 47 45N 59 15W
St. Andrews, *U.K.* **12 E6** 56 20N 2 47W
St-Anicet, *Canada* **79 A10** 45 8N 74 22W
St. Ann B., *Canada* **71 C7** 46 22N 60 25W
St. Ann's Bay, *Jamaica* **88 C4** 18 26N 77 15W
St. Anthony, *Canada* **71 B8** 51 22N 55 35W
St. Anthony, *U.S.A.* **82 E8** 43 58N 111 41W
St. Arnaud, *Australia* **63 F3** 36 40S 143 16 E
St. Arthur, *Canada* **71 C6** 47 33N 67 46W
St. Asaph, *U.K.* **10 D4** 53 15N 3 27W
St-Augustin-Saguenay, *Canada* **71 B8** 51 13N 58 38W
St. Augustine, *U.S.A.* **77 L5** 29 54N 81 19W
St. Austell, *U.K.* **11 G3** 50 20N 4 47W
St.-Barthélemy, I., *W. Indies* **89 C7** 17 50N 62 50W
St. Bees Hd., *U.K.* **10 C4** 54 31N 3 38W
St. Boniface, *Canada* **73 D9** 49 53N 97 5W
St. Bride's, *Canada* **71 C9** 46 56N 54 10W
St. Brides B., *U.K.* **11 F2** 51 49N 5 9W
St.-Brieuc, *France* **18 B2** 48 30N 2 46W
St. Catharines, *Canada* **70 D4** 43 10N 79 15W
St. Catherines I., *U.S.A.* **77 K5** 31 40N 81 10W
St. Catherine's Pt., *U.K.* **11 G6** 50 34N 1 18W
St.-Chamond, *France* **18 D6** 45 28N 4 31 E
St. Charles, *Ill., U.S.A.* **76 E1** 41 54N 88 19W
St. Charles, *Mo., U.S.A.* **80 F9** 38 47N 90 29W
St. Christopher = St. Kitts, *W. Indies* **89 C7** 17 20N 62 40W
St. Christopher-Nevis ■ = St. Kitts & Nevis ■, *W. Indies* **89 C7** 17 20N 62 40W
St. Clair, *Mich., U.S.A.* **78 D2** 42 50N 82 30W
St. Clair, *Pa., U.S.A.* **79 F8** 40 43N 76 12W
St. Clair, L., *Canada* **70 D3** 42 30N 82 45W
St. Clairsville, *U.S.A.* **78 F4** 40 5N 80 54W
St. Claude, *Canada* **73 D9** 49 40N 98 20W
St. Cloud, *Fla., U.S.A.* **77 L5** 28 15N 81 17W
St. Cloud, *Minn., U.S.A.* **80 C7** 45 34N 94 10W
St-Coeur de Marie, *Canada* **71 C5** 48 39N 71 43W
St. Cricq, C., *Australia* **61 E1** 25 17S 113 6 E
St. Croix, *Virgin Is.* **89 C7** 17 45N 64 45W
St. Croix →, *U.S.A.* **80 C8** 44 45N 92 48W
St. Croix Falls, *U.S.A.* **80 C8** 45 24N 92 38W
St. David's, *Canada* **71 C8** 48 12N 58 52W
St. David's, *U.K.* **11 F2** 51 53N 5 16W
St. David's Head, *U.K.* **11 F2** 51 54N 5 19W
St.-Denis, *France* **18 B5** 48 56N 2 22 E
St.-Dizier, *France* **18 B6** 48 38N 4 56 E
St. Elias, Mt., *U.S.A.* **68 B5** 60 18N 140 56W
St. Elias Mts., *Canada* **72 A1** 60 33N 139 28W
St.-Étienne, *France* **18 D6** 45 27N 4 22 E
St. Eugène, *Canada* **79 A10** 45 30N 74 28W
St. Eustatius, *W. Indies* **89 C7** 17 20N 63 0W
St-Félicien, *Canada* **70 C5** 48 40N 72 25W
St.-Flour, *France* **18 D5** 45 2N 3 6 E
St. Francis, *U.S.A.* **80 F4** 39 47N 101 48W
St. Francis →, *U.S.A.* **81 H9** 34 38N 90 36W
St. Francis, C., *S. Africa* **56 E3** 34 14S 24 49 E
St. Francisville, *U.S.A.* **81 K9** 30 47N 91 23W
St-François, L., *Canada* **79 A10** 45 10N 74 22W
St-Gabriel-de-Brandon, *Canada* **70 C5** 46 17N 73 24W
St. Gallen = Sankt Gallen, *Switz.* **16 E5** 47 26N 9 22 E
St.-Gaudens, *France* **18 E4** 43 6N 0 44 E
St. George, *Australia* **63 D4** 28 1S 148 30 E
St. George, *Canada* **71 C6** 45 11N 66 50W
St. George, *S.C., U.S.A.* **77 J5** 33 11N 80 35W
St. George, *Utah, U.S.A.* **83 H7** 37 6N 113 35W
St. George, C., *Canada* **71 C8** 48 30N 59 16W
St. George, C., *U.S.A.* **77 L3** 29 40N 85 5W
St. George Ra., *Australia* **60 C4** 18 40S 125 0 E
St-Georges, *Belgium* **15 D5** 50 37N 5 20 E
St. George's, *Canada* **71 C8** 48 26N 58 31W
St-Georges, *Canada* **71 C5** 46 8N 70 40W
St.-Georges, *Fr. Guiana* **93 C8** 4 0N 52 0W
St. George's, *Grenada* **89 D7** 12 5N 61 43W
St. George's B., *Canada* **71 C8** 48 24N 58 53W
St. Georges Basin, *Australia* **60 C4** 15 23S 125 2 E
St. George's Channel, *Europe* **13 E6** 52 0N 6 0W
St. Georges Hd., *Australia* **63 F5** 35 12S 150 42 E
St. Gotthard P. = San Gottardo, P. del, *Switz.* **16 E5** 46 33N 8 33 E
St. Helena, *U.S.A.* **82 G2** 38 30N 122 28W
St. Helena ■, *Atl. Oc.* **49 H3** 15 55S 5 44W
St. Helena, Mt., *U.S.A.* **84 G4** 38 40N 122 36W
St. Helena B., *S. Africa* **56 E2** 32 40S 18 10 E
St. Helens, *Australia* **62 G4** 41 20S 148 15 E
St. Helens, *U.K.* **10 D5** 53 27N 2 44W
St. Helens, *U.S.A.* **84 E4** 45 52N 122 48W
St. Helens, Mt., *U.S.A.* **84 D4** 46 12N 122 12W
St. Helier, *U.K.* **11 H5** 49 10N 2 7W
St-Hubert, *Belgium* **15 D5** 50 2N 5 23 E
St-Hyacinthe, *Canada* **70 C5** 45 40N 72 58W
St. Ignace, *U.S.A.* **76 C3** 45 52N 84 44W
St. Ignace I., *Canada* **70 C2** 48 45N 88 0W
St. Ignatius, *U.S.A.* **82 C6** 47 19N 114 6W
St. Ives, *Cambs., U.K.* **11 E7** 52 20N 0 4W
St. Ives, *Corn., U.K.* **11 G2** 50 12N 5 30W
St. James, *U.S.A.* **80 D7** 43 59N 94 38W
St-Jean, *Canada* **70 C5** 45 20N 73 20W
St-Jean →, *Canada* **71 B7** 50 17N 64 20W
St-Jean, L., *Canada* **71 C5** 48 40N 72 0W
St. Jean Baptiste, *Canada* **73 D9** 49 15N 97 20W
St-Jean-Port-Joli, *Canada* **71 C5** 47 15N 70 13W
St-Jérôme, *Qué., Canada* **70 C5** 45 47N 74 0W
St-Jérôme, *Qué., Canada* **71 C5** 48 26N 71 53W
St. John, *Canada* **71 C6** 45 20N 66 8W
St. John, *Kans., U.S.A.* **81 G5** 38 0N 98 46W
St. John, *N. Dak., U.S.A.* **80 A5** 48 57N 99 43W
St. John →, *U.S.A.* **71 C6** 45 12N 66 5W
St. John, C., *Canada* **71 B8** 50 0N 55 32W
St. John's, *Antigua* **89 C7** 17 6N 61 51W
St. John's, *Canada* **71 C9** 47 35N 52 40W
St. Johns, *Ariz., U.S.A.* **83 J9** 34 30N 109 22W
St. Johns, *Mich., U.S.A.* **76 D3** 43 0N 84 33W
St. Johns →, *U.S.A.* **77 K5** 30 24N 81 24W
St. Johnsbury, *U.S.A.* **79 B12** 44 25N 72 1W
St. Johnsville, *U.S.A.* **79 C10** 43 0N 74 43W
St. Joseph, *La., U.S.A.* **81 K9** 31 55N 91 14W
St. Joseph, *Mich., U.S.A.* **76 D2** 42 6N 86 29W
St. Joseph, *Mo., U.S.A.* **80 F7** 39 46N 94 50W
St. Joseph →, *U.S.A.* **76 D2** 42 7N 86 29W
St. Joseph, I., *Canada* **70 C3** 46 12N 83 58W
St. Joseph, L., *Canada* **70 B1** 51 10N 90 35W
St-Jovite, *Canada* **70 C5** 46 8N 74 38W
St. Kilda, *N.Z.* **59 L3** 45 53S 170 31 E
St. Kitts, *W. Indies* **89 C7** 17 20N 62 40W
St. Kitts & Nevis ■, *W. Indies* **89 C7** 17 20N 62 40W
St. Laurent, *Canada* **73 C9** 50 25N 97 58W
St.-Laurent, *Fr. Guiana* **93 B8** 5 29N 54 3W
St. Lawrence, *Australia* **62 C4** 22 16S 149 31 E
St. Lawrence, *Canada* **71 C8** 46 54N 55 23W
St. Lawrence →, *Canada* **71 C6** 49 30N 66 0W
St. Lawrence, Gulf of, *Canada* **71 C7** 48 25N 62 0W
St. Lawrence I., *U.S.A.* **68 B3** 63 30N 170 30W
St. Leonard, *Canada* **71 C6** 47 12N 67 58W
St. Lewis →, *Canada* **71 B8** 52 26N 56 11W
St-Lô, *France* **18 B3** 49 7N 1 5W
St-Louis, *Senegal* **50 E1** 16 8N 16 27W
St. Louis, *Mich., U.S.A.* **76 D3** 43 25N 84 36W
St. Louis, *Mo., U.S.A.* **80 F9** 38 37N 90 12W
St. Louis →, *U.S.A.* **80 B8** 47 15N 92 45W
St. Lucia ■, *W. Indies* **89 D7** 14 0N 60 50W
St. Lucia, L., *S. Africa* **57 D5** 28 5S 32 30 E
St. Lucia Channel, *W. Indies* **89 D7** 14 15N 61 0W
St. Lunaire-Griquet, *Canada* **71 B8** 51 31N 55 28W
St. Maarten, *W. Indies* **89 C7** 18 0N 63 5W
St.-Malo, *France* **18 B2** 48 39N 2 1W
St-Marc, *Haiti* **89 C5** 19 10N 72 41W
St. Maries, *U.S.A.* **82 C5** 47 19N 116 35W
St-Martin, *W. Indies* **89 C7** 18 0N 63 0W
St. Martin, L., *Canada* **73 C9** 51 40N 98 30W
St. Martins, *Canada* **71 C6** 45 22N 65 34W
St. Martinville, *U.S.A.* **81 K9** 30 7N 91 50W
St. Mary Pk., *Australia* **63 E2** 31 32S 138 34 E
St. Marys, *Australia* **62 G4** 41 35S 148 11 E
St. Marys, *Canada* **78 C3** 43 20N 81 10W
St. Mary's, *U.K.* **11 H1** 49 55N 6 18W
St. Marys, *U.S.A.* **78 E6** 41 26N 78 34W
St. Mary's, C., *Canada* **71 C9** 46 50N 54 12W
St. Mary's B., *Canada* **71 C9** 46 50N 53 50W
St. Marys Bay, *Canada* **71 D6** 44 25N 66 10W
St.-Mathieu, Pte., *France* **18 B1** 48 20N 4 45W
St. Matthews, I. = Zadetkyi Kyun, *Burma* **39 H2** 10 0N 98 25 E
St-Maurice →, *Canada* **70 C5** 46 21N 72 31W
St. Michael's Mount, *U.K.* **11 G2** 50 7N 5 29W
St.-Nazaire, *France* **18 C2** 47 17N 2 12W
St. Neots, *U.K.* **11 E7** 52 14N 0 15W
St. Niklass = Sint Niklaas, *Belgium* **15 C4** 51 10N 4 9 E
St.-Omer, *France* **18 A5** 50 45N 2 15 E
St-Pacome, *Canada* **71 C6** 47 24N 69 58W
St-Pamphile, *Canada* **71 C6** 46 58N 69 48W
St. Pascal, *Canada* **71 C6** 47 32N 69 48W
St. Paul, *Canada* **72 C6** 54 0N 111 17W
St. Paul, *Minn., U.S.A.* **80 C8** 44 57N 93 6W
St. Paul, *Nebr., U.S.A.* **80 E5** 41 13N 98 27W
St. Paul, I., *Ind. Oc.* **3 F13** 38 55S 77 34 E
St. Paul I., *Canada* **71 C7** 47 12N 60 9W
St. Peter, *U.S.A.* **80 C8** 44 20N 93 57W
St. Peter Port, *U.K.* **11 H5** 49 26N 2 33W
St. Peters, *N.S., Canada* **71 C7** 45 40N 60 53W
St. Peters, *P.E.I., Canada* **71 C7** 46 25N 62 35W
St. Petersburg = Sankt-Peterburg, *Russia* **24 C5** 59 55N 30 20 E
St. Petersburg, *U.S.A.* **77 M4** 27 46N 82 39W
St.-Pierre, *St- P. & M.* **71 C8** 46 46N 56 12W
St-Pierre, L., *Canada* **70 C5** 46 12N 72 52W
St.-Pierre et Miquelon □, *St- P. & M.* **71 C8** 46 55N 56 10W
St.-Quentin, *France* **18 B5** 49 50N 3 16 E
St. Regis, *U.S.A.* **82 C6** 47 18N 115 6W
St. Sebastien, Tanjon' i, *Madag.* **57 A8** 12 26S 48 44 E
St-Siméon, *Canada* **71 C6** 47 51N 69 54W
St. Stephen, *Canada* **71 C6** 45 16N 67 17W
St. Thomas, *Canada* **70 D3** 42 45N 81 10W
St. Thomas I., *Virgin Is.* **89 C7** 18 20N 64 55W
St-Tite, *Canada* **70 C5** 46 45N 72 34W
St.-Tropez, *France* **18 E7** 43 17N 6 38 E
St. Troud = Sint Truiden, *Belgium* **15 D5** 50 48N 5 10 E
St. Vincent, *W. Indies* **89 D7** 13 10N 61 10W
St. Vincent, G., *Australia* **63 F2** 35 0S 138 0 E
St. Vincent & the Grenadines ■, *W. Indies* **89 D7** 13 0N 61 10W
St. Vincent Passage, *W. Indies* **89 D7** 13 30N 61 0W
St-Vith, *Belgium* **15 D6** 50 17N 6 9 E
Ste-Agathe-des-Monts, *Canada* **70 C5** 46 3N 74 17W
Ste-Anne de Beaupré, *Canada* **71 C5** 47 2N 70 58W
Ste-Anne-des-Monts, *Canada* **71 C6** 49 8N 66 30W
Ste. Genevieve, *U.S.A.* **80 G9** 37 59N 90 2W
Ste-Marguerite →, *Canada* **71 B6** 50 9N 66 36W
Ste.-Marie, *Martinique* **89 D7** 14 48N 61 1W
Ste-Marie de la Madeleine, *Canada* **71 C5** 46 26N 71 0W
Ste.-Rose, *Guadeloupe* **89 C7** 16 20N 61 45W
Ste. Rose du Lac, *Canada* **73 C9** 51 4N 99 30W
Saintes, *France* **18 D3** 45 45N 0 37W
Saintes, I. des, *Guadeloupe* **89 C7** 15 50N 61 35W
Saintonge, *France* **18 D3** 45 40N 0 50W
Saipan, *Pac. Oc.* **64 F6** 15 12N 145 45 E
Sairang, *India* **41 H18** 23 50N 92 45 E
Sairecábur, Cerro, *Bolivia* **94 A2** 22 43S 67 54W
Saitama □, *Japan* **31 F9** 36 25N 139 30 E
Sajama, *Bolivia* **92 G5** 18 7S 69 0W
Sajó, *Hungary* **17 D11** 48 12N 20 44 E
Sajum, *India* **43 C8** 33 20N 79 0 E
Sak →, *S. Africa* **56 E3** 30 52S 20 25 E
Sakai, *Japan* **31 G7** 34 30N 135 30 E
Sakaide, *Japan* **31 G6** 34 15N 133 50 E
Sakaiminato, *Japan* **31 G6** 35 38N 133 11 E
Sakākah, *Si. Arabia* **44 D4** 30 0N 40 8 E
Sakakawea, L., *U.S.A.* **80 B3** 47 30N 101 25W
Sakami, L., *Canada* **70 B4** 53 15N 77 0W
Sakania, *Zaïre* **55 E2** 12 43S 28 30 E
Sakarya = Adapazarı, *Turkey* **25 F5** 40 48N 30 25 E
Sakarya →, *Turkey* **25 F5** 41 7N 30 39 E
Sakashima-Guntō, *Japan* **31 M2** 24 46N 124 0 E
Sakata, *Japan* **30 E9** 38 55N 139 50 E
Sakchu, *N. Korea* **35 D13** 40 23N 125 2 E
Sakeny →, *Madag.* **57 C8** 20 0S 45 25 E
Sakha □, *Russia* **27 C13** 62 0N 130 0 E
Sakhalin, *Russia* **27 D15** 51 0N 143 0 E
Sakhalinskiy Zaliv, *Russia* **27 D15** 54 0N 141 0 E
Šakiai, *Lithuania* **9 J20** 54 59N 23 0 E
Sakon Nakhon, *Thailand* **38 D5** 17 10N 104 9 E
Sakrand, *Pakistan* **42 F3** 26 10N 68 15 E
Sakrivier, *S. Africa* **56 E3** 30 54S 20 28 E
Sakuma, *Japan* **31 G8** 35 3N 137 49 E
Sakurai, *Japan* **31 G7** 34 30N 135 51 E
Sala, *Sweden* **9 G17** 59 58N 16 35 E
Sala Consilina, *Italy* **20 D6** 40 23N 15 36 E
Sala-y-Gómez, *Pac. Oc.* **65 K17** 26 28S 105 28W
Salaberry-de-Valleyfield, *Canada* **70 C5** 45 15N 74 8W
Saladas, *Argentina* **94 B4** 28 15S 58 40W
Saladillo, *Argentina* **94 D4** 35 40S 59 55W
Salado →, *Buenos Aires, Argentina* **94 D4** 35 44S 57 22W
Salado →, *La Pampa, Argentina* **96 D3** 37 30S 67 0W
Salado →, *Santa Fe, Argentina* **94 C3** 31 40S 60 41W
Salado →, *Mexico* **86 B5** 26 52N 99 19W
Salaga, *Ghana* **50 G4** 8 31N 0 31W
Sālah, *Syria* **47 C5** 32 40N 36 45 E
Sálakhos, *Greece* **23 C9** 36 17N 27 57 E
Salālah, *Oman* **46 D5** 16 56N 53 59 E
Salamanca, *Chile* **94 C1** 31 46S 70 59W
Salamanca, *Spain* **19 B3** 40 58N 5 39W
Salamanca, *U.S.A.* **78 D6** 42 10N 78 43W
Salāmatābād, *Iran* **44 C5** 35 39N 47 50 E
Salamis, *Cyprus* **23 D12** 35 11N 33 54 E
Salamís, *Greece* **21 F10** 37 56N 23 30 E
Salar de Atacama, *Chile* **94 A2** 23 30S 68 25W
Salar de Uyuni, *Bolivia* **92 H5** 20 30S 67 45W
Salatiga, *Indonesia* **37 G14** 7 19S 110 30 E
Salavat, *Russia* **24 D10** 53 21N 55 55 E
Salaverry, *Peru* **92 E3** 8 15S 79 0W
Salawati, *Indonesia* **37 E8** 1 7S 130 52 E
Salayar, *Indonesia* **37 F6** 6 7S 120 30 E
Salcombe, *U.K.* **11 G4** 50 14N 3 47W
Saldanha, *S. Africa* **56 E2** 33 0S 17 58 E
Saldanha B., *S. Africa* **56 E2** 33 6S 18 0 E
Saldus, *Latvia* **9 H20** 56 38N 22 30 E
Sale, *Australia* **63 F4** 38 6S 147 6 E
Salé, *Morocco* **50 B3** 34 3N 6 48W
Sale, *U.K.* **10 D5** 53 26N 2 19W
Salekhard, *Russia* **24 A12** 66 30N 66 35 E
Salem, *India* **40 P11** 11 40N 78 11 E
Salem, *Ind., U.S.A.* **76 F2** 38 36N 86 6W
Salem, *Mass., U.S.A.* **79 D14** 42 31N 70 53W
Salem, *Mo., U.S.A.* **81 G9** 37 39N 91 32W
Salem, *N.J., U.S.A.* **76 F8** 39 34N 75 28W
Salem, *Ohio, U.S.A.* **78 F4** 40 54N 80 52W
Salem, *Oreg., U.S.A.* **82 D2** 44 56N 123 2W
Salem, *S. Dak., U.S.A.* **80 D6** 43 44N 97 23W
Salem, *Va., U.S.A.* **76 G5** 37 18N 80 3W
Salerno, *Italy* **20 D6** 40 41N 14 47 E
Salford, *U.K.* **10 D5** 53 30N 2 18W
Salgótarján, *Hungary* **17 D10** 48 5N 19 47 E
Salida, *U.S.A.* **74 C5** 38 32N 106 0W
Salihli, *Turkey* **21 E13** 38 28N 28 8 E
Salihorsk, *Belarus* **17 B14** 52 51N 27 27 E
Salima, *Malawi* **53 G6** 13 47S 34 28 E
Salina, *Italy* **20 E6** 38 34N 14 50 E
Salina, *U.S.A.* **80 F6** 38 50N 97 37W
Salina Cruz, *Mexico* **87 D5** 16 10N 95 10W
Salinas, *Brazil* **93 G10** 16 10S 42 10W
Salinas, *Chile* **94 A2** 23 31S 69 29W
Salinas, *Ecuador* **92 D2** 2 10S 80 58W
Salinas, *U.S.A.* **83 H3** 36 40N 121 39W
Salinas →, *Guatemala* **87 D6** 16 28N 90 31W
Salinas →, *U.S.A.* **83 H3** 36 45N 121 48W
Salinas, B. de, *Nic.* **88 D2** 11 4N 85 45W
Salinas, C. de, *Spain* **22 B10** 39 16N 3 4 E
Salinas, Pampa de las, *Argentina* **94 C2** 31 58S 66 42W
Salinas Ambargasta, *Argentina* **94 B3** 29 0S 65 0W
Salinas de Hidalgo, *Mexico* **86 C4** 22 30N 101 40W
Salinas Grandes, *Argentina* **94 B2** 30 0S 65 0W
Saline →, *Ark., U.S.A.* **81 J8** 33 10N 92 8W
Saline →, *Kans., U.S.A.* **80 F6** 38 52N 97 30W
Salines, *Spain* **22 B10** 39 21N 3 3 E
Salinópolis, *Brazil* **93 D9** 0 40S 47 20W
Salisbury = Harare, *Zimbabwe* **55 F3** 17 43S 31 2 E
Salisbury, *Australia* **63 E2** 34 46S 138 40 E
Salisbury, *U.K.* **11 F6** 51 4N 1 47W
Salisbury, *Md., U.S.A.* **76 F8** 38 22N 75 36W
Salisbury, *N.C., U.S.A.* **77 H5** 35 40N 80 29W
Salisbury Plain, *U.K.* **11 F6** 51 14N 1 55W
Şalkhad, *Syria* **47 C5** 32 29N 36 43 E
Salla, *Finland* **8 C23** 66 50N 28 49 E
Sallisaw, *U.S.A.* **81 H7** 35 28N 94 47W
Salluit, *Canada* **69 B12** 62 14N 75 38W
Salmās, *Iran* **44 B5** 38 11N 44 47 E
Salmo, *Canada* **72 D5** 49 10N 117 20W
Salmon, *U.S.A.* **82 D7** 45 11N 113 54W
Salmon →, *Canada* **72 C4** 54 3N 122 40W
Salmon →, *U.S.A.* **82 D5** 45 51N 116 47W
Salmon Arm, *Canada* **72 C5** 50 40N 119 15W
Salmon Falls, *U.S.A.* **82 E6** 42 48N 114 59W
Salmon Gums, *Australia* **61 F3** 32 59S 121 38 E

Salmon Res., *Canada* **71 C8** 48 5N 56 0W
Salmon River Mts., *U.S.A.* **82 D6** 45 0N 114 30W
Salo, *Finland* **9 F20** 60 22N 23 10 E
Salome, *U.S.A.* **85 M13** 33 47N 113 37W
Salon-de-Provence, *France* **18 E6** 43 39N 5 6 E
Salonica = Thessaloníki, *Greece* **21 D10** 40 38N 22 58 E
Salonta, *Romania* **17 E11** 46 49N 21 42 E
Salpausselkä, *Finland* **9 F22** 61 0N 27 0 E
Salsacate, *Argentina* **94 C2** 31 20S 65 5W
Salsk, *Russia* **25 E7** 46 28N 41 30 E
Salso →, *Italy* **20 F5** 37 6N 13 57 E
Salt →, *Canada* **72 B6** 60 0N 112 25W
Salt →, *U.S.A.* **83 K7** 33 23N 112 19W
Salt Creek, *Australia* **63 F2** 36 8S 139 38 E
Salt Fork Arkansas →, *U.S.A.* **81 G6** 36 36N 97 3W
Salt Lake City, *U.S.A.* **82 F8** 40 45N 111 53W
Salt Range, *Pakistan* **42 C5** 32 30N 72 25 E
Salta, *Argentina* **94 A2** 24 57S 65 25W
Salta □, *Argentina* **94 A2** 24 48S 65 30W
Saltcoats, *U.K.* **12 F4** 55 38N 4 47W
Saltee Is., *Ireland* **13 D5** 52 7N 6 37W
Saltfjellet, *Norway* **8 C16** 66 40N 15 15 E
Saltfjorden, *Norway* **8 C16** 67 15N 14 10 E
Saltillo, *Mexico* **86 B4** 25 25N 101 0W
Salto, *Argentina* **94 C3** 34 20S 60 15W
Salto, *Uruguay* **94 C4** 31 27S 57 50W
Salto →, *Italy* **20 C5** 42 26N 12 25 E
Salton City, *U.S.A.* **85 M11** 33 29N 115 51W
Salton Sea, *U.S.A.* **85 M11** 33 15N 115 45W
Saltpond, *Ghana* **50 G4** 5 15N 1 3W
Saltville, *U.S.A.* **76 G5** 36 53N 81 46W
Saluda →, *U.S.A.* **77 H5** 34 1N 81 4W
Salûm, *Egypt* **51 B10** 31 31N 25 7 E
Salûm, Khâlig el, *Egypt* **51 B10** 31 35N 25 24 E
Salur, *India* **41 K13** 18 27N 83 18 E
Salvador, *Brazil* **93 F11** 13 0S 38 30W
Salvador, *Canada* **73 C7** 52 10N 109 32W
Salvador, L., *U.S.A.* **81 L9** 29 43N 90 15W
Salween →, *Burma* **41 L20** 16 31N 97 37 E
Salyan, *Azerbaijan* **25 G8** 39 33N 48 59 E
Salyersville, *U.S.A.* **76 G4** 37 45N 83 4W
Salzach →, *Austria* **16 D7** 48 12N 12 56 E
Salzburg, *Austria* **16 E7** 47 48N 13 2 E
Salzgitter, *Germany* **16 B6** 52 9N 10 19 E
Salzwedel, *Germany* **16 B6** 52 52N 11 10 E
Sam Neua, *Laos* **38 B5** 20 29N 104 5 E
Sam Ngao, *Thailand* **38 D2** 17 18N 99 0 E
Sam Rayburn Reservoir, *U.S.A.* **81 K7** 31 4N 94 5W
Sam Son, *Vietnam* **38 C5** 19 44N 105 54 E
Sam Teu, *Laos* **38 C5** 19 59N 104 38 E
Sama, *Russia* **26 C7** 60 12N 60 22 E
Sama de Langreo, *Spain* **19 A3** 43 18N 5 40W
Samagaltay, *Russia* **27 D10** 50 36N 95 3 E
Samales Group, *Phil.* **37 C6** 6 0N 122 0 E
Samana, *India* **42 D7** 30 10N 76 13 E
Samana Cay, *Bahamas* **89 B5** 23 3N 73 45W
Samanga, *Tanzania* **55 D4** 8 20S 39 13 E
Samangwa, *Zaïre* **54 C1** 4 23S 24 10 E
Samani, *Japan* **30 C11** 42 7N 142 56 E
Samar, *Phil.* **37 B7** 12 0N 125 0 E
Samara, *Russia* **24 D9** 53 8N 50 6 E
Samaria = Shōmrōn, *West Bank* **47 C4** 32 15N 35 13 E
Samariá, *Greece* **23 D5** 35 17N 23 58 E
Samarinda, *Indonesia* **36 E5** 0 30S 117 9 E
Samarkand = Samarqand, *Uzbekistan* **26 F7** 39 40N 66 55 E
Samarqand, *Uzbekistan* **26 F7** 39 40N 66 55 E
Sāmarrā, *Iraq* **44 C4** 34 12N 43 52 E
Samastipur, *India* **43 G11** 25 50N 85 50 E
Samba, *India* **43 C6** 32 32N 75 10 E
Samba, *Zaïre* **54 C2** 4 38S 26 22 E
Sambalpur, *India* **41 J14** 21 28N 84 4 E
Sambar, Tanjung, *Indonesia* **36 E4** 2 59S 110 19 E
Sambas, *Indonesia* **36 D3** 1 20N 109 20 E
Sambava, *Madag.* **57 A9** 14 16S 50 10 E
Sambawizi, *Zimbabwe* **55 F2** 18 24S 26 13 E
Sambhal, *India* **43 E8** 28 35N 78 37 E
Sambhar, *India* **42 F6** 26 52N 75 6 E
Sambiase, *Italy* **20 E7** 38 58N 16 17 E
Sambir, *Ukraine* **17 D12** 49 30N 23 10 E
Sambor, *Cambodia* **38 F6** 12 46N 106 0 E
Sambre →, *Europe* **15 D4** 50 27N 4 52 E
Samburu □, *Kenya* **54 B4** 1 10N 37 0 E
Samchŏk, *S. Korea* **35 F15** 37 30N 129 10 E
Samchonpo, *S. Korea* **35 G15** 35 0N 128 6 E
Same, *Tanzania* **54 C4** 4 2S 37 38 E
Samfya, *Zambia* **55 E2** 11 22S 29 31 E
Samnah, *Si. Arabia* **44 E3** 25 10N 37 15 E
Samo Alto, *Chile* **94 C1** 30 22S 71 0W
Samokov, *Bulgaria* **21 C10** 42 18N 23 35 E
Samoorombón, B., *Argentina* **94 D4** 36 5S 57 20W
Sámos, *Greece* **21 F12** 37 45N 26 50 E
Samothráki, *Évros, Greece* **21 D11** 40 28N 25 28 E
Samothráki, *Kérkira, Greece* **23 A3** 39 48N 19 31 E
Sampacho, *Argentina* **94 C3** 33 20S 64 50W
Sampang, *Indonesia* **37 G15** 7 11S 113 13 E
Sampit, *Indonesia* **36 E4** 2 34S 113 0 E
Sampit, Teluk, *Indonesia* **36 E4** 3 5S 113 3 E
Samrong, *Cambodia* **38 E4** 14 15N 103 30 E
Samrong, *Thailand* **38 E3** 15 10N 100 40 E
Samsø, *Denmark* **9 J14** 55 50N 10 35 E
Samsun, *Turkey* **25 F6** 41 15N 36 22 E
Samui, Ko, *Thailand* **39 H3** 9 30N 100 0 E
Samusole, *Zaïre* **55 E1** 10 2S 24 0 E
Samut Prakan, *Thailand* **38 F3** 13 32N 100 40 E
Samut Sakhon, *Thailand* **38 F3** 13 31N 100 13 E
Samut Songkhram →, *Thailand* **38 F3** 13 24N 100 1 E
Samwari, *Pakistan* **42 E2** 28 30N 66 46 E
San, *Mali* **50 F4** 13 15N 4 57W
San →, *Cambodia* **38 F5** 13 32N 105 57 E
San →, *Poland* **17 C11** 50 45N 21 51 E
San Agustin, C., *Phil.* **37 C7** 6 20N 126 13 E
San Agustín de Valle Fértil, *Argentina* **94 C2** 30 35S 67 30W
San Ambrosio, *Pac. Oc.* **90 F3** 26 28S 79 53W
San Andreas, *U.S.A.* **84 G6** 38 12N 120 41W
San Andrés, I. de, *Caribbean* **88 D3** 12 42N 81 46W
San Andres Mts., *U.S.A.* **83 K10** 33 0N 106 30W
San Andrés Tuxtla, *Mexico* **87 D5** 18 30N 95 20W
San Angelo, *U.S.A.* **81 K4** 31 28N 100 26W
San Anselmo, *U.S.A.* **84 H4** 37 59N 122 34W
San Antonio, *Belize* **87 D7** 16 15N 89 2W
San Antonio, *Chile* **94 C1** 33 40S 71 40W
San Antonio, *Spain* **22 C7** 38 59N 1 19 E
San Antonio, *N. Mex., U.S.A.* **83 K10** 33 55N 106 52W
San Antonio, *Tex., U.S.A.* **81 L5** 29 25N 98 30W
San Antonio →, *U.S.A.* **81 L6** 28 30N 96 54W
San Antonio, C., *Argentina* **94 D4** 36 15S 56 40W
San Antonio, C., *Cuba* **88 B3** 21 50N 84 57W
San Antonio, Mt., *U.S.A.* **85 L9** 34 17N 117 38W
San Antonio de los Baños, *Cuba* **88 B3** 22 54N 82 31W
San Antonio de los Cobres, *Argentina* **94 A2** 24 10S 66 17W
San Antonio Oeste, *Argentina* **96 E4** 40 40S 65 0W
San Ardo, *U.S.A.* **84 J6** 36 1N 120 54W
San Augustín, *Canary Is.* **22 G4** 27 47N 15 32W
San Augustine, *U.S.A.* **81 K7** 31 30N 94 7W
San Bartolomé, *Canary Is.* **22 F6** 28 59N 13 37W
San Bartolomé de Tirajana, *Canary Is.* **22 G4** 27 54N 15 34W
San Benedetto del Tronto, *Italy* **20 C5** 42 57N 13 53 E
San Benedicto, I., *Mexico* **86 D2** 19 18N 110 49W
San Benito, *U.S.A.* **81 M6** 26 8N 97 38W
San Benito →, *U.S.A.* **84 J5** 36 53N 121 34W
San Benito Mt., *U.S.A.* **84 J6** 36 22N 120 37W
San Bernardino, *U.S.A.* **85 L9** 34 7N 117 19W
San Bernardino Mts., *U.S.A.* **85 L10** 34 10N 116 45W
San Bernardino Str., *Phil.* **37 B6** 13 0N 125 0 E
San Bernardo, *Chile* **94 C1** 33 40S 70 50W
San Bernardo, I. de, *Colombia* **92 B3** 9 45N 75 50W
San Blas, *Mexico* **86 B3** 26 4N 108 46W
San Blas, Arch. de, *Panama* **88 E4** 9 50N 78 31W
San Blas, C., *U.S.A.* **77 L3** 29 40N 85 21W
San Borja, *Bolivia* **92 F5** 14 50S 66 52W
San Buenaventura, *Mexico* **86 B4** 27 5N 101 32W
San Carlos, *Argentina* **94 C2** 33 50S 69 0W
San Carlos, *Chile* **94 D1** 36 10S 72 0W
San Carlos, *Mexico* **86 B4** 29 0N 100 54W
San Carlos, *Nic.* **88 D3** 11 12N 84 50W
San Carlos, *Phil.* **37 B6** 10 29N 123 25 E
San Carlos, *Spain* **22 B8** 39 3N 1 34 E
San Carlos, *Uruguay* **95 C5** 34 46S 54 58W
San Carlos, *U.S.A.* **83 K8** 33 21N 110 27W
San Carlos, *Amazonas, Venezuela* **92 C5** 1 55N 67 4W
San Carlos, *Cojedes, Venezuela* **92 B5** 9 40N 68 36W
San Carlos de Bariloche, *Argentina* **96 E2** 41 10S 71 25W
San Carlos del Zulia, *Venezuela* **92 B4** 9 1N 71 55W
San Carlos L., *U.S.A.* **83 K8** 33 11N 110 32W
San Clemente, *Chile* **94 D1** 35 30S 71 29W
San Clemente, *U.S.A.* **85 M9** 33 26N 117 37W
San Clemente I., *U.S.A.* **85 N8** 32 53N 118 29W
San Cristóbal, *Argentina* **94 C3** 30 20S 61 10W
San Cristóbal, *Dom. Rep.* **89 C5** 18 25N 70 6W
San Cristóbal, *Mexico* **87 D6** 16 50N 92 33W
San Cristóbal, *Spain* **22 B11** 39 57N 4 3 E
San Cristóbal, *Venezuela* **92 B4** 7 46N 72 14W
San Diego, *Calif., U.S.A.* **85 N9** 32 43N 117 9W
San Diego, *Tex., U.S.A.* **81 M5** 27 46N 98 14W
San Diego, C., *Argentina* **96 G3** 54 40S 65 10W
San Diego de la Unión, *Mexico* **86 C4** 21 28N 100 52W
San Dimitri, Ras, *Malta* **23 C1** 36 4N 14 11 E
San Estanislao, *Paraguay* **94 A4** 24 39S 56 26W
San Felipe, *Chile* **94 C1** 32 43S 70 42W
San Felipe, *Mexico* **86 A2** 31 0N 114 52W
San Felipe, *Venezuela* **92 A5** 10 20N 68 44W
San Felipe →, *U.S.A.* **85 M11** 33 12N 115 49W
San Feliu de Guíxols, *Spain* **19 B7** 41 45N 3 1 E
San Félix, *Pac. Oc.* **90 F2** 26 23S 80 0W
San Fernando, *Chile* **94 C1** 34 30S 71 0W
San Fernando, *Mexico* **86 B1** 29 55N 115 10W
San Fernando, *La Union, Phil.* **37 A6** 16 40N 120 23 E
San Fernando, *Pampanga, Phil.* **37 A6** 15 5N 120 37 E
San Fernando, *Baleares, Spain* **22 C7** 38 42N 1 28 E
San Fernando, *Cádiz, Spain* **19 D2** 36 28N 6 17W
San Fernando, *Trin. & Tob.* **89 D7** 10 20N 61 30W
San Fernando, *U.S.A.* **85 L8** 34 17N 118 26W
San Fernando →, *Mexico* **86 C5** 24 55N 98 10W
San Fernando de Apure, *Venezuela* **92 B5** 7 54N 67 15W
San Fernando de Atabapo, *Venezuela* **92 C5** 4 3N 67 42W
San Francisco, *Argentina* **94 C3** 31 30S 62 5W
San Francisco, *U.S.A.* **83 H2** 37 47N 122 25W
San Francisco →, *U.S.A.* **83 K9** 32 59N 109 22W
San Francisco, Paso de, *S. Amer.* **94 B2** 27 0S 68 0W
San Francisco de Macorís, *Dom. Rep.* **89 C5** 19 19N 70 15W
San Francisco del Monte de Oro, *Argentina* **94 C2** 32 36S 66 8W
San Francisco del Oro, *Mexico* **86 B3** 26 52N 105 50W
San Francisco Javier, *Spain* **22 C7** 38 42N 1 26 E
San Francisco Solano, Pta., *Colombia* **90 C3** 6 18N 77 29W
San Gil, *Colombia* **92 B4** 6 33N 73 8W
San Gorgonio Mt., *U.S.A.* **85 L10** 34 7N 116 51W
San Gottardo, P. del, *Switz.* **16 E5** 46 33N 8 33 E
San Gregorio, *Uruguay* **95 C4** 32 37S 55 40W
San Gregorio, *U.S.A.* **84 H4** 37 20N 122 23W
San Ignacio, *Belize* **87 D7** 17 10N 89 0W
San Ignacio, *Bolivia* **92 G6** 16 20S 60 55W
San Ignacio, *Mexico* **86 B2** 27 27N 113 0W
San Ignacio, *Paraguay* **94 B4** 26 52S 57 3W
San Ignacio, L., *Mexico* **86 B2** 26 50N 113 11W
San Ildefonso, C., *Phil.* **37 A6** 16 0N 122 1 E
San Isidro, *Argentina* **94 C4** 34 29S 58 31W
San Jacinto, *U.S.A.* **85 M10** 33 47N 116 57W
San Jaime, *Spain* **22 B11** 39 54N 4 4 E
San Javier, *Misiones, Argentina* **95 B4** 27 55S 55 5W
San Javier, *Santa Fe, Argentina* **94 C4** 30 40S 59 55W
San Javier, *Bolivia* **92 G6** 16 18S 62 30W
San Javier, *Chile* **94 D1** 35 40S 71 45W
San Jeronimo Taviche, *Mexico* **87 D5** 16 38N 96 32W
San Joaquin, *U.S.A.* **84 J6** 36 36N 120 11W
San Joaquin →, *U.S.A.* **83 G3** 38 4N 121 51W
San Joaquin Valley, *U.S.A.* **84 J6** 37 20N 121 0W
San Jordi, *Spain* **22 B9** 39 33N 2 46 E
San Jorge, *Argentina* **94 C3** 31 54S 61 50W
San Jorge, *Spain* **22 C7** 38 54N 1 24 E
San Jorge, B. de, *Mexico* **86 A2** 31 20N 113 20W
San Jorge, G., *Argentina* **96 F3** 46 0S 66 0W
San Jorge, G. de, *Spain* **19 B6** 40 53N 1 2 E
San Jorge, G. of, *Argentina* **90 H4** 46 0S 66 0W
San José, *Bolivia* **92 G6** 17 53S 60 50W
San José, *Costa Rica* **88 E3** 9 55N 84 2W
San José, *Guatemala* **88 D1** 14 0N 90 50W
San José, *Mexico* **86 C2** 25 0N 110 50W
San Jose, *Phil.* **37 A6** 15 45N 120 55 E
San José, *Spain* **22 C7** 38 55N 1 18 E
San Jose, *U.S.A.* **83 H3** 37 20N 121 53W
San Jose →, *U.S.A.* **83 J10** 34 25N 106 45W
San Jose de Buenovista, *Phil.* **37 B6** 12 27N 121 4 E
San José de Feliciano, *Argentina* **94 C4** 30 26S 58 46W
San José de Jáchal, *Argentina* **94 C2** 30 15S 68 46W
San José de Mayo, *Uruguay* **94 C4** 34 27S 56 40W
San José de Ocune, *Colombia* **92 C4** 4 15N 70 20W
San José del Cabo, *Mexico* **86 C3** 23 0N 109 40W
San José del Guaviare, *Colombia* **92 C4** 2 35N 72 38W
San Juan, *Argentina* **94 C2** 31 30S 68 30W
San Juan, *Mexico* **86 C4** 21 20N 102 50W
San Juan, *Phil.* **37 C7** 8 25N 126 20 E
San Juan, *Puerto Rico* **89 C6** 18 28N 66 7W
San Juan □, *Argentina* **94 C2** 31 9S 69 0W
San Juan →, *Argentina* **94 C2** 32 20S 67 25W
San Juan →, *Nic.* **88 D3** 10 56N 83 42W
San Juan →, *U.S.A.* **83 H8** 37 16N 110 26W
San Juan, C., *Eq. Guin.* **52 D1** 1 5N 9 20 E
San Juan Bautista, *Paraguay* **94 B4** 26 37S 57 6W
San Juan Bautista, *Spain* **22 B8** 39 5N 1 31 E
San Juan Bautista, *U.S.A.* **83 H3** 36 51N 121 32W
San Juan Bautista Valle Nacional, *Mexico* **87 D5** 17 47N 96 19W
San Juan Capistrano, *U.S.A.* **85 M9** 33 30N 117 40W
San Juan Cr. →, *U.S.A.* **84 J5** 35 40N 120 22W
San Juan de Guadalupe, *Mexico* **86 C4** 24 38N 102 44W
San Juan de los Morros, *Venezuela* **92 B5** 9 55N 67 21W
San Juan del Norte, *Nic.* **88 D3** 10 58N 83 40W
San Juan del Norte, B. de, *Nic.* **88 D3** 11 0N 83 40W
San Juan del Rio, *Mexico* **87 C5** 20 25N 100 0W
San Juan del Sur, *Nic.* **88 D2** 11 20N 85 51W
San Juan I., *U.S.A.* **84 B3** 48 32N 123 5W
San Juan Mts., *U.S.A.* **83 H10** 37 30N 107 0W
San Julián, *Argentina* **96 F3** 49 15S 67 45W
San Justo, *Argentina* **94 C3** 30 47S 60 30W
San Kamphaeng, *Thailand* **38 C2** 18 45N 99 8 E
San Lázaro, C., *Mexico* **86 C2** 24 50N 112 18W
San Lázaro, Sa., *Mexico* **86 C3** 23 25N 110 0W
San Leandro, *U.S.A.* **83 H2** 37 44N 122 9W
San Lorenzo, *Argentina* **94 C3** 32 45S 60 45W
San Lorenzo, *Ecuador* **92 C3** 1 15N 78 50W
San Lorenzo, *Paraguay* **94 B4** 25 20S 57 32W
San Lorenzo, *Spain* **22 B10** 39 37N 3 17 E
San Lorenzo →, *Mexico* **86 C3** 24 15N 107 24W
San Lorenzo, I., *Mexico* **86 B2** 28 35N 112 50W
San Lorenzo, I., *Peru* **92 F3** 12 7S 77 15W
San Lorenzo, Mt., *Argentina* **96 F2** 47 40S 72 20W
San Lucas, *Bolivia* **92 H5** 20 5S 65 7W
San Lucas, *Baja Calif. S., Mexico* **86 C3** 22 53N 109 54W
San Lucas, *Baja Calif. S., Mexico* **86 B2** 27 10N 112 14W
San Lucas, *U.S.A.* **84 J5** 36 8N 121 1W
San Lucas, C., *Mexico* **86 C3** 22 50N 110 0W
San Luis, *Argentina* **94 C2** 33 20S 66 20W
San Luis, *Cuba* **88 B3** 22 17N 83 46W
San Luis, *Guatemala* **88 C2** 16 14N 89 27W
San Luis, *U.S.A.* **83 H11** 37 12N 105 25W
San Luis □, *Argentina* **94 C2** 34 0S 66 0W
San Luis, I., *Mexico* **86 B2** 29 58N 114 26W
San Luis, Sierra de, *Argentina* **94 C2** 32 30S 66 10W
San Luis de la Paz, *Mexico* **86 C4** 21 19N 100 32W
San Luis Obispo, *U.S.A.* **85 K6** 35 17N 120 40W
San Luis Potosí, *Mexico* **86 C4** 22 9N 100 59W
San Luis Potosí □, *Mexico* **86 C4** 22 10N 101 0W
San Luis Reservoir, *U.S.A.* **84 H5** 37 4N 121 5W
San Luis Río Colorado, *Mexico* **86 A2** 32 29N 114 58W
San Marcos, *Guatemala* **88 D1** 14 59N 91 52W
San Marcos, *Mexico* **86 B2** 27 13N 112 6W
San Marcos, *U.S.A.* **81 L6** 29 53N 97 56W
San Marino, *San Marino* **16 G7** 43 55N 12 30 E
San Marino ■, *Europe* **20 C5** 43 56N 12 25 E
San Martín, *Argentina* **94 C2** 33 5S 68 28W
San Martín, L., *Argentina* **96 F2** 48 50S 72 50W
San Mateo, *Spain* **22 B7** 39 3N 1 23 E
San Mateo, *U.S.A.* **83 H2** 37 34N 122 19W
San Matías, *Bolivia* **92 G7** 16 25S 58 20W
San Matías, G., *Argentina* **96 E4** 41 30S 64 0W
San Miguel, *El Salv.* **88 D2** 13 30N 88 12W
San Miguel, *Panama* **88 E4** 8 27N 78 55W
San Miguel, *Spain* **22 B7** 39 3N 1 26 E
San Miguel, *U.S.A.* **83 J3** 35 45N 120 42W
San Miguel →, *Bolivia* **92 F6** 13 52S 63 56W
San Miguel de Tucumán, *Argentina* **94 B2** 26 50S 65 20W
San Miguel del Monte, *Argentina* **94 D4** 35 23S 58 50W
San Miguel I., *U.S.A.* **85 L6** 34 2N 120 23W
San Narciso, *Phil.* **37 A6** 15 2N 120 3 E
San Nicolás, *Canary Is.* **22 G4** 27 58N 15 47W
San Nicolás de los Arroyas, *Argentina* **94 C3** 33 25S 60 10W
San Nicolas I., *U.S.A.* **85 M7** 33 15N 119 30W
San Onofre, *U.S.A.* **85 M9** 33 22N 117 34W
San Pablo, *Bolivia* **94 A2** 21 43S 66 38W
San Pedro, *Buenos Aires, Argentina* **95 B5** 26 30S 54 10W
San Pedro, *Jujuy, Argentina* **94 A3** 24 12S 64 55W
San-Pédro, *Ivory C.* **50 H3** 4 50N 6 33W
San Pedro, *Mexico* **86 C2** 23 55N 110 17W
San Pedro □, *Paraguay* **94 A4** 24 0S 57 0W
San Pedro →, *Chihuahua, Mexico* **86 B3** 28 20N 106 10W
San Pedro →, *Michoacan, Mexico* **86 D4** 19 23N 103 51W
San Pedro →, *Nayarit, Mexico* **86 C3** 21 45N 105 30W
San Pedro →, *U.S.A.* **83 K8** 32 59N 110 47W
San Pedro, Pta., *Chile* **94 B1** 25 30S 70 38W
San Pedro Channel, *U.S.A.* **85 M8** 33 30N 118 25W
San Pedro de Atacama, *Chile* **94 A2** 22 55S 68 15W
San Pedro de Jujuy, *Argentina* **94 A3** 24 12S 64 55W
San Pedro de las Colonias, *Mexico* **86 B4** 25 50N 102 59W
San Pedro de Lloc, *Peru* **92 E3** 7 15S 79 28W
San Pedro de Macorís, *Dom. Rep.* **89 C6** 18 30N 69 18W
San Pedro del Norte, *Nic.* **88 D3** 13 4N 84 33W
San Pedro del Paraná, *Paraguay* **94 B4** 26 43S 56 13W
San Pedro Mártir, Sierra, *Mexico* **86 A1** 31 0N 115 30W
San Pedro Mixtepec, *Mexico* **87 D5** 16 2N 97 7W
San Pedro Ocampo = Melchor Ocampo, *Mexico* **86 C4** 24 52N 101 40W
San Pedro Sula, *Honduras* **88 C2** 15 30N 88 0W
San Pieto, *Italy* **20 E3** 39 8N 8 17 E
San Quintín, *Mexico* **86 A1** 30 29N 115 57W
San Rafael, *Argentina* **94 C2** 34 40S 68 21W
San Rafael, *Calif., U.S.A.* **84 H4** 37 58N 122 32W
San Rafael, *N. Mex., U.S.A.* **83 J10** 35 7N 107 53W
San Rafael Mt., *U.S.A.* **85 L7** 34 41N 119 52W
San Rafael Mts., *U.S.A.* **85 L7** 34 40N 119 50W
San Ramón de la Nueva Orán, *Argentina* **94 A3** 23 10S 64 20W
San Remo, *Italy* **20 C2** 43 49N 7 46 E
San Roque, *Argentina* **94 B4** 28 25S 58 45W
San Roque, *Spain* **19 D3** 36 17N 5 21W
San Rosendo, *Chile* **94 D1** 37 16S 72 43W
San Saba, *U.S.A.* **81 K5** 31 12N 98 43W
San Salvador, *Bahamas* **89 B5** 24 0N 74 40W
San Salvador, *El Salv.* **88 D2** 13 40N 89 10W
San Salvador, *Spain* **22 B10** 39 27N 3 11 E
San Salvador de Jujuy, *Argentina* **94 A3** 24 10S 64 48W
San Salvador I., *Bahamas* **89 B5** 24 0N 74 32W
San Sebastián, *Argentina* **96 G3** 53 10S 68 30W
San Sebastián, *Spain* **19 A5** 43 17N 1 58W
San Sebastian de la Gomera, *Canary Is.* **22 F2** 28 5N 17 7W
San Serra, *Spain* **22 B10** 39 43N 3 13 E
San Severo, *Italy* **20 D6** 41 41N 15 23 E
San Simeon, *U.S.A.* **84 K5** 35 39N 121 11W
San Simon, *U.S.A.* **83 K9** 32 16N 109 14W
San Telmo, *Mexico* **86 A1** 30 58N 116 6W
San Telmo, *Spain* **22 B9** 39 35N 2 21 E
San Tiburcio, *Mexico* **86 C4** 24 8N 101 32W
San Valentin, Mte., *Chile* **96 F2** 46 30S 73 30W
San Vicente de la Barquera, *Spain* **19 A3** 43 23N 4 29W
San Ygnacio, *U.S.A.* **81 M5** 27 3N 99 26W
Sana', *Yemen* **46 D3** 15 27N 44 12 E
Sana →, *Bos.-H.* **16 F9** 45 3N 16 23 E
Sanaga →, *Cameroon* **50 H6** 3 35N 9 38 E
Sanaloa, Presa, *Mexico* **86 C3** 24 50N 107 20W
Sanana, *Indonesia* **37 E7** 2 4S 125 58 E
Sanand, *India* **42 H5** 22 59N 72 25 E
Sanandaj, *Iran* **44 C5** 35 18N 47 1 E
Sanandita, *Bolivia* **94 A3** 21 40S 63 45W
Sanawad, *India* **42 H7** 22 11N 76 5 E
Sancellas, *Spain* **22 B9** 39 39N 2 54 E
Sanchahe, *China* **35 B14** 44 50N 126 2 E
Sánchez, *Dom. Rep.* **89 C6** 19 15N 69 36W
Sanchor, *India* **42 G4** 24 45N 71 55 E
Sanco Pt., *Phil.* **37 C7** 8 15N 126 27 E
Sancti-Spíritus, *Cuba* **88 B4** 21 52N 79 33W
Sancy, Puy de, *France* **18 D5** 45 32N 2 50 E
Sand →, *S. Africa* **57 C5** 22 25S 30 5 E
Sand Springs, *U.S.A.* **81 G6** 36 9N 96 7W
Sanda, *Japan* **31 G7** 34 53N 135 14 E
Sandakan, *Malaysia* **36 C5** 5 53N 118 4 E
Sandan = Sambor, *Cambodia* **38 F6** 12 46N 106 0 E
Sandanski, *Bulgaria* **21 D10** 41 35N 23 16 E
Sanday, *U.K.* **12 B6** 59 16N 2 31W
Sandefjord, *Norway* **9 G14** 59 10N 10 15 E
Sanders, *U.S.A.* **83 J9** 35 13N 109 20W

Sanderson, U.S.A. 81 K3 30 9N 102 24W
Sandfly L., Canada 73 B7 55 43N 106 6W
Sandgate, Australia 63 D5 27 18S 153 3 E
Sandia, Peru 92 F5 14 10S 69 30W
Sandnes, Norway 9 G11 58 50N 5 45 E
Sandness, U.K. 12 A7 60 18N 1 40W
Sandnessjøen, Norway 8 C15 66 2N 12 38 E
Sandoa, Zaïre 52 F4 9 41S 23 0 E
Sandomierz, Poland 17 C11 50 40N 21 43 E
Sandover →, Australia 62 C2 21 43S 136 32 E
Sandoway, Burma 41 K19 18 20N 94 30 E
Sandoy, Færoe Is. 8 F9 61 52N 6 46W
Sandpoint, U.S.A. 82 B5 48 17N 116 33W
Sandringham, U.K. 10 E8 52 51N 0 31 E
Sandspit, Canada 72 C2 53 14N 131 49W
Sandstone, Australia 61 E2 27 59S 119 16 E
Sandusky, Mich., U.S.A. 78 C2 43 25N 82 50W
Sandusky, Ohio, U.S.A. 78 E2 41 27N 82 42W
Sandviken, Sweden 9 F17 60 38N 16 46 E
Sandwich, C., Australia 62 B4 18 14S 146 18 E
Sandwich B., Canada 71 B8 53 40N 57 15W
Sandwich B., Namibia 56 C1 23 25S 14 20 E
Sandwip Chan., Bangla. 41 H17 22 35N 91 35 E
Sandy, Nev., U.S.A. 85 K11 35 49N 115 36W
Sandy, Oreg., U.S.A. 84 E4 45 24N 122 16W
Sandy, Utah, U.S.A. 82 F8 40 35N 111 50W
Sandy Bight, Australia 61 F3 33 50S 123 20 E
Sandy C., Queens., Australia 62 C5 24 42S 153 15 E
Sandy C., Tas., Australia 62 G3 41 25S 144 45 E
Sandy Cay, Bahamas 89 B4 23 13N 75 18W
Sandy Cr. →, U.S.A. 82 F9 41 51N 109 47W
Sandy L., Canada 70 B1 53 2N 93 0W
Sandy Lake, Canada 70 B1 53 0N 93 15W
Sandy Narrows, Canada 73 B8 55 5N 103 4W
Sanford, Fla., U.S.A. 77 L5 28 48N 81 16W
Sanford, Maine, U.S.A. 79 C14 43 27N 70 47W
Sanford, N.C., U.S.A. 77 H6 35 29N 79 10W
Sanford →, Australia 61 E2 27 22S 115 53 E
Sanford, Mt., U.S.A. 68 B5 62 13N 144 8W
Sang-i-Masha, Afghan. 42 C2 33 8N 67 27 E
Sanga, Mozam. 55 E4 12 22S 35 21 E
Sanga →, Congo 52 E3 1 5S 17 0 E
Sanga-Tolon, Russia 27 C15 61 50N 149 40 E
Sangamner, India 40 K9 19 37N 74 15 E
Sangar, Afghan. 42 C1 32 56N 65 30 E
Sangar, Russia 27 C13 64 2N 127 31 E
Sangar Sarai, Afghan. 42 B4 34 27N 70 35 E
Sangasangadalam, Indonesia 36 E5 0 36S 117 13 E
Sange, Zaïre 54 D2 6 58S 28 21 E
Sangeang, Indonesia 37 F5 8 12S 119 6 E
Sanger, U.S.A. 83 H4 36 42N 119 33W
Sangerhausen, Germany 16 C6 51 28N 11 18 E
Sanggan He →, China 34 E9 38 12N 117 15 E
Sanggau, Indonesia 36 D4 0 5N 110 30 E
Sangihe, Kepulauan, Indonesia 37 D7 3 0N 126 0 E
Sangihe, P., Indonesia 37 D7 3 45N 125 30 E
Sangju, S. Korea 35 F15 36 25N 128 10 E
Sangkapura, Indonesia 36 F4 5 52S 112 40 E
Sangkhla, Thailand 38 E2 14 57N 98 28 E
Sangli, India 40 L9 16 55N 74 33 E
Sangmélina, Cameroon 52 D2 2 57N 12 1 E
Sangre de Cristo Mts., U.S.A. 81 G2 37 0N 105 0W
Sangudo, Canada 72 C6 53 50N 114 54W
Sanje, Uganda 54 C3 0 49S 31 30 E
Sanjo, Japan 30 F9 37 37N 138 57 E
Sankt Gallen, Switz. 16 E5 47 26N 9 22 E
Sankt Moritz, Switz. 16 E5 46 30N 9 50 E
Sankt-Peterburg, Russia 24 C5 59 55N 30 20 E
Sankt Pölten, Austria 16 D8 48 12N 15 38 E
Sankuru →, Zaïre 52 E4 4 17S 20 25 E
Sanliurfa, Turkey 25 G6 37 12N 38 50 E
Sanlúcar de Barrameda, Spain 19 D2 36 46N 6 21W
Sanmenxia, China 34 G6 34 47N 111 12 E
Sanming, China 33 D6 26 15N 117 40 E
Sannaspos, S. Africa 56 D4 29 6S 26 34 E
Sannicandro Gargánico, Italy 20 D6 41 50N 15 34 E
Sannieshof, S. Africa 56 D4 26 30S 25 47 E
Sannīn, J., Lebanon 47 B4 33 57N 35 52 E
Sanok, Poland 17 D12 49 35N 22 10 E
Sanquhar, U.K. 12 F5 55 22N 3 54W
Santa Ana, Bolivia 92 F5 13 50S 65 40W
Santa Ana, Ecuador 92 D2 1 16S 80 20W
Santa Ana, El Salv. 88 D2 14 0N 89 31W
Santa Ana, Mexico 86 A2 30 31N 111 8W
Santa Ana, U.S.A. 85 M9 33 46N 117 52W
Santa Barbara, Honduras 88 D2 14 53N 88 14W
Santa Bárbara, Mexico 86 B3 26 48N 105 50W
Santa Barbara, U.S.A. 85 L7 34 25N 119 42W
Santa Barbara Channel, U.S.A. 85 L7 34 15N 120 0W
Santa Barbara I., U.S.A. 85 M7 33 29N 119 2W
Santa Catalina, Mexico 86 B2 25 40N 110 50W
Santa Catalina, Gulf of, U.S.A. 85 N9 33 10N 117 50W
Santa Catalina I., U.S.A. 85 M8 33 23N 118 25W
Santa Catarina □, Brazil 95 B6 27 25S 48 30W
Santa Catarina, I. de, Brazil 95 B6 27 30S 48 40W
Santa Cecília, Brazil 95 B5 26 56S 50 18W
Santa Clara, Cuba 88 B4 22 20N 80 0W
Santa Clara, Calif., U.S.A. 83 H3 37 21N 121 57W
Santa Clara, Utah, U.S.A. 83 H7 37 8N 113 39W
Santa Clara de Olimar, Uruguay 95 C5 32 50S 54 54W
Santa Clotilde, Peru 92 D4 2 33S 73 45W
Santa Coloma de Gramanet, Spain 19 B7 41 27N 2 13 E
Santa Cruz, Argentina 96 G3 50 0S 68 32W
Santa Cruz, Bolivia 92 G6 17 43S 63 10W
Santa Cruz, Chile 94 C1 34 38S 71 27W
Santa Cruz, Costa Rica 88 D2 10 15N 85 35W
Santa Cruz, Madeira 22 D3 32 42N 16 46W
Santa Cruz, Phil. 37 B6 14 20N 121 24 E
Santa Cruz, U.S.A. 83 H2 36 58N 122 1W
Santa Cruz →, Argentina 96 G3 50 10S 68 20W
Santa Cruz de la Palma, Canary Is. 22 F2 28 41N 17 46W
Santa Cruz de Tenerife, Canary Is. 22 F3 28 28N 16 15W
Santa Cruz del Norte, Cuba 88 B3 23 9N 81 55W
Santa Cruz del Sur, Cuba 88 B4 20 44N 78 0W
Santa Cruz do Rio Pardo, Brazil 95 A6 22 54S 49 37W
Santa Cruz do Sul, Brazil 95 B5 29 42S 52 25W
Santa Cruz I., Solomon Is. 64 J8 10 30S 166 0 E
Santa Cruz I., U.S.A. 85 M7 34 1N 119 43W
Santa Domingo, Cay, Bahamas 88 B4 21 25N 75 15W
Santa Elena, Argentina 94 C4 30 58S 59 47W
Santa Elena, Ecuador 92 D2 2 16S 80 52W
Santa Elena, C., Costa Rica 88 D2 10 54N 85 56W
Santa Eugenia, Pta., Mexico 86 B1 27 50N 115 5W
Santa Eulalia, Spain 22 C8 38 59N 1 32 E
Santa Fe, Argentina 94 C3 31 35S 60 41W
Santa Fe, U.S.A. 83 J11 35 41N 105 57W
Santa Fé □, Argentina 94 C3 31 50S 60 55W
Santa Filomena, Brazil 93 E9 9 6S 45 50W
Santa Galdana, Spain 22 B10 39 56N 3 58 E
Santa Gertrudis, Spain 22 B7 39 0N 1 26 E
Santa Inés, Spain 22 B7 39 3N 1 21 E
Santa Inés, I., Chile 96 G2 54 0S 73 0W
Santa Isabel = Rey Malabo, Eq. Guin. 50 H6 3 45N 8 50 E
Santa Isabel, Argentina 94 D2 36 10S 66 54W
Santa Isabel, Brazil 93 F8 11 45S 51 30W
Santa Lucía, Corrientes, Argentina 94 B4 28 58S 59 5W
Santa Lucía, San Juan, Argentina 94 C2 31 30S 68 30W
Santa Lucía, Uruguay 94 C4 34 27S 56 24W
Santa Lucia Range, U.S.A. 83 J3 36 0N 121 20W
Santa Magdalena, I., Mexico 86 C2 24 40N 112 15W
Santa Margarita, Argentina 94 D3 38 28S 61 35W
Santa Margarita, Mexico 86 C2 24 30N 111 50W
Santa Margarita, Spain 22 B10 39 42N 3 6 E
Santa Margarita, U.S.A. 84 K6 35 23N 120 37W
Santa Margarita →, U.S.A. 85 M9 33 13N 117 23W
Santa María, Argentina 94 B2 26 40S 66 0W
Santa Maria, Brazil 95 B5 29 40S 53 48W
Santa María, Spain 22 B9 39 38N 2 47 E
Santa Maria, U.S.A. 85 L6 34 57N 120 26W
Santa María →, Mexico 86 A3 31 0N 107 14W
Santa María, B. de, Mexico 86 B3 25 10N 108 40W
Santa Maria da Vitória, Brazil 93 F10 13 24S 44 12W
Santa Maria di Leuca, C., Italy 21 E8 39 47N 18 22 E
Santa Marta, Colombia 92 A4 11 15N 74 13W
Santa Marta, Sierra Nevada de, Colombia 92 A4 10 55N 73 50W
Santa Marta Grande, C., Brazil 95 B6 28 43S 48 50W
Santa Maura = Levkás, Greece 21 E9 38 40N 20 43 E
Santa Monica, U.S.A. 85 M8 34 1N 118 29W
Santa Ponsa, Spain 22 B9 39 30N 2 28 E
Santa Rita, U.S.A. 83 K10 32 48N 108 4W
Santa Rosa, La Pampa, Argentina 94 D3 36 40S 64 17W
Santa Rosa, San Luis, Argentina 94 C2 32 21S 65 10W
Santa Rosa, Bolivia 92 F5 10 36S 67 20W
Santa Rosa, Brazil 95 B5 27 52S 54 29W
Santa Rosa, Calif., U.S.A. 84 G4 38 26N 122 43W
Santa Rosa, N. Mex., U.S.A. 81 H2 34 57N 104 41W
Santa Rosa de Copán, Honduras 88 D2 14 47N 88 46W
Santa Rosa de Río Primero, Argentina 94 C3 31 8S 63 20W
Santa Rosa I., Calif., U.S.A. 85 M6 33 58N 120 6W
Santa Rosa I., Fla., U.S.A. 77 K2 30 20N 86 50W
Santa Rosa Range, U.S.A. 82 F5 41 45N 117 40W
Santa Rosalía, Mexico 86 B2 27 20N 112 20W
Santa Sylvina, Argentina 94 B3 27 50S 61 10W
Santa Tecla = Nueva San Salvador, El Salv. 88 D2 13 40N 89 18W
Santa Teresa, Argentina 94 C3 33 25S 60 47W
Santa Teresa, Mexico 87 B5 25 17N 97 51W
Santa Vitória do Palmar, Brazil 95 C5 33 32S 53 25W
Santa Ynez →, U.S.A. 85 L6 35 41N 120 36W
Santa Ynez Mts., U.S.A. 85 L6 34 30N 120 0W
Santa Ysabel, U.S.A. 85 M10 33 7N 116 40W
Santai, China 32 C5 31 5N 104 58 E
Santana, Madeira 22 D3 32 48N 16 52W
Santana, Coxilha de, Brazil 95 C4 30 50S 55 35W
Santana do Livramento, Brazil 95 C4 30 55S 55 30W
Santanayi, Spain 22 B10 39 20N 3 5 E
Santander, Spain 19 A4 43 27N 3 51W
Santander Jiménez, Mexico 87 C5 24 11N 98 29W
Sant'Antioco, Italy 20 E3 39 4N 8 27 E
Santaquin, U.S.A. 82 G8 39 59N 111 47W
Santarém, Brazil 93 D8 2 25S 54 42W
Santarém, Portugal 19 C1 39 12N 8 42W
Santaren Channel, W. Indies 88 B4 24 0N 79 30W
Santee, U.S.A. 85 N10 32 50N 116 58W
Santiago, Brazil 95 B5 29 11S 54 52W
Santiago, Chile 94 C1 33 24S 70 40W
Santiago, Panama 88 E3 8 0N 81 0W
Santiago □, Chile 94 C1 33 30S 70 50W
Santiago →, Mexico 66 G9 25 11N 105 26W
Santiago →, Peru 92 D3 4 27S 77 38W
Santiago de Compostela, Spain 19 A1 42 52N 8 37W
Santiago de Cuba, Cuba 88 C4 20 0N 75 49W
Santiago de los Cabelleros, Dom. Rep. 89 C5 19 30N 70 40W
Santiago del Estero, Argentina 94 B3 27 50S 64 15W
Santiago del Estero □, Argentina 94 B3 27 40S 63 15W
Santiago del Teide, Canary Is. 22 F3 28 17N 16 48W
Santiago Ixcuintla, Mexico 86 C3 21 50N 105 11W
Santiago Papasquiaro, Mexico 86 B3 25 0N 105 20W
Santiaguillo, L. de, Mexico 86 C4 24 50N 104 50W
Santo Amaro, Brazil 93 F11 12 30S 38 43W
Santo Anastácio, Brazil 95 A5 21 58S 51 39W
Santo André, Brazil 95 A6 23 39S 46 29W
Santo Ângelo, Brazil 95 B5 28 15S 54 15W
Santo Antonio, Brazil 93 G7 15 50S 56 0W
Santo Corazón, Bolivia 92 G7 18 0S 58 45W
Santo Domingo, Dom. Rep. 89 C6 18 30N 69 59W
Santo Domingo, Baja Calif., Mexico 86 A1 30 43N 116 2W
Santo Domingo, Baja Calif. S., Mexico 86 B2 25 32N 112 2W
Santo Domingo, Nic. 88 D3 12 14N 84 59W
Santo Tomás, Mexico 86 A1 31 33N 116 24W
Santo Tomás, Peru 92 F4 14 26S 72 8W
Santo Tomé, Argentina 95 B4 28 40S 56 5W
Santo Tomé de Guayana = Ciudad Guayana, Venezuela 92 B6 8 0N 62 30W
Santoña, Spain 19 A4 43 29N 3 27W
Santorini = Thíra, Greece 21 F11 36 23N 25 27 E
Santos, Brazil 95 A6 24 0S 46 20W
Santos Dumont, Brazil 95 A7 22 55S 43 10W
Sanyuan, China 34 G5 34 35N 108 58 E
Sanza Pombo, Angola 52 F3 7 18S 15 56 E
São Anastácio, Brazil 95 A5 22 0S 51 40W
São Bernardo de Campo, Brazil 95 A6 23 45S 46 34W
São Borja, Brazil 95 B4 28 39S 56 0W
São Carlos, Brazil 95 A6 22 0S 47 50W
São Cristóvão, Brazil 93 F11 11 1S 37 15W
São Domingos, Brazil 93 F9 13 25S 46 19W
São Francisco, Brazil 93 G10 16 0S 44 50W
São Francisco →, Brazil 93 F11 10 30S 36 24W
São Francisco do Sul, Brazil 95 B6 26 15S 48 36W
São Gabriel, Brazil 95 C5 30 20S 54 20W
São Gonçalo, Brazil 95 A7 22 48S 43 5W
Sao Hill, Tanzania 55 D4 8 20S 35 12 E
São João da Boa Vista, Brazil 95 A6 22 0S 46 52W
São João da Madeira, Portugal 19 B1 40 54N 8 30W
São João del Rei, Brazil 95 A7 21 8S 44 15W
São João do Araguaia, Brazil 93 E9 5 23S 48 46W
São João do Piauí, Brazil 93 E10 8 21S 42 15W
São Jorge, Pta. de, Madeira 22 D3 32 50N 16 53W
São José do Rio Prêto, Brazil 95 A6 20 50S 49 20W
São José dos Campos, Brazil 95 A6 23 7S 45 52W
São Leopoldo, Brazil 95 B5 29 50S 51 10W
São Lourenço, Brazil 95 A6 22 7S 45 3W
São Lourenço →, Brazil 93 G7 17 53S 57 27W
São Lourenço, Pta. de, Madeira 22 D3 32 44N 16 39W
São Luís, Brazil 93 D10 2 39S 44 15W
São Luís Gonzaga, Brazil 95 B5 28 25S 55 0W
São Marcos →, Brazil 93 G9 18 15S 47 37W
São Marcos, B. de, Brazil 93 D10 2 0S 44 0W
São Mateus, Brazil 93 G11 18 44S 39 50W
São Paulo, Brazil 95 A6 23 32S 46 37W
São Paulo □, Brazil 95 A6 22 0S 49 0W
São Paulo, I., Atl. Oc. 2 D8 0 50N 31 40W
São Roque, Madeira 22 D3 32 46N 16 48W
São Roque, C. de, Brazil 93 E11 5 30S 35 16W
São Sebastião, I. de, Brazil 95 A6 23 50S 45 18W
São Sebastião do Paraíso, Brazil 95 A6 20 54S 46 59W
São Tomé, Atl. Oc. 48 F4 0 10N 6 39 E
São Tomé, C. de, Brazil 95 A7 22 0S 40 59W
São Tomé & Principe ■, Africa 49 F4 0 12N 6 39 E
São Vicente, Brazil 95 A6 23 57S 46 23W
São Vicente, Madeira 22 D2 32 48N 17 3W
São Vicente, C. de, Portugal 19 D1 37 0N 9 0W
Saona, I., Dom. Rep. 89 C6 18 10N 68 40W
Saône →, France 18 D6 45 44N 4 50 E
Saonek, Indonesia 37 E8 0 22S 130 55 E
Saparua, Indonesia 37 E7 3 33S 128 40 E
Sapele, Nigeria 50 G6 5 50N 5 40 E
Sapelo I., U.S.A. 77 K5 31 25N 81 12W
Saposoa, Peru 92 E3 6 55S 76 45W
Sappho, U.S.A. 84 B2 48 4N 124 16W
Sapporo, Japan 30 C10 43 0N 141 21 E
Sapudi, Indonesia 37 G16 7 6S 114 20 E
Sapulpa, U.S.A. 81 G7 35 59N 96 5W
Saqqez, Iran 44 B5 36 15N 46 20 E
Sar Dasht, Iran 45 C6 32 32N 48 52 E
Sar Gachineh, Iran 45 D6 30 31N 51 31 E
Sar Planina, Macedonia 21 C9 42 10N 21 0 E
Sara Buri, Thailand 38 E3 14 30N 100 55 E
Sarāb, Iran 44 B5 37 55N 47 40 E
Sarabadi, Iraq 44 C5 33 1N 44 48 E
Sarada →, India 41 F12 27 21N 81 23 E
Saragossa = Zaragoza, Spain 19 B5 41 39N 0 53W
Saraguro, Ecuador 92 D3 3 35S 79 16W
Sarajevo, Bos.-H. 21 C8 43 52N 18 26 E
Saran, G., Indonesia 36 E4 0 30S 111 25 E
Saranac Lake, U.S.A. 79 B10 44 20N 74 8W
Saranda, Tanzania 54 D3 5 45S 34 59 E
Sarandí del Yi, Uruguay 95 C4 33 18S 55 38W
Sarandí Grande, Uruguay 94 C4 33 44S 56 20W
Sarangani B., Phil. 37 C7 6 0N 125 13 E
Sarangani Is., Phil. 37 C7 5 25N 125 25 E
Sarangarh, India 41 J13 21 30N 83 5 E
Saransk, Russia 24 D8 54 10N 45 10 E
Sarapul, Russia 24 C9 56 28N 53 48 E
Sarasota, U.S.A. 77 M4 27 20N 82 32W
Saratoga, Calif., U.S.A. 84 H4 37 16N 122 2W
Saratoga, Wyo., U.S.A. 82 F10 41 27N 106 49W
Saratoga Springs, U.S.A. 79 C11 43 5N 73 47W
Saratov, Russia 24 D8 51 30N 46 2 E
Saravane, Laos 38 E6 15 43N 106 25 E
Sarawak □, Malaysia 36 D4 2 0N 113 0 E
Saray, Turkey 21 D12 41 26N 27 55 E
Sarayköy, Turkey 21 F13 37 55N 28 54 E
Sarbāz, Iran 45 E9 26 38N 61 19 E
Sarbīsheh, Iran 45 C8 32 30N 59 40 E
Sarda → = Sarada →, India 41 F12 27 21N 81 23 E
Sardalas, Libya 50 C7 25 50N 10 34 E
Sardarshahr, India 42 E6 28 30N 74 29 E
Sardegna □, Italy 20 D3 40 0N 9 0 E
Sardhana, India 42 E7 29 9N 77 39 E
Sardina, Pta., Canary Is. 22 F4 28 9N 15 44W
Sardinia = Sardegna □, Italy 20 D3 40 0N 9 0 E
Sardis, Turkey 21 E12 38 28N 28 2 E
Sārdūīyeh = Dar Mazār, Iran 45 D8 29 14N 57 20 E
Sargent, U.S.A. 80 E5 41 39N 99 22W
Sargodha, Pakistan 42 C5 32 10N 72 40 E
Sarh, Chad 51 G8 9 5N 18 23 E
Sārī, Iran 45 B7 36 30N 53 4 E
Sarıgöl, Turkey 21 E13 38 14N 28 41 E
Sarikei, Malaysia 36 D4 2 8N 111 30 E
Sarina, Australia 62 C4 21 22S 149 13 E
Sarita, U.S.A. 81 M6 27 13N 97 47W
Sariwŏn, N. Korea 35 E13 38 31N 125 46 E
Sark, U.K. 11 H5 49 25N 2 22W
Şarköy, Turkey 21 D12 40 36N 27 6 E
Sarlat-la-Canéda, France 18 D4 44 54N 1 13 E
Sarles, U.S.A. 80 A5 48 58N 99 0W
Sarmi, Indonesia 37 E9 1 49S 138 44 E
Sarmiento, Argentina 96 F3 45 35S 69 5W
Särna, Sweden 9 F15 61 41N 13 8 E
Sarnia, Canada 70 D3 42 58N 82 23W
Sarny, Ukraine 24 D4 51 17N 26 40 E
Sarolangun, Indonesia 36 E2 2 19S 102 42 E
Saronikós Kólpos, Greece 21 F10 37 45N 23 45 E
Saros Körfezi, Turkey 21 D12 40 30N 26 15 E
Sarpsborg, Norway 9 G14 59 16N 11 7 E
Sarre = Saar →, Europe 16 D4 49 41N 6 32 E
Sarreguemines, France 18 B7 49 5N 7 4 E
Sarro, Mali 50 F3 13 40N 5 15W
Sarthe →, France 18 C3 47 33N 0 31W
Sartynya, Russia 26 C7 63 22N 63 11 E
Sarvestān, Iran 45 D7 29 20N 53 10 E
Sary-Tash, Kyrgyzstan 26 F8 39 44N 73 15 E
Saryshagan, Kazakstan 26 E8 46 12N 73 38 E
Sasabeneh, Ethiopia 46 F3 7 59N 44 43 E
Sasaram, India 43 G11 24 57N 84 5 E
Sasebo, Japan 31 H4 33 10N 129 43 E
Saser, India 43 B7 34 50N 77 50 E
Saskatchewan □, Canada 73 C7 54 40N 106 0W
Saskatchewan →, Canada 73 C8 53 37N 100 40W
Saskatoon, Canada 73 C7 52 10N 106 38W
Saskylakh, Russia 27 B12 71 55N 114 1 E
Sasolburg, S. Africa 57 D4 26 46S 27 49 E
Sasovo, Russia 24 D7 54 25N 41 55 E
Sassandra, Ivory C. 50 H3 4 55N 6 8W
Sassandra →, Ivory C. 50 H3 4 58N 6 5W
Sássari, Italy 20 D3 40 43N 8 34 E
Sassnitz, Germany 16 A7 54 29N 13 39 E
Sassuolo, Italy 20 B4 44 33N 10 47 E
Sasumua Dam, Kenya 54 C4 0 45S 36 40 E
Sasyk, Ozero, Ukraine 17 F15 45 45N 29 20 E
Sata-Misaki, Japan 31 J5 31 0N 130 40 E
Satadougou, Mali 50 F2 12 25N 11 25W
Satakunta, Finland 9 F20 61 45N 23 0 E
Satanta, U.S.A. 81 G4 37 26N 100 59W
Satara, India 40 L8 17 44N 73 58 E
Satilla →, U.S.A. 77 K5 30 59N 81 29W
Satka, Russia 24 C10 55 3N 59 1 E
Satmala Hills, India 40 J9 20 15N 74 40 E
Satna, India 43 G9 24 35N 80 50 E
Sátoraljaújhely, Hungary 17 D11 48 25N 21 41 E
Satpura Ra., India 42 J7 21 25N 76 10 E
Satsuna-Shotō, Japan 31 K5 30 0N 130 0 E
Sattahip, Thailand 38 F3 12 41N 100 54 E
Satu Mare, Romania 17 E12 47 46N 22 55 E
Satui, Indonesia 36 E5 3 50S 115 27 E
Satun, Thailand 39 J3 6 43N 100 2 E
Saturnina →, Brazil 92 F7 12 15S 58 10W
Sauce, Argentina 94 C4 30 5S 58 46W
Sauceda, Mexico 86 B4 25 55N 101 18W
Saucillo, Mexico 86 B3 28 1N 105 17W
Sauda, Norway 9 G12 59 40N 6 20 E
Sauðarkrókur, Iceland 8 D4 65 45N 19 40W
Saudi Arabia ■, Asia 46 B3 26 0N 44 0 E
Sauer →, Germany 15 E6 49 44N 6 31 E
Sauerland, Germany 16 C4 51 12N 7 59 E
Saugeen →, Canada 78 B3 44 30N 81 22W
Saugerties, U.S.A. 79 D11 42 5N 73 57W
Sauk Centre, U.S.A. 80 C7 45 44N 94 57W
Sauk Rapids, U.S.A. 80 C7 45 35N 94 10W
Sault Ste. Marie, Canada 70 C3 46 30N 84 20W
Sault Ste. Marie, U.S.A. 76 B3 46 30N 84 21W
Saumlaki, Indonesia 37 F8 7 55S 131 20 E
Saumur, France 18 C3 47 15N 0 5W
Saunders C., N.Z. 59 L3 45 53S 170 45 E
Saunders I., Antarctica 5 B1 57 48S 26 28W
Saunders Point, Australia 61 E4 27 52S 125 38 E
Sauri, Nigeria 50 F6 11 42N 6 44 E
Saurimo, Angola 52 F4 9 40S 20 12 E
Sausalito, U.S.A. 84 H4 37 51N 122 29W
Savá, Honduras 88 C2 15 32N 86 15W
Sava →, Serbia, Yug. 21 B9 44 50N 20 26 E
Savage, U.S.A. 80 B2 47 27N 104 21W
Savage I. = Niue, Cook Is. 65 J11 19 2S 169 54W
Savai'i, W. Samoa 59 A12 13 28S 172 24W
Savalou, Benin 50 G5 7 57N 1 58 E
Savane, Mozam. 55 F4 19 37S 35 8 E
Savanna, U.S.A. 80 D9 42 5N 90 8W
Savanna la Mar, Jamaica 88 C4 18 10N 78 10W
Savannah, Ga., U.S.A. 77 J5 32 5N 81 6W
Savannah, Mo., U.S.A. 80 F7 39 56N 94 50W
Savannah, Tenn., U.S.A. 77 H1 35 14N 88 15W
Savannah →, U.S.A. 77 J5 32 2N 80 53W
Savannakhet, Laos 38 D5 16 30N 104 49 E
Savant L., Canada 70 B1 50 16N 90 44W
Savant Lake, Canada 70 B1 50 14N 90 40W

Savanur, India 40 M9 14 59N 75 21 E
Savé, Benin 50 G5 8 2N 2 29 E
Save →, Mozam. 57 C5 21 16S 34 0 E
Sāveh, Iran 45 C6 35 2N 50 20 E
Savelugu, Ghana 50 G4 9 38N 0 54W
Savo, Finland 8 E22 62 45N 27 30 E
Savoie □, France 18 D7 45 26N 6 25 E
Savona, Italy 20 B3 44 17N 8 30 E
Savonlinna, Finland 24 B4 61 52N 28 53 E
Sawahlunto, Indonesia 36 E2 0 40S 100 52 E
Sawai, Indonesia 37 E7 3 0S 129 5 E
Sawai Madhopur, India 42 F7 26 0N 76 25 E
Sawang Daen Din, Thailand 38 D4 17 28N 103 28 E
Sawankhalok, Thailand 38 D2 17 19N 99 50 E
Sawara, Japan 31 G10 35 55N 140 30 E
Sawatch Mts., U.S.A. 83 G10 38 30N 106 30W
Sawel Mt., U.K. 13 B4 54 50N 7 2W
Sawi, Thailand 39 G2 10 14N 99 5 E
Sawmills, Zimbabwe 55 F2 19 30S 28 2 E
Sawu, Indonesia 37 F6 9 35S 121 50 E
Sawu Sea, Indonesia 37 F6 9 30S 121 50 E
Saxby →, Australia 62 B3 18 25S 140 53 E
Saxony, Lower = Niedersachsen □, Germany 16 B5 53 8N 9 0 E
Saxton, U.S.A. 78 F6 40 13N 78 15W
Say, Niger 50 F5 13 8N 2 22 E
Sayabec, Canada 71 C6 48 35N 67 41W
Sayaboury, Laos 38 C3 19 15N 101 45 E
Sayán, Peru 92 F3 11 8S 77 12W
Sayan, Vostochnyy, Russia 27 D10 54 0N 96 0 E
Sayan, Zapadnyy, Russia 27 D10 52 30N 94 0 E
Saydā, Lebanon 47 B4 33 35N 35 25 E
Sayhan-Ovoo, Mongolia 34 B2 45 27N 103 54 E
Sayhandulaan, Mongolia 34 B5 44 40N 109 1 E
Sayḥut, Yemen 46 D5 15 12N 51 10 E
Saynshand, Mongolia 34 B6 44 55N 110 11 E
Sayre, Okla., U.S.A. 81 H5 35 18N 99 38W
Sayre, Pa., U.S.A. 79 E8 41 59N 76 32W
Sayula, Mexico 86 D4 19 50N 103 40W
Sazanit, Albania 21 D8 40 30N 19 20 E
Sázava →, Czech. 16 D8 49 53N 14 24 E
Sazin, Pakistan 43 B5 35 35N 73 30 E
Scafell Pike, U.K. 10 C4 54 27N 3 14W
Scalpay, U.K. 12 D2 57 52N 6 40W
Scandia, Canada 72 C6 50 20N 112 0W
Scandicci, Italy 20 C4 43 45N 11 11 E
Scandinavia, Europe 6 C8 64 0N 12 0 E
Scapa Flow, U.K. 12 C5 58 53N 3 3W
Scappoose, U.S.A. 84 E4 45 45N 122 53W
Scarborough, Trin. & Tob. 89 D7 11 11N 60 42W
Scarborough, U.K. 10 C7 54 17N 0 24W
Scebeli, Wabi →, Somali Rep. 46 G3 2 0N 44 0 E
Scenic, U.S.A. 80 D3 43 47N 102 33W
Schaffhausen, Switz. 16 E5 47 42N 8 39 E
Schagen, Neths. 15 B4 52 49N 4 48 E
Schefferville, Canada 71 B6 54 48N 66 50W
Schelde →, Belgium 15 C4 51 15N 4 16 E
Schell Creek Ra., U.S.A. 82 G6 39 15N 114 30W
Schenectady, U.S.A. 79 D11 42 49N 73 57W
Scheveningen, Neths. 15 B4 52 6N 4 16 E
Schiedam, Neths. 15 C4 51 55N 4 25 E
Schiermonnikoog, Neths. 15 A6 53 30N 6 15 E
Schio, Italy 20 B4 45 43N 11 21 E
Schleswig, Germany 16 A5 54 31N 9 34 E
Schleswig-Holstein □, Germany 16 A5 54 30N 9 30 E
Schofield, U.S.A. 80 C10 44 54N 89 36W
Scholls, U.S.A. 84 E4 45 24N 122 56W
Schouten I., Australia 62 G4 42 20S 148 20 E
Schouten Is. = Supiori, Indonesia 37 E9 1 0S 136 0 E
Schouwen, Neths. 15 C3 51 43N 3 45 E
Schreiber, Canada 70 C2 48 45N 87 20W
Schuler, Canada 73 C6 50 20N 110 6W
Schumacher, Canada 70 C3 48 30N 81 16W
Schurz, U.S.A. 82 G4 38 57N 118 49W
Schuyler, U.S.A. 80 E6 41 27N 97 4W
Schuylkill Haven, U.S.A. 79 F8 40 37N 76 11W
Schwäbische Alb, Germany 16 D5 48 20N 9 30 E
Schwaner, Pegunungan, Indonesia 36 E4 1 0S 112 30 E
Schwarzwald, Germany 16 D5 48 30N 8 20 E
Schwedt, Germany 16 B8 53 3N 14 16 E
Schweinfurt, Germany 16 C6 50 3N 10 14 E
Schweizer-Reneke, S. Africa 56 D4 27 11S 25 18 E
Schwenningen = Villingen-Schwenningen, Germany 16 D5 48 3N 8 26 E
Schwerin, Germany 16 B6 53 36N 11 22 E
Schwyz, Switz. 16 E5 47 2N 8 39 E
Sciacca, Italy 20 F5 37 31N 13 3 E
Scilla, Italy 20 E6 38 15N 15 43 E
Scilly, Isles of, U.K. 11 H1 49 56N 6 22W
Scioto →, U.S.A. 76 F4 38 44N 83 1W
Scobey, U.S.A. 80 A2 48 47N 105 25W
Scone, Australia 63 E5 32 5S 150 52 E
Scoresbysund, Greenland 4 B6 70 20N 23 0W
Scotia, Calif., U.S.A. 82 F1 40 29N 124 6W
Scotia, N.Y., U.S.A. 79 D11 42 50N 73 58W
Scotia Sea, Antarctica 5 B18 56 5S 56 0W
Scotland, U.S.A. 80 D6 43 9N 97 43W
Scotland □, U.K. 12 E5 57 0N 4 0W
Scotland Neck, U.S.A. 77 G7 36 8N 77 25W
Scott, C., Australia 60 B4 13 30S 129 49 E
Scott City, U.S.A. 80 F4 38 29N 100 54W
Scott Glacier, Antarctica 5 C8 66 15S 100 5 E
Scott I., Antarctica 5 C11 67 0S 179 0 E
Scott Inlet, Canada 69 A12 71 0N 71 0W
Scott Is., Canada 72 C3 50 48N 128 40W
Scott L., Canada 73 B7 59 55N 106 18W
Scott Reef, Australia 60 B3 14 0S 121 50 E
Scottburgh, S. Africa 57 E5 30 15S 30 47 E
Scottdale, U.S.A. 78 F5 40 6N 79 35W
Scottsbluff, U.S.A. 80 E3 41 52N 103 40W
Scottsboro, U.S.A. 77 H2 34 40N 86 2W
Scottsburg, U.S.A. 76 F3 38 41N 85 47W
Scottsdale, Australia 62 G4 41 9S 147 31 E
Scottsdale, U.S.A. 83 K7 33 29N 111 56W
Scottsville, Ky., U.S.A. 77 G2 36 45N 86 11W
Scottsville, N.Y., U.S.A. 78 C7 43 2N 77 47W
Scottville, U.S.A. 76 D2 43 58N 86 17W
Scranton, U.S.A. 79 E9 41 25N 75 40W
Scugog, L., Canada 78 B6 44 10N 78 55W
Scunthorpe, U.K. 10 D7 53 36N 0 39W
Scusciuban, Somali Rep. 46 E5 10 18N 50 12 E
Scutari = Üsküdar, Turkey 25 F4 41 0N 29 5 E
Seabrook, L., Australia 61 F2 30 55S 119 40 E
Seaford, U.S.A. 76 F8 38 39N 75 37W
Seaforth, Canada 70 D3 43 35N 81 25W
Seagraves, U.S.A. 81 J3 32 57N 102 34W
Seal →, Canada 73 B10 59 4N 94 48W
Seal Cove, Canada 71 C8 49 57N 56 22W
Seal L., Canada 71 B7 54 20N 61 30W
Sealy, U.S.A. 81 L6 29 47N 96 9W
Searchlight, U.S.A. 85 K12 35 28N 114 55W
Searcy, U.S.A. 81 H9 35 15N 91 44W
Searles L., U.S.A. 85 K9 35 44N 117 21W
Seaside, Calif., U.S.A. 84 J5 36 37N 121 50W
Seaside, Oreg., U.S.A. 84 E3 46 0N 123 56W
Seaspray, Australia 63 F4 38 25S 147 15 E
Seattle, U.S.A. 84 C4 47 36N 122 20W
Seaview Ra., Australia 62 B4 18 40S 145 45 E
Sebastián Vizcaíno, B., Mexico 86 B2 28 0N 114 30W
Sebastopol = Sevastopol, Ukraine 25 F5 44 35N 33 30 E
Sebastopol, U.S.A. 84 G4 38 24N 122 49W
Sebewaing, U.S.A. 76 D4 43 44N 83 27W
Sebha = Sabhah, Libya 51 C7 27 9N 14 29 E
Sebring, Fla., U.S.A. 77 M5 27 30N 81 27W
Sebring, Ohio, U.S.A. 78 F3 40 55N 81 2W
Sebringville, Canada 78 C3 43 24N 81 4W
Sebta = Ceuta, N. Afr. 19 E3 35 52N 5 18W
Sebuku, Indonesia 36 E5 3 30S 116 25 E
Sebuku, Teluk, Malaysia 36 D5 4 0N 118 10 E
Sechelt, Canada 72 D4 49 25N 123 42W
Sechura, Desierto de, Peru 92 E2 6 0S 80 30W
Secretary I., N.Z. 59 L1 45 15S 166 56 E
Secunderabad, India 40 L11 17 28N 78 30 E
Sedalia, U.S.A. 80 F8 38 42N 93 14W
Sedan, Australia 63 E2 34 34S 139 19 E
Sedan, France 18 B6 49 43N 4 57 E
Sedan, U.S.A. 81 G6 37 8N 96 11W
Seddon, N.Z. 59 J5 41 40S 174 7 E
Seddonville, N.Z. 59 J4 41 33S 172 1 E
Sedeh, Fārs, Iran 45 D7 30 45N 52 11 E
Sedeh, Khorāsān, Iran 45 C8 33 20N 59 14 E
Sederot, Israel 47 D3 31 32N 34 37 E
Sedgewick, Canada 72 C6 52 48N 111 41W
Sedhiou, Senegal 50 F1 12 44N 15 30W
Sedley, Canada 73 C8 50 10N 104 0W
Sedova, Pik, Russia 26 B6 73 29N 54 58 E
Sedro Woolley, U.S.A. 84 B4 48 30N 122 14W
Seeheim, Namibia 56 D2 26 50S 17 45 E
Seekoei →, S. Africa 56 E4 30 18S 25 1 E
Seferihisar, Turkey 21 E12 38 10N 26 50 E
Seg-ozero, Russia 24 B5 63 20N 33 46 E
Segamat, Malaysia 39 L4 2 30N 102 50 E
Segesta, Italy 20 F5 37 56N 12 50 E
Seget, Indonesia 37 E8 1 24S 130 58 E
Segezha, Russia 24 B5 63 44N 34 19 E
Ségou, Mali 50 F3 13 30N 6 16W
Segovia = Coco →, Cent. Amer. 88 D3 15 0N 83 8W
Segovia, Spain 19 B3 40 57N 4 10W
Segre →, Spain 19 B6 41 40N 0 43 E
Séguéla, Ivory C. 50 G3 7 55N 6 40W
Seguin, U.S.A. 81 L6 29 34N 97 58W
Segundo →, Argentina 94 C3 30 53S 62 44W
Segura →, Spain 19 C5 38 3N 0 44W
Seh Qal'eh, Iran 45 C8 33 40N 58 24 E
Sehitwa, Botswana 56 C3 20 30S 22 30 E
Sehore, India 42 H7 23 10N 77 5 E
Sehwan, Pakistan 42 F2 26 28N 67 53 E
Seiland, Norway 8 A20 70 25N 23 15 E
Seiling, U.S.A. 81 G5 36 9N 98 56W
Seinäjoki, Finland 9 E20 62 40N 22 51 E
Seine →, France 18 B4 49 26N 0 26 E
Seistan, Iran 45 D9 30 50N 61 0 E
Sekayu, Indonesia 36 E2 2 51S 103 51 E
Seke, Tanzania 54 C3 3 20S 33 31 E
Sekenke, Tanzania 54 C3 4 18S 34 11 E
Sekondi-Takoradi, Ghana 50 H4 4 58N 1 45W
Sekuma, Botswana 56 C3 24 36S 23 50 E
Selah, U.S.A. 82 C3 46 39N 120 32W
Selama, Malaysia 39 K3 5 12N 100 42 E
Selaru, Indonesia 37 F8 8 9S 131 0 E
Selby, U.K. 10 D6 53 47N 1 5W
Selby, U.S.A. 80 C4 45 31N 100 2W
Selçuk, Turkey 21 F12 37 56N 27 22 E
Selden, U.S.A. 80 F4 39 33N 100 34W
Sele →, Italy 20 D6 40 29N 14 56 E
Selemdzha →, Russia 27 D13 51 42N 128 53 E
Selenga = Selenge Mörön →, Asia 32 A5 52 16N 106 16 E
Selenge Mörön →, Asia 32 A5 52 16N 106 16 E
Seletan, Tg., Indonesia 36 E4 4 10S 114 40 E
Selfridge, U.S.A. 80 B4 46 2N 100 56W
Sélibabi, Mauritania 50 E2 15 10N 12 15W
Seligman, U.S.A. 83 J7 35 20N 112 53W
Selîma, El Wâhât el, Sudan 51 D10 21 22N 29 19 E
Selinda Spillway, Botswana 56 B3 18 35S 23 10 E
Selkirk, Canada 73 C9 50 10N 96 55W
Selkirk, U.K. 12 F6 55 33N 2 50W
Selkirk I., Canada 73 C9 53 20N 99 6W
Selkirk Mts., Canada 72 C5 51 15N 117 40W
Selliá, Greece 23 D6 35 12N 24 23 E
Sells, U.S.A. 83 L8 31 55N 111 53W
Selma, Ala., U.S.A. 77 J2 32 25N 87 1W
Selma, Calif., U.S.A. 83 H4 36 34N 119 37W
Selma, N.C., U.S.A. 77 H6 35 32N 78 17W
Selmer, U.S.A. 77 H1 35 10N 88 36W
Selowandoma Falls, Zimbabwe 55 G3 21 15S 31 50 E
Selpele, Indonesia 37 E8 0 1S 130 5 E
Selsey Bill, U.K. 11 G7 50 43N 0 47W
Selu, Indonesia 37 F8 7 32S 130 55 E
Selva, Argentina 94 B3 29 50S 62 0W
Selvas, Brazil 92 E5 6 30S 67 0W
Selwyn, Australia 62 C3 21 32S 140 30 E
Selwyn L., Canada 73 A8 60 0N 104 30W
Selwyn Ra., Australia 62 C3 21 10S 140 0 E
Semani →, Albania 21 D8 40 47N 19 30 E
Semarang, Indonesia 37 G14 7 0S 110 26 E
Semau, Indonesia 37 F6 10 13S 123 22 E
Sembabule, Uganda 54 C3 0 4S 31 25 E
Semeru, Indonesia 37 H15 8 4S 112 55 E
Semey, Kazakstan 26 D9 50 30N 80 10 E
Seminoe Reservoir, U.S.A. 82 E10 42 9N 106 55W
Seminole, Okla., U.S.A. 81 H6 35 14N 96 41W
Seminole, Tex., U.S.A. 81 J3 32 43N 102 39W
Semiozernoye, Kazakstan 26 D7 52 22N 64 8 E
Semipalatinsk = Semey, Kazakstan 26 D9 50 30N 80 10 E
Semirara Is., Phil. 37 B6 12 0N 121 20 E
Semisopochnoi I., U.S.A. 68 C2 51 55N 179 36 E
Semitau, Indonesia 36 D4 0 29N 111 57 E
Semiyarka, Kazakstan 26 D8 50 55N 78 23 E
Semiyarskoye = Semiyarka, Kazakstan 26 D8 50 55N 78 23 E
Semmering P., Austria 16 E8 47 41N 15 45 E
Semnān, Iran 45 C7 35 40N 53 23 E
Semnān □, Iran 45 C7 36 0N 54 0 E
Semois →, Europe 15 E4 49 53N 4 44 E
Semporna, Malaysia 37 D5 4 30N 118 33 E
Semuda, Indonesia 36 E4 2 51S 112 58 E
Senā, Iran 45 D6 28 27N 51 36 E
Sena, Mozam. 55 F3 17 25S 35 0 E
Sena Madureira, Brazil 92 E5 9 5S 68 45W
Senador Pompeu, Brazil 93 E11 5 40S 39 20W
Senaja, Malaysia 36 C5 6 45N 117 3 E
Senanga, Zambia 56 B3 16 2S 23 14 E
Senatobia, U.S.A. 81 H10 34 37N 89 58W
Sendai, Kagoshima, Japan 31 J5 31 50N 130 20 E
Sendai, Miyagi, Japan 30 E10 38 15N 140 53 E
Sendai-Wan, Japan 30 E10 38 15N 141 0 E
Seneca, Oreg., U.S.A. 82 D4 44 8N 118 58W
Seneca, S.C., U.S.A. 77 H4 34 41N 82 57W
Seneca Falls, U.S.A. 79 D8 42 55N 76 48W
Seneca L., U.S.A. 78 D8 42 40N 76 54W
Senegal ■, W. Afr. 50 F2 14 30N 14 30W
Senegal →, W. Afr. 50 E1 15 48N 16 32W
Senegambia, Africa 48 E2 12 45N 12 0W
Senekal, S. Africa 57 D4 28 20S 27 36 E
Senga Hill, Zambia 55 D3 9 19S 31 11 E
Senge Khambab = Indus →, Pakistan 42 G2 24 20N 67 47 E
Sengerema □, Tanzania 54 C3 2 10S 32 20 E
Sengkang, Indonesia 37 E6 4 8S 120 1 E
Sengua →, Zimbabwe 55 F2 17 7S 28 5 E
Senhor-do-Bonfim, Brazil 93 F10 10 30S 40 10W
Senigállia, Italy 20 C5 43 43N 13 13 E
Senj, Croatia 16 F8 45 0N 14 58 E
Senja, Norway 8 B17 69 25N 17 30 E
Senlis, France 18 B5 49 13N 2 35 E
Senmonorom, Cambodia 38 F6 12 27N 107 12 E
Sennâr, Sudan 51 F11 13 30N 33 35 E
Senneterre, Canada 70 C4 48 25N 77 15W
Seno, Laos 38 D5 16 35N 104 50 E
Sens, France 18 B5 48 11N 3 15 E
Senta, Serbia, Yug. 21 B9 45 55N 20 3 E
Sentani, Indonesia 37 E10 2 36S 140 37 E
Sentery, Zaïre 54 D2 5 17S 25 42 E
Sentinel, U.S.A. 83 K7 32 52N 113 13W
Sentolo, Indonesia 37 G14 7 55S 110 13 E
Seo de Urgel, Spain 19 A6 42 22N 1 23 E
Seohara, India 43 E8 29 15N 78 33 E
Seoni, India 43 H8 22 5N 79 30 E
Seoul = Sŏul, S. Korea 35 F14 37 31N 126 58 E
Separation Point, Canada 71 B8 53 37N 57 25W
Sepīdān, Iran 45 D7 30 20N 52 5 E
Sepo-ri, N. Korea 35 E14 38 57N 127 25 E
Sepone, Laos 38 D6 16 45N 106 13 E
Sept-Îles, Canada 71 B6 50 13N 66 22W
Sequim, U.S.A. 84 B3 48 5N 123 6W
Sequoia National Park, U.S.A. 83 H4 36 30N 118 30W
Seraing, Belgium 15 D5 50 35N 5 32 E
Seraja, Indonesia 39 L7 2 41N 108 35 E
Serakhis →, Cyprus 23 D11 35 13N 32 55 E
Seram, Indonesia 37 E7 3 10S 129 0 E
Seram Laut, Kepulauan, Indonesia 37 E8 4 5S 131 25 E
Seram Sea, Indonesia 37 E7 2 30S 128 30 E
Serang, Indonesia 37 G12 6 8S 106 10 E
Serasan, Indonesia 39 L7 2 29N 109 4 E
Serbia □, Yugoslavia 21 C9 43 30N 21 0 E
Serdobsk, Russia 24 D7 52 28N 44 10 E
Seremban, Malaysia 39 L3 2 43N 101 53 E
Serengeti □, Tanzania 54 C3 2 0S 34 30 E
Serengeti Plain, Tanzania 54 C3 2 40S 35 0 E
Serenje, Zambia 55 E3 13 14S 30 15 E
Sereth = Siret →, Romania 17 F14 45 24N 28 1 E
Sergipe □, Brazil 93 F11 10 30S 37 30W
Sergiyev Posad, Russia 24 C6 56 20N 38 10 E
Seria, Brunei 36 D4 4 37N 114 23 E
Serian, Malaysia 36 D4 1 10N 110 31 E
Seribu, Kepulauan, Indonesia 36 F3 5 36S 106 33 E
Sérifos, Greece 21 F11 37 9N 24 30 E
Seringapatam Reef, Australia 60 B3 13 38S 122 5 E
Sermata, Indonesia 37 F7 8 15S 128 50 E
Serny Zavod, Turkmenistan 26 F6 39 59N 58 50 E
Serov, Russia 24 C11 59 29N 60 35 E
Serowe, Botswana 56 C4 22 25S 26 43 E
Serpentine, Australia 61 F2 32 23S 115 58 E
Serpentine Lakes, Australia 61 E4 28 30S 129 10 E
Serpukhov, Russia 24 D6 54 55N 37 28 E
Sérrai, Greece 21 D10 41 5N 23 31 E
Serrezuela, Argentina 94 C2 30 40S 65 20W
Serrinha, Brazil 93 F11 11 39S 39 0W
Sertânia, Brazil 93 E11 8 5S 37 20W
Sertanópolis, Brazil 95 A5 23 4S 51 2W
Serua, Indonesia 37 F8 6 18S 130 1 E
Serui, Indonesia 37 E9 1 53S 136 10 E
Serule, Botswana 56 C4 21 57S 27 20 E
Sese Is., Uganda 54 C3 0 20S 32 20 E
Sesepe, Indonesia 37 E7 1 30S 127 59 E
Sesfontein, Namibia 56 B1 19 7S 13 39 E
Sesheke, Zambia 56 B3 17 29S 24 13 E
S'estañol, Spain 22 B9 39 22N 2 54 E
Setana, Japan 30 C9 42 26N 139 51 E
Sète, France 18 E5 43 25N 3 42 E
Sete Lagôas, Brazil 93 G10 19 27S 44 16W
Sétif, Algeria 50 A6 36 9N 5 26 E
Seto, Japan 31 G8 35 14N 137 6 E
Setonaikai, Japan 31 G6 34 20N 133 30 E
Settat, Morocco 50 B3 33 0N 7 40W
Setté-Cama, Gabon 52 E1 2 32S 9 45 E
Setting L., Canada 73 B9 55 0N 98 38W
Settle, U.K. 10 C5 54 5N 2 16W
Settlement Pt., Bahamas 77 M6 26 40N 79 0W
Setúbal, Portugal 19 C1 38 30N 8 58W
Setúbal, B. de, Portugal 19 C1 38 40N 8 56W
Seulimeum, Indonesia 36 C1 5 27N 95 15 E
Sevan, Ozero = Sevana Lich, Armenia 25 F8 40 30N 45 20 E
Sevana Lich, Armenia 25 F8 40 30N 45 20 E
Sevastopol, Ukraine 25 F5 44 35N 33 30 E
Seven Emu, Australia 62 B2 16 20S 137 8 E
Seven Sisters, Canada 72 C3 54 56N 128 10W
Severn →, Canada 70 A2 56 2N 87 36W
Severn →, U.K. 11 F5 51 35N 2 40W
Severn L., Canada 70 B1 53 54N 90 48W
Severnaya Zemlya, Russia 27 B10 79 0N 100 0 E
Severnyye Uvaly, Russia 24 C8 60 0N 50 0 E
Severo-Kurilsk, Russia 27 D16 50 40N 156 8 E
Severo-Yeniseyskiy, Russia 27 C10 60 22N 93 1 E
Severodvinsk, Russia 24 B6 64 27N 39 58 E
Severomorsk, Russia 24 A5 69 5N 33 27 E
Severouralsk, Russia 24 B10 60 9N 59 57 E
Sevier, U.S.A. 83 G7 38 39N 112 11W
Sevier →, U.S.A. 83 G7 39 4N 113 6W
Sevier L., U.S.A. 82 G7 38 54N 113 9W
Sevilla, Spain 19 D2 37 23N 6 0W
Seville = Sevilla, Spain 19 D2 37 23N 6 0W
Sevlievo, Bulgaria 21 C11 43 2N 25 3 E
Seward, Alaska, U.S.A. 68 B5 60 7N 149 27W
Seward, Nebr., U.S.A. 80 E6 40 55N 97 6W
Seward Pen., U.S.A. 68 B3 65 0N 164 0W
Sewell, Chile 94 C1 34 10S 70 23W
Sewer, Indonesia 37 F8 5 53S 134 40 E
Sewickley, U.S.A. 78 F4 40 32N 80 12W
Sexsmith, Canada 72 B5 55 21N 118 47W
Seychelles ■, Ind. Oc. 49 G9 5 0S 56 0 E
Seyðisfjörður, Iceland 8 D6 65 16N 13 57W
Seydvān, Iran 44 B5 38 34N 45 2 E
Seymchan, Russia 27 C16 62 54N 152 30 E
Seymour, Australia 63 F4 37 0S 145 10 E
Seymour, S. Africa 57 E4 32 33S 26 46 E
Seymour, Conn., U.S.A. 79 E11 41 24N 73 4W
Seymour, Ind., U.S.A. 76 F3 38 58N 85 53W
Seymour, Tex., U.S.A. 81 J5 33 35N 99 16W
Seymour, Wis., U.S.A. 76 C1 44 31N 88 20W
Sfax, Tunisia 51 B7 34 49N 10 48 E
Sfîntu Gheorghe, Romania 17 F13 45 52N 25 48 E
Shaanxi □, China 34 G5 35 0N 109 0 E
Shaba □, Zaïre 54 D2 8 0S 25 0 E
Shabunda, Zaïre 54 C2 2 40S 27 16 E
Shache, China 32 C2 38 20N 77 10 E
Shackleton Ice Shelf, Antarctica 5 C8 66 0S 100 0 E
Shackleton Inlet, Antarctica 5 E11 83 0S 160 0 E
Shādegān, Iran 45 D6 30 40N 48 38 E
Shadi, India 43 C7 33 24N 77 14 E
Shadrinsk, Russia 26 D7 56 5N 63 32 E
Shafter, Calif., U.S.A. 85 K7 35 30N 119 16W
Shafter, Tex., U.S.A. 81 L2 29 49N 104 18W
Shaftesbury, U.K. 11 F5 51 0N 2 11W
Shagram, Pakistan 43 A5 36 24N 72 20 E
Shah Bunder, Pakistan 42 G2 24 13N 67 56 E
Shahabad, Punjab, India 42 D7 30 10N 76 55 E
Shahabad, Raj., India 42 G7 25 15N 77 11 E
Shahabad, Ut. P., India 43 F8 27 36N 79 56 E
Shahadpur, Pakistan 42 G3 25 55N 68 35 E
Shahba, Syria 47 C5 32 52N 36 38 E
Shahdād, Iran 45 D8 30 30N 57 40 E
Shahdadkot, Pakistan 42 F2 27 50N 67 55 E
Shahe, China 34 F8 37 0N 114 32 E
Shahganj, India 43 F10 26 3N 82 44 E
Shahgarh, India 40 F6 27 15N 69 50 E
Shaḥḥāt, Libya 51 B9 32 48N 21 54 E
Shahjahanpur, India 43 F8 27 54N 79 57 E
Shahpur, India 42 H7 22 12N 77 58 E
Shahpur, Pakistan 42 E3 28 46N 68 27 E
Shahpura, India 43 H9 23 10N 80 45 E
Shahr Kord, Iran 45 C6 32 15N 50 55 E
Shāhrakht, Iran 45 C9 33 38N 60 16 E
Shahrig, Pakistan 42 D2 30 15N 67 40 E
Shahukou, China 34 D7 40 20N 112 18 E
Shaikhabad, Afghan. 42 B3 34 2N 68 45 E
Shajapur, India 42 H7 23 27N 76 21 E
Shakargarh, Pakistan 42 C6 32 17N 75 10 E
Shakawe, Botswana 56 B3 18 28S 21 49 E
Shaker Heights, U.S.A. 78 E3 41 29N 81 32W
Shakhty, Russia 25 E7 47 40N 40 16 E
Shakhunya, Russia 24 C8 57 40N 46 46 E
Shaki, Nigeria 50 G5 8 41N 3 21 E
Shala, L., Ethiopia 51 G12 7 30N 38 30 E
Shallow Lake, Canada 78 B3 44 36N 81 5W
Shalqar, Kazakstan 26 E6 47 48N 59 39 E
Shaluli Shan, China 32 C4 30 40N 99 55 E
Shām, Iran 45 E8 26 39N 57 21 E
Shamâl Kordofân □, Sudan 48 E6 15 0N 30 0 E
Shamanovo, Russia 27 C15 69 45N 147 20 E
Shamattawa, Canada 73 B10 55 51N 92 5W
Shamattawa →, Canada 70 A2 55 1N 85 23W
Shamīl, Iran 45 E8 27 30N 56 55 E
Shāmkūh, Iran 45 C8 35 47N 57 50 E
Shamli, India 42 E7 29 32N 77 18 E
Shamo = Gobi, Asia 34 C5 44 0N 111 0 E
Shamo, L., Ethiopia 51 G12 5 45N 37 30 E
Shamokin, U.S.A. 79 F8 40 47N 76 34W
Shamrock, U.S.A. 81 H4 35 13N 100 15W
Shamva, Zimbabwe 55 F3 17 20S 31 32 E
Shan □, Burma 41 J21 21 30N 98 30 E
Shan Xian, China 34 G9 34 50N 116 5 E
Shanchengzhen, China 35 C13 42 20N 125 20 E

Shāndak, Iran 45 D9 28 28N 60 27 E
Shandon, U.S.A. 84 K6 35 39N 120 23W
Shandong □, China 35 F10 36 0N 118 0 E
Shandong Bandao, China 35 F11 37 0N 121 0 E
Shang Xian, China 34 H5 33 50N 109 58 E
Shangalowe, Zaïre 55 E2 10 50S 26 30 E
Shangani →, Zimbabwe 55 F2 18 41S 27 10 E
Shangbancheng, China 35 D10 40 50N 118 1 E
Shangdu, China 34 D7 41 30N 113 30 E
Shanghai, China 33 C7 31 15N 121 26 E
Shanghe, China 35 F9 37 20N 117 10 E
Shangnan, China 34 H6 33 32N 110 50 E
Shangqiu, China 34 G8 34 26N 115 36 E
Shangrao, China 33 D6 28 25N 117 59 E
Shangshui, China 34 H8 33 42N 114 35 E
Shangzhi, China 35 B14 45 22N 127 56 E
Shanhetun, China 35 B14 44 33N 127 15 E
Shaniko, U.S.A. 82 D3 45 0N 120 45W
Shannon, N.Z. 59 J5 40 33S 175 25 E
Shannon →, Ireland 13 D2 52 35N 9 30W
Shansi = Shanxi □, China 34 F7 37 0N 112 0 E
Shantar, Ostrov Bolshoy, Russia 27 D14 55 9N 137 40 E
Shantipur, India 43 H13 23 17N 88 25 E
Shantou, China 33 D6 23 18N 116 40 E
Shantung = Shandong □, China 35 F10 36 0N 118 0 E
Shanxi □, China 34 F7 37 0N 112 0 E
Shanyang, China 34 H5 33 31N 109 55 E
Shanyin, China 34 E7 39 25N 112 56 E
Shaoguan, China 33 D6 24 48N 113 35 E
Shaoxing, China 33 C7 30 0N 120 35 E
Shaoyang, China 33 D6 27 14N 111 25 E
Shapinsay, U.K. 12 B6 59 3N 2 51W
Shaqra', Si. Arabia 44 E5 25 15N 45 16 E
Shaqrā', Yemen 46 E4 13 22N 45 44 E
Sharbot Lake, Canada 79 B8 44 46N 76 41W
Shari, Japan 30 C12 43 55N 144 40 E
Sharjah = Ash Shāriqah, U.A.E. 45 E7 25 23N 55 26 E
Shark B., Australia 61 E1 25 30S 113 32 E
Sharon, Mass., U.S.A. 79 D13 42 7N 71 11W
Sharon, Pa., U.S.A. 78 E4 41 14N 80 31W
Sharon Springs, U.S.A. 80 F4 38 54N 101 45W
Sharp Pt., Australia 62 A3 10 58S 142 43 E
Sharpe L., Canada 73 C10 54 24N 93 40W
Sharpsville, U.S.A. 78 E4 41 15N 80 29W
Sharya, Russia 24 C8 58 22N 45 20 E
Shashi, Botswana 57 C4 21 15S 27 27 E
Shashi, China 33 C6 30 25N 112 14 E
Shashi →, Africa 55 G2 21 14S 29 20 E
Shasta, Mt., U.S.A. 82 F2 41 25N 122 12W
Shasta L., U.S.A. 82 F2 40 43N 122 25W
Shatt al'Arab →, Iraq 45 D6 29 57N 48 34 E
Shattuck, U.S.A. 81 G5 36 16N 99 53W
Shaunavon, Canada 73 D7 49 35N 108 25W
Shaver L., U.S.A. 84 H7 37 9N 119 18W
Shaw →, Australia 60 D2 20 21S 119 17 E
Shaw I., Australia 62 C4 20 30S 149 2 E
Shawanaga, Canada 78 A4 45 31N 80 17W
Shawano, U.S.A. 76 C1 44 47N 88 36W
Shawinigan, Canada 70 C5 46 35N 72 50W
Shawnee, U.S.A. 81 H6 35 20N 96 55W
Shaybārā, Si. Arabia 44 E3 25 26N 36 47 E
Shaykh Sa'īd, Iraq 44 C5 32 34N 46 17 E
Shcherbakov = Rybinsk, Russia 24 C6 58 5N 38 50 E
Shchuchinsk, Kazakstan 26 D8 52 56N 70 12 E
She Xian, China 34 F7 36 30N 113 40 E
Shebele = Scebeli, Wabi →, Somali Rep. 46 G3 2 0N 44 0 E
Sheboygan, U.S.A. 76 D2 43 46N 87 45W
Shediac, Canada 71 C7 46 14N 64 32W
Sheelin, L., Ireland 13 C4 53 48N 7 20W
Sheep Haven, Ireland 13 A4 55 11N 7 52W
Sheerness, U.K. 11 F8 51 26N 0 47 E
Sheet Harbour, Canada 71 D7 44 56N 62 31W
Sheffield, U.K. 10 D6 53 23N 1 28W
Sheffield, Ala., U.S.A. 77 H2 34 46N 87 41W
Sheffield, Mass., U.S.A. 79 D11 42 5N 73 21W
Sheffield, Pa., U.S.A. 78 E5 41 42N 79 3W
Sheffield, Tex., U.S.A. 81 K4 30 41N 101 49W
Sheho, Canada 73 C8 51 35N 103 13W
Sheikhpura, India 43 G11 25 9N 85 53 E
Shekhupura, Pakistan 42 D5 31 42N 73 58 E
Shelburne, N.S., Canada 71 D6 43 47N 65 20W
Shelburne, Ont., Canada 70 D3 44 4N 80 15W
Shelburne, U.S.A. 79 B11 44 23N 73 14W
Shelburne B., Australia 62 A3 11 50S 142 50 E
Shelburne Falls, U.S.A. 79 D12 42 36N 72 45W
Shelby, Mich., U.S.A. 76 D2 43 37N 86 22W
Shelby, Mont., U.S.A. 82 B8 48 30N 111 51W
Shelby, N.C., U.S.A. 77 H5 35 17N 81 32W
Shelby, Ohio, U.S.A. 78 F2 40 53N 82 40W
Shelbyville, Ill., U.S.A. 80 F10 39 24N 88 48W
Shelbyville, Ind., U.S.A. 76 F3 39 31N 85 47W
Shelbyville, Tenn., U.S.A. 77 H2 35 29N 86 28W
Sheldon, U.S.A. 80 D7 43 11N 95 51W
Sheldrake, Canada 71 B7 50 20N 64 51W
Shelikhova, Zaliv, Russia 27 D16 59 30N 157 0 E
Shell Lake, Canada 73 C7 53 19N 107 2W
Shell Lakes, Australia 61 E4 29 20S 127 30 E
Shellbrook, Canada 73 C7 53 13N 106 24W
Shellharbour, Australia 63 E5 34 31S 150 51 E
Shelling Rocks, Ireland 13 E1 51 45N 10 35W
Shelton, Conn., U.S.A. 79 E11 41 19N 73 5W
Shelton, Wash., U.S.A. 84 C3 47 13N 123 6W
Shen Xian, China 34 F8 36 15N 115 40 E
Shenandoah, Iowa, U.S.A. 80 E7 40 46N 95 22W
Shenandoah, Pa., U.S.A. 79 F8 40 49N 76 12W
Shenandoah, Va., U.S.A. 76 F6 38 29N 78 37W
Shenandoah →, U.S.A. 76 F7 39 19N 77 44W
Shenchi, China 34 E7 39 8N 112 10 E
Shendam, Nigeria 50 G6 8 49N 9 30 E
Shendī, Sudan 51 E11 16 46N 33 22 E
Shengfang, China 34 E9 39 3N 116 42 E
Shenjingzi, China 35 B13 44 40N 124 30 E
Shenmu, China 34 E6 38 50N 110 29 E
Shenqiu, China 34 H8 33 25N 115 5 E
Shenqiucheng, China 34 H8 33 24N 115 2 E
Shensi = Shaanxi □, China 34 G5 35 0N 109 0 E
Shenyang, China 35 D12 41 48N 123 27 E
Sheopur Kalan, India 40 G10 25 40N 76 40 E
Shepetivka, Ukraine 17 C14 50 10N 27 10 E
Shepetovka = Shepetivka, Ukraine 17 C14 50 10N 27 10 E
Shepparton, Australia 63 F4 36 23S 145 26 E
Sheqi, China 34 H7 33 12N 112 57 E
Sher Qila, Pakistan 43 A6 36 7N 74 2 E
Sherborne, U.K. 11 G5 50 57N 2 31W
Sherbro I., S. Leone 50 G2 7 30N 12 40W
Sherbrooke, Canada 71 C5 45 28N 71 57W
Sheridan, Ark., U.S.A. 81 H8 34 19N 92 24W
Sheridan, Wyo., U.S.A. 82 D10 44 48N 106 58W
Sherkot, India 43 E8 29 22N 78 35 E
Sherman, U.S.A. 81 J6 33 40N 96 35W
Sherridon, Canada 73 B8 55 8N 101 5W
Sherwood, N. Dak., U.S.A. 80 A4 48 57N 101 38W
Sherwood, Tex., U.S.A. 81 K4 31 18N 100 45W
Sherwood Forest, U.K. 10 D6 53 6N 1 7W
Sheslay, Canada 72 B2 58 17N 131 52W
Sheslay →, Canada 72 B2 58 48N 132 5W
Shethanei L., Canada 73 B9 58 48N 97 50W
Shetland □, U.K. 12 A7 60 30N 1 30W
Shetland Is., U.K. 12 A7 60 30N 1 30W
Sheyenne, U.S.A. 80 B5 47 50N 99 7W
Sheyenne →, U.S.A. 80 B6 47 2N 96 50W
Shibām, Yemen 46 D4 16 0N 48 36 E
Shibata, Japan 30 F9 37 57N 139 20 E
Shibecha, Japan 30 C12 43 17N 144 36 E
Shibetsu, Japan 30 B11 44 10N 142 23 E
Shibogama L., Canada 70 B2 53 35N 88 15W
Shibushi, Japan 31 J5 31 25N 131 8 E
Shickshock Mts. = Chic-Chocs, Mts., Canada 71 C6 48 55N 66 0W
Shidao, China 35 F12 36 50N 122 25 E
Shido, Japan 31 G7 34 19N 134 10 E
Shiel, L., U.K. 12 E3 56 48N 5 34W
Shield, C., Australia 62 A2 13 20S 136 20 E
Shiga □, Japan 31 G8 35 20N 136 0 E
Shigaib, Sudan 51 E9 15 5N 23 35 E
Shiguaigou, China 34 D6 40 52N 110 15 E
Shihchiachuangi = Shijiazhuang, China 34 E8 38 2N 114 28 E
Shijiazhuang, China 34 E8 38 2N 114 28 E
Shikarpur, India 42 E8 28 17N 78 7 E
Shikarpur, Pakistan 42 F3 27 57N 68 39 E
Shikoku □, Japan 31 H6 33 30N 133 30 E
Shikoku-Sanchi, Japan 31 H6 33 30N 133 30 E
Shilabo, Ethiopia 46 F3 6 22N 44 32 E
Shiliguri, India 41 F16 26 45N 88 25 E
Shilka, Russia 27 D12 52 0N 115 55 E
Shilka →, Russia 27 D13 53 20N 121 26 E
Shillelagh, Ireland 13 D5 52 45N 6 32W
Shillong, India 41 G17 25 35N 91 53 E
Shilo, West Bank 47 C4 32 4N 35 18 E
Shilou, China 34 F6 37 0N 110 48 E
Shimabara, Japan 31 H5 32 48N 130 20 E
Shimada, Japan 31 G9 34 49N 138 10 E
Shimane □, Japan 31 G6 35 0N 132 30 E
Shimanovsk, Russia 27 D13 52 15N 127 30 E
Shimizu, Japan 31 G9 35 0N 138 30 E
Shimodate, Japan 31 F9 36 20N 139 55 E
Shimoga, India 40 N9 13 57N 75 32 E
Shimoni, Kenya 54 C4 4 38S 39 20 E
Shimonoseki, Japan 31 H5 33 58N 130 55 E
Shimpuru Rapids, Angola 56 B2 17 45S 19 55 E
Shin, L., U.K. 12 C4 58 5N 4 30W
Shin-Tone →, Japan 31 G10 35 44N 140 51 E
Shinano →, Japan 31 F9 36 50N 138 30 E
Shīndand, Afghan. 40 C3 33 12N 62 8 E
Shingleton, U.S.A. 70 C2 46 21N 86 28W
Shingū, Japan 31 H7 33 40N 135 55 E
Shinjō, Japan 30 E10 38 46N 140 18 E
Shinshār, Syria 47 A5 34 36N 36 43 E
Shinyanga, Tanzania 54 C3 3 45S 33 27 E
Shinyanga □, Tanzania 54 C3 3 50S 34 0 E
Shiogama, Japan 30 E10 38 19N 141 1 E
Shiojiri, Japan 31 F8 36 6N 137 58 E
Ship I., U.S.A. 81 K10 30 13N 88 55W
Shipehenski Prokhod, Bulgaria 21 C11 42 45N 25 15 E
Shiping, China 32 D5 23 45N 102 23 E
Shipki La, India 40 D11 31 45N 78 40 E
Shippegan, Canada 71 C7 47 45N 64 45W
Shippensburg, U.S.A. 78 F7 40 3N 77 31W
Shiprock, U.S.A. 83 H9 36 47N 108 41W
Shiqma, N. →, Israel 47 D3 31 37N 34 30 E
Shiquan, China 34 H5 33 5N 108 15 E
Shīr Kūh, Iran 45 D7 31 39N 54 3 E
Shiragami-Misaki, Japan 30 D10 41 24N 140 12 E
Shirakawa, Fukushima, Japan 31 F10 37 7N 140 13 E
Shirakawa, Gifu, Japan 31 F8 36 17N 136 56 E
Shirane-San, Gumma, Japan 31 F9 36 48N 139 22 E
Shirane-San, Yamanashi, Japan 31 G9 35 42N 138 9 E
Shiraoi, Japan 30 C10 42 33N 141 21 E
Shīrāz, Iran 45 D7 29 42N 52 30 E
Shire →, Africa 55 F4 17 42S 35 19 E
Shiretoko-Misaki, Japan 30 B12 44 21N 145 20 E
Shirinab →, Pakistan 42 D2 30 15N 66 28 E
Shiriya-Zaki, Japan 30 D10 41 25N 141 30 E
Shiroishi, Japan 30 E10 38 0N 140 37 E
Shīrvān, Iran 45 B8 37 30N 57 50 E
Shirwa, L. = Chilwa, L., Malawi 55 F4 15 15S 35 40 E
Shivpuri, India 42 G7 25 26N 77 42 E
Shixian, China 35 C15 43 5N 129 50 E
Shizuishan, China 34 E4 39 15N 106 50 E
Shizuoka, Japan 31 G9 34 57N 138 24 E
Shizuoka □, Japan 31 G9 35 15N 138 40 E
Shklov = Shklow, Belarus 17 A16 54 16N 30 15 E
Shklow, Belarus 17 A16 54 16N 30 15 E
Shkoder = Shkodra, Albania 21 C8 42 4N 19 32 E
Shkodra, Albania 21 C8 42 4N 19 32 E
Shkumbini →, Albania 21 D8 41 2N 19 31 E
Shmidta, O., Russia 27 A10 81 0N 91 0 E
Shō-Gawa →, Japan 31 F8 36 47N 137 4 E
Shoal Lake, Canada 73 C8 50 30N 100 35W
Shōdo-Shima, Japan 31 G7 34 30N 134 15 E
Shoeburyness, U.K. 11 F8 51 32N 0 49 E
Sholapur = Solapur, India 40 L9 17 43N 75 56 E
Shologontsy, Russia 27 C12 66 13N 114 0 E
Shōmrōn, West Bank 47 C4 32 15N 35 13 E
Shoshone, Calif., U.S.A. 85 K10 35 58N 116 16W
Shoshone, Idaho, U.S.A. 82 E6 42 56N 114 25W
Shoshone L., U.S.A. 82 D8 44 22N 110 43W
Shoshone Mts., U.S.A. 82 G5 39 20N 117 25W
Shoshong, Botswana 56 C4 22 56S 26 31 E
Shoshoni, U.S.A. 82 E9 43 14N 108 7W
Shouguang, China 35 F10 37 52N 118 45 E
Shouyang, China 34 F7 37 54N 113 8 E
Show Low, U.S.A. 83 J9 34 15N 110 2W
Shreveport, U.S.A. 81 J8 32 31N 93 45W
Shrewsbury, U.K. 10 E5 52 43N 2 45W
Shrirampur, India 43 H13 22 44N 88 21 E
Shropshire □, U.K. 11 E5 52 36N 2 45W
Shu, Kazakstan 26 E8 43 36N 73 42 E
Shu →, Kazakstan 28 E10 45 0N 67 44 E
Shuangcheng, China 35 B14 45 20N 126 15 E
Shuanggou, China 35 G9 34 2N 117 30 E
Shuangliao, China 35 C12 43 29N 123 30 E
Shuangshanzi, China 35 D10 40 20N 119 8 E
Shuangyang, China 35 C13 43 28N 125 40 E
Shuangyashan, China 33 B8 46 28N 131 5 E
Shuguri Falls, Tanzania 55 D4 8 33S 37 22 E
Shuiye, China 34 F8 36 7N 114 8 E
Shujalpur, India 42 H7 23 18N 76 46 E
Shukpa Kunzang, India 43 B8 34 22N 78 22 E
Shulan, China 35 B14 44 28N 127 0 E
Shule, China 32 C2 39 25N 76 3 E
Shumagin Is., U.S.A. 68 C4 55 7N 159 45W
Shumikha, Russia 26 D7 55 10N 63 15 E
Shungnak, U.S.A. 68 B4 66 52N 157 9W
Shuo Xian, China 34 E7 39 20N 112 33 E
Shūr →, Iran 45 D7 28 30N 55 0 E
Shūr Āb, Iran 45 C6 34 23N 51 11 E
Shūr Gaz, Iran 45 D8 29 10N 59 20 E
Shūrāb, Iran 45 C8 33 43N 56 29 E
Shūrjestān, Iran 45 D7 31 24N 52 25 E
Shurugwi, Zimbabwe 55 F3 19 40S 30 0 E
Shūsf, Iran 45 D9 31 50N 60 5 E
Shūshtar, Iran 45 D6 32 0N 48 50 E
Shuswap L., Canada 72 C5 50 55N 119 3W
Shuyang, China 35 G10 34 10N 118 42 E
Shūzū, Iran 45 D7 29 52N 54 30 E
Shwebo, Burma 41 H19 22 30N 95 45 E
Shwegu, Burma 41 G20 24 15N 96 26 E
Shweli →, Burma 41 H20 23 45N 96 45 E
Shymkent, Kazakstan 26 E7 42 18N 69 36 E
Shyok, India 43 B8 34 15N 78 12 E
Shyok →, Pakistan 43 B6 35 13N 75 53 E
Si Chon, Thailand 39 H2 9 0N 99 54 E
Si Kiang = Xi Jiang →, China 33 D6 22 5N 113 20 E
Si-ngan = Xi'an, China 34 G5 34 15N 109 0 E
Si Prachan, Thailand 38 E3 14 37N 100 9 E
Si Racha, Thailand 38 F3 13 10N 100 48 E
Si Xian, China 35 H9 33 30N 117 50 E
Siahan Range, Pakistan 40 F4 27 30N 64 40 E
Siaksriindrapura, Indonesia 36 D2 0 51N 102 0 E
Sialkot, Pakistan 42 C6 32 32N 74 30 E
Siam = Thailand ■, Asia 38 E4 16 0N 102 0 E
Siantan, P., Indonesia 39 L6 3 10N 106 15 E
Sīāreh, Iran 45 D9 28 5N 60 14 E
Siargao, Phil. 37 C7 9 52N 126 3 E
Siari, Pakistan 43 B7 34 55N 76 40 E
Siasi, Phil. 37 C6 5 34N 120 50 E
Siau, Indonesia 37 D7 2 50N 125 25 E
Šiauliai, Lithuania 9 J20 55 56N 23 15 E
Siaya □, Kenya 54 B3 0 0 34 20 E
Sibay, Russia 24 D10 52 42N 58 39 E
Sibayi, L., S. Africa 57 D5 27 20S 32 45 E
Šibenik, Croatia 20 C6 43 48N 15 54 E
Siberia, Russia 4 D13 60 0N 100 0 E
Siberut, Indonesia 36 E1 1 30S 99 0 E
Sibi, Pakistan 42 E2 29 30N 67 54 E
Sibil, Indonesia 37 E10 4 59S 140 35 E
Sibiti, Congo 52 E2 3 38S 13 19 E
Sibiu, Romania 17 F13 45 45N 24 9 E
Sibley, Iowa, U.S.A. 80 D7 43 24N 95 45W
Sibley, La., U.S.A. 81 J8 32 33N 93 18W
Sibolga, Indonesia 36 D1 1 42N 98 45 E
Sibsagar, India 41 F19 27 0N 94 36 E
Sibu, Malaysia 36 D4 2 18N 111 49 E
Sibuco, Phil. 37 C6 7 20N 122 10 E
Sibuguey B., Phil. 37 C6 7 50N 122 45 E
Sibut, C.A.R. 51 G8 5 46N 19 10 E
Sibutu, Phil. 37 D5 4 45N 119 30 E
Sibutu Passage, E. Indies 37 D5 4 50N 120 0 E
Sibuyan, Phil. 37 B6 12 25N 122 40 E
Sibuyan Sea, Phil. 37 B6 12 30N 122 20 E
Sicamous, Canada 72 C5 50 49N 119 0W
Sichuan □, China 32 C5 31 0N 104 0 E
Sicilia, Italy 20 F6 37 30N 14 30 E
Sicily = Sicilia, Italy 20 F6 37 30N 14 30 E
Sicuani, Peru 92 F4 14 21S 71 10W
Sidári, Greece 23 A3 39 47N 19 41 E
Siddhapur, India 42 H5 23 56N 72 25 E
Siddipet, India 40 K11 18 5N 78 51 E
Sidéradougou, Burkina Faso 50 F4 10 42N 4 12W
Sídheros, Ákra, Greece 23 D8 35 19N 26 19 E
Sīdī Barrānī, Egypt 51 B10 31 38N 25 58 E
Sidi-bel-Abbès, Algeria 50 A4 35 13N 0 39W
Sidlaw Hills, U.K. 12 E5 56 32N 3 2W
Sidley, Mt., Antarctica 5 D14 77 2S 126 2W
Sidmouth, U.K. 11 G4 50 40N 3 15W
Sidmouth, C., Australia 62 A3 13 25S 143 36 E
Sidney, Canada 72 D4 48 39N 123 24W
Sidney, Mont., U.S.A. 80 B2 47 43N 104 9W
Sidney, N.Y., U.S.A. 79 D9 42 19N 75 24W
Sidney, Nebr., U.S.A. 80 E3 41 8N 102 59W
Sidney, Ohio, U.S.A. 76 E3 40 17N 84 9W
Sidoarjo, Indonesia 37 G15 7 27S 112 43 E
Sidon = Saydā, Lebanon 47 B4 33 35N 35 25 E
Sidra, G. of = Surt, Khalīj, Libya 51 B8 31 40N 18 30 E
Sidra, G. of, Libya 48 C5 31 40N 18 30 E
Siedlce, Poland 17 B12 52 10N 22 20 E
Sieg →, Germany 16 C4 50 46N 7 6 E
Siegen, Germany 16 C5 50 51N 8 0 E
Siem Pang, Cambodia 38 E6 14 7N 106 23 E
Siem Reap, Cambodia 38 F4 13 20N 103 52 E
Siena, Italy 20 C4 43 19N 11 21 E
Sieradz, Poland 17 C10 51 37N 18 41 E
Sierra Blanca, U.S.A. 83 L11 31 11N 105 22W
Sierra Blanca Peak, U.S.A. 83 K11 33 23N 105 49W
Sierra City, U.S.A. 84 F6 39 34N 120 38W
Sierra Colorada, Argentina 96 E3 40 35S 67 50W
Sierra Gorda, Chile 94 A2 22 50S 69 15W
Sierra Leone ■, W. Afr. 50 G2 9 0N 12 0W
Sierra Madre, Mexico 87 D6 16 0N 93 0W
Sierra Mojada, Mexico 86 B4 27 19N 103 42W
Sierraville, U.S.A. 84 F6 39 36N 120 22W
Sífnos, Greece 21 F11 37 0N 24 45 E
Sifton, Canada 73 C8 51 21N 100 8W
Sifton Pass, Canada 72 B3 57 52N 126 15W
Sighetu-Marmatiei, Romania 17 E12 47 57N 23 52 E
Sighişoara, Romania 17 E13 46 12N 24 50 E
Sigli, Indonesia 36 C1 5 25N 96 0 E
Siglufjörður, Iceland 8 C4 66 12N 18 55W
Signal, U.S.A. 85 L13 34 30N 113 38W
Signal Pk., U.S.A. 85 M12 33 20N 114 2W
Sigsig, Ecuador 92 D3 3 0S 78 50W
Sigtuna, Sweden 9 G17 59 36N 17 44 E
Sigüenza, Spain 19 B4 41 3N 2 40W
Siguiri, Guinea 50 F3 11 31N 9 10W
Sigulda, Latvia 9 H21 57 10N 24 55 E
Sigurd, U.S.A. 83 G8 38 50N 111 58W
Sihanoukville = Kompong Som, Cambodia 39 G4 10 38N 103 30 E
Siikajoki →, Finland 8 D21 64 50N 24 43 E
Siilinjärvi, Finland 8 E22 63 4N 27 39 E
Sijarira Ra., Zimbabwe 55 F2 17 36S 27 45 E
Sikao, Thailand 39 J2 7 34N 99 21 E
Sikar, India 42 F6 27 33N 75 10 E
Sikasso, Mali 50 F3 11 18N 5 35W
Sikeston, U.S.A. 81 G10 36 53N 89 35W
Sikhote Alin, Khrebet, Russia 27 E14 45 0N 136 0 E
Sikhote Alin Ra. = Sikhote Alin, Khrebet, Russia 27 E14 45 0N 136 0 E
Síkinos, Greece 21 F11 36 40N 25 8 E
Sikkani Chief →, Canada 72 B4 57 47N 122 15W
Sikkim □, India 41 F16 27 50N 88 30 E
Sikotu-Ko, Japan 30 C10 42 45N 141 25 E
Sil →, Spain 19 A2 42 27N 7 43W
Silacayoapan, Mexico 87 D5 17 30N 98 9W
Silchar, India 41 G18 24 49N 92 48 E
Silcox, Canada 73 B10 57 12N 94 10W
Siler City, U.S.A. 77 H6 35 44N 79 28W
Silesia = Śląsk, Poland 16 C9 51 0N 16 30 E
Silgarhi Doti, Nepal 43 E9 29 15N 81 0 E
Silghat, India 41 F18 26 35N 93 0 E
Silifke, Turkey 25 G5 36 22N 33 58 E
Siliguri = Shiliguri, India 41 F16 26 45N 88 25 E
Siling Co, China 32 C3 31 50N 89 20 E
Silistra, Bulgaria 21 B12 44 6N 27 19 E
Silivri, Turkey 21 D13 41 4N 28 14 E
Siljan, Sweden 9 F16 60 55N 14 45 E
Silkeborg, Denmark 9 H13 56 10N 9 32 E
Sillajhuay, Cordillera, Chile 92 G5 19 46S 68 40W
Sillamäe, Estonia 9 G22 59 24N 27 45 E
Siloam Springs, U.S.A. 81 G7 36 11N 94 32W
Silsbee, U.S.A. 81 K7 30 21N 94 11W
Šilute, Lithuania 9 J19 55 21N 21 33 E
Silva Porto = Kuito, Angola 53 G3 12 22S 16 55 E
Silver City, N. Mex., U.S.A. 83 K9 32 46N 108 17W
Silver City, Nev., U.S.A. 82 G4 39 15N 119 48W
Silver Cr. →, U.S.A. 82 E4 43 16N 119 13W
Silver Creek, U.S.A. 78 D5 42 33N 79 10W
Silver L., Calif., U.S.A. 84 G6 38 39N 120 6W
Silver L., Calif., U.S.A. 85 K10 35 21N 116 7W
Silver Lake, U.S.A. 82 E3 43 8N 121 3W
Silver Streams, S. Africa 56 D3 28 20S 23 33 E
Silverton, Colo., U.S.A. 83 H10 37 49N 107 40W
Silverton, Tex., U.S.A. 81 H4 34 28N 101 19W
Silvies →, U.S.A. 82 E4 43 34N 119 2W
Simanggang, Malaysia 36 D4 1 15N 111 32 E
Simard, L., Canada 70 C4 47 40N 78 40W
Simav, Turkey 21 E13 39 4N 28 58 E
Simba, Tanzania 54 C4 2 10S 37 36 E
Simbirsk, Russia 24 D8 54 20N 48 25 E
Simbo, Tanzania 54 C2 4 51S 29 41 E
Simcoe, Canada 70 D3 42 50N 80 20W
Simcoe, L., Canada 70 D4 44 25N 79 20W
Simenga, Russia 27 C11 62 42N 108 25 E
Simeria, Romania 17 F12 45 51N 23 1 E
Simeulue, Indonesia 36 D1 2 45N 95 45 E
Simferopol, Ukraine 25 F5 44 55N 34 3 E
Sími, Greece 21 F12 36 35N 27 50 E
Simi Valley, U.S.A. 85 L8 34 16N 118 47W
Simikot, Nepal 43 E9 30 0N 81 50 E
Simla, India 42 D7 31 2N 77 9 E
Simmie, Canada 73 D7 49 56N 108 6W
Simmler, U.S.A. 85 K7 35 21N 119 59W
Simojoki →, Finland 8 D21 65 35N 25 1 E
Simojovel, Mexico 87 D6 17 12N 92 38W
Simonette →, Canada 72 B5 55 9N 118 15W
Simonstown, S. Africa 56 E2 34 14S 18 26 E
Simplon P., Switz. 16 E5 46 15N 8 3 E
Simpson Desert, Australia 62 D2 25 0S 137 0 E
Simpungdong, N. Korea 35 D15 40 56N 129 29 E
Simrishamn, Sweden 9 J16 55 33N 14 22 E
Simunjan, Malaysia 36 D4 1 25N 110 45 E
Simushir, Ostrov, Russia 27 E16 46 50N 152 30 E
Sinabang, Indonesia 36 D1 2 30N 96 24 E
Sinadogo, Somali Rep. 46 F4 5 50N 47 0 E
Sinai = Es Sînâ', Egypt 51 C11 29 0N 34 0 E
Sinai, Mt. = Mûsa, G., Egypt 51 C11 28 33N 33 59 E
Sinai Peninsula, Egypt 47 F2 29 30N 34 0 E
Sinaloa □, Mexico 86 C3 25 0N 107 30W
Sinaloa de Leyva, Mexico 86 B3 25 50N 108 20W
Sinarádhes, Greece 23 A3 39 34N 19 51 E
Sīnāwan, Libya 50 B7 31 0N 10 37 E
Sincelejo, Colombia 92 B3 9 18N 75 24W
Sinchang, N. Korea 35 D15 40 7N 128 28 E
Sinchang-ni, N. Korea 35 E14 39 24N 126 8 E
Sinclair, U.S.A. 82 F10 41 47N 107 7W
Sinclair Mills, Canada 72 C4 54 5N 121 40W
Sincorá, Serra do, Brazil 93 F10 13 30S 41 0W
Sind, Pakistan 42 G3 26 0N 68 30 E

Sind □, Pakistan 42 F3 26 0N 69 0 E
Sind →, India 43 B6 34 18N 74 45 E
Sind Sagar Doab, Pakistan 42 D4 32 0N 71 30 E
Sindangan, Phil. 37 C6 8 10N 123 5 E
Sindangbarang, Indonesia 37 G12 7 27S 107 1 E
Sinde, Zambia 55 F2 17 28S 25 51 E
Sines, Portugal 19 D1 37 56N 8 51W
Sines, C. de, Portugal 19 D1 37 58N 8 53W
Sineu, Spain 22 B10 39 38N 3 1 E
Sing Buri, Thailand 38 E3 14 53N 100 25 E
Singa, Sudan 51 F11 13 10N 33 57 E
Singapore ■, Asia 39 M4 1 17N 103 51 E
Singapore, Straits of, Asia 39 M5 1 15N 104 0 E
Singaraja, Indonesia 36 F5 8 6S 115 10 E
Singida, Tanzania 54 C3 4 49S 34 48 E
Singida □, Tanzania 54 D3 6 0S 34 30 E
Singitikós Kólpos, Greece 21 D11 40 6N 24 0 E
Singkaling Hkamti, Burma 41 G19 26 0N 95 39 E
Singkawang, Indonesia 36 D3 1 0N 108 57 E
Singleton, Australia 63 E5 32 33S 151 0 E
Singleton, Mt., N. Terr., Australia 60 D5 22 0S 130 46 E
Singleton, Mt., W. Austral., Australia 61 E2 29 27S 117 15 E
Singoli, India 42 G6 25 0N 75 22 E
Singora = Songkhla, Thailand 39 J3 7 13N 100 37 E
Singosan, N. Korea 35 E14 38 52N 127 25 E
Sinhung, N. Korea 35 D14 40 11N 127 34 E
Sînî □, Egypt 47 F2 30 0N 34 0 E
Sinjai, Indonesia 37 F6 5 7S 120 20 E
Sinjār, Iraq 44 B4 36 19N 41 52 E
Sinkat, Sudan 51 E12 18 55N 36 49 E
Sinkiang Uighur = Xinjiang Uygur Zizhiqu □, China 32 B3 42 0N 86 0 E
Sinmak, N. Korea 35 E14 38 25N 126 14 E
Sinni →, Italy 20 D7 40 8N 16 41 E
Sînnicolau Maré, Romania 17 E11 46 5N 20 39 E
Sinnuris, Egypt 51 C11 29 26N 30 31 E
Sinop, Turkey 25 F6 42 1N 35 11 E
Sinpo, N. Korea 35 E15 40 0N 128 13 E
Sinsk, Russia 27 C13 61 8N 126 48 E
Sint Eustatius, I., Neth. Ant. 89 C7 17 30N 62 59W
Sint Maarten, I., W. Indies 89 C7 18 4N 63 4W
Sint Niklaas, Belgium 15 C4 51 10N 4 9 E
Sint Truiden, Belgium 15 D5 50 48N 5 10 E
Sintang, Indonesia 36 D4 0 5N 111 35 E
Sinton, U.S.A. 81 L6 28 2N 97 31W
Sintra, Portugal 19 C1 38 47N 9 25W
Sinŭiju, N. Korea 35 D13 40 5N 124 24 E
Siocon, Phil. 37 C6 7 40N 122 10 E
Siófok, Hungary 17 E10 46 54N 18 3 E
Sioma, Zambia 56 B3 16 25S 23 28 E
Sion, Switz. 16 E4 46 14N 7 20 E
Sioux City, U.S.A. 80 D6 42 30N 96 24W
Sioux Falls, U.S.A. 80 D6 43 33N 96 44W
Sioux Lookout, Canada 70 B1 50 10N 91 50W
Siping, China 35 C13 43 8N 124 21 E
Sipiwesk L., Canada 73 B9 55 5N 97 35W
Sipura, Indonesia 36 E1 2 18S 99 40 E
Siquia →, Nic. 88 D3 12 10N 84 20W
Siquijor, Phil. 37 C6 9 12N 123 35 E
Siquirres, Costa Rica 88 D3 10 6N 83 30W
Sir Edward Pellew Group, Australia 62 B2 15 40S 137 10 E
Sir Graham Moore Is., Australia 60 B4 13 53S 126 34 E
Sira →, Norway 9 G12 58 23N 6 34 E
Siracusa, Italy 20 F6 37 4N 15 17 E
Sirajganj, Bangla. 43 G13 24 25N 89 47 E
Sirdān, Iran 45 B6 36 39N 49 12 E
Sirdaryo = Syrdarya →, Kazakstan 26 E7 46 3N 61 0 E
Sirer, Spain 22 C7 38 56N 1 22 E
Siret →, Romania 17 F14 45 24N 28 1 E
Sirohi, India 42 G5 24 52N 72 53 E
Sironj, India 42 G7 24 5N 77 39 E
Síros, Greece 21 F11 37 28N 24 57 E
Sirretta Pk., U.S.A. 85 K8 35 56N 118 19W
Sirsa, India 42 E6 29 33N 75 4 E
Sisak, Croatia 16 F9 45 30N 16 21 E
Sisaket, Thailand 38 E5 15 8N 104 23 E
Sishen, S. Africa 56 D3 27 47S 22 59 E
Sishui, Henan, China 34 G7 34 48N 113 15 E
Sishui, Shandong, China 35 G9 35 42N 117 18 E
Sisipuk L., Canada 73 B8 55 45N 101 50W
Sisophon, Cambodia 38 F4 13 38N 102 59 E
Sisseton, U.S.A. 80 C6 45 40N 97 3W
Sīstān va Balūchestān □, Iran 45 E9 27 0N 62 0 E
Sisters, U.S.A. 82 D3 44 18N 121 33W
Sitamarhi, India 43 F11 26 37N 85 30 E
Sitapur, India 43 F9 27 38N 80 45 E
Siteki, Swaziland 57 D5 26 32S 31 58 E
Sitges, Spain 19 B6 41 17N 1 47 E
Sitía, Greece 23 D8 35 13N 26 6 E
Sitka, U.S.A. 68 C6 57 3N 135 20W
Sitoti, Botswana 56 C3 23 15S 23 40 E
Sittang Myit →, Burma 41 L20 17 20N 96 45 E
Sittard, Neths. 15 C5 51 0N 5 52 E
Sittwe, Burma 41 J18 20 18N 92 45 E
Siuna, Nic. 88 D3 13 37N 84 45W
Siuri, India 43 H12 23 50N 87 34 E
Sivana, India 42 E8 28 37N 78 6 E
Sīvand, Iran 45 D7 30 5N 52 55 E
Sivas, Turkey 25 G6 39 43N 36 58 E
Sivomaskinskiy, Russia 24 A11 66 40N 62 35 E
Sivrihisar, Turkey 25 G5 39 30N 31 35 E
Sîwa, Egypt 51 C10 29 11N 25 31 E
Siwa Oasis, Egypt 48 D6 29 10N 25 30 E
Siwalik Range, Nepal 43 F10 28 0N 83 0 E
Siwan, India 43 F11 26 13N 84 21 E
Sizewell, U.K. 11 E9 52 12N 1 37 E
Siziwang Qi, China 34 D6 41 25N 111 40 E
Sjælland, Denmark 9 J14 55 30N 11 30 E
Sjumen = Šumen, Bulgaria 21 C12 43 18N 26 55 E
Skadarsko Jezero, Montenegro, Yug. 21 C8 42 10N 19 20 E
Skaftafell, Iceland 8 D5 64 1N 17 0W
Skagafjörður, Iceland 8 D4 65 54N 19 35W
Skagastølstindane, Norway 9 F12 61 28N 7 52 E
Skagaströnd, Iceland 8 D3 65 50N 20 19W
Skagen, Denmark 9 H14 57 43N 10 35 E
Skagerrak, Denmark 9 H13 57 30N 9 0 E
Skagit →, U.S.A. 84 B4 48 23N 122 22W
Skagway, U.S.A. 72 B1 59 28N 135 19W
Skala-Podilska, Ukraine 17 D14 48 50N 26 15 E
Skala Podolskaya = Skala-Podilska, Ukraine 17 D14 48 50N 26 15 E
Skalat, Ukraine 17 D13 49 23N 25 55 E
Skåne, Sweden 9 J15 55 59N 13 30 E
Skara, Sweden 9 G15 58 25N 13 30 E
Skardu, Pakistan 43 B6 35 20N 75 44 E
Skarzysko-Kamienna, Poland 17 C11 51 7N 20 52 E
Skeena →, Canada 72 C2 54 9N 130 5W
Skeena Mts., Canada 72 B3 56 40N 128 30W
Skegness, U.K. 10 D8 53 9N 0 20 E
Skeldon, Guyana 92 B7 5 55N 57 20W
Skellefte älv →, Sweden 8 D19 64 45N 21 10 E
Skellefteå, Sweden 8 D19 64 45N 20 50 E
Skelleftehamn, Sweden 8 D19 64 40N 21 9 E
Skerries, The, U.K. 10 D3 53 25N 4 36W
Ski, Norway 9 G14 59 43N 10 52 E
Skíathos, Greece 21 E10 39 12N 23 30 E
Skibbereen, Ireland 13 E2 51 33N 9 16W
Skiddaw, U.K. 10 C4 54 39N 3 9W
Skien, Norway 9 G13 59 12N 9 35 E
Skierniewice, Poland 17 C11 51 58N 20 10 E
Skikda, Algeria 50 A6 36 50N 6 58 E
Skilloura, Cyprus 23 D12 35 14N 33 10 E
Skipton, Australia 63 F3 37 39S 143 40 E
Skipton, U.K. 10 D5 53 58N 2 3W
Skirmish Pt., Australia 62 A1 11 59S 134 17 E
Skíros, Greece 21 E11 38 55N 24 34 E
Skive, Denmark 9 H13 56 33N 9 2 E
Skjálfandafljót →, Iceland 8 D5 65 59N 17 25W
Skjálfandi, Iceland 8 C5 66 5N 17 30W
Skoghall, Sweden 9 G15 59 20N 13 30 E
Skole, Ukraine 17 D12 49 3N 23 30 E
Skópelos, Greece 21 E10 39 9N 23 47 E
Skopí, Greece 23 D8 35 11N 26 2 E
Skopje, Macedonia 21 C9 42 1N 21 32 E
Skövde, Sweden 9 G15 58 24N 13 50 E
Skovorodino, Russia 27 D13 54 0N 124 0 E
Skowhegan, U.S.A. 71 D6 44 46N 69 43W
Skownan, Canada 73 C9 51 58N 99 35W
Skull, Ireland 13 E2 51 32N 9 34W
Skunk →, U.S.A. 80 E9 40 42N 91 7W
Skuodas, Lithuania 9 H19 56 16N 21 33 E
Skvyra, Ukraine 17 D15 49 44N 29 40 E
Skye, U.K. 12 D2 57 15N 6 10W
Skykomish, U.S.A. 82 C3 47 42N 121 22W
Skyros = Skíros, Greece 21 E11 38 55N 24 34 E
Slættaratindur, Færoe Is. 8 E9 62 18N 7 1W
Slagelse, Denmark 9 J14 55 23N 11 19 E
Slamet, Indonesia 36 F3 7 16S 109 8 E
Slaney →, Ireland 13 D5 52 26N 6 33W
Śląsk, Poland 16 C9 51 0N 16 30 E
Slate Is., Canada 70 C2 48 40N 87 0W
Slatina, Romania 17 F13 44 28N 24 22 E
Slaton, U.S.A. 81 J4 33 26N 101 39W
Slave →, Canada 72 A6 61 18N 113 39W
Slave Coast, W. Afr. 48 F4 6 0N 2 30 E
Slave Lake, Canada 72 B6 55 17N 114 43W
Slave Pt., Canada 72 A5 61 11N 115 56W
Slavgorod, Russia 26 D8 53 1N 78 37 E
Slavonski Brod, Croatia 21 B8 45 11N 18 0 E
Slavuta, Ukraine 17 C14 50 15N 27 2 E
Slavyanka, Russia 30 C5 42 53N 131 21 E
Slavyansk = Slovyansk, Ukraine 25 E6 48 55N 37 36 E
Slawharad, Belarus 17 B16 53 27N 31 0 E
Sleaford, U.K. 10 E7 53 0N 0 24W
Sleaford B., Australia 63 E2 34 55S 135 45 E
Sleat, Sd. of, U.K. 12 D3 57 5N 5 47W
Sleeper Is., Canada 69 C11 58 30N 81 0W
Sleepy Eye, U.S.A. 80 C7 44 18N 94 43W
Sleman, Indonesia 37 G14 7 40S 110 20 E
Slemon L., Canada 72 A5 63 13N 116 4W
Slidell, U.S.A. 81 K10 30 17N 89 47W
Sliedrecht, Neths. 15 C4 51 50N 4 45 E
Sliema, Malta 23 D2 35 54N 14 30 E
Slieve Aughty, Ireland 13 C3 53 4N 8 30W
Slieve Bloom, Ireland 13 C4 53 4N 7 40W
Slieve Donard, U.K. 13 B6 54 11N 5 55W
Slieve Gullion, U.K. 13 B5 54 7N 6 26W
Slieve Mish, Ireland 13 D2 52 12N 9 50W
Slievenamon, Ireland 13 D4 52 25N 7 34W
Sligeach = Sligo, Ireland 13 B3 54 16N 8 28W
Sligo, Ireland 13 B3 54 16N 8 28W
Sligo □, Ireland 13 B3 54 8N 8 42W
Sligo B., Ireland 13 B3 54 18N 8 40W
Slite, Sweden 9 H18 57 42N 18 48 E
Sliven, Bulgaria 21 C12 42 42N 26 19 E
Sloan, U.S.A. 85 K11 35 57N 115 13W
Sloansville, U.S.A. 79 D10 42 45N 74 22W
Slobodskoy, Russia 24 C9 58 40N 50 6 E
Slobozia, Romania 17 F14 44 34N 27 23 E
Slocan, Canada 72 D5 49 48N 117 28W
Slochteren, Neths. 15 A6 53 12N 6 48 E
Slonim, Belarus 17 B13 53 4N 25 19 E
Slough, U.K. 11 F7 51 30N 0 36W
Sloughhouse, U.S.A. 84 G5 38 26N 121 12W
Slovak Rep. ■, Europe 17 D10 48 30N 20 0 E
Slovakia = Slovak Rep. ■, Europe 17 D10 48 30N 20 0 E
Slovakian Ore Mts. = Slovenské Rudohorie, Slovak Rep. 17 D10 48 45N 20 0 E
Slovenia ■, Europe 16 F8 45 58N 14 30 E
Slovenija = Slovenia ■, Europe 16 F8 45 58N 14 30 E
Slovenska Republika = Slovak Rep. ■, Europe 17 D10 48 30N 20 0 E
Slovenské Rudohorie, Slovak Rep. 17 D10 48 45N 20 0 E
Slovyansk, Ukraine 25 E6 48 55N 37 36 E
Sluch →, Ukraine 17 C14 51 37N 26 38 E
Sluis, Neths. 15 C3 51 18N 3 23 E
Słupsk, Poland 17 A9 54 30N 17 3 E
Slurry, S. Africa 56 D4 25 49S 25 42 E
Slutsk, Belarus 17 B14 53 2N 27 31 E
Slyne Hd., Ireland 13 C1 53 25N 10 10W
Slyudyanka, Russia 27 D11 51 40N 103 40 E
Småland, Sweden 9 H16 57 15N 15 25 E
Smalltree L., Canada 73 A7 61 0N 105 0W
Smara, Morocco 50 B3 32 9N 8 16W
Smarhon, Belarus 17 A14 54 20N 26 24 E
Smartt Syndicate Dam, S. Africa 56 E3 30 45S 23 10 E
Smartville, U.S.A. 84 F5 39 13N 121 18W
Smeaton, Canada 73 C8 53 30N 104 49W
Smederevo, Serbia, Yug. 21 B9 44 40N 20 57 E
Smethport, U.S.A. 78 E6 41 49N 78 27W
Smidovich, Russia 27 E14 48 36N 133 49 E
Smiley, Canada 73 C7 51 38N 109 29W
Smith, Canada 72 B6 55 10N 114 0W
Smith →, Canada 72 B3 59 34N 126 30W
Smith Arm, Canada 68 B7 66 15N 123 0W
Smith Center, U.S.A. 80 F5 39 47N 98 47W
Smith Sund, Greenland 4 B4 78 30N 74 0W
Smithburne →, Australia 62 B3 17 3S 140 57 E
Smithers, Canada 72 C3 54 45N 127 10W
Smithfield, S. Africa 57 E4 30 9S 26 30 E
Smithfield, N.C., U.S.A. 77 H6 35 31N 78 21W
Smithfield, Utah, U.S.A. 82 F8 41 50N 111 50W
Smiths Falls, Canada 70 D4 44 55N 76 0W
Smithton, Australia 62 G4 40 53S 145 6 E
Smithtown, Australia 63 E5 30 58S 152 48 E
Smithville, Canada 78 C5 43 6N 79 33W
Smithville, U.S.A. 81 K6 30 1N 97 10W
Smoky →, Canada 72 B5 56 10N 117 21W
Smoky Bay, Australia 63 E1 32 22S 134 13 E
Smoky Falls, Canada 70 B3 50 4N 82 10W
Smoky Hill →, U.S.A. 80 F6 39 4N 96 48W
Smoky Lake, Canada 72 C6 54 10N 112 30W
Smøla, Norway 8 E13 63 23N 8 3 E
Smolensk, Russia 24 D5 54 45N 32 5 E
Smolikas, Óros, Greece 21 D9 40 9N 20 58 E
Smolyan, Bulgaria 21 D11 41 36N 24 38 E
Smooth Rock Falls, Canada 70 C3 49 17N 81 37W
Smoothstone L., Canada 73 C7 54 40N 106 50W
Smorgon = Smarhon, Belarus 17 A14 54 20N 26 24 E
Smyrna = İzmir, Turkey 21 E12 38 25N 27 8 E
Snæfell, Iceland 8 D6 64 48N 15 34W
Snaefell, U.K. 10 C3 54 16N 4 27W
Snæfellsjökull, Iceland 8 D2 64 49N 23 46W
Snake →, U.S.A. 82 C4 46 12N 119 2W
Snake I., Australia 63 F4 38 47S 146 33 E
Snake L., Canada 73 B7 55 32N 106 35W
Snake Range, U.S.A. 82 G6 39 0N 114 20W
Snake River Plain, U.S.A. 82 E7 42 50N 114 0W
Snåsavatnet, Norway 8 D14 64 12N 12 0 E
Sneek, Neths. 15 A5 53 2N 5 40 E
Sneeuberge, S. Africa 56 E3 31 46S 24 20 E
Snelling, U.S.A. 84 H6 37 31N 120 26W
Sněžka, Europe 16 C8 50 41N 15 50 E
Snizort, L., U.K. 12 D2 57 33N 6 28W
Snøhetta, Norway 9 E13 62 19N 9 16 E
Snohomish, U.S.A. 84 C4 47 55N 122 6W
Snoul, Cambodia 39 F6 12 4N 106 26 E
Snow Hill, U.S.A. 76 F8 38 11N 75 24W
Snow Lake, Canada 73 C8 54 52N 100 3W
Snow Mt., U.S.A. 84 F4 39 23N 122 45W
Snowbird L., Canada 73 A8 60 45N 103 0W
Snowdon, U.K. 10 D3 53 4N 4 5W
Snowdrift →, Canada 73 A6 62 24N 110 44W
Snowflake, U.S.A. 83 J8 34 30N 110 5W
Snowshoe Pk., U.S.A. 82 B6 48 13N 115 41W
Snowtown, Australia 63 E2 33 46S 138 14 E
Snowville, U.S.A. 82 F7 41 58N 112 43W
Snowy →, Australia 63 F4 37 46S 148 30 E
Snowy Mts., Australia 63 F4 36 30S 148 20 E
Snug Corner, Bahamas 89 B5 22 33N 73 52W
Snyatyn, Ukraine 17 D13 48 27N 25 38 E
Snyder, Okla., U.S.A. 81 H5 34 40N 98 57W
Snyder, Tex., U.S.A. 81 J4 32 44N 100 55W
Soahanina, Madag. 57 B7 18 42S 44 13 E
Soalala, Madag. 57 B8 16 6S 45 20 E
Soan →, Pakistan 42 C4 33 1N 71 44 E
Soanierana-Ivongo, Madag. 57 B8 16 55S 49 35 E
Soap Lake, U.S.A. 82 C4 47 23N 119 29W
Sobat, Nahr →, Sudan 51 G11 9 22N 31 33 E
Sobhapur, India 42 H8 22 47N 78 17 E
Sobolevo, Russia 27 D16 54 20N 155 30 E
Sobradinho, Reprêsa de, Brazil 93 E10 9 30S 42 0 E
Sobral, Brazil 93 D10 3 50S 40 20W
Soc Giang, Vietnam 38 A6 22 54N 106 1 E
Soc Trang, Vietnam 39 H5 9 37N 105 50 E
Soch'e = Shache, China 32 C2 38 20N 77 10 E
Sochi, Russia 25 F6 43 35N 39 40 E
Société, Is. de la, Pac. Oc. 65 J12 17 0S 151 0W
Society Is. = Société, Is. de la, Pac. Oc. 65 J12 17 0S 151 0W
Socompa, Portezuelo de, Chile 94 A2 24 27S 68 18W
Socorro, Colombia 92 B4 6 29N 73 16W
Socorro, U.S.A. 83 J10 34 4N 106 54W
Socorro, I., Mexico 86 D2 18 45N 110 58W
Socotra, Ind. Oc. 46 E5 12 30N 54 0 E
Soda L., U.S.A. 83 J5 35 10N 116 4W
Soda Plains, India 43 B8 35 30N 79 0 E
Soda Springs, U.S.A. 82 E8 42 39N 111 36W
Sodankylä, Finland 8 C22 67 29N 26 40 E
Söderhamn, Sweden 9 F17 61 18N 17 10 E
Söderköping, Sweden 9 G17 58 31N 16 20 E
Södermanland, Sweden 9 G17 58 56N 16 55 E
Södertälje, Sweden 9 G17 59 12N 17 39 E
Sodiri, Sudan 51 F10 14 27N 29 0 E
Sodo, Ethiopia 51 G12 7 0N 37 41 E
Sodus, U.S.A. 78 C7 43 14N 77 4W
Soekmekaar, S. Africa 57 C4 23 30S 29 55 E
Soest, Neths. 15 B5 52 9N 5 19 E
Sofia = Sofiya, Bulgaria 21 C10 42 45N 23 20 E
Sofia →, Madag. 57 B8 15 27S 47 23 E
Sofiya, Bulgaria 21 C10 42 45N 23 20 E
Sofiysk, Russia 27 D14 52 15N 133 59 E
Sōfu-Gan, Japan 31 K10 29 49N 140 21 E
Sogamoso, Colombia 92 B4 5 43N 72 56W
Sogār, Iran 45 E8 25 53N 58 6 E
Sogndalsfjøra, Norway 9 F12 61 14N 7 5 E
Søgne, Norway 9 G12 58 5N 7 48 E
Sognefjorden, Norway 9 F11 61 10N 5 50 E
Sŏgwi-po, S. Korea 35 H14 33 13N 126 34 E
Soh, Iran 45 C6 33 26N 51 27 E
Sohâg, Egypt 51 C11 26 33N 31 43 E
Sŏhori, N. Korea 35 D15 40 7N 128 23 E
Soignies, Belgium 15 D4 50 35N 4 5 E
Soissons, France 18 B5 49 25N 3 19 E
Sōja, Japan 31 G6 34 40N 133 45 E
Sojat, India 42 G5 25 55N 73 45 E
Sokal, Ukraine 17 C13 50 31N 24 15 E
Söke, Turkey 21 F12 37 48N 27 28 E
Sokelo, Zaïre 55 D1 9 55S 24 36 E
Sokhumi, Georgia 25 F7 43 0N 41 0 E
Sokodé, Togo 50 G5 9 0N 1 11 E
Sokol, Russia 24 C7 59 30N 40 5 E
Sokółka, Poland 17 B12 53 25N 23 30 E
Sokolo, Mali 50 F3 14 53N 6 8W
Sokołów Podlaski, Poland 17 B12 52 25N 22 15 E
Sokoto, Nigeria 50 F6 13 2N 5 16 E
Sol Iletsk, Russia 24 D10 51 10N 55 0 E
Solai, Kenya 54 B4 0 2N 36 12 E
Solano, Phil. 37 A6 16 31N 121 15 E
Solapur, India 40 L9 17 43N 75 56 E
Soléa □, Cyprus 23 D12 35 5N 33 4 E
Soledad, U.S.A. 83 H3 36 26N 121 20W
Soledad, Venezuela 92 B6 8 10N 63 34W
Solent, The, U.K. 11 G6 50 45N 1 25W
Solfonn, Norway 9 F12 60 2N 6 57 E
Soligalich, Russia 24 C7 59 5N 42 10 E
Soligorsk = Salihorsk, Belarus 17 B14 52 51N 27 27 E
Solikamsk, Russia 24 C10 59 38N 56 50 E
Solila, Madag. 57 C8 21 25S 46 37 E
Solimões = Amazonas →, S. Amer. 93 C9 0 5S 50 0W
Solingen, Germany 15 C7 51 10N 7 5 E
Sollefteå, Sweden 8 E17 63 12N 17 20 E
Sóller, Spain 22 B9 39 46N 2 43 E
Sologne, France 18 C4 47 40N 1 45 E
Solok, Indonesia 36 E2 0 45S 100 40 E
Sololá, Guatemala 88 D1 14 49N 91 10W
Solomon, N. Fork →, U.S.A. 80 F5 39 29N 98 26W
Solomon, S. Fork →, U.S.A. 80 F5 39 25N 99 12W
Solomon Is. ■, Pac. Oc. 64 H7 6 0S 155 0 E
Solon, China 33 B7 46 32N 121 10 E
Solon Springs, U.S.A. 80 B9 46 22N 91 49W
Solor, Indonesia 37 F6 8 27S 123 0 E
Solothurn, Switz. 16 E4 47 13N 7 32 E
Šolta, Croatia 20 C7 43 24N 16 15 E
Soltānābād, Khorāsān, Iran 45 C8 34 13N 59 58 E
Soltānābād, Khorāsān, Iran 45 B8 36 29N 58 5 E
Soltānābād, Markazī, Iran 45 C6 35 31N 51 10 E
Solunska Glava, Macedonia 21 D9 41 44N 21 31 E
Solvang, U.S.A. 85 L6 34 36N 120 8W
Solvay, U.S.A. 79 C8 43 3N 76 13W
Sölvesborg, Sweden 9 H16 56 5N 14 35 E
Solvychegodsk, Russia 24 B8 61 21N 46 56 E
Solway Firth, U.K. 10 C4 54 49N 3 35W
Solwezi, Zambia 55 E2 12 11S 26 21 E
Sōma, Japan 30 F10 37 40N 140 50 E
Soma, Turkey 21 E12 39 10N 27 35 E
Somali Pen., Africa 48 F8 7 0N 46 0 E
Somali Rep. ■, Africa 46 F4 7 0N 47 0 E
Somalia = Somali Rep. ■, Africa 46 F4 7 0N 47 0 E
Sombor, Serbia, Yug. 21 B8 45 46N 19 9 E
Sombra, Canada 78 D2 42 43N 82 29W
Sombrerete, Mexico 86 C4 23 40N 103 40W
Sombrero, Anguilla 89 C7 18 37N 63 30W
Somers, U.S.A. 82 B6 48 5N 114 13W
Somerset, Canada 73 D9 49 25N 98 39W
Somerset, Colo., U.S.A. 83 G10 38 56N 107 28W
Somerset, Ky., U.S.A. 76 G3 37 5N 84 36W
Somerset, Mass., U.S.A. 79 E13 41 47N 71 8W
Somerset, Pa., U.S.A. 78 F5 40 1N 79 5W
Somerset □, U.K. 11 F5 51 9N 3 0W
Somerset East, S. Africa 56 E4 32 42S 25 35 E
Somerset I., Canada 68 A10 73 30N 93 0W
Somerset West, S. Africa 56 E2 34 8S 18 50 E
Somerton, U.S.A. 83 K6 32 36N 114 43W
Somerville, U.S.A. 79 F10 40 35N 74 38W
Someş →, Romania 17 D12 47 49N 22 43 E
Sommariva, Australia 63 D4 26 24S 146 36 E
Somme →, France 18 A4 50 11N 1 38 E
Somosierra, Puerto de, Spain 19 B4 41 4N 3 35W
Somoto, Nic. 88 D2 13 28N 86 37W
Somport, Puerto de, Spain 18 E3 42 48N 0 31W
Son Ha, Vietnam 38 E7 15 3N 108 34 E
Son Hoa, Vietnam 38 F7 13 2N 108 58 E
Son La, Vietnam 38 B4 21 20N 103 50 E
Son Tay, Vietnam 38 B5 21 8N 105 30 E
Soná, Panama 88 E3 8 0N 81 20W
Sonamarg, India 43 B6 34 18N 75 21 E
Sonamukhi, India 43 H12 23 18N 87 27 E
Sŏnchŏn, N. Korea 35 E13 39 48N 124 55 E
Sondags →, S. Africa 56 E4 33 44S 25 51 E
Sondar, India 43 C6 33 28N 75 56 E
Sønderborg, Denmark 9 J13 54 55N 9 49 E
Søndre Strømfjord, Greenland 69 B14 66 59N 50 40W
Sóndrio, Italy 20 A3 46 10N 9 52 E
Sone, Mozam. 55 F3 17 23S 34 55 E
Sonepur, India 41 J13 20 55N 83 50 E
Song, Thailand 38 C3 18 28N 100 11 E
Song Cau, Vietnam 38 F7 13 27N 109 18 E
Song Xian, China 34 G7 34 12N 112 8 E
Songchŏn, N. Korea 35 E14 39 12N 126 15 E
Songea, Tanzania 55 E4 10 40S 35 40 E
Songea □, Tanzania 55 E4 10 30S 36 0 E
Songhua Hu, China 35 C14 43 35N 126 50 E
Songhua Jiang →, China 33 B8 47 45N 132 30 E
Songjin, N. Korea 35 D15 40 40N 129 10 E
Songjŏng-ni, S. Korea 35 G14 35 8N 126 47 E
Songkhla, Thailand 39 J3 7 13N 100 37 E
Songnim, N. Korea 35 E13 38 45N 125 39 E

Songpan, *China* **32 C5** 32 40N 103 30 E
Songwe, *Zaïre* **54 C2** 3 20S 26 16 E
Songwe →, *Africa* **55 D3** 9 44S 33 58 E
Sonid Youqi, *China* **34 C7** 42 45N 112 48 E
Sonipat, *India* **42 E7** 29 0N 77 5 E
Sonmiani, *Pakistan* **42 G2** 25 25N 66 40 E
Sono →, *Brazil* **93 E9** 9 58S 48 11W
Sonora, *Calif., U.S.A.* **83 H3** 37 59N 120 23W
Sonora, *Tex., U.S.A.* **81 K4** 30 34N 100 39W
Sonora □, *Mexico* **86 B2** 29 0N 111 0W
Sonora →, *Mexico* **86 B2** 28 50N 111 33W
Sonora Desert, *U.S.A.* **85 L12** 33 40N 114 15W
Sonoyta, *Mexico* **86 A2** 31 51N 112 50W
Sŏnsan, *S. Korea* **35 F15** 36 14N 128 17 E
Sonsonate, *El Salv.* **88 D2** 13 43N 89 44W
Soochow = Suzhou, *China* **33 C7** 31 19N 120 38 E
Sop Hao, *Laos* **38 B5** 20 33N 104 27 E
Sop Prap, *Thailand* **38 D2** 17 53N 99 20 E
Sopi, *Indonesia* **37 D7** 2 34N 128 28 E
Sopot, *Poland* **17 A10** 54 27N 18 31 E
Sopron, *Hungary* **17 E9** 47 45N 16 32 E
Sop's Arm, *Canada* **71 C8** 49 46N 56 56W
Sopur, *India* **43 B6** 34 18N 74 27 E
Sør-Rondane, *Antarctica* **5 D4** 72 0S 25 0 E
Sorah, *Pakistan* **42 F3** 27 13N 68 56 E
Sorata, *Bolivia* **92 G5** 15 50S 68 40W
Sorel, *Canada* **70 C5** 46 0N 73 10W
Soreq, N. →, *Israel* **47 D3** 31 57N 34 43 E
Sorgono, *Italy* **20 D3** 40 1N 9 6 E
Soria, *Spain* **19 B4** 41 43N 2 32W
Soriano, *Uruguay* **94 C4** 33 24S 58 19W
Sorkh, Kuh-e, *Iran* **45 C8** 35 40N 58 30 E
Soroca, *Moldova* **17 D15** 48 8N 28 12 E
Sorocaba, *Brazil* **95 A6** 23 31S 47 27W
Sorochinsk, *Russia* **24 D9** 52 26N 53 10 E
Soroki = Soroca, *Moldova* **17 D15** 48 8N 28 12 E
Soron, *India* **43 F8** 27 55N 78 45 E
Sorong, *Indonesia* **37 E8** 0 55S 131 15 E
Soroní, *Greece* **23 C10** 36 21N 28 1 E
Soroti, *Uganda* **54 B3** 1 43N 33 35 E
Sørøya, *Norway* **8 A20** 70 40N 22 30 E
Sørøysundet, *Norway* **8 A20** 70 25N 23 0 E
Sorrento, *Australia* **63 F3** 38 22S 144 47 E
Sorsele, *Sweden* **8 D17** 65 31N 17 30 E
Sorsogon, *Phil.* **37 B6** 13 0N 124 0 E
Sortavala, *Russia* **24 B5** 61 42N 30 41 E
Sortland, *Norway* **8 B16** 68 42N 15 25 E
Sŏsan, *S. Korea* **35 F14** 36 47N 126 27 E
Soscumica, L., *Canada* **70 B4** 50 15N 77 27W
Sosnogorsk, *Russia* **24 B9** 63 37N 53 51 E
Sosnovka, *Russia* **27 D11** 54 9N 109 35 E
Sosnowiec, *Poland* **17 C10** 50 20N 19 10 E
Sŏsura, *N. Korea* **35 C16** 42 16N 130 36 E
Sosva, *Russia* **24 C11** 59 10N 61 50 E
Sotkamo, *Finland* **8 D23** 64 8N 28 23 E
Soto la Marina →, *Mexico* **87 C5** 23 40N 97 40W
Sotuta, *Mexico* **87 C7** 20 29N 89 43W
Souanké, *Congo* **52 D2** 2 10N 14 3 E
Soúdha, *Greece* **23 D6** 35 29N 24 4 E
Soúdhas, Kólpos, *Greece* **23 D6** 35 25N 24 10 E
Soukhouma, *Laos* **38 E5** 14 38N 105 48 E
Sŏul, *S. Korea* **35 F14** 37 31N 126 58 E
Sound, The, *U.K.* **11 G3** 50 20N 4 10W
Sources, Mt. aux, *Lesotho* **57 D4** 28 45S 28 50 E
Soure, *Brazil* **93 D9** 0 35S 48 30W
Souris, *Man., Canada* **73 D8** 49 40N 100 20W
Souris, *P.E.I., Canada* **71 C7** 46 21N 62 15W
Souris →, *Canada* **80 A5** 49 40N 99 34W
Sousa, *Brazil* **93 E11** 6 45S 38 10W
Sousel, *Brazil* **93 D8** 2 38S 52 29W
Sousse, *Tunisia* **51 A7** 35 50N 10 38 E
South Africa ■, *Africa* **56 E3** 32 0S 23 0 E
South Atlantic Ocean **90 H7** 20 0S 10 0W
South Aulatsivik I., *Canada* **71 A7** 56 45N 61 30W
South Australia □, *Australia* **63 E2** 32 0S 139 0 E
South Ayrshire □, *U.K.* **12 F4** 55 18N 4 41W
South Baldy, *U.S.A.* **83 J10** 33 59N 107 11W
South Bend, *Ind., U.S.A.* **76 E2** 41 41N 86 15W
South Bend, *Wash., U.S.A.* **84 D3** 46 40N 123 48W
South Boston, *U.S.A.* **77 G6** 36 42N 78 54W
South Branch, *Canada* **71 C8** 47 55N 59 2W
South Brook, *Canada* **71 C8** 49 26N 56 5W
South Carolina □, *U.S.A.* **77 J5** 34 0N 81 0W
South Charleston, *U.S.A.* **76 F5** 38 22N 81 44W
South China Sea, *Asia* **36 C4** 10 0N 113 0 E
South Dakota □, *U.S.A.* **80 C5** 44 15N 100 0W
South Downs, *U.K.* **11 G7** 50 52N 0 25W
South East C., *Australia* **62 G4** 43 40S 146 50 E
South East Is., *Australia* **61 F3** 34 17S 123 30 E
South Esk →, *U.K.* **12 E5** 56 43N 2 31W
South Foreland, *U.K.* **11 F9** 51 8N 1 24 E
South Fork →, *U.S.A.* **82 C7** 47 54N 113 15W
South Fork, American →, *U.S.A.* **84 G5** 38 45N 121 5W
South Fork, Feather →, *U.S.A.* **84 F5** 39 17N 121 36W
South Georgia, *Antarctica* **5 B1** 54 30S 37 0W
South Gloucestershire □, *U.K.* **11 F5** 51 32N 2 28W
South Haven, *U.S.A.* **76 D2** 42 24N 86 16W
South Henik, L., *Canada* **73 A9** 61 30N 97 30W
South Honshu Ridge, *Pac. Oc.* **64 E6** 23 0N 143 0 E
South Horr, *Kenya* **54 B4** 2 12N 36 56 E
South I., *Kenya* **54 B4** 2 35N 36 35 E
South I., *N.Z.* **59 L3** 44 0S 170 0 E
South Invercargill, *N.Z.* **59 M2** 46 26S 168 23 E
South Knife →, *Canada* **73 B10** 58 55N 94 37W
South Korea ■, *Asia* **35 F15** 36 0N 128 0 E
South Lake Tahoe, *U.S.A.* **84 G6** 38 57N 119 59W
South Lanarkshire □, *U.K.* **12 F5** 55 37N 3 53W
South Loup →, *U.S.A.* **80 E5** 41 4N 98 39W
South Magnetic Pole, *Antarctica* **5 C9** 64 8S 138 8 E
South Milwaukee, *U.S.A.* **76 D2** 42 55N 87 52W
South Molton, *U.K.* **11 F4** 51 1N 3 51W
South Nahanni →, *Canada* **72 A4** 61 3N 123 21W
South Natuna Is. = Natuna Selatan, Kepulauan, *Indonesia* **39 L7** 2 45N 109 0 E
South Negril Pt., *Jamaica* **88 C4** 18 14N 78 30W
South Orkney Is., *Antarctica* **5 C18** 63 0S 45 0W
South Pagai, I. = Pagai Selatan, P., *Indonesia* **36 E2** 3 0S 100 15 E
South Pass, *U.S.A.* **82 E9** 42 20N 108 58W
South Pittsburg, *U.S.A.* **77 H3** 35 1N 85 42W
South Platte →, *U.S.A.* **80 E4** 41 7N 100 42W
South Pole, *Antarctica* **5 E** 90 0S 0 0 E
South Porcupine, *Canada* **70 C3** 48 30N 81 12W
South River, *Canada* **70 C4** 45 52N 79 23W
South River, *U.S.A.* **79 F10** 40 27N 74 23W
South Ronaldsay, *U.K.* **12 C6** 58 48N 2 58W
South Sandwich Is., *Antarctica* **5 B1** 57 0S 27 0W
South Saskatchewan →, *Canada* **73 C7** 53 15N 105 5W
South Seal →, *Canada* **73 B9** 58 48N 98 8W
South Shetland Is., *Antarctica* **5 C18** 62 0S 59 0W
South Shields, *U.K.* **10 C6** 55 0N 1 25W
South Sioux City, *U.S.A.* **80 D6** 42 28N 96 24W
South Taranaki Bight, *N.Z.* **59 H5** 39 40S 174 5 E
South Thompson →, *Canada* **72 C4** 50 40N 120 20W
South Twin I., *Canada* **70 B4** 53 7N 79 52W
South Tyne →, *U.K.* **10 C5** 54 59N 2 8W
South Uist, *U.K.* **12 D1** 57 20N 7 15W
South West Africa = Namibia ■, *Africa* **56 C2** 22 0S 18 9 E
South West C., *Australia* **62 G4** 43 34S 146 3 E
South Yorkshire □, *U.K.* **10 D6** 53 27N 1 36W
Southampton, *Canada* **70 D3** 44 30N 81 25W
Southampton, *U.K.* **11 G6** 50 54N 1 23W
Southampton, *U.S.A.* **79 F12** 40 53N 72 23W
Southampton I., *Canada* **69 B11** 64 30N 84 0W
Southbridge, *N.Z.* **59 K4** 43 48S 172 16 E
Southbridge, *U.S.A.* **79 D12** 42 5N 72 2W
Southend, *Canada* **73 B8** 56 19N 103 22W
Southend-on-Sea, *U.K.* **11 F8** 51 32N 0 44 E
Southern □, *Malawi* **55 F4** 15 0S 35 0 E
Southern □, *Uganda* **54 C3** 0 15S 31 30 E
Southern □, *Zambia* **55 F2** 16 20S 26 20 E
Southern Alps, *N.Z.* **59 K3** 43 41S 170 11 E
Southern Cross, *Australia* **61 F2** 31 12S 119 15 E
Southern Hills, *Australia* **61 F3** 32 15S 122 40 E
Southern Indian L., *Canada* **73 B9** 57 10N 98 30W
Southern Ocean, *Antarctica* **5 C6** 62 0S 60 0 E
Southern Pines, *U.S.A.* **77 H6** 35 11N 79 24W
Southern Uplands, *U.K.* **12 F5** 55 28N 3 52W
Southington, *U.S.A.* **79 E12** 41 36N 72 53W
Southold, *U.S.A.* **79 E12** 41 4N 72 26W
Southport, *Australia* **63 D5** 27 58S 153 25 E
Southport, *U.K.* **10 D4** 53 39N 3 0W
Southport, *U.S.A.* **77 J6** 33 55N 78 1W
Southwest C., *N.Z.* **59 M1** 47 17S 167 28 E
Southwold, *U.K.* **11 E9** 52 20N 1 41 E
Soutpansberg, *S. Africa* **57 C4** 23 0S 29 30 E
Sovetsk, *Kaliningd., Russia* **9 J19** 55 6N 21 50 E
Sovetsk, *Kirov, Russia* **24 C8** 57 38N 48 53 E
Sovetskaya Gavan, *Russia* **27 E15** 48 50N 140 5 E
Soweto, *S. Africa* **57 D4** 26 14S 27 54 E
Sōya-Kaikyō = La Perouse Str., *Asia* **30 B11** 45 40N 142 0 E
Sōya-Misaki, *Japan* **30 B10** 45 30N 141 55 E
Soyo, *Angola* **52 F2** 6 13S 12 20 E
Sozh →, *Belarus* **17 B16** 51 57N 30 48 E
Spa, *Belgium* **15 D5** 50 29N 5 53 E
Spain ■, *Europe* **19 B4** 39 0N 4 0W
Spalding, *Australia* **63 E2** 33 30S 138 37 E
Spalding, *U.K.* **10 E7** 52 48N 0 9W
Spalding, *U.S.A.* **80 E5** 41 42N 98 22W
Spangler, *U.S.A.* **78 F6** 40 39N 78 48W
Spaniard's Bay, *Canada* **71 C9** 47 38N 53 20W
Spanish, *Canada* **70 C3** 46 12N 82 20W
Spanish Fork, *U.S.A.* **82 F8** 40 7N 111 39W
Spanish Town, *Jamaica* **88 C4** 18 0N 76 57W
Sparks, *U.S.A.* **84 F7** 39 32N 119 45W
Sparta = Spárti, *Greece* **21 F10** 37 5N 22 25 E
Sparta, *Ga., U.S.A.* **77 J4** 33 17N 82 58W
Sparta, *Wis., U.S.A.* **80 D9** 43 56N 90 49W
Spartanburg, *U.S.A.* **77 H4** 34 56N 81 57W
Spartansburg, *U.S.A.* **78 E5** 41 49N 79 41W
Spárti, *Greece* **21 F10** 37 5N 22 25 E
Spartivento, C., *Calabria, Italy* **20 F7** 37 55N 16 4 E
Spartivento, C., *Sard., Italy* **20 E3** 38 53N 8 50 E
Spassk Dalniy, *Russia* **27 E14** 44 40N 132 48 E
Spátha, Ákra, *Greece* **23 D5** 35 42N 23 43 E
Spatsizi →, *Canada* **72 B3** 57 42N 128 7W
Spearfish, *U.S.A.* **80 C3** 44 30N 103 52W
Spearman, *U.S.A.* **81 G4** 36 12N 101 12W
Speers, *Canada* **73 C7** 52 43N 107 34W
Speightstown, *Barbados* **89 D8** 13 15N 59 39W
Speke Gulf, *Tanzania* **54 C3** 2 20S 32 50 E
Spencer, *Idaho, U.S.A.* **82 D7** 44 22N 112 11W
Spencer, *Iowa, U.S.A.* **80 D7** 43 9N 95 9W
Spencer, *N.Y., U.S.A.* **79 D8** 42 13N 76 30W
Spencer, *Nebr., U.S.A.* **80 D5** 42 53N 98 42W
Spencer, *W. Va., U.S.A.* **76 F5** 38 48N 81 21W
Spencer, C., *Australia* **63 F2** 35 20S 136 53 E
Spencer B., *Namibia* **56 D1** 25 30S 14 47 E
Spencer G., *Australia* **63 E2** 34 0S 137 20 E
Spencerville, *Canada* **79 B9** 44 51N 75 33W
Spences Bridge, *Canada* **72 C4** 50 25N 121 20W
Spenser Mts., *N.Z.* **59 K4** 42 15S 172 45 E
Sperrin Mts., *U.K.* **13 B5** 54 50N 7 0W
Spey →, *U.K.* **12 D5** 57 40N 3 6W
Speyer, *Germany* **16 D5** 49 29N 8 25 E
Spili, *Greece* **23 D6** 35 13N 24 31 E
Spin Baldak = Qala-i-Jadid, *Afghan.* **42 D2** 31 1N 66 25 E
Spinalónga, *Greece* **23 D7** 35 18N 25 44 E
Spirit Lake, *Idaho, U.S.A.* **82 C5** 47 58N 116 52W
Spirit Lake, *Wash., U.S.A.* **84 D4** 46 15N 122 9W
Spirit River, *Canada* **72 B5** 55 45N 118 50W
Spiritwood, *Canada* **73 C7** 53 24N 107 33W
Spithead, *U.K.* **11 G6** 50 45N 1 10W
Spitzbergen = Svalbard, *Arctic* **4 B8** 78 0N 17 0 E
Spjelkavik, *Norway* **9 E12** 62 28N 6 22 E
Split, *Croatia* **20 C7** 43 31N 16 26 E
Split L., *Canada* **73 B9** 56 8N 96 15W
Spofford, *U.S.A.* **81 L4** 29 10N 100 25W
Spokane, *U.S.A.* **82 C5** 47 40N 117 24W
Spoleto, *Italy* **20 C5** 42 44N 12 44 E
Spooner, *U.S.A.* **80 C9** 45 50N 91 53W
Sporyy Navolok, Mys, *Russia* **26 B7** 75 50N 68 40 E
Spragge, *Canada* **70 C3** 46 15N 82 40W
Sprague, *U.S.A.* **82 C5** 47 18N 117 59W
Sprague River, *U.S.A.* **82 E3** 42 27N 121 30W
Spratly Is., *S. China Sea* **36 C4** 8 20N 112 0 E
Spray, *U.S.A.* **82 D4** 44 50N 119 48W
Spree →, *Germany* **16 B7** 52 32N 13 13 E
Sprengisandur, *Iceland* **8 D5** 64 52N 18 7W
Spring City, *U.S.A.* **82 G8** 39 29N 111 30W
Spring Garden, *U.S.A.* **84 F6** 39 52N 120 47W
Spring Mts., *U.S.A.* **83 H6** 36 0N 115 45W
Spring Valley, *Calif., U.S.A.* **85 N10** 32 45N 117 5W
Spring Valley, *Minn., U.S.A.* **80 D8** 43 41N 92 23W
Springbok, *S. Africa* **56 D2** 29 42S 17 54 E
Springdale, *Canada* **71 C8** 49 30N 56 6W
Springdale, *Ark., U.S.A.* **81 G7** 36 11N 94 8W
Springdale, *Wash., U.S.A.* **82 B5** 48 4N 117 45W
Springer, *U.S.A.* **81 G2** 36 22N 104 36W
Springerville, *U.S.A.* **83 J9** 34 8N 109 17W
Springfield, *Canada* **78 D4** 42 50N 80 56W
Springfield, *N.Z.* **59 K3** 43 19S 171 56 E
Springfield, *Colo., U.S.A.* **81 G3** 37 24N 102 37W
Springfield, *Ill., U.S.A.* **80 F10** 39 48N 89 39W
Springfield, *Mass., U.S.A.* **79 D12** 42 6N 72 35W
Springfield, *Mo., U.S.A.* **81 G8** 37 13N 93 17W
Springfield, *Ohio, U.S.A.* **76 F4** 39 55N 83 49W
Springfield, *Oreg., U.S.A.* **82 D2** 44 3N 123 1W
Springfield, *Tenn., U.S.A.* **77 G2** 36 31N 86 53W
Springfield, *Vt., U.S.A.* **79 C12** 43 18N 72 29W
Springfontein, *S. Africa* **56 E4** 30 15S 25 40 E
Springhill, *Canada* **71 C7** 45 40N 64 4W
Springhouse, *Canada* **72 C4** 51 56N 122 7W
Springhurst, *Australia* **63 F4** 36 10S 146 31 E
Springs, *S. Africa* **57 D4** 26 13S 28 25 E
Springsure, *Australia* **62 C4** 24 8S 148 6 E
Springvale, *Queens., Australia* **62 C3** 23 33S 140 42 E
Springvale, *W. Austral., Australia* **60 C4** 17 48S 127 41 E
Springvale, *U.S.A.* **79 C14** 43 28N 70 48W
Springville, *Calif., U.S.A.* **84 J8** 36 8N 118 49W
Springville, *N.Y., U.S.A.* **78 D6** 42 31N 78 40W
Springville, *Utah, U.S.A.* **82 F8** 40 10N 111 37W
Springwater, *Canada* **73 C7** 51 58N 108 23W
Spruce-Creek, *U.S.A.* **78 F6** 40 36N 78 9W
Spur, *U.S.A.* **81 J4** 33 28N 100 52W
Spurn Hd., *U.K.* **10 D8** 53 35N 0 8 E
Spuzzum, *Canada* **72 D4** 49 37N 121 23W
Squam L., *U.S.A.* **79 C13** 43 45N 71 32W
Squamish, *Canada* **72 D4** 49 45N 123 10W
Square Islands, *Canada* **71 B8** 52 47N 55 47W
Squires, Mt., *Australia* **61 E4** 26 14S 127 28 E
Sragen, *Indonesia* **37 G14** 7 26S 111 2 E
Srbija = Serbia □, *Yugoslavia* **21 C9** 43 30N 21 0 E
Sre Khtum, *Cambodia* **39 F6** 12 10N 106 52 E
Sre Umbell, *Cambodia* **39 G4** 11 8N 103 46 E
Srebrnica, *Bos.-H.* **21 B8** 44 10N 19 18 E
Sredinny Ra. = Sredinnyy Khrebet, *Russia* **27 D16** 57 0N 160 0 E
Sredinnyy Khrebet, *Russia* **27 D16** 57 0N 160 0 E
Sredne Tambovskoye, *Russia* **27 D14** 50 55N 137 45 E
Srednekolymsk, *Russia* **27 C16** 67 27N 153 40 E
Srednevilyuysk, *Russia* **27 C13** 63 50N 123 5 E
Śrem, *Poland* **17 B9** 52 6N 17 2 E
Sremska Mitrovica, *Serbia, Yug.* **21 B8** 44 59N 19 33 E
Srepok →, *Cambodia* **38 F6** 13 33N 106 16 E
Sretensk, *Russia* **27 D12** 52 10N 117 40 E
Sri Lanka ■, *Asia* **40 R12** 7 30N 80 50 E
Srikakulam, *India* **41 K13** 18 14N 83 58 E
Srinagar, *India* **43 B6** 34 5N 74 50 E
Staaten →, *Australia* **62 B3** 16 24S 141 17 E
Stade, *Germany* **16 B5** 53 35N 9 29 E
Stadskanaal, *Neths.* **15 A6** 53 4N 6 55 E
Staffa, *U.K.* **12 E2** 56 27N 6 21W
Stafford, *U.K.* **10 E5** 52 49N 2 7W
Stafford, *U.S.A.* **81 G5** 37 58N 98 36W
Stafford Springs, *U.S.A.* **79 E12** 41 57N 72 18W
Staffordshire □, *U.K.* **10 E5** 52 53N 2 10W
Staines, *U.K.* **11 F7** 51 26N 0 29W
Stakhanov, *Ukraine* **25 E6** 48 35N 38 40 E
Stalingrad = Volgograd, *Russia* **25 E7** 48 40N 44 25 E
Staliniri = Tskhinvali, *Georgia* **25 F7** 42 14N 44 1 E
Stalino = Donetsk, *Ukraine* **25 E6** 48 0N 37 45 E
Stalinogorsk = Novomoskovsk, *Russia* **24 D6** 54 5N 38 15 E
Stalis, *Greece* **23 D7** 35 17N 25 25 E
Stalowa Wola, *Poland* **17 C12** 50 34N 22 3 E
Stalybridge, *U.K.* **10 D5** 53 28N 2 3W
Stamford, *Australia* **62 C3** 21 15S 143 46 E
Stamford, *U.K.* **11 E7** 52 39N 0 29W
Stamford, *Conn., U.S.A.* **79 E11** 41 3N 73 32W
Stamford, *Tex., U.S.A.* **81 J5** 32 57N 99 48W
Stamps, *U.S.A.* **81 J8** 33 22N 93 30W
Stanberry, *U.S.A.* **80 E7** 40 13N 94 35W
Standerton, *S. Africa* **57 D4** 26 55S 29 7 E
Standish, *U.S.A.* **76 D4** 43 59N 83 57W
Stanford, *U.S.A.* **82 C8** 47 9N 110 13W
Stanger, *S. Africa* **57 D5** 29 27S 31 14 E
Stanislaus →, *U.S.A.* **84 H5** 37 40N 121 14W
Stanislav = Ivano-Frankivsk, *Ukraine* **17 D13** 48 40N 24 40 E
Stanke Dimitrov, *Bulgaria* **21 C10** 42 17N 23 9 E
Stanley, *Australia* **62 G4** 40 46S 145 19 E
Stanley, *N.B., Canada* **71 C6** 46 20N 66 44W
Stanley, *Sask., Canada* **73 B8** 55 24N 104 22W
Stanley, *Falk. Is.* **96 G5** 51 40S 59 51W
Stanley, *Idaho, U.S.A.* **82 D6** 44 13N 114 56W
Stanley, *N. Dak., U.S.A.* **80 A3** 48 19N 102 23W
Stanley, *N.Y., U.S.A.* **78 D7** 42 48N 77 6W
Stanley, *Wis., U.S.A.* **80 C9** 44 58N 90 56W
Stanovoy Khrebet, *Russia* **27 D13** 55 0N 130 0 E
Stanovoy Ra. = Stanovoy Khrebet, *Russia* **27 D13** 55 0N 130 0 E
Stansmore Ra., *Australia* **60 D4** 21 23S 128 33 E
Stanthorpe, *Australia* **63 D5** 28 36S 151 59 E
Stanton, *U.S.A.* **81 J4** 32 8N 101 48W
Stanwood, *U.S.A.* **84 B4** 48 15N 122 23W
Staples, *U.S.A.* **80 B7** 46 21N 94 48W
Stapleton, *U.S.A.* **80 E4** 41 29N 100 31W
Star City, *Canada* **73 C8** 52 50N 104 20W
Stara Planina, *Bulgaria* **21 C10** 43 15N 23 0 E
Stara Zagora, *Bulgaria* **21 C11** 42 26N 25 39 E
Starachowice, *Poland* **17 C11** 51 3N 21 2 E
Staraya Russa, *Russia* **24 C5** 57 58N 31 23 E
Starbuck I., *Kiribati* **65 H12** 5 37S 155 55W
Stargard Szczeciński, *Poland* **16 B8** 53 20N 15 0 E
Staritsa, *Russia* **24 C5** 56 33N 34 55 E
Starke, *U.S.A.* **77 K4** 29 57N 82 7W
Starkville, *Colo., U.S.A.* **81 G2** 37 8N 104 30W
Starkville, *Miss., U.S.A.* **77 J1** 33 28N 88 49W
Starogard Gdański, *Poland* **17 B10** 53 59N 18 30 E
Starokonstantinov = Starokonstyantyniv, *Ukraine* **17 D14** 49 48N 27 10 E
Starokonstyantyniv, *Ukraine* **17 D14** 49 48N 27 10 E
Start Pt., *U.K.* **11 G4** 50 13N 3 39W
Staryy Chartoriysk, *Ukraine* **17 C13** 51 15N 25 54 E
Staryy Kheydzhan, *Russia* **27 C15** 60 0N 144 50 E
Staryy Oskol, *Russia* **24 D6** 51 19N 37 55 E
State College, *U.S.A.* **78 F7** 40 48N 77 52W
Stateline, *U.S.A.* **84 G7** 38 57N 119 56W
Staten, I. = Estados, I. de Los, *Argentina* **96 G4** 54 40S 64 30W
Staten, I., *Argentina* **90 J4** 54 40S 64 0W
Staten I., *U.S.A.* **79 F10** 40 35N 74 9W
Statesboro, *U.S.A.* **77 J5** 32 27N 81 47W
Statesville, *U.S.A.* **77 H5** 35 47N 80 53W
Stauffer, *U.S.A.* **85 L7** 34 45N 119 3W
Staunton, *Ill., U.S.A.* **80 F10** 39 1N 89 47W
Staunton, *Va., U.S.A.* **76 F6** 38 9N 79 4W
Stavanger, *Norway* **9 G11** 58 57N 5 40 E
Staveley, *N.Z.* **59 K3** 43 40S 171 32 E
Stavelot, *Belgium* **15 D5** 50 23N 5 55 E
Staveren, *Neths.* **15 B5** 52 53N 5 22 E
Stavern, *Norway* **9 G14** 59 0N 10 1 E
Stavropol, *Russia* **25 E7** 45 5N 42 0 E
Stavros, *Cyprus* **23 D11** 35 1N 32 38 E
Stavrós, *Greece* **23 D6** 35 12N 24 45 E
Stavros, Ákra, *Greece* **23 D6** 35 26N 24 58 E
Stawell, *Australia* **63 F3** 37 5S 142 47 E
Stawell →, *Australia* **62 C3** 20 20S 142 55 E
Stayner, *Canada* **78 B4** 44 25N 80 5W
Steamboat Springs, *U.S.A.* **82 F10** 40 29N 106 50W
Steele, *U.S.A.* **80 B5** 46 51N 99 55W
Steelton, *U.S.A.* **78 F8** 40 14N 76 50W
Steelville, *U.S.A.* **81 G9** 37 58N 91 22W
Steen River, *Canada* **72 B5** 59 40N 117 12W
Steenkool = Bintuni, *Indonesia* **37 E8** 2 7S 133 32 E
Steenwijk, *Neths.* **15 B6** 52 47N 6 7 E
Steep Pt., *Australia* **61 E1** 26 8S 113 8 E
Steep Rock, *Canada* **73 C9** 51 30N 98 48W
Stefanie L. = Chew Bahir, *Ethiopia* **51 H12** 4 40N 36 50 E
Stefansson Bay, *Antarctica* **5 C5** 67 20S 59 8 E
Steiermark □, *Austria* **16 E8** 47 26N 15 0 E
Steilacoom, *U.S.A.* **84 C4** 47 10N 122 36W
Steinbach, *Canada* **73 D9** 49 32N 96 40W
Steinfort, *Lux.* **15 E5** 49 39N 5 55 E
Steinkjer, *Norway* **8 D14** 64 1N 11 31 E
Steinkopf, *S. Africa* **56 D2** 29 18S 17 43 E
Stellarton, *Canada* **71 C7** 45 32N 62 30W
Stellenbosch, *S. Africa* **56 E2** 33 58S 18 50 E
Stendal, *Germany* **16 B6** 52 36N 11 53 E
Steornabhaigh = Stornoway, *U.K.* **12 C2** 58 13N 6 23W
Stepanakert = Xankändi, *Azerbaijan* **25 G8** 39 52N 46 49 E
Stephen, *U.S.A.* **80 A6** 48 27N 96 53W
Stephens Creek, *Australia* **63 E3** 31 50S 141 30 E
Stephens I., *Canada* **72 C2** 54 10N 130 45W
Stephenville, *Canada* **71 C8** 48 31N 58 35W
Stephenville, *U.S.A.* **81 J5** 32 13N 98 12W
Stepnoi = Elista, *Russia* **25 E7** 46 16N 44 14 E
Stepnyak, *Kazakstan* **26 D8** 52 50N 70 50 E
Steppe, *Asia* **28 D9** 50 0N 50 0 E
Sterkstroom, *S. Africa* **56 E4** 31 32S 26 32 E
Sterling, *Colo., U.S.A.* **80 E3** 40 37N 103 13W
Sterling, *Ill., U.S.A.* **80 E10** 41 48N 89 42W
Sterling, *Kans., U.S.A.* **80 F5** 38 13N 98 12W
Sterling City, *U.S.A.* **81 K4** 31 51N 101 0W
Sterling Run, *U.S.A.* **78 E6** 41 25N 78 12W
Sterlitamak, *Russia* **24 D10** 53 40N 56 0 E
Stérnes, *Greece* **23 D6** 35 30N 24 9 E
Stettin = Szczecin, *Poland* **16 B8** 53 27N 14 27 E
Stettiner Haff, *Germany* **16 B8** 53 47N 14 15 E
Stettler, *Canada* **72 C6** 52 19N 112 40W
Steubenville, *U.S.A.* **78 F4** 40 22N 80 37W
Stevens Point, *U.S.A.* **80 C10** 44 31N 89 34W
Stevenson, *U.S.A.* **84 E5** 45 42N 121 53W
Stevenson L., *Canada* **73 C9** 53 55N 96 0W
Stewart, *B.C., Canada* **72 B3** 55 56N 129 57W
Stewart, *N.W.T., Canada* **68 B6** 63 19N 139 26W
Stewart, *U.S.A.* **84 F7** 39 5N 119 46W
Stewart, C., *Australia* **62 A1** 11 57S 134 56 E
Stewart, I., *Chile* **96 G2** 54 50S 71 15W
Stewart I., *N.Z.* **59 M1** 46 58S 167 54 E
Stewarts Point, *U.S.A.* **84 G3** 38 39N 123 24W
Stewiacke, *Canada* **71 C7** 45 9N 63 22W
Steynsburg, *S. Africa* **56 E4** 31 15S 25 49 E
Steyr, *Austria* **16 D8** 48 3N 14 25 E
Steytlerville, *S. Africa* **56 E3** 33 17S 24 19 E
Stigler, *U.S.A.* **81 H7** 35 15N 95 8W
Stikine →, *Canada* **72 B2** 56 40N 132 30W
Stilfontein, *S. Africa* **56 D4** 26 51S 26 50 E
Stillwater, *N.Z.* **59 K3** 42 27S 171 20 E
Stillwater, *Minn., U.S.A.* **80 C8** 45 3N 92 49W
Stillwater, *N.Y., U.S.A.* **79 D11** 42 55N 73 41W

Stillwater, Okla., U.S.A. 81 G6 36 7N 97 4W
Stillwater Range, U.S.A. 82 G4 39 50N 118 5W
Stilwell, U.S.A. 81 H7 35 49N 94 38W
Štip, Macedonia 21 D10 41 42N 22 10 E
Stirling, Australia 62 B3 17 12S 141 35 E
Stirling, Canada 72 D6 49 30N 112 30W
Stirling, U.K. 12 E5 56 8N 3 57W
Stirling □, U.K. 12 E4 56 12N 4 18W
Stirling Ra., Australia 61 F2 34 23S 118 0 E
Stittsville, Canada 79 A9 45 15N 75 55W
Stjernøya, Norway 8 A20 70 20N 22 40 E
Stjørdalshalsen, Norway 8 E14 63 29N 10 51 E
Stockerau, Austria 16 D9 48 24N 16 12 E
Stockett, U.S.A. 82 C8 47 21N 111 10W
Stockholm, Sweden 9 G18 59 20N 18 3 E
Stockport, U.K. 10 D5 53 25N 2 9W
Stockton, Calif., U.S.A. 83 H3 37 58N 121 17W
Stockton, Kans., U.S.A. 80 F5 39 26N 99 16W
Stockton, Mo., U.S.A. 81 G8 37 42N 93 48W
Stockton-on-Tees, U.K. 10 C6 54 35N 1 19W
Stockton-on-Tees □, U.K. 10 C6 54 35N 1 19W
Stoke on Trent, U.K. 10 D5 53 1N 2 11W
Stokes Bay, Canada 70 C3 45 0N 81 28W
Stokes Pt., Australia 62 G3 40 10S 143 56 E
Stokes Ra., Australia 60 C5 15 50S 130 50 E
Stokksnes, Iceland 8 D6 64 14N 14 58W
Stokmarknes, Norway 8 B16 68 34N 14 54 E
Stolac, Bos.-H. 21 C7 43 8N 17 59 E
Stolbovaya, Russia 27 C16 64 50N 153 50 E
Stolbovoy, Ostrov, Russia 27 D17 74 44N 135 14 E
Stolbtsy = Stowbtsy, Belarus 17 B14 53 30N 26 43 E
Stolin, Belarus 17 C14 51 53N 26 50 E
Stomion, Greece 23 D5 35 21N 23 32 E
Stonehaven, U.K. 12 E6 56 59N 2 12W
Stonehenge, Australia 62 C3 24 22S 143 17 E
Stonewall, Canada 73 C9 50 10N 97 19W
Stony L., Man., Canada 73 B9 58 51N 98 40W
Stony L., Ont., Canada 78 B6 44 30N 78 5W
Stony Rapids, Canada 73 B7 59 16N 105 50W
Stony Tunguska = Podkamennaya Tunguska →, Russia 27 C10 61 50N 90 13 E
Stonyford, U.S.A. 84 F4 39 23N 122 33W
Stora Lulevatten, Sweden 8 C18 67 10N 19 30 E
Storavan, Sweden 8 D18 65 45N 18 10 E
Stord, Norway 9 G11 59 52N 5 23 E
Store Bælt, Denmark 9 J14 55 20N 11 0 E
Store Creek, Australia 63 E4 32 54S 149 6 E
Storm B., Australia 62 G4 43 10S 147 30 E
Storm Lake, U.S.A. 80 D7 42 39N 95 13W
Stormberge, S. Africa 56 E4 31 16S 26 17 E
Stormsrivier, S. Africa 56 E3 33 59S 23 52 E
Stornoway, U.K. 12 C2 58 13N 6 23W
Storozhinets = Storozhynets, Ukraine 17 D13 48 14N 25 45 E
Storozhynets, Ukraine 17 D13 48 14N 25 45 E
Storsjön, Sweden 8 E16 63 9N 14 30 E
Storuman, Sweden 8 D17 65 5N 17 10 E
Storuman, sjö, Sweden 8 D17 65 13N 16 50 E
Stoughton, Canada 73 D8 49 40N 103 0W
Stour →, Dorset, U.K. 11 G5 50 43N 1 47W
Stour →, Here. & Worcs., U.K. 11 E5 52 21N 2 17W
Stour →, Kent, U.K. 11 F9 51 18N 1 22 E
Stour →, Suffolk, U.K. 11 F9 51 57N 1 4 E
Stourbridge, U.K. 11 E5 52 28N 2 8W
Stout, L., Canada 73 C10 52 0N 94 40W
Stove Pipe Wells Village, U.S.A. 85 J9 36 35N 117 11W
Stowbtsy, Belarus 17 B14 53 30N 26 43 E
Stowmarket, U.K. 11 E9 52 12N 1 0 E
Strabane, U.K. 13 B4 54 50N 7 27W
Strabane □, U.K. 13 B4 54 45N 7 25W
Strahan, Australia 62 G4 42 9S 145 20 E
Stralsund, Germany 16 A7 54 18N 13 4 E
Strand, S. Africa 56 E2 34 9S 18 48 E
Stranda, Møre og Romsdal, Norway 9 E12 62 19N 6 58 E
Stranda, Nord-Trøndelag, Norway 8 E14 63 33N 10 14 E
Strangford L., U.K. 13 B6 54 30N 5 37W
Strangsville, U.S.A. 78 E3 41 19N 81 50W
Stranraer, U.K. 12 G3 54 54N 5 1W
Strasbourg, Canada 73 C8 51 4N 104 55W
Strasbourg, France 18 B7 48 35N 7 42 E
Strasburg, U.S.A. 80 B4 46 8N 100 10W
Stratford, Canada 70 D3 43 23N 81 0W
Stratford, N.Z. 59 H5 39 20S 174 19 E
Stratford, Calif., U.S.A. 83 H4 36 11N 119 49W
Stratford, Conn., U.S.A. 79 E11 41 12N 73 8W
Stratford, Tex., U.S.A. 81 G3 36 20N 102 4W
Stratford-upon-Avon, U.K. 11 E6 52 12N 1 42W
Strath Spey, U.K. 12 D5 57 9N 3 49W
Strathalbyn, Australia 63 F2 35 13S 138 53 E
Strathcona Prov. Park, Canada 72 D3 49 38N 125 40W
Strathmore, Australia 62 B3 17 50S 142 35 E
Strathmore, Canada 72 C6 51 5N 113 18W
Strathmore, U.K. 12 E5 56 37N 3 7W
Strathmore, U.S.A. 84 J7 36 9N 119 4W
Strathnaver, Canada 72 C4 53 20N 122 33W
Strathpeffer, U.K. 12 D4 57 35N 4 32W
Strathroy, Canada 70 D3 42 58N 81 38W
Strathy Pt., U.K. 12 C4 58 36N 4 1W
Stratton, U.S.A. 80 F3 39 19N 102 36W
Straubing, Germany 16 D7 48 52N 12 34 E
Straumnes, Iceland 8 C2 66 26N 23 8W
Strawberry Reservoir, U.S.A. 82 F8 40 8N 111 9W
Strawn, U.S.A. 81 J5 32 33N 98 30W
Streaky B., Australia 63 E1 32 48S 134 13 E
Streaky Bay, Australia 63 E1 32 51S 134 18 E
Streator, U.S.A. 80 E10 41 8N 88 50W
Streeter, U.S.A. 80 B5 46 39N 99 21W
Streetsville, Canada 78 C5 43 35N 79 42W
Strelka, Russia 27 D10 58 5N 93 3 E
Streng →, Cambodia 38 F4 13 12N 103 37 E
Streymoy, Færoe Is. 8 E9 62 8N 7 5W
Strezhevoy, Russia 26 C8 60 42N 77 34 E
Strimón →, Greece 21 D10 40 46N 23 51 E
Strimonikós Kólpos, Greece 21 D11 40 33N 24 0 E
Strómboli, Italy 20 E6 38 47N 15 13 E
Stromeferry, U.K. 12 D3 57 21N 5 33W
Stromness, U.K. 12 C5 58 58N 3 17W
Stromsburg, U.S.A. 80 E6 41 7N 97 36W
Strömstad, Sweden 9 G14 58 56N 11 10 E
Strömsund, Sweden 8 E16 63 51N 15 33 E
Stronsay, U.K. 12 B6 59 7N 2 35W
Stroud, U.K. 11 F5 51 45N 2 13W
Stroud Road, Australia 63 E5 32 18S 151 57 E
Stroudsburg, U.S.A. 79 F9 40 59N 75 12W
Stroumbi, Cyprus 23 E11 34 53N 32 29 E
Struer, Denmark 9 H13 56 30N 8 35 E
Strumica, Macedonia 21 D10 41 28N 22 41 E
Struthers, Canada 70 C2 48 41N 85 51W
Struthers, U.S.A. 78 E4 41 4N 80 39W
Stryker, U.S.A. 82 B6 48 41N 114 46W
Stryy, Ukraine 17 D12 49 16N 23 48 E
Strzelecki Cr. →, Australia 63 D2 29 37S 139 59 E
Stuart, Fla., U.S.A. 77 M5 27 12N 80 15W
Stuart, Nebr., U.S.A. 80 D5 42 36N 99 8W
Stuart →, Canada 72 C4 54 0N 123 35W
Stuart Bluff Ra., Australia 60 D5 22 50S 131 52 E
Stuart L., Canada 72 C4 54 30N 124 30W
Stuart Ra., Australia 63 D1 29 10S 134 56 E
Stull, L., Canada 70 B1 54 24N 92 34W
Stung Treng, Cambodia 38 F5 13 31N 105 58 E
Stupart →, Canada 73 B10 56 0N 93 25W
Sturgeon B., Canada 73 C9 52 0N 97 50W
Sturgeon Bay, U.S.A. 76 C2 44 50N 87 23W
Sturgeon Falls, Canada 70 C4 46 25N 79 57W
Sturgeon L., Alta., Canada 72 B5 55 6N 117 32W
Sturgeon L., Ont., Canada 70 B1 50 0N 90 45W
Sturgeon L., Ont., Canada 78 B6 44 28N 78 43W
Sturgis, Mich., U.S.A. 76 E3 41 48N 85 25W
Sturgis, S. Dak., U.S.A. 80 C3 44 25N 103 31W
Sturt Cr. →, Australia 60 C4 19 8S 127 50 E
Sturt Creek, Australia 60 C4 19 12S 128 8 E
Stutterheim, S. Africa 56 E4 32 33S 27 28 E
Stuttgart, Germany 16 D5 48 48N 9 11 E
Stuttgart, U.S.A. 81 H9 34 30N 91 33W
Stuyvesant, U.S.A. 79 D11 42 23N 73 45W
Stykkishólmur, Iceland 8 D2 65 2N 22 40W
Styria = Steiermark □, Austria 16 E8 47 26N 15 0 E
Su Xian, China 34 H9 33 41N 116 59 E
Suakin, Sudan 51 E12 19 8N 37 20 E
Suan, N. Korea 35 E14 38 42N 126 22 E
Suaqui, Mexico 86 B3 29 12N 109 41W
Subang, Indonesia 37 G12 6 34S 107 45 E
Subansiri →, India 41 F18 26 48N 93 50 E
Subayhah, Si. Arabia 44 D3 30 2N 38 50 E
Subi, Indonesia 39 L7 2 58N 108 50 E
Subotica, Serbia, Yug. 21 A8 46 6N 19 39 E
Success, Canada 73 C7 50 28N 108 6W
Suceava, Romania 17 E14 47 38N 26 16 E
Suchan, Russia 30 C6 43 8N 133 9 E
Suchitoto, El Salv. 88 D2 13 56N 89 0W
Suchou = Suzhou, China 33 C7 31 19N 120 38 E
Süchow = Xuzhou, China 35 G9 34 18N 117 10 E
Suck →, Ireland 13 C3 53 17N 8 3W
Sucre, Bolivia 92 G5 19 0S 65 15W
Sud, Pte., Canada 71 C7 49 3N 62 14W
Sud-Ouest, Pte. du, Canada 71 C7 49 23N 63 36W
Sudan, U.S.A. 81 H3 34 4N 102 32W
Sudan ■, Africa 51 E11 15 0N 30 0 E
Sudbury, Canada 70 C3 46 30N 81 0W
Sudbury, U.K. 11 E8 52 2N 0 45 E
Sûdd, Sudan 51 G11 8 20N 30 0 E
Sudeten Mts. = Sudety, Europe 17 C9 50 20N 16 45 E
Sudety, Europe 17 C9 50 20N 16 45 E
Suðuroy, Færoe Is. 8 F9 61 32N 6 50W
Sudi, Tanzania 55 E4 10 11S 39 57 E
Sudirman, Pegunungan, Indonesia 37 E9 4 30S 137 0 E
Sueca, Spain 19 C5 39 12N 0 21W
Suez = El Suweis, Egypt 51 C11 29 58N 32 31 E
Suez, G. of = Suweis, Khalîg el, Egypt 51 C11 28 40N 33 0 E
Suffield, Canada 73 C6 50 12N 111 10W
Suffolk, U.S.A. 76 G7 36 44N 76 35W
Suffolk □, U.K. 11 E9 52 16N 1 0 E
Sugar City, U.S.A. 80 F3 38 14N 103 40W
Sugluk = Salluit, Canada 69 B12 62 14N 75 38W
Suhār, Oman 45 E8 24 20N 56 40 E
Sühbaatar □, Mongolia 34 B8 45 30N 114 0 E
Suhl, Germany 16 C6 50 36N 10 42 E
Sui Xian, China 34 G8 34 25N 115 2 E
Suide, China 34 F6 37 30N 110 12 E
Suifenhe, China 35 B16 44 25N 131 10 E
Suihua, China 33 B7 46 32N 126 55 E
Suining, China 35 H9 33 56N 117 58 E
Suiping, China 34 H7 33 10N 113 59 E
Suir →, Ireland 13 D4 52 16N 7 9W
Suiyang, China 35 B16 44 30N 130 56 E
Suizhong, China 35 D11 40 21N 120 20 E
Sujangarh, India 42 F6 27 42N 74 31 E
Sukabumi, Indonesia 37 G12 6 56S 106 50 E
Sukadana, Kalimantan, Indonesia 36 E3 1 10S 110 0 E
Sukadana, Sumatera, Indonesia 36 F3 5 5S 105 33 E
Sukagawa, Japan 31 F10 37 17N 140 23 E
Sukaraja, Indonesia 36 E4 2 28S 110 25 E
Sukarnapura = Jayapura, Indonesia 37 E10 2 28S 140 38 E
Sukchŏn, N. Korea 35 E13 39 22N 125 35 E
Sukhona →, Russia 24 C6 61 15N 46 39 E
Sukhothai, Thailand 38 D2 17 1N 99 49 E
Sukhumi = Sokhumi, Georgia 25 F7 43 0N 41 0 E
Sukkur, Pakistan 42 F3 27 42N 68 54 E
Sukkur Barrage, Pakistan 42 F3 27 40N 68 50 E
Sukumo, Japan 31 H6 32 56N 132 44 E
Sukunka →, Canada 72 B4 55 45N 121 15W
Sula, Kepulauan, Indonesia 37 E7 1 45S 125 0 E
Sulaco →, Honduras 88 C2 15 2N 87 44W
Sulaiman Range, Pakistan 42 D3 30 30N 69 50 E
Sūlār, Iran 45 D6 31 53N 51 54 E
Sulawesi □, Indonesia 37 E6 2 0S 120 0 E
Sulawesi Sea = Celebes Sea, Indonesia 37 D6 3 0N 123 0 E
Sulima, S. Leone 50 G2 6 58N 11 32W
Sulina, Romania 17 F15 45 10N 29 40 E
Sulitjelma, Norway 8 C17 67 9N 16 3 E
Sullana, Peru 92 D2 4 52S 80 39W
Sullivan, Ill., U.S.A. 80 F10 39 36N 88 37W
Sullivan, Ind., U.S.A. 76 F2 39 6N 87 24W
Sullivan, Mo., U.S.A. 80 F9 38 13N 91 10W
Sullivan Bay, Canada 72 C3 50 55N 126 50W
Sulphur, La., U.S.A. 81 K8 30 14N 93 23W
Sulphur, Okla., U.S.A. 81 H6 34 31N 96 58W
Sulphur Pt., Canada 72 A6 60 56N 114 48W
Sulphur Springs, U.S.A. 81 J7 33 8N 95 36W
Sulphur Springs Draw →, U.S.A. 81 J4 32 12N 101 36W
Sultan, Canada 70 C3 47 36N 82 47W
Sultan, U.S.A. 84 C5 47 52N 121 49W
Sultanpur, India 43 F10 26 18N 82 4 E
Sultsa, Russia 24 B8 63 27N 46 2 E
Sulu Arch., Phil. 37 C6 6 0N 121 0 E
Sulu Sea, E. Indies 37 C6 8 0N 120 0 E
Suluq, Libya 51 B9 31 44N 20 14 E
Sulzberger Ice Shelf, Antarctica 5 D10 78 0S 150 0 E
Sumalata, Indonesia 37 D6 1 0N 122 31 E
Sumampa, Argentina 94 B3 29 25S 63 29W
Sumatera □, Indonesia 36 D2 0 40N 100 20 E
Sumatra = Sumatera □, Indonesia 36 D2 0 40N 100 20 E
Sumatra, U.S.A. 82 C10 46 37N 107 33W
Sumba, Indonesia 37 F5 9 45S 119 35 E
Sumba, Selat, Indonesia 37 F5 9 0S 118 40 E
Sumbawa, Indonesia 36 F5 8 26S 117 30 E
Sumbawa Besar, Indonesia 36 F5 8 30S 117 26 E
Sumbawanga □, Tanzania 54 D3 8 0S 31 30 E
Sumbe, Angola 52 G2 11 10S 13 48 E
Sumburgh Hd., U.K. 12 B7 59 52N 1 17W
Sumdo, India 43 B8 35 6N 78 41 E
Sumedang, Indonesia 37 G12 6 52S 107 55 E
Šumen, Bulgaria 21 C12 43 18N 26 55 E
Sumenep, Indonesia 37 G15 7 1S 113 52 E
Sumgait = Sumqayıt, Azerbaijan 25 F8 40 34N 49 38 E
Summer L., U.S.A. 82 E3 42 50N 120 45W
Summerland, Canada 72 D5 49 32N 119 41W
Summerside, Canada 71 C7 46 24N 63 47W
Summerville, Ga., U.S.A. 77 H3 34 29N 85 21W
Summerville, S.C., U.S.A. 77 J5 33 1N 80 11W
Summit Lake, Canada 72 C4 54 20N 122 40W
Summit Peak, U.S.A. 83 H10 37 21N 106 42W
Sumner, Iowa, U.S.A. 80 D8 42 51N 92 6W
Sumner, Wash., U.S.A. 84 C4 47 12N 122 14W
Sumoto, Japan 31 G7 34 21N 134 54 E
Sumperk, Czech. 17 D9 49 59N 17 0 E
Sumqayıt, Azerbaijan 25 F8 40 34N 49 38 E
Sumter, U.S.A. 77 J5 33 55N 80 21W
Sumy, Ukraine 25 D5 50 57N 34 50 E
Sun City, Ariz., U.S.A. 83 K7 33 36N 112 17W
Sun City, Calif., U.S.A. 85 M9 33 42N 117 11W
Sunagawa, Japan 30 C10 43 29N 141 55 E
Sunan, N. Korea 35 E13 39 15N 125 40 E
Sunart, L., U.K. 12 E3 56 42N 5 43W
Sunburst, U.S.A. 82 B8 48 53N 111 55W
Sunbury, Australia 63 F3 37 35S 144 44 E
Sunbury, U.S.A. 79 F8 40 52N 76 48W
Sunchales, Argentina 94 C3 30 58S 61 35W
Suncho Corral, Argentina 94 B3 27 55S 63 27W
Sunchon, S. Korea 35 G14 34 52N 127 31 E
Suncook, U.S.A. 79 C13 43 8N 71 27W
Sunda, Selat, Indonesia 36 F3 6 20S 105 30 E
Sunda Is., Indonesia 28 K14 5 0S 105 0 E
Sunda Str. = Sunda, Selat, Indonesia 36 F3 6 20S 105 30 E
Sundance, U.S.A. 80 C2 44 24N 104 23W
Sundarbans, The, Asia 41 J16 22 0N 89 0 E
Sundargarh, India 41 H14 22 4N 84 5 E
Sundays = Sondags →, S. Africa 56 E4 33 44S 25 51 E
Sunderland, Canada 78 B5 44 16N 79 4W
Sunderland, U.K. 10 C6 54 55N 1 23W
Sundre, Canada 72 C6 51 49N 114 38W
Sundridge, Canada 70 C4 45 45N 79 25W
Sundsvall, Sweden 9 E17 62 23N 17 17 E
Sung Hei, Vietnam 39 G6 10 20N 106 2 E
Sungai Kolok, Thailand 39 J3 6 2N 101 58 E
Sungai Lembing, Malaysia 39 L4 3 55N 103 3 E
Sungai Patani, Malaysia 39 K3 5 37N 100 30 E
Sungaigerong, Indonesia 36 E2 2 59S 104 52 E
Sungailiat, Indonesia 36 E3 1 51S 106 8 E
Sungaipakning, Indonesia 36 D2 1 19N 102 0 E
Sungaipenuh, Indonesia 36 E2 2 1S 101 20 E
Sungaitiram, Indonesia 36 E5 0 45S 117 8 E
Sungari = Songhua Jiang →, China 33 B8 47 45N 132 30 E
Sungguminasa, Indonesia 37 F5 5 17S 119 30 E
Sunghua Chiang = Songhua Jiang →, China 33 B8 47 45N 132 30 E
Sunndalsøra, Norway 9 E13 62 40N 8 33 E
Sunnyside, Utah, U.S.A. 82 G8 39 34N 110 23W
Sunnyside, Wash., U.S.A. 82 C3 46 20N 120 0W
Sunnyvale, U.S.A. 83 H2 37 23N 122 2W
Sunray, U.S.A. 81 G4 36 1N 101 49W
Suntar, Russia 27 C12 62 15N 117 30 E
Suomenselkä, Finland 8 E21 62 52N 24 0 E
Suomussalmi, Finland 8 D23 64 54N 29 10 E
Suoyarvi, Russia 24 B5 62 3N 32 20 E
Supai, U.S.A. 83 H7 36 15N 112 41W
Supaul, India 43 F12 26 10N 86 40 E
Superior, Ariz., U.S.A. 83 K8 33 18N 111 6W
Superior, Mont., U.S.A. 82 C6 47 12N 114 53W
Superior, Nebr., U.S.A. 80 E5 40 1N 98 4W
Superior, Wis., U.S.A. 80 B8 46 44N 92 6W
Superior, L., U.S.A. 70 C2 47 0N 87 0W
Suphan Buri, Thailand 38 E3 14 14N 100 10 E
Supiori, Indonesia 37 E9 1 0S 136 0 E
Supung Sk., China 35 D13 40 35N 124 50 E
Sūq Suwayq, Si. Arabia 44 E3 24 23N 38 27 E
Suqian, China 35 H10 33 54N 118 8 E
Sūr, Lebanon 47 B4 33 19N 35 16 E
Sur, Pt., U.S.A. 83 H3 36 18N 121 54W
Sura →, Russia 24 C8 56 6N 46 0 E
Surab, Pakistan 42 E2 28 25N 66 15 E
Surabaja = Surabaya, Indonesia 37 G15 7 17S 112 45 E
Surabaya, Indonesia 37 G15 7 17S 112 45 E
Surakarta, Indonesia 37 G14 7 35S 110 48 E
Surat, Australia 63 D4 27 10S 149 6 E
Surat, India 40 J8 21 12N 72 55 E
Surat Thani, Thailand 39 H2 9 6N 99 20 E
Suratgarh, India 42 E5 29 18N 73 55 E
Sûre = Sauer →, Germany 15 E6 49 44N 6 31 E
Surendranagar, India 42 H4 22 45N 71 40 E
Surf, U.S.A. 85 L6 34 41N 120 36W
Surgut, Russia 26 C8 61 14N 73 20 E
Suriapet, India 40 L11 17 10N 79 40 E
Surigao, Phil. 37 C7 9 47N 125 29 E
Surin, Thailand 38 E4 14 50N 103 34 E
Surin Nua, Ko, Thailand 39 H1 9 30N 97 55 E
Surinam ■, S. Amer. 93 C7 4 0N 56 0W
Suriname ■ = Surinam ■, S. Amer. 93 C7 4 0N 56 0W
Suriname →, Surinam 93 B7 5 50N 55 15W
Sūrmaq, Iran 45 D7 31 3N 52 48 E
Surprise L., Canada 72 B2 59 40N 133 15W
Surrey □, U.K. 11 F7 51 15N 0 31W
Surt, Libya 51 B8 31 11N 16 39 E
Surt, Khalīj, Libya 51 B8 31 40N 18 30 E
Surtsey, Iceland 8 E3 63 20N 20 30W
Suruga-Wan, Japan 31 G9 34 45N 138 30 E
Susaki, Japan 31 H6 33 22N 133 17 E
Süsangerd, Iran 45 D6 31 35N 48 6 E
Susanino, Russia 27 D15 52 50N 140 14 E
Susanville, U.S.A. 82 F3 40 25N 120 39W
Susquehanna →, U.S.A. 79 G8 39 33N 76 5W
Susquehanna Depot, U.S.A. 79 E9 41 57N 75 36W
Susques, Argentina 94 A2 23 35S 66 25W
Sussex, Canada 71 C6 45 45N 65 37W
Sussex, U.S.A. 79 E10 41 13N 74 37W
Sussex, E. □, U.K. 11 G8 51 0N 0 20 E
Sussex, W. □, U.K. 11 G7 51 0N 0 30W
Sustut →, Canada 72 B3 56 20N 127 30W
Susuman, Russia 27 C15 62 47N 148 10 E
Susunu, Indonesia 37 E8 3 20S 133 25 E
Susurluk, Turkey 21 E13 39 54N 28 8 E
Sutherland, S. Africa 56 E3 32 24S 20 40 E
Sutherland, U.S.A. 80 E4 41 10N 101 8W
Sutherland Falls, N.Z. 59 L1 44 48S 167 46 E
Sutherlin, U.S.A. 82 E2 43 23N 123 19W
Sutlej →, Pakistan 42 E4 29 23N 71 3 E
Sutter, U.S.A. 84 F5 39 10N 121 45W
Sutter Creek, U.S.A. 84 G6 38 24N 120 48W
Sutton, Canada 79 A12 45 6N 72 37W
Sutton, U.S.A. 80 E6 40 36N 97 52W
Sutton →, Canada 70 A3 55 15N 83 45W
Sutton in Ashfield, U.K. 10 D6 53 8N 1 16W
Suttor →, Australia 62 C4 21 36S 147 2 E
Suttsu, Japan 30 C10 42 48N 140 14 E
Suva, Fiji 59 D8 18 6S 178 30 E
Suva Planina, Serbia, Yug. 21 C10 43 10N 22 5 E
Suvorov Is. = Suwarrow Is., Cook Is. 65 J11 15 0S 163 0W
Suwałki, Poland 17 A12 54 8N 22 59 E
Suwannaphum, Thailand 38 E4 15 33N 103 47 E
Suwannee →, U.S.A. 77 L4 29 17N 83 10W
Suwanose-Jima, Japan 31 K4 29 38N 129 43 E
Suwarrow Is., Cook Is. 65 J11 15 0S 163 0W
Suwayq aş Şuqban, Iraq 44 D5 31 32N 46 7 E
Suweis, Khalîg el, Egypt 51 C11 28 40N 33 0 E
Suwŏn, S. Korea 35 F14 37 17N 127 1 E
Suzdal, Russia 24 C7 56 29N 40 26 E
Suzhou, China 33 C7 31 19N 120 38 E
Suzu, Japan 31 F8 37 25N 137 17 E
Suzu-Misaki, Japan 31 F8 37 31N 137 21 E
Suzuka, Japan 31 G8 34 55N 136 36 E
Svalbard, Arctic 4 B8 78 0N 17 0 E
Svappavaara, Sweden 8 C19 67 40N 21 3 E
Svartisen, Norway 8 C15 66 40N 13 50 E
Svay Chek, Cambodia 38 F4 13 48N 102 58 E
Svay Rieng, Cambodia 39 G5 11 5N 105 48 E
Svealand □, Sweden 9 G16 59 55N 15 0 E
Sveg, Sweden 9 E16 62 2N 14 21 E
Svendborg, Denmark 9 J14 55 4N 10 35 E
Sverdlovsk = Yekaterinburg, Russia 24 C11 56 50N 60 30 E
Sverdrup Is., Canada 4 B3 79 0N 97 0W
Svetlaya, Russia 30 A9 46 33N 138 18 E
Svetlogorsk = Svyetlahorsk, Belarus 17 B15 52 38N 29 46 E
Svetozarevo, Serbia, Yug. 21 C9 44 5N 21 15 E
Svir →, Russia 24 B5 60 30N 32 48 E
Svishtov, Bulgaria 21 C11 43 36N 25 23 E
Svislach, Belarus 17 B13 53 3N 24 2 E
Svobodnyy, Russia 27 D13 51 20N 128 0 E
Svolvær, Norway 8 B16 68 15N 14 34 E
Svyetlahorsk, Belarus 17 B15 52 38N 29 46 E
Swabian Alps = Schwäbische Alb, Germany 16 D5 48 20N 9 30 E
Swainsboro, U.S.A. 77 J4 32 36N 82 20W
Swakopmund, Namibia 56 C1 22 37S 14 30 E
Swale →, U.K. 10 C6 54 5N 1 20W
Swan Hill, Australia 63 F3 35 20S 143 33 E
Swan Hills, Canada 72 C5 54 42N 115 24W
Swan Is., W. Indies 88 C3 17 22N 83 57W
Swan L., Canada 73 C8 52 30N 100 40W
Swan River, Canada 73 C8 52 10N 101 16W
Swanage, U.K. 11 G6 50 36N 1 58W
Swansea, Australia 63 E5 33 3S 151 35 E
Swansea, U.K. 11 F4 51 37N 3 57W
Swansea □, U.K. 11 F3 51 38N 4 3W
Swar →, Pakistan 43 B5 34 40N 72 5 E
Swartberge, S. Africa 56 E3 33 20S 22 0 E
Swartmodder, S. Africa 56 D3 28 1S 20 32 E
Swartruggens, S. Africa 56 D4 25 39S 26 42 E
Swastika, Canada 70 C3 48 7N 80 6W
Swatow = Shantou, China 33 D6 23 18N 116 40 E
Swaziland ■, Africa 57 D5 26 30S 31 30 E
Sweden ■, Europe 9 G16 57 0N 15 0 E
Sweet Home, U.S.A. 82 D2 44 24N 122 44W
Sweetwater, Nev., U.S.A. 84 G7 38 27N 119 9W
Sweetwater, Tex., U.S.A. 81 J4 32 28N 100 25W

Sweetwater →, U.S.A. **82 E10** 42 31N 107 2W
Swellendam, S. Africa **56 E3** 34 1S 20 26 E
Świdnica, Poland **17 C9** 50 50N 16 30 E
Świdnik, Poland **17 C12** 51 13N 22 39 E
Świebodzin, Poland **16 B8** 52 15N 15 31 E
Świecie, Poland **17 B10** 53 25N 18 30 E
Swift Current, Canada **73 C7** 50 20N 107 45W
Swiftcurrent →, Canada **73 C7** 50 38N 107 44W
Swilly, L., Ireland **13 A4** 55 12N 7 33W
Swindle, I., Canada **72 C3** 52 30N 128 35W
Swindon, U.K. **11 F6** 51 34N 1 46W
Swinemünde = Świnoujście, Poland **16 B8** 53 54N 14 16 E
Świnoujście, Poland **16 B8** 53 54N 14 16 E
Switzerland ■, Europe **16 E5** 46 30N 8 0 E
Swords, Ireland **13 C5** 53 28N 6 13W
Sydney, Australia **63 E5** 33 53S 151 10 E
Sydney, Canada **71 C7** 46 7N 60 7W
Sydney Mines, Canada **71 C7** 46 18N 60 15W
Sydprøven, Greenland **4 C5** 60 30N 45 35W
Sydra, G. of = Surt, Khalīj, Libya **51 B8** 31 40N 18 30 E
Syktyvkar, Russia **24 B9** 61 45N 50 40 E
Sylacauga, U.S.A. **77 J2** 33 10N 86 15W
Sylarna, Sweden **8 E15** 63 2N 12 13 E
Sylhet, Bangla. **41 G17** 24 54N 91 52 E
Sylt, Germany **16 A5** 54 54N 8 22 E
Sylvan Lake, Canada **72 C6** 52 20N 114 3W
Sylvania, U.S.A. **77 J5** 32 45N 81 38W
Sylvester, U.S.A. **77 K4** 31 32N 83 50W
Sym, Russia **26 C9** 60 20N 88 18 E
Symón, Mexico **86 C4** 24 42N 102 35W
Synnott Ra., Australia **60 C4** 16 30S 125 20 E
Syracuse, Kans., U.S.A. **81 F4** 37 59N 101 45W
Syracuse, N.Y., U.S.A. **79 C8** 43 3N 76 9W
Syrdarya →, Kazakstan **26 E7** 46 3N 61 0 E
Syria ■, Asia **44 C3** 35 0N 38 0 E
Syrian Desert = Ash Shām, Bādiyat, Asia **28 F7** 32 0N 40 0 E
Syul'dzhyukyor, Russia **27 C12** 63 14N 113 32 E
Syzran, Russia **24 D8** 53 12N 48 30 E
Szczecin, Poland **16 B8** 53 27N 14 27 E
Szczecinek, Poland **17 B9** 53 43N 16 41 E
Szczytno, Poland **17 B11** 53 33N 21 0 E
Szechwan = Sichuan □, China **32 C5** 31 0N 104 0 E
Szeged, Hungary **17 E11** 46 16N 20 10 E
Székesfehérvár, Hungary **17 E10** 47 15N 18 25 E
Szekszárd, Hungary **17 E10** 46 22N 18 42 E
Szentes, Hungary **17 E11** 46 39N 20 21 E
Szolnok, Hungary **17 E11** 47 10N 20 15 E
Szombathely, Hungary **17 E9** 47 14N 16 38 E

T

Ta Khli Khok, Thailand **38 E3** 15 18N 100 20 E
Ta Lai, Vietnam **39 G6** 11 24N 107 23 E
Tabacal, Argentina **94 A3** 23 15S 64 15W
Tabaco, Phil. **37 B6** 13 22N 123 44 E
Ţābah, Si. Arabia **44 E4** 26 55N 42 38 E
Tabarka, Tunisia **50 A6** 36 56N 8 46 E
Ţabas, Khorāsān, Iran **45 C9** 32 48N 60 12 E
Ţabas, Khorāsān, Iran **45 C8** 33 35N 56 55 E
Tabasará, Serranía de, Panama **88 E3** 8 35N 81 40W
Tabasco □, Mexico **87 D6** 17 45N 93 30W
Tabatinga, Serra da, Brazil **93 F10** 10 30S 44 0W
Tabāzīn, Iran **45 D8** 31 12N 57 54 E
Taber, Canada **72 D6** 49 47N 112 8W
Tablas, Phil. **37 B6** 12 25N 122 2 E
Table B. = Tafelbaai, S. Africa **56 E2** 33 35S 18 25 E
Table B., Canada **71 B8** 53 40N 56 25W
Table Mt., S. Africa **56 E2** 34 0S 18 22 E
Tableland, Australia **60 C4** 17 16S 126 51 E
Tabletop, Mt., Australia **62 C4** 23 24S 147 11 E
Tábor, Czech. **16 D8** 49 25N 14 39 E
Tabora, Tanzania **54 D3** 5 2S 32 50 E
Tabora □, Tanzania **54 D3** 5 0S 33 0 E
Tabou, Ivory C. **50 H3** 4 30N 7 20W
Tabrīz, Iran **44 B5** 38 7N 46 20 E
Tabuaeran, Pac. Oc. **65 G12** 3 51N 159 22W
Tabūk, Si. Arabia **44 D3** 28 23N 36 36 E
Tacámbaro de Codallos, Mexico **86 D4** 19 14N 101 28W
Tacheng, China **32 B3** 46 40N 82 58 E
Tach'ing Shan = Daqing Shan, China **34 D6** 40 40N 111 0 E
Tacloban, Phil. **37 B6** 11 15N 124 58 E
Tacna, Peru **92 G4** 18 0S 70 20W
Tacoma, U.S.A. **84 C4** 47 14N 122 26W
Tacuarembó, Uruguay **95 C4** 31 45S 56 0W
Tademaït, Plateau du, Algeria **50 C5** 28 30N 2 30 E
Tadjoura, Djibouti **46 E3** 11 50N 42 55 E
Tadmor, N.Z. **59 J4** 41 27S 172 45 E
Tadoule, L., Canada **73 B9** 58 36N 98 20W
Tadoussac, Canada **71 C6** 48 11N 69 42W
Tadzhikistan = Tajikistan ■, Asia **26 F8** 38 30N 70 0 E
Taechŏn-ni, S. Korea **35 F14** 36 21N 126 36 E
Taegu, S. Korea **35 G15** 35 50N 128 37 E
Taegwan, N. Korea **35 D13** 40 13N 125 12 E
Taejŏn, S. Korea **35 F14** 36 20N 127 28 E
Tafalla, Spain **19 A5** 42 30N 1 41W
Tafelbaai, S. Africa **56 E2** 33 35S 18 25 E
Tafermaar, Indonesia **37 F8** 6 47S 134 10 E
Tafí Viejo, Argentina **94 B2** 26 43S 65 17W
Tafīhān, Iran **45 D7** 29 25N 52 39 E
Taft, Iran **45 D7** 31 45N 54 14 E
Taft, Phil. **37 B7** 11 57N 125 30 E
Taft, Calif., U.S.A. **85 K7** 35 8N 119 28W
Taft, Tex., U.S.A. **81 M6** 27 59N 97 24W
Taga Dzong, Bhutan **41 F16** 27 5N 89 55 E
Taganrog, Russia **25 E6** 47 12N 38 50 E
Tagbilaran, Phil. **37 C6** 9 39N 123 51 E
Tagish, Canada **72 A2** 60 19N 134 16W
Tagish L., Canada **72 A2** 60 10N 134 20W
Tagliamento →, Italy **20 B5** 45 38N 13 6 E
Tagomago, I. de, Spain **22 B8** 39 2N 1 39 E
Taguatinga, Brazil **93 F10** 12 16S 42 26W
Tagum, Phil. **37 C7** 7 33N 125 53 E
Tagus = Tejo →, Europe **19 C1** 38 40N 9 24W
Tahakopa, N.Z. **59 M2** 46 30S 169 23 E
Tahan, Gunong, Malaysia **39 K4** 4 34N 102 17 E
Tahat, Algeria **50 D6** 23 18N 5 33 E
Tāherī, Iran **45 E7** 27 43N 52 20 E
Tahiti, Pac. Oc. **65 J13** 17 37S 149 27W
Tahoe, L., U.S.A. **84 G6** 39 6N 120 2W
Tahoe City, U.S.A. **84 F6** 39 10N 120 9W
Taholah, U.S.A. **84 C2** 47 21N 124 17W
Tahoua, Niger **50 F6** 14 57N 5 16 E
Tahta, Egypt **51 C11** 26 44N 31 32 E
Tahulandang, Indonesia **37 D7** 2 27N 125 23 E
Tahuna, Indonesia **37 D7** 3 38N 125 30 E
Taï, Ivory C. **50 G3** 5 55N 7 30W
Tai Shan, China **35 F9** 36 25N 117 20 E
Tai'an, China **35 F9** 36 12N 117 8 E
Taibei = T'aipei, Taiwan **33 D7** 25 2N 121 30 E
Taibique, Canary Is. **22 G2** 27 42N 17 58W
Taibus Qi, China **34 D8** 41 54N 115 22 E
T'aichung, Taiwan **33 D7** 24 9N 120 37 E
Taieri →, N.Z. **59 M3** 46 3S 170 12 E
Taigu, China **34 F7** 37 28N 112 30 E
Taihang Shan, China **34 G7** 36 0N 113 30 E
Taihape, N.Z. **59 H5** 39 41S 175 48 E
Taihe, China **34 H8** 33 20N 115 42 E
Taikang, China **34 G8** 34 5N 114 50 E
Tailem Bend, Australia **63 F2** 35 12S 139 29 E
Taimyr Peninsula = Taymyr, Poluostrov, Russia **27 B11** 75 0N 100 0 E
Tain, U.K. **12 D4** 57 49N 4 4W
T'ainan, Taiwan **33 D7** 23 0N 120 10 E
Tainaron, Ákra, Greece **21 F10** 36 22N 22 27 E
T'aipei, Taiwan **33 D7** 25 2N 121 30 E
Taiping, Malaysia **39 K3** 4 51N 100 44 E
Taipingzhen, China **34 H6** 33 35N 111 42 E
Tairbeart = Tarbert, U.K. **12 D2** 57 54N 6 49W
Taita □, Kenya **54 C4** 4 0S 38 30 E
Taita Hills, Kenya **54 C4** 3 25S 38 15 E
Taitao, Pen. de, Chile **96 F2** 46 30S 75 0W
T'aitung, Taiwan **33 D7** 22 43N 121 4 E
Taivalkoski, Finland **8 D23** 65 33N 28 12 E
Taiwan ■, Asia **33 D7** 23 30N 121 0 E
Taïyetos Óros, Greece **21 F10** 37 0N 22 23 E
Taiyiba, Israel **47 C4** 32 36N 35 27 E
Taiyuan, China **34 F7** 37 52N 112 33 E
Taizhong = T'aichung, Taiwan **33 D7** 24 9N 120 37 E
Ta'izz, Yemen **46 E3** 13 35N 44 2 E
Tājābād, Iran **45 D7** 30 2N 54 24 E
Tajikistan ■, Asia **26 F8** 38 30N 70 0 E
Tajima, Japan **31 F9** 37 12N 139 46 E
Tajo = Tejo →, Europe **19 C1** 38 40N 9 24W
Tajrīsh, Iran **45 C6** 35 48N 51 25 E
Tājūrā, Libya **51 B7** 32 51N 13 21 E
Tak, Thailand **38 D2** 16 52N 99 8 E
Takāb, Iran **44 B5** 36 24N 47 7 E
Takachiho, Japan **31 H5** 32 42N 131 18 E
Takada, Japan **31 F9** 37 7N 138 15 E
Takahagi, Japan **31 F10** 36 43N 140 45 E
Takaka, N.Z. **59 J4** 40 51S 172 50 E
Takamatsu, Japan **31 G7** 34 20N 134 5 E
Takaoka, Japan **31 F8** 36 47N 137 0 E
Takapuna, N.Z. **59 G5** 36 47S 174 47 E
Takasaki, Japan **31 F9** 36 20N 139 0 E
Takatsuki, Japan **31 G7** 34 51N 135 37 E
Takaungu, Kenya **54 C4** 3 38S 39 52 E
Takayama, Japan **31 F8** 36 18N 137 11 E
Take-Shima, Japan **31 J5** 30 49N 130 26 E
Takefu, Japan **31 G8** 35 50N 136 10 E
Takengon, Indonesia **36 D1** 4 45N 96 50 E
Takeo, Cambodia **39 G5** 10 59N 104 47 E
Takeo, Japan **31 H5** 33 12N 130 1 E
Tākestān, Iran **45 C6** 36 0N 49 40 E
Taketa, Japan **31 H5** 32 58N 131 24 E
Takh, India **43 C7** 33 6N 77 32 E
Takhman, Cambodia **39 G5** 11 29N 104 57 E
Takikawa, Japan **30 C10** 43 33N 141 54 E
Takla L., Canada **72 B3** 55 15N 125 45W
Takla Landing, Canada **72 B3** 55 30N 125 50W
Takla Makan = Taklamakan Shamo, China **32 C3** 38 0N 83 0 E
Taklamakan Shamo, China **32 C3** 38 0N 83 0 E
Taku →, Canada **72 B2** 58 30N 133 50W
Takum, Nigeria **50 G6** 7 18N 9 36 E
Tal Halāl, Iran **45 D7** 28 54N 55 1 E
Tala, Uruguay **95 C4** 34 21S 55 46W
Talagante, Chile **94 C1** 33 40S 70 50W
Talamanca, Cordillera de, Cent. Amer. **88 E3** 9 20N 83 20W
Talara, Peru **92 D2** 4 38S 81 18W
Talas, Kyrgyzstan **26 E8** 42 30N 72 13 E
Talâta, Egypt **47 E1** 30 36N 32 20 E
Talaud, Kepulauan, Indonesia **37 D7** 4 30N 127 10 E
Talaud Is. = Talaud, Kepulauan, Indonesia **37 D7** 4 30N 127 10 E
Talavera de la Reina, Spain **19 C3** 39 55N 4 46W
Talawana, Australia **60 D3** 22 51S 121 9 E
Talayan, Phil. **37 C6** 6 52N 124 24 E
Talbot, C., Australia **60 B4** 13 48S 126 43 E
Talbragar →, Australia **63 E4** 32 12S 148 37 E
Talca, Chile **94 D1** 35 28S 71 40W
Talca □, Chile **94 D1** 35 20S 71 46W
Talcahuano, Chile **94 D1** 36 40S 73 10W
Talcher, India **41 J14** 21 0N 85 18 E
Taldy Kurgan = Taldyqorghan, Kazakstan **26 E8** 45 10N 78 45 E
Taldyqorghan, Kazakstan **26 E8** 45 10N 78 45 E
Tālesh, Iran **45 B6** 37 58N 48 58 E
Ţalesh, Kūhhā-ye, Iran **45 B6** 37 42N 48 55 E
Tali Post, Sudan **51 G11** 5 55N 30 44 E
Talibon, Phil. **37 B6** 10 9N 124 20 E
Talibong, Ko, Thailand **39 J2** 7 15N 99 23 E
Talihina, U.S.A. **81 H7** 34 45N 95 3W
Taliwang, Indonesia **36 F5** 8 50S 116 55 E
Tall 'Asūr, West Bank **47 D4** 31 59N 35 17 E
Tall Kalakh, Syria **47 A5** 34 41N 36 15 E
Talladega, U.S.A. **77 J2** 33 26N 86 6W
Tallahassee, U.S.A. **77 K3** 30 27N 84 17W
Tallangatta, Australia **63 F4** 36 15S 147 19 E
Tallarook, Australia **63 F4** 37 5S 145 6 E
Tallering Pk., Australia **61 E2** 28 6S 115 37 E
Tallinn, Estonia **9 G21** 59 22N 24 48 E
Tallulah, U.S.A. **81 J9** 32 25N 91 11W
Talodi, Sudan **51 F11** 10 35N 30 22 E
Taloyoak, Canada **68 B10** 69 32N 93 32W
Talpa de Allende, Mexico **86 C4** 20 23N 104 51W
Talsi, Latvia **9 H20** 57 10N 22 30 E
Taltal, Chile **94 B1** 25 23S 70 33W
Taltson →, Canada **72 A6** 61 24N 112 46W
Talwood, Australia **63 D4** 28 29S 149 29 E
Talyawalka Cr. →, Australia **63 E3** 32 28S 142 22 E
Tam Chau, Vietnam **39 G5** 10 48N 105 12 E
Tam Ky, Vietnam **38 E7** 15 34N 108 29 E
Tam Quan, Vietnam **38 E7** 14 35N 109 3 E
Tama, U.S.A. **80 E8** 41 58N 92 35W
Tamala, Australia **61 E1** 26 42S 113 47 E
Tamale, Ghana **50 G4** 9 22N 0 50W
Tamano, Japan **31 G6** 34 29N 133 59 E
Tamanrasset, Algeria **50 D6** 22 50N 5 30 E
Tamaqua, U.S.A. **79 F9** 40 48N 75 58W
Tamar →, U.K. **11 G3** 50 27N 4 15W
Tamarang, Australia **63 E5** 31 27S 150 5 E
Tamarinda, Spain **22 B10** 39 55N 3 49 E
Tamashima, Japan **31 G6** 34 32N 133 40 E
Tamaské, Niger **50 F6** 14 49N 5 43 E
Tamaulipas □, Mexico **87 C5** 24 0N 99 0W
Tamaulipas, Sierra de, Mexico **87 C5** 23 30N 98 20W
Tamazula, Mexico **86 C3** 24 55N 106 58W
Tamazunchale, Mexico **87 C5** 21 16N 98 47W
Tambacounda, Senegal **50 F2** 13 45N 13 40W
Tambelan, Kepulauan, Indonesia **36 D3** 1 0N 107 30 E
Tambellup, Australia **61 F2** 34 4S 117 37 E
Tambo, Australia **62 C4** 24 54S 146 14 E
Tambo de Mora, Peru **92 F3** 13 30S 76 8W
Tambohorano, Madag. **57 B7** 17 30S 43 58 E
Tambora, Indonesia **36 F5** 8 12S 118 5 E
Tambov, Russia **24 D7** 52 45N 41 28 E
Tambuku, Indonesia **37 G15** 7 8S 113 40 E
Tamburâ, Sudan **51 G10** 5 40N 27 25 E
Tâmchekket, Mauritania **50 E2** 17 25N 10 40W
Tamega →, Portugal **19 B1** 41 5N 8 21W
Tamenglong, India **41 G18** 25 0N 93 35 E
Tamgak, Mts., Niger **50 E6** 19 12N 8 35 E
Tamiahua, L. de, Mexico **87 C5** 21 30N 97 30W
Tamil Nadu □, India **40 P10** 11 0N 77 0 E
Tamluk, India **43 H12** 22 18N 87 58 E
Tammerfors = Tampere, Finland **9 F20** 61 30N 23 50 E
Tammisaari, Finland **9 F20** 60 0N 23 26 E
Tamo Abu, Pegunungan, Malaysia **36 D5** 3 10N 115 5 E
Tampa, U.S.A. **77 M4** 27 57N 82 27W
Tampa B., U.S.A. **77 M4** 27 50N 82 30W
Tampere, Finland **9 F20** 61 30N 23 50 E
Tampico, Mexico **87 C5** 22 20N 97 50W
Tampin, Malaysia **39 L4** 2 28N 102 13 E
Tamrida = Qādib, Yemen **46 E5** 12 37N 53 57 E
Tamu, Burma **41 G19** 24 13N 94 12 E
Tamworth, Australia **63 E5** 31 7S 150 58 E
Tamworth, U.K. **11 E6** 52 39N 1 41W
Tamyang, S. Korea **35 G14** 35 19N 126 59 E
Tan An, Vietnam **39 G6** 10 32N 106 25 E
Tana →, Kenya **54 C5** 2 32S 40 31 E
Tana →, Norway **8 A23** 70 30N 28 14 E
Tana, L., Ethiopia **51 F12** 13 5N 37 30 E
Tana River, Kenya **54 C4** 2 0S 39 30 E
Tanabe, Japan **31 H7** 33 44N 135 22 E
Tanafjorden, Norway **8 A23** 70 45N 28 25 E
Tanaga, Pta., Canary Is. **22 G1** 27 42N 18 10W
Tanahbala, Indonesia **36 E1** 0 30S 98 30 E
Tanahgrogot, Indonesia **36 E5** 1 55S 116 15 E
Tanahjampea, Indonesia **37 F6** 7 10S 120 35 E
Tanahmasa, Indonesia **36 E1** 0 12S 98 39 E
Tanahmerah, Indonesia **37 F10** 6 5S 140 16 E
Tanakura, Japan **31 F10** 37 10N 140 20 E
Tanami, Australia **60 C4** 19 59S 129 43 E
Tanami Desert, Australia **60 C5** 18 50S 132 0 E
Tanana, U.S.A. **68 B4** 65 10N 152 4W
Tanana →, U.S.A. **68 B4** 65 10N 151 58W
Tananarive = Antananarivo, Madag. **57 B8** 18 55S 47 31 E
Tánaro →, Italy **20 B3** 44 55N 8 40 E
Tanbar, Australia **62 D3** 25 51S 141 55 E
Tancheng, China **35 G10** 34 25N 118 20 E
Tanchŏn, N. Korea **35 D15** 40 27N 128 54 E
Tanda, Ut. P., India **43 F10** 26 33N 82 35 E
Tanda, Ut. P., India **43 E8** 28 57N 78 56 E
Tandag, Phil. **37 C7** 9 4N 126 9 E
Tandaia, Tanzania **55 D3** 9 25S 34 15 E
Tandaué, Angola **56 B2** 16 58S 18 5 E
Tandil, Argentina **94 D4** 37 15S 59 6W
Tandil, Sa. del, Argentina **94 D4** 37 30S 59 0W
Tandlianwala, Pakistan **42 D5** 31 3N 73 9 E
Tando Adam, Pakistan **42 G3** 25 45N 68 40 E
Tandou L., Australia **63 E3** 32 40S 142 5 E
Tane-ga-Shima, Japan **31 J5** 30 30N 131 0 E
Taneatua, N.Z. **59 H6** 38 4S 177 1 E
Tanen Tong Dan, Burma **38 D2** 16 30N 98 30 E
Tanezrouft, Algeria **50 D5** 23 9N 0 11 E
Tang, Koh, Cambodia **39 G4** 10 16N 103 7 E
Tang Krasang, Cambodia **38 F5** 12 34N 105 3 E
Tanga, Tanzania **54 D4** 5 5S 39 2 E
Tanga □, Tanzania **54 D4** 5 20S 38 0 E
Tanganyika, L., Africa **54 D2** 6 40S 30 0 E
Tanger = Tangier, Morocco **50 A3** 35 50N 5 49W
Tangerang, Indonesia **37 G12** 6 11S 106 37 E
Tanggu, China **35 E9** 39 2N 117 40 E
Tanggula Shan, China **32 C4** 32 40N 92 10 E
Tanghe, China **34 H7** 32 47N 112 50 E
Tangier, Morocco **50 A3** 35 50N 5 49W
Tangorin P.O., Australia **62 C3** 21 47S 144 12 E
Tangshan, China **35 E10** 39 38N 118 10 E
Tangtou, China **35 G10** 35 28N 118 30 E
Tanimbar, Kepulauan, Indonesia **37 F8** 7 30S 131 30 E
Tanimbar Is. = Tanimbar, Kepulauan, Indonesia **37 F8** 7 30S 131 30 E
Tanjay, Phil. **37 C6** 9 30N 123 5 E
Tanjong Malim, Malaysia **39 L3** 3 42N 101 31 E
Tanjore = Thanjavur, India **40 P11** 10 48N 79 12 E
Tanjung, Indonesia **36 E5** 2 10S 115 25 E
Tanjungbalai, Indonesia **36 D1** 2 55N 99 44 E
Tanjungbatu, Indonesia **36 D5** 2 23N 118 3 E
Tanjungkarang Telukbetung, Indonesia **36 F3** 5 20S 105 10 E
Tanjungpandan, Indonesia **36 E3** 2 43S 107 38 E
Tanjungpinang, Indonesia **36 D2** 1 5N 104 30 E
Tanjungpriok, Indonesia **37 G12** 6 8S 106 55 E
Tanjungredeb, Indonesia **36 D5** 2 9N 117 29 E
Tanjungselor, Indonesia **36 D5** 2 55N 117 25 E
Tank, Pakistan **42 C4** 32 14N 70 25 E
Tannu-Ola, Russia **27 D10** 51 0N 94 0 E
Tanout, Niger **50 F6** 14 50N 8 55 E
Tanta, Egypt **51 B11** 30 45N 30 57 E
Tantoyuca, Mexico **87 C5** 21 21N 98 10W
Tantung = Dandong, China **35 D13** 40 10N 124 20 E
Tanunda, Australia **63 E2** 34 30S 139 0 E
Tanzania ■, Africa **54 D3** 6 0S 34 0 E
Tanzilla →, Canada **72 B2** 58 8N 130 43W
Tao Ko, Thailand **39 G2** 10 5N 99 52 E
Tao'an, China **35 B12** 45 22N 122 40 E
Tao'er He →, China **35 B13** 45 45N 124 5 E
Taolanaro, Madag. **57 D8** 25 2S 47 0 E
Taole, China **34 E4** 38 48N 106 40 E
Taos, U.S.A. **83 H11** 36 24N 105 35W
Taoudenni, Mali **50 D4** 22 40N 3 55W
Taourirt, Morocco **50 B4** 34 25N 2 53W
Tapa, Estonia **9 G21** 59 15N 25 50 E
Tapa Shan = Daba Shan, China **33 C5** 32 0N 109 0 E
Tapachula, Mexico **87 E6** 14 54N 92 17W
Tapah, Malaysia **39 K3** 4 12N 101 15 E
Tapajós →, Brazil **93 D8** 2 24S 54 41W
Tapaktuan, Indonesia **36 D1** 3 15N 97 10 E
Tapanui, N.Z. **59 L2** 45 56S 169 18 E
Tapauá →, Brazil **92 E6** 5 40S 64 21W
Tapeta, Liberia **50 G3** 6 29N 8 52W
Taphan Hin, Thailand **38 D3** 16 13N 100 26 E
Tapi →, India **40 J8** 21 8N 72 41 E
Tapirapecó, Serra, Venezuela **92 C6** 1 10N 65 0W
Tappahannock, U.S.A. **76 G7** 37 56N 76 52W
Tapuaenuku, Mt., N.Z. **59 J4** 42 0S 173 39 E
Tapul Group, Phil. **37 C6** 5 35N 120 50 E
Taqīābād, Iran **45 C8** 35 33N 59 11 E
Ţaqţaq, Iraq **44 C5** 35 53N 44 35 E
Taquara, Brazil **95 B5** 29 36S 50 46W
Taquari →, Brazil **92 G7** 19 15S 57 17W
Tara, Australia **63 D5** 27 17S 150 31 E
Tara, Canada **78 B3** 44 28N 81 9W
Tara, Russia **26 D8** 56 55N 74 24 E
Tara, Zambia **55 F2** 16 58S 26 45 E
Tara →, Montenegro, Yug. **21 C8** 43 21N 18 51 E
Tara →, Russia **26 D8** 56 42N 74 36 E
Tarabagatay, Khrebet, Kazakstan **26 E9** 48 0N 83 0 E
Tarābulus, Lebanon **47 A4** 34 31N 35 50 E
Tarābulus, Libya **51 B7** 32 49N 13 7 E
Tarajalejo, Canary Is. **22 F5** 28 12N 14 7W
Tarakan, Indonesia **36 D5** 3 20N 117 35 E
Tarakit, Mt., Kenya **54 B4** 2 2N 35 10 E
Taralga, Australia **63 E4** 34 26S 149 52 E
Tarama-Jima, Japan **31 M2** 24 39N 124 42 E
Taran, Mys, Russia **9 J18** 54 56N 19 59 E
Taranagar, India **42 E6** 28 43N 74 50 E
Taranaki □, N.Z. **59 H5** 39 25S 174 30 E
Tarancón, Spain **19 B4** 40 1N 3 0W
Taranga, India **42 H5** 23 56N 72 43 E
Taranga Hill, India **42 H5** 24 0N 72 40 E
Táranto, Italy **20 D7** 40 28N 17 14 E
Táranto, G. di, Italy **20 D7** 40 8N 17 20 E
Tarapacá, Colombia **92 D5** 2 56S 69 46W
Tarapacá □, Chile **94 A2** 20 45S 69 30W
Tararua Ra., N.Z. **59 J5** 40 45S 175 25 E
Tarashcha, Ukraine **17 D16** 49 30N 30 31 E
Tarauacá, Brazil **92 E4** 8 6S 70 48W
Tarauacá →, Brazil **92 E5** 6 42S 69 48W
Tarawera, N.Z. **59 H6** 39 2S 176 36 E
Tarawera L., N.Z. **59 H6** 38 13S 176 27 E
Tarazona, Spain **19 B5** 41 55N 1 43W
Tarbat Ness, U.K. **12 D5** 57 52N 3 47W
Tarbela Dam, Pakistan **42 B5** 34 8N 72 52 E
Tarbert, Arg. & Bute, U.K. **12 F3** 55 52N 5 25W
Tarbert, W. Isles, U.K. **12 D2** 57 54N 6 49W
Tarbes, France **18 E4** 43 15N 0 3 E
Tarboro, U.S.A. **77 H7** 35 54N 77 32W
Tarbrax, Australia **62 C3** 21 7S 142 26 E
Tarcoola, Australia **63 E1** 30 44S 134 36 E
Tarcoon, Australia **63 E4** 30 15S 146 43 E
Taree, Australia **63 E5** 31 50S 152 30 E
Tarfaya, Morocco **50 C2** 27 55N 12 55W
Tarifa, Spain **19 D3** 36 1N 5 36W
Tarija, Bolivia **94 A3** 21 30S 64 40W
Tarija □, Bolivia **94 A3** 21 30S 63 30W
Tariku →, Indonesia **37 E9** 2 55S 138 26 E
Tarim Basin = Tarim Pendi, China **32 B3** 40 0N 84 0 E
Tarim He →, China **32 C3** 39 30N 88 30 E
Tarim Pendi, China **32 B3** 40 0N 84 0 E
Tarime □, Tanzania **54 C3** 1 15S 34 0 E
Taritatu →, Indonesia **37 E9** 2 54S 138 27 E
Tarka →, S. Africa **56 E4** 32 10S 26 0 E
Tarkastad, S. Africa **56 E4** 32 0S 26 16 E
Tarkhankut, Mys, Ukraine **25 E5** 45 25N 32 30 E
Tarko Sale, Russia **26 C8** 64 55N 77 50 E
Tarkwa, Ghana **50 G4** 5 20N 2 0W
Tarlac, Phil. **37 A6** 15 29N 120 35 E
Tarlton Downs, Australia **62 C2** 22 40S 136 45 E
Tarma, Peru **92 F3** 11 25S 75 45W
Tarn →, France **18 E4** 44 5N 1 6 E
Tarnobrzeg, Poland **17 C11** 50 35N 21 41 E
Tarnów, Poland **17 C11** 50 3N 21 0 E
Tarnowskie Góry, Poland **17 C10** 50 27N 18 54 E
Taroom, Australia **63 D4** 25 36S 149 48 E
Taroudannt, Morocco **50 B3** 30 30N 8 52W
Tarpon Springs, U.S.A. **77 L4** 28 9N 82 45W

Tarragona, *Spain* 19 B6 41 5N 1 17 E
Tarrasa, *Spain* 19 B7 41 34N 2 1 E
Tarrytown, *U.S.A.* 79 E11 41 4N 73 52W
Tarshiha = Me'ona, *Israel* 47 B4 33 1N 35 15 E
Tarso Emissi, *Chad* 51 D8 21 27N 18 36 E
Tarsus, *Turkey* 25 G5 36 58N 34 55 E
Tartagal, *Argentina* 94 A3 22 30S 63 50W
Tartu, *Estonia* 9 G22 58 20N 26 44 E
Tarṭūs, *Syria* 44 C2 34 55N 35 55 E
Tarumizu, *Japan* 31 J5 31 29N 130 42 E
Tarutao, Ko, *Thailand* 39 J2 6 33N 99 40 E
Tarutung, *Indonesia* 36 D1 2 0N 98 54 E
Tasāwah, *Libya* 51 C7 26 0N 13 30 E
Taschereau, *Canada* 70 C4 48 40N 78 40W
Taseko →, *Canada* 72 C4 52 8N 123 45W
Tash-Kömür, *Kyrgyzstan* 26 E8 41 40N 72 10 E
Tash-Kumyr = Tash-Kömür, *Kyrgyzstan* 26 E8 41 40N 72 10 E
Tashauz = Dashhowuz, *Turkmenistan* 26 E6 41 49N 59 58 E
Tashi Chho Dzong = Thimphu, *Bhutan* 41 F16 27 31N 89 45 E
Tashkent = Toshkent, *Uzbekistan* 26 E7 41 20N 69 10 E
Tashtagol, *Russia* 26 D9 52 47N 87 53 E
Tasikmalaya, *Indonesia* 37 G13 7 18S 108 12 E
Tåsjön, *Sweden* 8 D16 64 15N 15 40 E
Taskan, *Russia* 27 C16 62 59N 150 20 E
Tasman B., *N.Z.* 59 J4 40 59S 173 25 E
Tasman Mts., *N.Z.* 59 J4 41 3S 172 25 E
Tasman Pen., *Australia* 62 G4 43 10S 148 0 E
Tasman Sea, *Pac. Oc.* 64 L8 36 0S 160 0 E
Tasmania □, *Australia* 62 G4 42 0S 146 30 E
Tassili n-Ajjer, *Algeria* 48 D4 25 47N 8 1 E
Tasu Sd., *Canada* 72 C2 52 47N 132 2W
Tatabánya, *Hungary* 17 E10 47 32N 18 25 E
Tatar Republic □ = Tatarstan □, *Russia* 24 C9 55 30N 51 30 E
Tatarbunary, *Ukraine* 17 F15 45 50N 29 39 E
Tatarsk, *Russia* 26 D8 55 14N 76 0 E
Tatarstan □, *Russia* 24 C9 55 30N 51 30 E
Tateyama, *Japan* 31 G9 35 0N 139 50 E
Tathlina L., *Canada* 72 A5 60 33N 117 39W
Tathra, *Australia* 63 F4 36 44S 149 59 E
Tatinnai L., *Canada* 73 A9 60 55N 97 40W
Tatnam, C., *Canada* 73 B10 57 16N 91 0W
Tatra = Tatry, *Slovak Rep.* 17 D11 49 20N 20 0 E
Tatry, *Slovak Rep.* 17 D11 49 20N 20 0 E
Tatsuno, *Japan* 31 G7 34 52N 134 33 E
Tatta, *Pakistan* 42 G2 24 42N 67 55 E
Tatui, *Brazil* 95 A6 23 25S 47 53W
Tatum, *U.S.A.* 81 J3 33 16N 103 19W
Tat'ung = Datong, *China* 34 D7 40 6N 113 18 E
Tatvan, *Turkey* 25 G7 38 31N 42 15 E
Taubaté, *Brazil* 95 A6 23 0S 45 36W
Tauern, *Austria* 16 E7 47 15N 12 40 E
Taumarunui, *N.Z.* 59 H5 38 53S 175 15 E
Taumaturgo, *Brazil* 92 E4 8 54S 72 51W
Taung, *S. Africa* 56 D3 27 33S 24 47 E
Taungdwingyi, *Burma* 41 J19 20 1N 95 40 E
Taunggyi, *Burma* 41 J20 20 50N 97 0 E
Taungup, *Burma* 41 K19 18 51N 94 14 E
Taungup Pass, *Burma* 41 K19 18 40N 94 45 E
Taungup Taunggya, *Burma* 41 K18 18 20N 93 40 E
Taunsa Barrage, *Pakistan* 42 D4 30 42N 70 50 E
Taunton, *U.K.* 11 F4 51 1N 3 5W
Taunton, *U.S.A.* 79 E13 41 54N 71 6W
Taunus, *Germany* 16 C5 50 13N 8 34 E
Taupo, *N.Z.* 59 H6 38 41S 176 7 E
Taupo, L., *N.Z.* 59 H5 38 46S 175 55 E
Tauragė, *Lithuania* 9 J20 55 14N 22 16 E
Tauranga, *N.Z.* 59 G6 37 42S 176 11 E
Tauranga Harb., *N.Z.* 59 G6 37 30S 176 5 E
Taurianova, *Italy* 20 E7 38 21N 16 1 E
Taurus Mts. = Toros Dağlan, *Turkey* 25 G5 37 0N 32 30 E
Tavda, *Russia* 26 D7 58 7N 65 8 E
Tavda →, *Russia* 26 D7 57 47N 67 18 E
Taveta, *Tanzania* 54 C4 3 23S 37 37 E
Taveuni, *Fiji* 59 C9 16 51S 179 58W
Tavira, *Portugal* 19 D2 37 8N 7 40W
Tavistock, *Canada* 78 C4 43 19N 80 50W
Tavistock, *U.K.* 11 G3 50 33N 4 9W
Tavoy, *Burma* 38 E2 14 2N 98 12 E
Taw →, *U.K.* 11 F3 51 4N 4 4W
Tawas City, *U.S.A.* 76 C4 44 16N 83 31W
Tawau, *Malaysia* 36 D5 4 20N 117 55 E
Tawitawi, *Phil.* 37 B6 5 10N 120 0 E
Taxila, *Pakistan* 42 C5 33 42N 72 52 E
Tay →, *U.K.* 12 E5 56 37N 3 38W
Tay, Firth of, *U.K.* 12 E5 56 25N 3 8W
Tay, L., *Australia* 61 F3 32 55S 120 48 E
Tay, L., *U.K.* 12 E4 56 32N 4 8W
Tay Ninh, *Vietnam* 39 G6 11 20N 106 5 E
Tayabamba, *Peru* 92 E3 8 15S 77 16W
Taylakova, *Russia* 26 D8 59 13N 74 0 E
Taylakovy = Taylakova, *Russia* 26 D8 59 13N 74 0 E
Taylor, *Canada* 72 B4 56 13N 120 40W
Taylor, *Nebr., U.S.A.* 80 E5 41 46N 99 23W
Taylor, *Pa., U.S.A.* 79 E9 41 23N 75 43W
Taylor, *Tex., U.S.A.* 81 K6 30 34N 97 25W
Taylor, Mt., *U.S.A.* 83 J10 35 14N 107 37W
Taylorville, *U.S.A.* 80 F10 39 33N 89 18W
Taymā, *Si. Arabia* 44 E3 27 35N 38 45 E
Taymyr, Oz., *Russia* 27 B11 74 20N 102 0 E
Taymyr, Poluostrov, *Russia* 27 B11 75 0N 100 0 E
Tayport, *U.K.* 12 E6 56 27N 2 52W
Tayshet, *Russia* 27 D10 55 58N 98 1 E
Taytay, *Phil.* 37 B5 10 45N 119 30 E
Taz →, *Russia* 26 C8 67 32N 78 40 E
Taza, *Morocco* 50 B4 34 16N 4 6W
Tāzah Khurmātū, *Iraq* 44 C5 35 18N 44 20 E
Tazawa-Ko, *Japan* 30 E10 39 43N 140 40 E
Tazin L., *Canada* 73 B7 59 44N 108 42W
Tazovskiy, *Russia* 26 C8 67 30N 78 44 E
Tbilisi, *Georgia* 25 F7 41 43N 44 50 E
Tchad = Chad ■, *Africa* 51 E8 15 0N 17 15 E
Tchad, L., *Chad* 51 F7 13 30N 14 30 E
Tch'eng-tou = Chengdu, *China* 32 C5 30 38N 104 2 E
Tchentlo L., *Canada* 72 B4 55 15N 125 0W
Tchibanga, *Gabon* 52 E2 2 45S 11 0 E
Tch'ong-k'ing = Chongqing, *China* 32 D5 29 35N 106 25 E
Tczew, *Poland* 17 A10 54 8N 18 50 E
Te Anau, L., *N.Z.* 59 L1 45 15S 167 45 E
Te Aroha, *N.Z.* 59 G5 37 32S 175 44 E
Te Awamutu, *N.Z.* 59 H5 38 1S 175 20 E
Te Kuiti, *N.Z.* 59 H5 38 20S 175 11 E
Te Puke, *N.Z.* 59 G6 37 46S 176 22 E
Te Waewae B., *N.Z.* 59 M1 46 13S 167 33 E
Tea Tree, *Australia* 62 C1 22 5S 133 22 E
Teague, *U.S.A.* 81 K6 31 38N 96 17W
Teapa, *Mexico* 87 D6 18 35N 92 56W
Tebakang, *Malaysia* 36 D4 1 6N 110 30 E
Tébessa, *Algeria* 50 A6 35 22N 8 8 E
Tebicuary →, *Paraguay* 94 B4 26 36S 58 16W
Tebingtinggi, *Indonesia* 36 D1 3 20N 99 9 E
Tecate, *Mexico* 85 N10 32 34N 116 38W
Tecomán, *Mexico* 86 D4 18 55N 103 53W
Tecopa, *U.S.A.* 85 K10 35 51N 116 13W
Tecoripa, *Mexico* 86 B3 28 37N 109 57W
Tecuala, *Mexico* 86 C3 22 23N 105 27W
Tecuci, *Romania* 17 F14 45 51N 27 27 E
Tecumseh, *U.S.A.* 76 D4 42 0N 83 57W
Tedzhen = Tejen, *Turkmenistan* 26 F7 37 23N 60 31 E
Tees →, *U.K.* 10 C6 54 37N 1 10W
Teesside, *U.K.* 10 C6 54 36N 1 15W
Teeswater, *Canada* 78 C3 43 59N 81 17W
Tefé, *Brazil* 92 D6 3 25S 64 50W
Tegal, *Indonesia* 37 G13 6 52S 109 8 E
Tegelen, *Neths.* 15 C6 51 20N 6 9 E
Teghra, *India* 43 G11 25 30N 85 34 E
Tegid, L. = Bala, L., *U.K.* 10 E4 52 53N 3 37W
Tegina, *Nigeria* 50 F6 10 5N 6 11 E
Tegucigalpa, *Honduras* 88 D2 14 5N 87 14W
Tehachapi, *U.S.A.* 85 K8 35 8N 118 27W
Tehachapi Mts., *U.S.A.* 85 L8 35 0N 118 30W
Tehrān, *Iran* 45 C6 35 44N 51 30 E
Tehuacán, *Mexico* 87 D5 18 30N 97 30W
Tehuantepec, *Mexico* 87 D5 16 21N 95 13W
Tehuantepec, G. de, *Mexico* 87 D5 15 50N 95 12W
Tehuantepec, Istmo de, *Mexico* 87 D6 17 0N 94 30W
Teide, *Canary Is.* 22 F3 28 15N 16 38W
Teifi →, *U.K.* 11 E3 52 5N 4 41W
Teign →, *U.K.* 11 G4 50 32N 3 32W
Teignmouth, *U.K.* 11 G4 50 33N 3 31W
Tejen, *Turkmenistan* 26 F7 37 23N 60 31 E
Tejo →, *Europe* 19 C1 38 40N 9 24W
Tejon Pass, *U.S.A.* 85 L8 34 49N 118 53W
Tekamah, *U.S.A.* 80 E6 41 47N 96 13W
Tekapo, L., *N.Z.* 59 K3 43 53S 170 33 E
Tekax, *Mexico* 87 C7 20 11N 89 18W
Tekeli, *Kazakstan* 26 E8 44 50N 79 0 E
Tekirdağ, *Turkey* 21 D12 40 58N 27 30 E
Tekkali, *India* 41 K14 18 37N 84 15 E
Tekoa, *U.S.A.* 82 C5 47 14N 117 4W
Tel Aviv-Yafo, *Israel* 47 C3 32 4N 34 48 E
Tel Lakhish, *Israel* 47 D3 31 34N 34 51 E
Tel Megiddo, *Israel* 47 C4 32 35N 35 11 E
Tela, *Honduras* 88 C2 15 40N 87 28W
Telanaipura = Jambi, *Indonesia* 36 E2 1 38S 103 30 E
Telavi, *Georgia* 25 F8 42 0N 45 30 E
Telde, *Canary Is.* 22 G4 27 59N 15 25W
Telegraph Creek, *Canada* 72 B2 58 0N 131 10W
Telekhany = Tsyelyakhany, *Belarus* 17 B13 52 30N 25 46 E
Telemark, *Norway* 9 G12 59 15N 7 40 E
Telén, *Argentina* 94 D2 36 15S 65 31W
Teleng, *Iran* 45 E9 25 47N 61 3 E
Teles Pires →, *Brazil* 92 E7 7 21S 58 3W
Telescope Pk., *U.S.A.* 85 J9 36 10N 117 5W
Telford, *U.K.* 10 E5 52 40N 2 27W
Télimélé, *Guinea* 50 F2 10 54N 13 2W
Telkwa, *Canada* 72 C3 54 41N 127 5W
Tell City, *U.S.A.* 76 G2 37 57N 86 46W
Tellicherry, *India* 40 P9 11 45N 75 30 E
Telluride, *U.S.A.* 83 H10 37 56N 107 49W
Teloloapán, *Mexico* 87 D5 18 21N 99 51W
Telpos Iz, *Russia* 24 B10 63 16N 59 13 E
Telsen, *Argentina* 96 E3 42 30S 66 50W
Telšiai, *Lithuania* 9 H20 55 59N 22 14 E
Teluk Anson, *Malaysia* 39 K3 4 3N 101 0 E
Teluk Betung = Tanjungkarang Telukbetung, *Indonesia* 36 F3 5 20S 105 10 E
Teluk Intan = Teluk Anson, *Malaysia* 39 K3 4 3N 101 0 E
Telukbutun, *Indonesia* 39 K7 4 13N 108 12 E
Telukdalem, *Indonesia* 36 D1 0 33N 97 50 E
Tema, *Ghana* 50 G5 5 41N 0 0 E
Temanggung, *Indonesia* 37 G14 7 18S 110 10 E
Temapache, *Mexico* 87 C5 21 4N 97 38W
Temax, *Mexico* 87 C7 21 10N 88 50W
Temba, *S. Africa* 57 D4 25 20S 28 17 E
Tembe, *Zaïre* 54 C2 0 16S 28 14 E
Temblor Range, *U.S.A.* 85 K7 35 20N 119 50W
Teme →, *U.K.* 11 E5 52 11N 2 13W
Temecula, *U.S.A.* 85 M9 33 30N 117 9W
Temerloh, *Malaysia* 39 L4 3 27N 102 25 E
Temir, *Kazakstan* 26 E6 49 1N 57 14 E
Temirtau, *Kazakstan* 26 D8 50 5N 72 56 E
Temirtau, *Russia* 26 D9 53 10N 87 30 E
Témiscaming, *Canada* 70 C4 46 44N 79 5W
Temma, *Australia* 62 G3 41 12S 144 48 E
Temora, *Australia* 63 E4 34 30S 147 30 E
Temosachic, *Mexico* 86 B3 28 58N 107 50W
Tempe, *U.S.A.* 83 K8 33 25N 111 56W
Tempe Downs, *Australia* 60 D5 24 22S 132 24 E
Tempiute, *U.S.A.* 84 H11 37 39N 115 38W
Temple, *U.S.A.* 81 K6 31 6N 97 21W
Temple B., *Australia* 62 A3 12 15S 143 3 E
Templemore, *Ireland* 13 D4 52 47N 7 51W
Templeton, *U.S.A.* 84 K6 35 33N 120 42W
Templeton →, *Australia* 62 C2 21 0S 138 40 E
Tempoal, *Mexico* 87 C5 21 31N 98 23W
Temuco, *Chile* 96 D2 38 45S 72 40W
Temuka, *N.Z.* 59 L3 44 14S 171 17 E
Tenabo, *Mexico* 87 C6 20 2N 90 12W
Tenaha, *U.S.A.* 81 K7 31 57N 94 15W
Tenali, *India* 40 L12 16 15N 80 35 E
Tenancingo, *Mexico* 87 D5 19 0N 99 33W
Tenango, *Mexico* 87 D5 19 7N 99 33W
Tenasserim, *Burma* 39 F2 12 6N 99 3 E
Tenasserim □, *Burma* 38 F2 14 0N 98 30 E
Tenby, *U.K.* 11 F3 51 40N 4 42W
Tenda, Col di, *France* 20 B2 44 7N 7 36 E
Tendaho, *Ethiopia* 46 E3 11 48N 40 54 E
Tenerife, *Canary Is.* 22 F3 28 15N 16 35W
Tenerife, Pico, *Canary Is.* 22 G1 27 43N 18 1W
Teng Xian, *China* 35 G9 35 5N 117 10 E
Tengah □, *Indonesia* 37 E6 2 0S 122 0 E
Tengah Kepulauan, *Indonesia* 36 F5 7 5S 118 15 E
Tengchong, *China* 32 D4 25 0N 98 28 E
Tengchowfu = Penglai, *China* 35 F11 37 48N 120 42 E
Tenggara □, *Indonesia* 37 E6 3 0S 122 0 E
Tenggarong, *Indonesia* 36 E5 0 24S 116 58 E
Tenggol, P., *Malaysia* 39 K4 4 48N 103 41 E
Tengiz, Ozero, *Kazakstan* 26 D7 50 30N 69 0 E
Tenino, *U.S.A.* 84 D4 46 51N 122 51W
Tenkasi, *India* 40 Q10 8 55N 77 20 E
Tenke, *Shaba, Zaïre* 55 E2 11 22S 26 40 E
Tenke, *Shaba, Zaïre* 55 E2 10 32S 26 7 E
Tenkodogo, *Burkina Faso* 50 F4 11 54N 0 19W
Tennant Creek, *Australia* 62 B1 19 30S 134 15 E
Tennessee □, *U.S.A.* 77 H2 36 0N 86 30W
Tennessee →, *U.S.A.* 76 G1 37 4N 88 34W
Tennille, *U.S.A.* 77 J4 32 56N 82 48W
Teno, Pta. de, *Canary Is.* 22 F3 28 21N 16 55W
Tenom, *Malaysia* 36 C5 5 4N 115 57 E
Tenosique, *Mexico* 87 D6 17 30N 91 24W
Tenryū-Gawa →, *Japan* 31 G8 35 39N 137 48 E
Tent L., *Canada* 73 A7 62 25N 107 54W
Tenterfield, *Australia* 63 D5 29 0S 152 0 E
Teófilo Otoni, *Brazil* 93 G10 17 50S 41 30W
Teotihuacán, *Mexico* 87 D5 19 44N 98 50W
Tepa, *Indonesia* 37 F7 7 52S 129 31 E
Tepalcatepec →, *Mexico* 86 D4 18 35N 101 59W
Tepehuanes, *Mexico* 86 B3 25 21N 105 44W
Tepetongo, *Mexico* 86 C4 22 28N 103 9W
Tepic, *Mexico* 86 C4 21 30N 104 54W
Teplice, *Czech.* 16 C7 50 40N 13 48 E
Tepoca, C., *Mexico* 86 A2 30 20N 112 25W
Tequila, *Mexico* 86 C4 20 54N 103 47W
Ter →, *Spain* 19 A7 42 2N 3 12 E
Ter Apel, *Neths.* 15 B7 52 53N 7 5 E
Téra, *Niger* 50 F5 14 0N 0 45 E
Teraina, *Kiribati* 65 G11 4 43N 160 25W
Téramo, *Italy* 20 C5 42 39N 13 42 E
Terang, *Australia* 63 F3 38 15S 142 55 E
Tercero →, *Argentina* 94 C3 32 58S 61 47W
Terebovlya, *Ukraine* 17 D13 49 18N 25 44 E
Terek →, *Russia* 25 F8 44 0N 47 30 E
Teresina, *Brazil* 93 E10 5 9S 42 45W
Terewah, L., *Australia* 63 D4 29 52S 147 35 E
Terhazza, *Mali* 50 D3 23 38N 5 22W
Teridgerie Cr. →, *Australia* 63 E4 30 25S 148 50 E
Termez = Termiz, *Uzbekistan* 26 F7 37 15N 67 15 E
Términi Imerese, *Italy* 20 F5 37 59N 13 42 E
Términos, L. de, *Mexico* 87 D6 18 35N 91 30W
Termiz, *Uzbekistan* 26 F7 37 15N 67 15 E
Térmoli, *Italy* 20 C6 42 0N 15 0 E
Ternate, *Indonesia* 37 D7 0 45N 127 25 E
Terneuzen, *Neths.* 15 C3 51 20N 3 50 E
Terney, *Russia* 27 E14 45 3N 136 37 E
Terni, *Italy* 20 C5 42 34N 12 37 E
Ternopil, *Ukraine* 17 D13 49 30N 25 40 E
Ternopol = Ternopil, *Ukraine* 17 D13 49 30N 25 40 E
Terowie, *N.S.W., Australia* 63 E4 32 27S 147 52 E
Terowie, *S. Austral., Australia* 63 E2 33 8S 138 55 E
Terra Bella, *U.S.A.* 85 K7 35 58N 119 3W
Terrace, *Canada* 72 C3 54 30N 128 35W
Terrace Bay, *Canada* 70 C2 48 47N 87 5W
Terracina, *Italy* 20 D5 41 17N 13 15 E
Terralba, *Italy* 20 E3 39 43N 8 39 E
Terranova = Ólbia, *Italy* 20 D3 40 55N 9 31 E
Terrassa = Tarrasa, *Spain* 19 B7 41 34N 2 1 E
Terre Haute, *U.S.A.* 76 F2 39 28N 87 25W
Terrebonne B., *U.S.A.* 81 L9 29 5N 90 35W
Terrell, *U.S.A.* 81 J6 32 44N 96 17W
Terrenceville, *Canada* 71 C9 47 40N 54 44W
Terrick Terrick, *Australia* 62 C4 24 44S 145 5 E
Terry, *U.S.A.* 80 B2 46 47N 105 19W
Terschelling, *Neths.* 15 A5 53 25N 5 20 E
Teruel, *Spain* 19 B5 40 22N 1 8W
Tervola, *Finland* 8 C21 66 6N 24 49 E
Teryaweyna L., *Australia* 63 E3 32 18S 143 22 E
Teshio, *Japan* 30 B10 44 53N 141 44 E
Teshio-Gawa →, *Japan* 30 B10 44 53N 141 45 E
Tesiyn Gol →, *Mongolia* 32 A4 50 40N 93 20 E
Teslin, *Canada* 72 A2 60 10N 132 43W
Teslin →, *Canada* 72 A2 61 34N 134 35W
Teslin L., *Canada* 72 A2 60 15N 132 57W
Tessalit, *Mali* 50 D5 20 12N 1 0 E
Tessaoua, *Niger* 50 F6 13 47N 7 56 E
Test →, *U.K.* 11 F6 50 56N 1 29W
Tetachuck L., *Canada* 72 C3 53 18N 125 55W
Tetas, Pta., *Chile* 94 A1 23 31S 70 38W
Tete, *Mozam.* 55 F3 16 13S 33 33 E
Tete □, *Mozam.* 55 F3 15 15S 32 40 E
Teterev →, *Ukraine* 17 C16 51 1N 30 5 E
Teteven, *Bulgaria* 21 C11 42 58N 24 17 E
Tethul →, *Canada* 72 A6 60 35N 112 12W
Tetiyev, *Ukraine* 17 D15 49 22N 29 38 E
Teton →, *U.S.A.* 82 C8 47 56N 110 31W
Tétouan, *Morocco* 50 A3 35 35N 5 21W
Tetovo, *Macedonia* 21 C9 42 1N 21 2 E
Tetyukhe Pristan, *Russia* 30 B7 44 22N 135 48 E
Teuco →, *Argentina* 94 B3 25 35S 60 11W
Teulon, *Canada* 73 C9 50 23N 97 16W
Teun, *Indonesia* 37 F7 6 59S 129 8 E
Teutoburger Wald, *Germany* 16 B5 52 5N 8 22 E
Tévere →, *Italy* 20 D5 41 44N 12 14 E
Teverya, *Israel* 47 C4 32 47N 35 32 E
Teviot →, *U.K.* 12 F6 55 29N 2 38W
Tewantin, *Australia* 63 D5 26 27S 153 3 E
Tewkesbury, *U.K.* 11 F5 51 59N 2 9W
Texada I., *Canada* 72 D4 49 40N 124 25W
Texarkana, *Ark., U.S.A.* 81 J8 33 26N 94 2W
Texarkana, *Tex., U.S.A.* 81 J7 33 26N 94 3W
Texas, *Australia* 63 D5 28 49S 151 9 E
Texas □, *U.S.A.* 81 K5 31 40N 98 30W
Texas City, *U.S.A.* 81 L7 29 24N 94 54W
Texel, *Neths.* 15 A4 53 5N 4 50 E
Texhoma, *U.S.A.* 81 G4 36 30N 101 47W
Texline, *U.S.A.* 81 G3 36 23N 103 2W
Texoma, L., *U.S.A.* 81 J6 33 50N 96 34W
Tezin, *Afghan.* 42 B3 34 24N 69 30 E
Teziutlán, *Mexico* 87 D5 19 50N 97 22W
Tezpur, *India* 41 F18 26 40N 92 45 E
Tezzeron L., *Canada* 72 C4 54 43N 124 30W
Tha-anne →, *Canada* 73 A10 60 31N 94 37W
Tha Deua, *Laos* 38 D4 17 57N 102 53 E
Tha Deua, *Laos* 38 C3 19 26N 101 50 E
Tha Pla, *Thailand* 38 D3 17 48N 100 32 E
Tha Rua, *Thailand* 38 E3 14 34N 100 44 E
Tha Sala, *Thailand* 39 H2 8 40N 99 56 E
Tha Song Yang, *Thailand* 38 D1 17 34N 97 55 E
Thaba Putsoa, *Lesotho* 57 D4 29 45S 28 0 E
Thabana Ntlenyana, *Lesotho* 57 D4 29 30S 29 16 E
Thabazimbi, *S. Africa* 57 C4 24 40S 27 21 E
Thai Binh, *Vietnam* 38 B6 20 35N 106 1 E
Thai Hoa, *Vietnam* 38 C5 19 20N 105 20 E
Thai Muang, *Thailand* 39 H2 8 24N 98 16 E
Thai Nguyen, *Vietnam* 38 B5 21 35N 105 55 E
Thailand ■, *Asia* 38 E4 16 0N 102 0 E
Thailand, G. of, *Asia* 39 G3 11 30N 101 0 E
Thakhek, *Laos* 38 D5 17 25N 104 45 E
Thal, *Pakistan* 42 C4 33 28N 70 33 E
Thal Desert, *Pakistan* 42 D4 31 10N 71 30 E
Thala La, *Burma* 41 E20 28 25N 97 23 E
Thalabarivat, *Cambodia* 38 F5 13 33N 105 57 E
Thallon, *Australia* 63 D4 28 39S 148 49 E
Thame →, *U.K.* 11 F6 51 39N 1 9W
Thames, *N.Z.* 59 G5 37 7S 175 34 E
Thames →, *Canada* 70 D3 42 20N 82 25W
Thames →, *U.K.* 11 F8 51 29N 0 34 E
Thames →, *U.S.A.* 79 E12 41 18N 72 5W
Thamesdown □, *U.K.* 11 F6 51 33N 1 47W
Thamesford, *Canada* 78 C3 43 4N 81 0W
Thamesville, *Canada* 78 D3 42 33N 81 59W
Than Uyen, *Vietnam* 38 B4 22 0N 103 54 E
Thane, *India* 40 K8 19 12N 72 59 E
Thanesar, *India* 42 D7 30 1N 76 52 E
Thanet, I. of, *U.K.* 11 F9 51 21N 1 20 E
Thangoo, *Australia* 60 C3 18 10S 122 22 E
Thangool, *Australia* 62 C5 24 38S 150 42 E
Thanh Hoa, *Vietnam* 38 C5 19 48N 105 46 E
Thanh Hung, *Vietnam* 39 H5 9 55N 105 43 E
Thanh Pho Ho Chi Minh = Phanh Bho Ho Chi Minh, *Vietnam* 39 G6 10 58N 106 40 E
Thanh Thuy, *Vietnam* 38 A5 22 55N 104 51 E
Thanjavur, *India* 40 P11 10 48N 79 12 E
Thap Sakae, *Thailand* 39 G2 11 30N 99 37 E
Thap Than, *Thailand* 38 E2 15 27N 99 54 E
Thar Desert, *India* 42 F4 28 0N 72 0 E
Tharad, *India* 42 G4 24 30N 71 44 E
Thargomindah, *Australia* 63 D3 27 58S 143 46 E
Tharrawaddy, *Burma* 41 L19 17 38N 95 48 E
Tharthār, Mileh, *Iraq* 44 C4 34 0N 43 15 E
Tharthār, W. ath →, *Iraq* 44 C4 33 59N 43 12 E
Thásos, *Greece* 21 D11 40 40N 24 40 E
That Khe, *Vietnam* 38 A6 22 16N 106 28 E
Thatcher, *Ariz., U.S.A.* 83 K9 32 51N 109 46W
Thatcher, *Colo., U.S.A.* 81 G2 37 33N 104 7W
Thaton, *Burma* 41 L20 16 55N 97 22 E
Thaungdut, *Burma* 41 G19 24 30N 94 40 E
Thayer, *U.S.A.* 81 G9 36 31N 91 33W
Thayetmyo, *Burma* 41 K19 19 20N 95 10 E
Thazi, *Burma* 41 J20 21 0N 96 5 E
The Alberga →, *Australia* 63 D2 27 6S 135 33 E
The Bight, *Bahamas* 89 B4 24 19N 75 24W
The Coorong, *Australia* 63 F2 35 50S 139 20 E
The Dalles, *U.S.A.* 82 D3 45 36N 121 10W
The English Company's Is., *Australia* 62 A2 11 50S 136 32 E
The Frome →, *Australia* 63 D2 29 8S 137 54 E
The Grampians, *Australia* 63 F3 37 0S 142 20 E
The Great Divide = Great Dividing Ra., *Australia* 62 C4 23 0S 146 0 E
The Hague = 's-Gravenhage, *Neths.* 15 B4 52 7N 4 17 E
The Hamilton →, *Australia* 63 D2 26 40S 135 19 E
The Macumba →, *Australia* 63 D2 27 52S 137 12 E
The Neales →, *Australia* 63 D2 28 8S 136 47 E
The Officer →, *Australia* 61 E5 27 46S 132 30 E
The Pas, *Canada* 73 C8 53 45N 101 15W
The Range, *Zimbabwe* 55 F3 19 2S 31 2 E
The Rock, *Australia* 63 F4 35 15S 147 2 E
The Salt L., *Australia* 63 E3 30 6S 142 8 E
The Stevenson →, *Australia* 63 D2 27 6S 135 33 E
The Warburton →, *Australia* 63 D2 28 4S 137 28 E
Thebes = Thívai, *Greece* 21 E10 38 19N 23 19 E
Thedford, *Canada* 78 C3 43 9N 81 51W
Thedford, *U.S.A.* 80 E4 41 59N 100 35W
Theebine, *Australia* 63 D5 25 57S 152 34 E
Thekulthili L., *Canada* 73 A7 61 3N 110 0W
Thelon →, *Canada* 73 A8 62 35N 104 3W
Theodore, *Australia* 62 C5 24 55S 150 3 E
Thepha, *Thailand* 39 J3 6 52N 100 58 E
Theresa, *U.S.A.* 79 B9 44 13N 75 48W
Thermaïkós Kólpos, *Greece* 21 D10 40 15N 22 45 E
Thermopolis, *U.S.A.* 82 E9 43 39N 108 13W
Thermopylae P., *Greece* 21 E10 38 48N 22 35 E
Thessalon, *Canada* 70 C3 46 20N 83 30W
Thessaloníki, *Greece* 21 D10 40 38N 22 58 E
Thessaloniki, Gulf of = Thermaïkós Kólpos, *Greece* 21 D10 40 15N 22 45 E
Thetford, *U.K.* 11 E8 52 25N 0 45 E
Thetford Mines, *Canada* 71 C5 46 8N 71 18W
Theun →, *Laos* 38 C5 18 19N 104 0 E

Theunissen, *S. Africa* **56 D4** 28 26S 26 43 E
Thevenard, *Australia* **63 E1** 32 9S 133 38 E
Thibodaux, *U.S.A.* **81 L9** 29 48N 90 49W
Thicket Portage, *Canada* **73 B9** 55 19N 97 42W
Thief River Falls, *U.S.A.* **80 A6** 48 7N 96 10W
Thiel Mts., *Antarctica* **5 E16** 85 15S 91 0W
Thiers, *France* **18 D5** 45 52N 3 33 E
Thies, *Senegal* **50 F1** 14 50N 16 51W
Thika, *Kenya* **54 C4** 1 1S 37 5 E
Thikombia, *Fiji* **59 B9** 15 44S 179 55W
Thimphu, *Bhutan* **41 F16** 27 31N 89 45 E
þingvallavatn, *Iceland* **8 D3** 64 11N 21 9W
Thionville, *France* **18 B7** 49 20N 6 10 E
Thíra, *Greece* **21 F11** 36 23N 25 27 E
Thirsk, *U.K.* **10 C6** 54 14N 1 19W
Thisted, *Denmark* **9 H13** 56 58N 8 40 E
Thistle I., *Australia* **63 F2** 35 0S 136 8 E
Thívai, *Greece* **21 E10** 38 19N 23 19 E
þjórsá →, *Iceland* **8 E3** 63 47N 20 48W
Thlewiaza →, *Man., Canada* **73 B8** 59 43N 100 5W
Thlewiaza →, *N.W.T., Canada* **73 A10** 60 29N 94 40W
Thmar Puok, *Cambodia* **38 F4** 13 57N 103 4 E
Tho Vinh, *Vietnam* **38 C5** 19 16N 105 42 E
Thoa →, *Canada* **73 A7** 60 31N 109 47W
Thoen, *Thailand* **38 D2** 17 43N 99 12 E
Thoeng, *Thailand* **38 C3** 19 41N 100 12 E
Tholdi, *Pakistan* **43 B7** 35 5N 76 6 E
Thomas, *Okla., U.S.A.* **81 H5** 35 45N 98 45W
Thomas, *W. Va., U.S.A.* **76 F6** 39 9N 79 30W
Thomas, L., *Australia* **63 D2** 26 4S 137 58 E
Thomaston, *U.S.A.* **77 J3** 32 53N 84 20W
Thomasville, *Ala., U.S.A.* **77 K2** 31 55N 87 44W
Thomasville, *Ga., U.S.A.* **77 K3** 30 50N 83 59W
Thomasville, *N.C., U.S.A.* **77 H5** 35 53N 80 5W
Thompson, *Canada* **73 B9** 55 45N 97 52W
Thompson, *U.S.A.* **83 G9** 38 58N 109 43W
Thompson →, *Canada* **72 C4** 50 15N 121 24W
Thompson →, *U.S.A.* **80 F8** 39 46N 93 37W
Thompson Falls, *U.S.A.* **82 C6** 47 36N 115 21W
Thompson Landing, *Canada* **73 A6** 62 56N 110 40W
Thompson Pk., *U.S.A.* **82 F2** 41 0N 123 0W
Thomson's Falls = Nyahururu, *Kenya* **54 B4** 0 2N 36 27 E
Thon Buri, *Thailand* **39 F3** 13 43N 100 29 E
þórisvatn, *Iceland* **8 D4** 64 20N 18 55W
Thornaby on Tees, *U.K.* **10 C6** 54 33N 1 18W
Thornbury, *Canada* **78 B4** 44 34N 80 26W
Thorold, *Canada* **78 C5** 43 7N 79 12W
þórshöfn, *Iceland* **8 C6** 66 12N 15 20W
Thouin, C., *Australia* **60 D2** 20 20S 118 10 E
Thousand Oaks, *U.S.A.* **85 L8** 34 10N 118 50W
Thrace, *Turkey* **21 D12** 41 0N 27 0 E
Three Forks, *U.S.A.* **82 D8** 45 54N 111 33W
Three Hills, *Canada* **72 C6** 51 43N 113 15W
Three Hummock I., *Australia* **62 G3** 40 25S 144 55 E
Three Lakes, *U.S.A.* **80 C10** 45 48N 89 10W
Three Points, C., *Ghana* **50 H4** 4 42N 2 6W
Three Rivers, *Australia* **61 E2** 25 10S 119 5 E
Three Rivers, *Calif., U.S.A.* **84 J8** 36 26N 118 54W
Three Rivers, *Tex., U.S.A.* **81 L5** 28 28N 98 11W
Three Sisters, *U.S.A.* **82 D3** 44 4N 121 51W
Throssell, L., *Australia* **61 E3** 27 33S 124 10 E
Throssell Ra., *Australia* **60 D3** 22 3S 121 43 E
Thuan Hoa, *Vietnam* **39 H5** 8 58N 105 30 E
Thubun Lakes, *Canada* **73 A6** 61 30N 112 0W
Thuin, *Belgium* **15 D4** 50 20N 4 17 E
Thule, *Greenland* **4 B4** 77 40N 69 0W
Thun, *Switz.* **16 E4** 46 45N 7 38 E
Thundelarra, *Australia* **61 E2** 28 53S 117 7 E
Thunder B., *U.S.A.* **78 B1** 45 0N 83 20W
Thunder Bay, *Canada* **70 C2** 48 20N 89 15W
Thung Song, *Thailand* **39 H2** 8 10N 99 40 E
Thunkar, *Bhutan* **41 F17** 27 55N 91 0 E
Thuong Tra, *Vietnam* **38 D6** 16 2N 107 42 E
Thüringer Wald, *Germany* **16 C6** 50 35N 11 0 E
Thurles, *Ireland* **13 D4** 52 41N 7 49W
Thurloo Downs, *Australia* **63 D3** 29 15S 143 30 E
Thursday I., *Australia* **62 A3** 10 30S 142 3 E
Thurso, *Canada* **70 C4** 45 36N 75 15W
Thurso, *U.K.* **12 C5** 58 36N 3 32W
Thurston I., *Antarctica* **5 D16** 72 0S 100 0W
Thutade L., *Canada* **72 B3** 57 0N 126 55W
Thylungra, *Australia* **63 D3** 26 4S 143 28 E
Thyolo, *Malawi* **55 F4** 16 7S 35 5 E
Thysville = Mbanza Ngungu, *Zaïre* **52 F2** 5 12S 14 53 E
Tia, *Australia* **63 E5** 31 10S 151 50 E
Tian Shan, *Asia* **32 B3** 42 0N 76 0 E
Tianjin, *China* **35 E9** 39 8N 117 10 E
Tianshui, *China* **34 G3** 34 32N 105 40 E
Tianzhen, *China* **34 D8** 40 24N 114 5 E
Tianzhuangtai, *China* **35 D12** 40 43N 122 5 E
Tiaret, *Algeria* **50 A5** 35 20N 1 21 E
Tiassalé, *Ivory C.* **50 G4** 5 58N 4 57W
Tibagi, *Brazil* **95 A5** 24 30S 50 24W
Tibagi →, *Brazil* **95 A5** 22 47S 51 1W
Tibati, *Cameroon* **51 G7** 6 22N 12 30 E
Tiber = Tévere →, *Italy* **20 D5** 41 44N 12 14 E
Tiber Reservoir, *U.S.A.* **82 B8** 48 19N 111 6W
Tiberias = Teverya, *Israel* **47 C4** 32 47N 35 32 E
Tiberias, L. = Yam Kinneret, *Israel* **47 C4** 32 45N 35 35 E
Tibesti, *Chad* **51 D8** 21 0N 17 30 E
Tibet = Xizang □, *China* **32 C3** 32 0N 88 0 E
Tibet, Plateau of, *Asia* **28 F12** 32 0N 86 0 E
Tibni, *Syria* **44 C3** 35 36N 39 50 E
Tibooburra, *Australia* **63 D3** 29 26S 142 1 E
Tiburón, *Mexico* **86 B2** 29 0N 112 30W
Tîchît, *Mauritania* **50 E3** 18 21N 9 29W
Ticino →, *Italy* **20 B3** 45 9N 9 14 E
Ticonderoga, *U.S.A.* **79 C11** 43 51N 73 26W
Ticul, *Mexico* **87 C7** 20 20N 89 31W
Tidaholm, *Sweden* **9 G15** 58 12N 13 55 E
Tiddim, *Burma* **41 H18** 23 28N 93 45 E
Tidjikja, *Mauritania* **50 E2** 18 29N 11 35W
Tidore, *Indonesia* **37 D7** 0 40N 127 25 E
Tiel, *Neths.* **15 C5** 51 53N 5 26 E
Tiel, *Senegal* **50 F1** 14 55N 15 5W
Tieling, *China* **35 C12** 42 20N 123 55 E
Tielt, *Belgium* **15 D3** 51 0N 3 20 E
Tien Shan = Tian Shan, *Asia* **32 B3** 42 0N 76 0 E
Tien-tsin = Tianjin, *China* **35 E9** 39 8N 117 10 E
Tien Yen, *Vietnam* **38 B6** 21 20N 107 24 E
T'ienching = Tianjin, *China* **35 E9** 39 8N 117 10 E
Tienen, *Belgium* **15 D4** 50 48N 4 57 E
Tientsin = Tianjin, *China* **35 E9** 39 8N 117 10 E
Tierra Amarilla, *Chile* **94 B1** 27 28S 70 18W
Tierra Amarilla, *U.S.A.* **83 H10** 36 42N 106 33W
Tierra Colorada, *Mexico* **87 D5** 17 10N 99 35W
Tierra de Campos, *Spain* **19 A3** 42 10N 4 50W
Tierra del Fuego, I. Gr. de, *Argentina* **96 G3** 54 0S 69 0W
Tiétar →, *Spain* **19 C3** 39 50N 6 1W
Tieté →, *Brazil* **95 A5** 20 40S 51 35W
Tieyon, *Australia* **63 D1** 26 12S 133 52 E
Tiffin, *U.S.A.* **76 E4** 41 7N 83 11W
Tiflis = Tbilisi, *Georgia* **25 F7** 41 43N 44 50 E
Tifton, *U.S.A.* **77 K4** 31 27N 83 31W
Tifu, *Indonesia* **37 E7** 3 39S 126 24 E
Tighina, *Moldova* **17 E15** 46 50N 29 30 E
Tigil, *Russia* **27 D16** 57 49N 158 40 E
Tignish, *Canada* **71 C7** 46 58N 64 2W
Tigre →, *Peru* **92 D4** 4 30S 74 10W
Tigris = Dijlah, Nahr →, *Asia* **44 D5** 31 0N 47 25 E
Tigyaing, *Burma* **41 H20** 23 45N 96 10 E
Tîh, Gebel el, *Egypt* **51 C11** 29 32N 33 26 E
Tijuana, *Mexico* **85 N9** 32 30N 117 10W
Tikal, *Guatemala* **88 C2** 17 13N 89 24W
Tikamgarh, *India* **43 G8** 24 44N 78 50 E
Tikhoretsk, *Russia* **25 E7** 45 56N 40 5 E
Tikrīt, *Iraq* **44 C4** 34 35N 43 37 E
Tiksi, *Russia* **27 B13** 71 40N 128 45 E
Tilamuta, *Indonesia* **37 D6** 0 32N 122 23 E
Tilburg, *Neths.* **15 C5** 51 31N 5 6 E
Tilbury, *Canada* **70 D3** 42 17N 82 23W
Tilbury, *U.K.* **11 F8** 51 27N 0 22 E
Tilcara, *Argentina* **94 A2** 23 36S 65 23W
Tilden, *Nebr., U.S.A.* **80 D6** 42 3N 97 50W
Tilden, *Tex., U.S.A.* **81 L5** 28 28N 98 33W
Tilhar, *India* **43 F8** 28 0N 79 45 E
Tilichiki, *Russia* **27 C17** 60 27N 166 5 E
Tílissos, *Greece* **23 D7** 35 20N 25 1 E
Till →, *U.K.* **10 B5** 55 35N 2 3W
Tillabéri, *Niger* **50 F5** 14 28N 1 28 E
Tillamook, *U.S.A.* **82 D2** 45 27N 123 51W
Tillsonburg, *Canada* **70 D3** 42 53N 80 44W
Tillyeria □, *Cyprus* **23 D11** 35 6N 32 40 E
Tílos, *Greece* **21 F12** 36 27N 27 27 E
Tilpa, *Australia* **63 E3** 30 57S 144 24 E
Tilsit = Sovetsk, *Russia* **9 J19** 55 6N 21 50 E
Tilt →, *U.K.* **12 E5** 56 46N 3 51W
Tilton, *U.S.A.* **79 C13** 43 27N 71 36W
Timagami L., *Canada* **70 C3** 47 0N 80 10W
Timanskiy Kryazh, *Russia* **24 A9** 65 58N 50 5 E
Timaru, *N.Z.* **59 L3** 44 23S 171 14 E
Timau, *Kenya* **54 B4** 0 4N 37 15 E
Timbákion, *Greece* **23 D6** 35 4N 24 45 E
Timbedgha, *Mauritania* **50 E3** 16 17N 8 16W
Timber Lake, *U.S.A.* **80 C4** 45 26N 101 5W
Timber Mt., *U.S.A.* **84 H10** 37 6N 116 28W
Timboon, *Australia* **63 F3** 38 30S 142 58 E
Timbuktu = Tombouctou, *Mali* **50 E4** 16 50N 3 0W
Timi, *Cyprus* **23 E11** 34 44N 32 31 E
Timimoun, *Algeria* **50 C5** 29 14N 0 16 E
Timişoara, *Romania* **17 F11** 45 43N 21 15 E
Timmins, *Canada* **70 C3** 48 28N 81 25W
Timok →, *Serbia, Yug.* **21 B10** 44 10N 22 40 E
Timon, *Brazil* **93 E10** 5 8S 42 52W
Timor, *Indonesia* **37 F7** 9 0S 125 0 E
Timor □, *Indonesia* **37 F7** 9 0S 125 0 E
Timor Sea, *Ind. Oc.* **60 B4** 12 0S 127 0 E
Tin Mt., *U.S.A.* **84 J9** 36 50N 117 10W
Tinaca Pt., *Phil.* **37 C7** 5 30N 125 25 E
Tinajo, *Canary Is.* **22 E6** 29 4N 13 42W
Tindouf, *Algeria* **50 C3** 27 42N 8 10W
Tinggi, Pulau, *Malaysia* **39 L5** 2 18N 104 7 E
Tingo Maria, *Peru* **92 E3** 9 10S 75 54W
Tinh Bien, *Vietnam* **39 G5** 10 36N 104 57 E
Tinjoub, *Algeria* **50 C3** 29 45N 5 40W
Tinkurrin, *Australia* **61 F2** 32 59S 117 46 E
Tinnevelly = Tirunelveli, *India* **40 Q10** 8 45N 77 45 E
Tinogasta, *Argentina* **94 B2** 28 5S 67 32W
Tínos, *Greece* **21 F11** 37 33N 25 8 E
Tintina, *Argentina* **94 B3** 27 2S 62 45W
Tintinara, *Australia* **63 F3** 35 48S 140 2 E
Tioga, *U.S.A.* **78 E7** 41 55N 77 8W
Tioman, Pulau, *Malaysia* **39 L5** 2 50N 104 10 E
Tionesta, *U.S.A.* **78 E5** 41 30N 79 28W
Tipongpani, *India* **41 F19** 27 20N 95 55 E
Tipperary, *Ireland* **13 D3** 52 28N 8 10W
Tipperary □, *Ireland* **13 D4** 52 37N 7 55W
Tipton, *U.K.* **11 E5** 52 32N 2 4W
Tipton, *Calif., U.S.A.* **83 H4** 36 4N 119 19W
Tipton, *Ind., U.S.A.* **76 E2** 40 17N 86 2W
Tipton, *Iowa, U.S.A.* **80 E9** 41 46N 91 8W
Tipton Mt., *U.S.A.* **85 K12** 35 32N 114 12W
Tiptonville, *U.S.A.* **81 G10** 36 23N 89 29W
Tīrān, *Iran* **45 C6** 32 45N 51 8 E
Tirana, *Albania* **21 D8** 41 18N 19 49 E
Tiranë = Tirana, *Albania* **21 D8** 41 18N 19 49 E
Tiraspol, *Moldova* **17 E15** 46 55N 29 35 E
Tirat Karmel, *Israel* **47 C3** 32 46N 34 58 E
Tire, *Turkey* **21 E12** 38 5N 27 50 E
Tirebolu, *Turkey* **25 F6** 40 58N 38 45 E
Tiree, *U.K.* **12 E2** 56 31N 6 55W
Tîrgovişte, *Romania* **17 F13** 44 55N 25 27 E
Tîrgu-Jiu, *Romania* **17 F12** 45 5N 23 19 E
Tîrgu Mureş, *Romania* **17 E13** 46 31N 24 38 E
Tirich Mir, *Pakistan* **40 A7** 36 15N 71 55 E
Tîrnăveni, *Romania* **17 E13** 46 19N 24 13 E
Tírnavos, *Greece* **21 E10** 39 45N 22 18 E
Tirodi, *India* **40 J11** 21 40N 79 44 E
Tirol □, *Austria* **16 E6** 47 3N 10 43 E
Tirso →, *Italy* **20 E3** 39 53N 8 32 E
Tiruchchirappalli, *India* **40 P11** 10 45N 78 45 E
Tirunelveli, *India* **40 Q10** 8 45N 77 45 E
Tirupati, *India* **40 N11** 13 39N 79 25 E
Tiruppur, *India* **40 P10** 11 5N 77 22 E
Tiruvannamalai, *India* **40 N11** 12 15N 79 5 E
Tisa →, *Serbia, Yug.* **21 B9** 45 15N 20 17 E
Tisdale, *Canada* **73 C8** 52 50N 104 0W
Tishomingo, *U.S.A.* **81 H6** 34 14N 96 41W
Tisza = Tisa →, *Serbia, Yug.* **21 B9** 45 15N 20 17 E
Tit-Ary, *Russia* **27 B13** 71 55N 127 2 E
Tithwal, *Pakistan* **43 B5** 34 21N 73 50 E
Titicaca, L., *S. Amer.* **92 G5** 15 30S 69 30W
Titograd = Podgorica, *Montenegro, Yug.* **21 C8** 42 30N 19 19 E
Titov Veles, *Macedonia* **21 D9** 41 46N 21 47 E
Titova-Mitrovica, *Serbia, Yug.* **21 C9** 42 54N 20 52 E
Titovo Užice, *Serbia, Yug.* **21 C8** 43 55N 19 50 E
Titule, *Zaïre* **54 B2** 3 15N 25 31 E
Titusville, *Fla., U.S.A.* **77 L5** 28 37N 80 49W
Titusville, *Pa., U.S.A.* **78 E5** 41 38N 79 41W
Tivaouane, *Senegal* **50 F1** 14 56N 16 45W
Tiverton, *U.K.* **11 G4** 50 54N 3 29W
Tívoli, *Italy* **20 D5** 41 58N 12 45 E
Tizi-Ouzou, *Algeria* **50 A5** 36 42N 4 3 E
Tizimín, *Mexico* **87 C7** 21 0N 88 1W
Tiznit, *Morocco* **50 C3** 29 48N 9 45W
Tjeggelvas, *Sweden* **8 C17** 66 37N 17 45 E
Tjirebon = Cirebon, *Indonesia* **37 G13** 6 45S 108 32 E
Tjörn, *Sweden* **9 G14** 58 0N 11 35 E
Tlacotalpan, *Mexico* **87 D5** 18 37N 95 40W
Tlahualilo, *Mexico* **86 B4** 26 20N 103 30W
Tlaquepaque, *Mexico* **86 C4** 20 39N 103 19W
Tlaxcala, *Mexico* **87 D5** 19 20N 98 14W
Tlaxcala □, *Mexico* **87 D5** 19 30N 98 20W
Tlaxiaco, *Mexico* **87 D5** 17 18N 97 40W
Tlell, *Canada* **72 C2** 53 34N 131 56W
Tlemcen, *Algeria* **50 B4** 34 52N 1 21W
Tmassah, *Libya* **51 C8** 26 19N 15 51 E
To Bong, *Vietnam* **38 F7** 12 45N 109 16 E
Toad →, *Canada* **72 B4** 59 25N 124 57W
Toamasina, *Madag.* **57 B8** 18 10S 49 25 E
Toamasina □, *Madag.* **57 B8** 18 0S 49 0 E
Toay, *Argentina* **94 D3** 36 43S 64 38W
Toba, *Japan* **31 G8** 34 30N 136 51 E
Toba Kakar, *Pakistan* **42 D3** 31 30N 69 0 E
Toba Tek Singh, *Pakistan* **42 D5** 30 55N 72 25 E
Tobago, *W. Indies* **89 D7** 11 10N 60 30W
Tobelo, *Indonesia* **37 D7** 1 45N 127 56 E
Tobermorey, *Australia* **62 C2** 22 12S 138 0 E
Tobermory, *Canada* **70 C3** 45 12N 81 40W
Tobermory, *U.K.* **12 E2** 56 38N 6 5W
Tobin, *U.S.A.* **84 F5** 39 55N 121 19W
Tobin, L., *Australia* **60 D4** 21 45S 125 49 E
Tobin L., *Canada* **73 C8** 53 35N 103 30W
Toboali, *Indonesia* **36 E3** 3 0S 106 25 E
Tobol, *Kazakstan* **26 D7** 52 40N 62 39 E
Tobol →, *Russia* **26 D7** 58 10N 68 12 E
Toboli, *Indonesia* **37 E6** 0 38S 120 5 E
Tobolsk, *Russia* **26 D7** 58 15N 68 10 E
Tobruk = Tubruq, *Libya* **51 B9** 32 7N 23 55 E
Tobyhanna, *U.S.A.* **79 E9** 41 11N 75 25W
Tobyl = Tobol →, *Russia* **26 D7** 58 10N 68 12 E
Tocantinópolis, *Brazil* **93 E9** 6 20S 47 25W
Tocantins □, *Brazil* **93 F9** 10 0S 48 0W
Tocantins →, *Brazil* **93 D9** 1 45S 49 10W
Toccoa, *U.S.A.* **77 H4** 34 35N 83 19W
Tochigi, *Japan* **31 F9** 36 25N 139 45 E
Tochigi □, *Japan* **31 F9** 36 45N 139 45 E
Tocopilla, *Chile* **94 A1** 22 5S 70 10W
Tocumwal, *Australia* **63 F4** 35 51S 145 31 E
Tocuyo →, *Venezuela* **92 A5** 11 3N 68 23W
Todd →, *Australia* **62 C2** 24 52S 135 48 E
Todeli, *Indonesia* **37 E6** 1 38S 124 34 E
Todenyang, *Kenya* **54 B4** 4 35N 35 56 E
Todos os Santos, B. de, *Brazil* **93 F11** 12 48S 38 38W
Todos Santos, *Mexico* **86 C2** 23 27N 110 13W
Tofield, *Canada* **72 C6** 53 25N 112 40W
Tofino, *Canada* **72 D3** 49 11N 125 55W
Tofua, *Tonga* **59 D11** 19 45S 175 5W
Tōgane, *Japan* **31 G10** 35 33N 140 22 E
Togba, *Mauritania* **50 E2** 17 26N 10 12W
Togian, Kepulauan, *Indonesia* **37 E6** 0 20S 121 50 E
Togliatti, *Russia* **24 D8** 53 32N 49 24 E
Togo ■, *W. Afr.* **50 G5** 8 30N 1 35 E
Togtoh, *China* **34 D6** 40 15N 111 10 E
Tōhoku □, *Japan* **30 E10** 39 50N 141 45 E
Toinya, *Sudan* **51 G10** 6 17N 29 46 E
Tojikiston = Tajikistan ■, *Asia* **26 F8** 38 30N 70 0 E
Tojo, *Indonesia* **37 E6** 1 20S 121 15 E
Tōjō, *Japan* **31 G6** 34 53N 133 16 E
Tokachi-Dake, *Japan* **30 C11** 43 17N 142 5 E
Tokachi-Gawa →, *Japan* **30 C11** 42 44N 143 42 E
Tokala, *Indonesia* **37 E6** 1 30S 121 40 E
Tōkamachi, *Japan* **31 F9** 37 8N 138 43 E
Tokanui, *N.Z.* **59 M2** 46 34S 168 56 E
Tokar, *Sudan* **51 E12** 18 27N 37 56 E
Tokara-Rettō, *Japan* **31 K4** 29 37N 129 43 E
Tokarahi, *N.Z.* **59 L3** 44 56S 170 39 E
Tokashiki-Shima, *Japan* **31 L3** 26 11N 127 21 E
Tŏkchŏn, *N. Korea* **35 E14** 39 45N 126 18 E
Tokeland, *U.S.A.* **84 D3** 46 42N 123 59W
Tokelau Is., *Pac. Oc.* **64 H10** 9 0S 171 45W
Tokmak, *Kyrgyzstan* **26 E8** 42 49N 75 15 E
Toko Ra., *Australia* **62 C2** 23 5S 138 20 E
Tokoro-Gawa →, *Japan* **30 B12** 44 7N 144 5 E
Tokuno-Shima, *Japan* **31 L4** 27 56N 128 55 E
Tokushima, *Japan* **31 G7** 34 4N 134 34 E
Tokushima □, *Japan* **31 H7** 33 55N 134 0 E
Tokuyama, *Japan* **31 G5** 34 3N 131 50 E
Tōkyō, *Japan* **31 G9** 35 45N 139 45 E
Tolaga Bay, *N.Z.* **59 H7** 38 21S 178 20 E
Tolbukhin = Dobrich, *Bulgaria* **21 C12** 43 37N 27 49 E
Toledo, *Spain* **19 C3** 39 50N 4 2W
Toledo, *Ohio, U.S.A.* **76 E4** 41 39N 83 33W
Toledo, *Oreg., U.S.A.* **82 D2** 44 37N 123 56W
Toledo, *Wash., U.S.A.* **82 C2** 46 26N 122 51W
Toledo, Montes de, *Spain* **19 C3** 39 33N 4 20W
Tolga, *Algeria* **50 B6** 34 40N 5 22 E
Toliara, *Madag.* **57 C7** 23 21S 43 40 E
Toliara □, *Madag.* **57 C8** 21 0S 45 0 E
Tolima, *Colombia* **92 C3** 4 40N 75 19W
Tolitoli, *Indonesia* **37 D6** 1 5N 120 50 E
Tollhouse, *U.S.A.* **84 H7** 37 1N 119 24W
Tolo, *Zaïre* **52 E3** 2 55S 18 34 E
Tolo, Teluk, *Indonesia* **37 E6** 2 20S 122 10 E
Toluca, *Mexico* **87 D5** 19 20N 99 40W
Tom Burke, *S. Africa* **57 C4** 23 5S 28 0 E
Tom Price, *Australia* **60 D2** 22 40S 117 48 E
Tomah, *U.S.A.* **80 D9** 43 59N 90 30W
Tomahawk, *U.S.A.* **80 C10** 45 28N 89 44W
Tomakomai, *Japan* **30 C10** 42 38N 141 36 E
Tomales, *U.S.A.* **84 G4** 38 15N 122 53W
Tomales B., *U.S.A.* **84 G3** 38 15N 123 58W
Tomar, *Portugal* **19 C1** 39 36N 8 25W
Tomaszów Mazowiecki, *Poland* **17 C10** 51 30N 19 57 E
Tomatlán, *Mexico* **86 D3** 19 56N 105 15W
Tombé, *Sudan* **51 G11** 5 53N 31 40 E
Tombigbee →, *U.S.A.* **77 K2** 31 8N 87 57W
Tombouctou, *Mali* **50 E4** 16 50N 3 0W
Tombstone, *U.S.A.* **83 L8** 31 43N 110 4W
Tombua, *Angola* **56 B1** 15 55S 11 55 E
Tomé, *Chile* **94 D1** 36 36S 72 57W
Tomelloso, *Spain* **19 C4** 39 10N 3 2W
Tomingley, *Australia* **63 E4** 32 26S 148 16 E
Tomini, *Indonesia* **37 D6** 0 30N 120 30 E
Tomini, Teluk, *Indonesia* **37 E6** 0 10S 122 0 E
Tomkinson Ras., *Australia* **61 E4** 26 11S 129 5 E
Tommot, *Russia* **27 D13** 59 4N 126 20 E
Tomnavoulin, *U.K.* **12 D5** 57 19N 3 19W
Tomnop Ta Suos, *Cambodia* **39 G5** 11 20N 104 15 E
Tomorit, *Albania* **21 D9** 40 42N 20 11 E
Toms Place, *U.S.A.* **84 H8** 37 34N 118 41W
Toms River, *U.S.A.* **79 G10** 39 58N 74 12W
Tomsk, *Russia* **26 D9** 56 30N 85 5 E
Tonalá, *Mexico* **87 D6** 16 8N 93 41W
Tonalea, *U.S.A.* **83 H8** 36 19N 110 56W
Tonantins, *Brazil* **92 D5** 2 45S 67 45W
Tonasket, *U.S.A.* **82 B4** 48 42N 119 26W
Tonawanda, *U.S.A.* **78 D6** 43 1N 78 53W
Tonbridge, *U.K.* **11 F8** 51 11N 0 17 E
Tondano, *Indonesia* **37 D6** 1 35N 124 54 E
Tonekābon, *Iran* **45 B6** 36 45N 51 12 E
Tong Xian, *China* **34 E9** 39 55N 116 35 E
Tonga ■, *Pac. Oc.* **59 D11** 19 50S 174 30W
Tonga Trench, *Pac. Oc.* **64 J10** 18 0S 173 0W
Tongaat, *S. Africa* **57 D5** 29 33S 31 9 E
Tongareva, *Cook Is.* **65 H12** 9 0S 158 0W
Tongatapu, *Tonga* **59 E11** 21 10S 174 0W
Tongchŏn-ni, *N. Korea* **35 E14** 39 50N 127 25 E
Tongchuan, *China* **34 G5** 35 6N 109 3 E
Tongeren, *Belgium* **15 D5** 50 47N 5 28 E
Tongguan, *China* **34 G6** 34 40N 110 25 E
Tonghua, *China* **35 D13** 41 42N 125 58 E
Tongjosŏn Man, *N. Korea* **35 E14** 39 30N 128 0 E
Tongking, G. of = Tonkin, G. of, *Asia* **32 E5** 20 0N 108 0 E
Tongliao, *China* **35 C12** 43 38N 122 18 E
Tongnae, *S. Korea* **35 G15** 35 12N 129 5 E
Tongobory, *Madag.* **57 C7** 23 32S 44 20 E
Tongoy, *Chile* **94 C1** 30 16S 71 31W
Tongres = Tongeren, *Belgium* **15 D5** 50 47N 5 28 E
Tongsa Dzong, *Bhutan* **41 F17** 27 31N 90 31 E
Tongue, *U.K.* **12 C4** 58 29N 4 25W
Tongue →, *U.S.A.* **80 B2** 46 25N 105 52W
Tongwei, *China* **34 G3** 35 0N 105 5 E
Tongxin, *China* **34 F3** 36 59N 105 58 E
Tongyang, *N. Korea* **35 E14** 39 9N 126 53 E
Tongyu, *China* **35 B12** 44 45N 123 4 E
Tonk, *India* **42 F6** 26 6N 75 54 E
Tonkawa, *U.S.A.* **81 G6** 36 41N 97 18W
Tonkin = Bac Phan, *Vietnam* **38 B5** 22 0N 105 0 E
Tonkin, G. of, *Asia* **32 E5** 20 0N 108 0 E
Tonlé Sap, *Cambodia* **38 F4** 13 0N 104 0 E
Tono, *Japan* **30 E10** 39 19N 141 32 E
Tonopah, *U.S.A.* **83 G5** 38 4N 117 14W
Tonosí, *Panama* **88 E3** 7 20N 80 20W
Tønsberg, *Norway* **9 G14** 59 19N 10 25 E
Tooele, *U.S.A.* **82 F7** 40 32N 112 18W
Toompine, *Australia* **63 D3** 27 15S 144 19 E
Toonpan, *Australia* **62 B4** 19 28S 146 48 E
Toora, *Australia* **63 F4** 38 39S 146 23 E
Toora-Khem, *Russia* **27 D10** 52 28N 96 17 E
Toowoomba, *Australia* **63 D5** 27 32S 151 56 E
Top-ozero, *Russia* **24 A5** 65 35N 32 0 E
Topaz, *U.S.A.* **84 G7** 38 41N 119 30W
Topeka, *U.S.A.* **80 F7** 39 3N 95 40W
Topki, *Russia* **26 D9** 55 20N 85 35 E
Topley, *Canada* **72 C3** 54 49N 126 18W
Topocalma, Pta., *Chile* **94 C1** 34 10S 72 2W
Topock, *U.S.A.* **85 L12** 34 46N 114 29W
Topol'čany, *Slovak Rep.* **17 D10** 48 35N 18 12 E
Topolobampo, *Mexico* **86 B3** 25 40N 109 4W
Toppenish, *U.S.A.* **82 C3** 46 23N 120 19W
Toraka Vestale, *Madag.* **57 B7** 16 20S 43 58 E
Torata, *Peru* **92 G4** 17 23S 70 1W
Torbali, *Turkey* **21 E12** 38 10N 27 21 E
Torbay, *Canada* **71 C9** 47 40N 52 42W
Torbay, *U.K.* **11 G4** 50 26N 3 31W
Tordesillas, *Spain* **19 B3** 41 30N 5 0W
Torfaen □, *U.K.* **11 F4** 51 43N 3 3W
Torgau, *Germany* **16 C7** 51 34N 13 0 E
Torhout, *Belgium* **15 C3** 51 5N 3 7 E
Tori-Shima, *Japan* **31 J10** 30 29N 140 19 E
Torin, *Mexico* **86 B2** 27 33N 110 15W
Torino, *Italy* **20 B2** 45 3N 7 40 E
Torit, *Sudan* **51 H11** 4 27N 32 31 E
Tormes →, *Spain* **19 B2** 41 18N 6 29W
Tornado Mt., *Canada* **72 D6** 49 55N 114 40W
Torne älv →, *Sweden* **8 D21** 65 50N 24 12 E
Torneå = Tornio, *Finland* **8 D21** 65 50N 24 12 E
Torneträsk, *Sweden* **8 B18** 68 24N 19 15 E
Tornio, *Finland* **8 D21** 65 50N 24 12 E
Tornionjoki →, *Finland* **8 D21** 65 50N 24 12 E
Tornquist, *Argentina* **94 D3** 38 8S 62 15W
Toro, *Spain* **22 B11** 39 59N 4 8 E
Toro, Cerro del, *Chile* **94 B2** 29 10S 69 50W
Toro Pk., *U.S.A.* **85 M10** 33 34N 116 24W
Toroníios Kólpos, *Greece* **21 D10** 40 5N 23 30 E

Toronto, Australia 63 E5 33 0S 151 30 E
Toronto, Canada 70 D4 43 39N 79 20W
Toronto, U.S.A. 78 F4 40 28N 80 36W
Toropets, Russia 24 C5 56 30N 31 40 E
Tororo, Uganda 54 B3 0 45N 34 12 E
Toros Dağlan, Turkey 25 G5 37 0N 32 30 E
Torquay, Canada 73 D8 49 9N 103 30W
Torquay, U.K. 11 G4 50 27N 3 32W
Torrance, U.S.A. 85 M8 33 50N 118 19W
Tôrre de Moncorvo, Portugal 19 B2 41 12N 7 8W
Torre del Greco, Italy 20 D6 40 47N 14 22 E
Torrejón de Ardoz, Spain 19 B4 40 27N 3 29W
Torrelavega, Spain 19 A3 43 20N 4 5W
Torremolinos, Spain 19 D3 36 38N 4 30W
Torrens, L., Australia 63 E2 31 0S 137 50 E
Torrens Cr. →, Australia 62 C4 22 23S 145 9 E
Torrens Creek, Australia 62 C4 20 48S 145 3 E
Torrente, Spain 19 C5 39 27N 0 28W
Torreón, Mexico 86 B4 25 33N 103 26W
Torres, Mexico 86 B2 28 46N 110 47W
Torres Strait, Australia 64 H6 9 50S 142 20 E
Torres Vedras, Portugal 19 C1 39 5N 9 15W
Torrevieja, Spain 19 D5 37 59N 0 42W
Torrey, U.S.A. 83 G8 38 18N 111 25W
Torridge →, U.K. 11 G3 51 0N 4 13W
Torridon, L., U.K. 12 D3 57 35N 5 50W
Torrington, Conn., U.S.A. 79 E11 41 48N 73 7W
Torrington, Wyo., U.S.A. 80 D2 42 4N 104 11W
Tórshavn, Færoe Is. 8 E9 62 5N 6 56W
Tortola, Virgin Is. 89 C7 18 19N 64 45W
Tortosa, Spain 19 B6 40 49N 0 31 E
Tortosa, C. de, Spain 19 B6 40 41N 0 52 E
Tortue, I. de la, Haiti 89 B5 20 5N 72 57W
Torūd, Iran 45 C7 35 25N 55 5 E
Toruń, Poland 17 B10 53 2N 18 39 E
Tory I., Ireland 13 A3 55 16N 8 14W
Tosa, Japan 31 H6 33 24N 133 23 E
Tosa-Shimizu, Japan 31 H6 32 52N 132 58 E
Tosa-Wan, Japan 31 H6 33 15N 133 30 E
Toscana □, Italy 20 C4 43 25N 11 0 E
Toshkent, Uzbekistan 26 E7 41 20N 69 10 E
Tostado, Argentina 94 B3 29 15S 61 50W
Tostón, Pta. de, Canary Is. 22 F5 28 42N 14 2W
Tosu, Japan 31 H5 33 22N 130 31 E
Toteng, Botswana 56 C3 20 22S 22 58 E
Totma, Russia 24 C7 60 0N 42 40 E
Totnes, U.K. 11 G4 50 26N 3 42W
Totonicapán, Guatemala 88 D1 14 58N 91 12W
Totten Glacier, Antarctica 5 C8 66 45S 116 10 E
Tottenham, Australia 63 E4 32 14S 147 21 E
Tottenham, Canada 78 B5 44 1N 79 49W
Tottori, Japan 31 G7 35 30N 134 15 E
Tottori □, Japan 31 G7 35 30N 134 12 E
Touba, Ivory C. 50 G3 8 22N 7 40W
Toubkal, Djebel, Morocco 50 B3 31 0N 8 0W
Tougan, Burkina Faso 50 F4 13 11N 2 58W
Touggourt, Algeria 50 B6 33 6N 6 4 E
Tougué, Guinea 50 F2 11 25N 11 50W
Toul, France 18 B6 48 40N 5 53 E
Toulepleu, Ivory C. 50 G3 6 32N 8 24W
Toulon, France 18 E6 43 10N 5 55 E
Toulouse, France 18 E4 43 37N 1 27 E
Toummo, Niger 51 D7 22 45N 14 8 E
Toungoo, Burma 41 K20 19 0N 96 30 E
Touraine, France 18 C4 47 20N 0 30 E
Tourane = Da Nang, Vietnam 38 D7 16 4N 108 13 E
Tourcoing, France 18 A5 50 42N 3 10 E
Touriñán, C., Spain 19 A1 43 3N 9 18W
Tournai, Belgium 15 D3 50 35N 3 25 E
Tournon, France 18 D6 45 4N 4 50 E
Tours, France 18 C4 47 22N 0 40 E
Touwsrivier, S. Africa 56 E3 33 20S 20 2 E
Towada, Japan 30 D10 40 37N 141 13 E
Towada-Ko, Japan 30 D10 40 28N 140 55 E
Towamba, Australia 63 F4 37 6S 149 43 E
Towanda, U.S.A. 79 E8 41 46N 76 27W
Towang, India 41 F17 27 37N 91 50 E
Tower, U.S.A. 80 B8 47 48N 92 17W
Towerhill Cr. →, Australia 62 C3 22 28S 144 35 E
Towner, U.S.A. 80 A4 48 21N 100 25W
Townsend, U.S.A. 82 C8 46 19N 111 31W
Townshend I., Australia 62 C5 22 10S 150 31 E
Townsville, Australia 62 B4 19 15S 146 45 E
Towson, U.S.A. 76 F7 39 24N 76 36W
Toya-Ko, Japan 30 C10 42 35N 140 51 E
Toyah, U.S.A. 81 K3 31 19N 103 48W
Toyahvale, U.S.A. 81 K3 30 57N 103 47W
Toyama, Japan 31 F8 36 40N 137 15 E
Toyama □, Japan 31 F8 36 45N 137 30 E
Toyama-Wan, Japan 31 F8 37 0N 137 30 E
Toyohashi, Japan 31 G8 34 45N 137 25 E
Toyokawa, Japan 31 G8 34 48N 137 27 E
Toyonaka, Japan 31 G7 34 50N 135 28 E
Toyooka, Japan 31 G7 35 35N 134 48 E
Toyota, Japan 31 G8 35 3N 137 7 E
Tozeur, Tunisia 50 B6 33 56N 8 8 E
Trá Li = Tralee, Ireland 13 D2 52 16N 9 42W
Tra On, Vietnam 39 H5 9 58N 105 55 E
Trabzon, Turkey 25 F6 41 0N 39 45 E
Tracadie, Canada 71 C7 47 30N 64 55W
Tracy, Calif., U.S.A. 83 H3 37 44N 121 26W
Tracy, Minn., U.S.A. 80 C7 44 14N 95 37W
Trafalgar, C., Spain 19 D2 36 10N 6 2W
Trail, Canada 72 D5 49 5N 117 40W
Trainor L., Canada 72 A4 60 24N 120 17W
Trákhonas, Cyprus 23 D12 35 12N 33 21 E
Tralee, Ireland 13 D2 52 16N 9 42W
Tralee B., Ireland 13 D2 52 17N 9 55W
Tramore, Ireland 13 D4 52 10N 7 10W
Tran Ninh, Cao Nguyen, Laos 38 C4 19 30N 103 10 E
Tranås, Sweden 9 G16 58 3N 14 59 E
Trancas, Argentina 94 B2 26 11S 65 20W
Trang, Thailand 39 J2 7 33N 99 38 E
Trangahy, Madag. 57 B7 19 7S 44 31 E
Trangan, Indonesia 37 F8 6 40S 134 20 E
Trangie, Australia 63 E4 32 4S 148 0 E
Trani, Italy 20 D7 41 17N 16 25 E
Tranoroa, Madag. 57 C8 24 42S 45 4 E
Tranqueras, Uruguay 95 C4 31 13S 55 45W
Trans Nzoia □, Kenya 54 B3 1 0N 35 0 E
Transantarctic Mts., Antarctica 5 E12 85 0S 170 0W
Transcaucasia = Zakavkazye, Asia 25 F7 42 0N 44 0 E
Transcona, Canada 73 D9 49 55N 97 0W
Transilvania, Romania 17 E12 45 19N 25 0 E
Transilvanian Alps = Carpaţii Meridionali, Romania 17 F13 45 30N 25 0 E
Transylvania = Transilvania, Romania 17 E12 45 19N 25 0 E
Trápani, Italy 20 E5 38 1N 12 29 E
Trapper Pk., U.S.A. 82 D6 45 54N 114 18W
Traralgon, Australia 63 F4 38 12S 146 34 E
Trasimeno, L., Italy 20 C5 43 8N 12 6 E
Trat, Thailand 39 F4 12 14N 102 33 E
Traun, Austria 16 D8 48 14N 14 15 E
Traveller's L., Australia 63 E3 33 20S 142 0 E
Travemünde, Germany 16 B6 53 57N 10 52 E
Travers, Mt., N.Z. 59 K4 42 1S 172 45 E
Traverse City, U.S.A. 76 C3 44 46N 85 38W
Travnik, Bos.-H. 21 B7 44 17N 17 39 E
Trayning, Australia 61 F2 31 7S 117 40 E
Trébbia →, Italy 20 B3 45 4N 9 41 E
Třebíč, Czech. 16 D8 49 14N 15 55 E
Trebinje, Bos.-H. 21 C8 42 44N 18 22 E
Tredegar, U.K. 11 F4 51 47N 3 14W
Tregaron, U.K. 11 E4 52 14N 3 56W
Tregrosse Is., Australia 62 B5 17 41S 150 43 E
Treherne, Canada 73 D9 49 38N 98 42W
Treinta y Tres, Uruguay 95 C5 33 16S 54 17W
Trelew, Argentina 96 E3 43 10S 65 20W
Trelleborg, Sweden 9 J15 55 20N 13 10 E
Tremonton, U.S.A. 82 F7 41 43N 112 10W
Tremp, Spain 19 A6 42 10N 0 52 E
Trenche →, Canada 70 C5 47 46N 72 53W
Trenčín, Slovak Rep. 17 D10 48 52N 18 4 E
Trenggalek, Indonesia 37 H14 8 3S 111 43 E
Trenque Lauquen, Argentina 94 D3 36 5S 62 45W
Trent →, U.K. 10 D7 53 41N 0 42W
Trento, Italy 20 A4 46 4N 11 8 E
Trenton, Canada 70 D4 44 10N 77 34W
Trenton, Mo., U.S.A. 80 E8 40 5N 93 37W
Trenton, N.J., U.S.A. 79 F10 40 14N 74 46W
Trenton, Nebr., U.S.A. 80 E4 40 11N 101 1W
Trenton, Tenn., U.S.A. 81 H10 35 59N 88 56W
Trepassey, Canada 71 C9 46 43N 53 25W
Tres Arroyos, Argentina 94 D3 38 26S 60 20W
Três Corações, Brazil 95 A6 21 44S 45 15W
Três Lagoas, Brazil 93 H8 20 50S 51 43W
Tres Marías, Mexico 86 C3 21 25N 106 28W
Tres Montes, C., Chile 96 F1 46 50S 75 30W
Tres Pinos, U.S.A. 84 J5 36 48N 121 19W
Três Pontas, Brazil 95 A6 21 23S 45 29W
Tres Puentes, Chile 94 B1 27 50S 70 15W
Tres Puntas, C., Argentina 96 F3 47 0S 66 0W
Três Rios, Brazil 95 A7 22 6S 43 15W
Tres Valles, Mexico 87 D5 18 15N 96 8W
Treviso, Italy 20 B5 45 40N 12 15 E
Triabunna, Australia 62 G4 42 30S 147 55 E
Triánda, Greece 23 C10 36 25N 28 10 E
Triang, Malaysia 39 L4 3 15N 102 26 E
Tribulation, C., Australia 62 B4 16 5S 145 29 E
Tribune, U.S.A. 80 F4 38 28N 101 45W
Trichinopoly = Tiruchchirappalli, India 40 P11 10 45N 78 45 E
Trichur, India 40 P10 10 30N 76 18 E
Trida, Australia 63 E4 33 1S 145 1 E
Trier, Germany 16 D4 49 45N 6 38 E
Trieste, Italy 20 B5 45 40N 13 46 E
Triglav, Slovenia 16 E7 46 21N 13 50 E
Trikkala, Greece 21 E9 39 34N 21 47 E
Trikomo, Cyprus 23 D12 35 17N 33 52 E
Trikora, Puncak, Indonesia 37 E9 4 15S 138 45 E
Trim, Ireland 13 C5 53 33N 6 48W
Trincomalee, Sri Lanka 40 Q12 8 38N 81 15 E
Trindade, I., Atl. Oc. 2 F8 20 20S 29 50W
Trinidad, Bolivia 92 F6 14 46S 64 50W
Trinidad, Colombia 92 B4 5 25N 71 40W
Trinidad, Cuba 88 B3 21 48N 80 0W
Trinidad, Uruguay 94 C4 33 30S 56 50W
Trinidad, U.S.A. 81 G2 37 10N 104 31W
Trinidad, W. Indies 89 D7 10 30N 61 15W
Trinidad →, Mexico 87 D5 17 49N 95 9W
Trinidad, I., Argentina 96 D4 39 10S 62 0W
Trinidad & Tobago ■, W. Indies 89 D7 10 30N 61 20W
Trinity, Canada 71 C9 48 59N 53 55W
Trinity, U.S.A. 81 K7 30 57N 95 22W
Trinity →, Calif., U.S.A. 82 F2 41 11N 123 42W
Trinity →, Tex., U.S.A. 81 L7 29 45N 94 43W
Trinity B., Canada 71 C9 48 20N 53 10W
Trinity Range, U.S.A. 82 F4 40 15N 118 45W
Trinkitat, Sudan 51 E12 18 45N 37 51 E
Trion, U.S.A. 77 H3 34 33N 85 19W
Tripoli = Tarābulus, Lebanon 47 A4 34 31N 35 50 E
Tripoli = Tarābulus, Libya 51 B7 32 49N 13 7 E
Trípolis, Greece 21 F10 37 31N 22 25 E
Tripolitania, N. Afr. 48 C5 31 0N 13 0 E
Tripp, U.S.A. 80 D6 43 13N 97 58W
Tripura □, India 41 H17 24 0N 92 0 E
Tripylos, Cyprus 23 E11 34 59N 32 41 E
Tristan da Cunha, Atl. Oc. 49 K2 37 6S 12 20W
Trivandrum, India 40 Q10 8 41N 77 0 E
Trnava, Slovak Rep. 17 D9 48 23N 17 35 E
Trochu, Canada 72 C6 51 50N 113 13W
Trodely I., Canada 70 B4 52 15N 79 26W
Troglav, Croatia 20 C7 43 56N 16 36 E
Troilus, L., Canada 70 B5 50 50N 74 35W
Trois-Pistoles, Canada 71 C6 48 5N 69 10W
Trois-Rivières, Canada 70 C5 46 25N 72 34W
Troitsk, Russia 26 D7 54 10N 61 35 E
Troitsko Pechorsk, Russia 24 B10 62 40N 56 10 E
Trölladyngja, Iceland 8 D5 64 54N 17 16W
Trollhättan, Sweden 9 G15 58 17N 12 20 E
Trollheimen, Norway 8 E13 62 46N 9 1 E
Tromsø, Norway 8 B18 69 40N 18 56 E
Trona, U.S.A. 85 K9 35 46N 117 23W
Tronador, Argentina 96 E2 41 10S 71 50W
Trøndelag, Norway 8 D14 64 17N 11 50 E
Trondheim, Norway 8 E14 63 36N 10 25 E
Trondheimsfjorden, Norway 8 E14 63 35N 10 30 E
Troodos, Cyprus 23 E11 34 55N 32 52 E
Troon, U.K. 12 F4 55 33N 4 39W
Tropic, U.S.A. 83 H7 37 37N 112 5W
Trossachs, The, U.K. 12 E4 56 14N 4 24W
Trostan, U.K. 13 A5 55 3N 6 10W
Trotternish, U.K. 12 D2 57 32N 6 15W
Troup, U.S.A. 81 J7 32 9N 95 7W
Trout →, Canada 72 A5 61 19N 119 51W
Trout L., N.W.T., Canada 72 A4 60 40N 121 14W
Trout L., Ont., Canada 73 C10 51 20N 93 15W
Trout Lake, Mich., U.S.A. 70 C2 46 12N 85 1W
Trout Lake, Wash., U.S.A. 84 E5 46 0N 121 32W
Trout River, Canada 71 C8 49 29N 58 8W
Trouville-sur-Mer, France 18 B4 49 21N 0 5 E
Trowbridge, U.K. 11 F5 51 18N 2 12W
Troy, Turkey 21 E12 39 57N 26 12 E
Troy, Ala., U.S.A. 77 K3 31 48N 85 58W
Troy, Idaho, U.S.A. 82 C5 46 44N 116 46W
Troy, Kans., U.S.A. 80 F7 39 47N 95 5W
Troy, Mo., U.S.A. 80 F9 38 59N 90 59W
Troy, Mont., U.S.A. 82 B6 48 28N 115 53W
Troy, N.Y., U.S.A. 79 D11 42 44N 73 41W
Troy, Ohio, U.S.A. 76 E3 40 2N 84 12W
Troyes, France 18 B6 48 19N 4 3 E
Trucial States = United Arab Emirates ■, Asia 45 F7 23 50N 54 0 E
Truckee, U.S.A. 84 F6 39 20N 120 11W
Trudovoye, Russia 30 C6 43 17N 132 5 E
Trujillo, Honduras 88 C2 16 0N 86 0W
Trujillo, Peru 92 E3 8 6S 79 0W
Trujillo, Spain 19 C3 39 28N 5 55W
Trujillo, U.S.A. 81 H2 35 32N 104 42W
Trujillo, Venezuela 92 B4 9 22N 70 38W
Truk, Pac. Oc. 64 G7 7 25N 151 46 E
Trumann, U.S.A. 81 H9 35 41N 90 31W
Trumbull, Mt., U.S.A. 83 H7 36 25N 113 8W
Trundle, Australia 63 E4 32 53S 147 35 E
Trung-Phan, Vietnam 38 E7 16 0N 108 0 E
Truro, Canada 71 C7 45 21N 63 14W
Truro, U.K. 11 G2 50 16N 5 4W
Truskavets, Ukraine 17 D12 49 17N 23 30 E
Truslove, Australia 61 F3 33 20S 121 45 E
Truth or Consequences, U.S.A. 83 K10 33 8N 107 15W
Trutnov, Czech. 16 C8 50 37N 15 54 E
Tryon, U.S.A. 77 H4 35 13N 82 14W
Tryonville, U.S.A. 78 E5 41 42N 79 48W
Tsaratanana, Madag. 57 B8 16 47S 47 39 E
Tsaratanana, Mt. de, Madag. 57 A8 14 0S 49 0 E
Tsarevo = Michurin, Bulgaria 21 C12 42 9N 27 51 E
Tsau, Botswana 56 C3 20 8S 22 22 E
Tselinograd = Aqmola, Kazakstan 26 D8 51 10N 71 30 E
Tsetserleg, Mongolia 32 B5 47 36N 101 32 E
Tshabong, Botswana 56 D3 26 2S 22 29 E
Tshane, Botswana 56 C3 24 5S 21 54 E
Tshela, Zaïre 52 E2 4 57S 13 4 E
Tshesebe, Botswana 57 C4 21 51S 27 32 E
Tshibeke, Zaïre 54 C2 2 40S 28 35 E
Tshibinda, Zaïre 54 C2 2 23S 28 43 E
Tshikapa, Zaïre 52 F4 6 28S 20 48 E
Tshilenge, Zaïre 54 D1 6 17S 23 48 E
Tshinsenda, Zaïre 55 E2 12 20S 28 0 E
Tshofa, Zaïre 54 D2 5 13S 25 16 E
Tshwane, Botswana 56 C3 22 24S 22 1 E
Tsigara, Botswana 56 C4 20 22S 25 54 E
Tsihombe, Madag. 57 D8 25 10S 45 41 E
Tsimlyansk Res. = Tsimlyanskoye Vdkhr., Russia 25 E7 48 0N 43 0 E
Tsimlyanskoye Vdkhr., Russia 25 E7 48 0N 43 0 E
Tsinan = Jinan, China 34 F9 36 38N 117 1 E
Tsineng, S. Africa 56 D3 27 5S 23 5 E
Tsinghai = Qinghai □, China 32 C4 36 0N 98 0 E
Tsingtao = Qingdao, China 35 F11 36 5N 120 20 E
Tsinjomitondraka, Madag. 57 B8 15 40S 47 8 E
Tsiroanomandidy, Madag. 57 B8 18 46S 46 2 E
Tsivory, Madag. 57 C8 24 4S 46 5 E
Tskhinvali, Georgia 25 F7 42 14N 44 1 E
Tsna →, Russia 24 D7 54 55N 41 58 E
Tso Moriri, L., India 43 C8 32 50N 78 20 E
Tsodilo Hill, Botswana 56 B3 18 49S 21 43 E
Tsogttsetsiy, Mongolia 34 C3 43 43N 105 35 E
Tsolo, S. Africa 57 E4 31 18S 28 37 E
Tsomo, S. Africa 57 E4 32 0S 27 42 E
Tsu, Japan 31 G8 34 45N 136 25 E
Tsu L., Canada 72 A6 60 40N 111 52W
Tsuchiura, Japan 31 F10 36 5N 140 15 E
Tsugaru-Kaikyō, Japan 30 D10 41 35N 141 0 E
Tsumeb, Namibia 56 B2 19 9S 17 44 E
Tsumis, Namibia 56 C2 23 39S 17 29 E
Tsuruga, Japan 31 G8 35 45N 136 2 E
Tsurugi-San, Japan 31 H7 33 51N 134 6 E
Tsuruoka, Japan 30 E9 38 44N 139 50 E
Tsushima, Gifu, Japan 31 G8 35 10N 136 43 E
Tsushima, Nagasaki, Japan 31 G4 34 20N 129 20 E
Tsyelyakhany, Belarus 17 B13 52 30N 25 46 E
Tual, Indonesia 37 F8 5 38S 132 44 E
Tuam, Ireland 13 C3 53 31N 8 51W
Tuamotu Arch. = Tuamotu Is., Pac. Oc. 65 J13 17 0S 144 0W
Tuamotu Is., Pac. Oc. 65 J13 17 0S 144 0W
Tuamotu Ridge, Pac. Oc. 65 K14 20 0S 138 0W
Tuao, Phil. 37 A6 17 55N 121 22 E
Tuapse, Russia 25 F6 44 5N 39 10 E
Tuatapere, N.Z. 59 M1 46 8S 167 41 E
Tuba City, U.S.A. 83 H8 36 8N 111 14W
Tuban, Indonesia 37 G15 6 54S 112 3 E
Tubarão, Brazil 95 B6 28 30S 49 0W
Tūbās, West Bank 47 C4 32 20N 35 22 E
Tubau, Malaysia 36 D4 3 10N 113 40 E
Tübingen, Germany 16 D5 48 31N 9 4 E
Tubruq, Libya 51 B9 32 7N 23 55 E
Tubuai Is., Pac. Oc. 65 K12 25 0S 150 0W
Tuc Trung, Vietnam 39 G6 11 1N 107 12 E
Tucacas, Venezuela 92 A5 10 48N 68 19W
Tuchodi →, Canada 72 B4 58 17N 123 42W
Tucson, U.S.A. 83 K8 32 13N 110 58W
Tucumán □, Argentina 94 B2 26 48S 66 2W
Tucumcari, U.S.A. 81 H3 35 10N 103 44W
Tucupita, Venezuela 92 B6 9 2N 62 3W
Tucuruí, Brazil 93 D9 3 42S 49 44W
Tucuruí, Reprêsa de, Brazil 93 D9 4 0S 49 30W
Tudela, Spain 19 A5 42 4N 1 39W
Tudmur, Syria 44 C3 34 36N 38 15 E
Tudor, L., Canada 71 A6 55 50N 65 25W
Tuen, Australia 63 D4 28 33S 145 37 E
Tugela →, S. Africa 57 D5 29 14S 31 30 E
Tuguegarao, Phil. 37 A6 17 35N 121 42 E
Tugur, Russia 27 D14 53 44N 136 45 E
Tuineje, Canary Is. 22 F5 28 19N 14 3W
Tukangbesi, Kepulauan, Indonesia 37 F6 6 0S 124 0 E
Tukarak I., Canada 70 A4 56 15N 78 45W
Tukayyid, Iraq 44 D5 29 47N 45 36 E
Tūkrah, Libya 51 B9 32 30N 20 37 E
Tuktoyaktuk, Canada 68 B6 69 27N 133 2W
Tukums, Latvia 9 H20 57 2N 23 10 E
Tukuyu, Tanzania 55 D3 9 17S 33 35 E
Tula, Hidalgo, Mexico 87 C5 20 5N 99 20W
Tula, Tamaulipas, Mexico 87 C5 23 0N 99 40W
Tula, Russia 24 D6 54 13N 37 38 E
Tulancingo, Mexico 87 C5 20 5N 99 22W
Tulare, U.S.A. 83 H4 36 13N 119 21W
Tulare Lake Bed, U.S.A. 83 J4 36 0N 119 48W
Tularosa, U.S.A. 83 K10 33 5N 106 1W
Tulbagh, S. Africa 56 E2 33 16S 19 6 E
Tulcán, Ecuador 92 C3 0 48N 77 43W
Tulcea, Romania 17 F15 45 13N 28 46 E
Tulchyn, Ukraine 17 D15 48 41N 28 49 E
Tūleh, Iran 45 C7 34 35N 52 33 E
Tulemalu L., Canada 73 A9 62 58N 99 25W
Tuli, Indonesia 37 E6 1 24S 122 26 E
Tuli, Zimbabwe 55 G2 21 58S 29 13 E
Tulia, U.S.A. 81 H4 34 32N 101 46W
Tulita, Canada 68 B7 64 57N 125 30W
Ţūlkarm, West Bank 47 C4 32 19N 35 2 E
Tullahoma, U.S.A. 77 H2 35 22N 86 13W
Tullamore, Australia 63 E4 32 39S 147 36 E
Tullamore, Ireland 13 C4 53 16N 7 31W
Tulle, France 18 D4 45 16N 1 46 E
Tullibigeal, Australia 63 E4 33 25S 146 44 E
Tullow, Ireland 13 D4 52 49N 6 45W
Tully, Australia 62 B4 17 56S 145 55 E
Ţulmaythah, Libya 51 B9 32 40N 20 55 E
Tulmur, Australia 62 C3 22 40S 142 20 E
Tulsa, U.S.A. 81 G7 36 10N 95 55W
Tulsequah, Canada 72 B2 58 39N 133 35W
Tulua, Colombia 92 C3 4 6N 76 11W
Tulun, Russia 27 D11 54 32N 100 35 E
Tulungagung, Indonesia 36 F4 8 5S 111 54 E
Tum, Indonesia 37 E8 3 36S 130 21 E
Tuma →, Nic. 88 D3 13 6N 84 35W
Tumaco, Colombia 92 C3 1 50N 78 45W
Tumatumari, Guyana 92 B7 5 20N 58 55W
Tumba, Sweden 9 G17 59 12N 17 48 E
Tumba, L., Zaïre 52 E3 0 50S 18 0 E
Tumbarumba, Australia 63 F4 35 44S 148 0 E
Tumbaya, Argentina 94 A2 23 50S 65 26W
Túmbes, Peru 92 D2 3 37S 80 27W
Tumbwe, Zaïre 55 E2 11 25S 27 15 E
Tumby Bay, Australia 63 E2 34 21S 136 8 E
Tumd Youqi, China 34 D6 40 30N 110 30 E
Tumen, China 35 C15 43 0N 129 50 E
Tumen Jiang →, China 35 C16 42 20N 130 35 E
Tumeremo, Venezuela 92 B6 7 18N 61 30W
Tumkur, India 40 N10 13 18N 77 6 E
Tummel, L., U.K. 12 E5 56 43N 3 55W
Tump, Pakistan 40 F3 26 7N 62 16 E
Tumpat, Malaysia 39 J4 6 11N 102 10 E
Tumu, Ghana 50 F4 10 56N 1 56W
Tumucumaque, Serra, Brazil 93 C8 2 0N 55 0W
Tumut, Australia 63 F4 35 16S 148 13 E
Tumwater, U.S.A. 82 C2 47 1N 122 54W
Tunas de Zaza, Cuba 88 B4 21 39N 79 34W
Tunbridge Wells = Royal Tunbridge Wells, U.K. 11 F8 51 7N 0 16 E
Tuncurry, Australia 63 E5 32 17S 152 29 E
Tunduru, Tanzania 55 E4 11 8S 37 25 E
Tunduru □, Tanzania 55 E4 11 5S 37 22 E
Tundzha →, Bulgaria 21 C11 41 40N 26 35 E
Tunga Pass, India 41 E19 29 0N 94 14 E
Tungabhadra →, India 40 M11 15 57N 78 15 E
Tungaru, Sudan 51 F11 10 9N 30 52 E
Tungla, Nic. 88 D3 13 24N 84 21W
Tungsten, Canada 72 A3 61 57N 128 16W
Tunguska, Nizhnyaya →, Russia 27 C9 65 48N 88 4 E
Tunica, U.S.A. 81 H9 34 41N 90 23W
Tunis, Tunisia 50 A7 36 50N 10 11 E
Tunisia ■, Africa 50 B6 33 30N 9 10 E
Tunja, Colombia 92 B4 5 33N 73 25W
Tunkhannock, U.S.A. 79 E9 41 32N 75 57W
Tunliu, China 34 F7 36 13N 112 52 E
Tunnsjøen, Norway 8 D15 64 45N 13 25 E
Tunungayualok I., Canada 71 A7 56 0N 61 0W
Tunuyán, Argentina 94 C2 33 35S 69 0W
Tunuyán →, Argentina 94 C2 33 33S 67 30W
Tunxi, China 33 D6 29 42N 118 25 E
Tuolumne, U.S.A. 83 H3 37 58N 120 15W
Tuolumne →, U.S.A. 84 H5 37 36N 121 13W
Tuoy-Khaya, Russia 27 C12 62 32N 111 25 E
Tūp Āghāj, Iran 44 B5 36 3N 47 50 E
Tupã, Brazil 95 A5 21 57S 50 28W
Tupelo, U.S.A. 77 H1 34 16N 88 43W
Tupik, Russia 27 D12 54 26N 119 57 E
Tupinambaranas, Brazil 92 D7 3 0S 58 0W
Tupiza, Bolivia 94 A2 21 30S 65 40W
Tupman, U.S.A. 85 K7 35 18N 119 21W
Tupper, Canada 72 B4 55 32N 120 1W
Tupper Lake, U.S.A. 79 B10 44 14N 74 28W
Tupungato, Cerro, S. Amer. 94 C2 33 15S 69 50W
Tuquan, China 35 B11 45 18N 121 38 E
Túquerres, Colombia 92 C3 1 5N 77 37W
Tura, Russia 27 C11 64 20N 100 17 E

Turabah, *Si. Arabia* **44 D4** 28 20N 43 15 E
Tūrān, *Iran* **45 C8** 35 39N 56 42 E
Turan, *Russia* **27 D10** 51 55N 95 0 E
Turayf, *Si. Arabia* **44 D3** 31 41N 38 39 E
Turda, *Romania* **17 E12** 46 34N 23 47 E
Turek, *Poland* **17 B10** 52 3N 18 30 E
Turfan = Turpan, *China* **32 B3** 43 58N 89 10 E
Turfan Depression = Turpan Hami, *China* **28 E12** 42 40N 89 25 E
Tŭrgovishte, *Bulgaria* **21 C12** 43 17N 26 38 E
Turgutlu, *Turkey* **21 E12** 38 30N 27 48 E
Turia →, *Spain* **19 C5** 39 27N 0 19W
Turiaçu, *Brazil* **93 D9** 1 40S 45 19W
Turiaçu →, *Brazil* **93 D9** 1 36S 45 19W
Turin = Torino, *Italy* **20 B2** 45 3N 7 40 E
Turin, *Canada* **72 D6** 49 58N 112 31W
Turkana □, *Kenya* **54 B4** 3 0N 35 30 E
Turkana, L., *Africa* **54 B4** 3 30N 36 5 E
Turkestan = Türkistan, *Kazakstan* **26 E7** 43 17N 68 16 E
Turkey ■, *Eurasia* **25 G6** 39 0N 36 0 E
Turkey Creek, *Australia* **60 C4** 17 2S 128 12 E
Türkistan, *Kazakstan* **26 E7** 43 17N 68 16 E
Türkmenbashi, *Turkmenistan* **25 F9** 40 5N 53 5 E
Turkmenistan ■, *Asia* **26 F6** 39 0N 59 0 E
Turks & Caicos Is. ■, *W. Indies* **89 B5** 21 20N 71 20W
Turks Island Passage, *W. Indies* **89 B5** 21 30N 71 30W
Turku, *Finland* **9 F20** 60 30N 22 19 E
Turkwe →, *Kenya* **54 B4** 3 6N 36 6 E
Turlock, *U.S.A.* **83 H3** 37 30N 120 51W
Turnagain →, *Canada* **72 B3** 59 12N 127 35W
Turnagain, C., *N.Z.* **59 J6** 40 28S 176 38 E
Turneffe Is., *Belize* **87 D7** 17 20N 87 50W
Turner, *Australia* **60 C4** 17 52S 128 16 E
Turner, *U.S.A.* **82 B9** 48 51N 108 24W
Turner Pt., *Australia* **62 A1** 11 47S 133 32 E
Turner Valley, *Canada* **72 C6** 50 40N 114 17W
Turners Falls, *U.S.A.* **79 D12** 42 36N 72 33W
Turnhout, *Belgium* **15 C4** 51 19N 4 57 E
Turnor L., *Canada* **73 B7** 56 35N 108 35W
Tŭrnovo = Veliko Tŭrnovo, *Bulgaria* **21 C11** 43 5N 25 41 E
Turnu Măgurele, *Romania* **17 G13** 43 46N 24 56 E
Turnu Roşu, P., *Romania* **17 F13** 45 33N 24 17 E
Turon, *U.S.A.* **81 G5** 37 48N 98 26W
Turpan, *China* **32 B3** 43 58N 89 10 E
Turpan Hami, *China* **28 E12** 42 40N 89 25 E
Turriff, *U.K.* **12 D6** 57 32N 2 27W
Tursāq, *Iraq* **44 C5** 33 27N 45 47 E
Turtle Head I., *Australia* **62 A3** 10 56S 142 37 E
Turtle L., *Canada* **73 C7** 53 36N 108 38W
Turtle Lake, *N. Dak., U.S.A.* **80 B4** 47 31N 100 53W
Turtle Lake, *Wis., U.S.A.* **80 C8** 45 24N 92 8W
Turtleford, *Canada* **73 C7** 53 23N 108 57W
Turukhansk, *Russia* **27 C9** 65 21N 88 5 E
Tuscaloosa, *U.S.A.* **77 J2** 33 12N 87 34W
Tuscany = Toscana □, *Italy* **20 C4** 43 25N 11 0 E
Tuscola, *Ill., U.S.A.* **76 F1** 39 48N 88 17W
Tuscola, *Tex., U.S.A.* **81 J5** 32 12N 99 48W
Tuscumbia, *U.S.A.* **77 H2** 34 44N 87 42W
Tuskar Rock, *Ireland* **13 D5** 52 12N 6 10W
Tuskegee, *U.S.A.* **77 J3** 32 25N 85 42W
Tustin, *U.S.A.* **85 M9** 33 44N 117 49W
Tuticorin, *India* **40 Q11** 8 50N 78 12 E
Tutóia, *Brazil* **93 D10** 2 45S 42 20W
Tutong, *Brunei* **36 D4** 4 47N 114 40 E
Tutrakan, *Bulgaria* **21 B12** 44 2N 26 40 E
Tutshi L., *Canada* **72 B2** 59 56N 134 30W
Tuttle, *U.S.A.* **80 B5** 47 9N 100 0W
Tuttlingen, *Germany* **16 E5** 47 58N 8 48 E
Tutuala, *Indonesia* **37 F7** 8 25S 127 15 E
Tutuila, *Amer. Samoa* **59 B13** 14 19S 170 50W
Tututepec, *Mexico* **87 D5** 16 9N 97 38W
Tuva □, *Russia* **27 D10** 51 30N 95 0 E
Tuvalu ■, *Pac. Oc.* **64 H9** 8 0S 178 0 E
Tuxpan, *Mexico* **87 C5** 20 58N 97 23W
Tuxtla Gutiérrez, *Mexico* **87 D6** 16 50N 93 10W
Tuy, *Spain* **19 A1** 42 3N 8 39W
Tuy An, *Vietnam* **38 F7** 13 17N 109 16 E
Tuy Duc, *Vietnam* **39 F6** 12 15N 107 27 E
Tuy Hoa, *Vietnam* **38 F7** 13 5N 109 10 E
Tuy Phong, *Vietnam* **39 G7** 11 14N 108 43 E
Tuya L., *Canada* **72 B2** 59 7N 130 35W
Tuyen Hoa, *Vietnam* **38 D6** 17 50N 106 10 E
Tuyen Quang, *Vietnam* **38 B5** 21 50N 105 10 E
Tūysarkān, *Iran* **45 C6** 34 33N 48 27 E
Tuz Gölü, *Turkey* **25 G5** 38 42N 33 18 E
Ţūz Khurmātū, *Iraq* **44 C5** 34 56N 44 38 E
Tuzla, *Bos.-H.* **21 B8** 44 34N 18 41 E
Tver, *Russia* **24 C6** 56 55N 35 55 E
Twain, *U.S.A.* **84 E5** 40 1N 121 3W
Twain Harte, *U.S.A.* **84 G6** 38 2N 120 14W
Tweed, *Canada* **78 B7** 44 29N 77 19W
Tweed →, *U.K.* **12 F7** 55 45N 2 0W
Tweed Heads, *Australia* **63 D5** 28 10S 153 31 E
Tweedsmuir Prov. Park, *Canada* **72 C3** 53 0N 126 20W
Twentynine Palms, *U.S.A.* **85 L10** 34 8N 116 3W
Twillingate, *Canada* **71 C9** 49 42N 54 45W
Twin Bridges, *U.S.A.* **82 D7** 45 33N 112 20W
Twin Falls, *U.S.A.* **82 E6** 42 34N 114 28W
Twin Valley, *U.S.A.* **80 B6** 47 16N 96 16W
Twisp, *U.S.A.* **82 B3** 48 22N 120 7W
Two Harbors, *U.S.A.* **80 B9** 47 2N 91 40W
Two Hills, *Canada* **72 C6** 53 43N 111 52W
Two Rivers, *U.S.A.* **76 C2** 44 9N 87 34W
Twofold B., *Australia* **63 F4** 37 8S 149 59 E
Tyachiv, *Ukraine* **17 D12** 48 1N 23 35 E
Tychy, *Poland* **17 C10** 50 9N 18 59 E
Tyler, *U.S.A.* **75 D7** 32 18N 95 17W
Tyler, *Minn., U.S.A.* **80 C6** 44 18N 96 8W
Tyler, *Tex., U.S.A.* **81 J7** 32 21N 95 18W
Tynda, *Russia* **27 D13** 55 10N 124 43 E
Tyne →, *U.K.* **10 C6** 54 59N 1 32W
Tyne & Wear □, *U.K.* **10 C6** 55 6N 1 17W
Tynemouth, *U.K.* **10 B6** 55 1N 1 26W
Tyre = Sūr, *Lebanon* **47 B4** 33 19N 35 16 E
Tyrifjorden, *Norway* **9 F14** 60 2N 10 8 E
Tyrol = Tirol □, *Austria* **16 E6** 47 3N 10 43 E
Tyrone, *U.S.A.* **78 F6** 40 40N 78 14W
Tyrrell →, *Australia* **63 F3** 35 26S 142 51 E
Tyrrell, L., *Australia* **63 F3** 35 20S 142 50 E
Tyrrell Arm, *Canada* **73 A9** 62 27N 97 30W
Tyrrell L., *Canada* **73 A7** 63 7N 105 27W
Tyrrhenian Sea, *Medit. S.* **20 E5** 40 0N 12 30 E
Tysfjorden, *Norway* **8 B17** 68 7N 16 25 E
Tyulgan, *Russia* **24 D10** 52 22N 56 12 E
Tyumen, *Russia* **26 D7** 57 11N 65 29 E
Tywi →, *U.K.* **11 F3** 51 48N 4 21W
Tywyn, *U.K.* **11 E3** 52 35N 4 5W
Tzaneen, *S. Africa* **57 C5** 23 47S 30 9 E
Tzermiádhes, *Greece* **23 D7** 35 12N 25 29 E
Tzukong = Zigong, *China* **32 D5** 29 15N 104 48 E

U

U Taphao, *Thailand* **38 F3** 12 35N 101 0 E
U.S.A. = United States of America ■, *N. Amer.* **74 C7** 37 0N 96 0W
Uanda, *Australia* **62 C3** 21 37S 144 55 E
Uarsciek, *Somali Rep.* **46 G4** 2 28N 45 55 E
Uasin □, *Kenya* **54 B4** 0 30N 35 20 E
Uato-Udo, *Indonesia* **37 F7** 9 7S 125 36 E
Uatumã →, *Brazil* **92 D7** 2 26S 57 37W
Uaupés, *Brazil* **92 D5** 0 8S 67 5W
Uaupés →, *Brazil* **92 C5** 0 2N 67 16W
Uaxactún, *Guatemala* **88 C2** 17 25N 89 29W
Ubá, *Brazil* **95 A7** 21 8S 43 0W
Ubaitaba, *Brazil* **93 F11** 14 18S 39 20W
Ubangi = Oubangi →, *Zaïre* **52 E3** 0 30S 17 50 E
Ubauro, *Pakistan* **42 E3** 28 15N 69 45 E
Ube, *Japan* **31 H5** 33 56N 131 15 E
Ubeda, *Spain* **19 C4** 38 3N 3 23W
Uberaba, *Brazil* **93 G9** 19 50S 47 55W
Uberlândia, *Brazil* **93 G9** 19 0S 48 20W
Ubolratna Res., *Thailand* **38 D4** 16 45N 102 30 E
Ubombo, *S. Africa* **57 D5** 27 31S 32 4 E
Ubon Ratchathani, *Thailand* **38 E5** 15 15N 104 50 E
Ubondo, *Zaïre* **54 C2** 0 55S 25 42 E
Ubort →, *Belarus* **17 B15** 52 6N 28 30 E
Ubundu, *Zaïre* **54 C2** 0 22S 25 30 E
Ucayali →, *Peru* **92 D4** 4 30S 73 30W
Uchi Lake, *Canada* **73 C10** 51 5N 92 35W
Uchiura-Wan, *Japan* **30 C10** 42 25N 140 40 E
Uchur →, *Russia* **27 D14** 58 48N 130 35 E
Ucluelet, *Canada* **72 D3** 48 57N 125 32W
Uda →, *Russia* **27 D14** 54 42N 135 14 E
Udaipur, *India* **42 G5** 24 36N 73 44 E
Udaipur Garhi, *Nepal* **43 F12** 27 0N 86 35 E
Uddevalla, *Sweden* **9 G14** 58 21N 11 55 E
Uddjaur, *Sweden* **8 D17** 65 56N 17 49 E
Udgir, *India* **40 K10** 18 25N 77 5 E
Udhampur, *India* **43 C6** 33 0N 75 5 E
Udi, *Nigeria* **50 G6** 6 17N 7 21 E
Údine, *Italy* **20 A5** 46 3N 13 14 E
Udmurtia □, *Russia* **24 C9** 57 30N 52 30 E
Udon Thani, *Thailand* **38 D4** 17 29N 102 46 E
Udupi, *India* **40 N9** 13 25N 74 42 E
Udzungwa Range, *Tanzania* **55 D4** 9 30S 35 10 E
Ueda, *Japan* **31 F9** 36 24N 138 16 E
Uedineniya, Os., *Russia* **4 B12** 78 0N 85 0 E
Uele →, *Zaïre* **52 D4** 3 45N 24 45 E
Uelen, *Russia* **27 C19** 66 10N 170 0W
Uelzen, *Germany* **16 B6** 52 57N 10 32 E
Ufa, *Russia* **24 D10** 54 45N 55 55 E
Ufa →, *Russia* **24 D10** 54 40N 56 0 E
Ugab →, *Namibia* **56 C1** 20 55S 13 30 E
Ugalla →, *Tanzania* **54 D3** 5 8S 30 42 E
Uganda ■, *Africa* **54 B3** 2 0N 32 0 E
Ugie, *S. Africa* **57 E4** 31 10S 28 13 E
Uglegorsk, *Russia* **27 E15** 49 5N 142 2 E
Ugljane, *Croatia* **16 F8** 44 12N 14 56 E
Ugolyak, *Russia* **27 C13** 64 33N 120 30 E
Ugûn Mûsa, *Egypt* **47 F1** 29 53N 32 40 E
Uhrichsville, *U.S.A.* **78 F3** 40 24N 81 21W
Uibhist a Deas = South Uist, *U.K.* **12 D1** 57 20N 7 15W
Uibhist a Tuath = North Uist, *U.K.* **12 D1** 57 40N 7 15W
Uíge, *Angola* **52 F2** 7 30S 14 40 E
Uijŏngbu, *S. Korea* **35 F14** 37 48N 127 0 E
Ŭiju, *N. Korea* **35 D13** 40 15N 124 35 E
Uinta Mts., *U.S.A.* **82 F8** 40 45N 110 30W
Uitenhage, *S. Africa* **56 E4** 33 40S 25 28 E
Uithuizen, *Neths.* **15 A6** 53 24N 6 41 E
Ujhani, *India* **43 F8** 28 0N 79 6 E
Uji-guntō, *Japan* **31 J4** 31 15N 129 25 E
Ujjain, *India* **42 H6** 23 9N 75 43 E
Ujung Pandang, *Indonesia* **37 F5** 5 10S 119 20 E
Uka, *Russia* **27 D17** 57 50N 162 0 E
Ukara I., *Tanzania* **54 C3** 1 50S 33 0 E
Uke-Shima, *Japan* **31 K4** 28 2N 129 14 E
Ukerewe □, *Tanzania* **54 C3** 2 0S 32 30 E
Ukerewe I., *Tanzania* **54 C3** 2 0S 33 0 E
Ukhrul, *India* **41 G19** 25 10N 94 25 E
Ukhta, *Russia* **24 B9** 63 34N 53 41 E
Ukiah, *U.S.A.* **84 F3** 39 9N 123 13W
Ukki Fort, *India* **43 C7** 33 28N 76 54 E
Ukmerge, *Lithuania* **9 J21** 55 15N 24 45 E
Ukraine ■, *Europe* **25 E5** 49 0N 32 0 E
Ukwi, *Botswana* **56 C3** 23 29S 20 30 E
Ulaanbaatar, *Mongolia* **27 E11** 47 55N 106 53 E
Ulaangom, *Mongolia* **32 A4** 50 5N 92 10 E
Ulamba, *Zaïre* **55 D1** 9 3S 23 38 E
Ulan Bator = Ulaanbaatar, *Mongolia* **27 E11** 47 55N 106 53 E
Ulan Ude, *Russia* **27 D11** 51 45N 107 40 E
Ulanga □, *Tanzania* **55 D4** 8 40S 36 50 E
Ulaya, *Morogoro, Tanzania* **54 D4** 7 3S 36 55 E
Ulaya, *Tabora, Tanzania* **54 C3** 4 25S 33 30 E
Ulcinj, *Montenegro, Yug.* **21 D8** 41 58N 19 10 E
Ulco, *S. Africa* **56 D3** 28 21S 24 15 E
Ulefoss, *Norway* **9 G13** 59 17N 9 16 E
Ulhasnagar, *India* **40 K8** 19 15N 73 10 E
Ulladulla, *Australia* **63 F5** 35 21S 150 29 E
Ullapool, *U.K.* **12 D3** 57 54N 5 9W
Ullswater, *U.K.* **10 C5** 54 34N 2 52W
Ullung-do, *S. Korea* **35 F16** 37 30N 130 30 E
Ulm, *Germany* **16 D5** 48 23N 9 58 E
Ulmarra, *Australia* **63 D5** 29 37S 153 4 E
Ulonguè, *Mozam.* **55 E3** 14 37S 34 19 E
Ulricehamn, *Sweden* **9 H15** 57 46N 13 26 E
Ulsan, *S. Korea* **35 G15** 35 20N 129 15 E
Ulster □, *U.K.* **13 B5** 54 35N 6 30W
Ulubaria, *India* **43 H13** 22 31N 88 4 E
Ulubat Gölü, *Turkey* **21 D13** 40 9N 28 35 E
Uludağ, *Turkey* **21 D13** 40 4N 29 13 E
Uluguru Mts., *Tanzania* **54 D4** 7 15S 37 40 E
Ulungur He →, *China* **32 B3** 47 1N 87 24 E
Uluru = Ayers Rock, *Australia* **61 E5** 25 23S 131 5 E
Ulutau, *Kazakstan* **26 E7** 48 39N 67 1 E
Ulverston, *U.K.* **10 C4** 54 13N 3 5W
Ulverstone, *Australia* **62 G4** 41 11S 146 11 E
Ulya, *Russia* **27 D15** 59 10N 142 0 E
Ulyanovsk = Simbirsk, *Russia* **24 D8** 54 20N 48 25 E
Ulyasutay, *Mongolia* **32 B4** 47 56N 97 28 E
Ulysses, *U.S.A.* **81 G4** 37 35N 101 22W
Umala, *Bolivia* **92 G5** 17 25S 68 5W
Uman, *Ukraine* **17 D16** 48 40N 30 12 E
Umaria, *India* **41 H12** 23 35N 80 50 E
Umarkot, *Pakistan* **40 G6** 25 15N 69 40 E
Umatilla, *U.S.A.* **82 D4** 45 55N 119 21W
Umba, *Russia* **24 A5** 66 42N 34 11 E
Umbrella Mts., *N.Z.* **59 L2** 45 35S 169 5 E
Ume älv →, *Sweden* **8 E19** 63 45N 20 20 E
Umeå, *Sweden* **8 E19** 63 45N 20 20 E
Umera, *Indonesia* **37 E7** 0 12S 129 37 E
Umfuli →, *Zimbabwe* **55 F2** 17 30S 29 23 E
Umgusa, *Zimbabwe* **55 F2** 19 29S 27 52 E
Umkomaas, *S. Africa* **57 E5** 30 13S 30 48 E
Umm ad Daraj, J., *Jordan* **47 C4** 32 18N 35 48 E
Umm al Qaywayn, *U.A.E.* **45 E7** 25 30N 55 35 E
Umm al Qittayn, *Jordan* **47 C5** 32 18N 36 40 E
Umm Bāb, *Qatar* **45 E6** 25 12N 50 48 E
Umm Bel, *Sudan* **51 F10** 13 35N 28 0 E
Umm el Fahm, *Israel* **47 C4** 32 31N 35 9 E
Umm Lajj, *Si. Arabia* **44 E3** 25 0N 37 23 E
Umm Ruwaba, *Sudan* **51 F11** 12 50N 31 20 E
Umnak I., *U.S.A.* **68 C3** 53 15N 168 20W
Umniati →, *Zimbabwe* **55 F2** 16 49S 28 45 E
Umpqua →, *U.S.A.* **82 E1** 43 40N 124 12W
Umreth, *India* **42 H5** 22 41N 73 4 E
Umtata, *S. Africa* **57 E4** 31 36S 28 49 E
Umuarama, *Brazil* **95 A5** 23 45S 53 20W
Umvukwe Ra., *Zimbabwe* **55 F3** 16 45S 30 45 E
Umzimvubu = Port St. Johns, *S. Africa* **57 E4** 31 38S 29 33 E
Umzingwane →, *Zimbabwe* **55 G2** 22 12S 29 56 E
Umzinto, *S. Africa* **57 E5** 30 15S 30 45 E
Una, *India* **42 J4** 20 46N 71 8 E
Una →, *Bos.-H.* **16 F9** 45 0N 16 20 E
Unadilla, *U.S.A.* **79 D9** 42 20N 75 19W
Unalaska, *U.S.A.* **68 C3** 53 53N 166 32W
Uncía, *Bolivia* **92 G5** 18 25S 66 40W
Uncompahgre Peak, *U.S.A.* **83 G10** 38 4N 107 28W
Underbool, *Australia* **63 F3** 35 10S 141 51 E
Ungarie, *Australia* **63 E4** 33 38S 146 56 E
Ungarra, *Australia* **63 E2** 34 12S 136 2 E
Ungava B., *Canada* **69 C13** 59 30N 67 30W
Ungava Pen., *Canada* **69 C12** 60 0N 74 0W
Ungeny = Ungheni, *Moldova* **17 E14** 47 11N 27 51 E
Unggi, *N. Korea* **35 C16** 42 16N 130 28 E
Ungheni, *Moldova* **17 E14** 47 11N 27 51 E
União da Vitória, *Brazil* **95 B5** 26 13S 51 5W
Unimak I., *U.S.A.* **68 C3** 54 45N 164 0W
Union, *Miss., U.S.A.* **81 J10** 32 34N 89 7W
Union, *Mo., U.S.A.* **80 F9** 38 27N 91 0W
Union, *S.C., U.S.A.* **77 H5** 34 43N 81 37W
Union, Mt., *U.S.A.* **83 J7** 34 34N 112 21W
Union City, *Calif., U.S.A.* **84 H4** 37 36N 122 1W
Union City, *N.J., U.S.A.* **79 F10** 40 45N 74 2W
Union City, *Pa., U.S.A.* **78 E5** 41 54N 79 51W
Union City, *Tenn., U.S.A.* **81 G10** 36 26N 89 3W
Union Gap, *U.S.A.* **82 C3** 46 33N 120 28W
Union Springs, *U.S.A.* **77 J3** 32 9N 85 43W
Uniondale, *S. Africa* **56 E3** 33 39S 23 7 E
Uniontown, *U.S.A.* **76 F6** 39 54N 79 44W
Unionville, *U.S.A.* **80 E8** 40 29N 93 1W
United Arab Emirates ■, *Asia* **45 F7** 23 50N 54 0 E
United Kingdom ■, *Europe* **7 E5** 53 0N 2 0W
United States of America ■, *N. Amer.* **74 C7** 37 0N 96 0W
Unity, *Canada* **73 C7** 52 30N 109 5W
Unjha, *India* **42 H5** 23 46N 72 24 E
Unnao, *India* **43 F9** 26 35N 80 30 E
Unst, *U.K.* **12 A8** 60 44N 0 53W
Unuk →, *Canada* **72 B2** 56 5N 131 3W
Uozu, *Japan* **31 F8** 36 48N 137 24 E
Upata, *Venezuela* **92 B6** 8 1N 62 24W
Upemba, L., *Zaïre* **55 D2** 8 30S 26 20 E
Upernavik, *Greenland* **4 B5** 72 49N 56 20W
Upington, *S. Africa* **56 D3** 28 25S 21 15 E
Upleta, *India* **42 J4** 21 46N 70 16 E
Upolu, *W. Samoa* **59 A13** 13 58S 172 0W
Upper Alkali Lake, *U.S.A.* **82 F3** 41 47N 120 8W
Upper Arrow L., *Canada* **72 C5** 50 30N 117 50W
Upper Foster L., *Canada* **73 B7** 56 47N 105 20W
Upper Hutt, *N.Z.* **59 J5** 41 8S 175 5 E
Upper Klamath L., *U.S.A.* **82 E3** 42 25N 121 55W
Upper Lake, *U.S.A.* **84 F4** 39 10N 122 54W
Upper Musquodoboit, *Canada* **71 C7** 45 10N 62 58W
Upper Red L., *U.S.A.* **80 A7** 48 8N 94 45W
Upper Sandusky, *U.S.A.* **76 E4** 40 50N 83 17W
Upper Volta = Burkina Faso ■, *Africa* **50 F4** 12 0N 1 0W
Uppland, *Sweden* **9 F17** 59 59N 17 48 E
Uppsala, *Sweden* **9 G17** 59 53N 17 38 E
Upshi, *India* **43 C7** 33 48N 77 52 E
Upstart, C., *Australia* **62 B4** 19 41S 147 45 E
Upton, *U.S.A.* **80 C2** 44 6N 104 38W
Ur, *Iraq* **44 D5** 30 55N 46 25 E
Uracara, *Brazil* **92 D7** 2 20S 57 50W
Urad Qianqi, *China* **34 D5** 40 40N 108 30 E
Urakawa, *Japan* **30 C11** 42 9N 142 47 E
Ural = Zhayyq →, *Kazakstan* **25 E9** 47 0N 51 48 E
Ural, *Australia* **63 E4** 33 21S 146 12 E
Ural Mts. = Uralskie Gory, *Eurasia* **24 C10** 60 0N 59 0 E
Uralla, *Australia* **63 E5** 30 37S 151 29 E
Uralsk = Oral, *Kazakstan* **24 D9** 51 20N 51 20 E
Uralskie Gory, *Eurasia* **24 C10** 60 0N 59 0 E
Urambo, *Tanzania* **54 D3** 5 4S 32 0 E
Urambo □, *Tanzania* **54 D3** 5 0S 32 0 E
Urandangi, *Australia* **62 C2** 21 32S 138 14 E
Uranium City, *Canada* **73 B7** 59 34N 108 37W
Uranquinty, *Australia* **63 F4** 35 10S 147 12 E
Urawa, *Japan* **31 G9** 35 50N 139 40 E
Uray, *Russia* **26 C7** 60 5N 65 15 E
'Uray'irah, *Si. Arabia* **45 E6** 25 57N 48 53 E
Urbana, *Ill., U.S.A.* **76 E1** 40 7N 88 12W
Urbana, *Ohio, U.S.A.* **76 E4** 40 7N 83 45W
Urbino, *Italy* **20 C5** 43 43N 12 38 E
Urbión, Picos de, *Spain* **19 A4** 42 1N 2 52W
Urcos, *Peru* **92 F4** 13 40S 71 38W
Urda, *Kazakstan* **25 E8** 48 52N 47 23 E
Urdinarrain, *Argentina* **94 C4** 32 37S 58 52W
Urdzhar, *Kazakstan* **26 E9** 47 5N 81 38 E
Ure →, *U.K.* **10 C6** 54 19N 1 31W
Ures, *Mexico* **86 B2** 29 30N 110 30W
Urfa = Sanliurfa, *Turkey* **25 G6** 37 12N 38 50 E
Urganch, *Uzbekistan* **26 E7** 41 40N 60 41 E
Urgench = Urganch, *Uzbekistan* **26 E7** 41 40N 60 41 E
Uri, *India* **43 B6** 34 8N 74 2 E
Uribia, *Colombia* **92 A4** 11 43N 72 16W
Uriondo, *Bolivia* **94 A3** 21 41S 64 41W
Urique, *Mexico* **86 B3** 27 13N 107 55W
Urique →, *Mexico* **86 B3** 26 29N 107 58W
Urk, *Neths.* **15 B5** 52 39N 5 36 E
Urla, *Turkey* **21 E12** 38 20N 26 47 E
Urmia = Orūmīyeh, *Iran* **44 B5** 37 40N 45 0 E
Urmia, L. = Orūmīyeh, Daryācheh-ye, *Iran* **44 B5** 37 50N 45 30 E
Uroševac, *Serbia, Yug.* **21 C9** 42 23N 21 10 E
Uruana, *Brazil* **93 G9** 15 30S 49 41W
Uruapan, *Mexico* **86 D4** 19 30N 102 0W
Urubamba, *Peru* **92 F4** 13 20S 72 10W
Urubamba →, *Peru* **92 F4** 10 43S 73 48W
Uruçuí, *Brazil* **93 E10** 7 20S 44 28W
Uruguai →, *Brazil* **95 B5** 26 0S 53 30W
Uruguaiana, *Brazil* **94 B4** 29 50S 57 0W
Uruguay ■, *S. Amer.* **94 C4** 32 30S 56 30W
Uruguay →, *S. Amer.* **94 C4** 34 12S 58 18W
Urumchi = Ürümqi, *China* **26 E9** 43 45N 87 45 E
Ürümqi, *China* **26 E9** 43 45N 87 45 E
Urup, Os., *Russia* **27 E16** 46 0N 151 0 E
Usa →, *Russia* **24 A10** 66 16N 59 49 E
Uşak, *Turkey* **25 G4** 38 43N 29 28 E
Usakos, *Namibia* **56 C2** 21 54S 15 31 E
Usedom, *Germany* **16 B8** 53 55N 14 2 E
Ush-Tobe, *Kazakstan* **26 E8** 45 16N 78 0 E
Ushakova, Os., *Russia* **4 A12** 82 0N 80 0 E
Ushant = Ouessant, I. d', *France* **18 B1** 48 28N 5 6W
Ushashi, *Tanzania* **54 C3** 1 59S 33 57 E
Ushibuka, *Japan* **31 H5** 32 11N 130 1 E
Ushuaia, *Argentina* **96 G3** 54 50S 68 23W
Ushumun, *Russia* **27 D13** 52 47N 126 32 E
Usk →, *U.K.* **11 F5** 51 33N 2 58W
Üsküdar, *Turkey* **25 F4** 41 0N 29 5 E
Usman, *Russia* **24 D6** 52 5N 39 48 E
Usoke, *Tanzania* **54 D3** 5 8S 32 24 E
Usolye Sibirskoye, *Russia* **27 D11** 52 48N 103 40 E
Uspallata, P. de, *Argentina* **94 C2** 32 37S 69 22W
Uspenskiy, *Kazakstan* **26 E8** 48 41N 72 43 E
Ussuri →, *Asia* **30 A7** 48 27N 135 0 E
Ussuriysk, *Russia* **27 E14** 43 48N 131 59 E
Ussurka, *Russia* **30 B6** 45 12N 133 31 E
Ust-Aldan = Batamay, *Russia* **27 C13** 63 30N 129 15 E
Ust Amginskoye = Khandyga, *Russia* **27 C14** 62 42N 135 35 E
Ust-Bolsheretsk, *Russia* **27 D16** 52 50N 156 15 E
Ust Chaun, *Russia* **27 C18** 68 47N 170 30 E
Ust'-Ilga, *Russia* **27 D11** 55 5N 104 55 E
Ust Ilimpeya = Yukti, *Russia* **27 C11** 63 26N 105 42 E
Ust-Ilimsk, *Russia* **27 D11** 58 3N 102 39 E
Ust Ishim, *Russia* **26 D8** 57 45N 71 10 E
Ust-Kamchatsk, *Russia* **27 D17** 56 10N 162 28 E
Ust-Kamenogorsk = Öskemen, *Kazakstan* **26 E9** 50 0N 82 36 E
Ust-Karenga, *Russia* **27 D12** 54 25N 116 30 E
Ust Khayryuzovo, *Russia* **27 D16** 57 15N 156 45 E
Ust-Kut, *Russia* **27 D11** 56 50N 105 42 E
Ust Kuyga, *Russia* **27 B14** 70 1N 135 43 E
Ust Maya, *Russia* **27 C14** 60 30N 134 28 E
Ust-Mil, *Russia* **27 D14** 59 40N 133 11 E
Ust Muya, *Russia* **27 D12** 56 27N 115 50 E
Ust-Nera, *Russia* **27 C15** 64 35N 143 15 E
Ust-Nyukzha, *Russia* **27 D13** 56 34N 121 37 E
Ust Olenek, *Russia* **27 B12** 73 0N 120 5 E
Ust-Omchug, *Russia* **27 C15** 61 9N 149 38 E
Ust Port, *Russia* **26 C9** 69 40N 84 26 E
Ust Tsilma, *Russia* **24 A9** 65 28N 52 11 E
Ust-Tungir, *Russia* **27 D13** 55 25N 120 36 E
Ust Urt = Ustyurt, Plateau, *Asia* **26 E6** 44 0N 55 0 E
Ust Usa, *Russia* **24 A10** 66 2N 56 57 E
Ust Vorkuta, *Russia* **26 C7** 67 24N 64 0 E
Ústí nad Labem, *Czech.* **16 C8** 50 41N 14 3 E
Ustica, *Italy* **20 E5** 38 42N 13 11 E
Ustinov = Izhevsk, *Russia* **24 C9** 56 51N 53 14 E
Ustye, *Russia* **27 D10** 57 46N 94 37 E
Ustyurt, Plateau, *Asia* **26 E6** 44 0N 55 0 E
Usu, *China* **32 B3** 44 27N 84 40 E
Usuki, *Japan* **31 H5** 33 8N 131 49 E
Usulután, *El Salv.* **88 D2** 13 25N 88 28W
Usumacinta →, *Mexico* **87 D6** 17 0N 91 0W
Usumbura = Bujumbura, *Burundi* **54 C2** 3 16S 29 18 E

Usure, *Tanzania* .... **54 C3** 4 40S 34 22 E
Uta, *Indonesia* .... **37 E9** 4 33S 136 0 E
Utah □, *U.S.A.* **82 G8** 39 20N 111 30W
Utah, L., *U.S.A.* **82 F8** 40 10N 111 58W
Ute Creek →, *U.S.A.* ... **81 H3** 35 21N 103 50W
Utena, *Lithuania* .... **9 J21** 55 27N 25 40 E
Utete, *Tanzania* .... **54 D4** 8 0S 38 45 E
Uthai Thani, *Thailand* ... **38 E3** 15 22N 100 3 E
Uthal, *Pakistan* .... **42 G2** 25 44N 66 40 E
Utiariti, *Brazil* **92 F7** 13 0S 58 10W
Utica, *N.Y., U.S.A.* **79 C9** 43 6N 75 14W
Utica, *Ohio, U.S.A.* **78 F2** 40 14N 82 27W
Utik L., *Canada* **73 B9** 55 15N 96 0W
Utikuma L., *Canada* **72 B5** 55 50N 115 30W
Utrecht, *Neths.* .... **15 B5** 52 5N 5 8 E
Utrecht, *S. Africa* ...... **57 D5** 27 38S 30 20 E
Utrecht □, *Neths.* ...... **15 B5** 52 6N 5 7 E
Utrera, *Spain* **19 D3** 37 12N 5 48W
Utsjoki, *Finland* .... **8 B22** 69 51N 26 59 E
Utsunomiya, *Japan* ..... **31 F9** 36 30N 139 50 E
Uttar Pradesh □, *India* .. **43 F9** 27 0N 80 0 E
Uttaradit, *Thailand* ...... **38 D3** 17 36N 100 5 E
Uttoxeter, *U.K.* **10 E6** 52 54N 1 52W
Uummannarsuaq =
Farvel, Kap, *Greenland* **4 D5** 59 48N 43 55W
Uusikaarlepyy, *Finland* .. **8 E20** 63 32N 22 31 E
Uusikaupunki, *Finland* ... **9 F19** 60 47N 21 25 E
Uva, *Russia* .... **24 C9** 56 59N 52 13 E
Uvalde, *U.S.A.* **81 L5** 29 13N 99 47W
Uvat, *Russia* .... **26 D7** 59 5N 68 50 E
Uvinza, *Tanzania* .... **54 D3** 5 5S 30 24 E
Uvira, *Zaïre* .... **54 C2** 3 22S 29 3 E
Uvs Nuur, *Mongolia* ..... **32 A4** 50 20N 92 30 E
Uwajima, *Japan* .... **31 H6** 33 10N 132 35 E
Uxbridge, *Canada* **78 B5** 44 6N 79 7W
Uxin Qi, *China* .... **34 E5** 38 50N 109 5 E
Uxmal, *Mexico* **87 C7** 20 22N 89 46W
Uyandi, *Russia* .... **27 C15** 69 19N 141 0 E
Uyuni, *Bolivia* **92 H5** 20 28S 66 47W
Uzbekistan ■, *Asia* ..... **26 E7** 41 30N 65 0 E
Uzen, *Kazakstan* .... **25 F9** 43 29N 52 54 E
Uzerche, *France* .... **18 D4** 45 25N 1 34 E
Uzh →, *Ukraine* .... **17 C16** 51 15N 30 12 E
Uzhgorod = Uzhhorod,
Ukraine .... **17 D12** 48 36N 22 18 E
Uzhhorod, *Ukraine* ...... **17 D12** 48 36N 22 18 E
Uzunköprü, *Turkey* ..... **21 D12** 41 16N 26 43 E

V

Vaal →, *S. Africa* ...... **56 D3** 29 4S 23 38 E
Vaal Dam, *S. Africa* ..... **57 D4** 27 0S 28 14 E
Vaalwater, *S. Africa* ..... **57 C4** 24 15S 28 8 E
Vaasa, *Finland* .... **8 E19** 63 6N 21 38 E
Vác, *Hungary* .... **17 E10** 47 49N 19 10 E
Vacaria, *Brazil* **95 B5** 28 31S 50 52W
Vacaville, *U.S.A.* **84 G5** 38 21N 121 59W
Vach → = Vakh →,
Russia .... **26 C8** 60 45N 76 45 E
Vache, Î.-à-, *Haiti* **89 C5** 18 2N 73 35W
Vadnagar, *India* .... **42 H5** 23 47N 72 40 E
Vadodara, *India* .... **42 H5** 22 20N 73 10 E
Vadsø, *Norway* .... **8 A23** 70 3N 29 50 E
Vaduz, *Liech.* .... **16 E5** 47 8N 9 31 E
Værøy, *Norway* .... **8 C15** 67 40N 12 40 E
Vágar, *Færoe Is.* **8 E9** 62 5N 7 15W
Vågsfjorden, *Norway* .... **8 B17** 68 50N 16 50 E
Váh →, *Slovak Rep.* .... **17 D9** 47 43N 18 7 E
Vahsel B., *Antarctica* **5 D1** 75 0S 35 0W
Váï, *Greece* .... **23 D8** 35 15N 26 18 E
Vaigach, *Russia* .... **26 B6** 70 10N 59 0 E
Vakh →, *Russia* .... **26 C8** 60 45N 76 45 E
Val d'Or, *Canada* **70 C4** 48 7N 77 47W
Val Marie, *Canada* **73 D7** 49 15N 107 45W
Valahia, *Romania* .... **17 F13** 44 35N 25 0 E
Valandovo, *Macedonia* .. **21 D10** 41 19N 22 34 E
Valcheta, *Argentina* **96 E3** 40 40S 66 8W
Valdayskaya
Vozvyshennost, *Russia* **24 C5** 57 0N 33 30 E
Valdepeñas, *Spain* **19 C4** 38 43N 3 25W
Valdés, Pen., *Argentina* .. **96 E4** 42 30S 63 45W
Valdez, *U.S.A.* **68 B5** 61 7N 146 16W
Valdivia, *Chile* **96 D2** 39 50S 73 14W
Valdosta, *U.S.A.* **77 K4** 30 50N 83 17W
Valdres, *Norway* .... **9 F13** 61 5N 9 5 E
Vale, *U.S.A.* **82 E5** 43 59N 117 15W
Vale of Glamorgan □, *U.K.* **11 F4** 51 28N 3 25W
Valença, *Brazil* **93 F11** 13 20S 39 5W
Valença do Piauí, *Brazil* . **93 E10** 6 20S 41 45W
Valence, *France* .... **18 D6** 44 57N 4 54 E
Valencia, *Spain* **19 C5** 39 27N 0 23W
Valencia, *Venezuela* **92 A5** 10 11N 68 0W
Valencia □, *Spain* **19 C5** 39 20N 0 40W
Valencia, G. de, *Spain* ... **19 C6** 39 30N 0 20 E
Valencia de Alcántara,
Spain **19 C2** 39 25N 7 14W
Valencia Harbour, *Ireland* **13 E1** 51 56N 10 19W
Valencia I., *Ireland* **13 E1** 51 54N 10 22W
Valenciennes, *France* .... **18 A5** 50 20N 3 34 E
Valentim, Sa. do, *Brazil* .. **93 E10** 6 0S 43 30W
Valentin, *Russia* .... **30 C7** 43 8N 134 17 E
Valentine, *Nebr., U.S.A.* .. **80 D4** 42 52N 100 33W
Valentine, *Tex., U.S.A.* ... **81 K2** 30 35N 104 30W
Valera, *Venezuela* **92 B4** 9 19N 70 37W
Valga, *Estonia* .... **9 H22** 57 47N 26 2 E
Valier, *U.S.A.* **82 B7** 48 18N 112 16W
Valjevo, *Serbia, Yug.* ..... **21 B8** 44 18N 19 53 E
Valka, *Latvia* .... **9 H21** 57 42N 25 57 E
Valkeakoski, *Finland* .... **9 F20** 61 16N 24 2 E
Valkenswaard, *Neths.* .... **15 C5** 51 21N 5 29 E
Vall de Uxó, *Spain* **19 C5** 39 49N 0 15W
Valladolid, *Mexico* **87 C7** 20 40N 88 11W
Valladolid, *Spain* **19 B3** 41 38N 4 43W
Valldemosa, *Spain* ...... **22 B9** 39 43N 2 37 E
Valle de la Pascua,
Venezuela **92 B5** 9 13N 66 0W
Valle de las Palmas,
Mexico **85 N10** 32 20N 116 43W
Valle de Santiago, *Mexico* **86 C4** 20 25N 101 15W
Valle de Suchil, *Mexico* .. **86 C4** 23 38N 103 55W
Valle de Zaragoza, *Mexico* **86 B3** 27 28N 105 49W
Valle Fértil, Sierra del,
Argentina **94 C2** 30 20S 68 0W
Valle Hermoso, *Mexico* ... **87 B5** 25 35N 97 40W
Valledupar, *Colombia* ... **92 A4** 10 29N 73 15W
Vallehermoso, *Canary Is.* **22 F2** 28 10N 17 15W
Vallejo, *U.S.A.* **84 G4** 38 7N 122 14W
Vallenar, *Chile* **94 B1** 28 30S 70 50W
Valletta, *Malta* .... **23 D2** 35 54N 14 31 E
Valley Center, *U.S.A.* **85 M9** 33 13N 117 2W
Valley City, *U.S.A.* **80 B6** 46 55N 98 0W
Valley Falls, *U.S.A.* **82 E3** 42 29N 120 17W
Valley Springs, *U.S.A.* ... **84 G6** 38 12N 120 50W
Valley Wells, *U.S.A.* **85 K11** 35 27N 115 46W
Valleyview, *Canada* **72 B5** 55 5N 117 17W
Vallimanca, Arroyo,
Argentina **94 D4** 35 40S 59 10W
Valls, *Spain* .... **19 B6** 41 18N 1 15 E
Valmiera, *Latvia* .... **9 H21** 57 37N 25 29 E
Valognes, *France* **18 B3** 49 30N 1 28W
Valona = Vlóra, *Albania* . **21 D8** 40 32N 19 28 E
Valozhyn, *Belarus* ...... **17 A14** 54 3N 26 30 E
Valparaíso, *Chile* **94 C1** 33 2S 71 40W
Valparaíso, *Mexico* **86 C4** 22 50N 103 32W
Valparaiso, *U.S.A.* **76 E2** 41 28N 87 4W
Valparaíso □, *Chile* **94 C1** 33 2S 71 40W
Vals →, *S. Africa* ...... **56 D4** 27 23S 26 30 E
Vals, Tanjung, *Indonesia* . **37 F9** 8 26S 137 25 E
Valsad, *India* .... **40 J8** 20 40N 72 58 E
Valverde, *Canary Is.* **22 G2** 27 48N 17 55W
Valverde del Camino,
Spain **19 D2** 37 35N 6 47W
Vammala, *Finland* ...... **9 F20** 61 20N 22 54 E
Vámos, *Greece* .... **23 D6** 35 24N 24 13 E
Van, *Turkey* .... **25 G7** 38 30N 43 20 E
Van, L. = Van Gölü,
Turkey .... **25 G7** 38 30N 43 0 E
Van Alstyne, *U.S.A.* **81 J6** 33 25N 96 35W
Van Bruyssel, *Canada* ... **71 C5** 47 56N 72 9W
Van Buren, *Canada* **71 C6** 47 10N 67 55W
Van Buren, *Ark., U.S.A.* ... **81 H7** 35 26N 94 21W
Van Buren, *Maine, U.S.A.* **77 B11** 47 10N 67 58W
Van Buren, *Mo., U.S.A.* .. **81 G9** 37 0N 91 1W
Van Canh, *Vietnam* ..... **38 F7** 13 37N 109 0 E
Van Diemen, C., *N. Terr.,*
Australia .... **60 B5** 11 9S 130 24 E
Van Diemen, C., *Queens.,*
Australia .... **62 B2** 16 30S 139 46 E
Van Diemen G., *Australia* **60 B5** 11 45S 132 0 E
Van Gölü, *Turkey* .... **25 G7** 38 30N 43 0 E
Van Horn, *U.S.A.* **81 K2** 31 3N 104 50W
Van Ninh, *Vietnam* ..... **38 F7** 12 42N 109 14 E
Van Rees, Pegunungan,
Indonesia .... **37 E9** 2 35S 138 15 E
Van Tassell, *U.S.A.* **80 D2** 42 40N 104 5W
Van Wert, *U.S.A.* **76 E3** 40 52N 84 35W
Van Yen, *Vietnam* ...... **38 B5** 21 4N 104 42 E
Vanadzor, *Armenia* ..... **25 F7** 40 48N 44 30 E
Vanavara, *Russia* ...... **27 C11** 60 22N 102 16 E
Vancouver, *Canada* **72 D4** 49 15N 123 10W
Vancouver, *U.S.A.* **84 E4** 45 38N 122 40W
Vancouver, C., *Australia* . **61 G2** 35 2S 118 11 E
Vancouver I., *Canada* **72 D3** 49 50N 126 0W
Vandalia, *Ill., U.S.A.* **80 F10** 38 58N 89 6W
Vandalia, *Mo., U.S.A.* **80 F9** 39 19N 91 29W
Vandenburg, *U.S.A.* **85 L6** 34 35N 120 33W
Vanderbijlpark, *S. Africa* . **57 D4** 26 42S 27 54 E
Vandergrift, *U.S.A.* **78 F5** 40 36N 79 34W
Vanderhoof, *Canada* **72 C4** 54 0N 124 0W
Vanderkloof Dam,
S. Africa .... **56 E3** 30 4S 24 40 E
Vanderlin I., *Australia* ... **62 B2** 15 44S 137 2 E
Vandyke, *Australia* ..... **62 C4** 24 10S 147 51 E
Vänern, *Sweden* .... **9 G15** 58 47N 13 30 E
Vänersborg, *Sweden* .... **9 G15** 58 26N 12 19 E
Vang Vieng, *Laos* .... **38 C4** 18 58N 102 32 E
Vanga, *Kenya* .... **54 C4** 4 35S 39 12 E
Vangaindrano, *Madag.* ... **57 C8** 23 21S 47 36 E
Vanguard, *Canada* **73 D7** 49 55N 107 20W
Vanier, *Canada* **70 C4** 45 27N 75 40W
Vankleek Hill, *Canada* ... **70 C5** 45 32N 74 40W
Vanna, *Norway* .... **8 A18** 70 6N 19 50 E
Vännäs, *Sweden* .... **8 E18** 63 58N 19 48 E
Vannes, *France* **18 C2** 47 40N 2 47W
Vanrhynsdorp, *S. Africa* . **56 E2** 31 36S 18 44 E
Vanrook, *Australia* ...... **62 B3** 16 57S 141 57 E
Vansbro, *Sweden* .... **9 F16** 60 32N 14 15 E
Vansittart B., *Australia* .. **60 B4** 14 3S 126 17 E
Vantaa, *Finland* .... **9 F21** 60 18N 24 58 E
Vanthli, *India* .... **42 J4** 21 28N 70 25 E
Vanua Levu, *Fiji* .... **59 C8** 16 33S 179 15 E
Vanua Mbalavu, *Fiji* **59 C9** 17 40S 178 57W
Vanuatu ■, *Pac. Oc.* ..... **64 J8** 15 0S 168 0 E
Vanwyksvlei, *S. Africa* ... **56 E3** 30 18S 21 49 E
Vanzylsrus, *S. Africa* .... **56 D3** 26 52S 22 4 E
Vapnyarka, *Ukraine* ..... **17 D15** 48 32N 28 45 E
Varanasi, *India* .... **43 G10** 25 22N 83 0 E
Varanger-halvøya, *Norway* **8 A23** 70 25N 29 30 E
Varangerfjorden, *Norway* **8 A23** 70 3N 29 25 E
Varaždin, *Croatia* .... **16 E9** 46 20N 16 20 E
Varberg, *Sweden* .... **9 H15** 57 6N 12 20 E
Vardar → = Axiós →,
Greece .... **21 D10** 40 57N 22 35 E
Varde, *Denmark* .... **9 J13** 55 38N 8 29 E
Vardø, *Norway* .... **8 A24** 70 23N 31 5 E
Varella, Mui, *Vietnam* .... **38 F7** 12 54N 109 26 E
Varena, *Lithuania* .... **9 J21** 54 12N 24 30 E
Varese, *Italy* .... **20 B3** 45 48N 8 50 E
Varginha, *Brazil* **95 A6** 21 33S 45 25W
Variadero, *U.S.A.* **81 H2** 35 43N 104 17W
Varillas, *Chile* **94 A1** 24 0S 70 10W
Varkaus, *Finland* .... **9 E22** 62 19N 27 50 E
Varna, *Bulgaria* .... **21 C12** 43 13N 27 56 E
Värnamo, *Sweden* ...... **9 H16** 57 10N 14 3 E
Vars, *Canada* **79 A9** 45 21N 75 21W
Varzaneh, *Iran* .... **45 C7** 32 25N 52 40 E
Vasa Barris →, *Brazil* .. **93 F11** 11 10S 37 10W
Vascongadas = Pais
Vasco □, *Spain* **19 A4** 42 50N 2 45W
Vasht = Khāsh, *Iran* .... **40 E2** 28 15N 61 15 E
Vasilevichi, *Belarus* ..... **17 B15** 52 15N 29 50 E
Vasilkov = Vasylkiv,
Ukraine .... **17 C16** 50 7N 30 15 E
Vaslui, *Romania* .... **17 E14** 46 38N 27 42 E
Vassar, *Canada* **73 D9** 49 10N 95 55W
Vassar, *U.S.A.* **76 D4** 43 22N 83 35W
Västerås, *Sweden* ...... **9 G17** 59 37N 16 38 E
Västerbotten, *Sweden* ... **8 D18** 64 36N 20 4 E
Västerdalälven →,
Sweden .... **9 F16** 60 30N 14 7 E
Västervik, *Sweden* ...... **9 H17** 57 43N 16 33 E
Västmanland, *Sweden* .. **9 G16** 59 45N 16 20 E
Vasto, *Italy* .... **20 C6** 42 8N 14 40 E
Vasylkiv, *Ukraine* ...... **17 C16** 50 7N 30 15 E
Vatican City ■, *Europe* .. **20 D5** 41 54N 12 27 E
Vatili, *Cyprus* .... **23 D12** 35 6N 33 40 E
Vatnajökull, *Iceland* **8 D5** 64 30N 16 48W
Vatoa, *Fiji* **59 D9** 19 50S 178 13W
Vatólakkos, *Greece* ..... **23 D5** 35 27N 23 53 E
Vatoloha, *Madag.* .... **57 B8** 17 52S 47 48 E
Vatomandry, *Madag.* .... **57 B8** 19 20S 48 59 E
Vatra-Dornei, *Romania* .. **17 E13** 47 22N 25 22 E
Vättern, *Sweden* .... **9 G16** 58 25N 14 30 E
Vaughn, *Mont., U.S.A.* ... **82 C8** 47 33N 111 33W
Vaughn, *N. Mex., U.S.A.* . **83 J11** 34 36N 105 13W
Vaupés = Uaupés →,
Brazil **92 C5** 0 2N 67 16W
Vauxhall, *Canada* **72 C6** 50 5N 112 9W
Vava'u, *Tonga* **59 D11** 18 36S 174 0W
Vawkavysk, *Belarus* ..... **17 B13** 53 9N 24 30 E
Växjö, *Sweden* .... **9 H16** 56 52N 14 50 E
Vaygach, Ostrov, *Russia* . **26 C6** 70 0N 60 0 E
Váyia, Ákra, *Greece* ..... **23 C10** 36 15N 28 11 E
Vechte →, *Neths.* ...... **15 B6** 52 34N 6 6 E
Vedea →, *Romania* ..... **17 G13** 43 53N 25 59 E
Vedia, *Argentina* **94 C3** 34 30S 61 31W
Vedra, I. del, *Spain* ..... **22 C7** 38 52N 1 12 E
Veendam, *Neths.* .... **15 A6** 53 5N 6 52 E
Veenendaal, *Neths.* ..... **15 B5** 52 2N 5 34 E
Vefsna →, *Norway* ..... **8 D15** 65 48N 13 10 E
Vega, *Norway* .... **8 D14** 65 40N 11 55 E
Vega, *U.S.A.* **81 H3** 35 15N 102 26W
Veghel, *Neths.* .... **15 C5** 51 37N 5 32 E
Vegreville, *Canada* **72 C6** 53 30N 112 5W
Vejer de la Frontera, *Spain* **19 D3** 36 15N 5 59W
Vejle, *Denmark* .... **9 J13** 55 43N 9 30 E
Velas, C., *Costa Rica* **88 D2** 10 21N 85 52W
Velasco, Sierra de,
Argentina **94 B2** 29 20S 67 10W
Velddrif, *S. Africa* .... **56 E2** 32 42S 18 11 E
Velebit Planina, *Croatia* .. **16 F8** 44 50N 15 20 E
Vélez, *Colombia* **92 B4** 6 1N 73 41W
Vélez Málaga, *Spain* **19 D3** 36 48N 4 5W
Vélez Rubio, *Spain* **19 D4** 37 41N 2 5W
Velhas →, *Brazil* **93 G10** 17 13S 44 49W
Velika Kapela, *Croatia* ... **16 F8** 45 10N 15 5 E
Velikaya →, *Russia* .... **24 C4** 57 48N 28 10 E
Velikaya Kema, *Russia* .. **30 B8** 45 30N 137 12 E
Veliki Ustyug, *Russia* .... **24 B8** 60 47N 46 20 E
Velikiye Luki, *Russia* .... **24 C5** 56 25N 30 32 E
Veliko Tŭrnovo, *Bulgaria* **21 C11** 43 5N 25 41 E
Velikonda Range, *India* .. **40 M11** 14 45N 79 10 E
Velletri, *Italy* .... **20 D5** 41 41N 12 47 E
Vellore, *India* .... **40 N11** 12 57N 79 10 E
Velsen-Noord, *Neths.* .... **15 B4** 52 27N 4 40 E
Velsk, *Russia* .... **24 B7** 61 10N 42 5 E
Velva, *U.S.A.* **80 A4** 48 4N 100 56W
Venado Tuerto, *Argentina* **94 C3** 33 50S 62 0W
Vendée □, *France* **18 C3** 46 50N 1 35W
Vendôme, *France* ...... **18 C4** 47 47N 1 3 E
Venézia, *Italy* .... **20 B5** 45 27N 12 21 E
Venézia, G. di, *Italy* .... **20 B5** 45 15N 13 0 E
Venezuela ■, *S. Amer.* .. **92 B5** 8 0N 66 0W
Venezuela, G. de,
Venezuela **92 A4** 11 30N 71 0W
Vengurla, *India* .... **40 M8** 15 53N 73 45 E
Venice = Venézia, *Italy* .. **20 B5** 45 27N 12 21 E
Venkatapuram, *India* .... **41 K12** 18 20N 80 30 E
Venlo, *Neths.* .... **15 C6** 51 22N 6 11 E
Vennesla, *Norway* ...... **9 G12** 58 15N 8 0 E
Venraij, *Neths.* .... **15 C5** 51 31N 6 0 E
Ventana, Punta de la,
Mexico **86 C3** 24 4N 109 48W
Ventana, Sa. de la,
Argentina **94 D3** 38 0S 62 30W
Ventersburg, *S. Africa* ... **56 D4** 28 7S 27 9 E
Venterstad, *S. Africa* .... **56 E4** 30 47S 25 48 E
Ventnor, *U.K.* **11 G6** 50 36N 1 12W
Ventotene, *Italy* .... **20 D5** 40 47N 13 25 E
Ventoux, Mt., *France* .... **18 D6** 44 10N 5 17 E
Ventspils, *Latvia* .... **9 H19** 57 25N 21 32 E
Ventuarí →, *Venezuela* . **92 C5** 3 58N 67 2W
Ventucopa, *U.S.A.* **85 L7** 34 50N 119 29W
Ventura, *U.S.A.* **85 L7** 34 17N 119 18W
Venus B., *Australia* ..... **63 F4** 38 40S 145 42 E
Vera, *Argentina* **94 B3** 29 30S 60 20W
Vera, *Spain* **19 D5** 37 15N 1 51W
Veracruz, *Mexico* **87 D5** 19 10N 96 10W
Veracruz □, *Mexico* **87 D5** 19 0N 96 15W
Veraval, *India* .... **42 J4** 20 53N 70 27 E
Verbánia, *Italy* .... **20 B3** 45 56N 8 33 E
Vercelli, *Italy* .... **20 B3** 45 19N 8 25 E
Verdalsøra, *Norway* ..... **8 E14** 63 48N 11 30 E
Verde →, *Argentina* **96 E3** 41 56S 65 5W
Verde →, *Chihuahua,*
Mexico **86 B3** 26 29N 107 58W
Verde →, *Oaxaca,*
Mexico **87 D5** 15 59N 97 50W
Verde →, *Veracruz,*
Mexico **86 C4** 21 10N 102 50W
Verde →, *Paraguay* **94 A4** 23 9S 57 37W
Verde, Cay, *Bahamas* ... **88 B4** 23 0N 75 5W
Verden, *Germany* .... **16 B5** 52 55N 9 14 E
Verdi, *U.S.A.* **84 F7** 39 31N 119 59W
Verdigre, *U.S.A.* **80 D5** 42 36N 98 2W
Verdun, *France* .... **18 B6** 49 9N 5 24 E
Vereeniging, *S. Africa* ... **57 D4** 26 38S 27 57 E
Vérendrye, Parc Prov. de
la, *Canada* **70 C4** 47 20N 76 40W
Verga, C., *Guinea* **50 F2** 10 30N 14 10W
Vergemont, *Australia* ... **62 C3** 23 33S 143 1 E
Vergemont Cr. →,
Australia .... **62 C3** 24 16S 143 16 E
Vergennes, *U.S.A.* **79 B11** 44 10N 73 15W
Verin, *Spain* **19 B2** 41 57N 7 27W
Verkhnevilyuysk, *Russia* . **27 C13** 63 27N 120 18 E
Verkhneye Kalinino,
Russia .... **27 D11** 59 54N 108 8 E
Verkhniy Baskunchak,
Russia .... **25 E8** 48 14N 46 44 E
Verkhnyaya Amga, *Russia* **27 D13** 59 50N 127 0 E
Verkhoyansk, *Russia* .... **27 C14** 67 35N 133 25 E
Verkhoyansk Ra. =
Verkhoyanskiy Khrebet,
Russia .... **27 C13** 66 0N 129 0 E
Verkhoyanskiy Khrebet,
Russia .... **27 C13** 66 0N 129 0 E
Verlo, *Canada* **73 C7** 50 19N 108 35W
Vermilion, *Canada* **73 C6** 53 20N 110 50W
Vermilion →, *Alta.,*
Canada **73 C6** 53 22N 110 51W
Vermilion →, *Qué.,*
Canada **70 C5** 47 38N 72 56W
Vermilion, B., *U.S.A.* **81 L9** 29 45N 91 55W
Vermilion Bay, *Canada* .. **73 D10** 49 51N 93 34W
Vermilion Chutes, *Canada* **72 B6** 58 22N 114 51W
Vermilion L., *U.S.A.* **80 B8** 47 53N 92 26W
Vermillion, *U.S.A.* **80 D6** 42 47N 96 56W
Vermont □, *U.S.A.* **79 C12** 44 0N 73 0W
Vernal, *U.S.A.* **82 F9** 40 27N 109 32W
Vernalis, *U.S.A.* **84 H5** 37 36N 121 17W
Verner, *Canada* **70 C3** 46 25N 80 8W
Verneukpan, *S. Africa* ... **56 D3** 30 0S 21 0 E
Vernon, *Canada* **72 C5** 50 20N 119 15W
Vernon, *U.S.A.* **81 H5** 34 9N 99 17W
Vernonia, *U.S.A.* **84 E3** 45 52N 123 11W
Vero Beach, *U.S.A.* **77 M5** 27 38N 80 24W
Véroia, *Greece* .... **21 D10** 40 34N 22 12 E
Verona, *Italy* .... **20 B4** 45 27N 11 0 E
Versailles, *France* ...... **18 B5** 48 48N 2 8 E
Vert, C., *Senegal* **50 F1** 14 45N 17 30W
Verulam, *S. Africa* ...... **57 D5** 29 38S 31 2 E
Verviers, *Belgium* ...... **15 D5** 50 37N 5 52 E
Veselovskoye Vdkhr.,
Russia .... **25 E7** 46 58N 41 25 E
Vesoul, *France* .... **18 C7** 47 40N 6 11 E
Vesterålen, *Norway* ..... **8 B16** 68 45N 15 0 E
Vestfjorden, *Norway* .... **8 C15** 67 55N 14 0 E
Vestmannaeyjar, *Iceland* . **8 E3** 63 27N 20 15W
Vestspitsbergen, *Svalbard* **4 B8** 78 40N 17 0 E
Vestvågøy, *Norway* ..... **8 B15** 68 18N 13 50 E
Vesuvio, *Italy* .... **20 D6** 40 49N 14 26 E
Vesuvius, Mt. = Vesuvio,
Italy .... **20 D6** 40 49N 14 26 E
Veszprém, *Hungary* ..... **17 E9** 47 8N 17 57 E
Vetlanda, *Sweden* ...... **9 H16** 57 24N 15 3 E
Vetlugu →, *Russia* ..... **26 D5** 56 36N 46 4 E
Vettore, Mte., *Italy* ..... **20 C5** 42 49N 13 16 E
Veurne, *Belgium* .... **15 C2** 51 5N 2 40 E
Veys, *Iran* .... **45 D6** 31 30N 49 0 E
Vezhen, *Bulgaria* .... **21 C11** 42 50N 24 20 E
Vi Thanh, *Vietnam* ..... **39 H5** 9 42N 105 26 E
Viacha, *Bolivia* **92 G5** 16 39S 68 18W
Viamão, *Brazil* **95 C5** 30 5S 51 0W
Viana, *Brazil* **93 D10** 3 13S 44 55W
Viana do Alentejo,
Portugal **19 C2** 38 17N 7 59W
Viana do Castelo, *Portugal* **19 B1** 41 42N 8 50W
Vianópolis, *Brazil* **93 G9** 16 40S 48 35W
Viaréggio, *Italy* .... **20 C4** 43 52N 10 14 E
Vibank, *Canada* **73 C8** 50 20N 103 56W
Vibo Valéntia, *Italy* ..... **20 E7** 38 40N 16 6 E
Viborg, *Denmark* .... **9 H13** 56 27N 9 23 E
Vicenza, *Italy* .... **20 B4** 45 33N 11 33 E
Vich, *Spain* .... **19 B7** 41 58N 2 19 E
Vichy, *France* .... **18 C5** 46 9N 3 26 E
Vicksburg, *Ariz., U.S.A.* .. **85 M13** 33 45N 113 45W
Vicksburg, *Mich., U.S.A.* . **76 D3** 42 7N 85 32W
Vicksburg, *Miss., U.S.A.* . **81 J9** 32 21N 90 53W
Viçosa, *Brazil* **93 E11** 9 28S 36 14W
Victor, *India* .... **42 J4** 21 0N 71 30 E
Victor, *Colo., U.S.A.* **80 F2** 38 43N 105 9W
Victor, *N.Y., U.S.A.* **78 D7** 42 58N 77 24W
Victor Harbor, *Australia* . **63 F2** 35 30S 138 37 E
Victoria, *Argentina* **94 C3** 32 40S 60 10W
Victoria, *Canada* **72 D4** 48 30N 123 25W
Victoria, *Chile* **96 D2** 38 13S 72 20W
Victoria, *Guinea* **50 F2** 10 50N 14 32W
Victoria, *Malaysia* ...... **36 C5** 5 20N 115 14 E
Victoria, *Malta* .... **23 C1** 36 2N 14 14 E
Victoria, *Kans., U.S.A.* ... **80 F5** 38 52N 99 9W
Victoria, *Tex., U.S.A.* **81 L6** 28 48N 97 0W
Victoria □, *Australia* .... **63 F3** 37 0S 144 0 E
Victoria →, *Australia* ... **60 C4** 15 10S 129 40 E
Victoria, Grand L., *Canada* **70 C4** 47 31N 77 30W
Victoria, L., *Africa* ...... **54 C3** 1 0S 33 0 E
Victoria, L., *Australia* ... **63 E3** 33 57S 141 15 E
Victoria Beach, *Canada* . **73 C9** 50 40N 96 35W
Victoria de Durango,
Mexico **86 C4** 24 3N 104 39W
Victoria de las Tunas,
Cuba **88 B4** 20 58N 76 59W
Victoria Falls, *Zimbabwe* . **55 F2** 17 58S 25 52 E
Victoria Harbour, *Canada* **70 D4** 44 45N 79 45W
Victoria I., *Canada* **68 A8** 71 0N 111 0W
Victoria Ld., *Antarctica* .. **5 D11** 75 0S 160 0 E
Victoria Nile →, *Uganda* **54 B3** 2 14N 31 26 E
Victoria Res., *Canada* ... **71 C8** 48 20N 57 27W
Victoria River Downs,
Australia .... **60 C5** 16 25S 131 0 E
Victoria Taungdeik, *Burma* **41 J18** 21 15N 93 55 E
Victoria West, *S. Africa* . **56 E3** 31 25S 23 4 E
Victoriaville, *Canada* **71 C5** 46 4N 71 56W
Victorica, *Argentina* **94 D2** 36 20S 65 30W
Victorville, *U.S.A.* **85 L9** 34 32N 117 18W
Vicuña, *Chile* **94 C1** 30 0S 70 50W
Vicuña Mackenna,
Argentina **94 C3** 33 53S 64 25W
Vidal, *U.S.A.* **85 L12** 34 7N 114 31W
Vidal Junction, *U.S.A.* ... **85 L12** 34 11N 114 34W
Vidalia, *U.S.A.* **77 J4** 32 13N 82 25W
Vídho, *Greece* .... **23 A3** 39 38N 19 55 E
Vidin, *Bulgaria* .... **21 C10** 43 59N 22 50 E
Vidisha, *India* .... **42 H7** 23 28N 77 53 E
Vidzy, *Belarus* .... **9 J22** 55 23N 26 37 E
Viedma, *Argentina* **96 E4** 40 50S 63 0W
Viedma, L., *Argentina* ... **96 F2** 49 30S 72 30W
Vieng Pou Kha, *Laos* .... **38 B3** 20 41N 101 4 E
Vienna = Wien, *Austria* .. **16 D9** 48 12N 16 22 E

Vienna, U.S.A. 81 G10 37 25N 88 54W
Vienne, France 18 D6 45 31N 4 53 E
Vienne →, France 18 C4 47 13N 0 5 E
Vientiane, Laos 38 D4 17 58N 102 36 E
Vientos, Paso de los, Caribbean 89 C5 20 0N 74 0W
Vierzon, France 18 C5 47 13N 2 5 E
Vietnam ■, Asia 38 C5 19 0N 106 0 E
Vigan, Phil. 37 A6 17 35N 120 28 E
Vigévano, Italy 20 B3 45 19N 8 51 E
Vigia, Brazil 93 D9 0 50S 48 5W
Vigía Chico, Mexico 87 D7 19 46N 87 35W
Vígla, Ákra, Greece 23 D9 35 54N 27 51 E
Vigo, Spain 19 A1 42 12N 8 41W
Vijayawada, India 41 L12 16 31N 80 39 E
Vík, Iceland 8 E4 63 25N 19 1W
Vikeke, Indonesia 37 F7 8 52S 126 23 E
Viking, Canada 72 C6 53 7N 111 50W
Vikna, Norway 8 D14 64 55N 10 58 E
Vikulovo, Russia 26 D8 56 50N 70 40 E
Vila da Maganja, Mozam. 55 F4 17 18S 37 30 E
Vila de João Belo = Xai-Xai, Mozam. 57 D5 25 6S 33 31 E
Vila do Bispo, Portugal 19 D1 37 5N 8 53W
Vila do Chibuto, Mozam. 57 C5 24 40S 33 33 E
Vila Franca de Xira, Portugal 19 C1 38 57N 8 59W
Vila Gamito, Mozam. 55 E3 14 12S 33 0 E
Vila Gomes da Costa, Mozam. 57 C5 24 20S 33 37 E
Vila Machado, Mozam. 55 F3 19 15S 34 14 E
Vila Mouzinho, Mozam. 55 E3 14 48S 34 25 E
Vila Nova de Gaia, Portugal 19 B1 41 8N 8 37W
Vila Real, Portugal 19 B2 41 17N 7 48W
Vila Real de Santo António, Portugal 19 D2 37 10N 7 28W
Vila Vasco da Gama, Mozam. 55 E3 14 54S 32 14 E
Vila Velha, Brazil 95 A7 20 20S 40 17W
Vilaine →, France 18 C2 47 30N 2 27W
Vilanandro, Tanjona, Madag. 57 B7 16 11S 44 27 E
Vilanculos, Mozam. 57 C6 22 1S 35 17 E
Vileyka, Belarus 17 A14 54 30N 26 53 E
Vilhelmina, Sweden 8 D17 64 35N 16 39 E
Vilhena, Brazil 92 F6 12 40S 60 5W
Viliga, Russia 27 C16 61 36N 156 56 E
Viliya →, Lithuania 9 J21 55 8N 24 16 E
Viljandi, Estonia 9 G21 58 28N 25 30 E
Vilkitskogo, Proliv, Russia 27 B11 78 0N 103 0 E
Vilkovo = Vylkove, Ukraine 17 F15 45 28N 29 32 E
Villa Abecia, Bolivia 94 A2 21 0S 68 18W
Villa Ahumada, Mexico 86 A3 30 38N 106 30W
Villa Ana, Argentina 94 B4 28 28S 59 40W
Villa Ángela, Argentina 94 B3 27 34S 60 45W
Villa Bella, Bolivia 92 F5 10 25S 65 22W
Villa Bens = Tarfaya, Morocco 50 C2 27 55N 12 55W
Villa Cañás, Argentina 94 C3 34 0S 61 35W
Villa Carlos, Spain 22 B11 39 53N 4 17 E
Villa Cisneros = Dakhla, W. Sahara 50 D1 23 50N 15 53W
Villa Colón, Argentina 94 C2 31 38S 68 20W
Villa Constitución, Argentina 94 C3 33 15S 60 20W
Villa de María, Argentina 94 B3 29 55S 63 43W
Villa Dolores, Argentina 94 C2 31 58S 65 15W
Villa Frontera, Mexico 86 B4 26 56N 101 27W
Villa Guillermina, Argentina 94 B4 28 15S 59 29W
Villa Hayes, Paraguay 94 B4 25 5S 57 20W
Villa Iris, Argentina 94 D3 38 12S 63 12W
Villa Juárez, Mexico 86 B4 27 37N 100 44W
Villa María, Argentina 94 C3 32 20S 63 10W
Villa Mazán, Argentina 94 B2 28 40S 66 30W
Villa Montes, Bolivia 94 A3 21 10S 63 30W
Villa Ocampo, Argentina 94 B4 28 30S 59 20W
Villa Ocampo, Mexico 86 B3 26 29N 105 30W
Villa Ojo de Agua, Argentina 94 B3 29 30S 63 44W
Villa San José, Argentina 94 C4 32 12S 58 15W
Villa San Martín, Argentina 94 B3 28 15S 64 9W
Villa Unión, Mexico 86 C3 23 12N 106 14W
Villacarrillo, Spain 19 C4 38 7N 3 3W
Villach, Austria 16 E7 46 37N 13 51 E
Villafranca de los Caballeros, Spain 22 B10 39 34N 3 25 E
Villagarcía de Arosa, Spain 19 A1 42 34N 8 46W
Villagrán, Mexico 87 C5 24 29N 99 29W
Villaguay, Argentina 94 C4 32 0S 59 0W
Villahermosa, Mexico 87 D6 17 59N 92 55W
Villajoyosa, Spain 19 C5 38 30N 0 12W
Villalba, Spain 19 A2 43 26N 7 40W
Villanueva, U.S.A. 83 J11 35 16N 105 22W
Villanueva de la Serena, Spain 19 C3 38 59N 5 50W
Villanueva y Geltrú, Spain 19 B6 41 13N 1 40 E
Villarreal, Spain 19 C5 39 55N 0 3W
Villarrica, Chile 96 D2 39 15S 72 15W
Villarrica, Paraguay 94 B4 25 40S 56 30W
Villarrobledo, Spain 19 C4 39 18N 2 36W
Villavicencio, Argentina 94 C2 32 28S 69 0W
Villavicencio, Colombia 92 C4 4 9N 73 37W
Villaviciosa, Spain 19 A3 43 32N 5 27W
Villazón, Bolivia 94 A2 22 0S 65 35W
Ville-Marie, Canada 70 C4 47 20N 79 30W
Ville Platte, U.S.A. 81 K8 30 41N 92 17W
Villena, Spain 19 C5 38 39N 0 52W
Villeneuve-d'Ascq, France 18 A5 50 38N 3 9 E
Villeneuve-sur-Lot, France 18 D4 44 24N 0 42 E
Villiers, S. Africa 57 D4 27 2S 28 36 E
Villingen-Schwenningen, Germany 16 D5 48 3N 8 26 E
Villisca, U.S.A. 80 E7 40 56N 94 59W
Vilna, Canada 72 C6 54 7N 111 55W
Vilnius, Lithuania 9 J21 54 38N 25 19 E
Vilvoorde, Belgium 15 D4 50 56N 4 26 E
Vilyuy →, Russia 27 C13 64 24N 126 26 E
Vilyuysk, Russia 27 C13 63 40N 121 35 E
Viña del Mar, Chile 94 C1 33 0S 71 30W
Vinaroz, Spain 19 B6 40 30N 0 27 E
Vincennes, U.S.A. 76 F2 38 41N 87 32W
Vincent, U.S.A. 85 L8 34 33N 118 11W
Vinchina, Argentina 94 B2 28 45S 68 15W
Vindelälven →, Sweden 8 E18 63 55N 19 50 E
Vindeln, Sweden 8 D18 64 12N 19 43 E
Vindhya Ra., India 42 H7 22 50N 77 0 E
Vineland, U.S.A. 76 F8 39 29N 75 2W
Vinh, Vietnam 38 C5 18 45N 105 38 E
Vinh Linh, Vietnam 38 D6 17 4N 107 2 E
Vinh Long, Vietnam 39 G5 10 16N 105 57 E
Vinh Yen, Vietnam 38 B5 21 21N 105 35 E
Vinita, U.S.A. 81 G7 36 39N 95 9W
Vinkovci, Croatia 21 B8 45 19N 18 48 E
Vinnitsa = Vinnytsya, Ukraine 17 D15 49 15N 28 30 E
Vinnytsya, Ukraine 17 D15 49 15N 28 30 E
Vinton, Calif., U.S.A. 84 F6 39 48N 120 10W
Vinton, Iowa, U.S.A. 80 D8 42 10N 92 1W
Vinton, La., U.S.A. 81 K8 30 11N 93 35W
Virac, Phil. 37 B6 13 30N 124 20 E
Virachei, Cambodia 38 F6 13 59N 106 49 E
Virago Sd., Canada 72 C2 54 0N 132 30W
Viramgam, India 42 H5 23 5N 72 0 E
Virden, Canada 73 D8 49 50N 100 56W
Vire, France 18 B3 48 50N 0 53W
Virgenes, C., Argentina 96 G3 52 19S 68 21W
Virgin →, Canada 73 B7 57 2N 108 17W
Virgin →, U.S.A. 83 H6 36 28N 114 21W
Virgin Gorda, Virgin Is. 89 C7 18 30N 64 26W
Virgin Is. (British) ■, W. Indies 89 C7 18 30N 64 30W
Virgin Is. (U.S.) ■, W. Indies 89 C7 18 20N 65 0W
Virginia, S. Africa 56 D4 28 8S 26 55 E
Virginia, U.S.A. 80 B8 47 31N 92 32W
Virginia □, U.S.A. 76 G7 37 30N 78 45W
Virginia Beach, U.S.A. 76 G8 36 51N 75 59W
Virginia City, Mont., U.S.A. 82 D8 45 18N 111 56W
Virginia City, Nev., U.S.A. 84 F7 39 19N 119 39W
Virginia Falls, Canada 72 A3 61 38N 125 42W
Virginiatown, Canada 70 C4 48 9N 79 36W
Viroqua, U.S.A. 80 D9 43 34N 90 53W
Virovitica, Croatia 20 B7 45 51N 17 21 E
Virton, Belgium 15 E5 49 35N 5 32 E
Virudunagar, India 40 Q10 9 30N 77 58 E
Vis, Croatia 20 C7 43 4N 16 10 E
Visalia, U.S.A. 83 H4 36 20N 119 18W
Visayan Sea, Phil. 37 B6 11 30N 123 30 E
Visby, Sweden 9 H18 57 37N 18 18 E
Viscount Melville Sd., Canada 4 B2 74 10N 108 0W
Visé, Belgium 15 D5 50 44N 5 41 E
Višegrad, Bos.-H. 21 C8 43 47N 19 17 E
Viseu, Brazil 93 D9 1 10S 46 5W
Viseu, Portugal 19 B2 40 40N 7 55W
Vishakhapatnam, India 41 L13 17 45N 83 20 E
Visnagar, India 42 H5 23 45N 72 32 E
Viso, Mte., Italy 20 B2 44 38N 7 5 E
Visokoi I., Antarctica 5 B1 56 43S 27 15W
Vista, U.S.A. 85 M9 33 12N 117 14W
Vistula = Wisła →, Poland 17 A10 54 22N 18 55 E
Vitebsk = Vitsyebsk, Belarus 24 C5 55 10N 30 15 E
Viterbo, Italy 20 C5 42 25N 12 6 E
Viti Levu, Fiji 59 C7 17 30S 177 30 E
Vitigudino, Spain 19 B2 41 1N 6 26W
Vitim, Russia 27 D12 59 28N 112 35 E
Vitim →, Russia 27 D12 59 26N 112 34 E
Vitória, Brazil 93 H10 20 20S 40 22W
Vitoria, Spain 19 A4 42 50N 2 41W
Vitória da Conquista, Brazil 93 F10 14 51S 40 51W
Vitsyebsk, Belarus 24 C5 55 10N 30 15 E
Vittória, Italy 20 F6 36 57N 14 32 E
Vittório Véneto, Italy 20 B5 45 59N 12 18 E
Vivero, Spain 19 A2 43 39N 7 38W
Vizcaíno, Desierto de, Mexico 86 B2 27 40N 113 50W
Vizcaíno, Sierra, Mexico 86 B2 27 30N 114 0W
Vize, Turkey 21 D12 41 34N 27 45 E
Vizianagaram, India 41 K13 18 6N 83 30 E
Vjosa →, Albania 21 D8 40 37N 19 24 E
Vlaardingen, Neths. 15 C4 51 55N 4 21 E
Vladikavkaz, Russia 25 F7 43 0N 44 35 E
Vladimir, Russia 24 C7 56 15N 40 30 E
Vladimir Volynskiy = Volodymyr-Volynskyy, Ukraine 17 C13 50 50N 24 18 E
Vladivostok, Russia 27 E14 43 10N 131 53 E
Vlieland, Neths. 15 A4 53 16N 4 55 E
Vlissingen, Neths. 15 C3 51 26N 3 34 E
Vlóra, Albania 21 D8 40 32N 19 28 E
Vltava →, Czech. 16 D8 50 21N 14 30 E
Vo Dat, Vietnam 39 G6 11 9N 107 31 E
Vogelkop = Doberai, Jazirah, Indonesia 37 E8 1 25S 133 0 E
Vogelsberg, Germany 16 C5 50 31N 9 12 E
Voghera, Italy 20 B3 44 59N 9 1 E
Vohibinany, Madag. 57 B8 18 49S 49 4 E
Vohimarina, Madag. 57 A9 13 25S 50 0 E
Vohimena, Tanjon' i, Madag. 57 D8 25 36S 45 8 E
Vohipeno, Madag. 57 C8 22 22S 47 51 E
Voi, Kenya 54 C4 3 25S 38 32 E
Voiron, France 18 D6 45 22N 5 35 E
Voisey B., Canada 71 A7 56 15N 61 50W
Vojmsjön, Sweden 8 D17 64 55N 16 40 E
Vojvodina □, Serbia, Yug. 21 B9 45 20N 20 0 E
Volborg, U.S.A. 80 C2 45 51N 105 41W
Volcano Is. = Kazan-Rettō, Pac. Oc. 64 E6 25 0N 141 0 E
Volchayevka, Russia 27 E14 48 40N 134 30 E
Volda, Norway 9 E12 62 9N 6 5 E
Volga →, Russia 25 E8 46 0N 48 30 E
Volga Hts. = Privolzhskaya Vozvyshennost, Russia 25 D8 51 0N 46 0 E
Volgodonsk, Russia 25 E7 47 33N 42 5 E
Volgograd, Russia 25 E7 48 40N 44 25 E
Volgogradskoye Vdkhr., Russia 25 D8 50 0N 45 20 E
Volkhov →, Russia 24 B5 60 8N 32 20 E
Volkovysk = Vawkavysk, Belarus 17 B13 53 9N 24 30 E
Volksrust, S. Africa 57 D4 27 24S 29 53 E
Vollenhove, Neths. 15 B5 52 40N 5 58 E
Volochanka, Russia 27 B10 71 0N 94 28 E
Volodymyr-Volynskyy, Ukraine 17 C13 50 50N 24 18 E
Vologda, Russia 24 C6 59 10N 39 45 E
Vólos, Greece 21 E10 39 24N 22 59 E
Volovets, Ukraine 17 D12 48 43N 23 11 E
Volozhin = Valozhyn, Belarus 17 A14 54 3N 26 30 E
Volsk, Russia 24 D8 52 5N 47 22 E
Volta →, Ghana 50 5 46N 0 41 E
Volta, L., Ghana 50 G5 7 30N 0 15 E
Volta Redonda, Brazil 95 A7 22 31S 44 5W
Voltaire, C., Australia 60 B4 14 16S 125 35 E
Volterra, Italy 20 C4 43 24N 10 51 E
Volturno →, Italy 20 D5 41 1N 13 55 E
Volvo, Australia 63 E3 31 41S 143 57 E
Volzhskiy, Russia 25 E7 48 56N 44 46 E
Vondrozo, Madag. 57 C8 22 49S 47 20 E
Voorburg, Neths. 15 B4 52 5N 4 24 E
Vopnafjörður, Iceland 8 D6 65 45N 14 50W
Voríai Sporádhes, Greece 21 E10 39 15N 23 30 E
Vorkuta, Russia 24 A11 67 48N 64 20 E
Vormsi, Estonia 9 G20 59 1N 23 13 E
Voronezh, Russia 24 D6 51 40N 39 10 E
Voroshilovgrad = Luhansk, Ukraine 25 E6 48 38N 39 15 E
Voroshilovsk = Alchevsk, Ukraine 25 E6 48 30N 38 45 E
Vorovskoye, Russia 27 D16 54 30N 155 50 E
Võrts Järv, Estonia 9 G22 58 16N 26 3 E
Võru, Estonia 9 H22 57 48N 26 54 E
Vosges, France 18 B7 48 20N 7 10 E
Voss, Norway 9 F12 60 38N 6 26 E
Vostok I., Kiribati 65 J12 10 5S 152 23W
Votkinsk, Russia 24 C9 57 0N 53 55 E
Votkinskoye Vdkhr., Russia 24 C10 57 22N 55 12 E
Vouga →, Portugal 19 B1 40 41N 8 40W
Voúxa, Ákra, Greece 23 D5 35 37N 23 32 E
Vozhe Ozero, Russia 24 B6 60 45N 39 0 E
Voznesenka, Russia 27 D10 56 40N 95 3 E
Voznesensk, Ukraine 25 E5 47 35N 31 21 E
Voznesenye, Russia 24 B6 61 0N 35 28 E
Vrangelya, Ostrov, Russia 27 B19 71 0N 180 0 E
Vranje, Serbia, Yug. 21 C9 42 34N 21 54 E
Vratsa, Bulgaria 21 C10 43 13N 23 30 E
Vrbas →, Bos.-H. 20 B7 45 8N 17 29 E
Vrede, S. Africa 57 D4 27 24S 29 6 E
Vredefort, S. Africa 56 D4 27 0S 27 22 E
Vredenburg, S. Africa 56 E2 32 56S 18 0 E
Vredendal, S. Africa 56 E2 31 41S 18 35 E
Vrindavan, India 42 F7 27 37N 77 40 E
Vrises, Greece 23 D6 35 23N 24 13 E
Vršac, Serbia, Yug. 21 B9 45 8N 21 18 E
Vryburg, S. Africa 56 D3 26 55S 24 45 E
Vryheid, S. Africa 57 D5 27 45S 30 47 E
Vu Liet, Vietnam 38 C5 18 43N 105 23 E
Vught, Neths. 15 C5 51 38N 5 20 E
Vukovar, Croatia 21 B8 45 21N 18 59 E
Vulcan, Canada 72 C6 50 25N 113 15W
Vulcan, Romania 17 F12 45 23N 23 17 E
Vulcan, U.S.A. 76 C2 45 47N 87 53W
Vulcaneşti, Moldova 17 F15 45 35N 28 30 E
Vulcano, Italy 20 E6 38 24N 14 58 E
Vulkaneshty = Vulcaneşti, Moldova 17 F15 45 35N 28 30 E
Vunduzi →, Mozam. 55 F3 18 56S 34 1 E
Vung Tau, Vietnam 39 G6 10 21N 107 4 E
Vyatka = Kirov, Russia 24 C8 58 35N 49 40 E
Vyatka →, Russia 24 C9 55 37N 51 28 E
Vyatskiye Polyany, Russia 24 C9 56 14N 51 5 E
Vyazemskiy, Russia 27 E14 47 32N 134 45 E
Vyazma, Russia 24 C5 55 10N 34 15 E
Vyborg, Russia 24 B4 60 43N 28 47 E
Vychegda →, Russia 24 B8 61 18N 46 36 E
Vychodné Beskydy, Europe 17 D11 49 20N 22 0 E
Vyg-ozero, Russia 24 B5 63 47N 34 29 E
Vylkove, Ukraine 17 F15 45 28N 29 32 E
Vynohradiv, Ukraine 17 D12 48 9N 23 2 E
Vyrnwy, L., U.K. 10 E4 52 48N 3 31W
Vyshniy Volochek, Russia 24 C5 57 30N 34 30 E
Vyshzha = imeni 26 Bakinskikh Komissarov, Turkmenistan 25 G9 39 22N 54 10 E
Vyškov, Czech. 17 D9 49 17N 17 0 E
Vytegra, Russia 24 B6 61 0N 36 27 E

W

W.A.C. Bennett Dam, Canada 72 B4 56 2N 122 6W
Wa, Ghana 50 F4 10 7N 2 25W
Waal →, Neths. 15 C5 51 37N 5 0 E
Wabakimi L., Canada 70 B2 50 38N 89 45W
Wabana, Canada 71 C9 47 40N 53 0W
Wabasca, Canada 72 B6 55 57N 113 56W
Wabash, U.S.A. 76 E3 40 48N 85 49W
Wabash →, U.S.A. 76 G1 37 48N 88 2W
Wabeno, U.S.A. 76 C1 45 26N 88 39W
Wabigoon L., Canada 73 D10 49 44N 92 44W
Wabowden, Canada 73 C9 54 55N 98 38W
Wabuk Pt., Canada 70 A2 55 20N 85 5W
Wabush, Canada 71 B6 52 55N 66 52W
Wabuska, U.S.A. 82 G4 39 9N 119 11W
Waco, U.S.A. 81 K6 31 33N 97 9W
Waconichi, L., Canada 70 B5 50 8N 74 0W
Wad Banda, Sudan 51 F10 13 10N 27 56 E
Wad Hamid, Sudan 51 E11 16 30N 32 45 E
Wâd Medanî, Sudan 51 F11 14 28N 33 30 E
Wad Thana, Pakistan 42 F2 27 22N 66 23 E
Wadai, Africa 48 E5 12 0N 19 0 E
Wadayama, Japan 31 G7 35 19N 134 52 E
Waddeneilanden, Neths. 15 A5 53 25N 5 10 E
Waddenzee, Neths. 15 A5 53 6N 5 10 E
Wadderin Hill, Australia 61 F2 32 0S 118 25 E
Waddington, U.S.A. 79 B9 44 52N 75 12W
Waddington, Mt., Canada 72 C3 51 23N 125 15W
Waddy Pt., Australia 63 C5 24 58S 153 21 E
Wadena, Canada 73 C8 51 57N 103 47W
Wadena, U.S.A. 80 B7 46 26N 95 8W
Wadesboro, U.S.A. 77 H5 34 58N 80 5W
Wadhams, Canada 72 C3 51 30N 127 30W
Wādī as Sīr, Jordan 47 D4 31 56N 35 49 E
Wadi Halfa, Sudan 51 D11 21 53N 31 19 E
Wadsworth, U.S.A. 82 G4 39 38N 119 17W
Waegwan, S. Korea 35 G15 35 59N 128 23 E
Wafrah, Si. Arabia 44 D5 28 33N 47 56 E
Wageningen, Neths. 15 C5 51 58N 5 40 E
Wager B., Canada 69 B11 65 26N 88 40W
Wager Bay, Canada 69 B10 65 56N 90 49W
Wagga Wagga, Australia 63 F4 35 7S 147 24 E
Waghete, Indonesia 37 E9 4 10S 135 50 E
Wagin, Australia 61 F2 33 17S 117 25 E
Wagon Mound, U.S.A. 81 G2 36 1N 104 42W
Wagoner, U.S.A. 81 G7 35 58N 95 22W
Wah, Pakistan 42 C5 33 45N 72 40 E
Wahai, Indonesia 37 E7 2 48S 129 35 E
Wahiawa, U.S.A. 74 H15 21 30N 158 2W
Wâḥid, Egypt 47 E1 30 48N 32 21 E
Wahnai, Afghan. 42 C1 32 40N 65 50 E
Wahoo, U.S.A. 80 E6 41 13N 96 37W
Wahpeton, U.S.A. 80 B6 46 16N 96 36W
Wai, Koh, Cambodia 39 H4 9 55N 102 55 E
Waiau →, N.Z. 59 K4 42 47S 173 22 E
Waibeem, Indonesia 37 E8 0 30S 132 59 E
Waigeo, Indonesia 37 E8 0 20S 130 40 E
Waihi, N.Z. 59 G5 37 23S 175 52 E
Waihou →, N.Z. 59 G5 37 15S 175 40 E
Waika, Zaïre 54 C2 2 22S 25 42 E
Waikabubak, Indonesia 37 F5 9 45S 119 25 E
Waikari, N.Z. 59 K4 42 58S 172 41 E
Waikato →, N.Z. 59 G5 37 23S 174 43 E
Waikerie, Australia 63 E2 34 9S 140 0 E
Waikokopu, N.Z. 59 H6 39 3S 177 52 E
Waikouaiti, N.Z. 59 L3 45 36S 170 41 E
Waimakariri →, N.Z. 59 K4 43 24S 172 42 E
Waimate, N.Z. 59 L3 44 45S 171 3 E
Wainganga →, India 40 K11 18 50N 79 55 E
Waingapu, Indonesia 37 F6 9 35S 120 11 E
Wainwright, Canada 73 C6 52 50N 110 50W
Wainwright, U.S.A. 68 A3 70 38N 160 2W
Waiouru, N.Z. 59 H5 39 28S 175 41 E
Waipara, N.Z. 59 K4 43 3S 172 46 E
Waipawa, N.Z. 59 H6 39 56S 176 38 E
Waipiro, N.Z. 59 H7 38 2S 178 22 E
Waipu, N.Z. 59 F5 35 59S 174 29 E
Waipukurau, N.Z. 59 J6 40 1S 176 33 E
Wairakei, N.Z. 59 H6 38 37S 176 6 E
Wairarapa, L., N.Z. 59 J5 41 14S 175 15 E
Wairoa, N.Z. 59 H6 39 3S 177 25 E
Waitaki →, N.Z. 59 L3 44 56S 171 7 E
Waitara, N.Z. 59 H5 38 59S 174 15 E
Waitsburg, U.S.A. 82 C5 46 16N 118 9W
Waiuku, N.Z. 59 G5 37 15S 174 45 E
Wajima, Japan 31 F8 37 30N 137 0 E
Wajir, Kenya 54 B5 1 42N 40 5 E
Wajir □, Kenya 54 B5 1 42N 40 20 E
Wakasa, Japan 31 G7 35 20N 134 24 E
Wakasa-Wan, Japan 31 G7 35 40N 135 30 E
Wakatipu, L., N.Z. 59 L2 45 5S 168 33 E
Wakaw, Canada 73 C7 52 39N 105 44W
Wakayama, Japan 31 G7 34 15N 135 15 E
Wakayama-ken □, Japan 31 H7 33 50N 135 30 E
Wake Forest, U.S.A. 77 H6 35 59N 78 30W
Wake I., Pac. Oc. 64 F8 19 18N 166 36 E
Wakefield, N.Z. 59 J4 41 24S 173 5 E
Wakefield, U.K. 10 D6 53 41N 1 29W
Wakefield, Mass., U.S.A. 79 D13 42 30N 71 4W
Wakefield, Mich., U.S.A. 80 B10 46 29N 89 56W
Wakeham Bay = Maricourt, Canada 69 C12 56 34N 70 49W
Wakema, Burma 41 L19 16 30N 95 11 E
Wakkanai, Japan 30 B10 45 28N 141 35 E
Wakkerstroom, S. Africa 57 D5 27 24S 30 10 E
Wakool, Australia 63 F3 35 28S 144 23 E
Wakool →, Australia 63 F3 35 5S 143 33 E
Wakre, Indonesia 37 E8 0 19S 131 5 E
Wakuach L., Canada 71 A6 55 34N 67 32W
Walamba, Zambia 55 E2 13 30S 28 42 E
Wałbrzych, Poland 16 C9 50 45N 16 18 E
Walbury Hill, U.K. 11 F6 51 22N 1 28W
Walcha, Australia 63 E5 30 55S 151 31 E
Walcheren, Neths. 15 C3 51 30N 3 35 E
Walcott, U.S.A. 82 F10 41 46N 106 51W
Wałcz, Poland 16 B9 53 17N 16 27 E
Waldburg Ra., Australia 60 D2 24 40S 117 35 E
Walden, Colo., U.S.A. 82 F10 40 44N 106 17W
Walden, N.Y., U.S.A. 79 E10 41 34N 74 11W
Waldport, U.S.A. 82 D1 44 26N 124 4W
Waldron, U.S.A. 81 H7 34 54N 94 5W
Wales □, U.K. 11 E4 52 19N 4 43W
Walgett, Australia 63 E4 30 0S 148 5 E
Walgreen Coast, Antarctica 5 D15 75 15S 105 0W
Walhalla, Australia 63 F4 37 56S 146 29 E
Walhalla, U.S.A. 73 D9 48 55N 97 55W
Walker, U.S.A. 80 B7 47 6N 94 35W
Walker L., Man., Canada 73 C9 54 42N 95 57W
Walker L., Qué., Canada 71 B6 50 20N 67 11W
Walker L., U.S.A. 82 G4 38 42N 118 43W
Walkerston, Australia 62 C4 21 11S 149 8 E
Walkerton, Canada 78 B3 44 10N 81 10W
Wall, U.S.A. 80 C3 44 0N 102 8W
Walla Walla, U.S.A. 82 C4 46 4N 118 20W
Wallabadah, Australia 62 B3 17 57S 142 15 E
Wallace, Idaho, U.S.A. 82 C6 47 28N 115 56W
Wallace, N.C., U.S.A. 77 H7 34 44N 77 59W
Wallace, Nebr., U.S.A. 80 E4 40 50N 101 10W
Wallaceburg, Canada 70 D3 42 34N 82 23W
Wallachia = Valahia, Romania 17 F13 44 35N 25 0 E
Wallal, Australia 63 D4 26 32S 146 7 E
Wallal Downs, Australia 60 C3 19 47S 120 40 E
Wallambin, L., Australia 61 F2 30 57S 117 35 E
Wallaroo, Australia 63 E2 33 56S 137 39 E
Wallasey, U.K. 10 D4 53 25N 3 2W
Wallerawang, Australia 63 E5 33 25S 150 4 E
Wallhallow, Australia 62 B2 17 50S 135 50 E

Wallingford, U.S.A. **79 E12** 41 27N 72 50W
Wallis & Futuna, Is., Pac. Oc. **64 J10** 13 18S 176 10W
Wallowa, U.S.A. **82 D5** 45 34N 117 32W
Wallowa Mts., U.S.A. **82 D5** 45 20N 117 30W
Wallsend, Australia **63 E5** 32 55S 151 40 E
Wallsend, U.K. **10 C6** 54 59N 1 31W
Wallula, U.S.A. **82 C4** 46 5N 118 54W
Wallumbilla, Australia **63 D4** 26 33S 149 9 E
Walmsley, L., Canada **73 A7** 63 25N 108 36W
Walney, I. of, U.K. **10 C4** 54 6N 3 15W
Walnut Creek, U.S.A. **84 H4** 37 54N 122 4W
Walnut Ridge, U.S.A. **81 G9** 36 4N 90 57W
Walsall, U.K. **11 E6** 52 35N 1 58W
Walsenburg, U.S.A. **81 G2** 37 38N 104 47W
Walsh, U.S.A. **81 G3** 37 23N 102 17W
Walsh →, Australia **62 B3** 16 31S 143 42 E
Walsh P.O., Australia **62 B3** 16 40S 144 0 E
Walterboro, U.S.A. **77 J5** 32 55N 80 40W
Walters, U.S.A. **81 H5** 34 22N 98 19W
Waltham, U.S.A. **79 D13** 42 23N 71 14W
Waltham Station, Canada **70 C4** 45 57N 76 57W
Waltman, U.S.A. **82 E10** 43 4N 107 12W
Walton, U.S.A. **79 D9** 42 10N 75 8W
Walvisbaai, Namibia **56 C1** 23 0S 14 28 E
Wamba, Kenya **54 B4** 0 58N 37 19 E
Wamba, Zaïre **54 B2** 2 10N 27 57 E
Wamego, U.S.A. **80 F6** 39 12N 96 18W
Wamena, Indonesia **37 E9** 4 4S 138 57 E
Wamulan, Indonesia **37 E7** 3 27S 126 7 E
Wan Xian, China **34 E8** 38 47N 115 7 E
Wana, Pakistan **42 C3** 32 20N 69 32 E
Wanaaring, Australia **63 D3** 29 38S 144 9 E
Wanaka, N.Z. **59 L2** 44 42S 169 9 E
Wanaka L., N.Z. **59 L2** 44 33S 169 7 E
Wanapiri, Indonesia **37 E9** 4 30S 135 59 E
Wanapitei L., Canada **70 C3** 46 45N 80 40W
Wanbi, Australia **63 E3** 34 46S 140 17 E
Wandarrie, Australia **61 E2** 27 50S 117 52 E
Wandel Sea = McKinley Sea, Arctic **4 A7** 82 0N 0 0 E
Wanderer, Zimbabwe **55 F3** 19 36S 30 1 E
Wandoan, Australia **63 D4** 26 5S 149 55 E
Wanfu, China **35 D12** 40 8N 122 38 E
Wang →, Thailand **38 D2** 17 8N 99 2 E
Wang Noi, Thailand **38 E3** 14 13N 100 44 E
Wang Saphung, Thailand **38 D3** 17 18N 101 46 E
Wang Thong, Thailand **38 D3** 16 50N 100 26 E
Wanga, Zaïre **54 B2** 2 58N 29 12 E
Wangal, Indonesia **37 F8** 6 8S 134 9 E
Wanganella, Australia **63 F3** 35 6S 144 49 E
Wanganui, N.Z. **59 H5** 39 56S 175 3 E
Wangaratta, Australia **63 F4** 36 21S 146 19 E
Wangary, Australia **63 E2** 34 35S 135 29 E
Wangdu, China **34 E8** 38 40N 115 7 E
Wangerooge, Germany **16 B4** 53 47N 7 54 E
Wangi, Kenya **54 C5** 1 58S 40 58 E
Wangiwangi, Indonesia **37 F6** 5 22S 123 37 E
Wangqing, China **35 C15** 43 12N 129 42 E
Wankaner, India **42 H4** 22 35N 71 0 E
Wanless, Canada **73 C8** 54 11N 101 21W
Wanon Niwat, Thailand **38 D4** 17 38N 103 46 E
Wanquan, China **34 D8** 40 50N 114 40 E
Wanrong, China **34 G6** 35 25N 110 50 E
Wanxian, China **33 C5** 30 42N 108 20 E
Wapakoneta, U.S.A. **76 E3** 40 34N 84 12W
Wapato, U.S.A. **82 C3** 46 27N 120 25W
Wapawekka L., Canada **73 C8** 54 55N 104 40W
Wapikopa L., Canada **70 B2** 52 56N 87 53W
Wappingers Falls, U.S.A. **79 E11** 41 36N 73 55W
Wapsipinicon →, U.S.A. **80 E9** 41 44N 90 19W
Warangal, India **40 L11** 17 58N 79 35 E
Waratah, Australia **62 G4** 41 30S 145 30 E
Waratah B., Australia **63 F4** 38 54S 146 5 E
Warburton, Vic., Australia **63 F4** 37 47S 145 42 E
Warburton, W. Austral., Australia **61 E4** 26 8S 126 35 E
Warburton Ra., Australia **61 E4** 25 55S 126 28 E
Ward, N.Z. **59 J5** 41 49S 174 11 E
Ward →, Australia **63 D4** 26 28S 146 6 E
Ward Cove, U.S.A. **72 B2** 55 25N 132 43W
Ward Mt., U.S.A. **84 H8** 37 12N 118 54W
Warden, S. Africa **57 D4** 27 50S 29 0 E
Wardha, India **40 J11** 20 45N 78 39 E
Wardha →, India **40 K11** 19 57N 79 11 E
Wardlow, Canada **72 C6** 50 56N 111 31W
Ware, Canada **72 B3** 57 26N 125 41W
Ware, U.S.A. **79 D12** 42 16N 72 14W
Wareham, U.S.A. **79 E14** 41 46N 70 43W
Warialda, Australia **63 D5** 29 29S 150 33 E
Wariap, Indonesia **37 E8** 1 30S 134 5 E
Warin Chamrap, Thailand **38 E5** 15 12N 104 53 E
Warkopi, Indonesia **37 E8** 1 12S 134 9 E
Warley, U.K. **11 E6** 52 30N 1 59W
Warm Springs, U.S.A. **83 G5** 38 10N 116 20W
Warman, Canada **73 C7** 52 19N 106 30W
Warmbad, Namibia **56 D2** 28 25S 18 42 E
Warmbad, S. Africa **57 C4** 24 51S 28 19 E
Warnambool Downs, Australia **62 C3** 22 48S 142 52 E
Warner, Canada **72 D6** 49 17N 112 12W
Warner Mts., U.S.A. **82 F3** 41 40N 120 15W
Warner Robins, U.S.A. **77 J4** 32 37N 83 36W
Waroona, Australia **61 F2** 32 50S 115 58 E
Warracknabeal, Australia **63 F3** 36 9S 142 26 E
Warragul, Australia **63 F4** 38 10S 145 58 E
Warrawagine, Australia **60 D3** 20 51S 120 42 E
Warrego →, Australia **63 E4** 30 24S 145 21 E
Warrego Ra., Australia **62 C4** 24 58S 146 0 E
Warren, Australia **63 E4** 31 42S 147 51 E
Warren, Ark., U.S.A. **81 J8** 33 37N 92 4W
Warren, Mich., U.S.A. **76 D4** 42 30N 83 0W
Warren, Minn., U.S.A. **80 A6** 48 12N 96 46W
Warren, Ohio, U.S.A. **78 E4** 41 14N 80 49W
Warren, Pa., U.S.A. **78 E5** 41 51N 79 9W
Warrenpoint, U.K. **13 B5** 54 6N 6 15W
Warrensburg, U.S.A. **80 F8** 38 46N 93 44W
Warrenton, S. Africa **56 D3** 28 9S 24 47 E
Warrenton, U.S.A. **84 D3** 46 10N 123 56W
Warrenville, Australia **63 D4** 25 48S 147 22 E
Warri, Nigeria **50 G6** 5 30N 5 41 E
Warrina, Australia **63 D2** 28 12S 135 50 E
Warrington, U.K. **10 D5** 53 24N 2 35W
Warrington, U.S.A. **77 K2** 30 23N 87 17W
Warrnambool, Australia **63 F3** 38 25S 142 30 E
Warroad, U.S.A. **80 A7** 48 54N 95 19W
Warsa, Indonesia **37 E9** 0 47S 135 55 E
Warsaw = Warszawa, Poland **17 B11** 52 13N 21 0 E
Warsaw, Ind., U.S.A. **76 E3** 41 14N 85 51W
Warsaw, N.Y., U.S.A. **78 D6** 42 45N 78 8W
Warsaw, Ohio, U.S.A. **78 F2** 40 20N 82 0W
Warszawa, Poland **17 B11** 52 13N 21 0 E
Warta →, Poland **16 B8** 52 35N 14 39 E
Warthe = Warta →, Poland **16 B8** 52 35N 14 39 E
Waru, Indonesia **37 E8** 3 30S 130 36 E
Warwick, Australia **63 D5** 28 10S 152 1 E
Warwick, U.K. **11 E6** 52 18N 1 35W
Warwick, U.S.A. **79 E13** 41 42N 71 28W
Warwickshire □, U.K. **11 E6** 52 14N 1 38W
Wasaga Beach, Canada **78 B4** 44 31N 80 1W
Wasatch Ra., U.S.A. **82 F8** 40 30N 111 15W
Wasbank, S. Africa **57 D5** 28 15S 30 9 E
Wasco, Calif., U.S.A. **85 K7** 35 36N 119 20W
Wasco, Oreg., U.S.A. **82 D3** 45 36N 120 42W
Waseca, U.S.A. **80 C8** 44 5N 93 30W
Wasekamio L., Canada **73 B7** 56 45N 108 45W
Wash, The, U.K. **10 E8** 52 58N 0 20 E
Washago, Canada **78 B5** 44 45N 79 20W
Washburn, N. Dak., U.S.A. **80 B4** 47 17N 101 2W
Washburn, Wis., U.S.A. **80 B9** 46 40N 90 54W
Washim, India **40 J10** 20 3N 77 0 E
Washington, D.C., U.S.A. **76 F7** 38 54N 77 2W
Washington, Ga., U.S.A. **77 J4** 33 44N 82 44W
Washington, Ind., U.S.A. **76 F2** 38 40N 87 10W
Washington, Iowa, U.S.A. **80 E9** 41 18N 91 42W
Washington, Mo., U.S.A. **80 F9** 38 33N 91 1W
Washington, N.C., U.S.A. **77 H7** 35 33N 77 3W
Washington, N.J., U.S.A. **79 F10** 40 46N 74 59W
Washington, Pa., U.S.A. **78 F4** 40 10N 80 15W
Washington, Utah, U.S.A. **83 H7** 37 8N 113 31W
Washington □, U.S.A. **82 C3** 47 30N 120 30W
Washington, Mt., U.S.A. **79 B13** 44 16N 71 18W
Washington I., U.S.A. **76 C2** 45 23N 86 54W
Washougal, U.S.A. **84 E4** 45 35N 122 21W
Wasian, Indonesia **37 E8** 1 47S 133 19 E
Wasior, Indonesia **37 E8** 2 43S 134 30 E
Waskaganish, Canada **70 B4** 51 30N 78 40W
Waskaiowaka, L., Canada **73 B9** 56 33N 96 23W
Waskesiu Lake, Canada **73 C7** 53 55N 106 5W
Wassenaar, Neths. **15 B4** 52 8N 4 24 E
Wasserkuppe, Germany **16 C5** 50 29N 9 55 E
Waswanipi, Canada **70 C4** 49 40N 76 29W
Waswanipi, L., Canada **70 C4** 49 35N 76 40W
Watangpone, Indonesia **37 E6** 4 29S 120 25 E
Water Park Pt., Australia **62 C5** 22 56S 150 47 E
Water Valley, U.S.A. **81 H10** 34 10N 89 38W
Waterberge, S. Africa **57 C4** 24 10S 28 0 E
Waterbury, Conn., U.S.A. **79 E11** 41 33N 73 3W
Waterbury, Vt., U.S.A. **79 B12** 44 20N 72 46W
Waterbury L., Canada **73 B8** 58 10N 104 22W
Waterdown, Canada **78 C5** 43 20N 79 53W
Waterford, Canada **78 D4** 42 56N 80 17W
Waterford, Ireland **13 D4** 52 16N 7 8W
Waterford, U.S.A. **84 H6** 37 38N 120 46W
Waterford □, Ireland **13 D4** 52 10N 7 40W
Waterford Harbour, Ireland **13 D5** 52 8N 6 58W
Waterhen L., Man., Canada **73 C9** 52 10N 99 40W
Waterhen L., Sask., Canada **73 C7** 54 28N 108 25W
Waterloo, Belgium **15 D4** 50 43N 4 25 E
Waterloo, Ont., Canada **70 D3** 43 30N 80 32W
Waterloo, Qué., Canada **79 A12** 45 22N 72 32W
Waterloo, S. Leone **50 G2** 8 26N 13 8W
Waterloo, Ill., U.S.A. **80 F9** 38 20N 90 9W
Waterloo, Iowa, U.S.A. **80 D8** 42 30N 92 21W
Waterloo, N.Y., U.S.A. **78 D8** 42 54N 76 52W
Watersmeet, U.S.A. **80 B10** 46 16N 89 11W
Waterton-Glacier International Peace Park, U.S.A. **82 B7** 48 45N 115 0W
Watertown, Conn., U.S.A. **79 E11** 41 36N 73 7W
Watertown, N.Y., U.S.A. **79 C9** 43 59N 75 55W
Watertown, S. Dak., U.S.A. **80 C6** 44 54N 97 7W
Watertown, Wis., U.S.A. **80 D10** 43 12N 88 43W
Waterval-Boven, S. Africa **57 D5** 25 40S 30 18 E
Waterville, Canada **79 A13** 45 16N 71 54W
Waterville, Maine, U.S.A. **71 D6** 44 33N 69 38W
Waterville, N.Y., U.S.A. **79 D9** 42 56N 75 23W
Waterville, Pa., U.S.A. **78 E7** 41 19N 77 21W
Waterville, Wash., U.S.A. **82 C3** 47 39N 120 4W
Watervliet, U.S.A. **79 D11** 42 44N 73 42W
Wates, Indonesia **37 G14** 7 51S 110 10 E
Watford, Canada **78 D3** 42 57N 81 53W
Watford, U.K. **11 F7** 51 40N 0 24W
Watford City, U.S.A. **80 B3** 47 48N 103 17W
Wathaman →, Canada **73 B8** 57 16N 102 59W
Watheroo, Australia **61 F2** 30 15S 116 0 E
Wating, China **34 G4** 35 40N 106 38 E
Watkins Glen, U.S.A. **78 D8** 42 23N 76 52W
Watling I. = San Salvador, Bahamas **89 B5** 24 0N 74 40W
Watonga, U.S.A. **81 H5** 35 51N 98 25W
Watrous, Canada **73 C7** 51 40N 105 25W
Watrous, U.S.A. **81 H2** 35 48N 104 59W
Watsa, Zaïre **54 B2** 3 4N 29 30 E
Watseka, U.S.A. **76 E2** 40 47N 87 44W
Watson, Australia **61 F5** 30 29S 131 31 E
Watson, Canada **73 C8** 52 10N 104 30W
Watson Lake, Canada **72 A3** 60 6N 128 49W
Watsonville, U.S.A. **83 H3** 36 55N 121 45W
Wattiwarriganna Cr. →, Australia **63 D2** 28 57S 136 10 E
Watuata = Batuata, Indonesia **37 F6** 6 12S 122 42 E
Watubela, Kepulauan, Indonesia **37 E8** 4 28S 131 35 E
Watubela Is. = Watubela, Kepulauan, Indonesia **37 E8** 4 28S 131 35 E
Wau, Sudan **49 F6** 7 45N 28 1 E
Waubamik, Canada **78 A4** 45 27N 80 1W
Waubay, U.S.A. **80 C6** 45 20N 97 18W
Waubra, Australia **63 F3** 37 21S 143 39 E
Wauchope, Australia **63 E5** 31 28S 152 45 E
Wauchula, U.S.A. **77 M5** 27 33N 81 49W
Waugh, Canada **73 D9** 49 40N 95 11W
Waukarlycarly, L., Australia **60 D3** 21 18S 121 56 E
Waukegan, U.S.A. **76 D2** 42 22N 87 50W
Waukesha, U.S.A. **76 D1** 43 1N 88 14W
Waukon, U.S.A. **80 D9** 43 16N 91 29W
Wauneta, U.S.A. **80 E4** 40 25N 101 23W
Waupaca, U.S.A. **80 C10** 44 21N 89 5W
Waupun, U.S.A. **80 D10** 43 38N 88 44W
Waurika, U.S.A. **81 H6** 34 10N 98 0W
Wausau, U.S.A. **80 C10** 44 58N 89 38W
Wautoma, U.S.A. **80 C10** 44 4N 89 18W
Wauwatosa, U.S.A. **76 D2** 43 3N 88 0W
Wave Hill, Australia **60 C5** 17 32S 131 0 E
Waveney →, U.K. **11 E9** 52 35N 1 39 E
Waverley, N.Z. **59 H5** 39 46S 174 37 E
Waverly, Iowa, U.S.A. **80 D8** 42 44N 92 29W
Waverly, N.Y., U.S.A. **79 D8** 42 1N 76 32W
Wavre, Belgium **15 D4** 50 43N 4 38 E
Wâw, Sudan **51 G10** 7 45N 28 1 E
Wâw al Kabīr, Libya **51 C8** 25 20N 16 43 E
Wawa, Canada **70 C3** 47 59N 84 47W
Wawanesa, Canada **73 D9** 49 36N 99 40W
Wawona, U.S.A. **84 H7** 37 32N 119 39W
Waxahachie, U.S.A. **81 J6** 32 24N 96 51W
Way, L., Australia **61 E3** 26 45S 120 16 E
Wayabula Rau, Indonesia **37 D7** 2 29N 128 17 E
Wayatinah, Australia **62 G4** 42 19S 146 27 E
Waycross, U.S.A. **77 K4** 31 13N 82 21W
Wayne, Nebr., U.S.A. **80 D6** 42 14N 97 1W
Wayne, W. Va., U.S.A. **76 F4** 38 13N 82 27W
Waynesboro, Ga., U.S.A. **77 J4** 33 6N 82 1W
Waynesboro, Miss., U.S.A. **77 K1** 31 40N 88 39W
Waynesboro, Pa., U.S.A. **76 F7** 39 45N 77 35W
Waynesboro, Va., U.S.A. **76 F6** 38 4N 78 53W
Waynesburg, U.S.A. **76 F5** 39 54N 80 11W
Waynesville, U.S.A. **77 H4** 35 28N 82 58W
Waynoka, U.S.A. **81 G5** 36 35N 98 53W
Wazirabad, Pakistan **42 C6** 32 30N 74 8 E
We, Indonesia **36 C1** 5 51N 95 18 E
Weald, The, U.K. **11 F8** 51 4N 0 20 E
Wear →, U.K. **10 C6** 54 55N 1 23W
Weatherford, Okla., U.S.A. **81 H5** 35 32N 98 43W
Weatherford, Tex., U.S.A. **81 J6** 32 46N 97 48W
Weaverville, U.S.A. **82 F2** 40 44N 122 56W
Webb City, U.S.A. **81 G7** 37 9N 94 28W
Webster, Mass., U.S.A. **79 D13** 42 3N 71 53W
Webster, N.Y., U.S.A. **78 C7** 43 13N 77 26W
Webster, S. Dak., U.S.A. **80 C6** 45 20N 97 31W
Webster, Wis., U.S.A. **80 C8** 45 53N 92 22W
Webster City, U.S.A. **80 D8** 42 28N 93 49W
Webster Green, U.S.A. **80 F9** 38 38N 90 20W
Webster Springs, U.S.A. **76 F5** 38 29N 80 25W
Weda, Indonesia **37 D7** 0 21N 127 50 E
Weda, Teluk, Indonesia **37 D7** 0 30N 127 50 E
Weddell I., Falk. Is. **96 G4** 51 50S 61 0W
Weddell Sea, Antarctica **5 D1** 72 30S 40 0W
Wedderburn, Australia **63 F3** 36 26S 143 33 E
Wedgeport, Canada **71 D6** 43 44N 65 59W
Wedza, Zimbabwe **55 F3** 18 40S 31 33 E
Wee Waa, Australia **63 E4** 30 11S 149 26 E
Weed, U.S.A. **82 F2** 41 25N 122 23W
Weed Heights, U.S.A. **84 G7** 38 59N 119 13W
Weedsport, U.S.A. **79 C8** 43 3N 76 35W
Weedville, U.S.A. **78 E6** 41 17N 78 30W
Weemelah, Australia **63 D4** 29 2S 149 15 E
Weenen, S. Africa **57 D5** 28 48S 30 7 E
Weert, Neths. **15 C5** 51 15N 5 43 E
Wei He →, Hebei, China **34 F8** 36 10N 115 45 E
Wei He →, Shaanxi, China **34 G6** 34 38N 110 15 E
Weichang, China **35 D9** 41 58N 117 49 E
Weichuan, China **34 G7** 34 20N 113 59 E
Weiden, Germany **16 D7** 49 41N 12 10 E
Weifang, China **35 F10** 36 44N 119 7 E
Weihai, China **35 F12** 37 30N 122 6 E
Weimar, Germany **16 C6** 50 58N 11 19 E
Weinan, China **34 G5** 34 31N 109 29 E
Weipa, Australia **62 A3** 12 40S 141 50 E
Weir →, Australia **63 D4** 28 20S 149 50 E
Weir →, Canada **73 B10** 56 54N 93 21W
Weir River, Canada **73 B10** 56 49N 94 6W
Weirton, U.S.A. **78 F4** 40 24N 80 35W
Weiser, U.S.A. **82 D5** 44 10N 117 0W
Weishan, China **35 G9** 34 47N 117 5 E
Weiyuan, China **34 G3** 35 7N 104 10 E
Wejherowo, Poland **17 A10** 54 35N 18 12 E
Wekusko L., Canada **73 C9** 54 40N 99 50W
Welbourn Hill, Australia **63 D1** 27 21S 134 6 E
Welch, U.S.A. **76 G5** 37 26N 81 35W
Welkom, S. Africa **56 D4** 28 0S 26 46 E
Welland, Canada **70 D4** 43 0N 79 15W
Welland →, U.K. **10 E7** 52 51N 0 5W
Wellesley Is., Australia **62 B2** 16 42S 139 30 E
Wellin, Belgium **15 D5** 50 5N 5 6 E
Wellingborough, U.K. **11 E7** 52 19N 0 41W
Wellington, Australia **63 E4** 32 35S 148 59 E
Wellington, Canada **70 D4** 43 57N 77 20W
Wellington, N.Z. **59 J5** 41 19S 174 46 E
Wellington, S. Africa **56 E2** 33 38S 19 1 E
Wellington, Shrops., U.K. **10 E5** 52 42N 2 30W
Wellington, Somst., U.K. **11 G4** 50 58N 3 13W
Wellington, Colo., U.S.A. **80 E2** 40 42N 105 0W
Wellington, Kans., U.S.A. **81 G6** 37 16N 97 24W
Wellington, Nev., U.S.A. **84 G7** 38 45N 119 23W
Wellington, Ohio, U.S.A. **78 E2** 41 10N 82 13W
Wellington, Tex., U.S.A. **81 H4** 34 51N 100 13W
Wellington, I., Chile **96 F1** 49 30S 75 0W
Wellington, L., Australia **63 F4** 38 6S 147 20 E
Wells, U.K. **11 F5** 51 13N 2 39W
Wells, Maine, U.S.A. **79 C14** 43 20N 70 35W
Wells, Minn., U.S.A. **80 D8** 43 45N 93 44W
Wells, Nev., U.S.A. **82 F6** 41 7N 114 58W
Wells, L., Australia **61 E3** 26 44S 123 15 E
Wells Gray Prov. Park, Canada **72 C4** 52 30N 120 15W
Wells-next-the-Sea, U.K. **10 E8** 52 57N 0 51 E
Wells River, U.S.A. **79 B12** 44 9N 72 4W
Wellsboro, U.S.A. **78 E7** 41 45N 77 18W
Wellsburg, U.S.A. **78 F4** 40 16N 80 37W
Wellsville, Mo., U.S.A. **80 F9** 39 4N 91 34W
Wellsville, N.Y., U.S.A. **78 D7** 42 7N 77 57W
Wellsville, Ohio, U.S.A. **78 F4** 40 36N 80 39W
Wellsville, Utah, U.S.A. **82 F8** 41 38N 111 56W
Wellton, U.S.A. **83 K6** 32 40N 114 8W
Wels, Austria **16 D8** 48 9N 14 1 E
Welshpool, U.K. **11 E4** 52 39N 3 8W
Wem, U.K. **10 E5** 52 52N 2 44W
Wembere →, Tanzania **54 C3** 4 10S 34 15 E
Wemindji, Canada **70 B4** 53 0N 78 49W
Wen Xian, Gansu, China **34 H3** 32 43N 104 36 E
Wen Xian, Henan, China **34 G7** 34 55N 113 5 E
Wenatchee, U.S.A. **82 C3** 47 25N 120 19W
Wenchang, China **38 C8** 19 38N 110 42 E
Wenchi, Ghana **50 G4** 7 46N 2 8W
Wenchow = Wenzhou, China **33 D7** 28 0N 120 38 E
Wendell, U.S.A. **82 E6** 42 47N 114 42W
Wenden, U.S.A. **85 M13** 33 49N 113 33W
Wendeng, China **35 F12** 37 15N 122 5 E
Wendesi, Indonesia **37 E8** 2 30S 134 17 E
Wendover, U.S.A. **82 F6** 40 44N 114 2W
Wenlock →, Australia **62 A3** 12 2S 141 55 E
Wenshan, China **32 D5** 23 20N 104 18 E
Wenshang, China **34 G9** 35 45N 116 30 E
Wenshui, China **34 F7** 37 26N 112 1 E
Wensu, China **32 B3** 41 15N 80 10 E
Wentworth, Australia **63 E3** 34 2S 141 54 E
Wenut, Indonesia **37 E8** 3 11S 133 19 E
Wenxi, China **34 G6** 35 20N 111 10 E
Wenzhou, China **33 D7** 28 0N 120 38 E
Weott, U.S.A. **82 F2** 40 20N 123 55W
Wepener, S. Africa **56 D4** 29 42S 27 3 E
Werda, Botswana **56 D3** 25 24S 23 15 E
Werder, Ethiopia **46 F4** 6 58N 45 1 E
Weri, Indonesia **37 E8** 3 10S 132 38 E
Werra →, Germany **16 C5** 51 24N 9 39 E
Werribee, Australia **63 F3** 37 54S 144 40 E
Werrimull, Australia **63 E3** 34 25S 141 38 E
Werris Creek, Australia **63 E5** 31 18S 150 38 E
Wersar, Indonesia **37 E8** 1 30S 131 55 E
Weser →, Germany **16 B5** 53 36N 8 28 E
Wesiri, Indonesia **37 F7** 7 30S 126 30 E
Wesley Vale, U.S.A. **83 J10** 35 3N 106 2W
Wesleyville, Canada **71 C9** 49 8N 53 36W
Wesleyville, U.S.A. **78 D4** 42 9N 80 0W
Wessel, C., Australia **62 A2** 10 59S 136 46 E
Wessel Is., Australia **62 A2** 11 10S 136 45 E
Wessington, U.S.A. **80 C5** 44 27N 98 42W
Wessington Springs, U.S.A. **80 C5** 44 5N 98 34W
West, U.S.A. **81 K6** 31 48N 97 6W
West Allis, U.S.A. **76 D1** 43 1N 88 0W
West B., U.S.A. **81 L10** 29 3N 89 22W
West Baines →, Australia **60 C4** 15 38S 129 59 E
West Bank □, Asia **47 C4** 32 6N 35 13 E
West Bend, U.S.A. **76 D1** 43 25N 88 11W
West Bengal □, India **43 H12** 23 0N 88 0 E
West Beskids = Západné Beskydy, Europe **17 D10** 49 30N 19 0 E
West Branch, U.S.A. **76 C3** 44 17N 84 14W
West Bromwich, U.K. **11 E5** 52 32N 1 59W
West Cape Howe, Australia **61 G2** 35 8S 117 36 E
West Chazy, U.S.A. **79 B11** 44 49N 73 28W
West Chester, U.S.A. **76 F8** 39 58N 75 36W
West Columbia, U.S.A. **81 L7** 29 9N 95 39W
West Covina, U.S.A. **85 L9** 34 4N 117 54W
West Des Moines, U.S.A. **80 E8** 41 35N 93 43W
West Dunbartonshire □, U.K. **12 F4** 55 59N 4 30W
West End, Bahamas **88 A4** 26 41N 78 58W
West Falkland, Falk. Is. **96 G4** 51 40S 60 0W
West Fjord = Vestfjorden, Norway **8 C15** 67 55N 14 0 E
West Frankfort, U.S.A. **80 G10** 37 54N 88 55W
West Hartford, U.S.A. **79 E12** 41 45N 72 44W
West Haven, U.S.A. **79 E12** 41 17N 72 57W
West Helena, U.S.A. **81 H9** 34 33N 90 38W
West Ice Shelf, Antarctica **5 C7** 67 0S 85 0 E
West Indies, Cent. Amer. **89 C7** 15 0N 65 0W
West Lorne, Canada **78 D3** 42 36N 81 36W
West Lothian □, U.K. **12 F5** 55 54N 3 36W
West Lunga →, Zambia **55 E1** 13 6S 24 39 E
West Memphis, U.S.A. **81 H9** 35 9N 90 11W
West Midlands □, U.K. **11 E6** 52 26N 2 0W
West Mifflin, U.S.A. **78 F5** 40 22N 79 52W
West Monroe, U.S.A. **81 J8** 32 31N 92 9W
West Newton, U.S.A. **78 F5** 40 14N 79 46W
West Nicholson, Zimbabwe **55 G2** 21 2S 29 20 E
West Palm Beach, U.S.A. **77 M5** 26 43N 80 3W
West Plains, U.S.A. **81 G9** 36 44N 91 51W
West Point, Ga., U.S.A. **77 J3** 32 53N 85 11W
West Point, Miss., U.S.A. **77 J1** 33 36N 88 39W
West Point, Nebr., U.S.A. **80 E6** 41 51N 96 43W
West Point, Va., U.S.A. **76 G7** 37 32N 76 48W
West Pokot □, Kenya **54 B4** 1 30N 35 15 E
West Pt. = Ouest, Pte., Canada **71 C7** 49 52N 64 40W
West Pt., Australia **63 F2** 35 1S 135 56 E
West Road →, Canada **72 C4** 53 18N 122 53W
West Rutland, U.S.A. **79 C11** 43 38N 73 5W
West Schelde = Westerschelde →, Neths. **15 C3** 51 25N 3 25 E
West Seneca, U.S.A. **78 D6** 42 51N 78 48W
West Siberian Plain, Russia **28 C11** 62 0N 75 0 E
West Sussex □, U.K. **11 G7** 50 55N 0 30W
West-Terschelling, Neths. **15 A5** 53 22N 5 13 E
West Valley City, U.S.A. **82 F8** 40 42N 111 57W
West Virginia □, U.S.A. **76 F5** 38 45N 80 30W
West-Vlaanderen □, Belgium **15 D3** 51 0N 3 0 E
West Walker →, U.S.A. **84 G7** 38 54N 119 9W
West Wyalong, Australia **63 E4** 33 56S 147 10 E
West Yellowstone, U.S.A. **82 D8** 44 40N 111 6W
West Yorkshire □, U.K. **10 D6** 53 45N 1 40W
Westall Pt., Australia **63 E1** 32 55S 134 4 E
Westbrook, Maine, U.S.A. **77 D10** 43 41N 70 22W
Westbrook, Tex., U.S.A. **81 J4** 32 21N 101 1W
Westbury, Australia **62 G4** 41 30S 146 51 E
Westby, U.S.A. **80 A2** 48 52N 104 3W

Westend, *U.S.A.* **85 K9** 35 42N 117 24W
Westerland, *Germany* ... **9 J13** 54 54N 8 17 E
Western □, *Kenya* ...... **54 B3** 0 30N 34 30 E
Western □, *Uganda* ..... **54 B3** 1 45N 31 30 E
Western □, *Zambia* ..... **55 F1** 15 15S 24 30 E
Western Australia □, *Australia* .... **61 E2** 25 0S 118 0 E
Western Cape □, *S. Africa* **56 E3** 34 0S 20 0 E
Western Dvina = Daugava —►, *Latvia* .. **9 H21** 57 4N 24 3 E
Western Ghats, *India* .... **40 N9** 14 0N 75 0 E
Western Isles □, *U.K.* ... **12 D1** 57 30N 7 10W
Western Sahara ■, *Africa* **50 D2** 25 0N 13 0W
Western Samoa ■, *Pac. Oc.* **59 A13** 14 0S 172 0W
Westernport, *U.S.A.* **76 F6** 39 29N 79 3W
Westerschelde —►, *Neths.* **15 C3** 51 25N 3 25 E
Westerwald, *Germany* .... **16 C4** 50 38N 7 56 E
Westfield, *Mass., U.S.A.* . **79 D12** 42 7N 72 45W
Westfield, *N.Y., U.S.A.* .. **78 D5** 42 20N 79 35W
Westfield, *Pa., U.S.A.* ... **78 E7** 41 55N 77 32W
Westhope, *U.S.A.* **80 A4** 48 55N 101 1W
Westland Bight, *N.Z.* .... **59 K3** 42 55S 170 5 E
Westlock, *Canada* **72 C6** 54 9N 113 55W
Westmeath □, *Ireland* ... **13 C4** 53 33N 7 34W
Westminster, *U.S.A.* **76 F7** 39 34N 76 59W
Westmorland, *U.S.A.* **83 K6** 33 2N 115 37W
Weston, *Malaysia* ...... **36 C5** 5 10N 115 35 E
Weston, *Oreg., U.S.A.* ... **82 D4** 45 49N 118 26W
Weston, *W. Va., U.S.A.* .. **76 F5** 39 2N 80 28W
Weston I., *Canada* **70 B4** 52 33N 79 36W
Weston-super-Mare, *U.K.* **11 F5** 51 21N 2 58W
Westport, *Canada* **79 B8** 44 40N 76 25W
Westport, *Ireland* **13 C2** 53 48N 9 31W
Westport, *N.Z.* .... **59 J3** 41 46S 171 37 E
Westport, *Oreg., U.S.A.* .. **84 D3** 46 8N 123 23W
Westport, *Wash., U.S.A.* . **82 C1** 46 53N 124 6W
Westray, *Canada* **73 C8** 53 36N 101 24W
Westray, *U.K.* **12 B6** 59 18N 3 0W
Westree, *Canada* **70 C3** 47 26N 81 34W
Westville, *Calif., U.S.A.* .. **84 F6** 39 8N 120 42W
Westville, *Ill., U.S.A.* ... **76 E2** 40 2N 87 38W
Westville, *Okla., U.S.A.* .. **81 G7** 35 58N 94 40W
Westwood, *U.S.A.* **82 F3** 40 18N 121 0W
Wetar, *Indonesia* .... **37 F7** 7 30S 126 30 E
Wetaskiwin, *Canada* **72 C6** 52 55N 113 24W
Wethersfield, *U.S.A.* **79 E12** 41 42N 72 40W
Wetteren, *Belgium* ...... **15 D3** 51 0N 3 52 E
Wetzlar, *Germany* ...... **16 C5** 50 32N 8 31 E
Wewoka, *U.S.A.* **81 H6** 35 9N 96 30W
Wexford, *Ireland* **13 D5** 52 20N 6 28W
Wexford □, *Ireland* **13 D5** 52 20N 6 25W
Wexford Harbour, *Ireland* **13 D5** 52 20N 6 25W
Weyburn, *Canada* **73 D8** 49 40N 103 50W
Weyburn L., *Canada* **72 A5** 63 0N 117 59W
Weymouth, *Canada* **71 D6** 44 30N 66 1W
Weymouth, *U.K.* **11 G5** 50 37N 2 28W
Weymouth, *U.S.A.* **79 D14** 42 13N 70 58W
Weymouth, C., *Australia* . **62 A3** 12 37S 143 27 E
Wha Ti, *Canada* **68 B8** 63 8N 117 16W
Whakatane, *N.Z.* .... **59 G6** 37 57S 177 1 E
Whale —►, *Canada* **71 A6** 58 15N 67 40W
Whale Cove, *Canada* **73 A10** 62 11N 92 36W
Whales, B. of, *Antarctica* . **5 D12** 78 0S 165 0W
Whalsay, *U.K.* **12 A7** 60 22N 0 59W
Whangamomona, *N.Z.* .. **59 H5** 39 8S 174 44 E
Whangarei, *N.Z.* .... **59 F5** 35 43S 174 21 E
Whangarei Harb., *N.Z.* ... **59 F5** 35 45S 174 28 E
Wharfe —►, *U.K.* **10 D6** 53 51N 1 9W
Wharfedale, *U.K.* **10 C5** 54 6N 2 1W
Wharton, *N.J., U.S.A.* ... **79 F10** 40 54N 74 35W
Wharton, *Pa., U.S.A.* **78 E6** 41 31N 78 1W
Wharton, *Tex., U.S.A.* ... **81 L6** 29 19N 96 6W
Wheatland, *Calif., U.S.A.* . **84 F5** 39 1N 121 25W
Wheatland, *Wyo., U.S.A.* . **80 D2** 42 3N 104 58W
Wheatley, *Canada* **78 D2** 42 6N 82 27W
Wheaton, *U.S.A.* **80 C6** 45 48N 96 30W
Wheelbarrow Pk., *U.S.A.* . **84 H10** 37 26N 116 5W
Wheeler, *Oreg., U.S.A.* .. **82 D2** 45 41N 123 53W
Wheeler, *Tex., U.S.A.* ... **81 H4** 35 27N 100 16W
Wheeler —►, *Canada* ... **73 B7** 57 25N 105 30W
Wheeler Pk., *N. Mex., U.S.A.* **83 H11** 36 34N 105 25W
Wheeler Pk., *Nev., U.S.A.* **83 G6** 38 57N 114 15W
Wheeler Ridge, *U.S.A.* ... **85 L8** 35 0N 118 57W
Wheeling, *U.S.A.* **78 F4** 40 4N 80 43W
Whernside, *U.K.* **10 C5** 54 14N 2 24W
Whidbey I., *U.S.A.* **72 D4** 48 12N 122 17W
Whiskey Gap, *Canada* ... **72 D6** 49 0N 113 3W
Whiskey Jack L., *Canada* **73 B8** 58 23N 101 55W
Whistleduck Cr. —►, *Australia* .... **62 C2** 20 15S 135 18 E
Whitby, *Canada* **78 C6** 43 52N 78 56W
Whitby, *U.K.* **10 C7** 54 29N 0 37W
White —►, *Ark., U.S.A.* .. **81 J9** 33 57N 91 5W
White —►, *Ind., U.S.A.* .. **76 F2** 38 25N 87 45W
White —►, *S. Dak., U.S.A.* **80 D5** 43 42N 99 27W
White —►, *Utah, U.S.A.* . **82 F9** 40 4N 109 41W
White —►, *Wash., U.S.A.* **84 C4** 47 12N 122 15W
White, L., *Australia* ..... **60 D4** 21 9S 128 56 E
White B., *Canada* **71 B8** 50 0N 56 35W
White Bear Res., *Canada* **71 C8** 48 10N 57 5W
White Bird, *U.S.A.* **82 D5** 45 46N 116 18W
White Butte, *U.S.A.* **80 B3** 46 23N 103 18W
White City, *U.S.A.* **80 F6** 38 48N 96 44W
White Cliffs, *Australia* ... **63 E3** 30 50S 143 10 E
White Deer, *U.S.A.* **81 H4** 35 26N 101 10W
White Hall, *U.S.A.* **80 F9** 39 26N 90 24W
White Haven, *U.S.A.* **79 E9** 41 4N 75 47W
White Horse, Vale of, *U.K.* **11 F6** 51 37N 1 30W
White I., *N.Z.* .... **59 G6** 37 30S 177 13 E
White L., *Canada* **79 A8** 45 18N 76 31W
White L., *U.S.A.* **81 L8** 29 44N 92 30W
White Mts., *Calif., U.S.A.* **83 H4** 37 30N 118 15W
White Mts., *N.H., U.S.A.* . **75 B12** 44 15N 71 15W
White Nile = Nîl el Abyad —►, *Sudan* .... **51 E11** 15 38N 32 31 E
White Otter L., *Canada* .. **70 C1** 49 5N 91 55W
White Pass, *Canada* **72 B1** 59 40N 135 3W
White Pass, *U.S.A.* **84 D5** 46 38N 121 24W
White Plains, *U.S.A.* **79 E11** 41 2N 73 46W
White River, *Canada* **70 C2** 48 35N 85 20W
White River, *S. Africa* ... **57 D5** 25 20S 31 0 E
White River, *U.S.A.* **80 D4** 43 34N 100 45W
White Russia = Belarus ■, *Europe* .... **17 B14** 53 30N 27 0 E
White Sea = Beloye More, *Russia* .... **24 A6** 66 30N 38 0 E
White Sulphur Springs, *Mont., U.S.A.* **82 C8** 46 33N 110 54W
White Sulphur Springs, *W. Va., U.S.A.* **76 G5** 37 48N 80 18W
White Swan, *U.S.A.* **84 D6** 46 23N 120 44W
Whitecliffs, *N.Z.* .... **59 K3** 43 26S 171 55 E
Whitecourt, *Canada* **72 C5** 54 10N 115 45W
Whiteface, *U.S.A.* **81 J3** 33 36N 102 37W
Whitefield, *U.S.A.* **79 B13** 44 23N 71 37W
Whitefish, *U.S.A.* **82 B6** 48 25N 114 20W
Whitefish L., *Canada* **73 A7** 62 41N 106 48W
Whitefish Point, *U.S.A.* .. **76 B3** 46 45N 84 59W
Whitegull, L., *Canada* ... **71 A7** 55 27N 64 17W
Whitehall, *Mich., U.S.A.* .. **76 D2** 43 24N 86 21W
Whitehall, *Mont., U.S.A.* . **82 D7** 45 52N 112 6W
Whitehall, *N.Y., U.S.A.* .. **79 C11** 43 33N 73 24W
Whitehall, *Wis., U.S.A.* ... **80 C9** 44 22N 91 19W
Whitehaven, *U.K.* **10 C4** 54 33N 3 35W
Whitehorse, *Canada* **72 A1** 60 43N 135 3W
Whitemark, *Australia* .... **62 G4** 40 7S 148 3 E
Whitemouth, *Canada* **73 D9** 49 57N 95 58W
Whitesboro, *N.Y., U.S.A.* . **79 C9** 43 7N 75 18W
Whitesboro, *Tex., U.S.A.* . **81 J6** 33 39N 96 54W
Whiteshell Prov. Park, *Canada* **73 C9** 50 0N 95 40W
Whitetail, *U.S.A.* **80 A2** 48 54N 105 10W
Whiteville, *U.S.A.* **77 H6** 34 20N 78 42W
Whitewater, *U.S.A.* **76 D1** 42 50N 88 44W
Whitewater Baldy, *U.S.A.* **83 K9** 33 20N 108 39W
Whitewater L., *Canada* ... **70 B2** 50 50N 89 10W
Whitewood, *Australia* ... **62 C3** 21 28S 143 30 E
Whitewood, *Canada* **73 C8** 50 20N 102 20W
Whitfield, *Australia* ..... **63 F4** 36 42S 146 24 E
Whithorn, *U.K.* **12 G4** 54 44N 4 26W
Whitianga, *N.Z.* .... **59 G5** 36 47S 175 41 E
Whitman, *U.S.A.* **79 D14** 42 5N 70 56W
Whitmire, *U.S.A.* **77 H5** 34 30N 81 37W
Whitney, *Canada* **78 A6** 45 31N 78 14W
Whitney, Mt., *U.S.A.* **83 H4** 36 35N 118 18W
Whitney Point, *U.S.A.* ... **79 D9** 42 20N 75 58W
Whitstable, *U.K.* .... **11 F9** 51 21N 1 3 E
Whitsunday I., *Australia* . **62 C4** 20 15S 149 4 E
Whittier, *U.S.A.* **85 M8** 33 58N 118 3W
Whittlesea, *Australia* ... **63 F4** 37 27S 145 9 E
Whitwell, *U.S.A.* **77 H3** 35 12N 85 31W
Wholdaia L., *Canada* **73 A8** 60 43N 104 20W
Whyalla, *Australia* ...... **63 E2** 33 2S 137 30 E
Whyjonta, *Australia* ..... **63 D3** 29 41S 142 28 E
Wiarton, *Canada* **78 B3** 44 40N 81 10W
Wibaux, *U.S.A.* **80 B2** 46 59N 104 11W
Wichian Buri, *Thailand* ... **38 E3** 15 39N 101 7 E
Wichita, *U.S.A.* **81 G6** 37 42N 97 20W
Wichita Falls, *U.S.A.* **81 J5** 33 54N 98 30W
Wick, *U.K.* **12 C5** 58 26N 3 5W
Wickenburg, *U.S.A.* **83 K7** 33 58N 112 44W
Wickepin, *Australia* ..... **61 F2** 32 50S 117 30 E
Wickham, C., *Australia* .. **62 F3** 39 35S 143 57 E
Wickliffe, *U.S.A.* **78 E3** 41 36N 81 28W
Wicklow, *Ireland* **13 D5** 52 59N 6 3W
Wicklow □, *Ireland* **13 D5** 52 57N 6 25W
Wicklow Hd., *Ireland* **13 D5** 52 58N 6 0W
Widgeemooltha, *Australia* **61 F3** 31 30S 121 34 E
Widnes, *U.K.* **10 D5** 53 23N 2 45W
Wieluń, *Poland* .... **17 C10** 51 15N 18 34 E
Wien, *Austria* .... **16 D9** 48 12N 16 22 E
Wiener Neustadt, *Austria* **16 E9** 47 49N 16 16 E
Wierden, *Neths.* .... **15 B6** 52 22N 6 35 E
Wiesbaden, *Germany* .... **16 C5** 50 4N 8 14 E
Wigan, *U.K.* **10 D5** 53 33N 2 38W
Wiggins, *Colo., U.S.A.* ... **80 E2** 40 14N 104 4W
Wiggins, *Miss., U.S.A.* ... **81 K10** 30 51N 89 8W
Wight, I. of □, *U.K.* **11 G6** 50 40N 1 20W
Wigtown, *U.K.* **12 G4** 54 53N 4 27W
Wigtown B., *U.K.* **12 G4** 54 46N 4 15W
Wilber, *U.S.A.* **80 E6** 40 29N 96 58W
Wilberforce, *Canada* **78 A6** 45 2N 78 13W
Wilberforce, C., *Australia* **62 A2** 11 54S 136 35 E
Wilburton, *U.S.A.* **81 H7** 34 55N 95 19W
Wilcannia, *Australia* .... **63 E3** 31 30S 143 26 E
Wilcox, *U.S.A.* **78 E6** 41 35N 78 41W
Wildrose, *Calif., U.S.A.* .. **85 J9** 36 14N 117 11W
Wildrose, *N. Dak., U.S.A.* **80 A3** 48 38N 103 11W
Wildspitze, *Austria* ..... **16 E6** 46 53N 10 53 E
Wildwood, *U.S.A.* **76 F8** 38 59N 74 50W
Wilge —►, *S. Africa* .... **57 D4** 27 3S 28 20 E
Wilhelm II Coast, *Antarctica* .... **5 C7** 68 0S 90 0 E
Wilhelmshaven, *Germany* **16 B5** 53 31N 8 7 E
Wilhelmstal, *Namibia* .... **56 C2** 21 58S 16 21 E
Wilkes-Barre, *U.S.A.* **79 E9** 41 15N 75 53W
Wilkesboro, *U.S.A.* **77 G5** 36 9N 81 10W
Wilkie, *Canada* **73 C7** 52 27N 108 42W
Wilkinsburg, *U.S.A.* **78 F5** 40 26N 79 53W
Wilkinson Lakes, *Australia* **61 E5** 29 40S 132 39 E
Willamina, *U.S.A.* **82 D2** 45 5N 123 29W
Willandra Billabong Creek —►, *Australia* ... **63 E4** 33 22S 145 52 E
Willapa B., *U.S.A.* **82 C2** 46 40N 124 0W
Willapa Hills, *U.S.A.* **84 D3** 46 35N 123 25W
Willard, *N. Mex., U.S.A.* . **83 J10** 34 36N 106 2W
Willard, *Utah, U.S.A.* **82 F7** 41 25N 112 2W
Willcox, *U.S.A.* **83 K9** 32 15N 109 50W
Willemstad, *Neth. Ant.* ... **89 D6** 12 5N 69 0W
Willeroo, *Australia* ..... **60 C5** 15 14S 131 37 E
William —►, *Canada* **73 B7** 59 8N 109 19W
William Creek, *Australia* . **63 D2** 28 58S 136 22 E
Williambury, *Australia* ... **61 D2** 23 45S 115 12 E
Williams, *Australia* ..... **61 F2** 33 2S 116 52 E
Williams, *Ariz., U.S.A.* ... **83 J7** 35 15N 112 11W
Williams, *Calif., U.S.A.* .. **84 F4** 39 9N 122 9W
Williams Lake, *Canada* .. **72 C4** 52 10N 122 10W
Williamsburg, *Ky., U.S.A.* **77 G3** 36 44N 84 10W
Williamsburg, *Pa., U.S.A.* **78 F6** 40 28N 78 12W
Williamsburg, *Va., U.S.A.* **76 G7** 37 17N 76 44W
Williamson, *N.Y., U.S.A.* . **78 C7** 43 14N 77 11W
Williamson, *W. Va., U.S.A.* **76 G4** 37 41N 82 17W
Williamsport, *U.S.A.* **78 E7** 41 15N 77 0W
Williamston, *U.S.A.* **77 H7** 35 51N 77 4W
Williamstown, *Australia* . **63 F3** 37 51S 144 52 E
Williamstown, *Mass., U.S.A.* **79 D11** 42 41N 73 12W
Williamstown, *N.Y., U.S.A.* **79 C9** 43 26N 75 53W
Williamsville, *U.S.A.* **81 G9** 36 58N 90 33W
Willimantic, *U.S.A.* **79 E12** 41 43N 72 13W
Willis Group, *Australia* .. **62 B5** 16 18S 150 0 E
Williston, *S. Africa* ..... **56 E3** 31 20S 20 53 E
Williston, *Fla., U.S.A.* **77 L4** 29 23N 82 27W
Williston, *N. Dak., U.S.A.* **80 A3** 48 9N 103 37W
Williston L., *Canada* **72 B4** 56 0N 124 0W
Willits, *U.S.A.* **82 G2** 39 25N 123 21W
Willmar, *U.S.A.* **80 C7** 45 7N 95 3W
Willoughby, *U.S.A.* **78 E3** 41 39N 81 24W
Willow Bunch, *Canada* ... **73 D7** 49 20N 105 35W
Willow L., *Canada* **72 A5** 62 10N 119 8W
Willow Lake, *U.S.A.* **80 C6** 44 38N 97 38W
Willow Springs, *U.S.A.* ... **81 G8** 37 0N 91 58W
Willow Wall, The, *China* . **35 C12** 42 10N 122 0 E
Willowlake —►, *Canada* . **72 A4** 62 42N 123 8W
Willowmore, *S. Africa* ... **56 E3** 33 15S 23 30 E
Willows, *Australia* ...... **62 C4** 23 39S 147 25 E
Willows, *U.S.A.* **84 F4** 39 31N 122 12W
Willowvale = Gatyana, *S. Africa* .... **57 E4** 32 16S 28 31 E
Wills, L., *Australia* ...... **60 D4** 21 25S 128 51 E
Wills Cr. —►, *Australia* .. **62 C3** 22 43S 140 2 E
Wills Point, *U.S.A.* **81 J7** 32 43N 96 1W
Willunga, *Australia* ..... **63 F2** 35 15S 138 30 E
Wilmette, *U.S.A.* **76 D2** 42 5N 87 42W
Wilmington, *Australia* ... **63 E2** 32 39S 138 7 E
Wilmington, *Del., U.S.A.* . **76 F8** 39 45N 75 33W
Wilmington, *Ill., U.S.A.* .. **76 E1** 41 18N 88 9W
Wilmington, *N.C., U.S.A.* . **77 H7** 34 14N 77 55W
Wilmington, *Ohio, U.S.A.* **76 F4** 39 27N 83 50W
Wilpena Cr. —►, *Australia* **63 E2** 31 25S 139 29 E
Wilsall, *U.S.A.* **82 D8** 45 59N 110 38W
Wilson, *U.S.A.* **77 H7** 35 44N 77 55W
Wilson —►, *Queens., Australia* .... **63 D3** 27 38S 141 24 E
Wilson —►, *W. Austral., Australia* .... **60 C4** 16 48S 128 16 E
Wilson Bluff, *Australia* ... **61 F4** 31 41S 129 0 E
Wilsons Promontory, *Australia* .... **63 F4** 38 55S 146 25 E
Wilton, *U.K.* **11 F6** 51 5N 1 51W
Wilton, *U.S.A.* **80 B4** 47 10N 100 47W
Wilton —►, *Australia* .... **62 A1** 14 45S 134 33 E
Wiltshire □, *U.K.* **11 F6** 51 18N 1 53W
Wiltz, *Lux.* .... **15 E5** 49 57N 5 55 E
Wiluna, *Australia* .... **61 E3** 26 36S 120 14 E
Wimmera —►, *Australia* .. **63 F3** 36 8S 141 56 E
Winam G., *Kenya* .... **54 C3** 0 20S 34 15 E
Winburg, *S. Africa* ...... **56 D4** 28 30S 27 2 E
Winchendon, *U.S.A.* **79 D12** 42 41N 72 3W
Winchester, *U.K.* **11 F6** 51 4N 1 18W
Winchester, *Conn., U.S.A.* **79 E11** 41 53N 73 9W
Winchester, *Idaho, U.S.A.* **82 C5** 46 14N 116 38W
Winchester, *Ind., U.S.A.* . **76 E3** 40 10N 84 59W
Winchester, *Ky., U.S.A.* .. **76 G3** 38 0N 84 11W
Winchester, *N.H., U.S.A.* . **79 D12** 42 46N 72 23W
Winchester, *Nev., U.S.A.* . **85 J11** 36 6N 115 10W
Winchester, *Tenn., U.S.A.* **77 H2** 35 11N 86 7W
Winchester, *Va., U.S.A.* .. **76 F6** 39 11N 78 10W
Wind —►, *U.S.A.* **82 E9** 43 12N 108 12W
Wind River Range, *U.S.A.* **82 E9** 43 0N 109 30W
Windau = Ventspils, *Latvia* .... **9 H19** 57 25N 21 32 E
Windber, *U.S.A.* **78 F6** 40 14N 78 50W
Windermere, L., *U.K.* **10 C5** 54 22N 2 56W
Windfall, *Canada* **72 C5** 54 12N 116 13W
Windflower L., *Canada* .. **72 A5** 62 52N 118 30W
Windhoek, *Namibia* ..... **56 C2** 22 35S 17 4 E
Windom, *U.S.A.* **80 D7** 43 52N 95 7W
Windorah, *Australia* ..... **62 D3** 25 24S 142 36 E
Window Rock, *U.S.A.* ... **83 J9** 35 41N 109 3W
Windrush —►, *U.K.* **11 F6** 51 43N 1 24W
Windsor, *Australia* ...... **63 E5** 33 37S 150 50 E
Windsor, *N.S., Canada* ... **71 D7** 44 59N 64 5W
Windsor, *Nfld., Canada* .. **71 C8** 48 57N 55 40W
Windsor, *Ont., Canada* ... **70 D3** 42 18N 83 0W
Windsor, *U.K.* **11 F7** 51 29N 0 36W
Windsor, *Colo., U.S.A.* ... **80 E2** 40 29N 104 54W
Windsor, *Conn., U.S.A.* .. **79 E12** 41 50N 72 39W
Windsor, *Mo., U.S.A.* **80 F8** 38 32N 93 31W
Windsor, *N.Y., U.S.A.* ... **79 D9** 42 5N 75 37W
Windsor, *Vt., U.S.A.* **79 C12** 43 29N 72 24W
Windsorton, *S. Africa* ... **56 D3** 28 16S 24 44 E
Windward Is., *W. Indies* . **89 D7** 13 0N 61 0W
Windward Passage = Vientos, Paso de los, *Caribbean* **89 C5** 20 0N 74 0W
Windy L., *Canada* **73 A8** 60 20N 100 2W
Winefred L., *Canada* **73 B6** 55 30N 110 30W
Winfield, *U.S.A.* **81 G6** 37 15N 96 59W
Wingate Mts., *Australia* . **60 B5** 14 25S 130 40 E
Wingen, *Australia* ...... **63 E5** 31 54S 150 54 E
Wingham, *Australia* ..... **63 E5** 31 48S 152 22 E
Wingham, *Canada* **70 D3** 43 55N 81 20W
Winifred, *U.S.A.* **82 C9** 47 34N 109 23W
Winisk, *Canada* **70 A2** 55 20N 85 15W
Winisk —►, *Canada* **70 A2** 55 17N 85 5W
Winisk L., *Canada* **70 B2** 52 55N 87 22W
Wink, *U.S.A.* **81 K3** 31 45N 103 9W
Winkler, *Canada* **73 D9** 49 10N 97 56W
Winlock, *U.S.A.* **84 D4** 46 30N 122 56W
Winneba, *Ghana* **50 G4** 5 25N 0 36W
Winnebago, *U.S.A.* **80 D7** 43 46N 94 10W
Winnebago, L., *U.S.A.* ... **76 D1** 44 0N 88 26W
Winnecke Cr. —►, *Australia* .... **60 C5** 18 35S 131 34 E
Winnemucca, *U.S.A.* **82 F5** 40 58N 117 44W
Winnemucca L., *U.S.A.* .. **82 F4** 40 7N 119 21W
Winner, *U.S.A.* **80 D5** 43 22N 99 52W
Winnett, *U.S.A.* **82 C9** 47 0N 108 21W
Winnfield, *U.S.A.* **81 K8** 31 56N 92 38W
Winnibigoshish, L., *U.S.A.* **80 B7** 47 27N 94 13W
Winning, *Australia* ...... **60 D1** 23 9S 114 30 E
Winnipeg, *Canada* **73 D9** 49 54N 97 9W
Winnipeg —►, *Canada* ... **73 C9** 50 38N 96 19W
Winnipeg, L., *Canada* **73 C9** 52 0N 97 0W
Winnipeg Beach, *Canada* **73 C9** 50 30N 96 58W
Winnipegosis, *Canada* ... **73 C9** 51 39N 99 55W
Winnipegosis L., *Canada* **73 C9** 52 30N 100 0W
Winnipesaukee, L., *U.S.A.* **79 C13** 43 38N 71 21W
Winnsboro, *La., U.S.A.* .. **81 J9** 32 10N 91 43W
Winnsboro, *S.C., U.S.A.* . **77 H5** 34 23N 81 5W
Winnsboro, *Tex., U.S.A.* . **81 J7** 32 58N 95 17W
Winokapau, L., *Canada* .. **71 B7** 53 15N 62 50W
Winona, *Minn., U.S.A.* ... **80 C9** 44 3N 91 39W
Winona, *Miss., U.S.A.* ... **81 J10** 33 29N 89 44W
Winooski, *U.S.A.* **79 B11** 44 29N 73 11W
Winschoten, *Neths.* ..... **15 A7** 53 9N 7 3 E
Winslow, *Ariz., U.S.A.* ... **83 J8** 35 2N 110 42W
Winslow, *Wash., U.S.A.* . **84 C4** 47 38N 122 31W
Winsted, *U.S.A.* **79 E11** 41 55N 73 4W
Winston-Salem, *U.S.A.* .. **77 G5** 36 6N 80 15W
Winter Garden, *U.S.A.* ... **77 L5** 28 34N 81 35W
Winter Haven, *U.S.A.* ... **77 M5** 28 1N 81 44W
Winter Park, *U.S.A.* **77 L5** 28 36N 81 20W
Winterhaven, *U.S.A.* **85 N12** 32 47N 114 39W
Winters, *Calif., U.S.A.* ... **84 G5** 38 32N 121 58W
Winters, *Tex., U.S.A.* **81 K5** 31 58N 99 58W
Winterset, *U.S.A.* **80 E7** 41 20N 94 1W
Wintersville, *U.S.A.* **78 F4** 40 23N 80 42W
Winterswijk, *Neths.* ..... **15 C6** 51 58N 6 43 E
Winterthur, *Switz.* ...... **16 E5** 47 30N 8 44 E
Winthrop, *Minn., U.S.A.* . **80 C7** 44 32N 94 22W
Winthrop, *Wash., U.S.A.* . **82 B3** 48 28N 120 10W
Winton, *Australia* ...... **62 C3** 22 24S 143 3 E
Winton, *N.Z.* .... **59 M2** 46 8S 168 20 E
Winton, *U.S.A.* **77 G7** 36 24N 76 56W
Wirral, *U.K.* **10 D4** 53 25N 3 0W
Wirrulla, *Australia* ...... **63 E1** 32 24S 134 31 E
Wisbech, *U.K.* .... **10 E8** 52 41N 0 9 E
Wisconsin □, *U.S.A.* **80 C10** 44 45N 89 30W
Wisconsin —►, *U.S.A.* ... **80 D9** 43 0N 91 15W
Wisconsin Dells, *U.S.A.* . **80 D10** 43 38N 89 46W
Wisconsin Rapids, *U.S.A.* **80 C10** 44 23N 89 49W
Wisdom, *U.S.A.* **82 D7** 45 37N 113 27W
Wishaw, *U.K.* **12 F5** 55 46N 3 54W
Wishek, *U.S.A.* **80 B5** 46 16N 99 33W
Wisła —►, *Poland* ...... **17 A10** 54 22N 18 55 E
Wismar, *Germany* ...... **16 B6** 53 54N 11 29 E
Wisner, *U.S.A.* **80 E6** 41 59N 96 55W
Witbank, *S. Africa* ...... **57 D4** 25 51S 29 14 E
Witdraai, *S. Africa* ...... **56 D3** 26 58S 20 48 E
Witham —►, *U.K.* **10 D7** 52 59N 0 2W
Withernsea, *U.K.* .... **10 D8** 53 44N 0 1 E
Witney, *U.K.* **11 F6** 51 48N 1 28W
Witnossob —►, *Namibia* . **56 D3** 26 55S 20 37 E
Witten, *Germany* .... **15 C7** 51 26N 7 20 E
Wittenberg, *Germany* ... **16 C7** 51 53N 12 39 E
Wittenberge, *Germany* ... **16 B6** 53 0N 11 45 E
Wittenoom, *Australia* ... **60 D2** 22 15S 118 20 E
Wkra —►, *Poland* ...... **17 B11** 52 27N 20 44 E
Wlingi, *Indonesia* ...... **37 H15** 8 5S 112 25 E
Włocławek, *Poland* ..... **17 B10** 52 40N 19 3 E
Włodawa, *Poland* ...... **17 C12** 51 33N 23 31 E
Woburn, *U.S.A.* **79 D13** 42 29N 71 9W
Wodian, *China* .... **34 H7** 32 50N 112 35 E
Wodonga, *Australia* ..... **63 F4** 36 5S 146 50 E
Wokam, *Indonesia* ...... **37 F8** 5 45S 134 28 E
Wolf —►, *Canada* **72 A2** 60 17N 132 33W
Wolf Creek, *U.S.A.* **82 C7** 47 0N 112 4W
Wolf L., *Canada* **72 A2** 60 24N 131 40W
Wolf Point, *U.S.A.* **80 A2** 48 5N 105 39W
Wolfe I., *Canada* **70 D4** 44 7N 76 20W
Wolfsberg, *Austria* ..... **16 E8** 46 50N 14 52 E
Wolfsburg, *Germany* .... **16 B6** 52 25N 10 48 E
Wolin, *Poland* .... **16 B8** 53 50N 14 37 E
Wollaston, Is., *Chile* **96 H3** 55 40S 67 30W
Wollaston L., *Canada* ... **73 B8** 58 7N 103 10W
Wollaston Pen., *Canada* . **68 B8** 69 30N 115 0W
Wollogorang, *Australia* .. **62 B2** 17 13S 137 57 E
Wollongong, *Australia* ... **63 E5** 34 25S 150 54 E
Wolmaransstad, *S. Africa* **56 D4** 27 12S 25 59 E
Wolseley, *Australia* ..... **63 F3** 36 23S 140 54 E
Wolseley, *Canada* **73 C8** 50 25N 103 15W
Wolseley, *S. Africa* ..... **56 E2** 33 26S 19 7 E
Wolstenholme, C., *Canada* **66 C12** 62 35N 77 30W
Wolvega, *Neths.* .... **15 B6** 52 52N 6 0 E
Wolverhampton, *U.K.* **11 E5** 52 35N 2 7W
Wonarah, *Australia* ..... **62 B2** 19 55S 136 20 E
Wondai, *Australia* ...... **63 D5** 26 20S 151 49 E
Wongalarroo L., *Australia* **63 E3** 31 32S 144 0 E
Wongan Hills, *Australia* . **61 F2** 30 51S 116 37 E
Wongawol, *Australia* .... **61 E3** 26 5S 121 55 E
Wŏnju, *S. Korea* .... **35 F14** 37 22N 127 58 E
Wonosari, *Indonesia* .... **37 G14** 7 58S 110 36 E
Wŏnsan, *N. Korea* ...... **35 E14** 39 11N 127 27 E
Wonthaggi, *Australia* ... **63 F4** 38 37S 145 37 E
Woocalla, *Australia* ..... **63 E2** 31 42S 137 12 E
Wood Buffalo Nat. Park, *Canada* **72 B6** 59 0N 113 41W
Wood Is., *Australia* ..... **60 C3** 16 24S 123 19 E
Wood L., *Canada* **73 B8** 55 17N 103 17W
Wood Lake, *U.S.A.* **80 D4** 42 38N 100 14W
Woodah I., *Australia* .... **62 A2** 13 27S 136 10 E
Woodanilling, *Australia* .. **61 F2** 33 31S 117 24 E
Woodbridge, *Canada* **78 C5** 43 47N 79 36W
Woodburn, *Australia* .... **63 D5** 29 6S 153 23 E
Woodenbong, *Australia* .. **63 D5** 28 24S 152 39 E
Woodend, *Australia* .... **63 F3** 37 20S 144 33 E
Woodfords, *U.S.A.* **84 G7** 38 47N 119 50W
Woodgreen, *Australia* ... **62 C1** 22 26S 134 12 E
Woodlake, *U.S.A.* **84 J7** 36 25N 119 6W
Woodland, *U.S.A.* **84 G5** 38 41N 121 46W
Woodlands, *Australia* ... **60 D2** 24 46S 118 8 E
Woodpecker, *Canada* ... **72 C4** 53 30N 122 40W
Woodridge, *Canada* **73 D9** 49 20N 96 9W
Woodroffe, Mt., *Australia* **61 E5** 26 20S 131 45 E
Woodruff, *Ariz., U.S.A.* .. **83 J8** 34 51N 110 1W
Woodruff, *Utah, U.S.A.* .. **82 F8** 41 31N 111 10W
Woods, L., *Australia* .... **62 B1** 17 50S 133 30 E
Woods, L., *Canada* **71 B6** 54 30N 65 13W
Woods, L. of the, *Canada* **73 D10** 49 15N 94 45W
Woodstock, *Queens., Australia* .... **62 B4** 19 35S 146 50 E
Woodstock, *W. Austral., Australia* .... **60 D2** 21 41S 118 57 E
Woodstock, *N.B., Canada* **71 C6** 46 11N 67 37W
Woodstock, *Ont., Canada* **70 D3** 43 10N 80 45W
Woodstock, *U.K.* **11 F6** 51 51N 1 20W
Woodstock, *Ill., U.S.A.* .. **80 D10** 42 19N 88 27W
Woodstock, *Vt., U.S.A.* .. **79 C12** 43 37N 72 31W
Woodsville, *U.S.A.* **79 B13** 44 9N 72 2W
Woodville, *N.Z.* .... **59 J5** 40 20S 175 53 E

Woodville, *U.S.A.* **81 K7** 30 47N 94 25W
Woodward, *U.S.A.* **81 G5** 36 26N 99 24W
Woody, *U.S.A.* **85 K8** 35 42N 118 50W
Woolamai, C., *Australia* . **63 F4** 38 30S 145 23 E
Woolgoolga, *Australia* ... **63 E5** 30 6S 153 11 E
Woombye, *Australia* .... **63 D5** 26 40S 152 55 E
Woomera, *Australia* ..... **63 E2** 31 5S 136 50 E
Woonsocket, *R.I., U.S.A.* . **79 D13** 42 0N 71 31W
Woonsocket, *S. Dak., U.S.A.* **80 C5** 44 3N 98 17W
Wooramel, *Australia* .... **61 E1** 25 45S 114 17 E
Wooramel →, *Australia* **61 E1** 25 47S 114 10 E
Wooroloo, *Australia* .... **61 F2** 31 48S 116 18 E
Wooster, *U.S.A.* **78 F3** 40 48N 81 56W
Worcester, *S. Africa* ..... **56 E2** 33 39S 19 27 E
Worcester, *U.K.* **11 E5** 52 11N 2 12W
Worcester, *Mass., U.S.A.* **79 D13** 42 16N 71 48W
Worcester, *N.Y., U.S.A.* .. **79 D10** 42 36N 74 45W
Workington, *U.K.* **10 C4** 54 39N 3 33W
Worksop, *U.K.* **10 D6** 53 18N 1 7W
Workum, *Neths.* .... **15 B5** 52 59N 5 26 E
Worland, *U.S.A.* **82 D10** 44 1N 107 57W
Worms, *Germany* .... **16 D5** 49 37N 8 21 E
Wortham, *U.S.A.* **81 K6** 31 47N 96 28W
Worthing, *U.K.* **11 G7** 50 49N 0 21W
Worthington, *U.S.A.* **80 D7** 43 37N 95 36W
Wosi, *Indonesia* .... **37 E7** 0 15S 128 0 E
Wou-han = Wuhan, *China* **33 C6** 30 31N 114 18 E
Wour, *Chad* .... **51 D8** 21 14N 16 0 E
Wousi = Wuxi, *China* ... **33 C7** 31 33N 120 18 E
Wowoni, *Indonesia* ..... **37 E6** 4 5S 123 5 E
Woy Woy, *Australia* .... **63 E5** 33 30S 151 19 E
Wrangel I. = Vrangelya, Ostrov, *Russia* .... **27 B19** 71 0N 180 0 E
Wrangell, *U.S.A.* **68 C6** 56 28N 132 23W
Wrangell I., *U.S.A.* **72 B2** 56 16N 132 12W
Wrangell Mts., *U.S.A.* ... **68 B5** 61 30N 142 0W
Wrath, C., *U.K.* **12 C3** 58 38N 5 1W
Wray, *U.S.A.* **80 E3** 40 5N 102 13W
Wrekin, The, *U.K.* **10 E5** 52 41N 2 32W
Wrens, *U.S.A.* **77 J4** 33 12N 82 23W
Wrexham, *U.K.* **10 D4** 53 3N 3 0W
Wrexham □, *U.K.* **10 D5** 53 1N 2 58W
Wright, *Canada* **72 C4** 51 52N 121 40W
Wright, *Phil.* .... **37 B7** 11 42N 125 2 E
Wrightson Mt., *U.S.A.* ... **83 L8** 31 42N 110 51W
Wrightwood, *U.S.A.* **85 L9** 34 21N 117 38W
Wrigley, *Canada* **68 B7** 63 16N 123 37W
Wrocław, *Poland* .... **17 C9** 51 5N 17 5 E
Września, *Poland* .... **17 B9** 52 21N 17 36 E
Wu Jiang →, *China* ..... **32 D5** 29 40N 107 20 E
Wu'an, *China* .... **34 F8** 36 40N 114 15 E
Wubin, *Australia* .... **61 F2** 30 6S 116 37 E
Wubu, *China* .... **34 F6** 37 28N 110 42 E
Wuchang, *China* .... **35 B14** 44 55N 127 5 E
Wucheng, *China* .... **34 F9** 37 12N 116 20 E
Wuchuan, *China* .... **34 D6** 41 5N 111 28 E
Wudi, *China* .... **35 F9** 37 40N 117 35 E
Wuding He →, *China* .. **34 F6** 37 2N 110 23 E
Wudu, *China* .... **34 H3** 33 22N 104 54 E
Wuhan, *China* .... **33 C6** 30 31N 114 18 E
Wuhe, *China* .... **35 H9** 33 10N 117 50 E
Wuhsi = Wuxi, *China* ... **33 C7** 31 33N 120 18 E
Wuhu, *China* .... **33 C6** 31 22N 118 21 E
Wukari, *Nigeria* .... **50 G6** 7 51N 9 42 E
Wulajie, *China* .... **35 B14** 44 6N 126 33 E
Wulanbulang, *China* .... **34 D6** 41 5N 110 55 E
Wulian, *China* .... **35 G10** 35 40N 119 12 E
Wuliaru, *Indonesia* ...... **37 F8** 7 27S 131 0 E
Wuluk'omushih Ling, *China* .... **32 C3** 36 25N 87 25 E
Wulumuchi = Ürümqi, *China* .... **26 E9** 43 45N 87 45 E
Wum, *Cameroon* .... **50 G7** 6 24N 10 2 E
Wunnummin L., *Canada* . **70 B2** 52 55N 89 10W
Wuntho, *Burma* .... **41 H19** 23 55N 95 45 E
Wuppertal, *Germany* .... **16 C4** 51 16N 7 12 E
Wuppertal, *S. Africa* .... **56 E2** 32 13S 19 12 E
Wuqing, *China* .... **35 E9** 39 23N 117 4 E
Wurung, *Australia* ...... **62 B3** 19 13S 140 38 E
Würzburg, *Germany* .... **16 D5** 49 46N 9 55 E
Wushan, *China* .... **34 G3** 34 43N 104 53 E
Wusuli Jiang = Ussuri →, *Asia* ...... **30 A7** 48 27N 135 0 E
Wutai, *China* .... **34 E7** 38 40N 113 12 E
Wuting = Huimin, *China* . **35 F9** 37 27N 117 28 E
Wutonghaolai, *China* .... **35 C11** 42 50N 120 5 E
Wutongqiao, *China* ..... **32 D5** 29 22N 103 50 E
Wuwei, *China* .... **32 C5** 37 57N 102 34 E
Wuxi, *China* .... **33 C7** 31 33N 120 18 E
Wuxiang, *China* .... **34 F7** 36 49N 112 50 E
Wuxing, *China* .... **33 C7** 30 51N 120 8 E
Wuyang, *China* .... **34 H7** 33 25N 113 35 E
Wuyi, *China* .... **34 F8** 37 46N 115 56 E
Wuyi Shan, *China* ...... **33 D6** 27 0N 117 0 E
Wuyuan, *China* .... **34 D5** 41 2N 108 20 E
Wuzhai, *China* .... **34 E6** 38 54N 111 48 E
Wuzhi Shan, *China* ..... **38 C7** 18 45N 109 45 E
Wuzhong, *China* .... **34 E4** 38 2N 106 12 E
Wuzhou, *China* .... **33 D6** 23 30N 111 18 E
Wyaaba Cr. →, *Australia* **62 B3** 16 27S 141 35 E
Wyalkatchem, *Australia* . **61 F2** 31 8S 117 22 E
Wyalusing, *U.S.A.* **79 E8** 41 40N 76 16W
Wyandotte, *U.S.A.* **76 D4** 42 12N 83 9W
Wyandra, *Australia* ..... **63 D4** 27 12S 145 56 E
Wyangala Res., *Australia* **63 E4** 33 54S 149 0 E
Wyara, L., *Australia* ..... **63 D3** 28 42S 144 14 E
Wycheproof, *Australia* ... **63 F3** 36 5S 143 17 E
Wye →, *U.K.* **11 F5** 51 38N 2 40W
Wyemandoo, *Australia* .. **61 E2** 28 28S 118 29 E
Wymondham, *U.K.* **11 E7** 52 45N 0 43W
Wymore, *U.S.A.* **80 E6** 40 7N 96 40W
Wynbring, *Australia* .... **63 E1** 30 33S 133 32 E
Wyndham, *Australia* .... **60 C4** 15 33S 128 3 E
Wyndham, *N.Z.* .... **59 M2** 46 20S 168 51 E
Wyndmere, *U.S.A.* **80 B6** 46 16N 97 8W
Wynne, *U.S.A.* **81 H9** 35 14N 90 47W
Wynnum, *Australia* ..... **63 D5** 27 27S 153 9 E
Wynyard, *Australia* ..... **62 G4** 41 5S 145 44 E
Wynyard, *Canada* **73 C8** 51 45N 104 10W
Wyola, L., *Australia* ..... **61 E5** 29 8S 130 17 E
Wyoming □, *U.S.A.* **82 E10** 43 0N 107 30W
Wyong, *Australia* .... **63 E5** 33 14S 151 24 E
Wytheville, *U.S.A.* **76 G5** 36 57N 81 5W

X

Xai-Xai, *Mozam.* .... **57 D5** 25 6S 33 31 E
Xainza, *China* .... **32 C3** 30 58N 88 35 E
Xangongo, *Angola* ...... **56 B2** 16 45S 15 5 E
Xankändi, *Azerbaijan* ... **25 G8** 39 52N 46 49 E
Xánthi, *Greece* .... **21 D11** 41 10N 24 58 E
Xapuri, *Brazil* **92 F5** 10 35S 68 35W
Xar Moron He →, *China* **35 C11** 43 25N 120 35 E
Xau, L., *Botswana* ...... **56 C3** 21 15S 24 44 E
Xavantina, *Brazil* **95 A5** 21 15S 52 48W
Xenia, *U.S.A.* **76 F4** 39 41N 83 56W
Xeropotamos →, *Cyprus* **23 E11** 34 42N 32 33 E
Xhora, *S. Africa* .... **57 E4** 31 55S 28 38 E
Xhumo, *Botswana* ...... **56 C3** 21 7S 24 35 E
Xi Jiang →, *China* ..... **33 D6** 22 5N 113 20 E
Xi Xian, *China* .... **34 F6** 36 41N 110 58 E
Xia Xian, *China* .... **34 G6** 35 8N 111 12 E
Xiachengzi, *China* ...... **35 B16** 44 40N 130 18 E
Xiaguan, *China* .... **32 D5** 25 32N 100 16 E
Xiajin, *China* .... **34 F8** 36 56N 116 0 E
Xiamen, *China* .... **33 D6** 24 25N 118 4 E
Xi'an, *China* .... **34 G5** 34 15N 109 0 E
Xian Xian, *China* .... **34 E9** 38 12N 116 6 E
Xiang Jiang →, *China* .. **33 D6** 28 55N 112 50 E
Xiangcheng, *Henan, China* **34 H8** 33 29N 114 52 E
Xiangcheng, *Henan, China* **34 H7** 33 50N 113 27 E
Xiangfan, *China* .... **33 C6** 32 2N 112 8 E
Xianghuang Qi, *China* ... **34 C7** 42 2N 113 50 E
Xiangning, *China* .... **34 G6** 35 58N 110 50 E
Xiangquan, *China* ...... **34 F7** 36 30N 113 1 E
Xiangshui, *China* .... **35 G10** 34 12N 119 33 E
Xiangtan, *China* .... **33 D6** 27 51N 112 54 E
Xianyang, *China* .... **34 G5** 34 20N 108 40 E
Xiao Hinggan Ling, *China* **33 B7** 49 0N 127 0 E
Xiao Xian, *China* .... **34 G9** 34 15N 116 55 E
Xiaoyi, *China* .... **34 F6** 37 8N 111 48 E
Xiawa, *China* .... **35 C11** 42 35N 120 38 E
Xiayi, *China* .... **34 G9** 34 15N 116 10 E
Xichang, *China* .... **32 D5** 27 51N 102 19 E
Xichuan, *China* .... **34 H6** 33 0N 111 30 E
Xieng Khouang, *Laos* ... **38 C4** 19 17N 103 25 E
Xifei He →, *China* ..... **34 H9** 32 45N 116 40 E
Xifeng, *China* .... **35 C13** 42 42N 124 45 E
Xifengzhen, *China* ...... **34 G4** 35 40N 107 40 E
Xigazê, *China* .... **32 D3** 29 5N 88 45 E
Xihe, *China* .... **34 G3** 34 2N 105 20 E
Xihua, *China* .... **34 H8** 33 45N 114 30 E
Xiliao He →, *China* .... **35 C12** 43 32N 123 35 E
Xin Xian, *China* .... **34 E7** 38 22N 112 46 E
Xinavane, *Mozam.* ...... **57 D5** 25 2S 32 47 E
Xinbin, *China* .... **35 D13** 41 40N 125 2 E
Xing Xian, *China* .... **34 E6** 38 27N 111 7 E
Xing'an, *China* .... **33 D6** 25 38N 110 40 E
Xingcheng, *China* ...... **35 D11** 40 40N 120 45 E
Xinghe, *China* .... **34 D7** 40 55N 113 55 E
Xinghua, *China* .... **35 H10** 32 58N 119 48 E
Xinglong, *China* .... **35 D9** 40 25N 117 30 E
Xingping, *China* .... **34 G5** 34 20N 108 28 E
Xingtai, *China* .... **34 F8** 37 3N 114 32 E
Xingu →, *Brazil* **93 D8** 1 30S 51 53W
Xingyang, *China* .... **34 G7** 34 45N 112 52 E
Xinhe, *China* .... **34 F8** 37 30N 115 15 E
Xining, *China* .... **32 C5** 36 34N 101 40 E
Xinjiang, *China* .... **34 G6** 35 34N 111 11 E
Xinjiang Uygur Zizhiqu □, *China* .... **32 B3** 42 0N 86 0 E
Xinjin, *China* .... **35 E11** 39 25N 121 58 E
Xinkai He →, *China* .... **35 C12** 43 32N 123 35 E
Xinle, *China* .... **34 E8** 38 25N 114 40 E
Xinlitun, *China* .... **35 D12** 42 0N 122 8 E
Xinmin, *China* .... **35 D12** 41 59N 122 50 E
Xintai, *China* .... **35 G9** 35 55N 117 45 E
Xinxiang, *China* .... **34 G7** 35 18N 113 50 E
Xinzhan, *China* .... **35 C14** 43 50N 127 18 E
Xinzheng, *China* .... **34 G7** 34 20N 113 45 E
Xiong Xian, *China* ...... **34 E9** 38 59N 116 8 E
Xiongyuecheng, *China* ... **35 D12** 40 12N 122 5 E
Xiping, *Henan, China* ... **34 H8** 33 22N 114 5 E
Xiping, *Henan, China* ... **34 H6** 33 25N 111 8 E
Xique-Xique, *Brazil* **93 F10** 10 50S 42 40W
Xisha Qundao = Paracel Is., *S. China Sea* ...... **36 A4** 15 50N 112 0 E
Xiuyan, *China* .... **35 D12** 40 18N 123 11 E
Xixabangma Feng, *China* **41 E14** 28 20N 85 40 E
Xixia, *China* .... **34 H6** 33 25N 111 29 E
Xixiang, *China* .... **34 H4** 33 0N 107 44 E
Xiyang, *China* .... **34 F7** 37 38N 113 38 E
Xizang □, *China* .... **32 C3** 32 0N 88 0 E
Xlendi, *Malta* .... **23 C1** 36 1N 14 12 E
Xuan Loc, *Vietnam* ..... **39 G6** 10 56N 107 14 E
Xuanhua, *China* .... **34 D8** 40 40N 115 2 E
Xuchang, *China* .... **34 G7** 34 2N 113 48 E
Xun Xian, *China* .... **34 G8** 35 42N 114 33 E
Xunyang, *China* .... **34 H5** 32 48N 109 22 E
Xunyi, *China* .... **34 G5** 35 8N 108 20 E
Xushui, *China* .... **34 E8** 39 2N 115 40 E
Xuyen Moc, *Vietnam* .... **39 G6** 10 34N 107 25 E
Xuzhou, *China* .... **35 G9** 34 18N 117 10 E
Xylophagou, *Cyprus* .... **23 E12** 34 54N 33 51 E

Y

Ya Xian, *China* .... **38 C7** 18 14N 109 29 E
Yaamba, *Australia* ...... **62 C5** 23 8S 150 22 E
Yaapeet, *Australia* ...... **63 F3** 35 45S 142 3 E
Yabelo, *Ethiopia* .... **51 H12** 4 50N 38 8 E
Yablonovy Ra. = Yablonovyy Khrebet, *Russia* .... **27 D12** 53 0N 114 0 E
Yablonovyy Khrebet, *Russia* .... **27 D12** 53 0N 114 0 E
Yabrai Shan, *China* ..... **34 E2** 39 40N 103 0 E
Yabrūd, *Syria* .... **47 B5** 33 58N 36 39 E
Yacheng, *China* .... **33 E5** 18 22N 109 6 E
Yacuiba, *Bolivia* **94 A3** 22 0S 63 43W
Yadgir, *India* .... **40 L10** 16 45N 77 5 E
Yadkin →, *U.S.A.* **77 H5** 35 29N 80 9W
Yagodnoye, *Russia* ..... **27 C15** 62 33N 149 40 E
Yagoua, *Cameroon* ..... **52 B3** 10 20N 15 13 E
Yaha, *Thailand* .... **39 J3** 6 29N 101 8 E
Yahila, *Zaïre* .... **54 B1** 0 13N 24 28 E
Yahk, *Canada* **72 D5** 49 6N 116 10W
Yahuma, *Zaïre* .... **52 D4** 1 0N 23 10 E
Yaita, *Japan* .... **31 F9** 36 48N 139 56 E
Yaiza, *Canary Is.* **22 F6** 28 57N 13 46W
Yakima, *U.S.A.* **82 C3** 46 36N 120 31W
Yakima →, *U.S.A.* **82 C3** 47 0N 120 30W
Yakovlevka, *Russia* ..... **30 B6** 44 26N 133 28 E
Yaku-Shima, *Japan* ..... **31 J5** 30 20N 130 30 E
Yakutat, *U.S.A.* **68 C6** 59 33N 139 44W
Yakutia = Sakha □, *Russia* **27 C13** 62 0N 130 0 E
Yakutsk, *Russia* .... **27 C13** 62 5N 129 50 E
Yala, *Thailand* .... **39 J3** 6 33N 101 18 E
Yalbalgo, *Australia* ..... **61 E1** 25 10S 114 45 E
Yalboroo, *Australia* ..... **62 C4** 20 50S 148 40 E
Yale, *U.S.A.* **78 C2** 43 8N 82 48W
Yalgoo, *Australia* .... **61 E2** 28 16S 116 39 E
Yalinga, *C.A.R.* .... **51 G9** 6 33N 23 10 E
Yalkubul, Punta, *Mexico* . **87 C7** 21 32N 88 37W
Yalleroi, *Australia* ...... **62 C4** 24 3S 145 42 E
Yalobusha →, *U.S.A.* ... **81 J9** 33 33N 90 10W
Yalong Jiang →, *China* . **32 D5** 26 40N 101 55 E
Yalova, *Turkey* .... **21 D13** 40 41N 29 15 E
Yalta, *Ukraine* .... **25 F5** 44 30N 34 10 E
Yalu Jiang →, *China* ... **35 E13** 40 0N 124 22 E
Yalutorovsk, *Russia* ..... **26 D7** 56 41N 66 12 E
Yam Ha Melaḥ = Dead Sea, *Asia* .... **47 D4** 31 30N 35 30 E
Yam Kinneret, *Israel* .... **47 C4** 32 45N 35 35 E
Yamada, *Japan* .... **31 H5** 33 33N 130 49 E
Yamagata, *Japan* .... **30 E10** 38 15N 140 15 E
Yamagata □, *Japan* .... **30 E10** 38 30N 140 0 E
Yamaguchi, *Japan* ...... **31 G5** 34 10N 131 32 E
Yamaguchi □, *Japan* ... **31 G5** 34 20N 131 40 E
Yamal, Poluostrov, *Russia* **26 B8** 71 0N 70 0 E
Yamal Pen. = Yamal, Poluostrov, *Russia* .... **26 B8** 71 0N 70 0 E
Yamanashi □, *Japan* .... **31 G9** 35 40N 138 40 E
Yamantau, Gora, *Russia* . **24 D10** 54 15N 58 6 E
Yamba, *N.S.W., Australia* **63 D5** 29 26S 153 23 E
Yamba, *S. Austral., Australia* .... **63 E3** 34 10S 140 52 E
Yambah, *Australia* ...... **62 C1** 23 10S 133 50 E
Yambarran Ra., *Australia* **60 C5** 15 10S 130 25 E
Yâmbiô, *Sudan* .... **51 H10** 4 35N 28 16 E
Yambol, *Bulgaria* .... **21 C12** 42 30N 26 36 E
Yamdena, *Indonesia* .... **37 F8** 7 45S 131 20 E
Yame, *Japan* .... **31 H5** 33 13N 130 35 E
Yamethin, *Burma* .... **41 J20** 20 29N 96 18 E
Yamma-Yamma, L., *Australia* .... **63 D3** 26 16S 141 20 E
Yamoussoukro, *Ivory C.* . **50 G3** 6 49N 5 17W
Yampa →, *U.S.A.* **82 F9** 40 32N 108 59W
Yampi Sd., *Australia* .... **60 C3** 16 8S 123 38 E
Yampil, *Moldova* .... **17 D15** 48 15N 28 15 E
Yampol = Yampil, *Moldova* .... **17 D15** 48 15N 28 15 E
Yamuna →, *India* ...... **43 G9** 25 30N 81 53 E
Yamzho Yumco, *China* .. **32 D4** 28 48N 90 35 E
Yana →, *Russia* .... **27 B14** 71 30N 136 0 E
Yanac, *Australia* .... **63 F3** 36 8S 141 25 E
Yanagawa, *Japan* ...... **31 H5** 33 10N 130 24 E
Yanai, *Japan* .... **31 H6** 33 58N 132 7 E
Yan'an, *China* .... **34 F5** 36 35N 109 26 E
Yanaul, *Russia* .... **24 C10** 56 25N 55 0 E
Yanbu 'al Baḥr, *Si. Arabia* **44 F3** 24 0N 38 5 E
Yancannia, *Australia* .... **63 E3** 30 12S 142 35 E
Yanchang, *China* .... **34 F6** 36 43N 110 1 E
Yancheng, *Henan, China* . **34 H7** 33 35N 114 0 E
Yancheng, *Jiangsu, China* **35 H11** 33 23N 120 8 E
Yanchi, *China* .... **34 F4** 37 48N 107 20 E
Yanchuan, *China* .... **34 F6** 36 51N 110 10 E
Yanco Cr. →, *Australia* . **63 F4** 35 14S 145 35 E
Yandal, *Australia* .... **61 E3** 27 35S 121 10 E
Yandanooka, *Australia* .. **61 E2** 29 18S 115 29 E
Yandaran, *Australia* ..... **62 C5** 24 43S 152 6 E
Yandoon, *Burma* .... **41 L19** 17 0N 95 40 E
Yang Xian, *China* .... **34 H4** 33 15N 107 30 E
Yangambi, *Zaïre* .... **54 B1** 0 47N 24 20 E
Yangcheng, *China* ...... **34 G7** 35 28N 112 22 E
Yangch'ü = Taiyuan, *China* .... **34 F7** 37 52N 112 33 E
Yanggao, *China* .... **34 D7** 40 21N 113 55 E
Yanggu, *China* .... **34 F8** 36 8N 115 43 E
Yangliuqing, *China* ..... **35 E9** 39 2N 117 5 E
Yangon = Rangoon, *Burma* .... **41 L20** 16 45N 96 20 E
Yangpingguan, *China* ... **34 H4** 32 58N 106 5 E
Yangquan, *China* .... **34 F7** 37 58N 113 31 E
Yangtze Kiang = Chang Jiang →, *China* ..... **33 C7** 31 48N 121 10 E
Yangyang, *S. Korea* .... **35 E15** 38 4N 128 38 E
Yangyuan, *China* .... **34 D8** 40 1N 114 10 E
Yangzhou, *China* .... **33 C6** 32 21N 119 26 E
Yanji, *China* .... **35 C15** 42 59N 129 30 E
Yankton, *U.S.A.* **80 D6** 42 53N 97 23W
Yanna, *Australia* .... **63 D4** 26 58S 146 0 E
Yanonge, *Zaïre* .... **54 B1** 0 35N 24 38 E
Yanqi, *China* .... **32 B3** 42 5N 86 35 E
Yanqing, *China* .... **34 D8** 40 30N 115 58 E
Yanshan, *China* .... **35 E9** 38 4N 117 22 E
Yanshou, *China* .... **35 B15** 45 28N 128 22 E
Yantabulla, *Australia* ... **63 D4** 29 21S 145 0 E
Yantai, *China* .... **35 F11** 37 34N 121 22 E
Yanykurgan, *Kazakstan* .. **26 E7** 43 50N 68 48 E
Yanzhou, *China* .... **34 G9** 35 35N 116 49 E
Yao, *Chad* .... **51 F8** 12 56N 17 33 E
Yao Xian, *China* .... **34 G5** 34 55N 108 59 E
Yao Yai, Ko, *Thailand* ... **39 J2** 8 0N 98 35 E
Yaoundé, *Cameroon* ..... **50 H7** 3 50N 11 35 E
Yaowan, *China* .... **35 G10** 34 15N 118 3 E
Yap I., *Pac. Oc.* .... **64 G5** 9 30N 138 10 E
Yapen, *Indonesia* .... **37 E9** 1 50S 136 0 E
Yapen, Selat, *Indonesia* .. **37 E9** 1 20S 136 10 E
Yappar →, *Australia* ... **62 B3** 18 22S 141 16 E
Yaqui →, *Mexico* **86 B2** 27 37N 110 39W
Yar-Sale, *Russia* .... **26 C8** 66 50N 70 50 E
Yaraka, *Australia* ...... **62 C3** 24 53S 144 3 E
Yaransk, *Russia* .... **24 C8** 57 22N 47 49 E
Yardea P.O., *Australia* ... **63 E2** 32 23S 135 32 E
Yare →, *U.K.* .... **11 E9** 52 35N 1 38 E
Yaremcha, *Ukraine* ..... **17 D13** 48 27N 24 33 E
Yarensk, *Russia* .... **24 B8** 62 11N 49 15 E
Yarí →, *Colombia* **92 D4** 0 20S 72 20W
Yarkand = Shache, *China* **32 C2** 38 20N 77 10 E
Yarker, *Canada* **79 B8** 44 23N 76 46W
Yarkhun →, *Pakistan* ... **43 A5** 36 17N 72 30 E
Yarmouth, *Canada* **71 D6** 43 50N 66 7W
Yarmūk →, *Syria* ...... **47 C4** 32 42N 35 40 E
Yaroslavl, *Russia* .... **24 C6** 57 35N 39 55 E
Yarqa, W. →, *Egypt* .... **47 F2** 30 0N 33 49 E
Yarra Yarra Lakes, *Australia* .... **61 E2** 29 40S 115 45 E
Yarraden, *Australia* ..... **62 A3** 14 17S 143 15 E
Yarraloola, *Australia* .... **60 D2** 21 33S 115 52 E
Yarram, *Australia* .... **63 F4** 38 29S 146 39 E
Yarraman, *Australia* .... **63 D5** 26 50S 152 0 E
Yarranvale, *Australia* ... **63 D4** 26 50S 145 20 E
Yarras, *Australia* .... **63 E5** 31 25S 152 20 E
Yarrowmere, *Australia* .. **62 C4** 21 27S 145 53 E
Yartsevo, *Russia* .... **27 C10** 60 20N 90 0 E
Yasawa Group, *Fiji* ..... **59 C7** 17 0S 177 23 E
Yaselda, *Belarus* .... **17 B14** 52 7N 26 28 E
Yasin, *Pakistan* .... **43 A5** 36 24N 73 23 E
Yasinski, L., *Canada* **70 B4** 53 16N 77 35W
Yasinya, *Ukraine* .... **17 D13** 48 16N 24 21 E
Yasothon, *Thailand* ..... **38 E5** 15 50N 104 10 E
Yass, *Australia* .... **63 E4** 34 49S 148 54 E
Yatağan, *Turkey* .... **21 F13** 37 20N 28 10 E
Yates Center, *U.S.A.* **81 G7** 37 53N 95 44W
Yathkyed L., *Canada* **73 A9** 62 40N 98 0W
Yatsushiro, *Japan* ...... **31 H5** 32 30N 130 40 E
Yatta Plateau, *Kenya* ... **54 C4** 2 0S 38 0 E
Yauyos, *Peru* **92 F3** 12 19S 75 50W
Yavari →, *Peru* **92 D4** 4 21S 70 2W
Yavatmal, *India* .... **40 J11** 20 20N 78 15 E
Yavne, *Israel* .... **47 D3** 31 52N 34 45 E
Yavoriv, *Ukraine* .... **17 D12** 49 55N 23 20 E
Yavorov = Yavoriv, *Ukraine* .... **17 D12** 49 55N 23 20 E
Yawatahama, *Japan* ..... **31 H6** 33 27N 132 24 E
Yayama-Rettō, *Japan* ... **31 M1** 24 30N 123 40 E
Yazd, *Iran* .... **45 D7** 31 55N 54 27 E
Yazd □, *Iran* .... **45 D7** 32 0N 55 0 E
Yazoo →, *U.S.A.* **81 J9** 32 22N 90 54W
Yazoo City, *U.S.A.* **81 J9** 32 51N 90 25W
Yding Skovhøj, *Denmark* **9 J13** 55 59N 9 46 E
Ye Xian, *Henan, China* .. **34 H7** 33 35N 113 25 E
Ye Xian, *Shandong, China* **35 F10** 37 8N 119 57 E
Yealering, *Australia* .... **61 F2** 32 36S 117 36 E
Yebyu, *Burma* .... **41 M21** 14 15N 98 13 E
Yechŏn, *S. Korea* ...... **35 F15** 36 39N 128 27 E
Yecla, *Spain* **19 C5** 38 35N 1 5W
Yécora, *Mexico* **86 B3** 28 20N 108 58W
Yedintsy = Edinița, *Moldova* .... **17 D14** 48 9N 27 18 E
Yeeda, *Australia* .... **60 C3** 17 31S 123 38 E
Yeelanna, *Australia* ..... **63 E2** 34 9S 135 45 E
Yegros, *Paraguay* **94 B4** 26 20S 56 25W
Yehuda, Midbar, *Israel* .. **47 D4** 31 35N 35 15 E
Yei, *Sudan* .... **51 H11** 4 9N 30 40 E
Yekaterinburg, *Russia* ... **24 C11** 56 50N 60 30 E
Yekaterinodar = Krasnodar, *Russia* .... **25 E6** 45 5N 39 0 E
Yelanskoye, *Russia* ..... **27 C13** 61 25N 128 0 E
Yelarbon, *Australia* ..... **63 D5** 28 33S 150 38 E
Yelets, *Russia* .... **24 D6** 52 40N 38 30 E
Yelizavetgrad = Kirovohrad, *Ukraine* ... **25 E5** 48 35N 32 20 E
Yell, *U.K.* **12 A7** 60 35N 1 5W
Yell Sd., *U.K.* **12 A7** 60 33N 1 15W
Yellow Sea, *China* ...... **35 G12** 35 0N 123 0 E
Yellowhead Pass, *Canada* **72 C5** 52 53N 118 25W
Yellowknife, *Canada* **72 A6** 62 27N 114 29W
Yellowknife →, *Canada* . **72 A6** 62 31N 114 19W
Yellowstone →, *U.S.A.* . **80 B3** 47 59N 103 59W
Yellowstone L., *U.S.A.* ... **82 D8** 44 27N 110 22W
Yellowstone National Park, *U.S.A.* **82 D8** 44 40N 110 30W
Yellowtail Res., *U.S.A.* ... **82 D9** 45 6N 108 8W
Yelsk, *Belarus* .... **17 C15** 51 50N 29 10 E
Yelvertoft, *Australia* .... **62 C2** 20 13S 138 45 E
Yemen ■, *Asia* .... **46 E3** 15 0N 44 0 E
Yen Bai, *Vietnam* .... **38 B5** 21 42N 104 52 E
Yenangyaung, *Burma* ... **41 J19** 20 30N 95 0 E
Yenbo = Yanbu 'al Baḥr, *Si. Arabia* .... **44 F3** 24 0N 38 5 E
Yenda, *Australia* .... **63 E4** 34 13S 146 14 E
Yenice, *Turkey* .... **21 E12** 39 55N 27 17 E
Yenisey →, *Russia* ..... **26 B9** 71 50N 82 40 E
Yeniseysk, *Russia* ...... **27 D10** 58 27N 92 13 E
Yeniseyskiy Zaliv, *Russia* **26 B9** 72 20N 81 0 E
Yennádhi, *Greece* ...... **23 C9** 36 2N 27 56 E
Yenyuka, *Russia* .... **27 D13** 57 57N 121 15 E
Yeo, L., *Australia* ...... **61 E3** 28 0S 124 30 E
Yeola, *India* .... **40 J9** 20 2N 74 30 E
Yeoryioúpolis, *Greece* ... **23 D6** 35 20N 24 15 E
Yeovil, *U.K.* **11 G5** 50 57N 2 38W
Yeppoon, *Australia* ..... **62 C5** 23 5S 150 47 E
Yerbent, *Turkmenistan* .. **26 F6** 39 30N 58 50 E
Yerbogachen, *Russia* .... **27 C11** 61 16N 108 0 E
Yerevan, *Armenia* ...... **25 F7** 40 10N 44 31 E
Yerilla, *Australia* .... **61 E3** 29 24S 121 47 E
Yermak, *Kazakstan* ..... **26 D8** 52 2N 76 55 E
Yermakovo, *Russia* ..... **27 D13** 52 25N 126 20 E
Yermo, *U.S.A.* **85 L10** 34 54N 116 50W
Yerofey Pavlovich, *Russia* **27 D13** 54 0N 122 0 E
Yerólakkos, *Cyprus* ..... **23 D12** 35 11N 33 15 E
Yeropol, *Russia* .... **27 C17** 65 15N 168 40 E
Yeropótamos →, *Greece* **23 D6** 35 3N 24 50 E
Yeroskipos, *Cyprus* ..... **23 E11** 34 46N 32 28 E
Yershov, *Russia* .... **25 D8** 51 23N 48 27 E
Yerushalayim = Jerusalem, *Israel* ...... **47 D4** 31 47N 35 10 E
Yes Tor, *U.K.* **11 G4** 50 41N 4 0W
Yesan, *S. Korea* .... **35 F14** 36 41N 126 51 E
Yeso, *U.S.A.* **81 H2** 34 26N 104 37W
Yessey, *Russia* .... **27 C11** 68 29N 102 10 E
Yeu, I. d', *France* **18 C2** 46 42N 2 20W
Yevpatoriya, *Ukraine* .... **25 E5** 45 15N 33 20 E
Yeysk, *Russia* .... **25 E6** 46 40N 38 12 E
Yezd = Yazd, *Iran* ...... **45 D7** 31 55N 54 27 E
Yhati, *Paraguay* **94 B4** 25 0S 56 35W
Yhú, *Paraguay* **95 B4** 25 0S 56 0W
Yi →, *Uruguay* **94 C4** 33 7S 57 8W

Yi 'Allaq, G., *Egypt* ..... **47 E2** 30 22N 33 32 E
Yi He —►, *China* .... **35 G10** 34 10N 118 8 E
Yi Xian, *Hebei, China* ... **34 E8** 39 20N 115 30 E
Yi Xian, *Liaoning, China* . **35 D11** 41 30N 121 22 E
Yialiás —►, *Cyprus* ..... **23 D12** 35 9N 33 44 E
Yialousa, *Cyprus* .... **23 D13** 35 32N 34 10 E
Yianisádhes, *Greece* .... **23 D8** 35 20N 26 10 E
Yiannitsa, *Greece* .... **21 D10** 40 46N 22 24 E
Yibin, *China* .... **32 D5** 28 45N 104 32 E
Yichang, *China* .... **33 C6** 30 40N 111 20 E
Yicheng, *China* .... **34 G6** 35 42N 111 40 E
Yichuan, *China* .... **34 F6** 36 2N 110 10 E
Yichun, *China* .... **33 B7** 47 44N 128 52 E
Yidu, *China* .... **35 F10** 36 43N 118 28 E
Yijun, *China* .... **34 G5** 35 28N 109 8 E
Yilehuli Shan, *China* .... **33 A7** 51 20N 124 20 E
Yimianpo, *China* .... **35 B15** 45 7N 128 2 E
Yinchuan, *China* .... **34 E4** 38 30N 106 15 E
Yindarlgooda, L., *Australia* **61 F3** 30 40S 121 52 E
Ying He —►, *China* ..... **34 H9** 32 30N 116 30 E
Ying Xian, *China* .... **34 E7** 39 32N 113 10 E
Yingkou, *China* .... **35 D12** 40 37N 122 18 E
Yining, *China* .... **26 E9** 43 58N 81 10 E
Yinmabin, *Burma* .... **41 H19** 22 10N 94 55 E
Yinnietharra, *Australia* .. **60 D2** 24 39S 116 12 E
Yiofiros —►, *Greece* .... **23 D7** 35 20N 25 6 E
Yishan, *China* .... **32 D5** 24 28N 108 38 E
Yishui, *China* .... **35 G10** 35 47N 118 30 E
Yíthion, *Greece* .... **21 F10** 36 46N 22 34 E
Yitiaoshan, *China* .... **34 F3** 37 5N 104 2 E
Yitong, *China* .... **35 C13** 43 13N 125 20 E
Yiyang, *Henan, China* ... **34 G7** 34 27N 112 10 E
Yiyang, *Hunan, China* ... **33 D6** 28 35N 112 18 E
Yli-Kitka, *Finland* .... **8 C23** 66 8N 28 30 E
Ylitornio, *Finland* .... **8 C20** 66 19N 23 39 E
Ylivieska, *Finland* .... **8 D21** 64 4N 24 28 E
Ynykchanskiy, *Russia* ... **27 C14** 60 15N 137 35 E
Yoakum, *U.S.A.* **81 L6** 29 17N 97 9W
Yog Pt., *Phil.* .... **37 B6** 14 6N 124 12 E
Yogyakarta, *Indonesia* ... **37 G14** 7 49S 110 22 E
Yoho Nat. Park, *Canada* . **72 C5** 51 25N 116 30W
Yojoa, L. de, *Honduras* .. **88 D2** 14 53N 88 0W
Yŏju, *S. Korea* .... **35 F14** 37 20N 127 35 E
Yokadouma, *Cameroon* .. **52 D2** 3 26N 14 55 E
Yokkaichi, *Japan* .... **31 G8** 34 55N 136 38 E
Yoko, *Cameroon* .... **51 G7** 5 32N 12 20 E
Yokohama, *Japan* ...... **31 G9** 35 27N 139 28 E
Yokosuka, *Japan* ...... **31 G9** 35 20N 139 40 E
Yokote, *Japan* .... **30 E10** 39 20N 140 30 E
Yola, *Nigeria* .... **51 G7** 9 10N 12 29 E
Yolaina, Cordillera de, *Nic.* **88 D3** 11 30N 84 0W
Yonago, *Japan* .... **31 G6** 35 25N 133 19 E
Yonaguni-Jima, *Japan* .. **31 M1** 24 27N 123 0 E
Yŏnan, *N. Korea* .... **35 F14** 37 55N 126 11 E
Yonezawa, *Japan* .... **30 F10** 37 57N 140 4 E
Yong Peng, *Malaysia* .... **39 L4** 2 0N 103 3 E
Yong Sata, *Thailand* .... **39 J2** 7 8N 99 41 E
Yongampo, *N. Korea* .... **35 E13** 39 56N 124 23 E
Yongcheng, *China* ...... **34 H9** 33 55N 116 20 E
Yŏngchŏn, *S. Korea* .... **35 G15** 35 58N 128 56 E
Yongdeng, *China* .... **34 F2** 36 38N 103 25 E
Yŏngdŏk, *S. Korea* ..... **35 F15** 36 24N 129 22 E
Yŏngdŭngpo, *S. Korea* .. **35 F14** 37 31N 126 54 E
Yonghe, *China* .... **34 F6** 36 46N 110 38 E
Yŏnghŭng, *N. Korea* .... **35 E14** 39 31N 127 18 E
Yongji, *China* .... **34 G6** 34 52N 110 28 E
Yŏngju, *S. Korea* .... **35 F15** 36 50N 128 40 E
Yongnian, *China* .... **34 F8** 36 47N 114 29 E
Yongning, *China* .... **34 E4** 38 15N 106 14 E
Yongqing, *China* .... **34 E9** 39 25N 116 28 E
Yŏngwŏl, *S. Korea* ..... **35 F15** 37 11N 128 28 E
Yonibana, *S. Leone* **50 G2** 8 30N 12 19W
Yonkers, *U.S.A.* **79 F11** 40 56N 73 54W
Yonne —►, *France* ..... **18 B5** 48 23N 2 58 E
York, *Australia* .... **61 F2** 31 52S 116 47 E
York, *U.K.* **10 D6** 53 58N 1 6W
York, *Ala., U.S.A.* **77 J1** 32 29N 88 18W
York, *Nebr., U.S.A.* **80 E6** 40 52N 97 36W
York, *Pa., U.S.A.* **76 F7** 39 58N 76 44W
York □, *U.K.* **10 D6** 53 58N 1 6W
York, C., *Australia* ...... **62 A3** 10 42S 142 31 E
York, Kap, *Greenland* **4 B4** 75 55N 66 25W
York Sd., *Australia* ...... **60 B4** 15 0S 125 5 E
Yorke Pen., *Australia* .... **63 E2** 34 50S 137 40 E
Yorkshire Wolds, *U.K.* ... **10 D7** 54 8N 0 31W
Yorkton, *Canada* **73 C8** 51 11N 102 28W
Yorktown, *U.S.A.* **81 L6** 28 59N 97 30W
Yorkville, *U.S.A.* **84 G3** 38 52N 123 13W
Yornup, *Australia* ...... **61 F2** 34 2S 116 10 E
Yoro, *Honduras* **88 C2** 15 9N 87 7W
Yoron-Jima, *Japan* ..... **31 L4** 27 2N 128 26 E
Yos Sudarso, Pulau, *Indonesia* .... **37 F9** 8 0S 138 30 E
Yosemite National Park, *U.S.A.* **83 H4** 37 45N 119 40W
Yosemite Village, *U.S.A.* . **84 H7** 37 45N 119 35W
Yoshkar Ola, *Russia* ..... **24 C8** 56 38N 47 55 E
Yŏsu, *S. Korea* .... **35 G14** 34 47N 127 45 E
Yotvata, *Israel* .... **47 F4** 29 55N 35 2 E
Youbou, *Canada* **72 D4** 48 53N 124 13W
Youghal, *Ireland* **13 E4** 51 56N 7 52W
Youghal B., *Ireland* **13 E4** 51 55N 7 49W
Young, *Australia* .... **63 E4** 34 19S 148 18 E
Young, *Canada* **73 C7** 51 47N 105 45W
Young, *Uruguay* **94 C4** 32 44S 57 36W
Younghusband, L., *Australia* .... **63 E2** 30 50S 136 5 E
Younghusband Pen., *Australia* .... **63 F2** 36 0S 139 25 E
Youngstown, *Canada* ... **73 C6** 51 35N 111 10W
Youngstown, *N.Y., U.S.A.* **78 C5** 43 15N 79 3W
Youngstown, *Ohio, U.S.A.* **78 E4** 41 6N 80 39W
Youngsville, *U.S.A.* **78 E5** 41 51N 79 19W
Youyu, *China* .... **34 D7** 40 10N 112 20 E
Yoweragabbie, *Australia* . **61 E2** 28 14S 117 39 E
Yozgat, *Turkey* .... **25 G5** 39 51N 34 47 E
Ypané —►, *Paraguay* ... **94 A4** 23 29S 57 19W
Ypres = Ieper, *Belgium* .. **15 D2** 50 51N 2 53 E
Ypsilanti, *U.S.A.* **76 D4** 42 14N 83 37W
Yreka, *U.S.A.* **82 F2** 41 44N 122 38W
Ysleta, *U.S.A.* **83 L10** 31 45N 106 24W
Ystad, *Sweden* .... **9 J15** 55 26N 13 50 E
Ysyk-Köl, *Kyrgyzstan* ... **28 E11** 42 26N 76 12 E
Ysyk-Köl, Ozero, *Kyrgyzstan* .... **26 E8** 42 25N 77 15 E
Ythan —►, *U.K.* **12 D7** 57 19N 1 59W
Ytyk Kyuyel, *Russia* ..... **27 C14** 62 30N 133 45 E
Yu Jiang —►, *China* .... **33 D6** 23 22N 110 3 E
Yu Xian, *Hebei, China* ... **34 E8** 39 50N 114 35 E
Yu Xian, *Henan, China* ... **34 G7** 34 10N 113 28 E
Yu Xian, *Shanxi, China* .. **34 E7** 38 5N 113 20 E
Yuan Jiang —►, *China* .. **33 D6** 28 55N 111 50 E
Yuanqu, *China* .... **34 G6** 35 18N 111 40 E
Yuanyang, *China* .... **34 G7** 35 3N 113 58 E
Yuba —►, *U.S.A.* **84 F5** 39 8N 121 36W
Yuba City, *U.S.A.* **84 F5** 39 8N 121 37W
Yūbari, *Japan* .... **30 C10** 43 4N 141 59 E
Yūbetsu, *Japan* .... **30 B11** 44 13N 143 50 E
Yucatán □, *Mexico* **87 C7** 21 30N 86 30W
Yucatán, Canal de, *Caribbean* **88 B2** 22 0N 86 30W
Yucatán, Península de, *Mexico* **66 H11** 19 30N 89 0W
Yucatán Basin, *Cent. Amer.* **66 H11** 19 0N 86 0W
Yucatan Str. = Yucatán, Canal de, *Caribbean* ... **88 B2** 22 0N 86 30W
Yucca, *U.S.A.* **85 L12** 34 52N 114 9W
Yucca Valley, *U.S.A.* **85 L10** 34 8N 116 27W
Yucheng, *China* .... **34 F9** 36 55N 116 32 E
Yuci, *China* .... **34 F7** 37 42N 112 46 E
Yudino, *Russia* .... **26 D7** 55 10N 67 55 E
Yuendumu, *Australia* .... **60 D5** 22 16S 131 49 E
Yugoslavia ■, *Europe* .... **21 B9** 44 0N 20 0 E
Yukon —►, *U.S.A.* **68 B3** 62 32N 163 54W
Yukon Territory □, *Canada* **68 B6** 63 0N 135 0W
Yukti, *Russia* .... **27 C11** 63 26N 105 42 E
Yukuhashi, *Japan* .... **31 H5** 33 44N 130 59 E
Yule —►, *Australia* ..... **60 D2** 20 41S 118 17 E
Yulin, *China* .... **34 E5** 38 20N 109 30 E
Yuma, *Ariz., U.S.A.* **85 N12** 32 43N 114 37W
Yuma, *Colo., U.S.A.* **80 E3** 40 8N 102 43W
Yuma, B. de, *Dom. Rep.* . **89 C6** 18 20N 68 35W
Yumbe, *Uganda* .... **54 B3** 3 28N 31 15 E
Yumbi, *Zaïre* .... **54 C2** 1 12S 26 15 E
Yumen, *China* .... **32 C4** 39 50N 97 30 E
Yun Ho —►, *China* ..... **35 E9** 39 10N 117 10 E
Yuncheng, *Henan, China* **34 G8** 35 36N 115 57 E
Yuncheng, *Shanxi, China* **34 G6** 35 2N 111 0 E
Yundamindra, *Australia* . **61 E3** 29 15S 122 6 E
Yungas, *Bolivia* **92 G5** 17 0S 66 0W
Yungay, *Chile* **94 D1** 37 10S 72 5W
Yunnan □, *China* .... **32 D5** 25 0N 102 0 E
Yunta, *Australia* .... **63 E2** 32 34S 139 36 E
Yunxi, *China* .... **34 H6** 33 0N 110 22 E
Yupyongdong, *N. Korea* . **35 D15** 41 49N 128 53 E
Yur, *Russia* .... **27 D14** 59 52N 137 41 E
Yurgao, *Russia* .... **26 D9** 55 42N 84 51 E
Yuribei, *Russia* .... **26 B8** 71 8N 76 58 E
Yurimaguas, *Peru* **92 E3** 5 55S 76 7W
Yuryung Kaya, *Russia* ... **27 B12** 72 48N 113 23 E
Yuscarán, *Honduras* **88 D2** 13 58N 86 45W
Yushe, *China* .... **34 F7** 37 4N 112 58 E
Yushu, *Jilin, China* ..... **35 B14** 44 43N 126 38 E
Yushu, *Qinghai, China* .. **32 C4** 33 5N 96 55 E
Yutai, *China* .... **34 G9** 35 0N 116 45 E
Yutian, *China* .... **35 E9** 39 53N 117 45 E
Yuxi, *China* .... **32 D5** 24 30N 102 35 E
Yuzawa, *Japan* .... **30 E10** 39 10N 140 30 E
Yuzhno-Sakhalinsk, *Russia* **27 E15** 46 58N 142 45 E
Yvetot, *France* .... **18 B4** 49 37N 0 44 E

Z

Zaandam, *Neths.* .... **15 B4** 52 26N 4 49 E
Zabaykalsk, *Russia* ..... **27 E12** 49 40N 117 25 E
Zabid, *Yemen* .... **46 E3** 14 0N 43 10 E
Zābol, *Iran* .... **45 D9** 31 0N 61 32 E
Zāboli, *Iran* .... **45 E9** 27 10N 61 35 E
Zabrze, *Poland* .... **17 C10** 50 18N 18 50 E
Zacapa, *Guatemala* **88 D2** 14 59N 89 31W
Zacapu, *Mexico* **86 D4** 19 50N 101 43W
Zacatecas, *Mexico* **86 C4** 22 49N 102 34W
Zacatecas □, *Mexico* **86 C4** 23 30N 103 0W
Zacatecoluca, *El Salv.* ... **88 D2** 13 29N 88 51W
Zacoalco, *Mexico* **86 C4** 20 14N 103 33W
Zacualtipán, *Mexico* **87 C5** 20 39N 98 36W
Zadar, *Croatia* .... **16 F8** 44 8N 15 14 E
Zadetkyi Kyun, *Burma* .. **39 H2** 10 0N 98 25 E
Zafarqand, *Iran* .... **45 C7** 33 11N 52 29 E
Zafra, *Spain* **19 C2** 38 26N 6 30W
Żagań, *Poland* .... **16 C8** 51 39N 15 22 E
Zagazig, *Egypt* .... **51 B11** 30 40N 31 30 E
Zāgheh, *Iran* .... **45 C6** 33 30N 48 42 E
Zagorsk = Sergiyev Posad, *Russia* .... **24 C6** 56 20N 38 10 E
Zagreb, *Croatia* .... **16 F9** 45 50N 16 0 E
Zāgros, Kūhhā-ye, *Iran* .. **45 C6** 33 45N 48 5 E
Zagros Mts. = Zāgros, Kūhhā-ye, *Iran* ...... **45 C6** 33 45N 48 5 E
Zāhedān, *Fārs, Iran* ..... **45 D7** 28 46N 53 52 E
Zāhedān, *Sīstān va Balūchestān, Iran* .... **45 D9** 29 30N 60 50 E
Zahlah, *Lebanon* .... **47 B4** 33 52N 35 50 E
Zaïre ■, *Africa* .... **52 E4** 3 0S 23 0 E
Zaïre —►, *Africa* .... **52 F2** 6 4S 12 24 E
Zaječar, *Serbia, Yug.* .... **21 C10** 43 53N 22 18 E
Zakamensk, *Russia* ..... **27 D11** 50 23N 103 17 E
Zakavkazye, *Asia* .... **25 F7** 42 0N 44 0 E
Zakhodnaya Dzvina = Daugava —►, *Latvia* .. **9 H21** 57 4N 24 3 E
Zākhū, *Iraq* .... **44 B4** 37 10N 42 50 E
Zákinthos, *Greece* ...... **21 F9** 37 47N 20 57 E
Zakopane, *Poland* ...... **17 D10** 49 18N 19 57 E
Zákros, *Greece* .... **23 D8** 35 6N 26 10 E
Zalaegerszeg, *Hungary* .. **17 E9** 46 53N 16 47 E
Zalău, *Romania* .... **17 E12** 47 12N 23 3 E
Zaleshchiki = Zalishchyky, *Ukraine* .... **17 D13** 48 45N 25 45 E
Zalew Wiślany, *Poland* .. **17 A10** 54 20N 19 50 E
Zalingei, *Sudan* .... **51 F9** 12 51N 23 29 E
Zalishchyky, *Ukraine* .... **17 D13** 48 45N 25 45 E
Zambeke, *Zaïre* .... **54 B2** 2 8N 25 17 E
Zambeze —►, *Africa* .... **55 F4** 18 35S 36 20 E
Zambezi = Zambeze —►, *Africa* .... **55 F4** 18 35S 36 20 E
Zambezi, *Zambia* .... **53 G4** 13 30S 23 15 E
Zambezia □, *Mozam.* ... **55 F4** 16 15S 37 30 E
Zambia ■, *Africa* .... **55 E2** 15 0S 28 0 E
Zamboanga, *Phil.* .... **37 C6** 6 59N 122 3 E
Zamora, *Mexico* **86 C4** 20 0N 102 21W
Zamora, *Spain* **19 B3** 41 30N 5 45W
Zamość, *Poland* .... **17 C12** 50 43N 23 15 E
Zanaga, *Congo* .... **52 E2** 2 48S 13 48 E
Zandvoort, *Neths.* ...... **15 B4** 52 22N 4 32 E
Zanesville, *U.S.A.* **78 G2** 39 56N 82 1W
Zangābād, *Iran* .... **44 B5** 38 26N 46 44 E
Zangue —►, *Mozam.* .... **55 F4** 17 50S 35 21 E
Zanjān, *Iran* .... **45 B6** 36 40N 48 35 E
Zanjān □, *Iran* .... **45 B6** 37 20N 49 30 E
Zante = Zákinthos, *Greece* **21 F9** 37 47N 20 57 E
Zanthus, *Australia* ...... **61 F3** 31 2S 123 34 E
Zanzibar, *Tanzania* ..... **54 D4** 6 12S 39 12 E
Zaouiet El-Kala = Bordj Omar Driss, *Algeria* ... **50 C6** 28 10N 6 40 E
Zaouiet Reggane, *Algeria* **50 C5** 26 32N 0 3 E
Zaozhuang, *China* ...... **35 G9** 34 50N 117 35 E
Zap Suyu = Kabīr, Zab al —►, *Iraq* .... **44 C4** 36 1N 43 24 E
Zapadnaya Dvina, *Russia* **26 D4** 56 15N 32 3 E
Zapadnaya Dvina —► = Daugava —►, *Latvia* .. **9 H21** 57 4N 24 3 E
Západné Beskydy, *Europe* **17 D10** 49 30N 19 0 E
Zapala, *Argentina* **96 D2** 39 0S 70 5W
Zapaleri, Cerro, *Bolivia* .. **94 A2** 22 49S 67 11W
Zapata, *U.S.A.* **81 M5** 26 55N 99 16W
Zapolyarnyy, *Russia* .... **24 A5** 69 26N 30 51 E
Zaporizhzhya, *Ukraine* ... **25 E6** 47 50N 35 10 E
Zaporozhye = Zaporizhzhya, *Ukraine* . **25 E6** 47 50N 35 10 E
Zaragoza, *Coahuila, Mexico* **86 B4** 28 30N 101 0W
Zaragoza, *Nuevo León, Mexico* **87 C5** 24 0N 99 46W
Zaragoza, *Spain* **19 B5** 41 39N 0 53W
Zarand, *Kermān, Iran* ... **45 D8** 30 46N 56 34 E
Zarand, *Markazī, Iran* ... **45 C6** 35 18N 50 25 E
Zaranj, *Afghan.* .... **40 D2** 30 55N 61 55 E
Zarasai, *Lithuania* ...... **9 J22** 55 40N 26 20 E
Zarate, *Argentina* **94 C4** 34 7S 59 0W
Zāreh, *Iran* .... **45 C6** 35 7N 49 9 E
Zaremba I., *U.S.A.* **72 B2** 56 20N 132 50W
Zaria, *Nigeria* .... **50 F6** 11 0N 7 40 E
Zarneh, *Iran* .... **44 C5** 33 55N 46 10 E
Zarós, *Greece* .... **23 D6** 35 8N 24 54 E
Zarqā' —►, *Jordan* ..... **47 C4** 32 10N 35 37 E
Zarrīn, *Iran* .... **45 C7** 32 46N 54 37 E
Zaruma, *Ecuador* **92 D3** 3 40S 79 38W
Żary, *Poland* .... **16 C8** 51 37N 15 10 E
Zarzis, *Tunisia* .... **51 B7** 33 31N 11 2 E
Zashiversk, *Russia* ..... **27 C15** 67 25N 142 40 E
Zaskar —►, *India* ...... **43 B7** 34 13N 77 20 E
Zaskar Mts., *India* ...... **43 C7** 33 15N 77 30 E
Zastron, *S. Africa* ...... **56 E4** 30 18S 27 7 E
Zavāreh, *Iran* .... **45 C7** 33 29N 52 28 E
Zavitinsk, *Russia* .... **27 D13** 50 10N 129 20 E
Zavodovski, I., *Antarctica* **5 B1** 56 0S 27 45W
Zawiercie, *Poland* ...... **17 C10** 50 30N 19 24 E
Zāyā, *Iraq* .... **44 C5** 33 33N 44 13 E
Zayarsk, *Russia* .... **27 D11** 56 12N 102 55 E
Zaysan, *Kazakstan* ..... **26 E9** 47 28N 84 52 E
Zaysan, Oz., *Kazakstan* .. **26 E9** 48 0N 83 0 E
Zayü, *China* .... **32 D4** 28 48N 97 27 E
Zbarazh, *Ukraine* .... **17 D13** 49 43N 25 44 E
Zdolbuniv, *Ukraine* ..... **17 C14** 50 30N 26 15 E
Zduńska Wola, *Poland* .. **17 C10** 51 37N 18 59 E
Zeballos, *Canada* **72 D3** 49 59N 126 50W
Zebediela, *S. Africa* ..... **57 C4** 24 20S 29 17 E
Zeebrugge, *Belgium* ..... **15 C3** 51 19N 3 12 E
Zeehan, *Australia* .... **62 G4** 41 52S 145 25 E
Zeeland □, *Neths.* ...... **15 C3** 51 30N 3 50 E
Zeerust, *S. Africa* .... **56 D4** 25 31S 26 4 E
Zefat, *Israel* .... **47 C4** 32 58N 35 29 E
Zeil, Mt., *Australia* ..... **60 D5** 23 30S 132 23 E
Zeila, *Somali Rep.* ...... **46 E3** 11 21N 43 30 E
Zeist, *Neths.* .... **15 B5** 52 5N 5 15 E
Zeitz, *Germany* .... **16 C7** 51 2N 12 7 E
Zelenograd, *Russia* ..... **24 C6** 56 1N 37 12 E
Zelenogradsk, *Russia* .... **9 J19** 54 53N 20 29 E
Zelzate, *Belgium* .... **15 C3** 51 13N 3 47 E
Zémio, *C.A.R.* .... **54 A2** 5 2N 25 5 E
Zemun, *Serbia, Yug.* .... **21 B9** 44 51N 20 25 E
Zenica, *Bos.-H.* .... **21 B7** 44 10N 17 57 E
Žepce, *Bos.-H.* .... **21 B8** 44 28N 18 2 E
Zeya, *Russia* .... **27 D13** 53 48N 127 14 E
Zeya —►, *Russia* .... **27 D13** 51 42N 128 53 E
Zêzere —►, *Portugal* ... **19 C1** 39 28N 8 20W
Zghartā, *Lebanon* .... **47 A4** 34 21N 35 53 E
Zgorzelec, *Poland* ...... **16 C8** 51 10N 15 0 E
Zhabinka, *Belarus* ...... **17 B13** 52 13N 24 2 E
Zhailma, *Kazakstan* ..... **26 D7** 51 37N 61 33 E
Zhambyl, *Kazakstan* ..... **26 E8** 42 54N 71 22 E
Zhangaqazaly, *Kazakstan* **26 E7** 45 48N 62 6 E
Zhangbei, *China* .... **34 D8** 41 10N 114 45 E
Zhangguangcai Ling, *China* .... **35 B15** 45 0N 129 0 E
Zhangjiakou, *China* ..... **34 D8** 40 48N 114 55 E
Zhangwu, *China* .... **35 C12** 42 43N 123 52 E
Zhangye, *China* .... **32 C5** 38 50N 100 23 E
Zhangzhou, *China* ...... **33 D6** 24 30N 117 35 E
Zhanhua, *China* .... **35 F10** 37 40N 118 8 E
Zhanjiang, *China* .... **33 D6** 21 15N 110 20 E
Zhanyi, *China* .... **32 D5** 25 38N 103 48 E
Zhanyu, *China* .... **35 B12** 44 30N 122 30 E
Zhao Xian, *China* ...... **34 F8** 37 43N 114 45 E
Zhaocheng, *China* ...... **34 F6** 36 22N 111 38 E
Zhaotong, *China* .... **32 D5** 27 20N 103 44 E
Zhaoyuan, *Heilongjiang, China* .... **35 B13** 45 27N 125 0 E
Zhaoyuan, *Shandong, China* .... **35 F11** 37 20N 120 23 E
Zhashkiv, *Ukraine* ...... **17 D16** 49 15N 30 5 E
Zhashui, *China* .... **34 H5** 33 40N 109 8 E
Zhayyq —►, *Kazakstan* .. **25 E9** 47 0N 51 48 E
Zhdanov = Mariupol, *Ukraine* .... **25 E6** 47 5N 37 31 E
Zhecheng, *China* .... **34 G8** 34 7N 115 20 E
Zhejiang □, *China* ...... **33 D7** 29 0N 120 0 E
Zheleznodorozhnyy, *Russia* .... **24 B9** 62 35N 50 55 E
Zheleznogorsk-Ilimskiy, *Russia* .... **27 D11** 56 34N 104 8 E
Zhen'an, *China* .... **34 H5** 33 27N 109 9 E
Zhengding, *China* ...... **34 E8** 38 8N 114 32 E
Zhengzhou, *China* ...... **34 G7** 34 45N 113 34 E
Zhenlai, *China* .... **35 B12** 45 50N 123 5 E
Zhenping, *China* .... **34 H7** 33 10N 112 16 E
Zhenyuan, *China* .... **34 G4** 35 35N 107 30 E
Zhetiqara, *Kazakstan* .... **26 D7** 52 11N 61 12 E
Zhezqazghan, *Kazakstan* . **26 E7** 47 44N 67 40 E
Zhidan, *China* .... **34 F5** 36 48N 108 48 E
Zhigansk, *Russia* .... **27 C13** 66 48N 123 27 E
Zhilinda, *Russia* .... **27 C12** 70 0N 114 20 E
Zhitomir = Zhytomyr, *Ukraine* .... **17 C15** 50 20N 28 40 E
Zhlobin, *Belarus* .... **17 B16** 52 55N 30 0 E
Zhmerinka = Zhmerynka, *Ukraine* .... **17 D15** 49 2N 28 2 E
Zhmerynka, *Ukraine* ..... **17 D15** 49 2N 28 2 E
Zhodino = Zhodzina, *Belarus* .... **17 A15** 54 5N 28 17 E
Zhodzina, *Belarus* ...... **17 A15** 54 5N 28 17 E
Zhokhova, Ostrov, *Russia* **27 B16** 76 4N 152 40 E
Zhongdian, *China* ...... **32 D4** 27 48N 99 42 E
Zhongning, *China* ...... **34 F3** 37 29N 105 40 E
Zhongtiao Shan, *China* .. **34 G6** 35 0N 111 10 E
Zhongwei, *China* .... **34 F3** 37 30N 105 12 E
Zhongyang, *China* ...... **34 F6** 37 20N 111 11 E
Zhoucun, *China* .... **35 F9** 36 47N 117 48 E
Zhouzhi, *China* .... **34 G5** 34 10N 108 12 E
Zhuanghe, *China* .... **35 E12** 39 40N 123 0 E
Zhucheng, *China* .... **35 G10** 36 0N 119 27 E
Zhugqu, *China* .... **34 H3** 33 40N 104 30 E
Zhumadian, *China* ...... **34 H8** 32 59N 114 2 E
Zhuo Xian, *China* ...... **34 E8** 39 28N 115 58 E
Zhuolu, *China* .... **34 D8** 40 20N 115 12 E
Zhuozi, *China* .... **34 D7** 41 0N 112 25 E
Zhupanovo, *Russia* ..... **27 D16** 53 40N 159 52 E
Zhytomyr, *Ukraine* ...... **17 C15** 50 20N 28 40 E
Zīārān, *Iran* .... **45 B6** 36 7N 50 32 E
Ziarat, *Pakistan* .... **42 D2** 30 25N 67 49 E
Zibo, *China* .... **35 F10** 36 47N 118 3 E
Zichang, *China* .... **34 F5** 37 18N 109 40 E
Zielona Góra, *Poland* .... **16 C8** 51 57N 15 31 E
Zierikzee, *Neths.* .... **15 C3** 51 40N 3 55 E
Zigey, *Chad* .... **51 F8** 14 43N 15 50 E
Zigong, *China* .... **32 D5** 29 15N 104 48 E
Ziguinchor, *Senegal* **50 F1** 12 35N 16 20W
Zihuatanejo, *Mexico* **86 D4** 17 38N 101 33W
Žilina, *Slovak Rep.* ...... **17 D10** 49 12N 18 42 E
Zillah, *Libya* .... **51 C8** 28 30N 17 33 E
Zima, *Russia* .... **27 D11** 54 0N 102 5 E
Zimapán, *Mexico* **87 C5** 20 54N 99 20W
Zimba, *Zambia* .... **55 F2** 17 20S 26 11 E
Zimbabwe, *Zimbabwe* ... **55 G3** 20 16S 30 54 E
Zimbabwe ■, *Africa* .... **55 F2** 19 0S 30 0 E
Zimnicea, *Romania* ..... **17 G13** 43 40N 25 22 E
Zinder, *Niger* .... **50 F6** 13 48N 9 0 E
Zinga, *Tanzania* .... **55 D4** 9 16S 38 49 E
Zion National Park, *U.S.A.* **83 H7** 37 15N 113 5W
Zipaquirá, *Colombia* **92 C4** 5 0N 74 0W
Ziros, *Greece* .... **23 D8** 35 5N 26 8 E
Zitácuaro, *Mexico* **86 D4** 19 28N 100 21W
Zitundo, *Mozam.* .... **57 D5** 26 48S 32 47 E
Ziway, L., *Ethiopia* ..... **51 G12** 8 0N 38 50 E
Ziyang, *China* .... **34 H5** 32 32N 108 31 E
Zlatograd, *Bulgaria* .... **21 D11** 41 22N 25 7 E
Zlatoust, *Russia* .... **24 C10** 55 10N 59 40 E
Zlin, *Czech.* .... **17 D9** 49 14N 17 40 E
Zlītan, *Libya* .... **51 B7** 32 32N 14 35 E
Zmeinogorsk, *Kazakstan* . **26 D9** 51 10N 82 13 E
Znojmo, *Czech.* .... **16 D9** 48 50N 16 2 E
Zobeyrī, *Iran* .... **44 C5** 34 10N 46 40 E
Zobia, *Zaïre* .... **54 B2** 3 0N 25 59 E
Zolochev = Zolochiv, *Ukraine* .... **17 D13** 49 45N 24 51 E
Zolochiv, *Ukraine* ...... **17 D13** 49 45N 24 51 E
Zomba, *Malawi* .... **55 F4** 15 22S 35 19 E
Zongo, *Zaïre* .... **52 D3** 4 20N 18 35 E
Zonguldak, *Turkey* ...... **25 F5** 41 28N 31 50 E
Zonqor Pt., *Malta* .... **23 D2** 35 51N 14 34 E
Zorritos, *Peru* **92 D2** 3 43S 80 40W
Zou Xiang, *China* ...... **34 G9** 35 30N 116 58 E
Zouar, *Chad* .... **51 D8** 20 30N 16 32 E
Zouérate, *Mauritania* ... **50 D2** 22 44N 12 21W
Zoutkamp, *Neths.* ...... **15 A6** 53 20N 6 18 E
Zrenjanin, *Serbia, Yug.* .. **21 B9** 45 22N 20 23 E
Zuetina = Az Zuwaytīnah, *Libya* .... **51 B9** 30 58N 20 7 E
Zufar, *Oman* .... **46 D5** 17 40N 54 0 E
Zug, *Switz.* .... **16 E5** 47 10N 8 31 E
Zugspitze, *Germany* .... **16 E6** 47 25N 10 59 E
Zuid-Holland □, *Neths.* .. **15 C4** 52 0N 4 35 E
Zuidhorn, *Neths.* .... **15 A6** 53 15N 6 23 E
Zula, *Eritrea* .... **51 E12** 15 17N 39 40 E
Zumbo, *Mozam.* .... **55 F3** 15 35S 30 26 E
Zumpango, *Mexico* **87 D5** 19 48N 99 6W
Zungeru, *Nigeria* .... **50 G6** 9 48N 6 8 E
Zunhua, *China* .... **35 D9** 40 18N 117 58 E
Zuni, *U.S.A.* **83 J9** 35 4N 108 51W
Zunyi, *China* .... **32 D5** 27 42N 106 53 E
Zuoquan, *China* .... **34 F7** 37 5N 113 22 E
Zurbātīyah, *Iraq* .... **44 C5** 33 9N 46 3 E
Zürich, *Switz.* .... **16 E5** 47 22N 8 32 E
Zutphen, *Neths.* .... **15 B6** 52 9N 6 12 E
Zuwārah, *Libya* .... **51 B7** 32 58N 12 1 E
Zūzan, *Iran* .... **45 C8** 34 22N 59 53 E
Zverinogolovskoye, *Russia* **26 D7** 54 26N 64 50 E
Zvishavane, *Zimbabwe* .. **55 G3** 20 17S 30 2 E
Zvolen, *Slovak Rep.* .... **17 D10** 48 33N 19 10 E
Zwettl, *Austria* .... **16 D8** 48 35N 15 9 E
Zwickau, *Germany* ..... **16 C7** 50 44N 12 30 E
Zwolle, *Neths.* .... **15 B6** 52 31N 6 6 E
Zwolle, *U.S.A.* **81 K8** 31 38N 93 39W
Żymoetz —►, *Canada* ... **72 C3** 54 33N 128 31W
Żyrardów, *Poland* ...... **17 B11** 52 3N 20 28 E
Zyryan, *Kazakstan* ..... **26 E9** 49 43N 84 20 E
Zyryanka, *Russia* .... **27 C16** 65 45N 150 51 E
Zyryanovsk = Zyryan, *Kazakstan* .... **26 E9** 49 43N 84 20 E
Żywiec, *Poland* .... **17 D10** 49 42N 19 10 E
Zyyi, *Cyprus* .... **23 E12** 34 43N 33 20 E

REGIONS IN THE NEWS

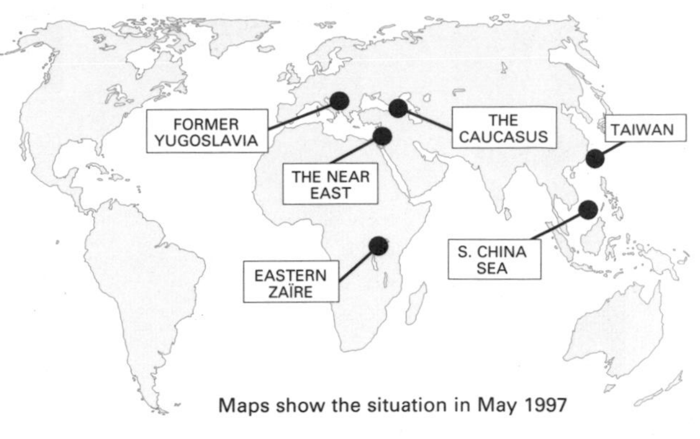

FORMER YUGOSLAVIA

0 50 100 miles

- International boundaries
- Republic boundaries
- Province boundaries
- Capital cities
- Dayton Peace Agreement Boundary
- Muslim-Croat Federation
- Bosnian Serb Republic

AUSTRIA HUNGARY ROMANIA ITALY SLOVENIA CROATIA BOSNIA HERZEGOVINA YUGOSLAVIA SERBIA Vojvodina MONTENEGRO Kosovo ALBANIA MACEDONIA BULGARIA GREECE ADRIATIC SEA

Maribor Ljubljana Zagreb Rijeka Osijek Vukovar Subotica Novi Sad Timişoara Beograd (Belgrade) Smederevo Bihać Banja Luka Tuzla Zenica Sarajevo Split Mostar Dubrovnik Kraljevo Niš Podgorica Priština Skopje Tirana Bitola

Drava Sava Danube Tisa Morava Ibar Kalina

THE CAUCASUS

0 50 100 miles

- International boundaries
- Republic boundaries
- Capital cities

RUSSIA CASPIAN SEA BLACK SEA GEORGIA ARMENIA AZERBAIJAN TURKEY IRAN ADYGEA KARACHEY-CHERKESSIA KABARDINO-BALKARIA NORTH OSSETIA INGUSHETIA CHECHENIA DAGESTAN ABKHAZIA South Ossetia AJARIA Nagorno-Karabakh NAXCIVAN Caucasus Mountains

Maykop Cherkessk Sochi Malgobek Groznyy Nalchik Vladikavkaz Makhachkala Sukhumi Tskhinvali Tbilisi Batumi Yerevan Baki (Baku) Xankändi Naxçivan Sevana Lich Kür Rud-e Aras

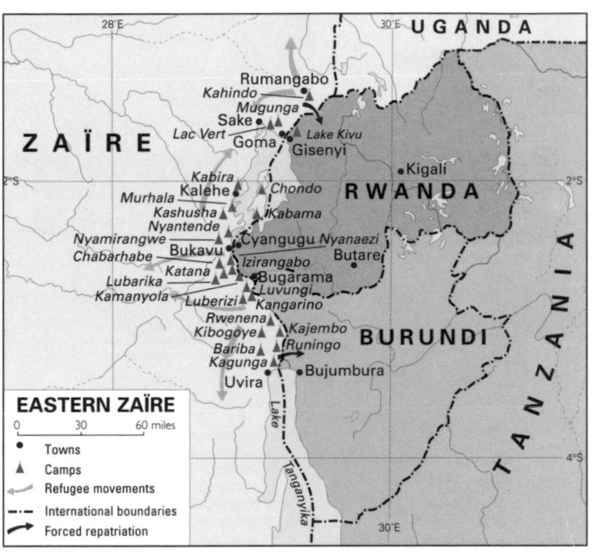

AFGHANISTAN

ALBANIA

ALGERIA

ANDORRA

ANGOLA

ANTIGUA & BARBUDA

ARGENTINA

BARBADOS

BELARUS

BELGIUM

BELIZE

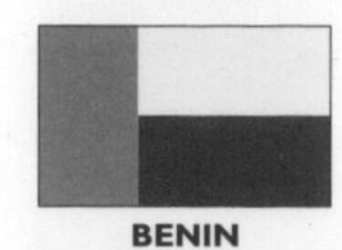
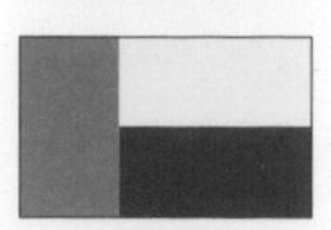
BENIN

BHUTAN

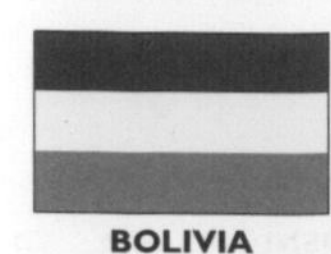
BOLIVIA

BURUNDI

CAMBODIA

CAMEROON

CANADA

CAPE VERDE

CENTRAL AFRICAN REP.

CHAD

CUBA

CYPRUS

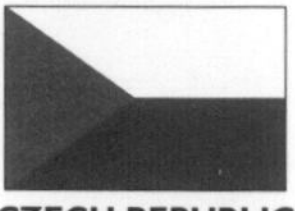
CZECH REPUBLIC

DENMARK

DJIBOUTI

DOMINICA

DOMINICAN REPUBLIC

FAROE ISLANDS

FIJI

FINLAND

FRANCE

GABON

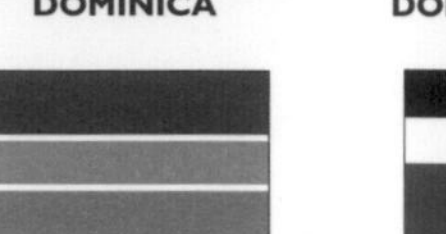
GAMBIA

GEORGIA

GUINEA-BISSAU

GUYANA

HAITI

HONDURAS

HONG KONG

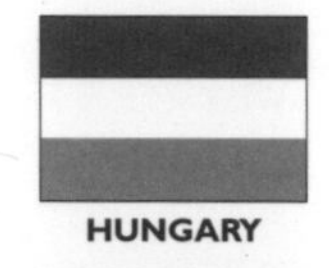
HUNGARY

ICELAND

IVORY COAST

JAMAICA

JAPAN

JORDAN

KAZAKSTAN

KENYA

KIRIBATI

LESOTHO

LIBERIA

LIBYA

LIECHTENSTEIN

LITHUANIA

LUXEMBOURG

MACAU

MAURITANIA

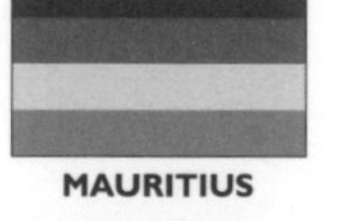

MAURITIUS

MEXICO

MICRONESIA

MOLDOVA

MONACO

MONGOLIA

NICARAGUA

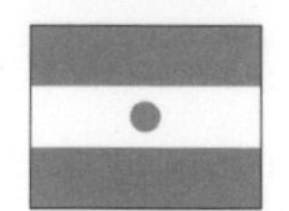
NIGER

NIGERIA

NORTHERN MARIANAS

NORWAY

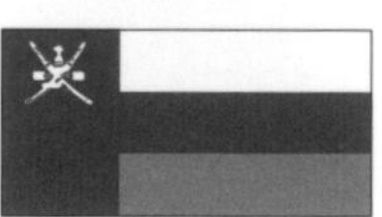
OMAN

PAKISTAN

PUERTO RICO

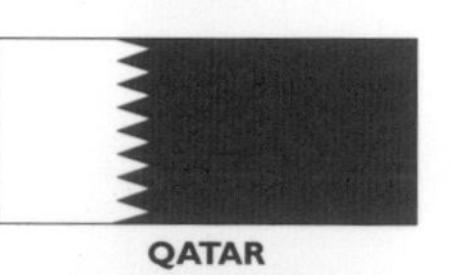
QATAR

ROMANIA

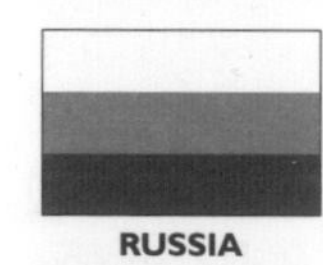
RUSSIA

RWANDA

SAN MARINO

SÃO TOMÉ & PRÍNCIPE

SOLOMON ISLANDS

SOMALIA

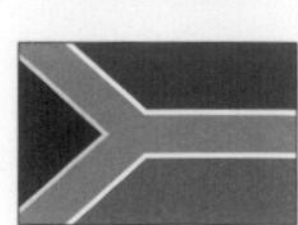
SOUTH AFRICA

SPAIN

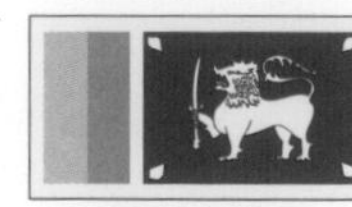
SRI LANKA

ST KITTS & NEVIS

ST LUCIA

TAIWAN

TAJIKISTAN

TANZANIA

THAILAND

TOGO

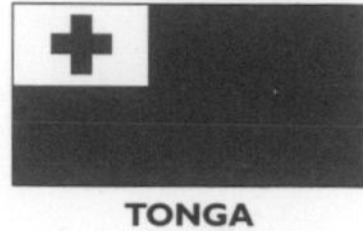
TONGA

TRINIDAD & TOBAGO

UNITED KINGDOM

UNITED STATES

URUGUAY

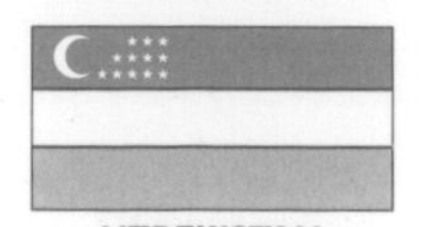
UZBEKISTAN

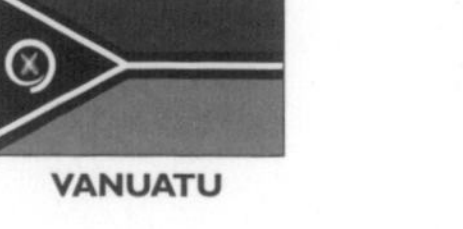

VANUATU

VATICAN CITY

VENEZUELA